ISBN 978-0-260-72013-9
PIBN 10622673

THE SCOTTISH GEOGRAPHICAL MAGAZINE

Authors are alone responsible for their respective statements.

THE SCOTTISH GEOGRAPHICAL MAGAZINE

PUBLISHED BY THE ROYAL SCOTTISH GEOGRAPHICAL SOCIETY

PROFESSOR JAMES GEIKIE, LL.D., D.C.L., F.R.S., HON. EDITOR

MARION I. NEWBIGIN, D.Sc., ACTING EDITOR

VOLUME XXIX: 1913

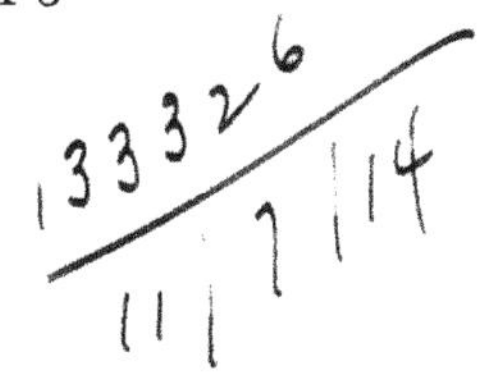

EDINBURGH

PRINTED BY T. AND A. CONSTABLE, PRINTERS TO HIS MAJESTY

AT THE UNIVERSITY PRESS

1913

ROYAL SCOTTISH GEOGRAPHICAL SOCIETY.

CONDITIONS AND PRIVILEGES OF MEMBERSHIP.

THE conditions of Ordinary Membership are:—Approval by the Council; the payment in the case of members in the Edinburgh district of an entrance fee of One Guinea; in the case of other members one of Half-a-Guinea; and the payment of the ordinary Annual Subscription or a composition for Life-Membership. The Annual Subscription is One Guinea and is payable *in advance* at the commencement of each Session.

A Member may compound for Life-Membership by payment as follows, viz.:—When under ten years' standing, £20; when over ten and under twenty years' standing, £15; when over twenty and under 30 years' standing, £10; when over thirty years' standing, £5.

The Official Year, or Session, of the Society is from October 1st to September 30th. New Members are required to pay the Subscription for the Session in which they join the Society, at whatever period, and they are entitled to receive the ordinary publications of that Session. Resignations, to take effect, must be lodged with the Secretary *before* the commencement of a new Session.

The privileges of Ordinary Membership include admission (with one Guest) to the Ordinary Meetings of the Society, and the use of the Library and Map-Room. Non-resident Members may borrow books from the Library, but they must defray the cost of transit both ways. Each Member is entitled to receive, free by post, the *Scottish Geographical Magazine*, which is published monthly by the Society.

TEACHER ASSOCIATE MEMBERSHIP.—With the object of helping to promote the teaching of Geography in Schools, "Teacher Associates" are admitted to certain privileges of the Society at a reduced Subscription of Half-a-Guinea. The privileges are limited to the use of the Society's Rooms, receipt of the monthly magazine, the right to borrow *one volume* from the Library, and one ticket (not transferable and admitting only one) for the Society's meetings. Only professional teachers are eligible for election, and the applicants must state the name of the school at which they are engaged.

Branches of the Society have been established in Glasgow, Dundee, and Aberdeen, where periodical Meetings are held.

CONTENTS.

VOL. XXIX: 1913.

No. I.—JANUARY.

No. II.—FEBRUARY.

No. III.—MARCH.

No. IV.—APRIL.

(*Map, Diagram and Illustrations.*)

No. V.—MAY.

(Port*rait, Maps, Illustrations and Diagrams.*)

No. VI.—JUNE.

No. VII.—JULY.

No. VIII.—AUGUST.

No. IX.—SEPTEMBER.

(*Plate.*)

No. X.—OCTOBER.

(*Maps.*)

No. XI.—NOVEMBER.

No. XII.—DECEMBER.

THE SCOTTISH GEOGRAPHICAL MAGAZINE.

THE NORWEGIAN SOUTH POLAR EXPEDITION.[1]

By Captain ROALD AMUNDSEN,
Livingstone Medallist of the Royal Scottish Geographical Society.

(With Map and Illustrations.)

WHILE the struggle for the North Pole covers hundreds of years, the struggle for the South Pole is of comparatively recent date. About 1900 we find several expeditions—English, German, French, Belgian, Scottish, and Swedish—working hand-in-hand in order to withdraw the veil and lay open the great mysteries of the Antarctic. The object of several of these expeditions was of course scientific, but I believe I may say that the Pole itself loomed behind as the ultimate goal.

My time to-night does not permit me to review all the expeditions that have contributed to increase our knowledge of the great unexplored section of the Antarctic continent. I shall mention only those which did work earlier in the region where we had to look for our starting-point. Our object being to reach the South Pole, we had first of all to push forward with the ship as far south as possible and there build our station. The sledge journey would be long enough, anyway. I knew that the English would go to their old winter quarters in M'Murdo Sound, South Victoria Land. The newspapers had stated that the Japanese had reserved King Edward VII. Land. Thus there was nothing else for us to do but to build our hut on the barrier itself, as far from these two expeditions as possible, in order not to be in their way.

The great Antarctic barrier, or the Ross Barrier as it is called, between South Victoria and King Edward VII. Land, has an extent of about 450 geographical miles. The first who met with this enormous glacial

[1] An address delivered before the Society in Edinburgh on Nov. 21, 1912.

formation was Sir James Clark Ross in 1841. He, naturally enough, did not take the risk of running his two sailing vessels, the *Erebus* and *Terror*, close under the mighty 100-feet high ice-wall, which barred his progress towards the south. But he examined it at a reasonable distance as well as circumstances would permit. These observations made it clear that the barrier was not a straight, steep ice-wall, but was broken at intervals by bights and small inlets. In Ross's chart we notice an imposing bay-formation in 164° W. and 78° 30′ S. The next expedition that sailed down to these regions was the Southern Cross expedition in 1900. It is interesting that this expedition found this bay in the same place where Ross saw it in 1841—sixty years earlier. As interesting is it, that this expedition succeeded in landing in a little bay—Balloon Bight—some miles to the eastward of the big one, and from there climbed up on the barrier, which up to this time had been considered inaccessible and an invincible hindrance to an advance towards the south.

In 1901 the *Discovery* steamed along the barrier and confirmed in every respect the observations of the Southern Cross expedition. It also succeeded in discovering land in the direction mentioned by Ross—King Edward VII. Land. Scott also landed in Balloon Bight and observed, like his predecessors, the big bay formation to the west.

In 1908 Shackleton in the *Nimrod*, like his predecessors, followed the barrier, and arrived at the conclusion that disturbances in the barrier had broken away the shore-line of Balloon Bight, merging that indentation into the bay to the west. To the big newly formed bay he gave the name of the "Bay of Whales." But his original plan of landing here was abandoned—the barrier in this place looking too dangerous as a foundation for winter quarters.

When the two charts were compared, it was not difficult to decide that the bay put down on the chart by Ross and the Bay of Whales was one and the same. Though some few pieces had broken off here and there, this bay had remained constant for about seventy years. It was an obvious conclusion that the bay was no casual formation, but owed its existence to subjacent land, banks, etc. This bay we selected as our basis of operation. It is 350 nautical miles (650 kiloms.) from the English station in M'Murdo Sound, and 100 nautical miles (185 kiloms.) from King Edward VII. Land. It seemed to us therefore that we were at a sufficient distance from the English sphere, and need not fear that we should come in their way. The reports in regard to the Japanese on King Edward Land were rather vague, and we thought that a distance of 100 miles from them was more than enough.

On board of Nansen's well-known polar ship, the *Fram*, we left Norway on August 9, 1910. We carried on board ninety-seven fine Eskimo dogs from Greenland, and provisions for two years. The first port we touched was Madeira, where we finally made everything ready for the long voyage to the Ross Barrier. It was no short distance we had to cover—about 16,000 nautical miles (29,600 kiloms.)—from Norway to the Bay of Whales. We had calculated that it would take us five months to make the trip. The *Fram*, which with good reason is said to be the most solid polar ship in the world, proved to be

The *Fram* in the Bay of Whales.

To ace page 3.]

(From *The South Pole*.

exceedingly seaworthy during this long voyage over pretty nearly all the oceans. Thus we sailed through the north-east and south-east trades, the roaring forties, the foggy fifties and the icy sixties without any mishap, and arrived at our sphere of work at the barrier on January 14, 1911. Everything had gone unusually well.

The ice in the Bay of Whales had just broken up, and to such an extent that we succeeded in sailing quite a distance further south than any of our predecessors, and found a cosy little corner behind a projecting ice-cape, from which we could, in comparative safety, bring our outfit up to the barrier. Another great advantage was that the barrier here sloped very gently down to the sea ice, and this gave us good ground for sledging. The first thing we did on our arrival was to climb the barrier to examine the nearest surroundings and to find a convenient place for the house we had brought from home. The supposition that this part of the barrier rested on subjacent land seemed to be confirmed at once by the surroundings. Instead of the plain, smooth surface that the outer barrier wall shows, we found the surface here greatly disturbed. Steep hills and crests, with intervening dales, filled with huge hummocks and pressure ridges, were seen everywhere. And these formations were not of recent date. It was easy to notice that they dated from a time far beyond the days of old father Ross. Our original plan was to build our station several miles from the barrier edge in order to guard us against an unwished-for sea-trip, in case the part of the barrier on which our house was built, should break off. But that was not necessary. The formations we met on our first examination were warrant enough for the barrier's stability in this region. In a little valley 2 nautical miles (3½ kiloms.) from the spot where we had made fast the ship, well sheltered against all winds, we selected a place for our winter quarters. On the following day we commenced to discharge the ship:—materials for housebuilding, outfit and provisions for nine men for several years. We were divided into two parties, the ship and the shore party. The first consisted of the master of the ship, Captain Nilsen, and nine men who would stay on board to navigate the *Fram* out of the ice and up to Buenos Aires. The other party consisted of those of us who were to go into winter quarters and march towards the south. It was the duty of the ship's party to unload everything from the ship on to the ice. There the shore party took it and drove it to the spot we had selected for our house. In the beginning we were a little unaccustomed to this work, untrained as we were after the long voyage. But it didn't take long before all of us were well trained, and then everything went along at a dizzy speed between the ship and our future home, Framheim, which daily increased in size. As soon as all the materials for the hut had been driven up, our experienced carpenters, Olav Bjaaland and Jörgen Stubberud, commenced to erect the house. It was a ready-built house, and now there was nothing to do but to put together the different parts, all of which were marked. That the house might be able to withstand all the storms we expected, the site was excavated to a depth of 4 feet below the barrier surface.

On January 28, fourteen days after our arrival, the home was ready and the provisions ashore. A gigantic piece of work had been done and everything promised well for the future. But time was precious, and it was our duty to make the most of it. The shore party was again divided into two. One would go on driving up from the ship the rest of the stores, outfit, and the like, while the other would prepare for a trip southward for the purpose of exploring our immediate surroundings and establishing a depôt. On February 10, the last-named party was off. We were four men, eighteen dogs and three sledges fully laden with provisions. How well I remember that morning when we for the first time made our way towards the south! It was calm and slightly overcast. Ahead of us was the vast, endless snow-plain; behind us the Bay of Whales with the great prominent ice-capes, and at the further end of the bay our dear *Fram*. The flag was hoisted, a last farewell from our comrades on board. Nobody knew when we should see them again. Most likely they would be gone when we returned, and then a year would elapse before we should see them. Another look behind, another farewell, and then southward. This first trip of ours on the barrier was an exciting one. What would the region be like? How about the sledging? Had we the proper outfit? Had we the right traction power? If our task was to be accomplished, everything had to be of the best. Our equipment was essentially different from that of our English competitors. We pinned all our faith to our dogs and ski.

We travelled fast on the smooth, flat snow-plain. On the 14th we reached 80° S., having travelled a distance of 85 nautical miles (160 kiloms.) and established a depôt, consisting chiefly of provisions to be used on our main march towards the south when spring came. The weight of the provisions was 1600 lbs. The return trip was made in two days. The first day we travelled 46 nautical miles (75 kiloms.), and the second day 50 nautical miles (93 kiloms.). On our arrival at the station, the *Fram* had sailed. The bay looked dreary and desolate. Seals and penguins had taken possession of the place. Our first trip southward, however short, was of great importance. We now knew to a certainty that our equipment and traction power was of the very best. No errors had been made in the selection of it. Now it was for us to use them in the best manner possible.

We did not stay long at home. On the 22nd we were once more ready to carry our depôts towards the south. The intention was to take them as far south as possible. We were eight men, seven sledges, and forty-two dogs. The cook alone stayed at home. On the 27th we passed the depôt in 80° S., where everything was in first-class order. On March 4, we made lat. 81° S., and deposited there 1050 lbs. of provisions. From here three men returned, while five men continued their way southward, and on March 8 reached 82° S., where 1250 lbs. of provisions were left. We then returned, and were home again on the 22nd. Once more before the winter set in we were in the field, and carried 2200 lbs. of fresh seal meat and 400 lbs. of other provisions to the depôt in 80° S. On April 11 the trip was completed, and all the depôt work had come to an end. Up to this time we had carried 6500 lbs. of provisions, distributing them at the three depôts.

The part of the barrier over which we had travelled had an average height of about 150 feet, and looked like a smooth plain, rising in great waves or undulations without characteristic marks of any kind. It has been the common opinion that depôts could not be laid out on such an endless plateau without an imminent probability of losing them. But if there was to be any chance of reaching our goal, we had to lay out our depôts, and that even on a large scale. We talked much over the question, and arrived at the conclusion that we must use signals athwart our course instead of along it as is commonly the case. Consequently, we put down a line of flags at right angles to our course, that is, in an east-west direction, with the depôt as central point. Each of the three depôts was marked in this manner, 5 nautical miles (9 kiloms.) on each side of the depôt, with half a mile (1 kilom.) between each flag. Besides, all the flags were marked, so that, wherever we met them, we were able to discover in which direction the depôt lay, and how far we were from it. This plan proved to be absolutely reliable, and even in the densest fog we succeeded in finding our depôts. Our compasses and distance meters were examined at the station, and we knew that we could depend upon them.

We had gained much on our depôt trips. Not only had we succeeded in carrying plenty of provisions southward, but we had acquired a lot of experience, which was possibly more important, and came in handy on our final dash for the Pole. The lowest temperature observed on these depôt trips was $-50°$ Fahr. Considering that it was still summer when this temperature was observed, it was a serious warning to us that we must have our equipment in good order. We had also seen that our solid, heavy sledges were too clumsy, and that without risk they could be lightened considerably. The same could be done with the greater part of our other outfit.

Some days more were spent on a seal hunt, before the sun disappeared. The total weight of the seals killed amounted to 120,000 lbs. Thus we had provisions in plenty for ourselves as well as for our dogs, now increased to 115. The first thing we did was to make them a shelter. We had brought with us ten very big tents, large enough to accommodate sixteen men. They were pitched on the barrier, after which the snow under each tent was dug out, 6 feet down, so that the ultimate height of these dog-houses became 18 feet. The diameter of the floor was 15 feet. Our intention in building these houses so large was to make them as airy as possible, in order to avoid the hoar frost so annoying to the dogs. We achieved our object, for even during the most severe period of the winter no frost could be noticed. The tents were always cosy and warm. Each tent had room for twelve dogs, and every man had his own team to look after.

Having thus cared for our dogs, our own turn came. Mother nature had stretched out a helping hand, and we were not slow in taking it. By April the house was completely covered with snow. In this newly drifted snow excavations were made in direct communication with the hut. Thus we got large and spacious rooms without buying or fetching materials. There were workshops, forge, sewing-room, packing-room, a space for coal, wood and oil, ordinary bath and steam bath. However cold and stormy the winter might be, it would not annoy us at all.

On April 21 the sun disappeared, and then began the longest night ever experienced by men in the Antarctic. We had no fear of meeting it. We had provisions enough for years, a cosy house well ventilated, well lighted, and well heated, with an excellent bath—a complete sanatorium indeed. As soon as all these buildings were finished we began to make preparations for the final journey in the spring. Our business was to improve our equipment and reduce its weight. Thus we condemned all our sledges. They were too heavy and clumsy for the smooth surface of the barrier. The weight of such a sledge was 150 lbs. Our ski and sledgemaker, Bjaaland, took care of the sledges and did all the necessary work connected with them, and when the spring came, a complete new sledge outfit was ready from his hand. These sledges weighed only one-third of the original ones. In a like proportion we succeeded in reducing the weight of everything. Of the utmost importance was the packing of the provisions selected for the trip. It was the work of Captain Johansen during the winter, and had to be done with care and attention. Of the 42,000 biscuits that were packed, each and every one was turned in the hand, before the right place for it was found. In this manner the winter passed quickly and comfortably. Every one had his hands full all the time; our house was warm and dry, light and airy; consequently the health of everybody was excellent. We had no physician, and we didn't need one.

Meteorological observations were taken all the time; the results were surprising. We believed that we should encounter unpleasant, stormy weather, but it proved to be contrary to expectations. During the whole year we stayed there we didn't have more than two moderate storms; for the rest calm and light breezes—mostly easterly. The atmospherical pressure was mostly very low, but steady. The temperature became very low, and it is probable that the mean temperature for the year, $-14^{\circ}\cdot 8$ Fahr. (-26° C.), which we observed, is the lowest mean temperature on record. In five months of the year we had temperatures below -58° Fahr. (-50° C.). On August 13, we had the lowest temperature observed, $-74^{\circ}\cdot 2$ Fahr. (-59° C.). The aurora australis was very frequent in all directions and shapes; it was extremely lively, but not very intense. There were, however, a few exceptions.

On August 24 the sun returned; the winter was over. Some days before we had all our things in full order, and when the sun peeped over the barrier everything was ready for a start. The dogs were in excellent condition, some of them too much so. From now on we watched the temperature daily. As long as the glass remained as low as -58° Fahr. (-50 C.), there could be no question of starting. In the first days of September there was every sign that it was going to rise, and we therefore resolved to push off as soon as possible. On September 8 we had a temperature of −31° Fahr., and off we were. But this trip was not to be one of long duration. On the following day the temperature began to go down rapidly, and within a few days we had −72°·4 Fahr. (−58° C.). We human beings might have kept going for some time in this temperature, well clothed as we were, but our dogs could not stand it very long. We were therefore satisfied to reach 80° S., and arriving there we laid all our provisions and outfit in the old depôt and returned to Framheim.

Now came a period of doubtful weather, the change from winter into spring, and we never knew what the next day would bring forth. Some frost-bitten heels from the last trip forced us to wait until we knew for certain that the spring had come in earnest. On September 24 the first obvious sign of spring appeared—the seals began to go up on the ice. This sign, and not least the fresh seal meat which Bjaaland brought that very day, was welcomed with rejoicing. The dogs also appreciated this sign of spring. They were especially fond of fresh blubber. On the 29th appeared another and more obvious sign—a flight of Antarctic petrels; they were flying round the house, to the delight of the men as well as of the dogs. The dogs were wild with joy and excitement, ran after the birds and stupidly counted on a delicate bird for dinner; the hunt resulted in a wild fight. At last on October 20 the weather had settled so much that we were able to start. The original plan, that all of us should march southward, had been changed during the interval. We understood that without risk we could divide into two parties, and in this manner do considerably more work. We had arrived at the decision that three men should go east to King Edward VII. Land and examine it, while the other five should carry out the main plan—the dash for the Pole.

October 20 was a fine day, clear and mild: temperature −1° Fahr. We were five men, fifty-two dogs, and four sledges. Our sledges were light, and the going was lively. It was not necessary to encourage the dogs, they were willing enough without. With our depôts in 80°, 81° and 82°, we had provisions enough for 120 days. Two days after our departure a serious accident nearly happened, Bjaaland's sledge falling down one of the many crevasses we had to pass over that day. He got assistance at the last moment, but it was in the nick of time, or his sledge with the thirteen dogs would have disappeared in the apparently bottomless pit.

On the fourth day we arrived at our depôt in 80° S. Here we rested two days and gave our dogs as much fresh seal meat as they could eat. Between 80° and 81°, in the direction we went, the barrier is smooth and level with the exception of a few low undulations, and there are no hidden dangers. It is quite different between 81° and 82°. During the first 15 miles we were in a perfect labyrinth of crevasses and pressure ridges, rendering the passage extremely dangerous. Big pieces of the surface have been broken off, and gaping abysses are met with everywhere. From these gulfs cracks are to be found in all directions, and the surface consequently is very unsafe. We passed this bit of road four times. The three first times it was such a dense fog, that we could not see many yards ahead of us. Only the fourth time did we get clear weather, and we saw then what difficulties we had escaped. On November 5 we reached the depot in 82° S., and found everything all right. For the last time our dogs could get a good rest and plenty of food. And they got it thoroughly during a two days' stay. In 80° we commenced to build snow mounds, intended to serve as track marks on the homeward trip. We erected one hundred and fifty mounds in all. Sixty snowblocks were needed for a single mound; thus 9000 blocks were cut for these marks. The mounds proved to answer expectations, as by them we returned by precisely the same route we had gone.

The barrier south of 82° was, if possible, still more smooth than to the north of that latitude, and we marched along at a good speed. We agreed to lay down a depôt at each whole degree of latitude on our way south. Undeniably we ran a risk in doing so, as there was no time for putting down cross-marks. But we had to be satisfied with the snow mounds and pin our faith to them. And, on the other hand, our sledges became so much lighter that they were never too heavy for our dogs. In 83° S. we sighted land in a south-westerly direction. It could be nothing else but South Victoria Land, and probably a continuation of the mountain range running in a south-easterly direction, as drawn by Shackleton in his chart. From day to day the land became more distinct, each peak being more magnificent than the next; they are from 10,000 to 15,000 feet in height, sharp cones and sharp needle-like spurs. I have never seen a landscape more beautiful, more wild, and more imposing. Here a weathered summit—dark and cold—there snow and ice glaciers pell-mell in terrible chaos. On November 11 we sighted land due south, and pretty soon ascertained that South Victoria Land in about 86° S. and 163° W. is met by a range trending east and north-east. This mountain range is considerably lower than South Victoria Land's mighty mountains. Summits from 2000 to 4000 feet were the highest. We were able to see the range to 84° S., where it disappeared on the horizon. On the 17th we arrived at the spot where the ice barrier and the land are joined. We had all the time been steering due south from our winter quarters. The latitude was 85°, and the longitude 165° W. The junction between barrier and land was not followed by any great disturbances, but only by a few large undulations broken off at intervals by crevasses. Nothing there could impede our progress. Our plan was to go due south from Framheim to the Pole, and not go out of the way unless natural obstacles should force us to do so. If we succeeded, we would thus be able to explore an absolutely unknown land, and do good geographical work.

The nearest ascent due south was between the mighty peaks of South Victoria Land. Obviously there were no serious difficulties in store for us. We might probably have found a less steep ascent if we had crossed over to the newly discovered range. But we had once taken the notion that due south was the shortest way to the goal, and then we had to take the chances. On this spot we established our main depôt and left provisions here for thirty days. On our four sledges we carried provisions for sixty days. And up we went for the plateau. The first part

Captain Amundsen in Polar dress.

To face page 8.] (From *The South Pole*)

of the ascent was over sloping, snow-covered mountain sides, in some places rather steep, although not bad enough to prevent each team managing to haul up its own sledge. Further up we met with some short but very steep glaciers, in fact, they were so steep that we had to harness twenty dogs to each sledge. But now they went briskly enough in spite of precipices which were so steep that we had the greatest difficulty in climbing them on our skis. The first night we pitched camp at a height of 2000 feet. The second day we climbed mostly up some small glaciers, and camped at a height of 4000 feet. The third day we unfortunately had to swallow a bitter pill and descend about 2000 feet, being surprised by a large glacier running east-west, which divided the mountains we had climbed from the higher peaks further south. Down hill the expedition went therefore at a dizzy speed, and in a very little time we were down on the before-mentioned mighty glacier, Axel Heiberg's Glacier. That night our camp was about 3000 feet above sea-level. The following day began the longest part of our climb, as we were obliged to follow Heiberg's Glacier. This glacier was in places filled with hummocky ice, the surface rising into hillocks and splitting into chasms, and we had to make detours in order to escape the broad chasms which open into great gullies. The latter were of course mostly filled, the glacier apparently having stopped every movement long ago. But we had to be very careful, not knowing how thick was the covering layer. Our camp that night lay in very picturesque surroundings, in an altitude of 5000 feet above sea-level. The glacier here was compressed between two 15,000-feet high mountains—Fridtjof Nansen and Don Pedro Christophersen mountains. To the west at the further end of the glacier rose Mount Ole Engelstad to a height of 13,000 feet. The glacier in this comparatively narrow passage was very hummocky and broken by huge crevasses, so that our progress very often seemed to be impeded. On the following day we reached a slightly sloping plateau, which was supposed to be the plateau described by Shackleton. Our dogs worked so well that day that their superiority must be admitted once for all. Added to the toil of the preceding weary days, they travelled this day 17 nautical miles (33 kiloms.), ascending 5600 feet. We camped that night in an altitude of 10,600 feet. Now the time had come when, unfortunately, we were obliged to kill some of our dogs. Twenty-four of our brave companions had to lay down their lives. We had to remain here for four days on account of bad weather. When we at last broke up camp on November 26, we had only ten carcasses left, and these we laid in depôt. Some fresh provisions on our return trip would not do any harm. During the following days the weather was stormy and the snowdrift dense, so we were not able to see any of the surroundings. But we did notice that we were going down hill very fast. Once in a while when the drift lifted we saw high mountains due east. In a dense snowdrift on November 28 we were close under two peculiar-looking crests of mountains, running north-south, the only two peaks we observed on our right side. These we called Helland Hansen's Mountains; they were wholly snow-clad and had a height of 9000 feet. They became later on an excellent landmark. The gale slackened the next day and the sun shone through. Then it appeared to us as if we were transported to an absolutely new country. In the direction of our course trended a huge glacier. On its eastern boundary was a range running south-east—north-west. To the west the fog was dense over the glacier and hid even the nearest surroundings. The hypsometer gave 8000 feet above sea-level at the foot of "The Devil's Glacier," which means that we had descended 2600 feet from the Butchery. It was no pleasant discovery. Without doubt we would have the same climb again and probably more. We established a six-days' depôt and continued our march. From our camp that night we had a splendid view of the eastern mountain range. There was the most peculiar-looking peak I have seen. It was 12,000 feet high, and the top was round in shape and covered by a torn glacier. It looked as if nature in a fit of anger had showered sharp ice blocks down on it. We called it Helmer Haussen Mountain, and it became our best landmark. And there were Oscar Wisting, Olav Bjaaland, Sverre Hassel mountains, glittering dark and red, glaring white and blue in the rays of the midnight sun. In the distance appeared a romantic mountain—enormous to behold through the heavy masses of clouds and fog, which from time to time drifted over, now and then exposing to our view its mighty peaks and broken glaciers. Thus appeared Thv. Nilsen Mountain for the first time. The height of it was 15,000 feet. It took us three days to climb the Devil's Glacier. On December 1 we had left behind us this crevassed glacier, so full of holes and bottomless chasms. Our altitude was 9100 feet. Ahead of us, and looking like a frozen sea in the fog and snowdrift, was a sloping ice-plateau studded with hummocks. The march over "The Devil's Dancing-room" was not entirely pleasant. Gales from south-east, followed by snowdrift, were of daily occurrence. We saw nothing, absolutely nothing. The ground below us was hollow, and it sounded as if we were walking on the bottom of empty barrels. We crossed this unpleasant and ugly place as quickly and as light of foot as possible, all the time with the unpleasant possibility of being engulfed.

On December 6 we reached our greatest height—10,750 feet above sea-level, according to hypsometer and aneroid. From here the main inland plateau didn't rise any more, but ran into an absolutely flat plain. The height was constant as far as 88° 25′, from where it began to slope down to the other side. In 88° 23′ we had reached Shackleton's furthest south, and camped in 88° 25′. Here we established our last depôt—depôt No. 10, and deposited 200 lbs. of provisions. Then we began to go very slowly down-hill. The state of the ground was excellent, absolutely flat without undulations, hills, or sastrugi. The sledging was ideal, and the weather beautiful. We covered daily 15 nautical miles (30 kiloms.). There was nothing to prevent us from making a good deal longer marches, but we had time and food enough, and considered it more prudent to save the dogs and not overwork them. Without adventures of any kind we made latitude 89° on December 11. It seemed as if we were arrived at a region with perpetually fine weather. The most obvious sign of constant, calm weather was the absolutely plain surface. We were able to thrust a tent-pole 6 feet down into the snow without being met with any resistance. This is a clear enough proof that the snow has fallen in the same kind of weather—calm or very light breeze. Varying weather conditions—calm and gale—would have formed layers of different compactness, which would soon have been felt when one stuck the pole through the snow.

Dead reckoning and observations had always given like results. The last eight days of our outward march we had sunshine all the time. Every day we stopped at noon to take a meridian latitude, and every evening we took an azimuth observation. On December 13 the latitude gave 89° 37′; dead reckoning 89° 38′. In 88° 25′ we got the last good azimuth observation. Later on they were of no use. As the last observations gave pretty near the same result, the variation being almost constant, we used the observation taken in 88° 25′. We made out that we would reach the goal on December 14. The 14th arrived. I have a feeling that we slept less, breakfasted at a greater speed, and started earlier this morning than the previous days. The day was fine as usual—brilliant sunshine with a very gentle breeze. We made good headway. We didn't talk much. Everybody was occupied with his own thoughts, I think. Or had probably all of us the same thought, which urged all of us to gaze fixedly towards the south over the endless plateau? Were we the first or?—Halt! It sounded like a sound of exultation. The distance was covered. The goal reached. Calm, so calm stretched the mighty plateau before us, unseen and untrod by the foot of man. No sign or mark in any direction. It was undeniably a moment of solemnity when all of us with our hands on the flag-staff planted the colours of our country on the geographical South Pole, on King Haakon the Seventh Plateau.

During the night—according to our time—three men encircled our camp, the length of the semidiameter being 10 nautical miles (18 kiloms.), putting down marks, while the two others remained at the tent, taking hourly observations of the sun. These gave 89° 55′. We might very well have been satisfied with the result, but we had plenty of time and the weather was fine, so why not try to observe the very Pole itself? On the 16th we therefore moved our tent the remaining 5 nautical miles (9 kiloms.) further south, and camped there. We made everything as comfortable and snug as possible in order to take a series of observations throughout the twenty-four hours of the day. The altitudes were observed every hour by four men with sextant and artificial horizon. The observations will be worked out at the Norwegian University. With this camp as a centre we drew a circle with a radius of 4½ nautical miles (8 kiloms.), and marks were put down. From this camp we went out for 4 miles in different directions. A little tent we had carried with us in order to mark the spot was pitched here, and the Norwegian flag with the *Fram* pendant hoisted on the top of the tent. This Norwegian home got the name "Polheim." Judging from the weather conditions, this tent may stay here many years to come. In the tent we left a letter addressed to H.M. King Haakon the Seventh, with information of what we had done. The next man will bring it home. Besides, we left some clothing, a sextant, an artificial horizon, and a hypsometer.

On December 17 we were ready to start on our return journey. The outward journey had, according to distance meters, covered a distance of 750 nautical miles (1400 kiloms.). The daily average speed had been 13 nautical miles (25 kiloms.). When we left the Pole we had two sledges and seventeen dogs. Now we enjoyed the great triumph of being able to increase our daily rations, unlike earlier expeditions, all of which were obliged to go on short commons—already at a much earlier moment of time. The rations were also increased for the dogs, who got from time to time one of their comrades as an extra. The fresh meat had a favourable effect on the dogs, and contributed no doubt to the good result.

A last look and farewell to Polheim and then off. We see the flag yet. It is still waving to us. It is gradually diminishing. Then it disappears; a last good-bye from the little Norway on the South Pole. We left King Haakon's Plateau as we had found it, bathed in sunlight. The mean temperature during our stay here was −13 Fahr. (−25° C.). It felt much milder.

I am not going to weary the audience with a detailed account of the return journey. I shall only mention a few incidents of interest. The beautiful weather we got on our homeward run exposed to our view the whole of the mighty mountain range, that is, the continuation of the two ranges joined in 86° S. The newly discovered range, trending in a south-easterly direction, was everywhere studded with peaks of a height of from 10,000 to 15,000 feet. In 88° S. the range disappears on the horizon. The whole of the newly discovered mountain ranges—about 460 nautical miles (850 kiloms.)—has been given the name "Queen Maud Ranges."

All the depôts—ten in all—were found, and the abundant provisions, of which we at last had plenty, were taken along down to 80°, where they were deposited. From 86° we didn't go on rations, but everybody could eat as much as he liked. On January 25 we arrived at our winter quarters after an absence of ninety-nine days. The distance home, 750 nautical miles (1400 kiloms.), was thus covered in thirty-nine days without a single day of rest. The daily average speed was 19·2 nautical miles (36 kiloms.). On our arrival we had two sledges and eleven dogs safe and sound. Not even for a moment had we helped the dogs to pull the sledges. Our provisions consisted of pemmican, biscuits, milk in powder, and chocolate. Not much of a variation, but a healthy, nutritious food which invigorated the body, just what it needed. The best proof was, that we always felt well, and were never raving about food, which has been so common a feature in all the longer sledge journeys, and is an infallible sign of deficient nourishment.

In the meantime Lieutenant Prestrud and his two companions had succeeded in doing excellent work to the east and in the neighbourhood of the Bay of Whales. They succeeded in reaching King Edward Land—discovered by Scott—and confirm what he had seen. Alexandra

Mountains appeared to be a wholly snow-covered crest—1200 feet high—stretching in a south-easterly direction as far as the eye could see, the northern boundary being two bare peaks—Scott's Nunataks—1700 feet high. This expedition's exploration of Framheim's surroundings is of great interest. It appears from their observations that the Bay of Whales is formed by underlying land still snow-covered.

At the same time as our work inshore was going on, Captain Nilsen with his companions on the *Fram* succeeded in doing work which, from a scientific point of view, probably will turn out to be the most valuable of the expedition. On an 8000 nautical miles' cruise from Buenos Aires to Africa and back, they established a series of oceanographic stations, sixty in all. Twice they circumnavigated the world, voyages full of dangers and toil. The voyage out of the ice in the autumn 1911 was of a very serious character. They were ten men all told. Through darkness and fog, cyclones and hurricanes, pack-ice and icebergs, it was their lot to beat their way out. Last but not least let me mention, that the same ten men, on February 15, 1911, hoisted the Norwegian flag further south than a ship has ever floated before.

A fine record in the century of records:—Farthest north, farthest south.

SOME NOTES ON MY 1912 EXPEDITION TO THE SIACHEN, OR ROSE GLACIER, EASTERN KARAKORAM.

By Mrs. Fanny Bullock Workman,

Officier de l'Instruction Publique, France; F.R.S.G.S.

As reported in 1911 in the geographical journals, Dr. W. Hunter Workman and I, after completing our exploration among the Hushe and Kondus systems of glaciers, crossed to the Saltoro valley and traversed the Bilaphon La, 18,900 feet, to the Siachen or Rose glacier on August 19. We examined it for some distance above the entrance of the Bilaphon branch, and visited two of its most important affluents, but it was far too late in the season to contemplate the survey of a glacier the most noteworthy points of which lie at altitudes above 17,000 feet.

The object of my 1912 expedition was to have as detailed a map as possible made of the whole glacier, and to ascend to, and examine myself, its apparently rather complex sources, and elucidate, if possible, the problem of their relations to regions beyond. Mr. C. Grant Peterkin, who had received the diploma of the Royal Geographical Society, and was kindly recommended to me by Mr. E. A. Reeves, F.R.A.S., agreed to act as surveyor. The Surveyor-General of India, Colonel Burrard, R.E., also kindly loaned a native plane tabler, Sarjan Singh, from the Survey to assist in the work. One Italian Alpine porter accompanied the surveyor's party, and four Italian guides and porters, Dr. Hunter Workman and myself.

A Parsee, T. Byramgi, from Srinagar, who had handled the transport part in 1911, undertook the work this season, and was assisted on the Siachen side by two sepoy reservist orderlies from the Pindi division, Royal Indian Army, who did very good work. Goma, in the Saltoro valley, was the base for supplies.

On reaching the top of the Bilaphon Pass on July 11, the weather being very fine, I decided to attempt an impressive rock and snow peak lying above an elevated snow plateau, north-west, which was likely to command a wide view toward the upper Siachen. We camped on the plateau at 19,000 feet, and the peak was climbed by myself and three guides on the 12th. It proved a difficult mountain, as the last 800 feet lay at a very sharp angle, and each step had to be cut in a surface of black ice.

At the top I found myself separated from the great double-summited K^{10}, 25,415 feet, which heads the Dong Dong glacier, only by the wide snow basin on its north-east side, which we traversed in ascending on that mountain to 20,000 feet last season. Its precipitous walls loomed above us with startling grandeur. North and south the vista of glaciers and peaks, some of which I was able to identify, was grand beyond description. Here I can only say that this ascent, being possible at the beginning of the journey, was of much value in enabling me, as it did, to get a comprehensive idea of the ice region about to be entered. Full light was also thrown upon certain points regarding the Dong Dong and Shorpigang glaciers explored by us in 1911. This peak works out from hypsometric observations available at about 21,000 feet, but heights here mentioned will be subject to change when compared with lower station readings obtained from Skardu.

On reaching the Siachen a sheep and burtsa-camp was set up at our old 1911 base, the Teram Shehr promontory, which abuts on the Siachen at the junction of the large north-east affluent. Sheep were kept here, there being no grass above that point, and burtsa for fuel was collected and sent to higher points. I have named this north-east glacier the Teram Shehr, which is almost the only native name connected with the Siachen of which the oldest inhabitants have a dim remembrance. A flour base-camp, with a baker in charge, was made on moraine a little further up the glacier.

All camps of my own party, except for a few days when we descended the glacier, were at or above 17,000 feet for five weeks, and were mostly either on moraine-covered ice or wholly on snow, for the Rose glacier offers no dainty plots of soil or grass to the tent-nomad visiting its precincts. The mountain flanks on both sides are so precipitous that it is dangerous to pitch tents on them.

Weather did not admit of our reaching the sources of the glacier at once. On the first attempt, which lasted two or three days, we camped far up the glacier at 18,000 feet, but were driven down in a stinging snow-storm early the next day. Two weeks of very fine weather, however, soon set in, and we were able to accomplish in this time what under the usual summer Himalayan climatic conditions might take six weeks of running backward and forward, for in order to visit what is to be seen at the Siachen sources, clear skies are essential.

Contrary to the usual conditions on this glacier, it was found possible to cut out terraces and camp on a mountain flank at 18,500 feet, just below the high snow-fields leading to the north and north-east heads. From here the north saddle or very apex of the Siachen was ascended. Various facts of much interest were observed there, but these are notes of where we went and not a narrative of what we saw. This saddle is at an altitude of 20,900 feet. The north-east col was ascended the next day. It is not so arduous as the north one, and is less high, about 19,300 feet. We discovered from here some very high peaks beyond the east Siachen boundary on the Chinese-Turkestan side. These are about seven miles north of and are higher than Teram Kangri; one is probably well over 25,000 feet, but it is futile to discuss heights of peaks unless one's data can be verified by careful triangulation.

After another climb to a ridge of 19,700 feet for observations and photography, Dr. Hunter Workman and I headed the caravan toward the glacier leading to the west Siachen source. We descended the main stream, crossing to the (orographical) right side. This glacier enters the Rose at about 17,400 feet. Ascending it for several hours we at last found a point free of crevasses where a camp on snow was pitched at 18,700. We passed three nights here, which made a serious impression on the coolies, there being no lakelets nor running streams such as are so abundant on the main glacier. A peerless white peak heads this glacier to the west, of (probably) 24,000 feet; the exact height will be known later. It is connected by a high col with a lower snow peak, and at the base of this lower peak is another col or depression.

The higher saddle was first ascended, height 19,700 feet, and from it the source of a large glacier was seen about 4000 feet below where we stood. The following day the lower depression was examined, and, as I hoped, it proved to be the main outlet from the west Siachen source, the summit of a rather difficult, but feasible pass for a loaded caravan to the glacier above mentioned lying west. Next an ascent was made to a wide snow plateau lying directly under the chief summit of the 24,000 foot peak. This plateau is at over 21,000 feet, and offered very fine views of the Hidden Peak, Gusherbrum, and other giants of the region.

Our stint of wonderful weather was now ended, and a snow-storm drove us down to the Siachen. I saw clearly though, that, after our work on the lower part was finished, a return would have to be made to this icy head and the Rose glacier finally left by the passage here discovered.

After we had descended the glacier the 11th August found us again on the high Siachen awaiting a chance to advance to the pass. Here fog and snow-storms held us prisoners for nine days. At the beginning of this detention I ordered forty coolies to return under two Europeans over the Bilaphon for flour, which was now running short. This brought on a strike which lasted eighteen hours. They refused to go for supplies, saying, if they went, they would not return. Their long visit to the sources had made a deep impression on their minds, and they declared they would not return there. Finally, upon the wazir explaining that, if they crossed the new pass, they would have only two days on snow and then be always on moraine down to villages, they gave in and no further complaint was heard. Flour arrived, and one bitterly cold morning under a leaden sky we started with sixty-six coolies laden with provisious for ten days.

Although it did not appear to be far enough north, we thought the pass might afford a passage to the Baltoro glacier. If not, we should find ourselves probably on the Kondus, which was surveyed in 1911 by our topographer. Either way would be new and interesting; the Kondus more novel because bearing no footprints of illustrious predecessors, and from the fact that our 1911 expedition had not pushed up to its extreme upper portion. We camped a few hundred feet below the summit of the pass in a freezing temperature, for by 3 P.M. it was snowing hard. By 6 P.M. it cleared. After a cool night under canvas, minimum temperature 3° Fahr., we ascended under a deep blue sky to the col. Here a boiling-point reading was taken and numerous photographs. The pass works out, pending later data, at 19,000 feet.

Kind indeed was the weather god, who had allowed us three weeks before under sunlit skies to discover this new passage, and again after dreary cold days of waiting drew back the curtain of mist for a day, permitting my caravan of sixty-six to pass safely to the other side under the fairest of skies. The descent to the glacier is of about 3000 feet. It was smooth going for, say, a quarter of the way down, but after that the surface was seamed from side to side with wide crevasses, some open abysses, others treacherously covered by soft snow. We reached moraine for camping below the source of the new glacier, which is at somewhat over 16,000 feet.

The next day it became evident we were on the upper Kondus, and the worst monster of a glacier to travel on we have met with in the Karakoram. After one short march from its head ice-bands and easy moraine end, and for four days moraine-strewn ice-hillocks, often four to five hundred feet high, must be steadily clambered over. Owing to the danger of stones falling from the perpendicular mountain walls on its sides no camps even to the end of its tongue can be pitched except on the tops of these depressing hillocks.

Sometime I shall have something to say as to whether the Siachen is a complement of the Baltoro and what are the chances of a passage from one to the other, likewise upon the mythical Kondus and Chogolisa saddles.

I am glad to have carried out with fair completeness the task, which has kept me two summers in India, and I think the work will, perhaps, be appreciated by future geographers in the days when the special examination and survey of more restricted unknown areas than vast polar continents becomes the vogue, as it must perforce in years to come. At least not having the pioneer exploring to do, they will be spared the expense, the ennui with irresponsible coolies, and the many obstacles which the Siachen offers as a region for exploration, such as being separated from all base of supplies by two glaciers and a snow-pass like the Bilaphon, and from wood by a distance of at least thirty miles.

Still, I am satisfied, and pay my deep salaams to the weather god, for, without his aid, the survey of so huge a glacier could not have been completed, nor could I, considering the perfidious actions of the only headmen available, and the prodigious pilferings of the Saltoro valley coolies, have had the rare privilege of first standing on two new points of the north-east Karakoram water-parting, of observing their relation to Chinese Turkestan, and of attaining other heights of geographical interest on this the largest and longest Himalayan glacier. Other geographical and glaciological features were studied, some relating to questions already discussed, but not settled, which must be reserved for detailed treatment. Mr. Grant Peterkin and his assistant made a full survey of the Siachen from its sources to the end of the tongue at the Nubra River, including its numerous tributaries.

I would add that one of my Italian porters, through momentary carelessness in not testing the ice with his axe, fell into a deep crevasse, carrying with him the only rope with us at the time. He remained there one and a half hours before ropes and the guides were available to extricate him, and died of shock and the effects of cold the same night. I was walking directly behind him, and, supposing him to be on the watch for crevasses, noticed nothing until I saw him disappear two steps in front of me. Fortunately I held up on the brink of the chasm and called to the caravan ascending a short distance below.

I mention this, because, owing to untrue stories being spread by Skardu coolies, garbled reports were published all over the world by the press. It is a source of satisfaction to me, that no coolies in my party were seriously injured or died on the glacier, for the Siachen is incomparably more difficult to move about on than any of the five great Karakoram glaciers, all of which, except the Baltoro, we know pretty thoroughly. One special danger for loaded coolies is water. In crossing the Siachen opposite Teram Shehr, a distance of about three miles, eleven glacial streams were met with this season. Three are at least ten feet wide, and so deep that it was impossible to ford them after 11 A.M.

THE SOUTH POLE: A REVIEW.[1]

THE expedition of the *Fram*, which Captain Roald Amundsen led to South Polar regions, was not designed as a scientific expedition. It made no pretensions on that score, and that must be borne in mind in estimating its scientific results. "Our object was to reach the Pole—everything else was secondary," writes Captain Amundsen. "On this little détour science would have to look after itself; but, of course, I knew very well that we could not reach the Pole by the route I had determined to take without enriching in a considerable degree several

[1] *The South Pole: An Account of the Norwegian Antarctic Expedition in the Fram, 1910-12.* By Roald Amundsen. Translated from the Norwegian by A. G. Chater. Maps and illustrations. Two vols. London: John Murray, 1912. *Price* 42*s. net.*

branches of science." The reasons which led Captain Amundsen to make this "détour" are a curious reflection on the spirit of the age. Amundsen's whole interests lay in North Polar regions; no doubt he must have heard the call of the North Pole, but he had never answered it, for all his inclination led to solid, useful, scientific exploration, and he is not the man to abandon his ambitions however strong the temptation. The *Fram* had been lent by the Norwegian Government to Captain Amundsen for a five years' drift across the North Polar basin, on the lines of Nansen's famous expedition, for magnetical, meteorological, and oceanographical work. The attainment of the North Pole was merely incidental, and no stress was laid upon it. Then in September 1909 came the news of the discovery of the North Pole. Amundsen's funds were far short of the amount required, and now the stimulus to contribute was gone. Once the Pole was reached, who would pay for a Polar expedition? "Just as rapidly as the message had travelled over the cables I decided on my change of front—to turn to the right-about and face to the south." There, by a piece of record-breaking, so alien to the custom of the man and his work, Amundsen felt he could put his projected Arctic expedition on its feet when otherwise he must fail for lack of funds. It was a strange use to find for the South Pole, but more than justifiable when it was the only means of arousing public support for scientific work. And it was a courageous decision, for failure to get to his southern goal would necessitate the abandonment of the northern explorations; but Amundsen did not think of failure. The possibility of it seems never to enter into the scheme of his plans. And indeed when once this remarkable man and his band of Norwegians had turned towards the Antarctic, any one who knew the nature of the task and the calibre of the men felt that the South Pole was as good as reached. There could be no failure: success was a foregone conclusion.

On December 14, 1911, when the Pole was reached—a date that will be memorable as long as deeds of daring are cherished for their own sakes—Amundsen's feelings were curiously mixed. He writes: "I cannot say—though I know it would sound much more effective—that the object of my life was attained. That would be romancing rather too barefacedly. I had better be honest and admit straight out that I have never known any man to be placed in such a diametrically opposite position to the goal of his desires as I was at that moment. The regions round the North Pole—well, yes, the North Pole itself—had attracted me from childhood, and here I was at the South Pole. Can anything more topsy-turvy be imagined?"

But he knew that the success of his cherished northern expedition was assured, and, though it meant little to this man, he had won imperishable fame. For we are men first and geographers after, in so much as, deprecate as we will a race to the Pole, the feat is one that fills us with admiration, especially when it is accomplished with the success and skill of Captain Amundsen's journey.

Captain Amundsen himself gives an account of his expedition in this issue of the *Magazine*, so that we need not recapitulate any of the details here. But he has made so light of the difficulties encountered, both in his lecture and in his book, that they tend to escape notice. It was a bold move to take an entirely new route to the Pole, but it was fortunate for geographical science. We now know the details of two tracks into the heart of Antarctica, that of Sir Ernest Shackleton and that of Captain Amundsen. The former route, which was to be used by Captain Scott, Amundsen considered out of bounds, and although it was always maintained that Captain Scott's expedition was for scientific exploration, and looked on the attainment of the Pole as quite incidental, Amundsen cabled him to New Zealand of his change of plans and his intended line of attack. For the same reason Edward Land was avoided, though, as later events proved, Captain Scott was unable to land a party there with a second base. The Norwegian base was therefore on the Ross Barrier in 78° 38′ S., 163° W., at the Bay of Whales, where Amundsen found the Barrier was aground, probably on reefs or low islands. From here to the Pole and back the course followed, which was very nearly along the meridian, was 1860 miles, and the time taken ninety-nine days. The success of this marvellous journey was not due to luck, but to the experience, capability, and endurance of the men, their utilisation of ski, in which they were all experts, and their employment of excellent Eskimo dogs, which they not only cared for above all other things, even themselves, but thoroughly understood how to handle. It is not a little curious that on this record sledge journey dogs should have been used, the earliest method of sledging, and that more modern expedients, motor or ponies, should have played no part. The dogs are the burden of this story on almost every page: "dogs first and dogs all the time" was the watchword of the expedition. And they served their masters well.

While one party of five men made the southern journey, another of three men went east to explore Edward Land. Both brought back important geographical results which earned for Amundsen the medal of this Society. These results we noted in a previous article (see this *Magazine*, xxviii. pp. 204-208), but the present volumes amplify the information then available.

The Ross Barrier, at least at the Bay of Whales, Amundsen believes is aground. That bight Amundsen believes to be the same, but for slight changes in outline, as was observed by Sir James Clark Ross. Sir Ernest Shackleton came to a different conclusion, but his visit to the Bay of Whales in 1908 was a hurried one. When he saw the ice in the bay breaking up, he abandoned the idea of wintering there. "Otherwise," says Amundsen, "the problem of the South Pole would probably have been solved long before December 1911. With his keen sight and sound judgment, it would not have taken him long to determine that the inner part of the bay does not consist of floating barrier, but that the barrier there rests upon a good solid foundation, probably in the form of small islands, skerries or shoals, and from this point he and his able companions would have disposed of the South Polar question once for all." Off the south-western end of the bay are a number of stranded ice-islands, between which Lieutenant Prestrud got depths of as much as 200 fathoms. The shallowest soundings are not recorded. This is about

200 miles from Edward Land, and from the irregularities in the ice-surface south of the Bay of Whales, it seems probable that the barrier on its eastern side has overridden low-lying land, or else is resting on a shelf not far below sea-level.

The mountains to the south of the Ross Sea, which Shackleton named the Commonwealth Range, are continued in Prince Olav's Mountains, to the south-west of which Amundsen ascended to the plateau by the Axel Heiberg glacier (see map at end). Amundsen made the most important discovery that the main line of peaks from there is continued not towards Graham Land, but towards Coats Land, in the range named after Queen Maud, while in about 86° S., 160° W., another range strikes away to the north-east: this bears no name on Amundsen's map, but in the text it is called Carmen Land. It was definitely established from 86° to 84° S., and on the return journey, when the party was in 81° 20′ S. they saw, to the south-east, high bare land running north-east and south-west with "two lofty white summits to the south-east, probably in about 82° S. Although what we have seen apparently justifies us in concluding that Carmen Land extends from 86° S. to this position—about 81° 30′ S.—and possibly further to the north-east, I have not ventured to lay it down thus on the map. I have contented myself with giving the name of Carmen Land to the land between 86° and 84° and have called the rest 'Appearance of Land.'" Such caution is admirable, but it will likely enough result in some explorer in the future taking unto himself the credit for the discovery of this land. "Appearance of Land" is only marked between 81° and 82° S.: Amundsen's chart is left blank between 82° and 84° S.

A party under Lieutenant Prestrud was sent to examine Edward Land since the *Terra Nova* had failed to land a party there the previous summer. The Alexandra Mountains of Scott were seen, snow-clad from end to end, and Scott's Nunatak (1700 feet) was climbed. Here were obtained the first rock specimens known from Edward Land, and they prove to be of the highest importance. They consist of granitic rocks and crystalline schists, and are identical with those brought by the southern party from Mount Betty beside Axel Heiberg glacier in 85° S. Moreover, they agree so closely with the rocks of South Victoria Land that we can now say that an identity of structure has been established all round the Ross Barrier. Edward Land undoubtedly seems to belong to the plateau formation of Victoria Land, and the presumption grows in strength that the Ross Sea is a rift valley. This exploration of Edward Land was, of course, not exhaustive, and we hope that Captain Scott will extend his earlier discoveries there, but it may be said now, with fair certainty, that the folded ranges of Graham Land do not reach the Ross Sea, and that the Ross Sea is a great bight and not—if indeed any one ever seriously believed so—the end of a transcontinental strait.

In addition to these geographical discoveries the Norwegian South Polar Expedition obtained an important series of hourly meteorological observations at their winter quarters from April 1911 to January 1912. These will enhance the value of Captain Scott's more extended observations to the west. The meteorological equipment was not as complete as it might have been, but every possible care was taken over the observations, and several of the party were practised observers.

The only other scientific work consisted of an oceanographical cruise of the *Fram* in South Atlantic. During July and August 1911, Captain Nilsen took temperatures and salinities at sixty stations between 15° and 35° S. from South America to Africa and back. At nearly all the stations the observations were taken at twelve depths down to 545 fathoms. They are discussed in an appendix to the book. Valuable as they are, we feel it somewhat disappointing that a ship like the *Fram* did not do this work in higher southern latitudes in the South Atlantic where the work is even more required, and where the ordinarily con-

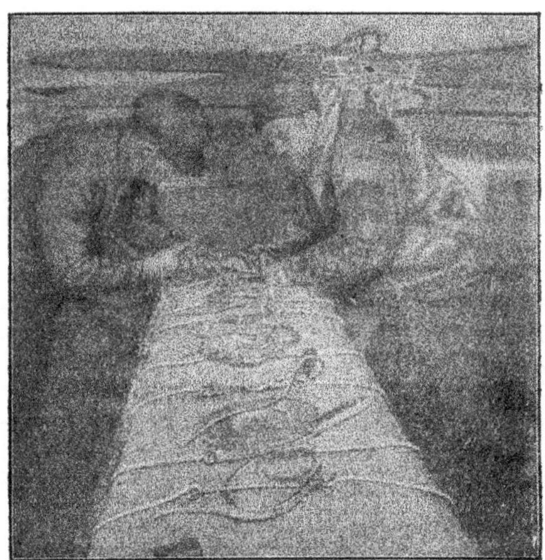

Packing sledges in the snow-house

structed ship could not work with the same safety and success as the *Fram*.

Other appendices deal with the construction of the *Fram*, by Commodore Christian Blom, and the astronomical observations taken at the Pole.

Some criticism has been levelled at Captain Amundsen for competing with Captain Scott for the Pole. It would serve no purpose to discuss that matter in these pages, and we consider that criticism most unfounded, and certain to be little to the liking of Captain Scott himself. Writing of the British expedition, Captain Amundsen says, "Our preparations were entirely different, and I doubt whether Captain Scott, with his great knowledge of Antarctic exploration, would have departed in any point from the experience he had gained and altered his equipment in accordance with that which I found it best to employ. For I came far short of Scott both in experience and means." There is one deed in the relationships of these two expeditions which is not recorded in these volumes, and speaks much for the goodwill of Captain Amundsen towards the British expedition. When the *Fram* was lying in the Bay of Whales in February 1911, unloading stores for winter quarters, the *Terra Nova* arrived from the east after failing to effect a landing on the difficult shores of Edward Land. Captain Amundsen suggested to Lieutenant Pennell, the commander of the *Terra Nova*, that Lieutenant Campbell and his party should land on the Barrier, and winter near the Norwegians, so that the British party could carry out their original plan of exploring Edward Land. In this case Captain Amundsen would not have sent a party eastward, but would have left Edward Land entirely to Lieutenant Campbell and his men.

To Sir Ernest Shackleton Captain Amundsen pays many tributes. "The name has a brisk sound. At its mere mention we see before us a man of indomitable will and boundless courage. He has shown us what the will and energy of a single man can perform." "Shackleton's exploit is the most brilliant incident in the history of Antarctic exploration." When they reached 88° 23′ S., Shackleton's farthest, we read, "We did not pass that spot without according our highest tribute of admiration to the man, who—together with his gallant companions—had planted his country's flag so infinitely nearer to the goal than any of his precursors."

Captain Amundsen writes well, in a straightforward style devoid of any affectations. The chapter on the winter life at *Framheim* is a fine piece of vivid writing (see illustration). We have seldom read a book of travel in which the personal note is less sounded; every one gets his due share of praise and notice, so that we make the acquaintance of all the expedition as we read the pages, and when we close the book we feel the richer for having met a band of remarkable men. The book will become the classic of Antarctic travel.

There is a brief preface by Dr. Fridtjof Nansen, and several chapters by Captain Th. Nilsen on the voyage of the *Fram*. The index is not perfect, and the maps are too meagre for so important a work. The translator's work is well done. The book is well illustrated and we are indebted to the publisher for the three illustrations in this issue and for the map.

R. N. Rudmose Brown.

A GEOGRAPHICAL DESCRIPTION OF EAST LOTHIAN.

By Charles M. Ewing.

(*Continued from* vol. xxviii. p. 641.)

VIII.—General Industries.[1]

East Lothian cannot now be termed an industrial county. A century or two ago there were quite a number of flourishing industries, but most of these, owing mainly to the increasing competition of other countries or other parts of Britain more favourably situated, have either ceased or are seriously on the decline.

One of the most ancient industries was salt manufacture; as early as the twelfth century it is recorded that the monks of Newbattle established ten pans at Prestonpans. The neighbouring town of Cockenzie, too, has from early times produced large quantities of salt from brine. Since the repeal of the salt duty in 1826, this industry has shrunk to insignificance. Numerous attempts were made in the late seventeenth, eighteenth, and nineteenth centuries to establish cloth, textile, and other allied industries. Thus the Scottish Cloth Manufacturing Company commenced work on a large scale at Haddington in 1681, but wound up in 1687; woollen mills were established in 1730 at Haddington and Dunbar; at Ormiston in 1730 Cockburn set up what seems to have been the first bleaching-fields in Scotland, fine linen having been previously sent over to Haarlem for this purpose; cotton and linen factories flourished at Dunbar between 1801 and 1821, as many as 400 to 500 hands being employed; while stocking weaving and dyeing, for which madder was grown at Aberlady, were formerly largely carried on at Haddington. Among other industries that have disappeared are distilling at East Linton, where prior to 1840 as many as 500,000 gallons of whisky were sometimes produced annually; the manufacture of roofing tiles, bricks, and drain-pipes at Dunbar, and of vitriol and sal-ammoniac at Prestonpans.

Now the chief industries at Haddington are the manufacture of Bermaline flour, of agricultural implements, of woollens and sacking, with brewing and tanning; while at Prestonpans there are soap-works, breweries, factories for the preparation of bone-dust and feeding stuffs, and extensive pottery, fireclay, and brick works. Agricultural implements and other farming accessories are turned out at Dunbar.

IX.—Fisheries and Shipping.[2]

About a mile from the Midlothian border is the harbour of Morrison's Haven. It seems to have been known as a harbour long before Leith, and is said to have been first used by the monks of Newbattle for the

---

1 Chief sources are the *Statistical Accounts*, Martine, Croall.

2 Chief authorities: *Fisheries Report; North Sea Pilot.*

export of salt and coal. It was not, however, till after 1526, when the harbour seems to have been reconstructed on a larger scale, that Morrison's Haven became an important commercial port; in fact we learn that, at this time, the trade of Morrison's Haven and Port Seton together rivalled that of Leith. The present harbour has an area of four acres, a depth of 13 to 14 feet at high water, an entrance 70 feet wide, and is connected by a goods line with the North British Railway. It is now the only port in the county where foreign trading vessels call with any regularity. Its chief exports are coal, earthenware, and pottery goods, bricks and tiles.

Prestonpans has no harbour, but possesses four fishing-boats, totalling 101 tons, and has 20 resident fishermen.

Further east are the two most prosperous fishing-ports in the county—Cockenzie and Port Seton. The fishermen here are very industrious and enterprising. Many of them follow the herring round the coasts of Britain, and are usually very successful. In 1910 the Fisherrow and Cockenzie fishermen conjointly employed 29 boats in the English, and 9 in the Irish fishing, and their total catch—£10,500 in value—amounted to 9635 crans.[1] Port Seton harbour is hollowed out of rocks of the Coal-Measures, and much of its masonry reposes on hard rocks of Millstone Grit. That of Cockenzie is well protected behind an east-and-west dolerite dyke. It has an area of three acres, an entrance 78 feet wide, and a depth of 12 to 13 feet at high-water, but its approach is rendered dangerous by the sunken Corsik Rock, about 250 feet offshore.[2] The oyster-beds of Cockenzie—now exhausted—were once celebrated, 20,000 oysters being sometimes obtained annually in the eighteenth century.[3] In 1910 the total number of fishing-boats at Port Seton and Cockenzie was 133 (76 with keels of upwards of 45 feet); 606 men and boys were employed in the fishing industry, and 12,856 cwts. of fish, realising £9881, were taken—mainly in the Firth of Forth and off the Isle of May. Aberlady was once the port of the county town, and, sixty to seventy years ago, considerable quantities of bark, linseed oil, and guano were here imported.[4] The channel of the Peffer stream, which joins the south side of the bay, admits of vessels drawing not more than 9 feet of water, but it is now seldom used.

There is a small harbour at North Berwick, one acre in area, with an entrance 25 feet wide and a depth at high tide from 10 to 11 feet.[5] Opening as it does to the west-south-west it is extremely difficult of access in a north-east gale. In 1910 there were six fishing-boats with a total tonnage of 11, and 15 fishermen and boys; 526 cwts. of fish (more than half shell-fish) were taken.

For long Belhaven was the port of Dunbar. In the seventeenth century Dunbar was the largest fishing centre in Scotland, and its importance was greatly enhanced when a harbour—now called the Old Harbour—towards the construction of which Cromwell gave £300, was erected at Dunbar itself shortly before the Restoration. It has an entrance of

[1] *Fisheries Report.*
[2] *North Sea Pilot.*
[3] *Statistical Account* (1791-99).
[4] *Abridged Statistical Account* (1845).
[5] *North Sea Pilot.*

40 feet, an enclosed space of 2½ acres, and a depth of 10 feet at high tide, but a sunken rock outside, called the Outer Buss, makes approach perilous in rough weather.[1] From 1752 to 1804 Dunbar was the headquarters of the prosperous East Lothian and Merse Whale Fishing Company. During the eighteenth century it carried on a considerable export trade, the chief articles shipped being corn, whisky, fish, wool, and hides. In 1792 it had 16 ships with a total tonnage of 1505, besides 2 Greenland vessels of 675 tons. In the first twenty years of the nineteenth century the fishing industry was especially flourishing, and we learn that in 1819 no fewer than 200 fishing-boats belonged to the port, supplying 30 to 60 crans of herrings per day in the season, while 35,000 barrels of herrings were cured.[2] A few years later a decline in trade set in, for while in 1830 as many as six Dunbar vessels were engaged in the timber and grain trade with the Baltic, and thirty in the coasting trade, there were only 11 ships altogether in 1851, with a total tonnage of 658. In 1862 the Victoria Harbour, for which an entrance had to be cut through the solid whinstone on which the castle stands, was constructed farther to the west. It has an entrance to the west 50 feet wide, a depth at high water of 12 feet,[3] and is four acres in extent. During fierce onshore gales the heavy swell within the harbour causes serious difficulty. Its construction, however, did not materially arrest the decline of Dunbar's general trade, and now it is little more than a fishing-port. In 1910 there were 29 fishing-boats; the fishing industry gave employment to 125 men and boys; and 7144 cwt. of fish—of a value of £2566 (£1318 representing shell-fish)—were taken.

X.—Mining.[4]

After agriculture, coal-mining is the chief industry of the county. Some of the mines in the Prestonpans-Tranent area were worked at least as early as the twelfth century, and are said to be the oldest in Scotland. Coal was mined at Penston in Gladsmuir parish in the fourteenth century, and at the time of the Commonwealth produced a revenue of £400.

The coalfield of East Lothian is some 30 miles in area. It is confined to a region bounded by a line drawn from Port Seton through Gladsmuir to the confluence of the Birns Water and Kinchie Burn on the east, by the latter stream on the south, by the county boundary on the west, and by the coast on the north. It occupies the eastern and smaller of two synclines separated by the anticlinal Roman Camp upland, the western forming the Midlothian coalfield. The gentleness of the dip of the coal-bearing strata and their insignificant depth in parts made them workable even in very early times, and account for the fact that in places the best seams are exhausted. The only coal-bearing formations of consequence in the county are those belonging to the Edge Coal Group, the middle member of the Carboniferous Limestone series. Nine different seams yield coal in quantities sufficient to be profitable; of

[1] *North Sea Pilot.* [2] *Abridged Statistical Account* (1845). [3] *North Sea Pilot.*
[4] *Report of Mines and Quarries; Geological Memoir* (Economics).

these the Great Seam (maximum thickness at Prestonpans, 7 feet) is the chief. According to the estimate of Hull, who had not fully realised the submarine possibilities of the coalfield, there were in 1880 still 86,849,000 tons of coal to be mined in East Lothian, above the 4000-feet limit. The following is a list of mines at present worked within the county:[1]

Name of Company.	Name of locality of Mine.	Persons Employed.		Kind of Coal and other Minerals.
		Under-ground.	Above-ground.	
Bankpark	Tranent	100	15	Household and steam.
Edinburgh Coll.	Bankton	255	34	Household, manufacturing, steam.
	Elphinstone	283	46	Household, manufacturing, steam.
	Northfield	...	...	...
	Preston Links	350	40	Household, manufacturing, steam.
	Tranent	240	33	Household, manufacturing, steam; fire-clay.
Ormiston	Meadow	60	14	Household, manufacturing, steam.
	Ormiston Station	158	33	Household, manufacturing, steam; fire-clay.
Riggonhead	Penston	195	45	Household, manufacturing, steam.
,,	Riggonhead	71	14	Household, manufacturing, steam.
Summerlee Iron	Prestongrange	759	152	Household, manufacturing, steam; fire-clay.
White and Co.	Tyneside	38	6	Household, manufacturing, steam.
Woodhall	Pencaitland	130	37	Household, manufacturing, steam.
	Total	2639	469	

Total output for county (1909), 1,048,499 tons.

One of the most prosperous mines in the county, the Preston Links Colliery, is partially submarine; it has, in fact, recently taken a new lease of life since the dolerite dyke, running from Port Seton to Cockenzie Harbour, and cutting right through the Great Seam, was pierced, and coal began to be worked to seaward of the dyke.

The future prosperity of East Lothian coal-mining will probably largely depend on the productivity of these beds extending beneath the Firth; and if the Coal-Measures which geologists believe to occur under the sea off the East Lothian coast can also be reached, it may be long before mining in Haddingtonshire can be called a declining industry.[2]

The only other materials at present worked in the county are blaes,

[1] From *Report of Mines and Quarries.*

[2] *Geological Memoir.*

a kind of shale, and fireclay. In 1909, 20,464 tons of the former, and 14,100 of the latter were obtained. The fireclay which is found associated with coal seams has given rise to the considerable manufacture of bricks, tiles, and pottery at Prestonpans.

At the old Dolphingstone colliery and at Penston, blackband ironstone (10 and 14 inches thick respectively) was worked up to 1880, and at one time as much as 100 tons a day was brought to the surface. Haematite was discovered on the Garleton Hills in 1866, but was mined for only a few years.[1] Ironstone of good quality exists on the farm of Garvald Grange; it is, however, doubtful whether the extension of the Gifford Railway would, as one writer opines, make the working of it practicable. Copper has been found in various parts of the Lammermuirs, but never in sufficient quantities to be worked.

XL—Quarries.

Thirty quarries are at present worked in East Lothian, and 100 people employed. In 1909 48,862 tons of stone were quarried, of which 18,611 were limestone, 3667 sandstone, and 26,584 igneous rock.[2]

The building stone used in the county is mainly sandstone or of igneous origin.[3] Much of Dunbar is built of a poor red sandstone obtained from the neighbouring Old Red Sandstone quarries. A common stone used for building purposes in East Linton district is porphyritic basalt, while much of Haddington burgh is built of porphyritic trachyte, obtained from Peppercraig quarry, on the southern slope of the Garleton Hills. The chief building material in North Berwick is the reddish phonolitetrachyte of North Berwick Law, which is somewhat decomposed and easily worked. Perhaps the best building stone in the county is a light-coloured, easily worked, yet durable sandstone belonging to the Calciferous Sandstone Series, and occurring in the south-east corner of the county in the neighbourhood of Dunglass.

For road metal the more basic igneous rocks are most suitable. The chief source of supply at present is the trachytic phonolite quarry of Traprain Law. This stone, however, is not ideal, as it is difficult to work, yet brittle, and offers a somewhat rough surface in summer. Better road-metal is obtained at the Kidlaw Quarry (three miles south-west of Gifford), at Millstone Neuk, near Dunbar, and at Gosford Bay, near Aberlady. The two latter are not, however, accessible at high tide. The only sedimentary rock used for road-metal in the county is the Silurian greywacke—a soft stone suitable only for by-roads.

Limestone is worked within the county for agricultural, smelting, and building purposes. The Skateraw Quarry, near Dunbar, was formerly extensively worked, as much as 4000 tons being in some years shipped to Devonshire; now 30 to 40 tons per day are obtained here, of which two-thirds are sent to the west of Scotland for smelting, and one-third is

[1] *Geological Memoir.* [2] *Report of Mines and Quarries.*

[3] *Geological Memoir* (Economics) is chief authority for rest of this section.

used locally for agricultural purposes. The stone of the Harelaw Quarry, near Longniddry, contains 96 per cent. of carbonate of lime, a fact which renders it useful alike in farming, in building and plastering, and in gas purifying.

XII.—Communications.

Just as the hills and uplands of East Lothian run in an east-and-west direction, so the main routes have developed along east-and-west lines. In most parts of the county north- and south-running roads have been more difficult to construct, and have had more difficult gradients to surmount. An aeroplane setting out for Berwick-on-Tweed from Edinburgh would proceed straight across the Lammermuirs, south-east of Gifford, but so difficult has it been found to cross these hills that practically all traffic bound for Berwick has been diverted to the west by way of Lauderdale, or to the east by way of the much more important route through Dunbar and Ayton.

[1] In early times there was but one fairly easy route from England to the Scottish capital, viz. that by Dunbar. To reach it from Carlisle the lofty southern uplands had to be traversed. From Norham, Coldstream, or places higher up the Tweed, the direct way was rendered difficult by the Moorfoots and Lammermuirs. Two routes across the latter, now rough and little frequented roads, must indeed have at one time been considerably used. One traverses the hills from the upper Whitadder valley to Garvald, and its former importance is shown by the fact that its northern exit from the hills was defended by two castles —Castle Moffat and White Castle. The other, farther west, also defended by forts called the Black and the Green Castles, and almost certainly the route taken by the Highlanders under Mackintosh of Borlum in 1715, crosses the hills from Longformacus to Gifford. But the most frequented route was that followed by the present Great North Road via Ayton and Dunbar; by this way Edward I. reached the Lothians in 1296 and 1298, Somerset before the battle of Pinkie in 1547, and Cromwell in 1650. Yet even this way was not without its difficulties; slightly south of Cockburnspath is the narrow gorge by which it cuts its way through the Eastern Lammermuirs; of it Cromwell said that "ten men to hinder were better than forty to make their way." Moreover, just to the south of Dunbar, invading armies could easily be brought to a standstill; here in 1296 Edward I. was opposed by the Scots, and on the same ground Leslie threatened Cromwell's retreat to Berwick in 1650.

West of Dunbar the Great North Road proceeds to East Linton, and thereafter, though a much easier route lies to the north, is carried over somewhat hilly country to reach the county town. From Haddington there is a long and tedious ascent to Gladsmuir (355 ft.), and only $1\frac{1}{2}$ miles west of Tranent does the road leave the upland. Now that traffic is

---

[1] Chief authority in this paragraph: Kermack, *A Geographical Factor in Scottish Independence* (*S.G.M.*, Jan. 1912).

returning to the roads, it is rather unfortunate that the only direct road of the first class between Dunbar and the capital should require to surmount these uplands; as the trade of Haddington itself is now of minor importance, a direct road from East Linton to Drem should be constructed, and the road thence to the capital improved.

The other roads of East Lothian are mainly used by local traffic. A good road skirts the base of the Lammermuirs, and connects Dunbar with Garvald, Gifford, Humbie, and Dalkeith in Midlothian. Another proceeds from Dunbar to North Berwick; formerly it passed through East Linton, but since 1750, when the new bridge over the Tyne at Tyninghame was constructed, its course has been farther to the east; from North Berwick it continues westward along the coast, and as it connects the rising watering-places along its course with the capital it has recently become one of the busiest roads of the county. The north-and-south roads are mainly connecting-links between the more important villages on the east-and-west high roads. There is, however, one exception. The burgh of Haddington, owing doubtless to the former importance of its local traffic, is a distinct centre on which no fewer than eight roads from neighbouring towns and villages converge. Of these perhaps the most interesting is the old, yet well-built road which was carried over the steep slope of the Garleton Hills (road summit 451 feet) to Aberlady, the former port of Haddington.

All the railways in East Lothian belong to the North British Railway Company. There is one mile of railway to every five square miles.

The chief railway of the district, opened in 1846, is a section of the East Coast route from Edinburgh to London. While it avoids the uplands traversed by the great high road, it unfortunately loses the considerable traffic of the populous coastal strip in the west by running quite a mile inland as far as Longniddry. Here two branch lines diverge, one to the north to the watering-places of Aberlady and Gullane, the other to the south-east, over a considerable incline to the county town. At Drem, 4½ miles farther on, a branch line proceeds to the north to North Berwick. A considerable saving in railway construction would have been effected had it been possible to continue the Gullane line direct to North Berwick, but long before a branch to the former was contemplated, the Drem-North Berwick Railway had been constructed. From Longniddry to Dunbar the main line follows the most level part of the county. Beyond Dunbar it runs through the fertile coastal strip, and just north of the Berwickshire boundary it ascends to 150 feet, its highest point throughout the county. At the time of the construction of the main line the people of Haddington were greatly opposed to its passing through this burgh; this, it is alleged, was the reason why it took a more northerly course, but it seems probable that the present course of the railway was decided upon by the much easier gradients and by the greater accessibility to the rising coastal resorts.

Just before Inveresk station in Midlothian, a branch railway leaves the main line, enters the county near Crossgatehall, and proceeds via Ormiston to Macmerry. It was constructed primarily as a mineral railway and must have paid better when iron was worked at Macmerry.

Four years ago a branch line was opened from Ormiston to Gifford, the main purpose of which was to provide a more direct outlet for the agricultural produce of south-west East Lothian. Though called the Gifford and Garvald Light Railway, it has yet to be extended to the latter place, it is doubtful whether the extension of the line to this thinly populated neighbourhood would prove a financial success. Between Ormiston and Gifford the country is irregular and hilly, and just beyond Humbie the line reaches an elevation of 521 feet.

From Joppa, three miles to the east of Edinburgh and connected with the capital by cable tramway, an electric tram-line, completed in 1909, runs through Musselburgh to Levenhall, just beyond which it reaches the Haddingtonshire border. Thereafter it skirts the shore of the Firth of Forth, and after passing through the thickly populated places called Morrison's Haven, Cuthill, Prestonpans, and Cockenzie, terminates at the fishing-village of Port Seton, ten miles distant from Edinburgh. As Morrison's Haven and Port Seton are two miles, and the other places mentioned at least one mile distant from the nearest station, this tramway has had the effect of diverting much of the passenger traffic from the railway, even travellers bound for Edinburgh making use of the tramway as far as Musselburgh station, which is connected by a cheap and frequent service of trains with the capital.

XIII.—Distribution and Statistics of Population.[1]

At present there are four areas in which the population of East Lothian has chiefly concentrated—(1) the Haddington district; (2) the Prestonpans, Tranent, Port Seton, Cockenzie, Ormiston and Pencaitland district; (3) the North Berwick, Gullane, and Aberlady coastal strip; (4) the Dunbar and East Linton district. The choice of Haddington itself by the monks who founded the abbey in 1178 can easily be understood. It lies in a fertile hollow, and on the banks of the Tyne, a river that supplied abundance of water for domestic and agricultural purposes as well as for the working of mills. It is sheltered from the north and east, and is well concealed on the south and east, a fact of considerable importance in times of English invasion. Though it has declined much in relative importance it is remarkable how little the county town has varied in size during the last three centuries. Its population is chiefly held together nowadays by the fact that it is the county town, a considerable agricultural centre, and has a number of industries. The dense population of district (2) is to be explained in several ways. Prestonpans owed its rise to the establishment of salt pans and coal mines in the twelfth century, and on the coal and other industries its inhabitants are still dependent. There are no natural advantages in the high, inhospitable site of Tranent, where until the recent water-supply from Pathhead, on the northwest slopes of the Lammermuirs, was installed, there had always been a scarcity of water; it is, as it seems always to have been, chiefly a coal town.

---

[1] Statistics derived from *Preliminary Census Report* (1911); *Detailed Census Report for Scotland*, 1901.

Cockenzie was primarily a place of salt manufacture, Port Seton a port for general shipping; their recent growth, however, has been due to the fishing industry and to their popularity as summer resorts. The Ormiston and Pencaitland populations are partly agricultural, partly mining.

A Cistercian Abbey founded by David I. seems to have formed the original nucleus of North Berwick, and the choice of its site was largely influenced by the presence of the natural stronghold, North Berwick Law, immediately to the south. In the last century North Berwick has entirely changed its character; thanks to its excellent links and bracing air it has become one of the most fashionable resorts in Scotland. Gullane, too, which two centuries ago was an abode of horse-rearers and rather extravagantly termed the "Newmarket of Scotland" by writers of the time, is now simply a golfing resort.

Dunbar grew up around its almost impregnable castle, once the most important fortress between Berwick and Edinburgh, a stout barrier to invaders from the south, and in fact sometimes termed the "key to the Lothians." Protected by this stronghold a considerable fishing and industrial population took up its residence here. It is now a fishing and agricultural centre, and attracts a considerable number of visitors in summer. East Linton, situated on the Tyne where it is crossed by the high road, grew up as a bridge-town. Its population was until half a century ago maintained by local industries, particularly distilling.

By far the least populous part of the county is the Lammermuir hill area, where human habitations are almost entirely confined to the more important valleys.

The population of East Lothian at the 1911 census amounted to 43,253, of which 21,446 were males and 21,807 females. This represents a density of 162 persons per square mile. There was an actual increase of 4582 in the last decade, or an increase per cent. of 11·9. In only five Scottish counties was the increase per cent. within this period greater, that of Dumbartonshire and also of Fifeshire amounting to as much as 22·3 per cent.

Decennial Increase or Decrease of Population in East Lothian.

1801 1811	1811 1821	1821 1831	1831 1841	1841 1851	1851 1861	1861 1871	1871 1881	1881 1891	1891 1901	1901 1911
1064	4077	1018	−259	500	1248	137	731	−1017	1180	4588
3·5%	13·1%	2·9%	0·7%	1·4%	3·4%	·4%	1·9%	2·6%	3·1%	11·9%

Only two of the past decades show a decrease in population, viz. between 1831 and 1841, and between 1881 and 1891, in the latter case to the extent of more than a thousand. After so recent a decrease it is satisfactory to find that the increase per cent. of the past ten years has only once been exceeded, viz. between 1811 and 1821, since the census began to be taken. Considerable light is thrown on this marked increase by the statistics for the various parishes :—

Name of Parish.	Population in 1911.	Increase or Decrease per cent. in last decade.
Aberlady	963	1·4
Athelstaneford	666	4·9
Bolton	256	-14·1
Dirleton	2064	14
Dunbar	4830	4
Garvald *	511	18·6
Gladsmuir	1433	-3·2
Haddington	5424	5·8
Humbie	679	5·7
Innerwick	676	-13·6
Norham	199	-1·0
North Berwick	3969	8·8
Oldhamstocks	404	-5·8
Ormiston	1598	34·3
Pencaitland	1273	14·5
Prestonkirk	1631	-2·5
Prestonpans	4722	39·6
Salton	386	-10·4
Spott	386	-8·7
Stenton	461	-9·8
Tranent	8677	41·9
Whitekirk and Tyninghame	776	-7·1
Whittinghame	523	·4
Yester	746	2·2

* Navvies at work here in 1901.

The four parishes which show the greatest increase are Tranent, Prestonpans, Ormiston, and Pencaitland, and in all these coal-mining is an important industry, though in Tranent the growing population dependent on summer visitors and the successful fishing industry, in Prestonpans on general industries, and in Ormiston on market-gardening, have all, doubtless, helped to swell the numbers. Notable increases are also recorded in the parishes of North Berwick and Dirleton, due certainly to the large increase of persons catering for summer visitors in North Berwick and Gullane. The only purely agricultural parish that has materially grown in the last decade is Athelstaneford. The most unsatisfactory results are furnished by the eastern group of parishes —Whitekirk, Innerwick, Stenton, Spott, Oldhamstocks, and Dunbar— every one of which shows a marked falling off. They, like Garvald, Bolton, and Salton, are, however, only sharing in that tendency to depletion manifested by so many agricultural parishes in Scotland.

The preponderance of males over females is greatest in the four mining parishes of Ormiston (53·9 per cent.), Pencaitland (52·9 per cent.), Tranent (52·8 per cent.), Prestonpans (52·5 per cent.), and in Salton (57·7 per cent).

There is little difference between the number of males and females in the agricultural parishes. The preponderance of females over males is most marked in the parishes of North Berwick (55·7 per cent.), Dunbar (53·4 per cent.), Dirleton (52 per cent.), Whitekirk (53·5 per cent.), and Haddington (51·8 per cent.), and in the case of the first three may be mainly attributed to the large amount of female labour required in the summer resorts.

The population of the following burghs within the county has increased in the last decade: Tranent, the largest town in the county (4369 inhabitants), by 58 per cent., Cockenzie and Port Seton (2400 inhabitants together) by 42·3 per cent., North Berwick (3247 inhabitants) by 11·9 per cent., Prestonpans (1923 inhabitants) by 11·7 per cent., and Haddington (4140 inhabitants) by 3·7 per cent. East Linton (population 877), on the other hand, and Dunbar (population 3346) have fallen off by 3·8 per cent. and 6·6 per cent. respectively. The decrease in the case of East Linton was quite to be expected, for while its industries have almost disappeared, it is too close to Dunbar on the one hand, and Haddington on the other, to hold its own as an agricultural centre. But the case of Dunbar is not so easily explained. Not only is it an important station on the main line, and is connected by a fast and frequent service of trains with Edinburgh and Berwick, the North of England and London itself, but it is a considerable summer resort. Perhaps its falling off is chiefly accounted for by the decline of its fishing industry and by the rivalry of the other rather more popular watering-places of Haddingtonshire.

The following table shows the percentage of the population of ten years and upwards engaged in each class of occupation at the 1901 census:—

	Total engaged in occupations.	Unoccupied or at present without employment.	Professional.	Commercial.	Domestic.	Mining.	Fishing.	Agriculture.	Other industries.	Miscellaneous occupations (mainly represented in columns described as "Conveyance of men, food, and messages," and "sale of food, tobacco, drink; lodging."
Males, . .	82·5	17·5	3·9	1·5	3·9	10·5	3·8	25·7	17·2	16·3
Females, .	31·5	68·5	1·7	·5	13·2	...	...	9·7	3·6	2·6

These figures show clearly that East Lothian is primarily an agricultural county, no fewer than 5276 persons being engaged in this pursuit. But while this figure has for some time back remained comparatively stationary, the number of persons engaged in mines and quarries has since 1901 almost doubled itself, as we see by comparing the number of persons thus employed in 1901, viz. 1542, with that given in the *Blue Book of Mines and Quarries* for 1910, viz. 3107. The percentage of males without occupation in the county is unusually high, higher even than that of Midlothian itself, which is the abode of retired people from all over Scotland. North Berwick, Dunbar, Gullane (as well as the county town) are, indeed, inhabited by considerable numbers of the "affluent-unoccupied" for the greater part of the year, attracted hither, no doubt, by the excellent golf and bracing sea air.

XIV.—Conclusion.

In very ancient times mountain-making movements folded and hardened the rocks of the Lammermuir area. Later sand and limestone formations were laid down to north-west and north-east of the great fault, and being much less affected by hardening agents yielded readily to erosion and form much of the lowlands to-day. At the same time volcanic action played an important part and left numerous minor hills, islands, and lava-beds as its memorials. Moreover, as the Forth wore its bed deeper and the land here subsided, access was given to the sea all along the north and north-west of the county. Late in physical history came the Great Ice Age, under whose influence the configuration and more particularly the drainage system was profoundly modified, and much of the county became deeply covered with glacial deposits. Thus in brief were upland, lowland, and the present coastline determined.

On relief and proximity to the sea the climate of the county depends. Being in the rain-shadow of the Lammermuirs, Haddingtonshire has a low rainfall, and being so much hemmed in by the sea it has a small range of temperature, but it is subjected to cold east winds from the North Sea.

The occupations of the inhabitants are mainly conditioned by the structure and relief of the county or by climate, and sometimes by both. Thus the Lammermuirs, too exposed for cultivation, are essentially a pastoral district. The lowlands are on the whole well suited for agriculture by reason of the favourable climate, and of the generally fertile boulder-clay and soils of igneous origin. The lengthy seaboard has encouraged the growth of a vigorous fishing class, whose efforts are more often favoured by a breezy climate than hindered by storms.

Sheep-rearing has given rise to a small woollen industry. Agriculture has called into being the manufacture of agricultural implements and drainage tiles, brewing and flour-milling. Owing to the luxuriant plant-life in Carboniferous times and the preservative qualities of the sediments deposited on it, there is a good supply of fossil-fuel in the west, and coal-mining has arisen, while a number of industries dependent on coal have grown up.

The chief factors affecting the growth of what may be termed the summer-resort industry are the invigorating coastal climate, the abundance of blown sand and raised beaches (for golf courses), and a varied and picturesque seaboard.

The prosperity of agriculture and market-gardening, of fishing and summer resorts, has been increased by the proximity of East Lothian to Edinburgh, and by the East Coast main line, which runs through the county.

The principal centres of population have grown up, either where the conditions for the various industries are most favourable, mining in the west, fishing and catering for summer visitors on the coast, agriculture in many lowland parts, or where water is most plentiful, as at Haddington and East Linton on the Tyne.

Most of the towns now receive their water-supplies from the Lammermuir upland, where, owing to the greater rainfall and the impervious nature of the underlying rocks, numerous reservoirs have been constructed.

The two most important factors in the growth of communications have been relief and population, though the latter has to some extent been a result of communications. The Lammermuirs are crossed by few roads, because travelling here is difficult and the number of inhabitants small. In the lowlands the gradients of east-and-west-running roads are usually easier than those running north and south, and the former are on the whole more frequented than the latter. As the Southern Uplands offer a considerable obstacle to communications between the Central Valley and the south, the easiest and probably the most frequented route from England to Edinburgh passes through East Lothian.

PROCEEDINGS OF THE ROYAL SCOTTISH GEOGRAPHICAL SOCIETY.

Lectures in January.

Captain Ejnar Mikkelsen will address the Society on his travels in North-Eastern Greenland, in Aberdeen, Dundee, Edinburgh and Glasgow, on the 21st, 22nd, 23rd, and 24th, respectively.

GEOGRAPHICAL NOTES.

Europe.

Nature Reserve in Norfolk.—We mentioned here in the last volume, p. 648, the fact that a National Park is being established in Switzerland. According to the *Times*, a small reserve of wild nature is similarly to be established in Norfolk, where an area of about 1000 acres has been placed in the hands of the National Trust. The area has been acquired by the generosity of the Fishmongers' Company and of some private individuals, and consists of a shingle spit and a system of sand dunes and salt marshes together forming Blakeney Point, on the north coast of Norfolk between Sheringham and Wells. From a physical point of view the area is interesting because of the contrast of shingle, dune and marsh, and because of the processes of silting up and so forth which are going on in the region. Again, the vegetation is interesting, including as it does characteristic salt-marsh plants, notably four species of sea-lavender, the oyster-plant (*Mertensia*), and the curious seablite (*Suæda fruticosa*), a plant of wide distribution which in Great Britain is confined to certain localities on the east and south coasts of England. The point is a favourite breeding-spot for sea birds, and these have been protected for some years, so that there is now a rich fauna; insects, molluscs and other invertebrates are well represented in addition to birds. It is hoped that the area may form one only of a series of reserves to be established within the British Isles.

The Variations of European Glaciers.—In the *Annales de Glaciologie* for September 1912 there appears the usual annual report on the condition of the glaciers of Europe, the report referring to the year 1911. In the Alps it is noticeable that the continuously warm and sunny summer of that year led to a marked decrease in the case of the majority of the glaciers. In the Swiss Alps in the preceding years although the majority of the glaciers were certainly retreating, yet certain glaciers, numbering eight in 1910, showed signs of advance, and two, one being the Lower Grindelwald, had been advancing for three years in succession. All these eight, however, responded to the summer of 1911 by distinct retreat, and of all the Swiss glaciers examined, three only showed an advance during 1911. Similar marked retreat showed itself in the other parts of the Alps. On the other hand, of the five glaciers studied in Sweden four showed advance, an advance which in some cases at least has been going on for some years. In Norway the conditions are very variable, both advance and retreat having taken place among the numerous glaciers measured. On the whole, however, retreat was best marked here.

Glacial Overdeepening in the Canton Ticino.—We have received a pamphlet with this title, by Dr. Hermann Lautensach, which forms one of the *Geographische Abhandlungen* of the University of Berlin. In his preface the author points out that Penck and Brückner's great work, *Die Alpen im Eiszeitalter*, covered such an extent of ground, and dealt with so many aspects of the Ice Age, that necessarily the treatment of some phenomena and of particular regions had to be summary. Since the completion of that memoir, also, a considerable number of criticisms of particular points have appeared. The author, therefore, believed that it would be profitable to undertake a detailed examination of a limited area, and investigate the problem of how far the appearances presented there could be explained on the view of glacial action adopted by Professors Penck and Brückner, and to what extent, if at all, on the hypotheses supported by others. In this pamphlet, therefore, he devotes himself to a particular problem, that of glacial overdeepening, and to a particular area. As to his general results we can only say here that his observations, in his opinion, support the views put forth in *Die Alpen im Eiszeitalter*, and are inexplicable on any other hypothesis. In the descriptive part of his paper the most interesting sections seem to us to relate to the great alpine passes, several of which of course lie within the region investigated. As an example we select the account of the San Bernardino Pass, which, it will be remembered, brings the traveller from Thusis and Splügen down to Mesocco and so to Bellinzona and Lugano. The pass is markedly asymmetrical. A very steep rise, traversed by the carriage road in great windings, leads up from the valley of the Hinterrhein to the broad, open, gently-sloping pass with its lake. From this summit a less steep slope leads southward, this showing a sudden steepening above and near the village of San Bernardino. The author explains these relations as follows:—Prior to the Ice Age, owing to the nature of the rock, there was a comparatively low watershed in the region of the present pass. At the maximum glaciation the Hinterrhein valley was filled so high with ice that part of this ice flowed ("transflowed") over the watershed and found its way into the Mesocco valley. This distributary eroded the summit of the old watershed, and produced the present broad open pass. Meantime the far larger mass of ice in the Hinterrhein valley was overdeepening its own valley, its eroding force being much greater than that of the small distributary. Hence the pass was undercut, and the very steep slope on its northern side produced. The sudden change of slope on the south side was produced by the confluence with the distributary of small glaciers from the side walls of the Mesocco valley, the consequent sudden increase in the amount of ice giving increased erosive power. Thus, Dr. Lautensach says, the two gigantic steps which bound the San Bernardino Pass are due the one (the northern) to the *transfluence*, the other (the southern) to the *confluence* of Ice Age glaciers.

The Course of the Upper Danube.—M. Ernest Fleury contributes to a recent issue of *La Géographie* a long note on the evolution of the course of the Danube above Vienna, a subject on which much research has been conducted lately. There is reason to believe that in essence the Danube is a very old river, which once flowed eastward towards the inland sea of Tertiary times, but its present course shows no apparent relation to the relief of the region through which it flows.

The first point of interest is in connection with the source. According to most geographers this is to be sought in the confluence of the two streamlets called Brigach and Breg, which descend from the flanks of the Black Forest in the Grand Duchy of Baden, and at Donaueschingen unite with a spring regarded by some as the true source. Afterwards the stream so formed flows south-eastward towards the basin of Rhine and the lake of Constance, till, at Gutmadingen, it takes a sharp turn to the north-east. All this part of the course is through the calcareous beds of the Swabian Jura, and, especially near Immendingen, part of the water flows into sinks and rifts. By means of colouring matter it has been shown that this subterranean water reaches the Rhine basin. As the basin of the Danube is here placed at a higher level than that of the lake of Constance there is every reason to expect that this leakage of Danube water will increase, and indeed, in 1876, several sinks near Immendingen enlarged suddenly, and led to a considerable loss of water, a loss which the manufacturers using the power tried to prevent by blocking-up the openings. Suess thinks that ultimately the whole of the Danube above Tuttlingen will probably be captured by the Rhine, and, as it is known that further down, near Friedlingen and Beuron, infiltered water reappears in the channel of the river, he thinks that the sources of the Danube should really be placed in Wurtemburg.

Below the gorge of Sigmaringen the river emerges on a wide depressed area, but here the glacial deposits, which are spread out on the Bavarian plain like a huge alluvial fan, drive the stream northwards in a great curve against the Franconian and Swabian Jura. In this depressed area there are many indications of changes in the river's course, fragments of dry valleys and of deserted meanders being obvious. Fixed westwards by the channel which it has cut through the Jura, and eastwards by the similar channel cut through the massif of Bohemia, in the intervening plain the river has swayed to and fro in response to conditions which have now largely disappeared or are disappearing. Part of the plain was once a plateau, some 1600 feet high, and it was the existence of this which enabled the river to cut through the hard rocks of Bohemia to the east.

The present northerly position of the river on the Bavarian plain, however, is a consequence of the above-mentioned glacial deposits of the plain. But it is possible to show that older channels still exist nearer the edge of the Jura than the present one, so that the Danube here has been creeping south within recent geological time. It is probable that this tendency will continue, in spite of man's efforts to prevent it.

Asia.

The Plants, Animals and Minerals of the Bible.—An exhibition to illustrate the plants, animals and minerals mentioned in the Bible has been arranged at the British Museum (Natural History) at South Kensington, and the trustees have had printed an interesting *Guide* to this collection which contains some facts of considerable geographical interest. We note one or two of the more striking points. It is pointed out that the common fowl is not mentioned in the Old Testament and that it was probably introduced into Palestine after the Roman Conquest. The "unicorn" of the Old Testament is stated to have probably been the Syrian aurochs, which is known from Assyrian sculptures to have been living in Asia Minor in Biblical times. The common assertion that behemoth was the hippopotamus appears to be negatived by the fact that there is no record of the species in Syria or Palestine in historical time. Some interesting notes are also given in regard to the cereals. Wheat was sown in November or December and was reaped in May, while barley came to maturity nearly a month carlier and was usually got in at the end of Passover. The "tares" of the parable are darnel grasses, whose seeds are poisonous. The "rose" of our translation is probably the narcissus, while the "lily" is the poppy anemone (*A. coronaria*) of our gardens—a very curious inversion of the facts!

AUSTRALASIA.

The John M'Douall Stuart Transcontinental Expedition, 1861-62.—The South Australian Branch of the Royal Geographical Society of Australasia has sent to the Society a copy in bronze of a Gold Medal presented by citizens of Adelaide, in July last, to Messrs. W. P. Auld, S. King, J. W. Billiatt, H. Nash, and J. McGorrery, the then surviving members of the above expedition, which returned fifty years ago.

COMMERCIAL GEOGRAPHY.

A Desert Rubber Plant.—Professor F. E. Lloyd gives, in the *Popular Scientific Monthly*, an interesting account of a composite plant called guayule, which grows in desert regions in Mexico, and is being used as a commercial source of rubber. The plant, whose scientific name is not given, is a low greenish-grey shrub found within the Chihuahuan desert, in the State of Zacatecas, Mexico. It has a very limited geographical distribution, and though it extends into Texas, grows there only in small amounts, and does not attain a large size. Both the plant itself and two allied forms, *Parthenium incanum* and *P. lyratum*, are used by the Mexicans as a source of rubber for balls, the bark of the plant being assiduously chewed and the rubber thus obtained in pellets. The ancient Mexicans also seem to have used guayule for this purpose. The rubber does not occur as latex in special vessels, but in minute droplets in the cells of the plant, hence the necessity for the process of mastication to rupture the cells and liberate the minute particles. In consequence of this method of occurrence tapping cannot be practised, and in the process of manufacture it is necessary first to tear the plant into fragments, and then grind these fragments finely in pebble mills with flints and water, by which means the rubber is expressed.

In consequence, the method of collection consists in pulling up the whole plant and conveying it to the mills in bales. As it grows naturally in rocky and desert places, far from water and supplies, the collecting is done by peons with mules. The loads are brought to a central camp, from which they are conveyed by wagon to the nearest railway station. In the first instance, the plants were, as suggested, simply pulled up, root and all, as the easiest method. This is obviously "robber economy" in its worst form, and as the plant, like desert plants in general, is very slow-growing, an industry based on such methods must have a very short existence. It is now found that if the plant be cut off near the ground, instead of being pulled up, reproduction takes place by shoots growing up from the lateral roots, as happens for instance in the lilac of gardens. Attempts are also being made to cultivate the plant, as it thrives in localities otherwise useless. A curious point which emerges from these experiments is that on irrigated land growth is much more rapid than in the wild state, but the rubber content of the plants is also much smaller. Though, however, the net rubber content of the tissues is small, yet the amount of tissue produced within a given time is so much greater than in the case of the wild plant, that the actual amount of rubber produced would seem to be nearly equal. As yet the cultivation experiments have not been carried on on a sufficiently large scale to render it possible to draw conclusions as to its profitableness or otherwise.

Sugar-Cane in Hawaii.—In vol. xxviii. p. 486 we discussed here the relative production of cane and beet-sugar, showing how cane-sugar is recovering lost ground. In a recent issue of the *Revue Générale des Sciences* Professor Henri Jumelle describes some of the new methods in use in the cane-sugar industry in Hawaii, and his article affords an interesting commentary on the figures previously given, by showing how it is that this industry is progressing. In Hawaii attention has been directed to two separate problems, methods of increasing the production of raw-sugar by better methods of extraction, and methods of increasing the yield of cane per acre, by attention to deep cultivation, manuring and irrigation. Owing to the success of the diffusion method with beet-sugar, it was suggested that this method should be applied to the cane. It has, however, many disadvantages in hot countries; it demands much skilled labour, much costly fuel, and yields a waste (bagasse) so watery that it burns badly. Latterly, therefore, attempts have been made rather to improve the crushing machinery, for the crushing method has not these disadvantages and yields a bagasse which burns readily. The attempts have been crowned with such success that with the newest machinery the yield of crude sugar has been raised from about 75 per cent. of that in the cane to 98·43 per cent.

No less successful have been the results obtained by steam-ploughing. With the elaborate machinery now used in Hawaii it is possible to cultivate the soil to a depth greater than two feet, as against about ten inches before. The result, in combination with improved methods of manuring, has been extraordinary. It is recorded that crops of 272 metric tons per hectare (2¼ acres) have been obtained, a yield some four times greater than that previously obtained. Indeed it is stated that in Réunion a yield of only 40 metric tons per hectare is usually reckoned on. Professor Jumelle points out that if such improvements can be effected with sugar-cane it is probable that such crops as rice and cotton would also respond to scientific methods by a largely increased yield.

General.

Origin and Reclamation of Sand-dune Areas.—Two recent publications of some size deal with the subject of sand-dunes. One, an official *Report on the Dune-areas of New Zealand, their Geology, Botany, and Reclamation*, by L. Cockayne, is essentially practical in nature, while the other—Dr. Gustav Braun's *Entwicklungsgeschichte Studien an Europaeischen Flachlandsküsten und ihren Dunen*, published by the Geographische Institut of Berlin, is a purely scientific study of the problem, and includes an attempt to apply the American developmental terminology to this particular kind of land form.

Dr. Cockayne points out that, while sand-dune areas occasionally occur in the interior of New Zealand, these are of most importance along the coast. In the North Island it is estimated that they cover an area of about 290,000 acres, and in the South Island about 24,000 acres. Their tendency to spread and overwhelm fertile land has been recognised, and the Department of Lands and Survey commissioned Dr. Cockayne in 1908 to make a thorough study of the subject. The present is the second report issued, and is especially full in regard to the question of reclamation. The paper is illustrated by a number of striking photographs, which give a good idea of the dune areas. In connection with methods of reclamation it is of interest to note that the marram grass, so widely distributed in the Northern Hemisphere generally, including Great Britain, is regarded as the most useful plant for this purpose, though the tree lupin of California, commonly grown in gardens here, is useful under certain conditions.

Dr. Braun describes in detail many of the dune coasts of Europe, including the German Baltic coast, and those of Gascony, Portugal, Languedoc, Catalonia, etc., and then discusses the origin of shores generally, noting the parts played respectively by the sea and rivers, with a special consideration of the characters, conditions of formation, etc., of dunes. His paper is thus of a much more general nature. It also contains some good illustrations.

EDUCATIONAL.

We have been furnished by Mr. T. S. Muir with the following particulars in regard to the recently-formed Association of Scottish Teachers of Geography:—

A meeting of those interested in the teaching of geography was held in the Society's Rooms, Synod Hall, Edinburgh, on December 7 last. The attendance was representative of several parts of Scotland. Mr. Geo. G. Chisholm occupied the chair. After full discussion the proposed Constitution as approved by the Council of the Society was unanimously adopted, and office-bearers and a committee were appointed.

Constitution.

1. This association shall be composed of those interested in the teaching of geography in Scotland, and its objects shall be to provide a means of discussing such problems as may arise in connection with the status of the subject, and generally to promote the teaching of geography on modern lines.

2. This association shall be under the auspices and shall form a section of the Royal Scottish Geographical Society.

3. All members and associates of the Royal Scottish Geographical Society who are engaged in or connected with the teaching of geography

shall be eligible for membership of this association without further subscription.

4. The subscription for those who are not members or associates of the Royal Scottish Geographical Society shall be 2s. 6d. per annum, payable to the Royal Scottish Geographical Society.

5. The office-bearers shall consist of a chairman, two vice-chairmen, and an honorary secretary. These, and five other elected members shall form the committee, three to be a quorum.

6. A general meeting for the election of office-bearers and committee and any other competent business shall be held each year not later than November 15. Other meetings shall be called at the discretion of the committee. The meetings shall be held in the rooms of the Royal Scottish Geographical Society, Edinburgh.

7. All resolutions passed at meetings of the association shall be laid before the Council of the Royal Scottish Geographical Society.

OFFICE-BEARERS FOR 1912-1913.

Chairman . . Mr. Geo. G. Chisholm.

Vice-Chairmen. . Mr. J. Cossar, Training College, Glasgow.
Mr. F. Spence, Training College, Edinburgh.

Hon. Secretary . Mr. T. S. Muir, Royal High School, Edinburgh.

Committee.—Mr. Findlay, George Watson's College, Edinburgh; Mr. Macgregor, Boroughmuir H. G. School, Edinburgh; Mr. Corrie, Kirkcaldy; Mr. Philip, Aberdeen; and Mr. Wilson, Dollar.

All teachers of geography in Scotland who are members or associates of the Royal Scottish Geographical Society are earnestly requested to send in their adhesion to the new association without delay. They are also expected to do their utmost to secure recruits so that the association may be thoroughly representative. It is hoped to hold a meeting early in 1913 when a paper dealing with methods of teaching geography will be read. Should a sufficient number of members join from any well-defined district local branches will be started.

NEW BOOKS.

EUROPE.

Celtic Place-Names in Aberdeenshire. With a Vocabulary of Gaelic Words not in Dictionaries. The Meaning and Etymology of the Gaelic Names of Places in Aberdeenshire. Written for the Committee of the Carnegie Trust. By JOHN MILNE, M.A., LL.D. Aberdeen: *Daily Journal* Office, 1912.

Gaelic Place-Names of the Lothians. By JOHN MILNE, M.A., LL.D. London: M'Dougall's Educational Co., 1912.

These are two exceedingly interesting books, containing many facts of great importance to the geographer. Dr. Milne points out that the process of corrupting the often highly appropriate Gaelic place-names—a process which has been going on for a prolonged period—has been greatly accelerated by the publication of the Ordnance Survey Sheets. Names on the sheets have been applied to a particular feature when another was meant; they have been taken down wrongly and then ingeniously corrected by persons at a distance; they have been doctored and improved and modified in a hundred ways till the original meaning has often quite disappeared in the printed form. This is a disadvantage, not only from the point of view of poetry and beauty, but also because the true names often throw a flood of light on vanished conditions of life. Real though these disadvantages are, however, we fear the process cannot be stopped. The old tongue with all its beauties is disappearing, and we doubt if, for example, the Scot of to-day can be persuaded to abandon the convenient, if foolish, *Gateside* for the appropriate but unfamiliar *Gaothach Suidhe*, windy place. If, however, his tongue prefers the familiar, he can nevertheless scarcely fail to be interested in the interpretations which Dr. Milne offers. Who that has descended from Lochnagar and found The Ladder far less formidable than the name suggests will not be interested to know that it is An Leitir, the hillside? Curious also is the story of Beinn Iutharn Bheag and Beinn Iutharn Mhor, little and big mountain of hell, an ingenious Ordnance Survey rendering of Ben Uarn, the phonetic spelling of Ben Bhearn, mountain of the gap, because of the gap between the two mountains. The two books are full of such curious facts, and may be cordially recommended to the geographer.

Rome, the Cradle of Western Civilisation: as Illustrated by Existing Monuments. By H. T. INMAN, M.A. London: Edward Stanford, 1912. Price *4s. net.*

This is a guide-book upon a somewhat new plan, an attempt being made to classify the monuments of Rome according to the time of their construction, and thus show by a continuous narrative the debt of the western world to Rome. The idea is doubtless good enough, but there are certain practical difficulties in carrying out the author's scheme. For example, the Mausoleum of Hadrian is described on p. 56 under the heading of The Campus Martius and Neighbourhood, while the fact that it is enclosed within the Castle of St. Angelo is not mentioned till p. 287, under the heading of Art Collections. Again, the tourist who, trusting in the author's classification, sets out to see a fourth-century church in Santa Maria Maggiore will, we fancy, get something of a shock when its façade opens before him. Nor can we quite see the logic of placing the very curious lower church of San Clemente under the twelfth century. In fact, as reverence for the buildings of a past epoch was never a Roman characteristic, we think the attempt to classify the monuments in the author's fashion is a task of very great difficulty. If, however, the book can hardly be recommended to the tourist visiting Rome for the first time, those already familiar with the objects of interest will doubtless find it very useful.

Les Alpes de Provence: Guide du Touriste, du Naturaliste et de l'Archéologue. Par GUSTAVE TARDIEU. Paris: Masson et Cie, 1912. Prix *4 fr.* 50.

This is another of the series of guides which are being published under the general editorship of Professor Marcellin Boule, and are remarkable for the amount of space given to various branches of science as illustrated in the different regions considered. The first part in this volume consists of a monograph on the Alps of Provence, the geological structure, topography, climate, flora and fauna, archæology, and generally the conditions of existence as they affect human life, being all carefully studied. This is followed by a second part, giving suggested itineraries, and the whole is illustrated by some striking views. The book may be recommended to the notice of those contemplating a visit to a region not so well known as it ought to be.

Norway, Sweden, and Denmark, with Excursions to Iceland and Spitsbergen. Handbook for Travellers. By KARL BAEDEKER. Tenth Edition. Revised and Augmented. Leipzig: Karl Baedeker. Price 8*s.* net.

The publication of this new edition speaks to the increasing popularity of Scandinavia and the neighbouring regions with tourists. Among the additions are several new maps, notably one of the Lofoten Islands, and an instructive one of the Jostedalsbræ, the largest glacier in Europe.

Russia. By SIR DONALD MACKENZIE WALLACE, K.C.I.E., K.C.V.O. Revised and enlarged edition. With Maps. London: Cassell and Co., 1912. Price 12*s.* 6*d.* net.

In vol. xxi. p. 504, we published a long review of the second edition of Sir Donald Wallace's *Russia.* That volume was a revised, supplemented and re-arranged version of a book originally published in 1877. Thus while a period of eighteen years intervened between the publication of the first and second editions there is only seven years between the issue of the second and third. Necessarily therefore the alterations have been of a less elaborate nature than before, and deal chiefly with the history of the Dumas and other recent events. In other respects also, however, the volume has been brought up to date, and remains a treasure-house of facts in regard to the great Russian empire.

Rambles in Ireland. By ROBERT LYND. London: Mills and Boon, 1912. Price 6*s.*

Mr. Lynd takes us here and there through Ireland in a delightfully unpremeditated fashion. We start in Galway of the Races, but come in our rambling to spots so different as Cashel of the Kings, and we find an account of an evening in Lisdoonvarna among priests at play, along with legends of Oisin and of many another hero who trod the soil of Erin in times past. We are also given the opportunity—though the book is not a professed study of Irish character and problems—of understanding a little better what manner of man the Irishman at home really is, for the book is full of anecdotes of Irish life in the various parts of the country.

The illustrations consist of a series of fine photographs and some coloured drawings by Mr. J. B. Yeats.

The Guadalquivir. By PAUL GWYNNE. London: Constable and Co., Ltd., 1912. Price 7*s.* 6*d.* net.

This is a delightful book to take up on a lazy, idle day. It contains a pleasantly written, if somewhat garrulous description of the country on each side of the Guadalquivir from its source to the sea, with more detailed and very interesting disquisitions on the two famous cities of Cordova and Seville. The author knows Andalusia well, and writes with generous enthusiasm of its former glories and its present attractions. He has something to tell us of the geology and ethnography of the tract; he makes skilful use of its brilliant history in the halcyon days of the Romans and the Moors; and he is a sturdy champion of Andalusian music, painting and art. He is at his best, however, in his amusing and shrewd delineations of the present inhabitants, which he successfully illustrates in many good anecdotes. He is not forgetful of, though he does not dwell upon, the economic conditions of Andalusia, and he inveighs against the excessive taxation and general mismanagement of the Government. We may add a word of commendation of the illustrations, a dozen of which are coloured.

My Parisian Year: A Woman's Point of View. By MAUDE ANNESLEY. With 20 Illustrations. London: Mills and Boon, Ltd., 1912. *Price* 10*s.* 6*d. net.*

Paris is the city inexhaustible. It will bear any amount of "writing up." In adding another to the many books about it Miss Annesley need not fear being accused of vain repetition, for, though she does not tell us much that is new, she has a light, happy touch, and her point of view is unhackneyed. As a resident in Paris she simply relates her experiences, not, indeed, in chronological order, but beginning with the more obvious and concluding with the more intimate features. We view the Parisians as they are, with all their faults and virtues, in a series of moving word pictures as vivid as they are truly realistic. We have only one complaint to make, and that is that in the chapter on "Money" the author does not refer to the absolutely scandalous way in which false coins are "planted" upon the unfortunate tourist. The result is that every one scrutinises his change in a manner which would give intolerable offence in this country.

The Building of the Alps. By T. G. BONNEY, Sc.D., F.R.S., etc. etc. With 48 Illustrations. London and Leipsic: T. Fisher Unwin, 1912. *Price* 12*s.* 6*d. net.*

Professor Bonney's is a name that has long ranked high in the affections of those who love the Alps, as it takes one back to the heroic days when almost every mountain expedition meant a new ascent, and when the sill of each valley was the threshold of the unknown. The mountaineer will turn with most interest to those concluding chapters in which the reminiscent vein is charmingly exploited, while even the most indefatigable of peak-baggers will examine with admiring envy the appendix in which Professor Bonney's numerous and almost ubiquitous excursions are detailed. The first chapter deals with the geographical distribution of Alpine rocks, and here the nature of the metamorphic rocks is very fully treated. Passing on to the materials of the Alps, their nature and origin, we find considerable space devoted to the age of these gneisses and schists, and Professor Bonney, in opposition to the Lyellian or Uniformitarian School, leans to the opinion that they are Archæan. Chapter III. gives some account of the growth of the Alps, with special reference to flat folds. The next chapter, entitled "Mountain Forms," is remarkable for the easy way in which the author cites examples from all parts of the Alps, the result of personal knowledge attained by perhaps only three others among Anglo-Saxons—Ball, Tuckett, and Coolidge. Chapter V. gives the distribution of snow-fields and glaciers. On page 116 it is stated that the Col de Fenêtre, at the head of the Val de Bagnes, cannot be crossed without traversing a glacier. This is not the case, and indeed it is contradicted on page 123. In Chapters VI, VII, and VIII Professor Bonney re-states his well-known opinions on the erosive power of ice. He declines to admit that glaciers have borne more than a very minor part in carving mountain systems. Only the smallest and loftiest of lake basins have been, in his opinion, excavated by ice. First the sea, as the dome-shaped mass slowly elevated itself, then running water and the ordinary atmospheric agencies have been the main factors in the creation of valleys and the isolation of peaks. Even cirques and steps, he maintains, have been formed chiefly by streams. The larger lakes, apart from those obviously formed by morainic dams, are due, he thinks, to differential movements in the beds of valleys.

The illustrations, chiefly from photographs, are really illustrative. The Index is curiously unequal, one of the references at least being almost freakish.

Edinburgh. By R. L. STEVENSON. With 24 illustrations in colour by James Heron. London: Seeley, Service and Co., 1912. Price 12*s.* 6*d. net.*

This is a very handsome edition of Robert Louis Stevenson's well-known book, and will doubtless find many purchasers among his admirers. Printing and get-up are admirable, and Stevenson's melodious prose may be here enjoyed under the most favourable conditions. Some of Mr. Heron's illustrations have come out well, others seem rather to lack definiteness, and display too great uniformity of colour.

The Geology of the Districts of Braemar, Ballater, and Glen Clova (Explanation of Sheet 65). *Memoirs of the Geological Survey of Scotland.* By GEORGE BARROW, E. H. CUNNINGHAM CRAIG, and L. W. HINXMAN. Edinburgh: H.M. Stationery Office, 1912. *Price* 2*s.* 6*d.*

Geographers will be grateful for this interesting Memoir, which deals with a most attractive district, including as it does the Dee Valley, the massif of Lochnagar, and the upper parts of several of the Forfarshire Glens. In addition to the detailed description of the geological map the *Memoir* contains, especially in its first chapter, a great deal of matter of direct geographical importance, the description of the physical features of the region being to our mind excellent. We do not, however, understand why the Dee on the sheet is described as consequent—surely it is subsequent, the Lee, Mark, and North and South Esk being remnants of the original consequents?

ASIA.

Snapshots in India. By JOHN WEAR BURTON. London: Elliot Stock, 1912. *Price* 5*s. net.*

This volume contains a number of sketches by a missionary whose proper field of work is in Fiji, who made a tour in India to see how missionary work was being carried on there. He "suffered nearly eight thousand miles of weary travelling" in India, and visited between forty and fifty mission stations, and "looked at things with the missionary eye." Remembering this limitation, the reader will enjoy a glance over these *snapshots*, which are pleasantly and even enthusiastically written by one who is evidently most earnest in his zeal for the missionary cause. The book is profusely illustrated with photographs of well-known places and scenes.

The Indian Scene. By J. A. SPENDER. London: Methuen and Co., Ltd. Price 3*s.* 6*d. net.*

This volume is a reprint of the articles contributed to the *Westminster Gazette* by its well-known editor, Mr. Spender, during a brief tour in India at the time of the Coronation Durbar. As might be expected from the writer of the Bagshot papers, the articles deserved and secured a good deal of attention at the time of their publication, and were as popular in India as they were at home. Mr. Spender recognises his own limitations, as practically a novice on Anglo-Indian affairs, but this does not prevent him from venturing now and again on debatable and contentious matter, on which his observations are always shrewd, incisive, and good-humoured, and not too dogmatic. Usually such articles as we have in this volume do not bear republication, but there are exceptional cases, and this is one of them.

The Chinese. By JOHN STUART THOMSON. London: T. Werner Laurie, 1912. *Price* 12*s.* 6*d. net.*

This is a collection of papers on things Chinese, ranging over a wide field, from the daily life of foreigners to Chinese humour, art, politics, and literature, and the book affords some entertaining reading. Mr. Thomson is an American, and when he essays to give his readers a picture of "the daily life of foreigners in China," he might have said Hongkong in place of China, for his first section, which bears the above heading, treats chiefly of life in our Crown Colony. But any one who judges that life from Mr. Thomson's account would be justified in believing that Hongkong is a very strange and fearsome place for the foreigner to live in, afflicted with a pestilential climate in which the exile from home drags out a precarious and miserable existence. It is, perhaps, not quite fair to assume that the author is wilfully distorting facts, and it may be needless to take him too seriously in his flights of fancy. But Hongkong largely owes its prosperity and its continued existence as a British colony to the energy of Scotsmen, and if the book is to be noticed at all here, it is not possible to refrain from challenging some of the grotesque and extravagant statements in which Mr. Thomson indulges with regard to Hongkong and South China. We take a few instances at random:—

"I never knew a foreigner in Southern Chinese ports who did not languish for nine months of his first two years in sickness." "This British colony takes vitality out of its citizens more than any port of the Orient. Its line of invalids and derelicts who have fallen back for repairs is a long one, and not all of them reach Glasgow, or even Chifu, Yokohama, or Colombo, before the chill ghost-order 'Halt' is all too willingly obeyed for ever." "The band who rove the East find their discoveries as melancholy as did the followers of Camoens' hero Da Gama, to whom 'a grave was the first and awful sight of every shore.' Certainly three quarters of those who adventure, float out on the tide again as dead culls" (*sic*). "The soldiers in the barracks of this garrison post (Hongkong), which is the strongest in the far East, lie all day on their backs and call to the punkah coolies to fan away their curses." In July "the pigs crawl to the gutter and become molten grease from their own and the sun's heat." And so on, *ad nauseam.*

Hongkong can be very unpleasant during the damp heat of early summer, but it is on the whole a very cheery and healthy community, and that most of the above is sheer nonsense, any old Hongkongite can testify. Many of the author's remarks on native life and customs are marred by similar misstatements, which detract a good deal from the value of what would otherwise be a very readable book. The text is illustrated with some good photographs.

China's Revolution. By EDWIN J. DINGLE. London: T. Fisher Unwin, 1912. *Price* 15*s. net.*

That part of the valley of the Yangtsze which stretches from Hankow to Nanking has been the scene of many struggles for the mastery of China, furnishing materials for many red pages in the history of that country. Not the least bloody of these records will be found in the story of the fighting and massacres which took place from October to December of 1911, when the cities of Hankow, Hanyang, and Wuchang were the scene of the outbreak of the revolution and of the struggle which took place between the revolutionaries and the imperial troops under the command of Yuan Shi Kai. Mr. Dingle, whose book bears the sub-title of "a historical and political record of the civil war," was in Hankow throughout the time of the fighting and the burning of that city by the imperialists, and he claims that as a personal friend of the revolution leader General Li Yuan Hung, and the repository of exclusive information, he is equipped to write of the main doings of the revolution. We have his own word for this, and we can say from a perusal of his book that it contains much which is of value in giving an idea of the beginnings of events which are fraught with wide-reaching possibilities for the future of China and of the world. The chronicle is written in instalments, bearing various dates, and there are not wanting signs of haste in its composition, so much so that it is not always easy to follow the story; while some of the writer's quotations "from an American daily," or "from a writer in a London journal," do not carry so much weight as they would if the source of authority were more definitely stated. However, after allowing for this, we have in this book a real contribution to the history of the outbreak of a civil war as to which not very much is known by the public in this country.

Mr. Dingle is of opinion that the new-born republic may encounter some of its greatest dangers from the numerous semi-independent aboriginal tribes who inhabit far Western China, and who are not yet animated by any enthusiasm for progress and reform. This is a side of the question which we do not remember to have seen stated elsewhere, but it is of minor importance when compared with the great problem as to what the body of young intellectuals and Chinese educated abroad who are now superseding the old mandarinate will make of the government of the country under republican forms. Mr. Dingle prints some interesting biographical notes concerning some of the leaders, such as Sun Yat Sen, Yuan Shi Kai, Li Yuan Hung, and others, which may help to a solution. That he looks for great things himself may be judged from the following quotation:—

"The Republicans of China, new-born into a life full of highest promise to mankind, now have free way. In them, if they are wise and good, as wise and good we believe them anxious to be, we shall soon see on the horizon of the East a nation whose power will be ultimately predominant on the earth, upon whose integrity will undeniably depend the peace of the world."

Mr. Dingle is a journalist who has recently travelled a great deal in the far interior of China, as is shown in his book, *Across China on Foot*, published about a year ago, and we have no intention of detracting from the value of his latest book when we say that his attitude is mainly that of a journalist.

From the Black Mountain to Waziristan. By Colonel H. C. Wylly, C.B. London: Macmillan and Co., Limited, 1912. Price 10*s.* 6*d. net.*

This is a book written by a soldier for soldiers, more especially for the Indian and European troops which are likely to be employed in what are commonly called Frontier punitive expeditions. But it has an interest far beyond that limited circle, and will be welcomed by statesmen, politicians, students of history, geography, and ethnography, and especially by all who have to study and deal with the complicated and delicate problems summed up in the expression, the North-Western Frontier of India. In 1897, Colonel Wylly had to take his regiment, the Sherwood Foresters, to the Tirah expedition, and he found that "few of us knew anything of the wild men against whom we were to fight, or of the equally wild country in which the operations were to be conducted." Recognising the want of a handy compendium of information on this subject, he has compiled from official and other sources this volume, which contains an account of "the more turbulent of the tribes beyond our Border, the countries they inhabit, and the campaigns which the Indian Government has undertaken against them during the last sixty-five years." Questions have frequently been asked in Parliament and out of it as to whether such expeditions are really necessary or useful. They are abhorred by the Government of India, and it is doubtful if they are popular in the army. To such questions Colonel Wylly's book furnishes a conclusive answer. He shows how the patience and forbearance of the Government of India have been always stretched to the breaking point, ere sanction was given for punitive measures, and that they were never prolonged one day longer than was necessary. Indeed, it would seem that the prompt withdrawal of our troops from the territories of the offending tribes too often gave rise to misunderstandings and doubts of our sincerity, and in a measure contributed to encouraging the offenders to repeat their offence. In the two years before the Mutiny the Mohmands of Pandiali, we are told, perpetrated thirty-six raids having plunder and murder for their objects, and this they did without reprisals, although Sir John Lawrence had urged the Government of India to take punitive measures. Again, between 1884 and 1890 almost every clan of the Orakzais were continually raiding the frontier; "one division committed forty-eight fresh offences in one year." In an appendix to this work Colonel Wylly gives a list of over sixty punitive expeditions which had to be undertaken against offending tribes within the last sixty years, and, considering the provocation, the number is small. The plan of the book is to take the principal tribes, one by one, to set forth the locality where it lives, to describe its character, and to give a brief history of how it has conducted itself in the past, with a more detailed account of any military operations undertaken against it. Colonel Wylly has many of the qualities which we look for in military writers, simplicity of style, brevity compatible with a just sense of proportion, impartiality of judgment, justice to the good qualities of the enemy; and he has made a careful and effective study of the authorities already existing on his subject. The book is completed with an ample supply of useful maps, and should be in every regimental library and military club. But, as we have already observed, it will also be found of much interest to many outside military circles.

In Abor Jungles. By ANGUS HAMILTON, F.R.G.S. London: Eveleigh Nash, 1912. Price 18*s.* *net.*

When we hear of a punitive expedition on the Indian frontier, our minds instinctively turn to the North-Western frontier, where raids and forays by the fanatical Mohmands, Afridis, Orakzais, etc., may be expected with a most unwelcome regularity. This work on the Abor expedition comes to remind us that there is a North-Eastern frontier as well as a North-Western one, on which it is necessary now and again to send out a punitive expedition. The North-Eastern frontier may be described roughly as the frontier between India and Tibet, and owing to a variety of circumstances much less is known of the inhabitants of the tracts along this frontier than of those on the North-West. The Abors, who are the subject-matter of this book, for example, were comparatively unknown until the murder of Mr. Williamson forced them into public notice.

The principal object of the expedition was, of course, punitive, but the interests of science were not forgotten, and so far as geography and ethnography are concerned the results constitute a substantial addition to our knowledge of the country and the people. The Indian Pundit Krishna had shown long ago that the Tsau-po and the Irrawaddy were different rivers, and his statement had been confirmed by Mr. Needham and Prince Henry of Orleans. But an interesting question remained unanswered as to how the waters of the Tsan-po reached the Brahmaputra in Assam—a descent of well over ten thousand feet. Was it by a magnificent waterfall or by a series of rapids? It will be remembered that Sir Frank Younghusband, at the conclusion of the Lhasa expedition, proposed a mission to investigate the question, but, most unfortunately for the interests of geography, his proposal was disallowed by the Home Government. At the conclusion of the Abor expedition Mr. Bentinck was deputed by General Bower on the same task, but again most unfortunately the mission failed. "Fogs, torrential rains, and deep snows" prevented it getting farther north than Singging. We need not however dilate farther on this subject, as doubtless it will be fully dealt with by Mr. Bentinck in his address to the Society on the geographical results of the expedition, promised for a later period of the session. A second survey party proceeded up the Yamne river and conducted operations as far as Peram, from which place they observed to the north-east a chain of high mountains forming one wall of the valley of the Tsan-po, which mingled its waters with those of the Dihang. They are in reality one river. The survey of the Yamne river was carried out "to a point twelve miles north of its junction with the Dihang, while the valley of the Dihang itself was surveyed for fifteen miles beyond the same junction, and its course sketched in for yet another twenty-five miles. The region had not been visited before by any mission, and it was found with surprise that it differed considerably in character from the country though which the main column had been passing." Other short expeditions were made to Damro, the chief village of the Kadam Abors, who were quite friendly and offered no opposition, and along the Siyon and Shimong rivers. Mr. Hamilton sums up the work of the survey officers thus: "An accurate series of triangulation emanating from the Assam longitudinal series of the great trigonometrical survey of India was carried over the outlying ranges to the latitude of Kebang, terminating in the base Sadup H. S. Namkam H. S. From this series, and an extension of reconnaissance triangulation to the latitude of Simong, several large snow peaks were fixed on what appeared to be the main Himalayan Divide. One of these was the 25,000-foot peak to which reference has already been made. Other peaks were fixed on the watershed between the Dihang and Subansiri rivers, which seems to be a prominent spur of the main water-parting. It was, of course, only possible to obtain a mere approximation of the topography of these snow-ranges; but, none the less, the results which were forthcoming will be of great future value. Three thousand five hundred square miles of country in all were correctly mapped on a scale of four miles to the inch. This area included the whole valley of the Dihang as far north as Singging, the whole of the Yamne valley, the whole of the Shimong valley, and a portion of the Siyon river." So much for the geographical results of the expedition.

Besides these scientific expeditions there were two political ones, the first to ascertain the truth as to the allegations of Chinese aggression in the North-Eastern frontier zone, and this mission returned, having failed to find either Chinese officials or soldiers within the zone. The second mission was to the Miri country to survey the Kamla Subansiri tracts, and according to Mr. Hamilton the mission was given an impossible task owing to the inexperience of the political officer in charge and the insufficiency of the means placed at his disposal by the Assam Government. But presumably Mr. Bentinck will have much to say on this point also. The ethnographical results of the expedition are not less considerable, but we must refer our readers for details to Mr. Hamilton's exceedingly interesting chapters on life in the Abor jungles. He has used his exceptional opportunities (he was the only war-correspondent with the force) to the best advantage, and has much new and valuable information to give us about the Abors, the Miris, the Mishmis, the Nagas, their subdivisions, relations to adjoining tribes, their laws, religion, customs, methods of warfare, family life, etc. Naturally the greater part of the book is taken up with an account of the inception, progress and results of the punitive expedition, and these are related with the easy grace and literary skill which distinguish Mr. Hamilton's other works, and which we expect from the practised journalist. An excellent map materially assists the reader in following the various military operations, and the book is illustrated by a profusion of good photographs.

AFRICA.

In French Africa. By Miss BETHAM-EDWARDS. London: Chapman and Hall, Ltd., 1912. *Price* 10*s.* 6*d. net.*

In this volume Miss Betham-Edwards gives us some bright and crisply-written sketches of ancient history in French Africa, for she tells of travels among conditions which prevailed when Napoleon III. and Eugénie reigned in the Tuilleries and MacMahon was Governor of Algiers. The sketches will be enjoyed by tourists and residents in Algiers now, as enabling them to appreciate the great change and improvement effected in Algeria by the French administration, to which reference was made in our notice of M. Victor Piquet's *La Colonisation Française dans L'Afrique du Nord* in this Magazine last September. Miss Betham-Edwards has no exciting adventures to relate, nothing worse than a snow-storm in a cedar forest, but she is an adept at description, and her pictures of the *fête* in honour of the Mahommedan saint Aissaoua, and of the effects of an invasion of locusts, and of an earthquake, are vividly realistic and sympathetic. Indeed sympathy with and admiration of the Arab population are marked features of this book. It is illustrated by a number of good photographs.

Au Pays de Salammbô. Par MARTIAL DOUËL. Paris: Fontemoing, 1912. *Prix* 3 *fr.* 50.

Since Flaubert visited Tunis and Carthage in 1858, and wrote *Salammbô*, many things have occurred there. M. René Cagnat, who writes a Preface to this volume, thinks the French occupation has destroyed much of the charm of the district by introducing railways, villas, hotels, and boulevards, but no one can survey the ruins of Carthage, or inspect the museum containing the antiquities discovered there, without feeling transported to another epoch. M. Douël evidently experienced this when he realised in thought "the sumptuous bustle of the city of Hamilcar, the splendour of its temples, the animation of its public places, and when these came tumultuously to his mind, and he enjoyed the penetrating illusion of reconstructing with Flaubert the proud and savage capital." He was also impressed, as all visitors are, with the enormous subterranean cisterns which furnished the water supply of Carthage, and which Flaubert only saw in ruins, but which were restored about 1862, and now supply the little town of La Goulette instead of the vast Phœnician metropolis.

AUSTRALASIA.

By Flood and Field: Adventures Ashore and Afloat in North Australia. By ALFRED SEARCY, author of *In Australian Tropics.* London: G. Bell and Sons, Ltd. Price 6*s. net.*

Beginning with a shipwreck, a massacre, and residence for some months among savages, the author takes us to Port Darwin, and narrates in a breezy manner his many and varied adventures, first as customs officer, and then as a member of the police. It is a fascinating story of life in Northern Australia, and can be cordially recommended to all those who have not outlived their love of tales of moving accident in flood or field.

GENERAL.

Cities Seen in East and West. By Mrs. WALTER TIBBITS. London: Hurst and Blackett, Ltd., 1912. Price 16*s. net.*

We may confess our belief at once, that this book does not lend itself readily to review in a geographical magazine. The author, a lady of Irish descent, is quite frank about herself. She tells us in as many words that she is "an orthodox Hindu and therefore an idolatress. Mahadev is the Lord and the lingam the idol." She has been "in the Outer Court of practical occultism for many years, and in constant association with high initiates," and she has at least eleven personal friends who are direct pupils of the Mahatmas. The book is divided into two parts, the second of which is presumably the more important as it is intended "for those men and women, forming a belt of fire all round the world, of every colour, creed, and clime, who have known and sickened of all experiences of the outer life and wish for definite knowledge of the World Unseen, but have not yet the power to penetrate it for themselves," and if any of our readers come under this category and have some idle hours to waste, they may read this part for themselves. We may add that, having read it, we find "definite knowledge" conspicuous by its absence. Mrs. Tibbits tells us that to some "it will appear naught but the imaginings of an unbalanced mind," and we are ready to admit that we are of that number. The first part of the book is more intelligible, and is dedicated "to the men and women met in all cities and seas who also have wished for all sensations." It consists for the most part of sketches of cities, a few in Europe and the majority in India, with which Mrs. Tibbits evinces a very peculiar and extensive acquaintance. Each city is ticketed off: Paris, the city of sin; Nice, the city of pleasure; Bombay, the city of destiny; Agra, the city of love, etc. She claims to be a great traveller. "I have," she says, "made eighteen long voyages and innumerable short ones. The tale of my travels must approximate two hundred thousand miles," and apparently she was the confidant and *chère amie* of princesses, duchesses, diplomats, and officials of all ranks in the West, and also in the East with the addition of maharajahs, ranis, gurus, and priests of every grade. She has considerable power of word-painting, and goes off into a rhapsody over a flower or a temple, etc., rather oftener than the ordinary reader cares for. As for her opinions on life in India, their value may be gauged from a quotation: "Simla, with its sordid scheming, its artifices, its made, forced conversation, its sensualism, its pettiness of passion, its social middle-classness, and its political immensities. . . . For the Simla season are required nerves of steel, Spartan self-control, the self-restraint of St. Antony, the patience of Piccadilly, the talents of a Crichton, the tact of Edward VII., the diplomacy of an empress, the courage of despair," and so on. Who of us, who knows Simla, can refrain from a smile at such a farrago of nonsense? At one place in her book Mrs. Tibbits says: "My experience, in many sorts of investigations, has led me to the conclusion that the world is inclined to be too incredulous." *Cities Seen in East and West* will not improve matters.

A Modern Pilgrim in Mecca. By A. J. B. WAVELL, F.R.G.S. London: Constable and Co., Ltd., 1912. *Price* 10*s.* 6*d.*

Apart altogether from the interest which this book arouses, it is quite an achievement to have written it. As the author points out, of the dozen or so Europeans who have been to Mecca, during the past hundred years, only four—of which he himself is one—were Englishmen. There is much to interest the reader in the description of the journey from London to Beyrout, the stay in Damascus, and the journey thence *via* the Hedjaz Railway to Medina. From this point the pilgrimage proper begins, and through many perils the author and his two faithful servants find themselves at last in the stronghold of Islam. One of the servants seems to have been a "lad o' pairts," as on more than one occasion he extricated his master from a difficult situation by declaring that he had uncles in exalted positions in Yemba and Jiddah, whether consanguinean or otherwise is not stated.

Arrived at Mecca we have a graphic account of the various religious performances, much of which is impressive, but much quite the reverse. The author's advice to intending pilgrims is to enter the country in disguise, but one would require to be of a very adventurous nature to attempt the pilgrimage at all.

The latter portion of the book should be of particular interest at the present time, as it gives the author's experiences in the war between the Turks and the Arabs in the Yemen. The chapters on his escape from Sanaa, where he was a prisoner, on his recapture and final release, are quite as exciting as anything of the kind in fiction. We are inclined to doubt the advisability of printing as an appendix the letters which passed between the author and the Foreign Office in regard to his arrest. There are a few interesting illustrations and a useful map.

The Sea-Trader: his Friends and his Enemies. By DAVID HANNAY. London and New York: Harper and Brothers, 1912. *Price* 15*s. net.*

Some time ago Mr. Hannay published an acceptable book on the *Royal Navy from 1217 to 1688.* In this substantial volume he has undertaken, and with conspicuous success, a much more formidable task. It is to give an account of the sea-faring man "in normal conditions going upon the sea on his 'lawful occasions,' and in port." The task was one involving much study, much patience, and much erudition, for, as the author points out, most writers expatiate on the romantic and exciting details of adventures and incidents, of which there is notoriously *un embarras de richesse*, and they do not care to take away from the interest of their narratives by prosaic details of commonplace life. And yet these details have an intense interest as throwing light on the conditions in which the trade, commerce and war of the past were carried on, and in which deeds of derring-do were done that changed the course of history and affected the fate of empires. Mr. Hannay has confined himself to what may now be called ancient history, for his work contains only one too brief chapter on the revolutionary changes produced in the nineteenth century by the use of iron in the construction of ships and steam in their propulsion. These were of course the principal factors in the change, but Mr. Hannay adduces others, *e.g.* the break-down of the Navigation Laws and the competition of the United States, on which he has some weighty observations to make. This book must have an intense interest to the "mariners of England" and to the whole English-speaking peoples now scattered all over the world; but its interest will not be confined to them, for the author has much to say about the conditions of sea-service among the Venetians, Spaniards, Portuguese, French, and Dutch. In passing, we may add that he is scrupulously impartial, and does full justice to the prowess of the sailors and explorers of other countries.

Malta and the Mediterranean Race. By R. N. BRADLEY. London: T. Fisher Unwin, 1912. *Price* 8*s*. 6*d*. *net.*

From time to time lately brochures have appeared written by Professor Zammit and Mr. Bradley himself, giving an account of the work of excavation in Malta and the adjacent Gozo. In a sense the present volume is a summing up of the work done to date, and a rendering into popular form of the methods and investigations of Sergi and Angelo Mosso. From discoveries in Crete and other parts it has been known that there existed a pre-Grecian and prehistoric race of high culture which has apparently entirely disappeared. Who were they? Professor Sergi of Rome, pursuing his own method of classifying skulls, deduces that this great race belonged to Neolithic interglacial man, who spread throughout North Africa and the whole of Europe, and were a Hamitic people, "probably fairly represented to-day by the light-skinned Berbers." They comprised the early Egyptians, migrating through Spain as the Iberians, through Sicily and Italy as the Ligurians, and throughout the Mediterranean basin generally. The Empires, he says, of Greece and Rome have disappeared, but have left this old population practically where they stood before. It will be seen how much that is new and controversial there is in all this. There is also a very interesting contribution on the underground dwellings, and their connection with cave dwellings and dolmens.

BOOKS RECEIVED.

Karakoram and Western Himalaya, 1909: An Account of the Expedition of H.R.H. Prince Luigi Amedeo of Savoy, The Duke of the Abruzzi. By Dr. FILIPPO DE FILIPPI, F.R.G.S., Author of *Ruwenzori.* With a Preface by H.R.H. THE DUKE OF THE ABRUZZI. Profusely Illustrated with Maps, Panoramas, and Photographs by VITTORIA SELLA. In 2 vols. Large 4to. Pp. xviii + 470. £3, 3*s*. *net.* London: Constable and Co., Ltd., 1912.

Greece of the Twentieth Century. By PERCY F. MARTIN, F.R.G.S. Demy 8vo. Pp. 391. *Price* 15*s*. *net.* London: T. Fisher Unwin, 1912.

The Conquest of New Granada. By Sir CLEMENTS MARKHAM, K.C.B., D.Sc. (Cam.). With a Map. Demy 8vo. Pp. xvii + 232. *Price* 6*s*. *net.* London: Smith, Elder and Co., 1912.

Recent Events and Present Policies in China. By J. O. P. BLAND. Illustrated. Royal 4to. Pp. xi + 482. *Price* 16*s*. *net.* London: William Heinemann, 1912.

The Little World of an Indian District Officer. By. R. CARSTAIRS. Demy 8vo. Pp. ix + 381. *Price* 8*s*. 6*d*. *net.* London: Macmillan and Co., Ltd., 1912.

The Sea and the Jungle. By H. M. TOMLINSON. Demy 8vo. Pp. 354. *Price* 7*s*. 6*d*. *net.* London: Duckworth and Co., 1912.

Religion und Zauberei auf Bismarck-Archipel. Von P. G. PEEKEL. Demy 8vo. Pp. iv + 135. *Preis M.* 6. Münster i W.: Aschendorffsche Verlagsbuchhandlung, 1910.

The South Pole: An Account of the Norwegian Antarctic Expedition in the "Fram," 1910-1912. By ROALD AMUNDSEN. Translated from the Norwegian by A. G. CHATER. With Maps and numerous Illustrations. In 2 vols. Royal 8vo.

Pp. vol. I. xxxv + 392; vol. II. x + 449. *Price* £2, 2*s. net.* London: John Murray, 1912.

Geographie des Atlantischen Ozeans. Von Prof. Dr. GERHARD SCHOTT. Mit 1 Titelbild, 28 Tafeln und 93 Textfiguren. 4to. Pp. xii + 330. *Preis M.* 23. Hamburg: C. Boysen, 1912.

Aux Sources du Nil par le Chemin de fer de l'Ouganda. Par JULES LECLERCQ. Avec 16 gravures hors texte et une carte. Crown 8vo. Pp. v + 302. *Prix* 4 *fr.* Paris: Plon-Nourrit et Cie, 1913.

The True History of the Conquest of New Spain. By BERNAL DIAZ DEL CASTELLO, one of its conquerors, from the exact copy made of the Original Manuscript. Edited and Published in Mexico by GENARO GARCIA. Translated into English, with Introduction and Notes by ALFRED PERCIVAL MAUDSLAY, M.A., D.Sc. Vol. IV. Demy 8vo. Pp. xiii + 395. London: Hakluyt Society, 1912.

A History of Geographical Discovery in the Seventeenth and Eighteenth Centuries. By EDWARD HEAWOOD, M.A. Demy 8vo. Pp. xii + 475. *Price* 12*s.* 6*d. net.* Cambridge: University Press, 1912.

South America: Observations and Impressions. By JAMES BRYCE. With Maps. Demy 8vo. Pp. xxiv + 611. *Price* 8*s.* 6*d.* London: Macmillan and Co., Ltd., 1912.

Weather Science. By R. G. K. LEMPFERT, M.A. Small Crown 8vo. Pp. 94. *Price* 6*d. net.* London: T. C. and E. C. Jack, 1912.

The Structure of the Earth. By F. G. BONNEY, Sc.D., F.R.S. Small Crown 8vo. Pp. 94. *Price* 6*d. net.* London: T. C. and E. C. Jack, 1912.

In the Shadow of the Bush. By P. AMAURY TALBOT. With Illustrations and a Map. Demy 8vo. Pp. xiv + 500. *Price* 18*s. net.* London: William Heinemann, 1912.

The Pagan Tribes of Borneo: A Description of their Physical, Moral and Intellectual Conditions, with some discussion of their Ethnic Relations. By CHARLES HOSE, D.Sc., and WILLIAM M'DOUGALL, M.B., F.R.S. Demy 8vo. Pp. vol. I. xvi + 283; vol. II. x + 374. *Price* 42*s. net.* London: Macmillan and Co., Ltd., 1912.

The Putumayo: The Devil's Paradise. Travels in the Peruvian Amazon Region, and an Account of the Atrocities committed upon the Indians therein. By W. E. HARDENBURG. Edited and with an Introduction by C. REGINALD ENOCK. Demy 8vo. Pp. 347. *Price* 10*s.* 6*d. net.* London: T. Fisher Unwin, 1912.

Linlithgow (Cambridge County Geographies). By T. S. MUIR, M.A., F.R.S.G.S. With Maps, Diagrams, and Illustrations. Crown 8vo. Pp. 142. *Price* 1*s.* 6*d.* Cambridge: University Press, 1912.

Ski-Runs in the High Alps. By F. F. ROGET, S.A.C. With 25 Illustrations by L. M. CRISP and 6 Maps. Demy 8vo. Pp. 312. *Price* 10*s.* 6*d. net.* London: T. Fisher Unwin, 1913.

Deuxième Expédition Antarctique Française (1908-1910) *commandée par le Dr. Jean Charcot.* Documents Scientifiques. 4 vols. Demy 4to. Paris: Masson et Cie, 1912.

Experimental Researches on the Specific Gravity and the Displacement of some Saline Solutions. By J. Y. BUCHANAN, M.A., F.R.S. Royal 4to. Pp. 227. *Price* 7*s.* 6*d. net.* Edinburgh: Neill and Co., Ltd., 1912.

Pionieri Italiani in Libia: Relazioni dei Delegati della Società Italiana di Esplorazioni Geografiche e Commerciali di Milano, 1880-1896. Con 140 Illustrazioni e Tavole. Royal 4to. Pp. xi + 403. *Prezzo Lire* 12. Milano: Dottor Francesco Vallardi, 1912.

With the Indians in the Rockies. By JAMES WILLARD SCHULTZ. Illustrated.

Demy 8vo. Pp. xii + 228. Price 4s. 6d. net. London: Constable and Co., 1912.

The "Wellcome" Photographic Exposure Record and Diary, 1913. Crown 8vo. Pp. 280. Price 1s. London: Burroughs, Wellcome and Co., 1912.

The Daily Mail Year Book for 1913. Edited by D. WILLIAMSON. Crown 8vo. Pp. xxxii + 314. Price 6d. London: Associated Newspapers, Ltd., 1912.

Die Erklärende Beschreibung der Landformen. Von WILLIAM MORRIS DAVIES. Deutsch bearbeitet von Dr. A. RÜHL. Mit 212 Abbildungen und 13 Tafeln. Small Royal 8vo. Pp. xvii + 565. Preis M. 11. Leipzig: B. G. Teubner, 1912.

Scottish National Antarctic Expedition: Report on the Scientific Results of the Voyage of S.Y. "Scotia" during the Years 1902, 1903, and 1904, under the Leadership of William S. Bruce, LL.D., F.R.S.E. Vol. VI. Zoology. 4to. Pp. viii + 353. Price 30s. Edinburgh: The Scottish Oceanographical Laboratory, 1912.

Bismaya, or the Lost City of Adad; A Story of Adventure, of Exploration, and of Excavation among the Ruins of the Oldest of the Buried Cities of Babylonia. By EDGAR JAMES BANKS, Ph.D. With 174 Illustrations. Demy 8vo. Pp. xxii + 455. Price 21s. net. New York: G. P. Putnam's Sons, 1912.

Die Portugiesen in Abessinien. Von KURT KRAUSE. Demy 8vo. Pp. 118. Dresden, 1912.

On the Tracks of the Abor. By POWELL MILLINGTON. Crown 8vo. Pp. xii + 218. Price 3s. 6d. net. London: Smith, Elder and Co., 1912.

From Pole to Pole: A Book for Young People. By SVEN HEDIN. Crown 8vo. Pp. xiv + 407. Price 7s. 6d. net. London: Macmillan and Co., Ltd., 1912.

Highways and Byways in Somerset. By EDWARD HUTTON. With Illustrations by Nelly Erichsen. Crown 8vo. Pp. xviii + 419. Price 5s. net. London: Macmillan and Co., Ltd., 1912.

Village Directory of the Presidency of Bengal, 45 volumes: *North-Western Provinces*, 37 volumes: *Rajputana Circle*, 5 volumes: *Province of Assam*, 7 volumes. *Presented by Dr. James Burgess, C.I.E.*

Postal Directory of the Bombay Circle. Compiled by H. E. M. JAMES, Esq., Bombay C. S. Bombay, 1879. *Presented by Dr. James Burgess, C.I.E.*

Who's Who? 1913. An Annual Biographical Dictionary, with which is incorporated *Men and Women of the Time.* Sixty-Fifth Year of Issue. Price 10s. net. London: Adam and Charles Black, 1912.

Colonial Import Duties, 1912. Price 3s. 3d. London, 1912.

Foreign Import Duties, 1912. Price 4s. 10d. London, 1912.

Diplomatic and Consular Reports:—

Trade of Valparaiso, 1910-11 (5007); Trade, etc., of Peru, 1910-11 (5008); Trade, etc., of Bosnia-Herzegovina, 1911 (5009); Trade of the Provinces of Seistan and Kain, 1911-12 (5010); Trade of Smyrna, 1911-12 (5011); Trade, etc., of Belgium, 1911-12 (5012); Trade, etc., of Hamburg, 1911 (5013); Trade, etc., of the Vilayet of Trebizond, 1910-11 (5014); Trade of the Adrianople Vilayet, 1911 (5015); Trade of Damascus, 1911 (5016); Trade of Salonica, 1911 (5017); Trade of Norway (Supplementary), 1911 (5018); Finances of the German Empire, 1912 (5019); Trade of Senggora, 1911-12 (5020); Trade and Commerce of Fiume, 1911 (5021); Trade of the Consular District of Riga, 1911 (5002); Trade and Commerce of Uruguay, 1911 (5024); Trade of Santa Marta, 1911 (5025); Trade and Navigation of Port of Dairen, 1911 (5023); Trade via Port Sudan, 1911 (5026).

Publishers forwarding books for review will greatly oblige by marking the **price** *in clear figures, especially in the case of foreign books.*

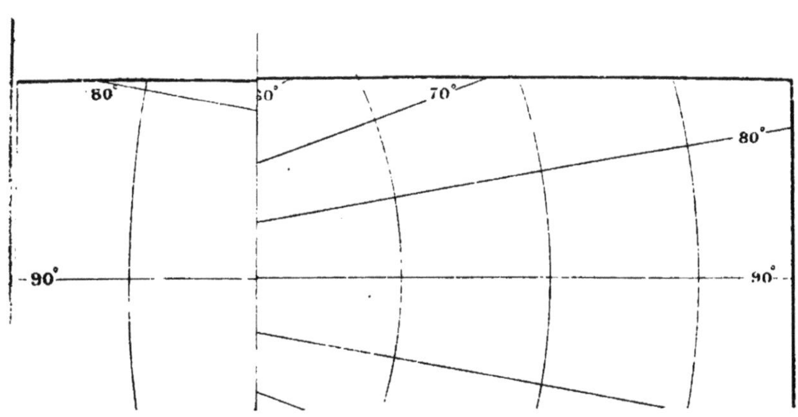

80°
60°
70°
80°
90°
90°

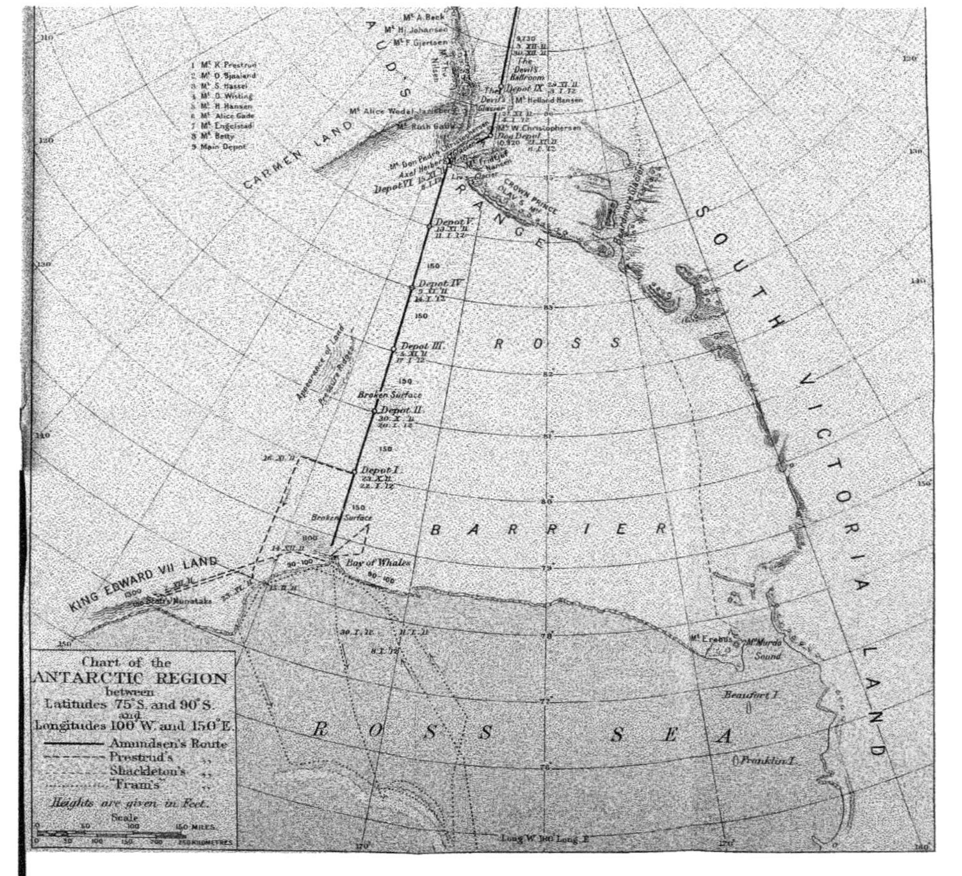

Chart of the
ANTARCTIC REGION
between
Latitudes 75°S. and 90°S.
and
Longitudes 100°W. and 150°E.
Amundsen's Route
Prestrud's "
Shackleton's "
"Fram's" "
Heights are given in Feet.
Scale
150 MILES
SOUTH VICTORIA LAND
KING EDWARD VII LAND
CARMEN LAND
ROSS BARRIER
ROSS SEA
Bay of Whales
Broken Surface
Depot I
Depot II
Depot III
Depot IV
Depot V
Depot VI
Depot IX
The Devil's Ballroom
The Devil's Glacier
Mt. A. Beck
Mt. H. Johansen
Mt. F. Gjertsen
Mt. Alice Wedel-Jarlsberg
Mt. Ruth Gade
Mt. Helland Hansen
Mt. W. Christophersen
Mt. Fridtjof Nansen
Mt. Don Pedro Christophersen
Axel Heiberg Glacier
Crown Prince Olav's Mts.
Beardmore Glacier
Appearance of Land
Pressure Ridges
Mt. Erebus
McMurdo Sound
Beaufort I.
Franklin I.
1 Mt. K. Prestrud
2 Mt. O. Sjaaland
3 Mt. S. Hassel
4 Mt. O. Wisting
5 Mt. H. Hansen
6 Mt. Alice Gade
7 Mt. Engelstad
8 Mt. Betty
9 Main Depot
Long. W. 180° Long. E.

THE SCOTTISH
GEOGRAPHICAL
MAGAZINE.

AHEAD OF CIVILISATION: SOME ASPECTS OF PIONEER WORK IN A NEW STATE.[1]

By Lieutenant-Colonel The Hon. Sir NEWTON J. MOORE, K.C.M.G., J.P., Agent-General for Western Australia.

(*With Illustrations.*)

LAST year, when I had the honour of addressing this Society in Aberdeen and Dundee,[2] I attempted to tell you something of the way in which the plan of Empire is filled in, and my remarks to-night may be regarded in the light of a sequel to that discourse. For I wish more particularly to deal in a more or less sketchy way with the main heads of the history of exploration in Western Australia—the chief landmarks on the path of geographical progress, as it were—and to give you some slight impression, founded partly on personal experience, of the life led and work done by the pioneers of civilisation in a new State.

Founded in 1829, the colony of Western Australia, comprising within its area nearly a million square miles of territory, or the equivalent of more than eight United Kingdoms, has produced a history which, if rather barren in economic results for the first sixty years, until the discovery of payable gold, is remarkably rich in exploration records.

Of all the States of Australia, the Western has been by far the most continuous and persistent in carrying on this work of filling in the map of the interior of the great island-continent. Hardly a year passed from 1829 to 1912 in which some one or more expeditions did

[1] An address delivered before the Society in Edinburgh on December 13, 1912.

[2] See this *Magazine*, vol. xxviii. p. 37.

not leave the boundaries of civilisation and settlement to penetrate into the unknown land, and to report fresh discoveries of pastoral land, rivers, mountains, plains, mineral deposits, examples of an ancient flora and fauna, and many curious and interesting features, sometimes productive ones, of a wide and varied Imperial domain.

Many years ago, the Rev. E. Jenison Woods, one of the most celebrated of Australian naturalists, wrote :—"Since the time of its first foundation, Western Australia has never given up the subject of exploration. Unlike the other colonies, which have always gone into the matter by fits and starts, there have always been continuous expeditions from Perth"; and on another occasion, he said :—"The first of the colonies to wake up again to the importance of examining the interior was, as usual, the indefatigable colony of Western Australia." And this activity in land exploration is of a piece with the earlier prominence of the western coast of the continent in maritime investigation, for its exploration by sea began more than 150 years before Captain Cook sailed into Botany Bay, on the eastern coast of Australia, and began his wonderfully accurate charting of that portion of the shore-line of the great island-continent.

The indefatigable Dutch seamen of Batavia were as consistent and continuous in their navigation of our western seas, as were afterwards our own countrymen, many of them Australian-born, in the mapping of that vast hinterland which is even now not completely explored. Perhaps to the uninitiated, to those who read of these matters in books, there is more of an air of romance about the doings of those who go down to the sea in ships, and meet with that great sensation so vividly pictured in Coleridge's *Ancient Mariner* :—

"We were the first that ever burst
Into that silent sea."

The adventures of the early navigators circling the world in 20-ton or 50-ton sailing vessels, little cockle-shells as they are often called, appeal more readily to the imagination, especially of an island sea-faring people, than does the greyer life of the bushman who, taking his life in his hand, sets out into an unknown interior, knowing not how he shall fare for water or what human enemies he may meet. But the courage of such men is of the highest order. No sea is so lonely as that great unknown land, where no white man is to be met, and where the blacks are often hostile and seldom, if ever—except those whom you take with you and who are half-civilised—of any assistance.

The ocean, in all its parts, is a great highway of traffic in these days, where you cannot proceed far without meeting somebody, and where it is impossible to hide yourself; even in the earliest times there was always the possibility of meeting other craft, and even in case of shipwreck, so long as you were not drowned, the chance of discovery and return always existed as a hope in the breast. But even at this day in Western or Central Australia, once you leave civilisation behind to proceed into unexplored regions, you have left it definitely, and burnt your boats. You must rely thereafter on Providence, on expert bushcraft, and on an alert imagination and keen instinct.

So the story of these explorations, great in number and rich in results as they have been in Western Australia, has an indefinable charm to those who have once experienced the fascination of the never-ending blue haze of sombre landscape, the eternal alternation of sun and stars, and the never-varying heavenly dome which in its serenity and clear expanse seems wider and vaster and more magnificent than the homelier and more varying skies of other lands.

Among the long roll of men who did pioneer work of this character in Western Australia are many distinguished names. One of the earliest of our explorers was Captain, afterwards Sir George Grey, known as the great pro-consul, who was sent out from England in 1837 by the Secretary of State for the Colonies, Lord Glenelg, after whom he named the Glenelg River, a large stream running into the Indian Ocean at a point on the extreme north coast.

Grey spent a considerable time in the exploration of a part of the surrounding country, but was eventually driven back by ill-health, reduced stores, mortality among his horses, and the generally unfavourable character of the season. He retired temporarily to the Isle of France, but in the following year did much valuable work in exploring the country between the Williams and Leschenault Rivers, and subsequently that between Shark Bay and Perth, the capital and centre of the Swan River Settlement, which he reached ultimately in 1839. While he did not penetrate very far inland, it may be said that Grey explored a great part of the long coastal stretch from Prince Regent's River to Perth, a distance of some 3000 miles. He discovered and named a number of rivers, mountain chains, and bays, and his reports were of high value and stimulative of further efforts to investigate the conditions of this vast new land.

But although Grey was one of the first he was only one among many, from whom the chief names alone can be selected for mention; these include Edward John Eyre, famous not only as an Australian explorer, but afterwards as a colonial governor; Captain James Stirling, who proclaimed the colony, and was its first administrator; A. C. Gregory; Frank T. Gregory, who from 1857 to 1861 conducted minute investigations of the north-western districts; Sir John Forrest, 1869 to 1883, who first succeeded in making his way from Perth to Adelaide, and explored the route of the trans-continental telegraph line; Mr. Alexander Forrest, Mr. Ernest Giles, and Mr. L. A. Wells, besides many others.

Eyre's journey from Fowler's Bay to Albany in 1841 is regarded as one of the greatest feats of human endurance on record. In an attempt, begun in 1840, to cross overland from Adelaide to Western Australia he had been foiled by want of water, but he sent back the majority of his party, and started out from Fowler's Bay, in South Australia, with one white companion, Baxter, a black boy, and two other natives, determined to reach King George's Sound (Albany) or perish in the undertaking.

One night the two natives murdered Baxter, and escaped with the greater part of the provisions, leaving Eyre and the black boy, Wylie, ill-equipped and provisioned, to complete the journey of some hundreds of miles on foot, and alone. They endured incredible hardships, and, when within about three weeks' march of Albany, fortunately fell in with the French whaler *Mississippi*, whose commander, Captain Rossiter, enabled them by his kind treatment to survive, when they were reduced to the last extremity of hunger, thirst, and fatigue.

The accounts of this arduous journey form a most thrilling narrative and demonstrate, with the history of the fatal expeditions of Leichardt, of Burke and Wills, and at a later date of Wells and Jones, that the annals of Australian exploration are not without their features of tragedy and mystery. But the work of these intrepid pioneers was to draw a few thin lines upon the map rather than to fill it in with ample details. After them follow the surveyor and prospector. The life of these men is simple, and their surroundings are such as to instil into them a spirit of reverence and an imagination which is seldom possessed by the town-dweller.

As I have led this life, some description of it may prove interesting. In the early days of surveying a survey-party would consist usually of the surveyor, his assistant, three or four hands, and a native boy, with perhaps twenty to thirty pack-horses. In later years the "ship of the desert" (the camel) has largely superseded the pack-horse, owing to the fact that in some parts of Western Australia there are long distances between supplies of water, and the camel, as is well known, can go without water for long periods. This animal has thus proved a most useful factor in surveying and prospecting work.

There are now also superior mechanical contrivances for carrying water on these journeys. In the old days any water that was carried was taken in small wooden barrels, but in later years these have been discarded, and perhaps one of the finest contrivances for the carriage of water in the interior has been substituted in the shape of the water-bag. Where camels are used, huge bags, capable of carrying from three to six gallons of water, are swung on either side of the camel, and similar ones are hung on the horses, or are carried by the men.

The life on survey is quite dissimilar to anything one might imagine in a country like this. The horses live on the natural herbage. In some places this is fairly luxuriant; in others, coarse and not particularly nutritious. At night hobbles are put round the forelegs of each horse, and such is the love of company in these animals that they rarely stray more than a mile or two from the camp-fire. Even if they do stray, the black boy, who is a wonderful "tracker," has no difficulty in finding them. As an instance of this remarkable faculty, I may mention that many years ago I had made a rendezvous with a fellow-surveyor at a point some sixty miles from where is now situated the famous town of Coolgardie. I had to go round two sides of a triangle to make this point, and he was to come direct across the hypotenuse, and we were to meet on a certain afternoon approximately at a given point. When within a mile or two of this place each was to signal by firing off his rifle. On arriving at the meeting-place I fired repeatedly without eliciting any response, and so I made back to water, where I thought my friend might be. But he was not there, so next day I set out to look for him, accompanied by a black boy, who eventually discovered his tracks, and tracked him without the slightest difficulty. He would even stop and pick up a twig to show where my friend had picked his teeth, and with this sure detective we found my friend next day, none the worse for being temporarily "bushed."

In the Kimberley district the aboriginal natives have some customs which are peculiar to themselves, and which are of considerable interest.

Camel teams leaving Carnarvon.

One peculiar fact about them that was noted by Dr. House, the naturalist in Mr. Fred. Brockman's exploration party of 1901, was the complete absence from their weapons of the boomerang or kylie (which is, more than any other implement, always associated with the Australian black); not one being seen amongst the weapons met with.

The most interesting point in regard to these natives is the custom they have of making drawings on suitable faces of rock. These are done in colour—in red, yellow, black, and white; the black (says Dr. House) being charcoal, and the other colours argillaceous earth, specimens of which were found by Mr. Brockman's party wrapped up in paper-bark parcels in deserted native camps. Most of the figures are of human heads, and are better drawn than would be expected from the primitive state in which these people live. Other drawings represent snakes devouring human beings.

That this rude art preceded the advent of the white man is shown by the fact that Sir George Grey discovered similar drawings on the Glenelg River in 1837, although it is singular that the figures have clothes on, making it appear that the natives must at least have seen shipwrecked mariners.

There are other evidences of considerable ingenuity on the part of these natives. According to Dr. House, they make very passable string out of the inner bark of a wattle which grows in abundance. Opossum, kangaroo, and human hair are also used for spinning into string, this being accomplished by rolling it on the thigh with one hand, the string as it is made being wound on a cross with two sticks bound together.

They live largely on fish, and their methods of obtaining this item in their food supply are both interesting and ingenious. As the rivers get low, they travel down from pool to pool. With a coarse grass and wattle-bark they make what looks like an enormous straw bottle, the inside of which they fill with bark obtained from the root of a shrub that grows on the banks of all these rivers, and is known as "Majalla." This straw bottle they then drag backwards and forwards along the pool, with the result that the fish are stupefied, probably by the "Majalla," rise to the surface, and are easily caught.

Besides fish, the natives in these regions use a great number of different edible roots, seeds, nuts, and reptiles as elements in their food supply. Large game they get occasionally, but seldom, and their life is one long nomadic hunt for food. Stones for pounding their food are found in all their camps, and the use of fire is known to them.

The surveyor literally "blazes" the track for the settler and miner, as by his well marked survey lines the prospector is able to make his way through the bush, and the information that is put down on the plans as a result of the surveys is of a most valuable character to the selector who goes out to look for a new home, or the squatter in search of new land on which to locate his station, or the prospector following up a line of mineral country, in the hope of striking minerals in paying quantities.

It is some twenty years ago—I think it was this month—that I was at Coolgardie, in the capacity of a junior surveyor of the Lands Department, and our work was to locate the recent find which had then been made by Arthur Bailey at Coolgardie. With that object in view a tie line was run from what was known as the slate well; and in November we arrived at Coolgardie, and then we were able to demonstrate that this field, which it was anticipated was lying within the concession of the Hampton Plains Syndicate, was twelve miles inside Government land. But I little anticipated when looking at that small hole in the ground, at that time something like eighteen inches deep—they were awaiting the arrival of the survey party, and no development at all had taken place—when I looked at that hole lying underneath a big quartz reef, little did I anticipate that twenty years afterwards I, as representative of the State of Western Australia, would be in a position to say, in this great centre of our Empire, that since that period gold to the value of one hundred and seven millions sterling had been produced by Western Australia, and that dividends amounting in all to something like twenty-three millions sterling have been paid by the various gold-mining companies operating in Western Australia. The mineral industry has also been the fount and origin of other industries. It has attracted population, the human power which, applied to the slumbering potentialities of our land, has awakened them into productive life.

And so it is that we have now entered upon the second phase or stage of our development as a productive State.

The explorer, the surveyor, the prospector (and these three are really one, all three functions being often combined in the one person)

Miners' Camp near Roebourne, in the West Pilbarra Goldfield.

are still doing their work in filling in the map, but first mineral and pastoral development resulted from their labours, and now we have come to the agricultural and horticultural stage. The precedent condition to all productive expansion is, as you will agree, the attraction of population, and for this the goldfields first acted as the magnet. Between the year 1892, the year of the discovery of Coolgardie, and the year 1906, the population of Western Australia increased from some 58,658 to 255,173 souls, or had more than quadrupled in fourteen years.

Then began a new phase; a return of prosperity visited the Eastern States, and the inflow of surplus population into Western Australia began to decrease; at the same time the expansion of mining became restricted. Manifestly Western Australia had to look for new sources of population and wealth. Of the result, if I speak ever so humbly, I can but speak eloquently, for the facts and figures themselves are eloquent.

Just at this period the people of the State began to turn their attention to agriculture, and the Government, of which I was the head, tried to devise means by which agricultural settlement could be promoted. The first arm of State assistance which we put forth was liberal terms of settlement—probably the most liberal in the British Empire. The second was the construction of agricultural railways by which the settler has been enabled to convey the produce of his land to the markets. From 1906 to the beginning of 1912, about 1000 miles of new railway line were constructed, consisting of extensions into the various new agricultural districts, as well as mining districts, extending from Nannine in the north to Hopetoun in the south, while the agricultural

A load of wool at Carnarvon, the collecting centre of the Gascoyne River district.

railways were practically confined to the south-western division, *i.e.* within a safe rainfall limit as far as agriculture was concerned. According to the Budget speech, which was only delivered a month ago, there are now under construction, or proposed, some 850 miles of line, of which 350 miles will be completed by the end of the current year. When all these new lines, now in contemplation or in course of construction, are completed, the State will own no less than 3450 miles of railways, or an average of one mile of railway for every ninety persons within the State. That, I believe, is a record as far as railway construction is concerned.

The third arm of State assistance is the Government Agricultural Bank, which lends money on approved holdings in sums ranging from £25 to £2000 for improvements, water conservation, stock, farming implements, and other approved purposes.

The success of this policy has been completely demonstrated, and when I say that a total sum of £2,000,000 has been advanced, and that the arrears at the present time only amount to £1000, you will realise that it is a tribute to the business capacity of the manager, as well as a tribute to the fertility of the land on which these advances have been made. Last year the profits made at the bank were some £8000, which brings the present reserves up to something like £35,000.

There are now 7101 accounts current on the books, representing as many customers, and the effect of the policy pursued has been to induce other banks and financial institutions to come into the field for the purpose of lending money on Western Australian agricultural land.

The results of this triple-armed policy of State assistance are apparent. In the strides taken by the wheat industry alone the rate of progress is very remarkable. In 1903, excluding the area cut for hay, there were 92,398 acres under wheat; this year, with the same exclusion, there are 750,000 acres, and from this area a total crop of 9,000,000 bushels—compared with 985,550 bushels in 1903—is expected. In other words where, ten years ago, we reckoned our wheat by the hundred thousand bushels, we reckon it now by the million.

That is the difference which the State policy of immigration, land settlement, railway construction, and financial aid, working on the sound basis of good land and effective human power, has made to Western Australia in wheat growing alone. Nor is this all. While we had only some 6000 acres of land under orchards in 1903, we have now 18,193 acres, and there is further a large annual increase in planting. There are also 2821 acres under vines. We exported in 1911 about 22,000 cases of fruit; this year we exported 65,000 cases; and from the advice that I have there is every possibility that the next year there will be considerably over 100,000 cases; so that for the last three years, so far as fruit is concerned, each succeeding year has doubled the yield of its predecessor. Our apples and grapes bring the highest prices in the market; they are ahead of all others by a considerable percentage, in the London and continental markets; and there is every prospect of an excellent trade opening up in India. To the policy of opening up to agriculture the fertile lands of the south-western district, and irrigating them in the summer season, the Government are now devoting attention, and quite recently a bill has been introduced with a view of giving effect to this policy.

There is a stretch of country 20 million acres in extent in the south-west, with an average assured rainfall of from 30 to 50 inches per annum, which is suitable for fruit-growing, mixed farming, and dairying. On that area the Government, assisted by the best expert advice, are concentrating their efforts at the present time. They have obtained the best possible experts they are able to get in Australia, and are looking forward in the very near future to some very big developments in this part of the State, a portion which, so far as agriculture and mixed farming is concerned, has been neglected to some extent.

It has been found that it was a so much simpler matter to get a prompt return in the far-eastern districts, that most of the new settlers have gone to the wheat areas; but, with the assistance that is now being offered, the cheaper cost of collecting and other inducements that are held out, we hope, in the very near future, that Western Australia will make a substantial advance not only in the fruit industry, but also in dairying.

The timber and wool industries antedated what I might call the second phase of Western Australia's economic development. But they

Karri (*Eucalyptus diversicolor*) forest in Western Australia. Karri is one of the important timber trees of the region.

have progressed exceedingly under the new conditions. The powerful stimulus of increased population has had its effect upon these great industries. The export of wool has gone up from a value of £443,000 odd in 1903 to over £925,000 last year, and in a more favourable season it rose to over a million sterling. The number of sheep was doubled in the ten years; and cattle and horses have nearly doubled during the same period. The export of timber has risen from about £500,000, a decade ago, to a million sterling annually, and orders are flowing in at the present time to the timber companies.

Twenty thousand emigrants whom we have sent out in about three years have all been absorbed. There is not a trace of them in the general community, and the leading newspaper of the State declares that we want, not only 10,000 a year, but 20,000 a year, and we can absorb them to our advantage and to theirs.

An evidence of the great progress of the State in recent years is supplied by the figures of imports indented from the United Kingdom by the Government alone.

It just gives some slight index when I say that while in 1908-9 the value of the Government indents was £120,000, in the year 1911-12 it has gone up to considerably over £530,000, or more than four times the sum in the former year. At the present time our indents amount to something like just under three-quarters of a million.

This is but one evidence of the great expansion in trade and industry that is taking place; and when I tell you that in a little community of just 300,000 souls, the total trade amounts to £19,727,292, made up of exports £10,443,570, imports £9,283,722, you will get some idea of the trade of Western Australia. That brings the average up, *per capita*, to no less a sum than £65, 5s. 3d.

Now, I have given you all this information about the second stage of productive development, partly because it is all part of our pioneer work, and partly because it is the direct result of exploration and discovery. The explorers who first went out with horses, then those with camels, and then the prospector, have, by their patient heroic work, prepared the way for the miner and the farmer.

It is a fact which will stand examining, that, had there been no explorer there would have been no farmer in Western Australia, and had there been no farmer, there would have been no trader, so that the great Dominion's trade with the United Kingdom, which is growing year by year, is the direct outcome of the work of the explorer. Thus it is that the science and labour of the geographer make empires in fact as well as in name, and that this restless desire which possesses the British people to penetrate unknown regions is the true source of their Imperial greatness.

A GEOGRAPHICAL DESCRIPTION OF FIFE, KINROSS AND CLACKMANNAN.[1]

By Laurance J. Saunders.

(*With Sketch-Maps.*)

I.—Physical Characteristics.

Position.—One of the most important features of the Scottish Lowlands is the ridge of harder rock which crosses it under various names from sea to sea. This runs north-east and south-west parallel to the crests of

[1] This essay was awarded the Silver Medal offered by the Council for competition among the students of geography at Edinburgh University during 1912.

the Grampians and the Southern Uplands, and so divides the Lowlands into two valleys. Between them communication is ensured by three important breaks in the dividing line. The Clyde draining to the south-west through a transverse valley from the Southern Uplands, gains the sea between the Renfrewshire heights and the Lennox Hills. The Teith flows south-eastwards from the highlands of West Perthshire in another transverse valley parallel to the Clyde, and then, joining the Forth near Stirling, separates the northern spurs of the Campsie Fells from the south-western extremities of the Ochil Hills. Below Stirling this river widens out into a firth which, through taking its name from the Forth, lies in a line with the Teith in its narrow enclosed upper basin. At Queensferry there is a change of direction; and the firth, now very much broader and more open to the influences of the sea, turns and trends to the north-east. The third interruption in the Mid-Lowland range is near Perth, where the Tay breaks seawards between the Ochils and the Sidlaws. Then, like the Teith-Forth, it widens out into a firth which stretches in a north-easterly direction between the eastern prolongation of the Ochils and the Sidlaw Hills. A peninsula is thus formed between the Firth of Tay and the Firth of Forth. Eastwards lies the sea; westwards, the Ochils run from the head of the one firth to the head of the other, and so cut off the peninsula from the rest of Scotland. The interior is occupied by the three counties of Fife, Kinross and Clackmannan.

Shape and Build.—Since this peninsula is marked out by parallel and transverse elevations and depressions, it is rhomboidal in shape, and lies with its longer and parallel sides running north-east and south-west. The northern of these longer lines is formed by the Ochils, which stretch from within a few miles of Stirling on the Forth in a north-easterly direction for thirty miles, meeting the Tay near Newburgh, eight miles below Perth. This line is prolonged in the same direction for twenty miles along the south shore of the Firth of Tay to where the Tay enters the North Sea at Tents Muir Point. Thence the eastern side of the rhomboid runs to Fife Ness, but the regularity of the coastline is here broken by the wide sweep of St. Andrews Bay, which adds ten to the fourteen miles of straight line from point to point. Fife Ness is the easternmost projection of the peninsula (56° 16′ N., 2° 25′ W.), and from it the coastline of the Firth of Forth runs south-west to the southern apex at Ferryhill. This line, roughly parallel to the Ochils, is cut into by the semi-circle of Largo Bay ($6\frac{5}{8}$ by $2\frac{1}{4}$ miles) which brings the length of the stretch of coast up to 55 miles. From Ferryhill to where the Forth almost washes the base of the Ochils near Stirling, the northern shore of the upper firth completes the rhomboid. The key to the elevation of the interior is the circular area of lowland lying round Loch Leven in the centre of the peninsula. Thence radiate three lowland strips separated the one from the other by uplands which occasionally rise above a thousand feet. To the north-east the valley of the Eden opens on to St. Andrews Bay; to the east, the valley of the Leven from Loch Leven opens on to Largo Bay; to the south-west the valley of the Lower Devon opens on to the head of the Upper Firth. All along the northern side of the depression formed by the Devon valley, the plain of Kinross and Stratheden, are the Ochil Hills and their eastern continuations, forming a natural boundary from firth to firth, to which the county marches approximate. Then between the Eden and the Leven rise the Lomonds and the highlands of the East Neuk running eastwards to end in Fife Ness. To the south of the pit of Loch Leven, and filling in the area between the Leven and the Devon, is another elevated region, its highest peaks, well over a thousand feet, bearing the names of the Cleish and Saline Hills.

The Ochils.—Among these features the oldest rocks are to be found in the Ochil system, which is composed of underlying andesites, igneous rocks of the Lower Old Red Sandstone Age, associated in some places with beds of sandstone and argillaceous shale. Over these sedimentary strata are poured the lavas of the Upper Old Red, which stretch along the entire system, highest in the west and sinking to low heights near Tayport in the east.

In the west the Ochils rise steeply above the gorges cut by the Allan Water while making its way southward along their flanks to join the Forth near Stirling. Thence the ground mounts to Blairdennon Hill (2027 feet), from which flow the headwaters of the Allan and the Devon, the former at first due north-east and parallel to its subsequent lower course through Strathallan, the latter eastwards through a longitudinal valley. In the southern of the two parallel ranges into which this valley cuts the Ochils are the highest peaks of the whole Mid-Lowland range, culminating in Bencleuch (2363 feet). Then about six miles in a direct line eastwards from Blairdennon, whence start the two ranges and the valley between, all three are cut by a great transverse which never rises above 900 feet. On entering this transverse the longitudinal valley ceases; the cut crossing the northern range drains to the Earn, and so the Devon finds its way through the southern part of the transverse (Glendevon) to the plain beneath. Ten miles due north-east of Glendevon comes another transverse, dropping to 480 feet. This is formed to the north by the little river Farg, which flows through Glenfarg to join the Earn as it enters the Tay; southwards, the valley opens on to the lowlands near the watershed between the Eden and those streams which flow to Loch Leven. With Glenfarg ends that part of the Ochils which, but for the proximity of the Grampians, might be called mountainous. In the line of twenty miles from Strathallan to Glenfarg, there are no less than five summits over 2000 feet in height, and only one well-marked depression, Glendevon; and even it rises to 900 feet from lowlands well below 500 feet. In addition, the breadth of the hills is on an average ten miles, so that it can easily be understood what an enormous obstacle to communication they form, and how important Glenfarg must become as the first low crossing-place east of Strathallan.

Eastwards of Glenfarg, the Ochils sink in height to under a thousand feet. They now come close to the Tay, leaving little or no level land along the northern coast of the peninsula. Then just where the range approaches the firth (seven miles from Glenfarg) occurs a third transverse well below 250 feet, dominated to the west by a conspicuous cliff of intrusive andesite. This valley opens southwards on to the part of Stratheden known as the Howe; to the north lies the Firth of Tay, but communication with the river lands to the west is maintained by a narrow strip of lowland running close to the river along the northern base of the Ochils. In the centre of this low transverse lies the small Loch of Lindores, whence another longitudinal valley runs eastwards again, dividing the hills into two parallel ranges. The northern range rises to over 900 feet in Normans Law, then sinks to wooded heights near Tayport, where it ends above the dunes of Tents Muir. The southern range rises to and terminates in Lucklaw Hill (625 feet), round the base of which the Motray Water carries the drainage of the central valley to the estuary of the Eden.

Stratheden.—To the south of this low prolongation of the Ochils lie the unconformable red-and-yellow sandstones of the Upper Old Red, found far underneath the surface drift of the Eden valley. Southwards they are bounded conformably by the carboniferous systems, the division between the two running in a fairly straight line from near the Eden mouth to Falkland Road Station, thence round the base of the intrusive igneous mass of the Lomonds into the plain of Kinross. Most of this area is occupied by the low-lying plain of the river Eden, which, formed by the junction of several small streams from the Ochils and the Lomonds, reaches the sea after a north-eastward course of 30 miles. Since one of its headwaters comes from the southern prolongation of Glenfarg, communication is thus maintained across the Ochils with the Tay valley and Strathmore, while the opening westwards between the Lomonds and the Ochils to the plain of Kinross is low and easily traversed—in fact only 400 feet high. This upper part of the valley of the Eden is constricted by the hills on either side to a breadth of less than two miles; but, by the north-eastward trend of the Ochils and the curve of the Lomonds, it widens out at Strathmiglo into what may once have been the bed of an old lake, now the Howe o' Fife, four miles broad and triangular in shape. To the north it is bounded by the Ochils, which, becoming broader to the east, again impinge on the Eden at Cupar. Across them the Pass of Lindores opens from the Tay valley just where the Howe is broadest, that is, opposite the southern apex of the triangle. The south-western side is formed by the slopes of the Lomond Hills, which curve to the southern apex, formed by a tongue of lowland running between the Lomonds and the highlands of the East Neuk. This depression, the Pass of Markinch, from a town at its southern extremity, connects the Howe with the river lands of the Leven. Further east the Howe narrows as the hills of the East Neuk press north, until at Cupar the Eden, which has pursued a meandering course over the dead level of the plain, is hemmed in between low spurs from the Ochils and these hills of the East Neuk. Thence to its estuary, six miles eastward, its valley is narrowed to less than two miles in breadth. Three miles below Cupar it receives its most important tributary, the Ceres Burn, which follows a parallel course north-eastward among the hills of the East Neuk and now cuts abruptly northwards through the famous Dura Den, where it exposes the underlying sandstone with its fossil treasures. The Eden finally reaches the sea through a winding muddy estuary between Tents Muir and the links of St. Andrews.

The Eastern Highlands.—With the exception of a portion of Kinross, the rest of the area to be described is composed of rocks which form a part of the great carboniferous systems of the Scottish rift valley. Eastwards of a line from Elie to St. Andrews lies a triangle of calciferous sandstone, the other two sides being St. Andrews Bay and the Firth of Forth, the eastern apex Fife Ness. From the northern apex an irregular tongue-like area of this same rock projects westward to underlie the basin of the Ceres Burn; otherwise the rest of the East Neuk is occupied by carboniferous limestone. Northwards are the sandstones of the Eden, while to the south a line of fault runs from Largo to near Leslie. In this area of the Neuk volcanic phenomena occur extensively. For example, to the east of Loch Leven rise the isolated Lomond Hills in three great peaks—West Lomond (1713 ft.), White Craigs (1492 ft.) and East Lomond (1471 ft.). The mass is circular in shape, and is formed of soft marls, sandstones, and limestones under a cap of basalt which has intruded through the sedimentary strata and so preserved them. Separated from the Ochils by the valley of the upper Eden, and, westwards, from the volcanic Cleish Hills by the Leven, the Lomonds are connected with the highlands of the East Neuk by a low ridge of carboniferous limestone which forms the depression of Markinch already noted. Eastwards of this the hills spread out intricately to occupy the peninsula between St. Andrews and Largo Bays, but the contours seem to have been determined by two lines of harder dolerite sills which diverge from Clatto Hill in the west. The northern ridge, lying between the Eden and the Ceres Burn, is continued eastwards by Walton Hill (622 ft.) and Ceres Moor. Here the Ceres Burn cuts down to the Eden, but the range reappears on the other side of the stream to terminate near St. Andrews. From Clatto Hill the southern line of heights—an easterly continuation of the volcanic rocks of the Lomonds—forms a broad watershed running eastwards to Kellie Law and Fife Ness, between the tributaries of the Ceres Burn, which drain the central valley, and the small streams which find their way independently to the Firth of Forth. Outwith these lines of high land stands the isolated Largo Law (948 ft.), an intrusive mass of igneous rock which ages of denudation have worn down into the Vesuvius-like cone so conspicuous a feature of the scenery of the coast resorts. These hills of the Neuk are by no means inaccessible; except where they rise above the lowlands of the Eden or where they form such rare irregular cones as Largo Law, they appear as swelling ridges above slightly inclined plateaus which sink to the sea in terraces and form low cliffs or grassy banks between the numerous small creeks from St. Andrews to Kincraig. At this latter point (near Elie) a small igneous intrusion forms a long cliff of a height of 200 feet, almost the only example in Fife of wild rock scenery.

The Leven Basin.—Between the Lomonds and the highlands which fill in the area between Kirkcaldy and the lower Devon, lies the valley of the Leven, comprising two of the four lowlands of the peninsula. The upper area, 400 feet above sea-level and circular in form, centres on Loch Leven (353 ft.). To the north and the north-west rise the Ochils; to the west the Lomonds. Between them a strip of lowland leads to the upper Eden valley, and so low is the watershed between the two systems that the transverse of Glenfarg, drained southwards by a head-water of the Eden, is accessible from Kinross and so connects with the Tay. Another strip of lowland leads west between the Ochils and the Cleish Hills to the Devon valley at Crook o' Devon, thence to the Glendevon transverse or the plains of the Forth. Then to the south of Loch Leven rise the basalt-capped Cleish Hills (Dunglow, 1241 ft.) and Benarty (1131 ft.), between which a depression slightly under 500 feet leads to the upper Ore valley; while between Benarty and the Lomonds the Leven finds its way from the loch through the narrowest gap of all to Largo Bay. Within this broken circle of hills lies the plain of Kinross, made up of the soft sandstones, marls, and conglomerates of the Old Red and so geologically connected with the Eden valley. To the south these rocks dip conformably beneath the harder carboniferous limestones of the hill area. The drainage of this circular depression collects in Loch Leven, whose waters occupy an irregularity of surface left by ice action. Roughly heart-shaped, it lies with its axis pointing north-west and south-east and measures 3⅝ miles by 2 with an area of 3406 acres—the largest loch of the Lowlands. Numerous sandy islets dot its surface, and in two pits its depth exceeds 10 fathoms. From its southern apex it discharges the Leven, which flows due east between the Lomonds and Benarty to reach Largo Bay after a course of 16 miles.

Three miles above its mouth the Leven receives the river Ore from the east. This stream rises at a height of nearly 1000 feet in the Cleish Hills and thence flows eastwards until it meets the Leven. It has several tributary streams, which, following the general trend of the land as left by the ice action, also flow east or north-east—the Fitty from Loch Fitty near Dunfermline, which meets the Ore near Lochgelly, and the Lochty from Benarty, which meets the Ore near Thornton. On the north side, the Leven receives one important tributary, the Conland, which rises in the Lomonds and flows past Markinch. Round the lower courses of these various rivers is formed a fan-shaped plain area. To the east it is bounded by the sweep of Largo Bay; to the north lie the Lomonds and the heights of the Neuk, between which the depression at Markinch allows fairly easy communication up the Conland to the Howe. Westwards the lowlands of the river Leven stretch in a narrow strip to the Kinross plain. To the south of this the land rises in long parallel ridges trending west or south-west, between which are the upper valleys of the Ore, Fitty, and Lochty. Along the coast to the south lies a smaller plain through which the little river Tiel flows eastwards to reach the sea at Kirkcaldy. This area is separated from the Leven riverlands by a gentle ridge running eastwards from the higher land to the west; southwards it ends where the volcanic formations of the Burntisland district rise close to the coast at Kinghorn.

The geology of this part of Fife is most important. To the north of the fault from Largo to Leslie lies the carboniferous limestone already described, but, southwards of this, the fault has preserved the more recent rocks which give the area its present economic value. These lie for the most part in the triangle marked out by the coast, the fault, and a line drawn from Leslie to Pathhead on the coast. Westwards of this the carboniferous limestone reappears; eastwards lies a narrow strip of millstone grit exposed in a shore section from Pathhead to Dysart. Above this to the east are the true coal measures, which, except where, along the coast from West Wemyss to Methil, a line of still more recent rock (Upper Red Sandstone series) occurs, fill up the rest of the triangle and are connected under the firth with the measures of Midlothian. Isolated from the larger coal area, a small field occurs near Kinglassie, which runs tongue-like up the Lochty valley from the millstone grit, with carboniferous limestones on either side.

The Western Highlands.—Westwards of this plain long elevated ridges run north-east and south-west, swell into hills and then sink to the coast lands of the upper firth or the river lands of the Devon valley. Most of the rocks belong to the carboniferous system, with igneous rocks contemporaneously erupted, not intruded as in the Neuk. Round Burntisland calciferous sandstone is present, and the coast sections there show remarkable examples of alternate sedimentary and igneous bedding; but the rest of the area as far west as Culross is composed of underlying carboniferous limestone with lavas capping the higher hills. Of these latter the best marked ridge runs from Benarty south-westwards, temporarily disappearing between that hill and the Cleish Hills in the depression already noted as connecting the Kinross plain with the upper Ore valley. Westwards of this rise the Cleish Hills, and a little to the south of them, and parallel, the two volcanic vents of the Saline and Knock Hills, 1178 and 1189 feet respectively. To the south of these runs the narrow valley of the Fitty, then another parallel ridge culminating in the Hill of Beath (705 feet), near Dunfermline. East of this, and between these northern heights and the igneous hills around Burntisland, is a wider valley drained northwards by the Ore and the Tiel, southwards and south-eastwards by the Lyne, on which stands Dunfermline, and by the Keithing falling into the Forth at Inverkeithing. Thus these hills around Burntisland stand somewhat isolated, with a shallow but fairly wide valley between them and the elevations to the north. The highest of these remarkable hills are Cullalo and the Binn (632 ft.), while similar formations project out into the firth to form the peninsula on which Pettycur stands and the islands of Inchkeith and Inchcolm. West of Inverkeithing the further presence of these rocks is shown by the peninsula and promontory of Ferryhill, which, jutting out over against Mons Hill in Linlithgowshire, almost encloses the Upper Firth.

The Carse lands.—Westwards of this, low spurs from the interior come down to the coast, with here and there a break through which some small stream flows to the firth. At Longannet Point these ridges begin to draw away from the shore, and running inland are divided by the Black Devon, which twists down from the Saline Hills through the town of Clackmannan. Further west a single low ridge runs parallel to the Forth, and between it and the Devon to the north. This latter stream emerges from Glendevon and, meeting the opposition of the Cleish Hills to progress due south, turns in an abrupt curve at "Crook o' Devon" away from Loch Leven and elects to flow westwards to the Forth through deep ravines and chasms until near Dollar it comes out on to the level. Between it and the Forth there now lies the low continuation of the western heights, which forces the river to flow along the base of the Ochils, parallel to, but in the opposite direction from, its upper course in the higher longitudinal valley. On its right bank it receives numerous burns which have ploughed deep picturesque glens down the steep and in some places precipitous face of the hills. But west of Menstrie these burns find their own way to the Forth, for, the ridge between the Devon and the Forth coming to an end at Tullibody, the Devon turns abruptly south and flows for two miles across the Carse of Clackmannan, a plain similar in physical characteristics to the Howe. The river thus reaches the Forth after a course of 33 miles, only 5 miles south of its source in Blairdennon Hill. The alluvial plain of the Carse now narrows westwards between the Ochils, lying here almost due east and west, and the river Forth, flowing south-eastwards until communication is almost stopped by the Abbey Craig, an igneous mass showing the crag-and-tail effect, which rises between the river and the hills some 2 miles west of Devonmouth. Just to the west of this the Allan Water reaches the Forth from the flanks of the Ochils.

The geology of this plain area of the Devon and the Forth is similar to that of Strathleven, and it is for the same reason important. A line of fault runs east and west along the base of the Ochils, cutting the lavas of the Old Red Sandstone; south of this lie the preserved carboniferous systems. The limestone which occupies the Saline Hills ends at Culross and is bounded to the west by a strip of millstone grit which in its turn bears, as in the Leven plain, the important true coal measures. These seem to be an extension northwards of the great Lanarkshire field, buried in East Stirlingshire and the Carse lands under the more recent deposits of the Forth.

Soil.—Despite the close connection which appears to exist between the volcanic areas and the higher lands there is nothing to indicate that volcanic activity directly determined the present configuration. This was more probably produced in its main outlines by epigene action during long geological ages, and it is really due to the greater resisting power of the igneous formations that they in many cases now stand out as the higher ground. Then the ice flood descended from the Highlands, mounted over the Ochils and turned eastwards into what is now the Firth of Forth; but it came upon an area the main features of which were substantially those of to-day, only higher, more rugged, and covered with the results of long erosion. The ice planed down the hills and swept away the surface accumulations to lower levels, where it mingled them with other deposits it had brought from the north and strewed the mixture to form a soil over the land. When it retreated, the hills had decreased in height; many, like the Abbey Craig or the May Isle, had assumed the crag-and-tail formation, their basaltic-topped precipices fronting westwards whence came the ice. Volcanic chimneys of solidified ash had been ground down to the regular cone-shape of Largo Law or Kellie Law. Long regular ridges now ran in an east-west direction, while round the coast were traces of former beaches at heights of (roughly) 25, 50 and 100 feet. More important still were the vast accumulations which the ice left behind to form the superficial covering of the land. Of these the most widely distributed is till or boulder clay, which, pushed along from the north-west, filled up the lower valleys, mounted up the higher hills to over a thousand feet, and eventually passed under the waters of the firth to form an impervious bed above the coal measures there. It now forms over most of the area the surface soils, though in some places, which it has failed to cover, the soil is directly derived from the disintegration of the underlying rock; in some other places the boulder clay is covered by still more recent deposits. Where this clay forms the soil—mainly over the carboniferous system—it is cold, stiff and retentive, supporting in its natural state vegetation of a rushy character; but when properly treated by modern methods it has been found to be good corn land. Its texture and appearance depend largely on the rock from which it is derived; if in great part from the sandstones, it is warm in colour, sandy yet stiff in texture; if from the carboniferous systems, blue or black. It is among the higher ground of the peninsula that this covering of clay is sometimes broken by underlying rock. A conspicuous instance of this is to be found in the Ochils, the peaks of which rise above the limit of the clay and are covered with thin soil composed of their own detritus and capable of supporting rough pasture and heather. The lower slopes are covered with boulder clay. Then in the hills of the East Neuk are other examples of soils derived from underlying rocks. These are rarely sedimentary owing to the comparative ease with which such rocks weather away; yet at one spot, called suggestively Ceres Moor, the sandstone crops out to form a poor, thin soil. More numerous and of greater importance are the rich loamy soils derived from exposures of igneous rock. These soils are warm and fertile and usually contain a plentiful supply of phosphate of lime owing to the presence of apatite. In marked contrast to the dark rushy vegetation of the boulder clay, they support bright green grass, and thus it often happens in the East Neuk that the soils of the higher lands, which are composed of rotted igneous rock, are covered with more flourishing vegetation than the lower slopes of boulder clay. Otherwise the soils of the Neuk, except in some important stretches to be afterwards noted, are cold and clayey. To the south of the Neuk, the Ore valley and the highlands near Dunfermline are covered with a subsoil of boulder clay over which sometimes lies a thin loam.

Above the boulder clay were deposited more recent sands and gravels, which are now found in Stratheden, Strathleven, the plain of Kinross, the Devon valley, and their transverse depressions. They also run rim-like round certain stretches of coast. These deposits seldom rise above 200 feet in the southern parishes of the Neuk, and in many places they run along the 100-foot beach. From below Largo Law they stretch inland as far as Colinsburgh, and then drop down to the 100-foot terrace along the coast of the East Neuk and round Fife Ness to Kingsbarns. Thence they rise up the valley of the Kenly Water to Kinaldy (400 ft.), then contract at Boarhills to the 100-foot beach until they reach the valley of the Eden. To the north of this they reappear near Leuchars and creep up Lucklaw Hill to Balmullo. Thence they run as a narrow strip along the whole length of the Motray Water, between the boulder clay on both sides, until they join the larger deposits of the Howe through the Pass of Lindores. Along the north coast they form only a very narrow rim of gravel, for here the hills press close to the shore and leave little room for deposition. The soils produced over these mixed sands and clays of the older marine deposits are generally good loams. In Kinross the soil is occasionally clayey, but more often light and sandy, as there is an extensive development of morainic gravel. Round the base of the Lomonds it is light but sharp and valuable for grass. The gentle inclines of the Howe, where the 100-foot beach is well developed, are very fertile and productive, but the centre of the Howe near Ladybank is shingly, and it would seem as if the coating of richer earth had been washed away by a current of water. Among the most fertile soils of the peninsula are those lying on the 100-foot beach already described as lining the coast from St. Andrews to Largo, where is found a deep rich loam. Along this coast are scattered here and there other deposits of marine clays, evidently laid down when the surrounding seas were ice-covered. These occur near Elie, where they form light and sandy soils, and again at St. Andrews. Thence they run up the Eden to above Cupar, forming sandy soils which pass eastwards into the dunes of Tents Muir and Pilmour Links, where a littoral vegetation flourishes.. On the coast strip from Leven to Inverkeithing the soil varies from light dry to rich clayey loam, the most fertile districts being the alluvium area of the Leven, the raised terraces near Kirkcaldy, and the patches of igneous soil in the Burntisland district. The coasts of the upper firth between Dunfermline and Culross are very fertile, since the soil is derived from an igneous outcrop and from the deposits of the raised beaches which appear near Torryburn. Westwards of this the soils of the carse lands of the Forth are alluvial, for the upper firth, not being open to the influences of the sea, is favourable to deposition. These soils are composed of the finest clays without stones but with occasional beds of marine shells. In most parts they have been stripped of their peaty covering—a significant indication of their original condition—and are very fertile under modern methods of lime manuring.

Climate.—The vegetation which flourishes on these soils is determined by the climate. But the peninsula is too small to be able to claim a distinctive climate of its own, and the actual figures of temperature and rainfall, which vary from place to place according to local conditions, are confined to certain limits by the general climatic conditions common to the whole east of Scotland. When compared with the figures for the West Coast the temperatures show a faint approximation to a continental type—slightly colder winters and slightly warmer summers. For while the broad outlines of the climate are fixed by proximity to the Atlantic, these are a little modified by a position on the leeward side of Scotland and a consequent exposure to continental influences which the shallow North Sea does little to temper. Hence one would expect a gradual intensification of the continental features from west to east, the maximum being attained as far east as possible, away from the Atlantic and yet not coming under the influences of the North Sea. That the presence of this sea has very little influence at all, in fact only on the coastal rim, is seen from the data of the temperature stations of St. Andrews and Feddinch Mains, the latter only three miles south-west of the former, and 300 feet high, with an eastern exposure. It has the distinction of experiencing the lowest winter temperatures and the greatest range of the Fifeshire stations—that is, the climate of Feddinch Mains is the nearest recorded approach to the continental type, though naturally in a rather mild form.

Temperature Records for Fife.

Station.	Years.	Mean Annual.	Mean Jan.	Mean July.	Range.
St. Andrews . .	1855-95	47·1° F.	38·2° F.	58·0° F.	19·8° F.
Feddinch Mains	1855-95	46·0° F.	36·6° F.	57·7° F.	21·1° F.

There are also marked differences in the annual rainfalls of east and west. The rain-bearing winds from the Atlantic have access to the west of Fife through the mid-Scotland valley; but the highlands of the peninsula are in the west, and the east thus tends to have its drier continental tendencies aggravated. Consequently there is a marked diminution of rainfall from west to east:—

Rainfall.

Station.	Height in feet.	Years.	Mean Annual Rainfall in inches.
Dollar	178	1866-90	43·39
Loch Leven Sluice . .	365	1866-90	36·10
St. Andrews	57	1866-90	29·41

The seasonal and local variations may be briefly summarised. The spring is, comparatively speaking, the dry season, owing to the prevalence of east winds from the continental anti-cyclonic area; temperatures are low and there are occasional late hoar frosts which sometimes damage the young crops. In summer the higher temperatures of the east encourage sudden showers of the thunderstorm type, which add considerably to the total rainfall of that part; both in the east and in the west, however, the wettest period is in the late autumn and the winter. The continental conditions in the east tend to raise the autumn temperatures slightly above those experienced in the west of the peninsula, while on the actual coastline proximity to the North Sea moderates the winter temperatures a little. Local conditions are affected by elevation and exposure; the higher lands are colder and the hill slopes facing the west are wetter. High winds, too, are aggravated by local configuration, being stronger in the funnel-shaped parallel valleys and on the unprotected slopes of the east; but, conversely, the Ochils protect the Devon valley from north winds and the Lomonds shelter the Markinch district to their south-east.

The climate is on the whole particularly favourable to agriculture. The peninsula enjoys all the advantages of the climate of the north-west European margin, and by its leeward situation in the east of Scotland it escapes the great disadvantage of excessive wet which has retarded and still retards the agricultural development of the otherwise fertile lands of the west. In the peninsula the climatic disadvantages are minor—rare late frosts or occasional thunderstorms. Neither the rainfall of the wetter west nor the dryness of the warmer east is excessive. And, moreover, full advantage of this favourable climate can be taken, for, particularly in the straths of the east and on the coastal rim—the most favoured climatic areas—exceptionally fertile soils occur. But all over there is little unproductive land, for the whole history of the area shows the gradual transformation of original undrained waste into profitable farming land. At the present day the percentage of cultivated land is very high—for Fife alone nearly 80, for Kinross above 60. The great agricultural areas are the eastern plains of Stratheden, Kinross, and the coastal rim, which the warmer and drier summers and autumns make particularly suitable for wheat. The interior of the East Neuk is also extensively and successfully cultivated, though the exposure and elevation may retard the harvests perhaps three weeks later than those of the favoured coastal strip. Modern scientific farming has changed the soils of the wetter west into good land. The pastures of the higher hills are of considerable economic importance, and even the dunes of the east coast, covered with coarse grass of little or no agricultural value, have been turned to account, and form as golf links a popular and paying asset to the numerous coast resorts.

II.—Economic Development to 1707.

After these considerations of the relief, geology, soil, and climate, it remains to show how they have affected the history and development of the peninsula, and to sketch its present condition.

Local Isolation.—One is prepared for local isolation from the nature of the relief. There were felt none of the centralising influences of a great river such as made Lanarkshire the basin of the Clyde, or Perthshire the basin of the Tay; nor were there converging lowlands such as meet on Ayr, and make it a county town. It is true there were lowlands capable of development and of supporting a population, and the cause of centralisation was helped by the fact that the lowlands were to some extent continuous, inasmuch as they diverged from a common centre, Kinross. But Kinross was far away in the interior and not near the sea, and between the lowlands lay belts of uplands, not without their own economic value in early times, but nevertheless obstacles to communication.

In addition, while physical relief alone would have prevented the early growth of an efficiently centralised county, its bounds coincident with the natural barrier of the Ochils, the tendency to disunion was emphasised by the once extensive presence of forest, marsh and lake. The forests have been cut down, most of the marshes have been drained, and many of the lakes have been reduced in size or drained away altogether—either by the work of man or by the emerging stream cutting its way down through the rocky barrier which formed the lake. But as late as 1662, when Timothy Pont published his map of Fifeshire in Blaeu's Atlas, the land was dotted with lochs and marshes which no longer exist. To the south-west of Tayport, Timothy shows a lake which the monks of Balmerino called the Swan Loch; it is now completely gone. The whole level of the Howe was probably once occupied by a lake which decreased in size as the Eden cut its way through the bar of higher land from Cupar to the sea. In historic times this had shrunk to what was nevertheless the largest lake in the area after Loch Leven, the now vanished Loch Rossie near Auchtermuchty. It must have been surrounded by an extensive marsh, for the monks called the site of the modern Ladybank, in the centre of the Howe, Our Lady's Bog, when they came there to cut peats for the Abbey of Lindores. Loch Leven itself was once larger by a quarter, and each of the various tributaries of the lower Leven flowed—some indeed still flow—through a lake or marsh. Near Mark*inch* the Couland drains a level strip of ground known by the suggestive name of the Myres. The site of the old Loch Ore is preserved in the name of a village and mansion at the foot of Benarty, and Blaeu's Atlas shows near Kinglassie quite an extensive lake on the course of the Lochty. The Fitty still flows through a small loch on the west of the Hill of Beath, while to the east the waters of the Tiel come from Moss Morran, a part of which still exists, though once it must have stretched right across the valley to the Burntisland Hills. The carse lands of the west were probably marshy, for the English cavalry found it impracticable to advance up the carse of Stirling on their way to Bannockburn; and the carse of Stirling is similar in formation and character to the smaller carse of Clackmannan on the north bank of the Forth.

Where the soil and climate would permit, some areas were covered with forests, oak on the richer soils, pines on the sandy and more exposed parts. How extensive these forests were it is difficult to estimate; but the presence of remains of the aurochs has been adduced to prove that they were once fairly extensive. At any rate, three are known to have existed and to have been of geographical importance in early times. On the hill slopes which stretch round the carse of Clackmannan stood Clackmannan "Forest," a district of heath, moss and thicket, which emphasised the isolation of the carse lands from the rest of the peninsula. Remains still exist in the district known as "the Forest," near Clackmannan town, and in Tulliallan Forest to the north of Kincardine. Then to the north of the Ochils was the Forest of Earnside, which covered the lower hill slopes of the Ochils, and ran from the Earn river lands to the Pass of Lindores, thus rendering more difficult an important—in mediæval times almost the only convenient—pass leading to Strathmore. The third forest lay on the slopes of the Lomonds, and the excellent hunting which it afforded attracted the Stuarts to Falkland. But it is fairly certain that the "forests" were by no means confined to these three areas, and that many districts were covered with marsh, heath and spinney similar to those mentioned.

The location of these extensive marshes and forests may help to explain the site of some of the antiquities of Fife. Crannogs have been found near Strathvithie, where there was an old lake, and at Collessie near the marshes of the Howe; and the choice of the situation of the ancient camp—Roman or otherwise—near Ballingry is understood when it is remembered that it was surrounded by the loch and marsh of the Ore. But of far greater geographical interest is the effect which forest, loch, and marsh had in dividing up the peninsula, and in diverting routes.

Peninsular Isolation.—Then to local isolation was added peninsular isolation. The peninsula forms a natural area. On three sides it is surrounded by sea or firth, on the landward side by hills of considerable height. And so if local conditions favoured local independence, isolation from the rest of Scotland might have tended to unity in the peninsula itself. To a certain extent this was achieved. All through its history peninsular unity has so far triumphed over local independence, that though there might be a multitude of separate shires and districts in the interior, an area *almost* coincident with the peninsula was usually regarded as a political entity of the rank of such old natural divisions of Scotland as Tweeddale, Lothian and Moray; its isolation was recognised, and it was named Fife long before the creation of such a division as a shire.

Yet it has been hinted at that the term Fife did not quite cover the whole peninsula. The exception is not Kinross, which is a comparatively artificial erection with the geographical justification (slight at that) of a ring of highland; it is Clackmannan. In early times that area had but little connection with Fife. Northwards the Ochils proved an almost impenetrable barrier, twelve miles broad and in some places precipitously steep above the river lands; eastwards lay the Saline Hills, and in addition the "Forest" of Clackmannan, which rendered difficult routes up the Devon valley and along the coast. There was really far more open communication with Stirling than with Fife, and this is seen in its history and in its very name, Clackmannan, "the Rock of Mannan." Mannan was an ill-defined district stretching from the moors above Falkirk, where the name appears in Slamannan, along the carse of Stirling, and up the valleys of the various rivers which converge on the town of that name. The area which eventually became Clackmannanshire was evidently included in this district, which formed a battle ground for the Pict, Scot, Angle, and Strathclyde Briton. When these races consolidated to form a Scottish nation, part of Mannan, the carse of Stirling, lay along the great north-south route from Lothian to Strathmore and Moray, and this part consolidated around Stirling. But the smaller carse lands to the north of Forth lay off this line of communication; they were separated from Stirling by the Abbey Craig between the Ochils and the river, and, lying as it were in a back water, they preserved their existence as a separate county independent of either Fife or Stirling, though they had much more intimate relations with the latter than with the former.

To return to Fife, the history of that peninsula early gives evidence of the influence of its geographical situation. It was nominally comprised in the so-called kingdom of Pictavia, but owing to its position it acquired an independence which was evidently unusual even for that nebulous state. It either preserved or assumed the title of kingdom, and when eventually such a state could no longer be allowed to exist, its mormaers had to be bribed with prominent court duties and positions. The attempts of a higher civilisation to enter the peninsula show what obstacles had to be overcome ere its isolation could be broken down. A new culture could not very well come from the land, for the Ochils, the river marshes, and the forests cut off communication with Strathmore or Monteith; it could much more easily cross the firths or come directly overseas. It probably came first of all from the latter, for there seems to have been a considerable immigration of Frisians contemporary with, and most probably a part of the so-called Anglo-Saxon settlement. The next successful invaders also came overseas—the Danes; and certainly the coast of Fife with its numerous small creeks and its rivers leading inland was just such as would tempt sea-rovers. Before the aggressive period of the Danish invasions, Northumbria had risen to greatness, and forcible incorporation with this kingdom was prevented only by the difficulty of crossing the firth. But the Anglian influences, which had ebbed after the defeat of Nectans Mere, returned when the Scottish kings acquired Lothian. Thence Anglian civilisation spread across the firths, and by a strange coincidence the event which rendered its triumph permanent was a landing from the sea, when Saxon Margaret disembarked at Rosyth on her way to King Malcolm's palace of Dunfermline.

The Religious Foundations.—With Margaret's elevation to power in Scotland the supremacy of the Roman over the Celtic Church was ensured, and by the time of David I. the newer church was erecting its first great abbeys and monasteries. These were more than religious institutions; they were civilising agencies which spread a superior knowledge of agriculture and stimulated commerce. As several very important foundations were built in the peninsula, their precise situation deserves some note (see sketch-map). Most are to be found on the coast; for here were good soils, an encouraging climate, advantages for fishing, and numerous harbours. But not all of the coast was equally favoured. The Ochils come close to the southern shore of the Firth of Tay, and if this was an advantage, in so far as the hills gave good pasture and the fall of land ensured a certain amount of natural drainage, the cultivable area was restricted to a very narrow rim along the coast. Moreover, though the navigable channel was near the southern shore of the Tay, there was a decided lack of harbours. And so along Tayside only two abbeys were founded. One was at the northern entrance to the fertile gap of Lindores, and bore that name. It lay between the fisheries of the Tay and the fresh-water loch of Lindores,

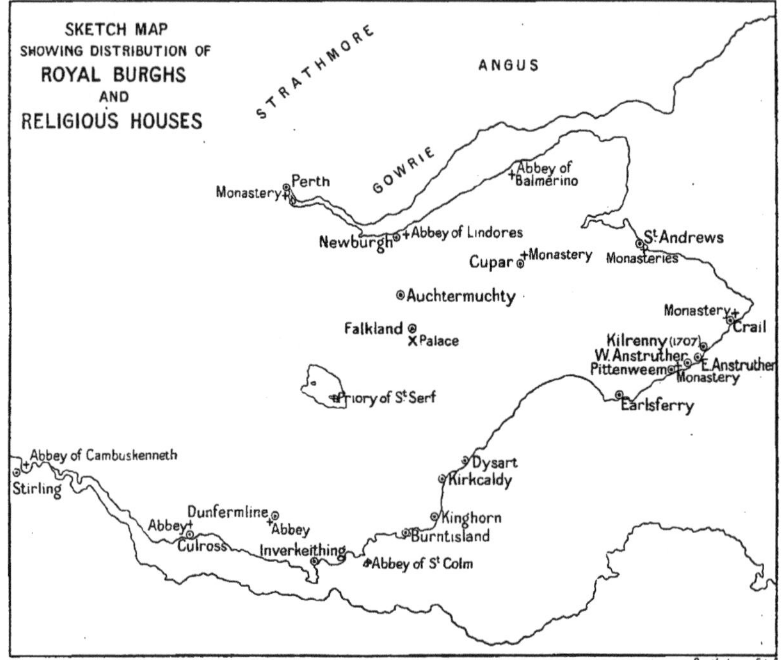

and below it grew up the port of Newburgh, described, in contrast to the decaying town of Abernethy, a few miles away in Perthshire, as the "novus burgus juxta monasterium de Lindores." Further east, where the hills decrease considerably in height, and a little curve on the coast could be used as a harbour, stood the Abbey of Balmerino (1227), in a fairly good situation for ferry traffic with Angus.

The east coast had even fewer attractions, for the coast of Tents Muir was without the slightest vestige of a harbour, the soil was unfit for cultivation, and further south, across the shallow and muddy estuary of the Eden, low shelving rocks ran seaward, causing dangerous eddies and difficult approaches. Yet here on the slopes, above the inconvenient harbour formed by the mouth of the little Kinness Burn, was fixed the ecclesiastical capital of Scotland, St. Andrews, with its cathedral and its three monasteries. It is difficult to find any geographical explanation of this importance. The situation was indeed protected from marauders, and this consideration does seem to have influenced the site of the primacy in its transit across Scotland from Iona to Dunkeld and Abernethy. But such a recommendation would apply to any situation east of the Ochils. Seemingly the determining attraction was the presence there of the only relics of an apostle which Scotland possessed, and the geographical factors of a good soil (except on the coast to the north) and a small harbour acted only in making the growth of a religious centre not impossible. Along the Firth of Forth, however, there was no lack of good soil or good harbours. In the east priories were founded at Crail and Pittenweem; then past the low sandy shores of Largo Bay the monks of the Abbey of St. Colm, situate on a small islet in the firth, farmed the fertile lands of Aberdour and Donibristle. On the productive strip near Torryburn the piety of St. Margaret founded an Abbey of Dunfermline at its eastern extremity, while the Cistercians raised another at Culross in the west. Then under the protection of the king's castle of Stirling was the Abbey of St. Mary of Cambuskenneth, where monks from Picardy, probably with a previous experience of marshy soil, drained the links of Forth. It was some time before the new agriculture penetrated inland, but though the clays of Strathore were not inviting, the more fertile soils of Kinross and Stratheden attracted attention. One monastery was founded at Cupar, and another displaced a Culdaic settlement on an island of Loch Leven, whence its monks tilled the lower slopes of the Lomonds and pastured their sheep on the upper.

Mediæval Commerce and the Royal Burghs.—The effect of these religious settlements was more than an improved agriculture; they stimulated commerce. Besides cultivating the land, the monks improved communication by building bridges; they developed the fisheries; they became the bankers of the time, and their demands for incense and articles of price brought foreign merchants to Scottish shores.

The situation of the peninsula was certainly advantageous for trade. Even if Scotland lay on the bounds of the known world it had considerable economic importance, for its wools and skins were as necessary to the manufacturing centres of Brabant and Flanders as the raw products of a non-manufacturing country are to the Britain of to-day. Moreover, in addition to the better climate, the east of Scotland had the advantage over the west of facing the continental trading towns, and so the fertile areas of Moray, Fife, Lothian and Tweeddale early carried on a considerable foreign commerce. But after the Wars of Independence, when the Scottish monarchy became a feeble distracted thing, Moray lay open to Highland raids, and Tweeddale to English invasions. From the Highlanders Fife was protected by the Ochils; from the English Lothian was protected by the Lammermuirs. Fife, however, might seem to have the advantage over Lothian as it had taken little part in the devastating Wars of Independence. These had raged up and down the routes from England to Moray, in Lothian, Stirling, and Strathmore; and Fife lay sheltered behind the Ochils. This advantage might seem to be increased by the superiority of the Fife ports on the Firth of Forth. They were sheltered from the north-east and, as the five-fathom line ran quite close to the shore, comparatively deep. To the north the ports of Angus shared the disadvantages of the exposed east coast of Fife—broad sands or dangerous rocks; and to the south the traffic of Lothian had to concentrate at Queensferry or Leith. But despite these advantages over the rest of Scotland, Fife was the centre of the kingdom only during the short period when the monarchy in its progress from Strathmore to Lothian halted mid-way at Dunfermline, as if, though it had broken with its Celtic subjects, it did not quite trust its new English ones, and thought the isolation of Fife rather an advantage. Lothian did eventually become the centre of the kingdom. It lay at the end of the mid-Scotland valley, and such of the products of the West Coast as did not find their way to Bordeaux, the only commercial centre with which Ayrshire could trade directly, turned eastwards to Lothian. From this traffic Fifeshire was separated by the upper firth. Indeed, if the Ochils and the firths protected Fife from the Highlanders and other invaders, they practically cut it off from participation in the trade of the interior. Communication had to be preserved with Strathmore by the Pass of Lindores, which led from the north, not to a port but into the Howe, and with Monteith by the carse of Clackmannan, which had really no well-developed connection with Fife at all. And so the produce of Strathmore was shipped directly from Perth or Dundee, while that of Monteith was shipped from the river ports of Stirling or Clackmannan. Yet Fife was second only to Lothian. Its position secured it from war or other disturbance—for example, it lay completely away from the Douglas wars—and its inhabitants could develop the natural riches under favourable conditions as far as they then could. In addition to its fertile coastal strips, with their rich abbeys, the extensive uplands of the interior and the Ochil Hills furnished abundant supplies of that most valuable of mediæval products, wool, and this was almost sufficient in itself to form a good commerce. Even if the peninsula was practically self-sufficing, its isolation enabled it to advance relatively faster than other parts of Scotland, and to support a safe and profitable trade. Hence one is prepared to find several important towns in mediæval Fife; but later, by the time of the early Stuarts, the number of royal burghs is surprising. There were no less than seventeen royal burghs in the peninsula alone out of about seventy in all Scotland, and their location leads to some interesting conclusions (see sketch-map, p. 82).

All except four were on the coast; and of these four inland burghs three (Cupar, Falkland, Auchtermuchty) were in Stratheden, and the fourth (Dunfermline) was only three miles from the coast. From Perth to Fife Ness, that is, on the coasts of the Firth of Tay and St. Andrews Bay, there were only two (Newburgh, St. Andrews); but in the short strip of coast from Fife Ness to Kincraig there were no less than five (Crail, Anstruther-Easter, Anstruther-Wester, Pittenweem, Earlsferry), and one (Kilrenny, 1707) was subsequently added—all these on a coast of thirteen miles. None were to be found on the low shores of Largo Bay from Kincraig to Dysart, but between Dysart and Ferryhill, fourteen miles in a direct line, there were five (Dysart, Kirkcaldy, Kinghorn, Burntisland, Inverkeithing). Westwards of this lay Dunfermline, and then Culross. There were no royal burghs in Clackmannanshire or Kinross, Strathore or Strathleven. Stratheden was the only district which contained royal burghs more than three miles from the coast. Of these seventeen royal burghs, six were in close proximity to religious foundations.

Here no doubt local isolation came into play. Trade could not be centralised at two or three great ports as in Angus or Lothian, for with the smaller vessels of those days any one of the numerous fairly deep creeks was convenient, and since the interior was cut up into independent valleys, each valley could have its own little port. But the distribution of these royal burghs may surely be taken as giving some idea of the balance of population. The conclusion then is, that Stratheden held the bulk of the population of the interior, that the coasts were as a whole much more developed than the interior, particularly the coasts from Fife Ness to Kincraig and from Dysart to Ferryhill. Are there any further geographical explanations of this?

Excepting Stratheden one presumes, from the absence of royal burghs in the interior, that it was not capable of supporting a centralised population, that it was still an undeveloped, raw material producing area. Probably there was little cultivation, for the river lands, before scientific drainage, would be swampy, and even on the slopes the clayey soils would not be very productive under mediæval methods. There are, indeed, places mentioned in the interior by various travellers, but they seem to have been very small, and noteworthy rather from their position. Thus Markinch, which was, be it noted, situated in the most favoured agricultural district of Strathleven, and was what might be called a pass town, is mentioned in the Chronicle of Edward I., but it had only a church and "iij houses." Kinross was called into existence by its position, where routes diverged down the various lowlands from Loch Leven; but though it occasionally figured as a royal residence it was never created a royal burgh. Nor were there any in Clackmannanshire, though a village, which eventually became the county town, lay round a tower built by the Bruce on a low spur running riverwards through the carse lands. Why then this development of Stratheden? It had a favourable climate, and, most important of all, the soils of its hill slopes above the marsh of the Howe were warm and fairly dry when most other soils tended to be cold and rather wet. Here then agriculture had evidently been spreading, for the three royal burghs of the Strath are all near the slopes, and one of them, Auchtermuchty, bears the significant seal of a sower, a horn of plenty, and the motto "Dum sero, spero." But in the cases of Falkland and Cupar there were additional reasons for importance. It seems probable that the Howe and Loch Rossie were difficult to cross. Falkland lay between them and the Lomonds, rising rapidly to over a thousand feet. Hence routes round the Lomonds, and from Perth to the ferry ports on the Firth of Forth, were constricted at Falkland, but it derived a more conspicuous importance from its royal residence near the Forest of Falkland and its hunting. Then the route from Perth to St. Andrews, two centres of supreme importance in mediæval Scotland, curved through the Pass of Lindores and round the foot of the marsh to

Cupar, where there was a convenient crossing of the Eden. Further, the route from the Forth ferry ports to Angus would naturally cross the Pass of Markinch, and keep the southern hill slopes above the Howe until a favourable opportunity of crossing the Eden presented itself. This was again at the foot of the marsh, and Cupar, at the junction of the two great routes across Fife, and conveniently situated equidistant from the Tay, St. Andrews Bay, and Largo Bay, naturally became the county town.

In discussing the distribution of the ecclesiastical foundations, the disadvantages of the coasts of the Firth of Tay and St. Andrews Bay have been noted—a narrow cultivable strip backed by hills which prevented expansion inland, sands or shelving rock, and a lack of harbours, or, at the best, inconvenient ones. It is easy then to see why traffic on this coast, except that of the ferry routes to Angus, should have centralised into the two burghs of Newburgh and St. Andrews, both beside religious foundations. Newburgh was the port of Lindores, but it was more; it was a pass town and an outlet for the produce of the slopes of the Howe, which were now becoming more important. And even though its little harbour was inconvenient, St. Andrews was the only possible outlet for lower Stratheden, and it, of course, derived additional importance as a religious centre. But up till this time (Charles I.) the two coastal strips of the Firth of Forth, from Fife Ness to Kincraig and from Dysart to Ferryhill, had always been the most important parts of the peninsula. Here were the great religious houses and the royal burghs, for they enjoyed the following advantages. The numerous creeks afforded excellent harbours sheltered from the north-east; behind them the soils of the igneous outcrops or of the older marine terraces were exceptionally fertile, and had been developed by the monks. Further inland were hills, those of the East Neuk or around Saline and Burntisland providing suitable pasture for sheep conveniently near the coast. In addition were the pursuits of fishing, relatively more important when all Europe was Catholic, and of salt-making, particularly in the western series of burghs, where for the first time the proximity to "sea-coal" became of economic importance in connection with this manufacture. Hence these favourably situated ports had trading relations with Flanders, Zeeland (Campvere), and France, and there still survive in many of the smaller burghs, particularly in Crail and Dysart, buildings of which the quaint Flemish or Dutch architecture recalls this profitable intercourse. Westwards of Inverkeithing lay the royal burgh of Culross, but, with the exception of fishing, the same industries as those in the eastern towns made Torry, Kincardine, and Alloa busy little river-side ports which had not as yet attained to the status of royal burgh.

Then followed a series of historical events which shattered the prosperity of the Fifeshire towns. With the Reformation and the fall of the old religion, St. Andrews and Dunfermline dwindled down to mere agricultural villages. The removal of the court to London left Falkland desolate. The fishermen of Anstruther and Kirkcaldy were slaughtered by the hundred on the Covenanting side at Kilsyth. The Dutch wars closed the European ports, and the Navigation Act the

English and Colonial. The saltworks and the coal mines ceased to supply any but local needs, and agriculture declined. With the Revolution, when secular interests predominated, the ruin of the Darien Scheme closed the prospect of a revival, and the Union extinguished the last vigorous industry, for the heavy duties on salt ruined the fishing. At the beginning of the eighteenth century the peninsula was in a state of industrial collapse.

(*To be continued.*)

THE DUKE OF THE ABRUZZI IN THE KARAKORAM.[1]

IN the *Scottish Geographical Magazine*, vol. xxvi. p. 204, we gave a brief account of the progress and results of the Duke of the Abruzzi's expedition of 1909 in the Karakoram range and Western Himalayas, and this renders it unnecessary to recapitulate to our readers what they will find in great detail in the sumptuous and beautifully illustrated volumes now before us. The objective of the expedition was the Titanic peak now well known as K^2, reputed to be the second highest mountain in the world, with an ascertained altitude of 28,250 feet above sea-level. We say "reputed," because according to present measurements Kinchinjunga is lower only by less than a hundred feet, and revised and more accurate measurements may easily change the figures and establish the greater altitude of Kinchinjunga. Up to a certain stage in his approach to K^2 the Duke was going over fairly well-known routes, which had been used by Godwin Austen, Younghusband, Conway, Eckenstein, Pfanul, Guillarmod and others. Guillarmod's account of his expedition in 1902, *Six Mois dans l'Himalaya, le Kara-Korum, et l'Hindu-Kush ; voyages et explorations aux plus hautes montagnes du monde* justly attracted much attention and was noticed at some length in this *Magazine*, vol. xx. p. 584. But since his day the science of mountain-climbing has made great advances. H.R.H. the Duke of the Abruzzi has a well-established reputation as an experienced and courageous mountaineer, and he brought to the service of this expedition every appliance that experience, wealth, and ingenuity could suggest. In his determined effort to ascend the K^2 giant he made approaches from the south, the west and the east, and he secured photographs of the north, a side which was at once seen to be unclimbable; but all in vain. "The mountain is a quadrangular pyramid, the corners being formed by four main crests meeting at right angles—the south-west and north-east, the north-west and south-east. The first two are prolonged in long and powerful buttresses, proportionate in size to the mass which they sustain. The other two are cut off short and precipitously . . . all its sides are equally fortified with the most formidable defences against

---

[1] *Karakoram and Western Himalaya, 1909: An Account of the Expedition of H.R.H. Prince Luigi Amedeo of Savoy, Duke of the Abruzzi.* By Filippo de Filippi, F.R.G.S. With a preface by H.R.H. the Duke of the Abruzzi. London: Constable and Company, Ltd., 1912. Two vols. Price £3, 3s. net.

the mountain-climber." Having most reluctantly abandoned the attempt on K², the Duke turned his attention to Bride Peak, some twenty miles to the south. Its altitude is 25,119 feet above sea-level, but when he had reached the altitude of 24,583 feet he had to retrace his steps owing to bad weather. In her "Notes on My 1912 Expedition to the Siachen or Rose Glacier, Eastern Karakoram," published in the January issue of this *Magazine* (p. 13), Mrs. Bullock Workman says, "Sometime I shall have something to say as to whether the Siachen is a complement of the Baltoro and what are the chances of a passage from one to the other, likewise upon the mythical Kondus and Chogolisa saddles." But the Duke does not seem to have found the Kondus and Chogolisa saddles mythical, for he encamped on the latter for several days at an altitude of 20,784 feet, and he had excellent views of the former, which is a little lower than Chogolisa, and at the foot of the Golden Throne peak.

Turning to the scientific results of the expedition, one of the most remarkable is in connection with the effect of the rarity of the air on the mountain-climber. Contrary to what might have been expected, "Twelve Europeans and fifteen coolies lived for about two months at above 17,000 feet of altitude, working regularly and not showing a single case of illness, even of the most fleeting character, attributable to mountain sickness. At the end of our campaign, seven Europeans spent nine days at a height of more than 20,700 feet, during which time four of them encamped for the night at 21,673 and 22,483 feet, and this without the inconvenience of sleeplessness. They likewise made two steep ascents, through deep soft snow, to 23,458 and 24,600 feet, without exhaustion, without lowering of morale, without exaggerated difficulty of breathing, palpitation or irregularity of the pulse, and with no symptoms of headache, nausea or the like. . . . The reason doubtless lies in the development and perfecting of the equipment, and in the gradual increase of knowledge as to the best plan of life and work under conditions of high altitude." Another very remarkable phenomenon of life in high altitudes was the loss of appetite. "It was possible for the lack of appetite to increase and become almost absolute repugnance to food, if after its appearance one moved and established oneself at a greater height. Thus, at Chogolisa Camp, the Duke and the guides had given up meat and lived on soups, coffee, tea, chocolate and biscuits. In the two ascents above 23,000 feet, their only food all day was a little chocolate, although they suffered no nausea or other unpleasant sensations. Of course, in the long run, this insufficient nourishment would cause a lowering of vitality, loss of flesh, and a certain amount of anæmia. However, the process is so slow that we were still, at the end of two months, in condition to make long marches without experiencing excessive fatigue." Throughout the expedition, the Duke and the members of his party eschewed the use of alcohol. The author has a good word to say for the Baltis, who with kind and considerate treatment turned out most courageous and effective transport-bearers.

In some respects the appendices are the most valuable part of this work. They furnish ample and most interesting details regarding the photogrammetic survey and the meteorology and altimetric calculations recorded during the period of the expedition, and also regarding the geology and botany of the regions through which it passed.

To appreciate and enjoy this work properly, the reader must have before him the maps and panoramas by Cavaliere Vittorio Sella, of which it is no exaggeration to say that they are the finest and best which have yet appeared. Since the publication of Guillarmod's exceptionally well-illustrated work, the sciences of cartography and photography have made great advances, and the artistic and scientific photographers who accompanied this expedition have availed themselves of them to the full. They did not do their work without considerable trouble and personal risk from time to time, but the results are admirable. The beauty of the illustrations alone will induce many an admirer of scenery to embellish his library with a copy of this work. It remains to be said that Signore de Filippi is an excellent narrator, clear, precise and enthusiastic, with a due sense of proportion and sound judgment.

Addendum: A Geographical Description of East Lothian.—In connection with this paper (cf. p. 23)—proof of which could not be submitted to the author owing to his absence abroad—it should have been stated that the work of preparation was done at the School of Geography of the University of Oxford.—Ed. *S.G.M.*

OBITUARY

Colonel F. Bailey, R.E., LL.D.

By the death of Colonel Bailey, which took place on Saturday, December 21st, 1912, the Society and the Council lost one of their most energetic members, and a former Secretary.

Colonel Bailey spent his early manhood in India, where several years were devoted to military duty, including active service in Bhutan (1864-5), for which he obtained a medal. Subsequently, in 1871, he was attached to the India Forest Service, and was soon afterwards intrusted with the formation and superintendence of the Survey Branch of that Department. He continued to hold this charge until 1884, when a very large area of forest-land, partly in the hills and partly in the plains, had been surveyed and mapped. In 1878 he was appointed a Conservator of Forests, and organised the Central Forest School, of which he was the first Director. He held this appointment until 1884, in addition to that of Superintendent of Forest Surveys. When on furlough in England, in 1884, he was appointed by the Secretary of State for India to the charge of the English students at the French Forest School at Nancy. He visited forests in various parts of France, as well as in Switzerland, Germany, and Hungary, and published notes on them in the *Transactions* of the Royal Scottish Arboricultural Society (Gold Medal) and of the Botanical Society of Edinburgh. In acknowledgment of his services to the cause of forestry, he was, in 1887, awarded the Cross for *Mérite Agricole* by the French Government.

On his return to India, in 1887, he was appointed temporarily to the Conservatorship of the Punjab forests, and was shortly afterwards selected by the Government of India to act as Inspector-General of Forests. In 1889 he was selected a second time to fill the office of Inspector-General, but was prevented from taking up the appointment by a temporary illness, which obliged him to return to this country.

THE LATE COL. F. BAILEY.

After his return he became Lecturer on Forestry in the University of Edinburgh, and, in 1892, was appointed Secretary of the Royal Scottish Geographical Society in succession to Mr. Silva White. This position he occupied for twelve years, until his resignation in 1903, which was due to failing health. The Society's appreciation of his devoted services during these years is expressed in the Reports of Council and of the Annual Business Meeting for that year (see the *Magazine*, vol. xix. pp. 668 and 669).

In spite of his failing health, Colonel Bailey was able to carry on his work as a Lecturer in Forestry in the University of Edinburgh until 1907, while other branches of his manifold activities, such as his editing of the *Transactions* of the Royal Scottish Aboricultural Society, he continued until the last.

As this brief account will indicate, Colonel Bailey's life-work lay mostly outside the activities of the Royal Scottish Geographical Society, but he gave to his work there that punctilious devotion which was his most marked characteristic. He possessed to the fullest extent the military virtues of promptness, exactness, and conscientiousness, and, to those who worked with him in his later years, it was pathetic to note his extreme unwillingness to delegate even minor duties, although his health was obviously making it difficult for him to discharge them. His interest in the Society's work did not cease with his resignation of the secretaryship. Not only did he take an active part in the work of the council during the years following his resignation, but during the temporary absence of the secretary, Mr. George G. Chisholm, in America during the past autumn, Colonel Bailey kindly undertook the duties of the office.

Colonel Bailey's work in connection with the Forestry Department of the University of Edinburgh received recognition during the past summer, when the degree of LL.D. was bestowed upon him at the July graduation ceremony. In addition to his association with the Royal

Scottish Geographical Society and the Royal Scottish Arboricultural Society, he was a Fellow of the Royal Geographical Society of London, and of the Royal Society of Edinburgh, and a past president of the Botanical Society of Edinburgh. Of his two sons, one, Captain Bailey, has already made important contributions to geographical science by his travels in SE. Tibet and the Mishmi Hills, of which some account appeared in the *Scottish Geographical Magazine*, vol. xxviii. p. 189.

We regret also to record the death of Léon Philippe Teisserenc de Bort, the distinguished French meteorologist, which took place in the early part of January. His name is associated with the investigation of the upper air by means of kites and *ballons sondes*. As a result of these investigations it was proved that the atmosphere is divided into two shells, the dividing surface lying at a height of about 10 kilometres, just above the highest clouds. The upper layer or *stratosphere* shows practically no change of temperature in a vertical column, while the lower layer or *troposphere* is the region of vertical temperature gradient and convection. Many of Teisserenc de Bort's investigations were carried on in conjunction with the American meteorologist Rotch, the founder of the Blue Hill Observatory, who died last year.

GEOGRAPHICAL NOTES.

EUROPE.

Drainage Conditions in the Jura.—A curious phenomenon, apparently of economic as well as of geographical importance, has taken place in connection with the piercing of the new tunnel through the Mont d'Or in the Swiss Jura. The engineers have encountered considerable difficulties, owing to outbreaks of water, and, according to the Paris correspondent of the *Times*, during the month of December water flowed from the tunnel for a few hours at a rate of 1100 gallons per second, reduced later to 900 gallons per second. At the same time the source of the Bief Rouge, a tributary of the Doubs, which runs through French territory, dried up, suggesting that the tunnel works had tapped the underground water which ultimately forms this stream. It appears probable, according to Professor Fournier of Besançon, that a permanent change in the watershed will take place owing to the construction of the tunnel, leading to an increase of water at Vallorbe on the Swiss side, and an impoverishment of the supplies of the French villages in the vicinity of Pontarlier. This means not only difficulties in regard to water supply on the French side, but also a great loss of power.

AFRICA.

Sleeping Sickness in Rhodesia.—Dr. Warrington York, director of the Runcorn Research Laboratories of the Liverpool School of

Tropical Medicine, delivered a lecture in Liverpool recently, in the course of which he gave an account of the conclusions arrived at by the Commission which has been studying the distribution of sleeping sickness in North-West Rhodesia. The most alarming result of the Commission's work has been to show that in this region it is not the fly *Glossina palpalis* which carries the disease, as in Uganda, but the species known as *G. morsitans*, which also acts as the carrier of fly-disease in domestic animals. The alarming feature of this discovery is that while the former fly has a restricted range in the areas where it occurs, never being found far from water, the common tsetse is very abundant, has a very wide distribution, and shows no tendency to follow watercourses. Thus the remedy for sleeping sickness which has been employed in Uganda, that of the removal of the inhabitants from the comparatively narrow area infected by *G. palpalis*, is inapplicable to Rhodesia. As yet no means is known whereby an area may be cleared of *G. morsitans*. The only possibility that remains is, therefore, the question whether the reservoir of the parasite can be got rid of. This reservoir is certainly constituted by the large game. We are glad to note, however, that Dr. York does not advocate the drastic remedy, which has been proposed elsewhere, of slaughtering the animals. He suggests, however, that an attempt might be made to drive them back from an experimental area, the area selected being one with a considerable human population, with the object of observing the effect upon the disease. In the newspaper account of the lecture from which the above is taken no mention is made of the question as to the migratory powers of the tsetses, that is of the extent of ground which it would be necessary to clear, or of the methods to be adopted to keep the big game from the area under observation.

POLAR.

The German Antarctic Expedition.—The *Deutschland*, the ship of the German Antarctic expedition, which left Buenos Ayres in October 1911 (cf. this *Magazine*, vol. xxvii. p. 661), returned there in the beginning of January in the present year. The leader, Lieutenant Filchner, has sent a short telegraphic message to Europe, stating that, after passing through 1200 nautical miles of pack-ice, he discovered new land in lat. 76° 35′ S. lat. and long. 30° W., which he has named Prinz Regent Luitpoldland, after the late regent of Bavaria. The position suggests that the new land is a continuation of Coats Land (cf. the map showing Dr. Bruce's results in vol. xxi.), but the telegram goes on to say that the land is bounded to the *south* in lat. 78° by the Weddell Sea, which is a little puzzling. Westward it is stated to be fringed by a barrier, which has been named the Kaiser Wilhelm barrier, while it continues (or was followed?) to 79° S. Lieutenant Filchner expresses his determination to return to the Antarctic in December to complete his explorations.

GENERAL.

The Royal Geographical Society and Women Fellows.—At a meeting of Fellows of the Royal Geographical Society of London, held on January 15, it was proposed by the President, Lord Curzon, that women should be eligible as Fellows. On a vote the motion was passed by 130 votes to 51. A previous referendum on the subject had given a majority of three to one. Henceforth, therefore, women will be eligible for the fellowship on the same terms as men.

Livingstone Centenary in Glasgow.—A meeting of a representative committee formed in Glasgow in connection with the celebration of the coming centenary of the birth of Dr. David Livingstone, was held in the City Chambers, Glasgow, on January 15 last. The Lord Provost presided, aud among those present were Principal Sir Donald MacAlister, K.C.B.; Professor Gregory; Professor Glaister; Mr. Paul Rottenburg, LL.D.; Mr. Robert Gourlay, LL.D.; and other representatives of the various public bodies interested.

On the motion of Professor Glaister, seconded by the Rev. David Watson, it was unanimously agreed that a fund be inaugurated for the purpose of establishing a permanent memorial of Dr. Livingstone. After some discussion the following became the finding of the meeting:—

"That a Livingstone Centenary Memorial Fund be established for the promotion (1) of medical missionary work in Central Africa, especially in connection with the London Missionary Society, the Livingstonia Mission of the United Free Church, and the Blantyre Mission of the Church of Scotland, and towards the establishment of a Livingstone scholarship in Anderson's College Medical School for the training of medical missionaries; and (2) for the promotion of the scientific study of Geography in the University of Glasgow by the endowment of a Livingstone lectureship or professorship."

The Committee arranged that an appeal be issued to the public in the terms of this resolution, intending subscribers to be asked to indicate to which section of the memorial they desired their subscriptions to be allocated, and that where no such indication was given the committee be authorised to exercise their own discretion.

Among other arrangements which are being made in Glasgow for the celebration of the Centenary, it is announced that an afternoon gathering will be held in the University on Tuesday, March 18, at which an appreciation of Livingstone will be delivered by Professor Gregory. The Very Rev. Dr. M'Adam Muir has arranged also to hold a public service in the Cathedral on Wednesday afternoon, March 19, and the Corporation will give a civic reception in the City Chambers the same evening.

Society for Promotion of Nature Reserves.—In connection with the note in our last issue (p. 35) on the formation of a Nature Reserve at Blakeney Point in Norfolk, we notice that it is announced in the *Times* that a new society, called the Society for the Promotion of Nature Reserves, has just been founded with the following objects:—

1. To collect and collate information as to areas of land in the United Kingdom which retain their primitive conditions, and contain rare and local species liable

to extinction owing to building, drainage, and disafforestation, or in consequence of the cupidity of collectors. All such information to be treated as strictly confidential. 2. To prepare a scheme showing which areas should be secured. 3. To obtain these areas and hand them over to the National Trust, under such conditions as may be necessary. 4. To preserve for posterity as a national possession some part at least of our native land, its fauna, flora, and geological features. 5. To encourage the love of Nature, and to educate public opinion to a better knowledge of the value of Nature study.

The society exacts no subscription ; members are formally elected by invitation of the Executive Committee (marked with * below), and all interested are invited to communicate with the secretaries. The control of the society's affairs is in the hands of a representative council consisting at present of the following :—

President—The Right Hon. J. W. Lowther, M.P.; Dr. I. Bayley Balfour, F.R.S.; Sir E. H. Busk ; Francis Darwin, F.R.S. ; Dr. F. D. Drewitt ; * G. Claridge Druce ; Professor J. Bretland Farmer, F.R.S. ; L. Fletcher, F.R.S. ; The Right Hon. Sir Edward Grey, Bt., K.G., M.P. ; The Right Hon L. V. Harcourt, M.P. ; * Sir Robert Hunter, K.C.B. ; Lord Lucas ; * E. G. B. Meade-Waldo ; * The Hon. E. S. Montagu, M.P. ; The Earl of Plymouth, C.B. ; Professor E. B. Poulton, F.R.S. ; Sir David Prain, F.R.S. ; * The Hon. N. C. Rothschild ; * W. H. St. Quintin ; Dr. R. F. Scharff; W. M. Webb. *Ex officio*—Hon. Treasurer, * C. E. Fagan, I.S.O. ; Hon. Secretaries * W. R. Ogilvie-Grant and the Hon. F. R. Henley.

The trustees of the British Museum have kindly given permission to the committee to use the Natural History Museum, Cromwell Road, London, S.W., as the temporary address of the society.

The formation of this society is part of a movement which has of late years being spreading widely in the civilised world, the attention of all those interested in nature having been drawn to the fact that many plants and animals are being wiped out, and many interesting natural phenomena destroyed, sometimes wantonly, sometimes inadvertently with the advance of "civilisation," the result being that the world is constantly growing poorer. Both in America and in many parts of the Continent efforts are being made to check this process by, among other methods, the formation of reserves. The wider objects of the movement may be gathered from the list given above of the objects of the new society.

New Type of Palæolithic Skull.—The discovery is announced, in a gravel deposit on Piltdown Common, Fletching, Sussex, of a human skull and mandible, apparently of Lower Pleistocene age, differing in many respects from the Neanderthal type. The find was described by Mr. Charles Dawson and Dr. Smith Woodward before the Geological Society in London in December last. In addition to the human remains the gravel contained "eoliths" of the same type as those found in Kent, two fragments of the molar teeth of a Pliocene elephant, and a part of the molar of a mastodon, together with teeth of hippopotamus, beaver, and horse, and early Palæolithic implements. The skull is remarkable for the thickness of the bones, the feeble development of the brow ridges, and the relatively high forehead. The back of the skull and the mandible are ape-like in character. The point of interest is thus that, while the skull is of a primitive type, the low forehead and heavy brow ridges of Neanderthal men are absent. The idea that modern man rose directly from the Neanderthal type has always presented certain difficulties, and Dr. Smith Woodward inclines to the view that Neanderthal man was a degenerate offshoot of primitive man and became extinct early, while the new remains may be those of a direct ancestor of modern man.

The Location of Icebergs.—In vol. xxviii. p. 546, we gave some account of Professor Howard Barnes' observations on the changes in temperature in sea water produced by the approach of icebergs. In a letter to *Nature* of 12th December last Professor Barnes describes his further observations during the past summer, as a result of which his previous conclusions have been slightly modified. He spent three weeks making observations in the Straits of Belle Isle, the ss. *Montcalm* having been placed at his disposal by the Canadian Government, and found, contrary to the earlier observations, that there was always a slight rise in temperature of the water near the iceberg, the fall previously observed close to the ice being absent. He is of opinion that this fall, when present, is due to cold arctic water in which the ice is floating, and not to ice-melting. Professor Barnes now holds that only two of the three Pettersson currents are caused by the melting of icebergs—the one a downward current which carries off the cold melt-water, and the other a horizontal one which brings warm surface water towards the ice. This surface water is warm, because it is withdrawn from the ordinary vertical circulation which cools sea water. Thus an iceberg automatically causes its own destruction by drawing warm water towards itself. The proximity of land, according to Professor Barnes, causes a fall of temperature, owing to the way in which the land turns up the colder under-water.

It will be seen from the above that Professor Barnes has somewhat modified his earlier views, and certain difficulties still remain. A long letter, pointing out some of these difficulties, and describing some laboratory experiments in ice-melting, is contributed to *Nature* of January 9, by Dr. John Aitkin. The subject is clearly one in regard to which further investigation is required.

Proposed Institute of Geography at Paris.—It is announced that the Marquise Arconati-Visconti, who has already presented a sum of £20,000 to the University of Paris, to be devoted "to the benefit of the Faculties of Science and Letters," has supplemented this gift by another of like amount. The Council of the University has resolved to use the money for the erection of an Institute of Geography to be built by the side of the Oceanographical Institute, which, it will be remembered, was endowed by the Prince of Monaco.

Geography and the India Police Force.—It is announced that the Secretary of State for India in Council has decided to modify the scheme of the competitive examination, held under the direction of the Civil Service Commissioners, for admission to the India Police Force, by requiring candidates to take up English history and geography as a compulsory subject. The change will come into operation for the examination of 1914.

Medical Science among Primitive Peoples.—In connection with the International Medical Congress, which is to meet in London in the summer of this year, Mr. Henry Wellcome is organising a Historical Medical Museum, intended to illustrate the development of medicine, surgery, pharmacy, and the allied sciences. A special feature is to be the exhibition of charms, amulets and talismans connected with the art of healing among primitive and savage races, as well as specimens of their surgical instruments, and exhibits of plants, etc., used in medicine. All those who have objects of this kind in their possession, and are willing to lend them, are requested to communicate with the Secretary, 54A Wigmore Street, London. W.

EDUCATIONAL.

The School Atlas.—The late Professor Karl Hüttl contributes to the *Kartographische u. Schulgeographische Zeitschrift* a critical article upon the atlases in use in Austrian schools, with a discussion of the ideal school atlas. The analyses of the various atlases have, of course, little significance for us, but the general discussion is of considerable interest. The author is of opinion that the school atlas should illustrate what he calls the three parts of geography—the astronomical, the physical, and the political, the first two being far more important than the last. He states, and the statement is probably true of our school atlases also, that the first branch of the subject is for the most part very inadequately treated. In some cases astronomical maps are entirely absent, in others the scale is too small, and the facts are represented in too unsystematic a manner to be of any use to the scholar. On the other hand, many sides of physical geography are fully treated, a statement borne out by the analyses given of the Austrian school atlases. We gather that in these respects the atlases are in advance of our own, for such plates as those illustrating contrasts between glaciated and non-glaciated mountains, karst phenomena, onset of spring in Europe, etc., etc., must be at least rare in our school atlases. But while recognising the merits of the national atlases from the physical standpoint, the author makes an energetic appeal for a combination of pictures with the plates of the atlas. Hachuring, he says, is absolutely useless for the representation of land forms for children, and he would have all physical maps of every kind accompanied by pictures, profiles, sketches, etc., in order to render the phenomena represented real and vivid. The political maps should be restricted in number as far as possible, and be drawn with a prudent limitation of detail. On the other hand regions of special importance, such as the Rhine, the Ardennes, Upper Italy, the part of the North Sea upon which England, the Netherlands, Belgium and France abut, California, etc., etc., should be illustrated by large scale maps. There should also be good ethnographic maps, accompanied by pictures showing the different peoples, their natural surroundings and their cultural conditions. Finally, a concluding map showing the great commercial routes over land and sea should be inserted.

In the arrangement of the atlas the standpoint and the needs of the adult should be definitely abandoned, and the whole arranged progressively, so as, first, to give the scholar vivid impressions of the chief facts of geography, and, second, to train him gradually to use and understand the highly abstract map. Thus no means of rendering the representations vivid should be neglected.

The article—though we do not know to what extent its recommendations can be carried out—seems to us of interest in that it is an attempt to apply to the atlas those principles which have now become a commonplace in regard to the geographical text-book. The old school-book of geography, it is not too much to say, was at best an abbreviated and summarised gazetteer; it contained solid facts to be assimilated by the scholar as he could, or, if not assimilated, at least swallowed. The old school atlas (and to some extent the existing one also) was a reduced and cheapened library atlas. Professor Hüttl, however, says openly and boldly that the needs of the traveller, the engineer, the soldier, the scholar, are not those of the school child, and the school atlas must be an entirely independent compilation.

Geographical Section of the Royal Philosophical Society of Glasgow.—This section has entered upon another session of successful work. Before Christmas the two following papers were delivered: "The Geography, Natural History and People of Southern Nigeria" (illustrated by views, native songs and choruses), by Mr. A. E. Kitson, F.G.S., F.R.G.S.; "The Uses of Place Names in the Teaching of Geography," by Mr. Jas. A. Ramsay, M.A.; while the other four items on the syllabus are as follows: *Thursday, 23rd January*—"The Roman Conquest and Occupation of Britain, Geographically Considered," by Mr. S. M. Miller, M.A. *Wednesday, 12th February* (Meeting of the Society)—"Scottish Scenery as shown in Ordnance Survey Maps," by Miss M. I. Newbigin, D.Sc. *Thursday, 6th March*—"The Whangie and its Origin," by Mr. G. W. Tyrell, M.A., B.Sc.; and "Civilisation as determined by Geographical Conditions," by Mr. Thos. W. M'Cance, M.A., B.Sc. *Thursday, 24th April*—"British Warfare in North America," by Mr. Thos. M'Michael, M.A., B.Sc. A conducted excursion is also to be arranged for a Saturday in May.

NEW BOOKS.

EUROPE.

The Gateway of Scotland; or East Lothian, Lammermoor, and the Merse. By A. G. BRADLEY. Illustrated with 8 coloured plates and numerous line drawings by A. L. COLLINS. London: Constable and Co., Ltd., 1912. *Price 10s. 6d. net.*

When one examines the list of the author's previously published works, one is led to expect a "made" book, but one does not read very far into *The Gateway of Scotland* before the half-formed prejudice is entirely dispelled. There are throughout abundant signs that Mr. Bradley has not only studied his district thoroughly, but is imbued with its peculiar spirit, and is inspired with a genuine affection for his subject. The explanation seems to be, as he himself hints, that East Lothian is his "calf" country, or at least that he knew it well in his youth, and has in his later survey retained much of that early glamour. The reminiscent, pleasantly garrulous vein which runs throughout must indeed charm readers—and we hope there are still many—who like to wander in by-paths. Yet that is not all. Mr. Bradley has a real sense of the value of a landscape, and the power of conveying to his audience an essentially true picture. This cannot be done by methods of compilation alone, but by loving study of the ground itself. For example, we may refer to the admirable treatment of East Lothian agriculture. Messrs. Constable are to be congratulated upon the general appearance of the book, and especially for their enterprise in illustrating it with line-drawings and the beautiful and suggestive sketches in colour by Mr. Collins.

Through Holland in the "Vivette." By E. KEBLE CHATTERTON. London: Seeley, Service and Co., Limited, 1913. *Price 6s. net.*

The friendly reception which the public gave to Mr. Chatterton's *Down Channel in the "Vivette"* (*S.G.M.*, vol. xxvii. p. 45), has induced him to publish the volume now before us, in which he tells of another cruise in the *Vivette* from Southampton to Calais, thence to Flushing, and on through the interior of Holland to Amsterdam, where the *Vivette* was laid up for the winter. Mr. Chatterton and Mr. Norman S. Carr, who had been at once the mate, artist, and ship's cook of the expedition, returned to England. They rejoined the *Vivette* in July of the following year, and sailed back from Amsterdam *via* Haarlem and Brassemer Meer to Gowre, where they picked up and retraced the route of the previous year to Flushing and across to Calais and Dover. They had a most enjoyable time, and received much courtesy and assistance from fellow yachtsmen and the inhabitants of the places where they halted. The only dangers they encountered were from bad weather and carelessness, which might at times have resulted in a collision, in which their four-ton yacht was certain to have come to grief. This breezily and crisply written book appeals especially to yachtsmen, and has been cleverly illustrated by Mr. Carr.

The Brenner Pass: Tirol from Kufstein to Riva. By CONSTANCE LEIGH CLARE. Illustrations by J. F. Leigh Clare. London: J. and J. Bennett, Ltd., 1912. *Price 6s. net.*

The Brenner Pass has from its nature and geographical position always been one of the great thoroughfares from Northern Europe into Italy, both in peace and war. Indeed, almost within the memory of living persons, before the advent of railways, travellers from Venice on their way north used to post through the Brenner Pass. Historically, therefore, it is of supreme interest. Tourists who are in the habit of hurrying through to the Süd Tirol would do well to linger on the route so ably described by the author. One of the most interesting chapters describes the stirring and pathetic episode of Andreas Hofer, a struggle which excited the sympathy of Great Britain at the time, and which will never be forgotten. That part of the country which is sacred to his memory, the Passeierthal, is also unusually attractive for many reasons. For many centuries the great trade route ran from Sterzing through this pass to Meran and so to Bozen, and the vestiges of this traffic remain to this day, both in the buildings and in the physique of the people. There is one point which will be new to some readers. The picturesque processions of "Banner wavers," so familiar to those who visit Siena, have their counterpart in the Burggrafenamt as the *Fahnenschwingen*. The illustrations are excellent and in keeping with the text.

ASIA.

Une Colonie Modèle, La Birmanie. Par JOSEPH DAUTREMER, Consul de France. Paris : E. Guilmoto, N.D. Prix 6 *francs.*

M. Dautremer was French consul at Rangoon from 1904 to 1908, and in this interesting work he gives us his impressions of the general results of the British occupation of Burma. It was formally and finally annexed by Lord Dufferin in 1885, not much more than quarter of a century ago ; then it was a semi-civilised country cursed by all the miseries resulting from the vagaries and eccentricities of a more or less insane and irresponsible despot ; now it is peaceful, contented, and rich, in a word, a model colony, which compares favourably with any province in the Indian empire. In his exposition of this transformation M. Dautremer, after a brief sketch of the previous history of Burma, enters upon a description of the various races which inhabit it, its geography, administration, communications, resources, products, trade, etc., and we may remark, in passing, that, unlike some other French authors, he has been careful to avail himself of the latest statistics. From various passages in the book we gather that one object on the part of the author is to arouse reflection in France, and to evoke comparison with the French administration in the adjoining Cochin-China. M. Dautremer is quite frank in his exposition of the errors which he finds in the French colony. They have been often pointed out, *e.g.* the continuous and unnecessary changes of officials, their consequent ignorance of the language and the people, the mistaken parsimony of the home government which does not expend the capital necessary to develop the resources of the country, and M. Dautremer adds another on which he lays considerable stress, viz., the difference in the personal habits of the administrators. "L'Anglais," he says, "n'est ni plus intelligent, ni plus travailleur que les autres, mais il est doué d'un merveilleux esprit d'organisation, lequel, accompagné d'une grande audace en affaires, lui assure des résultats abondants et solides là où d'autres ne récoltent la plupart du temps que des déboires. Je dis que l'Anglais n'est pas plus travailleur que les autres, mais son travail est fait consciencieusement, et depuis le moindre des agents de la couronne jusqu'au gouverneur, depuis le moindre employé de bureau jusqu'au chef de maison, tout le monde accomplit sa tâche quotidienne avec la plus grande sincérité et la plus scrupuleuse exactitude ; organisation bien comprise, travail méthodique, patience à toute épreuve, énergie remarquable dans les à-coups qui se produisent ici comme ailleurs, tel est le secret de la puissance anglaise dans le monde qu'elle enserre de mille liens divers." Burma is indeed, in his estimation, a model colony, but it is by no means perfect, and he sets forth fearlessly from time to time what he thinks flaws or deficiencies in the administration. For this reason it is to be hoped that his work will find many readers in Rangoon and Simla as well as in Paris. We commend it to our readers.

The Holy City (Benares). With 58 illustrations and a map. By RAJANI RANJAN SEN. Chittagong: 1912. *Price Rs.* 3.

Smiling Benares. Containing a sketch from the Vedic days to the Modern Times. With many illustrations and a map of Benares and its Environs. By K. S. MUTHIAH AND CO. Madras: 1911.

These books give two Indian views of the sacred city, and despite the difficulties of a language foreign to the authors, both succeed in giving a vivid presentation of its beauty and charm. It is the spiritual rather than the material which appeals to Mr. Sen, and though his book is written in the form of a guide he is at his best in his accounts of the legends which cluster round the holy men of the city. Some of the legends which he re-tells are full of charm and poetry, and he closes his chapter on Myths and Annals with this delightful little touch:—"And even in the deep hours of night, pious people are met with going along the river-bank with loads of eatables in search of the hungry poor who might need them, and calling out 'Koi bhuka hai?'—Is there any one hungry?"

Smiling Benares is a "Coronation Souvenir," and gives a detailed account of the city, illustrated by many photographs.

Stewart's Tourist's Guide to the Far East: From Southampton to Tokio. London: Raphael's Ltd., 1912. *Price* 2*s.* 6*d.*

This book, which is copiously interleaved with advertisements, gives a detailed account of the sea journey from Southampton to Ceylon, followed by brief descriptions of the parts of the far eastern countries most frequently visited. A considerable amount of practical information is given.

AFRICA.

Among Congo Cannibals: Experiences, Impressions, and Adventures during a Thirty Years' Sojourn amongst the Boloki and other Congo Tribes, with a Description of their curious Habits, Customs, Religion, and Laws. By JOHN H. WEEKS. London: Seeley, Service and Co., 1912. *Price* 16*s. net.*

In 1890 the Baptist Missionary Society had three stations on the Upper Congo,—one at Bolobo, about 200 miles above Stanley Pool, another at Lukolele, a little over 100 miles further on, and a third at Bopoto, more than 400 miles beyond Lukolele. It was thought advisable to plant a new station somewhere midway between Lukolele and Bopoto, among the Bangalas, a Bantu tribe, reputed to be a race of cannibals, strong, warlike, and cruel; and Monsembe was the district eventually selected. It is of the Bangalas of this district, whom the author prefers to call Boloki, that the book under review gives an account. The author is a missionary. He deals, however, not with missionary effort, but with the manners and customs of the people amongst whom he has resided for the last fifteen years; and we may say without further preface that he has produced a work of great interest and value.

The natives, on being assured that the new-comers were unarmed, received them willingly: and their permanent goodwill was gained by the distribution of a few presents. The authorities of the Congo Free State had given their sanction to the occupation by the missionaries of any plot of land suitable for their purposes. The latter, however, thought it better to treat with the actual owners; and accordingly they purchased a site and built a house. Gradually they overcame the difficulties of the language. Gradually, too, by their good sense and calm courage they dissipated the distrust of the natives.

The Boloki are a fine race. Their physical powers are highly developed. They can carry heavy loads across broken country, or paddle a heavy canoe hour after hour without much sign of fatigue. They have well-trained memories, and are keen and minute observers. They can adapt themselves to their surroundings, and possess a remarkable faculty for imitation. But they lack power of concentration or initiative. They will not, metaphorically speaking, follow a line of thought "for twenty yards." "Up to the age of fourteen or fifteen the boys and girls—especially the boys—are very receptive, and are easily taught; but after that age comparatively few make real advance in learning." The author suggests, as an explanation of this characteristic, that by that age the Congo boy has learned all that his father has to teach, and is content to keep to the old ways of parental practice or custom. A further explanation of their mental stagnation seems to be supplied by what is said of the influence of the "witch-doctor." "Before our arrival the 'witch-doctor,' by threats of 'witchcraft,' killed every aspiration of the people and smothered every sign of inventive genius that exhibited itself. To make anything out of the ordinary—any new article—was to be regarded as a 'witch,' and trouble was sure to follow any suspicions of that kind."

The Boloki man is sociable and kindly. He will attend to his sick relative or friend and nurse him faithfully. At the same time he is subject to outbursts of passion, during which he is capable of reckless cruelty. His affections are inconstant. He is neither a good lover nor a good hater. He is not without gratitude, but fears to show it for fear of being asked for favours. He is scrupulously fair in his division of meat or cloth among those who are entitled to a share, but he is niggardly to those who have no such right. He is innately selfish and gives only that he may receive the more in return. He does not consider it disgraceful to lie or to thieve, unless the lie or the theft be discovered. At the same time he is proud—"Touch his pride and he will act as though he were *un grand seigneur*." At the time of the missionaries' arrival the people were cannibals, and the horrors attending the practice were brought within their presence on several occasions. Widows were buried alive with their deceased husbands, and slaves were killed and interred with their dead masters.

The Boloki is a born trader, and, as a boy, he accompanies his father on all his trading expeditions. He learns in his village the relative values of things, and thoroughly enjoys the business of barter. Yet, strange to say, there are no markets in the country. If a Boloki has something to sell he hawks it through the town. If the owner of an animal wishes to sell its flesh, he cuts it up leaving a piece of skin on each portion so that the intending purchaser can tell whether it is taboo to him or not; and the man who has caught a fish tabooed to him seeks to barter it for something he can eat. Every kind of food is taboo to some one, and this prohibition may be due to various causes. The witch-doctor may have forbidden his patient to eat of a certain sort of food; or a family may not eat of a certain animal because it is their totem; or a permanent taboo may be put upon a particular kind of animal or fish or fruit.

In this connection it may be observed that when a free woman marries she observes not only her own but her husband's totem; and, when a child is born of the marriage, it takes the totems of both parents, until a family council, composed of members of both sides of the house, decides which it shall observe for the future. In the author's opinion Totemism is generally dying out on the Congo.

Artificial relationship in the form of milk-brotherhood and milk-sisterhood is recognised among this people ; and, while it does not constitute a marriage bar, public opinion regards the union of milk-brother with milk-sister as very irregular. Peace is concluded by "making brothers"—the tie being strengthened by the imposition of taboos, a disregard of which would be to court some great disaster or even death.

The avoidance by a man of his mother-in-law, and by a woman of her father-in-law, is a recognised practice.

What is said of the native songs is of great interest. They may be divided into three classes—topical songs, sung in canoes for the purpose of distributing news ; songs at funeral ceremonies, in which the praises of the dead are sung ; and songs in which the qualities of the men and women of a town or village are published to all and sundry. The canoe songs serve to warn a village of the approach of friends, and recall to us the signals of approach given in many parts of the world by means of blast of trumpet, or beat of drum, or ceremonial coughing. The songs which record the characters of men and women, their bravery or cowardice, generosity or meanness, etc., etc., remind us, in a way, of the Eskimo Nith-songs. They exercise a great popular influence ; for there is nothing which affects the African more than being praised or ridiculed in a song.

The book contains many interesting chapters regarding the arts of the Boloki, their marriage and burial customs, their methods of hunting and fishing and making war, their mythology, and their religious notions. In regard to the latter it may be said that while they have some vague idea of a Supreme Being, their system of belief is based on their fear of malignant spirits, infinitely numerous, which seek to compass their sickness, misfortune, or death, and which can be appeased or cajoled or forced only by means of witch-doctor and fetish.

A serviceable map and index are included in the volume.

We have read Mr. Week's book with unflagging interest, and take leave of it with a feeling of gratitude not only for the information which it contains, but for the admirable manner in which that information is conveyed.

Livingstone and the Exploration of Central Africa. By Sir H. H. Johnston. With numerous illustrations. London : G. Philip and Son, N.D. *Price* 1*s. net.*

This is a cheap reprint of the volume which Sir Harry Johnston contributed to "The World's Great Explorers" series, and has been prepared in connection with the approaching centenary of the birth of David Livingstone.

My Sudan Year. By E. S. Stevens. London : Mills and Boon, Ltd., 1912. *Price* 10*s.* 6*d. net.*

Few of us have even comparatively accurate ideas of the size and resources of the Sudan. We are told it is about two-thirds of the size of the Chinese empire, or that it has an area of more than a million square miles, but it has to be confessed that such a description does not go far to enlighten the ordinary reader. But it will suffice to explain, how, in her Sudan year (which was in reality by no means a whole year), Miss Stevens was able to see only a small part of the Sudan, but what she did see she describes in attractive and graphic style. We have, of course, descriptions of Khartoum and Omdurman with ever-exciting reminiscences, clustering around the names of Gordon, Kitchener, Slatin Pasha, and Father Ohrwalder, but the greater part of the book is taken up with the story of a sail from Khartoum up the Nile to the Sudd and back, rather a monotonous journey, in which almost the only distractions were the insects and occasional short expeditions for sport. But Miss Stevens is a careful and watchful observer, and her narrative is full of shrewd and instructive observations of the conditions, manners, customs, religion, folklore, legends, superstitions, dress (and want of dress), etc., of the many tribes whom she saw *en route*. A perusal of this book will go far to explain to us the difficulties and delicacy of the task undertaken by Great Britain in the civilisation of the Sudan, which, it has to be confessed, are in some ways greater than those we had to encounter when we annexed province after province of India. A particularly interesting chapter is one in which Miss Stevens gives translations of several characteristic Sudanese songs, and mention should also be made of her interesting notes on the bird and plant life of the Nile. Like others who have travelled in the Sudan, she takes a sanguine view of its future. It requires a large influx of capital to ensure irrigation, and of imported labour to till the soil, imported because the native population is evidently most disinclined to undertake regular field work. But, given irrigation and labour, the Sudan could produce an enormous amount of excellent cotton, equal to any we now get from India or America. The Sudan is a vast untilled garden. A word of commendation is due to the photographic illustrations.

AMERICA.

Florida Trails as seen from Jacksonville to Key West and from November to April inclusive. By WINTHROP PACKARD. Illustrated from photographs by the Author and others. London: Frank Palmer, 1912. *Price* 7*s.* 6*d. net.*

To visit such a country as Florida must be a pleasure in any circumstances, but to go in the company of Mr. Packard is something to live for. For besides being a poet by instinct, he is a trained botanist and naturalist, and is able to make one see not only what he himself sees, but the fairylike interpretations which he dreams. Human beings do not interest him, he "has no use" for them, but with the forest and its birds and flowers he is thoroughly at home. From the time the steamer leaves the north the birds follow in their hundreds, "far exceeding the paying passenger list," an escort varying in character from the northern gull to the pelican in the south, and all perfectly at home. There is a tale of a Myrtle Warbler which settled on the hat of a negro, and laid an egg there, "Yes, sir, it done did." Certainly one wonders why our American friends, with a paradise like Florida so near them, should ever stray across the Atlantic. In one touch of human nature he does mention Mrs. Beecher Stowe, who lived there, and did so much for freedom. Now there is nothing left of her dwelling, but freedom is in the air, for not only the negro but the wild bird is free. It seems that in Florida it is forbidden to cage wild birds. It would have been helpful to British readers had there been some sort of a glossary of the local popular names of the plants and birds. The illustrations are very beautiful.

AUSTRALASIA.

Travels in Australasia. By WANDANDIAN. Birmingham: Cornish Brothers, Ltd., Publishers to the University, 1912. *Price* 5*s. net.*

The author conceals his identity under the pseudonym of "Wandandian," a native Australian word meaning "a home for lost lovers," and applied to an inlet of St. George's Lake, in New South Wales. His profession or hobby is that of collector of insectivora to which he added later that of birds. In these pursuits he spent considerable periods at various places, and during his four years' sojourn became acquainted with many regions from South Australia to North Queensland, including Tasmania and New Guinea. A brief visit was paid to the Solomon Islands, Tonga, Samoa, and Fiji. A chapter is devoted to a "few Australian problems," notably that of "White Australia," on which the author is not afraid to express decided opinions. We presume it is useless to protest against his arrogant assumption to his own home political party of the sole possession of patriotism. The author is obviously a trained observer, and his descriptions provide a very illuminating series of pictures of Australian conditions. The printing of the book is distinctly unworthy of a firm which publishes to a University. On page 63 a line is partially duplicated, and another is omitted, the spacing throughout is very irregular, and towards the end the lines are closer together than elsewhere. On page 386 there are thirty-eight lines and on page 387 there are forty-one. There is no index.

EDUCATIONAL.

Map Projections. By Arthur R. Hinks, M.A. Cambridge: University Press, 1912. *Price 5s. net.*

This book may be said to occupy a place intermediate to the works of Hughes and Morrison, which are more or less popular discussions on the subject of Map Projections, and the more mathematical article of Clarke in the *Encyclopædia Britannica*. It is intended principally for map-users, yet map-makers will also find it serviceable. In six chapters the author gives a general description of the principal projections, while in the eighth the mathematics necessary for their construction are considered. The seventh chapter is perhaps the most interesting, dealing, as it does, not with construction, but with such matters as projections employed in actual maps, and their identification and choice.

While several books and articles on Map Projections have appeared within the last few years, and in spite of the fact that the subject appeals to but a limited number, Mr. Hinks's book amply justifies its publication. One matter alone leaves something to be desired—the tables at the end. Suitable as these no doubt are for general purposes, when considered from the map-maker's standpoint a regret may be expressed that the author has not made them fuller and more complete. Otherwise the book could scarcely be improved on, and even in regard to this it must be remembered that, as has been said, the book is primarily intended for map-users, for whom the tables are an improvement on what has appeared in former popular text-books. Altogether, as a book suitable for general use, Mr. Hinks's *Map Projections* is probably the best that has yet appeared in English.

Highroads of Geography: Book IV. The Continent of Europe; Book V. Britain Overseas. London: Thomas Nelson and Sons, 1912. *Price 1s. 6d. net each.*

These are reading-books of the journey type, copiously illustrated with coloured and uncoloured pictures. Book V. has no author's name on its title-page, while many authors have combined to produce Book V. under the general editorship of Sir J. Parrott. We presume, from internal evidence, that the authors in this latter book have been elected because of their own detailed acquaintance with the separate regions described, but the result is that far more details are supplied than can be assimilated by the scholar. This is especially the case when, as often happens, the details given are illustrated neither on the relief maps nor in the pictures, and we do not envy the task of the conscientious teacher, who will require many Baedekers, a continental Bradshaw, with maps and guide-books galore, to explain or illustrate the casual allusions of the text. Further, it seems to us that, in many cases, tourist fashion, minor incidents and details are given the prominence which should have been reserved for facts of genuine geographic interest. Book V. is devoted to a rather journalistic description of the British Empire.

Business Geography. By J. HAMILTON BIRRELL. London: Ralph Holland and Co., 1912. *Price* 1*s.* 6*d. net.*

A useful though greatly condensed little book, intended to serve as an introduction to commercial geography. It has a number of diagrams and simple sketch-maps, and gives references which will enable those interested to pursue the subject further.

Regional Geography: The World. By J. B. REYNOLDS, B.A. London: A. and C. Black, 1912. *Price* 3*s.* 6*d.*

This is an excellent and comprehensive book, illustrated by a number of clear and useful maps and diagrams. It is arranged progressively, the later sections being somewhat more advanced than the earlier ones.

The Elementary Geography: Vol. VI. The Three Southern Continents. *Price* 1*s.* 9*d.* *Vol. V. North and Central America and the West Indies.* *Price* 1*s.* 6*d.* By F. D. HERBERTSON. The Oxford Geographies. Edited by A. J. HERBERTSON. Oxford: At the Clarendon Press, 1912.

Mrs. Herbertson's *Elementary Geography* has already found its public, and these two additional volumes will be cordially welcomed. The carefully annotated illustrations are an excellent feature.

GENERAL.

Lehrbuch der Dänischen Sprache für den selbsten Unterricht. Von J. C. POESTION. Wien und Leipzig: Hartleben, 1912. *Price M.* 2.

A useful little book on the Danish language, in German, intended for use without a teacher. The usefulness is shown by the fact that this is the third edition, and that the book has been extensively used in university teaching. The present issue has been thoroughly revised and modified. It is noted as a curious fact that recently the Danish language has shown signs of undergoing modification in the direction of a *rapprochement* between the popular and the literary forms of speech, which were, even till the end of last century, markedly contrasted with each other.

Missions: Their Rise and Development. By LOUISE CREIGHTON. Home University Library. London: Williams and Norgate, 1912. *Price* 1*s.*

The editor of this series has done well to entrust the volume on missions to the widow of the late Bishop of London, herself a member of the Continuation Committee of the World Missionary Conference. Her useful work surveys the whole history of Christian missions, including to some extent Roman Catholic missions, from their commencement, and is replete with information and sound views on the subject. Withal it is well written. We learn with interest that the first recorded donation in England to the cause of foreign missions was Sir Walter Raleigh's £100 given for work in Virginia.

Hazell's Annual, 1913. Edited by HAMMOND HALL. London: Hazell, Watson and Viney, 1913. *Price* 3*s*. 6*d*. *net*.

Whitaker's Almanack, 1913. By JOSEPH WHITAKER. London. *Price* 2*s*. 6*d*.

The International Whitaker. London, 1912. *Price* 2*s*. *net*.

Whitaker's Peerage, Baronetage, Knightage, and Companionage, 1913. London. *Price* 5*s*. *net*.

Daily Mail Year-Book, 1913. Edited by DAVID WILLIAMSON. London: Associated Newspapers. 6*d*. *net*.

The beginning of the year brings the usual crop of annuals, etc. Among the newcomers we notice the *International Whitaker*, which is addressed "to the 200,000,000 English-reading people of the world," and treats the different countries on more or less uniform lines. *Hazell's Annual*—a useful production—contains as usual summaries of scientific progress during the year, good accounts being given of the various Arctic and Antarctic expeditions, while the meeting of the British Association at Dundee is briefly described. In the very cheap *Daily Mail Year-Book* we have what it is customary to call a marvel of journalistic enterprise in which the fastidious will see various points which might be criticised, though the book has evidently its public. To *Whitaker's Almanack* Mr. Edward Heawood contributes a valuable article, illustrated by sketch maps, on "Geographical Progress and Territorial Changes, 1911-12," and the volume shows its other usual features, having new articles on matters of current interest.

BOOKS RECEIVED.

Lost in the Arctic: Being the Story of the Alabama *Expedition 1909-1912.* By EJNAR MIKKELSEN, with numerous Illustrations and a Map. Crown 4to. Pp. xviii + 395. *Price* 18*s*. *net*. London: William Heinemann, 1913.

The Land of Zinj: Being an Account of British East Africa, its Ancient History and Present Inhabitants. By Captain G. H. STIGAND. Demy 8vo. Pp. xii + 351. *Price* 15*s*. *net*. London: Constable and Co., Ltd., 1913.

Greater Rome and Greater Britain. By Sir C. P. LUCAS, K.C.B., K.C.M.G. Demy 8vo. Pp. 184. *Price* 3*s*. 6*d*. *net*. Oxford: Clarendon Press, 1912.

A New Account of East India and Persia: Being Nine Years' Travels, 1672-1681. By JOHN FRYER. Edited, with Notes and an Introduction, by WILLIAM CROOKE, B.A. Vol. II. Demy 8vo. Pp. 371. London: Hakluyt Society, 1912.

Travellers' Practical Manual of Conversation, No. 2, in English, French, German and Dutch (Holland and the Rhine). Subjects arranged alphabetically. Pp. 144. *Price* 1*s*. 6*d*. London: E. Marlborough and Co., 1912.

Norwegian Self-Taught, with Phonetic Pronunciation. By C. A. THIMM. Revised and Enlarged by P. TH. HANSSEN. Fifth Edition. Crown 8vo. Pp. 128. *Price* 2*s*. 6*d*.

Portuguese Self-Taught (Thimm's System), with Phonetic Pronunciation. By E. DA CUNHA. Second Edition, Revised. Crown 8vo. Pp. 120. *Price* 2*s*. 6*d*. London: E. Marlborough and Co., 1912.

Northern Italy, including Leghorn, Florence, Ravenna and Routes through France, Switzerland, and Austria. Handbook for Travellers. By KARL BAEDEKER. With 36 Maps, 45 Plans, and a Panorama. Fourteenth Remodelled

Edition. Crown 8vo. Pp. lxviii + 698. Price 8s. net. Leipzig: Karl Baedeker, 1913.

An Almanack for the Year of the Lord 1913. By JOSEPH WHITAKER, F.S.A. Crown 8vo. Pp. v + 1034. Price 2s. 6d. London: Joseph Whitaker, 1912.

The International Whitaker. Crown 8vo. Pp. xlviii + 527. Price 2s. London: Joseph Whitaker, 1912.

Whitaker's Peerage, Baronetage, Knightage, and Companionage for the Year 1913. Crown 8vo. Pp. xvi + 854. Price 5s. net. London: Joseph Whitaker, 1912.

Hazell's Annual for 1913: A Record of the Men and Movements of the Time. Edited by HAMMOND HALL. Twenty-Eighth Year of Issue. Crown 8vo. Pp. lxvi + 592. Price 3s. 6d. net. London: Hazell, Watson and Viney, Ltd., 1913.

Monumental Java. By J. T. SCHELTEMA, M.A. With Illustrations, and Vignettes after Drawings of Javanese Chandi Ornament by the Author. Demy 8vo. Pp. xviii + 302. Price 12s. 6d. net. London: Macmillan and Co., Ltd. 1912.

The Fighting Spirit of Japan, and other Studies. By E. J. HARRISON. With 35 Illustrations. Demy 8vo. Pp. 352. Price 12s. 6d. London: T. Fisher Unwin, 1912.

Die Sunda—Expedition des Vereins für Geographie und Statistik zu Frankfurt am Main. Von Dr. JOHANNES ELBERT, Leiter der Expedition. Demy 4to. Pp. xv + 373. Frankfurt am Main: Hermann Minjon, 1912.

The River of London. By HILAIRE BELLOC. Crown 8vo. Pp. 145. Price 5s. net. Edinburgh: T. N. Foulis, 1912.

Latin America: Its Rise and Progress. By T. GARCIA CALDERON. With a Preface by RAYMOND POINCARÉ. Translated by BERNARD MIALL. With a Map and 34 Illustrations. Demy 8vo. Pp. 406. Price 10s. 6d. net. London: T. Fisher Unwin, 1912.

Twenty-Five Years in Qua Iboe: The Story of a Missionary Effort in Nigeria. By ROBERT L. M'KEOWN. Crown 8vo. Pp. vii + 170. Price 2s. 6d. London: Morgan and Scott, Ltd., 1912.

The Empire of India. By Sir BAMPFYLDE FULLER, K.C.S.I., C.I.E. Demy 8vo. Pp. x + 393. Price 7s. 6d. net. London: Sir Isaac Pitman and Sons, Ltd., 1913.

The Guide to South and East Africa, for the use of Tourists, Sportsmen, Invalids and Settlers. With Coloured Maps, Plans, and Diagrams. Edited annually by A. SAMLER BROWN and G. GORDON BROWN, 1913. Nineteenth Edition. Crown 8vo. Pp. liv + 695. Price 1s. net. London: Sampson Low, Marston and Co., Ltd., 1913.

Commodore Sir John Hayes: His Voyage and Life (1767-1831); with some Account of Admiral D'Entrecasteaux's Voyage of 1792-3. By IDA LEE (Mrs. CHARLES BRUCE MARRIOTT). With Illustrations. Demy 8vo. Pp. xvi + 340. Price 7s. 6d. net. London: Longmans, Green and Co., 1912.

The Russian Year-Book for 1913. Compiled and edited by HOWARD P. KENNARD, M.D., assisted by NETTA PEACOCK. Crown 8vo. Pp. xx + 810. Price 10s. 6d. net. London: Eyre and Spottiswoode, Ltd., 1913.

Common Sense in Foreign Policy. By Sir HARRY JOHNSTON, G.C.M.G., K.C.B., D.Sc. Illustrated with Eight Maps by the Author and Dr. J. G. Bartholomew of the Geographical Institute, Edinburgh. Demy 8vo. Pp. x + 119. Price 2s. 6d. net. London: Smith, Elder and Co., 1913.

Wissenschaftliche Ergebnisse der Expedition Filchner nach China und Tibet, 1903-1905. II. Band. Bilder aus Kan-su. Von WILHELM FILCHNER. Bearbeitet von HERBERT MUELLER. Crown 4to. Pp. x + 157. Preis *M.* 20. Berlin: Ernst Siegfried Mittler und Sohn, 1912.

The Sea West of Spitsbergen: the Oceanographic Observations of the Isachsen Spitsbergen Expedition in 1910. By BJØRN HELLAND-HANSEN and FRIDTJOF NANSEN. With 6 Plates. Royal 8vo. Pp. 89. Christiania: Jacob Dybwad, 1912.

Memoirs of the Geological Survey, Scotland: the Geology of Ben Wyvis, Carn Chuinneag, Inchbae and the Surrounding Country, including Garve, Evanton, Alness and Kincardine. Explanation of Sheet 93. By B. N. PEACH, LL.D., F.R.S., the late W. GUNN, etc. With Petrological Contributions by J. S. FLETT, M.A., D.Sc. Price 4*s.* Edinburgh: Geological Survey, 1912.

Thinking Black: Twenty-Two Years without a Break in the Long Grass of Central Africa. By D. CRAWFORD, F.R.G.S. Demy 8vo. Pp. xvi + 503. Price 7*s.* 6*d. net.* London: Morgan and Scott, Ltd., 1912.

Doreen Coasting: with Some Account of the Places she saw and the People she encountered. Edited by ALYS LOWTH. With 125 Illustrations. Demy 8vo. Pp. xix + 294. Price 10*s.* 6*d. net.* London: Longmans, Green and Co., 1912.

Echoes from the Hills: being the Reminiscences of the late John Hyslop, J.P., Langholm. Edited by his son Robert Hyslop, F.S.A. (Scot.). Crown 8vo. Pp. xv + 357. *Price* 2*s.* 6*d. net.* Sunderland: Hill and Co., 1912.

Northern Germany as far as the Bavarian and Austrian Frontiers. Handbook for Travellers. By KARL BAEDEKER. With 54 Maps and 101 Plans. Sixteenth Revised Edition. Crown 8vo. Pp. xxxviii + 439. *Price* 8*s. net.* Leipzig: Karl Baedeker, 1913.

My Russian Year. By ROTHAY REYNOLDS. With 28 Illustrations. Demy 8vo. Pp. xii + 304. Price 10*s.* 6*d. net.* London: Mills and Boon Limited, 1913.

D'Alger à Tombouctou; Des rives de la Loire aux rives du Niger. Par Comte RENÉ LE MORE. Avec une carte. Deuxième Edition. Crown 8vo. Pp. 260. Prix 3 *fr.* 50. Paris: Plon-Nourrit et Cie, 1913.

Statistical Abstract for the British Empire in each Year from 1896 to 1910. Price 1*s.* 3*d.* London, 1912.

East India (Trade): Review of the Trade of India in 1910-11. Price 1*s.* 1*d.* London, 1911.

Sudan Almanac, 1913. Price 1*s.* London: His Majesty's Stationery Office, 1913.

Statistical Abstract for the Several British Self-governing Dominions, Crown Colonies, Possessions and Protectorates in each Year from 1897 to 1911. Forty-ninth Number. Price 1*s.* 10*d.* London, 1912.

Annual Statement of the Trade of the United Kingdom with Foreign Countries and British Possessions, 1911. Supplement to Volumes I. and II. Price 3*s.* 1*d.* London, 1912.

Verhandlungen des achtzehnten Deutschen Geographentages zu Innsbruck, vom 28 mai bis 2 Juni 1912. Herausgegeben von GEORG KOLLON. Mit 15 Abbildungen im Text. *Preis M.* 8. Berlin: Dietrich Reimer, 1912.

Publishers forwarding books for review will greatly oblige by marking the price *in clear figures, especially in the case of foreign books.*

NEW MAPS.

EUROPE.

ORDNANCE SURVEY OF SCOTLAND.—The following publications were issued from 1st to 31st October 1912 :—

Six-inch Maps (Revised). Quarter Sheets, with contours in blue. Price 1s. each. *Lanarkshire.*—21 NE., 21 SW., 21 SE., 38 SE., 49 SE. (52 NE. and 52 SE.), 53 SW.

1 : 2500 Scale Maps (Revised), with houses stippled, and with areas. Price 3s. each. *Lanarkshire.*—Sheets VIII. 2, 7, 8, 12, 14, 15, 16. Sheet XVII. 12. Sheet XIX. 1.

Note.—There are no coloured editions of these Sheets, and the unrevised impressions have been withdrawn from sale.

Special Enlargements from 1 : 2500 to 1 : 1250 Scale. Prepared for the Land Valuation Department, Inland Revenue. Price 2s. 6d. each. *Aberdeenshire.*—Sheet XLVI. 5 NE., SE. *Argyllshire.*—Sheet XCVIII. 11 NE. *Ayrshire.*—Sheets VIII. 8 SE. ; 9 NW. ; 12 NE. ; 15 NW. ; XII. 4 NE. ; 7 SW. ; XVI. 16 NE. ; XVII. 15 NW. ; 16 NW. ; XVIII. 15 SW. ; XIX. 13 SE. ; XXI. 16 SE. ; XXII. 7 SW. ; 14 NW. ; 16 NW. ; XXIII. 3 NW. ; XXVII. 7 SW. ; XXVIII. 12 NE. ; XXIX. 10 NW. ; 11 NW. ; XXXIII. 3 SW. ; 14 SW. ; XXXIV. 9 SE. ; 13 NE. ; XXXV. 3 SW. ; 8 SW. ; XXXVIII. 6 NE. ; XXXIX. 10 NW. ; XLII. 9 NE., SW., SE. ; XLIV. 10 SW., SE. ; XLV. 6 SE. ; 10 SW. ; 14 NW. ; 16 SW. ; XLVI. 1 NE. ; 6 SE. ; L. 2 NE. ; LI. 1 SW. ; LV. 8 NW. ; LVI. 12 SW. ; LXV. 12 NW., SW. ; LXVI. 2 NE. ; 3 NW. ; LXVII. 10 NE. *Dumbartonshire.*—Sheets XVII. 2 SW. ; 5 NW., NE. ; XVIII. 5 NE., SE. ; 6 NW., SW. ; 13 NE., SE. ; 14 NW. ; XXII. 2 SW., SE. ; 5 NE., SE. ; XXIII. 13 SW. *Dumfriesshire.*—Sheets VI. 10 NE., SE. ; VII. 2 SW. ; 6 NE. ; XIII. 4 NW. ; XIV. 11 SE. ; XXII. 10 NE., SE. ; XXX. 8 SW. ; 10 SE. ; 15 NW. ; 16 NW. ; XXXI. ; 1 SE. ; 6 SW. *Edinburghshire.*—Sheets III. 8 NW., NE., SW., SE. ; IV. 5 NE., SE. ; 6 NW., SW., SE. ; 11 NE. ; VII. 12 SE. ; VIII. 6 NE., SE. ; 7 NW., NE., SW. ; 9 NW., NE., SW., SE. *Elginshire.*—Sheets V. 12 NE. ; IX. 5 NW. ; 9 NW. *Fifeshire.*—Sheets VIII. 10 NW. ; XII. 4 SE. ; 8 NE. ; XIII. 13 NE., SE. ; XIV. 15 NE. ; 16 NW. ; XIX. 4 SW., SE. ; 8 NE. ; XX. 15 NE., SE. ; XXII. 9 NW., SW. ; 13 SW. ; 14 SE. ; 16 NW. ; XXVI. 16 NE., SE. ; XXVII. 4 NE. ; 14 SW. ; XXVIII. 2 SW., SE. ; 3 NW. ; 5 NE. ; XXX. 2 NW., NE. ; 5 NW. ; XXXIV. 4 NE., SE. ; 8 NE., SW., SE. ; XXXV. 5 NW., SW. *Haddingtonshire.*—Sheet X. 6 NE. *Inverness-shire.*—Sheets IV. 13 NW., NE., SW., SE. ; 14 SW., SE. ; XII. 1 NE., SW., SE. ; 2 NW., SW. ; 5 NE., SE. ; 6 NW., SW. *Kincardineshire.*—Sheets XXV. 15 NW., SW. ; XXVIII. 13 NE. *Kirkcudbrightshire.*—Sheets XVIII. 10 NE. ; 16 NW. ; XXVI. 3 NW. ; XXXV. 11 SW., SE. ; XXXIX. 3 SW. ; XLVII. 1 NW., SW. ; 9 SE. *Lanarkshire.*—Sheets I. 14 SE. ; VI. 13 NW. ; VII. 12 NE. ; 14 NE. ; VIII. 5 SW., SE. ; 6 SW., SE. ; 9 NW., NE., SW., SE. ; 13 NW. ; X. 3 SW. ; 7 SE. ; 8 NW., NE., SW., SE. ; 14 SE. ; XI. 3 SW. ; 5 NW., SW., SE. ; 7 NW., NE. ; 8 NE., SE. ; 11 NE., SW., SE. ; 14 NE., SE. ; 15 NW., NE., SW., SE. ; 16 SW. ; XII. 10 SW. ; XVII. 4 NW., NE., SW., SE. ; XVIII. 2 NE. ; XXIII. 15 NE. ; XXIV. 6 NW. ; XXXI. 8 SE. *Linlithgowshire.*—Sheets I. 8 NW., SW. ; V. 3 NW., NE., SW., SE. ; IX. 6 NW., NE., SW., SE. *Orkney.*—Sheets CVIII. 3 NW., NE., SW., SE. *Perthshire.*—Sheets LXIV. 13 NE. ; XCVIII. 1 SW., SE. ; 2 NW. ; 5 NW. ; NE., SW., SE. ; 6 SW. ; 9 NW., NE., SW., SE. ; XCIX. 7 SW. ; CXV. 13 SW. *Ross and Cromarty.*—Sheets LXXVI. 15 SW., SE. ; LXXXVIII. 3 NW., NE., SE. *Roxburghshire.*—Sheets IX. 4 SE. ; 8 NE. *Stirlingshire.*—Sheets XVII. 12 NW.,

NE., SW., SE. ; xxx. 2 NE., SE. *Wigtownshire.*—Sheets xxx. 7 SE. ; 11 NE. ; *Zetland.*—Sheets LIII. 9 SE. ; 14 NW., SW.

The following publications were issued from 1st to 30th November 1912 :—

One-inch and Smaller Scale Maps.—One-inch Map. Third edition ; engraved in outline. Sheets CXXVIII. Price 1s. 6d. each.

Six-inch and Larger Scale Maps.—Six-inch Maps (Revised). Quarter Sheets, with contours in blue. Price 1s. each. *Lanarkshire.*—20 SW. ; 26 N.E. ; 29 SE. ; 32 N.E.; 36 NW. ; 37 NW. ; 36 NW. ; 42 SW. ; 45 SE.

1 : 2500 Scale Maps (Revised), with houses stippled and with areas. Price 3s. each. *Lanarkshire.*—Sheets I. (7 and 6), 8, (10 and 9) ; VI. 16 ; VII. 5, 10, 16 ; VIII. 1, 3, 6, 11, 13 ; XI. 2, 3, 4, 10, 12, 14 ; XII. 1, 5, 9, 11, 12.

Note.—There are no coloured editions of these Sheets, and the unrevised impressions have been withdrawn from sale.

Special Enlargement from 1 : 2500 to 1 : 1250 scale. Prepared for the Land Valuation Department, Inland Revenue. Price 2s. 6d. each. *Ayrshire.*—Sheets XXII. 16 SW. ; XLVI. 6 NE. *Banffshire.*—Sheet VI. 9 NW. *Berwickshire.*—Sheets XVI. 7 SW., SE. *Dumbartonshire.*—Sheet XVII. 1 NE. ; XVIII. 10 NW. ; XXII. 1 SE. *Elginshire.*—Sheet IX. 5 SW. *Fifeshire.*—Sheets I. 11 SE. ; III. 12 NW., NE. ; VIII. 3 SE. ; 7 NE. ; 12 SW. ; 14 SE. ; 15 SW. ; 16 NW. ; XIII. 1 NW., NE. ; 7 SW., SE. ; 9 SE. ; 12 NE. ; 14 NE. ; 15 NW. ; XIV. 5 NW., SW., SE. XV. 4 NE. ; XVI. 5 NE., SE. ; XX. 1 NE., SE. ; XXI. 10 SE. ; 12 SE. ; XXVII. 10 NE. ; 13 NE. ; XXVIII. 3 SW., SE. ; 4 NE., SW. ; 7 NW., NE. ; 8 NE., SW. ; XXX. 1 SW., SE. ; 5 NE ; XXXIV. 11 NE., SE. ; XXXV. 2 NW., SW. ; XL. 8 SW.; XLIII. 2 NW., SW., SE. *Forfarshire.*—Sheet LIV. 1 SW. *Kirkcudbrightshire.*—Sheet VII. 12 SW. *Lanarkshire.*—Sheets VI. 8 NW., NE., SW. ; VII. 9 SW., SE. ; 13 NW., NE., SW. *Perthshire.*—Sheets XCVII. 8 NE., SE. ; XCVIII. 6 NW. *Selkirkshire.*—Sheets VIII. 2 NW., SW., SE. *Stirlingshire.*—Sheets XXX. 1 NW., SW. ; XXXI. 5 NW., NE., SW., SE. ; XXXV. 2 SW., SE.

The following publications were issued from 1st to 31st December 1912 :—

One-inch and Smaller Scale Maps.—Two-mile Map. Printed in colours and hills shown by layers, folded in cover or flat in sheets. Sheets 13, 14, 15. Price, on paper, 1s. 6d. Mounted on linen, 2s. Mounted in sections, 2s. 6d. each.

Six-inch and Larger Scale Maps.—Six-inch Maps (Revised). Quarter Sheets, with contours in blue. Price 1s. each. *Lanarkshire.*—20 SE. ; 21 NW. ; 22 SE. ; 28 NW. ; 29 NE. ; 30 NE. ; 31 NW., 31 NE. ; 31 SE. ; 32 SE. ; 33 NW.; 34 NE. ; 36 NE. ; 38 NE. ; 41 NW. ; 41 NE. (41 SW. and 41A SE) ; 42 NW. ; 43 NE. ; (45 NW. and 45A NE), 45 SW. ; 46 SW. ; 49 NW. ; 53 NW. ; 53 SE.

1 : 2500 Scale Maps (Revised), with houses stippled and with areas. Price 3s. each. *Lanarkshire.*—Sheets I. 11, 12, 14, 15, 16 ; II. 13 ; VI. 4, 8, 12, 15 ; VII. 1, 6, 8, 9, 11, 13, 15 ; VIII. 5, 9, 10 ; XI. 1, 7, 8, 9, 11, 15, 16 ; XII. 13 ; XVIII. 12 ; Sheet (II. 5; I. 4 and II. 1). Price 1s. 6d. *Renfrewshire.*—Sheets VII. 3, 6 ; XI. 7 ; XII. 11, 16 ; XVII. 1, 2.

Note.—There are no coloured editions of these Sheets, and the unrevised impressions have been withdrawn from sale.

Special Enlargements from 1 : 2500 to 1 : 1250 Scale. Prepared for the Land Valuation Department, Inland Revenue. Price 2s. 6d. each. *Caithness-shire.*—Sheets V. 11 NW., NE., SE. ; XXV. 5 NW., NE., SW. *Dumbartonshire.*—Sheets XVII. 1 SE. ; XXV. NW., SW., SE. *Fifeshire.*—Sheets XXVII. 13 SE. ; XXVIII. 4 SE. ; XXXV. 8 NE., SW. ; XLIII. 2 NE.; LIV. 1 NW. *Forfarshire.*—Sheets LI. 16 NW., NE. ; LIV. 1 NW., SE. ; LV. 1 NE., SW., SE. *Kinross-shire.*—Sheets XVIII. 10 NW., SW. *Lanarkshire.*—Sheet VI. 8 SE. *Linlithgowshire.*—Sheets I. 7 NW., NE., SW., SE.

GEOLOGICAL SURVEY OF SCOTLAND.—The following publications were issued from 1st to 31st December 1912 :—

One-inch Map. New Series.. Colour Printed, solid and drift editions. Alness. Sheet XCIII. Price 2s. 6d. net. The solid is shown in colour, the drift by symbols in black.

Memoirs of the Geological Survey. The geology of Ben Wyvis, Carn Chuinneag, Inchbae, and the surrounding country, including Garve, Evanton, Alness, and Kincardine. (Explanation of Sheet XCIII). By B. N. Peach, LL.D., F.R.S., the late W. Gunn, C. T. Clough, M.A., L. W. Hinxman, B.A., F.R.S.E., C. B. Crampton, M.B., C.M., and E. M. Anderson, M.A., B.Sc. ; with Petrological contributions by J. S. Flett, M.A., D.Sc. Price 4s.

The Oil Shales of the Lothians. Part I. The Geology of the Oil-Shale Fields. By R. G. Carruthers, based on the work of H. M. Cadell and J. S. Grant Wilson. Part II. Methods of working the Oil-Shales. By W. Caldwell. Part III. The Chemistry of the Oil-Shales. By D. R. Steuart, F.I.C. Second Edition. Price 2s. 6d.

ADMIRALTY CHARTS, SCOTLAND:—

St. Abb's Head to Aberdeen. New edition, September 1912. Number 1407. Price 3s.

Aberdeen to Banff. New edition, November 1912. Number 1409. Price 2s.

Island of Islay. New edition, November 1912. Number 3116. Price 3s.

Loch Alsh and Kyle Rhea. New edition, September 1912. Number 3292. Price 4s. *Admiralty Office, London.*

BALKAN PENINSULA.—Map to illustrate the New Eastern Question. Scale 1 : 2,533,000. Price 1s.
W. and A. K. Johnston, Limited, London and Edinburgh.

BALKAN PENINSULA.—Philip's Panorama Map. Price 6d.
George Philip and Son, Limited, London.

The above two maps are issued in connection with the war in the Balkans.

AFRICA.

AÏR.—Carte dressée par le Capitaine Cortier et l'Adjudant Malroux. Echelle de 1 : 500,000. 1912. 2 feuilles. Prix 2 francs (each sheet).
Service Géographique des Colonies, Paris.

The most complete general map of this great oasis yet published.

EAST AFRICAN PROTECTORATE.—Scale 1 : 125,000, or about 2 miles to an inch. Sheet—Fort Hall. 1912. Price 2s. net.
Geographical Section, General Staff, London.

SOUTHERN NIGERIA.—Scale 1 : 125,000, or about 2 miles to an inch. Sheet—Badagri. 1912. Price 1s. 6d. net. Published by the Director of Surveys, Lagos. *Sold by W. and A. K. Johnston, Limited, London and Edinburgh.*

KAMERUN.—Karte in 31 Blatter und 3 Ansatzstücken im Massstabe von 1 : 300,000, bearbeitet unter Leitung von Max Moisel. Im Auftrage und mit Unterstutzung des Reichs-Kolonialamts, 1912. Blatt B 4, Kusseri ; Blatt C 4, Marua ; Blatt D 2, Schebschi-Gebirge ; Blatt D 3, Garua ; Blatt E 3, Ngaundere. Price M. 2 (each sheet).
Dietrich Reimer (Ernst Vohsen), Berlin.

ASIA.

INDIAN GOVERNMENT SURVEYS :—

India and Adjacent Countries. 1912. Scale 1 : 8,110,080. Price 2 Rupees, coloured.

India, showing Railways. 1911. Scale 1 : 4,055,040. Price 1 Rupee.

India, Railway, Canal, and Road Map. 1911. In 6 sheets. Scale 1 : 2,027,520. Price 12 Rupees, coloured.

India and Adjacent Countries. Scale 1 : 1,000,000, or 16 miles to an inch. 1910-1912. Price 1 Rupee each sheet.

Sheets—30, 34, 35, 36, 46, 48, 63, 66.

This is a very clear and excellent new map on a most useful scale for general reference purposes.

Afghanistan and Frontier Province. 1911. Scale 4 miles to an inch. Sheet 38 N. Price 1 Rupee.

Bengal. 1911-1912. Scale 4 miles to an inch, 1 : 253,440. Sheets 79 A, 79 B, 79 E, 79 F, 79 I, 79 J, 79 M. Price 1 Rupee each sheet.

Bihar and Orissa. 1911-1912. Scale 4 miles to an inch, 1 : 253,440. Sheets —73 C, 73 F, 73 G, 73 L and P, 73 N, 73 O. Price 1 Rupee each sheet.

Burma. 1910. Scale 4 miles to an inch. Sheet 93 L. Price 1 Rupee.

Survey of India Office, Calcutta.

INDIA.—Philip's Travelling Map. Scale 1 : 5,000,000. With Index. In two sheets, mounted on cloth in case. Price 7s. net.

George Philip and Son, Limited, London.

This is a more compact new edition of the excellent map by Mr. E. G. Ravenstein. It appears to be carefully revised to date.

NEW ATLASES, ETC.

GROSZER DEUTSCHER KOLONIAL ATLAS.—Bearbeitet von Paul Sprigade und Max Moisel. Herausgegeben von Reichs-Kolonialamt. Lieferung, 8, Kamerun, No. 8a. Mbaiki (1 : 1,000,000), No. 8d. Bonga (1 : 1,000,000). Deutsch-Ostafrika, No. 17 Muansa (1 : 1,000,000). Namen-Verzeichnis für die Karte von Deutsch-Ostafrika. 1912. Price M. 2.

Dietrich Reimer (Ernst Vohsen), Berlin.

STANDARD ATLAS AND CHRONOLOGICAL HISTORY OF THE WORLD.—Arranged by Charles Leonard-Stuart. 1912. Price $1.50.

Syndicate Publishing Company, New York.

This volume contains much varied information about the World in general and the United States in particular. It includes a fairly comprehensive Historical Chronology, a Gazetteer of the Cities of the World, a useful review of the various departments of the U.S. Federal Government, together with Census Statistics and other information. The atlas contains 90 plates, including maps of each of the United States, but they are not distinguished by clearness or legibility.

THE MEDITERRANEAN LANDS.—Bacon's Excelsior Map. Scale 1 : 4,067,712. Price 16s., mounted on cloth, rollers, and varnished.

G. W. Bacon and Co., Limited, London.

This is an admirable school wall-map with bathy-orographical colouring similar to the other maps in this popular series.

"ADMIRALTY" TELEGRAPH CHART OF THE WORLD.—In 3 sheets. Numbers 3778, 3779, 3780. Showing Submarine Cables, Principal Land Lines, and Wireless Telegraph Stations open for Commercial Purposes. Compiled in the Hydrographic Department of the Admiralty. 1912. Price 3s. each sheet.

Admiralty Office, London.

A completely new chart on a much larger scale than the old one, which was on a single sheet. This enlarged size is quite necessary to show at all adequately the various cables, and to distinguish their ownership.

THE SCOTTISH

GEOGRAPHICAL

MAGAZINE.

A NEW VIEW OF WEST AFRICA.[1]

By Mrs. MARY GAUNT.

(With Illustrations.)

You have, some of you, always heard West Africa spoken of as the land of miasma and fever, of swamp, heat, and mosquito, a land in fact to which no man would go willingly, a land to which he is driven by sheer necessity. That is the way I thought of West Africa before I went. It presents itself to me now in many pictures, sometimes lovely and entrancing, sometimes cruel and forbidding, but for the coast I think there will always come into my mind one, that of palms and sea and sky. I wish I could give you the blue of the sky, the deeper blue of the sea, the yellow sand, the white foam of the surf, the green of the graceful palms; could make you hear the whisper of the wind in the palm fronds, the crash of the surf as it meets the sand; could make you feel the freshness of the early morning, the languor and the heat of the midday, the glory of the tropical nights. It is a mistake to think that African nights are hot. They are not. So long as you remain in the open air they are simply delightful, and I have never, so long as I slept where the air could have free access, as on a verandah or out in the open, found one single night when I did not require to draw up a blanket over me before morning. Wherefore have I come to the conclusion that some of the ill-health in West Africa is due to a too strict adherence to British ways. "Pull down the blinds, light the lamp, and draw round the fire," is the British standard of comfort, and a very delightful standard it is in these colder latitudes, but the man who is going to try and apply it in a country like West Africa is courting

[1] An address delivered before the Royal Scottish Geographical Society in Aberdeen on December 3, 1912.

disaster. What he wants to enjoy there is the beauty and glory of the tropical night, and if he will only learn to appreciate that properly he will turn to good account the sixteen hours of pleasant temperature that are always his if he does not shut himself in too much. Then he will find that the eight hours of heat will trouble him far less. You must no more seek British modes of comfort in West Africa than you must expect British service. When you land on her coasts you have done with much of the mechanism to which you have been accustomed, and all your belongings are handled by man. The carrier is the most important factor in life on the coast. The time will come, I suppose, when we shall do away with him, but I am glad I have met him. Not once all along that coast did I meet a white man on the march, but there was traffic for all that, for certainly in the western half of the colony the seashore is the king's highway, and here the people, men, women and children, pass to and fro from one village to another bearing all their merchandise on their heads. Everything is carried on the head by these stalwart sons and daughters of Anak.

The people along the shore live by fishing, so that the launching and drawing up of a surf boat were everyday sights. Sometimes I would come across a great catch of fish, and then, though I could not speak their tongue, and was an object of great curiosity, the people always made the most of the occasion, and rummaged over the fish to show me those they thought the most remarkable. I could not but feel friendly when a gentleman, black, of course, and lightly and suitably clad in a scrap of rag, raised high over his head the thorny tail of a flat fish that was flapping about on the sand, and the others stood round pointing at it with long "ohs" and "ahs." Of course all the shore was dotted with the fishing-villages, and very dirty indeed they were, and dirtier and dirtier they grew as I travelled eastward, till I reached German territory, where their surprising cleanliness astonished me. As long as those villages are filthy I shall never be surprised that any town on the Gold Coast is quarantined for yellow fever. They must be the most excellent breeding-places for the yellow fever mosquito. They are every one within short water distance of the various towns, and we all know that a benevolent Government is very loath to interfere with the personal liberty of the native, even if it be the liberty to spread disease and death. It is a pity, for this is going to be a great country, though for all the years that we have held West Africa I think it is only in this century that we are beginning to recognise its immense possibilities. The head of the station at Tarkwa, looking out at the gold mines round him, declared emphatically to me that the agricultural wealth of the country was far in excess of any gold output, though he acknowledged that was by no means to be despised, and he took me round and showed me the glories and the beginnings of the garden over which he ruled. The rubber plantations are very picturesque, with their dark, almost gloomy foliaged trees, with the strong tropical sunlight streaming down the aisles between. The black man, left to himself, will gather rubber till the tree is exhausted and dies; so much experiment was entered into to find the best way of scoring the tree. These rubber plantations ought to interest us deeply, for they are the beginning of what in the future I am convinced will be a very great industry.

People have said to me, "Of course you were well in West Africa, we all know the tourist never catches diseases," and then I, though they did not, asked myself "Why?" and of course found an answer easily

Tapping a rubber tree at Tarkwa.

enough. Providence does not make a tourist immune simply because he is a tourist. I think it is probably his state of mind, that of being interested and amused, that makes for health, so if we can interest and amuse the man who has to live in Africa we go far towards keeping him in good health,—but this is only by the way.

There are many other valuable products besides rubber that are grown at the Tarkwa Agricultural Station, and may be grown profitably all over the Gold Coast. Palm oil and cocoa and mahogany are staple products, and also pineapples. It is necessary to keep the remembrance of the amazing fertility of the land in your thoughts when one looks at the native town of Tarkwa, for anything more hot and dreary and ugly it would be difficult to find. Though I am bound to admit the negro generally has a shade tree or two in his town if left to himself, he seems on the whole to have no idea of the beauty of growing green things, and I am sorry to say that many of the men who rule over him in the British colony do not seem to have any idea of it either. Bare ugly native houses, bare ugly shops with all the wares displayed outside, seem to be accounted good enough for the natives of a country of surpassingly beautiful vegetation, and I never shall forget being taken by the doctor who was superintending its buildings to see the new village that was going up for the miners of the Prestea mine. It was clean, that is all I can say for it. The streets were absolutely bare of shade trees in a land where for at least eight hours of the day nearly all the year round shade is the great desideratum. This, I was told, was done in the interests of health, because trees brought the mosquito. I would rather die of fever than sunstroke any day, and in the new mining village of Prestea sunstroke has been carefully arranged for. He was an energetic man, that doctor. In the European settlement he had converted, in his war against mosquitoes, a beautiful hill into one great raw red scar, where not even a blade of grass grew, and he had crowned it with corrugated iron-roofed buildings. Do you wonder that in the tropical glare and heat every white woman had fallen sick and been sent home? Add to this the fact that every house was carefully fenced in, windows and doors, with mosquito-proof wire netting—and mosquito-proof wire netting, when it is not kept clean, and it never is in Africa, keeps out the air so efficaciously that to go into one of these houses is positive pain. Remember, the air can never be changed, and you go in and gasp. I am inclined to think quite as many men suffer from vitiated air in West Africa as from ailments peculiar to the country. The vitiated air that they live in for certainly a good third of their time must lower their vitality, and must give every other tropical disease a chance. A mosquito-net you must have. Let it be a clean cotton net, put it up in some breezy place, sleep under it out in the open air if possible, abstain from doubtful water and too many gin cocktails that are by no means doubtful, be interested in the life, and I feel sure that more than half the battle is won.

Mines, of course, do not make for beauty. I was born and brought up on a goldfield myself, so nobody knows that better than I do, but some of these West African goldfields have the most beautiful setting. Tarkwa is a town set in a great gully among hills clothed in vivid green. The mines burn, I believe, about a hundred trees a day, so that the forest is being rapidly cleared away. I grieve for that, for these paths through the forest are very fascinating. But all the forest must go, for the mines want fuel, and civilisation naturally does not require heavy forest. It is better, so the manager of another great mine, Obuasi, assured me, to clear the land and plant farms, bananas, plaintains, cassada and cocoa, so that ten men may live where not one lived before. He is right, of course, but presently I am going to put in a plea for the wise saving of some of the grand trees.

I have heard many opprobrious epithets applied to West Africa, but

A path in the forest.

no one had ever told me, before I went there, that it was a beautiful land, and yet a beautiful land it is, and one of the most beautiful rivers in this beautiful land is the Volta. It is the most lovely river I have ever seen. Not perhaps at Addah at the mouth. For to Addah comes all the trade of the river, palm oil and cocoa and other products. Here in the old days the Danes had a fort, and here was a great slaving station. But it is a place of swamps, a forbidding place, a place of strong winds too, luckily, for the mosquito here is a wild beast that makes you feel you must go about armed, and if you would live in even partial comfort, the wire netting I abominate is an absolute necessity. Luckily the sea-winds force their way through the meshes and keep the air from becoming vitiated, and my hostess was one of those clever capable women—I am proud to say she was born in my country—of whom there are all too few in West Africa, who made it her business to see to her husband's comfort, and incidentally had her netting carefully cleaned every day. I repeat, if you would live in a mosquito-proof room and keep your health, this is an absolute necessity. Addah was desolate, but it had a charm of its own. The weird swamp warned the traveller that this was the country accursed for the white man, and the gorgeous sunrises and lovely sunsets seemed to say that, if you only look for it, there is something lovely to be found everywhere.

The real interest of the Volta began for me sixty-five miles up, at Amedika, which is really the port of Akuse, a great trading town of the Krobo country. Here come two steamers a week from Addah, bringing mails and taking produce to the sea, but, when I was there, only the ordinary canoe marked it as a port. I was going up the Volta, so I engaged a canoe which was just a little out of the ordinary, in that it had a roof over the centre to shade me from sun and rain. That journey will remain in my mind as one of the most wonderful I have ever taken, for the river was entrancingly lovely, but if any woman is fired with a desire to do as I did and go there by herself I strongly advise her not to go. The half-civilised black men of whom the preventive service is made up are not, I think, safe for a white woman to go among alone. I think it was only surprise at my audacity that saved me; the next woman might not be so fortunate. I was told the other night by the District Commissioner from Akuse that I was warned not to go, so I suppose I was foolish, but I was warned against so many things when I was in West Africa, that I got into the habit of disregarding all warnings. On the whole, it was just as well, for I should have seen nothing had I paid attention to all of them. And even if I was unwise to go up the Volta by myself, it certainly was well worth seeing. No picture that I could take can give you any idea of the loveliness of the colouring. The blue sky was reflected in the water, the shores were a panorama of hills clothed to their peaks in verdure in every shade of glorious green, and every now and again would be a patch made by the thatched roofs of the numberless villages. Occasionally we came to a greater town, and at one of these, Kpong, I spent the first night (see illustration). It is entirely a native town, having about 4000 inhabitants. In the background is a peak, Yogaga, the long woman, which is the great feature of that stretch of the river. The river winds about, and above all the other greenery stands Yogaga, sometimes to one's left, sometimes to one's right, sometimes ahead, sometimes behind, but always vivid green against the vivid blue, as if proclaiming the enormous agricultural wealth of the country.

Kpong on the Volta, with Yogaga in the background.

The Volta is a difficult river to navigate, as I soon found. This is because it is a series of enchanting lovely reaches barred from one another by huge barriers of rock, and I am bound to say many a time when we were negotiating those barriers, and I found the canoe broadside on with the white water foaming all round her, I really thought to myself it looked decidedly as if I and my belongings were going to the bottom. The Senchi Rapids—the worst of these rapids—raise the river thirty-four feet in an incredibly short distance. It looked to me like a wild waste of boiling white water, and how we ever got up it I do not know. That we did was due to the skill of my canoe men.

There was plenty of life along the banks, men fishing in canoes, women selling cassada or maize meal along the banks, and every now and again we came to preventive service stations where I spent the night. There is a curious anomaly in the Gold Coast, two different customs duties being paid along the banks of the great river. On the east bank four per cent., the same as in German Togoland, is paid, and on the west it is ten per cent., the customs duty of the Gold Coast Colony. The temptation therefore is to smuggle from one side to the other, hence these preventive stations.

At Labolabo I left the river, with great reluctance, and turned inland, spending Easter at the British Cotton Growing Experimental Farm with its kind manager. The British Cotton Growing Farm lies, if I judge by my feelings, for I walked there in all the heat of a blazing tropical afternoon, about ten miles from the river, but subsequent investigation has convinced me it is really only about a mile and a half away. It is a picturesque red bungalow, and is set in the embrace of the green hills, looking out over acres of land laid out in cotton, away to where through the hills and greenery you can just see the silver of the river. Odysseus once saw the sea like a silver shield, and so in the blazing sunshine saw I the Volta from the verandah of the bungalow. The Cotton Growing Farm is a failure, for the native having discovered that he can grow cocoa at much less cost and at far greater profit naturally declines to grow cotton, but such a lovely site and such rich soil must be worth a great deal, and some day I hope to see it put to more use. It was there I learned how utterly lonely a man can be, cut off from his kind, and that beautiful surroundings do not save a man from the brooding terror of the African bush. Fear lived there. "There is nothing to fear," said I, full of the change and the delight of new scenes. "No," said my host doubtfully, "there is never anything to fear in Africa till the time when there is something and then it is too late." Oh, it was lonely! I found myself looking down the avenue between the grape-fruit tree and the shaddock at the front entrance and wondering what awful thing might come out of those deserted cotton-fields, and when a tornado on Easter Sunday swept down from the surrounding hills, blotting out everything in a sheet of grey mist, I found that Fear was in the howling wind and in the shuddering rain. Nothing made me realise, as that did, that it was the conditions of life and not the climate that made West Africa a word of ill omen to the white man. Anything more lovely than the surroundings of that cotton farm I defy any one to find anywhere in the world, yet the man who lived there by himself led perforce a most unwholesome life, a life that by no fault of his own was enough to drive any sane man mad. Up and down the verandah I used to hear him walking, up and down, "the lonely man's walk," said he; "you see those two boards, well, when I can't see them, I know that I'm drunk and I go to bed." But I don't believe by his looks he ever did get drunk, though it would have been quite excusable if he had, for his life was tragic.

From Labolabo I went up to Anum Mountain, where many years ago the Basel Mission established a station; so long ago, that in 1869 the fierce Ashanti warriors raided it and carried off the missionary and his wife, Mr. Ramseyer, and their child, into a captivity that lasted many years. Some years afterwards the Basel Mission came back to their old site, and as, with true German forethought, they have planted trees and made gardens most carefully, the natural beauties of the place have been enhanced. Here emphatically the German Basel Mission people have made a home, a home in the very best sense of the word, a beautiful garden, a comfortable house, a happy family, and presently I suspect they will bring up children for the mission field, here in the heart of savage Africa. On the other side of the mountain, looking away to the north, is the Government Rest House, looking out over far, far vistas, with the river winding its way across the Afram plain. I only looked at that plain from the top of the mountain, and I wondered what the Government were doing to neglect such an obvious health resort, for here on the top of the mountain, though the sun was hot enough at midday, it was quite cool at night. It was eerie too, for a white mist came up and enveloped everything, so that I could not see beyond the verandah, a white mist by which I knew the moon was shining somewhere beyond, but as I had no carriers about me, and did not even know where my boy had gone, I grew mightily afraid, of what I do not know, —of the stillness and the night, I suppose. Anyhow, I put in a bad time, there in a house that was doubly desolate because the white man had been there once and was there no longer, and the night passed slowly, and I was hungry, because there is a ju-ju over Anum Mountain and there was no possibility of getting food there. "This bush country no good, no can get chop here," announced my boy, and as I had been unprepared for it I dined and breakfasted frugally off tea and biscuits, which was all I had. However, before the mists had cleared away there came a chattering and a shouting, and the carriers who were to take me to Pekki Blengo had arrived.

But at Pekki Blengo I was really in a dilemma. There is a very steep, rugged—and for a woman who is not suited for mountain climbing—difficult range of hills to be crossed here. My helper was a very charming, educated black man named Olympia, one of Messrs. Swanzy's agents, and he explained that over that range I would have to walk. Therefore I decided to do it at earliest dawn. Unluckily I reckoned without my carriers. They came at 5.30, picked up my gear and laid it down again and walked away, saying in the vernacular what my carriers seemed to have said at intervals all along the coast, "We no be fit, Ma.' It was eleven o'clock before by strenuous efforts I succeeded in replacing them, and then when I had started them all off I found that my hammock boys had gone back to their cocoa farms, and I had not even a drink of water left, for of course I would not have dared to drink the native water, however thirsty I had been. They told me I could hope for no hammock boys till next day, and I felt bad about it because I did not know how I was going to sleep without a bed. Therefore I said I would walk, and proceeded to agitate for two carriers, men or women, to carry my empty hammock. As this would have thrown at least six men out of a good job, about one o'clock eight hammock boys appeared on the scene, deciding to accept four shillings each for the journey to Ho in German territory, and we started.

In a very short time we reached the Eveto Range and the trouble began. I had had nothing to eat or drink since five o'clock in the morning, the day was blazing, and the way—no words of mine can adequately describe that way. It was steep and it was rugged, and it was full of holes and roots and rocks, and I climbed and sat down, and climbed and lay down, and climbed and thought I would die, and did not care if I did, and wondered what joy people found in Alpine climbing, and when life seemed to have resolved itself into an eternity of effort to which there was no end I found I was at the top, looking out over a wilderness of verdure-clad hills, range upon range, dappled with shadow and evening sunlight, such a view as would be glorious in any land. My thoughts were confused; joy at my attainment and the unexpected beauty of the view, for very seldom does any one tell you you may expect beauty in West Africa, and horror at the thought of what a slave-driver I had become, for those wretched carriers were going over exactly the same ground as I had covered with so much difficulty, and all those luckless men and women had loads upon their heads which they agreed to carry to Tsito, at the foot of the range, for sums varying from sixpence to ninepence a head. I felt as I started down the Eveto Range that Legree could not hold a candle to me, and when, after a struggle nearly as difficult as going up, we arrived at Tsito and I was shown to my sleeping quarters, Swanzy's cocoa store, I called them in and, very unwisely to ease my conscience, paid them double.

I put it on record that that struggle did not do me any harm whatever. I had an evening meal of eggs and tea and biscuit. I looked at the small drop of gin I had left, and decided to keep it for greater emergencies, and the population at Tsito stood round the store and discussed my personal appearance aloud. At least I suppose they did, for they looked and talked a great deal, though I could not tell what they said, then presently I put out the lamp, had my bath in the dark, and went peaceably to bed, and as I dropped asleep I could hear the audience dispersing, still discussing the show at the top of their voices. I had no means of knowing whether they found it satisfactory, but anyhow it was cheap. The next morning was glorious, and I felt as if the world belonged to me. I started off my carriers for German territory, and went round taking photographs.

Swanzy's Store at Tsito was only a humble place, ruled over by a black man who bought cocoa and cotton from the farmers, and retailed them kerosene, the commoner sort of tinned fish, cotton cloths, and a little gin, but it is interesting as showing how trade and the desire for something beyond the mere necessaries of life is spreading into the remoter corners of these West African colonies.

I had some difficulty in getting away owing to my foolishness in paying my carriers extra the night before. When I turned to my hammock my boys were not forthcoming, and the head of the store, who was the only person about who could communicate with me, announced with negro pomposity, "There is trouble. They say they no can carry you; Ho be far, and you not give enough money." I was furious, for most of the day before they had only carried themselves and the empty hammock, and I had given them two bottles of rum. If I had thought about it I would have found the situation desperate, for I was entirely at their mercy, had not even my servant with me; he had gone with my carriers, and I did not know the road to Ho. Luckily I was too angry to think about that. I simply marched back to the store and said, "Tell them they can go home, and I will pay them nothing at all." What I proposed to do, if they had taken me at my word, I am sure I do not know, but evidently my position did not strike them as so helpless as subsequent reflection convinced me it might have been, for they came back promptly to work, and carried my hammock through orchard bush country where the road was so bad that most of the time I had to walk; and the butterflies, blue and purple and orange and white, were so gorgeous that the remembrance of them is with me still. I was troubled a little because I could not believe that this was really the main road to Ho, and I wondered if it was possible that these men were taking me out into the bush; and when we walked, much to their amusement, I made them all walk in front of me. I need not have worried: they were simple, honest peasant farmers, and I was worth a deal more to them safely at my destination than any other way, for after what seemed to me a very long and tiring march I arrived suddenly at the border. It is a purely imaginary line, orchard bush country on one side, orchard bush on the other, and the only difference is that on the British colony the road is barely marked from the surrounding bush, while in German Togo the road is broad and rolled, and well kept as a garden path. The carriers heaved a sigh of relief when they found themselves upon it, and we fairly raced to the first German village. It was my first introduction to German thoroughness, and I was amazed at the neatness and tidiness of the little town. It was swept and garnished, there were no ruts and holes, no waste water to breed mosquitoes, no garbage or litter lying about, and above all there were the most beautiful shade trees with wooden seats ranged under them, so that the villagers might take their noonday rest there comfortably. I don't know whether the native is happier under German rule, but I am quite certain he is cleaner and healthier. I met no mosquitoes in Togo, and since I was astonished and delighted at the way the authorities had conserved and planted trees, I cannot think that the trees are responsible for the mosquito, as I have so often been told. The natives prefer British rule, partly, I think, because the Briton has intuitively a better way with the native than the German, and partly because it gives him freedom to live exactly as he likes; it does not insist on such drastic cleanliness, and the mosquito has evidently plumped for British rule because the Briton, whether from a sense of fairness or not, certainly gives it a better chance of life.

I have been accused of unduly praising the Germans. But I have only spoken of what I have seen, and you would not surely care to come and listen to me if I bore false witness. It may be, of course, that a tremendous lot of money has been spent in Togo, that I do not know. I only know that last year it paid, and that if foundations well and truly laid mean anything, it will go on paying. To begin with, the Germans have learned what we so sadly need to learn, the value of continuous service. The German Commissioner's bungalow at Ho is on top of a range exactly like the appalling Eveto Range, but the road is gently graded, and is as easy of ascent as a path in Hyde Park. The Commissioner has been there for the last six years, and the place is laid out as a beautiful garden, all the natural beauties having been preserved. Behind the house is a magnificent plantation of teak-trees, and from the verandahs and garden paths is a view that takes in miles of lovely tropical country, with, in the far distance, the highest mountain in Togo.

The Commissioner had his wife with him, a smiling, rosy-cheeked girl who had been out fourteen months, and was the picture of happy contentment. Would you have had me ignore this prosperity and contentment? Is it not rather wisdom to seek out its reason? I say emphatically the nation that can take its women with it is going to hold a country far more firmly, and far more satisfactorily, than that one that leaves its women at home. And in this matter the English fall far behind the French and Germans. Sometimes this is the fault of the women. I have seen in West Africa women so dependent on civilisation, the intense civilisation of a British town, for their comfort, so absolutely without resources in themselves, that they were just a burden and a care to the unlucky man they married. They became ill, and blamed the country and the climate when they should rather have blamed their own helplessness and stupidity, and I am more convinced than ever that if Government or private firms would see to the proper housing of the women, and instead of either forbidding them to come, and penalising them if they do come, as they so often do, offered a bonus to every woman who stayed out her full term without being very sick, they would find a very great difference in the health and prosperity of the colonies.

If Britain is going to keep her pride of place in the Tropics, she must encourage the women to go out with their husbands. There are men who say they would not expose their wives to risks, and that is kindly, but unwise; there are quite as many risks in staying behind as in going with their husbands, and those other risks that come from divided homes and divided interests ought to be considered, for they are far more insidious and dangerous to a nation's welfare than the African climate. Of course where there is open warfare a woman would be in the way, but there are very few places now on the coast where there is any danger of that sort, and anywhere else a woman should be a comfort and a help to her husband. Some are, I know. If you have read Major Tremearne's book you will find what he says about the helpful resident's wife who gave cooking lessons to her bachelor friend's servants in Nigeria, and if you have read Captain Haywood's book you will see what he says about the French women in Senegal, where a man apparently takes not only his wife, but his daughters. I know, of course, there is in some quarters strong opposition to women going. I remember hearing one man say openly that when white women made their appearance life was spoilt, it was not so free. That made me more sure than ever that what West Africa wanted was good sensible wives to raise the tone and the health of the country. There is another thing to be remembered. The native finds it hard to bring himself to respect a man who has no wife, therefore to the man who brings out his wife, he at least will give full measure of honour. These things must be known to the authorities, and yet I am told that in Northern Nigeria, while a man may live in Government quarters rent free, the moment he brings out his wife he is charged £30 a year for his quarters, though he have the very same he occupied as a bachelor. I could hardly have believed this, did I not know there were most curious anomalies in the treatment of women on the coast. No one will deny that the nursing sisters at the hospitals must have made an enormous difference to the man going to the coast. Yet this is what happens to them. The hospitals are Government property, the medical officers are Government servants, but any officer who has a private patient too ill to remain in his own quarters, transfers him to the hospital, where he very properly pays a certain sum for his keep and attendance, which sum is divided between the medical officer and the hospital. The nurse, also a Government servant, nurses these private patients, but it has never occurred to any one that she has a right to one penny for extra work that may, perhaps, keep her up day and night for days on end. Who cares what happens to a nursing sister?

The farther I went, the more I was convinced of these two things, the necessity for bringing out white women, and the fact that in Togo at least the Germans were making a very good show. Understand, I only speak of what I saw, and I saw that the Germans were certainly running their colony with thoroughness.

The first town I stopped at after leaving Ho was Palime. In the market the women were cooking and selling maize-meal balls, fried a nice crisp brown in palm oil, and making and selling cassada porridge. Cassada is the root from which our tapioca comes, and this thin porridge forms the black man's first meal, at any rate in these parts, early in the morning. It is warm and wet, I suppose, but it looks uninterestingly like paste. Palime is railhead; and when I arrived there late one night I knew my bush journeying was done for a little. It is one of the cleanest towns I have ever seen, and I slept undisturbed, without mosquito curtains, a testimony to the thoroughness of the

Germans. The roads in Togo are excellent, and there are six hundred miles of them. When the German Commissioner told me that, I murmured "Forced labour," but he indignantly denied it. "There is a tax," said he, "of six shillings a year, and if the man does not pay, he must give a fortnight's labour. It is only fair, for if we have no roads we have no trade." And I, thinking of the twenty-five per cent. of the cocoa harvest left up the Afram creek, "because we no be fit to tote," had perforce to agree with him. Why should not the man to whom the roads are of paramount importance contribute to their upkeep? The country round Palime is like the country on the west of the border, with this difference, the excellent roads. Unless there were good roads, the carts heavily laden with merchandise could not get about the country, drawn as they are by Kroo boys. They pull these heavily laden carts through the most beautiful country, along hillsides, where the ground falls away from the roadway, and drops down hundreds of feet into the valley below, and away, far away, in the blue distance on the other side of the valley, rises another picturesque hill.

Misahohe, nestling on the hillside amidst luxuriant tropical greenery, but for the palms and the strong sunshine might have been a village on a Swiss hillside, and I went up away beyond, because I was bound for Mount Klutow and the sleeping-sickness camp the Germans have established there. It has to be on a mountain top away from all vegetation, because vegetation harbours *Glossina palpalis*, the fly that, when infected, is the cause of the evil. I went up there one hot Sunday in May 1911, and Dr. Van Raven, the German doctor in charge, received me so kindly and told me so much about it, though I was an outsider and a foreigner, that I almost felt myself competent to grapple with sleeping-sickness in all its forms. It is curable, he believes, though it is as ghastly as consumption, and often quite as hopeless. There was a time when the natives feared and dreaded the Germans, and could only be got to the sleeping-sickness camp by main force, but that time has passed away, and they now come eagerly from far distances; the sad part is that they so often come too late.

From Palime I came down by rail to Lome. Lome is one of the most charmingly laid out towns on the coast. The bungalows provided for the officials are like palaces, the gardens are things of beauty, and all the streets are nicely planted with avenues of trees, some tiny, just planted, and some four years old, which stretch their branches right across the street.

From Lome I made my way along the coast to Keta, passing across a narrow strip of sand that runs for many miles between a lagoon to the north-west and the ocean to the south-east. It is a very narrow strip of sand, and it is very hot, but thereon are many cocoa-nut palms from which it is expected in the future to make much copra. The villages, in contrast to the German ones, struck me as peculiarly filthy, and the reek of them was overpowering. It was as if all the natives who had desired the freedom and resulting dirt of the lax British rule had come across the border and settled on this strip of sand, just out of reach of the German scrubbing-brush and pail. I am only stating facts, those villages are filthy. It is for you to say whether the German methods are better. They are certainly more thorough. In Keta, where I arrived after a long day's tramp, I should say the people are decidedly happy and entirely contented. The sandbank here is possibly a quarter of a mile wide, and there is a great native town that carries on a big trade in poultry, chickens, turkeys and ducks, palm-oil from across the lagoon, and does much weaving. The negro woman is a born trader; from her cradle to her grave it seems to me she is buying and selling; and here in Keta, because I stayed with the Bremen sisters who spoke the vernacular, I got more closely in touch with the people, especially the women, and was deeply interested. There were dozens of little traders all down the streets, beginning with the woman who at earliest dawn stationed herself at the corner outside the fort with her little fire and her little pot to sell porridge to the early breakfasters, and ending up with the women who sat in the streets in the evening with big platters of native-made sweets at their feet. The regular market where the authorised buying and selling was done was held on a flat, featureless piece of ground alongside the lagoon, and one had to seek for beauty in the quaint folk and the varied assortment of goods laid out on the ground or on the stalls for sale.

These markets in the native towns, save for the heat and the dress and colouring of the people, are to all intents and purposes exactly like the markets we see held in country towns and in the poorer streets of the great cities of Britain. Human nature has points of resemblance all the world over. Certainly in England the daughters of the nobles do not sell beer for profit like the daughters of the chief here, but it is laid upon all the women of Keta to be self-supporting, and I think they are all the better for it. The lagoon is close to the market-place, and here the canoes take the goods that arrive from the sea to the interior and bring back the produce, palm-oil and cocoa I think mostly, that must be shipped to Europe. The houses of the people are comfortable enough in their way, for these are not savages, they are simply a peasant people. The sister took me to pay a round of calls upon the chief's wives. There were, I think, sixty of them. They were most hospitable, and many of them gave my hostess small sums of money varying from sixpence to half a crown, so that she might buy something to make a little feast for me.

In a courtyard of the house of one of the wives of the chief, a prosperous, contented woman, with children of her own, we saw two or three girls she had taken in to instruct in the art of bead rubbing and other necessary feminine accomplishments, just as in the old days in England the wives of the powerful nobles were surrounded by the daughters of the smaller gentry. I must confess I do not understand how in a community where apparently men and women are about equally divided, a man can take to himself sixty wives without seriously upsetting the balance of society. Here in Keta this powerful chief has most certainly sixty women who are nominally at all events his wives, and who live, some in houses in small compounds attached to his house, some scattered over the town, and some away in the country. I visited another household of smaller dimensions, the man having only four wives. He was a clerk attached to the Bremen Mission Factory, had once been a communicant and pillar of the church, and still wore a stiff collar, a straw hat, and tailor-made clothes, but the rules of the church had been too much for him, he could not deny himself the dignity which attaches to a man with many wives.

Weaving is a great industry at Keta. Why weaving should have gone ahead on this little strip of sand, which looks to me as if the first great storm would overwhelm it, I do not know, nor do I know who first taught the weavers, but Keta clothes are in demand all over the coast, and rightly so, for they are well woven and the colours skilfully blended. It

Lagoon at Christiansborg.

is only when the negro comes into contact with European wares and tries to ape them that he loses his sense of proportion. The thing that is entirely native is by no means ugly even if it may be quaint.

All along the shore about Accra and Christiansborg you come across ruins which are reminders of the old strenuous days when men built not only for a living place, but for defence. Some day I should like to tell you of the mighty castles now falling into decay, put up by the Danes when they landed and held these coasts in the wars when nations took little count of them. All the tribes seem to have been warlike fighting men in past days, and this is a matter we might well take to heart when thinking of West Africa. Long has it been known as the White Man's Grave, but this same place that for years has been stigmatised as having the most deadly climate in the world, produced a race of men and women so strong, so virile, so adaptable, and I suppose we must add so stupid, that the civilised world for at least three centuries came regularly here for its labour. This heat and damp produced mighty thews and sinews, and, so long as the native does not imitate too closely European modes of life, it produces them still. Both men and women carry heavy weights on their heads. The men are fine to look at, though you may not admire the negro type of feature. The young women, before hard work has worn them out, are things of beauty, like bronze statues, and walk and carry themselves like queens. No effete race is this, dying out before the arrival of the white man, but one strong and virile, and the country that produces it cannot deserve all the epithets that have been hurled at it. We know that the health of the coast is improving every day, and I look forward to the day when we shall find out the reason why the land of the stalwart negro makes the white man anæmic. Maybe the remedy will be as simply and as easily applied as the knowledge that fresh air kills tubercle. Of course nine months of the coast did not make me anæmic. I never had better health in my life; so possibly I am a little prejudiced in its favour.

The great race of the West Coast, which has terrorised over the rest for centuries, is that of the Ashantis, and, as I longed to see the forest country, I went by train to Kumasi and there prepared for a journey through the forest to Sunyani, a station on the outskirts of the colony, five days' march to the north. Ashanti, as you all know, is conquered country, and twelve years ago was at grips with Britain. But it is most effectually conquered now, and in the places that ran with the blood of human sacrifices are now going up large handsome buildings, the stores of the great African merchants. It is a charming place, Kumasi; I have but one fault to find with it; in a place where plenty of fresh air is the great desideratum, an absolute necessity, the African merchant princes are putting up their houses as close together almost as if they were in Liverpool itself, and very often they make their clerks live in cubicles that are little more than death-traps. The town is well laid out and beautifully planted with trees, but what interested me most were the big African chiefs who, with umbrellas twirling (the umbrella is the sign of rank) marched about the streets. I am thankful to say that the good taste of the people forbids the chief to appear in anything but native dress.

In Kumasi the transport men kindly provided me with carriers, and I started north through the great rubber forest. The forest remains in my mind as the most wonderful and awe-inspiring of my experiences. It was difficult to take photographs, for the trees day after day stood in close phalanxes, straight as Nelson's column, never breaking into branches till they were possibly 150 feet from the ground, and between the trunks was a dense undergrowth of fern and creeper, and jungle of all sorts, so closely twined that it was impossible to step a foot from the path, which must have been kept open by an infinity of pains. Never in the world have I seen such mighty trees.

The Ashanti houses are simply four verandahs built round a courtyard open to the sky, extremely healthy, I should think, and I for one am sorry to think they are passing away in favour of European houses entirely unsuited to the climate. What can be pleasanter on a hot tropical night than to sleep with the open air all round you? I confess that when first I arrived at the rest house I was a little dismayed to find it full of Hausa and Wangard traders, and I wondered where I was going to get the necessary privacy. But the lady who owned the house—she was simply clad in a scanty blue cloth and had her head shaven—cleared out the traders promptly, and having had a little tent made of my rugs for my bath, I found the rest simply delightful, and I slept. I slept like the dead and awoke refreshed, feeling as if the world was mine. People say to me, "Oh yes, it was all very well for you, you were interested," and, as I said before, in that I suspect lies the crux of the matter. I wish I could give you a detailed account of that forest journey, show you the green of it, so pale that it was gold, so dark that it was black, let you smell the rich sensuous scent of the orchids, see overhead the narrow strip of hard blue sky so far away that only for a short time at midday did the sun reach the doorway.

Potsikrom, my next stopping-place, is built in European fashion, and the consequence is that the two little rooms are unbearably hot and stuffy. The only thing was to have my bed placed right in the doorway, and this had its drawbacks. In an enclosed courtyard you feel safe if you are alone, but these rest houses are generally the last houses in the village, right on the outskirts, and when every living being had retired to the safety of the village and left me all alone, in the doorway of a house on the edge of the forest, with only a mosquito curtain between me and possible danger, it was generally rather terrifying. There was nothing to be afraid of but the trees, but African trees have a personality of their own, and I have put in some few nights wondering what they would do to me if they could catch me. I assure you I listened more than I liked in Potsikrom to the sounds of the night. Those fears were no light thing, but the journey was well worth it.

The farther north we went the denser the forest became till, north of the Tano river, we had left even the road behind and travelled in a pathway that wound beneath the great trees, travelled mostly in gloom, for neither sun nor rain could penetrate, and yet I assure you it had a wonderful charm. I do so hope and pray the British will not rashly cut down and destroy that forest. A mahogany-tree is just as well worth preserving as an oak, and though it takes two hundred and fifty years to its full growth it is often rudely cut down and destroyed in a single day. Opinion on the coast seems to be against that wonderful forest, that cathedral not builded with hands, and I shall not soon forget the delight of the medical officer who was with me when we came to a place where perhaps for two hundred yards back the trees had been ruthlessly felled, not one being left standing, though the climate cries out for shade, and the young man himself wore a very large helmet and a spinal pad, as he explained, to protect him from the sun. Roads must be made of course, and the forests must be cleared, but to you Geographical Societies who have any influence I appeal for some discretion in the clearing. Do not let Britain make a sun-scorched wilderness of Ashanti and think she has done her duty. I know there is a forestry department in the Gold Coast now, and some of the members are very keen on their job, so I am hoping great things for the future, but I would have every man who goes into the forest remember that a tree,

Entrance to an Ashanti village.

specially these African trees, mahogany and kaku and other great hard-wood trees that take hundreds of years to their growth, cannot be replaced in our time or our children's time, or our children's children's time. The faith that nothing matters in an outland will not do for Britain now that other nations are in the colonising world, eager to make their colonies things of profit and beauty.

And so through the forest I came at last to Sunyani, where six white men hold and rule the place for Britain. The official bungalows are built of mud without doors and windows and with steep pitched thatched roofs and mud floors, and because they are very spacious and wide, with openings where those same doors and windows should be, seem to me ideal places of residence for the climate. They are built, of course, away from the town, and, close at hand, a sort of reminder of the brooding fear of West Africa, is the little fort where the occupants could retire should there be necessity—the provincial commissioner says there will never be necessity again. The stakes round the wall are bound together with a tangle of barbed wire. It is all very simple but very suggestive. The sentry-box is at the only entrance, and there is a company of black

Ashanti warriors.

soldiers drawn from a distant part of the colony on guard. Half of the white population is composed of their officers. But those who know are emphatic that this is a phase that is passing away. The negro is a keen trader. He would rather make money than fight any day, there are stores growing up in the town, and the post-office just in the shelter of the fort is beginning to pay.

From Sunyani I went on five miles further to Odumase, where the last Ashanti war first broke out. As a punishment for that war it was almost destroyed, and though it is rapidly being rebuilt there is really not much to see there, but to me that journey was full of interest. I was the first white woman who had ever been there, as I was the first white woman along all the track north of Kumasi. The chiefs, in all the glory of umbrellas, silken robes of many colours, and numerous attendants, turned out to meet me, and everywhere in the streets crowds came to look on, but the arrival was the most exciting of all. To be fired at out of long Dane guns by warriors with their cloths girt about their waists and their powder-flasks slung under their arms, is the greatest honour an Ashanti can confer upon you, and the Ashanti warriors of Odumase conferred that great honour upon the first white woman to visit them. Luckily I had been warned, for I assure you it is quite a nervous thing to be carried along a forest path and suddenly from the greenery to find half a score of armed warriors dashing out at you and guns going off over your hammock, under it, before you, and all round about you, to smell the acrid smell of the smoke, listen to the barbaric yells, and then to see half-clad dark figures fleeing along so that at the next turning the fusillade may begin again.

Now, have I, I wonder, carried out my promise and given you a new view of West Africa? It is tropical, of course, but I do honestly believe it is no more difficult to live in than any other tropical climate. Improve the conditions and you will find that the climate will improve. Once more I say it, and I say it emphatically, if you want to improve this land which should be one of the richest jewels in the British crown, you must improve the conditions, you must encourage every man to take out his wife. Already there is a beginning and things move slowly. Already ten women go out where one went before, and even though some come home saying the land is hateful, some have a higher conception of their duty and stay and do their best to make it more comfortable for their husbands and those around them. Every one who stays makes it easier for the next woman, and the more white people, with their higher standards of comfort, in a place, the better and more habitable that place is bound to become. I assure you West Africa, the land of opportunities, one of the richest lands in the world, is well worth a most mighty effort.

A GEOGRAPHICAL DESCRIPTION OF FIFE, KINROSS AND CLACKMANNAN.

By LAURANCE J. SAUNDERS.

(*With Sketch-Map.*)

(*Continued from* p. 87.)

III.—MODERN FIFE.

Origin of Modern Fife.—Mediæval Fife was a beggar's mantle fringed with gold; the interior, save Stratheden, being unimportant and thinly peopled, the coasts, especially those of the Firth of Forth, strung with numerous royal burghs betokening considerable wealth and commerce. The presence of these burghs surely implies a certain degree of centralisation, and geographical influences had carried this centralisation a little further, inasmuch as the burghs were grouped along two particular stretches of coast. But the number of burghs also implies isolation within these limits. Each little port had its own fairs, its own ships, its own salt-pans; and after the Reformation had robbed St. Andrews and Dunfermline of their ecclesiastical importance, no one burgh was predominantly more wealthy or more populous than its neighbours. Indeed the commercial and industrial conditions were such that, along the favoured coasts, geographical isolation did not destroy commerce or industry. Each little creek had a burgh at its head, and each burgh had the same advantages as its neighbours. Perhaps the beginning of a different order of things may be surmised when the little ports of the upper firth began to use sea-coal in their salt manufactories, but whatever new distribution of population and wealth this factor could have made in the seventeenth century, was postponed by the great decline of the Restoration and Revolution periods.

Effects of the Industrial Revolution.—But after the Union of 1707 the markets of America and England were thrown open to Scottish enterprise, and there commenced a new industrial development which led up to the Industrial Revolution and the modern factory system. Contemporary with this was a continuous, and in the eighteenth century a very rapid, improvement in agriculture, and then, last of all, new methods of conveyance by land and sea. As the result of these advances the relative positions of the East and West Coasts of Scotland were somewhat altered. Round the western end of the Rift Valley, that is, in Clydesdale and Ayrshire, now gathered dense populations, attracted at first by the convenience of the situation for the new American trade, and then finding surer prosperity in the adjacent mineral wealth which now became almost an essential to industrial greatness. But though the East of Scotland was shut out from predominance in this western trade, its initial advantages of a good climate and accessibility from the Continent secured it from decay. It had its own coalfields and its own peculiar industries, and while it nowhere presented such an intense concentration of population as around Glasgow, its larger area suitable for development enabled it to support several important centres, such as Edinburgh, Dundee, Perth and Aberdeen. Thus immediately outside Fife the great centres remained almost the same, for the Industrial Revolution acted only in what was already the most populous part of Scotland, the Rift Valley, and with the exception of the rise of Glasgow (which was important even in mediæval Scotland), there was no great shifting of population as in England. Thus on the borders of the peninsula, Edinburgh, Stirling, Perth, and Dundee, places which had determined the routes to and from Fife in the Middle Ages, still retained their importance.

In the peninsula itself, however, there was considerable change. The east had long been supreme in wealth and population; but now, while the east did not languish, the determining factor of convenient supplies of coal drew industry and commerce to the west where are now the great centres of population like Kirkcaldy, Dunfermline, and Cowdenbeath. Moreover, when the commerce of the peninsula, like that of the rest of Scotland, with the increasing size of vessels employed, centred more and more at one or two points where the volume of trade encouraged the construction of large harbours, the small creeks of the east were deserted, and in the west arose ports like Charlestown and Methil whose greatness the Middle Ages had not known. Lastly, with the increasing use of railways, the long firths penetrating on either side of the peninsula to Stirling and Perth proved barriers to direct communication. Between Glasgow and Perth, Dundee, and Aberdeen, the old route by Stirling and Strathallan, up which the English invasions had gone, was still the most convenient; but between Edinburgh and the now populous North-East of England and the northern centres, long detours round the head of the firths or else inconvenient railway ferries proved both expensive and annoying. The difficulty of a more direct communication between the south-east and the north-east was solved by the construction of the Forth Bridge (1890), across the firth at the natural stepping-place of Ferryhill, and the Tay Bridge (1877 and 1887) constructed opposite Dundee. When it thus became a link in a great trunk railway system, the peninsula ceased to be so completely isolated as previously. These changes and advances prepare one for some corresponding change in the internal place relations.

For a more detailed description, the peninsula can now be divided into three areas, distinct to some extent by physical feature but more by diversity of industry—(1) the county of Clackmannan; (2) the west of Fife and in addition the small county of Kinross, that is, the whole Leven basin and the highlands to the south of Loch Leven as far west as the old Forest of Clackmannan; (3) the east of Fife, comprising the peninsula east of Glenfarg, the Lomonds, and a line following the height of land between the Lomonds and the hills of the East Neuk to strike the coast at Leven.

1. *Clackmannan: Industries and Communications.*—The boundaries of this county, the smallest in Britain, have already been described as formed by natural features—the Ochils, the Forth, and to the east, the Forest. In the west where there is no well-marked feature *large* enough to form a boundary between the Devon and the Allan, the line from Blairdennon Hill to the Forth follows the convenient course of small hill burns to the latter stream. Of this small area of 35,000 acres, the northern half is occupied by the Ochil Hills; the southern half by fertile flat carse-lands across which a low clay ridge runs westwards to divide the Devon and the Forth; underneath this lower plain is the small bed of true coal measures preserved by the fault from Dollar to Menstrie. All this diversity of surface and its various riches lie within easy distance of the Forth. Hence each new industrial development found Clackmannanshire able to profit by it. The Ochils were suitable for pasture, the carse-lands for cultivation. The agricultural progress after the Union improved the sheep, and new methods of draining and manuring made the lower lands particularly suitable for cereals and beans. From the produce of the hill area arose

a manufacture of woollens; from the produce of the carse-lands, distilling and brewing. And in pre-railway days, when transport was dear and, unless under exceptional circumstances, an industry could not flourish away from such cheap means of communication as a navigable river or the sea afforded, the close proximity of the Forth and the convenience of the port of Alloa enabled these industries of Clackmannanshire to advance beyond the supplying of only local needs. Nor did the first stage of the Industrial Revolution cause any change, for since the new machines were driven by water power, what was more convenient than the power supplied by the numerous mountain burns which plunge down the southern face of the Ochils to the Devon? And so a string of small towns arose along the base of the hills, in some instances, as at Glenochil (Menstrie[1]), employed in distilling but more often manufacturing woollens, as at Alva,[2] Tillicoultry,[3] and Dollar,[4] and exporting the finished product through the port of Alloa, around which grew up the most considerable town[5] of all, naturally situated as a centre of converging routes, and combining both the distilling and the woollen industries. Nor did the application of steam power disturb those industries, for part of the plain was underlain by coal measures, and there was little distance between the mines and the consuming industries. When railways were built they brought about no shifting of the population. It was hopeless to attempt any railway construction across the Ochils, for not even a carriage road crosses this range between Strathallan and Glendevon. But there was little workable mineral wealth in the hills to make this disadvantage felt, and so the railways of the shire are confined to the carse-lands where construction was relatively easy, and the population dense. Alloa is the centre; westwards a line runs to Stirling, one of the great junctions of central Scotland, with a branch from Cambus, a brewing village at the mouth of the Devon, across the Carse to the Hillfoots, where it taps the distilling and woollen centres of Menstrie and Alva. From the south a line enters Alloa from Larbert, crossing the Forth by a long narrow bridge, which thus avoids an inconvenient detour by Stirling and gives Alloa direct communication with Glasgow. This line from Glasgow is continued to Perth across the clay ridge behind Alloa to such centres as Tillicoultry and Dollar and up the Devon valley to Kinross. Another line cuts due eastwards through the thinly peopled Forest to Dunfermline and Edinburgh, while yet another diverges at the town of Clackmannan, now displaced by Alloa as the county town, to the coast of the firth at Kincardine. With such a variety of industry carried on by a population confined to a narrow plain some three miles broad, it is little to be wondered at that this little county of Clackmannan should form one of the most populous areas of Scotland.

2. *Western Fife:* (a) *Historical Progress.*—The second area into which the peninsula has been divided is also of great importance. It comprises the plateau lying around Loch Leven, the lower plain of the Leven,

---

[1] Pop. (1911) 918. [2] Pop. 4332. [3] Pop. 3105.
[4] Pop. 1497. [5] Pop. (1901) 11,417, (1911) 11,893.

and the alternate valley-and-hill region between these lowlands and the Devon. This area was neglected during the Middle Ages. Except along the coastal strip where were situate the abbeys and the royal burghs, it was mostly covered with cold and retentive clays, and was in addition high and exposed. Yet this area is now the most important in the whole peninsula. It contains four of the five towns which have a population of over ten thousand souls. Such a population as seems here foreshadowed could not have been attracted by agriculture alone, for although the agrarian improvements following the Union did something to break in the clays, the soil was not such as could be turned into really superior land, and with, in addition, large tracts unfit for anything but rough pasture, the area never was, and is not yet, conspicuous for its agriculture. In Kinross, the most favoured part, where farming is still the predominant industry on the morainic gravels, the largest town [1] has a population of under three thousand.

There was certainly considerable industrial progress subsequent to the Union, when the English and Colonial markets were opened, and the opportunity was taken to revive the linen manufacture, which had before been extensively carried on as a domestic industry, using local supplies of flax to satisfy local needs. At first very widely distributed over Scotland, the industry gradually centred in the east, which was most conveniently situated for the importation of foreign flax (particularly from Russia), and so, while the bulk of the industry settled in Angus, several very important branches flourished in Fife, both east and west. Its progress in the east is discussed later; in the west as a domestic industry two influences determined its location: the cost of transport and the advantages resulting from proximity to a good bleaching and dyeing water. Hence it was pursued along or near the coast at Kinghorn, Kirkcaldy and Dunfermline, the prosperity of the last being secured by the introduction [2] of the manufacture of damask cloth, learned surreptitiously at Drumsheugh, and by the establishment of the British Linen Company (1746). In the interior, the water of the Ore, coming from mosses and mixing with coal water, was unsuitable for bleaching and dyeing, but the peculiar excellence of the water of the Leven and its northern tributaries overcame the disadvantages of an interior situation, so that such villages as Markinch, Kennoway, Leslie, and Kinross, all in the Leven basin, became thriving weaving centres, importing their raw material through the rising harbour of Leven at the mouth of the river. Nor was there any further movement of the industry when the later stages of the Industrial Revolution brought in the extensive use of coal and machinery.

The true coal measures of this area have already been described as lying between Leslie and Pathhead, and passing under the Firth of Forth to the smaller Midlothian field. But the rest of the area (except the plain of Kinross, which, composed of underlying Old Red Sandstone, has no chance of bearing coal measures) is occupied by the carboniferous limestone, in the middle series of which are to be found extensive coal developments. This coal area lies, roughly, between Benarty and

---

[1] Kinross, pop. (1911) 2618. [2] 1718.

Burntisland, and runs from the newer measures of Strathleven on the east, south-westwards to the shores of the upper firth, near Torryburn and Culross. But, as in the carboniferous limestone of the East Neuk, where volcanic intrusion has either so destroyed or so broken up the measures there present as to render their working almost profitless, such volcanic areas as the Burntisland hills are by no means productive. The richer coal-bearing area of the interior is thus restricted to the wider, shallow valleys of the Ore and the Lyne, where volcanic action is least in evidence. At first development was confined to the stretches of coast where the measures are naturally exposed, at Wemyss and Torry, where, it will be remembered, the salters of Stuart times had begun to use sea-coal. After the Union, under the same transport conditions, there was little advance inland, but even this development, restricted as it was to the coast, secured the prosperity of the industries settled there when the application of steam power became general. The coarser linen manufactures of Kirkcaldy lay only two miles away from the pits at Dysart, the southern apex of the Wemyss field; the northern limit of the field touched Markinch and Leven, and was within convenient distance of Leslie. Kinghorn, further away to the south-west, did not derive such conspicuous benefits in these pre-railway days of expensive transport, but Dunfermline lay near the coastal field of the upper firth, of which the growing prosperity may be indicated by the construction, in 1761, just one year after the Carron Ironworks on the opposite side of the firth had created a brisk demand for coal, of the port of Charlestown, which served as an outlet for the products, mineral and textile, of Dunfermline and the Earl of Elgin's estate of Broomhall. Development of the interior was hindered by the transport problem, which could not be solved by canals owing to the nature of the surface. It was later solved by railways, which, running from the coast through the parallel valleys, opened up such a rich field in the valleys of the Ore and Lyne that towns like Cowdenbeath and Lochgelly sprang up with American rapidity. Lastly, the construction of the Forth Bridge made this western portion of Fife a link in the great trunk routes from the south to Perth and Dundee.

(b) *Population, Industries and Communications.*—This industrial progress has determined the present distribution of population. During the Middle Ages, population centred round the coast, particularly on the rim from Dysart to Ferryhill, where convenient harbours and a wool-producing hinterland encouraged the rise of numerous royal burghs. But in the modern development the determining factor was not convenient harbours or abundance of wool, but proximity to coal. From this it has resulted that the population has spread inland from those strips of coast where exposed coal measures occur, along the valleys towards the interior, but not so much up the highlands, which, being mostly of igneous rock, are of little economic value. In the valleys between the Burntisland and Saline hills, where the coal measures are least disturbed by igneous intrusions, a dense mining population exists. At the western end of this area is Dunfermline; at the eastern and more open end a string of coal ports, from Kirkcaldy to

Leven. To the south of this, the Burntisland hills form a sparsely peopled corner; to the north, a belt of similar character is formed by Benarty, the Cleish and Saline Hills. Then to the north of this belt of highland, and between it and the Ochils—another area incapable of supporting a population, and offering little attraction to mineral exploitation—lie the Devon valley and the plain of Kinross, where the better soils allow the growth of an agricultural population, gathering into small towns like Kinross and Milnathort,[1] where routes diverge and converge, and where are carried on such small industries as naturally arise in an almost purely agricultural area.

In this coal area of the west of Fife, which has attracted to itself the bulk of the population, lie the great towns of Fife. Of these, the two greatest, Dunfermline[2] and Kirkcaldy,[3] were already comparatively important before the Industrial Revolution, when proximity to the sea for cheap carriage and to water for bleaching, were essentials to the prosperity of their industries. Since then their situation near or on a coalfield has enabled their specialised branches, damask-weaving in Dunfermline and the manufacture of the coarser linens and oilcloth and linoleum in Kirkcaldy, to expand to enormous proportions. Neither can be said to owe very much to its position as a centre of diverging routes, for though Kirkcaldy stood at the fork of the two great coach routes from Pettycur Ferry (near Kinghorn), one branch going north to Cupar, the other east to the Neuk, the corresponding railway junction has moved north to Thornton; and Dunfermline is a rather artificial railway centre, certainly important but by no means inevitable. Kirkcaldy has, however, a fairly good port, which was nearest to the Strathleven field at the time when the smaller creeks had been largely deserted owing to the increasing size of vessels, when Leven was blocked by sand and Methil as yet not constructed; and though the whale fishing which Carlyle mentions in his *Reminiscences* is a thing of the past, the port is still among the three chief of Fife. Dunfermline has only recently reached the coast, when the burgh boundaries were extended to include the site of the new town expected to grow up round the naval base of Rosyth. The geographical recommendations for such a situation are obvious. From the time when Queen Margaret landed here under stress of weather, the little bay, sheltered by Ferryhill from east winds, has formed a harbour of refuge. In 1711 Sir Robert Sibbald wrote thus in his *History of Fife and Kinross*:—"To the east of Rosyth is St. Margaret's Bay, separated by a small neck of land from the bay of Inverkeithing: which if cutt would make the hill above the North Ferry an island, and this hill, which has a promontory stretching south into the firth over against Inch-garvie (a small island), if it were fortified and Inch-garvie and the south shoar opposite to it, it would secure all the western parts of the firth above that, and give great opportunity for docks, for building and repairing ships, and that with safety . . . and for laying up vessels of the greatest force and burden during the winter season." And now in these latter

---

[1] Milnathort, pop. 1052.

[2] Dunfermline, (1901) 25,500, (1911) 28,103.

[3] Kirkcaldy, (1901) 34,064, (1911) 39,601.

days, the political reason of the shifting of the centre of sea-power from the Mediterranean to the North Sea has drawn attention to this, the only natural site for a naval base on the East Coast, with convenient supplies of coal from Fife and oil from across the firth in Linlithgowshire. As yet the full effect of the new settlements has not been felt. Certainly the population of Inverkeithing increased by the influx of workers 80·9 per cent. in the period 1901-11,[1] and the authorities of Dunfermline have been warned that a population of 30,000 may have assembled within twenty years. A rapid increase in the population and importance of this enlarged burgh of Dunfermline is thus almost sure.

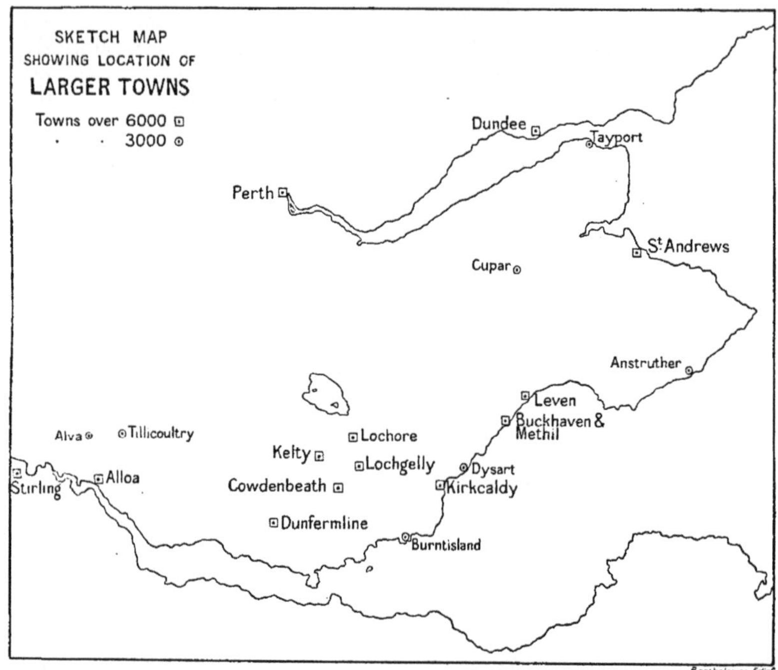

Most of the other royal burghs were not so favourably situated as Dunfermline and Kirkcaldy for the taking advantage of the new conditions. They were now on a stretch of coast which had not a very productive hinderland, they lay away from the coalfields, hence their industries rather survived than flourished. Their little creeks, which had proved sufficient in the days of small vessels, have now lost importance. And so Kinghorn[2] and Culross[3] are no longer great trading centres or even ferry-ports, but rather quiet coastal villages, and the same description could have been applied to Inverkeithing had not the new naval base wakened it to life. But there were at least two other royal burghs which did derive

---

1 Inverkeithing, (1901) 1504, (1911) 2806.

2 Kinghorn: pop. 1550.

3 Culross: pop. 456.

some advantage from their position. Dysart[1] lay on the edge of the coalfield, and while its harbour was not sufficiently good to overcome the advantages of Kirkcaldy, it has been so far improved as to serve modern needs. In addition to being a mining centre, Dysart is enabled to support flax and linen works, and still preserves a vigorous burghal life independent of its larger neighbour. The other burgh, Burntisland,[2] is a curious example of a port with no immediately productive hinderland. It has an excellent natural harbour, and rose to importance as a railway ferry-port from Granton before the Forth Bridge was constructed. The company, in whose system it formed an indispensable link, spent something like £150,000 in the carrying out of improvements, with the result that it became a great coal-port, and the harbour revenue rose in twenty years (from 1860 to 1880) from £197 to £14,785. At this time, it must be remembered, Methil had not been built, and Leven was barred by sand. Thus, in addition to the ferry-traffic, Burntisland was as yet the nearest approach to a convenient coal-port, better even than Kirkcaldy. But it had its disadvantages. The coal-bearing area was in the central valleys, and though the port lay only some five or six miles away from such a mining centre as Cowdenbeath, the Binn lay between and prevented direct railway communication. The mineral lines had therefore to curve round the extremities of the hills by Dunfermline (where Charlestown could handle only a certain amount of the output of coal) from the west, or else down the Tiel valleys and round by Kinghorn—both routes involving fairly long detours. The excellence of the harbour overcame these disadvantages until (1) the construction of the Forth Bridge deprived Burntisland of its railway ferry; and (2) the increasing output from the north-eastern fringe of the coal area, Lochore, Auchterderran, and Lochgelly, forced the little haven of Methil into becoming a great port (1887). Burntisland is still a great coal-port,[3] but it is among the three burghs in Fife which showed a decrease of population[4] between 1901 and 1911.

With the exception of Dunfermline and Kirkcaldy these old royal burghs have not become great centres of population under the new conditions, and "new towns" have arisen as the coal mining gradually advanced inland. These lie in definite areas. To the north-west of Dunfermline a phenomenal increase[5] of population has been observed in

---

1 Dysart: pop. 4197. 2 Burntisland: pop. 4707.

3 Exports for 1911:—

	Tons	Value
Coal, etc.,	1,781,275	£853,221
All other articles,	...	501
		£853,722

4 2·9%.

5 Parish of Beath:—

1801— 613	
1851— 1252	
1881— 5442	
1891— 8296	
1901—15,812	increase—54·0%
1911—24,351	

Parish of Ballingry:—

1801— 269	
1851— 568	
1881—1065	
1891—2275	
1901—4156	increase 121 7%
1911—9214	

Parish of Auchterderran:—1901— 8626 } increase—103·4%
1911—17,547

the series of coal-producing valleys between the Cleish and Burntisland Hills, and farther east where these come down to the coast between Dysart and Leven.[1] Here arose the port of Methil. Formerly, the eastern port of the area had been Leven, which is still an important town[2] with its mines, its rope-walks, its sawmills, and creosote works; but its once busy harbour has been rendered useless by a bar of sand, and a new port had to be constructed about a mile away to the south, off the mouth of the Leven. It was opened in 1887, when 219,884 tons of coal were shipped; by 1911 this had risen to 2,564,869 tons of a value of over a million pounds. The population of the combined burgh of Buckhaven and Methil increased at the same rapid rate.[3] Inland from Methil at this seaward end of the central valleys lies a string of new towns, which sprang up with extraordinary speed as the coal-mining developed. Where the line of rail running inland from Methil crosses the trunk line from Edinburgh to Dundee, is the important junction of Thornton,[4] whence a line runs up the Ore to Dunfermline, passing such mining towns as Auchterderran (station Cardenden), Lochgelly[5] and Cowdenbeath,[6] centres of the huge population which has grown up within the last few decades between Leven and Dunfermline.

The routes are determined by the physical features and the population. The centres of the railway system are Dunfermline, Thornton and Kinross, and lines run along the parallel valleys, east and west, with two trunks from the Forth Bridge using the transverses to Perth and Dundee. The line to Perth twists up to Dunfermline, where it meets two parallel lines, one directly from Alloa through the thinly peopled Forest, the other along the coast by Culross. Then the line runs up the central valleys between the Burntisland and Cleish Hills to Cowdenbeath, until the break in the northern range between Benarty and the Cleish Hills allows entrance to the agricultural plain of Kinross. On the southern side of this pass stands the mining town of Kelty,[7] a terminus of the West of Fife electric tramways, which run thence to Cowdenbeath and Dunfermline, with extensions to Lochgelly and (in construction) to Lochore.

At Kinross lines branch off down the Devon valley to Alloa, and eastwards to Strathleven, while the main line turns northward through the conveniently situated pass of Glenfarg to Perth.

After leaving the Forth Bridge the Dundee line is kept close to the shore by the Burntisland Hills as far as Kirkcaldy; it then cuts across Strathleven to the pass of Markinch, sending out numerous branches.

---

1 Parish of Wemyss:—1901—15,031 / 1911—23,104 } increase—53·7%.

2 Leven—6559.

3 Combined burgh of Buckhaven and Methil—
1861— 2824
1901— 8621 / 1911—15,149 } increase—75·7%

4 Thornton: 1147.

5 Lochgelly—1901—5472 / 1911—9076 } increase—67·5%.

6 Cowdenbeath—1861— 1148
1901— 8329 / 1911—14,029 } increase—68·4%

7 Kelty—7502.

At Thornton the great coal-line which left the Perth trunk at Cowdenbeath, and runs down the Ore valley, crosses the northern line on its way to Methil, and from Thornton another line runs to Leven and the East Neuk. North of this a short line from Markinch reaches Leslie, where the peculiar excellence of the water, which attracted linen bleaching, etc., has now brought the manufacture of paper. It is to be noted that no line runs from the Largo Bay coast to Lochleven, probably because such a route would connect no great industrial or mining centres with the coast.

In addition to these various branches, the coastal rim of Largo Bay is served by an extensive tramway system from Kirkcaldy through Wemyss and Buckhaven to Methil and Leven.

A population due to a rapid development of mining is liable to sudden fluctuations, but as yet, supposing the present economic conditions to continue, there is little reason to foresee any sudden decline in the population or prosperity of the West of Fife; indeed, rather the reverse is to be expected from the report of the Royal Commission of 1905:—"The County of Fife, with the smaller quantities in Kinross, takes the leading position in Scotland in the matter of its coal resources. Besides the coal in these counties, probably two-thirds of that under the Firth of Forth will be worked by collieries in Fifeshire, so that the available resources at less than 4000 feet deep will amount to something like 5,700,000,000 tons or sufficient to maintain the present (1905) output for 930 years. While the output from the Firth of Forth will not rapidly increase for a great many years, when so much coal is available under the land, the output from Fifeshire is certain to advance till it occupies a leading position in the Scotch coal trade." The output of 1905 was 7,241,439 tons; in 1887, the date of the opening of the Methil dock, it was 2,459,395 tons, and by 1911, it had risen to 9,037,790 tons, most of it shipped across the North Sea to Scandinavia and the Baltic countries.

3. *Eastern Fife:* (a) *Historical Progress.*—North of the pass of Markinch, and eastwards of the Lomonds, lies the remainder of the peninsula, its physical features arranged in alternate belts of high and low ground running east and west. To the north the Ochils, now low hills cut by numerous transverse valleys and by one important longitudinal valley, are divided by Stratheden and the Howe from the Lomonds and the hills of the East Neuk, separated by the depression to the north of Markinch. To the south lies the coastal rim.

This area was most prosperous and wealthy during the Middle Ages, when its facilities for trade and its advantages of climate and soil attracted a considerable population. In it were eleven of the seventeen royal burghs of the peninsula. Even after the Union its commercial development kept pace with that of the West.

With its good climate and soil it derived most benefit from the modern improvements in agriculture. There was very little absolutely unproductive land. The soil of the hills, especially in the East Neuk with its volcanic outcrops, could be extensively cultivated despite exposure and elevation. The new drainage turned the swampy haughs of the river valleys into good farming land. More particularly this took place in the Howe, where Loch Rossie disappeared about 1760, and the once marshy Howe is now successfully cultivated. With its dry summers and warm autumns the area was specially suited for wheat; and the improvement in the roads, the favourable market for cereals when the foreign supplies stopped with the outbreak of the Revolutionary Wars, the opening of the Forth and Clyde canal whereby the industrial population of the west could be reached—these causes all combined to make the early nineteenth century a time of great agricultural prosperity in the east. In addition it had never before been so industrially prosperous, for, in the early stages of the development of the linen industry, it had the same advantages as the west, the same facilities for importation of the raw material, numerous ports and, particularly in the Eden basin, excellent water. Hence the weaving was carried on as a domestic industry in the numerous small burghs and villages which lie around the slopes of the Howe—Freuchie, Falkland, Strathmiglo, Auchtermuchty—where the disadvantages of a comparatively interior situation were overcome by the advantage of superior water, and the convenience of the little port of Newburgh through the pass of Lindores. In the east, the ferry-ports of Newport and Tayport (Ferry-port-on-craig) opposite Dundee, were suitable centres for the importing of flax, to supply their own needs and the needs of the towns and villages of the Lower Eden, Cupar, Ceres and Dairsie. Then the burghs on the Firth of Forth, which had been almost ruined by the heavy duties on salt and the disappearance of the herring from their shores, revived when the duties were rescinded and the herring reappeared. Thus up to the time (about 1810) when steam-power was applied to the linen industry and the presence of a coalfield became an essential to industrial prosperity, the east had profited equally with the west from the new industrial conditions. But it had little coal, for the Ochils and Stratheden were composed of underlying sandstone, and the coal present in the limestone of the east Neuk was either so destroyed or broken up by igneous action as not to repay working on an extensive scale. While the west with its accessible coalfields drew to itself industry and population, the progress of the east slackened. It certainly did not stop altogether, for agriculture was pursued more and more scientifically and, on the whole, successfully, despite less advantageous markets after the importation of grain from abroad assumed larger proportions; the fishing flourished, and the linen industry was saved from extinction by the supreme excellence of the Eden water for dyeing and bleaching, and the accessibility of the interior to railway penetration. But there was no chance of great industrial development.

(b) *Population, Industries and Communications.*—Thus the distribution of population is unlike that of the west. There are nowhere presented the contrasts of the west, where an industrial area shows a dense population clustering in large towns along a valley, and, a few miles distant, an almost uninhabited mountain. In the east the hills, capable for the most part of some agricultural development, support a fair population, though, of course, the larger towns and the more thickly peopled parts are in the valleys, particularly in Stratheden and the coastal rim. But there are no great centres of population, no towns of over ten thousand inhabitants, and very many little villages and burghs of a thousand or under, agricultural centres with perhaps a bleaching-field or a paper-mill, or, if on the coast, a few fishing-boats. For even yet the isolation which produced so many royal burghs in the Middle Ages still acts sufficiently to keep the lesser villages from absolute decay.

The position of these towns of the east on the coast and in the valleys secures for them a convenient railway system. The trunk line, which has been traced above from Thornton Junction to the pass of Markinch, cuts across the level of the Howe to Ladybank,[1] one of the two examples in the east of a new town. On the slopes round the Howe stand old burghs and villages like Falkland,[2] important when the Howe was a marsh and Loch Rossie a reality. Routes then curved through these villages, but now the railway cuts straight north through the level and leaves Falkland isolated. This new town rose in the centre of the Howe, where various lines break away down the valleys, enabling it by its central position to become a small agricultural centre. Eastwards a line runs to Kinross between the Ochils and the Lomonds, through little burghs like Auchtermuchty[3] and Strathmiglo,[4] now agricultural centres with sawmills or bleachfields. Northwards from Ladybank a line which, before the construction of the Glenfarg route, was the main one between Burntisland (for Edinburgh) and Perth, cuts through the Pass of Lindores to Newburgh,[5] a declining river port, and so to Perth. Westwards the trunk line to Dundee runs up Stratheden to Cupar, which, though the county town and one of the three greatest centres of population in the east, has a population of only 4112. It used to be an important route-centre at the foot of the marsh, but now routes diverge above and below it at Ladybank and Leuchars, and Cupar is only an important agricultural centre with one or two linen works. At Leuchars Junction, near the estuary of the Eden, one route goes southwards to St. Andrews, another to Tayport, the great railway ferry of the Tay before the Tay Bridge was constructed. The main line cuts through a gap in the swelling hills which now represent the Ochils, to the great bridge and Dundee. It receives two branches, one from Tayport and Newport, practically residential suburbs of Dundee, with jute and linen mills, the other from Lindores and Newburgh, through the longitudinal valley of the Motray, opening up an agricultural district with nothing more than a few hamlets.

Probably the most important line of the east of Fife is that running round the coast from Leuchars to Leven and Thornton. At Guardbridge, between Leuchars and St. Andrews, are important papermills, using the water of the Eden and the Motray, and the esparto grass imported at the nearest convenient port, Tayport. Then comes St. Andrews,[6] and strangely enough, with all its associations of priest and

[1] Ladybank (1861) 576, (1891) 1178, (1911) 1266.
[2] Falkland (1911) 830.
[3] Auchtermuchty (1911) 1396.
[4] Strathmiglo (1901) 966.
[5] Newburgh (1871) 2777, (1911) 1977.
[6] St. Andrews (1911) 7851.

presbyter, it must be classed as a "new town." After the Reformation it sank to the status of an agricultural village, its cathedral fell into decay, its university was threatened with removal to Perth. After the Union few manufactures could be introduced, for the harbour is inconvenient and the situation of the town unsuitable. Only a little fishing is carried on. The revival dates from the first years of the nineteenth century, when the littoral vegetation of Pilmuir provided excellent golf links, which have rapidly grown in popularity. Its other industry is education; it has a small but important university and several large secondary and boarding schools. To the south of St. Andrews the railway runs through an agricultural area where, as of old, the lack of harbours on an exposed coast has prevented the rise of fishing-villages, to Crail. From Crail to Earlsferry lie the once predominant royal burghs. Of these, the most important is Anstruther,[1] with the advantages of a large modern härbour and convenient access to the markets by rail. It is now the great fishing centre of Fife, but the same industry is carried on to a lesser extent by the smaller burghs, which have equal facilities for transport, but smaller harbours, not quite equal to modern demands. Behind them is a fertile agricultural region. They are best known, however, as summer resorts frequented by visitors from Dundee and Edinburgh. After passing Elie, the most popular of these little towns, the line runs past Largo to Leven and Thornton.

Conclusion.—From this description it will be seen that, compared with the Middle Ages, modern development has centralised and specialised industry. The determining factor of the coal area confines the great industrial centres to the west. By the economy of nature, agriculture as a predominant industry is confined to the east. Shipping has centralised in one or two ports like Burntisland, Kirkcaldy, and Methil. There is no longer a string of little burghs, each engaging in foreign commerce, each with its own little fleet. The density and relative distribution of population has changed. There is not one town of over 10,000 inhabitants in the east, and yet two of over 20,000 in the west. While the agricultural parishes and villages of the east either decrease or remain practically constant, the populations of the western parishes show rapid increases unusual in Scottish statistics. But the aggregate populations of the industrial counties show a continued increase since the taking of the first census in 1801. The inhabitants of Clackmannan county in 1801 numbered 10,858; by 1911 this had increased to 31,121. Almost the same ratio of increase is to be found in Fifeshire—in 1801, 93,743; in 1901, 218,840; in 1911, 267,734, where for the decade 1901-1911 the intercensal increase of 48,899, or 22·8 per cent., is surpassed in actual amount only by Lanark. In Kinross-shire, almost purely agricultural, the population has remained at the same figure. In 1801 it was 6725, by 1851, when the industries introduced into the county town were as prosperous as they ever were, it had mounted up to 8924, by 1911 it had fallen to 7528.

---

[1] The three royal burghs of Anstruther Easter, Anstruther Wester, and Kilrenny practically form one town with a population of 4454.

But whatever degree of centralisation of industry Fife may have attained, there are abundant evidences that the old influences tending to decentralisation are still working. There are in the county of Fife twenty-eight burghs; this is more than in any other of the Scottish counties. The difficulties of the physical features have been evaded rather than overcome. The towns still lie string-like along the valleys, and the railway system resembles nothing more than the lines of a chess-board, with junctions innumerable. The old royal burghs, though deserted by the greater industries, are enabled by their position to preserve a certain degree of life as centres of a small local trade. The peninsular isolation is still in existence. The two bridges across the Forth below Stirling have not superseded but only supplemented the various ferries. Only three important railway routes connect Fife with the rest of Scotland by land, and all three—by the Carse of Clackmannan, by Glenfarg, or by the pass of Lindores—but follow old mediæval routes. The Tay ferries still ply briskly. In the interior the great lines of traffic are still east and west, not up and down the trunk lines, but across them.

STATISTICS, ETC.

I.—*Population of Fife.*

	1901.	1911.	Increase.
Burghal, .	136,087—62·2 %	163,407—61 %	27,320—20·1 %
Extra-burghal,	82,753—37·8 %	104,332—39 %	21,579—26·1 %

II.—*Industries.*

		1901.	1911.	Increase.
Males:	(1) mining, ...	15,583	26,634	70·9 %
	(2) agriculture, ...		6,858	
	(3) building trades, ...		5,352	
	(4) floorcloth making, . . .		2,416	
	(5) flax and linen manufacture, . .		2,235	
	(6) paper making, ...		772	
Females:	(1) flax and linen manufacture, .		8,112	
	(2) domestic servants, . . .		5,781	
	(3) mining (above ground), . .		919	

III.—*Distribution, etc.*

Mining (males)—		Flax and linen manufactures (females)—	
Dunfermline, .	1,956	Dunfermline, ...	2,520
Kirkcaldy, . .	1,594	Kirkcaldy, ...	2,290
Rest of County, .	23,084	Rest of County, . .	3,302

IV.—*Agriculture of Fife, Clackmannan and Kinross*—1911.

	Clackmannan.	Kinross.	Fife.
Total area—acres, ...	34,927	52,410	322,844
Total acreage (under crops and grass),	15,561	34,698	251,486
Arable land, ...	9,242	22,266	176,273
Permanent grass, ...	6,319	12,432	75,213
Wheat, ...	375	90	12,992
Barley, ...	240	272	16,898
Oats, ...	3,001	6,483	39,888
Potatoes, ...	399	888	16,342
Turnips, ...	784	2,428	22,569
Horses—number, ...	727	1,172	10,696
Cattle, ...	3,623	6,655	47,412
Sheep, ...	15,177	34,040	107,594

V.—*Progress of Agriculture in Fife.*

	1856.	1911.
Wheat—acres, . .	34,099	12,992
Barley, ...	22,856	16,898
Oats, ...	42,328	39,888
Potatoes, . . .	17,269	16,342
Turnips, ...	29,739	22,569
Sheep—number, . .	57,306	107,594
Cattle, ...	40,611	47,412
Horses, ...	12,258	10,696

THE LATE CAPTAIN ROBERT FALCON SCOTT, K.C.B., R.N., C.V.O., D.Sc.,

Livingstone Gold Medallist of the Royal Scottish Geographical Society.

AN APPRECIATION: By Dr. WILLIAM S. BRUCE.

ANTARCTIC exploration has been singularly free of disaster; so much so, indeed, that the general public have come to look upon these many recent expeditions to the South Pole and South Polar regions as comfortable excursions, and have concluded that it was much easier to get to the South Pole than the North Pole. The tragic end of Captain Scott and his companions has come, therefore, as all the greater shock. It is wonderful how all the early explorers came through almost unscathed. Neither Cook, Bellinghausen, Weddell, Biscoe, Wilkes, d'Urville, nor Ross had many or very serious mishaps, though many narrow escapes. Possibly of all these Biscoe fared the worst, though it is more than probable that Ross, had he succeeded in landing and wintering in M'Murdo Sound as he wished to, might have suffered in a way that would have been appalling. The reason, however, is that,

THE LATE CAPTAIN ROBERT FALCON SCOTT
K.C.B., R.N., C.V.O.
Commander of the British Antarctic Expedition.

since the time of Ross, Antarctic exploration was severely left alone for fifty years, because, with the exception of the *Challenger*, no further expeditions visited the Antarctic regions until the Scottish whalers, with an artist and men of science on board, worked and explored there, namely, 1892-93.

Consequently 1892 marks the date of modern Antarctic exploration. It was not until 1898 that anybody had experienced an Antarctic winter, and we were entirely ignorant of the conditions in the Antarctic regions up to that time except for a few summer months. It was the Belgians who first told us what an Antarctic winter was, and the following year a British expedition at Cape Adare was the first to winter on the Antarctic land. In contrast to this, more than three centuries have been spent in the exploration of the North Polar regions, and Europeans first wintered in the Arctic in the winter of 1630-31, when a British ship missed their companions in a fog and were compelled to winter in Bell Sound, Spitsbergen. Consequently Antarctic exploration began after three centuries' experience of exploration in the Arctic regions, during which time all the chief preliminary experiments had been made, resulting in a marvellous development in equipment and food supply. During the past twenty-five years especially an immense advance has been made in our knowledge of how to equip a Polar expedition, and especially in sledging, how to reduce weight of gear, and how to concentrate food supply into the smallest possible weight and bulk with the best possible nutrient power. The absence of disaster is consequently due to the sum of experience gained during that long period in Arctic regions, which have, as it were, been the training-ground of subsequent Antarctic work. That such a disaster has occurred at the present time emphasises the fact that the exploration of the Antarctic regions requires not only courage, perseverance, and foresight, but also experience.

Robert Falcon Scott had these qualifications. His first expedition was in the *Discovery* during the years 1901-1904. He was appointed to the command of that expedition, when it was decided that the leader was to be a naval man, because of his distinguished career in the service. He was at the time first lieutenant of the *Majestic*, flag-ship of the Channel squadron, and was appointed "Commander" on the eve of the departure of the expedition. The result of the work that he did then is given in his well-known narrative entitled *The Voyage of the "Discovery,"* and in the important scientific reports of the expedition which have been published by the British Museum, the Royal Society, and the National Physical Laboratory. The fact that there were at least three other expeditions working in the Antarctic regions at the same time, and that they were co-operating with him, greatly enhanced the value of the work done by Scott and his companions. The full significance of that work will perhaps be appreciated even more fully in years to come than it is at the present time.

It is a curious fact that it is seamen who have taken the lead in the exploration of the interior of the Antarctic continent, whilst landsmen have been paying more particular attention to the sea, and Scott was the first of these land workers to do extensive work in that direction. It was he who first attempted a sledge journey on the Ross Barrier to a high southern latitude, attaining a position of 82° 17′ S., and it was he who followed Armitage's track up the Western Glacier and succeeded in reaching 146° 33′ E. on the inland plateau at an altitude of about 9000 feet. These two journeys of his were the two most extensive journeys that had been taken into the interior of Antarctica at that time. Besides Armitage other colleagues of his did extensive sledge work into the interior. Royds travelled south-east over the Ross Barrier, and Barne and Koettlitz travelled to the westward of Scott's southern track. These sledge journeys elucidated the topography and geology of these lands, and at the same time an extensive series of valuable observations were being carried out at the base station at M'Murdo Sound. With the ship Scott sailed along and closely investigated the whole of the Ross Barrier and confirmed the "Appearance of Land" mentioned by Ross at its eastern extremity, naming it after King Edward. He cruised along the whole of the coast of Victoria Land, visited Cape Adare, and pushed westward along the coast to about 166° W. He then made further investigations with the *Discovery* in the region of the Balleny Islands and the eastern extremity of Wilkes Land, altogether making the most extensive investigation of this part of the Antarctic regions for sixty years, namely, since the time of his predecessor, Sir James Clarke Ross.

But a few years elapsed and Scott was once more successful in securing sufficient funds to enable him to sail to the Antarctic. I say, sufficient funds, but I bear in mind that the funds were made sufficient by personal sacrifice on the part of his wife and himself, and even then he would scarcely have succeeded in getting away had he not induced the Treasury to vote him £20,000 and the Admiralty to give him considerable assistance.

There have been various statements as to what the object of Scott's recent expedition really was. It has been refuted that he was attempting to reach the South Pole. It is best to quote his own words where he says:—

"During the winter, great preparations will be made for a great effort to reach the South Pole in the following season. By that time we shall know what reliance can be placed respectively on the ponies, the dogs, and the motor sledges. But in any case a large party of men will be detailed for the southern party. Some of the scientific staff will remain at the wintering station through the summer. A small party will act independently in the western mountains for geological purposes; but at least sixteen, and possibly more, men will accompany the main transport agents on the road to the south."

Scott considered the achievement of the South Pole by no means a certainty. He was not without hope that either the ponies, the dogs, or the motor sledges might traverse the disturbed regions of the glacier, and that if this was possible the difficulty of the journey would be greatly diminished, but he was prepared to make the whole of the journey, once off the barrier, with the unaided efforts of men alone.

He considered that the only manner in which such a record as

Shackleton's could be beaten was by taking a larger party of men and sending sections of them back at intervals. He calculated that a party with the aid of a supporting party of quick members could hope to achieve a distance greater than it would have done had it been without a supporting party. All these careful considerations, which he discusses in the July *Geographical Journal* for 1910, make it clear that the objective of the southern journey was to attain the South Pole, and undoubtedly with this objective view he had the wish to plant the British flag at the Pole before any other. He himself stated that his plans were not altered by the presence of Amundsen in the field, and personally I do not believe that Scott, once he knew that a band of expert Norwegian ski-runners were in the field, would attempt anything in the way of a race. If the Norwegians were to reach the Pole at all it was practically an athletic certainty that they would reach it first. This they did, and in Scott's words "they deserved their luck."

From a geographical standpoint it might have been more profitable to have done the same journey over new ground, for Scott himself and Shackleton, his former colleague, had covered the whole road except one hundred miles. But little as we know at present of the scientific results of this expedition, they appear to be of great value. Along the whole route observations of pressure, temperature, wind, and so forth must have been taken. These will be of wide interest when worked out in conjunction with the meteorological observations taken, not only at Scott's own base stations at M'Murdo Sound and Cape Adare, but also with those taken by Amundsen, Mawson, Filchner, and the Argentines. It seems likely also that Scott would have taken in some form or other certain magnetic observations which would again be of great interest. Besides this, we learn that at Buckley Island, on the Beardmore Glacier, and on the side of the Cloud Maker, Wilson and Bowers made collections of fossils. The fossils at the former locality are well-preserved plants, probably of Upper Palæozoic or early Mesozoic age. Those from the latter are described as "corals of a primitive form typical of the early Palæozoic age," which have been found in Cambrian limestone. If so, there is proof that this range of mountains is actually continuous with those of Graham Land and South America. But whatever may be the result of the examination of these fossils, they will give us an important clue to the geology of that region. Amundsen secured rocks further east in approximately the same latitude, and also in Edward Land. Consequently, the collections of Scott's expedition will be an important factor in clearing up the geological problems of the whole of the region traversed by Amundsen and himself Topographically, undoubtedly Scott would confirm and correct the results of his own work and Shackleton's of previous years, and would undoubtedly fill in many gaps that there would be in such a survey on account of weather, or unavoidable personal error in observation. We can say, therefore, that the results of this great southern journey seem likely to be of great value to science, but

> "The best laid schemes o' mice an' men
> Gang aft agley,"

and regretfully it must be admitted that for some reason or other Scott's carefully made plans failed to bring him and his companions back again.

We mourn their loss, and our deepest sympathy goes out from our hearts to the widows and fatherless, and to those others to whom they were nearest and dearest.

It is easy for any one of us, sitting comfortably at home, to be wise after the event, to point out that by a different arrangement the great tragedy that has happened might have been averted. But, as one who has been in the field himself, I know how even the most expert and kindly critic at home may just fail to realise some crucial condition which would upset all his calculations. Scott himself has criticised no one, therefore why should we! He says:

"We took risks—we know we took them. Things have come out against us, and therefore we have no reason for complaint, but bow the head to the will of Providence, determined still to do our best."

"Wha does the utmost that he can
Will whyles do mair."

And Scott and his companions have done more, and what they have done has sent a thrill through this island Empire of ours that has been re-echoed through the whole civilised world. They have made us proud to think that we are their fellow-countrymen. They have set us an example of devotion to duty, of stern resolution to accomplish what they set themselves to do, and of self-sacrifice. Those rough notes and their dead bodies have indeed told the tale.

At the meeting of the Society, held in Edinburgh on February 14, Dr. Horne, Chairman of Council, who was in the chair, made the following remarks in regard to Captain Scott's Antarctic expedition:—

"In the unavoidable absence of Lord Stair it is my first duty to-night to express our profound sorrow at the tragic disaster that has befallen Captain Scott and his brave comrades in the Antarctic region. Many of you will remember the graphic account of his first successful expedition which he gave in this hall. On his departure as leader of the recent expedition Dr. Bartholomew said to him, 'We expect you to open the session of the Royal Scottish Geographical Society in November 1913, and you will be sure of a most cordial reception.' He replied, waving his hand, 'I'll be there. I will always remember my last reception in Edinburgh.' But our hopes have been sadly shattered.

"Geographers are aware that the recent expedition was not organised to make a dash for the Pole. It was based on a comprehensive plan, elaborately constructed, to obtain all available information regarding the evolution of that part of the continent which the members of the expedition had to traverse. The despatches sent home last year showed that so far their efforts had been crowned with success. The messages cabled within the last few days from Commander Evans clearly prove that not only had the South Pole been located, but materials of the highest scientific value had been obtained, which will throw light on the age and structural relations of the mountain chains near the Beardmore Glacier. And yet, when nearing their goal, Captain Scott and his brave comrades had to face the grim reality that their lives had to be sacrificed, although success had been achieved. No more tragic situation could have arisen in the annals of Polar research than when Captain Scott sat down to pen the message to his countrymen which is inspired by the best traditions of the service to which he belonged. Every line of it vibrates with stern devotion to duty, unselfish loyalty to the cause he had at heart, and unflinching courage in the face of overwhelming odds. His message will endure. It adds glory to the race to which we are proud to belong. Need I express the hope that his pathetic appeal on behalf of the dependents of those who perished will be responded to by all parts of the British Empire. Several members of the Council of this Society think that the announcement should now be made that the Treasurer of the Society will be glad to receive subscriptions on behalf of those who have laid down their lives in the interests of geographical research."

PROCEEDINGS OF THE ROYAL SCOTTISH GEOGRAPHICAL SOCIETY.

At a Meeting of Council held on Tuesday, February 18, the following were elected Ordinary Members of the Society:—

Miss Penelope M. Ker.
Miss E. M. M. Thomson.
Mrs. A. L. Gemmell.
George Harding-Edgar.
W. A. Verel.
Lieut.-Col. C. de C. Etheridge, D.S.O.
Charles Matheson, M.A.

The following were elected Associate Members, viz.:—

Miss M. P. Gott.
Miss I. H. Brown.
Wilfrith Elstob, B.A.
S. Hyslop, M.A.
John M. Short, M.A.
William Byers.
Harry C. Crowden, M.A.
F. W. Taylor, B.A.
C. M. Lawrence.
John Cuthbert.
L. J. Wild, M.A.
Miss A. Corbett.
Miss E. G. Ball.

Diploma of Fellowship.

The ordinary Diploma of Fellowship was conferred upon the Rev. T. H. Walker, Uddingston, subject to his compliance with the prescribed conditions.

The late Colonel Bailey.

The following Minute relating to the late Colonel Bailey was unanimously adopted, and the Secretary was instructed to send a copy to Mrs. Bailey:—

"The Council expresses its profound regret at the death of Lieutenant-Colonel Frederick Bailey, R.E., LL.D., and its sense of the great loss that this Society has thereby sustained.

"Appointed Secretary to this Society in November 1892, Colonel Bailey brought to its service the benefit of the varied administrative experience he had had in India and elsewhere, and devoted himself to the work of this Society with the industry and enthusiasm which he had already displayed first in the service of his country, more particularly in connection with the Indian Forest Department, and afterwards in the University of Edinburgh, in which he was the first Lecturer on Forestry. After resigning the Secretaryship in 1903 he still showed his interest in the Society by acting as a Member of the Council and of various Committees. His services were continued to within a short time of his death, and his ready offer only a few months before his death, on the occasion of the Secretary's absence for two or three months from his duties, to undertake for that time the work of the secretarial office is an illustration of the kindly zeal and punctilious devotion with which those services were rendered. Of the value of those services the Council desires to place on record its high appreciation."

The Disaster to Captain Scott's Antarctic Expedition.

The Council unanimously adopted the following Minute with reference to the disaster to the Antarctic Expedition under the command of the late Captain Scott. The Secretary was instructed to send a copy to the Secretary of the Royal Geographical Society, to Mrs. Scott, and to other relatives of those who had met their death with Captain Scott:—

"The Council of the Royal Scottish Geographical Society, while placing on record an expression of their profound sorrow at the disaster that has befallen Captain Scott and his comrades, Captain Oates, Dr. F. A. Wilson, Lieutenant H. R. Bowers, and seaman Edgar Evans, on their return from their successful journey to the South Pole, desire to add an expression of their pride in thinking that the circumstances of their death, in strict and stern devotion to duty in the pursuit of geographical research, were such as to add glory to the race to which they belonged."

Bust of David Livingstone.

A letter expressing the cordial thanks of the Council was instructed to be sent to Captain Livingstone Bruce for the loan of a marble bust of his grandfather Dr. Livingstone.

Gaelic Place-Names.

It was agreed to revive the committee on the spelling of Gaelic place-names.

Congress Delegates.

Dr. J. Horne and Mr. H. M. Cadell were appointed delegates of the Society to the International Geological Congress to be held at

Toronto from August 7 to August 14. The Editor and the Secretary were appointed delegates to the International Geographical Congress to be held at Rome from March 27 to April 3.

Lectures in March.

Professor D'Arcy Thompson, C.B., will address the Society in Aberdeen, Edinburgh and Glasgow on the 11th, 13th and 14th respectively. The subject of his lecture will be "The North Sea."

Mr. George G. Chisholm, M.A., B.Sc., Secretary of the Society, will address the Dundee Centre on March 12. "The Transcontinental Excursion of the American Geographical Society of New York" will be the subject of his address.

Meetings to commemorate the centenary of the birth of David Livingstone will be held on the 26th and 27th in Glasgow and Edinburgh respectively, when an address will be delivered by Sir Harry Johnston, G.C.M.G., K.C.B.

Antarctic Exhibition.

Mr. Burn Murdoch and Dr. W. S. Bruce have arranged an Exhibition of Antarctic paintings and animals, together with clothing, equipment, etc., for Polar work, for the benefit of the Scott Memorial Fund. The Exhibition, which is under the auspices of the Royal Scottish Geographical Society, was opened on Wednesday, February 26, by the Lord Provost, and will remain open for about ten days.

GEOGRAPHICAL NOTES.

Africa.

Plant Geography of Algiers.—In a recent issue of the *Vierteljahrsschrift d. Naturforschenden Gesellschaft* in Zurich, there is an elaborately illustrated article by Messrs. Rikli and Schröter, with the collaboration of other authors, upon an excursion to Algiers and the margin of the Algerian Sahara, which is of interest from various points of view. The excursion was planned in connection with the Technische Hochschule in Zurich, but the party, which consisted of forty-two members, included, in addition to lecturers and students, a number of teachers from Swiss schools and scientists of other nationalities. The excursion lasted a little more than five weeks, and, though primarily botanical, the inclusion of a number of specialists in various branches of science allowed for the making of a great variety of observations, and a large number of characteristic photographs were also taken.

Polar.

The British Antarctic Expedition.—We allude elsewhere in the present issue to the tragedy which has clouded the close of this expedition. At a later period some account of the scientific results obtained will be given; meantime we desire merely to place upon record the bare facts of the narrative as they have reached the civilised world in successive telegrams.

On February 10 the *Terra Nova* returned to New Zealand bringing with her the surviving members of the party. Commander Evans reported that Captain Scott and his companions, Lieutenant Bowers, Captain Oates, Dr. Wilson, and Petty Officer Evans, failed to return from the Polar trip in March 1912, as expected, and a party sent to meet them returned from One Ton Depôt without achieving this object. In November 1912 another party, well provisioned, found, eleven miles beyond the depôt, and 155 miles from the base camp, a tent containing the bodies of Captain Scott, Lieutenant Bowers, and Dr. Wilson, with all their records and many geological specimens. From these records it appears that the party reached the Pole on January 18, 1912, about one month after Captain Amundsen, and found the tent, etc., left by Amundsen. The return journey was difficult from the first owing to the lateness of the season, the bad weather, the heavy going and deficient food and fuel. On the Beardmore Glacier Petty Officer Evans broke down in health owing to the continued strain, and died on February 17, his death being accelerated by concussion of the brain produced by an injury received while sledging over the heavy ice.

Soon afterward Captain Oates also fell ill, both feet and hands being badly frostbitten, and on March 17 he left the tent in a state of extreme exhaustion, in spite of his comrades' protests. He was not seen again, and his heroic act was prompted by the desire not to keep back his companions on their journey. The others struggled on, but were overtaken by a blizzard, which seems to have raged for many days. The last entry in Captain Scott's diary—a singularly pathetic appeal to the nation—is dated March 25. Some echoes of the emotions which have been roused everywhere by the tragic end of the brave explorers will be found, as has been said, elsewhere in this issue.

The northern party, under Lieutenant Campbell, though happily none were lost, suffered also extreme hardship, spending the winter of 1912 in a snow hut at Terra Nova Bay with practically no food but seal meat, and not reaching the base camp till November 7.

General.

Ice Dangers in the North Atlantic.—It is of interest to note that the *Scotia*, the vessel formerly employed by the Scottish National Antarctic Expedition, has been chartered for special work in the North Atlantic during the coming spring, in connection with ice and risks to navigation. The *Scotia* is to be stationed off the east coast of North America, to the north of the usual shipping routes, to watch the break-up of the ice, and to report upon its movements towards the shipping routes. The vessel has been fitted up at Dundee, and has a Marconi wireless installation of long range, which will enable it to keep in touch with the wireless stations in Newfoundland and Labrador. The cost of the expedition is to be divided between the British Government and the principal shipping lines, and there will be three scientific observers on board, who hope to make oceanographical and meteorological observations in addition to those on the movements of the ice. It is expected that the vessel will be ready to leave Dundee at the beginning of March, and her period of active service will include the spring months during which ice is a menace to navigation in the North Atlantic.

The Association of American Geographers.—Professor Isaiah Bowman informs us that this Association held its ninth annual meeting at New Haven, Connecticut, on December 27 and 28, 1912. "The sessions were held in Lampson Hall, Yale University, and an informal meeting took place on Friday evening at the Graduates' Club. In the absence of the President (Professor Salisbury), Mr. M. R. Campbell, the first Vice-President, presided. About thirty members attended.

"It is gratifying to the members to see the increasing number of papers on anthropogeography, regional geography and climatology that deal with human relations, a feature less prominent in the earlier programmes of the Association. Seven purely physiographic papers were presented out of a total of sixteen. Great interest is manifested in the *Annals* of the Association since the appearance of the first volume during the past year. The publication committee has performed a distinct service to geographic science in securing papers of high quality and a volume of excellent appearance.

"The newly elected officers for 1913 are as follows:—President, Henry G. Bryant; First Vice-President, Ellsworth Huntington; Second Vice-President, Charles C. Adams; Secretary, A. P. Brigham; Treasurer, F. E. Matthes; Councillor for three years, R. DeC. Ward. The publication committee appointed for two years (1913 and 1914) consists of R. E. Dodge, Editor, and Alfred H. Brooks, H. E. Gregory, and H. H. Barrows."

Commercial Geography.

Cotton-growing in the Sudan.—It was announced recently by the Prime Minister, to a deputation of the British Cotton-Growing Association, that a bill is to be drafted next session to authorise the Treasury to guarantee the payment of interest on a loan to be raised by the Government of the Sudan to the extent of three millions. The loan is intended to improve the condition of the land so that cotton can be grown in the Sudan, and particularly in the Gezira Plain. In this region there are some 5,000,000 acres of first-class cotton soil, of which only 2000 acres are now under irrigation. It is proposed to construct a dam across the Blue Nile at Sennar, and by means of canals to irrigate the plain. If, as preliminary experiments seem to show, cotton can be grown here as a winter crop, the irrigation works should lead to a great increase in the cotton supplies of England.

EDUCATIONAL.

A MEETING of the Association of Teachers of Geography in Scotland was held on Saturday, February 1, at eleven o'clock, in Sciennes School, Mr. Chisholm in the chair.

There was a good attendance, and a paper was read by Mr. T. L. Millar, the teacher of geography in the school, on the practical methods adopted by him for pupils in the supplementary stage. A brief account of Mr. Millar's methods will be given here, but is unavoidably held over this month owing to want of space.

A vote of thanks was unanimously passed to Mr. Millar on the motion of Mr. Cossar, and the Chairman expressed the thanks of the meeting to Mr. Crockett, the Headmaster of Sciennes School, who was present at the meeting, for his courtesy in allowing the use of the room for the holding of the meeting.

The next meeting of the Association is to be held in the Society's rooms on Saturday, March 8, at 11 A.M., when a paper upon "A Syllabus of Geography for Elementary Classes" will be read by Mr. James Cossar, M.A., Training College, Glasgow.

NEW BOOKS.

EUROPE.

The Botany of Iceland. Edited by L. KOLDERUP ROSENVINGE, Ph.D., and EUG. WARMING, Ph.D., Sc.D. Part I. *The Marine Algal Vegetation.* By HELGI JÓNSSON, Ph.D. London: John Wheldon and Co., 1912.

When completed, *The Botany of Iceland* will no doubt have a similar scope to *The Botany of the Færoes,* and judging from the volume now under review will be also a model of its kind. The present volume, entitled *The Marine Algal Vegetation,* by Dr. Helgi Jónsson, is a magnificent piece of work, the discussion of the relation of the algal flora to the temperature, salinity, etc., of sea-water being of great interest.

The coast of Iceland consists partly of rock and partly of sand, the former supporting a rich algal vegetation, while the latter is virtually a "desert." Fiords are numerous on the south-west, north-west, north, and east coasts, and vary greatly in size, some being about 10 by 12, while others are 18 by 10 geographical miles. The rocks of the coast consist mainly of basalt, but it is not so much the rock itself which influences the distribution of algæ as the nature of the surface. The movements of the ocean, such as tides, waves, and currents are naturally of great importance to the life of the algæ. Currents especially play a large part in determining the distribution of species. The warm water of the Atlantic Ocean here meets the cold water of the Arctic Ocean. The Gulf Atlantic Drift washes the south coast of the country, and sends an arm northwards along South-West and North-West Iceland. This, turning south along the east coast, mixes with cold water from the East Iceland polar current. From this brief indication it will be seen that complexities in the currents will be set up, but so far observations are too few to form opinions on their variation. Many interesting tables are given correlating depth of water, temperature, and salinity, and the following conclusions have been arrived at :—

"At the south coast warm pure Atlantic water of a high (above 35°/₀₀) and somewhat varying salinity occurs; at South-West Iceland there is a somewhat similar sea; at North-West and North Iceland there is Atlantic water mixed with cold water of low salinity from the East Greenland polar current; and, lastly, at East Iceland arctic water occurs (with a temperature of 0°–2° C., and salinity from 34·6°/₀₀ to 34·9°/₀₀), the East Iceland polar current mixing with water from the Atlantic current. The change of temperature in the surface-layers of the water, the cooling process during winter and the heating process during summer reaches down almost as deep as the algal vegetation, and is consequently of no slight importance to the latter." Regarding the temperature of water in the interior of small fiords, it is found that it is nearly the same from the surface downwards to the bottom, but in the larger fiords the temperature decreases regularly with the depth. In the fiords the salinity varies according to the amount of fresh water, and, where low, excludes certain species of algæ; generally it increases with depth.

Cambridge County Geographies. Forfarshire. By E. S. VALENTINE, M.A. Cambridge University Press, 1912. Price 1*s.* 6*d.*

In this—one of the latest volumes—the scope is similar to that of other books in the same series. Interesting statistics are given regarding climate, rainfall, agriculture, shipping, etc., which cannot fail to be of use to teachers of practical geography in Scotland, and the latest rainfall map of Scotland is reproduced. To one unacquainted with the Forfarshire dialect the section dealing with people, race, and dialect should prove of great interest. As in other members of the series the illustrations are excellent.

Handbook to Belgium, including the Ardennes and Luxemburg. London: Ward, Lock and Co., 1912. Price 2*s.* 6*d.*

This is the sixth edition of a comprehensive and moderately priced guide-book which is well illustrated by photographs and has a number of town plans, etc., as well as a general map on a small scale.

Southern England: Coast and Countryside. Edited by PRESCOTT ROW and ARTHUR HENRY ANDERSON. London: The Homeland Association, 1912. Price 1*s.*

A useful little compilation, forming No. 5 of the Homeland Reference Books, and containing articles on the Cathedrals of Southern England, on motoring, golf, etc., with alphabetically arranged notes on the towns, health resorts, etc., of the district. It is illustrated by some good photographs, and contains a list of books dealing with the region considered. Among the practical details given in regard to the towns we find in many cases the average rents and the rates, both points of considerable interest alike to the geographer and intending resident; the very heavy rates in many places are remarkable.

ASIA.

Recent Events and Present Policies in China. By J. O. P. BLAND. London: William Heinemann, 1912. *Price* 16*s.*

It is not possible to notice here in detail Mr. Bland's exhaustive account of the causes which have led to the upheaval in China and his criticism of the policy pursued by the Great Powers in that country in the last few years, although his book is without doubt one of the most important contributions which have been made to the literature of the subject; but those who have read his *China Under the Empress Dowager*, which he wrote in collaboration with Mr. E. Backhouse, will not require any further recommendation to induce them to give serious attention to Mr. Bland's views, as they are set forth in the large volume before us. He is not in agreement with Dr. Morrison, the former correspondent of the *Times* in Peking, and now the political adviser to the Republic, who sees in the revolution every cause for hope that a regenerate China will arise in the near future. Rather is Mr. Bland one of those who can see no good prospect of a settled state of affairs issuing from a so-called Republic, under the influence of a large body of half-educated youths of the student class who have climbed into office on the strength of the "Western learning" acquired by them in Japan, Europe, or America, and who now go to swell the crowd of expectant office-seekers. He sees in this class an element which will go to increase and prolong the prevailing unrest. A number of these youths of the Young China party have adopted the profession of arms, and, since the revolution, "hold rank to which neither their age nor their achievements entitle them." Mr. Bland regards the Republic as an accident—a result of the revolt against the Manchu régime which broke out at Hankow and other cities in the Yangtsze valley, in October 1911. He already sees throughout the country, and especially north of the Yangtsze, signs of a revolution against the new order and the band of southern hot-heads who lead and maintain it.—"The Republic is the offspring of unexpected opportunity, out of sudden chaos; accidental in its birth and foredoomed to early demise," while "Young China, as at present constituted, will pass, the shadowy fabric of a restless dream. An inevitable reaction will restore the ancient ways, the vital Confucian morality, and that enduring social structure whose apex is the Dragon Throne. But Young China, at its passing, will not have been in vain. Something of the Utopia of its visions will remain, to renovate and modify that ancient structure." Economic causes, rather than political, go to produce that unrest which is, and always has been, more or less prevalent in a country like China, in which the chronic condition is that of a struggle for life, unequalled in any other part of the world. The main factor in producing this condition is, as Mr. Bland well expresses it, "the procreative recklessness of the race, that blind frenzy of man-making, born of ancestor worship, which, despite plague, pestilence, and famine, battle, murder, and sudden death, persistently swells the numbers of the population up to, and beyond, the visible means of subsistence." To the great mass of such a population, immersed in the grim fight for bare subsistence, forms of government, be they monarchical or republican, can have but little importance. In estimating the factors which tell against permanence of the Young China régime Mr. Bland also gives prominence to the absence of religious inspiration, to the want of an authoritative aristocracy, and, most of all, to the lack of common honesty in public affairs. So that we are not unprepared for his conclusion that China is only likely to find her salvation in the coming of the strong man who can set up an autocracy. Whether the present President, Yuan Shih-k'ai, will prove to be that man time only can show.

Mr. Bland gives his readers some interesting character sketches of Young China's leaders:—of Wu Ting-fang (a barrister of Lincoln's Inn and until lately Chinese Minister to the United States), of Wen Tsung-yao, Tang Shao-yi, and, last, but not least, of Sun Yat-sen, "the perambulating Conspirator-in-Chief of the Radical Republicans"; but these sketches are too long to quote here.

The long protracted negotiations in connection with international loans and railway finance are detailed with great minuteness, and the history of them is brought up to last October. Mr. Bland has scant belief in the honesty of many of the negotiators, either foreign or Chinese, and he criticises severely the policy of the British Foreign Office during the last few years, which he thinks has resulted in a loss of position and prestige. He regards it as inevitable that Russia and Japan will eventually absorb Mongolia and Manchuria, in spite of the treaties to maintain the integrity of China; and, as for the Yellow Peril, he sums up his well-reasoned conclusions with these words: "If there be any menace to Europe in Cathay, it lies in the fierce struggle for life of three hundred million men who are ready to labour unceasingly for wages on which most white men must inevitably starve."

The vexed question of the Opium Traffic, and of Britain's connection therewith, come in for a special chapter, in which the pros and cons of the controversy are stated both with fairness and knowledge; and it is to be noted that what the author describes as Young China's deliberate intention to violate the new opium agreement with Great Britain is already taking shape as we write.

The Malay Peninsula, A Record of British Progress in the Middle East. By ARNOLD WRIGHT and THOMAS H. REID. London: T. Fisher Unwin, 1912. *Price 10s. 6d. net.*

To use their own words, the aim of the authors is to present "a comprehensive account of the development of British influence in the Middle East from the earliest times to the present day." They have not spared either labour or research in the accomplishment of their task, and we may congratulate them on the production of a very readable and succinct story of the growth of empire in the Malay Peninsula.

The beginnings of British power date from the occupation of the port of Bencoolen, on the south-west coast of Sumatra, by the East India Company in June 1685. The main attraction to Bencoolen lay in its nearness to the pepper-growing districts of the island, and the company hoped by opening a station there to be able to wrest a share of the trade from their Dutch rivals. Owing to the swampy nature of the soil, and the coral reefs which render it difficult of access from the sea, Bencoolen never was a success as a settlement. The early records quoted by the authors furnish a dismal story of losses by sickness—("'Ffeavours and flux' were the chief causes of death"),—and of dissensions amongst the Company's officers. It is interesting to learn that the company made a trial of sending out German emigrants, men and women, to Bencoolen about the year 1780; but most of these poor people speedily fell victims to the climate. The port was ceded to Holland by the London treaty of 1825, and so ended British rule in Sumatra.

The real foundation of the Straits Settlements dates from the concession of the island of Penang, made by the Sultan of Kedah to Francis Light, who was acting for the Government of Madras. The consideration paid to the Sultan took the shape of an annual subsidy of 9000 Mexican dollars, which seems a trifling figure, viewed in the light of the later importance of the place, and the British flag was hoisted on 11th August 1786. Better known to fame than the name of Francis Light is that of Sir Stamford Raffles, although the nation also owes a deep debt of gratitude to the former. Raffles was sent out to Penang by the East India Company in 1805, and, as is well known, it is to him we owe the possession of Singapore. How he raised himself to front rank by his own native merit, and was made Lieut.-Governor of Java when that island was taken from the Dutch in 1811, are matters of history; but it is worth recalling that he was brought into touch with John Leyden, when the Border poet and orientalist went to Java in the train of the Earl of Minto. A visit to Calcutta, which he made in 1818, gave Raffles the chance to induce the Marquess of Hastings, the then Governor-General, to found a new station to the eastward of Penang, in order to maintain British supremacy at the eastern end of the Straits of Malacca; and on the 29th January

1819, Raffles had the satisfaction of hoisting the Union Jack at Singapore. This far-seeing man set to work on founding the new settlement, constituting it a free port from the start; and a free port it has remained, to the infinite advantage of British trade in Malaya. The names of Francis Light and Stamford Raffles, belittled in their lives, now stand on the roll of fame as among the greatest of empire builders.

The history of British development in the Malay States is given in some detail, and we would draw special attention to the account of the states of Kelantan and Trengganu, which have so recently come under British protection, but both of which are already in a fair way to become flourishing territories. There has been so little known hitherto of these two states that the authors' description of them will be novel to many. We can only refer here in passing to the dramatic history of the rise of the Federated Malay States of Perak, Selangor, Pahang, and Negri Sembilan. In the latter part of the volume will be found interesting chapters on the development of the rubber industry in Malaya, on mining, and other matters, and we can cordially recommend this book to the attention of our readers.

Things Seen in Palestine. By A. GOODRICH-FREER, F.R.G.S. With fifty Illustrations. London: Seeley, Service and Co., 1913. *Price 2s. net.*

This chatty, delightful series of "Things Seen" has received an interesting and useful addition in *Things Seen in Palestine.* Mrs. Goodrich-Freer has made the country her own by her works, and she describes with a vivid pen the things seen in a land that never fails to sharpen the pencil point of every traveller. The glimpses she affords of the desert, the village, the town, the social and the religious life are captivating. She speaks with a fulness of knowledge that makes every sentence tell.

AFRICA.

South African Geology. By G. H. L. SCHWARZ. London: Blackie and Son, 1912. *Price 3s. 6d. net.*

This is an interesting little book, whose title, to our mind, does not quite adequately describe its contents. It is really an introduction to geology for the use of South Africans, *i.e.* the examples and illustrations are mostly taken from that region, and the section on stratigraphical geology is practically limited to South Africa. The book thus differs from the ordinary run of small text-books in containing a mass of material not found in them, and in omitting much that they include. The figures also depict conditions unfamiliar to most geologists, and have the same air of freshness as the rest of the book. Professor Schwarz is not afraid to state vigorously his own opinion on subjects which to many geologists seem as yet unsettled, and has therefore produced a very attractive little book, in places much condensed.

Life of a South African Tribe. By HENRI A. JUNOD. Vol. I. *The Social Life.* London: David Nutt, 1912. *Price 3s. 6d. net.*

The author was attached to the Swiss Romande Mission, and spent fifteen years with the Thonga Clans, a Bantu race of the East African Coast. (Their country extends roughly from 22° S. lat. to St. Lucia Bay, and from 32° East long. to the sea.) Most of the book is a minute literal transcription of the ceremonies, tabus, laws and customs of these clans as taken down from the reports of four or five natives in whom the author has confidence. It is, therefore, a human document of great importance to anthropologists, and full of interesting information regarding the clan system, the relationships, language, and especially the rites of birth, puberty, marriage, burial, war and chieftainship. The peculiar lubricity of Bantu life is here given with an accuracy which one might think unnecessary were it not that this, the governing factor in negro nature, is always forgotten by most missionaries, and the public generally.

There are many interesting points in the book. The graves described resemble those in certain cemeteries in Crete. The occasional use of human flesh as food is explained by the author as a mystic ceremony. The patriarchal government by the chief, who is in a sense the tribe, is carefully described. There are many misprints, *e.g.* "synonimous," "consaquinity," p. 242, "proferential," p. 247, and occasionally unusual expressions which, however, cannot but be expected if the author is writing in a foreign language.

Handbook of British East Africa, 1912-13. Illustrated and with Two Maps. Compiled by H. F. WARD and J. W. MILLIGAN. Second Impression. London: Sifton Praed and Co., Ltd. *Price* 4*s.* 6*d. net.*

This handbook should prove exceedingly useful to intending settlers and also to visitors to the country. It gives a considerable amount of detail as to the various crops which are being grown in the region as well as in regard to cattle-rearing, ostrich-farming, and so forth, and also much information for sportsmen. We note that sisal hemp is doing well, and is expected to become an important crop in the highlands, while there are also good prospects for rubber. On the whole, however, stock-farming seems the most profitable industry, but it demands considerable initial capital.

AMERICA.

A Holiday Trip to Canada. By MARY J. SANSON. London: The St. Catherine Press, 1912. *Price* 2*s.*

This is an unpretentious little book giving a very detailed account of a short visit to Canada, even the meals on board ship being fully described. It will appeal chiefly to the unsophisticated.

Joseph Pennell's Pictures of the Panama Canal. London: William Heinemann, 1912. *Price* 5*s. net.*

This is a collection of photographs of the Panama Canal made by Mr. Pennell for the *Century Magazine* and the *Illustrated London News*, and its main interest is that it gives us a vivid idea of many parts of the canal during the process of construction. When the canal is completed the appearance of the places, which now resemble busy work-yards, crowded by workmen and machinery, will be inevitably changed. Mr. Pennell has an introductory chapter in which he cordially acknowledges the welcome he everywhere received, and he naturally writes with pardonable enthusiasm and effusive patriotism of the splendid work, both engineering and sanitary, which is being achieved by his fellow-countrymen. To each picture there is appended an explanatory note.

The Flowing Road: Adventuring on the Great Rivers of South America. By CASPAR WHITNEY. London: William Heinemann, 1912. *Price* 12*s.* 6*d.*

This volume gives an account of five separate expeditions in South America, largely by canoe, and chiefly on streams more or less connected, hence the title. The author shows himself an indomitable and intrepid traveller, capable of surmounting many difficulties which might well have daunted a more timid soul. Though some of the journeys had a special purpose, as for instance the hunting of the jaguar, the author states frankly that for the most part the impelling motive was merely a desire to "see things," and he certainly succeeded in seeing a good many, and conveys many vivid impressions to his readers.

The book is illustrated by a series of photographs for whose deficiencies the author apologises, but which, nevertheless, give an excellent idea of the regions traversed.

With the Indians in the Rockies. By JAMES WILLARD SCHULTZ. London : Constable and Co., 1912. *Price* 4*s*. 6*d*. *net*.

This book contains the early recollections of Thomas Fox, who travelled to Fort Benton in 1856. He was plundered by the Indians in the Rockies, and had to pass the winter in the mountains with an Indian friend, where they suffered great hardships. The book is full of exciting adventures and hairbreadth escapes from avalanches, bears, Indians, etc. The conditions of frontier travel in 1856-8 are well described.

Les États-Unis du Mexique. By Comte MAURICE DE PÉRIGNY, Chargé de missions. With a Preface by Professor MARCEL DUBOIS. Librairie Orientale et Américaine. Paris : E. Guilmoto, 1911. *Price* 5 *fr*. 50.

The author of this book first deals with the history of ancient Mexico and passes briefly on to the Spanish Conquest and in chapter ii. he outlines the history of the Mexican Republic. The laws and constitution, public instructions, army finances, are next taken up, and in chapter vi. the commerce and industry are treated of. From chapter viii. to chapter xvii. (the last chapter) the book describes the different states and districts in order, and much attention is given to the mining industry in each state. To French readers who wish to have a general sketch of the country mainly from the tourist's point of view, the book may be interesting. It is without an index, and does not profess to deal profoundly with that immense country in which the author's countrymen had such unfortunate political experiences a generation ago.

GENERAL.

A Book of the Wilderness and Jungle. Edited by F. G. AFLALO, F.R.G.S. London : S. W. Partridge and Co., Ltd. *Price* 6*s*. *net*.

Here is another book on wild beasts and sport at the hands of the well-known writer, Mr. Aflalo, who is a recognised authority on this subject. It purports to be "a kind of nature-study book on the larger scale, and introduction to the study of big game in our overseas possessions," written for the "thousands of young fellows (who go) every year to India, Africa, Canada, or those farther colonies that lie on the other side of the world." In it the reader will find abundance of information regarding the big game of every part of the world, their localities and habits, the difficulties and the dangers of hunting them. The book is copiously illustrated with good stories of thrilling and exciting adventures experienced by such doughty shikaris as Selous, Lee Warner, de la Poer and Seton Karr. Mr. Aflalo is, of course, a real sportsman, and writes wisely and earnestly on the necessity of protecting efficiently, ere it be too late, those kinds of wild animals and birds which now run the risk of extinction, and in this matter he denounces those who kill game merely to be able to brag about the size of their bag or to establish a record. In this respect even Mr. Roosevelt comes in for censure on account of some of his African exploits. The inexperienced sportsman will do well to ponder the pages of the chapter on "the vengeance of the wild," in which are mentioned a number of the incidents in which the most experienced and fearless of sportsmen have paid for their sport by severe wounds or death. The illustrations of this book in colour and black and white are by Mr. E. F. Caldwell.

Palæolithic Man and Terramara Settlements in Europe. By ROBERT MUNRO, M.A., M.D., LL.D. Edinburgh: Oliver and Boyd, 1912. *Price* 16*s. net.*

This handsome volume will be gratefully received by all who are interested in the history of mankind, and is indeed a noteworthy addition to British Anthropology. The plates are not only numerous, attractive, and beautifully clear, but many of them will be new to British readers, which is rather an unusual circumstance in illustrations of prehistoric men and beasts. The first part consists, with the exception of a few changes in phraseology, of the first course of Munro Lectures in 1912. The second part practically forms the Dalrymple Lectures on Archæology, given in Glasgow in 1911.

The author begins with an interesting discussion on man's place in the organic world, and passes on to the story of the Ice Ages and chronological problems. Most of the first part, however, is devoted to a detailed and careful description of all the various stations in Britain, Belgium, France, Italy and other countries where fossil man and his artifacts have been discovered. In enumerating all this evidence, Dr. Munro includes many finds which, as he shows, are not now considered to be of palæolithic age. This information is, of course, of the greatest possible value to all students of the subject. Then follow chapters on the palæolithic races, and on the supposed hiatus between the palæolithic and neolithic periods. In the second part, the terramara settlements are described, and full lists are given of the implements, bones, etc., found in each of them. The final chapters deal with the culture of the Terramaricoli, the lake dwellings in the Po Valley, neolithic huts, the terpen and other pileworkers of Holland, Bosnia, and a few other countries.

It will be seen that the arrangement is strictly geographical, and one cannot help a regret that Dr. Munro has not given more space to his discussion of the general questions involved. During the last two or three years Dr. Penck and Professor Sollas in this country have suggested definite chronologies for the Ice Ages. The Acheulean period is dated by the latter as towards the culmination of the third glacial episode at about 34,000 B.C. Even if these dates should be taken as provisional, there is great advantage for the student, and especially for the man in the busy streets of to-day, in a definite working hypothesis. Moreover, owing to the works of Montelius and others, the chronology of the bronze and iron ages in Italy is surely fixed at any rate provisionally. Still, the beautiful illustrations, full lists and excellent bibliography of the Terramara settlements given in this volume are of great value, even though some of the information has already been supplied to British readers by Mr. Peet's book published in 1909. Indeed, prehistoric and proto-historic man in Italy has now, thanks to this volume, been most carefully attended to, and one wishes that the same could be said of his British contemporaries.

The volume is in every way beautifully turned out, and misprints are not at all evident. On page 369 a beam is described as 24½ *inches* long. Probably this should be feet, not inches.

From Pole to Pole. A Book for Young People. By Sven Hedin. London: Macmillan and Co., 1912. *Price 7s. 6d. net.*

Dr. Hedin's book has been translated, abridged, and edited for the use of young people, and with its numerous illustrations and its vivid descriptions should admirably fulfil its purpose of arousing interest in travel and exploration, and in the manners and customs of other lands. There are numerous sketch maps, though the scale of these is very small. Dr. Hedin's own travels are fully described, and in addition to accounts of journeys through the civilised regions of the world we have tales of exploration of all ages retold, so that the book contains both history and geography, and is full of human interest.

The Origin and Evolution of Primitive Man. By Albert Churchyard. London: George Allen and Co., 1912. *Price 5s. net.*

In this book Dr. Churchyard, with the aid of many italics, puts forward a view of human evolution which he prefaces with the observation that "the evidence in favour of my contentions is critically correct." Some of the points dealt with seem to us of interest, but the author rather spoils his chance of a hearing by his somewhat intemperate language in regard to other investigators.

EDUCATIONAL.

Dent's Historical and Economic Geographies. By Horace Piggott and Robert J. Finch. Book I. *World Studies.* London: Dent and Co., 1912. *Price 3s. 6d.*

This is the first volume of a new geography which is to be issued in six books, the remaining volumes treating of the continents and of the British Isles and the British Empire in more detail. Well and elaborately illustrated, the present volume shows in its text some remarkable deviation from tradition. Thus we have a considerable amount of geology, and a very full treatment of prehistoric man, illustrated by pictures taken from *From Nebula to Man* and other sources. From these latter chapters, mostly based, we think, upon popular or semi-popular works, the student passes on to a consideration of climate along more or less the usual lines. We are not satisfied that this is a wholly desirable method, on account of the great uncertainty which exists as to the interpretation of the evidence at present available in regard to the early records of man. The position of Pithecanthropos and of Neandertal man in the evolution of the human race, the significance of eoliths, these with many related problems are still being fiercely debated, and we feel that it is undesirable that subjects of this kind should be discussed in school-books under circumstances which make it impossible to indicate clearly the large element of uncertainty which exists. The authors' purpose could, we think, have been better attained by a brief discussion of the various forms of human societies which exist at the present day, and this could be done with certainty and in detail.

The book, however, contains a great deal of interesting matter, and many of the pictures are attractive.

Our Own and Other Lands: The World. London: M'Dougall's Educational Co., 1912. *Price 2s.*

This is a geography of the reading-book type, illustrated by a very admirable series of pictures and some coloured maps. It includes some elementary history in addition to geography. The book on the whole is written in a clear and interesting style, though we object to malaria being called a "poisonous emanation."

The absence of the author's name, also, is to our mind a real blemish. Otherwise the book should prove useful as a general summary of world geography.

BOOKS RECEIVED.

Lippincott's New Gazetteer: A Complete Pronouncing Gazetteer or Geographical Dictionary of the World. Edited by ANGELO HEILPRIN and LOUIS HEILPRIN. With a Conspectus of the Thirteenth Census of the United States. Royal 8vo. Pp. x + 2105. *Price £2, 2s. net.* London: J. B. Lippincott Co., 1912.

A Scientific Geography. Vol. VIII. *South America.* By ELLIS W. HEATON, B.Sc., F.G.S. Crown 8vo. Pp. 90. *Price 1s. net.* London: Ralph, Holland and Co., 1913.

A Pronouncing Vocabulary of Geographical Names, with Notes on Spelling and Pronunciation and Explanatory Lists and Derivations. By GEO. G. CHISHOLM, M.A., B.Sc., F.R.G.S. Crown 8vo. Pp. vii + 102. *Price 1s. net.* Glasgow: Blackie and Son, Ltd., 1913.

With the Conquered Turk: the Story of a Latter-Day Adventure. By LIONEL JAMES. Crown 8vo. Pp. 370. *Price 2s. net.* London and Edinburgh: T. Nelson and Sons, 1913.

Panama Canal: What it is; What it means. By JOHN BARRETT. Demy 8vo. Pp. 120. *Price $1 net.* Washington, D.C.: Pan-American Union, 1913.

Scientific Papers. By J. Y. BUCHANAN, M.A., F.R.S. Vol. I. Demy 8vo. Pp. xii + 314. *Price 10s. 6d. net.* Cambridge: University Press, 1913.

Deuxième Expédition Antarctique Française, 1908-1910. Commandée par le Dr. JEAN CHARCOT. *Étude sur les Marées.* Par R. E. GODFROY. *La Flore Algologique.* Par L. GAIN. Cartes. Three Volumes. Demy Folio. Paris: Masson et Cie, 1912.

A Handbook of Geography. By A. J. HERBERTSON, M.A., Ph.D. Vol. II. *Asia, Australasia, Africa and America.* Crown 8vo. Pp. xii + 681. *Price 4s. 6d.* London and Edinburgh: T. Nelson and Sons, 1913.

With the Turks in Thrace. By ELLIS ASHMEAD-BARTLETT, in collaboration with SEABURY ASHMEAD-BARTLETT. Illustrated. Demy 8vo. Pp. x + 335. *Price 10s. net.* London: William Heinemann, 1913.

With the Victorious Bulgarians. By Lieutenant HERMENEGILD WAGNER. With fifty-five Illustrations and six Maps. Demy 8vo. Pp. ix + 295. *Price 7s. 6d. net.* London: Constable and Co., Ltd., 1913.

How England Saved China. By J. MACGOWAN. With 38 Illustrations. Demy 8vo. Pp. v + 319. *Price 10s. 6d. net.* London: T. Fisher Unwin, 1913.

Aborigines of South America. By the late Colonel GEORGE EARL CHURCH. Edited by CLEMENTS R. MARKHAM, K.C.B. Demy 8vo. Pp. xxi + 314. *Price 10s. 6d. net.* London: Chapman and Hall, Ltd., 1912.

Les Sociétés Primitives de L'Afrique Équatoriale. Par le Docteur ADOLPHE CUREAU. Un vol. in-8° écu. Pp. xii + 420. *Prix 6 frs.* Paris: Librairie Armand Colin, 1912.

Three Years in the Libyan Desert: Travels, Discoveries and Excavations of the Menas Expedition. By J. C. EWALD FALLS. Translated by ELIZABETH LEE. With 61 Illustrations. Demy 8vo. Pp. xii + 356. *Price 15s. net.* London: T. Fisher Unwin, 1913.

A War Photographer in Thrace: An Account of Personal Experiences during the Turko-Balkan War, 1912. By HERBERT F. BALDWIN. With 36 Illustrations from Photographs by the Author. Demy 8vo. Pp. 311. Price 5s. net. London: T. Fisher Unwin, 1913.

The Story of the Forth. By H. M. CADELL, D.L., B.Sc., F.R.S.E., M.Inst.M.E. With 75 Illustrations and 8 Maps. Medium 8vo. Pp. xvii + 299. Price 16s. net. Glasgow: J. Maclehose and Sons, 1913.

Quebec, the Laurentian Province. By BECKLES WILSON. Fully illustrated. Demy 8vo. Pp. xii + 271. Price 10s. 6d. net. London: Constable and Co., Ltd., 1913.

The European in India. By H. HERVEY. Demy 8vo. Pp. viii + 312. Price 12s. 6d. net. London: Stanley Paul and Co., 1913.

Popular Guide for the Use of the Aix-la-Chapelle Waters. By A. LIEVEN, M.D. Fcap. 8vo. Pp. 90. Aachen: Barth'schen Buchhandlung, 1913.

Questions and Exercises in Geography. (Based on Heaton's *Scientific Geographies.*) IV. *North America: with Statistical Appendices.* By ROBERT J. FINCH, F.R.G.S. Crown 8vo. Pp. 48. Price 4d. net. London: Ralph, Holland and Co., 1913.

Dent's Practical Note-books of Regional Geography. By HORACE PIGGOT, M.A., Ph.D., and ROBERT J. FINCH, F.R.G.S. Book I. *The Americas.* Crown 4to. Pp. 64. Price 6d. net. London: Dent and Sons, Ltd., 1913.

Burma Gazetteers: Myitkyina, Lower Chindwin, Bhamo, Sandoway; Arakan and Tharrawady Districts. Rangoon, 1912.

Eastern Bengal District Gazetteers: Dacca and Dinajpur Districts. Allahabad, 1912.

Year-Book and Record of the Royal Geographical Society. London, 1912.

Iron and Steel, 1911. Board of Trade Report, 1913.

Catalogue of the War Office Library. Part III. Subject-Index. Compiled by F. J. HUDLESTON. London: War Office, 1912.

Census of India, 1911. Vol. 7, Bombay, Part I. Report; Part II. Imperial Tables. Vol. 8, Bombay (Town and Tables), Parts I. and II. Report and Tables. Vol. 21, Mysore, Part I. Report; Part II. Tables. Bombay and Bangalore, 1912.

Monographia Sobre a Industria da Borracha da Mangabeira e da Maniçóba No Estado da Parahyba do Notre (Brazil). Parahyba: Muzeu Commercial do Rio de Janeiro, 1912.

Diplomatic and Consular Reports.—Trade of Dunkirk, 1911 (5027); Trade of the Consular District of Buenos Ayres, 1911 (5029); Russian Budget, 1912 (5030); Trade and Commerce of Denmark, 1911 (5031); Trade and Commerce of Lingah, 1911-12 (5032); Trade and Commerce of Bunder Abbas, 1911-12 (5033); Trade of Bangkok, 1911-12 (5034).

Colonial Reports—Annual.—St. Helena, 1911 (714); Weihaiwei, 1911 (715); Ceylon, 1911 (716); Gibraltar, 1911 (717); Gambia, 1911 (718); Ashanti, 1911 (719); Falkland Islands, 1911 (720); Seychelles, 1911 (721); Northern Territories of the Gold Coast, 1911 (722); Hong-Kong, 1911 (723); Sierra Leone, 1911 (724); Gold Coast, 1911 (725); Turk and Caicos Islands, 1911 (726); Fiji, 1911 (727): Bechuanaland Protectorate, 1911-12 (728); Basutoland, 1911-12 (729); the Surveys of British Africa, Ceylon, Cyprus, Fiji, Jamaica, Trinidad, British Honduras (730); Malta, 1911-12 (731); Nyasaland, 1911-12 (732); British Honduras, 1911 (733); Gilbert and Ellice Islands Protectorate, 1910 (734); Southern Nigeria, 1911 (735); Somaliland, 1911-12 (736); Report on the Work of the Imperial Institute, 1911 (737); Northern Nigeria, 1911 (738); Jamaica, 1911-12 (739); Swaziland, 1911-12 (740); Bahamas, 1911-12 (741); Grenada. 1911-12 (742); Uganda, 1911-12 (743); Mauritius, 1911 (744); Trinidad and Tobago, 1911-12 (745); British Guiana, 1911-12 (746); Barbados, 1911-12 (748).

Publishers forwarding books for review will greatly oblige by marking the price *in clear figures, especially in the case of foreign books.*

THE SCOTTISH

GEOGRAPHICAL

MAGAZINE.

THE PLANT ECOLOGY OF BEN ARMINE (SUTHERLANDSHIRE).

By C. B. CRAMPTON, M.B., C.M., and M. MACGREGOR, B.Sc.

(*With Map and Diagram.*)

I. THE PHYSIOGRAPHY AND DISTRIBUTION OF THE VEGETATION.

THE ecological survey of the Ben Armine district, the results of which are given in the following paper, was carried out at intervals during the summers of 1910-12. The accompanying map has been reduced to two-thirds from the six-inch scale, and includes the greater part of Ben Armine and its immediate surroundings. The three principal summits of Ben Armine surmount a ridge extending over six miles in a north-north-west direction, and lying approximately twenty miles distant from the north coast, twenty-four miles from the east coast, and thirty-two miles from the west coast of Sutherland. Ben Armine is only separated from Ben Clibrick by the long, narrow, and steep-sided hollow which holds the beautiful lake known as Loch a Choire. This hollow is open to the north-east, but the high ground of the two mountains is connected by a narrow crest bounding the south-west end of the loch, and on its west side overlooking the point on the Tongue road known as the Crask.

Clibrick is 3184 feet in height, almost 1000 feet higher than Ben Armine, and the mass formed by these mountains is isolated on the Sutherland plateau. The nearest mountains have the same abrupt outstanding isolated character. The Morven group (Morven, 2313 feet) lies sixteen miles to the east; the two Ben Griams (1936 feet and 1903 feet) ten miles to the north-east; Ben Loyal (2504 feet) twelve miles to the north-west; Ben Hope (3040 feet) fourteen miles to the north-west;

and Ben Hee (2864 feet) sixteen miles to the west. All these mountains were mainly carved out during a past cycle of erosion, and have little relation to the headwaters of the present drainage system. They are (relict) types of mountains, known to geologists as monadnocks or mountains of circumdenudation. The string of mountains further to the west, between Ben Hee and the west coast, is fully twenty-four miles distant from Ben Armine. It intercepts some of the moisture brought by the Atlantic winds.

The following data are taken from Bartholomew's Meteorological Atlas: The annual rainfall is over 40 inches, *i.e.* more than on the Caithness high plateau. The wettest month is October, and the driest May. The monthly rainfall averages are: more than 4 inches from October to January; more than 3 inches during February and March; more than 2 inches from April to June; and more than 3 inches from July to September. The mean annual sunshine is less than 25 hours per month. The annual range of temperature is about 18°. The mean annual temperature varies from 40° on the higher ground to 47° on the lower part of the plateau.

The following table is constructed from the monthly isotherms for the plateau region:—

January,	37°	July,	56°
February,	38°	August,	56°
March,	39°	September,	53°
April,	44°	October,	47°
May,	49°	November,	41°
June,	53°	December,	39°.

The rocks of the Bèn Armine country are chiefly mica-schists with some hornblendic, felspathic and siliceous schists, but all are so impregnated with material of a granitic nature as to form coarsely granitoid gneisses having certain resemblances to the Lewisian Gneiss of the North-West Highlands.[1] These hard, massive, crystalline rocks have imposing joint planes and form great cliff-faces and immense blocks and boulders, but under the prolonged influence of frost, have weathered, on the mountain-tops, into a debris of stones, grit, and sand.

A glacial drift swathes the gentler slopes and the hollows of the plateau. It is of a loose, gravelly, and sandy nature, full of blocks and fragments of the country rocks, but is rarely exposed, except in burn sections, since a great mantle of peat is spread far and wide on the face of the county. The whole of the great Sutherland plateau is enveloped in this peat, which stretches as far as the eye can see into Caithness and Ross, and reaches to the cliffs of the north coast. The contrast afforded by this great stretch of peat and the widely spaced features of the mountains, as seen from one of the heights on a clear day, forms an impressive picture pregnant of the reality and immensity of the great time changes.

The higher slopes of Clibrick rise above this peaty mantle, but peat covers the whole of the plateau, rises to the slopes and ridge of

[1] See *Summaries of Progress of Geological Survey*, 1910-1911, pp. 41 and 37.

Ben Armine, and is absent only on small areas of the summits, or other places where the physiography and exposure to wind, or other geological agents of denudation, have prevented its growth or have been inconsistent with its preservation.

The Ben Armine ridge is divided by saddles, or cols, into a series of gently rounded, plateau-like summits, three of which, Creag Mhor, Creag a Choire Ghlais, and an unnamed top at the northern end of the mountain, exceed the 2300-feet contour. There is a fourth small top called Creag Bheag, 1750 feet, at the southern end of the ridge. The plateau on which the mountain rests averages 800 feet in height, but has small elevations reaching the 1000-feet contour. The lowest col on the ridge does not descend to 1500 feet.

At its northern end the summit is broad, flat, and peat-covered, but small areas forming the summits of Creag a Choire Ghlais, Creag Mhor and Creag Bheag are free of peat, though surrounded by it on all sides except where they directly overlie abrupt crags on the eastern face of the mountain.

Ben Armine has a very different aspect as seen from the west or the east. The western slopes rise gradually to the summits, and are completely peat-covered. The eastern face, on the other hand, is for the most part abrupt, and is marked by a series of five corries, three of which, the Coire a Saidhe Duibhe, and the two corries whose crags give their names to Creag Mhor and Creag Bheag, are backed by precipitous cliffs; whilst a fourth, the Coire Ghlais, is a "hanging-corrie" with very steep, rocky slopes. The fifth corrie, the Coire nan Eas, is wide, low, and open, backed by small scattered crags, breaking the lowest col in the centre of the ridge, where peat passes almost uninterruptedly from the western to the eastern slopes of the mountain.

The Ben Armine country is drained by a branch of the Brora River, the Black Water, whose tributaries take their rise to the east and west of Ben Armine, and flow in a south-east direction to join at a considerable distance beyond the confines of the map.

By far the greater part of the water from Ben Armine is derived from the peat, and flows away to the west. The streams descending the eastern flank (excepting the two traversing the Coire a Saidhe Duibhe and the Coire nan Eas, which have their origin in hollows lying to the west of the ridge) have no permanence and little volume, and the head-waters on this side of the mountain arise, for the most part, in the peat of the plateau.

The amount and direction of the drainage of Ben Armine is thus correlated with the physiography, the orientation of the mountain mass, and the distribution of peat. The westerly flowing drainage is held up by the peat, and is, comparatively speaking, somewhat sluggish and peaty in character; while that rising on the eastern flank is practically ephemeral and almost negligible in volume, but torrential in character, except where it issues as small springs from beneath the rock debris or peat near the tops of the corries. Much of the drainage of the eastern face, however, is also peaty in nature, since very little of the mountain mass rises above the great peat mantle. In the early part of the year some is derived from the melting of snow lying in the higher parts of the corries.

Apart from the numerous peat holes, or Dubh-lochans, there are only two lochs in the area, Gorm Loch Mhor and Gorm Loch Bheag, both of glacial origin. They lie close under the eastern face of the mountain, are deep in relation to their superficial area, and have rock or moraine-dammed banks. The small amount of drainage entering the lochs is derived from the adjacent corries, and the waters are clear, and show but slight fluctuation in level. The lochs, therefore, show little sign of silting up or lowering through erosion at the outlets, and very little effect from rise and fall of the waters or from sand-drift, but their shores are subject to wave influence. They are, in fact, typical mountain tarns.

Apart from those leaving these lochs, the smaller streams all arise in the peat, their channels, in many cases, being mere peat-walled gullies or pipe-like drains beneath the peat. The numerous narrow alluvial tracts flanking the streams are formed from undercut peat surfaces let down within the flood zone, or consist of sand and gravel resulting from erosion of the sandy drift.

Ben Armine differs remarkably from the Ben Griams further east, and also from Morven, Scaraben, and the other mountains of Caithness, both in the greater altitude up to which numerous springs are found and in the development of corries having precipitous sides (these two features being, no doubt, correlated). The corrie development, however, is chiefly a result of the persistence of localised ice erosion during the later phases of glaciation. The steep corries on the eastern face of Ben Armine are fronted by semicircular groups of terminal moraines of the corrie-glacier phase, the moraines still standing out as sharply as if the glaciation were but a recent phenomenon. The only precipitous corrie in the other mountains mentioned above is the one on the north side of Ben Griam Bheag, and this also is accompanied by well-marked terminal moraines of the corrie-glacier phase.

Though Ben Armine receives a greater rainfall than the other mountains, there is little doubt that the frequency of occurrence and distribution of springs is less to be attributed to this than to differences in the shape of the mountains, and in the nature of the disintegration undergone by the rocks. The pyramidal masses of Morven and the Ben Griams are of small bulk relative to their height, and the conglomerates and sandstones of which they are chiefly composed break up to form gravels or screes of large blocks, into which water percolates to a considerable depth. In these respects Morven is a more extreme case than the Ben Griams, and the plant associations are of a correspondingly more xerophytic type. The similar relations existing between the quartzite of Scaraben and the great depth of porous angular debris in which the mountain is completely buried have been discussed in *The Vegetation of Caithness.*[1]

---

[1] C. B Crampton.—*The Vegetation of Caithness. considered in relation to the Geology*, 1911.

The massive, felspathic gneisses forming the bulk of Ben Armine show somewhat different relations in regard to weathering and the retention of water near the surface. The debris on the summits is less porous, and the results of decomposition are more evident than in the other cases. Considered in relation to its height, the plateau-like mass of Ben Armine further increases the water-holding powers of the mountain-top debris. The base of the mountain-top debris is rarely exposed on Morven, Scaraben, and the Griams, but is deeply undercut by the steep corries on the eastern face of Ben Armine, and the water which accumulates in the debris appears as springs on the rocky walls of the corries. But of more importance still is the cloak of peat, the inherited time effect of all these factors, combined with the wetter climate resulting from the geographical position of Ben Armine and its close association with Ben Clibrick. This cloak of peat holds up the water like a sponge, and supplies it to all the lower slopes and crags as small ephemeral streams, or oozing trickling springs, bathing the cracks and surfaces of the corrie walls.

This effect increases with the dominance of peat on the summit of the ridge as we pass from south to north along the corries on the eastern face of Ben Armine, the complete peat covering at the northern end of the ridge affording a more constant and larger supply of water to the corrie walls, with a corresponding difference in the type and development of the vegetation.

The peat covering the plateau and the gentler slopes averages five feet, but is often much thicker. On most of the slopes it is gradually shrinking, cracking, and slipping down hill, and this has led to undermining of the peat by water channels, and the formation of sinks and "swallow holes." Further stages in erosion produce a network of steep-sided gullies and peat hags.

The feeders of the streams run for considerable distances underground, in tunnels in the loose sandy drift beneath the peat, and the roofs of these tunnels gradually collapse to form rows of sinks, and eventually open channels flanked by high banks of peat.

Even the permanent water-courses, apart from those of the true valley streams, show, by their close relations to deeply cut sections in the peat, that they are only a further stage of evolution in the same process. They occupy ground which was formerly peat-covered, and have been originally derived from underground channels beneath the peat.

Everything points, indeed, to comparatively recent conditions when the whole district was completely clad in unbroken peat, which enveloped the slopes and even the summit of Ben Armine, filled all the hollows, covered the wider ledges on the crags, and occupied the positions of all the existing smaller water-courses. The drainage was held up by the spongy nature of a luxuriant moorland vegetation, continually soaked and subject to no such spells of drought as characterise the present climate. Growth of peat was at a maximum, and erosion limited to those streams that could keep open channels in spite of the rapid growth of moorland vegetation. The alternating layers of buried forests and moorland peat encountered in all the sections, except in the most wind-swept places on the summit ridges, show that the phase of moorland plant activity was but a later one amongst others alternating with phases when the climatic conditions were different.[1]

The layers in the peat are the same as those described for the Caithness peat mosses.[2] At the base of the peat a "sub-arctic layer" similar to those discovered by Dr. Lewis, so widely spread in our Scottish peat mosses, has been noticed at a number of places in this area, and where it is wanting the base of the peat generally consists of layers of birch twigs and bark.

Higher up in the peat a layer of pine stumps is commonly met with throughout the district. The peat sections further show that since the death and decay of the pine forest a prolonged moorland phase has intervened. Though the great size of the stumps, and their widespread dominance, points to the trees having existed under fairly favourable conditions of drainage, it is quite evident that the succeeding moorland phase completely effaced such conditions, the growth of peat having been so rapid as to hold up the drainage and obliterate the smaller branches of the streams.

The plateau at that time must have exhibited a practically unbroken and soaking spongy mantle, dissected only by the larger stream courses. The slopes likewise were cloaked in wet peat whose surface drainage percolated slowly, without producing erosion, into the valley streams. The summits now bare of peat had then a thick humus covering, and even the most abrupt crags, where alone the peat could find no resting-place must have been continuously bathed in humous waters dripping from peaty accumulations on the ledges.

This we conceive as a picture of the last moorland phase, but the present conditions show that a great change supervened, such as can only be satisfactorily accounted for by a falling off in the amount and constancy of the precipitation.

---

[1] Professor James Geikie suggested as long ago as 1867 that the buried forests in our peat mosses indicated changes of climate ("On the Buried Forests and Peat Mosses of Scotland, and the Changes of Climate which they Indicate."—*Trans. Roy. Soc. Edin.*, xxiv., 1867). This view has been supported by Blytt in Norway, and Sernander and the Uppsala School in Sweden. It is, however, rejected by a number of investigators of peat bogs in Norway, Sweden, and Denmark (see Gunnar Andersson, *The Climate of Sweden in the Late Quaternary Period*, 1909).

Dr. Francis Lewis's recent work on the Peat Mosses of Scotland has established the presence of a layer of subarctic plants intercalated in the peat between the forest beds in certain districts, and this has afforded additional support for Professor Geikie's view. As far as Central Sutherland and Caithness are concerned no subarctic bed has yet been detected except at the base of the peat deposits. This is believed by Dr. Lewis to correspond with his second subarctic layer overlying the lower forest bed in Shetland, and in the Southern Uplands of Scotland. See the following papers by Dr. Lewis:—"The Plant Remains in the Scottish Peat Mosses," Pts. 1-4—*Transactions of the Royal Society of Edinburgh*, 1905-1911; "The History of the Scottish Peat Mosses and their relation to the Glacial Period."—*Scottish Geographical Magazine*, vol. xxii., 1906; "The Sequence of Plant Remains in the British Peat Mosses."—*Science Progress*, No. 6, October 1907.

[2] C. B. Crampton.—*The Vegetation of Caithness, considered in relation to the Geology*, 1911.

The summits are now bare of peat wherever the physiography has favoured rapid drainage and erosion by wind. All stages in the denudation of the tops can be observed. Some still retain a thick peat covering, while on others the peat is gradually undergoing removal. The peat on such summits and on exposed cols has shrunk, and forms a network of gullies and peat hags. The slopes and peat-filled hollows are dotted with sink holes and areas of subsidence due to undermining and slipping of the peat. The stream courses are gradually stealing up the slopes, and are invading the hollows which formerly had an unbroken peat covering. Crags are reappearing from beneath the peaty mantle, and the thick peat of the plateau is shrinking, cracking, and gradually gravitating in the direction of the newly formed water-courses.

Marking the centres of the areas now defined by the water-courses, Dubh-lochans have appeared as a result of the disturbance of the surface and the drainage. Two types of Dubh-lochans are represented, as in Caithness: (*a*) Those scattered in floating peat, (*b*) those associated with sinks and underground drainage channels.[1] Many of the latter type have unfortunately been drained, and now form wide areas of black exposed peat.

From the ecological standpoint this great peat covering must be looked upon as an inheritance of the present vegetation. In spite of its degradation and partial removal the thick peat still dominates the surface, and acts as a barrier to all plants that cannot grow on an acid humus soil.[1] The effects of the dynamic agents of retrogression are, moreover, somewhat diverse in their operations. In the one direction they tend to exacerbate the barren condition, and to strengthen the barrier against immigration and competition by plants other than moorland species. By the erosion of thick peat and the baring of its surface, by stagnancy resulting from inequalities of the drainage, and by the liberation at the surface of vast quantities of previously stored and buried toxic humous compounds,[2] conditions are set up which are even intolerable to some of the moorland species prevalent under wetter climatic conditions. Toxic, humous conditions are superadded to periodical physical drought, and to a greater degree of stagnancy than can have ever prevailed during the wetter climate of the preceding moorland phase, when the plant covering of the peat was unbroken and continuously bathed by fresh precipitation.

On the other hand, by erosion leading to exposure of rock and drift, which in the Ben Armine district sometimes contain a higher percentage of lime than is usual in our highland schists and gneisses, by a periodical flushing and drying of slopes and alluvial surfaces, an amelioration of the barren conditions has been locally brought about. Plants, which during the previous moorland phase were doubtless restricted to the stream alluvia of the lower country where peaty waters were less in evidence, have advanced far into the moorland region, and have taken possession of such alluvial surfaces, rainwashed slopes, and exposed surfaces of

---

[1] See *The Vegetation of Caithness.*

[2] B. E. Livingston.—"The Physiological Properties of Bog Water."—*Botanical Gazette*, 39, 1904.

rock and drift as are not still subjected to the most peaty type of drainage.

Further, an alpine flora of a meagre kind has reinvaded those crags and summits from which the peat has been removed. Some of the alpine species doubtless found refuge throughout the late moorland phase on the cliffs of the corries, but the species having the present majority of individuals are such as can flourish on peat surfaces when they are unbroken and subject to frequent precipitation. These belong to a group which everywhere is most evident on the alpine boundary of the moorland, and is capable of living either on leached alpine rocks and debris, or on peat, provided the competition is not adverse to them, and the surface is not too toxic with humous compounds.

Apart from critical forms it is probable that we have no endemic species in our limited alpine flora. Most of our alpine species, in fact, have a very wide range on the mountains of Europe and Asia, and many are circumpolar in distribution on the tundras, or are also found on some of the North American mountains; they are therefore often classed as arctic-alpines. The group of alpines referred to above,[1] however, appears to be separable from the rest of our alpines by its powers of withstanding the competition and edaphic conditions accompanying the rapid accumulation of moorland peat, since its species are not only commonly met with at the present time on thick peat, but must formerly have dominated wide areas of the plateau peat mosses during the climax conditions of the wettest moorland phases, and are represented in the history of the peat mosses by Dr. Lewis's subarctic layers.[2] [3]

The species referred to the "peat-alpines" are still, perhaps, more abundant in this area and the neighbouring districts of Sutherland than in any other part of Britain, while other alpine species are poorly represented. This numerical relation has doubtless resulted from the intensity and all-pervading nature of the moorland phases in these districts, the widely enveloping peat growth favouring the peat-dwelling arctic-alpines and leading to an elimination of the others.

In reference to these two types of alpine species the following notes are of importance. Ben Clibrick rises considerably above the peat mantle, and supports several species not found on Ben Armine, including *Sibbaldia procumbens* (Mountain Potentilla), *Cherleria sedoides* (Cyphel), *Luzula spicata* (Spiked Woodrush), and *Gnaphalium supinum* (Dwarf Cudweed).

The following are far more abundant on Clibrick than on Ben

---

[1] Other arctic-alpine species have only been found as yet in the basal deposits of the peat.

[2] C. B. Crampton.—*The Vegetation of Caithness, considered in relation to the Geology*, 1911.

[3] F. Lewis.—"The Plant Remains in the Scottish Peat Mosses," Pts. 1-4.—*Transactions of the Royal Society of Edinburgh*, 1905-1911.

—— "The History of the Scottish Peat Mosses and their relation to the Glacial Period."—*Scottish Geographical Magazine*, vol. xxii., May 1906.

—— "The Sequence of Plant Remains in the British Peat Mosses."—*Science Progress*, No. 6, October 1907.

Armine: *Polygonum viviparum*, *Juncus trifidus*, *Silene acaulis* (Moss Campion), *Saxifraga oppositifolia*, and *Antennaria dioica* (Mountain Everlasting). The following are confined to crags on both mountains: *Saussurea alpina*, *Hieraceum* sp. (Hawkweeds), *Oxyria digyna*, *Cystopteris fragilis* (Bladderfern), *Sedum Rhodiola* (Roseroot), *Salix lapponum*, *Asplenium viride*, *Polypodium Dryopteris* (Oakfern), *P. Phegopteris* (Beechfern).

Others are confined to alpine springs or wet cracks, as—*Epilobium alpinum*, *Saxifraga stellaris*, *S. aizoides*. Others again are chiefly in the alpine grassland, nursed by flushes, as—*Alchemilla alpina*, *Thalictrum alpinum* (Alpine Meadow rue), *Festuca vivipara*, *Aira cæspitosa*, var. *brevifolia*.

Certain species border on the "peat-alpines," as—*Azalea procumbens*, *Salix herbacea*, and *Carex rigida*, and specially characterise the smooth parts of the mountain-top formerly covered by peat, and now chiefly occupied by *Calluna* (Heather) mat, or *Rhacomitrium* carpet. They invade smooth, peat-covered surfaces greatly influenced by wind exposure.

The true "peat-alpines"—*Arctostaphylos alpina*, *A. Uva-ursi* (Bearberry), *Rubus Chamæmorus* (Cloudberry), *Salix repens* and allied creeping species, *Betula nana* (Dwarf Birch), *Lycopodium Selago*, *L. alpinum*, *Empetrum nigrum* (Crowberry), *Vaccinium Vitis-idæa* (Cowberry), are locally replaced or joined by others more susceptible to wind exposure and the chilling action of a constantly saturated surface, and usually found on well-drained and aerated peat. Such are—*Vaccinium Myrtillus* (Blaeberry), *V. uliginosum*, *Melampyrum montanum*, *Cornus suecica* (Dwarf Cornel). These, however, are chiefly found on well-drained, sheltered peat surfaces on the higher slopes and ledges of the corries, where they are sometimes found attempting to dispute the ground with the ubiquitous *Calluna* (Heather), but usually only in places where this is straggling, and would probably under quite natural conditions have been succeeded by scrub birch and Rowan, or a shrubby growth of *Betula nana*, *Vaccinium*, and *Salix*. They are, therefore, in all probability, to be classed as species locally dependent on an alpine dwarf scrub on a peat substratum, and are sometimes accompanied by one or other of the following: *Blechnum Spicant* (Hardfern), *Pyrola secunda*, *P. minor*, *Listera cordata*, *Lastræa montana*, *Trientalis europea*.

As a comparison with the nearest mountains it may be added that *Dryas octopetala* and *Carex capillaris* occur on Ben Griam Mhor, *Statice maritima* and *Gnaphalium supinum* on Ben Loyal, but none of these are found on Ben Armine.

It is therefore noteworthy that a number of alpine species of the Central Highland mountains are wanting on Ben Armine, and that Ben Clibrick and the nearest mountains only add seven species to the Ben Armine list, whilst at the same time the "peat-alpines" are abundant and vigorous in Central Sutherland, and on Ben Armine in particular. It appears from this that the greater altitude and more precipitous orography of the Central Highlands afforded better sanctuaries for the more characteristic arctic-alpine species during past periods of great peat growth, which caused an invasion and narrowing of their domain, but favoured the spread of "peat-alpines."

It may be further remarked that an exposed, highly glaciated, plateau and roches moutonnées topography, such as that of Central Sutherland and Western Caithness, is peculiarly favourable to a peat-moss or to a peat-alpine and tundra vegetation, according to the severity of the climate, while the steep-walled cirques, arêtes and over-steepened valley sides of a glaciated mountain complex like that of the Central and Western Highlands, provide conditions of soil, drainage and protection from wind, better suited for the more delicate alpine species and subalpine woodlands.

When the peat swept over the summit of Ben Armine, Clibrick still stood out as an isolated alpine sanctuary, and the Central Highland mountains afforded larger and more continuous areas of precipitous mountain crag and steep debris-clad slopes, where the alpine plants were isolated in cold and humid surroundings. During the drier continental phases the alpine plants would find the summer and winter conditions more favourable for them, provided they were sufficiently rigorous to limit incursions by the subalpine forest. Owing to their isolated geographical position, our arctic-alpines were evidently recruited with difficulty from other alpine or arctic centres. The species which are continental-alpine, as differentiated from arctic-alpine, are very few in Britain, and perhaps were never well represented in our alpine flora.

We have already indicated the present dominance of moorland plant associations, and their probable greater extension at a former time. Limits were, however, always set by the migratory agents of geological change, wind, gravity, and running water, and it is in places most affected by these agents that the greatest retreat of the moorland associations is found at the present time.[1]

The migratory plant associations take possession of these places, compete for them with the stable moorland associations, and gain ground at their expense, or are forced to retreat before them, according to (*a*) the phases of retreat or advance of the moorland climatic province, or (*b*) the advance or retreat of the zones of activity of the migratory agents of surface change.

The plant associations may therefore be classed into two groups as follows: [2]—

(I.) Stable associations chiefly owing their inheritance to the rigorous conditions imposed by the late glaciation and the post-glacial climate, and, apart from man's interference, attaining to the full stabilisation allowed by this inheritance under the present conditions of climate and physiography.

(II.) Migratory associations chiefly owing their present distribution to the migratory agents of surface change. The limits and nature of their

---

[1] C. B. Crampton.—*The Vegetation of Caithness, considered in relation to the Geology*, 1911.

[2] C. B. Crampton.—"The Geological Relations of Stable and Migratory Plant Formations."—*Scottish Botanical Review*, vol. i., pts. 1, 2, 3, Jan.-July 1912.

stabilisation accord with the changes produced by these agents and the effects of animal interference.

(I.) *The Stable Associations* are of three main kinds :—

(*A*) Associations occupying places where topographic factors still preserve habitats approaching those which must have been widespread and dominant during the late climax moorland phase. Such factors provide habitats for (1) the "alpine peat mosses," where an unbroken cloak of wet peat still covers the gentle slopes of the mountain summits, in positions subject to the maximum of present atmospheric moisture : (2) "Sphagnum mosses" in circumscribed areas of the plateau, where the surface of the peat is floating and still retains plant associations in which *Sphagnum* takes a controlling part.

(*B*) Plant associations occupying areas of peat where it has undergone shrinkage, slipping, undermining, and disturbance of the surface, leading either to drainage or bogginess. This has naturally brought about a redistribution of species in reference to their dominance, allowing certain species like *Calluna* (Heather), which must formerly have only covered well-drained areas, and *Rhacomitrium*, which specially affected wind-swept places, to invade the moorland far and wide, and to form new associations characteristic of the various retrogressive conditions now widespread and dominant. Under these conditions we have :—

(1) Retrogressive types with good drainage.
 - (i) *Calluna* moor (Heather moor).
 - (ii) Alpine *Calluna* moor.

(2) Retrogressive types with bad drainage.
 - (i) *Rhacomitrium* bogs.
 - (ii) *Scirpus-Eriophorum* bogs (Deer's Hair and Cottongrass bogs).

(*C*) Progressive plant associations, colonising steep banks and alluvial terraces, etc., which have been deserted by the geological agents of surface change. Amongst these we have :—

(1) Grass heaths, the degenerate representatives of associations passing into birch scrub by the process of succession. Such associations had former periods of dominance when the birch scrub occupied wider areas of the plateau, but at the present time they only assert themselves under the limited topographic conditions resulting from late erosion, and are but poorly represented in the Ben Armine district. Under a wetter climate they would probably succeed *Calluna* moor over limited areas.

(2) Sphagneta and grass moors, on more recent growths of peat, on deserted stream terraces, and elsewhere, under conditions where slipping and breaking of the peat are not entailed. There are indications in the plant succession of the moorland that the dry phase, which succeeded the last great moorland phase, and led to the widespread retrogression of our moorland associations, has probably been on the wane for some time past, and that some of these progressive types of moorland associations are beginning to reassert their claims, under climatic conditions more favourable to their growth.

These three types : (*a*) the climax, (*b*) the retrogressive, (*c*) the progressive, might be looked upon as three plant formations, regionally successive to one another, but existing side by side at the present time owing to topographic factors; or they might be classed as three sub-formations of a moorland formation, each in turn becoming dominant during (*a*) climax phases, (*b*) retrogressive phases, (*c*) progressive phases. The sphagneta, grass heaths, grass moors and birch scrub, characterise the initial and progressive stages (*c*); the alpine peat mosses, the climax phases (*a*); and the present dominant bogs and *Calluna* moor (together with the pine forests, unfortunately extinct so far as this area is concerned) the retrogressive phases (*b*).

But topographic differences may supply the needs of the various associations during stages other than those in which they are dominant, and thus we may find them all co-existing at the same time, where the surface is sufficiently diversified to afford places of refuge for those that are recessive. Such sanctuaries form centres of dispersal when the climatic conditions again become favourable for one or the other type that may be, for the time being, recessive.

In sections of peat mosses we may therefore expect layers pointing to such progressive, climax and retrogressive stages, marking both secular changes (regional successions), and the effects induced by the migratory geological agents of surface change (topographic successions).[1]

Before assuming climatic change it is necessary to weigh carefully the evidence as to the cause of the succession, and if climatic change is inferred, the evidence as to whether changes in temperature or differences in the amount and distribution of the atmospheric precipitations are chiefly responsible. It is probable that a climate favouring the spread of peat mosses tends to limit the distribution of many arctic-alpine species, though it may favour those referred to as "peat-alpines," whereas one leading to their retrogression, in the absence of higher temperatures, may have the reverse effect.

In regard to the former distribution of species in the moorland region, the barrier to plant migration resulting from the gradually accumulated peat deposits probably merits greater attention than has been given to it. In discussions on the former more northerly or southerly limits to the distribution of various tree species in north-west Europe the summer temperatures have been considered the determinating factor, but plants like the Hazel, that only grow on thick peat under exceptional conditions,[2] might well be gradually eliminated from areas where they formerly grew, by the increasing depth of moorland peat, whilst others like the pine might become more and more restricted to such areas by competition with other trees.

In this connection it is noteworthy that the buried forests in this country rarely contain Hazel except they be basal in position, while the pine stools are almost invariably found overlying thick peat.

Both buried forests and subarctic layers need, therefore, separate consideration as regards their basal position or their intercalation in

---

[1] C. B. Crampton.—"The Geological Relations of Stable and Migratory Plant Formations."—*Scottish Botanical Review*, vol. i., pts. 1, 2, 3, Jan.-July 1912.

[2] *E.g.* where the soil waters are alkaline.

thick peat. A number of forest or arctic-alpine species, according to the climate, may well have had a wide range before the thick peat accumulated, but their later migrations in the moorland region would become more and more restricted except in the case of those species like the Pine, Birch and the "peat-alpines" which could tolerate the competition and edaphic conditions encountered in the vegetation on thick peat accumulations.

(II.) *The Migratory Associations.*—The plant associations of migratory formations are chiefly found in relation to the stream belt and in the rock-bound corries and on the mountain tops. Thus we have—

(*A*) Plant associations limited to parts of the stream belt where the agents of geological change involve flushing of the surface of the peat, erosion of peat or drift, and the formation of rainwash or alluvial deposits. These form habitats which encourage the growth of plants debarred from the more stable areas of peat surfaces. The plants composing the associations may be—

(i) Species naturally associated with moorland vegetation and rarely found save in its vicinity. These include in particular those of moorland springs and wet moorland flushes, where certain bryophytes and so-called "bog-plants" have their centres of distribution.

(ii) Species with centres of distribution beyond the moorland districts, but encouraged to invade them through the conditions afforded by the migratory agents. These include grasses, marsh or other plants from the lower alluvial stretches, and bushland species following the stream alluvia and steep banks of the water courses.

(iii) Species from the alpine region also invade the moorland stream courses.

The alpine migrations appear to be mainly the result of plants or seeds carried down stream, while the lowland migrants seem to be chiefly transported by animals. Thus moorland, lowland, and alpine species contest the ground, and are found as pure associations, or forming various combinations, stabilised to the extent and in the direction allowed by the migratory agents, as represented by geological changes and those induced by herbivorous mammals.

The plant associations may be grouped as follows :—

1. Those of springs with alpine species.
2. Those of springs and wet flushes with moorland species.
3. Those of flushes with moorland and lowland species.
4. Grasslands of dry flushes with alpine species (alpine grasslands).
5. Grasslands of moorland alluvia with moorland species dominant.
6. Grasslands of sandy river alluvia with lowland immigrants.
7. Open associations of sandy and gravelly alluvia with alpines.

(*B*) Plant associations that are normal to rock surfaces as found in the rocky stream channel, gorges and waterfalls, and on the steep crags which back the corries. A study of such rock formations shows that several factors have to be taken into consideration, namely—

1. The age of the plant succession.

2. The history of the rock surface exposed.
3. Its physiographic and edaphic relations as a station for plant growth.

Many of the rock surfaces at present exposed to plant colonisation have been but recently, or are even now, undergoing denudation from a peaty covering, or are so situated in reference to the thick peat mantle as to be dominated by a peaty drainage. Whole cliff faces may in this manner be subject to drip from overlying thick peat and show plant associations which, though differing from normal moorland associations, yet have a large share in their ecological inheritance.

Some owe their exposure to late glacial conditions and are so situated physiographically that their features have little relation to the present migratory conditions of geological change. Many of these were recently covered with thick peat or are still partly covered by it and undergoing denudation, and their conditions are the same as those which have been noticed above. Others are more or less horizontal surfaces or large glacial boulders denuded of peat, and yet only subject to the slow processes of disintegration by frost under conditions which may be termed alpine or subalpine, as the case may be. These are slowly acquiring a vegetation in which the degree of stabilisation is only limited by the barren and slowly decaying nature of the substratum. These usually pass into those noted in the next section, where the extent of stabilisation is controlled by migratory factors of exposure to wind and surface creep.

In contrast with the above are those rock surfaces where the degree and kind of stabilisation is limited by certain presently acting geological factors, such as—

1. Instability of position due to physiography. The ledges, crevices and rock faces of the corrie walls have but a limited existence as a substratum for plant growth. The plants of such stations may be separated into lithophytes or rock-surface plants, chasmophytes or crevice dwellers, and ledge chomophytes or dwellers on soil accumulations on ledges.[1] [2] [3]

2. The flushing of rock surfaces by springs, which may be comparatively permanent or markedly periodic and migratory in their influence, deserting one part of the rock surface for another as time progresses. The habitats induced by such conditions may be separated as "rock flushes."[4]

3. Exposure of rock surfaces to the permanent or periodic flooding of streams, in waterfalls or on the floors and walls of rock chasms and gorges, or on boulders large enough to be stationary for long periods. These constitute the "rocky stream channel."

The rock plants are therefore a very varied assemblage and live under very varied conditions. Those of the rocky stream channel are aquatic or amphibious lithophytes, algae and bryophytes, capable of withstanding a considerable force of water and scour without being dislodged. They

---

[1] A. F. W. Schimper.—*Plant Geography on a Physiological Basis*, p. 178.

[2] M. E. Oettli.—*Beiträge z. Okologie der Felsflora*, Zurich, 1905.

[3] Members of the Central Committee for the Survey and Study of British Vegetation.—*Types of British Vegetation*, 1911.

[4] C. B. Crampton.—"The Geological Relations of Stable and Migratory Plant Formations."—*Scottish Botanical Review*, vol. i., pts. 1, 2, 3, Jan.-July 1912.

are strongly attached to the rock and float freely with the water currents, or lie prostrate with a special curvature, orientation and imbrication of the leaves and branches, protective against scour and dislocation.

Those of the rock flushes are amphibious and more often intricately matted and soil-accumulating, but sometimes show similar devices against erosion.

The subaerial lithophytes vary according to the orientation and nature of the rock face, and the degree of exposure to sun, wind and atmospheric precipitation. Crustaceous lichens preponderate and are accompanied by densely cushioned mosses, often hoary and drought-resisting. The frequently loosely attached, foliaceous and fruticose lichens affect less inclined surfaces with a tendency to soil accumulation. The strongly attached, umbilicate *Gyrophoreæ*, on the other hand, chiefly affect horizontal wind-swept and grit-scoured rock surfaces, which suggests that their habit may be of service to them against sand blast.

The chasmophytes are the most characteristic of the phanerogamic rock vegetation, since the isolation they procure from the nature of their habitat limits competition. The most characteristically alpine species are therefore usually chasmophytes, or are found on the narrowest rock ledges where the moorland plants fail to obtain a footing.

(*C*) Plant associations that take possession of—

1. The rock debris of the flatter summits of the mountains.
2. Exposed smooth crests of gravelly morainic accumulations.
3. The more or less horizontal broken rock surfaces of roches moutonnées on the plateau, where denuded of peat and exposed to frost and wind.

The rock debris covering the high mountain plateau is believed to have resulted from the combined action of frost and wind since late glacial times. The material is subject to creep away from the crests of the mountain ridges, and its surface is usually highly porous owing to deflation by wind. Many of the small roches moutonnées at much lower elevations on the Sutherland plateau (800-1000 feet) have been long subjected to the action of frost and wind, and are covered by a mantle of debris of a similar character. The whale-backed surfaces of moraines in this region also often consist of wind-deflated debris of small stones.

These habitats differ somewhat in altitude and the nature of the surface, but have in common the smooth, rounded contours originally due to glaciation, and the relatively elevated and whale-backed shape, porosity of surface, and exposure, which results in rapid drainage and wind deflation.

These influences have led to their escape or early emergence from the peat mantle that lately covered every surface compatible with its formation. These debris-strewn surfaces and the alpine peat mosses are the special habitats in Sutherland of that group of arctic-alpine species already referred to as capable of growing on thick peat. *Arctostaphylos alpina*, *A. Uva-ursi*, *Betula nana*, and *Vaccinium Vitis-idæa*, appear to be equally at home on either type of surface, while *Azalea procumbens* rather avoids the peat and is almost peculiar to the habitats under discussion.

The important point is that these surfaces form places of refuge for certain alpines during stages of shrinking and retrogression of the peat mosses, and again centres of dispersal during the wettest stages or climax moorland phases.

Certain plants like *Azalea procumbens* are rarely found on them except at the higher elevations, presumably because during the moorland phases these positions withstood longest the encroachment of peat and were the earliest to emerge from it. Species like *Arctostaphylos alpina* have found in them a place of refuge during stages of moorland retrogression, while *Rubus Chamæmorus*, which always occurs on peat, will persist or become extinct according to the fate of the alpine peat mosses. The associations in possession of these surfaces therefore consist of plants capable of withstanding the inhospitable conditions of exposure they afford, and are either xerophytic alpines like *Azalea procumbens*, which can only compete with moorland species under very restricted conditions, species like *Arctostaphylos alpina*, which compete with moorland species under less restricted conditions, or moorland species like *Rhacomitrium lanuginosum*, or forms of moorland species like the mat type of *Calluna*, which were capable of adapting themselves to the altered conditions when peat was denuded from the surface.

II. The Moorland Plant Associations.

The moorland plant associations of this area for the most part closely resemble those of Caithness, and their general relations to the physiography are similar so far as the plateau is concerned.

They may be tabulated as follows:—

Stable moorland.	*A*. Climax types but recessive and relict.	1. Alpine peat mosses. 2. *Sphagnum* aureoles.
	B. Retrogressive types. (*a*) Badly drained.	3. *Rhacomitrium* bogs. 4. *Scirpus-Eriophorum* bogs.
	(*b*) Well drained.	5. *Calluna* moor. 6. Alpine *Calluna* moor.
	C. Initiative types, progressive but local.	7. Grass heaths. 8. Alpine moss heaths. 9. Sphagneta.
Migratory types.	*D*. Moorland flushes, etc.	See Section on flushes.

Badly drained foci with Dubh-lochans, *Sphagnum* aureoles and *Rhacomitrium* bogs, occupy central areas furthest removed from the stream courses, whilst zones of *Scirpus-Eriophorum* bogs and *Calluna* moor are met with in succession as we approach the better drained belts along the stream courses or the slopes of the hills. It is in the development of the alpine peat mosses that the Ben Armine district excels.

Reference to the map will show that the alpine peat mosses form a zone on the highest slopes encircling the associations of the mountain-top debris, and that the zone is best developed on the north-west side of the summits. This is partly due to the gentler slopes in this direction, but undoubtedly has also close relations to the greater exposure to wind and atmospheric precipitation coming from this quarter. The surfaces of the alpine peat mosses are smooth and the underlying peat is unbroken. The plant association abruptly changes to others when slipping or undermining of the peat has taken place.

The plant association is as follows:—

Mosses and lichens forming the smooth carpet:—*Sphagna* ssp. (chiefly acutifolia), *Hylocomium splendens*, *Hylocomium loreum*, *Hypnum Schreberi*, *Rhacomitrium lanuginosum* (patches), *Cladina sylvatica*, *Cetraria islandica* (Iceland Moss).

Tips of shoots showing:—*Vaccinium Vitis-idæa* (Cowberry) (common), *V. Myrtillus* (Blaeberry) (less often), *Erica Tetralix* (Crossleaved Heath) (frequent), *Calluna vulgaris* (Heather).

Espalier growth:—*Arctostaphylos alpina* (characteristic), *Arctostaphylos Uva-ursi* (Bearberry) (occasional), *Empetrum nigrum* (Crowberry) (common).

Juncus squarrosus (common), *Carex rigida* (local), *Rubus Chamæmorus* (Cloudberry) (characteristic), *Eriophorum vaginatum* (Cotton Grass) (scattered not tufted), *Eriophorum polystachion* (scarce), *Cornus suecica* (Dwarf Cornel) (local), *Salix repens* (Creeping Willow) (local), *Betula nana* (Dwarf Birch) (local), *Pinguicula vulgaris* (Butterwort) (frequent).

The most noticeable vegetative characters are the luxuriance and even surface of the moss carpet and the small development of the epigeal in comparison to the underground parts of the vascular plants. The underground stems of several of the species have a strong general resemblance in their more or less horizontal position and greatly elongated, tough wiry nature, and generally pale, yellowish colour with little development of cortex. The roots are few, whitish and relatively thick.

The shoots may form rosettes like *Juncus squarrosus* and *Carex rigida*, barely emergent leafy branches like the *Vaccinia* and *Erica*, or appear for short distances creeping on the surface like *Arctostaphylos* and *Empetrum*. None of the plants except *Eriophorum* usually bear flowers or fruit in abundance. The type of vegetative growth appears to be related to the combined effects of an abundant water supply in the moss and the frequent evaporation and chilling by strong winds.

The alpine peat mosses are generally abruptly limited above by the mountain-top debris. On the north and west flanks of the summits, where the alpine peat mosses usually give place above to the *Rhacomitrium* carpet association, the passage may be gradual owing to the *Rhacomitrium* being dominant in the alpine peat mosses near the boundary, but the latter seem to be always on peat, while the *Rhacomitrium* carpet lies directly on the rock debris and is of comparatively recent origin. The junction suggests that the alpine peat mosses are spreading locally and succeed the *Rhacomitrium* carpet association.

On the southern flanks the alpine peat mosses come more frequently into contact with the *Calluna* mat association, and here the junction suggests a retreat of the alpine peat mosses. (See section on the mountain-top debris).

The *Sphagnum* aureoles, *Rhacomitrium* bogs, *Scirpus-Eriophorum* bogs, and *Calluna* moor are practically identical plant associations with those of

Caithness, and there is at present little to add to the account of them given in *The Vegetation of Caithness.*

In this part of Sutherland intermediate stages between *Sphagnum* aureoles and *Rhacomitrium* bogs are more frequently met with, but it is not yet certain whether these mark degenerative stages of *Sphagnum* aureoles, or their recent advance at the expense of the *Rhacomitrium* bogs. *Vaccinium oxycoccos* and *Carex pauciflora* are generally present in the *Sphagnum* aureoles.

Other noticeable points are the absence of Bog Myrtle (*Myrica Gale*) in the bogs on the higher Sutherland plateau (800′), and the common occurrence of *Betula nana* even in those bogs which have undergone retrogression.

The Bog Myrtle is abundant in the lower-lying bogs north and south of this district (in the hollows of the Sutherland plateau and on the Caithness plain), and its distribution in Sutherland is suggestive of a comparatively recent advance from low levels. The persistence of *Betula nana* may point to the plateau bogs of Central Sutherland having entered into the late phase of retrogression at a later date than the other bogs in the north-east of Scotland, or of their having come more recently under conditions of artificial interference affecting the flora.

A parallel case is found in the *Calluna* moor of the Ben Armine district being chiefly of the type of *Calluna* moor with alpines. In comparison with other districts of the north of Scotland, *Calluna* moor is poorly developed on the Sutherland plateau and especially in the Ben Armine district. It forms only small areas on the steeper slopes, and narrowly fringes the streams on parts of the peat which have slipped and have become drained without undergoing erosion or flooding. It also commonly forms small patches round sinks in the peat.

On descending the stream courses to the points where the drift is undergoing erosion the peat alpines disappear, and the banks deserted by the streams show a plant succession progressive to *Calluna* grass-heath. This consists of bushy *Calluna* with *Hylocomium splendens, Hypnum Schreberi, H. cupressiforme* var. *ericetorum, Dicranum scoparium, D. majus, Aira flexuosa, Luzula pilosa, Galium saxatile, Potentilla sylvestris, Viola Riviniana, Blechnum Spicant*, and *Vaccinium Myrtillus.*

This association would probably advance to birch scrub under completely natural conditions. Rowan seedlings are fairly common, but those of birch very rare. Birch is abundant at Loch Choire, but appears to have been destroyed in the Ben Armine Forest.

The alpine *Calluna* moor has a more straggling growth of *Calluna* than *Calluna* moor, with a more luxuriant and wetter type of moss growth, often including patches of *Sphagnum* with abundant *Hylocomium loreum, Diplophyllum albicans, Polytrichum alpinum*, etc., and the alpines *Cornus suecica, Vaccinium Vitis-idæa, V. Myrtillus, V. uliginosum, Rubus Chamæmorus*, and sometimes *Betula nana.*

Calluna heath, or moor, and alpine *Calluna* moor may sometimes be seen on the opposing banks of the upper stream courses, where one is more subject to drought and exposure to the sun's heat, and the other damper and more shaded, showing that in the former case these factors have much effect in encouraging the growth of *Calluna* and its associates and suppressing that of the alpines.

On the higher ledges of the corries and sheltered, steeper slopes of the mountain-top debris, the alpine *Calluna* moor gives place to an alpine *Vaccinium* heath. Here any peat that occurs is thin and of recent formation, and *Calluna* is almost absent, the association consisting of luxuriant masses of *Hylocomium* (chiefly *loreum*) with *Sphagnum*, numerous hepatics, *Dicranum sp.*, and abundant *Vaccinium Myrtillus*, *V. Vitis-idœa*, *V. uliginosum*, *Cornus suecica*, etc.

The progressive types of sphagneta are chiefly found round bog springs and in continuously wet moorland flushes, but also on deserted terraces of the stream alluvia when they have become subject to flushing by peaty ferruginous drainage from the adjacent bogs. An iron pan is then generally formed and the surface becomes stagnant. Under the former circumstances they depend on migratory factors for their appearance, and the species of *Sphagna* are generally cuspidata or subsecunda and have *Juncus effusus* and *Polytrichum commune* as associates.

On some of the terraces there has been a considerable formation of peat of apparently recent origin dominated by *Sphagna* (*cymbifolia* and *acutifolia*) with *Eriophorum*, *Erica Tetralix*, *Drosera* and other forms, and in these we have a progressive stable type of moorland association.

III. The Plant Associations of the Debris of the Mountain Plateaux.

The debris covering those parts of the mountain tops bare of peat consists chiefly of small subangular or more or less rounded fragments of the underlying granitoid gneisses. The surface is all stones, but much sand and fine grit is usually found a few inches below, since the gneiss decays and becomes disintegrated, and the action of the wind causes considerable attrition and rounding of the stones. Layers of humus also occur buried in the debris.

The plant associations are almost limited to small areas of the smooth, rounded, almost plateau-like summits of Creag a Choire Ghlais, Creag Mhor and Creag Bheag. These areas are shown on the map as occupied by two principal associations, (1) *Rhacomitrium* carpet,[1 2 3]; and (2) *Calluna* mat; but under certain circumstances to be mentioned below other associations may supervene. The nature and composition of these associations are as follows:—

In the *Rhacomitrium* carpet *Rhacomitrium lanuginosum* is dominant, forming a continuous smooth carpet varying from a few inches to a foot or more in thickness. All the other plants grow in this carpet, and are dependent on it for protection and moisture. The surface is greyish

[1] C. B. Crampton.—*The Vegetation of Caithness, considered in relation to Geology*, 1911.

[2] Members of the Central Committee for the Survey and Study of British vegetation. *Types of British Vegetation*, 1911.

[3] Compare *Grimmia* Heath of Ostenfeld.—*The Land Vegetation of the Faroes.*

green and pale when dry, owing to the woolly nature of the leaf tips of the moss. The lower parts are shades of olive-green and yellowish brown, and the appearance of this colour on the surface denotes recent wind erosion. The other plants growing in the moss generally only expose the tips of their shoots, but the extent to which they compete with the *Rhacomitrium* for surface area depends on the degree of protection from the wind. *Aira flexuosa*, *Carex rigida*, and *Salix herbacea* are generally abundant even in the most exposed places. They creep in the moss layers and usually are only to be detected by emergent leaves. Elsewhere there is also a thin weaving of *Galium saxatile*, *Empetrum* forms mats, and *Vaccinium Vitis-idæa* and *V. Myrtillus* appear either as scattered single or bunched shoots from stems creeping beneath the moss. Lichens are scarce, but there is often a thin weaving of *Cetraria islandica*. Dwarf plants of *Solidago Virg-aurea* (Golden rod), 1 to 3 inches, are scattered. As one walks from the exposed summits of the more sheltered easterly slopes above the crags, the *Rhacomitrium* carpet passes through stages where the *Vaccinia* become more abundant, into an alpine *Vaccinium* moss-heath. with *Hylocomium loreum*, *H. splendens*, *Polytrichum alpinum*, *Hypnum Schreberi* and patches of *Sphagna* (*acutifolia*), *Cladina*, *Dicranum uncinatum*, *Leptoscyphus Taylori*. These and other species gradually displace the *Rhacomitrium*. The *Vaccinia* become bushy in habit but dwarf, and are accompanied by *Cornus suecica*, *Rubus Chamæmorus* and *Vaccinium uliginosum*.

The *Rhacomitrium* carpet and the alpine *Vaccinium* moss-heath are progressive associations, but the former from its wind-exposed position is liable to be destroyed by the wind ripping it from the underlying debris, rolling it up and carrying it bodily away. The Alpine *Vaccinium* moss-heath is more stable, but from its protected position is more subject to snow lie, which sometimes leads to its destruction in places where peat has accumulated, by causing creep of the surface, the peat gradually slipping into the corries beneath. The central areas of these snow lies, whether on alpine *Vaccinium* moss-heath, or *Rhacomitrium* carpet, are always marked by a growth of *Nardus stricta* (Mat Grass).

On the northerly and westerly slopes the *Rhacomitrium* carpet may abut directly upon the alpine peat-mosses if the slope is gentle. In such cases the mosses appear to succeed the *Rhacomitrium* carpet by an invasion of *Sphagnum*, *Eriophorum*, *Juncus squarrosus*, *Arctostaphylos alpina*, etc. But for the most part the *Rhacomitrium* carpet passes into *Calluna* mat; and this is especially so on the southern and south-western aspects, where the exposure to wind and sun is greater in comparison with the atmospheric precipitation. As we pass outward from the *Rhacomitrium* carpet, the *Calluna* mat first appears as small patches in the carpet. These soon coalesce. until the *Rhacomitrium* only forms patches in the *Calluna* mat. In either case the patches are usually elongated, with a parallel orientation at right angles to the direction of the prevalent wind. The surfaces beyond, occupied only by *Calluna* mat, show a wave and trough-like arrangement of patches of *Calluna* mat and bare debris, which also have a similar orientation. The kind of passage at the junction of the two associations therefore suggests that the *Rhacomitrium* has locally advanced from above, and has captured the bare surfaces between the wave-like strips of the *Calluna* mat, and has then spread from these as centre. In *The Vegetation of Caithness* it was stated that the *Calluna* mat occupies surfaces which the *Rhacomitrium* was forced to vacate owing to the severity of wind erosion, and this may be so in some cases, but a study of the passage between the associations on Ben Armine rather supports the view given above, since the type of erosion and growth of the *Calluna* mat usually results in the wave-like type of formation, while wind erosion of the *Rhacomitrium* carpet is usually more irregular, and often results in complete removal.

The *Calluna* mat association differs according to the station studied. When considered in relation to each individual mountain summit its

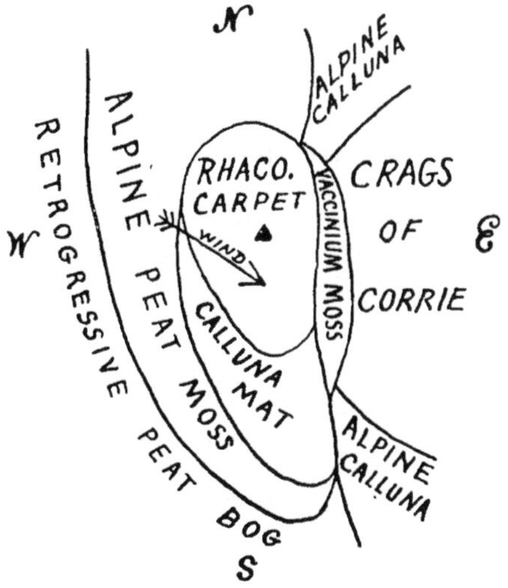

FIG. 1.—The Relations of the Summit Plant Associations.

outer margin is generally in contact with alpine peat-mosses, while its inner margin, if present, shows a passage into *Rhacomitrium* carpet as described above.

Diagrammatically the plant associations of a summit may be shown as in Fig. 1.

The differences in the *Calluna* mat association depend on (1) The approach of the alpine peat-mosses, (2) the approach of the *Rhacomitrium* carpet, (3) the degree of wind erosion, and (4) the degree of surface creep of the mountain-top debris.

As we pass from the alpine peat to the *Calluna* mat we find successive zones somewhat as follows:—

(1) Alpine peat mosses with *Sphagnum*, etc.

(2) *Juncus squarrosus* dominant, *Calluna* mat-like (closed), *Rhacomitrium*, *Cladina*.

(3) *Calluna* mat-like (closed), *Carex rigida* (abundant), *Rhacomitrium*, *Cladina*.

(4) *Calluna* mat-like (closed), *Cladina*, *Empetrum*, *Lycopodium alpinum*.

(5) *Calluna* mat-like in patches or wave-like from wind erosion, *Azalea procumbens* (mats), *Antennaria dioica* (dwarf), *Lycopodium*, *Selago* (dwarf), *Carex rigida* (dwarf), *Lotus corniculatus* (dwarf hairy), *Aira flexuosa* (prostrate), (mat), *Cladina* (in *Calluna* mat), *Hypnum Schreberi* (in *Calluna* mat).

All the plants are prostrate or very dwarf.

Where the slope is greater there is a marked creep of the debris to form a series of terraces and short abrupt vertical steps (from a few inches to two feet high), running parallel with the contours of the slope. In such places we find the surfaces of the exposed terraces have a plant association much like that described above (5). The little steep banks, on the other hand, show an association in which *Alchemilla alpina*, *Thymus Serpyllum* (Thyme), *Festuca vivipara*, *Carex rigida*, *Lotus corniculatus*, *Arctostaphylos alpina* and *Juncus trifidus* are prominent.[1] Where the slipping of the debris is a marked feature the surface is often covered with patches of *Arctostaphylos alpina*, *A. Uva-ursi*, *Thymus Serpyllum*, *Erica cinerea* (Bell Heather) (dwarf, prostrate or cushion-like), *Vaccinium Myrtillus* (prostrate) and *Calluna* (scattered and cushion-like).

We thus see that the *Vaccinia*, *Erica*, and *Thymus Serpyllum* rather avoid the wind-swept places in the *Calluna* mat, unless there is much slipping of the surface.

The *Calluna* is found in all stations, but varies in habit according to the conditions imposed by migratory factors. In the wind-eroded places it is completely prostrate in the direction of the prevailing wind, and is wave-like in patches. Nearest the peat mosses it forms a closed prostrate mat; further away it generally becomes open, and forms parallel wave-like patches arranged at right angles to the direction of the prevailing wind.

On pulling up patches of the *Calluna* mat we find that the younger branches send down fine thread-like roots into the ground, while many of the older twigs and often the main stem have succumbed from exposure of their roots. The young shoots are all arranged in a close fastigiate manner, and their tips are usually turned downward towards the ground. The racemes of flowers are markedly secund and are often twisted and may face the ground. This type is evidently an extreme case of the form of *Calluna* named *Ericæ*.[2]

When *Erica cinerea* accompanies the *Calluna* it assumes the same prostrate growth, and produces fine adventitious roots from its shoots. Examination of the conditions under which this takes place shows that the layering of the shoots by fine wind-driven debris is an important factor. Where the debris is subject to greater disturbance and movement through creep there is a more rapid accumulation of sand and fine

---

[1] Compare *Types of British Vegetation*, p. 313.

[2] Dr. C. E. Moss has confirmed this identification.

debris, wherever the vegetation temporarily arrests erosion, and the *Calluna* and *Erica* then assume another type of growth, that of more or less rounded cushions with close-pressed vertical fastigiate branches. On cutting into these cushions we find that there is an even greater development of fine adventitious roots from the branches than in the mat-type of growth. Sections of the cushions, subjected to prolonged beating to remove the sand, appear as a shaggy mass of pale brown fine capillary roots. It is therefore probable that it is the accumulation of debris and burying of the shoots that acts as a stimulus to the production of the adventitious roots in both cases. The root systems thus evolved are large in proportion to any normal shoot system that can survive in face of the severity of the exposure to winds, and this disproportion between the root and shoot systems may act as a stimulus to the production of the close fastigiate branching met with in the mat and cushion types of growth.

Essentially similar conditions with cushion types of *Calluna* are encountered in blown sand, in places where rabbits prevent the elongation of the shoots, and here also the larger adventitious root system is balanced by the shoot system by a close fastigiate type of branching. But cases also occur on the blown sand near lochs isolated in the moorland, where the cushion type evolves purely from sand accumulation and exposure to wind.

A somewhat close fastigiate type of *Calluna* growth is also found on wind-swept parts of the moorland in parts of Sutherland, and especially on roches moutonnées which have become peat-covered, and are isolated as islands in lochs. The surface of the *Calluna* growth is perfectly smooth and rounded, and humus accumulates between the stems which send down adventitious roots. The cushion and mat types of *Calluna* growth would therefore appear to be due to causes which at the same time limit vertical elongation of the shoots and stimulate the adventitious production of roots. It is possible that other cushion plants have similar conditions regulating their growth habit, but many appear to have had this growth habit fixed by inheritance.

In the case of the cushion type on the mountain slopes the effects of creep of the debris is a more potent factor than wind erosion. The creep is doubtless partly due to the wind eroding the debris, but chiefly to gravity and to the formation of gliding-planes where water accumulates beneath the surface. By digging beneath the debris at places where such slopes abut on the alpine peat-mosses forming their lower boundary, we find that the debris often overlies peat of some thickness, showing that the debris and its plant associations have advanced over the peat-mosses in these directions. Locally two or three layers of debris were noted in the peat, pointing to short phases of advance and retreat of the peat-mosses along the boundary. But peaty layers also occur in the debris at a distance from the margin on some of the slopes examined.

On some of the flatter parts of these tops the alpine peat-mosses are drying up and undergoing wind erosion, and patches of the superficial layers of the peat are undercut and let down on the bared surface of the debris. The *Calluna* in these patches then acquires the type of growth of the *Calluna* mat association.

The present evidence therefore points to the following conclusions:—

(1) That the rounded summits of Ben Armine were at one time covered by alpine peat-mosses which gradually underwent shrinkage and wind erosion, the process beginning at the highest and most exposed places.

(2) That the *Calluna* mat has evolved from the peat mosses through such erosion, and therefore marks a retrogressive stage from the peat mosses. Its constitution and vegetative type are, however, the direct results of migratory factors of surface change.

(3) That as the process continued the debris became exposed and underwent wind erosion and creep, gradually gravitating and invading parts of the peat-mosses covering the slopes beneath it.

(4) That the process has not been uniformly continuous, since the alpine peat-mosses have sometimes advanced over the front of the debris.

(5) That the *Rhacomitrium* carpet is a progressive type of association which has taken possession of the bared tops wherever it could obtain a hold.

(6) That it is generally successive to the *Calluna* mat along their boundaries, but is very unstable if exposed to drought as well as strong winds. It is therefore largely controlled by migratory factors of surface change.

(7) That ring-like patches and crescents of *Calluna* sometimes occur in the *Rhacomitrium* carpet, but it is doubtful if the *Calluna* mat ever replaces the *Rhacomitrium* carpet when this has been destroyed by wind.

(8) That the *Rhacomitrium* carpet is succeeded by more stable types of vegetation, by alpine peat-mosses on exposed places, and by alpine *Vaccinium* moss heaths on protected slopes.

(9) That a wetter climate would cause an advance of these associations on the *Rhacomitrium* carpet, and an advance of the latter on the *Calluna* mat.

(10) That there is evidence pointing to local advance of the *Calluna* mat at the expense of the peat-mosses, and that this would probably increase under a drier climate.

(11) That an advance of the *Calluna* mat at the expense of the peat mosses has been effective during a late phase of retrogression in the moorland, but that an advance of the *Rhacomitrium* carpet has perhaps been of more importance of late years, and points to a new phase of progression.

(*To be continued.*)

The *Alabama's* last harbour,

LOST IN THE ARCTIC: A REVIEW.[1]

(*With Illustrations.*)

THE expedition of Captain Ejnar Mikkelsen in 1909 was a sequel to that of Mylius Erichsen in the *Danmark* to north-east Greenland in 1906. How Mylius Erichsen, with Hφeg Hagen and the Eskimo Brönlund, all perished after completing their exploration of the unknown coastline from 78° N. to 83° 30′ N., is a story that will never be forgotten. Brönlund, the last survivor, dying of cold and starvation, dragged himself to a place where he knew that search must eventually find his body, and with it the records of success at the cost of three lives. Lieut. A. Trolle told that story in the pages of this *Magazine* (vol. xxv. pp. 57-70). But the diaries and maps of Erichsen and Hagen were still to be found before the story of their work could be completed. It was with this object that Captain Ejnar Mikkelsen organised the Danish expedition in the *Alabama* in 1909. The expedition was not one of exploration, nor did it carry a scientific staff. It was a search expedition which could only hope incidentally to add to our knowledge, so well had the work of Mylius Erichsen and his brave companions been done. But that in no wise detracts from, in fact it rather adds to, the marvellous record of this latest Danish expedition.

The sloop *Alabama*, of forty tons, with seven men on board, left Copenhagen in June 1909. She called at the Danish settlement of Angmasalik, on the east coast of Greenland, to collect dogs in place of the unsatisfactory ones which had been shipped to the Faroes to meet her. The *Alabama* then made northwards, and after many difficulties with ice on this unapproachable coast, anchored off Shannon Island in the end of August. This proved to be her last voyage. A month later Mikkelsen and five companions set out northward on a sledge journey to Lambert's Land to try and find the bodies of the three lost explorers. They travelled over the sea ice and met with many difficulties, but ultimately found Brönlund's body. Those of the others they failed to find. This trip revealed nothing new, for Captain Koch of the *Danmark* had found in 1908 all that there was to discover. Within a week of Christmas the sledge party returned to winter quarters in the *Alabama*. On March 4, 1909, the great journey was begun. The party consisted only of Mikkelsen and Iver Iversen, and their destination was Danmark Fjord, or beyond if necessary, with the object of finding all the missing records. They started over the sea ice to Dove Bay, where the inland ice was ascended and followed in continual gales and blizzards to the head of Danmark Fjord. There they followed the north coast of the fjord and came on Erichsen's camping-places, and recovered two important dispatches of his. The first cast considerable doubt on whether the three

[1] *Lost in the Arctic: Being the Story of the "Alabama" Expedition*, 1909-12. By Ejnar Mikkelsen. Illustrations and a map. London: William Heinemann, 1913. Price 18s. net.

men in their starvation march had really crossed the inland ice to Lambert's Land, or whether they had not followed the coast in their retreat to the ship at Cape Bismarck. The second contained word of an important discovery, namely, that Peary Channel between Peary Land and the mainland of Greenland does not exist. Navy Cliff and Heilprin Land are connected by land. Erichsen had reached Peary's Cape Glacier, and made this discovery, and rightly changed the name of Independence Bay to Independence Fjord. This news upset Mikkelsen's plan of penetrating Peary Channel to Robeson Channel (here wrongly called Robertson Channel), and so getting in touch with the West Greenland settlements and home by a less difficult route than he had come. He had not provisions for the journey round Peary Land, and could not risk crossing the high inland ice. Finally, at Cape Rigsdagen, at the south of the entrance to Independence Fjord, he and Iversen decided, on May 27, to turn back, and none too soon if they were to reach the ship before winter set in.

The return journey was to be round the coast in order to examine depôts in the hope of finding records. And now began the most arduous sledge journey on record, if we take into account the conditions and season of travelling and the scanty rations on which most of it was performed. It was a longer route than the outward journey, and the difficulties were immeasurably greater. The only compensation was the depôts of food along the route left by the *Danmark* expedition, and, as luck would have it, most of these had been used by that expedition or contained food that was scarcely fit to eat after several years' exposure. The two men had a journey of at least 750 miles before them, and they started it with twelve days' provision for the dogs and forty-five for themselves, and two and a half months in which to do it. Mikkelsen's orders to the *Alabama* were that she was not to leave Greenland before August 15 unless absolutely necessary. If they did not return by then, the ship was to leave for Denmark, and they would trust to being picked up by a sealer, several of which touch at Shannon Island every year. There was scarcely a chance that they could return in time, but they meant to try. Almost at the very start their misfortunes began, and then followed one after another in endless repetition such a series of difficulties and misfortunes as surely no polar traveller has ever lived to record. Rough and treacherous ice, slush, open water, steep ravines, glacier rivers, rocky coasts, crevasses, bad weather, want of food, absence of game, empty depôts, scurvy, death of dogs—the list is endless—no conceivable misfortune but befell these two tireless men, yet still they held on and never gave up hope. It is a story of marvellous tenacity of purpose and great endurance, and the crowning wonder of all is that they lived to tell the tale. A hundred times they were near death, and escaped with a laugh. Even forty hours on a cluster of rocks in the middle of Skaerfjord in a raging storm did not dismay them; nor did a hundred miles on practically nothing to eat. At length they had to abandon their tent and sleeping sacks and sledge, and struggle onward eighty miles to Danmark Haven, where they knew they would find food, and then at last they reached the *Alabama* on November 25,

Sledging in an ice valley.

Difficult going.

only to find that she had foundered and been abandoned. The two men were doomed to winter alone, after their nine months' isolation on this terrible journey. In the spring they went north again to Skaerfjord to pick up their abandoned note-books. The greater part of Mikkelsen's diary had been eaten by a bear, but fortunately Iversen's was complete. On their return to Shannon Island they passed the summer anxiously awaiting a ship, but in the end had to face a third winter, which they decided to spend at Bass Rock, some forty-five miles south of their previous quarters. Presumably this was decided on as putting them more on the track of sealers the following summer. Their disappointment must have been intense when they found there a note to say that on July 25 a sealer had called and sailed away again. That was when they were encamped on the north-east end of Shannon Island, only sixteen miles away, looking expectantly for a sail. But they made the best of things, and settled down for the winter, trying to keep themselves occupied to avoid their none too cheerful thoughts. There were stores enough, and their ammunition lasted out, so that fresh meat was procurable. But the inaction and anxiety was telling on them, and they decided that in spring they would make a desperate attempt to sledge southward to Angmasalik, and so make good their escape. When the time came the attempt was made, but had to be abandoned, for their remaining means of transport were quite inadequate for this journey of over eight hundred miles. Then at length, late in the summer, a sealer calls and brings deliverance to these hard-tried explorers. Ten days later, in August 1912, they are in Aalesund on their way home.

Mikkelsen's companions had left Shannon Island in a sealer shortly after the foundering of the *Alabama*, salving everything possible for the use of him and Iversen before they left.

The greater part of the volume is occupied with the record of the long sledge journey, but despite all the detail introduced, the story never flags for an moment. One reads with eager interest from beginning to end, divided between admiration for the men and horror at their seemingly hopeless position.

The two final winters and intervening summer are briefly sketched, but after all there can have been little to record. It must have been a time of almost uniform monotony.

A number of illustrations (some of which we are enabled to reproduce here) and a good map add to the value of a volume which we are glad to see, and which adds another glowing page to the record of Denmark's work in Greenland. The book is all the more important since it is the first in English to record the striking explorations of Mylius Erichsen and his comrades. Mikkelsen's and Iversen's three years' work was a worthy tribute to the memory of those men.

R. N. RUDMOSE BROWN.

THE THULE OF TACITUS.

By E. M. Horsburgh.

What place was meant by the Thule of Tacitus is a question which has often been asked and to which many different replies have been given. The Shetlands, the Faroes, Norway, Iceland, and even Greenland have all been suggested, but no very conclusive reasons have been advanced for the selection of any one of these places.

In the world-map of Ptolemy, about 160 A. D., Scotland is represented as pointing in a due easterly direction, almost at right angles to England, which is shown fairly correctly. One theory which has been advanced in explanation of this is that the Roman troops of Agricola marched possibly to the north of Scotland, and that the fleet, battling against the prevailing westerly winds, took so long to make the passage and rejoin the army, that, on comparing the itinerary of the troops with the log of the fleet, geographers arrived at the conclusion that Caledonia must taper far to the east, forming a kind of peninsula.

From the writings of Tacitus, however, it seems that the main army did not march to the extreme north of Scotland. During the great campaign which culminated in the battle of Mons Grampius, the fleet and army operated together, the army marching up the east coast, while the fleet, which would be used to convey stores and munitions of war, coasted along in touch with the troops. This, it is interesting to remark, was the method by which all the great subsequent invasions of Scotland were carried out, from the days of Edward I. to the time when David Leslie, though defeated at Dunbar, had yet out-manœuvred Cromwell.

After the battle of the Grampians, Agricola led his army into the territory of the Boresti, and apparently fitted out a smaller naval expedition to explore the north of the island. Though the context is obscure, and though minute details of the geography of a remote and barbarous spot would not appeal to so vivid a word-painter as Tacitus, yet he makes it clear that this expedition at all events reached the extreme north of Scotland. Agricola was probably not with this division, but he seems to have given the command to a high official, the admiral of the fleet *(praefectus classis)*. When the Roman ships had rounded Duncansby Head and viewed the coastline stretching west towards Dunnet Head and Cape Wrath, it was evident that the extreme northern end of Caledonia had been reached. The fleet and army would probably form a camp at Scrabster Roads—the most suitable anchorage in these parts—preparatory to crossing the Pentland Firth to explore and annex the Orkney Islands.

The wild speculations which insist that the fleet made a Columbus-like voyage from this point to Norway, Iceland or Greenland seem quite untenable, contradictory to the words of Tacitus, and a violation of common sense. Is it to be expected that a fleet which probably carried the food supplies and general stores of the army, and which would be necessary for the safety of the expedition, would have been allowed to stray aimlessly all over the waters of the North Sea and Arctic Ocean? Winter with the dreaded storms of the north was also approaching, and unnecessary risks could not be taken. If the fleet had actually cruised to the north of the Orkneys, could we possibly have had the passage which explains that the great Northern Ocean is so sluggish that its waves are not raised high by storms as in other seas? Assuredly the Roman fleet had not been so far north as the Roost of Sumburgh.

Where then did it go? Their orders as we read were distinct—to explore the Orcades, and to go no further (*quia hactenus jussum*). The giver of these orders could not have been Agricola, if he were at this time leading his victorious troops towards their winter quarters, perhaps at Rosyth, unless indeed he were actually on this enterprise, a suggestion which seems extremely unlikely. They were probably issued to a small exploring division by the commander of the expedition from his camp near Scrabster.

So far then the ships had only coasted along the eastern shores of Scotland, and that too in the fine weather of summer. They had been under the shelter of the land, at least during the prevailing westerly winds. But on a first visit to the Orcadian waters they would now encounter many dangers. They would have their first experience of ten-knots tides, and of the "Swelkie," where later King Haco lost a "long keel," and would find themselves among some of the most dangerous skerries and tide races of the world.

Once among the islands they might be swept out past the sunken rocks of Hoy Sound, and back through the boiling roost of Eynhallow, adventuring through many perils along these uncharted shores as far as North Ronaldshay. By that time their orders were carried out. Not only was it contrary to instructions to go further, but what could so practical a people possibly gain by faring forth on the open ocean with the approaching danger of the winter storms? (*Et hiems appetebat.*) The safety of the whole expedition was involved in the peril of the fleet, and there seems little doubt that the Romans would not go beyond North Ronaldshay.

From this point then as they gazed over the great turbulent northern ocean they would see a horizon unbroken by any line of coast, and in the far distance one speck of land, a rugged solitary giant pillar of rock—not Sumburgh Head in Shetland—but what we know as Fair Island. This outlier, the halfway post on the long journey from Orkney to Shetland must have been the northern limit of the Roman world of Tacitus. *Dispecta est et Thule!*

SCOTTISH GEOGRAPHY IN SCHOOL TEXT-BOOKS.

By John Murray, M.A.

I have before me seven comparatively recent text-books of geography,[1] some of general application, others dealing with restricted areas, but all intended for school use. In a perusal of the volumes I have paid special attention to the sections on Scotland, and have marked what seem to me inaccuracies, mis-statements, and terms or passages which might lead to misapprehension. The grammar in several of these text-books is rather slipshod, but I do not wish to weary my reader with lists of grammatical mistakes. I desire rather to appeal to teachers of geography.

Of the authors of the books in question, two are termed on the title-page English masters, one is called a Geography master, two are assistant-masters in English schools, and two (apart from teaching) are recognised experts in geography. So far as I know, six of the seven are either English or in English schools. Of the publishing firms, two or three have no Scottish house.

Most of the writers give us a few items from Scottish history. These items are intended apparently for our Southern neighbours. The following sentence is refreshing:—"Scotland, unable to cope single-handed with her aggressive and wealthy neighbour, was compelled to seek the aid of France, and the treaty of alliance with France made by John Balliol in 1295 lasted for 300 years" (3). No doubt Scottish historians will be glad to know what tangible assistance the French gave us up to and at the date 1314! Teachers in the Highlands are told that "Culloden, to the north-east of Inverness, is where the Young Pretender *turned sullenly* at bay in 1746" (3). The phrase italicised is specially good. Another note says—"Culloden Moor, where the Young Pretender was defeated in *1745*" (4). Then, for the further benefit of the much-maligned Celts, let me quote the following:—"Whisky distilling is a *typical* Highland industry"; "we find some (Highlanders) speaking their native Gaelic language and a few wearing a *peculiar dress*"; in the Highlands "fishing is the mainstay of the people, and fish and oatmeal the chief foods" (by the way, what is the verb to the nouns "fish" and "oatmeal"?); "deer and grouse on the moors, . . . and salmon in the streams fare better than men."

To come to geography proper. The geography teacher who has tried to eradicate certain time-honoured myths from his pupils' minds will read with astonishment that "the Atlantic Ocean has its waters heated by the Gulf Stream, which sweeps across it in a north-easterly

[1] (1) *The Clarendon Geography*, vol. i. (Herbertson).
(2) *A Text Book of Geography* (Fry).
(3) *Europe* (Snape).
(4) *A Regional Geography of the British Isles* (Parkinson).
(5) *A Systematic Geography of the British Isles* (Webb).
(6) *A Geography of the World* (Evans).
(7) *A Junior Geography of Scotland* (Frew).

direction from the Gulf of Mexico, and strikes the north of Scotland" (5). Perthshire teachers will rejoice in that they are now made aware that the Tay "passes the *towns* of *Dunkeld*, *Scone*, and Perth" (5). To them also the following sentences may be of some use :—"Among other lakes noted for their beauty are Loch Awe in Argyllshire, and Lochs Tay, Rannoch, Ericht, and Earn, in Perthshire. The last two are connected with Loch Rannoch, as their surplus waters fall into it" (5). The latter sentence is delightful. Apparently the geography of Perthshire is confusing, for another author informs us that "Callander (is) on the *Earn*, and Crieff on the *Teith*" (1).

Then, apart from orographical features, Harlaw is proof that the Aberdonians would resent their city being called "the largest town in the *Highlands*" (3). The mention of Aberdeen may serve to introduce this statement—"Peterhead has long been famous as a port for the whale fishery" (5). The geography teacher might forgive this if the author changed "has been" to "was," but Peterhead's herring-fishery remains unmentioned by the same writer, who also astounds us when he says that "Wick is the *chief seat* of the herring-fishery, and Thurso and Stornoway (*Kincardine*) have a similar trade."

Then from what I know of the county, I am disinclined to believe that granite is quarried in Dumfriesshire (6). Are not the granite quarries in Kirkcudbrightshire, as see p. 79 of (6)? Again, is Tinto Hill ever called Mount Tinto (7), is Ben Cleuch (or Bencleuch) written Ben Cleugh (7), is Kinnaird Head styled Kinnard's Head (7), is Duncansby Head usually termed Duncansbay Head (3) and (6)?

The Scottish railways have come in for a fair share of bad treatment. Quite a common error in the Railway Maps of the text-books is to mark the portion of the Caledonian Railway from a few miles north of Montrose to Aberdeen as North British, *e.g.* (5). The same mistake occurs in a recent atlas produced by one of the foremost firms of cartographers in the United Kingdom. The North British Railway has running powers only over the portion mentioned. In the Railway Map of one text-book (2), the Great North of Scotland Railway is entirely omitted, and in that of a second (4) it is shown as passing from Aberdeen round by the coast through Peterhead, Fraserburgh, Banff, and Elgin. Curiously enough, in direct opposition to this second map, which shows correctly the course of the Highland Railway, the text states that the Great North of Scotland Railway's route is "along the coast to Inverness, Thurso, and Wick, running a branch across country to Strome Ferry, near the Island of Skye." The same map (4) shows no direct and main Caledonian Railway route from Perth *viâ* Forfar and Stonehaven to Aberdeen, and indicates a Joint Caledonian and North British Railway line from Aberdeen to Dundee! Further, is Portpatrick the port for Ireland as indicated, and is Strome Ferry the terminus of the "Dingwall" branch of the Highland Railway?

In a "Commercial and Industrial" Map contained in one of the text-books (7), Bervie is marked—"Woollens, Chemicals." I am sure that the inhabitants of the small fishing burgh will rejoice at the importance of their place, all the more since Montrose has attached to it only

"Fishing." Then Banchory is inserted, while Stonehaven is omitted. Dyeing is not given as a Perth industry; and Dumfries gets the credit of ironworks, while the Sanquhar coalfield is not marked.

I append the table given under. It is based on the six books which deal "regionally" with England and Wales, Scotland, and Ireland as separate sections, and may prove both interesting and instructive. The pages include maps, diagrams, and figures.

	England and Wales.	Scotland.	Ireland.
(1)	51	16	23
(2)	32	13	9
(3)	26	9	6
(4)	103	39	28
(5)	43	16	16
(6)	39	12	10

It is outwith the scope and purpose of this brief article to attempt to lay down or indicate at any length the lines on which Scotland ought to be treated in school text-books, but I may be allowed, on the plea that mere negative criticism is often of little avail, to make a few suggestions. In the first place, teachers of geography desire to see text-books well, or at least clearly, written and free from grammatical mistakes. From one of the seven I could make a list of a score of errors in grammar, most of them due undoubtedly to hasty revision on the part of the author. It is aggravating to a teacher to see an otherwise good book marred by small mistakes, and it is harmful for a pupil to use such a book. In the second place, while fulness of detail cannot and need not be insisted upon, accuracy of detail is altogether desirable. Many of the inaccuracies pointed out above could have been corrected from reliable maps and reference works. Of the latter, the finest are Geikie's *Scenery of Scotland* and Mackinder's *Britain and the British Seas*. In the third place, the treatment of Scotland ought to follow the three well-recognised geological and orographical divisions of Southern Uplands, Central Lowlands or Midlands, and Northern Highlands. In each division, the differences—climatic and industrial—between east and west should be carefully noted. In a consideration of the railways, the main point is their relation to and dependence upon physical features. The routes linking Edinburgh and Glasgow with England are of especial interest. Lastly, in the case of the towns, the reasons for their locations must be dwelt upon. Obvious and convenient examples are Edinburgh, Glasgow, Perth, Stirling, Aberdeen, Dumfries, and Inverness.

Erratum.—Geographical Description of Fife, Kinross and Clackmannan:—On page 145 of this article, line 19 from top, the word *eastwards* should read *westwards*, and in line 25 *westwards* should read *eastwards*.

GEOGRAPHICAL NOTES.

EUROPE.

Geographical Causes of Italian Emigration.—Dr. Alfred Rühl contributes to the *Zeitschrift der Gesellschaft f. Erdkunde z. Berlin* (9. 1912) an article with this title. It will be remembered that we published here (vol. xxvi., p. 337) a paper upon Italian Emigration by our corresponding member, Mr. Victor Dingelstedt. This fact renders it unnecessary to give the figures contained in Dr. Rühl's paper, and we shall confine ourselves to his discussion of the causes of the emigration. He begins by pointing out that the Italian annexation of Tripoli and the war there must be considered in relation to the process of emigration which is depopulating parts of Italy. When the Government sees the constant stream of emigrants quitting its shores it is natural that it should seek to acquire a Mediterranean colony to which the stream may perchance be directed, with the result that the emigrants will not be lost to the motherland. In earlier days Italian politicians were wont to console themselves with the thought that, heavy as was the emigration rate, the birth-rate more than kept pace with it. As Dr. Rühl points out, however, this overlooks the fact that the emigration is not equally distributed over the country, so that in certain regions, especially in the south, the population is steadily diminishing, with economic results of much importance. To say that the cause of the emigration is over-population seems inexact, in view of the fact that it is the thinly peopled regions which furnish the most emigrants, and the country districts rather than the towns. Nor is it accurate to say that poverty is the prime cause, for from the poorest regions, *e.g.* Sardinia, there is little emigration, and in Sicily it is noted that the communes with the highest rate are those where wages are neither low nor high. While among the contributing causes are certainly the large estates, the high taxes, the want of technical education, and so on, Dr. Rühl emphasises other causes of a purely geographical nature.

Among these are to be reckoned the large percentage of mountainous land in Italy, the elevation being often such that even with elaborate terracing the more delicate and profitable crops cannot be cultivated. It is noticeable that the mountainous regions have a high percentage of emigrants. Further, though the plains are often fertile, yet, with the exception of the great plain of the Po, they are liable to malaria. This is especially true in the south, where, for example in Cotrone (Calabria), one-third of the inhabitants are known to be infected. The malarious regions are again regions with a high percentage of emigrants.

As is well known, the Apennines are young mountains, and among the soft rocks which occur in them clays are frequent (cf. vol. xxvii. p. 541). These predominate in the North Apennines, and with irrigation form a fertile soil. Further south, limestones occur, forming in the Abruzzi, Molise and Basilicata highly infertile areas. Here the emigration rate, low to the north, again becomes high. To the infertility of the soil in these regions is added, as a factor in promoting depopulation, the scarcity of water. Apulia especially suffers from this, and water is actually carried there in special trains from Naples to be sold. A scheme for supplying water in a great aqueduct has been proposed, and work is already begun, but it will be long before it is completed. Apart from the presence of soft rocks in the Apennines, with the consequent frequent slips and falls of earth, their geological youth brings great difficulties in the way of intercommunication, owing to the steep gradients, narrow valleys, and so forth. The result is that railways can only be constructed at great expense, and even road construction is prohibited by the expense in the poorer regions, which must thus continue poor. The author instances Bova in Calabria, which has 4600 inhabitants, as a place to which access is only obtainable first by a torrent bed, and afterwards by a dangerous mule path. In Sicily there are many communes which are without driving roads, and some have no proper access at all. Another disadvantage of the recent uplift is, of course, the constant liability to earthquake shock. The danger of volcanic eruption, on the other hand, is probably counterbalanced by the fertility of the volcanic soil, as indicated by the dense population round Naples.

The climate also is disadvantageous in that rain is deficient when it is most wanted, and comes in devastating downpours at other seasons. Devastation by swollen torrents often reaches serious dimensions, and the danger is increased by the deforestation which goes on ceaselessly, though its dangers are now well known. There is no doubt that the superabundant rainfall could be stored for irrigation, and also used as a means for generating electric power. It is in this white coal that the author sees the greatest hope for the future. It is being now utilised to some extent round Naples, chiefly with foreign capital, and also in Sicily. Italy must, however, he believes, remain chiefly an agricultural nation, and it is in the improvement of agriculture that her greatest hope lies.

Spanish Emigration.—To the above account we may add a brief note on Spanish emigration, considered by M. Albert Girard in the *Annales de Géographie* for November last. Though far from reaching the dimensions attained in Italy, emigration from Spain, after a temporary set-back due to the Spanish-American war, is steadily rising, and now probably amounts to about 200,000 persons per annum, out of a population of less than 20,000,000, or one per cent. in a thinly peopled country. The loss to the Mother Country is here chiefly permanent, for the Spaniard expatriates himself more willingly than the Italian. As in Italy the emigrants are chiefly agriculturists, and the agricultural population of Spain is but 29 per cent. of the total. Curiously enough, while in Andalusia, as in Italy generally, the size of the estates is given as a cause of emigration, in Galicia, Asturias, and a part of Leon, it is the excessive subdivision of the land which is said to be the cause. In this case however, the land is not freehold, and will not support the dues to the overlord in addition to the heavy taxes. The unequal distribution of the taxes, which press heavily upon the rural population, and the uncertainty of the rains are other predisposing causes.

The chief regions to which the emigrants direct themselves are the Antilles, especially Cuba, South America, especially Brazil and the Argentine, and (now to a diminishing extent) Algiers.

The two articles summarised show admirably the incapacity of the peasants of Southern Europe to compete with the farmers of the new lands across the sea, at least so long as the standard of expenditure of their various Governments is set by that of the industrial nations to the north.

ASIA.

Geography of North-West Mongolia.—Dr. J. G. Granö, of Helsingfors, contributes to a recent issue of the *Zeitschrift für Erdkunde zu Berlin* what he describes as a geographical sketch of North-west Mongolia, based upon his own travels in the region. From the paper, which is a somewhat lengthy one, we pick out a few facts of interest. One striking feature of the region is the abrupt way in which the mountains rise from the plains, the approximately level surface of the plains abutting suddenly upon the mountain slopes, the absence or scarcity of vegetation rendering the contrast all the sharper. The plains can be divided into two sets, the high-lying peneplains, occurring among the mountains and having the characters sometimes of desolate steppes and at others of alpine meadows, and the low-lying plains. The latter are sometimes steppes or deserts alternating with meadow land in the damp regions, and may be subdivided into two, those lying along the banks of the rivers, and the basins of internal drainage in whose deepest part lies a salt lake. The climate varies, of course, greatly with elevation, but generally the winters are cold, clear and still, the summer windy and rainy. The northern mountain region and the Dzungarian slopes of the Altai are moist, while the level regions are dry, though the author doubts if the high plains are so dry as has been believed. Autumn is everywhere the pleasantest time, for though the nights are cold the days are sunny and dry, and the nocturnal cold causes the insect pests to disappear.

The climatic variations are reflected in the vegetation. In the Syansk, the Tannu-ola, the northern Khangai, and on the western slopes of the Altai, there are extensive woods of larch, pine, Siberian pine, fir, aspens, and birches. Elsewhere the woods only occur on the highest slopes, especially the northern ones, and are there thin. Above the tree limit are Alpine meadows rich in flowers; such plants as globe flower, monkshood, larkspur, poppy, and so forth, occur below, with alpine forms like dryas, gentian, etc., higher up. Below the woods also meadows occur with many gaily coloured members of the ranunculus family. Further down come the steppes, which in the dry regions also ascend high up the slopes. These steppes are scantily covered with grass, sometimes mingled with edelweiss to such an extent that they may be called edelweiss steppes. In the river valleys there are sometimes light woods of poplar and birch, or these trees may be replaced by willow and sea-buckthorn. In the dry regions bushes of the Siberian pea-tree (*Caragana*) predominate, and there are few regions wholly devoid of vegetation.

To this general description we may add a brief note on two contrasted regions traversed by the author. In the west Syansk mountains he found very dense wood, of great extension, which was predominantly coniferous, though aspens and poplars also occurred. So dense is the wood that it is difficult to find grazing for the transport animals and travelling is difficult. Rain and mist are frequent, the paths are often nearly impassable, and insect pests are numerous. Further, the rivers have to be crossed at fords, and they are often so swollen with rain that it may be necessary to wait days before the crossing can be made. The original occupants of this region, the Soiots, who are nomad hunters, are now rarely met with, though game is abundant. On the Russian side of the mountains there is a settled population only where gold washing is carried on, but parties of fishers may sometimes be met with on their way to the river Bei-kem, which is rich in fish. Quite different are the conditions in the Khangai region. Here woods are absent save on the highest summits, pasture is abundant, and the slopes are so gentle and the passes so easy, that it is possible to take carts from Uliassutai to Urga over the mountains. On account of the abundant pasture the region is well peopled, there being great groups of yurts, with here and there a pagoda or a convent and Russian or Chinese factories. Caravans are constantly encountered on the march, and consist either of camels or oxen. The author travelled with an ox caravan. The animals were harnessed to carts with creaking wooden wheels, the carts being loaded with tea, wool, and Chinese artistic goods, and drawn slowly over hill and dale. The escort in such cases rides on horseback, and in front of the caravan marches a flock of sheep which are slaughtered during the journey to supply food for the party. The dung of the animals is used as fuel, no wood being obtainable. The caravan halts for several hours in the middle of the day as the oxen are very sensitive to heat, but marching is carried on early and late. The escort, however, is far from remaining with the caravan all the time. Often the whole is left to the guardianship of a single individual, the other members of the party spending their time in visiting neighbouring yurts, where they are welcomed as news-bringers. Besides the caravans, many pilgrims, rosary in hand, are to be found upon the road, and lamas in their yellow mantles. Generally, it may be said that this region is prosperous and advanced, and the ruins show that through historic time it has been a centre of Mongolian culture.

Polar.

Mr. Stefansson's Expedition to the Arctic.—It is announced that the Canadian Government is to supply funds for Mr. Stefansson's projected expedition to Arctic North America. Mr. Stefansson, who will take with him some Canadian students, and whose expedition is to be under the Canadian Geological Survey, is to leave Victoria, British Columbia, in June, and will proceed through Bering Strait to the hitherto unexplored region in the far north of Western Canada. The party will devote the four seasons during which the expedition is to last to the exploration of the islands and mainland of the Arctic coast, making their base either on the new land which it is hoped to reach, or on Land's End in Prince Patrick's Island. From this base the ship is expected to return to civilisation every year. Mr. Stefansson has done exploring work in this region for several years, in company with Mr. Leffingwell. In his last season's work he discovered, in Victoria Island, an Eskimo tribe of very light colour, and with marked European features. Among their weapons are copper knives with horn handles, such as those used by the old Norsemen, and Mr. Stefansson has suggested that the tribe may be descended from the old Norse settlers in Greenland. In his new expedition he will devote attention to the question of the origin of this interesting people. The party will include ten men of science, and the whaler *Kamluk* has been selected as the ship of the expedition.

The Australian Antarctic Expedition.—Not long after the news of the Scott disaster, word reached Melbourne by wireless from Macquarie Island of the loss of two members of the Mawson Antarctic expedition. It appears that in December, Dr. Mawson, Lieutenant Ninnis, and Dr. Mertz were exploring new land 300 miles south-east of the winter quarters at Commonwealth Bay, in the west of Adélie Land. During the course of the journey Lieutenant Ninnis, with a dog team, and almost all the food, disappeared in an unfathomable crevasse. Dr. Mawson and Dr. Mertz had thus to start back to the hut with insufficient food and six starving dogs. They experienced bad weather and were obliged to live chiefly upon the dogs. On January 17, Dr. Mertz succumbed to the hardships of the journey, leaving Dr. Mawson to struggle on alone to the hut, which he reached on February 8, after a very difficult journey. On his arrival he found that the *Aurora* had left a few hours before, having been unable to wait any longer, but had left behind a search party of six men. On the return journey the *Aurora* took off Mr. Wild's party of eight. Mr. Wild reports the survey of land trending east and west, which has been named King George v. Land, and stretches from Kaiser Wilhelm Land to long. 101° 30′ and lat. 67° 30′.

General.

The Meeting of the British Association.—We regret to announce that the President-elect of the British Association, Sir William H. White, F.R S., died suddenly on February 27, in the middle of his activities in connection with the Birmingham meeting, which is to last from September 10 to 17. Sir Oliver Lodge is to take his place as President. The Sectional Presidents have now been appointed.

Professor H. N. Dickson is to preside over E (Geography), and of the related sections Professor Garwood is to preside over C (Geology), and Sir Richard Temple over H (Anthropology).

EDUCATIONAL.

The Cinematograph in the Teaching of Geography.—A proposal has been adopted by the Education Committee of the London County Council, advocating the expenditure of a sum of money to permit of the employment of the cinematograph in the teaching of elementary school children in London. In connection with this proposal, so far as it affects geography, Mr. H. J. Mackinder contributes the following letter to the *Times*:—

"As chairman of the Council of the Geographical Association, a body which consists of some nine hundred expert teachers of geography in all parts of the country, I desire by your courtesy to place on record the views of my council in regard to the proposed adoption of the cinematograph for educational purposes, to which attention is drawn in your issue of to-day by the debate in the Education Committee of the London County Council. At our last meeting it was unanimously agreed that we should protest against any large expenditure of public money for this purpose, unless after considerable experiment and adequate discussion by those who are practically experienced in education. We realise that there are some things which only the cinematograph can do, but in our opinion there is grave danger of its abuse, especially in connection with the teaching of geography. It is the constant aim of modern teachers of that subject to compel their pupils to visualise accurately and at will. Obviously effort must be called for on the part of the pupils, and we fear that the use of the cinematograph, except on rare occasions, will tend to decrease this effort, and thus to weaken the imaginative power instead of strengthening it. My council would not deprecate such a small experiment as is proposed by the Education Committee of the London County Council, but they feel that a warning is at the same time necessary, because large commercial interests are now involved in the manufacture of cinematograph films and apparatus, and we understand that an exhibition of such apparatus is shortly to be held, one of the sections of which will be educational. It is, of course, good business for the cinematograph companies to create an atmosphere among the public which shall lead to the expenditure of public money upon their wares. My association asks, however, for caution in such expenditure, not merely from motives of economy, but, as we believe, in the interests of the children and of the nation."

Summer School of Geography at Aberystwyth.—We have received a programme of a summer school of geography which it is proposed to hold at University College, Aberystwyth, during the present year, from July 28 to August 16. Five theoretical and five practical courses are offered, the latter including simple survey work.

The classes are to be conducted by the following gentlemen:—Professor H. J. Fleure (Human Geography, and Wales and its Borders); Professor Lewis (Civic History); Mr. W. E. Whitehouse (Climatology and Trade Routes); Mr. E. S. Price (Land Forms and Natural Regions). The fee for the course is two pounds, and further information can be obtained by application to the University College of Wales, Aberystwyth.

NEW BOOKS.

EUROPE.

Biarritz and the Basque Country (Joanne Guide-Books). 4 Plans, 4 Maps, 40 Illustrations. Hachette and Co., 1912. *Price* 3*s*.

This is an admirable little guide-book to a region that is drawing every year an increasing number of tourists and visitors. In brief compass, compacted in small type, it provides almost everything one needs to know. To the sojourner at Biarritz, a town that to its 14,000 inhabitants adds every year a clientèle of about 40,000, to enjoy its breezes and its pleasures, and the interesting and beautiful region of the Basque country, this little work is invaluable. It is of handy size and moderate price.

Warfare in England. By HILAIRE BELLOC. (Home University Library.) London: Williams and Norgate, 1912. *Price* 1*s*.

Although this is only a shilling book, it deserves attention on the score both of its subject and its distinguished author. Mr. Belloc sets forth, to begin with, the geographical features of England which have affected the strategy of campaigns. These are, in particular, its penetrating rivers with their estuaries, though the Severn for certain reasons is not a river of importance, the great ridge of the Pennines with the bar to northward advance made by the Mersey and the Aire on either side, and the "great obstacle, depôt and nodal point of London." Most of the campaigns in England of which we have detailed record are of the nature of civil wars, and local and accidental considerations play a prominent part in them. But in the campaign of 1066 and the following years the author is able to illustrate clearly the influence of the geographical factors he has premised.

At the end of the book is a chapter on "The Scotch Wars," in which his erudition has somewhat failed him. He lays it down that "An English army marches north by the easiest road of invasion, that along the eastern coast: it makes for Stirling, the nodal point of the south: or a Scotch army marches south by the same road." "To this use of the eastern route there is no true exception." The only exceptions admitted by Mr. Belloc are the advances into England of 1651, 1715, and 1745. But these, as occurring after the Union of the Crowns, are treated as mere rebellions.

This comprehensive statement requires large qualifications. For instance, there were at least ten invasions of Scotland by Edward I., and in only a minority of them was the eastern route followed. One may instance, as specially important, the invasion of 1301, when a division of the English Army under the Prince of Wales entered by Dumfries, while another under the King marched by way of Peebles. His successor, who used the coast route in the Bannockburn campaign for special reasons, made his last big effort, in 1322, by way of Lauderdale. On the Scots side of the invasions of 1346 (Neville's Cross), 1513 (Flodden), and the contemplated invasion of 1542 (Solway Moss), none took the eastern route. It is perhaps worth while to mention two slips. Flodden was not fought by James VI., and Prestonpans is more than two hours' march from the capital. An index, as in other volumes of the series, would have been useful.

We wonder if Mr. Belloc is aware that a high-lying pass in the Roxburghshire hills still recalls the advance of a portion of Prince Charlie's forces in the name "The Not of the Gate" (Note of the gait or going). Enumerators are said to have been stationed there. Readers who wish a more extended view of the same subject are referred to an article in this magazine, vol. xvi. p. 138.

Arabic Spain: Sidelights on her History and Art. By BERNHARD and ELLEN M. WHISHAW. London: Smith, Elder and Co., 1912. *Price* 10*s*. 6*d*. *net*.

The period of Arab ascendency in Sicily has lately received attention, and this careful investigation of the contemporary field in Spain will stimulate other workers to explore this almost virgin soil. That intellectual but mysterious people, the Yemenite Arabs, brought learning and civilisation to a high eminence even before the ninth century, and the records here revealed are full of interest and romance. Their artistic remains, as in architecture and embroidery, receive full treatment, and the Greek, Roman, Coptic, and Persian sources in which they found their models carry the origins of their art into remote antiquity.

The authors are inclined to apologise for their ignorance of Arabic, and the handicap which this has imposed. But the defect seems to have been largely made good by the assistance they have obtained. Indeed they have been singularly successful in securing information not readily accessible. The eight years they have devoted to the study of the subject have been well spent. The book appeals to the eye from the clearness of its type and illustrations.

Voyage en Dalmatie, Bosnie-Herzégovine et Monténégro (*L'Europe en Automobile*). Par PIERRE MARGE. Paris: Plon-Nourrit et Cie, 1912. *Price* 3 *fr*. 50.

We have already reviewed here previous books by M. Marge, describing motor tours in other parts of Europe. Like his other works, this is much more than the ordinary motorist's notebook, for the author has not only made several visits to the regions described, but has got up the literature carefully, and has produced a book which will not only be useful to those who contemplate following his example, but also to the geographer proper. He refrains carefully from those elaborate descriptions of the petty incidents of travel which sometimes bulk so largely in accounts of journeys, and we are also spared would-be amusing accounts of his travelling companions. In short, his book is a serious piece of work, giving careful accounts of regions to which attention is being directed at the present time, and in regard to which accurate information is not too plentiful. Among the sections which we have read with particular interest are those discussing the *bora* at Zara, the author being himself witness of a violent tempest of this wind; the account of the Roman remains at Spalato, and of the curious contrasts of scenery and vegetation presented by the karst country. Readers of the book will probably feel a strong desire to follow the author's example, and tour through this interesting region, but he notes that it is for the most part accessible only to motor traffic, for railways are few, and the steep roads and the length of the stages make it difficult to use horse-drawn vehicles.

The book is well illustrated, but there is no map. It possesses—still a rare merit among French books—an index in addition to an analytical table of contents. Details as to the roads are relegated to footnotes, and the text is light and easy reading—altogether we have greatly enjoyed the book.

Linlithgowshire (Cambridge County Geographies). By T. S. Muir, M.A., F.R.S.G.S. Cambridge: University Press, 1912. *Price* 1*s*. 6*d*.

The county of West Lothian or Linlithgow is not more than 120 square miles in area, and there are only two smaller counties in Scotland—Kinross and Clackmannan. But while it is small superficially, it is large in other ways, and although in point of area its place is only thirty-first among the thirty-three counties of Scotland, in population it ranks twelfth, and in density of population third. At last census the population was 79,486, and it is interesting to note that while many other rural communities have diminished, West Lothian has continued to grow steadily during the last twenty years at the rate of more than 1200 persons per annum. This satisfactory state of matters is due mainly to the growth of the mineral oil industry and to coal mining near Bo'ness and Bathgate. In this connection Mr. Muir makes one or two misleading statements which should certainly be corrected in future editions of his most useful and on the whole very excellent book. At page 68 he says, "It is more than probable that in the neighbourhood of Linlithgow coal lies beneath the igneous rocks, and as the more easily obtainable fuel becomes exhausted this deposit will be attacked." Unfortunately for this hopeful view the evidence from recent deep borings and other geological observations is all the other way. Again at page 69 we read, "The total quantity of proven coal in the county is over 200 million tons, of which one-fifth has been worked. Added to this is an estimated quantity of unproven coal amounting to about the same, and 1000 million tons are supposed to lie under the Firth of Forth, making a grand total of 1400 million tons." In the first place, the figures given of proven and unproven coal are probably greatly overstated, and the estimate of unproven coal is largely hypothetical. Unfortunately the supplies of coal are not nearly so large as was once supposed, and while there may be reason to believe that coal exists at certain places, it may be too deep or too poor in quality to be workable, and is therefore of no appreciable value. But the author makes a further mistake when he adds to the supposed four hundred million tons of unworked coal under the county a thousand million tons of coal under the Firth of Forth, making a grand total of fourteen hundred million tons belonging to the West Lothian coalfield still to work. In Dron's book on the coalfields of Scotland, the author refers to the coal under the *whole* of the Firth of Forth between Fife and the three Lothians as amounting to one thousand million tons, and does not suggest, as Mr. Muir would lead us to suppose, that this prodigious quantity lies hidden in the very small part of the underground coalfield pertaining to West Lothian alone.

At page 11 there is a statement that no river flows or ever did flow along the valley at Linlithgow. If Mr. Muir will refer to the *Scottish Geographical Magazine*, vol. xxviii. page 250, and examine the map at page 258, he will see that there is reason to suppose that a large river did flow north-eastwards along this valley before the Glacial Period.

At the top of page 19 a theory is given to explain the SW. and NE. trend of the British rock formations which does not seem to be in accordance with well-established geological facts. The theory of the formation of good anthracite coal is also misleading, as much coal found near volcanic rock is not anthracitic, and most good anthracite both in Wales and in America is not found near volcanic rocks at all.

At top of page 110 the date 1790 should, to be strictly correct, read 1788.

At page 118 it is stated that the Union Canal, like the Forth and Clyde Canal, is the joint property of the North British and Caledonian Railway Companies. This is a mistake, the Union Canal belongs to the North British and the Forth and Clyde to the Caledonian Railway Company. The writer has apparently no great hope for the future growth of Bo'ness, situated as the town is at the foot of a steep slope, with but little flat ground for public works and expansion along the shore. The flat ground has mostly been produced by reclamation of the shallow foreshore, and he refers to a proposal to reclaim an additional area of 2000 acres lying between Bo'ness and Grangemouth. No doubt such a proposal would have important geographical results, both for Bo'ness and Grangemouth. Bo'ness, however, being nearer the high ground, where there are excellent building sites, and also being only five miles from Rosyth, may yet develop into a residential town, and has considerable advantages from the garden city point of view, as well as deeper water at low tide than exists at Grangemouth.

We have noted these points of possible criticism in an otherwise comprehensive and very useful little text-book of local geography, carefully drawn up, and generally accurate, in order that future editions may be amended. There are numerous good illustrations, and three maps which are not so good. The first, or topographical map, does not show the Grangemouth new docks, which entirely altered the coastline at the mouth of the Carron, when constructed ten years ago. It is also badly coloured bathymetrically, as it makes no distinction in tint from the line of high-water to the five-fathoms line below low-water, and thus gives the reader no idea of the extensive and reclaimable foreshores in the estuary. The map of place-names is capable of some improvement and not quite so complete as might be expected. The geological map also might have been better made without additional cost. On the explanation of colours there is one mistake; "Calciferous Limestone" should read "Calciferous Sandstone Series."

ASIA.

The Lushai Kuki Clans. By Lieut.-Colonel J. SHAKESPEARE. London: Macmillan and Co., 1912. *Price* 10*s. net.*

The people described in this monograph inhabit the Lushai Hills and certain hill-tracts of Manipur. They are of Tibetan or Burman stock, and are divided into Lushai, Old Kuki, Thado and Lakher or Mara clans, as well as into several smaller groups more or less influenced by the Lushais. They are scattered over 25,000 square miles, and speak many different dialects.

The author has succeeded in giving a very full and careful description of the people, their habits, customs, and beliefs, as well as interesting specimens of their folklore, and has two chapters on the language, grammar and vocabulary. The Mongolian type of countenance prevails; and the men are short and sturdy, with broad faces, high cheek bones, almond eyes and wide nostrils. Before the arrival of the British they were continually at war, and were much mixed up through the disintegration of small and weak tribes, whose members were obliged to take refuge with the more powerful clans. The chief crop is rice (also maize, tobacco, cotton, etc.), and many keep bison or mitzan, which are used in their sacrificial feasts. They cut down and burn stretches of jungle on which crops are grown for two years at most, and, in the case of the Lushais, are continually migrating to obtain forest to destroy. The author describes fully their marriage rites, tabus, death and burial ceremonies. The bachelors live in special houses, but each family has a house of its own. The power of the village chief seems to depend very much on his personal influence, but there are interesting transitional states between the free man and the slave. Head-hunting seems to have existed at no very distant date. One curious legend deals with the origin of mankind. The ancestors of the tribe emerged from a hole in the ground as soon as an inquisitive monkey raised a stone which had closed them in. There is an interesting map of Hades, in which a valley is set aside for the Christians. The carved stone monolith to the memory of Mangkasa is of great interest as occurring amongst these tribes. The author alludes to many points of resemblance with Chins, Nagas, Garos and Mikirs, but many of the customs referred to are very widely spread and could be paralleled in Burma, Borneo, and occasionally in Madagascar.

It is perhaps a little unfortunate that the book in places inevitably recalls a Government Report, and it would have been more interesting and of more scientific value if the author had allowed himself to enunciate theories and make deductions. But it is full of important ethnological facts which should attract the attention not only of anthropologists, but of the general reader.

The Land that is Desolate. By Sir FREDERICK TREVES, Bart., G.C.V.O., C.B., LL.D. London: Smith, Elder, and Co., 1912. *Price 9s. net.*

This last publication of Sir Frederick Treves is an account of a recent tour in Palestine, in which he and Lady Treves did not leave the beaten track; but the well-known story of what the devout, or, it may be, the merely curious, traveller has to find in the Holy Land is retold with all the verve and sparkle as well as the accuracy and directness, which characterises Sir Frederick's other works. He went to Palestine with no illusions and he found none there. It is a land not only desolate but steeped in sordid squalor and full of unabashed greed. Thanks to the ferocious iconoclasm of the Mohammedans on the one hand and the perverted zeal of Christians of all varieties of creed on the other, it is practically impossible now to fix down with anything like certainty the sites in the Holy Land of the scenes and incidents recorded in the Bible, and this, notwithstanding the learned and pious endeavours of such distinguished savants as Stanley, George Adam Smith, Conder, etc. The devout pilgrim is also grievously harassed both in mind and body by the people and conditions under which he strives to conjure up the occurrences of thousands of years ago. All this, and much more, is set forth in crisp and sparkling style by Sir Frederick, who is endowed with a happy sense of humour, which brings into prominence the comic aspect of things but never degenerates into irreverence. The sadder side of the story, such as is exemplified by the wailing of the Jews at the wall of the Temple, appeals to him powerfully, and evokes his genuine and profound sympathy. He is an adept at depicting the strange and varied scenery through which he passed, and he verifies it with apt illustrations from history. Altogether this is a delightful book, but it leaves on us the feeling, that Palestine is not only a land that is desolate, but one where the sacred and reverent impressions which are now a treasured possession would receive the severest shock and could hardly survive.

Troy: A Study in Homeric Geography. By WALTER LEAF, Litt.D., Hon.D. Litt. (Oxon), Sometime Fellow of Trinity College, Cambridge. With Maps, Plans, and Illustrations. London: Macmillan and Co., Ltd., 1912. *Price 12s. net.*

"Overtaxed reviewers," says Dr. Leaf in the preface, "need hardly consider themselves bound to begin before page 310 or thereabouts, but should kindly read as far as page 330; in this compass they will find the conclusions with which I am mainly concerned." A reviewer would require to be very overtaxed indeed, and besides be totally devoid of a taste for learning, who would not read the whole of this fascinating volume. The first four chapters are confessedly a synopsis of the standard work *Troja und Ilion* by the great scholar Dörpfeld, but a synopsis of great value and urgent necessity, since the German work is bulky, expensive, and without an index. It is also based on personal knowledge, for Dr. Leaf himself visited and carefully studied the Troad. We may take it therefore that the hill of Hissarlik is beyond question the site of the castle (it was never a city until Roman times) of Troy, and that the Trojan War as related in the *Iliad* actually took place. For details of the reasons which have led Dr. Leaf to this conclusion we must refer the reader to the book itself, but broadly stated they come to this, that from its geographical position Troy was a great emporium for goods from Thrace, the Euxine, and Asia Minor, that it held complete command of the Dardanelles, and that the growing commerce of the Greeks combined with the rise of their national spirit drove them to dispute this supremacy and by razing Troy to open the way direct to the markets and regions of supply.

Agamemnon, Odysseus, Achilles, and Priam may have been actual persons, the abduction of Helen may have really taken place, but whatever the immediate cause of the war may have been, the ultimate cause was economic, and was founded upon geographical conditions. The book is admirably produced. Maps, plans, and illustrations combine to elucidate as well as to adorn the text. Several appendices and an adequate index are invaluable for reference. Altogether the work is a credit to British scholarship and to the traditional method (derived from Scottish geologists we believe) of studying things on the ground.

The Little World of an Indian District Officer. By ROBERT CARSTAIRS. London: Macmillan and Co., Limited, 1912. Price 8s. 6d. net.

It was the late Lord Ripon who remarked that, in India, when anything goes right or anything goes wrong, *e.g.* if there is a copious fall of rain, or if there is a drought, there are only two men blamed or praised, viz., the Governor-General and the District Officer. His Excellency was quite right in his remark so far as the three hundred millions of natives are concerned. For ninety-nine out of the hundred of them know little or nothing of the system under which they are governed, and without hesitation refer everything connected with their prosperity or adversity to the *ikbal*, *i.e.*, the auspices of the Governor-General, a distant, nebulous, but unquestionably supreme authority, but still more often to the *ikbal* of the District Officer, with whom they have more immediately to do. In our notice of Mr. Fraser's *India, under Curzon and After* (*S.G.M.*, vol. xxvii. p. 46), we pointed out, that that book gave the reader an excellent idea of the multifarious duties which fall to the lot of the Governor-General. In the interesting and instructive volume now before us our readers will find a compendious but fairly full account of the multifarious duties of a District Officer. For each and all of these the Government and the people hold the District Officer responsible; and Mr. Carstairs' object in writing this book is to set forth how he dealt with these duties in the various places where he was stationed during his Indian service. Here and there throughout the book there is a vein of disappointment, natural and inevitable to all officers of the enthusiastic temperament of Mr. Carstairs, who find their best thought-out schemes and projects ruthlessly nipped in the bud or coldly neglected at headquarters. Like most Anglo-Indian officers he has no complaints to make against the Lieutenant-Governor or the Governor-General.

It is the Secretariat that is the District Officer's *bête noire* in other provinces as well as Bengal. The Anglo-Indian civilian is often twitted with want of sympathy with the natives, an unjust taunt in most instances; and with regard to this Mr. Carstairs has some wise and shrewd remarks on the difficulties and dangers incidental to free and cordial intercourse between the official and the governed classes. There is, however, another and a much more bright and cheering side to this book, for Mr. Carstairs has to tell us of administrative experiments skilfully contrived and carried out for the amelioration of the conditions of the people, and improvements wisely initiated and successfully effected, often under very discouraging circumstances. The whole story is one of a strenuous, energetic, and conscientious career, chequered undoubtedly with disappointments, but eminently happy and successful. At a time when it is said that the career of the Indian civilian is not presenting to the youth of the United Kingdom the attractions it once had, the publication of this book is most opportune, for a perusal of it will show that nowhere in the length and breadth of the empire is there a career so full of opportunities of distinction and usefulness for an ambitious, earnest, and conscientious young man as there is in the Indian Civil Service of the present day.

The Progress and Arrest of Islam in Sumatra. By GOTTFRIED SIMON. London, Edinburgh, and New York: Marshall Brothers, Ltd., 1912. *Price* 6*s. net.*

The aggressive power of Islam has been manifesting itself in various parts of the world of late, and causing anxiety to those who are convinced of its baleful effect on heathen peoples. This is seen in Dr. Karl W. Kumm's note on "The Spread of Islam in Africa" in a recent issue of this magazine. In the Malaysian Archipelago, where Mohammedanism planted its foot six centuries ago, it was nowhere so successful as in Sumatra. The advent of the capture of the island by Holland, a Christian power, did not affect its activities, as it was allowed to pursue its way, the new Government using it for its own purposes. Of a population of four millions, it claimed three millions and a half among its converts. But Christian missions found their way to the island, and, in spite of the fanaticism Moslemism seems always to engender, they assuredly have won their most striking triumphs over Islam in that island. Mr. Gottfried Simon has spent eleven years in Sumatra, and has examined the situation as presented by the three faiths of the people, the Animistic, the Moslem, and the Christian, with great minuteness and care. He has canvassed the highest authorities on this subject, such as Snouck, Hurgronje, and Poensen. He presents the results of his study and observation in this work, under three divisions, in the first of which he deals with the factors in the progress of Islam in Sumatra, which he attributes to its active propagandism and neutrality of colonial governments. In the second division of the work the social and religious conditions of the Pagans who have become Mohammedans are dealt with, and here he seriously questions if the native has really advanced at all under the Islamic faith. In the third division, he shows what has been the outcome of the impact of Christianity on Islam, arresting it, and leading the people to a purer and fuller life. He holds that Islam does not affect the inward man; "he remains the same as ever he was, at best he becomes haughty, fanatical, and more indifferent to much other evil-doing which even his heathen conscience branded as sin," and that Islam "has not proved able really to raise the peoples of the Dutch East Indies." Christianity goes to the heart of man's failure, and supplies him with energy and strength of will to lift him to a higher life. The author shows a prolonged study of the situation, and a masterly grasp of the questions involved. He has made a valuable contribution to this increasingly pressing problem of Islam and Christianity. His publication, however, sadly lacks two things—an index and a map. The latter is desirable, the former is indispensable.

AFRICA.

In the Shadow of the Bush. By P. AMAURY TALBOT, of the Nigerian Political Service. London: William Heinemann, 1912. *Price 18s. net.*

The Ekoi are a numerically insignificant tribe, springing from a Bantu stock, who now inhabit part of the Oban district of Nigeria and a slice of the German Cameroons adjacent to it. The whole Ekoi population both in British and German territory numbers only some 32,000, almost two-thirds of which live in British territory, into which there is a steady flow of immigration of Ekoi from the Cameroons. Mr. Talbot is the Deputy Commissioner of that part of Nigeria where the Ekoi dwell, and this highly interesting and beautifully illustrated work describes in simple and lucid style the manners, customs, religion, costumes, ceremonies, laws, etc. of this primitive and, as yet, unsophisticated race, gentle and kindly, but still steeped in superstition and not yet quite reclaimed from savagery. It is an excellent example of how Great Britain understands and fulfils the duty known as bearing the white man's burden, and how wisely and skilfully it adapts its form of government to local circumstances and conditions. The Ekoi people are practically allowed to govern themselves according to their own ideas of right and wrong, subject to the proviso that their courts shall do nothing contrary to the British sense of justice. In every town there is a council of elders which may be summoned by any one, even by a child. This council hears and decides all complaints. There is an appeal to a higher court consisting of picked representatives from other towns, and again there is a final appeal to the Commissioner.

A delightful feature of the book is an admirable and copious selection of amusing and quaint tales, mostly in the shape of folklore, by which the author has, so to speak, allowed the Ekoi to describe themselves at least in the infancy of their mental development. The numerous and quaint illustrations deserve special notice. The photographs of scenery are excellent, but even more remarkable and interesting are the photographs of individuals and incidents, illustrating the Ekoi religion, magic, domestic occurrences, costumes, etc.

Aux Sources du Nil. By JULES LECLERCQ. Paris: Plon-Nourrit et Cie, 1913. *Price 4 francs.*

The author travelled by the Uganda railway and the usual steamers on Lake Victoria. There is a full description of the usual incidents of the voyage (including the inevitable natives who dive and pick up money), of the Mombasa tramway, and of the carriages on the Uganda railway. His luggage was nearly carried off to Mombasa when he was changing steamers, and it is of such innocent adventures that the book is composed.

But the author makes statements which we cannot allow to pass unchallenged. Forced labour is not inflicted in Uganda simply to provide wood at a cheap rate for the railway. The natives are not oppressed, and are not discontented under British control. The book is full of carping criticism of everything British.

Moreover, some of the author's informants are not without a sense of humour which is a little unkind, for there are some people with whom it is very unsafe to joke. Thus, "Et l'on éprouve un petit frisson à l'idée que c'est dans ces mêmes eaux que se tient le gigantesque serpent d'eau dont parlent avec terreur les riverains du Nyanza et qui est si peu, comme on l'a cru longtemps, un produit de leur imagination que le monstre vint un jour dérouler ses six mètres d'anneaux visqueux sur le pont du Winnifred (sic) dont le capitaine put le photographier."

We hope that this and other instances will prevent our French friends from taking this work as a serious account of Uganda.

British Somaliland. By RALPH E. DRAKE-BROCKMAN, M.R.C.S.Eng., L.R.C.P. Lond., F.R.S., etc., etc. With 74 Illustrations from Photographs on art paper, other Illustrations, and a Map. London: Hurst and Blacket, Ltd., 1912. *Price* 12*s.* 6*d. net.*

This volume will undoubtedly take rank as a standard work on British Somaliland, its history, geography, ethnology, flora and fauna. After an account, going back to early Egyptian days, of its relations with the outside world, the author in four chapters describes the country as it is at present. The next three chapters are devoted to a careful study of the Somali race. Then the "Mad" Mullah comes under review, and the reader is given more than a glimpse of that remarkable man. Politics are not touched upon, except that the author gives his opinion that the abandonment of the whole interior was the only course open to the Government. Chapters follow on live stock, the outcast tribes, flora and fauna, and products of commercial value. Much interesting information is given as to the frankincense and myrrh and other characteristic exports. The illustrations are well chosen, while of the map we would only say that it imparts to the country a rather delusive appearance of being abundantly supplied with rivers.

Thinking Black. By D. CRAWFORD, F.R.G.S. Pp. xii + 485. With Illustrations. London: Morgan and Scott, 1912. *Price* 7*s.* 6*d. net.*

The author, who is a pioneer missionary, spent, as shown by the second title of his book, "twenty-two years without a break in the long grass of Central Africa." He landed at Benquella in 1889, and after a long delay at the coast, started inland on a tedious and difficult journey more or less along the twelfth parallel of latitude to Bunkeya, the capital of King Mushidi. After a long residence, mostly without any European companion, in these terrible marshes, he again started eastward, and, crossing Lake Mweru, established a mission station on the further side. The cool resolution and patient endurance of the author can perhaps only be realised by those who know the conditions which prevailed in Central Africa at that time. He managed to make a friend of the brutal Mushidi, and seems to have kept on good terms with both Portuguese and Belgian officials, although he speaks plainly regarding San Thomé slavery and the cannibal soldiery of the Congo Free State.

The chief interest of the book consists in the exact but sometimes lurid picture of negro life as it was led under a native king when the Arab slave-traders were still in full activity, and before efficient European control was brought about. There is much that will attract every one in the author's luminous descriptions of the country, animal and plant life, and especially of the wily tortuous workings of the black man's mind. Indeed both geographers and anthropologists will find the general impression left by the author of real value, although he does not attempt accurate scientific detail. The photographs (some by Mr. Kidd) are quite delicious, and the four coloured sketches are excellent.

But the reader must not be put off by certain peculiarities of the book which ought to be pointed out. At the beginning of each chapter there are three or four little texts in red which seem unnecessary. In this "rubric" one finds, *e.g.* Jeremiah, Kipling, Mr. Punch, the author himself, and many others. It is also unfortunate that the author indulges in really atrocious puns and in too familiar quotations. There is, of course, always something new out of Africa, and there are at least three "curious tales." One refers to an elephant which seized the tail of a crocodile by its trunk, and threw it thirty yards away. Another is about the honey ratel which "has struck the idea of making his own distillery! This he does by digging a pot-like hole with his three middle claws on the wet bank of a river. Now he rams home the honey with a careful admixture of water, and off he goes to await the issue of the loyal law of fermentation. In two days or so, back he comes to find his alcohol foaming up out of the hole, and in a few minutes Mr. Badger is rolling drunk. The liquor swimming through his veins like a glorious fire makes him mad for murder." In fact the natives apparently told Mr. Crawford that one of his native boys was killed by an intoxicated honey ratel.

Twenty-five Years in Qua Iboe. By ROBERT L. M'KEOWN. London: Morgan and Scott, Ltd., 1912. *Price* 2*s.* 6*d.*

This little volume contains a pleasantly written account of the foundation, progress, struggles, and success of a missionary enterprise commenced in 1887, under the auspices of the Irish Presbyterian Church, in Southern Nigeria under circumstances which in many ways were most unpromising. The climate was deadly for Europeans; the inhabitants were sunk in barbarism and superstition and addicted to every sort of cruelty and vice; tribal war and slave-driving were universal. Out of this chaos, thanks to the patient devotion and strenuous exertions of the missionaries, there have been evolved peace, prosperity, and brotherly kindness. Such a story hardly lends itself to the pages of a scientific magazine, but we may say that the author has much to tell us of the primitive manners, customs, superstitions, and practices of the indigenous peoples living on both sides of the Qua Iboe river, and the book is a useful contribution to the anthropology of that region. It is written strictly from the missionary point of view, but not the less will it be found interesting and instructive to the general reader.

AMERICA.

South America: Observations and Impressions. By JAMES BRYCE. London: Macmillan and Co., Limited. New York: The Macmillan Company, 1912. *Price* 8*s.* 6*d. net.*

In this thoughtful and instructive volume we have a record of the observations made, and impressions received by Mr. Bryce, the well-known publicist, who was lately our Ambassador at Washington, during a four months' tour in South America, in which he travelled "through western and southern South America from Panama to Argentina and Brazil, via the Straits of Magellan," and visited seven of the republics. It is needless to say, that any such work coming from the pen of so distinguished a statesman will be universally received with a cordial welcome. If we look at it simply as a literary work, it charms by the purity, lucidity, and grace of its style and language, all of which we naturally expect in another work by the author of *The Holy Roman Empire.* But it has many additional attractions. Of late years we have had many works, able, learned and practical, dealing with the history and conditions of several of the South American republics. But in this book we have a review of their condition as a whole, and as related to each other and to the outer world, profound, serious and discriminating, such as we can hope for only from an authority eminent as a historian, a jurist, a diplomat, and a statesman. The tour was brief, as Mr. Bryce himself points out, but, thanks to the courtesy and assistance he everywhere received, he had exceptional opportunities of seeing and appreciating the circumstances of the republics which he visited, and he was evidently familiar with their previous history. His survey then of South America is of peculiar value, more especially at this juncture when the great South American republics, having enjoyed a good spell of peace and prosperity, are preparing to take an honourable and important part in the affairs of the world. This work, welcome as it is on this side of the Atlantic, will probably receive greater attention in the United States, where, for example, the Monroe doctrine is still the acknowledged national policy. Mr. Bryce shows clearly, that however justifiable and suitable that doctrine once was, and even yet may be, for the smaller and less stable republics, it is now unsuitable, because unnecessary, for the great southern republics, where it is now generally resented and detested. This is only one example from a score of profound, intricate, and perplexing problems, political, racial, commercial, etc., connected with South America, which are calmly and impartially discussed by Mr. Bryce. For the general reader this book has many attractions. We have already mentioned the grace and lucidity of the style, and the interest and importance of its subject matter, and we may add, that the romance inseparably connected with such names as Pizarro, Bolivar, San Martin and Artigas, is freely used to adorn and enliven its pages.

The Spell of the Rockies. By Enos A. Mills. With Illustrations from Photographs by the Author. London: Constable and Co., 1912. Price 6*s. net.*

This book consists of a series of sketches mostly reprinted from various magazines published in the United States, and now presented in a very attractive volume, beautifully illustrated. The papers are not entirely devoted to the scenery of the Rockies, but are interspersed with interesting and instructive chapters on the denizens of the forest and streams. The chapter entitled "A Rainy Day" is, strange as it may seem, an eloquent appeal for afforestation, the benefits of which the author sets forth very clearly.

There is a breezy atmosphere about this collection of sketches which makes the book very readable, although of geography, in the strict sense it does not contain much.

AUSTRALASIA.

Papua or British New Guinea. By J. H. P. Murray, Lieutenant-Governor and Chief Judicial Officer, Papua. With an Introduction by Sir William MacGregor, G.C.M.G., C.B., D.Sc., LL.D. London: T. Fisher Unwin, 1912. Price 15*s. net.*

The story of the British occupation of Papua is one in which all concerned may feel legitimate pride. Never has the policy of "peaceful penetration" been more faithfully carried out, or the rights of the natives more scrupulously respected. This is largely owing to the outstanding qualifications of the various officials and lieutenant-governors responsible. This volume by the present lieutenant-governor gives a most complete account of Papua, its geography, history, ethnology, administration, exploration, and development. Several most amusing tales are related illustrative of native modes of thought. There are many excellent photographs, but the map unfortunately does not give all the places mentioned in the book.

The Pagan Tribes of Borneo. By CHARLES HOSE, D.Sc., and WILLIAM M'DOUGALL, M.B., F.R.S. With an Appendix by A. C. HADDON, D.Sc., F.R.S. Two volumes. London: Macmillan, 1912. *Price* 42*s. net.*

In these two handsome volumes Dr. Hose summarises the experiences of twenty-one years spent in North Borneo, where he was formerly Divisional Resident and member of the Supreme Council of Sarawak. Dr. W. M'Dougall has also spent a year in Borneo, but only claims to have assisted in the preparation and "delivery" of the book.

This work should surely be in future the standard book of reference, not only on the anthropology of the island but on Bornean affairs generally. But it is far more than this, for the general reader will surely appreciate this vivid picture of the daily life and, up to a certain point, of the motives and beliefs of a very interesting group of aborigines.

We do not think that any one who even glances at this work can lay it down without looking at every illustration. These are not only supplied in generous proportion, but most of the 211 plates have an unusual charm which will appeal to every reader. It is impossible to describe artistic excellence, but we must especially mention the Kayan charging a sacrificed pig with a message to the god, the elderly Punan Headman, various young women of prepossessing appearance, and the plates dealing with the evolution of decorative design in textiles, woodcarving, and on the human skin.

The authors begin with chapters on the geography, history, ethnology, and the material condition of the Pagan tribes. The social system, agriculture, and ordinary daily life in the communal houses, on the rivers and in the jungles are then described with much detail. War, handicrafts, and especially decorative art are then dealt with; the latter subject is treated in full, and forms a valuable addition to the history of art, and is especially worth the attention of those who are interested in the evolution of conventional design. The second volume treats of the spiritual beliefs, of the soul, burial customs, animistic ideas, magic, myths, childhood rites, and the moral and intellectual peculiarities. There are also chapters on the nomad hunters, on the present Government, and especially one on the ethnology of Borneo. It is this last which forms probably the most important part of the book.

The problem of the origin of these races is exceedingly difficult to solve. Most of the tribes live in large communal houses on the banks of great rivers which, from their mountain sources to the sea, are almost everywhere bordered by dense tropical jungle. All, except the nomad hunters, cultivate rice to a greater or less extent, but every stage of culture seems to be represented, from the wandering hunter to the more advanced Sea Dayaks, who have usually a large surplus of rice to dispose of and who weave cloth and are great traders and boatbuilders. There are resemblances due to environment, to similar stages of culture, and to borrowings of usages and superstitions. There have also been many invasions and immigrations.

The ethnology of these tribes is therefore exceedingly intricate. Dr. Hose has adopted and developed a suggestion made by Mr. J. R. Logan in 1850, and has succeeded in giving a simple and adequate explanation. He suggests that when

Borneo, Sumatra, and Java were still part of the mainland, they were inhabited by a race formed of two distinct elements, namely, a Caucasic or Aryan people from Bengal and the advanced guard of the Mongols from the north. When Borneo was separated from the mainland, this original Caucasic-Mongoloid stock was divided, and the various parts were subsequently isolated by further invasions from the north. One branch remained in Borneo; another, which contained pre-Dravidian as well as other Mongol elements, formed the Kayans, originally of the Irawadi basin, and, entering Borneo by the south coast rivers, began to penetrate up their valleys about seven hundred years ago. The Iban or Sea Dayaks are Proto-Malays from Sumatra who may have seized the river mouths not more than two hundred years ago when the Malays became, under Arab leaders, an aggressive power. Whilst in the present condition of ethnology one can only be thankful for this clear and attractive theory, there are certain points which require further investigation. The authors doubt if there ever were negritos in Borneo. But there are said to be cave drawings of black-skinned people of short stature with curly hair and very broad, flat noses. It seems doubtful if all this can be explained by the existence of a few African slaves. Then again Kayan "flaki," hawk, *may be* from the same root as "falco" or "falcon," and "aman" may be connected with "omen," but it requires courage to accept this derivation, even though the methods of divination by pig's livers and the flight of birds do resemble suspiciously those practised by the Romans. The practice of head-hunting has also, we think, a much wider distribution and an older history than the authors' suspect. Dr. Haddon's Appendix contains the measurements, as made by himself, Dr. M'Dougall, Mr. R. Shelford, and includes some of the figures given by Niewenhuis and others.

The anthropological information is very full, but is very condensed. On page 326 the author states that of *seven* malohs, *three* are low brachycephalic, yet goes on to say the "cephalic index is essentially dolichocephalic."

GENERAL.

Voyages and Wanderings in Far-off Seas and Lands. By J. Inches Thomson. Illustrated. London: Headley Brothers, 1912. *Price* 3*s.* 6*d. net.*

As the title indicates this is an account of journeys and adventures in many parts of the world. Among the author's adventures was included shipwreck on Macquarie Island, where he and a party spent four months before they were taken off. The book also includes many tales of the South Sea Islands in the earlier days when "blackbirding" was practised. It is illustrated by some good photographs, and gives an interesting picture of the seaman's life fifty or so years ago.

The Story of Jerusalem. By Col. Sir C. M. Watson, K.C.M.S., C.B., M.A.
The Story of Santiago de Compostela. By C. Gasquoine Hartley (Mrs. Wallis M. Gallichan). (Mediæval Town Series.) London: J. M. Dent and Co., 1912. *Price* 4*s.* 6*d. each net.*

These two books treat their respective subjects from a somewhat different point of view. In the case of Santiago Mrs. Hartley gives first a general account of Galicia—the background of her city,—then devotes three chapters to its history, and gives most of the rest of the book to a detailed account of the architecture of the city, with a final chapter of general information for the visitor. The book will thus serve as a good introduction to the charms of Galicia, the chapter entitled "The Way to Santiago" giving some account of the province. Sir Charles Watson's book is almost purely historical, and the earlier chapters contain an oddly uncritical summary of the statements of Josephus and of the books of the Old Testament. Even the narratives of the books of Kings and Chronicles are merely quoted, without comment, as though the Higher Criticism had never been. This complete neglect of the work of the critics means, as it seems to us, a great loss to the book. Further, as surely no visitor to Jerusalem would omit to put a Bible in his box, it seems a little needless to summarise its historical portions. In the final chapter some description is given of present-day Jerusalem for the use of the tourist.

La Sismologie Moderne: Les Tremblements de Terre. Par le Comte de MONTESSUS DE BALLORE. Paris: Armand Colin, 1911. *Prix* 4 *fr.*

This is an excellent and well-illustrated little book whose object is to give a general account of earthquakes and the associated phenomena, in language sufficiently simple to be understood by the non-specialist. The author is the director of the Seismological Service of the Republic of Chili, and is thus a seismologist by profession, already known by his larger and more technical works on the same subject. His book should prove of great use to the geographer.

BOOKS RECEIVED.

Marlborough's Travellers' Practical Manual of Conversation in Four Languages: English, French, German, and Italian. Third Edition. Revised. Pp. 144. London: E. Marlborough and Co., 1912. *Price* 1*s.* 6*d.*

Notre France d'Extrême Orient. Par le DUC DE MONTPENSIER. Préface de M. LE MYRE DE VILERS. Ouvrage orné de 18 Gravures d'après les Photographies de l'auteur. Cr. 8vo. Pp. iii + 348. Paris: Perrin et Cie. *Prix* 5 *fr.*

Half-Hours in the Levant: Personal Impressions of Cities and Peoples of the Near East. By ARCHIBALD B. SPENS. With 32 Illustrations in Half-tone on Art Paper. Cr. 8vo. Pp. 107. London: Stanley Paul and Co., 1913. *Price* 1*s. net.*

Asia: A Geography Reader. By ELLSWORTH HUNTINGTON, Assistant Professor of Geography in Yale University. With an Introduction by RICHARD E. DODGE, Professor of Geography, Teachers' College, Columbia University. Cr. 8vo. Pp. xxvi + 344. Chicago: Rand, M'Nally and Co., 1913.

An Elementary Historical Geography of the British Isles. By M. S. ELLIOTT, M.A. Containing 60 Illustrations, Maps, and Views. Cr. 8vo. Pp. x + 172. London: A. and C. Black, 1913.

American History and its Geographic Conditions. By ELLEN CHURCHILL SEMPLE. With Maps. Demy 8vo. Pp. viii + 466. London: Constable and Co., Ltd., 1913. *Price* 12*s.* 6*d. net.*

The Duab of Turkestan: A Physiographic Sketch and Account of Some Travels. By W. RICKMER RICKMERS. 4to. Pp. xvi + 564. Cambridge: University Press, 1913. *Price* 30*s. net.*

The Republic of Chile: A popular description of the Country; its People, and its Customs. By DAVID W. CADDICK. Cr. 8vo. Pp. 64. London: A. H. Stockwell, 1913. *Price* 1*s. net.*

Ordnance Survey Maps; Their Meaning and Use. With Descriptions of Typical Sheets. By MARION I. NEWBIGIN, D.Sc. (Lond.). Cr. 8vo. Pp. 126. Edinburgh: W. and A. K. Johnston, 1913. *Price* 1*s. net.*

Rambles in Holland. By EDWIN and MARION SHARPE GREW. Illustrated. Cr. 8vo. Pp. xii + 339. London: Mills and Boon, Ltd., 1913. *Price* 6*s.*

Cambridge County Geographies:—

Middlesex. By G. F. ROSWORTH. Pp. x + 166.

Radnorshire. By LEWIS DAVIS. Pp. x + 156.

Rutland. By G. PHILIPS. Pp. vii + 171.

With Maps, Diagrams and Illustrations. Crown 8vo. Cambridge: University Press, 1912. Price 1s. 6d. net.

The Syrian Goddess: Being a Translation of Lucian's "De Dea Syria," with a Life of Lucian. By Professor HERBERT A. STRONG, M.A., LL.D. Edited with Notes and an Introduction by JOHN GARSTANG, M.A., D.Sc. With Illustrations. Cr. 8vo. Pp. xiv + 111. London: Constable and Co., Ltd., 1913. Price 4s. net.

Les Deux Congo. Par Baron JEHAN DE WITTE. Un volume in-16 avec gravures et cartes. Pp. xii + 408. Paris: Plon-Nourrit et Cie, 1913. Prix 4 fr.

Eastern Bolivia: Description of its General Features and Resources. By Dr. GUILLERMO VELASCO. Demy 8vo. Pp. 20. London, 1912.

The Lord Wardens of the Marches of England and Scotland: Being a brief History of the Marches, the laws of March, and the Marchmen, together with some account of the ancient feud between England and Scotland. By HOWARD PEASE, M.A., F.S.A. Royal 8vo. Pp. xvi + 255. London: Constable and Co. Ltd., 1913. Price 10s. 6d. net.

A Wayfarer in China: Impressions of a Trip across West China and Mongolia. By ELIZABETH KENDALL. With Illustrations. Demy 8vo. Pp. xiv + 338. London: Constable and Co., Ltd. Price 10s. 6d. net.

Mexico and Her People of To-day: An Account of the Customs, Characteristics, Amusements, History and Advancement of the Mexicans, and the Development and Resources of their Country. By NEVIN O. WINTER. Illustrated from photographs. Demy 8vo. Pp. viii + 492. London: Cassell and Co., 1913. Price 7s. 6d.

Indian Pages and Pictures: Rajputana, Sikkim, The Punjab and Kashmir. By MICHAEL M. SHOEMAKER. With 63 Illustrations. Crown 8vo. Pp. xxii + 475. London: G. P. Putnam's Sons, 1912. Price 10s. 6d. net.

Philip's Geo-Graph Book: A Geographical Observation Note-Book for Climatic, Astronomical, and other Records. By J. H. HACK, M.A., Part I. London: George Philip and Son, 1913. Price 3d. net.

Vascular Plants of West Greenland. Between 71° and 73° N. Lat. By MORTEN P. PORSILD. Royal 8vo. Pp. 38. Kφbenhavn: Bianes Lunos Bogtrykkeri, 1912.

Wissenschaftliche Ergebnisse der Expedition Filchner nach China und Tibet 1903-1905. Von Dr. WILHELM FILCHNER. Demy 8vo. Berlin: Mittler und Sohn, 1913. Band IV. Pp. viii + 148. Preis 10 m. Band V., Bilder. Preis 20 m.

A Practical and Experimental Geography. By FREDERICK MORROW, B.A., F.R.G.S., and ERNEST LAMBERT, L.C.P. With 112 Maps and Diagrams and 254 Exercises. Cr. 8vo. Pp. xiv + 239. London: Meiklejohn and Son, 1913. Price 2s. 6d. net.

Ancient Babylonia. By C. H. W. JOHNS, Litt.D. Cr. 8vo. Pp. vii + 148. Cambridge: University Press, 1913. Price 1s. net.

The Earth: Its Shape, Size, Weight, and Spin. By J. H. POYNTING, Sc.D., F.R.S. Cr. 8vo. Pp. 141. Cambridge University Press, 1913. Price 1s. net.

The Atmosphere. By A. J. BARRY, M.A. Cr. 8vo. Pp. 146. Cambridge: University Press, 1913. Price 1s. net.

Official Year-Book of the Scientific and Learned Societies. Compiled from Official Sources. Twenty-ninth Annual Issue. Demy 8vo. Pp. vi + 373. Griffin and Co., Ltd., 1912. Price 7s. 6d.

The People of the Polar North. By KNUD RASMUSSEN. Compiled from the

Danish Originals, and edited by G. HERRING. Illustrations by Count HAROLD MOLTKE. Royal 8vo. Pp. xi + 358. London: Kegan Paul, Trench, Trübner and Co., Ltd., 1908. *Price* 21*s. net.*

Panama and What it Means. By JOHN FOSTER FRASER. With a Map and 48 Plates from Photographs. Cr. 8vo. Pp. ix + 291. London: Cassell and Co., Ltd., 1913. *Price* 6*s.*

Adventures in the Alps. By ARCHIBALD CAMPBELL KNOWLES. Illustrated. Cr. 8vo. Pp. xii + 176. London: Skeffington and Son, 1913. *Price* 3*s.* 6*d. net.*

Reminiscences of a South African Pioneer. By WILLIAM CHARLES SCALLY. With 16 Illustrations. Demy 8vo. Pp. 320. London: T. Fisher Unwin, 1913. *Price* 10*s.* 6*d. net.*

The Life of Philibert Commerson, D.M., Naturaliste Du Roi: An Old World Story of French Travel and Science in the Days of Linnæus. By the late Captain S. PASFIELD OLIVER, R.A., and edited by G. F. SCOTT ELLIOT, F.L.S., F.R.G.S. With Illustrations. Demy 8vo. Pp. xvi + 242. London: J. Murray, 1909. *Price* 10*s.* 6*d. net.* *Presented by G. F. Scott Elliot*, F.L.S., F.R.G.S.

What I saw in Russia. By the Hon. MAURICE BARING. Foolscap 8vo. Pp. vii + 381. London: T. Nelson and Sons, 1913. *Price* 1*s.*

A Guide for Laboratory Geography Teaching. By O. D. VON ENGELN, Ph.D. For use in connection with a *Laboratory Manual of Physical and Commercial Geography.* By the late Prof. R. S. Tarr and O. D. Von Engeln, Ph.D. Crown 4to. Pp. 20. New York: Macmillan and Co., 1913. *Price* 1*s.*

Diplomatic and Consular Reports.—Trade of Paraguay (5040); Trade of Vladivostock (5041).

Statistical Abstract relating to British India from 1901-2 to 1910-11. London, 1913.

Publishers forwarding books for review will greatly oblige by marking the price *in clear figures, especially in the case of foreign books.*

NEW MAPS.

EUROPE.

ORDNANCE SURVEY OF SCOTLAND.—The following publications were issued from 1st to 31st January 1913:—

Six-inch and Larger Scale Maps.—Six-inch Maps (Revised). Quarter Sheets, with contours in blue. Price 1s. each. *Lanarkshire.*—23 SW., 54 NE.

1:2500 Scale Maps (Revised), with houses stippled, and with areas. Price 3s. each. *Lanarkshire.*—Sheets VI. 6; VII. 7. *Renfrewshire.*—Sheets VII. 7, 9; VIII. 5; XI. 2, 6, 9; XII. 12. Sheet XIII. 1. Price 1s. 6d.

Note.—There are no coloured editions of these Sheets, and the unrevised impressions have been withdrawn from sale.

Special Enlargements from 1:2500 to 1:1250 Scale. Prepared for the Land Valuation Department, Inland Revenue. Price 2s. 6d. each. *Caithness*—Sheets V. 11 SW.; XXV. 1 SE.; 5 SE.; 6 NW., SW. *Dumbartonshire.*—Sheets XVII. 6 NW., SW. *Fifeshire.*—Sheets XXXV. 8 NW., SE. *Forfarshire.*—Sheets LIV. 7 NW.; 8 NW., NE., SW., SE. *Perthshire.*—Sheets LXIV. 9 SE.; LXXIV. 13 NW. *Renfrewshire.*—Sheets VIII. 11 NE., SE.; XI. 2 NE.; 8 NE., SW.; XII. 2 NW., SE.; 3 NW.; 4 NW., SW., SE.; 6 NW., NE., SW.; 15 SW.; XIII. 9 NE., SE.; 13 NW., NE.; 14 NW.; XVII. 2 SE. *Stirlingshire.*—Sheets XXV. 9 SW., SE.

The following publications were issued from 1st to 28th February 1913:—

One-inch and Smaller Scale Maps.—One-inch Map. Third Edition; printed in colours and folded in cover, or flat in sheets. Sheets 83, 93. Price, on paper, 1s. 6d.; mounted on linen, 2s.; mounted in sections, 2s. 6d. each.

Six-inch and Larger Scale Maps. Six-inch Maps (Revised). Quarter Sheets, with contours in blue. Price 1s. each. *Lanarkshire.*—19 SE.

1 : 2500 Scale Maps (Revised), with houses stippled, and with areas. Price 3s. each. *Lanarkshire.*—Sheets I. 13; II. 9; V. 12 (V. 16 and X*a*. 4); VI. 2, 3, 5, 7, 9, 10, 11, 13, 14; X. 1, 2, 3, 4, 6, 7, 8; XI. 5. Sheet V. 7. Price 1s. 6d. *Lanarkshire (Glasgow and its Environs).*—Sheet X. 5. *Renfrewshire.*—Sheets II. 12; VI. 8; VII. 1, 5, 8, 10; VIII. 10, 15, 16; XI. 3, 13; XII. 6, 7, 8, 15; XIII. 5, 9, 13, 14. Sheet XIII. 15. Price 1s. 6d.

Note.—There are no coloured editions of these Sheets, and the unrevised impressions have been withdrawn from sale.

Special Enlargements from 1 : 2500 to 1 : 1250 Scale. Prepared for the Land Valuation Department, Inland Revenue. Price 2s. 6d. each. *Aberdeenshire.*—Sheets LIV. 4 SW., SE. *Dumbartonshire.*—Sheet XXIII. 13 SE. *Fifeshire.*—Sheets I. 14 NE., SW.; IV. 1 NE., SW., SE.; XVI. 14 SW., SE.; XX. 2 NW.; XXII. 8 SE.; XXIII. 2 NW., NE.; 5 SW.; XXXV. 14 NW.; XL. 6 SE.; 10 NE. *Forfarshire.*—Sheet LIV. 7 NE. *Haddingtonshire.*—Sheets X. 2 SE.; 6 NW. *Peeblesshire.*—Sheets XIII. 6 NW., NE., SW., SE. *Perthshire.*—Sheets LXIII. 3 SE.; LXXIV. 13 SW. *Renfrewshire.*—Sheets II. 11 NW., NE., SE.; 12 SW.; VII. 5 NW., NE., SE.; VIII. 7 NE., SE.; 8 SW., SE.; 11 NW., SW.; 12 NW., NE.; XI. 4 NE.; 7 NW., NE., SE.; 8 NW.; 13 SW.; XII. 2 NE., SW.; 3 SW.; 4 NE.; 6 SE.; 15 NW., NE., SE.; XIII. 13 SE.; 14 NE., SW.; XVI. 3 NW., NE.; XVII. 2 NW., NE.; 6 NE. *Ross and Cromarty.*—Sheets XLI. 4 SE.; 8 NE. *Roxburghshire.*—Sheets X. 1 SW.; XXI. 5 NE., SE.; XXV. 3 SW., SE.; 4 NW., SW.; 7 NW., NE., SW., SE. *Selkirkshire.*—Sheets VIII. 1 NE., SE.; 6 NW., NE.; XII. 1 SW., SE.; 5 NW., NE., SW., SE.

ADMIRALTY CHART, SCOTLAND.—West Coast. Sheet IV. Ardnamurchan to Summer Isles, including the Inner Channel and part of the Minch. New Edition, January 1913. Number 2475. Price 3s.

Admiralty Office, London.

FORTH AND TAY.—Bathy-orographical Wall Map. Scale 1 : 126,720, or two miles to an inch. Constructed and engraved by W. and A. K. Johnston, Ltd. Price 12s., mounted on cloth, rollers, and varnished.

W. and A. K. Johnston, Ltd., London and Edinburgh.

This is an effective map of Eastern Central Scotland, drawn on a large enough scale to show the physical and political features with admirable clearness. It ought to prove of great value in teaching local geography.

AFRICA.

SIERRA LEONE.—Scale 1 : 250,000, or about 4 miles to an inch. Sheets—Ronietta, Panguma, Sherbro. 1912. Price 1s. 6d. net each sheet.

Geographical Section, General Staff, London.

EAST AFRICAN PROTECTORATE.—Scale 1 : 250,000, or about 4 miles to an inch. Sheet—Kenya. 1912. Price 1s. 6d. net.

Geographical Section, General Staff, London.

SOUTH AFRICA.—The Standard Commercial and Educational Map of the Union of South Africa and Southern Rhodesia, specially compiled for the Standard Publishing Co. of South Africa. 1913. Scale 1 : 2,500,000, or about 40 miles to an inch. With Index to towns.

John Bartholomew and Co., Edinburgh.

This new map gives the country from the Cape to the Zambesi with as much detail as the scale will permit. It is compiled from the latest surveys, and shows all new railways up to date. There are inset maps giving the environs of the principal towns.

AMERICA.

ONTARIO.—Topographic Map. Scale 1 : 63,360, or 1 mile to an inch. Sheet XXXIII.—Hamilton. Sheet XLV.—Romney. (Department of Militia and Defence.) Price 2s. net each sheet. 1912.
Geographical Section, General Staff, London.

ONTARIO AND QUEBEC.—Scale 1 : 126,720, or 2 miles to an inch. Cornwall Sheet. 1912. *Price 2s.* *Geographical Section, General Staff, London.*

PANAMA CANAL.—*Daily Mail* Chart. A Sheet of Maps and Diagrams. Price 1s. net. *George Philip and Son, Ltd., London.*

In this useful Sheet are comprised a Plan of the Canal, a Bird's Eye View, a Map of the Isthmus, Diagrams showing Comparison with other Ship Canals, a World Map showing new Routes, a Comparative Table of Routes, and Plans of Culebra Cut and Gatun Dam.

NEW ATLASES, ETC.

PHYSICAL AND POLITICAL SCHOOL ATLAS. By J. G. Bartholomew, LL.D., F.R.G.S. Demy 4to. 1913. Price 1s. net.
Oxford University Press, London.

This new school atlas shows political features on a physical basis. It consists of 32 plates of coloured maps and 16 plates of uncoloured maps and diagrams. The maps are clear and legible and the colouring is effective.

PHILIPS' VISUAL CONTOUR ATLAS.—49 Maps and Diagrams. With Index. Price 6d. net. *George Philip and Son, Ltd., London.*

A series of maps prepared to meet the want of a sixpenny elementary atlas. It embodies the latest educational requirements and is well adapted for elementary teaching.

ATLAS OF THE WORLD.—By J. Bartholomew, F.R.G.S. Price 6d. net.
T. C. and E. C. Jack, London and Edinburgh.

Designed as a reference atlas for the series entitled "The People's Books," this little volume contains 56 pages of coloured maps illustrating the general geography of the world at the present day.

BASE MAPS.—Prepared by Professor Paul Goode. A series of 32 outline maps. In two sizes, 8 × 10 in. and 15 × 10 in. Prices 1 cent and 3 cents each.
The University of Chicago Press, Chicago.

These maps are admirably adapted for the purpose for which they are designed. They are clear, free from detail, and printed on tough paper suitable for pen and coloured work.

TERRESTRIAL GLOBE.—Philips' Panama Canal Route Globe. Six inches diameter. Price 2s. 6d. net. *George Philip and Son, Ltd., London.*

This little globe enables one, as nothing else could, to realise the world-wide changes in international routes which will be brought about by the opening of the Panama Canal.

EXPLAN

Rha
Car

Call
Mat

Vac
Mos

Alpi
Pe

Scir
Eric
Bog

Rha
Bog

Alpi
Hea

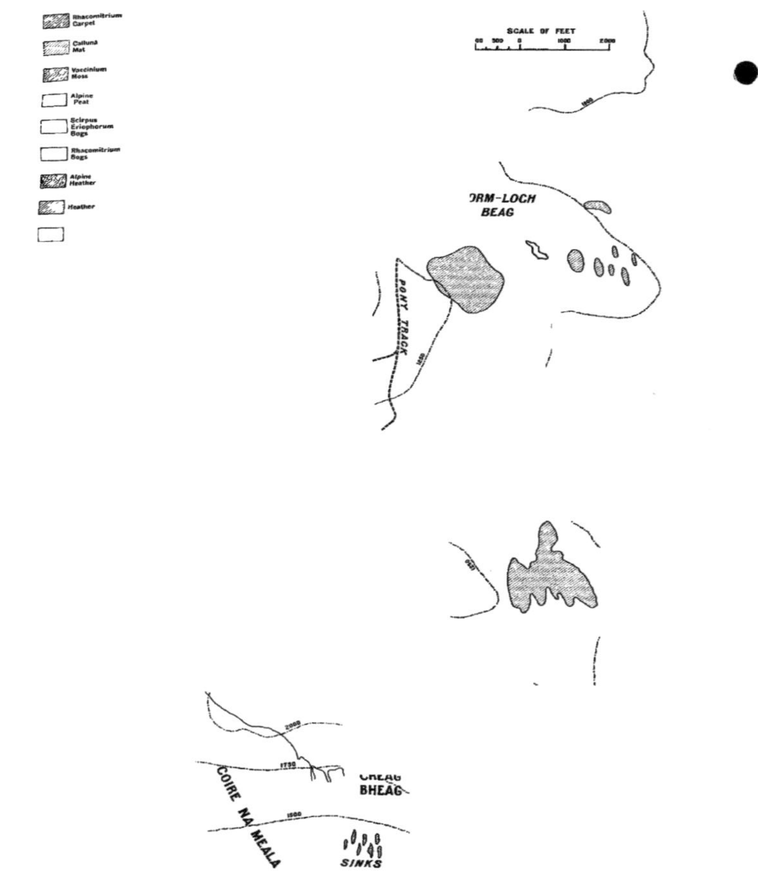

Rhacomitrium Carpet
Calluna Mat
Vaccinium Moss
Alpine Peat
Scirpus Eriophorum Bogs
Rhacomitrium Bogs
Alpine Heather
Heather
SCALE OF FEET
500
0
1000
2000
1800
ORM-LOCH BEAG
PONY TRACK
2000
1750
CREAG BHEAG
COIRE NA MEALA
1800
SINKS

DAVID LIVINGSTONE, 1813-1873

THE SCOTTISH

GEOGRAPHICAL

MAGAZINE.

LIVINGSTONE AS AN EXPLORER: AN APPRECIATION.[1]

By J. W. GREGORY, D.Sc., F.R.S.,
Professor of Geology in the University of Glasgow.

(*With Maps.*)

AT the dawn of the nineteenth century European ignorance of the interior of Africa was at its nadir. In the sixteenth and seventeenth centuries Portuguese missionaries and travellers had penetrated far inland, and their observations were supplemented by information collected from Arab traders. The maps compiled by European cartographers from the Portuguese reports were full of information about the geography of Central Africa. Thus the map published by Sanson of Abbeville in 1635 truly represented tropical Africa as a land rich in rivers and lakes: Tanganyika, the Victoria and Albert Nyanzas, Lakes Rudolf, Moero and Chad, can all be recognised.[2] Hence, in spite of their blunders, these seventeenth-century maps contain so many of the main facts of Central African geography that they must have been based on actual observation.

But the men who had made these discoveries were not trained explorers; they kept no regular journals; they made no astronomical observations to determine their positions; and on their return they could only spin such yarns as those on which Defoe, in 1720, founded his story of Captain Singleton's journey across Africa. As European geographers became more critical and scientific, and as these Portuguese pioneers were gradually forgotten, the knowledge they had gathered

[1] Delivered in the University of Glasgow on the occasion of the Centenary of David Livingstone, 18th March 1913, and published here in slightly abbreviated form.

[2] The chief errors in this map are that Tanganyika, "Zaire Lacus," is united with Nyasa, "Zembre Lacus"; that Tanganyika is given an outlet to the Nile as well as to the Congo; and that the rivers in various parts of the map are represented anastomosing like those of low alluvial plains.

was either lost or dismissed as fabulous. The lakes, rivers, goldfields, mountains and Portuguese stations were omitted from the maps, and the interior of tropical Africa was represented as one vast uninhabitable desert; there remained two fabulous mountain chains, "the Mountains of the Moon" on the north and the "Spine of the World" ("the Mountains of Lupata") on the east.

The maps of the interior of Africa issued between 1800 and 1850 contained less information and were less accurate regarding the interior of Africa than those of the sixteenth century. By the end of the nineteenth century the map of Africa had been again filled in; the deserts had been reduced to their true proportions; every important river had been tracked from source to sea; the highest mountains had been climbed; steamers were plying on the lakes and rivers; railways were superseding the slow and costly caravans of porters; and instead of only the Portuguese, Dutch and Turks holding extensive possessions in Africa, almost the whole continent had been partitioned between seven European states.

This revolution in the condition of Africa was mainly due to the great Scottish explorer whose memory we are assembled to honour.

David Livingstone's world-influencing achievements were the result of his own personal efforts and character. He had no social advantages, and his influence was due to his dogged persistence, utter fearlessness,[1] noble ambition, and his genius for travel. For he left school at the age of ten to begin work in the Blantyre cotton mills; and his education was continued at a night school after the day's mill-work was done. At the age of nineteen he was promoted to the grade of cotton spinner; he then earned enough in the summer, with some help from his brothers, to support himself, during the winter, as a student at this University and at Anderson's College.

That he was a student at the University has been denied; but his own testimony is conclusive. He referred in a letter to Murchison, printed in the *Journal of the Royal Geographical Society* (vol. xxvii. 1857, p. 386), to the encouragement he had received from "my former instructors in Glasgow University." In a speech[2] at the University Union in 1858, he told the students that "I remember well when I was among you," and he reminded them that "When Sir Robert Peel addressed us here in 1837, he told us that it was not genius that ensured success but hard, earnest working." Livingstone, therefore, attended that address as one of the Lord Rector's constituents, and he was clearly then a student of the University. Hence, on the evidence of Livingstone's statements, apart from the abundant testimony of his friends,[3] he was a son of this University.

---

[1] His superb courage was one of his most striking personal characteristics.

[2] For reference to the report of this speech, in the *Glasgow Herald* of 26th February 1858, I am indebted to Mr. E. G. Hawke.

[3] *E.g.* Prof. Adam Sedgwick, in his introduction to *Livingstone's Cambridge Lectures*, said that Livingstone had attended three courses of Lectures at the University of Glasgow, 1860, p. 55. The reference to Sir D. K. Sandford on p. 16 of the *Missionary Travels* supports the tradition that Livingstone attended the Greek class at the University.

We may also reflect with pleasure that the University of Glasgow was the first to recognise Livingstone's geographical work; for in 1854, while he was making his journey across Africa, the University conferred on him, "free of cost," the honorary degree of LL.D., "in high appreciation of his services in the cause of science and Christian philanthropy." Letters announcing this fact on his arrival on the east coast were especially welcome and contrasted strangely with a communication from the Missionary Society which he received at the same place.[1]

Livingstone's father was a deacon of the Congregational Church at Hamilton, and when he resolved to become a missionary, he naturally offered his services to the London Missionary Society, and was one of the many illustrious men whom the Congregational denomination sent into the mission-field. He was provisionally accepted and went to London in 1838. He was never a fluent public speaker, and narrowly escaped rejection; but after an extension of his period on probation he was accepted and continued his medical studies at Charing Cross Hospital. He returned to Glasgow in November 1840, to pass his final examination and receive his medical diploma; and he sailed the following month for South Africa. It was symptomatic of Livingstone's interests that during the three months' voyage to the Cape he learnt from the captain of the ship how to determine geographical positions; and this training, combined with subsequent instruction from Maclear, the Government Astronomer at Cape Town, enabled Livingstone to fix his routes with the precision of an experienced navigator.

Livingstone arrived at the Cape during a critical time in South African history. There, the ideal of the missionary party was to develop the country as a group of native states ruled by native chiefs, who were to be guided by missionary advisers—states, in fact, like Madagascar before its conquest by France.

The Reform Bill of 1832 had placed political power in the hands of the aggressive philanthropists; and under their influence the British Government adopted a policy in South Africa, which events, and the almost unanimous verdict of historians, have emphatically condemned. The Dutch farmers at the Cape found their position intolerable; and they sought freedom by emigration to the north of the British territories. There they proclaimed a free republic. The feelings of these emigrants towards the missionaries can be judged from one article of the constitution which was adopted by the Boers at Winburg on 6th June 1837; for it decreed that no man should join the republic unless he would take an oath to have no connection with the London Missionary Society or with any of its agents.

The success of a free Boer state to the north of the Cape would have been fatal to the missionary policy; so the British Government was persuaded to establish a chain of native states across South Africa, extending from Natal to the deserts north of the Orange River. This chain of states was completed in 1844. But the scheme proved a fiasco;

---

[1] Livingstone forgave but could never forget this letter, which probably helped his severance from Congregationalism.

the native states collapsed, and in 1852 the British Government acknowledged the independence of the South African Republic.

Livingstone arrived in South Africa at the time when the effort to establish missionary supremacy to the north of the Cape was being actively pursued; and he was sent to Bechuanaland to help the development of the northern outposts. He went at first to Lattakoo, or Kuruman, the most northern of the mission stations, then under the charge of the famous missionary Robert Moffat, whose daughter Livingstone married in 1844.

From 1841 to 1849 Livingstone was engaged in mission work on the south-eastern border of the Kalahari Desert in the district to the north of Mafeking. The history of this period of Livingstone's career has never been adequately written. His relations with many of his fellow-missionaries were not always harmonious. Livingstone had more practical insight than many of his colleagues, and he expressed his opinions of them and of the results of missionary enterprise in South Africa with ruthless candour.[1] Owing to the smallness of the native population, he was disappointed with the field which the London Missionary Society was then tending so diligently.

Livingstone had meanwhile abandoned medical work, a course which seems to have been partly due to his dialectic defeat on the question of rain-making. He endeavoured to convince the people of their folly in trying to make rain; he insisted that there was no visible connection between their medicines and the clouds; the natives replied that there was none between his medicines and the diseases they were given to cure; he argued that as their medicines often failed, the rain that sometimes fell after their use was not due to them; they retaliated that his medicines sometimes failed, yet he attributed to them whatever good results followed; and they concluded that as he persevered with his medicines, although his patients sometimes died, there was no reason why they should cease using theirs, because their methods were not always successful. Livingstone recognised that he had lost the argument, and gave up medical practice, except on special occasions, so that it might not hamper his spiritual work.[2]

In 1848 Livingstone settled at Kolobeng, the village of a chief Sechele, who was converted next year. After Sechele's abandonment of polygamy, Livingstone baptized him, amid his weeping people, who were terrified at the loss of their best rain-maker. With deplorable ill-fortune, the year following the country was stricken with so severe a drought,

---

[1] For example, the letter published in Sir H. H. Johnston's *Livingstone*, 1891, pp. 64-69. In an unpublished letter (now in the British Museum, No. 36525 f. 9), written from Tette in 1856 to his colleague, Rev. J. Moore, Livingstone remarked, "The natives too behaved liberally to us except when we came near to this. We Christians have not given the heathen fair play with our glorious Christianity. They believed my statements of all my goods being expended until we came into the vicinity of the Christians. It is not so only among R. Catholic Christians but everywhere. I often feel my own share of the guilt to be great."

[2] Thus he wrote to Cecil, his tutor at Ongar: "I did not at first intend to give up all attention to medicine and the treatment of disease, but now I feel it to be my duty to have as little to do with it as possible. I shall attend to none but severe cases in future, and my reasons for this determination are, I think, good."

that the men had to scatter far and wide to hunt, while the women and children tried to avert starvation by collecting locusts,[1] and later on Kolobeng had to be removed owing to the failure of water.

While Livingstone had been gradually growing less satisfied with his mission-field, the fascination of the unknown land beside him had captivated his mind. Livingstone was one of the idealists described in Kipling's lines:

> "We were dreamers, dreaming greatly, in the man-stifled town;
> We yearned beyond the sky-line, where the strange roads go down."

The fine imagination which changed the mill-hand at Blantyre into the missionary to the Bechuanas was leading Livingstone to another change in his career. He daily looked out into the wild waste of the Kalahari and yearned to follow the strange roads that went down into it, in the hope that they might lead to a better watered, more fertile, more populous country, where missionary work might be more profitable than among the small poverty-stricken tribes hemmed in between the Boers and the desert.

In the year of his worst troubles at Kolobeng his chance came. Two famous hunters, Cotton Oswell and Murray, visited his station and made the offer that, if Livingstone would find guides across the desert to the lake reported to exist on the other side, they would pay the expenses of the expedition. The evidence is contradictory as to who proposed this journey. According to the magazine of the London Missionary Society, Livingstone was afforded an opportunity for his "long-cherished purpose, by the visit of two benevolent travellers, Messrs. Murray and Oswell, who requested his co-operation in attempting to cross the Desert and exploring the unknown regions to the north. This overture Mr. Livingstone gladly embraced, unintimidated by the hardships or dangers of the undertaking."[2]

The party left Kolobeng on the 1st June, and after a journey of two months reached Lake Ngami on 1st August 1849. The country around the lake proved malarial and unpromising, but the rivers that discharged into it indicated the existence of higher and therefore probably more salubrious land to the north. The travellers endeavoured to reach this better land, but they were stopped by a large river. They resolved to try again; so Oswell went to the Cape to buy a boat, while Livingstone awaited him at Kolobeng.

This journey to Lake Ngami settled Livingstone's future. It com-

---

[1] Livingstone notes in his journal, with characteristic resignation, the local failure of the rains and their continuance in adjacent districts.

[2] *Miss. Mag. and Chron.*, vol. xiv., 1850, p. 34. This version is supported by Sir Samuel Baker, Sir Francis Galton, and the correspondence published in the *Life of Oswell*. According to Livingstone, he proposed the expedition and Murray and Oswell came from England to be present at the discovery (*Miss. Mag. and Chron.*, vol. xiv., 1850, pp. 35-37). Oswell appears certainly entitled to more credit in connection with this expedition than he has generally received. He seems to have deliberately effaced his share in the undertaking, as he knew that the reputation would be far more useful to his friend than it could be to himself. He sacrificed geographical fame, but gained an enduring reputation for chivalrous generosity.

pleted his conversion from a missionary into an explorer, a change which had been in progress throughout his residence in South Africa. Livingstone never again settled down to live on a mission-station. He wrote to his Directors: "I hope to be permitted to work, so long as I live, beyond other men's line of things, and plant the seed of the Gospel where others have not planted."[1] Livingstone had enlisted for life in that pioneer regiment whose mission is

To "preach in advance of the Army,"
To "skirmish ahead of the Church."

Livingstone heard at Lake Ngami of people who had come from the west coast and were probably Portuguese, and he saw there a coat which, he remarks, "we believe to be of Portuguese manufacture." It seemed probable that there was a practicable route from Lake Ngami to the west coast; and Livingstone was confirmed in this conclusion shortly after his return to Kolobeng by some envoys who brought him an invitation to visit Sebituane,[2] the famous chief of the Makololo. Some of this party had been to the west coast, and they described to Livingstone how they had seen ships there, "and called out to them, 'Hey, come and tell us the news,'" a request to which these indifferent ships paid no attention.

Livingstone was delighted to accept this invitation. His convert, Sechele, purchased a wagon, and together they started off on this journey into the unknown north. The effort failed, but it was renewed next year with Oswell, who had come back from the Cape, and it was then successful. Livingstone and Oswell reached Sebituane, who gave them a most friendly reception—but died a few days after their arrival. He was succeeded by his daughter, who subsequently resigned the chieftainship to her brother Sekeletu.

Sebituane had invited Livingstone because he needed the help of a British traveller, and he could not have made a better selection. Linyanti, his capital, was situated on the Chobe River, near its junction with the Zambesi, in a district equidistant from the eastern and western coasts. The country abounded in elephants, that were killed by the bold native hunters, and Sebituane wished to establish commercial relations for the sale of his ivory with the towns on the west coast. The road southward to the Cape skirted the Kalahari desert, and was always difficult and sometimes closed by war. The route to the east coast was also precarious, for impis of rebellious Zulu had settled in positions which commanded the lower Zambesi. Some slave traders from Benguella had recently visited Sebituane and exchanged guns for slaves, and as the way to the west coast was therefore practicable Sebituane wanted to use it independently of the Portuguese traders.

The scheme was delayed by Sebituane's death and Livingstone's preliminary arrangements. In order to send his wife and children to

---

[1] *Miss. Mag. and Chron.*, vol. xiv., 1850, p. 36.

[2] The spelling of most of the personal and place names is that adopted by Livingstone in his two chief books.

England, he returned with them to Cape Town, where he found that he had overdrawn his salary, and a further advance was refused. From this strait he was relieved by the wealthy and generous Oswell, who was fortunately then at the Cape. Livingstone returned northward to find that his house at Kolobeng had been destroyed by a party of Boers, who had also attacked and burned the adjacent native settlement. Livingstone wrote an indignant protest and predicted the extermination of the natives unless the Boers were controlled; and Sechele went to Cape Town to demand redress. Neither of them gained any satisfaction from the British officials, who were no longer in political co-operation with the missionaries: according to Dr. Moffat, the authorities at Cape Town paid Sechele "no attention, nor would even regard his tale of woe." [1]

Sir George Cathcart, the Governor and High Commissioner at the Cape, thought the incident only what Livingstone might have expected, and what Sechele deserved. He declared, in reference to Livingstone, "that the losses and inconveniences sustained do not amount to more than the ordinary occurrences incidental to a state of war, or to which those who live in remote regions beyond Her Majesty's dominions must be frequently liable." [2]

Livingstone reported that he found the tribes in the interior "just as anxious to have a path to the seaboard as I was to open a communication with the interior." [3] Hence he had little trouble in persuading Sekeletu, the new chief of the Makololo, to adopt his father's policy. Sekeletu fitted out a caravan, provided ivory for the expenses in Angola, and sent instructions throughout his extensive sphere of influence that the travellers were not to feel hungry. Livingstone left Linyanti on the 11th November 1853, and travelling to the north of the route used by the Portuguese traders, he arrived at Loanda on the 31st May 1854. He was received with the greatest kindness by the Portuguese authorities. "The Portugese have been amazingly kind," he wrote of them.[4]

After a residence of some months at Loanda, in a house which is now in ruins, but which the Governor-General has arranged shall be marked with a memorial tablet, he left on 20th September for Linyanti. Not contented with the great feat of having crossed from the middle of the continent to the Atlantic, he resolved to continue down the Zambesi valley to the east coast. The chief provided him with a still larger escort and fresh supplies, and thus re-equipped, Livingstone left Linyanti on 3rd November 1855, and reached Quilimane on the east coast on 22nd May 1856. He was thus the first white man, other than Portuguese traders, to cross tropical Africa from shore to shore.

The previous trans-African journeys of the Portuguese have been often ignored or discredited; but they may be fully admitted without detracting from Livingstone's merit. The evidence appears conclusive that Portuguese and Portuguese half-castes had occasionally crossed from the Portuguese

---

[1] *Miss. Mag. and Chron.*, vol. xviii., 1854, p. 75.

[2] "Further Correspondence Relative to the State of the Orange River Territory," *Parl. Pap.*, 1854, vol. xliii., p. 5.

[3] *Proc. R. Geog. Soc.*, vol. ii. 1858, p. 126.

[4] *Miss. Mag. and Chron.*, vol. xx., 1856, p. 196.

West African to their East African colonies. This journey was made both ways at the beginning of the nineteenth century by two men, P. J. Baptista and A. José Francisco, who were probably half-castes.[1]

Livingstone returned home and was rightly welcomed as a national hero, for his long journey, with its chain of well-fixed positions, is generally regarded as the most important contribution to the geography of tropical Africa that has ever been made. The feat was all the more magnificent as it was achieved by help enlisted from a negro chief. It showed that both Africa and its people had been unjustly maligned. The power of one man to persuade a Central African chief to equip so great an expedition aroused high hopes of what might be done in Africa by adequate use of native help.

Public opinion was well guided in its estimate of Livingstone's work by leading geographical experts. Sir Roderick Murchison, then president of the Royal Geographical Society, proclaimed the greatness of Livingstone's achievements. The Geographical Society gave the traveller his first public welcome. Livingstone gratefully dedicated his *Missionary Travels* to Murchison,[2] whose death he deplored as the loss of the truest friend he had ever had. Sir Thomas Maclear, the Government Astronomer of the Cape, testified to the unusual accuracy with which Livingstone had determined his positions. Honours and honorary degrees were showered upon him. It was soon recognised that he was not only in the front rank of explorers, but that he had the practical insight of a statesman, and was animated by lofty philanthropic ambitions. He was appointed British Consul for the East African coast, and "the independent districts in the interior," and went back to East Africa as head of a well-equipped Government expedition.

The master idea of all the later part of his life was the suppression of the slave-trade by the introduction of European commerce and colonisation.

That the object of the Zambesi expedition was mainly geographical

---

[1] These "Pombeiros" have been generally described as blacks and as slaves, but this view is not borne out by the documents connected with the expedition. The term Pombeiros is rendered as "bondsmen" by Beadle, who translated Baptista's journal for the volume issued by the Royal Geographical Society under the title *The Lands of Cazembe* (London, 1873, pp. vii., 271, and map). The Pombeiros are referred to as "home born slaves" in the list of contents to that work. The men were probably half-castes. They do not appear to have been ordinary slaves. Thus the Governor of Sena, in his letter in reference to this journey, speaks of slaves having journeyed to Cazembe; but he refers to the two Pombeiros as "persons," and twice refers to them as "explorers." He remarks, "I do not find a large amount of intelligence in these explorers; but, at the same time, I admit that, according to their capabilities, they did a great deal."

The style of Baptista's journal resembles that of an illiterate European rather than that of a negro. He constantly refers to "blacks" in a way which shows that he certainly did not regard himself as a black. He and his companions are referred to in the journal as white men. Thus on their introduction to the chief Cazembe, the guide said, "I bring you some white men here" (p. 187). Again (p. 211) some people at a farm answered, "it was very fortunate to see white people, whom they call Muzungos, coming from Angola" (Muzungu, plural Wazungu, is the Suahili word for European. See *e.g.* Krapf, *Suahili Dictionary*, 1882, p. 271).

[2] He wished to dedicate also to Murchison his book on the Zambesi expedition.

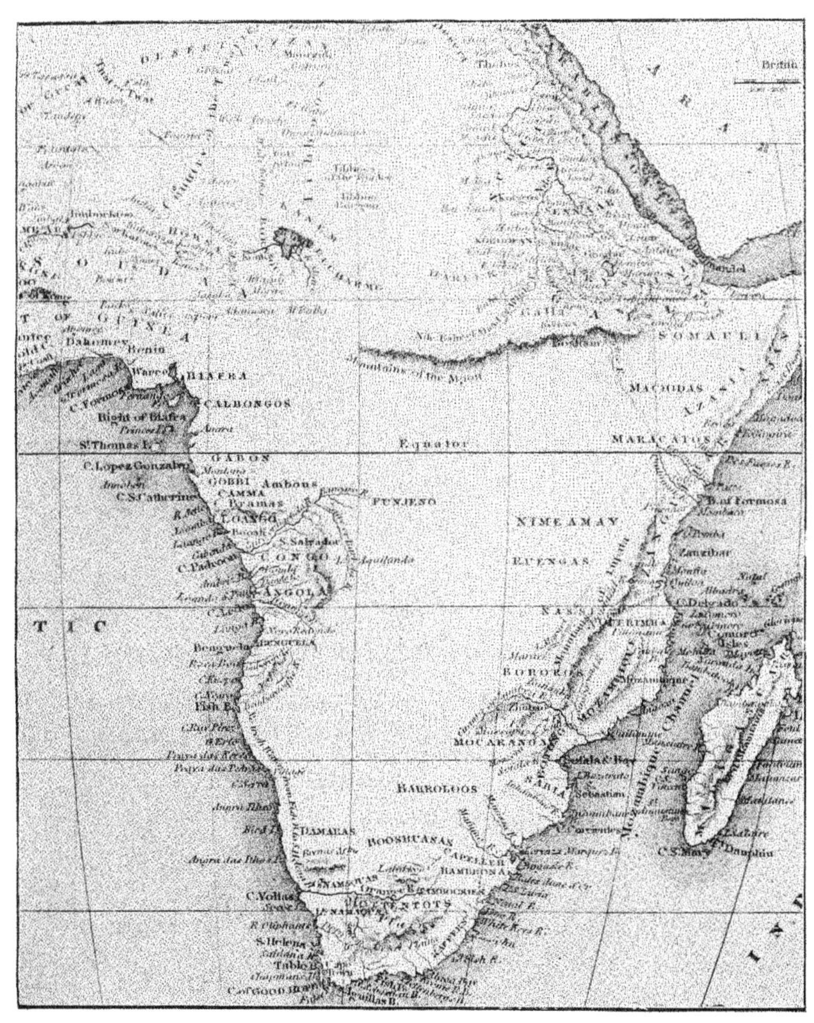

THE MAP OF AFRICA IN 1823.

(Hofland, *Africa Described.*)

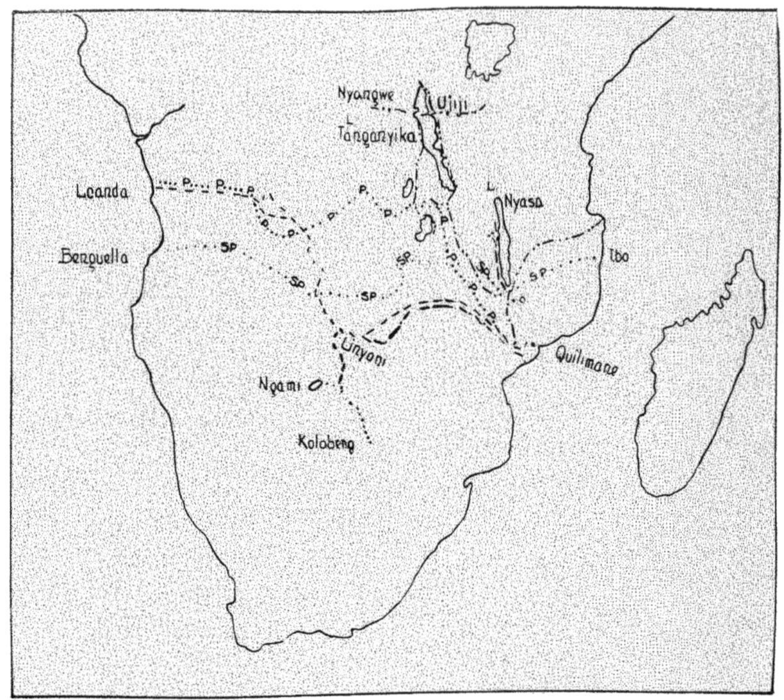

SKETCH-MAP OF LIvINGSTONE'S MAIN ROUTES OF EXPLORATION.

. . . . . . Expedition to Ngami.	.·. . . P . . . . P . . . . Approximate route of the Pombeiros, 1802-1811.
------ Trans-African Expedition.	. . . . . . SP . . . . . . Reported route of Silva Porto, 1853-1854.
-·-·-· Zambesi Expedition.	
-··-··- Last Expedition.	

has been clearly stated in the beginning of Livingstone's book upon its work.[1] The mission appealed to him, for he trusted that it might strike a great blow at the slave-trade throughout the world. He was convinced from the excellent results of the Portuguese administration in West Africa, where he described the slave-trade as practically dead, that, if the Africans were allowed to develop their keen instincts for trade, slavery would prove unprofitable and would collapse. He believed that the Lower Zambesi contained the best cotton-growing land in the world. "I think," he wrote, "the most important part of the discoveries I was privileged to make is, that there is an immense extent of country where sugar and cotton might be cultivated."[2] Livingstone believed the country could be so developed as to grow all the cotton and sugar required by the British markets, and thus render them independent of the slave-grown products of America. Slavery on both sides of the Atlantic might be slain with one blow.

The expedition, with its official position, ample funds, and large expert staff, appeared to give Livingstone the chance of his life. Its geographical results, though important, were less striking than those from his two other main journeys. Livingstone was away on this expedition from 1858 to 1864, and during that time surveyed the course of the Zambesi from the Victoria Falls to the sea; he explored its tributary the Shire, discovered Lake Shirwa, and marched up the western coast of Lake Nyasa nearly to its northern end.[3] The most important discovery was that the region around Lake Nyasa has a dense population, a fertile soil, large tracts of healthy highlands, and is near extensive coalfields. The work of the expedition thus led to the foundation of the British Protectorate of Nyasaland. The botanical collections made by Sir John Kirk, the medical officer of the expedition, were a most valuable contribution to African botany. In spite of all these results, the expedition was generally judged a comparative failure. Murchison expressed the current view by remarking in a letter to Livingstone, "the little success attending your last mission."[3]

Success in the main object of the expedition was, however, at the time impossible. Its work was hampered by three difficulties. The boat, the *Ma-Robert*, which had been especially built for the expedition, proved a failure. Livingstone denounced its builder as a worse thief than any of the Africans who had robbed him; but as the boat was

1 "The main object of the Zambesi expedition, as our instructions from Her Majesty's Government explicitly stated, was to extend the knowledge already attained of the geography and mineral and agricultural resources of Eastern and Central Africa—to improve our acquaintance with the inhabitants, and to endeavour to engage them to apply themselves to industrial pursuits and to the cultivation of their lands, with a view to the production of raw material to be exported to England in return for British manufactures; and it was hoped, that, by encouraging the natives to occupy themselves in the development of the resources of the country, a considerable advance might be made towards the extinction of the slave-trade, as they would not be long in discovering that the former would eventually be a more certain source of profit than the latter" (*Expedition to Zambesi*, 1865, p. 9).

2 *Proc. R. Geog. Soc.*, vol. ii., 1858, p. 58.

3 See map, p. 248.

4 Blaikie, *Personal Life of Livingstone*, p. 297.

passed by the Admiralty, it was probably well built, and Murchison expressly exonerated the builder from blame. The trouble apparently came from overloading a boat which had been designed to have a draft of sixteen inches till it drew two and a half feet. No wonder it was slow and leaky.[1] Livingstone, at his own expense, replaced the *Asthmatic*, as he nicknamed the first boat, by the *Lady Nyasa*. It was built in sections for portage past the Murchison Falls to Lake Nyasa; but before it could be used there the expedition was recalled. He sailed the *Lady Nyasa* to Bombay, where she was sold; and, with his usual financial ill-luck, he placed the money in a bank which failed, and he lost it all.

The second trouble was the despatch in 1861 of a Church of England mission in the wake of the expedition. Murchison strenuously objected to this proceeding, for there was ample room elsewhere in Africa for missionary enterprise; and he pointed out that the establishment of a Protestant mission, in territory claimed by Portugal, must lead to trouble between the expedition and the Portuguese.[2] Livingstone, however, welcomed the mission, and was deeply disappointed at its failure, for which he blamed the bishop in charge.

Murchison's objections proved only too well justified, and Livingstone became involved in a bitter feud with the Portuguese. He still referred with sympathy to their administration of Angola, but he denounced that in East Africa with characteristic vehemence. As the British Government recognised that it was impossible for a British mission to suppress the slave-trade, in an area that was isolated from the coast by a wide tract of Portuguese territory, the expedition was recalled. The justice of this decision was sadly admitted by Livingstone.

On Livingstone's return home he suffered a second period of lionisation; from this he escaped in August 1865, when he left England on the expedition that lasted till his death in 1873. Its object was clearly and repeatedly explained by Murchison. It was to determine between the view that Tanganyika was the source of the Nile as held by Beke, Findlay, and Burton, and the contrary view indicated on Speke's earlier map. Livingstone inclined from the first to believe that Tanganyika discharged to the Nile.

Though Livingstone retained his official position as British Consul, he did so without salary or claim to pension. Most of the expense of this expedition he contributed himself or obtained from personal friends. The largest donation, £1000, was contributed by his devoted Glasgow friend, James Young, founder of the Scottish oil industry.

Livingstone often declared that he would only go to Africa as a missionary. In the introduction to his book on the Zambesi Expedition, Livingstone has clearly explained the scope of his investigations during the trans-African journey described in his *Missionary Travels*. "In

---

1 This overloading appears to have been first due to the officer in charge of the boat, whom Livingstone described as his "naval donkey" (*Life of Oswell*, vol. ii. p. 59).

2 The Prince Consort's refusal to become patron of the Universities' Mission may have been due to the same view.

our exploration the chief object in view was not to discover objects of nine days' wonder, to gaze and be gazed at by barbarians; but to note the climate, the natural productions, the local diseases, the natives, and their relation to the rest of the world." Since that journey his geographical interests had been steadily growing; and his diary during his last expedition shows that he had become intensely interested in the problem of the Nile, and was anxious to crown his African career by settling that most prolonged of geographical controversies. "The object of my expedition," he wrote to his son Tom, in explanation of his last journey, "is the discovery of the sources of the Nile."[1] He also welcomed the mission owing to its new opportunities for furthering his schemes of African development; and owing to his quarrel with the Portuguese he was glad to explore a new road into the interior north of their dominions, and yet south of the region where an arid belt intervenes between the coast and the fertile interior. The last six years of the expedition was devoted exclusively to the Nile problem.

The zone of Africa north of the Zambesi is geographically more interesting than the country to the south; but the conditions of travel there are more difficult and annoying. Large tracts of South Africa were occupied by comparatively organised tribes. A traveller who had secured the friendship of a leading chief could explore safely far and wide under his authority; but north of the Zambesi there were no great chiefs. The country was occupied by small, independent tribes, who maintained perpetual blood feuds, so that inter-tribal intercourse was restricted. Travel was precarious, and in some districts impossible except to large and well-armed caravans. The natives were ignorant of the power of the white man. They had had no experience of punitive expeditions, and the massacre of a caravan usually secured rich loot and no punishment. For the conditions of travel south of the Zambesi, Livingstone was ideally suited; but in the region of his last expedition his very virtues told against him. He was bored by the continual wrangles with village elders, and in his later years he appears to have been too kind-hearted, or perhaps too indolent, to discipline his caravan.

The geographical results of Livingstone's last and longest expedition were most important, but the story how its great discoveries were made, is one of the most pathetic in the annals of African exploration.

On the Zambesi expedition Livingstone's relations with his European comrades had not been happy, so this time he wisely went alone. He started with a caravan of fifty-seven men and boys; he trusted for transport to camels and buffaloes, as he hoped that they would resist tsetse disease. His drivers were a party of sepoys, who soon lost heart, and tried to secure the return of the expedition by killing the transport animals. Livingstone, in his journal and still more in private letters, complained of these men's brutality. He said he had to hurry on ahead of the caravan so that he might not hear the groans of his baggage animals. Some of them, he says, were deliberately beaten to death, and others died through the torturing of their wounds. His experiment in

---

[1] Blaikie, *Personal Life of Livingstone*, p. 332.

transport had, therefore, no chance of success. The beasts of burden were soon dead and the men out of control.

The expedition began by the ascent of the Rovuma River, across what is now German East Africa. Livingstone's most direct route to his special field of work would have been past the northern end of Lake Nyasa and thence to Tanganyika along the route followed by the Stevenson road; but Livingstone was gradually pressed southward by the timidity of his men. He passed around the southern end of Lake Nyasa and began the exploration of what is now north-eastern Rhodesia.

Shortly after leaving Lake Nyasa the sepoys deserted in a body; and to excuse their return and secure their pay, they told a dramatic story of how Livingstone had been killed in a fight, after he had heroically slain several of his opponents. The two most competent authorities at Zanzibar, Dr. Seward, the Consul, and Sir John Kirk, the Vice-Consul, believed the men's story, which was also accepted by Sir Samuel Baker. It was scouted by Murchison and Oswell, and finally disproved by a search party under Young, which, though it failed to reach Livingstone, obtained conclusive evidence that he was alive after the date of his reported murder.

After the desertion of the sepoys Livingstone met some Arab traders, and the happiest part of the expedition was during his journeys with them. Livingstone wrote from Bangweolo (8th July 1868), "The Arabs have all been overflowing in kindness."[1] He had lost his stores and had only a few men left. The Arabs at once supplied him with provisions, cloth and beads; and "showed," said Livingstone, "the greatest kindness and anxiety for my safety and success."[2]

Livingstone still loathed the slave-trade, but he had come to recognise that a certain measure of domestic slavery was inevitable in that stage of African development; and he drew pleasing pictures of the kindly relations between his Arab friends and their slaves. "I was glad," he wrote, "to see the mode of ivory and slave trading of these men, it formed such a perfect contrast to that of the ruffians from Kilwa, and to the ways of the atrocious Portuguese from Tette."[3] At Nyangwe, however, he again saw the full horrors of the slave-trade.

Livingstone's work during this part of the expedition disproved the supposed connection between Lakes Tanganyika and Nyasa; he discovered Lakes Bangweolo and Moero, and evidence which convinced him that the rivers of that district were the head streams of the Nile. "I have found what I believe to be the sources of the Nile between 10° and 12° S.," he wrote joyfully home.[4] In one of his waking dreams he speculated as to the possibility of Lake Moero being the Meroe where, according to ancient tradition, Moses lived for some time with his Egyptian foster-mother, the Princess Merr.[5]

In a journey westward from Tanganyika, he reached the Lualaba River at Nyangwe, and from its vast volume feared that it might be the head stream of the Congo, as Stanley afterwards proved to be the case.

---

[1] *Proc. R. Geog. Soc.*, vol. xiv., 1870, p. 8. [2] *Ibid.* p. 10. [3] *Ibid.* p. 10.
[4] *Ibid.* p. 8. [5] *Last Journals*, vol. ii. p. 59.

Any water that flowed to the Congo had no interest for Livingstone. In one entry in his journal, he described himself as "oppressed with the apprehension that after all it may turn out that I have been following the Congo; and who would risk being put into a cannibal pot, and converted into black man for it." [1] He clung tenaciously to the idea that the Lualaba discharged to the Nile, and in a letter to Oswell, written in October 1869, said: "there can be little doubt that such is its destination." He wrote to Murchison: "the sources of the Nile are undoubtedly between 10° and 12° S." [2]

He was, however, unable to trace the Lualaba northward in order to determine whither it flowed. He returned to Ujiji, where, to his bitter disappointment, he found that his stores had been sold by an Arab. Another of the Ujiji Arabs at once called on him with the offer to sell some ivory and replace the stolen goods. Livingstone temporarily declined this generous offer, and a few days later his wants were relieved by the arrival of H. M. Stanley.

Livingstone's long stay to the west of the lakes had naturally occasioned great anxiety at home, for it had been thought that the expedition need not spend more than a year in the field, and the contribution from the Geographical Society had been given on that estimate. The news of the death of his transport animals, and the desertion of his sepoy escort, revealed his difficulties and danger, and repeated rumours of his death reached the coast.

After his long absence, and the failure to obtain news of him, Mr. Gordon Bennett of the *New York Herald* sent Stanley to find the lost explorer. The authorities regarded the quest as hopeless. Stanley reached East Africa in 1871, when Livingstone had a five years' start, and it was thought by some of his friends that he would certainly cross Africa to the west coast. But Stanley's pursuit was swift; he crashed through every obstacle which could not be avoided. He left his two European comrades buried beside his path, and after a rapid march of five months found Livingstone at Ujiji, the Arab settlement on the eastern shore of Tanganyika.

Stanley was then a young and adventurous journalist, and appears to have had no special interest in Livingstone, or sympathy with geographical research. Stanley found a man surprisingly unlike his expectations. The two men must have been extraordinarily different in original disposition, and Livingstone's thirty years' residence in Africa had left its mark on his temperament. In spite of his many trials there, he doubtless much preferred life in Africa to that in Europe. Sir Francis Galton, in explanation of Livingstone's long absence, remarked that "there was no ground for crediting Livingstone with any excessive home-sickness. He was as much at home in Africa as in England; . . . therefore, when he received his supplies, if he had more work to do, no doubt he would remain." [3] But in spite of Stanley's first disappointment with Livingstone, he gradually fell under his influence during their four

[1] *Last Journals*, vol. ii. p. 188.
[2] *Proc. R. Geog. Soc.*, vol. xiv., 1870, p. 12.
[3] *Ibid.* vol. xv., 1871, p. 209.

months' residence together. Livingstone was anxious to determine whether the river at the northern end of Tanganyika flowed into the lake or out of it, for, according to the Royal Geographical Society, that was "the great object of Livingstone's journey."[1] The Arabs who had seen the river said it flowed into the lake. Livingstone expected it to flow out of the lake. Stanley probably thought the difference immaterial, but that if Livingstone really wished to know he might as well satisfy this whim. So he made the necessary arrangements to visit the locality, and invited Livingstone to go as his guest. They found the Arabs were right; but the journey settled far more than that question.

Around the camp fires at night, Livingstone told Stanley the story of his life and wanderings, and explained his theories of the Central African river system. He seems to have laid in Stanley's mind the germ of the belief that knowledge is worth something for its own sake alone. Stanley, having supplied Livingstone's immediate needs, returned to the coast, and sent him back ample supplies, with a caravan of carefully selected and thoroughly reliable Zanzibari. Thus re-equipped, Livingstone started west again in 1872. He travelled slowly, for his usual deliberation was increased by growing weakness. He struggled on to Lake Bangweolo, where he died between the 1st and 4th of May 1873. To prove his death, the men embalmed the body, and, after many difficulties, carried it back to the coast, and it was buried in Westminster Abbey.

The geographical results of this long expedition were most important. Though they did not conclusively settle the problem which Livingstone had started to solve, they showed that Tanganyika had no northern outlet, and was not connected with Lake Nyasa. The freshness of the water of Tanganyika indicated that it had an outlet somewhere, which was subsequently found by Stanley. Livingstone proved that the river known as the Chambesi was not, as had been thought from its name, a tributary of the Zambesi, but that it flowed through Lake Bangweolo and thence discharged northward; but whether the northward-flowing rivers to the west of Tanganyika joined the Congo, or whether, as Livingstone both thought and hoped, they were tributaries of the Nile, was not settled until Stanley's epoch-making expedition, described in *Through the Dark Continent.* Nevertheless, Livingstone's exploration of the region to the south and west of Lake Tanganyika was a contribution of fundamental importance to African geography.

In the above attempt to summarise Livingstone's thirty years' geographical work in Africa in less than thirty minutes it has been only possible to refer to the general outline of his travels. His results were so great that Livingstone is universally accepted as one of the world's greatest explorers.

His fame rests mainly on the first of his three chief expeditions, owing to the great extent of new country it explored, the unexpected nature and practical value of its discoveries, and its achievement at the cost of a native chieftain. Sir Roderick Murchison proclaimed this

---

[1] *Proc. R. Geog. Soc.*, vol. xi., 1867, p. 144.

journey as "the greatest triumph in geographical research which has been effected in our times."

The outstanding features of Livingstone's explorations were the great length of his new routes and the precision with which he determined his positions. He travelled about 29,000 miles in Africa, and traversed so much new ground that, to quote the authoritative opinion of Dr. Scott Keltie, "no single explorer ever did as much for African geography as Livingstone." "No explorer on record has determined his path with the precision you have accomplished," wrote Sir Thomas Maclear to Livingstone.

The accuracy of his work may be shown by comparison of some of the positions he determined in Angola with those adopted by the official Portuguese map of the area published in 1910; his results, as shown by the following list, are extraordinarily accurate, considering the conditions under which he worked.

		Portuguese Map, 1910.
Katema's, near Lake Dilolo,	11° 35′ 49″ S.	11° 20′ S.
	22° 27′ E.	22° 2′ E.
Golungo Alto, . . .	9° 8′ 30″ S.	9° 4′ S.
Ambaca, ...	9° 16′ 35″ S.	9° 14′ S.
Probably, .	15° 23′ E.	15° 12′ E.
Pungo Andongo, . .	9° 42′ 14″ S.	9° 40′ S.
	15° 30′ E.	15° 34′ E.

His first journey, with its chain of accurately determined stations, gave the first reliable section across tropical Africa, and represents the greatest single contribution to African geography which has ever been made. His account of the country was a revelation to European geographers. The Kalahari had been regarded as part of a vast desert, which was believed to occupy the whole interior of Southern Africa, and had been called the Southern Sahara, from the belief that it corresponded in size to the Sahara of Northern Africa. Livingstone's discovery that southern tropical Africa was watered by large rivers, contained many great lakes, and a fertile soil which grew excellent cotton, sugar, indigo, and other tropical produce, and that its inhabitants were keen traders, expert agriculturists, and skilled craftsmen, was a discovery as momentous as it was unexpected.

Livingstone therefore discovered that southern tropical Africa, instead of being a useless desert, is a land of incalculable commercial possibilities. Europe was given an interest in Africa, which has never ceased to grow; Livingstone was thus the most influential pioneer in opening tropical Africa to civilising influences.

Livingstone's views as to the geographical structure of Equatorial Africa were of less merit than his practical achievements; but his lack of training in geography is responsible for any deficiencies in his theoretical conclusions. He represented Africa as a plateau bordered by two mountain chains which run north and south and are separated by a great central depression. This view had been advocated by Murchison in 1852, and Livingstone independently arrived at the same conclusion; and as a simple generalised statement of the structure of southern tropical Africa it is essentially correct.[1] Sometimes, however, probably under Murchison's influence, Livingstone overestimated the depth and importance of the central depression, and imagined that it had been filled by one vast fresh-water sea, which had been drained by the formation of cracks through the mountain ranges on each side. The gorge of the Zambesi below the Victoria Falls, one of his most famous discoveries, he regarded as the most conspicuous of these cracks. The existence of this sea has not been confirmed, and the gorges on the margins of the plateau were doubtless cut by the rivers that flow through them.[2]

In some of the statements of his theory Livingstone represented Africa as consisting of a low, malarial plain, enclosed by a rim of mountains which he described as "perfect sanatoria." Such a conception of the structure of Africa is only partially true. But Livingstone clearly recognised that Equatorial Africa is essentially an ancient plateau, of which the jagged sides look from below like mountain ranges. His account of the descent from the eastern plateau of Angola to the western coast shows that he understood its real geographical structure;[3] and his conclusion: "The continent seems to be an elevated tableland sloping chiefly towards the east" was quite correct for the zone he was describing.

Livingstone's views as to the structure of Africa were a great advance on the ideas of most of his contemporaries. Thus the *Encyclopædia Metropolitana*, in its article on Africa, in 1845, adopted a classification of African mountains which was very crude in comparison with Livingstone's explanation of their nature; and it described as the most striking feature of Africa, "the immensity of its deserts." In addition to the Sahara, it says (p. 195), the continent is "everywhere intersected with deserts of an inferior, but still of great extent; and these are to be found even in the southern parts, towards the European settlements. There is, probably, a wide wilderness of this nature, between the east and west ranges of mountains, pervaded by the race of people called Jagas, who sometimes are said to wander into the vicinity of the Cape."

Regarding the sources of the Nile, the problem to which Livingstone devoted the last six years of his life, his theoretical conclusions were less correct than those on the structure of the continent. The facts that he discovered during his last expedition would, no doubt, have led him to recognise, if he had been spared to write a connected summary of his results, that his view was practically impossible. But while in the field he was under the conviction, which almost amounted to an obsession, that he was extending southward the basin of the Nile. This view was probably a reaction from the supposed former limitation

---

1 According to Dr. Scott Keltie, in the article on Livingstone in the last edition of the *Encyclopædia Britannica*, vol. xvi., 1911, p. 841, "The conclusions he came to have been essentially confirmed by subsequent observations"

2 The formation of the gorge by the cutting back of the Victoria Falls has been proved by Mr. G. W. Lamplugh (*Quart. Journ. Geol. Soc.*, vol. lxiii., 1907, pp. 162-214).

3 The essential accuracy of Livingstone's interpretation of the mountains of Angola I had the pleasure of confirming last summer by a journey along a parallel line to the south of his route.

of the Nile to the mountains north of the Equator; Speke's work had extended the Nile basin south-eastward by the inclusion of the Victoria Nyanza, and Livingstone hoped to extend it still further south and south-westward to include the basin of Tanganyika. When his visit with Stanley disproved the supposed northern outlet from Tanganyika to the Albert Nyanza, he clung to the hope that the waters on the highlands west of Tanganyika flowed north-eastward into the Nile; and he probably died all the happier from the belief that he had been mapping the head waters and adding an enormous area to the basin of the Nile.

Livingstone is to be judged by his achievements as an explorer and a pioneer of civilisation rather than by his views as a geographer. The spiteful criticism that, as an explorer, he stands in the highest rank and as a geographer in the very lowest, it is only worth quoting as an illustration of the undoubted superiority of his exploration to his theoretical geography. Livingstone began his work with no training in the general principles of geography, but he showed such skill in surveying and such keen insight that it is difficult to realise how much the world has lost by Livingstone's lack of geographical education. The chief defects of his geographical writings are his neglect of his predecessors and his irritation at unfavourable criticism. Cooley, learned in African lore, was offended by neglect of this subject, and severely denounced what he described as Livingstone's strong dislike of preliminary information. Livingstone's books should, however, be judged as records of his own personal observations and opinions. After he left Britain for South Africa he spent less than two and a half years at home, and during his two visits he was so overpressed with engagements that he had no time to study the literature of African exploration.

It is more important in an appreciation of Livingstone to recognise his relations to his successors than to his predecessors; he was one of those independent pioneers who owe little to either forerunners or contemporaries; but the impression he made on his successors has profoundly influenced the politics of both Africa and Europe.

The year of Livingstone's death, 1873, marked a great turning-point in African history. It closed the work of the dreamers. Men of that type still went to Africa; they go there still; but henceforward they were unimportant compared with those who went to help the administration and the commercial development of the country. Lord Houghton referred, in some verses on Livingstone's funeral, to the promise of a new era in Africa:

> "Morning o'er that weird continent
> Is slowly breaking;
> Europe her sullen self-restraint
> Forsaking."

The dawn which Lord Houghton discerned broke on Africa with dazzling swiftness. Many influences helped its advance. Livingstone's death aroused Europe to a sense of its responsibilities to Africa and the conviction that we are our black brothers' keepers. Gordon had just reached Khartoum to continue Baker's attempt to rescue the Soudan from the raid of the slave-trader and the misrule of the Pashas. France was beginning to develop her west coast colonies so as to gain in Africa the supremacy she had lost in Europe. The greatest influence of all, however, was that which Livingstone exerted through his pupil Stanley, in whom were growing the interests which Livingstone had planted. Before the end of the year of Livingtone's funeral Stanley was back on the east coast, his life dedicated to the service in which his teacher had fallen.

Stanley's first journey across Africa solved the problems for which Livingstone died; and it led to the foundation of the Congo Free State. That State was founded with high ideals. "I am charged," wrote Stanley, "to open and keep open, if possible, all such districts and countries as I may explore, for the benefit of the commercial world. The mission is supported by a philanthropic society, which numbers noble-minded men of several nations. It is not a religious society, but my instructions are entirely of that spirit. No violence must be used, and wherever rejected, the mission must withdraw to seek another field." And the success of the undertaking which was begun with these motives led to the rapid growth of a keener European interest in Africa, and the subdivision of the continent among the States of Europe. The history of that change has been marred by many crimes and blunders, but there can be no doubt that the Africans are immeasurably safer and happier to-day than they were forty years ago.

Livingstone appears to have died under the impression that his life had been a failure. With the malicious irony of fate the only immediate commercial result of his journey from the Upper Zambesi to the west coast was the opening of the worst slave road and, by the help of his friend, Sekeletu, the most active slave market in Equatorial Africa. The three missions—Episcopalian, Congregational, and Presbyterian—sent to the Zambesi valley in answer to his appeal had all withdrawn; and the work during the later years of his last journey had thrown doubt on the confident conclusions he had announced from its first discoveries. But, forty years onward, we can better realize how great was the work that he accomplished; for the beneficent revolution that has taken place in Africa was due both to Livingtone's discovery of its vast possibilities and to Livingstone's influence on the men who established civilisation where he entered as the heroic pioneer.

THE LIVINGSTONE CENTENARY LOAN EXHIBITION AT THE ROYAL SCOTTISH MUSEUM.

By Alexander Galt, D.Sc., F.R.S.E., Keeper of the Technological Department, the Royal Scottish Museum.

(With Illustrations.)

About eighteen months ago it occurred to the writer that, as the hundredth anniversary of the birth of David Livingstone would fall this year, the occasion might be appropriately marked by the holding of a Livingstone Loan Exhibition in the Royal Scottish Museum. Official sanction was readily given to the proposal, and there has now been gathered together a highly interesting and complete collection of exhibits, largely of a personal nature, illustrating the life and work of Livingstone. For these the Museum is indebted to many lenders, particularly to the various members of the Livingstone family, all of whom were so kind as to put their treasured possessions at our disposal for a time. Special thanks are due to Mrs. Livingstone Wilson, the youngest and sole surviving member of Dr. Livingstone's own family, and to Captain Livingstone Bruce, a grandson of Dr. Livingstone.

The exhibition was opened on the evening of Monday, March 17, when the Lord Provost, Magistrates, and Council of the city gave a reception in the Museum. The exhibits have evidently aroused great public interest, many thousands of visitors having already come to see them. It is proposed to keep the exhibition open for six months.

The life of David Livingstone has recently been so fully described in the press and on the platform that only those particular aspects of it which are illustrated by the exhibits need be referred to here. Livingstone was born in the village of Blantyre, near Glasgow, on March 19, 1813, and when ten years of age he left school and began work in a cotton mill there. In those days the working hours were long—from six o'clock in the morning till eight at night; and yet, by attending a night school and by self-education, Livingstone contrived to gather a great store of sound learning during the second ten years of his life. The only relic of this period in the exhibition is an Arithmetic Book with Livingstone's signature scribbled all over the inside covers: the date (1826) shows that he was then thirteen years of age. It is recorded that at the age of sixteen he was well versed in Latin, and had read a great part of the writings of Virgil and Horace. His knowledge of the works of standard English and American authors was extensive, and he had made a special study of the subjects of botany, zoology, and geology.

When he had reached his twenty-first year Livingstone decided to prepare himself to become a medical missionary abroad, and accordingly two years later, at the beginning of November 1836, he went to Glasgow for his first session at College. He attended the classes of Chemistry and Anatomy, including lectures and practical work in both subjects, at the Andersonian University, now Anderson's College Medical School. Livingstone's Class Certificates, signed by the professors, are shown in a case labelled "Dr. Livingstone as a Medical Man." He was privileged to study under Dr. Graham, the distinguished Professor of Chemistry, and Dr. Graham's assistant, Mr. James Young, who afterwards gained fame and fortune as the originator of the shale-oil industry in Scotland. In Mr. Young's private laboratory Livingstone received a good training in the use of tools, a training which must have been of great service to him in after years in Africa, and here he met William Thomson (afterwards Lord Kelvin), who as a boy often visited the laboratory. This early friendship between Livingstone and James Young became a most intimate one, and it continued throughout the whole of Livingstone's subsequent career. In fact, Young's liberality in supporting Livingstone's work of missionary exploration in Africa was very great, culminating in 1872 in the fitting out of a relief expedition at his own expense. A Portrait of James Young is hung on the wall in the gallery. During the succeeding winter, 1837-8, Livingstone again attended classes at the Andersonian University, the subjects being Chemistry and Surgery (lectures and practical classes in both), and Materia Medica, the Class Certificates for which are also shown. Finally, in 1838-9, Livingstone again attended the Anatomy class. During these years he continued to work in the mill in the summer six months.

Meantime Livingstone had offered himself to the London Missionary Society, and he was invited in 1838 to go up to London to undergo, at Ongar in Essex, a few months' training for foreign mission work. Two of Livingstone's Letters belong to this period, and these, along with many others, are exhibited in chronological order. The small handwriting of the early letters is noteworthy, when comparison is made with the large bold style of handwriting adopted in later years.

In 1839 Livingstone began his final medical studies at Charing Cross Hospital, London, under the guidance of Dr. (afterwards Sir) J. Risdon Bennett, whose lectures on the Practice of Medicine he attended. For nearly eighteen months Livingstone gained much valuable training and experience in the practice of medicine and surgery in the wards of the hospital and in Aldersgate Street dispensary. A Certificate of attendance at the dispensary, signed by Dr. Bennett and other physicians and surgeons, may be noted. It was at this period that Livingstone met in London the famous missionary, the Rev. Robert Moffat of Kuruman (who afterwards became his father-in-law), through whose advice it was now arranged that Livingstone should go to the South African mission field. A fine Portrait of Dr. Moffat, at a considerably later period of his life, is included in the collections. Finally, Livingstone went back to Glasgow in November 1840 to submit himself for examination by the Faculty of Physicians and Surgeons. His Medical Diploma from the Faculty is exhibited along with a copy of the Oath, in his own handwriting, required from all new Licentiates. In 1857 the Faculty recognised his eminence by making him an Honorary Fellow. A fine Portrait in oil of Dr. Livingstone has been lent to the exhibition by the Faculty. We may also note here that the Dental and Surgical Instruments used by Dr. Livingstone in Africa are placed beside the diploma and the other medical exhibits.

In an adjacent case may be seen numerous Geological Specimens, collected by Dr. Livingstone during his first great journey, which he presented to the Museum in 1858 through Dr. George Wilson, the first Director, who had been a fellow-student with Livingstone in London. Above these may be noted a splendid Microscope with accessories, in two fine brass-bound boxes, presented by Miss Angela (afterwards the Baroness) Burdett Coutts. The inscription on the boxes is as follows: "To David Livingstone, LL.D., D.C.L., etc. A token of remembrance from Angela G. Burdett Coutts. With her earnest wish that his path in difficulties may be clear like as this Microscope clears that which the natural eye cannot see. London, January 25, 1858."

Dr. Livingstone sailed for Cape Town in December 1840, and during his long voyage out by sailing-ship, *viâ* Rio de Janeiro, he seems to have been on very friendly terms with the captain, who gave him a sound course of practical instruction in the use of the sextant and in the methods of determining the ship's position at sea. After a short stay at Cape Town, Dr. Livingstone proceeded to Port Elizabeth, from which place he started in May 1841 for Kuruman in Bechuanaland, the most distant missionary station; and during the two months' journey of 700 miles up country by ox-wagon he had his first opportunity of observing the life and customs of the natives. The next two years were spent in making long journeys of exploration into the unknown regions to the north of Kuruman, with the view of making the acquaintance of the different Bechuana tribes, and noting the suitability of the country for missionary stations. He lived alone with one of the tribes for some months to give himself an opportunity of mastering the language, and some years later he wrote a book on the *Analysis of the Bechuana Language*, the autograph manuscript of which is exhibited in a case marked "Dr. Livingstone's Literary and Scientific Work."

In August 1843 he founded a new mission station at Mabotsé, in the territory of the Bakhatla tribe, 250 miles beyond Kuruman. The district was found to be infested with lions, from one of which he had a very narrow escape from death. The lion's teeth had crunched through his upper left arm bone, and as Livingstone was unable to set it properly by himself, that arm was always weak for the rest of his life. It may be stated here that, when Dr. Livingstone's body was brought home from Africa in 1874, the false-jointed arm bone formed one of the principal means of identification of his remains. A Model of this Bone is exhibited in the collections.

A highly important event in Dr. Livingstone's career was his marriage to a daughter of Dr. Moffat, towards the close of the year 1844. Their first home was at Mabotsé, from which they shortly removed to Chonuane in the country of the Bakwain tribe, whose young chief, Setchelé, afterwards became a Christian and a great friend of Dr. Livingstone. A Photograph of Setchelé, taken many years later, is on view. On account of the lack of water, the whole tribe removed with the Livingstones in 1847 to the banks of the river Kolobeng, 40 miles distant, where, for the third time in three years, Dr. Livingstone had to build a home, and here they remained for five years, the last and almost the only home they ever had. Photographs of Mrs. Livingstone and of a family group (including Dr. and Mrs. Livingstone, their elder daughter Agnes, and their three sons), both belonging to a later period, are exhibited. There are also two excellent Busts of Dr. Livingstone in the exhibition.

Dr. Livingstone was not content to continue to live a placid existence as an ordinary medical missionary at Kolobeng, while the vast regions beyond remained in spiritual darkness. Progress westward was barred by the Kalahari desert and to the east by the Boers, who were determinedly opposed to missionary enterprise. It was therefore decided, in 1849, to make a journey of exploration further north. He was accompanied by Mr. Oswell, a British army officer who was on a hunting expedition, and after travelling about 600 miles Lake Ngami was reached, the first of Dr. Livingstone's many geographical discoveries. A Map of the route followed, constructed by the explorers, is exhibited. The journey to the lake was repeated in 1850, Dr. Livingstone's companions this time being his wife and family. A large Oil Painting, showing the whole family at the lake, has been lent to the exhibition. In 1851 Mr. Oswell and the Livingstones made a third and more extended journey to the north, where they met Sebituane, the chief of the Makololo tribe.

It was now clear to Dr. Livingstone that if the country still further north, of which he had heard great accounts from Sebituane and his people, were to be opened up for missionary enterprise and trade, a new and much shorter and easier route to the sea must be found, as land transport from the Cape was prohibitive on account of the distance (2000 miles). He therefore decided, early in 1852, to return to Cape Town, taking his wife and family with him that they might sail for home, while he remained at the Cape for a few months making preparations for the first, and what turned out to be the greatest, of three remarkable journeys. He had already obtained considerable experience in the use of the sextant and in surveying and mapping out new country, but he wisely placed himself in the hands of Sir Thomas Maclear, Astronomer-Royal, who was so kind as to give him a thorough course of instruction in the more advanced methods of making astronomical and other observations and calculations. Two Sextants used by Dr. Livingstone during his travels in Africa are exhibited in a case headed "Dr. Livingstone's Scientific Instruments." One of them has an inscription in Dr. Livingstone's handwriting: "Sextant much valued as an old companion, D. L." We may also note in the same case his Compass, Chronometer Watch, Artificial Horizon, Aneroid Barometer, Hypsometer, Thermometer, Hygrometer, Tape Measure, Foot-Rule, Drawing Instruments, Parallel Rulers, and Field-Glasses.

Dr. Livingstone left Cape Town on his first great journey in June 1852. He had with him, for the first time, a magic lantern, presented by Mr. Oswell's companion, Mr. Murray, and he used it to illustrate his Bible lessons to the natives. Dr. Livingstone's Lantern may be seen in one of the wall cases along with two Bibles, one of which is inscribed "The Bible which went with me in all my wanderings in Africa. David Livingstone, Feb. 1858." The other was presented by the Rev. Mr. Stewart in 1865, and it was used during the last great journey. Mr. Stewart had been sent out to Africa in 1861, by the then Free Church of Scotland, to meet with Dr. Livingstone and obtain information as to the possibility of founding a mission. Dr. Stewart afterwards became the distinguished head of the Lovedale Mission.

When Dr. Livingstone arrived at Kuruman, three months after leaving Cape Town, he heard that his home at Kolobeng had been destroyed by the Boers and that all his journals were lost. Linyanti, the capital of the Makololo, was reached in June 1853, and some months later, accompanied by a band of Makololo guides, he started for St. Paul de Loanda on the west coast, where he arrived at the end of May 1854. Great difficulties and dangers were experienced on the way. Dr. Livingstone had also suffered much from fever and dysentery, but by the kind attention of friends at Loanda he gradually recovered, and after a stay

Some of the Instruments used by Dr. Livingstone in Africa.

of a few months he began his wonderful journey right across the African continent, a deviation southwards being made to Linyanti, where he engaged a new and much larger body of Makololo guides for the last portion of the journey. The east coast was reached at Quilimane in May 1856, almost four years after leaving Cape Town. Near Linyanti he had discovered the great and now well-known waterfalls on the Zambesi, which he named after Queen Victoria. A large part of the route from Linyanti to Quilimane was down the Zambesi, and a Boat's

Compass used by Dr. Livingstone oı he voyage is exhibited. A Photograph of the last survivor of Dr. L ingstone's Makololo guides, taken many years afterwards, is on view.

Speaking at a public meeting at (pe Town in 1856, the Astronomer-Royal, to whom Dr. Livingstone ha forwarded copies of his observations so that accurate maps of the hi erto unknown regions explored by him might be constructed, said that, as far as he knew, Livingstone was the first traveller who had made acciate observations, and that, though Timbuctoo was so well known, it coı l not be reached by the aid of any recorded observations, as its position as only known in a vague manner. On the other hand, you could go to ay point across the entire continent along Livingstone's track and feel ceı ıin of your position. I say, what that man has done is unprecedented.' Certainly the labour involved in making the many thousands of astroı nical, meteorological, and surveying observations, measurements, and ılculations during all his travels must have been enormous, and our ıdmiration for the man is not lessened when we remember that, dur.g two of his three great journeys, Dr. Livingstone travelled alone with l native carriers and had thus to take his own observations, that thes must often have been made in periods of great weakness through illn s or of extreme personal danger, and that his equipment was of the sleı erest description. His scientific observations were generally recorded t Dr. Livingstone in special books, one of which has been placed on view the case labelled "Dr. Livingstone's Literary and Scientific Wo ." Dr. Livingstone's Pocket Dictionary, signed and dated, is also sl wn.

The Museum is fortunate in beir able to show fifteen or manuscript maps drawn by Dr. Livingone to illustrate the va of country explored by him during l his exploratory j Africa. That chosen for reproductioı here was kindly William Stanford.

After Dr. Livingstone's return ho , i were showered upon him. The Royal had already, in 1849, awarded him Twe of Lake Ngami; in the following year and in 1855 the Gold Medal of the made a Foreign Corresponding M also had awarded him, in 18 later he received a similar the Encouragement of A Medal of Honour." Dr of London, Edinburgh, Glasgow presented hir Africa gave him a S King of Italy award ferred upon him th Oxford University ı Documents, and Mec exhibited in a case stone Gold Medal has

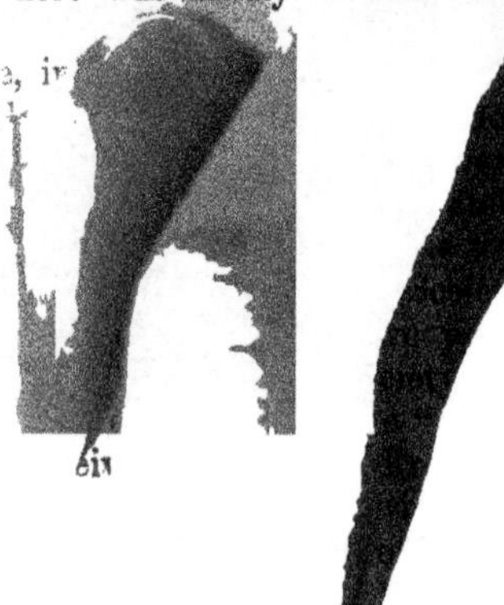

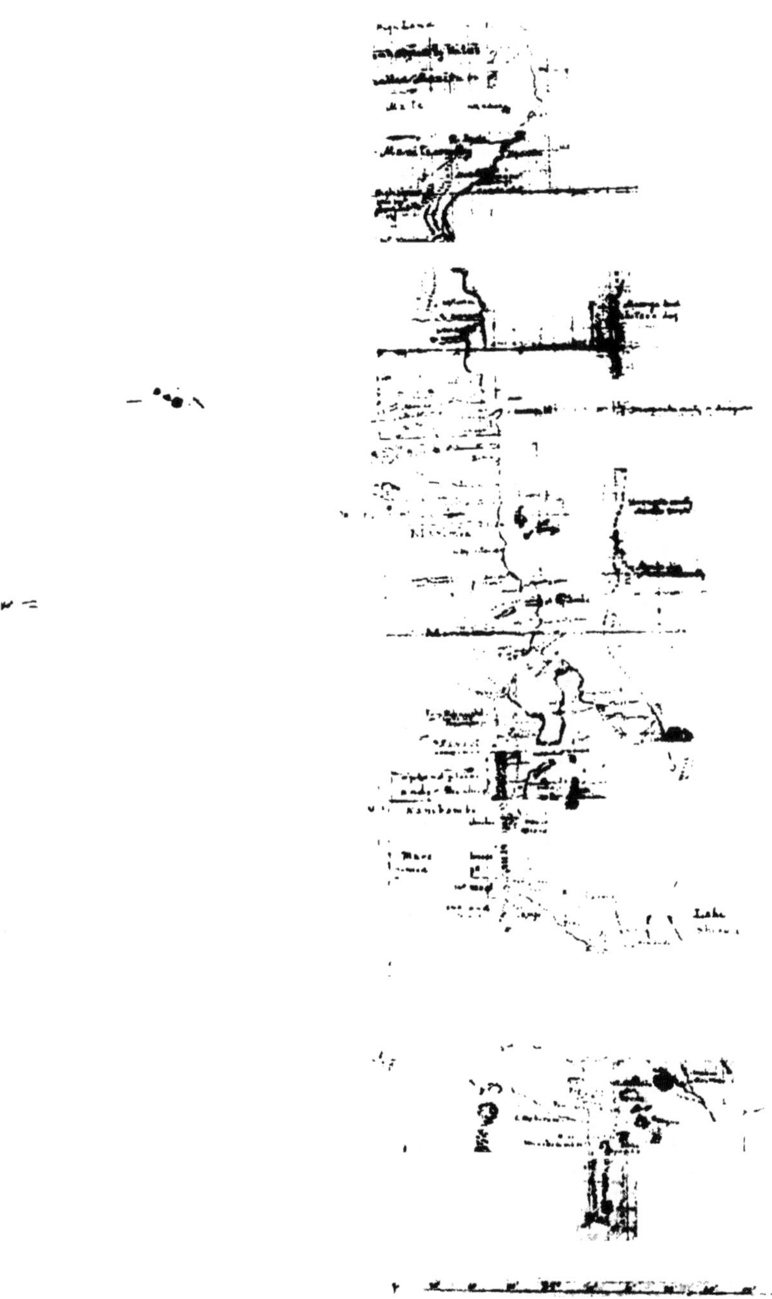

ʳt of La yasa, drawn about 1859 by David Livingstone.

Compass used by Dr. Livingstone on the voyage is exhibited. A Photograph of the last survivor of Dr. Livingstone's Makololo guides, taken many years afterwards, is on view.

Speaking at a public meeting at Cape Town in 1856, the Astronomer-Royal, to whom Dr. Livingstone had forwarded copies of his observations so that accurate maps of the hitherto unknown regions explored by him might be constructed, said that, "as far as he knew, Livingstone was the first traveller who had made accurate observations, and that, though Timbuctoo was so well known, it could not be reached by the aid of any recorded observations, as its position was only known in a vague manner. On the other hand, you could go to any point across the entire continent along Livingstone's track and feel certain of your position. I say, what that man has done is unprecedented." Certainly the labour involved in making the many thousands of astronomical, meteorological, and surveying observations, measurements, and calculations during all his travels must have been enormous, and our admiration for the man is not lessened when we remember that, during two of his three great journeys, Dr. Livingstone travelled alone with his native carriers and had thus to take his own observations, that these must often have been made in periods of great weakness through illness or of extreme personal danger, and that his equipment was of the slenderest description. His scientific observations were generally recorded by Dr. Livingstone in special books, one of which has been placed on view in the case labelled "Dr. Livingstone's Literary and Scientific Work." Dr. Livingstone's Pocket Dictionary, signed and dated, is also shown.

The Museum is fortunate in being able to show fifteen original manuscript maps drawn by Dr. Livingstone to illustrate the vast tracts of country explored by him during all his exploratory journeys in Africa. That chosen for reproduction here was kindly lent by Mr. William Stanford.

After Dr. Livingstone's return home, in December 1856, honours were showered upon him. The Royal Geographical Society of London had already, in 1849, awarded him Twenty-five Guineas for his discovery of Lake Ngami; in the following year he had been given a Silver Watch, and in 1855 the Gold Medal of the Society, and now, in 1857, he was made a Foreign Corresponding Member. The Paris Geographical Society also had awarded him, in 1849, its large Silver Medal, and five years later he received a similar Medal in Gold. In 1857 the Society for the Encouragement of Arts and Industries gave him its bronze "Prize Medal of Honour." Dr. Livingstone also received in 1857 the Freedom of London, Edinburgh, Glasgow, Dundee, and Hamilton; the citizens of Glasgow presented him with £2000, and in 1858 the people of South Africa gave him a Silver Casket and 800 guineas, while in 1872 the King of Italy awarded him a Gold Medal. Glasgow University conferred upon him the honorary degree of Doctor of Laws in 1854, and Oxford University made him a D.C.L. in 1857. Nearly all the Caskets, Documents, and Medals, and the Silver Watch, etc., just referred to, are exhibited in a case labelled "Dr. Livingstone's Honours." The Livingstone Gold Medal has also been placed beside the Honours. This medal

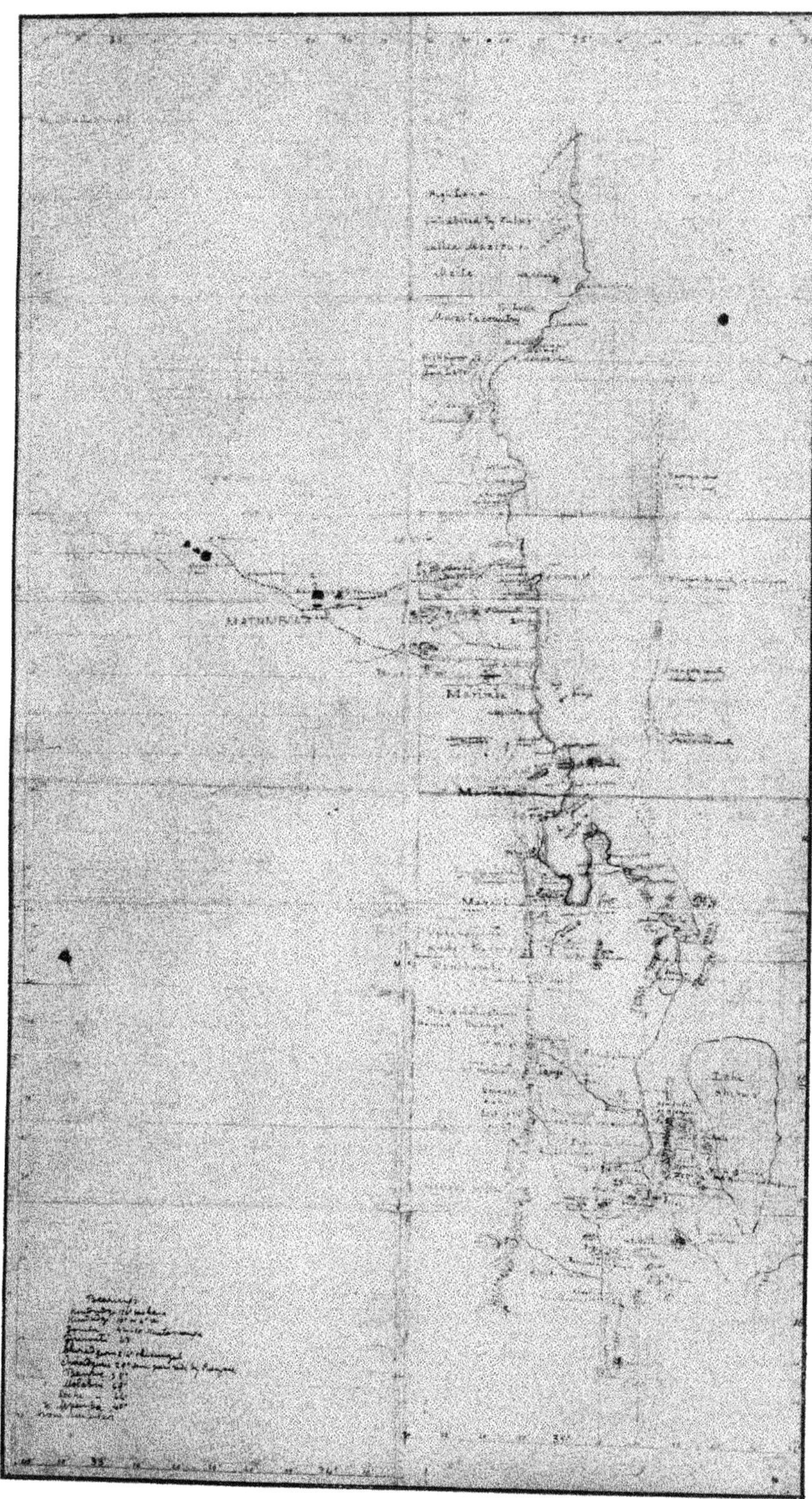

Lake Shirwa and part of Lake Nyasa, drawn about 1859 by David Livingstone.

was founded by the late Mrs. A. L. Bruce (Dr. Livingstone's elder daughter), and it is awarded annually by the Royal Scottish Geographical Society for eminence in geographical research.

Much of Dr. Livingstone's time in 1857 was spent in writing his first book, *Missionary Travels*, the Manuscript of which is exhibited. Mention may be made here of the Manuscript of his second book, *The Zambesi and its Tributaries*, descriptive of his second great journey, which is also shown, as well as Dr. Livingstone's Last Journal, sent home with Stanley in 1872. These, with the contents of Dr. Livingstone's many pocket and other Note Books, which may also be seen, formed the basis of the Rev. Mr. Waller's book, *The Last Journals of David Livingstone*. Numerous copies of the first of these three books are shown, one being Dr. Livingstone's own copy, with signature and date, and others with most interesting autograph inscriptions by Dr. Livingstone. Stanley presented a copy of his book, *How I Found Livingstone*, to Dr. Livingstone's elder daughter, and a pleasing autograph inscription by him, with signature and date, may be noted on the fly-leaf.

Dr. Livingstone had interviews with Queen Victoria and the Prince Consort and their children, and before he sailed again for Cape Town in March 1858, accompanied by Mrs. Livingstone and a staff of expert officers, who were to accompany him on his second great journey, he had been made a British Consul in Africa. His Consular Commission, signed by the Queen and by Earl Clarendon, Foreign Secretary, his Consular Cap (so frequently illustrated in books on Dr. Livingstone), and his full-dress Consular Uniform, form interesting exhibits in the collections. In the same wall case may be noted a large pocket knife inscribed "Dr. Livingstone. From his friend Lady Franklin." Dr. Livingstone and his staff proceeded from Cape Town up the east coast to the Zambesi, while Mrs. Livingstone went inland to pay a visit to her parents at Kuruman, where her youngest child was born towards the end of the year. Dr. (afterwards Sir) John Kirk, who is still with us at the ripe age of eighty, was one of the principal members of Dr. Livingstone's party, and he was British Consular Agent at Zanzibar during Dr. Livingstone's last journey. A comparatively recent Portrait of Sir John Kirk is on exhibition.

The second great journey was an unfortunate one in many respects. Difficulties occurred with some members of the staff; another died. Dr. Livingstone himself was repeatedly an eye-witness of some of the worst horrors of the slave-trade, and his indignation against the Portuguese was such that, in his despatches to our Foreign Office at home, he denounced them in the most vigorous language, leading to somewhat strained relations between the British and Portuguese governments. Many of these and other despatches, which have never been published, had been copied by Dr. Livingstone into his Journals, one of which may be seen open at the last page of one despatch and the first page of the next. Dr. Livingstone had promised the Makololo, who had accompanied him in the last part of his first great journey, from Linyanti towards the east coast, that he would return from England and convoy them home. This promise he now fulfilled, but he was dismayed to learn on nearing Linyanti that a mission station, established there by the London Missionary Society, had been abandoned owing to the death of the two leaders. A Universities' Mission from England, formed under Dr. Livingstone's guidance, next came to an untimely end, and in April 1862 Mrs. Livingstone, who had not long rejoined her husband, died at Shupanga, owing to protracted delays in the unhealthy lower regions of the Zambesi. A Water Colour Landscape Drawing, showing Mrs. Livingstone's grave near the banks of the Zambesi, and photographs of Bishop Mackenzie and the Rev. Mr. Waller of the Universities' Mission, may be observed on the wall of the exhibition gallery.

Still, the geographical and economic results of this expedition were, or have since become, of the highest importance. The Zambesi and some of its tributaries, especially the Shiré, were thoroughly explored, and Lakes Shirwa and Nyasa were discovered and surveyed. (See map facing p. 248.)

The expedition was recalled by the Government in 1863, and in the following year Dr. Livingstone sailed his little steamboat, the *Lady Nyasa*, across the Indian Ocean from Zanzibar to Bombay, to prevent its falling into the hands of the slave-dealers on the African coast. This voyage was one of the most remarkable incidents in the whole of his career.

Dr. Livingstone reached London in July 1864, and after little more than a year at home, spent mainly at Newstead Abbey, by the kind invitation of Mr. and Mrs. Webb, in writing his Second Book, *The Zambesi and its Tributaries*, he left in September 1865 for Bombay, where he disposed of the *Lady Nyasa*. Proceeding to Zanzibar early in 1866, he began his third and last great journey from the mouth of the Rovuma, 250 miles to the south of Zanzibar. Rumours of Dr. Livingstone's death reached the coast a year later, and the Royal Geographical Society sent out a search expedition under Mr. E. D. Young, a member of his staff on the previous expedition, who, though he did not find Livingstone, ascertained that the rumours were untrue. Bad as Dr. Livingstone's experiences were in his former journeys, they were as nothing compared with what happened now. Almost wherever he went he was a witness of the slave-trade and its dreadful evils. For long periods his own circumstances were so critical that but for the help of an Arab slave-dealer he must have died. He was even compelled to travel in the company of slave gangs, and while making one of his long journeys in this manner to Ujiji for letters and supplies he experienced another of his many narrow escapes from death, a hostile native having thrown at him a spear which just grazed his head. This Spear is shown in a wall case, along with small pocket Pistols and Powder Flasks and a double-barrelled breech-loading Rifle, used by Dr. Livingstone to obtain game for food. He had also run out of writing-paper and ink, and he had to continue the record of his voluminous daily notes and scientific observations on journal books made from Newspapers and the edgings of Pamphlets, etc., with "ink" of his own manufacture. Examples of these singular journals are on view.

The Arab traders between the great lakes and the east coast had

followers place the arrows in the bowstring, but stood in mute amazement while the guns mowed them down in great numbers. They use long spears in the thick vegetation of their country with great dexterity & they have told me frankly what is self-evident that but for the firearms not one of the Zanzibar slaves or half-castes would ever leave their country. There is not a single great chief in all Manyema. No matter what name the different divisions of people bear, Manyema, Balegga, Babire, Bazire, Bakos, there is no political cohesion - not one king or kingdom. Each head man is independent of every other. The people are industrious and most of them cultivate the soil largely. We found them everywhere very honest. When detained at Bambarre we had to send our goats & fowls to the Manyema villages to prevent them being all stolen by the Zanzibar slaves. The slave owners had to do the same. Manyema land is the only country in central Africa I have seen where cotton is not cultivated - spun and woven. The clothing is that known in Madagascar as "Lambas" or grass cloth made from the leaves of the "Muale" palm. They call the good spirit above "Ngulu" or the great one, and the spirit of evil who resides in the deep "Mulambu". A hot fountain near Bambarre is supposed to belong to this being - the author of death by drowning and other misfortunes.

Your Lordship's obedient
humble servant
David Livingstone
H M Consul Inner Africa

A true copy
by H M Stanley &
David Livingstone

A true copy
Henry M. Stanley

Part of a Despatch sent by Dr. Livingstone to the Foreign Office, describing his journeys in the Manyema country in search of the sources of the Nile.

deliberately foiled for years all his attempts to gain new supplies from, or to get into communication with, Zanzibar and home; and the loss of his medicine chest early in his travels, owing to the carelessness of a native carrier, resulted in almost constant and very severe illness, so that when Mr. H. M. Stanley, who had gone to Africa on behalf of the *New York Herald* to try and find Livingstone, actually did find him at Ujiji, on Lake Tanganyika, in October 1871, he was in sore straits indeed. A very interesting exhibit is Stanley's own Map constructed by himself in mapping the country traversed by him between Zanzibar and Ujiji, a distance of 800 miles.

Notwithstanding his terrible experiences, Dr. Livingstone had again been able to explore immense tracts of new country during the intervening five and a half years. He had travelled in a south-westerly direction from the Rovuma till the south end of Lake Nyasa was rounded, and then marching to the north-west he finally reached Lake Tanganyika, which he surveyed and mapped. The Map of Lake Tanganyika is one of the most interesting of Dr. Livingstone's manuscript maps. Lakes Mweru and Bangweolo were discovered and their boundaries determined, and the upper waters of the river Lualaba, which he at first thought might be the beginnings of the Nile but which really belonged to the basin of the Congo, were traced. It is interesting to note that Dr. Livingstone had with him a copy, evidently drawn by himself, of an old map of the northern half of Africa, in which the Nile is shown as rising in the "Mountains of the Moon." This Map is also on view.

After staying with Dr. Livingstone for a few months, Stanley left for Zanzibar. Dr. Livingstone waited at Unyanyembe, 500 miles distant from Zanzibar, for further stores and an excellent band of native helpers which Stanley obtained for him at Zanzibar, and then in August 1872 he marched towards the heart of Africa for the last time. But he was now unable to stand the strain, and on May 1, 1873, he died at Chitambo's village near the southern shore of Lake Bangweolo. His heart was buried under a tree near by, and the body was embalmed and carried, with his effects, by his faithful native followers, Chuma, Susi, Wainwright and others to Zanzibar, a distance of 1500 miles. Nine months were required for the performance of this self-imposed task of love and loyalty. From Zanzibar Dr. Livingstone's remains were conveyed to England and were buried in Westminster Abbey. A Model of the Hut in which Dr. Livingstone died, constructed by the native attendants after their arrival in England, is on view. Photographs of Wainwright, Chuma, and Susi may be noted, as well as a Picture showing all three taking an inventory of Dr. Livingstone's effects after his death. A fine water-colour Drawing of the interior of Westminster Abbey, indicating the position of Dr. Livingstone's grave, has been placed on the wall in the exhibition near the British Flag which Dr. Livingstone had carried during his travels in Africa.

Any appreciation of Livingstone's life and work would be incomplete if his eminence as a naturalist were not referred to. As a medical missionary and as an explorer his fame is known to all. But students of natural science who may read Dr. Livingstone's published writings cannot fail to be impressed by the scientific method shown when he pauses in his narrative at times to describe and discuss observations in such subjects as Botany, Zoology, Geology, Meteorology, etc. It is understood that a great mass of unpublished material in various branches of science exists in Dr. Livingstone's manuscript journals. Probably he had hoped to prepare it for publication at home when his work of African exploration was completed. Should not this duty be now undertaken by others?

As illustrative of Dr. Livingstone's abilities in Natural Science, we may take, for example, his observations on the tsetse fly, recorded on pages 80-83, 337, and 352 of his first book, *Missionary Travels.* This fly appears to have been indigenous to Africa for many centuries: the descriptions given in the Bible at Exodus (chapters 8 and 9) and Isaiah (chapter 7), clearly show that the tsetse fly is referred to. Yet Dr. Livingstone was evidently the first in modern times to give (in 1843) accurate authentic information regarding this African scourge, though Major Vardon, who, like Mr. Oswell, had gone to Africa on a hunting expedition and had met Dr. Livingstone, was the first to bring home (in 1848) specimens of the fly. During his early travels, Dr. Livingstone had also sent home specimens, for the identification of which the writer is indebted to the Natural History department of the Museum, and they are now lent to the exhibition.

It may be added that, with the view of illustrating native life and environment in the different regions which Dr. Livingstone traversed, many ethnographical and other specimens, drawn mainly from the permanent collections of the Museum, have been temporarily placed on view in the exhibition gallery.

DAVID LIVINGSTONE AND HIS WORK IN AFRICA.[1]

By Sir Harry Johnston, G.C.M.G., K.C.B.,
Livingstone Gold Medallist, Royal Scottish Geographical Society.

Before I venture to speak to you on the subject of Livingstone's journeys in South and Central Africa, his researches and discoveries, and the results which sprang from them, I should like to address you, as I have already done other members of this Geographical Society in Glasgow, *more especially as a Scottish audience*, as fellow-countrymen of Livingstone, and consequently, presumably, even more keenly interested in the celebration of his life's work, more jealous that he shall receive his due recognition, more anxious that the full effect of his researches

---

[1] This article contains Sir Harry Johnston's Introductory Remarks before the delivery of his Livingstone Centenary Address in Edinburgh on March 27 last. The Address will appear in the next issue of the *Magazine*.

should be given to the world than those of us who live south of the Border. Not only was Livingstone one of the greatest amongst the amazing explorers of Scotland sent out to make known the unknown,—and on your wonderful roll of honour you have already had to inscribe William Lithgow, who explored North Africa in the seventeenth century; the heroes of Darien; James Bruce; Sir Alexander Mackenzie; Mungo Park; and Robert Moffat—but the work of Livingstone was materially assisted, furthered, and brought to fruition by other Scots, by Sir John Kirk, Edward Rae (a Highlander whose name was originally spelt Reach), Joseph Thomson, John and Frederick Moir (sons of an Edinburgh doctor), Dr. Stewart of Lovedale, Dr. Robert Laws of Nyasaland, Henry Henderson, John Buchanan, David Ruffele-Scott, Duncan Fraser, and Alexander Hetherick—to mention only a few of those who have left their mark on South and Central Africa; while south of the Zambesi, a Scotsman, John Mackenzie, was the real creator, in succession to Livingstone, of British Bechuanaland, was the man who really secured the British road north from the Orange River and the Limpopo to the Zambesi.

For a nation which still numbers under five millions of people,—and was about half as numerous at the time when these great deeds were done, deeds which have changed the history of the world—this is an amazing record, exceeding that of Portugal, Norway, or Switzerland, the only European countries which being mountainous and of restricted agricultural area, have turned the energies of their adventurous people to regions far beyond their own frontiers.

But the question I wish to put to you to-night is "how much" are you proud of Livingstone and those who came before and followed after? Not much, I fear, so far as the expenditure of money and unproductive and disinterested effort are concerned. If, for example, Glasgow's pride in Livingstone, who was born in what is now virtually a suburb of Glasgow, were on a par with the wealth and enterprise of that mighty city—practically the second in the British dominions in point of population—the house in which Livingstone first saw the light, and in which he spent his boyhood, would not be as it is at this moment, a place of singular squalor. I visited and photographed this block of houses and its surroundings in 1890, and I have my photographs to prove that since that date not only has the neighbourhood "gone down," but it has been allowed by the civic authorities of Blantyre and of Glasgow to become actually disgusting. Yet this little group of two or three stone houses is far from being unpicturesque, nor are they at all discreditable examples of late eighteenth-century domestic architecture in Scotland. If Glasgow did its duty by the memory of Livingstone it would raise a public subscription, and its municipality would vote a sum out of the rates; something further might be squeezed from Blantyre and the neighbourhood; and the money thus produced would go to purchase this group of houses and the strip of land between them and the banks of the Clyde. Some of the rubbishy tenements which now crowd up to them could be demolished, and a sufficient space of land round them might be judiciously planted with trees to regain the picturesque surroundings of Livingstone's boyhood. The houses—at any rate the house wherein Livingstone lived—might be furnished as far as possible according to the period; and the whole should constitute a very interesting little museum of itself, illustrating domestic life amongst the educated poor in Scotland at the beginning of the nineteenth century. Many domestic relics of Livingstone might there be housed.

Then Americans, Englishmen, Germans, Scandinavians, Frenchmen, Portuguese, who are now beginning to make Blantyre a place of pilgrimage when on a visit to Scotland, would not recoil from it with something like loathing and rightly appraise the very scanty interest which Glasgow really takes in the name and fame of Livingstone.

Now: what can Edinburgh do? Edinburgh can put forth its whole strength as the royal city of Scotland to secure for permanent exhibition in its National Museum the collection we have all examined with so much interest and profit, which is at present being shown there—an exhibition which reflects great credit on Dr. Galt and Mr. D. J. Vallance, who have organised it, and on the descendants and friends of Livingstone who have contributed to it. I feel that a most strenuous appeal must be made to Scotland, in particular, that this collection thus brought together should never again be dissipated into space. A Livingstone Museum illustrative of a most interesting phase in the history of Africa should be a permanent section of the National Museum of Scotland. Of course, with the cheap way in which we have all become accustomed to deal with great explorers and men of science, this will seem to be an object easy of attainment: the various lenders of exhibits would be asked to present them to the nation, and if they declined, most unkind things would be said about them. Now this is not my idea at all. Livingstone was most scurvily treated by the Government of Great Britain. He served that Government for fifteen years as a consul in East-Central Africa. For six out of those fifteen years he received salary at the rate of £500 a year, but a good proportion of that had to go in meeting necessary expenses of the expedition, which the Foreign Office would not defray. In fact as against the £3000 in all which Livingstone received from the Foreign Office between 1858 and 1864 he actually expended on the work of this Government expedition in Zambezia nearly £7000 of his own money. This £7000 and all, in fact, that he was able to make and set aside for pursuing his disinterested objects, rearing and educating his children, was derived entirely from his work as an author and from his generous publisher, John Murray. But for Livingstone's three books he would have died a very poor man, leaving little or nothing to his children. As it was he left them comparatively little. It will seem almost incredible to future generations that after his death not one single penny was found by public subscription or obtained by direct grant from the British Government for the assistance of Livingstone's surviving sons and daughters. One of the sons struggled for years to maintain himself, his wife and family, though stricken with phthisis; one of the daughters, early left a widow, had similar difficulties in maintaining herself and educating her children; and had it not been for the fact that another daughter had the good fortune to marry one who was not only a man of means but a man of great generosity, the circumstances of Livingstone's descendants would have not been far removed from penury. Therefore I hold that some slight reparation should be made even late in the day, and if Scotland desires to be worthy of having engendered Livingstone, she should at once provide out of her revenues an endowment fund for this Livingstone section of the National Museum, and out of this purchase from all the lenders the exhibits which I regard as being of exceptional interest, not merely as illustrating in a very vivid fashion the adventures, resources, and talents of Livingstone, but as a unique contribution to the history of Africa. Such a Livingstone section will draw an appreciably large number of pilgrims annually to Edinburgh. Therefore, I think the great railway companies and the great hotels might at any rate send a small contribution to the Livingstone Exhibition fund.

So much for Scotland. Now what can Great Britain do unitedly to the memory of Livingstone and for the making of full and practical use of all Livingstone's work in Africa? The full force of public opinion must be brought to bear on the Foreign Office to induce that Department to publish as fully as possible all Livingstone's Government despatches between 1858 and 1873. Mere tedious detail and a few criticisms and personal remarks might be left out, but the mass of his despatches will be found to be of very great interest as part of the history of East Africa. If this is too much trouble for a Government Department to take, then it might authorise some person in whom it had confidence to have access to these despatches—many or all of which are in what is known as confidential print, though after this lapse of time there is of course nothing confidential about them. Next, the Government of Cape Colony should be approached with a view to the careful printing of all Livingstone's manuscript vocabularies, said to be stored in the Grey Library.

Then there are still surviving several noteworthy people who have known Livingstone. These should be encouraged to send in their reminiscences. For the most part they are silent folk averse from writing but not averse from criticism. They are very ready to find fault with all extant books and articles on Livingstone, declaring such and such a detail is inaccurate, but they do not themselves publish their reminiscences, though these, so far as I have seen them in private letters, are sometimes of great interest and obvious accuracy.

In short, what we want and what Scotland should make it a national duty to obtain, and if necessary to subsidise, is a new and complete Life of Livingstone, which shall contain all the unpublished scientific material now in the possession of Captain Livingstone Bruce, the manuscript vocabularies (of immense interest to students of the Bantu languages) from the Grey Library, the gist of the despatches to the Foreign Office, and all reminiscences regarding Livingstone's life in Scotland and in England which are not mere twaddle. I would suggest that before I leave these indiscretions and deliver my geographical discourse, that such a work as this might be edited by a *Committee* rather than by a single individual, and this committee should consist of *Captain Livingstone Bruce* (grandson), *Mrs. Livingstone Wilson* (daughter), *Captain John Kirk* (as representing his father, Sir John Kirk, who has a great mass of interesting correspondence and many original photographs taken by himself in Livingstone's company), *Mrs. Fraser* (the daughter of William Webb), *Miss Alice Werner*, the well-known student of Bantu languages, and one who has traversed herself nearly all the scenes of Livingstone's life in South and Central Africa, and *Mr. Ralph Durand*, whose recently published articles in the *African Mail* show him to be singularly well fitted to deal with the contemporary geographical criticism or acknowledgments of Livingstone's discoveries.

This, I think you will admit, is a concrete and practical proposal, and if adopted—and it could be adopted and carried through if Scotland willed—it would give to the world such a revelation of Livingstone's work as would ensure for him far greater recognition as an African discoverer than he has yet been accorded.

I make no appeal to private generosity. Private generosity has all along had to pay the bill in regard to Livingstone. I wish to urge all Scots, men and women, to demand that this national reparation to Livingstone, this full fructification of his work as Government official and Scottish scientific explorer shall be paid for—the total cost will not be great—out of Scottish national and civic funds, aided also by grants from a few great corporations which are virtually departments of State. None of the money raised is to be spent on philanthropic or missionary purposes (except that all expenditure on education *is* a contribution to Christian philanthropy). The objects to be attained are three: the preservation of the site of Livingstone's birth at Blantyre as a national monument and museum; the permanent instalment of a Livingstone Memorial Exhibition in the National Museum at Edinburgh; and the compilation of a new Life of Livingstone, which shall include the gist of his despatches to the Foreign Office, the unpublished scientific material in the possession of Captain Livingstone Bruce, and the manuscripts hidden in the Grey Library at Cape Town.

THE PLANT ECOLOGY OF BEN ARMINE (SUTHERLANDSHIRE).

By C. B. Crampton, M.B., C.M., and M. Macgregor, B.Sc.

(*With Diagrams.*)

(*Continued from page* 192.)

IV.—The Vegetation of the Corries and Crag Plant Associations.

The vegetation in the corries varies rapidly from point to point, and indeed is kaleidoscopic in its changes. Without studying the influence of the physiography any attempt to describe the ecology would be hopeless.

The walls of the corries are formed of highly inclined faces of rock, rarely vertical or overhanging, broken by ledges, and overlooking slopes formed chiefly of fallen blocks and fans of debris more or less peat-covered, or exposing ice-worn rock surfaces. This we find in the centre of the corries immediately underlying the mountain-top debris, while the marginal areas are generally far less precipitous, with glaciated rocks and scattered crags, separated and overhung by slopes covered with slipping peat.

The steep rock faces of the central parts of the corries were the areas of greatest rock erosion during the corrie glacier phase, and are still the areas most subject to change under the influence of frost and gravity. The lateral crags and slopes, on the other hand, are in many cases only now emerging from a widely enveloping mantle of peat.

We may therefore distinguish, from the outset, certain areas of steep crags and slopes exposed to atmospheric influences only, or to the water from springs percolating the mountain-top debris or the joints of the rocks, from others where the surface run off is apt to be more contaminated by waters from peat undergoing erosion. We have also to consider the physiography as it affects each ledge, rock surface, fallen block, or slope of debris. Most of these may be found under one or another of five conditions which to some extent merge into one another. They are that the surfaces are—

1. Subject only to atmospheric precipitation and leaching.
2. Little subject to atmospheric precipitation or to flushing with drainage waters.
3. Chasmophyte or chomophyte vegetation fed by mineralised waters draining from above.
4. Vegetation fed by waters receiving manure of animals, and especially beneath eyries.
5. Surfaces fed or flushed by waters contaminated with peat.

1. In the first class the rock faces become covered with flat growths of lichens, species of *Lecidea*, *Lecanora*, etc., and patches of *Frullania* and hair-like tresses of *Alectoria*, and dotted over with small close cushions of *Grimmia* and other mosses. The tops of spurs, crests and fallen blocks show peltate discs of *Umbilicaria*, or fruticose or foliaceous growths of *Cladonia*, *Sphærophoron*, *Platysma*, etc., with cushions of *Rhacomitrium*, *Hedwigia*, etc., and in the humus that accumulates, *Empetrum*, *Arctostaphylos*, *Vaccinium Vitis-idæa*, or even seedling Rowans (*Pyrus Aucuparia*) may take up their residence. The ridges and isolated prominences on slopes of debris usually support *Calluna*, and are generally more or less peat covered.

2. Old established overhanging faces, subject neither to leaching nor definite flushing, are sometimes very dry, at other times damp from condensation of moisture. In the latter cases large surfaces may be orange with the alga *Trentepohlia aurea*, and little damp cracks have swelling cushions of *Bartramia ithyphylla*, *Weissia rupestris* or *Blindia acuta*. Drier crevices frequently support cushions of *Grimmia torquata* or *G. funalis*.

3. In the third class we find debris-covered ledges with a chomophyte,

and cracks with chasmophyte vegetation of a more varied and attractive nature.

The wet cracks especially favour *Sedum Rhodiola*, *Saxifraga stellaris*, *Oxyria reniformis*, etc. Small ledges have *Saxifraga oppositifolia*, *Thalictrum alpinum*, *Saussurea alpina*, *Silene acaulis* and masses of various mosses and hepatics. Wider ledges support a more vigorous vegetation, especially *Vaccinium Myrtillus*, with species of *Salix*, sometimes *S. lapponum*, but if sloping and with much rainwash, grass-land with *Alchemilla alpina* usually takes possession. Shaded crevices show *Cystopteris fragilis*, and sometimes *Asplenium viride*, *Polypodium Dryopteris* or *Phegopteris polypioides*.

Flushed rock surfaces are sharply separable from those receiving only atmospheric precipitation and usually support algal or algal-bryophyte associations to the exclusion of lichens. They are referred to in another section. The waters descending the crags spread over the slopes of debris beneath, covering them with rainwash and forming channels of erosion on their surface. Those flushed surfaces are avoided by *Calluna* and most other peat-loving plants, and form patches of alpine grass-land or damp hollows with *Carex* or other types of flushes.

4. The fourth class are related by degrees to the last, but become specially prominent under crags with eagles' eyries or resting-places. The vegetation is very rank. *Luzula sylvatica* often forms dense mats, and certain mosses are very luxuriant, forming wide sheets and hanging masses. *Aira cæspitosa* appears in its tufted lowland form, and Valerian, *Spiræa Ulmaria*, *Geum urbanum*, *Carduus palustris* (Marsh Thistle) and other coarse plants are conspicuous. The Globe Flower (*Trollius*) is often very abundant in these places, but is also seen on most of the ledges with strong vegetation. Nettles (*Urtica*) are not uncommon beneath eagles' eyries in the Highlands, but were not noticed on Ben Armine.

5. In the fifth class we find ledges, rock faces and slopes subject to peaty drainage. The ledges accumulate peat and support a moorland flora, often masses of *Sphagnum* with *Pinguicula*, etc., when very wet, and *Calluna* when the drainage temporarily or permanently deserts the surface. Rock faces with a peaty drainage often support lichens, distinguishing them from most other flushed rocks, except those subject to salt spray. These lichens are chiefly species of *Cladonia*, *Verrucaria*, *Platysma*, *Stereocaulon*, and accompany patches of *Sphagnum*, black masses of *Nardia*, *Marsupella*, *Pleurozia*, *Rhacomitrium* or *Campylopus*. The slopes at these places are invariably peat covered and support *Calluna* moor or moorland flushes, according to the nature of the surface.

The apparent confusion in the vegetation of the corries is therefore in reality a nicely adjusted arrangement of chomophyte, chasmophyte, lithophyte and flush associations, resulting from the past and present conditions of physiography as affecting snow-lie, rock-erosion, exposure to light and wind, and the direction and the nature of the water supply and drainage.

The chomophyte associations are either outliers of moorland vegetation or peculiar to the station according to these influences. The chasmophyte species more closely approach the endemic. The lithophytes vary chiefly with the age, and orientation of the rock face and the nature of the drainage, and certain of them will need further consideration in the section on flushes.

Each ledge overlies a rock face, and this in turn another ledge or slope of talus. The drainage from each overlying member affects, or fails to affect, the vegetation of that beneath. Thus a ledge is liable to leaching or is fed with waters from the rock face above, and beneath all the larger crags we frequently find a slope covered with alpine grass-land.

As to the relative preponderance of these various associations in the corries of Ben Armine it must be kept in mind that the moorland plants still dominate all places in Central Sutherland where the migratory factors of change are not in full possession. On peat-covered mountains like Ben Armine alpine associations are very limited in their range, and the lithophyte and other associations are chiefly such as can withstand the rigorous conditions imposed by isolation in, and close contact with, moorland peat.

V. The Plant Associations of Springs and Flushes.

These formations, geologically speaking, belong to the drainage system, but the flow of water is so gentle and circumscribed or ephemeral in its action, or so different from that of streams in its physical and chemical properties, that the plant associations, though more closely related to those of the stream belt than any others, have peculiarities of their own which require examination. The springs of this area range from alpine springs with a very low temperature to springs rising in the moorland peat containing much humus and ferruginous substances, but there appear to be many stages connecting these types, and local differences are also perceptible, depending upon other factors which have not as yet been sufficiently studied to afford satisfactory explanations.

The alpine springs rise from beneath the rock debris in the higher parts of the corries, and often in positions that are subject to snow-lie in the early part of the year. They ooze up to the surface along the line where the junction of the mountain-top debris with the solid rock is exposed on the steep slopes. The water is thus derived from the debris, and is cold and but little contaminated with peaty substances. The alpine spring associations comprise characteristically smooth cushions of mosses of beautifully varied green and red tints, with various low plants growing on the surface and round the margins. The following are characteristic species :—

Philonotis fontana, *Dicranella squarrosa*, *Scapana dentata*, *Marsupella emarginata*, *Hypnum sarmentosum*, *Bryum Duvalii*, *Mnium punctatum* var. *elatum*, *Epilobium alpinum*, *Saxifraga stellaris*, *Viola palustris* (dwarf), *Rumex Acetosa* (dwarf), *Montia fontana*, *Selaginella selaginoides*.[1]

The bog springs break out in thick peat where it has cracked from movements, and generally on or near the base of gentle slopes. They

[1] C. H. Ostenfeld.—*The Land Vegetation of the Faroes*, 1908, p. 843. See also *Types of British Vegetation*, p. 326.

form spongy surfaces, flat or tumid, floating on water, and sometimes of sufficient size and depth to drown animals unwary enough to venture on to them. The floating masses consist of species of *Sphagna* (cuspidata) or *Hypnum revolvens*, or smooth fastigiate growths of *Hypnum stramineum*. *Stellaria uliginosa* grows on the surface, and *Juncus bulbosus* (*forma fluitans*) is generally present.

Springs with limy waters are common enough in the Ben Griam and Ben Loyal districts further north, and also in Caithness, where the Old Red Sandstone rocks and drift often contain much lime. In the Ben Armine district some of the hornblende gneisses have lime, but the drift is rarely very calcareous. Where gentle oozing springs break out from crevices and ledges of hornblende gneisses we find tumid cushions or sheets of one or other of the following mosses:—*Hypnum commutatum*, *H. scorpioides*, *H. molluscum*, *Breutelia arcuata*, *Hypnum filicinum*, *H. falcatum*, with *Saxifraga aizoides*, *Thalictrum alpinum*, *Saxifraga oppositifolia*, *Galium boreale*, etc.

Boggy calcareous flushes from springs have some of the above mosses, with *Saxifraga aizoides*, *Schœnus nigricans*, *Carex flava*, *C. flacca*, *C. pulicaris*, *Euphrasia* sp., *Linum catharticum*, *Pedicularis palustris* (Lousewort), *Scirpus pauciflorus*. They are generally margined with *Hypnum molluscum* and *Thalictrum alpinum*.

The most important flushes of this area are, however:—

1. Rock flushes.
2. Flushes forming alpine grass-lands.
3. Moorland flushes.

The two former are practically confined to the steep slopes and corries of the eastern face of Ben Armine, but the last named are widespread.

Rock flushes must be separated from other rock habitats, and also from other types of flushes on soils capable of easy erosion. They may be classified as rock surfaces subject to a gentle trickling stream of water, either constant in its action or sufficiently frequent in its occurrence to cause a special type of vegetation, different from neighbouring uninfluenced rock surfaces.

Vertical or nearly vertical faces with a comparatively constant trickle of water support an algal vegetation only, whilst others only periodically flushed form the station of very defined associations of bryophytes and algæ, varying from open to closed formations, and these again to open formations, with the results of progression and retrogression and reinitiation of plant successions induced by an increasing erosive power of the water as soil accumulates, and a tendency to migrate from point to point, deserting one part of the rock surface for another.[1]

The rock flushes of the Ben Armine corries have associations which have been looked upon as characteristic of positions of snow-lie, and have been described as such in Switzerland and in this country.[2] From the

---

[1] C. B. Crampton.—"The Geological Relations of Stable and Migratory Plant Formations."—*Scottish Botanical Review*, vol. i., pts. 1, 2, 3, Jan.-July 1912.

[2] W. G. Smith.—"Anthelia: an Arctic Alpine-Plant Association."—*Scottish Botanical Review*, vol. i., No. 2, April 1912, p. 81.

wide distribution of this type of association in the corries there seems, however, to be some doubt as to what extent snow waters are essential for their formation. They are always found on rock surfaces or coarse rock debris on which other types of flush plants can gain no footing. The probabilities seem to be that the snow-lie is the cause of the baring of rocks owing to any other soil than peat becoming slush, and slipping off the face of the rocks. The snow-lie may also act as a deterrent to other species, and the trapped dust and scrapings from the rocks may have some peculiar nutritive influence, as has been suggested. Even lichens are denied any rest at places where snow lies long on sloping rock faces. Parts of the corries that have well-defined snow-lie on a peat surface show an entirely different association, in which *Nardus stricta* is the dominant species.

In the rock flushes in the Ben Armine corries the plant succession is as follows:—

1. Blue-green algæ forming a thin slimy mat, and consisting of *Stigonema* and *Schizonema* closely interwoven.

2. *Anthelia* (*juliacea*?) and *Marsupella emarginata* (forma)[1] forming whitish green and purplish brown to black cushions and mats. The *Marsupella* occupies the parts liable to the greatest flushing.

3. Large black cushions of *Campylopus atrovirens* and patches of *Andræa petrophila* and *Gymnomitrium* (two sp.?) invade the station.

At this stage the association generally degenerates, either through invasion by various moorland species, including *Pleurozia*, *Pinguicula*, *Nardus*, etc., or through erosion or drying up, the flush migrating elsewhere.

If erosion takes place the succession may begin all over again, but when the flush dries up, most of the accumulated material gradually vanishes, and the rock is left scattered over with little patches of *Andræa* and *Gymnomitrium*. These gradually become confined to damp crevices, and lichens appropriate the general surface, and are joined by drought-resisting species of *Grimmia* and *Rhacomitrium*. Owing to the migratory nature of the formation large surfaces are left dotted over with *Andræa* and *Gymnomitrium*. Every stage between arrest by invasion and that by drying up may be seen. The order of succession is not quite constant, but the climax stage appears to be that of *Anthelia* and *Marsupella*.

Other types of flushes occur in the corries where the water from springs or surface run-off carries rain-wash over the slopes beneath the crags. They are partly on fans of debris and partly on peat, and show gradations between moorland flushes and alpine grass-land. Thus we have

1. Wet flushes with *Nardus*, *Juncus squarrosus*, etc.
2. Wet flushes with *Carex Goodenovii*, *C. echinata*, *Pinguicula*, *Viola palustris*, etc.
3. Wet flushes with *Carex pulicaris*, *C. flava*, *C. flacca*, *C. dioica*, *Pedicularis palustris*, *Linum catharticum*, etc., indicating the neutralising effect of lime in the waters.

[1] This species is perhaps *Pearsoni*. See MacVicar's *Student's Handbook*, p. 116.

4. Grassy patches with *Agrostis*, *Anthoxanthum*, *Carex pilulifera*, *Luzula multiflora*, *Galium saxatile*, *Polygala*, *Potentilla erecta*, *Hylocomium squarrosum*, etc.
5. Larger areas of alpine grass-land on the steeper slopes beneath large crags. *Festuca vivipara*, *Anthoxanthum*, *Aira cæspitosa*, var. *brevifolia*, *Luzula sylvatica*, *Vaccinium Myrtillus*, *Alchemilla alpina*, *Thalictrum alpinum*, etc.[1]

The moorland flushes on the plateau are generally due to the overflow from bog springs. The wetter types have a central belt of *Sphagna* (chiefly cuspidata and subsecunda) margined by *Juncus effusus* with *Carex Goodenovii*, *C. canescens*, *C. echinatus*, etc. Drier channels have a thick carpet of *Polytrichum commune*.

The large *Molinia* flushes, so common in the Baddanloch hollow a few miles further north and in the central hollow of Sutherland to the south-west, are rarely met with on the higher part of the plateau.

VI. The Plant Associations of the Stream Belt.

The Rocky Stream Channel may be considered to include all rock surfaces liable to submergence in the stream, but, comparatively speaking, immovable under conditions of flood.

There is a well-marked zonation of plants in the rocky stream channel of this area. First we have a zonation of bryophytes having relations to the frequency of submergence.

1. Associations beyond the ordinary limits of flooding—*Rhacomitrium heterostichum*, *R. fasciculare*, and lichens generally abundant.
2. Submerged only during floods—*Rhacomitrium aciculare*.
3. Always partly submerged (larger streams)—*Hypnum palustre*, *Brachythecium rivulare*, and in the higher-level streams with less erosive powers *Scapania dentata*, *Marsupella aquatica*.
4. High-level shallow streams, peaty waters—*Scapania undulata*.
5. Always completely submerged except in prolonged drought—*Fontinalis antipyretica*.

There is also a zonation relative to aeration and rapidity of the waters.

1. *Sacheria fluviatilis* (torrents only).
2. *Brachythecium rivulare*, *Hypnum ochraceum* (waterfalls particularly).
3. *Hypnum palustre*, *H. eugyrium* (rapid waters only).
4. *Scapania undulata*.
5. *Fontinalis antipyretica* (quiet waters).

Fontinalis squamosa takes the place of *F. antipyretica* in the granite bound pools of the Allt Coire na Fearna, where the waters are purer, but with much sand and grit.

Where the streams are full of boulders, each stone that is big enough to project from the water shows this zonation, and the up-stream side of

[1] The alpine grass-lands were first noted by Robert Smith in this country; further accounts of them will be found in Crampton, *The Vegetation of Caithness*, *The Geological Relations of Stable and Migratory Plant Formations*, and in *Types of British Vegetation*.

the larger boulders subject to a rapid current supports a different association to that which faces the quiet pool with eddies down-stream.

A few shaded rock gorges occur in the vicinity of Ben Armine, but outside the limits of the area shown on the map. In these places the zonation is different to that of the exposed waterfalls, and the following are noticeable :—*Porotrichum alopecurum*, *Hyocomium flagellare*, and *Eurynchium rusciforme*, the last named forming the lowest zone partly submerged at low water. The stems of nearly all the mosses in submerged habitats become black, hard, and fibrous, and denuded of leaves below.

Banks of Stream Erosion and Alluvial Surfaces.—When the bank within the flood zone consists of partly consolidated sand and gravel, or washed peat, we find the following zonation :—

1. Zone beyond the reach of ordinary floods, moorland species, *Hylocomium splendens* abundant.

2. At limits of ordinary floods, *Polytrichum commune*, *Hylocomium squarrosum*, *Sphagna* (cuspidata), *Viola palustris*.

3. *Bryum pallens*, *Philonotis fontana*, *Sphagnum squarrosum*, *Sagina procumbens*. Overhanging shady banks—*Pellia sp.*, *Mnium hornum*.

Where the banks are low, zones 2 or 3 form flood shingles or terraces of sand and gravel. Thorough sorting and grading of these materials implies constant movement, and the surfaces are then very barren, but when the material is ill-sorted, consisting of gravelly sand and boulders, an open association of certain plants in the above two zones appears. The boulders are fringed with *Rhacomitrium aciculare*, and the protected sandy surfaces have patches with *Hylocomium squarrosum*, *Bryum pallens*, *Dichodontium pellucidum* (occasionally), *Philonotis fontana*, *Rhacomitrium ericoides*, small forms of *Polytrichum*, *Sagina procumbens*, *Rumex Acetosa*, *Viola palustris*, and the alpines *Saxifraga stellaris*, *Alchemilla alpina*, *Epilobium alpinum*.

The limits of flooding and scour in these places is usually marked by a belt of *Nardus stricta*.

On ascending the burns till the waters are very peaty *Fontinalis* and *Scapania undulata* persist so long as there is a rocky substratum, but most of the other bryophytes disappear. *Fontinalis* occupies the pools, *Scapania* the better-lighted and aerated shallows.

Highly ferruginous springs often emerge from beneath the peat, and in some cases render the stream so irony that the ordinary aquatic vegetation is prevented from growing. The stones in the stream-bed are then sometimes completely covered with the beautiful little alga *Drapanauldia*, which looks blue-green against the red flocculent irony deposits, and waves continually in the stream like a matted spider-web. Every stone, and even the eroded peat surfaces have a slimy covering of the alga. As the waters become less ferruginous down stream the alga becomes displaced by mosses, especially *Fontinalis antipyretica*, which is the least exacting so far as the purity of the water is concerned.

The alluvial deposits consist of gravel and sand, derived from the sandy glacial drift, or of peat.

The alluvial surfaces fringing the stream channels nearest the points where the streams emerge from tunnels beneath the peat are formed of large strips of the upper part of the peat, undermined, and lowered bodily *in situ*. They are generally smooth, but often have pools in cracks or sink holes in the surface.

The following sketches show the relations of these alluvial surfaces to the surrounding peat.

The plants originally occupying these undermined and lowered peat surfaces are soon replaced by associations in which either *Carex Goodenovii* or *Nardus stricta* are dominant, along with *Polytrichum commune* and *Viola palustris*. These are replaced later by grass-land, with *Anthoxanthum*, *Luzula multiflora*, *Agrostis tenuis*, and *Festuca ovina*, with *Potentilla erecta*, and *Hylocomium squarrosum*, but *Juncus squarrosus* is often dominant.

FIG. 2. – Stream channel in peat bog with slipped peat alluvium. The channel is still underground at one place.

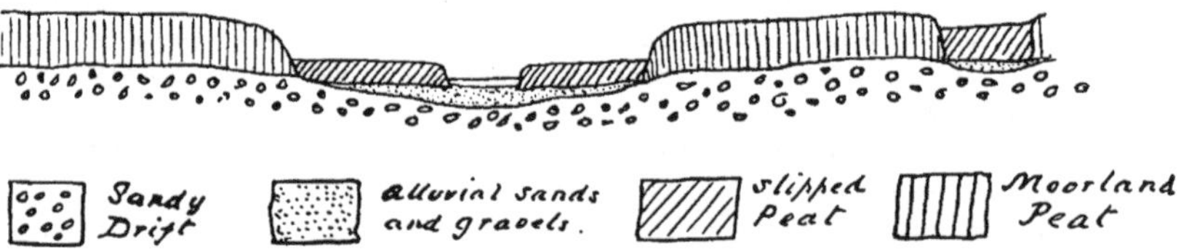

FIG. 3.—Diagrammatic horizontal section across a stream channel to show relations of moorland peat and slipped peat alluvia.

If the burn cuts deep enough to prevent flooding, the plants from the surrounding moorland may resume occupation.

Further down, where the stream channels are wider, we find, locally, extensive stretches of flood shingles which remain barren if the materials are well sorted and mobile. These flood shingles sometimes become colonised through the dumping of peat sods brought down the stream in flood and stranded on the shingle banks. The areas of shingle protected by sods thus become colonised by various species of the mosses, alpine plants, etc., already mentioned, and the sods become centres of dispersal for *Nardus stricta*, *Carex Goodenovii*, *Molinia cœrulea* and *Polytrichum commune*. By a further accumulation of sods, and through the silting and levelling influence of floods the surface gradually becomes more even and eventually similar to those of the peat alluvia already described. *Viola palustris*,

Rumex Acetosa, *Galium saxatile*, and *Potentilla erecta* (Tormentil) are almost always present on these peat alluvia.

This erosion of peat, undermining and lowering of its surface, and dumping of sods, is constantly going on in the moorland streams. The eroded and scoured surfaces of peat along the flood margins always invite colonisation by *Nardus stricta*, while the lowering and dumping of peat rather favours *Carex Goodenovii.*

The more normal types of moorland stream alluvia, consisting of sand and gravel, are barely represented within this area. They carry subxerophytic types of grass-land similar to those described for Caithness. Those furthest removed from the lowlands have *Festuca vivipara* dominant along with *Anthoxanthum* and *Aira cæspitosa*, var. *brevifolia.* A form of *Rumex Acetosa* (Sorrel Dock), *Taraxicum palustre* (Dandelion), *Alchemilla alpina*, and *Carduus palustris* are generally present, and these types show an approach to the alpine grass-lands of the flushed mountain slopes. Those in the peat of the plateau have associations more like those described for the peat alluvia, but generally have a greater variety of species.

Those nearest the present region of cultivation show *Veronica Chamædrys*, *Bellis perennis* (Daisy), *Holcus lanatus* (Yorkshire Fog), and other migrants that are advancing up stream.

Such species are generally absent from the moorland stream alluvia, except where ponies or cattle have afforded a means of transference. Daisies are often quite abundant near the stables used for "stalking ponies" in the heart of the moorland, and also round springs where these ponies are taken to water.

VII.—Lochs and Dubh-lochans.

The shore of Gorm Loch Mòr plunges rapidly into deep water. In a few places where gravelly shallows occur, *Litorella* and *Isoetes* grow on the bottom, and *Nitella opaca* occurs in rather deeper water. The larger stones are covered with *Nostoc* and *Rivularia* colonies, and sometimes *Drapanauldia.* Boulders on the shore of the lake have numerous mosses, including *Rhacomitrium fasciculare*, *Grimmia apocarpa*, *Ptilidium ciliare* (form approaching *pulcherima*), *Pterigynandrum filiforme*, *Hypnum cupressiforme*, var. *tectorum*, *Brachythecum* sp. etc.

A few tow-nettings taken near the shore across several of the smaller inlets, in the month of June, showed hardly any plankton, very few algæ, some cyprids and daphnids, and floating colonies of flagellates.

Several of the corrie lochans in neighbouring areas show a sudden summer development of great quantities of a species of *Nostoc* just visible to the naked eye. It accumulates as a brilliant peacock-green paint-like covering on the stones, or forms a scum on the surface of the water on the lee shores. This development takes place annually in July, but was not observed in Gorm Loch Mòr. It only lasts a few days, and appears to mark a process of propagation and dispersal.

There have been numberless dubh-lochans in the peat of the Ben Armine district, but many have, unfortunately, been artificially drained, with deplorable results so far as the beauty of the country and the distribution of the golden plover and dunlin are concerned.

The origin of these lochans was discussed in *The Vegetation of Caithness*.

In the Ben Armine district, as in Caithness, the dubh-lochans are often very barren of macroscopic flora, especially when they are subject to much wave action. Bogbean, *Juncus bulbosus* (floating type) and *Eriophorum polystachium* are commonly present, and *Carex rostrata* is occasionally encountered, but the last species is far more abundant in the low-lying dubh-lochans in the Baddanloch hollow to the north of the Ben Armine country, where it is sometimes associated with *Carex filiformis* and *Phragmites*, but in these cases the lochans occur in *Molinia* moors, where the waters contain more lime than on the high plateau.

The exposed dubh-lochans often show a sudden development of a brilliant, pale-green microscopic alga *Inefflegiata neglecta* which accumulates in quantities on the lee shores.

Those rather less exposed to wave action generally have the alga *Batrachospermum* growing in patches in the peat mud of the bottom. Small lochans may have flaccid masses of *Sphagnum*, but not floating types, including plumose forms of *S. cuspidatum* and a tumid form of *S. subsecundum* (var. *obesum*).

Squeezings from these mosses generally show an abundant microflora and fauna, including test-bearing rhizopods, rotifers, tardigrades, daphnids, ciliate infusoria, green turbellarians, and numerous desmids, and blue-green algæ, such as species of Chroococcus, Pediastrum, Glœocystis, Merismopedia, Netrium, Gymnozyga, Arthrodesmus, Cosmarium, Euastrum, Staurastrum, Docidium, Gonatozygon, Tetnamorus, etc. (at least thirty species or forms were detected in a few samples subjected to microscopic examination). Diatoms are very scarce.

THE STORY OF THE FORTH.[1]

HAPPY is the country that has no history ; but fortunate is the district which has such a varied and interesting history as that of the Forth, and so sympathetic and competent a historian as the author of this handsome volume.

Mr. H. M. Cadell knows the Forth in all its phases, geological, geographical, and industrial, and he has himself made important original contributions to the elucidation of its geological history. In this volume he deals with some sections of its physical geology and its applied geography. He writes on the physical history of the Forth with the authority of an official expert, for he was engaged in its investigation as a member of the staff of the Geological Survey of Scotland. He has, moreover, travelled extensively, and the series of

---

[1] *The Story of the Forth.* By H. M. Cadell, M.A., B.Sc. Pp. xvii+299. 75 Illustrations and 8 Maps. Glasgow, 1913. Price 16s. net.

memoirs which he has contributed to the *Scottish Geographical Magazine* shows that he has been an alert and thoughtful observer of geographical processes in other lands.

The first chapters deal with the geological history of the Forth basin and with the evolution of the river, and they are mainly reprinted from Mr. Cadell's recent paper in this Magazine. The author here describes the evidence for his view that the Forth was formerly a much greater river which originally rose near Loch Fyne, and that the drainage of the upper part of the basin has been diverted to the Clyde, which has been steadily enlarging its area at the expense of the old Forth. This view was advanced in Mr. Cadell's well-known and suggestive paper on the Dumbartonshire Highlands, which initiated the discussion on the relations of the Clyde, Forth, and Tweed. The main points in the author's early conclusion have been generally admitted, though there is still some room for doubt as to some of the secondary questions. Thus a more complex history is assigned to the Forth by Drs. Peach and Horne; but even where the author's views have not been universally accepted, there is much to be said in their favour. The summary in Chapter VI. of the buried river valleys of the Forth basin is a valuable compilation of this scattered but instructive evidence. The last chapter on the physical geography relates to the old lochs around Edinburgh. The author describes several lochs which are no longer in existence, and explains why they have disappeared.

The latter part of the work is devoted to some branches of the economic history of the district, including a most interesting account of the Carron Company with its long and eventful career in which Mr. Cadell's ancestors played an important part. One chapter is devoted to the Scottish mineral oil industry. Mr. Cadell knows this subject intimately, as he mapped much of the oil shale field for the Geological Survey, and prepared the first edition of the official memoir on these oil fields. Mr. Cadell eulogises the scientific and industrial skill by which the oil shale industry has been developed, in spite of severe foreign competition, until its present annual output of waste shale exceeds, in bulk, the largest of the Egyptian pyramids. The industry is flourishing, but Mr. Cadell does not expect any considerable further extension of it "until a change takes place in the public attitude to the fiscal question, and more common sense with less of party politics comes into vogue," (p. 222). Mr. Cadell is very severe on our fiscal policy and regards its adherents as simple fanatics. The Standard Oil Company, the best known competitor of the Scottish oil industry, the author condemns as an "unscrupulous and relentless combination," without any recognition of the great debt which the oil industry and the public owe to that great corporation.

The last chapters in the work are devoted to land reclamation in the Forth valley. The author has for years taken a very keen interest in this subject, and gave important evidence before the Royal Commission on coast erosion. He then advocated the reclamation of nearly 3000 acres of land between Bo'ness and Grangemouth. He convinced the Commissioners that his scheme was worthy of further consideration and expert investigation. His conclusions were opposed by Mr. Meik on both general and financial grounds. Mr. Meik objected that so extensive a reclamation would injuriously affect the Forth by diminishing the tidal scour. To these criticisms Mr. Cadell replies at length. His original scheme was to reclaim 2837 acres, but he now advances a smaller scheme for the reclamation of 1000 acres above Grangemouth. He feels, however, that there is little chance of such schemes being undertaken while land owners are injured "by unsympathetic or even positively malignant politicians."

The last chapter deals with the reclamation of moss lands. He describes Lord Kames' spirited action in clearing Blair-Drummond Moss by scouring off the peat, and thus exposing the good land below. The results of these early efforts appear to have been financially profitable, but land values are now so much lower, that the clearing of the land under present conditions might not be remunerative. Mr. Cadell makes the interesting suggestion that the effort to reclaim the land should be combined with utilisation of the peat. He points out that hitherto the attempts to use peat in Scotland, except as a local fuel, have been unprofitable; but he suggests that if the peat were so removed as to leave the ground ready for agriculture, then the value of the land gained might make the peat-working profitable.

Mr. Cadell's work is a very valuable contribution to Scottish geography, for it collects a great deal of scattered and useful information. He illustrates his views by many excellent maps, and sketches which are artistic, ingenious and instructive, and are beautifully reproduced. The book states the conclusions of an experienced geographer on many important problems to which its author has devoted years of careful observation and thoughtful study.

GEOGRAPHICAL NOTES.

EUROPE.

Place Names in Scotland.—Mr. J. Mathieson, the Convener of the recently revived Place-names Committee, sends us the following:—"The Committee held their second meeting in the Society's rooms on March the 28th, Dr. W. J. Watson in the chair. The chief business was the correct form and derivation of Beinu Iutharn, a hill on the watershed between the counties of Aberdeen and Perth, which Dr. J. Milne (*Place-names in Aberdeenshire*) derives from 'Bearn,' *a gap*, i.e. 'Beinn a' Bheairn,' *mountain of the gap*. The Committee concluded that there was no authority for Dr. Milne's derivation, but that the name was pronounced in Perthshire 'Iùrn' and on the Invereyside 'Iùrnan.' They also found that 'faobhar,' *an edge*, applied specially tc the sloping part of a hill as seen against the sky, is pronounced in Perthshire 'fieur,' and they were on the whole of the opinion that the name means 'edge hill,' and should from a philological point of view be written Beinn Fhiùbharain, the 'ain' being a terminal suffix, or it may be a plural termination. They considered the form as it appears on the O. S. Map, viz., Beinn Iutharn, a fair representation of the local pronunciation."

Rise and Fall of Turkish Power in Europe.—In connection with recent events in the Balkan Peninsula, it is of interest to note an article by Professor Nikolaus Jorga of Bukarest, which appears in *Petermann's Mitteilungen* for January, under the above title. The article is illustrated by an interesting series of maps, showing not only the varying boundaries of Turkey in Europe, since its greatest extension in the seventeenth century, but also the Bulgarian and Servian kingdoms at the period of their greatest extension. The accompanying article is historical in nature.

Asia.

Central and Northern Yunnan.—M. A. F. Legendre contributes to the December issue of *La Géographie* an account of his explorations in western China and Tibet, which includes some interesting facts in regard to the mode of life of the native and Chinese inhabitants of Yunnan. Both, it appears, subsist chiefly on maize, a curious fact as the cultivation of this cereal here cannot be of very ancient date. Red mountain rice is also grown, but on a small scale, and the other plants chiefly cultivated are buckwheat, millet, sorghum, tobacco and hemp. The millet and sorghum are not of great importance as food, their chief use being, at least so far as the Chinese are concerned, for the manufacture of spirit. On the other hand, tobacco and hemp bulk large in the estimation of the native. With the hemp fibre he makes a textile which, as it is loose-meshed and undyed, must present a close resemblance to sacking. It forms nevertheless the only native cloth, cotton, though highly prized, being a luxury. The Chinese and half-breeds, on the other hand, always use cotton, and have a great contempt for the native cloth.

In addition to the plants named various peas and beans are grown, including soya beans, while another leguminous plant, a species of Pachyrhizus, is cultivated for the sake of its tuberous root which is rich in starch. The ordinary "vegetables" of Europe, as spinach, cabbages, onions, carrots, turnips, etc., grow freely as winter crops, but winter cultivation is checked by the scarcity of rain then. Sugar cane and earth-nuts are also grown by the Chinese, and there is a certain production of fruit, such as kaki plums, oranges, the fruit of the jujube tree, pears, and a few chestnuts, but the stocks are poor and unimproved, and the people seem too inert to practise improved methods of agriculture or horticulture. Further, the Chinese have deforested the country almost completely, and surface erosion has consequently greatly increased, with a corresponding loss of soil on the slopes. Natural pasture is widespread, but though buffaloes, cattle, horses, asses, mules, goats and sheep are all kept, the number is small, the breeds poor, and the flocks produce little. Their milk is not utilised in any fashion.

The general lack of energy and initiative suggested by the methods of cultivation and the nature of the stocks both of plants and animals is apparently largely due to the prevalence of goitre. In the high valleys and plateaux dwarfs are frequent, and have been regarded as a pigmy breed, but, according to the author, they are merely crétins. Below about 6500 feet malaria also occurs, and the two diseases, with the enervating climate, probably account for the general listlessness.

The natives are cultivators and herdsmen and have practically no commerce or industries. The Chinese, on the other hand, in addition to these occupations, extract sugar from the sugar cane and oil from the earth-nuts; manufacture brandy; dye and weave the small amount of cotton grown; act as petty merchants; exploit the scanty remnants of the forests; and lend money to the cultivators. They have the greatest contempt for the natives, whom they characterise as "without faith and without law."

Currents in the Sea of Japan.—According to a note in *Nature*, the existence of circular currents in the Sea of Japan has recently been proved by Dr. Wada, the meteorologist of the Korean Government-General. The discovery is regarded in Japan as of great importance, both in connection with the distribution of marine life, and even with human migrations in East Asia. Dr. Wada's investigations were based upon the movements of mines laid in Vladivostock Bay during the Russo-Japanese war, and also upon the drift of 120 bottles, thrown from a steamer for the purpose. He believes that the results show that the Liman current, which runs down the coast of Siberia, flows southward past Kang-won and Ham-gyong provinces, Korea, as far as Cape Duroch. It then sweeps across to the coast of Echizen, Japan, and travels northward along the coast in company with the Tsushima current. One branch then goes out to the Pacific through the Tsugaru Strait, and another continues northward to Tartar Strait, where it joins the Liman current, thus completing the circle.

General.

The Royal Geographical Society's Annual Awards.—The following are the awards made for the current year by this Society. The Founder's Medal will not be awarded, but a casket, with a suitable inscription, will be presented to Lady Scott to contain the Patron's Medal and the special Antarctic Medal awarded to the late Captain Scott in 1904. The Patron's Medal is awarded to the late Dr. E. A. Wilson, while Lieut. Campbell is to receive a gold watch in recognition of the ability and good leadership displayed by him as commander of the Northern Party.

The Victoria Research Medal goes to Colonel S. Burrard, Surveyor-General of India, the Murchison Award to Major H. D. Pearson, the Gill Memorial to Miss Gertrude Lowthian Bell, the Cuthbert Peek Grant to Dr. Felix Oswald, and the Back Bequest to Mr. William S. Barclay.

International Geological Congress.—At this Congress, which is to be held at Toronto from August 7-14 of the present year, the principal subject for discussion is to be "The Coal Resources of the World," and a large monograph on the subject is being prepared by the committee. It is hoped that this monograph, which will consist of two quarto volumes and a folio atlas, will be ready for the meetings. Among the other subjects for discussion is—"To what extent was the ice-age broken by interglacial periods?" A fine series of excursions has been arranged in connection with the Congress, some of which will precede the meeting while others will follow it, and shorter excursions will also take place during the Congress week.

EDUCATIONAL.

Historical Development of Geographical Teaching in the United States.—Mr. Charles Dryer contributes to a recent issue of *The Journal of Geography* an interesting note on the development of secondary school geography in the States during the past century. In the earlier part of the nineteenth century the subject was in what he calls the "Gazeteer Stage," the books used giving mere descriptions of the salient features of the earth, with no attempt at scientific treatment except in the case of mathematical geography, taught by means of globes. Later this was replaced by the "Wonder Book" stage, when the books used in schools laid special stress upon striking and spectacular phenomena, such as earthquakes, volcanoes, etc. Such books were pious in tone, and found their natural development in Guyot's work (1873), which was an elaboration from the geographical side of Paley's argument from design. Such a sentence as that "the continents are made for human societies, as the body is made for the soul" sufficiently indicates the style. As the date of publication shows, Darwin had long before, in the *Origin of Species* (1859), undermined the basis of Paley's argument, and many of Guyot's purely geographical generalisations have as little permanent value as his philosophy. The new science and new philosophy found expression in Huxley's *Physiography*, which made the subject a general introduction to science, and paved the way for many books of similar type. But in the States the numerous striking discoveries of geologists, especially those connected with the Geological Survey, give a great impetus to physiography, considered as the science of land forms, and thus in a much narrower sense than Huxley's. A new terminology was elaborated, text-books multiplied and this specialised physiography apparently took hold of school teaching in a way which it has not done here. But as it has virtually become physical geology its study in the schoolroom is obviously beset with great difficulties, which even the fitting up of costly laboratories does not eliminate. Thus, within the last few years, this type of physical geography, in Mr. Dryer's opinion, is yielding before what he calls biogeography, or the relation of living creatures to their environment. This involves the throwing overboard of a good deal both of the specialised physical geography of the American text-books, and even of physical geography in a less specialised sense. Mr. Dryer believes that stress is being more and more laid upon climates as a more important factor than relief in determining distribution, and also upon plant ecology as throwing light upon climate; economic geography is also being reorganised. Finally, in this new development stress is being laid upon human geography in a very wide sense, and this involves a development of regional geography on a basis of natural rather than political divisions. The new geography will thus be a blend of natural, technical, and social science.

NEW BOOKS.

EUROPE.

Baedeker's Northern Italy. Fourteenth Remodelled Edition.
Baedeker's Northern Germany. Sixteenth Revised Edition. Leipzig: Karl Baedeker, 1913. Price 8 *marks each.*

In these new editions the usual revision has taken place without, so far as we notice, any great modification. Both are stated to have the newer arrangement by which the book can be cut up into sections for those who do not care to carry the entire volume about. Personally we have not found the arrangement very satisfactory, and think that still further improvements along the same lines are required, though doubtless the technical difficulties are considerable. Otherwise the books show the long familiar and valuable features of their predecessors.

The Early Norman Castles of the British Isles. By ELLA S. ARMITAGE. Numerous Plans and Illustrations. London: John Murray, 1912. Price 15*s. net.*

The title of this work may seem to cover only a small subject, yet the mere enumeration of the English castles of the period with short notes occupies 157 pages, and in truth the castle-building labours of the Normans were on an enormous scale. With them it was a matter of deliberate policy, and, as a recent writer has remarked, "round these castles all the subsequent history of warfare in Britain turns for 400 years." But though this fact may be familiar to the general reader of history, it is not always realised that these places of strength of our Norman conquerors were merely erections of earth and wood. There were probably not more than four castles of stone erected by them in the eleventh century,—the Tower of London, Colchester, Pevensey and Bramber. The other constructions consisted of an earthen mound, most often artificial, surmounted by a palisade and tower of wood, and with a separate enclosure or bailey, the plan corresponding exactly with that of the stone castles which immediately succeeded them.

These castles are termed *motte*-castles by the author, from the Norman-French word, motte, meaning the hillock of a castle—well known to us in Scotland in the form moat. The term appears also as moot, signifying judgment (*cf.* Muthill or Moot-hill in Perthshire), since the site of these ancient castles continued often to be the seat of feudal jurisdiction. Miss Armitage has expended many years of fruitful research in gleaning all that can be gathered concerning the design, purpose and history of these vanished fortifications, and the present work will receive the cordial encomiums of all students of the period, on which it throws fresh light. We can only hope that she may be enabled to give further attention to the castles of Scotland, in regard to which she and Dr. George Neilson have already done notable work, showing incidentally the extent of the Anglo-Norman occupation of the Lowlands. The book is a monument of erudition and at the same time most interesting reading.

Echoes from the Border Hills: being the Reminiscences of the late John Hyslop, J.P., Langholm. Edited by his son ROBERT HYSLOP, F.S.A. (Scot.). Sunderland: Hills and Co.; Langholm: Robert Scott and Walter Wilson, 1912. *Price 2s. 6d. net.*

In our notice of *Langholm as It Was* in December last (*S.G.M.*, vol. xxviii. page 660), we expressed a hope that Mr. Robert Hyslop might be able to carry out his intention of publishing another volume of his father's reminiscences. He has now done so by the publication of *Echoes from the Border Hills*, which will probably receive as enthusiastic a welcome as the former volume. It contains a series of reminiscences illustrating the conditions of life in Eskdale before the days of railways and penny post, not to speak of motors and wireless telegraphy. The stories are pleasantly told in familiar and often humorous language, with no straining after effect, and will appeal to a wide circle of readers.

ASIA.

The Empire of India. By Sir BAMPFYLDE FULLER, K.C.S.I., C.I.E. London: Sir Isaac Pitman and Sons, Ltd., 1913. *Price 7s. 6d. net.*

The publishers of the "All-Red Series" were fortunate in securing for their volume on India the services of Sir Bampfylde Fuller, whose experience of India is both recent and extensive. He has set about his arduous task with great care and assiduity, and has produced a very interesting volume of something less than 400 pages, which may fairly be entitled "Enquire within about everything Indian." It is a model of laborious and judicious condensation, and will, we trust, serve to whet the reader's appetite for other and more detailed works of which there are plenty, treating fully what Sir Bampfylde is only able to summarise. For example, the history of India from the most ancient down to recent times is sketched in 22 pages; to agriculture, and the vast majority of the 315 million inhabitants of India, are assigned only 19 pages; to commerce 20, and so on. The book is an encyclopedia, necessarily on a small scale, but packed with well-digested information and up to date; and incidentally it states some of the many perplexing and complicated problems connected with our Indian Empire and gives the opinions regarding them of one who by his training and experience is a very competent judge.

On the Track of the Abor. By POWELL MILLINGTON. London: Smith, Elder and Co., 1912. *Price 3s. 6d. net.*

This little volume deals with the same tract of country and the same tribe of people as Mr. Hamilton's book reviewed on p. 49. It is racily written and can easily be read at a sitting. It adds considerably to our knowledge of the hardships and inconveniences which a punitive expedition has to undergo on the remote North-Eastern Frontier of India, and gives an interesting account of the Abor's methods of cultivation. In the appendices come rough sketch-maps of Aborland, and a note on the course of the Tsanpo is given.

Bismaya; or The Lost City of Adab. By EDGAR JAMES BANKS. New York and London: G. P. Putnam's Sons, 1912. *Price* 21*s. net.*

Dr. Banks was the leader of an expedition sent by the University of Chicago at the end of 1899 to explore Babylonian ruins, the objects found being destined for the Smithsonian Institution. Great difficulty was experienced in obtaining the necessary permission from the Turks, and after Dr. Banks had spent two and a half years in Constantinople his committee abandoned hope and were disbanded. Not long afterwards, however, permission was obtained to excavate, not at the site first chosen, but at Bismaya in the heart of Mesopotamia, and, financial backing being obtained, a start for the desert was made in October 1903, nearly three years after Dr. Banks' original departure from the States. A journey of six weeks brought him to the mounds of Bismaya, where excavation was begun on Christmas day. The rest of the book gives an account of the objects found, the progress of the excavations, and of the author's journeys in Babylonia.

The mounds were found to cover the ruins of Adab, a city already known from tablets, and the most striking object obtained was a white marble statue, some 32 inches high, with an inscription interpreted as meaning that it represents a king of the city. The author believes that there was here first a pre-Sumerian city, which was replaced by a fortified, temple-containing Sumerian city. This city was sacked by the Semitic king Sargon I., but came again into the hands of the Sumerians before its final destruction. Large numbers of objects of all kinds, including clay tablets, gold, copper and ivory objects, as well as remains of temples, palaces and private houses were unearthed, and of these the author gives interesting accounts. Incidentally also he casts much light upon present conditions of life in Mesopotamia. The book is profusely illustrated, and is written for the general reader.

Monumental Java. By J. F. SCHELTEMA, M.A. London: Macmillan and Co., 1912. *Price* 12*s.* 6*d.*

The author seems to have lived in Java at intervals from 1874 to 1903, and he has endeavoured in this book to describe the monuments in their proper setting, their relations to natural scenery and native civilisation. As he himself points out in a rather long preface, he "sets up no pretence at completeness; there is no full enumeration of all the Hindu and Buddhist temples known by their remains; there are no measurements; no technical details, and no statistics."

Even in 412 or 413 when the Chinese pilgrim Fa Hiem visited Java, Brahmanism seems to have made considerable progress in the island, though Buddhism was not much in evidence. From 778 to 928 A.D. (according to the author), the fine Buddhist temples of Boro Budoor, Mendoot, and others seem to have been erected. About 1400 A.D., the Arabs introduced "the Islām" (*sic*) into Java, and most of the island seems to have been converted or enthralled by 1490 A.D.

After a short introductory chapter on the country, the people and their work, the author describes a few of the monuments in West, Central, and East Java, but the last three chapters deal with the Boro Budoor temple, of which there are several photographs.

The author is clearly an enthusiast, and there are passages in the book which are not only beautiful, but reveal an unusual sympathy with the devout spirit and inner meaning of the ancient Javanese builders. But there are several things which interfere with a comfortable enjoyment of the book. The author is always "dwelling on the tenebrous general aspect of governmental archæology in the past." We hold no brief for the Dutch government, but we suspect that Mohammédans had more to do with the destruction of the monuments than he supposes and in any case one coherent chapter of scathing invective would have pleased the reader better than perpetual complaints.

The incessant use of Hindu or Javan expressions, and the abundance of quotations (in Latin, Greek, French, German, Italian, Spanish, etc.), tend to disturb the main argument. The author's spelling is also eccentric (Sooltan for Sultan, Islām). Why also should a volcano be so often or always described as a fire-mountain?

These are unfortunate defects, for any one visiting Java who has already a working knowledge of Buddhism might find great pleasure in the author's descriptions.

AFRICA.

The Adventures of an Elephant-Hunter. By JAMES SUTHERLAND. London: Macmillan and Co., Limited, 1912. Price 7*s.* 6*d.* *net.*

Mr. Sutherland has certainly had a varied and adventurous career. He went out to Africa in 1896, and, since then, has roamed about the Transvaal, Matabeleland, Mashonaland, British Central Africa, Lake Tanganyika, the Congo, Portuguese and German East Africa, earning a living by "nigger-bossing," contracting for railways, acting as agent for trading companies, keeping native stores, prize-fighting, and volunteering with the Germans to put down an insurrection. But his heart was in none of these occupations. He is a sportsman, and during ten years has shot 447 bull elephants, besides females, for which he claims a world's record, which we imagine no one will care to dispute with him. In this volume he gives us a series of thrilling and exciting stories of his adventures in pursuit of elephants, man-eating lions, etc. He had no European companions, and had to trust to his own skill and prowess in many a tight place. He scouts at the idea of a speedy extermination of the big game in Africa, telling us that in addition to the many game reservations "there are thousands of square miles of quite uninhabited country in which there are hundreds of thousands of elephants and every other kind of game, with the exception, perhaps, of giraffes." He was equally unorthodox as to his potations during his expeditions, drinking a stiff whisky and soda after his day's work, half a bottle of port at dinner, and one or two strong whisky pegs ere he went to sleep.

Dawn in Darkest Africa. By JOHN H. HARRIS. London: Smith, Elder and Co., 1912. Price 10*s.* 6*d.* *net.*

Before he published this very interesting book, Mr. Harris did a very astute thing. He selected his own reviewer, and he made the happy selection of Lord Cromer, whose authority as a diplomat, a statesman, and a man of affairs is indisputable. Lord Cromer has contributed an introduction, which is neither more nor less than an able and elaborate review, from the conclusions of which it would be rash and risky to differ. The Dawn in Africa, which is the subject-matter of the book, is not the dawn in political or economic history of centuries ago. It is the dawn which is of quite recent origin, confined practically to the area of the Congo and colonies round the Gulf of Guinea, and the result of the lurid light thrown on the condition of the natives from various sources, notably missionaries, who, by exposing the horrors of the Congo and the Portuguese protectorates, have compelled the attention of Europe to Equatorial Africa, and have thus inaugurated a more hopeful and beneficent era for vast regions of territory, the populations of which were rapidly approaching extinction. Mr. Harris describes the miseries of the inhabitants of the Congo under the Belgian régime and those of the inhabitants of Angola, San Thomé, and Principe under the Portuguese administration, with the vivacity and earnestness of him who has a personal, recent, and extensive experience of them, and at the same time he does not conceal the few good points in these administrations, nor does he evade the very serious difficulties inevitably incidental to a change of system. He has many shrewd observations to make on the British, French, and German administrations, to each of which he plays the invidious and at all times unpleasant rôle of the candid friend. In this respect he has, indeed, little or nothing to tell us that is new, for the faults and deficiencies of administration in their African territories are notorious and are frankly acknowledged by French and German writers. He does not hesitate to plunge into *la haute politique*, and suggest, if he does not propose, a rearrangement of the map of Equatorial Africa, a proceeding in which it is needless to say he does not receive much encouragement or assistance from the experienced and cautious diplomat Lord Cromer. Much more useful and instructive are his notes and observations on the natural resources of the territories, *e.g.* oil palms, rubber trees, cocoa cultivation, and the like, which well merit the earnest attention and study of statesmen as well as of merchants. Mr. Harris is a missionary, obviously profoundly convinced that the true hope of Africa lies in the spread of Christianity and in the raising of the level of character both of Europeans and natives, but he sees the serious hindrances and difficulties, and in no way minimises them. Indeed, his remarks on some of them, notably on polygamy, show a candour, liberality of thought and charity which we do not often meet with in missionary writings. We cordially recommend this book to the perusal of our readers, agreeing with the high encomium of Lord Cromer, that no one has ever brought a more evenly balanced mind to bear on the numerous problems which perplex the African administrator than Mr. Harris, whose "enthusiasm is tempered by reason and by a solid appreciation of the differences between the ideal and the practical."

AMERICA.

The Putumayo: The Devil's Paradise. By W. E. Hardenburg. London: T. Fisher Unwin, 1912. *Price* 10*s.* 6*d. net.*

It seems but a short time ago since the moral sense of the civilised world was shocked by the disclosure of atrocities and horrors connected with the rubber traffic in the Belgian Congo, and now, in the volume now before us, we have a disclosure of prolonged brutality and cruelty in South America, rivalling, but certainly not exceeded by, anything which occurred under the Leopoldian régime. A deplorable feature of the affair is that the atrocities were committed by the agents of a company among the directors of which are English gentlemen high in rank in the social and commercial world. As we now write, their personal innocence or guilt is the subject-matter of a formal trial, so it would be most improper to express any opinion on that subject. Suffice it to say, that no condemnation and no punishment could be too severe for a government or a company who were aware that their agents were committing one tithe of the crime which is laid to their charge in this book and in the documents by which it is supported, crime by which a peaceful, docile, amiable race has been almost extirpated under circumstances of indescribable brutality. The geographical interest in the book lies in the description it gives of a very adventurous, and, in the end, disastrous journey over the Andes down to the Putumayo River by W. E. Hardenburg and his companion, W. A. Perkins, who, apparently, were prospecting for rubber and were prepared to invest money in any concern that promised good profits. Hardenburg has a pitiful story to tell of his misfortunes and bad treatment, when he reached the regions which were being exploited by the Peruvian Amazon Company, for the details of which we must refer the reader to the book. Incidentally Hardenburg has much interesting information to give us as to the conditions of life in the forest, the flora and fauna he found there, the Indian tribes, and more especially the Huitotos, their houses, dress, weapons, superstitions, domestic customs, means of livelihood, and the incredibly inhuman treatment they have been receiving at the hands of the agents of the Peruvian Amazon Company, owing to which a population estimated to have numbered some forty to fifty thousand a few years ago has been reduced to under ten thousand. The Putumayo is an affluent of the mighty Amazon. Its source is near the town of Pasto in Southern Columbia, and it has a course of over a thousand miles ere it enters the Amazon. The region described by Mr. Hardenburg lies for the most part between the Putumayo River and its effluent the Igara-Parana, and Mr. Reginald Enoch, an indisputable authority, who contributes an elaborate and incisive introduction to this book, describes it as "an extremely outlying part of Peru with corresponding difficulties of access and governance," and he confirms from his own knowledge much of what Mr. Hardenburg describes. The object of the book is undoubtedly an indictment against the Peruvian Government and the Peruvian Amazon Company, but in addition it is a distinct contribution to our knowledge of the natural history, anthropology, and geography of a little known but interesting tract of South America.

The Conquest of New Granada. By Sir Clements Markham, K.C.B., D.Sc. (Cam.). London: Smith, Elder and Co., 1912. *Price 6s. net.*

Mr. Bryce in his recent book on South America says, "When the Spaniards came to the New World, they came mainly for the sake of gold. Neither the extension of trade, the hope of which prompted the Dutch, nor the acquisition of lands to be settled and cultivated, thereby extending the dominion of their crowns, which moved most of the English and French, nor yet the desire of freedom to worship God in their own way, which sent out the Pilgrims and Puritans of New England—none of these things were uppermost in the minds of the companions of Columbus and Ponce de Leon, of Vasco Nuñez and Cortes and Pizarro. No doubt they also desired to propagate the faith, but their spiritual aims were never suffered to interfere with their secular enterprises." Probably, it is no exaggeration to say, that were we to search the whole course of the history of Spanish doings in America, we would not find a more striking illustration of the truth of Mr. Bryce's remarks than in the episode so gracefully and spiritedly told by Sir Clements Markham in the volume now before us. The conquest of New Granada, *i.e.* the modern republic of Columbia, took place in the first half of the sixteenth century. Several Spanish *conquistadores* took part in it, but the actual subjugation of the Chibchas will go down in history as the work of Gonzalo Jimenes de Quesada (who started life as an advocate in Granada in Spain) in his expedition in the years 1536-8. Sir Clements makes out the best case he can for Quesada, but he is fain to confess that "Quesada had arrived in the country of the Chibchas and found wide plains and beautiful valleys, thickly peopled by an industrious and intelligent race. He found an advancing civilisation guided by two sovereigns of ancient lineage, with a third sacred personage acting as arbitrator and peace-maker. He found chiefs and people happy and contented. When he departed all was changed. There was confusion and terror, cultivation neglected, some of the people in flight, others forced to work as slaves. He had killed two sovereigns, tortured another to death. Destruction had come upon Chibcha civilisation, and desolation brooded over the once prosperous land. True: but Quesada was taking home a box containing 758 emeralds for the emperor Charles v." A Spanish annalist, referred to by Mr. Bryce, states that within twenty years after the arrival of the Spaniards, the Chibchas, who had numbered nearly a million, were almost annihilated. Sir Clements has done good work in collecting together what is known of the Chibcha civilisation, their religion, language, calendar, system of government, etc., but he indicates that his work is by no means exhaustive, and he hopes that some other author will tackle the subject again. We may be permitted to echo the hope, and to add our hope that the author of the future will be endowed with the same patience, industry, lucidity and good judgment as the veteran president of the Hakluyt Society.

GENERAL.

Dans l'Atlantique. Par Henri Dehérain. Paris: Hachette et Cie, 1912. *Price* 3 *fr.* 50.

St. Helena has been immortalised as the prison of Napoleon, but in this book M. Dehérain describes the island as it was during the seventeenth and eighteenth centuries. No European Power occupied it till 1659, when it was taken possession of by the English East India Company, which received in 1661 a royal charter from Charles II. This conquest was resented by the Dutch East India Company, which, profiting by the war between the United Provinces and Charles II., the ally of Louis XIV., sent a fleet from the Cape of Good Hope in 1672 and captured the island. Next year, however, the island was retaken by a British fleet under Richard Munden, who received a reward of £2500, and was knighted by Charles II. A second royal charter was granted to the English East India Company, which administered the island for the next one hundred and sixty-one years.

A French vine grower named Etienne Poirier lived for thirty years in St. Helena, being employed by the company in cultivating the vine and producing wine and brandy. Owing to a revolt, during which the governor was murdered, Poirier, who had shown great courage, was appointed lieutenant, and in 1697 governor of the island. A Huguenot, he proved a strict administrator, repressing drunkenness and insisting on Sabbath observance. In 1706, under his very eyes, two of the company's ships were seized by two French men-of-war commanded by Des Augiers, and carried off to Brest. Poirier was greatly blamed by the company, and died in the following year. Slaves formed a large proportion of the population of the island, as many as fifteen hundred and forty being there in 1817. They were subject to such severe rules and punishments, that they often committed suicide or fled to the interior of the island or escaped by sea. Gentler treatment began in the eighteenth century, and at last, in 1833, the emancipation of slaves was decreed throughout the British empire.

The remainder of the volume (which is illustrated by maps) is devoted to descriptions of Tristan Da Cunha (the correct spelling), the coasts of South Africa, and the voyages of Auguste Broussonet to Morocco and the Canaries. The author mentions that Captain Dugald Carmichael, R.N., was the first European to describe the central peak of Tristan Da Cunha, which he ascended in 1817, afterwards contributing a scientific paper on the island to the Linnean Society of London. The mountain he ascended is the nesting-place of the great sea-birds of the Atlantic.

BOOKS RECEIVED.

The Age of the Earth. By ARTHUR HOLMES, B.Sc., A.R.C.S. Illustrated with 20 Figures and Diagrams. Fcap. 8vo. Pp. xii + 196. London: Harper Brothers, 1913. Price 2*s.* 6*d. net.*

Wayfaring in France: From Auvergne to the Bay of Biscay. By EDWARD HARRISON BARKER. 8vo. Pp. xv + 540. London: Macmillan and Co., Ltd., 1913. *Price* 7*s.* 6*d. net.*

The Continents and their People. Asia: A Supplementary Geography. By JAMES FRANKLIN CHAMBERLAIN, ED.B., S.B., and ARTHUR HENRY CHAMBERLAIN, B.S., A.M. Cr. 8vo. Pp. ix + 198. Macmillan and Co., Ltd., 1913. Price 3*s.*

The Land of the Peaks and the Pampas: South America of Yesterday and To-day. By JESSE PAGE, F.R.G.S. With a Map and 18 Illustrations from Photographs. Cr. 8vo. Pp. xv + 368. London: The Religious Tract Society, 1913. Price 3*s.* 6*d. net.*

Modern Chile. By W. H. KOEBEL. With Illustrations and Map. Demy 8vo. Pp. x + 278. G. Bell and Sons, Ltd., 1913. Price 10*s.* 6*d. net.*

A History of the Colonisation of Africa by Alien Races. By Sir HARRY JOHNSTON, G.C.M.G., K.C.B. With 8 Maps. New Edition, revised throughout and considerably enlarged. Cr. 8vo. Pp. xvi + 505. Cambridge University Press, 1913. Price 8*s. net.*

Changing Russia. By STEPHEN GRAHAM. With 15 Illustrations and a Map. Demy 8vo. Pp. ix + 309. London: John Lane, 1913. *Price* 7*s.* 6*d. net.*

The Land of the New Guinea Pygmies: An Account of the Story of a Pioneer Journey of Exploration into the Heart of New Guinea. By Captain C. G. RAWLING, C.I.E., F.R.G.S. With 48 Illustrations and a Map. Demy 8vo. Pp. xvi + 366. London: Seeley, Service and Co., Ltd., 1913. Price 16*s. net.*

The Norfolk Coast and The Suffolk Coast. By W. A. DUTT. Illustrated. Cr. 8vo. Pp. 413. London: T. Fisher Unwin, 1909. *Price* 6*s. net.*

The Cornwall Coast. By ARTHUR L. SALMON. Illustrated. Crown 8vo. Pp. 384. London: T. Fisher Unwin, 1910. Price 6*s. net.*

The Geography and Geology of South-Eastern Egypt. By JOHN BALL, Ph.D., D.Sc. Demy 8vo. Pp. xii + 394. Cairo: Survey Department, 1912.

Anthropological Report on the Ibo-speaking Peoples of Nigeria. By NORTHCOTE W. THOMAS, M.A., F.R.A.I. Demy 8vo. Part I.: *Law and Custom of the Ibo of the Awka Neighbourhood, S. Nigeria.* Pp. 161. Part II.: *English-Ibo and Ibo-English Dictionary.* Pp. viii + 391. Part III.: *Proverbs, Narratives, Vocabularies, and Grammar.* Pp. vi + 199. London: Harrison and Sons, 1913.

Mozambique: Its Agriculture and Development. By ROBERT NUNEZ LYNE, F.L.S., F.R.G.S. Illustrated. Demy 8vo. Pp. 352. London: T. Fisher Unwin, 1913. Price 12*s.* 6*d. net.*

The North Pole and Bradley Land. By EDWIN SWIFT BALCH. Royal 8vo. Pp. 91. Philadelphia: Campion and Co., 1913.

Are the Planets Inhabited? By E. WALTER MAUNDER, F.R.A.S. Fcap. 8vo. Pp. 166. London: Harper Brothers, 1913. *Price 2s. 6d. net.*

Twentieth Century Jamaica. By H. G. DE LISSER. With Illustrations. Demy 8vo. Pp. 208. Kingston: Jamaica Times, Ltd., 1913.

Le Pérou Économique. Par PAUL WALLE. Préface de M. PAUL LABBÉ. Un fort vol. in-8°. Illustr. et carte hors texte, broché. Pp. xvi + 388. Paris: E. Guilmoto, 1913. *Prix 9 fr.*

Ravenna: A Study. By EDWARD HUTTON. Illustrated in Colour and Line. By HARALD SUND. Demy 8vo. Pp. xii + 300. London: Dent and Sons, Ltd., 1913. *Price 10s. 6d. net.*

Big Game Shooting in India, Burma, and Somaliland. By Col. V. M. STOCKLEY. Royal 8vo. Pp. xii + 282. London: Horace Cox, 1913. *Price 21s. net.*

The Berwick and Lothian Coasts. By IAN C. HANNAH. With 65 Illustrations. By EDITH BRAND HANNAH. Demy 8vo. Pp. 368. London: T. Fisher Unwin, 1913. *Price 6s. net.*

The Passing of the Turkish Empire in Europe. By Captain B. GRANVILLE BAKER. With 33 Illustrations and a Map. Demy 8vo. Pp. 335. London: Seeley, Service and Co., Ltd., 1913. *Price 16s. net.*

Camp and Tramp in African Wilds: A Record of Adventure, Impressions, and Experiences during many years spent among the Savage Tribes round Lake Tanganyika and in Central Africa. With a description of Native Life, Character, and Customs. By E. TORDAY. With 45 Illustrations and a Map. Demy 8vo. Pp. xvi + 316. London: Seeley, Service and Co., Ltd., 1913. *Price 16s. net.*

Geography of Missouri. By FREDERICK V. EMERSON. Demy 8vo. Pp. 74. Columbia, 1912.

The Timber, Agricultural, and Industrial Resources of Vancouver Island, British Columbia. Pp. 44.

Preliminary Report on the Field Experiment upon Cotton carried out at Talbia, Egypt, 1912. Pp. 10. Cairo, 1912.

Report on the Administration of Coorg, 1911-12. Mercara, 1912.

Report on the Administration of Bangalore, 1911-12. Bangalore, 1912.

Report on the Administration of the Punjab and its Dependencies, 1911-12. Lahore, 1913.

Report on the Administration of the United Provinces of Agra and Oudh, 1911-12. Allahabad, 1912.

Diplomatic and Consular Reports.—Report on a Journey in the South of Brazil (684); Report on the Leased Territory of Liaotung Peninsula (5042); Report on the Trade and Resources of the Congo, 1911 (5043); Report of Trade of Amsterdam, 1912 (5044); Trade of Constantinople, 1912 (5045); Trade of Goa, 1911-12 (5046); Trade of Swatow, 1912 (5051); Trade of Rio de Janeiro, 1911-12 (5049); Trade of Ispahan, 1912 (5048).

Colonial Reports.—St. Lucia, Report for 1911-12 (752).

Publishers forwarding books for review will greatly oblige by marking the price *in clear figures, especially in the case of foreign books.*

THE SCOTTISH GEOGRAPHICAL MAGAZINE.

DAVID LIVINGSTONE:

A Review of his Work as Explorer and Man of Science.[1]

By Sir Harry Johnston, G.C.M.G., K.C.B.,
Livingstone Gold Medallist, Royal Scottish Geographical Society.

(With Map.)

David Livingstone, it is scarcely necessary to remind you, was of Highland descent, his grandfather having been a crofter on the little island of Ulva, off the west coast of the larger island, Mull. In appearance he showed clearly that the predominant strain in his ancestry was what we call Iberian for want of a more definite word. That is to say, that he was of that very old racial strain still existing in Western Scotland, Western Ireland, Wales, and Cornwall, which has apparently some kinship in origin with the peoples of the Mediterranean, and especially of Spain and Portugal. Indeed, according to such descriptions as we have of him (notably that by the late Duke of Argyll), and such portraits as illustrate his appearance, he was not unlike a Spaniard, especially in youth and early middle age. His height scarcely reached to 5 feet 7 inches, his hair and moustache, until they were whitened with premature old age, were black, his eyes hazel, his complexion much tanned by the African sun, but at all times inclining to sallow. He possessed a natural dignity of aspect, however, which never failed to make the requisite impression on Africans and Europeans alike. Bubbling over with sly humour, with world-wide sympathies, and entirely free from any narrowness of outlook, he possessed a very strong measure of

[1] The Livingstone Centenary Address, delivered before the Society in Edinburgh on March 27, 1913.

self-respect, coupled with a quiet, intense obstinacy of purpose. In earlier life he was so eager to advance the bounds of knowledge, and so certain that he was a predestined and appointed agent to accomplish great purposes, that he may have been slightly arrogant and contemptuous towards fools and palterers.

Livingstone succeeded in a measure to the work of exploration already commenced in the first quarter of the nineteenth century by men like Campbell, and Moffat, Dr. Philip, Edwards, Archbell and Melvill. Of such pioneers, including Livingstone himself, James Chapman, one of the greatest and most widely travelled of South African explorers, subsequently wrote in 1855: "Their labours were difficult, their trials many, their earthly reward was a bare subsistence. I believe that the real causes of dislike to the missionaries in South Africa are the avarice of trade, and jealousy of the influence they possess, and the check they are upon those who would like to exercise an arbitrary and unjust authority over the natives. I could say a great deal more on this subject; but the missionaries are a class of men, generally speaking, so irreproachable, that the scandals of the unprincipled cannot affect them with well-thinking men; nor do their characters require any further defence by me."

As an estimate which is one of unmitigated praise generally defeats its object and provokes a reaction of dislike, I have sought diligently to record all the aspects and details of the character and acts of David Livingstone which could be gathered from the remembrance of contemporaries (European, Arab, and Negro), or could be found in books and letters, and in my conclusions, based on this evidence, I have been careful to mention any points in disparagement of Livingstone for which there was any foundation. But, as a matter of fact, this research leaves me unable to quote anything of importance which could be regarded as serious dispraise of this remarkable man, whose life was saintly, whose disposition was naturally noble, yet full of attractive charm. On the other hand, a frequently repeated reading of his works leave me increasingly astonished at his achievements with the means that he possessed, and more than ever convinced that he was so far the greatest of African explorers, judged not only by his actual achievements, but by his character, disposition, and mental capacity.[1] He wrote things, he expressed ideas in the forties, fifties, and sixties of the last century which seem to those who read them to-day singularly modern as conceptions, conclusions, and lines of profitable study. For instance, apart from his boyish passion for geology and the records of the rocks, and his feeling that

---

[1] Besides getting himself taught on board ship, and later by Sir Thomas Maclear, to take with great accuracy astronomical observations for fixing latitude and longitude, besides acquainting himself with botany and geology, with patristic literature and Egyptology, Livingstone was an excellent mechanic, a steersman and a mariner. His resourcefulness was at all times remarkable. When he was hard up for fuel on his first steamer journey up the river Shiré he landed in the Elephant marsh. Here no trees existed and no fuel was obtainable, but his men found many bones of slaughtered elephants. Livingstone at once took the bones on board, burnt them in the furnace of the *Ma-robert*, and so continued his journey.

here lay before us a new and much vaster Bible, he had only just attained manhood when by dint of reading he begins to express his conviction that Christian missionaries were going to produce not only the awakening but the renaissance of China, an eventuality which has now come to pass. Scarcely landed in South Africa, he conceives the idea, barely formulated then, of the far-spreading affinities of the Bantu peoples, and the possibility through this community of language of carrying British missionary work and British political influence up through the centre of Africa to Abyssinia. He also, fifteen years afterwards, grasped the important fact before any other explorer of Africa, that the part of the continent white men should make for in their settlements was the high plateau region of the interior rather than the banks of the great rivers or the seaboard.

Indeed, it requires very little accentuation of his opinions expressed in private letters in 1841, to formulate the phrase, since so potent, of "The Cape to Cairo." He never lost sight of this ideal, and during his last years speculated on its ultimate achievement through the work of Sir Samuel Baker on the Mountain Nile and the Albert Nyanza. It was only when Stanley chilled these anticipations by informing him that Great Britain had lost her influence in African problems, and that it was perhaps the United States which was going to re-organise Egypt through the loan of American officers, that Livingstone's ideals now transcended the limitations of national politics. In his journal on May 1, 1872, just one year before his death, he wrote the celebrated words which have been recorded on his tombstone, "All I can add in my loneliness is, may Heaven's rich blessing come down on every one, American, Englishman, or Turk, who would help to heal the open sore of the world."

Yet he was under few illusions about the negro and his inherent weakness as a self-governing race: "The evils inflicted by the Arabs are enormous, but probably not greater than the people (the negroes) inflict on each other," is one of his mature conclusions. Of all the tribes with whom he came into contact, he was most indulgent towards the Makololo—the Basuto governing caste in Upper Zambezi. This, no doubt, was at first due to his personal admiration for Sebituane, the great Makololo conqueror, who may be said to have thrown open Zambezia to Livingstone's penetration. Secondly, Livingstone was deeply grateful to his Makololo followers for so faithfully escorting him, first to Angola and later to Tete. Yet it is evident that the Makololo had made themselves detested by the natives of the Zambezi Basin. "If the people have cattle," it was said, "the Makololo kill them for their cattle's sake; if they have not, they kill them for their children (to be sold as slaves); if they have neither, they kill merely to kill." And according to James Chapman, who visited this region whilst Livingstone was making his great journey from the Upper Zambezi to Angola, the Makololo, after Sebituane's death, were not so much impressed by Livingstone as a missionary teacher as by his skill in medicine and surgery.

"Livingstone," wrote Chapman, "being a doctor, has also the reputation of being a wizard. This makes him either feared or admired, and gives him a certain influence. They give him credit for being a good doctor, and say he has cured many, but killed some natives. They do not believe in natural deaths; when a man dies, he has been killed. By all accounts the doctor's preaching is barely tolerated by the chief, who is at heart highly displeased at his doctrines concerning rain and polygamy. The people say that Dr. Livingstone has promised them all the good things of this earth—rain, corn, cattle, etc.—if they would believe in God and refrain from polygamy, slavery, and other malpractices; that they have waited a long time for these good things; and that they would wait another year to see if the Good Man he talked about helped them nicely (*tusa sintle*). While they were relating these things, and conversation grew slack, the councillor Punoani was observed sitting with a piece of newspaper upside down, mimicking the doctor singing a hymn, and, observing that he had attracted our attention, he rolled over on his back, threw his feet into the air, and exclaimed, bursting out into a loud laugh of ridicule, 'Minari' (a corruption of the Dutch mynheer, generally applied to missionaries). Such is the sort of impression as yet made on these barbarians. It is to be hoped that in time better success will attend missionary efforts."

Livingstone's last seven years' explorations of South-Central Africa brought home to him the devastation of Africa by the negro peoples and their internecine wars. He records in 1866 his impression that a good deal of the Yao country between the Ruvuma and Lake Nyasa once supported a prodigious iron-smelting and grain-growing population. The land was marked with the ridges in which they formerly planted their crops and from which they drained off the too abundant moisture of the rains, while the ground was strewn with pieces of broken pots. Internecine wars had led to famine and depopulation, yet the surviving Yao tribes had invaded the lands of the A-nyanga and Alolo, and (wrote Livingstone) Yao raids in the middle nineteenth century for the supply of Arab caravans almost depopulated the fertile tracts to the south-east of Lake Nyasa. The evidence of Livingstone and other travellers of the fifties, sixties, and seventies, brings home to us the widespread devastation caused by bands of Angoni-Zulus. These Zulu raids over East-Central Africa during the nineteenth century were one of the greatest disasters of its history. They had their origins in the convulsions caused in Natal and Zululand by the conquests of Chaka the Destroyer, and their efforts long remained written on the surface of Nyasaland, North-east Rhodesia and German East Africa.

"It was wearisome to see the skulls and bones scattered about everywhere; one would fain not notice, but they are so striking that they cannot be avoided," is an extract from Livingstone's journal as he comes in contact with the Angoni raids in South-west Nyasaland. As he begins to leave the basin of the Upper Luangwa for the unknown Bemba regions beyond, he notices the uninhabited condition of the country due to the slave raids of the Awemba; a fresh factor in African history. The Awemba or Aba-emba did not come from the south like the Zulus, but from Congoland, and their irruption into South-Central Africa was one of the results of great tribal disturbances there due to the conquests of the Sudanese Bushongo. Livingstone writes in December, 1866: "I shall make this beautiful land (North-east Rhodesia) better known, which is an essential part of the process by which it will become the pleasant haunts of men. It is impossible to describe its rich luxuriance, but most of it is running to waste through the slave-trading and eternal wars."

Yet while condoling with the Mañanja survivors from the Yao slave-raids in Nyasaland, he was asked by the men amongst them for guns and powder, not to defend themselves only, but so that they might imitate the Yao and go slave-raiding; and he noted in 1866 that the much-harried A-chewa people, instead of loathing the Mazitu, or "wild-beast" Angoni Zulus for their raids, admired them and strove to dress-up their young men like them.

Livingstone's description of the horrors of the slave-raids and the slave-trade in Eastern Congoland are too well known to need quotation in order to convince you that the Central Africa he knew merited his pity and his appeal for intervention. Of course not all the Arabs were ruthless murderers and slave-raiders. Tipu-tipu (whose acquaintance he made in 1867), and some others were only ivory-traders, though they made use of slave-porters. Livingstone's attitude towards slavery and the slave-trade we now know to have been a perfectly reasonable one, based quite as much on a far-sighted appreciation of the economic importance of free labour in Africa as on an equally far-sighted instinct of philanthropy. He criticised severely, but not unjustly, the Dutch-speaking colonists of South Africa for their treatment of the Bechuana and Bushmen; the Portuguese for their acquiescence or participation in the Central African slave-trade; his own Government for lack of zeal in regard to the repression of the slave-trade at Zanzibar (I am referring to the sixties of the last century); and the Arabs—whose nobility of disposition he was well able to appreciate where it was manifest—for the misuse of the power they had acquired in the region of the great lakes.

In the middle of the nineteenth century we find him writing quietly to deprecate the butchery of big game which was beginning to rage in Central South Africa. Some of these gentlemen-hunters he denounced as "itinerant butchers," and he points out that so far from their reckless slaughter of wild beasts earning for them a high position in the regard of the natives, they were often regarded with a certain amount of contempt.[1] Yet if it were necessary for the protection of human life, or other reasonable human needs, such as hunger, or for the enlargement of scientific knowledge, he could handle a rifle and a shot-gun nearly as well as any one of his sportsmen friends; and that he was a thorough sportsman in

[1] Many sportsmen-travellers in Africa (wrote Livingstone) have recorded with satisfaction in their books what a high position they occupied in the regard of the natives for their prowess in shooting big game, yet Livingstone, who so often heard and understood exactly what the Bechuana people of these South-Central African hunting-grounds were saying, adds that the great hunters were regarded with a certain amount of contempt by the negroes whom they so lavishly supplied with meat. "Why do these men, who are rich and could slaughter oxen every day of their lives at home, come to our country and endure so much thirst for the sake of this dry meat, none of which is equal to beef? You say it is for play! But your friends are fools."

the best definitions of the word—what the Americans would call "a real white man"—is evident from the unqualified regard with which all the great sportsmen-naturalists and pioneers of South Africa expressed for him. Despite all rivalries in exploration, all attempts on the part of the malicious and the envious to sow discord between them, William Cotton Oswell, throughout his own unblemished career wrote and spoke of Livingstone in the highest terms, without any qualification whatever. The same was the case with Frank Vardon, Gordon Cumming (even though he may have winced at the reproof of his butchery), William Webb, Thomas Steele, and James Chapman.

But if he blamed and criticised, he was far readier to praise and thank. Those who have been so willing to underline his criticism of the Portuguese have omitted to place alongside it much that he recorded in their favour and many cordial expressions of thanks which he tendered them for their frequent assistance. "May God remember them in their day of need," he wrote in deepest gratitude for Portuguese hospitality and kindness at the crisis of his journey on the frontiers of Angola; and in another place, "The universal hospitality of the Portuguese was most gratifying as it was most unexpected. And even now as I copy my journal I remember it all with a glow of gratitude."

In very truth, but for the action of the Portuguese towards him when he reached the Kwango river from the Upper Zambezi, he might have perished and have been scarcely heard of in the history of Africa. He had only a small escort of scared Makololo, already frightened at their own boldness in travelling so far away from home; he was almost entirely without trade goods or provisions, ill and weak with semi-starvation and dysentery. The truculent natives on the east bank of the Kwango refused him a passage across the river, and avowed their intentions of seizing and enslaving his Makololo, while at him they had begun to fire their muskets. He would almost certainly have died from one cause or another at this juncture, but for the intervention of a Portuguese sergeant of militia on the west bank of the Kwango, who prevailed on the natives to ferry him across; and, once he had got him as a guest, treated him with every kindness and hospitality and sent him forward, safe and well provisioned, to finish his journey to the Atlantic coast. The Acting-Governor of Angola in those days was the Bishop of Angola, whose sentiments on the subject of religion, as recorded by Livingstone, are broad-minded enough not to seem out of date for an advanced Review of 1913. This Bishop-Governor gave him an excellent riding-horse, which was of material use to him on his return expedition to the Zambezi, besides helping him in every possible way not only to return to the Zambezi, but to cross Africa to the Portuguese dominions on the other side. The Portuguese residents in London subscribed to a fund for fitting out Livingstone anew for crossing the continent to the mouth of the Zambezi.

Livingstone, or more likely his brother Charles, wrote harshly of the Portuguese on the Lower Zambezi, and the want of good-will that they showed to his expedition of 1858-64. Yet in the account given of that six years of martyrdom one is struck over and over again with the forbearance, the unwearied kindness and hospitality, and often the acts of material help afforded by the Portuguese officials. The fact is that both parties were in a false position. The despatch by the British Government of the second Zambezi expedition was an act characteristic at times of our foreign and colonial policy—an attempt to shirk responsibility, to get some one else to pull the chestnuts out of the fire, and to avoid recourse to diplomacy.[1] Judged by modern lights, the Portuguese claim to lock up the Zambezi and the access to Lake Nyasa was unjustifiable, just as we cannot any longer defend in theory the attempt made by the British some years ago to constitute the commerce of Nigeria as the monopoly of a British company, or many other temporary or long-standing practices in our colonial policy. But the claim of the Portuguese was an historic and an uncontested one of 350 years' standing, when Livingstone started as a British Consul on a very indefinite mission to open up Zambezi to British commerce, settlement, and missionary work. The British Government should not have sent him on this enterprise without first of all coming to terms with Portugal; and in all probability the Portuguese might have been just as susceptible to reason as they proved to be when the matter was seriously tackled eighteen years after Livingstone's death. As it was, Livingstone and the members of his staff were allowed to do pretty much as they pleased in the Zambezi delta, on the Zambezi itself, and on the Shiré, but it was made quite clear to them that if their efforts resulted in success all British commerce in those regions must pass through a Portuguese customs house.

---

[1] Livingstone was also to blame for having leapt too hastily to a wrong conclusion about the navigability of the Zambezi in 1856, and thus having led the British Government completely astray. He followed the course of the Zambezi from the Victoria Falls to Tete mainly by land, and, of course, whole sections of the river escaped his observation. He thus emerged on the coast of the Indian Ocean at Quilimane with the conviction that, presuming the Zambezi river could be entered by shipping through one of the mouths of its delta—a possibility not concealed from him by the Portuguese—British ships or launches could steam up it to the Makololo country, in the heart of South-Central Africa. He exaggerated in his optimism the amount of sugar-cane and cotton grown in Central Zambezi at that time; and so, in short, the whole of the second Zambezi expedition was intended not to poach in Portuguese preserves, but to reach the heart of South Africa by the Zambezi channel, and there to carry on a profitable trade with a very enterprising and powerful negro people.

In the year 1860, however, Livingstone and Kirk made the journey from Tete to Sesheke and back mainly by water, and realised the impossibility of continuous navigation of the Zambezi channel between the Quebra Baço rapids and the Victoria Falls. There are several impassable barriers in between, such as the Morumbua or Quebra Baço, the Kakalole, and still more the Kariba gorge.

Here came the first great disappointment. Livingstone next chose to regard Nyasaland as his goal. In course of time proof of the difficulties of the Shiré navigation disheartened the British Government from giving him further support. Lastly, the Ruvuma route proved impracticable, and when this was made clear, the expedition was recalled. That the "six years of martyrdom" were not wasted we now thankfully realise in the existence of the prosperous colony of Nyasaland, and the steady progress of Northern Rhodesia. And it is a happy thing for an important alliance to be able to relate that Portuguese rule and commerce on the eastern Zambezi are better established now than they were in Livingstone's day. In fact, his ideals and his aspirations—even in regard to the Portuguese—have been fulfilled to the letter.

However that may be, the breakdown of this second Zambezi expedition was not really due to unwillingness on the part of the Portuguese to negotiate, but to the gigantic and unforeseen difficulties of the task on which Livingstone had embarked. The Chinde mouth of the Zambezi delta either did not exist in those days as a navigable passage, or was unknown or unrevealed. The Kongone mouth, practically discovered by Livingstone's expedition, was one with a dangerous and shallow bar; and although occasionally, by marvellous feats of seamanship, small steamers were got across it, the Kongone bar was one of the first causes of disappointment, loss, failure, and threatened disaster. In one instance, the disaster would have been almost complete but for the efforts and kindness of the Portuguese. Then the Zambezi and the Shiré proved full of difficulties for navigation. Above Tete were the almost impassable Quebra Baço rapids, while the navigation of the Central Zambezi between Quebra Baço and Victoria Falls was likewise beset with difficulties from the banks of the river on both sides being ranged over by truculent tribes and greedy chieftains.[1]

The Makololo power on the Upper Zambezi was rapidly decaying, yet owing to the great impression made on the chief and people by Livingstone's lieutenant, Sir John Kirk, the Makololo would have thrown their remaining energies into the support of a British settlement and administration of their country, and offered considerable grants of healthy, unoccupied land for that purpose: in fact, they wished to forestall the work which long since has been done by the British South Africa Company. The tract offered in 1860 by Sekeletu, the son of Sebituane, was in the country of the Batonga, north of the Zambezi. Here Livingstone had thought formerly that there were indications of a rather ancient negro civilisation, evidenced by the big, long-horned cattle of the Central Zambezi valley, and the dwarf cattle of the Batonga highlands, which were scarcely more than three feet in height. The Batonga also possessed a breed of small black and white goats and of small and very prolific fowls, sometimes known as "Makololo fowls." The hens of these pretty little bantams were said to lay an egg every day.

But Sekeletu's proposals remained a dead letter owing to the inaccessibility of his country. Advance up the Shiré to Lake Nyasa was temporarily stopped by the impassable Murchison Falls. The land route across the Shiré Highlands, across what is now one of the most peaceful, prosperous, and best governed parts of Africa, was arrested by the slave raids of the Muhammadan Yaos. But Lake Nyasa, in spite of all difficulties, was discovered by Livingstone, accompanied by Kirk, Charles Livingstone, and Edward Rae, on September 16, 1859.[2]

---

[1] Even this overland journey up the Zambezi valley to the Makololo country could hardly have been undertaken in 1860 but for the active help of the Portuguese. Most of the Makololo men whom Livingstone had left behind from his first expedition at Tete had been demoralised by long residence there and refused to carry loads or even to march. The Portuguese lent the expedition porters and donkeys.

[2] It is only fair to say that Livingstone himself supplies us with information pointing to the obvious fact that he was not the first white man to stand on the shores of Lake Nyasa. A Portuguese magistrate or judge in native affairs at Tete, Senhor Candido de Costa

Twice did Livingstone, after this date, return to Lake Nyasa; in 1861 and in 1863. With a sailing-boat he and Kirk explored its shores (chiefly on the west) as far north as the Tumbuka or Northern Atonga country, and he himself marched through Western Angoniland and the Arab settlements of Kotakota. In 1863 he was obliged to reach the lake by marching overland, boats having been lost or destroyed and steamers proved to draw too much water for the depleted Shiré river. Once more he sought the hospitality of Kotakota, and from this Arab town he marched with desperate eagerness to the north-westward, hoping to accomplish something remarkable in geographical discovery which might enlarge our knowledge of Central Africa. But the knowledge that his expedition had been publicly cancelled, obliged him to stop within ten days' march of Lake Bangweulu and return painfully to England.

His last seven years of African exploration were somewhat meagrely assisted by a grant of £500 from the Royal Geographical Society and a grant of £500 from the Foreign Office. He was allowed to retain the title of Consul, but accorded no pay, and further warned that he must expect no pension. He might have been unable to continue his work at all—the work for which he was so clearly cut out—had it not been for the private generosity of Mr. James Young, the great chemist of Glasgow, William Webb of Newstead Abbey, and, perhaps one should add, Mr. John Murray the publisher. Livingstone—few of us seem to realise—had himself borne a considerable share of the expense of this six years' government expedition to Zambezia. He had spent £6000 of his own money on the *Lady Nyasa*, a steamer specially designed for him. Gathering together, therefore, such funds as he could set aside from the provision he had made for the education and maintenance of his four children, and combining with this the money subscribed by private friends and public bodies, he embarked in 1866 on his self-imposed mission of inquiry into the hydrography of Central Africa.

When resting at Linyanti on the Chobe river in 1855, he had heard from an intelligent Arab of Zanzibar, who had just reached that place overland from the east coast, of Lake Tanganyika and the great river Lualaba of Central Africa. He was at that time puzzled by the contradictory Portuguese and Arab stories of the Chambezi, which on account of the similarity of name was declared by armchair geographers to be the real Zambezi and the head stream of the Kafue. In 1864 he asked himself to what system belonged this mysterious Lake Sbuia or Bemba (which he afterwards called Bangweulu)? To that of the Zambezi? If not, to what other? He realised in 1855 that the Kasai, which he had discovered in South Congoland, must be a very important affluent of

Cardoso, had made a journey about 1846 to the south-west gulf of Lake Nyasa, a very shallow inlet of water, and crossed this in canoes, probably to the Livingstone peninsula. Mr. Ralph Durand, in a series of articles on Livingstone, recently published in the African *Mail* (essays remarkable for their acumen and research) casts doubts on this achievement of Costa Cardoso; but the geographical details given to Livingstone in 1856 could scarcely have been quoted from one who had not made that journey. But it was clearly the shallow south-west bay of Nyasa which was visited.

the Congo, if not a main stream of that system. But he evidently did not conceive it possible till near the time of his death, that the Congo basin could extend so far as it does to the south-east and be fed by the Chambezi and the Lualaba. In his mind there could only be one alternative to the Zambezi as the recipient of these reported lakes and rivers of the Bemba country; and that was the Nile.

The animosity between Burton and Speke, the pardonable vanity as an explorer of Sir Samuel Baker, and other factors, had induced in the sixties of the last century very incorrect ideas about the ultimate sources of the Nile. The Victoria Nyanza had been split up into a number of separate lakelets or swamps, and the size and length of the Albert Nyanza exaggerated to an extraordinary degree. With regard to this last factor, we now know that Sir Samuel Baker's speculations were not so entirely unjustifiable. His gaze had penetrated south sufficiently far to have realised the general outlines of the Semliki valley. He was deceived by the blue slopes of Ruwenzori on the east and the lofty plateaus of Mboga and Bukonjo on the west of the Semliki, into imagining that the Albert lake[1] had an indefinite extension towards the south between vast mountain walls. In fact, if he could have seen a little farther or have heard stories of Lakes Edward and Kivu, he might have been still more positive on this score.[2] Then again, neither Burton nor Speke had properly examined the north end of Lake Tanganyika to ascertain whether water flowed out of it or into it. The rumoured Rusizi river might be an effluent, and Tanganyika be the farthest southern source of the Nile waters. Or Tanganyika might feed the Great River which since 1855 Livingstone knew to be dimly rumoured to emerge from a cluster of lakes and flow northward through the heart of Central Africa. Or it might be a locked basin independent of the Nile system; in which case the "Luapura" would prove to be the upper Albertine Nile.[3]

It was to the solution of this problem that Livingstone, without a thought for anything else, and yet regarding it as a mission divinely inspired, devoted all his remaining energies; but in the middle of this

---

[1] Lake Chowambe, as Livingstone calls it, following a name current amongst Zanzibar Arabs.

[2] In August 1870, Livingstone offered one of the Arabs about £270 in rupees and goods to leave the ivory trade "which is at present like gold digging," and convey him down the Lualaba "to see where it went, and back again to its western branching."

On February 25, 1871, he wrote, "I had to suspend my judgment so as to be prepared to find the Lualaba after all perhaps the Congo."

Livingstone heard, in 1871, of the Lomami and of other rivers of the Congo system farther west, and believed he might even get thus into touch with the French settlements and the Gaboon, and so prepared despatches to send home that way. He evidently felt by then that though the Lualaba might not be the Upper Congo, it could not be far from the Congo system.

[3] The geographical names, Luapula, Lualaba, Lufira, and Tanganyika were probably first inscribed on the map of Africa by Livingstone in 1856. "Zanganyika" is mentioned by Krapf as the name of a great trading-place in Central Africa, a short time before Burton and Speke discovered the lake in 1857. Livingstone derived his first knowledge of Tanganyika from the Arab, Ben Habib, whom he met at Linyanti in 1855; and he heard of the Luapula also from this Arab as well as from Portuguese reports. He derived much information about

task he was forced to realise the appalling devastation of Central Africa which was now resulting from the Arab slave raids. From about 1869 he had two objects ever before him: one was to solve the Nile problem, and the other to rouse the conscience of the world in regard to the Central African slave-trade.

Let us briefly consider his achievements as a geographical discoverer. He directly inspired the search for Lake Ngami, and was the main agent in carrying South African exploration beyond the arid plateaus of Bechuanaland and the Kalahari desert into what is really the Zambezi basin. Oswell and Murray contributed to the cost of his journeys, but he by his influence found the guides and secured the friendship or the neutrality of the native chiefs. He acted as interpreter-in-chief, and, thanks to the mastery he had acquired over the Sechuana language, was able to converse fully and freely with the natives of South-Central Africa. He also picked up a considerable knowledge of other dialects. He served diligently and skilfully as physician and surgeon all who were connected with these journeys. But his own predilections were for botany, zoology, and the study of man. It was the impression that native reports of his character had made on Sebituane, the Makololo conqueror of the Upper Zambezi, and the resultant protection afforded which made it so easy for Livingstone and Oswell to reach the Chobe river and the Upper Zambezi in 1851.

Between 1852 and 1856, Livingstone traced the main course of the Zambezi from its confluence with the Chobe northwards to near the sources of the Liba, and from this point westwards he was the first scientific geographer to lay down correctly the position of the upper Kasai and Kwango affluents of the Congo.

Livingstone may be quoted as the discoverer of the great Kasai (perhaps the principal among the Congo affluents for volume and for the extent of drainage area). At first it would seem probable that the Pombeiros, at the beginning of the nineteenth century, must have crossed the Kasai in order to reach the court of the Mwata Yanvo. But they appear to have deflected their route southwards, after leaving the upper Kwango, so that they passed round the sources of the Kasai, leaving them to the north. Ladislaus Magyar, the Hungarian explorer and trader (who married a negress of Bihé and travelled over Angola between 1849 and 1864), penetrated about 1851 to the upper Kwango

---

the geography of southern Angola and South-Central Africa from *viva voce* information and the quoted records of the Portuguese; thus from them in 1855-6 he drew pretty accurately the course of the Kunene river and recorded the existence of the great Kubango three years before Andersson discovered it.

The names Lualaba ("Guarava") and Luapula ("Guapula") are first mentioned in history by Dr. Lacerda in 1798. He derived his information from the Pereiras, the Goanese slave- and ivory-traders whose explorations north of Tete caused them to be chosen as guides for the great Portuguese expedition of 1798-9. "Lualaba" and "Luapura" appear in the records of travel of the negro Pombeiros, who crossed Africa from Angola to Tete in 1806-11. The Pombeiros do not appear actually to have crossed the Lualaba, but to have skirted it near its source and passed to the south of it. Under the name of "Lualap" the upper Congo near Nyangwe is mentioned in the story of a Kanyoka slave woman told to the missionary-philologist Koelle at Sierra Leone about 1849.

and the north-west limits of the Zambezi basin, and may have seen the infant Kasai in 1855, a few months before or after Livingstone passed by. But he did not communicate the information to the world until after Livingstone's journey, and never, I think, specifically mentioned the Kasai, at any rate, before the publication of Livingstone's book. Moreover, he was no trained geographer or taker of observations for fixing points of latitude and longitude. Silva Porto, a Portuguese trader of Bihé, reached the upper Zambezi and South Congoland in the fifties and sixties, but his wanderings resulted in no addition to the map of Africa.

It is, indeed, remarkable what Livingstone's predecessors missed rather than what they found. Dr. Lacerda reached to little Lake Mofwe, an isolated lagoon about twenty miles south of Mweru and a short distance east of the Luapula. Yet apparently neither he nor any member of his expedition, before or after his death, had the curiosity to penetrate northwards one day's journey and discover Lake Mweru, or visit the banks of the Luapula. Going through the Bisa country they heard of a lake—"Lake Chuia," or Shuia, a short distance to the westward, and knew that the Chambezi flowed into it. This was Livingstone's Bangweulu [1] (named, as he tells us, from one of its islands). But the Portuguese of Lacerda's mission, like those of the Monteire-Gamitto expedition of 1831-2, made no effort to locate Bangweulu and place it definitely on the map. Lake Nyasa was heard of (as "Nyanja") by the Portuguese of the eighteenth century and early nineteenth; but it was not till 1846 that its waters—so far as historical records go—were actually seen by a Portuguese (Candido de Costa Cardoso). Gasparo de Bocarro passed near to Lake Nyasa in 1616 on his way to Kilwa and Mombasa, but seems to have crossed Lake Malombe or the upper Shiré only, and not actually to have seen Lake Nyasa.

Returning from Angola to the Chobe river, he discovered the Victoria Falls, and followed the Zambezi more or less closely down to its delta emerging on the sea-coast at Quilimane.

On his second Zambezi expedition he revealed to the world Lake Nyasa, Lake Chilwa (mis-written Shirwa), the high mountains of the Shiré region, and the course of the Shiré river, the Luangwa river to the west of Lake Nyasa, most of the northern confluents of the Zambezi in their lower courses, and the Butonga highlands. This second expedition was also the means of effecting a great increase in our knowledge of the Zambezi delta.

On his third great African journey he renewed previous explorations in the direction of the Ruvuma, and traced a good deal of the course of that East African river. He was practically the first European to explore West Nyasaland and the northern Bemba or Awemba country; he discovered the south end of Tanganyika, and made a shrewd guess

---

[1] Probably we do not yet know the correct native name for the open water of Bangweulu. It seems sometimes to be known as the Nyanja ya Lubemba or Luemba; whence "Bemba." Bangweulu, which Livingstone spells Bangweolo, in his propensity to turn all u's into o's after the fashion of the Bechuana peoples, is, or was, the name of one of its islands. The root *-emba* in "Bemba," "Awemba," "Liemba," "Luemba" seems to mean "lake."

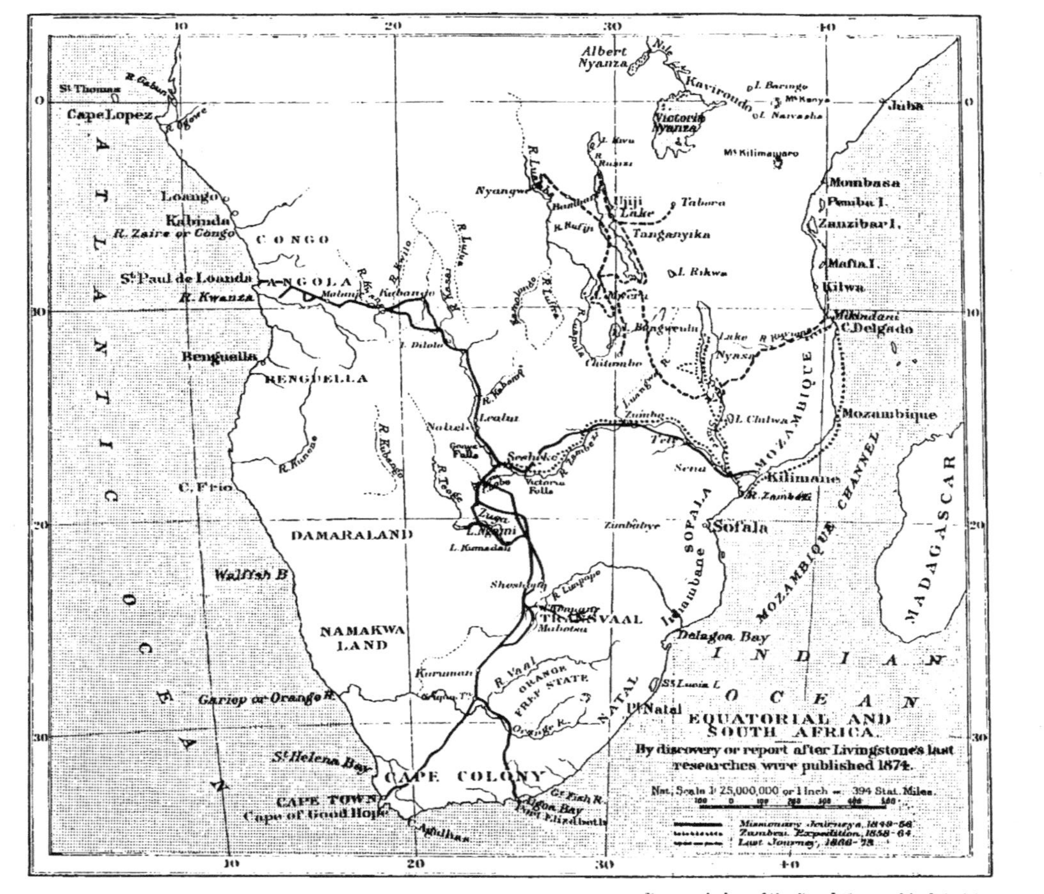

By permission of the Royal Geographical Society.

at its outlet through the Rukuga (which river he styled the Longumba).[1] He first revealed the great Mweru swamp or Chisera. "Elephants, buffaloes, and zebras grazed in large numbers on the long sloping banks of a river or marsh called Chisera." (This considerable extent of alternate swamps or shallow water was afterwards re-discovered by Sir Alfred Sharpe.) Livingstone made known to us Lakes Mweru and Bangweulu and the connecting Luapula river, and the course of the great Lualaba or upper Congo at Nyangwe. He also recorded the existence of the upper Lualaba or Kamolondo.[2] He was the first European to penetrate as far north as S. lat. 3° 30′ near the Elila river, and to describe the Manyuema[3] forests with the large chimpanzis and pygmy elephants found in them. He mentions for the first time the Lomami river, and is the first explorer to hear of the country of Katanga, its mineral wealth and its—as yet—unexplored, inhabited caverns of vast size. "A month to the westward of Kazembe's country lies Katanga, where the people smelt copper ore (malachite) into large ingots shaped like the capital letter I, weighing from fifty to a hundred pounds. The natives draw the copper into wire for armlets and leglets. Gold is also found at Katanga." Livingstone was the first writer to mention the possible existence of Lake Kivu; of Kavirondo gulf (Victoria Nyanza); and of Lake Naivasha: from Arab information, of course.

He was the first to record the existence of drilled stones in the country to the south-west of Tanganyika, which seemed to be evidence of the existence of a people of ancient bushmen culture in that direction,[4] and his remarks generally on the Stone Age in Africa, on the possible existence of undiscovered ancient types of mammals and of mammalian fossils, all show an enlightenment in speculative scientific

---

1 "It may be that the Loñgumba is the outlet of Tanganyika."—*Last Journals*, November 29, 1871. He states that in its lower course the Loñgumba is known as the Luamo. This is correct. "Lukuga" or "Rukuga," the name recorded by Joseph Thomson, is only a term applied to the sluggish, swampy leakage from Tanganyika which connected that lake with the Luamo affluent of the Congo.

2 He wished to name the Luvua-Luapula (Eastern Lualaba) after his great friend William Webb, and the Western or Kamolondo-Lualaba after James Young, the Lufira (an important affluent of the Kamolondo-Lualaba) after Sir Bartle Frere; and to give the name of Abraham Lincoln to a supposed lake which he believed to lie in the course of the western Lualaba. It was while being detained for a long time in the Ulungu country, in August 1867, that Livingstone first heard of the main Lualaba or Upper Congo river, "about fifteen days west of Tanganyika, said to be 10 miles broad and known as the Logarawa, flowing northwards." "Kamolondo" is an unrecognisable native name only retained for convenience of distinction; the better term would be Lualaba, as contrasted with Lufira and Luapula; and "Upper Congo" should be applied to the mighty river which is formed by the junction of all three.

3 The Manyuema is the southernmost portion of that vast equatorial forest belt which lies to the *north* and *east* of the main Congo, and extends with scarcely any interruption from South Kamerun to Ruwenzori and round to north-west Tanganyika and the Manyuema country. This region has a remarkable mammalian fauna, including the okapi, the gorilla, chimpanzi, great forest pig, bongo tragelaph, etc.

4 On August 1, 1867, Livingstone describes a perforated stone which had been picked up and placed on one of the poles of an Arab stockade. It was oblong and showed evidence of the boring process in rings, the diameter of the hole in the middle being an inch and a half. The stone was of hard porphyry "and resembled somewhat the weight of a digging-stick which I saw in 1841 in the hands of a Bushman."

imagination greatly in advance of his times. He was also in all probability the first writer since the Portuguese chroniclers of the sixteenth century to allude to the remarkable ruins of stone-built forts, villages, and cities in South-east Africa. He derived his information from natives, and perhaps also from Boer hunters. He also mentions the coins found in excavating the shore of Zanzibar island, with Kufic inscriptions, and perhaps dating back to the ninth or tenth century A.D. (Sir John Kirk confirms this statement, and adds that some of these coins were of Harun-ar-rashid's reign, and bore the name of his viziers, Yahya or Fadl).

His biblical studies drew him into Egyptology, and one of his incentives to the exploration of the Nile sources was the conviction that Moses when living in Egypt had taken a great interest in Nile exploration. Livingstone half hoped that in discovering the ultimate sources of the Nile he might come across archæological traces of Egyptian influence. He was not pursuing in this direction an absolute chimera.

The physical appearance of so many of the Bantu tribes between Lunda on the south-west and Manyuema, Bambare, and Buguha on the north-east, constantly suggested to Livingstone's mind the idea of an immigration of Egyptians into Central Africa. Had he lived to penetrate to the countries north of Tanganyika to see the Hima or Tusi aristocracy on the highlands of Equatorial Africa, he would have been still more convinced of the ancient inflow of Egyptian influence into these regions, though it is a theory which it is very unsafe to pursue on the scanty evidence we possess at the present time.

When travelling from Tanganyika to Mweru in 1869, he remarks on the appearance of the chief and people of Itawa.[1] "Nsama, the chief, was an old man with a head and face like those sculptured on the Assyrian monuments. . . . His people were particularly handsome, many of the Itawa men with as beautiful heads as one could find in an assembly of Europeans. Their bodies were well shaped, with small hands and feet —none of the West Coast ugliness—no prognathous jaws or lark heels."

There is another entry in his journal derived from Arab information which bears on this theory of the Hamitic permeation of Negro Africa.

"The royal house of Merere of the Basango" (north-east Nyasaland) "is said to have been founded by a light-coloured" (Hamitic?) "adventurer, who arrived in the country with six companions of the same race. Their descendants for a long time had straight noses, pale skins, and long hair."

His journeys into southern Congoland threw a very interesting light on a native kingdom made famous by the earlier Portuguese exploration—that of the Kazembe of Lunda, whose capital was between Lakes Mweru and Bangweulu.

In the early seventeenth century a great negro empire had arisen in

---

[1] The Itawa country, he remarks, between Tanganyika and Mweru had many traces of plutonic activity. Earthquakes were by no means rare, and there were hot springs, some of which were used to boil the natives' food.

southern Congoland, partly due, no doubt, to the arms and trade goods derived from the Portuguese, but partly also to the after-effects of the Sudanese civilisation of Central Congoland under the Bushongo dynasty. This Empire of Lunda ruled over all the south of Congoland and a small part of northern Zambezia.

In the early eighteenth century a member of the family of the Lunda emperor, or "Mwata Yanvo," moved to the south of Lake Mweru and founded a feudatory kingdom there. He received the title of Kazembe, or "lieutenant."

Kazembe's capital was by the side of a little lake called Mofwe. Livingstone approached it along a path as broad as a carriage road one mile long, the chief's residence being enclosed by a wall of reeds 8 or 9 feet high and 300 yards square. The innermost gateway was decorated by about sixty human skulls, and had a cannon, dressed in gaudy colours, placed under a shed before it. This, no doubt, was a gift from the Portuguese. Kazembe himself had a heavy, uninteresting countenance, without beard or whiskers, somewhat of the Chinese type, his eyes with an outward squint. He smiled but once during the day, yet that was pleasant enough, though the cropped ears of his courtiers and the human skulls at the gate made Livingstone indisposed to look on him with favour. Kazembe was usually attended by his executioner, who wore a broad Lunda sword under his arm, and a scissor-like instrument at his neck for cropping ears. This was the punishment inflicted on all who incurred the Kazembe's displeasure.

Kazembe sat before his hut on a square seat placed on lion and leopard skins. "He was clothed in white Manchester print and a red baize petticoat so as to look like a crinoline put on wrong side foremost. His arms, legs, and head were covered with ornaments, and a cap made of various coloured beads in neat patterns. A crown of yellow feathers surmounted his cap. His head men came forward, shaded by a huge ill-made umbrella and followed by dependants . . ." This Central African monarch (whose descendant was finally deposed for cruelties by the British Government) bore an evil reputation; yet he was a good friend to Livingstone and put no obstacle in his path; though he politely told him that lakes and rivers only consisted of water, and that to ascertain this fact by ocular inspection would not repay him for his fatigues and outlay in trade goods!

Livingstone from boyhood had taken a great interest in botany and in the appearance of trees and flowers in the landscape. His observant glance led him to note all the more salient features of the African flora from the Cape to the equatorial forests of Manyuema. His books are full of little word-pictures of the strange, stately or beautiful trees and plants he encounters. He records in his journal the spectacle of the Crinum "lilies" of the Luangwa valley, which in the first rains "flower so profusely that they almost mask the rich, dark, red colour of the loamy soil, and form a covering of pure white where the land has been cleared by the hoe." The weird stone- or pebble-like Mesembryanthemums of the Kalahari Desert, and the gouty, leafless geraniums and vividly coloured pumpkins and gourds of the same region arrest his attention; the Bauhinia bushes with their golden or bluish tinted, bifid leaves, and the scale insects on them exuding a sweet manna; the noble giraffe-acacia trees, the euphorbias of very diverse modes of growth, the *Strophanthus* creepers whose seeds possess medicinal or violently poisonous qualities, the borassus and hyphæne fan-palms, the wild date, and the "noble raphias," the pandanus and dracænas of the Zambezi delta or of inner Congoland, the innumerable forest trees of northern Zambezia and southern Congoland: all are illustrated in his pages by well-chosen words and sometimes explanatory drawings; and most are correctly named, in contrast with the very unscientific nomenclature of the generality of travellers in his day.

Livingstone notices as he descends the slopes of the mountains towards the Chambezi the abundance of the fig-tree which yields the bark-cloth, so that the natives cared little for the cotton cloths of Europe and India. He also in this region observed green mushrooms, which, on being peeled, revealed a pink fleshy inside (the *Visimba* of the natives). Only one or two of these mushrooms were put into a wooden mortar to flavour other and much larger kinds, the whole being pounded up into a savoury mess, which was then cooked and eaten. But in Livingstone's experience this mushroom diet "only produced dreams of the by-gone days, so that the saliva ran from the mouth in these dreams and wetted the pillow." The country on the Chambezi slope of these Muchinga mountains was devoid of game, the game having been killed out by far-reaching and long-continued drives through the hopo fences into pitfalls.

He reached the Chambezi first on January 28, 1867. It was flooded with clear water as it wended its way westwards. Crossing that river and passing northwards through almost trackless, dripping forests and across oozing bogs, through a region which furnishes the most important and the ultimate sources of the Congo (he notices the pretty grasses in the oozes, with pink seed-stalks and yellow seeds), he reached the village of the great Awemba chieftain, Chitapangwa. He gave Livingstone a tusk and a cow. At this village he found a small party of black Arab slave-traders from the Zanzibar coast, and to one of these men, who was returning to Zanzibar, Livingstone entrusted a packet of despatches, which actually did reach Zanzibar and England.

Chitapangwa gave him a great deal of trouble haggling for a large present and wishing to have one or more of Livingstone's invaluable boxes, threatening if a suitable present was not given to bar the way to further exploration. With tact and good-humour, however, on the doctor's part, friendly relations were maintained. The chief's heart was especially won by a piece of red serge, which was a vestige of a handsome present of serge of this taking colour given to Livingstone by the Baroness Burdett-Coutts (as she afterwards became) before he started on his second Zambezi expedition in 1858. At intervals through his nine years' subsequent experiences, this serge had stood him in good stead.

Arrived at the coast of Tanganyika, early in 1869, after his discovery of Mweru, Bangweulu, and the Luapula, he was put into a canoe by the Arabs, and made his way northwards along the coast till he reached certain islands, on one of which a young Maskat Arab received the party and fed them sumptuously. There were seventeen islands in this Kasanga group, and on the principal one was a breed of very large fowls, which Livingstone compares to the Cochin China variety. There were also numbers of Muscovy ducks.

He notes the decidedly brackish taste of Tanganyika water near the shore, but out in the middle the lake-water was quite sweet.

At last a favourable opportunity as regards the weather offered for the bold passage across the width of the Lake, a passage extremely dangerous to canoes when the wind raises the waves. And he reached Ujiji on the north-east coast of Tanganyika in March 1869, shattered in health and craving for letters and stores. It was to this place that he had directed his supplies and mails to be sent by caravans bound inland from Zanzibar, but little care seems to have been taken by the Arabs entrusted with the conveyance of these goods, and to Livingstone's intense disappointment he found that many of his stores had been robbed from him, whilst others had been left behind in Unyamwezi. Sixty-two out of eighty pieces of cloth had been stolen, and most of his best beads. His new supply of medicine, wine, and cheese had been left in Unyamwezi. The buffaloes which he had wished to import from India, in the hope that they might serve for transport, were all dead. Fortunately, there had come through a little flannel and a supply of tea. The flannel was particularly grateful.

He recrossed Tanganyika to resume his search for the Lualaba in July 1869. In his journal he recorded the abundance of pandanus screw pines off the west coast of that lake. As he travelled through the Guha and Manyuema countries he entered "the land of grey parrots with red tails" ["to play with grey parrots was the great amusement of the Manyuema people"]. The Manyuema country he describes as "surprisingly beautiful, palms crowning the highest heights of the mountains, and climbers of cable size in great numbers hanging among the gigantic trees." Strange birds and monkeys were everywhere to be seen. The women went innocently naked; and the Adams of this Eden wore nothing but a small piece of bark cloth. Both sexes atoned for their absence of clothing by having their bodies tattooed with full moons, stars, crocodiles, "and devices recalling Egyptian hieroglyphics." Yet although their country—prior to the Arab raids—seemed an earthly paradise smallpox came every three or four years to Manyuemaland and killed many of the people.

It was in the Manyuema country that he came into contact with the large chimpanzi (*Troglodytes schweinfurthi*) of eastern Equatorial Africa, whose range extends from the Welle-Muhangi river and Unyoro to the eastern bend of the Upper Congo and the west coast of Tanganyika.

The Soko, as he called this large chimpanzi, always tried to bite off the ends of the fingers and toes of the men with whom it fought, not otherwise doing them any harm. It made nests, which Livingstone described as poor contrivances with no architectural skill.

The Manyuema told him, however, that the flesh of the Soko was delicious; and Livingstone thinks that through devouring this ape they may have been led into cannibalism. The Sokos gave tongue like fox-hounds; this was their nearest approach to speech. They also laughed when in play, and in their relations with the natives were quite as often playful as ill-tempered. The lion, which seemingly existed in the Manyuema country in spite of the forest, was said to attack and kill the Soko, but never to eat him. The Sokos lived in monogamous communities of about ten. Intruders from other camps were beaten off with fists and loud yells. If one tried to seize the female of another, the remainder of the party united to box and bite him. The male often carried his child, relieving the mother occasionally of her burden.

Rhinoceroses were shot in the Manyuema country. He also alludes to the pygmy elephant of Congoland, "a small variety, only 5 feet 8 inches high at the withers, yet with tusks 6 feet 8 inches in length"; and notes the killing of an elephant with three tusks, one of them growing out through the base of the trunk.

[The pygmy elephant (*Elephas africanus pumilus*) of the equatorial Kamerun-Congo forests was only rediscovered in the early part of the twentieth century.]

Livingstone reached Ujiji for the second time at the end of October 1871. "I was now reduced to a skeleton . . . But I hoped that food and rest would soon restore me." He found that an Arab at that place had recently sequestrated and sold all his goods, not leaving even a single yard of calico or a string of beads out of an enormous quantity. "I now felt miserable, having to wait in beggary, which was what I never contemplated. . . . I felt in my destitution as if I were the man who went down to Jerusalem from Jericho and fell among thieves." Yet the Arabs were not all bad, for one of them came to him and said, "This is the first time we have been together. I have no goods but I have ivory. Let me, I pray you, sell some ivory and give the goods to you. . . . . But when my spirits were at their lowest ebb, the good Samaritan was close at hand, for one morning Susi came running at the top of his speed and gasped out, 'An Englishman, I see him,' and off he darted to meet him. The marine flag at the head of the caravan told of the nationality of the stranger. . . It was Henry Moreland Stanley, the travelling correspondent of the *New York Herald*, sent by James Gordon Bennett . . . at the expense of more than £10,000 to obtain accurate information of Dr. Livingstone, if living, and if dead, to bring home his bones." Livingstone also learned that the British Government had not forgotten him, but had voted £1000 for supplies (so, at least, he writes in his journal, but this is probably an exaggeration).

He then went off at the request of the Royal Geographical Society to explore the north end of Tanganyika and lay down its geography definitely. This was done in company with Stanley. It was now shown by actual observation that the Lake had no outlet on the north, but that the Lusizi flowed into Tanganyika as the overflow of the waters of a rumoured Lake Kivu, not actually discovered till some twenty years afterwards.

Stanley was often ill in January and February 1872, and had to be nursed by Livingstone. But in between his attacks of fever, Stanley recovered and shot a great deal of big game for the sustenance of the expedition. On December 27, 1871, Livingstone and Stanley had marched away together for Unyamwezi. Livingstone intended to pick up fresh supplies of trade goods and other porters. They reached Unyamwezi on February 18, and here Livingstone found letters awaiting him from friends and relations, pressing him with bland assurance to make a complete work of exploration of the sources of the Nile before he retired from the country. Livingstone, therefore, decided to return from Unyamwezi through Fipa to the south end of Tanganyika, then cross the Chambezi, pass south of Lake Bangweulu and due west to the "ancient fountains of the Nile and the underground excavations . . . of Katanga. This route will serve to certify that no other sources of the Nile can come from the south without being seen by me. No one will cut me out after this exploration has been accomplished, and may the good Lord of all help me to show myself one of his stout-hearted servants, an honour to my children and perhaps to my country."

Livingstone was almost an expert in geology and petrology. He felt the keenest interest in the records of the rocks, and fully realised the importance of palæo-botany. When descending the valley of the Central Zambezi in 1856, he discovered fossil remains of *Araucaria*, conifers now confined to South America and Australasia; and fully realised what his discovery meant in regard to ancient land connections between South Africa, India, and South America. He was much impressed with the probable coal-bearing strata of sandstone throughout the Ruvuma valley. A great many pieces or blocks of silicified wood appeared on the surface of the soil at the bottom of the slope up the plateaus. "This," he wrote, "in Africa is a sure indication of the presence of coal beneath." In the sands of some of the rivers pieces of coal were quite common. He originated the theory of the rift valley of Lake Nyasa. "It looks as though a sudden rent had been made so as to form the lake and tilt all these rocks nearly over" (namely, in the direction of Ruvuma). His observations would seem to show that the level of Lake Nyasa was once about 55 feet above its present high-water mark. It is possible that at this high level its overflow of waters first of all passed into the basin of Lake Chilwa, and then flowed northwards into the Ruvuma system.

Here follow a few word-pictures of Central African scenery selected from his journals. On January 9, 1867, he had ascended a hardened sandstone range (of what have since been called the Sharpe mountains[1]), with very beautiful valleys having the appearance of well-kept English parks; but they were in fact full of water to overflowing, immense sponges, covered with close, short, green turf. Then followed a march through mountains which he describes as being of delicately-tinted pink and white dolomite. In the ravines there were noble Raphia palms. He ascended this northern part of the dividing range between the Zambezi system and the Chambezi, till he reached a height of about 5380 feet

---

[1] See my maps of Central Africa in *George Grenfell and the Congo.*

above sea-level, the mountains further rising above that to nearly 7000 feet.

He thus describes the south end of Lake Tanganyika in the western part of Ulungu. "From altitude of nearly 6000 feet above sea-level one descends 2000 feet to the lake shores, and still the surface of the waters is upwards of 2500 feet above sea-level. The sides of its basin are very steep, sometimes the rocks run a sheer 2000 feet down to the water. Nowhere is there 3 miles of level land from the foot of the cliffs to the shore. Top, sides, and bottom of this tableland are covered with well-grown forest and rich grass, except where the bare rocks protrude. The scenery is extremely beautiful."

"The Aisi, a stream of 15 yards broad, and thigh-deep, came down alongside our precipitous path, forming cascades by leaping 300 feet at a time. The bright red of the schists among the green sward made the dullest of my attendants pause and remark with wonder. Antelopes, buffaloes, and elephants abound on the steep slopes, and hippopotami, crocodiles, and fish swarm in the water. One elephant got out of our way to a comparatively level spot, and then stood and roared at us. . . . The first village we came to on the banks of the lake had a grove of oil palms and other trees around it . . . not the dwarf species seen on Lake Nyasa, but one with fruit quite as large as those on the west coast. After being a fortnight at this lake (Tanganyika) it still appears one of surpassing loveliness. Its peacefulness is remarkable, though at times it is said to be lashed up by storms. It lies in a deep basin, whose sides are nearly perpendicular, but covered well with trees; the rocks which appear are bright-red argillaceous schist; the trees at present all green; down some of these rocks come beautiful cascades, and buffaloes, elephants, and antelopes wander and graze on the more level spots; while lions roar all night. The level place below is not 2 miles from the perpendicular heights. The village (Pambete), at which we first touched the lake, is surrounded by oil-palm trees—not the stunted ones of Lake Nyasa, but the real West Coast oil-palm tree, requiring two men to carry a bunch of the ripe fruit. In the morning and evening huge crocodiles may be observed quietly making their way to their feeding-grounds; hippopotami snort by night and at early morning."

This is how he describes the route and the way of travel with the Arabs from Tanganyika to Mweru through the Itawa country.

"The valleys along which we travelled at the base of a range of low granite mountains were beautiful with their green grass and their clumps of trees of a great variety of form, creating that park-like scenery so characteristic of tropical Africa. The long line of slaves and carriers brought up by their Arab employers added life to the scene. The great caravan went in three bodies and numbered four hundred and fifty persons in all, each body had a guide and a flag, and when that was planted all the company of this section stopped till it was lifted and a drum was beaten and a kudu's horn sounded. Each of the three parties was headed by about a dozen leaders, or *wenyi-para*, dressed with a fantastic head-gear of feathers and beads, red cloth on their bodies, and skins cut into strips and twisted. These took their places in line, the drum beat, the horn sounded harshly, and all fell in to resume the march. The female slaves walked bravely along carrying loads on their heads, but the actual wives of the Arabs were usually covered with a fine white shawl and wore ornaments of gold and silver on their heads, and many pounds' weight of fine copper leglets above the ankles. As soon as the slaves and wives arrived at the camping-place they began to cook, showing in this art a great deal of expertness, and making savoury dishes for their masters out of wild fruits and quite unlikely materials."

On March 29, 1871, he reached the outlying villages of Nyangwe on the Upper Congo. The country even at that date was open and dotted with trees, chiefly a species of *Bauhinia* that resists the annual grass-burnings. There were many Manyuema villages, each with a host of pigs. The altitudes seemed to be about 2000 feet above sea-level. The upper Congo or main Lualaba was narrower here than higher up its course in the south, but still a mighty river, at least 3000 yards broad, always deep, and quite impossible to ford. The current was about two miles an hour flowing north. The pigs at this place must have an interesting history as regards their origin. They could not have been brought thither by the Arabs on account of Muhammadan prejudice. They could not have come from the north, because the domestic pig of the *Sus scrofa* type is absolutely unknown in the interior of Equatorial Africa. They must have reached the Lualaba through the Rua countries, which in turn received them from Lunda, and that empire from Angola, and the Portuguese. From the same direction, perhaps, had come the Brazilian musk-ducks which Livingstone found in such abundance on the islands off the west coast of Tanganyika. The pine-apple also was just penetrating these Congo forest countries from the Atlantic seaboard.

This great explorer started on his earliest African journeys with a sound constitution; but the first shock to his system was the crunching of his left arm by the lion at Mabotsa. He did not, however, suffer much from malarial fever till he reached the upper Zambezi in Barotseland in 1854. The journey thence through South-west Congoland during the rainy season brought on severe attacks of dysentery, and these—alternating with malarial fever and rheumatism—followed all through Angola, so that he was seldom well for a week until he regained the bracing climate of South Africa at Linyanti in 1855. His rest at this place restored him to comparative health, a cure made more complete by the sea voyage home.

The six years spent in the exploration of Zambezia, Nyasaland, and the Ruvuma were marked by severe attacks of blackwater fever (as we now know it to have been) and by exhausting dysentery. He never quite regained his old strength and resiliency after that; even though he spent two years—1864-6—in England and Scotland. Moreover, during this time he had no tonic from the consciousness of success, and no complete freedom from monetary anxieties on behalf of his children's and his own future. He was in semi-disgrace, still holding a vague commission as a consul without a consulate, a salary, or any prospects of a pension. He would indeed have been in desperate straits had it not been for the previous and continuing generosity of his publisher and the faithfulness of his friends, William Webb and James Young. Such as these—not forgetting Oswell—would have combined to place him quite beyond the reach of monetary embarrassments had he not been too proud of his independence to accept such help. But, at any rate, they subscribed towards his last great expedition in search of the Nile sources.

When the early summer of 1866 found him once more on African soil free from all entanglements, free to search as he pleased and where he pleased for the mysterious lakes and rivers of innermost Central Africa, his sense of elation long prevailed to counterbalance disappointments from a badly-selected staff of India sepoys. Extracts like the following appeared in his journal as he approaches Lake Nyasa from the east:—

"The mere animal pleasure of travelling in a wild, unexplored country is very great. When on lands of a couple of thousand feet elevation, brisk exercise gives health, circulates the blood, and the mind works well; the eye is clear, the step is firm, and a day's exercise always makes the evening's repose thoroughly enjoyable."

In the region west of Lake Nyasa, however, a most serious loss occurred to him. Two of his Yao porters deserted. They had been very faithful to him all the way from Lake Nyasa, taking his part in every difficulty with the natives, and preventing many disputes by their knowledge of the languages. Yet these men "of uniform good conduct" were guilty not only of desertion but of the cruellest robbery. They took with them the load which contained his medicine box, merely because wrapped up with it were five large cloths and the clothing and beads of one of the coast porters. In addition, they took all Livingstone's dishes, a large box of gun-powder, all the flour which he had purchased to last him as far as the Chambezi river, the carpenter's tools, and two guns. "I felt as if I had now received sentence of death, like poor Bishop Mackenzie." All the other goods Livingstone had divided in case of loss or desertion, but he had never dreamed of losing the precious quinine and other remedies. "It is difficult to say from the heart 'Thy will be done,' but I shall try." He then goes on in his diary to put forward all the excuses that he could for these wicked Yao people, who had been, and were for long afterwards, the curse of Nyasaland, though now they are one of the mainstays of the administration, and their soldiers in British uniforms have gone far and wide over Africa to Ashanti and Somaliland.

From this time onward entries like the following are of frequent occurrence in his journal:—

"I am excessively weak—cannot walk without tottering, and have constant singing in the head, but the Highest will lead me further."

On the night of March 20, 1867, he was terribly bitten from head to foot by the driver ants which invaded his hut. "The more they are disturbed the more vicious are their bites. They become quite insolent." A few months before his death he was similarly attacked, and at last driven from shelter to shelter till he stood nearly naked and bleeding in the pouring rain.

On December 22, 1867, Livingstone (who wrote in his journal at that time "I am always ill when not working"), having left Kazembe's court in the terrible rains at the height of the rainy season, says of himself:—

"Every step I take jars in the chest, and in my very weak state I can scarcely keep up the march, though formerly I was always first, and had to keep in my pace. . . . I had a loud singing in the ears, and could scarcely hear the tick of the chronometers."

"After I had been a few days here (near Bangweulu), I had a fit of insensibility, which shows the power of fever without medicine. I found myself floundering outside my hut, and unable to get in. I tried to lift myself from my back by laying hold of two posts at the entrance, but when I got nearly upright I let them go, and fell back heavily on my head on a box. The boys had seen the wretched state I was in, and hung a blanket at the entrance of the hut, that no stranger might see my helplessness; some hours elapsed before I could recognise where I was."

During the winter or rainy season of 1868-9 Livingstone was very ill. He had been wet, times without number, and suffered from terrible pains in the chest and pneumonia. He was often semi-delirious and subject to delusions, such as that the bark of the trees was covered with figures and faces of men. He thought often of his children and friends, and his thoughts seemed almost to conjure them up before him. For the first time in his life he was being carried, and could not raise himself to a sitting position. The Arabs were very kind to him in his extreme weakness, but the vertical sun, blistering any part of the skin exposed to it, tired him sorely in the day marches. He also extracted twenty maggots from his emaciated body, due to a species of stinging fly, which inserts its eggs into the puncture. As the grubs grew they formed exceedingly painful pimples on his legs. In July 1870, his feet were almost consumed with irritable, eating ulcers, pulsating with pain and constantly discharging matter. These sores were obviously communicated by mosquitoes from the blood of the wretched slaves, who were tortured with them. Livingstone could fall asleep when he wished, at the shortest notice. A mat, and a shady tree under which to spread it, would at any time afford him a refreshing sleep. But in his last years of travel sleep was often made sad by the realistic dreams of happy English life from which he wakened, to find himself ill and consumed with anxiety that he might not live to complete his mission.

After 1869 he suffered much from the results of the decay and loss of his molar teeth, so that imperfect mastication of rough African food induced severe dyspepsia, and his bodily strength weakened under a condition of permanent mal-nutrition. Stanley, by relieving him when he did, gave him at least two more years of life, a certain measure of happiness, and the sweet consolation that he was not forgotten, and that the magnitude of his discoveries was appreciated. In this brief sunset glow of his life he turned his face once more towards Lake Bangweulu in order to trace the course of the Luapula to Mweru, and its junction with the Lualaba, half hoping that he might then travel down the broad stream till he entered the Bahr-al-Ghazal or the Albert Nyanza; but, although he now possessed comforts he had long lacked, and faithful, comparatively disciplined men, his strength gave out under constant exposure to rain, and to soakings in crossing rivers and marshes. Severe hæmorrhage set in from the bowels, and he died of exhaustion at Chitambo's village in the swamps near the south shore of Bangweulu on May 1, 1873.

NOTE.—I desire to add these remarks to the foregoing text of my address. In my preface to that address (see *S.G.M.*, p. 252), I animadverted with some emphasis on the treatment accorded to Livingstone by the Government of his day, more especially between 1863 and 1873. I see no reason to modify these passages. But in adding, as I did on March 27, that "after his death not one single penny was found by public subscription" (p. 254), I unwittingly erred. Some £3000 was publicly subscribed (chiefly in Scotland), which went to the relief, mainly, of Livingstone's two sisters at Hamilton. A larger amount would certainly have been raised but that the Misses Livingstone arrested the public generosity. Also the British Government awarded a pension for life to each of Livingstone's two daughters. H. H. J.

THE SPECIFIC CHARACTERISTICS AND COMPLEX CHARACTER OF THE SUBJECT-MATTER OF HUMAN GEOGRAPHY.[1]

By Professor JEAN BRUNHES (translated by E. S. BATES).

IT has been said by Taine: "Give a glance at a map. Greece is a triangular-shaped peninsula. Based on Turkey-in-Europe, it breaks away southwards, plunges into the sea, narrows down into the Isthmus of Corinth, and then continues southwards, forming a second peninsula, the Peloponnesus, which resembles a mulberry leaf, united as it is with the mainland by means of a sort of slender stalk. Add thereto a hundred or so islands, and likewise that part of the coast of Asia which confronts it—a fringe of little districts tacked on to huge barbarian continents, and bordering a blue sea bestrewn with islands—such is the country which has nursed and fashioned so precocious, so keen-witted a race. It was peculiarly well adapted for such a purpose. . . . A people fashioned by a climate like this would develop more quickly and more harmoniously than another; man is neither overwhelmed nor enervated by excessive heat; nor does cold constrain him to be stiff and stubborn. He is not condemned to dreamy indolence, nor to ceaseless exertion: he is not hampered by the meditation of the mystic nor by the brutality of barbarism."[2]

[1] An Inaugural Lecture delivered by M. Jean Brunhes, Professor of Human Geography at the Collége de France, Dec. 9, 1912; published in the *Annales de Géographie*, Jan. 1913. As the translation is not always word for word, it is as well to note that the variations occur with the author's approval.—E. S. B.

[2] *Philosophie de l'Art*, 4th ed., Paris, 1885, pp. 102, 104, 105. He pursues the thesis thus: "Compare a Neapolitan or a Provençal with a Breton, a Dutchman with a Hindoo; you will feel how gentleness and moderation on the part of Nature infuse vivacity and balance

It has also been said by Hegel, as if in reply: "I trust no one will come and talk to me of the climate of Greece, seeing that where once the Greeks dwelt, Turks are dwelling now: I hope that that will be ruled out of the discussion, and that we shall be left in peace."[1]

There the two opinions are face to face. Nevertheless, we need not treat the *Philosophie de l'Art* as summed up in that one geographical explanation, and we shall have occasion to return to it; but we are aware how Taine's determinist preconceptions led him into thinking of the laws of human knowledge as identical with those of Natural Science: "The philosophy of history faithfully reflects the philosophy of Natural history."[2]

To others, on the contrary, Man is wholly independent of his physical surroundings. Sometimes the secret of all progress among the various groups of human beings, economic, historical, and geographical, is detected in continuity of mentality, sometimes in race. Gobineau has asserted: "Even if the white group dwelt right amidst Polar ice or beneath the fiery rays of the Equator, thither would the world's intellectuality trend. There would be the focussing point of all ideas, all movements, all effort; and there would not be any natural obstacles equal to preventing the commodities and the products of the farthest distant countries from arriving there, across seas, rivers, and mountains."

Before we can claim, or rather, before we can substantiate what is the due of Human Geography, we must first free ourselves from the attractions of Taine's magnificent style and of his over-simplified, over-systematised, analysis, and likewise from the doctrines of specialists in "ethnic influence."

Our primary need is to confront and to controvert these extreme views concerning the modalities of human activities on the surface of the globe, in spite of the fact that these views bear the endorsement of such great names. You will recognise too that everything we shall attempt to define during this course of Human Geography is not so evident as our account of it may seem to suggest; and, further, that our business at present is to restate three problems with the help of a few plain facts:—(1) how Geography is affected by History; (2) how History

---

into the soul so as to be able to assist a mind that is alert and adaptable along the paths of thinking and acting" (*ibid.* p. 105). And farther on: "There you have the physical environment which, from the earliest times, has tended to stimulate the mental faculties. This race may be compared with a swarm of bees which, born where the heavens are kind but the earth inhospitable, makes the most of those aerial routes whereof it can take advantage. They reap, they plunder, they swarm, for their defence they have their wits and their stings; they construct fairy-like dwelling-places, they manufacture delicious honey; ever hunting, ever busy, humming away amidst the stolid creatures who live hard by, and who know no better than to feed under a master and to rage blindly against one another" (*ibid.* p. 111).

[1] This quotation and that from Gobineau which occurs later are given and discussed by L. Gumplowicz, *Der Rassenkampf*, Innsbruck, 1883, p. 15 *et seq.* W. Bagehot has said: "You could not show that the natural obstacles opposing human life much differed between Sparta and Athens . . . and yet Spartans and Athenians differ essentially" (*Physics and Politics*. 4th ed., 1876, p. 84).

[2] *Essais de critique et d'histoire*, p. 26, 2nd ed., 1866

is affected by Geography; (3) how far, and in what ways, human beings are geographical agents?

After having thus determined the specific characteristics of Human Geography, we shall have recourse to a practical experimental demonstration; that is, shall select some section of the terrestrial surface whereon it will be possible to distinguish very definitely what is essentially Human Geography (properly so called) from what is termed political, or historical, Geography. These considerations will enable us to add a third part, wherein we shall endeavour to give a threefold reason why Human Geography is so complex a subject of study: because each fact (1) contains and implies a social problem; (2) necessitates and implies a statistical problem; (3) conceals and implies a psychological problem.

Starting, therefore, with a sort of introductory meditation, and continuing with a monographic illustration, we shall lead to a series of deductions.

I. The Specific Characteristics of the Subject-Matter of Human Geography.

(1) *The Influence of History on Geography.*

As men go on making history on the earth, they also make geography; using the word history in the widest sense of the term, to include all history, agricultural and industrial, military and technical, economic and social, scientific and religious. Archæologists and epigraphists make it their business to reconstruct vanished civilisations with the help of those fragments of temples and stones which I should be inclined to term the geographical dust of history.

Dr. Capitan, in his course last year at the Collége de France dealing with the great civilisations of Mexico, based his lectures on all these more or less obliterated features, more particularly on that embankment-work which serves as a memorial of those who once lived there. He likewise examined and classified all the innumerable and tremendous earthworks and tumuli, from the conical mounds on to those modelled in the likeness of animals, which the famous "Mound-builders" have scattered all over the east and south of the territory which now belongs to the United States.

So too the paved roads and the Great Wall of China; the Roman roads of our Gaul; the Napoleonic roads of our Europe; the trans-continental routes of the United States; the network of English cables, are not only illustrations, allegories even, of history, but also, and far more emphatically, adequate expressions, local or world-wide, of so many different types of imperial rule.

History reveals both co-operation and government becoming more and more effective agents in human life, and it becomes the business of geography to illustrate this fact.

A town, especially an embryo town, has often been conditioned by a well, a spring, a market, etc., and has crystallised around cross-roads, or along the side of a highway; certain villages, indeed, are so clearly the creation of the road that the Germans have invented the well-justified term of Gassendorf or Strassendorf. In proportion as the agglomeration of houses continues, do the means of inter-communication grow more complex. And though the house, whether in the form of blocks of workmen's dwellings or of the twenty-five story hotel, does typify the collective character of the immense modern town, yet it is the means of intercommunication which typifies this best. Without this latter characteristic the great modern agglomeration could not exist. It is in evidence on the surface, forming a perfect web in its complexity, not only with its suburban lines and stations, railways and waterways, but also by means of arteries and veins which condition and regulate its internal life. Just for a moment let us try to visualise Parisian means of inter-communication. What a mass of lines, high and low, crossing and recrossing, various in levels and infinite in direction! A girdle of railways, a network of street trains, horse 'buses and motor 'buses, the subterranean and aerial network of the Métropolitain and the Nord-Sud, telegraph and telephone wires, pneumatic tubes, conductors of electric power and light, water pipes, gas pipes, and sewers, etc., whithersoever we turn or walk, or go down or go up, we pass through a maze of tubes and wires which hem us in, overshadow us, undermine us, and all to put us in touch with other human beings. Did the whole tangle lie visible before us in its entirety, it would seem to us like a vast, unsymmetrical tissue of steel, iron, copper and lead, more complex and more deliberate than the most intricate of spiders' webs; and therein we should see the supreme expression—so far as matter can express it—of all the mutual understanding, the ties, the sense of solidarity that have as yet come into being amongst that swarm of individual and domestic entities which, when crushed together on to a single point of space, create that excrescence, that uncouth blot, that species of geographical tumour, that we term—a town.

Others have pointed out the increasing attention which modern historians tend to give to all these surface-features, houses and roads, tillage and breeding, quarries and mines, which are, so to speak, the projection of the will of Man on the superficies of the Earth's crust. And M. Camille Jullian, the historian of Gaul, who is continually becoming better qualified to claim the proud title of historian of the national soil, lectured last year on the "Topographical Formation of Towns." He was inquiring into the origins of cities, and he dealt not only with the fortifications and with the sanctuary, but also with the spring and with the market; and, basing his researches on historical documents and map in hand, he examined the influence of the towns on the routes, and of the routes or havens on the towns. So also with M. Georges Renard, the Professor of the History of Labour; when he came to outline the course of the evolution of agriculture during the last 150 years, he did not merely note the advance in methods of culture due to the advance in chemical and biological knowledge; he indicated the result, namely, "the stretches of land wrested from the sea, from the marsh, from the sand-banks, from the desert, and from the bush," and further, "the migrations and variations of plants which Man finds useful or attractive."

All the great civilisations of Asia, the mother of our modern civilisations, have found their expression in magnificent and complex systems of irrigation, or else in innumerable flocks and herds, or else—and it was this last kind that proved the strongest—in oases of agriculture farmed in co-operation with brave men who derived huge profits from a pastoral life on immense grassy steppes.[1]

As in great things, so in small. Fortified churches, such as that of Royat in Auvergne, or that of Hunawihr in Upper Alsace, are, so to speak, documents which summarise a whole period of history. Round about Trier and throughout the schistose massif of the Rhine, the Prussian Government has replanted the districts which had been deforested by their former rulers or by the armies of the great revolutionary period; the country-folk call these trees *Preussenbäume*. Districts which once grew cereals would, under present economic conditions, be condemned to an ever increasing inferiority; but they have been successfully transformed into districts where pasture predominates: in the villages which are the natural centres for these districts, the former "halle au grain," or "grenette," is shut up or pulled down, at Saint-Flour, *e.g.*, or at Lausanne. Again, political powers who find themselves forced into developing navies, create harbours at any cost; Trieste and Fiume are towns which, but for the theory and practice of "free ports," could never have come into existence.

Some day we shall be studying, in some detail, the geographical effects of the Continental Blockade. We shall note the part played by the Hanseatic towns and by those, Gothenburg, Stralsund, Oldenburg, which constituted veritable cleavage-points in the blockade system. We shall inquire into the more or less lasting results of that daring endeavour, not only on the towns and on channels of commerce, but even on the general economy of our tillage; there is not one amongst us whose daily food is not still, by reason of the quality or quantity of sugar consumed, modified by the political plan devised in opposition to England by the extraordinary will-power and individuality of Napoleon.

Thus and thus is our daily life interwoven with a thousand trifling facts which symbolise one or other of the most decisive episodes of world-history. A Yankee passes a negro in the streets of New

---

[1] "At every period of history at which we find Mesopotamia under the control of a race who knew how to defend it against the nomads, and how to administer it and to manage the water wisely, abundant prosperity is manifest; the generous earth gives a hundred-fold return for the seed sown. When the Turk is the ruler, anarchy comes in his train, the canals become choked up, the Arabs of the desert make the laws, and the country falls into the state of insecurity, wretchedness, and unproductiveness, in which we see it to-day. A firm hand and a restoration of the irrigation on the lines proposed by the engineer Willcocks, would be sufficient to enable this country, which has seen Babylon, Nineveh, Ctesiphon, and Bagdad, to become once again one of the most fertile of the Earth. Thus profoundly does Man's action affect the geographical aspect of one and the same country; sometimes it is that of a conqueror, sometimes of a protector, sometimes of a destroyer." (René Pinon, "La Géographie Humaine," in the *Revue Hebdomadaire*, 1911, p. 180.

York: if he condescended to notice him, the line of thought would take him back to the black African chieftain to whom the contractors applied for the manual labour needed by the cotton and sugar planters: what an amount of history in that woolly hair, in those thick lips, in that country, in that latitude! The maize which the low-lying plains of the Save and of the Danube produce in such abundance; the potatoes which occupy so many an acre in Ireland, in France, and in Germany, convert vast tracts of land in this Old World of ours into permanent memorials of the discovery of America.

The plain truth is that the outstanding facts of Geography at the present time are not the discoveries of the North Pole and of the South Pole. For the bold and learned pioneers of Polar exploration we have the most enthusiastic admiration; we have just received the intrepid Roald Amundsen with the hearty welcome that is his due; and we, as Frenchmen, are proud to remind ourselves in this connection of the glorious name of Charcot. But whatever may be the most heroic facts in the history of the world, the most important are events such as the realisation of the Suez and Panama Canals. And if all the history of the world, political, economic, and social, has been, or is about to be, subjected to fundamental changes in consequence of human initiative taking directions like those to which I have just referred, it is precisely because human beings are transforming the surface of the globe itself, and so are factors in, indeed, incarnations of, Geography.

(2) *How History is affected by Geography.*

If facts which seem so tremendous from a human point of view and so trifling from the point of view of the Earth, bring about and set in motion, as they do, fundamental and widely-differing changes in the lives of human beings and in their mutual relations, we may deduce from them to what extent humanity, individually and collectively, directly and indirectly, is affected by the distribution of land and sea, by the conformation of continents, by the nature of the channels of inter-communication (*e.g.*, sea or river)—in a word, by situation. I cannot to-day enter into detail to the extent needful to make that idea of situation come home to you afresh, and to tell you how far all of us, individuals, municipalities, states, races, civilisations, are functions of the point of space whereon we have been born, functions of the point of space wherein we live, functions of situation.

A master-historian, Gabriel Hanotaux, speaking of Sicily, summing up its history in that terse and vivid way of his which sums up life, wrote: "Palermo looks towards Europe, Agrigentum towards Africa, Syracuse towards Greece: thus does Trinacria fulfil her formula, her threefold orientation, and her threefold fate." [1]

Routes have contributed their large share to history according as they have more or less conditioned the relative values of situations. Routes undoubtedly have brought men together, in the fullest sense of

---

[1] *En Méditerranée: La Paix Latine.* Paris, n.d., p. 175.

the phrase, for they have done work among groups of human beings analogous to that which would have been effected had stretches of land or water been obliterated, a reduction of stubborn distances. That is the idea—a true idea—which underlies the words: "Do we not know that the quest of routes—amber routes, tin routes, ivory-, silk-, and spice routes—lies at the bottom of the chief events that have happened to humanity? Consider what Suez, Bagdad, the Alpine tunnels, and Panama, have meant to history, and mean to-day."[1]

But what is it that creates a route? Several preliminary conditions must be satisfied; products must be in demand, products which can be transported; a sufficient number of people must demand those products; conditions must be favourable to inter-communication. One of the lines of communication which Nature has most unmistakably provided for mankind is that "transportation-belt" which runs from New York to Lake Erie, where the railway lines follow it, first along by the Hudson and afterwards westwards from Mohawk Pass. The districts either side of this line contain 30 per cent. of the land in the State suitable for tillage, 77 per cent. of the population, all the towns of more than 100,000 inhabitants—New York, Syracuse, Rochester, Buffalo,—and, with hardly an exception, all those of more than 50,000. That way, too, lies the means of communication by water between the sea and the lakes; and it possesses the double privilege of rich alluvial soils and a temperate winter climate. What else could happen than that such a district should sooner or later take precedence of those that border it?[2]

At a later date we shall have occasion to analyse the geographical intuitions of Montesquieu, Turgot, Daunou, and Heeren; but the first historian to feel and to assert, in any really striking way, that History is in some degree dependent on Geography was Michelet: "Without a geographical base, the people, the actors in the historical drama, seem to walk in the air, as they do in Chinese pictures from which the ground is omitted. And notice that the soil is not only their stage; it influences them in a hundred ways, through their food, the climate, etc."[3]

But difficulties arise immediately we attempt to define the share that the facts of Nature have in deciding human destinies. We shall endeavour to make clear how History had been little by little getting into touch with the earth previous to the appearance of P. Vidal de la Blache's "Tableau de la Géographie de la France,"[4] the most perfect example of scientific explanation that exists. And please allow me to tell you, to-day and here, how proud I am to say that I have been, and still am, the pupil of such a master; it is more than ideas that I owe him; I owe him inspiration, initiation, the sense of all the finer shades of the sub-aerial interplay between Earth and Man; in short, the passion for Geography.

---

[1] H. Hauser, in the *Revue Historique*, cviii. p. 170.

[2] Cf. "The Distribution of Population in the United States," by Albert Perry Brigham, in the *Geog. Journ.*, xxxii., 1908, p. 384.

[3] *Histoire de France*. preface of 1869, i. p. v.

[4] The introductory chapter in Lavisse's *Histoire de France*.

We must, then, make up our minds to put aside generalities, and vague analogies between nature and man. We must make it our business to search for facts of interaction.

Now if the facts of Nature react on the faculties and on the occupations of groups of human beings, it is because there exist intermediary facts, of fishing and hunting, of tillage, of diseases, etc.; and these throw light on points of contact between terrestrial and human activities, and suffice to reveal the influence of the former on the latter. We may truthfully say with Napoleon that the policy of States lies in their geography,[1] but that is not enough. In order to make it clear to you how desirable it is to orientate historico-geographical researches, I will quote some lines which illustrate with far greater completeness one of the material modes of political power; it was written in 1902, concerning Madagascar: "Every day proves the truth of the formula—he who is master of the cattle is master of the country. However refractory a tribe may be, it returns to its senses directly its cattle are seized."[2]

When you come from Paris and have reached the summit of the last of those circles of heights which form one of the essential features of the Parisian Basin, you see, silhouetted against the further horizon, the hill and town of Laon; a little plateau which, thanks to a stubborn stratum of twelve yards of coarse limestone, stands firm, both level and high, above the sandy strata around it. It looks like a sort of circumflex accent, cut short all round by sheer cliffs, and in former times the district which it dominates was unhealthy marshland, an excellent defence. All the history of Laon forces itself on the memory; how it was an outwork of Christianity in Northern Gaul, an islet fortress which the Vandals left alone, the refuge of the later Carolingians. Was not this exceptional conformation predestined to be the base of a stronghold, and that a most efficient one? Let us go further still. Is it not a matter of cause and effect that so clearly defined a township as this, at whose very gates there began a steep slope splendidly suitable for defensive purposes, should have been the first to summon up sufficient hardihood and sufficient consciousness of its own collective actuality to rebel against its episcopal overlord? Is not this an admirable example of history as influenced not only by situation, but also by the structure, by the conformation, of the ground?

And yet, and that even in this very example, while men are clearly predestined by the Earth, they are likewise predestined by what they themselves have created on their atom of space. Once a site has been chosen by its earliest inhabitants, their successors are subjected to two influences, Geography as Nature wrought it, and Geography as modified by man. To calculate the inter-relations of the two is no easy matter; but our inquiry drives home this moral, that men become the allies of their geographical surroundings not only by virtue of the inherent fitness of those surroundings, but also by virtue of facts of

---

1 *Correspondance de Napoléon*, letter of Nov. 10, 1804, addressed to the King of Prussia (4to ed., x. p. 59, No. 8170).

2 Cf. *Annales de Géographie*, xi., Chronique du 15 mai 1902, p. 285.

Human Geography which result from their deliberate choice and activities.[1]

So, too, with climate. Everybody later than Montesquieu has accepted the idea that the modes of human activity are conditioned by facts of climate. But what are the real conditions imposed? How does the climate influence us? It acts directly on our organism; it acts indirectly by means of other organisms whose growth it fosters, and which are the intermediaries for the propagation of this and that disease; it acts, further, through vegetation and the food which that vegetation supplies. And it is just as necessary here, as elsewhere, perhaps more necessary, to rejuvenate current assumptions by analysing them, for they are far too slipshod and superficial. In a given year certain relations exist between the sum and succession of the details of the weather, and the quality and quantity of the crops. But there is more than one kind of corn, and each kind requires different conditions as regards soil and climate, which latter will make almost another plant of it; certainly another plant so far as Human Geography is concerned. There was a time when Americans had never heard of hard wheat: Cherson wheat was only discovered about four or five years ago. This Cherson district is beyond the black earth limit; it is dry prairie: Cherson wheat has meant the gain of 100 miles from the Great Desert. To-day the United States are producing 60,000,000 bushels of this wheat annually.[2]

(3) *Men as Geographical Agents.*

We see the surface of the earth as partly rugged continent and partly level ocean, and we see, too, earth-waves as well as water-waves, the latter mobile, eternally vanishing and re-appearing, and the former, of stone, motionless; then there are sand-waves also, varying as water varies. Glaciers, likewise, and torrents, rivers, the phenomena of eruption—all these imply some marking or other of the Earth's surface; and more, they are incessantly at work changing and carving, and, within limits, remodelling it. But, side by side with, and amidst all these factors, whose activities, when scientifically analysed or explained, constitute the subject-matter of physical geography, co-exists yet another, the most potent of all those that are modifying the terrestrial surface—Humanity, the sixteen hundred millions who are always at work, and always increasing in numbers. What renders them the most potent agent is not superiority in strength, but universality of range; not a deadly certainty in method, but adaptability of method to surroundings; not a concentration of tremendous force on a given

---

1 By this means we come to replace the unreal "heterogeneous determinism," such as Taine's, by a determinism which may be "homogeneous." I am borrowing these ingenious and accurate expressions from M. Paolo Arcari ("The classification of the content of literary work from the point of view of the comparative history of Literatures," in the *Germanisch-Romanische Monatschrift*, 1910, p. 142).

2 For this information I am indebted to Dr. Aaronsohn, the specialist who so recently discovered wild wheat.

point, but the endless multiplication and synthesis of a series of small achievements whose sum-total is capable of transforming the appearance of continents.

In short, there is a very large part of the surface of the globe which may be termed a "human surface." The simplest and clearest definition of Human Geography is—the study of this "human," or, still better, "humanised" surface of our planet.

One of the most ardent students and admirers of this humanised surface is Mr. Albert Kahn: he believes that in the study of it lies one of the principal means, not only of the furtherance of co-operation among men of science, but also of the cause of Peace. His desire is that everybody should be enabled to go and see for themselves and appreciate what everybody else is doing, and, being wholly unsatisfied with fine phrases, and in the habit of combining the most idealistic of dreams with the practice of what is practicable, there has occurred to him, as you are aware, the ingenious idea of founding scholarships for world-travel, whereby young men may be enabled to gain a preliminary acquaintance with the inhabited world in the hope that they may thereby acquire the taste, in fact, a passion, for getting to know it better. One day Mr. Albert Kahn learnt that it was under consideration whether or no a place should be found for Human Geography at the Collége de France, and it occurred to him that he might give effect to what had so long been an earnest wish of his by endowing a new lectureship at this great school of scientific work. With that breadth of mind, therefore, which knows no yielding, and yet knows how to make the best of what is to hand, he has requested that, at this college and from this chair, instruction should be given which should utilise all the best means yet devised for recording facts, and should set before itself the aim of revealing human achievement in relation to the earth with complete scientific accuracy and, so far as possible, in all its inherent and absolute beauty. The founder of the world-travel scholarships has skilfully preserved his anonymity for years, and would not agree to be present here this evening, but he cannot prevent our thanks reaching him, the personal gratitude which I feel myself as well as that which I am commissioned to express as from the chief of the Collége de France, and from the college itself as a whole.

To-day, thanks to the advance in kindred branches of knowledge—geology, meteorology, botany, biology—physical geography is on a sound basis. And history, archaeology, prehistory, anthropology, ethnology, and the economic and statistical sciences, are in their turn becoming richer and richer, both in quantity and in quality, in genuine evidence, and are becoming masters in their own houses. They are all realising, and stating, that the isolated fact is susceptible of such contradictory interpretations that it is inadvisable to study it in isolation; it is necessary to restore it to its place in the current of the life which gave it birth. It is a link in a chain; and the welding of the chain is the work of that unknown quantity—Life.

Quite recently M. Émile Boutroux commented very truly: "A

famous philologian, M. Michel Bréal, in his celebrated *Essai sur la Sémantique*, has explained that we must avoid thinking of speech as of a thing that is of spontaneous growth in accordance with laws not made by Man. He considers that as regards the evolution of speech, all external conditions rank as no more than secondary and incidental causes; the only effective cause is human intelligence and volition."[1] Now Human Geography is in the same case, only still more so. All the phenomena of Life are dependent on their surroundings, but these latter are always in a state of evolution, and always will be. If we bear this idea in mind, and are intent on being something different from mere collectors of notes—are, in fact, among those who aim at bringing about what Michelet termed the wonderful "resurrection" in every department of research—then the sciences which depend on observation, whether economic, moral or social, are bound to become studies of surroundings, surroundings which Life is incessantly modifying and transforming.

For the purpose of throwing light generally on the sciences which are concerned with Life, more particularly with human life, recourse ought to be had to geographical researches which bear directly on human life. But such researches will have to begin with the most unequivocal observation; exactness in observation must needs be the primary and essential guarantee for the possibility of accuracy in the explanation that is to come.

One of the most genuine of contemporary geographers, Professor Woeikof, of St. Petersburg, very truly remarked that Man's power over the Earth is mainly dependent on his manipulation of its movables.[2] The remark can be insisted on to its furthest application. One of the chief aspects of human activity consists in controlling drops of water, *i.e.*, in determining the destination of raindrops, damming channels, constructing canals, directing water (whether running or stagnant), drop by drop, on to the surface of the fields for irrigation, or down below into drains to relieve excess of moisture. So too with immovables. Real cultivation always implies a more or less thorough reorganisation of the upper strata of the soil; constructing houses, roads, or mines, implies, to begin with, cutting through masses of earth and forests, and removing, we might almost say turning into movables, quantities of minerals and tree-trunks and branches which are fixed to the soil by their roots. On all sides we find men breaking up sections of the surface, sometimes amounting to hills and mountains; and dissecting, displacing, and stacking pebbles and blocks of stone. Wherever they are, they are giving stability to particles of sand and reducing rocks to dust, and, on the other hand, are at work on the dust, consolidating it all over again. Minerals, they melt down in order to make ingots. In short, the whole work of civilisation in relation to matter consists of subdividing, amalgamating, condensing.

[1] From his address delivered to the constitutive Assembly of the French Institute in the United States, June 14, 1911.

[2] A. Woeikof, "De l'influence de l'homme sur la terre" (*Annales de Géographie*, x., 1901, p. 98).

I have already made an effort, in a book of mine, to meet the elementary demands of observation by outlining a definite, though in nowise dogmatic, system of classification. I may be allowed to summarise the ideas underlying it.

It seems to me that all human activities on the terrestrial surface may be reduced to six essential kinds:

(1) Houses, and (2) Roads. These two are always co-existent, and combine to form not only villages and towns, but also, as we have already seen, the outward and visible expressions of more complex political entities, States and Empires. These two kinds of achievement form a first group which may be described as "facts of the unproductive occupation of the soil." Next come—

(3) Gardens and farm-land, and (4) Beasts of burden, flocks and herds, domestic animals, cultivated plants. These form the second group,—"facts of conquest, vegetable and animal." Lastly, men destroy, or carry off, without any intention of replacing, organic and inorganic products of both land and sea wherever their means of destruction enable them to do so. The third and last group, therefore, consists of—

(5) Devastation in relation to vegetable and animal products, and (6) Exploitation of minerals;—"facts of destructive economy."

Unproductive occupation of the soil, conquest of the vegetable and animal kingdoms, destructive economy: might we not state them more simply and with clearer antithesis as three forms of occupation, unproductive, productive, and destructive?[1] Destructive occupation is, as a rule, a sign that human beings are only just installing themselves on that point of the globe; productive occupation will follow if they are to win a livelihood there permanently; and unproductive occupation is the final development and obviously implies the most stable and the most characteristic stage of their activity. I have hitherto avoided making use of these expressions in my writings lest I might seem to have sought out formulae which were neater than they had any right to be, and which were verbal rather than real. Let us look at realities first, and that with eyes wide open; let us endeavour to analyse and to classify; the right phrases will come in due course.

And I myself will admit, with the greatest readiness, and with that frankness of criticism which ought to characterise every effort towards the reality of science, and which it is a man's first duty to apply to himself, that I am not very well satisfied with the designation of the two first essential "facts" in the above classification—houses and roads. The words "unproductive occupation of the soil" are accurate so far as the earth is concerned, seeing that, in both cases, it remains unproductive in consequence of men's action; but it is an ambiguous term as regards Man's efforts, inasmuch as, in relation to himself and to civilisation, they have a very real productive value. If you discover a better

[1] Some careful readers of my *Géographie Humaine* have already thought of these simpler formulae which I had omitted intentionally. Cf. the very interesting article by Ph. Gidel in *L'Enseignement secondaire*, July 1912, p. 211.

term I shall be the first to rejoice thereat: we will be fellow workers, without spending too much time on difficulties over words. If we were satisfied with ourselves, if we had abandoned that fruitful discontentedness which is the beginning of scientific wisdom, neither the one nor the other of us would be likely to be found here. I am bound to add, too, with the same frankness, that, as a matter of fact, this classification is the first of its kind.

It is put forth, therefore, without dogmatism, as a means to an end, as a convenient instrument to assist us in our purpose of unravelling the tangled skein of the web that human effort has woven on the surface of the earth. But what I do maintain, with all the zeal that conviction implies, is the principle which inspires this classification, and likewise the result it aims at. Readers of the second edition of my *Géographie Humaine* will at once concede that I have never tried to confine Human Geography to the single species of "visible and photographicable facts"; but I have said, and I repeat, that facts of unproductive occupation of the soil, facts of conquest, vegetable and animal, and facts of destructive economy, are, on the one hand, the intermediaries and the symbols which consecrate to the service of humanity all the rest of the subject-matter of physical geography, while, on the other hand, they form points of departure, are outward and visible signs of every thing which is both within the pale of human activities and also truly geographical. All the moral and social sciences take the same complexes of human activity for their subject-matter; but ethnologists and statisticians do not work on the same lines as historians, nor historians on those of ethnologists or statisticians, however much each may assist the other. Then geographers likewise should commit no trespasses: they ought to have their own prescribed domain for inquiry and analysis. Human Geography lies, as it were, where roads meet, whither come many facts, and those from many quarters; but for all that it has no call to turn itself into a bazaar in which everything is retailed. It is, and it ought to remain, a highly specialised department, admittance whereto shall only be vouchsafed to such facts as have a right to be there.

In order to make my meaning quite clear, I have judged it best to have recourse to the method which I have employed before, *i.e.*, an exhaustive monograph. What I want most of all is this, that both you and I should possess the clearest of ideas as to what it is that we are working at together; and since there is no process that reveals truths so explicitly as that of taking facts and trying to draw inferences from them, I intend to use part of my time now, and also on future Mondays, to considerations respecting Bosnia-Herzegovina.

II. A Monographic Illustration: Bosnia-Herzegovina.

Well, then, I have been to Bosnia-Herzegovina. I set out before the declarations of war; I was intent on the study of matters scientific, which seemed to me particularly well suited to lend point to this new course of mine; but circumstances, over which neither you nor I had any control, have imparted a remarkable degree of actuality to these wanderings of mine on Servian soil.

In Bosnia and in Herzegovina Turkish rule endured without mitigation from the fifteenth century right on to 1878, at which date a neighbouring Power was commissioned by the Congress of Berlin to administer these territories under the suzerainty of Turkey. This Power, Austria-Hungary, colonised — I use the word deliberately — this variant of Algeria and Tunis, and has subsequently reaped the benefit of prolonged and systematic effort in proclaiming the definitive annexation of the two provinces on October 5, 1908.

As soon as you forsake the Austrian half of the town of Brod and cross the broad Save by the great iron bridge, which serves both for railway and for all other traffic, you see outlined against the horizon the minaret of a mosque, the belfry of a Catholic church, and the belfry of an Orthodox church; it is the village of Bosna-Brod, the Bosnian half of Brod. You are entering Bosnia. Bosnia and Herzegovina are the most northerly of European countries in which Mohammedanism is recognised as a State religion, and the mosque at Brod is nothing exceptional: it is but the first of the thousand mosques which confront the traveller who enters from the north, and which are scattered throughout the 50,000 square miles of the territory of Bosnia-Herzegovina. They are likewise the only countries wherein the three faiths are found co-existent to this extent; the Christian churches are fewer than the mosques, but this is the result of Bosnia-Herzegovina having escaped from four centuries of Turkish rule only as lately as the date of the Congress of Berlin; but it is a district where Christianity dates very far back, where ancient churches are still standing, besides the ruins of many others, as well as besides the brand new ones—generally very ugly, it must be confessed —which bear witness to the newly won freedom. In Croatia and in Dalmatia there are no Mohammedans: in Servia and in Montenegro hardly any who are not Orthodox (98 and 90 per cent.). In Albania, strictly so called, most are Mohammedan, Catholics and Orthodox alike forming minorities. In Bosnia-Herzegovina, on the contrary, we find three groups, all very strong, though unequal in numbers; any two will outnumber the remaining one. In 1908, out of 1,900,000 inhabitants, 824,000 were Orthodox, 612,000 Mohammedan, and 433,000 Roman Catholics.

If you ask any one in the street, or in a railway carriage, in either province, what he is, he will no longer answer—"Mohammedan," or "Orthodox," or "Catholic": sometimes, indeed, he will not understand your question at first. Catholics describe themselves as "Croats," Orthodox as "Serbs," Mohammedans as "Turks." In other words, when they are questioned as to their belief, the term they use connotes a "nation," or even a "race." And yet this is a mere delusion; they are only one people, divided into three parties by creed alone. After the conquest of Bosnia by Mohammed II., in the fifteenth century, the aristocracy forsook its faith and passed over to Mohammedanism in order to retain its domains and privileges; ever since which period the contests between Christians and Mohammedans, and even, we are bound to add,

between the three creeds, have made up the whole history of Bosnia-Herzegovina, and are still at the root of all political questions, national and international. Behind the political and administrative façade erected by Austria-Hungary, symbolised by railway and telegraph lines, all either military or militarised, by fortifications and by barracks, etc., and personified by Austrian and Hungarian officials, we find, then, that there are three groups in Bosnia-Herzegovina: Turks, Serbs, and Croats. I am intentionally leaving on one side the two diminutive minorities of Jews (known as *Spagnoli* by reason of their Mediterranean origin), and Tsigans. These religious differences are clearly visible in the costume of the men, and still more in that of the women; the Mohammedan women, of course, are only seen in the streets with their faces veiled, and their whole figure enveloped in a loose robe. But a close observer will notice that the fez is worn as frequently by Christians as by Turks; and so also with the sash round the waist; and Serbs and Croats wind white linen bands round their heads just after the fashion of Mohammedan turbans. In fact, the two provinces, and even districts within them, can show distinctions which are very nearly as obvious as those external ones that are occasioned by creed. It is particularly necessary to attend to the physical types of the locality, and of these you will be able to form an opinion from the illustrations which I shall place before you in future lectures. It is most striking to find to what extent Mohammedans, Serbs, and Croats resemble each other in profile, in bearing, and in all the main features of physique: between the girl children of the Mohammedans and those of Serbs and Croats I defy you to establish any distinction, physical or racial. Yet what differences exist, on the contrary, between those minorities to which I have just referred, the Spagnoli and the Tsigans.

The truth is that it is really one and the same race that is here represented by three religious groups; all belong to the great Servian stock which, with the inhabitants of Servia, Montenegro, Dalmatia, Croatia, and the Diasporas constitutes a homogeneous whole of about ten millions of individuals. In Bosnia-Herzegovina there are three religions, but only one people. So too are there two alphabets, the Cyrillic and the Latin, but only one language, the Serbo-Croat, which is no other than Servian. The Mohammedans themselves, who call themselves Turks, are not only not Turks, but do not know a word of Turkish; of the Arabian language they know equally little, and they are incapable of reading or understanding the Koran; all of them speak Servian.

Now, this unity of the Servian inhabitants of Bosnia-Herzegovina is borne witness to by the "essential facts" of Human Geography. Croats, Serbs, and Turks cultivate fruit-trees, especially plum-trees, in the same way, and also the two cereals, wheat and maize, which yield them the major part of their nutriment; this is especially the case with the more important of the two, maize; and they make and eat their "johnny-cakes" in the same way, too. They use the same plough and the same breeds of cattle and sheep; also (Mohammedans, of course, excepted) of swine. In short, they lead the same lives, and at the great fair at Jajce in October I was able to verify how far the three groups do business with each other without distinction of religion. Everything, therefore, points to community of race and origin.

Moreover—and here we come upon the complementary entity which throws into relief the specific characteristics of the fundamentals of Human Geography—these essential facts vary, without differentiation due to religious or ethnical causes, according as the geographical surroundings vary.

In a general way, and without any exact correspondence between the political and the natural boundaries, Bosnia is a country of Flysch hills still thickly wooded, more than half its surface being covered with forest: it resembles its neighbour Servia, which long bore the name of Choumadia, from *chouma*, a forest. It comprises the three valleys of the Vrbas, the Bosna, and the Drina, which run from south to north and fall into the Save; whereas through Herzegovina runs the essential portion of the valley of the Narenta, which, descending from the north-east to south-west, leads to the Adriatic, in other words, to the Mediterranean. Herzegovina is, to the south, a continuation of the Karst, and is a far hotter district; a country of limestone mountain masses, deforested and bare; characteristic Mediterranean scenery, hemmed in by grim grey rocks. Now to the wooded district belong the wooden Bosnian house, so dainty and so original, and likewise the wooden mill, the wattle fencing, the mosque with roof and minaret of wood. To the treeless, rocky district belong the stone house, roofed with broad and thick limestone slabs, the ancient Turkish fortified house, called a *kouba*, the stone mill, walls of hewn stone, and the minaret of stone, tapering with needle-like fineness.

These "facts of observation" are obvious, and it seems a very simple matter to have noticed them. And yet, if you consult a conscientious compilation such as that which the *Revue Générale des Sciences* has put together with the help of a selection of learned men of the highest standing, or else a more specialised, but very detailed, book like the *Geologischer Führer* of Katzer, you will find no mention of such things.[1] Not only are they not accorded the place in the foreground to which they are entitled, but they are not even touched on.

The distinctive characteristic of a fact of Human Geography lies in its being a fact of the terrestrial surface which is seen to be a link between the facts of physics and the facts of human will. But it has been maintained that, whereas the attempt has been made to base Human Geography primarily on the observation of facts like the above, it abandons the intention of proceeding further or, if I may be allowed to phrase it so, to higher things. What a perversion! My idea is that in entering a house it is necessary to pass through the hall and ground floor, and that to gain access by way of the fifth, or even the fifteenth, floor is not really a convenient way. Bosnia-Herzegovina will show us, by means of facts of Human Geography, which are genuinely

1 "Études scientifiques de la Revue Générale des Sciences en Bosnie-Herzégovine" (*Rev. Gén. des Sc.*, xi. 1900, pp. 269-402, 419-555); Fr. Katzer, *Geologischer Führer durch Bosnien und Hercegovina* . . ., Sarajevo, 1903.

geographical, what degree of distinction can be traced between Human Geography and political geography; and that is the first demonstration which I wished to put before you.

In subsequent lectures I shall analyse still further, and in detail, life as it is lived in Bosnia-Herzegovina, inasmuch as this country is, at present, typical. For the time being, allow me to anticipate matters, and indicate how the specific subject-matter of Human Geography leads on to the study of certain other subjects which, so far as their root-principles are concerned form, within limits, integral parts of Human Geography.

"Human Geography certainly consists rather in the study of the material achievements of humanity than in the study of the races of humanity." Yes, these material achievements of humanity form the medium through which the geographer should see and study human beings. That is the special point of view which he ought to make his own. Does not this hasty and very brief indication that I have just given of certain facts of culture and of certain facts of habitation as regards Bosnia-Herzegovina put us in a better position to understand the racial unity and the community of method of living that exist among this tripartite Servian people than would any amount of phrases? While it obviously cannot serve as a description, it nevertheless throws light, geographically speaking. And once more, is it not the same sum-total of facts of the surface which constitutes the subject-matter of sciences so diverse as geography, ethnology, statistics, and sociology? Human beings and their achievements—those always, only the aspects vary, the aspects under which they are perceived and appraised. True as it is that every investigator in one of these branches of science cannot afford to ignore the chief results attained in kindred branches, it is no less true that we shall not be able to advance and prosper in the work that we do otherwise than by means of drastic specialisation and abolition of the "vaulting ambition" which tries to do everything at once.

Again, some of these essential facts serve to illustrate highly important events in diplomatic, military, and political history.

Ever since Austria-Hungary gained a hold in Bosnia-Herzegovina her efforts have been mainly concerned with the hemming in of Servia, and rendering it commercially dependent on the dual monarchy. Some of the clearest evidence of this tendency may be found in the south of Bosnia-Herzegovina, where the great natural highway runs from the Servian boundary at Mokra-Gora past Visegrad down to Foca and Bilec, reaching the Adriatic at Ragusa. By this route the Serbs' cattle could reach the sea in four or five days; it was a main road for exportation purposes. It has been closed. The closing of it has been rendered ever more and more effective, and for ten years or more Servia has been trying every means to regain her economic independence. That is how diplomats phrase it, but we should rather speak simply of a *via pecuaria*, a road which affords freedom of transit for her animals, and more particularly for her herds of swine. She has left no means untried; has endeavoured to send her animals across Hungary by rail; but Austria has increased the duties, and at times shut them out altogether at the frontier under pretext of epizootic diseases. She has also tried by way of the Danube; but the Iron Gates are badly managed and, there too, the guardian is Austria again, who imposes such taxes that the traffic is rendered impossible.[1] All the Balkan crises of the last fifteen years have turned partly on questions of routes, and Servian politics in particular must be looked at from that point of view. An attempt has been made to "bottle up" Servia: the great natural highway which links up the valley of the Morava with the Adriatic has been padlocked, and it is only a question of time before the Servian armies will put forth all their strength in order to attain the coast, and re-open this route.

Now what are the relations between these economic maladies and these far-reaching realities of geography? And what relation do the former, perceived as they are more or less dimly and felt more or less keenly, bear to the formulae which designate them in the often hazy, and even inconsistent, language of rulers and diplomats? That is a problem for historians to solve; but it is for us geographers to say to the historian, "But for the closing of the road through Bosnia-Herzegovina along which it was traditional for Servian live-stock to be driven to the sea, things would undoubtedly be happening differently to-day." Geography has its revenge on politics; it takes time because its life is longer; but it often has the last word.

Had I time to spare, I would go on (always, however, remaining on Servian soil) to the Sanjak of Novi Pazar, a mountainous district which, in relation to the above subject, has been treated as a buffer between Montenegro and Servia, and which remains what it is, *i.e.*, a sort of natural barrier. Germanism has had the desire to penetrate this district; has dreamed of continuing across this Sanjak the road from north to south which ought, in theory, to terminate at Salonica. But there is a problem to face here, how to turn a barrier into a route.[2] The Austrian staff have seriously considered geographical impossibilities, bearing in mind a war with Turkey, and I think I am in a position to state that the Austrian plans for invasion, from this point of view, would only allot a very small rôle to the Sanjak of Novi Pazar, even supposing the narrow-gauge line from Uvac to Mitrovica to have been completed; the chief military route would continue to be that of the great natural highway across Servia, Morava-Vardar.

The history of art ought likewise to take geography into account. Bosnia-Herzegovina here too affords a typical example of this kind of dependence on geography, a dependence which has its limits, but is nevertheless real. A wooden minaret cannot take the same shape as a stone minaret. The wooden one broadens out to a noticeable extent at the height of the balcony, inasmuch as the latter, being of wood, must be protected; and the top is capped by a wooden shelter whose supports rest on the balcony: whereas the slender-stone spire can rise higher without any break in its tapering outline, for the balcony half-way up seems merely a girdle.

---

[1] Cf. J. Cvijic, *L'Annexion de la Bosnie et la Question Serbe*, Paris, 1909, pp. 4-16.

[2] Cf. "Le Sandzak de Novi Pazar," by Gaston Gravier, in the *Annales de Géographie*, Jan. 1913, pp. 41-67.

Nor can social history, the history of all organisations of societies, ignore the essential facts of Human Geography. In the later lectures we shall inquire into that primitive organisation of property in Bosnia known as the "zadruga," and we shall see, further, how all modern agricultural progress has combined to dissipate the setting in which that ancient mode of collective ownership existed, and that the latter is consequently on the wane.

(To be continued.)

SCOTTISH PLACE NAMES: REVISION OF ORDNANCE SURVEY MAPS.

In connection with the revision of the names of places appearing on the Ordnance Survey maps, there was appointed in 1891 a small Committee, under the auspices of the Royal Scottish Geographical Society, whose function was to advise on the forms to be adopted for the names of Gaelic origin. This Committee, presided over by Dr. James Burgess, C.S.I., met and worked from 1891 to 1899, when it lapsed.

Early in this present year, on the suggestion of Colonel C. F. Close, R.E., Director-General of the Ordnance Survey, the Committee was revived and reconstituted with Dr. W. J. Watson as Chairman, and Mr. John Mathieson, F.R.S.G.S., as Honorary Secretary.

The Committee, having considered its method of procedure, has decided to invite the co-operation of all interested in the subject of Place-names. For map purposes the Celtic names fall into two classes—(1) names presented on the map in Gaelic spelling; (2) names which, while of Celtic origin, are given on the map in Anglicised form. The latter class is large, even in districts where traditional Gaelic forms of the names are well known and in common use, and includes practically all postal addresses. For example, the Lewis name which appears on maps as Ardroil is in Gaelic *Eadar-da-fhaodhail*, meaning "between two sea-fords." It is important to note that for commercial reasons names of this class cannot, *as a rule*, be altered on the map.

(1) The Committee in the first place desires to have its attention drawn to names of the former class that are given incorrectly on the map, *i.e.* where the map spelling does not accurately represent the local pronunciation in Gaelic. When on investigation the Committee is satisfied as to the true form it will advise the Ordnance Survey authorities.

(2) With regard to the important class of Anglicised names the Committee is of opinion that although, as a rule, the names must be regarded for ordinary map purposes as stereotyped, it is nevertheless of great importance for philological and historical purposes that the genuine traditional forms as pronounced in Gaelic should be ascertained and preserved, and information on this subject is therefore invited. It may still be possible, for instance, to recover the traditional pronunciation in Gaelic of such names as Forteviot, Renfrew, Birnam, as well as names in

the counties of Kincardine, Forfar, Fife, Aberdeen, Banff, Elgin, Nairn, Dumbarton, and Stirling, which counties, though they have ceased in whole or in part to be Gaelic-speaking, yet adjoin or contain districts where the language is still in use.

(3) In addition, information is invited as to names omitted from the map or given imperfectly thereon, whether names of natural features, of divisions of land now merged in large holdings, of wells, graveyards, and artificial structures, or of districts the ancient names of which still survive in Gaelic, though they do not appear on maps. As an example of the last-named may be taken *An Tòiseachd*, "the Thanedom" of Glen Lyon.

(4) With regard to the very large number of names of Norse origin occurring in or near Gaelic-speaking districts the Committee, recognising the importance for philological purposes of the forms these names have assumed in Gaelic, desires to obtain information here also as to the actual Gaelic pronunciation.

The gift or loan of papers and handbooks, guides, etc., embodying the results of investigations already made in any of the above departments will be welcomed.

In addition to communicating with the Ordnance Survey Department, the Committee have arranged for the publication of all decisions and authenticated information in the *Scottish Geographical Magazine*. Communications will be received by the Honorary Secretary, Mr. John Mathieson, addressed to the Royal Scottish Geographical Society, Synod Hall, Castle Terrace, Edinburgh, from whom also forms suited to the various departments of the inquiry may be obtained.

GEOGRAPHICAL NOTES.

AMERICA.

The Peninsula of Yucatan.—Mr. Ellsworth Huntington contributed to the *Bulletin* of the American Geographical Society for November last an exceedingly suggestive article with this title. The peninsula shows typical tropical conditions, and on a small scale illustrates most of the problems found in tropical countries. One special feature, discussed here already (vol. xxvi. p. 491), is the presence of limestone rock, giving rise to a characteristic karst type of country.

Taking the peninsula as a whole we find a northern belt clothed in scanty savanna forest (jungle), and a southern belt with true tropical forest. The difference Mr. Huntington believes is due not only to the marked difference in rainfall, which is scanty to the north and heavier to the south, but also to the severity of the seasonal drought in the north. The porous limestone of the north is also a contributing factor. The savanna region is well peopled, and prosperous, while the forest belt is scantily peopled by wild Indians. The difference, as in other parts of the tropics, seems to be due to the greater ease with which primitive agriculture can be carried on in the drier north. Here the well-marked dry season leads to a periodic check to plant growth, and permits of the burning off of the natural vegetation without difficulty. Maize is the staple crop, the grains being inserted into holes made with a pointed stick. Small amounts of beans and pumpkins are also grown, while the prosperity of the region depends upon the plantations of henequen or sisal hemp, exported largely to the wheat-growing regions of North America. The people, whether of Indian or Spanish blood, show a great lack of initiative and energy, but all, the Indians especially, are exceedingly clean and neat, the towns being stated to be models of cleanliness and order. The languor is apparently due to climatic causes, for it is noticeable that Spaniards who have recently entered the country are far more energetic than those who have been there for some generations. Further, it is commonly stated that labourers work harder than usual when the cool north wind blows, and Mr. Huntington believes that it is the uniformity of temperature throughout the year which has the debilitating effect. This is an interesting point in view of recent physiological observations on the lowering effect of unvarying temperatures in heated rooms and buildings. Another possible cause is malaria, which, especially in the forest region, is prevalent, and probably almost always affects the children in a mild form, which yet leaves permanent effects.

Mr. Huntington points out, further, that the present prosperity of Yucatan depends upon the demand, in the United States and elsewhere, for its henequen, a product which can be produced with relative ease. But the region formerly enjoyed a high degree of indigenous civilisation, the Maya Indians in the past having displayed great skill alike in obtaining water, in constructing temples and great buildings, and in cultivating the land. No reason has as yet been given to explain the development of such a civilisation in a region whose inhabitants are now so inert, and in a land which seems to lead to the degeneration of immigrant stocks.

Australasia.

Meeting of the Australian Association for the Advancement of Science.—This Association met at Melbourne in January last when several papers of geographical interest were read. Professor David discussed the Australian climate, with special reference to the meteorological observations made by the various Antarctic expeditions, notably those of Scott and Mawson. The observations obtained by wireless telegraphy from Macquarie Island, through members of Mawson's expeditions, show, he believes, that there is an intimate connection between the weather of Australia and that of the subantarctic region. It is hoped that an arrangement will be come to between Australia and New Zealand to insure the establishment of a permanent meteorological station on that island. Professor David believes that it is the existence of the lofty continent of Antarctica which produces the coldness of the Antarctic climate, and thus the marked contrast between south polar and equatorial temperatures. This again increases the rapidity of the air circulation in the southern hemisphere, and accounts for the periodic fierce outrushes of blizzard winds, which accompany the development of Antarctic low-pressures, and profoundly affect Australian weather.

Another interesting address was delivered by Professor Baldwin Spencer on the Northern Territory and its Aborigines. Professor Baldwin Spencer had just returned from a special mission to the region, lasting a year. The territory is four and a half times as large as Great Britain, but excluding aborigines the total population does not reach 4000, and the aborigines probably do not number more than 40,000. The climate is certainly trying, especially in the wet season, which lasts from March to September, but it is less trying in the interior than on the coastal belt, and the fact that there is a distinct lowering of temperature in the "winter" season, makes the region differ from other tropical countries. The fact that the natives are totally ignorant of agriculture makes the problem of dealing with them very difficult, but the Commonwealth Government, which now has control of the region, is endeavouring to develop the country, and to find solutions for the various problems involved.

Tasmania and Antarctica.—Professor T. W. Edgeworth Davis contributes to the May number of the *Geographical Journal* a note on certain soundings taken by Captain J. K. Davis of Dr. Mawson's Antarctic ship *Aurora* on his voyage southwards last November and December. These soundings prove the existence of a submarine bank lying to the south of Tasmania. For 100 miles south of the island the bottom deepens steadily to 2082 fathoms, then it begins to rise again to the crest of a long ridge, at least 150 miles in length. So far as is yet known the shallowest part of the ridge has a depth of 545 fathoms, but as the ocean in adjacent areas to the east and west sinks to depths of from 2450 to 2700 fathoms, this speaks to a rise of at least 11,000 feet from the ocean floor. The ridge seems to be at least 100 miles in width, and has a rocky bottom. Professor Davis believes that it is a fragment of a lost continent which once connected Tasmania and Antarctica, and compares the deep trench which separates it from Tasmania to the smaller trench which forms Bass strait, and separates Australia from Tasmania.

Another bank was discovered about 60 miles north of Macquarie Island, this bank rising from a depth of 1750 fathoms to within about 570 fathoms of the surface.

POLAR.

Captain Amundsen's North Polar Expedition. — Captain Amundsen has now obtained sufficient funds to insure the success of his proposed North Polar expedition. The Norwegian Geographical Society have contributed £8000, and the National Geographical Society of America, £4000. Captain Amundsen expects to start next spring from San Francisco or Seattle, and to be absent five years. He proposes to sail through Bering Strait, then allow the *Fram* to be frozen in the ice, and to drift across the Polar Basin to the Atlantic.

The American Crocker Land Expedition.—This expedition, which was postponed last year (cf. vol. xxviii. p. 479 *et antea*) on account of the death of the leader, has been reorganised and expects to start for the north during the coming month (July). Mr. Donald B. Macmillan is to be leader, Ensign Fitzhugh Green, U.S. Navy, is to have charge of the mapping, electrical work, terrestrial magnetism and seismology, while the other members of the party are to be Messrs. W. Elmer Ekblaw (geologist and ornithologist), and M. C. Tanquary (zoology, especially invertebrate zoology). The object of the expedition is the scientific exploration of the land supposed to lie north-west of the line of islands stretching from Grant Land to Prince Patrick Land. In addition to the mapping of the new land and of the uncharted coast lines in the vicinity of Grant Land and Axel Heiberg Land, the party will make investigations in the meteorology, terrestrial magnetism, seismology, geology, botany, zoology, ethnology, etc., of the regions visited.

General.

A Tropical University.—Some correspondence has taken place lately in the press on the advisability of establishing a University in the tropics to investigate the many problems which present themselves there, and also to serve as a training school for those who propose to spend a part of their lives in the tropics. Various localities have been suggested, but Mr. Kirkham, of Nairobi Government Laboratory, communicates to *Nature* for April 24, a letter in which he puts forward a strong case for Nairobi, and in the course of his letter sums up in an interesting fashion some of the characteristic features of British East Africa, a region with many peculiarities. He points out that the region is dissected by the equator, rises from sea-level to a plateau region with an average height exceeding 8000 feet, and includes a mountain of 17,000 feet in height, rising far above the snow line. The climatic conditions vary greatly: there are regions of heavy rainfall and almost rainless deserts; healthy districts and others rendered uninhabitable by deadly disease. Again the soils vary greatly, and the crops range from coffee, rubber, cocoanuts, to cotton, maize and wheat. The country contains a rich fauna and flora and a large native population, and is within easy reach of India. Nairobi itself has a healthy climate, is placed within a hundred miles of the equator, and is about equidistant by rail from the tropical coast belt and the Victoria Nyanzá and Uganda.

Origin of certain Submarine Valleys—Mr. Cyril Crossland contributes to *Nature* an interesting note on certain submarine valleys in the Red Sea and equatorial East Africa. He states that such harbours as Port Sudan, Suakin, the desert harbours of the Red Sea, no less than those of Mombasa, Kilindini, Tanga, Wasin, etc., are really the high parts of submarine fault valleys, and not, as might be supposed, drowned river valleys. More than this, the numerous fiords on the west coast of the island of Pemba are, he states, similarly rift valleys, for the island has undergone recent elevation, not depression. Off both Pemba and Zanzibar, also, there occur barrier coral reefs which owe their origin to abrasion alone, and not, as is usually assumed with barrier reefs, to subsidence. There is of course nothing intrinsically improbable in these statements, for if, as is admitted, faulting may produce steep-sided valleys on land, as in the case of the Great Rift valley of Africa, there is no reason why it should not also produce steep-sided submarine valleys, but the precise statements made are of considerable interest, and, if well founded, show the danger of assuming subsidence wherever submarine valleys or barrier reefs occur.

The Piltdown Skull.—In the February issue (p. 94) we noted the discovery of a new type of human skull at Piltdown, Sussex. A full report of the proceedings at the meeting of the Geological Society when the skull was described appears in Part I. of the current volume of the *Quarterly Journal* of that Society, and is accompanied by numerous illustrations. Those interested in the matter are referred to that journal for fuller details than can be given here. We may note, however, that Dr. Smith Woodward regards the skull as sufficiently peculiar to justify the erection of a new genus, and believes that it forms a link between the higher apes and man, for it presents certain features shown in the earlier stages of the growth of the skull of the living apes. It is probable that these features occurred in Mid-Tertiary apes, from which man and the living apes doubtless arose.

Geological Structure of Continental Shelf.—Under the rather startling title of "The Geology of the Sea-bottom," Dr. Lemoine contributes to the *Annales de Géographie* for November last an interesting summary of some recent researches on rocks dredged from the sea-bottom off the coast of Europe. These recent investigations have shown that rock specimens of cretaceous and eocene age lie over the floor of the English Channel in numbers sufficient to justify the conclusion that they are fragments of rocks *in situ*. The observations are of interest as they serve to confirm conclusions arrived at on theoretical grounds in regard to the probable structure of the Channel region. Off the western coastline of Europe specimens of eruptive rocks have been found in sufficient numbers, according to Dr. Lemoine, to lead to the conclusion that a line of fracture, marked by eruptive rocks, runs from the south of Portugal to Rockhall, and forms the western boundary of the continent of Europe. These observations, of course refer only to the Continental Shelf, regarded by all geologists as structurally a part of the continent of Europe. Dr. Lemoine, however, believes that it may not be impossible to obtain rock specimens from the great ocean depths, and thus investigate the geological structure of the sea-floor beneath the covering of ooze.

Commercial Geography.

Agricultural Developments in Uganda.—Mr. P. H. Lamb contributes an article with this title, based upon personal experience, to the *Bulletin* of the Imperial Institute. He points out that the cotton and most of the other products grown for export are cultivated in Uganda at an elevation of 3000 feet and over, in a plateau region close to the equator, with a rainfall of 40 to 60 inches per annum, and a remarkable equable temperature. The population comprises about 3,000,000 persons, and the Baganda, the principal tribe, are remarkably progressive, and take naturally to trade.

The cotton industry is especially interesting. The Agricultural Department has been making strenuous attempts to improve the yield, and found that occasional European inspection did little to eliminate native errors of cultivation. A number of native cotton instructors have therefore been trained and appointed, and, with the co-operation of the chiefs, these have done excellent work, and as their districts are small they are able to visit the various villages frequently and see that their instructions are carried out. Cotton cultivation is practically all in the hands of the natives, the crop being collected and purchased by small Indian traders, and ginned and exported by European firms. Seed can only be obtained through the Department, and the plant is grown as an annual in order to check insect pests, and prevent stained and dirty cotton being put upon the market. The crop is grown by the natives as a "money crop," to give them the means of buying wives and cattle, which to them constitute wealth. The crop is exceedingly popular with the natives, and its further development is only checked by difficulties of transport, the local demand being as yet unlimited. Ground-nuts and sesamum are also important crops, and are encouraged by the Department as yielding a concentrated food-supply, useful in case of famine or a sudden demand for extra labour for public works. Wheat is also grown on the foot-hills of Ruwenzori, and, though the yield is small, twelve bushels to the acre, it is believed that in certain regions the crop is capable of further development.

Elephant Grass and Paper-making.—A note in the *Bulletin* of the Imperial Institute, 1, suggests the possibility of the establishment of a new industry in Uganda, owing to the use of elephant grass for paper pulp making. Elephant grass (*Pennisetum purpureum*) is a member of the millet genus, and grows wild throughout a wide zone in tropical Africa, extending from 10° N. lat. to 9° S. lat. on the west coast, and from 9° to 20° S. lat. on the east coast. It occurs chiefly along water-courses and marshy depressions, and reaches a height of 6 to 10 feet, or occasionally even 20 feet. To agriculturists it is a great nuisance, in spite of the fact that it can be used as a fodder plant both for horses and cattle, owing to the rapid way in which it grows after the aërial shoots have been burnt or cut down. A bundle was recently sent to the Imperial Institute with a view to testing the grass's possibilities as a pulp-maker. The results were fairly satisfactory, but it is pointed out that the grass is too bulky in proportion to its probable market value to permit of profitable shipping to Europe. If, on the other hand, it could be converted into pulp by the soda process on the spot, it is possible that a remunerative industry might be carried on. In an article in the *Times* it is pointed out that there are immense deposits of soda in east Africa, so that there is a possibility that a pulp industry might be successfully started in Uganda.

EDUCATIONAL.

Vacation Courses at Hamburg.—We have received a programme giving full details of vacation courses to be held at Hamburg during the present summer from July 24 to August 6. The programme is a very elaborate one, and in addition to courses which touch more or less upon the province of geography, we notice the following which are of very special interest to geographers:—Professor E. V. Drygalski is to lecture on "Probleme der Eiszeit in den Polargebieten und in den Alpen"; Professor W. Branca on "Gegenwärtiger Stand, Aufgaben u. Ziele vulkanologischer Forschung; Dr. Ehrenbaum on "Die Internationale Meeresforschung"; Professor Meinardus on "Probleme des Kreislaufs des Wassers"; Professor Passarge on "Geomorphologische Probleme u. Streitfragen"; Professor Pfeffer on "Die grossen tiergeographischen Probleme"; Professor Schott on "Ergebnisse, Probleme u. zukünftige Aufgaben der Morphologie der Meeresräume, and so on.

In addition to the various courses of lectures, special practical courses in German for foreigners have been arranged, and will be held for three to four hours daily from June 16 to July 26. These will give students an opportunity of studying the language according to the most approved modern methods. The fee for the lecture courses is 25 marks, with a supplement of 50 marks for those taking the practical German course. Applications should be sent to Geschäftsstelle der Akademischen Ferienkurse zu Hamburg, Martinistrasse 52, Hamburg 20.

NEW BOOKS.

EUROPE.

Ski-Runs in the High Alps. By F. F. Roget, S.A.C., Honorary Member of the Alpine Ski Club, etc. With 25 Illustrations by L. M. Crisp, and 6 Maps. London and Leipzig: T. Fisher Unwin, 1913. *Price* 10*s.* 6*d. net.*

Let us say at once, before beginning to find fault, that we have read Professor Roget's book with pleasure, and have derived from it some instruction and much entertainment. We especially enjoyed the accounts of the ascents of the Finster-aarhorn, the Grand Combin, and of the "high-level route" to Zermatt. In pleasantly discursive fashion hints are given to beginners, the mechanics of ski-bindings are discussed, and a few notes are added on Swiss winter resorts. The illustrations are curiously unequal, and present the appearance of having been drawn from photographs. The maps are reproduced from those of the Swiss Topographical Survey, and are on such a small scale as to be practically worthless. There is no index, but the chapters have good detailed headings. We must emphatically protest against Professor Roget's pronunciation of the word "ski."

He uses no *s* in the plural, which is reasonable enough, but he insists on calling it "skee," on the ground that it is so written and also that it is good Norse. Now both in Norwegian and Swedish *k* is like *h* before the letter *i*. May we ask Professor Roget if he knows how the name of the Swedish explorer Nordenskiöld is pronounced?

The Story of Lucca. By JANET ROSS and NELLY ERICHSEN. Illustrated by NELLY ERICHSEN. London: J. M. Dent and Sons, Limited, 1912. *Price* 4*s*. 6*d*. *net*.

This is a further addition to the Mediæval Town Series, and will be quite as acceptable as any of its predecessors. Lucca, as the authors say, is too little known to the public. For one thing, it is slightly out of the beaten track down the coast from Genoa, and it is overshadowed by its old rival, Pisa. And yet it ought to have a claim on English-speaking travellers. Historically it has always had a connection with Great Britain, and many of our famous men of letters have resorted thither for the baths. Students of Dante cannot go far without hearing of the wars between Pisa and Lucca, the grim memory of Ugolino della Gherardesca, and the more pleasant reference to Nino Visconti. Lucca has been rather more famed for its religious tendencies than for its art, and yet there is much of the highest interest in the buildings and their associations. William the Conqueror, it is said, when he was planning his invasion of England, sent ambassadors to the Pope for his blessing, and it was at Lucca that they met him.

As has been said, the high standard of these publications and of the illustrations has been fully kept up.

The Cities of Lombardy. By EDWARD HUTTON. With 12 Illustrations by MAXWELL ARMFIELD. London: Methuen and Co., Limited. *Price* 6*s*.

In order rightly to understand the history of Italy itself it is necessary to give the Plains of Lombardy and their inhabitants their due measure of importance. Two points are brought out in the introductory chapter which should be borne in mind. The people inhabiting the valley of the Po are not, ethnologically speaking, Italian at all. Their country was known as Cisalpine Gaul; they were and are Gauls rather than Italian. These Gauls in the dawn of history poured into Italy by the various passes which cross the Alps. A brave and warlike race, they, having settled down, acted as a buffer state to protect Italy proper from foes from the north, and although wave after wave of invasion passed over them to Rome, the barbarian was finally driven out by another great Gaul, Charlemagne; and the people remain Gauls to this day, constituting probably the cream of the army. The other point on which the author enlarges is that it was when Julius Cæsar finally took Gaul, both Cisalpine and Transalpine, that the *Pax Romana* was established which made the civilisation of Europe, as we have it to-day, possible. The first four centuries of our era were the great and indestructible foundations of all that is worth having in the world. The descriptions of the various cities of Lombardy are as thoughtful as Mr. Hutton's always are. The illustrations are uncommonly good, and give an excellent idea of the colour of the country.

The Russian Year-Book for 1913. Compiled and edited by HOWARD P. KENNARD, M.D., assisted by NETTA PEACOCK. London: Eyre and Spottiswoode. *Price* 10*s*. 6*d*. *net*.

This edition takes a much more handy and compact form than last year's volume. The pages are only slightly fewer, but the paper is much thinner, and the book altogether much easier to work with. The general arrangement remains the same, but there are various minor additions, space being found *e.g.* by the omission of the Russian vocabulary (oddly enough, this remains in the index but not in the text), of the scheme for the transliteration of Russian, and so on. We think the author is right in these omissions and in the slight reduction of size, and hope that in future he will limit the contents strictly to those matters which are naturally looked for in a Year-Book and cannot easily be found elsewhere. The book is a wonderful storehouse of facts about an empire in regard to which details have hitherto not been easy to obtain. Among the additions to the 1913 edition are some instructive diagrams showing the distribution and quality of cereal crops in European Russia in 1912. To those teachers whose scholars still regard Siberia as an unpeopled waste we may recommend the study of the population statistics, which shows that its population exceeds that of Canada by one million.

Greece of the Twentieth Century. By PERCY F. MARTIN, F.R.G.S. London: T. Fisher Unwin, 1913. Price 15*s. net.*

Certainly the reader who peruses this book will endorse the encomium of the learned Dr. Andréadès who contributes the introduction, and will admire the patience and industry of Mr. Martin in its compilation. We trust we are not unjust to it when we describe it as a glorified Whittaker combined with a strong dash of Baedeker. It gives detailed information about the Royal Family of Greece, government, army, navy, finance, government offices, railways, agriculture, shipping, mining, etc., going into such minutiæ as the careers of every member of the Cabinet, the numbers and distribution of the officers and rank and file of the army, primary schools, bishops, monasteries, etc. In the chapters on railways we have details such as the gauge, bridges, tunnels, and rolling stock of each line. In domestic affairs we have the wages paid to men, women, and children in factories, the pay of domestic servants, their dress, and so on. Abundance of statistics are given as to population, economic productions, prices, imports, exports, etc. Even the prices of seats in the theatres, tram-cars, retail groceries, and refreshments are duly set down. In brief, the reader can inquire within this book on every subject, important and unimportant, connected with the Greece of to-day and be sure of finding an up-to-date answer.

ASIA.

Through Shên-Kan. By ROBERT STERLING CLARK and ARTHUR DE C. SOWERBY. Edited by Major C. H. CHEPWALL. London: T. Fisher Unwin, 1912. Price 25*s. net.*

This handsome and beautifully illustrated volume contains an interesting, if somewhat belated, account of an expedition through a part of the Chinese provinces of Shansi, Shensi, and Kansu in the years 1908-9. The expedition was projected by Mr. Clark of New York. He proposed to start from the terminus of the Pekin railway, T'ai-yüan Fu in Shansi, to traverse Shensi and Kansu, skirt the Tibetan border to Ch'êng-tu-Fu in Suchwan, descend the Min river to Sui-Fu, and return to Shanghai down the Yangtse river. In the course of the expedition it was intended to survey the route, and to take astronomical observations of latitude and longitude of the important towns as well as daily meteorological observations. Mr. Clark secured for the expedition a staff of competent workers, the most important of whom was Mr. Sowerby of the Smithsonian Museum, whose previous experience of China, and more especially of Shensi, marked him as peculiarly qualified for the work. The expedition started in September 1908 and proceeded due west to Yu-lin Fu by the Great Wall of China, thence south to Fu-Chou, and thence again west to Lan-Chou. Up to the middle of June 1909 the party had met with no misadventure, and the work was making satisfactory progress, but on the 21st of that month Hazrat Ali, the surveyor of the expedition, was treacherously murdered in the neighbourhood of Lan-Chou, under circumstances which are still more or less shrouded in mystery. The expedition was then summarily brought to a close, and the party returned to Pekin. The most valuable parts of the report are the chapters dealing with the biology, meteorology, and geology of the tract. There is also a valuable route map, which cannot fail to be of much service to future explorers. To many readers the most interesting feature of the book will be the many beautiful illustrations, some of them in colour. The story of the expedition is another illustration of the danger of travelling in Northern China without a sufficient escort.

A New Account of East India and Persia. By JOHN FRYER. Edited with Notes and an Introduction by WILLIAM CROOKE, B.A. Vol. II. London: Printed for the Hakluyt Society, 1912.

In the *Scottish Geographical Magazine*, vol. xxvi. p. 211, we gave some account of the life and career of John Fryer and of the first part of his travels in East India and Persia in 1672 to 1681. In the volume now before us we have a continuation of his narrative, dealing with the Carnatic in the days of Sivaji, the Mahratta freebooter who was a thorn in the side of the fanatic Moghal Emperor Aurangzeb, and with Persia, where Fryer travelled from Bandar Abbas to Ispahan, via Shiraz, and back. Not the least important chapter in the book is one which describes the coins, weights, and precious stones found in the places trading with the East India Company. Fryer is an acute and accurate observer and collector of information, and what he records has been confirmed by other travellers. As in the first volume, Mr. Crooke's editorial notes are copious and illuminating.

AFRICA.

The Gambia. By HENRY FENWICK REEVE. With Illustrations and Maps. London: Smith, Elder and Co., 1912. *Price* 10*s.* 6*d. net.*

This is a curious book, whose object is stated by Mr. Reeve in the Preface to be "to give his own people and Government a true perspective of the Colony of the Gambia, its value to the British Empire, and, above all, a clear view of their obligations as a Great Nation, towards those who have been our loyal fellow-subjects for more than three centuries." The author, whose love of metaphor runs away with him at times, "claims the indulgence of those who wield the facile pen and the purple pencil." His book is divided into three parts—(1) historical; (2) geographical, geological. and ethnical; (3) natural history. Under geology we learn that "the ocean is the source of the power that has in the past laid down the sedimentary series of rocks, and afterwards acted upon them with erosive force to form the valley of to-day and change it to its present conditions." Then we have a chapter headed "The Amphibians of the Gambia," which begins with the statement that "of the amphibians of the Gambia the hippopotamus takes the pride of place, as the largest and perhaps the most interesting." As the chapter goes on to describe the manatee and a porpoise, crocodiles, turtles, a "species of the otter tribe," and so on, it is evident that the author's statement that he speaks from a "popular point of view" is fully borne out.

AMERICA.

The Sea and the Jungle. By H. M. TOMLINSON. London: Duckworth and Co., 1912. Price 7*s.* 6*d. net.*

In December 1909 Mr. Tomlinson, who seems to have got tired, temporarily at least, of the conventionalities of home life, signed on as purser in a "tramp" steamer carrying coals from Swansea to Porto Velho on the Rio Madeira, a tributary of the Amazon, and in this well-written volume we have an account of the voyage, and of his companions at sea and at Porto Velho, where he made one or two excursions into the interior. Porto Velho is the spot on the Rio Madeira where supplies and materials are taken for the construction of the Madeira-Mamore railway, which is intended to open up Bolivia and provide an outlet for its products to the Atlantic via the Amazon. Two abortive attempts have already been made to construct this railway. In both cases the enterprise failed owing to the inability of the employees to resist the fatal malaria of the tropical forest in which they had to work. If we understand Mr. Tomlinson aright, it is to be feared that this third attempt will not be more successful. The interest of the book consists in the powerful description it gives of the scenery of the Amazon and its tributaries, and of the impenetrable forest on both sides of the rivers. There are also some cleverly written sketches of the strange types of men with whom Mr. Tomlinson chanced to come into contact here and there in the cruise, and we have vivid and thrilling descriptions of the terrors of the tropical forest, and especially of its insect life. Mr. Tomlinson has a keen eye for animal life, and has much to say on the beauty of the butterflies and some of the birds in these regions. On the return journey he left the "tramp" at Tampa, in Florida, and hurried back to England. We heartily commend this book to the perusal of our readers.

The True History of the Conquest of New Spain. By BERNAL DIAZ DEL CASTILLO. Edited and published in Mexico by GENARO GARCÍA. Translated into English with Introduction and Notes by ALFRED PERCIVAL MAUDSLAY M.A., D.Sc., London. Printed for the Hakluyt Society, 1912.

In this, the fourth volume of his *True History of the Conquest of New Spain*, Bernal Diaz brings his story down to the siege and capture of Mexico, and the subsequent proceedings of Cortes for the purpose of subduing and settling the adjacent territories. The artless but candid old soldier indulges in somewhat wearisome details, but his narrative throws valuable, although at times lurid, light on the doings of Cortes and his companions, and on the endless intrigues and counter-intrigues which prevailed at the court of the Spanish Emperor. Diaz acknowledges the valour and generalship of Cortes, but repeats with illustrations accusations against his personal capacity and neglect of the comrades who had contributed so much to his success. As in the former volumes, Mr. Maudslay's notes are valuable and illuminating.

GENERAL.

Commodore Sir John Hayes. By IDA LEE. London: Longmans, Green and Co., 1912. Price 7*s.* 6*d. net.*

Ida Lee has done good service in rescuing from the unmerited oblivion of the Bombay Secretariat the story of the life of Sir John Hayes. His career, especially in the early stages, was one which may well be the envy of every young officer in the British Navy to-day. Born in 1767, he joined the Bombay Marine at the age of thirteen, and served in the first war against Tippoo Sahib, the son of the redoubtable Haidar Ali, in 1783. After serving in various ships from Bussora, in the Persian Gulf, to Canton, he took part in the second and decisive war which annihilated the power of Tippoo Sahib. These were the palmy days when John Company had no objections to his servants doing a bit of trading on their own account; and Hayes, who had heard of a chance from a friend, John M'Cluer, formed a syndicate for an expedition to New Guinea, where M'Cluer had seen plenty round nutmegs, which then commanded high prices in various markets. The syndicate chartered two small ships, which started under Hayes' command in 1793. Contrary winds compelled them to go round by the west coast of Australia to Tasmania. In this course Hayes was unwittingly following the track of Bligh and the French Admiral D'Entrecasteaux, who was searching for traces of La Perouse. From Tasmania he sailed on to New Caledonia and the Louisiade Archipelago, and arrived at New Guinea, where he coolly took possession of the whole island in the name of King George III., built a fort, and settled a number of his followers so as to make the occupation effective. He then sailed for Calcutta, but had to make for Canton in order to escape from a French privateer. In Canton he sold his nutmegs to great advantage. On his return to Calcutta he reported his annexation of New Guinea, but to his great chagrin the Governor-General, Sir John Shore, repudiated his action, as New Guinea was too far away from India to be of much profit to the Company. Hayes rejoined the Indian Marine, and saw active service against the French and the pirates who infested the Gulf of Bombay and the Arabian seas. He also took an active part in the Dutch War, ended by the siege and capture of Ternate in 1801. Two years later he was put in command of a squadron to protect the Roads and Bay of Bengal. After a short furlough to England he returned to Calcutta as Deputy-Master Attendant of the port, but was sent with the expedition to Java in 1811, where he took part in the siege and capture of Batavia. After this he had a long spell of rest, during which as Master Attendant of the port of Calcutta he introduced many improvements in its administration. When war was declared with Burma in 1824 Hayes went with the expedition, and did good service at the siege and capture of Arakan. For these services he was knighted, and he continued to reside in Calcutta, where his health gradually gave way. In 1831, in quest of good health, he went on a voyage to New South Wales, but had to be landed on one of the Cocos Islands, where he died, finishing peacefully an adventurous, gallant, and useful career, fully justifying the eulogistic stanzas of Agnes Strickland, which appear for the first time in this volume. Ida Lee has done her part well. She has made a patient and conscientious study of all the sources of information available to her, and she uses her materials skilfully and judiciously, and with the generous enthusiasm and admiration which her hero justly evokes. This book will be welcomed certainly by every sailor, and also by all who are proud of the glories of the British Navy.

Lippincott's New Gazetteer: A Complete Pronouncing Gazetteer or Geographical Dictionary of the World. Edited by ANGELO HEILPRIN and LOUIS HEILPRIN. London: J. B. Lippincott Co., 1912. *Price £2, 2s. net.*

This new issue of a well-known work of reference differs from its predecessor of 1905 only in the addition of an appendix containing the results of the Thirteenth Census of the United States. We regret that space was not also found for the results of the 1911 census in Canada, which brought to light so many striking changes in the distribution of population.

Half-Hours in the Levant: Personal Impressions of Cities and Peoples of the Near East. By ARCHIBALD B. SPENS. London: Stanley Paul and Co., 1913. Price 1*s. net.*

This is a detailed diary of various pleasure cruises in the Mediterranean, illustrated by numerous photographs. It scarcely appears to us to have deserved publication.

EDUCATIONAL.

Ordnance Survey Maps: their Meaning and Use. By MARION I. NEWBIGIN, D.Sc. (Lond.). Edinburgh and London: W. and A. K. Johnston, 1912. Price 1*s. net.*

The object of this little book—the first of its kind published in this country—is to explain how the one-inch Ordnance Map can be used to the best advantage. The author shows how the geological features of a country can be discerned, such as, consequent, subsequent, and obsequent, rivers; the effects of ice, the erosion of corries, or cirques, and the general shaping of the land by the various agencies at work. How to measure the gradients from the contours is explained, and how the human geography of the country can be studied. For the purpose of illustration four sheets are selected in Scotland, and four in England. The Scottish sheets explained are—Haddington, Oban with Ben Cruachan, Balmoral district, and Ullapool district, in Ross-shire. No better selection could be made as typical examples of the scenery of Scotland. The English sheets are equally well chosen. It is regrettable that, owing to the cost, only one small diagram (Ben Cruachan) and one photograph are given. Let us hope that the Government departments concerned—the Ordnance Survey and the Geological Survey—will imitate the American and German departments, who, we understand, supply the plates for such publications.

The delineation of physical features on the one-inch O.S. map was observed by Sir A. Geikie, who says: "The varying forms of the surface are so faithfully delineated as frequently to indicate, to a trained observer, the nature of the rocks and the geological structure of the ground. The individual characteristics of schist, of granite, of quartz rock, of slate are often well depicted." A careful study of this little volume, together with the maps, will enable the student and the amateur to see, to some extent, what the trained geologist sees in these maps. The author complains that the O.S. maps are only contoured at 250 feet intervals above the 1000 foot line, but she also states that this is to be remedied. This is obviously a defect; for many of the minor peaks are not indicated except so far as the hachures suggest them. A glance at sheet 5 of Scotland, where the contours are given at 100 ft. intervals to the summits, will show the advantage of adopting a uniform interval. Attention is also called to the fact that while the contours for the sea-bed are shown, there is nothing to indicate the depths and contours of the inland lakes. This also leaves the map somewhat incomplete, especially for those who wish to use it for physical geography. For instance, to know that Loch Morar—the surface of which is only 30 ft. above sea-level—is 1017 ft. deep—deeper than any part of the sea within 100 miles, and that Lochs Ness and Lomond are respectively 751 and 623 ft. deep—both much below any portion of the adjacent sea, would greatly assist the student in constructing the original valleys.

During recent years, however, much has been done to make the maps more useful both to the scientist and the general user. They are published both in colour and engraved. The class of every road, the position of every post office and railway station are given. Woods are indicated, both coniferous and deciduous; ornamental, pasture, and rough pasture are clearly shown; and under the present policy of the Department the executive would doubtless be willing to consider any suggestions for the further improvement of these maps.

To many the chapters dealing with the human geography will be of most interest. The sparse population must not always be taken as due to the want of fertility of the land. In Scotland there is very little cultivation above the 800 ft. level, except such small patches as one meets with at Leadhills and Wanlockhead, 1300 ft., at Tomintoul, where there are crofts at 1400 ft., and a small patch at Sir John Stirling Maxwell's Lodge on Rannoch Moor, 1730 ft. There are, however, many fertile valleys in the Highlands of Scotland now devoted to deer, which supported a considerable population a hundred years ago.

The title "Ordnance Survey Maps" is slightly misleading, for the author deals only with the one-inch map; but a reference to the catalogue of the O.S. Maps will show that the Department publishes maps on about a dozen different scales, varying from 16 miles to an inch, to 10 ft. to a mile.

BOOKS RECEIVED.

The Battlefields of Scotland: Their Legend and Story. By T. C. F. BROTCHIE, F.S.A. (Scot.). With sixty drawings by the author. Demy 8vo. Pp. ix + 242. Edinburgh: T. C. and E. C. Jack, 1913. *Price 5s. net.*

The War of Quito: by Pedro de Cieza de Leon and Inca Documents. Translated and edited by Sir CLEMENTS R. MARKHAM, K.C.B. Demy 8vo. Pp. xii + 212. Cambridge University Press: John Clay, M.A., 1913.

The Life of a South African Tribe: II. The Psychic Life. By HENRI A. JUNOD. Demy 8vo. Pp. 574. London: Macmillan and Co. Ltd., 1913. *Price 15s. net.*

A Naturalist in Cannibal Land. By A. S. MEEK. Edited by FRANK FOX. With an Introduction by The Hon. WALTER ROTHSCHILD, and thirty-six Illustrations. Demy 8vo. Pp. xi + 238. London: T. Fisher Unwin, 1913. *Price 10s. 6d. net.*

Panama: The Creation, Destruction, and Resurrection. By PHILIPPE BUNAU-VARILLA. Royal 8vo. Pp. xvii + 568. London: Constable and Co. Ltd., 1913. *Price 12s. 6d. net.*

The Fringe of the East: A Journey Through Past and Present Provinces of Turkey. By HARRY CHARLES LUKACH. Demy 8vo. Pp. xiii + 273. London: Macmillan and Co. Ltd., 1913. *Price 12s. net.*

The Story of Belfast and its Surroundings. By MARY LOWRY. With over thirty Illustrations. Demy 8vo. Pp. 191. London: Headley Brothers, 1913. *Price 7s. 6d. net.*

Things as they are in Panama. By HARRY A. FRANCK. Illustrated. Demy 8vo. Pp. iv + 314. London: T. Fisher Unwin, 1913. *Price 7s. 6d. net.*

A Text-Book of Geography: Practical and Physical. By RONALD M. MUNRO, M.A. Crown 8vo. Pp. 480. Edinburgh: John Cormack, 1913. *Price 3s. 6d. net.*

Dent's Practical Notebooks of Regional Geography. By HORACE PIGGOTT, M.A., Ph.D., and ROBERT J. FINCH, F.R.G.S. Book II. *Asia.* Crown 4to. Pp. 64. Book III. *Africa.* Crown 4to. Pp. 48. London: J. M. Dent and Sons, Ltd., 1913. *Price 6d. net each.*

Handbook of Baltic and White Sea Loading Ports, including Denmark. By

J. F. MYHRE. Revised edition. Pp. 576. Demy 8vo. London: Rider and Son, Ltd., 1913. *Price* 21*s. net.*

Rouen: Étude d'une agglomération urbaine. Par J. LEVAINVILLE. Avec 24 figures dans le texte, 1 carte, 1 plan de Rouen. Demy 8vo. Pp. 418. Paris: Librairie Armand Colin, 1913. *Prix* 7 *fr.* 50.

Svenska Turist-Foreningens Arsskrift 1913. Med Trehundratjugoåtta Illustrationer Och Fyra Kartskisser. Demy 8vo. Pp. viii + 404. Stockholm: Wahlström and Widstrand, 1913. Pr. 4 *Kr.*

Burma Under British Rule. By JOSEPH DAUTREMER. Translated, and with an introduction, by Sir GEORGE SCOTT, K.C.I.E. With 24 Illustrations. Demy 8vo. Pp. 391. London: T. Fisher Unwin, 1913. *Price* 15*s. net.*

A Commercial Geography of the World. By FREDERICK MORT, M.A., B.Sc., F.G.S. With numerous Maps and Diagrams. Crown 8vo. Pp. viii + 392. Edinburgh: Oliver and Boyd, 1913. *Price* 2*s.* 6*d.*

L'Eritrea economica. By FERDINANDO MARTINI and others. Illustrated. Royal 8vo. Pp. xiv + 542. Novara—Roma: Istituto Geografico de Agostini, 1913. *Prezzo L.* 16.

A Turkish Woman's European Impressions. By ZEYNEB HANOUM. Edited and with an Introduction by GRACE ELLISON. Illustrated. Crown 8vo. Pp. xx + 246. London: Seeley, Service and Co., Ltd., 1913. *Price* 6*s. net.*

Naples and Environs: Mount Vesuvius, Pompeii, Sorrento, Amalfi, Ischia, and Capri. A Practical Guide (Grieben's Guide Books). Vol. 166. With 5 Maps and 3 Ground Plans. London: Williams and Norgate, 1913. *Price* 1*s.* 6*d. net.*

Athènes (Les Villes d'Art célèbres). Par GUSTAV FOUGÈRES. Crown 4to. Pp. 204. Paris: Librairie Renouard, 1912.

The Highways and Byways of England: Their History and Romance. By T. W. WILKINSON. Crown 8vo. Pp. xxiii + 270. London: Iliffe and Sons, Ltd., 1913. *Price* 4*s.* 6*d. net.*

The British Empire with its world setting. By J. B. REYNOLDS, B.A. Small Crown 8vo. Pp. viii + 200. London: A. and C. Black, 1913. *Price* 1*s.* 4*d.*

Spain and Portugal. By KARL BAEDEKER. With 20 Maps and 59 Plans. Fourth Edition. Leipzig: Karl Baedeker, 1913. *Price* 15*s. net.*

The Statesman's Year-Book: Statistical and Historical Annual of the States of the World, for the Year 1913. Edited by J. SCOTT KELTIE, LL.D. Fiftieth Annual Publication. Revised after Official Returns. Crown 8vo. Pp. xcvi + 1452. London: Macmillan and Co. Ltd., 1913. *Price* 10*s.* 6*d. net.*

Walch's Tasmanian Almanac for 1913. With Map of Tasmania, Fifty-first Year of Publication. Crown 8vo. Pp. 432. J. Walch and Son, Ltd.

Victorian Year-Book for 1911-12. By A. M. LAUGHTON, F.I.A., F.F.A., F.S.S. Thirty-second Issue. Melbourne, 1913.

Trade and Customs and Excise Revenue of the Commonwealth of Australia for the Year 1911. Melbourne, 1912.

Australia: Its Land, Conditions and Prospects. The Observations and Experiences of the Scottish Agricultural Commission of 1910-11. With numerous Illustrations. Edinburgh: Blackwood and Sons, 1911. *Price* 1*s. net.*

Australia To-Day, 1913. Illustrated. Melbourne, 1913. *Price* 1*s.* 6*d. net.*

Statistics of the State of Tasmania for the Year 1910-11. Hobart, 1911.

Tasmania: The Island State of the Commonwealth. Hobart, 1911.

Handbook of Tasmania: Orcharding, Dairy, Poultry, and General Farming, Information for Immigrants, etc. Hobart, 1912.

Tasmania: Crown Lands Guide, 1912. Hobart, 1912.

Tasmania's Wonderland. Eighty-one Views. Sydney, 1913.

Census of India, 1911. Vol. III., *Assam* ; Vol. XXII., *Rajputana-Merwara* ; Vol. XV., *United Provinces of Agra and Oudh.* Calcutta, 1912.

The Gold of the Klondike. By J. B. TYRRELL, F.G.S. Ottawa, 1912.

The Pocket Queensland, containing General Information regarding the Great North-Eastern State. Revised Edition. Brisbane, 1912.

The Year Book of Queensland, 1913. Brisbane, 1913.

The Official Year-Book of New South Wales, 1911. Sydney, 1912.

The Britannica Year-Book, 1913: A Survey of the World's Progress since the Completion in 1910 of the Encyclopædia Britannica. Eleventh Edition. Edited by HUGH CHISHOLM, M.A. Oxon. Demy 8vo. Pp. xliii+1226. London: The Encyclopædia Britannica Company, Ltd., 1913. *Price* 10*s*.

South Australia: Handbook of Information. By V. H. RYAN. Demy 8vo. Pp. 160. Adelaide: R. E. E. Rogers, 1913.

The Year-Book of South Australia, 1912. Demy 8vo. Pp. 192. London, 1913.

Statistical Register of the State of South Australia. Adelaide, 1911.

Catalogue of the Hocken Library, Dunedin, New Zealand. Compiled under instructions from the Hocken Trustees by the Librarian, W. H. TRIMBLE. With a Preface by WM. DOWNIE STEWART. Dunedin, 1912.

Diplomatic and Consular Reports.—Trade, etc. of Shasi, 1912 (5050); Hawaii, 1911-12 (5047); Bahrein Islands (5056); Iquitos, 1912 (5054); Mosul, 1912 (5055); District of Patras, 1912 (5065); District of Colima, 1912 (5064); Cape Verde Islands, 1912 (5063); Kiukiang, 1912 (5062); District of Kuingchow, 1912 (5061); St. Pierre and Miquelon, 1912 (5060); Pakhoi, 1912 (5059); Port of Antwerp, 1912 (5056).

Publishers forwarding books for review will greatly oblige by marking the price *in clear figures, especially in the case of foreign books.*

NEW MAPS.

EUROPE.

ORDNANCE SURVEY OF SCOTLAND.—The following publications were issued from 1st to 31st March 1913:—

One-inch Map. Third edition; printed in colours and folded in cover, or flat in sheets. Sheets 84, 94, 102, 103, 109, 110. Price—on paper, 1s. 6d.; mounted on linen, 2s.; mounted in sections, 2s. 6d. each.

Six-inch and Larger Scale Maps.—Six-inch Maps (Revised). Quarter Sheets, with contours in blue. Price 1s. each. *Lanarkshire.*—25 SW.; 36 SW.; 51 SW.

1 : 2500 Scale Maps (Revised), with houses stippled and with areas. Price 3s. each. *Lanarkshire.*—Sheets V. 4, 8; VI. 1; VII. 12; IX. 7. *Renfrewshire.*—Sheets II. 11; VIII. 11, 12; XII. 4; XVI. 3, 4; XVII. 5, 6.

Note.—There are no coloured editions of these Sheets, and the unrevised impressions have been withdrawn from sale.

Special Enlargements from 1 : 2500 to 1 : 1250 Scale. Prepared for the Land Valuation Department, Inland Revenue. Price 2s. 6d. each. *Dumbartonshire.*—Sheets XVIII. 10 SW.; XXV. 6 NE.; 7 NE. *Fifeshire.*—Sheets I. 14 SE.; IV. 2 NW., NE.; XX. 9 NW.; XXI. 4 NW.; 14 NW., NE.; XXVI. 12 SW.; XXXIV. 15 NW., NE., SW., SE.; XXXV. 7 NE.; 10 SW.; 16 NE.; XXXVI. 2 NE.; 5

NW., NE., SW., SE.; xxxix. 4 NW.; xl. 7 SW. *Fifeshire and Kinross-shire.*—Sheet xxvi. 12 SE. *Lanarkshire.*—Sheets xlix. 3 NE., SE.; 7 NE., SE.; 8 NW. *Renfrewshire.*—Sheets ii. 10 NE.; viii. 14 NE., SW., SE.; xii. 5 NW., NE.; xiii. 10 NW., NE., SE.; 11 SW. *Selkirkshire.*—Sheet iv. 13 SE. *Stirlingshire.*—Sheets x. 11 SW., SE.; 15 SW., SE.; xxviii. 9 NW., SW.

The following publications were issued from 1st to 30th April 1913 :—

Six-inch and Larger Scale Maps.—1 : 2500 Scale Maps (Revised), with houses stippled and with areas. Price 3s. each. *Lanarkshire.*—Sheets vii. 2, 3, 4, 14; ix. 6, 9, 10, 11, 13; xi. 6; xvi. 3, 4, 8, 12. *Renfrewshire.*—Sheets viii. (7 and 3), 8, 14; xi. 4, 8; xii. 2, 3.

Note.—There are no coloured editions of these Sheets, and the unrevised impressions have been withdrawn from sale.

Special Enlargements from 1 : 2500 to 1 : 1250 scale. Prepared for the Land Valuation Department, Inland Revenue. Price 2s. 6d. each. *Renfrewshire.*—Sheets xiii. 6 SW., 10 SW., 11 NW. *Roxburghshire.*—Sheet x. 5 NW. *Selkirkshire.*—Sheet iv. 13 SW. *Stirlingshire.*—Sheets xviii. 5 NE., SE.; 13 NW.; 16 SW.; xxix. 7 NW., NE. *Stirlingshire with Dumbartonshire* (Det.).—Sheets xxix. 7 SW., SE.

GEOLOGICAL SURVEY OF SCOTLAND.—The following publications were issued from 1st to 31st March 1913 :—

One-inch Map. New Series, colour printed; solid and drift editions. Sheet 64. Price 2s. 6d. net. The Solid is shown in colour. The Drift by symbols in black.

Memoirs of the Geological Survey.—The Geology of Upper Strathspey, Gaick, and the Forest of Atholl; Explanation of Sheet 64. By George Barrow, F.G.S., Lionel Hinxman, B.A., F.R.S.E., and E. H. Cunningham Craig, B.A.; with contributions by H. Kynaston, B.A. Price 2s.

ADMIRALTY CHART, SCOTLAND.—River Forth, Port Edgar to Carron River. Scale 1 : 18,260. New edition, March 1913. Number 114c. Price 4s.
Admiralty Office, London.

EDINBURGH DISTRICT.—Bartholomew's Motoring and Cycling Map. Reduced from Ordnance Survey to scale of 2 miles to an inch. Showing roads and distances. New edition, 1913. Price 1s. Mounted on cloth in case.
John Bartholomew and Co., Edinburgh.

A new edition with improved contour colouring and revised roads.

AFRICA.

CAPE OF GOOD HOPE.—Scale 1 : 250,000, or about 4 miles to an inch. Sheets—Fraserburg, Kenhardt, Schnit Drift, Van Wyk's Vlei. 1913. Price 1s. 6d. net, each sheet.
Geographical Section, General Staff, London.

EAST AFRICA PROTECTORATE.—Scale 1 : 250,000, or about 4 miles to an inch. Sheets—Mumoni (Provisional), Meru (Provisional). 1912. Scale 1 : 125,000, or about 2 miles to an inch. Sheet—Nyeri. 1913. Price 1s. 6d. net, each sheet. *Geographical Section, General Staff, London.*

AMERICA.

ONTARIO.—Topographic Map. Scale 1 : 63,360, or 1 mile to an inch. Sheets—32, Tillsonburg; 34, Toronto; 35, Brampton; 39, Bothwell; 40, Sarnia;

42, St. Clair Flats; 43, Ridgetown; 47, Belle River. (Department of Militia and Defence, 1910-12.) Price 2s. net each sheet.

Geographical Section, General Staff, London.

One is glad to see that the publication of this most valuable and beautiful map is now making more rapid progress.

NEW BRUNSWICK.—Scale 1 : 500,000, or about 8 miles to an inch. 1912. J. E. Chalifour, Chief Geographer. *Department of the Interior, Ottawa.*

MANITOBA.—Map showing Disposition of Lands. 1913. Scale 1 : 792,000, or 12½ miles to an inch. J. E. Chalifour, Chief Geographer.

SASKATCHEWAN.—Map showing Disposition of Lands. 1913. Scale 1 : 792,000, or 12½ miles to an inch. J. E. Chalifour, Chief Geographer.

Department of the Interior, Ottawa.

These interesting maps show the extent of lands already settled, also the lands applied for and available for settlers.

MANITOBA, SASKATCHEWAN AND ALBERTA.—Map showing Elevators, for storage of corn, situated on various railways. 1913. Scale 25 miles to an inch.

Department of the Interior, Ottawa.

BARBADOS.—Island of, on scale of 1 inch to a mile—coloured to show parishes. Price 3s. Mounted on cloth in case.

G. W. Bacon and Co., Ltd., London.

PANAMA CANAL.—*Daily Mail* Chart. School edition, with notes for teachers. Price 3s. net. Mounted on cloth.

George Philip and Son, Ltd., London.

This map has already been noticed. The present edition is mounted for school use.

PANAMA CANAL.—Aeronautical View with Statistics.

F. D. Graves, New York.

A somewhat unsuccessful attempt at a pictorial map of the canal.

NEW ATLASES, ETC.

CALENDARIO-ATLANTE DE AGOSTINI.—Anno x., 1913. Con note geografico-statische del Prof. Dr. A. Machetto. Price, lire 1,00.

Istituto Geografico de Agostini, Novara.

The 1913 edition of this neat little atlas contains two new maps and amplified text, with calendar and statistical tables revised to date.

ATLANTE GEOGRAFICO MUTO.—Fisico politico a colori ed albo di esercitazioni cartografiche. G. de Agostini.

Fascicolo Primo—Tavole 17. Price, lire 1,50.

Fascicolo Secondo—Tavole 11. Price, lire, 1,00.

Istituto Geografico de Agostini, Novara.

These maps are beautifully executed and admirably adapted for their purpose.

WIRELESS MAP OF THE WORLD.—Compiled by Marconi's Wireless Telegraph Company, Limited. 1913. Price 2s. 6d. net.

George Philip and Son, Ltd., London.

Shows the "Wireless" Telegraph Stations, existing and projected, all over the world.

THE SCOTTISH GEOGRAPHICAL MAGAZINE.

WANDERINGS OF A NATURALIST IN TIBET AND WESTERN CHINA.

By F. KINGDON WARD, F.R.G.S.[1]

(*With Illustrations.*)

Burma.—We will pick up the thread of travel at the little village of Bhamo in Upper Burma, one of those outposts of Empire where, in the market, one may jostle with the men of twenty or thirty races, and nearly as many creeds.

It is winter, the dry season in the monsoon area, and through the blur of hot air which blankets the Irrawaddy valley, the distant hills are seen indistinctly; but presently we are trotting down the long white road that leads to China, a road associated with such well-known names as Margary, Baber and Morrison. Soon we are up in the hills, the sound of the torrent floating up faintly from the gorge far below, the deep silence of the forest broken only by the jingle of the mule bells as the caravan swings along at a steady pace, or by the cry of the wild-fowl. There were monkeys to watch by day, and at dusk we listened for the barking deer, or the rustle of the sambur in the marshes. Then out of the dark forest to the bare hills on the threshold of the Yunnan plateau; now we were on the frontier of two Empires; behind us lay the golden land of Burma, before us, stretching its vast plains and mountain ranges far out towards the rising sun, the magic land of China.

Eight stages from Bhamo we reached the city of T'eng-yueh, where I spent ten days, receiving invaluable advice and help from the late consul there, Mr. Archibald Rose.

Yunnan Plateau.—And now onwards we went over the well-forested

[1] An Address delivered before the Royal Scottish Geographical Society, in Edinburgh, on February 14, 1913.

plateau, with abrupt descents into the deep gutters which have scored the face of south-western China from north to south, where flow the Shweli, Salween, and Mekong rivers, crossed by chain suspension bridges which sway from side to side as the mules are led across one at a time. The plateau is already ablaze with crimson rhododendrons and pink camellias; in the thickets lurk strange pheasants of gorgeous plumage, and it is delightful to leave the caravan and wander all day over the wild hills, gun in hand, picking up the road again towards evening.

At the city of Tali, twelve days' journey over the plateau, once the headquarters of Mohammedan power in Western China, another halt is called.

The Yangtse.—From Tali we turn northwards, soon striking the

Fig. 1.—Lake on the Mekong-Yangtse watershed, 16,000 ft.

great Yangtse River at the end of its long southward journey, just as it swings away to the east to flow for over two thousand miles across China. Thus we come to the edge of that marvellous region of parallel rivers which, rising in Tibet and cutting their way due south, are hemmed in between China on the one hand and Upper Burma and Assam on the other, leaving between them towering walls of rock which, covered with the most wonderful alpine flowers, may be called the garden of Asia.

A few days later we turned back westwards, recrossing the Yangtse-Mekong watershed at an altitude of 12,000 feet. On the way over the mountains I left the caravan for an hour in order to shoot pheasants, and lost myself for two days and a night in consequence—an unpleasant experience, for the weather was cold and wet, and I had to subsist on what green leaves I could find; they gave me dreadful indigestion. The only pheasant I saw was late on the second day, when I was too weary to put up my gun. I shot an innocent little finch instead and ate that.

We now descended to the city of Wei-hsi, situated 8000 feet above sea-level, the last Chinese city we were destined to see for some time.

Numerous Tibetan and Moso traders came in to the markets, and there is a lively bustle about the place. During a three days' rest I made friends with the military mandarin, who sent me a present of bacon and chickens, to which I retaliated with a box of scented soap. Now I began to experience the difficulties of photographing tribesmen, most of whom ran away when I pointed a camera at them; but the Tibetans, reckless fellows, would always stand up to it for the sheer

FIG. 2.—An oasis in the arid region of the Mekong valley, with terraced wheat-fields.

adventure of the thing. On April 25 we continued down to the Mekong.

The Mekong River.—We followed the Mekong northwards for eight days in glorious weather, the valley, at first clad with roses and emerald-green rice fields, and flanked by snow-clad mountains, presently giving place to the barren gorges of the arid regions.

Since leaving Wei-hsi we had been amongst the Moso, a curious folk, whose women are extraordinarily handsome and dress with the most becoming taste, having anticipated their European sisters in the discovery of the box-pleated skirt for instance; but now in the arid region we found only Tibetans. We were now right in the middle of the area where the Salween, Mekong and Yangtse rivers flow for 200 miles within fifty miles of each other. The deep gorges of the Mekong soon grow very dry and lifeless, small scattered villages standing out as oases in the wilderness of rock. The river is quite unnavigable, and is crossed by means of bamboo ropes sloping steeply from one bank to the other, from which the victim is suspended by leather thongs running on top of the rope. Thus there is a pair of ropes at each crossing, one sloping in either direction.

A-tun-tsi.—A-tun-tsi, which I selected as a base camp for six months' rambling, is a small Chino-Tibetan trading village, through which many of the lama caravans pass on their way to Lhasa. The houses are of mud, two stories high, with flat roofs; in the foreground are the ruins of the monastery destroyed by the Chinese troops in 1905.

The shops and inns are kept by Chinamen; the Tibetans till the slopes, and throughout the brief summer tend their flocks of sheep and yak on the high alpine pastures. Excellent milk and butter can be bought in the village, besides mutton, bacon, chickens, eggs, and barley flour. In addition, the Mohammedans sometimes kill a yak or a cow, so it will readily be understood that the village serves excellently as a base; moreover, rice, sugar, and other luxuries are imported, and though expensive, can generally be obtained. It is advisable to draw the milk into one's own utensils, as otherwise, owing to the dirty condition of Tibetan milk pails, it is miraculously transformed into a sort of sour cream cheese within half an hour; this is not unpalatable, but is limited in its application to cookery. Tibetan butter, too, suffers from being made by the simple expedient of kicking milk about in a skin bag, with the hair inside. However, these are minor details.

The village is 11,500 feet above sea-level, situated on a spur of the Mekong-Yangtse divide, on to which I could step from the roof of my house. The Tibetans I found cheerful and resourceful people to work with; they can stand anything, except soap and water, which they never make the acquaintance of, though this is chiefly a matter of climate. It is better to be dirty than cold, as I found when winter set in.

Journey to the Salween.—And now we will pass on to my first journey from A-tun-tsi westwards over the Mekong-Salween divide to the Lutzu country. It was impossible to get permission from the Chinese authorities to go into Tibet, so I collected men at a quiet village two days' journey distant, and went without.

It rained most of the time, and indeed my guide said it was always raining in the Salween valley. This guide of mine was a most intelligent fellow, who, in addition to Tibetan, spoke Chinese, Lutzu, and other tribal languages; he brightened the journey by pointing out things of interest—plants eaten by the Tibetans, or used as medicines; by collecting jungle produce, including toadstools, bamboo shoots, and fried mice for my supper; and by singing songs. I had also a Tibetan dog to guard my tent at night—a fierce furry animal who became much attached to me. In the night I would sometimes wake suddenly to hear him barking furiously as some denizen of the forest prowled by.

After crossing two passes about 14,000 feet high we had a view of the snow-clad Salween-Irrawaddy divide in the west, and a vast extent of unexplored country spread out in front of us, crossed by Prince Henry of Orleans in the south, and by Captain Bailey in the north. On the tenth day we at last saw the Salween far below us, and descended to a Lutzu village. The Lutzu are a quiet agricultural people, very friendly, rather good-looking, and nicely dressed; they use the cross-bow with poisoned arrows, and drink vast quantities of a thick soupy liquid made from fermented maize. I was besieged for foreign medicine, and treated half the people in the village for simple maladies, but I did not wait to see the results. The houses consist of a single room, often built on piles, the space underneath being used as a cattle byre; on one occasion we slept fifteen in a room, my own party of nine and the original occupants, not counting various domestic animals.

Next day we crossed the river in dug-outs, and continued up the left bank to T'sam-pu-tong, from where a road is said to strike westwards to Assam. Here the Chinese official requested me to return to A-tun-tsi,

FIG. 3.—Clouds rising up from the Salween valley. The view is taken at the tree-limit, about 13,000 ft.

and gave me two soldiers as escort, but we lost the soldiers next day, and continued up the Salween, the rainy jungle region suddenly giving place to terrific arid gorges.

It was a memorable sight to see that great river smashing its way through the mountains, the endless gorges growing more and more lifeless as we continued northwards. A scorching wind raged through them by day; the glare from the hot pale-coloured rocks grew intolerable; vast screes, sometimes smoking with the dust of falling rocks, rose on either hand; and there was always the booming of the restless river. Then came a voyage in a dug-out, there being fourteen of us on board, squatting on our haunches in single file, besides the dog and luggage. Our gunwale was often awash in the rough water, and I thought we must capsize, but the tribesmen brought us through safely. They had changed a good deal in the last day or two, being now much dirtier.

They never wash, but are eminently hospitable. The children, when they dress at all, wear the skins of goats inside out; but the adults wear hempen dressing-gowns, like the Tibetans.

Finally we got amongst the Tibetans again, and reached Menkong, the capital of Tsa-ring, a few days after Captain Bailey and Mr. Edgar. The former had started for India only three days before our arrival, and of his subsequent remarkable journey to India you have all heard.

The return trip was made over the sacred mountain of Doker-la in appalling weather. When at last we reached the Mekong, we learnt that on the previous day two soldiers had arrived in search of us with orders to bring us back, and as soon as we arrived at A-tun-tsi the official came round to see me in genuine alarm, saying that I might have been killed!

Journey to Batang.—A month later I reaped the whirlwind unconsciously raised by Captain Bailey's exploit, for a whisper sprang up that the English had taken Tibet, an officer having gone in on secret service from China; and I was advised to fly. I heard the news one evening at the end of July, and early next morning we were on the road again, not south to Burma, but northwards along the Tibetan frontier, the journey to Batang, by a great effort, including two days of twelve hours in the saddle, being accomplished in six days. However, at Batang we could hear nothing of the startling rumour which had sent us post-haste to the telegraph office, and after a few days' rest I obtained permission from the official, through the good offices of Mr. Edgar, an English missionary who had accompanied Captain Bailey to Meukong, to travel westwards into Tibet by the Lhasa road, provided I signed a paper absolving the Chinese Government from all responsibility.

Next day we started, voyaging twenty miles down the Yangtse in a light skin coracle, made of three ox-hides sewn together. There were five of us and the luggage on board, and when we stepped ashore the navigator picked up our craft, put it on his head like a gigantic hat, and set off home with it. But if you really want some fun in a Tibetan coracle, you must get a native to take you through bad water when he is thoroughly drunk; he fears nothing then, and if you live to tell the story you will have something to boast about all your life.

The Tibetan Plateau.—We were soon up on the great grassland plateau, where it rains incessantly for six months in the summer, and we reached Gartok without incident, only to learn that the Chinese and Tibetans were fighting further north. We therefore left the Lhasa road, and continued westwards as far as the village of Samba-dhuka.

Never shall I forget that evening after four days of rain and cold on the plateau. It was dusk when we climbed the last spur, and looked away down into a vast trench from which floated up, like a far away song, a murmur of raging water; it was the Mekong thudding southwards through the gorges. The Mekong is undoubtedly the smallest of the three rivers, an exaggerated mountain torrent; yet, when at sunset I saw that streak of light pouring out of the deep sword-cut with which it has gashed the face of south-eastern Tibet, I thought it the grandest river of them all.

At this time I had with me a Tibetan soldier as escort, a most bellicose fellow who wanted to pound everybody with brickbats in his zeal to make them accomplish my will at a word. He nearly got me

FIG. 4.—View on the Tibetan plateau at about 13,000 ft.

into trouble at one village by laying out a man with a rock as big as a football; his zeal outran his discretion, and my patience.

The return to A-tun-tsi was made by a different route over the plateau to Y'a-k'a-lo, where we found more trouble, for the official had been roundly abused by the Viceroy for not stopping Captain Bailey, and the sudden unauthorised advent of another European out of Tibet only added to his concern. However, he allowed me to proceed by the small road on condition that I signed a paper saying that he was not responsible for consequences, which I gladly did, not being able to read a word of Chinese! There were no consequences except a midnight assault on a Tibetan house in the course of which I was nearly eaten by a large watch-dog, and we reached A-tun-tsi four days later.

Journey to the Yangtse.—Except for a week's journey to the Yangtse and back, *via* the monastery of Tung-chu-ling, September was spent in A-tun-tsi, and the time passed pleasantly enough. We climbed great precipices in search of plants, the nicest of which always select the most abominable situations, and one day in a mist I was pursued by a bull yak, and on another occasion in the forest I found myself face to face with a black bear. Then we had a wedding, and the whole village got gloriously drunk; also a funeral, and they got drunker.

With such a mixed population there was no lack of festivals, which

FIG. 5.—The home of *Meconopsis speciosa*, at a height of 17,000 ft. on the Mekong-Yangtse watershed.

occurred about once a month. They were like a combination of May Day and Guy Fawkes' Day—that is to say, there were flowers and bonfires and noise and things to eat. Tibetan children are keen on dancing, and when their annual picnic came round, the girls danced all the evening in the mule square, playing a game which may be described as a parody on "Here we go gathering nuts and may," though the words, which I did not understand, were probably less ingenuous than in that innocent ditty.

In October we journeyed north-east to the Yangtse, crossing the Run-tsi-la, a pass 18,000 feet above sea-level. This trip was marked by a fine eclipse of the sun, which, however, impressed the Tibetans far less than it did me; they treated it as an almost daily occurrence, though it was dusk at eleven o'clock in the morning.

Journey to the Salween.—When back in A-tun-tsi again at the end of October we heard rumours of the revolution, and the missionaries from Batang passed through on their way south. Living in England, where one gets letters and telegrams several times a day from all over the world, it is difficult to realise the sense of isolation in such a country, where one has no idea of what is going on fifty miles away. I received letters about once in six weeks, but, for the time being, the hustle of the west seemed such a small thing in the world of nature that it had little interest for me. We held a council of war, but nothing could be done while we had no news, and one by one the missionaries left for Burma.

On November 1 our work was finished and we set out for the south, having been warned not to go into Tibet again, though before leaving I presented the official, who throughout had treated me courteously, with a silver-topped glass bottle. Two days south of the village we bribed one of the soldiers to come with us as interpreter, and dashed across the mountains again to the Salween. However, things had to be done hurriedly, in case of official interference, and in the rush, the soldier and the tents got left behind, and we had to go without either.

The weather was better, but the nights in the open forest, with snow falling, were bitterly cold. From the pass, 13,000 feet, a fine view was obtained over the Salween-Irrawaddy divide.

Through the Land of the Crossbow.—On our return to the Mekong we continued southwards to Wei-hsi, and it was now that I congratulated myself on having made friends with the official there; for owing to the disturbed state of the country it seemed a good opportunity to vary the route home by going due south through the tribal country. The caravan, however, would be safer on the main road, and I therefore proposed to part company for the time and meet at T'eng-yueh in three weeks.

The official demurred, saying that the country was unsafe; happily, he had a severe cold, and on my sending him a little medicine, he came round to my point of view; thus I started south down the Mekong valley with two porters, while the mules set out by the main road.

We passed through Lissu, Moso, Minchia, and other tribes, and on the sixth day fell in with the revolutionists. The leader, however, treated me courteously, giving me a special passport, on which it was set forth, to all whom it might concern, that I was to be well treated; and wherever I presented this passport I was courteously received.

The journey was not without incident, however, for one evening I got into trouble with some muleteers, who were about to hit me over the head with benches; however, I calmed them down, and jokingly asked whether they wished to kill me, adding that for the life of one Englishman the Chinese officials would sacrifice all four muleteers! One fellow laughed it off, and added, with true Chinese politeness, that they were not worthy to kill me—a very ambiguous compliment.

Crossing the Mekong we climbed the dividing ridge and so reached the Salween, now finding ourselves amongst the Shans, a people closely akin to the Siamese. Meanwhile the native chiefs of the villages were very friendly, inviting me in to meals, and taking great interest in my property. An ejector gun was a source of considerable amusement to them, for I often pretended that by whistling I could make the cartridge hop out on to my lap at will—a droll performance that elicited roars of mirth.

On two subsequent occasions I met bodies of tribesmen on the march northwards—fierce-featured men, armed with dâh and crossbow; but I was never molested, and we reached the summit of the Salween-Shweli divide without incident; that night we saw the sun set in a blaze of glory over the golden land of Burma.

T'eng-yueh.—Three days later we were back on the small T'eng-yueh plain, the first piece of flat ground I had seen for six months, only to find that the country was seething, there had been heavy fighting on the main road, and my caravan had not arrived.

However, the consul did everything that was possible, and five days later we received news from the revolutionist leader at Tali saying that my caravan was safe and on its way; it arrived shortly after Christmas, and we at once set out for Burma, getting back to Bhamo just after the New Year.

EXPLORATION IN THE ROCKY MOUNTAINS NORTH OF THE YELLOWHEAD PASS.[1]

By Professor J. NORMAN COLLIE.

UP to only a very few years ago little was known about the country that lies along the watershed of the Rocky Mountains in Canada. The first man to cross the continent in what is now Canada was Sir Alexander Mackenzie in 1793. After that date and up to 1857 the only people who visited that "Great Lone Land" were the hunters and trappers, employed by the Hudson Bay or the North West Companies. In 1857 an expedition was sent by the British Government under Captain Palliser to explore the southern portion of the Canadian Rockies and to report if it was possible for a road to be made over the mountains so as to connect Eastern with Western Canada. The report was unfavourable.

In 1871 a survey was started for a trans-continental railway, and in 1886 the Canadian Pacific Railway was opened. This rendered it possible for people to get to the Rockies without difficulty, but still, even ten years later, large portions of the mountainous country, quite near to where the railway passes over the divide at the Kicking Horse Pass, were unknown and unexplored.

During the summers of 1897, 1898, 1900, and 1902 the author explored the country for about one hundred and twenty miles north of the Kicking Horse Pass, discovered many peaks, passes, and glaciers, the most important being a large central snow-field that was the source

[1] An Address delivered before the Society in Edinburgh on February 20, 1913.

of the three largest rivers in Western Canada. It was named the Columbia ice-field; from it flowed the Columbia, the Saskatchewan, and the Athabasca rivers, that drain respectively into the Pacific, the Atlantic, and the Arctic oceans. A map was made of the country thus explored.

But there still remained much more country further north about which practically nothing was known. Dr. Dawson, the late head of the Canadian Survey, wrote only twenty years ago, "In Canada there are 3,470,000 square miles, of which 954,000 square miles (exclusive of the inhospitable Arctic portions) is for all practicable purposes entirely unknown."

It was in 1910 that the first section of the Grand Trunk Pacific Railway was opened almost up to the foothills of the Rockies on the east. This made it possible to reach the Rockies at a point about one hundred miles to the north of the ground explored during 1897-1902, and by following the Athabasca River reach the Yellowhead Pass. This pass leads over the main range from the valley of the Athabasca to that of the Fraser River in British Columbia.

It was in 1910 that the author with Mr. A. L. Mumm made use of the railway; the railhead was then at Wolf Creek. From there to the Yellowhead Pass was roughly one hundred and fifty miles. With the party came five men together with about twenty horses, the latter carrying all the provisions and tents of the party.

We started on July 17, and on the 27th we crossed the Athabasca, reaching the Yellowhead Pass on the 29th. After crossing the pass we descended about fifteen miles, then struck north up the Moose River, crossed another pass, arriving under Mount Robson on August 9. Mount Robson is the highest measured mountain in the Canadian Rockies (13,700 ft.), and has been known ever since the Yellowhead Pass was first used by white men, for it can be seen from the trail leading down from the Pass to Tête Jaune Cache. It was not till 1907 that it was first visited by Dr. Coleman of Toronto. It has once been climbed—Messrs. Kinney and Phillips reaching the top in 1909.

We intended to remain under Mount Robson for some time, and, if the weather were favourable, climb it. But after waiting eleven days the weather, which had been bad most of the time, became worse, and our whole camp, after a snow-storm lasting two days and two nights, was buried in snow. This finally prevented us even attempting the high peak, but we ascended several smaller mountains. One on the north-west of our camp gave us a good view to the north, into an absolutely unknown and unexplored country. We also went to the head of the great glacier on the east of Mount Robson, but the snow was so deep that we were unable to climb any peak.

As the weather had become too bad for the ascents of high mountains, we determined to work north down the valley of a river, the Smoky, so as to get, if possible, into the country we had seen from the peak we climbed on the north-west of our camp near Mount Robson. The weather, however, continued wretched. We discovered a new pass over the divide well below tree limit at about six thousand one hundred feet. An old

Indian trail crosses this pass, but it had not been used for years. To the north of the pass was a fine snow peak and glacier, and one of our men named it Mount Bess.

It was now time for us to return to civilisation. We had heard from some Indians on the Athabasca that a pass existed somewhere leading from the Smoky River to the Stoney River, and the latter flows ultimately into the Athabasca below Jasper House. Obviously, if we could find this pass, it would be a comparatively short route out of the mountains. But, again, if we went wrong and by mistake crossed a pass to the Sulphur River, it would take us a hundred miles away to the north to the Smoky River again, and it would be two months before we got back to civilisation instead of two weeks.

Fortunately, as it eventually turned out, we found the right pass, but for a week we were uncertain whether we were on the Smoky or the Sulphur River. The following year the same party started back for the Smoky River, our newly discovered trail of the year before being a short cut into the unknown country north of Mount Robson.

This time, as I was collecting material for a map, it was necessary to occasionally climb some mountain sufficiently high to enable me to find out where we were with regard to Mount Robson and the high peaks we had seen to the north of Mount Robson. The first peak climbed was one on the east side of the pass that led from the Stoney River to the Smoky River. From it we had a magnificent view in all directions, including Mount Bess, and to the north of Mount Bess, where a splendid group of snow mountains could be seen. This peak we named Hoodoo Peak, after a bulldog that insisted on climbing the peak with his master, who was one of the men who came with us.

On our way from the pass down to the Smoky River we passed a fine lake. Having reached the Smoky River we crossed it and struck up a valley that came down from the west, for from the top of Mount Hoodoo we had seen a fine snow mountain that lay at the head of this side valley.

At the head of this valley we camped near the foot of a splendid glacier, and underneath several fine snow peaks.

This glacier is probably the biggest and certainly the finest I have seen in the Canadian Rocky Mountains. The glacier that comes from Mount Columbia to the headwaters of the Bush Valley must be of great size, but I have never seen it at close quarters. The glacier under Mount Robson is not so large, nor does it possess ice scenery to compare with this new glacier.

Later we found that it had its source in a large snow-field, and is about two miles wide at the point it issues from the snow-field. It then drops a thousand feet in a magnificent cascade of ice pinnacles. Then there is a nearly flat stretch of a mile and then a second drop occurs. It then narrows and finally ends with a third drop to the floor of the valley. We explored during the next few days the country round the bottom of the glacier. Finally we started for an exploration of the upper part of the glacier and with the intention of climbing a snow peak towards the west. The chief difficulty lay in surmounting the great upper ice fall. It was impossible to cut a way up it, our only chance lay in climbing up the corner of the ice-fall where it abutted on to the rock precipice on the north side of the glacier. The route was not free from danger, as it was overhung by ice walls that occasionally sent down avalanches. In the early morning we were able to pass this spot safely, and after climbing up some steep snow slopes arrived on the edge of the great snow-field. It stretched for miles to the north-west, and about four miles away to the west at the other side lay the peak we wished to climb. Fortunately it was very easy walking over the névé, and, soon reaching a col on the south side of the peak, we were able to ascend an easy ridge to the top of the peak, 11,300 ft.

From the summit we had a fine view. South-east of us was the highest peak of the group, probably 11,500 ft. high, to the south was Mount Bess, the same height as the peak we were on, and further away we could see Mount Robson and his attendant satellites. To the west lay an unknown land, and our peak fell away steeply for thousands of feet down to a beautiful valley that ran parallel with the main range to the north. Far away to the south-west and beyond the Fraser River we could see the unknown Cariboo Mountains.

It is curious that no one has ever mentioned these excessively fine peaks. They are probably bigger and finer than the Selkirk Mountains, although they do not stretch over so much ground. Two peaks we could see were especially fine, one a snow peak and the other sharper and with more rock on it.

Now, however, that the trans-continental railway, the Grand Trunk Pacific, is going down the Fraser Valley, we shall probably hear a good deal more of this unknown mountain land.

We next attempted to force our way further north along the valley of the Smoky River that runs parallel with the main range. The burnt timber and difficulty of getting our pack horses through soon made us turn back; and as we had some few days left we determined to visit the country round Mount Bess and if possible climb that mountain. This we were fortunately able to do as the weather was favourable. We ascended from the south side, and found the last two thousand feet very steep, partly ice and partly snow. The top is dome-shaped, and again we had a magnificent view from the summit.

We then made an excursion over to the west side of Mount Bess into British Columbia and were able to see the main range from the western side. And then, as our time was up, we had to turn our backs on this happy hunting-ground and hurry hot foot on the homeward trail. We should soon have to change the camp life for that of the hotel, and our small world was to be ended for the time being. But if one has once wandered in such a land one is always hearing the call to come back again; the sombre forests, the rushing rivers, the beautiful quiet lakes, and the snow-white mountains, they all call and call again, and the memories of one's old friends on the trail and the free life rise up, and back one has to go to those valleys amongst the mountains and wander once more.

SOME ASPECTS OF GEOGRAPHICAL WORK IN THE SUPPLEMENTARY STAGE.[1]

By T. L. MILLAR, M.A.

THE remarks made, and the whole of the work set out for inspection, emphasised and illustrated the now well-accepted educational idea, that practice lies at the foundation of theory, and that it is by actual doing, and by personal observation, that the child best comprehends geographical problems, and creates his standards of comparison. Below we give a few of the practical methods adopted in Sciennes School, together with a brief description of the instruments used, all of which have been made in the School workshop by pupils between 12 and 14 years of age.

I. To find the Latitude and Longitude of the school.—

Latitude.—The instrument used consists of a pole five feet high, standing on a cross-base. Inserted in the top of the pole is a skeleton quadrant made of three-ply wood. Behind, and projecting to the right of this quadrant, is a small vertical mirror, four inches by three inches, in a frame hinged to a small platform at right angles to the pole, and across the centre of the mirror is drawn a horizontal line to represent the artificial horizon. Screwed to the nose of the quadrant is a thin piece of hard wood with smoked glass on the inside, the whole at right angles to the plane of the quadrant. The sighting-hole in this piece of wood is in the plane of the artificial horizon; while the pointer, a fine banjo-string stretched on a frame resembling the front sighting-vane of a prismatic compass, is screwed to the mirror frame on a line with the artificial horizon. The pupil stands with his back to the sun, and allows the mirror to fall back, thus raising the pointer, until he gets the lower limb of the sun on his artificial horizon. The angle is then read off.

With this must be used the data given on the first two pages for each month in the *Nautical Almanac*—*i.e.*, declination, equation of time, and semi-diameter—and of course wedges and plumb-line for adjusting the pole on sloping ground.

Example—

Altitude of sun at noon (lower limb)	=	33° 10′ 0″
Semi-diameter		16′ 5″
∴ True altitude	=	33° 26′ 5″
∴ Zenith distance	=	90° − 33° 26′ 5″
	=	56° 33′ 55″
Declination (minus)	=	34′ 57″
∴ Latitude	=	55° 58′ 58″
Actual latitude	=	55° 57′ 23″

[1] An Abstract of a paper delivered before the Association of Teachers of Geography in Scotland on February 1, 1913 (see p. 158).

This is the best result the instrument has given, and results have varied between 55° 30′ and 56′ 32″. It forms a good exercise to get the pupil to estimate his error in miles.

1° Lat. at 56° N. = 69·19 mls.
Here error = 1′ 35″ +
∴ Distance too far N. = 1·826 mls.

Longitude.—The watch used must show Greenwich Time. In practice the watch is corrected on day of observation by Castle one o'clock gun.

(*a*) Shortly after 10 a.m. the sun's altitude is observed and the exact time and altitude noted.

(*b*) After the lunch-hour the instrument is set to the morning observation, and the time noted when the sun is again at the same altitude.

(*c*) Division of the sum of these times by 2 gives local noon in Greenwich Mean Time, but, since "apparent time" is wanted, to this result must be added or subtracted the equation of time as indicated in the *Nautical Almanac.*

(*d*) The difference between Greenwich apparent time and 12 o'clock = difference in longitude between Greenwich Meridian and School Meridian in time.

(*e*) Dividing this by 4, as 4″ = 1° Long., we obtain the longitude of the observation station in angular measure.

Example.—Morning observation = 10 hrs. 12 mins.
Afternoon " = 14 " 4 "
24 hrs. 16 mins.
Watch 1 min. fast ∴ 2 "
24 hrs. 14 mins.

$$\therefore \text{ Noon "G.M.T."} = \frac{24 \text{ hrs. } 14 \text{ mins.}}{2}$$

= 12 hrs. 7 mins. 0 secs.
Equation of time + = 5 " 27·8 "
12 hrs. 12 mins. 27·8 secs.
∴ Longitude of school = 12 min. 27·8 secs. W.

$$= \frac{12 \text{ " } 27{\cdot}8 \text{ secs.}}{4} \text{ W.}$$

= 3° 7′ W.
Actual longitude = 3° 10′ W.

Again results have varied, being usually slightly higher, and the worst result gave 3° 40′ W. Again the pupils might calculate error in miles and time. 1° Long. at 56° N. = 38·77 miles.

The instrument can also be used to calculate heights by reading elevations at two ends of a measured base line, and then setting out the results to scale on squared paper with ruler and protractor.

Example.—Base line 36 feet 6 inches.
Elevations 34° and 42°.

This gave the height of the school weather-vane as

102 feet + 5 feet for pole = 107 feet.

Actual height from plans = 106 feet.

The instrument might easily be mounted on a tripod, and would then be much more serviceable and portable.

II. The following experiment is used in connection with a graph showing actual and possible hours of sunshine per week at Edinburgh and London—the actual hours data being taken from *The Scotsman* every Saturday. It at once helps the child to appreciate the varying positions of sun and earth relative to each other, and the reason for prolonged hours of summer daylight in high latitudes. Having been shown how to work the experiment, and how to use the *Nautical Almanac*, pupils readily acquire the knack of it, and need practically no supervision.

Experiment.—To find from the globe with the aid of an inch-tape or quadrant the possible hours of sunshine at Edinburgh on March 26, 1912—

(1) When on the observer's meridian the sun's rays reach 90° on a great circle in every direction, . . . = 90° 0′ 0″
N. declination on 26.3.1912, ... 2° 57′ 35″

92° 57′ 35″

i.e. the rays at noon are 2° 57′ 35″ on the side of the N. pole remote from the sun.

(Note that when declination is S. from September 23 to March 21 the declination must be subtracted.)

(2) Place the tape across meridian 0° at right angles (*i.e.* inch mark coincides with meridian), and with the edge nearer Edinburgh at 92° 57′ 35″—*i.e.* at 87° 2′ 25″ N. lat. on side of globe remote from Edinburgh.

(3) Stretch the tape flat across the globe and calculate along latitude of Edinburgh, 56° N., the number of degrees of longitude E. and W. of meridian 0°, intercepted between that meridian and edge of tape touching 92° 57′ 35″—in this case

95° E. + 95° W., = 190° long.

(4) Divide by 15 to obtain hours $= \frac{190}{15} =$ 12 hrs. 40 mins.

(5) From *Murray's Time-Table* :

Sun rose 5 hrs. 57 min. a.m.
Sun set 6 hrs. 40 min. p.m.
∴ hours of sunshine = 12 hrs. 43 min.
∴ error = 3 min.

In this experiment a small allowance ought to be made for refraction, and more frequently than not by adding some 8 mins. to his result the pupil gets an absolutely correct answer.

III. *Contours.*—The following model was devised to explain how contours are laid down on Ordnance Survey Maps, and shows steep and gentle slopes, a valley, a section, and is of course used to elucidate the interpretation of such contours generally. From the use of the instrument described above a pupil readily appreciates how heights are estimated, and that contour lines join points at the same altitude above "Ordnance Datum."

The large model consists of two blackboards hinged together, and painted on the lower board are lines to represent the vertical and horizontal scales of a section. Let in along the centre of the upper board is a detachable section through the two hills with steep and gentle slopes and a valley between. In front, behind, and down the middle of these sections run vertical rails, on which are built up the eight, inch-thick, horizontal sections of the hills while the boards are horizontal. The pupil outlines the lowest contours, withdraws the lowest horizontal sections, and the whole model slides down the rails, and of course falls within the drawn contours. The other contours are treated similarly until the whole model has been contoured.

The centre section-board is then removed and the boards hung vertically to exhibit the contours. The line of insertion of the centre-board shows a line of section, and from points of intersection perpendiculars are dropped on to the scale on the lower board. The ends of these are then joined, and by laying the centre-board against the drawing the pupil soon grasps the significance of such a section and what it really shows.

IV. *Synoptic Charts.*—In connection with the study of "Synoptic Charts" a record of daily weather observations is kept by each pupil. To sustain interest in these observations a weather-pole has been constructed on which are shown, in coloured flags made by the girls, the main weather observations over a month.

This is a sixteen-sided column, 5 feet high, with 31 holes down each face, and standing on a cross-base on which are painted in various colours the cardinal points. Inserted in the top of the column is a rectangular frame, 19 inches × 10 inches, carrying three vertical columns on which can be read by means of sliding pointers the rainfall for the month to hundredths of an inch. Round this frame and in its edge are bored 31 holes with dates opposite for the insertion of brown or white flags to mark days of rain or snow respectively, while the strength and nature of the wind for the day are shown on grey-coloured flags with words printed across them, inserted in the upper corners of the frame. The scheme of flag-colour was chosen thus:—

N. *Red*—cold winds, red noses, often red skies towards evening.

S. *Yellow*—sun moves round by south.

E. *Black*—cold, raw, uncomfortable, dirty weather in Edinburgh.

W. *Blue*—the region of clearing showers—blue sky.

Intermediate flags are of various colours, *e.g.* SW. = upper half *yellow*; lower half *blue*. ENE. = ⅓ *black*; middle ⅓ *red*; lower ⅓ *black*. At the end of the month the results are set out in graphic form and in colour, corresponding to the flags, *e.g.* the "wind-star" pointers converge towards the centre of a large circle, and show to scale for any day the direction, strength and nature of the wind. From these graphs the lessons of the month are deduced and examined in the light of the "Synoptic Charts," and the necessary explanations of high and low pressures and atmospheric physics are given.

Excursions.—During the spring and summer months a series of geographical excursions are organised on the principle that the same ground is covered twice.

(*a*) In the first attention is given to phenomena relating to physical geography, and with this is associated such historical and economic connections as may arise en route.

(*b*) During the second ramble plant geography only is taken, special attention being given to climatic and edaphic factors, and in this connection a chart of life histories, with illustrative specimens of the common district trees, is filled in throughout the year.

On these rambles survey maps are carried, and heights such as Arthur's Seat are calculated by boiling water at the "Bench-Mark" nearest the base and at the summit, and by pocket aneroid. When on the top (a whole day excursion) a shadow stick is erected and read every fifteen minutes, and the results graphed later give a good basis for the interpretation of longitude. Here also a lesson is given on how to find direction—

(*a*) By compass.
(*b*) By watch.
(*c*) By shortest shadow.

Having nailed direction tapes across the summit the pupils erect a table and construct a "Direction Chart" by drawing rays from the origin (Edinburgh) of N. and S., and E. and W. co-ordinates, to towns, hills, islands, etc., in view on both sides of and in the Firth of Forth. Direct distances are taken from large scale maps. These distances are then compared with rail and road distances, and the chart is made the basis of many valuable lessons on how Nature aids or impedes man's movements, and how man circumvents natural obstacles.

THE SPECIFIC CHARACTERISTICS AND COMPLEX CHARACTER OF THE SUBJECT-MATTER OF HUMAN GEOGRAPHY.

By Professor JEAN BRUNHES (translated by E. S. BATES).

(*Continued from p.* 322.)

III.—THE COMPLEX CHARACTER OF THE SUBJECT-MATTER OF HUMAN GEOGRAPHY.

THERE exist, then, facts which have affinity with Politics, with History, with Art, Linguistic Science, Ethnography, Political Economy, and yet belong to none of these studies. These are the facts whose specific character we have just defined, and which ought to be the primary objects of study among anthropo-geographers. But these, using as they must subject-matter of this description, are obliged—as Bosnia-Herzegovina has just shown to us—to take into account more complex entities, entities which are both tremendous and yet elusive; but facts which are strictly geographical in nature, realities of the most practical kind, lead us on to the study of these. In proportion as they do so, we must study such facts.

(1) *Each Fact of Human Geography contains and implies a Social Problem.*

Let us bear in mind, in point of fact, that when geographers wish to show the connection between particular instances and general ideas, they should not merely draw upon the results of ethnography, or history, or statistics; they must go further. I maintain that the ultimate analysis of any fact of Human Geography reveals a problem which is not only an economic problem, but a social one too. If we ascertain the numbers of a drove of horses or of camels, if we go down a copper- or a coal-mine, if we examine the yield of a Fang hunt or the Bergen fish-market, we cannot prevent deductions relating to social states ensuing from our observations or our analysis. It is never a case of mere juxtaposition of the respective facts; the deductions are part and parcel of the whole material reality to such an extent that it is an impossibility to get to the bottom of the reality without perceiving the social fact which is part and parcel of it too.

Show me your plough, and how you drive your furrow, and I will tell you whether your share, which barely breaks up the surface, is that of a nomad shepherd whose main source of livelihood is his flock, and who comes in haste, almost by stealth, to sow where he will come again only to reap; or whether that wooden share, whose furrow is also no deep one (as well may be the case since the mud it cleaves is so impregnated with salts that it is risky to trench too deeply if plenty of water is not available wherewith to irrigate freely enough to dissolve the salts), belongs to an Egyptian *fellah* and co-exists with a given type of irrigation-methods. I will tell you whether you are a peasant from these fields of ours that have been cornfields time out of mind, from the districts of Beauce and Brie; where the character of the furrows, a tradition in themselves which dates back hundreds of years, at once reveals individual ownership, whether on a small or large scale, and also one of the most persistent and most perfect forms of conquest by culture. Or, at sight of the plough, I can say if you are engaged in trenching chiefly rich earth, in order to sink the beetroots with which the syndicate owning the neighbouring sugar factory provides you, sugar-producing plants which are all bought and sold in advance: I will tell you, finally, when you show me the row of ploughs—a whole battery adapted to steam or electric force—which are simultaneously at work beside your seed-drills—that you belong to a region where labourers are few, or almost unobtainable, for you must belong either to the vast tracts where the New World practises its rapid and highly specialised methods of culture, or else to the Old World's model farms.

The inhabitants of Utah and Colorado are identical in nature,

Mormons mainly (one of the most zealous apostles of "dry farming," Widtsoe, hails from Utah), but there is a wide difference between the two States as regards cultivation, Colorado being by far the more prosperous, for the reason that the land has been distributed there with greater discretion. Between the two States exists the same difference as between Tunis and Algeria. In Colorado land has not been allotted, as a matter of local politics, to people who frequently lack the requisite physical strength. The authorities waited until candidates presented themselves who were possessed of sufficient will-power, incentive, and means to tackle virgin soil effectually.

Our essential facts, then, separately or collectively, are concrete illustrations of a series of social facts. In the course of every immigration or emigration movement how varied must be the social phenomena implied, in the old home, and in the new! Some day or other I shall certainly endeavour to show you how the truly geographical factors in ethnographical problems come to light, in Human Geography, as kinds of social problems.

(2) *Every Fact of Human Geography necessitates and implies a Statistical Problem.*

A fact of Human Geography, however interesting it may be, has not become known to us in the fulness of its import as a scientific discovery until we have recognised and can appreciate its coefficient in statistical values.

How many travellers and observers exaggerate the importance of some isolated fact and thereby falsify their glimpse of reality. They base their idea of a race on careful measurements and photographs of a single individual; they represent a whole nation as hospitable because they were well treated in a particular place by some few of its inhabitants; or, on the other hand, they vilify some other whole nation because an unfortunate mood of their own, or some failure in tactfulness, has occasioned one unpleasant experience. These things are examples of the moral order, but they afford us some insight into the number of errors which can accumulate behind the magic veil of an unquestionably authentic photograph or of first-hand observation of persons or of places.

It cannot be repeated too often that the concern of Geography is with the normal, not the abnormal; not with the weight of the exceptional ear of wheat, but with the yield of bushels per acre; not with the extraordinary specimen of a mineral, but with average percentage. "Normal yield," "average value"—do not these imply a sober and rational statistical estimate of facts noted one by one, such as will attribute to each its due importance as a factor in relation to a just calculation of the generalities of the subject. In exact proportion to the worth of the data and of the methods of statistical calculations will be the worth of the inferences drawn.[1]

---

[1] I may use as a reference pp. 623 *et seq.* of the 2nd ed. of my *Géographie Humaine*, Paris, 1912.

But while conceding this much as regards statistical valuations whose real values we sometimes find to be not quite great enough, we are bound to recognise that Human Geography ought always to retain the assistance of this most useful auxiliary. The point of view of the detailed, meticulous monograph and that of the generalising epitome may often seem to be irreconcilable; but each acts as a check on the other. Nevertheless, let us not go away with the idea that this discordance is only an imaginary one. While we must, and ought to, pass from the one to the other, and while any conception is incomplete which fails to co-ordinate the two, they are nevertheless so alien to one another, so sharply contrasted, that a great deal of knowledge, and a great deal of faith, are required before we can get accustomed to the transition from either perspective to the other.

That master whom we all regret, who taught Economic Geography at the Collége de France over so long a period, and who has left us the memory and the high example of a lifetime of unwearied toil and strict integrity, may be cited as witness to this; he has delivered to us the theory underlying these connecting links between geography and statistics, subtle and complex though they be. In the opening lecture of his course, on the history of economic and social achievements, M. Marion ably summarised Levasseur's work. Allow the occupant of the chair of Human Geography likewise to utter his ready and heartfelt homage to him who, throughout his life, right on to the final days of his vigorous old age, remained geographer, historian, and statistician.

One fine attribute of Émile Levasseur's was that of retaining that feeling for living realities with which he began his studies; he never slackened in his vigilance in verifying average and abstract values by confronting them with the practical facts of the terrestrial surface. In his very remarkable *Introduction sur la Statistique*, which forms the preface to his work on the population of France, he wrote:—"Statistics supply figures which labour under the defect of being abstractions." But these abstractions correspond to realities of such a kind that we cannot be prevented from treating them almost on a footing with facts. The averages of births, deaths, marriages, etc., the average yields per acre of wheat or rye, or barley, etc., harvests over a given area, have both in their variations, and in spite of their variations, so much continuity that we end by being struck chiefly by the universality, and even, in a measure, by the inevitability, of the causes which seem to govern human societies and activities; and the temptation arises to ignore causes which are accidental, local, or special, which throw light on genuine characteristics of groups of individuals, or of this and that individual, just as the same temptation arises with reference to the reasons which vary the conditions of production in this or that field, and in this or that district, despite proximity. We are fascinated by that somewhat super-real harmony which figures set before us, and then we grow less appreciative of the analytic detail which is so specially valuable a product of studies such as history and geography. The difficulty was recognised by Levasseur; probably it is recognised by everybody who passes from first-hand observation of details to generalised estimates based on summaries. Is it not a fact that the general laws which govern the modalities and caprices of individuals are so rigid as to lessen the interest which ought to be a factor in inquiry into those unsophisticated expedients hazarded by individuals and by localities? Does not the curve sometimes seem unable to bend? Does not the practice of (mathematical) interpolation suggest a sort of implied protest against the "infinite variety" of geography? But that is what statistics are for—to legitimise this plan of interpolation, and to underline that seeming, or at any rate relative, inflexibility of curves.

On the other hand, have not the most regular of curves been diverted, or even shattered, by well concerted geographical conditions, just as much so as by the strength of will of certain individuals in history? We must therefore learn how to combine the use of statistics with inquiry into the evolutionary processes with which history acquaints us, and also with a critical and comparative study of all the variety of combinations existing in space. This trinity of entities delimited the whole of Levasseur's learned labours, whose aim was to define the interrelations of these subjects, so vast in their scope, sometimes at war with each other, and sometimes co-operating. He knew how to modify the calculations of statistics by means of historical research and geographical investigations. And note that this geographer-statistician of the population of France has been a hard fighter on behalf of reform in the teaching of geography, and also the historian of the working classes; the issue of the second volume of his splendid *Histoire du Commerce de la France*, the posthumous volume for which we are indebted to the loyal efforts of Auguste Deschamps, occurs very aptly to illustrate what I say. Levasseur's clear insight and enlightened common-sense enabled him to find common ground for these irreconcilables. Therein lies the profound and illuminative significance of his varied and complex labours; it has ensured its merit and will ensure its durability.

In short, statistics serve a purpose; in their own way they organise and group, using words to explain numbers and numbers to elucidate ideas, and both in reference to those same modes and realities of dynamic energy which come within the scope of scientific vision once again, albeit from another angle, both in history and in geography. But in so far as our "essential facts" of the terrestrial surface are concerned, what are statistics, what can they be, without Human Geography, which both creates and modifies them?

We shall have to put our trust in tact and use the utmost discretion in estimating the relative values of the facts (necessarily few in number), which we have seen with our own eyes, for the purpose of these generalising calculations which, when considered simply in themselves, so often prove misleading. The truth will be ascertainable only by means of comparison; and even then no geometrical truth will ensue, but merely what is relatively true, whose value will be the joint product of the value of the observations and of the degree of critical faculty employed in determining the compromise between the things seen and arithmetical deductions, whether totals or averages. I do not hesitate to state definitely that every truth concerning the relations between natural surroundings and human activities can never be anything but approximate: to represent it as something more exact than that is to falsify it, is to become anti-scientific in the highest degree. For this reason I shall demonstrate later how radically vitiated have been those systems which have exaggerated man's dependence on his surroundings, and which have represented the evolution of history and of civilisation as a matter which logic might have prophesied.

Speaking of the truth of a fact is a curious instance of the misuse of words. A fact possesses dimensions, colour, duration; but it does not possess truth; it is in our perception of the fact that truth or falsehood lies; it is the criticism that we pass on it that possesses more or less accuracy. Science consists of nothing but the relations we establish between facts. Now all scientific truth is inherently similar, in fact more or less identical, with what we have just been terming a truth of Human Geography. And therefore let us avoid demanding of statistics, as do so many superficial minds, the illusion of pseudo-arithmetical precision (which would imply the illusion of a pseudo precision of deduction), but simply ask for a just estimate of objective truth—assuredly a very real thing—the basis whereof we shall recognise, implicitly or explicitly, to be an estimate of probabilities.

One of the most remarkable mathematicians in this country and of our age, Émile Borel, has written several articles which deserve geographers' special attention: "Le calcul des probabilités et la méthode des majorités"[1]; "Le calcul des probabilités et la mentalité individualiste"[2]; "Un paradoxe économique, le sophisme du tas de blé et les vérités statistiques."[3] Once statisticians cease to rest content with making computations, once they endeavour to predict, they formulate truths which acquire a peculiar, but none the less real, value from the fact of their being based on the theory of probabilities.

"The idea which I am trying to isolate," says Émile Borel, "is this, that the answer for a mathematician to give to many a practical question consists of a coefficient of probability. Such an answer will seem unsatisfactory to many persons, who expect mathematics to yield certainties; a very lamentable tendency. It is most unfortunate that the education of the public should be, from that point of view, so very backward. It is doubtless the result of the computation of probabilities being so almost universally ignored, in spite of the process entering with ever increasing frequency into everybody's life (in the shape of the various forms of insurance, friendly societies, pensions, etc.). A coefficient of probability constitutes a perfectly clear answer, which is in accordance with a perfectly tangible reality. Some people will affirm that they 'prefer' certainties; perhaps they would also 'prefer' that 2 and 2 should make 5.

"If the conception of statistical truth came to be familiar to all who write or speak on questions wherein statistical truth is the only kind of truth available, many sophisms and many paradoxes would be avoided."[4]

---

[1] *L'Année Psychologique*, xiv. 1908, pp. 125-151.
[2] *La Revue du Mois*, vi. 1908, pp. 641-650.
[3] *La Revue du Mois*, iv. 1907, pp. 688-699.
[4] "Un paradoxe économique," article quoted, p. 698.

He goes on to say: "Many persons, acute enough in other ways, imagine that there are no truths except those particular truths to which they give the name of facts. As I came off the boat at Dover I noticed three Englishmen, each of whom was over six feet high; that is a fact; but it is a fact which does not possess any importance; whereas the proposition that Englishmen's average height is less than six feet is not a statement of fact, but an average struck from a multitude of facts; yet it is a genuine scientific truth for all that."

In another article Émile Borel writes further: "The development of modern physical theories renders it clearer every day that the idea, which Maxwell was the first to foreshadow, namely, that most, if not all, physical laws are truths of statistics, is a true one; in other words, they embody a net result in relation to phenomena which are too numerous and too complex for detailed analysis of them to be practicable. If these statistical laws are more definite than the laws of demographical statistics, it is merely because the number of entities concerned (molecules or electrons) cannot be compared with the number of human individuals, and we know that the accuracy of a statistical estimate is roughly in proportion to the square root of the number of its constituent entities. Thus a statistician can foretell how many births will occur in Paris next week to within 10 per cent. or thereabouts, whereas a physicist, on being given a cubic measurement, a temperature, and a pressure, will know the corresponding content of gas to within 1 or ·01 per cent. . . ." And, after expressing a well-founded regret at the persistence with which criticism is directed, in the name of absolute truth, against many a method or institution which cannot lay claim to anything more than "statistical truth," M. Borel concludes: "The social sciences, using the term in the widest sense, often submit to the use of statisticians' methods in the absence of any other available methods, and they prove so serviceable as to leave no room for lamentation at the frequent failure to discover some more scientific method. And this is true not only in relation to the innumerable questions where social procedure is at stake, but also in sciences which are more exclusively intellectual and less immediately practical, such as philology, linguistic science, the history of art, or of literature, etc."[1] To these we may add Human Geography.

All biological relations, all œcological truths, are, and can be, nothing more than statistical truths. An example taken from botany, a recollection of the mountains of that splendid and hospitable country associated in my mind with sixteen years of work and Alpine excursions, may serve to make clear how far an approximate truth is a truth, what kind of a truth it is, and how a readiness to over-define its applicability tends to subvert its nature. As you climb higher on the slopes of the Alps you find changes taking place in the flora. And the most ordinary, the most significant, the most clearly marked of the characteristics of the flora of higher altitudes may be summed up as follows:—(1) diminution of the organs exposed to the air, so much so that in many cases dwarf

[5] "La théorie des probabilités et l'éducation" (in *Hommage à Louis Olivier*, Paris, 1911, pp. 37-40).

types appear; (2) subterranean organs, roots and root-stocks, proportionately more highly developed; (3) various arrangements tending to retard transpiration and to minimise the risks run by the plant in connection with that part of it which is exposed to the air, whether through frost, or nocturnal radiation, or too great heat by day (villosity, carnosity, increased thickness of the epidermis, narrower diameter of the cells, greater power of osmotic attraction in the cellular fluid); (4) flowers which shoot up to a greater height and display a more intense and deeper colour. It is when our eyes are held by the velvet-like colouring of the gentians of the higher altitudes, that penetrating, full-toned, vivid blue, that the fulness of the truth about these transformations of vegetation dawns upon us. But have we here a series of transformations which obey a uniform, rigid law? Certainly not. As we go higher, the vegetation is seen to be undergoing alteration little by little, without sudden and thoroughgoing transitions; there exists no boundaries at which these phenomena begin or end. Yet no less certain is it that these strange and fitful modifications, when considered as a whole, constitute one of the most striking cases of a transformation-scene in Nature that could be discovered. Here we find a statistical truth in connection with the physiology and the geography of vegetation which exactly corresponds with what we term truths of Human Geography when we are considering connecting links.

(3) *Every fact of Human Geography conceals and implies a Psychological Problem.*

With these principles clearly before us, we shall go on collecting truths of Human Geography and the harvest will be a great one. And the outcome of it all will be nothing less than light thrown on the most puzzling and unfathomable of problems, that which is dimly outlined where even imagination fails to admit us, in the background which is common to history and prehistory and ethnology and sociology; namely, the problem of the conditions of the successive implantation of humanity on this earth and of the part that humanity has played in the transformation of the planet: the earliest forms of tillage, of alloys, of industry, of townships. But we shall be aware just how much truth these truths contain; we shall not be credulous, we shall not be dupes; for nothing within these limits is absolute or stable. All connecting links between human activities and Nature (and, consequently, all facts of Human Geography, being, as they must be, effects and symbols of those connecting links) are dependent upon that factor which is ever at work, whether at the bidding of free-will or of fate,—the human being.

For the purposes of objective study we have, in *La Géographie Humaine*, presupposed that we have risen above the earth's surface as if in a balloon, endeavouring to discern and to classify all "visible and photographicable" facts which are due to humanity's existence thereon. But as soon as we have landed and have followed up our elementary efforts at first-hand observation and definite classification by a practical recognition of our duty to go on to an analysis of causation, everything combines to make us aware that the facts of Human Geography derive their primary specific characteristics—I do not say their predominant characteristics, nor their essential features—from a fact of human life evolved either from the physiological necessities of our bodies, or from our desires, or from our concepts, or just from our ephemeral illusions.

Let me remind you of some well-known facts—too well known, and chosen because they are so. Tea and coffee only possess their economic value because we like them; we like them because they have been brought to our notice, and because we have become convinced that they are good to drink. The original and the abiding geographical causes of the cultivation of tea and of coffee on a large scale are within ourselves.

So far as Man is concerned coal was non-existent until he came to understand it; in other words, until he became aware that this black rock, unlike basalts and black sandstone, would burn and formed an incomparable source of latent energy. From that time onward he has been the industrial slave of the coalfield; has built factories, or transferred them, near the mouths of the pits; but, here too, if coal is his master, it is he himself who has made it so. If it lies with the coal supply to create or to withhold certain modes of his activities, the initiative was his by virtue of his knowing how to use it and, so to speak, domesticate it; and determining to do so.

As with these elementary facts, so too with the greatest and most complex facts of political geography; conquering a country implies becoming more dependent on it; but this result is a result of conquest.[1] Let us try and identify the human "points of departure," whence radiates all the bio-geography of man.

To begin with, there are those fundamental physiological needs which have received detailed treatment in *La Géographie Humaine*; the need for food, with all that pertains thereto; the need to drink, the curious physiological desire for salt, etc.; the need to sleep, which begets the need to erect a shelter, a house; the need for protection against changes, sudden or extreme, of temperature, which brings in its train a need for clothing. And the exigencies of these elementary needs grow ever more and more manifold and insistent; their iron rule is responsible for the creation and for the continually increasing development of the major part of terrestrial economics.

We have been told how different geographical horizons exist for different groups of human beings, and that these horizons expand in proportion as each group becomes more civilised and more effective; "Räumliche Anschauung, enge oder weite Horizonte."[2] We might well speak of an horizon of free choice for each little group. When a peasant

[1] It is in this sense that we must interpret Marcel Dubois' bold assertion: "Nature promises nothing; everything must be wrested from her by means of work and knowledge" (*La crise maritime*, p. 21).

[2] Cf. Fr. Ratzel, *Anthropogeographie*. I. Theil, 3 Aufl., Stuttgart, 1909, pp. 148 *et seqq.*; II. Theil, 2 Aufl., Stuttgart, 1912, pp. 29 *et seqq.*

decides on a site for his house, he considers the most favourable situation in relation to sun, wind, and water; but the limits within which he may choose are usually very narrow; his horizon of free choice is close around him. Now when the earliest settlers were seeking settlements among the chalky plateaux of Normandy, they found better shelter from the wind, more springs, greater proximity to the sea, whence came their food, in the "valleuses"; everywhere, therefore, we find the villages established in these hollows. This neglect of the shelterless higher ground in favour of the little *thalwegs* abutting on a stretch of sand implies some more or less conscious survey of the whole of a district which is of considerable extent; and we cannot do otherwise than assume so systematic a crystallisation of humanity to be the work of men whose horizon of free choice was likewise of considerable extent. In so far as the origin of a given town was due to a founder—whether overlord, bishop, or conqueror—who intended to be a founder, so far has its site been selected for it by virtue of a comparison of sites ranging over what is a territory in the fullest sense of the word. The farther history advances, and the higher the degree of perfection attained by human groups in the way of culture, so much the nearer does our horizon of free choice approximate to the farthest limits of habitable land; all progress in means of communication creates its equivalent in greater freedom of choice for each group, and often, indeed, for single individuals; a twofold choice, not only where to live, but also what to do. When the archduke Ludwig Salvator had finally satisfied his peripatetic instincts and his observant eye by journeyings along most of the coastline of the Mediterranean, he took up his abode halfway up the glorious slopes on the north-east side of Majorca, and there founded the Miramar of the Balearic islands. And a whole State, the Commonwealth of Australia, undertakes a prolonged comparative inquiry for the purpose of deciding on the site of its future capital, and institutes a competition between the architects of two continents concerning the planning of the city that is to be. The real advantage that a certain number of Anglo-Saxons have secured in practical life is doubtless due to their having taken into account this horizon of free choice, which is the gift of modern civilisation and of their having tested the benefits of the variety of the opportunities it offers more thoroughly than others have done.

The outstanding psychological fact, then, is the antithesis of a rigid fatalistic determination of human acts by climate and by soil. It is this—that natural surroundings, whether as a whole or in detail, react on us just so far, and just in such a way, as we adopt them; in other words, according to our interpretation of them.

A river, a mountain, forms a frontier only in so far as we form such and such ideas, economic or political, concerning frontiers; ideas which undergo modification in the course of history. Mont Blanc, for instance, is a supreme example of a majestic and decisive boundary to those to whom the idea of watersheds represents an abiding truth; yet, not only is that idea less than two centuries old, but at this very moment Mont Blanc and its setting are in reality so far from constituting a boundary that one and the same language, French, maintains its hold in every valley on every side of it, just as firmly in the valley of Aosta, which belongs to Italy, as in the Swiss Bas Valais and the valleys of Savoy. The fact is that the two routes of the Great and the Little St. Bernard have remained so much used and so important as to uphold the natural unity of that great entity in spite of all the claims and chances of politics.

In Nature there exist no frontiers but those of our theories. Instead of pursuing the mirage of a classification which distinguishes between natural and artificial frontiers, instead of succumbing to the temptation to bemuse ourselves with the still more factitious distinctions between "Naturgrenzen und natürliche Grenzen," let us keep this clearly in mind, that time and place have caused the same features of Nature both to be, and to cease to be, boundaries: *frontier antinomies.*

Among islands which are on the same footing as regards physical and climatic conditions, some are over-populated, like Java, while others, such as Sumatra and Borneo, are more or less uninhabited: *insular antinomies.*

Places seemingly predestined to be the sites of great capitals have been abandoned, whereas Pekin and Madrid have come into being on barren wilds: *urban antinomies.*

Confronted with the manifold, but clearly defined, possibilities, which every district and every homestead lay before us, men are bound, as P. Vidal de la Blache has phrased it, to take sides. And, seeing that their achievements on the surface of the globe tend to transform the original conditions of their natural surroundings in ways that, though imperceptible, are nevertheless unceasing, we need not be surprised at what seems to be contradictory in historical and geographical life. They suddenly take it into their heads to give effect to some side or other of reality which they had been ignoring. The Portuguese turned themselves into sailors at the shortest notice, once America had been discovered. The Dutchmen who had been sailors and town-dwellers and accustomed to co-operation on a grand scale, became shepherds and individualists as soon as their natural surroundings were changed. And when the Highlanders of Scotland were hunted from the agricultural holdings, which their inhuman masters ravaged with fire, they took refuge by the waterside and became fishermen: *racial and social antinomies.*

What splendid suggestions for study are these! Just think of the new fields of research they open up for us!

Thence too, we geographers may infer how to visualise the vicissitudes, both great and small, of economic history. The damp and flinty lands at the heart of the district of Léon in Brittany once produced nothing but mediocre cereals; to-day, thanks to the pastoral industry, it has become a land of beautiful green meadows. In the Vannetais and Cornouaille the railway has led to apples being grown for cider, apples which are utilised by a Würtemburg factory more than 1000 kilometres away.[1] And the same method in observing and meditating

[1] Cf. H. Hauser, "La géographie humaine et l'histoire économique" (*La Revue du Mois*, i. 1906, p. 209).

will assuredly enable us to perceive the minutest, but equally real, contradictions, such as those to which our attention has been drawn by A. Demangeon (in relation to dwellings)[1] and by R. Blanchard (in relation to tillage).[2]

In two weighty articles which M. P. Vidal de la Blache has published in the *Annales de Géographie* he has thrown much light on the theory and on the consequences of what he terms "genres de vie."[3] We may summarise his argument as follows:—

The life of a district, as we see it to-day, is divided up in such a way that differing modes of life co-exist quite separately, albeit side by side. If a highly perfected agricultural civilisation has clearly gained the upper hand in that vast oasis of the Nile which we know as Egypt, "the victory of agriculture is, nevertheless, not immune from reaction and checks. Twenty or thirty kilometres from the sea, where occurs the classic delta of the Nile, the gradient decreases almost to non-existence, thereby hindering the outflow, a consequence of which is that the salts are brought to the surface by capillarity, thus forming the salt marshes which are a continuation of the desert."[4] Within this region the lagoon fishermen and the nomad Bedouin replace the fellah. So too round the marshes near Kerbela intó which what used to be the western branch of the Euphrates now disappears, thanks to men's negligence. All the vicissitudes that characterise the life of the river thus find their reflection in the modes of human life. A vivid epitome of this affinity between the degradation of modes of existence and the pathological phases of water-courses in arid wastes is given by Sven Hedin.[5] On passing away from the great oases of Yarkand and Kashgar we find but a scanty population of shepherds among the groves of willows and poplars which fringe the Tarim as it wends its way amidst the sand, only to vanish utterly in the end in immense jungles of reeds, in the intervals between which a few tribes derive a bare existence from fishing. The same story would apply, almost word for word, to the Shari during its course towards Lake Chad.

Moreover, in the same territory, or even in a territory which possesses some sort of unity, either physical or political, kinds of occupation which are akin to "modes of life," but which are wholly distinct from one another, are not only to be found existing in alternation, but are harmonised and supplement each other. Thus when the shepherds lead their flocks to graze in the stubble on the great corn-plains of the Beauce, the pastoral industry is found in intimate relation with agriculture. In all the sheep-rearing district, indeed, an organic relation between the pastoral and agricultural industries exists. Far as we are

---

1 "La Montagne en Limousin: Étude de géographie humaine" (*Annales de Géographie*, xx. 1911, p. 329).

2 "L'habitation en Queyras" (*La Géographie*, xxii. 1910, p. 324).

3 "Les genres de vie dans la géographie humaine" (*Annales de Géographie*, xx. 1911, pp. 193-212, 289-304).

4 Jean Brunhes, *L'irrigation*, Paris, 1902, pp. 324 *et seqq.*

5 *Scientific Results of a Journey in Central Asia*, 1899-1902, Vol. II. *Lop-Nor.* Stockholm, London and Leipzig, 1905, pp. 609 *et seqq.*, with map, pl. 63.

from the over-simplified classifications of certain schools of sociologists (vanishing now, it is hoped) which told men off into irreducible categories, shepherds, agriculturists, fishermen, hunters, etc., it is nevertheless clear that groups do exist which have had, and have still, but one form of occupation; but yet, as geographical evidence has shown, a more normal state is that in which one form, say, tillage or hunting, is no more than a predominant one, while one or several other forms are in use, albeit subordinate. The Fan or Pahuin of equatorial Africa is primarily a hunter; but he also fishes, and, furthermore, cultivates bananas.

Our predecessors on the age-old surface of this Europe of ours, in the course of their work of deforestation, whether in relation to forests proper or to fen districts, have frequently left more or less circular thickets standing; Ordnance maps of France and of Germany still indicate their outlines. Now what were these men doing? While they were doing away with the forest, they continued to exploit it—a gleaning industry; above all they cultivated the soil and grew the indispensable cereals for food,—an agricultural industry; and, lastly, they reared some horses, or cattle, or swine, for which they found grazing ground along the faintly defined, or rather, variable, edge of the forest,—a pastoral industry.

Then too there are cases in which human occupations vary according to the seasons, and thereby bring into being a very complex mode of life, which we may term that of seasonal recurrence. The majority of our trees in these climates are those with deciduous foliage, that is, they multiply their organs and surfaces of transpiration by means of leaves during the warmer months, and shed these organs during the colder period, when they would prove useless and even dangerous; both changes adapting them to climatic conditions. They are, in fact, plants which are of one physiological type in summer, and of another in winter; hydrophytes in summer, they become xerophytes in winter. In the same way one may say of many groups of peasants in the Jura that they are agriculturists or herdsmen in summer, whereas in winter they turn into true artisans, wood-workers, or even watchmakers or jewellers. Elsewhere, in those parts of France that still remain forest-land, men who in spring, summer and autumn are highly specialised field-workers, in winter-time go to live in the forest and turn wood-cutters.

But what will be the result of the irresistible tendency of industrialism? It will tend to specialise not only the various occupations but also the processes related thereto, to suppress more and more those complex types, to introduce vigorous and persistent specialisation even in the depths of the mountains. But at present there still exist groups of men, homogeneous, energetic, living a life of their own, just as do, to recur to our former simile, beeches, oaks, and apple-trees, with a mode of life which combines heterogeneous modalities in the course of a year.

More striking still, men whose occupations and whose means of subsistence are the same year in and year out adopt quite distinct modalities of family and social life season by season. We are referring more particularly to the Eskimo, the northerners who have not domesticated the reindeer and who live by hunting and fishing. In summer, these Eskimo disperse into tents; one tent, one family; whereas in winter they reunite in houses which are more or less subterranean and contain several apartments—what may be termed family cells. My proposal is that we should devote part of our time to an examination of this Eskimo life, by way of a monograph of the Human Geography of primitive man. I shall rely on the indispensable works of Rink, on Nansen's charming and keen-witted book, *Eskimoleben*, and, above all, on that perfect example of how to study social morphology which M. Mauss has given us.[1]

All the essential facts begin and end in facts of psychology. A fresh instance of this, and a typical one, will provide us with a short-cut to the point towards which we are hastening. We have heard of the "dry-farming" methods[2] which are destined to promote agriculture in arid or semi-arid regions, the antithesis of those of wet-farming, *i.e.* of cultivation by irrigation. In all ages dry-farming has been in use in the exacting regions bordering on the Mediterranean; but during the last twenty years the chief centre has been the United States, where it has been investigated, extolled and preached. The idea underlying dry-farming is this:—the more the earth is worked, the riper it becomes; the greater the extent to which the uppermost layer of the soil remains fine soil, the less does it settle and create that hard crust which interferes with the infiltration of rain-water and facilitates evaporation. The aim of dry-farming is to counteract shortage of rain by ensuring that the soil shall benefit to the utmost by all precipitation; the earth must be ready to imbibe the raindrops directly they fall and to hoard them up as far as needful from the surface. In these dry regions the rainfall is irregular and capricious in the extreme; if the soil is to be always ready the tillers must be always at work; plough, harrow and spade must be continually turning and re-turning, and that without any faith in the coming of the rain. How toilsome is such continual physical effort, how high a standard of moral perseverance are implied by such a method of cultivation! But, like man, the earth responds to training; it becomes easier and easier to work; it acquires the habit of remaining free instead of caking. We might almost say that the raindrops get out of the habit of stopping on the surface, and work their way down into the soil to do their utmost among the roots of the sown plants. In short, after ten years of such labour the same annual rainfall produces twice the effect it did. Man's persistent energy has obtained results equivalent to a change of climate.

There we see a psychological fact in the act of achievement—concentration of attention; orientation, thought-out and followed-up, of human will-power. If we interpreted all the facts of Human Geography

---

[1] M. Mauss and H. Beuchat, "Essai sur les variations saisonnières des Sociétés Eskimos" (*L'Année Sociologique*, 1904-5, pp. 39-130).

[2] Cf. Augustin Bernard, "Le 'Dry-farming' et ses applications dans l'Afrique du Nord" (*Annales de Géographie*, xx., 1911, pp. 411-430).

in a dry-farming region solely in the light of natural conditions, we should never succeed in understanding them; the data of the rain-gauge have been modified by men's co-ordinated efforts. The true rain-gauge there is human will-power.

On the other hand, men undoubtedly cannot work in this obstinate fashion for ten years on end without being markedly affected thereby; their moral temperament will be different from what it was ten years back. At the close of these ten years, then, a transformation has been effected in both the human beings and in their natural surroundings; they are new men on a new earth.

Such are the elements of the psychological interaction which lies at the root of Human Geography. The characteristics of the soil, its chemical composition, the configuration of localities, the trend of the relief, the articulation of the coast, the meteorological conditions of a country, the régime of its watercourses have, and always will have, their effect. How far-reaching that influence is we shall find as we go on; we shall always be assigning it a share. While it is the business of geographers to be aware to what extent natural surroundings are responsible for our earthly achievements, it is clear that their share is more or less called into existence by our own acts. The man is there, and the flint is there; but it is the man who makes the spark fly.

It is obvious then that men succeed in modifying natural conditions. It is not the districts whose natural fertility is greatest, or whose water-supply is most copious, that yield the richest harvests; that plain common-sense idea ought to be elevated into a fundamental truth. We may go further. Extreme cases occur in which unfavourable natural conditions have forced men into making better-planned and more persevering efforts, and, by prohibiting life on easy terms, have attuned, so to speak, human will-power to a higher pitch, and thereby have perfected his agricultural activities.

In the heart of the Sahara there are tracts of sand which are absolutely barren, and yet the Soafas have succeeded in raising the best yielding date-palm plantations of the desert. Without fountains and without streams, they discovered how to utilise subterranean lakes for the cultivation of these date-palms, which rank among the most delicate of trees. A little farther westwards, still in this same Sahara, lies the white, stern, bare, inhospitable stretch of the Sebkha; scattered about it are the oases of M'Zab, growing cereals, fruit trees, and date-palms; the water required for these is laboriously drawn from wells 120, 150, 200 feet deep. They are marvels of cultivation, and the cultivation is done by poor men as a hobby.

In the Balearic Islands, the two little villages wherein the fields which rise one above another in terraces are cultivated with the highest degree of perfection, are Estallenchs and Bañalbufar, in Majorca. Now, in the ordinary way, the fishermen and the tillers of the villages hereabouts form two distinct groups, in fact, they often live in separate villages some kilometres apart; but in the two villages in question the men who tend the vines are the same as those who fish; the olive trees on the terraces are cared for by men who are also sailors and coasting traders.

The greater the pressure brought to bear on men to win a living, the higher rises their productive faculty, and often the standard of quality too.

Over the tracts which human beings have wrested, with infinite trouble, from the sea, from the marsh, from the bog, tillage is spreading which is exceptionally deserving of admiration.[1] We shall immediately be studying together one of the instances in which these methods have been applied on the greatest scale and with the most signal success in a setting in which Nature is hostile. The Finns are near neighbours of ours, but they live on the northern limits of habitable land in Europe. For all that, they have achieved such success as not only to increase their own resources substantially, but even to turn their country into an exporter of food products; Finland butter is at this moment competing in London with Danish, Normandy, and Irish butter.

Without quitting the surface of our Earth—geographers should always remain unhesitating realists—we shall become conscious to what extent we must turn to ideas for enlightenment as to co-operation between Earth and Humanity. The case of Finland will prove specially valuable, inasmuch as it will enable us to see how beliefs which are strongly held by many succeed in communicating, albeit by channels but dimly perceptible, efficiency and quality even to the most material products of that group's collective activities. On this Earth of ours there are regions, and in the lives of human societies there occur moments, wherein some essential fact of Human Geography—the maintenance of a route, the driving of a furrow, the grafting of a tree, the blow of a pickaxe in a quarry, the casting of a net in the sea—wins its way through a series of commonplace, almost unthinking efforts to a state of perfection of striking world-wide importance. And why? Because, at that point and at that moment, a collective psychic impulse—the cohesion of a nation that is being born or being attacked, the pride of a people which is making history or is endeavouring to do so, a tradition of bounden duty, apostolic zeal for the triumph of a creed—reinforces the feeling that all are working for each other, multiplying thereby the potentialities of every individual act, however trivial. To the psychology of crowds, therefore, we shall sometimes have to look, and to that alone, for the explanation of exceptional fruitfulness in a given region.

Man's eye and brain convey certain impressions to him concerning the Universe, impressions which are to some extent his own creation. Not only does the surface of the Earth suffer changes at our hands, but our vision, our interpretation of it, undergo changes too. Widely different ideas from ours did our predecessors entertain concerning that strip of land that was theirs and is ours. And in so far as our vision is a new vision, the reality is likewise a new reality. That is the principal fact; change is upon it even when it is only we who are changing. But then, too,—as we have already pointed out—fundamental material changes

[1] Cf. E. Coquidé, *Recherches sur les propriétés des sols tourbeux de la Picardie*, Paris, 1912.

do take place—North America and South America become islands because human beings will it so, and that for all the geography of intercommunication of the future. It does alter, and that through us.

Of these two kinds of alterations, the former is frequently subconscious or unconscious, and consequently more or less collective; while the latter is far more conscious, and often has been predetermined by individual initiative previous to entering on its collective phase. The essence and aim of Human Geography, then, amount to this:—observation, analysis, and explanation of these twofold and never-ceasing transformations of the inhabited surface of our planet, which is successively, or often simultaneously, becoming *different to us and different by us.* "Successively" and "simultaneously"; what characterises the ever-recurring and limitless interaction of either kind of transformation upon the other? Will it ever be practicable to define this with any certainty? A State ensues from a co-ordination of individuals and of groups which cannot be conceived without a material co-ordination between the various parts of the surface on which these individuals and groups exist. What is the minimum of connection necessary between a State, say, France or Switzerland, and its domain? What is the necessary connection between a city, a collection of citizens, and a town, a collection of houses and streets?

How far is spiritual co-ordination a function of material co-ordination, and how far does material co-ordination necessitate spiritual co-ordination? There we have the whole problem, and there, too, we have its geographical formula.

Can the mutual relations of these two kinds of co-ordination be ascertained more amply than in relation to particular cases, sufficiently so to enable us to look forward to a time when we may formulate their "law," in the fullest sense of the term? Is there a harmony, not antecedent but evolved, foreshadowing collective domination of the Earth, domination at once fruitful, peaceful, stable; a harmony capable of welding, if not into one group, at any rate into analogous or homologous groups, all peoples, all the historical aggregation of races and human societies? I do not know; as yet I can say nothing on the subject. That is what you and I will search after, in all humility and patience, in all the years that are to come.

MOZAMBIQUE: ITS AGRICULTURE AND DEVELOPMENT.[1]

(*With Illustrations.*)

The author of this book, Mr. Robert Nunez Lyne, F.L.S., F.R.G.S., has already written on "Zanzibar in Contemporary Times," he having been Director of Agriculture at Zanzibar. At present he is Director of Agriculture in Ceylon, but he was also for some time Director of

[1] *Mozambique: Its Agriculture and Development.* By Robert N. Lyne. London: T. Fisher Unwin, 1913. Price 12s. 6d. net.

Agriculture in the Province of Mozambique. His latest work, proceeding as it does from an agricultural specialist, may thus be regarded as a volume of much practical value. He tells us in his Preface that "with

A good Coconut Palm.

the exception of the case of the clove industry of Zanzibar, which has been carried on by the Arabs for nearly a hundred years, agriculture on the East coast of Africa is in its infancy. This is especially so in Portuguese East Africa. Hence, no handbook of the agriculture of that country, with any pretence to authority, could as yet be written, and this book lays no claim to such. It is an epitome of the conclusions

American Tobacco in East Africa.

come to after an eighteen months' examination of the territory and its agricultural resources and prospects, and is an attempt to reveal Portuguese East Africa to the investor."

Mozambique possesses several natural advantages. Mr. Lyne points out that the Province is particularly favoured in its navigable rivers, and its enormous stretch of seaboard, and that it is also well placed as regards markets. While its rubber plantations compare favourably with those of German East Africa, the River Zambezi is pre-eminently associated with sugar-planting, cotton, cattle and tobacco. There are seven sugar factories at work in the Province of Mozambique. There is good coconut land in the delta of the Zambezi between the Chinde River and the main stream. There are probably a million hectares of Landolphia forests in the Province carrying rubber of the value, at £10 per hectare, of £10,000,000. Sisal hemp is being grown in several parts of the Province, notably in Quilimane, where there are close upon three million plants in cultivation.

The most profitable planting industry in East Africa at present is probably that of tobacco-planting in Nyasaland, and there is no reason why it should not also succeed in Portuguese East Africa. Cotton is also making headway in Nyasaland, and it is safe to predict a future for it in Mozambique. In the oil-producing *Trichilia emetica* tree, the country of the lower Limpopo and on to Lourenzo Marques possesses an enormous source of wealth as yet practically untapped. The cashew-nut tree ought to furnish Inhambane with another extensive means of wealth.

Cattle-breeding and dairying are also making good progress in Mozambique, but there is only a limited future for fruit-farming. Many miles of river bank at Lourenzo Marques could be put under lucerne, which is an irrigation crop. The cultivation of vanilla, nutmegs, cinnamon and other spices will probably attract attention when settlers begin to enter the country, but timber will never become an important article of export because of the cost of transport.

One of the difficulties of developing East Africa is caused by the scarcity of labour arising from three circumstances, viz. (1) a sparse population; (2) the indolent disposition of the natives; and (3) the inexperience of white employers. Africans, taken as a whole, dislike regular work. The natives often also show a predilection for intoxicants. In Mozambique it is generally Portuguese wine, in Nigeria gin.

The volume concludes with "Notes on Ceara Rubber in East Africa," and a translation of "The Land Law for the Province of Mozambique," approved by decree of 9th July 1909. It has a good map of the Province, and index, and should be indispensable for every emigrant to Mozambique or investor in its industries. By the courtesy of the publishers we are enabled to reproduce here two of the illustrations.

GEOGRAPHICAL NOTES.

Australasia.

Dr. Wollaston's Expedition to New Guinea.—Dr. Wollaston returned last year to New Guinea with a new expedition to the Snow Mountains, in order to make another attempt on Mount Carstensz, which the last expedition did not succeed in ascending. The present expedition was more successful, though the difficulties proved very great.

After proceeding to Batavia to confer with the Dutch authorities, Dr. Wollaston went to Borneo, and spent eight weeks collecting Dyaks before returning to Batavia. Here he was joined by Mr. C. B. Kloss, curator of the Kuala Lumpor Museum, and the two, with five native collectors and seventy-five Dyaks, proceeded in a Dutch Government ship to the south coast of Dutch New Guinea. They were escorted by forty Dutch soldiers and eighty convicts from Batavia under a Dutch officer.

Last September the party disembarked at the mouth of the Utakwa River, which had been ascended by Dutch travellers two years previously, and appeared to be the best route into the unknown interior. From the deck of the ship could be seen the snow-capped peaks of Carstensz. A motor boat that had been built in England and all the stores and equipment were landed, and a base camp made twenty miles up this river. All this region was quite uninhabited, and the expedition had to carry all its own food. Canoes were made by the Dyaks, and the river was ascended for two days beyond the base, but thereafter the expedition travelled by land. Depôts were established three days' travel apart, the first being three days' march up the foothills of the Snow Range. From the fourth depôt the ascent was made to the snow line. Progress was very slow, the ridges being appallingly steep and the track rough.

In the high mountains the sun was never visible except for an hour in the morning, and the travellers were always in the clouds. At about five thousand feet the expedition met some curious but friendly folk of smallish stature, who showed the travellers their track, and helped them. They were not, however, pigmies. They are a rather small people of the Papuan type, and are of a dark chocolate-brown colour. They wear no clothing, notwithstanding that at the altitude at which they live it is very cold at night. In some respects they are more intelligent than the people on the coast. They are a race of cultivators, growing sweet potatoes, tobacco, and sugar-cane. They carry bows and arrows, and when they travel they sling over their shoulders bags containing apparatus for producing fire, their tobacco, knives, spoons, and so forth. Sweet potatoes, which they roast in a fire, form their staple food. They dig out the inside of the potato with a spoon made of the shoulder-blade of a pig or other animal. Their knives are made of stone of a hard, slaty variety which takes a very fine edge, and is made so sharp as to enable the user to cut a bamboo with it. A stone axe, which they also employ, is likewise susceptible of receiving a sharp cutting edge.

As regards their means of subsistence, they live from hand to mouth, killing rats, mice, and any other small animals they can find, and shooting birds with their bows and arrows, thus eking out the small resources derived from cultivation.

The highest point (15,000 feet) was reached after five days' march from the last base. The rain descended in a continuous torrent, and although Mount Carstensz is close to the equator, the fog-laden air was freezingly cold. During the ascent a fine panorama was observed, but the mist again closed in, and when the party were within a very short distance of the top the steep surface and dense fog necessitated a retreat. Two attempts to reach the actual summit were made, but eventually food gave out As the last load was being taken to the base camp the canoe containing Dr. Wollaston and six Dyaks struck a snag in a swirling torrent and capsized. Dr. Wollaston was carried a long way down stream, and was almost completely exhausted. Much valuable property was lost, including maps, cameras, instruments and three months' diaries. The expedition secured an important collection of birds and plants.

POLAR.

The Australian Antarctic Expedition.—In a short note in our April issue (p. 205) we noted the fact that Captain Davis, in the *Aurora*, took off Mr. Wild's party of eight. Mr. Wild has recently communicated to Reuter's agency an account of the experiences of this party, from which the following is taken :—

"The party under my command were G. H. Dovers, surveyor; C. Harrison, artist and biologist; A. C. Hoadley, geologist; B. E. Jones, surgeon; A. L. Kennedy, magnetician; J. H. Hoyes, meteorologist; and A. D. Watson, geologist. We left Mawson at his base in Adélie Land on January 19 of last year with orders to form a second base on Sabrina Land or Knox Land. The former we soon ascertained did not exist, and the impenetrable pack prevented us from getting within sixty miles of Knox Land, with the result that instead of 400 miles we cruised for 1300 miles and still found no chance of landing. On February 11 we sighted a glacier which had probably been mistaken by Wilkes for Termination Land, and on 15th found a landing. This being Shackleton's birthday, we named it Shackleton Glacier. It looked an impossible spot. It was clearly a moving glacier, and its terrible cliffs, a hundred feet high, were badly broken and crevassed. Landing our hut, stores, etc., and hoisting them up the dangerous cliff was a long and difficult business, and our next care was to move them from the broken edge to a spot 640 yards distant, where we erected our hut. The temperature varied 40 degrees, dropping as low as $-15°$ Fahr. We covered 105 miles drawing stores between the glacier edge and the hut. We next made preparations for sledging, but were detained until the middle of March by blizzards and snow-drifts 15 feet in depth.

"In the meantime, all sea ice blew away, leaving us with a perpendicular glacier edge, up which it was impossible for penguins or seals to reach, and for five months we had to depend entirely on tinned foods.

As soon as the weather permitted, a party of six left the hut to lay out a depôt on land, which we could see to the south 17 miles distant from our glacier. Kennedy and Watson were left at the base. Although in a direct line the land lay 17 miles off, our first journey to it was 88 miles in length. After eight days' travelling we reached a spot 35 miles inland at an altitude of 2200 feet. Crevasses abounded, and from March 21 for a period of nine days we were kept in camp by the same blizzard which proved fatal to Scott and his gallant companions. We soon found it impossible to go on, and turned back for home. Carrying only 50 lb. per man, the going was so hard that we only covered a mile and a quarter in eight hours down hill and sinking three feet in snow. When two miles from our hut, another blizzard held us up. One tent collapsed, and its three occupants were unable to move or get food for thirty-six hours.

"The days were now becoming too short and the weather too uncertain for extensive sledge work, and we made preparations for winter. In August we again made preparations for sledging, one party of three going eastward and another to west. The latter surveyed all the coastline to the point reached by the German Expedition of 1902. The western party did most of its travelling on land at an altitude of 2000 to 3000 feet. On one trip it did 510 miles at that altitude. This party discovered the largest Emperor penguin rookery ever recorded. This was on an island sixty-five miles west of our glacier hut, and here were congregated some seven thousand young Emperor birds, in addition to innumerable ordinary penguins. The eastern party surveyed as far as 101° east longitude, and the west inland for fifty miles, reaching an altitude of 4500 feet. The blizzards were very severe; one exceptionally bad one split a tent and caused the others to collapse. We were thus without shelter in a hundred-miles-an-hour wind. For five days we lay in a covered hole twelve feet by six feet by three feet. At intervals awful avalanches crashed down from a 600-feet cliff 400 yards from us, while giant boulders of ice, weighing twenty tons, came to within a hundred yards of our hole, which three months later was itself engulfed.

"On Christmas Day we formally took possession of Queen Mary's Land, and hoisted the Union Jack. I called my companions together to witness the act, as we took the land in the name of the Australian expedition for King George V. The land is a continuation of King Edward VII. plateau, and has a coastline of 350 miles, and ascends gradually, probably to the Pole itself."

Mr. Stefansson's Expedition to the Arctic.—Mr. Stefansson's expedition, which includes among its members Mr. James Murray, who, it will be remembered, was a member of the Scottish Loch Survey, and also accompanied the British Antarctic expedition of 1907-9, has made a start from Esquimalt. The whaler *Karluk* (not *Kamluk*, as stated in our previous note, p. 204) left that port on June 17 for Nome (Alaska). Mr. Stefansson, Dr. Anderson, and Mr. James Murray will join the ship there at the end of the first week of July. At this port the ship will put in only long enough to replenish stores, and thence proceed to Point Barrow, which is to be reached about July 25, the earliest possible date on account of the ice. If wind is easterly the explorers will sail into the Beaufort Sea. If, in turn, the Beaufort Sea is open and the *Karluk* gets to Herschel safely, eight men under the commander, Dr. Anderson, comprising the southern section, will disembark and sail eastward in the auxiliary vessel with stores and scientific equipment to Victoria Island. There a second base will be established.

Mr. Stefansson in the *Karluk* will go northward to Patrick Land, where he will make his first base, and then by boat or dog-sledge explore as far north as possible the unknown Arctic region. The *Karluk* will return to Victoria in November, and sail back to the Arctic in the summer of 1914 with fresh stores and mails for the expedition.

General.

The Birmingham Meeting of the British Association.—The arrangements for this meeting, which is to open on Wednesday, September 10, are now far advanced, and the attendance promises to be large. Though it has been found impossible to house the whole of the sections and the offices of the association in one building, the buildings utilised all lie within easy reach of one another. Eight of the thirteen sections are to be housed in Mason College, the remaining five finding quarters elsewhere. Geography is to find a home in the lecture theatre of the Midland Institute.

Of the evening discourses one will be delivered by Dr. Smith Woodward, of the British Museum, and will be on missing links among extinct animals. A handbook and an excursion guidebook are being prepared, the former containing some geological and topographical maps prepared under the guidance and help of Professor Lapworth, which are described as marking an epoch in map-making. Numerous excursions have been arranged for, and a considerable number of entertainments will also be provided.

Commercial Geography.

Mineral Wealth of Canada and United States.—A note in a recent issue of *Science* gives some interesting figures comparing the produce of the mines of Canada and the United States. In 1911 the total value of the mineral produce of Canada was 103,220,994 dollars, or 14·42 dollars per head; during the same year the United States produced minerals valued at 1,918,326,253 dollars. Among the provinces of Canada, Ontario led with a total value of nearly 43,000,000 dollars, British Columbia following with a value of 21,000,000. The most valuable Canadian product was coal, 26½ million dollars. This is true also of the United States where, however, coal to the value of 327 million dollars was obtained. Only as regards nickel does any Canadian product exceed in value those of the United States. The Canadian produce of nickel was in 1911 worth over 10 million dollars, while that of the United States was very insignificant. Of the total Canadian produce, exports amounting to over 33 million dollars went to the United States, while under 7 million dollars' worth went to the United Kingdom. It is to be noted, however, that there is considerable interchange of mineral produce across the United States frontier. Coal especially is both exported to the States and imported from them, according to local supplies.

The preliminary figures for 1912 show a great increase in the production of minerals in Canada, the total value being 133 million dollars, as compared with 103 million in 1911.

The Lötschberg Tunnel.—We have recorded here from time to time the progress of this tunnel, and it is of interest to note that the opening of the new direct route to Italy through the tunnel is announced for July 15. There will be three expresses daily each way, and the fastest of these will accomplish the distance between Paris and Milan in 16¼ hours.

Personal.—Mr. A. R. Hinks, F.R.S., chief assistant at the Cambridge Observatory, and University lecturer in surveying and cartography, has been appointed Assistant Secretary of the Royal Geographical Society.

EDUCATIONAL.

We are informed that a Yorkshire Summer School of Geography has been instituted by the Universities of Leeds and Sheffield, in co-operation with Armstrong College, Newcastle-upon-Tyne, and with the help of the Education Committees of the County Councils of the North, West, and East Ridings of Yorkshire, and of the County Boroughs of Bradford, Huddersfield, Hull, Leeds, Middlesbrough, Sheffield, and York. For the present year, the work of organisation has been undertaken by the University of Leeds.

The headquarters of the School will, this year, be the County School, Whitby, the school buildings having been kindly lent by the Governors for this purpose.

The course will consist of lectures, laboratory-work, field-work, and demonstrations. Lectures will begin on Monday morning, August 4, and the course will end on Saturday, August 23. Two lectures will be given on each of five mornings in each week of the course. There will also be occasional evening lectures and discussions. The laboratory-work will include map-reading, methods of map-enlargement, and the making of relief-maps. The field-work will comprise plane-table and contour work. All the apparatus used will be simple and inexpensive. Methods applicable to school-work will be adopted. There will be three or four whole-day excursions, and eight afternoon excursions for field-work. The staff of teachers will be large in order that in laboratory-work the students may receive individual attention.

Lectures will be given on the following subjects :—The Geological Structure of Yorkshire, with especial reference to Relief and Mineral Distribution; The Historical Geography of Yorkshire, including pre-historic, Roman, Saxon, Danish, Medieval and Military Geography; Language and Place-names of Yorkshire; Sites of Towns; Architecture; The Vegetation and Agriculture of Yorkshire; The General Economic Geography of the Yorkshire area; Yorkshire Mining, past and present; The Textile and Iron and Steel Industries of Yorkshire; Meteorology; The Teaching of Geography.

Among the lecturers will be Professor Kendall, M.Sc., F.G.S. (Professor of Geology in the University of Leeds); Professor F. W. Moorman, B.A., Ph.D. (Professor of English Language); Mr. A. Gilligan, B.Sc., F.G.S. (Assistant Lecturer and Demonstrator in Geology); Mr. L. Rodwell Jones, B.Sc. (Assistant Lecturer in Geography); Dr. W. G. Smith, Ph.D. (Lecturer at the Edinburgh and East of Scotland College of Agriculture); Mr. W. P. Welpton, B.Sc. (Lecturer in Education and Master of Method in the University of Leeds); Mr. P. W. Dodd, B.A. (Assistant Lecturer in Classics). Certificates will be granted at the end of the course to students who have made a complete attendance at the lectures, practical classes, etc.

The charge for admission to the whole course of instruction is £3. It is proposed to limit the numbers to about two hundred. Applications for tickets should be made, with remittance, to the Secretary of the Yorkshire Summer School of Geography, The University, Leeds.

In the *Report* for the year 1912-13 of the Royal Scottish Museum, Edinburgh, we note that an Advising Committee on Education has now been appointed in connection with the Museum, with the objects of encouraging teachers and pupils to make greater use of the Museum, and of shaping the exhibits in the School Gallery to direct educational needs. It is proposed to issue a list with short descriptions of the objects which may best be employed by teachers as illustrations, and the Committee has also favourably considered the proposal to purchase certain Educational models for the Museum, notably a planetarium. This will show the sun and planets with their satellites, the planets and their satellites revolving with a velocity proportionate to their true velocity. The whole will be surrounded by a glass sphere about five feet in diameter, on which the stars of the best-known constellations are indicated. The whole mechanism will be worked by electricity and will be so arranged that it can be set in motion by pressing a button.

The *Report* gives also some account of the very popular lantern demonstrations which were held in the Museum during last winter. As the list given shows many of these had a more or less direct geographical bearing, and the Museum itself of course contains many objects of great interest to the geographical teacher. The efforts of Sir T. Carlaw Martin and those associated with him to render the Museum more useful and more accessible to teachers and pupils will therefore receive grateful acknowledgment from all those interested in geographical education.

NEW BOOKS.

EUROPE.

My Russian Year. By ROTHAY REYNOLDS. London: Mills and Boon, Ltd., 1913. Price 10*s.* 6*d. net.*

In this cleverly written and entertaining book we have a series of sketches portraying all sorts and conditions of men in Russia, and the general impression it conveys is that Russia is a land of most extraordinary contrasts, intellect and learning side by side with stupidity and ignorance, piety and religion with superstition and bigotry, charity and gentleness with hatred and cruelty, opulence and wealth with poverty and degradation, tyranny and arbitrariness with a paternal Government solicitous to undertake everything for its subjects. We do not forget that similar contrasts may be observed in many other countries, but it is a matter of proportion, and in Russia the contrasts are, according to the writer, unusually sharp, so much so that it may be called the land of paradox. The author has evidently travelled much in the interior as well as in the cities, and has been able to win the confidence of revolutionists and reactionaries both male and female, and not the least interesting part of this book is the insight it gives us into the train of thought, we might even call it the philosophy of these classes. There is an amusing as well as instructive chapter on prison-life, the anomalies and peculiarities of which are strange to our British ideas. No good account of the Russia of to-day could omit to accentuate the extraordinary influence of the religion of the Orthodox Greek Church, which permeates more or less through every grade in society in one shape or another, and this feature of the country receives ample attention in this book. The chapters on the Black and White Clergy are excellent, and so also are the two quaint tales told by Dimitri, which throw a brilliant as well as a humorous light on the religious intellectuality of the peasants. *My Russian Year* gives us a good impression of the profoundly perplexing and intricate problems which constitute the difficulty now confronting the Russian Government. They are the result of the huge area of the country, its geography, variety of races, religion, history, education, etc., combined with an antiquated system of government which insists on keeping its subjects in leading strings, thereby evoking more or less general discontent which takes every form of expressing itself from sullen despair to the ready use of dynamite. The graver interest of the book is lightened and indeed sustained by many anecdotes and conversations told with much vivacity and literary skill.

Rambles in Holland. By EDWIN and MARION SHARPE GREW. London: Mills and Boon, Ltd., 1913. Price 6*s.*

In our February number we noticed a cleverly written book *Through Holland in the Vivette*, which describes one way of seeing Holland, viz., from the deck of a yacht. In this book we have an excellent description of the country as seen from the railway and motor-car. It so happens that this year thousands of people will be flocking to Holland. Not only is there the special attraction of the formal inauguration of the Palace of Peace at the Hague, but in almost every place of importance there are to be exhibitions and festivities in honour of the centenary of the liberation of the Dutch from Napoleonic rule. Instead of concentrating their activities on one spot, the Dutch have decided to hold local exhibitions wherever local industries and circumstances indicate a good chance of success. Thus there will be a maritime exhibition at Amsterdam, an agricultural and horticultural exhibition at the Hague, a rose exhibition at Boskoop, a bulb exhibition at Haarlem, and so on, and it is needless to add that the exhibitions of Dutch fine arts at Utrecht, etc., will be well worthy of the traditions of the nation. The volume now before us can be confidently commended as well worth study both beforehand and on the spot by any of our readers who think of going to Holland. It is a most interesting and judicious *mélange.* The reader will find many well-told anecdotes and episodes from the romantic and glorious history of the nation; much useful information regarding local industries and economic conditions, both past and present, without an excessive use of statistics; many wise and discriminating observations about the fine arts, more especially architecture, both secular and ecclesiastical; and plenty of hints and suggestions, attention to which will save much time and trouble, and greatly facilitate the traveller's movements. We may add that there are a number of very good photographic illustrations.

ASIA.

The Duab of Turkestan. By W. RICKMER RICKMERS. Cambridge: At the University Press, 1913. Price 30*s. net.*

Mr. Rickmers has had a long and varied experience of Bokhara, and we trust that the members of this Society have not forgotten, that so long ago as the year 1900 he lectured on Bokhara in Edinburgh and at the branches. In this handsome and beautifully illustrated volume we have the account of his later wanderings, explorations and mountaineering in the territory between the Jaxartes and the Oxus, the Sir-Darya and the Amu-Darya, the Duab of Turkestan. The work is essentially a study of physiography and geology, but the author goes far afield, and has many acute and instructive observations to offer on the climate, zoology and botany of the tract, and also on the political and economic conditions of its inhabitants. Cordially received as old friends, he and his party were welcomed wherever they went, alike by the Russian authorities, and by the local chiefs. Except on one occasion, they were never in danger of personal violence, but in the course of their wanderings and climbing they fell in with many difficulties, and it is very evident that a cool head and wary movement are quite as necessary in the mountains of Turkestan as in the Alps or the Andes. Mr. Rickmers has had much experience in mountaineering in other countries, and not the least valuable part of this work is the instructive comparisons which are drawn from time to time between the conditions and methods of similar expeditions elsewhere. For example, his remarks on the well-known phenomenon of mountain nausea are well worth comparing with the accounts we have of it in the Andes by Sir Martin Conway, and in the Karakorams by Mrs. Bullock Workman and the historian of the Duke of the Abruzzi's expedition. The work is, as we have said, principally a study in physiography, and the narrative does not maintain a chronological sequence. For instance, we have an interesting account of an expedition to Karshi in 1898 followed a few pages afterwards by an account of an ascent of the Kemkutan peak near Samarkand in 1907. Other expeditions to the mountains of Urgat, the Fan or Hazret Sultan, the range of Peter the Great, etc., are described with great wealth of details bearing on the physiography and geology of the tract. Vivid and sparkling descriptions of important towns such as Samarkand and Bokhara lend a pleasing variety to the severer features of the book. It was at the camp at Landan, in the Chapdara mountains, that Mr. Rickmers and his wife—a granddaughter of the well-known Dr. Alexander Duff, who read an interesting paper on her travels with her husband and Dr. Krafft in Eastern Bokhara to the British Association in 1898, and who is the authoress of the standard work on the *Chronology of India*—were for a brief space of time in danger of their lives from robbers. Great prominence is given to the Zarafshan river, the 400 miles course of which is minutely traced from its source in a glacier near the Igol peak in the Alai range to where it disappears in the sands south-west of the city of Bokhara, and Mr. Rickmers succeeds in investing the river with an unwonted variety of interest.

Excellent maps and indices greatly facilitate the reader in enjoying this book, but the numerous illustrations deserve special notice. Many of them considered merely as representations of landscapes are exceedingly beautiful, and many of them are exceptionally illuminative as well as beautiful, because they assist the reader materially in visualising to himself the scenery described under varied and strange climatic and physiological conditions, the representation of which is the chief object of the book.

Notre France d'Extrême-Orient. Par le Duc de Montpensier. Paris: Perrin et Cie, 1913. *Prix 5 fr.*

By the publication of this work the Duc de Montpensier concludes his exposition of his study of French Indo-China during the five years which he spent there. In his former volumes he gave us the lighter side of the subject, and in this volume we have his views on more serious topics, ethnographical, political, economic, etc. They are set forth in crisp and graceful language, lucid and vivid, such as we find in good French authors, modest also and undogmatic, which impresses the reader with a sense of the author's impartiality and accuracy, and as they are the result of patient careful study of the conditions of Indo-China by an able and dispassionate observer they are well worth the attention of the French Colonial Office. The Duc without unwise optimism or equally unwise chauvinism anticipates a great future for the French empire in the East. He sets forth in much detail—too much detail perhaps for an English reader, but not too much for a French one making a serious study of the subject—its natural resources, its capabilities, the distribution of the many different races which inhabit Indo-China, an estimate of their social conditions, and the causes which contribute to or retard material progress. French Indo-China, like French Equatorial Africa, has suffered from the ill-timed parsimony and neglect of the home authorities, but its prospects are brighter than those of French Equatorial Africa, for it appears that the home Government is now making a serious effort to effect what is a primary necessity of development, viz., a great increase and improvement in the various methods of communication between the principal towns and villages and to connect them with the sea. Indo-China has had the good fortune to have had as Governors-General some excellent administrators, and it is noticeable that the Duc de Montpensier does not animadvert on the frequent changes of officials, which have been so often the subject of just complaint in other colonies. One of the most interesting chapters is that in which he describes the system of government, which in some respects resembles the system established in British India while differing from it in others; another chapter, and one much too short, treats of the local literature, and relates two or three charmingly quaint fairy tales. In their Indo-China empire the French are wisely recognising that they have to deal with many races differing from each other widely in origin, civilisation, development, and progress, and are adapting their methods of government to existing conditions. A Government working on these sagacious and kindly principles is sure of success, although it must be a work of patience and time—success which, one day, will fulfil the patriotic hope of the author that Indo-China will be "un des joyaux les plus précieux du collier glorieux de notre belle France."

By Desert Ways to Baghdad. By LOUISA JEBB. London, Edinburgh, Dublin and New York: Thomas Nelson and Sons, 1912. *Price* 1*s. net.*

This volume contains a pleasantly written series of sketches of a journey from Brusa in Asia Minor to Bagdad, Damascus, and Palmyra. The travellers were two plucky English ladies, who, at first at any rate, knew no Arabic. They had, however, good luck, and came in for no worse mishaps than rainstorms and malarial fever; on the other hand, they had often to rough it in many ways, all of which they seem to have enjoyed heartily. Perhaps the best chapters in the book are those describing the descent of the Tigris on a raft from Diarbekir to Bagdad.

AFRICA.

D'Alger à Tombouctou. Par Comte RENÉ LE MORE. Paris: Plon-Nourrit et Cie, 1913. *Prix* 3 *fr.* 50.

This volume contains an account of a very plucky and successful journey of fifteen months' duration from Algiers to Timbuctu and back, right across the desert. On his way to Timbuctu Count René Le More proceeded from station to station occupied by French troops, but on his return journey he, for a part of the way, took a short cut through a more dangerous, because less known tract. He had various difficulties to contend with, such as malarial fever, and the chance of finding himself and his party without water, but above all, his real danger was from roving bands of Tuaregs, who may be justly called the pirates of the desert. Under any circumstances the journey is an arduous one, and it tested the Count's physique pretty severely. The narrative is very simply and modestly written, and gives useful hints on many points, mostly military, and some commercial, in connection with the various places passed en route. The Count seems anxious to establish an aeroplane service between Algiers and Timbuctu, a project which, some years ago, might have emanated from the fertile brain of Jules Verne, but here we have it calmly worked out in detail, and the Count is ready to make the first aerial journey. The advantages, on which he relies to recommend his project to the authorities, are that the journey through the air will be much quicker and healthier than the present toilsome and unhealthy one across the desert on camels, and that the service once established would put an end to Tuareg raids, and so stimulate commerce, and effect considerable reductions in the military budget. The French now lead the way in aerial science, and it would be rash to say that such a service as the Count contemplates is impossible.

Les Deux Congo. Par Baron JEHAN DE WITTE. Paris: Librairie Plon-Nourrit et Cie, 1913. *Prix* 4 *fr.*

The two Congos which are the subject-matter of this book are the French and the Belgian. The story of the French Congo is treated from the missionary point of view, and may be described as an enthusiastic eulogistic account of the rise and spread of Christianity under the auspices of the Roman Catholic Church. The author follows the missionaries from their arrival in 1843 up to the present day, laying much, but not too much, stress on the special work of Père Augouard, who by his labours of thirty-five years has justly earned the title of the Apostle of the Congo. In the course of his career he encountered perils and dangers such as rarely fall to the lot of men, but he persevered and succeeded in founding and fostering missions, churches, schools, and even farms in the midst of savage cannibal tribes under conditions which might well have daunted the most adventurous spirit. It is no vain boast that in French Congo the missionaries have done more to open up the country than any other agency. For the details of the story we must refer our readers to the book itself. Incidentally it throws much light on the condition of the inhabitants and the gross mismanagement of the country owing to the neglect of the Government to attend to the elementary principles of good administration. Attention has often been directed to the constant changes of officials, obviously so detrimental to good government (cf. *L'Afrique Équatoriale Française*, par Maurice Boudet-Saint, *S.G.M.*, vol. xxviii. p. 491). Baron de Witte gives us some examples of this, on which comment is needless; there were 10 administrators in Brazzaville in two years, 7 in Cap-Lopez in one year, and 14 governors in Gabon in fifteen years, "fonctionnaires éphémères" as he justly calls them. The conclusion of this part of the book is a wail against the action of M. Caillard, who, after the affair of Agadir, ceded to Germany 275,000 square kilometres of fertile land on the east of the Cameroons in exchange for 14,000 square kilometres of swamp on the north. The second part, dealing with the Belgian Congo, is practically a *résumé* of the action of King Leopold II. from 1876 till his death. The Baron does full justice to the astuteness and tenacity of purpose of the king, but denounces his duplicity and greed. Into this subject we need not follow him. Dislike of England and distrust of Germany are both very apparent throughout the work.

The Land of Zinj. By Captain C. H. STIGAND. London: Constable and Co., Ltd., 1913. Price 15*s. net.*

Captain Stigand, of the King's African Rifles, is well known as a mighty Nimrod, who has seen much service in British East Africa and Egypt, and some years ago made an adventurous expedition from Uganda by Lake Rudolph to Abyssinia (*S.G.M.*, vol. xxvi. p. 500). In this volume he has brought together the notes and observations of his long service regarding the many tribes and peoples with whom he has come in contact, and in so doing he has made a notable contribution to our knowledge of the anthropology of the Protectorate. He introduces his subject "The Land of Zinj," or "the land of the blacks," by a brief but fairly complete compilation of what is known of its history from various documents and authors beginning with the days of King Solomon. Amidst a mass of mixed material it is often hard to distinguish what is trustworthy from what is merely legendary or mythical. He was not allowed access to the original Swahili records, which are apparently kept studiously concealed, but his information was received first-hand from a direct descendant of the Pate Sultans, and taken down by Captain Stigand to his dictation. The greater and most instructive part of the book consists of a detailed description of the geography of the tract and of the distribution of the inhabitants, with copious notes on their appearance, dress, habits, laws, ceremonies, religion, economic conditions, character, etc. He has spared no pains to expiscate the minute differences in the customs, etc., of the tribes, and apparently he was able to ingratiate himself with them, so that they have not concealed from him anything he wished to ascertain. When it is remembered that he has had experience of more than a hundred tribes, it will be readily understood that this book is a valuable contribution to our knowledge of the anthropology of the tract. In the last chapters Captain Stigand deals with the thorny problem of the African and his future. He accepts as beyond controversy that for good or evil the European has taken over Africa, and is responsible for the African's future. But he earnestly deprecates over-haste in the process of civilisation, urging that in the case of many tribes the native is richer, happier, and better than he would be semi-civilised. He suggests that the most promising tribes should be placed in reserved tracts, where they would develop on their own lines under the paternal supervision of European officers, but preserved from the contamination inevitable when their country is open to traders of any race. He denounces the present system of education, viz., teaching the natives to read and write English, and he would substitute education in cleanliness, sanitation, and agriculture. He would relegate "the indolent and worthless" tribes to be the hewers of wood and drawers of water for the Europeans and rest of the community. It is obviously beyond our province to enter on a discussion of such a contentious subject, but we think we have said enough to indicate that *The Land of Zinj* is an exceedingly interesting and instructive work, and one well worthy of the perusal of our readers. We must add that it is illustrated by many good photographs, and is equipped with an excellent map.

AMERICA.

Latin America. By F. GARCIA CALDERON. London : T. Fisher Unwin. Price 10*s.* 6*d. net.*

In this volume a young Peruvian diplomatist has essayed what is wellnigh an impossible task, viz. to give in the compass of 400 pages an appreciation of the rise and progress of the score of large and small republics on the other side of the Atlantic which he includes in Latin America. He proposes to himself to lay before his readers "a balance-sheet" of these republics, sketching their tangled history, the races from which their populations are drawn, their condition, intellectual, economic, political, religious, ethical, etc., the evolution by which they have reached their present condition, and their probable future. The mere statement of such a subject indicates the width of its scope and the difficulties surrounding it. On the whole M. Calderon has done ample justice to his subject from his own point of view, subject to the limitations of space imposed on him. He points to the broad fact that, notwithstanding the diversity of races and geographical and economic conditions within the various republics, their histories present at least one feature in common, viz. that after the struggle for independence of a century ago they pass through a more or less lengthy period of what may be called anarchy. Colombia, for example, has had twenty-seven civil wars. "Uruguay," says Mr. Bryce in his *South America*, "saw more incessant fighting from 1810 to 1876 than any other part of the world has seen for the last hundred years, and even since then risings and conflicts have been frequent." The rule of military dictators lasts for a greater or a shorter period, and is succeeded by "civilism," which, according to M. Calderon, is fatal to clericalism ; civilism is succeeded by plutocracy ; and after a lapse of nearly a century only five states, Argentina, Brazil, Chili, Peru, and Bolivia, have made serious attempts to institute constitutional government. It would be interesting to contrast the views of the South American diplomatist, M. Calderon, with those of the Anglo-Saxon diplomatist, Mr. Bryce, but limitations of space preclude us from entering upon so fascinating a comparison. It is well known that the Latin republics of America are at various stages of political, industrial and economic development, and M. Calderon has many shrewd and judicious observations as to the causes of progress or delay in their national development. His observations on the effects of the climate, geography, intermixture of races, religion, education, etc., and on the possibilities of foreign interference are well worth serious study, and here again a comparison with Mr. Bryce's observations might profitably be made. As to the future, M. Calderon lays much stress on the inevitable rearrangement of the trade and commerce of the world which will result from the opening of the Panama Canal. We observe that he takes a very gloomy view of the future of England. "England has reached the zenith of her industrial period, the maximum of her political development; the figures of the birth-rate in the industrial towns are diminishing, and emigration has almost ceased. The State is becoming the protector of a demagogic and decadent crowd."

We trust we have said enough to show to our readers that this volume is one of intense and varied interest, all the more so because readers, politicians or economists, on both sides of the Atlantic will find much in it that they are inclined to dispute. To many readers the most interesting part of the book will be the preface, which has been contributed by M. Poincaré, now President of the French Republic. The preface is, in fact, a thoughtful and appreciative review.

The Panama Canal. By JOHN BARRETT. Washington, D.C., U.S.A.: Pan-American Union, 1913. *Price* $1.00.

Mr. Barrett is Director General of the Pan-American Union, which is an "international organisation and office maintained by the twenty-one American republics . . . and devoted to the development of commerce, friendly intercourse, good understanding, and peace among all the American republics," and it will be readily admitted that in at least a good many of these republics, there is ample scope for the further development of these essentials of prosperity and stability. This little book is designed to answer two important questions about the Panama Canal, viz., what it is and what it means. It bristles with facts and statistics of intense interest, not only to those who live on the other side of the Atlantic, but to practically the whole civilised world. Panama and its wonderful Canal have been much in evidence of late years, and have received ample attention from this Society. In 1910 (*S.G.M.*, vol. xxvi. p. 148), we had an interesting lecture from Mr. Cornish, describing the engineering features of the work while it was in progress: in 1911 (*S.G.M.*, vol. xxvii. page 17), we had an instructive article by Mr. Peddie on the various former projects to cut through the Isthmus, some of them dating from the early years of the sixteenth century; in 1912 (*S.G.M.*, vol. xxviii. p. 330), there is a review of a clever book by Arthur Brellard (Albert Edwards) describing the Canal two years after Mr. Cornish had seen it; now, this book by the Director of the Pan-American Union brings the story up to date. It seems primarily intended for the use of tourists who go to see the Canal, and it is also an advertisement of how to go, what to see, and where to stay: it describes the local engineering and sanitary problems, and how they have been overcome, and it attempts to furnish an answer to the difficult question of what will be the results to the world when the Canal is ready. Mr. Barrett writes with exceptional knowledge, for he has been United States Ambassador in Panama, and he has produced an excellent compendium of information sufficient to furnish an answer to almost any question. The book is profusely illustrated, and has several excellent maps.

GENERAL.

Doreen Coasting. Edited by ALYS LOWTH. London: Longmans, Green and Co., 1912. *Price* 10*s.* 6*d.*

Doreen, being seized with a sudden desire, such as comes to many of us at times, to see the world, took boat for Africa, and this book is an account of her travels. The life on board, with its dances and its gossip, its quarrels and reconciliations, is recounted in a pleasant enough fashion, and numerous photographs may give those who must stay at home some idea of the places Doreen saw in her "coasting."

Weather Science. By R. G. K. LEMPFERT, M.A., Superintendent of the Forecast Division of the Meteorological Office. ("The People's Books.") Pp. 94. London: T. C. and E. C. Jack, 1912. *Price* 6*d.*

This little monograph can only be properly described as a masterly piece of work. Written out of the fulness of knowledge by one who is himself well known for his own original researches, the key to the whole is struck at the very outset, where the physical processes involved in atmospheric changes are discussed in the simplest possible terms. There is everywhere an appeal to these physical processes, so that the book forms a coherent whole. Considerable space is devoted to synoptic meteorology and the problems of forecasting weather by means of synoptic charts, and Mr. Lempfert is equally happy in dealing with the comparatively local problem of a "line squall" and in describing the great seasonal variations of pressure over the surface of the globe and their relation to variations in the seasonal incidence of rainfall. A short chapter deals with recent researches in the Upper Air, and it is pointed out that "the changes met with above are quite as great in magnitude and as sudden in their occurrence as those with which we are familiar at ground-level."

There is a generous supply of maps and diagrams, and we rejoice that the book is published at a price which will ensure for it a very wide circulation. It would form an admirable text-book, to be supplemented by such statistics as might be found desirable.

Portuguese Self-Taught. By E. DA CUNHA. Second Edition. Revised.
Norwegian Self-Taught. By C. A. THIMM. Revised and Enlarged by P. HANSSEN. Fifth Edition. London: E. Marlborough and Co., 1912. *Price* 2*s.* 6*d. each.*

Both of these are new editions of books which have already proved their usefulness to travellers and others, and will be welcomed by the considerable public which is in the habit of visiting one or other of the two countries concerned.

The Syrian Goddess: being a translation of Lucian's "De Dea Syria." By Professor HERBERT A. STRONG, M.A., LL.D. Edited by JOHN GARSTANG, M.A., D.Sc. London: Constable and Co., 1913. *Price* 4*s. net.*

This little volume is an interesting contribution to the study of Oriental religions. It contains a translation of Lucian's treatise on the Syrian goddess Astarte, by Professor Strong, with abundance of learned notes and comments by Dr. Garstang, the greatest living authority on the Hittites. To this has been added a brief sketch of the life of Lucian with an estimate of his opportunities of forming a correct opinion of the subject of his treatise. The book will be valued by all students of ancient cults and their mysteries.

Greater Rome and Greater Britain. By Sir C. P. Lucas, K.C.B., K.C.M.G. Oxford: At the Clarendon Press, 1912. *Price* 3*s.* 6*d. net.*

This intensely interesting and most instructive essay is the work of a scholar, a student, an administrator and a statesman. It contains an exhaustive and thoughtful comparison between the Roman and the British empires, setting forth how each was acquired, held and administered, and it enlarges on how each has been affected by permanent as well as transient causes, such as space, distance, water, medical science, and the like. It contrasts the genius of Roman rule with that of Britain, showing how in the case of the former the ideal of rule was to withdraw freedom from those ruled, while in the latter the ideal is to increase freedom. The Roman empire was in the main a Mediterranean empire, "all or nearly all within the temperate zone, not concerned with lands of great cold, not concerned with the Tropics, not concerned to any appreciable extent with coloured races." But "England has overflowed into temperate zones like her own; she has found them wholly outside Europe, and very largely in the most remote part of the world, in the far south. The contrast between the Roman and the British Empires is illustrated and emphasised, if it is borne in mind that they are in the main geographically exclusive of each other." And now the British Empire is in reality two Empires, one the self-governing dominions, and the other, India, the Crown Colonies and Protectorates. It is in his wise discrimination of the methods which should be adopted for the development and administration of these two empires that Sir Charles shows his mature and skilful statesmanship. He recognises the complexities and difficulties of the many problems, which are incidental and inevitable to the work, and he states many of them with the lucidity and impartiality of the experienced administrator. He recognises that, as at present constituted, there are conflicting interests in various parts of the Empire which need the most careful handling, and may entail on the Colonial Office inconsistency in the methods of administration. He is frankly an opportunist. "The British present," he says, "has grown up on no definite plan. So far from being logical, it is a unity of contradictions, absolutely impossible on paper, but working very comfortably in fact. To anything like an orderly ground-plan of the future, British instinct, which constitutes British genius, is opposed. There is only one sure guide to the future, and that is the race instinct which represents day to day opportunism." Sir Charles comes to the conclusion that if England is to hold her own as a nation she must keep the self-governing dominions with her, and that Imperial Preference is not too high a price to pay for this: he believes it to be inevitable that in course of time, when the self-governing dominions have come to maturity, they must find outlets beyond their present boundaries, and the natural as well as the most profitable outlets are the Crown Colonies and the Protectorates. The self-governing dominions know this, and so will be most reluctant to break away from the Mother Country. This, however, is not the place for a political discussion. In conclusion, we can assure our readers that there is not a chapter in this book which is not replete with the results of wise, thoughtful, and enlightened study, presented in clear and undogmatic language which commands admiration and repays attention.

EDUCATIONAL.

Asia: A Geography Reader. By ELLSWORTH HUNTINGTON. Chicago: Rand M'Nally and Co., 1913.

This is an exceedingly interesting little book, illustrated by many photographs and some rather crude maps. Professor Huntington's detailed knowledge of the regions described enables him to add many little human touches to complete the picture. We have enjoyed especially the account of the journey by the Transcaspian Railway, with its description of the boiling samovars at the stations, surrounded by a crowd of eager passengers, each with his empty teapot ready to be filled. Teachers here will perhaps regard the American spelling as a disadvantage, otherwise the book may be warmly recommended.

A Practical and Experimental Geography. By FREDERICK MORROW, B.A., F.R.G.S., and ERNEST LAMBERT, L.C.P. London: Meiklejohn and Son, 1913. *Price 2s. 6d. net.*

This book must prove very useful in such schools as devote considerable time to the so-called practical side of geography. It may be doubted, of course, whether the calculation of distances, field surveying, etc., are after all really geography. In our opinion geography proper does not commence till Chapter VII. is reached. From this point onwards the book is very good and would acquire even great ervalue if Chapter VIII. were doubled in length. The whole get-up of the volume—printing, binding, etc.—is excellent; the diagrams are simple and clear; and the explanations easy to follow. Messrs. Morrow and Lambert's book will be a formidable rival to time-honoured Simmons and Richardson, and to the more recent works by Unstead, and by Fairgrieve and Young.

A Scientific Geography. Book VIII.: South America. By ELLIS W. HEATON, B.Sc., F.G.S. London: Ralph, Holland and Co., 1913. *Price 1s. net.*

This small volume completes the series of books which comprise "A Scientific Geography." Mr. Heaton's experience in writing the companion volumes has made Book VIII. the best of the series, though it is followed closely by Book II., The British Isles, and Book VII., The British Empire. Of the present volume Part II. is decidedly better than Part I., and of the separate chapters, Nos. II., IV., and VI. may be specially mentioned. The diagrams are to the point, and the maps are clear and not over-crowded with names. The most useful figures are 6, 10, 12, 13, and 18. We would wish to have had more about the Trans-Andine Railway and the Panama Canal. Further, there is, as in the other volumes of the series, a slight tendency to make too much use of the short paragraph. Connection of ideas often demands the linking up of certain paragraphs, *e.g.* on p. 14. A few minor defects, mainly linguistic, do not detract from the value of the book considered strictly geographically. For instance, the spelling of Barranquilla and Buenaventura (p. 38) should be corrected, and *contrast* (p. 40) and *series* (p. 69) should be given a singular verb.

BOOKS RECEIVED.

The Moselle. By CHARLES TOWER. With Illustrations in colour and black and white by LIONEL EDWARDS. Demy 8vo. Pp. x + 332. London: Constable and Co., Ltd., 1913. *Price 7s. 6d. net.*

Siam. By PIERRE LOTI. Translated from the French by W. P. BAINES. Illustrated. Demy 8vo. Pp. xi + 182. London: T. Werner Laurie, Ltd., 1913. *Price 7s. 6d. net.*

Dent's Practical Notebooks of Regional Geography. By HORACE PIGGOTT, M.A., Ph.D., and ROBERT J. FINCH, F.R.G.S. *Book VI.: The British Empire in America and Asia.* With a general survey of the Empire. Crown 4to. Pp. 64. London: J. M. Dent and Sons, Ltd., 1913. *Price 6d. net.*

The Framework of Union: A Comparison of Some Union Constitutions. With a Sketch of the Development of Union in Canada, Australia, and Germany, and the Text of the Constitutions of the United States, Canada, Germany, Switzerland, and Australia. Medium 8vo. Pp. viii + 207 + cxviii. Cape Town: Cape Times, Ltd., 1908.

The Log of a Rolling Stone. By HENRY ARTHUR BROOME. Illustrated from Water-colour Drawings by the Author. Demy 8vo. Pp. xv + 325. London: T. Werner Laurie, Ltd., 1913. *Price 12s. 6d. net.*

The Car Road-Book and Guide: An Encyclopædia of Motoring. With large folding road-map of England, Scotland, and Ireland. Edited by Lord MONTAGU. Crown 8vo. Pp. xvi + 623. London: The Car Illustrated, Ltd., 1913. *Price 12s. 6d. net.*

The Loire: The Record of a Pilgrimage from Gerbier de Joncs to St. Nazaire. By DOUGLAS GOLDRING. With Illustrations in colour and black and white by A. L. COLLINS. Demy 8vo. Pp. xxii + 332. London: Constable and Co., Ltd., 1913. *Price 7s. 6d. net.*

With Camera and Rucksack in the Oberland and Valais. By REGINALD A. MALBY, F.R.P.S., F.R.H.S. With over seventy photographic studies by the Author, in colour, photogravure, and half-tone. Demy 8vo. Pp. 316. London: Headley Brothers, 1913. *Price 10s. 6d. net.*

The Elements of Geography. By ROLLIN D. SALISBURY, HARLAN H. BARROWS, and WALTER S. TOWER. Crown 8vo. Pp. ix + 616. New York: Holt and Co., 1913.

The "Times" Shipping Number, 1912. Reprinted from the *Times*, Friday, December 13, 1912. Large 4to. Pp. xiv + 286.

The "Times" American Railway Number, 1912. Reprinted from the *Times*, Friday, June 28, 1912. Large 4to. Pp. liv + 245.

The Japanese Empire. A reprint of the *Times* Japanese edition, July 19, 1910. Large 4to. Pp. 439. London: The *Times* Office.

Maps and Survey. By ARTHUR R. HINCKS, M.A., F.R.S. Demy 8vo. Pp. xvi + 206. Cambridge: University Press, 1913. *Price 6s. net.*

Senior Geography of North America. By G. C. FRY, M.Sc. Crown 8vo. Pp. 44. London: University Tutorial Press, Ltd., 1913. *Price 1s.*

Senior Geography of Asia. By G. C. FRY, M.Sc. Crown 8vo. Pp. 36. London: University Tutorial Press, Ltd., 1913. *Price 1s.*

The Government of South Africa: Two vols. Crown 4to. Pp. xviii + 890. South Africa: The Central News Agency, Ltd., 1908.

The New World of the South: Australia in the Making. By W. H. FITCHETT, B.A., LL.D. Crown 8vo. Pp. xiv + 402. London: Smith Elder and Co., 1913. *Price 6s.*

The Argentine in the Twentieth Century. By ALBERT B. MARTINEZ, Under Secretary of State, and MAURICE LEWANDOWSKI, Doctor in Law. With a Map. Demy 8vo. Pp. liii + 376. London: T. Fisher Unwin, 1911. *Price 12s. 6d. net.*

Presented by the Vice-Consul for Argentina, Leith.

Trans-Himalaya: Discoveries and Adventures in Tibet. By SVEN HEDIN. With 156 Illustrations from Photographs, Water-colour Sketches, and Drawings by the Author, and 4 Maps. Vol. III. Demy 8vo. Pp. xv + 426. London: Macmillan and Co. Ltd., 1913. *Price* 15*s. net.*

A Busy Time in Mexico: An Unconventional Record of Mexican Incident. By HUGH B. C. POLLARD. Demy 8vo. Pp. vii + 243. London: Constable and Co., 1913. *Price* 8*s.* 6*d. net.*

An Introduction to Plant Geography. By M. E. HARDY, D.Sc. ("The Oxford Geographies." Edited by A. J. HERBERTSON.) Crown 8vo. Pp. 192. Oxford: Clarendon Press, 1913. *Price* 2*s.* 6*d.*

Opportunities in Canada, 1913: Official Information. Furnished by the Provincial Governments and the Boards of Trade in Canada. Editors: ERNEST HEATON, B.A. (Oxon.), and J. BEVERLEY ROBINSON. Crown 8vo. Pp. xxvi + 366. Toronto: Heaton's Agency, 1913.

Travels in the Pyrenees: including Andorra and the Coast from Barcelona to Carcassonne. By V. C. SCOTT O'CONNOR. With four coloured Plates and 158 other Illustrations and a Map. Demy 8vo. Pp. xx + 348. London: J. Long, Ltd., 1913. *Price* 10*s.* 6*d. net.*

The Year-Book of Wireless Telegraphy and Telephony, 1913. Crown 8vo. Pp. lxxv + 563. London: St. Catherine's Press, 1913. *Price* 2*s.* 6*d. net.*

The Land of the Blue Poppy: Travels of a Naturalist in Eastern Tibet. By F. KINGDON WARD, B.A., F.R.G.S. Medium 8vo. Pp. xii + 283. Cambridge: University Press, 1913. *Price* 12*s.* 6*d. net.*

Eine Geographische Studienreise durch das Westliche Europa. Von W. HANNS, A. RUHL, H. SPETHMANN, H. WALDBAUR. Mit einer Einleitung von W. M. DAVIS. Herausgegeben von Verein der Geographen an der Universität Leipzig. Demy 8vo. Pp. iv + 75. Leipzig: B. G. Teubner, 1913. *Preis M.* 2.40.

La Géographie de l'Afrique Centrale dans l'Antiquité et au moyen âge. Par TH. SIMAR. Demy 8vo. Pp. 134. Bruxelles: Vromant et Cie, 1913.

Les Wangata (Tribu du Congo Belge): Etude Ethnographique. Par le Lieutenant ENGELS. Demy 8vo. Pp. 104. Bruxelles: Vromant et Cie, 1912.

Les Pyrénées méditerranéennes: Etude de géographie biologique. Par M. MAXIMILIEN SORRE, Professeur à l'École Normale de Montpellier, docteur ès lettres. Un vol. in-8° raisin, 41 figures dans le texte, 11 planches de phototypies et une carte en couleur *hors texte.* Pp. 508. Paris: Librairie Armand Colin. *Prix* 12 *fr.*

Le Monde Polaire. Par OTTO NORDENSKJÖLD. Traduit du Suédois par Georges Parmentier et Maurice Zimmermann. Préface du Docteur JEAN CHARCOT. Un volume in-18 jésus, avec 30 planches de cartes et de gravures hors texte Pp. xi + 324. Paris: Librairie Armand Colin. *Prix* 5 *fr.*

Les Etats-Unis d'Amérique. Par D'ESTOURNELLES DE CONSTANT. Un volume in-16. Pp. x + 536. Paris: Librairie Armand Colin, 1913. *Prix* 5 *fr.*

La Région du Haut Tell en Tunisie: Essai de Monographie Géographique. Par CH. MONCHICOURT. Un volume in-8 raisin, 14 cartes dont une en couleur hors texte, 4 figures et 12 planches de photogravures. Pp. xv + 488. Paris: Librairie Armand Colin, 1913. *Prix* 12 *fr.*

Mittelmeerbilder. Gesammelte Abhandlungen zur Kunde der Mittelmeerländer. Von Dr. THEOBALD FISCHER. Zweite Auflage besorgt von Dr. ALFRED RÜHL. Demy 8vo. Pp. 472. Leipzig: B. G. Teubner, 1913. *Preis M.* 7.

Die Vegetation der Erde. Herausgegeben von A. ENGLER und O. DRUDE. Leipzig: Wilhelm Engelmann.

Vol. I. *Grundzüge der Pflanzenverbreitung au der Iberischen Halbinsel.*

Von MORITZ WILLKOMM. Mit 21 Textfiguren, 2 Heliogravüren und 2 Karten. Royal 8vo. Pp. xvi + 395. 1896. *Preis M.* 10.

VOL. II. Band i. *Grundzüge der Pflanzenverbreitung in den Karpathen.* Von F. PAX. Mit 9 Textfiguren, 3 Heliogravüren und 1 Karte. Royal 8vo. Pp. xii + 320. 1898. *Preis M.* 11.

Vol. V. *Die Heide Norddeutschlands und die sich anschliessenden Formationen in biologischer Betrachtung.* Von P. GRAEBNER. Mit einer Karte. Royal 8vo. Pp. xii + 320. 1901. *Preis M.* 16.

Vol. X. Band ii. *Gründzüge der Pflanzenverbreitung in den Karpathen.* Von F. PAX. Mit 29 Textfiguren und 1 Karte. Royal 8vo. Pp. viii + 322. 1909. *Preis M.* 11.

Vol. XII. *Die Pflanzenwelt der Peruanischen Anden in ihren Grundzügen dargestellt.* Von Prof. Dr. A. WEBERBAUER. Mit 40 Vollbildern, 63 Textfiguren und 2 Karten. Royal 8vo. Pp. xii + 355. 1911. *Preis M.* 20.

Vol. XIII. *Phytogeographic Survey of North America, including Mexico, Central America, and the West Indies, together with the Evolution of North American Plant Distribution.* By JOHN W. HARSHBERGER, A.B., B.S., Ph.D. With 1 Map, 18 Plates, and 32 Figures in Text. Royal 8vo. Pp. lxiv + 790. 1911. *Preis M.* 40.

Memoirs of the Geological Survey of Scotland:—Explanations of Sheets 5 (Kirkcudbrightshire); 34 (Eastern Berwickshire); 36 (Seaboard of Mid-Argyll); 45 (Oban and Dalmally); 64 (Upper Strathspey, Gaick, and the Forest of Atholl).

Memoirs of the Geological Survey of Great Britain:—Monograph on the Higher Crustacea of the Carboniferous Rocks of Scotland. By B. N. PEACH, LL.D., F.R.S.

Western Australia. Selectors' Guide for 1912. Published under the direction of the Minister for Lands. Sydney, 1912.

Tours in Sweden, 1913. Railway Guide. Stockholm, 1913.

Guide-Annuaire du Gouvernement Général de Madagascar et Dépendances, 1913. Tananarive, 1913.

Census of India, 1911. Vol. xiii., North-West Frontier Province; Vol. ii., Andaman and Nicobar Islands; Vol. xvii., Central India Agency; Vol. x., Central Provinces and Berar; Vol. xiv., Punjab. Calcutta, 1912.

Diplomatic and Consular Reports.—Trade of Ichang, 1912 (5058); Trade, etc., of Hayti, 1912 (5057); Trade of Coquimbo, 1912 (5053); Trade of St. Thomas and St. Croix, 1912 (5070); Trade of Chefoo, 1912 (5071); Trade, etc., of Canary Islands, 1912 (5073); Trade, etc., of Hungary, 1912 (5074); Trade of Yucatan, 1912 (5075); Trade, etc., of Bordeaux, 1912 (5081); Trade, etc., of Norway, 1912 (5081); Trade, etc., of San Francisco, 1912 (5068); Trade, etc., of Dresden, 1912 (5072); Trade of Bahia, 1912 (5076); Trade, etc., of Brest, 1912 (5077); Trade of Batoum, 1912 (5078); Trade, etc., of Java, Sumatra, etc., 1912 (5083); Trade, etc., of Havre, 1912 (5084); Trade, etc., of Ciudad Bolivar, 1912 (5085); Trade, etc., of the Dominican Republic, 1912 (5066); Trade, etc., of Rotterdam, 1912 (5082); Trade of Shimonoseki, 1912 (5086); Trade of St. Louis, 1912 (5087); Trade of Azerbaijan, 1912 (5088); Trade, etc., of the Philippine Islands, 1912 (5089); Trade of Savannah, 1912 (5091); Trade of Germany, 1912 (5092); Trade, etc., of Philadelphia, 1912 (5090); Trade of Bushire, 1911-12 (5093); Trade of Amoy, 1912 (5094); Commerce, etc., of Dalmatia, 1912 (5069); Trade, etc., of Trieste (5079).

Publishers forwarding books for review will greatly oblige by marking the **price** *in clear figures, especially in the case of foreign books.*

THE SCOTTISH GEOGRAPHICAL MAGAZINE.

CEYLON IN 1913.

By A. L. Cross, F.R.C.I.

(*With Illustrations.*)

During a short visit to the island of Ceylon in the spring of this year I was greatly impressed with the progress made recently, and my impressions may be worth recording.

Ceylon has made enormous progress in many directions, notably in connection with what was once almost the exclusive product of Brazil—rubber. Great stretches of the low country of Ceylon, which was formerly jungle or scrub, are now covered with European plantations, mingled with the native coco-nut estates. The rubber plantations are well equipped with factories and bungalows of the most up-to-date description. Railways and roads are not far off, and the rubber can be transported to Colombo for sale locally, or shipped to London or other parts of the world direct. Millions of pounds of rubber are either sold in Colombo, or shipped direct without being sold, and the business is rapidly extending as new plantations come into bearing, which occurs after six or seven years from the time of planting. The Hevea (Para) rubber-tree is unlike any other tropical tree in that in Ceylon in January it winters. The leaves become a russet hue, and then gradually drop off, and the old trees stand under bare poles, so that it might be assumed that they were dying rapidly. But this is not so, for in a month or six weeks the new leaves begin to make their appearance, and by the end of March the trees are clothed with bright green foliage.

Tea planting succeeded coffee, which was killed out by the mysterious leaf disease, *Hemelia Vastatrix*, and is still a great industry, but tea flourishes best in the hill country at elevations of from 2000 to 7000 feet above sea-level. Strange as it may appear, the highest cultivated tea in the world is said to be found in Ceylon on the Nuwara Eliya Tea Estates, at an elevation of from 6200 to 7000 feet, the highest estate—"Excelsior"—which I visited, running up to 7300 feet. I understand that at Darjeeling, on the borders of the Himalayas, scarcely any of the tea estates run much over 6000 feet. The tea estates in the

A coolie woman tapping a rubber-tree.

Gampola, Maskeliya, Dimbula, Nuwara Eliya and Uda Pussellawa districts are well served by both railway and road communication. The finest tea plantations in the world probably are to be seen in the Agrapatana district of the great Dimbula Valley, the true source of Ceylon's great river, the Mahavillāganga. It is a wonderful sight to look on this valley of lovely green bushes in the very highest state of cultivation, and all at an elevation of from 4500 to 5000 feet. The above named are only a portion of the tea districts. There are the great districts of Ouvah and Haputale in the eastern province, and many others, too numerous to mention, in other parts of the island at lower elevations, but not of so much importance. Many of the tea factories turn out over 1,000,000 lbs. of tea per annum. In contrast with Assam there is no isolated life in the districts, for the estates all adjoin each

The interior of a rubber factory, where the rolling takes place. The white masses are the sheets of rolled rubber.

ther, and each district possesses its own Planters' Association, with hairman and secretary, where the wants of the district are discussed. uch questions as coolie wages, medical wants, hospitals, diseases affect. ng tea, coco-nuts, or rubber, roads, railway transport, arrack and toddy tadi) taverns are discussed, and action taken if necessary or feasible. hese local associations are all affiliated to the parent Planters' Associa. ion, which has its permanent headquarters in Kandy, the mountain apital, and meets annually in February of each year, and sometimes oftener, if any serious question arises. It is a great power in the country, and keeps an eye on all Government measures introduced into the Legislative Council. In fact, no measures affecting the planting community are ever passed without consultation with that body. The chairman and secretary are the most able men that can be found. The chairman often passes on to become the representative of the Rural Community in the Legislative Council, after his term of office has expired.

Each district has its sports club, with tennis, and in some cases cricket grounds, and in the districts of Kandy, Dickoya, and Nuwara Eliya golf-courses as well.

The revenue of the colony for last year (1912), according to the report of the late Governor, Sir Henry M'Callum, amounted to

A plantation bungalow on a tea and rubber estate.

£3,149,748, the largest ever attained. This year there is a probability that this amount may even be exceeded. Rubber exports have risen from 300 tons in 1907 to over 4000 tons last year (1912), and they will no doubt become a gradually increasing quantity as the young rubber plantations come into bearing.

The tea exported now averages some 186 millions of pounds. In 1881 it reached only 4000 to 5000 pounds. In the low country, however, where in recent years rubber has been interplanted with the tea, the shade cast by the rubber will render the tea not worth cultivating, and it will have to be eradicated.

Another great industry which has hitherto been, to a great extent, in the hands of the natives of Ceylon, like the cultivation of rice, is that of the coco-nut; but it is now occupying the attention of Europeans as well. There are several English companies in Ceylon engaged in this industry, and it is likely to be very profitable. The coco-nut palm in

Ceylon is practically cultivated along the west coast from Hambantota in the extreme south, to Manaar in the north-west, and also in parts of the east coast about Trincomali and Batticaloa. In visiting the Kelani Valley I was surprised to see what an immense tract of low-country land all along the Kelani Valley Railway, and inland from it, had been planted with coco-nuts, not yet in bearing, and probably not likely to be for several years to come, as the coco-nut even in favourable situations takes seven or eight years to come into bearing.

The great harbour of Trincomali has practically been abandoned as a naval station, and the buildings long occupied by the naval authorities, some artillery, and a detachment of the regiment garrisoning Ceylon,

A street in Colombo.

have been allowed to fall into disrepair. This seems decidedly retrograde, as now that a motor service has been established in connection with the northern railway from Anuradhapura to Trincomali there was the less necessity for abandoning this splendid naval harbour. Colombo is, of course, the seat of Government, and a great harbour, where the largest mail and other steamers of all nations call, and where all the shipping of produce takes place.

The great scheme for uniting Ceylon to India, which took shape some years ago in Mr. Joseph Chamberlain's time as Colonial Secretary of State, is now almost a *fait accompli*. The northern railway, which runs to Jaffna in the extreme north of Ceylon, was completed some five years ago, and the branch line to Manaar to connect with Adam's Bridge and India, some thirty-five miles in length, is expected to be opened in

August or September, communication with India being carried on by steam ferry till such time as the bridges and causeways across the islands are finished. The Indian Government has already constructed their portion of the line as far as the island of Rameswaram (celebrated for its Hindu temple), to connect with the Ceylon railway system, so that in the near future one may travel by rail from Colombo to Quetta without a break. This is no mere stretch of imagination, but will soon be an accomplished fact.

Everywhere throughout Ceylon there are marks of great prosperity. The natives are building better houses in the coco-nut districts.

The golf-course at Nuwara Eliya.

Wealthy natives are buying up all the house property in Colombo, and even in the sanitarium—Nuwara Eliya—it is quite the fashion among them to seek the higher regions in the hot-weather season. Colombo is a place of 200,000 inhabitants, and has many fine buildings. The Galle Face Hotel and the Grand Oriental are palatial buildings, with dining-rooms of 120 by 40 feet, and they can accommodate hundreds of passengers. When a big P. and O. mail steamer comes in, on the way to or from Australia, she will probably land 400 to 500 passengers; and others, German, French, or Bibby liners, in proportion.

Nuwara Eliya, 6200 feet above sea level, has railway communication with Colombo all the way. It has grown considerably within the last few years, and is still growing. There is a beautiful public park along the banks of the Nanuoya stream, a good racecourse and polo grounds, a large building called the United Club, with reading-room and library and assembly rooms, large tennis-courts and croquet-ground, and a nine hole golf-course for ladies. There is a full eighteen-hole course for men of a very sporting character, the Nanuoya river having to be crossed some eight times. There is a beautiful artificial lake some three miles in circumference, made by placing a bund across the Nanuoya where it leaves the Nuwara plain, which is stocked with carp and trout, though the latter stick more to the river. There are also miles of fine driving roads round the lake and Moon Plains, and to the Ramboda Pass, and a road and light railway to the beautiful tea district of Uda Pussellawa. On the Badulla road to the eastern province, about six miles from Nuwara Eliya, are the beautiful Government gardens of Hakgalla, well worth a visit, especially to botanists.

The Galle Face Drive in Colombo, with the military barracks in the background.

For any one desirous of having a panorama of the mountain system of Ceylon, there is nothing finer than the view from the top of Pedrotallagalla on a fine day in March. Though this mountain is 8300 feet above sea-level, it is only 2100 feet above Nuwara Eliya, and the path being good and sheltered by the jungle very nearly to the top, it can be ascended in a little over an hour or so. The writer was favoured with a splendid day, and from the top the whole main mountain system of the island, from Adam's Peak in the south-west right round to Pedro again, was distinctly visible. Outside the mountains there was a complete circle of snow-white clouds, like gigantic fleeces, but each separate from the other, and perfectly still, as there was no wind to move them.

Though there are practically no unexplored parts of Ceylon now, there are many great stretches of country covered with forest or scrub where wild elephants and other animals, such as deer and buffalo, roam about. These parts are seldom visited except by hunting expeditions. Wild elephants, however, unless rogues, cannot be shot without a licence from the Government.

The ancient fresh-water tanks in the neighbourhood of Anuradhapura and Polonnaruwa owe their origin to the old Buddhist kings of Ceylon.

Coco-nut trees on the seashore. The photo was taken during the monsoon.

Some of them were constructed 500 and 400 B.C., others 400 A.D. Many of them have been restored, but the land underneath them is not as much used by the Sinhalese inhabitants as it might be. Some of these old tanks were twenty and forty miles in circumference, with enormous long bunds. The Kalawewa tank, constructed A.D. 437, had a bund of solid stone twelve miles in length which remains to this day, and forms what Turnour designates as "perhaps one of the most stupendous monuments of misapplied human labour in the island." The so-called "buried cities,"—buried, that is to say, under forest trees—of Anuradhapura and Polonnaruwa, in existence about the time when these tanks were constructed, have been partly freed from the dense jungle that has covered them, and many interesting ruins laid bare. The Anuradhapura pagodas look in the distance like islands standing out of a level sea of jungle. Eight miles from Anuradhapura is a hill called Mihintale, covered with ruined small pagodas, and a tank. It is said that the Buddhist religion was first proclaimed from this hill long before the Christian era. The *Mahawanso*, discovered by George Turnour of the Ceylon civil service in 1826, and translated by him, gives a history of Ceylon for twenty-three centuries, from 543 B.C. to 1758 A.D., "presenting a connected history of the island itself" from these remote centuries. From this wonderful hill Mihintale there is a splendid view of the surrounding country, all forest, with open glades and tanks, stretching as far as the eye can reach in every direction. No doubt some day this fine country will be available for cotton and other kinds of tropical cultivation. Meantime it is only used for the rearing of great herds of red cattle to feed the towns of Ceylon and the planting districts.

Ceylon is a most interesting country, and the wilder parts may still repay the enterprising traveller who does not mind roughing it. There is plenty of scope on the east coast from Trincomali to Point Pedro in the extreme north for such investigations, if carried out in the healthy season, from the middle of February to the middle of April. At other times it is apt to be malarious.

THE TENTH INTERNATIONAL GEOGRAPHICAL CONGRESS.

THE much postponed Tenth International Geographical Congress was opened at Rome on the morning of March 27. The opening ceremony, which was graced by the presence of the King of Italy, took place in the Sala degli Orazi e Curiazi in the Palazzo dei Conservatori, on the Campidoglio, and was of an impressive nature. The Mayor of Rome, Signor Nathan, welcomed the Congress in the name of the city, and was followed by the president of the Congress, the Marchese Raffaele Cappelli, who is also president of the Royal Italian Geographical Society. He delivered a speech of much eloquence, in the course of which he spoke of the last International Geographical Congress held in Italy, that of Venice in 1881. He also noted that at the Geneva Congress the president, the late Dr. Arthur de Claparède, announced that Peary had just prior to the Congress set out on his journey to the North Pole. Admiral Peary was present at the Roman Congress with all the honours of his success, and since the previous Congress the South Pole has also been conquered.

The Marchese was followed by the Minister of Public Instruction, who, in the name of the King, formally declared the Congress open. On behalf of the delegates Professor Otto Nordenskiöld returned thanks for the speeches of welcome, and the meeting, which was quite a short one, closed. Most of the members of the audience grouped themselves round the entrance to see the picturesque departure of the King from Michelangelo's beautiful piazza in the spring sunshine, and then the strangers betook themselves for the most part to the museums of the Campidoglio. The Congress tickets gave right of free access to all the museums and collections belonging to the city, and this great privilege undoubtedly proved a temptation to many of the members, who found Rome's treasures, ancient and mediæval, more attractive than the meetings of the sections at least. This is perhaps a necessary consequence of the choosing of so attractive a city as Rome for the Congress, but its effect was the more to be regretted in that the total number of participants was small. The list of members and associates present published on March 28 only contains some three hundred names in all, of which only two hundred were members, and though later lists added nearly a hundred more the total was never large, and the attendance at the meetings of sections was frequently reduced to a minimum. Indeed, it soon became necessary to hold joint sectional meetings owing to the small afternoon attendance.

The small attendance was no doubt partially due to the double postponement, and especially to the fact that the original postponement took place so short a time before the proposed meeting in October 1911. Further, it was unfortunate that the highly successful International Zoological Congress at Monaco should have overlapped the Geographical Congress by a day or two, for this no doubt further reduced the number of members. Whatever the cause it was noticeable that, as compared with Geneva, the number of British members was very small, and of these the majority were from Scotland.

The real business of the Congress began on Friday, March 28, the meetings being held in the University, a centrally-situated building well fitted for the purpose. The office arrangements proved very good, and the excellent *Diario*, which appeared promptly every morning, contained —a most useful innovation—a short account of the proceedings of the previous day. The Geographical Society also, together with various other geographical bodies, offered to the members and associates a large amount of valuable literature, including some very useful maps, and the Istituto Geografico de' Agostini supplied every one with their handy little guide to Rome, with its excellent plans. Various invitations to visit exhibitions, institutions, etc., also reached the members.

Sunday was devoted to the excursions offered by the Organising Committee. These tended to be archæological rather than geographical in the strict sense, and even that which was most strictly geographical, the visit to the Alban Hills and Lake Nemi, was a little disappointing from this aspect in that opportunities for observation were very limited, no expert guidance was provided, and generally of geography in the strict sense there was but little. Regarded as a pleasure excursion, however, the day was a great success, and those who were present will not soon forget the beauty of the blue anemones which clothed the banks of the lovely lake.

Of the entertainments mention may be made especially of the reception given by the town in the palace of the Campidoglio.

Meetings were held from Friday, March 28, to Thursday, April 3, and on Friday the long excursions began. These included expeditions both to the north of Italy (Po valley, etc.) and also to the south, to Naples and Sicily.

Some account of the proceedings at the meetings may now be given.

At the first general meeting, which, as already stated, took place on Friday morning, March 28, in addition to various minor pieces of business, three papers were read. These were first one, copiously illustrated by lantern slides and maps, by Dr. L. von Loczy of Budapest, on the survey of Lake Balaton in Hungary. Dr. von Loczy himself initiated this survey some twenty years ago, and, as the readers of this *Magazine* know, the results have been appearing for many years in the form of elaborate monographs. The interest of the lake lies in its large area and its shallowness, and the variety of types of scenery, depending upon variations in the surface rocks, which appear upon its banks.

Dr. von Loczy was followed by Dr. C. van Overbergh of Brussels, whose paper was entitled "Human Geography." It was, however, devoted to some account of the work of the International Institute at Brussels, which has undertaken a very elaborate study of the negro race, having collected together four hundred thousand references in regard to this race. The material so collected has been most elaborately classified and arranged, with the result that information on practically any point connected with the black races can be furnished to inquirers. The task has involved the collaboration of six hundred persons, and has taken seven years. If a similar bibliography could be compiled for other races a veritable encyclopædia of human and social geography would be constructed. In the course of his address Dr. van Overbergh made frequent allusion to the work of the International Institute of Agriculture at Rome, an institution which the members had later an opportunity of visiting, and expressed the wish that the Brussels Institute might evolve along similar lines, and be of similar utility to the world.

Dr. O. J. Skattum then gave an account of the Norwegian explorations in Spitsbergen, especially from the point of view of geology. The paper was illustrated by a fine series of lantern views.

The proceedings closed with the formal presentation of reports upon the progress of geography in the different countries of Europe, since 1889. The report for Scotland will be published here at a later date.

The sectional meetings in the afternoon were mostly somewhat sparsely attended, and a number of papers on the programme of the different sections were not read owing to the absence of the authors. Most interest perhaps attaches to the proceedings of Section VIII. (Methodology), where representatives of the different countries presented reports on the state of geographical teaching in their respective countries, that for England being presented by Mr. F. Grant Ogilvie. The speakers for the most part confined themselves to a short summary of their reports, and the publication of the complete papers in the *Proceedings* of the Congress will be looked forward to with interest.

Another interesting paper was one by Professor Descombes (Section III.) on the work of the "Association Centrale pour l'Aménagement des Montagnes." The Association has its seat at Bordeaux, and has for objects the afforestation of mountain regions, the prevention of excessive grazing, the regulation of mountain torrents, and so forth. The association has already done excellent work in the Pyrenees. General Schokalsky gave an account (Section IIA.) of the work done by the officers of the Russian Navy in studying the Russian seas, and presented hydrographical maps of the Arctic Ocean, the Baltic, the Black Sea, the Caspian, and the northern Pacific. In the Economic section (V.) Professor Zimmerer of Regensburg read a paper on the Danube as a commercial highway, and showed that commerce along this stream has been carried on uninterruptedly for a very prolonged period.

On Saturday, March 29, the greater part of the General Meeting was devoted to a discussion of the 1 : 1,000,000 map of the World. The proceedings opened with a paper by Colonel Close, who briefly summarised the steps which had been taken in connection with this map since the Geneva Congress. The substance of this paper has been already reported here. Colonel Close was followed by Colonel Shidzouma, delegate of the Japanese Government, who recalled the fact that in November 1909 it had been resolved that the maps of Asia and China should be undertaken by France and Germany. Subsequently, however, Japan joined the Convention in charge of the map, and as she has had in preparation for a number of years a map of eastern China, on the 1 : 1,000,000 scale, which, with little alteration, could be made to conform to the International regulations, she desires that at least a part of the map of this region should be confided to her. A number of other speakers followed, and in the course of their remarks it became apparent that a number of minor points in regard to the map still remain unsettled, or open to criticism. It was therefore proposed by Professor Penck that a new International Official Conference should be held to examine the sheets already published, and decide whether any modification of the previous resolutions had become necessary.

After the close of this discussion Professor Lallemand read a paper on a new International map on the scale of 1 : 200,000 for aviators. He discussed in a very interesting fashion the aviator's difficulty in orienting himself, and also the kind of information which an aeronautical map should contain. The Aero Club of France has already published a dozen sheets of an aviator's map of Central France, which has proved very useful. Professor Lallemand concluded by proposing that a new International Conference should be arranged for the purpose of settling upon the desirable symbols for an aeronautical map on the 1 : 200,000 scale, to be based upon the 1 : 1,000,000 map.

The session concluded with a paper by Professor Helbronner on his survey work in the French Alps. He has carried on this survey for ten years, and the first instalment of his results has appeared as vol. i. of a work entitled *Description Géometrique détaillée des Alpes françaises.*

In the sectional meetings in the afternoon an interesting paper was read by Professor von Cholnoky on the influence of the Asiatic monsoons on the climate of Europe (Sections II. and VI. united). Apparently a lowering of temperature of one degree at the beginning of June, which occurs from Sweden through Germany to the Balkan States, is due to the effect of the monsoons. In speaking on this paper Professor Woeikof pointed out that both meteorologically and climatologically Europe can be regarded as a peninsula of Asia.

Professor Sapper spoke (same Sections) of "Bodenflüsse," that is, of those masses of earth which owing to their water-content are capable of movement, and of spreading out upon the lower grounds. Such earth streams occur in the tropics, in temperate latitudes, and in polar regions, and have great influence upon topography. The paper gave rise to an interesting discussion in which Professor Supan took part. In Section VIII. Captain Giannitrapani proposed that the 1 : 1,000,000 map be supplemented by the issue of a Universal Geography, and that the International Commission for that map should draw up a scheme for its execution, this scheme to be presented to the next Congress.

The General Meeting on the morning of Monday, March 31, was a brief one, as the session was interrupted at half past ten to permit the members to take advantage of the Marchese Raffaele Capelli's invitation to visit the International Institute of Agriculture. The meeting was, however, rendered interesting by the appearance of Admiral Peary, who was very warmly received by the members, and, at the invitation of the president, took his seat on the platform, and spoke a few words of thanks.

Professor Emile Chaix opened the proceedings by a short account of the *Atlas of Erosion*, which, as readers of this *Magazine* already know, has made much progress, and promises to be an extremely fine and very useful work. Professor Oberhummer then spoke on the question of the reproduction of ancient maps. He pointed out that many of the existing reproductions were made before mechanical methods had reached their present state of advancement, and that they do not satisfy the needs of students. He regarded it as unlikely that governments could be induced to take up the matter, but much might be done by the geographical societies of the different countries. The question aroused considerable interest, and the discussion was ultimately referred to Section VII., which, in the afternoon, accepted a resolution that a Commission should be formed to draw up a catalogue of the ancient maps in the various countries of the world.

The General Meeting was resumed in the afternoon, when General Schokalsky read an interesting paper on Russian investigations during the last fifteen years in the Arctic Ocean. He was followed by Mr. Stefansson, who gave an illustrated address on his explorations in Arctic America, the pictures, especially of Eskimo life, being very attractive.

No outstanding papers were read in the afternoon before the Sections, the chief incident of the afternoon being a private meeting of delegates of the different countries interested in the production of the 1 : 1,000,000 map. After discussion the following resolutions were passed at this meeting:—

(1) "It is desirable that another official Conference should be held to consider questions affecting the International Map of the World on the scale of 1 : 1,000,000, in the capital of a State which has already undertaken the preparation of sheets of the map; and it is thought that it would be convenient to all concerned if this capital were Paris."

(2) "In view of the fact that the general principles governing the construction of the map are already settled and adopted, the new Conference would be asked to consider questions of detail only, such as the size of the lettering, conventional signs for railways, etc."

(3) "It is desirable that all civilised States should be invited to send delegates to the proposed Conference."

(4) "It would be convenient if the date of the propcsed Conference were towards the end of the present year."

(5) "London (Geographical Section of the General Staff, War Office) remains the centre of the undertaking until the meeting of the proposed Conference, and communications connected with the proposed map should be addressed to that office. Also it is desirable that a set of not less than fifty copies of a sheet already printed should be sent by each country which has produced a sheet or sheets to the above office at an early date. These sheets will be distributed to those Governments invited to take part in the new Conference, and to recognised private authorities."

In the morning session on April 1, M. Semenow-Tianchanskij opened the proceedings with a paper on a proposed "Index nominum geographicorum universalis." Such a work should, he thought, be undertaken by the great geographical societies of the world, and should be followed by periodical supplements, containing additions and corrections. It should be published at so cheap a rate as to ensure its maximum circulation.

Professor Schott of Hamburg, and Commander Drechsel, General Secretary of the International Committee for the Exploration of the Sea, both spoke on various problems connected with investigations in the Atlantic, a subject which was the subject of one of the resolutions of the Geneva Congress. After their speeches the two members united in presenting the following resolution, which was accepted:—

"The Tenth International Congress of Geography recommends:—

"(1) That, in accordance with the decision arrived at by the Geneva Congress (1908), and the meeting of the Monaco Commission (1910), the first problem for the international exploration of the Atlantic be regarded as the sending out of preliminary expeditions to the North Atlantic to investigate the amount, extension in space, and nature of the periodic variations in the layers of water down to a depth of 100 metres, and the determination of the extent to which a single observation represents the mean condition of the great depths.

"(2) That at the same time the observations on the temperature and salinity of the surface, which have been already carried out for many years by the different nations, should be continued on a larger scale, and that the Congress also recommends that floats in large number should be liberated in order to study the ocean currents.

"(3) That the President of the Congress be requested to communicate this resolution to the Governments of the States interested in navigation and in fishing in the North Atlantic, and particularly to those States which are at present united for the International Exploration of the Sea, in order that they may consider whether the above directions can be carried out."

Professor Ole Olufsen then spoke of the labours of the Committee appointed at Geneva with the object of attempting to obtain a closer union between the chief geographical societies of the world. As yet not much has been done in this direction, and he suggested that the Congress should express the desire that the secretaries of the chief geographical societies of the world should meet in Copenhagen not later than 1914, in order to consider the matter.

Mr. Bridgman of New York then spoke of Peary's Arctic work, illustrating by some fine lantern slides, and was followed by Commander Roncagli, the general secretary of the Congress, who spoke on Peary's observations for latitude. The session closed after some minor business.

The most interesting papers in the afternoon sections were those by M. de Quervain (Sections II. and VI. united) on his crossing of Greenland in 1912, and by Professor Nordenskiöld on the Inland Ice of the Arctic and Antarctic regions. Professor Nordenskiöld stated that inland ice only occurs in continental regions with a polar climate. The exceptions, *i.e.* the inland ice of South Greenland, are to be explained as the result of an earlier period of greater cold.

In the morning session on April 2, Dr. W. S. Bruce gave a full account, illustrated by lantern slides, of the proposed second Scottish Antarctic Expedition. The project was subsequently warmly supported by Admiral Peary, by Mr. G. G. Chisholm on behalf of the Royal Scottish Geographical Society, and by Professor Penck. Subsequently Mr. Stefansson described the plans of the Canadian Arctic Expedition, to which frequent reference has already been made here. Admiral Peary and Dr. Bruce congratulated Mr. Stefansson on having obtained the support of the Canadian Government, and expressed warm approval of the details of the expedition.

After the transaction of various minor pieces of business, Professor Ricchieri of Milan spoke of the labours of the Committee on the question of the transcription of geographical names. The subject is recognised on all sides as a very difficult one, and after a discussion in which a number of delegates took part, it was resolved that the question of how best to obtain a uniform method of writing and pronouncing geographical names should be referred to the new Conference which is to meet in connection with the map of 1 : 1,000,000.

In the sections the most interesting paper was one by Professor Supan on the Europe of geographers (Section VI.). As at present defined, he considers that Europe is neither a peninsula of Asia nor an independent entity. It is a purely historical concept with no true geographical value. If, however, Russia be regarded as a part of Asia, then the remainder of Europe is a true peninsula, whose morphological characters are in striking contrast with those of Asia. The point is of considerable interest, for it is another indication of the fact which we have frequently emphasised here lately, that the new geography will have to discard, sooner or later, the nomenclature which an earlier geography imposed upon it. The immediate problem, however, is when will teachers be bold enough to discard this outworn terminology? Must they, for example, always continue to define Europe and then explain that the name has no geographical value?

On Thursday morning the delegates met to consider the next place of meeting, the question of adding Spanish to the official languages of the Congress, and the resolutions proposed during the meeting.

As regards the first point it was unanimously agreed that the next Congress should be held at St. Petersburg in 1916.

The second point gave rise to a lively discussion, but when General Schokalsky asked if, supposing Spanish to be accepted as an official language, Russian would similarly be accepted, the proposal was doomed, and was finally withdrawn.

All the resolutions submitted to the delegates were accepted, with only minor alterations. Most of these have been already mentioned, but for convenience' sake, they may be summed up as follows:—

(1) Professor Lallemand's resolution in regard to an international aeronautical map (see p. 408).

(2) Resolution in regard to the 1 : 1,000,000 map (see p. 409).

(3) A recommendation in regard to the Atlas of Terrestrial Relief (see p. 409).

(4) Commander Drechsel and Professor Schott's resolution in regard to the study of the North Atlantic (see p. 410).

(5) Professor Olufsen's proposal in regard to measures to be taken to form a world-wide Union of Geographical Societies (see p. 411).

(6) A motion by General Schokalsky emphasising the necessity that maps accompanying scientific memoirs, as well as isolated maps, should always show the scale, the projection used, and the source of the information utilised. This is a reaffirmation of a recommendation adopted at Berlin in 1899.

(7) The proposal that the International Conference for the map of the world should consider the question of the unification of geographical nomenclature (see p. 411).

(8) Professor Oberhummer's proposal that a committee should be appointed to further the reproduction of ancient maps, and to prepare catalogues of the ancient maps hitherto reproduced, these catalogues to be published before the next Congress.

(9) A resolution by Dr. Erödi on questions connected with geographical teaching, which included a recommendation that the history of geographical discovery should be included in the teaching of geography; that geographical excursions to foreign countries should be arranged for the pupils of higher schools; that summer courses in geography should be arranged; and that an International Geographical Institute should be established.

(10) A resolution proposed by M. Helbronner that a commission should be appointed to settle disputed questions of nomenclature along mountain frontiers, and especially to attempt to get rid of duplicate names.

(11) Captain Giannitrapani's resolution in regard to the drawing up of an official universal geography (see p. 409).

In the afternoon the closing meeting was held, and put an end to a successful if not largely attended Congress.

THE ARMENIANS OR HAIKANS: AN ETHNOGRAPHICAL SKETCH.

By V. DINGELSTEDT, Corresponding Member, Royal Scottish Geographical Society.

ARMENIANS are one of the very few old nations, who, having lost their autonomy, being deprived of their national rights, crushed by centuries of oppression, have yet preserved their individuality, their national creed, their customs and habits, their patriotism and their belief in a national resurrection. Though dispersed all over the earth, a considerable number of them still occupy their fatherland, a beautiful, fertile and healthy mountainous country, a high plateau (Hayasduni or Haikh) strewn with the ruins of their ancient cities, commanding an important strategical position in connection with intercommunication between the central plateau of Asia and Anatolia, and including in its frontier the origin of four famous historical rivers: the Kur with the Aras, the Tigris and the Euphrates.

Armenians have serious claims to the attention of geographers. They have had a glorious history, and have their own national character, their peculiar feelings and traditions, their own mental characteristics, creed, moral habits and usages, which enrich humanity and are particularly interesting on account of their ancient origin. It has not always been big nations, but sometimes small ones, such as the Etruscans, Greeks, Jews, etc., who have made the biggest contribution to human enlightenment. The Armenians are not Greeks, nor has their native country any resemblance to Hellas, yet they have also solid if not brilliant virtues.

The acceptance of Christian doctrines condemned as heresies by other Christian nations has greatly influenced the destiny of Armenians, excluding them from the benefit of religious intercourse. Even in our enlightened times they are not regarded as true Christians deserving full protection from intolerance and persecution. The repudiation of the Armenians as a true Christian nation has had an effect analogous to the condemnation of the Jews as enemies of Christ; the two people become thus associated, not indeed as friends, but as rivals condemned to seek in money the means to fight against the injustices of the world.

There is a resemblance and not improbably a blood parentage between Armenians and Jews. This idea was first announced by F. v. Luschow who, in his *Ethnological Studies*, reached the conclusion that both nations are descended from the old Hittites or Khita (Héthéens) mentioned in the history of Rameses II., on the monuments of Egypt, as having founded before the Phœnicians a powerful empire in Asia Minor. A little later another German ethnologist, L. Sofer, came to the same conclusion.[1]

[1] "Armenien und Juden," *Zeitschrift Demogr. Stat. Juden*, 1907.

Both nations have the same commercial acumen, the same greed for gold, the same characteristic elongated features, with a well developed, prominent nose, black hair, and heavy yet often handsome physiognomies. Armenians are not an heroic, perhaps not even a chivalrous nation, heroism and chivalry having been lost in long subjection to foreign rule, but they are not without many civic and manly virtues. Intelligent, industrious, hard-working, persevering, greedy, shrewd, cunning, sordid, hard-hearted, fond of knowledge, inquisitive, insinuating—they are the Jews of the East.

But though an Oriental nation they are capable of being inoculated with Occidental civilisation, for in this race are mingled many Eastern and Western vices and virtues. In their Oriental soul there are admirable resources of patience. An Armenian understands almost equally well Oriental and Occidental man—whence their superiority as business men. It is through such a race of men that a reconciliation between Eastern and Western ideas and mode of life may be attained.

Armenians are mainly known to the general public by the appalling and frequent massacres perpetrated on them under the auspices of the Turkish Government. Being generally of a peaceful character, Armenians submit to oppression, but they are yet capable of energetic action when roused to high indignation by the iniquities of their foes, and when called upon to defend the honour of their wives. A peaceful peasant, merchant or citizen may then suddenly turn into a terrible bloodthirsty *haiduc*, animated with a blind fury of revenge and not stopping before even the most cruel deeds. Yet, during centuries of oppression, they have lost the martial virtues which play such a vital part in the health and honour of a free nation.

Geographical Description.—Armenia (Hayasduni or Haikh) is an extensive country of Western Asia, extending from the Caucasus in the north to the mountains of Kurdistan in the south, from the Caspian Sea in the east to Asia Minor in the west. There were two Armenias—major and minor. Armenia major—Medz Hayotz,—or Armenia proper, to the east of the Euphrates, is an elevated tableland, about 7000 feet in height, culminating in the peaks of Ararat and sinking towards the plains of Iran and the valley of the Aras on the east and south, and to those of Asia Minor in the west, while Armenia minor (Phokhr Hayotz) extends to the west of the Euphrates.

Armenia proper is traversed by many mountains, some of which are of volcanic origin; it is broken by glens and valleys, and contains within its borders the sources of mighty rivers—the Euphrates, Tigris, Aras, Kur—and a great number of lakes—Van or Aghtamar (3690 square miles), Sevanga or Khabodan (1280 square miles), Urmia or Vieyhem.

West of Ararat, along the Russo-Turkish frontier, extends a high chain separating the Aras from the Murad (the Eastern Euphrates), called the *Aghri Dagh*. It extends towards Erzerum, where it joins the high (water-parting) ridge separating the river-basins of the Black Sea, the Caspian and the Persian Gulf. Towards the sources of the Aras the chain rises to an altitude of 3600 metres, and is called Bingol Dagh. Hence radiate the mountain chains which separate the waters of the Murad from the Kara-su, the two uniting somewhat farther south to form the Euphrates. In consequence of the irregular topographical character of this country some Armenian authors have found an analogy between the tormented course of their national history and the tormented character of their national soil. The mean altitude of the Armenian plateau is 1500 metres.

The volcanic mountains of Ararat—Arm. *Masis*—(16,925 feet) have an imposing aspect; they enjoy in the eyes of Armenians a sacred character and constitute a centre of their patriotic affection. Armenia enjoys a healthy but, in winter, rather a cold climate, and has luxuriant pastures, romantic scenery and a naturally fertile soil, producing grain, tobacco, grapes, cotton, rice, ricinus, hemp, etc. Unluckily law and order scarcely exist, wars and troubles are frequent, and consequently there are numerous ruins of once flourishing and important towns, such as Armavir to the north of Ararat, Artaxata, Ervandashat, Pakharij, Varzahan, Ani and others.

The Armenian mountains are treeless, showing only rocks and vast pasturage, and the inhabitants use as fuel dry cow-dung mixed with chopped straw.

A description of the geographical features of the high plateau of Greater Armenia, enclosed in the quadrilateral Erzerum, Mush, Bashkala and Bayazid, and now called Turkish Armenia, has been given in this *Magazine* by Captain F. R. Maunsell.[1] With the actual and promised extensions of the railway, the Armenian plateau will certainly attract before long the attention of the ruling nations. It is situated within an area which opens on the south-east to the Persian Gulf; on the west it has access to the Mediterranean, on the north to the Black Sea, and on the north-east it commands the Caspian and the routes of Northern Persia.

Religion.—Creed, language and traditions of past woes and glories are the triple ties which, besides the otherwise powerful natural ties of racial unity, bind together the people of Haikh.

Like the Jews, the Armenians are an ancient nation at whose birth a creed presided. Their religious organisation has survived their national independence and political ruin, and it is mainly thanks to its faith that the Armenian nation, though persecuted, dispersed and brought under foreign yoke, has continued to maintain itself through centuries, and still exercises a not quite negligible importance in the industrial and commercial movement of the world. The great majority of Armenians continue to belong to the Gregorian Church. About a quarter of them have retired from this church, and under the name of Armenians-United have become adherents of Roman Catholicism. They call themselves Catholics, and are divided into two groups, a majority recognising the Pope as its chief, and a minority forming "Old Catholics." A not inconsiderable number of Armenians have also become Lutherans, and some are Greek-Armenians. Considerations of safety or some material advantage have caused a considerable number

---

[1] Vol. xii. p. 225.

of Armenians in Turkey and Persia to become Mussulmans. The Armenian Church claims an older than apostolic foundation; for according to legend this Church was already founded in 34 A.D. The historical founder was, however, St. Gregory, called the Illuminator, who was the first Armenian bishop (302 A.D.).

Armenians have adopted the teaching of Arius, the founder of Arianism, according to which Christ was a mere creature and a work of the deity, not of the same substance with the Father. These doctrines were condemned at the Council of Nicæa in Bithynia (325 A.D.). Arianism was virtually abolished in the Roman Empire about 379-395, and has gradually lapsed into Unitarianism.

The Armeno-Gregorian Church has maintained a strong ecclesiastical organisation. The supreme chief of the Church is a catholicos (at present Mattheos II.) residing in Etchmiadzin, a little trans-Caucasian town at the foot of Ararat. His humble palace is far from having the splendour of the Vatican, but he bears the pompous title of "His Holiness the supreme Patriarch and Catholicos of all Armenians." He is chosen by a council consisting of two delegates from each of the eparchies, directly depending upon the catholicos, eight members of the synod of Etchmiadzin, and seven delegates from the congregation of Etchmiadzin. He is confirmed in his powers by the decree of the Emperor of Russia, who has his very powerful "procureur" in the synod. There are eight eparchies directly depending upon the catholicos; six of them are in Russia, viz.: that of Erivan, comprising within its jurisdiction the towns and the districts of Erivan, Kars, Alexandropol, Nakhidjevan, Kagisman and Olty. Almost one-half of the Russian Armenians belong to this bishopric; the other five are those of Tiflis, Shusha, Shemakha, Astrakhan and Kicheneff (Bessarabia). The two eparchies directly depending on the catholicos outside Russia are that of Azerbeidjan, with its centre of Tabriz, and that of Nor Djoulfa, near Ispahan, its jurisdiction extending to South-Persia, India and the Indian archipelago.

There are forty-five Armenian eparchies in Turkey, under the superior guidance of the Patriarch in Constantinople who governs through a mixed ecclesiastical and lay assembly, called the National Armenian-Assembly. There are besides three high ecclesiastical dignitaries: a patriarch in Jerusalem, whose jurisdiction extends to the Armenians of Syria and Cyprus; a catholicos at Sis, who is hierarchically second to the catholicos of Etchmiadzin; he ordains his own bishops and bears the title of "*Catholicos de la Maison de Cilicie*"; a catholicos of Akthamar, residing in the isle of this name in the lake of Van, and governing the mountains of Kurdistan. Outside Turkey there is an eparchy in Egypt, and another in Bulgaria, governing the Austrian and Rumanian Armenians. In the United States there is an Armenian bishop of Worcester.

The Armenians do not keep up their churches with much care, and in everyday life they are far from being devout. Their churches and monasteries have mostly their own glebes, cultivated lands, vineyards, orchards, gardens, shops, droves of cattle, flocks of sheep, etc.

They let their lands to peasants, undertake commerce and work to increase their worldly goods. To most of the temples and monasteries are attached schools. Armenians have many saints, mostly among ancient patriarchs. They like making pilgrimages to the tombs of these saints, to monasteries famous for some "miraculous" image and to some tumulus or stone or tree which has become sacred in the eyes of superstitious people. Those objects are sometimes equally venerated both by Armenians and Turks, some former Christian sanctuaries having become Turkish.

Tree worship exists at Tzugrut, near Akhaltsikh; pilgrimages towards the sacred tree are organised, wax tapers are lighted, incense burned, small animals sacrificed, and prayers said. The tree is placed under the safeguard of the public. Sick persons tear off rags from their garments, and attach them to the tree, believing that they can thus get rid of their complaint.

Armenians worship old temples in ruins, as also ancient MS. Both form objects of pilgrimages and religious practices. In some places they also worship serpents. Armenian peasants do not kill serpents found in a house; these are home-serpents and bring fortune to the house. In Artvin, the power of curing some diseases is attributed to the serpent. There are black and white serpents, regarded as good and evil. Epileptic women (called *enkaror*) are considered by the people as being struck by some saint, and they then regard themselves as servants of this saint. In the popular feasts one may see these women prostrated in ecstasy before church doors, sacred trees or rocks, and articulating some mandates of the saint. That occurs mainly in time of some popular calamity.[1]

NUMBERS.—According to Colonel Mark Bell, there are about four million Armenians,[2] that is a little more than Swiss (3,741,971). Probably more than half of them are actually in Russia, one quarter in Turkey, some hundred thousands in Persia, some thousands in Austria-Hungary, and the rest scattered from the Gulf of Iskandarun to Lake Urmia, over Austria, the Balkan States, India, etc.

More than half of the Russian Armenians are in Transcaucasus. They were formerly more numerous in Turkey, but the terrible reign of Abdul Hamid forced many of them to leave the country. Armenians are known in most of European great cities, and their numbers are rapidly increasing, their women being prolific.

Armenians yield a variety of types. The purest type is to be found in the province of Astrakhan; the people here are tall, well-built, but inclined to obesity. They have large, black, deeply set eyes, their forehead is low, the nose of great length, prominent and aquiline; the oval of the face is even longer than that of the Persians. All Armenians are brachycephalic, their cephalic index, according to Khanikov, being about 85. The beard is abundant and of precocious development; one may meet in the streets of Tiflis boys of 13-14 already bearded.

---

1 Fr. Macler, *Rapport sur une mission scientifique* Paris, 1900.
2 *S.G.M.*, vol. 1890.

A valiant and courageous type of Armenian inhabits the high valleys in the north-west of Ispahan. There are also Armenian Gypsies called Magagorse, who inhabit mainly the villayet of Sivas. They are to be found also in Mersivan. At the beginning of spring, they spread to seek agricultural work, and return for the winter to their respective hovels. They speak good Armenian, and are zealous for religion. The highest ambition of the men is to make a pilgrimage to Jerusalem, and return hence with the new title of Magdessi (or Chadji). They do not know how to write or read, but enjoy some prosperity, and are decently clad. The true Armenians, however, do not marry with them, and between both parties there are endless and ruinous disputes in the tribunals.

The Armenian highlanders in the Caucasus, and elsewhere, differ widely from the Armenians in the lower valleys and plains.

Occupations.—Unlike Jews, the majority of the Armenians are agriculturalists, peasants, field labourers, cattle-breeders, working hard and leading a very simple, patriarchal, mostly miserable life. But like Jews they are also an eminently commercial and industrial people, good financiers, excellent artisans, competing successfully not only with such relatively weak commercial nations as Georgians, Tartars and Turks, but also with astute people like Greeks, Jews and even European nations. Though suffering from injustice, their situation in commerce in Turkey remains on a satisfactory level. When they do not export themselves, they remain necessary intermediaries in the delivery to European markets of many of the products of Anatolia—carpets, wool, cotton, silk, cereals. Armenians have known also how to maintain an honourable place in the trade in the importation of textile fabrics, metals, hardwares, building materials, watches and jewellery. Armenian artisans are known for their intelligence and skill in small industries such as metalwork, weaving, carpet manufacture or embroidering, printing, dyeing, etc.

Armenians as merchants and artisans are far superior to the indolent Turks; they value education, are more open-minded, have more initiative and are intelligent, insinuating, cunning and artful. Armenians have developed a harder and stronger type of mind than that of their rather effeminate, lazy, improvident and ignorant neighbours, the Persians, Georgians and Tartars. They not seldom abuse their superior knowledge and craftiness in their dealings even with such warlike people as the Kurds, brave riders, excellent herdsmen, but very children of nature, completely ignorant of the laws and usages governing modern civilised communities. A Kurd brings his sheep to be sold at a town bazaar, but Armenians will not allow him to proceed as far as the market place; they surround him before he gets there and know how, with shameless hypocrisy and insistence, to persuade him to sell his sheep under the market price. The poor Kurd, incapable of resisting the pressure, abandons his wares with a bleeding heart. Commerce in the Caucasus is in the hands of the Armenians, and the range of their commercial activity in Tiflis, Baku, Constantinople, Sivas, in the Black Sea ports, and even in Moscow and Astrakhan, is considerable. Some merchants in Sivas send their agents to Bokhara, Samarkand and other large Asiatic cities.

RUSSIAN ARMENIANS.—Russian Armenians do not suffer from any legal disabilities, for they are placed before the law on an equal footing with orthodox Russians. In consequence they thrive in Russia and are continually growing in number and prosperity. A number of them have attained a high situation in civil and military state service. This state of things attracts numerous immigrants from Turkey. Thanks to their industry, intelligence and great commercial acumen the Armenians have become the most important ethnic element in the Caucasus and Transcaucasia, and are the true economic masters of these beautiful and most promising countries. The principal Armenian towns in the Caucasus and Transcaucasia are Tiflis, Erivan, Kars, Alexandropol, Nakhidjevan, Kagisman, Olty, Elisabetpol, Shusha, Shemakha, Batum, Baku, etc.

The Armenians in Tiflis, from eighty to one hundred thousand strong, constitute the masterful majority of the population. The local industries and almost the whole commerce are in their hands, and besides they own most of the houses and landed property, which their former proprietors, the Georgians, have mortgaged and were not able to redeem. The Armenians constitute three-quarters of the municipal council of Tiflis and they develop great activity. They have many influential papers: literary, political, agrarian, ethnographical, humoristic; numbers of benevolent, scientific and dramatic societies, and their own theatres, where are represented the works of such renowned authors as Gabriel Sundukiantz, Chirvanzadê, Abelian. Besides Tiflis, the Armenians are also very influential in Batum, and in Baku, where they own some of the naphtha wells, and in other towns.

Outside of the Caucasus there are numerous Armenian colonies in the town of Astrakhan, in ancient Bessarabia, in Crimea, Moscow,[1] St. Petersburg, Odessa, and Nakhidjevan on the Don, near Rostoff, which is an important commercial centre of a number of Armenian villages.

Armenians in Russia, as in Turkey, do live, but exceptionally, in compact national communities. The well-to-do Armenians wear a characteristic high conical headgear of dressed lamb-skin (*merlushka*); their dress is a long under-tunic (*beshmet*), and above it a tunic with long sleeves cut through all their length to be flung away over the shoulders. But their greatest pride is the embroidered girdle and the dagger.

DISPROPORTION OF SEXES.—The number of women among the Armenian population of Russia is far less than that of men, the proportion being on the average as 100 : 85; in more favourable cases as 100 : 90 (province of Terek and Armenian Catholics in the government of Tiflis); and in the less favourable cases only as 100 : 66 (Kutais) and even as 100 : 60 (Baku). There is no exception to this phenomenon; everywhere among Armenians in Russia the women are less numerous than men. Also Armenians attach far greater importance to the birth of a boy than to that of a girl, and certainly they bestow more care on the education of a boy than on that of a girl.

TURKISH ARMENIANS.—The lot of Turkish Armenians is most to be

[1] *S.G.M.*, 1885, p. 321.

pitied. They suffer from the oppression and iniquities of their own government, the barbarous incursions and raids of their neighbours, the Kurdish and Cherkess tribes, and the disdainful attitude towards them of the Mussulman Turkish population. Many of them have been themselves compelled to become Mussulmans. Many Armenians have been expelled from their native towns, and been established by force on the lands near Isnik, Brusa, Angora, and elsewhere.

The Armenians in Turkey suffered greatly from the terrible massacres organised by the government, or at least countenanced by it, in the year 1895-96. The horrors of the slaughter, where at least two hundred thousand victims were cruelly murdered and many hundred villages destroyed, defy description.[1] The lands of the victims have been confiscated or stolen by Kurds. The Armenians who fled after the massacre are now returning and attempting to re-obtain their property.

Out of twenty-five Turkish vilayets, there are eleven partly inhabited by Armenians, but the last are to be found also outside Armenia proper, especially in Constantinople. The most important Armenian towns are Erzerum, Van, Mouch, Bitlis, Kharputt, Diarbekr, Egin, Erzinggan, Sivas, Trebizond. The plain of Erzerum is one of the richest and best cultivated in the whole Ottoman Empire. One of the sanjaks of the Erzerum vilayet, Bajazid, at the southern foot of Ararat, is exclusively Armenian. In the remaining three sanjaks—Erzerum, Erzinjan and Baiburt—the Armenians are mainly concentrated on the two rivers (Kara-su and Murad) forming the Euphrates. They form about half of the whole population of Erzerum. Southwards of Erzerum three small vilayets, Van, Bitlis and Kharputt, constitute the central region of Turkish Armenians.

Van, a town on the lake of the same name, as also Kharputt, have been converted to Islam, like many other villages such as Sandukht, Argo, Agrorik, Shabanos, etc. In Kharputt a number of Armenians have been wrongfully dispossessed of their lands. There are many Armenian Protestants here, who live in peace with their Gregorian co-nationalists. It is a large city with four temples, is the seat of a bishop, and has renowned cloth manufactories. In the vicinity there are fourteen Armenian villages, having temples and schools. There is no difference in customs and habits between the citizens and the country people. Being on the main route to Bagdad, Kharputt has a great future before it.

On the southern shore of the Black Sea are to be mentioned Trebizond, Samsun, Sinope, Ordu, Kerasun, Bafra, Ineboli. Trebizond, inhabited by 8000 Armenians, of whom 6000 are Gregorians, 1500 Catholics, and 500 Protestants,[2] is the most important Turkish town on the coast. It is now declining, but was once a splendid city on the great through route between Persia, Syria, and Mesopotamia on one side, and Caucasus and Turkey on the other.

---

[1] Ed. Noguères, *Arménie*. Genève, 1897.

[2] Fr. Macler, *Rapport sur une mission scientifique*. The other numbers of Armenians are taken from the same document.

Samsun is an important commercial port with about 4000 Armenians, partly Gregorians, partly Catholics and Protestants, out of the total population of 17,000. It extends in a hemicycle along the circular harbour. One principal artery is well paved, the streets are neat, and planted with trees. The Armenians inhabit their special quarter; they are mainly occupied with the transport of imported and exported goods, and they cultivate tobacco.

Kerasun is a nice town dominated by an Acropolis. Picturesque hillocks run along the shore. There are ten per cent. of Armenians among the 10,000 Turks and Greeks which form the population. The Greeks are renowned for their weakness for resorting to the tribunals on the most trifling occasions, whereas Armenians prefer to arrange their quarrels by recourse to their notables. Kerasun, like Ordu, exports to Marseilles very considerable quantities of hazel-nuts.

Ordu includes about 2500 Armenians, occupied like their other co-religionaries on the Black Sea with commerce and tobacco growing. The town extends along the shore, and the hills that surround it are covered with vines. Like some towns in Sicily, Ordu has low square houses covered with red tiles.

Bafra or Balira has above 1000 Gregorian and Protestant Armenians, and some 300 Armenian Tzygan, or Magagorse, called also Bosha.

Sinub or Sinope, famous in Russian naval history, has about 1000 Armenians suffering much from incursions of Cherkess and Chetchen brigands.

South-west of Mount Karabel is situated the vilayet Sivas with its chief town of the same name, one of the most graceful and picturesque cities in Asia Minor. Besides Sivas, the Armenian quarter of which—Ochdar—is separated from the town by some ploughed lands, there are in this vilayet about forty Armenian villages with a total Armenian population of more than 20,000, Gregorian and Protestant. In the Ochdar was born Mhitar, founder of the celebrated brotherhood whose chief seat is on an isle near Venice. Gack-Medresse in Sivas is believed to be an ancient palace. The streets and squares in Sivas are dirty, but the inner courts of the houses are kept with relative cleanliness. In the town is shown a cemetery called "Black Soil," where were buried thousands of children trampled down by the cavalry of Timur-Lenk. In another cemetery called "Forty Children" are buried 4000 Armenians killed by the same conqueror. Almost all the tradesmen in Sivas are Armenian, especially the smiths and armourers. They have all great fear of the Turks, who treat them harshly, and use them ill, and, according to Armenian statements, in order to conceal their fortune, even rich men are clad as mendicants, and live in miserable hovels.[1] In the vilayet of Sivas, the main centre of Armenian population, there are some noteworthy Armenian towns and villages, such as Tokat, Amasia, Sherek, Niksar, Turik or Divric.

---

[1] Archimandrite Garegin Srvandstianz, *Toros Achpar: An account of the situation of Turkish Armenians.*

Tokat, the seat of an Armenian eparchy, is an old stronghold now in ruins. There are above 10,000 Armenians, Catholics, Protestants, and Greeks, besides a majority of Gregorians. The main industries of the Armenians are copperwork, copper being extracted from the Argpy and Kopan mines, dyeing and painting of cloth, and the cultivation of tobacco and the poppy. In the villages surrounding Tokat there are above 10,000 Armenians, and thirteen Armenian temples.

Amasia with 6000 Armenians, of whom 100 are Protestants, is an old stronghold supposed to have been founded by Alexander the Great, and having attained its highest glory at the time of Mithridates.

Not far from Amasia, in the lower basin of the Iris, is the town of Mersivan or Mersifoun, the theatre of violent struggles between Armenians, Gregorians and Protestants. The town possesses a splendid Armenian temple. All her artisans are Armenians. They have founded a national league for the defence of their religion.

Turic or Divric (ancient stronghold) is renowned for the strong constitution of its women, who not seldom bring forth twin boys. In the bishopric of Egin are numbers of Greek-Armenians. They speak and write Armenian, and their priests speak Armenian only; their headgear, priestly vestments, and ritual, are Greek. In the Church the Gospel, missal, and breviary, are read in Armenian, but the mass is said in Greek, with a translation into Armenian. Some prayers are a mixture of Greek and Armenian. Some of the temples are divided into two parts, thus in the vilayet Vank one of the oratories is Greek, the other Armenian, in the Mushehka one of the chancels and a font belong to the Armenians, and the two other chancels to Greek-Armenians.

There are more Armenians in Constantinople than in Tiflis; both these cities are great centres of Armenian activity. Here they are divided into three unequal groups: (1) Stambuliots or Armenians who have been settled in the capital for many generations, and speak a special idiom, mingled with Turkish words, with a peculiar pronunciation; before the massacres of 1895-96 these were about 100,000 strong; (2) Provincials from different centres of Armenia and Asia Minor, Caesarea, Angora, and Brusa, who have arrived with their families, and have mostly forgotten their native language; (3) Temporary immigrants, both young and married people, who have come to seek work as porters, navvies, masons, servants, labourers, workmen, citizens, and petty merchants. The two last groups are each from 20,000 to 40,000 strong. The Turks do not easily accept Armenians as State officials; there are scarcely half-a-hundred of them as such in a total of about 10,000 State servants in Constantinople, and the proportion is even less favourable in the provinces. It is only in the administration of the Ottoman public debt that Armenians are freely admitted. Since the proclamation of the Constitution their lot has, however, been a little bettered. An Armenian has even become a Secretary of State for Foreign Affairs. Armenians now occupy fairly honourable places in international and private financial and commercial institutions, such as banks, railway and tramway companies, mines, water and gas corporations, etc. They occupy also a very honourable place in the import trade of textile fabrics, metals, ironmongery, construction materials, watches and jewellery.

PERSIAN AND OTHER ARMENIANS.—Armenians in Persia are steadily decreasing; they emigrate to seek for work in Russia or in Constantinople, India, Java, or China. They are indigenous only in the province of Azerbaijan, and concentrate in Tabriz, Salmas, and Urumia; outside this province they are immigrants. They are mostly commercial people, and constitute a class not much above poor, miserable peasants, not safe from exactions, nor the bastinado, at least it was so before the famous Constitution. Some once-rich Armenian colonies, as for instance one established by Shah Abbas I., are now decayed and find themselves, like so many things in Persia, in miserable circumstances.

In Austro-Hungary Armenians are pretty numerous, probably above 10,000. The Armenians in Vienna are divided into two distinct groups; a small secular colony, and the congregation of P. P. Mekhitharistes. The first, about 60 to 70 in number, frequent the church of Mekhitharistes, but for their religious requirements address themselves to the Greek clergy. The Mekhithariste arrived from Trieste in 1810, and have been received by the Imperial family and the Viennese clergy with great favour. They study science and philosophy, possess an excellent printing establishment, and direct a seminary with about thirty pupils destined for ecclesiastical careers. They have a rich collection of Armenian money.

Hungarian-Armenians, mostly Catholic, dwell in Transylvania and Bukovina. In the former province there are two Armenian commercial colonies: Szamos-Ujvar or Armenopolis and Ebesfalva or Elisabetopolis. These colonies were assigned to them in the seventeenth century, but they are being gradually absorbed by the Magyars, and only a few of these Armenians remember their native language. They are mostly commercial travellers, but are not nearly so ubiquitous as their commercial rivals the Jews; they have conserved most of the former traditions in commercial affairs. Armenians settled in Bukovina as long ago as the eleventh century; they have increased since and constitute now in Czernowitz, Suczawa, and other towns prosperous communities. More or less prosperous colonies are to be found also in Venice, Trieste, Marseilles, London, Amsterdam, Malacca, Singapore and Shanghai.

CUSTOMS AND HABITS.—Being a very old and conservative nation, Armenians, especially the country-people, have preserved ancient customs and usages which as such have a considerable interest. These illustrate patriarchal life of, perhaps, more than two thousand years ago. There is, of course, a great diversity of customs in different lands and in different social strata, for there are educated Armenians of high social standing, and very ignorant, poor, crushed and bigoted people.

Armenians appreciate social distinctions, rank, culture, learning, piety. Before all things they are fond of money, and in the art of making money they are not inferior to any other nation. They are unrivalled in the economic struggle with the nations they come in contact with, and easily get the upper hand. Most Armenians lead a true Asiatic life. Their women envelop themselves from head to feet in a veil; in Turkey many women as well as men go bare-footed. Women do not speak with men and seek to avoid them. Men wear, usually, long dresses. There is no furniture in the house, neither chairs nor tables; the people sit on carpets or low divans, and for meals they gather around brazen dishes put on the floor. They write sitting down on their heels, keeping the paper on their knees.

They are hospitable. If any one has a guest in a village, his neighbours send him victuals—milk, butter, honey, and wine—and come themselves to partake of the meal. There being few distractions, family festivals play a very important rôle.

Birth.—A new-born infant is sprinkled with salt and is not bathed till the second day. Soon after the birth a *terter* (priest) is brought into the room to say prayers, sprinkle the chamber with holy water, and put the Gospel under the young mother's pillow. The parents and their friends drink wine, and manifest much joy at the birth of a boy, but it is rather a disappointment when a girl is born.

Baptism takes place on a Sunday about forty days after birth. There is a godfather but no godmother. The child, carried by the nurse, is plunged into water, and then left for a week without a bath. Whilst the godfather takes the child from the font, the priest twists threads of white cotton and red silk to make a string which he binds round the neck of the infant, as a symbol of the blood and tears shed by Jesus crucified. After the baptism the child receives the sacrament, and as early as possible it is brought to the church to receive the sacrament once more, and so on every important holy day. On the evening of the day of baptism there are festivities in the house of the parents, accompanied with much drinking and noisy manifestations of joy. To preside at the banquet, a *tulumbash* (chairman) is chosen, who is invested with discretionary powers to propose the health of each person present in turn, according to their rank, to see that at every toast the full cups are emptied, and to punish the refractory by obliging him—to the great joy of those present—to drink twice as much. After each toast the whole assembly sings in chorus thrice *mravoljavia* (the German Hoch).

Betrothal.—Armenians are not a sentimental people. A young man desiring to take a wife is not guided in the choice of his future companion by affection. Wives are scarcely to be considered as true companions of their husbands, their situation is quite subordinate; they have to serve their husbands and obey them.

The Armenians do not recognise divorce *a mensà et thoro*, but husband and wife may be divorced, and the guilty party has no right to marry again. An Armenian desiring to take a wife sends, without previous courting, one of his friends or relations to the parents of the young person to ask in his name for her hand. When the parents are not disinclined to accept the offer, they appoint the day and the hour when the intended may meet and see his future bride; the place of rendezvous being usually a public garden or a church during the service hours. It is not seldom the first time that the wooer has the opportunity of examining silently his intended wife. If the examination is satisfactory and the wooer declares that the young person pleases him, then, as a token of the seriousness of his intentions, he offers a small present. There is no question of the consent of the bride, who has to obey the will of her parents. A week after this first meeting, the time being employed in obtaining information as to the fortune of the interested parties, the wooer, accompanied by a priest and friends, calls at the house of his intended. The priest consecrates the betrothal, and the bridegroom this time presents his bride with a gift of some value. Armenians keep their head cool, they seldom depart from a certain mistrust, and are always rigidly calculating. There are stories of beautiful brides replaced at the decisive moment by others less attractive. If all goes well, the elders do not delay any longer in appointing the wedding-day. Before taking leave of his bachelor's state, an Armenian needs to sow his wild oats; for some days he leads a disorderly life in company with his friends.

On the wedding day the bridegroom sends his friends to the parents of the betrothed in order to bring home the promised trousseau. The proceedings are noisy, for the joyful messengers are accompanied by musicians, and the procession of street-porters (*musha*) charged with chests of drawers, tables, looking-glasses, beds, cushions, carpets, and articles of clothing, naturally excites general curiosity. The conveyance of the trousseau, in fact, rouses the whole neighbourhood, and provokes on the part of the audience a thousand comments. On his side the bridegroom sends on this occasion to the parents of his bride some bread and a sheep. On the evening of the same day, the bridegroom calls at the house of his bride, offers her a new wedding gift, and presents to his future parents upon a tray four wax-tapers and two sugar-loaves. The wedding takes place at midnight. The betrothed are accompanied to the church by their sponsors. After the nuptial ceremony, during which the priest reads the duties of man and wife, the new married couple are conducted to the home of the man's parents, and at each door they pass their friends cross daggers above their heads. On approaching their home they are met by a near relative, who offers them a cup of water to drink. In crossing the threshold of his own house the new husband is expected to break a plate placed in his way.

These are the usual proceedings in Caucasian villages. The marriage ceremonies in Turkey are somewhat different. The young man rides to seek his betrothed on horseback, and she awaits him also mounted. She is veiled in white and keeps in her hands the end of a long belt, the other end of which is left loose, and is to be grasped by the bridegroom. The parents and friends, all mounted, follow the couple to the temple, bearing wax-tapers in their hands. To poor people, the necessary horses are readily lent by their rich neighbours. At the foot of the altar, the betrothed join their foreheads, the priest places the Gospel on them and reads the passages concerning the duties of married people. He asks the man whether he is willing to be master and patron of his wife, and he asks the wife whether she is willing to be docile and devoted to her man. Then he joins their hands and the ceremony is over. The nuptials are always accompanied by many days of feasting to which the relatives and friends of the young couple are invited. These feasts cost a lot of money, and are very noisy. Any place is suitable for the gathering: an inn, a wine cellar, a garden, a flat roof, the main condition required being abundance of food and drink. Armenians are generally sober, but on these special occasions they allow themselves to drink to excess, and manifest their joy with great exuberance.

Armenians have kept up the patriarchal habits of their ancestors; the newly married couple remains dependent on their parents; the young husband brings his wife to dwell under the parental roof, and she is expected to obey not only her husband, but also her mother-in-law and all the other older female relatives of her new family. Before she gives a child to her husband she remains extremely docile and submissive, with tightly closed mouth. With a child on her bosom she gains in respect and consideration, but even then and for a long time an Armenian woman appears in the street with uncovered face, but with the mouth and lower part of the chin closed by a white bandage. It is the tongue of women, not so much their faces, that the Armenians consider most dangerous to their domestic peace.

Armenians respect the family, the *thônir* or *tandur* (hearth) is a sacred thing; the children have great respect for the head of the family, and do not dare to take a seat in his presence without permission. Armenian mothers are known for the deep affection they cherish for their children. There may be not much poetry in the married life of Armenians, but there is not much infidelity.

The young women have all an ardent desire to have children, and if they do not get them they have recourse to numerous superstitious practices in the hope of obtaining them. Thus in the neighbourhood of Etchmiadzin is a village which organises every year at the beginning of summer a famous feast in the honour of S. Messrop. Women desirous of having children betake themselves to the saint and steal some objects in the temple. If later they become pregnant the object is restored with numerous gifts.

Funerals.—In funerals as in other family ceremonies Armenians like noise and publicity. Persons of some notoriety are buried on Sundays to get more people at their funerals; those of common rank are buried on the third day after death. The washing, dressing, and laying in a coffin or a bag of new cloth, as also the prayers are usually performed by a much detested but useful man called *mordichon* (French *croque-mort*). He is a humble personage who gets no other recompense for his sad work than the clothes and ornaments that may be on the person of the deceased. At the news of somebody's death, his relatives and friends, especially women, betake themselves to the mortuary house to express their condolences. The mourning is noisy. The female members of the household squat in one corner of their room and there receive the visits of condolence. At the view of each new visitor there is a renewed burst of lamentation, gradually appeased. The visitors are in no hurry, it is not becoming to leave the mourners alone with their grief, and the house remains full of people, who expect to be entertained. But there is no fire kindled so long as the body of the deceased remains in the house. The preparations for the great post-mortem entertainment begin only after the removal of the dead. The body is laid in the coffin as soon as this is brought, and the coffin is put on a table in the middle of the room and surrounded by tapers. All the while a sacristan or mordichon reads prayers. On the third day after the death the body is brought to church for the mass and then later to the cemetery. Men alone accompany the body to the grave. The coffin of a distinguished person is borne to the cemetery on the shoulders of his friends. Those present do not form a procession but gather around the coffin in a crowd. Priests taking part in the funerals as well as singers wear sacramental dresses of light colour. The coffin usually remains open. Sometimes the procession is accompanied by sacred images and banners, and the funeral car is adorned with wreaths and flowers. There is no calculated ceremonial deliberateness in the proceeding, but on the contrary some haste. The clergy precede the procession chanting in a sing-song manner the *Sarakans.* At the cemetery the coffin is lowered into the grave and the mourners hurry to gather around the great repast prepared meanwhile by the parents of the deceased. When the bereaved family is rich those feasts last quite a week and are offered even to strangers. Armenian funerals are expensive and are certainly characteristic of a nation otherwise so sober and economical. The excesses they indulge in on these occasions may be a natural reaction against their otherwise dull life.

ARTS AND LITERATURE.—Works of art and literature are the best testimonies of a nation's spiritual grandeur. The Armenians have of old distinguished themselves by architectural art, theological and historical works, as also by translations from Syriac, Arabic, etc.

According to Professor Patkanoff the Armenian language constitutes a testimony of the great antiquity of the people speaking it; it occupies a middle place between the Iranian and the Slav-Lithuanian groups, and represents an independent branch of a lost family of languages.[1] There are twelve dialects.

The ancient Armenian works have mostly if not entirely perished, but there remains innumerable churches and convents dispersed all over Asia Minor, Caucasus, Persia, that are testimonies of high art and remarkable for the difficulties surmounted in their construction on the summits of mountains or in narrow mountain gorges. There are churches hewn out in rock. Many of them contain precious collections of MSS., mostly copies of Gospels, themselves remarkable objects of fine art. Even in our days, the extensive ruins of Ani, on the Arpa Chai, with their cathedrals, chapels, castles, of a high style of architecture, afford evidence of Armenian art.

The most famous of the still existing sanctuaries in the eyes of Russian Armenians are the cathedral of Etchmiadzin and the chapels in its vicinity belonging to the saints Gayiane, Ripsime and Solakath;

---

[1] Patkanoff, *Izsledovanie o sostare Armianskago iasyka.*—Moskva.

whilst in the eyes of Turkish Armenians the most esteemed sanctuary is the temple of Surb Karapet in Mush.

As objects of fine art may be regarded the Armenian manuscripts dispersed in a number of libraries and convents, such as Paris, Vienna, Venice, Etchmiadzin, Constantinople, etc. They have considerable value for Christian theologians, and have been the objects of the studious researches of scientific missions from France and Germany.

A considerable number of MSS. were made by pious men, who desired to copy the Holy Scriptures to ornament convents and churches. In their dedication, they implored the faithful of the community to pray for the safety of their souls and strongly condemned those who under any pretext should remove these MSS. from the church.

Some MSS. contain fragments of the history of Armenia. Many of them are coloured and illustrated; the illustrations represent usually religious scenes and portraits of saints. They are written in very old *erkathagir* or *bolorgir* upon parchment, with golden or purple ink. The copies of the Gospel deposited in the convents are often ornamented with miniatures, rich stuff, pearls and other precious stones.[1]

Literature.—The first impulse given to modern Armenian literature came from Venice and Vienna in the eighteenth century, and a little later from Constantinople and Smyrna (at about 1820-1830). Armenians have produced many newspapers, and have translated some French and German romantic works. The Armenian language has acquired in the absence of national unity quite a remarkable degree of development. Printing presses have been established in most of the cities where Armenians are numerous. In Russia the Armenian press has grown in importance since 1828, that is since the annexation of the Ararat part of Armenia. The Armenians have now in Tiflis and Baku a number of renowned literary, dramatic and scientific authors. They have two daily papers in Armenian, and one in the Russian language, which exercise influence also in the Turkish and Persian parts of Armenia. There are besides two weekly literary reviews, one agronomic, one humoristic, and two or three monthly, and one ethnographical publication, quite at the level of similar publications in Europe.

Notwithstanding the rigours of the censorship the Armenian press in Constantinople is by far the most important among the polyglot publications of the Turkish capital. There are now published fourteen periodicals, of which seven are daily, five weekly, one bi-weekly, and one monthly. These journals and reviews are political, literary, scientific, economic, sociological and satiric. There are also Armenian publications dealing with history, biography, philology, pedagogy, social economy and philosophy, both originals, and also in translation from the French, English, German, Italian and Russian languages.

An Armenian writes in a figurative, passionate, high-flown style; he would say, for instance: "The sea of life boils in my soul, its troubled and poisonous waves gnaw my breast as one buried alive gnaws the lid of his coffin."[2]

---

[1] *Nouvelles archives des missions scientifiques et littéraires.* Paris, 1900.

[2] *Avetis Aharonian.*

BIBLIOGRAPHY.—Documents for the study of Armenians both in their own and in other European languages are not wanting. The following are some of those in English:—H. F. B. Lynch: *Armenia: Travels and Studies* (London, 1901) (see *S.G.M.*, 1902). Curzon: *Armenia* (London, 1854). H. Barkley: *A Ride through Asia Minor and Armenia* (London, 1891). Rolin Jacquemyns: *Armenia, the Armenians and the Treaties* (London, 1891). Mark Bell: *Around and about Armenia* (see *S.G.M.*, 1890). Anglo-Armenian Association: *The Case for the Armenians* (London, 1893). Dr. Felix Oswald: *Explorations in Armenia.*

THE USE OF PLACE-NAMES IN THE TEACHING OF GEOGRAPHY.[1]

By JAMES A. RAMSAY, M.A.

PRESENT-DAY ideas concerning Geography have shown us how very comprehensive the subject is. The matter dealt with and the methods adopted in teaching the subject have also shown that Geography draws largely upon many other sciences, including Geology, Physics, Astronomy, and Botany, and to a small extent Ethnology. There is also, however, another "side-issue" which might be treated, namely, place-names, The aspect of the subject has, so far, been almost entirely left in the hands of a few enthusiastic philologists; it has not yet entered to any appreciable degree into the teaching of Geography, and it is because I have found place-names, their meanings and their "story" so interesting, that I venture to lay my arguments before you.

Isaac Taylor in his *Words and Places* says: "The names which still remain upon our maps are able to supply us with traces of the history of nations which have left us with no other monument. . . . The knowledge of the history and the migrations of such tribes must be recovered from the study of the names of the places which they once inhabited, but which now know them no more—from the names of the hills which they fortified, of the rivers by which they dwelt, of the distant mountains upon which they gazed. . . . Language adheres to the soil when the race by which it was spoken has been swept from off the earth, or when its remnants have been driven from the plains which they once peopled into the fastnesses of the surrounding mountains."

To study place-names in detail, and follow the movements of all nations on the earth as traced through place-names, would involve a task beyond the scope of the teacher of geography, and far beyond the grasp of his pupils. The teacher will mention in his teaching the meanings and significance of such names only as he thinks his class will comprehend thoroughly. Hence his use of Etymological Geography is of necessity limited in extent, but, however few be the names he deals with, he can by their means make his subject more vivid and realistic. The imagination of his pupils will be stirred, and the results will be an

[1] A paper read at a meeting of the Geographical Section of the Royal Philosophical Society of Glasgow on 5th December 1912.

awakened interest and a more intelligent view of the world and of mankind in the history of the world.

It has been argued that the study of place-names lies within the province of the historian rather than the geographer, and place-names are undoubtedly a very useful adjunct to the study of history. The movements of peoples, their conquests, their settlements, may all be followed out by studying names. Geography, however, deals with the conditions under which men live, and these same conditions may be the cause of their migrations. Therefore I contend that in so far as the circumstances under which or owing to which a people moves and forms new settlements with names of their own are geographical, it lies within the province of the geographer to deal with these names.

But even if it be granted that the story of place-names lies in the province of history, still the teacher of history makes little or no use of these names in order to show colonial expansion or any movement of nations. His work in schools is usually confined to our own islands and Europe. The rest of the world's history is a closed book to the pupil. Why, then, should not the teacher of geography take up this line of study? It is a branch of education of great value; the child can be told to look for names in a particular language on the map. Such names indicate the presence of a particular people in the country at some period in history. Take, for example, Spanish, French, and English names on the map of North America.

The main point, however, lies in the fact that geography and history are in certain important respects closely related. As I have already said, geographical environment may be the cause of historical events; the configuration of a country determines the sites of battlefields, fortresses, etc. No teacher can study the geography of a country, say one of our colonies, without indicating to a certain extent the facts which led to its discovery or exploration. The motive may have been—and it frequently was—a commercial one. Commerce, as we know, depends largely on geographical circumstances. The crops and products generally —vegetable, animal, or mineral—depend on certain conditions; the observation of these conditions, principally in connection with vegetation and animal life, and the showing how they are related to these products, form a large and important part of the domain of geographical teaching. Plants, as we know, are very susceptible to differences in temperature and amount of moisture, etc.; the finding of lands suitable for their growth, the colonisation and the eventual commercial activity of these lands, must therefore be within the sphere of geographical teaching.

One of the best examples of this close connection between geographical environment and colonial expansion may be instanced in the case of the rise of the Portuguese navigators in the fifteenth century. The motive of their movements was purely commercial, and was due to the then important spice trade. The spices, as is well known, came from the East—India and the East Indies—by the overland route, a route which was beset with difficulties and dangers which led to the spices becoming very expensive. The Portuguese, under their famous prince, Henry the Navigator, had studied the maps of the world as it then was supposed to be. They determined to find a way to the Indies by sea, and after a long and arduous struggle they succeeded. All along the course of their voyages are to be found place-names in the Portuguese language. The names Madeira, Canary, Cape Blanco, Cape Verde, Lagos, Guinea, are of Portuguese derivation. The Dutch followed up the Portuguese after the defeat of the latter country at the hands of Spain; hence the Dutch settlements in Africa with their characteristic Dutch names—Orange, Vaal, etc.—and the Dutch East Indies, the main source of Holland's wealth to-day.

The stories of the discovery and development of new lands are of immense value to geographical teaching. The motive, the course of the voyages, the maps, the products obtained—all form part of the material at the disposal of the geographer, and so, also, do the *names*.

There are, of course, other circumstances which may have induced movement on the part of a people—the sterility of a country, as in the case of the Persians and Arabs; in these cases there was a struggle for existence, a struggle which eventually led to invasion and the forming of vast empires. Look how the names in Southern Spain indicate the presence of the Arabs once upon a time; the teacher or geographer cannot pass this point when dealing with this country. In the case of the Phœnicians there was an inhospitable hinterland. Expansion was impossible in that direction. The finding of a particular shellfish which produced a valuable purple dye led to the formation of settlements where such shellfish were found. Here they could extract the dye and leave the dead-weight behind; this was more suitable than taking the shellfish to their homes at Tyre and Sidon.

Then we may mention the use of copper and tin by the Phœnicians, their journeys in search of these metals, and the routes they followed—the valley of the Rhone, and thence by the Seine to the shores of France and Britain. Traces of their settlements may still be found in the place-names in certain parts of Italy, Sicily, and Spain.

The sterility of the land may be given as the cause of the invasions and consequent settlement by the Franks, the Longobards, the Vandals, the Huns, and the Turks, and it is interesting to note the names of the portions of Europe occupied by these peoples; they are named after the people who settled there (France, Lombardy, Andalusia, Hungary Turkey).

The question may now be raised—can the significance and meaning of place-names be dealt with by pupils of all ages? There most certainly must be a limit somewhere; very young pupils are practically out of the question altogether. But in elementary schools the subject can be discussed with a measure of success; the amount of success or of interest excited will depend on the teacher. He must use his own judgment concerning which names can be explained to his class; the ages and intelligence of the pupils must be considered. Names are frequently descriptive in character, indicating size, colour, or some other outstanding characteristic, and the lesson can be made vivid, the imagination of the pupils can be awakened by the description of the place showing why the name was given.

In higher classes or in Secondary Schools the task is much easier. Here the education of the pupils is farther advanced, they have a knowledge of languages which is of immense value, and the work of the teacher is easier.

It would be a serious mistake, however, to exaggerate the importance of place-names and their meanings in relation to the other work dealt with in geographical teaching: there are many factors of greater importance in geography than names and their import, *e.g.* climate, productions, occupations of the people. In certain classes, where the field of work is extensive, the time spent in dealing with names must be of necessity very limited. There is no reason, however, why in certain higher classes a special study of this branch of geography should not be undertaken.

If no special examination be in view, there is a splendid opportunity offered to the teacher of geography to take up some special department of geography for study in these classes, and the story of place-names might form a very useful and interesting study.

The results of a course of study of place-names, however scanty be the amount of work done in this department, are of great and far-reaching importance. I have already mentioned the voyages of the Portuguese in their search for the sea-road to India. The story of this famous exploit would form a most interesting lesson, the names dealt with would give an added interest to the narrative, and no doubt the lesson would create a desire to learn more concerning these early navigators. There is, therefore, a distinct effect on the reading of the pupils. Lives of great navigators and explorers form most interesting and instructive reading, and the very mention of Frobisher, Parry, Franklin, Magellan, Cook, as recalled by parts of the world's surface commemorated by their names, should act as an incentive to a healthy outlook on life.

I have examined at various times copies of text-books of geography used long ago, and in quite a number of cases there were given lists of names and their meanings, showing that our grandparents saw some important facts in mere names, and made use of these in the teaching of the subject. It seems as if I am really attempting to revive an old interest; it has been lying hidden away among the dust of the past, and if this interest be reawakened and put to practical use in our schools in connection with the teaching of geography, I am sure the results will be all for the benefit of education and the training of those faculties which tend to make a boy a good citizen.

Note.—Among the books which the teacher may use are:—A. J. Bury, *The Association of History and Geography* (Library of Pedagogics); J. B. Johnston, *Place-names of Scotland* (1903); Sir Herbert Maxwell, *Scottish Land-Names* (1894).

GEOGRAPHICAL NOTES.

Asia.

The Physiographical Divisions of China.—Professor Eliot Blackwelder contributes to a recent issue of *The Popular Science Monthly* an interesting article on China, in which he recognises the following typical physiographical divisions:—(1) The north-eastern mountain region, typified by the province of Shantung, with broad valleys connected by easy passes over the low, isolated mountains. Roads are plentiful and easily made, so that carts and wheelbarrows are extensively used, but on the other hand the streams are of little use for navigation. The people are clustered in the broad valleys. (2) To the west comes the broad plain of the Hoang-ho River, floored by thick deposits of silt, densely peopled and intensively cultivated. There is not sufficient water for rice, but the dry-land grains, together with cabbages and potatoes, are grown. The numerous streams and their shifting courses make road construction difficult, but the extensive canal system allows for great freedom of intercommunication over the plain, and makes the people more homogeneous in customs and languages than in the other parts of China. The great drawbacks are the absence of fuel and fodder, and every scrap of straw or stubble is used up for one of these purposes, so that the fields are absolutely bare in autumn and give rise to great clouds of dust. (3) To the west and north-west lie the rugged plateaus and mountains of the loess region. Here there is little water for irrigation, but the fertility of the loess leads to its being most elaborately terraced on the hill sides. Communication by water is not possible, and in the more mountainous parts road construction is difficult, and pack animals are largely used. Coal is extensively produced here, but the difficulties of communication make it impossible to transport it to any great distance. (4) To the south-west lie the Central Ranges, with a very rugged surface, and isolated patches of cultivated land. Pack animals and coolies are the only available means of transport in the more remote regions. In consequence, despite the demand for fuel in other parts of China, it is almost impossible to bring the lumber of this mountain region to the plains. Rice is cultivated where possible, and the opium poppy, though not in local demand, used to be a favourite crop owing to the fact that the product is of high value and small bulk and will thus pay the cost of transport. This region of the Central Ranges is traversed by the great Yangtse River, which, despite the difficulties presented by its gorges enables the products of (5) the rich province of Szechuan to be carried eastwards. This broad basin, with its floor of soft sandstone beds, and its varied topography, produces a great variety of crops, such as silk, oil, lacquer, rice, with many other grains and garden vegetables. Protected by mountains to the west the province is one of the most stable and progressive in China. Finally, to the south-west comes (6) the mountain region of Southern China, with a very rugged surface, great difficulties of communication, and settlement practically confined to the few level valleys.

America.

Meteorological Observations in the Sierra Nevada.—We have received two pamphlets giving interesting accounts of observations made at Mount Rose Observatory, a high level station in the Sierra Nevada placed on the summit of the mountain named at a height of 10,800 feet above sea-level. This is the highest meteorological station in the United States, and owing to the position of Mount Rose at the western edge of the Great Plateau is well fitted for the study of mountain and desert meteorology. Two other stations at lower levels carry on observations simultaneously with the summit station, and in addition very detailed observations are being made with the object of predicting frost, and also of determining the limits within which fruit-raising may be profitably carried on on the lower ground. Observations spread over a wide area in the basin of the Truckee River show that where the minimum temperature is never less than 28° F., fruit-raising is highly profitable. Where the minimum temperature lies between 24° and 27° F., heating of the orchards must be resorted to in frost, hence the necessity for frost warnings from the summit station. Where the minimum temperature lies between 18° and 23° F. fruit-raising ceases to be profitable.

Another point to which attention is being given, and which is of great importance in this irrigated region, is the relation of mountains and forests to snow conservation. It is upon the mountain snow that the irrigator must often depend to fill his canals, and it is therefore very necessary to find out what kind of forest is most effective as a snow trap. It is found that belts of timber on the tops of cliffs and lee slopes increase snow accumulation at the foot of these, and also by checking wind diminish the rate of evaporation of the snow. On the other hand unbroken forests catch falling snow in proportion to their openness, but conserve it afterwards in proportion to their density, *i.e.* in proportion to the extent to which they shelter the ground from wind and sun. Forests should therefore be light, with numerous glades to permit of snow accumulation, but the area of the glades must bear some proportion to the height of the surrounding trees. At high levels, 8000 feet and higher, the mountain hemlock is found to be the best tree, for its tapering form allows the snow to slip down to the ground below, while the dense foliage afterwards protects the fallen snow from evaporation. Thus the planting and management of forests in catchment basins which in winter are snow-covered, requires a considerable amount of care and consideration.

Glacial Excursion of the Geological Congress.—Several of the excursions in connection with the twelfth International Geological Congress, which is being held in Canada this summer, will go from Toronto to Vancouver. Then an excursion (C8., August 29 to September 22), under the leadership of R. G. M'Connell, and with the guidance of R. W. Brock, D. D. Cairnes, and W. W. Leach, will traverse the fiords of British Columbia, ascend the Skeena River valley from Prince Rupert to Aldermere by rail, visiting the silver-lead mines and coal mines, and continuing to Skagway by steamer. There will be stops at the copper mines on Portland Canal and the Treadwell gold mine on the Gastineau fiord at Juneau. The excursion will then cross the Canadian Coast Range by the White Pass and Yukon Railway to Whitehorse, stopping at the copper deposits there and the coal mines at Tantalus, descending the Yukon River to Dawson and the Klondyke gold field in the driftless interior plateau near latitude 64° north.

After the return to Skagway an excursion, under the direction of Lawrence Martin of the University of Wisconsin, will be made on a special steamer to the Malaspina Glacier, Yakutat Bay, and Muir Glacier. This glacial excursion will last five days, with a possibility of two days more in case of cloudy weather.

The first day will afford an opportunity of seeing the Fairweather and St. Elias Ranges, 16,000 to 18,000 feet high, and covered by snow-fields and glaciers. These ice tongues include the La Perouse, Malaspina, and many smaller glaciers. The front of the great piedmont ice sheet of Malaspina Glacier will be followed, affording an opportunity of seeing the tidal ice front of the Guyot lobe west of Yahtse River, the moraine-veneered ice cliff of the Seward lobe at Sitkagi Bluffs, and the forest-covered terminus of the Marvine lobe near Point Manby.

On the second day something will be seen of the eastern border of Malaspina Glacier in Yakutat Bay, and the forested terminal moraine of the Yakutat Foreland. Landings will be made in Disenchantment Bay in connection with various glacial phenomena such as the shrub-covered ablation moraine upon the ice of Variegated Glacier, the streams engaged in carrying and depositing outwash gravels, the calving of icebergs from Hubbard and Turner glaciers, the cirque vacated by a fallen glacier, and the beaches, rock benches, sea cliffs and islands which were uplifted from 7 to 47⅓ feet during the earthquakes of September 1899.

The third day will be spent on and near the Nunatak Glacier in Russell Fiord. Here the hanging valleys, the till-veneered, overridden outwash gravels, and the tidal, land-ending and cascading glaciers will be visited and studied, as well as the phenomena of glacial erosion in the barren area from which the ice has recently retreated and of fault scarps made during the 1899 earthquake. Some of these scarps are vertical and are 4½ to 8 feet high.

The fourth day will afford an opportunity of seeing the morainic and glacio-fluviatile phenomena about the terminus of the Hidden Glacier, which advanced 2 miles between 1906 to 1909, as a result of the earthquake avalanching in 1899, which has subsequently caused nine ice tongues of Yakutat Bay to move forward. After this landing something will be seen of a fiord with submerged hanging valleys, submarine moraines, buried forests, shorelines depressed in 1899, and the high strand lines of a former glacial lake.

Part of the fifth day will be devoted to Glacier Bay, where there has been a recession of 8¾ miles at Muir Glacier from 1899 to 1911. A landing will be made in Muir Inlet to see the buried forests, the vertical ablation of over 1200 feet of ice in twelve years, and many other phenomena. The rapid recession of Grand Pacific Glacier in Reid Inlet at the head of Glacier Bay now places part of this fiord in Canada.

The glacier melted back 5000 to 7400 feet in two months during the summer of 1912, as was determined by N. J. Ogilvie of the Canadian Boundary Survey. At the International Boundary there is now dry land and open fiord where the ice was at least 1750 feet thick as recently as 1894. Sixty miles of Glacier Bay have been opened to the ocean by glacier recession since 1794, making an arm of the sea as long as Hardanger fiord in Norway

Progress in Brazil.—A recent *Report* on the Consular District of Rio de Janeiro, by Mr. Acting Consul-General Hambloch, contains some interesting facts both in regard to the city of Rio de Janeiro and to Brazil in general. It is noticed that the importation of Argentine cattle into Brazil has almost ceased, and that not only does Brazil practically supply its own market, but the question is being raised whether an export trade in cattle and meat cannot be started. The quality of the meat is not very high, and attempts to improve the breed by importing European stock are rendered difficult owing to the liability of European cattle to Texas fever. Cold storage plants are, however, to be erected, and a company is being started with the object of exporting frozen beef to Liverpool. The matter is of considerable theoretical interest, as the question as to whether tropical countries will in course of time become stock-raising areas has often been debated by geographers, without any definite decision being arrived at.

Brazil is also endeavouring to develop its own fisheries, with as yet apparently but little success. At present dried cod is imported to the extent of about one million pounds' worth yearly, about half the total coming from Newfoundland. The export of fruit is increasing, though the total export is still insignificant. Attempts are being made to stimulate wheat production, though as yet with but little result, but on the other hand the levy of a customs tax on imported rice has considerably increased the home production, with a corresponding decrease in the imports, of which India is the chief source. On the other hand the effect of high taxation may be seen in the cost of living, estimated as three times as great, in the city of Rio, as in Europe, and in the growth of trusts. Another disquieting feature is the employment, without regulation, of female and child labour in factories. In 1911 a measure to prevent the employment of children under ten years of age was proposed but dropped.

The cattle industry is carried on chiefly in Minas Geraes, which also produces for export rice, coffee, potatoes, beans, tobacco, maize, etc., and has very large and apparently very valuable deposits of iron ore, as yet not worked on a large scale owing to the difficulties of communication in this hilly state. Of the population of four and a half millions in this state some three and a half millions are estimated to be engaged in agriculture and the remainder in the mines.

General.

Winds in the Free Air.—Mr. Charles Cave recently delivered before the Royal Institution an address on this subject which is printed in *Nature* for May 22. The address is based largely upon results obtained by Mr. Cave from the sending up of balloons containing a meteorograph from Ditcham on the South Downs, and for the most part these results are of somewhat too detailed a nature to lend themselves to a summary. Thus a very full account is given of the different currents of air met with by the balloons in ascending through the troposphere. From the discussion we extract two points of general interest. The first is an explanation of the phenomenon—probably more frequent at Edinburgh than at Ditcham!—of continuous rain with a north-easterly wind. It is difficult to see how such a wind, especially in south-eastern England, can be saturated, in view of the fact that it has crossed only a narrow sea. Mr. Cave finds evidence that the rainfall is due to the presence in the upper air of a south-westerly wind bringing moisture from the Atlantic. So sure is he of this that, though living only twelve miles from the Channel, he does not hesitate to send up balloons even in this type of weather, being sure that they will be caught in the upper current, and so carried westward and not out to sea. In point of fact they generally descend in the Midlands or in the eastern counties.

The other point of special interest is that Mr. Cave's results confirm the suggestion, made by Dr. Shaw, that the changes of pressure to which the weather at the surface owes its origin, take place, not near the surface of the earth, but just below the level of the stratosphere, at a height of about 9 kilometres. This suggestion is considered in some detail, various types of observed pressure and winds at different levels being considered in relation to weather changes. For details, reference should be made to the original paper.

Commercial Geography.

The Cape Verde Islands.—A recent *Diplomatic Report* gives some interesting details in regard to these islands, which are said to be badly in need of capital for their further development. It is stated that there is plenty of subterranean water which could be raised and used for drinking and agriculture; meantime drought is a serious drawback. The two most important articles exported are coffee, which practically all goes to Lisbon, and Purgueira seeds, which go mostly to Lisbon, but partly to Marseilles. Purgueira (*Jatropha curcas*) is a tropical shrub belonging to the *Euphorbia* family, and offering certain resemblances in the characters of its seeds to the castor oil plant, with which it has been confused. In addition to its purgative properties the oil expressed from the seeds is of use as an illuminant, and the husk of the seed contains much potash, hence the use of the seeds in soap-making. The shrub demands practically no care in cultivation, and is very resistant to drought and strong winds. From time to time the Government has attempted to use it to afforest the waste lands of the archipelago, but the ravages of goats and the fact that the inhabitants prize the plant as fuel, have led to the destruction of these plantations. The output of seed is, however, increasing.

Among the other products of the islands are fruits, which could be exported to a considerable extent if communications were better.

The small exports (about £60,000 in value) go chiefly to Portugal, the mother country. On the other hand the imports, chiefly coal, textiles and provisions, come largely from the United Kingdom, but all come *via* Lisbon, where they are transhipped to avoid the import duties on foreign goods.

NEW BOOKS.

EUROPE.

The Lord Wardens of the Marches of England and Scotland. By Howard Pease, M.A., F.S.A. London: Constable and Co., Ltd., 1913. *Price 10s. 6d. net.*

Mr. Pease claims for his book, which, by the way, is dedicated to the Master and Fellows of Balliol, that it was written on the only true principle—to please the author. We believe, however, that it will fulfil another that is, commercially, at least as important, and please its readers. After the rubbish that so often passes for history nowadays, it is a most welcome achievement. The Scottish and English Border is a subject which would lend itself very easily to a sentimental, superficial treatment. Mr. Pease's book is nothing of this sort. It is written with a sound historical sense, and based on careful and extensive reading, as well as, obviously, knowledge of the Marches themselves, which is, after all, the best way to study history. The account of the English Marches is rather fuller than that of the Scottish side; Mr. Pease, for example, does not give the limits of the Scottish Eastern and Middle Marches, though he deals in detail with their English counterparts. This fuller English treatment may perhaps be due to more available authorities. But, taken as a whole, the subject is most completely dealt with. Many of the chapters are, of necessity, purely historical. Such are those dealing with the Border Laws, the Courts and Jurisdiction of the Lord Wardens, the origin of the office of Warden, and such kindred details. But much of the treatment is bound to be geographical. It is suggested, for instance, that the richer soil of the Scottish side of the Border supported a proportionately larger population than the English, but that this advantage was lessened by the bitter family feuds in Scotland, and the fact that this very fertility meant an easier and a richer booty for the raider. Then there are the routes taken by the raiders; fairly numerous, evidently, as the whole line of the English Borderland was systematically watched night and day for half the year. In the discussion of this part of his subject, Mr. Pease, it seems to us, has slightly misunderstood one of his authorities. He speaks (page 162) of a pamphlet of the Historical Association on the historical geography of Northumberland as giving a "Raider's Line" from Carham on Tweed by Harbottle to Chollerford, and seems to understand this "Line" as a common route of Scottish moss-troopers. But what is really meant by speaking of this as a "Raider's Line" is not that it was a common line for raids, but that, once the Northern English shires had come to some measure of prosperity, this line, which roughly coincides with the 600 feet contour-line, was *never* crossed by mere raids, but only by forces of considerable strength. A few such slips are probably inevitable in any book where many authorities are drawn on, and Mr. Pease's method—a very successful and interesting one—is to give the words of his authorities wherever possible, rather than paraphrase them in his own. There are interesting reproductions of old maps of the Marches from Blaeu's Atlas and elsewhere, but in addition reference is necessary to a modern large scale orographical map, if the geographical points are to be properly understood.

Wayfaring in France: From Auvergne to the Bay of Biscay. By EDWARD HARRISON BARKER. Illustrated. London: Macmillan and Co., Ltd., 1913. *Price 7s. 6d. net.*

The title correctly describes the author's wanderings, which were made afoot. In this way he was enabled to penetrate into regions denied to the motorist, and gained a personal acquaintance with the mind and manners of the peasantry. The disadvantage occurred when he had to share their mode of life, and the tale of his experiences is not always encouraging. His course lay from the headwaters of the Dordogne, with its deep-cut gorges, down to its meanderings in the Plains of Périgord, and by the valleys of its right bank tributaries, the Vézère, the Isle and the Dronne. He then crossed to the Garonne and the Adour, terminating his wanderings at Biarritz. Whilst not disregarding history, architecture, etc., his main interest is devoted to descriptions of scenery, and to personal experiences. A striking feature in these unchanging parts is the tradition of the *temps des Anglais* which seems to be credited in the native mind with too large a share of the destruction which the centuries have wrought.

The Oxford Country: Its Attractions and Associations described by several Authors. Collected and arranged by R. T. GÜNTHER. London: John Murray, 1912. *Price 7s. 6d. net.*

This anthology of the Oxford district comprises thirty-six papers, gathered mostly from scattered magazine articles, and not likely to be known to those who were not previously in touch with Oxford. But most of us are familiar with "The Scouring of the White Horse," and will be glad to renew acquaintance with the extract given. The articles are written with spirit and point, and wild life, hunting, geology, literary associations and history, all have their share.

It is not Oxford itself, with its Dons and Colleges, which is commemorated, but the country centred round it, and all likely to visit the country will do well to dip into this cheerful miscellany. The illustrations are in keeping with the text.

The Norfolk and Suffolk Coast. ("The County Coast Series.") By W. A. DUTT. Illustrated. London: T. Fisher Unwin, 1909. *Price 6s.*

It was natural for the editor of the series to turn to Mr. Dutt for this volume, and the choice has been justified. It is a capital book. Dealing only with a portion of the counties, it treats its subject from all points of view, but possibly local personalities and legends and the wild life of the district have first place. Those of us who hold as an article of faith that the great German descent will be made upon the East Anglian coast some month of July, will be interested in the old saying—

"He who would old England win,
Must at Weybourn Hoop begin."

Weybourn lies west of Cromer. A specialty of the county used to be the flocks of great bustards, which continued well into last century.

Probably no volume of the series will have such a tale to tell of the age-old combat between land and sea. It was the Romans who first successfully dyked out the waters—under Cromwell another stand was made—and Great Yarmouth has arisen in what was once a wide estuary. Yet the sea can claim to have engulfed at least a score of parishes, and the Report of the recent Royal Commission does not tempt us to think that its ravages have ceased. Many incidents of the struggle are recorded in the book. Two small criticisms we have to make. The map does not give nearly all the places mentioned in the text. Mr. Dutt has failed to avoid the common misquotation—*Tempora mutantur, et nos*, etc.

The Passing of the Turkish Empire in Europe. By Captain B. GRANVILLE BAKER. London: Seeley, Service, and Co., Ltd., 1913. Price 16*s. net.*

Captain Baker tells us that he wrote this volume in Constantinople while "the sound of firing was borne on the westerly wind" from the lines of Chatalja, and "a mighty empire was tottering to its fall"; and this may account for two things, one, an air of depression which runs through the book, and the other, the marks of haste in its composition. It is an ambitious work, for besides giving a very detailed account of Constantinople during the anxious days before the signing of the armistice on the 4th December last, it gives us a rapid survey of the origin and history of the Turks in Europe, of the modern Greeks, Bulgarians, Servians, and Montenegrins, and short notices of the Armenians, Albanians, and Vlacks who inhabit the Balkans. This is compressed into 328 pages, in which occur some repetitions. But not the less has Captain Baker produced an intensely interesting and instructive book. The history of the Turks is a sad one, *i.e.*, that of a once martial race, which rose rapidly to the possession of a wide and magnificent empire, and lost it by sheer incapacity to keep abreast of civilisation, the incapacity being mainly the result of the stagnating religion of Islam; but it is replete with thrilling episodes of romantic adventure and prowess on the side both of conqueror and conquered. Captain Baker's style is sometimes somewhat artificial and stilted, but he has made an honest study of the authorities on his subject: his narrative runs smoothly, and the interest never flags. He does not conceal his admiration of the Balkan allies, the deliberate prudence with which they made efficient preparations for their grand *coup*, and the heavy sacrifices they unhesitatingly made. Evidently he is of opinion that Great Britain has much to learn from the lessons of the war, but into this complicated and contentious question we do not need to follow him. We heartily commend the book to our readers.

A War Photographer in Thrace. By HERBERT F. BALDWIN. London: T. FISHER UNWIN. Price 5*s. net.*

With the Conquered Turk. By LIONEL JAMES. London and Edinburgh: Thomas Nelson and Sons, 1913. *Price* 2*s. net.*

With the Turks in Thrace. By ELLIS ASHMEAD-BARTLETT. London: William Heinemann, 1913. *Price* 10*s. net.*

With the Victorious Bulgarians. By Lieutenant HERMENEGILD WAGNER. London: Constable and Co., Ltd., 1913. Price 7*s.* 6*d. net.*

We have now before us a number of books written by war correspondents in the late campaign in the Balkans. Three are by Englishmen, worthy successors to William Russell and Arthur Forbes, and one by an Austrian officer. The whole campaign lasted only a few weeks, so there is a certain amount of sameness in the story told, and in the observations and criticisms of the correspondents. One and all agree in denouncing the obstacles thrown in the way of their seeing the facts, and especially the fighting for themselves, and for preventing them sending accurate news to London; and one and all had no hesitation in exerting their ingenuity to the utmost to outwit and evade the regulations imposed on them. In this they were very successful, but they did not succeed without running much personal risk. To be a successful war correspondent one has to have an iron physique, great personal courage, much fertility of resource, plenty of money, a working acquaintance with several languages, a facile pen, a fairly elastic conscience, and a capacity to live and work hard for two or three days on cigarettes, a few sardines, or a stick of chocolate. The first and foremost consideration is the interest of the paper the correspondent represents: when that is secured, the correspondent can be, and is, a good comrade and a kind friend to all and sundry, most especially to a brother correspondent in a fix. All these traits of character are well brought out in the volume before us, and we may add that the English writers have already won their literary spurs by various works which met with general acceptance.

The first of the four volumes is by Mr. Baldwin, the war photographer of the *Central News*, who got as far as Seidler and Karashtiran on the way to Lule Burgas, but was prevented from going farther. He gives a thrilling and life-like picture of the disastrous flight of the inhabitants and retreat of the Turkish army after the severe defeat of Lule Burgas. From Karashtiran he had an exciting journey to Chorlu, and thence to Constantinople. Here he and another war correspondent became possessed of "an interesting document" in "a curious way," which they prefer not to describe, and by means of it they managed to work their way back to Hademkeui, in the Chatalja lines and have an interview with the ill-fated Nizam Pasha, the commander-in-chief of the Turks. The story is well told. The illustrations of the book are very good, and there are many shrewd hints and suggestions which will be found very useful to photographers in similar expeditions in the future.

In his volume on the *Conquered Turk* Mr. Lionel James, an old campaigner and war correspondent of the *Times*, goes over a good deal of the same ground as Mr. Baldwin, and he has many caustic observations to make on the raw material of the troops sent to the front at first by the Turks, and the many and grave faults and deficiencies in the arrangements and administration, at least at the beginning of the campaign. He is not sparing of his criticism of the allies, who, according to his view, did not take full advantage of their first successes. He saw a good deal of the fighting, as he had a motor car which enabled him to get over the ground rather more quickly than some of the other war correspondents. Like Mr. Baldwin he was able to ingratiate himself with the Turkish officers, and his *savoir faire* and good-humoured audacity stood him in good stead in several tight places. The illustrations in this book are taken from the *Illustrated London News.*

The third volume—*With the Turks in Thrace*—is by Mr. Ashmead-Bartlett, the war correspondent of the *Daily Telegraph.* He was fortunate and skilful enough to outwit the Turkish authorities, and thus to have a good view of the all-important battle of Lule Burgas, of which he gives an excellent description. He also saw some of the fighting at the now famous lines of Chatalja, and at the attack on Rodosto. The book is considerably padded with disquisitions on a variety of subjects, *e.g.* the military history of the Turks, their unpreparedness for war, their lack of organisation, and so on; but it is all very interesting reading, and the personal adventures and escapades are narrated with a happy swing and verve, so that although there is a certain amount of sameness in them the attention of the reader never flags. Mr. Ashmead-Bartlett also comments on the fact that the victorious Bulgarian army did not take full advantage of the victory of Lule Burgas, which he attributes to their want of cavalry, and to exhaustion after the tremendous exertions of the previous week. This volume has many good illustrations from photographs by the author, and also from the *Daily Mirror*.

The fourth volume is by Lieutenant Wagner, the war correspondent of the Vienna *Reichspost*, who was deputed to watch and report the course of events at the Bulgarian headquarters. Although born in Vienna, Lieutenant Wagner grew up among the Balkans, speaks Bulgarian, Servian, and Bosnian fluently, is a *persona grata* in influential Bulgarian circles, and has had some experience of war conditions. This volume, like the others, is heavily padded. It contains only 285 pages, but it is page 125 before we reach the declaration of war. However, Lieutenant Wagner tells his story well, and lays his finger with unerring judgment on the deficiencies, weaknesses, and errors of the Turkish army and administration, which were apparently well known to the Bulgarian authorities long before the war began, and were counted upon as certain to affect gravely the issue of the campaign. Lieutenant Wagner, like his English *confrères*, gave the slip to those who were herding the war correspondents (and the hindrances and regulations in the Bulgarian camp seem to have been quite as irksome and rigorous as those in the Turkish, and indeed rather more difficult to evade), and he saw a good deal of the fighting. The accuracy of his telegrams to the *Reichspost* has been questioned, especially by Mr. Ashmead-Bartlett in his *With the Turks in Thrace*. But the narrative now before us is full of life and interest, and we heartily commend it to our readers. It is profusely illustrated with photographs, of which those of the leading Bulgarian statesmen and officers are particularly good, and there are several useful and instructive maps.

The Berwick and Lothian Coasts. By IAN C. HANNAH. London: T. Fisher Unwin, 1913. *Price* 6*s. net.*

Written by the grandson of Dr. John Hannah, Rector of the Edinburgh Academy from 1847 to 1854, this book is full of interesting and valuable information, evidently lovingly collected. It opens with a description of Berwick-on-Tweed, and passes along the coast by St. Abb's Head, Cockburnspath, and Dunbar to North Berwick, Haddington, Portobello, and Craigmillar, and thus arrives at Edinburgh. It is difficult to say anything new of Edinburgh, but the author tells us that he "has spoken to sailors who had walked the streets and listened to the bands of the Eternal City herself without finding in her anything more remarkable than they might have discovered in Birmingham, over whom nevertheless Edinburgh had not failed to cast its spell." There is nothing of importance in the city which the author does not refer to with care and describe with erudition. He is also critical, as when he points out that St. Mary's Cathedral, although a noble Gothic pile, does not "harmonise with the spirit of the New Town" so much as the parish church of St. George, which is, however, "architecturally deplorable (even its dome is not seen within)." He remarks that "the New Town of Edinburgh is one of the extremely few instances in which the British have consented to lay out a new city on any definite plan, instead of reflecting in the crude irregularity of uninteresting streets, something of the policy of Drift that so possesses the national mind. At Washington we may get some idea of what London might have been had Wren's plan for rebuilding after the Great Fire got any further than paper." Descriptions of Leith, Queensferry, Linlithgow, and Borrowstounness, and sixty-five excellent illustrations by Mrs. Hannah throughout the volume, complete a work for which the dwellers in Berwickshire and the Lothians owe gratitude to the author.

The River of London. By HILAIRE BELLOC. London and Edinburgh: T. N. Foulis, 1912. Price 5*s. net.*

The River of London is the seafaring name for that part of the Thames by which seamen approach the great city, and Mr. Belloc essays to show in his book how the Thames has made London, and that London owes its very existence through the ages to the fact that it occupies the point nearest to the sea at which, under primitive conditions, it has been possible to span the river with a bridge, To the eastward of London the banks of the river are mud flats or marshes, on one side or other, and hence London became the collecting and distributing centre upon which roads converged from all the points of the compass. We thus have the geography of the lower Thames expressed in its sociological aspect; and the only parallel which Mr. Belloc allows to London and the Thames is that of Antwerp and the Scheldt, though he thinks that the Thames is "probably more destined to endure in its functions and commerce." The subject is in itself an attractive one, and it is treated by Mr. Belloc in such a way that the reader is impressed with the important part which the Thames has played in the making of London and of the southern counties of England. The influence of the lower Thames in early warfare has a chapter to itself. In his text Mr. Belloc, as we have said above, deals only with the lower Thames, but the illustrations, which are very beautiful reproductions of oil paintings by Mr. John Muirhead, take us as far upstream as Teddington and Pangbourne, and they add greatly to the charm of the book as a whole.

What I saw in Russia. By the Honourable MAURICE BARING. London: Thomas Nelson and Sons, 1913. Price 1*s.*

Mr. Baring was the correspondent of the *Morning Post* in the Russo-Japanese War. In the first half of this little volume he gives us some vivid and stirring sketches of scenes in the war, more especially of the famous battles of Liaoyang and Shaho, of which he was a spectator. He bears eloquent testimony to the bravery of the Russians and the humanity of the Cossacks. The other half of the book has interesting sketches of life in Russia. He was present in Moscow during a part of the famous strike and during the religious ceremonies of Holy Week. He attended several sittings of the Duma in St. Petersburg and made acquaintance with several of the members, who apparently had no hesitation in speaking their minds to him very freely. There is also a good description of the well-known Nijni-Novgorod Fair and various scenes in the life of the peasantry. Mr. Baring is proficient in the Russian language, and this stood him in good stead everywhere. The volume is well worth reading.

Changing Russia. By STEPHEN GRAHAM. London: John Lane, 1913. Price 7*s.* 6*d. net.*

In the *Scottish Geographical Magazine*, vol. xxviii. p. 151, we noticed Mr. Graham's *Undiscovered Russia*, and on p. 665 his *Tramp Sketches.* In the volume now before us we have another work on much the same lines, a charming description of a tramp along the shores of the Black Sea with excursions into the Ural Mountains and the Crimea. Mr. Graham is a delightful companion on such expeditions, having a keen eye for the beauties of nature, a marvellous knack of falling in with strange and interesting people, much skill in drawing out from them their inmost characteristics, and a graphic and lucid pen to portray the whole scene for the delectation of his readers. But in this volume he seems unable to refrain from politics, and his invective against commercialism and the bourgeoisie is both bitter and extravagant. He is apparently obsessed with the idea that the one way to keep peace and prosperity in Russia is to preserve the peasantry in their present condition. "The hope," he says, "lies in the Tsar and his advisers, who are all conservatives, that they may truly conserve and keep the peasantry living simply and sweetly on the land, that they will not make any more commercial concessions when once the present pecuniary needs are satisfied. Of course if the Tsar and his advisers are not wise enough to save their people from commercialism, they will certainly bring ruin on their own heads. Every peasant brought into a factory or a mine or a railway is one man subtracted from the forces of the Tsar, and one added to the social revolutionary party." And of the bourgeoisie, or what he calls the "Lower Intelligentia," after a scathing description of their vulgarity and degradation, part of which it appears is their "miserable imitation of modern English life," he sums up his indictment in these words, "the curse of Russia, and, as the years go on, the increasing curse, is the bourgeoisie, the lower middle class, aware of itself articulately as the lower intelligentia. It is forming everywhere in the towns as a result of the commercial development of the nation. They are worse than the English middle class, worse than the Forsytes, because they wish to be considered national. They are of the race who 'limerick' and treasure-hunt, but are not occupied so innocently. They are unwilling to sacrifice anything, or to take any risks for political ends. Through them the *revolution* failed; they would have liked the revolution to have succeeded, but as they had not the faith of the true revolutionaries they waited to see who would win; selfish as it is possible to be, crass, heavy, ugly, unfaithful in marriage, unclean, impure, incapable apparently of understanding the good and the true in their neighbours and in life—such is the Russian bourgeois." No allowance is made for the inevitable results of the spread of education, and the fact that civilisation, with all that it implies, is penetrating gradually but surely throughout the Russian Empire, so that it is impossible for the peasantry and the middle classes to preserve the *status quo*. Moreover, if he gave a thought to the two countries where commercialism is most highly developed, and the middle classes, the bourgeoisie, is most powerful, Belgium and England, he would be fain to confess that the direful consequences which he so eloquently deprecates for Russia have not ensued, and are not in the least likely to ensue. Apart from these sinister and gloomy prognostications, this book will be welcomed by all who can enjoy a fresh breezy description of out-of-door life, in a strangely interesting country, exceedingly well told by an enthusiast who evidently enjoyed himself thoroughly, and succeeds admirably in visualising his tramp and his companions to his readers.

ASIA.

The European in India. By H. HARVEY. London: Stanley Paul and Co., 1913. *Price* 12*s.* 6*d. net.*

This volume is divided into three sections, in each of which we have twenty sketches of men, women, and functions or scenes in ordinary Indian life. This way of portraying Europeans in India may be made very amusing and quite effective, and may convey a great deal of information and instruction alike to the European in India as to the European at home. But in this respect this volume completely fails: it deals with only a few phases of the least important part of life in India, and does not treat them with anything like adequacy. Twenty different types of men and as many of women are disposed of with an average of five pages to each, and the descriptions are misleading, because so defective.

For example, one would never gather from the descriptions of "the Military man," "the Uncovenanted Civilian," "the Railway man," and so on, that men have to work in India quite as hard as they do at home. The headings of the chapters on women are a sufficient indication of how they are depicted: "the Fast Married Woman," "the Scorpion," "the Passée Belle," and so on. As for the descriptions of Indian functions and station life, our readers will be wise to consult some of the scores of good Indian novels which may be found in any circulating library rather than the crude caricatures of this book.

A Wayfarer in China: Impressions of a Trip across West China and Mongolia. By ELIZABETH KENDALL. London: Constable and Co. *Price* 10*s*. 6*d*. *net.*

In the midst of the perennial supply of ordinary globe-trotter books dealing with China, it is refreshing to come upon one like Miss Kendall's which belongs to a different category, for in it we have an account of a very genuine piece of travel. Miss Kendall tells her story simply and unaffectedly, and we are sure that the reader will follow it with unflagging interest, from the author's start in Tongking in the extreme south, up through Western Yunnan and Szechuan, following the Tibetan border, down the Yangtsze from Chengtu to Hankow, thence by rail to Peking and Kalgan, and on across the great Mongolian desert until her arrival at the Trans-Siberian railway. The distance is, approximately, three thousand miles, much of it over country unknown to any European save the wandering missionary or explorer, and this Miss Kendall covered alone, and without escort beyond what was afforded by her faithful coolies, cook, and "boy," and by her Irish terrier, Jack, who appears to be nearly as great a "brick" as his mistress. Miss Kendall is not a novice at travel of this kind, and she knew how to face the discomforts inseparable from native inns and curious crowds, but she met with no serious difficulties and no adventures, and, at the end of her book, she pays a tribute to the good manners of the Chinese from whom she says she never met anything but courtesy and consideration. In another place she says "I have met bad manners in the Flowery Kingdom, but not among the natives." The journey was made in the last quiet months preceding the outbreak of the revolution in China, and, though Miss Kendall may not have made any fresh additions to geographical science, she has written a very charming book which we have no hesitation in commending to the attention of our readers.

AFRICA.

Three Years in the Libyan Desert: Travels, Discoveries, and Excavations of the Menas Expedition. (*Kaufmann Expedition.*) By J. C. EWALD FALLS, Member of the Expedition. Translated by ELIZABETH LEE. London: T. Fisher Unwin, 1913. *Price* 15*s*. *net.*

The writer of this narrative is a cousin of Father Kaufmann, the results of whose expedition to the Libyan desert, mainly the discovery of the buried city of Menas, are well known. While giving a general account of the treasures unearthed, the author devotes himself more particularly to a description of the organisation of the party, the engaging and victualling of the army of workmen, the incidents of camp life, the manners and customs of the Beduins, and various excursions from headquarters. It is rather curious that but for a casual reference here and there one would scarcely know that Egypt was in British hands. The illustrations are excellent, and there is an index, but one feels the want of a map. In the concluding chapter the author gives a favourable opinion on the possibility of the reclamation of a large part of the desert of Libya by means of irrigation.

Camp and Tramp in African Wilds: A Record of Adventure, Impressions, and Experiences during many years spent among the Savage Tribes round Lake Tanganyika and in Central Africa. With a Description of Native Life, Character, and Customs. By E. TORDAY, Member of the Council of the Royal Anthropological Institute, etc. With 45 Illustrations and a Map. London: Seeley, Service and Co., Ltd., 1913. *Price 16s. net.*

Works on the Congo region are numerous, but this volume fills a niche of its own. As indicated in the title, the author has spent many years in Tropical Africa, and has had time to become intimately acquainted with the daily routine and customs of many savage tribes. It is clear that his character is such as not only to have commanded respect, but also to have invited confidences from his black neighbours, and his great influence over them, as well as the prestige his name enjoyed, are abundantly evident from several adventures modestly but graphically related. There are many excellent illustrations from photographs, but instead of the small scale map we should have preferred one of the author's own region giving all the place-names he uses in describing his journeys. The account of the customs of the various tribes is of interest, and will be valuable to the ethnologist.

A History of the Colonization of Africa by Alien Races. By Sir HARRY H. JOHNSTON, G.C.M.G., K.C.B., Hon. Sc.D. Cantab. Cambridge: At the University Press. *Price 8s. net.*

In a prefatory note to this book Sir Harry Johnston recalls to us that in 1898 he contributed to the Cambridge Historical Series a work on the History of African Colonisation which was reprinted with additions in 1905. But there has been much making of history in Africa since 1898, and even since 1905, and this more than justifies the decision of the Cambridge University Press that the whole book should be rewritten from beginning to end. How thoroughly this has been done may be gathered from the fact that the first edition contained only 288 pages, while the volume now before us with the appendices contains 471. The former maps have necessarily been thoroughly revised; a map of German Africa has been added; while a map showing the extent of the now practically extinct slave trade has been withdrawn. A comparison of the maps of these two editions is particularly interesting and instructive, as they show at a glance the important political changes which have taken place in the distribution of territory in Africa between the years 1898 and 1912. Of this new edition we have only to repeat what we said of the first edition (*S.G.M.*, vol. xv. p. 274) viz., that Sir Harry "has accomplished his task with thoroughness, and yet with clearness, and the reader is impressed with the well-balanced acumen and judiciousness with which he has selected what is really important, and rejected what is immaterial, in a huge mass of detail, which in undisciplined or incautious hands could hardly have failed to be unmanageable and obscure." As in the first edition, to some readers the most interesting chapters will be those on the slave trade, on Christian missions, and on the legion of distinguished explorers. The statesman, the politician, the philanthropist, and the economist will remark that, notwithstanding the many and important changes effected in Africa during the last fourteen years, Sir Harry has not seen sufficient grounds for departing much from the general conclusions as to the future of Africa which he formed in 1898. We most heartily commend this book to the perusal of our readers.

Reminiscences of a South African Pioneer. By WILLIAM CHARLES SCULLY. London: T. Fisher Unwin, 1913. *Price* 10*s.* 6*d. net.*

In November last year (*S.G.M.*, vol. xxviii. p. 601) we noticed a book by Mr. Scully on Johannesburg. In the volume now before us we have the first instalment of his autobiography. He was born in Dublin in 1855, and when he was twelve years old his father decided to emigrate to Cape Colony, where he tried to make a living by farming, but without much success. Apparently the writer of this book began to carve out his own career at the early age of fourteen, when, after a futile attempt at sheep-farming he joined in the rush for the diamond fields at Kimberley, where on the whole he had somewhat bad luck, and just missed discovering the famous Kimberley mine. Here he met many men who were to figure prominently in the history of South Africa, such as Mr. Merriman, Mr. Becher, Mr. Barry, and the three brothers Rhodes, of whom some interesting anecdotes are given. Mr. Scully seems to have got disheartened with the diamond fields, and so, in 1873, he started for the gold fields in Lydenberg and Pilgrim's Creek and the adjacent lands, but here also he was not very successful. In 1874 he joined an expedition to Delagoa Bay, in the course of which he had some exciting adventures; and, in 1875, he with a Scottish friend, the claimant of a peerage, started off for Swaziland in search of a party of Australian explorers, regarding whom contradictory rumours of success and disaster had become prevalent. The two reached Swaziland and had various adventures with the native tribes, who were hospitable and kind to them, but very much the reverse to their two porters, who ultimately deserted them. This necessitated a return to Pilgrim's Rest. There Mr. Scully had a turn at working in a store, at which he was as unsuccessful as formerly, whereupon he resumed his prospecting for gold and succeeded in finding a good place, but his usual bad luck reappeared; the water supply failed, and he had to abandon what is now the Theta mine. Early in 1876 he started for Natal, passing Lydenberg, Laing's Nek, Majuba Hill, and Ladysmith to Durban, whence he got a passage to East London. Here he found most uncongenial employment in the service of a boating company, from which he was rescued by receiving an appointment as clerk to a resident magistrate. This commenced a career in the public service which lasted thirty-six years, and is to be the subject of a second volume of reminiscences, which should be every whit as interesting as the first one, which is well worth perusal as giving a vivid and graphic account of South Africa in the stirring days when the diamond and gold mines first attracted attention, and when roads and communications were few.

AMERICA.

Aborigines of South America. By Colonel GEORGE EARL CHURCH. Edited by an old friend, CLEMENTS R. MARKHAM, K.C.B. London: Chapman and Hall, Ltd., 1912. *Price* 10*s.* 6*d. net.*

All who read this erudite and interesting book will heartily agree with the regret expressed by the Editor, Sir Clements Markham, that it is unfinished, and at the same time they will be grateful to Colonel Church's widow, who has authorised the publication of as much as now appears. Colonel Church was well known to geographers on both sides of the Atlantic, and, as we learn from the biographical notice contributed by Sir Clements Markham, he was the first foreigner, not an English citizen, who was appointed Vice-President of the Royal Geographical Society. The story of his life has been condensed into the compass of a few pages, but they suffice to outline his brilliant career as a soldier, a savant, a geographer, and a man of affairs. In this volume we have an account of the Aboriginal tribes of South America, of whom he made a patient study, for which he had exceptional opportunities. In connection with this he describes the localities in which they live now, and in which they have been driven to and fro by cruel and ruthless oppressors, of whom decidedly the worst were the Spaniards. He had made himself familiar with the history of these localities, and has enlivened this work with historical anecdotes and episodes, which throw a bright, and often a lurid, light on bygone times. The general impression produced is a sad one, and the sadness is not relieved by any certainty that oppression and cruelty are now things of the past. The story of the Putumayo, still being slowly and painfully expiscated, precludes any such feeling. Colonel Church's book is a very valuable contribution to our knowledge of the geography and ethnology of South America, and we heartily commend it to our readers.

Mexico and Her People of To-Day. By NEVIN O. WINTER. London: Cassell and Co., Ltd., 1913. *Price 7s. 6d. net.*

Mr. Winter has produced a very interesting book, in which he depicts the customs and characteristics of the modern Mexican, and gives an account of the mineral, industrial, and agricultural development of the country, in all of which branches of human activity he is convinced there is still scope for further improvement. He tells us in the preface that he has read all the leading books on Mexico and Mexican history, and has supplemented the knowledge thereby gained by traversing the country from north to south, and from the Atlantic to the Pacific, so that his information and advice to travellers contemplating a visit to that most romantic land are presumably reliable. The book is admirably illustrated.

Panama and What it Means. By JOHN FOSTER FRASER. London: Cassell and Co., Ltd., 1913.

This is a very interesting book, and gives to the general and non-technical reader a very comprehensive and graphic description of the Panama Ship Canal, which will very shortly, it is expected, be completed and open for traffic. The author deals fully with the inception and history of the undertaking, he describes vividly the life and work of those engaged in its construction, the difficulties that have been encountered and been overcome, such as disease due to bad climate, labour troubles, turbulent rivers, landslips, etc. Specially interesting and instructive is the account of the difficulties that are being met with in making the deepest cutting on the line of the canal, the Culebra cut. Unfortunately, this cut is in comparatively soft material, and hence is subject to tremendous land slides, necessitating much of the excavation being done over and over again, adding greatly to the cost.

The author's visit to the country appears to have been made at an interesting period, when the works were nearing completion, and before the water was admitted to the canal, after which the nature and magnitude of the operations will be less in evidence. The illustrations, which are numerous and good, are not the least instructive part of the book.

Modern Chile. By W. H. KOEBEL. London: G. Bell and Sons, Ltd., 1913. *Price 10s. 6d. net.*

Mr. Koebel is a recognised authority on Argentina, and in this volume he sets forth his observations and impressions of its former enemy and present rival,

Chile. The two countries present very marked contrasts. Separated from Argentina by the Andes, Chile on the map looks like a strip of coast, and as a fact it has some 2600 miles of coast, and its greatest length inland is somewhere about 250 miles. But it is exceedingly rich both in agricultural and mineral wealth, and above all in the nitrates which it distributes to the marked benefit of farmers all over the world. The demand for these nitrates has been so great that the question was raised whether or not the supply might fail, and it is satisfactory to learn from Mr. Koebel that this is in the last degree unlikely. The real want in Chile is population, the present figure being only a little over three millions, and we must add there is also much want of great improvements in roads and means of communications generally. Some time ago we should have had to say that the first need of the country is peace, but happily this need no longer be said. Since the suicide of Balmaceda more than twenty years ago, there has been no constitutional crisis, and now that there remains no disputed boundary with Argentina, the chances of war seem very remote. Yet it is doubtful if the commercial development of the country will make much or rapid progress unless it is taken in hand by foreigners. The native Chilian is intensely conservative, and disinclined to the policy of what our American cousins call "hustle"; but foreigners are making their appearance all over the country, and its development, though tardy, is sure to come. As it is, the statistics of imports and exports given by Mr. Koebel indicate a satisfactorily steady increase of trade. Mr. Koebel gives several graphic sketches of the principal towns, Valparaiso, Santiago, Concepcion, Valdivia, etc., and of the life and manners of the inhabitants, and also some excellent descriptions of country life and typical estates. The opening of the Panama Canal cannot fail to make a great difference to Chilian trade and commerce. This book will be found of great interest to merchants and engineers as well as to tourists. It is illustrated with photographs and has a good map.

The Land of the Peaks and the Pampas. By JESSE PAGE, F.R.G.S. London: The Religious Tract Society, 1913. Price 3*s.* 6*d. net.*

This book does not lend itself readily to review in a scientific geographical magazine. Its object is to make an earnest appeal to the Protestants of Europe and North America, to set before them the present spiritual destitution of the inhabitants of South America, to point out the dangers, social, religious, political and economical, which are incidental to such a situation, and to set forth what, in the opinion of the writer, is the only efficient remedy, viz., renewed and increased Protestant missionary efforts and propaganda. From beginning to end it is written from the Protestant point of view.

AUSTRALASIA.

The Land of the New Guinea Pygmies. By C. G. RAWLING, C.I.E., F.R.G.S. London: Seeley, Service and Co., Ltd. Price 16*s. net.*

So little of the earth's surface remains a *terra incognita* that the reader turns with a fresh interest to Captain Rawling's account of his explorations in Dutch New Guinea and of its natives, who are still in the Stone Age. The expedition was organised by the British Ornithologists' Union, and the Royal Geographical Society of London contributed liberally to the funds for exploration and survey work. Mr. Walter Goodfellow was selected as leader of the party, Captain Rawling being put in charge of the survey work; and, later, when Mr. Goodfellow was invalided home, Captain Rawling took the command. The provisioning of the expedition appears to have been faulty at the start, a large quantity of the stores having been taken over from the surplus of the Shackleton expedition, which, however suitable they may have been for the Antarctic, were quite unfit for use in New Guinea; and a long period elapsed and much suffering was entailed before suitable stores were sent from England. The Dutch government afforded generous assistance, and an escort of forty Javanese troops, under the command of Lieutenant Cramer, was attached to the expedition. In addition to these, ten Gurkas were enlisted in India, as guards, and a large number of convicts from Java, as well as coolies, were taken to act as carriers. The sufferings endured in the forests and swamps told cruelly on the carriers, and as many as eighty or ninety per cent. of them died or were invalided. As Captain Rawling remarks, "this is the darkest side of the expedition, and it may be said of all journeys undertaken in New Guinea. . . . Of the four hundred Europeans and natives enlisted, only eleven lasted out till the end, a total period of fifteen months. Of the survivors four were Europeans, four were Gurkas, two were soldiers, and one a convict."

The forest land which lies between the sea and the foot of the mountains is nothing but a malarial bog, soaked by constant torrential rains, and swept by floods, and the coast natives are restricted to narrow strips of land bordering on the rivers. This part of New Guinea appears to merit the author's description of it as "a horrible country." The natives seem to have treated the members of the expedition better than might have been expected from such a low type of savages. For example, Captain Rawling tells us, "every one says the natives of New Guinea are blood-thirsty savages; perhaps they are, but they were decent enough to us"; and it is worthy of note that we find no record in this book of any serious attempt to attack the members of this expedition. This may have been due to the conciliatory attitude of the leaders, for there were frequent occasions when the lives of Captain Rawling and his companions were at the mercy of the savages. It is also noticeable that no mention is made of any instance of cannibalism.

Unfortunately, the expedition chose for their original landing-place the mouth of the crooked, shallow Mimika river, which, as later experience showed, takes its rise in the swamps of the jungle, in place of the Kamura river to the east, which, with its tributary the Wataikwa, proved to be navigable almost to the foot of the mountains. The mistake is not surprising, considering that this part of New Guinea was absolutely unknown; but it cost the expedition very dear, in time, in life, and in suffering. However, Captain Rawling had his compensation in his great discovery of the pygmies, which he could not have made but for his attempt to penetrate to the hills from the west. He was able to visit the pygmies in one of their villages, and to get a number of pictures, together with notes on this interesting people, which form a valuable contribution to anthropology. A separate monograph on the pygmies from the pen of Mr. H. S. Harrison, B.Sc., F.R.A.I., forms one of the chapters in the book. Mr. Harrison considers them to be "true aborigines, primitive man," who have been driven to the fastnesses of the hills by invading tribes from the coast. The explorers saw the male pygmies only. Later, a special journey was made by two of the party, Dr. Marshall and Mr. Woolaston, to the village of Wamberini, in the hope of getting a sight of the women; but the ladies themselves were too shy, or their men too jealous, and no bribe proved large enough to induce the pygmies to show their women-folk, so the party had to return disappointed.

After thirteen months of privation and toil, and after many obstacles to progress had been overcome, including the bridging of a dangerous river, with the help of a gallant Gurka havildar, who crossed at the peril of his life, Captain

Rawling and his companions succeeded in getting a view of the mountains which are mapped in the present volume. The explorers failed to reach the snows, but, apart from this, all the objects of the expedition were at length accomplished. A tribute of admiration is due to the author and his company for the courage and perseverance which enabled them to remain so long among the pestilential swamps of Dutch New Guinea, and to do so much for science. The help so generously rendered by the Netherlands government, who provided not only guards and bearers, but a gun vessel to carry the people to and fro, is gratefully acknowledged. Captain Rawling has given us a book of absorbing interest, and we can confidently recommend it to our readers.

GENERAL.

A History of Geographical Discovery in the Seventeenth and Eighteenth Centuries. By EDWARD HEAWOOD, M.A. Cambridge: At the University Press, 1912. *Price* 12*s.* 6*d. net.*

A perusal of this the latest addition to the Cambridge Geographical Series cannot fail to impress the reader with admiration for the patience, industry, and erudition, which the learned Librarian to the Royal Geographical Society has brought to his task. Into a compass of less than four hundred pages he has condensed the story of exploration and research over the whole known world throughout the seventeenth and eighteenth centuries. In such a task he was compelled to weigh and select from his plethora of material; and it would be easy to criticise and cavil at details, *e.g.*, the space accorded to some particular explorers or events as compared with others. The author evinces a keen desire to do justice to all who have contributed, and indeed to those who have attempted to contribute to geographical knowledge, and for this cause he has made mention of expeditions, etc., more or less unfruitful, silence with regard to which would scarcely have detracted from the value of his review. This book is indeed a model of patient and judicious condensation, but the author has so handled his subject that the interest never flags, and the reader has only a feeling of regret, that limitations of space compel such short notices of famous explorers and travellers as Champlain, La Salle, Tasman, La Perouse, Cook, James Bruce, to name but a few. It is greatly to be hoped that a perusal of this book will send its readers to the study of the more detailed and classic authorities of whom there is no lack. The book is profusely illustrated with reproductions of quaint old maps and prints, very few of which are usually accessible to the ordinary reader.

Big Game Shooting in India, Burma, and Somaliland. By Colonel V. M. STOCKLEY, late 16th Cavalry, Indian Army. London: HORACE COX, "Field" Office, 1913. *Price* 21*s. net.*

The sporting reminiscences of which this book is the record are spread over a period of nearly forty years, and they relate to many and varied climes, from the eternal snows of the high Himalayan passes to the redhot plains of Somaliland. The book is interesting as the record of the performances of a keen and successful sportsman, who evidently thoroughly enjoys fighting his battles over again, and is not given to boasting. Of course in such a varied experience there were plenty thrilling and exciting adventures and escapades, the excitement of which is not lost by the way they are described. On the other hand, similar adventures may be found in abundance in many of the scores of books which have been written on big game sport in all parts of the world within the last ten or twenty years. Colonel

Stockley is a keen observer of the habits of animals, and he gives interesting details of the measurements of his trophies. A novice, who is contemplating a shooting expedition to any of the places where the colonel has shot, would do well to go carefully over this book, in which he will find plenty useful hints and information which will serve him in good stead in his preparations, and also when he gets to his ground.

Common Sense in Foreign Policy. By Sir HARRY JOHNSTON, G.C.M.G. London: Smith, Elder, and Co., 1913. *Price* 2*s.* 6*d.*

It is not often that geographers have the privilege of seeing the world, not as it is, but as it ought to be. In this volume Sir Harry Johnston tells us what he considers "common sense," commends, and illustrates his opinions with eight maps of the Future drawn by himself and Dr. Bartholomew. The reason he gives for writing this book is "because—in comparison with our gigantic stake in the wise conduct of our foreign affairs—we are *in the mass* so lacking in a just appreciation of the lands and peoples outside the scope of the British Islands." At the same time, he admits that travel and colonisation and the discussion of foreign affairs in our newspapers are increasing our knowledge, so that "for ten Members of Parliament considered qualified thirty years ago for the discussion of foreign affairs in the House of Commons, there are now at least three hundred able to speak on the subject with acumen and representing the very decided views and interests of millions behind them."

In so original a work as this we must be prepared to be startled by some suggestions, such as that Germany might surrender her claims on Luxemburg and give up Metz in order to receive the French Congo and a predominant partnership in the administration of the Turkish empire. The author makes short work of existing European alliances, for he anticipates France's withdrawal "from the useless and unprofitable Russian alliance," and union with Italy and Spain in a Latin confederation, while a Central European alliance would include Germany, Austria, and the Balkan kingdoms, a suggestion regardless of the racial antipathy between Slav and Teuton.

The author thinks that Britain should cede Walfisch Bay to Germany, who would then undertake to withdraw from direct access to the Upper Zambezi. He reminds Germany, however, that she has done little as yet to make use of her African colonies. As to Europe, he suggests that Austria might let Italy have the Trentino, and Germany Trieste, Austria herself receiving Salonika in exchange, a suggestion which the results of the Balkan War seem absolutely to negative. Although the Magyars have ruled Hungary for a thousand years, the author declares that they "are an incongruity in Central Europe." He thinks Hungary should be "a practically independent country," but as Slavs, Rumanians, Germans, and Jews outnumber the Magyars in Hungary, its independence would not assure racial agreement. The author's suggestion that the Kingdom of Poland might be re-created as an independent state "to become a close ally of Germany and Austria," ignores the antagonism between German and Pole. Similarly, his settlement of the Finnish question by Russia granting independence to North and West Finland, and obtaining power to fortify the Aland Islands, means that Russia would be brought closer to Stockholm than the Swedes might like.

While suggestions such as these may be open to criticism, no one can fail to appreciate the wealth of historical, ethnical, and geographical knowledge displayed in this book and expressed with the lucidity and vivacity which mark all Sir Harry Johnston's writings.

THE SCOTTISH GEOGRAPHICAL MAGAZINE.

MALTHUS AND SOME RECENT CENSUS RETURNS.[1]

By GEORGE G. CHISHOLM, M.A., B.Sc.,
Lecturer on Geography in the University of Edinburgh.

(*With Plate.*)

WHEN you did me the honour to invite me to lecture to your Section of the Royal Philosophical Society of Glasgow, I suggested as a subject the "Scottish Census of 1911," because I happened to be then engaged in the examination of the preliminary report of that census with a view to the preparation of some brief notes to accompany a map of the density of the population of Scotland, compiled by the Edinburgh Geographical Institute to illustrate the returns. Your council was good enough to accept that suggestion, but on proceeding to prepare my paper I found that I had planned it on so large a scale that all I could get within the limits of an hour is more suited to the title now adopted. The brief notes referred to appeared in the September number of the *Scottish Geographical Magazine* for 1911. The details there given showed that during the last intercensal period the growth of population in different parts of Scotland had been due principally to the influence of mining and fisheries, and that almost the only other causes of growth indicated were commerce and manufactures, and the attractions of health and pleasure resorts. The total population showed a reduction in the rate of increase as compared with the previous census from 11·1 to 6·4 per cent.

[1] A lecture delivered to the Royal Philosophical Society of Glasgow, and reprinted by permission from the *Bulletin* of the Geographical Society of Philadelphia, January 1913.

Some of these results, especially the diminished rate of increase of the total population, when first published, were received with a certain amount of disappointment and even dismay, not unmingled with hints that somebody or some class was to blame. And I propose, in the remainder of this paper, to consider how far these feelings are justified.

The cause of the disappointment is, no doubt, to be found in the fact that economic prosperity is almost universally believed to be an unquestionable good, and that a rapidly increasing population is an unmistakable sign of economic prosperity. This latter is an idea not confined to the man in the street, but occurs again and again in the writings of those who have considered the subject of population, from Malthus downwards. In the first edition of his *Essay on the Principle of Population*, Malthus says :[1] "There is not a truer criterion of the happiness and innocence of the people than the rapidity of their increase" (a statement that certainly goes far beyond the evidence that he has there adduced). And again : "The happiness of a country does not depend, absolutely, upon its poverty, or its riches, upon its youth, or its age, upon its being thinly, or fully inhabited, but upon the rapidity with which it is increasing."[2] And in our own times we have Sir Athelstane Baines stating in the article on population in the new edition of the *Encyclopædia Britannica* that "an increasing population is one of the most certain signs of the well-being of a community."

I am myself disposed to think, however, that if these statements are to be accepted, a very peculiar meaning must be given to the term "prosperous." It seems to be generally admitted that Ireland is, at the present time, one of the most prosperous parts of the British Isles, and yet there the population is still decreasing, and the birth-rate is the lowest of the four countries making up the United Kingdom. On the other hand, Russia seems to be the most striking example of a country with a rapidly increasing population. Taking the estimates of population for Russia given in the *Statesman's Year Book* for 1907, and comparing those with the Census returns of 1897, we find that population seems to have increased at the average rate of 1.67 per cent. per annum. And yet a Russian writer tells us "the general impoverishment of the Russian peasantry leaps to the eyes. If we ask ourselves how the peasant is fed, we find that the diet of himself and his family includes neither flesh, nor milk, nor eggs ; he eats only rye bread, and often not even that, with the addition of, at most, cabbage soup and weak brick tea."[3] All that I can admit to be certainly shown by an increasing population is that the conditions are such as to allow of the population multiplying in the circumstances in which they are willing to live and increase, which, I grant, is not admitting very much. It is an admission on all-fours with the statement that opium sends to sleep because it has a soporific virtue.

And with reference to the first of the authorities I have cited as to this idea of the relation between prosperity and an increasing population,

---

[1] P. 108. [2] P. 137.

[3] Quoted in Andree's *Geographie des Welthandels*, 2nd ed., vol. i. p. 880, from Annenski's *Needs of the Village Communes.*

Malthus, I must point out that, so far as I am aware, the statements quoted belong only to the first edition of his *Essay*, and reconsideration led to their disappearance. In the third edition the paragraph, containing the first of the passages quoted from him, disappears altogether and the second is modified into this, "Other circumstances being the same it may be affirmed, that countries are populous according to the quantity of human food which they produce, or can acquire; and happy, according to the liberality with which this food is divided, or the quantity which a day's labour will purchase,"[1] a statement that seems to me much more satisfactory. I should add, too, that even the first edition shows the interpretation which was to be put on the statement made therein as to the happiness of a country depending upon the rapidity with which it is increasing inasmuch as he adds as an equivalent, "upon the degree in which the yearly increase of food approaches to the yearly increase of unrestricted population." That is, he looked upon the rapid increase of population merely as implying a rapid increase in the food supply, and this he took as a sign of the happiness of the population. In the sixth edition, we have this view given more explicitly, where he asks "if anything could be more desirable than the most rapid increase of population unaccompanied by vice and misery."[2] Obviously, Malthus, referring to "happiness," means the presence of the material conditions necessary for happiness, which, he had no need to be reminded, is a very different thing from happiness itself.

Holding this view, therefore, Malthus was under no temptation to regard an increase of population under all circumstances as a sign of "happiness," and, as is well known, he did not so regard it, either at the time that he was writing the first or any of the subsequent editions of his well-known *Essay*, which has evoked as much subsidiary literature as Adam Smith's *Wealth of Nations* or Darwin's *Origin of Species*. That literature is still growing in amount, and it is still worth while to inquire whether the views which he actually held are sound, and if so, whether they warrant us in regarding with disappointment the returns of the last Scottish Census, and, in that case, how far, and in what manner the feeling is well grounded.

Now, there is one statement which Malthus regarded throughout as fundamental and to which, accordingly, he appears to have consistently adhered amid all the changes of the various editions, which, during his lifetime, followed the first *Essay* of 1798; the statement that, while population tended to increase in a geometrical, food tended to increase in only an arithmetical ratio. Yet this statement I cannot but look upon as purely fanciful, one to which, indeed, it is hardly possible to attach any definite meaning consistent with facts that Malthus well knew. What is the meaning of "tendency"? Language is full of figurative expressions and the word "tend" in English is entirely figurative, and the figure, in this case, is so clear as to seem to leave no possibility of mistake. "Tend" is literally "to stretch," and "stretch" suggests or even implies a resisting force. Any prevailing movement

[1] Vol. i. p. 71. [2] Vol. ii. p. 450.

liable to counteraction thus naturally suggests the use of this metaphor. We do not say that the earth "tends" to move round the sun in an ellipse, but that it *does* so move; but we say that the snow on a mountain top *tends* to move downhill either as snow or ice, or that the dust of the desert *tends* to accumulate most deeply in the hollows, because, however much or often the wind may blow snow or dust upwards, gravity is always pulling it downwards. But, while the figure here is quite clear, I cannot imagine how any statement could be ventured on as to the precise degree or rate at which snow tends downwards or desert dust accumulates in hollows. We may also say that the earth, while it moves round the sun in an ellipse, "tends" to fall into the sun, and is only prevented from doing so by the counteracting tendency at each instant to fly off at a tangent; and, in this case, we may add that the tendency is to fall towards the sun at a rate corresponding to the law that the two masses attract one another with a force directly proportional to the mass, and inversely proportional to the square of the distance. Here we have an intelligible measure of a "tendency." What is meant is quite clear. It implies the statement of a universal law, whose operation can be fully overcome only by a force equal to that described as operating. But what parallel to this can we get in the statement of the law of population? Malthus shows us that, in fact, population increases in various cases at various rates—that, under very favourable circumstances, it can be shown to have doubled itself in even fifteen years; yet he does not take that as the rate at which it will double itself "if unchecked." He says we may safely assume from experience in the early years of the North American Colonies that, if unchecked, it will do so in twenty-five years—in any case, in a geometrical ratio, which must mean some definite ratio.

Well, let it be so. But what about food? Here we find no proviso "if unchecked," when he tells us that it will increase (of course, in any case, only through the labour of men) only in an arithmetical ratio. Yet he himself suggests a case in which it might be supposed that food would go on increasing at a geometrical ratio. He says, "when acre has been added to acre till all the fertile land is occupied"[1] a diminished rate of production must ensue, showing that, until then, he does not venture to make that assertion. Whether, therefore, it be population or food, we can imagine that, if the increase is unchecked (in which case we must take the term "unchecked," as applied to food production, to mean "not counteracted by greater hindrances to human labour") it might go on increasing at the same geometrical ratio.

But this objection to what Malthus puts forth as a fundamental fact, is, as regards the burden of his *Essay* merely trivial, superficial, inessential. The supposed fact about which he is so emphatic is not really fundamental. To make good his case, he was not required to show that population tends to increase in a geometrical and food in an arithmetical ratio. That way of putting the facts makes them, indeed, very plainly alarming, but his case would have been equally well estab-

[1] Third ed., vol. i. p. 8.

lished, if he could have shown that, while both population and food tend to increase at a geometrical ratio, population tends to increase at the rate of 2·5 per cent. per annum, food only at 2·4 per cent. per annum. The essential fact on which his case is built up is, indeed, stated in what follows after the words already quoted about the adding of acre to acre till all the fertile land is occupied. "The yearly increase of food must depend upon the melioration of the land already in possession. This is a stream, which, from the nature of all soils, instead of increasing, must be gradually diminishing. But population, could it be supplied with food, would go on with unexhausted vigour." This is nothing else than a statement of the law of diminishing returns, and it would have been better if Malthus had stated his case so, for his own statement of his case has led some people to suppose that he has been answered when some of the assertions made by him by way of corollary from his own way of stating his case have been proved to be wrong; as when it is shown that in certain cases, even in old countries, food supplies increase at an even more rapid rate than population. For instance, in a recent article on Japan, in the *Journal of the Royal Statistical Society*, there is a table showing that, if in the period 1888-92, population, and the production of rice, barley, wheat, and rye in the aggregate, in that country, be both taken at 100, in 1903-07 population had risen to 117·8, whereas the production of the cereals mentioned had increased to 120·8. It is notorious too that, in the latter part of the nineteenth century, at any rate, the food supply of the civilised countries of western Europe did actually increase at a more rapid rate than the population.

The fact to which Malthus draws attention had been recognised by others, as he well knew, again and again in many places from ancient times downwards. It was, in many cases, too obvious to be overlooked that increasing numbers might gradually increase the difficulty of providing food for the population. The merit of Malthus, and it seems to me to be a merit that cannot be questioned, was to call attention to and insist upon the universal importance of the relation between numbers and food supply to the wellbeing of the people. He adduced ample evidence to show that almost everywhere and at all times, population was pressing upon certain barriers, by the removal of which, it would be sure to increase at a more rapid rate than that actually observed. And he pointed out the nature of those barriers and showed how they operated in different cases.

But, in doing this, he came to differ from himself. While what seems to me the futile contention that population tends to increase in a geometrical and food only in an arithmetical ratio, runs through all the editions of the *Essay*, the first edition, as is well known, differs in a very important point from all the subsequent ones, with the result that the later editions take a quite opposite view of the possibilities of ameliorating the condition of the poorer classes of the people to that taken in the first. The first edition takes the view that all the checks that keep the actual population within the limits of the food supply, which, of course, it cannot pass beyond, may be reduced to the two heads "vice" and "misery," and that hence any permanent melioration of the condition of the poorer classes, that is, of the great bulk of the community, is impossible. In the second and subsequent editions, however, he enumerates three checks to the undue increase of population, vice, misery, and moral restraint, and much space is devoted in these later editions to insisting on the importance which ought to be given to this last check, and to picture the improvements that might ensue from the more extended operation of this check as compared with the other two.

It is true that in the first edition he did not altogether overlook this third check. As he himself says, he had observed that, if some check to population must exist, it was better that this check should arise from a foresight of the difficulties attending a family and the fear of impending poverty, than from the actual presence of want and sickness.[1] But, in the first edition, he had classed this check, rather curiously perhaps, under the head of "misery."[2]

It is right, however, that we should take the second thoughts of Malthus on this subject and, if we do so, it seems to me that the whole of human history since his day has been such as to confirm his doctrines and to bring into relief their supreme importance,—the doctrines that population in its increase is always held back by certain barriers, that the checks which keep the population within those barriers may be classed as "vice," "misery," and "moral restraint," that population increases with greater or less rapidity when these barriers are pushed back, and with very remarkable rapidity, when they are almost entirely removed[3] (of their entire removal there is probably no example in history); that with like rapidity even old states recover from the ravages of war, pestilence, famine, or the convulsions of nature,[4] that the operations of these checks is not deferred till the world is filled up—a contingency that all writers on population recognise as an inevitable, though remote event—but "has existed in most countries ever since we have had any histories of mankind, and continues to exist at the present moment";[5] and that the only hope of the permanent amelioration of the condition of the bulk of mankind lies in the extension of the operation of the checks which come under the head of moral restraint. Some of his corollaries are also strikingly confirmed, as, that it is impossible to argue any future rate of increase from a present rate; that, as to America, we may be "perfectly sure, that population will not long continue to increase with the same rapidity as it did then"[6] that, even in a well-governed state, disastrous results may ensue from a sudden cessation of an extraordinary stimulus to wealth and population;[7] that those who live in the most frugal way are the nearest to a great disaster, and that where the great mass of the industry of a country is directed to the land, there may be, according to circumstances, instances in which the poor are in the best state, as well as others in which they

---

[1] Sixth ed., vol. ii. p. 256, apparently referring to first ed., p. 62.
[2] P. 108. [3] First ed., p. 101. [4] *Ibid.*, p. 109.
[5] Sixth ed., vol. ii. p. 7. [6] First ed., p. 343.
[7] Sixth ed., vol. ii. pp. 329-31.

are in the worst state.[1] But, in spite of all these dismal forecasts, you may recognise that he was also right in foreseeing the possibility of a great increase of population in Great Britain, with greater comfort than was enjoyed by the mass of the people in his time.[2]

Let us consider some of these propositions individually. Malthus tells us that "under a government constructed upon the best and purest principles, and executed by men of the highest talents and integrity, the most squalid poverty and wretchedness might universally prevail from an inattention to the prudential check to population."[3] One knows not where to find a government which will meet the conditions which Malthus here sets down, but we may, I think, take the testimony of various enlightened foreign critics as proving that the British government of India is, at least, of a relatively high degree of excellence; and it is true that in quite recent years we have heard little of famine in India. Sir Theodore Morison, indeed, tells us that "the term 'famine,' which is still applied in India to a harvest failure, is now an anachronism and a misnomer. The true meaning of the word 'famine' according to the Oxford Dictionary is 'extreme and general scarcity of food.' This phenomenon has entirely passed away. Widespread death from starvation, which this word may be held to connote, has also ceased. . . . 'Famine' now means a prolonged period of unemployment, accompanied by dear food, and this is undoubtedly an economic calamity, which inflicts great hardship upon the working classes in India, as it would in any country."[4] Still, if this picture represents the truth now, we must remember that it is not very long since it was not true.

TABLE I.

POPULATION OF INDIA.

	Population, Millions.		Variation, per Cent	
	1901.	1911.	1901-1911.	Per Ann.
Bengal,	50·7	52·7	+ 3·8	+0·38
Bombay,	15·3	16·1	+ 5·2	+0·51
Burma,	10·5	12·1	+14·9	+1·40
Central Province and Berar,	12·0	13·9	+16·3	+1·52
Madras,	38·2	41·4	+ 8·3	+0·80
North-west Frontier Province,	2·0	2·2	+ 7·5	+0·73
Punjab,	20·3	20·0	− 1·8	−0·18
Sind,	3·2	3·5	+ 9·4	+0·91
Agra,	34·8	34·6	− 0·7	−0·07
Oudh,	12·	12·6	− 2·1	−0·21
Central India Agency,	8·5	9·4	+10·0	+0·96
Central Provinces States,	1·6	2·1	+29·8	+2·64

[1] Sixth ed., vol. ii. p. 119. [2] *Ibid.*, vol. ii. p. 293. [3] *Ibid.*, vol. ii. p. 323.
[4] *The Economic Transition in India*, pp. 120-1.

Even the figures of the last census, when examined in the light of recent history, bear testimony to the efficiency of famine, with its attendants, disease and pestilence, in putting a check on the growth of population, and then afterwards allowing it to grow again with renewed rapidity. The accompanying table gives some examples of the differences in the movement of population in some of the provinces of India, in all cases according to the present area of those provinces, between 1901-11; and the ratio of increase, it will be observed, differed very greatly, there being, in some cases, actual decrease. The decrease in the Punjab is worth inquiry inasmuch as it is a large and populous province, and one in which the growth of population has been favoured in recent years by the construction of irrigation canals.

"The Chenab and Jehlum canals, by rendering cultivable vast areas of waste, have been of incalculable help in reducing the pressure on the soil in the most thickly populated districts, and in increasing the productive power of the province. . . .

"Of recent years the immediate effects of scarcity on the population of the province have been practically negligible. The famine of 1899-1900, the most severe since annexation, affected the health of the people, so that many were unable to withstand disease which under more favourable circumstances might not have proved fatal. . . .

"Whether it will ever be possible to render the Punjab free from liability to famine is a difficult question at present to answer. . . .

"During famine cholera is most to be feared; but when famine ceases, after a plentiful monsoon, malaria, acting on a people whose vitality has been reduced by privation, claims a long tale of victims." [1]

And from the last "Moral and Material Progress of India," we learn that the average death-rate in the province in 1908 was 50·7, against a birth-rate of 41·8, followed in 1909 by a decline in the birth-rate to 35·1, with the explanation that this decline in the birth-rate was due to the fact that the people suffered severely from an epidemic of malaria in the autumn of 1908.

By way of contrast we may take the population of the Central Provinces and Berar, which showed the largest increase in population between the dates of the last two censuses. On inquiry into this case, we find that the high rate of increase expressed, in a large measure, rapid recovery from previous decline due to famine and disease. The decrease in Berar, shown in the census of 1901, was due to the famines of 1896-1897 and 1899-1900, and to the abnormally high mortality from disease in 1894-97, and in 1900. In the ten years preceding 1901, there was but one year which could be described as very favourable, and even then the rabi crops partially failed. We are also told that, between 1881-91, the population deduced from the vital statistics of the Central Provinces, when compared with those of the census returns, differed by only 50,000, while, in 1901, the corresponding difference amounted to 450,000. This, it is added, may be partly accounted for by emigration, but was mainly due to the deficient reporting of deaths

[1] *Imperial Gazetteer of India*, new edition, vol. xx. pp. 330-31.

in famine years. Other similar testimony is given in the *Gazetteer*. The population of these territories had decreased from 13·06 millions in 1891, to 11·99 millions (a decline of almost exactly a million) in 1901, when it was brought back to exactly the same figures as in 1881, so that the average rate of increase between 1891-1911 is equal to only 0·32 per cent. per annum.

On the other hand, the large increase in the case of Burma is quite normal. For Burma, as a whole, we cannot make comparisons going far back, but the rate of increase of population for 1901-11, given in the table, corresponds to an average increase of 1·42 per cent. per annum, which may be compared with the successive rates for Lower Burma according to the results of the censuses of 1872, 1881, 1891, and 1901, which are 3·48, 2·43, and 1·94 per cent. per annum, respectively. But here we have to do with a province which suffered from hundreds of years of misgovernment before it came into British possession, in consequence of which, when good government was established, there was plenty of cheap land to be had and the vast resources of the country came to be developed with great rapidity.

Let us now consider another country, the conditions of which are more similar to those in Scotland. The accompanying table shows for successive, but unequal periods, the rate of increase in Germany, and that of Scotland as a whole, the figures for Germany being, even in 1840, those for the present area of the German Empire. To make the figures for the two countries comparable, the rate of increase has, in every case, been expressed as the average rate of increase per cent. per annum.

TABLE II.

POPULATION OF GERMANY IN MILLIONS. INCREASE OF POPULATION PER CENT. PER ANNUM IN GERMANY AND SCOTLAND.

Germany.			Scotland.	
Year.	Population.	Increase.	Year.	Increase.
			1801-11	1·14
			1811-21	1·48
			1821-31	1·23
			1831-41	1·03
1840	32·8		1841-51	0·95
			1851-61	0·60
1871	41·1	0·72	1861-71	0·93
1885	46·9	0·95	1871-81	1·07
1890	49·4	1·06	1881-91	0·75
1900	56·4	1·32	1891-1901	1·06
1910	64·9	1·43	1901-11	0·62

With reference to these tables, I may point out first, that either of them may be taken as proving the observation of Malthus that it is impossible to argue the future rate of increase from the present, for both show that the rate fluctuates from period to period. This may seem too familiar to need pointing out at all, and yet we find a distinguished French statistician, Mr. A. de Foville, comparing France with Germany, stating without qualification that it is probable that at the end of the present century, Germany will have a population of 120 millions or more, while France will not have 60 millions! If this forecast of Mr. de Foville as to the population of Germany at the end of the twentieth century proves to be fulfilled, it will be one of the most surprising things that that century will have to show. It is all the more remarkable that Mr. de Foville should have made this observation inasmuch as he makes it in an article[1] in which he is reviewing, not with disfavour, a forecast, by a Danish statistician, Professor Westergaard, of a totally different nature, Professor Westergaard anticipating that a conspicuous decline of the birth-rate will, by and by, appear in all civilised countries, that the growth of large European towns which was so marked a feature of the nineteenth century will be followed by a cessation of their growth, and an increase of population in the rural districts and the smaller towns of those districts.

Next, one may note that it is only since the latter part of last century that Germany has come to show any high rate of increase as compared with that of Scotland. Down to 1901, the rate of increase of population in Scotland was only once lower than the average in Germany between 1840-1871. It is true that the latter census was taken in a year when the increase of population must have been checked considerably by war; but, even making allowance for that, the calculated rate for the long period 1840-1871 must still have been a low one. I would lay special emphasis on this fact. It seems to show that the population in Germany was then pressing pretty severely against some barrier to its increase, and no very recondite investigation is required to reveal what that barrier was. At that time, the agriculture of the country was in a very backward condition, and the means of communication, especially by rail, very defective, and for the improvement of agricultural production better means of communication were absolutely necessary. But this was a barrier very easy to push back, once the modern methods, already widely in use in other countries, came to be applied, and they were pushed back all the more rapidly in consequence of the fact that Germany is rich in material resources for industrial development. Then the agricultural and manufacturing industries stimulated one another to such a degree that the average rate of increase has not merely been higher than Scotland has ever recorded, but higher for a decennial period than England (exclusive of Wales) has ever reached since 1821-31. Even for a quinquennial period, it may be mentioned, Germany has never attained the average reached by England (exclusive of Wales) in 1811-21 (1·68 per cent. per annum), and latterly the quinquennial censuses have been showing a declining rate.

And, unquestionably, the population in Germany, as in every other country, has throughout been more or less kept down by misery and vice, but, when one considers that the period for which quinquennial averages are given is one in which the death-rate had been going down,

---

[1] *Économiste français*, 30 November, 1907.

I think that we are safe in assuming that these checks have not been operating in an increasing proportion, and that the checks that have determined the fluctuations, have been chiefly those which have come under the head of "moral restraint." And what is true of Germany is true also of Scotland, with this geographical difference, that the growth of Scottish mining and manufactures is to a much less extent a stimulus to Scottish agriculture, than corresponding conditions are to that of Germany.

TABLE III.

GERMANY. INCREASE OF POPULATION PER CENT. PER ANNUM.

	1840-71.	1871-85.	1885-90.	1890-1900.	1900-10.
Prussia	0·87	1·00	1·11	1·98	1·54
East Prussia	0·89	0·52	−0·02	0·19	0·23
Brandenburg	1·40	1·77	2·41	1·95	2·12
Schleswig-Holstein	0·60	1·04	1·18	1·30	1·56
Westphalia	0·81	1·57	1·93	2·76	2·61
Hesse-Nassau	0·32	0·93	0·87	1·32	1·58
Rhine Province	1·03	1·40	1·61	2·03	2·14
Bavaria	0·36	0·80	0·62	1·00	1·08
Saxony	1·30	1·59	1·91	1·84	1·35
Würtemberg	0·32	0·67	0·40	0·63	1·17
Baden	0·39	0·66	0·68	1·20	1·38
Hesse	0·30	0·83	0·73	1·21	1·36
Mecklenburg-Schwerin	0·39	0·23	0·09	0·51	0·51
Alsace-Lorraine	0·10	0·07	0·48	0·69	0·86
German Empire	0·72	0·95	1·06	1·32	1·43

Similar results are reached when we look into details. Here are some rates of increase of population in recent periods for different parts of Germany. Geographical conditions in relation to the present state of industry here also obviously have an important influence on the rate of increase, and the differences are not unlike those which we find in Scotland. The lowest rates of increase, at least in recent years, are shown in the agricultural provinces of East Prussia and Mecklenburg-Schwerin. In one of these, you will notice, the rate is quite insignificant, and in one period there was even a decline. Notwithstanding the great improvement that has gone on in German agriculture, and a great increase in the number of the population employed in agriculture, there has been, as in Scotland and England, an actual decline of population in many of the rural districts of Germany. What is called a "landflight" in Germany is known there just as it is here. It is not very wonderful, for in old countries improvement in agriculture means in a large measure the substitution of machinery for men, as in a new country it means in a large measure the addition of men to machinery. And in this connection, returning again to Mr. de Foville's anticipation as to the population of Germany at the end of the twentieth century, I may mention that, while a slackening of the birth-rate in the towns of Germany has, for some years, been subject for remark, the diminished fruitfulness of marriages in the country districts of Germany is now beginning to attract attention. It is the subject of an article in the *Zeitschrift fur Socialwissenschaft* for December of last year. The really high rates of increase, amounting to a maximum of 2·76, are in the mining and manufacturing provinces of Westphalia and the Rhine. The other great manufacturing region, the kingdom of Saxony, it will be noted, has never shown so rapid a rate of increase as the two others have since 1890, and Saxony since the period 1885-90 has shown a declining rate of increase.

In comparing Scottish counties with German provinces we are indeed comparing small things with great, still the comparison is not uninstructive. To facilitate that comparison, the accompanying diagram, for the idea, as well as for the preparation of which, I am indebted to my university assistant, Miss A. B. Lennie, M.A., B.Sc., has been drawn up. Looking at this diagram, we may notice a regular decline in the rate of increase of population in most of the non-manufacturing counties since 1871, but the great majority of those counties reached their maximum rate in the period 1861-71 or before it. But some of the non-manufacturing counties, Elgin, Banff, and Kincardineshire, have maintained the same increase to the end, probably, as already stated, through the development of the fishing industry. Special comment may be made on two counties, Argyllshire and Perthshire, whose decline began specially early, in both cases in the period 1831-41, and two, which show specially notable advances, Fife, in which there was on the whole a declining rate from 1851 to 1861, when the rate was only ·08 per cent. and then a steady, and, even latterly, rapid rise to 22·3 per cent. in 1901-11, and Selkirk, which showed a rapid increase in the two periods, 1861-71 and 1871-81. In the case of Argyllshire, there probably can be little doubt that what brought about the early decline of population was the rapid growth of the city of Glasgow, which, of course, attracted population from many parts, and no county was more favourably situated for supplying a population than the county of Argyll. In Perthshire, we probably see the result of a similar attraction due to the growth of the manufacturing towns of Dundee and Arbroath. As for the peculiarly rapid growth of Selkirkshire, culminating in the period 1871-81, it is necessary to note that this growth is almost entirely confined to the manufacturing towns, Galashiels accounting for the greater part of it. In the table in the census report, the increase of population for that county is represented as having been as much as 82½ per cent., but this arises from the inclusion of the Roxburghshire portion of Galashiels in the county of Selkirk in the population of 1881, but not in the population of 1871. In the diagram, for the sake of making the comparison more instructive, the population of that part of Galashiels that lies upon the left bank of the Gala, has been included in the Selkirkshire portion and not that of the county of Roxburgh, from 1861 onwards. Even then, a rapid increase is noted in the two periods, to which attention has just been called; and the explanation of that rapid increase is not to be found in this county any more than in any other county in any growth of the rural population, but solely in the fact that there was a very steady advance in the woollen industries of the Scottish

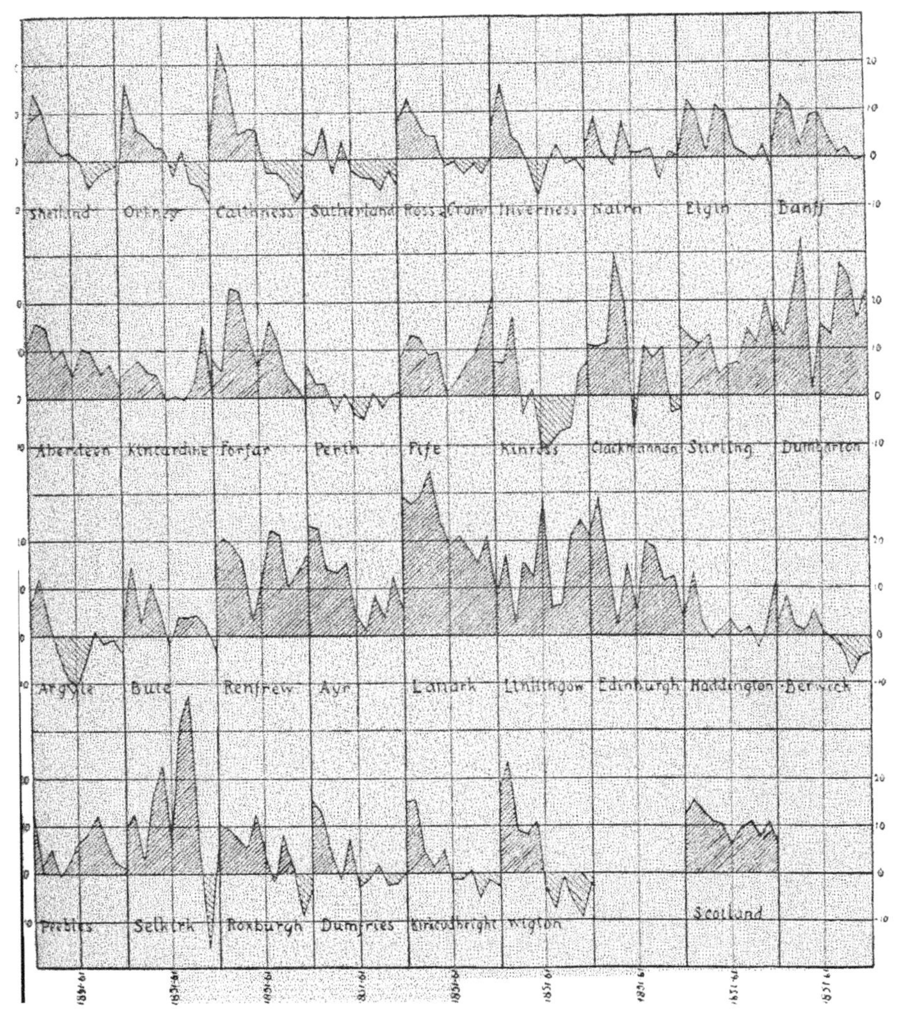

Diagram showing changes in Rate of Increase of Population in Counties of Scotland.

borders generally, down to the date 1872, which was the culminating year as regards the quantity of goods produced. There can be little doubt that the Franco-German war, which interfered with the industry of two of the most important woollen manufacturing countries on the continent of Europe, gave excessive temporary stimulus to that industry in this country, and, therefore, may be looked upon as a contributory cause of the slackening of the rate of increase and the decline which ultimately followed. The rapid and growing rate of increase of population in Fife since 1861 is a sign amongst other things of one way in which the recent prosperity of Germany has influenced this country. The growth of the mining industry in Fife is promoted in a large measure by the rapidly rising demand for coal in the Baltic as well as in other parts of Europe.

But with all that Malthus foresaw, there were certain important facts bearing upon the growth of population since his time which Malthus did not foresee, and could not possibly have foreseen. No more than Adam Smith did Malthus foresee the enormous revolution that was going to be wrought by railways. Unlike Adam Smith, he lived into the railway era. The last edition of his *Essay* was published in 1826, the year after the opening of the Stockton-Darlington line, which may be called the pioneer railway of the world. But neither railways nor steamships are mentioned by him, and, in any case, it was not to be expected that he would be able to foresee the marvellous degree to which railways have cheapened the carriage of produce as bulky as the most essential of our food-stuffs. He understood the manner in which commerce may put into the hands of the people in one country command of labour in other countries, but he could not foresee the extraordinary extent to which the accumulation of capital, and its investment in foreign countries, would give the inhabitants of one little country command of the labour of millions in the most distant regions of the world. But, if he could not foresee the way in which railways could make the plains of America available for the food-supplies of this country, it may be contended that he did foresee what would happen in America in such an event. "In a country," he says, "where there is an abundance of good land, where there are no difficulties in the way of its purchase and distribution and where there is a foreign vent for raw produce, both the profits of stock and the wages of labour will be high. These high profits and high wages, if habits of economy pretty generally prevail, will furnish the means of a rapid accumulation of capital, and a great and continued demand for labour, while the rapid increase of population which will ensue will maintain undiminished the demand for produce, and check the fall of profits. . . . America affords a practical instance of the agricultural system in a state, the most favourable to the condition of the labouring classes. . . . They have been able to command an unusual quantity of the necessaries of life, and the progress of population has been unusually rapid."[1] It is enough, perhaps, by way of comment upon this, to draw attention to the manner in which that is illustrated by the accompanying

[1] Sixth ed., vol. ii. pp. 119-21

table, showing the rates of increase of population in different parts of the United States at different periods.

TABLE V.

UNITED STATES. POPULATION (IN MILLIONS) AND INCREASE PER CENT. PER ANNUM.

Division.	1790.	1840.	1860.	1890.	1900.	1910.
1. N. Atlantic Div.	2·0	6·8	10·6	17·4	21·0	25·9
Rate of Incr.		2·50	2·27	1·67	1·92	2·08
2. S. Atlantic Div.	1·9	3·9	5·4	8·9	10·4	12·2
Rate of Incr.		1·52	1·57	1·69	1·66	1·56
3. N. Central Div.	—	3·3	9·1	22·4	26·3	29·9
Rate of Incr.		—	5·12	3·04	1·63	1·27
4. S. Central Div.	0·1	3·0	5·8	11·0	14·1	17·2
Rate of Incr.		—	—	2·17	2·34	2·02
5. Western Div.	—	—	0·6	3·0	4·1	6·8
Rate of Incr.		—	—	5·43	2·81	5·25
Total[1] . . .	3·9	17·1	31·4	62·6	76·0	92·0
Rate of Increase .		—	—	2·32	1·90	1·93

Even in Malthus's day this rapid increase of population depended upon the small amount of labour with which increased supplies of food could be produced, and, since his time, the application of machinery to food production has still further diminished, and that, in a great degree, the amount of labour required for production on the fields, while the improved means of transport has been reducing that amount as required for its production, in the sense in which that word is used by economists, on the markets of old countries across the ocean.

So far has this proceeded that there is a tendency to suppose that food production does not, in any way, act as a check upon the increase of population in old countries with an advanced industry; that for them there is sure to be enough, whatever countries go short. But to those who hold this view, Malthus would have pointed out, first, that home industry can give us the power of purchasing from abroad only in so far as we are able to sell abroad, and the extent of our sales, is, in the last resort, dependent upon the food supply somewhere. In years of famine, India can buy from us fewer cottons, and, as is well known, this great British industry, which, on the balance of years, is a very profitable one, is one that is characterised by marked fluctuations of profit and loss. A highly organised industry in an advanced manufacturing country gives employment to numbers of people with various degrees of ability and training and character, but, in the long run, the number of people for whom food, with a small surplus, is afforded in India, China, and other countries, is an important factor in determining how many of each grade can be employed in this country.

But, secondly, Malthus would have pointed out that, however favour-

[1] Exclusive of Alaska.

able conditions may be at any given time to the more extended utilisation of the food-producing areas of the world, these areas are, after all, as distinctly limited as those of any one country. The world is larger than our little island, but no more extensible. We are justified in arguing from what we see going on around us, that a time must come when the increase of food production all over the world will involve more labour, as it has come to do in one country after another. No doubt this result may be prevented by a variety of possibilities which we have no means of foreseeing, but it is only experience that can offer us probabilities.

A more plausible consideration is that the extension of the area of food supply has this effect at least, that so long as it continues to extend, it does not matter where industry is carried on. The products of industry will always ensure the supply of food and other necessaries. Railways and steamships, we are told, have annihilated distance. But this epigrammatic way of expressing an unquestionably remarkable development of the means of transport in modern times is apt to be taken too literally. The truth is, that the local market is still a powerful factor in stimulating the growth of nearly all industries. It is, indeed, a truth that is apt to be overlooked at present owing to the way in which the growth of the world, as a whole, tends to conceal the degree in which our hold on foreign or rather external markets is getting encroached on by the more rapid growth of the home industries in those markets. In new countries, without doubt, there are great difficulties in the way of establishing manufacturing industries, and especially those which involve a high degree of organisation. The population is too scattered to form a sufficiently large local market; the supply of skilled labour is inadequate; capital is dear. But skilled labour becomes more plentiful and capital becomes cheaper, and the demand of the local market, which from the first counts for something, becomes steadily greater as population increases from natural causes. Besides, if it is of no moment where industry is carried on, provided people will only work, why have Distress Committees in Great Britain to make such reports as that "there are still, however, a very considerable number of willing and able workers who are unable to procure situations"?

In new countries, or at least in the new parts of new countries, this difficulty does not present itself. There the abundance of free land gives a value to mere manual labour which ensures to it a high degree of mobility, at least if the labourer is endowed with only a modicum of foresight. As a Winnipeg waiter, who, before he went out to Canada, earned fifteen shillings a week in one of the eastern counties of England, expressed it to me, "Now I can go where I like, east, west, north, or south," and the buoyancy of feeling revealed by this statement is what seems to be common to all the young and strong throughout the north-west of Canada.

It was considerations such as those above indicated that induced Malthus to declare himself no very determined friend to "trade and manufactures." But, in the later editions, while he elaborated, even more fully, his views as to the evils of manufactures, he lays great stress on the fact that "manufactures, by inspiring a taste for comforts, tend to promote a favourable change in the [habits of the labouring poor], and in this way, perhaps, counterbalance all other disadvantages."[1] This is the point to which he again and again returns as that which is most likely to render generally operative the check of moral restraint, by which alone he looks forward to a general improvement in the condition of humanity as at least a possibility if not a probability. In one enumeration of the peculiar advantages of Great Britain, after mentioning the excellence of her soil, the comparative steadiness of her climate, the happiness of her insular situation, he adds as the last of those advantages, "above all, throughout a very large class of people, a decided taste for the conveniences and comforts of life, a strong desire of bettering their condition (that master-spring of public prosperity) and, in consequence, a most laudable spirit of industry and foresight are observed to prevail."[2] And with England he contrasts France, as to which he says, "With all her advantages of situation and climate, the tendency to population is so great, and the want of foresight among the lower classes of the people so remarkable, that, if poor-laws were established, the landed property would soon sink under the burden, and the wretchedness of the people at the same time but increase"[3]—another reminder of the impossibility of foreseeing a future rate of increase of population from the rate at any given time.

If Malthus had lived at the present day he would probably, in considering the unquestionable increase of economic wellbeing in the great bulk of the people in this country, have pointed out, that the tendency of legislation also had been, in some respects, to retard the rate of increase of the population. It is, perhaps, not without significance that Malthus speaks of a "standard of wretchedness"[4] as placing a limit on the increase of population. Our phrase for the same thing is the "standard of living," apparently a recognition of the fact that such a standard has come to operate above the standard of wretchedness. Among the causes which have tended to raise that standard may be included, I think, the Factory Acts, as to the operation of which it is not impossible that both the supporters and opponents of those Acts were, in a measure, right, the supporters, in thinking that they would tend to raise the economic level of the workers, the opponents in thinking they would check, to some extent, the growth of British industries—a supposition not disproved by the very rapid expansion of those industries in spite of the Factory Acts.

The removal of the restrictions on the action of trade unions may be credited with the same tendencies, for, so far as union action is able to maintain a higher level of wages or better conditions of another kind, than would otherwise have obtained, it can only be, on the one hand, by restricting the supply of labour in a particular industry and, on the other hand, by compelling the leaders of industry to restrict their employment of capital to those branches of the industry in which higher

---

[1] Third ed., vol. ii. p. 206.
[2] Sixth ed., vol. ii. p. 346.
[3] *Ibid.*, p. 348.
[4] *Ibid.*, p. 359.

wages can be earned. The endeavours of trade unions to restrict the number of apprentices, and to drive non-unionists out of employment, are also conscious or unconscious efforts to keep up the standard by keeping down the numbers. Industries or branches of industry that do not come up to the British standard of living tend to die out. Probably the dominating tendency of the Education Acts in this connection also is to retard the growth of population through its effect on the standard of living, even though their effect in promoting industry and population, their necessity as a defence of industry in competition with other countries, can hardly be called in question. In making the suggestion of a possible retarding effect, ascribable, on the balance of tendencies, to our Education Acts, I shall probably be told, as we are so often nowadays told, whether appropriately or inappropriately, to look to the example of Germany. Germany, it will be said, is one of the best educated countries in the world, and has been in that position longer, perhaps, than any other country, and yet its population is still increasing with striking rapidity. True, but if I look at Germany, I should like to see, if I could, all the relevant facts, and I think, at least, that it is not irrelevant to refer once more to what I believe to be the fact, that one essential condition of the remarkably rapid advance that Germany has made since the early seventies of last century was the very backward state of German agriculture at that time.

I have wandered far away from Scotland, but, if I now return to our own country, I trust I shall be able to show that I have not wandered without reason. We have seen that elsewhere population increases or diminishes in different parts of the country, at varying rates, and that these rates depend, in the long run, on the extent of the food supply somewhere, and, directly, in civilised countries, to a large extent on the standard of living which the different classes endeavour to maintain. If the maintenance of this standard involves a diminished rate of increase of population, the fact that this is so is no doubt matter for regret, but it may be a matter for which no one is responsible. And, if it is so, I cannot say that I see anything to regret in the fact that that degree of moral restraint is observed which is necessary to maintain the standard. As things are, it would probably be easy enough for Scotland as a whole to recover the rate of increase which she showed between 1891 and 1901. Looking to the growing importance of the mining industry, one may say that a willingness on the part of the miners to accept 3d. a ton less wages would go a long way to bring that about, provided, of course, that there was a corresponding willingness in other branches of British industry, without which there could really be no such reduction in that of mining. There would then soon be an increased demand for Scottish coal, and an increased demand for labour, and that demand would, no doubt, speedily be supplied. But, if it were to be supplied in this way, I wonder whether there is any one here who would look upon that mode of increasing population with any degree of satisfaction. But there may be more wholesome means of increasing the population, possibly by the improvement of Scottish agriculture, and all, I should think, will be agreed that, if that could be done, without lowering the standard of living for the classes concerned, it is much to be desired. Still, even with reference to this, it is important to bear in mind the fact mentioned by Malthus, that it is possible for an agricultural population to be in the worst state as well as in the best, and I may here point out that Malthus was one of those who foresaw the possibility of disaster in Ireland from an unrestrained increase of an agricultural population long before the famine took place. Referring to Arthur Young's project for improving the condition of the people by the more extended use of milk and potatoes in their diet, he says, "when, from the increasing population, and diminishing sources of subsistence, the average growth of potatoes was not more than the average consumption, a scarcity of potatoes would be, in every respect, as probable as a scarcity of wheat at present; and, when it did arrive, it would be beyond all comparison more dreadful"[1]—a Cassandra prophecy uttered in 1826, leaving twenty years for scorn before its fulfilment. Indirectly, even more than directly, it may be also possible to promote the growth of a numerous, strong, and healthy population, by a scientific cultivation of forestry in Scotland, but, in any case, it appears to me that for the present day, as for the time when Malthus wrote, the sound conclusion is as he puts it: "It is clearly the duty of each individual not to marry till he has a prospect of supporting his children"; while, however, we add with him, that it is "at the same time to be wished that he should retain, undiminished, his desire of marriage, in order that he may exert himself to realize this prospect, and be stimulated to make provision for the support of greater numbers."[2]

Still, we cannot overlook the fact that, if the economic conditions, to which Malthus calls attention, do tend to bring about a healthy degree of moral restraint in the great bulk of the community, there is still a large section of the population in whom the motives that bring about moral restraint do not and, perhaps, in the present circumstances, cannot be expected to act. They are born and brought up in conditions which give them nothing to strive for, except, perhaps on the part of those who, even in such surroundings, may be endowed with exceptional force of character. With regard, also, to this section of the population, it seems to me that the problem is still as it was stated by Malthus: "How to provide for those who are in want, in such a manner as to prevent a continual increase of their numbers, and of the proportion which they bear to the whole society"[3]—not "how to provide in the cheapest and best manner for a given number of people. If this had been the sole question it would never have taken so many hundred years to resolve."[4] And, probably, it is the most important result of the labours of that keen thinker, whose sympathy with the poor haunted him like a passion and governed the labour of his whole life, that that idea has, more or less, entered the minds of nearly all social reformers, whether they accept the doctrines of Malthus, or profess to denounce them. Whatever means can be taken that will answer to the "touchstone" of Malthus, that they shall tend to increase the foresight of the classes

---

[1] Sixth ed., vol. ii. p. 388. [2] *Ibid.*, p. 269. [3] *Ibid.*, p. 395. [4] *Ibid.*, p. 392.

concerned will be looked on with approval by all, but there still remains a large class, whom motives cannot reach, the feeble-minded, the hopelessly degenerate, the criminally idle; but with regard to these classes I will merely say that it is very significant that from so many sides, representing different tendencies of thought, the demand is daily becoming stronger that these classes should be isolated and not allowed to multiply to the detriment of the community.

GEOGRAPHY IN SCOTLAND SINCE 1889.

A REPORT PRESENTED TO THE TENTH INTERNATIONAL GEOGRAPHICAL CONGRESS AT ROME, MARCH-APRIL 1913.

By MARION I. NEWBIGIN, D.Sc., Editor of the *Scottish Geographical Magazine.*

IN the year 1889, Mr. A. Silva White, then Secretary of the Royal Scottish Geographical Society, and Editor of its *Magazine*, presented to the Fourth International Geographical Congress at Paris a *Report* on the Achievements of Scotsmen during the Nineteenth Century in the Fields of Geographical Exploration and Research (see *S.G.M.*, vol. v., p. 480 *et seq.*). This Report, covering as it did a period of nearly ninety years, was of a very comprehensive nature, and summarised the work of Scotsmen both at home and abroad. The present Report, which refers only to a period of some twenty-three years, and brings the previous one up to date, must necessarily be shorter.

We may note in the first instance that during this period geography has followed in Scotland much the same lines as in other countries. As, in broad outline, the surface of the globe, at least in lower latitudes, was tolerably well known before the beginning of the period, there has been, naturally, everywhere a diminution in the amount of exploration done. More attention has therefore been devoted to the detailed study of the homeland, and to the investigation of special problems relating to the geography of Scotland. Thus we have had, as will be noted directly, investigations of the Scottish Lochs, of the vegetation of the country, and so forth. In devoting special attention to such detailed problems Scotland has followed the example of other countries within recent years.

In the second place, in Scotland, as elsewhere, the fact that the land surfaces of the earth are now tolerably well known has led to an increasing concentration of attention upon the ocean, with the result that many papers have been published and many investigations conducted on problems connected with Oceanography.

Finally, as high latitudes constitute the least-known part of the surface, we find that Scotland has taken her full share in Polar Exploration during the last two decades.

Again, in Scotland, as elsewhere, great interest has been taken in recent years in methods of improving Geographical Education, and in this, as in the fields of research, considerable progress has been made.

I. INVESTIGATIONS IN SCOTLAND.—After this brief introduction we may take up the different subjects in order, beginning with research in the homeland. Of such investigations the most elaborate has been the *Survey of Scottish Lochs*, conducted under the direction of Sir John Murray and Mr. Laurence Pullar. This Survey has not only yielded a great mass of scientific material, but it has also served as a training school for a number of the younger investigators, and has thus had a direct as well as an indirect importance in connection with geography in Scotland.

The inception of the scheme may be briefly noted. In 1883 the Royal Society of Edinburgh wrote a letter to the Secretary of the Treasury, pointing out the desirability of having a bathymetrical survey made of at least some of the more interesting of the Scottish Lochs. This letter was followed by one from the Royal Society at London, and by other efforts to bring pressure upon the Government of the day. These efforts were ineffective, and in the year 1900 Sir John Murray and Mr. Fred Pullar published a first article containing observations made by them on the depth, temperature, life and so forth of the Lochs of the Trossachs and Callander District. This article was copiously illustrated with maps and figures, and was based upon the personal observations of the two gentlemen named; geological notes by Messrs. Horne and Peach, of the Geological Survey, being added, together with a geological map. Two further instalments were published during the following year, but the work was interrupted by the tragic death of Mr. Fred Pullar, who was drowned while endeavouring to save the lives of others, in the early spring of 1901. The early investigations had owed much to Mr. Pullar's mechanical skill as shown in the invention and modification of sounding-apparatus, and his father, Mr. Laurence Pullar, resolved that the Survey should be completed as a memorial to his son. The successive papers appeared in the *Geographical Journal* and in the *Scottish Geographical Magazine*, before their final issue in six volumes under the title of *Bathymetrical Survey of the Scottish Fresh-water Lochs* (1910). These volumes, in addition to containing the detailed results, soundings, temperatures, and so forth, and also papers on the fauna and flora of the lakes, and on the geology of the regions in which they lie, include also a number of general papers which render the whole monograph to a large extent a summary of our present knowledge of the subject of lake study. The beautiful maps which accompany it are the work of Dr. J. G. Bartholomew.

As already suggested, the Survey also stimulated research in a number of directions. One of the most important of the subsidiary problems raised was that of seiches and other oscillations of lake surfaces. The late Prof. Chrystal of Edinburgh University and Mr. E. M. Wedderburn undertook the investigation of this subject in the Scottish Lochs, a very detailed research being carried on upon Loch Earn. As a result some interesting new conclusions were arrived at, and the Scottish investigators have the satisfaction of knowing that their work has led to a greatly increased interest being taken in seiches throughout the civilised world.

The second important line of investigation which has been pursued in Scotland during the last twenty-two years is the *Botanical Survey* of the country, and this investigation also, sadly enough, had its inception marked by the loss of a young and promising life. Mr. Robert Smith, a student of University College, Dundee, by the advice of Prof. Geddes, went to continue his studies at Montpellier under Prof. Flahault, and was there initiated into the new methods of botanical survey. On his return he began to apply these methods to Scotland, and in 1900 published two sheets of his survey, the Edinburgh District and the North Perthshire District, with accompanying letterpress. These two sheets were to be but the first instalment of a very elaborate scheme, but, only a month or two after their publication, Mr. Smith succumbed suddenly to a severe attack of illness, leaving his work begun merely.

But it has not been allowed to remain untouched. The new methods aroused much enthusiasm in Great Britain generally, and a number of separate investigations have been carried on, a Committee having been appointed to ensure uniformity of method, and to avoid overlapping. In Scotland Mr. Robert Smith's work has been continued by his brother, Dr. W. G. Smith, who has published (1904-5) sheets for Forfar and Fife, with illustrative letterpress. Dr. Marcel Hardy, also associated with Dundee College, has published (1906) a very beautiful vegetation map of the whole of the northern part of Scotland, with an accompanying article. A number of less elaborate articles have also appeared; we may note as illustrating the way in which the work is being continued the article on the vegetation of a small region which appeared in the issue of the *Scottish Geographical Magazine* for September 1911, and the more elaborate study of a limited region which is to appear in the April issue of the present year (1913).

A third line of investigation which has been receiving a considerable amount of attention is that of *Regional Surveys*. A scheme for such surveys, based upon the 1-inch maps of the Ordnance Survey, was first suggested by Dr. H. R. Mill in 1896, and he subsequently prepared a specimen memoir based on sheets 317 and 332 of the Map of England. In connection with the application of the method to Scotland my own article on the Kingussie district (1906) may be mentioned. Surveys along similar lines are also being undertaken in connection with the work of the recently established lecturership in Geography in Edinburgh University. A number of the papers prepared by the students here have already appeared, and in the course of time the whole of Scotland will no doubt be studied in this detailed fashion. Further, the publication, at Cambridge, of a series of volumes dealing with the counties of the United Kingdom (*The Cambridge County Geographies*), should be noticed. The volumes on the Scottish counties have been contributed by Scottish geographers and others, and attest the increasing interest in Regional geography. Another indication of this is the attention which is being paid to the study of Ordnance Survey maps in schools.

In connection with the subject of Regional Survey we may note that in 1908 a Royal Commission was appointed to report upon the ancient and historical monuments and constructions of Scotland. Three sections of this Report, dealing with Berwick and Sutherland and Caithness respectively, have now been prepared. Much archæological work has also been done, especially in connection with Roman antiquities, by Dr. David Christison, Mr. James and others.

To this account of progress which has been made in the investigation of detailed problems in Scotland, we must add a note in regard to the High-Level Meteorological Observatory on Ben Nevis, to which allusion was made in Mr. Silva White's Report. Here we have to record not progress but retrogression. In October 1904 the High Level Observatory, and also the Observatory at Fort William carried on in connection with it, were both closed for want of funds to continue their work.

The Observatory at the top of Ben Nevis was commenced in 1883 and completed in 1884, and in the autumn of 1890 a fully-equipped observatory, almost at sea-level, was opened in Fort William. For the summit of the mountain a complete set of hourly observations of the various meteorological elements is available for a period of fully twenty years, and for the last fourteen years of that period there are simultaneous hourly observations for Fort William. From 1884 to 1890 there is a useful Fort William record of observations made at fixed hours several times each day. The entire series of observations has been published in detail, along with numerous scientific papers, in vols. 36, 42, 43, and 46 of the *Transactions* of the Royal Society of Edinburgh, and these volumes form a splendid record of the enterprise of the Scottish Meteorological Society, which, with only the scantiest help from Government, raised the extensive funds necessary for such a great undertaking. Imperishably associated with the history of these Observatories are the names of Mr. R. T. Omond and the late Dr. Buchan. Great regret was expressed at the time of the closing of the Observatories, and it is hoped that they may be re-opened at some future date. Both were of great importance in connection with Scottish geography, not only from the information which the records supplied, but also from the way in which they afforded training to many young investigators, some of whom are now occupying important posts in different parts of the world, and have already done much useful work.

Dr. John Horne, F.R.S., has supplied the following account of the progress made by the Geological Survey :—

"The chief work done by the Geological Survey in Scotland since 1889 may be grouped under the following heads: (1) the Highlands and Western Isles; (2) the Southern Uplands; and (3) the Midland Valley.

"The mapping of the belt of complicated ground in the north-west Highlands, comprising the Lewisian Gneiss, the Torridon Sandstone, Cambrian strata and Eastern Schists, which was begun in 1883, was completed about 1896. Nearly all the maps embracing this area have been published, and the results of the work are embodied in an important memoir on *The Geological Structure of the North-West Highlands of Scotland*, published in 1907. Since the publication of this memoir the north-west of Scotland has become a classic region for the study of those great movements by which the crust of the earth has been affected in mountain building.

"East of the great post-Cambrian displacements large areas of the Eastern Schists in the counties of Sutherland, Ross and Inverness have been surveyed. The conclusion has been reached by some of the members of the staff that the Moine Series represents a succession of altered sediments which rest unconformably on inliers of Lewisian Gneiss. Of special interest is the evidence obtained in Ross-shire between Loch Luichart and the river Carron, which drains into the Kyle of Sutherland, proving the manufacture of crystalline schists and gneisses out of sediments and intrusive igneous rocks by movements after the consolidation of the igneous material.

"The greater portion of the area occupied by metamorphic rocks between the Great Glen and the Highland border has been surveyed. Various maps, with accompanying memoirs, comprising parts of this wide region, have been published since 1889, of which the following may be mentioned: Sheet 93, Ben Wyvis; Sheet 85, Lower Strathspey; Sheet 65, Braemar and Ballater; Sheet 55, the district of Blair Atholl and Pitlochry; Sheet 45, Oban and Dalmally; Sheet 36, Mid-Argyll; Sheet 37, Upper Loch Fyne; Sheet 28, Knapdale and Jura; Sheet 29, Cowal. The islands of Islay and Colonsay have been surveyed, and the maps 19, 27, and 35, with the memoirs, have been published. While much valuable information has been obtained regarding the metamorphic rocks of this region, it must be admitted that the order of succession of the strata has not been satisfactorily determined. Various theories have been advanced to explain the tectonics; the latest put forward by one of the officers of the staff postulates great recumbent folds and displacement of the rocks.

"Progress has been made with the mapping of the Jurassic sediments and the plutonic and volcanic rocks of the Western Isles. The central and south-eastern parts of Skye and the islands of Raasay, Scalpay, Rum, Canna, Eigg and Muck have been surveyed, and sheets 70, 71 and 60, with descriptive memoirs, have been published. Special reference ought to be made to the valuable memoir on *The Tertiary Igneous Rocks of Skye*, published in 1904, which deals specially with the petrology of the rocks of that region.

"The revision of the Southern Uplands in accordance with the sequence of the graptolites, as determined by Professor Lapworth, was begun in 1888 and completed about 1896. The results of this work are set forth in a large memoir on *The Silurian Rocks of Scotland*, published in 1899, which deals with the stratigraphy, the palæontology, and the petrography of the region. In the course of this revision an important zone of Radiolarian Cherts of Lower Llandeilo and Arenig age was detected, underlain by Arenig volcanic rocks, which appear along anticlines throughout the Uplands.

"Regarding the Midland Valley progress has been made with the revision of the coalfields. In 1900 a memoir on *The Geology of Central and Western Fife* was published, and in 1902 a volume appeared, descriptive of the geology of Eastern Fife—both by Sir Archibald Geikie.

"The detailed revision of the coalfields was begun in 1902, and since that date the Carboniferous areas in the Lothians and large parts of the Lanarkshire and Stirlingshire fields have been re-examined. New editions of sheet 32 (Midlothian) and sheet 33 (East Lothian), with accompanying memoirs, were published in 1910. A special map of the Glasgow district with a descriptive memoir was also published in 1911. The characteristic feature of these memoirs is the large additions to our knowledge of the economics of these fields, and of the petrology of the igneous rocks associated with the Carboniferous strata. A special economic memoir on *The Oil Shales of the Lothians* was issued in 1906, for which there was a steady demand, and a second edition was published in 1912.

"Colour-printed geological maps on the scale of four miles to one inch are now being issued, chiefly for educational purposes. Four sheets have been published.

"In 1910 colour-printed geological maps on the scale of one inch to a mile were issued for the first time in the history of the survey in Scotland. These maps show both the solid geology and superficial deposits.

"Since 1889 a photographic department has been organised in connection with the Geological Survey. Nearly 3000 photographs have been taken, illustrating the geological features of different parts of Scotland. A brief descriptive catalogue of about 2000 of these photographs has been issued for the benefit of educational institutions.

"In the course of the detailed mapping of the geological structure of the country much valuable information has been acquired regarding the evolution of mountain forms and the various systems of drainage."

II. OCEANOGRAPHY.—The second subject with which Scottish geographers have been especially busied during the last twenty-two years is that of Oceanography. An interest in this subject is easily accounted for in a country which has so large a coastline, and which in certain regions depends more upon its fisheries than upon the products of the often barren land.

It will be recollected that the publication of the *Challenger Reports* took place from Edinburgh, and the Challenger Office, under the Directorship of Sir John Murray, was an active centre of scientific work in Scotland for more than twenty years. Sir John Murray took over the Directorship in 1882, on the death of Sir C. Wyville Thomson, and the concluding volumes of the Reports were not issued till 1895. Even after their completion the Challenger Office remained, and still remains a centre of active life; from it the Reports of the Loch Survey were issued in 1910.

The two concluding volumes of the *Challenger Reports*, entitled Summary and Results, include what is virtually a treatise on Oceanography by Sir John Murray. Since the date of their publication the author's continuous interest in the subject is attested not only by many separate papers and memoirs, but also by the fact that in 1911 he supplied the funds for the *Michael Sars* expedition to the North Atlantic, an expedition which has yielded results of high value for oceanographical science, and led to the publication of a very important book, *The Depths of the Ocean* (1912), by Sir John Murray and Dr. Johann Hjort.

Further, the North Sea Fisheries Investigation Committee, acting in co-operation with the International Council for the Exploration of the Sea, has for some time been carrying on detailed investigations both of practical and theoretical problems. Prof. Darcy W. Thompson, of Dundee College, together with his assistants, has already been able to draw from these investigations various conclusions of great interest in connection with hydrographical problems.

Under the heading of Oceanography also we should include much of the work done by Dr. W. S. Bruce, although Dr. Bruce has mostly carried on oceanographical research in high latitudes, so that his work falls under the heading of Polar Exploration—but the general subject is one which he has enriched by many pieces of research. Dr. Bruce has also founded, in Edinburgh, the Scottish Oceanographical Laboratory, where a considerable number of expeditions have been organised.

III. Polar Exploration.—The mention of expeditions leads us to the third subject in which great interest has been displayed in Scotland in recent years, this being Arctic and especially Antarctic exploration. In 1892-3 Dr. Bruce, in company with some other Edinburgh gentlemen, visited the Antarctic regions on board a Dundee whaling expedition. Some of the results of this expedition were laid before the British Association in the autumn of 1893, and before the Royal Scottish Geographical Society in the early part of the following year. To the British Association also Dr. Bruce expressed his strong desire to return to the Antarctic, a desire which was not fulfilled till some years later.

In 1898 the Royal Scottish Geographical Society issued a special Antarctic number of their *Magazine* with the special object of enlisting sympathy with work in this region, and of obtaining support for a British Antarctic Expedition. This number contained a stirring appeal from Sir John Murray, and also an article by the same author setting forth the advantages to science which might be expected to result from a well-equipped Antarctic expedition. There were also included in addition articles on the History of Antarctic Discovery, on the Flora and Fauna of the Antarctic, and an Antarctic Bibliography.

These efforts no doubt played their part in determining the British Government to give a grant in aid of the proposed British Expedition, but they did not exhaust Scottish interest in Antarctic exploration. In 1900, Dr. W. S. Bruce put forward his plans, the result of long consideration, for a Scottish National Expedition to the Weddell Sea. In the following year Dr. Bruce published a further article indicating the final form of his matured plans, and at the beginning of November 1902 the *Scotia* set out from Troon for the Far South with the members of the expedition on board.

Work was carried on in the south for two years, and not till July 1904 did the *Scotia* return to Scottish waters. During their first summer, in March 1903, the members of the expedition set up a Meteorological Station in the South Orkneys. This station was subsequently taken over by the Argentine Government, and is still in operation, so that it is satisfactory to think that one result of the Scottish Expedition has been the permanent establishment of a Meteorological Station in the Far South.

Mr. R. C. Mossman, the meteorologist of the expedition, remained behind in the South Orkneys for another year, in the service of the Argentine Government, and he has since become permanently attached to the Meteorological Office of that country, and has published a number of articles on meteorological subjects.

The *Results* of the *Scotia* expedition are still in course of publication, but we may note as especially important the rich zoological collections obtained, and the extensive soundings which were made in the South Atlantic Ocean and the Weddell Sea, which permitted Dr. Bruce to construct a bathymetrical map of this part of the ocean. It is interesting to note that Lieut. Filchner's recent explorations in the Antarctic have led to the discovery of a coastline which is undoubtedly a prolongation of Coats Land, discovered by Dr. Bruce. As already suggested, the meteorological observations made were also extremely valuable, and have thrown much light upon the atmospheric conditions in the Weddell Sea.

Dr. Bruce proposes to return to the Antarctic area with another expedition, but in the meantime he has taken a number of cruises to the north, notably to Prince Charles Foreland, Spitsbergen, of which he has made a special survey. Other members of his scientific staff also have carried on various Polar investigations. For example, Mr. R. C. Mossman made two visits to the Greenland Sea, which permitted him to publish a valuable paper on the Summer Climate of that Sea and the Ice Distribution there. We may perhaps also note that the *Scotia*, the ship in which Dr. Bruce made his cruise, has recently been chartered by the Government for ice observation in the Atlantic, with a view to preventing a repetition of the *Titanic* catastrophe.

IV. General Travel and Exploration.—As already noted, the day for sensational discoveries in this field, save in the case of Polar regions, is almost past, but during the last twenty years a number of Scotsmen have advanced geographical knowledge in various regions of the globe by their exploring work. Among others, we may mention, in addition to those already referred to, the names of Lieut. Boyd Alexander and his brother Lieut. Claude Alexander, whose African explorations, so sadly cut short, disclosed many facts of great interest; of the Hon. D. W. Carnegie, who during his short life undertook three expeditions into the desert interior of Western Australia; of Mr. G. F. Scott Elliot, who has done exploring work in the Ruwenzori region of Africa and in South America; of Dr. H. O. Forbes, who has carried on investigations in various parts of the world; of Prof. J. W. Gregory, now of Glasgow, who has done exploring work in Africa and Australia; of Col. J. R. L. Macdonald, who has added much to our knowledge of Uganda; of Prof. Sir W. H. Ramsay, whose researches in Asia Minor, though primarily archæological, touch geography on various sides.

V. Geographical Education.—In regard to this progress has been rapid during the last twenty-two years. To the Oxford School of Geography, established in 1899, Scotland has sent two successive Readers, Mr. Mackinder, and Dr. A. J. Herbertson, both of whom have promoted geographical education greatly both by teaching and by their text-books. The Universities of Scotland were later than those of England in establishing Departments of Geography, and not until 1908 were the efforts made by the Royal Scottish Geographical Society to establish geographical lecturerships in Scottish Universities crowned with success. In January of that year Mr. Geo. G. Chisholm, M.A., B.Sc., the author of the well-known *Handbook of Commercial Geography* and other books, was appointed Lecturer on Geography in the University of Edinburgh. In the following year, Captain H. G. Lyons, Director-General of the Survey Department of Egypt, was appointed Lecturer in the same subject in the University of Glasgow, and on his resignation in 1911, Dr. J. D. Falconer, late principal officer of the Mineral Survey of Northern Nigeria, took his place.

Considerable progress has also been made in regard to the teaching of geography in schools, which is now beginning to follow distinctively modern lines.

VI. The Work of the Royal Scottish Geographical Society.—In Mr. Silva White's *Report* allusion was made to the foundation of this Society, and to the work which it had done at the time of the publication of the article. Since this date the membership has been doubled, the Society has received an annual Government Grant of £200, and all its activities have greatly increased. On the 31st October last there were 1898 Ordinary Members, and 166 Teacher Associates. In connection with the Society an Association of Scottish Teachers of Geography has been formed, with the object of promoting the teaching of geography on modern lines.

VII. Geographical Publications.—The number of geographical text-books has increased so greatly since 1889 that it would be impossible to give a list of all those written by Scottish authors, or dealing with Scottish subjects, while it would be invidious to make a selection. But three atlases of outstanding importance must at least be mentioned. The first of these is the Royal Scottish Geographical Society's *Atlas of Scotland*, which was designed and prepared under the direction of Dr. J. G. Bartholomew, with the assistance of a number of specialists, and was published in 1895.

In 1899 there appeared as the first instalment of *The Royal Physical Atlas*, the *Atlas of Meteorology*, which was prepared by Drs. J. G. Bartholomew and A. J. Herbertson, under the general editorship of the late Dr. Buchan. In the beginning of the present year another instalment of the same atlas appeared in the shape of the *Atlas of Zoogeography*, prepared by Dr. Bartholomew and Messrs. Eagle Clarke and Grimshaw. All these are beautiful pieces of work, almost invaluable to the student.

In the above summary account attention has been limited to geographers who have had a definite association with Scotland. It would have been easy to extent the list of achievements considerably by including the names of those of Scottish birth or descent, whose work has not been carried on in close association with Scotland; but it has been thought better to leave these to be discussed among English explorers.

THE GEOGRAPHICAL FACTOR IN CIVILISATION.[1]

By THOS. W. M'CANCE, M.A.

Geography and History Related.—There are probably few departments of knowledge in which greater progress has been made within recent years than in those of History and of Geography ; in History, as showing the progress achieved by man ; and in Geography as indicating in a general way the places most suited for man's habitation, those which assist or hinder his life, his movements, and his means of communication. A few years ago our knowledge of History was obtained chiefly from books with records of man's life, and from coins and inscriptions ; but, nowadays, the student of History is forced to some extent to take account of climate, soil, and other geographical conditions, in order to understand thoroughly the life and progress of any people. Both History and Geography are now, as it were, Natural Sciences perfecting their methods of investigation.

Geographical conditions have remained in general fairly stationary for the last few thousand years ; and yet, in that time, man has made enormous progress, especially in Western Europe. His progress, therefore, is not due chiefly to altered geographical conditions. A difference in conditions between one place and another helps to explain a difference in progress between one people and another ; but it does not necessitate this difference. Let us compare Britain, Russia, and China in this respect.

Probably few factors have contributed more to human progress than the greatly extended use of coal in the nineteenth century. Coal has existed, where it now lies, for long ages in China, in Russia, and in Britain ; but it became important in national development only when man had learned how to use it and how to obtain it easily. Russia has only recently begun to realise her great mineral wealth, while China has scarcely touched her coalfields, although she possesses the largest known coalfield in the world, in which some of the seams are forty feet thick. That is, between the fact of the existence of coal in our island and our industrial supremacy in the last century, there was no relation of cause and effect—only a geographical condition, without which it would have been almost impossible for our country to occupy its present position among the nations.

To take another example, Carlyle tells us in his essay on the state of German literature that, "The three great elements of modern civilisation were gunpowder, printing, and the Protestant religion." Now, before man could use gunpowder he must have known how to make it, and this knowledge was gained, probably, from the accidental discovery of the explosive nature of nitre (found so plentifully in the soil of China) when mixed with the charcoal of some wood fire. We know that saltpetre (or

[1] A paper read before the Geographical Section of the Royal Philosophical Society of Glasgow, March 6, 1913.

nitre) was early known as "Chinese snow," and gunpowder was used by the Chinese long before it was used in Britain. The knowledge of the process of manufacturing gunpowder travelled west, and was brought by the Arabs into Spain; the explosive was used by the Saracens in their wars in Eastern Europe and by the Moors in Western Europe. This knowledge of gunpowder, so easily and so early gained in China, has not given China a great place among the nations, any more than the possession of coalfields; because the Chinese did not know of the existence of the one or how to develop fully their knowledge of the other. That is, geography supplies conditions, not causes of progress: in other words, the geographical factor is mainly passive.

The importance of favourable conditions varies directly, of unfavourable conditions inversely, with the character and the stage of civilisation of the people who may act upon them.

Difficulty in knowing very early Conditions.—But before one can understand the connection between Civilisation and Geography one must try to reconstruct the earth as it appeared in ages long past. One is forced to do so in order to know what influences worked on early man, and here one is confronted by an almost insuperable difficulty. At present we know practically nothing of these early conditions; and yet, it is all the more important to know these conditions, because primitive man was so comparatively powerless to overcome adverse surroundings. It was only after long ages of struggle against nature that he learned how to choose the places with the most favourable conditions, and how to overcome some of the more unfavourable conditions in other places. An important point in connection with our difficulty in understanding the physical conditions which acted on primitive man, is the relation of our earth to the sun and to other heavenly bodies. At the present moment astronomers are baffled by the behaviour of certain satellites of other planets and by the movements of certain stars, which, it seems, cannot in every case be reasonably attributed to gravitation. In the same way we cannot be certain that we know all the forces which have ever acted on our earth, or that the forces at present acting have always acted with the same relative degree, nor can we know all the changes through which our earth has passed. The seasons may have been, and probably were, different from what we now experience. If, as was probably at one time the case, in December the axis of our earth in northern latitudes was inclined *towards*, instead of *away from*, the sun, as at present, there would be a great difference in climate; and climate is a determining factor in the existence of most plants and of most animals, and is therefore for that reason alone, if for no other, supremely important to man. We know that the Arctic Regions must have had at some one or more times a much more temperate climate than they now possess.

Again, in past times, there have been enormous changes in configuration. Geology proves that continents now submerged once made complete land communication possible from South Eastern Asia to New Guinea and Australia, and thence by way of Antarctica to South America; and from India by way of Madagascar to Africa. We have not less strong proof that Africa was at one time united to Europe where the Strait of Gibraltar now separates these continents. The remains of a hearth with bones of the hippopotamus have been found in Malta, and also bones of pygmy elephants. Connect the existence of these animals with the great valleys in Malta formed by erosion, and it is almost certain that the island at one time formed part of the African continent, if it was not indeed a link completing a land ridge between Africa and the mainland of Europe.

The Sahara was not always so broad as it is now, nor was the South-Eastern Mediterranean region so barren as at present. The Western Hemisphere was connected with Europe by Greenland, Iceland, the Faroes and Britain on the east; and by Behring Strait to Asia on its west, if not at other places as well. Geology informs us that the peopling of our earth took place in inter-glacial, if not in pre-glacial ages, when the climate was much milder in our Northern Hemisphere, before the Bosporus was formed and before the great sea now represented by the Caspian Sea, the Sea of Aral, and Lake Balkash had shrunk to its present condition.

Ethnologists believe that all men are descended from the same common stock. Thus, an anatomist may fairly easily distinguish the bones of the arm of a man from those of the fore limb of a gorilla; but it would puzzle him to tell whether those same bones had belonged to a Chinaman, a Bushman or an Englishman. We do not know where the parent tribe lived nor when it appeared. Probably the excavations at present being carried on at Troy, in Palestine, in Egypt, in Crete, in Italy and in other places may throw some light on these questions.

In this paper I do not intend to discuss the question of civilised man's former state as a savage. There is too much uncertainty and supposition here also. All peoples do not pass through Morgan's successive stages. We have almost no knowledge of when our primitive ancestors learned how to create fire, to make the bow and arrow, or to fashion a clay pot. We do not even know when they first domesticated the dog and other animals. The bones of the dog are found amongst very early remains of man.

In order to see some of the relations between Civilisation and Natural Conditions I purpose dealing chiefly with the early Mediterranean region as best showing the continuous progress of man; and, then, since the development of minerals is also the history of man's progress, and the use of a few minerals as weapons or as tools was probably the determining factor which gave one people the distinct advantage over another, I shall consider briefly the search for minerals and more particularly the use of copper and of clay; with the consequent progress made by peoples brought into contact with each other in this region.

The Mediterranean most suitable for Progress.—Civilisation, as we know it, was nourished for thousands of years in the Mediterranean countries, the "Orbis terrarum" of the Romans, because this region is a huge junction zone, more favourable to continuous progress than any other part of the earth. Consider the shape of the sea, its length, very great compared with its breadth; its peninsulas and its islands, all tending to facilitate crossing at a time when man had no mariner's compass. Consider the varieties of soil in the surrounding countries, the grass lands and wheat lands, the sloping mountains, giving as many varieties of climate and of vegetation as may be found between the Sahara and the Arctic Circle, and one must say that, until man had learned boldly to face the stormier ocean, it was the best nursing ground for early progressive human life; and this because the Great Sea afforded such easy communication between different peoples striving to "live well" under different natural and social conditions.

Let us now see some of the efforts of these early Mediterranean peoples to "live well." To "live well" is a comparative term, and with different peoples inhabiting a great region stretching from Spain to the Persian Gulf, and being more and more gradually brought into contact with each other, the standard of living was likely to rise as better methods and better tools were used and attainable luxuries became more common.

Neolithic Trade.—Professor Angelo Mosso in his *Dawn of Mediterranean Civilisation*, says that in Neolithic times (which he places from about 7000 B.C. till 4500 B.C. for Southern Europe) man was sailing on the Mediterranean, in a cabined boat propelled by wind and oars, which argues a certain proficiency in navigation; and with one of the obsidian knives used by Neolithic man more than six thousand years ago, the author was able to sharpen his own pencil while taking notes of excavations in Crete. Obsidian is a kind of volcanic glass originally found abundantly on the island of Melos in the Cyclades. In the Neolithic age this glass was sent all over the borders of the Aegean and found its way into Troy, and even into Egypt, where it was commonly used for razors. Again it is known there was considerable traffic in the Eastern Mediterranean in henna, Gr. κύπρος, a plant used in perfumery, toilet purposes, and in some of the religious services. The fact that the Greek name for this plant is the same as that for the island of Cyprus, points to Cyprus having been a chief source of supply of this plant in Neolithic times.

Silver, to which I now refer, reveals still further the extent of Mediterranean commerce. This metal has been found in the most ancient tombs in Crete; and also along with copper at the close of the Neolithic period in Egypt. Where did it come from? probably from Spain or from Sardinia, in both of which places it is found pure. It might have come from Attica, in Greece, where silver is still worked; but in Attica it has to be extracted from lead ore, and though the Greeks, of say 500 B.C., seem to have understood partly the process of extraction, it is improbable that Neolithic man had any knowledge of the art. We thus see that in the search for obsidian, henna, and silver the Neolithic navigators of the Mediterranean had penetrated east and west, and also north and south.

With regard to copper, we know it was imported into Egypt from Crete in very early times. Chrysocamina (oven of gold), near Gournia, in Crete, bears evidence of the very early use of copper in that island.

Here has been found a prehistoric copper mine. The early inhabitants of Crete seem to have been the first in the Mediterranean to work in copper. Professor Mosso says: "The bronze age in Crete shows such perfect work in the art of casting, that nothing can be found equal to it in artistic worth, even in the north of Europe. The Phœnicians, who came after the Minoan civilisation, and were believed to be the artists who diffused the art of casting in bronze, remained inferior to their Cretan masters in the technique of metal work and in the plastic of bronze."[1]

No copper ores or veins have been found in Crete nor yet in Cyprus, so that it is probable that in both these islands, the trade in copper, and later, with extended navigation, in bronze, arose from copper being found on the surface of the earth. This substance mixed with carbonate was collected and was the origin of Cretan and Cyprian trade in copper weapons and tools. Copper and bronze relics of Minoan stamp have been found in Sardinia, in Italy, and in Egypt.

Let us now look shortly at iron. Iron relics of this very early period are rare. Mr. Chisholm tells us in his *Handbook of Commercial Geography* that the most ancient relic of an iron implement brought to light is a piece discovered in one side of the great Pyramid of Gizeh. That carries us back five thousand years. He then explains the rarity of remains of ancient iron implements as compared with those of bronze, by the fact that, "Under the influence of air and moisture iron is eaten away so rapidly that its preservation for a long period is possible only under very exceptional conditions. So liable is it to disappear that, of all the numerous articles of iron that must have existed in ancient Egypt, the remnants which have been discovered do not weigh in all more than half a pound, and this in a country with a dry climate specially suited for the preservation of such articles." At Knossos, in Crete, in Neolithic strata a small piece of magnetic iron has been found, weighing half a kilogramme. It seems to have been a sort of charm or to have been connected with worship, and was evidently never used as a hammer.

Mount Sinai is rich in copper and iron, but the ores are too deep seated to have been worked by early man. The Homeric poems represent Greece as beginning an iron age about 1200 B.C., and we know the Greeks obtained their supplies of iron from Chalybes to the west of Trebizond on the Black Sea. Iron was not worked in Elba till after the arrival of the Etruscans in Italy about 1100 B.C. Strabo, the geographer and historian, speaks of iron being produced in very early times in Calabria, though no longer worked in his day, *i.e.* about the time of Christ. The Romans knew of the rich ores of Bilbao, in Spain, but their methods of working were very defective and very expensive.

The iron industry, costly as it was, gave the Romans weapons and tools superior to those of most of their contemporaries, and, in conjunction with the situation of Rome at the geographical centre of the

[1] Professor Mosso's views on the Cretan origin of our culture seem to be shared by Professor Myres of Oxford. In the *Dawn of History* we find: "It is here (in Crete) that man *first* achieved an artistic style which was naturalist and idealist in one; . . . wonderfully skilled in the art of the potter, painter, gem-engraver, and goldsmith."

Mediterranean, assisted in no small measure the spread of Roman civilisation. We thus see that, though iron is not scarce in the Mediterranean region, man was long in learning how to obtain it; but we know that once he had learned its importance, the knowledge would easily be spread and no efforts spared to obtain the valuable mineral.

Early Pottery.—Now with regard to clay, we find that the early Cretans were also clever workers in pottery, and the manufacture of earthenware reached a very high level in Crete in Neolithic times. There would seem to have been factories at that time, and potters placed their own private marks on the pottery they had made. Some fine examples of this predynastic pottery, showing the potters' private marks, may be inspected in the Royal Scottish Museum, Edinburgh. Pots with Minoan marks have been found in Hungary, in Egypt, and elsewhere, indicating the extent of Cretan trade.

In the *Dawn of History* Professor Myres says there is no evidence that the potter's art arose in the Nile Valley, adding that the raw Nile mud has not the qualities of a good pot clay.

"The new art appears rather suddenly in one well-defined district of Upper Egypt at a high level of technical skill." It is a good thing for the historian that clay was plentiful in the places most suitable for early man, that is the clayey wheat-producing soils of Egypt and of Mesopotamia. When the potter out of soft shapeless clay makes a definite shape, we have the creative art. The clay pot once formed was exposed to fire or dried in the sun, and thus made brittle. Hence it had a brief existence, so brief, that pottery, the manufacture of common household pots, was as regular, though not, of course, so frequent an occupation with the Neolithic housewife as baking. Pottery is more important to us, from the historical point of view, than Cretan or Egyptian or Babylonian weaving; because cloth is easily consumed, but the broken pot cast among the refuse remains, and reveals to later ages the life of the potter and the history of his times; and, as Professor Stewart Macalister says, "Pottery is the key to chronology."

Maritime Origin of Writing.—But clay is not only useful for pots in daily use, and for ornamental pots developing still further the creative and imaginative arts; clay soon became useful as a means of recording and transmitting signs. These signs or seals seem to have been first used on goods being shipped. A piece of soft clay was clapped on some particular part of a package and stamped with the seal or distinguishing mark of the vessel conveying it, as well as with the seal of the sender. The earliest Cretan seals are rude representations of ships. These seals soon changed into particular signs, and probably lost their original significance. Additional signs were used, and were also applied to other purposes, and with a sea-going people such as the Cretans, capable of artistic expression, and feeling the necessity of transmitting messages, with abundance of clay easily sun-dried in which to fix or record these messages, we can readily understand how the rudiments of writing once used were soon disseminated all over the Eastern Mediterranean region.

Of course it may be objected that the necessities of trade alone would have produced the art of writing; but in no other place were so many favourable conditions conjoined in early times to produce this art as in Crete. Wallace found rudiments of picture writing in the Amazon valley; but the conditions there were unfavourable to further development. The art of writing arose in China under less favourable conditions, and probably at a later period. At any rate our script was little influenced by Eastern Asiatic man. The origin of writing, as we know it, was probably in maritime trade and, so far as we can learn, took place in Crete, whence it spread to Egypt and all round the Eastern Mediterranean.

The necessity of organised labour to fight the rising Nile, and to irrigate the fertile Nile valley, the expeditions against covetous marauding tribes, the highly centralised despotic government with messages and commands to be conveyed to great distances, led to many changes in devices stamped on clay, or recorded in other ways, which gradually developed into Egyptian hieroglyphics.

Again we learn that the ancient name of Gaza in Palestine was Minoa, from Minos, king of Crete. David, king of Israel, had a bodyguard composed of Cretans and Philistines, showing a close connection between Crete and Philistia in his time. The Bible speaks of the skill of the Philistines in archery, and the Cretans were famous among the Greeks as archers. In the time of the Ptolemies, say 250 B.C., the Egyptians believed that the Phœnicians were Cretans. Phœnician φοινικός, red man; connect this derivation with the fact that the Cretans painted in red; and, considering all these points, one is forced to conclude that both the Phœnicians and, later on, the Philistines came from Crete. Hence the Phœnician phonic alphabet also took its origin in Crete.

This art of writing was carried east by traders, and also gave rise to the Babylonian syllabic alphabet. Traces of hieroglyphic characters, older than the wedge-shaped or cuneiform, will be found in the inscription on the statue of Gudea, No. 74, Royal Scottish Museum, Edinburgh. Again it may be objected that these three forms of script, Egyptian, Phœnician and Babylonian, are different; but Dr. Tylor remarks in the *Early History of Mankind* that "One people got the art of writing from another, though not the same characters for particular letters." Cadmus is said to have introduced into Greece, from Phœnicia or from Egypt, the Greek alphabet of sixteen letters, and the Phœnician script was adopted by the Semites. Thus there is a strong probability that all the various scripts of the Eastern Mediterranean region, and therefore, of our own Northern European languages, which are all derived from them, were Cretan in origin, though differing in particular letters; and the ancient script of Crete, not yet fully deciphered, began in maritime trade. That is, the art of writing, man's most potent means of furthering progress, was called into existence by the fortunate combination of geographical conditions which determined the occupations of the earliest navigators of the Mediterranean, and was by them spread throughout the whole region.

Colonisation in the Mediterranean.—Let us now look shortly at the progress of a few of the peoples on this Mediterranean Sea. As indicated above, in connection with the art of writing, the Cretans formed settlements on the mainland to the east of the Mediterranean, probably driven by trouble in their own densely populated home, "The Island of a hundred cities," according to Homer; probably also, attracted by the central position of Palestine among the then civilised countries.

They established themselves about 2700 B.C. in Phœnicia to the north, and much later, about 1500 B.C., in Philistia to the south. We might ask, What took the Cretans to Philistia? As compared with Egypt and some other regions round the Mediterranean, this country is now comparatively barren, though considered a land "flowing with milk and honey" by the Israelites who took possession of a part of the same land about a century later. Dean Stanley in *Sinai and Palestine* states that the flora of the district has been much changed within the last three thousand years. Probably the climate has become drier; but, in any case, unfavourable though the soil may be in the eyes of an Egyptian, to the Israelite coming from the desert country east of Jordan it appeared all that was good. The Cretans were traders and probably wished to capture some of the through trade between the surrounding civilised peoples, as their kinsmen, the Phœnicians, had done so successfully further to the north.

If the Philistine Cretans were not successful, we know that the Phœnician Cretans were eminently so; for Tyre existed as a trading centre of great importance for more than three thousand years, until it finally decayed under the combined influences of Turkish misrule and altered natural conditions. Different conquerors overran it in turn; but though man might destroy the commercial prosperity of Tyre and Sidon for a short time, while the geographical conditions which produced these cities remained unchanged, the fortunes of Phœnicia were assured.

The Phœnicians were traders, and in pushing their business everywhere, both by land and by sea, they distributed to the world the wares of Egypt, of Babylon and of other places. To the great powers around, Phœnician ships and Phœnician sailors were indispensable. Sennacherib, Xerxes, and Alexander the Great, each in turn employed them for their transports, and even for naval warfare. One might say the Battle of Salamis, 480 B.C., possibly the most decisive in the history of European civilisation, was not so much a struggle between Greek and Persian, for the Persians were not sailors, as rather a naval contest between the sea-powers of Greece and of Phœnicia. To facilitate trade the Phœnicians had established trading stations or factories along various sea routes. One of these, Kart Hadjat (the New City), better known as Carthage, in later times the great rival of Rome, was founded in 813 B.C. by a fugitive princess from Tyre; another further west was Cadiz, the Tarshish of the Bible, which had long been used by Phœnicians and Carthaginians as a trading centre for tin, silver and amber.

This station or outpost was in 501 B.C. definitely occupied by the Carthaginians, who were attracted by its large natural harbour, its profitable fisheries, and its rich metalliferous hinterland; and probably, in order better to compete, by the Atlantic route, with Cumæ, north-west of Naples, founded about 1000 B.C. by Greek Chalcidians from Chalcis (from which χαλκός, which was the Greek word for copper) to the west of Negropont, where copper was now exhausted. Tin was much in demand for the manufacture of bronze for weapons, tools or ornaments; and the chief source of supply was Cornwall and the Scilly Isles, the Cassiterides (tin islands). The Phœnicians held the sea route, the easier route; and the Greeks wanted to hold the overland route, *via* the Loire and the Rhone. About 560 B.C. another Greek colony was founded by Phocæans from north-west of Smyrna in Asia Minor, copper and tin workers driven out of Phocæa by Persian pressure. This colony was Marseilles, nearer the overland tin route than Cumæ; and soon there was war between Marseilles and Cumæ, the result of commercial rivalry. Indeed rivalry in trade was both a blessing and a curse to Greece. It has little hinterland, and the various town states were separated by mountain chains, so that communication was easier by water.

As happened often in the history of Phœnicia, the Greek towns fought against each other or against their own colonies, so that the physical conditions which assisted in developing Greek civilisation also tended to destroy these states as political powers. Poverty produced emigration, which was further increased by their skill on the sea; hence the enterprising successful Greek colonies. But sailors, having to share the same dangers, are intolerant of tyranny; and the spirit of independence thus engendered, further stimulated by their success as colonists, produced commercial rivalry among the Greeks.

After the rise of Rome, Greece conquered her conqueror in another way, for Constantinople remained essentially Greek in culture for over a thousand years; and, on the occupation of that city by the victorious Turks, Greek culture was disseminated over the whole of Western Europe.

Thus the search for minerals was a determining factor in the development of Mediterranean civilisation, just as in our own day the search for minerals has developed Australia, Eastern Alaska, California, Chili and the Transvaal. Indeed, we may say that mineral wealth and superiority in mining methods account in no small measure for the present position of Britain and of Germany among the nations. Man can increase his agricultural riches even in unfavourable regions; but the supply of minerals is limited. He cannot manufacture coal or zinc.

With the extended use of the mariner's compass and the discovery of other commercial routes, the Mediterranean fell behind the more northern regions facing the Atlantic; and even the Suez Canal and the tunnels through the Alps, though assisting the region to a certain extent, have not restored to it anything like its former comparative importance, chiefly because of the almost entire absence of fuel, though electricity generated by cheap water-power may do something for the junction zones at the foot of the snow-capped mountains. Here again the conditions which were favourable to progress in early ages are comparatively unfavourable at the present time.

Geographical Hindrances to Progress.—As intercourse and easy routes of communication are of vast importance in the progress of any people, so isolation tends to prevent development. We see this in the condition of the native inhabitants of Australia at the beginning of last century; in the present condition of Spain, where the parallel ranges of high mountains have formed so many isolated valleys; and in the condition of many tribes in the great equatorial forests of Central Africa and of the Amazon. China from her natural and artificial barriers has not made progress commensurate with her advantages in soil and climate; but we know not what the future will do for her. Signs of her awakening are becoming apparent. Within a few years Pekin will be connected by rail with Canton *via* Hankow, and another line of rail is already projected between Hankow and Rangoon *via* Chungking and Yunnan. The vast infant Republic is at last adopting Western civilisation.

Dr. Gregory has urged that white man may live in the tropics. Poisonous insects and malaria can be overcome, swampy regions can be reclaimed and barren lands rendered comparatively fertile. Yet with advancing civilisation man's dependence on geographical conditions is not destroyed, it is merely modified. He cannot entirely overcome difficulties in climate, in soil, or in environment; his life is so much bound up with that of animals and of plants. The effect of environment on animals is seen in the case of horses taken to the Falkland Islands; which, according to Darwin, have during successive generations become smaller and weaker, as in the case of the pygmy elephants of Malta: while horses running wild on the Pampas have acquired larger and coarser heads. The effect of natural surroundings is seen on plants, in the case of Sea Island cotton which degenerates in India, whilst Indian cotton improves in the South Eastern States round Florida. The English miller cannot produce Hungarian flour, even from Hungarian wheat, and the spinners of Pilsen and of Brünn need to charge their mills with artificial vapour to compete successfully in the production of the same kind of fabrics produced in the moist atmosphere of Britain.

That physical conditions affect national temperament, national morality, and national character is undoubted; though the influences are subtle, they are real, but are very difficult to determine, because different peoples are affected by the same kind of surroundings in different manners.

Britain, France, Italy, Germany and the United States of America differ in language, or in history, or in government, or in geographical conditions; yet their civilisations are practically the same. Freedom in communication has almost levelled down differences in progress arising from different conditions or from peculiar ethnical qualities.

Importance of Geographical Conditions.—Other factors in civilisation are as important as the geographical. But, nevertheless, geographical conditions, though rarely active, are enormously important in the progress of man. He is affected by these conditions at every point; in bodily frame, in physical health, in character and in material prosperity; and, though man may overcome many adverse conditions he can never rise very far above them. Rather, his success as a rule is possible only in modifying their effects and in working out his advancement along the lines permitted by his geographical surroundings.

PROCEEDINGS OF THE ROYAL SCOTTISH GEOGRAPHICAL SOCIETY.

At a meeting of Council held in the Society's Rooms on July 28, it was resolved that the Livingstone Medal for the coming session be presented to Commander Evans, of the British Antarctic Expedition, and the Society's Gold Medal to Professor Albrecht Penck for his signal services to geographical science.

OBITUARY

DR. JOHN WATSON M'CRINDLE.

Dr. John Watson M'Crindle, an original member of the Royal Scottish Geographical Society, died at Westcliff-on-Sea on July 16, in his 89th year. Dr. M'Crindle was born at Maybole and educated there and at the Edinburgh University, where he graduated with distinction in 1854. For some years he taught classics in Edinburgh, but in 1859 he went to Calcutta as Principal of the Duff College in succession to Dr. George Smith. Seven years later he entered the Bengal Education Department, and after serving as a professor in Krishnagar College he became first Principal of the Government College at Patna; there he remained until 1880, when he retired. During his stay in Patna, M'Crindle, along with his wife, founded the first high-school for native girls. While in India he contributed a series of learned articles to the *Indian Antiquary*, chiefly annotated translations of classical works relating to ancient India. These were subsequently amplified and collected into book form, and have become widely known as a nearly complete collection and translation of everything relating to India to be found in Greek or Roman literature. He also edited for the Hakluyt Society *The Christian Topography* of Cosmas Indicopleustes, an Egyptian monk, which appeared in 1891. Shortly after M'Crindle left India he settled in Edinburgh, where he joined the Hellenic Club in the days of its glory—the days of Blackie and Butcher, of Hutchison Stirling and Walter Smith. In India, M'Crindle was honoured with the fellowship of the University of Calcutta, and at home, in 1898, with the honorary LL.D. of the University of Edinburgh. When the Scottish Geographical Society was founded in 1884 he took enthusiastic interest in its work, and in the following year he was elected to the Council of which he remained a member for sixteen years. It was not in the Council, however, that his influence on the Society's fortunes in its early days was felt so much as in the Geographical Club, a little band of those who first projected the Society, and who periodically came together in quiet convivial meeting to discuss geographical affairs, to plan out articles for the Magazine, and frequently to entertain distinguished explorers with simple tavern refreshment. At these unostentatious reunions there was much inspiring intercourse, and many an anecdote of thrilling adventure which had not been told in public lecture was recounted by travellers from Central Africa or Asia or Arctic regions or islands of the Pacific. By the few who survive, M'Crindle's happy personality at these meetings will ever be remembered; his genial countenance would beam unalloyed friendliness and diffuse a feeling of contented satisfaction among all who were present. His knowledge, his experience, his scholarship were a source of strength, and on occasions his perhaps somewhat lingering wit would flash out in unexpected sallies, which delighted his associates. He was of great assistance to the editors in the early days, and contributed much to the pages of the Society's Magazine.

Dr. M'Crindle retired from the Society's Council in 1902, when the advancing infirmities of age forced him to give up active work. A distressing disorder of the eyes rendered him nearly blind, and he left Scotland to be with relatives in the south of England. There he died, deeply regretted by many old friends. W. B. B.

GEOGRAPHICAL NOTES.

EUROPE.

The Population of Scotland.—At the beginning of August, vol. xi. of the twelfth Census of Scotland was issued as a blue-book, and we note the following interesting points in regard to its contents:—The population of the country is given at 4,760,904 (2,308,839 males and 2,452,065 females). Compared with the census of 1901 there was an increase of 288,801, or 6·5 per cent. The present population is fully one million greater than the population of 1881, fully two millions more than that of 1841, nearly three millions more than that of 1811, and is 3,152,484 more than that of 1801. The ratio of females to males in Scotland at the present census was 106·2 to 100. At the last census this ratio was 105·7 to 100.

The 33 counties in Scotland vary in amount of population from 1,447,034 in Lanark, 507,666 in Edinburgh, 314,552 in Renfrew, and 312,177 in Aberdeen, to 7,527 in Kinross, 9,319 in Nairn, 15,258 in Peebles, and 18,186 in Bute.

Housing conditions are dealt with at some length, and it is shown that of the population of Scotland 2,077,277, or 43·6 per cent., are living more than two in a room; 1,005,991, or 21·2 per cent., more than three in a room; and 397,262, or 8·3 per cent., more than four in a room. Comparison with the figures of the earlier censuses shows that the proportion of the population living in the less crowded conditions has increased, and the proportion of those living in the more crowded conditions has decreased.

The enumeration of Gaelic-speaking persons in Scotland was first instituted in the census of 1881. The census now under review gives the number of Gaelic speakers at 202,398, as against 230,806 in 1901. Speakers of Gaelic and English in the year 1901 numbered 202,700, and of Gaelic only 28,107, the former being 7,977, or 3·8 per cent., fewer, and the latter 15,632, or 35·7 per cent., fewer than in 1891. By this census speakers of Gaelic and English are found to number 183,998, and to be 18,702, or 9·2 per cent., fewer than in 1901, and 26,679, or 12·7 per cent., fewer than in 1891; and speakers of Gaelic only are found to number 18,400, and to be 9,706, or 34·5 per cent., fewer than in 1901, and 25,338, or 57·9 per cent., fewer than in 1891.

Fishery Research in the North Sea.—The research vessel S.Y. *Hiawatha*, chartered for fishery research in the North Sea, left the Tyne on August 4 for the purpose of making certain practically continuous hydrographic observations, at a fixed position, during the first fortnight of August. She was to take part in a co-ordinated research into the movements of the great water masses in the North Sea, and for this purpose was to drop her anchor about 150 miles "E. by N. ½ N." of Shields. Her labours were to be identical in aim with researches simultaneously carried out on board eight other vessels, also at anchor. Two of these other vessels were to be research vessels, acting on behalf of Sweden and Scotland, the Swedish vessel working in the Skagerak, the Scottish well to the north-east of Aberdeen. The remaining vessels are light vessels, two acting for Holland and the other four for the English department.

The observations were to consist of current measurements made near both surface and bottom every hour night and day throughout the fortnight, and in fine weather at other intermediate depths. Special attention was to be paid to the submarine waves which, it is expected, are to be met with at the depth at which the heavier bottom water and the lighter surface water are in contact. Specially devised current meters are used in this work. The temperature and salinity of the various layers of the sea were also to be ascertained, special water-bottles being employed to secure samples of the sea from any desired depth. Samples of the minute floating organisms which, directly or indirectly, constitute the food of all our food fishes were also to be taken at various depths and at the extremes of the tide. It is expected that some 8000 independent current measurements would be made from the English vessels alone.

The operations have been planned by a special committee of the International Council for the Exploration of the Sea, it is stated, because a knowledge of the constitution and movements of the sea water is essential to the understanding of the movements and of the abundance of the fishes upon which the fishing industry depends. For instance, the abundance or scarcity of the herring of the Kattegat and Skagerak has been found to be connected directly with the amount of water which enters the Baltic from the North Sea, and other fisheries of Southern Sweden have been shown to change with the ebb and flow of this layer of cold, salt water.

The Scottish Zoological Park.—As a matter of local interest we may note here the opening of the Zoological Park, Corstorphine, Edinburgh, during the present summer. The Zoological Society of Scotland are to be congratulated both on the choice of so excellent a site for a zoological garden, and on the progress which has been already made. As already stated here it is proposed, so far as possible, to keep the animals in open dens, on the Hagenbeck system, and those who have watched, *e.g.* the bears living in their open, water-surrounded enclosure, will have no doubt of the educational value of the system. The difference in the behaviour of the animals after they had been turned out of their confined temporary cages into the open was very marked, and it is now possible to make observations of great interest on the differences between the brown and polar bear, in relation to the natural surroundings of both animals. The park has been opened to the public, although the process of laying-out is just begun, and thus the gradual development of a modern type of zoological garden can be watched in detail day by day.

AFRICA.

Insect-borne Diseases in Africa.—We have discussed here from time to time the various investigations and observations which have been made in Africa on the deadly diseases caused in man and domestic animals by the minute organisms known as trypanosomes, which are transmitted by tsetse flies. We have also spoken of the proposals made in various quarters that the larger game animals of Africa should be slaughtered on the ground that they serve as reservoirs of the parasites. In this connection it is interesting to note that the Secretary of State for the Colonies has nominated a Committee to report:—

1. Upon the present knowledge available on the question of the parts played by wild animals and tsetse flies in Africa in the maintenance and spread of trypanosome infections of man and stock. 2. Whether it is necessary and feasible to carry out an experiment of game destruction in a localised area in order to gain further knowledge on these questions, and, if so, to decide the locality, probable cost, and other details of such an experiment, and to provide a scheme for its conduct. 3. Whether it is advisable to attempt the extermination of wild animals, either generally or locally, with a view of checking the trypanosome diseases of man and stock. 4. Whether any other measures should be taken in order to obtain means of controlling these diseases.

The Committee is a large and representative one, and the inclusion upon it of some zoologists, notably of Dr. Chalmers Mitchell, will reassure those who fear that hasty and irrevocable steps may be taken with regard to those beautiful mammals which are already disappearing too fast from the surface of Africa.

POLAR.

Arctic Expeditions.—It is announced that Captain Koch (cf. this Magazine, vol. xxvii. p. 656) has successfully accomplished his journey across the inland ice of Greenland, for he arrived at Proeven, near Upernivik, on the west coast, in the middle of July, after having left the east coast on April 20 of the present year. Horses were used for traction purposes across the ice, but had finally to be killed for lack of fodder.

We learn further from the *Times* that M. Jules de Payer left Havre on August 10 on an expedition to Franz Josef Land, the little known north-eastern corner of which will be explored.

The American Crocker Land Expedition.—This expedition (cf. p. 326) left New York in the beginning of July, on board the steam sealer *Diana*, and, after putting in at Boston, reached Sydney, N.S., on July 9. Here some additional equipment was taken on board, and the boat left on July 13 for Battle Harbour, Labrador. From Labrador the vessel was to make for the west coast of Greenland, the objective being Cape York, where walrus and seal hunting were to be carried on. Much of the cargo may be landed at Payer Harbour, Pim Island, but the main headquarters of the expedition are to be at Flagler Bay, on the south side of Bache Peninsula.

The objects of the expedition, as finally arranged, are detailed fully in *Science* as follows:—"(1.) To reach, map the coastline, and explore Crocker Land, the mountainous tops of which were seen across the Polar Sea by Rear-Admiral Peary in 1906. (2.) To search for other lands in the unexplored region west and south-west of Axel Heiberg Land, and north of the Parry Islands. (3.) To penetrate into the interior of Greenland at its widest part, between the 77th and 79th parallels of north latitude, studying meteorological and glaciological conditions on the summit of the great ice cap. (4.) To study the geology, geography, glaciology, meteorology, terrestrial magnetism, electrical phenomena, seismology, zoology (both vertebrate and invertebrate), botany, oceanography, ethnology, and archæology throughout the extensive region which is to be traversed—all of it lying above the 77th parallel."

Though the original intention was to spend three summers and two winters in the north, sufficient supplies have been taken to permit of a longer stay should this seem desirable. The expedition is supplied with a powerful wireless telegraphy apparatus, and it is hoped to keep the base station in communication with the wireless stations of Canada, and so with Washington.

General.

The Scott Fund.—It is announced that £75,000 has been raised for the purposes of the fund. In the allocation of the money ample provision has been made for the publication of the results, which are to appear under the editorship of Captain Lyons, F.R.S. A total sum of £17,500 has been set apart for this purpose, which includes, besides the cost of publication, an allowance for the services of three biologists, three geologists, two physicists, and other specialists, in addition to a draughtsman, while a sum of £800 has been earmarked for maps and charts. Further, a sum of £10,000 is to be set aside to form a trust fund for the endowment of Polar research in the future. It is also proposed that a group of statuary should be erected in Hyde Park facing the Royal Geographical Society's house, and there is also to be a tablet in St. Paul's Cathedral. Provision for the relatives of the dead explorers has also been made.

Communications in Mountain Regions.—It is a commonplace of geography that mountains offer obstacles to intercommunication. Hence it is natural to conclude that roads and other means of communication are less abundant in such regions than elsewhere. In a short note in a recent issue of the *Annales de Géographie*, however, M. Charles Biermann of Lausanne points out that, for parts of Western Europe at least, such a generalisation is not borne out by the facts of observation. Many of the mountain regions of Western Europe offer rich pasturage, and therefore the pastoral populations in their vicinity practise that form of nomadism called in French *transhumance*, a term for which we have no English equivalent, although it has been stated recently that there is evidence that the phenomenon once occurred in Highland Scotland. The migrating flocks leave their traces on the mountains in the shape of numerous tracks and roads, now often utilised by tourists. In the Swiss Alps, as we have had often occasion to point out here, nomadism is not confined to the pastoral folk, for indeed purely pastoral peoples hardly exist. The cultivator also moves up and down the mountain slopes according to the season, sowing and reaping his scanty crops where he may. In consequence the hill-sides are scored by tracks, not the roads of the plains, but paths suited to the conditions. Three types of these, M. Biermann points out, can be recognised:—the zig-zag mule track or mere path, used by pack mule or human porter; the straighter, steeper track, traversed by the winter sledges on the upward journey; and the very steep and straight short cut by which the descent is made. In addition, as means of communication to the geographer, must be recognised the wood slides, the small irrigating channels (whose banks, by the way, are often used by lightly loaded persons as short cuts), and finally the erratic paths marked out by the migrating beasts.

Even so in most cases the total mileage of road and track in the mountain is small in proportion to the area, but this, says the author, is to be ascribed to the usual poverty and scanty population of the area, not to the difficulty of construction. As an example he takes the region round Montreux, which has always been populous because the local climate is especially suited for the vine, as well as for fruit-trees and maize, and which has in recent years become one of the great centres of the tourist industry. Here three types of mountain road exist, representing as many periods of social evolution. The first, the ancient road, is straight and steep, sometimes paved, sometimes covered with sand, interrupted at intervals by logs of wood which both diminish the slope by forming steps, and also throw the drainage water to the side. These roads were meant only for mules or pedestrians, and are sometimes replaced by genuine staircases built of stone. Second come the nineteenth-century roads, which zig-zag up the slopes, but are too steep for anything but light, wheeled vehicles. The twentieth century has found this method of transport too slow and costly, and has "corrected" the roads, giving them a wider sweep and gentler slopes, so that heavily loaded vehicles can be pulled up. The old roads are, however, still kept up and often serve for the down journey. The consequence of the co-existence of the three types has been to give the Montreux region a great wealth of roads, a total of some sixty-two miles for an area of twelve and a half square miles.

The railways have followed a somewhat similar process of evolution in their constantly increasing sacrifice of directness. The first was the Territet-Glion funicular, as straight as one of the primitive mule tracks. But the cable in such cases cannot be indefinitely extended—the principle is only available over short distances. Thus the next railway, that from Glion to Naye, is on the rack and pinion principle, and is zig-zag, not direct. Finally, in the last route, that from Montreux to the Oberland, the rack and pinion, which greatly diminishes speed, has been abandoned, and the slope is circumvented by huge loops, which increase the distance sevenfold. More curious still is the construction of a "corrected" line from Montreux to Glion, which lies alongside the funicular but is four and a half times longer and permits of the transport of merchandise. The total result is to give the Montreux district ten thousand yards (or not far from two miles) of rail to every square mile of area. The conclusion is that in connection with means of communication, as with some other geographical facts, the activity displayed by man is not in direct, but in inverse ratio to the facilities given by nature—the more difficult it is to make roads, the more roads are built.

Tropical Diseases and the Colonisation of Tropical Lands.—Dr. M. Alsberg contributes to a recent issue of the *Geographische Zeitschrift* an article on sanitation in tropical lands, which, although it does not contain anything new, is of interest as a concise summary of recent work. He points out that till recently it was universally assumed by medical men that the tropical climate had a directly injurious effect on the white man. So much was this the case that such diseases as "tropical heart," "tropical liver," and so forth, were described. Recently, however, with the discovery of the organisms associated with malaria in all its forms, with sleeping sickness, etc., this tendency has been entirely reversed, and tropical diseases in general are regarded as having specific causes, like many of the diseases of temperate lands. The problem which remains, and it is one which Dr. Alsberg scarcely considers, is—if such diseases can be eliminated, and the gross tropical death rate thus greatly reduced, will Europeans be able to retain permanently their efficiency in tropical lands?

Dr. Alsberg limits himself to three great tropical diseases—malaria, with blackwater fever, etc., now known to be disseminated by the mosquito *Anopheles*; sleeping sickness, associated chiefly with the tsetse called *Glossina palpalis*, though apparently with other species also; yellow fever, associated with the mosquito *Stegomyia*. As far as the first and the last kinds of insect are concerned, there is little doubt that great reduction in the total number can be accomplished with comparative ease, and that an associated reduction of the two types of disease is merely a matter of an efficient sanitary service. In regard to sleeping sickness the matter is more difficult, and Dr. Alsberg seems to take rather a more optimistic view of the beneficial effect of clearing away the bush from the streams, and thus of destroying the lurking-places of the tsetse, than is justified by experience. Further, he does not touch upon the difficult question of how the numbers of *Glossina morsitans* are to be reduced if, as now seems certain (see p. 91), this fly also can serve as a carrier of the disease. He makes, however, the rather curious suggestion that the European Powers interested in tropical Africa should combine to limit native trading, and thus diminish the risk of transference of disease from one area to another, as well as checking the trade in strong drink and in gunpowder.

Programme of the British Association.—It is announced that Dr. H. N. Dickson's presidential address to Section E at the Birmingham meeting of the British Association will deal with the increasing recognition of the importance of human geography in the study of social and economic questions. In regard to the programme of the Section, we note first that there are to be two discussions. One, to be held jointly with Section A, is to deal with mathematical geography; the other is to be on the natural regions of the world, and is to be opened by Prof. A. J. Herbertson of Oxford. It is expected that a number of geographers will take part in this latter discussion, which is on a subject to which much attention is being given at the present time. In connection with the mathematical discussion, Capt. Winterbotham is expected to read a paper on the accuracy of the principal triangulation of Great Britain. Among other papers promised are one on Australia by Prof. J. W. Gregory, and one by Mr. I. N. Dracopoli on his recent travels in Jubaland, British East Africa. In the Economic Section Prof. A. W. Kirkaldy is to speak on the economic effects of the opening of the Panama Canal, while in the Zoological Section Prof. Minchin is to lecture on the sleeping sickness problem, both subjects which interest the geographer greatly.

There seems to be at present a prospect that the meeting will prove a great success, and it is hoped by the local committee that the attendance will reach 3000. A very comprehensive programme of excursions has been arranged.

Commercial Geography.

Suggested Cross-Channel Train Ferry.—Recently the proposal to construct a tunnel across the English Channel has been revived, and the Prime Minister, in answer to a deputation, stated a short time ago that the subject was receiving the attention of the Government. It seems probable, however, from the tone of his reply that the proposal, as before, will be rejected on strategical grounds. In these circumstances other proposals for establishing continuous rail communication between England and France have been receiving attention. One of these is the establishment of a train ferry, and the *Times* has an interesting article on the various train ferries at present in action. None now exists in the United Kingdom, but it will be recalled that till the opening of the Forth Bridge a train ferry existed between Burntisland and Granton, and was not only established at a small cost, but effected a considerable saving of time as compared with ship transport. At present in Europe the most interesting train ferries are those which exist in connection with the Danish State railways. Among the more important of these are that from Korsoer to Nyborg across the Storabelt, at a point where the water is sixteen miles wide; that from Copenhagen to Malmo, a distance of nineteen miles; and that from Warnemunde to Gjedser, a distance of twenty-six miles.

It is this last crossing which offers most analogy to the conditions which would prevail on a cross-channel ferry. It crosses the Baltic at a point where rough water is not infrequent, it connects two different countries, and is part of an international line of communication. The conditions are, however, somewhat simpler in that only a small tidal range has to be allowed for. Of more recent type than the Danish ferries are those which connect the German and Swedish State railways, crossing the Baltic from Trelleborg to Sassnitz. This service was opened in 1909, and has proved a great success; the distance of fifty-eight miles is covered in three and a half hours, and the boats are large and well-appointed.

In America, especially in the region of the Great Lakes, train ferries exist on a much larger scale, and have been found economical and capable of being worked without special difficulties. In some of the American ferries, *e.g.* in New York Bay, considerable tidal range has to be allowed for.

NEW BOOKS.

EUROPE.

Switzerland and the Adjacent Portions of Italy, Savoy, and Tyrol. Handbook for Travellers. By KARL BAEDEKER. Twenty-fifth Edition. London: T. Fisher Unwin, 1913. *Price* 8*s*.

The superior person is fond of asserting that Switzerland has become so vulgarised as to be impossible, but the fact that the twenty-fifth edition of Baedeker's Guide has just appeared shows that the superior person does not count for very much. Indeed, as one turns over the pages and maps of the book, it is difficult to avoid feeling that no other region of the Alps can vie with Switzerland in the splendour of its scenery. Tyrol, it is true, is simpler, more unspoiled, and certainly cheaper, but its mountains for the most part shrink into insignificance beside the central Alps, and the colouring of its strange Dolomites perhaps hardly compensates for the absence of the magnificent glaciers which encircle places like Zermatt.

The new edition of Baedeker is, as usual, thoroughly up to date. We note, for instance, that the Lotschberg line takes its place without remark among the other mountain railways, as if it had existed since time was, instead of being opened, so to speak, yesterday. It is stated that sleeping-cars are run on the night trains, though we believe that, in point of fact, great difficulties have been experienced in running a night service through the tunnel, difficulties connected especially with the electric traction. The recently opened line from Bevers to Schuls-Tarasp is also described, and generally all notable changes are recorded. The book, in short, shows the usual combination of accuracy and completeness.

An Elementary Historical Geography of the British Isles. By M. S. ELLIOTT, B.A. London: Adam and Charles Black, 1913. *Price* 1*s.* 6*d.*

This is an interesting little book, intended to be used by such pupils in secondary schools as have already an elementary acquaintance with British history and geography. It is very fully illustrated, some of the little sketch-maps being especially attractive. In the preface the author states that historical geography as a subject gives great opportunity for teaching sympathy with other nations and other types of civilisation, and the book seems admirably fitted for this purpose in the way it describes the interrelations of the nations in times of peace—an admirable contrast with that excessive stress upon war which used to be the great blemish of school histories. We recommend the book to the notice of teachers.

The Story of Belfast and its Surroundings. By MARY LOWRY. London: Headley Brothers, 1913. *Price* 7*s.* 6*d. net.*

This is a delightful book, which appears at an opportune moment, and will appeal to a much wider circle than those for whom the talented author says it is written, viz. the rising generation of Belfast. To them it will have an irresistible fascination, but it will also attract every student of the history of what is, at least for the time, the most interesting part of an interesting country. The author has imbued herself with the spirit of the locality, and has made a patient and exhaustive study of its history, municipal, political, and ecclesiastical, from which she has chosen a series of most interesting and amusing incidents, told in crisp, fresh, attractive language. She had, of course, an *embarras de richesse* in the shape of materials from which to chose, as the city of Belfast is first mentioned so long ago as the seventh century, although it was not till the twelfth century that it came into prominence. Early in the sixteenth century a Lord Mayor was appointed with the grandiloquent title of the Sovereign; one of his duties was "to attend the Corporation Church every Sunday followed by the burgesses and freemen in state, and every person over thirteen years of age was obliged also to go to church under a penalty of being fined in the sum of sixpence or up to five shillings, which was a serious matter in 1632." His successors undertook other work, as appears from an advertisement in 1767, which announces that "The Company at the Mill Gate will give a benefit to the Poor. Pit and Gallery, 2s. 6d. each. The Sovereign will attend to take the tickets." The condition of the postal arrangements may be gathered from a notice put up by the postmaster in 1795, "that merchants were requested not to send little boys for letters, but to send proper persons, as little boys behaved very badly and gave great annoyance to the postmaster, who could not be accountable for their conduct." Sir Walter Raleigh, it is true, introduced tobacco into England in 1585, but smoking is shown to have been common in Ireland at least some centuries before that date, for "there is a recumbent figure on a monument erected to the memory of Donogh O'Brien, king of Thomond, who was killed in the year 1227, and was buried in the Abbey. He is represented lying in the usual position, with the short pipe or 'dudeen' of the Irish in his mouth." In 1800 "two masons struck work and were sent to gaol for three months; and six shoemakers combined to have their wages raised, and they were at once sent to Carrickfergus Gaol, the judge remarking, 'How could trade go on, or trade improve, if such actions were permitted?'" The penalty for breaking or extinguishing a street lamp was six months' imprisonment, and for stealing any part of a lamp transportation for seven years, or public whipping. The book is a delightful treasury of quaint archæological lore, of which we refer to only one more incident, viz., the sacrilegious act of one James Cleland, who, not quite a century ago, introduced six snakes into St. Patrick's country, but happily the shameful crime was abortive. The second half of the book treats in the same way Armagh, Carrickfergus, Bangor, and several other well-known places in the vicinity of Belfast, and is quite as interesting as the first half. It is needless to add that the element of romance is not wanting throughout the book. There are also a great many illustrations from old pictures and prints which are not easily accessible to the ordinary reader. We heartily commend this book to the perusal of our readers.

The Tourists' Russia. By RUTH KEDZIE WOOD. London: Andrew Melrose, 1912. *Price* 6*s. net.*

This book seems to have been written for American as well as British tourists. It contains a series of very brief sketches of the principal cities and places in Russia, with information as to routes, cost of travel, objects to see, and the like; in brief, a modified Baedeker.

ASIA.

Indian Pages and Pictures. By MICHAEL MYERS SHOEMAKER. New York: E. P. Putnam's Sons; and London: The Knickerbocker Press, 1912. *Price* 10*s.* 6*d. net.*

Mr. Shoemaker is a voluminous writer, and has published many books. In this last one he takes his readers to India, including Kashmir, with regard to which, however, it has to be confessed he has little or nothing fresh to say, fresh at least to British readers. Apparently he is a man of considerable means who is able to tour about in great comfort, and having good introductions he saw to great advantage whatever there is to be seen. His experience as a writer stands him in good stead, and he is never dull. The book is profusely illustrated with photographs of average merit, most of them taken by members of the party.

The Fighting Spirit of Japan. By E. J. HARRISON. London: T. Fisher Unwin. *Price* 12*s.* 6*d. net.*

Mr. Harrison's aim in this book is to impart to his readers something of the history and rationale of *Judo*, more familiarly known as Jiujitsu, and he states that he, together with one other foreigner in Japan, has the distinction of holding the rank of *Shodan*, the highest as yet attained by any alien. Mr. Harrison thus brings to his task, not only keen enthusiasm for his subject, but also special knowledge, and the result is one of the most interesting books on Japan that it has recently fallen to our lot to read. Apparently a bout of Judo, as it is practised by the devotees of the cult in Tokio, imposes the severest demands upon the nervous and physical forces of any form of athletics, and we can recommend any one who cares to study the subject to turn to Mr. Harrison's pages for information. There will be found a very full account of the practice of Judo by the best Japanese experts. We may mention, as a small sample of the contents of the book, the chapter on the art of strangulation, *i.e.* of choking an adversary into a state of insensibility. Arguing from his own experience of being choked in this way, Mr. Harrison concludes that the actual physical sensations felt by the victims of capital punishment by hanging are, on the whole, rather pleasant than otherwise; which is a comforting doctrine for the humanitarian! We find a good deal about the rôle played by transcendentalism in fighting, and, in connection with this part of his subject, the author tells some rather tall stories; as, for instance, that the Japanese Bujin, or fighting man of feudal days, had the power of doing three things. First, he could render himself invisible to his adversary at the moment of combat; second, arrest his opponent's weapon in mid-air; and, third, convert the latter from rage into a condition of fatuous good humour. Some of this savours of the claims which the Chinese Boxers made to miraculous powers, and, probably, in common with most of the arts of Japan, it comes from a similar Chinese origin.

Mr. Harrison tells a good deal about the esoteric aspect of Judo, and he admits that he is himself a convert to occultism. Otherwise one may fairly suppose that he is sometimes laughing in his sleeve at the expense of his readers. The practice of deep-breathing and the development of the abdominal muscles are inculcated, and the notes on these aspects of Judo will be found to be of real value. The rest of the book is made up of chapters on the Japanese equivalent of the Indian Yogi, on the Stage, and other matters. The photographic illustrations are excellent.

How England saved China. By J. MACGOWAN. London: T. Fisher Unwin, 1913. *Price 10s. 6d. net.*

This book describes the labour of English missionaries, especially medical missionaries, in China, and the effect of their work in suppressing infanticide, foot-binding, and other barbarous customs. It is written with much fervour of conviction, and contains many photographs illustrating the habits and customs of the people.

GENERAL.

Das Gesetz der Wüstenbildung in Gegenwart und Vorzeit. Von JOHANNES WALTHER. Leipzig: Quelle u. Meyer, 1912. *Preis 12 marks.*

Professor Walther tells us in his preface that his interest in deserts dates back for twenty-five years, and began with his study of those barren rocks, transitional in date between Palaeozoic and Mesozoic time, which point to desert conditions in the Germany of an earlier period. The first edition of his book, based on desert journeyings in Africa and elsewhere, appeared in 1900, and gave rise to much discussion and some criticism. His desire to reinvestigate certain of the problems connected with deserts led to the planning of a new journey to Egypt and the vicinity, but this plan was not carried out till 1911, and upon that journey this new edition is largely based. It is not a general treatise on deserts in the wide sense, for the author expressly limits himself to the phenomena which he has studied personally, but his observations have been sufficiently widespread to give his book considerable breadth of treatment. His special interest, as indicated above, is in the problem of the extent to which existing deserts throw light upon the deserts of earlier periods of the earth's history. The book, which has been divided up into a greater number of chapters than the previous issue, consists of four sections: the general characters of deserts; erosion in deserts; deposition; deserts in geological time. It is illustrated by a fine series of photographs, taken from very varied parts of the world, and will be of great value to the physical geographer.

Die Erklärende Beschreibung der Landformen. Von WILLIAM MORRIS DAVIS. Deutsch bearbeitet von Dr. A. RÜHL. Leipzig u. Berlin: B. G. Teubner, 1912. *Preis* 11 *Marks.*

In the winter of 1908-9 Professor Davis exchanged professorships with Professor Penck of Berlin, and the present volume is based upon the course of lectures which he delivered at that university. As the title indicates, the book is an elaboration of the thesis, put forward by Professor Davis on a number of occasions, that it is possible to devise a nomenclature for land forms which shall indicate their origin as well as their present characters. This of course involves the further assumption that the agents which model the surface act in a certain limited number of ways, and that any particular land form can be classified as occupying a particular position in a given "cycle of erosion." Though a certain number of the terms introduced or employed by Professor Davis have already won acceptance, many geographers shrink from the terminology in its entirety as having in it something too simple and mechanical for the complexity of nature, and as not being without disadvantages and risks. Both those who are fully and those who are only partially sympathetic will welcome in this book a full exposition of the scheme.

Further, the student will find in Professor Davis's book an instructive account of certain aspects of physical geography, illustrated by examples taken from very varied parts of the earth's surface.

A Handbook of Geography. By A. J. HERBERTSON, M.A., Ph.D. Vol. I., *General Geography, The British Isles and Europe*; Vol. II., *Asia, Australasia, Africa, and America.* London and Edinburgh: Thomas Nelson and Sons, 1911 and 1913. *Price* 4*s.* 6*d. net, each vol.*

This comprehensive and well-illustrated book is stated in the preface to be written for students and teachers. It includes a somewhat full treatment of the general principles of geography, including physical geography in the narrower sense, biogeography, economic geography, and so forth, followed by detailed discussions of the different regions of the earth. In the regional sections the physical feature of the continents are first fully discussed, and then the political divisions are described. This is of course progress, but we feel that the problem as to how a logical division into natural regions can be reconciled with the utilitarian need of considering political divisions will have to be squarely faced, sooner or later, by geographers. Mr. Dryer, in a text-book recently reviewed here, adopts what seems to us a more logical though less utilitarian standpoint. Some position between his and Dr. Herbertson's will, we think, have to be found, for the student of the latter's book will be disposed to ask why he should trouble, for example, to learn about Laurentian Uplands and Appalachian Highlands, if he is to pass on next to a consideration of provinces and states on traditional lines without reference to such natural regions.

Dr. Herbertson states in his preface that his book was written some time ago, but has been delayed in the press, and has been brought up to date by a pupil. We are sorry, however, to be obliged to indicate that the revision is often partial and imperfect, and in a number of cases seems to consist in the addition of new figures without removal or alteration of those previously present. In the section relating to Canada and the States this is specially noticeable, owing to the recent censuses, and in the former case to recent modification of provincial boundaries. Thus we note that three different figures are given for the population of Canada, two dated and one undated. Two figures for area are given, with a very considerable difference between them, and the indications as to the areas of the provinces given in the text do not agree with those given in the table on page 496. The census figures of the United States for 1900 are given in two different forms on two pages facing one another (pp. 458 and 459), while still another figure for the same census is given in the text on page 512.

Scientific Papers. By J. Y. BUCHANAN, M.A., F.R.S. Vol. I. Cambridge: At the University Press, 1913. *Price* 10*s*. 6*d*. *net.*

Mr. Buchanan reprints here, without alteration, the papers on Oceanographical subjects which he has contributed to various journals, including the *Scottish Geographical Magazine.* Some of these papers deal with matters of great interest, and many will be glad to have them collected together in a convenient form.

The Year-Book of the Scientific and Learned Societies of Great Britain and Ireland. Compiled from official sources. Twenty-ninth Annual Issue. London: Charles Griffiths and Co., 1912. *Price* 7*s*. 6*d*. *net.*

This very useful compilation gives as usual full accounts of the activities of the various learned societies and bodies, with detailed records of the work accomplished during the session 1911-12.

BOOKS RECEIVED.

La Scandinavie (Norvège-Suède-Danemark-Islande). Par CHARLES RABOT. Crown Folio. Pp. 68. Illustrated. Paris: Bong et Cie, 1913.

Borrowstounness and District: being Historical Sketches of Kinneil, Carriden, and Bo'ness, 1550-1850. By THOMAS J. SALMON. With Illustrations and Maps. Demy 8vo. Pp. x + 476. *Price* 6*s*. *net.* Edinburgh: Wm. Hodge and Co., 1913.

Bennett's Norway: with Through Routes to Sweden and Denmark. Thirtieth edition, revised. With Route Map and four Plans. Crown 8vo. Pp. 256. *Price* 2*s*. 6*d*. London: Simpkin, Marshall and Co., 1913.

Animal Geography: the Faunas of the Natural Regions of the Globe. By MARION I. NEWBIGIN, D.Sc. (The Oxford Geographies.) Crown 8vo. Pp. 238. *Price* 4*s*. 6*d*. Oxford: Clarendon Press, 1913.

The Land of Footprints: Adventures in East Africa. By S. E. WHITE, F.R.G.S. Crown 8vo. Pp. 462. *Price* 2*s*. *net.* Edinburgh: T. Nelson and Sons, 1913.

The Panama Canal: A History and Description of the Enterprise. By J. SAXON MILLS, M.A. With Map and Illustrations. Crown 8vo. Pp. 344. *Price* 2*s*. *net.* Edinburgh: T. Nelson and Sons, 1913.

The Southland of North America: Rambles and Observations in Central America during the Year 1912. By GEORGE PALMER PUTNAM. With 96 Illustrations from Photographs by the Author, and a Map. Demy 8vo. Pp. xiv + 425. *Price* 10*s*. 6*d*. *net.* London: G. P. Putnam's Sons, 1913.

Switzerland and the Adjacent Portions of Italy, Savoy and Tyrol. Handbook for Travellers. By KARL BAEDEKER. With 77 Maps, 21 Plans, and 14 Panoramas.

Crown 8vo. Pp. xl + 604. *Price* 8*s*. *net*. Twenty-fifth Edition. Leipzig: Karl Baedeker, 1913.

Trails and Tramps in Alaska and Newfoundland. By WILLIAM S. THOMAS. With 147 Illustrations from Original Photographs. Demy 8vo. Pp. xvi + 330. *Price* 7*s*. 6*d*. *net*. London: G. P. Putnam's Sons, 1913.

The Atlas Geographies: A New Visual Atlas and Geography Combined. Part I. *Physical Geography*. By THOMAS FRANKLIN and E. D. GRIFFITHS, B.Sc., F.R.G.S., F.C.S. Crown 4to. Pp. vi + 88. *Price* 1*s*. 6*d*. *net*. Edinburgh W. and A. K. Johnston, Ltd., 1913.

La Libia. By Prof. GIUSEPPE RICCHIERI. Con Illustrazioni, due carte geografiche in nero e due a colori. Crown 8vo. Pp. 144. Milan: Federazione Italiana delle Biblioteche Popolari, 1913. *Price L*. 1.50.

The Cathedrals of Southern France. By T. FRANCIS BUMPUS. Crown 8vo. Pp. viii + 224. *Price* 6*s*. *net*. London: T. Werner Laurie, 1913.

Bradshaw's Through Routes to the Chief Cities and Bathing and Health Resorts of the World. Edited by EUSTACE REYNOLDS BALL, F.R.G.S. Maps, Plans, and Vocabularies. Pp. lxviii + 708. *Price* 6*s*. *net*. London: H. Blacklock and Co., Ltd., 1913.

From a Punjab Pomegranate Grove. By C. C. DYSON. With 14 Illustrations. Demy 8vo. Pp. vii + 289. *Price* 10*s*. 6*d*. *net*. London: Mills and Boon, Ltd., 1913.

Résultats du Voyage du S.Y. Belgica en 1897-1898-1899, sous le commandement de A. De Gerlache De Gomery. Rapports Scientifiques. 2 Parts. Anvers, 1912.

General Description of the Argentine Republic. Demy 8vo. Pp. 66. Buenos Aires, 1913.

Allgemeine Verkehrsgeographie. Von Dr. KURT HASSERT, Professor der Geographie an der Handels-Hochschule, Köln. Mit 12 Karten und graphischen Darstellungen. 8vo. Pp. viii + 494. *Preis M*. 10. Leipzig: G. J. Göschen, 1913.

Deuxième Expédition Antarctique Française (1908-1910), commandée par le Dr. Jean Charcot. Sciences Naturelles: Documents Scientifiques. 2 Parts. Large 4to. Paris: Masson et Cie, 1913.

Aux Confins de l'Europe et de l'Asie. Par MAURICE RONDET-SAINT. Préface de M. ANDRÉ LEBON. Deuxième édition; un volume en-16. Pp. vi + 334. *Prix* 3 *fr*. 50. Paris: Plon Nourrit et Cie, 1913.

Die Cordillerenstaaten. Von Dr. WILHELM SIEVERS. Erster Band: Bolivia und Peru. Zweiter Band: Ecuador, Colombia, und Venezuela. 12 mo. Pp. 148 und 128. *Preis* 90 *Pfennig*. Berlin: G. J. Göschen, 1913.

The Geography of South Australia: Including the Northern Territory: Historical, Physical, Political, and Commercial. By WALTER HOWCHIN. With an Introduction by Prof. J. W. GREGORY, D.Sc., F.R.S. Crown 8vo. Pp. 320. *Price* 3*s*. 6*d*. Melbourne: Whitcombe and Tombs, Ltd., 1913.

The Geography of Victoria: Historical, Physical, and Commercial. By J. W. GREGORY, D.Sc., F.R.S. New and revised edition. Crown 8vo. Pp. 306. *Price* 3*s*. 6*d*. Melbourne: Whitcombe and Tombs, Ltd., 1913.

Making the Most of the Land. By JAMES LONG. Crown 8vo. Pp. xiv + 282. *Price* 5*s*. *net*. London: Hodder and Stoughton, 1913.

Plant Life. By J. BRETLAND FARMER, M.A., D.Sc., F.R.S. 12mo. Pp. viii + 255. *Price* 1*s*. *net*. London: Williams and Norgate, 1913.

Burma Gazetteers. 12 Vols. Rangoon, 1913.

Census of India, 1911. Vol. VI., City of Calcutta. Part I. Report; Part II. Tables. Vol. IV. Baluchistan. Part I. Report; Part II. Tables. Calcutta, 1913. Vol. XX., Kashmir. Part I. Report; Part II. Tables. Lucknow, 1912.

The Currents in the Entrance to the St. Lawrence. From Investigations of the Tidal and Current Survey in the Seasons of 1895, 1911, and 1912. Pp. 50. Ottawa : Department of the Naval Service, 1913.

Chile of To-day: Its Commerce, its Production, and its Resources. By A. Ortúzar. Royal 8vo. Pp. 525. New York: The Tribune Association, 1907.

Diplomatic and Consular Reports.—Trade, etc., of Chicago, 1912 (5095); Trade, etc., of the Netherlands, 1912 (5096); Trade, etc., of Vera Cruz, 1912 (5100); Report on the Swedish Budget, 1914 (5101); Trade of Corea, 1912 (5104); Trade, etc., of Bosnia-Herzegovina, 1912 (5067); Trade, etc., of Alexandria, 1912 (5097); Trade of Genoa, 1912 (5098); Trade, etc., of Roumania, 1911-12 (5102); Trade of Baltimore, 1912 (5103); Trade, etc., of Azores, 1912 (5105); Trade, etc., of Sicily, 1912 (5106); Trade, etc., of Jerusalem, 1912 (5107); Trade, etc., of Lyons, 1912 (5108); Trade, etc., of Corunna, 1912 (5109); Trade of Corsica, 1912 (5111); Trade, etc., of Bolivia, 1912 (5110); Trade of Antofagasta, 1912 (5112); Trade of Wuchow and Nanning, 1912 (5113); Trade of Odessa, 1912 (5114); Trade, etc., of Boston, 1912 (5116); Trade of Hankow, 1912 (5120); Trade, etc., of Dairen, 1912 (5121); Trade, etc., of Chungking, 1912 (5115); Trade of Wuhu, 1912 (5117); Trade of Changsha, 1912 (5119); Trade of Foochow, 1912 (5122); Trade of Harbin, 1912 (5123); Trade of Nanking, 1912 (5124); Trade, etc., of Portland, Oregon, 1912 (5118); Trade, etc., of Rio Grande, 1912 (5130); Trade of Calais, 1912 (5125); Trade of Chinkiang, 1912 (5126); Trade of Malaga, 1912 (5127); Trade of Teng Yueh, 1912 (5128); Trade, etc., of Peru, 1911-12 (5129); Trade, etc., of New York, 1912 (5131); Trade, etc., of Seville, 1912 (5132); Trade of Mengtsz, 1912 (5134); Trade, etc., of the Cyclades, 1912 (5133); Trade of Iquique, 1912 (5135); Trade of France, 1912 (5137); Trade of Lisbon, 1912 (5138); Trade of Réunion, 1912 (5139); Trade, etc., of Rouen, 1912 (5140); Trade of German South-West Africa, 1912 (5141); Trade, etc., of Portugal, 1912 (5142); Trade of Angola, 1912 (5143); Trade of Mannheim, 1912 (5144); Industries and Economic Condition of Turin, 1912 (5148); Budget for 1913 and Finances of Paraguay (5150); Trade, etc., of Porto Rico, 1912 (5154); Trade of Antioquia, 1912 (5146); Trade of Oporto, 1912 (5149); Trade of Hakodate, 1912 (5151); Trade, etc., of Madeira, 1912 (5152); Trade, etc., of the Society Islands, 1912 (5153); Trade, etc., of Tampico, 1912 (5155); Trade of Rome, 1912 (5157); Trade of Salina Cruz, 1912 (5158); Trade of Erzeroum, 1912 (5159); Trade of the State of São Paulo, 1912 (5160); Trade of Basra, 1912 (5168); Trade, etc., of Costa Rica, 1912 (5147); Trade of Venezuela and Caracas, 1911-12 (5156); Trade of Japan, 1912 (5161); Finances of Italy, 1912 (5162); Trade of Abyssinia, 1911-12 (5163); Trade, etc., of Stettin, 1912 (5164); Trade, etc., of the Vilayet of Trebrizond, 1912 (5166); Trade of the Aleppo Vilayet, 1912 (5167).

Publishers forwarding books for review will greatly oblige by marking the price *in clear figures, especially in the case of foreign books.*

NEW MAPS.

EUROPE.

Ordnance Survey of Scotland.—The following publications were issued from 1st to 31st May 1913 :—

Six-inch and Larger Scale Maps.—Six-inch Maps (Revised). Quarter Sheets with contours in blue. Price 1s. each. *Lanarkshire.*—23 SE., 26 SW. With contours in red. Price 1s. each. *Lanarkshire.*—15 SW.

1:2500 Scale Maps (Revised), with houses stippled, and with areas. Price 3s. each. *Dumbartonshire.*—Sheet xxv. 4. *Lanarkshire.*—Sheets xvi. 7, 11.

Note.—There are no coloured editions of these Sheets, and the unrevised impressions have been withdrawn from sale.

Special Enlargements from 1 : 2500 to 1 : 1250 Scale. Prepared for the Land Valuation Department, Inland Revenue. Price 2s. 6d. each. *Clackmannanshire.*—Sheets cxxxiii. 16 NE., SE.; cxxxiv. 5 SW., SE.; 13 NE.; cxl. 5 NW., NE., SW., SE. *Fifeshire.*—Sheets xxii. 12 SW., 15 NE., SW., SE.; xxiii. 9 NW., SW.; xxviii. 14 SE.; xxxvi. 1 SE. *Lanarkshire.*—Sheets v. 12 SE.; vi. 13 SW., SE.; x. 4 NE., SE. *Perthshire.*—Sheets cxxv. 13 SE.; cxxx. 1 NW., SW. *Renfrewshire.*—Sheets ix. 5 NW., SW. *Roxburghshire.*—Sheet xxv. 3 NE. *Stirlingshire.*—Sheets x. 11 NW., NE.; xviii. 13 NE.; xxiv. 1 NW., 4 NW.; 11 NW.; 12 SW.; 16 NW., NE., SW., SE.; xxvii. 6 NW., SW.; xxix. 4 NE., SW., SE.; 8 NW., SW.; xxx. 8 NE., SE.; 12 SW.; xxxi. 9 SE.; 13 SW., SE.

Publications issued from 1st to 30th June 1913 :—

One-inch Map; third edition. Printed in colours and folded in cover, or flat in sheets. Sheets 115, 116. Price on paper 1s. 6d. Mounted on linen, 2s. Mounted in sections, 2s. 6d. each. Printed in colours and folded in cover, or flat in sheets, Special Map of Aberdeen District. Price on paper 1s. 6d. Mounted on linen 2s. Mounted in sections 2s. 6d.

Six-inch and Larger Scale Maps :—

1 : 2500 Scale Maps (Revised), with houses stippled, and with areas. Price 3s. each. *Lanarkshire.*—Sheets ii. 10, 15; iii. 11, 16; iv. 13; viii. 4; ix. 1, 5. Sheet ix. 2. Price 1s. 6d.

Note.—There are no coloured editions of these Sheets, and the unrevised impressions have been withdrawn from sale.

Special Enlargements from 1 : 2500 to 1 : 1250 Scale. Prepared for the Land Valuation Department, Inland Revenue. Price 3s. each.* *Fifeshire.*—Sheets xxii. 12 NE., SE.; xxviii. 8 NW.; 11 SW.; xxxiv. 16 NW., NE., SW., SE.; xxxv. 12 NE., SW., SE.; xxxvi. 1 SW. *Stirlingshire.*—Sheets xxxi.—9 NW., NE.

* Note.—The price of all 1 : 1250 Scale Enlargements is raised from 2s. 6d. to 3s.

Publications issued from 1st to 31st July 1913 :—

Two-mile Map; printed in colours, and hills shown by layers, folded in cover, or flat in sheets. Sheet 12. Price, on paper 1s. 6d.; mounted on linen, 2s.; mounted in sections, 2s. 6d.

Six-inch and Larger Scale Maps.—Six-inch Maps (Revised). Quarter Sheets, with contours in blue. Price 1s. each. *Lanarkshire.*—24 NE.; 25 NE.; 26 NW.; 36 SE.; 39 SE.

1 : 2500 Scale Maps (Revised), with houses stippled and with areas. Price 3s. each. *Lanarkshire.*—Sheets ii. 12, 16; iii. 5, 6, 8, 9, 10, 13, 14, 15; xvi. 2. *Renfrewshire.*—Sheet xii. 10.

Note.—There are no coloured editions of these Sheets, and the unrevised impressions have been withdrawn from sale.

Special Enlargements from 1 : 2500 to 1 : 1250 Scale. Prepared for the Land Valuation Department, Inland Revenue. Price 3s. each. *Clackmannanshire.*—Sheets cxxxiii. 10 NE.; cxxxix. 2 NE.; 3 NW. *Renfrewshire.*—Sheets ii. 10 NW.

GEOLOGICAL SURVEY OF SCOTLAND.—One-inch Map. New Series, colour printed. Sheet 70. Minginish. 1913. Price 2s. 6d. The Solid shown in Colour. The Drift by symbols in Black.

Ordnance Survey Office, Southampton.

This interesting new sheet contains the Cuillin Hills.

ADMIRALTY CHARTS, SCOTLAND. New Editions.

Firth of Clyde, Western Approaches. 1913. No. 1577. Price 3s.

Firth of Forth, St. Abb's Head to Edinburgh. 1913. No. 114*a*. Price 3s.

River Tay. 1913. No. 1481. Price 3s.

North Sea. 1913. No. 2339. Price 3s.

St. Abb's Head to Aberdeen. 1913. No. 1407. Price 3s.

British Islands. 1913. No. 2. Price 3s. *Admiralty Office, London.*

GLASGOW DISTRICT.—Bartholomew's Motoring and Cycling Map. Reduced from the Ordnance Survey to scale of 2 miles to an inch. Showing roads and distances, with contour colouring. New Edition, 1913. Price 1s. net. Mounted on cloth in case. *John Bartholomew and Co., Edinburgh.*

BRITISH ISLES.—Bartholomew's Contour Motoring Map, showing the best touring roads, with heights and distances. Scale 16 miles to an inch. New edition, 1913. Price 5s. net. Mounted on cloth, dissected.

John Bartholomew and Co., Edinburgh.

NORTH YORKSHIRE.—Graded Road Map on scale of half-an-inch to a mile. Price 1s. net, folded in case. *Gall and Inglis, Edinburgh.*

The roads on this series are revised from special surveys by Mr. Harry R. G. Inglis, and appear to be carefully done.

ITALY.—Testo-Atlante delle Ferrovie e Tramvie Italiane. Leonida Leoni. Prefazione dell' Ing. Pietro Lanino. 1913. Prezzo L. 5.

Istituto Geografico de Agostini, Novara.

This useful little atlas contains provincial maps showing railways and tramways; there are also enlarged maps for the principal towns, together with statistical text.

ITALY.—Carta amministrativa stradale. Scala di 1 : 250.000.

Provincia di Alessandria, Prezzo L. 0.60
„ Milano, „ L. 0.50
„ Padova, „ L. 0.50
„ Torino, „ L. 1.20

Istituto Geografico de Agostini, Novara.

These maps are on a uniform scale of 4 miles to an inch. They are clear and well adapted for travelling uses. The hills are printed in brown, rivers in blue, and railways in black, whilst the character of the roads is distinguished by red and green lines.

AFRICA

SOUTHERN NIGERIA.—Scale 1 : 125,000, or about 2 miles to an inch. Sheets—Akande, Shaki-West. 1913. Price 1s. 6d. each.

Sold by W. and A. K. Johnston, Limited, Edinburgh and London.

KAMERUN.—Karte in 31 Blätter und 3 Ansatzstücken im Massstabe von 1:300,000, bearbeitet von Max Moisel. Im Auftrage und mit Unterstutzung des Reichskolonialamts. 1913. Blatt A 4, Tschad; Blatt B 3, Dikoa; Blatt C 3, Mubi. Price M. 2 each sheet. *Dietrich Reimer (Ernst Vohsen), Berlin.*

AMERICA.

SOUTHERN ALBERTA.—Map showing Disposition of Lands. 1913. Scale 1 : 792,000 or 12½ miles to an inch. J. E. Chalifour, Chief Geographer.
Department of the Interior, Ottawa.

NOVA SCOTIA AND NEW BRUNSWICK.—Standard Topographical Map. Scale 1 : 250,000 or about 4 miles to an inch. Moncton Street, 14 SW. J. E. Chalifour, Chief Geographer. *Department of the Interior, Ottawa.*

ONTARIO.—Topographic Map. Scale 1 : 63,360 or 1 mile to an inch. Sheets—37, St. Thomas; 41, Wallaceburg; 38, Strathroy. (Department of Militia and Defence, 1913.) Price 2s. net each sheet.
Geographical Section, General Staff, London.

UNITED STATES GEOLOGICAL SURVEY.—Geologic Atlas Folios. No. 186, Apishapa, Colorado; No. 183, Llano-Burnet, Texas; No. 184, Kenova, Kentucky. Issued 1912. George Otis Smith, Director.
United States Geological Survey, Washington, D.C.

AUSTRALIA.

NEW SOUTH WALES.—Agricultural Map. 1909.
Issued by the Immigration and Tourist Bureau, Sydney.

NEW SOUTH WALES.—Maps indicating Eastern, Central and Western Divisions, 1911, and Railways, 1911. *Department of Lands Office, Sydney.*

QUEENSLAND.—Geological Sketch Map showing Mineral Localities. Scale 40 miles to an inch. Prepared under the supervision of B. Dunstan, F.G.S., Government Geologist. *Geological Survey Office, Brisbane.*

QUEENSLAND.—Railway Map revised to April 1912.
Office of Government Railways, Brisbane.

SOUTH AUSTRALIA.—Southern Portion. 1912. Scale about 16 miles to an inch. Showing Government Lands. *Surveyor General's Office, Adelaide.*

TASMANIA.—Scale about 12 miles to an inch. 1913. Showing Government Lands. With inset showing Timber Areas.
Surveyor General's Office, Hobart.

VICTORIA.—Maps showing Water Supply and Agriculture, also Railways, Schools, and Temperature. Scale 22 miles to an inch.
Issued by the Intelligence Bureau, Melbourne.

SCHOOL WALL MAPS.

WALES.—Bacon's New Contour Map. Scale 1 : 278,784 or about 4½ miles to an inch. (In Welsh.) Edited by Professor Timothy Lewis, M.A. 1913. Price 7s. 6d. Mounted on cloth, rollers and varnished.
G. W. Bacon and Co., Ltd., London.

NEAR AND MIDDLE EAST.—Bacon's New Contour Map. Scale 1 : 6,000,000. With inset Map of Palestine on scale 1 : 685,000. Price 7s. 6d. Mounted on cloth, rollers and varnished. *G. W. Bacon and Co., Ltd., London.*

The above maps are produced in effective orographical colouring, and are admirably adapted for teaching purposes.

THE SCOTTISH

GEOGRAPHICAL

MAGAZINE.

THE WORLD'S RESOURCES AND THE DISTRIBUTION OF MANKIND.[1]

By Professor H. N. DICKSON, M.A., D.Sc.

SINCE the last meeting of this Section the tragic fate of Captain Scott's party, after its successful journey to the South Pole, has become known; and our hopes of welcoming a great leader, after great achievement, have been disappointed. There is no need to repeat here the narrative of events, or to dwell upon the lessons afforded by the skill, and resource, and heroic persistence, which endured to the end. All these have been, or will be, placed upon permanent record. But it is right that we should add our word of appreciation, and proffer our sympathy to those who have suffered loss. It is for us also to take note that this last of the great Antarctic expeditions has not merely reached the Pole, as another has done, but has added, to an extent that few successful exploratory undertakings have ever been able to do, to the sum of scientific geographical knowledge. As the materials secured are worked out it will, I believe, become more and more apparent that few of the physical and biological sciences have not received contributions, and important contributions, of new facts; and also that problems concerning the distribution of the different groups of phenomena and their action and reaction upon one another—the problems which are specially within the domain of the geographer—have not merely been extended in their scope but have been helped towards their solution.

The reaching of the two Poles of the earth brings to a close a long and brilliant chapter in the story of geographical exploration. There is still before us a vista of arduous research in geography, bewildering almost in its extent, in such a degree indeed that "the scope of geography"

[1] Presidential Address to Section E (Geography), at the Birmingham Meeting of the British Association, September 1913.

is in itself a subject of perennial interest But the days of great pioneer discoveries in topography have definitely drawn to their close. We know the size and shape of the earth, at least to a first approximation, and as the map fills up we know that there can be no new continents and no new oceans to discover, although all are still, in a sense, to conquer. Looking back, we find that the qualities of human enterprise and endurance have shown no change; we need no list of names to prove that they were alike in the days of the earliest explorations, of the discovery of the New World or of the sea route to India, of the "Principall Navigations," or of this final attainment of the Poles. The love of adventure and the gifts of courage and endurance have remained the same: the order of discovery has been determined rather by the play of imagination upon accumulated knowledge, suggesting new methods and developing appropriate inventions. Men have dared to do risky things with inadequate appliances, and in doing so have shown how the appliances may be improved and how new enterprises may become possible as well as old ones easier and safer. As we come to the end of these "great explorations," and are restricted more and more to investigations of a less striking sort, it is well to remember that in geography, as in all other sciences, research continues to make as great demands as ever upon those same qualities, and that the same recognition is due to those who continue in patient labour.

When we look into the future of geographical study, it appears that for some time to come we shall still be largely dependent upon work similar to that of the pioneer type to which I have referred, the work of perfecting the geographer's principal weapon, the map. There are many parts of the world about which we can say little except that we know they exist; even the topographical map, or the material for making it, is wanting; and of only a few regions are there really adequate distributional maps of any kind. These matters have been brought before this Section and discussed very fully in recent years, so I need say no more about them, except perhaps to express the hope and belief that the production of topographical maps of difficult regions may soon be greatly facilitated and accelerated with the help of the new art of flying.

I wish to-day rather to ask your attention for a short time to a phase of pioneer exploration which has excited an increasing amount of interest in recent years. Civilised man is, or ought to be, beginning to realise that in reducing more and more of the available surface of the earth to what he considers a habitable condition he is making so much progress, and making it so rapidly, that the problem of finding suitable accommodation for his increasing numbers must become urgent in a few generations. We are getting into the position of the merchant whose trade is constantly expanding and who foresees that his premises will shortly be too small for him. In our case removal to more commodious premises elsewhere seems impossible—we are not likely to find a means of migrating to another planet—so we are driven to consider means of rebuilding on the old site, and so making the best of what we have, that our business may not suffer.

In the type of civilisation with which we are most familiar there are two fundamental elements—supplies of food energy, and supplies of mechanical energy. Since at present, partly because of geographical conditions, these do not necessarily (or even in general) occur together, there is a third essential factor, the line of transport. It may be of interest to glance, in the cursory manner which is possible upon such occasions, at some geographical points concerning each of these factors, and to hazard some speculations as to the probable course of events in the future.

In his Presidential Address to the British Association at its meeting at Bristol in 1898, Sir William Crookes gave some valuable estimates of the world's supply of wheat, which, as he pointed out, is "the most sustaining food-grain of the great Caucasian race." Founding upon these estimates, he made a forecast of the relations between the probable rates of increase of supply and demand, and concluded that "Should all the wheat-growing countries add to their (producing) area to the utmost capacity, on the most careful calculation the yield would give us only an addition of some 100,000,000 acres, supplying, at the average world-yield of 12·7 bushels to the acre, 1,270,000,000 bushels, just enough to supply the increase of population among bread-eaters till the year 1931." The president then added, "Thirty years is but a day in the life of a nation. Those present who may attend the meeting of the British Association thirty years hence will judge how far my forecasts are justified."

Half the allotted span has now elapsed, and it may be useful to inquire how things are going. Fortunately this can be easily done, up to a certain point at any rate, by reference to a paper published recently by Dr. J. F. Unstead,[1] in which comparisons are given for the decades 1881-90, 1891-1900, and 1901-10. Dr. Unstead shows that the total wheat harvest for the world may be estimated at 2258 million bushels for the first of these periods, 2575 million for the second, and 3233 million for the third, increases of 14 per cent. and 25 per cent. respectively. He points out that the increases were due "mainly to an increased acreage," the areas being 192, 211, and 242 million acres, but also "to some extent (about 8 per cent.) to an increased average yield per acre, for while in the first two periods this was 12 bushels, in the third period it rose to 13 bushels per acre."

If we take the period 1891-1900, as nearly corresponding to Sir William Crookes' initial date, we find that the succeeding period shows an increase of 658 million bushels, or about half the estimated increase required by 1931, and that attained chiefly by "increased acreage."

But signs are not wanting that increase in this way will not go on indefinitely. We note (also from Dr. Unstead's paper) that in the two later periods the percentage of total wheat produced which was exported from the United States fell from 32 to 19, the yield per acre showing an increase meanwhile to 14 bushels. In the Russian Empire the percentage fell from 26 to 23, and only in the youngest of the new countries—Canada, Australia, and the Argentine—do we find large proportional increases. Again, it is significant that in the United Kingdom, which is, and always has been, the most sensitive of all wheat-producing

[1] *Geographical Journal*, August and September 1913.

countries to variations in the floating supply, the *rate* of falling-off of home production shows marked if irregular diminution.

Looking at it in another way, we find (still from Dr. Unstead's figures) that the total amount sent out by the great exporting countries averaged, in 1881-90 295 million bushels, 1891-1900 402 million, 1901-10 532 million. These quantities represent respectively 13·0, 15·6, and 16·1 per cent. of the total production, and it would appear that the percentage available for export from these regions is, for the time at least, approaching its limit, *i.e.* that only about one-sixth of the wheat produced is available from surpluses in the regions of production for making good deficiencies elsewhere.

There is, on the other hand, abundant evidence that improved agriculture is beginning to raise the yield per acre over a large part of the producing area. Between the periods 1881-90 and 1901-10 the average in the United States rose from 12 to 14 bushels; in Russia from 8 to 10; in Australia from 8 to 10. It is likely that, in these last two cases at least, a part of the increase is due merely to more active occupatiou of fresh lands as well as to the use of more suitable varieties of seed, and the effect of improvements in methods of cultivation alone is more apparent in the older countries. During the same period the average yield increased in the United Kingdom from 28 to 32 bushels, in France from 17 to 20, Holland 27 to 33, Belgium 30 to 35, and it is most marked in the German Empire, for which the figures are 19 and 29.

In another important paper[1] Dr. Unstead has shown that the production of wheat in North America may still, in all likelihood, be very largely increased by merely increasing the area under cultivation, and the reasoning by which he justifies this conclusion certainly holds good over large districts elsewhere. It is of course impossible, in the present crude state of our knowledge of our own planet, to form any accurate estimate of the area which may, by the use of suitable seeds or otherwise, become available for extensive cultivation. But I think it is clear that the available proportion of the total supply from "extensive" sources has reached, or almost reached, its maximum, and that we must depend more and more upon intensive farming, with its greater demands for labour.

The average total area under wheat is estimated by Dr. Unstead as 192 million acres for 1881-90, 211 million acres for 1891-1900, and 242 million acres for 1901-10. Making the guess, for we can make nothing better, that this area may be increased to 300 million acres, and that under ordinary agriculture the average yield may eventually be increased to 20 bushels over the whole, we get an average harvest of 6000 million bushels of wheat. The average wheat-eater consumes, according to Sir William Crookes' figures, about four and a half bushels per annum; but the amount tends to increase. It is as much (according to Dr. Unstead) as six bushels in the United Kingdom and eight bushels in France. Let us take the British figure, and it appears that on a liberal estimate

[1] *Geographical Journal*, April and May 1912.

the earth may in the end be able to feed permanently 1000 million wheat-eaters. If prophecies based on population statistics are trustworthy, the crisis will be upon us before the end of this century. After that we must either depend upon some substitute to reduce the consumption per head of the staple foodstuff, or we must take to intensive farming of the most strenuous sort, absorbing enormous quantities of labour and introducing, sooner or later, serious difficulties connected with plant-food. We leave the possibility of diminishing the rate of increase in the number of bread-eaters out of account.

We gather, then, that the estimates formed in 1898 are in the main correct, and the wheat problem must become one of urgency at no distant date, although actual shortage of food is a long way off. What is of more immediate significance to the geographer is the element of change, of return to earlier conditions, which is emerging even at the present time. If we admit, as I think we must do, that the days of increase of extensive farming on new land are drawing to a close, then we admit that the assignment of special areas for the production of the food-supply of other distant areas is also coming to its end. The opening up of such areas, in which a sparse population produces food in quantities largely in excess of its own needs, has been the characteristic of our time, but it must give place to a more uniform distribution of things, tending always to the condition of a moderately dense population, more uniformly distributed over large areas, capable of providing the increased labour necessary for the higher type of cultivation, and self-supporting in respect of grain-food at least. We observe in passing that the colonial system of our time only became possible on the large scale with the invention of the steam locomotive, and that the introduction of railway systems in the appropriate regions, and the first tapping of nearly all such regions on the globe, has taken less than a century.

Concentration in special areas of settlement, formerly chiefly effected for military reasons, has in modern times been determined more and more by the distribution of supplies of energy. The position of the manufacturing district is primarily determined by the supply of coal. Other forms of energy are, no doubt, available, but, as Sir William Ramsay showed in his Presidential Address at the Portsmouth meeting in 1911, we must in all probability look to coal as being the chief permanent source.

In the early days of manufacturing industries the main difficulties arose from defective land transport. The first growth of the industrial system, therefore, took place where sea transport was relatively easy; raw material produced in a region near a coast was carried to a coalfield also near a coast, just as in the times when military power was chiefly a matter of "natural defences," the centre of power and the food-producing colony had to be mutually accessible. Hence the Atlantic took the place of the Mediterranean, Great Britain eventually succeeded Rome, and eastern North America became the counterpart of Northern Africa. It is to this, perhaps more than to anything else, that we in Britain owe our tremendous start amongst the industrial nations, and we observe that we used it to provide less favoured nations with the means of improving their system of land transport, as well as actually to manufacture imported raw material and redistribute the products.

But there is, of course, this difference between the supply of foodstuff (or even military power) and mechanical energy, that in the case of coal at least it is necessary to live entirely upon capital; the storing up of energy in new coalfields goes on so slowly in comparison with our rate of expenditure that it may be altogether neglected. Now in this country we began to use coal on a large scale a little more than a century ago. Our present yearly consumption is of the order of 300 millions of tons, and it is computed[1] that at the present rate of increase "the whole of our available supply will be exhausted in 170 years." With regard to the rest of the world we cannot, from lack of *data*, make even the broad assumptions that were possible in the case of wheat supply, and for that and other reasons it is therefore impossible even to guess at the time which must elapse before a universal dearth of coal becomes imminent; it is perhaps sufficient to observe that to the best of our knowledge and belief one of the world's largest groups of coalfields (our own) is not likely to last three centuries in all.

Here again the present interest lies rather in the phases of change which are actually with us. During the first stages of the manufacturing period energy in any form was exceedingly difficult to transport, and this led to intense concentration. Coal was taken from the most accessible coalfield and used, as far as possible, on the spot. It was chiefly converted into mechanical energy by means of the steam engine, an extremely wasteful apparatus in small units, hence still further concentration; thus the steam engine is responsible in part for the factory system in its worst aspect. The less accessible coalfields were neglected. Also, the only other really available source of energy—water-power—remained unused, because the difficulties in the way of utilising movements of large quantities of water through small vertical distances (as in tidal movements) are enormous; the only easily applied source occurs where comparatively small quantities of water fall through considerable vertical distances, as in the case of waterfalls. But, arising from the geographical conditions, waterfalls (with rare exceptions such as Niagara) occur in the "torrential" part of the typical river-course, perhaps far from the sea, almost certainly in a region too broken in surface to allow of easy communication or even of industrial settlement of any kind.

However accessible a coalfield may be to begin with, it sooner or later becomes inaccessible in another way, as the coal near the surface is exhausted and the workings get deeper. No doubt the evil day is postponed for a time by improvements in methods of mining—a sort of intensive cultivation—but as we can put nothing back the end must be the same, and successful competition with more remote but more superficial deposits becomes impossible. And every improvement in land transport favours the geographically less accessible coalfields.

From this point of view it is impossible to overestimate the import-

[1] *General Report of the Royal Commission on Coal Supplies*, 1906.

ance of what is to all intents and purposes a new departure of the same order of magnitude as the discovery of the art of smelting iron with coal, or the invention of the steam engine, or of the steam locomotive. I mean the conversion of energy into electricity, and its transmission in that form (at small cost and with small loss) through great distances. First we have the immediately increased availability of the great sources of cheap power in waterfalls. The energy may be transmitted through comparatively small distances and converted into heat in the electric furnace, making it possible to smelt economically the most refractory ores, as those of aluminium, and converting such unlikely places as the coast of Norway or the West Highlands of Scotland into manufacturing districts. Or it may be transmitted through greater distances to regions producing quantities of raw materials, distributed there widespread to manufacturing centres, and re-converted into mechanical energy. The Plain of Lombardy produces raw material in abundance, but Italy has no coal supply. The waterfalls of the Alps yield much energy, and this transmitted in the form of electricity, in some cases for great distances, is converting Northern Italy into one of the world's great industrial regions. Chisholm gives an estimate of a possible supply of power amounting to 3,000,000 horse-power, and says that of this about one-tenth was already being utilised in the year 1900.

But assuming again, with Sir William Ramsay, that coal must continue to be the chief source of energy, it is clear that the question of accessibility now wears an entirely different aspect. It is not altogether beyond reason to imagine that the necessity for mining, as such, might entirely disappear, the coal being burnt *in situ* and energy converted directly into electricity. In this way some coalfields might conceivably be exhausted to their last pound without serious increase in the cost of getting. But for the present it is enough to note that, however inaccessible any coalfield may be from supplies of raw material, it is only necessary to establish generating stations at the pit's mouth and transport the energy to where it can be used. One may imagine, for example, vast manufactures carried on in what are now the immense agricultural regions of China, worked by power supplied from the great coal deposits of Shan-si.

There is, however, another peculiarity of electrical power which will exercise increasing influence upon the geographical distribution of industries. The small electric motor is a much more efficient apparatus than the small steam engine. We are, accordingly, already becoming familiar with the great factory in which, instead of all tools being huddled together to save loss through shafting and belting, and all kept running all the time, whether busy or not (because the main engine must be run), each tool stands by itself and is worked by its own motor, and that only when it is wanted. Another of the causes of concentration of manufacturing industry is therefore reduced in importance. We may expect to see the effects of this becoming more and more marked as time goes on, and other forces working towards uniform distribution make themselves more felt.

The points to be emphasised so far, then, are, first, that the time when the available areas whence food supply, as represented by wheat, is derived are likely to be taxed to their full capacity within a period of about the same length as that during which the modern colonial system has been developing in the past; secondly, that cheap supplies of energy may continue for a longer time, although eventually they must greatly diminish; and, thirdly, there must begin in the near future a great equalisation in the distribution of population. This equalisation must arise from a number of causes. More intensive cultivation will increase the amount of labour required in agriculture, and there will be less difference in the cost of production and yield due to differences of soil and climate. Manufacturing industries will be more uniformly distributed, because energy, obtained from a larger number of sources in the less accessible places, will be distributed over an increased number of centres. The distinction between agricultural and industrial regions will tend to become less and less clearly marked, and will eventually almost disappear in many parts of the world.

The effect of this upon the third element is of first-rate importance. It is clear that as the process of equalisation goes on the relative amount of long-distance transport will diminish, for each district will tend more and more to produce its own supply of staple food and carry on its own principal manufactures. This result will naturally be most marked in what we may call the "east-and-west" transport, for as climatic controls primarily follow the parallels of latitude, the great *quantitative* trade, the flow of food-stuffs and manufactured articles to and fro between peoples of like habits and modes of life, runs primarily east and west. Thus the transcontinental functions of the great North American and Eurasian railways, the east-and-west systems of the inland waterways of the two continents, and the connecting-links furnished by the great ocean ferries, must become of relatively less importance.

The various stages may be represented, perhaps, in some such manner as this. If **I** is the cost of producing a thing locally at a place A by intensive cultivation or what corresponds to it, if **E** is the cost of producing the same thing at a distant place B, and **T** the cost of transporting it to A, then at A we may at some point of time have a more or less close approximation to

$$\mathbf{I}=\mathbf{E}+\mathbf{T}.$$

We have seen that in this country, for example, **I** has been greater than **E**+**T** for wheat ever since, say, the introduction of railways in North America, that the excess tends steadily to diminish, and that however much it may be possible to reduce **T** either by devising cheaper modes of transport or by shortening the distance through which wheat is transported, **E**+**T** must become greater than **I**, and it will pay us to grow all or most of our own wheat. Conversely, in the seventies of last century **I** was greater than **E**+**T** in North America and Germany for such things as steel rails and rolling-stock, which we in this country were cultivating "extensively" at the time on more accessible coalfields, with more skilled labour and better organisation than could be found elsewhere. In many cases the positions are now, as we know, reversed, but geographically **I** must win all round in the long run.

In the case of transport between points in different latitudes, the conditions are, of course, altogether dissimilar, for in this case commodities consist of foodstuffs, or raw materials, or manufactured articles, which may be termed luxuries, in the sense that their use is scarcely known until cheap transport makes them easily accessible, when they rapidly become "necessaries of life." Of these the most familiar examples are tea, coffee, cocoa, and bananas, indiarubber and manufactured cotton goods. There is here, of course, always the possibility that wheat as a staple might be replaced by a foodstuff produced in the tropics, and it would be extremely interesting to study the geographical consequences of such an event as one-half of the surface of the earth suddenly coming to help in feeding the two quarters on either side; but for many reasons, which I need not go into here, such a consummation is exceedingly unlikely. What seems more probable is that the trade between different latitudes will continue to be characterised specially by its variety, the variety doubtless increasing, and the quantity increasing in still larger measure. The chief modification in the future may perhaps be looked for in the occasional transference of manufactures of raw materials produced in the tropics to places within the tropics, especially when the manufactured article is itself largely consumed near regions of production. The necessary condition here is a region, such as, *e.g.*, the monsoon region, in which there is sufficient variation in the seasons to make the native population laborious; for then, and apparently only then, is it possible to secure sufficient industry and skill by training, and therefore to be able to yield to the ever-growing pressure in more temperate latitudes due to increased cost of labour. The best examples of this to-day are probably the familiar ones of cotton and jute manufacture in India. With certain limitations, manufacturing trade of this kind is, however, likely to continue between temperate and strictly tropical regions, where the climate is so uniform throughout the year that the native has no incentive to work. There the collection of the raw material is as much as, or even more than can be looked for—as in the case of mahogany or wild rubber. Where raw material has to be cultivated—as cotton, cultivated rubber, etc.—the raw material has to be produced in regions more of the monsoon type, but it will probably—perhaps as much for economic as geographical reasons—be manufactured at some centre in the temperate zones, and the finished product transported thence, when necessary, to the point of consumption in the tropics.

We are here, however, specially liable to grave disturbances of distribution arising from invention of new machinery or new chemical methods; one need only mention the production of sugar or indigo. Another aspect of this which is not without importance may perhaps be referred to here, although it means the transference of certain industries to more accessible regions merely, rather than a definite change of such an element as latitude. I have in mind the sudden conversion of an industry in which much labour is expended on a small amount of raw material into one where much raw material is consumed, and by the application of power-driven machinery the labour required is greatly diminished. One remembers when a fifty-shilling Swiss watch, although then still by tradition regarded as sufficiently valuable to deserve enclosure in a case constructed of a precious metal, was considered a marvel of cheapness. American machine-made watches, produced by the ton, are now encased in the baser metals and sold at some five shillings each, and the watch-making industry has ceased to be specially suited to mountainous districts.

In considering the differences which seem likely to arise in what we may call the regional pressures of one kind and another, pressures which are relieved or adjusted by and along certain lines of transport, I have made a primary distinction between "east-and-west" and "north-and-south" types, because both in matters of food-supply and in the modes of life which control the nature of the demand for manufactured articles climate is eventually the dominant factor; and, as I have said, climate varies primarily with latitude. This is true specially of atmospheric temperature; but temperature varies also with altitude, or height above the level of the sea. To a less extent rainfall, the other great element of climate, varies with altitude, but the variation is much more irregular. More important in this case is the influence of the distribution of land and sea, and more especially the configuration of the land surface, the tendency here being sometimes to strengthen the latitude effect where a continuous ridge is interposed, as in Asia, practically cutting off "north-and-south" communication altogether along a certain line, emphasising the parallel-strip arrangement running east and west to the north of the line, and inducing the quite special conditions of the monsoon region to the south of it. We may contrast this with the effect of a "north-and-south" structure, which (in temperate latitudes especially) tends to swing what we may call the regional lines round till they cross the parallels of latitude obliquely. This is typically illustrated in North America, where the angle is locally sometimes nearly a right angle. It follows, therefore, that the contrast of "east-and-west" and "north-and-south" lines, which I have here used for purposes of illustration, is necessarily extremely crude, and one of the most pressing duties of geographers at the present moment is to elaborate a more satisfactory method of classification. I am very glad that we are to have a discussion on "Natural Regions" at one of our sederunts. Perhaps I may be permitted to express the hope that we shall concern ourselves with the types of region we want, their structure or "grain," and their relative positions, rather than with the precise delimitation of their boundaries, to which I think we have sometimes been inclined, for educational purposes, to give a little too much attention.

Before leaving this I should like to add, speaking still in terms of "east-and-west" and "north-and-south," one word more about the essentially east-and-west structure of the Old World. I have already referred to the great central axis of Asia. This axis is prolonged westward through Europe, but it is cut through and broken to such an extent that we may include the Mediterranean region with the area lying further north, to which indeed it geographically belongs, in any discussion of this sort. But the Mediterranean region is bounded on the other side by the Sahara, and none of our modern inventions facilitating transport has made any impression upon the dry desert; nor does it seem likely that such a desert will ever become a less formidable barrier than a great mountain mass or range. We may conclude, then, that in so far as the Old World is concerned, the "north-and-south" transport can never be carried on as freely as it may in the New, but only through certain weak points, or "round the ends," *i.e.* by sea. It may be further pointed out that the land areas in the southern hemisphere are so narrow that they will scarcely enter into the "east-and-west" category at all—the transcontinental railway as understood in the northern hemisphere cannot exist; it is scarcely a pioneer system, but rather comes into existence as a later by-product of local east-and-west lines, as in Africa.

These geographical facts must exercise a profound influence upon the future of the British Isles. Trade south of the great dividing line must always be to a large extent of the "north-and-south" type, and the British Isles stand practically at the western end of the great natural barrier. From their position the British Isles will always be a centre of immense importance in *entrepôt* trade, importing commodities from "south" and distributing "east and west," and similarly in the reverse direction. This movement will be permanent, and will increase in volume long after the present type of purely "east-and-west" trade has become relatively less important than it is now, and long after the British Isles have ceased to have any of the special advantages for manufacturing industries which are due to their own resources either in the way of energy or of raw material. We can well imagine, however, that this permanent advantage of position will react favourably, if indirectly, upon certain types of our manufactures, at least for a very long time to come.

Reverting briefly to the equalisation of the distribution of population in the wheat-producing areas and the causes which are now at work in this direction, it is interesting to inquire how geographical conditions are likely to influence this on the smaller scale. We may suppose that the production of staple foodstuffs must always be more uniformly distributed than the manufacture of raw materials, or the production of the raw materials themselves, for the most important raw materials of vegetable origin (as cotton, rubber, etc.) demand special climatic conditions, and, apart from the distribution of energy, manufacturing industries are strongly influenced by the distribution of mineral deposits, providing metals for machinery, and so on. It may, however, be remarked that the useful metals, such as iron, are widely distributed in or near regions which are not as a rule unfavourable to agriculture. Nevertheless, the fact remains that while a more uniform distribution is necessary and inevitable in the case of agriculture, many of the conditions of industrial and social life are in favour of concentration; the electrical transmission of energy removes, in whole or in part, only one or two of the centripetal forces. The general result might be an approximation to the conditions occurring in many parts of the monsoon areas—a number of fairly large towns pretty evenly distributed over a given agricultural area, and each drawing its main food supplies from the region surrounding it. The positions of such towns would be determined much more by industrial conditions, and less by military conditions, than in the past (military power being in these days mobile, and not fixed); but the result would on a larger scale be of the same type as was developed in the central counties of England, which, as Mackinder has pointed out, are of almost equal size and take the name of the county town. Concentration within the towns would, of course, be less severe than in the early days of manufacturing industry. Each town would require a very elaborate and highly organised system of local transport, touching all points of its agricultural area, in addition to lines of communication with other towns and with the great "north-and-south" lines of world-wide commerce, but these outside lines would be relatively of less importance than they are now. We note that the more perfect the system of local transport the less the need for points of intermediate exchange. The village and the local market-town will be "sleepy" or decadent as they are now, but for a different reason; the symptoms are at present visible mainly because the country round about such local centres is overwhelmed by the great lines of transport which pass through them; they will survive for a time through inertia and the ease of foreign investment of capital. The effect of this influence is already apparent since the advent of the "commercial motor," but up to the present it has been more in the direction of distributing from the towns than collecting to them, producing a kind of "suburbanisation" which throws things still further out of balance. The importance of the road motor in relation to the future development of the food-producing area is incalculable. It has long been clear that the railway of the type required for the great through lines of fast transport is ill-adapted for the detailed work of a small district, and the "light" railway solves little and introduces many complications. The problem of determining the direction and capacity of a system of roads adequate to any particular region is at this stage one of extraordinary difficulty; experiments are exceedingly costly, and we have as yet little experience of a satisfactory kind to guide us. The geographer, if he will, can here be of considerable service to the engineer.

In the same connexion, the development of the agricultural area supplying an industrial centre offers many difficult problems in relation to what may be called accessory products, more especially those of a perishable nature, such as meat and milk. In the case of meat the present position is that much land which may eventually become available for grain crops is used for grazing, or cattle are fed on some grain, like maize, which is difficult to transport or is not satisfactory for bread-making. The meat is then temporarily deprived of its perishable property by refrigeration, and does not suffer in transport. Modern refrigerating machinery is elaborate and complicated, and more suited to use on board ship than on any kind of land transport. Hence the most convenient regions for producing meat for export are those near the sea-coast, such as occur in the Argentine or the Canterbury plains of New Zealand. The case is similar to that of the "accessible" coalfield. Possibly the preserving processes may be simplified and cheapened, making overland transport easier, but the fact that it usually takes a good deal of land to produce a comparatively small quantity of meat will make the difficulty greater as land becomes more valuable. Cow's milk, which in modern times has become a "necessary of life" in most parts of the civilised world, is in much the same category as meat, except that difficulties of preservation, and therefore of transport, are even greater. That the problem has not become acute is largely due to the growth of the long-transport system available for wheat, which has enabled land round the great centres of population to be devoted to dairy produce. If we are right in supposing that this state of things cannot be permanent the difficulty of milk supply must increase, although relieved somewhat by the less intense concentration in the towns; unless, as seems not unlikely, a wholly successful method of permanent preservation is devised.

In determining the positions of the main centres, or rather, in subdividing the larger areas for the distribution of towns with their supporting and dependent districts, water supply must be one of the chief factors in the future, as it has been in the past; and in the case of industrial centres the quality as well as the quantity of water has to be considered. A fundamental division here would probably be into districts having a natural local supply, probably of hard water, and districts in which the supply must be obtained from a distance. In the latter case engineering works of great magnitude must often be involved, and the question of total resources available in one district for the supply of another must be much more fully investigated than it has been. In many cases, as in this country, the protection of such resources pending investigation is already much needed. It is worth noting that the question may often be closely related to the development and transmission of electrical energy from waterfalls, and the two problems might in such cases be dealt with together. Much may be learned about the relation of water supply to distribution of population from a study of history, and a more active prosecution of combined historical and geographical research would, I believe, furnish useful material in this connexion, besides throwing interesting light on many historical questions.

Continued exchange of the "north-and-south" type and at least a part of that described as "east-and-west" gives permanence to a certain number of points where, so far as can be seen, there must always be a change in the mode of transport. It is not likely that we shall have heavy freight-carrying monsters in the air for a long time to come, and until we have the aerial "tramp" transport must be effected on the surfaces of land and sea. However much we may improve and cheapen land transport it cannot in the nature of things become as cheap as transport by sea. For on land the essential idea is always that of a prepared road of some kind, and, as Chisholm has pointed out, no road can carry more than a certain amount; traffic beyond a certain quantity constantly requires the construction of new roads. It follows, then, that no device is likely to provide transport indifferently over land and sea, and the seaport has in consequence inherent elements of permanence. Improved and cheapened land transport increases the economy arising from the employment of large ships rather than small ones, for not only does transport inland become relatively more important, but distribution along a coast from one large seaport becomes as easy as from a number of small coastal towns. Hence the conditions are in favour of the growth of a comparatively small number of immense seaport cities like London and New York, in which there must be great concentration not merely of work directly connected with shipping, but of commercial and financial interests of all sorts. The seaport is, in fact, the type of great city which seems likely to increase continually in size, and provision for its needs cannot in general be made from the region immediately surrounding it, as in the case of towns of other kinds. In special cases there is also, no doubt, permanent need of large inland centres of the type of the "railway creation," but under severe geographic control these must depend very much on the nature and efficiency of the systems of land transport. It is not too much to say (for we possess some evidence of it already) that the number of distinct geographical causes which give rise to the establishment and maintenance of individual great cities is steadily diminishing, but that the large seaport is a permanent and increasing necessity. It follows that aggregations of the type of London and Liverpool, Glasgow and Belfast will always be amongst the chief things to be reckoned with in these islands, irrespective of local coal supply or accessory manufacturing industries, which may decay through exhaustion.

I have attempted in what precedes to draw attention once more to certain matters for which it seems strangely difficult to get a hearing. What it amounts to is this, that as far as our information goes the development of the steamship and the railway, and the universal introduction of machinery which has arisen from it, have so increased the demand made by man upon the earth's resources that in less than a century they will have become fully taxed. When colonisation and settlement in a new country proceeded slowly and laboriously, extending centrifugally from one or two favourable spots on the coast, it took a matter of four centuries to open up a region the size of England. Now we do as much for a continent like North America in about as many decades. In the first case it was not worth troubling about the exhaustion of resources, for they were scarcely more than touched, and even if they were exhausted there were other whole continents to conquer. But now, so far as our information goes, we are already making serious inroads upon the resources of the whole earth. One has no desire to sound an unduly alarmist note, or to suggest that we are in imminent danger of starvation, but surely it would be well, even on the suspicion, to see if our information is adequate and reliable and if our conclusions are correct; and not merely to drift in a manner which was justifiable enough in Saxon times, but which, at the rate things are going now, may land us unexpectedly in difficulties of appalling magnitude.

What is wanted is that we should seriously address ourselves to a stock-taking of our resources. A beginning has been made with a great map on the scale of one to a million, but that is not sufficient; we should vigorously proceed with the collection and discussion of geographical *data* of all kinds, so that the major natural distributions shall be adequately known, and not merely those parts which commend themselves, for one reason or another, to special national or private enterprises. The method of

Government survey, employed in most civilised countries for the construction of maps, the examination of geological structure or the observation of weather and climate, is satisfactory as far as it goes, but it should go further, and be made to include such things as vegetation, water supply, supplies of energy of all kinds, and, what is quite as important, the bearings of one element upon others under different conditions. Much, if not most, of the work of collecting *data* would naturally be done as it is now by experts in the special branches of knowledge, but it is essential that there should be a definite plan of a *geographical* survey as a whole, in order that the regional or distributional aspect should never be lost sight of. I may venture to suggest that a committee formed jointly by the great national geographical societies, or by the International Geographical Congress, might be entrusted with the work of formulating some such uniform plan and suggesting practicable methods of carrying it out. It should not be impossible to secure international co-operation, for there is no need to investigate too closely the secrets of any one's particular private vineyard--it is merely a question of doing thoroughly and systematically what is already done in some regions, sometimes thoroughly, but not systematically. We should thus arrive eventually at uniform methods of stock-taking, and the actual operations could be carried on as opportunity offered and indifference or opposition was overcome by the increasing need for information. Eventually we shall find that "country-planning" will become as important as town-planning, but it will be a more complex business, and it will not be possible to get the facts together in a hurry. And in the meanwhile increased geographical knowledge will yield scientific results of much significance about such matters as distribution of populations and industries, and the degree of adjustment to new conditions which occurs or is possible in different regions and amongst different peoples. Primary surveys on the large scale are specially important in new regions, but the best methods of developing such areas and of adjusting distributions in old areas to new economic conditions are to be discovered by extending the detailed surveys of small districts. An example of how this may be done has been given by Dr. Mill in his *Fragment of the Geography of Sussex*. Dr. Mill's methods have been successfully applied by individual investigators to other districts, but a definitely organised system, marked out on a carefully matured uniform plan, is necessary if the results are to be fully comparable. The schools of geography in this country have already done a good deal of local geography of this type, and could give much valuable assistance if the work were organised beforehand on an adequate scale.

But in whatever way and on whatever scale the work is done, it must be clearly understood that no partial study from the physical, or biological, or historical, or economic point of view will ever suffice. The urgent matters are questions of distribution upon the surface of the earth, and their elucidation is not the special business of the physicist, or the biologist, or the historian, or the economist, but of the geographer.

THE DOMINION OF CANADA: A STUDY IN REGIONAL GEOGRAPHY.[1]

By Arthur Silva White, Hon. F.R.S.G.S.

(With Maps.)

PART I. THE COUNTRY.

Unveiling of the New World.

The geographical characteristics of the American continents, and the unveiling of these to the Old World, present an instructive and fascinating study. In physical structure, in relief and drainage, and in general configuration the resemblance of the Americas is close; in climatic conditions, the contrast is no less sharp. Pear-shaped in outline, North and South America resemble each other, as well as the land-masses of Africa and India, by having their broadest lands in the north. Both continents are ridged up on their western sides by a massive backbone of mountain ranges and highlands, which differentiate the climates of the Atlantic and Pacific coastlands; and, in each, two vast river-systems discharge south and east into the Atlantic basin by long and gentle slopes. Their unveiling, like their climates—and, indeed, largely because of these—may be contrasted broadly: North America was opened up from the south and east; South America, mainly from the north and west. The early voyagers, explorers, and settlers acted each according to their kind: the English, clinging to the coasts, occupied islands and peninsular lands; the Dutch entrenched themselves on the Hudson at New Amsterdam (New York); the French boldly penetrated the great arteries of North America—the St. Lawrence and the Mississippi; while the Portuguese and Spaniards, in the Eastern and Western worlds of the Papal award, sought wealth in warmer climes and richer lands. Their common objective, however, was the fabulous wealth of the Indies, search for which was pushed forward in every direction. Thus it came about that the barren lands of Africa were passed by and the uninviting lands of North America received less attention than the West India Islands and tropical South America. Before the first quarter of the sixteenth century had passed, the early voyagers had revealed the great length of the Atlantic coastline of the Americas, while the tropical lands of Central and South America were being opened up, somewhat roughly, to the policy of Europe.

Historical Retrospect.

The rise of Canada from its obscurity, half a century ago, to its position as the Premier Colony of the British Empire is, perhaps, the most notable achievement in British colonial expansion. The disunity of the early days—due to mutual misunderstandings between the scattered

Provinces, separate Customs, and the neglect of communications—has been overcome by co-operation and by loyalty to English ideals of life and government. A supreme faith in herself and in the development of her natural resources has been, and remains, the guiding spirit of Canada.

The voyages of Columbus, Vasco da Gama, and Magellan stimulated others to seek for a northern passage to Asia, both eastward and westward; and scientific exploration of the Arctic Regions became eventually a great national undertaking, in which the British Navy co-operated. John Cabot, in 1497, was the pioneer. His search for the North-West Passage led to the discovery of Newfoundland and the continent of North America; and he was followed by the Portuguese voyager, Corte Real. In 1576, the quest was taken up by Martin Frobisher, an English navigator and explorer, who sailed westward until brought up in Frobisher Bay (Baffin Land), thereby narrowly missing the discovery of Hudson Strait. His return and subsequent attempt to found a settlement there yielded no result. Other English navigators followed, including John Davis and William Baffin; but it was left to the ill-fated Henry Hudson, in 1610, to enter the strait that bears his name, and to outline the southern limit of Hudson Bay. Considering the small vessels and the scanty crews that took part in this search for a North-West Passage, it is surprising that the early navigators accomplished so much.

The Portuguese were equally active in seeking a short-cut to their Eastern colonies. France, too, whose navigators played the leading rôle in the sixteenth century, made determined attempts to penetrate westwards. Several expeditions left Norman and Breton seaports for the New Lands, one of which (under Denys of Honfleur) is said to have reached the St. Lawrence as early as 1506. The true discoverer of Canada—who named it New France—was Jacques Cartier, a Breton sailor. In 1534 and 1535, he entered the St. Lawrence and was well received by the Indians at Quebec and Montreal. Subsequently, in 1541, he and De Roberval made the first serious attempt at colonisation, at Cap Rouge; but it failed ignominiously. More than fifty years passed before permanent settlement began under the brilliant leadership of Champlain, who founded Quebec in 1608.

England—particularly Devonshire—in the sixteenth century produced men who were filled with the spirit of adventure, and who boldly faced the limitless horizon in search of it. Of such was Sir Francis Drake, the master-spirit in the overthrow of the Armada and the first Englishman to circumnavigate the globe. Drake, from the heights above Panama, "stared on the Pacific"; and his vow "to sail an English ship in those seas" was nobly redeemed on his voyage of discovery, which lasted nearly three years.[1]

Among other Elizabethan navigators, none stand out with greater distinction than Sir Humphrey Gilbert and Sir Walter Raleigh—both pioneers in the foundation of empire in British North America—who

[1] Mrs. Nuttall, an American archæologist, made a communication to the Hakluyt Society (Oct. 9, 1912), based on some recently discovered documents, which would appear to confirm the discovery of Cape Horn by Drake, and to throw new light on many incidents in his career.

made serious attempts at colonisation. In 1578, Gilbert obtained a patent "for the inhabiting and planting of our people in America," and, in virtue of this, he proclaimed the Queen's sovereignty over Newfoundland. He, however, was pursued by misfortune. Returning home, his small ship (the *Squirrel,* 10 tons) foundered in mid-Atlantic. "We are as near heaven by sea as by land" was his last utterance, so characteristic of the man and of his times.

Raleigh's attempts at colonisation began, a year after Gilbert's loss, in Virginia and ended ingloriously in Guiana. But, when war with Spain defeated the first attempt to colonise Virginia, freebooting became the order of the day. Private enterprise, under Royal licence, entered too much into these early colonial adventures, to lead to any permanent results. Later, when commercial companies were formed, more systematic and thorough methods were adopted, the most conspicuous example of which was the Hudson's Bay Company, founded in 1670.

The long struggle, lasting over a century and a half, between France and England for supremacy in the North American continent was determined largely by geographical conditions. The French on the St. Lawrence, with Quebec and Montreal for their rallying points, and the British on the Hudson, with New York and Boston for their base, were in constant conflict, together with their Indian allies—the Algonkin and the Iroquois, respectively. The peninsular lands between the St. Lawrence and the sea, deeply afforested and peopled with warlike Indians, were the scene of fierce and ruthless encounters. The policy of France was to surround the English maritime settlements, and to join forces with the French settlements in the Mississippi basin—the vast province of Louisiana—but it was a task beyond her strength. The British settlements, north and south of New York, occupied an interior strategic position based on the coast. Thus, access to the interior being blocked, attention was given to agriculture rather than to conquest, New England and Virginia being relatively populous.

At the beginning of the eighteenth century, New York and Quebec —the one on an island, the other on a fortified height: so characteristic of the two peoples—were the keys to the political situation, in which aristocratic and paternal Quebec strove in vain against the democratic and individualistic elements of New York and Boston. Quebec stood at the head of inland navigation (which was interrupted at Montreal by La Chine Rapids) the vast extent of which created an object of rivalry between New England and New France. French settlements, of no great importance, were formed in the maritime provinces (Acadia: *i.e.* Nova Scotia, New Brunswick, and Prince Edward Island); and when Nova Scotia fell to England, by the Treaty of Utrecht, in 1713, the British secured a foothold in New France which led by degrees to its final conquest in 1759, when Quebec surrendered to the heroic Wolfe. Montreal capitulated in the following year; and in 1763, by the Peace of Paris, which ended the Seven Years' War, all the French Possessions in North America became British, although in St. Pierre and Miquelon Isles and on the west coast of Newfoundland French claims survived for nearly a century.

Illustrating Mr Silva White's Article.

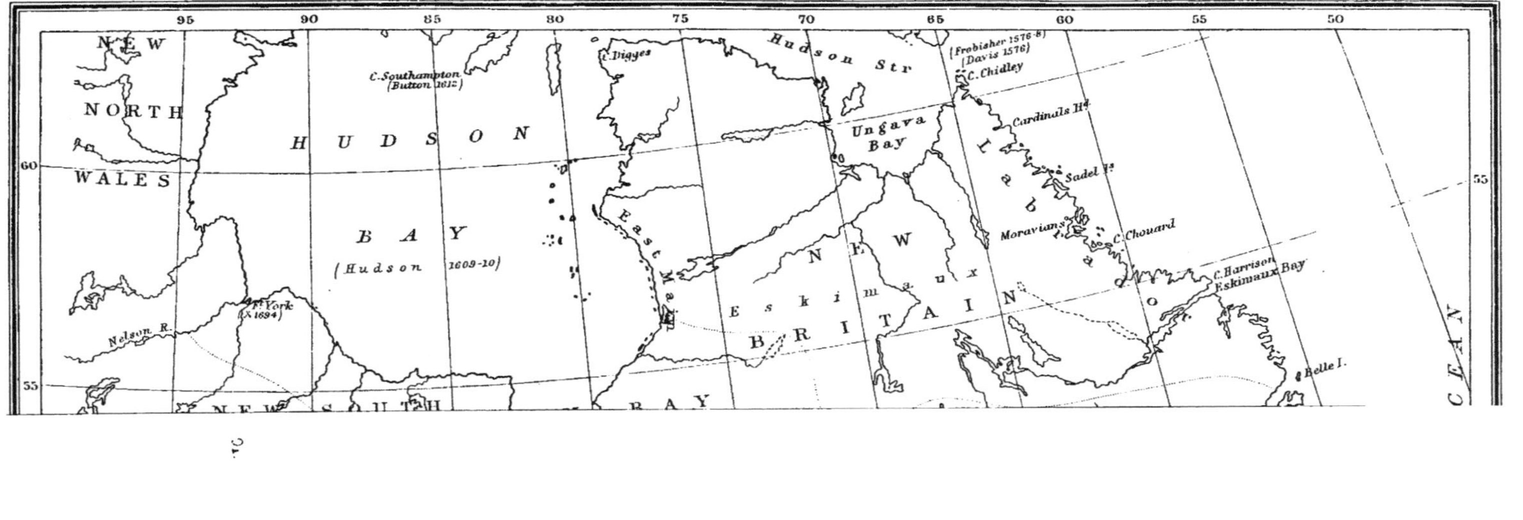

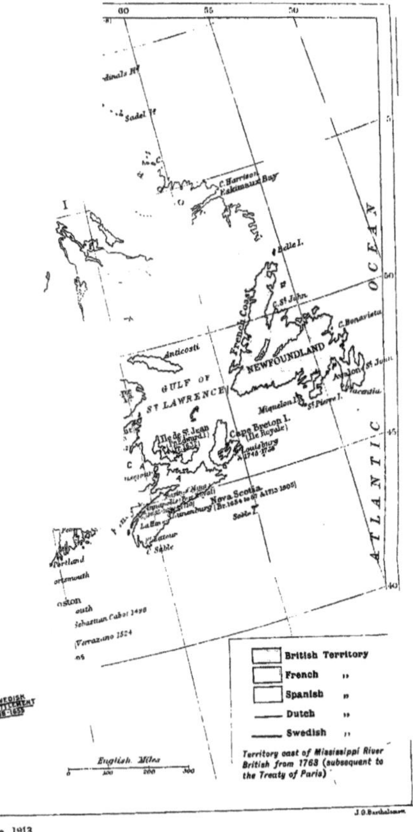

Scottish Geographical Magazine, 1913

Twenty years later, the Thirteen Colonies, which had declared their independence of the Mother Country in 1776, became the United States; but they were defeated in their attack upon Quebec. The United Empire Loyalists, however, to the number of 40,000, preferring British rule, migrated in 1784 to Ontario, Nova Scotia, and New Brunswick, settling mainly on the southern shore of the St. Lawrence. Canada was again invaded in 1812, when war broke out between Great Britain and the United States; but the handful of Loyalists and French Canadians repulsed the invaders and held their homes inviolate until the end of the war in 1814. West of Lake Superior—or, more precisely, Lake of the Woods—an artificial frontier, the forty-ninth parallel (created by the Oregon treaty of 1846) separates Canada from the United States; and this frontier is now crossed yearly by a hundred thousand peaceful emigrants from the American Union, who find new homes in the prairie lands.

In 1791, when Upper Canada (now Ontario) and Lower Canada (now Quebec) were formed by Act of Parliament into two provinces, the population of the country numbered about 160,000; and all the colonies had representative institutions. At the close of the eighteenth century, British settlements in Canada—apart from the Hudson Bay traders and Newfoundlanders—were still confined to the Maritime Provinces, the shorelands of the St. Lawrence and of the Great Lakes; while the Prairies to the west were left to the vast herds of bison, or buffalo, and Indian hunters.

After the war between the United States and Great Britain (1812-1814), under the industrial depression and social unrest in the Mother Country which followed her supreme effort at Waterloo, Canada received a steady stream of English and Scottish emigrants, and began to open up the western waterways. The Hudson's Bay Company and the North-West Company, engaged in friendly rivalry, pushed their search for furs farther afield; and, in 1812, the Scottish and Irish settlement on the Red River was founded by Lord Selkirk.

The long-sustained conflict between the executive and legislative authorities and the struggle for supremacy between the French and British elements of the population led to the outbreak of a rebellion, in both Upper and Lower Canada, and, consequently, to the suspension of the constitution of the latter. But the rebels were easily dispersed, and constitutional grievances were redressed by the Act of 1840, which re-united the Provinces. The new Constitution of 1841 did not, however, find full expression until the appointment of Lord Elgin, as Governor-General, in 1847. A few years later, responsible government was granted to the Maritime Provinces. The repeal of the navigation laws in 1849 threw open the St. Lawrence to the merchant shipping of the world.

Canada—and particularly Ontario—developed rapidly in the next quarter-century, greatly improving her internal and external communications. In 1867 she received her *Magna Carta*, in the British North America Act, which united Ontario, Quebec, New Brunswick and Nova Scotia under a Federal Dominion which left room for union with the other Provinces. The population of the Dominion then numbered about 3,400,000.

West of the Woodland Region, on the wide plains which extend to the foot of the Cordilleras, or Rocky Mountains, settlement lagged behind. The Hudson's Bay Company, which absorbed the North-West Company in 1821, had its principal post at Winnipeg—then called the Red River Settlement—and carried on an extensive trade in furs, little realising the gold-mine that lay in the rich soil of the prairies. In 1869, the Canadian Government acquired, by purchase, the territorial rights of the Company over the North-West Territories; but on attempting to take possession, the half-breeds, who constituted the main population of the Red River Settlement, rebelled, and, under the leadership of Riel, instituted a provisional government. No resistance, however, was offered to the expedition sent out under Sir Garnet Wolseley, which restored order. In 1870, the Red River district received responsible government and was admitted to the Dominion as the Province of Manitoba.

The rapid development of Manitoba and the opening up of the Prairie Provinces—in which, it is estimated, over two hundred million acres are available for cultivation—are unsurpassed achievements in Colonial settlement. To this success the Canadian Pacific Railway, which was opened for traffic in 1886, largely contributed, side by side with the linking up of Ontario and Quebec by the Inter-Colonial Railway. Under a liberal land policy, emigrants poured into the Prairie Provinces; and the flood of migration is still unabated.

Prince Edward Island joined the Federation in 1873. Other unions followed; and before the close of 1905, the Prairie Lands entered the Dominion as the Provinces of Alberta and Saskatchewan.

On the Pacific Coast, with Vancouver as a centre, settlement lagged until discoveries of gold in the Yukon district attracted a large floating population. In 1898, British Columbia was constituted into a separate territory; and its progress since then has been rapid.

The general election of 1911, turning on the question of Reciprocity with the United States, was an epoch-making event, not only for Canada but for the Empire at large. In the words of Mr. Borden, Canada then "determined that for her there shall be no parting of the ways, but that she will continue in the old path of Canadian unity, Canadian manhood, and the British connection: she has emphasised the ties that bind her to the Empire."

With these few landmarks in the history of Canada to guide us, we shall enter upon our survey of the great Dominion.[1] Its past history,

---

[1] The political history of the country has been divided, by a Canadian historian, into five fairly distinct periods:

1. The period of French rule, from 1608 to 1760, or the régime of absolute government.
2. The period of a Crown Colony from 1760 to 1791, when representative and legislative institutions were established.
3. The period from 1791 to 1840, when representative institutions were slowly developing into responsible or complete self-government.
4. The period from 1840 to 1867, during which responsible government was enjoyed in

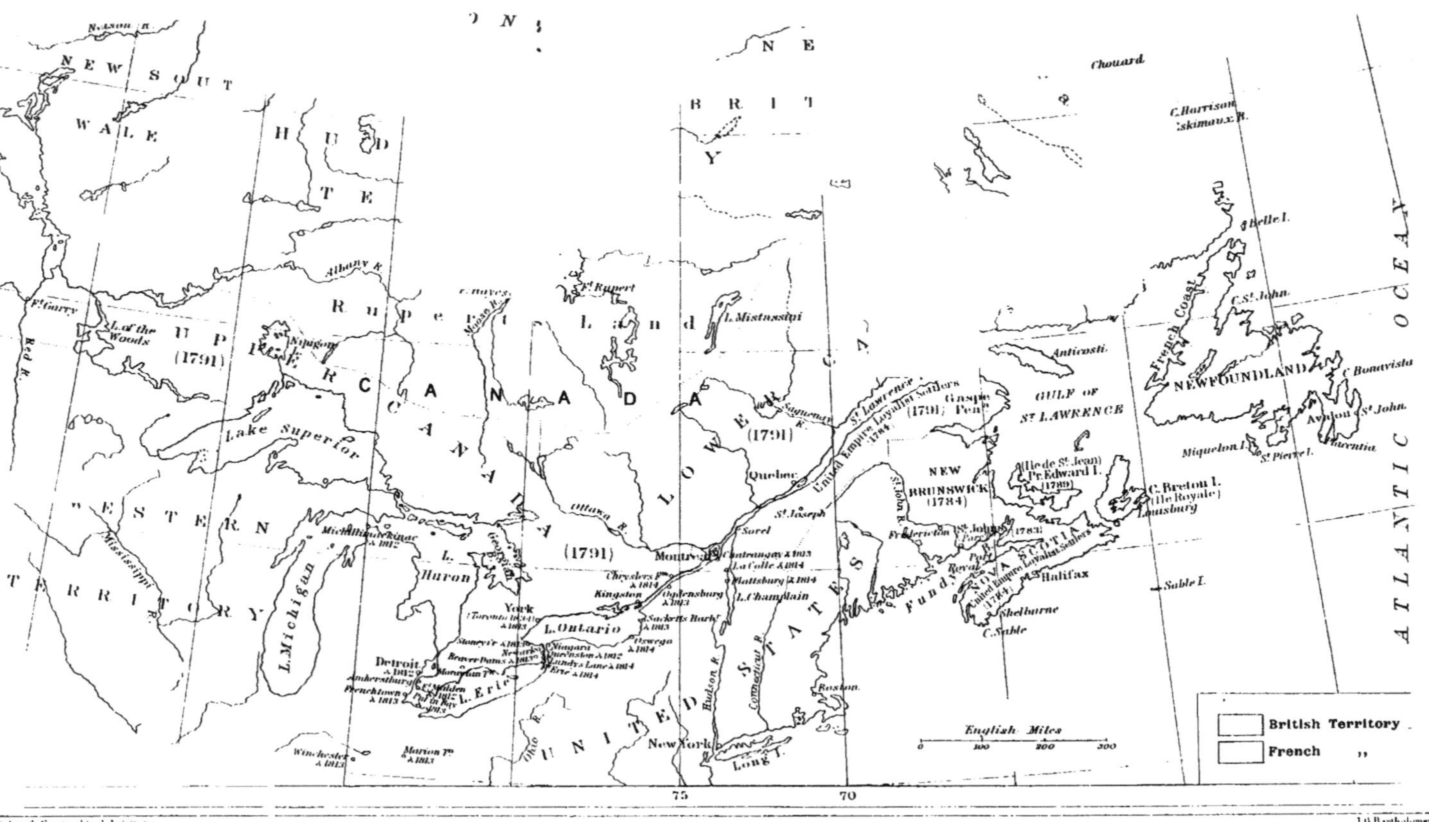

The Edinburgh Geographical Institute

J. G. Bartholomew.

which has been signalised by the implanting and cultivation of English traditions, although beginning in conquest by force of arms, illustrates for us the capacity of two great colonising peoples to pursue the arts of peace, side by side, in friendly rivalry and towards a common goal.

AREA AND SETTLEMENT.

The Dominion of Canada covers an area greatly exceeding that of any other portion or political aggregate of the Empire, and, with the exception of the United Kingdom and India, has a larger population: nevertheless, the population of Canada in 1911 was less than that of "Greater London."[1] Compared with the United States, Canada occupies a much larger area—nearly a million square miles—but has only one-thirteenth of its total population. This disparity in numbers is accounted for by the vast extent of Arctic or virtually unprospected lands (covering two-thirds of the total area of Canada) in the Northern Territories and the corresponding absence of natural resources to support a large, and in the main agricultural, population.

Inclusive of the aggregate area (125,755 square miles) of lakes and rivers within the territorial limits of the Dominion, the total area of Canada is 3,729,665 square miles, or over thirty times that of the United Kingdom. Consequently, nearly one-thirteenth of the surface of the Dominion is occupied by water. Nothing could be more fortunate, since the waterways of Canada offer a magnificent—perhaps an unrivalled—system of inter-communication between east and west, along the main line of her natural development. The settlement of this area is, of course, exclusively in the south, contiguous to the coastlands of the Maritime Provinces, the shorelands of the St. Lawrence and Great Lakes, and along the trans-continental, natural and artificial highways. Facility of communication has, in short, largely determined the settlement of Canada, influenced by a general tendency of concentration towards the United States frontier.

PHYSICAL STRUCTURE.

The physical structure of the North American continent is both simple and massive. The land-masses are carved out on bold lines.

---

the fullest sense of the term, and the federal union was finally accomplished as the natural result of the extended liberties of the people.

5. The existing period—the period of confederation—in which the political system has been adapted to the circumstances of the country.—*Canadian Life in Town and Country* (p. 43), by H. J. Morgan and L. J. Burpee.

[1] The population of Registration London (of the "Outer Ring") and of "Greater London" (or the Metropolitan and City Police district) according to the census of 1911, was:—

Registration London, .	4,522,961
"Outer Ring," . .	2,730,002
"Greater London," . .	7.252,963

Stateman's Year Book, 1912.

The catchment-basins of the Atlantic and Pacific drainage-areas are divided on the western side, where the Cordilleras, or Rocky Mountains, run the entire length of the continent. The main chain is accompanied, in various places, by parallel ridges, some of which have distinctive names and throw off spurs in all directions. This continental uplift, uniting somewhat abruptly with the Great Central plain, is drained towards the south and east by the Missouri-Mississippi river-system, towards the north by the Yukon and Mackenzie systems, and to the west by numerous rivers of no great length. The great central area is drained towards Hudson Bay, but its two largest rivers—Mackenzie and St. Lawrence—have separate watersheds. It is separated from the Great Lakes by The Height of Land. The Appalachian Highlands intervene between the catchment-basin of the St. Lawrence system and the Atlantic. The Rocky Mountain system in the west may be said to constitute the terrene of British Columbia, the area of which is nearly three times the size of Great Britain and Ireland.

The drainage-areas of Canada are as follows :—

	Sq. Miles.
Atlantic Basin (exclusive of Hudson Bay),	554,000
Hudson Bay Basin,	1,486,000
Pacific Basin,	387,300
Arctic Basin,	1,290,000

The relative areas of Canadian territory drained by the St. Lawrence Basin are :—

	Sq. Miles.
Above the mouth (Pt. de Monts),	498,500
,, Montreal,	368,900
,, Kingston,	302,300
,, Niagara,	267,100
,, Windsor,	230,500
,, Sault St. Marie,	82,890

The areas of the Great Lakes are :—

	Sq. Miles.
Lake Superior,	31,800
,, Huron,	23,200
,, Erie,	10,000
Ontario,	7,260
and	
,, Michigan (in U.S. territory),	22,400

Lake Superior is the largest body of fresh water in the world. Its area is greater than that of Scotland.

The general strike of the country is along the spread of its coastline. The Rocky Mountain system attains its widest expansion in about latitude 40°, between the Great Basin on the west and the High Western Prairie on the east. The Great Plains slope eastwards from the High Western Prairie, and are broadest between latitudes 35° and 55°. The Labrador Peninsula slopes from south to north, the highest elevations (1000 to 2000 ft.) being in the south and east. There is no natural frontier between Canada and the United States, save those provided by the St. Lawrence system: on the contrary, the Great Lakes unite, and there is a co-mergence of physical structure along the general strike of the country. This physical co-unity is of political significance.

Hudson Bay lies in a cup-shaped depression, with a deep channel outlet (Hudson Strait) to the Atlantic, which is blocked by icefields during the greater part of the year; otherwise it would attract the mercantile marine of the world into the very heart of the Dominion: in other words, climate has condemned this region to sterility. The Hudson Bay basin is built up of Archæan rocks, between the Mackenzie and the St. Lawrence river-systems, surrounded by a zone of Palæozoic rocks. This core, or nucleus, of ancient rock-structure—which occurs again in many mountain ranges in North America—is separated from the Palæozoic and Mesozoic masses of the mountainous western range by broad plains and prairies, which are underlain by more recent formations. The vast ice-sheet, which in the Great Ice Age filled the lowlands of Canada, left behind deposits of boulder clay that cover the underlying formations and impede the more direct flow of the rivers, thus creating shallow and sluggish waterways in the Great North-West which are open to navigation by canoe.

The Appalachian Highlands represent the vestiges of a geologically ancient mountain range, so eroded and denuded that only moderate heights of hard rock remain, the chief of which—White and Black Mountains—lie in New Hampshire and North Carolina. Granite rocks are widely distributed in the Maritime Provinces.

On the other side of the continent, the Rocky Mountain system (Cordilleras), with precipitous slopes and escarpments on its eastern borders, prevails, in a series of ridges, throughout the length of Western Canada—almost everywhere up to the coastline—and it is very varied in structure and form. A high-plateau occupies the central parts. Its general elevation falls within the two-to-five thousand feet contour-line; but there is a well-defined backbone carrying elevations of from five-to-ten thousand feet, between which and the Pacific coast there are numerous *enclaves* or *massifs* of the same height. It is a "sea of mountains." There are peaks of from twelve to fourteen thousand feet, the highest (recorded) elevations being reached in Alaska by Mount St. Elias (18,024 ft.) and Mount Logan (19,534 ft.). The Alpine mountains of the Selkirk Range, traversed by the Canadian Pacific Railway, contain many snowfields and glaciers.

CLIMATE.

The North American continent ranges through every category of climate, from sub-tropical to Polar. In the main, the climate is continental in character, being subject to the greatest extremes of seasonal changes of temperature in the broad lands of the north. In the south, the heat-equator (indicated by the mean annual temperature) carries an isotherm of 70° Fahr. far into the temperate zone. Inter-seasonal changes are gentle on the Pacific Coast, but abrupt in the interior and on the Middle Atlantic border.

The highest mean annual rainfall[1] (from 50 to 75 inches) occurs round the Gulf of Mexico and on the coast of British Columbia, where in places it exceeds 75 inches. The warm Kuro Siwo, or Japanese current, splits between Vancouver Island and Queen Charlotte Island—one arm going north, the other south—thus modifying the climate, which is mild and humid on the coast.

The Rockies and the cold waters of Hudson Bay carry the isotherms for the year further south on either side of Canada. The isotherms for summer show the great northern "loop"—or, penetration into the interior—of the 55° isotherm, which, together with other influences, makes the cultivation of cereals possible, even in the valley of the Mackenzie River. In the northern portion of Canada there is considerably more sunshine in the summer than in the southern; and, owing to the clearness of the skies, the average duration of actual sunshine is greater in northern Canada than in the eastern portion. In consequence of the moderating effect of the Atlantic Ocean and the Great Lakes in eastern Canada and of the Japanese current on the Pacific coast, temperatures at Quebec average 290 days in the year above the freezing point (32°), and at Victoria, B.C., as many as 363 days. The high summer temperature of towns in southern Canada may be indicated by a line passing through Nova Scotia, Ottawa, the extreme southern portion of Alberta and Saskatchewan, along which the temperature is above 50° during 200 days in the year. In south-western Nova Scotia and in eastern Quebec, temperatures exceed 70° on only fifty days in the year. The Ottawa valley, in the vicinity of Ottawa city, has upwards of a hundred days above 70°; and Winnipeg has 70 days or more. These figures are based on observations extending over a series of years.[2]

Following Dr. G. M. Dawson,[3] we may, in general terms, divide the whole country into three climatic areas—the Eastern, the Inland, and the Pacific Coast regions, to which, however, we must add a fourth: the sub-Arctic lands of the Northern Territories. The Eastern region, characterised by great range of temperature and ample rainfall, coincides with the Southern Forest area, which includes all the old provinces, and extends westwards almost to Winnipeg. The Inland region, stretching still further westward to within a short distance of the Pacific Coast, is characterised by a still greater range of temperature—hot in the summer and cold in the winter—but has a moderate rainfall. The Pacific Coast region, with an oceanic climate—small range of temperature, heavy rainfall, and high humidity—does not include the whole of the Pacific slope. but only the coastal zone of the western mountain range of the Cordilleras, which enjoys a climate that is comparable to

---

[1] Precipitation includes snowfall as well as rainfall—ten inches of snow counting as one inch of rain, on a rough average.

[2] Cf. *Atlas of Canada*: Department of the Interior, 1906.

[3] Mill's *International Geography*, p. 682.

that of England at its best. The exhilarating atmosphere of Canada mitigates seasonal severities; and the four or five months of winter, with it's mantle of snow, prepares the land for a rich yield of cereals.

FLORA.

The floras of the Maritime Provinces, Quebec and Eastern Ontario, are, generally speaking, much the same; in Western Ontario, trees, shrubs, and herbaceous plants not found in the eastern parts become common. In New Brunswick, the western flora appears in conjunction with some southern immigrants, and in Quebec the flora is very varied. Beyond the forest country of Ontario, in Manitoba and the former North-West Territories, the eastern flora penetrates the ravines, but succumbs to the prairie, which prevails everywhere between the Red River and the Rocky Mountains, except in wooded and damp localities. In Saskatchewan, the flora of the forest and that of the prairie intermingle. The flora of the forest belt of the North-West Territories resembles that of northern Ontario. In the mountainous country of British Columbia, the flora varies in accordance with climatic conditions: in some of the valleys open to the south, it is partly peculiar to the American desert; near the Pacific coast, the woods and open spaces are filled with flowers and shrubs.[1]

SURFACE FEATURES.

Canada is widely afforested, in deep zones extending from south to north, and most densely in the Cordilleran forest of British Columbia. The southern forest, between Manitoba and the Atlantic, which has been largely cleared in the Maritime Provinces and along the shores of the St. Lawrence, merges into the densely wooded zone of the northern forest in a line drawn irregularly between Winnipeg and the mouth of the St. Lawrence. The great Northern Forest extends in a deep zone right across the continent, getting thinner and thinner as the sub-Arctic lands are approached. Finally, prairie lands extend from Winnipeg to the Rockies, merging in the north, along the settled area, into mixed prairie and woodland country. There are national parks and timber reserves in many parts of the Dominion.

For the purposes of our survey, it will be convenient to divide up the country into the following natural regions:—(1) *The Woodland Region*, covered by the Southern Forest, including the Maritime Provinces, Ontario, and Quebec; (2) *The Prairie Provinces* of Manitoba, Saskatchewan, and Alberta (partially cleared in the south, and opened up to agriculture); (3) *The Sub-Arctic Lands* of the Northern Territories, including Hudson Bay, between Bering Sea and Labrador Coast; and (4) *The Pacific Slope* of British Columbia.

From west to east, bordering the United States frontier, the Cordilleran forest region extends over a distance of 500 miles; the prairie lands, including mixed prairie and woodland, cover a zone at least 800

---

[1] *Encyclopædia Britannica*, 11th Ed., vol. v., p. 147.

miles in length and from 300 to 400 miles in depth; and the woodlands of the Southern Forest, bordering the 50th parallel, extend over a distance of 1500 miles, to the mouth of the St. Lawrence River. The northern limit of forest approximates to the isotherm of mean summer temperature: 50° F.

THE WOODLAND REGION:

ONTARIO, QUEBEC, NEW BRUNSWICK, PRINCE EDWARD ISLAND, NOVA SCOTIA, CAPE BRETON ISLAND.

Surface Features.—The Woodland Region lies between the Great Lakes and Hudson Bay, and extends into the Maritime Provinces. It is pre-eminently a region of unrivalled inland waterways, in which agriculture and woodcraft vie with mixed farming and mineral production. The most populous and important provinces of Canada and the seat of the Dominion Government are included in its area.

A magnificent system of waterways opens up from the Gulf of St. Lawrence, which carries tide-water for 700 miles—or, from the mouth of the St. Lawrence to Three Rivers, 330 miles inland from the sea—and extends to the head of Lake Superior, at the western extremity of Ontario, over a distance of 2384 miles. Combined with the canals—eight in number, with fifty-four locks—a continuous waterway, nowhere less than 14 ft. deep, is thus provided between Lake Superior and the Atlantic. The canal system begins at La Chine Rapids, above Montreal, which is the first obstruction to free navigation from the sea; and at Montreal—the principal emporium of trade and nodal point of communications—the first railway bridge crosses the river. Montreal is thus the gateway of Canada, from whence railways radiate in all directions.

Above Montreal and La Chine, the St. Lawrence River—still a mighty stream, carrying numerous islands in its bed—connects with the chain of fresh-water Lakes—Ontario, Erie, Huron and Michigan (united on the same level), and Lake Superior—each a little larger and higher as one proceeds from the one to the other. Between Lakes Ontario and Erie, the famous Falls occur on the Niagara River, about half-way; a navigable river connects Lakes Erie and Huron; and between Lakes Superior and Huron the connecting stream is obstructed by the Sault Rapid—or Leap of St. Mary—locally known as the Soo, past which a canal with only one lock carries ships drawing 20 ft. of water.

Navigation on the St. Lawrence to Quebec and Montreal—both ice-bound ports—is usually open from the end of April to the end of November. Above Lake Superior, navigation, with interruptions, may be continued westwards to almost the foot of the Rockies by Rainy Lake and River, Winnipeg Lake and River, and the North Saskatchewan River.

Natural Resources: Minerals.—Mining, although only in its infancy, has become a leading industry of Canada, second only to agriculture. Coal and iron in proximity appear in many parts of the Maritime Provinces; and lignite (brown coal) deposits are found on the southern border of James Bay (Hudson Bay). Ontario, Nova Scotia, and Quebec are the only provinces of Canada in which at present iron ores are smelted and converted into iron and steel.

Gold is obtained in Nova Scotia from saddle reefs and veins in fissures associated with them. Auriferous veins are found also in eastern Ontario. Other gold districts are the Beuce, near Quebec, and a considerable area of the country between the Lake of the Woods and Lake Superior, which contains also Atikokan and Hunter Is. iron districts. Silver is mined at Thunder Bay, near Port Arthur, where it occurs in the form of the sulphide, argentite; and silver-bearing galena is found also in New Brunswick and Nova Scotia. In 1910, Canada took third rank in the world's production of silver. Copper deposits are widely distributed, the bulk of the production coming from Quebec and Ontario.

Canada supplies the greater portion of the world's output of nickel (chiefly to the United States) from the deposits at Sudbury, Ontario, near Lake Superior; a small proportion of cobalt and traces of platinum are present, in addition to the nickel. Nickel occurs also in the ores of cobalt near Lake Tenniskaming, in Huronian rocks. At the present time, too, about 90 per cent. of the world's output of asbestos is derived from the Thetford and Black Lake region of the eastern townships of Quebec, south of the River St. Lawrence, where the mineral is worked in open quarries. Canada supplies over 85 per cent. of the world's total production of corundum, the principal deposits being in Renfrew County; and she has become one of the chief mica-producing countries, mainly for electrical purposes.

Graphite of fair quality occurs at Cape Breton and at other localities in Nova Scotia, as well as near St. John, New Brunswick. The deposits which, up to the present, have been worked chiefly are in the crystalline limestones and gneisses of Quebec, north of Ottawa. Gypsum is quarried in large quantities in Nova Scotia and New Brunswick, where it occurs in beds of Carboniferous age; whilst in Ontario it is found in Silurian beds.

The mineral wealth of Ontario is both varied and extensive: there are, perhaps, few regions of the world in which more promising prospects are to be found. Petroleum occurs in large quantities in the Ontario peninsula; but the Province has no coal (apart from the deposits of lignite, referred to) and the development of the mining industry is thereby handicapped.

Of the principal metals and minerals produced by Canada, in 1911, coal took the lead in the value of the year's output; and was followed by silver, nickel, and gold. The coal production of Canada may be divided into four parts:[1] Nova Scotia, British Columbia, and Alberta, in the rank indicated, contributed the bulk of the output, a small residue coming from Saskatchewan, New Brunswick, and Yukon. In Nova Scotia—by far the most important area—coal has been mined for over two centuries. The coals, belonging to the middle measures of the Carboniferous forma-

---

[1] *Commission of Conservation* (Canada), 1911: Lands, Fisheries and Game, Minerals, p. 433.

tion, are found in several parts of the Province but chiefly at Sydney, where it is easily and economically mined. Practically, all the Nova Scotia coal-fields have the advantage of being on tide-water.

As the coal deposits of Canada are located only in the eastern and western portions of the country—omitting the rapidly growing industry in Alberta and Saskatchewan, where the coal mined is retained for home consumption—it is necessary for Ontario, Quebec and Manitoba to import coal from the United States. Consequently, in many parts of Canada coal has not yet replaced wood as a fuel. Natural gas wells are productive in six of the counties of Ontario—the main source of Canada's supply.

Water-Power.—Water-power is not usually included among the natural resources of a country; but in Canada it is one of special significance, constituting an important national asset. The Hon. Clifford Sifton, chairman of the Conservation Commission, claims that Canada possesses nearly one-half of the total available water-power of the globe.[1] But only a tithe of this has been tapped. The total water-power developed in Canada during 1910 amounted to 1,016,521 h.p., of which the bulk was electrical energy, and the principal industry using it was the paper and pulp mills. Ontario takes the lead in the development of its water resources, the Ottawa river and its tributaries forming a power district of signal importance. The power stations of the Niagara Falls constitutes the most important hydro-electric power site in the world. Quebec, too, is well served: it is calculated that 80 per cent. of the power used in the Province is water-power. The Maritime Provinces, having no large watersheds—although watershed area is by no means correlative with water-power—rely mainly on steam-power for economic requirements. In Manitoba, the largest water-power resources are on the Winnipeg River. In British Columbia, Yukon Territory, and the vast North-West, water-power, though little used so far, is practically unlimited.

Agriculture and Farming.—Although mining is attracting an increasing amount of attention, and some day may rival agriculture, the latter must be regarded as Canada's chief national asset. The resources of agriculture tend to settle and develop the land as no other industry can do. Moreover, these resources create interests that foster the best elements of national life and have an important bearing upon the fortunes of the nation.

Agriculture is the chief industry of the Maritime Provinces, in which soil and climate combine to offer the most favourable productive conditions. The forests have been cleared extensively along the coastal zones, where the fertile soil is capable of supporting a relatively large population, particularly in Prince Edward Island—"the Garden Province of Canada"—the greater portion of which is occupied by farms and lots. Even in Nova Scotia, which has so many varied resources—mines, forests, and fisheries—mixed farming, dairying, and apple-growing are carried on with satisfactory, if somewhat halting, results: the apples of the salt marshlands of Annapolis valley are famous, but only ten per

---

[1] *Commission of Conservation* (Canada), 1911.

cent. of the land is under orchard cultivation. New Brunswick, larger and less densely populated than the other Maritime Provinces, enjoys every advantage as an agricultural and fruit-growing country, and has a better water-supply. The great majority of the people in New Brunswick are farmers; but stock-raising and dairy farming have been found to be more profitable than grain crops or fruit-growing.

Ontario has large areas with soil and climate suitable for fruit-growing; and practically all the fruits of the temperate zone can be produced in the Province. The centres of the industry are the Niagara Peninsula and the regions round Lake Erie, where the climate is particularly mild, favouring the cultivation of grapes and peaches in addition to hardier fruits. Ontario is said to produce 75 per cent. of all the fruit grown in Canada. In agriculture, too, Ontario enjoys a leading position; but the Province has found dairying more profitable than grain-growing. In particular, the cheese industry—both in Ontario and Quebec—is of great importance to Canada, which exports every year, chiefly to Great Britain, quantities approximating five million pounds sterling in value.

Agriculture, too, is the principal industry of Quebec Province, in which mixed farming generally is carried on by the French-Canadians. The St. Lawrence Plain is well suited for agriculture; and it is fully settled. In addition to the ordinary farm crops of temperate countries, tobacco is grown on a limited scale. Dairy produce and market gardening find a ready market.

The Federal and Provincial Governments foster the dairy industry in various ways. All the provincial departments of agriculture, except that of Nova Scotia, where dairy farming is not so important, have organised dairy divisions. Dairy schools are maintained in Ontario, Quebec, and New Brunswick, as well as in Manitoba. Experts are engaged to visit creameries and factories during the working season in order to give instruction and advice. Usually, Government inspection is carried out; and the commercial interests of the producer are, together with other matters, cared for by the Dairy and Cold Storage Commissioner and Staff.

An important factor in the practical and scientific development of the agricultural resources of Canada has been the invaluable assistance given by the experimental farms. The farms have a large staff of skilled investigators; and experiments are made as to the best methods of preparing the land for crops, maintaining its fertility, and disposing of the produce to the best advantage. The Central Experimental Farm is at Ottawa, which is close to the boundary between Ontario and Quebec. A branch at Nappau, in Nova Scotia, serves the three Maritime Provinces.

Forest Products.—The eastern Provinces were at one time covered with forests, and, in the early days of settlement, lumbering was the principal occupation of the people. It still is one of the most important industries of Canada. The forest area of the Woodlands Region is officially estimated[1] as under:—

[1] *Statesman's Year Book*, 1912, p. 254.

	Thousand Acres.
Quebec,	120,000
Ontario,	40,000
New Brunswick,	7,500
Nova Scotia,	5,500
Total,	173,000

This area is nearly equal to the forest areas of British Columbia and of Manitoba, Saskatchewan, Alberta and Territories. The Crown Forests of the Woodlands Region belong to the Provincial Governments.

In valuable species of timber, the Eastern Forest is the richest in Canada. It contains a monopoly—or what is now left after clearance and forest fires—of the hardwoods, of white and red pine, and rich stocks of spruce, cedar, balsam fir, and larch. But hardwoods are now scarce, and have to be imported in excess of the local supply. The Eastern Forest produces about four-fifths of the lumber cut in Canada.

Maple-sugar is made generally from the hard maple, which is a common tree from Nova Scotia westwards to Lake Superior, and reaches its greatest size in SW. Ontario.

Fisheries.—It is claimed that the Canadian territorial waters contain the principal food-fishes of commerce in greater abundance than the waters of any other part of the world. The value of the catches for 1909-10 ranged between seven and one million dollars—in the order given—for salmon, cod, lobsters, herring, halibut, and haddock. Canadian salmon ranks, nearly commensurately in value, with the export of dried codfish from Newfoundland. As in Newfoundland, the coastlands of the Atlantic Provinces, from the Bay of Fundy to the Straits of Belle Isle, are highly incised and indented; and this coastline, measuring over five thousand miles, contains innumerable natural harbours and coves. Deep-sea fishing is pursued on the banks lying from 20 to 90 miles from the Canadian coast, where cod, haddock, hake, and halibut abound; and the inshore fishery includes these food-fishes as well as herring, mackerel, smelt, flounder and sardine. A most extensive lobster fishery is carried on along this coast; and there are good oyster beds in many parts of the Gulf of St. Lawrence. In the inland lake fisheries, the principal commercial fishes caught are whitefish, trout, pike, sturgeon, and, in the Great Lakes of Ontario, fresh-water herring.[1]

The fisheries of the Gulf of St. Lawrence and the Bay of Fundy are the oldest in Canada, and rank among the great fisheries of the world: cod, mackerel, halibut, salmon and herring are taken in great numbers. The lobster and oyster industries of Prince Edward Island also are of considerable magnitude. According to official returns (1911), Nova Scotia is the leading fish-producing Province of Canada, followed by British Columbia, New Brunswick, Ontario, Quebec, and Prince Edward Island.

---

[1] *Commission of Conservation* (Canada), 1911: Fisheries and Game.

Game, and Animal Products.[1]—Moose and caribou abound in many parts of the forests of Quebec, while Virginian or red deer have increased, in some parts to an inconvenient extent. The greater part of the country about The Height of Land between Lake Edward and Kiskisink is full of game. Moose, caribou and deer are found in the Ottawa district, where partridges, wild ducks and wild geese abound in their seasons. In the country north of the Ottawa, the moose have been thinned out. On the south side of the St. Lawrence, red deer are plentiful in almost every part of the Eastern Townships where any wild forest lands remain. Almost the whole southern portion of the Province of Quebec, adjacent to the State of Maine—an unsettled wilderness—is full of big game. The caribou have disappeared from Prince Edward Island and from the greater part of the Province of Nova Scotia: on Cape Breton Island they are found in considerable numbers.

The birds of Canada are, for the most part, migratory. Nearly all the sea-birds of the British Isles are found in Canadian waters or are represented by allied species. In the Gulf of St. Lawrence, the gannet is very abundant. There are several varieties of geese and grouse—among the latter, the ruffled grouse, perhaps the most valuable of the game birds—and in certain parts of Ontario the wild turkey is found occasionally. New Brunswick is the best game province, ranking next to British Columbia in that respect; the extent of the hunting ground covers about four-fifths of its area.

Animals and their products come next to agricultural products and manufactures among leading exports from Canada; but we have no statistics to show the relative share of the Woodlands in that respect. Ontario and Quebec, however, possess by far the largest share of live stock (horses, horned cattle, sheep and pigs), of the provinces of Canada.

THE PRAIRIE PROVINCES:

MANITOBA, SASKATCHEWAN, ALBERTA.

Development.—To a large extent, the Prairie Provinces have been carved out of the more or less densely wooded Northern Forest. The treeless prairie is not so dominating as is commonly supposed, except in the settled area.[2] The settled area (prior to the new boundary arrangements) embraced the greater portion of the Provinces of Saskatchewan and Alberta and all Manitoba. The population of Manitoba is now exceeded by that of Saskatchewan, and Alberta is not far behind.

Although Canada is among the important grain-producing regions of the world and cereals are cultivated over a large part of the Dominion, the rapid development of Western Canada during recent

---

[1] *Commission of Conservation* (Canada), 1911; Fisheries and Game.

[2] Cf. *Atlas of Canada*, by James White (Department of the Interior): Forests—Plate 8.

years has been due in great measure to its natural capacity for the successful production of wheat and oats. The undulating prairie lands, which extend westwards from the Great Lakes to the foot of the Rocky Mountains, embracing Manitoba—the central province of the Dominion—Saskatchewan and Alberta, produce wheat of very high quality, in addition to oats, barley and rye. "It is not unwise to predict," says Mr. W. P. Rutter, in *Wheat Growing in Canada*, "that future wheat exporting on a large scale from America will be with Canada rather than with the United States or even Argentina. No other conclusions are possible than that Canada has greater possibilities of an immediate and rapid increase in wheat production than any other country in America, and that, though it is not possible to gauge all her potentialities as a wheat producer, yet her possible future wheat lands would seem to be greater than those of the United States and Argentina" (p. 292). During the twelve months ended July 1911, Canada exported more wheat than any other country in the world.

Surface Features.—The Prairie Provinces are built up in three steppe-like zones along the eastern flank of the Rocky Mountains, with a general strike corresponding to that of the main axis; north-west and south-east. These zones represent natural areas, in striking contrast to the Provincial boundaries, which partition the country after the manner of town-lots. In general terms, Manitoba may be said to occupy the lowest prairie-steppe (500 to 1000 ft., *s.m.*); Alberta, the highest (2000 to 5000 ft.); and Saskatchewan, the middle region (1000 to 2000 ft.) The northern parts of Alberta and Saskatchewan are drained through the Athabaska and Peace Rivers into Great Slave Lake, from whence the Mackenzie issues in its long course to the Arctic Ocean. The greater portion of the Province of Manitoba lies in the dry bed of a geologically ancient glacial lake, the soil of which—a rich clay, or black lake-silt, for the most part—is the most fertile in the world for growing wheat. The basin is cut up into lake, marsh, and dry land—the marshlands overlying a rock floor of Silurian or Devonian limestone, which protrudes in places. Lake Winnipeg and the other lakes, all inter-communicating, are expanses of shallow water which, while reproducing, *in petto*, some of the features of the Great Lakes, are mere lagoons broken up by marshlands. This inter-lake-land, which drains into Hudson Bay by the unnavigable Nelson River, receives on its southern border the waters of the navigable Red River and its affluent, the Assiniboine. The watershed between the Red River and the Lake of the Woods—the last true lake, from east to west—is imperceptible.

The Manitoba escarpment in the west leads up, in steep slopes, to the middle steppe of Prairie-land, which contains hills and deep river valleys, draining into Hudson Bay. The scarp has no rocks, and consists of shale, sand, clay, and marl of the uppermost Secondary Age: the deep blue clays of Cretaceous formations rival in fertility the best lands of Manitoba. The slope westwards is gradual and wooded in parts, for 350 miles—a region of "limitless prairie," with hills and trees everywhere. The North Saskatchewan River has a permanent river-bed and follows the straightest course of all the meandering streams of Prairie-land: it is navigable between May and August practically throughout its length, but interrupted by rapids, from the head of the Grand Rapids, near Lake Winnipeg, to Edmonton and beyond.

The uppermost steppe of the High Western Prairie is drier, barer, and more hilly than the other two we have been considering:. it resembles, in some degree, the arid lands of the United States; but a strip between the dry tract and the Rockies receives the warm Chinook winds from the Pacific. Between the dry tract and the rolling prairies of the Peace River district, the prairie land is interrupted by swampy forests round the head-waters of the Athabaska. The Peace and Athabaska Rivers flow northwards into the Mackenzie; and the North and South Saskatchewan drain east into Lake Winnipeg, watering prairie lands that have confused watersheds.

Valuable power sites are located on the Athabaska and Peace Rivers.

Natural Resources: Minerals.—Coal is found in many parts of the third steppe country; natural gas is tapped from Devonian formations; oil occurs in the north and in the south; bitumen is found in the Athabaska districts; salt-springs are worked on the south-west shores of Lake Winnipegosis and at Swan Lake. Coal and lignitic coal are the principal economic minerals found in this great central plain.

The wealth of Manitoba lies in its soil. Minerals are not worked to any extent, at present. Iron ore occurs; and some gold-bearing Huronian rocks have been located on the east shore of Lake of the Woods. The Alberta Coal District—the most extensive in Canada—is estimated to cover an area of nearly twenty thousand square miles; and there is a lignite-yielding area in Manitoba.

Fisheries.—The fisheries of Manitoba are important; those of Saskatchewan and Alberta, relatively insignificant. The principal catches are whitefish, pickerel, and pike.

Agriculture and Ranching.—The Prairie Provinces have a total area of three hundred and seventy million acres, of which over twelve million are water.[1] While all are fertile, the best lands carry soils very rich in the constituents of plant-food. Wheat of the finest quality is raised on the prairie lands of Manitoba and Saskatchewan. Alberta, with a milder winter climate, is devoted chiefly to horse and cattle ranching. Manitoba—the earliest of the Prairie Provinces to be settled—is passing through a stage of transition from wheat-growing to mixed farming.

The total wheat acreage of the Prairie Provinces, in 1911, was over ten million acres—Alberta having five million and Manitoba only three. In Saskatchewan, the wheat yield (over a million and a half acres, in 1911) has decreased. In oats, too—the next most valuable crop—Alberta leads the way, and Saskatchewan has a large share. Barley and flax also are valuable crops, the latter coming mainly from Alberta. In Alberta, dry-farming is practised, and agriculture is keeping pace with the development of communications.

There is ample room for extension in the cultivation of wheat and

---

[1] This estimate does not cover the new provincial boundaries. *Vide*, p. 546.

the manufacture and exportation of flour in the Prairie Provinces. Canadian wheats and flour command the highest prices in the markets.

Cordilleran Belt.

A strip of territory in the south-western portion of Alberta abuts on the eastern watershed of the Rocky Mountains. Near the height-of-land between Alberta and British Columbia, many peaks rise from 10,000 feet to 12,000 feet above sea-level, Mount Robson being 13,700 feet. The outer ranges in Alberta have steep cliffs towards the north-east; along the dip of the beds, towards the south-west, the slopes are gentler.

Pacific Slope.

Physical Features.—Proceeding on our survey from east to west, we may enter British Columbia from Alberta by several gateways: among others, by the Yellowhead Pass, which lies 3738 feet above sea-level, or by the still higher passes—the Athabaska, the Kicking Horse, or the Crow's Nest. The watershed is confused, giving rise on the steep western slopes, to tortuous river-beds. West of the Rockies, high mountains, divided by deep valleys, prevail in five or more ranges running parallel to the coastline, each of which is distinguished by its physical features and mineral yield.

The coast ranges in most parts do not exceed an elevation of 9000 ft.; but mountains everywhere—a "sea of mountains"—bar the way to freedom of intercourse between east and west. Our view of British Columbia, therefore, may best be taken from the Pacific, to which the Province belongs climatically and structurally.

The great rivers are navigable—except, of course, in the gorges—throughout their course. The Columbia and the Fraser—the chief rivers of British Columbia—take their rise in the longest and most characteristic valley of the Province, which extends north-west and south-east for 800 miles, and gives rise to other rivers.

The Fraser River, unlike the Columbia, flows throughout in Canadian territory, and, bursting slantwise through the Coast range, its waters, reinforced by the Thompsons, reach the ocean near New Westminster. Five hundred miles further north-west, the Skeena breaks through the coast range on its course to the sea. From the mouth of the Skeena, nearly opposite Queen Charlotte Islands, proceeding northwards for forty miles, the only other large river of Canada to reach the sea is the Nass, which empties itself into a fiord, from which the Portland Canal, separating Dominion territory from Alaska, is reached.

Four hundred miles of United States territory constitute a barrier between British Columbia and the Pacific, in the fiords that extend north-westward to Yukon territory. The United States, therefore, holds the main entrance to the mineral district of the Yukon, which is entered by way of Lynn Canal and the White Pass. There are other entrances in Canadian territory, but not of equal value: and the White Pass, leading to Bennett, in British Columbia, is commonly used.

All the main river-valleys of British Columbia rise, tier above tier, in terraces; and these are afforested by equally characteristic trees, marking off this province as distinct from the remainder of Canada: the giant Douglas fir and the great girth of the red cedar are well-known features. Resembling these, respectively, are the white spruce and yellow cedar. Trees 300 ft. high are common.

The Cordilleran forest region is a zone, densely wooded, four hundred miles deep in the southern part, which narrows and thins out as more northerly latitudes are reached, until the sea is touched near Mount St. Elias.

Owing to the general configuration of the country, by which the coastal ranges intercept the rain-bearing winds of the Pacific, dry and wet strips of land are another characteristic of the Province. Farms are found as high as 2500 ft., and cattle graze up to another thousand feet, above sea-level.

Natural Resources.—The natural resources of British Columbia are diverse in character and almost limitless in extent. They comprise fisheries, mines, lumber, agricultural and fruit lands.

Minerals.—The production of various metals in the Province shows progressive growth during the last two decades, except for the years 1902-4 and 1908-9, in which there was a decrease of output. In this production gold (placer and lode) ranks first, followed by coal (including coke), copper, silver and lead. Although, in respect of these minerals, the Cordilleran belt is recognised as one of the great mining regions of the world, the total mineral output of British Columbia is still at an early stage of economic development; but the gold production of Canada comes largely from Yukon and British Columbia. Since the discovery of the rich placers in 1897, Yukon has produced over 136½ million dollars' worth of gold, and its yield is increasing. The known resources of placer gold in the Klondyke alone have been estimated at a hundred million dollars.[1]

British Columbia and Ontario are practically the only Provinces of Canada producing silver, but of this output the share of British Columbia is only ten per cent., the remainder being derived from the cobalt-silver ores of Cobalt, Ontario.

At present, the reserves of copper ore in Canada are greater than that of any other metallic mineral, excepting iron; and British Columbia contains at least three copper districts of importance: the Rossland, Boundary, and Coast districts. The copper is produced from sulphide ores containing gold and silver; and only part of the production is consumed in Canada.

The production of lead in the Dominion is obtained almost entirely from the Province of British Columbia, chiefly (since 1908) from Fort Steele Mining Division.

The Pacific Coast and the Western Mountains Division, containing the semi-anthracite and bituminous coal-fields of Vancouver Island, at

---

[1] *Commission of Conservation* (Canada): Minerals, p. 409. The Report furnishes us with material under this sub-section.

Nanaimo and Comox, the bituminous coal-fields of the interior, and the lignites of Yukon are important assets in the coal resources of the Dominion. The coal-fields of British Columbia are situated in the Eastern Rocky Mountain area, the Southern Interior basin, the Northern Interior basin, and the Coast District. The formations in the areas known to contain coal are the Cretaceous and the Tertiary: no rocks of Carboniferous age have been found, so far, to be coal-bearing. The series of mineral fuels range from excellent anthracites to lignites. Whereas in Alberta, with its rich coal-field, coal exclusively is used on the farms, about eighty-five per cent. of the fuel used on the farms of British Columbia is, of course, wood, of which the Province should have an inexhaustible supply, if adequately conserved.

Coal has been for many years one of the chief items in the natural wealth of British Columbia, and the importance of its extensive coal-fields, even in a country otherwise so rich in minerals, cannot be over-estimated. When located close to mining centres, as at Crow's Nest, the beds are of still greater importance, for smelting and other purposes.

Agriculture.—In spite of the total absence of prairie lands, British Columbia, with its multiplicity of land-forms and with every variety of climate, enjoys many advantages for mixed farming and fruit culture. Agriculture is restricted to the coastal plains, the valley bottoms, and the inland plateau. The whole of the province, south of latitude 52°, and east of the Coast Range, is a grazing country up to 3500 ft., and, where irrigation is possible, a farming country up to 2500 ft.

Vancouver Island, with its genial oceanic climate, took the lead in agriculture and fruit-growing, in the early days, when farming settlements were established in the Saanich Peninsula; and its orchards and homesteads now are comparable to those of Kent and Devonshire. Nanaimo, in addition to its coal resources, has good garden-land. The coastal plains are so heavily wooded that clearing is a costly preparation to the laying-out of farms; but good lands in the valleys of the Thompson and Fraser Rivers have been for a long time under settlement, especially round New Westminster and in the Lower Fraser valley. The most extensive farms and the largest cattle ranches are found in the Okanagan valley; and the Kootenay district, opened up by the C. P. R., contains excellent farming and ranching lands. Oats are the principal grain crop.

Food products for the mining population and fruit-growing, partly for export, are growing industries. Hogs, in small farming, are the most profitable of live stock.

Forests.—The vast extent of forest and woodland in British Columbia, covering an area of over 182 million acres, represents one of its chief natural resources. The coast is heavily timbered as far north as Alaska. The most valuable and widely distributed tree is the Douglas fir (Oregon pine), which grows as far north as latitude 55°, where it is supplanted by the cypress, or yellow cedar, red cedar, hemlock, and spruce. Firs—the staple of commerce, owing to their durability and strength—are widely distributed, from the coast to the Rocky Mountains; the best average trees are 150 ft. high, and measure

5 to 6 ft. in diameter. The bulk of this timber is found on Vancouver Island, on the coast, and in the Selkirk and Gold Mountains. The red and yellow cedars also are of marketable value on account of their straight grain and fine growth. Red cedar shingles are the standard in the markets of Eastern Canada; and there are many other trees of commerce which are manufactured into lumber, including white pine, tamarac, balsam, yew, maple, and cottonwood.

The province is unsurpassed as a field for the manufacture of paper-pulp and paper.

Game.—The greater portion of British Columbia being unsettled, and to some extent unexplored, game abounds everywhere. Moose show no sign of decrease—rather the reverse. Deer and bear are killed even close to Vancouver. Excepting antelope and musk ox, British Columbia possesses every species of big game indigenous to North America, even the rarest. In addition to wild fowl and game birds—some acclimatised—there is the finest trout and salmon fishing to attract the sportsman.

The best game district is Cassiar, the least accessible of all. The game to be obtained there includes Stone's mountain sheep (*Ovis Stonei*), and mountain goats are found everywhere in abundance. Black and grizzly bear also are to be found, while caribou and moose are plentiful. There still are some wapiti left on Vancouver Island.

Other good hunting districts are Savona, for wildfowl shooting; Bridge River and Chilcotin, for *Ovis Montana* and other game; East Kootenay, for the greatest variety of sport; West Kootenay and Kamloops, for fishing. The big-horn mountain sheep (*Ovis Canadensis*) reaches its highest development in the Rockies, and still is plentiful in south-east Kootenay.

Fisheries.—The coastline of British Columbia, extending from the 49th to the 55th parallel, between United States territory, is estimated to measure over 7000 miles, exclusive of Vancouver and Queen Charlotte Islands. Being highly indented and fringed with islets, the Pacific Coast offer a very favourable ground for inshore fisheries. The principal take is salmon, which are very plentiful at the mouths of the rivers Fraser and Skeena during the spawning season; and salmon canning is one of the leading industries of the Province.

According to the 45th *Annual Report* (1911-12) of the Department of Marine and Fisheries, British Columbia contributed by far the leading share (13 out of 34 parts) of the total yield of Canadian Fisheries, the bulk of which was salmon, herring and halibut ranking next in value. This was a record year. The salmon are of five varieties—sockeye, spring or tyee, cohoe, humpback, and dog.

Whales are fairly plentiful along the coast and in Bering Sea. Seals are decreasing, and sealing has fallen off since the Bering Sea Award, which imposed restrictions on their catch. The average number for five years, ended 1903, was 26,300 skins, as compared with an average of 62,000 for the previous five-year period. The headquarters of the sealing industry are at Victoria.

From this brief description, but more particularly if we consider economic conditions, it will be seen that British Columbia is essentially a Pacific State, whose interests "lie on the water": interests that will be affected profoundly by the opening of the Panama Canal. Its mines, its forests, its fisheries are all served best from the ocean, though it may continue to import some of its supplies through the back doors, so to speak, of the mountain passes.

Yukon Territory.

The territory of Yukon, between the frontiers of Alaska and the watershed of the Mackenzie River, is conterminous with the provincial boundary of British Columbia along the 60th parallel. It contains the gold districts of Klondyke and Kluane, and abuts on that of Atlin, B.C.

The physical features of the country approximate to those already given in respect of the Rocky Mountains, of which Yukon Territory is an extension, but the ranges are less distinct and regular. Nearly all the rivers are tributaries of the Yukon, which is navigable by river-steamers from Teslin Lake—one of its sources—to Bering Sea, a distance of 2400 miles. Except in the northern tundra, the lowlands are wooded—in parts covered only by poor scrub—and hardy crops can be raised.

Yukon Territory is essentially a mining country, somewhat inaccessible. In spite of its northern lands being situated within the Arctic circle, the climate is less severe than that of the North-West Territories, in consequence of the prevalent westerly winds. Most of its area falls within the Pacific basin, to which it naturally belongs.

Northern Territories.

The recent re-arrangement of Provincial frontiers under which Quebec, Ontario, and Manitoba reach the shores of Hudson Bay— Quebec absorbing Ungava, Ontario and Manitoba dividing Keewatin, up to the 60th parallel—leaves only the Mackenzie district to represent the North-West Territories, between the Yukon and Hudson Bay. But the barren and unprospected lands of Canada, whatever value they may be found to have in the future, are now better grouped under the title of Northern Territories, in order to convey a clear meaning in Regional Geography. The term "North-West Territories," as employed in the past, marked simply a point of view—that of the seat of Government at Ottawa. Thus, the Prairie Provinces originally were called North-West Territories—with Provincial capitals, first at Winnipeg, then at Regina; and, as the districts became organised and opened up to settlement they took on distinctive names. This will happen again; but, in the meantime, ignoring the expansion of Provincial frontiers, since we are restricting our survey to natural regions, we may return to the *status quo ante* in our geographical survey.

The Northern Territories comprise the unsettled lands of Canada, not previously surveyed by us, merging into the tundra and frozen wastes of the High North which, commonly coloured red on British maps, make so imposing a display of territory on Mercator's projection. These lands may well be styled Arctic Canada. Annexed to the Crown by an Imperial Order in Council, in 1880, and then transferred to the Dominion Parliament for purposes of administration, they are visited periodically by the North-West Mounted Police, who collect dues at specified posts.

The most striking feature in the structure of this region is the Archæan protaxis covering about two million square miles of territory. It includes Labrador, Ungava and most of Quebec in the east, Northern Ontario in the south, and the Great Central Plain in the west extending from the Lake of the Woods north-westwards to the Arctic Ocean, near the mouth of the Mackenzie River. Most of this vast basin is made up of barren Laurentian gneiss and granite, and it bears evidence of glaciation (*roches moutonnées*). Woodlands occur south of the 60th parallel, in scattered clumps of black spruce, white spruce, and with larch growing in the swampy tracts. North of this parallel, the country is almost treeless or only lightly timbered, with short grasses and sedges covering the undulating, stony plain. The timber line is reached near Churchill, on the west shore of Hudson Bay, but it does not extend so high north in Labrador. To the north of this are the "barren grounds," overrun by caribou and musk ox.

Except on the Atlantic border—in Labrador, where summits of the Nachvak Mountains are said to attain 6000 ft. in elevation—the plateau is covered with hills of uniform level; while the lowlands round Hudson Bay—the Northern Plain—extend to a great distance outwards. The northern parts of Alberta, Saskatchewan, and, to some extent, British Columbia, are drained by the Athabaska and Peace Rivers into Lake Athabaska, out of which the Slave River drains into Great Slave Lake—the main source of the Mackenzie River. From the head of the Peace River to the Arctic Ocean, the Mackenzie River is over two thousand miles in length, half of which is navigable by stern-wheel steamers.

Arctic Canada, except for a few fur-traders, is inhabited only by Indians and Eskimo. The winters are long and severe; but in the autumn, after a short spell of warm weather, the "Indian summer," with a bracing and exhilarating atmosphere, is said to be delightful. Hudson Strait is blocked with ice during eight or nine months in the year, when the North-West Passages all but preclude navigation: thus, the magnificent open waters of Hudson Bay, penetrating deep into British North America, are to some extent a *mare clausum*. Nevertheless a railway to Port Nelson is now under construction.

Northern Canada has been styled "the last great fur reserve of the world." The management of the fur trade is in the hands chiefly of the Hudson's Bay Company, whose posts are widely distributed throughout the Northern Territories. The principal skins brought in by the trappers are musquash, marten, mink, beaver, lynx, ermine, red and white fox. The most valuable fur is the silver fox, of which relatively few skins are taken.

(*Part II., "The People," to follow.*)

GEOGRAPHICAL NOTES.

EUROPE.

Discovery of Oil-shale in Skye.—It is announced that the Geological Survey officers have discovered a seam of oil-shale, 11 feet in thickness, in the island of Skye, the seam extending over a considerable distance. The shale is not of such good quality as those worked in the Lothians, but it is believed it will prove of sufficient value to be worked commercially.

AFRICA.

The Algerian Sahara.—We have published here at various times illustrated accounts of the investigations carried on at the Desert Laboratory at Tucson, Arizona, and it is interesting to note that a member of the staff of that laboratory, Dr. W. A. Cannon, contributes to the July issue of the *Bulletin* of the American Geographical Society an article on his travels in the Algerian Sahara. The article is illustrated by a fine series of photographs, and the incidental comparisons with conditions in Arizona are of much interest. Dr. Cannon includes in his paper a good many notes on the vegetation, and lays especial stress on the association in the Sahara of the jujube-tree (Zizyphus) with an unnamed species of Pistacia. He explains the association as due not only to adaptation to similar conditions of life, but also to a protective effect of the jujube on the young Pistacia plants. The jujube has small leathery leaves and spines and is not attacked by grazing animals; the Pistacia, with more abundant foliage, is greedily devoured by such animals. According to Dr. Cannon the Pistacia seedlings are only able to survive if the seed sprouts within reach of the branches of a jujube bush, whose spines protect the young plants until the crowns are high enough to be beyond the reach of the herbivores. Elsewhere, however, he notes that the small shrub Haloxylon, one of the Chenopodiaceae, is eaten by sheep, goats, and camels as well as by the wild gazelle, and is yet abundant and widespread in the desert. We are disposed to ask, if the Haloxylon can survive without special protection, why not the Pistacia?

Dr. Cannon states that he has noticed as a curious paradox that the intensely arid regions of the desert, when the soil is thin, have more plants to the square yard than slightly less arid areas with deeper soil. He explains this by stating that in the latter area the plants are larger and better developed, and thus each individual plant makes greater demands upon the water supply, and so prevents the growth of other plants near at hand. On the other hand, the plants in arid areas with shallow soils are so dwarfed and stunted that each plant requires less water, and thus sterilises, as it were, a smaller area around it. The areas with deep soil have, however, a flora richer in species than those with shallow soil.

General.

Australian Meeting of the British Association.—According to an article in *Nature*, in addition to the various excursions which are being arranged in connection with the meetings in the different capital cities of Australia during the visit of the British Association in 1914, two or more special excursions are being planned with the object of permitting selected members to see parts of the continent which will not otherwise be touched.

The first of these special excursions is to be in Western Australia, and is to be for a party of twenty-five to thirty geologists, zoologists, anthropologists and botanists. The Government of Western Australia and the committee in charge are prepared to arrange for railway facilities and hospitality for a period of from a week to a fortnight, and the regions to be visited lie in various directions from Perth.

The second tour would begin after the last meeting in Brisbane, and would be limited to four to five persons, selected as representative. The tour would give fine opportunities for the study of botany, geology and agriculture, and especially of the questions connected with the possibility of white settlement in the tropics. The trip would occupy about one month, and expenses would be defrayed by the Government of Queensland and the Administration of the Northern Territory.

The proposed itinerary of this tour is as follows:—Brisbane *viâ* Rockhampton to Longreach by rail, coach to Winton, rail to Hughenden and Cloncurry, motor to Croydon, and rail to Normanton. The party would then be taken to the mouth of the Norman river, and be met by the steamer belonging to the Administration of the Northern Territory, and conveyed across the Gulf of Carpentaria, and about one hundred miles up the Roper river. They would proceed through the territory by motor-car (there are no roads) to Pine Creek, thence by rail to Port Darwin, where the steamer for Colombo would be met.

"Freezing Caverns."—Those who take an interest in the bypaths of physical geography are probably acquainted with the phenomenon of "freezing caves." These are cavities and passages which contain ice in summer, though the temperature of the surrounding region is far above freezing-point. Such cavities are not very uncommon in limestone districts, and a controversy has always existed as to whether the ice is actually formed in summer, or whether that found at this season is merely a remnant of a previous accumulation. In a recent article in the *Popular Science Monthly* Mr. M. O. Andrews gives a very interesting account of an "Ice-mine" in the Sweden valley in Pennsylvania, and supports strongly the theory that the ice forms in summer. According to him there is an indraught of cold air into the "mine" in winter, with the result that very cold air is stored in the fissures at the bottom of the pit. In summer, on the other hand, especially on hot days, there is an updraught from the cavern to the free air, and the rising air is cold enough to freeze percolating water, and thus form deposits of ice on the sides of the pit. So long as the supply of cold air within the fissures lasts this effect continues.

In an article in *Science* Mr. Arthur Miller vigorously attacks Mr. Andrews' views. Apparently he accepts the general outline of the explanation offered for ice-caves in general, but he denies that the formation of new ice goes on in summer, though it may continue after the external air has risen above freezing-point.

Commercial Geography.

Cotton-Growing in the Sudan.—We noted here (p. 157) the announcement that the British Government have guaranteed to pay the interest on a loan to be raised by the Government of the Sudan to the extent of £3,000,000, the money to be used for irrigation and other schemes for the promotion of cotton-growing. A recent issue of the *Bulletin* of the Imperial Institute gives some account of the existing position of affairs in regard to cotton-growing in the Sudan.

At present the chief areas where cotton is cultivated are the Tokar district of the Red Sea province, parts of the Kassala province, and of the Gezira plain, and along the Nile between Khartum and the Atbara river, principally in Berber province. Only in the last region is irrigation meantime practised systematically. As, however, it is proposed to use a part of the loan mentioned above for irrigation works in the Gezira plain, experiments on a large scale are now being carried on there. At present considerable amounts of rain-grown cotton are produced in this region in good years, but the uncertainty of the precipitation causes great fluctuation in the amount produced. In the Tokar district cotton is grown on land flooded by the annual overflow of the Baraka river. Until recently the cotton produced here was of poor quality, but under Government supervision an improvement is taking place. In the Berber province the yield is poor and the industry is not very thriving, while the Kassala yield is mostly used locally or sent to Abyssinia.

The following figures indicate both the fluctuations in the total yield and the progress which is being made. Total export of cotton, ginned and unginned, from the Sudan from 1908-1911 inclusive:—

	Weight.	Value.
1908, . . .	3477 tons	£82,762.
1909, . . .	2481 „	57,707.
1910, . . .	6893 „	230,551.
1911, . . .	5141 „	241,988.

In first nine months of 1912 there was a very large decrease both in weight and value as compared with the corresponding months of 1911, owing to a shortage in the amount of rain and flood-grown cotton, but the complete figures are not yet available. The figures given show, however, that the industry cannot be regarded as stable when climatic variations produce so marked an effect, and suggest that the future in the Sudan must lie with cotton grown on irrigated land.

Tobacco-Growing in England.—One of the crops which the visitor from North America or the continent of Europe misses in Great Britain is the tobacco plant. Attempts are now, however, being made to cultivate this plant in England, and small experiments are also being carried on in Scotland and Wales. Ireland has now a considerable area (about 150 acres) under tobacco. In England experiments are being carried on at Byfleet in Surrey on a sandy soil, with shelter belts of artichokes; at Fleet in Hampshire, on hop land consisting of a silty loam; at Methvold in Norfolk and at Elveden Hall, Suffolk, as well as in Lincolnshire and Worcestershire. The Hampshire experiment is being carried on on the largest scale, and here a large curing barn has been built. In Norfolk tobacco is being grown on small holdings in patches, and the crop is stated to look well. The experiments are being carried out by the Tobacco Growers' Society with the aid of advances from the Development Commission, and under the supervision of an expert from Rothamstead, and it is intended that they should be continued for a period of five years.

Railways in Nigeria.—The construction of a new railway line in Nigeria has been sanctioned by the Imperial Government, and the line will be started without delay, and pushed forward as rapidly as possible.

The terminal point of the line, which will be some 400 miles in length and will probably take three years to construct, has been fixed at the head of the Bonny estuary, and will be called Port Harcourt. From Port Harcourt the line, which is called the New Eastern Railway, will run through the Central Province, traversing a rich and populous district, to the coalfields near Udi, a distance of about 120 miles. Thence it will run to the Benue River, crossing that stream by means of an important bridge a little below Abinsi. It will next run to the neighbourhood of Jemaa, on the tin-fields (to which a branch line may be subsequently constructed), and thence to the point at which the existing line crosses the Kaduna River. There a junction with the Kano-Lagos-Baro system will be effected. The bar at Bonny is said to be the best in British West Africa, affording a depth of 23 ft. at high water, and any vessel that can enter the Bonny estuary can reach Port Harcourt.

POLAR.

British Expedition to King Edward's Land.—It is announced that Mr. J. Foster Stackhouse, F.R.G.S., F.R.S.G.S., who was associated with Captain Scott in organising the voyage of the *Terra Nova*, is arranging an expedition to King Edward the Seventh's Land.

It is proposed that the members of the expedition sail from the Thames about the middle of August 1914, in the steam yacht *Polaris*, a ship especially built for ice navigation in accordance with designs approved by an international committee of explorers, including Charcot, de Gerlache, and Nansen. Captain Scott saw the plans of construction and made suggestions concerning them. The *Polaris* was built at

Sandefjord, in Norway, and her trial trips have been quite satisfactory. The expedition will, it is expected, be away for twenty months or more.

EDUCATIONAL.

The Teaching of Plant Geography.—Mr. N. Miller Johnson sends us the following note on this subject:—"There are many books dealing with Plant Geography; there are few treating the subject from an educational point of view. Of these latter may be mentioned the section on this subject in *An Introduction to Practical Geography*,[1] and a more recent book by Dr. Hardy.[2] In addition to these may be mentioned various exercises which occur here and there in the ordinary geographical text-books.

"It appears to the present writer that in the case of older pupils who have already done some Elementary Botanical Nature Study a course in Plant Geography would give definiteness to nature knowledge, especially in those schools where microscopic work is almost an impossibility. The object of the present paper is to make suggestions for such a course. The suggestions, if they have no other merit, have been actually carried out, and are therefore based upon actual experience:—

"I. The effect of warmth, light, and moisture on plant life. Recapitulation of former experimental work, together with outdoor observations. The influence of altitude. A study of vegetation survey work (see papers by Dr. W. G. Smith and others).[3] Alpine plants.

"II. The distribution of cereals considered as a question of climatic control. The preparation of maps (see Economic Atlas).[4] Wheat and oats in Great Britain.[5] Examination of Map of British Isles, showing average annual rainfall, highest summer and lowest winter temperatures. Correlation of distribution with these factors. Preparation of diagrams respecting comparative areas under oats, wheat, turnips, etc.

"III. Soil and Flora. A study of the climate of the home district. A comparison of rainfall, temperature, etc., with other parts of Scotland. A study of tables correlating vegetation zone, altitude, and rainfall.[3] A study of the geological formation of the home district. Classification of plants according to water supply. Xerophytes, Mesophytes, Hydrophytes. The occurrence of different plants on sharply defined areas, such as peat moor, sand dune, deciduous wood, etc. Plant associations.

---

1 Simmons, A. T., and Richardson, H.—*An Introduction to Practical Geography.* Macmillan, 1906.

2 Hardy, M. E.—*An Introduction to Plant Geography.* Oxford: At the Clarendon Press, 1913.

3 Smith, R., and Smith, W. G.—"Botanical Survey of Scotland." *Scot. Geog. Mag.*, 1900, 1904, 1905.

4 Bartholomew, J. G.—*Economic Atlas.* Introduction by Lyde. Oxford: At the Clarendon Press, 1911.

5 Bridges, J. S., and Dicks, A. J.—*Plant Study in School, Field, and Garden.* Ralph, Holland and Co., 1908.

"IV. Botanical Mapping. The mapping of dominant plant associations in fields, woodlands, etc. Correlation of distribution, with amount of humus, water content, kind of soil, etc. Mapping in colours of distribution of crops on 6-inch ordnance survey map of home district. The effect of geological and climatic conditions upon this distribution. The connection between the vegetation of a district and the occupations of the people.

"V. General survey of main plant landscapes of the world. Selvas, mangroves, savanas, deserts, tundra, mountains, etc."

NEW BOOKS.

EUROPE.

Adventures in the Alps. By ARCHIBALD CAMPBELL KNOWLES, author of "Jocelyn Vernon," etc. Illustrated. London: Skeffington and Co., 1913. *Price* 3*s*. 6*d*. *net*.

The title of this book is misleading, for the only "adventure" we have been able to find is that of being caught on the Gemmi path in a summer thunderstorm, while the author shows little knowledge of the mountains and glaciers usually connoted with most people by the term "Alps." But if there are few adventures there are many "reflections" on Alpine and other subjects.

The Highways and Byways of England: Their History and Romance. By T. W. WILKINSON. London: Iliffe and Sons, Limited, 1913. *Price* 4*s*. 6*d*. *net*.

This is not a volume of the well-known green-covered series, but deals with the highways of England as a whole.

Many roads date from the time when the country was one-third forest, and the remainder largely swamp. Naturally roads, like cultivation, kept to the higher levels, since the low ground was impracticable. Pack-horse traffic settled the lines of some routes, while others owe their origin to the turnpike system—in which this country was unique. Adam Smith and authorities of his era attribute our trade pre-eminence partly to the superiority of our means of communication—and probably with justice. But the motorist of to-day sees no benefit in having to follow a route originally designed to save pack-horses from the danger of being lost in bogs. Hence the Development Act is due quite as much to the hoary antiquity of our roads as to the invention of the motor.

The book is full of interesting lore, but why is there no index? The wanderer, say, in Hampshire or in Yorkshire, wishes to know whether there is anything in it about his particular district, and failing references to information will probably leave the book alone. This is the more to be regretted as Mr. Wilkinson is a real authority on his subject, and has presented here not only a history of the development of the national highways, but much local information of value. There are excellent illustrations.

Athènes. Par GUSTAVE FOUGÈRES. ("Les Villes d'Art Célèbres.") Paris: H. Laurens, 1912.

This is a beautifully illustrated book, giving a full account of ancient and modern Athens, with some description of the interesting sites and monuments in the vicinity of the town. Two introductory chapters describe, the one the scenery of Attica, and the other, in brief outline, its history. There is a short bibliography and an index. The author is full of enthusiasm for his subject, and the book may be cordially recommended to those who contemplate a visit to the "place where perfection exists," or who have already seen its splendours.

Travels in the Pyrenees, including Andorra and the Coast from Barcelona to Carcassonne. By V. C. SCOTT O'CONNOR. London: John Long, 1913. *Price* 10*s.* 6*d. net.*

To those who echo Kipling's line, "France beloved of every soul that loves or serves its kind," this volume will be welcome. It is devoted to the Eastern Pyrenees, a tract of country which has not been so long known to the British traveller as the Atlantic Pyrenees from St. Jean de Luz to Luchon. First of all we have that fascinating mountain eyry, Andorra, a republic "which has dexterously preserved its existence for a thousand years by dividing its allegiance between France and Spain." Then we have what the author styles "Vernet of the English," with the Canigou, regarding whose charms Kipling wrote a glowing letter, while at Vernet Lord Roberts laid the foundation stone of an English Church. Then follow the Abbey of St. Martin du Canigou, Villefranche, Perpignan, and Gerona, the starting-place for Barcelona, where, "if anywhere in Spain, resides the spirit of modernity, the faculty of organisation, the power of accumulating wealth by trade." The volume closes with a chapter on Carcassonne, whose ancient Cité is perhaps the most unique of all the sights of France, while historically none can be more interesting, for this magnificent natural stronghold was held successively by the Gaul, the Roman, the Visigoth, the Spaniard, the Moslem, and the Frank. The author might have dwelt more fully on the exquisite Cathedral of Saint Nazaire at Carcassonne, but as his book relates chiefly to the Pyrenees, Carcassonne can be regarded as only a distant excursion made more easily from Toulouse. The volume contains 4 plates, 158 other illustrations, and a sketch-map of the Eastern Pyrenees.

The Battlefields of Scotland: Their Legend and Story. By T. C. F. BROTCHIE, F.S.A.Scot. Illustrated. Edinburgh: T. C. and E. C. Jack, 1913. *Price* 5*s. net.*

Mr. Brotchie has connected his "Battlefields" with explanations of the causes which led up to them, so that his book becomes in effect a sketch of the history of Scotland in which prominence is given to the episodes of war. He makes good use of the early chronicles, and is to be commended for his verbatim quotations. His footnotes, too, are interesting. But his style is something of a mosaic, and such effusions as "an unperishable nocturne silhouetted against the mists of eld" are not to be commended.

The purpose of the book is not to deal with topographical but with historical questions. For instance, the events immediately preceding Bannockburn occupy fourteen pages, while the four or five devoted to the fight itself give no consideration to the questions existing as to the disposition of the Scottish army on those two eventful days, or the movements of their opponents. Dunbar, where there are also interesting points of topography, is dismissed with a mere mention. As flaws in detail we may note that on p. 23 Shakespeare's "Gentle Duncan" should be "Gracious Duncan," and on p. 89 the date of Otterburn was 1388, and not 1338. Still the book is a painstaking and well-informed work. It is pitched in a warmly patriotic key.

ASIA.

Burma under British Rule. By JOSEPH DAUTREMER. Translated and with an introduction by Sir JAMES GEORGE SCOTT, K.C.I.E. London and Leipsic: T. Fisher Unwin, 1913. *Price* 15*s. net.*

In the issue of this *Magazine* for last February we noticed at some length M. Dautremer's excellent book on Burma. We have now before us a translation of the work by Sir James George Scott, K.C.I.E., a distinguished political officer who retired from the service some three years ago. He contributes to the work an introduction in which he inveighs in refreshingly plain language against what he considers the unjust and unwise treatment of Burma by the Indian authorities, more especially in the matter of construction of means of communication. In one passage he compares the Government of India to "a man who would go on selling matches in the street, because he can make a living out of it, and would never dare to dream of the possibility of owning a match-factory for himself, and sending consignments to the ends of the earth"; in another to "a rich uncle who is asked to get a favourite niece a *bonbonnière*, silk and hand-painted, or preferably silver, and puts her off with a couple of bananas or a water-ice instead"; in a third he says "the administrative view is that of the parish beadle, and the enterprise that of the country-carrier with a light cart instead of a motor-van." To those who cannot read M. Dautremer in the original this translation will be very welcome, and the plain speaking of the introduction will give ample grounds for reflection to the Indian authorities, as well as to the reader.

The Land of the Blue Poppy. By F. KINGDON WARD, B.A., F.R.S.E. Cambridge: At the University Press, 1913. *Price* 12*s. net.*

The members of this Society who had the good fortune to hear Mr. Ward's address in Edinburgh and Glasgow in February last will welcome the publication of this volume, and it will appeal to a much wider circle, viz., to the ethnologist, the geologist, the geographer, and, above all, to the botanist. The scene of Mr. Ward's wanderings and studies was a corner of the borderland between Western China and South-Eastern Tibet, a tract of which very little is known, although here and there Mr. Ward found outposts of French and English missionaries. It has also been visited by Captain Bailey. Mr. Ward selected the village of A-tun-tsi as the basis of his operations, as the inhabitants were friendly, and there was no difficulty in procuring supplies. From thence he made journeys in pursuit of plants and seeds, hither and thither over the lofty mountain passes which separate the great rivers of the Yangtse, the Mekong, and the Salween. In the course of these expeditions he had several remarkable adventures, not the least exciting of which was an encounter with a band of revolutionaries from whom he received much courtesy and assistance. He came across all sorts and conditions of men, from pompous, pedantic Chinese officials down to semi-savage Mosos and Lutzus, whose favourite weapon is the cross-bow, with which they are fairly efficient marksmen. As in the case of Sir Sven Hedin and other travellers, the Chinese officials seemed bent on thwarting his plans at every turn, and sending him back whence he came, but Mr. Ward displayed the same good-humoured diplomacy as Sir Sven, and outwitted his opponents whenever it was expedient to do so. At the hands of the ordinary villagers he found much kindness and hospitality. As usual he was hailed everywhere as a first-class physician, and his medicine case was often brought into requisition. The book abounds with indications of the skill and tact with which he managed to ingratiate himself with the inhabitants of these uncivilised tracts, and his observations on their habits, customs, dress, houses, agriculture, and manner of life are of profound interest to the ethnologist. Not less interesting are his observations on the geology of the tract; but the supreme interest of the book lies in the results of his botanical studies and labours. In an appendix he gives a list of two hundred plants, of which twenty-two are new species. Seeds were obtained of seventy-six species, which are now being grown in Cheshire, and it is hoped will ere long be placed on the market. Mr. Ward is master of a lucid, graceful, and easy style; he is never dogmatic or wearisome, and, although graphic and even eloquent when occasion arises, he never gives the impression of exaggeration or of writing for effect. The book is illustrated by a large number of very beautiful photographs, all of which were taken by the author. We cordially commend it to the perusal of our readers.

AFRICA.

Life of a South African Tribe. Vol. II. By HENRI A. JUNOD. With numerous Illustrations. London: Macmillan and Co., Ltd., 1913. *Price* 15*s. net.*

The first volume of this work has been already reviewed in this Journal. The second volume shows the same painstaking carefulness, and contains an enormous mass of information, much of which is of the first importance to anthropologists. There are chapers on Land Tenure, Agriculture, Food and Drink, Clothing and Ornament, Houses, Pottery, Boatbuilding, Metallurgy, and Trade. The major part of the volume deals, however, with the characteristics of the Bantu Intellect as revealed in their language, poetry and folklore (of which many examples are given), and especially with their religious life and superstitions. The author describes in exact detail all sorts of rites and ceremonies, the black and white magic, and especially the customs of divination by bones (astragalus) and taboo.

It is impossible in a brief review to refer to the many interesting questions which are involved. There are *e.g.* curious similarities to the beliefs prevalent in classical times; an interesting parallel in the method of burial to the interment of the palæolithic *Homo mousteriensis*, and a full account of certain sacred woods which are the burial-places of old-time chieftains and are now inhabited by malignant and murderous ghosts.

The author's account of Bantu religion is of great value. At first, like most Europeans who have learnt a native language, he supposed that their religious ideas consisted only of a sort of ancestor-worship. But he discovered (almost accidentally) a second set of religious intuitions which he describes as a "Deistic Conception of Heaven." This confused belief in a Supreme Being is in the author's opinion general throughout the Bantu race, and is perhaps the original form of their religion. Heaven sends thunder and lightning, protects some individuals, and punishes others, especially those who steal. There are also in this volume suggestions regarding negro education and religious instruction which should be carefully considered by all who are interested in the native question.

AMERICA.

Things as they are in Panama. By HARRY A. FRANCK. London: T. Fisher Unwin, N.D. *Price* 7*s.* 6*d. net.*

From internal evidence in this book it would appear that Mr. Franck is very much a wanderer over the face of the earth, and in 1912 he betook himself to

Panama with a view to joining the police. Being proficient in Spanish and a number of other languages he was at once employed in the census department, and the first part of this book is devoted to his quaint experiences in it. The American zone, through which the canal runs, is some 50 miles long by 10 wide, and was found to contain a population of 62,810 souls, representatives of no less than 72 separate states and dependencies. Some of the curious scenes and incidents which occurred during the six weeks of the census are very well described. After that Mr. Franck became "Zone Policeman 88." Sometimes he worked as a private detective, sometimes in the ordinary line. The whole force consisted of some 250 rank and file, a very mixed lot, but on the whole "good-hearted, well-set-up, young Americans, almost all of military training," and practically guiltless of "graft." Mr. Franck's experiences as a policeman were also quaint and varied, but we must refer our readers to his book for details. It is illustrated with a number of good photographs.

Quebec: The Laurentian Province. By BECKLES WILLSON. London: Constable and Co., 1913. *Price* 10*s.* 6*d. net.*

Previous books on Canada from the pen of Mr. Beckles Willson have served to establish his reputation as an authority on the Dominion, and in this, his latest book, he gives his readers a comprehensive view of the vast province of Quebec, of its rivers and lakes and towns and villages, of its people and their lives. Lying on both sides of the St. Lawrence and stretching from above Montreal on the west to Labrador on the east, the area of the province is greater than that of France and Germany combined, while its population of two millions is three-fourths of French origin. The prevailing stock of Quebec to-day is Norman by descent, and in their religion and folklore the *habitants* still preserve traces of much that was best in old France of the seventeenth century. The fecundity of the habitants is well known, families of twenty being common enough; and it is thus the less surprising that from 69,000 souls in 1763, the time of the conquest, the French Canadians now number over a million and a half.

Mr. Willson lays stress on the power and predominance of the Roman Catholic Church: "The Church—it cannot be repeated too often—is the ruling factor in the life of this people. The priest, or the shadow of the priest, is always by the habitant's side." The Catholic Church in Quebec is really a State Church, and the clergy enjoy the right by Act of Parliament to collect the revenues formerly coming to them before the conquest by the British. On the whole, the rule of the clergy, from the village *curé* to the bishop, appears to be beneficent, and Mr. Willson quotes many instances of valuable technical teaching given in church schools and colleges. He draws a striking picture of the religious life of the city of Quebec, of the services, and of the processions on a holy day. The devotion of the Quebecquois to the French language amounts to a passion. For example, among the mottoes on the walls of the drill hall in Quebec during the sessions of the great Congrès du Parler Français are the following, "Il n'est pas de plus grande gloire que de combattre pour la langue de la patrie," and "C'est un crime de lèse-majesté d'abandonner le langage de son pays." This goes side by side with a loyal devotion to the British flag which preserves to the people the right to maintain their own language and church and usages.

It is impossible even to refer here to the many interesting details which Mr. Willson gives of the various places he visited in his tour through the province, but we may say they are enough to inspire the reader with a desire to go and see for himself.

Having regard to the desolation which might be worked in the country by wholesale destruction of its forests, it is comforting to know that "the Government has recently established an efficient forestry service, headed by forestry engineers of the highest standing, whose staff is to be recruited from the students of a forestry school recently endowed by the Province."

Mr. Willson has also something to say of the sporting features of the province in the shape of hunting and fishing, and it is worth noting that, like the English in the Scottish Highlands, rich Americans have monopolised many of the preserves for salmon and trout in the more accessible of the lakes and rivers.

We have no hesitation in recommending Mr. Willson's book as a mine of information about a very attractive part of Canada.

The War of Quito, by Pedro de Ciaza de Leon, and Inca documents. Translated and edited by Sir Clements R. Markham, K.C.B., Vice-President of the Hakluyt Society. London: Printed for the Hakluyt Society, 1913.

In this, the latest of the Hakluyt publications, we have translations by the veteran Sir Clements Markham of a number of documents, which throw a lurid light on one episode in the tangled history of Spanish misgovernment of South America in the middle of the sixteenth century. Apparently the misery and oppression of his Indian subjects had at last come to the knowledge and touched the heart of the Emperor Charles, and he sent out Blasco Nunez as viceroy, with orders to put an end to slavery and to introduce other reforms. However just and praiseworthy the emperor's intentions were, they were confided for execution to an absolutely incompetent and tactless administrator, and were met with determined opposition in all quarters from the Spaniards in America; and this was quite natural, for the wholesale reforms, if suddenly and immediately introduced, meant ruin to them. The head and brains of the rebels—for as a fact the opposition amounted to rebellion—was Gonzalo Pizarro, the brother of the conqueror of Peru, who did not hesitate to collect troops and lead them against the viceroy. For the details of the war we must refer our readers to the book itself. The narrative, somewhat garrulously told by a trustworthy simple old soldier, who was an eye-witness of what he relates, is well worthy of perusal. So also are the other documents which form part of the book. Sir Clements Markham has contributed a too brief sequel which completes the story to the death of the ill-starred viceroy. It is hardly necessary to add that the work has been ably translated and edited, and that the notes appended to the narrative, etc., are interesting and illuminating.

A Busy Time in Mexico. By Hugh B. C. Pollard. London: Constable and Co., Ltd., 1913. Price 8*s*. 6*d*. *net.*

"For the traveller," Mr. Pollard assures us, "Mexico is a charming country, and offers boundless possibilities to the artist and pleasure-seeker, archæologist, or tourist, and to people in search for something entirely different to everywhere else. Adventure and romance are still to be found, and the primitive is within a day's journey of the railroad. In Mexico there is still the spirit of the early days of the West and places little changed since the days of Cortes." In an appendix to this book he adds, "Mexico needs saving from the Mexicans and Mexican government is a tragic farce." For a significant and convincing proof of these observations our readers could desire no better evidence than the story related by Mr. Pollard. Romance and adventure are to be found in almost every chapter, and more especially in his vivid and graphic details of the revolution of 1911, which ended in the fall of the famous President Porfirio Diaz. Mr. Pollard was an eye-witness of the occurrences in the city of Mexico at that time, and made his escape to Vera Cruz in a train in front of the one in which the fallen President fled from his capital. But besides a graphic account of the social and political conditions of the country, Mr. Pollard has much useful information to give which deserves the careful attention of any one thinking of making a career in Mexico. Attention to his advice, based on personal experience, will save an emigrant many a disappointment and much expense ; to sportsmen and tourists his advice is invaluable. Not less valuable and cogent is his warning to "casuals" against going to Mexico on the chance of picking up a plum. "To succeed in Mexico you need special knowledge of a trade or profession, a good knowledge of Spanish, a sufficiency of capital for your enterprise, and at least a year's experience of the country before you invest a penny of it." The events of the last three years cause Mr. Pollard to doubt whether the development of Mexico, "a great country rich and teeming with possibilities," is possible or even probable so long as the Mexicans are left to govern themselves, and his solution of the problem points to a protectorate by the United States, which could then organise and enforce a stable government, under which an era of internal peace and prosperity would commence. Taking its past history and present conditions into consideration a true friend of Mexico would come to the conclusion that annexation to the United States in some form or other would probably be the best fate for Mexico in the immediate, if not the remote, future.

The Panama Canal. By J. SAXON MILLS. Edinburgh : Thomas Nelson and Sons, 1913. *Price 2s. net.*

This volume is intended as a popular history and description of the Panama Canal from the time of the first project for a canal, some four hundred years ago, down to the present date. It gives an interesting and lucid account of the greatest work of civil engineering in modern times, and indicates the probable economic and other changes that will follow when the canal is opened. It has good illustrations of the Culebra Cut and the Gatun and Pedro Miguel locks, an excellent bird's-eye view of the canal and the route it traverses, and portraits of Colonel Goethals, the Chief Engineer, and Colonel Gorgas, the head of the Department of Sanitation. The book will be read with interest and appreciation by those who have neither time nor inclination to wade through a lengthy and scientific treatise on the subject.

Panama*: The Creation, Destruction and Resurrection.* By PHILIPPE BUNAU-VARILLA. London : Constable and Co., Limited, 1913. *Price 12s. 6d. net.*

This volume, which is written by the eminent French engineer and diplomatist who played so prominent a part in the drama which the volume records, is the largest and most exhaustive treatise on the Panama Canal that has yet appeared. The author's object is to vindicate French genius, conception and enterprise alike in the eyes of his own countrymen and of the world in general, to exonerate the French pioneers of the great project from the stigma of incompetence and blundering wastefulness and peculation, which the intermediate collapse of the project unjustly attached to them, and to justify his own ideas, schemes and methods in carrying out the work ; as well as to furnish a permanent and reliable history of the project from its inception to the present day, when it stands on the verge of successful completion. He deals with the project in its three stages, viz., "The Creation, The Destruction, and The Resurrection." The first heading embraces under it the project in all aspects and stages from the earliest scheme in 1513, until the financial wreck of the company in 1888. The period of "Destruction" lasted until 1899, during which time all efforts at resuscitation signally failed, and the project appeared hopelessly doomed. The "Resurrection" began in 1899. There was a fierce and protracted controversy, or "battle of the routes," Panama versus Nicaragua, and another battle over the question of a high-level lock canal, or a sea-level canal, ending in the victory of the Panama route and the high-level project. On the heels of this followed the secession of Panama from Colombia, the details and justification of which the author enters into at considerable length. The actual work of construction was then begun, and has since been so vigorously and successfully carried on in face of seemingly almost insuperable difficulties, that it is fully expected that the first vessel will sail through the canal in October, although it will be some months later before the waterway is opened for general traffic. The volume closes with "Supplementary Chapters," dealing with "The Key of the Secret of the Straits," about which the author holds very pronounced views; "The Panama Tolls," "The Fortification of the Straits," etc. It contains a number of excellent illustrations of the work done by the French engineers, and photos of prominent persons connected with the project. Engineers, contractors, financiers, and others interested in the carrying out of great public works will find the book a veritable mine of valuable information.

GENERAL.

Notes and Queries on Anthropology. Fourth Edition. Edited for the British Association for the Advancement of Science. By BARBARA FREIRE-MARRECO and JOHN LINTON MYRES. London: The Royal Anthropological Institute, 1912. *Price* 5*s*.

This is a new edition of a booklet first published in 1874, with the object of promoting accurate anthropological observations on the part of travellers. The last edition was published in 1899, and the science of anthropology is making such rapid progress that even in this relatively short interval considerable changes have occurred. Despite its avowedly utilitarian object the present edition is full of interest, and the sections relating to religion, morals, government, and so forth show especially with what care and precautions the study of primitive man is now being carried on. We have drifted as far from the "noble savage" standpoint as from that of the "soulless brute." Indeed, we feel disposed to wish that some of the statements and recommendations of those *Notes and Queries* might reach a larger public than that for which the book is intended. The instructions as to the carefully qualified acceptance which should alone be given to the statements of "earnest politicians" among savage peoples, might well be repeated to a larger public, and with a wider reference!

A Turkish Woman's European Impressions. By ZEYNEB HANOUM. Edited, and with an Introduction, by GRACE ELLISON. London: Seeley, Service and Co., Ltd., 1913. *Price* 6*s*.

Zeyneb Hanoum is one of the heroines of Pierre Loti's novel *Les Désenchantées*, which is of much interest in that it gives an extraordinarily vivid account of Constantinople and of the life of its educated women at a time of great social change. The present book describes the effect upon such a woman of experience at close quarters of the customs and manners of Western Europe. Perhaps naturally Zeyneb Hanoum's sojourn in the West ended in further disenchantment and a return to

Turkey. The interest of the book lies in its presentation of Western ways by an Eastern.

EDUCATIONAL.

A Commercial Geography of the World. By FREDERICK MORT, M.A., B.Sc., F.G.S. Edinburgh and London: Oliver and Boyd, 1913. *Price* 2*s.* 6*d.*

This is an admirable text-book and ought to take its place as the best of the smaller volumes on commercial geography which have appeared within the last few years. Its lucid style, well-chosen matter, and careful arrangement are all to be commended. A large number of figures renders the book most useful, and mention should be made of the series of excellent graphs. Of special value are Figs. 1, 2, 5, 8, 9, 12, 14, 17, 22, 30, 31, 34, 39, 65, 70.

The introductory chapter is sound and sane, and might be read with profit by others as well as teachers or students of commercial geography. The same might be said with truth regarding the chapter on "Geographical Factors that influence Commerce." In the description of commercial commodities are to be found very valuable paragraphs on wheat, coal, and cotton. It may be doubted, however, whether the technical paragraphs on the making of steel can in strictness be said to deal with commercial geography. The treatment of the separate countries is fully causal.

A few minor inaccuracies have crept into the text. Aberdeen is not mentioned as one of the chief trawling ports, p. 86; Kashmir is spelt correctly on the map, p. 137, but as Cashmere, p. 135, and as Kashmere, p. 140 (the word, by the way, does not appear in the Index); we prefer Trans-Atlantic to Transatlantic, p. 107 (cf. transatlantic, p. 203, and Trans-Caspian, Trans-Caucasia, Trans-Siberian in Index); Trondhiem has become Tronhiem, p. 218; Kiev has become Khiev, p. 244; Triest, map, p. 234, and text, p. 235, has become Trieste, map, p. 236; the names L. Victoria Nyanza, L. Albert Nyanza, and L. Albert Edward Nyanza, Fig. 85, are surely out of the date; it is somewhat misleading to speak of "Indian" labour, p. 313. These may seem trifling errors, but we expect accuracy from one of Mr. Mort's standing.

A Text-Book of Geography: Practical and Physical. By R. M. MUNRO, M.A. Edinburgh: John Cormack, 1913. *Price* 3*s.* 6*d. net.*

The name of the writer is new to us, but Mr. Munro has now made his mark in Geography. The main defect of the book, and it is a defect at which we need not cavil overmuch, is that it is too encyclopædic in its scope, with the result that causal connections are somewhat lost sight of, and that the short paragraph style is too much in evidence. Mr. Munro's bent seems to lie to the mathematical side of Geography, for chapters i., ii., iii. and xvii. are exceptionally well done. We note, for instance, that the volume under review is among the very few text-books which deal satisfactorily with the problem of the frequency of eclipses, a matter which often exercises the minds of young students. Due praise must also be given to the account of the Means of Communication.

The paragraphs on those Heavenly Bodies, the influence in Geography of which is practically *nil*, might well be omitted. Then again, when the author makes a quotation, he might leave the name of the writer of the passage to be discovered by the intelligence of the teacher or student (pp. 40, 255, and 305). When he deals with the Ocean Currents Mr. Munro might have given us more about those of the North Atlantic, which are after all the most important for us. *Kuro Siwo* is so called on Fig. 59, but it is spelt *Kuro Sivo* on Fig. 61 and in the text. Then a study of Shaw's recent meteorological work might alter somewhat the notes on anticyclones. Further, in Climate, the reasons for the Horse Latitude High Pressures are not adequately given. Is it true that *roches moutonnées* are so called from their resemblance to a sheep's back? Lastly, it might be necessary to state that the Antarctic icebergs do not float according to Fig. 257, and text p. 276.

The general appearance of the book reflects credit on the enterprising publishing house which issues the work, but it might have been improved by a wider margin, more artistic colouring in the maps, and by the omission of such Figs. as Nos. 140 and 142-152.

The book ought to be specially helpful to students who are preparing themselves for the many public examinations in which Geography is a set subject, and for the busy teacher it will provide a *vade-mecum* to assist him in obtaining material for his geography lessons.

Asia: A Supplementary Geography. By J. F. and A. H. CHAMBERLAIN. New York: The Macmillan Co., 1913.

This volume, the third in "The Continents and their People" Series, is intended to supplement, not to replace, the ordinary school text-book. Stress is laid on the social conditions of the various Asiatic countries, and thereby the human element in geography is brought into prominence. The accounts are eminently readable, and the information is conveyed in a highly pleasant form. A large number of excellent photographic reproductions embellishes the text. Of these Figs. 4, 8, 11, 22, 42, 68, and 76 seem to us specially interesting. While the three maps given are scarcely up to the standard of similar examples of British cartography, we might well follow the Americans in the art of binding.

The Elements of Geography. By R. D. SALISBURY, H. H. BARROWS, and W. S. TOWER. New York: Henry Holt and Co., 1913.

The three authors who have collaborated in this treatise belong to the Department of Geography in the University of Chicago. The preface to the book makes no mention of special sections assigned to the several writers, but the name of Professor Salisbury, whose *Physiography* is one of the finest works of its kind ever produced, is a guarantee of the soundness of the treatment given to geography. Naturally the principles enunciated in the preface have been applied most fully to the home country, but students on this side of the Atlantic will find much that is suggestive, much that can be applied equally well to Britain. We wish to direct attention especially to the admirable sections on Climate. No fewer than eight of the twenty-one chapters of the book deal more or less directly with this all-important subject, while the treatment of surface-features and their modifications serves as an excellent counterpart. At chapter xvi. we reach Human Geography proper, a subject to which American geographers, following the lead given by Miss Semple, have paid special attention. Mountains, plateaus, the various types of plains, and coastlines are all dealt with, and that fully and attractively, as parts of man's environment. In the Economic chapter the writers have confined their attention to the United States—in fact for the British student the whole book is a commentary from the points of view of Physiography and Ontography on the geography of the United States.

A good index, an extensive series of serviceable figures, a list of plates at the end, and full references at the close of each chapter, are additional merits in a book which must take its place as one of the finest and most exhaustive of the more recent general works of a moderate size dealing with geography.

BOOKS RECEIVED.

Colombia. By PHANOR JAMES EDER. With 40 Illustrations and 2 Maps. Demy 8vo. Pp. xxiv + 312. Price 10*s.* 6*d. net.* London : T. Fisher Unwin, 1913.

Further Reminiscences of a South African Pioneer. By WILLIAM CHARLES SCULLY. With 16 Illustrations. Demy 8vo. Pp. 384. Price 10*s.* 6*d. net.* London : T. Fisher Unwin, 1913.

Outback in Australia; or Three Australian Overlanders: Being an Account of the longest overlanding Journey ever attempted in Australia with a single horse, and including chapters on Various Phases of Outback Life. By WALTER KILROY HARRIS, F.R.G.S., F.R.C.I. With Map and 29 Illustrations from Photographs. Demy 8vo. Pp. x + 224. Price 5*s. net.* Letchworth : Garden City Press, 1913.

A Motor Tour in Belgium and Holland. By TOM R. XENIER Illustrated. Demy 8vo. Pp. viii + 311. Price 10*s.* 6*d. net.* London : Mills and Boon, Ltd, 1913.

Italy in North Africa: An Account of the Tripoli Enterprise. By W. K. M'CLURE. With Illustrations and Maps. Demy 8vo. Pp. xi + 328. Price 10*s.* 6*d. net.* London : Constable and Co., Ltd., 1913.

The Republics of Central and South America: Their Resources, Industries, Sociology, and Future. By G. REGINALD ENOCK, F.R.G.S. With 16 Illustrations and 9 Maps. Demy 8vo. Pp. 544. Price 10*s.* 6*d. net.* London : J. M. Dent and Sons, Ltd., 1913.

Questions and Exercises in Geography. Based on Heaton's "Scientific Geographies." By R. J. FINCH, F.R.G.S. Crown 8vo. Pp. 334. Price 2*s.* 6*d. net.* London : Ralph Holland and Co., 1913.

The Carolina Mountains. By MARGARET W. MORLEY. With Illustrations. Demy 8vo. Pp. viii + 397. Price 8*s.* 6*d. net.* London : Constable and Co. Ltd., 1913.

Pemba: The Spice Island of Zanzibar. By Captain J. E. E. CRASTER. With 30 Illustrations. Demy 8vo. Pp. 358. Price 12*s.* 6*d. net.* London : T. Fisher Unwin, 1913.

With the Russian Pilgrims to Jerusalem. By STEPHEN GRAHAM. With 38 Illustrations from Photographs and a Map. Demy 8vo. Pp. x + 306. Price 7*s.* 6*d. net.* London : Macmillan and Co. Ltd., 1913.

The Battlefields around Stirling. By JOHN E. SHEARER, F.S.A.Scot. Demy 8vo. Pp. viii + 96. Price 3*s. net.* Stirling : R. S. Shearer and Son, 1913.

Turistruter-paa Island. Ved DANIEL BRUUN. Pp. xxiv + 254. Kφbenhavn. Gyldendalske Boghandel, 1912-13.

Les Nègres d'Afrique (Géographie Humaine). Par CYR. VAN OVERBERGH. Royal 8vo. Pp. xii + 276. Bruxelles : Albert Dewit, 1913.

The Awakening of the Desert. By JULIUS C. BIRGE. With 25 Photographic Illustrations. Crown 8vo. Pp. 429. Price 7*s.* 6*d. net.* London : Heath, Cranton and Ouseley, Ltd., 1913.

A Comparative Geography of the Six Continents. By ELLIS W. HEATON, B.Sc., F.G.S. Crown 8vo. Pp. 219. Price 1*s.* 9*d. net.* London . Ralph Holland and Co., 1913.

Sketches of North and West Ireland. By MARY RIDSDALE WILSON. Illustrated by JESSIE S. WILSON. Crown 8vo. Pp. 77. Price 2*s. net.* London : A. H. Stockwell, 1913.

Contours and Maps: Explained and Illustrated with Six Full-Page Maps and Forty Diagrams. By FREDERICK MORROW, B.A., F.R.G.S. Crown 8vo. Pp. viii + 116. Price 1*s.* 6*d. net.* London : Meiklejohn and Son, 1913.

The Natural History of the Toronto Region: Ontario, Canada. Edited by J. H. FAULL, B.A., Ph.D. Crown 8vo. Pp. 419. Price 2 *dollars.* Toronto: The Canadian Institute, 1913.

Madrolle's Guide Books: Northern China; The Valley of the Blue River; Korea. 43 Maps and Plans. Crown 8vo. Pp. 471. Price 15*s.* London: Hachette and Co., 1912.

Visual Geography: A Practical Pictorial Method of Teaching Introductory Geography. By AGNES NIGHTINGALE. Book II.: *Continents and Countries.* With 22 page Outline Picture Maps for colouring and completing. Crown 4to. Pp. 48. Price 8*d.* London: A. and C. Black, 1913.

The Atlas Geographies: Preparatory. By THOMAS FRANKLIN and E. D. GRIFFITHS. 2 Parts. *British Isles* and *Europe.* Crown 4to. Pages 33 and 64. Price 6*d. each.* Edinburgh: W. and A. K. Johnston, Ltd., 1913.

National Antarctic Expedition, 1901-1904. Meteorology. Part II. Prepared under the superintendence of M. W. CAMPBELL HEPWORTH, C.B., R.D., Commander R.N.R. Large 4to. London: Royal Society, 1913.

General Report on the Operations of the Survey of India, 1911-12. Prepared under the Direction of Colonel S. G. BURRARD, C.S.I., R.E., F.R.S. Calcutta: Survey of India, 1913.

Report on the Administration of Bengal, 1911-12. Calcutta, 1913.

Burma Gazetteers; Pakokku, Upper Chindwin, and Insein Districts. Rangoon, 1913.

Punjab District Gazetteers. Vols. 2, 6, 8, 10, 14, 19, 25, 32. Lahore, 1912.

Memoirs of Geological Survey. Summary of Progress of the Geological Survey of Great Britain and the Museum of Practical Geology for 1912. London, 1913.

Statement Exhibiting the Moral and Material Progress and Condition of India during the year 1911-12 and the nine preceding years. London, 1913.

Census of India, 1911. Vol. V.: Bihar and Orissa; Part III. Tables. Vol. XXII.: Travancore; Part I. Report; Part II. Imperial Tables. Vol. XXII.: Rajputana and Ajmer=Merwara; Part I. Report. Calcutta, 1913.

Board of Agriculture for Scotland: Agricultural Statistics. Vol. I. Part I. Acreage and Live Stock returns of Scotland. London, 1913.

Edinburgh University Calendar, 1913-14. Edinburgh, 1913.

Diplomatic and Consular Reports.—Trade, etc., of Chinde, 1912 (5136); Trade of Katanga, 1912 (5165); Trade of Newchwang, 1912 (5169); Trade of Yokohama, 1912 (5170); Trade, etc., of German East Africa, 1909-12 (5171); Trade of Moscow, 1912 (5172); Trade of Marseilles, 1912 (5174); Trade, etc., of Saigon, 1912 (5177); Trade, etc., of Nice, 1912 (5178); Industries and Commerce of Spain, 1912 (5179); Trade of Harrar, 1912 (5180); Trade, etc., of Barcelona, 1912 (5181); Trade of Naples, 1912 (5182); Trade of Finland, 1912 (5183); Trade of Pernambuco, 1912 (5173); Trade, etc., of Zanzibar, 1911-12 (5176); Trade of Mexico, 1912 (5175); Trade, etc., of Beirut and the Coast of Syria, 1912 (5184); Trade of Madagascar, 1912 (5186); Trade of Switzerland, 1912 (5187); Trade, etc., of the Faroe Islands and Iceland, 1912 (5189); Trade, etc., of the Island of Cuba, 1912 (5190); Trade of Port Said, 1912 (5185); Trade of the Piræus and District, 1912 (5188); Trade, etc., of Kobe, 1912 (5191); Trade of Tientsin, 1912 (5192); Trade of Hanchow, 1912 (5193).

Publishers forwarding books for review will greatly oblige by marking the price *in clear figures, especially in the case of foreign books.*

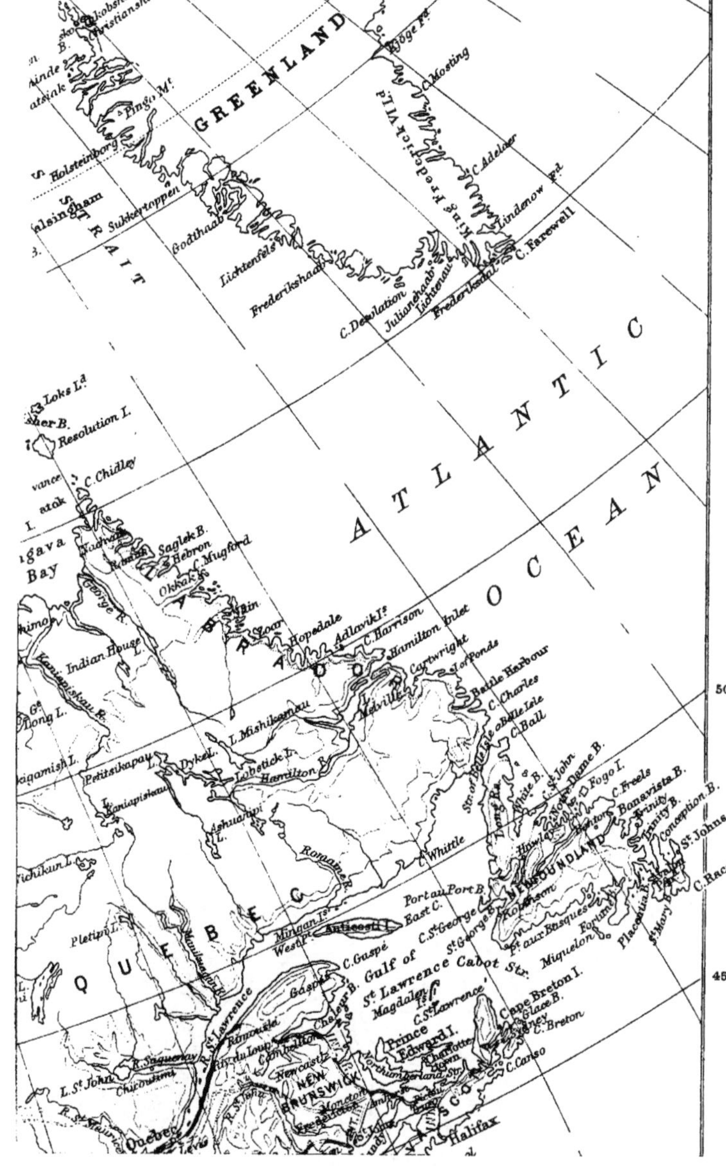

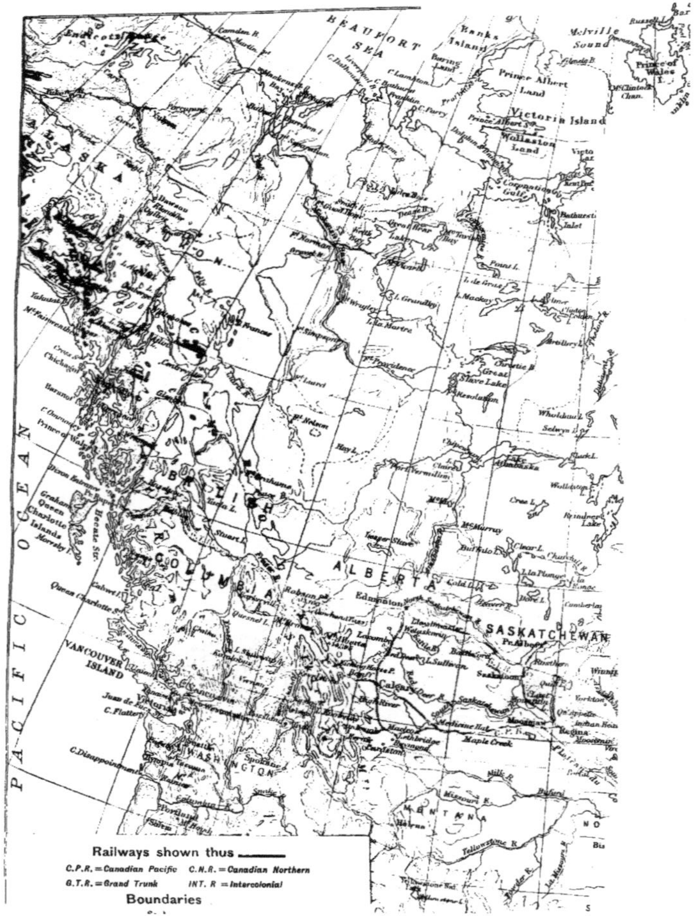

BEAUFORT SEA

ALASKA

YUKON

BRITISH COLUMBIA

ALBERTA

SASKATCHEWAN

WASHINGTON

MONTANA

VANCOUVER ISLAND

Queen Charlotte Islands

Hecate Str.

Banks Island

Victoria Island

Prince Albert Land

Wollaston Land

Great Slave Lake

Edmonton

Calgary

Medicine Hat

Maple Creek

Regina

Saskatoon

PACIFIC OCEAN

Railways shown thus ______

C.P.R. = Canadian Pacific C.N.R. = Canadian Northern

G.T.R. = Grand Trunk INT. R = Intercolonial

Boundaries

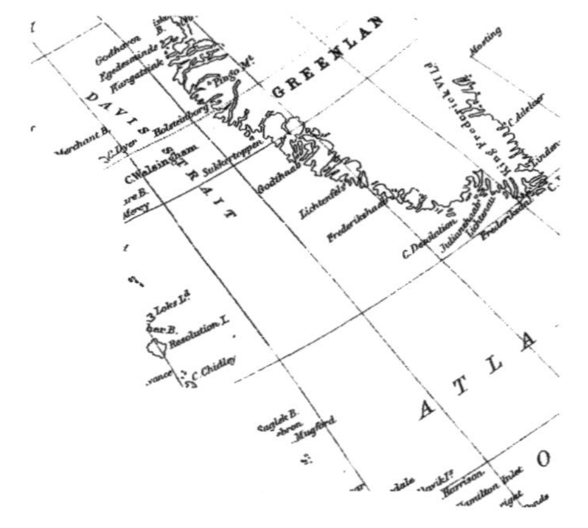

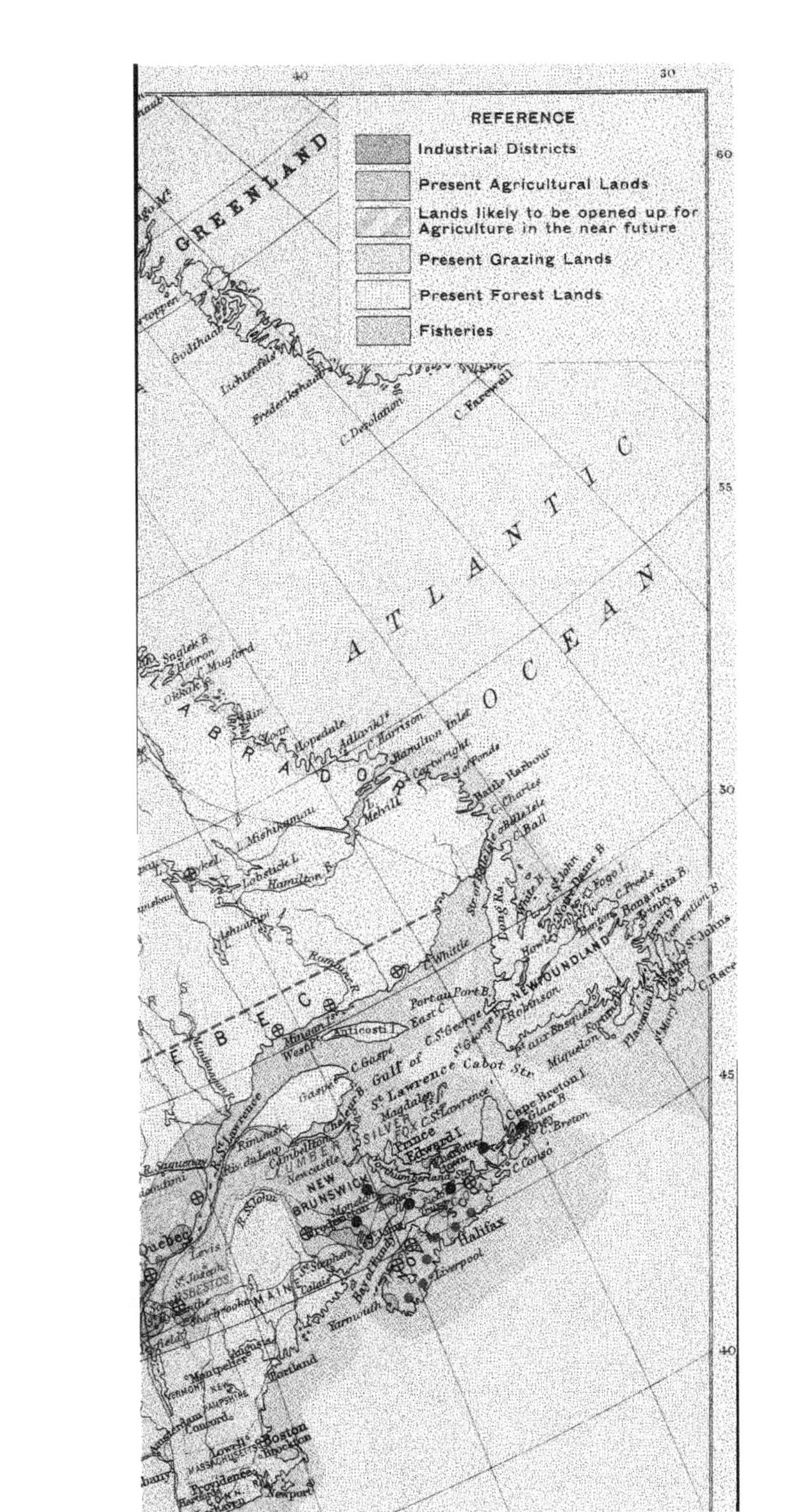
REFERENCE
Industrial Districts
Present Agricultural Lands
Lands likely to be opened up for Agriculture in the near future
Present Grazing Lands
Present Forest Lands
Fisheries
GREENLAND
ATLANTIC
OCEAN
LABRADOR
NEWFOUNDLAND
NEW BRUNSWICK
MAINE
VERMONT
MASSACHUSETTS
C. Farewell
C. Desolation
Hamilton Inlet
Battle Harbour
Anticosti I.
Gulf of St Lawrence
Cabot Str.
Prince Edward I.
Cape Breton I.
Halifax
Liverpool
Yarmouth
Quebec
Portland
Boston
Concord
Providence
Newport
C. Race
St Johns
Miquelon
Magdalen Is.
C. Gaspe
Hopedale
Cartwright
C. Harrison
Port au Port B.
C. St George
C. Canso
Sherbrooke
Montpelier
Lowell
Brockton

Illustrating Mr Silva White's Article.

CANADA—ECONOMIC DEVELOPMENT

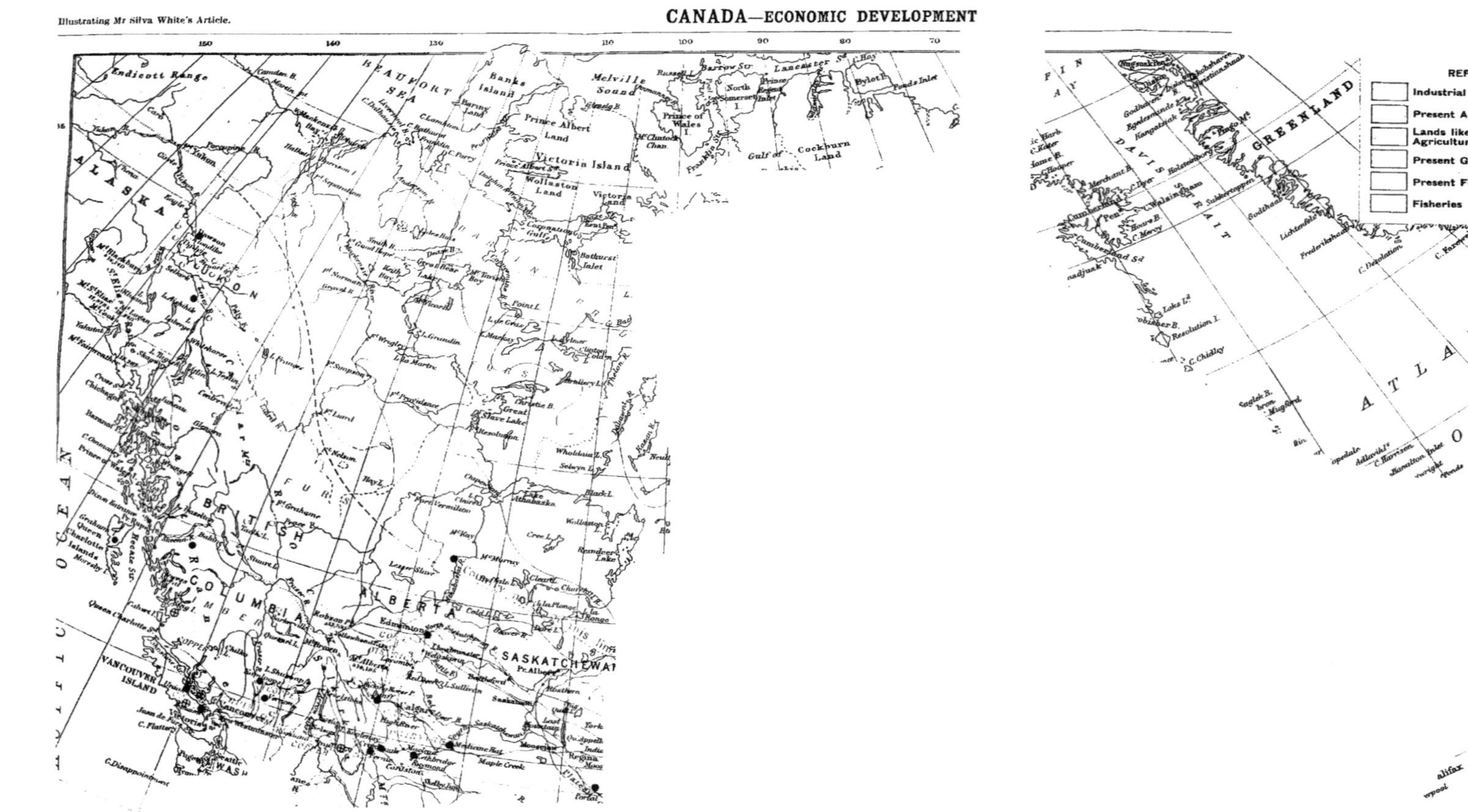

THE SCOTTISH GEOGRAPHICAL MAGAZINE.

THE DOMINION OF CANADA: A STUDY IN REGIONAL GEOGRAPHY.

By ARTHUR SILVA WHITE, Hon. F.R.S.G.S.

(With Map.)

(Continued from p. 547.)

PART II.—THE PEOPLE.

POPULATION.

THE following figures show the increase, except in Prince Edward Island, of population during the decade 1901-1911.[1]

	Provinces.	1911.	1901.
	Ontario, . .	2,523,208	2,182,947
	Quebec, . .	2,002,712	1,648,898
Maritime	Nova Scotia, .	492,338	459,574
	New Brunswick, .	351,889	331,120
	Prince Edward Isl.	93,728	103,259
Prairie	Manitoba, . .	455,614	255,211
	Saskatchewan, .	492,432	91,279
	Alberta, . .	374,663	73,022
	British Columbia,	392,480	178,657
	Other Territories (including Yukon),	15,463	47,348
DOMINION OF CANADA, . .		7,204,527	5,371,315

Decrease in the population of Prince Edward Island—the most densely occupied Province of the Dominion—is probably accounted for

[1] *Parliamentary Paper*: Dominions, 1912. Cd. 6091.

by migration to the Prairie and other Provinces; nearly all are Canada-born on the Island. In general, the movement of the population is, naturally, from the Eastern to the Western Provinces; Alberta and Saskatchewan, in particular, show the highest relative increase.

The British element predominates in all the Provinces except Quebec, which is mainly French in origin; and Protestant denominations are everywhere in a majority save in Quebec. Roman Catholics in the Dominion exceed 2¼ millions. The proportion of foreign-born (American, Russian, German, Austro-Hungarian, Chinese, and others) is largest in the Western Provinces and in the unorganised Territories. In the latter, it is estimated that there is not more than one inhabitant to fifty square miles. In 1910 there were 110,597 Indians and Eskimo (in Arctic lands) representing the aboriginal population. The Athabaskan and Algonkin tribes are the most widely distributed. The Indians are grouped in reservations—as small and as wide apart as possible—and give little or no trouble. In Eastern Canada they follow industrial pursuits, and the franchise has been extended to them.

Following is the population (1911) of the principal cities and towns:

	Inhabitants.
Montreal,	470,480
Toronto,	376,240
Winnipeg,	135,430
Vancouver,	100,333
,, (North and South),	23,802
Victoria,	31,620
Ottawa,	86,340
Quebec,	78,190

Immigration into Canada, in 1911, was very active, totalling 351,072 persons. Of these, 130,102 came from the United States, and 175,000 from the United Kingdom. English immigrants have increased by 20 per cent., and Scottish by 30 per cent. The total increase, as compared with 1910, was 40,000. There was a relative increase in the number of British over American immigrants, and a relative decline in the influx from other countries. During the twelve months from April 1, 1911, to March 31, 1912, emigration to Canada was even higher:[1]

From Great Britain,	138,121
,, the United States,	133,710
,, other Countries,	82,406
Total,	354,237

CHARACTERISTICS.

The emergence of a national type out of the two good stocks on which it is founded, in Canada, has been necessarily affected by the proximity and close intercourse of the United States, no less than by

[1] Statement by Mr. Buxton in the House of Commons, Oct. 21, 1912.

the influx in large numbers of Scotsmen in Nova Scotia and of the United Empire Loyalists into the Maritime Provinces. But the blend marks off Canada, in its social and national characteristics, from the American Union, and carries on the traditions of the Mother Country, except that there is no leisured class in the Dominion.

The Americanisation of Canada, introduced through the newspaper press rather than by the flood of emigrants to the Dominion, does not appear to have affected materially the more conservative and idealistic elements of the population, based on its Northern ancestry. The born Canadian is suspicious of talkers and hustlers, and his farm is more his home than his place of business: he is not for ever on the tiptoe of expectation. But his ignorance of the Mother Country, in its modern aspects, is great, though not so great as in the reverse direction: in both cases, ideas of each other are based on information some generations out of date.

Constitution and Government.[1]

Canada is a Self-Governing Dominion of the British Empire. The Governor-General is appointed by the King. The Executive Government is vested in the Crown, and is exercised by the Governor-General, assisted by a Privy Council chosen and summoned by himself. The Cabinet is a Committee of the Privy Council, consisting at present of fifteen of the principal members of the Dominion Government, or Ministers.

Supreme legislative power is vested in a Parliament consisting of the King, a Senate, and a House of Commons. The *Governor-General in Council* appoints Lieutenant-Governors for the Provinces, and, in certain circumstances, removes them; appoints also officers for the effectual execution of the Constitution. Each Province has a Legislative Assembly; in Quebec and Nova Scotia there is also a Legislative Council, forming a Second Chamber. Unlike the Constitution of the United States, that of Canada vests the Dominion Parliament with exclusive legislative power in all matters except those specifically delegated by the Constitution to the Provincial Legislatures, which on their side possess the power of altering their own Constitutions. Territory not comprised within any Province is governed by a Commissioner and a Council of four appointed by the Governor-General in Council.

The Judges are appointed by the Governor-General in the Superior, District, and County Courts throughout the Dominion, except in the Probate Courts of Nova Scotia and New Brunswick. His Excellency has the right of pardon and reprieve.

As regards bills passed by the Provincial Legislatures, the Governor-General is in the same position as the King in Council for Self-Governing Colonies. His salary (£10,000 a year) is paid out of the Consolidated Revenue Fund of Canada.

---

[1] *Colonial Office List*, 1912; *Analysis of the System of Government throughout the British Empire* (Anon.), p. 79.

The Lieutenant-Governor of a Province is assisted by an Executive Council composed usually of the chief officials who possess the confidence of the Provincial Assembly. Manhood Suffrage prevails to a very considerable extent.

The Central Government, which assumed the Provincial Debts existing at the time of the Federal Union, pays to each Province an annual allowance for purposes of government and an annual subsidy, *per capita.* Divergence of interests—inevitable in a young and rapidly expanding Colony—there has been in the past between the Central and Provincial administrations; but a tendency towards consolidation of interests, rather than the exercise of restraint by the Central Government, is markedly supporting a true national and imperial policy, and is giving effect to that material unity which Canada has sought to attain, in all earnestness and sincerity.

Local self-government in municipal affairs is general, and it has been developed to the fullest extent.

Ottawa is the seat of the Dominion Government, which is represented in the United Kingdom by a High Commissioner, resident in London.

ATLANTIC PROVINCES.

The United Empire Loyalists and emigrants from the Highlands of Scotland, who found new homes in a New Scotland, encountered familiar surroundings in the bold coastlands of Nova Scotia; and the Nova Scotian Assembly is the prototype of our Colonial Parliaments. Nova Scotia—itself all but an island—together with Cape Breton and Prince Edward Islands, reproduce many of the physical features of the British Isles, and, with New Brunswick, represent a stage of transition from oceanic to continental conditions. But the earliest colonists were French, who called the country L'Acadie.

Nova Scotia.—The capital city, Halifax, with its magnificent harbour on the Atlantic coast, was—and is, though not actively—an important naval base for the British fleet. Since the Boer war, Halifax has been taken over, for naval purposes, by the Dominion Government. It is a thriving seaport, open, unlike the ports of the St. Lawrence, to the world's shipping all the year round; and, like the port of London, it serves as a distributing centre—at least in the winter months—for Canadian produce and foreign imports. The city of Halifax has a University (Dalhousie), and is the centre of educational as well as of industrial life. West of Halifax county, on the south coast, is Lunenburg county, peopled by the descendants of German settlers who emigrated from Hanover in 1751.

Although agriculture and mixed farming—chiefly fruit-growing and dairy farming—are the principal occupations of the people, the fisheries of Nova Scotia give employment to about 14,000 men. Thousands of men, too, are employed during the winter in the forests, which, though greatly thinned out, yield good supplies of pulp-wood. The "marsh" lands, formed by the phenomenally high tides of the Bay of Fundy, are of peculiar value to the dairy farmer, and the land reclaimed is of inexhaustible fertility. We have referred already to the mining industry, which gives employment to a considerable section of the population.

The vast majority of the people are of British stock.

Prince Edward Island, the smallest Province of the Dominion, is also the most densely populated: 47 inhabitants to the square mile, the mean for the whole Dominion being 1·9. Scotsmen predominate, followed by English and Irish, with an increasing number of French Canadians. Here the forest has been cleared to make room for farms, which cover practically the whole Island: "the million-acre farm," it is called. Charlottetown—the only city in the Island—is well situated on a bay in the middle of the south shore.

A considerable portion of the population are employed in industrial pursuits, but the bulk are farmers. The Provincial Government, as in other Provinces, for many years has carried on a stock farm, and has imported also a number of pure-bred stock, besides establishing model orchards; while the farmers themselves have organised co-operative agricultural associations. Many of the smaller farmers engage also in the "fisheries," chiefly lobsters and oysters.

Education is free; and for higher education the Prince of Wales College is maintained at Charlottetown. There are many churches on the Island, the majority being Protestant; and there is also a strong Roman Catholic community. Social life, to a large extent, centres round the churches.

New Brunswick, though nearly as large as Scotland, has relatively only one-tenth of the population. Originally the most densely afforested of the Maritime Provinces, the coastal lands have been cleared and deep inroads made in the central parts for farming. Most of the pine timber has gone, but there remain spruce and a great variety of hardwood, the products of which are exported in large quantities, besides being sent to other parts of Canada and to the United States.

The forest and the fisheries, to a certain extent, have distracted attention from farming; but agriculture flourishes, not only in the broad valleys and on the undulating plain, but along the banks of many streams running through parts of the country still devoted chiefly to lumbering. The county of Northumberland, in the north-east, contains the largest area of wild land. In the valley of the St. John River—on which Fredericton, the capital, is situated—and in the valley of the Miramichi River—which divides the county of Northumberland—there are good farm lands and many farmers. The establishment of cheese and butter factories all over the country has raised this branch of agriculture to the first place.

The Provincial Government imports cattle, sheep, and pigs of the best breeds, and sells these by auction at prices below cost, on condition that the animals be kept in the Province; the farmers also co-operate among themselves. A dairy school is maintained at Sussex.

The bulk of the population—scattered over the country districts—are of British stock, descended mostly from the United Empire Loyalists. There are three universities in the Province, one being a State institution at the capital. St. John, at the mouth of the river of that name is a much larger town than Fredericton. It is a seaport, railway terminus, and manufacturing centre.

River and Lake Provinces.[1]

The settlement and development of Quebec and Ontario Provinces have been determined largely by the well-defined geographical conditions that make of this region one of the most favoured in America for economic growth. The extensive system of waterways, great variety and wealth of natural resources, and a people welded together out of the descendants of two great colonising powers, combine to give the River and Lake Provinces of Canada advantages of unique value. In Ontario the English, and in Quebec the French elements predominate; but in the Dominion Government at Ottawa harmony and identity of interests may be said to prevail over local and racial prejudice. The absorption of New France into the body politic of our premier Colony is one of the most notable achievements of British Colonial polity, and a favourable augury for the experiment now being worked out in South Africa between British and Dutch elements. Even the bi-lingual impediment, which in South Africa retards complete fusion in the present generation, has been overcome to a great extent in Canada by the logic of events that could not be denied a prosperous people working towards a common end—the unity and consolidation of the Dominion.

Quebec.—In the Province of Quebec many of the customs, institutions, and picturesque trappings of Old France are enshrined and displayed by a people who, while cherishing their birthright as the first conquerors and settlers, are as loyal to the British connection as any other in the Empire. When the British Colonists further south rebelled against the rule of King George, the French-Canadians refused to join them, and themselves helped the British troops to repel the American invasion in 1775, and again in 1812. The subsequent revolt of the Colonists, both British and French, were ebullitions of patriotism which won freedom for their country by the destruction of the feudal system.

English-speaking people outside the city of Montreal are found chiefly in the southern section, bordering the United States. In the rest of the settled districts of the Province they are scattered very thinly among the French-Canadians. The French spoken is no patois, but the language of Middle France—except in accent—in which some English words and corruptions are traceable.

Like the French in Europe, the French-Canadian is light-hearted, hospitable, industrious and thrifty; and he is fond of politics. Almost all are Roman Catholic, and devoted to the Church, to which they pay a percent-

---

[1] The Dominion Government have come to a decision as to the settlement of the question of the extension of the boundaries of Ontario and Quebec. Quebec obtains the territory known as Ungava, and Ontario that part of the North-West Territories which lies south of a line drawn approximately from the north-east corner of Manitoba to a point fifty miles south-east of Port Nelson, the remainder of the country south of the 60th parallel being allotted to Manitoba. The Provincial Legislatures concerned in this matter have accorded their assent.—*Dominions Report*, No. 12, for 1911-12 (Cd. 6091), page 64.

age of their produce for maintenance charges. Although about, or nearly, half a million of the people live in Montreal—the commercial capital of the Province—almost two-thirds of the entire population are found in the country districts, where the farmers nearly all own their own land and homesteads. The distribution of the population coincides mainly with the valleys of the St. Lawrence and its tributary, the Ottawa River, which it shares with Ontario; and flourishing homesteads are found in the south, right up to the United States frontier.

Quebec, the capital city, beautifully situated on the St. Lawrence, and most picturesque in itself, has extensive docks and important manufacturing industries. It is the seat of Provincial legislation and administration, and has one of the oldest universities in Canada, named after the first bishop, Laval.

Montreal—the Liverpool of Canada—with rapidly developing interests in shipping and manufactures, is a busy centre in which a large and influential section of the community is English. The city lies a thousand miles inland from the open Atlantic, but the St. Lawrence River is here a mile and a half wide. It is the headquarters of the Canadian Pacific Railway, which crosses Canada to the Pacific Coast, and also of the Grand Trunk Railway; other lines radiate in all directions, linking up this important distributing centre with the American Continent.

Montreal is rich in churches and cathedrals—so rich, that Mark Twain once said he could not throw a stone without breaking a church window. The largest are the Roman Catholic cathedral, built on the plan of St. Peter's at Rome, and the ancient parish church. Christ Church is the Anglican cathedral.

The educational system of the Province is adapted to the requirements of the people. Roman Catholic institutions are both many and large. The Church of England maintains a university college at Lennoxville. As an educational centre Montreal ranks high, its McGill University being comparable to Harvard and Yale in the United States; and there is a famous agricultural college, created by Sir William Macdonald, at St. Anne's.

Other urban centres are the old French settlement of Three Rivers, between Montreal and Quebec; the city of Hull on the Ottawa River, and the industrial town of Valleyfield, with its cotton and paper mills. Sherbrooke is the capital city of the eastern townships, charmingly situated on the banks of the St. Francis River. St. Hyacinthe is a centre of the leather industry. Northwards of the St. Lawrence, a manufacturing centre is growing up on the banks of the St. Maurice River, where the Shawinigan Falls are utilised as a power station.

When winter comes, the lumbermen invade the densely wooded areas lying north of the cultivated valleys of the St. Lawrence and Ottawa Rivers. Much of the land being of good quality, the farmer is making inroads into the forest; and it is also a favoured region for sport and residential estates.

Ontario.—The leading Province of the Dominion was discovered and opened up by a band of famous explorers—Champlain, founder of Canada; Joliet, discoverer of the Mississippi; and La Salle, who navigated that mighty river to its mouth in the Gulf of Mexico—and missionaries who, together with the intrepid "runners of the wood," braved perils by land and water in untracked Ontario, the record of which is among the most romantic in pioneer literature. But the solid and enduring foundations of the structure we know to-day were built up by the Loyalists and others of British stock, as in the Maritime Provinces. Ottawa, the seat of the Dominion Government, is situated in Ontario, on the right bank of the Ottawa River, which flows eastwards for one hundred miles to its union with the St. Lawrence near Montreal. Ottawa is very much smaller than Toronto.

Toronto, on the north-west shore of Lake Ontario, is second only to Montreal in population and importance. Nearly half of the Canadian banks have their headquarters in this city, in which there are numerous factories, department stores, and wholesale warehouses. Its university is the largest in the Dominion; and it is perhaps the leading centre in Canada for the newspaper press. Toronto is the political capital of the Provincial Government.

Ontario leads the way in manufacturing activity. In towns too numerous to mention, mills and factories are busy. Iron and its products, from tin-tacks to locomotives—and, in particular, agricultural machinery—support a large and increasing industry. In the south-western peninsula—surrounded by Lakes Huron, Erie, and Ontario—which is covered by a relatively dense population, this activity is at its highest. The great coalfields of Pennsylvania lie on the other side of Lake Erie, and Niagara Falls are near by to supply power. A great iron and steel industry is being created at Sault St. Marie, where the waters of Lake Superior spill into Lake Huron.

The Provincial Government has set aside some ten million acres of wild land in northern and western Ontario as a forest reserve; and in the peninsula the Algonkin National Park provides a game and fish sanctuary. Allusion has been made already to the forest and mineral resources of Ontario. Neither these nor manufactures constitute the principal wealth of the Province. Ontario is essentially a farmer's country. The whole of the peninsula, between the Ottawa River and Detroit River, at the United States frontier, is an agricultural country. Wheat and oats are raised in very large quantities, as well as maize, barley, peas, roots and hay; but a large proportion of these field-crops are consumed as food for the live stock used in dairying. Ontario makes more cheese than any other Province. Fruit-growing, too, especially in the south, round Lake Erie, is a profitable industry: in all this favoured district of orchard land the peach and vine grow to perfection. At Guelph there is a well-known agricultural college; and the Federal Government maintains experimental farms and orchards in many parts, including the central one at Ottawa, at which emigrants and others receive training.

In the eastern counties, as might be expected, French-Canadians have settled in considerable numbers; but the Province as a whole is essentially British in settlement. In the county of Waterloo there is a strong German contingent; and Negroes, Indians, and half-breeds are scattered throughout.

There are about six thousand free public schools, many High Schools and Collegiate Institutes, besides the Universities of Toronto and Ottawa. The Province is mainly Protestant; but, although denominational grouping prevails, as elsewhere in Eastern Canada, no Church is privileged.

Inland Provinces.

Manitoba and the great North-West is indissolubly connected with the name of "The Company"—the famous Hudson's Bay Company, with Prince Rupert at its head, which received from Charles the Second a grant of some two and a half million square miles of territory round Hudson Bay at an annual rental of "2 elks and 2 black beavers." Its fortified posts—chief of which was Fort Garry, south of Lake Winnipeg—became trading centres, to which the Indian and other trappers brought their pelts; and its treatment of the natives was proverbially good. One of the first acts of the Federal Government was to buy out the monopoly of the Company; and, when settlers began to come in to the "North-West," the Red River district was organised and Fort Garry became Winnipeg by name.

The phenomenal growth of this Prairie City and the rapid development of the vast area of wheat-lands to the west are among the leading achievements in the settlement of British North America. Other branches of agriculture have proved no less successful: and dairy farming has made great progress. The Dominion Government maintains an experimental farm at Brandon.

About eighty-five per cent. of Manitobans are English-speaking; and by far the greater number of these were born either in the Province itself or are immigrants from Ontario. Waves of immigration have swept in from south-eastern and north-western Europe, leaving settlers from many lands. While a large number of the Roman Catholics are the descendants of the old French fur-traders, Presbyterians predominate.

Elementary education is supported out of public funds. The Province does not lack High Schools and Colleges. At Winnipeg the University of Manitoba is well equipped.

Saskatchewan.—The physical features of Manitoba—a gently rolling prairie, mostly bare of trees—are continued in the southern borders of the newer Province of Saskatchewan, but the further one proceeds west the drier the country becomes. This dry patch carries good wheat-lands in the south-east and south-west, with an agricultural and stock-ranching population that is rapidly increasing. In the well-watered park-lands to the north, mixed and dairy farming have better opportunities.

The people are of various origin, having come in with a vast stream of migration that has increased the population of Saskatchewan and Alberta over five times within the last decade. Most of the people are English-speaking immigrants from other parts of Canada, the Home Country, and the United States; while many of the emigrants from the States are Old-Canadians. Galicians and Scandinavians are numerous.

Regina, the Provincial capital, is the headquarters of the famous North-West Mounted Police. It has a university college.

Alberta.—The twin-province of Alberta resembles Saskatchewan in its population and resources. The vast high-plateau in the south, which formerly was the home of the cattle-king and the cowboy, is, in spite of a light rainfall, suitable for autumn-sown wheat; and land is being brought under cultivation. Irrigation has been introduced in southern Alberta, where the land slopes down from the mountains; and here beets are grown for sugar. Cattle-ranching still flourishes in among the foot-hills of the Rockies and elsewhere, and horse-ranching supplies local requirements for the farm.

Railways are opening up the country rapidly, and settlers follow in their track. In the northern half of the Province, beyond the reach of the railway at present, the well-watered forest-lands are the home of fur-traders, who bring in their skins to the Hudson's Bay Company's stores at Edmonton.

The growth of Edmonton, the Provincial capital, has been very rapid within the last decade: it is now a city with fine public buildings and a university. Edmonton, too, like Calgary—which rivals it—is an important railway centre.

Pacific Province.

The discovery of gold, in 1858, let in a rush of population. Thirty-three thousand men from the Californian mining camps came up in one summer to prospect the valley of the Fraser River and its tributaries; but they disappeared as suddenly, with disappointed hopes. This influx, however, led to the organisation of the Province. Gold-mining continued, until the discovery of rich deposits in the mountainous Caribou district brought on another invasion of gold-seekers from California, Australia, and other parts.

British Columbia, on joining the Federation, stipulated for a trans-continental railway to link up its rich mineral lands with the Eastern Provinces. Thus, the Canadian Pacific Railway was carried to, and created, Vancouver—now the largest city of the Province and the fourth largest of the Dominion. Victoria, the Provincial capital, stands at the southern end of Vancouver Island. The railway terminus and seaport at Prince Rupert are, in like manner, the creation of the Grand Trunk Pacific Railway.

Chinese and Japanese labour was drawn upon largely in the early days of settlement, with a result that their exclusion or strict limitation nowadays are topics that closely concern the Provincial Legislature, which is anxious to keep British Columbia essentially a white man's country. The bulk of the population is British in origin—mainly by direct emigration—and includes a large contingent of Canadians from other parts of the Dominion. The influx from the United States has been very considerable, and Scandinavians also are numerous. Protestant denominations predominate.

The Territories.

East of the Yukon, to which reference has been made, lies Mackenzie Territory, which comes under the authority of the Crown exercised by the Royal North-West Mounted Police. Its northern shore, washed by the Arctic Ocean, is visited by these smart troopers and by whalers from Bering Sea. Both in Mackenzie and in Keewatin, which includes the western shore of Hudson Bay, there are some missionaries and fur-traders, but otherwise no white population. Ungava, too—between the eastern shore of Hudson Bay and Labrador Coast—is little visited.

There are over one hundred thousand Indians and half-breeds scattered over the Dominion, and their numbers are not increasing. A majority of these are found in Manitoba and the North-West Territories; but over twenty thousand remain in each of the other regional areas surveyed by us: British Columbia, Ontario, Quebec, and the Maritime Provinces. Dr. Bryce, of Winnipeg, classifies the aborigines under (1) Algonkin, (2) Dakotas, or Sioux, (3) Chippewayans, or Athabaskans, (4) Indians of British Columbia (Iroquois in the north, Salish and others in the south), and (5) Eskimo. The great Algonkin family is the connecting link between East and West; but the origin of by far the most remarkable of these tribes, the Iroquois, is lost in hopeless obscurity. The great reformer and lawgiver of the Iroquois, Hiawatha, who rose to power in the middle of the fifteenth century, constructed a federal system which revived the Five Nations and survives to this day, although his scheme for a confederation of all the tribes fell short of fruition.

Trade and Commerce.

An appreciation of Canadian literature and art should, properly, precede the consideration of trade and commerce; but exigencies of space preclude more than this passing reference.[1] Of trade and commerce, too, to which incidental references have been made already, little need be said in this place. The oversea trade of the Dominion is expanding, the gross figures for the year 1912 exceeding, for the first time, 200 million pounds sterling. In 1912-13 the proportion of imports from the United Kingdom to total imports was 20·3 per cent., whereas that of the United States was 65·5 per cent.

The economic development of Canada is indicated by the stages through which she has passed in her experimental search for a fiscal policy. Under the French *régime* she was ground down by monopolies set up and supported by a ruthless administration of military autocrats. "Let us beware," wrote Montcalm, "how we allow the establishment of manufactures in Canada; she would become proud and mutinous like the English. . . . So long as France is a nursery to Canada, let not the Canadians be allowed to trade, but kept to their wandering, laborious life with the savages, and to their military exercises."[2] Cession of the country to

[1] Cf. *Canadian Life in Town and Country*, by H. J. Morgan and L. J. Burpee (p. 192); and *British America*, vol. iii. of the "British Empire Series," in which Sir J. G. Bourinot gives a short review of Canadian literature.

[2] Quoted in *Canadian Life*, *op. cit.* (p. 63).

England brought some relief; but the prohibition of foreign trade was maintained. The Navigation Laws restricted external commerce to the Mother Country and the Thirteen Colonies, and thus set up a preferential policy which was met in England by substantial reductions in the tariff. But this growing trade collapsed on the abolition of the Corn Laws. It revived, however, between 1855 and 1866, when a Reciprocity Treaty was in force between Canada and the United States. The treaty was abrogated by the Republic, but not before Canadians had sown their wild oats and settled down to business: it helped materially to break down antagonistic interests and hostile tariffs between Province and Province, leading eventually to Federation and the pooling of economic interests.

This was not accomplished without much friction between the Conservative and Liberal parties, the final issue of which was fought out at the General Election of 1911, when Reciprocity with the United States was the question at issue. When Parliament was dissolved, the Liberals held 133 seats and the Conservatives held 88 seats. When again it met, the Conservatives held 131 seats and the Liberals were reduced to 87 seats: a complete landslide, which destroyed the Reciprocity Agreement by which Canada would have become "an adjunct" of the Republic, and definitely confirmed the adherence of Canada to "the old house." While controlling her own fiscal policy, Canada, through the Underwood Tariff, will derive nearly all the benefits suggested by the Reciprocity Agreement.

The Customs tariff of Canada is protective, but preferential towards the United Kingdom and most of the British Colonies. It is urged by Tariff reformers in the United Kingdom that Reciprocity with Canada and the other Dominions should replace—where these exist—Free Trade restrictions at present prevailing.

The banking system of Canada, though marked by efficient Government control, is very accommodating, the banks being allowed to increase their note circulation when the needs of commerce demand it: this is important for Canada, whose trade expands and contracts with the seasons. The value of English money is fixed by law, a sovereign being worth $4{\cdot}86\frac{2}{3}$ dollars, and a crown being worth 1·2 dollars.

A branch of the Royal Mint is established in Ottawa: sovereigns and half-sovereigns coined there are legal tender in every country under the British flag. Only five coins are used in Canada: the 50-cent piece (half-dollar), the 25-cent piece (quarter-dollar), corresponding to the shilling, the 10- and 5-cent pieces, and the copper cent, about equal to a halfpenny. Paper money, issued by the Government and the banks, is used universally for all sums from one dollar upwards.

Consular Service.—The British Consular Service is open to Canada and the other Dominions. The Canadian Trade Department is furnished by our Foreign Office with Consular Reports, as required; and any Canadian firm may apply for information direct to British Consuls, who will report thereon. Canadian Trade Commissioners (resident in Europe, etc.) are at liberty to apply to British Consuls for advice and assistance. Members of the Canadian Commercial Service, who may be accommodated with office-room—if available—in British Consulates, are eligible for the

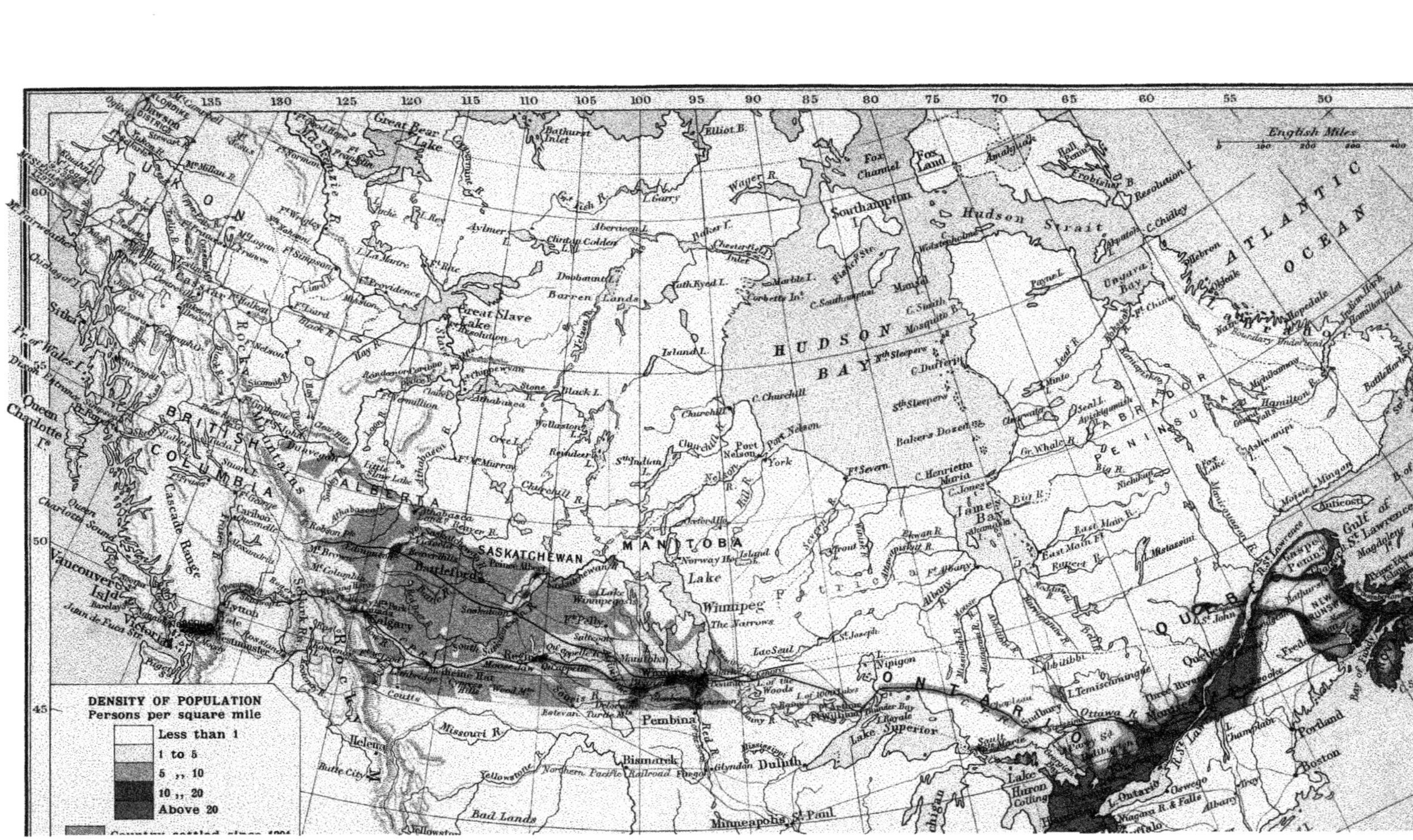

DENSITY OF POPULATION
Persons per square mile
Less than 1
1 to 5
5 ,, 10
10 ,, 20
Above 20
English Miles
HUDSON BAY
ATLANTIC OCEAN
Hudson Strait
LABRADOR PENINSULA
BRITISH COLUMBIA
ALBERTA
SASKATCHEWAN
MANITOBA
ONTARIO
YUKON
Rocky Mountains
Cascade Range
Great Bear Lake
Great Slave Lake
Lake Winnipeg
Lake Superior
James Bay
Gulf of St. Lawrence
Vancouver Isl.
Queen Charlotte Is.
Southampton I.
Barren Lands
Battleford
Calgary
Regina
Pembina
Duluth
Bismarck
Minneapolis St. Paul
Helena
Portland
Boston

British Commercial Service, under an arrangement that has been come to between our Foreign Secretary and the Canadian Minister of Trade and Commerce.[1]

External Relations.

We now have reached a stage at which we can report progress. We have seen the bold Elizabethan navigators and adventurers opening up the horizon of the New World. We have followed in the path of conquest and empire, and we have witnessed the flow of settlement from east to west across British North America—deepest in the Eastern Provinces and thinning out to a few rivulets passing through the Rockies to the Pacific Province. All that we have experienced, so far, has been enacted well within the North Temperate zone—dipping, indeed, into the Arctic—under conditions not dissimilar, save in degree, to Northern Europe; but enacted in so vast a theatre, that in many parts the human element is all but lost to view. Finally, we have to realise that in the people of Canada we find a definite grouping, but not too disproportionate a share in numbers,[2] between the British and French elements of national life, in which English ideals prevail so strenuously that, like Aaron's rod, all other elements are thereby absorbed. It is an English country from end to end, save in Quebec, which, though not imperial-minded, is thoroughly loyal and constant to the British cause—which, indeed, is their cause, inherited through their Norman ancestry. What Canada wants is men and money—more men and more capital—to develop her boundless resources and to foster her national life. Sir Wilfrid Laurier, the eminent French-Canadian, characterised Canada as "the gem of the British Crown." Rather, we would regard the premier Colony of the Empire as Britain beyond the Inland Seas, to whom is entrusted the task of carrying on English traditions and European culture. The women of Canada, too, are more identified with industrial and social progress than are their cousins in the Home Country. They have inspired many of the most vital social reforms and educational advances. Woman, in Canada, is the bulwark of religion, and reigns supreme in the home, upon which the whole social structure rests—and this, more particularly, is the case in the Province of Quebec.

In our view, Canada is a country which the Dominion Government holds—as in the case of the other Dominion Governments—*in trust for the Empire*: it is not exclusively a country to be used for its own purposes by the favoured few who have their homes there. It has duties to the Empire, as such—duties of British citizenship, so to speak—and these constitute a redemption of the great moral debt which was built up overseas by the blood and treasure of the Motherland.

Defence.—The Imperial garrisons at Halifax and Esquimault were

---

1 Statement by Mr. Ackland in the House of Commons, November 26, 1912.

2 In 1911, English-speaking persons numbered 3,721,944; and French-speaking, 1,649,371. Moreover, French-Canadians claim a million compatriots in the neighbouring Republic, some of whom are returning to the home of their ancestors. Direct emigration from France has ceased long ago.

withdrawn, owing to the patriotic offer of Canada during the Boer War to maintain these prime naval bases at her own expense. The control of Admiralty property at these stations was handed over finally in 1910 to the Dominion Government, on the understanding that H.M.'s ships of war shall be given facilities, and that the naval dockyards be kept up.

In regard to naval defence, towards which Canada makes no direct contribution, the proposal of the Borden ministry to provide for the construction—as an emergency measure—of three battleships or battle cruisers of the latest type, at a cost of thirty-five million dollars, has not yet been sanctioned by the Dominion Parliament. In the Home Country, Canada is entitled to representation at special meetings of the Committee of Imperial Defence, together with the other Dominions, who have accepted in principle the proposal—brought forward at the Imperial Conference of 1911—that a Defence Committee be established in each Dominion.

The number of men who were trained in the Militia in 1910-11 was 39,512. A scheme has been drawn up whereby the organisation of the Militia in Eastern Canada will be the same in peace as it would be in war. In March 1911 there were 263 gazetted Cadet Corps, the success of which movement has an important bearing on the efficiency of the Militia, which at present is crippled by lack of qualified officers and N.C.O.'s, as well as by the restricted term (twelve to sixteen days) of annual training. The general plan of the Department of Militia and Defence of the Dominion is to organise 100,000 men in first line and 100,000 in second line. The first line, or Active Militia, is making fair progress; the second, or Reserve Militia, is unorganised.

Communications.

Internal.—The Canadian Pacific Railway, 2906 miles in length, traversing the continent, linking up centres of settlement and penetrating new fields for conquest, always has stood in the van of progress. Its work and its policy have been broadly Imperial, no less than vital to the Dominion in its early days of development. Other transcontinental railways have come into existence; and two of these will reach the Pacific by 1914, when the Panama Canal will be open for traffic. The two new lines are the National Transcontinental, or Grand Trunk Pacific, Railway to Prince Rupert, and the Canadian Northern Railway to Vancouver. The Mackenzie and Mann system also is constructing transcontinental connections.

A network of recently constructed railways criss-crosses the Prairie Provinces in their southern parts; and the River and Lake Provinces are fully served by railroads as well as by waterways. The same may be said of the Maritime Provinces, which also have the Intercolonial Railway.

In June, 1911, the railway mileage in operation throughout Canada was 25,400 miles, or over 2000 miles more than in the United Kingdom in the same year; and another 1600 miles of railway were in partial operation.

The Federal Government have undertaken, by 1914, to construct a railway to Hudson Bay—to the tide-water terminus at Port Nelson—in connection with which it is contemplated to run a line of boats across James Bay to the Nottaway River, where a railway line could be carried to Labrador or Quebec. The road starts from the crossing of the Saskatchewan River at the Pas Mission. The distance from the wheat-fields to Hudson Bay is practically the same as to Fort William, at the head of Lake Superior; likewise, the ocean voyage from the Bay, under favourable conditions, equals approximately that from the St. Lawrence during the period of safe navigation: thus, under favourable conditions, the economic saving would be equivalent to the cost of transportation from Fort William to Montreal. The period of safe navigation is reckoned at four to five months (July to November) for Hudson Bay, and a month less for Hudson Strait, where ice-breakers could be used to keep open the fairway and to signal ice-conditions by radio-telegraphy.

External.—The increase in the popularity of the St. Lawrence route, for immigrants and tourists, is indicated by the following figures of traffic by the principal trans-Atlantic lines operating direct between British and Canadian ports during 1911:[1]

Steamship Line.	Westbound.	Eastbound.	Total.
Allan, . . .	61,709	19,330	81,039
Canadian Pacific Railway,	40,200	18,750	58,950
White Star Dominion, .	29,162	12,985	42,147
Royal (Bristol), . .	13,204	6,351	19,555
Donaldson, . . .	9,911	1,771	11,682
Cunard, . . .	5,972	1,080	7,052

The contract for the conveyance of the direct weekly mail (Liverpool-Halifax in the winter months, Liverpool-Quebec in the summer) is held by the Allan Line. Supplemental routes also are used.

The British Government pays £20,000, and the Dominion Government pays £25,000, per annum, to the Canadian Pacific Steamship Company for the trans-Pacific mail service between Vancouver and Hong-Kong, via Yokohama: this is the Empress Line, with two 18-knot steamships, of 15,000 tons gross, and six others. The service is three-weekly from each terminal port; and the contract time, from Liverpool to Hong-Kong, is 818 hours via Quebec, and 853 hours via St. John, N.B. In 1911 the Dominion Government contracted afresh with the Union Steamship Company of New Zealand for a five years' service, with monthly sailings. Other mail contracts include Lines to Havre, South Africa, Central American and Mexican ports. The Holt Line (Liverpool) operates to Vancouver via the Suez Canal and Asiatic ports.

The Great Lakes are well lighted; and cargo vessels from Lake ports reach the Atlantic without breaking bulk. In 1910 over 36,000 vessels, aggregating nearly 43 million tons, passed through the Canadian canals. Freights were chiefly in grain, iron ore, timber and coal; and the preponderance of traffic is from west to east. But nearly 80 per cent. of this traffic is American. A comprehensive scheme for the improvement of the navigation on the Great Lakes and the Welland and St. Lawrence canals is in course of construction.

[1] *Canada To-day* (Annual), 1912.

Cables.—There are thirteen cables between the United Kingdom and Canada, all of which are under American administration. The only completely British-owned cable is from Halifax to the West Indies, via Bermuda. It is, however, proposed—as complementary to the All-British Pacific Link—to lay an Atlantic Imperial cable from Blacksod Bay, on the west coast of Ireland, to Halifax.

Improved steamship and cable services, in support of the Dominion Government's trade agreements with the British West Indies, have now been arranged.

Radio-telegraph, or "wireless," stations, open to ship and shore communications, are established along the Atlantic and Pacific coasts, on the St. Lawrence, and round the shores of the Great Lakes. At Glace Bay (Cape Breton Is.) there is a high-power public service station, for trans-Atlantic messages.

ITALIAN EMIGRATION TO SCOTLAND.

By Ralph Richardson, F.R.S.E.

An Italian visiting Scotland must be struck with the fact that, go where he will, he will almost certainly find a compatriot engaged in business. There is scarcely a village in Scotland without its "Italian Ice Cream" shop, a gaily decorated place of business, with the Italian National flag painted outside, plenty of mirrors inside, and, to Scotsmen, a strange foreign name on its signboard. At one time the Italian emigration to Scotland took the form of men selling stucco casts of figures from the antique, often of considerable beauty; at another time, Italians were, and still are sometimes, conspicuous as organ-grinders or travelling musicians. Now, however, Italian emigration has followed a more important and more lucrative channel, and we have throughout Scotland Italians occupying shops for the sale of ice-cream, confectionery, and other refreshments, taking part in Scottish trades and hotels, and becoming regular, resident, Scottish citizens.

One would imagine that Italians suffering from poverty would hesitate about a voyage to distant Scotland, with a different language, climate, and manners to their own. It is, however, their very poverty which first propelled Italians there, although latterly their business is so widely and solidly established that Scotland draws to itself Italians of the hardworking, but by no means impecunious class. While well-to-do Italians seldom visit Scotland, working Italians emigrate there in large numbers, and apparently are quite satisfied with the financial results. They know that they go not merely to a wealthy country, but also to a people who have always regarded Italians as friends, and who enthusiastically admired Garibaldi—

"A name that earth will not forget
Till earth has roll'd her latest year."[1]

---

[1] Tennyson on Garibaldi, who planted a tree in the poet's garden.

There is also a fellow-feeling between Italians and Scots, for they both went through a "War of Independence" in the course of their history.

By the kindness of Signor Lagorara, Italian Consul-General for Scotland, my attention was drawn to an interesting article on "The Italians of Glasgow," which appeared in the *Weekly Mail* (special) of October 21, 1911. This article states that "the first real invasion" of Scotland by Italians took place "about forty years ago, when there arrived in Glasgow from Tuscany a few sellers of statuettes, and about the same time there came over some Swiss-Italians, who, in consequence of a landslip in their native Canton, had to depart into the world to seek their fortune. Some of the latter found refuge in Italy, some went to London (among them being the now well-known Gatti and Monaco), while others came to Glasgow." There they noticed that on holidays and Sundays the shops were shut, so some of the Italians supplied on those days "various refreshing drinks and a species of iced refreshment, popularly known as ice-cream." The success of these ice-cream sellers induced other Italians to emigrate, and in 1911 there were in Glasgow no fewer than 300 ice-cream shops served by 900 of the 2000 comprising the total Italian population of the city. It may be remarked, with regard to such shops being kept open on Sundays, that Italians and foreigners generally are accustomed in their own country to regard Sunday as the great holiday of the week, and a day to be largely devoted to recreation. I have observed this even in Calvinistic Holland, much more in Lutheran Germany.

Signor Lagorara, in a letter with which he favoured me, dated December 22, 1911, gave the approximate number of Italians actively employed in business in Glasgow at that date as follows:—

Business.	Shops.	Employees.
Ice-cream, etc., ...	300	900
Restaurants, ...	80	240
Barbers, ...	40	160
Mosaic workers, ...	...	120
Waiters, ...	...	60
Organ-grinders, ...	...	20
Ironworkers, etc., ...	...	100
Produce merchants, . . .	...	20
		1620

The *Weekly Mail* states that the Italian immigrants are chiefly "of the crofting class," having been brought up on farms, and know nothing about the ice-cream trade until they arrive in this country. When they return to Italy "they generally resume their old occupation on the farm." With regard to their character, the same newspaper states that, "the Italian, as we know him, is generally hardworking, thrifty, and honest. Other good traits in his character are his sobriety and morality. In a recent year's statistics of Glasgow's police, it is shown that out of

37,707 persons convicted of offences, only forty-five were Italians; twenty-one of these offences were on technical points connected with the conduct of their shops, and only two were serious cases."

The Lord Provost of Glasgow, in the course of an address in 1908, characterised the Italians in Glasgow as "sober, hardworking, and honest, and an example to his fellow-countrymen." The Italian women are excellent housewives, thrifty and economical and good cooks. The majority of the Italians are Roman Catholics.

True to the motto of New Italy, "Italia farà da se," the Italians in Glasgow have established various societies to assist Italian immigrants, morally and materially. Their principal society is the "Mutuo Soccorso" (mutual help), which attends to cases of sickness or necessity. It was established in 1891, and is in a sound financial position, having at its credit in 1911 a sum of £784, 18s., of which £650 is in Corporation loans.

There is also the "Società Italiana di Beneficenza" devoted entirely to charitable purposes, its principal object being the relief of indigent Italians so that they do not become a burden on the local authorities, or have recourse to begging. This society likewise enables Italians seeking employment to travel to other parts of the country; and, in cases of absolute poverty, especially of women and children who have been left destitute, assists them to return to Italy.

We find also in Glasgow a Branch of the National Dante Alighieri Society, with a Consiglio Direttivo (Executive Committee), of which Mr. Ernest B. Thomson has been President from the beginning. He favoured me with a letter, dated 23rd August 1913, in which he stated that "our aim is to provide a number of lectures in Italian and on Italian subjects for the benefit of members and friends. The lectures were given, and meetings, until a few months ago, were held in one of the rooms of the Glasgow Athenæum, but an arrangement was recently made with the Italian Society 'Mutuo Soccorso' for the use of their hall at 430 Argyll Street for all meetings and lectures, and also for a small circulating library we are endeavouring to form. Our society has not yet many members, owing to the fact that most of them are shopkeepers or men employed in shops which are open at the time our lectures are given. We also have considerable difficulty in procuring lectures in Italian in Glasgow, where visits of cultured men from Italy are somewhat rare. On the whole, our members have reason to be satisfied with what our Comitato has been able to accomplish hitherto, in comparison with Comitati which are more favourably placed."

Mr. Thomson also sent me the List of Office-bearers and the Laws of his Comitato, which was established in Glasgow on 26th August 1909, and was the first Comitato of the National Dante Alighieri Society instituted in Great Britain. The Honorary Presidents in Glasgow are the Lord Provost of the city and the Principal of the University, while its President is Professor Phillimore of the University of Glasgow, and its Vice-President is Mr. Francis H. Newbery. The Secretary of the Glasgow Executive is Signor Carmine Alfieri, and its Treasurer is Mr. Robert A. Buchanan, Vice-Consul for Italy, 21 Bothwell Street, Glasgow, which is also the official address of the Secretary. Such a society deserves the support of every one interested in Italy, its literature, its history, and its people, and Glasgow may well be proud of being the first in Britain to establish a Comitato.

According to Signor Lagorara, the Italian Consul-General for Scotland, in his letter to me of December 22, 1911, the total Italian population in Glasgow and suburbs was about 2000, and the total for Scotland about 5000. The Italian Consul for Edinburgh and Leith (Sir Richard Mackie) favoured me with a letter dated December 30, 1911, setting forth the number and condition of the Italians there. He stated that "there are close on 950 Italians in Edinburgh and Leith—men, women, and children. Their occupations are nearly all connected with the ice-cream and confectionery trade and fried fish and potato restaurants. There are a few who go round with organs, but not nearly so many as there were a few years ago. There is only one Mutual Society in Edinburgh, of which Mr. Giovanni Dagostino, 121 Leith Street, is President."

While, as far as the Consul is aware, there has never been any article in the newspapers on the Italians of Edinburgh and Leith, and he has no printed matter or reports regarding them, he adds, "But I may tell you that, as far as my experience goes, and I know them pretty well, they are a most sober and industrious class of people."

The last statement comes with peculiar force from Sir Richard Mackie, as he was for some years Provost and Chief Magistrate of the burgh of Leith, and had thus, from his official position, a very intimate acquaintance with its population. His testimony to the high character of the Italians resident in Edinburgh and Leith coincides with that already expressed regarding those resident in Glasgow by the Lord Provost of that city in 1908.

As the majority of the Italians are Roman Catholics, the opinion of a leading clergyman of that Church is valuable. Not long ago, the Very Reverend Canon Stuart, a member of Edinburgh School Board, bore testimony from personal knowledge to the excellent character and industrious habits of the Italians in Edinburgh and Leith.

Although there is a prejudice in some minds against Italians, and indeed against foreigners generally, especially as immigrants, a broader view is taken by well-informed authorities such as the Editor of the *Spectator*, who, in an editorial note (May 10, 1913), assured a correspondent that "we needed no proof of the assertion so often made in these columns, that the Italian peasant is an admirable immigrant, and that a new land in want of settlers should be glad to get him and his wholesome blood, large family, and saving habits. As has been proved over and over again in many generations, the mixture of English and Italian strains is excellent." While Scotland is anything but a "new land," the character of her immigrants is as important to her as it is to any newly settled country.

We have seen that, beginning some forty years ago, the emigration of Italians to Scotland has gradually increased, until now there are about 5000 scattered throughout Scotland, occupying in many cases positions of commercial importance and acquiring considerable wealth. Unlike their compatriots in the United States, they prefer to start shops in towns, and do not care much to work as railway navvies or as labourers, although, in special cases, they act as mosaic workers, of whom there are 120 in Glasgow alone. I was told that, in laying the caissons of the Forth Bridge, Italians were employed, as their magnificent lung development (so noticeable in their public singers) enabled them to remain below water longer than men of other nationalities.

Just as Italians are increasingly arriving in Scotland, the Scots are leaving it in numbers which are causing national alarm. Seeing that during the first half of 1913 some 35,000 Scots sailed for Canada or the United States, which will henceforth be their homes, those who depended upon them in Scotland for their support—whether intellectual, moral, financial, or as labourers—may well look with foreboding on the future of Scotland. For our Scottish emigrants are usually men and women in the prime of life, full of health and energy, otherwise they would not, and could not, be emigrants. By an Act passed on February 20, 1907, the United States accepts as immigrants only aliens who are sound in mind, body, and morals.

It is therefore a drain of blood of the highest quality which is pouring from our native land into one of the Britains overseas or into the United States. It has affected the Scottish labour market—railway and farm servants, foundry and factory hands being now scarce; indeed at Brechin feeing market it was "stated that, owing to the scarcity of labour, more than one farmer has been entirely depleted of workers."—*Scotsman*, June 4, 1913. It has even affected the religious atmosphere of Scotland and occasioned the lament of Sir Colin Macrae, in the General Assembly of the Church of Scotland (May 23, 1913), that "it was probable, nay, almost certain, that the decrease of 3000 in the number of scholars attending that Church's schools was due to the large emigration which had taken place from our shores during the last few years and which seemed to be rather increasing than diminishing. The birthrate in Scotland was steadily falling, and, in addition, there was the vast stream of emigration draining away the strongest youth of the country."

No one denies that for the "individual Scotsman," as the *Spectator* lately put it, it may be a good thing "to determine for himself whether he will continue to live under the British flag in a difficult climate in the north of Britain, or whether he will transfer himself and his home to another country under the same flag, where he will find ampler space and the opportunities for a larger life." This may be a good thing for the individual Scotsman and for the British Empire; but is it an altogether good thing for Scotland, whose teeming industries cannot be carried on if Scotland is permanently deprived of labour on a large scale, and whose many agencies for the advancement of the best interests of the people of Scotland will be crippled if the young and vigorous are, in increasing numbers, carried away for ever from Scotland, to distant shores?

The arrival of a multitude of Italian immigrants will not, from the point of view of the predominant Scottish Churches, better the situation, for the Italians will be Roman Catholics; but they will supply for Scotland, as they have long done for France, a counterpoise to a decreasing native population, and will furnish a laborious, thrifty, and sober class of tradesmen and operatives which no country would undervalue. Already the great Scottish mining industry has been indebted to Poles for much of its development; and, if Scotland has thus profited by the labour of Slavs, it may also benefit by the industry of members of the Latin race.

SOME OF THE ECONOMIC EFFECTS OF THE PANAMA CANAL.[1]

By Professor A. W. KIRKALDY.

IN estimating what the economic effects of the opening of the Panama Canal will be, it should be noted that from the broad business point of view there are two spheres—(1) the local, (2) that of world trade, in which it is likely to have considerable, though different effects.

Beginning with the second of these, it has been realised, though perhaps not so clearly as it should have been, that there will be other factors affecting the situation in addition to the saving of distance. But until the new route is fully available for commerce it will hardly be possible to realise every consideration that will weigh with importers and shippers. Some light, however, may be thrown upon the subject by a consideration of a few of the factors which have been and will continue to be taken into account, either directly or indirectly, in deciding (*a*) who shall supply certain markets with either manufactured goods, raw materials or food stuffs; and (*b*) what routes shall be taken by the shipping which performs the necessary work of transport. No route can possibly furnish the whole of these factors, and as here as in other spheres the less important has to give way to the more important, the final decision will depend on the balance of advantages.

The principal of these factors, arranged rather in their natural order than in order of importance, are:—

(1) Distance.
(2) Tolls.
(3) Freights and the possibility of continuous freight earning.
(4) Fuel stations—coal or oil.
(5) Insurance rates.
(6) The political factor.
(7) Rates of exchange.

---

[1] A paper read before Section F (Economics) at the Birmingham meeting of the British Association. The paper is excerpted from a book by the author on *British Shipping*, which will shortly be published by Messrs. Kegan Paul, Trench & Co., who have kindly permitted this paper to be printed.

(8) Invested capital and banking facilities.

(9) The human factor (manufacturing and commercial ability, experience of trade and markets, present possession).

1. *Distance.*—The lessening of an ocean voyage by a few hundred miles may have a very direct effect in decreasing the cost of transporting goods, and so should favourably affect freights in the interest of the importer. A modern steamer represents a large amount of capital; there is connected with the working of such a ship considerable expense each day for wages, food, fuel, and running expenses generally; and then in addition to this there is interest on the capital, depreciation, and other items inseparably connected with the business of shipowning. Thus mileage run in the case of cargo transported by sea is a serious factor with which shipowners have to reckon. In this shipping offers a great contrast to railways. When railway trucks are once loaded, length of haul has comparatively little effect on the cost of the service. Thus superficially one would expect that a route saving about 2500 miles, as will be the case via Panama on the voyage between England and New Zealand, and New York and New Zealand, would very considerably benefit the eastern manufacturing States of America at the expense of British manufacturers, and have a very marked effect in assisting the progress of the American mercantile marine at the expense of the British shipowner. Even though the ships of both countries pay the same rate of toll, the saving of distance promises to give a great advantage to both the American shipowner and the American manufacturer.

The keenest competitors for the supply of manufactured goods in the world's markets at the present moment are the United Kingdom, the United States of America, and Germany. All three of these must be affected to some extent by the opening of the Panama Canal. So far as the mere saving of distance is concerned the United Kingdom and Germany will suffer or benefit together, for the short distances between the exporting ports of North-Western Europe are practically negligible in the long sea voyage to the Far East, or even to South American ports. Thus when making a comparison between London and New York, from this point of view, one is really considering the advantages or disadvantages accruing to the manufacturers of Western Europe, as compared with those of the Eastern States of America.

In trying to estimate the possible effects of the opening of the Panama Canal from this point of view, the chief foreign markets of the world demanding imported manufactured goods may be roughly divided into three classes:—

(1) In the first class come all those countries which are in close, or fairly close, proximity to the Canal. These include all the States of North and South America and the adjacent islands. So far as some of these are concerned the transport services should pass through the Canal. From Valparaiso northward, all trade to or from the Eastern States of America and Europe; and from Pernambuco northward, all trade to and from ports on the Pacific coast of North America should almost of necessity pass through the Canal. The effect of the Canal on this trade,

so far as distance is concerned, will be very considerable. From London to San Francisco round Cape Horn is only about three hundred miles further than from New York to San Francisco. The Canal will affect both voyages, but while it shortens the distance from London by 5500 miles, it brings New York nearer to San Francisco by 8000 miles; an advantage of over two thousand miles, which, with the possibility of freedom from tolls for coasting vessels belonging to the United States, will give the manufacturers of the Eastern States a very great advantage in any open markets. The Western States of America will also benefit in another way. It is practically certain that the opening of the Canal will give a great impetus to the immigration of Europeans. Thus a rapid development of States which have hitherto been comparatively unprogressive may be looked for.

(2) In the second class come Australasia and the Far East. At present the shipping supplying these markets has a choice of routes, for some ports there is a choice of three routes, via the Suez Canal, via the Cape of Good Hope, and via Cape Horn; for others only two of these. From the moment that the Isthmus of Panama is pierced, there will be another possibility. This second class shades off gradually into the third class, (3) which comprises ports which will not be directly affected by the new route. It is mainly to the ports comprised in the second class that attention is being directed when an attempt is made to estimate the importance of the Panama Canal to world trade.

There is a parallel which so far as actual steaming or sailing distance is concerned is about equidistant from London via the Suez Canal, and from New York via the Panama Canal. The point of this parallel on the south coast of Australia is represented very nearly by Port Lincoln, South Australia; Adelaide being the nearest of the great ports of Australia. Most of the Asiatic trading ports will continue to be nearer to London than to New York, for though by going via Panama instead of via Suez the distance between New York and Shanghai is reduced from 12,321 to 11,240 nautical miles, London via Suez is nearer by 799 miles, the distance being 10,441 miles. All ports west of Shanghai will remain nearer to London via the Suez Canal than from New York via Panama. For instance, the voyage to Manila will be 2000 miles less from London via Suez than from New York via Panama; indeed, and this makes the point perhaps clearer, it will save something like 200 miles when steaming from New York to Manila to go via Suez rather than by the new route.

On the other hand, all Japanese and New Zealand ports and Australian ports east of Port Lincoln, will be nearer to New York via Panama than to London by any route. For instance, Yokohama will be 892 miles nearer New York than London, Melbourne 831 miles, Sydney 1612 miles, Brisbane 2933 miles, Auckland 3660 miles, Wellington 3717 miles, and Dunedin 3137 miles. Thus, with the opening of the Panama Canal, the chief Australasian ports from Melbourne eastwards, which have hitherto been, roughly speaking, from 900 to 1800 miles nearer to London than to New York, will be nearer to New York by several hundred miles. If it be correct to reckon the normal cost of transporting a ton of goods 1000 miles at about two shillings, then, even though the Suez and Panama tolls be of equal amount, this saving of distance will give the manufacturers of the Eastern States of America an advantage of from two shillings to seven shillings and sixpence a ton on any goods they may export to Australian and New Zealand ports between Melbourne and Wellington. Here then is the case so far as mere distance to be traversed is concerned; on the face of it one might conclude that the advantage given to American manufacturers would be such that in certain markets Europe would no longer be able to compete efficiently.

At this point, however, other factors emerge, and in considering them it becomes apparent that distance saved is but one of several considerations which weigh with both the importer and the exporter.

2. *Tolls.*—When the Suez Canal was opened to traffic, the great saving of distance between Europe and the Far East operated at once. The only alternative route entailed steaming a distance of 4000 miles further than that via the Canal, and this was sufficient to give the Canal a practical monopoly. So far as tolls are concerned, a diminution in the rate charged, whilst giving satisfaction to the users of the Canal, cannot appreciably increase the tonnage passing through, and on the other hand, an increase in the charges might cause grumbling, but, up to a certain point, would not decrease business. With the Panama Canal the circumstances are somewhat different; there are alternative routes which much of the traffic likely to pass through the Canal might take. Thus the fixing of regulations and tolls is a very serious business. There has been a great amount of discussion on this subject from the political as well as from the commercial point of view. Does the United States Government wish the Canal to be a paying concern like the Suez Company; will it be content if running expenses are covered, together with a small surplus in addition with which to form a sinking fund to pay off gradually the original cost of construction; or will the tolls be but nominal, the American Government shouldering the burden almost entirely? The question was by no means set at rest by the publication in November 1912 of the conditions on which the Canal may be used, and the tolls to be charged. That proclamation, according to some authorities, contravenes treaties made both with Great Britain and the State of Panama. International lawyers and diplomacy must decide as to this. To the plain man the situation is that the United States have constructed the Canal, and think it a good opportunity to administer a check to the monopolistic policy of their own trans-continental railways, and at the same time give an advantage to American-owned tonnage which might give just that amount of impetus necessary to enable it to regain the position it enjoyed seventy years ago.

The main points about the schedule of tolls issued in November 1912 are, that merchant ships carrying either passengers or cargo will pay a toll of $1.20 cents per net ton. Ships in ballast, without passengers or cargo, will pay 72 cents per ton. Men of war will pay 50 cents on each ton of displacement, and naval and army transports will be on the same footing as merchant ships.

So far good, but unfortunately there is another proviso, namely, that ships belonging to the United States register and employed in the coasting trade, except those belonging to railway companies, shall pass toll free.

Against this provision a serious protest has been raised by Great Britain, supported by Germany, whose interests in that part of the world are great and growing. Should the United States Government maintain freedom of tolls for American ships engaged in the coasting trade a number of problems will arise. The press has already discussed this matter very fully, and some of the principal points at issue may be set down. Great Britain contends that under the treaty it was agreed by the United States that the Canal should be open to the commerce of the world on equal terms. Apparently the Americans do not altogether dispute this, but it is urged on their side that when the Hay-Pauncefote treaty was negotiated, the proposal was to construct a Canal through foreign territory, namely, through the State of Colombia. But since then the revolt of Panama has created a new situation, for that State made over to the United States Government, practically in full sovereignty, a strip of country ten miles wide across the Isthmus. The construction of a ship canal through their own territory is, it is contended, a very different thing from constructing one through a foreign State. This may raise one of those nice points of international law which rejoice the heart of a lawyer, but Monsieur Bunau-Varilla, who negotiated the treaty between the United States and Panama, points out that it was specially stipulated in that treaty that the clauses of the Hay-Pauncefote treaty guaranteeing equality of tolls to the ships of all nations should be incorporated in the treaty between the United States and Panama, and that these clauses still stand.

The United States Government may not have been aiming altogether at giving an undue advantage to the American coasting trade in drawing up the scheme of tolls. Indeed it is undoubtedly true that one great consideration weighing with the authorities was the possibility through the Canal of striking a blow against the monopoly selfishly exercised by the trans-continental railways. For ships belonging to these companies are stringently excluded from this benefit. One object of this is to create healthy competition between sea and land transport agencies, and obtain for the community the benefit of a state of competition rather than allow the present unsatisfactory system of monopoly to continue. Perhaps this might have been accomplished without raising international complications, possibly the difficulty may yet be arranged by the United States discriminating in favour of their coasting vessels, or against vessels owned by the trans-continental railway companies, and thus make the dispute domestic instead of international. But if the immunity of tolls for American coasting ships be maintained in spite of protests, or of arbitration on the subject, several interesting points will emerge. Doubtless a full and frank definition of a coasting voyage will be framed by the Government. It would certainly be interesting to see such a definition. A voyage between any ports belonging to the United States is counted a coasting voyage, even the voyage from New York to Manila is euphemistically so called. It has been hinted that the whole coastline of both North and South America might possibly be included, so that any American ships trading between the ports of the United States and Canada, or of any South American States, would be entitled to traverse the Canal toll free. This would be stretching matters too far, and there is hardly a possibility that it would be seriously considered. But even as the term coasting voyage is at present understood, either the coasting ships themselves will find their operations restricted in a very galling manner, or it may be the strict letter of the regulations will be disregarded. For instance, as at present understood, an American coasting vessel would be barred from touching at the West Indian Islands, or at any foreign port on the coast; for by doing so she would render herself liable to the exaction of the toll.

There would be many difficulties in enforcing full and literal compliance with the regulations. Take, for example, a ship sailing from New York to San Francisco via the Canal. It would require an immense amount of supervision to prevent such a ship proceeding either northwards to Canadian ports, or southwards to ports on the coast of South America, and this difficulty would be intensified in the case of ships trading with the Philippines. The United States enjoy a growing commerce with the Far East. The saving of rather more than a dollar per net ton would give a distinct advantage to American traders. How would it be possible to prevent ships and goods ostensibly going to Manila from going still further east when once clear of American waters? It would clearly not be to the advantage of the American Government to exercise supervision in this matter, and experience shows the hopelessness of getting justice in such a case through diplomatic channels.

There is also another consideration which may be serious indeed to foreign shipping. The nominal toll has been fixed at $1.20 per net ton. It has been estimated that altogether ten and a half million tons of shipping will pass through the Canal during the first twelve months. Accepting this figure for the moment, the total amount produced by the tolls, if all ships contributed, would be something less than eleven million dollars. It is estimated that the cost of upkeep and the running expenses will be about two million dollars a year. An allowance of three per cent. on the cost of construction, viz., $400,000,000, will be twelve million dollars. These taken together amount to a sum of fourteen million dollars per annum. Thus at the outset, even though all ships passing through the Canal without distinction pay tolls, there will be an annual loss on this calculation of at least three million dollars. But if American coasting vessels pass through toll free, this loss will be still greater. How long will the Americans be content to subsidise shipping to this extent? And be it remembered these are the commercial figures only, nothing has been allowed for the amount which must be spent on defensive works and their maintenance. Is the American Government prepared to adopt a policy which will place it in the position of public benefactor, in the first instance to their own coasting vessels, and then to a lesser extent to the shipping of all nations? It does not seem a likely proposition. The most that can be expected is that the Canal be self-supporting, that is, the tolls received must pay working expenses, upkeep and maintenance, together with the interest on the capital expended. If the sum of fourteen million dollars be accepted as covering these, and then one allows, that of the tonnage passing through one-third will be able to claim exemption of tolls under the coasting vessels clause, then the tonnage on which tolls may be levied would be reduced from ten and a half million to seven million, and it would require not a toll of $1.20 per net ton, but a toll of $2 to produce the required amount. This will be a heavier toll than that charged by the Suez Canal Company, which has been reduced to 6 francs 75 centimes ($1.40) per net ton. Moreover, there are two important considerations connected with the question of tolls at Panama that must be taken into account. Every addition made to the tolls will decrease the amount of tonnage making use of the Canal. Here Panama differs from Suez, as has already been pointed out. Thus the ten and a half million tons calculated on the basis of a toll of $1.20 would shrink considerably were a toll of anything like $2 imposed. The other point is this: the foreigner may not consent to be a dumb beast of burden in this matter. Already Japanese shipping interests have outlined an ingenious policy by means of which the Canal might be rendered almost entirely unremunerative. It has been threatened that in the event of American coasting ships being allowed to pass through the Canal free, Japanese ships bound from the Far East to European or Eastern American ports, will tranship their cargo at a Pacific port whence it may be carried through the Canal free by an American ship, and transhipped again at an Atlantic port, being carried to its final destination in a Japanese ship. It has been calculated that this transhipment would cost something like $2.50 per ton, which on the face of it would be a loss to the Japanese, but if all foreign shipping adopted this suggestion, how long would the American Government maintain the exemption from tolls of the coasting vessels?

Trade and commerce frequently take an unexpected course. It may be that before the Canal is finally opened to traffic the Government will see the wisdom of simplifying the toll system, charging the ships of all nations, including American ships, on one basis. This will settle the international question, and will very considerably free the hands of the Government. Then as to the machinery by means of which coasting ships may be benefited as against those of the trans-continental railways, or even American shipping against that of any other nation using the Canal, this could be more easily arranged by a system of bounties—a system already adequately understood on the other side of the Atlantic.

3. *Freights and the possibility of continuous freight earning.*—For the Panama route to give an advantage to American manufacturers in the shape of low freights, and to American shipping in the form of full cargoes and continuous demand for services (and these two really go hand in hand), there must not only be a large amount of freight from America to Australasia and the Far East, but there must be an

equally large amount of return freight and the possibility of picking up freight along the route. At present the demands of Australasia for manufactured goods are mainly supplied by Europe. The ships transporting these goods go either by way of the Cape of Good Hope, or through the Suez Canal. By the latter route a steamer passes, and may touch at, a number of trading ports. Mails, passengers, or cargo can be picked up, or landed by the regular liners. This does not, to so great an extent, affect cargo steamers, at any rate so far as intermediate freight is concerned; but the regular trading requirements have resulted in providing ample, and on the whole, economical, coaling and other facilities all along this route. Under different conditions the same is true of the Cape route. South Africa demands more bulk than it gives in exchange. Gold and diamonds are not exported in ship loads, nor do ostrich feathers help to ballast a large steamer. Thus the regular liners return to England "flying light," as sailors say. But the route from England via the Cape is exceptionally well supplied with coaling facilities, and it is this, together with the invention of marine engines having a comparatively small coal consumption, which enables so many steamers in the Australasian trade to utilise the Cape route both out and home, instead of going through the Suez Canal, unless this be contrary to the charter-party.

It has been suggested that when the Panama route is available, all-round-the-world services will be organised by the leading shipping companies. This may be the explanation of the attempt made a few months ago by a well-known shipping combination to absorb one of the oldest British companies trading in the Far East. Had this attempt succeeded, it would have been possible (and as there is nothing to prevent the attempt being repeated, it may yet be carried through) to inaugurate in the early days of 1915 a service of steamers from the United Kingdom via the Suez Canal to Far Eastern and Australasian ports, making a circuit of the globe via Panama. This combination would at the same time provide a very complete system by means of which the trade of West, South, and East Africa, both sides of South America, Central America and the West Indies, would all be co-ordinated as integral parts of a remarkably complete network of shipping routes, giving facilities under one management for handling, or at any rate competing for, under peculiarly advantageous circumstances, and with the benefit of long-tried experience and knowledge, the shipping business of some of the greatest markets in the world. American shipowners would find that a combination of this strength would be extremely difficult to compete with, its resources and elasticity would enable it to command existing freight, and to be first in the field whenever and wherever new trades might be opened up.

4. *Fuel-Stations—Coal or Oil.*—The United Kingdom produces large quantities of coal, and this can be carried very cheaply to distant parts of the world where coal is lacking. By loading a cargo of coal, a ship avoids the necessity of going in ballast to certain ports, and any freight carried in this way is money gained. In transport services if either ships or railway wagons have to go empty in one direction, the rates

charged must be considerably greater than if there be a return load, and thus any rate of freight, for the possibly empty journey, is an advantage. It is owing to this that there are low transport rates in certain directions both on shore and at sea. Of the freight carried by shipping rather over one quarter of the weight consists of coal, and English coal, transported mainly by English ships, is to be found almost all over the world.

There are signs that the United States fully realise the advantage that the coal trade is to Great Britain, and undoubtedly in connection with the opening of the Panama Canal a great effort will be made to increase the demand for American coal, not only at the terminal ports of the Canal itself, but at all the coaling stations connected either directly or indirectly with the new routes, and no doubt it is hoped eventually to gain for American shipping all those advantages which now and for so long have benefited the British shipowner. To the extent to which it is found possible to displace British coal in foreign markets and coaling stations, American trade should in the first instance experience a proportional gain, and on the strength of this, it is no doubt hoped, would be able with comparative ease to take up a position of greater importance in the sphere of ocean transport. Thus in the immediate future there will undoubtedly be a period of very keen competition in the various coal markets of the world, and it is of the greatest importance to the commercial and industrial interests of this country that our position be maintained. British colliery proprietors and the miners' unions will be on their trial, and the issue at stake will be the future wellbeing of the whole country. The position should be made perfectly plain to all who have a voice in shaping the relations between capital and labour in this great industry, in order that they may realise to the full their responsibility.

The United Kingdom exports between sixty and seventy million tons of coal each year, and in addition to this sends another twenty million tons abroad for the use of her ships engaged in foreign trade. This gives an immense volume of steady freight which is a great advantage to British shipping. This trade, however, is passing through a stage of transition. Nearly every country is looking seriously into its fuel resources, and in the near future there will be a great fight for supremacy between coal and oil. Coal itself will be subject to modification, and some experts are of opinion that before many years are over it will be exceptional to find coal in use in its raw state away from the district in which it is mined; gases and oils of various grades will be the main products of the colliery. Will America as a fuel-producing and exporting country be able to oust Great Britain from her present position in foreign markets? The solution of this question will have a very marked effect on the future commercial history of the world. The United Kingdom, in spite of its great coal resources, is small compared with America, but the British Empire has vast districts rich in coal and oil resources which are hardly yet known, and for the production of which, so far, only the surface has been lightly scratched. With increasing commerce the United Kingdom may, and probably will, hold her own, and it may be from the Dominions beyond the seas that America will meet the keenest competition in this other respect. Australia, India, British Africa, and, most important of all, Canada, have the possibility of competing for the supply of fuel for the mercantile marine of the world.

Hitherto British coal has been practically free from competition on the Suez route; from the moment that the Panama Canal is opened a very keen and growing competition will be experienced. The effects of American competition have before this been felt in other industries, nor has the result always been to the disadvantage of the United Kingdom. This threatened competition in the overseas coal trade may have a very much needed effect in making some of our collieries put their house in order. The coal industry of this country may be very effectually awakened by American rivalry, and the whole question of coal-getting, the organisation of the industry, royalties, and the facilities for transport, storage, and loading, may be overhauled and revised with useful results.

5. *Insurance Rates.*—Will probably be the same on both routes.

6. *The Political Factor.*—The working of the Imperial ideal in Great Britain, America, and Germany should be noted.

Preferences granted by the Dominions have materially assisted British trade. The possession of the Philippines has displaced Spain from the position of chief trader there in favour of America. The importance of this factor can be traced in the case of Japan; and China, when settled government comes, will be another notable instance. But it is impossible to elaborate fully this factor here.

7. *Rates of Exchange.*—China has a silver, Europe and America a gold, standard. Rates of exchange affect trading relations. The whole question should be carefully studied. About seven years ago when a Chinese merchant could get exchange on the west coast of America at the rate of 119 taels for $100 gold, it paid him to import thence timber and flour; but at the present rates, namely, 160 taels for $100 gold, this ceases to be profitable business, and he can trade to greater advantage locally. This factor works independently of trade routes.

8. *Finance, Banking, and Investment.*—It has been estimated that the people of Great Britain have a sum of about £2,500,000,000 invested abroad. This vast sum has been advanced partly to our own Colonies and Dominions, and partly also to new or needy countries in every part of the world. Wherever there is fairly good security and the chance of a reasonable return, Englishmen are usually willing to supply capital. In making advances of this nature, it is usual to stipulate that the interest shall be paid in gold, but the merest tyro in these matters knows that it would not be possible either for the creditor nation to export golden sovereigns to the extent of the sum advanced, or for the debtor States to ship golden sovereigns in payment of the interest on their debts. The original sum advanced leaves the shores of the United Kingdom in the form of goods, railway equipment, constructional material, textiles, and other manufactures. The interest, too, comes to these shores in the form of goods, the goods in both cases being covered by bills of exchange

in terms of sterling. The State needing capital thus obtains what is necessary for its development, and when the interest becomes due it can be collected by the creditor in gold, through banking establishments. In this way there have been built up and strengthened trading relations which will exist not only so long as the indebtedness continues, but, thanks to good relations on both sides, are likely to be permanent. To illustrate the importance of this factor in international trading relations, one has but to refer to the States of South America.

Conclusion.—In conclusion there can be no doubt that the opening of the Panama Canal will effect much for the New World. Development on the west coast of America, both North and South, will receive a great impetus. Hitherto emigrants from Europe have felt that the distance and journey to these parts made the western less eligible than the eastern side of North America. The improved sea voyage should do much to remove this prejudice. Local trade too should be greatly increased; facilities in transport doing much to develop trading relations between the east and west coasts.

Both England and Germany will feel growing competition in the supply of manufactured goods in South America, especially in the Republics on the west coast.

Very great benefit will undoubtedly accrue to the West Indies. The West Indian Islands include Bermuda, the Bahamas, and the Greater and Lesser Antilles. Of these Bermuda and the Bahamas belong to Great Britain. Jamaica is the chief of her possessions in the Greater Antilles, whilst the principal islands of the Lesser Antilles are all British. In the same region Great Britain has British Honduras and British Guiana, both on the mainland. All these, but especially the islands, must very materially benefit when the Isthmus of Panama becomes a thoroughfare.

Since Mr. Chamberlain appointed a Royal Commission in the year 1896 to visit the West Indies and report thoroughly as to their existing condition and prospects, these, the oldest of Great Britain's colonial possessions, have grown in public interest, and owing to the impetus given by Mr. Chamberlain's policy, considerable progress has resulted. Still these possessions were situated, as it were, in an ocean *cul-de-sac.* The trade has in the past depended to a great extent on local demands, and local industry, in producing food stuffs and raw materials for other parts of the world. The sugar industry, once the great backbone of West Indian trade, has passed through a long period of depression, and gives but little indication of regaining its old importance. Raw cotton, fruit, cocoa, with perhaps some coffee and less tea, are the main agricultural products. Of recent years, however, mineral oil has been worked at Trinidad and some of the other islands. The oil trade may become for the West Indies what the sugar industry at one time was. Especially does this promise to be the case when, instead of being situated towards the end of a *cul-de-sac*, these islands are on one of the great highways of the world's trade. A wise policy can do much to make these the fuel stations for a great part of the shipping making use of the Canal route. It should be possible to supply American coal on the islands at the same rates as at the coaling ports connected with the Canal, unless the American Government, in the supposed interests of the Canal, arrange for an artificially cheap rate of coal at their own depots. Further, later on the islands, owing to the local supplies, should be able to attract a large number of, if not all, those ships which either use oil for fuel or have internal-combustion engines. The political factor will doubtless make its influence felt here, but British and many foreign ships will, owing to the convenience of situation and the growing trading possibilities, undoubtedly make use of the islands, especially if they are equipped with up-to-date facilities, not only for trade, but for dry-docking repairs and bunkering. Taking a comprehensive view of the new situation created by the Canal, it would appear that the islands must grow both in commercial and in strategical importance, attractive alike both to population and to capital. So much briefly for the local effects of the Canal.

The effects on world trade can easily be exaggerated; indeed, in many of the reports and estimates drawn up this has already been done. If the new régime in the United States should succeed in freeing her trade and industries from a protection under which shipping is severely handicapped, it might come to pass that if it were found practicable to supply suitable coaling stations along the routes with good cheap coal, a heavy blow would be dealt, not only to the Suez and Cape routes, but English and German shipping would find competition so keen that they would have to fight for very existence. A survey, however, of commercial history tends to reassure one: a position which is the result of centuries of growth can only cease to be effective when it is attacked not only from without, but is subject also to internal decay. At the present moment neither British manufactures nor shipping show any serious signs of internal weakening. On the contrary, the most reliable indications warrant the assertion that at no previous time was the nation more fitted to occupy and enlarge the great position that is the heritage of many generations.

It must be remembered that there is the great question of the awakening of the Far East to be considered in connection with this new route. At the moment it is true that China is in a state of political unrest which is a serious handicap to trade developments. But this state of affairs can only be temporary. In a few years' time one may expect to find China settled down under an orderly system of government. Already the Chinaman has shown his capability in the heavy industries. Iron ore and coal are apparently abundant in China. High-class ore is produced within forty miles of Hankow, and can be put on railway trucks by Chinese labour at a contract price of 5d. per ton. Chinese pig-iron can be produced at a cost which permits it being sold f. o. b. at Hankow at about £2, 10s. per ton. America has already been importing this pig-iron in large quantities. Chinese labour is stated to be within ten per cent. of an equality with the best white labour, and its cost is in comparison very cheap: one report says that wages are one-fifteenth of those paid to white labour at Pittsburg. In a state of political peace, the Chinese should be able to develop enormously their manufactures; and the time may come when not only Chinese pig-iron, but Chinese structural steel will invade the American markets. The Panama route will assist in this, but one doubts whether the Chinese exports will be transported in American bottoms. The Japanese are being forced to be a commercial nation, and it is rather to be expected that it will be Japanese shipping, steaming via Panama, that will perform services of transport between the Far East and the North American east coast. That the above forecast is not without some foundation the following extract published in a London daily paper last May (1913) seems to show :—

"The President of the American Steel Trust stated before a Government Commission yesterday, that Indian pig-iron is being imported into the United States at a total cost (including manufacture, freight, and duty) of a little more than half the price of the same material produced by the Trust.

"At present the Indian pig-iron can only be landed at this cost on the western coast of the United States, but the opening of the Panama Canal towards the end of next year will bring the Indian producer into direct competition with the manufacturers in the eastern American States.

"NEW YORK, *Tuesday*.

"Mr. J. A. Farrell, President of the United States Steel Trust, continued his evidence to-day in the suit by the Government for the dissolution of the Steel Corporation.

"Mr. Farrell stated that pig-iron could be manufactured in India and laid down in Calcutta at 24s. 6d. per ton. There was now under way from Calcutta to San Francisco the first cargo of Indian pig-iron ever brought into the United States. The freight rate was 22s. 11d. per ton, and under the new tariff the duty would be 5½d. per ton.

"Thus pig-iron could be laid down in San Francisco at a cost of about 47s. 11d. per ton, and Chinese pig-iron could be similarly supplied at 44s. 11d. per ton, while the present market price of pig-iron on the Pacific coast was 89s. 7d."

The world is on the eve of great things full of great possibilities, probably the greatest being the awakening of the Oriental. At the moment that this awakening is in progress America, with splendid energy, regardless of money cost, and really regardless of the possible ultimate effects upon world commerce, has taken in hand the cutting of the Isthmus of Panama. This act is awakening the Western world, for it is already making the leading commercial nations consider not only new possibilities, but review old-established ways and methods. It may well be that the greatest effect of the great engineering feat now being rapidly brought to a successful issue by American genius, will be the indirect one of awakening the white man. Thus in an age that can rightly call itself progressive there is to be more progress, new spheres will be developed, whilst the old also will be reviewed. The result of the present transition which is affecting both Occidental and Oriental must be the raising of the world as a whole to a higher plane of civilisation, the West and East, in closer contact, acting and reacting upon each other, bringing in a new era of peace and prosperity, and the enlarging of life in its highest spheres.

COMPLETION OF THE MAP OF PRINCE CHARLES FORELAND, SPITSBERGEN.[1]

By WILLIAM S. BRUCE, LL.D., F.R.S.E.

ON two previous occasions I have reported to the British Association the progress of the work in connection with the charting of Prince Charles Foreland. I have now to report not only the completion of the survey, but also during the past few weeks the publication of the map.

The work has been conducted by Mr. John Mathieson, F.R.S.G.S., of H.M. Ordnance Survey (Scotland), and myself, assisted by Mr. J. V. Burn Murdoch, Dr. R. N. Rudmose Brown, Mr. E. A. Miller, the late Mr. Angus Peach, Mr. Alastair Geddes, and Mr. Gilbert Kerr.

The field work was carried on during the summers of 1906, 1907, and 1909, and I may mention that Mr. Gilbert Kerr, who was one of my shipmates on board the *Scotia*, accompanied me on all three expeditions to Prince Charles Foreland. I much appreciated his excellent services, for whilst joining all these four expeditions as "piper," he also showed great aptitude as a naturalist, and was a very able assistant in actual survey work, besides being the strong and handy man of each of the expeditions. Mr. Burn Murdoch and Mr. Miller accompanied me on two of the expeditions, and both gave much skilled assistance in the detailed survey of the Foreland. Dr. Rudmose Brown, who was also botanist of the *Scotia*, the late Mr. Peach, and Mr. Geddes accompanied me on the expedition of 1909, and each gave valuable skilled assistance towards actual survey work. Dr. Brown made a special study and has reported upon the botany of Prince Charles Foreland, and Mr. Peach, whose lamentable death occurred subsequently, while carrying on geological investigations in Egypt, left behind him excellent geological field notes, which are being put together by his distinguished father, Dr. Benjamin Peach, and will be published during the coming winter.

I must also make mention of the services of Captain Napier and Mr. Sword, master and mate of the *Conqueror* respectively, whose excellent help during 1909 materially aided the completion of the chart, and last, but not least, I must thank H.S.H. The Prince of Monaco for his much appreciated help by conveying my party to and from the Foreland on board his yacht the *Princesse Alice* in 1906, for paying almost the whole of the expenses of that season, and for subscribing substantially to the expeditions of 1907 and 1909. To the Prince also I am indebted for having reproduced, entirely at his own expense and under his auspices, the final results of the work of Mr. Mathieson and myself and our assistants. It is this map which I now have pleasure in showing to the members of the British Association.

The plotting has been done during many long night watches in the Scottish Oceanographical Laboratory by Mr. Mathieson and myself.

---

[1] Read at the Meeting of the British Association (Section E), Birmingham, September 1913.

The island of Prince Charles Foreland extends along the west of Spitsbergen from lat. 78° 12′ to 78° 54′, and is therefore about 50 miles long, and varies in width from 3½ to 7 miles. At the south end the coast is fissured with fiords which are not more than 200 feet wide and extend for half a mile inland. Then comes the mass of Saddle Mount, and beyond that the island is flat from sea to sea for a distance of about 10 miles. Rising precipitously from this flat there is a group of mountains extending for about 6 miles and cut off by Scotia Glen. To the north of this glen lies the most important and in many ways the most conspicuous group of mountains in the whole of Spitsbergen. These vary in height from 1600 to nearly 4000 feet, the most prominent being Mount Monaco, Mount Jessie and Mount Charles, and the Devil's Thumb; the last was seen by us at a distance of over 80 miles. Owing to the inaccessibility of most of the mountain peaks the method adopted for the survey was to measure two base lines, one at the north end 12,000 feet, and another across the flat land at the south end 20,146 feet. These were measured by a Chesterman steel tape, and from their extremities a network of triangles was spread over the whole of the island, more than 80 summits being observed.

The coastline was surveyed by a measured traverse which was connected with the triangulation at every available point.

Many parts of the coast offered peculiar difficulties, such as the fiords at the south end where measurements had to be taken by subtense method, and along the 20 miles of crevassed glaciers on the east side, where trigonometrical points had to be fixed at suitable intervals as actual measurements were out of the question.

The heights and horizontal angles were observed by six-inch theodolites, and nearly all the heights were observed from at least two independent points, *e.g.*, The Devil's Thumb was observed from more than fifteen points, and heights obtained varying from 2590 to 3610 feet, giving a mean of 2602 feet. Similarly Saddle Mount was observed from over 20 stations, and the heights obtained varied from 1400 to 1410 feet, giving a mean of 1406 feet. This mountain was stated by early explorers to be 600 feet, but later it rose to 800, then to 1000, and latterly to 1200 feet.

Some of the peaks offered exceptional facilities for accurate observation, *e g.* Mount Charles (3200 feet) is surmounted by a pillar which is over 30 feet in height, and only a few feet in diameter, and Mount Jessie (3351 feet), as seen from the south, terminates in a sharp cone. These and many others could be observed with almost as much accuracy as if a pole had been fixed on their summits.

The depth of the sea was taken in about forty places in the Foreland Sound, which varied from 137 to 5 fathoms.

The magnetic variation was found to be 14° 40′ in 1909, and the dip 73° 23′.

The latitude was observed in five different places. The area of the island is about 250 square miles, half of which is below the 100 feet contour, and one-fifth covered by glaciers; the remainder consists of high mountain peaks and extensive moraines.

GEOGRAPHICAL NOTES.

Europe.

Proposed New Highland Road (*With Map*).—The accompanying sketch-map shows the course of the proposed new road from Deeside to Speyside, which it is believed will be undertaken shortly. At present, as is well known, only two tracks, both possible only for robust pedestrians, lead from the head of the Dee valley to the Spey. One, the Lairig Ghru, ascends to nearly 3000 feet. The other, that which it is proposed to replace by a carriage road, follows the Geldie Burn, crosses the low watershed between the Geldie and Feshie valleys at a height of about 1830 feet, and reaches the Feshie at its junction with the Eidart. The Feshie is then followed down to Achlean, where a track turns off to the left to join the road on the right bank of the Spey near Tromie Bridge. At present, between the end of the Deeside road at Linn of Dee and the Speyside road, a distance of about 23 miles intervenes, over which the new road would require to be constructed. The gradients are moderate, but bridges would be required at Linn of Dee to replace the present White Bridge, over the Eidart, and over the Feshie. The sum of £40,000 has been mentioned as the possible cost.

Asia.

New Land North of Siberia.—Press telegrams state that Captain Wilkitzky, an officer in command of two Russian vessels engaged in surveying the coast of North Siberia, has discovered new land to the north of Cape Chelyuskin. The new land is stated to extend as far north as lat. 81° in long. 96° E., and thus lies to the south-east of Franz Josef Land. We hope to give fuller details in a future issue.

Italian Expedition to the Karakoram.—Dr. Filippi is to lead an expedition to the Himalayas next summer. The explorer intends to spend the present autumn in Chinese Turkestan, carry on observations into Russian Turkestan, winter in Scardo in Baltistan, and early next spring travel to Leh by the inner Indus valley. From Leh the expedition will travel to the Karakoram to survey and map the unknown portion of the range between the Karakoram Pass and the Siachen glacier. The Government of India has subscribed £1000 to the funds, and Major Woods of the Trigonometrical Survey will accompany the expedition.

Exploration of Central Asia.—It is announced that Sir Aurel Stein, K.C.I.E., Superintendent of the Frontier Circle of the Archæological Survey of India, has been deputed by the Government of India, with the sanction of the Secretary of State, to resume his archæological and geographical explorations in Central Asia and westernmost China, in continuation of the work he carried out between 1906 and 1908. For

THE PROJECTED NEW ROAD FROM SPEYSIDE TO DEESIDE

his journey to the border of Chinese Turkestan on the Pamirs he is taking on this occasion the route which leads through the Darel and Tangir territories, which have not been previously visited by a European. Sir Aurel Stein will be accompanied by his old travel companion, Rai Bahadur Lal Singh, who, with a second surveyor, will assist him in topographical work.

AFRICA.

An Animal Reserve in Tunisia.—We learn from the *Times* that a reserve is to be established in Tunisia for the wild animals which are being so rapidly exterminated there. For this purpose a wild mountainous stretch of 4000 acres, with an adjoining marsh of 5000 acres, has been secured near Bizerta and offers peculiarly advantageous conditions. There are already inhabiting this virgin district wild boar, hyenas, jackals, foxes, lynx, civet cat, porcupines, eagles, vultures, etc., besides many kinds of waterfowl, including a number of migratory species.

The object is to isolate, so far as possible, this area, and reintroduce those species of animals which, through the spread of European civilisation, have either been exterminated or driven beyond the frontier. The district will be maintained as a permanent "reserve."

AMERICA.

Legal Time in Brazil.—According to a note in *Science* a bill has been passed in Brazil fixing time zones there on the basis of the meridian of Greenwich. Four zones are laid down: (*a*) Archipelago of Fernando de Noronha and Trinidad, Greenwich time less two hours; (*b*) the whole of the coast, and the States of the interior (except Matto Grosso and Amazonas) and the part of Para east of a line starting from Mount Grevaux in French Guiana, following the Rio Pecuary to the Javary, then east to the Amazon, and southward along the Rio Xingu to the boundary of Matto Grosso, Greenwich time less three hours; (*c*) State of Para west of the line given above, Matto Grosso, State of Amazonas east of a line drawn on a great circle from Tabatinga to Porto Acre, Greenwich time less four hours; (*d*) Territory of Acre and district west of (*c*), Greenwich time less five hours.

GENERAL.

The Future of the International Geographical Congress.—Prof. G. Braun contributed to the May issue of *Petermann's Mitteilungen* a very outspoken criticism of the International Geographical Congress at Rome, and, following upon this article, there appears in the September issue of the same periodical an international discussion of the changes which, in the opinion of the various writers, should be made in the form and constitution of the meetings. A number of eminent geographers take part in the discussion, and the chief point which emerges is that there is a general consensus of opinion that it is time to define the geographer, no less than geography, more strictly. In other words, the meetings of the Congress tend to suffer from the presence, in too great numbers, of the untrained person. Prof. de Martonne puts the matter concisely by saying that "The Geographical Societies, who are the organisers of the Geographical Congress, are not scientific societies comparable to Geological Societies. The number of their members is so large that the specialists included are more or less submerged (*noyés*)." He adds that a congress is successful only when "géographes de métier" play the predominant part in its inception, and this can only happen when they predominate in the organising societies. He would therefore strictly limit the meeting-place to those towns where this condition is fulfilled. Prof. Davis would keep the congress as quiet and small as possible, so that its members would tend to be only those with a genuine interest in the subject, not persons attracted by the entertainments, excursions, etc. Prof. Eugen v. Cholnoky proposes a division of members into two groups, ordinary members and extraordinary members, the former to be geographers according to a very strict definition, the latter to be members of the general public who are to pay a higher subscription, enjoy diminished privileges, and to be excluded from certain meetings—a proposal which we fear presents practical difficulties.

Many of the writers speak of the excursions which, from the scientific point of view, were certainly disappointing at Rome, and various detailed recommendations in regard to them are made. Thus various contributors emphasise the need of making papers on phenomena exemplified in the local district bulk large on the agenda, and giving strangers every opportunity—practical and theoretical—of learning what has been done, and is being done, in connection with local geography. Finally, various detailed suggestions in regard to the conduct of the business are made, which will no doubt be considered by the organising committee of the St. Petersburg Congress.

Commercial Geography.

The Mont d'Or Tunnel.—We have spoken here several times of this tunnel, which is nearly four miles long, and is on the new Frasne-Vallorbe line. Its construction has been greatly impeded by the tapping of springs of water (cf. p. 91), but it was finally pierced in the beginning of October. The new line of which the tunnel forms a part shortens the Paris-Lausanne route by about 10½ miles, but more important than the actual shortening is the avoidance of a portion of the existing track where the gradients are very heavy, and the liability to snow-block in winter considerable. The construction of the new line has been very arduous, but it is hoped that when completed it will divert to Lausanne traffic which is now reaching the Simplon *via* the new Lötschberg tunnel.

Personal.

The late Professor Arminius Vambéry.—Professor Vambéry, the great orientalist, died at Budapest on September 14 last, in his eighty-second year. While chiefly a linguist with a special knowledge of oriental languages, Professor Vambéry's expedition to the interior of Asia in the early sixties brought him to the notice of geographers. In 1862 he started on a journey to Persia, disguised as a dervish, and ultimately visited Khiva, Samarkand, Herat, Meshed, etc., enduring great privations especially on the way from Khiva to Bokhara. After his journey Professor Vambéry paid his first visit to this country, a visit which was subsequently repeated many times. In 1885 he lectured before the Royal Scottish Geographical Society on *Herat and its Environs*, and his paper appears in vol. i. of the *Magazine*.

EDUCATIONAL.

Geographical Teaching in Scotland.—We noted here last year (vol. xxviii. p. 587) the appointment of a British Association Committee at the Dundee Meeting "to inquire into and report upon the present position of geographical education in Scotland." This Committee presented its report to the Birmingham Meeting of the British Association, and from the report we take the following extracts:—

"I. It is clear that in very many Elementary schools, especially in country districts, Geography teaching is still on the old lines. To quote from one correspondent: 'All elementary teachers (in my district) are of the old school—*i.e.* text-book, map, and memory. I find great difficulty, more in this than in any other subject, in getting them to teach Geography in a reasonable and attractive way.' The causes of this unfortunate state of matters are: (1) lack of knowledge on the part of the teachers, due to the extremely limited opportunities for acquiring instruction; (2) the reluctance, frequently the refusal, of School Boards to supply modern equipment, even such essentials as proper text-books and physical wall-maps. Some Boards issue admirable lists of approved text-books, etc., but, rightly or wrongly, many teachers are of opinion that the smaller their annual bill for such things is the more favourably they are looked upon by their employers. Teachers here and there exist who, at the expenditure of their own time and labour, construct wall maps and simple instruments; but the ordinary elementary teacher who has to undertake many subjects has little, if any, leisure to devote to special work in one of these subjects. Nor can it reasonably be expected of him.

"The 'Memorandum on the Teaching of Geography in Scottish Primary Schools' issued by the Scotch Education Department in 1912, in spite of some defects which need not be mentioned here, as they have already been noticed in several geographical reviews, undoubtedly marks a great advance, and will promote the setting up of a higher standard than before in Elementary schools.

"It is advisable that classes for teachers be held in suitable centres all over Scotland. The experiment has been tried in at least one place with considerable success, and correspondents indicate that the demand for such instruction is both strong and widespread. Secondly, pressure should be brought to bear upon School Boards by inspectors or by other means to equip their schools with at least modern text-books and physical wall-maps.

"II. In the Intermediate stage (ages 12 to 15) a higher standard of teaching is maintained. The Intermediate Certificate examination is here the end in view. Geography is compulsory, but is counted as part of English on the basis of 100 marks to English, and 50 to Geography. No time allowance for teaching is prescribed, but one and a half hours per week is recommended. Needless to say, that allowance is rarely exceeded. Six schools only reported an allowance of more than two hours, one of them giving three hours.

"The Committee notes with satisfaction the recent improvement in the type of paper set in the Intermediate examination, but thinks it capable of improvement. . . . In the compulsory sections it would be advantageous that some further choice be afforded to the candidates, that greater opportunity be given for displaying knowledge of places associated with current events, and that a more reasonable proportion of the marks be allotted to that part of the paper which exercises the intelligence of the candidate.

"The written examination is supposed to be supplemented by an oral examination conducted by an inspector, but this is usually perfunctory, and in some schools the inspector pays no attention whatever to Geography. Throughout the Intermediate course a compulsory minimum time allowance of three and a quarter hours per week would be very beneficial, and inspectors of Geography might encourage attempts at a higher standard of teaching.

"III. In the Post-Intermediate stage Geography is no longer a compulsory subject, except in the case of junior students, who, however, are at no time examined as to their knowledge of Geography. In the Leaving Certificate examination Geography is separated from English, which is the only compulsory subject, is put on a level with other optional subjects, and is allotted 100 marks. It may form one of a 'group,' but the curriculum must then be submitted to the Scotch Education Department for its specific approval. This is not required if a school commits itself to English, Mathematics, and French; or to English, Mathematics, and Latin. It is distinctly laid down that Geography is on the same level with, for example, languages, and that a candidate must spend upon it an adequate amount of time. The Committee finds that the average amount of time spent upon languages at this stage is seven hours per week. The following tabular statement will help to make matters clear:—

	1912.	1913.
No. of candidates for Group Leaving Certificate	2,202	2,290
Successful	1,711	1,739
No. of candidates with Geography as part of Group	195	146
Successful	155	92
No. of candidates who sat Geography examination	319	212
Successful	227	116

"From these figures it will be seen that while in 1912 Geography candidates formed only about 9 per cent. of the total number of Leaving Certificate candidates, in 1913 even that small proportion was reduced to a little more than 6 per cent. The presentations for Geography as a separate paper also fell from 124 to 66; while the number of candidates with Geography as part of their group fell from 195 to 146.

"The Committee has received some information regarding 1912. It is aware of 70 candidates who were accepted by the Department, and who had had an hour and a half, or less, teaching per week. It is also aware of 35 of these candidates who passed. This last piece of information was not asked for in the circular sent out, but some correspondents gave it voluntarily. These figures speak for themselves.

"From the returns received it is definitely proved that the making an optional subject of Geography has practically killed it in the Post-Intermediate stage. Only seventy-two schools sent in information on this point. The average time allowance was just over an hour and a half per week; in nine schools it was more than two hours. Some give an hour or so for the first year, then drop it entirely. Twelve have dropped it altogether. The Committee is aware of some others which have made no returns, but have also dropped Geography. It is a fair inference that a complete census would reveal many more. In only eleven schools has the time allowance been recently increased, and in most cases this increase has been from a totally inadequate to but a slightly less inadequate amount. The Committee is of opinion that this is a very serious matter. It finds that in many Secondary schools, some of them the largest and most important in the country, situated in great educational centres, the pupils cease to study Geography at the age of fifteen. Further, that the average time devoted to Geography up to that age is only an hour and a half per week. Now the time up to the close of the Intermediate stage should be devoted to providing that foundation of fact which is the basis of scientific Geography, and it is only in the Post-Intermediate stage that a pupil is mentally fitted to build upon that foundation by studying Political and Economic Geography—in other words, how man adapts himself to his environment, and how that environment reacts upon man. It is not considered necessary to emphasise the value of Geography as an educational subject beyond expressing the opinion that after a knowledge of the English language there is nothing more essential to the mental equipment of the modern Briton than a thorough grounding in Geography. This is impossible of achievement under the present regulations. It seems only reasonable that Geography be made a compulsory subject throughout the Post-Intermediate stage, and that in this stage also a minimum time allowance of three hours and a quarter per week be fixed.

"IV. *Training Colleges.*—It may be explained that students preparing for the Teaching Profession in Scotland may either receive their training at the Training Colleges, where the course extends for two years, or may continue their professional training with a University course, or may first complete their graduation and then devote one year to their professional training under the auspices of the Provincial Committees for the Training of Teachers established in the four centres—Aberdeen, Edinburgh, Glasgow, and St. Andrews.

"The University students in training at Edinburgh or Glasgow may include Geography among the subjects required for graduation at the University, but this is not possible at the other centres, where so far there is no University teaching of the subject.

"The position of the subject varies considerably at the different centres. At the Training Colleges of Edinburgh, Glasgow, and St. Andrews, lecturers in Geography have been appointed, and at these centres instruction in Geography forms an integral part of the curriculum for all Training College students.

"At Aberdeen Training College the previous training of the students and their knowledge of the subject are regarded as satisfactory, so that there is now no special instruction in the subject, and attention is confined to the methods of teaching Geography. The classes consist of thirty to forty students, and thirty periods are devoted to the methods of teaching Mathematics, Nature Study, and Geography, so that if the time is equally divided Geography can receive only ten periods.

"At Edinburgh at least thirty periods are given to the study of Geography, and the classes consist of forty to fifty students. At Glasgow the Geography course extends to thirty periods, and for lectures the classes average eighty students, while for practical work they are reduced to twenty-seven. At St. Andrews sixty periods are devoted to the study of Geography, and the classes number twenty-five students.

"The University students in training at the Edinburgh centre receive no instruction in Geography unless they elect to include the subject in their graduation course at the University; a considerable number do so, but the larger number, who do not, are being sent out each year—many to teach in Secondary schools—without any equipment to teach the subject so far as the Training College is concerned.

"At Glasgow the subject has been dropped from the curriculum of the University students in training, and attention is now confined to methods in teaching Geography, in spite of the fact that in very many cases the previous study of the subject has been quite insufficient.

"Finally, recent legislation by the Scotch Education Department, and local conditions at several of the Training Centres, now make it quite possible for students who may have ceased the study of Geography after obtaining the Intermediate Certificate to complete their professional training without much, if any, further instruction in the subject.

"In the opinion of the Committee it should be rendered necessary for all University students in training to have obtained the Leaving Certificate in Geography unless adequate instruction in the subject is provided in their professional course, or unless they include the subject in their graduation course at the University.

"V. *Universities.*—Geography was first recognised by the Scottish Universities in 1908, when a lecturer was appointed as head of a new department in that subject in the University of Edinburgh. The lecturer has an ordinary class extending over the whole session (three terms) and two advanced classes, each of which is confined to a single term. From the first the ordinary class has qualified for admission to the M.A. examination, Geography being now one of the optional subjects in that degree. One of the advanced classes is a non-graduation class. The other, which is devoted especially to Economic Geography, is the qualifying class for an optional paper for the degree of M.A. with honours in Economic Science. In the five years during which the ordinary class has been held, the attendance has been 48, 40, 116, 132, 98. The attendance at the advanced class varies from 5 to 10.

"The only other Scottish University which so far recognises Geography is Glasgow, where the lecturer was appointed on similar conditions to those in Edinburgh in 1909. There Geography may now be taken as a subject for either the M.A. or the B.Sc. degree. The ordinary class is the qualifying class for the M.A., and the advanced class for the honours degree of B.Sc., and was held for the first time last winter. The attendance at the ordinary class for the four years during which the lectureship has been in existence has been about 30, 65, 73, 94.

"It should be added that under a recent regulation, which comes into force next year, the position of Geography in the preliminary examination for admission to the Arts and Science Faculties of Edinburgh University has been seriously prejudiced. Down to 1913 Geography was one of the branches under the head of English, which is a compulsory subject in the preliminary examination, but from 1914 onwards the only recognition of Geography is in connection with the history of the British people, one of the subjects included in the English syllabus. 'Candidates will be expected to show acquaintance with the social as well as the political history of the British people and the relevant geography.'

"In conclusion the Committee is of opinion that while the worst result of the present regulations for the Post-Intermediate stage is that pupils leave school with a very imperfect and one-sided educational equipment, a subsidiary result of nearly as much importance may soon appear. It is that the majority of the pupils who intend to become teachers will not care to take up the study of Geography again after the lapse of two or three years. Thus the supply of capable teachers will diminish, and once more, as in the past, even in the Intermediate stage, Geography will be entrusted to the 'general' teacher, and it will fall back into its old position of memory work, unintelligent and uncomprehended."

NEW BOOKS.

EUROPE.

With Camera and Rücksack in the Oberland and Valais. By REGINALD A. MALBY. London: Headley Brothers, 1913. *Price* 10*s.* 6*d. net.*

Mr. Malby is an enthusiast about Alpine rock-plants, and is already known to garden lovers as the author of *The Story of my Rock Garden.* In the present volume he resumes this theme and tells of an expedition which he made to

Switzerland in pursuit of his favourite study. He has made very good use of his camera, and, in a number of colour-plates and monochromes, many of which are marvels of painstaking photography, we get beautiful representations of Alpine flowers blooming among their native rocks. The text of the book gives an idea of the difficulties, and even dangers, which attended the taking of these pictures, and, though he did but little in the way of real mountain-climbing, Mr. Malby is able to give an entertaining account of his experiences among the Alps. We would draw attention to his interesting theory as to the formation of soil for plant life above the tree level (stated on pages 233 to 235), and we can recommend the book to all who intend to take a holiday in Switzerland. Some of the camera pictures of Alpine scenery are very fine.

The Cornish Coast and Moors. By A. G. FOLLIOTT-STOKES. London: Greening and Co., Ltd., 1912. *Price* 12*s.* 6*d. net.*

In this book the author has attempted a somewhat formidable task, viz., to take his readers a walk with him round the whole of the Cornish coast, varied by occasional excursions to points of interest inland. He performs this task with considerable success, although it has to be admitted that there is a good deal of repetition in the scenery and monotony in the course of the road. The author, however, has a keen eye for the endless beauties of the scenery and the flora, which, in this district, are exceedingly luxuriant and varied, and the story of the course of the ramble is enlivened with anecdotes and historical reminiscences of not a little interest. The geological and antiquarian features of the locality also receive due attention. We must add a special word of commendation of the numerous photographic illustrations.

ASIA.

Siam. By PIERRE LOTI. Translated from the French by W. P. BAINES. London: T. Werner Laurie, Ltd., 1913. *Price* 7*s.* 6*d. net.*

In its palmy days the temple of Angkor Wat in Cambodia must have been one of the wonders of the world. A competent authority, Mr. Campbell, says of it, "The comparatively few European travellers who have visited this temple all unite in declaring it the most colossally stupendous as well as architecturally beautiful structure they have ever beheld, so that while it rivals or eclipses the Egyptian pyramids in one respect, it hardly falls short of the highest Hellenic standard as regards artistic detail in the other. The huge building, which is between two and three miles in circumference, contains a multitude of courts, colonnades and chambers. There are twelve superb staircases, the four in the middle having from forty to fifty steps, each step a single slab, and over five thousand columns, while everywhere the stones are fitted together in a manner so perfect that the joinings are not easy to find. The walls and portals are covered with sculptures, the exterior of the temple being ornamented with bas-reliefs of scenes from the Ramayana, the great Sanscrit epic poem, with vast processions of warriors, horses and chariots, and animals of all sorts, both real and mythical. Angkor Wat was certainly commenced as a Brahminical temple, but before its completion Buddhism had become the religion of the land, and so it is that we find here, as in the temple of Boroboddo in Java, artistic representations of both the religions." We can easily understand how these magnificent buildings in their present state of decay and ruin would appeal to the imagination and emotions of the famous French traveller Pierre Loti, and in this volume we have a description of his feelings and sensations in seeing them and their surroundings in the winter of 1901. One brief quotation will let our readers understand what to expect. "I mount the steps of the temple for the last time. No rain has fallen since last night to refresh the suspended plants, or moisten the heaps of stones, and an intolerable heat, as of glowing coal, now emanates from the terraces, the walls and statuary, on which the sun has been blazing all day long. But the divine Apsaras (*i.e.* celestial dancers), who have been used for centuries to be thus burnt with rays, smile at me by way of adieu, without losing their ease or customary gracious irony. As I took leave of them I little thought that within a few hours, by the lavish caprice of the king of Pnom-Penh, I should see them again, one night, at the evocation of the sound of the old music of their times, see them no longer dead, with these fixed smiles of stone, but in the fulness of life and youth, no longer with these breasts of rigid sandstone, but with palpitating breasts of flesh, and coifed in veritable tiaras of gold, and sparkling with veritable jewels." We trust we have said enough to show that Pierre Loti's subject in this book is magnificent, and his treatment of it characteristic.

From a Punjaub Pomegranate Grove. By C. C. DYSON. London: Mills and Boon, Limited, 1913. *Price* 10*s.* 6*d. net.*

In this book Mrs. Dyson, the wife of a junior civilian in the Punjaub Commission, gives a bright and chatty description of a brief sojourn in India, where she saw and heard a good deal in British and in native territory. She evidently enjoyed herself thoroughly, and is well qualified to give an excellent account of what she saw and heard. She was delighted with the country and with the cordial reception she received alike from the natives and the Europeans. Indeed her admiration of the former goes the length of making her exclaim—"as to the Indian aristocracy, where will you find more courtesy, more generosity, more dignity, more of all the qualities that constitute 'a gentleman,' than among the noblemen and high-class Indian people?" She visited a good many places and attended many functions, mixed with all sorts of people, kept her eyes and ears wide open, put the best construction on everything, and had a real good time. The book contains nothing new, but is most readable, and so we commend it to the perusal of our readers. Some of the illustrations are good.

AFRICA.

La Région du Haut Tell. Par CH. MONCHICOURT. Paris: Librairie Armand Colin, 1913. *Prix* 12 *fr.*

In this erudite and exhaustive monograph Dr. Monchicourt gives us a minute and detailed description of a very interesting tract in Tunisia known as the Haut Tell. The word "tell" signifies a peculiar kind of soil, "tout sol argilleux recevant chaque année assez de pluie pour en être pénétré au point d'accumuler en lui des réserves aqueuses assurant constamment le complet développement normal des plantes annuelles, céréales et herbages." Thus from its very definition we gather that the area of the Tell is not and cannot be fixed, as it varies with the rainfall of the locality. Dr. Monchicourt has made a patient and detailed study of the tract, and has embodied in this book a luminous account of its geography and history, more particularly during the eighty years during which it has come under the influence of France. He describes from personal observation and a careful study of the literature of the subject its natural features, its economic condition, the distribution and movements of its population, its resources, agricultural and mineral, its means of communication, etc. The book will appeal to but a small circle of readers in this country, but it deserves and will doubtless receive a hearty welcome from the wide circle of readers in France who are justly interested in everything that relates to their oldest and most prosperous colony in the north of Africa.

AMERICA.

Le Pérou Economique. Par PAUL WALLE. Paris: E. Guilmoto, 1913. *Prix 9 francs.*

We have now before us the fourth edition of a very useful work by M. Walle, in which he sets forth in considerable detail the natural resources and present condition of Peru. His object is stated quite frankly. In his travels in South America in general, and in Peru in particular, he was distressed to notice that France had not taken the definite and commanding position with regard to commerce which he thinks she should have taken. In the race for the trade of Peru she is far behind Germany and England, and this thoughtful and instructive volume has been compiled in the hope that the commercial world of France will rise to the occasion, will not accept the present situation as hopeless or irremediable, but will reform its methods, adapt them to the circumstances of the case, and remembering the old maxim, "fas est ab hoste doceri," will not hesitate to take several leaves out of the German book, and reconstruct their trade with Peru on a new basis and on new lines. This reconstruction is all the more necessary, as M. Walle clearly sees, in the prospect of a more or less early opening of the Panama Canal, which cannot fail to have prodigious and permanent effects on the distribution of commerce between Europe and America and the world generally. This book is written for commercial men and intending emigrants, and thus M. Walle has deliberately refrained from much that might make his work attractive to the general reader. There are no descriptions of magnificent scenery, no thrilling narratives of personal adventure, no notice of the wondrous archæological remains which are still a puzzle and enigma to the antiquarian, no excursions into the history of a country which is replete with romance and excitement. But for business men and emigrants it will be most valuable and interesting, although it has a fault not uncommon in similar works, viz., the statistics are very often not up to date. Each province and each of the principal towns is treated separately, and ample details are given of its conditions and resources so far as they are known, and it has to be remembered that M. Walle writes from personal knowledge and experience. Considerations of space prevent us from following him into details, but we may remark he is eminently judicious and cautious both in his judgment of the resources of the districts and in the advice he gives to emigrants, to investors, and to commercial firms considering the question of starting agencies or businesses in Peru. The situation, it has to be confessed, although much better than it was, leaves much to be desired. The republic is not yet a hundred years old, and it has already had forty Presidents, so that, although there has been peace and order for a few years of late, capitalists may reasonably hesitate before assuming that revolutions are now things of the past. Then an undesirable and most serious drawback to the development of the undoubted resources of the country is the want of good internal communications of all sorts, a want which has now received the serious attention of the Government with a view to earnest and speedy reform. There are other difficulties which may be mentioned, such as those of the climate, the diversity of races which inhabit the country, the absence of local enterprise and labour, abuses and venality in commerce and justice, etc.; on all such matters M. Walle's opinions and observations are well worthy the attention of English commercial men as well as French.

Les États-Unis d'Amérique. Par P. d'Estournelles de Constant. Paris : Librairie Armand Colin, 1913. *Prix 5 fr.*

The author of this thoughtful and instructive book has evidently spent many happy days in the United States during the last ten or twelve years. A French Senator and distinguished traveller and educationalist, he was invited to be present at various functions across the Atlantic, and there he was invariably received with the effusive hospitality and genuine hearty welcome which are characteristic of our American cousins. On his part he lost no opportunities; he made detailed and careful observations and serious study of all that he saw and heard, and he does not hesitate to set forth his views in vigorous and at times dogmatic language, which we fear will not be always received with general approval in the United States. The volume now before us is the latest of a series which the author has been publishing during the last fifteen years. The first half of the volume contains a series of sketches of things worth noting in several of the great cities which the Senator visited in 1911, into which are introduced excursions (we had almost said lectures) on miscellaneous topics, such as votes for women, a possible war with Japan, the influence of Germany, the general ignorance of French, and so on. In the latter half the Senator plunges freely into "la haute politique" in a very wide sense of the word; he discusses education, religion, the race question, economics, including commercial competition, internal and foreign; he denounces emphatically the present Panama policy, that of excessive tariffs, and the assumption by the United States of the part of a colonising power and its consequent entry into foreign politics. We cannot, of course, follow him into any of these thorny subjects, but his obvious earnestness and enthusiasm, his grasp of his subject, and his wide experience all entitle him to a dispassionate hearing. We must add that he writes with all the sparkling vivacity and spontaneity which we expect in an accomplished and practised French author. Even when dogmatic and professorial he is never wearisome or dull. In Scotland there are prevalent a good many theories regarding Paradise. We have to thank M. de Constant for a new one, viz., that Paradise is now on the other side of the Forth. "Que n'ai-je le temps de parler de ce beau domaine écossais que Andrew Carnegie a acheté à Dumfermline, son pays natal, transformé en parc royal réservé aux générations venues après lui ; un parc ? un paradis plutôt."

Trails and Tramps in Alaska and Newfoundland. By William S. Thomas, author of "Hunting Big Game with Gun and Kodak." With 147 illustrations from original photographs. New York and London: G. P. Putnam's Sons, The Knickerbocker Press, 1913. *Price 7s. 6d. net.*

Experiences during camping-out in southern Alaska, a short trip to Newfoundland, and excursions round the author's home in Pennsylvania, are here pleasantly narrated by Mr. Thomas. He seems to be as great an expert with the camera as with rod and gun, as the numerous charming photographs of wild life prove. The description of a glacier (pp. 32-34), especially of the origin of crevasses, is inaccurate, but that is a small blemish. There is a good index, and the "get-up" of the book is excellent.

American History and its Geographic Conditions. By Ellen Churchill Semple. With Maps. London : Constable and Company, Ltd., 1913. *Price 12s. 6d. net.*

This is a re-issue of a book first published in 1903. We welcome it more especially for the sake of the new generation of geography students that has arisen during the last decade. Miss Semple's style is singularly lucid, and it gives one the greatest of pleasure to follow her arguments, even if one does not go with her the whole way to her conclusions. We may draw particular attention to the concluding chapters—"The American Mediterranean," and "America as a Pacific Ocean Power"—as gaining fresh value in the light of current events.

AUSTRALASIA.

Outback in Australia; or Three Australian Overlanders. Being an account of the longest overlanding journey ever attempted in Australia with a single horse. and including chapters on various phases of Outback Life. By WALTER KILROY HARRIS, F.R.G.S., F.R.C.I. With Map and 29 Illustrations from Photographs. Letchworth: Garden City Press, Ltd., 1913. *Price* 5*s. net.*

The title of this book is sufficiently descriptive of its contents, except that the journey was from Newcastle in New South Wales by way of Melbourne to Adelaide, returning by the Murray River and the western plains of New South Wales. The average distance covered on each travelling day was 22 miles on the outward, and 28 miles on the homeward journey. The author praises Australian hospitality very highly, but, judging from his narrative, not more highly than it deserves. The most interesting parts of the book are those where the author interrupts his tale to enlarge upon some side-issue which may have emerged during the day's run—*e.g.* "The Outback Goat," "The Rabbit in Australia," "The Outback Parson"—the chapters on the Murray irrigation schemes, and the account of the transformation of the so-called ninety-mile desert between the Victoria border and the Murray. We cordially recommend this book to those interested in the modern progress of Australia.

A Naturalist in Cannibal Land. By A. S. MEEK. Edited by FRANK FOX. With an Introduction by the Hon. WALTER ROTHSCHILD, and 36 Illustrations. London and Leipsic: T. Fisher Unwin, 1913. *Price* 10*s.* 6*d. net.*

There are in this book 231 pages, and we could easily have put up with double the number. Mr. Meek is one who knows neither hardship nor danger when engaged in the pursuit of "specimens." In Papua, the Solomon Islands, and the adjacent seas he has spent nearly twenty years, and has gathered a rich harvest, not only of birds and insects, but of adventures. All his collections were destined for the Tring Museum, whose owner testifies in a graceful introduction to the author's work. Mr. Meek seems to have managed the somewhat treacherous natives with great skill, chiefly by means of timely firmness. One does not need to be a naturalist in order to enjoy this volume, yet a naturalist will find in it much that is of interest. There is also something for the geographer and the ethnologist, and indeed for every one who enjoys reading about real things. We trust Mr. Meek during his next leisure-spell will be encouraged to give us more. Mr. Fox, the Australian friend who acted as editor, must have had a pleasant task. There are many good illustrations, an index, and an adequate map.

The New World of the South. By W. H. FITCHETT, B.A., LL.D. London: Smith, Elder and Co., 1913. *Price* 6*s.*

This is an excellent book which should be put into the hands of every boy of the English-speaking nations. It contains a series of charmingly written sketches describing the discovery, exploration, and colonisation of Australia, and in this connection we have brief but attractive notices of famous men, *e.g.*, Cook, Flinders,

Bligh, Oxley, Hume, and Sturt. The narratives are calculated to stir the imagination and fire the spirit of adventure in the rising generation. To this we have to add that Dr. Fitchett depicts in glowing but honest language the miserable condition of the convicts who were the first settlers in Australia, for whose unhappy fate he has a sincere and kindly sympathy. He has also a kindly word for the aborigines, who seem doomed to extinction before the advance of the white man. Interspersed in the narratives are wise and thoughtful observations, which cannot fail to arrest the attention and evoke serious reflection from even a youthful reader. A perusal of this book will set before him in a most attractive form some of the profoundest lessons of history, and will insensibly but convincingly demonstrate the beauty and the power of the manly virtues, such as perseverance, generosity, tenacity of purpose, gentleness, and loyalty to the calls of duty.

GENERAL.

The Log of a Rolling Stone. By HENRY ARTHUR BROOME. London: T. Werner Laurie, Ltd., 1913. *Price* 12*s*. 6*d*. *net*.

Mr. Broome has unquestionably selected an excellent title for this remarkable work. In it we have an autobiography of some forty years of his life, during which he roamed over a great part of the known world, *e.g.* South Africa, India, New Zealand, Australia, Chili, Argentina, Brazil, Patagonia, besides minor excursions into Belgium, Holland, France, Italy and Switzerland; and amongst his numerous avocations we may mention a few, viz. mounted policeman, wood-engraver, journalist, ordinary seaman, railway contractor, convict guard, resident magistrate, whaler, quarry contractor, and stores-clerk. Such an unusually varied life and career could not fail to be intensely interesting, and Mr. Broome tells his story well. He writes with the easy skill of a practised journalist, is not prone to exaggerated language, and although it is the story of his own life, he is not egotistical or over-self-conscious. The greater part of the matter refers to South Africa, to which he returned more than once. At one time he lived there for a continuous period of nearly twelve years, during which he rose in the Civil Service from being a convict guard to being a resident magistrate; but at the end of the Boer War he apparently thoroughly disapproved of the policy and proceedings of the Government and his fellow-countrymen, and he found it expedient to resign and to start off on a new career. It was inevitable that in the course of such a life he should come across some strange characters and have many exciting adventures, and thus the interest of the book is well maintained. Some of the illustrations are from water-colour paintings by the author. We commend the book to the perusal of our readers.

Are the Planets Inhabited? By E. WALTER MAUNDER, F.R.A.S.
The Age of the Earth. By ARTHUR HOLMES, B.Sc., A.R.C.S.

"Harper's Library of Living Thought." London and New York: Harper and Brothers, Limited, 1913. *Price* 2*s*. 6*d*. *net* each volume.

These volumes are excellent additions to this series. Mr. Maunder comes to the conclusion on what seems to be sufficient data that the earth is the only member of the solar system which is fitted to be the abode of life. Since the conditions which determine the existence of life are so numerous and so stringent it is even conceivable that our planet is unique among the hosts of stars in being the home of organised beings. Mr. Holmes deals with the fascinating problem of the earth's age, a problem which has excited acute controversy between the geologist and the physicist, as well as among geologists themselves. His conclusion is that in the radioactive minerals will be found the means of reconciling the widely divergent time-estimates.

The Structure of the Earth. By T. G. BONNEY, Sc.D., F.R.S. "The People's Books." London, Edinburgh, and New York: T. C. and E. C. Jack, 1912. Price 6*d. net.*

Professor Bonney has given in this booklet an admirable summary of geology which cannot but be helpful to the beginner, and interesting to the general reader. The list of reference books at the end provides a useful guide to future study.

The Statesman's Year-book, 1913. Fiftieth Annual Publication. Edited by J. SCOTT KELTIE, LL.D. London: Macmillan and Co., Limited, 1913. *Price* 10*s.* 6*d. net.*

The chief feature of interest about this edition is the fine series of maps showing the progress of the world since the first issue of the annual in 1863. These maps show the enormous increase in the number of railways within this period, especially, of course, in America, Africa and Asia, but also in Russia and other parts of Europe. Very striking are the contrasts between the maps of Africa and Australia in 1863 and in 1913. In the introductory matter, also, special stress has been laid upon the contrasts between then and now. Here again some of the differences in population figures are very remarkable.

Maps and Survey. By ARTHUR R. HINKS, M.A., F.R.S. Cambridge: University Press, 1913. *Price* 6*s. net.*

While the present work treats of map-making and surveying, the author writes neither for the map-maker nor the surveyor. In this sense the book is not a text-book, for Mr. Hinks writes for a wider public. There is little that is new in the book, but the established principles of survey work in the field and in the office are explained in simple language. There are chapters dealing with maps, map-analyses, route traversing, simple land survey, compass and plane-table sketching, topographical survey, geodetic survey, and survey instruments. In his treatment under these headings the author has steered clear of a too technical handling of his subject, and by the aid of many well-chosen illustrations has produced a work which ought to be read by students in the higher-grade geography class. It is for such that the book has been written, but all interested in geography, especially in the means of representing geographical facts on the map, will find much to interest them.

BOOKS RECEIVED.

Our Villa in Italy. By J. LUCAS. Crown 8vo. Pp. 200. *Price* 5*s. net.* London: T. Fisher Unwin, 1913.

Historical Geography of Scotland. By W. R. KERMACK, B.A. (Oxon). Demy 8vo. Pp. 134. *Price* 2*s.* 6*d. net.* Edinburgh: W. and A. K. Johnston, Ltd., 1913.

Brief Biography and Popular Account of Unparalleled Discoveries of T. J. J. See. By W. L. WEBB. Demy 8vo. Pp. xii + 298. Mass., U.S.A.: Thos. P. Nichols and Son.

The Realm of Nature. Second Edition. An outline of Physiography. By HUGH ROBERT MILL, D.Sc., LL.D. Crown 8vo. Pp. xii + 404. *Price* 5*s.* London: John Murray, 1913.

Preliminary Geography. By E. G. Hodgkison, B.A. (Lond.), F.R.G.S. Crown 8vo. Pp. xvi + 225. *Price* 1s. 6d. London: University Tutorial Press, Ltd., 1913.

Australia from a Woman's Point of View. By Jessie Ackermann, F.R.S.G.S. With Sixty-four Plates. Crown 8vo. Pp. xiv + 317. *Price* 6s. London: Cassell and Co., Ltd., 1913.

Catalogue of the Mammals of Western Europe in the Collection of the British Museum. By Gerrit S. Miller. Demy 8vo. Pp. xv + 1019. London: The British Museum, 1912.

Catalogue of the Heads and Horns of Indian Big Game, bequeathed by A. O. Hume, C.B., to the British Museum (Natural History). By R. Lydekker, F.R.S. Demy 8vo. Pp. xvi + 45. London: The British Museum, 1913.

Rambles in the North Yorkshire Dales. By J. E. Buckrose. Illustrated. Crown 8vo. Pp. x + 192. *Price* 3s. 6d. *net.* London: Mills and Boon, Ltd., 1913.

Social Welfare in New Zealand: The Result of Twenty Years of Progressive Social Legislation, and its Significance for the United States and other Countries. By Hugh H. Lusk. Crown 8vo. Pp. viii + 287. *Price* 6s. *net.* London: William Heinemann, 1913.

Japan's Inheritance: The Country, its People and their Destiny. By E. Bruce Mitford, F.R.G.S. With Twelve Maps and Plans, and Seventy-five Illustrations from Photographs. Demy 8vo. Pp. 384. *Price* 10s. 6d. *net.* London: T. Fisher Unwin, 1913.

My Cosmopolitan Year. By the Author of *Mastering Flame.* With nineteen Illustrations. Demy 8vo. Pp. xi + 289. *Price* 10s. 6d. *net.* London: Mills and Boon, Ltd., 1913.

Handbook for Travellers in Scotland. Edited by Scott Moncrieff Penney, M.A. Ninth Edition. With Fifty-seven Travelling Maps and Plans. Crown 8vo. Pp. lvi + 539. *Price* 10s. 6d. London: Edward Stanford, Ltd., 1913.

Wild Animals of Yesterday and To-day. By Frank Finn, B.A. (Oxon), F.Z.S. Illustrated in Colour and Black and White. Demy 8vo. Pp. 382. *Price* 6s. *net.* London: S. W. Partridge and Co., Ltd., 1913.

The Continent of Europe. By Lionel W. Lyde, M.A., F.R.G.S. Demy 8vo. Pp. xv + 446. *Price* 7s. 6d. *net.* London: Macmillan and Co., Ltd., 1913.

Camp Fire Yarns of the Lost Legion. By Colonel G. Hamilton-Browne. Demy 8vo. Pp. xi + 301. *Price* 12s. 6d. *net.* London: T. Werner Laurie, Ltd., 1913.

Astronomy Simplified. By the Rev. Alex. C. Henderson, B.D., F.R.A.S. With Twelve Engravings, Copious Notes and an Index. Demy 8vo. Pp. 152. *Price* 2s. 6d. *net.* London: J. Clarke and Co., 1913.

A Commercial Geography of the British Isles. By Frederick Mort, M.A., B.Sc., F.G.S. With numerous Maps and Diagrams. Demy 8vo. Pp. viii + 160. *Price* 1s. Edinburgh: Oliver and Boyd.

The Nature and Origin of Fiords. By J. W. Gregory, F.R.S., D.Sc. With Diagrams and Illustrations. Demy 8vo. Pp. xvi + 542. *Price* 16s. *net. cash.* London: John Murray, 1913.

Histoire de l'État Indépendant du Congo. Par Fritz Masoin. 2 Vols. Pp. 382 + 444. *Prix* 8 *fr.* Namur: Picard-Balon, 1913.

A Geography of the British Empire. By W. L. Bunting, M.A., and H. L. Collen, M.A. Medium 8vo. Pp. 159. Cambridge University Press, 1913.

The Holy Land. By Robert Hichens. Crown 8vo. Pp. viii + 311. *Price* 6s. London: Hodder and Stoughton, 1913.

Wild Life Across the World. Written and Illustrated by Cherry Kearton.

Introduction by THEODORE ROOSEVELT. Medium 8vo. Pp. xxvii + 286. Price 20*s. net.* London: Hodder and Stoughton, 1913.

Luxembourg: The Grand Duchy and its People. By GEORGE RENWICK, F.R.G.S. With Thirty-four Illustrations and a Map. Demy 8vo. Pp. 320. Price 10*s.* 6*d. net.* London: T. Fisher Unwin, 1913.

Official Year-Book of the Commonwealth of Australia, 1901-12. No. 6. By G. H. KNIBBS. Melbourne 1913.

Round the British Empire. By ALEX. HILL. With Twenty-four full-page Illustrations. Crown 8vo. Pp. 383. *Price* 2*s.* 6*d. net.* London: Herbert Jenkins, Ltd., 1913.

A Handbook for Travellers in India, Burma and Ceylon. Ninth Edition. With Twenty-nine Maps and Plans. Crown 8vo. Pp. clxviii + 664. Price 20*s.* London: John Murray, 1913.

Highways and Byways in the Border. By ANDREW LANG and JOHN LANG. With Illustrations by HUGH THOMSON. Demy 8vo. Pp. xvi + 439. Price 5*s. net.* London: Macmillan and Co., Ltd., 1913.

Burma Gazetteers: Meiktila, Kyaukse, Shwebo, Bhamo, Minbu, Upper and Lower Chindwin, Rangoon, Prome, *Toungoo, Amherst, Mandalay, Sagaing, and Myingyan Districts.* Rangoon, 1913.

Statistical Abstract relating to British India from 1902-3 to 1911-12. Forty-seventh Number. Price 1*s.* 3*d.* London, 1913.

Annual Statement of the Trade of the United Kingdom with Foreign Countries and British Possessions, 1912. Vol. ii. Price 4*s.* 1*d.* London, 1913.

East India (Trade): Review of the Trade of India in 1911-12 and 1912-13. Price 1*s.* 1*d.* London, 1913.

Statistical Abstract for the United Kingdom in each of the last Fifteen Years, from 1898 to 1912. Price 1*s.* 11*d.* London, 1913.

Report on the Administration of the Territories now included in the Province of Bihar and Orissa, 1911-12. Price 7*s.* 6*d.* Patna, 1913.

Punjab District Gazetteers. Vols. 3, 4, 7, 12, 17, 21, 23, 26, 28, 29, 34, 35, 36. Lahore, 1913.

Annual Report of the Secretary for Mines for Victoria for the Year 1912. Melbourne, 1913.

Diplomatic and Consular Reports.—Trade, etc., of Bavaria, 1912, and part of 1913 (5194); Trade of Bilbao, 1912 (5196); Trade of Muscat, 1912-13 (5198); Trade of Canton, 1912 (5199); Trade, etc., of Chiengmai, 1912 (5200); Trade of Ningpo, 1912 (5201); Trade, etc., of Gothenburg, 1912 (5202); Foreign Trade of Austria-Hungary, 1912 (5205); Trade, etc., of Brindisi, 1912 (5203); Trade of Kermanshah, 1913 (5204); Trade, etc., of New Orleans, 1912 (5206); Trade of Shanghai, 1912 (5207); Trade, etc., of Fiume, 1912 (5208); Trade of Khorasan, 1912-13 (5211); Trade, etc., of Hamburg, 1912 (5212); Trade, etc., of Curacoa, 1912 (5195); Trade of Panama, 1912 (5214); Foreign Trade of China, 1912 (5216); Trade of Nagasaki, 1912 (5209); Trade, etc., of Leghorn, 1912 (5213).

Colonial Reports (Annual).—Gilbert and Ellice Islands Protectorate, 1911 (753); Cayman Islands, 1911-12 (754); Gibraltar, 1912 (755); St. Helena, 1912 (756); Weihaiwei, 1912 (757); Ceylon, 1911-12 (758); Sierra Leone, 1912 (759); Seychelles, 1912 (760); Somaliland, 1912-13 (761); Hong-Kong, 1912 (762); Turks and Caicos Islands, 1912 (763); Falkland Islands, 1912 (764); Northern

Territories of the Gold Coast, 1912 (765) ; Bahamas, 1912-13 (766) ; Gambia, 1912 (767) ; Fiji, 1912 (768).

Colonial Reports (*Miscellaneous*).—Southern Nigeria, 1911 (85).

Publishers forwarding books for review will greatly oblige by marking the price *in clear figures, especially in the case of foreign books.*

NEW MAPS.

EUROPE.

ORDNANCE SURVEY OF SCOTLAND.—The following publications were issued from 1st to 31st August 1913 :—

Two-mile Map, printed in Colours, and hills shown by layers, folded in cover or flat in sheets. Sheets 8, 9, 11. Price—on paper, 1s. 6d. ; mounted on linen, 2s. ; mounted in sections, 2s. 6d. each.

Six-inch and Larger Scale Maps.—Six-inch Maps (Revised). Quarter Sheets, with contours in blue. Price 1s. each. *Lanarkshire.*—24 SW. ; 24 SE. ; 25 NW. ; 30 SE. ; 35 NE.

1 : 2500 Scale Maps (Revised), with houses stippled, and with areas. Price 3s. each. *Lanarkshire.*—Sheets II. 14 ; III. 12 ; XVI. 6. Sheet XVI. 10, price 1s. 6d. *Renfrewshire.*—Sheets XVI. 6, 7, 10, 11, 12 ; XVII. 9, 10, 14. Sheet XVII. 15. Price 1s. 6d.

Note.—There are no coloured editions of these Sheets, and the unrevised impressions have been withdrawn from sale.

Special Enlargements from 1 : 2500 to 1 : 1250 Scale. Prepared for the Land Valuation Department, Inland Revenue. Price 3s. each. *Stirlingshire.*—Sheets X. 15 NW., NE. ; 16 NW., SW.

The following publications were issued from 1st to 30th September 1913 :—

Six-inch and Larger Scale Maps.—Six-inch Maps (Revised). Quarter Sheets, with contours in blue. Price 1s. each. *Lanarkshire.*—19 SW ; 20 NW. ; 23 NW. ; 37 SW. With contours in red. Price 1s. each. *Lanarkshire.*—17 SW. ; 18 NE.

1 : 2500 Scale Maps (Revised), with houses stippled, and with areas. Price 3s. each. *Lanarkshire.*—Sheets II. 11 and 7 ; IV. 14 and 10. Sheets II. 8 ; III. 7 ; IV. 9. Price 1s. 6d. each. *Renfrewshire.*—Sheets XVI. 15 ; XVII. 13 ; XVIII. 2, 3.

Note.—There are no coloured editions of these Sheets, and the unrevised impressions have been withdrawn from sale.

Special Enlargements from 1 : 2500 to 1 : 1250 Scale. Prepared for the Land Valuation Department, Inland Revenue. Price 3s. each. *Fifeshire.*—Sheets XXVIII. 9 NW., SW. ; 11 NE., SE. ; 12 NW., SW. ; XXIX. 1 NW., SW. *Stirlingshire.*—Sheets XVII. 4 SW. ; XXVIII. 8 SE. ; 13 NE. ; 14 NW. *Stirlingshire* with *Dumbartonshire* (*Det.*).—Sheet XXVIII. 12 S.E.

GEOLOGICAL SURVEY OF UNITED KINGDOM.—General Memoir. Summary of Progress of the Geological Survey of Great Britain and the Museum of Practical Geology for 1912. Price 1s.

AFRICA.

ADRAR DES IFOGHAS.—(French Sahara) Carte dressée par le Capitaine Cortier et l'Adjutant Malroux. Echelle de 1 : 500,000. Feuilles, No. 1, et No. 2, 1912. Prix 2 francs (each sheet).

Service Géographique des Colonies, Paris.

EAST AFRICA PROTECTORATE.—Scale 1 : 125,000 or about 2 miles to an inch. Sheet—Elgon, Nakuru-Nyeri, Nasin-Gishu. 1913. Price 2s. 6d. each.
Geographical Section, General Staff, London.

DEUTSCH-SÜDWESTAFRIKA.—Karte des Sperrgebiets. Massstab 1 : 100,000. Im Auftrage der Deutschen Diamanten-Gesellschaft, M.B.H., bearbeitet von Paul Sprigade und Dr. U. Lotz. 1913. 10 Blätter. Price M. 8 each sheet.
Dietrich Reimer (Ernst Vohsen), Berlin.

KAMERUN mit **TOGO.**—Massstab 1 : 2,000,000. Bearbeitet von Max Moisel. 1913. Price M. 5. *Dietrich Reimer (Ernst Vohsen), Berlin.*

This is the latest and best general map of the Kamerun Territory.

KAMERUN.—Karte in 31 Blatter und 3 Ansatzstucken im Massstabe von 1 : 300,000, bearbeitet unter Leitung von Max Moisel. Im Auftrage und mit Unterstützung des Reichs-Kolonialamts. 1913. Blatt E 2, Banjo; Blatt F 2, Fumban; Blatt G 2, Jaunde. Price M. 2 each sheet.
Dietrich Reimer (Ernst Vohsen), Berlin.

SOUTHERN NIGERIA.—Scale 1 : 125,000 or about 2 miles to an inch. Sheets—Ibadan, Igangan, Igangan West, Ijebu-Ode, Shaki-West. 1913. Price 1s. 6d. each.
Sold by W. and A. K. Johnston, Limited, Edinburgh and London.

SOUTH AFRICA.—Bartholomew's Reduced Survey Map, coloured to show height of land. Scale 1 : 2,500,000 or 39·4 miles to an inch. With inset plans of Capetown, Port Elizabeth and Durban. New Edition. 1913. Price 3s. mounted on cloth. *John Bartholomew and Co., Edinburgh.*

NORTH AMERICA.

NORTH AMERICA.—General Sketch Map in one sheet, Scale 1 : 10,000,000. Compiled by U.S. Geological Survey. George Otis Smith, Director. 1912.
United States Geological Survey, Washington, D.C.

NORTH AMERICA, INTERNATIONAL MAP.—Scale 1 : 1,000,000 or 15·78 miles to an inch. Sheet North K 19, Boston. Compiled, engraved and published by the U.S. Geological Survey. George Otis Smith, Director. 1912.
United States Geological Survey, Washington, D.C.

We welcome this first contribution from the United States to the International Map of the World. It fully realises the prescribed standard of cartography, both in drawing and execution.

ALBERTA AND SASKATCHEWAN.—Pitner's New Map on Scale of 1 : 1,300,560 or 21 miles to an inch. Compiled by G. W. Bacon and Co., Ltd. 1913. Price 21s. *G. W. Bacon and Co., Ltd., London.*

A clear, up-to-date office reference map on a scale adequate to show all places of importance. Railways, existing and in construction, are distinguished according to different companies. An index to towns and villages is printed on the margin.

MANITOBA, SASKATCHEWAN AND ALBERTA.—Scale 35 miles to an inch. Showing the number of Quarter-Sections available for homestead entry in each Township, also the pre-emption and purchased homestead area as defined by the Dominion Lands Act, 1908. Corrected to May 1913.
Department of the Interior, Ottawa.

UNITED STATES SURVEY.—George Otis Smith, Director.

(*The figures after the name of each State indicate the number of Sheets received.*)

Topographic Sheets on Scale of 1 : 12,000 or about 5¼ inches to a mile. (Special Scale for Mining Districts) Arizona, 2; Colorado, 1; New Mexico, 1.

Topographic Sheets on Scale of 1 : 31,680 or about 2 inches to a mile. California, 18; Louisiana, 2; Mississippi, 4; Texas, 3.

Topographic Sheets on Scale of 1 : 62,500 or about 1 inch to a mile. Arizona, 1; Arkansas, 1; California, 2; Georgia, 1; Illinois, 3; Iowa, 2; Kentucky, 3; Maine, 2; Maryland, 2; Michigan, 2; Minnesota, 3; Missouri, 2; Montana, 4; New York, 8; Ohio, 7; Oregon, 1; Pennsylvania, 3; Tennessee, 1; Washington, 2; West Virginia, 10.

Topographic Sheets on Scale of 1 : 125,000 or about 2 miles to an inch. Arizona, 2; California, 5; Colorado, 1; Montana, 5; New Mexico, 1; Ohio, 1; Tennessee, 1; Wyoming, 2; Yellowstone Park, 1.

Hawaii, Island of Kauai.—Scale 1 : 62,500 or about 1 inch to a mile.

United States Geological Survey, Washington, D.C.

THE HUDSON, MOHAWK GAP.—Contoured Outline-Map on scale of 1 : 1,250,000, by B. B. Dickinson. Prepared for London School of Economics.

Sifton, Praed and Co., Ltd., London.

ARCTIC AND ASIA.

ARCTIC REGIONS.—Scale 1 : 6,300,000 or 100 miles to an inch. Projected and drawn by A. Briesemeister, 1912.

American Geographical Society, New York.

The special feature of this map is the introduction of black gores, at every 10° of longitude, which take up the distortion caused by the representation of the spherical surface on a plane.

SPITSBERGEN.—Map of Prince Charles Foreland from Surveys by W. S. Bruce, LL.D., and J. Mathieson, F.R.S.G.S. Scale 1 : 140,000. Surveyed 1906-7-9. Published 1913.

This map is the subject of a special article in this number, pp. 598-9.

HONG KONG, and Part of the Leased Territory of Kowloon. 1913. Scale 2⅔ inches to a mile. Two sheets. Price 2s. each.

Geographical Section, General Staff, London.

The latest and best map of the Hong Kong Territory, showing full topographical detail, with height of land and depth of sea.

ATLAS.

ATLAS OF COMMERCIAL GEOGRAPHY.—Compiled by Fawcett Allen, Assistant Map Curator, R.G.S., with an Introduction by D. A. Jones, Assistant Librarian R.G.S. Price 3s. 6d. net. *The University Press, Cambridge.*

A series of forty-eight plates of maps, with a general index. The maps illustrate physical, political and economic features, with special reference to Commercial Geography. The Atlas is primarily designed to accompany the *Elementary Commercial Geography*, also issued by the University Press.

THE SCOTTISH

GEOGRAPHICAL

MAGAZINE.

THE BRITISH ANTARCTIC EXPEDITION, 1910-13.[1]

By Commander E. R. G. R. Evans, C.B., R.N.,
Livingstone Medallist of the Royal Scottish Geographical Society.

(*With Maps.*)

So much has been published concerning the British Antarctic Expedition, the tragic loss of its gallant leader and his four brave companions, whose names we know so well, that there is no need to preface the story by telling you at length how Captain Scott made his preparations. His organisation was complete, his equipment splendid, and no expedition ever left our shores with a better outfit or a more enthusiastic and determined personnel. Thanks to Captain Scott's fine organisation our expedition remains self-contained even after his death.

On June 1, 1910, the *Terra Nova* left London with most of the members of the expedition. She finally left New Zealand on November 29. Captain Scott had with him fifty-nine officers, scientists, and seamen. The *Terra Nova* left New Zealand a very full ship; besides four hundred tons of coal she carried provisions for three years, two huts, forty sledges, fur sleeping bags, bales of clothing, all kinds of instruments, and the hundreds of little items of equipment necessary to a Polar expedition with an ambitious scientific programme. Besides these things, which filled our ship's holds and the between deck spaces, we carried nineteen Siberian ponies, thirty-four dogs, three motor sledges, 2500 gallons of petrol, and our paraffin on the upper deck. The animals were under the charge of Mr. Cecil Meares, who with Lieut. Bruce had brought them down from Siberia. The ponies after we left New Zealand were taken charge of by Captain Oates, of the Inniskilling Dragoons.

The first exciting incident on the southward voyage occurred on

1 An Address delivered before the Society in Edinburgh, Nov. 19, 1913. Reprinted by permission from the *Geographical Journal*.

December 2, when we encountered a gale which, in the deeply laden condition of the ship, nearly caused the loss of the expedition. First the engine-room choked, and then the hand-pumps. Heavy seas washed over the vessel, and fires had to be extinguished as the engine-room was feet deep in water. While the pump suctions were being cleared the after-guard formed a bucket discharge party, and baled the ship out continuously for twenty-four hours. At the end of this time the gale abated, and we proceeded southward, having come through with no loss save two ponies, and one dog which was drowned.

Proceeding south on the meridian of 179° W., the first ice was seen in lat. 64°. The ship passed all kinds of icebergs, from huge tabular to little weathered water-worn bergs. The Antarctic pack was reached on December 9, in lat. 65° S., and the ship boldly pushed through for some 200 miles under steam and sail, when her progress was retarded to such an extent that, to save coal, engines were stopped, sail was furled, and the ship lay under banked fires for some days. We spent three weeks in the pack, and emerged on December 30, after pushing through 380 miles of ice. The time was not wasted: magnetic observations, deep-sea soundings, and serial sea-temperatures were obtained. The zoologists and marine biologists secured valuable specimens. Once in open water we proceeded full speed to Cape Crozier, as Dr. Wilson wished to study the embryology of the Emperor penguins during the winter season. Captain Scott was quite prepared to make Cape Crozier our base, if a suitable landing-place was to be found. As no good place was to be seen, we rounded Cape Bird at midnight, and entered M'Murdo Sound. It was remarkably clear of ice. We passed Shackleton's winter quarters, and noticed his hut at Cape Royds looking quite new and fresh. Six miles further south the ship brought up against the fast ice, which extended right across the Sound.

On January 4, 1911, thirty-six days out from New Zealand, Captain Scott, Wilson, and myself went across the ice and visited a little cape which looked, and subsequently proved to be, an ideal spot for wintering. This place Captain Scott named Cape Evans. Immediately the winter quarters were selected, out came the stores and transport. Lieut. Pennell took charge of the ship, Lieut. Campbell of the transport over the mile and a half of sea-ice; the charge of the base was given to me, while Captain Scott supervised, planned, and improved.

Meares' dogs, Oates' ponies, and Day's motors supplemented by man-hauling parties bustled between ship and shore, transporting stores over the frozen sea. At the cape, Davis, the carpenter, with his willing crew, put up the tent. In less than a week the main party had their equipment ashore.

We will now follow Captain Scott and his companions at the principal base. The weather was so hot when first we landed that the ice melted, and we could wash in fresh water, and even draw our drinking-water from a cascade. We built ice-caves to stow our fresh mutton in, and for magnetic observations. Outside the hut we soon had fine stables. Directly the construction of the base station was assured, away went every available man to lay a depôt. We said good-bye to the ship, and on January 24, 1911, Captain Scott and eleven companions left with two dog teams and eight ponies to lay out a depôt of foodstuffs before the Antarctic winter set in. Nearly one ton of provisions was taken out to a point 144 miles from our base. This spot was named One Ton Depôt. The party for the return journey was split up into three detachments. Captain Scott with Meares, Wilson, and Cherry Garrard, came home with the dogs. Scott and Meares had the misfortune to run along the snow bridge of a crevasse. The bridge gave way, and all the dogs but

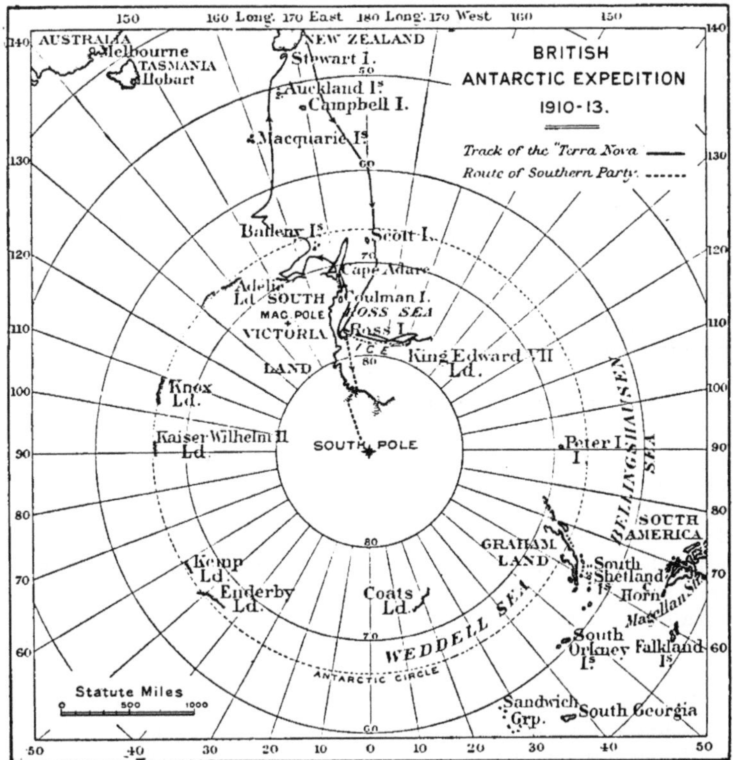

Osman, the leader, and the two rear animals, disappeared down a yawning chasm. With the greatest difficulty the dogs were rescued. Scott and Meares were lowered by Wilson and Cherry Garrard into the crevasse. They found the dogs twisting round, suspended by the harness, fighting, howling, and snapping. One by one they were freed from the trace, and hauled up on to solid ice; as each animal regained safety he lay down and slept. It was an anxious period for all concerned. Captain Scott spoke most highly of Wilson, Meares, and Cherry Garrard's behaviour and resource on this occasion.

One party, consisting of the second in command and two seamen, returned from the depôt journey with the three oldest and weakest ponies—Blossom, Blucher, and James Pig. The ponies were in very poor condition, and Oates, their master, expected all three to give out on their return march. They were christened by the seamen "The Baltic Fleet." Two of them died owing to the severe weather conditions that obtained at the end of February, but the third pony, James Pig, was a plucky little animal, and he survived. Lieut. Bowers, in charge of the detachment which built up "One Ton Depôt," returned after the other two parties. He had with him Cherry Garrard and Crean, when, on March 1, he was sent across the sea-ice to reach Hut Point. The ponies were tired and listless after their hard journey and in bad condition, and they had to be frequently rested. As they advanced towards Hut Point cracks in the ice became apparent, and when the party reached a crack which showed the ice to be actually on the move, they turned and hastened back—but the ice was drifting out to sea! The ponies behaved splendidly, jumping the ever-widening cracks with extraordinary sagacity. Bowers, Cherry Garrard, and Crean launched the sledges back over the cracks in order not to risk the ponies' legs. Eventually they reached what looked like a safe place. Men and ponies were thoroughly exhausted. Camp was pitched, and the weary party soon fell asleep. Bowers soon awoke, hearing a strange noise. He found the party in a dreadful plight—the ice had again commenced to break up, and they were surrounded by water. One of their four ponies had disappeared in the sea. Camp was again struck, and for five hours this noble little party fought their way over three-quarters of a mile of drifting ice. They never thought of abandoning their charge, realising that Scott's Polar plans might be ruined if four more ponies were lost with their sledges and equipment. Crean, with great gallantry, went for support, clambering with difficulty over the ice. He jumped from floe to floe, and at last climbed up the face of the Barrier from a piece of ice which touched the ice-cliff at the right moment. Cherry Garrard stayed with Bowers at his request, for little Bowers would never give up his charge while a gleam of hope remained. For a whole day these two were afloat, and eventually Captain Scott, Oates, Gran, and Crean appeared on the Barrier edge, and on seeing them Bowers and Cherry Garrard jumped some floes till they reached a piece of ice resting against the Barrier face, thanks to the return of the tide. Bowers and Cherry Garrard were rescued, and after a further piece of manœuvring, a pony and all the sledges were recovered. The other three ponies were drowned. During this trying time Killer whales were about almost continuously, blowing and snorting in the intervening water spaces. Only those who have served in the Antarctic can realise fully what Bowers' party, and also Scott's own rescue party, went through.

By March 4 all the depôt parties were safely, if not comfortably, housed at Hut Point, with the two dog teams and the two remaining ponies. We were unable to return to Cape Evans for six weeks, as the sea would not freeze over properly on account of persistent high winds. We lived in the old hut left by the *Discovery*, and our existence was rather primitive. Meares and Oates perfected a blubber stove. We killed seals, and thus obtained food and fuel. Although rather short of luxuries, such as sugar, we were never in any great want of good plain food, and the time passed agreeably enough. On March 14, the Depôt party was joined by Griffith Taylor, Debenham, Wright, and Petty-Officer Evans.

Taylor's party had been landed by the *Terra Nova* on January 27, after the start of the Depôt party, to make a geological reconnaissance. They traversed the Ferrar glacier, and then came down a new glacier, which Scott named after Taylor, and descended into Dry Valley, so called because it was entirely free from snow. Their way led over a deep freshwater lake four miles long, which was only surface frozen. This lake was full of algæ. The gravels below—a promising region of limestones, rich in garnets—were washed for gold, but only magnetite was found. When Taylor had thoroughly explored and examined this region, his party retraced their footsteps and proceeded southward to examine the Koettlitz glacier. They returned from the Koettlitz glacier along the edge of the almost impenetrable pinnacle ice, and part of their journey actually led them through an extraordinary and difficult ice-field. It took two days to negotiate 6 miles of this surface; the party were then able to get back on to sea-ice, and without mishap marched to Hut Point. We now numbered sixteen at this congested station, and 15 miles of open water separated us from Cape Evans. The gales were so bad that spray dashed over the hut sometimes, and all round the low-lying parts of the coast spray ridges of ice formed. But at last ice formed which was not blown out, first in little pancakes, which cemented together and formed floes, these in their turn were frozen together, and at last a party of nine made the passage over the new sea-ice to Cape Evans, and, on April 13, 1911, they marched into the hut at our main base, dirty but cheerful. The cook soon had all kinds of luxuries prepared for us. Captain Scott was delighted at the progress made by those left in our hut under Dr. Simpson.

And now that communication was established between Hut Point and Cape Evans, we settled down for the winter. Thanks to Ponting, our photographic artist, we have a magnificent pictorial record of events. Ponting went everywhere with his camera and kinematograph machine. Even when we came South in the ship he kinematographed the bow of the *Terra Nova* breaking the ice. If a sledge-party set out, a penguin appeared, or a pony "played up," or even if a dog broke adrift, Ponting was there with his artillery, ready for action. He even had a galloping carriage with his quick-firing cameras drawn by dogs. He would get seals to pose for him if he wished, by his persuasive methods, or by exciting their curiosity. Ponting never missed an opportunity of making an artistic photograph.

We must now hurry through four months' darkness. The first winter seemed to pass very quickly. Every one was busy at his special subject. Dr. Wilson, the chief of our scientific staff, helped us all. He was our Solomon. To "Uncle Bill" we all went for sound practical advice. Wilson was a friend and companion to Captain Scott, and, indeed, to all in the Expedition.

During the winter months holes were made in the sea-ice through which we lowered a wire fish-tray. ' By this means we caught a number of Notothenia. When Atkinson, the helminthologist, had examined these fish, they were handed to the cook who served them up for breakfast. These fish were a great delicacy.

A small hut was erected some 50 yards from the main station to contain the magnetic observatory under Dr. G. C. Simpson of Simla. His place was at the base station; his important work as physicist and meteorologist prevented him from taking an active part in our sledge journeys. When he was recalled to Simla in 1912, his work was ably continued by Wright, the Canadian chemist, who made a special study of ice structure and glaciation.

On June 27 Dr. Wilson, with Bowers and Cherry Garrard, started on a remarkable journey to Cape Crozier. Their object was to observe the incubation of the Emperor penguins at their rookery. During this first Antarctic mid-winter journey the temperatures were seldom above 60°, and they actually fell to 77° below zero, that is 109° of frost. The party took a fortnight to reach Cape Crozier, meeting with good weather —that is, calm weather—but bad surfaces, which handicapped them severely. After rounding Cape Mackay they reached a wind-swept area, and experienced a series of blizzards. Their best light was moonlight, and they were denied this practically by overcast skies. Picture their hardships—frozen bags to sleep in, frozen finneskoe to put their feet in every time they struck camp. They scarcely slept at all. And when they reached Cape Crozier, only about one hundred Emperor penguins could be seen. In the *Discovery* days this rookery was found to contain two or three thousand birds. Possibly the early date accounted for the absence of Emperors. However, half a dozen eggs were collected, and three of these are now in our possession. Wilson on his return told us that he picked up rounded pieces of ice which the stupid birds had been cherishing, fondly imagining they were eggs. The maternal instinct of the penguin is very strong.

At Cape Crozier, Wilson's party had built a stone hut behind a land ridge on the slopes of Mount Terror. This hut was roofed with canvas. The same night that the eggs were collected a terrific storm arose. One of the hurricane gusts of wind swept the roof of the hut away, and for two days the unfortunate party lay in their bags half smothered with fine drifting snow. The second day was Dr. Wilson's birthday. He told me afterwards that had the gale not abated, then they must all three have perished. They dared not stir out of the meagre shelter afforded by their bags. Wilson prayed hard that they might be spared. His prayer was answered; but, as you know, two of this courageous little band lost their lives later on in their eager thirst for scientific knowledge. When the three men crept out of their bags into the dull winter gloom they groped about and searched for their tent, which had blown away from its pitch near the stone hut. By an extraordinary piece of good fortune it was recovered, scarcely damaged, a quarter of a mile away. Wilson, Bowers, and Cherry Garrard started home the next day. They were caught by another blizzard, which imprisoned them in their tent for forty-eight hours. After a very rough march, full of horrible hardships and discomforts, the little band won through and reached Cape Evans on August 1, having faced the dreadful winter weather conditions on the great barrier for five weeks. Of the three only Cherry Garrard survives. He was Wilson's special friend amongst the younger members. On their return they wanted bread, butter, and jam most, and loaves disappeared with extraordinary speed. They were suffering from want of sleep, but were all right in a few days. A remarkable feature of this journey was the increase of weights due to ice collecting in the sleeping-bags, tent, and clothing. The three sleeping-bags weighed 47 lbs. at the start,

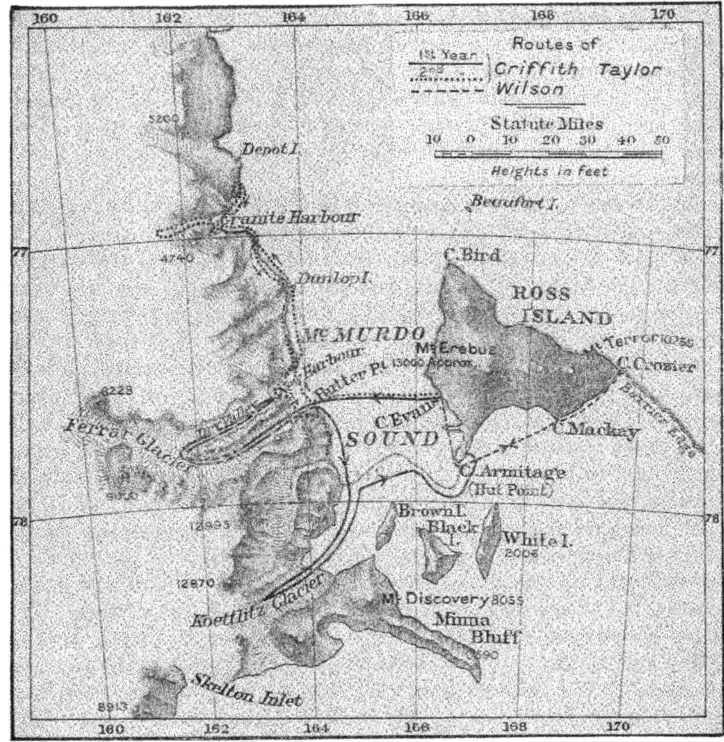

and 118 lbs. on their return. Other weights increased in the same proportion, and their sledge had dragged very heavily in consequence. The three men when they arrived in the hut were almost encased with ice. I well remember undressing poor Wilson in the cubicle he and I shared. His clothes had almost to be cut off him.

From this journey we derived additional experience in the matter of sledging rations. Thanks to the experiments made, we arrived at the most suitable ration. This was for the colder weather expected during the second half of the forthcoming Polar journey. It was to consist of 16 ozs. biscuit, 12 ozs. pemmican, 3 ozs. sugar, 2 ozs. butter, 0·7 oz. tea, 0·6 oz. cocoa—equals 34·4 ozs. food daily. This is one man's food per day. No one could possibly eat this in a temperate climate; it was a fine filling ration even for the Antarctic. The pemmican consists of beef extract with 60 per cent. pure fat.

No casualties occurred during the winter, but Dr. Atkinson had a severely frostbitten hand. He had gone out to read a thermometer on the sea-ice 800 yards from our hut. It was blowing and drifting, and Atkinson lost his way in a blizzard. He was adrift for eight hours, but luckily found his way back during a lull in the weather.

During the second half of the winter we were all busy preparing for the sledge expedition to the Pole. Food rations had to be prepared, instruments calibrated, sledges specially fitted to carry the travelling equipment, and our own clothes adapted for sledging according to experience gained in the depôt and winter journeys. Meares and the second in command took parties out and laid depôts during the early spring, and Captain Scott made a coastal journey to the west. These spring journeys were all interesting in their way, but cannot now be dwelt on owing to want of time.

On October 24 the advance guard of the Southern party, consisting of Day, Lashly, Hooper and myself, left with two motor sledges. We had with us three tons of stores, pony food and petrol, carried on six sledges. The object of sending forward such a weight of stores was to save the ponies' legs over the variable sea-ice, which was in some places hummocky, and in others too slippery to stand on. The first 30 miles of Barrier was known to be bad travelling. The motor party had rather trying experiences, owing to the frequent over-heating of the air-cooled engines. Directly the engines became too hot we had to stop, and by the time they were reasonably cooled the carburetter would refuse duty—it had often to be warmed up with a blow-lamp. Day and Lashly, the engineers, had great trouble in starting the motor-sledges. We all four would heave on the spans of the towing-sledges, to ease the starting strain; the engines would generally give a few sniffs, and then stop. It is true that the motors advanced the necessaries for the Southern journey 51 miles, but at the expense of the men who had charge of them. The engineers continually got their fingers frost-bitten tinkering with the engines and replacing big end brasses, which several times gave out. But although the temperatures were low, we were all very happy, and Day was most keen to bring the motors through with credit. They were abandoned a mile south of Corner Camp, but had advanced their weights in turn over rough, slippery and crevassed ice, and thus given the ponies a chance to march light.

The first 30 miles of Barrier surface led over very deep soft snow, and, in fairness to the despised motors, they went better over soft snow than any other part of our transport. The man-hauling party, as we now became, marching for a fortnight, covered nearly 180 miles, and halted at a rendezvous on November 15 in lat. 80° 32′ south. We waited here six days, and built an enormous snow cairn 15 feet high. We called this rendezvous Mount Hooper, after our youngest member.

On November 21, Captain Scott arrived with eleven men, ten ponies and two dog teams. We heard that they had been delayed partly by bad

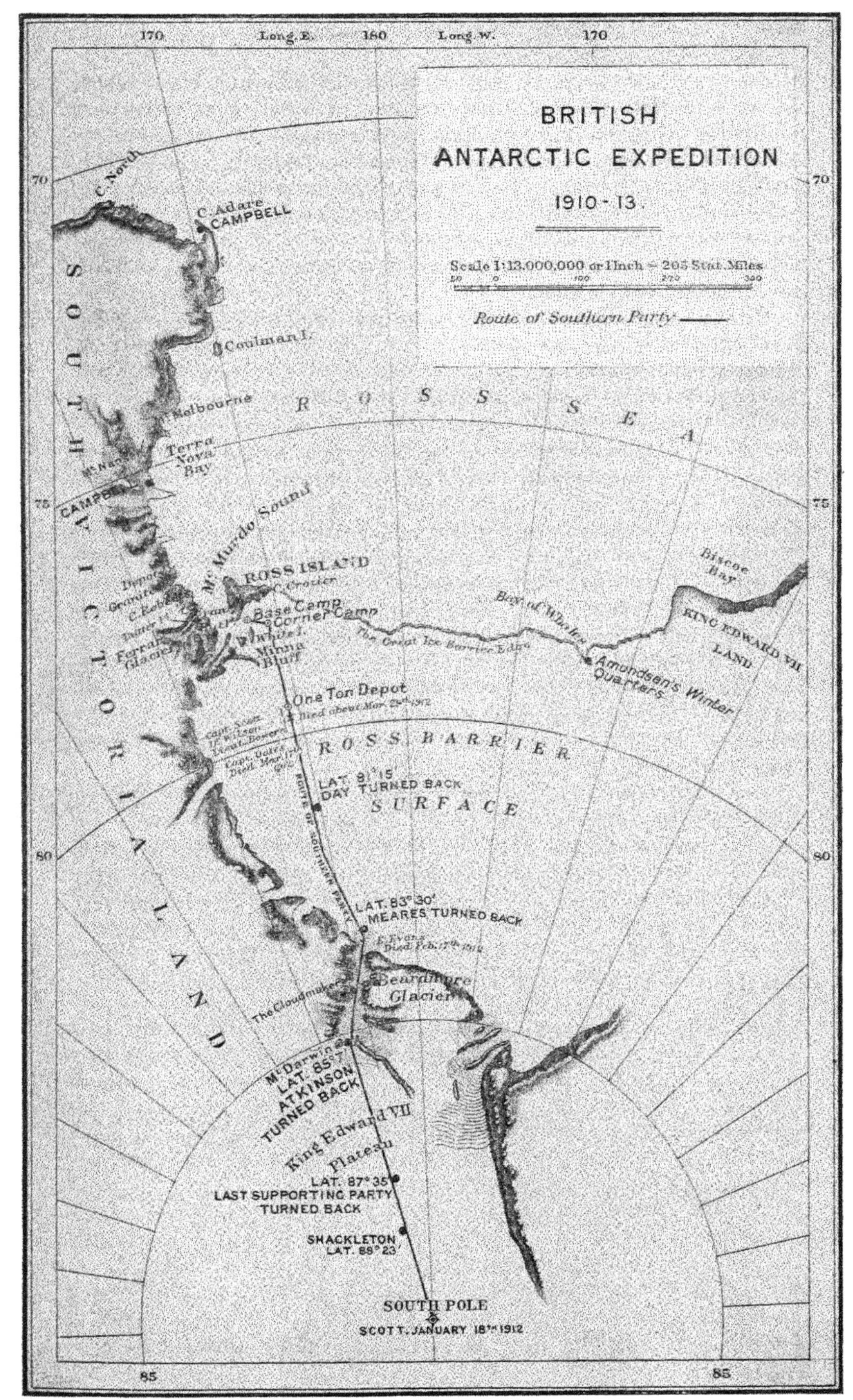

BRITISH
ANTARCTIC EXPEDITION
1910-13.
Scale 1:13,000,000 or 1 Inch = 205 Stat. Miles
Route of Southern Party
Long. E.
Long. W.
C. North
C. Adare
CAMPBELL
Coulman I.
R O S S S E A
Terra Nova Bay
CAMPBELL
SOUTH VICTORIA LAND
McMurdo Sound
ROSS ISLAND
C. Crozier
Base Camp
Corner Camp
Ferrar Glacier
White I.
Minna Bluff
Biscoe Bay
Bay of Whales
KING EDWARD VII LAND
The Great Ice Barrier Edge
Amundsen's Winter Quarters
One Ton Depot
ROSS BARRIER SURFACE
LAT. 81° 15'
DAY TURNED BACK
LAT. 83° 30'
MEARES TURNED BACK
The Cloudmaker
Beardmore Glacier
Mt Darwin
LAT. 85° 7'
ATKINSON TURNED BACK
King Edward VII Plateau
LAT. 87° 35'
LAST SUPPORTING PARTY TURNED BACK
SHACKLETON LAT. 88° 23'
SOUTH POLE
SCOTT. JANUARY 18th 1912

weather, and had purposely kept down the marches to give the weaker animals a chance. However, every one was well and eager to advance southward. Captain Scott ordered us to continue to go forward in advance of the dogs and ponies. We marched exactly 15 miles daily, erecting cairns at certain pre-arranged distances, surveying, navigating, and selecting the camping site. The ponies marching by night were able to rest when the sun was high and the air warmer. Meares' dogs would bring up the rear; they started some hours after the ponies, as their speed was so much greater.

Captain Scott's plans worked easily and well. The ponies pulled splendidly, and their masters vied with each other in their care and management. Oates always kept a very careful look-out on his charges. The tough little beasts pulled about 650 lbs. each, and were fed daily on 10 lbs. of oats and 3 lbs. of oilcake. On camping, large walls would be erected by the pony leaders to shelter their animals from the wind, and while this was being done the cook of each tent would prepare the supper hoosh. We were all so happy and full of life on the march over the great ice barrier that we often would wrestle and skylark at the end of the day. We had our good and bad weather, and we had our turns of snow blindness. This ailment was common to the ponies and dogs as well as to ourselves. Depôts were made every 65 miles. They were marked by big black flags, and we saw one of them, Mount Hooper, 9 miles away. Each depôt contained one week's rations for every returning unit. That outward Barrier march will long be remembered—it was so full of life, health and hope. Our sad days came when the ponies were killed one by one. But hunger soon defeated sentiment, and we used to relish our pony meat, which made the hoosh more solid and satisfactory.

Day and Hooper were the first to return, their places being taken in the man-hauling party by Atkinson and Wright, whose ponies, Jehu and Chinaman, were first shot. With two invalid dogs, Day and Hooper left us at 81° 15′ and marched back to the base. With only two they had great difficulty in pulling their sledge home, so cut it in half and saved themselves considerable labour. On December 4 we arrived within 12 miles of Shackleton's Gap or Southern Gateway. We could see the outflow of the Beardmore glacier stretching away to our left as we advanced southward that day. Hopes ran high, for we still had the dogs and five ponies to help us. Captain Scott expected to camp on the Beardmore itself after the next march. Luck was against us. On December 5 we encountered a blizzard which lasted four whole days. The temperature rose to 35° Fahrenheit, and the drift was very bad indeed. The snow was in big flakes, driving from the SSE., but as the gale took its course snow was succeeded by sleet, and even rain. The Barrier surface was covered with 18 inches of slush. The poor ponies had continually to be dug out from the snowdrifts, which accumulated behind their walls. The dogs suffered less, but they themselves looked like wet rats when Meares and Demitri went to feed them. All our tents, clothing and sleeping-bags were soaked. On December 9 the blizzard was over, and all hands dug out sledges and stores. We wallowed sometimes thigh deep in this Antarctic morass, and after marching for fourteen days on end the remaining five ponies were shot, as no food was left for them. Poor things! they did their job well, and I believe every pony leader gave half his biscuits to his own animal, so they had some little reward for their last march.

As arranged, three teams of four, pulling 170 lbs. per man, now advanced up the glacier. Meares and his Russian dog-boy came along with us for two marches, and then turned homeward. To help us Meares had travelled further south than his return rations allowed for, and for the 450-mile northward march to Cape Evans he and his companion Demitri went short one meal a day, rather than deplete the depôts. It is a dreadful thing on an Antarctic sledge journey to forfeit a whole meal daily, and Meares' generosity should not be forgotten. The advance of the twelve men up the Beardmore was retarded considerably by the soft wet snow which had accumulated in the lower reaches of the glacier. Panting and sweating, we could only make 4-mile marches until the 15th. But after that the surfaces were better, and we were far less tired in doing more than twice the distances.

On December 16 we reached Blue Ice, 3000 feet above the Barrier, and, with the exception of little delays caused by people falling into crevasses, our progress was not impeded. Wilson did a large amount of sketching on the Beardmore. His sketches, besides being wonderful works of art, helped us very much in our surveys. We had fine weather generally, and with twelve men the Beardmore glacier was overcome without great difficulty. Of course, we had Shackleton's charts, diaries, and experience to help us. We often discussed Shackleton's journey, and were amazed at his fine performance. We always had full rations, which Shackleton's party never enjoyed at this stage. Our marches from December 16 worked up from 13 to 23 miles a day.

On December 21 we were on the plateau in lat. 85° 7′ S., 6800 feet above the Barrier, fit and ready to go forward. Here we established the Upper Glacier depôt. The third supporting party, consisting of Atkinson, Wright, Cherry Garrard, and Keohane, left us the next day, and marched home 584 miles. They spent Christmas Day collecting geological specimens, and reached Cape Evans on January 28, after a strenuous journey of 1164 miles. They had some sickness in the shape of enteritis and scurvy. But Dr. Atkinson's care and medical knowledge brought them through safely.

Captain Scott with his two sledge teams now pushed forward, keeping an average speed of 15 miles a day with full loads of 190 lbs. a man. We steered south-west for the first two days after leaving the Beardmore, to avoid the great pressure ridges and icefalls which were plainly visible to the south. On December 23 we came across enormous crevasses which were as big as Regent Street. They were nearly all well bridged with snow, but we took them at the rush and had no serious falls. The dangerous part is at the edge of the snow bridge, and we frequently fell through up to our armpits, just stepping on to or leaving the bridge. We experienced on this plateau the same tingling southerly wind that Shackleton speaks of, and men's noses were frequently frost-bitten. On Christmas Eve we were 8000 feet above the Barrier, and we imagined we were clear of crevasses and pressure ridges. We now felt the cold far more, when marching, than we had done on the Beardmore. The wind all the time turned our breath into cakes of ice on our beards. Taking sights when we stopped was a bitterly cold job, fingers had to be bared to work the little theodolite screws, and in the biting wind one's finger-tips soon went. On Christmas Day we marched 17¼ miles, and during the forenoon again crossed a badly crevassed area. Lashly celebrated his forty-fourth birthday by falling into a crevasse 8 feet wide. The laden sledge just bridged the chasm, and poor Lashly was suspended below spinning round with 80 feet of clear space beneath him. We had great difficulty in hauling him up on account of his being directly under the sledge. When he reached the surface, one of the party wished him a happy Christmas and another many happy returns. I will not tell you what Lashly's reply was.

At 7.30 we camped and had our Christmas dinner—extra thick pemmican with pony meat in it, a chocolate and biscuit hoosh, plum pudding, cocoa, ginger, and caramels, and a mug of water each. We were all so full that we could hardly shift our foot-gear, and although the temperature was well below zero, we lay on our sleeping-bags unable to muster the energy to get into them. The 87th parallel was reached on New Year's Eve, after a short march; we made a depôt here, and the seamen of the party converted the 12-foot sledges into 10-foot ones by the spare short runners we had brought along. This took nine hours, but the reduction in bearing surface was worth it. We saw the New Year in that night with a fine feed of pemmican and a stick of chocolate which Bowers had kept for the occasion.

On January 3, Captain Scott came into our tent and told us that he was sure he could reach the Pole if my party gave up one man, and made the homeward journey short-handed. Of course we consented, and Bowers was taken into the Polar party. On January 4 the last supporting party, consisting of Lashly, Crean, and myself, marched south to lat. 87° 34′ with the Polar party, and seeing that they were travelling rapidly, yet easily, we halted, shook hands all round, and said good-bye.

Up to this time no traces of the successful Norwegians had been seen, and we all fondly imagined that our flag would be the first to fly at the South Pole. We gave three huge cheers for the Southern party as they stepped off, and then turned our sledge and commenced our homeward march of nearly 800 miles. We frequently looked back until we saw the last of Captain Scott and his four companions, a tiny black speck on the horizon, and little did we think that we were the last to see them alive, that our three cheers were the last appreciation they would ever know.

The return of the last supporting party nearly ended in disaster. After the first day's homeward march, we found that we could not do the necessary distances in the nine hours, owing to being a man short. This was serious, and in order not to make my seamen companions anxious, that night I handspiked my watch, putting the hands on one hour, so that we therefore turned out about 4 a.m., making from ten to twelve hours a day. On January 8 we were overtaken by a blizzard which continued for three days. We dared not stop our marches, and thanks to the wind being with us we were able to push on. But the soft snow spoilt the surface, and the outlook was so bad we cut off a big corner and saved two days' march, by shaping course direct for the Upper Glacier depôt under Mount Darwin. This led us over Shackleton's ice-falls at the head of the Beardmore glacier. We descended many hundred feet, mostly riding on the sledge; we had frequent capsizes and broke the bow of the sledge. Crean had the misfortune to catch his trousers somehow in our headlong flight and had them torn to shreds. We reached the Upper Glacier depôt the same day, however, and reclothed Crean, who had left a pair of Mandleberg wind-proof trousers in the depôt cairn with some of his tobacco wrapped up in them.

Returning down the Beardmore, we had some misty weather which hid the land, and we were embarrassed by getting into a mass of ice-falls, pressure ridges, and crevasses. We fell about a great deal, and were two days getting clear. We had no food left, and when we reached the next depôt under Cloudmaker mountain we had marched 17 miles without anything to eat except one biscuit and a mug of tea. To make things worse, I developed scurvy about January 17, when we had 500 miles to go. My condition became daily more serious until I entirely lost the use of my legs. But I could not afford to give up as I was the only one in the party who knew anything about navigation, and I had to keep them marching until they could see Mount Erebus, or some known landmark. When 75 miles from Hut Point I ordered Crean and Lashly to leave me with my sleeping-bag and some food, and go on, sending out relief if possible. They refused to do this, and strapping me on the sledge, dragged me 40 miles in four days, helped by a southerly wind. When 35 miles from Hut Point we had a heavy snowfall, which made it impossible for Lashly and Crean to move the sledge.

Crean then left us on February 19, and marched for eighteen hours with nothing to eat but a few biscuits. He plodded on stolidly through the soft snow, and eventually reached Hut Point utterly exhausted and numbed with cold; but he gave our whereabouts to Dr. Atkinson, who was there with Demitri and the dog-teams, and they came out and rescued us. Lashly undoubtedly saved my life by his careful nursing. It was very brave of him to stay with me, as he only had three meals left, and if relief had not come in time he could never have walked in without food, as he himself was very done after hauling in my sledge-team for over 1500 miles. Crean volunteered to come out again with Atkinson, but was, of course, not allowed to.

And now we will turn again to the Polar party itself. They covered the 145 geographical miles that remained in a fortnight. Captain Scott came across Amundsen's dog-tracks soon after lat. 88°, and followed them to the Polar area. Scott, Wilson, Oates, Bowers, and Seaman Evans reached the South Pole on January 17, 1912. They fixed the exact spot by means of a 4-inch theodolite, and the result of their careful observations located the Pole at a point which only differed from Amundsen's by half a mile, as shown by his flag. This difference actually meant that the British and Norwegian observers differed by one scale division on the theodolite, which was graduated to half a minute of arc. Experts in navigation and surveying will always look on this splendidly accurate determination as a fine piece of work by our own people as well as by the Norwegian expedition.

Lady Scott has remarked on the magnificent spirit shown by her husband and his four specially selected tent-mates, when they knew that Queen Alexandra's little silk Union Jack had been anticipated by the flag of another nation. Scott and his companions had done their best, and never from one of them came an uncharitable remark. On January 19 the homeward march was commenced; the party had before them a distance of over 900 miles. They came back at a fine pace over the ice-capped plateau. A blizzard stopped them from travelling on January 25, but otherwise their progress was not retarded materially. Seaman Evans was causing anxiety, and his condition naturally worried Captain Scott and his comrades. But, however great their anxieties, they looked after Evans most carefully, and hoped to pull him through. He was rested on the Beardmore glacier, Oates looking after him while the others made a halt for geographical investigation by the Cloudmaker depôt. But Evans also sustained a serious concussion through falling and hitting his head, and then the party were greatly hampered. They were so delayed that the surplus foodstuffs rapidly diminished, and the outlook became serious. Bad weather was encountered, and near the foot of the Beardmore poor Seaman Evans died. He was a man of enormous strength, a tried sledger, and a veteran in Antarctic experience. Captain Scott had the highest opinion of this British seaman. He was the sledge-master, and to Evans we owed the splendid fitting of our travelling equipment, every detail of which came under his charge.

Seaman Evans' death took place on February 17, and then the bereaved little band pushed northward with fine perseverance, although they must have known by their gradually shortening marches that little hope of reaching their winter quarters remained. Their best march on the Barrier was only 9 miles, and in the later stages their marches dropped to 3 miles. The depôts were 65 miles apart, and contained six weeks' provisions; they knew their slow progress was not good enough, but they could not increase their speed over such bad surfaces. The temperatures fell as they advanced, instead of rising as expected, and we find them recording a temperature of $-46{\cdot}2°$ one night.

Poor Oates' feet and hands were badly frost-bitten—he constantly appealed to Wilson for advice. What should he do, what could he do? Poor gallant soldier, we thought such worlds of him. Wilson could only answer, "Slog on—just slog on." On March 17, which was Oates' birthday, he walked out to his death in a noble endeavour to save his three comrades beset with hardships, and as our dead leader wrote, "*It was the act of a brave man and an English gentleman.*"

Scott, Wilson, and Bowers fought on until March 21, only doing about 20 miles in the four days, and then they were forced to camp 11 miles south of One Ton Depôt. They were kept here by a blizzard which was too violent to permit them to move, and on March 25 Captain Scott wrote his great message to the public.

Thanks to Atkinson and the search party we have all the records of these brave men, and so the surviving members of the expedition can work on them, and for Scott's and Wilson's sakes particularly let us hope justice will be done to these same records.

I must now take you right away from the main party to give you an insight into Lieutenant Victor Campbell's work. Campbell's party consisted of Surgeon Levick, Raymond Priestley, geologist, and Seamen Abbott, Browning, and Dickason. Lieutenant Pennell, who now commanded the *Terra Nova*, took this expedition along the Barrier to King Edward's Land in the beginning of February 1911. They got within 10 miles of Cape Colbeck, but the most formidable pack-ice yet seen lay between them and the ice cliffs of this inhospitable-looking land. It was out of the question for Campbell to put his hut and gear out on to sea-ice, with no certain prospect of being able to climb the cliffs of King Edward VII. Land. So he and Pennell reluctantly returned to seek a landing elsewhere.

Coal was short, and the season drawing on. The *Terra Nova* steamed back along the face of the Great Ice Barrier, and in the Bay of Whales sighted the *Fram*. The two ships' companies soon made friends, and the commanding officers exchanged calls. Amundsen was anxious for Campbell to winter alongside of him, but Campbell decided to make his winter quarters in another region—it being undesirable to have two expeditions wintering at the same base. Campbell eventually landed at Cape Adare, after vainly searching for a more profitable wintering place. He was most handicapped by the shortage of coal in the *Terra Nova*, which limited the radius of their search.

Campbell and his party did excellent meteorological, geological, and magnetic work, and he himself made some very good surveys. Levick made a special study of the penguins, and Priestley, with his previous Antarctic knowledge, made himself invaluable, apart from his own scientific work. Campbell was loud in his praise of the seamen in this party.

Lieutenant Pennell, in the *Terra Nova*, revisited Cape Adare after the first winter, took off the party and their collections, and landed them again on January 8 at the Terra Nova Bay, to sledge round Mount Melbourne to Wood Bay, and examine this part of Victoria Land. Campbell and his crew returned, after a month's sledge journey to Terra Nova bay, on February 6. They had found garnets and many excellent fossils on this trip. Campbell did some very good work surveying, and has added a good deal to the existing maps.

On February 17, the party began to look for the *Terra Nova*, but as time went on and she did not put in an appearance, Campbell prepared to winter. Pennell, who had brought the *Terra Nova* back to pick this sledge team up, met with ice conditions that were insuperable, and never got within 30 miles of Terra Nova Bay. Pennell, Rennick, and Bruce did all that any men could do to work their ship through, but communication was impossible that season, and so Campbell was left with only four weeks' sledging provisions to face an Antarctic winter. His party could not have been better chosen to help him through this ordeal. Campbell knew his men absolutely, and they themselves were lucky in having such a resourceful and determined officer in charge.

On March 1 Campbell selected a hard snow slope for their winter home, and into this they cut and burrowed until they had constructed an igloo or snow-house, 13 feet by 9 feet. This they insulated with blocks of snow and seaweed. A trench roofed with sealskins and snow formed the entrance, and at the sides of this passage they had their store-rooms and larder.

All the time this house was under construction a party was employed killing penguins and seals, for which they kept a constant look-out. By March 15 their larder contained 120 penguins and 11 seals. After this date gale succeeded gale, and the winter set in with a long run of bad weather.

Campbell and his companions led a very primitive existence here for six and a half months. They only had their light summer sledging clothes to wear, and these soon became saturated with blubber; their hair and beards grew, and they were soon recognisable only by their voices. Some idea of their discomforts will be gleaned by a description of their diet. Owing to their prospective journey to Cape Evans, Campbell had to first reduce the biscuit supply from eight to two biscuits a day, and then to one.

Generally their diet consisted of one mug of pemmican and seal hoosh and one biscuit for breakfast. Nothing for lunch. One and a half mugs of seal, one biscuit and three-quarters of a pint of thin cocoa for supper. On Sunday, weak tea was substituted for cocoa; this they reboiled for Monday's supper, and they used the dried tea-leaves for tobacco on Tuesday. Their only luxuries were a piece of chocolate and twelve lumps of sugar weekly. They sometimes used tea-leaves and wood shavings for tobacco. They kept twenty-five raisins each for birthdays. One lucky find was thirty-six fish in the stomach of a seal, which, fried in blubber, proved excellent. The biscuit ration had to be stopped entirely from July to September.

The six men cooked their food in sea-water as they had no salt, and seaweed was used as a vegetable. Priestley did not like it, and no wonder, for it had probably rotted in the sun for years, and the penguins had trampled it all down, etc.

Campbell kept a wonderful discipline in his party, and as they were sometimes confined to the igloo for days, Swedish drill was introduced to keep them healthy. A glance at their weather record shows how necessary this was. We find one day snowing hard, next day blowing hard, and the third day blowing and snowing hard, nearly all through the winter. But there was never a complaint. On Sunday divine service was performed. This consisted of Campbell reading a chapter of the Bible, followed by hymns. They had no hymn-book, but Priestley remembered several, while Abbot and Browning and Dickason had all been at some time or other in a choir.

To add to their discomforts, owing to the state of their clothing and meagre food supply, they were very susceptible to frost-bites, and Jack

Frost made havoc with feet, fingers, and faces. Then sickness set in, in the shape of enteritis—Browning suffering dreadfully, but always remaining cheerful. The sickness was undoubtedly due to the meat diet, and its ravages weakened the party sadly.

On May 6 Campbell's party sustained a severe disappointment, for they saw what appeared to be four men coming towards them. Immediately they jumped to the conclusion that the ship had been frozen in and this was a search party. The four figures turned out to be Emperor penguins, and although disappointing in one way they served to replenish the larder, and so had their use.

Campbell and his five companions started for Cape Evans on September 30. Progress was slow and the party weak, but thanks to their grit and to Campbell's splendid leadership, this party all got through to the winter quarters alive. Browning had to be carried on the sledge part of the way, but fortunately they picked up one of Griffith Taylor's depôts, and the biscuit found here quite altered Browning's condition.

It seems a pity that full justice cannot be done to all the parties who went forth sledging in various directions, but a single lecture does not permit very full descriptions. Griffith Taylor, the Australian physiographer, with Debenham, Gran, and Seaman Forde, made a most valuable journey along the coast of Victoria Land for geological and surveying purposes. I hope Taylor will prepare a paper on this expedition at some future date. The work of the *Terra Nova* is also worthy of a special lecture; and here I would like to say that Lieutenant Pennell, her commander, Lieutenants Rennick and Bruce, Mr. Drake and Mr. Lillie, have worked incessantly in the ship and on the less frequented coasts of New Zealand for nearly three years. They have been ably and loyally assisted by the seamen and stokers of the *Terra Nova*—worthy fellows, whose by-word has been, "*Play the game.*"

GEOGRAPHY AT THE BRITISH ASSOCIATION.

THE Birmingham Meeting of the British Association opened on Wednesday, September 10, with the delivery of the presidential address by Sir Oliver Lodge.

The Sections began work on the following morning, Section E being conveniently housed in the Midland Institute, close to the Reception Room. Professor Dickson's presidential address appeared in the October issue. He was followed by Dr. W. S. Bruce, whose paper we published last month (pp. 598-9).

Dr. C. A. Hill then described the exploration of Gaping Gyll, a limestone shaft in the south-eastern flank of Ingleborough, by the members of the Yorkshire Ramblers' Club. During the course of the Survey some interesting facts in regard to underground circulation of water have emerged.

Afterwards Mr. I. N. Dracopoli read a paper on his journey across

Southern Jubaland from the Coast to Mt. Kenia, of which the following abstract was supplied :

The immediate object of this journey was to elucidate the hydrographical problems presented by the disappearance of the river Uaso Nyiro into the Lorian Swamp, and to map as much as possible of the unknown country lying between the latter place and the sea, which had not been previously visited by a white man.

A start was made from Kismayu, behind which, and parallel to the coast, lies a low range of sandhills covered with dense bush, above which stand out numerous conifers (*Juniperus procera*). The general slope of the land is from the north-west towards the south-east. The country consists of a series of broad, shallow valleys almost imperceptible to the eye, for the most part overgrown with dense bush and forest, and running in the same general direction. Down the centre of these valleys there are dry river-beds, mostly sandy and densely covered with jungle. As they draw near the sea these valleys and low, rounded ridges disappear, giving place in the north to a level arid plain, which is bounded on the east by the above-mentioned sandhills. In the south-east the country consists of a densely wooded plateau of slight elevation, drained by the rivers Durnford and Arnolé. Near the coast in the Biskayia district are several mangrove swamps infested with the tsetse-fly. The main watershed in Southern Jubaland is that which divides the valley of the Lak Dera from that of the Guranlagga. The latter stream rises in the district of Kurde and flows almost due east, a very different course from that marked on existing maps. There is an important swamp containing a large amount of water in the district of Gulola. The country in the interior alternates between impenetrable acacia scrub and open park-like glades which afford admirable pasturage for cattle, goats, and sheep, but the districts of Rama Gudi and Arroga are arid in the extreme.

Southern Jubaland is inhabited by the Ogaden section of the Darod Somali, and by a fast-diminishing tribe called the Waboni, who live in a state of semi-slavery The Galla, who originally inhabited the country, have been driven southwards and westwards by the victorious Somali. Of the latter, the most important sub-tribes are the Mohammed Zubheir, the Abd Wak, the Aulehan, the Abdulla, and the Maghábul, of which the latter are the only ones to profess friendship for the white man. Near the coast are found some Herti Somali. The wealth of the Somali consists of vast herds of camels, cattle, goats, and sheep, while they depend for food chiefly on milk and ghee. They are strict Mohammedans of the Shujai sect.

The Lorian district differs in many essential points from Southern Jubaland. The Uaso Nyiro River, which rises in the high tableland to the NW. of Mount Kenia, attains its greatest development near the remarkable volcanic plateau called Marti by the natives. It then gradually diminishes in volume, flowing slowly through gently sloping alluvial plains until, on entering the main Lorian swamp, it is scarcely thirty feet broad and two deep. This swamp is roughly oval in shape, with its long axis NW. and SE., and it is very roughly fifty miles in circumference. Much water is here lost by evaporation and percolation. The river, however, emerges in a still further attenuated form, and flows between high banks for five miles before entering a second and smaller swamp. On emerging once again, it gradually dwindles until permanent water ceases in a series of pools known to the natives as Madolé. There is, however, a distinct stream bed that runs in a shallow valley towards the east until Afmadu is reached, when it turns southwards, finally joining the Juba River by means of the Deshek Wama. In an exceptionally wet season the overflow from the Lorian Swamp runs down this stream bed, which is known as the Lak Dera, and into the sea, by way of the Juba

River. The alluvial plains of the Lorian district are remarkably fertile, and are eminently suited for agriculture.

Friday was devoted to a number of papers of local interest. Miss C. A. Simpson spoke of the Upper Basin of the Warwick Avon, pointing out that as a whole it is a longitudinal river, but very few streams in its basin, above Warwick, are actually strike streams, though there is a line of detached valleys at the foot of the escarpment. The main streams flow obliquely across parallel outcrops of rock, ranging from Inferior Oolites in the SE., through bands of lias and triassic marls, to an inlier of Permian sandstone, which is bounded on the NE. by the Nuneaton coalfield.

The northern and western boundary of the Upper Avon basin crosses the uplands of this formation; and on the south the boundary follows the Oolitic escarpment. On the north-east is a very narrow water-parting between the Avon and the Welland—at a comparatively low level. Within the Avon basin, the two almost equally important valleys of the Upper Avon, and of the Leam continued by Rainsbrook Valley, have defined the plateau on which Rugby stands, and so have further dissected the Great Midland valley.

Above Rugby population is comparatively scanty in the Avon valley, and the country districts are tending to decrease in population.

The Upper Avon basin is crossed by several important lines of communication, mainly from SE. to N. or NW. The Watling Street and the Fosse Way are no longer used as main roads, but, in addition to these, certain main roads cross the district. They run mainly across uplands, in contrast to the railways and canals, which often follow valleys.

Professor W. W. Watts discussed the Geography of Shropshire, and Mr. P. E. Martineau The Midland Plateau and its Influence on the English Settlement of Britain. Professor Watts described the geomorphology and physical features of Shropshire, laying especial stress on the fact that the country is divided into two parts by the Severn, whose course within the county is roughly from north-west to south-east. The two parts differ much from each other, for while the southern is a continuation of the Welsh mountains, the northern, save for a few minor prolongations of the ranges, consists of comparatively flat plains, with marsh lands near the rivers. This difference in relief naturally affects the course of the streams, which are straight and rapid in the south, but tortuous to the north, and has much influence on the lines of communication. This was illustrated by a description of the roads, etc., from Roman times onwards, of the distribution of the population, the sites of the towns and villages, and the chief facts connected with agriculture and industries.

Mr. Martineau said that the Plateau lying between the Severn, Avon, and Trent, was of oval form, and its area was almost exactly 1000 square miles. It was part of the Triassic plain of Middle England, and owed its resistance to denudation, and its consequent elevation above the general level of the plain to a rim of hard rocks by which it is enclosed. The outward escarpment was marked on the map by the 300-feet and 400-feet contour lines, which were close together. The 500-feet line was generally close by, and on the southern side there were two points over 1000 feet and many of 900 and 800. The south edge of the plateau was thus not only steep, but of considerable height, and was a serious obstacle to travel, as was shown by the Midland Railways Blackwell bank of 1 in 37·5 and the Great Western's Old Hill bank of 1 in 50. During the fifth and sixth centuries the English (or Anglo-Saxon) settlement of Britain was gradually completed, and the southern edge of the plateau marked the meeting of the two main waves of colonists, who might be described for convenience of reference as the Humber (or English) and the Southampton (or Saxon).

Mr. W. H. Foxhall, in discussing the growth and development of Birmingham, said that the early history of Birmingham was somewhat scantily recorded. Situated on the outskirts of Arden Forest, the hamlet lay considerably off the main Roman roads, but a connecting link between Watling Street and the southern part of the Fosse Way was in close proximity. The determination of the town's position was due wholly to geographical conditions, for peninsulas of high ground lay between the valleys of the Tame, the Rea, and the Cole, and in each valley there existed extensive and almost impassable marsh-land which could only be crossed at certain well-defined points, and here ran the ancient ways which connected the towns of the north and west with those of the south and east. The existence of Birmingham was recorded in *Domesday Book*, but it was seldom mentioned in the historical records of the country previous to the time of the Stuarts; yet during that period the place was growing in size and importance. From the surrounding centres of population already established, trackways converged upon this locality, which thus became a centre—a meeting-place for individuals and a market for the exchange of commodities. In the reign of Henry II. the lord of the manor secured rights to establish a market. Small as this was, compared with our present-day idea of a market, it was sufficient to give the place an advantage over other localities; for the possession of a favoured and established market not only attracted traders, but also necessitated the provision of accommodation for those traders and their wares. It was from such small beginnings that the trade and, later, the local manufactures of the town developed. Practically every writer on old Birmingham referred to the town in this twofold aspect. With the development of the coal and iron industry in South Staffordshire the business of Birmingham increased, for, from being the market town, it became also the distributing centre; so that the one expanded in proportion as the development of the other progressed. The construction of canals which radiated from Birmingham, the improvements to the steam engine, followed later by the introduction of railways, gave an impetus to the town's manufactures and to the industries of the Black Country, and further established Birmingham as a trading and distributing centre. The passing of Acts for the improvement of roads, buildings, markets, etc., in the early part of the nineteenth century, marked the beginning of the transformation of Birmingham from an overgrown market-town into the chief midland commercial centre. Its later progress was due to the industrial skill and inventive genius of its inhabitants, and to that commercial enterprise which had recognised and utilised the advantages that depend on geographical situation.

Mr. Henry Kay contributed a paper on The Black Country and its Borderlands. Its population, he said, approached 1,750,000, and was perhaps denser than that of any area of equal size outside London. The presence of so many people was due to the abundance of valuable minerals near the surface and to the industries which had arisen in consequence. The mineral wealth of the Black Country consisted of unrivalled stores of coal and ironstone, together with limestone, fire-clay, and brick-clay. One seam of coal was ten yards thick. The mines of the central portion were now largely exhausted, or flooded by water, and fresh supplies were being developed east and west. Limestone was got from underground workings at Dudley, the Wren's Nest, and Walsall, and magnificent caverns had been excavated at those places. Fire-clay was obtained from the Stour Valley, and brick-clay was abundant everywhere.

On Monday morning Mr. C. B. Fawcett read a paper on the Expansion of the Fiord Peoples and its Geographical Conditions, taking Norway, the North-west Coast of North America and Magellanes as types. We quote his abstract as follows :

Each of these is a narrow and fragmentary strip of coast backed by barren highland, with only small and scattered patches of lowland. Their climates are all of the same type, wet, cool, and equable, unfavourable to agriculture, but with open sea at all seasons. The peoples depend mainly on the sea for their food and means of communication, and therefore their expansion has been along the waterways.

Each fjord coast is on the mountainous western edge of a continent in high latitudes, but in other respects their positions are very different. Magallanes is at an end of the inhabited earth, with desert coasts to the north, and a wide ocean on all other sides. The NW. Coast is exposed to Asiatic and Polynesian influences, and is less completely shut off from its hinderland than the other fjordlands. The south of Norway borders the "Narrow Seas," and thence had communication with Europe and the Mediterranean.

In each of these lands there has been a mingling of races, but in each the uniformity of the local conditions and the separation from other peoples has made the inhabitants one people in their modes of life. Their relative social development has been mainly influenced by their skill in navigation. This directly determined the range and security of their food supply, and the growth of population and organisation, and therefore the power of expansion. The development of navigation was local, and limited, in Magallanes. On the NW. Coast only dug-out vessels were employed. The Norse ships were influenced by those of Europe and the Mediterranean, and here navigation very early reached a high stage of development.

The expansion of the NW. Amerinds was limited to the region bounded by the mountains and the ocean east and west, the desert, and Arctic coasts south and north. The only practicable route was up the rivers ; and that demanded a complete change in their mode of life. The limits to expansion from Magallanes were similar, but even narrower.

The Norsemen were not checked by any such barriers, and they spread along three chief routes: (1) NE. to Finmark and the White Sea, (2) westward to Iceland and Greenland, and (3) SW. to the British Isles. The fourth route, to the SE. was blocked by the Danes and Swedes. The form of the expansion in each case was determined by the social condition of the people visited, and the causes of the early success and later rapid decline of the Viking Power are to be found in the state of affairs in Norway and Western Europe at the time and the limited resources of Norway.

After the reading of the paper Professor Myres delivered an interesting speech, in the course of which he referred especially to the social conditions prevailing among fiord peoples.

Dr. W. S. Bruce gave an address on Spitsbergen economically considered, in which he urged that the time had come for Britain to annex Spitsbergen in order to protect the interests of British subjects there. Mr. Alan G. Ogilvie discussed the Physical Geography of the Inverness Firth, in a paper which we shall publish here later. Mr. Ogilvie proposes to continue his work in this region, and received a grant from the funds of the Association for the purpose.

Subsequently Professor J. W. Gregory gave an account, illustrated by a series of very fine lantern slides, of Australia, in view of the British Association's visit to that continent next year.

On Tuesday the Section divided into two parts, one uniting with Section A (Mathematical and Physical Science) to hear various papers connected with Survey work, while the other took part in a discussion on Natural Regions opened by Professor Herbertson, to whom is due the original conception of Natural Regions as applied to geography.

Professor A. J. Herbertson began the proceedings by reading a paper of which the following abstract was supplied:

The idea of Natural Regions is not a new one. It has been used, consciously or unconsciously, by every traveller and geographical student with insight. What we are attempting now, is (i) to divide the World into its Natural Regions, taking into account all the elements composing them, (ii) to recognise and group Natural Regions into different classes and orders, and (iii) to trace the consequences of the recognition of Natural Regions as entities.

I. *The Natural Region.*

It is unnecessary to review the attempts to divide the Earth's surface into zones or regions each possessing some special property—land, water, igneous rocks, little or no rainfall, forests, a roundheaded population, etc. etc. In this raw material the geographer finds certain relations between different elements, certain laws of combination. The elements and the combinations are far more complex than those of organic chemistry, but this complexity is no reason for not applying scientific methods to their investigation, nor for doubting that substantial results can be gained by their use. It is a reason for not consigning the study of Geography to the least experienced, but for giving it to the man who knows most. The only apology the geographer has to make is not for his subject but for his ignorance.

The Natural Region is a vital unit as well as a physical one, a symbiosis on a vast scale. It is more than an association of plants, or of animals, or of men. It is a symbiotic association of all these, indissolubly bound up with certain structures and forms of the land, possessing a definite water circulation, and subjected to a seasonal climatic rhythm. As each element in a region has its own history, and as each varies in its rate of change, so the evolution of the region is highly complex.

II. *Types and Orders of Natural Regions.*

The advantage of classifying Natural Regions into types is obvious. Some of the larger ones with common morphological or climatic characters have long been recognised, *e.g.*, Mountainous Lands, Plains, Monsoon Lands, Deserts, the Mediterranean Type of Region, etc. The systematic analysis and classification of all types are more recent efforts of geographers, who encounter two difficulties at the outset. One of these is to fix the limit of a Natural region: where does the Sahara begin and the Sudan end? The other is to distinguish properly between different orders or classes of natural regions: are there to be two, or three, or four, or more? For instance, is it sufficient to divide the Monsoon Lands of Asia into great river basins and each of them again into their minor basins, and these again into valleys and plains? Are the classes the same in rainy and rainless lands, in mountains and plains? Can we arrive at any grouping of order of natural regions as little objectionable as the biologist's organism, organ, tissue, cell?

Let us consider some of the simpler natural regions. In the Upper Thames Plain there are belts of (*a*) flowing water bordered by (*b*) meadow flood plains, (*c*) land rising above the flood level. Such belts are also found in the valleys of the adjacent Cotswolds; but here all three are more complex, (*a*) the flowing water is not confined to the surface but permeates (*c*) which rises in steep banks above the irregular (*b*) flood plains and forms a gently undulating and sloping surface very different from that of (*c*) in the Plain. Further, while the varieties of (*c*) in the Plain are merely (1) clay and (2) gravel-capped clay, the varieties of (*c*) in the Cotswolds are (*a*) the relatively flat land, (1) very fertile cornbrash, (2) less fertile belts of Oolitic limestone, (3) cappings of gravel, (4) cappings of clay, and (*b*) the steep slopes (1) of the valleys, and (2) of the escarpment. There are also considerable differences in climate differentiating the Plain from the Scarped Ridges, and the climatic conditions within the natural region itself vary much more in the hilly region than they do in the plain. The Upper Thames Vale is then clearly a different natural region from the Cotswold Scarped Ridge. There are, however, vales and vales, scarped ridges and scarped ridges. The Upper Thames Vale is a much simpler variety of the same species of natural region than the Vale of Aylesbury, and the chalk-scarped ridges are a variety of the species differing from the oolitic limestone variety.

This grouping according to orographical conditions has its advantages, but others suggest themselves—more particularly the division into regions with a common drainage, *e.g.* the basins of Cherwell, Evenlode, Windrush, etc. Probably it is best to combine the two and see in the Cotswolds each incised river-valley as a sub-division of the natural region and the relatively flat land between as another sub-division. In plains the land between the rivers becomes by far the largest and most important sub-division.

We might look on the geological structure as the tissue of a region; but this geological structure does not suffice for a complete natural region. The water circulation must also be taken into account and the surface forms it has carved. In a mountainous area the valleys seem the natural units, divided by crests; in the plains the land between the rivers is the dominant element, and the rivers divide the natural regions. There are many intermediate forms between these extremes of mountain and plain, of which the Cotswolds are a good example.

Individual valleys, plateaus, and plains form simple geographical units.[1] Groups of different varieties of one or more of these constitute a new and more complex geographical order. These in their turn may be combined to form something more complex, such as the English Scarplands, the West Highlands, etc. Yet more complex are associations of different natural regions such as form the British Isles or Iberia or Scandinavia.

There are therefore various orders of natural region, which we may roughly compare with the species, genera, orders, etc., of the biologist. For these we may have no definite names. Tentatively we may speak of species, genera, orders, classes of natural regions, even though this suggests biological analogies, which may mislead the beginner. From this point of view valley, plain, and plateau are chorographical species each with many varieties. These themselves may be grouped into different genera of mountainous-, plateau-, and plain-lands, and these again into different orders of country, compounded of different combinations of these, which finally grouped together compose the continents.

This morphological or topographical classification is not enough. Climate must be taken into account in the larger divisions. The need for this is most easily recognised in the vast plains, where vegetation affords an index of climatic and edaphic influence. The botanist's classification into plant societies, associations and formations suggests new groupings, which have to be considered in making natural regions.

Beginning with the continents we might recognise major natural regions, each with its own climate rhythm, and this acting on different genera of land forms. These topographical genera enable us to determine the geographical genera, which may themselves be sub-divided into different geographical species, each of which may have a number of varieties.

I venture to suggest as a convenient nomenclature for the geographical natural regions that we should consider each continent as composed of countries containing different regions divisible into districts each with its various localities or neighbourhoods. This would not interfere with such political terms as empire, state, province, county and parish.

Some of the attempts to divide the world into natural regions will be reviewed, if time permits.

Hitherto nothing has been said about Man. The different natural regions exist, whether he is part of them or not. Even in the most complex class, the continent, he may not live at all, *e.g.* in Antarctica. In some natural regions he counts for no more than other animals. In others he has so profoundly altered the surface that it is necessary to consider him and his works in any classification. This is not merely the case in such districts as the Fens or the Lancashire coal-field, but even in such complex countries as China or those which border the Norland Seas of Europe, where much of the original forest has been converted into cultivated land. The concentration of man in the cities obviously alters a locality.

---

[1] One other class—the cone, usually volcanic—may be mentioned. Only rainy regions are considered for the moment.

We must also differentiate between two natural regions originally of the same type, in one of which man has settled in farms, villages and towns and added a new characteristic to it; and we must also distinguish between a natural region in its present condition from the same region in an earlier state when man did not play the part in it that he does at present, *e.g.* Egypt or Mesopotamia, for we have also to consider reversion.

Men in the natural region may be compared with nerve-cells in an animal. Some have a highly developed, others a very simple undifferentiated nervous (social) system.

III. *The Significance of Natural Regions.*

There are many important consequences which cannot be examined in detail now. It is, however, desirable to point out that while it is possible to study men in families, races, etc., without taking into account that they form but a part of a more comprehensive entity, to do so is to deal with less than the whole problem. It does not suffice to give once and for all a brief description of the physical conditions of a country as a prelude to the account of the history of its inhabitants. The history is incomplete which does not consider both together at all stages. Neither is passive, neither remains unchanged when man becomes important enough to have a recorded history. The changes are not usually sharp and dramatic as in the American Prairies or in South Wales, but slow and subtle as in most countries of Europe. We are accustomed to think of the changes in the men, we rarely consider the changes in the region, which are also of vital significance.

The entity higher than the individual is not the family nor the race, nor any association of men alone, but the more complex association of the Natural Region. This has always been so, and until we recognise the fact neither the history of man nor his relation to the universe, nor even his very pressing present problems, economic or political, can be properly understood. This is much too vast a subject to examine in detail on the present occasion; but I should like to add a note of warning to those who imagine that any such conclusions do more than show that the problems of the universe are somewhat more complex than is commonly supposed. It is not meant that the geographer can solve them; only that they cannot be solved without taking these geographical considerations into account.

The discussion was taken part in by Mr. Geo. G. Chisholm, Dr. M. I. Newbigin, Rev. W. J. Barton, Mr. Alan Ogilvie and others, but was felt on the whole to be a little disappointing. The audience was small throughout, and there was a lack of definiteness in the terms used, which was felt by many to keep the discussion in the air rather than on firm ground.

After the discussion Professor Herbertson showed, on behalf of Mr. W. G. Kendrew, a series of slides illustrating the rainfall of China, and incorporating a mass of material collected and arranged by Father Louis Froc. This new material has rendered necessary some modifications in the conclusions previously arrived at in regard to the seasonal precipitation.

While one half of the Section was thus engaged, another part was listening to the following papers in the joint meeting with Section A. Captain Winterbotham discussed the Accuracy of the Principal Triangulation of the United Kingdom, in a paper in which he described some recent measurements undertaken in the Lossiemouth region, with the object of determining whether any considerable error was present. The results of these measurements have been very satisfactory, and a full report will be published by the Ordnance Survey. Captain Lyons then spoke on the need of a more precise terminology in Higher Surveying, and two papers on Survey Work in Egypt were communicated.

Two Reports of Committees were presented, one on Geographical Teaching in Scotland, which we published in our last issue, and another on School Atlases. In connection with the latter committee a short address was given by Colonel Close in the Committee Room on Monday, followed by a discussion, and an exhibition of foreign school atlases was also arranged by Dr. Herbertson. The School Atlas Committee asked for re-appointment, and also for a grant to prepare specimen sheets of maps. The draft report and the discussion both showed not only that improvement in school atlases is highly desirable, but that many difficult questions of detail arise in connection with cost, and so forth. The final report of the Committee will be looked forward to with interest.

The meeting as a whole was a great success, and was largely attended. Next year the Association is to visit Australia, as has been already announced here.

THE GEOGRAPHY OF THE ATLANTIC OCEAN.[1]

(With Illustrations.)

THIS is a very fine book, full of varied learning, and written with singular grace and skill. Dr. Schott has tried to write for the traveller, for the scientific student of geography, and not least for the mariner, whether on his hurried passage across the great Atlantic ferry, or on his long voyage out and home from the Pacific, aboard of some splendid full-rigged ship, relic of the old fleets of sail. He has indeed found the art of weaving together the knowledge of sailor, of student, of trader, of man of the world; of describing complicated things with simplicity and ease; of pressing much learning into little space; of giving us, in short, within this small and modest volume, a comprehensive and adequate account of the great ocean with which it deals.

In his first chapter, Dr. Schott gives us a brief review of the older and the newer history of Atlantic exploration. Passing quickly over the old voyages of priest and viking, of which Nansen tells us so much, he reminds us that Columbus did not sail from Cadiz into an utterly unknown sea; the Canaries had been visited two hundred years before

---

[1] *Geographie des Atlantischen Ozeans.* Von Dr. Gerhard Schott. 28 Charts and 90 Text-figures. Hamburg: C. Boysen, 1912. *Price* 23 *M.*

We are indebted to the publishers for the loan of the three blocks used in illustrating this article.—ED. *S.G.M.*

his day, and for a hundred and fifty years, not only those islands, but also Madeira, and even the Azores, had been embodied in the geographical knowledge of the age. Among the many old maps which Dr. Schott reproduces, are Wagner's restoration of the lost chart of Toscanelli (1474), which shows "Antilia" and "Zipangu," lying in a western ocean, on whose farther side was far "Cathay"; the first chart of America (1500) by Juan de la Cosa, the Basque, Columbus's fellow-navigator; and the wonderful Atlantic chart of 1529, by Diego Ribero, in which, just thirty-six years after Columbus's first home-coming, we find well depicted the whole Atlantic coastline, from Norway to the Promontorium Bonae Spei, from

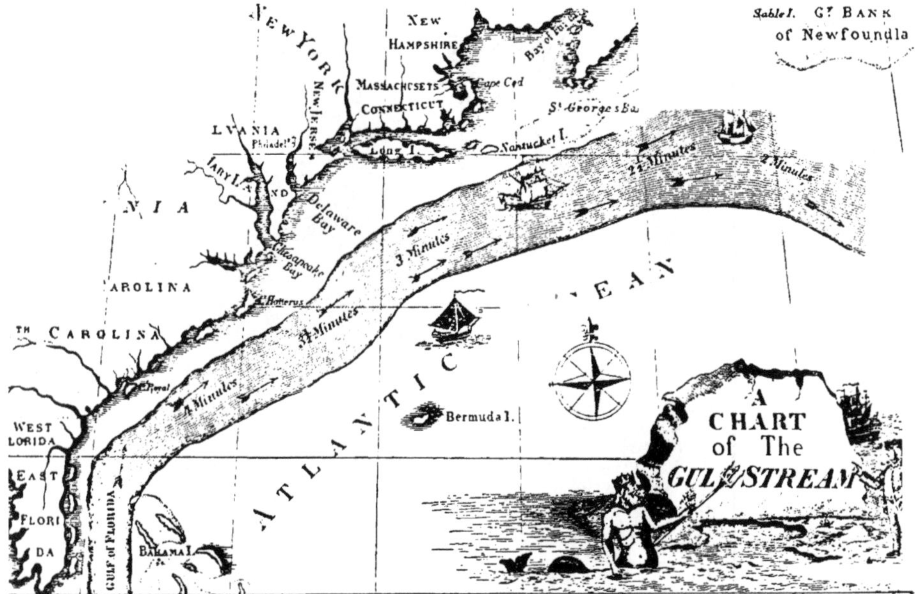

Benjamin Franklin's Map of the Gulf Stream.

Labrador to the Tiera de Fernañ de Magelläes. Dr. Schott adds an instructive chart showing the routes of the great navigators of the fifteenth and sixteenth centuries, and the stretches of Atlantic coastline which the several great exploring nations had contributed to the map.

Next, Dr. Schott traces out for us the growth of knowledge from the sixteenth century onwards, as it is shown forth in the spirit and the work of the mariner and of the scientific geographer. This "dualism," as Dr. Schott calls it, is already apparent when we compare, for instance, Juan de la Cosa's chart of 1500 with the great geographer Ortelius' map of 1570; and so, through the succeeding centuries, the same two streams of work and thought flowed on, until at last Maury drew them together in his monumental book: for it was he who showed first, as Dr. Schott now shows us again, how much the seaman has to gain from the theoretical knowledge of the oceanographer. Here, then, we have shown us the first chart of Atlantic currents, drawn by that strange old Encyclopaedist of the Jesuits, Athanasius Kircher, in 1678; Halley's

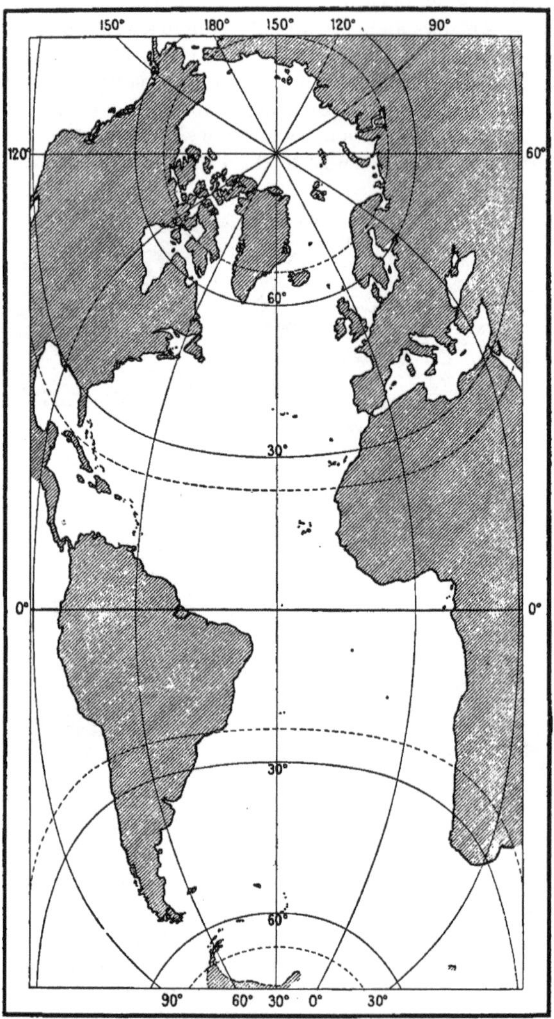

The Atlantic Ocean from Bering Strait to the Antarctic Region; Lambert's Equivalent Cylindrical Projection 1:160,000,000. (*After a Map by M. Grolls.*)

first chart of the North Atlantic winds, published ten years later; and Benjamin Franklin's Gulf-stream chart of 1770. And so we come to Maury, to that "new chapter," the chapter of Marine Meteorology, which he "opened in the volume of Nature," and on which are based those great epitomes of meteorological and hydrographical knowledge which constitute the "Sailing Directions" for the sea-faring peoples of the world. Dr. Schott's sketch of historical geography closes with an account of the great exploring expeditions of our own and the last generation, beginning with that of the *Challenger*, the value of which is "unsurpassed even to the present day."

The next brief chapter is wholly geographical. Among its contents may be mentioned an interesting series of charts of the Atlantic, drawn under various projections, and a reproduction of a curious map by Jean de Windt, showing the zones of distance from the nearest land. Here we see that only one small spot of ocean, half-way between the Bermudas and the Cape Verdes, lies as much as 2000 kilometres, or say 1200 miles, from adjacent land.

We pass to a geological chapter, full of interesting matter, where Dr. Schott describes the various speculations of Frech, Neumayr and others, as to the probable extent and contour of the Atlantic, so far back as Carboniferous and Jurassic times. We come much nearer to the region of certainty in the Pliocene period, when the Wyville Thomson ridge extended its great solid barrier from the Scottish coast to Greenland, and isolated from the tempering warmth of the Atlantic a great Arctic sea. The great changes within historic times upon the Dutch and North German coasts are here also touched upon. And lastly, a comprehensive chart shows, for the whole Atlantic coastline, the extensive regions where the land is known to be gaining on the sea, and the comparatively scanty portions in the North Sea, and in the Eastern United States, where the continental level is apparently sinking, and where the sea gains upon the land.

The next three chapters deal with the soundings of the Atlantic basin, with the physical properties, colour, temperature, and salinity of its waters, and with the various climates and other meteorological phenomena. In the first of these chapters we find a capital chart, based on the work of Sir John Murray and of Dr. E. Philippi, of the bottom deposits of the Atlantic, showing the wide distribution of Globigerina ooze over the whole basin, the manner in which it extends far northward with the branches of the warm "Gulf-stream," the curiously local patches of Pteropod ooze, and the great extent of the central patches of Red Clay. The Telegraph Plateau is carefully described; and a very ingenious illustration shows us a great modern ocean liner, in relation first to the shallow waters of the North Sea, whose depth is but a fraction of the vessel's length: again to the depth of the English Channel, which scarcely equals it; and lastly to the depths of the Atlantic, which after all, at least on the passage to New York, are only equal to some twenty times the depth of the vessel. The Atlantic is very deep, but a great ship floating on its surface is (comparing the length of the one with the depth of the other) very much like a cork floating in a wash-tub. Here also Dr. Schott gives us a picturesque account of the Atlantic banks and islands, and showing us, by the way, a view of Rockall, reminds us that this is but a tiny rock, a few hundred square yards in surface, but that it is the sole visible peak of an immense bank, some 2500 square miles of which are within about 100 fathoms depth.

Dr. Schott, as he has already shown us in the reports of the Valdivia Expedition, is a past-master in the art of drawing hydrographical charts; and we have in this book a whole series of admirable coloured maps, in which are set forth the distribution of the various tints of blue and green in the surface waters of the ocean, and of its temperature, salinity, and density. Furthermore, we are shown charts of the air-temperatures, both mean and seasonal, of the difference between air and water temperature, of barometric pressures and of winds, of cloud and fog, of rainfall, and of the course of the great ocean-currents. Were it for these charts alone, we should be heartily grateful for the book.

One interesting diagram shows the barograph records of a swift mail-steamer out and home from New York, in stormy weather. On the outward passage the barograph shows a succession of strong maxima and minima: on the homeward route the curve is smooth and all but level. For the ship had been running against the track of the barometric minima in the one case, and had been keeping pace with them on the other.

A chapter follows on the natural history of the ocean, dealing mainly with the distribution of its birds and fishes, its seals and whales. In one chart the ránge and habitat of the sea-birds is shown, for instance the southern limits of the eider and the auks and guillemots, the range of the commoner gulls, the limits of the penguins, the albatross and the frigate birds. In another chart we have the great northern fisheries of herring, cod and sardine, the summer and winter limits of the flying fish, the mid-Atlantic breeding places of the eel, the boundaries of the Sargasso Sea, and the distribution of coral reefs in the West Indian and Brazilian regions. A third chart, based on the work of Dr. J. Hjort and Mr. A. H. Clark, shows us the distribution of the seals and walrus, the fur seals and the whales. A naturalist may be inclined to carp a little at a few statements and some omissions in this chapter. For instance, Dr. Guldberg's papers on the migrations of the various whales might well have been referred to. The steady catch of sperm-whales, in not inconsiderable numbers, by the Scoto-Norwegian whale-fishers off Rockall and still nearer to the Scottish coast, shows the statement to be inaccurate that the northern limit of that species lies "with very rare exceptions" in 40° N. lat.

The last chapter of the book deals with the commercial geography of the Atlantic, the trade-routes of steamship and sailing ship, the submarine cable-lines, and briefly and finally the network of wireless telegraphy. As we might expect and count upon from a member of the Deutsche Seewarte, where so much care is taken to help and instruct the captains of the German Mercantile Marine, this chapter is admirably done, and in particular that section of it which deals with the courses of the great sailing ships. Dr. Schott reminds us that, narrowed as is the field for these fine vessels, they still play an important part especially in France and Germany, and especially in the transport of wheat and nitre from the Pacific Coast. But few and narrow are the ocean-tracks on which nowadays the sailing ship is found, and her last great highway is that old one round the Horn. Dr. Schott gives us a striking diagram to show how, and why, captains outward-bound now give a wide berth to the Horn, taking a far southerly course into seas that were feared and avoided a generation ago: the simple reason being that the great cyclonic systems are constantly passing to the southward of the Horn, where their direction of rotation is clock-wise, or the opposite of that in the Northern hemisphere. So it is on their far southern border that the skipper hopes to sail westward; to come home again, still with a favouring wind, close under the Horn, just where the older mariners, outward bound and beating to the westward, fought for days and weeks against the baffling westerly winds (see figure below).

A pair of charts is then given to show, by means of contour lines or "isochrones," the mean number of days spent by sailing ships on their

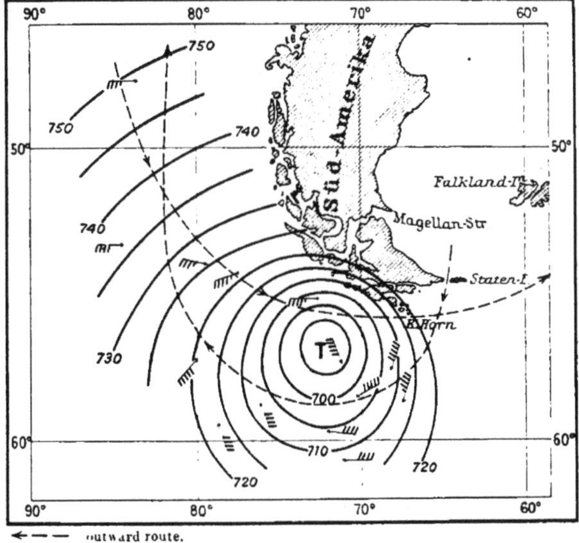

Pressure and wind in the Cape Horn Region on April 21, 1896. The arrows show the mean direction of sailing-ship courses on the outward and return journey.

outward and homeward passage between the Lizard and distant parts of the world. The curious fact appears that the homeward passage is on the average considerably the longer, the difference amounting to about five days, or say eight per cent., in a voyage to the Cape of Good Hope or Cape Horn. While indicating some meteorological reasons connected with this discrepancy, Dr. Schott does not let us forget that the greater part of the difference is to be accounted for by the simple fact that the ship goes out clean and comes home foul. The chart of sailing-routes is full of interest. One must not only know something of Great-circle Sailing, but a good deal of the inner mysteries of Meteorology, if one is to comprehend the many curious facts which it reveals. Only then may one hope to understand, for instance, why two ships, bound for the Horn and for Cape Town, should begin by keeping the same course all the way to the neighbourhood of Pernambuco or Bahia.

But we must take leave of this most instructive and delightful book. The great shipping firms of Hamburg, as the author tells us, have contributed generously towards its publication, and in so doing they have given yet another proof of the sympathy between commerce and learning in that great Free City of Trade. We feel most heartily that the book does credit not only to its author, but also to the great Institute, the Deutsche Seewarte, to which he has the honour to belong.

D'ARCY W. THOMPSON.

PROCEEDINGS OF THE ROYAL SCOTTISH GEOGRAPHICAL SOCIETY.

AT a meeting of Council held on Tuesday, 18th November, the following ladies and gentlemen were elected Ordinary Members of the Society :—

The Rt. Hon. Lord Ruthven.
David M. Thomson, South Nigeria.
Sir Alex. Lyon, Aberdeen.
Charles D. Gairdner.
Robert Goudie.
Thomas Wilson.
James Kay Small.
Thomas C. Bramah.
Donald Macdonald.
Mrs. Margaret Cathcart.
Rev. John Martin.
James A. Aikman.
Harvey M. Anthony.
Joseph G. M. Bannerman.
Miss I. W. Cree.
Miss E. P. Hogg.
J. M. Anderson.
Miss Louise Buchanan.
James Ross.
Peter Howie, M.B., C.M.
Alexander Anderson.
John Boyd, Jr., C.A.
William Arnot.
A. W. M. Beveridge.
Miss Sarah A. Cowan.
Miss J. E. Gray-Buchanan.
John Tait.
Samuel R. Skilling, M.A.
James Adam.
C. P. Ross.
Admiral A. W. Chisholm Batten, M.V.O.
W. J. Watson, LL.D.
Robert Abernethy, M.D., F.R.C.P.E.
William Cowan, W.S.
Arthur H. H. Sinclair, M.D., F.R.C.P.E.
Mrs. W. M. Smith.
Mrs. Wm. Gibson.
Miss S. M'L. M'Vean.
Robert R. Mathers, M.D.
Mrs. Greig.
P. Neill Fraser.
Mrs. E. M. Strang.
James Shaw Nowery.
John Stewart.
C. Weddall Holroyd.
Miss M. Boog.
Miss C. K. Wilkie.
George Dickson.
Robert F. Alexander.
William Menelaws.
Miss Mary Dakers.
C. D. Geddes, M.Inst. C.E.
J. Dall, M.A.
Wm. F. Scrimgeour.
Charles C. Duncan.
Gibson C. Ferrier.
Mrs. Gibson C. Ferrier.
William Parker.
J. Noel Johnston.
James Milne, J.P.
W. Leslie Mackenzie, M.D., D.P.H., F.R.C.P.E., LL.D.
James B. Goold.
Thomas E. Watson.
Mrs. Donald.
G. M. Wood, W.S.
Alexander M'Dougall, M.A., B.Sc.
Miss A. Taylor.
Miss Gladys M. Younger.
Thomas Wilson, Jr.
James W. Telford.
James Tuke.
Mrs. Bailey-Duncan.
Thomas Law.
John S. Galbraith, B.L., J.P.
Miss Grace Ramsay.
Joseph Williams.
Mrs. Albert Harvey.
R. C. Greig.
Miss Janet Fraser.
W. L. Pattullo.
W. Frain.
John Nicoll.
Frederick Thomson, J.P.
John Rose.
G. Andrew, M.A., F.R.S.E.
J. A. Briggs.
David Hyne.
G. F. Whyte, M.B., C.M.
James Ross.
W. B. Dickie, Solicitor.
James Young, W.S.
Andrew Dunlop.
Miss E. M. Edwards.
Harry G. Crowden, M.A.
Mrs. J. G. Christie.
J. F. Flint.
J. B. Clark.
A. C. Halley.
Charles Parker.
T. L. Miller.
Edward Cowan.
Robert C. Thomson.
William Henry.
Rev. Alexander Macdonald, M.A., B.D.
J. C. Robertson, C.A.
George Bonar.
John Anderson.
John M. Nairn.
Prof. J. Cossar Ewart, M.D., F.R.S.E.
Edwin A. Watson.
K. Johnston, M.B., C.M.
William Kinnear, M.D.
J. Ernest Cox.
James E. Erskine.

W. Clarke Read.
J. C. Thompson, B.L.
William Gibson.
William Byers, M.A.
Andrew F. Donald.
R. W. Pentland.
Miss E. M. S. Jackson.
John Y. Braidwood.
Miss M. S. Brodie.
J. N. Ferguson.
O. A. C. Alexander.
Miss Annie B. Crow.
J. C. Robertson, J.P.
A. J. Brown.
J. P. Cuthbert.
Alexander Gilroy.
J. F. Anderson.
William J. Brown.
Norman Anderson, M.D., C.M.
John Murray, M.A.
Douglas D. Taylor.
D. K. Symington.
Mrs. J. M. Bell.
Alexander Guthrie.
George Ure Thomson.
George C. Spence.
Charles Corsar.
A. M. Henderson.
Mrs. Richardson.
Albert O. Knoblauch.
Robert W. Bow.
Charles Ogilvy.
John Corrie.
J. G. Gerrard.
W. M. Anderson.
John Macdonald.

The following were elected "Teacher Associate" Members.

Miss M. S. Cormack.
Miss Helen Littlejohn.
Miss Helen Clark.
Miss Chrissy Gray.
Miss Margaret M. Clark.
Miss G. A. Polgreen.
Miss Charlotte Buchan.
Miss J. R. Stewart.
Miss Janet M. Rae.
Miss E. J. Lindsay.
William P. Reid.
Miss Annie H. Dickie.
Miss Margaret R. Cruden.
Miss J. P. Morrison.
R. M. Munro, M.A.
Miss Ann A. Rae.
Miss H. A. Hamilton.
Miss Margaret Hay.
Miss E. Crawford.
Miss J. Crawford.
Miss C. R. Allan.
John Andrew.
Miss Jean Ferguson.
Miss M. B. Wilson.
Miss H. J. Kinghorn.
Miss E. G. Young.
Miss A. Scrymgeour.
John Thomson.
Miss A. H. Watson.
Miss A. Tait Sloan.
Miss A. B. Irving.
John S. Gilchrist, F.E.I.S.
William Baird, M.A.
William Lee, M.A.
Miss Kate Fraser.
Miss Annie Kinloch.
Miss Margaret Johnston.
Miss I. M. Ruxton.
Miss I. Lambert.
Miss C. C. Ritchie.
Miss E. Jenkins.
Miss R. M. H. Hutton.
Miss I. Donald.
Miss M. T. M'Kissock.
Miss Jeanie Ferguson
Miss E. E. Fletcher.
Miss E. S. Hamilton.
Thomas Duncan, M.A.
James B. Richardson.
James Gall, M.A.
Miss M. MacKail.
E. Wakefield.
Miss M. W. Crear.
Miss J. I. Cochrane.
Miss E. M. Pickett.
Miss I. L. Morrison.
William S. Anderson
H. F. Hunter.
Miss M. M. H. Cram.
Miss N. P. Hogarth.
Miss M. R. Smith.

Diploma of Fellowship.

The ordinary Diploma of Fellowship was conferred on David M. Thomson, Northern Nigeria; J. K. Small, Loanhead: Charles Ogilvy, Edinburgh; and Miss G. H. Picken, Glasgow, subject to the prescribed conditions being complied with.

Dr. W. S. Bruce.

Dr. W. S. Bruce, Leader of the Scottish National Antarctic Expedition, 1902-4, was made an Honorary Life Member of the Society.

The Annual Business Meeting.

The Annual Business Meeting was held on the 18th November in the Society's Hall, the Rt. Hon. the Earl of Stair, President of the Society, being in the Chair.

The Council of the Society submitted its report for session 1912-1913, the twenty-ninth session of the Society. This Report appears on pp. 672-674.

On the motion of Mr. H. M. Rush, seconded by Mr. D. MacRitchie, the report was adopted.

The list of office-bearers recommended by the Council for election for the current session was also submitted to the meeting, and on the motion of Mr. T. Clendinnen, seconded by Mr. W. Lee, the gentlemen recommended were duly elected.

The following are the appointments so made:—

The Rt. Hon. the Earl of Stair was re-appointed President. Dr. Paul Rottenburg, Glasgow, and Dr. James Burgess, C.I.E., were elected Vice-Presidents.

The Very Rev. G. Adam Smith, D.D., LL.D., Principal of Aberdeen University, was elected Chairman of the Aberdeen Centre.

Mr. J. C. Buist was elected Chairman of the Dundee Centre.

The following members of Council who retire by rotation were re-elected:—W. B. Blaikie, LL.D.; H. B. Finlay; D. F. Lowe, LL.D.; H. M. Cadell, B.Sc.; Major-General P. C. Dalmahoy; Colonel T. Cadell, V.C.; Sir James A. Russell, LL.D.; William B. Wilson, W.S. (Edinburgh).

The following members of the Society were elected to fill vacancies on the Council:—James Cornwall, John Harrison, Charles E. Price, M.P., and W. J. Watson, LL.D. (*Edinburgh*); M. Pearce Campbell (*Glasgow*); J. Bentley Philip, M.A. (*Aberdeen*); R. Polack, LL.D. (*Dundee*).

Lectures in December.

Mr. F. W. Christian, B.A. (Oxon.), will lecture before the Aberdeen, Dundee, Glasgow, and Edinburgh Centres on the 2nd, 3rd, 4th, and 5th respectively. The subject of his address will be "New Zealand: Its Past History, Present Prosperity, and Future Commercial Importance." The lecture will be illustrated with lantern views.

The Christmas lecture will be delivered by Mr. H. Charles Woods, F.R.G.S., in Aberdeen on the 15th, Edinburgh on the 16th, Glasgow on the 17th, and Dundee on the 18th December. The title of his lecture will be "Travels in the Balkan Peninsula," and it will be illustrated with lantern views.

GEOGRAPHICAL NOTES.

Europe.

The Oil-shale of Raasay and Skye.—We published here on p. 548 a note recording the discovery of oil-shale in the island of Skye. The Director of the Geological Survey of Great Britain communicates to a recent issue of *Nature* an authoritative account of the deposits, written by Dr. C. W. Lee, who made the discovery, and illustrated by the accompanying figures, which we are enabled to reproduce by the kind permission of the editor of *Nature*.

The sketch-map shows the outcrop of the shale in the islands of Raasay and Skye. In Raasay, it will be noted, the bed appears at the surface between Dùn Caan and the boundary fault which throws Mesozoic rocks against the Torridonian. The distance between these points is three miles, and the field has an average width of seven-eighths of a mile, diminishing towards the south. The beds are not folded, and have a dip of about 10 degrees to the west.

In Skye, in the Portree region, a very extensive field once existed now largely destroyed by the contact action of intrusive rocks. The outcrop has been traced from Ollach, five miles south of Portree, to the

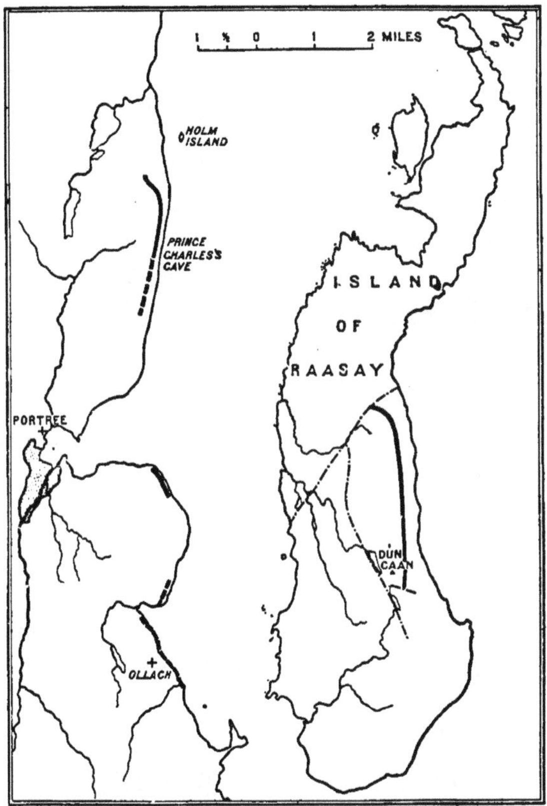

Fig. 1.—Sketch-map showing the outcrop of the Oil Shale in Skye and Raasay.
- - - - - - - Oil Shale, where burnt by contact with igneous rocks.
——— Oil Shale, unaltered by igneous rocks.
—·—·—·— Faults.

Holm burn, five miles north of Portree; but, as the map shows, north of Portree only the part north of Prince Charles's cave has escaped the action of the igneous rocks. South of Portree also, only a small part has escaped alteration. Inland no exposure of the oil-shale horizon has been found, the strata being everywhere covered by higher beds; but, since the dip is low, the shale should be within reach for some distance inland. The oil-shale seam appears to vary in thickness from seven to ten feet. The analyses so far made have been of weathered portions of the shale, but indicate that the deposit is worth investigation from the commercial point

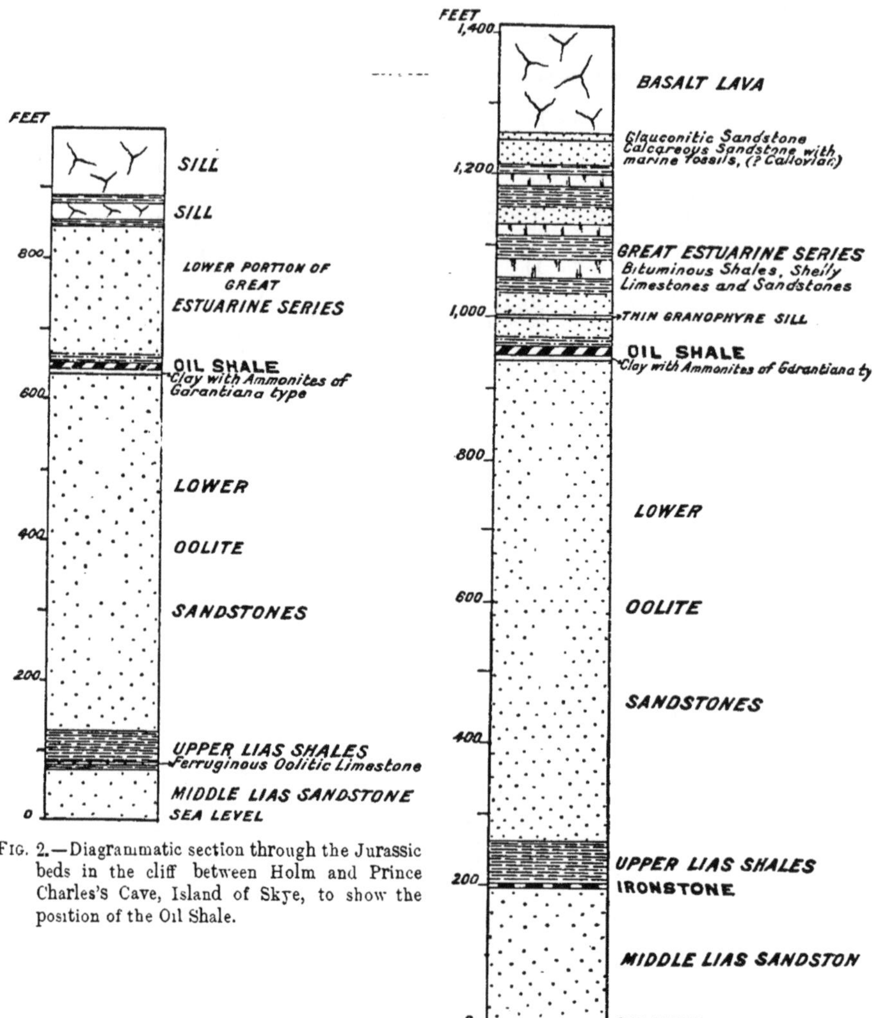

FIG. 2.—Diagrammatic section through the Jurassic beds in the cliff between Holm and Prince Charles's Cave, Island of Skye, to show the position of the Oil Shale.

FIG. 3.—Diagrammatic section to show the relati of the Jurassic rocks below Dùn Caan, in Island of Raasay, and the position of the Shale.

of view. The two diagrammatic sections show the position of the oil-shale beds in the Jurassic beds of Skye and Raasay.

Asia.

Exploration of the Brahmaputra.—Captains Bailey and Morshead have returned to India after an arduous journey of exploration in the Sanpo region. They have proved conclusively that the Sanpo and the Brahmaputra are the same river, but at the same time find that the supposed falls of the Sanpo do not exist. In their journey through the inaccessible Abor country the two officers were without an escort, and were accompanied by ten coolies only. Details in regard to this important journey will be awaited with much interest; it has finally settled one of the most actively debated questions of Asiatic geography.

Africa.

The Climate of the Mariout District.—In connection with the much-debated question of the possibility of notable climatic change occurring within the human period, a paper by Mr. A. L. P. Weedon in *The Cairo Scientific Journal* is of much interest. Mr. Weedon has studied the Mariout district, a strip of land lying west of Alexandria, which in ancient times was famous for its fertility. Now the tract is largely barren and waste, growing only some barley in certain parts. This crop depends upon the winter rains, which are precarious and seem to be about the same in average amount as at Alexandria, *i.e.* about 8·7 inches per annum. The possibility therefore presents itself that the former greater fertility depended upon a greater precipitation. But the rainfall depends upon the temperature of the Mediterranean and the general atmospheric circulation, and Mr. Weedon brings forward evidence which in his opinion shows that neither these nor the rainfall have altered since Roman times. He believes that even at present more profitable crops could be produced by greater skill on the part of the cultivators, and the use of scientific methods in conserving the soil water, or supplying more water.

America.

Forests and Rainfall in the United States.—Dr. Raphael Zon contributes to *Science* of July 18 a long article on the relation of the forests in the Atlantic plain of the United States to the humidity of the central States and the prairie region. The thesis maintained in this article is that forests may exercise a marked effect on the rainfall of distant regions, owing to the fact that winds blowing over a forested area take up more moisture than when blowing over cultivated land, and much more than when blowing over waste land. The thesis involves the assumption that moisture-carrying winds do not, as is usually assumed, take up the greater part of their moisture from the surface of the ocean, but that much is taken up from land surfaces.

Dr. Zon begins by pointing out the special features of the American continent as regards configuration and climate. The fact that the mountain chains run along meridians and not along parallels is of course fundamental, resulting as it does in a free circulation of air over a large area in latitude. Another peculiar feature is the rapid decrease of temperature from south to north. Thus between Florida and Labrador there is a decrease of temperature in January of 2·9° F. for every degree of latitude, in July of 1·08° F., and for the whole year of 1·7° F. In the same latitudes in Europe the decrease is less than half. Thus between the Canary Islands and northern Scotland the decrease in the mean annual temperature is only 0·8° per degree of latitude.

Again, owing to the position of the mountain chains the whole of the part of the North American continent east of the hundredth meridian forms one climatic province. Throughout this region in winter, and partially in autumn and spring, the prevailing winds are north and north-west, and bring cold dry air from the interior of the continent. When such winds prevail in late spring and summer they are hot and dry. In summer the prevailing winds in Texas are from the south-east, while further north and east they come from the south and south-west. These southerly winds of summer are humid and bring most of the rainfall, and this is especially true of the rainfall of the prairie States and the central region. When southerly winds are frequent during summer these regions produce abundant crops. But seasons of frequent northerly winds are seasons of drought. Dr. Zon is of opinion that the existing forests in the Atlantic plain and Appalachian region help to supply these southerly winds with moisture, and that any destruction of them would diminish the amount of moisture carried to the interior of the continent. He considers therefore that a forest cover should be carefully maintained in the Atlantic plain and the southern Appalachians, especially (1) on moist soils, swamps, which lose water by run-off, being drained to permit of more forest growth; (2) on sandy soils, for otherwise the water percolates to underground channels and thus local surface evaporation is diminished; (3) on steep slopes and rocky places, for here the forest cover prevents rapid run-off and loss of surface water.

Nomenclature of the Amazonian Forest.—Mr. J. C. Branner contributes to *Science* a letter protesting against the common use of the Portuguese word Selva to designate the dense equatorial forest of the Amazon basin. According to him the Brazilians call this, as well as other dense forests, *Mattas*, and Selva is a poetic word infrequently used in Portuguese, and having no special application to the Amazonian forest.

AUSTRALASIA.

Exploration in Papua.—The Rev. J. Bryant sends us the following note on this subject:—Mr. G. H. Baker, Resident Magistrate of the Western Division of British Papua, has reported on the expedition he recently led into the Fly River districts. He found a considerable portion of the country to consist of swamps, but also a very large area suitable for occupation. The cocoanuts were the finest he had ever seen; tobacco was growing freely; and the sugar cane was excellent.

Mr. Baker selected a spot fifteen miles up the Strickland River as the site for a new control station. The position was suitable in every way except for the crowds of mosquitoes, which on a sunny afternoon were simply maddening.

The expedition started from Daru on June 2nd in two launches and a whale boat. A month was occupied in exploring the Fly and Strickland Rivers. After ascending the Strickland about twenty miles the expedition entered a noble lake. Mr. Baker reports it as the finest he has seen in Papua, and judges its perimeter to be fully one hundred miles. It has been named Lake Murray. The natives proved troublesome and treacherous, and of this stage of the journey he writes:—

"We were inveigled ashore by friendly unarmed natives in canoes, but on landing and proceeding towards the main house, we were surrounded by a host of painted, fully armed warriors, whose intentions were unmistakable. The steadiness of the police, combined with a slow backward movement to the whale-boat, saved the situation.

"On another occasion steaming up the river we ran into the arms of about a hundred and fifty warriors in canoes, fresh from a head-hunting expedition. It required a volley from the police to clear the way. Shortly alterwards we came upon the result of their vile work in the shape of bodies, headless and armless, as well as partly skinned trunks of women, lying on the shore, at what was apparently a peaceful sago-making camp.

"On the way back to the base camp about eight hundred warriors attempted to stop us, but thanks to our powerful glasses the ruse was discovered in time. Two or three unarmed natives stood on the beach and made peaceful signs, while about four hundred warriors with their canoes ready lay hidden in the grass and as many more lay behind the village. As soon as they saw we were going on without taking any notice of them they showed themselves, and the whole place was alive with them. I put a good many miles between us before camping that night."

General.

Pastoral Peoples in Asia and Africa.—Dr. F. Hahn contributes to two recent issues of the *Geographische Zeitschrift* a long article on this subject, in which he attacks vigorously the old theory that hunter, herdsman, and farmer represents three successive stages in human evolution. He says that this exploded theory is not only still current among non-specialists, but is tacitly accepted by those who ought to know better, leads to all sorts of erroneous deductions, and prevents the development of an accurate terminology.

According to him pastoral groups have always been derived from agricultural ones, and are never self-sufficing. Even the Bedouins, he says, do not live upon the produce of their flocks, but by preference consume cereals, coffee, etc., obtained by barter or trade or by stealing. He holds that the Asiatic peoples in the region round Babylon domesticated the ox as a draught animal, and also domesticated sheep and goats. The oxen were more or less tied to the settlements owing to their use in ploughing, but the sheep and goats were taken by herdsmen afield to feed upon the periodical abundance of the steppe. The herdsmen were thus a part of the agricultural community. The next step, he believes, was the domestication of the ass, which the Asiatics obtained from Africa. The ass was used as a transport animal, and greatly facilitated the movements of the herdsmen. But though used to carry the women and children, the aged and infirm, and also holy men, it was never a riding animal in the strict sense. The horse was in his opinion first used as a draught animal to pull chariots of war, and only later did the pastoral peoples begin to use it as a riding animal. With the acquisition of this habit they acquired great mobility, and the evolution of the warlike hordes of Central Asia was rendered possible.

In Africa, Dr. Hahn believes, the evolution of pastoral peoples took place independently. Cattle were obtained from Asia, and since the negroes do not use the plough, in negro Africa the cow could not acquire there a specially sacred character owing to a connection with agriculture. But he believes that when cattle reached Africa it was with this sacred character already attached, and in his opinion it is impossible to understand the attitude of those negro races who have herds of cattle to their flocks without making allowance for this fact. Cattle in negro Africa are not valued for their usefulness, for sometimes they have little direct use, but because they are the most highly prized possession within reach of the negro, and are comparable to family treasures like linen or silver, often equally unused. Sometimes the negro does not eat the flesh of his herds at all, at other times its use is limited to great occasions, or to the chiefs. Even so purely a pastoral group as the Masai are, he believes, in all probability a recent derivative of an agricultural stock.

The Geographical Society of Lisbon.—We are informed by the President of this Society that it has recently established a Permanent Committee for the protection of the aborigines of the Portuguese Colonies, whose aims are at once to assist the natives and to endeavour to civilise them. This step has been taken on account of the criticism which has been directed against the administration of certain of the colonies, "by those who have forgotten the services to the world rendered by the Portuguese navigators."

Commercial Geography.

Fox-farming in British North America.—A recent article in the *Times* gives an account of a new industry which is apparently becoming of some importance in Prince Edward Island, Newfoundland, Nova Scotia, New Brunswick, and other forested parts of eastern North America. This is the rearing of silver foxes for their coats, which form a costly fur, especially that of the variety called the black fox. The industry has been established in Prince Edward Island on a small scale for about twenty-five years, but it is now spreading and taking a firm hold, especially in Newfoundland where wild foxes are still abundant.

EDUCATIONAL.

Association of Scottish Teachers of Geography.—Mr. T. S. Muir, Hon. Secretary of this Association, furnishes us with the following note:—

"A meeting of the Association was held in the Society's Rooms on Saturday, Nov. 15. The following office-bearers were elected for the year 1913-14:—*Chairman*, Mr. Geo. G. Chisholm; *Vice-chairmen*, Mr. Cossar and Mr. Philip; *Secretary*, Mr. T. S. Muir, Royal High School, Edinburgh; *Committee*, Messrs. Findlay, Macgregor, Corrie, Millar, and Miss Lennie.

"It was resolved to make up a roll of members and to ask all teachers who are Members or Associates of the Society to intimate to the Secretary whether they wish also to be Members of the Association. Notices of meetings will continue to appear in the *Magazine*, but they will also be sent specially to members of the Association.

"An interesting discussion followed on the Report to the British Association on the teaching of geography in Scotland, an abstract of which appeared in last month's *Magazine*."

We may call attention here to a very fine series of views of Mt. Blanc which appears in the *National Geographic Magazine* for August last. The accompanying article gives some account of the first ascents of Mt. Blanc and also of the conditions under which the climb is made at the present day. The same issue contains a number of striking views of other parts of Switzerland, and also of the Italian lakes.

NEW BOOKS.

EUROPE.

Borrowstounness and District. By THOMAS JAMES SALMON. Edinburgh and London: William Hodge and Company, 1913. *Price 6s. net.*

Of late years we have had many monographs dealing with particular counties and places in Scotland, some of them of conspicuous merit. In this volume we have a monograph on Borrowstounness and its vicinity, of which we can say at once that it takes an honourable place among similar works. The author has conscientiously prepared himself for his task by a patient and minute study of the historical and ecclesiastical documents which bear upon it. He has also studied the family histories of the men and women who, connected with Bo'ness, have made their mark in the annals of Scotland. It is very obvious that he suffered at times from an *embarras de richesses* in his materials, and had to condense, and even omit, much matter that would be of great interest to the student of history and of archæology. In these pages we have most interesting and vivid pictures of the appearance and life of the locality from the middle of the sixteenth century, when it emerges into the light of history: trustworthy, because they are vouched for by contemporaneous documents, and lifelike, because Mr. Salmon has most judiciously made no effort to overstate or exaggerate the facts, and has wisely trusted to the intrinsic interest which they are sure to arouse. He is equally interesting and successful in portraying the crafty and tortuous Hamiltons, and the bigoted but sincere Covenanting ministers. All who are interested in the rise and progress of municipal life and government will find much well worth their study and consideration in the accounts of the various experiments in local self-government in this petty and obscure, but adventurous and stout-hearted village. It is very evident that to Mr. Salmon the preparation of this book has been a labour of love, and he writes with the pardonable enthusiasm and patriotism of one who is justly proud of the place of his birth and its history. We cordially commend this work to our readers.

Handbook of Baltic and White Sea Loading Ports, including Denmark. Revised Edition. By J. F. MYHRE. London: William Rider and Son, 1913. *Price* 21*s. net.*

This is the second edition of a useful work of reference, originally published in two parts, but now issued in one volume with the addition of Denmark, which was previously omitted. As well as supplying the practical needs of shipowners and others, the mass of detail contained in it makes the book of importance to the geographer, who will find here facts otherwise not easy to obtain.

Bennett's Handbook for Travellers in Norway. Thirteenth Edition, revised. London: Simpkin, Marshall, Hamilton, Kent and Co., 1913. *Price* 2*s.* 6*d.*

This is a handy little book, published under the auspices of a tourist office in Christiania, and giving practical hints for the traveller in Norway. The route map at the end of the volume does not give much detail, and is similar in type to the maps which usually accompany railway guides.

Sketches of North and West Ireland. By MARY REDSDALE WILSON. Illustrated by JESSIE S. WILSON. London: Arthur Stockwell, 1913. *Price* 2*s. net.*

A simple little account of wanderings in North and West Ireland, illustrated by a number of pleasing sketches. The text includes notes on the more interesting plants seen in the regions visited.

Mittelmeerbilder. Von Dr. THEOBALD FISCHER. Zweite Auflage besorgt von Dr. ALFRED RÜHL. Leipzig: B. G. Teubner, 1913. *Price* 7 *marks.*

Before his death Professor Fischer had revised his book to a certain extent, and had prepared various notes and additions. His colleague, Dr. Rühl, was entrusted with the task of preparing the work for the press, and it now appears with a portrait of the late author. The alterations and additions are only of a minor character, the book remaining in essentials the same as before. In its new form it will be welcomed by all those to whom the Mediterranean region makes an appeal.

Spain and Portugal: Handbook for Travellers. By KARL BAEDEKER. Fourth Edition. London: T. Fisher Unwin, 1913. *Price* 15*s.*

The price of this volume indicates that visitors to Spain and Portugal are not yet large in numbers; and the introduction, which is perhaps needlessly pessimistic in tone, gives incidentally many reasons for this reluctance of the ordinary tourist to venture across the Pyrenees. This is unfortunate, for a glance through the pages of the book shows how much of interest, both to the geographer and the artist, the peninsula contains; and that very indifference to the northerner's point of view, which the book chronicles so often, means that here, far more than in, for example, Italy, can the traveller come in contact with an unspoilt civilisation differing in many respects from his own, and therefore all the more instructive. We hope that the appearance of this new edition will stimulate interest in the region, and encourage readers to use it, as it is meant to be used, within the country itself.

Naples and Environs: Capri. Grieben's Guide-Books. London: Williams and Norgate, 1913. *Price* 1*s.* 6*d.*

This is another of the handy little Grieben series, containing some useful maps and plans, but no great wealth of detail in the text. The translation is moderately well done. Oddly enough, in a book of German origin, the fact that German is far more commonly spoken round Naples than English is not alluded to.

Photographic Supplement to Stanford's Geological Atlas of Great Britain and Ireland. Arranged and edited by HORACE B. WOODWARD, F.R.S., F.G.S. London: Edward Stanford, 1913. *Price* 4*s. net.*

This little book consists of a fine series of photographs, predominantly of beach and quarry scenes, intended to illustrate the rock features and scenery of the principal formations of Great Britain, and thus to serve as a companion to the *Geological Atlas.* The photographs have been collected from a variety of sources, and we note that the first two reproduced are fine examples from the collection of those taken by the Geological Survey of Scotland. Each figure is described, and references are given to publications where fuller details can be obtained.

The Cathedrals of Southern France. By T. FRANCIS BUMPUS. London: T. Werner Laurie, Ltd., 1913. *Price* 6*s. net.*

When a new and restricted definition of geography is pronounced it is possible that architecture will not be included. If so, the exclusion would be regrettable, for architecture reveals in permanent and visible form the ideas of its period, and often it affords a clue to the outside influences at work in the formation of those ideas. At the lowest it indicates the local industries and resources. On these grounds it seems to us that geography should welcome such a scholarly and comprehensive work as this is.

Mr. Bumpus has already published seven volumes on Continental ecclesiology and the latest product of his pen is an epitome of the writings of many authors on a favourite and inexhaustible subject. He does himself some injustice in the title, for his scheme includes not merely Southern, but Central and Western France, including Brittany. An intending visitor to Poitiers for example might not think of consulting this book. Yet the account of the cathedral there is well worth reading. The same may be said of many of the other descriptions, and there is further much information on cognate subjects. The illustrations are numerous and good, and we may instance specially the porch at St. Gilles, where the relief effect is most happily rendered. The index is scarcely as full as we could have wished.

AFRICA.

The Land of Footprints. By STEWART EDWARD WHITE, F.R.G.S. London: Thomas Nelson and Sons, 1913. *Price* 2*s. net.*

In this volume we have an account of an expedition for big game in East Equatorial Africa. The party, consisting of three American gentlemen and a lady, had a very good time and bagged plenty of lions, rhinoceros and buffaloes, not to mention smaller game, and they saw some of the now familiar episodes of native life in Uganda. The narrative is very similar to what may be found in a score of books recently published on the same subject, but is pleasantly and crisply written, and will doubtless find acceptance with many readers. Mr. White's experience of the natives with whom he came in contact did not differ much from that of other sportsmen, and there is nothing novel in what he describes. It appears that he is preparing a second volume describing the continuation of the hunting expedition.

Les Sociétés Primitives de l'Afrique Equatoriale. With 9 Figures, 18 Plates, and Map. By Dr. Ad. Cureau. Paris: Armand Colin, 1912. *Price* 6 *fr.*

The author spent twenty years amongst the natives of the Congo, during which time he travelled from the Atlantic to the Bahr-el-Ghazal. In this book he endeavours to give a sociological and psychological study of the negro race considered as a whole, and traces the evolution of the family, of the village, and of slavery. The village is studied under the following heads: être physique, être vivant, être sensible, être organisé, and être moral. He describes the environment, the overwhelming vegetation of the forest, and the great African rivers with their still smooth surfaces sometimes broken with savage rapids and waterfalls. He lays stress on the effect of the "l'écrasant soleil de midi" with its "impitoyable ardeur" upon the soul of the black man. From these factors of the environment he endeavours to deduce the character of the various natives and their institutions. Naturally, this is a supremely difficult undertaking. The author points out the hopelessness of the negro in face of the exuberance of vegetation, the swarms of insects which make life almost unbearable ("la vraie bête féroce de l'Afrique équatoriale c'est l'insecte"), and the exhaustion due to tropical heat and the effects of disease. Opinions will differ as to whether these feelings, which certainly oppress the European traveller, are experienced with equal force by the negro.

One cannot help sometimes thinking that the author is too didactive in his conclusions, and that he does not fully appreciate other points of view. Thus he hardly refers to the effect of slave-raiding carried on probably for centuries, and to the general insecurity of life and property which has always weighed upon the African negro.

AMERICA.

The Southland of North America. By George Palmer Putnam. New York and London: G. P. Putnam's Sons, The Knickerbocker Press, 1913. *Price* $2.50 *net.*

When in the near future the Panama Canal is opened to the commerce of the world, there is little doubt but that the republics of Central America will require and will receive greater attention than they get now, and for this reason the publication of this volume is opportune and welcome. The author visited half a dozen of these republics in 1912, and he gives us in a pleasant genial way his impressions of their present condition. It has to be admitted at once that, however attractive they are from the beauty of the scenery, the fertility of the soil, their archæological remains, their racial problems, etc., the republics are not, on the whole, very likely to attract many tourists, as they are an outstanding example of how regions, which under good government might vie with any other in the tropics, are paralysed and wellnigh ruined by chronic misgovernment and corruption; much less likely are they to attract commercial enterprise which would involve the risk of expending capital on their development. So long as the Monroe doctrine holds good (and there are no signs of its being abandoned), the only remedy for the misgovernment of these republics would be action on the part of the United States, but this, at least in the meantime, is out of the range of practical politics. The first attempt of the United States to dabble in foreign affairs was far from successful. The writer of this book sums it up as, "Cuba perhaps a failure, and the Philippines a perpetual grief." Moreover, nearer to the United States is the more perplexing problem of Mexico, a country at the present time in a state of revolution. The Mexican question must be settled before the United States can deal with the hornets' nests farther south. Thus, everything seems to indicate that, for some time at least, the central republics will be left to stew in their own juice. The writer of this book seems to have had exceptional facilities for seeing everything that is worth seeing in them, and he relates his impressions in a series of sketches which are amusing and interesting. There are a number of anecdotes and personal adventures and experiences, some of which involved personal danger, and the book is profusely illustrated with good photographs.

Colombia. By Phanor James Eder. London: T. Fisher Unwin, 1913. Price 10*s.* 6*d. net.*

This volume is the latest of Mr. Fisher Unwin's excellent South American series, and Mr. Eder, who by birth, friendships and business relations with the country is well qualified for his task, endeavours to present a "true, fair and sympathetic picture of present-day commercial and industrial conditions in Colombia." A glance at the map will show that the chief geographical feature of Colombia is its great Andean mountain system, forming the three "cordilleras" or main ranges of the Western, Central, and Eastern Cordilleras. Of the three, the central Cordillera is the most important, constituting the backbone of the country, while the table-land of Bogotá forms the heart of the central zone. Here, at an elevation of from 7000 to 8000 feet, an equable climate prevails, and the conditions are more favourable to civilised life than in the low belts by the river valleys and on the sea coasts. Boundary questions are the most fruitful source of trouble between Colombia and its neighbouring states, the frontiers not having been delimited towards Ecuador, Venezuela, and Peru, although the limits with Brazil on the west are now defined under a recent treaty. Fears of aggression on the part of the United States are the dominant note in a movement towards an alliance among the Spanish-American countries, and no one can say that such fears are devoid of foundation when one remembers the ruthless seizure of the department of Panama by President Roosevelt as a canal zone. As Mr. Eder puts it, "the Colombians have already felt the talons of the Eagle; they have a hysterical dread that the voracious bird will again swoop down upon their country. Hysterical is the only word. Suspicion of the designs of the American Government is carried to an absurd limit." This admission is the more significant coming from a citizen of the United States.

Mr. Eder deals mainly with the industrial and financial conditions of Colombia, and this aspect is of especial interest in view of the large concessions recently granted to a British firm, or syndicate, but there is much in the book which cannot fail to be of interest to all who are attracted by the boundless possibilities of a rich but imperfectly developed state. The author's opinion is that the railway building which has been attempted, at present resulting in only six hundred miles of "provisional tracks," will have to be done over again, the sooner the better; and that, consequently, "the country offers an almost virgin field for the railway operator and financier." Concessions are likely to be the order of the day, as Colombia itself cannot find the necessary capital, and, if nothing untoward happens, British engineers and others will find an outlet for their energies in the near future. The sections of the book which deal with Commerce, Agriculture, Mines and Forests contain some valuable hints.

Mr. Eder has eliminated the personal incidents of travel which speak home to the general reader, but he finds space for one picture of the inns on the Quindo road, between the Cauca and Magdalena valleys, which shows what the traveller may have to put up with. However, there are many compensations to be found in the fine air and magnificent scenery of the Andean Uplands. A curious point is that the little group of islands off the Colombian coast known as San André and Providencia, are populated by blacks and mulattoes who are English-speaking, and mainly Protestant, though belonging to a Spanish State.

A bibliography and a large and clear map help to make this an excellent book of reference on matters Colombian, which we can confidently recommend to our readers.

Twentieth-Century Jamaica. By H. G. De Lisser. Kingston: *The Jamaica Times*, 1913.

This unpretending little book gives an account of the social, economic, and political conditions which exist in Jamaica at the present time. The introductory chapter, which deals with Jamaica's future, discusses the three possibilities before the island—that it should remain attached to the mother country as at present, that it should be taken over by Canada or by the United States. The author seems to consider that the balance of probabilities lies in the last-named direction. Thus he points out that in 1911 the United States purchased Jamaica's products to the value of £1,825,000, the United Kingdom to £438,000 only, and Canada to the value of £253,000. Thus whatever Canada may be in a position to do in the future, at present as a purchaser of Jamaican products she is almost negligible. The banana trade, now the most important in the island, is in American hands, and the Americans are introducing improved business methods in connection with the export of other Jamaican products also. In the supply of manufactured goods the United Kingdom and the United States bulk almost equally (United Kingdom, £1,291,000; United States, £1,200,000), but the imports from Canada just about balance the exports.

Very interesting also is the account of the evolution of a distinctively Jamaican type, in which colour counts for very little, and of the growing dissatisfaction of the people with the present mode of government. The book is soberly written, and is well worth the attention of the geographer.

POLAR.

Le Monde Polaire, par Otto Nordenskjöld. Traduit du Suédois par Georges Parmentier et Maurice Zimmermann. Paris: Librairie Armand Colin, 1913. Prix 5 *fr.*

In this volume we have a translation in French of an exceedingly interesting work on the Polar world by Professor Otto Nordenskjöld. The writer deals with his subject mainly from a geological point of view, but he has added much valuable information and speculation as to the origin and distribution of the inhabitants of the Arctic circle. The subject is a large one, and considerations of space have compelled Dr. Nordenskjöld to condense ruthlessly, and as a result we have what may be described as a series of sketches or essays, no one of which is the last word that is to be said on its subject-matter. Indeed, a perusal of this book impresses the reader with the feeling that even now comparatively little is known of the conditions of the Polar world, and that its former history is still a matter of doubt and conjecture. This is mainly the result of the modesty of the writer, who, although an acknowledged authority and expert, carefully refrains from dogmatising, but advances his theories and states his opinions and conclusions with great caution, and is the first to admit that we have much to learn yet about the Polar world. As a summary of what is known of the geological conditions of the Arctic and Antarctic regions this work is excellent, and it also points out the directions in which further investigation is necessary.

GENERAL.

Animal Geography. By MARION I. NEWBIGIN, D.Sc. Oxford: The Clarendon Press, 1913. *Price* 4*s.* 6*d.*

We welcome the appearance of this useful study of "the faunas of the natural regions of the globe." The complaint in the Preface that the relationship existing between fauna and natural environment has not been given its due share of treatment in modern text-books of geography is, we think, well founded, but Miss Newbigin's book has removed the reproach.

The natural regions chosen are distinctly phytogeographical, for, as the introductory chapter points out, "all animals depend ultimately for their food upon plants." Following upon a general discussion of the natural regions come detailed accounts of these—their physical and climatic conditions and the animals living under the conditions. In our opinion, chapter i., The Tundra and its Fauna; chapter iii., Steppe Faunas; chapter vi., Tropical Savanas and Deserts; and chapter viii., The Distribution of Animal Life in the Sea, are most valuable. If one single chapter can be singled out as specially geographical, it is the last mentioned. A general sketch of the recognised zoo-geographical regions occupies chapter ix., while an outline classification of animals is given as an Appendix.

A distinguishing feature is the excellent plan of giving lists of references at the close of the various chapters. For those who wish to pursue the study further, the lists will serve as guide-posts. Close on fifty well-chosen illustrations enhance the value of the book, which altogether is a credit to both writer and publishers.

Making the most of the Land. By JAMES LONG. London, New York, Toronto: Hodder and Stoughton, 1913. *Price* 5*s. net.*

This work, the production of a well-known authority on agriculture, does not lend itself very readily to notice in the pages of this magazine. It is frankly political. It deals with the important but highly controversial subject of how to make the most of the agricultural land of Great Britain, and the writer in no way conceals his approval of the policy of practically unlimited creation of small holdings and of a new system of rating, which would press heavily on owners of uncultivated land. He treats his subject with much knowledge, experience and study, and under these circumstances his views, supported as they are by wealth of statistics and illustrations taken from home and foreign sources, are well worth serious attention, however unpalatable they may be found by many.

A perusal of the book will open the eyes of many to the present serious condition of British agriculture in its continual and severe competition with foreign nations, and may convince some that the greatest hindrance to agricultural success in these islands is the dogged conservatism and invincible obstinacy of the average farmer.

The "Welcome": Photographic Exposure Record and Diary, 1914. London: Borroughs, Wellcome and Co. *Price* 1*s.*

This useful little pocket-book appears as usual with great promptness, and we note as special features of the present issue that it contains reproductions of photographs taken by the Scott Expedition, and also a striking war-plane view of the river Maritza. The book does not now require recommendation to the amateur photographer, to whom its merits are well known.

"The Car" Road-Book and Guide. An Encyclopædia of Motoring. Edited and revised by Lord Montagu. London: *The Car Illustrated*, Ltd., 1913. *Price* 12*s.* 6*d. net.*

We have already reviewed here an earlier edition of this useful compilation with its excellent map. The present issue has been brought up to date, but shows no very novel features.

"The Britannica" Year-Book, 1913. A Survey of the World's Progress since the completion in 1910 of the *Encyclopædia Britannica*, eleventh edition. Edited by Hugh Chisholm. London: The *Encyclopædia Britannica* Co., 1913. *Price* 10*s.*

This is a year-book of a somewhat new type, whose scope is sufficiently indicated by the title. It contains in a convenient form many facts of great importance to the geographer.

Bradshaw's Through Routes to the Chief Cities, Bathing and Health Resorts of the World. Edited by Eustace Reynolds Bell. London: Henry Blacklock and Co., 1913. *Price* 6*s. net.*

The present edition of this book has been revised and enlarged, and gives in a compact form a great mass of information for the intending traveller. In addition to the traveller, however, the teacher will find many little facts and details which will help to give reality to his lessons. There are a number of useful maps, plans, and charts.

EDUCATIONAL.

Dent's Practical Notebooks of Regional Geography. By Horace Piggott and Robert Finch. Book I. The Americas; II. Asia; III. Africa; IV. The British Empire in America and Asia. London: Dent and Sons, 1913. *Price* 6*d. net each.*

These notebooks contain a series of blank maps, etc., facing pages intended for notes, each map having a list of instructions giving the exercises for which it can be used. The actual facts required are intended to be sought for in an atlas or geography book. Most of the ruled sheets intended for graphs seem to us very trying for the eyes, the ruling being in black and the squares small, but the exercises on the map should prove of great use if carried out carefully.

Questions and Exercises in Geography. By R. J. FINCH, F.R.G.S. London: Ralph Holland and Co., 1913. *Price 2s. 6d. net.*

To the busy teacher this must prove a most valuable source for obtaining examination questions. The wise selection, careful arrangement, and skilful grading should do much to make for the author's aim, "to fit the student for dealing with geographical problems in the examination room." By no means the least valuable portion of the book is the able introduction where typical answers are given to typical questions. Especially useful is the figure showing the geographical significance of the site of Buda-Pesth.

A Comparative Geography of the Six Continents. By ELLIS W. HEATON, B.Sc., F.G.S. London: Ralph Holland and Co., 1913. *Price 1s. 9d. net.*

Mr. Heaton's "Scientific Geographies" are already well known to and much appreciated by teachers, but it is doubtful whether he has yet done anything more strictly geographical on comparative lines than the little work before us. Environmental, more especially climatic, conditions are never lost sight of, and the book has a great deal to say which is most suggestive. Several of the diagrams must prove very useful; we refer especially to figs. 12, 13, 14, 15, 16a, 16b, 27, 34, 58, 60, 61, 62, 66, 67, and 68. A new feature in Mr. Heaton's books is the insertion of photographic illustrations.

The British Empire with its World Setting. By J. B. REYNOLDS, B.A. London: A. and C. Black, 1913. *Price 1s. 4d.*

For the needs of the pupils for which this book is specially written, viz., those of "the upper classes of elementary schools," the little work by Miss Reynolds is eminently suitable. Its neat appearance, clear print, and well-selected illustrations must make this Junior Regional Geography pleasing to the pupils. The "Questions and Suggestions" at the end of the sections are useful, and the material of the text is well arranged and happily chosen.

BOOKS RECEIVED.

Lands and their Stories: Asia, Africa, and America, outside the British Empire. By A. W. PALMER. With Appendices by David Frew, B.A. Crown 8vo. Pp. 304. *Price 1s. 9d.* Blackie and Son, Ltd., 1913.

Camping in Crete, with Notes upon the Animal and Plant Life of the Island. By AUBYN TREVOR-BATTYE, M.A., F.L.S., F.Z.S., F.R.G.S., etc. Including a Description of certain Caves and their Ancient Deposits. By DOROTHEA M. A. BATE, M.B.O.U. With 32 Plates and a Map. Demy 8vo. Pp. xxi + 308. *Price 10s. 6d. net.* London: Witherby and Co., 1913.

Ten Years near the Arctic Circle. By J. J. ARMISTEAD. Crown 8vo. Pp. 252. *Price 3s. 6d. net.* London: Headley Brothers, 1913.

My Somali Book: A Record of Two Shooting Trips. By Captain A. H. E. MOSSE, F.Z.S. Illustrated with Photographs by the Author, and with Sketches and Cover Design by Lieut. D. D. HASKARD, R.A. With an Introduction by Colonel H. G. C. SWAYNE, R.E., F.Z.S. Demy 8vo. Pp. xxv + 314. *Price 12s. 6d. net.* London: Sampson Low, Marston and Co., Ltd., 1913.

Principles and Methods of Teaching Geography. By FREDERICK L. HOLTZ, A.M. Crown 8vo. Pp. xii + 359. *Price 5s. net.* New York: The Macmillan Co., 1913.

Dix mille Kilomètres à travers Le Mexique, 1909-1910. Par VITOLD DE

SZYSZLO. Avec 22 gravures hors texte. Crown 8vo. Pp. iv + 343. Prix 4 *francs*. Paris : Plon-Nourrit et Cie, 1913.

Mountains: Their Origin, Growth and Decay. By JAMES GEIKIE, LL.D., D.C.L., F.R.S. (L. and E). Demy 8vo. Pp. xix + 311. Price 12*s*. 6*d*. *net*. Edinburgh : Oliver and Boyd, 1913.

A Text-Book of Geography. By A. W. ANDREWS, F.R.G.S., F.R.S.G.S. Illustrated. Crown 8vo. Pp. xii + 655. Price 5*s*. London : Edward Arnold, 1913.

History of Geography. By J. SCOTT KELTIE, LL.D. and O. J. R. HOWARTH, M.A. Crown 8vo. Pp. ix + 154. Price 1*s*. *net*. London : Watts and Co., 1913.

Die Sorge-Bai: Aus den Schicksalstagen der Schröder-Stranz Expedition. Von Dr. HERMANN RÜDIGER. Mit 46 Bildern im Text und 5 Tafeln nach Zeichnungen u. photographischen Aufnahmen des Marinemalers Christopher Rave, sowie einer Übersichtskarte. Medium 8vo. Pp. xii + 215. Preis 5 *Marks*. Berlin : Georg Reimer, 1913.

Things Seen in Oxford. By NORMAN J. DAVIDSON, B.A. (Oxon). With 50 Illustrations. Foolscap 8vo. Pp. xv + 258. Price 2*s*. *net*. London : Seeley, Service and Co., Ltd., 1913.

Winning a Primitive People: Sixteen Years' Work among the Warlike Tribe of the Ngoni and the Senga and Tumbuka Peoples of Central Africa. By DONALD FRASER. With an Introduction by JOHN R. MOTT, LL.D., F.R.G.S. With 27 Illustrations and 2 Maps. Crown 8vo. Pp. 320. Price 5*s*. *net*. London : Seeley, Service and Co., Ltd., 1913.

A Church in the Wilds: The Remarkable Story of the Establishment of the South American Mission amongst the hitherto Savage and Intractable Natives of the Paraguayan Chaco. By W. BARBROOKE GRUBB. Edited by H. T. MORREY JONES, M.A. (Oxon.). With 23 Illustrations and 2 Maps. Crown 8vo. Pp. xv + 287. Price 5*s*. London : Seeley, Service and Co., Ltd., 1913.

Through the Heart of Canada. By FRANK YEIGH. With 38 Illustrations. Demy 8vo. Pp. 319. Price 5*s*. London : T. Fisher Unwin, 1913.

Chambers's Commercial Handbooks: Commercial Geography of the World. Part II. : *Outside the British Isles*. By A. J. HERBERTSON, M.A., Ph.D. Revised by JAMES COSSAR, M.A. Second Edition. Crown 8vo. Pp. 383. *Price* 2*s*. 6*d*. London : W. and R. Chambers, Ltd., 1913.

Northumberland: Yesterday and To-day. By JEAN F. TERRY, LL.A. Crown 8vo. Pp. 235. Price 1*s*. 6*d*. *net*. Newcastle-upon-Tyne : A. Reid and Co., Ltd., 1913.

To Norway and the North Cape in "Blue Dragon II.," 1911-12. By C. C. LYNAM, M.A. Illustrated in Colour, and with Photograph Sketches and Maps. Medium 8vo. Pp. xxiv + 232. *Price* 6*s*. *net*. London : Sidgwick and Jackson, Ltd., 1913.

Il Genere umano morirà di Fame? By Gustavo Coen. Crown 8vo. Pp. 132. Livorno : S. Belforte e Ci.

The Ocean: A General Account of the Science of the Sea. By Sir JOHN MURRAY, K.C.B., F.R.S., LL.D., D.Sc., Ph.D. Foolscap 8vo. Pp. 256. *Price* 1*s*. *net*. London : Williams and Norgate, 1913.

The Humour and Pathos of the Australian Desert. By JOHN BEUKERS. Crown 8vo. Pp. 249. Price 4*s*. *net*. London : A. H. Stockwell, 1913.

The Conquest of the Desert. By WM. MACDONALD, M.S. Agr., Sc.D., Ph.D., D.Sc. With 50 Illustrations. Demy 8vo. Pp. xii + 197. Price 7*s*. 6*d*. *net*. London : T. Werner Laurie, Ltd., 1913.

An Artist in Italy. By WALTER TYNDALE, R.I. Royal 4to. Pp. 307. Price 20*s*. *net*. London : Hodder and Stoughton, 1913.

Islands grosster Vulkan: Die Dyngjufjöll mit der Askja. Von Dr. HANS SPETHMANN. Mit 36 Abbildungen im Text. Demy 8vo. Pp. vii + 143. *Preis 6 Marks.* Leipzig: Veit und Co., 1913.

Two Years with the Natives in the Western Pacific. By Dr. FELIX SPEISER. Demy 8vo. Pp. xii + 291. *Price* 10*s.* 6*d. net.* London: Mills and Boon, Ltd., 1913.

Au Yunnan et dans le Massif du Kin-Ho (Fleuve d'Or). Par le Dr. A. F. LEGENDRE. Avec 20 gravures et une carte. Crown 8vo. Pp. xii + 434. Paris: Plon Nourrit et Cie, 1913.

Guatemala and the States of Central America. By CHARLES W. DOMVILLE-FIFE. Illustrated. Demy 8vo. Pp. 310. *Price* 12*s.* 6*d. net.* London: Francis Griffiths, 1913.

La Bolivie et ses Mines. Par PAUL WALLE. Soixante et une Illustrations hors texte et quatre Cartes. Demy 8vo. Pp. xvi + 444. *Prix* 7 *francs* 50. Paris: E. Guilmoto, 1913.

Across Unknown South America. By A. HENRY SAVAGE-LANDOR. With 2 Maps, 8 Coloured Plates and 260 Illustrations from Photographs by the Author. In 2 Vols. Pp. xxiv + 432 and xvi + 504. Medium 8vo. *Price* 30*s. net.* London: Hodder and Stoughton, 1913.

Scott's Last Expedition. In 2 Volumes. Vol. I. Being the Journals of Captain R. F. Scott, R.N., C.V.O. Vol. II. Being the Reports of the Journeys and the Scientific Work undertaken by Dr. E. A. Wilson and the Surviving Members of the Expedition. Arranged by LEONARD HUXLEY. With a preface by SIR CLEMENTS R. MARKHAM, K.C.B., F.R.S. Illustrated. Demy 8vo. Pp. I. xxvi + 633; II. xiv + 524. *Price* 42*s. net.* London: Smith, Elder and Co., 1913.

The Waters of the North-Eastern North Atlantic: Investigations made during the Cruise of the "Frithjof," of the Norwegian Royal Navy in July 1910. By FRIDTJOF NANSEN. With 52 Text-Figures and 17 Plates. Medium 8vo. Pp. 139. Leipzig: Dr. Werner Klinkhardt, 1913.

A Report on the Land Settlement of the Gezina (Mesellemia District). By H. St. G. PEACOCK, Esq. Royal 8vo. Pp. 68. London: Sifton Praed and Co., Ltd., 1913.

Memoirs of the Geological Survey, Scotland: The Geology of the Fannich Mountains and the Country around Upper Loch Maree and Strath Broom. (Explanation of Sheet 92.) By B. N. PEACH, LL.D., F.R.S.; J. HORNE, LL.D., F.R.S.; the late W. GUNN; C. T. CLOUGH, M.A., F.G.S., and E. GREENLY, F.G.S. *Price* 2*s.* 6*d.* Edinburgh, 1913.

Diplomatic and Consular Reports.—Trade, etc., of the Portuguese Possessions in East Africa, 1912 (5210); Trade of Casablanca, 1912 (5217); Trade of Milan, 1912 (5218); Foreign Trade of Italy, 1912 (5219); Finances of Greece, 1912-13 (5224); Foreign Commerce of Russia and Trade of the Consular District of St. Petersburg, 1912 (5197); Trade, etc., of Denmark, 1912 (5221); Trade, etc., of Düsseldorf, 1912 (5222); Commerce and Industry of the Atlantic Coast of Honduras, 1912 (5215); Trade of Tsinan and Tsingtau, 1912 (5223).

Colonial Reports (Annual).—Basutoland, 1912-13 (769); Gold Coast, 1912 (770); Nyasaland, 1912-13 (772).

Diplomatic and Consular Reports (Miscellaneous).—Report on the Supply of Electricity in Germany by the chief works in which private concerns and public bodies are jointly interested (685).

Publishers forwarding books for review will greatly oblige by marking the price *in clear figures, especially in the case of foreign books.*

ROYAL SCOTTISH GEOGRAPHICAL SOCIETY.

REPORT OF COUNCIL.

TWENTY-NINTH SESSION, 1912-1913.

The Council has the honour to submit the following Report:—

ORDINARY MEMBERSHIP.

The changes which occurred during the Session in the number of members were as follows:—

Number on 1st October 1912,		1898
New Members added,		50
		1948
Deduct by Death,	54	
„ Resignation,	147	
„ Arrears,	20	
		221
Number of ordinary members remaining on roll on 30th September 1913,		1727

Of this number, 920 are on the Edinburgh list, 422 are on the Glasgow list, 149 and 113 on the Dundee and Aberdeen lists respectively. In addition to those on the lists named, 35 members reside abroad, and 88 reside in England. The number of life members in the total given above is 222.

TEACHER ASSOCIATE MEMBERS.

Number on 1st October 1912,		166
New Members added,		34
		200
Deduct by Death,	1	
„ Resignation,	36	
		37
Number of Teacher Associates remaining on roll on 30th September 1913,		163

CORRESPONDING MEMBERS.

The corresponding members number forty-six.

MEETINGS OF THE SOCIETY.

The Society's Anniversary Meeting was addressed by Captain Roald Amundsen. Thirty-one ordinary meetings were held, seven in Edinburgh and eight in Glasgow, Dundee, Aberdeen. These meetings were addressed by Miss Ellen Churchill Semple; the Hon. Sir Newton J. Moore, K.C.M.G., J.P., M.L.A.; Captain Ejnar Mikkelsen; Mr. A. Bentinck, F.R.G.S.; Mr. F. Kingdon Ward, F.R.G.S.; Prof. J. Norman Collie; Prof. D'Arcy Thompson, C.B.; Sir Harry Johnston, G.C.M.G., K.C.B.; Mr. Harry de Windt, F.R.G.S.; Mrs. Mary Gaunt; Dr. Grieve; and Mr. Geo. G. Chisholm, M.A., B.Sc.

Medals Awarded.

The Livingstone Gold Medal was conferred on Captain Roald Amundsen for his valuable geographical discoveries in connection with his recent expedition to the South Pole.

The Society's Silver Medal was awarded to Captain Ejnar Mikkelsen for his explorations in North-Eastern Greenland.

The Silver Medal offered by the Society in connection with the Lectureship in Geography at Edinburgh University was awarded to Mr. Laurance J. Saunders for his essay entitled "A Geographical Description of Fife, Kinross, and Clackmannan."

The Late Colonel Bailey, R.E.

The Council have to express their regret at the loss which the Society has sustained through the death of Colonel Bailey, a minute with regard to which has already been published in the *Magazine* (p. 153).

The Disaster to Captain Scott's Antarctic Expedition.

On the instruction of the Council the Secretary of the Royal Scottish Geographical Society sent a message of condolence, which has already been published in the *Magazine* (p. 154), to Lady Scott.

Gaelic Place-Names.

A request having been received from the Director-General of the Ordnance Survey asking for the co-operation of the Society in the settling of queries in the spelling of Gaelic Place-Names in Scotland, the Council decided to revive the place-names committee. The committee is not confined to members of the Society, and has Dr. John Watson, Rector of the Royal High School, as chairman, the convener being Mr. John Mathieson. The committee have also secured the services of gentlemen as corresponding members in the counties of Argyll, Aberdeen, Inverness, Perth, Ross and Sutherland. Six meetings have been held since their appointment in March last, and over forty cases have been submitted to them for their opinion.

Recently the Survey authorities submitted to them for criticism a glossary of Gaelic and corrupt forms of Gaelic and Norse words, which is to appear in the pocket editions of the Ordnance Survey maps, and the Director-General of the Survey has thanked them for their assistance. The committee received considerable assistance from the publication in the *Magazine* of their letter in June last, and they have now issued forms for the use of correspondents.

Livingstone Centenary.

Meetings commemorative of the centenary of the birth of Dr. David Livingstone were held by the Society in Edinburgh and Glasgow. Sir Harry Johnston, G.C.M.G., K.C.B., addressed the meetings.

Bust of Dr. Livingstone.

Captain Livingtone Bruce has very kindly lent to the Society a marble bust of his grandfather, Dr. David Livingstone.

Association of Scottish Teachers of Geography.

The Association of Teachers of Geography in Scotland, under the auspices of the Society, has held several meetings, and although so far small in numbers it promises to perform useful work. It has been the means of attracting additional teacher associates to the Society, and as the Association becomes better known it will probably continue to do so more largely, besides securing members of the

Association outside the Society. There is some prospect of a branch being started in Aberdeen. The Association has made a beginning with discussing the scope of geography in the elementary and supplementary classes in schools, and it is hoped that this side of the Association's work will be further developed next session.

Diploma of Fellowship.

There are at present six Honorary Members, forty-seven Honorary Fellows, and one hundred and six Ordinary Fellows of the Society.

During the session the ordinary Diploma of Fellowship was awarded to the Rev. S. B. Rohold, Toronto.

The Society's Magazine.

The *Scottish Geographical Magazine* has, as usual, been published throughout the past session monthly, with maps and illustrations.

The Council is glad to acknowledge its obligations to the contributors of articles, and to the following gentlemen who have rendered valuable assistance to the editors:—Dr. J. G. Bartholomew; Mr. E. S. Bates; Dr. R. N. Rudmose Brown; Dr. W. S. Bruce; Mr. H. M. Cadell; Mr. Geo. G. Chisholm; Dr. A. S. Cumming; Rev. A. K. Dallas; Mr. G. F. Scott Elliot; Mr. H. B. Finlay; Prof. J. W. Gregory; Sir Philip Hamilton Grierson; Mr. N. Miller Johnston; Mr. W. R. Kermack; Rev. R. Mackenzie; Mr. J. Mathieson; Mr. J. D. Monro; Mr. T. S. Muir; Mr. J. Murray; Mr. H. J. Peddie; Mr. D. Pryde; Mr. Ralph Richardson; Mr. Andrew Watt; Mr. R. Wilkie; Mr. W. B. Wilson.

Library and Map Department.

During the past session 393 books, 57 pamphlets, 402 reports, 9 atlases, 394 map-sheets and charts have been added to the library. The Transactions and periodicals received regularly number 175. The number of volumes borrowed by members was 4262, and the library was, as usual, much consulted by non-members in search of geographical information.

The Council desires to record its thanks to foreign and colonial governments for the official publications they have presented to the library; to the Treasury, for the revised Ordnance Survey Maps of Scotland, both in outline and colour, as each of the revisions now in progress is published; to the Geological Survey of Scotland; to the Geographical Section, General Staff, War Office; to the Admiralty; to the British Museum (Natural History), the Edinburgh University Library, the Royal Society of Edinburgh, and the Survey of India; and also to the undermentioned private donors of books and maps, viz.:—H.R.H. The Duke of the Abruzzi; Dr. J. G. Bartholomew; Mr. Charles Boog-Watson; Dr. W. S. Bruce; Mr. J. Y. Buchanan, M.A.; Dr. James Burgess, C.I.E.; Mr. W. Caldwell Crawford; Dr. Jean Charcot; Mr. C. B. Crampton, M.B., C.M.; Mr. G. F. Scott Elliot; Sir Sven Hedin, K.C.I.E., LL.D.; Mr. Henri A. Junod; Mr. John Menzies; Captain Ejnar Mikkelsen; Prof. E. Oberhummer; Dr. Albrecht Penck; Major-General J. de Schokalsky; Mr. Lewis Spence; Mr. W. Stevenson; Mr. J. B. Tyrrell, F.G.S.; and Prof. Eug. Warming.

Glasgow, Dundee, and Aberdeen Centres.

The Council has again pleasure in acknowledging the services rendered by the honorary officials of the Glasgow, Dundee, and Aberdeen centres for their continued successful conduct of the business of the Society.

Finance.

The Council begs to submit the Annual Financial Statement.

ABSTRACT OF THE ACCOUNTS OF THE ROYAL SCOTTISH [illegible]

From 1st October 1912 *to* 30*th September* 1913.

Funds at the Close of last Account:—			
£1000 Glasgow and South-Western Railway Company 4 per cent. Debenture Stock, at cost price, as at 29th May 1885,			£1102 19 9
£533, 6s. 8d. North British Railway Company 3 per cent. Consolidated Lien Stock, at cost price, as at 30th December 1890,			496 2 0
£200 Great Central Railway Company 4½ per cent. Debenture Stock, at cost price, as at 31st December 1898,			312 2 9
£240 North British Railway Company 4 per cent. Consolidated Preference Stock, No. 1, at cost price, as at 11th December 1902,			306 8 9
£400 Dominion of Canada 3½ per cent. Registered Stock, 1930-50, at cost price, as at 9th March 1911,			405 1 0
			£2622 14 3
Balance due by Banks:—			
On Account Current,		£83 7 10	
On Deposit Receipt,		304 19 5	
On Do. do. (Glasgow Equipment Fund)		63 8 8	
Due by Treasurer,		4 3 9	
			455 19 8
			£3078 13 11
Less—Subscriptions received applicable to Sessions 1912-13 and 1913-14,			12 12 0
			£3066 1 11
Subscriptions from Ordinary Members,		£1694 14 0	
Less—Paid in Advance,		18 18 0	
		£1675 16 0	
Subscriptions from Associate Members,	£91 7 0		
Less—Paid in Advance,	0 10 6		
		90 16 6	
Fees received for Society's Diploma,		3 3 0	
Dividends and Interests,		110 0 4	
Government Grant,		200 0 0	
Miscellaneous,		5 0 8	
Entrance Fees from new Members,	£33 1 6		
Less—Paid in Advance,	2 12 6		
		30 9 0	
			2115 5

Magazine—Expenses of Publication, etc., for Nos. 10 to 12 of Vol. XXVIII., and Nos. 1 to 9 of Vol. XXIX.:—			
Printing and other charges,			£614 1 0
Illustrations,			42 19 2
Maps,			17 8 1
			£674 8 3
Less—Magazines sold,		£70 8 2	
Advertising Receipts, less Payments,		77 18 2	
			148 6 4
			£526 1 11
General Printing,			24 2 0
Books and other furnishings for Library,			43 17 0
Expenses in connection with Lectures—*Less* Lecture Tickets sold,			541 2 7
Glasgow Branch, Rent of Room,			15 0 0
Rent of Council Room, Taxes, Gas, Repairs, Insurance, etc.,			232 19 1
Salaries,			572 9 2
Miscellaneous, including Stationery, General Expenses, and Postages,			101 1 8
Medals,			3 4 0
			£2059 17 5
Funds at the Close of this Account:—			
£1000 Glasgow and South-Western Railway Company 4 per cent. Debenture Stock at cost price, as at 29th May 1885,		£1102 19 9	
£533, 6s. 8d. North British Railway Company 3 per cent. Consolidated Lien Stock, at cost price, as at 30th December 1890,		496 2 0	
£200 Great Central Railway Co. 4½ per cent. Debenture Stock, at cost price, as at 31st December 1898,		312 2 9	
£240 North British Railway Company 4 per cent. Consolidated Preference Stock, No. 1, at cost price, as at 11th December 1902,		306 8 9	
£400 Dominion of Canada 3½ per cent. Registered Stock, 1930-50, at cost price, as at 9th March 1911,		405 1 0	
		£2622 14 3	
Balance due by Banks:—			
On Account Current,	£49 0 4		
On Deposit Receipt,	404 19 5		
On do. do. (Glasg. Equipment Fund),	63 8 8		
Due by Treasurer,	3 8 4		
		520 16 9	
		£3143 11 0	
Less—Subscriptions received in advance:—			
Ordinary Members,	£18 18 0		
[illegible]ssociate Members	0 10 6		

LIVINGSTONE MEDAL FUND.

FUNDS AT THE CLOSE OF LAST ACCOUNT :—		
£770 North British Railway Company 4 per cent. Preference Stock No. 1, at cost price,	£999 11 6	
Deposit Receipt,	50 13 5	
	£1050 4 11	
INCOME RECEIVED—		
Dividend on £770 North British Railway Company 4 per cent. Preference Stock No. 1— From 31st January 1912 to 30th June 1913,		41 1 4
Bank Interest,		1 12 3
Income-tax recovered,		1 13 2
		£1094 11 8

Paid Messrs. Alexander Kirkwood & Sons for Medal,		£30 0 0
FUNDS AT THE CLOSE OF THIS ACCOUNT—		
£770 North British Railway Company 4 per cent. Preference Stock No. 1, at cost price,	£999 11 6	
Deposit Receipts,	65 0 2	
		1064 11 8
		£1094 11 8

Edinburgh, 21*st October* 1913.—Examined and found correct.

P. C. ROBERTSON, M.A., C.A., Auditor.

INDEX: VOL. XXIX.

In the following Index the ALPHABETICAL ORDER *is adhered to throughout. Titles of Papers are in deeper type. Contraction,* rev. = *Review in the Magazine.*

ILLUSTRATIONS.

ASIA.

AFRICA.

AUSTRALASIA.

ARCTIC.

ANTARCTIC.

GENERAL.

PORTRAITS.

MAPS AND DIAGRAMS.

END OF VOLUME XXIX.

Printed by T. and A. Constable, Printers to His Majesty
at the Edinburgh University Press

Lightning Source UK Ltd.
Milton Keynes UK
UKHW011224061118
331795UK00010B/1408/P

9 780260 720139

C000010336

EXPLORE AUSTRALIA

EXPLORE AUSTRALIA 2012

CONTENTS

EXPLORE AUSTRALIA

Australia is breathtaking in beauty, daunting in size and rich in diversity. There is so much to explore as you travel the length and breadth of this extraordinary country. For centuries 'The Great South Land' remained undiscovered and unexplored by Europeans, and it is this relative 'newness' that has helped protect so many of its secrets. As a visitor you have an astonishing 7 686 850 square kilometres of land to cover – so plan your trip wisely!

For more than 50 000 years before European settlement, Aboriginal people lived on the continent, occupying country across the landscape, including its driest deserts. Living in great affinity with the land on which they were dependent, they established rich, diverse and highly spiritual cultures, with sacred places, Dreaming tracks and art sites. With permission, much of this can be explored by travellers who are prepared to sit, listen and learn from Indigenous Australians.

Australians live and work right across this vast continent. Graziers, farmers, jack and jillaroos, and miners work the land and there are plenty of outback rural towns where the warm greeting you'll receive will be as welcome as the cold drink served at the local pub.

However, most Australians live in coastal cities. The expanding east-coast cities of Sydney, Melbourne and Brisbane are the three largest in the country and are representative of people from around the world. They have thriving arts, music, sport and dining cultures. The other capital and major cities also have unique flavours and heritage, and are worth visiting in their own right.

A visit to Australia means you must travel widely, taste boldly and be tempted to find out what's around that next corner. If you're lucky (you are travelling in the lucky country after all!), you might just uncover another of Australia's secrets.

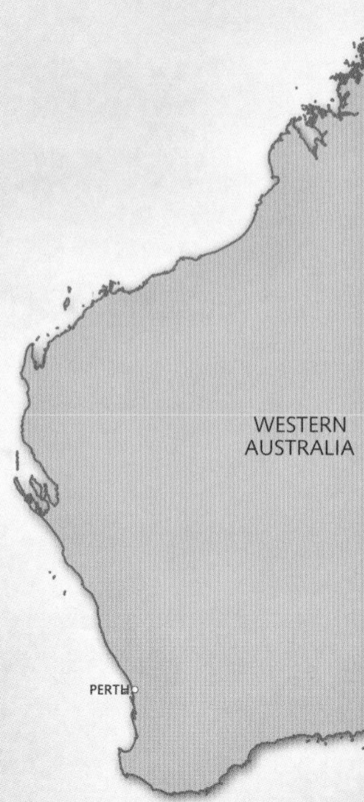

INDIGENOUS AUSTRALIANS

For an estimated 50–65 000 years, Aboriginal people have lived in Australia. They are believed to have occupied the country from the north, reaching Tasmania about 35 000 years ago. Aboriginal people were traditionally separated into some 600 different societies, most with completely different languages from those around them. Rather than settling in one spot, they lived and moved around their 'country'.

There are close to half a million Aboriginal people in Australia today, and although much of the cultures have been destroyed or damaged by European settlement, many Aboriginal communities still proudly carry on rich traditions, and have some of the oldest surviving cultural practices in the world.

Must-see Indigenous places and experiences:

- Uluṟu-Kata Tjuṯa National Park, Northern Territory
- Kimberley rock art sites, Western Australia
- Kakadu National Park/Arnhem Land, Northern Territory
- Thursday Island, Torres Strait
- Burrup Peninsula, Western Australia
- Bunya Mountains, Queensland
- Tjapukai Aboriginal Cultural Park, Queensland

[ABORIGINAL DANCERS AT THE BARUNGA FESTIVAL, NORTHERN TERRITORY]

WORLD HERITAGE SITES

[PORT ARTHUR HISTORIC SITE, TASMANIA]

Australia has 18 listings on UNESCO's World Heritage register. Most are areas of extreme natural beauty and hold exceptional conservation value. However, the latest additions placed on the list in 2010 included 11 penal sites, such as Port Arthur in Tasmania and Fremantle Prison in Western Australia. There are two other Australian sites listed exclusively for cultural reasons: the Sydney Opera House and Melbourne's Royal Exhibition Building in Carlton Gardens.

The other listings are:

- Australian Fossil Mammal Sites: Riversleigh, Lawn Hill National Park, Queensland and Naracoorte Caves, South Australia
- Fraser Island, Queensland
- Gondwana Rainforests, Queensland and northern NSW
- Great Barrier Reef, Queensland
- Greater Blue Mountains Area, New South Wales
- Heard and McDonald Islands, sub-Antarctic islands
- Kakadu National Park, Northern Territory
- Lord Howe Island Group, New South Wales
- Macquarie Island, sub-Antarctic island
- Purnululu National Park, Western Australia
- Shark Bay, Western Australia
- Tasmanian Wilderness
- Uluṟu-Kata Tjuṯa National Park, Northern Territory
- Wet Tropics of Queensland
- Willandra Lakes Region, New South Wales

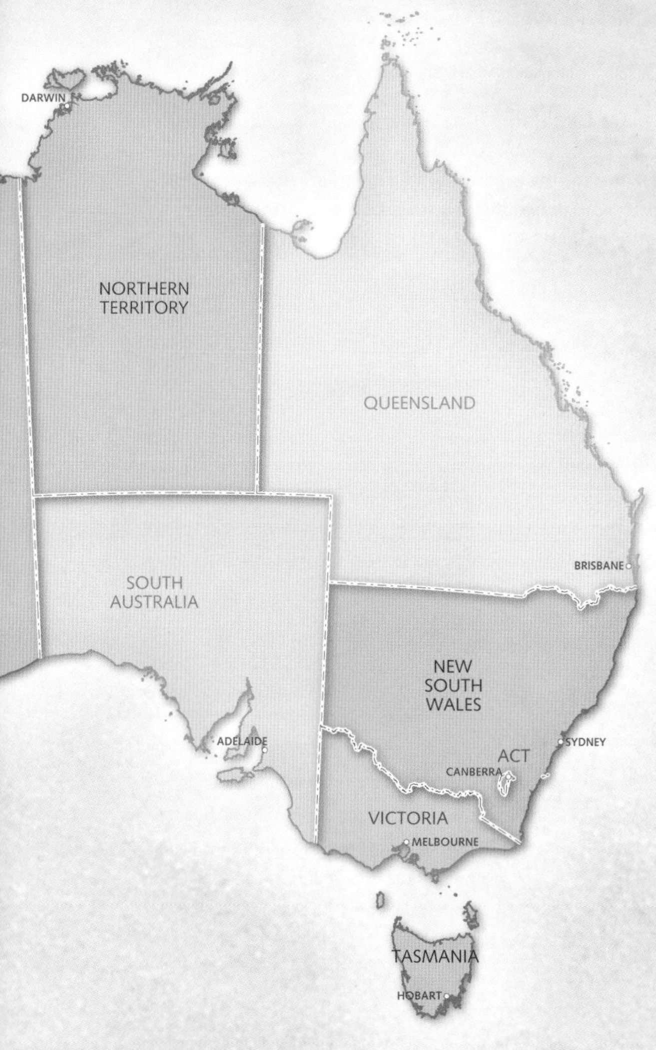

NATURAL WONDERS

One of the great attractions of travelling around Australia is the unique plant and animal life. You'll see startlingly white ghost gums set against red rocks in Central Australia, and in contrast, lime-green cushion plants in the Tasmanian highlands. Australia is exceedingly rich in diversity, in fact, south-western Australia is one of the world's major biodiversity hotspots.

Australia holds more than 18 000 flowering plant species, as well as grasses and innumerable fungi and lichens, and thousands of these species are unique to this country.

Our animal life includes the largest collection of marsupials on earth, as well as more than 860 bird species (about half of which are only found here), a similar number of reptile species, and more than 200 species of frog. In the water you can also find more than 4400 fish species, and 400 coral species on the Great Barrier Reef alone.

Australia is truly a land of natural wonders.

[SNORKELLING IN THE GREAT BARRIER REEF, QUEENSLAND]

BUILDING A STYLE

[FEDERATION SQUARE, MELBOURNE, VICTORIA]

As an eclectic mix of cultures, Australia has a rich diversity of styles on which to build its architecture. From the Aboriginal humpy or shelter, to the world-famous white sails of the Sydney Opera House, architects have much to admire and ponder.

Much of suburbia is dominated by production-line 'McMansions', but occasionally hints of Australia peak through: bits of corrugated iron; large, airy verandahs; the classic 'Queenslander' up on stilts; and in historic or official buildings, the prolific use of granite or local sandstone.

Increasingly, Australian architecture is focusing on sustainability. This can mean everything from solar technology to the use of natural heating and cooling, and the return of sustainably harvested timbers – each making use of our resourceful land.

FOOD AND WINE

[VINEYARDS IN THE BAROSSA VALLEY, SOUTH AUSTRALIA]

Australia is one of the world's major food producers. Wherever you travel, food production is apparent on the landscape, from the vast inland sheep and cattle stations, to the canefields of New South Wales and Queensland. Coastal towns abound in fresh seafood, and increasingly there is an interest in other foods indigenous to Australia, such as bush plums, nuts, herbs and kangaroo. Visit the many farm gate or food trails to access this local produce, or pick up fresh bargains at farmers' markets.

Australia is consistently in the world's top ten producers of wine, making more than 1.4 billion litres a year. Australia's major wine growing regions are nearly all tourist attractions in their own right, with delightful B&Bs, historic places to explore and cellar doors galore.

Don't miss these experiences:

- Honey from Ligurian bees on Kangaroo Island, South Australia
- Margaret River's chocolate, dairy and wines, Western Australia
- Sydney Fish Markets, New South Wales
- Tasmania's pinot noir, cheese and berries
- Queensland's mangoes and other tropical fruits
- Historic wine regions such as Rutherglen, Victoria
- Barramundi and crocodile, Northern Territory

TRAVEL THROUGH HISTORY

Although still relatively young, Australia has an interesting and rich history since Europeans 'discovered' her, and many of the most important sites can be visited and experienced by travellers today. Wars on home soil have been mercifully limited, but there have been no end of other challenges and conflicts, including floods, droughts, fires and plagues.

Through this turbid history, the legend of the 'Australian spirit' has been forged: easygoing outdoor-loving people who will barrack for the underdog, stick by their mates, challenge authority, and do it all with a sense of humour that's as dry as the Simpson Desert.

~50 000 YEARS AGO

First people arrive in northern Australia.

1629

[THE VOYAGE OF THE BATAVIA]

The Dutch ship *Batavia*, one of several to have explored the Great South Land, is wrecked on the Houtman Abrolhos, WA. See timbers from the wreck at the Western Australian Museum – Maritime, at Fremantle, and discover more about our early maritime explorers.

1770

[JAMES COOK]

Englishman Captain James Cook, in the *Endeavour*, sights the east coast, landing several times. Visit Botany Bay, in Sydney, and the town of Seventeen Seventy, 400 km north of Brisbane.

1788

The First Fleet arrive, establishing the first settlement at Botany Bay, then Sydney Cove.

1803

A settlement is established on the Derwent River in Tasmania – Hobart is born.

1813

Blaxland, Lawson and Wentworth find a way through the Blue Mountains, New South Wales, one of the barriers to settling further west. See the legendary tree they are believed to have blazed near Katoomba.

1824

The Moreton Bay Penal Settlement is established near what will become Brisbane.

1830s

Port Arthur in Tasmania becomes one of the prime penal colonies in Australia.

1851

[ZEALOUS GOLD DIGGERS IN BENDIGO]

Gold is discovered in New South Wales and Victoria, leading to mass gold rushes. Relive this lustrous history in Victorian towns such as Clunes, Bendigo and Ballarat.

1860–61

[RETURN OF BURKE AND WILLS TO COOPER CREEK]

Explorers Burke and Wills cross the continent from south to north, but then die on the return, tragic journey. See the infamous Dig Tree and death sites along Cooper Creek near Innamincka, in the far north-east corner of South Australia.

1872

The 3200 km Overland Telegraph line between Darwin and Port Augusta, South Australia, is completed, allowing fast communication between Australia and the rest of the world. The repeater station at Alice Springs has been partially restored.

1880

[PORTRAIT OF NED KELLY]

Australia's best-known bushranger, Ned Kelly, is captured at Glenrowan, in Victoria, and hung at Melbourne Gaol.

1891

Shearers across Australia go on strike for better pay and conditions. The protests are centred around Barcaldine in Queensland.

MID-1890s

Gold is discovered in remote Western Australian fields, particularly around Kalgoorlie and Coolgardie, prompting another gold rush.

1901

[MELBOURNE'S ROYAL EXHIBITION BUILDING]

The six Australian colonies form the Commonwealth of Australia in a grand celebration at the Royal Exhibition Building in Melbourne's Carlton Gardens. The building is now listed on the World Heritage register.

1908

Land known as Canberra, at the foothills of the Australian Alps, is chosen as the site of the national capital.

1914–18

World War I: Australia fights against Germany and its allies. Vessels carrying ANZAC soldiers bound for Gallipoli depart from Albany in Western Australia. Learn more of this history at the Australian War Memorial in Canberra.

1927

The former Parliament House opens in Canberra. It now houses the Museum of Australian Democracy.

1939–45

[AUSTRALIAN SOLDIERS DURING WORLD WAR II]

World War II: Australia fights in Europe and the Pacific. Darwin is bombed and Sydney attacked by Japanese submarines. There are still plenty of old WW II airstrips hidden in the dense jungles and scrub of the Top End.

1949–74

Australia's largest engineering project to date, the Snowy Mountains Hydro-Electricity Scheme, is built.

1955–63

Large atomic bombs are detonated for testing at Maralinga in western South Australia. Visit the ground zero site at Emu Junction, or the museum at Woomera.

1965

Australia joins the Vietnam War, last combat troops come home in 1972.

1973

[GOUGH WHITLAM AT THE NATIONAL PRESS CLUB]

Legal end of the white Australia policy with the Whitlam government.

1974

Cyclone Tracy devastates Darwin on Christmas Eve.

1988

[CANBERRA'S NEW PARLIAMENT HOUSE]

At the bicentenary of settlement, Australia's new Parliament House opens in Canberra.

1992

Native land title is recognised after a decade of litigation. This is known as the 'Mabo' Judgement.

2000

Olympic Games held in Sydney. Tour the Olympic site at Homebush.

2008

On February 13, Prime Minister Kevin Rudd apologises for the hurt caused by decades of state-sponsored ill-treatment of Indigenous Australians.

2009

In February, Victoria suffers the most catastrophic bushfire disaster in history, with the loss of 173 lives. Saturday, February 7, becomes known as Black Saturday. See towns that have rebuilt in places like Kinglake and Marysville.

2011

In January, three-quarters of Queensland is declared a disaster zone as a result of flooding. Highly affected populated areas included Toowoomba, Brisbane and Ipswich. In February severe tropical Cyclone Yasi devastates Queensland again.

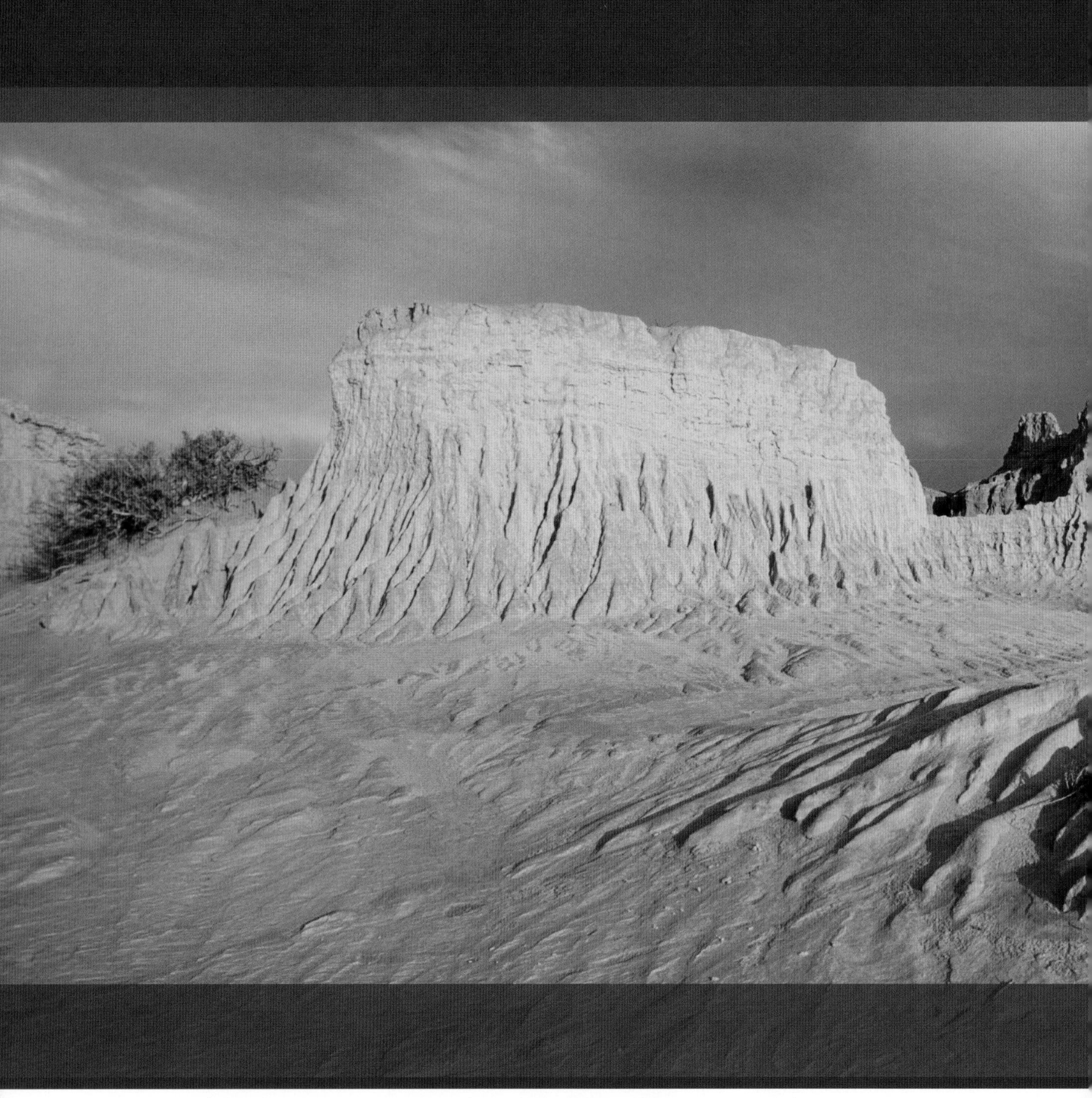

NEW SOUTH WALES IS...

Bodysurfing at BONDI / Viewing the Three Sisters rock formation in the BLUE MOUNTAINS / Skiing at Thredbo in KOSCIUSZKO NATIONAL PARK / Tasting wines at cellar doors in the HUNTER VALLEY / A country music experience in TAMWORTH / Seeing weather-eroded sand formations and skeletons of

NEW SOUTH WALES

[THE WALLS OF CHINA, MUNGO NATIONAL PARK]

megafauna in MUNGO NATIONAL PARK / Walking along Byron Bay's beaches to CAPE BYRON LIGHTHOUSE, Australia's most easterly point / A whale-watching boat tour in MERIMBULA / Strolling through the historic town of BERRIMA, visiting antique shops along the way / Fossicking for opals at LIGHTNING RIDGE

NEW SOUTH WALES is a land of contrasts. Lush rainforests, pristine beaches, snowfields and the rugged beauty of the outback all vie for visitors' attention.

Beaches are a clear drawcard, with those at Bondi and Byron Bay among the most popular. Surfing, swimming and whale-watching can be enjoyed almost anywhere along the coast, but Hyams Beach in Jervis Bay National Park is home to the whitest sand in the world.

The discovery of Mungo Man and Woman, the miraculously preserved remains of two ancient Aboriginal people found in a dune over three decades ago, prove that civilisation existed here 40 000 years ago. Numerous Aboriginal nations have called the state home, and still do. Well-preserved fish traps in Brewarrina are thought by some to be the oldest man-made structures in the world.

After the American War of Independence spelt the end for British penal settlements in North America, the recently annexed New South Wales was an obvious solution to the problem of overcrowded prisons. Conditions were harsh; the first inmates of Maitland Gaol, who

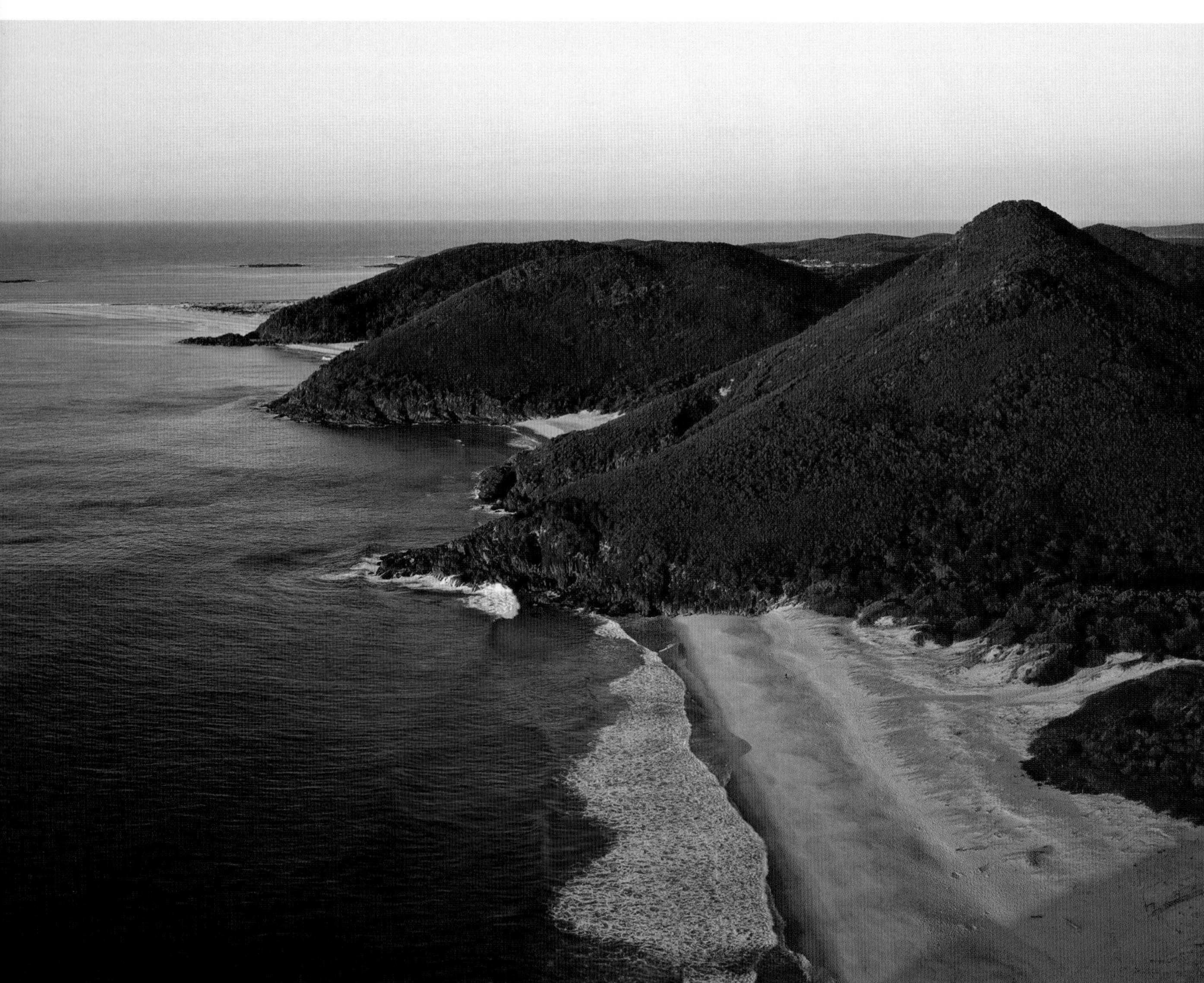

[ZENITH BEACH AT SUNRISE, TOMAREE NATIONAL PARK]

included many children, were forced to march the 6 kilometres from the wharf at Morpeth to the prison in shackles and chains.

Harsh conditions were not limited to the prisons. In 1845, due to the pitiless terrain explorer Charles Sturt lost his second-in-command and was stranded for six months in the outback near Milparinka. Today remote Silverton stands as a reminder of outback isolation, with its buildings and stark surrounds featuring in Australian films such as *Mad Max II* and *The Adventures of Priscilla, Queen of the Desert.*

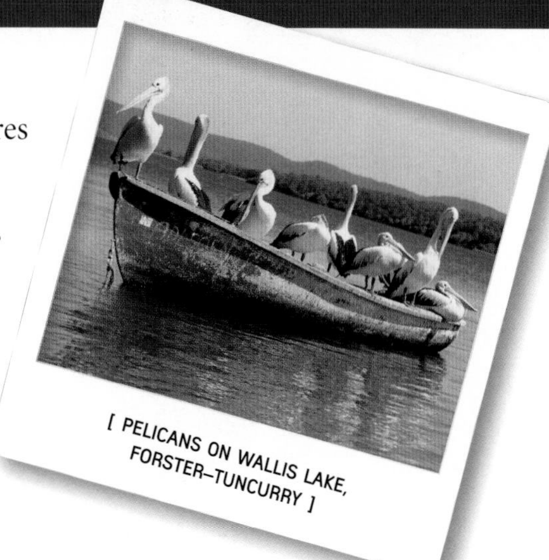

[PELICANS ON WALLIS LAKE, FORSTER–TUNCURRY]

In stunning contrast, Sydney's bright lights and sophistication sit beside the sparkling waters of the largest natural harbour in the world. The iconic Harbour Bridge and Opera House, along with the successes of the 2000 Olympics and the popular Gay and Lesbian Mardi Gras, have ensured Sydney and New South Wales a place on the world stage.

fact file

Population 7 238 800
Total land area 800 628 square kilometres
People per square kilometre 8.5
Sheep per square kilometre 48
Length of coastline 2007 kilometres
Number of islands 109
Longest river Darling River (1390 kilometres)
Largest lake Lake Eucumbene (dam), (145 square kilometres)
Highest mountain Mount Kosciuszko (2228 metres)
Highest waterfall Wollomombi Falls (220 metres), Oxley Wild Rivers National Park
Highest town Cabramurra (1488 metres)
Hottest place Bourke (average 35.6°C in summer)
Coldest place Charlotte Pass (average 2.6°C in winter)
Wettest place Dorrigo (average 2004 millimetres per year)
Most remote town Tibooburra
Strangest place name Come-by-Chance
Most famous person Nicole Kidman
Quirkiest festival Stroud International Brick and Rolling Pin Throwing Competition
Number of 'big things' 49
Most scenic road Lawrence Hargrave Drive, Royal National Park
Favourite food Sydney rock oysters
Local beer Tooheys
Interesting fact The Stockton Sand Dunes, 32 kilometres long, 2 kilometres wide and up to 30 metres high, form the largest moving coastal sand mass in the Southern Hemisphere

gift ideas

Cookies (Byron Bay Cookie Company, Byron Bay) Drool-worthy cookies in flavours like triple choc fudge or white choc chunk and macadamia nut. See Byron Bay p. 53, 557 O3

Australian country music CD (Big Golden Guitar Tourist Centre, Tamworth) Bring back a music sample by one of Australia's talented country music legends. See Tamworth p. 96, 556 H9

Mead (Dutton's Meadery, Manilla) Manilla is home to one of only two meaderies in Australia, producing this ancient alcoholic beverage from fermented local honey and water. Purchase some bottles and see the private museum. Barraba St, Manilla. See Manilla p. 79, 556 H8

Replica of Sydney Opera House (Opera House Shop, Sydney) Take home a tiny replica of this icon in gleaming pewter or sparkling Waterford crystal. See Sydney Opera House p. 11, 8 C2

Sydney Harbour Bridge coathanger (Pylon Lookout Shop, Sydney) A quirky – and useful – variation on the old 'Coathanger' itself. See Sydney Harbour Bridge p. 13, 8 C1

Beach gear (Bondi Beach) Sun hats, beach towels and unique swimwear that sport the name of this iconic beach in bold letters are available from various stores. See Bondi Beach p. 17, 6 D6, C5

Arts and crafts (Paddington Markets) Local artisans sell fabulous artworks, jewellery, clothing and collectibles. See Markets p. 14, 6 C5

Gumnut products (Nutcote, Neutral Bay) The home of May Gibbs sells beautiful books, postcards, CDs, DVDs, tea towels and more, featuring the enchanting lives of Snugglepot, Cuddlepie and the Banksia Men. See Nutcote p. 18, 6 C3

Wool products (The Big Merino, Goulburn) Stylish woollen clothing, slippers lined with lambs wool and beauty products made from lanolin. See Goulburn p. 65, 565 E3

Wine (Hunter Valley) Try the shiraz and chardonnay from one of Australia's most renowned wine-producing districts. See Hunter Valley & Coast p. 26, 543 D1

SYDNEY IS...

Bodysurfing at **BONDI** / Wandering through the lanes and alleyways of **THE ROCKS** / Views from **SYDNEY TOWER** / A trip on the **MANLY FERRY** / Fish and chips at **WATSONS BAY** / Soaking up the atmosphere of **KINGS CROSS** / A picnic at **TARONGA ZOO** / Harbour views from **MRS MACQUARIES CHAIR** / A performance at the **SYDNEY OPERA HOUSE** / Climbing the **SYDNEY HARBOUR BRIDGE** / A stroll through **BALMAIN** / Bargain hunting at **PADDY'S MARKETS** / Following the Games Trail at **OLYMPIC PARK** / Visiting **ELIZABETH FARM** in Parramatta

VISITOR INFORMATION

Sydney Visitor Centre
→ Level 1, cnr Argyle and Playfair sts, The Rocks
→ Palm Grove between Cockle Bay Wharf and Harbourside, Darling Harbour
(02) 9240 8788 or 1800 067 676
www.sydneyvisitorcentre.com

Australia's largest city stretches from the shores of the Tasman Sea to the foot of the Blue Mountains. Along with outstanding natural assets – stunning beaches, extensive parklands and the vast expanse of the harbour – Sydney boasts an impressive list of urban attractions, including world-class shopping and a host of superb restaurants and nightclubs.

Sydney began life in 1788 as a penal colony, a fact long considered a taint on the city's character. Today, echoes of those bygone days remain in areas such as The Rocks, Macquarie Street and the western suburb of Parramatta.

Since those early days, the one-time prison settlement has become one of the world's great cities. Home to two of Australia's most famous icons, the Sydney Harbour Bridge and the Sydney Opera House, Sydney attracts more than two million international visitors a year. For a true Sydney experience, try watching a Rugby League Grand Final at ANZ Stadium with a crowd of 80 000 cheering fans. Or if good food and fine wine are more your style, sample the waterfront dining at Circular Quay or Darling Harbour, and multicultural flavours in inner-city Darlinghurst.

With a population of 4 500 000, Sydney offers a multitude of activities. Surf the breakers at Bondi Beach or jump on a Manly ferry and see the harbour sights. Whatever you do, Sydney is a great place to explore.

Macquarie Park
Lane Cove NP
North Ryde
Chatswood West
Chatswood Chase Shopping Centre
Chatswood Plaza
Chatswood
North Willoughby
Castle Cove
H C Press Park
Middle Cove
Harold Reid Reserve
Sugarloaf Point
Garigal National Park
Manly Warringah War Memorial Park
Manly Reservoir
Manly Vale
North Balgowlah
Seaforth
Manly West Organic Market (Sat)
Balgowlah
Fairlight
Lane Cove North
Lane Cove West
Artarmon
Artarmon Reserve
Willoughby
Castlecrag
The Pinnacle
The Spit Bridge
The Spit
Clontarf
Balgowlah Heights
Sydney Harbour NP
Lane Cove
Northbridge
Northbridge Golf Course
Beauty Point
Fig Tree Point
Pearl Bay
Naremburn
Riverview
Linley Point
Fig Tree Bridge
Longueville
Osborne Park
Crows Nest
Cammeray
Long Bay
Balmoral
Chinamans Beach
Wy-ar-gine Point
Grotto Point
Dobroyd Head
The Sound
Hunters Hill
Northwood
St Leonards
Burlington Bar and Dining
Greenwich
Cremorne
The Hayden Orpheum Picture Palace
Mosman
Rocky Point
Hunters Bay
Balmoral Beach
HMAS Penguin Naval Base
Middle Head
Wollstonecraft
Woolwich
Gladesville Bridge
Waverton
Mary MacKillop Place Museum
Greenwich
Neutral Bay
Georges Heights
Sydney Harbour
Parramatta River
Cockatoo Island
Spectacle Island
North Sydney
Nutcote
Cremorne Point
Clifton Gardens
Taronga Zoo
Milsons Point
McMahons Point
Birkenhead Point Shopping Centre
Drummoyne
Birchgrove
Kirribilli
Little Sirius Point
Kurraba Point
Chowder Bay
Chowder Head
Jackson
Lady Bay
Camp Cove
Green Point
Port
Balmain Market (Sat)
Balmain East
Goat Island
Sydney Harbour Bridge
Iron Cove Bridge
Rodd Island
Balmain
Bridge Hotel
Fort Denison
Robertsons Point
Bradleys Head
Watsons Bay
Vaucluse Point
Village Point
Steel Point
Rozelle
Royal Australian Navy Heritage Centre
Shark Island
Vaucluse
Vaucluse House
Lilyfield
Pyrmont
Garden Island
Potts Point
Sydney Harbour
Clark Island
Darling Point
Point Piper
Woolahra Point
Sydney
SEE PAGE 8
Elizabeth Bay
Elizabeth Bay House
Sydney Fish Market
Bicentennial Park
Leichhardt
Annandale
Glebe
Rushcutters Bay
Double Bay
Darling Point
Cosmopolitan Centre
Norton Plaza
Glebe Markets (Sat)
Broadway Shopping Centre
Ultimo
Haymarket
Darlinghurst
Surry Hills
Edgecliff Centre
Edgecliff
Bellevue Hill
Rose Bay
Royal Sydney Golf Course
Dover Heights
Annandale Hotel
Camperdown
The University of Sydney
Paddington
Strawberry Hills Hotel
Brett Whiteley Studio
Victoria Barracks Museum
Chauvel Cinema
Australian Centre for Photography
Paddington Markets (Sat)
Woollahra
Stanmore
Petersham
Darlington
Dendy Newtown
Vanguard
CarriageWorks
Belvoir St Theatre
Sydney Football Stadium
SCG
North Bondi
Newtown
Eveleigh
Redfern
Entertainment Quarter & EQ Village Markets (Sat–Sun)
The Forum
Cinema Paris
Westfield Bondi Junction
Bondi Beach
Bondi Markets (Sun)
Enmore
Enmore Theatre
Erskineville
Waterloo
Moore Park
Centennial Parklands Equestrian Centre
Centennial Park
Eastgate Shopping Centre
Bondi Junction
Bondi
Bondi Beach
St Peters
Marrickville Metro Shopping Centre
Sydney Park
Supa Centa Shopping Centre
Moore Park
Queens Park
Waverley
Tamarama
Mackenzies Point
Marrickville
Sydenham
Alexandria
Zetland
Bronte
Bronte Park
Beaconsfield
Randwick Racecourse
Randwick
Waverley Cemetery
Kensington
Clovelly
Tempe
Rosebery
Australian Golf Course
Royal Randwick Shopping Centre
Mascot
Shark Point
Cooks River Bridge
Giovanni Brunetti Bridge
Sydney Airport
The Lakes Golf Club
Eastlakes
Daceyville
Kingsford
Coogee
Coogee Bay Hotel
Wedding Cake Island
Astrolabe Park
Pagewood
South Coogee
Bondi to Coogee Walk

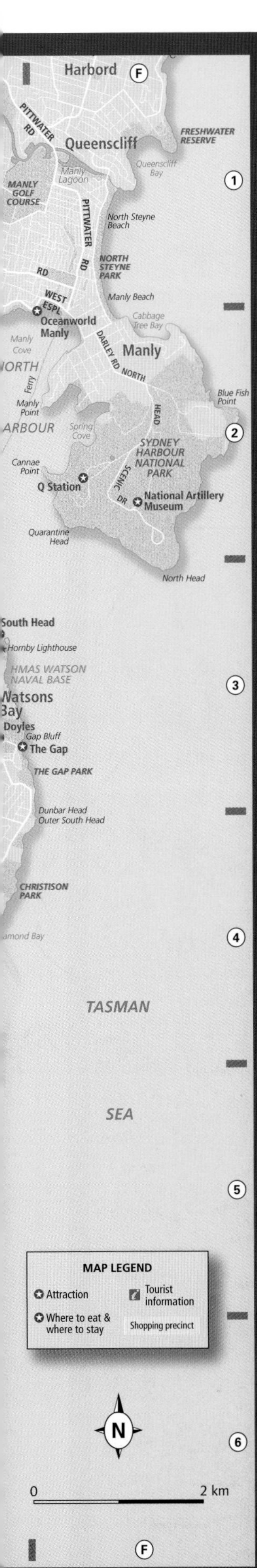

City Centre

Taking in the area between Martin Place, Bathurst Street, Elizabeth Street and George Street, this is primarily a retail district, with Pitt Street Mall, the main shopping precinct, situated between King and Market streets, and the exclusive shops of Castlereagh Street to the east.

Martin Place 8 B4

A sweeping vista of buildings in the High Victorian and Art Deco styles line Martin Place all the way to Macquarie Street. Chief among these notable buildings is the old **GPO** on the corner of Martin Place and George Street, which was designed by colonial architect James Barnet and has been gloriously transformed into the **Westin Sydney**, a five-star hotel with stunning interiors. The lower ground floor contains an up-market food hall along with a carefully preserved part of the old **Tank Stream**, which until recently was thought to be irretrievably lost below the streets of Sydney. Once a major source of water for the Eora people, it was also a deciding factor in the choice of Sydney Cove as a settlement site.

Pitt Street Mall 8 B4

If you head south along Pitt Street from Martin Place you will come to a busy pedestrian precinct. This is Pitt Street Mall, home of the Westfield Sydney Shopping Centre and the heart of the CBD's retail area, which is connected to a network of overhead walkways, small arcades and underground tunnels that lead to places as far away as David Jones and the QVB (the Queen Victoria Building). The mall houses beautiful department stores, stylish boutiques and vast emporiums selling music and books. Here too is the lovely **Strand**, the last of the old arcades in what was once a city of arcades.

Sydney Tower 8 B4

A visit to Sydney Tower is a must for any visitor. Your ticket grants you access to the 250-metre-high observation tower, which commands superb views of Sydney, all the way from the Blue Mountains to the Pacific. Access the tower by Westfield Sydney in Pitt Street Mall. Included in your ticket price is OzTrek, a virtual tour of Australia and the outback. For those who like to live dangerously, there's the new Skywalk, an outdoor walk over the roof of the Sydney Tower. *100 Market St; tickets www.myfun.com.au; open 9am–10.30pm daily.*

State Theatre 8 B4

An extravagant mix of Art Deco, Italianate and Gothic architecture, the State Theatre is the final word in opulence. Built in 1929 as a 'Palace of Dreams' with marble columns, mosaic floors and plush furnishings, it boasts a beautiful chandelier, the Koh-I-Noor, and several paintings by well-known Australian artists. Classified by the National Trust, it is still a working theatre and has seen performances by artists such as Bette Midler and Rudolf Nureyev. Self-guide tours of the building are available, but it's best to ring first to check times. *49 Market St; (02) 9373 6655.*

Queen Victoria Building 8 B4

It's astonishing to realise now that the QVB was once in danger of becoming a multistorey carpark, but this was indeed the case. Built in 1898 to replace the old Sydney markets, it was later used for a number of purposes, at various times housing a concert hall and the city library before being restored in 1984 to its former splendour. Now one of Sydney's most cherished landmarks, it is considered by some to be the most beautiful shopping centre in the world, with three levels of stylish shops and cafes. The QVB features elaborate stained-glass windows, intricate tiled floors, arches, pillars, balustrades and a mighty central dome. The building's best-kept secret is the old ballroom on the third floor, now used as the supremely elegant Tea Room. *455 George St.*

Sydney Town Hall 8 B4

Immediately to the south of the QVB stands the Sydney Town Hall, a wildly extravagant piece of Victoriana, now the seat of city government. Erected in 1869, it was built on the site of a convict burial ground – as recently evidenced by the accidental discovery of an old brick tomb. While you're here, slip inside for a look at the Grand Organ, which was installed in the Concert Hall in 1890. Free lunchtime concerts are held occasionally. *483 George St; (02) 9265 9189.*

climate

Sydney is blessed with a warm, sunny climate, often described as Mediterranean, making it possible to enjoy outdoor activities all year-round. The driest time of year occurs in spring, with autumn being the wettest season. Summers are hot and humid, with temperatures often in the mid-30s, while winters are cool and mostly dry, with temperatures usually around 15–17°C.

J	F	M	A	M	J	J	A	S	O	N	D	
25	25	24	22	19	16	16	17	19	22	23	25	MAX °C
18	18	17	14	11	9	8	8	11	13	15	17	MIN °C
102	116	130	125	122	128	97	81	69	77	83	77	RAIN MM
8	8	9	8	8	8	7	7	7	8	8	8	RAIN DAYS

SYDNEY HARBOUR NATIONAL PARK
Goat Island
Simmons Point
Blues Point Reserve
Blues Point
Milsons Point
Kirribilli
Wudyong Point
Kirribilli Point
Sydney Harbour Bridge
Walsh Bay
Dawes Point
Millers Point
Thornton Park
Balmain East
Illoura Reserve
Peacock Point
Clyne Reserve
Wharf Theatres
Pylon Lookout
Dawes Point Park
SYDNEY HARBOUR
Bennelong Point
Sydney Opera House
Sydney Opera House Markets (Sun)
Campbell's Storehouse
The Rocks
Sydney Harbour Bridge Visitor Centre & BridgeClimb
Sydney Theatre at Walsh Bay
The Rocks Markets (Sat–Sun)
Garrison Church
Foundation Park
Rocks Square
Bel Mondo
Quay
Puppet Shop
Argyle Cut
Argyle Stores
Cadmans Cottage
Observatory Park
Sydney Observatory Museum
Susannah Place Museum
The Rocks Discovery Museum
Museum of Contemporary Art
Sydney Cove
Dendy Opera Quays
Government House
Darling Harbour
Observatory Hotel
S.H. Ervin Gallery
Shangri-La Hotel
Circular Quay
Customs House
Justice & Police Museum
The Basement
Macquarie Place Park
Establishment Hotel
Department of Planning
Museum of Sydney
Conservatorium of Music
Farm Cove
Mrs Macquaries Point
Mrs Macquaries Chair
The Domain
Royal Botanic Gardens
Andrew (Boy) Charlton Pool
Woolloomooloo Bay
Darling Point
Star City
Pyrmont Bay
The Good Living Growers' Market
Wynyard
Sydney
City Recital Hall
Cenotaph
Martin Place
Westin Sydney
State Library of NSW
Parliament of NSW
Sydney Hospital
The Domain
Finger Wharf
King Street Wharf
Pyrmont
Australian National Maritime Museum
Sydney Wildlife World
Sydney Aquarium
Supreme Court
The Mint
St James Church
Strand Arcade
Sydney Tower
Art Gallery of NSW
Hyde Park Barracks Museum
BLUE Sydney
Potts Point
Harry's Cafe de Wheels
To Sydney Fish Market
Harbourside
Pyrmont Bridge
Darling Park
Queen Victoria Building
State Theatre
City Centre
David Jones
Archibald Fountain
St James
St Mary's Cathedral
Woolloomooloo
Cockle Bay Wharf
Convention
Sydney Convention Centre
The Galeries Victoria
Hyde Park
The Great Synagogue
Yurong Water Garden
Sydney Town Hall
IMAX Theatre
Galeries Victoria
St Andrew's Cathedral
Town Hall
Cook and Phillip Aquatic & Fitness Centre
Cook and Phillip Park
Fitzroy Gardens
Sydney Exhibition Centre
Tumbalong Park
Event Cinemas George Street
Pool of Reflection
Australian Museum
Tetsuya's
The Metro
Kings Cross
Exhibition Centre
Chinese Garden of Friendship
Chinatown
World Square
Museum
ANZAC Memorial
Powerhouse Museum
Sydney Entertainment Centre
Darlinghurst
Ultimo
Paddy's Markets
Paddy's Markets (Wed–Sun)
Capitol Square
Medusa
a tavola
Spice I Am
Sydney Jewish Museum
Old Darlinghurst Gaol
Green Park
Haymarket
Belmore Park
Central
Chippendale
Surry Hills
Fringe Bar Markets (Sat)
Paddington
0 300 m
N

Hyde Park & Macquarie Street

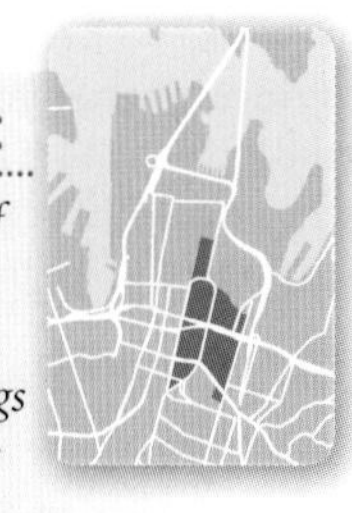

One of the most historically significant areas of Sydney, this district contains one of the oldest parks in the city, Australia's first museum, a Gothic cathedral and the beautiful old buildings of Macquarie Street, once the heart of Sydney's fashionable society.

Hyde Park 8 C4, C5

Originally laid out as the city's first racecourse, Hyde Park is now a place of sunny lawns and wide avenues shaded by spreading trees. At the quiet end of the park, near Liverpool Street, you will find the beautiful Art Deco **ANZAC Memorial** and the **Pool of Reflection.** In the offices below the memorial there's an excellent ongoing exhibition, 'Spirit of ANZAC'. At the busy end of the park stands the gorgeously kitsch **Archibald Fountain,** which commemorates the association of France and Australia during World War I.

Australian Museum 8 C5

On the corner of William and College streets, you will find the excellent Australian Museum. Established in 1827, the present complex is an intriguing mix of Victorian museum and 21st-century educational centre. It houses several unique natural history collections as well as a superb display of Indigenous Australian culture, with lots of hands-on activities and weekly sessions of Aboriginal music and dance. *6 College St; (02) 9320 6000; open 9.30am–5pm daily.*

Cook and Phillip Park 8 C5

For a change of pace, do as the locals do and go for a swim at the **Cook and Phillip Park Aquatic & Fitness Centre,** which is located across from the Australian Museum at the southern end of the park. The complex offers a full range of swimming and recreational activities, including a wave pool, but it's worth visiting just to see the mural that graces the western wall of the Olympic pool. Inspired by the life of Australian swimming champion Annette Kellerman, it consists of eight painted panels depicting scenes from a long and colourful career. Also worth a look is the **Yurong Water Garden** near the northern end of the park. *(02) 9326 0444; open Mon-Fri 6am-10pm and Sat-Sun 7am-8pm.*

St Mary's Cathedral 8 C4

The cathedral is located on the east side of Hyde Park North, and was designed by William Wardell in a soaring Gothic Revival style that recalls the cathedrals of medieval Europe. Work on the cathedral began in 1868 and was finished in 1928, leaving the twin towers in the southern facade without their spires. The completed spires were added to the cathedral in 2000. A particular highlight is the crypt beneath the nave, which features a stunning terrazzo mosaic floor. *College St, facing Hyde Park; (02) 9220 0400.*

St James Church 8 C4

This fine sandstone church, with its elegant tower and copper-sheathed spire, is the oldest ecclesiastical building in the city. The commemorative tablets on the walls read like a page from a history of early Australia. Don't miss the little children's chapel either, which is located in the crypt. The chapel is decorated with an enchanting mural inspired by the Christmas carol 'I Saw Three Ships Come Sailing In', which depicts the land and seascapes of Sydney Harbour. Regular lunchtime concerts. *173 King St, opposite Hyde Park; (02) 8227 1300.*

Macquarie Street 8 C4

Named after one of Sydney's most dynamic and far-seeing governors, Macquarie Street was a thriving centre of upper-class society during the 19th century, evidence of which can still be seen in the magnificent buildings that line the eastern side of

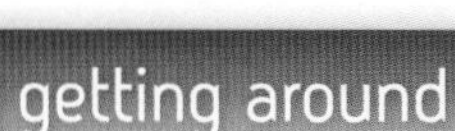

getting around

Sydney has an extensive network of rail, bus and ferry services. When negotiating the inner city, buses are probably best, with regular services on George and Elizabeth streets, between Park Street and Circular Quay, including the free 555 CBD Shuttle. The Red Explorer bus covers city attractions, and the Blue Explorer bus focuses on the eastern beach and harbourside suburbs. Trains are another option, with the City Circle line, which runs in a loop between Central Station and Circular Quay. All bus travel in the CBD and many surrounding areas is Pre Pay only 7am-7pm Monday to Friday. Buy tickets from newsagents, convenience stores or information kiosks.

The monorail, an elevated ride through the streets of Sydney, is an experience in itself. It runs in a circle that includes the north, west and south sides of Darling Harbour, and Liverpool, Pitt and Market streets.

The light rail service runs from Central Station to the inner-west suburb of Lilyfield. As Sydney's only tram service, it is particularly useful for accessing places such as Star City and the Sydney Fish Market.

Ferries are also a great way to travel, with services to many locations on the inner and outer harbour (*see Getting around on ferries, p. 10*). Inquire about TravelTen, MyMulti Day Pass and weekly tickets, as these can considerably reduce the cost of your trip.

If you're driving, an up-to-the-minute road map or GPS is essential. There are ten tollways in Sydney, including the Harbour Bridge and the Cross-City Tunnel. Tolls can sometimes be paid on the spot, but the Cross-City Tunnel, Sydney Harbour Bridge, Sydney Harbour Tunnel, Lane Cove Tunnel, Falcon Street Gateway and M7 will only accept payment by E-tag, an electronic device obtained through the Roads and Traffic Authority (RTA). www.myrta.com

Public transport Train, bus and ferry information line 13 1500.

Airport rail service Airport Link (02) 8337 8417.

Tollways Roads and Traffic Authority (RTA) 13 2213.

Motoring organisation NRMA 13 1122.

Car rental Avis 13 6333; Bayswater Car Rental (02) 9360 3622; Budget 13 2727; Hertz 13 3039; Thrifty 1300 367 227.

Specialty trips Monorail and Metro Light Rail (02) 8584 5288.

Harbour cruises Sydney Harbour Ferries 13 1500; Captain Cook Cruises (02) 9206 1111 or 1800 804 843.

Taxis ABC Taxis 13 2522; Legion Cabs 13 1451; Manly Warringah Cabs 13 1668; Premier Cabs 13 1017.

Water taxis Water Taxis Combined 1300 666 484; Yellow Water Taxis 1300 138 840.

Tourist bus Red Explorer, Blue Explorer 13 1500.

Bicycle hire Centennial Park Cycles (02) 9398 5027; Bonza Bike Tours (02) 9247 8800.

getting around on ferries

Ferries are a great way to get about and see the harbour. Sydney Ferries and private operators run daily services from Circular Quay to more than 30 locations around the harbour and Parramatta River. Timetables, network maps and information about link tickets (combining a ferry fare with admission to various tourist attractions) can be obtained from the Sydney Ferries Information Centre at Circular Quay. Matilda Catamarans and Captain Cook Cruises are the main private ferry operators. They run ferry cruises and some express services to various points around Sydney Harbour, including some not serviced by Sydney Ferries.

Sydney Ferries from Circular Quay

Manly Ferry Departs Wharf 3.

Taronga Zoo Ferry Departs Wharf 2.

Watsons Bay Ferry Darling Point (Mon–Fri only), Double Bay, Rose Bay and Watsons Bay, departs Wharf 4.

Mosman Ferry Mosman and Cremorne, departs Wharf 4 (Mon–Sat) and Wharf 2 (Sun).

North Sydney Ferry Kirribilli, North Sydney, Neutral Bay and Kurraba Point, departs Wharf 4.

Woolwich Ferry North Shore, Balmain, Cockatoo Island and Drummoyne, departs Wharf 5 (Mon–Sat) and Wharf 4 (Sun).

Parramatta RiverCat Express service to Parramatta, departs Wharf 5.

Rydalmere RiverCat North Sydney, Balmain, Darling Harbour and the Parramatta River to Rydalmere, departs Wharf 5.

Birkenhead Point Ferry North Sydney, Balmain and Birkenhead Point, departs Wharf 5 (Mon–Fri).

Darling Harbour Ferry North Sydney, Balmain and Darling Harbour, departs Wharf 5.

Private services from Darling Harbour and Circular Quay

Manly Fast Ferry Express service to Manly from in front of Museum of Contemporary Art at Circular Quay. Tickets may be purchased on board.

Matilda Catamaran City Loop Service Darling Harbour to Circular Quay via Luna Park. Tickets may be purchased on board.

Matilda Catamaran Cruise + Attraction Combines Matilda Harbour Express ferry ticket with entry to one of the following harbourside attractions: Taronga Zoo, Shark Island or Fort Denison. Bookings on (02) 9264 7377.

Captain Cook Cruises Hop-on and hop-off cruise departs Darling Harbour, stopping at Luna Park, Circular Quay, Taronga Zoo, Shark Island, Fort Denison and Watsons Bay. Bookings on 1800 804 843.

Sydney Harbour Dreaming On this Aboriginal cruise, you'll experience traditional performances, taste bush tucker and hear indigenous history and stories of the area. Departs Circular Quay and King Street Wharf. (02) 9518 7813

the street. In keeping with its old-world character, Macquarie Street is also home to a number of statues. Up near Hyde Park, for example, you will find **Queen Victoria** and her royal consort, **Prince Albert**. While you're in this part of Macquarie Street, take the lift to the 14th floor of the **Supreme Court** building and visit the **Buena Vista** cafe. You can enjoy one of the best harbour views in Sydney, all for the price of a latte.

Hyde Park Barracks Museum 8 C4

Every elegant line and delicate arch of Hyde Park Barracks, one of the loveliest of Sydney's older buildings, bears the stamp of its convict architect, Francis Greenway. Built in 1819, the barracks have provided accommodation for a wide range of individuals including convicts. Various ongoing exhibitions reveal the many layers of this building's rich social history, with poignant displays of artefacts gleaned from recent excavations. *Cnr College and Macquarie sts; (02) 8239 2311; open 9.30am–5pm daily.*

The Mint 8 C4

Once the South Wing of the old Rum Hospital, the site of the colony's first mint is now the headquarters of the Historic Houses Trust. Besides a pleasant reading room there's a good cafe on the upper floor with balcony seating and fine views of Macquarie Street and Hyde Park. The mint artefacts are now housed in the Powerhouse Museum (*see p. 14*). *10 Macquarie St; (02) 8239 2288; open 9am–5pm Mon–Fri; general admission free.*

Sydney Hospital 8 C4

Now housing both the Sydney Hospital and the Sydney Eye Hospital, this imposing complex of sandstone buildings occupies the original site of the centre wing of the old Rum Hospital. Tours of the hospital's historic buildings are available (bookings essential). The little cobbled walkways that lead to the rear of the complex bring you to the oldest building, the Nightingale Wing, which houses the **Lucy Osburn-Nightingale Foundation Museum**. Among other items, you can see the sewing basket used by Florence Nightingale in the Crimea. *Museum open 10am–3pm Tues; tours (02) 9382 7111.*

Before you move on, don't forget to pay a visit to '**Il Porcellino**', a favourite photo opportunity with tourists and a 'collector' of money for the hospital. It is considered lucky to rub the statue's nose, then toss a coin in the fountain and make a wish.

Parliament of New South Wales 8 C4

Between the Sydney Hospital and the State Library stands the northern wing of the old Rum Hospital, now the seat of the Parliament of New South Wales. If it is a non-sitting day, individuals can see inside both chambers of parliament, after speaking to the assistants at the front desk. There is also a public tour every Thursday at 1 pm. *(02) 9230 3444; open daily; general admission free.*

State Library of New South Wales 8 C3

Facing the Royal Botanic Gardens, on the corner of Macquarie Street and Shakespeare Place, the state library houses a remarkable collection of Australian books, records, personal papers, drawings, paintings and photographs, most of which are regularly displayed in the library's five public galleries. The magnificent Mitchell Library Reading Room is a highlight, as is the exquisite mosaic on the floor of the lobby, which depicts Abel Tasman's map of Australia. Just outside the library, check out the statues of **Matthew Flinders** and his beloved cat **Trim**. Stolen at least four times, Trim now perches on a sandstone ledge, well beyond the reach of souvenir hunters. *Open 9am–8pm Mon–Thurs, 9am–5pm Fri, 10am–5pm Sat–Sun.*

Gardens & Domain

One of the loveliest parts of the city, this area consists of extensive parkland, much of which was once part of the property surrounding the first Government House. It is thanks to the wisdom of governors Phillip and Macquarie that so much of this land was preserved, saving it from 200 years of ferocious development.

Royal Botanic Gardens 8 C3

A landscaped oasis on the edge of the harbour, the gardens are a wonderful place in which to stroll or relax with a picnic. Aside from its sweeping parklands, there are several formal gardens, including the Aboriginal garden, Cadi Jam Ora, and a stunning rose garden. Stock up on bush tucker, take in the botanical drawings at the Red Box Gallery, see a film by moonlight (summer only), or even adopt a tree. Ask at the Gardens Shop for details. *Mrs Macquaries Rd; (02) 9231 8125.*

Government House 8 C2

Government House is located in the north-west corner of the Botanic Gardens, close to Macquarie's old stables, now the **Conservatorium of Music** *(see Grand old buildings, p. 16)*. Built in 1845, Government House is an elaborate example of the Gothic Revival style, with extensive gardens and harbour views. Free guided tours of the state apartments are available, and there are 'Behind the Scenes' tours of the servants' halls, cellars, kitchens and dairies run by the Historic Houses Trust requiring an admission fee. Contact Trust headquarters at the Mint *(see The Mint, facing page)* for details.

The Domain 8 C4

Separated from the Botanic Gardens by the Cahill Expressway, the Domain falls into two distinct parts. South of the expressway, it's a wide green park where soapbox orators and an attendant crowd of hecklers once gathered each Sunday to debate the issues of the day. Now a place where office workers come to soak up the sun and play some lunchtime soccer, this area comes into its own in January when it hosts popular jazz, opera and symphony concerts. North of the Cahill, the rest of the Domain runs along the promontory to **Mrs Macquaries Chair**, a bench that was carved out of the sandstone bluff specifically so that Elizabeth Macquarie could sit in comfort as she watched for ships arriving from England with longed-for letters from home.

Art Gallery of New South Wales 8 C4

The gallery is situated to the south of the Cahill Expressway, opposite the South Domain. Built in an imposing Classical Revival style with ultra-modern additions, it houses an impressive collection of both Australian and international artworks, including a large permanent collection of Aboriginal art and a superb Asian collection. *Art Gallery Rd, The Domain; open 10am–5pm Thurs–Tues, 10am–9pm Wed; general admission free; 1800 679 278 or (02) 9225 1744.*

Andrew (Boy) Charlton Pool 8 D3

A sensational place for a quick dip, a swimming lesson or a leisurely bite at the harbourside cafe. Boy Charlton was just 14 when he won a major swimming championship and became a national idol. His status was confirmed when he beat the Swedish world record holder Arne Borg three times in early 1924, and he went on to compete in three Olympic Games (1924, 1928 and 1932). Mrs Macquaries Rd; (02) 9358 6686; open 6–8pm Sep–May.

[SYDNEY OPERA HOUSE AT NIGHT]

Circular Quay 8 C3

Since the moment that Sydney was declared a settlement, Circular Quay has been where it all happens. With many journeys beginning and ending here, it's a major junction for bus, rail and ferry transport *(see Getting around, p. 9)*, and a natural meeting place. There's entertainment on the quay itself in the form of buskers and street performers, while the many excellent cafes and bars offer stunning views of the harbour.

Sydney Opera House 8 C2

One of the great buildings of the 20th century, the Sydney Opera House stands at the far end of East Circular Quay, breathtaking in its beauty against the backdrop of Sydney Harbour. Thought by many to echo the sails seen in the harbour, it was in fact the natural fall of a segmented orange that inspired Jørn Utzon's design, although Utzon never saw the completed building. After resigning in anger over changes made to the interior design, he left, swearing never to return. The recent renovations to the

Around Circular Quay

Bounded by the 19th-century streetscape of Bridge Street and including the promenades of East and West Circular Quay, this area contains one of Sydney's most famous icons. It also houses two must-see museums and boasts one of the few corners of the CBD that has remained unaltered for almost 200 years.

top events

Sydney Festival A celebration of the city, this includes cultural events at Sydney's most stunning indoor and outdoor venues. January.

Gay and Lesbian Mardi Gras A two-week cultural festival culminating in a spectacular street parade. February–March.

Royal Easter Show The country comes to the city in the Great Australian Muster. Easter.

Archibald Prize The Archibald national portrait prize and exhibition is one of Sydney's most controversial events. March–May.

Sydney Comedy Festival Three weeks of local and international comedy acts at various venues around Sydney. April–May.

Sydney Film Festival A showcase for the newest offerings in cinema. June.

City to Surf A 14-kilometre fun run, from Hyde Park to Bondi. August.

Crave Sydney International Food Festival A month-long festival of food and outdoor art, including the Sydney International Food Festival, Darling Harbour Fiesta and Art & About. October.

Rugby League Grand Final The leaders of the football competition compete for the title in the final match of the season. October.

Sydney to Hobart Yacht Race Classic blue-water sailing event. Begins 1pm 26 December.

museums

Brett Whiteley Studio Paintings and sculptures in the former studio and home of this great Australian artist. 2 Raper St, Surry Hills; (02) 9225 1881; open 10am–4pm Sat–Sun.

Mary MacKillop Place Museum A tribute to this remarkable woman who brought education to the children of the bush. 7 Mount St, North Sydney; (02) 8912 4878; open 10am–4pm daily.

Royal Australian Navy Heritage Centre Over 100 years of navy history in the harbourside setting of Garden Island. Take Watsons Bay ferry from Circular Quay to Garden Island; (02) 9359 2003; open 9am-4pm; admission free.

Sydney Jewish Museum A history of the Jewish people in Australia, along with a poignant tribute to the victims of the Holocaust. Cnr Darlinghurst Rd and Burton St, Darlinghurst; open 10am–4pm Sun–Thu, 10am–2pm Fri; free admission on the first Sun of each month.

Justice and Police Museum Located in the old Water Police Station, with exhibitions on crime and punishment in Sydney, including the city's most notorious cases. Cnr Albert and Phillip sts, Circular Quay; (02) 9252 1144; open 10am–5pm.

Victoria Barracks Museum Military pride in a colonial setting, with a stirring flag-raising ceremony, followed by a guided tour of the barracks and the museum. Oxford St, Paddington; (02) 8335 5330; open 10am–1pm Thurs, 10am–4pm Sun; admission free.

Sydney Tramway Museum Historic trams from Sydney, Nagasaki, Berlin and San Francisco. Entry fee includes unlimited rides on the trams. Cnr Pitt St and Princes Hwy, Loftus; (02) 9542 3646; open 10am–3pm Wed, 10am–5pm Sun, and weekdays during some school holidays.

The Rocks Discovery Museum A fascinating glimpse into the chequered past of The Rocks, featuring both interactive technology and archaeological artefacts. 2–8 Kendall La, The Rocks; (02) 9240 8680 open 10am–5pm daily; admission free.

Australian Centre for Photography Exhibitions of work by the world's best art, fashion and documentary photographers. 257 Oxford St, Paddington; (02) 9332 1455; open 12–7pm Tues–Fri, 10am–6pm Sat–Sun; admission free.

See also Australian Museum, p. 9, Hyde Park Barracks Museum, p. 10, Museum of Sydney, this page, Museum of Contemporary Art, this page, Susannah Place Museum, p. 13, Australian National Maritime Museum, p. 14, Powerhouse Museum, p. 14, La Perouse Museum, p. 18

interior were carried out under his direction prior to his death in November 2008. With nearly 1000 rooms that include theatres and concert halls, rehearsal studios and dressing-rooms, a guided tour is recommended (bookings essential). *Tours (02) 9250 7250.*

Museum of Sydney 8 C3

A gem of a museum built of sandstone, steel and glass, the Museum of Sydney is situated on the site of the first Government House, the original foundations of which can still be seen. Through a series of intriguing displays, exhibitions and films, the many layers that have gone into the making of Sydney are revealed. *Cnr Phillip and Bridge sts; (02) 9251 5988; open 9.30am–5pm daily.*

Macquarie Place Park 8 C3

On the corner of Loftus and Bridge streets stands a remnant of the old Government House garden, now known as Macquarie Place Park. The elegant sandstone obelisk near the south-east corner of the park marks the place from which all distances in the colony were once measured. The nearby anchor and small gun belonged to Arthur Phillip's flagship *Sirius*. More of this ship has since been recovered, and can be viewed at the Australian National Maritime Museum *(see p. 14)*.

The Rocks

Once the haunt of pickpockets, prostitutes and sailors, The Rocks contains some of the city's most important historic sites, and is one of Sydney's most treasured attractions. Divided into two parts by the approaches to the Harbour Bridge, the area on the eastern side of the Argyle Cut is centred on the lanes north of George Street, while the area to the west takes in one of Sydney's most beautiful parks and the gracious 19th-century houses of Lower Fort Street.

Sydney Visitor Centre 8 B2

A visit to The Rocks should begin with a visit to the excellent Sydney Visitor Centre, where you'll find all the detailed information you'll need regarding accommodation, dining, shopping and sightseeing in the area. Of special interest are the many guided tours available, including several walking tours, a ghost tour and a tour of the area's historic pubs. *Cnr Argyle and Playfair sts; (02) 9240 8500; open 9.30am–5.30pm daily.*

George Street – north 8 B2, C2

Much of The Rocks' previous character as a notorious seaport rookery can still be glimpsed in the winding streets and tiny lanes that run behind George Street. There are quiet courtyard cafes and some unusual shops to be found here, including the enchanting **Puppet Shop**, with hundreds of exquisitely handcrafted marionettes hanging from the ceilings of four rooms. Located in the sandstone cellars of 77 George Street, it's a must-see, and not just for children; *(02) 9247 9137.*

Museum of Contemporary Art 8 C2

The museum occupies the old Maritime Services building – a brooding Art Deco structure that dominates the western side of Circular Quay. But once inside, the large, open white rooms offer an ideal setting to showcase works by Australian and international contemporary artists. Visitors can expect an ever-changing display of innovative sculptures, paintings, photographs and video installations. Make sure to visit the giftshop. *(02) 9245 2400; open 10am–5pm daily; general admission free.*

Cadmans Cottage 8 C2

This charming little cottage has the distinction of being Sydney's oldest surviving residence. Built in 1816, for many years it was the home of John Cadman, an ex-convict and boatman to Governor Macquarie. Today it is the Sydney Harbour National Park Information Centre and the starting point for tours of the harbour islands *(see Harbour islands, p. 20). 110 George St; (02) 9247 5033; open 9.30am–4.30pm Mon–Fri and 10am–4.30pm Sat–Sun.*

Playfair Street 8 B2

The heart and soul of The Rocks, Playfair Street is the place to be, particularly on the weekends. There's corn-on-the-cob, wandering street performers, endless live entertainment, and a fantastic vibe from the crowd that is constantly on the move. Find a seat in **Rocks Square** and soak it all up. Not far away is **Argyle Stores**, a converted warehouse that now houses a collection of open-plan shops specialising in beautiful and unusual clothes. The main entrance to the stores is on Argyle Street, by way of Mary Reibey's old bond stores. As you pass through the arched gateway, spare a thought for this remarkable woman who arrived in Australia in 1792 at the age of 14 and is today remembered as one of the most successful businesswomen in the colony of New South Wales. Her portrait appears on the 20-dollar note.

Foundation Park 8 B2

Hidden behind a row of souvenir shops in Playfair Street, the quirky and charming Foundation Park occupies the almost vertical site of three former dwellings that were built into the face of the sandstone escarpment. A front door was situated at the top of the cliff, and a back door halfway down. All that remains now are a few scattered foundations among the grassy terraces of the park and a steep stairway or two leading nowhere, but here and there you can find a chair, a fireplace, a clock or a table, recalling those long-vanished homes.

Susannah Place Museum 8 B2

From the top of Foundation Park, this wonderful museum is just up the Argyle Stairs and around the corner in Gloucester Street. Occupying four terrace houses, it affords a glimpse of what it was like for working-class people living in The Rocks at varying stages in its history. The museum is extremely people-friendly, with visitors encouraged to linger among the exhibits and try the piano in the parlour of No. 64. Take a moment to look around the little shop before you go; re-created in turn-of-the-century style, it sells toys, sweets, soft drinks and other goods from that era. *58–64 Gloucester St; (02) 9241 1893; open 2–6pm Mon–Fri, (winter 2–5pm) 10am–5pm Sat–Sun and school holidays.*

Sydney Harbour Bridge 8 C1

With almost every view in The Rocks dominated by its soaring arch, it's hard to ignore the presence of Sydney's magnificent harbour bridge. With its longest span at 503 metres, it was completed and opened to traffic in 1932. One of the best views of Sydney can be had from the **Pylon Lookout**, which contains an excellent exhibition detailing the bridge's history. The **Sydney Harbour Bridge Visitor Centre** also contains an exhibition and cinemas showing how the bridge was built. The same entrance leads to **BridgeClimb Sydney**, offering unique and unforgettable tours to the very top of the span *(see Walks & tours, p. 15).* *3 Cumberland St; (02) 8274 7777; open 8.30am–5pm Mon–Fri, 9am–5pm Sat–Sun.*

Argyle Cut 8 B2

Work on the cutting connecting the eastern and western sides of The Rocks commenced in 1843, with chained convicts doing most of the hard labour. Initially much narrower than it is now, it was once the haunt of 'pushes' (larrikin youths who specialised in gang warfare and rolling the lone passer-by). In the heyday of the pushes, even the police went through the Argyle Cut in pairs.

Observatory Park 8 B2

High above The Rocks stands what is perhaps the loveliest park in Sydney. It has an old-world ambience and stunning views of the western harbour. For stargazers, there's the **Sydney Observatory Museum**, which hosts exhibitions on astronomy as well as talks, films and viewings of the night sky (bookings essential). *Bookings (02) 9921 3485; open 10am–10pm daily; day admission free.*

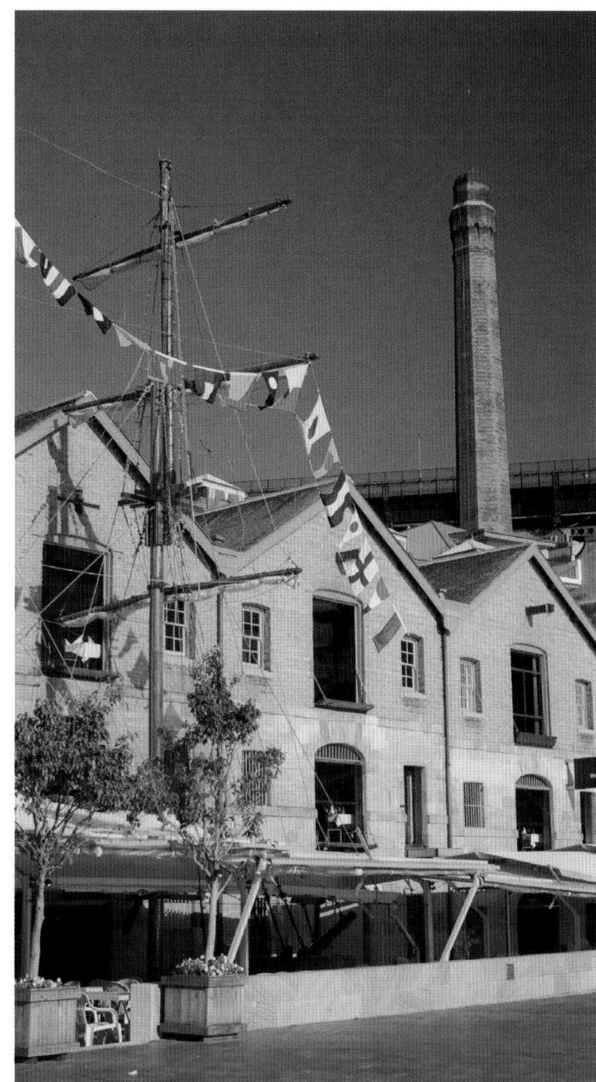

[THE HISTORIC ROCKS PRECINCT]

shopping

Pitt Street Mall, City Sydney's major shopping area.

Castlereagh Street, City Sheer indulgence with some of the world's leading designer labels.

The Galeries Victoria, City A dazzling array of top-quality fashion and lifestyle brands.

The Rocks The place to go for top-quality Australian art, jewellery and clothing.

Oxford Street, Darlinghurst Up-to-the-minute street fashion and funky alternative clothing.

Oxford Street, Paddington Cutting-edge designers and a mecca for antique hunters.

Double Bay Sydney's most exclusive shopping suburb.

Birkenhead Point, Drummoyne Designer shopping at bargain prices in a historic venue.

Military Road, Mosman Classy shopping in a village atmosphere.

Darling Harbour

Located to the immediate west of the CBD and easily accessed by ferry, light rail and monorail, Darling Harbour is the focus of much of the city's culture and entertainment. In the inner-west area, families will delight in Australia's largest and most popular museum, the Powerhouse. There's also the glitz of Star City casino, and food lovers should head to the Sydney Fish Market, the largest market of its kind in the Southern Hemisphere.

Not far from the observatory, hidden by a curve of the Cahill Expressway, stands the old Fort Street School for Girls, now the headquarters for the National Trust. This is where you will find the **S. H. Ervin Gallery**, which is renowned for its innovative and unusual art exhibitions. *Watson Rd, Observatory Hill; (02) 9258 0173; open 11am–5pm Tue–Sun.*

markets

Paddy's Markets Sydney's most famous markets, with superb fresh produce and bargains in clothing, souvenirs, toys and gifts. Haymarket; Wed–Sun. 8 A5

The Rocks Markets Classic street market, Fri 10am–4pm; Weekend market 10am–5pm. 8 C2

Sydney Opera House Markets Australian souvenirs, arts and crafts in a glorious harbour setting; Sun. 8 C2

Paddington Markets Fabulous mix of fashion, artworks, jewellery and collectibles in one of Sydney's trendiest suburbs. 395 Oxford St; (02) 9331 2923; Sat. 6 C5

Sydney Markets Vast 42-hectare market, includes Sydney Produce Market, Sydney Growers Market and Sydney Flower Market. 250-318 Parramatta Rd, Flemington; (02) 9325 6200; Mon–Sat from 6am. 545 K9

EQ Village Markets Gourmet food, produce and coffee in the village-like atmosphere of the Showring. Moore Park; Wed and Sat. 6 C5

Balmain Market Jewellery and leather goods, arts and craft in the grounds of an old sandstone church. St Andrews Church; cnr Darling St and Curtis Rd; (02) 9555 1791; Sat. 6 B3

Bondi Markets Clothes, jewellery, a range of new and second-hand collectibles, and a lively beachside atmosphere. Bondi Beach Public School, Campbell Pde; (02) 9315 8988; Sun. 6 E5

The Good Living Growers' Market The gourmet's choice, with superb breads, cheeses, fruit and vegetables, close to Darling Harbour. Pyrmont Bay Park; 1st Sat each month. 8 A3

Glebe Markets Decorative homewares, arts and crafts, new and second-hand clothing, with a background of live music. Glebe Public School, cnr Derby Pl and Glebe Point Rd; 0419 291 449; Sat. 6 B4

Eveleigh Market Farmers Market every Sat, Artisans' Market first Sun each month. 243 Wilson St, Darlington; (02) 9209 4735. 6 B5

Fringe Bar Markets Trendy new fashions and vintage clothing. Paddington; Fringe Bar, 106 Oxford St, Paddington; (02) 9360 5443; Sat. 8 D6

Manly West Organic Market Stock up on exclusively organic produce at this market at Manly West Public School. Hill St, Balgowlah; 0431 532 378; Sat. 6 E1

North Sydney Produce Market Foodies market Miller St, North Sydney, between Ridge and McLaren sts; (02) 9922 2299; 3rd Sat of month.

See also Sydney Fish Market, facing page

Darling Harbour East 8 A4

Once very much a working harbour, Darling Harbour has a maritime past reflected in the structures that surround its landscaped promenades and parks. One good example is the National Trust–classified **Pyrmont Bridge**, which spans the harbour from east to west. North-east of the bridge are **King Street Wharf** and **Cockle Bay Wharf**, renowned for exclusive shopping and dining. South-east of Pyrmont Bridge, the ultra-modern **Imax Theatre** boasts the largest cinema screen in the world and offers films in both 2D and 3D format. *(02) 9281 3300.*

Sydney Aquarium 8 A4

The Sydney Aquarium rates high among the world's aquariums for sheer spectacle-value. Highlights include the fabulous underwater tunnels that enable you to walk with the stingrays and stroll with the sharks, a Great Barrier Reef exhibition that's pure magic, two of only five dugongs in captivity in the world, and the Seal Sanctuary, where you can watch the seals flirt and frolic all around you. *Aquarium Pier; (02) 8251 7800; open 9am–8pm daily.*

Sydney Wildlife World 8 A4

Situated right next door to the Aquarium, Sydney Wildlife World takes you deep into the heart of the Australian bush, from the rainforests of the tropical north to the deserts of the Red Centre. Over 250 different Australian species live within their own natural habitats and ecosystems. *Aquarium Pier; (02) 9333 9288; open 9am–5pm daily.*

Chinese Garden of Friendship 8 A5

Tucked away to the south is this garden, where airy pavilions and tiny arched bridges stand reflected in tranquil lakes, surrounded by weeping willows and graceful bamboo. There's an elegant teahouse that serves traditional Chinese teas, as well as a tiny pavilion where for a small fee you can dress up in costumes from the Peking Opera. *(02) 9240 8888; open 9.30am–5.30pm daily.*

Australian National Maritime Museum 8 A4

Located to the north of Pyrmont Bridge, this museum highlights Australia's multifaceted relationship with the sea, from the days of convict transports and immigrant ships, to the beach culture of today. Special exhibitions include tours of the museum's fleet. *2 Murray St, Darling Harbour; (02) 9298 3777; open 9.30am–5pm daily; general admission free.*

South of the harbour 8 A5, B5

Just past the **Sydney Entertainment Centre**, host to the world's biggest music stars, you'll find **Paddy's Markets** *(see Markets, this page)*. Opposite the Thomas Street entrance to the market is Dixon Street, the main thoroughfare of **Chinatown.** There are colourful shops offering everything from inexpensive souvenirs to beautiful and costly Chinese items, as well as superb Asian cuisine to suit every taste.

Powerhouse Museum 8 A5

South-west of Darling Harbour, the museum is best accessed by monorail. One of Sydney's most fascinating museums, it houses an extraordinary collection of oddments and treasures. Highlights include the Hall of Transport, with its fleet of aeroplanes suspended from the ceiling, and the tiny 1930s-style cinema with

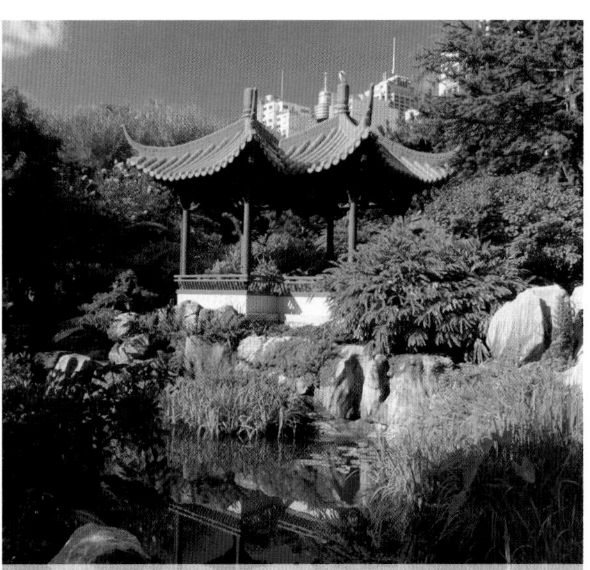

[CHINESE GARDEN OF FRIENDSHIP, DARLING HARBOUR]

its program of old newsreels and documentaries. Free guided tours are available as well as a broad range of daily activities. *500 Harris St, Ultimo; (02) 9217 0111; open 10am–5pm daily.*

Star City 8 A3

Star City, Sydney's only official casino, can be found to the north-west of Darling Harbour. Star City is also home to two of Sydney's finest theatres and several restaurants that offer fine dining with panoramic views of Darling Harbour. *80 Pyrmont St, Pyrmont; (02) 9777 9000; open 24 hours.*

Sydney Fish Market 6 B4

The Sydney Fish Market is still within walking distance of Darling Harbour, but is probably best accessed by light rail. Fast-paced and vibrant, with a good bakery, an excellent deli and some of the best and freshest seafood you're ever likely to find, the market is a great place to visit. There are guided tours that enable you to watch a Dutch auction (where the price actually *drops* every minute), but only for the early risers. *Bank St, Pyrmont; tours (02) 9004 1143; open 7am–4pm daily.*

walks & tours

The Rocks Walking Tours The history of The Rocks and its many colourful characters are brought to life during this 90-minute tour. Bookings on (02) 9247 6678.

Harbour Circle Walk For a comprehensive overview of Sydney's harbour, starting and ending at Observatory Hill, this walk covers 26 kilometres and could take you the whole day. Download a map from planning.nsw.gov.au/harbour

BridgeClimb Sydney Various tours of the Sydney Harbour Bridge that involve climbing to the top of the span clad in protective clothing and secured with a harness. Bookings on (02) 8274 7777.

Sculpture by the Sea During November, this free outdoor exhibition on the Bondi to Tamarama coastal walk draws large crowds viewing over 100 modern art pieces with the sea as their backdrop.

Sydney Ferry Harbour Walks Discover some of Sydney's more out-of-the-way places on foot. Pick up a brochure at the Sydney Ferries Information Centre at Circular Quay, take a ferry to any of the listed destinations, and start walking.

The Rocks Pub Tour Tales of scandal and intrigue unfold on this pub crawl with a difference, as you knock back a schooner or two at The Rocks' most famous hotels. Bookings on 1300 458 437.

Oz Jet Boating Tours Tour the harbour in a high-performance V8 jet boat. Bookings on (02) 9808 3700.

Destiny Tours Stories of sex, scandal and the supernatural. Explore Sydney's darker side on this night-time minibus tour. Bookings on (02) 9487 2895.

Easyrider Motorbike Tours See Sydney from the back of a Harley-Davidson. Bookings on 1300 882 065.

Gourmet Safaris Choose from a range of foodie tours that uncover Sydney's multicultural world of gourmet food. Experienced chefs will take you to local shops and invite you to try traditional Greek, Italian, Lebanese, Portuguese, Turkish or Vietnamese cuisine. Bookings on (02) 8969 6555.

City Sightseeing Sydney and Bondi Tours Cover the city and the eastern suburbs in a double-decker, open-topped bus, hopping on and off as often as you like. No need to book. (02) 9567 8400.

Chocolate Espresso Tours Tour the city's CBD and shopping districts, with a focus on either coffee or chocolate. Bookings on 0417 167 766.

Inner West

The inner west starts at the beginning of Parramatta Road near Sydney University and takes in the historic suburbs of Glebe and Balmain. Glebe has retained many of its grand Victorian homes, Federation houses and modest workers' cottages, and has a reputation as an alternative suburb. With its arty flair, in complete contrast to its previous population of dockland workers, Balmain sits on a peninsula in Port Jackson, adjacent to the suburbs of Rozelle to the south-west, Birchgrove to the north-west and Balmain East.

Glebe 6 B4

Once a predominantly working-class suburb, Glebe has become more refined over the years. With Sydney University close by, you can find excellent bookshops, lively cafes and some great weekend markets here *(see Markets, facing page)*. Its leafy streets lined with old weatherboard houses, Victorian terraces and the occasional mansion, Glebe is a fabulous area to explore on foot. At the far end of Glebe Point Road is **Glebe Park**, which features landscaped walks with views of Blackwattle Bay and Anzac Bridge.

Balmain 6 B4

Tucked away on its own little peninsula, Balmain is another suburb of quaint houses, stepped lanes and harbour views. The main shopping centre is in Darling Street, a lively area with cafes, boutiques and top-quality shops. Quickly and easily reached by ferry, this is another suburb that is best seen on foot.

Inner East

The inner-eastern suburbs begin just beyond the Domain and include Kings Cross, the old maritime suburb of Woolloomooloo, Paddington and Darlinghurst, once the haunt of some of the city's most notorious gangsters and now home to some very trendy cafes and restaurants.

grand old buildings

Customs House Elegant sandstone building designed by colonial architect James Barnet in the Classical Revival style. Alfred St, Circular Quay.

Department of Planning This building is particularly noteworthy for the statues of famous explorers and legislators that grace the exterior. 22–33 Bridge St.

Conservatorium of Music Much altered, but the castellated facade still recalls Macquarie's fancy Government House Stables. Macquarie St.

Cadmans Cottage Sydney's oldest surviving residence, now a NPWS office. 110 George St, The Rocks.

Campbell's Storehouse Built from bricks made by convicts. Hickson Rd, The Rocks.

The Garrison Church View the red-cedar pulpit and the beautiful stained-glass window. Lower Fort St, Millers Point.

The Great Synagogue Exotic and remarkable in its originality, with a gorgeous mix of Byzantine and Gothic architecture, and sumptuous interiors. Elizabeth St.

St Andrew's Cathedral With twin towers that recall York Minster, it is best seen in November through a cloud of purple jacaranda. Cnr Bathurst and George sts.

Old Darlinghurst Gaol An impressive early Victorian sandstone prison with an imposing entrance, it now houses the National Art School. Cnr Burton and Forbes sts.

The University of Sydney Landscaped grounds and historic sandstone buildings. Parramatta Rd, Broadway.

See also Queen Victoria Building, p. 7, Sydney Town Hall, p. 7, Hyde Park Barracks Museum, p. 10, Government House, p. 11

Woolloomooloo 8 D4

While there are buses that service this area, the best way to get to Woolloomooloo is to take the stairs near the Art Gallery in Mrs Macquaries Road. These will bring you down close to Woolloomooloo's most recent and controversial development, the **Finger Wharf**. A remnant of the days when the 'Loo had a reputation for toughness and lawlessness, the Finger Wharf is now one of Sydney's most exclusive addresses. For a traditional Sydney culinary experience, try **Harry's Cafe de Wheels**, a pie-wagon and long-time Sydney icon on the eastern side of the bay.

Kings Cross 8 D5

The Cross is one of the most fascinating parts of Sydney. At night, when the strip clubs of Darlinghurst Road swing into action, the vibe is raw and edgy. But by daylight it retains the charm of its bohemian past, with gracious tree-lined streets and pretty sidewalk cafes. Look for the **El Alamein Fountain**, the most photographed fountain in Sydney.

Elizabeth Bay House 6 C4

Designed by John Verge as a home for Alexander Macleay (whose family history is displayed on the walls of St James Church in Queens Square), the house is particularly famous for the stunning proportions of its elegant oval saloon with a sky-lit dome, and beautiful curving staircase. After spending many years as a shabby boarding house and artists' squat, it is now a Historic Houses Trust museum, and its rooms have been restored to their former 19th-century graciousness. *7 Onslow Ave, Elizabeth Bay; (02) 9356 3022; open 9.30am–4pm Fri–Sun.*

Paddington 6 C5

Considered an eyesore until quite recently, Paddington is now one of Sydney's most beautiful suburbs. Its backstreets are lined with rows of iron-lace-trimmed terraces decked with flowers, and its main thoroughfare, Oxford Street, is home to dozens of boutiques, galleries and cafes. Always busy, it is at its most vibrant on weekends, when the **Paddington Markets** are in full swing *(see Markets, p. 14)*. Paddington is easily accessed by bus – you can disembark on Oxford Street and walk around.

East

Sydney's east is dominated by the city's green lungs, Centennial Park. Stretching south of Paddington, the park is a magnet for joggers, sun-worshippers, picnickers, tai-chi exponents, horseriders and lovers of the great outdoors.

Centennial Park 6 C5

South of Paddington lies Centennial Park, a massive complex of parklands, playing fields, bridle paths and riding tracks. It's a great place to get away from it all and relax with a picnic. Once a huge catchment of creeks, swamps, springs, sand dunes and ponds fed by groundwater, it was traditionally home to the Gadi people. The distinctive Federation Pavilion commemorates the site of the inauguration of Australian Federation on 1 January 1901. *(02) 9339 6699.*

Inner South

The Moore Park precinct in the city's inner south takes in the Entertainment Quarter, one of Sydney's premier leisure playgrounds. It was the home of the Royal Agricultural Society and the Sydney Royal Easter Show until January 1998, when they were relocated to Sydney Olympic Park.

Entertainment Quarter 6 C5

One of Sydney's newer entertainment centres, the Entertainment Quarter occupies the site of the old showgrounds and is located next to the **Sydney Cricket Ground** at Moore Park. While film studios and sound sets occupy much of the site, the Showring is now a vast village green, adjacent to a vibrant pedestrian precinct lined with top-quality fashion and homewares outlets. The Entertainment Quarter also boasts two cinema complexes and more than a dozen cafes, restaurants and bars. Always lively, the village atmosphere is most noticeable when the weekly markets set up their stalls in the Showring *(see Markets, p. 14)*.

Equestrian Centre 6 C5

There are five riding schools at the **Centennial Parklands Equestrian Centre**, offering park rides for those who have experience and riding lessons for those who do not. The Equestrian Centre is located on the corner of Lang and Cook roads in Moore Park. If riding a bike is more your style, try one of the bicycle-hire shops in Clovelly Road, on the south side of the park *(see Getting around, p. 9)*.

Eastern Beachside

Vaucluse is one of Sydney's most exclusive suburbs, and old and new money abounds in the lavish homes gracing the harbour foreshore. The suburb's nearby cousin, Watsons Bay, on the southern head of the harbour entrance, is recognised as Australia's oldest fishing village. It was Governor Phillip's first landing point after discovering the harbour in 1788.

Vaucluse House 6 E4

In the heart of Vaucluse, not far from the harbour, this early 19th-century house was once owned by the flamboyant William Charles Wentworth. Now a Historic Houses Trust museum, it is a beautifully preserved example of an early Victorian well-to-do household. The gardens are open to the public daily; guided tours of the house are available on request. *Wentworth Rd; (02) 9388 7922; open 9.30am–4pm Fri–Sun.*

Watsons Bay 7 E3

All roads in the eastern suburbs lead to Watsons Bay. This is where you will find the spectacular ocean cliffs known as **The Gap** and **Doyles**, a famous seafood restaurant with superb views of the city. Nearby is **Camp Cove**, a popular family beach and the starting point of a 1.5-kilometre walking track, which takes you past Sydney's first nude-bathing beach, **Lady Bay**, to the windswept promontory of **South Head**. South Head boasts magnificent views across the harbour to Manly and a poignant memorial to the crew and passengers of the *Dunbar* who were all lost in 1853 when the ill-fated ship ran aground and sank just outside the heads.

Bondi Beach 6 E5

On a hot summer's day, Bondi is about as iconic as it gets – a sweep of pale sand covered with a rainbow of towels and umbrellas, encircling the blue-and-white breaks of the bay. Interestingly, Bondi Beach was privately owned until 1856, when the government purchased it for the 'pleasure of the people'. The suburb was a bohemian and immigrant enclave from the 1950s onwards, and even today with soaring real estate prices, it retains a diverse mix of professionals, artists, students and surfers. The beach stretches 800 metres between a set of headlands, from the **Icebergs** sea baths in the south to a rockpool at the northern end. In the south is where you'll find sizeable, if a little inconsistent, surf, while the more sheltered north is suitable for families.

Bondi to Coogee walk 6 D6, C5

Though all these suburbs are easily accessed by bus, the best way to view them is by taking the walking track that starts at the southern end of **Bondi Beach**, known for its excellent weekend market *(see Markets, p. 14)* and chic sidewalk cafes. From Bondi, the walking track winds south along the cliffs through the tiny boutique beach of **Tamarama**. In November, this section of the track becomes crowded with people viewing the Sculpture by the Sea exhibition *(see Walks & tours, p. 15)*. **Bronte**, a lovely beach with a natural-rock swimming pool known as the Bogey Hole, completes the track. For those who want to explore further, the walk extends through **Waverley Cemetery** (where you can find the grave of Henry Lawson), past **Clovelly** (a popular swimming place), to **Coogee**. Along with the shops and cafes of Arden Street, you can visit the **Coogee Bay Hotel**, a lively pub with a sunny waterfront beer garden, which lies directly opposite the beach.

[DOYLES RESTAURANT, WATSONS BAY]

South-eastern Beachside

The city of Botany Bay, on the northern shores of Botany Bay, is Australia's largest municipality. Botany Bay National Park straddles the bay's headlands, and at its entrance is the suburb of La Perouse, home to an intriguing combination of natural and cultural heritage.

La Perouse 545 M10

La Perouse is situated on the northern head of Botany Bay. Originally the home of the Muru-ora-dial people, it was named after the Comte de La Perouse, a French navigator who arrived in Botany Bay around the same time as the First Fleet, as part of a competing French contingent. With some beautiful beaches and interesting walks, it also contains an excellent museum *(see p. 18)*.

where to eat

a tavola Authentic Italian in an intimate dining room setting in the heart of Sydney's trendy inner east. 348 Victoria St, Darlinghurst; (02) 9331 7871; open Fri for lunch and Mon–Sat for dinner. 8 D5

Bel Mondo Modern Sydney cuisine served with delicious glimpses of the Harbour Bridge and Opera House. Gloucester Walk, The Rocks; (02) 9241 3700; open Thurs–Fri lunch and Tues–Sat dinner. 8 C2

Berowra Waters Inn Spectacular regional location and cuisine make this star definitely worth the journey. Via Public Wharf; (02) 9456 1027; open Fri–Sun for lunch and Fri–Sat for dinner. 454 L5

Burlington Bar and Dining Dining excellence without the hefty price tag. Experience what locals are raving about. 6 Burlington St, Crows Nest; (02) 9439 7888; open Mon–Fri for lunch and Mon–Sat for dinner. 6 C2

Quay Right at the top of everyone's list for first-class dining. The location – directly opposite the Opera House – isn't bad either. Upper Level, Overseas Passenger Terminal, Circular Quay West; (02) 9251 5600; open Tues–Fri for lunch and daily for dinner. 8 C2

Spice I Am It might not look like much, but Spice I Am has the best Thai food in Sydney and at affordable prices. 90 Wentworth Ave, Surry Hills; (02) 9280 0928; open Tues–Sun for lunch and dinner. 8 B5

Tetsuya's Experience a genius of Japanese-French fusion and individuality with the incredible degustation at Tetsuya's. 529 Kent St; (02) 9267 2900; open Sat for lunch and Tues–Sat for dinner. 8 B5

[BEACHGOERS, COOGEE]

entertainment

Cinema Located in George Street, between Bathurst and Liverpool streets, the 17-screen Event Cinemas George Street complex is the major cinema centre in the CBD. Arthouse cinemas include the Chauvel at Paddington Town Hall; the Dendy at Newtown and Circular Quay; and the Cinema Paris at the Entertainment Quarter. For a unique cinema experience, try the Hayden Orpheum Picture Palace in Cremorne, famous for its Art Deco interior and Wurlitzer pipe organ. See the newspapers or websites for details of films being shown.

Live music Sydney has always had a strong live music scene with some excellent venues throughout the city. Apart from the larger, more formal places for live bands that include the Enmore Theatre, the Metro in George Street and the Forum at the Entertainment Quarter, the city's pubs are the main focus for live music. Try the Annandale Hotel, a popular venue for local indie bands; or the legendary Bridge Hotel in Rozelle, which specialises in blues and pub rock. For jazz lovers, there's the Vanguard in Newtown and the famous Basement at Circular Quay. Check newspaper lift-outs such as 'Metro' for what's on, or get hold of one of the free magazines such as *3D World* or *Drum Media* for details.

Classical music and performing arts The ultimate venue for theatre, dance and classical music, the Sydney Opera House plays host to companies such as the Sydney Symphony Orchestra, the Sydney Theatre Company, Opera Australia and the Australian Ballet. It's worth checking out some of the smaller venues, such as the City Recital Hall in Angel Place or the Conservatorium of Music. Other venues for excellent live theatre include the Belvoir St Theatre in Surry Hills, and the Wharf Theatres and the Sydney Theatre, both at Walsh Bay. If you are interested in dance, you can catch the Sydney Dance Company and the Bangarra Aboriginal dance group between tours at the Sydney Theatre or at the Sydney Opera House. For details, check the *Sydney Morning Herald*'s Friday lift-out, 'Metro', or *Time Out Sydney* magazine.

Bare Island 545 M10

From the earliest days of European settlement, La Perouse was considered to be crucial to the defence of the colony. Governor Macquarie built the sandstone tower that stands at the highest point of the promontory, and the fortifications at Bare Island were added in 1885. Guided tours are the only way to view these buildings (which movie fans will recognise from *Mission: Impossible 2*). *Accessed via a footbridge from Anzac Pde; (02) 9247 5033; guided tours Sun 1.30pm, 2.30pm and 3.30pm.*

La Perouse Museum 545 M10

This highly recommended museum occupies the old Cable Station, and presents the rich and varied history of La Perouse – it was once the site of an Aboriginal mission station and a Depression-era shanty town. The real focus of the museum is the life and times of La Perouse himself, with galleries devoted to the history of Pacific exploration, the voyage to Botany Bay and the eventual wreck and loss of the entire expedition. Not far from the museum is a monument to La Perouse near the grave of Father Receveur, the chaplain on the expedition. *Anzac Pde, La Perouse; (02) 9247 5033; open 10am–4pm Wed–Sun.*

Inner North

Sydney's inner north includes some of the city's most sought-after suburbs, including Neutral Bay, Mosman and the harbour beach suburb of Balmoral. The main shopping districts of Neutral Bay and Mosman sit along Military Road, the only connection to the Spit Bridge spanning Mosman and Seaforth on Middle Harbour.

Nutcote 6 C3

Nutcote is a must-see for anyone who remembers the characters of Snugglepot, Cuddlepie and the Banksia Men. Once home to May Gibbs, one of Australia's best-known authors and illustrators of children's books, Nutcote is situated in the exclusive harbourside suburb of Neutral Bay. It has been restored and furnished as it would have been in the 1930s, when May Gibbs lived there. It is now a centre for children's literature, the arts and the environment. *5 Wallaringa Ave; (02) 9953 4453; open 11am–3pm Wed–Sun.*

Taronga Zoo 6 D3

Located at the end of Bradleys Head Road in Mosman, Taronga Zoo is best accessed by ferry from Circular Quay. A world-class institution, it has an impressive reputation in the fields of conservation, and the care and management of rare and endangered species. It houses over 2000 animals, most of which enjoy stunning views of the harbour and city. Visitors can enjoy animal feeding, keeper talks and displays. The zoo boasts plenty of places for picnics and barbecues. It is also the venue for the ever-popular *Twilight at Taronga* open-air concerts, which take place in February and March. *(02) 9969 2777; open 9am–5pm daily.*

Balmoral 6 D2

Some say Balmoral, with its curving promenade, shady trees and elegant bridge connecting a rocky outcrop to the mainland, is reminiscent of Edwardian Sydney. Extremely popular in summer, Balmoral has a lovely beach with a fenced-in pool, pleasant foreshore and some excellent restaurants.

North-east

Manly 7 F2

A ferry company once promoted Manly as 'Seven miles from Sydney, and a thousand miles from care'. Today it still retains much of the holiday atmosphere of a seaside village. The Manly ferry, which travels to and from Circular Quay, is an experience in itself, and the most pleasant way of getting there. With two beaches, an aquarium *(see next entry)*, the Old Quarantine Station *(see entry on this page)* and the carnival atmosphere of Manly Corso, a visit to Manly is a must.

Oceanworld Manly 7 F2

Situated to the west of the ferry terminal, Oceanworld is an exciting bottom-of-the-sea experience. With colourful displays of marine life that include a touch pool and an underwater tunnel, it also reveals some of the scarier aspects of the continent's wildlife. There are shark-feedings daily, exhibitions that involve some of Australia's deadliest snakes and spiders, plus the chance to dive with sharks. *West Esplanade, Manly; (02) 8251 7877; open 10am–5.30pm daily.*

Q Station 7 F2

On the west side of North Head, up the road from Manly, the Old Quarantine Station has been transformed into the Q Station, a unique resort with accommodation, a restaurant, interactive tours and a theatre experience within Sydney National Park. Built to house incoming passengers and crew suspected of carrying contagious diseases from the 1830s to 1984, there's an extensive collection of 65 heritage buildings as well as some fascinating 19th-century rock carvings that were made by the crew and passengers interned here. Access is via car or water shuttle from The Rocks and Manly, and tour bookings are essential. *North Head Scenic Dr, Manly; (02) 9466 1551.*

West

Sydney Olympic Park, Homebush Bay 545 K8

The site of the 2000 Olympic Games, the park now hosts Sydney's yearly Royal Easter Show and is home to several other attractions and events. It can be accessed by Parramatta Road, by rail or

sport

Sydneysiders have always been passionate about their sport. As a city that hosted the Olympic Games 12 years ago, Sydney is home to some of the best sporting facilities in the world.

Football, cricket and racing dominate the sporting scene. Although Sydney does have an **AFL** (Australian Football League) team – the Sydney Swans – Rugby League and Rugby Union hold more sway here, with the season for both codes beginning in March. Key games throughout the **Rugby League** season are played at the Sydney Football Stadium at Moore Park, with the Grand Final taking place in September at ANZ Stadium, in Olympic Park. A particular highlight is the **State of Origin** competition, which showcases the cream of Rugby League talent in a series of three matches between Queensland and New South Wales. These take place in the middle of the season, and the New South Wales matches are played at ANZ Stadium.

The Waratahs are the New South Wales side in the Super 12s, the **Rugby Union** competition in which local and overseas teams go head to head. These games are played at Sydney Football Stadium, while the **Bledisloe Cup** games (between Australia's Wallabies and the New Zealand All Blacks) are played at ANZ Stadium and attract up to 80 000 spectators.

In summer, **cricket** takes centre stage. The highlight is the New Year's Day Test, followed by the One-Day Internationals, all of which are played at the Sydney Cricket Ground (SCG). Sydney's official **soccer** (football) team is Sydney FC, which competes in the Hyundai A-League at Sydney Football Stadium.

Other sporting highlights include the **Spring and Autumn Racing Carnivals**, with the world's richest horserace for two-year-olds, the Golden Slipper, being held just before Easter.

where to stay

BLUE Sydney Stunning Sydney waterfront accommodation nestled at one of Sydney's premier landmarks: Woolloomooloo Wharf. The Wharf at Woolloomooloo, 6 Cowper Wharf Rd; (02) 9331 9000. 8 D4

Establishment Hotel Experience fabulous luxury in the heart of the city at this celebrity-studded jewel. 5 Bridge La; (02) 9240 3100. 8 B3

Medusa Boutique Hotel Eclectic bohemian colour and life burst from this unique boutique hotel cradled in one of Sydney's funkiest districts. 267 Darlinghurst Rd, Darlinghurst; (02) 9331 1000. 8 D5

Observatory Hotel Timeless elegance and luxury saturate this old-world treasure tucked away at The Rocks. 89–113 Kent St; (02) 9256 2222. 8 B2

Shangri-La Hotel Guests' concept of contemporary comfort is catapulted sky high at this statuesque beauty towering over Sydney Harbour. 176 Cumberland St; (02) 9250 6000. 8 B3

other suburbs

Surry Hills Where old Sydney meets new urban chic, with 19th-century streetscapes, trendy cafes, bars and clusters of fashion warehouses close to Central Station. 8 B6

Newtown A suburb with a funky, alternative feel, Newtown seems to sleep late and party late. It is best visited in the afternoon and early evening, when the shops and cafes of King Street come alive. 6 A5

Leichhardt Sydney's little Italy, with some of the city's best Italian restaurants, superb shopping in Norton Street and the unique shopping and dining precinct of Italian Forum. 6 A4

Cabramatta The heart of the Vietnamese community, with a vibrant shopping strip specialising in good-quality fabrics, fresh Asian produce and fabulous pho (rice noodle soup). 545 J9

Avalon Pretty beachside village with a stunning backdrop of bush-clad hills, and a laid-back shopping precinct crammed with delis, cafes and some very up-market clothing and homewares stores. 545 N6

Penrith Close to the Nepean River, this is a paradise for watersports enthusiasts, with the Cables Waterski Park and the Whitewater Stadium located here. There is a pleasant shopping precinct with some good eateries, and an excellent regional gallery. 544 H6

Camden One of Sydney's most far-flung suburbs, with a pleasant rural atmosphere, pretty, old-world streetscapes and good cafes. 544 G11

Cronulla With some interesting shops and cafes in the main street and a superb beach close by, this is a good starting point for trips to Royal National Park and the important historical site of Kurnell (Captain Cook's landing site and the birthplace of modern Australia). 545 L11

harbour islands

Sydney Harbour is dotted with islands, but only five of them are open to the public. Four come under the authority of the National Parks & Wildlife Service, which charges a $7 landing fee per person. Visits must be prebooked and prepaid, so the information and tour booking office located in Cadmans Cottage, George St, The Rocks, is the starting point for any tour or visit to the islands (except Cockatoo Island). (02) 9247 5033

Fort Denison Crime, punishment and the defence of Sydney Harbour are all part of Fort Denison's past. Now it plays a vital role in assessing and predicting the tides, and is the site of the One O'clock Gun. Access is by guided tour only, and there is a restaurant on the island. 6 C3

Shark Island Sandy beaches, shaded grassy areas and superb views of the harbour make this the perfect place for a picnic. Matilda Catamarans runs a daily ferry service from Darling Harbour and Circular Quay to Shark Island, with the fare including the landing fee. 6 D4

Clark Island Named for Ralph Clark, an officer of the First Fleet who once planted a vegetable garden here, Clark Island is now a place of unspoiled bushland and pleasant grassy areas, and is popular with picnicking families. Access by private vessel or water taxi. 6 D4

Rodd Island Another favourite picnic place, Rodd Island has a colonial-style hall, which dates back to 1889, and 1920s summer houses that shelter long tables, making the island suitable for picnicking in all seasons. Access by private vessel or water taxi. 6 A4

Cockatoo Island A former prison and shipyard, Cockatoo Island has abandoned workshops and wharves, a camping ground and cafe, and hosts art and concert events. Tours are available. It's free to visit and accessible by regular ferries from Circular Quay Wharf 5 or 4. 6 A3

by RiverCat from Circular Quay. Take a lift to the Observation Deck on the 17th floor of the Novotel Hotel to see fantastic views of the entire park. There are various activities available, as well as **Bicentennial Park**, with its extensive wetlands and bird sanctuaries, and the fabulous **Aquatic Centre**.

Outer West

Elizabeth Farm, Rosehill 21 D3

Elizabeth Macarthur was one of the more fascinating and unusual characters in early Australian history. With her husband, John, she gave the Australian wool industry a kick-start, running their large merino farm while John was away in England. The lovely sandstone building surrounded by gardens is now a museum run by the Historic Houses Trust. Visitors are encouraged to wander the rooms, touch the furniture, try the featherbeds and generally behave as if they are guests of the family. *Alice St; (02) 9635 9488; open 9.30am–4pm Fri–Sun.*

Hambledon Cottage, Harris Park 21 C3

Not far from Elizabeth Farm is the charming Hambledon Cottage, built for a Miss Penelope Lucas, governess of the Macarthur children. Small but elegant, it has been restored and furnished in a style that reflects the early reign of Queen Victoria. It is surrounded by trees said to have been planted by John Macarthur himself. *Hassall St; (02) 9635 6924; open 11am–3.30pm, Wed, Thurs, Sat, Sun.*

Parramatta 21 B2

The quickest way to get to Parramatta is probably by train, but the most pleasant and relaxing way to travel is by RiverCat. The area was discovered not long after the arrival of the First Fleet, and Arthur Phillip immediately recognised its farming potential. The colony's first private farm was established here in November 1788, making this outer suburb almost as old as Sydney itself. With places of historical interest around every corner and the lively commercial atmosphere of a busy regional centre, Parramatta is a fascinating spot to visit.

Once you arrive in Parramatta, your first stop should be the **Parramatta Heritage Centre**. Located alongside the Parramatta River, the centre presents a number of exhibitions highlighting the experiences of those who helped to shape this part of Sydney. *346A Church St; (02) 8839 3311; open 9am–5pm daily.*

The nearby **Riverside Walk** takes on a whole new meaning when you follow the 800-metre painted path. A combination of paintings, interpretive plaques and native gardens help to reveal the history of this area and its inhabitants, all from the perspective of the Aboriginal people. The walk culminates in a moving soundscape of music and spoken words – it is an experience that should not be missed.

Nearby in the sweeping grounds of Parramatta Regional Park, **Old Government House** stands as one of the oldest public buildings in Australia. Built between 1799 and 1818, chiefly by Governors Hunter and Macquarie, it is very much Macquarie's house. It has been restored to reflect the life and times of the Macquarie family, and includes a fine collection of their own furniture. Guided tours are available and a ghost tour runs on the third Friday of each month; bookings essential. *(02) 9635 8149; open 10am–4pm Tues–Fri, 10.30am–4pm Sat–Sun.*

parramatta

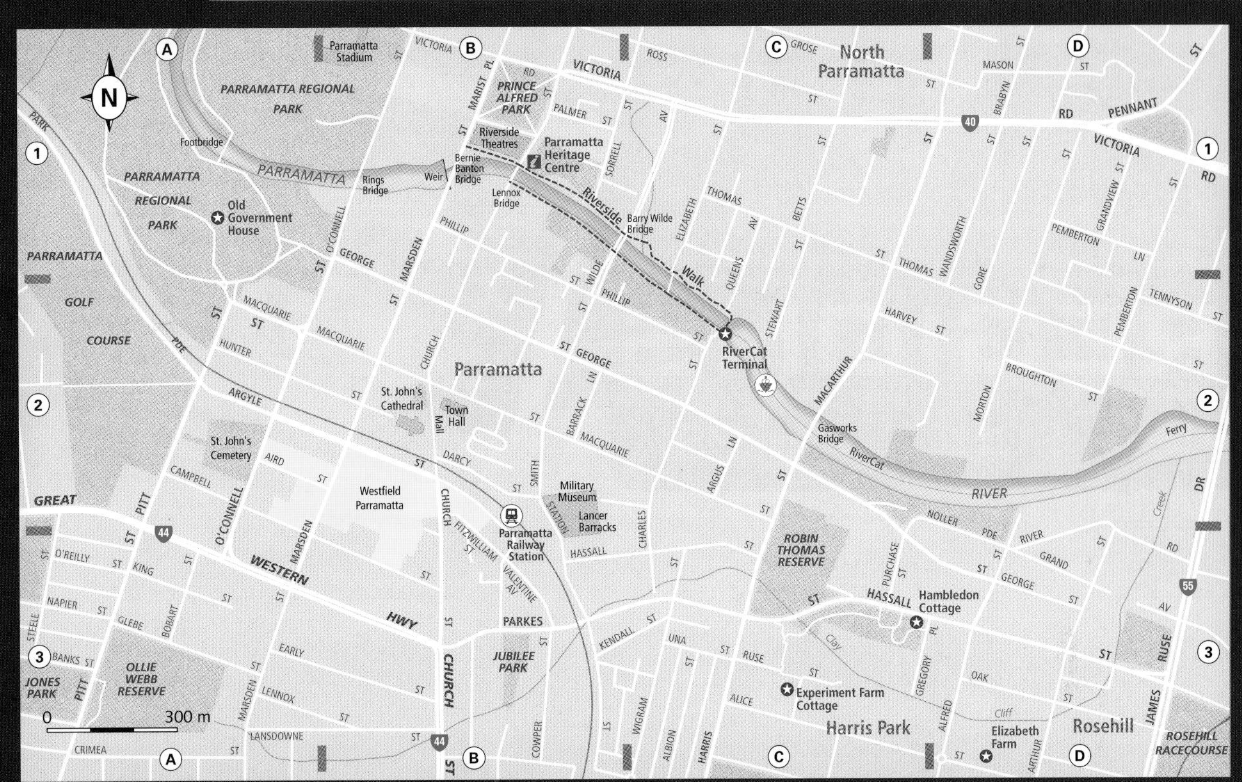

A visit to Parramatta would not be complete without visiting the **Experiment Farm Cottage**. Built in 1798 and named because it was built on the site of Australia's first private farm, the cottage is now run by the National Trust. It features lovely gardens as well as an excellent collection of 1830s furniture. If you are fortunate, your visit may coincide with one of the National Trust exhibitions that are held there from time to time. *Ruse St; (02) 9635 5655; open 10.30am–3.30pm Tues–Fri, 11am–3.30pm Sat–Sun.*

Far Outer West

Featherdale Wildlife Park, Doonside 545 I7

This is the place to get up close and personal with Australian wildlife. Over 2000 animals live in a beautiful bushland setting, and there are opportunities to cuddle koalas and handfeed kangaroos. It is also a perfect venue for picnics and barbecues. *217 Kildare Rd; (02) 9622 1644; open 9am–5pm daily.*

day tours

Blue Mountains Only 100 kilometres from Sydney, the brooding sandstone cliffs and deep, tree-lined gorges of the Blue Mountains provide a superb natural retreat. Aboriginal cave art, cool-climate gardens, charming mountain villages, walking trails and adventure activities are among the many attractions. For more details see p. 23

Southern Highlands Nestled into the folds and hills of the Great Dividing Range, the Southern Highlands offer pretty rural scenery, historic townships with superb European-style gardens, a variety of festivals, and wonderful guesthouses and restaurants. For more details see p. 24

Central Coast & Hawkesbury Stretching north from Sydney Harbour to Broken Bay, the northern beaches are one of Sydney's loveliest natural assets. Pittwater Road takes you from Manly through the beach suburbs of Dee Why, Long Reef, Collaroy and Narrabeen. Barrenjoey Road leads to more exclusive beaches, culminating in posh Whale Beach and Palm Beach. The Upper Hawkesbury, north-west of Sydney, encompasses a scenic river landscape dotted with charming Georgian villages. It is Australia's most historic rural area. For more details see p. 25

Ku-ring-gai Chase North of Sydney, Ku-ring-gai Chase National Park encloses a magnificent stretch of bushland, set around the glittering waters of the Hawkesbury River and Broken Bay. Fishing, river cruises and bushwalking are popular activities here. For more details see p. 25

South along the coast Abutting Sydney's southern suburbs, Royal National Park encloses a landscape of sandstone outcrops, wild heathland, rainforest, plunging cliffs and secluded beaches – perfect for walks, fishing, wildlife-watching and camping. Beyond the park you'll find the city of Wollongong and a magnificent stretch of surf coast dotted with pleasant resort towns. For more details see p. 24

Hunter Valley Located about 160 kilometres north-west of Sydney, the Lower Hunter area is Australia's oldest winegrowing district, with over 60 wineries radiating from the town of Cessnock. Take a tour of the wineries, starting with a visit to the wine centre in town, or book into one of the many excellent restaurants. For more details see p. 26

[NEDS BEACH, LORD HOWE ISLAND]

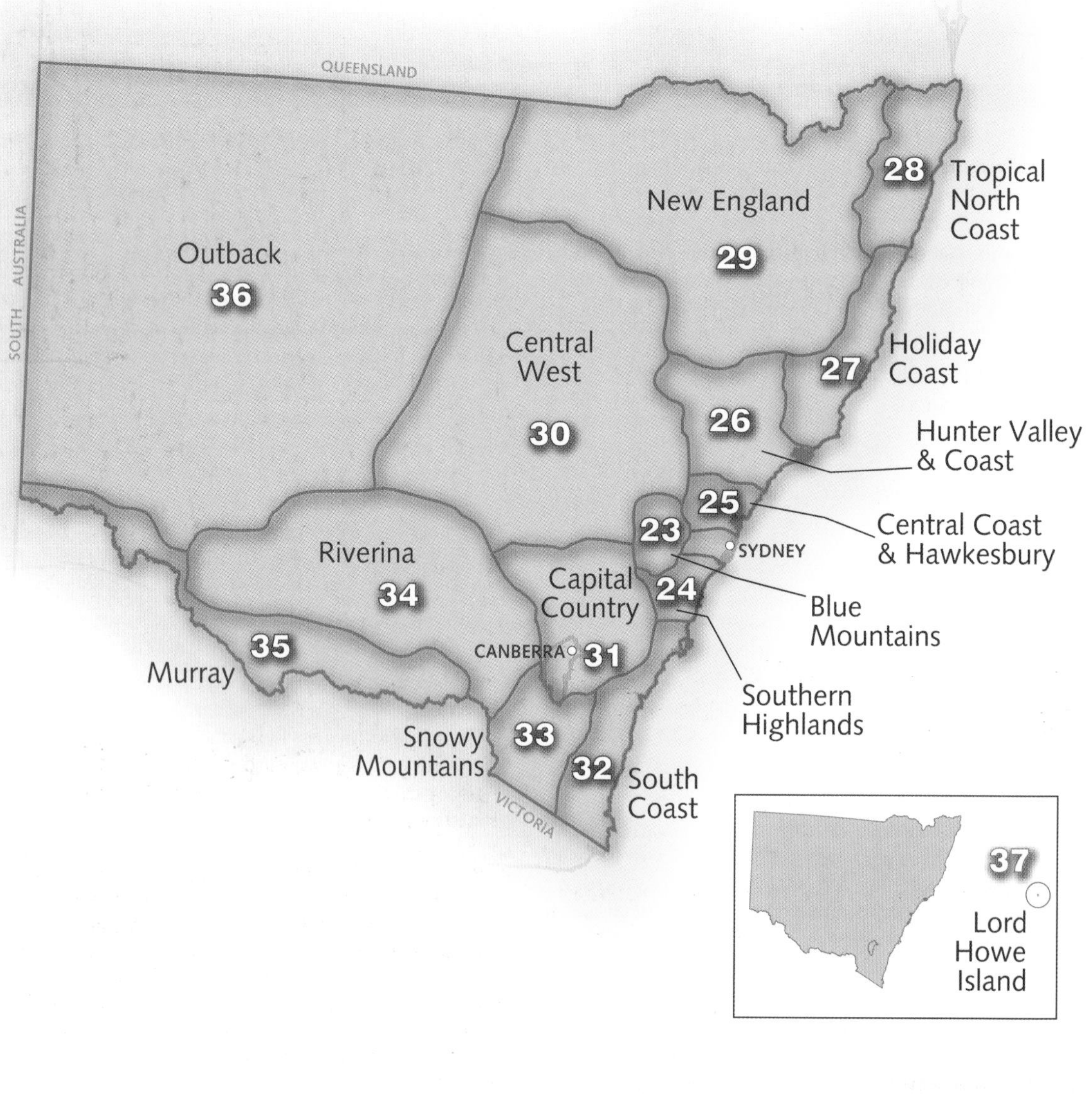

[SKIING AT THREDBO, SNOWY MOUNTAINS]

REGIONS
new south wales

QUEENSLAND
SOUTH AUSTRALIA
VICTORIA

Outback 36
New England 29
Tropical North Coast 28
Central West 30
Holiday Coast 27
26 Hunter Valley & Coast
25 Central Coast & Hawkesbury
23 Blue Mountains
24 Southern Highlands
SYDNEY
Riverina 34
Capital Country 31
CANBERRA
Murray 35
Snowy Mountains 33
32 South Coast
37 Lord Howe Island

BLUE MOUNTAINS

The misty, bush-clad cliffs and valleys of the Blue Mountains provide a beautiful nature retreat for nearby Sydney residents.

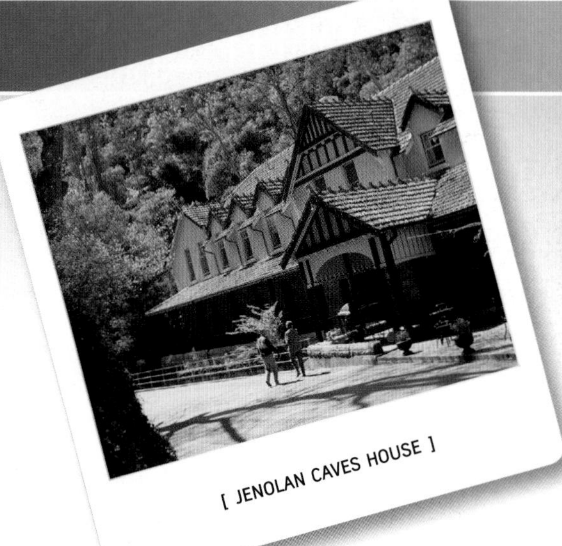

[JENOLAN CAVES HOUSE]

Zig Zag Railway When it was built in 1869, the Zig Zag Railway was a revolutionary engineering achievement, opening up Blue Mountains towns such as Lithgow, which had previously been isolated from the coast. Today it operates as a tourist railway, passing through beautiful mountain scenery, with trips departing from Clarence (10 kilometres east of Lithgow) daily at 11am, 1pm and 3pm. Kids will love the special Friends of Thomas days, when Thomas the Tank Engine himself heads the train.

Jenolan Caves Formed 400 million years ago, the Jenolan Caves make up one of the most extensive and complex underground limestone cave systems in the world. Of the 300 or so 'rooms', nine are open to the public – by tour only.

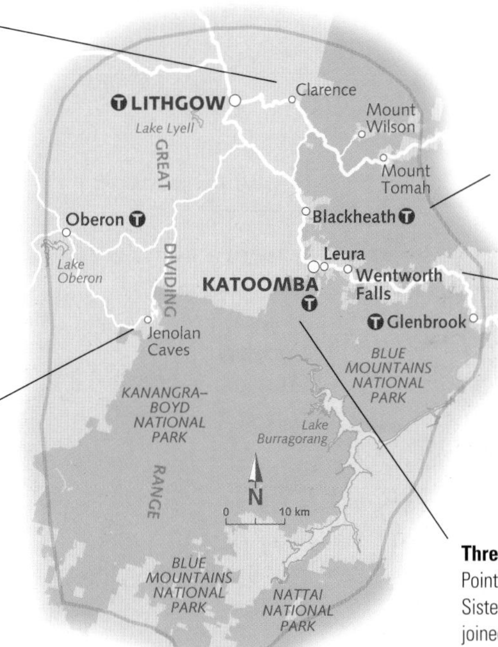

Heritage Centre The Blue Mountains Heritage Centre east of Blackheath has a wealth of knowledge on the geology, history, flora and fauna of the mountains, and also on the network of walking trails leading into their heart. Nearby Govetts Leap Lookout, high above the Grose Valley, is the starting point for a handful of trails.

Wentworth Falls One of the most beautiful towns in the Blue Mountains, Wentworth Falls offers bushwalks with phenomenal views. A corridor of trees leads to picturesque Wentworth Falls Lake, a popular picnic spot, and the massive waterfall after which the town was named.

Three Sisters and Echo Point The lookout at Echo Point is the best vantage point to view the famous Three Sisters. Echo Point and the Three Sisters were once joined, but over time great blocks of rock broke off and fell away into the Jamison Valley. Visitors can ride on the Scenic Railway, the steepest railway in the world, or enjoy views from the Scenic Skyway.

TOP EVENTS

APR	Ironfest (Lithgow)
JUNE	Winter Magic Festival (Katoomba)
JUNE–AUG	Yulefest (throughout region)
SEPT	Daffodil Festival (Oberon)
OCT	Leura Gardens Festival (Leura)
NOV	Rhododendron Festival (Blackheath)

CLIMATE

KATOOMBA

J	F	M	A	M	J	J	A	S	O	N	D	
23	22	20	17	13	10	9	11	14	18	20	22	MAX °C
13	13	11	9	6	4	3	3	5	8	10	12	MIN °C
160	174	170	123	104	117	85	82	73	90	103	124	RAIN MM
13	13	13	11	10	10	9	9	9	11	11	12	RAIN DAYS

GARDENS

Volcanic soil and cool-climate conditions have made the Blue Mountains one of the best-known gardening regions in Australia. Visit Everglades near Leura, a 6-hectare classically designed garden that melds with the surrounding bush. Mount Wilson is a tiny village of grand, historic estates, nearly all with large gardens of formal lawns, cool-climate plantings and tall European trees; many properties are open to the public. Mount Tomah Botanic Garden is the cool-climate annexe of Sydney's Royal Botanic Gardens. Here specialist displays bring together thousands of rare species from around the world, with a focus on those from the Southern Hemisphere. One feature is the grove of young Wollemi pines – this species was only discovered in 1994 in Wollemi National Park.

→ For more detail see maps 544–5, 546–7 & 554–5. For descriptions of Ⓣ towns see Towns from A–Z (p. 38).

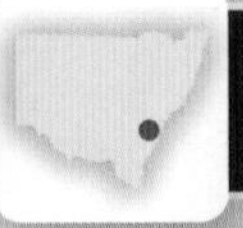

SOUTHERN HIGHLANDS

This region offers European-style rural scenery, combined with a stretch of typically Australian coastline – both are great settings for holiday-makers.

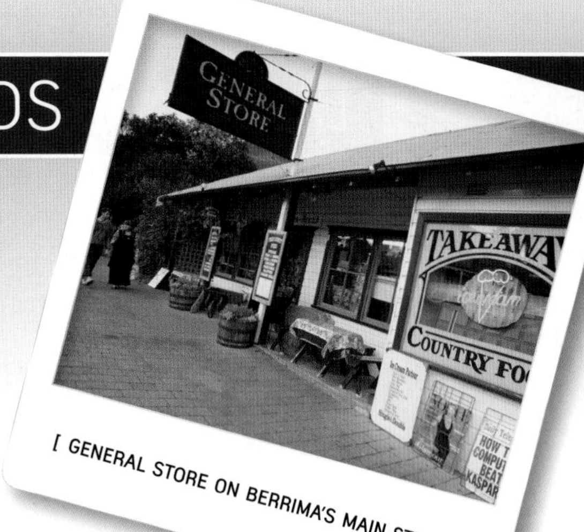

[GENERAL STORE ON BERRIMA'S MAIN STREET]

Wollongong 'Australia's most liveable regional city' is set along 17 surf beaches and surrounded by fantastic mountain scenery. Visit the city's boat harbour, fish co-op and restaurants on Flagstaff Point; drive north through the old coalmines for coastal views; and see Nan Tien Temple, the largest Buddhist temple in the Southern Hemisphere.

Minnamurra Falls Minnamurra Falls, in Budderoo National Park, is one of several waterfalls formed by the massive sandstone escarpment that defines the edge of the Southern Highlands. Nearby is the Minnamurra Rainforest Centre, which incorporates a raised walkway leading into pockets of dense temperate and subtropical rainforest.

Morton National Park The northern reaches of Morton National Park were once known as the Bundanoon Gullies; in autumn, mists roll out and engulf the town, giving it the feeling of Brigadoon. Explore the sandstone cliffs, wooded valleys and waterfalls of the park on one of the various walking tracks from the town.

Kangaroo Valley This valley, known for its lovely combination of rural and native scenery, can be explored on the scenic route leading from the highlands to the coast (from Moss Vale to Nowra). The route takes in the 80-metre-high Fitzroy Falls, crosses Australia's oldest suspension bridge, and passes through the historic township of Kangaroo Valley.

TOP EVENTS

JAN	Illawarra Folk Festival (Bulli)
MAR	Jazz and Blues Festival (Kiama)
APR	Bundanoon is Brigadoon Highland Gathering (Bundanoon)
SEPT	Festival in the Forest (Shellharbour)
SEPT–OCT	Tulip Time Festival (Bowral)
NOV	Festival of Fisher's Ghost (Campbelltown)

CLIMATE

BOWRAL

J	F	M	A	M	J	J	A	S	O	N	D	
25	25	22	19	15	12	12	13	16	19	21	24	MAX °C
13	13	11	8	5	3	2	3	5	7	9	11	MIN °C
87	91	97	86	77	82	46	66	59	75	92	71	RAIN MM
14	13	14	11	13	11	10	10	11	12	13	12	RAIN DAYS

TOWNS OF THE HIGHLANDS

Bowral, with its historic streetscapes, restaurants and cafes, guesthouses and superb gardens, is the centre of the highlands. Historic Berrima, once the commercial heart of the district, now serves as a timepiece of colonial Australia. Nearby Mittagong boasts lovely gardens as well as good cafes and interesting shopfront architecture. Sutton Forest and Moss Vale are pretty towns with an air of the English countryside, while Bundanoon, further south, is known for its excellent guesthouses, health resort and views across Morton National Park. Away from the main tourist route are Robertson and Burrawang, peaceful settlements steeped in history. Don't miss Burrawang's general store (c. 1875), still with a sign advertising the *Sydney Morning Herald* for a penny. The rolling hills of this district were the backdrop for the Australian film *Babe* (1995). To the south is Berry, set in the dairy country of the Shoalhaven River district and boasting charming heritage buildings, galleries, antique shops and guesthouses.

→ For more detail see maps 542, 544–5, 552, 553 & 565. For descriptions of Ⓣ towns see Towns from A–Z (p. 38).

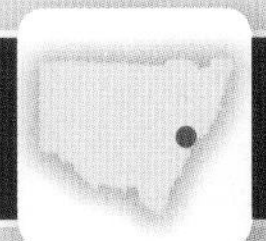

CENTRAL COAST & HAWKESBURY

Discover rugged wilderness, Australia's most historic farming district, and the glittering waters of one of the country's most popular national parks.

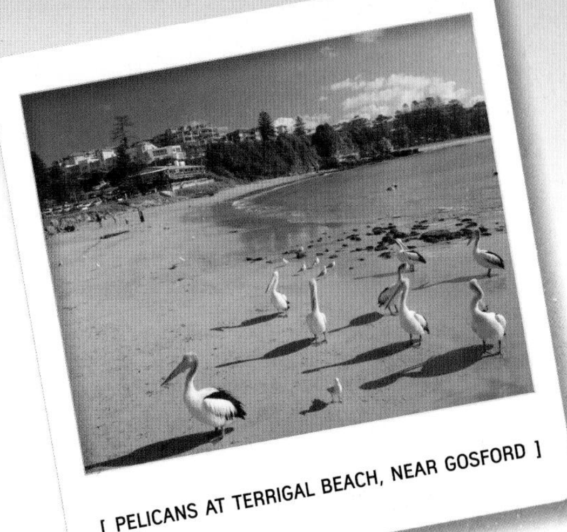

[PELICANS AT TERRIGAL BEACH, NEAR GOSFORD]

Wollemi National Park This 500 000-hectare wilderness includes areas still unmapped. In 1994 the discovery of a new tree species – the Wollemi pine – in a rainforest gully was compared to finding a living dinosaur. See Newnes' historic ruins, walk to the Glow Worm Tunnel, or visit the secluded beaches along the Colo River.

Pelican feeding at The Entrance Every day at 3.30pm sharp, dozens of entertaining pelicans wander in to the pelican pavilion on The Entrance's foreshore for a feed, and every year tens of thousands of people turn up to watch them. What has become one of the Central Coast's most popular tourist attractions started out by accident over 20 years ago, when someone working at a local fish 'n' chip shop threw out some scraps ...

Lower Hawkesbury The mouth of the Hawkesbury River opens up into a dazzling network of calm, bush-lined waterways, which provides the perfect setting for a houseboat holiday or a day of boating and fishing. Hop aboard Australia's last river postal run: the Riverboat Postman departs from Brooklyn at 9.30am on weekdays.

Macquarie towns In 1810 Governor Macquarie established a number of towns on fertile river plains north-west of Sydney. The area now preserves some of Australia's oldest buildings and sites. In Windsor is a cemetery containing the graves of some of the pioneers who arrived in the First Fleet, and to the north in Ebenezer is Australia's oldest church, built in 1809.

Ku-ring-gai Chase National Park This scenic 15 000-hectare bush, river and sandstone landscape forms a border on the edge of Sydney's northern suburbs. Hidden coves, sheltered beaches and panoramic lookouts combine with a rich Aboriginal heritage. See the Guringgai rock engravings on the Basin Trail at West Head.

TOP EVENTS

MAY	Bridge to Bridge Powerboat Classic (Windsor)
JULY	Terrigal Food and Wine Festival (Terrigal)
SEP	CoastFest (Gosford)
SEPT–NOV	Fruits of the Hawkesbury Festival (Windsor and Richmond)
DEC	Tuggerah Lakes Mardi Gras Festival (The Entrance)

CLIMATE

GOSFORD

J	F	M	A	M	J	J	A	S	O	N	D	
27	27	26	24	20	18	17	19	21	24	25	27	MAX °C
17	17	15	12	8	6	5	5	8	11	13	15	MIN °C
139	148	150	136	119	128	79	76	69	83	92	102	RAIN MM
11	11	11	11	9	10	8	8	8	9	10	10	RAIN DAYS

FISHING

Fishing is one of the area's great attractions. There are plenty of boat ramps and boat-hire outlets, and no shortage of local bait-and-tackle suppliers. The trouble-free waters of the Hawkesbury are good for bream, luderick, mulloway and flathead, with bass in the upper reaches. Bream and whiting are a possibility for anglers in the calm stretches of Brisbane Water and Tuggerah Lake. On the coast, beach fishing yields bream, tailor and mulloway. There are rock platforms, particularly around Terrigal, where anglers try for some of the big ocean fish, including tuna and kingfish.

→ For more detail see maps 542, 543, 544–5 & 555. For descriptions of T towns see Towns from A–Z (p. 38).

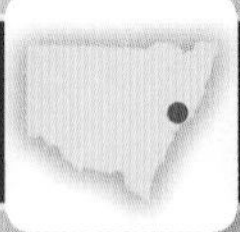

HUNTER VALLEY & COAST

Taste wines at cellar doors in the Hunter Valley, one of Australia's premier wine-tourism destinations.

[PEPPER TREE WINES CELLAR DOOR, HUNTER VALLEY]

Barrington Tops National Park The World Heritage–listed Barrington Tops are situated on one of the highest points of the Great Dividing Range (1600 metres). This landscape of rugged basalt cliffs, rainforests, gorges and waterfalls, with a touch of snow in winter, is a mecca for walkers, campers and climbers.

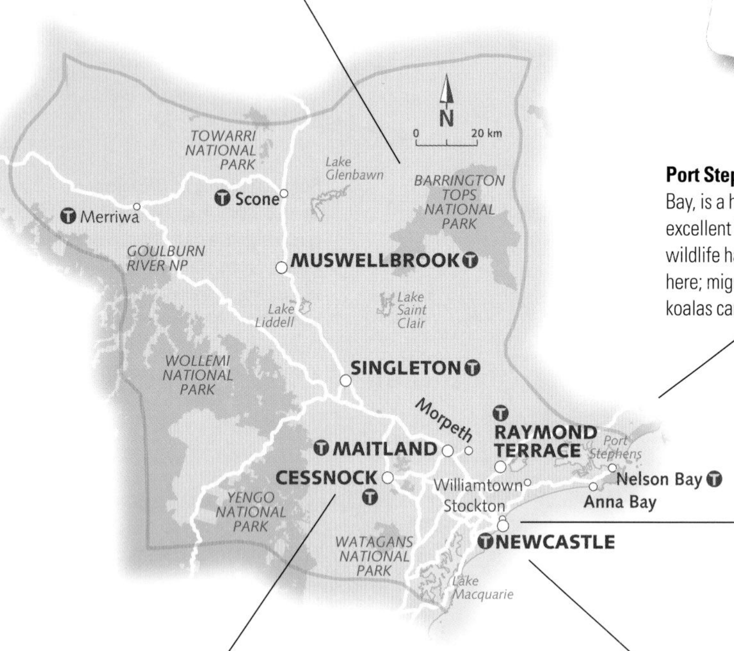

Port Stephens Port Stephens, reached via the township of Nelson Bay, is a haven of calm blue waters and sandy beaches, offering excellent boating, fishing and swimming. It is also something of a wildlife haven: over 100 bottlenose dolphins are permanent residents here; migrating whales can be seen in season on a boat cruise; and koalas can be spotted at Tilligerry Habitat.

Stockton Sand Dunes This huge dune area, the largest moving coastal sand mass in the Southern Hemisphere, extends for 32 kilometres, rises to a height of 30 metres and is, understandably, a popular place for sand boarding. Deep within its midst, you'd be forgiven for thinking you were in the Sahara, not coastal New South Wales. Access is from Anna Bay or Williamtown by four-wheel drive or safari (book at the Port Stephens visitor centre).

Lower Hunter The Lower Hunter is Australia's oldest and best-known wine-producing district. Around 3000 hectares of vines and over 60 wineries sprawl across the foothills of the Broken Back Range. A day tour should start at the Wine and Visitors Centre in Cessnock. Excellent food and accommodation are available throughout the region.

Newcastle Australia's second oldest city was founded as a penal colony in 1804. Newcastle rises up the surrounding hills from a spectacular surf coastline, its buildings a pleasant chaos of architectural styles from different periods. The city boasts a range of attractions including cosmopolitan restaurants, a premier regional art gallery and many historic sites.

TOP EVENTS

MAR	Hamilton Music, Food and Wine Festival (Hamilton)
MAR–APR	Harvest Festival (throughout region)
SEPT	Jazz Festival (Morpeth)
OCT	Jazz in the Vines (Cessnock)
OCT	Opera in the Vineyards (Cessnock)
DEC	King Street Fair (Newcastle)

CLIMATE

CESSNOCK

J	F	M	A	M	J	J	A	S	O	N	D	
30	29	27	25	21	18	18	20	23	25	27	30	MAX °C
18	18	15	12	9	6	5	5	8	11	14	16	MIN °C
91	100	90	58	57	50	33	38	41	58	69	69	RAIN MM
10	10	11	9	9	9	7	8	8	9	11	9	RAIN DAYS

WINES

Over 130 wineries – with over 100 cellar doors – can be found in the Hunter Valley. The region built its reputation on outstanding semillon, shiraz, chardonnay and verdelho, but exciting new wines are emerging, including merlot, chambourcin and pinot noir. The wine industry here is huge, taking up over 4000 hectares and making around $200 million worth of wine each year. The red earth found on the hillsides produces some of the country's great reds, including Tyrell's Shiraz, Lake's Folly Cabernet and McWilliams Rosehill Shiraz. The alluvial soils of the river flats produce excellent semillons, many coming from Tyrell's and McWilliams Lovedale. To complete a day of tasting, enjoy a night in one of the valley's gourmet restaurants or boutique hotels.

→ For more detail see maps 548, 549 & 555. For descriptions of towns see Towns from A–Z (p. 38).

HOLIDAY COAST

[DIAMOND HEAD, CROWDY BAY NATIONAL PARK]

This is classic Australian holiday territory: miles of perfect beaches, friendly seaside towns, areas of pristine wilderness and a near-perfect subtropical climate.

Dorrigo National Park This World Heritage–listed park preserves rugged escarpment country and the continent's most accessible area of temperate rainforest. There are excellent walking tracks, lookouts and picnic areas, and opportunities to see lyrebirds and brush turkeys. The popular Skywalk is an elevated walkway high above the rainforest canopy.

Bellingen Bat Island This 3-hectare island, in the centre of Bellingen, is home to a colony of around 40 000 grey-headed flying foxes (fruit bats). At dusk you can witness the bats setting off in search of food, filling the sky to spectacular effect in the process. The best time to see them is between September and March; access is via Bellingen Caravan Park in Dowle Street. Don't forget your hat.

Dorrigo
DORRIGO NP
COFFS HARBOUR
BONGIL BONGIL NATIONAL PARK
Bellingen
Urunga
Nambucca Heads
N
0 20 km
Macksville
YARRAHAPPINI NP
South West Rocks
HAT HEAD NATIONAL PARK
KUMBATINE NATIONAL PARK
Kempsey
Crescent Head
PORT MACQUARIE
Wauchope
TAPIN TOPS NATIONAL PARK
Laurieton
CROWDY BAY NATIONAL PARK
Wingham
Gloucester
TAREE
TASMAN SEA
FORSTER–TUNCURRY
Wallis Lake
Stroud
Myall Lake
Elizabeth Beach
MYALL LAKES NATIONAL PARK
Bulahdelah
Boolambayte Lake

Coffs Harbour During the holiday season the population of Coffs quadruples. Attractions include the Big Banana, marking one of the area's largest industries; the Pet Porpoise Pool, which features trained sea mammals; and Muttonbird Island, the breeding ground of thousands of wedge-tailed shearwaters, which you can access via a short walk along the sea wall.

Timbertown This re-created 1880s sawmilling village near Wauchope is a step back in time to the days of horsedrawn wagons, bullock teams and blacksmiths. Features include a wood-turning workshop, a craft gallery and a leather goods outlet; the activities range from rides on a restored steam train to sleeper-cutting demonstrations. Australian bush songs, roast meats and damper can be enjoyed in the authentic 1880s hotel.

Myall Lakes The 'Murmuring Myalls' are 10 000 hectares of connected lakes protected from the South Pacific by a long line of windswept dunes. Hire a houseboat and explore the calm waters, or enjoy fishing, windsurfing or canoeing. There are 40 kilometres of spectacular beaches and lookouts along the adjoining coast.

TOP EVENTS

JAN	Golden Lure Tournament (Port Macquarie)
EASTER	Gaol Break Swim (South West Rocks, near Kempsey)
MAY	Shakespeare Festival (Gloucester)
JULY	International Brick and Rolling Pin Throwing Competition (Stroud)
OCT	Global Carnival (Bellingen)
OCT–NOV	Country Music Festival (Kempsey)

CLIMATE

PORT MACQUARIE

J	F	M	A	M	J	J	A	S	O	N	D	
26	26	25	23	21	19	18	19	20	22	23	25	MAX °C
18	18	17	14	11	9	7	8	10	13	15	17	MIN °C
153	177	176	170	147	132	98	83	83	94	102	127	RAIN MM
12	13	14	13	11	10	9	9	9	11	11	11	RAIN DAYS

BEACHES

Just like the towns here, the beaches of this region cover both ends of the scale; you can choose a stretch of sand with a lively social scene, a secluded inlet, or one of the many options in between. Elizabeth Beach (near Forster) is calm, seasonally patrolled and popular with families. Solitude seekers and nature lovers should explore the bays and coves protected by Crowdy Bay National Park. Further north, top spots include Crescent Head (popular with surfers, particularly longboard riders), South West Rocks, the beaches of Hat Head National Park and those around Coffs Harbour.

→ For more detail see maps 550, 551, 555 & 557. For descriptions of towns see Towns from A–Z (p. 38).

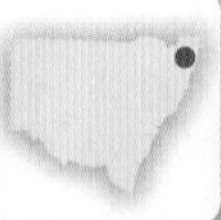

TROPICAL NORTH COAST

Exquisite beaches, wide rivers and World Heritage rainforest feature in this area, which is also famous for attracting alternative-lifestylers.

[JACARANDA TREES, GRAFTON]

Mount Warning National Park The summit of Mount Warning is the first place in Australia to be lit by the morning sun, being the highest point along Australia's eastern fringe. A bushwalking track leads up from the carpark through pockets of subtropical and warm–temperate rainforest. The park is World Heritage–listed and the mountain itself is the plug of one of the world's oldest volcanoes, which stretches as far as Cape Byron.

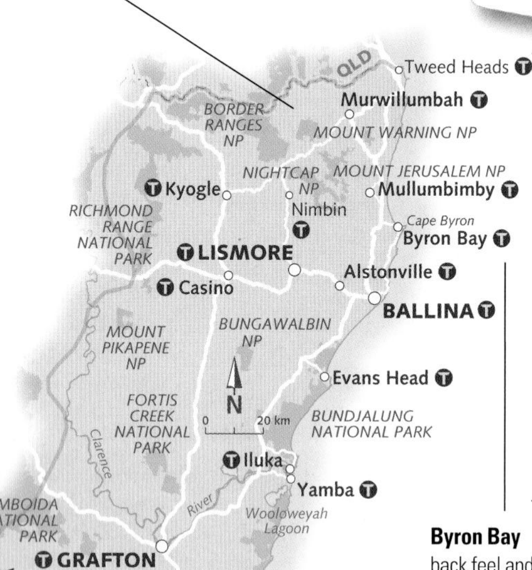

Holiday coast The combination of excellent weather and pristine beaches makes this region popular with visitors year-round. Visit family-friendly Tweed Heads on the Queensland border, or the fishing towns of Iluka and Yamba, great for their fresh seafood. Iluka is also worth a visit for the World Heritage–listed Iluka Nature Reserve, which contains the largest area of coastal rainforest in New South Wales and is great for birdwatching and walking.

Grafton This picturesque rural town, with a number of 19th-century buildings, is best known for its beautiful civic landscaping, particularly for the mature jacaranda trees with their vivid purple springtime blossom. Located on the Clarence River, Grafton is also a busy centre for watersports, including whitewater rafting and canoeing.

Byron Bay Byron Bay's excellent beaches, laid-back feel and great weather have long made it a hideaway for surfers, backpackers and alternative-lifestylers. The town centre has a vast array of restaurants, cafes, New Age shops, galleries, day spas, yoga centres and natural therapy parlours. Byron also hosts many great events, such as the blues festival over Easter. You can stand at Australia's most easterly point at the tip of Cape Byron and, during the cooler months, watch for migrating humpback whales offshore.

TOP EVENTS

EASTER	East Coast Blues and Roots Festival (Byron Bay)
JUNE	Wintersun Carnival (Tweed Heads)
AUG	Tweed Valley Banana Festival and Harvest Week (Murwillumbah)
SEPT	Chincogan Fiesta (Mullumbimby)
OCT	Rainforest Week (Tweed Heads)
OCT–NOV	Jacaranda Festival (Grafton)
NOV	Fairymount Festival (Kyogle)

CLIMATE

BYRON BAY

J	F	M	A	M	J	J	A	S	O	N	D	
28	28	27	25	22	20	19	20	22	23	25	26	MAX °C
21	21	20	17	15	12	12	13	14	16	18	20	MIN °C
161	192	215	185	188	158	100	92	66	102	118	143	RAIN MM
15	16	17	15	15	12	10	9	9	11	12	13	RAIN DAYS

RAINFOREST

The region's tropical heat and humidity create perfect conditions for the lush rainforest that is the largest remaining area of rainforest in New South Wales. With rainforest comes abundant wildlife, especially native reptiles and mammals, and tours can be arranged through the National Parks & Wildlife Service. Kyogle is known as the 'gateway to the rainforests' as it is almost completely surrounded by them; nearby Border Ranges National Park is a forest full of gorges, creeks and waterfalls. The stunning swimming hole of Wanganui Gorge, west of Mullumbimby, is surrounded by rainforest and strangler figs, and Rotary Rainforest Reserve is 6 hectares of rainforest in the centre of Lismore.

→ For more detail see map 551. For descriptions of T towns see Towns from A–Z (p. 38).

NEW ENGLAND

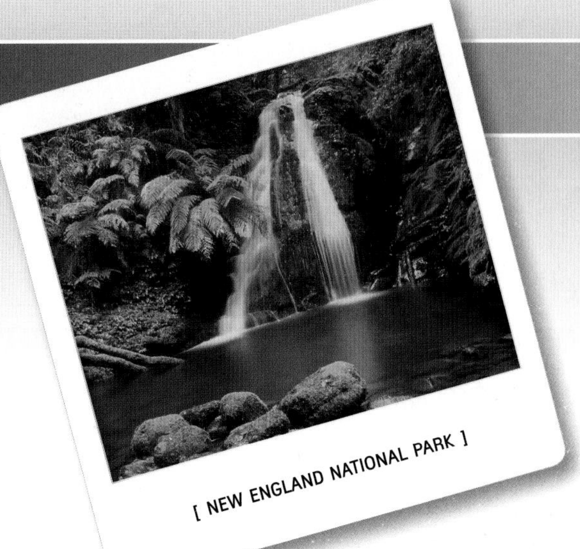

[NEW ENGLAND NATIONAL PARK]

As its name suggests, this region features chilly winters, golden autumns and heritage buildings; it is also famous for gem fossicking.

Lightning Ridge This opal-mining town is one of the few places in the world, and the only place in Australia, where black opal can be found. Local Aboriginal legend says that the opals were created when a huge wheel of fire fell to earth. Visitors can tour the opal mines and museums, and those who feel lucky can fossick for their own gems.

Tenterfield Saddler Tenterfield is perhaps best known from Peter Allen's 1970s song 'Tenterfield Saddler', which he wrote about his grandfather, George Woolnough. You can visit the famous saddlery, which is still open for business in Tenterfield's High Street and classified by the National Trust of Australia. Since the shop's inception in 1870, its customers have included none other than the illustrious Banjo Paterson.

Armidale Armidale is a sophisticated town with over 30 National Trust–listed buildings, two cathedrals, and the first Australian university established outside a capital city, all set in gracious tree-lined streets. Visit the New England Regional Art Museum, featuring the multimillion-dollar Hinton and Coventry collections, Australia's most significant provincial art holdings.

Tamworth Tamworth is Australia's country music capital, hosting the huge Tamworth Country Music Festival each January. Beneath the 12-metre Golden Guitar you will find the Gallery of Stars Wax Museum, with replicas of favourite country artists, and the Country Music Roll of Renown – Australia's highest country music honour. The Hands of Fame Park contains the handprints of over 200 country music stars.

Oxley Wild Rivers National Park The centrepiece of this World Heritage–listed rainforest park is the stunning Wollomombi Falls – at 220 metres, they are one of the highest falls in Australia. The ruggedly beautiful Dangars Falls drop 120 metres into a spectacular gorge. Activities in the park include camping, canoeing, walking and horseriding.

TOP EVENTS

JAN	Tamworth Country Music Festival
MAR	Bavarian Beer Fest (Tenterfield)
APR–MAY	Australian Celtic Festival (Glen Innes)
OCT	Gourmet in the Glen (Glen Innes)
NOV	Land of the Beardies Bush Festival (Glen Innes)
DEC	Great Inland Fishing Festival (Inverell)

CLIMATE

ARMIDALE

J	F	M	A	M	J	J	A	S	O	N	D	
27	26	24	21	16	13	12	14	18	21	24	27	MAX °C
13	13	11	8	4	2	0	1	4	7	10	12	MIN °C
105	87	65	46	44	57	49	48	52	68	80	89	RAIN MM
10	10	10	8	8	10	9	9	8	9	9	10	RAIN DAYS

FOSSICKING

The New England district is a fossicker's paradise. Quartz, jasper, serpentine and crystal are common finds, while sapphires, diamonds and gold present more of a challenge. Fossickers Way is a well-signposted route that introduces visitors to the district, beginning at Nundle and travelling north as far as Glen Innes. It passes through a number of towns including Inverell, the world's largest producer of sapphires. In the far west of the district, Lightning Ridge is known as a source of the famed black opal.

→ For more detail see maps 556–7. For descriptions of T towns see Towns from A–Z (p. 38).

CENTRAL WEST

[PARKES CSIRO RADIO TELESCOPE]

The open spaces and clear skies are ideal spots for the two observatories here, while natural features include the Warrumbungles rock formations.

Western Plains Zoo Five kilometres from Dubbo, this excellent open-range zoo covers 300 hectares and is home to animals representing five continents. The zoo is renowned throughout the world for its breeding programs, particularly of endangered species. Visitors can use their own cars, hire bikes or walk along the trails to all areas.

Warrumbungle National Park The Warrumbungles are extraordinary rock formations created by ancient volcanic activity. Best known is The Breadknife, which juts savagely out of the surrounding bushland. The 21 000-hectare national park marks the area where the flora and fauna of the Western Plains merge with that of the Great Dividing Range.

Siding Springs Observatory Located west of Coonabarabran on the road to Warrumbungle National Park, Siding Springs Observatory is home to Australia's largest optical telescope; in all there are 12 telescopes dotted around the site. Unfortunately there is no night viewing available; however, the observatory is open to the public during the day, and its Exploratory Centre has hands-on activities and a great astronomy exhibition.

Wellington Caves One of the largest stalagmites in the world (with a circumference of 32 metres) can be viewed in the limestone caves west of Wellington. The caves also contain rare cave coral and for thousands of years have acted as natural animal traps. Fossils of a diprotodon and a giant kangaroo have been found here.

Bathurst This is Australia's oldest inland settlement. Founded in 1815, Bathurst is noted for its colonial and Victorian architecture, including Miss Traill's House (1845), open to the public. Also of interest are Ben Chifley's cottage, the excellent regional art gallery and Mount Panorama, venue for the V8 Supercars 1000 (Bathurst 1000).

TOP EVENTS

EASTER	Western Plains Country Music Festival (Dubbo)
SEPT	Red Ochre Festival (Dubbo)
SEPT	Mudgee Wine Festival (Mudgee)
SEPT–OCT	Sakura Matsuri (cherry blossom festival, Cowra)
OCT	Festival of the Stars (Coonabarabran)
OCT	V8 Supercars 1000 (Bathurst)

CLIMATE

DUBBO

J	F	M	A	M	J	J	A	S	O	N	D	
33	32	29	25	20	16	15	17	21	25	29	32	MAX °C
18	18	15	11	7	4	3	4	6	10	13	16	MIN °C
61	54	48	45	48	49	45	45	44	49	51	50	RAIN MM
6	5	5	5	6	8	8	8	7	7	6	6	RAIN DAYS

MUDGEE REGION WINES

German settler Adam Roth planted vines at Mudgee in the 1850s. Thirteen wineries were established by 1890, but just three survived the Depression later in the decade. The red-wine boom of the 1960s saw many new vines planted, and today the Mudgee region, stretching from Dunedoo in the north down through Gulgong, Mudgee, Rylstone and Kandos, has over 40 vineyards, most of which offer tastings. Warm summers favour the production of full-bodied shiraz and chardonnay. Try Huntington's shiraz; the organic, preservative-free wines of Botobolar; Craigmoor's chardonnay; or the cabernet sauvignon from Thistle Hill.

→ For more detail see maps 554–5, 556, 559 & 561. For descriptions of towns see Towns from A–Z (p. 38).

CAPITAL COUNTRY

Best known for its colonial history, this region also has boutique cool-climate wineries and beautiful areas of natural bushland.

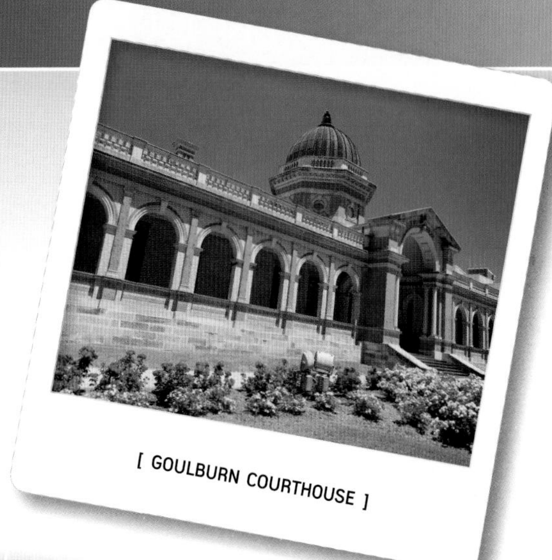

[GOULBURN COURTHOUSE]

Murrumbateman wineries
The first vines in this hilly terrain were planted in 1971. Many of the original winemakers had no experience in winemaking and came from government or science jobs in the ACT. Their wine was slow to take off, but the industry has seen a reversal of fortunes now. The region produces a great range of styles, including riesling and shiraz viognier. Cellar doors include Clonakilla, Helm and Jeir Creek. Stop for lunch at Poachers' Pantry in Hall and pick up some gourmet produce to enjoy with your wine purchases.

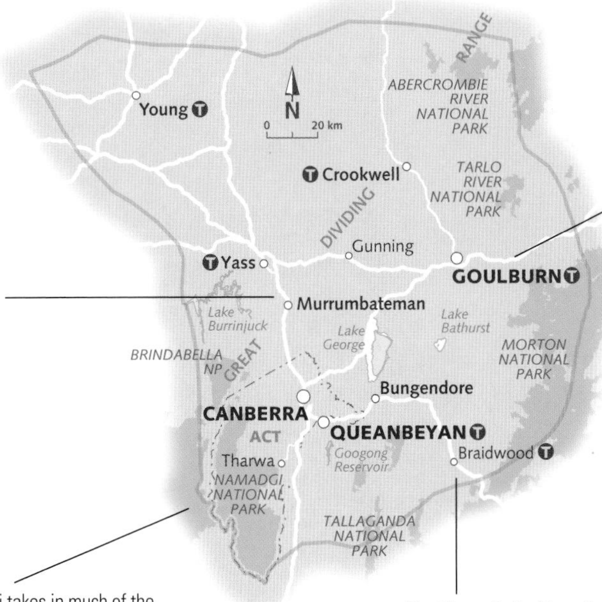

Goulburn Established in 1833, this town displays elaborate 19th-century architecture, a legacy of early wool-growing wealth. On the town's outskirts, the Big Merino sells wool products, has an educational display on the history of wool, and provides a lookout with sweeping views of Goulburn and surrounds.

Namadgi National Park Namadgi takes in much of the Brindabella Range, covering almost half of the ACT. It boasts significant Aboriginal rock art and a beautiful environment of mountains, valleys and bush, regenerating after the 2003 bushfires. Camping and bushwalking are popular. The excellent Namadgi Visitor Centre is just south of Tharwa.

Braidwood Braidwood's elegant Georgian buildings recall early agricultural settlement, while ornate Victorian structures mark the town's goldmining boom. It was once the haunt of bushrangers such as Ben Hall, and today is the haunt of people making movies about bushrangers (and other subjects). Films made here include *Ned Kelly* in 1969, and *On Our Selection* in 1994.

TOP EVENTS

FEB	Australian Blues Music Festival (Goulburn)
MAR	Celebration of Heritage and Roses (Goulburn)
MAY	Heritage Festival (Braidwood)
OCT	Lilac City Festival (Goulburn)
NOV	Festival of Wine, Roses and all that Jazz (throughout Murrumbateman district)
NOV–DEC	National Cherry Festival (Young)

CLIMATE

GOULBURN

J	F	M	A	M	J	J	A	S	O	N	D	
27	26	24	20	16	12	12	13	16	19	22	26	MAX °C
13	14	11	8	5	2	1	2	5	7	9	12	MIN °C
61	59	57	53	49	47	45	58	51	56	66	55	RAIN MM
10	9	9	9	11	11	12	12	11	11	12	9	RAIN DAYS

HISTORY

The first inhabitants of the area were the Ngunnawal, whose ancestors may have arrived anywhere from 75 000 to 12 000 years ago according to archaeological evidence from Lake George. Europeans sighted the district in the late 1790s and the Goulburn plains were named in 1818. Settlers arrived between the 1820s and the 1850s, first attracted by the rich grazing land and later by the discovery of gold. Today a number of towns, including Yass, Young, Gunning, Bungendore and Braidwood, features heritage buildings and charming streetscapes. The district as a whole provides an evocative glimpse of 19th-century life in rural Australia.

→ For more detail see maps 554, 562, 563, 564, 565 & 566–7. For descriptions of T towns see Towns from A–Z (p. 38).

SOUTH COAST

This region of sandy beaches, fishing villages and native forests is popular with holiday-makers in summer, yet never loses its laid-back feel.

[BOURNDA LAGOON, BOURNDA NATIONAL PARK]

Jervis Bay Part of this area is in Jervis Bay Territory (the third and least-known mainland territory of Australia) and includes the national capital's seaport. The bay, partially flanked by Jervis Bay National Park, is known for its dramatic underwater landscapes and its dolphins; diving and dolphin cruises are available.

Central Tilba Classified by the National Trust as an 'unusual mountain village', picturesque Central Tilba is a showcase for late-19th-century town architecture, set in a magnificent rural landscape. New buildings are required to meet National Trust specifications, maintaining the original charm of the streetscapes. Central Tilba was founded in 1895 as a goldmining town, but now caters for tourists with cafes, galleries and art and craft shops.

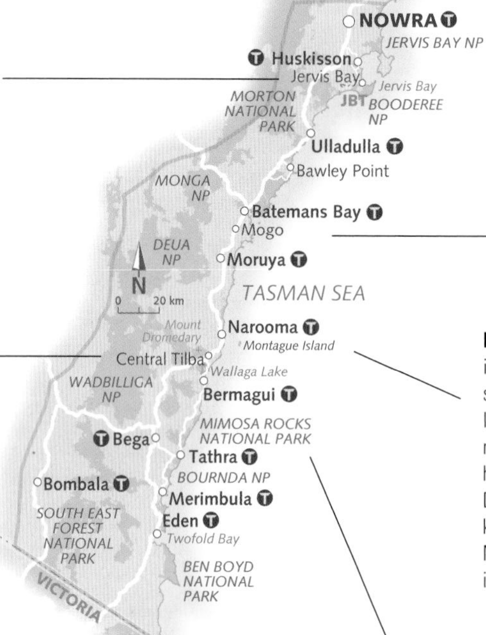

Mogo Mogo is an old goldmining town that has been restored and turned into a tourist destination, with several craft shops and a zoo that specialises in raising endangered species. Visitors can pan for gold and take a mine tour at Goldfields Park, visit the reconstructed 19th-century goldmining village of Old Mogo Town or enjoy the sclerophyll forest of the Mogo Bushwalk.

Montague Island Nature Reserve This isolated island, accessed only by guided tour, is a major shearwater breeding site and home to 8000 pairs of little penguins and 600 Australian fur seals. Cruise from nearby Narooma to tour the island and learn about its history as a fertile hunting ground for the Walbanga and Djiringanj tribes, or you can see migrating humpback and killer whales off the island's coast from September to November on a whale-watching cruise. Recently, stays in the lighthouse keeper's quarters were introduced.

Mimosa Rocks National Park The park, named after a wrecked paddlesteamer, crosses a landscape of beaches, sea caves, cliffs, forests and wetlands. Rocks tumble across the beaches and form impressive sculptural stacks. There are several secluded campsites, and swimming, walking, diving, fishing and birdwatching are among the activities on offer.

TOP EVENTS

JAN	Blue Water Fishing Classic (Bermagui)
EASTER	Tilba Festival (Central Tilba)
OCT	Great Southern Blues and Rockabilly Festival (Narooma)
OCT–NOV	Eden Whale Festival (Eden)
NOV	Riverside Festival (Bombala)
NOV	Country Music Festival (Merimbula)

CLIMATE

MERIMBULA

J	F	M	A	M	J	J	A	S	O	N	D	
24	25	23	21	19	16	16	17	18	20	21	23	MAX °C
15	15	14	11	8	6	4	5	7	9	12	14	MIN °C
77	79	91	83	65	69	41	42	56	69	88	78	RAIN MM
10	10	10	9	9	9	7	9	11	11	13	11	RAIN DAYS

ABORIGINAL CULTURE

Before European colonisation, the Yuin occupied the area from Jervis Bay to Twofold Bay, sustained by the produce of coast and rivers. Today the area remains steeped in Yuin history. At Wallaga Lake the Umbarra Cultural Centre offers tours, including one to the summit of Mount Dromedary (Gulaga) where, according to legend, the great creation spirit, Daramulun, ascended to the sky. Booderee National Park, in Jervis Bay Territory, is once again Yuin land after a successful 1995 land claim. Jointly managed by the Wreck Bay Aboriginal Community and Parks Australia, Booderee has middens and other significant sites, and an art and craft centre. The Murramarang Aboriginal Area, near Bawley Point, offers a self-guide interpretive walk.

→ For more detail see maps 553, 554–5, 565 & 567. For descriptions of towns see Towns from A–Z (p. 38).

SNOWY MOUNTAINS

[SNOW GUMS, KOSCIUSZKO NATIONAL PARK]

This region features Kosciuszko National Park, which offers skiing in winter and bushwalking, mountain-bike riding and watersports in summer.

Yarrangobilly Caves Among myriad natural wonders in Kosciuszko National Park, Yarrangobilly Caves stand out as a world-class attraction. Of the 70 caves formed some 2 million years ago, six are open to the public, with highlights including underground pools, frozen waterfalls and a bizarre web of limestone formations. Visitors can marvel at the radiant white, orange and grey limestone flows, stalactites and stalagmites, or take a dip in the naturally formed thermal pool offering year-round swimming.

Alpine Way Stretching 111 kilometres from Jindabyne to Khancoban, this spectacular route winds around the Thredbo slopes, passes through Dead Horse Gap, and crosses the valley of the Murray headwaters. It is best driven during spring and summer, although the winter scenery is superb.

Snowy River power This once mighty river was dammed and diverted for the Snowy Mountains Scheme. Some 100 000 men from 30 countries worked for 25 years on the largest engineering project of its kind in Australia. Drop in at the Snowy Mountains Authority Information Centre in Cooma, or visit the power stations near Khancoban.

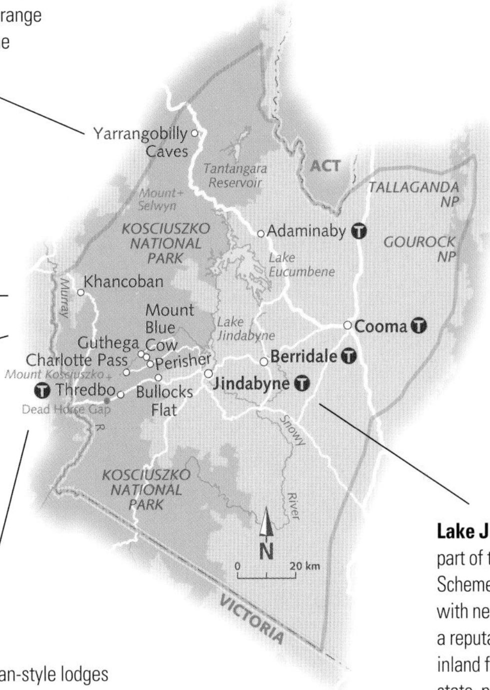

Thredbo This charming alpine village with its European-style lodges and hotels makes for an unusual sight in the Australian landscape. Packed and brimming with life during winter, it is also popular in summer with wildflower enthusiasts, anglers, mountain-bike riders and bushwalkers wanting to tackle Mount Kosciuszko. A chairlift drops people at the beginning of a 6-kilometre walk to the summit.

Lake Jindabyne Created as part of the Snowy Mountains Scheme, this huge lake, along with nearby Lake Eucumbene, has a reputation as one of the best inland fishing destinations in the state, particularly for trout. Sailing, windsurfing and waterskiing are popular in summer.

TOP EVENTS

JAN	Blues Festival (Thredbo)
FEB	Rodeo (Cooma)
MAY	Jazz Festival (Thredbo)
NOV	Cooma Street Fair (Cooma)
NOV	Snowy Mountains Trout Festival (throughout region)

CLIMATE

THREDBO

J	F	M	A	M	J	J	A	S	O	N	D	
21	21	18	14	10	6	5	6	10	13	16	19	MAX °C
7	7	4	2	0	−3	−4	−2	−1	1	4	5	MIN °C
108	87	112	114	165	166	158	192	212	206	155	117	RAIN MM
11	10	11	13	15	16	17	17	18	16	15	12	RAIN DAYS

SKI RESORTS

The Snowy Mountains resorts are well equipped in terms of lessons, lifts, ski hire, transport, food, accommodation and entertainment. Perisher, Smiggin Holes, Guthega and Mount Blue Cow (near Guthega) are collectively known as Perisher Blue, the largest ski resort in Australia, with 50 lifts and a variety of slopes. A good range of accommodation is available at Perisher and Smiggin Holes, but overnight parking is limited (many visitors leave their cars at Bullocks Flat and take the Skitube). Accommodation is limited at Guthega and not available at Blue Cow. Thredbo, the main village, has excellent skiing and tourist facilities. Charlotte Pass provides access to some of the region's most spectacular runs for experienced skiers. In the north of the park, Mount Selwyn is a good place for families and beginners, and is one of the main centres for cross-country skiing.

→ For more detail see maps 565 & 566. For descriptions of Ⓣ towns see Towns from A–Z (p. 38).

RIVERINA

[RESTORED HISTORIC BUILDING, COOTAMUNDRA]

A landscape of far horizons and clear skies makes touring these fertile plains a pleasure; regional towns feature heritage buildings and good restaurants.

Griffith The main township of the Riverina, Griffith developed in the early days of irrigation and was designed by Walter Burley Griffin, architect of Canberra. It is surrounded by farms and vineyards, but nearby, in Cocoparra National Park, there is also an opportunity to glimpse the landscape as it once was.

Riverina wineries The Riverina is responsible for 60 per cent of the grapes grown in New South Wales. There are 14 wineries in the district, mostly around Griffith, including the De Bortoli and McWilliams estates. This region is best known for its rich botrytised semillon (semillon improved by flavour-enhancing fungus).

Hay Hay lies at the centre of a huge stretch of semi-arid grazing country known as the Hay Plains. Established in 1859, the town boasts an interesting collection of late-19th-century buildings. The POW Internment Camp Interpretive Centre tells the fascinating and sometimes inspirational stories of Hay's role as a prison camp in World War II.

Wagga Wagga On the banks of the Murrumbidgee, Wagga Wagga is the state's largest inland city and a major centre for commerce, agriculture and education. Visit the Museum of the Riverina (incorporating the Sporting Hall of Fame), the Botanic Gardens and the Regional Art Gallery, home of the National Art Glass Collection. River walks and cruises are also on offer.

Dog on the Tuckerbox Originally mentioned in the poem *Bill the Bullocky* by Bowyang Yorke, this monument to pioneer teamsters and their dogs, situated on the highway 8 kilometres north of Gundagai, is recognised throughout the nation as an Australian icon. The faithful bronze dog was unveiled in 1932 by former prime minister Joseph Lyons and celebrated in the song 'Where the Dog Sits on the Tuckerbox' by Jack O'Hagan (who also penned the famous 'Along the Road to Gundagai').

TOP EVENTS

FEB	Tumbafest (food and wine, Tumbarumba)
MAR	John O'Brien Bush Festival (Narrandera)
AUG	Wattle Time Festival (Cootamundra)
OCT	Apple Blossom Festival (Batlow)
NOV	Dog on the Tuckerbox Festival and Snake Gully Cup (Gundagai)
NOV	Festival of Gardens (Griffith)

CLIMATE

TEMORA

J	F	M	A	M	J	J	A	S	O	N	D	
31	31	28	23	18	14	13	15	18	22	26	30	MAX °C
16	16	13	9	6	3	2	3	5	8	11	14	MIN °C
48	37	41	43	45	43	47	45	42	54	45	41	RAIN MM
5	5	5	6	9	10	13	12	9	9	7	6	RAIN DAYS

REGIONAL PRODUCE

The Riverina produces vast quantities of rice, citrus and stone fruit, grapes, poultry and vegetables. To learn about the rice industry, visit the Sunrice Country Visitors Centre in Leeton. For fruit products, tour the Catania Fruit Salad Farm at Hanwood (near Griffith). For the gourmet, Mick's Bakehouse in Leeton makes excellent breads and award-winning pies. Drop in to Charles Sturt University Winery and Cheese Factory in Wagga Wagga for local products.

→ For more detail see maps 554, 558–9 & 565. For descriptions of T towns see Towns from A–Z (p. 38).

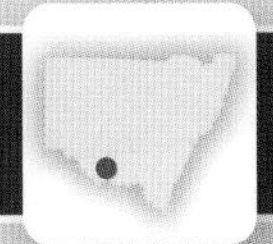

MURRAY

Australia's most important river supports this region's agriculture, but is also a holiday destination with activities including golf and watersports.

[RED GUM TREES ALONG THE MURRAY RIVER]

Perry Sandhills The vast orange dunes of Perry Sandhills near Wentworth cover 10 hectares and were formed after an ice age around 40 000 years ago. They hold the remains of megafauna such as kangaroos, wombats, emus and lions. Today the dunes are often used as a backdrop in film and television production.

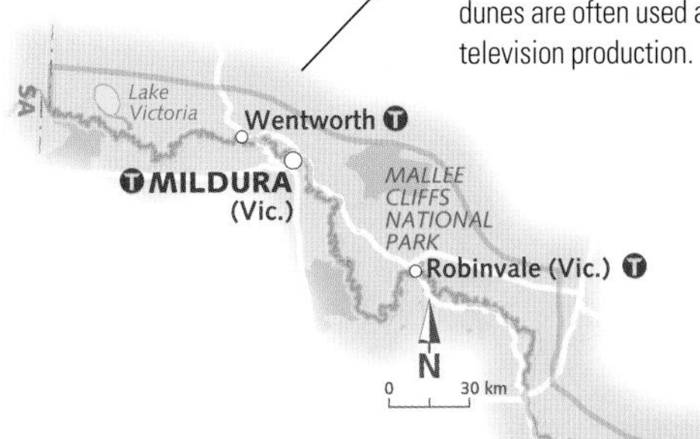

Corowa Corowa is the quintessential Australian river town, with wide streets and turn-of-the-century architecture. Federation got a jump-start here in 1893 at the Corowa Federation Conference, now commemorated in the Federation Museum. In 1889 Tom Roberts completed his iconic painting *Shearing the Rams* at a sheep station nearby.

Cobram Barooga Golf Club There are many renowned golf courses along the banks of the Murray River, and the Cobram Barooga Golf Club has two of the best. The two 18-hole, championship-standard golf courses, good for pros and novices alike, are consistently rated in Australia's top resort courses. With many holes surrounded by beautiful red gums and pines, and the odd kangaroo looking on, this is a great spot to have a swing.

Lake Mulwala The construction of the Yarrawonga Weir in 1939 created this 6000-hectare artificial lake, around which the town of Mulwala has grown. The lake is now a premier destination for watersports, offering yachting, sailboarding, swimming, canoeing and fishing, and the largest waterskiing school in the world.

Albury Once a meeting place for the local Aboriginal tribes, today Albury is a large regional centre familiar to motorists who travel the busy Hume Highway. Attractions include interesting heritage buildings, a large regional art gallery and good restaurants. Nearby Lake Hume, one of Australia's biggest artificial lakes, is another popular spot for watersports. Wineries around Albury can be explored on a camel trek.

TOP EVENTS

JAN	Sun Festival (Deniliquin)
JAN	Federation Festival (Corowa)
FEB	Border Flywheelers Vintage Engine Rally (Barham)
MAR–APR	Albury Gold Cup Racing Carnival (Albury)
SEPT–OCT	Play on the Plains Festival and Ute Muster (Deniliquin)
SEPT–OCT	Food and Wine Festival (Albury)

CLIMATE

ALBURY

J	F	M	A	M	J	J	A	S	O	N	D	
31	31	27	23	18	14	13	15	18	21	25	29	MAX °C
15	15	12	9	6	4	3	4	6	8	11	13	MIN °C
55	36	43	49	64	71	87	89	72	72	53	51	RAIN MM
6	6	6	7	11	14	16	15	13	11	9	7	RAIN DAYS

ALONG THE RIVER

The Murray River provides endless opportunities for swimming, boating, watersports and fishing. Perch, catfish and yabbies are plentiful, and Murray cod is the catch that almost every angler seeks. The parklands of Albury feature pleasant riverside walks and picnic facilities, and at Corowa Yabby Farm visitors can catch and cook their own yabbies. Lake Mulwala was formed by damming the Murray and is now a haven for watersports. The Rocks of Tocumwal, on the banks of the river, change colour according to the weather and, as Aboriginal legend has it, the mysterious Blowhole nearby was the home of a giant Murray cod that liked to eat young children. The Barham Lakes consist of four artificial lakes stocked with yabbies and fish, with walking trails through the surrounding bushland.

→ For more detail see maps 558–9. For descriptions of towns see Towns from A–Z (p. 38).

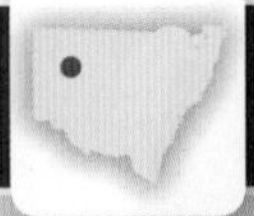

OUTBACK

A series of national parks preserve this sparsely populated region's stunning landscapes, natural heritage and rich Indigenous history.

[OUTBACK FARMHOUSE]

Sturt National Park Occupying 310 000 hectares of Corner Country – the point where three states meet – this park offers a varied landscape of hills, rocks and plains, congregations of native birds and animals, and wonderful wildflowers after rain. Camping is available; check in at the park office in Tibooburra and take advantage of their tours.

White Cliffs The area around White Cliffs has been rendered a moonscape, thanks to 5000 abandoned opal digs. Owing to soaring temperatures in summer much of the town has been built underground, where the temperature remains a steady 27°C. Opals are still mined here, and nearby is spectacular Mutawintji National Park with its sandstone cliffs, river red gums, gorges, rockpools, desert plains and some of the state's best Aboriginal rock art.

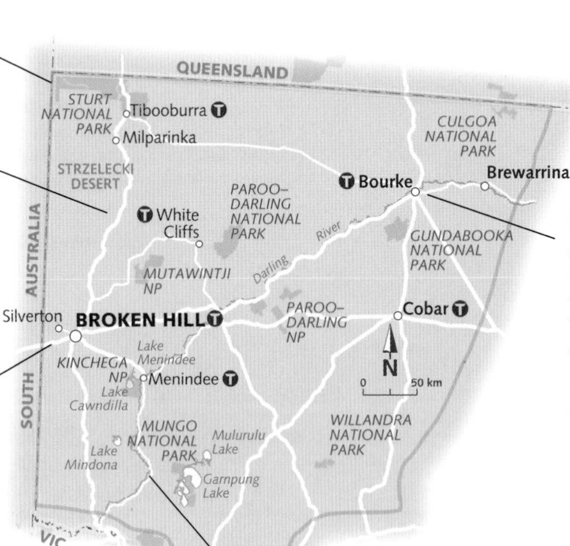

Bourke 'If you know Bourke you know Australia,' said Henry Lawson. This Darling River town has become synonymous with the outback and is rich in heritage sites. The Back O'Bourke Exhibition Centre is a fascinating and important modern facility set among river red gums. It tells the story of the river and the outback from the Dreamtime until now.

Broken Hill Broken Hill was established in the 1880s to service the mining of massive deposits of silver, lead and zinc in the Barrier Ranges. With its historic buildings, 30 or so art galleries and decent eating options, it is an oasis of civilisation in a sparse landscape. Nearby Silverton is a quintessential outback town, made famous in films such as *Mad Max II* (1981).

Mungo National Park The focal point of the 240 000–hectare Willandra Lakes World Heritage Area, Mungo National Park's weather-eroded, sculptural sand formations have become one of outback Australia's signature sights. Around the area's dry lakes astounding remains have been uncovered, including ancient Aboriginal artefacts and Mungo Man, a full male skeleton estimated to be around 40 000 years old. The undoubted highlight is the huge Walls of China crescent-shaped dune, which you can visit on a 70-kilometre self-guide drive through the park.

TOP EVENTS

OCT	Festival of the Miner's Ghost (Cobar)
OCT	Red Desert Live! (Broken Hill)

CLIMATE

BROKEN HILL

J	F	M	A	M	J	J	A	S	O	N	D	
33	32	29	24	19	16	15	17	21	25	29	31	MAX °C
18	18	16	12	9	6	5	6	9	12	15	17	MIN °C
23	25	20	18	23	21	19	19	21	25	20	22	RAIN MM
3	3	3	3	5	5	6	5	4	5	4	3	RAIN DAYS

RUGGED REPUTATION

Charles Sturt was marooned north-west of Milparinka for six months while waiting for rain to replenish his party's drinking water in 1845. Although some towns have been successfully established in this harsh but beautiful environment, unpredictable conditions suggest that it will never be tamed. The land of Broken Hill has brought forth valuable minerals, which can be viewed at White's Mineral Art and Mining Museum. White Cliffs receives the most solar radiation in New South Wales and so was the natural choice for Australia's first solar-power station. Sturt National Park is home to the red sands and reptiles of the Strzelecki Desert; here day temperatures are often over 40°C and nights can be below freezing.

→ For more detail see maps 558–9 & 560–1. For descriptions of Ⓣ towns see Towns from A–Z (p. 38).

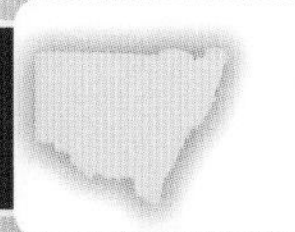

LORD HOWE ISLAND

This World Heritage–listed site has great natural beauty and some of the rarest flora and fauna on earth.

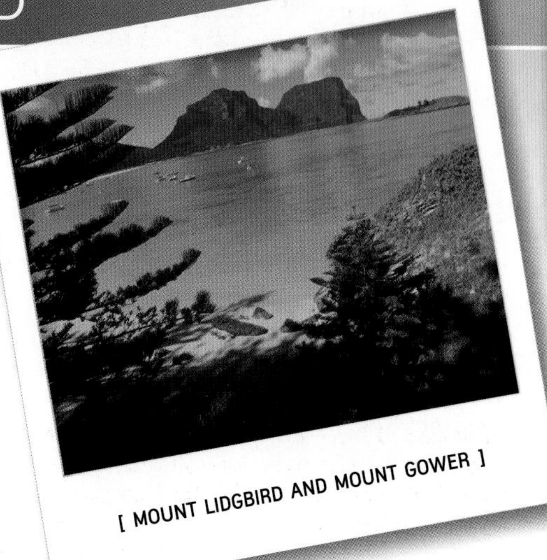

[MOUNT LIDGBIRD AND MOUNT GOWER]

Dining at Arajilla Widely regarded as 'the' place to dine on Lord Howe Island, Arajilla exudes casual elegance and lashings of style from its coveted forest location. Seafood is the focus of the contemporary menu, with the restaurant host personally reeling in the catch of the day brought fresh to your plate. The wine list features an impressive array of Australian and New Zealand labels, making it easy to linger over your meal under the beautiful canopy of trees.

Neds Beach Lush forests of kentia and banyan trees, two of the island's most prolific and distinctive species, fringe this beautiful surf beach. Handfeed the tropical fish, swim, surf, snorkel or take a 45-minute walk up Malabar Hill on the northern headland for superb island views.

The Lagoon A coral reef encloses the crystal waters of this 6-kilometre long lagoon on the western side of the island. Spend the day relaxing and swimming at Old Settlement Beach or snorkel at Escotts Hole, about 1 kilometre out. Glass-bottomed boat and snorkelling tours are available.

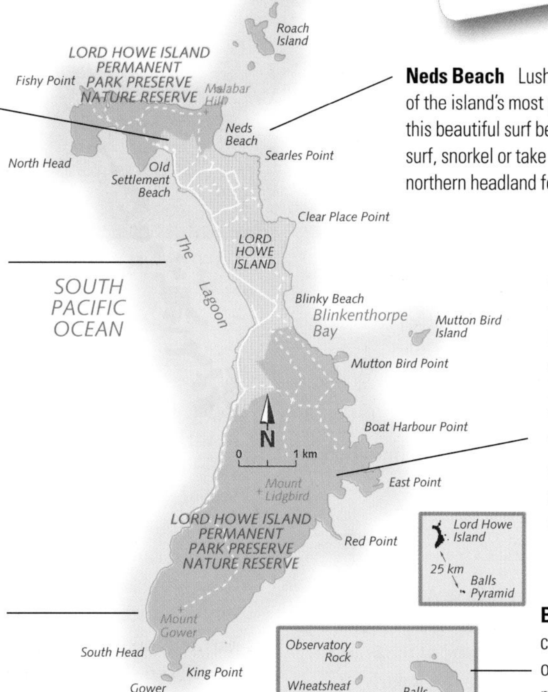

Golf at Mount Lidgbird Situated at the base of imposing Mount Lidgbird, Lord Howe Island's golf course is one of Australia's most scenic – its nine holes incorporate lush vegetation and open fairways with superb coastal views. Visitors are welcome to tee off at any time on this World Heritage turf, and you can even participate in the annual Lord Howe Open Golf Tournament held in November.

Mount Gower This 875-metre mountain is the island's highest point. Take a nine-hour guided walk to the summit through areas of stunted rainforest – a fairytale world of gnarled trees, orchids and moss-covered basalt outcrops. Mists permitting, the panoramic view from the summit is spectacular.

Balls Pyramid This extraordinary cathedral-shaped rock rises 551 metres out of the sea 25 kilometres south-east of the main island. Once nearly 6 kilometres wide, the rock's width has been eroded to only 400 metres. It is a major breeding ground for seabirds and can be seen by air charter or boat cruise. Deep-sea fishing in the vicinity is excellent.

TOP EVENTS

FEB	Discovery Day
OCT–NOV	Gosford to Lord Howe Island Yacht Race
NOV	Lord Howe Island Golf Open

CLIMATE

LORD HOWE ISLAND

J	F	M	A	M	J	J	A	S	O	N	D	
25	26	25	23	21	19	19	19	20	21	22	24	MAX °C
20	20	20	18	16	14	13	13	14	15	17	19	MIN °C
108	114	122	149	160	177	178	141	135	127	116	117	RAIN MM
11	13	15	18	21	22	23	21	17	14	12	12	RAIN DAYS

A WORLD OF ITS OWN

The Lord Howe Group of Islands comprises a series of seven-million-year-old volcanic formations. Because of its isolation and the absence of humans until recent times, Lord Howe has a unique natural history. Fifty-seven of the islands' 180 flowering plants and 54 fern species are not found anywhere else. At settlement (1834) there were 15 species of land birds, 14 of which were unique. Five species survive today. Hundreds of thousands of seabirds roost on the islands. These include sooty terns, brown noddies, several shearwater species, the world's largest colony of red-tailed tropic birds and the world's only breeding colony of providence petrels.

TOWNS A–Z

new south wales

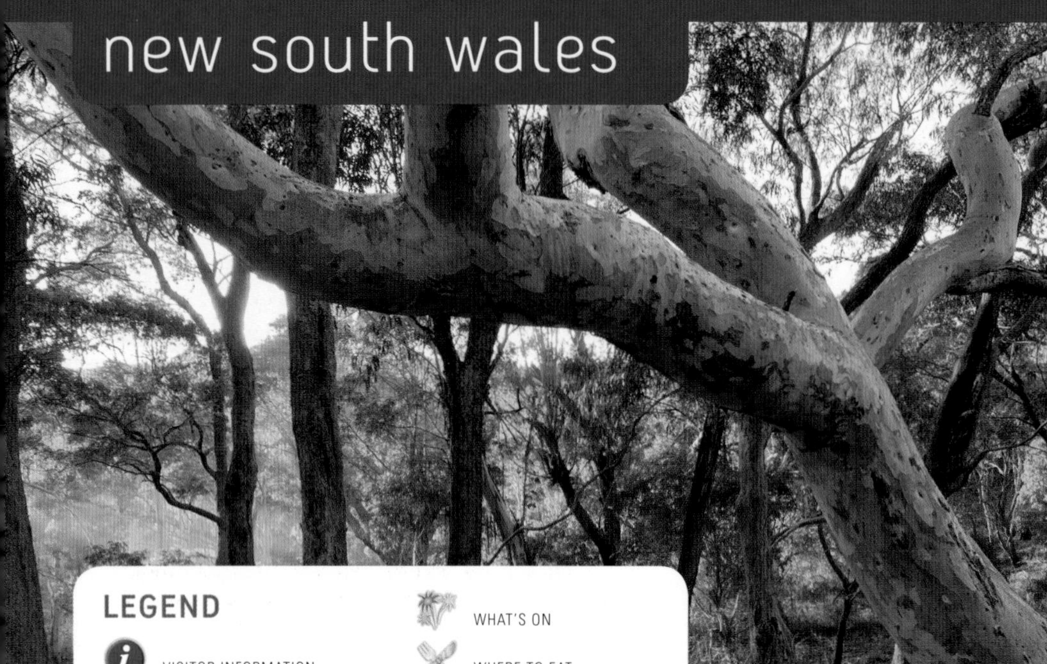

LEGEND

- VISITOR INFORMATION
- RADIO STATIONS
- IN TOWN
- WHAT'S ON
- WHERE TO EAT
- WHERE TO STAY
- NEARBY

Food and accommodation listings in town are ordered alphabetically with places nearby listed at the end

[SNOW GUM WOODLANDS AT SUNRISE, MORTON NATIONAL PARK]

Adaminaby

Pop. 235
Map ref. 554 E12 | 565 B6 | 566 E7 | 581 D2

Snowy Region Visitor Centre, Kosciuszko Rd, Jindabyne; (02) 6450 5600 or 1800 004 439; www.snowymountains.com.au

97.7 Snow FM, 1602 AM ABC Local

Over 100 buildings were moved in the 1950s to Adaminaby's current town site. The remaining town and surrounding valley were flooded to create Lake Eucumbene as part of the Snowy Mountains Hydro-Electric Scheme. The lake is regularly restocked with trout and Adaminaby is now a haven for anglers. Due to its proximity to Selwyn Snowfields, Adaminaby is also a popular base in winter for skiers.

 **The Big Trout** The world's largest fibreglass rainbow trout was erected after a local angler, attempting to drink a gallon of Guinness while fishing, was pulled into the water by a large trout and almost drowned. Legend has it the man then managed to finish the Guinness, but the 10 m high trout stands as a tribute to 'the one that got away'. Lions Club Park at town entrance.

Leigh Stewart Gallery: displays historical information including pictures and films; Main St. ***Historic buildings:*** several buildings, including 2 churches that were moved from Adaminaby's original site; details from visitor centre.

San Michele Resort: secluded contemporary accommodation; 1512 Caddigat Rd; (02) 6454 2229. ***Snowy Mountains Alpine Cabins:*** self-contained mountain cabins; 6078 Snowy Mountains Hwy; (02) 6454 1120.

Kosciuszko National Park On the road to Tumut is the historic goldmining site of Kiandra. North of the road, via Long Plain Rd, is historic Coolamine Homestead. All of New South Wales' ski fields exist inside Kosciuszko National Park; the focal point is Mt Kosciuszko, Australia's highest mountain. The summit can be reached easily via the Kosciuszko Express Chairlift (operating all year), which drops you at the beginning of a 13 km return walk. Among a myriad of natural wonders in the park, the Yarrangobilly Caves are a highlight. The string of 70 limestone caves, some 2 million years old, comprise marine sediments dating back 440 million years. Six caves are open to the public, featuring underground pools, frozen waterfalls, a bizarre web of limestone formations and a naturally formed thermal pool offering year-round swimming. For guided and self-guide tours, call (02) 6454 9597; more details from visitor centre.

Lake Eucumbene It is said that anyone can catch a trout in Lake Eucumbene, the largest of the Snowy Mountains artificial lakes. Its abundance of rainbow trout, brown trout and Atlantic salmon make it a popular spot for anglers. In recent years droughts have lowered the water level in the dam, revealing ruins of the old town. The Snowy Mountains Trout Festival draws hundreds of anglers every November. Fishing boats can be hired at Old Adaminaby and for fly-fishing tours, contact the visitor centre. Access via Old Adaminaby. 9 km sw.

Old Adaminaby Racetrack: featured in the film *Phar Lap* (1984); on the road to Rosedale, Cooma side of town. ***Power stations:*** tours and interactive displays; details from visitor centre.

See also SNOWY MOUNTAINS, p. 33

Adelong

Pop. 827
Map ref. 554 D10 | 565 A4 | 566 B2 | 585 P2

Tumut Region Visitor Centre, 5 Adelong Rd, Tumut; (02) 6947 7025; www.tumut.nsw.gov.au

96.3 FM Sounds of the Mountain, 99.5 FM ABC Radio National

Adelong was established and thrived as a goldmining town in the late 19th century. William Williams discovered gold and prospectors flocked to Adelong to seek their fortunes. Legend has it that Williams bought a mining claim for £40 000, only to sell it later the same day for £75 000. By World War I, over a million ounces of gold had been extracted from the mines, leaving little behind. The people began to disappear immediately. What is left is a charming rural village with a turn-of-the-century feel.

Historic buildings Many of the beautifully preserved buildings in Adelong have been classified by the National Trust. Take a stroll through Adelong's streets to discover banks, hotels and churches of the gold-rush era.

Coat of Arms Kaffeehaus Restaurant: authentic European; Beaufort Guest House, 77 Tumut St; (02) 6946 2273.

Beaufort House: heritage-listed B&B; 77 Tumut St; (02) 6946 2273.

Adelong Falls Reserve Richie's Gold Battery was one of the foremost gold-processing and quartz-crushing facilities in the country. See the ruins of its reefer machine, including water wheels and a red-brick chimney. Three clearly signposted walks explore the falls, and other ruins in the reserve. 1 km N.

See also RIVERINA, p. 34

Albury

Pop. 43 784
Map ref. 554 A12 | 559 O12 | 585 K6 | 587 I2

Gateway Visitor Information, Gateway Island, Hume Highway, Wodonga; 1300 796 222; www.destinationalburywodonga.com.au

105.7 FM The River, 990 AM ABC Radio National

The twin towns of Albury–Wodonga are 7 kilometres apart on opposite sides of the Murray River, which is also the New South Wales–Victoria border. Originally inhabited by Aboriginal people, the Albury area was 'discovered' in 1824 by explorers Hume and Hovell, who carved their comments into the trunks of two trees. Hume's tree was destroyed by fire, but Hovell's still stands today in Hovell Tree Park.

Albury Library Museum Exhibits include a display on one of Australia's largest postwar migrant centres, which existed at nearby Bonegilla. Free wireless is also available. Cnr Kiewa and Swift sts; (02) 6023 8333.

Botanical Gardens: array of native and exotic plants and signposted rainforest and heritage walks; cnr Wodonga Pl and Dean St. ***The Parklands:*** comprises Hovell Tree, Noreuil and Australia parks. Enjoy riverside walks, swimming, kiosk and picnic areas; Wodonga Pl. ***Monument Hill:*** spectacular views of town and alps; Dean St. ***Albury Regional Art Centre:*** extensive Russell Drysdale collection; Dean St; (02) 6023 8187. ***Wonga Wetlands:*** rehabilitated wetland along the Murray River, home to the black cormorant; Riverina Hwy.

Rotary Community Market: Townsend St; Sun. ***Opera in the Alps:*** Jan. ***Kinross Country Muster:*** Feb. ***Albury Gold Cup Racing Carnival:*** Mar/Apr. ***Food and Wine Festival:*** Sept/Oct. ***Mungabareena Ngan-Girra (Festival of the Bogong Moth):*** Nov.

sourcedining: contemporary Australian; 664 Dean St; (02) 6041 1288. ***Green Zebra:*** classic cafe fare; 484 Dean St; (02) 6023 1100.

Chifley Albury: 140 studio rooms; cnr Dean and Elizabeth sts; (02) 6021 5366. ***Quest Albury:*** 104 serviced apartments; 550 Kiewa St; (02) 6058 0900 or 1800 334 033. ***Table Top Mountain:*** self-contained bushland cottages; Bells Rd, Gregory; (02) 6026 0529.

Ettamogah A tiny but famous town in Australia, which is home to the original Ettamogah Pub, inspired by cartoonist Ken Maynard. 12 km NE. The OZE Wildlife Resort is worth a visit and Cooper's Ettamogah Winery is 3 km further along.

Lake Hume: watersports, camping and spectacular dam wall; 14 km E. ***Hume Weir Trout Farm:*** handfeeding and fishing for rainbow trout; Lake Hume; 14 km E. ***Jindera Pioneer Museum:*** originally a German settlement featuring a general store, a slab hut and a wattle and daub cottage; 14 km NW. ***Albury–Wodonga Trail System:*** walking trails in the footsteps of Hume and Hovell; maps available from visitor centre. ***Hume and Hovell Walking Track:*** 23-day, 440 km trek from Albury to Yass. For a kit, including maps, contact Department of Lands, Sydney (02) 6937 2700.

See also MURRAY, p. 35

Alstonville

Pop. 5001
Map ref. 551 G5 | 557 O3 | 653 N12

Ballina Visitor Information Centre, cnr Las Balsas Plaza and River St, Ballina; (02) 6686 3484 or 1800 777 666; www.discoverballina.com

101.9 Paradise FM, 738 AM ABC Local

Alstonville is between Lismore and Ballina, surrounded by macadamia and avocado plantations. It's known for its immaculate gardens and purple tibouchina trees that blossom in March, as well as quirky antique and gift shops.

Lumley Park: walk-through reserve of native plants, flying fox colony and open-air pioneer transport museum; Bruxner Hwy. ***Budgen Avenue:*** several shops and galleries with local art and craft. ***Elizabeth Ann Brown Park:*** rainforest park with picnic facilities; Main St.

E.S.P Espresso Bar: excellent coffee and food; Shop 1A/76 Main St; (02) 6628 3433.

Hume's Hovell B&B: B&B; Dalwood Rd; (02) 6629 5371. ***Tallaringa Views:*** self-contained cottages; 1344 Eltham Rd; (02) 6628 5005. ***Deux Belettes:*** French-style guesthouse; Victoria Park Rd, Dalwood; (02) 6629 5377.

Victoria Park Nature Reserve This remarkable rainforest reserve contains 68 species of trees in only 17.5 ha, 8 ha of which remain largely untouched. The area is also home to red-legged pademelons, potoroos, water rats and possums. There is a boardwalk, clearly marked walking trails and a spectacular lookout taking in the surrounding countryside. 8 km S.

Summerland House with No Steps: nursery, avocado and macadamia orchard, garden, crafts, fruit-processing plant, kids water park and Devonshire tea. Completely run by people with disabilities; (02) 6628 0610; 3 km S.

See also TROPICAL NORTH COAST, p. 28

Armidale

Pop. 25 494
Map ref. 550 B1 | 557 J8

82 Marsh St; (02) 6770 3888 or 1800 627 736; www.armidaletourism.com.au

 92.1 2ARM FM, 720 AM ABC Radio National

Armidale is the largest town of the New England district. It is home to New England University, the first university in Australia established outside a capital city. The transplanted birch, poplar and ash trees that line the streets make Armidale seem like an English village. It is one of those rare towns in Australia that enjoy four distinct seasons, with autumn turning the leaves stunning shades of crimson and gold. National parks in the area offer breathtaking forests, gorges and waterfalls.

New England Regional Art Museum The museum (closed Mon) has over 40 000 visitors each year and 8 gallery spaces, an audiovisual theatre, artist studio and cafe. The Howard Hinton and Chandler Coventry collections are among the most important and extensive regional collections, and include works by legendary Australian artists such as Arthur Streeton, Tom Roberts, Margaret Preston and John Coburn. There is also a separate Museum of Printing (open Thurs–Sun) featuring the F. T. Wimble & Co collection. Kentucky St; (02) 6772 5255.

Armidale Heritage Tour: includes Railway Museum, St Peter's Anglican Cathedral (built with 'Armidale blues' bricks) and University of New England; departs daily from visitor centre. ***Aboriginal Cultural Centre and Keeping Place:*** includes museum, education centre and craft displays; Kentucky St. ***Armidale Folk Museum:*** National Trust–classified building with comprehensive collection of pioneer artefacts from the region including toys and buggies; open 1–4pm daily; Cnr Faulkner and Rusden sts; (02) 6770 3536. ***Self-guide heritage walk and heritage drive:*** 3 km walk and 25 km drive provide history and points of interest in and around the town; maps from visitor centre.

Armidale Markets: Beardy St; last Sun each month (3rd Sun in Dec). ***Autumn Festival:*** Mar.

Archie's on the Park: modern Australian in historic homestead; Moore Park Inn, New England Hwy; (02) 6772 2358. ***Caffiends on Marsh:*** great coffee and food; Shop 1, 110 Marsh St; (02) 6771 3178. ***Split Dining:*** modern Australian; 1/117 Beardy St; (02) 6772 8313.

Annie's B&B: cosy, charming, self-contained; 37 Ellowera Rd; (02) 6772 5335. ***Claremont Crossing:*** self-contained farmhouse; 353 Claremont Rd; (02) 6775 1012. ***Petersens Armidale Winery & Guesthouse:*** luxury country manor; Dangarsleigh Rd; (02) 6772 0422.

Oxley Wild Rivers National Park World Heritage–listed with the largest area of dry rainforest in NSW, this park includes Dangars Falls (21 km SE), a 120 m waterfall in a spectacular gorge setting, and Wollomombi Falls (40 km E), which at 220 m is one of the highest falls in the state.

University of New England: features Booloominbah Homestead, Antiquities Museum, Zoology Museum and kangaroo and deer park; 5 km NW. ***Saumarez Homestead:*** National Trust–owned house offering tours (closed mid-June to end Sept); 5 km S. ***Dumaresq Dam:*** walking trails, boating, swimming and trout fishing (Oct–June); 15 km NE. ***Hillgrove:*** former mining town with Rural Life and History Museum featuring goldmining equipment (open Fri–Mon) and self-guide walk through old town site; brochure at visitor centre; 31 km E.

See also NEW ENGLAND, p. 29

Ballina

Pop. 14 675
Map ref. 551 G5 | 557 O4 | 653 N12

Cnr Las Balsas Plaza and River St; (02) 6686 3484 or 1800 777 666; www.discoverballina.com

 101.9 Paradise FM, 738 AM ABC Local

Ballina sits on an island at the mouth of the Richmond River in northern New South Wales surrounded by the Pacific Ocean and nearby fields of sugarcane. The sandy beaches, clear water and warm weather make the area popular. Ballina's name comes from the Aboriginal word 'bullenah', which is said to mean 'place where oysters are plentiful'. This is still the case, with fresh seafood readily available in many seaside restaurants.

Shelly Beach A superb spot for the whole family. Dolphins can be seen frolicking in the waves all year-round and humpback whales migrate through these waters June–July and Sept–Oct. The beach itself has rockpools, a wading pool for toddlers and a beachside cafe. Off Shelly Beach Rd.

Naval and Maritime Museum: features a restored Las Balsas Expedition raft that sailed from South America in 1973; Regatta Ave; (02) 6681 1002. ***Kerry Saxby Walkway:*** from behind the visitor centre to the river mouth with great river and ocean views. ***The Big Prawn:*** art and craft and fresh seafood; Pacific Hwy. ***Richmond Princess and Bennelong:*** river cruises; bookings at visitor centre. ***Shaws Bay:*** swimming and picnic area; off Compton Dr. ***Ballina Water Slide:*** River St.

Ballina Markets: Canal Drive; 3rd Sun each month. ***Ballina Cup:*** horserace; Sept.

Ballina Manor Boutique Hotel: modern Australian in Edwardian manor; 25 Norton St; (02) 6681 5888. ***Pelican 181:*** waterfront seafood, eat in or takeaway; 12–24 Fawcett St; (02) 6686 9181. ***Sandbar and Restaurant:*** modern European, tapas; 23 Compton Dr; (02) 6686 6602.

Ballina Beach Resort: incredible beachfront location; Compton Dr; (02) 6686 8888. ***Ballina Manor Boutique Hotel:*** award-winning, heritage-listed manor; 25 Norton St; (02) 6681 5888.

Lennox Head A beachside town with a good market on the 2nd and 5th Sun each month on the shores of Lake Ainsworth (also a popular spot for windsurfing). The lake has been nicknamed the Coca-Cola lake due to coloration from surrounding tea trees. Pat Morton Lookout affords excellent views along the coast, with whale-watching June–July and Sept–Oct. Below is The Point, a world-renowned surf beach. The outskirts of town offer scenic rainforest walks. 10 km N.

Thursday Plantation: tea-tree plantation with product sales and maze; (02) 6620 5150; 3 km W. ***Macadamia Castle:*** features macadamia products, industry displays, minigolf and children's wildlife park; (02) 6687 8432; 15 km N. ***Whale-watching:*** bookings at visitor centre.

See also TROPICAL NORTH COAST, p. 28

Balranald

Pop. 1217
Map ref. 558 H7 | 589 M9

Heritage Park, Market St; (03) 5020 1599; www.balranald.nsw.gov.au

 93.1 FM ABC Radio National, 102.1 FM ABC Local

Balranald is the oldest town on the lower part of the Murrumbidgee River. Situated on saltbush and mallee plains, the area now embraces the viticulture, horticulture and tourism industries. A string of dry lake beds stretches to the north of Balranald, the most famous of which are preserved in Mungo National Park. Recently, the oldest human footprints in Australia were found in the park, estimated to be up to 23 000 years old.

Heritage Park Investigate the old gaol, the Murray pine schoolhouse, local history displays and a historical museum. There are also picnic and barbecue facilities. Market St.

Art gallery: exhibitions by local artists, housed in 1880s Masonic Lodge; Mayall St. ***Balranald Weir:*** barbecues, picnics, fishing. ***Memorial Drive:*** great views. ***Frog Sculptures:*** 14 throughout town. ***Self-guide town walk:*** historically significant buildings in the town; maps available from visitor centre.

Balranald Cup: horserace; Feb/Mar.

Riverview on the Edward: B&B situated on a working sheep and cattle farm; Balpool Rd, Moulamein; (03) 5887 5241.

Mungo National Park This park is the focal point of the Willandra Lakes World Heritage Area, a 240 000 ha region dotted with 17 dry lakes. These lakes display astounding evidence of ancient Aboriginal life and of creatures that existed during the last Ice Age. The highlight is the 33 km crescent-shaped dune on the eastern edge of Lake Mungo, called the Walls of China. The park contains some remarkable animal remains, as well as ancient fireplaces, artefacts and tools. The Mungo Visitor Centre delves into the heritage of the Willandra Lakes, and has a replica of the diprotodon, a massive wombat-like marsupial; Arumpo Rd; (03) 5021 8900 or 1300 361 967. A 70 km self-guide drive tour, suitable for all vehicles, takes in the Walls of China, old tanks and wells and an old homestead site. Accommodation includes the old Mungo Shearers' Quarters and a campground. 100 km N.

Moulamein The oldest town in the Riverina, Moulamein has fascinating historic structures to explore including its restored courthouse (1845) and Old Wharf (1850s). There are picnic areas by the Edward River and Lake Moulamein. 99 km SE.

Yanga Lake: fishing and watersports; 7 km SE. ***Homebush Hotel:*** built in 1878 as a Cobb & Co station, the hotel now provides meals and accommodation; (03) 5020 6803; 25 km N. ***Kyalite:*** home to Australia's largest commercial pistachio nut farm and popular with campers and anglers; 36 km S. ***Redbank Weir:*** barbecues and picnics; Homebush–Oxley Rd; 58 km N.

See also RIVERINA, p. 34

Barham

Pop. 1131
Map ref. 559 I10 | 584 A2 | 591 O4

Golden Rivers Tourism, 15 Murray St; (03) 5453 3100.

107.7/102.5 Mix FM, 594 AM ABC Local

Barham and its twin town, Koondrook, sit beside the Murray River and the New South Wales–Victoria border. Barham is known as the southern gateway to Golden Rivers country. The Murray River makes Barham a great place for anglers with Murray cod, golden perch, catfish and yabbies in abundance. Barham Bridge is one of the oldest bridges on the Murray and was lifted manually until 1997.

Barham Lakes Complex The complex is popular with locals and visitors alike. It has 4 artificial lakes stocked with fish and yabbies, grasslands with hundreds of native plants, a walking track and barbecue facilities. Murray St.

Border Flywheelers Vintage Engine Rally: Feb. ***The Country Music Stampede:*** Feb and Aug. ***Jazz Festival:*** June. ***Barham Produce and Food Festival:*** Aug. ***Golden Rivers Red Gum Forest to Furniture Showcase:*** Oct.

Barham Caravan and Tourist Park: fantastic family park on Murray River; 1 Noorong St; (03) 5453 2553.

Koondrook State Forest Koondrook State Forest is 31 000 ha of native bushland that is perfect for birdwatchers and nature enthusiasts. The forest has over 100 bird species, kangaroos, emus and wild pigs. Forest drives winding through the park are well signposted. 12 km NE.

Koondrook: old sawmilling town and river port with historic buildings and tramway; 5 km SW. ***Murrabit:*** largest country markets in the region; 1st Sat each month; Murrabit Rd; 24 km NW.

See also MURRAY, p. 35

Barraba

Pop. 1163
Map ref. 556 H7

112 Queen St; (02) 6782 1255.

99.1 FM ABC Local, 648 AM ABC Local

The tree-lined streets of Barraba lie in the valley of the Manilla River. Surrounded by the Nandewar Ranges, Horton Valley and undulating tablelands, Barraba is a quiet and idyllic town. The area was once busy with mining and, although some mines still operate, the main industries today centre around sheep and wool.

Heritage walk The walk takes in a heritage-listed organ and historic buildings such as the courthouse, church, clock tower and the visitor centre itself. The Commercial Hotel on Queen St was once a Cobb & Co changing station.

Clay Pan and Fuller Gallery: exhibits art, craft and pottery; Queen St. ***The Playhouse:*** accommodation including a theatre and exhibition space; Queen St; (02) 6782 1109.

Market: 1st Sat each month; Queen St. ***Australia's Smallest Country Music Festival:*** Jan. ***Barraba Agricultural Show and Rodeo:*** Feb/Mar. ***Australia's Smallest Jazz and Blues Festival:*** Easter. ***Frost Over Barraba:*** art show; July. ***BarrArbor Festival:*** celebration of culture; Nov. ***Horton Valley Rodeo:*** Dec.

Barraba Motel: traditional, charming motel accommodation; 17 Edward St; (02) 6782 1555.
Playhouse Hotel: comfortable accommodation adjacent to theatre; 121–123 Queen St; (02) 6782 1109.

Mt Kaputar National Park This park is excellent for hiking, rising as high as 1200 m, and is the site of the now extinct Nandewar Volcano. The diverse vegetation ranges from semi-arid woodland to wet sclerophyll forest and alpine growth. Wildlife is abundant, especially bats, birds and quolls. Access to the park from Barraba is by foot only, although permission for 4WD access can be granted by the visitor centre. 48 km W.

Adams Lookout: panoramic views of the town and countryside; 5 km NE. ***Millie Park Vineyard:*** organic wine cellar-door tastings and sales; 5 km N. ***Glen Riddle Recreation Reserve:*** on Manilla River north of Split Rock Dam, for boating, fishing

continued on p. 43

BATEMANS BAY

Pop. 10 843

Map ref. 554 H12 | 565 F6 | 567 L6

Cnr Princes Hwy and Beach Rd; (02) 4472 6900 or 1800 802 528; www.eurobodalla.com.au

103.5 FM ABC South East, 104.3 Power FM

Batemans Bay is a popular town for holiday-makers at the mouth of the Clyde River. It has something for everyone with rolling surf beaches, quiet coves and rockpools, and wonderful views upriver to the hinterland mountains and out to sea to the islands on the horizon. Batemans Bay is the home of the famous Clyde River oyster and other excellent fresh seafood.

Birdland Animal Park: native wildlife and rainforest trail; Beach Rd. ***River cruises:*** daily cruises on the *Merinda* to historic Nelligen depart from the wharf behind The Boatshed; contact visitor centre for times. ***Houseboat hire and fishing charters:*** bookings at visitor centre. ***Corn Trail:*** a picturesque but challenging 12.5 km walking and horseriding track through Buckenboura State Forest, following a pioneer route from the top of Clyde Mountain down the valley to the coast; maps available from visitor centre. ***Joy-flights over the Clyde River:*** scenic flights or skydiving over the coast and Clyde River from Moruya Airport; inquiries 1800 802 528 or at visitor centre. ***Scenic lookouts:*** panoramic views of coast and hinterland from Big Bit Lookout or Folders Hill Lookout (north of town), Holmes Lookout (west) and Round Hill Lookout (south).

Batemans Bay Market: Museum Place, 1st, 2nd and 4th Sat each month; high school, 3rd Sun each month.

Briars Restaurant: modern Australian; Princes Hwy; (02) 4472 9200. ***On the Pier:*** waterfront seafood; 2 Old Punt Rd; (02) 4472 6405. ***Starfish Deli Restaurant:*** relaxed dining; The Promenade, 2 Clyde St; (02) 4472 4880.

Chalet Swisse Spa at Surf Beach Retreat: bushland and rainforest setting; 676 The Ridge Rd; (02) 4471 3671. ***Mariners on the Waterfront:*** resort-style accommodation; Orient St; (02) 4472 6222. ***The Bower at Broulee:*** romantic self-contained eco cabins; George Bass Dr, Broulee; (02) 4471 8666.

Mogo Alive with a quaint village atmosphere, Old Mogo Town re-creates a 19th-century goldmining town. Take a guided tour to experience gold-rush life, bushranger legends, goldmine workings and workshops. Wildlife enthusiasts will enjoy Mogo Zoo – it specialises in raising endangered species such as tigers, white lions, snow leopards and red pandas. Mogo offers fabulous village-style shopping and cafes, including antique and collectibles stores, art galleries and specialty shops. Mogo State Forest is popular with birdwatchers and bushwalkers, and is home to lorikeets, kookaburras, rosellas and cockatoos. 8 km S.

Eurobodalla Native Botanic Gardens: native plants, walking tracks, nursery and picnic area; open Wed–Sun; 5 km S. ***Murramarang National Park:*** undisturbed coastline and abundant kangaroos, particularly at Pebble Beach; good swimming, excellent surf beaches and plenty of picnic spots; rock and beach fishing are popular; 10 km NE. ***Murramarang Aboriginal Area:*** signposted walk through 12 000-year-old Aboriginal sites, including the largest complex of shell middens and stone tools on the south coast; just north of Murramarang National Park. ***Nelligen:*** an important Clyde River port in pioneer days, when goods were shipped from here down to Sydney or sent into the hinterland; nowadays holiday-makers use the historic town as a base for waterskiing, fishing and houseboat vacations; 10 km NW. ***Malua Bay:*** excellent surfing; 14 km SE. ***Durras Lake:*** fishing, kayaking and swimming; 16 km NE. ***Tomakin:*** coastal holiday village by the Tomaga River with long stretches of pristine family-friendly beaches and forested hinterland. 15 km S.

See also SOUTH COAST, p. 32

[BODY SURFERS CATCHING A WAVE AT MALUA BAY]

and picnicking; 15 km SE. ***Ironbark Creek:*** gold and mineral fossicking, with ruins of old village; 18 km E. ***Horton River Falls:*** 83 m waterfall, swimming and bushwalking; 38 km W. ***Birdwatching trails:*** the 165 species in the area include the rare regent honeyeater. Guides are available from the visitor centre.

See also NEW ENGLAND, **p. 29**

Bathurst

Pop. 35 269
Map ref. 554 H5

Kendall Ave; (02) 6332 1444 or 1800 681 000; www.visitbathurst.com.au

 96.7 FM ABC Radio National, 99.3 B-Rock FM

Bathurst, on the western side of the Great Dividing Range, is Australia's oldest inland settlement. Originally occupied by the Wiradjuri people, it was the site of enormous conflict in 1824 between its original inhabitants and the European settlers. Since then, Bathurst has become known as the birthplace of Ben Chifley, Australian prime minister 1945–49, and for its magnificent Georgian and Victorian architecture. Today it is best known for its motor racing circuit, Mount Panorama.

Miss Traill's House Ida Traill (1889–1976) was a fourth-generation descendant of pioneers William Lee and Thomas Kite who came to Bathurst in 1818. Her house, a colonial Georgian bungalow filled with artefacts and bequeathed to the National Trust, was built in 1845, making it one of the oldest houses in Bathurst. The 19th-century cottage garden is particularly charming in spring. Russell St; (02) 6332 4232.

Bathurst District Historical Museum: features notable local Aboriginal artefacts in the east wing of the Neoclassical Bathurst Courthouse; Russell St; (02) 6330 8455. ***Australian Fossil and Mineral Museum:*** exhibits close to 2000 fossil and mineral specimens, including rare and unique displays, housed in the 1876 public school building; Howick St; (02) 6331 5511. ***Bathurst Regional Art Gallery:*** focuses on Australian art after 1955, with frequently changing exibitions; Keppel St; (02) 6333 6555. ***Machattie Park:*** Victorian-era park in the heart of the city, with a begonia house full of blooms Feb–Apr; Keppel, William and George sts. ***Self-guide historical walking tour and self-drive tour:*** takes in Bathurst Gaol, the courthouse and historic homes, including Ben Chifley's house; map from visitor centre.

Gold Crown Festival: harness racing; Mar. ***Autumn Colours:*** variety of events celebrating autumn; Mar–May. ***V8 Supercars 1000:*** Oct.

Cobblestone Lane: modern Australian; 2/173–179 George St; (02) 6331 2202. ***The Church Bar:*** woodfired pizzas; 1 Ribbon Gang La; (02) 6334 2300.

Bishop's Court Estate Boutique Hotel: Victorian hilltop guesthouse; 226 Seymour St; (02) 6332 4447. ***Encore Apartments:*** contemporary self-contained apartments; 187 Piper St; (02) 6333 6000.

Mt Panorama The inaugural Bathurst 1000 was held here in 1960 and has since become an Australian institution. The 6.2 km scenic circuit is open year-round, and while the lap record is 124.08 seconds (over 170 km/h), visitors are limited to 60 km/h. The National Motor Racing Museum at the circuit displays race cars, trophies, memorabilia and special exhibits; (02) 6332 1872. Also at Mt Panorama are the Bathurst Goldfields, a reconstruction of a historic goldmining area, and McPhillamy Park, which has great views over Bathurst, especially at sunrise and sunset. 2 km S.

Abercrombie House: impressive 1870s baronial-style Gothic mansion; Ophir Rd; 6 km W. ***Bathurst Sheep and Cattle Drome:*** visitors can milk a cow and see shearing and sheepdog demonstrations; Limekilns Rd; 6 km NE. ***Bathurst Observatory:*** located at the Bathurst Goldfields; program varies throughout the year. The visitor centre features mineral, fossil and space displays. The complex is closed in inclement weather; 12 km NE. ***Wallaby Rocks:*** wall of rock rising from the Turon River and a popular spot for kangaroos and wallabies. Also an ideal swimming and picnic spot; 40 km N. ***Sofala:*** historic gold town and the setting for scenes from the films *The Cars That Ate Paris* (1974) and *Sirens* (1994); 42 km N. ***Hill End Historic Site:*** former goldfield with many original buildings. The area has inspired painters Russell Drysdale, Donald Friend, John Olsen and Brett Whiteley. There is a National Parks & Wildlife Service visitor centre in old Hill End Hospital, which has a historical display and information on panning and fossicking, with equipment for hire; (02) 6337 8206. Old gold towns nearby include Peel, Wattle Flat, Rockley, O'Connell and Trunkey; 86 km NW.

See also CENTRAL WEST, **p. 30**

Batlow

Pop. 997
Map ref. 554 D11 | 565 A5 | 566 C4 | 585 P3

Tumut Region Visitor Centre, 5 Adelong Rd, Tumut; (02) 6947 7025; www.tumut.nsw.gov.au

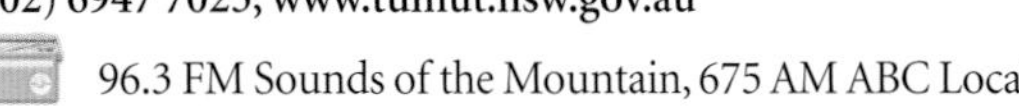 96.3 FM Sounds of the Mountain, 675 AM ABC Local Radio

In the 19th-century gold rush prospectors converged on nearby Reedy Creek, which sparked a sudden demand for fresh produce. The resulting orchards and farms became the town of Batlow. Set in the low-lying mountains of the state's south-west slopes, Batlow is a picturesque town still surrounded by orchards of delicious apples, pears, berries, cherries and stone fruit.

Batlow Woodworks: discover a range of art, pottery, woodworks and crafts, serving morning and afternoon tea; Pioneer St; (02) 6949 1265. ***Weemala Lookout:*** breathtaking views of town and Snowy Mountains; H. V. Smith Dr.

Apple Blossom Festival: Oct.

Mountain View Cafe: popular cafe; Pioneer St; (02) 6949 1110.

Batlow House: self-contained Federation cottage; 5 Pioneer St; 0439 475 383.

Hume and Hovell Lookout: great views over Blowering Valley and Blowering Reservoir, with picnic area at the site where explorers rested in 1824; 6 km E. ***Tumut Rd:*** fresh fruit along the road. Springfield Orchard, 6 km N, grows 16 apple varieties and has picnic and barbecue facilities. ***Kosciuszko National Park:*** this alpine park to the east includes nearby Bowering Reservoir and Buddong Falls; *for more details see Adaminaby*. ***Hume and Hovell Walking Track:*** access to short sections of the 440 km track via Tumut Rd. Maps available from visitor centre.

See also RIVERINA, **p. 34**

Bega

Pop. 4536
Map ref. 565 E9 | 567 J12 | 581 H6

 **Lagoon St; (02) 6491 7645; www.sapphirecoast.com.au**

 102.5 Power FM, 810 AM ABC Local

It is possible to ski and surf on the same day around Bega, set in a fertile valley with the mountains of the Kosciuszko snow resorts to the west and breathtaking coastline to the east. Bega is best known for its dairy industry, particularly cheese-making.

Bega Family Museum: houses town memorabilia including silverware, ball gowns, farm machinery and photographs; open Mon–Sat Sept–May, Tues and Fri June–Aug; Cnr Bega and Auckland sts; (02) 6492 1453.

 ***Bega Valley Art Awards:*** Oct.

Bega Downs Restaurant: modern Australian; Bega Downs Motor Inn, cnr Princes Hwy and High St; (02) 6492 2944. ***Pepperberry Restaurant:*** friendly cafe; Shop 1, Ayres Walkway; (02) 6492 0361.

Rock Lily Cottages: self-contained cottages; 864 Warrigal Range Rd, Brogo via Bega; (02) 6492 7364.

Biamanga National Park Now a popular spot for swimming, bushwalking and picnics, Biamanga National Park has long been a sacred site to the Yuin people. Mumbulla Mountain was an initiation site for young men and Mumbulla Creek was used to wash off ceremonial ochre. Visitors can now enjoy the rockpools, natural water slides, boardwalks, viewing platforms and picnic sites of this culturally significant area. 19 km NE.

Bega Cheese Heritage Centre: restored cheese factory with cheese-tasting and displays of cheese-making equipment; (02) 6491 7762; 3 km N. ***Lookouts:*** excellent views at Bega Valley Lookout (3 km N) and Dr George Lookout (8 km NE). ***Candelo:*** charming and peaceful village with market on 1st Sun each month; 24 km SW. ***Brogo Dam:*** haven for native birdlife such as sea eagles and azure kingfishers. Also popular for bass fishing, swimming, picnicking, boating and canoeing (canoe hire on-site; (02) 6492 7328); 30 km NW.

See also **South Coast, p. 32**

Bellingen

Pop. 2878
Map ref. 550 G1 | 551 D12 | 557 M8

 **Hyde St; (02) 6655 1522; www.bellingermagic.com**

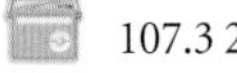 107.3 2BBB FM, 738 AM ABC Local

Bellingen is an attractive tree-lined town on the banks of the Bellinger River, surrounded by rich pasturelands. Traditionally serving dairy farmers and timber cutters, Bellingen is now a haven for urban folk fleeing the big cities, attracted to the relaxed and alternative lifestyle on offer. City touches can be found in shops and cafes, but the town retains its laid-back feel. The area is the setting for Peter Carey's novel *Oscar and Lucinda*.

Bellingen Bat Island This 3 ha island is home to a colony of up to 40 000 grey-headed flying foxes (fruit bats). At dusk the flying foxes set off in search of food, filling the sky. The best time to visit is Sept–Mar. Access is via Dowle St (north of the river).

Horse-drawn carriage tours: guided tours of town and river picnics (operate during holidays and by appt); 0423 671 581. ***Bellingen Museum:*** features extensive photo collection of early pioneer life and early transportation; Hyde St; (02) 6655 1259. ***Hammond and Wheatley Emporium:*** the first concrete block construction in Australia, the emporium has been magnificently restored – including a grand staircase leading to a mezzanine floor – and is now home to boutiques, homewares retailers and jewellery galleries; Hyde St. ***Local art and craft:*** galleries thoughout the town including The Yellow Shed, Cnr Hyde and Prince sts, and The Old Butter Factory, Doepel St.

Bellingen Markets: Bellingen Park, Church St; 3rd Sat each month. ***Jazz Festival:*** Aug. ***Global Carnival:*** world music; Oct.

Elite Espresso Gallery: great coffee and food; 62 Hyde St; (02) 6655 0509. ***Lodge 241 Gallery Cafe:*** historic riverside cafe; 117–212 Hyde St; (02) 6655 2470. ***No 2 Oak Street:*** modern Australian; 2 Oak St; (02) 6655 9000.

Casa Belle: Tuscan-style guesthouse; 90 Gleniffer Rd; (02) 6655 9311. ***Kumbaingiri Retreat:*** luxurious self-contained retreat; 184 Kalang Rd; 0434 074 075. ***Promised Land Cottages:*** self-contained cottages; 934 Promised Land Rd, Promised Land; (02) 6655 9578.

Raleigh Vineyard and Winery: tastings available; (02) 6655 4388; 11 km E. ***Walking, cycling, horseriding and canoeing:*** along the Bellinger River and in forest areas; information and maps from visitor centre. ***Scenic drive:*** north-east through wooded valleys and farmlands, across Never Never Creek to Promised Land; map available from visitor centre.

See also **Holiday Coast, p. 27**

Bermagui

Pop. 1298
Map ref. 565 E8 | 567 K11

Bunga St; (02) 6493 3054 or 1800 645 808; www.sapphirecoast.com.au

 102.5 Power FM, 105.9 FM 2EC, 810 AM ABC Local

Bermagui is a charming and sleepy coastal village. The continental shelf is at its closest to the mainland off Bermagui and this results in excellent fishing for marlin, tuna and shark. Zane Grey was a famous visitor in the 1930s and the town featured in two of his books. Bermagui is also the centre of a mystery involving a geologist, Lamont Young, who was sent to investigate goldfields in 1880. When he decided to head north to investigate further, he and his assistant were offered passage on a small boat with three men. All five disappeared en route. When their boat was discovered, it was found to have five bags of clothing, Young's books and papers, and a bullet in the starboard side. Despite extensive searches and media attention, no trace of the men was ever found.

Fish Co-op: freshly caught fish and prawns; Fishermans Wharf, harbourside. ***Blue Pool:*** large and attractive saltwater rockpool offering an unusual swimming experience; off Scenic Dr. ***Horseshoe Bay Beach:*** safe swimming spot. ***Good surfing beaches:*** Beares, Mooreheads, Cuttagee and Haywards beaches; maps from visitor centre. ***Gamefishing, deep-sea fishing and reef-fishing:*** bookings at visitor centre.

Craft Market: Dickinson Park; last Sun each month. ***Blue Water Fishing Classic:*** Jan. ***Seaside Fair:*** Mar. ***Tag and Release Gamefishing Tournament:*** Mar. ***Four Winds Festival:*** even-numbered years, Easter. ***Victorian Southern Gamefish Challenge:*** Easter.

Bermagui Mudworks & Cafe: cafe with Balinese feel; 23 Alexander Dr; (02) 6493 4661. ***Saltwater:*** seafood cafe; 59 Lamont St; (02) 6493 4328.

Bermagui Beach House: contemporary beach house; address provided on inquiry; 0402 337 396. ***Bimbimbi House:*** modern self-contained beach house; 62 Nutleys Creek Rd; (02) 6493 4456. ***Wu Wei House:*** luxurious, Japanese-inspired holiday home; address provided on inquiry; 0410 081 599.

Gulaga National Park The Wallaga Lakes area has an 8 km coastal walk through wetland flora and fauna reserves and remnants of the Montreal Goldfield north to Wallaga Lake. It passes Camel Rock, an unusual rock formation in the general shape of a camel. The park is hilly with steep gullies, so it is best explored by boat (available for hire from Regatta Pt and Beauty Pt). Guided tours are available through Montreal Goldfield (bookings at visitor centre). There are good walking trails, including one to the summit of Mt Dromedary. The park is excellent for boating, fishing, swimming, picnicking and bushwalking. The Yuin people run Aboriginal cultural tours from the Umballa Cultural Centre, sharing Dreamtime stories that have never been recorded on paper. Activities include ochre painting, bark-hut building and boomerang throwing. Bookings and information (02) 4473 7232; 6 km N.

Mimosa Rocks National Park: 17 km S; *see Tathra*. ***Mystery Bay:*** the site of the discovery of Lamont Young's abandoned boat and a memorial; 17 km N. ***Cobargo:*** historic working village with art galleries, wood and leather crafts, antiques, pottery and tearooms. A country market on 4th Sat each month at RSL hall grounds; 20 km W.

See also SOUTH COAST, p. 32

Berridale

Pop. 844
Map ref. 565 C8 | 566 F10 | 581 D4

Snowy Region Visitor Centre, Kosciuszko Rd, Jindabyne; (02) 6450 5600 or 1800 004 439; www.snowymountains.com.au

 97.7 Snow FM, 810 AM ABC Local

This charming small town calls itself the 'Crossroads of the Snowy' and is a popular stopover point in winter between Cooma and the snowfields. In the 1860s and 1870s it was known as Gegedzerick, but later changed its name to Berridale, the name of a local property. The main street is lined with poplars that provide a striking show in autumn. The trees were planted about 100 years ago by children from Berridale School.

Ray Killen Gallery: landscape photographs from around Australia; Jindabyne Rd. ***Historic buildings:*** St Mary's (1860), Mary St; Berridale School (1883), Oliver St; Berridale Inn (1863), Exchange Sq; Berridale Store (1863), Exchange Sq. ***Boulders:*** unique granite boulders near to the main road were formed from crystallised magma 400 million years ago.

Snowy River Winery: charming restaurant, wine tastings; Jindabyne Rd; (02) 6456 4015.

Snowy Mountains Coach and Motor Inn: resort-style accommodation; Jindabyne Rd; (02) 6456 3283. ***The Range:*** lodge-style accommodation; Kosciuszko Rd; (02) 6456 3368.

Dalgety: small town featuring historic Buckley's Crossing Hotel, marking the spot where cattle crossed the Snowy River; 18 km S. ***Eucumbene Trout Farm:*** sales, horseriding and tours; (02) 6456 8866; 19 km N. ***Snowy River Ag Barn and Fibre Centre:*** museum, craft and fibre shop, animals and restaurant; 21 km S.

See also SNOWY MOUNTAINS, p. 33

Berrima

Pop. 867
Map ref. 553 B5 | 555 I8 | 565 G2

Berrima Courthouse, cnr Wilshire and Argyle sts; (02) 4877 1505; www.berrimavillage.com.au

 102.9 FM 2ST, 675 AM ABC Local

A superbly preserved 1830s village, Berrima is nestled in a valley next to the Wingecarribee River. The National Heritage Council declared the village a historic precinct in the 1960s. Many buildings have been restored as antique shops, restaurants and galleries.

Berrima Courthouse The courthouse was the scene in 1841 of Australia's first trial by jury, in which Lucretia Dunkley and Martin Beech were accused of having an affair and tried for murdering Lucretia's much older husband, Henry, with an axe. They were both found guilty and hanged. The building, said to be the finest in town, now houses displays on the trial and early Berrima. Cnr Wilshire and Argyle sts; (02) 4877 1505.

Berrima District Historical Museum: displays focus on colonial settlement and the struggles of pioneer days; Market Pl; (02) 4877 1130. ***Harpers Mansion:*** Georgian house built in 1834, now owned by the National Trust; Wilkinson St. ***Australian Alpaca Centre:*** sales of knitwear and toys; Market Pl. ***The Surveyor General:*** built in 1835, Australia's oldest continually licensed hotel; Old Hume Hwy. ***Berrima Gaol:*** Bushranger Paddy Curran was the first man hanged there in 1842 and Lucretia Dunkley was the first and only woman executed there; Argyle St.

Market: 2nd Sun each month; school grounds; Oxley St.

Eschalot: modern Australian; 24 Old Hume Hwy; (02) 4877 1977. ***Josh's Cafe:*** popular eatery with Middle Eastern influences; 2/9 Old Hume Hwy; (02) 4877 2200.

Jellore Cottage and Settlers Hut: luxury self-contained cottages; 16 and 16A Jellore St; 0404 951 884. ***Surveyor General Inn:*** colonial-style pub; Old Hume Hwy; (02) 4877 1226.

Wineries: numerous in the area with cellar-door tastings including Southern Highland Wines in Sutton Forest ((02) 4868 2300); 10 km S; and Mundrakoona Estate in Mittagong ((02) 4872 1311); 10 km NE; maps from visitor centre.

See also SOUTHERN HIGHLANDS, p. 24

Berry

Pop. 1485
Map ref. 553 D8 | 555 J9 | 565 G3

Shoalhaven Visitors Centre, cnr Princes Hwy and Pleasant Way, Nowra; (02) 4421 0778 or 1300 662 808; www.berry.org.au

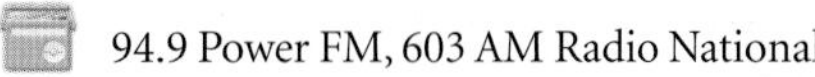 94.9 Power FM, 603 AM Radio National

The local chamber of commerce named Berry 'The Town of Trees' because of the extensive stands of English oaks, elms and beech trees planted by settlers in the 1800s. Berry is a popular weekend destination for Sydneysiders searching for bargains in the antique and craft shops and looking to enjoy the laid-back atmosphere. With the Cambewarra Ranges as its backdrop, Berry is the first truly rural town south of Sydney.

Berry Historical Museum: records and photographs of early settlement; open 11am–2pm Sat, 11am–3pm Sun, daily during school holidays; Queen St; (02) 4464 3097. ***Precinct Galleries:*** local contemporary art, craft and design; Alexandra St. ***Great Warrior Aboriginal Art Gallery:*** contemporary and traditional Aboriginal art, weapons, artefacts and didgeridoos; Queen St. ***Antique and craft shops:*** contact visitor centre for details.

Country Fair Markets: showground, cnr Alexandra and Victoria sts, 1st Sun each month (except Feb); Great Southern Hotel, Queen St, 3rd Sun each month. ***Musicale festival:*** June. ***Garden Festival:*** Oct.

Hungry Duck: modern Asian/Australian; 85 Queen St; (02) 4464 2323. ***The Posthouse:*** modern Australian; 137 Queen St; (02) 4464 2444.

Sojourn at Far Meadow: contemporary self-contained cabins; 47 Bryces Rd; (02) 4448 5497. ***Crystal Creek Meadows:*** eco-friendly retreat; 1655 Kangaroo Valley Rd, Kangaroo Valley; (02) 4465 1406.

Coolangatta: convict-built cottages, winery (open for tastings) and accommodation on site of first European settlement in area; 11 km SE. ***Other wineries in area:*** open for tastings and sales; map from visitor centre. ***Mild to Wild Tours:*** adventure tours including sea-kayaking with dolphins, rock climbing, moonlight canoeing and mountain-biking; bookings (02) 4464 2211.

See also SOUTHERN HIGHLANDS, p. 24

Bingara

Pop. 1205
Map ref. 556 H5

Roxy Theatre Building, Maitland St; (02) 6724 0066 or 1300 659 919; www.bingara.com.au

102.9 Gem FM, 648 AM ABC Local

Located in the centre of an area known as Fossickers Way, Bingara is an old gold- and diamond-mining town in the Gwydir River Valley. Gold was discovered here in 1852. Prospectors have been attracted to the town ever since, drawn by the chance of discovering their own fortunes in gold, tourmaline, sapphires and garnets, and by the peaceful cypress-covered mountain surrounds.

Orange Tree Memorial Orange trees along Finch St and Gwydir Oval stand as a memorial to those who have fallen in war. It is a town tradition that during the Orange Festival, Bingara's children pick the fruit and present it to hospital patients and the elderly.

All Nations Goldmine: a stamper battery is the only visible remnant; Hill St. ***Bingara Historical Museum:*** slab building (1860) displays gems and minerals and 19th-century furniture and photographs; Maitland St. ***Visitor Information Centre:*** set in the beautifully restored 1930s Art Deco Roxy Theatre complex; Maitland St. ***Gwydir River Rides:*** trail rides; Keera St. ***Gwydir River:*** walking track along the bank, and reportedly the best Murray cod fishing in NSW. ***Self-guide historical/scenic town walk and drive:*** contact visitor centre for maps.

Bingara Cup Race Meeting: Feb. ***Bingara Fishing Competition:*** Easter. ***Campdraft***: Apr. ***Orange Festival:*** Aug.

Bingara Fossickers Way Motel: friendly, comfortable accommodation; 2 Finch St; (02) 6724 1373.

Three Creeks Goldmine: working mine open to the public for gold panning, crystal fossicking and bushwalking; 24 km S. ***Myall Creek Memorial:*** monument to 28 Aboriginal men, women and children killed in the massacre of 1838; Delungra–Bingara Rd; 27 km NE. ***Rocky Creek glacial area:*** unusual conglomerate rock formations; 37 km SW. ***Sawn Rocks:*** pipe-shaped volcanic rock formations; 70 km SW. ***Birdwatching and fossicking:*** maps available from visitor centre.

See also NEW ENGLAND, p. 29

Blackheath

Pop. 4178
Map ref. 544 C4 | 546 F6 | 555 I6

Heritage Centre, Govetts Leap Rd; (02) 4787 8877; www.visitbluemountains.com.au

89.1 BLU FM, 675 AM ABC Local

This pretty resort town, the highest in the Blue Mountains, has breathtaking views. Known for its guesthouses, gardens and bushwalks, Blackheath is in an ideal location at the edge of Blue Mountains National Park. It is also known as 'Rhododendron Town' for the myriad varieties that bloom every November.

National Parks & Wildlife Heritage Centre More than just an information resource, the centre features an interactive display on the geology, wildlife, Aboriginal and European history of the area and offers historical tours and guided walks. It is the starting point for the Fairfax Heritage Walk, a gentle bushwalk with wheelchair access and facilities for the visually impaired. The 4 km return trail goes to Govetts Leap Lookout for views across several waterfalls and the Grose Valley. Govetts Leap Rd.

Govett statue: commemorates the bushranger known as Govett, said to have ridden his horse over a cliff rather than be captured by police; centre of town.

Markets: Growers Market, Community Hall, Great Western Hwy, 2nd Sun each month; Community Market, School Grounds, 1st Sun each month. ***Blue Mountains Food and Wine Fair:*** Apr. ***Rhododendron Festival:*** Nov.

Ashcrofts: modern Australian; 18 Govetts Leap Rd; (02) 4787 8297. ***Vulcan's:*** modern Australian; 33 Govetts Leap Rd; (02) 4787 6899.

Lavender Cottage and Manor: self-contained cottages; 117 Evans Lookout Rd; (02) 4782 2385. ***Secrets Hideaway:*** self-contained cottages; 173 Evans Lookout Rd; (02) 4787 8453. ***Woolshed Cabins:*** modern, spacious, self-contained cabins; Kanimbla Dr; (02) 4787 8199.

Mount Victoria National Trust–classified Mount Victoria is the westernmost township of the Blue Mountains. There are buildings from the 1870s, including the Imperial Hotel, St Peter's Church of England and The Manor House. Also in town are craft shops, a museum at the train station and the Mount Vic Flicks historic cinema (open Thurs–Sun and school holidays). Mount Victoria has wonderful views of the mountains and many picnic spots and walking trails. 6 km NW.

Blue Mountains National Park In the area east of Blackheath are various waterfalls and lookout points, including the much photographed Hanging Rock (around 9 km N via Ridgewell Rd). The Blue Mountains have been home to Aboriginal people for at least 22 000 years – the Gundungurra people in the north, the Dharug people in the south and the Wiradjuri people in the west. It is an area rich not only with Dreamtime stories but also with over 700 heritage sites that descendants of the original inhabitants continue to protect today. *For other parts of the park see Katoomba and Glenbrook.*

Bacchante Gardens: rhododendrons and azaleas; 1.5 km N. ***Mermaid Cave:*** picturesque rock cave where parts of *Mad Max III* (1985) were filmed; Megalong Rd; 4 km S. ***Shipley Gallery:*** art exhibitions (open weekends); Shipley Rd; 4.6 km S. ***Pulpit Rock Reserve and Lookout:*** sweeping views of Mt Banks and Grose Valley; 6 km E. ***Hargraves Lookout:*** overlooking Megalong Valley; Panorama Point Rd; 7.4 km SW via Shipley Gallery. ***Mt Blackheath Lookout:*** views of Kanimbla Valley; Mount Blackheath Rd; 8.2 km N via Shipley Gallery. ***Megalong Australian Heritage Centre:*** horseriding, adventure tours and tourist farm; (02) 4787 8188; 9 km S. ***Werriberri Trail Rides:*** various trail rides on horseback; bookings (02) 4787 9171.

See also **Blue Mountains, p. 23**

Blayney

Pop. 2748
Map ref. 554 G5

 **97 Adelaide St; (02) 6368 3534; www.blayney.nsw.gov.au**

105.9 Star FM, 576 AM ABC Radio National

Blayney is a farming town in the central tablelands. National Trust–classified buildings and avenues of deciduous trees add a touch of charm to the town, particularly in autumn.

Heritage Park: small wetland area, barbecue facilities and tennis courts; Adelaide St. ***Local craft shops:*** contact visitor centre for details. ***Self-guide heritage walk:*** includes churches and the courthouse; brochure from visitor centre.

Mams Coffee Shoppe: rustic cafe; 129B Adelaide St; (02) 6368 3262.

Blayney Goldfields Motor Inn: comfortable accommodation option; 48 Martha St; (02) 6368 2000.

Carcoar The National Trust–classified town of Carcoar, surrounded by oak trees on the banks of the Belubula River, has a wealth of historic buildings. In early settlement days, convicts and bushrangers caused a lot of trouble in the town. Johnny Gilbert and John O'Meally committed Australia's first daylight bank robbery in 1863 at the Commercial Bank (still standing) on Belubula St. The hold-up was unsuccessful and the robbers fled when a teller fired a shot into the ceiling. 14 km SW.

Millthorpe: National Trust–classified village with quaint shopfronts, art and craft shops, historic churches and a museum with blacksmith's shop and old-style kitchen. Also visit the Golden Memories museum for displays of life in the 19th century; 11 km NW. ***Wind farm:*** with 15 turbines and an interpretive centre, the largest farm of its kind in Australia; 11 km SW. ***Carcoar Dam:*** watersports and camping with picnic and barbecue facilities; 12 km SW. ***Newbridge:*** historic buildings and craft shops; 20 km E. ***Abercrombie Caves:*** cave system in a 220 ha reserve that features the largest natural limestone arch in the Southern Hemisphere. Carols in the Caves is held here in Dec. Various tours are available; contact visitor centre for details. 50 km SE.

See also **Central West, p. 30**

Bombala

Pop. 1204
Map ref. 565 C9 | 581 F7

Platypus Country Tourist Information Centre, Railway Park, Monaro Hwy; (02) 6458 4622; www.platypuscountry.org.au

 87.6 Bombala FM, 103.7 Monaro FM, 810 AM Radio Local

Situated halfway between the Snowy Mountains and the Sapphire Coast, this charming small town has remained largely untouched by time. It is the centre for the surrounding wool, beef, lamb, vegetable and timber industries. The Bombala River is well known as a platypus habitat and for trout fishing. Bombala was considered as a site for Australia's capital city but missed out due to lack of water.

Railway Park There is much to see at Railway Park, including the historic engine shed (open by appt) and the museum of local artefacts and farm implements. The most unusual, though, is Lavender House, home to the oldest lavender association in the country. Lavender House has education facilities, displays on distillation and an array of lavender products such as lavender jams, soaps and oils. Monaro Hwy.

Endeavour Reserve: features a 2 km return walking track to a lookout with views over town; Caveat St. ***Bicentennial Park:*** wetlands and a pleasant river walk; Mahratta St. ***Self-guide historical walk:*** 1 hr walk includes courthouse (1882) and School of Art (1871); leaflet available from visitor centre.

Market: Imperial Hotel, Cnr Forbes and Maybe sts; 1st Sat each month. ***Wool and Wood Festival:*** Jan. ***Celebrate Lavender:*** Jan. ***Celebration of Motorcycles:*** Nov. ***Riverside Festival:*** celebration of diversity of regional crafts and products; Nov. ***Historic Engine Shed Engine Rally:*** even-numbered years, Nov.

Cosmo Cafe: country-style cafe; 133 Maybe St; (02) 6458 3510. ***The Heritage Guest House:*** eatery in heritage house; 121 Maybe St; (02) 6458 4464.

The Heritage Guest House: historic guesthouse; 121 Maybe St; (02) 6458 4464.

Platypus Sanctuary Bombala has one of the densest populations of platypus in NSW. They can be seen here in their natural environment from the Platypus Reserve Viewing Platform. The best times for viewing are at dawn and dusk, when platypus are at their most active, but they can be seen at any time of day. Off Monaro Hwy on the road to Delegate; 3 km S.

Cathcart: charming township with historical town walk and Cathcart Collectables, a fascinating collection of Monaro history; brochure from visitor centre; 14 km NE. ***Myanba Gorge:*** boardwalk and bushwalks through old-growth eucalypt forest with spectacular views of waterfalls, granite boulders and Towamba Valley. Enjoy a picnic or barbecue at the gorge. South East Forest National Park; 20 km SE. ***Delegate:*** scenic town with Early Settlers Hut, believed to be the first dwelling on the Monaro plains, and Platypus Walk and River Walk (leaflets from visitor centre); 36 km SW. ***Scenic drive:*** gold fossicking en route to Bendoc Mines in Victoria; 57 km SW. ***Fly-fishing and trout fishing:*** maps from visitor centre. ***Mountain-biking:*** there are many trails in nearby state forest areas; maps from visitor centre.

See also **South Coast, p. 32**

Bourke

Pop. 2145
Map ref. 561 M5

Exhibition Centre, Kidman Way; (02) 6872 1321; www.backobourke.com.au

 106.5 FM, 585 AM TWB

The saying 'back o' Bourke' has come to mean the middle of nowhere, which is why Bourke is known as the gateway to the real outback. Bourke is a prosperous country town in the centre of thriving wool, cotton and citrus areas on the Darling River. It wasn't always so. Charles Sturt described it as 'unlikely to become the haunt of civilised man'. In 1835 Sir Thomas Mitchell came to the area and, thinking that the local Aboriginal people were a great threat, built himself a sturdy fort out of logs. Fort Bourke, as it became known, encouraged permanent settlement. Bourke quickly became a bustling major town, with Henry Lawson describing it as 'the metropolis of great scrubs'.

Back O' Bourke Exhibition Centre This fascinating modern facility is set among river red gums on the Darling River. It tells the story of the river and the outback from the Dreamtime to 100 years into the future. Visitors walk through colourful displays that re-create the past: paddleboats, early settlers and pastoralists, Afghan cameleers, Cobb & Co coaches, the history of unionism and Aboriginal heritage. The centre also looks at the sustainability of agriculture and the social structures of the outback. Kidman Way; (02) 6872 1321.

Fred Hollows' Grave and Memorial: the eye surgeon and famous humanitarian is buried in the cemetery; Cobar Rd. ***Historic wharf replica:*** reminder of days when Bourke was a busy paddlesteamer port. Take a paddlesteamer ride on the river; Sturt St. ***Mateship Country Tours:*** include historic buildings of Bourke (such as the Carriers Arms Inn, frequented by Henry Lawson) and surrounding citrus and grape farms; bookings at visitor centre; 0428 465 464.

My Litle Cafe: pasta, sandwiches; 27 Oxley St; (02) 6872 1701. ***Morralls Bakery and Cafe:*** sandwiches, pies; 37 Mitchell St; (02) 6872 2086. ***All Rounder Outback Cafe:*** sandwiches; 77 Mertin St; (02) 6872 2086.

Bourke Riverside Motel: heritage-listed accommodation; 3 Mitchell St; (02) 6872 2539. ***Kidman's Camp:*** riverside tourist park; Kidman Way, North Bourke; (02) 6872 1612.

Gundabooka National Park The park is a woodland haven for wildlife. There are over 130 species of bird, including the endangered pink cockatoo, pied honeyeater and painted honeyeater. Kangaroos, euros and endangered bats also make their homes here. Mt Gundabooka offers great walking tracks and a spectacular lookout. The Ngemba people have a history of ceremonial gatherings in the area and their art can be seen in some caves (to book tours, contact National Parks & Wildlife Service (02) 6872 2744). The park also provides excellent camping and barbecue facilities. 74 km s.

Fort Bourke Stockade replica: memorial to Sir Thomas Mitchell, who built the original fort; 20 km sw. ***Mt Oxley:*** home to wedge-tailed eagles and with views of plains from the summit. A key is needed for access to the mountain (collect from visitor centre); 40 km se. ***Comeroo Camel Station:*** working sheep and cattle station with artesian spa and outback experiences including a camel-drawn wagon; accommodation available; (02) 6874 7735; 150 km nw.

See also OUTBACK, p. 36

Bowral

Pop. 11 496
Map ref. 553 C5 | 555 I8 | 565 G2

Southern Highlands Visitor Information Centre, 62–70 Main St, Mittagong; (02) 4871 2888 or 1300 657 559; www.southern-highlands.com.au

102.9 2ST FM, 1431 AM ABC Radio National

Bowral is best known as the home town of 'the boy from Bowral', cricketing legend Sir Donald Bradman. There are Bradman tours, a sporting ground and a museum. Now an up-market tourist town and the commercial centre of the Southern Highlands, Bowral's close proximity to Sydney made it a popular retreat for the wealthy in earlier times. This is still evident today in the magnificent mansions and gardens around town.

Bradman Museum A comprehensive history of cricket is on display, including an oak bat from the 1750s. The Don Bradman memorabilia collection includes the bat he used to score 304 at Headingley in 1934. A cinema plays Bradman footage and newsreels. The Bradman Walk through town takes in significant sites including the Don's two family homes. A leaflet is available from the museum. Bradman Oval; St Jude St; (02) 4862 1247.

The Milk Factory Gallery: art and design exhibition centre and cafe; Station St. ***Bong Bong Street:*** specialty shopping including books and antiques. ***Historic buildings:*** mostly in Wingecarribee and Bendooley sts; leaflet available from visitor centre.

Produce Market: Bowral Public School, Bendooley St; 2nd Sat each month. ***Autumn Gardens in the Southern Highlands:*** throughout region; Apr. ***Tulip Time Festival:*** Sept–Oct. ***Bong Bong Races:*** Nov.

Centennial Vineyards Restaurant: local produce, vineyard views; Woodside, Centennial Rd; (02) 4861 8701. ***Hordern's Restaurant:*** modern Australian; Milton Park Country House Hotel, Horderns Rd; (02) 4861 1522. ***Onesta Cucina:*** modern Italian; Shop 2, The Penders, cnr Station and Wingecarribee sts; (02) 4861 6620.

Milton Park Country House: historic resort-style accommodation; Horderns Rd; (02) 4861 8100. ***Strathburn Cottage:*** award-winning historic bed and breakfast: 85 Bowral St; (02) 4862 3391.

Mittagong This small and appealing town has historic cemeteries and buildings. Lake Alexandra in Queen St is artificial and great for birdwatching and walking. There is a market at the Uniting Church hall on the 3rd Sat each month. 8 km ne.

Mt Gibraltar: bushwalking trails and lookout over Bowral and Mittagong; 2 km n. ***Box Vale Mine walking track:*** begins at the northern end of Welby, passes through old railway tunnel; 12 km n. ***Nattai National Park:*** protects landforms, geological features, catchments and biodiversity in the Sydney Basin. Only low-impact activities are encouraged and there is a 3 km exclusion zone around Lake Burragorang; via Hilltop; 19 km n.

See also SOUTHERN HIGHLANDS, p. 24

Braidwood

Pop. 1108
Map ref. 554 H11 | 565 E5 | 567 K4

National Theatre, Wallace St; (02) 4842 1144; www.visitbraidwood.com.au

94.5 Braidwood FM, 103.5 FM ABC Local

Braidwood is an old gold-rush area that has been declared a historic town by the National Trust. Gold was plentiful here in the 1800s, the largest gold discovery being 170 kilograms in 1869. With the discovery of gold came bushrangers, such as the Clarke Gang and Ben Hall, and Braidwood became one of the most infamous and dangerous towns in the region. The 19th-century buildings have been carefully maintained and restored, and are still in use. The town appears to be from a bygone era, which has come in handy for film producers, who have found Braidwood a perfect setting for movies such as *Ned Kelly* (1969), *The Year My Voice Broke* (1986) and *On Our Selection* (1994).

Braidwood Museum Built of local granite and originally the Royal Mail Hotel, the museum houses over 2100 artefacts and 900 photographs. On display are exhibits of Aboriginal history, goldmining, the armour worn by Mick Jagger in *Ned Kelly*, a machinery shed and a library of local records, newspapers and family histories. An unusual collection is from the Namchong family, who came here from China during the gold rush and became traders in town from the 1870s to the 1990s. Open Fri–Mon, daily during school holidays; Wallace St; (02) 4842 2310.

Galleries and craft and antique shops: details from visitor centre. ***Tallaganda Heritage Trail and scenic drive:*** tours of historic buildings such as the Royal Mail Hotel and St Andrew's Church; leaflet from visitor centre.

Picnic Race Meeting: Feb. ***Braidwood Heritage Festival:*** May. ***Music at the Creek:*** Nov. ***The Quilt Event:*** Nov.

Braidwood Bakery: pies, sandwiches; 99 Wallace St; (02) 4842 2541. ***Braidwood Deli:*** sandwiches, pies; Shop 1, 91 Wallace St; (02) 4842 1201.

Araluen Old Courthouse: fascinating historical B&B; The Old Courthouse, Araluen Rd; (02) 4846 4053. ***Mona:*** romantic retreat; 140 Little River Rd; (02) 4842 1288.

Monga National Park This park features a boardwalk through rainforest areas dating back to the ancient Gondwana period. Penance Grove, a small pocket of rainforest, is filled with ancient plumwood trees and tree ferns. Maps at visitor centre. Access via Kings Hwy; 20 km SE.

The Big Hole and Marble Arch The Big Hole is thought to have formed when overlying sandstone collapsed into a subterranean limestone cavern creating an impressive chasm 96 m deep and 50 m wide. Wildlife in the area includes native birds, echidnas, wallabies, wombats and tiger quolls. Marble Arch is a narrow canyon 3–4 m wide and 25 m deep. It is over 1 km in length and bands of marble are visible along the walls. There are caves along the way, but special permission is required to enter some of them. Some are very dark and require a torch, so it is best to check with the NSW National Parks & Wildlife Service if you intend to explore. Inquiries (02) 4887 7270; near Gundillion; 45 km S.

Scenic drives: rugged countryside; brochure from visitor centre. ***Fishing:*** good trout fishing, especially in the Mongarlowe and Shoalhaven rivers; details from visitor centre.

See also CAPITAL COUNTRY, p. 31

Brewarrina

Pop. 1123
Map ref. 561 O5

Bathurst St; (02) 6830 5152; www.breshire.com

106.5 2CUZ FM, 657 AM ABC Local

This charming outback town on the banks of the Barwon River is affectionately known as Bre. It was developed in the 1860s as a river crossing for stock, but later thrived because of its position on a Cobb & Co route. Brewarrina was once a meeting place for Aboriginal tribes, with sacred sites including burial and ceremonial grounds, pointing to a culture that revolved around the river. By far the most impressive relics are the ancient stone fish traps of the Barwon River estimated to be 40 000 years old – among the oldest constructions in the world.

Aboriginal fish traps The traditional Aboriginal story states that the traps were built during a drought. Gurrungga, the water hole at Brewarrina, dried up and this opportunity was used to build the traps in the dry bed. The Ngemba people, who were facing famine, were never hungry again. Anthropologists claim the traps are impressive evidence of early engineering, river hydrology and knowledge of fish biology. Around 500 m long, these traps relied on currents to sweep fish inside, where they would be confined when the water level dropped. Thousands of years ago, the traps formed the centrepiece of a seasonal festival, attended by up to 50 000 people from Aboriginal groups along the east coast. At night they held corroborees and shared stories around campfires. The traps preserve the memory of these ancient times. Guided tours are available from visitor centre.

Aboriginal Cultural Museum: displays on aspects of Aboriginal life, from tales of the Dreamtime to the present; open Mon–Fri; Bathurst St. ***Barwon Bridge:*** one of two surviving examples of the first series of lift span bridges in the state (1889); Bridge Rd. ***Wildlife park:*** native fauna in bush setting; Doyle St. ***Self-guide drive:*** 19th-century buildings; brochure from visitor centre.

Barwon River Rodeo: Easter. ***Brewarrina Races:*** May.

Bokhara Hutz: self-contained accommodation; Bokhara Plains; (02) 6874 4921.

Narran Lake Nature Reserve: wetlands and a breeding ground for native and migratory birds; access permits from visitor centre; 50 km NE. ***Culgoa National Park:*** wildlife unique to the western flood plains including falcons, striped-faced dunnat and pied bats. Information from visitor centre; 100 km N. ***Fishing:*** plentiful Murray cod in the Barwon River. ***Start of Darling River Run:*** self-drive tour; brochure from visitor centre.

See also OUTBACK, p. 36

Broken Hill

see inset box on next page

Bulahdelah

Pop. 1092
Map ref. 549 H4 | 550 C10 | 555 N3

Cnr Pacific Hwy and Crawford St; (02) 4997 4981; www.greatlakes.org.au

101.5 FM Great Lakes, 1233 AM ABC Local

Bulahdelah is a pretty town at the foot of Bulahdelah Mountain (known to locals as Alum Mountain because of the alunite that was mined here). Surrounded by rainforests and the beautiful Myall Lakes, it is a popular destination for bushwalkers and watersports enthusiasts.

Bulahdelah Mountain Park A park of contrasts, with meandering walking trails taking in tall forest, rare orchids in spring and the remains of mining machinery. There are picnic and barbecue facilities and a lookout over Bulahdelah and the Myall Lakes. Meade St.

Bulahdelah Court House: museum featuring Bulahdelah's logging past, with cells out the back; open Sat mornings or by appt; Cnr Crawford and Anne sts; bookings on (02) 6597 4838.

Market: beside visitor centre, Crawford St; 1st Sat each month. ***Bulahdelah Music Festival:*** Jan. ***The Bass Bash:*** fishing festival; Feb. ***Junior Rodeo:*** July. ***Bulahdelah Show:*** campdraft and rodeo; Sept.

continued on p. 51

BROKEN HILL

Pop. 18 856

Map ref. 560 B10 | 603 P1 | 605 O8

Cnr Blende and Bromide sts; (08) 8080 3560; www.visitbrokenhill.com.au

106.9 Hill FM, 999 AM ABC Local

In the vast, arid lands of far-western New South Wales, Broken Hill was first an intermittent home for the Willyama people; its lack of water made it impossible to settle permanently. When Charles Sturt encountered the area while searching for an inland sea, he described it as some of the most barren and desolate land he had ever seen. Enthusiasm was soon generated, however, with the discovery of silver and Broken Hill was born. A syndicate of seven men quickly bought much of the land and in 1885 they discovered the world's largest silver-lead-zinc lodes. Later that same year they decided to form a company and float shares. That company was Broken Hill Proprietary (BHP), now BHP Billiton, the largest mining company in the world. Referred to as the 'Silver City', Broken Hill is also the centre of the 16-million-hectare West Darling pastoral industry, which has 1.75 million merino sheep surrounded by a 600-kilometre dog-proof fence. As you would expect of a hot, arid mining town, Broken Hill has many pubs. Note that Broken Hill operates on Central Standard Time, half an hour behind the rest of New South Wales.

Line of Lode Miners Memorial and Cafe: perched on top of the mullock heap at the centre of town, and delving into Broken Hill's mining heritage; Federation Way. ***Railway, Mineral and Train Museum:*** displays on old mining and rail services. Also incorporates the Hospital Museum and Migrant Museum; open 10am–3pm; Bromide St. ***Albert Kersten Mining & Minerals Museum:*** displays of minerals, mining specimens and a silver tree; Cnr Crystal and Bromide sts; (08) 8080 3560. ***Photographic Recollections:*** more than 600 photos telling Broken Hill's history; Eyre St. ***White's Mineral Art and Mining Museum:*** walk-in mine and mining models; Allendale St. ***Joe Keenan's Lookout:*** view of town and mining dumps; Marks St. ***Muslim Mosque:*** built by Afghan community in 1891; open 2–4pm Sun or by appt; Buck St. ***Zinc Twin Lakes:*** popular picnic spot at lakes used as a water source for mines; off Wentworth Rd, South Broken Hill. ***Art galleries:*** over 30 in town including Pro Hart Gallery, Wyman St; Jack Absalom Gallery, Chapple St; and Broken Hill Regional Art Gallery, Argent St, the oldest regional gallery in the state. ***Broken Hill Walk and Silver Trail self-guide historical town drive:*** leaflets from visitor centre.

St Patrick's Race Day: horseraces; Mar/Apr. ***Outback 4x4 Challenge:*** 4WD rally; May. ***Red Desert Live!:*** music festival; Oct.

Bells Milk Bar: diner serving delicious milkshakes; 160 Patton St; (08) 8087 5380. ***Broken Hill Musicians Club:*** cafe, community club; 276 Crystal St; (08) 8088 1777.

Royal Exchange Hotel: Art Deco building; 320 Argent St; (08) 8087 2308 or 1800 670 160. ***The Imperial Fine Accommodation:*** luxury heritage-listed retreat; 88 Oxide St; (08) 8087 7444.

Silverton This National Trust–classified town was established when silver chloride was found 27 km NW of Broken Hill in 1883.

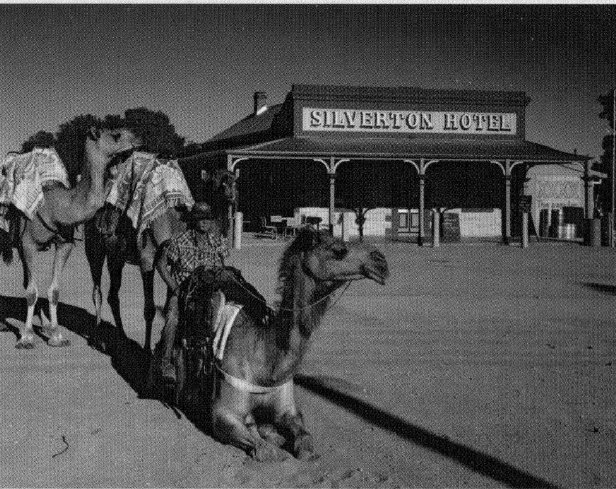

[CAMELS OUTSIDE THE SILVERTON HOTEL]

It now has less than 100 inhabitants and is surrounded by stark, arid plains, making it popular with filmmakers wanting an outback setting. *Mad Max II* (1981), *Razorback* (1983), *Young Einstein* (1988), *The Adventures of Priscilla, Queen of the Desert* (1994) and *Dirty Deeds* (2001) have all been filmed here, and the Silverton Hotel displays photographs of the film sets on its walls. In a tin shed, up the hill from the Silverton Hotel, a Mad Max Museum was opened in 2010 by Adrian Bennett – Mad Max number one fan. Take the Silverton Heritage Walking Trail (leaflet from visitor centre) and visit Silverton Gaol Museum. Silverton Camel Farm offers 15 min rides around the outskirts of town and 3-day safaris. Mundi Mundi Plain Lookout, a further 10 km N, affords views of the desolate yet awe-inspiring landscape. Daydream Mine, 13 km NE, operated in the 1880s and is now open for 1 hr guided tours, 10am–3.30pm daily.

Mutawintji National Park Bushwalks lead through a rugged terrain of colourful gorges, rockpools and creek beds, while Aboriginal rock engravings and paintings tell stories of creation. The land was returned to its traditional owners in 1998, and a historic site in the centre containing a vast gallery of rock art is accessed by tour only. There are three tour companies: Corner Country Adventures; (08) 8087 6956. Silver City Tours; (08) 8087 6956. Tri-State Safaris; (08) 8088 2389; 130 km NE.

Royal Flying Doctor Service base and visitor centre: headquarters, radio room and aircraft hangar open to visitors; at airport; 10 km S. ***Living Desert:*** magnificent sandstone sculptures set on a hillside (particularly striking at sunrise and sunset) and walking trails through the mulga-dotted landscape; leaflet available from visitor centre; Nine Mile Rd; 11 km N. ***Dingo Fence:*** longest fence in the world at 5300 km from Jimbour in Queensland to the Great Australian Bight. Originally constructed in the 1880s to halt rabbit invasion, now maintained to keep dingoes out of sheep grazing areas; 150 km NW. ***Tours:*** for 4WD, camel safaris, outback excursions and scenic flights; bookings at visitor centre.

See also **OUTBACK, p. 36**

Detours Cafe: homemade hamburgers; 82 Stroud St; (02) 4997 4755. ***Myalla Cafe:*** traditional cafe fare; 84 Stroud St; (02) 4997 4900.

Bombah Point Eco Cottages: luxurious self-contained cottages; 969 Bombah Point Rd; (02) 4997 4401. ***Peacehaven Country Cottages:*** self-contained cottages; 353 Upper Myall Rd; (02) 4997 8247.

Myall Lakes National Park Unspoiled coastal lakes and 40 km of beaches make this national park one of the most visited in the state. It is ideal for all types of watersports – canoe and houseboat hire is available, and Broughton Island, 2 km offshore, is a popular spot for diving. Enjoy the bushwalks and campsites set in the rainforest, heathlands and eucalypt forest. 12 km E.

Bulahdelah State Forest: with scenic picnic area and walking trails along old mining trolley lines, and one of the tallest trees in NSW – the 84 m flooded gum (*Eucalyptus grandis*); off The Lakes Way; 14 km N. ***Wootton:*** charming small town with a 6 hr rainforest walk along an old timber railway; 15 km N. ***Sugar Creek Toymakers:*** fine wooden toys and hand-painted dolls; (02) 4997 6142; 31 km E. ***Seal Rocks:*** fishing village with seals sometimes resting on the offshore rocks. Grey nurse sharks breed in underwater caves and whales pass by June–Aug. Sugarloaf Point Lighthouse (1875) has a lookout tower, and there are pleasant beaches and camping areas; 40 km E. ***Wallingat National Park:*** walking trails and picnic facilities. Stop at Whoota Whoota Lookout for sweeping views of forest, coast and lakes; 43 km NE.

See also HOLIDAY COAST, p. 27

Bundanoon

Pop. 2035
Map ref. 553 A6 | 555 I9 | 565 G3

Southern Highlands Visitor Information Centre, 62–70 Main St, Mittagong; (02) 4871 2888 or 1300 657 559; www.southern-highlands.com.au

97.3 FM ABC Local, 107.1 Highlands FM

Bundanoon is a quiet village with a European feel thanks to its green, tree-lined avenues. The first European to investigate the district was ex-convict John Wilson, who was sent by Governor Hunter to collect information on the area that would discourage Sydney convicts from trying to escape in this direction. Today the sleepy town is a popular yet unspoiled tourist destination with many delightful guesthouses and a health resort.

Craft shops and art galleries: several featuring local work; contact visitor centre for details. ***Drive and walk to several lookouts:*** map from visitor centre.

Market: Memorial Hall, Railway Ave; 1st Sun each month. ***Bundanoon is Brigadoon:*** highland gathering; Apr. ***Village Garden Ramble:*** Oct.

Ye Olde Bicycle Shoppe: popular cafe and bike hire; 9 Church St; (02) 4883 6043.

Tree Tops Country Guesthouse: Edwardian-style guesthouse; 101 Railway Ave; (02) 4883 6372.

Morton National Park, Bundanoon section This section of the park has stunning views, lookouts and walking trails. The park consists mainly of rainforest and eucalypts and is home to myriad native fauna including wallabies, potoroos and bush rats. See glow worms at night in the remarkable Glow Worm Glen. 1 km S. *For other sections of the park see Nowra, Robertson and Ulladulla.*

Exeter: quaint village with an English feel; 7 km N.

See also SOUTHERN HIGHLANDS, p. 24

Byron Bay

see inset box on next page

Camden

Pop. 3179
Map ref. 544 G11 | 553 E1 | 555 J7 | 565 H1

John Oxley Cottage, Camden Valley Way; (02) 4658 1370; www.visitcamden.com.au

C91.3 FM Campbelltown, 684 AM ABC Local

Camden, in a picturesque setting on the Nepean River just south-west of Sydney, was once a hunting ground of the Gundungurra people, who called it 'Benkennie', meaning 'dry land'. Governor Macquarie sent men to kill or imprison the Aboriginal people in 1816, and although records are poor, the brutal mission had some success. European settlement began after eight cattle wandered off four months after the First Fleet landed. They were not seen again until 1795 when it was discovered their number had grown to more than 40. The site on which they were found was named Cowpasture Plains, but was later changed to Camden. Camden was home to John and Elizabeth Macarthur, pioneers of the Australian wool industry, and also the first in Australia to grow tobacco, use mechanical irrigation, produce wine of respectable quality and quantity and make brandy. The Macarthurs sent thousands of vines to the Barossa Valley and are thereby credited with helping to start South Australia's wine industry.

Self-guide walk and scenic drive: includes historic buildings such as the Macarthur Camden Estate, St John the Evangelist Church and Kirkham Stables; brochure from visitor centre.

Produce Market: Cnr Exeter and Mitchell sts; 2nd and 4th Sat each month. ***Craft and Fine Food Market:*** Camden Showground; 3rd Sat each month (except Jan). ***Food, Wine and Music Festival:*** Sept. ***Camden House Open Weekend:*** Nov.

Enzo Italian Restaurant: modern Italian; 39 John St; (02) 4655 9260. ***Impassion:*** modern Australian; Level 1, 100 Argyle St; (02) 4655 8163.

Camden Valley Inn: stylish, comfortable accommodation; Remembrance Dr (Old Hume Hwy); (02) 4655 8413.

Camden Museum of Aviation This privately owned museum has the largest specialist aircraft collection in Australia. Where possible, the aircraft have been painstakingly restored (to a standard allowing them to taxi along a runway but not necessarily fly), with accurate wartime markings and camouflage colours carefully researched through service records and photographs. Open Sun and public holidays; (02) 4648 2419; 3 km NE.

Struggletown Fine Arts Complex: gallery featuring stained glass, pottery, traditional art and a restaurant; 3 km N. ***Belgenny Farm:*** includes Belgenny Cottage (1820) and the oldest surviving collection of farm buildings in Australia; (02) 4655 9651; 6 km SE. ***Camden Aerodrome:*** ballooning, gliding and scenic flights with vintage aircraft on display; (02) 4655 8064; 3 km NW. ***Cobbity:*** historic rural village with market 1st Sat each month; 11 km NW. ***The Oaks:*** small town in open countryside featuring the slab-built St Matthew's Church and Wollondilly Heritage Centre, a social history museum; 16 km W. ***Burragorang Lookout:*** views over Lake Burragorang; 24 km w. ***Yerranderie:*** this fascinating old silver-mining town can be reached by normal vehicle only in dry conditions, otherwise only by 4WD or plane; 40 km w. ***Wineries:*** several in the area; contact visitor centre for map; Gledswood (10 km N) features a working colonial farm.

See also SOUTHERN HIGHLANDS, p. 24

Campbelltown

Pop. 147 460
Map ref. 544 H11 | 553 F1 | 555 J7 | 565 H1

15 Old Menangle Rd; (02) 4645 4921; www.visitmacarthur.com.au

C91.3 FM Campbelltown, 603 AM ABC Radio National

Campbelltown was founded in 1820 by Governor Macquarie and named after his wife, Elizabeth Campbell. While the town is being engulfed by the urban sprawl of Sydney, it manages to combine the best of two worlds, enjoying the convenience of city living with the rustic charm of 19th-century buildings. It is also the location of the legend of Fisher's ghost. In 1826 an ex-convict, Frederick Fisher, disappeared. Another ex-convict, George Worrell, claimed that Fisher had left town, leaving him in charge of Fisher's farm. A farmer claimed to have seen the ghost of Fisher pointing at the creek bank where his body was subsequently found. Worrell was tried and hanged for Fisher's murder.

Campbelltown Arts Centre Visitors can see, explore and participate in art-creation at this interactive centre. Exhibitions are diverse and include local, regional, national and international shows of art and craft. Behind the gallery is a sculpture garden established in 2001 as a Centenary of Federation project. New permanent sculptures are added to the garden on a regular basis. Adjacent to the gallery is the Koshigaya-tei Japanese Teahouse and Garden, a bicentennial gift to the people of Campbelltown from its sister city, Koshigaya. The garden is a peaceful area with a waterfall, koi pond and timber bridge, perfect for picnics and tranquil contemplation. Art Gallery Rd; (02) 4645 4100.

Campbelltown Visitor Information Centre: formerly St Patrick's, the first Catholic school in Australia, displays include early world maps, desks, inkwells, canes and curriculums (1840); Old Menangle Rd. ***Stables Museum:*** display of historic farm equipment and household goods on 1st, 3rd and 5th Sun each month; Lithgow St. ***Self-guide heritage walks:*** take in numerous historic buildings including St Peter's Church (1823), Emily Cottage (1840) and Fisher's Ghost restaurant, formerly Kendall's Millhouse (1844); leaflet from visitor centre.

Ingleburn Festival: Mar. ***Autumn Harvest Food and Wine Fair:*** Mar/Apr. ***Festival of Fisher's Ghost:*** Nov.

The Barn Restaurant: modern Australian; 12 Queen St; (02) 4625 4521.

Campbelltown Colonial Motor Inn: heritage-listed sandstone motel; 20 Queen St; (02) 4625 2345. ***Rydges Hotel and Resort:*** contemporary hotel; Old Menangle Rd; (02) 4645 0500.

Mt Annan Botanic Garden Australia's largest botanic garden is a striking 400 ha garden with 20 km of walking trails. Attractions include two ornamental lakes with picnic areas, a nursery, an arboretum, themed gardens, the rare and endangered plants garden and the banksia garden. The botanic garden is a haven to over 160 bird species and mammals such as the wallaroo and swamp wallaby. The human sundial allows visitors to tell the time by standing in its centre and raising their arms. Guided tours are available; (02) 4648 2477; 3 km w.

Sugarloaf Horse Centre: heritage-listed site with amazing views and trail-rides for all skill levels; (02) 4625 9565; 2.2 km sw. ***Macarthur Centre for Sustainable Living:*** showcasing sustainable homes, gardens and lifestyles. Regular workshops are also held throughout the year; (02) 4647 9828; 3 km w. ***Steam and Machinery Museum:*** history of Australia's working past, including interactive displays; Menangle Rd; 5 km sw. ***Menangle:*** small town featuring a historic homestead (1834) and The Store (1904), an old-style country store with everything from antiques to ice-creams, and the Menangle Railway Bridge (1863), the colony's first iron bridge; 9 km sw. ***Eschol Park House:*** grand colonial home (1820) set in landscaped gardens; 15 km N. ***Appin:*** historic coalmining town with a monument to Hume and Hovell. Also weekend markets in 10 locations (leaflet available from visitor centre) and a celebration of Scottish links through the Highland Gathering and Pioneer Festival each Nov; 16 km s.

See also SOUTHERN HIGHLANDS, p. 24

Canowindra

Pop. 1501
Map ref. 554 F5

Age of Fishes Museum, Gaskill St; (02) 6344 1008; www.ageoffishes.org.au

104.3 FM ABC Radio National, 105.1 2GZ FM

Canowindra, meaning 'home' in the Wiradjuri language, is in the Lachlan Valley with sandstone mountains to the west and the old volcano, Mount Candobolas, to the north-east. Ben Hall and his gang struck in the town twice in 1863. During the first visit they robbed two homesteads and then forced residents and local police into Robinson's Inn (now the Royal Hotel), where they held an impromptu and compulsory two-day party. Two weeks later they returned and held a similar three-day party, reportedly at their own expense. Today Canowindra is still a genuine old-style country town with a National Trust–classified main street (Gaskill St) that follows the crooked path of an old bullock track. It is now known for balloons, fossils and boutique cellar doors. Canowindra calls itself the 'balloon capital of Australia' because there are more hot-air balloon flights here than anywhere else in the country. In 1956, 3500 fish fossils over 360 million years old were found in the area. Another major dig took place in 1993.

Age of Fishes Museum Long before dinosaurs walked the earth bizarre fish populated local rivers, including fish with armoured shells, fish with lungs and fish with jaws like crocodiles. The museum displays many of the fossils from the Devonian era found during the 1956 and 1993 digs along with information about the digs. There are also live aquarium displays and re-creations of life in the Devonian period. Gaskill St.

Historical museum: local history displays and agricultural equipment; weekends or by appt; Gaskill St. ***Hot-air balloon rides:*** over picturesque Lachlan Valley; Mar–Nov (weather permitting); details at visitor centre. ***Historical tourist drive and riverbank self-guide walks:*** include historic buildings of Gaskill St; brochure available from visitor centre.

Springfest: wine and art; Oct.

Taste Canowindra: lunches and cakes; 42 Ferguson St; (02) 6344 2332. ***Tom's Waterhole Winery:*** ploughman's lunches; 'Felton', Longs Corner Rd; (02) 6344 1819.

Everview Retreat: luxury self-contained cottages; 72 Cultowa La; (02) 6344 3116. ***Falls Vineyard Retreat:*** luxury homestead guesthouse; Belubula Way; (02) 6344 1293.

continued on p. 54

BYRON BAY

Pop. 4978

Map ref. 551 H4 | 557 O3 | 653 N12

80 Jonson St; (02) 6680 8558; www.visitbyronbay.com

92.2 FM North Coast Radio, 720 AM ABC Local

Byron Bay's excellent beaches, laid-back feel and great weather have made it a long-time destination for surfers, backpackers and alternative-lifestylers. A forward-thinking council moved early to ban all drive-in takeaway food outlets and all buildings more than three storeys high. Now people from all walks of life, including celebrities, flock to Byron Bay for whale-watching, surfing, swimming and relaxing. The town centre has an array of restaurants and cafes, shops selling discount surfboards, New Age products and interesting clothing, and galleries with everything from handmade jewellery to timber furniture. Day spas, yoga centres and natural therapy parlours offer expert pampering.

Tours and activities: everything from diving and sea-kayaking to surf lessons, skydiving and gliding; contact visitor centre for details. ***Beach Hotel:*** well-established venue for live music and dining, with a beer garden that has ocean views; cnr Jonson and Bay sts; (02) 6685 6402. ***Beaches:*** safe swimming at Main Beach (patrolled in summer school holidays and at Easter), and surfing at Clarkes Beach and the Pass; look out for 'sandologist' Steve Machell and his fabulous sand sculptures on Main or Clarkes beaches; there's more surfing at Wategos, child-friendly swimming at secluded Little Wategos, dogs are welcome at Belongil and Brunswick Heads, or go horseriding or beach-fishing at Seven Mile Beach stretching towards Lennox Head. ***Health and wellbeing:*** relax and unwind with a massage at Byron's numerous spas, including the tropical day spa at Buddha Gardens Balinese Spa; 15 Gordon St; (02) 6680 7844. Try massage therapy and float tanks Relax Haven at Belongil Beachhouse, 25 Childe St; (02) 6685 7868. ***Byron Bay Arts & Industry Park:*** arts precinct 3 km from town with artists', jewellers' and sculptors' studios, galleries, cafes and restaurants; Artist Trail guide from visitor centre.

Farmers Market: Thurs mornings; Butler St Reserve. ***East Coast Blues and Roots Festival:*** Easter. ***Byron Bay Writers Festival:*** July/Aug. ***Taste of Byron Food Fest:*** Sept. ***Buzz–Byron Bay Film Fest:*** Oct.

Byron Beach Cafe: modern Australian; Clarkes Beach, Lawson St; (02) 6685 8400. ***Fishmongers Cafe:*** gourmet fish and chips; Bay La; (02) 6680 8080. ***Italian at the Pacific:*** contemporary Italian with great views; Bay St; (02) 6680 7055. ***Red Ginger:*** eat dumplings in the store's front window; 2/111 Jonson St; (02) 6680 9779. ***The Balcony Bar & Restaurant:*** bohemian charm, world menu; cnr Lawson and Jonson sts; (02) 6680 9666. ***Twisted Sista Cafe:*** light meals and desserts; 1/4 Lawson St; (02) 6685 6810.

Amber Gardens Guesthouse: small, reasonably priced guesthouse; 66 Plantation Dr, Ewingsdale; (02) 6684 8215. ***Byron Retreat:*** contemporary self-contained accommodation; Ewingsdale (address given on inquiry); 1300 660 422. ***Rae's on Watego's:*** iconic Byron Bay resort; 8 Marine Pde, Wategos Beach; (02) 6685 5366. ***The Garden Burées:*** Balinese-inspired bungalows and chalets; 17 Gordon St; (02) 6685 5390.

Cape Byron This headland forms part of the world's oldest caldera – the rim of an enormous extinct volcano (the centre is Mt Warning). It is the easternmost point on the mainland and provides breathtaking views up and down the coast. Dolphins can be seen year-round and humpback whales migrate up the coast June–July and back down Sept–Oct. Cape Byron Lighthouse, the 22 m structure completed in 1901, houses a visitor centre with displays of the area's cultural and natural history. 3 km SE.

Julian Rocks Aquatic Reserve: protects 450 underwater species and is great for diving; 3 km S. ***Broken Head Nature Reserve:*** rainforest, secluded beaches and dolphin-watching; (02) 6627 0200; 9 km S. ***Bangalow:*** scenic village with antique shops, arts and crafts, walking tracks and a popular market on 4th Sun each month; 10 km SW.

See also TROPICAL NORTH COAST, p. 28

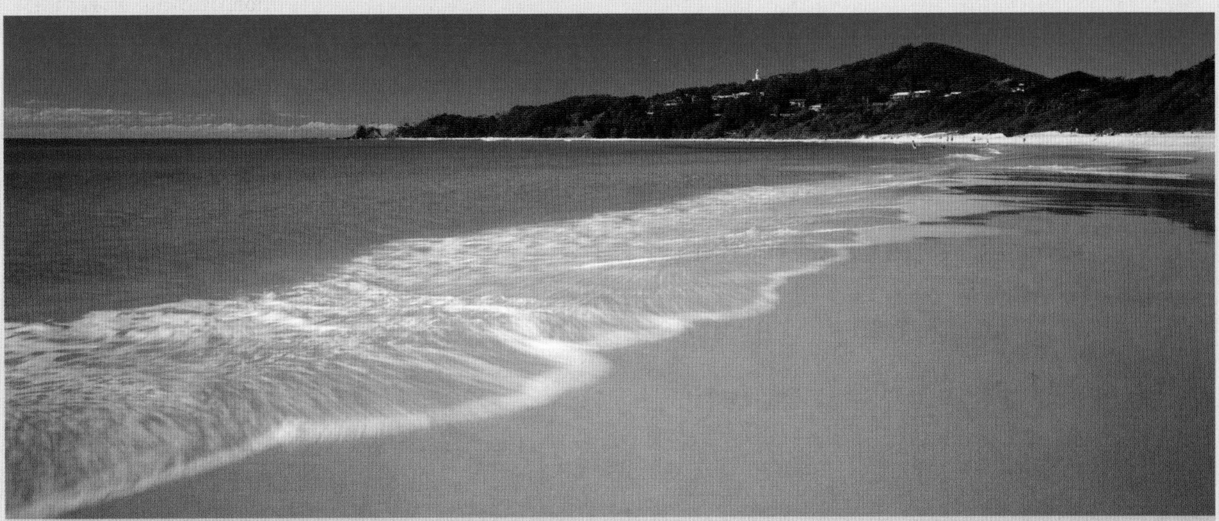

[BYRON BAY'S MAIN BEACH]

CESSNOCK

Pop. 18 318

Map ref. 543 E2 | 549 A8 | 555 L4

455 Wine Country Dr, Pokolbin; (02) 4990 0900; www.winecountry.com.au

96.5 CHR-FM Community Radio, 1044 AM ABC Local

Cessnock was once a coalmining town, but that was before the Hunter Valley, of which Cessnock is at the southern end, began producing some of Australia's best wines. A visit to Cessnock almost guarantees relaxation as it focuses on fresh produce and wine in the spectacular scenery of rainforests and green valleys.

Galleries and antique and craft shops: brochures from visitor centre.

Harvest Festival: throughout region; Mar–Apr. ***Hunter Semillon and Seafood Festival:*** Apr. ***Jazz in the Vines:*** Oct. ***Opera in the Vineyards:*** Oct.

Swill N Grill: international; The Australia Hotel, 136 Wollombi Rd; (02) 4990 1256. ***The Kurrajong Restaurant:*** pub bistro; Cessnock Hotel, 234 Wollombi Rd; (02) 4990 1002. ***Esca Bimbadgen:*** modern Australian; Bimbadgen Estate, 790 McDonalds Rd, Pokolbin; (02) 4998 7585.

Cessnock Heritage Inn: comfortable, affordable B&B; 167 Vincent St; (02) 4991 2744. ***Greta Main Pay Office B&B:*** guesthouse in historic surrounds; 988 Wollombi Rd, Greta Main; (02) 4998 1703. ***Peppers Guest House:*** 48 deluxe rooms; Ekerts Rd, Pokolbin; (02) 4993 8999.

Hunter Valley Wine Region Over 120 wineries, most open for cellar-door tastings. Older establishments include Tyrrell's (Broke Rd, Pokolbin; (02) 9889 4450), while newcomers include Tulloch Wines (Cnr McDonalds and DeBeyers rds, Pokolbin; (02) 4998 7580). Bimbadgen Estate is a lovely place to stop for lunch (790 McDonalds Rd, Pokolbin; (02) 4998 7585). Further details from visitor centre.

Watagan National Park: this park features several lookout points over mountains and valleys, including the lookout at Gap Creek Falls that reveals rainforest gullies of magnificent red cedar and Illawarra flame trees. There are many scenic rainforest walks along the creek, which is ideal for swimming. Some walks lead to picnic and barbecue facilities at Heaton, Hunter and McLean's lookouts and the serene Boarding House Dam picnic area; 33 km SE. ***Hunter Valley Zoo:*** hands-on zoo where most animals (native and introduced) can be patted and fed; 7 km NW. ***Lovedale:*** home of the Long Lunch (food, wine and music), held at various vineyards in May; 10 km N. ***Bimbadeen Lookout:*** spectacular views over Hunter Valley; 10 km E. ***Hunter Valley Cheese Factory:*** factory tours, tastings and sales; McGuigan Cellars Complex, McDonalds Rd; 13 km NW. ***Richmond Vale Railway Museum:*** rail and mining museum with steam-train rides and John Brown's Richmond Main Colliery, once the largest shaft mine in the Southern Hemisphere; 17 km NE. ***Wollombi:*** picturesque village with a wealth of historic sandstone buildings. Tour the Aboriginal cave paintings (inquire at general store). Visit Undercliff Winery and Studio (for etchings); 29 km SW. ***Galleries and craft shops:*** arts and crafts by local artists; maps from visitor centre.

[OAK BARRELS AT ALLANDALE WINERY, LOWER HUNTER VALLEY]

See also HUNTER VALLEY & COAST, p. 26

Gondwana Dreaming Historical Fossil Digs Tours of 1–6 days can be arranged to go on real archaeological digs led by palaeontologists to learn about and possibly find fossils. The tours promote hands-on learning and focus on the area as a whole, including local flora and fauna. This program funds ongoing scientific research. Bookings on (02) 6285 1872.

Wineries and vineyards: cellar-door tastings and tours; maps from visitor centre.

See also CENTRAL WEST, p. 30

Casino

Pop. 9398

Map ref. 551 E5 | 557 N3 | 653 M12

86 Centre St; (02) 6662 3566; www.richmondvalley.nsw.gov.au

107.9 COW FM, 738 AM ABC Local

Beside the Richmond River in the north-east of the state, Casino is a town of grand old buildings and magnificent parklands. Casino is named after the beautiful Italian town of Monte Cassino and is known as the beef capital of Australia. More than 12 000 cattle are sold each year at the Casino Livestock Selling Centre.

Jabiru Geneebeinga Wetlands These parklands have picnic facilities and are home to native bird species and wildlife including the jabiru (black-necked stork), egret and black swan. The park is circled by a mini-railway that operates each Sun. West St.

Casino Folk Museum: locally significant documents and photographs; open Mon, Wed afternoons and Sun mornings; Walker St. ***Self-guide heritage and scenic walks and drives:*** include Bicentennial Mural, St Mark's Church of England and Cecil Hotel; maps available from visitor centre.

Flower and Art Show: Mar. ***Beef Week Festival:*** May. ***Gold Cup:*** horseracing; May. ***Primex:*** primary industry exhibition; June.

Clydesdale Motel and Steak Barn: steakhouse; Bruxner Hwy; (02) 6662 5982.

Clydesdale Motel and Steak Barn: unpretentious country accommodation; Bruxner Hwy; (02) 6662 5982.

Fossicking: gold, labradorite and quartz (both smoky and clear types); maps from visitor centre. ***Freshwater fishing:*** Cookes Weir and Richmond River are popular fishing spots; maps from visitor centre.

See also Tropical North Coast, p. 28

Cobar

Pop. 4127
Map ref. 561 L9

Great Cobar Heritage Centre, Barrier Hwy; (02) 6836 2448; www.cobar.nsw.gov.au

103.7 Zoo FM, 810 AM ABC Local

Cobar got its name from the Aboriginal word 'kubbur', meaning 'burnt earth'. The story goes that three European settlers were camping at Kubbur watering hole near where Cobar is today when they noticed the unusual colour of the water. They took samples and showed them to Mrs Kruge, the publican at the Gilgunnia Pub, who identified them as copper ore. From this discovery, the Cobar mining industry was born. The mines made Cobar so prosperous that at one point the town had a population of 20 000 and its own stock exchange. Mines still operate, including the CSA Copper Mine which is the deepest in Australia at 3 km. Today Cobar is a surprisingly green and picturesque outback town.

Great Cobar Outback Heritage Centre This centre has displays on the local mining of copper, gold and silver-lead-zinc, an authentic re-creation of a local woolshed and displays on Aboriginal culture. Learn about the chronic water shortages in the early days of European settlement and the bush skills the settlers needed to survive the harsh environment. The Centenary of Federation Walking Track begins here and is a 2 hr scenic walk past mines and a slag dump. Barrier Hwy; (02) 6836 2448.

Commonwealth Meteorological Station: visitors can view the radar tracking process and the launching of weather balloons at 9.15am and 3.15pm daily (Eastern Standard Time). Louth Rd; inquiries (02) 6836 2149. ***Golden Walk:*** tour the operating Peak Gold Mine or view from observation deck; Kidman Way; (02) 6830 2265. ***Self-guide heritage walks and heritage bus tours:*** historic buildings including the courthouse and the Great Western Hotel (with the longest iron-lace verandah in NSW), and mining and agricultural sites around town; brochure from visitor centre.

Festival of the Miner's Ghost: Oct.

Copper City Restaurant: modern Australian; 40 Lewis St; (02) 6836 1022. ***Empire Hotel:*** steaks, pub food; 6 Barton St; (02) 6836 2725.

Mt Grenfell Historic Site The 5 km Ngiyambaa Walkabout leads visitors on a scenic tour with breathtaking views of the Cobar area. There are hundreds of Aboriginal stencils and paintings of great cultural significance in spectacular reds, yellows and ochres on rock overhangs along the trail. Picnic and barbecue facilities are available. Off Barrier Hwy; 67 km NW.

The Old Reservoir and Devil's Rock: Devil's Rock was a site for ceremonial rites for the Ngemba people. Good swimming and watersports at the reservoir; 3 km N. ***Mount Drysdale:*** deserted mining town where visitors can investigate old mine shafts and remains of the town. Features historic Aboriginal sites including rock wells; permission required (02) 6836 3462; 34 km N.

See also Outback, p. 36

Coffs Harbour

see inset box on next page

Cooma

Pop. 6587
Map ref. 565 C7 | 566 G9 | 581 E3

119 Sharp St; (02) 6450 1742 or 1800 636 525; www.visitcooma.com.au

 92.1 FM 2XL, 97.7 Snow FM, 810 AM ABC Local

The regional centre of the Snowy Mountains, Cooma was once dubbed Australia's most cosmopolitan city, thanks to the thousands of migrants who flocked to the region to work on the Snowy Mountains Hydro-Electric Scheme. It is a charming and bustling centre with visitors coming for the snow in winter and the greenery and crisp mountain air in summer. Motorists are advised to stop here to check tyres and stock up on petrol and provisions before heading into the alpine country.

Centennial Park Originally a swamp, Centennial Park was established in 1890. During WW II slit trenches were dug here in case of air attacks. The Avenue of Flags was constructed in 1959 to commemorate the 10th anniversary of the Snowy Mountains Hydro-Electric Scheme with one flag for each of the 27 nationalities of the workers. The Time Walk depicts the district's history in 40 ceramic mosaics laid below the flags. There is also a sculpture of Banjo Paterson's famous 'Man from Snowy River'. Sharp St.

Snowy Mountains Hydro-Information and Education Centre: interactive displays, photographs, models and films on the scheme. A memorial next door commemorates the 121 people killed while working on it. Monaro Hwy; inquiries 1800 623 776. ***NSW Corrective Services Museum:*** features unique displays of over 200 years of history, from convicts through to the modern prison system in the old Cooma Gaol; Vale Street; (02) 6452 5974. ***Southern Cloud Park:*** features Southern Cloud Memorial, a display of remains of the aircraft *Southern Cloud*, which crashed in the region in 1931 and was found in 1958. ***Bike path:*** picturesque path following Cooma Creek between Lambie St and Rotary Oval. ***Historic railcar:*** trips from Cooma to Bunyan on weekends; bookings at visitor centre. ***Lambie Town self-guide walk:*** designed in 1985, with over 5 km of easy walking. The tour incorporates three National Trust heritage areas including Lambie St, lined with huge oaks, pines and elms, and St Paul's Church, constructed with local alpine ash and granite and with striking stained-glass windows; brochure available from visitor centre.

Market: Centennial Park; 3rd Sun each month. ***Rodeo:*** Feb. ***Back to Earth Festival:*** Apr. ***Cooma Festival:*** Oct. ***Cooma Street Fair:*** celebrating Cooma's multicultural society; Nov. ***Cooma Races and Sundowner Cup:*** Dec.

The Lott Food Store: excellent cafe and bakery; 178–180 Sharp St; (02) 6452 1414.

Feathers B&B: romantic retreat; 12 Cosgrove St, Bredbo; (02) 6454 4151.

Tuross Falls Part of Wadbilliga National Park, the stunning Tuross Falls are a drop of 35 m. A picturesque 2 km walk from the camping area at Cascades leads visitors to the lookout platform, with views of the falls and the Tuross River Gorge. 30 km E.

Kosciuszko Memorial: donated in 1988 by the Polish government to commemorate Tadeuz Kosciuszko, after whom Australia's highest mountain is named; 2.5 km N. ***Mt Gladstone Lookout:*** impressive views, mountain-bike trails and Austrian teahouse; 6.5 km W. ***Transylvania Winery:*** cellar-door tastings of cool-climate organic wines; Monaro Hwy; 14 km N. ***Whitewater rafting:*** Snowy River tours from half-day to 2-day overnight trips; pick-up from Cooma, Jindabyne or Thredbo; bookings on 1800 677 179.

See also SNOWY MOUNTAINS, p. 33

Coonabarabran

Pop. 2605
Map ref. 556 E9

Newell Hwy; (02) 6849 2144 or 1800 242 881; www.warrumbungleregion.com.au

99.5 FM 2WCR, 107.1 FM Western Plains Community Radio

Coonabarabran is located on the Castlereagh River and is nestled in the foothills of the stunning Warrumbungle Mountain Range. It is known as the 'Astronomy Capital of Australia' as it has some of the clearest skies in the country, and is the gateway to the striking scenery of the Warrumbungles.

Visitor information centre The Australian Museum worked with Warrumbungle Shire Council to produce the unique Diprotodon Display. The diprotodon is the largest marsupial that ever lived and the skeleton on show was found in a creek bed 40 km east of town in 1979. Newell Hwy.

Crystal Kingdom: unique collection of minerals, including zeolite crystals and fossils, from the Warrumbungle Range; Newell Hwy; (02) 6842 1927. ***Newcastle Hats:*** hat factory outlet store; Ulan St.

Market: Dalgamo St; fourth Sun each month. ***Bunny Bazaar:*** includes markets; Easter. ***Festival of the Stars:*** includes Coona Cup Racing Carnival; Oct.

Blue Wren Cafe: cafe and pottery shop; Pilliga Pottery, Dandry Rd; (02) 6842 2239. ***Gecko Red:*** family restaurant; Imperial Hotel, John St; (02) 6842 1023.

Acacia Motor Lodge: contemporary, spacious accommodation; Newell Hwy; (02) 6842 1922. ***Amber Court:*** 21 ground-floor units; Newell Hwy; (02) 6842 1188. ***El Paso Motel:*** convenient ground-floor suites; Newell Hwy; (02) 6842 1722. ***Pilliga Pottery Old Schoolhouse:*** unique, spacious retreat; Pilliga Pottery, Dandry Rd; (02) 6842 2239.

Warrumbungle National Park Forested ridges, rocky spires and deep gorges coupled with excellent camping and visitor facilities have made this one of the state's most popular parks. Highlights include the Breadknife, a 90 m high rock wall, and the Grand High Tops walking trail with fabulous views of ancient volcanic remains. The park is outstanding for rock climbing, but climbing is prohibited on the Breadknife and permits are required for other areas (contact visitor centre). The Crooked Mountain Concert takes place here in Nov. Guided nature walks are conducted during school holidays or by appt. 35 km W.

Warrumbungle Observatory: night-sky viewing of the stars through computerised telescopes; National Park Road; 0488 425 112; 10 km W. ***Siding Spring Observatory:*** Australia's largest optical telescope, with a hands-on exhibition, science shop and cafe; National Park Rd; (02) 6842 6211; 27 km W. ***Pilliga Pottery:*** terracotta pottery, showrooms and tearooms in an attractive bushland setting; off Newell Hwy; (02) 6842 2239; 34 km NW. ***Barkala Bird Tours:*** guided birdwatching tours; off Newell Highway; (02) 6842 2239, 34 km NW. ***Sandstone Caves:*** formed by natural erosion of sandstone, these impressive caves are not signposted, so visitors are advised to seek directions from the visitor centre; 35 km N. ***Pilliga Forest:*** 450 000 ha of white cypress and ironbark trees, with plains of dense heath and scrub. There are sculptures in the scrub which are a series of Aborigianl sculptures overlooking a gorge. It is an excellent habitat for koalas, which can be spotted from signposted viewing areas. Also scenic forest drives and walking trails; maps are available from Pilliga Forest Discovery Centre in Baradine; (02) 6843 4011; 44 km NW. ***Local wineries:*** open for tastings and cellar-door sales; contact visitor centre for details.

See also CENTRAL WEST, p. 30

Coonamble

Pop. 2550
Map ref. 556 C8

26 Castlereagh St; (02) 6822 4532; www.coonamble.org

91.9 MTM FM, 648 AM ABC Local

Coonamble owes its existence to the discovery of artesian water in the area in the 1890s. On the Great Inland Way, the route via Coonamble provides an alternative to the more commonly taken coastal route between Queensland and the southern states. This town lost many of its buildings in the great fire of 1929. Castlereagh Street had to be rebuilt, so most constructions are relatively modern. Coonamble is also the birthplace of Sir Edward Hallstrom, the pioneer of refrigeration.

Historical Museum Housed in the old police station, the museum outlines the rich Aboriginal and pastoral history of Coonamble through photographs, household items and stables. Behind the museum is an authentic Cobb & Co coach and stables. Open by appt (02) 6822 4532; Aberford St.

Warrana Creek Weir: swimming, boating and fishing; southern outskirts of town. ***Self-guide town walk:*** takes in historic sites; brochure from visitor centre.

Rodeo and Campdraft: largest combined event in Southern Hemisphere; June. ***Gold Cup Race Meeting:*** Oct.

Sons of the Soil Hotel: bistro-style meals; 54 Castlereagh St; (02) 6822 1009.

Castlereagh Lodge Motel: comfortable, affordable accommodation; 81 Aberford St; (02) 6822 1999. ***Coonamble Motel:*** comfortable riverside accommodation; 86 Castlereagh St; (02) 6822 1400.

Quambone Locally known as the gateway to Macquarie Marshes, Quambone also has Australia's smallest library and the Marthaguy Picnic Races in Sept. 55 km W.

Gulargambone: a small town with a restored steam train in Memorial Park; 45 km S. ***Macquarie Marshes:*** 80 km W; *see Nyngan.*

See also CENTRAL WEST, p. 30

COFFS HARBOUR

Pop. 47 710

Map ref. 550 H1 | 551 E12 | 557 N8

[COFFS HARBOUR'S HISTORIC JETTY]

Cnr Pacific Hwy and McLean St; (02) 6648 4990 or 1300 369 070; www.coffscoast.com.au

107.9 2AIR FM, 819 AM ABC Local

Coffs Harbour, a subtropical holiday town on Coffs Coast, is known for its banana plantations (and the iconic Big Banana) and for its great fishing. The combination of great weather, stunning hinterland forests, sandy beaches and a growing cosmopolitan centre make Coffs Harbour a popular spot for tourists seeking fun and relaxation.

Muttonbird Island Nature Reserve Visitors can get an up-close look at the life cycle of one of Australia's most interesting migratory birds. The wedge-tailed shearwaters (muttonbirds) fly thousands of kilometres from South-East Asia each August, with large numbers settling at Muttonbird Island to breed. A walking trail winds through the burrows of the birds, which can be seen Aug–Apr. Muttonbird Island is also a vantage point for whale-watching June–Nov and is a great place for fishing and picnics. Access is via a 500 m walk along the sea wall from the harbour.

Bunker Cartoon Gallery: largest private collection of cartoons in the Southern Hemisphere, housed in a WW II bunker; City Hill; (02) 6651 7343. ***Legends Surf Museum:*** displays of classic photography, videos and equipment; Gaudrons Rd; (02) 6653 6536. ***Coffs Harbour Regional Gallery:*** varied program of contemporary art exhibitions; Cnr Coff and Duke sts; (02) 6648 4863; open Tue–Sat 10am–4pm. ***Coffs Harbour International Marina:*** departure point for fishing charters, scuba diving and whale-watching trips (June–Nov); Marina Dr; (02) 6651 4222. ***North Coast Regional Botanical Garden:*** rainforest, mangrove boardwalks, herbarium and diverse birdlife; Hardacre St; (02) 6648 4188. ***Pet Porpoise Pool:*** performing dolphins and seals with research and nursery facilities; Orlando St; (02) 6659 1900; open 9am–4pm every day. ***Self-guide walks:*** include Jetty Walk and Coffs Creek Walk; maps from visitor centre.

Growers market: Harbour Dr; Thurs Nov–Mar. ***Market:*** jetty, Harbour Dr; Sun. ***Uptown Market:*** Vernon St; Sun. ***Pittwater and Coffs Harbour Offshore Series:*** yachting; Jan. ***Sawtell Chilli Festival:*** July. ***International Buskers and Comedy Festival:*** Sept–Oct. ***Food and Wine Festival:*** Oct.

Caffe Fiasco: northern Italian; 22 Orlando St; (02) 6651 2006. ***Ocean Front Brasserie:*** spectacular views and seafood platters; Coffs Harbour Deep Sea Fishing Club, 1 Jordan Espl; (02) 6651 2819. ***Shearwater Restaurant:*** seafood; 321 Harbour Dr; (02) 6651 6053.

Breakfree Aanuka Beach: absolute beachfront accommodation; 11 Firman Dr, Diggers Beach; (02) 6652 7555. ***Sante Fe Luxury B&B:*** casually elegant accommodation; 235 The Mountain Way; (02) 6653 7700.

Big Banana Large banana-shaped landmark with displays on the banana industry, giant water slides and toboggan rides. 351 Pacific Hwy; (02) 6652 4355; open 9am–4.30 every day; 4 km N.

Bindarri National Park Not for the unseasoned bushwalker, Bindarri National Park is a largely untouched forest without facilities, but amazing views reward those who make the effort. The headwaters of the Urumbilum River form breathtaking waterfalls in a remote and rugged setting. Pockets of old-growth forest are scattered across the plateau and rich rainforest protects the steeper slopes. While there are no campgrounds, backpack camping is allowed and there are bushwalking trails to follow; (02) 6652 0900. 20 km W.

Clog Barn: Dutch village with clogmaking; Pacific Hwy; 2 km N; (02) 6652 4633. ***Bruxner Park Flora Reserve:*** dense tropical jungle area of vines, ferns and orchids with bushwalking trails, picnic area and Sealy Lookout; Bruxner Park Rd, Korora; (02) 6652 8900; 9 km NW. ***Butterfly House:*** enclosed subtropical garden with live native and exotic butterflies; 5 Strouds Rd, Bonville; (02) 6653 4766; open 9am–4pm daily; 9 km S. ***Adventure tours:*** include whitewater rafting, canoeing, reef-fishing, diving, horseriding through rainforest, surf rafting, skydiving, helicopter flights, go-karting and surf schools; see visitor centre for brochures and bookings.

See also HOLIDAY COAST, p. 27

Cootamundra

Pop. 5565
Map ref. 554 D8 | 565 A2

Railway station, Hovell St; (02) 6942 4212; www.cootamundra.nsw.gov.au

 107.7 Star FM, 675 AM ABC Local

This prosperous rural service centre and major junction on the railway line between Sydney and Melbourne prides itself on being the birthplace of cricketing legend Sir Donald Bradman. Much reference is made to him and to cricket around the town. Cootamundra also lends its name to the famous Cootamundra wattle (*Acacia baileyana*), which blooms in the area each July and August, and Cootamundra Gold, the locally produced canola oil.

Pioneer Park This natural bushland reserve on the northern outskirts of town has a scenic 1.3 km walking trail to the top of Mt Slippery. At the summit there are panoramic views. The park also has excellent picnic sites. Backbrawlin St.

Bradman's birthplace: restored cottage where 'the greatest batsman the world has ever known' was born. Contains memorabilia from cricket and his life in the Cootamundra district; Adams St. ***Arts Centre:*** community arts space with exhibitions, workshops, art sales and a theatre; Wallendoon St. ***Memorabilia Cottage:*** displays local history memorabilia and bric-a-brac; Adams St. ***Captains Walk:*** bronze sculptures of Australia's past cricket captains; Jubilee Park, Wallendoon St. ***Heritage Centre:*** local memorabilia including an Olympic cauldron and war relics; railway station, Hovell St. ***Self-guide 'Two Foot Tour':*** includes Sir Donald Bradman's birthplace and the town's historic buildings; brochure from visitor centre. ***Local crafts:*** at visitor centre and at Art and Craft Centre, Hovell St.

Markets: Fisher Park, 2nd Sun each month; Wallendbeen, 1st Sun each month. ***Wattle Time Festival:*** Aug. ***Rose Show:*** Nov.

Country Cuisine: cafe with homemade sweet treats; 265 Parker St; (02) 6942 1788. ***Helen's Coffee Lounge:*** cafe-style fare or hearty mains; 248 Parker St; (02) 6942 7400.

Heritage Motel: executive motel accommodation; 94 Hurley St; 1300 221 000. ***The White Ibis:*** elegant boutique B&B; 21 Wallendoon St; (02) 6942 1850.

Murrumburrah This small rural community has the Harden–Murrumburrah Historical Museum, which is open weekends and features pioneer artefacts, an old chemist shop exhibit and early Australian kitchenwares. Also in town are local craft shops and some outstanding picnic spots. 35 km NE. Stocks Native Nursery in Harden, 2 km E on Simmonds Rd, features 1.5 ha of native bush garden with a scenic walking trail and billabong. The Picnic Races are held here in Nov.

Green Tree Indigenous Food Gardens: Australian bush garden divided into different zones providing education on how indigenous people survived on bush tucker; tours available; (02) 6943 2628; 8 km N. ***Migurra Reserve:*** bushland walking trail with birdwatching and 5 species of wattles; 15 km SW. ***The Milestones:*** cast-concrete sculptures representing the importance of wheat to the area; 19 km NE. ***Bethungra:*** dam ideal for canoeing and sailing. The rail spiral is an unusual engineering feat; 23 km SW. ***Kamilaroi Cottage Violets:*** violet farm with tours by appt; bookings (02) 6943 2207; 25 km N. ***Illabo:*** charming town with impressive clock museum; 33 km SW. ***Cellar doors:*** in the Harden area; winery brochure from visitor centre.

See also RIVERINA, p. 34

Corowa

Pop. 5628
Map ref. 559 M12 | 585 I5 | 586 F1

88 Sanger St; (02) 6033 3221 or 1800 814 054; www.visitcorowashire.com.au

 105.7 FM The River, 675 AM ABC Local

Corowa has been known for its goldmining, winemaking, timber milling and as the 'birthplace of Federation'. Traders in the 19th century had to pay taxes both in New South Wales and Victoria when taking goods over the border, which caused much agitation. It was argued that free trade would benefit everyone and the Border Federation League was formed in Corowa, which led to the 1893 Corowa Federation Conference. In 1895 the proposals put forth at the conference were acted upon and on 1 January 1901 the Commonwealth of Australia was born.

Federation Museum This museum focuses on the reasons behind Federation and Corowa's involvement in it. Also on display are local Aboriginal artefacts, Tommy McRae sketches, horse-drawn vehicles and saddlery, and antique agricultural implements. Open Sat and Sun afternoons; Queen St.

Murray Bank Yabby Farm: catch and cook yabbies, go canoeing and enjoy a picnic or barbecue; Federation Ave; (02) 6033 2922. ***Self-guide historical town walk:*** includes Sanger St, Corowa's historic main street with its century-old verandahed buildings. Guide available for groups; brochure from visitor centre.

Market: Bangerang Park; 1st Sun each month (except Feb). ***Federation Festival:*** Jan. ***Billycart Championships:*** Easter.

Easdown House Restaurant: modern Australian; 1 Sanger St; (02) 6033 4077.

Rivergum Holiday Retreat: family resort-style accommodation; 386 Honour Ave; (02) 6033 1990. ***Savernake Farmstay Cottage:*** self-catering cottage; Mulwala–Savernake Rd; (02) 6035 9415.

All Saints Estate Situated in the respected Rutherglen district, All Saints is a winery like no other. Behind the hedge fence and imposing set of gates lies an enormous medieval castle built by the original owner, George Smith, based on the Castle of Mey in Scotland. Now owned and operated by Peter Brown (of the famous Brown Brothers), All Saints offers a large cellar-door operation, a renowned restaurant, The Keg Factory, Indigo Cheese and beautiful gardens. Inquiries (02) 6035 2222; 5 km SW.

Corowa Jump Shak: skydiving and gliding weekends, weather permitting; off Redlands Rd; (02) 6033 2435. ***Savernake Station:*** offers eco-heritage tours of their 400 ha woodland including 120 bird species, woodshed and shearers quarters (1912) and cooks museum (1930) and store; inquiries (02) 6035 9415; 50 km NW.

See also MURRAY, p. 35

Cowra

Pop. 8426
Map ref. 554 F6

Olympic Park, Mid Western Hwy; (02) 6342 4333; www.cowratourism.com.au

 99.5 Star FM, 549 AM ABC Local

Cowra is nestled in the Lachlan Valley, where you can enjoy local food and wine. The peaceful air of this town on the Lachlan River belies its dramatic history. The Cowra Breakout is an infamous World War II incident in Australia when 1104 Japanese POWs staged a mass breakout that was the biggest in British and

Australian war history. While it remains what Cowra is most famous for, the town has moved forward.

Australia's World Peace Bell Each country has only one peace bell and it is normally located in the nation's capital, but Cowra was awarded Australia's Peace Bell owing to local efforts for peace. The bell is a replica of the United Nations World Peace Bell in New York City and was made by melting down coins donated from 103 member countries of the United Nations. It is rung each year during the Festival of International Understanding. Darling St.

Japanese Garden This garden, opened in 1979, is complete with a cultural centre (with a collection of Japanese artwork and artefacts), a traditional teahouse, a bonsai house and a pottery. The garden itself represents the landscape of Japan, with mountain, river and sea re-created. From here gracious Sakura Ave, lined with cherry trees that blossom in spring, leads to the site of the POW camp and to the Australian and Japanese cemeteries. The camp includes the original foundations and replica guardtower, with photo displays and signage; audio tours available. Off Binni Creek Rd; (02) 6341 2233.

Olympic Park: information centre with a fascinating interpretive POW display and theatre. Also here is Cowra Rose Garden with over 1000 rose bushes in over 100 varieties; Mid Western Hwy. ***Cowra–Italy Friendship Monument:*** in recognition of Italians who died in WW II (Italian POWs interned at Cowra formed a strong friendship with the town); Kendal St. ***Lachlan Valley Railway Museum:*** displays and train rides; Campbell St. ***Cowra Mill Winery:*** winery in former flour mill (1861) with cellar-door tastings and restaurant; Vaux St; (02) 6341 4141. ***Aboriginal murals:*** by local artist Kym Freeman on pylons of bridge over the Lachlan River. ***Cowra Heritage Walk:*** Federation, colonial and Victorian buildings including the town's first hotel and oldest home; map from visitor centre.

Farmers market: showgrounds; 3rd Sat each month. ***Festival of International Understanding:*** Mar. ***Picnic Races:*** July. Cowra Wine Show: July. ***Sakura Matsuri:*** cherry blossom festival; Sept/Oct.

Neila: modern Australian; 5 Kendal St; (02) 6341 2188.

Country Gardens Motor Inn: spacious, comfortable accommodation; 75 Grenfell Rd; (02) 6341 1100. ***Old Milburn Schoolhouse B&B:*** award-winning B&B with great food; Nangaree Reg Hailstone Way; (02) 6345 1276. ***Tinnies at Back Creek Vineyard:*** self-contained cottage; Chiverton Rd; (02) 6342 9251.

Cowra museums: war, rail and rural museums all in one complex; Sydney Rd; 5 km E. ***Darby Falls Observatory:*** one of the largest telescopes accessible to the public. Check opening times; (02) 6345 1900 or 0417 461 162; Observatory Rd; 25 km SE. ***Conimbla National Park:*** known for its wildflowers, rock ledges, waterfalls, bushwalks and picnics; 27 km W. ***Lake Wyangala and Grabine Lakeside State Park:*** ideal for watersports and fishing; 40 km SE. ***Self-guide drives:*** through countryside including a wine-lovers' drive; leaflets from visitor centre. ***Local wineries:*** open for cellar-door tastings; leaflets from visitor centre.

See also CENTRAL WEST, p. 30

Crookwell

Pop. 1994
Map ref. 554 G8 | 565 E2

106 Goulburn St; (02) 4832 1988; www.upperlachlantourism.com

106.1 2GN FM, 846 AM ABC Radio National

This picturesque tree-lined township is a service centre to the local agricultural and pastoral district, and enjoys a cool climate and lush gardens. Australia's first grid-connected wind farm was opened here in 1998 and is capable of supplying electricity to 3500 homes. The Country Women's Association was formed here in 1922 and has since spread nationwide.

Crookwell Wind Farm: viewing platform and information board; Goulburn Rd.

Market: Uniting Church, Goulburn St; 1st Sat each month. ***Crookwell Country Festival:*** traditional country festival with music, markets, sports and a parade; Mar. ***Open Gardens weekends:*** spring and autumn (dates from visitor centre).

Paul's Cafe: sandwiches, burgers, fish and chips; 102 Goulburn St; (02) 4832 1745.

Gundowringa Homestead: charming country manor; Goulburn Rd; (02) 4848 1212.

Wombeyan Caves There are 5 caves open to the public including Figtree Cave, widely regarded as the best self-guide cave in NSW. Junction Cave has a colourful underground river; Wollondilly Cave has 5 main chambers with outstanding formations; Mulwarree Cave is intimate, with delicate formations; and Kooringa Cave is huge and majestic. Wombeyan Gorge is made of marble, providing an unusual swimming experience. There are several campgrounds and walking trails in the area. 60 km E.

Redground Lookout: excellent views of surrounding area; 8 km NW. ***Willow Vale Mill:*** restored flour mill with restaurant and accommodation; Laggan; (02) 4837 3319; 9 km NE. ***Lake Wyangala and Grabine Lakeside State Park:*** upper reaches ideal for waterskiing, picnicking, fishing, bushwalking and camping; 65 km NW. ***Bike riding:*** the area surrounding Crookwell is popular for bike riding; trail maps from visitor centre. ***Historic villages:*** associated with goldmining, coppermining and bushrangers, these villages include Tuena, Peelwood, Laggan, Bigga, Binda (all north) and Roslyn (south), the birthplace of poet Dame Mary Gilmore; maps from visitor centre. ***Historical and scenic drives:*** explore sites and countryside frequented by bushrangers such as Ben Hall; brochure from visitor centre.

See also CAPITAL COUNTRY, p. 31

Culcairn

Pop. 1118
Map ref. 554 B11 | 559 O11 | 585 L4

Greater Hume Shire Visitor Information Centre, 15 Wallace St, Holbrook; (02) 6036 2422; www.greaterhume.nsw.gov.au/tourism.html

93.1 Star FM, 675 AM ABC Local

Culcairn is located at the heart of 'Morgan country' where Dan 'Mad Dog' Morgan terrorised the district between 1862 and 1865. This peaceful town owes its tree-lined streets and lush green parks to an underground water supply discovered in 1926.

Stationmaster's Residence: beautifully restored museum (1883) reflects the importance of the railway; just across railway line. ***Billabong Creek:*** good fishing, one of the longest creeks in the Southern Hemisphere. ***National Trust–classified buildings:*** includes historic Culcairn Hotel (1891), still operating; Railway Pde and Olympic Way.

Henty This historic pastoral town has the Headlie Taylor Header Memorial, a tribute to the mechanical header harvester that revolutionised the grain industry. The nearby Sergeant Smith Memorial Stone marks the spot where Morgan fatally wounded a police officer, and the adjacent Doodle Cooma Swamp is 2000 ha of breeding area for waterbirds. 24 km N.

John McLean's grave: McLean was shot by Mad Dog Morgan; 3 km E. ***Premier Yabby Farm:*** tours through yabby-related displays and open ponds; also fishing, picnicking and barbecue facilities; (02) 6029 8351; 6 km SW. ***Round Hill Station:*** where Morgan committed his first hold-up in the area; Holbrook Rd; 15 km E. ***Walla Walla:*** old schoolhouse (1875) and the largest Lutheran church in NSW (1924); 18 km SW. ***Morgan's Lookout:*** granite outcrop on otherwise flat land, allegedly used by Morgan to look for approaching victims and police; 18 km NW.

See also MURRAY, p. 35

Deniliquin

Pop. 7433
Map ref. 559 J10 | 584 D2

Peppin Heritage Centre, George St; (03) 5898 3120 or 1800 650 712; www.denitourism.com.au

99.3 FM ABC Radio National, 102.5 Classic Rock FM

Deniliquin, at the centre of Australia's largest irrigation system, lies on the Edward River, part of the Murray River and formed by a fault in the earth. Situated next to the world's largest red-gum forest, birdlife and wildlife remain abudant, despite the drought currently crippling the town's rice-growing industry. 'Deni' proclaims itself the 'ute capital of the world', holding the official record for most utes mustered in one place in 2007.

Peppin Heritage Centre This museum is dedicated to George Hall Peppin and his sons' development of the merino sheep industry. Dissatisfied with the quality and yield of the wool from merino sheep, they developed a new breed, the peppin, that was better adapted to the harsh Australian conditions. Peppin sheep now predominate among flocks in New Zealand, South Africa and South America. The museum is housed in the National Trust–classified Old George Street Public School (1879), which still has an intact classroom on display. There is also a lock-up gaol from Wanganella and a 1920s thatched ram shed. George St.

Island Sanctuary: features kangaroos and birdlife, and the burial site of 'Old Jack', a member of the Melville gang who visited Deniliquin in 1851; off Cressy St footbridge. ***Ute on a Pole:*** confirms Deniliquin's status as 'ute capital of the world'; near National Bridge. ***Waring Gardens:*** originally a chain of lagoons, the park was established in the 1880s; Cressy St. ***Pioneer Steam Museum:*** private collection of steam engines and pumps; Hay Rd. ***Long Paddock River Walk:*** old stock route, includes interpretive panels; from Heritage Centre to Island Sanctuary. ***Self-guide walks:*** historical and nature walks taking in National Trust–classified buildings and town gardens; brochure from visitor centre.

Market: Waring Gardens; 4th Sat each month. ***Sun Festival:*** includes gala parade and international food and entertainment; Jan. ***RSL Fishing Classic:*** Jan. ***Play on the Plains Festival and Ute Muster:*** celebration of music and cars; Sept/Oct.

Deniliquin Bakery: coffee and snacks; 69 Davidson St; (03) 5881 2278. ***The Federal Hotel:*** pub bistro; 46 Napier St; (03) 5881 1260.

Centrepoint Motel: affordable 4-star accommodation; 399 Cressy St; (03) 5881 3544.

Pioneer Tourist Park A modern caravan park including charming features of the past (open to the public), with an antique steam and pump display and a blacksmith shop. 2 km N.

Irrigation works: at Lawsons Syphon (7 km E) and Stevens Weir (25 km W). ***Clancy's Winery:*** cellar-door tastings and sales; 18 km N. ***Conargo Pub:*** authentic bush pub with photo gallery depicting history of merino wool in the area; 25 km NE. ***Bird Observatory Tower:*** excellent vantage point for birdwatching; Mathoura; 34 km S.

See also MURRAY, p. 35

Dorrigo

Pop. 970
Map ref. 550 G1 | 551 C12 | 557 M8

Hickory St; (02) 6657 2486; www.bellingermagic.com

105.5 Star FM, 738 AM ABC Local

Dorrigo is known as 'Australia's national park capital'. It is entirely surrounded by national parks including Dorrigo and Cathedral Rock. The Dorrigo Plateau provides crisp, clean air and wonderful views in all directions. The town is small enough to be friendly, but popular enough to provide excellent facilities.

Historical museum: memorabilia, documents and photographs detailing the history of Dorrigo and surrounding national parks; Cudgery St. ***Local crafts:*** at Pinnata Gallery; Hickory St. ***Wood-fired bakery:*** produces popular products with local produce; Hickory St. ***Waterfall Way Winery:*** unique fruit wines and fortified wines; tastings available; Hickory St.

Market: showground, Armidale Rd; 1st Sat each month. ***Arts and Crafts Exhibition:*** Easter. ***Bluegrass Festival:*** Oct.

Lick the Spoon: cafe and provedore; 51–53 Hickory St; (02) 6657 1373. ***Tallawalla Teahouse:*** cafe in stunning surrounds; 113 Old Coramba Rd; (02) 6657 2315.

Fernbrook Lodge B&B: charming country guesthouse; 4705 Waterfall Way; (02) 6657 2573. ***Gracemere Grange:*** laid-back country guesthouse; 325 Dome Rd; (02) 6657 2630. ***Moss Grove B&B:*** English-style traditional B&B; 589 Old Coast Rd; (02) 6657 5388.

Dorrigo National Park This park takes in World Heritage–listed rainforest and offers plenty for visitors to see and do. Attractions include spectacular waterfalls and a variety of birds such as bowerbirds and lyrebirds. The Rainforest Centre has picnic facilities, a cafe, a video theatre and exhibitions. There is also the Skywalk, a boardwalk offering views over the canopy of the rainforest, and the Walk with the Birds boardwalk. 3 km E.

Cathedral Rock National Park Giant boulders, sculpted rock, distinctive granite hills and wedge-tailed eagles make Cathedral Rock spectacular viewing and popular among photographers. Walks include a 3 hr circuit walk to the summit of Cathedral Rock for amazing 360-degree views of the tableland. 56 km SW.

Trout fishing: in streams on the Dorrigo Plateau (between Dorrigo and Urunga); contact visitor centre for locations. ***Dangar Falls:*** viewing platform over beautiful 30 m waterfall; 2 km N. ***Griffiths Lookout:*** sweeping views of the mountains;

6 km s. ***Guy Fawkes River National Park:*** rugged and scenic surrounds with limited facilities, but worth the effort for experienced bushwalkers. Ebor Falls has cliff-top viewing platforms above and there is also good canoeing and fishing; 40 km w. ***L. P. Dutton Trout Hatchery:*** educational visitor centre and trout feeding; 63 km sw. ***Point Lookout:*** in New England National Park for spectacular panoramic views of Bellinger Valley and across to the ocean; 74 km sw.

See also HOLIDAY COAST, p. 27

Dubbo

Pop. 34 318
Map ref. 554 F1 | 556 C12

Cnr Newell Hwy and Macquarie St; (02) 6801 4450 or 1800 674 443; www.dubbotourism.com.au

 92.7 Zoo FM, 549 AM ABC Local

One of Australia's fastest growing inland cities, Dubbo is most famous for its world-class open-range zoo. The city, on the banks of the Macquarie River, is thriving and prosperous with more than half a million visitors each year. Dubbo prides itself on city standards with a country smile.

Western Plains Cultural Centre The centre includes the Dubbo Regional Gallery – the Armati Bequest, Dubbo Regional Museum and the Community Arts Centre. It exhibits local and national visual arts, heritage and social history. In 2007, the gallery received a generous gift from the Armati Family including Michael Riley's celebrated 'Cloud' series of photographs. To acknowledge this, the gallery was renamed the Armati Bequest. It specialises in the theme of Animals in Art, collecting works in a broad range of media and styles by artists from all areas. The museum, housed in the original Dubbo High School building, features a permanent space devoted to the story of Dubbo entitled 'People Places Possessions' and a temporary exhibition space. Closed Tues; Wingewarra Street; (02) 6801 4444.

Old Dubbo Gaol Closed as a penal institution in 1966, Old Dubbo Gaol now offers a glimpse at convict life. See the original gallows (where 8 men were hanged for murder) and solitary confinement cells, or walk along the watchtower. An amazing animatronic robot tells historical tales. There are also holograms and theatrical enactments. Macquarie St; (02) 6801 4460.

Shoyoen Sister City Garden: Japanese garden and teahouse designed and built with the support of Dubbo's sister city, Minokamo; Coronation Dr East. ***Traintasia:*** detailed operating model railway display with spectator interaction; Yarrandale Road; (02) 6884 9944.

Markets: Macquarie Lions Park, 1st and 3rd Sat each month; Dubbo Showground, Wingewarra St, 2nd Sun each month. ***Western Plains Country Music Championships:*** Easter. ***Jazz Festival:*** Aug. ***Red Ochre Festival:*** Sept.

Rose Garden Thai Restaurant: takeaway available; 208 Brisbane St; (02) 6882 8322. ***Two Doors Tapas & Wine Bar:*** Spanish; 215B Macquarie St; (02) 6885 2333.

No 95 Dubbo: stylish, good-value motel; 95 Cobra St; (02) 6882 7888. ***Taronga Western Plains Zoo:*** lodge accommodation or camping at the zoo; Zoofari Lodge (02) 6881 1488, Roar and Snore (02) 6881 1405.

Taronga Western Plains Zoo Australia's first open-range zoo, with over 1000 animals from 5 continents, is set on more than 300 ha of bushland. The zoo is renowned for its breeding programs (especially with endangered species), conservation programs and education facilities and exhibits. There are talks by the keepers and early morning walks, as well as accommodation at Zoofari Lodge. The Tracker Riley Cycleway paves the 5 km from Dubbo to the zoo and bicycles and maps are available from the visitor centre. Inquiries (02) 9969 2777; 5 km s.

Dubbo Observatory: explore the skies via Schmidt Cassegrain telescopes. Open nightly; Camp Road; (02) 6885 3022; 5.5 km s. ***Dundullimal Homestead:*** an 1840s restored squatter's slab-style homestead with working saddler, blacksmith and farm animals; Obley Rd; 7 km se. ***Terramungamine Rock Grooves:*** the 150 rock grooves were created by the Tubbagah people; Burraway Rd via Brocklehurst; 10 km n. ***Narromine:*** agricultural centre well known for gliding and an outstanding aviation museum. The Air Pageant and Evolution of Flight Festival is held here each Sept/Oct; 40 km e. ***Heritage drives and river cruises:*** brochures from visitor centre. ***Wineries:*** several in region offering cellar-door tastings; brochure from visitor centre.

See also CENTRAL WEST, p. 30

Eden

Pop. 3010
Map ref. 565 E10 | 581 H3

Eden Gateway Centre, Princes Hwy; (02) 6496 1953; www.sapphirecoast.com.au

 102.5 Power FM, 810 AM ABC Local

Situated on the Sapphire Coast, the aptly named Eden is an idyllic and peaceful town on Twofold Bay. The location is excellent, with national park to the north and south, water to the east and woodland to the west. The beautiful bay is rimmed with mountains. Originally settled by whalers, it is now a fishing port and a popular, but relatively undeveloped, tourist town.

Aslings Beach: surf beach with rockpools and excellent platforms for whale-watching Oct–Nov; Aslings Beach Rd. ***Snug Cove:*** working fishing port with plenty of restaurants and cafes. ***Eden Killer Whale Museum:*** fascinating displays on the history of the local whaling industry including the skeleton of 'Old Tom' the killer whale; Imlay St; (02) 6496 2094.

Market: Calle Calle St; 1st Sat each month. ***Eden Seafood and Arts Festival:*** Easter. ***Eden Whale Festival:*** Oct/Nov.

Eden Fishermen's Club: steak and seafood; 217 Imlay St; (02) 6496 1577. ***Essentially Eden:*** modern Australian; The Great Southern Inn, 158 Imlay St; (02) 6496 1515.

Crown & Anchor: traditional B&B in 1840s inn; 239 Imlay St; (02) 6496 1017. ***Quarantine Bay Beach Cottages:*** self-contained cottages; Princes Hwy; (02) 6496 1483.

Ben Boyd National Park This park's scenery includes rugged stretches of coastline, unique rock formations, heaths and banksia forest. The area is excellent for fishing, swimming, wreck diving, bushwalking and camping. Boyd's Tower at Red Point, 32 km se, was originally built for whale-spotting. Cape Green Lighthouse, 45 km se, is the first cast-concrete lighthouse in Australia and the second tallest in NSW; tours by appt; bookings (02) 6495 5000. The Pinnacles, 8 km n, are an unusual earth formation with red gravel atop white sand cliffs.

Jiggamy Farm: Aboriginal cultural and bush tucker experience; 9 km N. ***Boydtown:*** former rival settlement on the shores of Twofold Bay with convict-built Seahorse Inn (still licensed), safe beach and good fishing; 9 km S. ***Davidson Whaling Station Historic Site:*** provides unique insight into the lives of 19th-century whalers; Kiah Inlet; 30 km SE. ***South East Fibre Exports Visitors Centre:*** logging and milling displays; Jews Head; bookings essential (02) 6496 0222; 34 km SE. ***Nadgee Nature Reserve:*** walking track, access via Wonboyn Lake; 35 km SE. ***Wonboyn Lake:*** scenic area with good fishing and 4WD tracks; 40 km S.

See also **South Coast, p. 32**

Eugowra

Pop. 532
Map ref. 554 E5

Cnr Byng and Peisley sts; (02) 6393 8226 or 1800 069 466; www.orange.nsw.gov.au

 105.9 Star FM, 549 AM ABC Local

Situated on the rich basin of the Lachlan River, Eugowra is a tiny country town known for its crafts. Nearby on the Orange–Forbes road, the famous Gold Escort Robbery took place in 1862.

Eugowra Museum and Bushranger Centre: displays on pioneer life, a pistol used in the Gold Escort Robbery, gemstones, early farm equipment, wagons and Aboriginal artefacts; open Wed–Sun; (02) 6859 2214. ***Local craft shops:*** leaflet from visitor centre. ***Self-guide bushranging tour:*** maps from visitor centre.

Escort Rock Cafe: hearty country meals; 61 Nanima St; (02) 6859 2727. ***The Lady Bushranger:*** quaint cafe; Shop 2, 51 Nanima St; (02) 6859 2900.

Nangar National Park The horseshoe-shaped red cliffs of the Nangar–Murga Range stand out against the central west's plains. Nangar National Park's flowering shrubs and timbered hills provide an important wildlife refuge among mostly cleared land. Rocky slopes and pretty creeks make it a scenic site for bushwalks and popular for rock climbing. The park does not have facilities, so visitors are advised to take water and provisions with them and to give friends or family their itinerary. 10 km E.

Escort Rock: where bushranger Frank Gardiner and gang (including Ben Hall) hid before ambushing the Forbes gold escort. A plaque on the road gives details; 3 km E. ***Nanami Lane Lavender Farm:*** products and plants for sale and workshops on growing lavender; 19 km SE.

See also **Central West, p. 30**

Evans Head

Pop. 2629
Map ref. 551 G6 | 557 O4

Ballina Visitor Information Centre, cnr Las Balsas Plaza and River St, Ballina; (02) 6686 3484 or 1800 777 666; www.tropicalnsw.com.au

 92.9 FM North Coast Radio, 738 AM ABC Local

Evans Head is located at the mouth of the Evans River. It was the first prawning port in Australia and is still predominantly a fishing village, but with 6 kilometres of safe surfing beaches, sandy river flats and coastal scenery, it is also a tourist town. There is excellent rock, beach and ocean fishing.

Goanna Headland: site of great mythical importance to the Bundjalung people and favourite spot of serious surfers. ***Razorback Lookout:*** views up and down the coast. On a clear day, Cape Byron Lighthouse can be seen to the north; Ocean Dr.

Market: Cnr Oak and Park sts; 4th Sat each month. ***Fishing Classic:*** July. ***Evans Head Flower Show:*** Sept. ***Evans Head Longboard Invitational:*** Sept.

New Italy A monument and remains are all that are left of this settlement that was the result of the ill-fated Marquis de Rays expedition in 1880. The Marquis tricked 340 Italians into purchasing nonexistent property in a Pacific paradise. Disaster struck several times for the emigrants, as they travelled first to Papua New Guinea and then to New Caledonia. Eventually Sir Henry Parkes arranged for their passage to Australia, where the 217 survivors built this village. Also in the New Italy Complex, Guuragai Aboriginal Arts and Crafts offers quality works and information on Aboriginal culture. 23 km SW.

Bundjalung National Park: Aboriginal relics, fishing, swimming and bushwalking; 2 km S. ***Broadwater National Park:*** bushwalking, birdwatching, fishing and swimming; 5 km N. ***Woodburn:*** friendly town on the Richmond River with great spots for picnicking, swimming, fishing and boating; flower show in Aug; 11 km NW.

See also **Tropical North Coast, p. 28**

Finley

Pop. 2053
Map ref. 559 L11 | 584 F3

Tocumwal Visitor Centre, 41 Deniliquin St, Tocumwal; (03) 5874 2131 or 1800 677 271.

 102.5 FM Classic Rock, 675 AM ABC Local

This town, on the Newell Highway and close to the Victorian border, is a tidy and peaceful spot. It is the centre of the Berriquin Irrigation Area. The main street spans Mulwala Canal, the largest irrigation channel in Australia.

Mary Lawson Wayside Rest Features a log cabin that is an authentic replica of a pioneer home. It houses the Finley and District Historical Museum with displays of antique pumping equipment and machinery. Newell Hwy.

Finley Lake: popular boating, sailboarding and picnic area; Newell Hwy. ***Finley Livestock Exchange:*** experience a cattle sale; Fri mornings.

 ***Rodeo:*** Jan.

Berrigan: charming historic town known for its connections to horseracing; 22 km E. ***Sojourn Station Art Studio:*** spacious rural property providing accommodation for visiting artists; 25 km SE via Berrigan. ***Grassleigh Woodturning and Crafts:*** displays of woodturning in action and wood products; 37 km NE via Berrigan.

See also **Murray, p. 35**

Forbes

Pop. 6954
Map ref. 554 D4

Railway station, Union St; (02) 6852 4155; www.forbes.nsw.gov.au

 97.9 Valley FM, 104.3 FM Radio National

When John Oxley passed through in 1817, he was so unimpressed by the area's clay soil, poor timber and swamps that he claimed 'it is impossible to imagine a worse country'. Today Forbes is a pleasant spot bisected by Lake Forbes, a large lagoon in the middle of town. It was the discovery of gold that caused the town to be built and the legends of old bushrangers that keep it buzzing today.

Historical Museum: features relics associated with bushranger Ben Hall, a vintage colonial kitchen and antique farm machinery; open 2–4pm; Cross St; (02) 6851 6600. ***Cemetery:*** graves of Ben Hall, Kate Foster (Ned Kelly's sister), Rebecca Shields (Captain Cook's niece) and French author Paul Wenz; Bogan Gate Rd. ***King George V Park:*** memorial where 'German Harry' discovered gold in 1861, and a pleasant spot for picnics and barbecues; Lawler St. ***Dowling Street Park:*** memorial marks the spot where John Oxley first passed through in 1817; Dowling St. ***Lake Forbes:*** picnic spots, barbecue facilities, fishing, and a walking and cycling track; off Gordon Duff Dr. ***Historical town walk:*** includes the post office (1862) and the town hall (1861) where Dame Nellie Melba performed in 1909; map from visitor centre. ***Local arts and crafts:*** brochure from visitor centre.

Jazz Festival: Jan. ***Banderra Estate Vineyard winetasting:*** McFeeters Motor Museum, Newell Hwy; daily.

Wattle Cafe: light lunches; 85 Rankin St; (02) 6852 4310.

Lake Forbes Motel: affordable accommodation, picturesque views; 8 Junction St; (02) 6852 2922.

Gum Swamp Sanctuary: birdlife and other fauna, best seen at sunrise or sunset; 4 km s. ***Banderra Estate Vineyard:*** French winemaker with cellar-door tastings; off Orange Rd; 5 km E. ***Chateau Champsaur:*** oldest winery in the area (1886), with cellar-door tastings; (02) 6852 3908; 5 km SE. ***Ben Hall's Place:*** marks the site where the bushranger was shot dead by policemen; 20 km W. ***Jemalong Weir:*** with parklands by the Lachan River, good spot for fishing and picnicking; 24 km s.

See also CENTRAL WEST, p. 30

Forster–Tuncurry

Pop. 18 374
Map ref. 550 E9 | 555 O2

Little St, Forster; (02) 6554 8799 or 1800 802 692; www.greatlakes.org.au

 95.5 FM ABC Local, 107.3 Max FM

Located in the Great Lakes district, Forster is connected to its twin town, Tuncurry, by a concrete bridge across Wallis Lake, forming one large resort town. The area has an excellent reputation for its fishing and seafood, particularly its oysters.

Forster Arts and Crafts Centre: the largest working craft centre in NSW; Breese Pde. ***Tobwabba Art Studio:*** specialises in urban coastal Aboriginal art; cnr Breckenridge and Little sts, Forster. ***Pebbly Beach Bicentennial Walk:*** gentle and scenic 2 km walk to Bennetts Head, beginning at baths off North St, Forster. ***Wallis Lake Fishermen's Co-op:*** fresh and cooked oysters and ocean fish; Wharf St, Tuncurry. ***Dolphin-spotting and lake cruises:*** bookings at visitor centre. Dolphins can also be seen from Tuncurry Breakwall and Bennetts Head.

Forster Market: Town Park; 2nd Sun each month. ***Tuncurry Market:*** John Wright Park; 4th Sat each month.

Ripples: modern Australian; Shop 6, 24–30 Memorial Dr, Forster; (02) 6555 3949.

Tokelau Guest House: Federation guesthouse; 2 Manning St, Tuncurry; (02) 6557 5157.

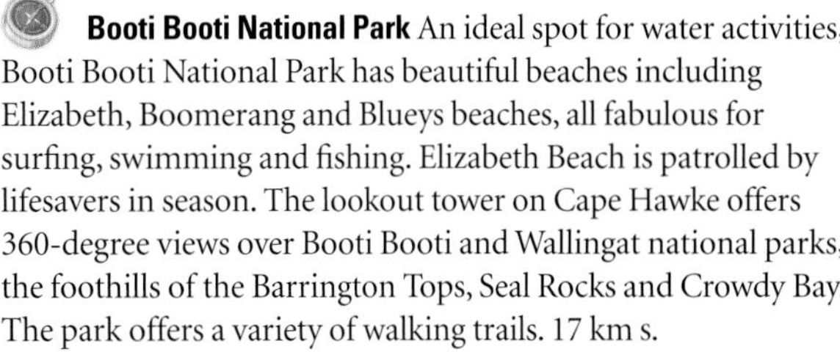

Booti Booti National Park An ideal spot for water activities, Booti Booti National Park has beautiful beaches including Elizabeth, Boomerang and Blueys beaches, all fabulous for surfing, swimming and fishing. Elizabeth Beach is patrolled by lifesavers in season. The lookout tower on Cape Hawke offers 360-degree views over Booti Booti and Wallingat national parks, the foothills of the Barrington Tops, Seal Rocks and Crowdy Bay. The park offers a variety of walking trails. 17 km s.

Cape Hawke: steep 400 m track to summit for views of Wallis Lake, Seal Rocks and inland to Great Dividing Range; 8 km s. ***The Green Cathedral:*** open-air church with pews and altar, sheltered by cabbage palm canopy; Tiona, on the shores of Wallis Lake; 13 km s. ***Smiths Lake:*** sheltered lake for safe swimming; 30 km s. ***Tours of Great Lakes area:*** kayak, 4WD, nature and eco-tours, including bushwalks; brochure from visitor centre.

See also HOLIDAY COAST, p. 27

Gilgandra

Pop. 2679
Map ref. 556 C10

Coo-ee Heritage Centre, Coo-ee March Memorial Park, Newell Hwy; (02) 6817 8700; www.gilgandra.nsw.gov.au

101.3 Star FM, 549 AM ABC Local

A historic town at the junction of three highways, 'Gil' is the centre for the surrounding wool and farming country. The 1915 Coo-ee March in which 35 men, given no support from the army, marched the 500 kilometres to Sydney to enlist for World War I left from here. Along the way they recruited over 200 men, announcing their arrival with a call of 'coo-ee!' The march sparked seven other such marches from country towns.

Coo-ee Heritage Centre Memorabilia from the 1915 Coo-ee March, and items relating to the Breelong Massacre, which took place after an Aboriginal man was insulted for marrying a white woman, and on which Thomas Keneally's *The Chant of Jimmy Blacksmith* was based. Coo-ee Memorial Park, Newell Hwy.

Rural Museum: vast collection of agricultural artefacts including antique farm machinery and early model tractors on display; Newell Hwy. ***Hitchen House Museum:*** the home of the Hitchen brothers, who initiated the Coo-ee March. The museum has memorabilia from WW I, WW II and Vietnam; Miller St. ***Orana Cactus World:*** almost 1000 different cacti on display collected over 40 years; open most weekends and by appt; Newell Hwy; bookings (02) 6847 0566. ***Gilgandra Observatory:*** Newtonian reflector and refractor telescopes and a sundial; open 7–10pm (8.30–10pm during daylight saving) Mon–Sat; Cnr Wamboin and Willie sts; (02) 6847 2646. ***Tourist drives:*** around town; brochure from visitor centre.

Rodeo: Oct. ***Coo-ee Festival:*** Oct.

Cafe Country Style: home-style meals; 44 Miller St; (02) 6487 1571. ***Holland's Family Diner:*** classic cafe; 11 Castlereagh St; (02) 6847 1199.

Anna's Place B&B: friendly, comfortable accommodation; 13 Morris St; (02) 6847 2790.

Warren Its location on the Macquarie River makes Warren a popular spot with anglers. For a stroll along the riverbank take the River Red Gum Walk, for birdwatching go to Tiger Bay Wildlife Reserve, and for a day of picnicking and swimming visit

Warren Weir. The racecourse is known as the 'Randwick of the west' and hosts some fantastic race days. 85 km W.

Gilgandra Flora Reserve: 8.5 ha of bushland, perfect for picnics and barbecues. Most plants flower in spring, making the park particularly spectacular Sept–Nov; 14 km NE. ***Emu Farm:*** raised for oil, leather and meat, guided tours; Tooraweenah; 40 km NW.

See also CENTRAL WEST, p. 30

Glen Innes

Pop. 5944
Map ref. 557 K5

152 Church St; (02) 6730 2400; www.gleninnestourism.com

106.7 Gem FM, 819 AM ABC Local

This beautiful town, set among rolling hills on the northern tablelands of New South Wales at an elevation of 1075 metres, is known for its fine parks, which are especially striking in autumn. It was the scene of many bushranging exploits in the 19th century, including some by the infamous Captain Thunderbolt. In the 1830s, two particularly hairy convict stockmen, Chandler and Duval, advised and guided settlers to new land where they settled stations. Because of Chandler and Duval, people came to know Glen Innes as the 'land of the beardies', a nickname that has stuck to this day. The town is undeniably proud of its Celtic beginnings (the first settlers were predominantly Scots), as illustrated by its attractions and festivals.

Centennial Parklands The site of Celtic monument 'Australian Standing Stones', which was built with 38 giant granite monoliths in recognition of the contribution made in Australia by people of Celtic origin. A full explanation of the stones can be read at Crofters Cottage, which also sells Celtic food and gifts. St Martin's Lookout provides superb views. Meade St.

Land of the Beardies History House: folk museum in the town's first hospital building; it has a reconstructed slab hut, period room settings and pioneer relics; Cnr Ferguson St and West Ave; (02) 6732 1035. ***Self-guide walks:*** past historic public buildings, especially on Grey St; brochure from visitor centre.

Market: Grey St; 2nd Sun each month. ***Minerama Gem Festival:*** world-class gems on display and guided fossicking trips to unique locations; Mar. ***Australian Celtic Festival:*** Apr/May. ***Gourmet in the Glen:*** Oct. ***Land of the Beardies Festival:*** Nov.

Ramona's Restaurant: modern Australian; 160 Church St; (02) 6732 2922. ***The Hereford Steakhouse:*** award-winning steakhouse; Rest Point Motel, New England Hwy; (02) 6732 2255.

Tudor House Boutique B&B: excellent boutique accommodation; 139–141 Church St; (02) 6732 3884 or 1800 139 149.

Gibraltar Range National Park and Washpool National Park These adjoining parks were World Heritage–listed in 1986 because of their ancient and isolated remnants of rainforest and their great variety of plant and animal species. Gibraltar Range is known for its scenic creeks and cascades and its unusual granite formations, The Needles and Anvil Rock. Gibraltar Range also contains over 100 km of excellent walking trails. 70 km NE. Washpool has the largest remaining stand of coachwood trees in the world and a unique array of eucalypt woods and rainforest. It has some of the least disturbed forest in NSW. 75 km NE.

Stonehenge: unusual balancing rock formations; 18 km S. ***Emmaville:*** the Australian beginnings of St John Ambulance occurred here. Includes a mining museum; 39 km NW. ***Deepwater:*** good fishing for trout, perch and cod with regular fishing safaris; bookings at visitor centre; 40 km N. ***Mann River Nature Reserve:*** popular camping spot due to fantastic swimming holes; 40 km NW. ***Torrington:*** gem fossicking, bushwalks and unusual rock formations; 66 km NW. ***Convict-carved tunnel:*** road tunnel halfway between Glen Innes and Grafton; Old Grafton Rd; 72 km W. ***Horse treks:*** accommodation at historic pubs; bookings at visitor centre.

See also NEW ENGLAND, p. 29

Glenbrook

Pop. 5138
Map ref. 544 G6 | 547 L10 | 555 J6

Blue Mountains Visitor Information Centre, Great Western Hwy; 1300 653 408; www.visitbluemountains.com.au

89.1 BLU FM, 576 AM ABC Radio National

Glenbrook is a picturesque village on the edge of the Blue Mountains. It was originally known as Watertank because it was used for the storage of water for local steam trains. Today it is a charming town with a large lagoon, close to the impressive Red Hands Cave that lies in Blue Mountains National Park.

Lapstone Zig Zag Walking Track The track follows the 3 km path of the original Lapstone Zig Zag Railway. The track includes convict-built Lennox Bridge (the oldest surviving bridge on the mainland), the abandoned Lucasville Station and numerous lookouts with views of Penrith and the Cumberland Plain. Nearby there is a monument to John Whitton, a pioneer in railway development. Starts in Knapsack St.

Glenbrook Lagoon: filled with ducks, and a perfect picnic spot with walking trails.

Market: Infants School, Ross St; 3rd Sat each month. ***Spring Festival:*** Nov.

Mash Cafe Restaurant: modern Australian, vegetarian and gluten-free options; 19 Ross St; (02) 4739 5908.

Lorikeet Cottage: self-contained cottage; 7 Powell St, Blaxland; 0418 602 690.

Faulconbridge This scenic town features the Corridor of Oaks, a line of trees, each one planted by an Australian prime minister. There is also the grave of Sir Henry Parkes, 'the father of Federation', and the stone cottage where Norman Lindsay lived, now a gallery and museum dedicated to his life and work. Lindsay was the author of Australian classics such as *The Magic Pudding* and was the subject of the film *Sirens* (1994). 16 km NW.

Wascoe Siding Miniature Railway: 300 m of steam and motor railway plus picnic and barbecue facilities. Trains operate 1st Sun each month; off Great Western Hwy; 2.5 km W. ***Blue Mountains National Park:*** Red Hands Cave is accessed by a 6 km return walk. The cave features hand stencils (mostly red, although some are white or orange) that were created between 500 and 1600 years ago. The artists created the stencils by placing their hands against the cave wall and blowing a mixture of ochre and water from their mouths. Euroka Clearing, 4 km S, is a popular camping spot home to many kangaroos. *For other parts of the park see Katoomba and Blackheath.* ***Springwood:*** galleries and craft and antique shops. Also home to the Ivy Markets; 2nd Sat each month (except Jan); civic centre, Macquarie Rd; 12 km NW. ***Linden:*** impressive Kings Cave with Caleys Repulse Cairn nearby commemorating early surveyor George Caley; 20 km W.

See also BLUE MOUNTAINS, p. 23

Gloucester

Pop. 2446
Map ref. 550 C8 | 555 M1 | 557 K12

27 Denison St; (02) 6558 1408; www.gloucester.org.au

97.7 Breeze FM, 100.9 FM ABC Local

At the foot of the impressive monolithic hills of the Bucketts Range, Gloucester calls itself the base camp to the Barrington Tops. This green and peaceful town is known for its top-quality produce including Barrington beef and perch.

Minimbah Aboriginal Native Gardens: bush-tucker gardens; Gloucester District Park. ***Folk Museum:*** pioneer household relics, toys, and gemstones and rocks; open 10am–2pm Thurs and Sat, 11am–3pm Sun; Church St; (02) 6558 9989. ***Town heritage walk:*** includes Lostrochs Cordial Factory and Gloucester Powerhouse; brochure from visitor centre. ***Walking trails:*** through nearby parks, brochure from visitor centre.

Shakespeare Festival: May. ***Mountain Man Triathlon:*** kayaking, mountain-biking and running; Sept.

Perenti: chic cafe with local produce; 69 Church St; (02) 6558 9219.

Altamira Retreat: nestled in scenic countryside; 1507 Bakers Creek Rd; (02) 6550 6558. ***Arrowee House B&B:*** traditional B&B; 152 Thunderbolts Way; (02) 6558 2050.

Barrington Tops National Park This World Heritage–listed rainforest is enormous and has a great variety of landscapes, flora and fauna. There are some good walking trails, beautiful forest drives and breathtaking views from Mt Allyn, 1100 m above sea level. The Barrington Tops Forest Dr from Gloucester to Scone has rainforest walks and picnic spots en route; brochure from visitor centre. 60 km W.

Copeland State Conservation Area Easily accessible dry rainforest, well known for its gold production in the 1870s and large stands of red cedar. Walking trails that utilise the old wagon and logging tracks are open to the public; 17km W. ***The Bucketts Walk:*** 90 min return with great views of town; Bucketts Rd; 2 km W. ***Lookouts:*** amazing views of the national park, town and surrounding hills at Kia-ora Lookout (4 km N), Mograni Lookout (5 km E) and Berrico Trig Station (14 km W). ***Goldtown:*** former site of the Mountain Maid Goldmine (1876), now mostly covered with rainforest. Also a historical museum, gold panning and underground mine tours; 16 km W.

See also HOLIDAY COAST, p. 27

Gosford

Pop. 166 626
Map ref. 543 D8 | 545 O3 | 555 L5

200 Mann St; (02) 4343 2353 or 1300 130 708; www.visitcentralcoast.com.au

92.5 FM ABC Local, 101.3 Sea FM

Part of the idyllic Central Coast region, Gosford is surrounded by national parks, steep hills and valleys, rainforest, lakes and ocean beaches. Understandably, Gosford continues to increase in popularity and has grown into a bustling city known for its high standard of tourism and its orchards and seafood.

Art galleries and craft and antique shops: many in town; contact visitor centre for details.

Gosford Country Show: May. ***Springtime Flora Festival:*** Sept. ***CoastFest:*** arts festival; Oct. ***Gosford to Lord Howe Island Yacht Race:*** Oct/Nov.

BodyFuel: fresh, healthy cafe meals; 1/9 William Court; (02) 4323 6669. ***Upper Deck:*** seafood, great views; 61 Masons Pde; (02) 4324 6705. ***Flair:*** modern Australian; 1/488 The Entrance Rd, Erina Heights; (02) 4365 2777.

Wombats B&B: traditional B&B; 144 Brisbane Water Dr; (02) 4325 5633. ***Claddagh House B&B:*** romantic retreat; 421 Avoca Dr, Green Point; (02) 4369 5251.

Ku-ring-gai Chase National Park Here the Hawkesbury River meets the sea with winding creeks, attractive beaches, hidden coves and clear water. Highlights are Resolute Track, with Aboriginal rock engravings and hand stencils; Bobbin Head, with a visitor centre and marina; and West Head Lookout, with panoramic views over the water. 33 km SW.

Gosford City Arts Centre: local art and craft and Japanese garden; 3 km E. ***Henry Kendall Cottage:*** museum in the poet's sandstone home. Also picnic and barbecue facilities in the grounds; 3 km SW. ***Australian Rainforest Sanctuary:*** walking trails through peaceful rainforest. The Firefly Festival is held here Nov–Dec; 14 km NW. ***Firescreek Fruit Wines:*** Holgate; (02) 4365 0768; 10 km NE. ***Australian Reptile Park:*** snakes, spiders and Galapagos tortoises. See shows throughout the day. Somersby Falls nearby provide an ideal picnic spot; Somersby; 15 km NW. ***Australian Walkabout Wildlife Park:*** native forest with 2 km of walking trails. Animals extinct in the area have been re-introduced with success thanks to the fence that keeps out feral animals; Calga; 20 km NW. ***Brooklyn:*** access to lower Hawkesbury for houseboating, fishing and river cruises. Historic Riverboat Postman ferry leaves Brooklyn weekdays for cruises and postal deliveries; 32 km S.

See also CENTRAL COAST & HAWKESBURY, p. 25

Goulburn

Pop. 20 131
Map ref. 554 H9 | 565 E3

201 Sloane St; (02) 4823 4492 or 1800 353 646; www.igoulburn.com

93.5 Eagle FM, 103.3 FM Community Radio, 1368 2GN Forever Classic Goulburn

Goulburn is at the junction of the Wollondilly and Mulwaree rivers, at the centre of a wealthy farming district. It was one of Australia's first inland settlements and the last proclaimed city in the British Empire. It is now known for its merino wool industry.

The Big Merino Even though it was only built in 1985, the Big Merino is an instantly recognisable landmark associated with Goulburn and its thriving merino wool industry. The 15 m high and 18 m long sculptured sheep has 3 floors, with a souvenir shop and an educational display on the history of wool in the area. Cnr Hume and Sowerby sts; (02) 4822 8013.

Old Goulburn Brewery: weekend tours of Australia's oldest brewery designed by colonial architect Francis Greenway. Brewing to original recipe; Bungonia Rd; (02) 4821 6071. ***Goulburn Historic Waterworks:*** displays antique waterworks engines, beside attractive parkland with picnic and barbecue facilities on Marsden Weir; off Fitzroy St. ***Rocky Hill War Memorial and Museum:*** erected in 1925 as a tribute to the

Goulburn men and women who served during WW I, it offers outstanding views across the city. Museum open weekends and holidays; Memorial Dr. ***Self-guide tour:*** historic buildings include Goulburn Courthouse and St Saviour's Cathedral; brochure from visitor centre.

Australian Blues Music Festival: Feb. ***Celebration of Heritage and Roses:*** Mar. ***Lilac City Festival:*** Oct.

Bungonia National Park Popular for adventurers with perfect terrain for canyoning, caving and canoeing. Walking trails offer fantastic river and canyon views. One walk passes through the spectacular Bungonia Gorge. 35 km E.

Pelican Sheep Station: farm tours, shearing and sheepdog demonstrations by appt; bookings (02) 4821 4668. Accommodation is available; 10 km S. ***Wombeyan Caves:*** Daily guided and self-guide cave tours through ancient and spectacular limestone cave system; inquiries (02) 4843 5976; 70 km N.

See also CAPITAL COUNTRY, p. 31

Grafton

Pop. 17 499
Map ref. 551 D9 | 557 N6

Clarence River Visitor Information Centre, Pacific Hwy, South Grafton; (02) 6642 4677; www.clarencetourism.com

 104.7 FM Clarence Coast, 738 AM ABC Local

With over 6500 trees in 24 parks, Grafton is known for its riverbank parks and jacaranda trees. Its city centre adds to the charm with wide streets, elegant Victorian buildings and the Clarence River passing through. Water lovers are spoiled for choice with everything from whitewater adventures, waterskiing and fishing.

Susan Island This rainforest recreation reserve in the Clarence River is home to a large fruit bat colony. Dusk is the time to visit to watch the bats flying off in search of food (wearing a hat is advisable if visiting at this time). During the day the island is a good spot for rainforest walks, barbecues and picnics. Access is via hired boat, skippered cruise or Clarence Islander ferry.

Grafton Regional Gallery: rated as one of the most outstanding regional galleries in Australia, the Grafton Regional Gallery inside Prentice House has permanent exhibitions such as the Jacaranda Art Society and Contemporary Australian Drawing collections; closed Mon; Fitzroy St; (02) 6642 3177. ***National Trust–classified buildings:*** include Schaeffer House, home of the Clarence River Historical Society, Christ Church Cathedral and the notorious Grafton Gaol; heritage trail brochure from visitor centre. ***Local art and craft shops:*** brochure from visitor centre.

Markets: Lawrence Rd, last Sat each month; Prince St, Grafton Showgrounds, 3rd Sat each month; Armidale Rd, South Grafton, every Sat morning. ***Autumn Artsfest:*** Apr. ***Grafton Cup:*** horserace; July. ***Grafton to Inverell Cycling Classic:*** Sept. ***Bridge to Bridge Ski Race:*** Oct. ***Jacaranda Festival:*** Oct–Nov.

Georgie's: modern Australian and art gallery; Grafton Regional Gallery, 158 Fitzroy St; (02) 6642 6996. ***Moos Restaurant:*** steakhouse; Pacific Hwy; (02) 6642 2833. ***Victoria's Restaurant:*** international; Quality Inn Grafton, 51 Fitzroy St; (02) 6640 9100.

Blooms Cottage: self-contained cottage; 6 Queen St; 0407 297 764. ***Ivy Cottage:*** B&B; 130 Centenary Dr; (02) 6642 3202.

Museum of Interesting Things Australian actor Russell Crowe chose Nymboida's coaching station for the home of his amazing collection of film memorabilia (mostly from Crowe's films like *Gladiator* and *L.A. Confidential*), vintage cars, motorcycles, a giant Cobb & Co coach and other 'interesting things'. Open 11am–3pm Mon–Fri, 10am–5pm Sat, 10am–3pm Sun; 3970 Armidale Rd, Nymboida; (02) 6649 4126. 47 km SW.

Nymboida: Waters pumped by the hydro-electric power station from the Nymboida River into Goolang Creek provide a high-standard canoe course that hosts competitions throughout the year. Canoe hire and lessons are on offer and facilities range from the learners' pond for beginners to grade III rapids for the experienced, thrill-seeking canoeist. The beautiful rainforest surrounds are excellent for bushwalking, abseiling, trail rides and platypus viewing. 47 km SW.

Ulmarra Village: National Trust–classified turn-of-the-century river port with exceptional galleries, craft shops and studios where you can watch artists at work; 12 km NE. ***Yuraygir National Park:*** highlights include Wooli for unspoiled surf beaches and Minnie Water for walking trails, secluded beaches, camping and abundant wildlife (especially the very friendly wallabies). Minnie Lagoon is a popular swimming, picnicking and boating spot; 50 km SE. ***Fishing:*** river fishing for saltwater or freshwater fish, depending on time of year and rainfall; details at visitor centre.

See also TROPICAL NORTH COAST, p. 28

Grenfell

Pop. 1991
Map ref. 554 D6

CWA Craft Shop and Visitors Centre, 68 Main St; (02) 6343 1612; www.grenfell.org.au

 99.5 Star FM, 549 AM ABC Local

Nestled at the foot of the Weddin Mountains, Grenfell is best known as the birthplace of writer Henry Lawson. The wealth appropriated during the days of the gold rush is evident in the opulent original buildings on Main Street. Originally named Emu Creek, the town was renamed after Gold Commissioner John Granville Grenfell, who was gunned down by bushrangers.

Henry Lawson Obelisk: memorial on the site of the house where the poet is believed to have been born; next to Lawson Park on the road to Young. ***Grenfell Museum:*** local relics (and their stories) from world wars, the gold rush, Henry Lawson and bushrangers; open weekends 2–4pm; Camp St. ***O'Brien's Reef Lookout:*** views of the town on a gold-discovery site with walkway and picnic facilities; access from O'Brien St. ***Weddin Bird Trails:*** unique Grenfell birdlife; maps from visitors centre. ***Historic buildings:*** walk and drive tours; brochures from visitor centre.

Guinea Pig Races: Easter and June. ***Henry Lawson Festival of Arts:*** June. ***Amateur Country Music Festival:*** Sept. ***Open Day:*** Oct.

Happy Inn Chinese Restaurant: traditional Chinese; Main St; (02) 6343 1366.

Weddin Mountains National Park The park is a rugged crescent of cliffs and gullies providing superb bushwalking, camping and picnicking spots. Two of the highlights of the bushwalks are Ben Hall's Cave, where the bushranger hid from the police, and Seaton's Farm, a historic homestead set on beautiful parkland. The bush is also rich with fauna including wedge-tailed eagles, honeyeaters and wallabies. 18 km SW.

Company Dam Nature Reserve: excellent bushwalking area; 1 km NW. ***Site of Ben Hall's farmhouse and stockyards:*** memorial; Sandy Creek Rd, off Mid Western Hwy; 25 km W.

See also CENTRAL WEST, p. 30

Griffith

Pop. 16 185
Map ref. 559 M7

Cnr Banna and Jondaryan avenues; (02) 6962 4145 or 1800 681 141; www.griffith.com.au

 99.7 Star FM, 549 AM ABC Local

Griffith is surrounded by low hills and fragrant citrus orchards in the heart of the Murrumbidgee Irrigation Area. It was designed by Walter Burley Griffin and named after Sir Arthur Griffith, the first Minister for Public Works in New South Wales. It is one of the largest vegetable-production regions and produces more than 60 per cent of the state's wine.

Hermits Cave and Sir Dudley de Chair's Lookout Hermits Cave is located down a path below the lookout. The cave is named because it was once home to Valerio Ricetti, an Italian miner from Broken Hill. After being jilted, he left his home and job and became a hermit in this cave. After many years of solitude he fell and broke his leg, and when he was hospitalised he was recognised by people who had known him in Broken Hill. In later years he became ill and local citizens collected money to send him back home to Italy, where he died three months later. Scenic Dr.

Pioneer Park Museum: the 11 ha of bushland features 40 replica and restored buildings, early 20th-century memorabilia and re-created Bagtown Village; Remembrance Dr. ***Italian Museum:*** exhibiting the lives of the early Italian migrants to Griffith including collections of memorabilia and stories from local families; Remembrance Dr. ***Griffith Regional Art Gallery:*** exhibition program of international and Australian artists that changes monthly; Banna Ave. ***Griffith Cottage Gallery:*** local paintings, pottery and handicrafts with exhibitions at various times throughout the year; Bridge Rd. ***Two Foot Tour and self-drive tour:*** feature the city's historic buildings and surrounding pastures; brochure from visitor centre.

Markets: Griffith Showground; Cnr Jondaryan Ave and Canal St; each Sun morning. ***La Festa:*** food and wine festival celebrating cultural diversity; Easter. ***UnWined in the Riverina:*** food and wine festival; June. ***Festival of Gardens:*** Oct. ***Catonia Fruit Salad Farm:*** guided tours Mon–Sat Mar–Nov at 1.30pm; bookings essential Dec–Feb and all Sundays.

The Clock Restaurant & Wine Lounge: modern Australian; 239–242 Banna Ave; (02) 6962 7111. ***La Scala:*** traditional Italian; 455B Banna Ave; (02) 6962 4322.

Ingleden Park Cottages: self-contained cottages; Coghlan Rd; (02) 6963 6527. ***Kidman Wayside Inn:*** comfortable motel; 58–72 Jondaryan Ave; (02) 6964 5666. ***Wilga Park Cottage:*** self-contained cottage; 6 Condon Rd, via Bilpul; (02) 6968 1661.

Cocoparra National Park Original Riverina forest full of wattles, orchids and ironbarks, the park is spectacular in spring, when the wildflowers bloom. The site is ideal for bushwalking, camping, birdwatching and picnicking, and the rugged terrain and vivid colours also make it popular with photographers. 25 km NE.

Altina Wildlife Park: range of exotic and native wildlife, on the banks of the Murrumbidgee River, horserides and cart rides; inquiries 0412 060 342; 35 km S. ***Catania Fruit Salad Farm:*** horticultural farm with tours at 1.30pm daily; Cox Rd, Hanwood; (02) 6963 0219; 8 km S. ***Lake Wyangan:*** good spot for variety of watersports and picnicking; 10 km NW. ***Many wineries:*** include De Bortoli and McWilliams, open for cellar-door tastings; map from visitor centre.

See also RIVERINA, p. 34

Gulgong

Pop. 1904
Map ref. 554 H2

109 Herbert St; (02) 6374 1202; www.gulgong.net

93.1 Real FM, 107.1 FM ABC Local

Gulgong was named by the Wiradjuri people (the name means 'deep waterhole'.) The town did not excite European interest until gold was discovered in 1866. By 1872 there were 20 000 people living in the area. By the end of the decade 15 000 kilograms of gold had been unearthed, the prospectors had gone and almost all of the local Aboriginal people had been slaughtered. Today, the town stands visually almost unchanged from these times. The narrow, winding streets follow the paths of the original bullock tracks past iron-lace verandahs, horse troughs and hitching rails.

Henry Lawson Centre Housed in the Salvation Army Hall, which was built in 1922, the year Lawson died, the centre has the largest collection of Lawson memorabilia outside Sydney's Mitchell Library. It includes original manuscripts, artefacts, photographs, paintings and an extensive collection of rare first editions. 'A Walk Through Lawson's Life' is an exhibition that uses Lawson's words to illustrate the poverty, family disintegration, deafness and alcoholism that shaped his life, as well as the causes he was passionate about such as republicanism, unionism and votes for women. Mayne St; (02) 6374 2049.

Pioneers Museum: illustrates every era of Gulgong's history. Exhibits include a replica of a classroom from the 1880s, period clothing and rare antique crockery; Cnr Herbert and Bayly sts. ***Red Hill:*** site of the town's original gold strike, featuring restored stamper mill, poppet head and memorial to Henry Lawson; off White St. ***Mayne Street Symbols:*** inscribed in the pavement by a local artist, to depict the 'language of the road' used by diggers to advise their mates who may have followed them from other goldfields. ***Town trail:*** self-guide walking tour of historic buildings such as Prince of Wales Opera House and Ten Dollar Town Motel; brochure from visitor centre.

Folk Festival: Jan. ***Henry Lawson Festival:*** June.

Butcher Shop Cafe: relaxed atmosphere, delicious food; 113 Mayne St; (02) 6374 2622. ***Phoebe's Licensed Restaurant:*** modern Australian; cnr Mayne and Medley sts; (02) 6374 1204.

Green Gables B&B: quaint B&B; 8 Tallawang Rd; (02) 6374 2855.

Goulburn River National Park The park follows approximately 90 km of the Goulburn River with sandy riverbanks making easy walking trails and beautiful camping sites. Rare and threatened plants abound here, as do wombats, eastern grey kangaroos, emus and birds. Highlights include the Drip, 50 m curtains of water dripping through the rocks alongside the Goulburn River, sandstone cliffs honeycombed with caves, and over 300 significant Aboriginal sites. 30 km NE.

Ulan: Ulan Coal Mine has viewing areas overlooking a large open-cut mine. Also here is Hands on the Rock, a prime example of Aboriginal rock art; 22 km NE. ***Talbragar Fossil Fish Beds:*** one of the few Jurassic-period fossil deposits in Australia; 35 km NE. ***Wineries:*** cellar-door tastings; brochure from visitor centre.

See also CENTRAL WEST, p. 30

Gundagai

Pop. 1999
Map ref. 554 D9 | 565 A3 | 566 C1 | 585 P1

249 Sheridan St; (02) 6944 0250; www.gundagai.local-e.nsw.gov.au

94.3 FM Sounds of the Mountain, 549 AM ABC Local

Gundagai is a tiny town on the Murrumbidgee River at the foot of Mount Parnassus. The town and the nearby Dog on the Tuckerbox statue have been celebrated through song and verse for many years. Banjo Paterson, C. J. Dennis and Henry Lawson all included the town in their works. It was also the scene in 1852 of Australia's worst flood disaster when 89 of the 250 townsfolk died. The count could have been worse but for a local Aboriginal man, Yarri, who paddled his bark canoe throughout the night to rescue stranded victims. Gundagai was moved to higher ground soon after, and there are monuments celebrating Yarri's efforts. Inside the visitor centre are statues of Dad, Dave, Mum and Mabel (characters from the writings of Steele Rudd).

Marble Masterpiece In an amazing display of patience and determination, local sculptor Frank Rusconi, who is also responsible for the Dog on the Tuckerbox statue, worked to create a cathedral in miniature. He built it in his spare time over 28 years, hand-turning and polishing the 20 948 individual pieces required to build it. Visitor centre, Sheridan St.

Gabriel Gallery: outstanding collection of photographs, letters and possessions illustrating Gundagai's unique history; Sheridan St. ***Gundagai Museum:*** relics include Phar Lap's saddle, Frank Rusconi's tools, and artefacts from the horse and buggy era; Homer St. ***Lookouts:*** excellent views of the town and surrounding green valleys from the Mt Parnassus Lookout in Hanley St and the Rotary Lookout in Luke St; South Gundagai. ***Historical town walk:*** includes the National Trust–classified Prince Alfred Bridge and St John's Anglican Church; leaflet from visitor centre.

Turing Wave Festival: celebrating Irish heritage; Sept. ***Spring Flower Show:*** Oct. ***Dog on the Tuckerbox Festival:*** Nov. ***Snake Gully Cup Carnival:*** horserace; Nov. ***Rodeo:*** Dec.

The Poets' Recall Motel & Licensed Restaurant: country restaurant; Cnr Pinch and West sts; (02) 6944 1777.

Bengarralong B&B: riverside homestead; Tarrabandra Rd; (02) 6944 1175. ***Gundagai Historic Cottages:*** heritage self-contained exclusive-use accommodation; 51 Sheridan St; (02) 6944 2385. ***Gundagai Tourist Park:*** camping, basic cabins, caravan sites; 1 Nangus Rd; (02) 6944 4440 or 1300 722 906.

The Dog on the Tuckerbox Originally mentioned in the poem *Bill the Bullocky* by Bowyang Yorke, this monument to pioneer teamsters and their dogs is recognised throughout the nation as an Australian icon. It was celebrated in the song 'Where the Dog Sits on the Tuckerbox' by Jack O'Hagan (the songwriter responsible for 'Along the Road to Gundagai'). The dog was unveiled in 1932 by Prime Minister Joseph Lyons. 8.5 km N.

See also RIVERINA, p. 34

Gunnedah

Pop. 7541
Map ref. 556 G9

 **Anzac Park, South St; (02) 6740 2230 or 1800 562 527; www.infogunnedah.com.au**

1080 AM 2MO

At the heart of the Namoi Valley, Gunnedah is instantly recognisable by the grain silos that tower over the town. The area is abundant with native wildlife, especially koalas. Gunnedah claims to be the koala capital of the world, with one of the largest koala populations in the country; they are often seen wandering around town. A large centre, Gunnedah still manages to keep a laid-back atmosphere and has been home to famous Australians such as Dorothea Mackellar and Breaker Morant.

Anzac Park The Water Tower Museum, housed in the town's main water tower, has a mural and display of early explorers, memorabilia from several wars and schools, and an Aboriginal history display. Dorothea Mackellar, the renowned Australian poet responsible for *My Country,* has a memorial statue in the park. Memorabilia of her life and of the annual national school poetry competition in her name can be viewed at the park's visitor centre; open 10am–2pm; South St.

Rural Museum: early agricultural machinery and the largest privately owned firearm collection in the country; Mullaley Rd. ***Red Chief Memorial:*** to Aboriginal warrior Cumbo Gunnerah, of the Gunn-e-dar people of the Kamilaroi tribe; State Office building, Abbott St. ***Old Bank Gallery:*** local art and craft; Conadilly St. ***Bicentennial Creative Arts Centre:*** art and pottery display and the watercolour series 'My Country' by Jean Isherwood; open 10am–4pm Thurs–Sun; Chandos St. ***Plains of Plenty:*** local craft and produce; Barber St. ***Eighth Division Memorial Avenue:*** with 45 flowering gums, each with a plaque in memory of men who served in the Eighth Division in WW II; Memorial Ave. ***Breaker Morant Drive:*** a plaque tells the story of Henry Morant, known as 'the breaker' because of his skill with horses; Kitchener Park. ***Poets Drive:*** celebration of Australian Poetry, the Poets Drive is a self-guide drive tour inspired by Gunnedah's iconic landmarks and local heroes. ***Bindea walking track and town walk:*** memorials, koala and kangaroo sites, lookouts and porcupine reserve; brochure from visitor centre.

Market: Wolseley Park, Conadilly St; 3rd Sat each month. ***National Tomato Competition:*** search for Australia's biggest tomato, and related celebrations; Jan. ***Week of Speed:*** festival includes go-karts, cars, bikes and athletics; Mar. ***Gunnedah Bird Expo and Sale:*** Apr. ***Ag Quip:*** agricultural field days; Aug. ***North-West Swap Meet:*** vintage cars; Sept.

Two Rivers Brasserie: hearty bistro meals; 313 Conadilly St; (02) 6742 0400. ***The Verdict Coffee:*** modern cafe; 147 Conadilly St; (02) 6742 0310.

Alyn Motel: comfortable contemporary accommodation; 351 Conadilly St; (02) 6742 5028.

Lake Keepit This lake is great for watersports, fishing and boating, and there is even a children's pool. If you want to stay a while longer, there is the Keepit Country Campout. The Campout provides all the facilities needed to camp without roughing it too much, including tents, a kitchen and showers and toilets, and there is a campfire amphitheatre for evening entertainment. For the daytime, there is all the equipment needed for canoeing, kayaking, rock climbing, gliding and bushwalking. 34 km NE.

Porcupine Lookout: views over town and surrounding agricultural area; 3 km SE. ***Waterways Wildlife Park:*** abundant with native animals such as kangaroos, koalas, wombats and emus; Mullaley Rd; 7 km W. ***150° East Time Meridian:*** the basis of Eastern Standard Time, crossing the Oxley Hwy; 28 km W.

See also NEW ENGLAND, p. 29

Guyra

Pop. 1755
Map ref. 557 J7

Rafters of Guyra Restaurant and Visitor Information Centre, New England Highway; (02) 6779 1876.

 100.3 FM, 720 AM ABC Radio National

Guyra is the highest town on the New England tablelands at an altitude of 1320 metres on the watershed of the Great Dividing Range. Snow is not unusual in winter and at other times the town is crisp and green.

Mother of Ducks Lagoon The reserve is a rare high-country wetland and home to hundreds of waterbirds. The migratory Japanese snipe is known to stop here and it is a nesting site for swans. There is a viewing platform with an identification board covering dozens of different birds. McKie Pde.

Blush Tomatoes: largest greenhouse growing tomatoes in the Southern Hemisphere. It has boosted Guyra's population and economy enormously over the last 3 years; Elm St. ***Historical Society Museum:*** themed room displaying town memorabilia and the story of the Guyra ghost; open by appt; Bradley St; bookings (02) 6779 2132. ***Railway Station:*** large display of antique machinery, rail train rides; Bradley St.

Lamb and Potato Festival: includes Hydrangea Festival; Jan.

Rafters Seafood: hearty fare; New England Hwy; (02) 6779 1876.

George's Creek Cottages: self-contained cottages on working sheep and cattle farm; 'Cabarfeidh', Wandsworth via Guyra; (02) 6779 4235. ***Top of the Range Retreat:*** luxury self-contained cottages; 93 Ollera St; (02) 6779 2336.

Thunderbolt's Cave: picturesque and secluded cave, rumoured to be where the bushranger Captain Thunderbolt hid from police; 10 km S. ***Handcraft Hall – The Pink Stop:*** hand-knitted garments, paintings, pottery and Devonshire teas; 10 km N. ***Chandler's Peak:*** spectacular views of the tablelands from an altitude of 1471 m; 20 km E.

See also NEW ENGLAND, p. 29

Hay

Pop. 2632
Map ref. 559 J7

407 Moppett St; (02) 6993 4045; www.visithay.com.au

92.1 Hay FM, 100.9 FM ABC Radio National

Located in the heart of the Riverina, the most striking thing about Hay is the incredibly flat plains in which it sits. The saltbush flats afford amazing views across the land, especially at sunrise and sunset, and the terrain makes bicycles a popular and easy mode of transport for residents. American travel writer Bill Bryson described Hay as 'a modest splat' and 'extremely likeable'.

POW Internment Camp Interpretive Centre Housed in Hay's magnificent restored railway station, the centre documents the WW II internment in Hay of over 3000 prisoners of war. The first internees were known as the 'Dunera boys', Jewish intellectuals who had fled Germany and Austria. The camp established a garrison band and a newspaper and printed camp money. The Dunera boys even ran their own 'university', teaching subjects such as atomic research and classical Greek. The Dunera boys held a reunion in Hay in 1990 and there is a memorial on Showground Rd. Murray St; inquiries (02) 6993 4045.

Shear Outback Centre – The Australian Shearers Hall of Fame: interactive experiences and shearing deomonstrations with sheep dogs, historic Murray Downs Woolshed and exhibitions; Sturt Hwy; inquiries (02) 6993 4000. ***Witcombe Fountain:*** ornate drinking fountain presented to the people of Hay by mayor John Witcombe in 1883; Lachlan St. ***Coach house:*** features an 1886 Cobb & Co coach, which travelled the Deniliquin–Hay–Wilcannia route until 1901; Lachlan St. ***Hay Gaol Museum:*** contains memorabilia and photographs of the town, and the building's history from 1878 as a gaol, maternity hospital, hospital for the insane and POW compound; Church St; inquiries (02) 6993 4045. ***War Memorial High School Museum:*** built in recognition of those who served in WW I, with war memorabilia and an honour roll. The building still operates as a school, so call for opening times (02) 6993 1408; Pine St. ***Bishop's Lodge:*** restored 1888 iron house, now a museum and gallery with a unique and remarkable collection of heritage roses. Holds a market 3rd Sun in Oct. Open 2–4.30pm Mon–Sat; Cnr Roset St and Sturt Hwy. ***Ruberto's Winery:*** cellar-door tastings; Sturt Hwy. ***Hay Wetlands:*** especially spectacular in spring, the land is a breeding ground for over 60 inland bird species with a breeding island and tree plantation; north-western edge of town; brochure from visitor centre. ***Hay Park:*** pleasant picnic spot with a nature walk along the banks of the river; off Brunker St. ***Murrumbidgee River:*** excellent sandy river beaches and calm water, perfect for waterskiing, canoeing, swimming and picnics. Enjoy excellent freshwater fishing for Murray cod, yellow-belly perch and redfin. A licence is required; available from outlets in town, including visitor centre. ***Heritage walk and scenic drive:*** walk includes city structures built for the harsh outback such as the beautifully restored courthouse (1892) on Moppett St and the shire office (1877) on Lachlan St. The drive takes in the parklands, river and surrounding saltbush plains; brochure from visitor centre.

Sheep Show: June. ***Rodeo:*** Oct. ***Hay Races:*** Nov.

Cumquats Cuisine: popular cafe; 161 Lachlan St; (02) 6993 4399. ***Jolly Jumbuck Bistro:*** relaxed bistro; 148 Lachlan St; (02) 6993 1137.

BIG 4 Hay Plains Holiday Park: basic cabins, camping sites; 4 Nailor St; (02) 6993 1875 or 1800 251 974. ***Saltbush Motor Inn:*** contemporary motel; 193 Lachlan St; (02) 6993 4555. ***The Bank:*** heritage-listed, centrally located; 86 Lachlan St; (02) 6993 1730.

Booligal In an area known as the 'devil's claypan', this hot and dusty sheep- and cattle-town is mentioned in Banjo Paterson's poem *Hay and Hell and Booligal.* The poem says that a visit to Booligal is a fate worse than hell with topics of complaint including heat, flies, dust, rabbits, mosquitoes and snakes. On the plus side, the atmosphere is relaxed and friendly. It is off the beaten tourist track and there is a memorial to John Oxley, the first European in the area, in the shape of a giant theodolite (surveyor's tool). 78 km N. Halfway to Booligal, look out for the lonely ruins of One Tree Hotel.

Hay Weir: on the Murrumbidgee River, excellent for picnics, barbecues and Murray cod fishing; 12 km w. ***Sunset viewing area:*** the vast plains provide amazingly broad and spectacular sunsets; Booligal Rd; 16 km N. ***Maude Weir:*** surprisingly green and lush oasis, ideal for picnics and barbecues; 53 km w. ***Goonawarra Nature Reserve:*** no facilities for visitors, but the flood plains with river red gum forests and black box woodlands are still worth a visit – waterfowl in the billabongs, plenty of Murray cod in the Lachlan River and kangaroos and emus on the plains; 59 km N. ***Oxley:*** tiny town with river red gums and prolific wildlife (best seen at dusk); 87 km NW.

See also Riverina, p. 34

Holbrook

Pop. 1339
Map ref. 554 B11 | 559 P11 | 585 M4

Greater Hume Shire Visitor Information Centre, 15 Wallace St; (02) 6036 2422; www.holbrook.nsw.au

93.1 Star FM, 990 AM ABC Radio National

Holbrook is a well-known stock-breeding centre rich with history. Originally called The Germans because of its first European settlers, the name was later changed to Germanton. During World War I, with the allies fighting the Germans, even this name became unacceptable and a new name had to be found. British Commander Norman Holbrook was a war hero and had been awarded the Victoria Cross and the French Legion of Honour, so it was decided the town would be named after him.

Otway Submarine The 30 m vessel, once under the command of Norman Holbrook, was decommissioned in 1995. The town was given the fin of the submarine by the Royal Australian Navy and was busily trying to raise funds to purchase the full piece of history, when a gift of $100 000 from Commander Holbrook's widow made the purchase possible. Mrs Holbrook was the guest of honour at the unveiling in 1996. The Submarine Museum next door features submariner memorabilia, including a control room with working periscope, and Commander Holbrook Room. Hume Hwy.

Bronze statue: Commander Holbrook and his submarine, a scale model of the one in which Holbrook won the VC in WW I; Holbrook Park, Hume Hwy. ***Woolpack Inn Museum:*** 22 rooms furnished in turn-of-the-century style, horse-drawn vehicles and farm equipment surrounded by lovely gardens; Albury St. ***Ten Mile Creek Gardens:*** attractive gardens, excellent for picnics. Also features a miniature railway, operating on the 2nd and 4th Sun each month, and every Sat during holidays; behind museum. ***National Museum of Australian Pottery:*** extensive range of 19th- and early-20th-century domestic pottery, and photographs; closed Wed, and Aug; Albury St. ***Ian Geddes Walk:*** through tranquil bushland, following Ten Mile Creek. Begins behind Grimwood's Craft Shop; Hume Hwy.

Planes, Trains and Submarines Festival: Mar.

Holbrook Airfield: ultralight flights over town and surrounds available; 3 km N. ***Hume and Hovell Walking Track:*** access to short sections of the 440 km track from Albury to Yass via Woomargama; 15 km s. For a kit, including maps, contact Department of Lands, Sydney (02) 9228 6666.

See also Murray, p. 35

Huskisson

Pop. 3391
Map ref. 553 D11 | 555 J10 | 565 G4 | 567 O2

Shoalhaven Visitors Centre, cnr Princes Hwy and Pleasant Way, Nowra; (02) 4421 0778 or 1300 662 808; www.shoalhaven.nsw.gov.au

94.9 Power FM, 603 AM Radio National

Huskisson is a sleepy holiday resort and fishing port on Jervis (pronounced 'Jarvis') Bay. It was named after British politician William Huskisson, secretary for the colonies and leader of the House of Commons, who was killed by a train in 1830 while talking to the Duke of Wellington at a railway opening. The idyllic bay is renowned for its white sand and clear water, and there are usually several pods of dolphins living in the bay, making the area ideal for cruises and diving.

Lady Denman Heritage Complex The centre features displays on the history of wooden shipbuilding at Huskisson. There is also Laddie Timbery's Aboriginal Art and Craft Centre, with bush tucker demonstrations and talks on request, and the Museum of Jervis Bay Science and the Sea with fine maritime and surveying collections. Woollamia Rd; inquiries (02) 4441 5675.

Market: Huskisson Sporting Ground; 2nd Sun each month. ***White Sands Carnival:*** stalls, music and entertainment; Easter.

Huskisson Bakery and Cafe: sandwiches, homemade pies; 11 Currambene St; (02) 4441 5015. ***Locavore:*** cafe with organic and Fair Trade products; 2/66 Owen St; (02) 4441 5464. ***Seagrass Brasserie:*** seafood; 13 Currambene St; (02) 4441 6124. ***The Gunyah Restaurant:*** modern Australian, stunning treetop views; Paperbark Camp, 571 Woollamia Rd, Woollamia; (02) 4441 7299.

Sandholme Guesthouse: romantic boutique coastal retreat; 2 Jervis St; (02) 4441 8855. ***Paperbark Camp:*** luxury safari-style tree houses; 571 Woollamia Rd, Woollamia; (02) 4441 6066 or 1300 668 167.

Booderee National Park This park is jointly managed by the Wreck Bay Aboriginal Community and Environment Australia. Highlights include Aboriginal sites, the Cape George Lighthouse, Booderee Botanic Gardens and magnificent beaches. Barry's Bush Tucker Tours offers guided walks drawing on Aboriginal food and culture. Bookings on (02) 4442 1168; 11 km s.

Jervis Bay National Park: walking tracks, mangrove boardwalk and amazing beaches (Hyams Beach claims the whitest sand in the world); north and south of town. ***Jervis Bay Marine Park:*** the clear waters, reefs and deep-water cliffs with caves offer superb diving. Bookings are taken at the visitor centre for dolphin-watching cruises.

See also South Coast, p. 32

Iluka

Pop. 1739
Map ref. 551 F7

Lower Clarence Visitor Centre, Ferry Park, Pacific Hwy, Maclean; (02) 6645 4121; www.clarencetourism.com

 104.7 FM Clarence Coast's FM, 738 AM ABC Local

Located at the mouth of the Clarence River on the north coast, Iluka is a relatively uncommercial fishing and holiday village. Its attractiveness is evident in the long stretches of sandy white beaches and rare and accessible rainforest. Iluka Nature Reserve has the largest remnant of littoral rainforest (trees obtaining water via filtration through coastal sand and nutrients from airborne particles) in New South Wales.

River cruises: day cruises and evening barbecue cruises (Wed only); Wed, Fri and Sun; from the Boatshed. ***Passenger ferry:*** travelling daily to Yamba; from the Boatshed. ***Iluka Fish Co-op:*** fresh catches on sale from 9am; adjacent to the Boatshed. ***Walking track:*** picturesque coastline walk; access via Iluka Bluff to the north and Long St to the south.

Makuti Cafe: waterfront cafe; 2A Charles St; (02) 6646 5212.

Iluka Nature Reserve This area, nestled on the narrow peninsula where the Clarence River meets the ocean, was World Heritage–listed in 1986. It happens to be the largest remaining coastal rainforest in NSW. It is rich with birdlife and is a beautiful spot for activities such as fishing, swimming, surfing, canoeing, walking and camping. 1 km N.

Iluka Bluff Beach: safe swimming beach with good surf and a whale-watching lookout; brochure from visitors centre; 1 km N. ***Bundjalung National Park:*** protects ancient rainforest and the Esk River, the largest untouched coastal river system on the north coast. Woody Head has rare rainforest with campground, fishing and swimming; 4 km N. ***Woombah Coffee Plantation:*** world's southernmost coffee plantation; tours by appt; bookings (02) 6646 4121; 12 km NW.

See also TROPICAL NORTH COAST, p. 28

Inverell

Pop. 9748
Map ref. 557 I5

Campbell St; (02) 6728 8161 or 1800 067 626; www.inverell-online.com.au

95.1 Gem FM, 738 AM ABC Local

This town on the Macintyre River at the centre of the New England tablelands is known as the 'Sapphire City'. It is also rich in other mineral deposits, including zircons, industrial diamonds and tin. The country here has lush farm and grazing land and excellent weather conditions with cool nights and warm sunny days. A Scottish immigrant gave Inverell its name, which means 'meeting place of the swans' in Gaelic.

Pioneer Village This collection of homes and buildings dating from 1840 was moved from its original site to form a 'village of yesteryear'. Attractions include Grove homestead, Paddy's Pub and Mt Drummond Woolshed. Gooda Cottage has an impressive collection of gems and minerals. Tea and damper are served by prior arrangement. Tingha Rd; (02) 6722 1717.

Visitor Centre: local and imported gems, a static display on the local mining industry, photographs and a video of local attractions; Campbell St. ***Inverell Art Centre:*** paintings, pottery and craft; Evans St. ***Gem Centre:*** visitors can see local stones being processed; Byron St. ***National Transport Museum:*** over 200 vehicles on display with an impressive collection of rarities; Taylor Ave. ***Town Stroll:*** includes sites such as the National Trust–classified courthouse and the CBC Bank building with stables at the rear; brochure from visitor centre. ***Arts, crafts and wood-turning:*** work by local artists; brochure from visitor centre.

Market: Campbell Park: 1st (except Jan) and 3rd Sun each month. ***Grafton to Inverell Cycling Classic:*** Sept/Oct. ***Sapphire City Festival:*** Oct. ***Antique Machinery Rally:*** Oct. ***Great Inland Fishing Festival:*** Dec. ***Inverell Jockey Club Boxing Day Meeting:*** Dec.

The Royal Club Hotel: international cuisine; 260 Byron St; (02) 6722 2811.

Blair Athol Estate: luxury country guesthouse; Warialda Rd; (02) 6722 4288. ***Fossickers Rest Tourist Park:*** lovely holiday park; Lake Inverell Dr; (02) 6722 2261.

Kwiambal National Park The Macintyre River flows through gorges and plunge pools to Macintyre Falls and then leads into the Severn River. The park is rich with protected woodlands of white cypress pine, box and ironbark. Bat nurseries can be viewed with a torch in the remarkable Ashford Caves, which until the 1960s were mined for guano (bat droppings) to be used as fertiliser on local farms. The park makes a serene site for swimming, bushwalking and camping. Encounters with kangaroos, emus and koalas are common. 90 km N.

McIlveen Lookout: excellent views of town and surrounding pastures and nature reserve; 2 km W. ***Lake Inverell Reserve:*** 100 ha of unique aquatic sanctuary for birds and wildlife. Also an excellent site for picnics, bushwalking, birdwatching and fishing; 3 km E. ***Goonoowigall Bushland Reserve:*** rough granite country rich with birdlife and marsupials, it offers superb birdwatching, remains of a Chinese settlement, bushwalking trails and picnic areas; 5 km S. ***Copeton Dam State Recreation Area:*** perfect for boating, waterskiing, swimming, fishing, bushwalking and rock climbing. It also has adventure playgrounds, and picnic and barbecue facilities. Kangaroos graze on the golf course at dusk; inquiries (02) 6723 6269; 40 km S. ***DeJon Sapphire Centre:*** no longer a working mine, but has sapphires on display and for sale; Glen Innes Rd; (02) 6723 2222; 19 km E. ***Green Valley Farm:*** working sheep property with zoo, accommodation, extensive gardens, playground and picnic and barbecue facilities. The highlight is Smith's Mining and Natural History Museum, with a rare collection of gems and minerals, local Aboriginal artefacts, antiques and period clothing; inquiries (02) 6723 3370; 36 km SE. ***Pindari Dam:*** fishing, swimming, camping and picnic facilities; 58 km N. ***Gwydir River:*** one of the best whitewater rafting locations in the country during the summer months; brochure from visitor centre. ***Cool-weather wineries:*** cellar-door tastings; brochure from visitor centre. ***Fossicking sites:*** great spots for searching for tin, sapphires, quartz or even diamonds; maps at visitor centre.

See also NEW ENGLAND, p. 29

Jamberoo

Pop. 935
Map ref. 553 E7 | 555 J9 | 565 H3

Jamberoo Newsagency Visitor Centre, Shop 2, 18 Allowrie St; (02) 4236 0100; www.kiama.com.au

98 FM, 603 AM ABC Radio National

One of the most picturesque areas of the New South Wales coast, Jamberoo was once tropical forest. The town is surrounded by nature reserves and national parks, but is now situated on the lush green dairy pastures that have made Jamberoo prosperous as a dairy farming region. The surrounding forests are popular with bushwalkers and birdwatchers.

Jamberoo Hotel: charming 1857 building with meals and Sun afternoon entertainment; Allowrie St.

Market: Kevin Walsh Oval; last Sun each month.

Ben Ricketts Environmental Preserve: basic self-contained cabins; 774 Jamberoo Mountain Rd; (02) 4236 0208. ***Cedar Mists Cottage:*** rustic rainforest cottage; 130 Misty La; (02) 4237 6199. ***The Birdhouse:*** contemporary self-contained cottage; location provided with booking; 0432 445 229.

Budderoo National Park This park offers views from a plateau across sandstone country, heathlands and rainforest. There are excellent walking trails, including one that is accessible by wheelchair, and there are 3 lookouts with views of Carrington Falls. The Minnamurra Rainforest Centre is the highlight with an elevated boardwalk through rainforest and a steep paved walkway to Minnamurra Falls. 4 km W.

Jamberoo Action Park: family fun park with water slides, speedboats, racing cars and bobsleds; (02) 4236 0114; 3 km N. ***Jerrara Dam:*** a picturesque reserve on the banks of a 9 ha dam that was once the town's main water supply. Picnic area surrounded by remnant rainforest and freshwater wetland; 4 km SE. ***Saddleback Lookout:*** 180-degree views of the coast and the starting point for Hoddles Trail, a 1 hr walk with beautiful views to Barren Grounds escarpment; 7 km S. ***Barren Grounds Nature Reserve:*** this 1750 ha heathland plateau on the Illawarra Escarpment protects over 450 species of plant and 150 species of bird, including the rare ground parrot and eastern bristlebird. It has fabulous bushwalking and birdwatching; guided tours available (bookings (02) 4423 2170); park details from Fitzroy Falls Visitor Centre (02) 4887 7270; 10 km SW.

See also SOUTHERN HIGHLANDS, p. 24

Jerilderie

Pop. 769
Map ref. 559 L10 | 584 G1

The Willows Museum, 11 Powell St; (03) 5886 1666.

94.1 FM ABC Radio National, Classic Rock 102.5 FM

Jerilderie is an important merino stud area, but is better known for its links with the Kelly Gang. In 1879 the gang captured the police, held the townspeople hostage for two days, cut the telegraph wires and robbed the bank. It was here that Kelly handed over the famous Jerilderie Letter, justifying his actions and voicing his disrespect for police, whom he called 'a parcel of big ugly fat-necked wombat-headed, big-bellied, magpie-legged, narrow-hipped, splay-footed sons of Irish bailiffs or English landlords'. The town of Jerilderie is now the gateway to the Kidman Way, an 800-kilometre outback highway.

Telegraph Office and The Willows Museum The well-preserved Telegraph Office is where the Kelly Gang cut the telegraph wires in 1879. The Willows Museum next door houses photographs, including some of the Kelly Gang, documents of local historical significance and the cell door from the old police station. Samples of local craft are also on display and Devonshire teas are available. Powell St.

St Luke Park: features Steel Wings, one of the largest windmills in the Southern Hemisphere. The park runs along the bank of Lake Jerilderie, which is popular for all watersports, especially waterskiing. ***Mini Heritage Steam Rail:*** runs along charming Billabong Creek, which also features the 1.8 km Horgans Walk. Entry behind The Willows Museum; runs 2nd and 5th Sun each month. ***Ned Kelly Heritage Trail:*** retraces the gangs' visit; brochure from visitor centre.

Coleambally Officially opened in 1968, Coleambally is NSW's newest town and is at the centre of the Coleambally Irrigation Area. It features the Wineglass Water Tower and a dragline excavator used in the irrigation scheme. The excavator is still in working order and can be viewed in the Lions Park at the town's entrance. Tours of the rice mill and of the farms are possible. The area is a haven for birdlife and kangaroos. 62 km N.

See also MURRAY, p. 35

Jindabyne

Pop. 1902
Map ref. 565 B8 | 566 E10 | 581 C4

Snowy Region Visitor Centre, Kosciuszko Rd; (02) 6450 5600; www.snowymountains.com.au

97.1 FM Radio National, 97.7 Snow FM

Jindabyne, adjacent to the south-eastern section of Kosciuszko National Park (*see Adaminaby and Thredbo*) and just below the snowline, was relocated from its original site on the banks of the Snowy River to make way for the Snowy Mountains Hydro-Electric Scheme. The original town was flooded, but a few of the buildings were moved to their current site beside Lake Jindabyne. The area is popular with skiers in the winter and with anglers, bushwalkers and whitewater rafters in the summer.

Walkway and cycleway: around Lake Jindabyne's foreshore, from Banjo Paterson Park on Kosciuszko Rd to Snowline Caravan Park. ***Winter shuttle bus service:*** several operators depart from various spots in town to Bullocks Flat and Thredbo; contact visitor centre for more information.

Speed and Marathon Championships: Lake Jindabyne; Jan. ***Flowing Festival:*** music and dragon-boat races: Feb. ***Lake Light Sculpture:*** Apr. ***Snowy Mountains Trout Festival:*** fishing competition throughout region; Nov. ***Boxing Day Rodeo:*** Dec.

Mario's Mineshaft Restaurant and Bar: modern Italian; Lakeview Plaza Motel, 2 Snowy River Ave; (02) 6456 2727.

Jindy Inn: simple accommodation, gorgeous lake views; 18 Clyde St; (02) 6456 1957. ***Wanderers Retreat:*** B&B or self-contained units; 54 Gippsland St; (02) 6456 2091.

Perisher The largest ski resort in Australia with an impressive 1250 ha of skiing area, Perisher caters for all levels of skier and snowboarder. State-of-the-art equipment includes many high-quality snow guns and Australia's first 8-seater chairlifts. The resort consists of slopes, accommodation, restaurants, bars and all the facilities and equipment hire needed to enjoy winter sports at Perisher and the nearby skiing areas of Smiggin Holes, Mt Blue Cow and Guthega. 30 km W.

Lake Jindabyne: well stocked with rainbow trout and ideal for boating, waterskiing and other watersports. When the water level is low, remains of the submerged town can be seen; western edge of town. ***Kunama Gallery:*** over 200 paintings by local artists and work by acclaimed water colourist Alan Grosvenor. Also panoramic views of Lake Jindabyne; (02) 6456 1100; 7 km NE. ***Snowy Valley Lookout:*** stunning views of Lake Jindabyne; 8 km N. ***Gaden Trout Hatchery:*** daily tours and barbecues along Thredbo River; (02) 6451 3400; 10 km NW. ***Wildbrumby Schnapps Distillery:*** boutique schnapps made from seasonal fruits, with cafe and cellar door; Wollondibby Rd; (02) 6457 1447; 11 km W. ***Crackenback Cottage:*** local craft, maze, restaurant and guesthouse; (02) 6456 2198; 12 km SW. ***Sawpit Creek:*** the Kosciuszko Education Centre can be found here. The start of the Palliabo (walking) Track is at the Sawpit Creek picnic area; 14 km NW. ***Bullocks Flat:*** terminal for Skitube, a European-style alpine train to Perisher ski resort; operates daily during ski season; 20 km sw. ***Wallace Craigie Lookout:*** views of Snowy River Valley; 40 km sw. ***Charlotte Pass:*** highest ski resort in Australia with challenging slopes for experienced skiers. Magnificent 24 km walking track past Blue Lake in summer; 45 km w. ***Scenic walks:*** varying lengths; brochures from visitor centre. ***Alpine Way:*** 111 km road through mountains to Khancoban provides superb scenic touring in summer; chains frequently required in winter.

See also Snowy Mountains, p. 33

Katoomba

see inset box on next page

Kempsey

Pop. 8139
Map ref. 550 G4 | 557 M10

Cultural Centre, Pacific Hwy, South Kempsey; (02) 6563 1555 or 1800 642 480; www.macleayvalleycoast.com.au

105.1 ROX FM, 684 AM ABC Local

Kempsey is an attractive town in the Macleay River Valley on the mid-north coast, with white sandy beaches and an unspoiled hinterland. It claims two quintessential Aussies as its own: singer Slim Dusty was born here in 1927 (while he died in 2003, he remains one of the country's best-loved country singers) and the Akubra hat has been made here since 1974.

Wigay Aboriginal Cultural Park: Aboriginal cultural experience, including introduction to bush tucker, learning about the use of plants and throwing a boomerang; Sea St. ***Cultural Centre:*** incorporates the Macleay River Historical Society Museum and visitor centre, a settlers cottage, displays on Akubra hats and Slim Dusty, and a working model of a timber mill; Pacific Hwy, South Kempsey. ***Historical walks:*** carefully restored historic buildings include the courthouse, post office and West Kempsey Hotel; brochure from visitor centre.

Markets: showground, 1st Sat each month; South West Rocks, 2nd Sat each month. ***Gaol Break Swim:*** 2.7 km race along Trial Bay beach to South West Rocks; Easter. ***Kempsey Cup:*** horserace; May. ***Akubra Classic Motorcycle Championships:*** June. ***Truck and Ute Show:*** Oct. ***Kempsey Country Music Festival:*** Oct/Nov.

Netherby Cafe: tranquil cafe; 5 Little Rudder St; (02) 6563 1777.

Netherby House B&B: tranquil retreat; 5 Little Rudder St, East Kempsey; (02) 6563 1777. ***Grass Trees Escape:*** self-contained cottages; Fern Tree Close, Arakoon; (02) 9907 1440.

Limeburners Creek Nature Reserve Showcases a beautiful coastline of heathlands, banksia and blackbutt forests. The rare rainforest of Big Hill is home to the threatened ground parrot. The beach is popular for swimming, surfing and fishing and there are camping areas with varying levels of facilities. 33 km se.

South West Rocks Attractions include a pristine white beach, a maritime history display at the restored Boatmans Cottage and the opportunity to handfeed fish at the Everglades Aquarium. The area is excellent for watersports, diving, camping and boating. The nearby Trial Bay Gaol (1886) was a public works prison until 1903 and reopened to hold 'enemy aliens' in WW I. Smoky Cape has a lighthouse offering tours and accommodation and clear views up and down the coast. Fish Rock Cave, just off the cape, is well known for its excellent diving. 37 km ne.

Frederickton: 'Fredo' has beautiful views of river flats, and its award-winning pie shop boasts 148 varieties; 8 km ne. ***Kundabung:*** tiny village with Australasian bullriding titles in Oct; 14 km s. ***Gladstone:*** fishing, antiques and crafts; 15 km ne. ***Crescent Head:*** this seaside holiday town has good surfing. There is an Aboriginal bora ring (ceremonial ground) just to the north. The Sky Show with kites, flying displays and fireworks is held each June and the Crescent Head Malibu Longboard Classic is in May (the largest longboard competition in Australia); 20 km se. ***Barnett's Rainbow Beach Oyster Barn:*** direct purchases and viewing of oyster processing; 31 km ne. ***Hat Head National Park:*** magnificent dunes and unspoiled beaches, popular for birdwatching, snorkelling, swimming and walking. Korogoro Point is a fabulous spot for whale-watching May–July and Sept–Oct; 32 km e. ***Bellbrook:*** classified by the National Trust as a significant example of a turn-of-the-century hamlet; 50 km nw. ***Walks and self-guide drives:*** nature-reserve walks and historical and scenic drives; details from visitor centre.

See also Holiday Coast, p. 27

Kiama

Pop. 12 290
Map ref. 553 F7 | 555 J9 | 565 H3

Blowhole Point Rd; (02) 4232 3322 or 1300 654 262; www.kiama.com.au

 98 FM, 1431 AM ABC Radio National

This popular holiday town, hosting one million visitors each year, is best known as the home of the Kiama Blowhole. The rocky coastline, sandy beaches and appealing harbour provide an attractive contrast to the green rolling hills of the lush dairy pastures of the hinterland.

Kiama Blowhole The spectacular blowhole sprays to heights of 60 m and is floodlit at night. Beside the blowhole is a constructed rockpool and a cafe. Pilots Cottage Historical Museum has displays on the blowhole, early settlement, the dairy industry and shipping. Blowhole Pt; (02) 4232 1001.

Family History Centre: world-wide collection of records for compiling and tracing family history; Railway Pde; (02) 4233 1122. ***Heritage walk:*** includes terraced houses in Collins St and Pilots Cottage; leaflet from visitor centre. ***Specialty and craft shops:*** several in town showcasing local work; leaflet from visitor centre. ***Beaches:*** perfect for surfing, swimming and fishing.

Craft market: Black Beach; 3rd Sun each month. ***Produce market:*** Black Beach; 4th Sat each month. ***Rotary Antique Fair:*** Jan. ***Jazz and Blues Festival:*** Mar. ***Big Fish Classic:*** game-fishing competition; Apr.

55 on Collins: modern Australian; 55 Collins St; (02) 4232 2811. ***Cargo's Wharf Restaurant:*** seafood; Kiama Wharf; (02) 4233 2771.

Seashells Kiama: self-contained family accommodation; 72 Bong Bong St; (02) 4232 2504. ***Spring Creek Retreat:*** resort-style accommodation; 41 Jerrara Rd; (02) 4232 2700. ***Bellachara Boutique Hotel:*** chic boutique hotel; 1 Fern St, Gerringong; (02) 4234 1359.

Gerringong A coastal town with a renowned heritage museum featuring remarkable scale models of the Illawarra coast. Gerringong's name comes from the Wodi Wodi language and is said to mean 'place of peril'. It is unclear where the peril lies, however, with safe beaches ideal for surfing, swimming and fishing. Heavy rainfall means the hinterland is lush and green. 10 km S.

Little Blowhole: smaller but more active than the Kiama Blowhole; off Tingira Cres; 2 km S. ***Bombo Headland:*** blue-metal quarrying in the 1880s left an eerie 'moonscape' of basalt walls and columns, which have been used in commercials and video clips; 2.5 km N. ***Cathedral Rocks:*** scenic rocky outcrop best viewed at dawn; Jones Beach; 3 km N. ***Kingsford Smith Memorial and Lookout:*** site of Charles Kingsford Smith's 1933 take-off in the *Southern Cross*, with panoramic views; 14 km S. ***Seven Mile Beach National Park:*** surrounded by sand dunes, this low forest is inhabited by birds and small marsupials. It makes a pretty spot for picnics and barbecues, beach fishing and swimming; 17 km S. ***Scenic drives:*** in all directions to visit beaches, rock formations, cemeteries and craft shops; brochure from visitor centre.

See also **SOUTHERN HIGHLANDS, p. 24**

Kyogle

Pop. 2730
Map ref. 551 E4 | 557 N3 | 653 M12

Summerland Way; (02) 6632 2700; www.tropicalnsw.com.au

 104.3 FM 2LM, 738 AM ABC Local

Located at the upper reaches of the Richmond River and the base of the charmingly named Fairy Mountain, Kyogle is known as the 'gateway to the rainforests'. It is almost completely surrounded by the largest remaining areas of rainforest in New South Wales.

Captain Cook Memorial Lookout: at the top of Mt Fairy, the lookout provides stunning views of the town and surrounding countryside; Fairy St. ***Botanical Gardens:*** combination of formal gardens and revegetated creek environments on the banks of Fawcetts Creek; Summerland Way.

Campdraft: July. ***Fairymount Festival – Hell on Hooves:*** bullriding; July. ***Gateway to the Rainforest Motorcycle Rally:*** Sept. ***Fairymount Festival – Remembering Yesterday, Dreaming Tomorrow***: Nov.

Ripples on the Creek: popular restaurant/cafe; Lions Road Tourist Drive, Gradys Creek; (02) 6636 6132.

Border Ranges National Park World Heritage–listed, this 30 000 ha park has walking tracks, camping, swimming, rock climbing and fantastic views of Mt Warning and the Tweed Valley. Sheepstation Creek is an attractive picnic spot. The Tweed Range Scenic Drive (64 km) is a breathtaking journey of rainforest, deep gorges and waterfalls. Brochure from visitor centre. 27 km N.

Wiangaree: rural community with rodeo in Mar; 15 km N. ***Roseberry Forest Park:*** picturesque picnic spot; 23 km N. ***Moore Park Nature Reserve:*** tiny reserve with the most important example of black bean rainforest in NSW; 26 km NW. ***Toonumbar Dam:*** built from earth and rocks, this offers scenic bushwalking with picnic and barbecue facilities. Nearby Bells Bay is known for its bass fishing and has a campsite; 31 km W. ***Toonumbar National Park:*** contains 2 World Heritage–listed rainforests and the volcanic remnants of Edinburgh Castle, Dome Mountain and Mt Lindesay; 35 km W. ***Richmond Range National Park:*** protected rainforest perfect for camping, birdwatching, picnics and barbecues, with good bushwalking on a 2 km or 6 km track; 40 km W. ***Scenic forest drive:*** via Mt Lindesay (45 km NW), offers magnificent views of both the rainforest and the countryside; brochure from visitor centre.

See also **TROPICAL NORTH COAST, p. 28**

Lake Cargelligo

Pop. 1149
Map ref. 554 A4 | 559 N4

1 Foster St; (02) 6898 1501; www.heartofnsw.com.au

 99.7 Star FM, 549 AM ABC Local

Lake Cargelligo is a surprisingly attractive small town in the heart of the wide, brown Riverina plains. Built beside the lake of the same name, the town's activities revolve around the water with fishing and regular lake festivals.

Lake Cargelligo The lake dominates the town and is popular for fishing (silver perch, golden perch and redfin), boating, sailing, waterskiing and swimming. It is also appealing to birdwatchers, being home to many bird species including the rare black cockatoo. There is a historic walkway and bicycle track.

Wagon Rides and Working Horse Demonstrations: can be arranged by appt on (02) 6898 1384. ***Information centre:*** houses a large gem collection and carved stone butterflies; Foster St. ***Kejole Koori Studio:*** Aboriginal art, jewellery and didgeridoos; Grace St.

Blue Waters Art Exhibition and Competition: June. ***Lake Show:*** Sept. ***Lake Cargelligo Fisharama***: Oct.

Willandra National Park Once one-eighth of a huge merino sheep station, the 20 000 ha park features a restored homestead (offering accommodation), stables, a shearing complex and men's quarters. The buildings house a display of pastoral and natural history of the area. Plains, wetlands and Willandra Creek make up the rest of the park, with a walking track that is best at dawn or dusk to see the myriad waterbirds, kangaroos and emus. The creek is popular for canoeing and fishing. 163 km W.

Murrin Bridge Vineyard: Australia's first Aboriginal wine producer; 13 km N. ***Cockies Shed Lavender Farm and Crafty Corner:*** Tullibegeal; 13 km E. ***Lake Brewster:*** 1500 ha birdwatcher's paradise with fishing and picnic area. No guns, dogs or boats; 41 km W. ***Nombinnie Nature Reserve:*** birdwatching, bushwalking and abundant spring wildflowers Sept–Dec; 45 km N. ***Hillston:*** main street lined with palms, thanks to its situation on top of a large artesian basin. Hillston Lake is popular for watersports and picnics, and a swinging bridge provides access to a nature reserve and walking trail; 93 km SW.

See also **CENTRAL WEST, p. 30**

Laurieton

Pop. 2088
Map ref. 550 F7 | 555 O1 | 557 M12

Greater Port Macquarie Tourism, cnr Gordon and Gore sts, Port Macquarie; (02) 6581 8000 or 1300 303 155; www.portmacquarieinfo.com.au

 100.7 2MC FM, 684 AM ABC Local

Laurieton is on an attractive tidal inlet at the base of North Brother Mountain. The mountain, which was named by Captain Cook in 1770, provides Laurieton with shelter from the wind, so the weather is mild all year-round. There are spectacular views and bushwalks on the mountain and the inlet is popular for estuary fishing.

Historical museum: documents history of the town in old post office; open by appt; Laurie St; bookings (02) 6559 9096. ***Armstrong Oysters:*** oyster farm on the river open to the public for fresh oysters; Short St.

Riverwalk Market: Cnr Tunis and Short sts; 3rd Sun each month. ***Camden Haven Music Festival:*** internationally

continued on p. 76

KATOOMBA

Pop. 7923

Map ref. 544 D5 | 546 F9 | 555 I6

Echo Point Rd; 1300 653 408; www.visitbluemountains.com.au

89.1 BLU FM, 97.3 FM ABC Local

Originally named Crushers but renamed a year later, Katoomba stands at an elevation of 1017 metres and is the region's principal tourist destination. Blue Mountains National Park lies to the north and south of town. The explanation for why these mountains look blue lies in the eucalyptus trees covering them; they disperse eucalyptus oil into the atmosphere, highlighting the sun's rays of blue light.

The Edge Cinema: daily screenings of *The Edge* (images of Blue Mountains) on a 6-storey screen; Great Western Hwy (access through Civic Pl); programs (02) 4782 8900. ***The Carrington Hotel:*** Katoomba's first hotel; Katoomba St.

Blue Mountains Festival of Folk, Roots and Blues: Mar. ***Six Foot Track Festival:*** Mar. ***Winter Magic Festival:*** June. ***Yulefest:*** throughout region; June–Aug. ***Spring Gardens Festival:*** throughout region; Sept–Nov.

Darley's Restaurant: modern Australian; Lilianfels Blue Mountains Resort & Spa, Lilianfels Ave; (02) 4780 1200. ***Echoes:*** modern Australian, great location; 3 Lilianfels Ave; (02) 4782 1966. ***Ashcrofts Licensed Restaurant:*** contemporary cuisine; 18 Govetts Leap Rd, Blackheath; (02) 4787 8297. ***The rooster:*** European; Jamison Guesthouse, 48 Merriwa Street; (02) 4782 1206. ***Hominy Bakery:*** best bread in the Blue Mountains; 185 Katoomba St; (02) 4782 9816. ***The Conservation Hut Cafe:*** modern Australian; Fletcher St, Wentworth Falls; (02) 4757 3827. ***Seven:*** Italian; 7 Station St, Wentworth Falls; (02) 4757 4997. ***Silks Brasserie:*** modern Australian; 128 The Mall, Leura; (02) 4784 2534. ***Solitary:*** French-influenced menu, great views; 90 Cliff Dr, Leura; (02) 4782 1164.

Echoes Boutique Hotel: boutique hotel; 3 Lilianfels Ave; (02) 4782 1966. ***Lilianfels Blue Mountains Resort & Spa:*** luxury resort and spa; Lilianfels Ave; (02) 4780 1200. ***Whispering Pines:*** heritage-listed self-contained suites; 178–186 Falls Rd, Wentworth Falls; (02) 4757 1449. ***The Falls:*** luxury self-contained accommodation; The Avenue (off Falls Rd), Wentworth Falls; (02) 4757 8801.

Leura Considered the most urbane and sophisticated village in the Blue Mountains, Leura has a beautiful tree-lined main street with impressive gardens, specialty shops, galleries and restaurants. Everglades Gardens (Everglades Ave) is a celebrated 1930s garden with a gallery devoted to its creator, Paul Sorensen. Leuralla (Olympian Pde) is a historic Art Deco mansion with a major collection of toys, dolls, trains and railway memorabilia. There are spectacular mountain views from Sublime Pt and Cliff Dr. The Leura Gardens Festival is held in Oct. Lyrebird Dell, near Leura, is an Aboriginal campsite estimated to be 12 000 years old. Relics can be found throughout the area, including rock engravings, axe-grinding grooves and cave paintings. All Aboriginal sites, discovered or undiscovered, are protected and are not to be disturbed by visitors. 3 km E.

[THREE SISTERS, ECHO POINT]

Blue Mountains National Park Echo Pt is the best place to view the Three Sisters, which are floodlit at night. Aboriginal Dreamtime legend tells of three beautiful sisters, Meehni, Wimlah and Gunnedoo, who lived in the Jamison Valley with the Katoomba people. The girls fell in love with three brothers from the Nepean people, but tribal law forbade their marriages. The brothers would not accept this and attempted to capture the sisters, causing a major battle. A witchdoctor from the Katoomba people feared that the girls were in danger, so he turned them to stone to protect them, intending to return them to their true forms when the battle was over. This was not to be, however, as the witchdoctor was killed in battle and the three sisters, at 922 m, 918 m and 906 m high, remain trapped as a magnificent rock formation. Nearby Orphan Rock is thought to be the witchdoctor. Also at Echo Pt is the Giant Stairway: 800 steps leading to the valley floor. Near Wentworth Falls, 7 km E, is an eco-designed cafe with great views, and the Valley of the Waters picnic area. *For other parts of the park see Blackheath and Glenbrook.*

Blue Mountains Scenic World Located near Katoomba Falls, the Three Sisters and Echo Pt, Scenic World takes in some of the best scenery in the national park. Take a ride on the Scenic Railway, which was originally built to transport coal and miners and is the world's steepest railway, or on the Scenic Skyway, a 7 km ride in a cable car high up over the Jamison Valley. There is also Sceniscender, which descends into the heart of the valley, and plenty of other attractions to explore on foot; (02) 4780 0200, 1300 759 929; South of town.

Explorers Tree: blackbutt tree reportedly carved with initials of Blaxland, Wentworth and Lawson (there is some question about whether this was done by the explorers or by early tourism operators); west of town, off hwy. ***Hazelbrook:*** small village with Selwood Science and Puzzles featuring a puzzle room, science kits, bookshop and local artwork. Hazelbrook hosts Regatta Day in Feb on Wentworth Falls Lake; 17 km E.

See also BLUE MOUNTAINS, p. 23

acclaimed music artists from all genres; Apr. ***Watermark Literary Muster:*** biennial event attracting scholars, authors and writers from around the world; odd-numbered years, Oct.

Rock Lobster Cafe: cafe favourites and gelato; Shop 3, 80 Bold St; (02) 6559 8322.

Gypsy Falls Retreat: secluded self-contained cottages; Tipperary Rd, Lorne; (02) 6556 9702.

Crowdy Bay National Park The park is known for its prolific birdlife and magnificent ocean beach. Diamond Head is an interesting sculpted rock formation and the hut beneath is where Kylie Tennant wrote *The Man and the Headland.* A headland walking track offers stunning views and the area is also popular for fishing, birdwatching and the abundant wildlife. Campsites are at Diamond Head, Indian Head and Kylies Beach. 5 km s.

Dunbogan: this seaside village borders the river and ocean. A fisherman's co-op offers the best fish and chips in the region. River cruises and patrolled swimming beach; 2 km SE. ***Kattang Nature Reserve:*** in spring the Flower Bowl Circuit leads through stunning wildflowers. Enjoy good coastal views all year-round from sharp cliffs jutting into the ocean; 5 km E. ***Dooragan National Park:*** according to local Aboriginal legend, Dooragan (North Brother Mountain) was the youngest of three brothers who avenged his brothers' deaths at the hands of a witch by killing the witch and then killing himself. Blackbutt and subtropical forest is home to gliders, bats and koalas. Viewing platforms on North Brother Mountain provide some of the best views anywhere on the NSW coast; 6 km w. ***North Haven:*** riverside dining, boutique gift shops and patrolled swimming beach; 6 km NE. ***Queens Lake Picnic Area:*** a beautiful reserve with St Peter the Fisherman Church nearby. Kayak access to the river; 6 km w. ***Kendall:*** poets walk, art and craft galleries. Also market 1st Sun each month on Logans Crossing Rd; 10 km w. ***Big Fella Gum Tree:*** 67 m flooded gum tree in Middle Brother State Forest; 18 km sw. ***Lorne Valley Macadamia Farm:*** tours at 11am and 2pm, products for sale and cafe; open Sat–Thurs; 23 km w.

See also HOLIDAY COAST, p. 27

Leeton

Pop. 6829
Map ref. 559 N7

10 Yanco Ave; (02) 6953 6481; www.leetontourism.com.au

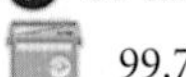
99.7 Star FM, 549 AM ABC Local

Leeton was designed by Walter Burley Griffin, the American architect who designed Canberra. Like Canberra, the town is built in a circular fashion with streets radiating from the centre. The Murrumbidgee Irrigation Area brought fertility to the dry plains of the Riverina and now Leeton has 102 hectares of public parks and reserves and a thriving primary industry.

Visitor centre: beautifully restored building with photographic displays, local artwork and a heritage garden; Yanco Ave. ***Sunrice Centre:*** product displays and tastings, with 'Paddy to Plate' video presentation at 9.30am and 2.45pm on weekdays; Calrose St. ***Mick's Bakehouse:*** home of award-winning pies; Pine Ave. ***Art Deco streetscape:*** includes Roxy Theatre and historic Hydro Motor Inn; Chelmsford Pl.

Bidgee Classic Fishing Competition: Mar. ***Sunrice Festival:*** even-numbered years, Easter. ***Picnic Races:*** May. ***Australian Birdfair:*** Nov. ***Light up Leeton:*** Dec.

Pages on Pine Restaurant: modern Australian; 119B Pine Ave; (02) 6953 7300.

Historic Hydro Guest House: landmark heritage-listed guesthouse; Chelmsford Pl; (02) 6953 1555. ***Leeton Heritage Motor Inn:*** traditional motel with rustic twist; 439 Yanco Ave; (02) 6953 4100.

Yanco This town is the site where Sir Samuel McCaughey developed the irrigation scheme that led to the establishment of the Murrumbidgee Irrigation Area. Attractions in town include McCaughey Park, the Powerhouse Museum and a miniature train that runs on market days. Village Markets are held at Yanco Hall on the last Sun each month. 8 km s.

Fivebough Wetlands: 400 ha home to over 150 species of waterbird with interpretive centre, walking trails and viewing hides; 2 km N. ***Brobenah Airfield:*** gliding and hot-air ballooning; 9 km N. ***McCaughey's Mansion:*** 1899 mansion with stained-glass windows and attractive gardens, now an agricultural high school, but drive-through inspections welcome. Yanco Agricultural Institute nearby is open to the public and provides farmer-training facilities, research and advisory services; 11 km s. ***Murrumbidgee State Forest:*** scenic drives; brochure from visitor centre; 12 km s. ***Whitton Historical Museum:*** housed in old courthouse and gaol with photographs, documents and early farming equipment; 23 km w. ***Gogeldrie Weir:*** pleasant spot for fishing, picnics and camping; 23 km sw. ***Wineries:*** cellar-door tastings and tours at Toorak and Lillypilly Estate; brochure at visitor centre.

See also RIVERINA, p. 34

Lightning Ridge

Pop. 2598
Map ref. 556 B3 | 652 E12

Lions Park, Morilla St; (02) 6829 1670; www.lightningridge.info.com.au

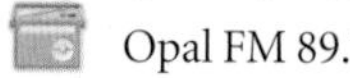
Opal FM 89.7

This famous opal-mining town is the only place in Australia where true black opals are found. Miners on the opal fields in most cases need to provide their own electricity and catch or cart their own water as these services are not available, making residents pretty creative and resilient. The otherwise desolate area receives 50 000 visitors each year and has modern facilities in an otherwise minimalist town. Famous finds in the region include Big Ben and the Flame Queen, which was sold for £80 because the miner who found it had not eaten properly in three weeks. Local Aboriginal legend explains the opals by saying that a huge wheel of fire fell to earth, spraying the land with brilliant stones. Lightning Ridge is notoriously hot in summer but boasts ideal weather in winter, and has lots of native marsupials and birdlife.

Bottle House Museum: collection of bottles, minerals and mining relics; originally a miner's camp; Opal St. ***Big Opal:*** opal-cutting demonstrations and daily underground working-mine tours; Three Mile Rd. ***John Murray Art:*** exclusive outlet for original paintings, limited-edition prints, postcards and posters; Opal St. ***Goondee Aboriginal Keeping Place:*** Aboriginal artefacts and educational tours of the premises; open by appt; Pandora St. ***Chambers of the Black Hand:*** unique underground sculptures and carvings in sandstone walls of an old mine; also underground shop; Three Mile Rd. ***Displays of art and craft:*** including beautiful displays of black opal in the many opal showrooms; several locations; leaflet from visitor centre. ***Black Queen's Legacy:*** outback theatrical experience based on true stories; Mar–Nov; (02) 6829 0980. ***Car Door Explorer Tours:*** follow painted car doors to find attractions; brochure from visitor centre.

Local craft market: Morilla St; every Fri and most Sun. ***Great Goat Race:*** Easter. ***Rodeo:*** Easter. ***Opal Open Pistol Shoot:*** June. ***Opal and Gem Festival:*** July.

Bowling Club: coffee shop, bistro and restaurant; Morilla St; (02) 6829 0408. ***Chats on Opal:*** homemade cakes, gourmet sandwiches; Shop 1, 5 Opal St; (02) 6829 4228. ***Morilla's Cafe:*** light meals, cakes; 2 Morilla St; (02) 6829 0066. ***Nobbies Restaurant:*** bistro; cnr Onyx and Morilla sts; (02) 6829 0304.

Chasin' Opal Holiday Park: cabins in centre of town; 10 Morilla St; (02) 6829 0448. ***Sonja's B&B:*** friendly, unpretentious accommodation; 60 Butterfly Ave; (02) 6829 2010.

Walk-In Mine: working mine with easy access and tours on demand. Also a cactus nursery nearby; off Bald Hill Rd; 2 km N. ***Hot artesian bore baths:*** open baths with average temperature 42°C; 2 km NE. ***Kangaroo Hill Tourist Complex:*** displays of antiques, bottles, shells, rocks and minerals, and mining memorabilia with a fossicking area outside; 3 km S. ***Opal fields:*** Grawin (65 km W) and Sheepyards (76 km W); brochure from visitor centre. ***Designated fossicking areas:*** maps from visitor centre.

See also NEW ENGLAND, **p. 29**

Lismore

Pop. 30 088
Map ref. 551 F5 | 557 O3 | 653 N12

Cnr Ballina and Molesworth sts; (02) 6626 0100 or 1300 369 795; www.visitlismore.com.au

92.9 FM North Coast Radio, 738 AM ABC Local

This regional centre of the Northern Rivers district is on the banks of the Wilsons River. European settlement came in 1840 when John Brown broke an axle near a small chain of ponds and stopped to have a look around. He liked what he saw and decided to settle. Called Browns Water Hole until 1853, it was changed to Lismore after a town in Ireland. The town is now known for its ecotourism.

Rotary Rainforest Reserve There are 6 ha of original tropical rainforest in the middle of the city, but this is only a small remnant of the original 'Big Scrub' that stood here before European settlement. Over 3 km of paths, including a boardwalk, lead visitors past hoop pines and giant figs, with rare species of labelled rainforest plants. Rotary Dr.

Visitor centre: indoor rainforest walk, historical displays, and local art and craft. The surrounding Heritage Park has pleasant picnic areas and a mini steam train offering rides (10am–2pm Thurs and public holidays, 10am–4pm Sat, Sun and school holidays). Cnr Ballina and Molesworth sts. ***Richmond River Historical Museum:*** geological specimens, Aboriginal artefacts and pioneer clothing, implements, furniture and handiwork; Molesworth St. ***Lismore Regional Art Gallery:*** permanent collection of paintings, pottery and ceramics with changing exhibitions by local and touring artists; Molesworth St. ***Robinson's Lookout:*** views south across the river to South Lismore and north to the mountains; Robinson Ave. ***Claude Riley Memorial Lookout:*** views over Lismore city; New Ballina Rd. ***Wilsons Park:*** contains original rainforest with labelled trees; Wyrallah St, East Lismore. ***Riverside Walk:*** along the banks of the Wilsons River from the town centre to Spinks Park, where there are picnic and barbecue facilities. ***Koala Care Centre:*** looks after injured and orphaned koalas. Guided tours 10am and 2pm Mon–Fri, and 10am Sat; Rifle Range Rd. ***Cafe and Culture Trail:*** self-guide walk through sites of historical significance and cafes in the town centre. ***Heritage walk:*** historic buildings and churches; brochure from visitor centre.

Car Boot Markets: Shopping Sq, Uralba St; 1st and 3rd Sun each month. ***Rainbow Region Organic Markets:*** Lismore Showground, Tues morning until 11am; Alexander Pde. ***Lismore Gemfest:*** May. ***Lantern Parade:*** June. ***Cup Day:*** horserace; Sept.

Fire in the Belly: modern Italian; 109 Dawson St; (02) 6621 4899. ***Paupiettes:*** French; 56 Ballina St; (02) 6621 6135. ***The Left Bank:*** seafood; 133 Molesworth St; (02) 6622 2338.

Elindale House B&B: rustic B&B; 34 Second Ave; (02) 6622 2533. ***Melville House:*** Art Deco B&B; 267 Ballina St; (02) 6621 5778. ***The Lismore Wilson Motel:*** heritage-listed motel; 119 Ballina Rd; (02) 6622 3383.

Boatharbour Reserve: 17 ha of rainforest, wildlife sanctuary, picnic area and walking tracks; maps from visitor centre; 6 km E. ***Tucki Tucki Koala Reserve:*** woodland planted by local residents to protect the diminishing koala population, with walking track and Aboriginal bora ring nearby; 15 km S. ***Rocky Creek Dam:*** between Nightcap National Park and Whian Whian State Forest, includes spectacular views of the lake, boardwalks, walking trails, platypus-viewing platform, barbecues and a playground; 18 km N.

See also TROPICAL NORTH COAST, **p. 28**

Lithgow

Pop. 11 298
Map ref. 544 B2 | 546 D2 | 555 I5

Great Western Hwy; 1300 760 276; www.tourism.lithgow.com

107.9 2ICE FM, 1395 AM ABC Local

Lithgow promotes itself as being 'surrounded by nature'. Easy access to several of the state's finest national parks and the city's charm make it worth visiting this isolated but staggeringly beautiful region. It was isolated from the coastal cities until the revolutionary Zig Zag Railway, built with gently sloping ramps to cut through the mountains, opened it up in 1869.

Eskbank House Museum Built in 1842, this sandstone Georgian mansion houses an extensive collection of Lithgow pottery, memorabilia and photographs. The front 4 rooms are authentically furnished with Regency and Victorian furniture. In the gardens are a stone stable, coach house and picnic area. Open Wed–Sun or by appt; (02) 6351 3557; Bennett St.

Blast Furnace Park: ruins of Australia's first blast furnace complex (1886) with a pleasant walk around adjacent Lake Pillans Wetland; off Inch St. ***State Mine Railway Heritage Park:*** mining and railway equipment and historic mining buildings; State Mine Gully Rd. ***Small Arms Museum:*** established in 1912, some argue that this is the birthplace of modern manufacturing in Australia. Displays range from firearms to sewing machines. Open 9.30am–2.30pm Tues and Thurs, 10am–4pm Sat, Sun, public and school holidays; Methven St.

Ironfest: cultural festival celebrating metal; Apr. ***Rally of Lithgow:*** May. ***Celebrate Lithgow:*** Nov.

Secret Creek Cafe and Restaurant: modern/native Australian; 35 Crane Rd; (02) 6352 1133. ***Lochiel House:*** modern Australian; 1259 Bells Line of Road, Kurrajong Heights; (02) 4567 7754.

The Bowen Inn Motel: contemporary motel; 5 Col Drewe Dr; (02) 6352 5111 ***Rustic Spirit:*** award-winning cottages; 23 Glenara Rd, Kurrajong Heights; (02) 4567 7170.

Wollemi National Park This is the largest wilderness area in NSW and is a breathtaking display of canyons, cliffs and undisturbed forest. Highlights include historic ruins at Newnes, the beaches of the Colo Gorge and the glow worms in a disused rail tunnel. The park also includes Glen Davis, home to more species of birds than anywhere else in the Southern Hemisphere. Mt Wilson, surrounded by the park, is a 19th-century village with large homes and superb gardens, many open to the public. Via the town is the Cathedral of Ferns; begins 16 km E.

Blue Mountains SpaRadise: Japanese bathhouse; Bowenfels; 1 km N. ***Hassans Walls Lookout:*** spectacular views of the Blue Mountains and Hartley Valley; via Hassans Walls Rd; 5 km S. ***Lake Lyell:*** stunning lake in mountain setting, popular for activities such as power-boating, waterskiing, trout fishing and picnics. Also canoe and boat hire available on-site; 9 km W. ***Zig Zag Railway:*** built in 1869, and later restored, it offers train trips of 1 hr 40 min return, departing 11am, 1pm and 3pm daily; via Bells Line of Road, Clarence; 10 km E. ***Lake Wallace:*** sailing and trout fishing; 11 km NW. ***Jannei Goat Dairy:*** produces cheeses, yoghurt and milk and is open to visitors, with free cheese tastings; 11 km NW. ***Hartley:*** became obsolete after the construction of the Great Western Railway in 1887. Explore 17 historical buildings administered by the National Parks & Wildlife Service; 14 km SE. ***Portland:*** charming town with a power station offering tours and interactive exhibits, a museum with much Australian memorabilia and several pleasant picnic areas; 17 km NW. ***Mt Piper Power Station:*** hands-on exhibits in the information centre and daily tours at 11am; 21 km NW. ***Gardens of Stone National Park:*** fascinating pagoda rock formations, sandstone escarpments and beehive-shaped domes caused by erosion. This is a great spot for rock climbing and picnics; 30 km N. ***Mount Tomah Botanic Garden:*** 5000 species of cool-climate plants at over 1000 m above sea level. Also award-winning restaurant; 35 km E.

See also **BLUE MOUNTAINS, p. 23**

Macksville

Pop. 2658
Map ref. 550 G2 | 557 M9

Cnr Pacific Hwy and Riverside Dr, Nambucca Heads; (02) 6568 6954 or 1800 646 587; www.nambuccatourism.com

105.9 FM Radio Nambucca, 738 AM ABC Local

Macksville is a fishing and oyster-farming town on the banks of the Nambucca River. There is an abundance of water-based activities, with picnic areas and riverside parks from which dolphins can often be observed. The town bustles with activities and festivals including the country's second oldest footrace, the Macksville Gift, each November.

Mary Boulton Pioneer Cottage: replica of a pioneer home and farm buildings with horse-drawn vehicles. Open by appt 2–4pm Wed–Sat; River St; (02) 6568 1280. ***Hotels:*** Star Hotel, River St, and Nambucca Hotel, Cooper St, both heritage buildings from late 1800s with many original features; both offer meals and accommodation. ***Nambucca MacNuts Factory:*** macadamia products at wholesale prices; Yarrawonga St.

Market: Scout Hall, Partridge St; 4th Sat each month. ***Patchwork and Quilt Display:*** Easter. ***Rusty Iron Rally:*** heritage machinery show; Sept. ***Macksville Gift:*** Nov.

Dangerous Dan's Butchery: excellent butchery; 13 Princess St; (02) 6568 1036. ***Short Order Cafe:*** light meals, cakes; Shop 1, 10 Princess St; (02) 6568 4550.

Jacaranda Country Lodge: charming B&B; 292 Wilson Rd; (02) 6568 2737.

Bowraville This unspoiled town has much to see with a National Trust–classified main street, the Bowra Folk Museum, Frank Partridge VC Military Museum, Bowra Art Gallery and many craft galleries. There are markets on 3rd Sun each month, regular races at the racecourse and the Bowraville Hinterland Festival each Oct. 16 km NW.

Mt Yarahappini Lookout: the highest point in Nambucca Valley with fabulous 360-degree views, in Yarriabini National Park; 10 km S. ***Scotts Head:*** coastal town with good beaches for surfing, swimming, fishing and dolphin-watching; 18 km SE. ***Pub with No Beer:*** built in 1896, this is the hotel that was made famous in Slim Dusty's song 'The Pub with No Beer', so named because it would often run out of beer before the next quota arrived. It is still largely in its original form and offers meals; 26 km SW. ***Bakers Creek Station:*** old cattle station, now an impressive resort offering horseriding, fishing, rainforest walking, canoeing and picnicking as well as accommodation; (02) 6564 2165; 30 km W. ***Local craft:*** featured in several shops and galleries in the area; leaflet from visitor centre.

See also **HOLIDAY COAST, p. 27**

Maitland

Pop. 61 431
Map ref. 543 F1 | 549 C7 | 550 A11 | 555 L3

Ministers Park, cnr New England Hwy and High St; (02) 4931 2800; www.maitlandhuntervalley.com.au

106.9 NX FM, 549 AM ABC Local

Maitland is in the heart of the world-class Hunter Valley wine region, on the Hunter River. Built on flood plains, it has suffered 15 major floods since settlement. The city has a significant Polish contingent as a result of immigration after World War II.

Maitland Gaol The gaol was built in 1844 and served as a maximum security prison for 154 years. The first inmates were convicts, including some children, and were forced to march 6 km from the wharf at Morpeth in shackles and chains. It has been home to some of Australia's most notorious and dangerous criminals and is now said to be the most haunted gaol in the country. Audio tours are available or there are guided tours with ex-inmates and ex-officers. For the extremely brave, there are also overnight stays. John St, East Maitland; bookings (02) 4936 6482.

Grossman House: National Trust–classified Georgian-style house, now a museum with pioneer silverware, porcelain and handmade clothing; Church St. ***Maitland Regional Gallery:*** city's art collection specialising in local works; High St. ***Self-guide heritage walks:*** showcasing Maitland, Morpeth, Lorn and East Maitland; brochures from visitor centre.

Market: showground; 1st Sun each month (except Jan). ***Hunter Valley Steamfest:*** celebration of steam trains; Apr. ***Garden Ramble:*** Sept. ***Bitter & Twisted International Boutique Beer Festival:*** Nov.

The Angels Inn Restaurant: modern Australian; Molly Morgan Motor Inn, New England Hwy; (02) 4933 5422. ***The Old George & Dragon Restaurant:*** modern European; 48 Melbourne St; (02) 4933 7272. ***Organic Feast:*** excellent provedore; 10–12 William St, East Maitland; (02) 4934 7351.

The Old George & Dragon Guesthouse: English-style romantic retreat; 50 Melbourne St; (02) 4934 6080. ***Pindari House:*** luxury guesthouse; 78 Winders La, Lochinvar; (02) 4930 7480.

Morpeth This riverside village has been classified by the National Trust. The town can be explored on foot and features magnificent old sandstone buildings such as St James Church (1830s), and antique and craft shops. There is the Weird and Wonderful Novelty Teapot Exhibition in Aug. A self-guide heritage walk brochure is available from the visitor centre. 5 km NE.

Walka: a scenic drive from Maitland, this former pumping station, now an excellent recreation area, is popular for picnics and bushwalks; 3 km N. ***Tocal Homestead:*** historic Georgian homestead; open by appt; bookings (02) 4939 8888; 14 km N. ***Paterson:*** signposted scenic drive leads to this charming hamlet on the Paterson River; 16 km N.

See also HUNTER VALLEY & COAST, p. 26

Manilla

Pop. 2081
Map ref. 556 H8

 **197 Manilla St; (02) 6785 1207; www.manillamuseum.org.au**

 92.9 FM, 648 AM ABC Local

This rural town is located at the junction of the Manilla and Namoi rivers and surrounded by attractive rural countryside. Its location between Lake Keepit and Split Rock Dam has made it a popular setting for myriad outdoor activities, especially aerosports. Nearby Mount Borah hosted the world's paragliding championships in 2007. Manilla is also known for its production of mead (an alcoholic drink made from fermented honey and water). It is home to one of only two meaderies in the country.

Dutton's Meadery: tastings and sales of honey and mead; Barraba St. ***Manilla Heritage Museum:*** incorporates Royce Cottage Collection and exhibits pioneer items such as clothing and furniture, and a bakery. Also has displays on platypus, which are often found in the area. ***Manilla St:*** antique and coffee shops.

National and international paragliding competitions: Nov–Apr.

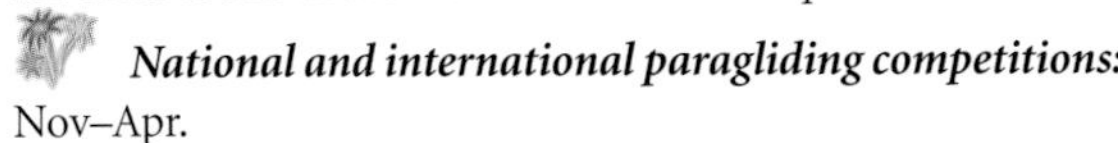

Oakhampton Homestead: homely farmstay accommodation; Oakhampton Rd; (02) 6785 6517.

Warrabah National Park At this peaceful riverside retreat you'll find enormous granite boulders sitting above still valley pools and rapids suitable for experienced canoeists. Activities include swimming and fishing in the Namoi and Manilla rivers and rock climbing on the cliffs. 40 km NE.

Manilla Paragliding: offers tandem flights; a 2-day introduction, and a 9-day 'live in' licensing course with equipment supplied; Mt Borah; (02) 6785 6545; 12 km N. ***Split Rock Dam:*** watersports, boating, camping and fishing for species such as Murray cod and golden perch; turn-off 15 km N. ***Manilla Ski Gardens:*** area at the northern end of Lake Keepit with waterskiing, fishing and swimming; 20 km SW. *For more details on Lake Keepit see Gunnedah.*

See also NEW ENGLAND, p. 29

Menindee

Pop. 331
Map ref. 560 D11

Yartla St; (08) 8091 4274.

95.7 FM ABC Radio National, 97.3 FM ABC Western Plains

Menindee is like an oasis in the middle of a desert. Although located in the state's arid inland plains, its immediate surrounds comprise fertile land, thanks to the 20 lakes in the area fed by the Darling River and a dam constructed in the 1960s. The orchards and vegetable farms provide a stark contrast to the vast freshwater lakes full of dead trees, and surrounding saltbush and red soil, on the way into this tiny settlement.

Maiden's Menindee Hotel: Burke and Wills lodged here in 1860. Meals and accommodation are available; Yartla St. ***Ah Chung's Bakehouse Gallery:*** William Ah Chung established one of the first market gardens in town. His bakery (c. 1880) now houses a gallery featuring local artists. Open for groups by appt from visitor centre; Menindee St. ***Heritage trail:*** through town; maps from visitor centre.

National and international paragliding competitions: Nov–Apr.

Bindara Station: working heritage homestead farm; Bindara Station, via Broken Hill; (08) 8091 7412.

Kinchega National Park When full, the lakes support waterbirds such as egrets, cormorants, black swans and spoonbills, and numerous other wildlife. There are giant river red gums growing along the banks of the Darling River, and campsites along the river in the north. Attractions include the wreck of paddlesteamer *Providence*, the old homestead and woolshed from Kinchega Station, and accommodation in the restored shearers' quarters. There is also a cemetery near the homestead. Activities include swimming, fishing and canoeing. 1 km W.

Menindee Lakes: in dry times only the upper lakes have water, making them the most reliable for fishing, swimming, birdwatching and watersports, and the best place for camping; details from visitor centre. ***Menindee Lake Lookout:*** good views of the lake; 10 km N. ***Copi Hollow:*** great spot for waterskiers, swimmers and powerboat enthusiasts, with campsites on the waterfront; 18 km N.

See also OUTBACK, p. 36

Merimbula

Pop. 3851
Map ref. 565 E10 | 581 H7

Beach St; (02) 6495 1129; www.merimbulatourism.com.au

105.5 2EC FM, 810 AM ABC South East

Merimbula is a modern seaside town known for its surfing, fishing and oyster farming. Middens found in the area indicate that oysters were gathered here by Aboriginal people well before the arrival of Europeans. The town began as a private village belonging to the Twofold Bay Pastoral Association, who opened it as a port in 1855.

Merimbula Aquarium Twenty-seven tanks here present a wide range of sea life and an oceanarium showcases large ocean fish including sharks. Fish feeding time is 11.30am Mon, Wed and Fri (Mon–Fri in school holidays). Excellent seafood restaurant on-site. Merimbula Wharf, Lake St.

Old School Museum: town history displayed in excellent collection of photos, documents and memorabilia; Main St.

NEW SOUTH WALES

Scenic flights: view the Sapphire Coast from the air. Bookings through Merimbula Air Services; (02) 6495 1074. ***Trike Tours:*** enjoy the beauty of the coast on three wheels; (02) 6495 2300.

Seaside Markets: Ford Oval; 3rd Sun each month. ***Jazz Festival:*** June. ***Country Music Festival:*** Nov.

Poppy's Courtyard Cafe: peaceful cafe; 15 The Plaza; (02) 6495 1110. ***Zanzibar Cafe:*** modern Australian; Cnr Main and Market sts; (02) 6495 3636.

Robyns Nest: absolute waterfront accommodation; 188 Merimbula Dr; (02) 6495 4956.

Pambula This historic sister village of Merimbula has excellent fishing on the Pambula River and a market 2nd Sun each month. Pambula Beach has a scenic walking track and lookout, with kangaroos and wallabies gathering on the foreshore at dawn and dusk; 7 km sw.

Magic Mountain Family Recreation Park: rollercoaster, water slides, minigolf and picnic area; Sapphire Coast Dr; 5 km N. ***Tura Beach:*** resort town with excellent beach for surfing; 5 km NE. ***Yellow Pinch Wildlife Park:*** peaceful bushland setting with array of native animals, birds and reptiles; 5 km E. ***Oakland Farm Trail Rides:*** scenic trail rides to the Pambula River; Princes Hwy, South Pambula; 0428 957 257; 10 km sw. ***Whale-watching (Oct–Nov), boat cruises and boat hire:*** bookings at visitor centre.

See also SOUTH COAST, p. 32

Merriwa

Pop. 944
Map ref. 555 J1 | 556 G12

Vennacher St; (02) 6548 2607.

101.9 FM ABC Local, 102.7 Power FM

This small town in the western Hunter region beside the Merriwa River is known for its majestic early colonial buildings. It is the centre of a vast farming district of cattle, sheep, horses, wheat and olive trees. People converge on the town each year for the Festival of the Fleeces, which includes shearing competitions, yard dog trials and a woolshed dance.

Historical Museum: in stone cottage (1857) with documented history of the region and the belongings of European pioneers; Bettington St. ***Bottle Museum:*** over 5000 bottles of all shapes and sizes; open Mon–Fri; visitor centre. ***Self-guide historical walk:*** early school buildings, Holy Trinity Anglican Church (1875) and the Fitzroy Hotel (1892); brochure from visitor centre.

Polocrosse Carnival: June. ***Festival of the Fleeces:*** June. ***Merriwa Motorcycle River Rally:*** Oct.

Merriwa Cakes and Pastries: bakery; 147 Bettington St; (02) 6548 2851.

Hidden Valley Escapes: B&B, farmstay; 3615 Willow Tree Rd; (02) 6548 8588 or 1800

Coolah Tops National Park The plateaus at high altitude in this park provide wonderful lookouts over the Liverpool plains and some spectacular waterfalls. Vegetation consists of giant grass trees and tall open forests of snow gums, providing a home for wallabies, gliders, eagles and rare owls. There are superb campsites, walking trails and picnic spots. 107 km NW.

Cassilis: tiny village with historic sandstone buildings including St Columba's Anglican Church (1899) and the courthouse/ police station (1858). The main streets have been declared an urban conservation area; 25 km NW. ***Flags Rd:*** old convict-built road leading to Gungal; 25 km sw. ***Gem-fossicking area:*** open to the public; 27 km sw. ***Goulburn River National Park:*** mostly sandstone walking tracks along the Goulburn River, honeycombed with caves; good rafting and access for boats; 35 km s.

See also HUNTER VALLEY & COAST, p. 26

Moree

Pop. 8085
Map ref. 556 G4

Lyle Houlihan Park, cnr Newell and Gwydir hwys; (02) 6757 3350; www.moreetourism.com

98.3 NOW FM, 819 AM ABC New England North West

Moree sits at the junction of the Mehi and Gwydir rivers and is the centre of the thriving local farming district. Thanks to its rich black-soil plains, one local claimed, 'You could put a matchstick in the ground overnight and get a walking-stick in the morning'. The town is also known for its artesian spas, with therapeutic qualities said to cure arthritis and rheumatism.

Spa complex These spas were discovered accidentally when settlers were searching for reliable irrigation water. A bore was sunk into the Great Artesian Basin and the water that emerged was 41°C. The complex also has an outdoor heated pool and an array of leisure activities that attract 300 000 visitors each year. Cnr Anne and Gosport sts.

Moree Plains Regional Gallery: contemporary Aboriginal art and artefacts and changing exhibitions; Frome St. ***The Big Plane:*** DC3 transport plane with tours available at Amaroo Tavern; Amaroo Dr. ***Dhiiyaan Indigenous Centre:*** located in the town's library, this was the first Aboriginal genealogy centre where Indigenous people could access historical family information and photographs; Cnr Balo and Albert sts. ***Barry Roberts Historical Walk:*** self-guide tour includes the courthouse and the Moree Lands Office, which was restored after a fire in 1982; brochure from visitor centre.

Market: Jellicoe Park; 1st Sun each month (not Jan). ***Opera in the Paddock:*** odd-numbered years; Mar. ***Picnic Races:*** May. ***Moree on a Plate:*** food and wine festival; May. ***Australian Cotton Trade Show:*** May. ***Harmony on the Plains Multicultural Festival:*** Aug. ***Golden Grain Festival:*** Nov.

Dragon & Phoenix Palace: Chinese; 361 Frome St; (02) 6752 4444.

Artesian Spa Motor Inn: comfortable accommodation; 2 Webb Ave, Newell Hwy; (02) 6752 2466. ***Moree Spa Motor Inn:*** traditional motel accommodation; Cnr Alice and Gosport sts; (02) 6752 3455.

Trewalla Pecan Farm: largest orchard in the Southern Hemisphere yielding 95 per cent of Australia's pecans; tour bookings at visitor centre; 35 km E. ***Cotton Gins:*** inspections during harvest; Apr–July; details at visitor centre. ***Birdwatching:*** several excellent sites in the area; brochure from visitor centre.

See also NEW ENGLAND, p. 29

Moruya

Pop. 2433
Map ref. 565 F7 | 567 L7

Vulcan St; (02) 4474 1345 or 1800 802 528; www.eurobodalla.com.au

104.3 Power FM, 105.1 FM ABC Radio National

This town on the Moruya River was once a gateway to local goldfields, but is now a riverside township. It is known for dairying and oyster farming, and also for its granite, which can be seen in some of the older buildings in town and was also used to build the pylons of the Sydney Harbour Bridge.

Eurobodalla Historic Museum: depicts gold discovery at Mogo and the district history of shipping, dairying and goldmining; town centre. ***South Head:*** beautiful views across the river mouth.

Market: Main St; Sat. ***Jazz Festival:*** Oct. ***Rodeo:*** New Year's Day.

The River Moruya: modern European, fantastic views; 16B Church St; (02) 4474 5505.

Blue Hills B&B: romantic getaway; 59 Hawdon St; (02) 4474 0028. ***Mogendoura Farm Holidays:*** self-catering cottages; 189 Hawdons Rd; (02) 4474 2057 ***The Knoll:*** self-contained accommodation; 182 South Head Rd; 0408 629 752.

Deua National Park A wilderness of rugged mountain ranges, plateaus, gentle and wild rivers and a magnificent limestone belt, the area is popular for canyoning and caving. The rivers are a base for most water activities, including swimming, fishing and canoeing. There are scenic walking and 4WD tracks and 4 main campsites to choose from. 20 km w.

Broulee: great surfing and swimming; 12 km NE. ***Bodalla:*** All Saints Church, built from local granite, is of historical significance; 24 km s. ***Comans Mine:*** visit the historic tramway and stamper battery used to crush ore and separate the gold; 40 km sw. ***Nerrigundah:*** former goldmining town with a monument to Miles O'Grady, who was killed here in a battle with the Clarke bushranging gang; 44 km sw.

See also SOUTH COAST, p. 32

Moss Vale

Pop. 6725
Map ref. 553 B6 | 555 I9

Southern Highlands Visitor Centre, 62–70 Main St, Mittagong; (02) 4871 2888 or 1300 657 559; www.southern-highlands.com.au

102.9 FM 2ST, 1431 AM ABC Radio National

Moss Vale is the industrial and agricultural centre of Wingecarribee Shire and the Southern Highlands. Once it was home to the Dharawal people, but by the 1870s they had all been driven off or killed. The town stands on part of the 1000 acres (approximately 400 hectares) of land granted to explorer Charles Throsby by Governor Macquarie in 1819. For most of the last century Moss Vale was a railway town and is dominated by the architecture of the Victorian railway station.

Leighton Gardens: picturesque area popular for picnics; Main St. ***Historical walk:*** includes Aurora College (formerly Dominican Convent) and Kalourgan, believed to have been a residence of Mary MacKillop; brochure from visitor centre.

Southern Highlands Country Fair: showgrounds; 4th Sun each month. ***Autumn Gardens in the Southern Highlands:*** throughout region; Apr. ***Tulip Time Festival:*** throughout region; Sept–Oct.

Katers: modern Australian; Peppers Manor House, 52 Kater Rd, Sutton Forest; (02) 4860 3111.

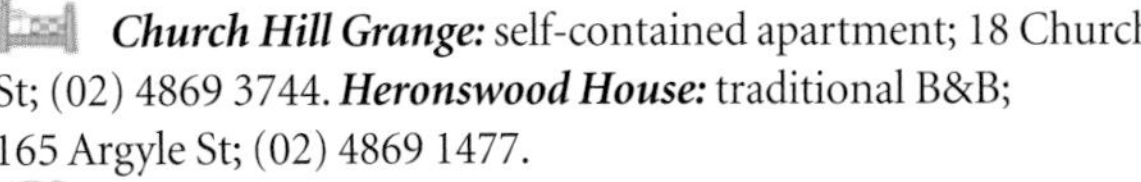

Church Hill Grange: self-contained apartment; 18 Church St; (02) 4869 3744. ***Heronswood House:*** traditional B&B; 165 Argyle St; (02) 4869 1477.

Sutton Forest Set among green hills, this tiny town has a shop called A Little Piece of Scotland for all things Scottish, and The Everything Store, c. 1859. Hillview House, just north, was the official residence of NSW governors 1882–1958. 6 km sw.

Throsby Park Historic Site: owned for 150 years by the Throsby family. Buildings that depict early settlement life include original stables, former barn, flour mill, Gundagai Cottage and Christ Church. Access is by tour only; bookings (02) 4887 7270; 1.5 km E. ***Cecil Hoskins Nature Reserve:*** tranquil wetland with over 90 bird species, one-third of which are waterfowl; 3 km NE.

See also SOUTHERN HIGHLANDS, p. 24

Mudgee

Pop. 8248
Map ref. 554 H2

84 Market St; (02) 6372 1020; www.visitmudgeeregion.com.au

93.1 Real FM, 549 AM ABC Central West

Mudgee derives its name from the Wiradjuri word 'moothi', meaning 'nest in the hills'. The name is apt as the town is situated among green and blue hills in the Cudgegong River Valley. Mudgee is graced with wide streets and historic Victorian buildings and is the centre of the Mudgee Wine Region, one of the largest winegrowing regions in Australia. Local produce features heavily in town and includes yabbies, lamb, asparagus, summer berries, peaches and hazelnuts.

Colonial Inn Museum: local history in photographs, documents, machinery, dolls and agricultural implements; check opening times (02) 6372 3365; Market St. ***Honey Haven:*** honey, jam and mustard tastings, and bees under glass; Cnr Castlereagh Hwy. ***Mudgee Gourmet:*** a gourmand's delight where you can enjoy the local produce, tastings, local art and craft; Cnr Henry Lawson Dr and Ula Rd. ***Lawson Park:*** home to possums, water rats and tortoises. Includes a playground, barbecues and duck pond; Short St. ***Mandurah at the Railway:*** local art and craft cooperative at the historic railway station; Cnr Inglis and Church sts. ***Mudgee Brewing Company:*** enjoy some pale ale at this home-grown brewery; Church St. ***Roth's Wine Bar:*** oldest wine bar in NSW with displays of wine history in the region and a wide selection of local wines; Market St. ***Mudgee Observatory:*** astronomical wonders of the NSW sky. Day and night sessions available; Old Grattai Rd. ***Town trail:*** self-guide walk taking in National Trust buildings including St John's Church of England (1860) and the Regent Theatre; brochure available from visitor centre.

Markets: St John's Anglican Church, 1st Sat each month; Lawson Park, 2nd Sat each month; St Mary's Catholic Church, 3rd Sat each month; Railway Station, 4th Sat each month. ***Small Farm Field Days:*** July. ***Mudgee Food and Wine Fair:*** Sept. ***Mudgee Wine Festival:*** Sept. ***Mudgee Cup:*** Dec.

Deeb's Kitchen: Lebanese; Cnr Buckeroo La and Cassilis Rd; (02) 6373 3133. ***Eltons Brasserie:*** modern Australian; 81 Market St; (02) 6372 0772.

Lauralla Guest House: Victorian guesthouse; 25 Lewis St; (02) 6372 4480. ***Wildwood Guesthouse:*** secluded B&B; Henry

 RADIO STATIONS   IN TOWN  WHAT'S ON   WHERE TO EAT  WHERE TO STAY  NEARBY

Lawson Dr; (02) 6373 3701. ***Wombadah and Tierney House:*** luxury B&B; 46 Tierney La; (02) 6373 3176.

Munghorn Gap Nature Reserve Over 160 bird species have been identified here, including the rare regent honeyeater. For bushwalkers, the Castle Rock walking trail is an 8 km journey with stunning views from sandstone outcrops. Camping, barbecue and picnic facilities are available; 34 km NE.

Site of Old Bark School: attended by Henry Lawson and made famous in several of his poems. Eurunderee Provisional School is nearby with historical displays of school life; 6 km N. ***Windermere Dam:*** watersports, trout fishing and camping facilities; 24 km SE. ***Hargraves:*** old goldmining town where Kerr's Hundredweight was discovered in 1851, yielding 1272 oz of gold. Ask at general store for gold-panning tours; 39 km SW. ***Wineries:*** over 40 in the area (many offering cellar-door tastings) including Robert Oatley, Huntington Estate and Botobolar; self-guide drives brochure from visitor centre.

See also CENTRAL WEST, p. 30

Mullumbimby

Pop. 3130
Map ref. 551 G4 | 557 O3 | 653 N12

Byron Visitor Centre, 80 Jonson St, Byron Bay; (02) 6680 8558; www.tropicalnsw.com.au

103.5 Radio 97 FM, 720 AM ABC North Coast

When Mullumbimby's economy, based on local agriculture, started to flag in the late 1960s, the town was saved by becoming an alternative-lifestyle centre. Drawn by lush subtropical countryside and excellent weather conditions, people settled at the foot of Mount Chincogan. The town still has a delightful laid-back feel today with all the facilities of a mature tourist town. A great place to rewind and reflect.

Brunswick Valley Historical Museum The museum covers local history in detail, including timber-getters, dairy farmers, pioneers and local government. It is in a pleasant park on the banks of Saltwater Creek. Outdoor displays include horse-drawn agricultural equipment and a pioneer slab cottage. Opening times vary so check with museum; Stuart St; (02) 6685 1385.

Mullumbimby Art Gallery: changing exhibitions of paintings, sculptures and prints; Cnr Burringbar and Stuart sts. ***Cedar House:*** National Trust–classified building housing an antiques gallery; Dalley St. ***Brunswick Valley Heritage Park:*** over 200 rainforest plants, including palms, and a 2 km park and river walk; Tyagarah St.

Market: Stuart St; 3rd Sat each month. ***Chincogan Fiesta:*** community celebration with stalls, parade and mountain footrace; Sept.

Milk and Honey: Mediterranean/Italian; Shop 5, 59A Station St; (02) 6684 1422. ***Poinciana:*** funky 'global' cafe; 55 Station St; (02) 6684 4036.

Mooyabil Farm Holidays: affordable self-contained accommodation; 448 Left Bank Rd; (02) 6684 1128. ***Figgy Byron Bay:*** luxury boutique accommodation; 1 Fig Tree La, Myocum; (02) 6684 7977.

Brunswick Heads This town on the Brunswick River estuary is a charming mix of quiet holiday retreat and large commercial fishing town. Despite having some truly beautiful beaches, Brunswick Heads has managed to remain remarkably serene and unassuming. Enjoy the excellent seafood in town. Highlights during the year include a wood-chopping festival in Jan and the Kite and Bike Festival in Mar. 7 km W.

Crystal Castle: spectacular natural crystal display, jewellery and gifts; 7 km SW. ***Tyagarah Airstrip:*** skydiving and paragliding; Pacific Hwy; 13 km SW. ***Wanganui Gorge:*** scenic 4 km bushwalk through the gorge with rainforest trees, enormous strangler figs and a pretty swimming hole; 20 km W. ***Crystal Creek Miniatures:*** enjoy the amazing world of minature animals; 40 km NW.

See also TROPICAL NORTH COAST, p. 28

Mulwala

Pop. 1629
Map ref. 559 M12 | 584 H5 | 586 D1

Irvine Pde, Yarrawonga; (03) 5744 1989 or 1800 062 260; www.yarrawongamulwala.com.au

96.9 Sun FM, 675 ABC AM Riverina

Mulwala and Yarrawonga (in Victoria) are twin towns sitting astride the Murray River. Mulwala prides itself on being an 'inland aquatic paradise', with plenty of water-based activities for visitors to enjoy. It is surrounded by forests and vineyards.

Lake Mulwala This artificial lake was formed by the 1939 damming of the Murray River at Yarrawonga Weir and is now home to myriad birdlife. The eastern end has river red gums up to 600 years old. The lake is popular for yachting, sailboarding, canoeing, swimming and fishing (especially for Murray cod). The Mulwala Water Ski Club is the largest in the world with 6000 members; it offers lessons and equipment hire for skiing, wakeboarding, and banana and tube rides. Day and evening cruises can be booked at the visitor centre.

Linley Park Animal Farm: working farm with horse and pony rides, and opportunity to handfeed native and exotic animals. Open weekends and school holidays; Corowa Rd. ***Pioneer Museum:*** historic farming exhibits, photographs and local artefacts; open Wed–Sun; Melbourne St. ***Tunzafun Amusement Park:*** minigolf, mini-train and dodgem cars; Melbourne St.

E.C. Griffiths Cup: powerboat racing; Apr.

La Porchetta: Italian; Mulwala Waterski Club, Melbourne St; (03) 5744 1507. ***Yarrawonga & Border Golf Club:*** lively country bistro; Gulai Rd; (03) 5744 1911.

Club Mulwala: motel-style rooms; 271 Melbourne St; (03) 5744 2331. ***Lake Mulwala Holiday Park:*** includes 18-hole golf course; Melbourne St; (03) 5744 1050. ***Lake Mulwala Hotel Motel:*** centrally located affordable accommodation; 88 Melbourne St; (03) 5744 2499. ***Mulwala Paradise Palms Motel:*** modern, comfortable motel; 121 Melbourne St; (03) 5743 2555. ***Yarrawonga & Border Golf Club:*** apartments, suites and cabins; Gulai Rd; (03) 5744 1911.

Savenake Station Woolshed: 1930s-style woolshed in working order producing merino wool; open by appt; bookings (02) 6035 9415; 28 km N. ***Everglade and swamp tours:*** to waterbird rookeries and native animal habitats; bookings at visitor centre. ***Local wineries:*** several in area offering cellar-door tastings; brochure from visitor centre.

See also MURRAY, p. 35

Murrurundi

Pop. 804
Map ref. 556 H11

113 Mayne St; (02) 6546 6446.

96.6 FM ABC Upper Hunter, 1044 AM ABC Local

Murrurundi (pronounced 'Murrurund-eye') is a rural town set in the lush Pages River Valley at the foot of the Liverpool Ranges. It is a well-preserved, quiet town and any changes have been gradual, thanks to the lack of heavy industry in the region. The main street has been declared an urban conservation area.

Paradise Park This horseshoe-shaped park lies at the base of a steep hill. Behind the park take a walk through the 'Edge of the Needle', a small gap in the rocks that opens to a path leading to the top of the hill and fantastic views. Paradise Rd.

St Joseph's Catholic Church: 1000-piece Italian marble altar; Polding St. ***Self-guide heritage walk:*** National Trust–classified sites; brochure from visitor centre.

Murrurundi Markets: last Sun of the month; outside visitor centre. ***Australia Day Carnival:*** markets, parade and activities; Jan. ***King of the Ranges Stockman's Challenge:*** May. ***Bushmans Rodeo and Campdraft Carnival:*** Oct. ***Murrurundi Stampede:*** Dec.

Cafe Telegraph: cafe with wine cellar; 155 Mayne St; (02) 6546 6733.

Combadello Cottage: luxury self-contained cottage; Pages River Rd; 0400 461 363.

Wallabadah Rock The second largest monolith in the Southern Hemisphere, the rock is a large plug (959 m high) of an extinct volcano. There are spectacular flowering orchids in Oct. The rock is on private property, with a good view from the road. Access is possible with the owner's permission (02) 6546 6881. 26 km NE.

Chilcotts Creek: diprotodon remains were found here (now in Australian Museum, Sydney); 15 km N. ***Burning Mountain:*** deep coal seam that has been smouldering for at least 5000 years; 20 km S.

See also NEW ENGLAND, p. 29

Murwillumbah

Pop. 7954
Map ref. 551 G2 | 557 O2 | 645 F12 | 653 N11

World Heritage Rainforest Centre, cnr Tweed Valley Way and Alma St; (02) 6672 1340 or 1800 674 414; www.tweedtourism.com.au

 101.3 Tweed Coast Country, 720 AM ABC Local

Murwillumbah is located on the banks of the Tweed River near the Queensland border. It is a centre for sugarcane, banana and cattle farms. In 1907 Murwillumbah was almost completely wiped out by fire. The town was rebuilt and many of those buildings can still be seen on the main street today.

Tweed River Regional Art Gallery This gallery displays a variety of paintings, portraits, glasswork, pottery, ceramics and photography. The two main themes are Australian portraits (nationwide subjects in all mediums), and depictions of the local area by regional artists. It is also home to the Doug Moran National Portrait Prize, the richest portrait prize in the world. Past winners are on display along with changing exhibitions. Mistral Rd; (02) 6670 2790.

Escape Gallery: sculpture, glass and fine arts showcasing local and regional artists, exhibitions changing constantly; Brisbane St. ***World Heritage Rainforest Centre:*** visitor centre with displays on local vegetation and wildlife, local Aboriginal and European history, and World Heritage regions; Cnr Pacific Hwy and Alma St. ***Tweed River Regional Museum:*** war memorabilia, genealogy documents, and domestic items and clothing through the ages; open Wed, Fri and 4th Sun each month, or by appt on (02) 6672 1865; Cnr Queensland Rd and Bent St.

Market: Knox Park, 1st and 3rd Sat each month; showground, 4th Sun each month. ***Tweed Valley Banana Festival and Harvest Week:*** Aug. ***Speed on Tweed:*** classic car rally; Sept.

Escape Gallery & Coffee Lounge: cafe and fine art gallery; 1 Brisbane St; (02) 6672 2433. ***The Modern Grocer:*** provedore; Shop 3, 1 Wollumbin St; (02) 6672 5007.

Ripples on Tweed: absolute waterfront B&B; 21 Tweed Valley Way; (02) 6672 2515. ***Ecoasis:*** eco-friendly accommodation; 55 Tateywan St, Uki; (02) 6679 5959.

Mt Warning National Park World Heritage–listed Mt Warning is the rhyolite plug of a massive ancient volcano left behind after surrounding basalt eroded away. The local Bundjalung nation calls the mountain Wollumbin. It is a traditional place of cultural law, initiation and spiritual education, so visitors are requested not to climb the mountain. For those who ignore this advice, it is steep in places and the return trip takes 4–5 hours, so take plenty of water and make sure there is enough time to return before sunset. The National Parks & Wildlife Service conducts regular tours in the park; contact visitor centre for bookings. 17 km SW.

Lisnagar: historic homestead in lush surrounds; northern edge of town just past showgrounds; open Sun. ***Stokers Siding:*** historic village at the foot of the Burringbar Ranges, includes pottery gallery with resident potter; 8 km S. ***Madura Tea Estates:*** tea plantation with tastings and tours by appt; bookings (02) 6677 7215; 12 km NE. ***Banana Cabana:*** garden and shop with over 20 bush-tucker species and exotic fruits; Chillingham; 12 km NW. ***Mooball:*** small town with a cow theme that has painted almost anything that stands still in the style of a black and white cow, including buildings, cars and electricity poles; 19 km SE.

See also TROPICAL NORTH COAST, p. 28

Muswellbrook

Pop. 10 225
Map ref. 555 K2

87 Hill St; (02) 6541 4050; www.muswellbrook.nsw.gov.au

98.1 Power FM, 1044 AM ABC Upper Hunter

Muswellbrook (the 'w' is silent) is in the Upper Hunter Valley and prides itself on being 'blue heeler country'. Here cattle farmers developed the blue heeler dog by crossing dingoes with Northumberland Blue Merles to produce a working dog that thrives in Australia's harsh conditions. The blue heeler is now in demand all over the world. There are several open-cut coal mines in the local area and the Upper Hunter Valley has many fine wineries.

Muswellbrook Art Gallery: in the restored town hall and School of the Arts building, its centrepiece is the Max Watters collection, which displays pieces from renowned Australian artists in paintings, drawings, ceramics and sculptures; Bridge St. ***Upper Hunter Wine Centre:*** displays on the local wine industry and information on Upper Hunter wineries; Loxton House, Bridge St. ***Historical town walk:*** 4.5 km walk featuring St Alban's Church, the police station and the town hall; map from visitor centre.

Muswellbrook Cup: Apr. ***Spring Festival:*** Aug–Nov.

Hunter Belle Cheese: cheese factory with cheese tastings; Purple Olive, 75 Aberdeen St, New England Hwy; (02) 6541 5066. ***Pukara Estate:*** olive grove with tastings; 1440 Denman Rd; (02) 6547 1055. ***SSS BBQ Barn:*** steakhouse; Cnr Hill Street and New England Highway; (02) 6541 4044.

Red Cedar Motel: simple, comfortable motel; 12 Maitland St (New England Hwy); (02) 6543 2852.

Aberdeen This small town is famous for its prize-winning beef cattle. There are markets at St Joseph's High School on the 3rd Sun each month and the famous Aberdeen Highland Games are held each July. The Scottish festivities including Scottish food and music, Highland dancing, caber tossing, a jousting tournament, a warriors competition and a kilted dash. 12 km N.

Bayswater Power Station: massive electricity source with coal-fired boilers and cooling towers; 16 km S. ***Sandy Hollow:*** picturesque village surrounded by horse studs and vineyards, with Bush Ride in Apr; 36 km SW. ***Local wineries:*** those open for cellar-door tastings include Arrowfield Wines (28 km S) and Rosemount Estate (35 km SW); brochure from visitor centre.

See also HUNTER VALLEY & COAST, p. 26

Nambucca Heads

Pop. 5874
Map ref. 550 H2 | 557 M9

Cnr Pacific Hwy and Riverside Dr; (02) 6568 6954 or 1800 646 587; www.nambuccatourism.com

105.9 2NVR FM, 684 AM ABC Mid North Coast

Located at the mouth of the Nambucca River, this is a beautiful coastal holiday town. The stunning long white beaches offer perfect conditions for fishing, swimming, boating and surfing.

V-Wall Breakwater Also known as the Graffiti Gallery, this rock wall gives visitors the opportunity to paint their own postcards on a rock. Mementos from all over the world are on display, including cartoons, paintings and poetry. Wellington Dr.

Headland Historical Museum: photographic history of the town and its residents, historic documents, antique farming implements and household tools; Headland Reserve. ***Model Train Display:*** miniature display models, including the Ghan and the Indian Pacific; Pelican Cres. ***Stringer Art Gallery:*** showcases local art and craft; Ridge St. ***Valley Community Art:*** cooperative of local artists and crafters with work for sale; Bowra St. ***Mosaic sculpture:*** the history of the town portrayed in a mosaic wrapped around a corner of the police station; Bowra St. ***Gordon Park Rainforest:*** unique walking trails through rainforest in the middle of urban development; between town centre and Inner Harbour. ***Foreshore Walk:*** 5 km pathway from Pacific Hwy to V-Wall with storyboards on shipbuilding yards and mills.

Market: Nambucca Plaza; 2nd Sun each month. ***Country Music Jamboree:*** Easter. ***Breakfast by the River:*** Apr. ***VW Spectacular:*** odd-numbered years, July–Aug. ***Show 'n' Shine Hot Rod Exhibition:*** Oct. ***Hinterland Spring Festival:*** Oct.

Matilda's: seafood and steak; 6 Wellington Dr; (02) 6568 6024.

White Albatross Holiday Centre: waterfront, family-friendly accommodation; 1 Wellington Dr; (02) 6568 6468 or 1800 152 505. ***Scared Mountain Retreat:*** luxurious romantic retreat; Burkes La, Valla; (02) 6569 5026.

Valla Beach Apart from secluded beaches and rainforest surrounds, Valla Beach is a hive of activity. Attractions include an art and craft gallery, the Valla Smokehouse (specialising in smoked products), the Australiana Workshop and the Gallery of Hidden Treasures. The Valla Beach Fair is held each Jan; 10 km N.

Swiss Toymaker: visitors can view wooden toys being crafted; closed Sun; 5 km N.

See also HOLIDAY COAST, p. 27

Narooma

Pop. 3100
Map ref. 565 F8 | 567 L10

Princes Hwy; (02) 4476 2881 or 1800 240 003; www.eurobodalla.com.au

 104.3 Power FM, 810 AM ABC Local

Narooma is a tranquil resort and fishing town at the mouth of Wagonga Inlet, well known for its natural beauty. The stunning beaches and waterways continue to draw people back to enjoy the excellent boating, aquatic sports and big-game fishing. Excellent fresh local seafood is a specialty in many of the restaurants.

Wagonga Princess This environmentally friendly, electronically powered boat is a converted huon pine ferry offering scenic cruises most days, taking in mangroves, forests and birdlife. The tour includes Devonshire tea. Commentary and tales (both tall and true) of local history, flora and fauna are provided by a third-generation local. Bookings (02) 4476 2665.

Walking tracks: several walking tracks incorporate places such as the Mill Bay Boardwalk, Australia Rock and lookout and numerous foreshore paths; maps from visitor centre. ***Whale-watching cruises:*** humpback and killer whales can be seen migrating, often with calves, Sept–Nov; bookings at visitor centre. ***Scuba-diving cruises:*** to shipwrecks; bookings at visitor centre.

Narooma Oyster Festival: May. ***Great Southern Blues and Rockabilly Festival:*** Oct.

Casey's Wholefood Cafe: healthy snacks, great coastal views; Cnr Wagonga and Canty sts; (02) 4476 1241. ***Quarterdeck Cafe:*** retro cafe; Riverside Dr; (02) 4476 2723. ***The Whale Restaurant:*** modern Australian; Whale Motor Inn, 104 Wagonga St; (02) 4476 2411. ***Anton's at Kianga:*** seaside cafe; 65 Dalmeny Dr, Kianga; (02) 4476 1802.

Anchors Aweigh B&B: homely 4-star accommodation; 5 Tilba St; (02) 4476 4000. ***Whale Motor Inn:*** stunning views, comfortable accommodation; 104 Wagonga St; (02) 4476 2411.

Montague Island Nature Reserve This isolated island, with access only by guided tours (bookings at visitor centre), is a major shearwater breeding site and home to little penguins and Australian and New Zealand fur seals. Whales can be viewed off the coast Sept–Nov. The tour includes historic buildings such as the Montague Lighthouse, which was first lit in 1881, but is now fully automated. Guides also explain the history of the island (known as Barunguba) as a fertile hunting ground for the Walbanga and Djiringanj tribes. 9 km SE.

Central Tilba Classified as an 'unusual mountain village' by the National Trust, Central Tilba was founded in 1895 and has many quality arts and craft shops. It has several old buildings worth a visit, including the ABC Cheese Factory in original 19th-century condition. The Tilba Festival is held here each Easter. 17 km SW. Tilba Tilba, a further 2 km S, features Foxglove Spires, a historic cottage surrounded by a beautiful 3.5 ha garden.

Mystery Bay: popular spot with strange-looking stones, snorkelling and access to Eurobodalla National Park; 17 km S.

See also SOUTH COAST, p. 32

Narrabri

Pop. 6104
Map ref. 556 F7

Newell Hwy; (02) 6799 6760 or 1800 659 931; www.narrabri.nsw.gov.au

91.3 MAX FM, 648 AM ABC New England North West

The fledgling town of Narrabri was devastated by a flood in 1864, but was rebuilt and grew in regional importance from 1865 when the newly constructed courthouse took over local services from Wee Waa. Cotton was introduced to the area in 1962 and the region now enjoys one of Australia's largest yields. This success has brought prosperity to Narrabri and surrounding towns. The town is located between the Nandewar Range and Pilliga scrub country in the Namoi River Valley.

Narrabri Old Gaol and Museum: historic museum with local artefacts. Night tours of the gaol on offer; Barwan St. ***Riverside park:*** pleasant surroundings next to the Namoi River with barbecue and picnic facilities; Tibbereena St. ***Self-guide town walk:*** historic buildings including the original courthouse (1865) and police residence (1879); leaflet from visitor centre.

Rodeo: Mar. ***Nosh on the Namoi:*** food and wine festival; Mar.

The Bush Coffee House: relaxed cafe; Shop 5, 111 Maitland St; (02) 6792 5747.

Dawsons Spring Cabins: basic self-contained cabins; Mt Kaputar National Park; (02) 6792 7300. ***Nandewar Motor Inn:*** central, traditional motel accommodation; Newell Hwy; (02) 6792 1155.

Paul Wild Observatory Here you'll find an impressive line of 22 m diameter antennas all facing the sky. They are connected on a rail track and are moved around to get full coverage. A sixth antenna lies 3 km away, and all 6 are sometimes connected with telescopes at Coonabarabran and Parkes. The visitor centre features a video, displays and an opportunity to view the telescope; open Mon–Fri (weekends during school holidays); 25 km W.

Yarrie Lake: birdwatching, waterskiing and windsurfing; 32 km W. ***Scenic drive:*** includes Mount Kaputar National Park (*see Barraba*) and cotton fields; leaflet from visitor centre.

See also NEW ENGLAND, p. 29

Narrandera

Pop. 3960
Map ref. 554 A8 | 559 N8

Narrandera Park, Newell Hwy; (02) 6959 1766 or 1800 672 392; www.narrandera.nsw.gov.au

92.3 FM, 549 AM ABC Central West

This historic town on the Murrumbidgee River in the Riverina district is an urban conservation area with several National Trust–classified buildings. It has been home to two Australian writers. Local magistrate Thomas Alexander Browne used the nom de plume Rolf Boldrewood to write early Australian novels such as *Robbery Under Arms*. Father Patrick Hartigan, parish priest of St Mel's Catholic Church, was better known as poet John O'Brien.

Parkside Cottage Museum Displays include the scarlet Macarthur Opera Cloak, made from the first bale of merino wool the Macarthur family sent to England in 1816. Also on display are a snow shoe and ski from Scott's Antarctic expedition, a valuable collection of shells from around the world and a set of silver ingots commemorating 1000 years of the British monarchy. Open 2–5pm daily; Newell Hwy.

Lake Talbot: boating, waterskiing, fishing and canoeing. Also scenic walking trails around the lake; Lake Dr. ***Lake Talbot Holiday Complex:*** water slides, swimming and barbecue facilities. Aquatic facilities open Oct–Mar; Lake Dr. ***NSW Forestry Tree Nursery:*** seedlings of a huge range of native trees for sale; open Mon–Fri; Broad St. ***Narrandera Park and Tiger Moth Memorial:*** beautiful park that houses the restored DN82 Tiger Moth commemorating the WW II pilots who trained in the district; Cadell St. ***Big Playable Guitar:*** built in 1988, the guitar is 5.82 m long. Largest in the Southern hemisphere; Visitor Centre Narrandera Park. ***Lavender Farm:*** tours and product sales; Bells Rd. ***Two-foot town heritage tour:*** sights include the Royal Mail Hotel (1868) and the former police station (c. 1870); brochure from visitor centre. ***Antique Corner and Objects d'Art:*** beautifully restored historic home which now houses fine antiques; Larmer St. ***Bundidgerry Walking Track:*** track passes through the koala regeneration reserve; best viewing time is at dawn; brochure from visitor centre. ***Blue Arrow scenic drive:*** historic sites, cemetery and lake; brochure from visitor centre. Eco Tours are available of the wetlands.

Rodeo: Jan. ***John O'Brien Festival:*** Mar. ***Hot Rod Rally:*** Easter. ***National Cavy Show:*** guinea pig show; Aug. ***National Model Aeroplane Championships:*** Apr. ***Camellia Show:*** Aug. ***Narrandera Cup:*** horse racing; Aug.

Narrandera Bakery: fantastic bakery; 108 East St; (02) 6959 3677.

Farrellee House: luxury 1920s guesthouse; 53 Douglas St; 0427 533 682. ***Midtown Motor Inn:*** comfortable, affordable, centrally located; 110 Larmer St; (02) 6959 2122. ***Narrandera Club Motor Inn:*** budget-friendly accommodation; 38 Bolton St; (02) 6959 3123 or 1800 422 225

John Lake Centre: fisheries visitor centre with live exhibits, audiovisual presentations and guided tours; open Mon–Fri; 6 km SE. ***Craigtop Deer Farm:*** deer raised for venison and velvet with presentation on deer-farming and a tour of the deer-handling facility; 8 km NW. ***Robertsons Gladioli Farm:*** produces flowers for Sydney and regional markets; 8 km W. ***Berembed Weir:*** picnicking, fishing and boating; 40 km SE.

See also RIVERINA, p. 34

Nelson Bay

Pop. 8153
Map ref. 549 G7 | 550 C12 | 555 N4

Port Stephens Visitor Information Centre, Victoria Pde; (02) 4980 6900 or 1800 808 900; www.portstephens.org.au

1233 AM ABC Newcastle

Nelson Bay is a coastal tourist centre that has remained small enough to maintain its charm. With outstanding white beaches and gentle waters, it is a superb spot for all aquatic activities and enjoys close proximity to Tomaree National Park. The attractive bay is the main anchorage of Port Stephens.

Inner Lighthouse This 1872 lighthouse, originally lit with 4 kerosene lamps, has been restored by the National Trust and is now completely automated. The adjacent museum features a display of the area's early history, souvenirs and a teahouse. Views of Nelson Bay are stunning. Nelson Head.

continued on p. 87

 RADIO STATIONS  IN TOWN  WHAT'S ON   WHERE TO EAT  WHERE TO STAY  NEARBY

NEWCASTLE

Pop. 498 000

Map ref. 543 H3 | 548 | 549 D9 | 550 B12 | 555 M4

3 Honeysuckle Dr; 1800 654 558; www.visitnewcastle.com.au

105.3 New FM, 1233 AM ABC Local

Newcastle began as a penal settlement and coalmining town, with a shipment of coal to Bengal in 1799 noted as Australia's first export. It soon became an industrial city, known for its steel works and port. As the steel industry is phased out, Newcastle is developing a reputation for being an elegant and cosmopolitan seaside city. The spectacular harbour is the largest export harbour in the Commonwealth and the town is bordered by some of the world's finest surfing beaches.

Fort Scratchley This fascinating fort was built in 1882 amid fears of Russian attack. Soldiers' barracks and officers' residences were built in 1886. It is one of the few gun installations to have fired on the Japanese in WW II and it remains in excellent condition. Explore networks of tunnels, gun emplacements and fascinating military and maritime museums, all perched high above Newcastle Harbour; Nobbys Rd; (02) 4929 3066; open 10am–4pm Wed–Mon.

Queens Wharf: centre point of foreshore redevelopment with restaurants, boutique brewery and observation tower linked by a walkway to Hunter Street Mall. ***Newcastle Region Art Gallery:*** broad collection of Australian art including works by Arthur Streeton, Brett Whiteley, William Dobell, Sidney Nolan and Russell Drysdale, as well as changing exhibitions; closed Mon; Laman St; (02) 4974 5100. ***Darby Street:*** Newcastle's bohemian enclave has vibrant restaurants and cafes, giftshops, galleries and young designer boutiques. ***King Edward Park:*** waterfront recreation reserve since 1863 featuring sunken gardens, ocean views, band rotunda (1898), Soldiers Baths (public pool) and Bogey Hole, a hole cut in rocks by convicts; Shortland Espl. ***Merewether Baths:*** largest ocean baths in the Southern Hemisphere; Scenic Dr. ***Blackbutt Reserve:*** 182 ha of bushland with duck ponds, native animal enclosures, walking trails, and picnic and barbecue facilities; New Lambton, off Carnley Ave; (02) 4904 3344. ***Hunter Wetlands Centre:*** 45 ha wetlands reserve 10 min from the city centre, with bike trails, playground, treasure hunt, guided eco-tours, canoe hire, picnic areas and cafe; Sandgate Rd, Shortland; (02) 4951 6466. ***Self-guide walks:*** include Town Walk and Shipwreck Walk; maps from visitor centre. ***Cruises:*** on the river and harbour; bookings at visitor centre.

Surfest: Mar. ***Newcastle Show:*** Mar. ***Offshore Superboat Championships***: Apr.

Bacchus: modern European; 141 King St; (02) 4927 1332. ***Queens Wharf Brewery:*** seafood; 150 Wharf Rd; (02) 4929 6333. ***Restaurant II:*** modern Australian; 8 Bolton St; (02) 4929 1233. ***Silo Restaurant & Lounge:*** modern Australian, harbourside; 18/1 Honeysuckle Dr, The Boardwalk; (02) 4926 2828.

Ashiana: Federation B&B; 8 Helen St, Merewether; (02) 4929 4979. ***Cooks Hill Cottage:*** luxury self-contained house; 102 Dawson St; 0401 269 863. ***Crowne Plaza Newcastle:*** chic, contemporary hotel; cnr Merewether St and Wharf Rd; (02) 4907 5000.

Lake Macquarie The enormous saltwater lake provides a huge aquatic playground with secluded bays and coves, sandy beaches and well-maintained parks lining its foreshore. Lake cruises leave from Toronto Wharf and Belmont Public Wharf. Dobell House, on the shore at Wangi Wangi, was the home of artist Sir William Dobell and has a collection of his work and memorabilia. Open to the public Sat and Sun 1–4pm and public holidays; (02) 4975 4115. The Lake Macquarie Heritage Afloat Festival is held here every Apr. 20 km s.

Munmorah State Conservation Area: coastal wilderness with great walking, picnicking, camping, swimming, surfing and fishing; (02) 4972 9000; 15 km w. ***Eraring Power Station:*** regular tours include access to cooling towers and a simulator used to train operators; bookings essential (02) 4973 0700; 22 km s. ***Swansea:*** modern resort town enjoying both lake and ocean exposure. It is popular with anglers and has excellent surf beaches; 24 km s. ***Surf beaches:*** many with world-class breaks, including Newcastle, Merewether and Nobbys; details from visitor centre.

See also HUNTER VALLEY & COAST, p. 26

[SCENIC VIEW OF FORT SCRATCHLEY]

Community Art Centre: oil and watercolour paintings, pottery china and quilting; Shoal Bay Rd. ***Self-guide heritage walk:*** from Dutchmans Bay to Little Beach; brochure from visitor centre. ***Cruises:*** dolphin-watching (all year-round) and whale-watching (May–July and Sept–Nov); on the harbour, Myall River and to Broughton Island. Also dive charters; bookings at visitor centre. ***4WD tours:*** along coastal dunes; bookings at visitor centre.

Craft markets: Neil Carroll Park, Shoal Bay Rd, 1st and 3rd Sun each month; Lutheran Church grounds, Anna Bay, 1st Sat each month. ***Tomaree Markets:*** Tomaree Sports Complex, Nelson Bay Rd, Salamander Bay; 2nd and 4th Sun each month. ***Medowie market:*** Bull 'n' Bush Hotel; 2nd Sat each month. ***Jazz Wine and Food Festival:*** Sept. ***Port Stephens Whale Festival:*** Sept–Oct. ***Tastes of the Bay Food, Wine and Jazz Festival:*** Nov.

Fishermen's Wharf Seafoods: fish and chips; 1 Teramby Rd; (02) 4984 3330. ***Ritual Organics:*** modern Australian; Shops 1 and 2, Austral St Shopping Village, Austral St; (02) 4981 5514. ***Zest Restaurant:*** modern Australian; 16 Stockton St; (02) 4984 2211.

Oaks Lure: self-contained apartments, motel rooms; 20 Tomaree St; (02) 4980 4888. ***The Landmark Nelson Bay:*** resort accommodation; Dowling St; (02) 4984 4633. ***Thurlows at the Bay:*** luxury guesthouse; 60 Thurlow Ave; (02) 4984 3722.

Tomaree National Park This park consists of bushland, sand dunes, heathland, native forest and over 20 km of rocky coastline and beaches. There is a signposted walk around the headland and another up to Fort Tomaree Lookout for breathtaking 360-degree views. Yacaaba Lookout across the bay (70 km by road) also offers great views. The park is a popular spot for bushwalking, swimming, surfing, snorkelling, fishing and picnicking. The park stretches from Shoal Bay (3 km NE) to Anna Bay (10 km SW).

Little Beach: white beach with native flora reserve behind; 1 km E. ***Gan Gan Lookout:*** spectacular views south to Newcastle and north to Myall Lakes; Nelson Bay Rd; 2 km S. ***Shoal Bay:*** popular and protected bay with spa resort; 3 km NE. ***Shell Museum:*** diverse display of shells, some rare; Sandy Point Rd; 3 km SW. ***Toboggan Hill Park:*** toboggan runs, minigolf and indoor wall-climbing; 5 km SW. ***Oakvale Farm and Fauna World:*** 150 species of native and farm animals, with visitor activities and feeding shows; 16 km SW. ***Tomago House:*** 1843 sandstone villa with family chapel and 19th-century gardens; open 11am–3pm Sun; Tomago Rd; 30 km SW. ***Stockton Sand Dunes:*** huge dune area popular for sand-boarding and whale-watching. Access is from Anna Bay or Williamtown by 4WD or safari (bookings at visitor centre); 38 km SW. ***Port Stephens wineries:*** several in area, with Stephens Winery featuring Jazz at the Winery in Mar; brochure from visitor centre.

See also **HUNTER VALLEY & COAST, p. 26**

Nimbin

Pop. 351
Map ref. 551 F4 | 557 O3

3/46 Cullen St; (02) 6689 1388; www.visitnimbin.com.au

102.3 NIM FM, 738 AM ABC Local

Situated in a beautiful valley, Nimbin is a place of healing and initiation for the Bundjalung people. When its original industry of timber faltered, the cleared land was put to use for dairy and banana farming, but Nimbin hit a depression in the late 1960s when the dairy industry collapsed. It took the 1973 Aquarius Festival to establish Nimbin as the alternative-culture capital of Australia and resurrect its fortunes. Today Nimbin is a blend of hippie counter-culture, artists' community and cottage industries amid a backdrop of farmland and forests.

Nimbin Museum: dedicated to hippie culture and Aboriginal heritage; Cullen St; (02) 6689 1123. ***Nimbin Candle Factory:*** produces stunning, environmentally friendly candles sold throughout Australia and the world; Butter Factory, Cullen St; (02) 6686 6433. ***The Rainbow Cafe:*** infamous cafe with gorgeous courtyard and colourful paintings; Cullen St. ***Town hall:*** features mural of Aboriginal art; Cullen St. ***Rainbow Power Company:*** take a tour of this alternative power supplier, now exporting power products such as solar water pumps and hydro-generators to over 20 countries; Alternative Way; (02) 6689 1430. ***Local art, craft and psychedelia:*** brochure from visitor centre.

Market: local craft; Nimbin Community Centre, Cullen St; 3rd and 5th Sun each month. ***Rainbow Lane Markets:*** next to museum, Sat mornings. ***Artist Gallery Autumn Extravaganza:*** Mar/Apr. ***Mardi Grass Festival:*** organised by the Nimbin HEMP (Help End Marijuana Prohibition) Embassy; May.

Rainbow Cafe: healthy and delicious food; 70 Cullen St; (02) 6689 1997.

Black Sheep Farm Rainforest Guest House: cosy, contemporary rainforest retreat; 449A Gungas Rd; (02) 6689 1095.

Nightcap National Park This lush World Heritage–listed forest offers signposted bushwalks from easy to very difficult, requiring a map and compass. The dramatic Protestors Falls is the site of the 1979 anti-logging protest that led to the area being gazetted as a national park. Rocky Creek Dam has a platypus-viewing platform and views of Mt Warning. 5 km NE.

Nimbin Rocks: spectacular remnants of an ancient volcano overlooking the town. This is a sacred initiation site for the Bundjalung people, so viewing is from the road only; Lismore Rd; 3 km S. ***The Channon:*** town featuring an alternative-craft market, 2nd Sun each month; Opera at the Channon in Aug; and Music Bowl Live Band Concert in Nov. 15 km SE.

See also **TROPICAL NORTH COAST, p. 28**

Nowra

Pop. 30 953
Map ref. 553 D9 | 555 I10 | 565 G4 | 567 O1

Shoalhaven Visitors Centre, cnr Princes Hwy and Pleasant Way, Nowra; (02) 4421 0778 or 1300 662 808; www.shoalhavenholidays.com.au

99 AM 2ST, 94.9 Power FM, 104.5 Triple U FM, 104.5 Shoalhaven FM, 603 AM ABC Radio National

Nowra is the principal town in the Shoalhaven district and is popular with tourists for its attractive river and water activities. The racehorse Archer began his famous 550-mile (880-kilometre) walk from Terara, on the outskirts of Nowra, to compete in the Melbourne Cup. He was led and ridden by his jockey, Dave Power. The two went on to be the first winners of the Melbourne Cup in 1861 and returned for a second win in 1862.

Meroogal Said to be the most intact 19th-century home in NSW, this 1885 property was passed down through 4 generations of women. Furniture, household objects, diaries, letters, scrapbooks, photographs and even clothes have been saved so visitors can see relics from each generation of its occupation. Open Sat afternoon and Sun; Cnr Worrigee and West sts; (02) 4421 8150.

Shoalhaven Historical Museum: old police station exhibiting the history of the town in records, photographs, household items and tools; Cnr Plunkett and Kinghorne sts; (02) 4421 8150. ***Shoalhaven River:*** fishing, waterskiing, canoeing and sailing. ***Shoalhaven River Cruises:*** departing from Nowra Wharf, Riverview Rd; bookings and times 0429 981 007. ***Hanging Rock:*** 46 m above the river with scenic views; off Junction St. ***Nowra Animal Park:*** native animals, reptiles and birds plus pony rides and opportunities to pat a koala; Rockhill Rd. ***Scenic walks:*** include Bens Walk along the river and Bomaderry Creek Walk from Bomaderry; leaflets from visitor centre.

Market: Nowra Greyhound Track; 4th Sun each month.

River Deli: excellent cafe and provedore; 84 Kinghorne St; (02) 4423 1344. ***The Boatshed:*** modern Australian, beautiful setting; 10 Wharf Rd; (02) 4421 2419.

The White House Heritage Guest House: rustic, contemporary comfort; 30 Junction St; (02) 4421 2084.

Kangaroo Valley This town of historic buildings has the National Trust–classified Friendly Inn, a Pioneer Settlement Reserve (a reconstruction of an 1880s dairy farm) and the Hampden Bridge, built in 1898 and is the oldest suspension bridge in Australia. Kangaroo Valley Fruit World is a working fruit farm open to visitors. Canoeing and kayaking safaris to Kangaroo River and Shoalhaven Gorge can be booked at the visitor centre. There is also beautiful rural scenery along Nowra Rd. 23 km NW.

Fleet Air Arm Museum Australia's largest aviation museum with displays and planes over 6000 sq m. The Wings Over Water exhibition tells the story of Australian naval aviation and the Royal Australian Navy's Fleet Air Arm; (02) 4424 1999; 8 km SW.

Marayong Park Emu Farm: group tours include the emu incubation room, brooder area, chick-rearing shed and breeding pen. Chicks hatch Aug–Oct; bookings required (02) 4447 8505; 11 km S. ***Cambewarra Lookout:*** spectacular views of the Shoalhaven River and Kangaroo Valley; 12 km NW. ***Greenwell Point:*** fresh fish and oyster sales; 14 km E. ***Bundanon:*** National Estate–listed homestead donated to the nation by artist Arthur Boyd and his wife Yvonne. The Bundanon collection and Boyd's studio are open Sun; (02) 4423 5999; 21 km W. ***Culburra:*** nearby Lake Wollumboola and coastal beaches are good for surfing, swimming, prawning and fishing. Culburra also holds the Open Fishing Carnival each Jan; 21 km SE. ***Morton National Park:*** the Nowra section features the Tallowa Dam water-catchment area, a popular spot for picnics; via Kangaroo Valley; 38 km NW. *For the Ulladulla section of the park see Ulladulla.* ***Beaches:*** many beautiful beaches in the vicinity for swimming and surfing; maps at visitor centre. ***Wineries:*** several in region offering cellar-door tastings; brochure from visitor centre.

See also SOUTH COAST, p. 32

Nundle

Pop. 290
Map ref. 557 I10

Fossickers Tourist Park, Jenkins St; (02) 6769 3355.

92.9 2TTT FM, 648 AM ABC New England North West

A thriving gold town in the 1850s, Nundle drew prospectors from California, Jamaica, China and Europe. Today the town is a quiet place nestled between the Great Dividing Range and the Peel River, but traces of gold can still be found and it is known as 'the town in the hills of gold'. The Peel River is well known for its fishing, with yellow-belly, trout and catfish being common catches.

Woollen Mill: Australia's famous wool mill offers a fascinating insight into one of the country's largest industries. View the inner workings of the mill from the observation deck or take the factory tour. Retail store also on site; Oakenville St. ***Courthouse Museum:*** built in 1880, now housing a history of Nundle and the gold-rush era; Jenkins St. ***Peel Inn:*** 1860s pub with meals and accommodation; Jenkins St. ***Mount Misery Gold Mine:*** walk back in time through a re-created goldmine evoking life on the Nundle goldfields 150 years ago; Gill St.

Go for Gold Chinese Festival: Easter. ***Nundle Dog Race:*** May.

Nundle Country Cafe: tranquil cafe; 96 Jenkins St; (02) 6789 3158.

The Rose and Willow Guest House: charming self-contained cottage; 53 Jenkins St; (02) 6765 7433.

Hanging Rock The area is popular for mineral fossicking, with good samples of scheelite and an excellent site for gold panning on the Peel River. Sheba Dams Reserve is home to numerous birds and animals. Activities include picnicking, bushwalking, camping and fishing, with regular stockings of trout and salmon. 11 km W.

Chaffey Reservoir: enjoy good swimming, fishing, sailing and picnicking. Dulegal Arboretum, an attractive garden of native trees and shrubs, is on the foreshore; 11 km N. ***Fossicker's Way tour:*** through scenic New England countryside; brochure from visitor centre.

See also NEW ENGLAND, p. 29

Nyngan

Pop. 1975
Map ref. 561 O10

Nyngan Leisure and Van Park, 12 Old Warren Rd; (02) 6832 2366; www.nyngan.com

95.1 FM ABC Local, 96.7 Rebel FM, 100.7 FM Outback Radio

Nyngan is a pleasant country town on the Bogan River, on the edge of the outback. It was largely unknown to the rest of the country until 1990, when the worst floods of the century struck here, doing damage worth $50 million. A helicopter was called in to airlift 2000 people – almost the whole town – to safety. Today the town is at the centre of a sheep, wheat and wool district.

Nyngan Museum: local memorabilia, photographs, an audio room with local stories, an 1800s kitchen and remnants from the 1990 flood; at railway station, Railway Sq. ***Mid-state Shearing Shed:*** informative displays of the continuing importance of shearing to the region, with work of local artists in murals; Mitchell Hwy. ***Historical town drive and Levee Tour:*** includes historic buildings in Cobar and Pangee sts, the Bicentennial Mural Wall and the heritage-listed railway overbridge with a lookout over town. The levee was built after the 1990 floods; brochure from visitor centre.

ANZAC Day Race Meeting: horserace; Apr. ***Nyngan Show:*** highlights the produce of the area; May.

Beancounters House: B&B; 103 Pangee St; (02) 6832 1610. ***Country Manor Motor Inn:*** comfortable, affordable accommodation; 145 Pangee St; (02) 6832 1447 or 1800 819 913.

Macquarie Marshes This mosaic of semi-permanent wetlands includes two major areas: the south marsh and the north marsh. The wetlands expand and contract, depending on recent rainfall, and provide a waterbird sanctuary and breeding ground. It is thought that the Macquarie Marshes contributed to the early myth of an inland sea, which led explorers – most notably Charles Sturt – on many ill-fated journeys. NSW National Parks & Wildlife Service has discovery rangers available to take visitors on tours of the reserve. Bookings in advance (02) 6842 1311. Marsh Meanders offers a range of activities including kayaking and bushwalking. Inquiries (02) 6824 2070. Both tours are seasonally dependant on water levels. 64 km N.

Cairn: marking the geographic centre of NSW. It is on private property but visible from the road; 65 km S. ***Richard Cunningham's grave:*** botanist with explorer Major Mitchell's party, Cunningham was killed by Aboriginal people in 1835 and buried here; 70 km S.

See also CENTRAL WEST, p. 30

Oberon

Pop. 2475
Map ref. 554 H6

137–139 Oberon St; (02) 6336 0666; www.oberonweb.com

 549 AM ABC Central West

This picturesque farming town is 1113 metres above sea level, which gives Oberon a mountain climate of cool summers, crisp winters and occasional snow. The town was named after the king in *A Midsummer Night's Dream* at the suggestion of a local Shakespeare enthusiast, after it was decided that the original name Glyndwr was unpleasant to the ear.

Oberon Museum Almost 1 ha of displays including early farming equipment, a fully furnished early settlers' house, a blacksmith shop and a functioning forge. A wide collection of artefacts and memorabilia are housed in the town's original 1920s railway station. Open 2–5pm Sat or by appt; Lowes Mount Rd.

Lake Oberon: good spot for trout fishing (both brown and rainbow). Boats and swimming are not permitted, but there are barbecue and picnic facilities; Jenolan St. ***The Common:*** green park with a small lake and picnic facilities; Edith Rd. ***Reef Reserve:*** natural bushland with access to the lake foreshore; Reef Rd. ***Cobweb Craft Shop:*** 8 tapestries on show depicting the town's landscape and buildings; Oberon St.

Market: Anglican Church; 1st Sat each month. ***Rodeo:*** Feb. ***Kowmung Music Festival:*** chamber music in caves; Mar. ***Sheep Show:*** Aug. ***Daffodil Festival:*** Sept.

McKeown's Rest: contemporary B&B; 5194 Jenolan Caves Rd; (02) 6335 6252.

Kanangra–Boyd National Park This is a rugged and dramatic piece of Australia with vast gorges, spectacular lookouts and scenic rivers. Sandstone formations of Thurat Spires, Kanangra Walls and Mt Cloudmaker are breathtaking and the park is excellent for bushwalking, rock climbing and camping. The Jenolan Caves are just outside the park border and are justifiably the country's best known cluster of caves. Tours of the majestic caverns feature fascinating flowstone deposits, helictites, columns and lakes. On the south-east edge of the park is Yerranderie, a restored silver-mining town with accommodation and walking trails. Inquiries (02) 6336 1972; Jenolan Caves 30 km SE; Kanangra Walls 52 km SE; Yerranderie 85 km SE.

Evans Crown Nature Reserve: bushwalking area with diverse flora and fauna and granite tors. Crown Rock was an initiation and corroboree site for the Wiradjuri people and is now popular for abseiling; 21 km N. ***Abercrombie River National Park:*** low eucalypt forest ideal for bushwalks, with kangaroos, wallaroos and wallabies. Abercrombie River, Retreat River and Silent Creek are havens for platypus and great for fishing, swimming and canoeing; (02) 6336 1972; 40 km S. ***Driving tours:*** routes taking in caves, national parks and surrounding towns; brochure from visitor centre. ***Wood mushrooms:*** delicacies that grow in Jenolan, Vulcan and Gurnang state forests Jan–early May. Mushrooms should be correctly identified before picking. Brochure from visitor centre.

See also BLUE MOUNTAINS, p. 23

Orange

Pop. 35 338
Map ref. 554 G4

Cnr Byng and Peisley sts; (02) 6393 8226 or 1800 069 466; www.orange.nsw.gov.au

 105.9 Star FM, 549 AM ABC Central West

Before European occupation this area was home to the Wiradjuri people, who thrived on the plentiful bush tucker resulting from the fertile volcanic soil and the abundant kangaroos and wallabies. The town was named by explorer Sir Thomas Mitchell after the Dutch Prince of Orange – they had fought together in a war in Spain. Today the prosperous 'colour city' on the slopes of Mount Canobolas enjoys a reputation for excellent food, wine, parks and gardens. It is also known for its goldmining history and as the birthplace of renowned Australian poet A. B. (Banjo) Paterson.

Civic Square This is Orange's cultural hub and also the first stop for visitors, with the visitor centre (incorporating a wine-tasting facility), the City Library, the Civic Theatre and a monument to Banjo Paterson. The Orange Regional Gallery (open Tues–Sun) is one of the busiest in the country. It has touring exhibitions as well as permanent collections that focus on jewellery, ceramics and clothing. Cnr Byng and Peisley sts.

Cook Park: colourful in any season with a begonia house (flowers Feb–May), duck pond, fernery, native bird aviary, Cook Park Guildry (for arts and crafts) and a picnic area; Summer St. ***Botanic Gardens:*** 20 ha parklands with an impressive exotic and native plant collection and a signposted walk through billabongs, rose gardens, orchards and woodlands; Kearneys Dr. ***Banjo Paterson Memorial Park:*** remains of Narambla Homestead, Paterson's birthplace, and a memorial obelisk; Ophir Rd. ***Self-guide historical walk:*** 90 min stroll past historic homes and buildings; brochure from visitor centre.

Market: Kmart carpark; Sun. ***Farmers market:*** showgrounds; 2nd Sat each month. ***Slow Food Festival:*** Feb. ***Orange Cup:*** horseracing; Mar. ***FOOD (Food of Orange District) Week:*** Apr. ***Music Week:*** Aug. ***Orange Region Winefest:*** Oct.

Blue Wren Restaurant: modern Australian, located at beautiful winery; 433 Cassilis Rd; (02) 6372 6205. ***Lolli Redini:*** modern Australian; 48 Sale St; (02) 6361 7748.

Tonic Resaurant: modern Australian; cnr Pym and Victoria sts, Millthorpe; (02) 6366 3811.

Arancia B&B: luxury retreat; 69 Moulder St; (02) 6365 3305 ***Abbey Lodge:*** romantic retreat; 224 Strathnook La; (02) 6365 1231. ***de Russie Suites:*** boutique hotel; 72 Hill St; (02) 6360 0973.

Ophir goldfields This was the site of the first discovery of payable gold in Australia (1851). The 1850s saw an influx of immigrants from Britain, Germany and China, all hoping to strike it rich. Features today include a fossicking centre, picnic area, walking trails to historic tunnels and tours of a working goldmine. There is still plenty of gold to be found in the area; the gold medals at the 2000 Sydney Olympic Games were made of Ophir gold. Brochure from visitor centre; 27 km N.

Campbell's Corner: cool-climate gardens and popular picnic spot; Pinnacle Rd; 8 km S. ***Lake Canobolas Reserve:*** recreation area with trout fishing in the lake, diving pontoons, children's playground and picnic and barbecue facilities; 9 km SW. ***Lucknow:*** old goldmining town and site of Australia's second gold discovery, now with historic bluestone buildings and craft shops; 10 km SE. ***Borenore Caves:*** undeveloped caves with evidence of fossils. Outside are walking trails, and picnic and barbecue facilities. Torch required if entering the caves; brochure from visitor centre; 22 km W. ***Cadia Mines:*** largest goldmine and coppermine in NSW; check with visitor centre for open days; 25 km SE. ***Mitchell's Monument:*** site of Sir Thomas Mitchell's base camp; 33 km W. ***Molong:*** charming rural town with Yarn Market, Craft Cottage and Coach House Gallery; 35 km NW. Grave of Yuranigh, Mitchell's Aboriginal guide, lies 2 km E of Molong. ***Wineries:*** over 30 cellar doors can be found in surrounding wine region; brochure from visitor centre.

See also CENTRAL WEST, p. 30

Parkes

Pop. 9825
Map ref. 554 E4

Henry Parkes Centre, Pioneer Park, Newell Hwy; (02) 6863 8860; www.visitparkes.com.au

 95.5 ROK FM, 549 AM ABC Local

Parkes is most famous for its huge telescope. It originated as a tent city, which grew into a town named Bushmans, built almost overnight when gold was found in the area in 1862. The name of the town changed following visits from New South Wales Colonial Secretary Henry Parkes in 1873. The main street, Clarinda Street, was named after Mrs Parkes the following year.

Pioneer Park Museum Set in a historic school and church, displays include early farm machinery and transport. The museum incorporates the collection of the previously separate Henry Parkes Historical Museum, which specialises in memorabilia from the gold rush, and includes the fascinating 1000-volume personal library of Sir Henry Parkes. Pioneer St, North Parkes; (02) 6862 3509.

Motor Museum: displays of vintage and veteran vehicles and local art and craft; Pioneer Park, Newell Hwy. ***Memorial Hill:*** excellent views of the town and surrounds; Bushman St, North Parkes. ***Kelly Reserve:*** playground and picnic and barbecue facilities in bush setting; Newell Hwy, North Parkes. ***Bushmans Hill Reserve:*** take the walking trail to a lookout, passing mining relics and a memorial to those who lost their lives in local mines; Newell Hwy, North Parkes. ***Self-guide historical town walk and drive:*** highlights include the police station (1875), post office (c. 1880) and Balmoral, one of the town's oldest homes, noted for its iron lace, Italian marble and stained-glass windows; brochure from visitor centre.

Elvis Festival: Jan. ***Parkes National Marbles Championships:*** Mar. ***Parkes Astrofest:*** July. ***Trundle Bush Tucker Day:*** Sept. ***Country Music Spectacular:*** Oct.

Dish Cafe: homemade lunches; Parkes Radio Telescope; (02) 6862 1566. ***Parkes International Restaurant:*** diverse menu; Parkes International Comfort Inn, Newell Hwy; (02) 6862 5222. ***Station Hotel:*** modern Australian; 82 Peak Hill Rd; (02) 6862 8444.

Kadina B&B: friendly, traditional B&B; 22 Mengarvie Rd; (02) 6862 3995 or 0412 444 452. ***The Old Parkes Convent:*** rustic B&B; 33 Currajong St; (02) 6862 5385.

Parkes CSIRO Radio Telescope Commissioned in 1961, the telescope is the largest and oldest of the 8 antennae making up the Australian Telescope National Facility. It has been used for globally important work such as identifying the first quasar in 1963, mapping important regions of the Milky Way, and tracking the NASA Apollo moon missions. It was most famously instrumental in transmitting images of Neil Armstrong's first steps on the moon to the world. The story of the events on the ground at Parkes is portrayed in the film *The Dish* (2000). The visitor centre explains the uses of the telescope and has 3D displays. Inquiries (02) 6861 1777; 23 km N.

Condobolin This country town is where the first *Australian Idol* runner-up Shannon Noll has his roots. A lookout on Reservoir Hill gives views over the town, which is at the junction of 2 rivers and surrounded by red-soil plains. Mt Tilga, 8 km N, is said to be the geographical centre of NSW, and Gum Bend Lake, 5 km W, is a good spot for fishing and watersports. 95 km W.

Macusani Alpaca Farm and Shop: includes workshops throughout the year; Tichborne, 10 km S. ***Peak Hill:*** working goldmine with lookout offering views of Parkes; 48 km N. ***Utes in the Paddock:*** 18 utes make up a quirky Australian exhibition; Ootha, 70 km W.

See also CENTRAL WEST, p. 30

Picton

Pop. 3025
Map ref. 544 F12 | 553 D2 | 555 J7 | 565 G1

Old Post Office, cnr Argyle and Menangle sts; (02) 4677 8313; www.visitwollondilly.com.au

C91.3 FM Campbelltown, 603 AM ABC Radio National

Located in the foothills of the Southern Highlands, Picton was once a thriving town, but since the re-routing of the Hume Highway it has become a peaceful and well-preserved village. Originally gazetted as Stonequarry, the town was renamed after Thomas Picton, one of Wellington's generals at the battle of Waterloo.

George IV Inn and Scharer's Brewery This is one of Australia's oldest operating inns. It only serves beer from its own brewery made from its own original German recipe (one of only a few pub breweries in the country) and has meals and regular entertainment. Argyle St; (02) 4677 1415.

Picton Botanical Gardens: quiet rural park with views over the farmland, barbecues and picnic facilities; Regreme Rd. ***Self-guide historical walk:*** includes the splendid railway viaduct (1862) over

Stonequarry Creek, and St Mark's Church (1848). ***Ghost tours:*** local pioneer ghost tour and tales with supper or 2-course meal. Bookings essential (02) 4677 2044.

 ***Brush with the Bush:*** art festival; Oct.

Dreamcatcher Lodge B&B: B&B plus self-contained cottage; 2330 Remembrance Dr; (02) 4683 3232. ***White Waratah Retreat:*** country guesthouse; 1665 Remembrance Dr; (02) 4677 2121.

Thirlmere This quiet and attractive town is best known for its NSW Rail Transport Museum. The complex offers steam-train rides on the 1st and 3rd Sun of each month (except in summer when the diesel trains run). The Festival of Steam is held here in Mar, and includes market stalls and street parade. 5 km SW. Thirlmere Lakes National Park (a further 3 km SW) protects 5 reed-fringed freshwater lakes that are home to waterbirds and other wildlife. This is a great place for swimming, picnicking and canoeing, and there is a scenic walk around the lakes.

Jarvisfield Homestead: 1865 home of pioneer landowners, now the clubhouse of Antill Park Golf Club; Remembrance Dr; 2 km N. ***Sydney Skydiving Centre:*** catering for beginners and experienced skydivers with video-viewing facilities and a picnic and barbecue area; (02) 9791 9155; 5 km E. ***Maldon Suspension Bridge:*** bungee jumping; 5 km SE. ***Wirrimbirra Sanctuary:*** native flora and fauna including Dingo Sanctuary, with regular events and twilight tours. Overnight cabins are available. Closed Mon; inquiries (02) 4684 1112; 13 km SW.

See also SOUTHERN HIGHLANDS, p. 24

Port Macquarie

see inset box on next page

Queanbeyan

Pop. 38 593

Map ref. 554 F11 | 563 G7 | 564 F6 | 565 D5 | 566 H3

Cnr Farrer Pl and Lowe St; (02) 6299 7307; www.visitqueanbeyan.com.au

 106.3 Mix FM, 549 AM ABC Local

Queanbeyan is a growing city adjoining Canberra. Even though most of the city is in New South Wales, the outskirts sprawl into the ACT. Queanbeyan takes its name from a squat ex-convict Timothy Beard inhabited near the Molonglo River. He called 'Quinbean' after an Aboriginal word meaning 'clear waters'.

History Museum: documented history of the city in restored police sergeant's residence; open 1–4pm Sat and Sun; Farrer Pl. ***Queanbeyan Printing Museum:*** includes memorabilia from the first newspaper in Queanbeyan; open 2pm–4pm Sat and Sun; Farrer Pl. ***Queanbeyan Art Society Inc:*** exhibits local art and craft; Trinculo Pl. ***Railway Historical Society:*** steam-train rides depart from station in Henderson St; check times (02) 6284 2790. ***Self-guide town walks:*** include Byrne's Mill (1883), now a restaurant, and St Benedict's Convent, built in the 1800s for the Sisters of the Good Samaritan; now home to an art and bead gallery; brochure available from visitor centre.

Queanbeyan Gift: footrace; Nov.

Basil at Byrne's Mill Restaurant: modern Australian; 55 Collett St; (02) 6297 8283.

Hamiltons Townhouse Motel: comfortable contemporary accommodation; 53 Tharwa Rd, cnr Canberra Ave; (02) 6297 1877. ***Quality Inn Country Plaza:*** well-appointed country motel; 147 Uriarra Rd; (02) 6297 1211.

Bungendore This historic country village is set in a picturesque valley near Lake George and still consists of old stone, brick and timber buildings that have been there since the 19th century. The town square contains charming colonial-style shops selling crafts and antiques, and there are several hobby farms in the area. There is a Country Muster in Feb and a rodeo in Oct. 26 km NE. Lark Hill Winery, 7 km N of Bungendore, has cellar-door tastings and sales.

Molonglo Gorge: scenic drive and 3 km walking trail provide spectacular views of Molonglo River; 2 km N. ***Googong Dam:*** fishing, bushwalking and picnicking; 10 km S. ***London Bridge Woolshed and Shearers' Quarters:*** visual history of turn-of-the-century farming and settlement life. Take the easy 1 km walk to a remarkable limestone arch. 24 km S. ***Captains Flat:*** tiny mining town, ideal for walking around, with historic buildings; 45 km S.

See also CAPITAL COUNTRY, p. 31

Raymond Terrace

Pop. 12 700

Map ref. 543 H2 | 549 D7 | 550 A12 | 555 M3

Shop 7, 42 Williams St; (02) 4987 5276; www.portstephens.org.au

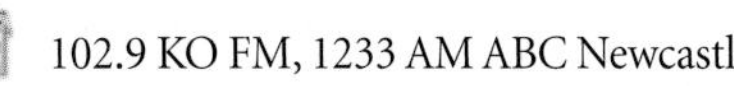 102.9 KO FM, 1233 AM ABC Newcastle

Situated on the banks of the Hunter and William rivers just outside Newcastle, Raymond Terrace was an important wool-shipping area in the 1840s. Many historic buildings from that era remain today. The town is in the middle of a koala corridor thanks to the vast remaining eucalypt forests in the region.

Sketchley Cottage: built in 1840 and rebuilt after being destroyed by fire in 1857. Displays include early Australian farming equipment, wine casks, furniture, handicrafts and photography; Pacific Hwy. ***Self-guide historical town walk:*** includes courthouse (1838) and an 1830s Anglican church built of hand-hewn sandstone; map from visitor centre.

Golden Terrace Chinese Restaurant: takeaway available; Pacific Gardens Caravan Park, 278 Pacific Hwy; (02) 4983 1515.

Motto Farm Motel: spacious budget-friendly accommodation; 2285 Pacific Hwy; (02) 4987 1211.

Tanilba House Home to the first white settler in Tanilba Bay, Lieutenant Caswell, Tanilba House was convict-built in 1831. Features of the house include decorative quoins defining the building edge, door and window openings, and high ceilings, archways and large rooms. There is said to be a resident ghost, thought to be an 1830s governess; 36 km NE.

Hunter Region Botanic Gardens: over 2000 native plants and several theme gardens; Pacific Hwy; 3 km S. ***Grahamstown Lake:*** beautiful serene lake with picnic facilities; 12 km N. ***Fighter World:*** hands-on displays of old fighter planes, engines and equipment; RAAF base, Williamtown; 16 km NE. ***Clarence Town:*** this historic village was one of the first European settlements in Australia; 27 km NW. ***Tilligerry Habitat:*** ecotourism centre with art and craft and guided walks to see koalas; 34 km NE. ***Koala Reserve:*** boardwalk through koala colony; 40 km N.

See also HUNTER VALLEY & COAST, p. 26

Richmond

Pop. 25 011
Map ref. 544 H4 | 547 O7 | 555 J6

Hawkesbury Visitor Information Centre, Ham Common, Hawkesbury Valley Way, Clarendon; (02) 4578 0233 or 1300 362 874; www.hawkesburytourism.com.au

 89.9 FM Hawkesbury Radio, 94.5 FM ABC Local

Richmond provides a peaceful country atmosphere on the Hawkesbury River, but is close enough to Sydney to enjoy the best of both worlds. Many residents commute to the city each day. The township was settled in 1794 because of the rich Hawkesbury River flats. Richmond was soon being used as a granary to supply half of Sydney's grain and is still important agriculturally. There are magnificent views of the Blue Mountains.

Bowman's Cottage: restored c. 1815 cottage, now National Parks & Wildlife office; Windsor St. ***Pugh's Lagoon:*** pleasant picnic spot with plentiful waterbirds; Kurrajong Rd. ***Self-guide historical town walks:*** features many heritage buildings including St Peter's Church (1841) and adjacent pioneer graves; brochure from visitor centre.

Bellbird Craft Markets: March St; 2nd Sat each month. ***Fruits of the Hawkesbury Festival:*** throughout region; Sept–Nov.

Red Rum Italian Restaurant: modern Italian; 39 Bells Line of Road, North Richmond; (02) 4571 3148.

Bilpin This tiny town is known for its apples and apple juice. It was originally named Belpin after Archibald Bell Jnr, who was the first European to cross the mountains from Richmond (Bells Line of Road is also named after him). The fact that he did this with the help of local Aboriginal people, who had been doing it for thousands of years, did not seem to detract from the achievement. Bilpin now has many orchards that are part of the Hawkesbury Farm Gate Trail. Brochures are available from the visitor centre. In keeping with the Australian penchant for 'big' attractions, Bilpin has the Big Bowl of Fruit. There are markets every Sat. 31 km NW.

RAAF base: oldest Air Force establishment in Australia, used for civilian flying from 1915; Windsor–Richmond Rd; 3 km E. ***Kurrajong:*** quaint mountain village that's home to shops, cafes and galleries, and the Scarecrow Festival held in Oct. Just off Bells Line of Rd; 10 km NW. ***Bellbird Hill Lookout:*** clear views across to Sydney skyline; Kurrajong Heights; 13 km NW. ***Hawkesbury Lookout:*** great views of Sydney over the Cumberland plain; 15 km SW. ***Avoca Lookout:*** stunning views over Grose Valley; 20 km W. ***Mountain Lagoon:*** mountain bushland setting with walking trails leading down to the pristine Colo River; brochure from visitor centre; 40 km NW.

See also CENTRAL COAST & HAWKESBURY, p. 25

Robertson

Pop. 1207
Map ref. 553 D6 | 555 J9 | 565 G3

Southern Highlands Visitor Information Centre, 62–70 Main St, Mittagong; (02) 4871 2888 or 1300 657 559; www.southern-highlands.com.au

 102.9 2ST FM, 603 AM ABC Radio National

Robertson sits high atop Macquarie Pass with some points in town enjoying spectacular views all the way across to the Pacific Ocean. The region's rich red soil has made Robertson the centre of the largest potato-growing district in New South Wales and there is a 'big potato' on the main street, although it is not signposted and visitors could be forgiven for thinking it is merely a large brown cylinder. The undulating hills create a picturesque setting and were featured in the film *Babe* (1996).

Kev Neel's Old Time Music Machines This music museum displays antique gramophones (some still in working condition) and music memorabilia from as early as the 1800s. Outside are deer, birds and views of the Illawarra Escarpment. Tour the grounds and use the picnic facilities. Illawarra Hwy.

Cockatoo Run Heritage Railway: steam train (when available) to Port Kembla. Robertson Railway Station. For running times call 1300 653 801. ***Art and craft shops:*** several in town featuring local work; details from visitor centre.

Market: Robertson School of Arts; 2nd Sun each month. ***Springtime Festival:*** Oct.

Pizzas in the Mist: Italian/modern Australian; 42 Hoddle St; (02) 4885 1799.

Greetham B&B: traditional B&B; 2 Meryla St; (02) 4885 1220. ***Sentinel Farm Cottage:*** large self-contained cottage; 116 Illawarra Hwy; (02) 4885 1571.

Morton National Park, Robertson section This section of the park has two attractive features: Belmore Falls (10 km SW) and Fitzroy Falls (15 km SW). Belmore Falls plunges into 2 separate rockpools, which then cascade down to the valley below. The area also features walking tracks and pleasant picnic facilities. At Fitzroy Falls is the National Parks & Wildlife Service visitor centre, which has maps and information about the entire national park and offers guided tours. The falls drop 80 m over sandstone cliffs onto black rocks and then another 40 m into the valley below. The walking trail around the falls has excellent lookouts. *For other sections of the park see Bundanoon, Nowra and Ulladulla.*

Illawarra Fly Tree Top Walk Visitors can experience the Southern Highlands' native flora and fauna from 25 m above ground. The 600 m walkway has 2 cantilevered arms stretching to the forest fall line, with expansive views of Lake Illawarra and the South Pacific Ocean. For those who want to get even higher, a spiralling tower reaches 45 m above ground. The environmentally friendly visitor centre has a cafe. Knights Hill Rd; 1300 362 881; 15 km SE.

Robertson Rainforest: 5 ha portion of what was the 2500 ha Yarrawah Brush. It is home to abundant birdlife and features an attractive bushwalk; 2 km S. ***Fountaindale Grand Manor:*** c. 1924 with over 80 rooms and uninterrupted views of the surrounding hills and valley. It is now a guesthouse and function centre; Illawarra Hwy; (02) 4885 1111; 3 km E. ***Burrawang:*** 19th-century village with an excellent historic pub; 6 km W. ***Macquarie Pass National Park:*** preserved section of the Illawarra Escarpment with bushwalks through eucalypt forest and picnic facilities; 10 km E. ***Budderoo National Park:*** Robertson section features Carrington Falls, a 50 m waterfall with adjacent walking tracks, lookouts and picnic facilities; 10 km SE. ***Manning Lookout:*** views over Kangaroo Valley; 16 km SW.

See also SOUTHERN HIGHLANDS, p. 24

Rylstone

Pop. 616
Map ref. 555 I3

Council offices, Louee St; (02) 6379 4318; www.rylstone.com

98.7 KRR FM, 549 AM ABC Central West

Visitors are drawn to Rylstone for its rural tranquillity. This old stone village is on the Cudgegong River and is a popular spot for birdwatching and fishing.

continued on p. 94

PORT MACQUARIE

Pop. 39 508

Map ref. 550 G6 | 557 M11

[SUNSET OVER PORT MACQUARIE]

Cnr Clarence and Hay sts at the Glasshouse; (02) 6581 8000 or 1300 303 155; www.portmacquarieinfo.com.au

102.3 Star FM, 684 AM ABC Local

Port Macquarie, one of the oldest towns in the state, was established in 1821 as a self-sufficient penal settlement. Convicts chosen for their skills and good behaviour maintained the fledgling town, doing everything from farming, boatbuilding and blacksmithing to teaching, baking and clerical duties. Today the city is a major holiday resort at the mouth of the Hastings River. It provides a fascinating history and features historic buildings, nature reserves, excellent surf and fishing beaches, an outstanding museum and scenic coastal walking tracks.

Historical Society Museum This award-winning museum has over 20 000 items in its ever-increasing collection. It specialises in letters, photographs and documents covering convict and free settlement and the evolution of the town from penal colony to coastal metropolis. The museum is in one of the town's beautifully restored older buildings (c. 1836). Clarence St; (02) 6583 1108.

St Thomas' Church: third oldest surviving church in Australia (1824–28), designed by convict architect Thomas Owen; Hay St. ***Port Macquarie Historic Courthouse:*** built in 1869, it served the community for 117 years; tours of the restored building by appt; closed Sun; Cnr Hay and Clarence sts; (02) 6584 1818. ***Mid-north Coast Maritime Museum:*** shipwreck relics, model ships and early photographs; William St; (02) 6583 1866. ***Kooloonbung Creek Nature Reserve:*** 50 ha of nature reserve with boardwalks and picnic area. Visit the historic cemetery nearby dating from 1842; Gordon St. ***Koala Hospital:*** Koala rehabilitation and adoption centre, the only one of its kind in NSW; Lord St; (02) 6584 1522; open 8am–4pm, feeding at 8am and 3pm. ***Roto House and Macquarie Nature Reserve:*** adjacent to the koala hospital and built in 1890, classified by the National Trust; off Lord St; (02) 6584 2180; open 10am–4pm Mon–Fri, 9am–1pm Sat. ***Port Macquarie Observatory:*** planetarium, telescope and solar system display; call for opening times (02) 6582 2397; William St. ***Billabong Koala and Wildlife Park:*** wide variety of Australian and exotic wildlife with kangaroo-feeding and koala photo sessions each day; Billabong Dr; (02) 6585 1060. ***Town beach:*** patrolled swimming and surfing beach with sheltered coves at one end. ***Cruises:*** depart daily; bookings at visitor centre. ***Scenic walks:*** 9 km coastal walk from Westport Park in the town centre to Tacking Point Lighthouse; traverses beaches and subtropical rainforest with picnic spots, interpretative signs and viewing platforms along the trail; look out for the dolphins of Hastings River and a variety of birdlife; walk takes 3 hrs 30 min one way; brochure from visitor centre.

Market: Findlay Ave; 2nd and 4th Sun each month. ***Golden Lure Tournament***: deep-sea fishing; Jan. ***Australian Ironman Triathlon***: May. ***Heritage Festival:*** Apr. ***FreshArt:*** youth arts festival; June. ***Australian Surfing Festival:*** Aug ***Aquasculpture:*** odd-numbered years, Oct. ***Tastings of the Hastings:*** food and wine festival; Oct. ***Festival of the Sun:*** music festival; Dec.

Bliss Restaurant: modern Australian; Level 1, 74 Clarence St; (02) 6584 1422. ***The Restaurant at Cassegrain:*** French cuisine; 764 Fernbank Creek Rd; (02) 6582 8320. ***The Breakwall:*** modern Australian, relaxed ambience; The Observatory Resort, 40 William St; (02) 6583 9300.

Quality Inn HW Boutique: beautiful accommodation, stunning location; 1 Stewart St; (02) 6583 1200. ***Azura B&B:*** seaside B&B; 109 Pacific Dr; (02) 6582 2700.

Lake Innes Nature Reserve This reserve is home to koalas, kangaroos and bats, but was once the location of the grand Lake Innes House. Unfortunately, the house was left to decay, and the ruins are all that remain. Guided tours are available; (02) 6588 5555. 7 km sw.

Sea Acres Rainforest Centre: elevated 1.3 km boardwalk through canopy; tours available; Pacific Dr; (02) 6582 3355; 4 km s. ***Lighthouse Beach:*** 16 km of white sand with camel rides, dolphin-watching from shore and breathtaking views from the grounds of Tacking Point Lighthouse at the northern end of the beach (lighthouse not open to public); 10 km s. ***Lake Cathie:*** holiday town between surf beach and tidal lake for swimming and fishing; 16 km s. ***Wineries:*** several in the area offering cellar-door tastings including Lake Innes Vineyard (7 km w) and Cassegrain Winery (13 km w). ***Skydiving, sea plane flights, trike tours, golf:*** brochures from visitor centre.

See also HOLIDAY COAST, p. 27

Jack Tindale Park This park is a pleasant green reserve, perfect for swimming, picnics and barbecues. Platypus are sometimes spotted in the water here and in the river below the showground. Cox St.

Self-guide historical walk: includes the Bridge View Inn (restaurant, formerly a bank) and the post office; brochure from visitor centre. ***Art and craft outlets:*** several featuring local work; details at visitor centre.

Agricultural Show: Feb. ***Great Escapade Bike Ride:*** Apr. ***StreetFeast:*** highlight is the Long Lunch, with gourmet regional food served to 350 people on long tables; Nov.

Jessie's Steakhouse: steaks, pub bistro; Rylstone Hotel, 62 Louee St; (02) 6379 1118.

Above the Clouds: self-contained accommodation; 1710 Nullo Mountain Rd; (02) 6379 6253.

Kandos: industrial town known for its cement. It features the Bicentennial Industrial Museum (open weekends) and holds the Kandos Street Machine and Hot Rod Show each Jan; 3 km S. ***Fern Tree Gully:*** tree ferns in subtropical forest with walking trails and lookouts; 16 km N. ***Dunn's Swamp:*** camping, fishing, bushwalking; 18 km E. ***Windermere Dam:*** watersports, fishing, camping, picnic and barbecue facilities. Also home of fishing competition each Easter; 19 km W. ***Military Vehicle Museum:*** vehicles from WW II and the Korean and Vietnam wars; 20 km N. ***Turon Technology Museum:*** power museum with restored steam engines; 20 km S. ***Glen Davis:*** this fascinating shale oil ghost town is at the eastern end of the Capertee Valley. The valley is almost 30 km across and is surrounded by sheer sandstone cliffs, which makes it the largest enclosed valley in the Southern Hemisphere; 56 km SE. ***Wineries:*** several in the area, including De Beaurepaire and Louee Winery; brochure from visitor centre.

See also CENTRAL WEST, p. 30

Scone

Pop. 4628
Map ref. 555 K1 | 556 H12

Cnr Susan and Kelly sts; (02) 6545 1526; www.upperhuntertourism.com.au

98.1 Power FM, 549 AM ABC Local

Set among rolling green hills in the Hunter Valley, Scone (rhymes with stone) is a pleasant rural town with tree-lined streets. It is known as the 'horse capital of Australia' and is actually the world's second largest thoroughbred and horse-breeding centre.

Australian Stock Horse Museum: photographs and displays on stockhorse history, and the headquarters of the Australian Stock Horse Society; open Mon–Fri; Kelly St. ***Mare and Foal:*** life-size sculpture by Gabriel Sterk; Kelly St. ***Historical Society Museum:*** large collection of local photographs and household furniture and appliances in an old lock-up (1870); open 9.30am–2.30pm Wed, 2.30–4.30pm Sun; Kingdon St.

Horse Festival: Apr/May.

Kerv Espresso Bar: excellent coffee, light meals; 108 Liverpool St; (02) 6545 3111. ***The Larda:*** gourmet cafe plus homewares; 122 Kelly St; (02) 6545 9533.

Belltrees Country House: working ranch retreat; Gundy Rd; (02) 6546 1123. ***Segenhoe View B&B:*** beautiful French Provincial–style B&B; Pages Row, 429 Glenbawn Rd; (02) 6545 2081. ***Strathearn Park Lodge:*** elegant guesthouse; New England Hwy; (02) 6545 3200.

Moonan Flat This small town sits at the base of the Barrington Tops. It has a beautiful suspension bridge, a small post office and the Victoria Hotel (1856). The hotel was a Cobb & Co coach stop during the gold-rush era and was reputedly patronised by bushranger Captain Thunderbolt. It is small but friendly, and has accommodation and an adjoining restaurant. 50 km NE. *For Barrington Tops National Park, see Gloucester.*

Lake Glenbawn: watersports, bass fishing, picnic, barbecue and camping facilities and a rural life museum; 15 km E. ***Burning Mountain:*** deep coal seam that has been smouldering for at least 5000 years. Take the 4.6 km track through the bush; 20 km N. ***Wineries:*** several offering cellar-door tastings; brochure from visitor centre. ***Trail rides:*** throughout area, and mustering opportunities for experienced riders; bookings at visitor centre.

See also HUNTER VALLEY & COAST, p. 26

Shellharbour

Pop. 64 296
Map ref. 553 F6 | 555 J9 | 565 H3

Lamerton House, Lamerton Cres; (02) 4221 6169; www.tourismshellharbour.com.au

97.3 FM ABC Illawarra, 99.8 Wave FM

Shellharbour was a thriving port in the 1830s when development could not keep up with demand. The first shops did not appear until the 1850s and the courthouse and gaol were erected in 1877. Prior to this the local constable had to tie felons to a tree. Today the town is an attractive holiday resort close to Lake Illawarra and one of the oldest settlements on the South Coast.

Illawarra Light Railway Museum The museum offers tram rides, displays of steam trains and vintage carriages and a miniature railway. The ticket office and kiosk are in an original 1890s rail terminus and the volunteer staff are knowledgeable. Open Tues, Thurs and Sat. Steam-train rides 2nd Sun each month and Sun during public holiday weekends. Russell St.

Historical walk: take in the historical buildings of the town, beginning at the Steampacket Inn, parts of which date back to 1856. Contact visitor centre for more information. ***Snorkelling and scuba diving:*** at Bushranger's Bay; details from visitor centre or Shellharbour Scuba Centre; (02) 4296 4266.

Craft market: Shellharbour Public School, Mary St; 2nd Sun each month. ***Gamefishing Tournament:*** Feb. ***City Festival:*** 3-day local sports extravaganza; Mar. ***Festival in the Forest:*** Sept.

Relish on Addison: modern Australian; Shop 1, 31A Addison St; (02) 4295 5191.

Joylaine Beach House: luxury self-contained beach house; 5 Darley St; (02) 9550 5525. ***Seascape Beach House:*** large self-contained house; Addison St; (02) 4297 3574.

Lake Illawarra This large tidal estuary was once a valuable source of food for the Wadi Wadi people. It is home to waterbirds such as black swans, pelicans and royal spoonbills, and has picnic and barbecue areas. The lake is excellent for boating, swimming, waterskiing, windsurfing, fishing and prawning. 7 km N.

Blackbutt Forest Reserve: remnant of coastal plain forest in urban area. Walking trails offer views of Lake Illawarra and Illawarra Escarpment; 2 km W. ***Killalea Recreation Park:*** foreshore picnic area with an ideal beach for surfing, diving, snorkelling and fishing; 3 km S. ***Bass Point Aquatic and Marine Reserve:*** top spot for scuba diving, snorkelling, fishing and surfing, with a nice picnic area on the shore; 5 km SE.

Tongarra Museum: shows area's history through photographs, maps and sketches; 11 km W. ***Crooked River Wines:*** specialty is Chardonnay White Port, with gorgeous restaurant also on-site; 20 km S.

See also SOUTHERN HIGHLANDS, p. 24

Singleton

Pop. 13 665
Map ref. 555 K3

39 George St; (02) 6571 5888 or 1800 449 888; www.singleton.nsw.gov.au

98.1 Power FM, 1044 AM ABC Upper Hunter

Singleton is a pleasant and sleepy town set next to the Hunter River among beautiful pasturelands, mountains and national parks. It is the geographical heart of the Hunter Valley and is known for its excellent wines, with several famous vineyards.

Singleton Mercy Convent The Sisters of Mercy arrived from Ireland in 1875 and set up this convent. Set in manicured gardens is the prominent convent, a chapel with an impressive marble altar, and the Sisters of Mercy Museum in an old Georgian cottage. Tours are conducted by the sisters at 2pm on weekends Mar–Nov; Queen St.

James Cook Park: riverside park with picnic facilities and the largest monolithic sundial in the Southern Hemisphere; Ryan Ave. ***Singleton Historical Museum:*** memorabilia in Singleton's first courthouse and gaol from the town's pioneer days; Burdekin Park, New England Hwy. ***Town walk:*** enjoyable walk passing a historic Anglican church and lush parklands; brochure from visitor centre.

Market: Burdekin Park, New England Hwy; 4th Sun each month. ***Countryfest:*** includes the Australian Wife Carrying Race; Apr.

Charades Restaurant and Bar: modern Australian; Quality Inn Charbonnier Hallmark, 44 Maitland Rd; (02) 6572 2333. ***Henri's Brasserie:*** modern Australian; Level 1, 85 John St; (02) 6571 3566. ***Roberts:*** modern Australian, in National Trust cottage; Halls Rd, Pokolbin; (02) 4998 7330. ***Terroir:*** winery restaurant; Hungerford Hill, 1 Broke Rd, Pokolbin; (02) 4990 0711.

Glen-Nevis B&B: Victorian guesthouse; 399 Westbrook Rd; (02) 6577 5612.

Yengo National Park Mt Yengo is of cultural significance to local Aboriginal communities and there are extensive carvings and paintings in the area. The park is a rugged area of steep gorges and rocky ridges with several walking tracks and lookouts. Old Great North Rd, along the south-east boundary, is an intact example of early 19th-century convict roadbuilding. There are picnic and barbecue areas and campsites throughout the park. Inquiries (02) 6574 5555; 15 km S.

Royal Australian Military Corps Museum: traces history of the infantry corps in Australia; 5 km S. ***Wollemi National Park:*** Singleton section features picturesque walking trails, lookouts and campsites; (02) 6372 7199; 15 km SW. *For further details on the park see Lithgow.* ***Hillside Orange Orchard:*** pick your own oranges; Windsor–Putty Rd; 25 km SW. ***Lake St Clair:*** extensive recreational and waterway facilities and you can camp onshore. Nearby lookouts offer magnificent views of Mount Royal Range; 25 km N. ***Broke:*** tiny township with breathtaking national park views and village fair in Sept; 26 km S. ***Mt Royal National Park:*** rainforest area with scenic walking tracks and lookouts with spectacular 360-degree views from Mt Royal over the entire region; (02) 6574 5555; 32 km N. ***Local Hunter Valley wineries:*** wineries with cellar-door tastings including Wyndham Estate, Australia's oldest winery, and Cockfighters Ghost; tours available; brochure from visitor centre.

See also HUNTER VALLEY & COAST, p. 26

Stroud

Pop. 669
Map ref. 549 F4 | 550 B10 | 555 M3

Forster Visitor Information Centre, Little St; Forster; (02) 6554 8799 or 1800 802 692; www.greatlakes.org.au

93.1 Breeze FM, 1512 AM ABC Radio National

This delightful town is nestled in the green Karuah Valley and seems to be from another era. The absence of tourist facilities combined with the plethora of historic buildings gives Stroud an unaffected charm. The annual International Brick and Rolling Pin Throwing Competition sees residents competing against towns called Stroud in the United States, England and Canada.

St John's Anglican Church This convict-built church was made with bricks of local clay in 1833 and features beautiful stained-glass windows and original cedar furnishings. The church is noted as the place where bushranger Captain Thunderbolt married Mary Ann Bugg. Cowper St.

Underground silo: one of 8 brick-lined silos built in 1841 for grain storage, it can be inspected by descending a steel ladder; Silo Hill Reserve, off Broadway St. ***Self-guide town walk:*** covers 32 historic sites including Orchard Cottage (1830s) and St Columbanus Catholic Church (1857), which is still in original condition; brochure from visitor centre.

International Brick and Rolling Pin Throwing Competition: July. ***Rodeo:*** Sept/Oct.

Lizzie's Cafe Brasserie: golf course brasserie; Bucketts Way; (02) 4994 5113. ***Terra Cottage Gallery and Cafe:*** great coffee and desserts; 17 Bucketts Way; (02) 4994 5338.

Orchard Cottage Accommodation: cosy self-contained accommodation; 3 Broadway St; (02) 4994 5608. ***Telegherry Estate:*** huge estate, beautiful accommodation options; 524 Bucketts Way; (02) 4994 5100.

Dungog In 1838 this town, nestled in the Williams River valley, was established as a military outpost to prevent bushranging by local villains such as Captain Thunderbolt. North of Dungog is Chichester Dam, with its blue-gum surrounds, and east of the dam is Chichester State Forest. In the foothills of the Barrington Tops, the forest has picnic spots, camping, lookouts and walking trails. 22 km W.

See also HOLIDAY COAST, p. 27

Tamworth

Pop. 42 496
Map ref. 556 H9

561 Peel St; (02) 6755 4300; www.visittamworth.com

92.9 FM Tamworth's Hit Music Station, 648 AM ABC New England North West, 1011 AM The Goanna

Tamworth is a prosperous city and the self-proclaimed country music capital of Australia, an image that has been carefully cultivated since the late 1960s. There is no question that Tamworth has increased country music's credibility and acceptance in Australia, with local events helping to launch the international careers of several stars. Thousands of fans flock to the Tamworth Country Music Festival every year in January.

Walk a Country Mile Interpretive Centre From the air this building is guitar-shaped, and it features various displays including one that cleverly documents the history of country music through lyrics. This is also Tamworth's visitor centre, and the first port of call for information on the festival. Peel St.

Hands of Fame Park: Country Music Hands of Fame Cornerstone features handprints of over 200 country music stars; Cnr New England Hwy and Kable Ave. Here you will find the outdoor Country Music Roll of Renown, which is said to be Australia's highest honour in country music. There are special tributes to, among others, Tex Morton, Smoky Dawson and Slim Dusty. 6 km s. ***Australian Country Music Foundation:*** features the Legends of Australian Country Music exhibition, a display on the Country Music Awards and a theatrette playing films and documentaries; Brisbane St. ***Calala Cottage:*** National Trust–classified home of Tamworth's first mayor with antique household items and original shepherd's slab hut; Denson St. ***Tamworth Regional Gallery:*** houses over 700 works including some by Hans Heysen and Will Ashton, and the National Fibre Collection; closed Mon; Peel St. ***Powerhouse Motorcycle Museum:*** collection of immaculate motorbikes from the 1950s through to the 1980s; Armidale Rd. ***Marsupial Park:*** sanctuary for kangaroos and other marsupials, with picnic and barbecue facilities; top Brisbane St. ***Oxley Lookout:*** views of the city and beautiful Peel Valley. It is also the starting point for the Kamilaroi walking track (6.2 km); brochure from visitor centre; top of White St. ***Powerstation Museum:*** traces Tamworth's history as the first city in the Southern Hemisphere to have electric street lighting (installed in 1888); Peel St. ***Joe Maguire's Pub:*** features Noses of Fame, nose imprints of country music stars; Peel St. ***Anzac Park:*** attractive picnic and barbecue spot with playground; bordered by Brisbane, Napier, Fitzroy and Upper sts. ***Bicentennial Park:*** fountains, granite sculptures and period lighting; Kable Ave. ***Regional Botanic Gardens:*** 28 ha of native flora and exotic displays; top of Piper St. ***Line dancing:*** various venues; lessons offered; brochure from visitor centre. ***Historical town walks:*** two available of 90 min each, visiting churches, theatres and hotels; brochure from visitor centre. ***Art and craft:*** several shops and galleries; brochure from visitor centre.

Market: Showground Pavilion; 2nd Sun each month. ***Main St Market:*** Peel St Blvd; 3rd Sun each month. ***Tamworth Country Music Festival:*** Jan. ***National Pro Rodeo:*** Jan. ***Gold Cup Race Meeting:*** horserace; Apr/May. ***Australian Line Dance Festival:*** May. ***Hats Off to Country Festival:*** July. ***Bush Poets and Balladeers:*** traditional Australian variety show with poets and country singers: July. ***National Cutting Horse Association Futurity:*** competition and entertainment; Sept–Oct. ***North-West Craft Expo:*** Dec.

Bellepoque Restaurante: Mediterranean; Cnr Darling and Marius sts; (02) 6766 3495. ***Golden Guitar Coffee Shop:*** cafe and tourist attraction; The Big Golden Guitar Tourist Centre, 2 Ringers Rd; (02) 6765 2688.

Jacaranda Cottage B&B: secluded retreat; 105 Carthage St; (02) 6766 4281. ***Paradise Tourist Park:*** excellent facilities; 575 Peel St; (02) 6766 3120 or 1800 330 133. ***The Retreat at Froog-Moore Park:*** luxurious unique retreat; 78 Bligh St; (02) 6766 3353.

Big Golden Guitar Tourist Centre The 12 m golden guitar is a giant replica of the country music award and an Australian icon. Inside the complex is the Gallery of Stars Wax Museum, which features wax models of Australian country music legends alongside current stars.

Oxley anchor: the original anchor from John Oxley's ship marks the point where he crossed the Peel River on his expedition to the coast; 9 km NW. ***Birdwatching routes:*** great birdwatching walks and drives in and around Tamworth; brochure from visitor centre.

See also NEW ENGLAND, p. 29

Taree

Pop. 16 519
Map ref. 550 E8 | 555 O1 | 557 L12

Manning Valley Visitor Information Centre, 21 Manning River Dr, Taree North; (02) 6592 5444 or 1800 182 733; www.gtcc.nsw.gov.au

 107.3 MAX FM, 684 AM ABC Local

Taree is a big modern town on the Manning River and the commercial hub of the Manning River district. It is well known for its handicrafts and its beautiful parklands and nature reserves.

Fotheringham Park and Queen Elizabeth Park These parklands make an ideal riverside picnic spot to watch the boats go by. To mark the bicentenary in 1988, an unusual herb and sculpture garden was established – the herbs are available to locals for cooking. There are also several memorials throughout the park. Between Pacific Hwy and Manning River.

Taree Craft Centre: huge craft centre featuring local work and picnic facilities; Manning River Dr, Taree North; (02) 6551 5766. ***Manning Regional Art Gallery:*** changing exhibitions always include some local works; Macquarie St; (02) 6592 5455. ***Self-guide historical walks:*** through eastern and western sections of town; brochure from visitor centre. ***Manning Valley River cruises:*** offering a variety of cruises on the Manning River; bookings at visitor centre.

Weekly markets: at various venues in the region; contact visitor centre for details. ***Manning River Summer Festival:*** Jan. ***Powerboat racing:*** Easter. ***Envirofair:*** June. ***Manning Valley Festival of the Arts:*** even-numbered years, June.

Raw Sugar Cafe: traditional cafe favourites; 214 Victoria St; (02) 6550 0137. ***Rio's Churrascaria:*** Brazilian, waterfront setting; Best Western Taree Motor Inn, 1 Commerce St; (02) 6552 3511.

Nundoobah Retreat: luxury romantic retreat; 200 Woola Rd; (02) 6552 2818. ***Kiwarrak Country Retreat:*** self-contained accommodation; 239 Half Chain Rd, Old Bar; (02) 6553 7391.

Coorabakh National Park The park features the volcanic plug outcrops of Big Nellie, Flat Nellie and Little Nellie. The Lansdowne escarpment is made up of sandstone cliffs and also has spectacular views. From Newbys Lookout you might see sea eagles and wedge-tailed eagles. 20 km NE.

Ellenborough Falls: One of the longest single-drop waterfalls in the Southern Hemisphere at 200 m. Viewing platforms and a boardwalk to the bottom of the falls. Bulga Plateau, via Elands; 50 km NW. ***Joy-flights and tandem skydiving:*** flights over the Manning Valley depart from the airport on the northern outskirts of town; Lansdowne Rd; (02) 6551 7776. ***Deep Water Shark Gallery:*** Aboriginal art and craft; Peverill St; 8 km SW. ***Ghinni Wines:*** boutique winery; Pacific Hwy, Ghinni Ghinni; 10 km NE. ***The Big Buzz Funpark:*** toboggan run, water slides and go-karts; Lakes Way; (02) 6553 6000; 15 km S. ***Beaches:*** excellent surfing conditions; 16 km E. ***Hallidays Point:*** features a rainforest nature walk; brochure from visitor centre; 25 km SE. ***Manning River:*** 150 km of navigable waterway with beaches, good fishing and holiday spots. ***Art and craft galleries:*** several in the area; brochure from visitor centre. ***Nature reserves:*** numerous in the area with abundant wildlife in rainforest settings and walking trails; map from visitor centre.

See also HOLIDAY COAST, p. 27

Tathra

Pop. 1622
Map ref. 565 E9

Andy Poole Dr; (02) 6494 1436; www.sapphirecoast.com.au

87.6 Hot Country FM, 810 AM ABC South East

Tathra is an idyllic family holiday location with a 3-kilometre surf beach, frequented by dolphins, that is safe for swimming and excellent for fishing. The town started as a small jetty that served as a shipping outlet for a group of local farmers. It is now the only sea wharf on the east coast. The region is abundant with prawns from November to May.

Sea wharf Deterioration of the 1860s wharf led to a demolition order in 1973. Only strenuous local action and the intervention of the National Trust saved the wharf. It has always been a popular fishing platform and there is also a seafood cafe. Above the wharf is the Maritime Museum, which traces the history of the wharf and steam shipping in the area and has replicas of early vessels. Fur seals and little penguins can often be seen.

Tathra Beach: 3 km patrolled beach with excellent surfing conditions. ***Fishing spots:*** several good spots for salmon and tailor; map from visitor centre.

Wharf to Waves Weekend: includes 1200 m swim; Jan.

Tathra Beachhouse Apartments: award-winning contemporary accommodation; 57 Andy Poole Dr; (02) 6499 9900.

Mimosa Rocks National Park This beautifully rugged coastal park features surf beaches, caves, offshore rock stacks, lagoons, patches of rainforest and incredible volcanic sculptures. It is excellent for snorkelling, surfing, bushwalking, birdwatching and foreshore fossicking. The name of the park comes from the steamship *Mimosa*, which was wrecked on volcanic rock in 1863; (02) 4476 2888; 17 km N.

Kianinny Bay: known fossil site with steep cliffs and rugged rocks, and diving and deep-sea fishing charters available. The 9 km Kangarutha track follows the coast with spectacular scenery; 1 km S. ***Mogareeka Inlet:*** safe swimming ideal for small children; northern end of Tathra Beach; 2 km N. ***Bournda National Park:*** picturesque conservation area for great camping and bushwalking. Wallagoot Lake has a wetland area with birdwatching, fishing, prawning, swimming, watersports and boat hire; (02) 6495 5000; 11 km S.

See also SOUTH COAST, p. 32

Temora

Pop. 4082
Map ref. 554 C7 | 559 P7

294–296 Hoskins St; (02) 6977 1511; www.temora.com.au

89.1 FM ABC Radio National, 549 AM ABC Local

In 1879 gold was discovered in the area and in 1880 the town site was chosen. By 1881 the Temora district was producing half of the state's gold. Of course this could not be maintained, and the population of around 20 000 dwindled quickly. What is left now is a quiet rural Riverina town with several historic buildings. It is also a harness-racing centre with numerous studs in the district.

Temora Rural Museum This award-winning museum has several impressive displays on rural life. There are fashions from the mid-1800s ranging from baby clothes to wedding dresses, and a replica flour mill and display explaining the history of wheat since 3000 BC. Don Bradman's first home, a hardwood slab cottage, has been moved to the grounds from Cootamundra. There is also an impressive rock and mineral collection with an emphasis on the local gold industry. Wagga Rd; (02) 6977 1291.

Skydive Centre: instruction and adventure jumps; weekends; Aerodrome Rd. ***Aviation Museum:*** This claims to be the world's finest collection of flying historic aircraft. The museum is home to the country's only 2 flying spitfires, the oldest Tiger Moth still flying in Australia, a WW II Hudson, the only flying Gloster Meteor F.8 in the world and many more. The museum holds regular flying weekends throughout the year; Menzies St; (02) 6977 1088. ***Heritage walk and drives:*** include Edwardian and Federation buildings around town; brochure from visitor centre.

Quota Markets: Pale Face Park; last Sat each month. ***Temora Rural Museum Exhibition Day:*** Mar.

Waratah Cafe: takeaway/burgers; 222 Hoskins St; (02) 6977 2054.

Courthouse Cottage B&B: restored Federation B&B; 158 Deboos St; (02) 6978 0418.

Lake Centenary: boating, swimming and picnicking; 4 km N. ***Ingalba State Forest:*** 10 000 ha of state forest featuring flora and fauna native to the area; 10 km W. ***Paragon Goldmine:*** working mine until 1996; 15 km N. ***Ariah Park:*** town known as 'Wowsers, Bowsers and Peppercorn Trees', with beautiful historic streetscape lined with peppercorn trees. Hosts the Mary Gilmore Country Music Festival in Oct; 35 km W.

See also RIVERINA, p. 34

Tenterfield

Pop. 3129
Map ref. 551 A5 | 557 L4 | 653 L12

157 Rouse St; (02) 6736 1082; www.tenterfieldtourism.com.au

89.7 Ten FM

Tenterfield is a town of four seasons with many deciduous trees making it particularly spectacular in autumn. It is perhaps best known from Peter Allen's song 'Tenterfield Saddler', which he wrote about his grandfather George Woolnough. But Tenterfield is also the self-proclaimed 'birthplace of the nation', as it is where Sir Henry Parkes delivered his famous Federation speech in 1889.

Sir Henry Parkes Museum Sir Henry Parkes made his Federation speech in this National Trust–classified building built in 1876. Today it stands as a monument to Parkes. Memorabilia includes a life-size portrait by Julian Ashton and Parkes' scrimshaw walking-stick made of whale ivory and baleen. Guided tours available. Cnr Manners and Rouse sts.

Centenary Cottage: 1871 home with local history collection. Open Wed–Sun; Logan St. ***Railway Museum:*** railway memorabilia in a beautifully restored station; Railway Ave. ***Tenterfield Saddler:*** handmade saddles at the place that inspired the Peter Allen song; High St. ***Stannum House:*** stately mansion built in 1888 for John Holmes Reid, a tin-mining magnate. Homestay and group tours; Rouse St. ***Self-guide historical town walk:*** includes early residential buildings in Logan St, St Stephens Presbyterian (now Anglican) Church where Banjo Paterson married Alice Walker in 1903 and the grand National Trust–classified post office; brochure from visitor centre.

Railway Market: railway station; 1st Sat every second month. ***Tenterfield Show:*** Feb. ***Bavarian Beerfest:*** odd-numbered years, Mar. ***Oracles of the Bush:*** Australian culture and bush poetry festival; Apr. ***Food and Wine Affair:*** Nov.

Kurrajong Downs Wines: international, wine tastings; Casino Rd; (02) 6736 4590.

Best Western Henry Parkes Motel: stylish 4-star accommodation; 144 Rouse St; (02) 6736 1066. ***Rover Park:*** family tourist park; Bruxner Hwy; (02) 6737 6862. ***Wangrah Wilderness Lodge:*** vast mountain retreat; Bluff River Rd, Sandy Flat via Tenterfield; (02) 6737 3665.

Bald Rock National Park There are excellent 360-degree views from the summit of Bald Rock, the largest granite monolith in Australia. The park is full of canyons and stone arches, and kangaroos abound. 35 km N.

Mt McKenzie Granite Drive: 30 km circuit from Molesworth St in town where you can see Bluff Rock, an unusual granite outcrop; 10 km S. ***Thunderbolt's Hideout:*** reputed haunt of bushranger Captain Thunderbolt; 11 km NE. ***Drake:*** old goldmining town now popular for fossicking and fishing; 31 km NE. ***Boonoo Boonoo National Park:*** several bushwalks include an easy 30 min stroll to the spectacular 210 m Boonoo Boonoo Falls. Pleasant swimming area above the falls; (02) 6736 4298; 32 km NE. ***Wineries:*** several offering cellar-door tastings; brochure and bookings at visitor centre.

See also NEW ENGLAND, p. 29

Terrigal

Pop. 9746
Map ref. 543 E8 | 545 P3 | 555 L5

Gosford Visitor Information, 200 Mann St; Gosford; (02) 4343 4444 or 1300 130 708; www.visitcentralcoast.com.au

101.3 Sea FM, 1512 AM ABC Radio National

Terrigal is a scenic and peaceful coastal town well known for its outstanding beaches, which are popular for surfing, swimming and surf-fishing. The Norfolk pines along the beachfront add to the relaxed feel and the boutique shops and restaurants add a sophisticated touch.

Rotary Park: pleasant for picnics and barbecues, backing onto Terrigal Lagoon, a good family swimming spot; Terrigal Dr.

Artisans Market: Rotary Park; 2nd Sat each month. ***Terrigal Food and Wine Festival:*** July.

Onda Ristorante Italiano: Italian; 150 Terrigal Dr; (02) 4384 5554. ***Seasalt:*** seafood and stunning views; Crowne Plaza Terrigal, Pine Tree La; (02) 4384 9133. ***The Reef Restaurant:*** modern Australian; The Haven; (02) 4385 3222. ***Blue Bar and Restaurant:*** modern Australian; 3/85 Avoca Dr, Avoca; (02) 4381 0707. ***Lamiche:*** modern Australian; 80 Oceanview Dr, Wamberal; (02) 4384 2044.

Beach Hut Terrigal: Balinese-style guesthouse; 6 Ash St; (02) 4385 3950. ***Clan Lakeside Lodge:*** absolute waterfront accommodation; 1 Ocean View Dr; (02) 4384 1566. ***Terrigal Pacific Resort:*** relaxing resort; 224 Terrigal Dr; (02) 4385 1555.

Bouddi National Park This park ranges from secluded beaches beneath steep cliffs to lush pockets of rainforest, with several signposted bushwalks. Maitland Bay is at the heart of a 300 ha marine park extension to protect marine life, one of the first in NSW, and contains the wreck of the PS *Maitland*. Fishing is allowed in all other areas. Putty Beach is safe for swimming and Maitland Bay is good for snorkelling. Tallow Beach is not patrolled and is recommended for strong swimmers only. 17 km S.

The Skillion: headland offering excellent coastal views; 3 km SE. ***Erina:*** pretty town with St Fiacre Distillery; 4 km W. ***Ken Duncan Gallery:*** largest privately owned photographic collection in Australia; 8 km NW. ***Several excellent beaches:*** Wamberal Beach, a safe family beach with rockpools (3 km N), Avoca Beach (7.5 km S) and Shelly Beach (13 km N), both popular for surfing. ***Sea-kayaking tours:*** various routes available; bookings (02) 4342 2222.

See also CENTRAL COAST & HAWKESBURY, p. 25

The Entrance

Pop. 2632
Map ref. 543 E7 | 545 P2 | 555 L5

Memorial Park, Marine Pde; (02) 4334 4444 or 1300 130 708; www.theentrance.org

101.3 Star FM, 1233 AM ABC Local

This immaculate seaside and lakeside town is named for the narrow channel that connects Tuggerah Lake to the Pacific Ocean. Given its proximity to Sydney and Newcastle, it has become a popular aquatic playground for residents of both cities.

Memorial Park: pelican-feeding with informative commentary at 3.30pm daily; Marine Pde. ***The Waterfront:*** town mall with shops, pavement eateries and children's playground.

Markets: Waterfront Plaza, Sat; Bayview Ave, Sun. ***Art and craft market:*** Marine Pde; Sun. ***Central Coast Country Music Festival:*** Mar. ***Tuggerah Lakes Mardi Gras Festival:*** Dec.

Ocean Restaurant: seafood; 102 Ocean Pde, Blue Bay; (02) 4334 4600.

Lavender House By The Sea: traditional B&B; 66 Dening St; (02) 4332 2234. ***El Lago Waters Resort:*** comfortable, affordable accommodation; 41 The Entrance; 1300 664 554. ***Al Mare Beachfront Retreat:*** luxury getaway for couples; 42 Werrina Pde, Blue Bay; (02) 4333 7979.

Wyrrabalong National Park With sections lying north and south of town, this park conserves the last significant coastal rainforest on the Central Coast. Signposted walking tracks lead along rocky cliffs and beaches with lookouts and picnic spots along the way providing stunning coastal views. 5 km N and S.

Shell Museum: extensive shell collection; Dunleith Caravan Park; 1 km N. ***Crabneck Point Lookout:*** magnificent coastal views;

THREDBO

Pop. 477

Map ref. 565 A8 | 566 C10 | 581 B4

Thredbo Resort Centre, Friday Flat Dr, 1800 020 589; or Snowy Region Visitor Centre, Kosciuszko Rd, Jindabyne, (02) 6450 5600; www.thredbo.com.au

88.9 FM ABC Local, 97.7 Snow FM

Thredbo, a mountain village in Kosciuszko National Park (*see Adaminaby*), is a unique year-round resort, with some of Australia's best skiing and winter sports in the colder months and angling, bushwalking and mountain-biking in summer.

Ski fields Thredbo has 480 ha of skiing terrain and the longest ski runs in Australia (up to 5.9 km). Night skiing is a feature in July and Aug. There are slopes for beginners to advanced skiers and snowboarders, with lessons and equipment hire available. The ultimate skiing and scenic experience for intermediate skiers is the Dead Horse Gap run. If skiing's not your game, plunge down the 700 m luge-style track on a bobsled, or go in-line skating along paths threading through town. Thredbo Snowsports Outdoor Adventures offers off-piste skiing, freeheeling, cross-country and snowshoeing lessons and excursions; bookings (02) 6459 4044.

Thredbo Alpine bobsled: 700 m luge-style track; adjacent to ski lifts; closed in winter. ***Australian Institute of Sport Alpine Training Centre:*** quality sporting facilities used by athletes for high-altitude training; northern end of the village. ***Thredbo River:*** excellent trout fishing. ***Village walks:*** include the Meadows Nature Walk through tea trees and the Thredbo Village Walk for diversity of alpine architecture; brochure from visitor centre. ***Mountain-biking:*** several tracks including the Village Bike Track; all bike and equipment hire available from Thredbo Service Station; bookings (02) 6457 6282.

Blues Festival: Jan. ***Jazz Festival:*** May. ***SNOWYfest: film festival;*** June. ***Top to Bottom Ski Race:*** July/Aug. ***Food and Wine Festival:*** Sept.

Credo: modern Australian; Riverside Cabins, Diggings Tce; (02) 6457 6844. ***Santé:*** modern Australian bistro, bar, live music; Shop 4, Squatter Run, Village Square; (02) 6457 6083. ***Terrace:*** modern Australian; Denman Hotel, Diggings Tce; (02) 6457 6222. ***Crackenback Cottage Restaurant:*** modern Australian; 902 Alpine Way, Crackenback; (02) 6456 2198.

[CARVING UP THREDBO'S SLOPES]

Alpine Habitats: luxury eco-retreat; cnr Alpine Way and Wollondibby Rd; (02) 6457 2228. ***Squatter's Run:*** luxurious self-contained accommodation; Diggings Tce; (02) 6457 6066. ***Crackenback Cottage:*** award-winning guesthouse; 902 Alpine Way, Crackenback; (02) 6456 2601.

Pilot Lookout: the magnificent view is dominated by the Pilot (1828 m) and the Cobberas (1883 m) in Victoria; 10 km SE. ***Skitube:*** access to Perisher and Mt Blue Cow ski fields via this European-style train from Bullocks Flat. This winter-only service passes through Australia's largest train tunnel; 15 km NE. ***Charlotte Pass:*** tucked away past Perisher and accessible in winter over snow only, Australia's highest snow resort is perfect for beginner and intermediate skiers; it's popular with families who enjoy being able to ski in and out of their lodgings. 69 km NE. ***Tom Groggin Station:*** horseriding and accommodation at the historic station where Banjo Paterson met Irish brumby-hunter and horse-breaker Jack Riley, the inspiration for *The Man from Snowy River*; (02) 6076 9510; 24 km SW. ***Mt Kosciuszko:*** Australia's highest mountain, with easy access via chairlift from Thredbo and 13 km walk; contact visitor centre for more details.

See also SNOWY MOUNTAINS, p. 33

6 km S. ***Nora Head Lighthouse:*** attractive automated lighthouse built in 1903 after several ships were wrecked on the coast; 8 km N. ***Toukley:*** unspoiled coastal hamlet with breathtaking scenery. Holds markets each Sun in the shopping centre carpark and the Gathering of the Clans (Scottish festival) in Sept; 11 km N. ***Munmorah State Recreation Area:*** signposted bushwalking trails with magnificent coastal scenery; 21 km N. ***Lakes:*** 80 sq km lake system, on average less than 2 m deep and shark-free. The linked Tuggerah Lake, Budgewoi Lake and Lake Munmorah all empty into the ocean at The Entrance and are fabulous for fishing and prawning (in summer), as well as watersports.

See also CENTRAL COAST & HAWKESBURY, p. 25

Tibooburra

Pop. 161

Map ref. 560 D2 | 662 H11

National Parks and Wildlife Service, Briscoe St; (08) 8091 3308.

999 AM ABC Local

Tibooburra is one of the hottest and most isolated towns in New South Wales. Its name means 'heaps of rocks' in the local Aboriginal language and refers to the 450-million-year-old granite tors that surround the town. In a similar tale to the creation of the Blue Mountains' Three Sisters, three brothers were turned to stone after marrying women from another tribe, creating three large rocks (only one remains today). Gold was discovered in 1881 but a poor yield, outbreaks of typhoid and dysentery and a lack of water meant the population explosion did not last.

Pioneer Park: features a replica of the whaleboat Charles Sturt carried with him on his 1844–46 expedition to find an inland sea; Briscoe St. ***Courthouse Museum:*** history of the region told with photographs, relics and documents in the restored 1887 courthouse; Briscoe St. ***Tibooburra Aboriginal Land Council Keeping Place:*** photographs and Indigenous artefacts on display include a cockatoo-feather headdress; check opening times (08) 8091 3435; Briscoe St. ***School of the Air:*** most remote school in NSW servicing students of Tibooburra and the Cameron Corner region. Tours during school terms; Briscoe St. ***Family Hotel:*** pub walls have been painted on by artists including Russell Drysdale, Clifton Pugh and Rick Amor; Briscoe St.

Gymkhana and Rodeo: Oct.

Sturt National Park This semi-desert park that begins on the edge of town is noted for its wildlife – wedge-tailed eagles, kangaroos and myriad reptiles. The landscape is diverse, ranging from ephemeral lakes to jump-ups, grassy plains and the rolling dunes of the Strzelecki Desert. Temperatures range from well over 40°C in summer to below 0°C at night in winter. Lake Pinaroo in the west is the site where Charles Sturt once built a fort to protect his party's supplies and sheep. There is an outdoor pastoralist museum and camping and homestead accommodation at Mt Wood, and short walking trails that lead from here and the park's three other campsites. Details from visitor centre.

Milparinka The Albert Hotel continues to do business in this small settlement, now almost a ghost town; (08) 8091 3863. Historic buildings include a restored courthouse, the remains of an old police station, a bank, a general store and a post office; 40 km S. Depot Glen Billabong, 14 km NW of Milparinka, is where Charles Sturt was marooned for 6 months in 1845 while searching for an inland sea. One of the worst droughts in Australia's history kept the party there due to diminishing water supplies. The grave of James Poole, Sturt's second-in-command who died of scurvy, is 1 km further east under a grevillea tree. Poole's initials and the year of his death were carved into the tree and can still be seen. Poole's Cairn, commemorating the disastrous expedition, is located at Mt Poole, 7 km N of Depot Glen.

Cameron Corner: where Queensland, NSW and SA meet. The Dog Fence, the longest fence in the world, runs through here from Jimbour in Queensland to the Great Australian Bight; 133 km NW.

See also OUTBACK, p. 36

Tocumwal

Pop. 1863
Map ref. 559 L11 | 584 F4

41 Deniliquin St; (03) 5874 2131 or 1800 677 271; www.toconthemurray. com.au

102.5 Classic Rock FM, 675 AM ABC Riverina

This picturesque town ('Toc' to the locals) is on the northern bank of the Murray River. The region is a popular holiday spot due to its pleasant river beaches and laid-back lifestyle.

Foreshore Park: peaceful green park shaded by tall gum trees, and featuring a large fibreglass Murray cod. Foreshore markets are held 11 times during the year (dates from visitor centre); Deniliquin Rd. ***River cruises, walks, drives and bike tracks:*** self-guide and guided tours of town and the river; brochures from visitor centre. ***Art and craft shops:*** several in town featuring local work; brochure from visitor centre.

Foreshore market: Foreshore park; 3rd Sat each month (not July). ***Farmers market:*** 3rd Sat each month. ***Tocumwal Classic:*** fishing competition; Jan. ***Horseraces:*** Easter.

The Big Strawberry: cafe with strawberry picking; Goulburn Valley Hwy, Koonoomoo; (03) 5871 1300.

Kingswood Motel & Apartments: budget-friendly, traditional motel; cnr Barooga Rd and Kelly St; (03) 5874 2444.

Blowhole and the Rocks This area is sacred to the Ulupna and Bangaragn people. The Rocks change colour according to weather conditions, and the Blowhole is a 25 m deep hole that legend says was home to a giant Murray cod that ate young children who fell into it. One young boy escaped the cod and was chased into the crevice, only to emerge in the Murray, suggesting that the Blowhole and the river are linked. Strangely, water has been known to flow from the Blowhole in times of drought. Adjacent to it is a working granite quarry. Rocks Rd; 8.5 km NE.

Tocumwal Aerodrome: largest RAAF base in Australia during WW II, now home to international Sportavia Soaring Centre (glider joy-flights and learn-to-glide packages); 5 km NE. ***Beaches:*** around 25 attractive river beaches in the vicinity, some with picnic areas; map from visitor centre. ***Wineries:*** there are several wineries in the area; brochure from visitor centre.

See also MURRAY, p. 35

Tumbarumba

Pop. 1487
Map ref. 554 D12 | 565 A6 | 566 B5 | 585 P5

10 Bridge St; (02) 6948 3333; www.visittumbashire.com.au

107.7 FM Radio Upper Murray, 675 AM ABC Local

This former goldmining town in the foothills of the Snowy Mountains remains seemingly untouched by the modern world with old-style charm and well-preserved buildings. This has been helped by the fact that it has been bypassed by major road and rail routes. It experiences four distinct seasons and enjoys European-style vistas of snow-capped mountains, forested hills, rolling green pastures and a crystal-clear creek. Tumbarumba's name comes from the Wiradjuri language and is thought to mean 'sounding ground'. This relates to the suggestion that there are places in the region where the ground sounds hollow.

Bicentennial Botanic Gardens: mix of native and exotic trees, especially striking in autumn; Prince St. ***Artists on Parade Gallery:*** run by local artists, exhibitions change regularly; The Parade. ***Tumbarumba Museum and Information Centre:*** includes working model of a water-powered timber mill; Bridge St.

Campdraft: Jan. ***Tumbafest:*** food, wine and music festival; Feb. ***Heritage Week:*** Nov. ***Christmas Street Carnival:*** Dec.

Four Bears Cafe and Accommodation: cabin-style guest rooms; 32 The Parade; (02) 6948 3228. ***Sunnyside Lodge:*** heritage-listed homestead; Albury Close; (02) 6948 3200. ***Lazy Dog B&B:*** boutique guesthouse, country location; 235 Moody's Hill Rd (off Elliott Way); (02) 6948 3664.

Site of Old Union Jack Mining Area: memorial to the students of the Union Jack school who died in WW I; 3 km N. ***Henry Angel Trackhead:*** starting point for a 12 km section of the Hume and Hovell Walking Track along Burra Creek. It includes waterfalls and the place where Hume and Hovell first saw the Snowy Mountains. The full walking trail is a 23-day, 440 km trek from Albury to Yass. For a kit (including maps), contact Department of Lands, Sydney (02) 6937 2700; Tooma Rd; 7 km SE. ***Pioneer Women's Hut:*** fascinating domestic and rural museum focusing on women's stories. The National Quilt Register was an initiative of the women who run this museum. Open Wed, Sat and Sun; Wagga Rd; (02) 6948 2635; 8 km NW. ***Paddy's River Falls:*** the waterfall cascades over a 60 m drop in a beautiful bush setting with a scenic walking track and picnic area. A concreted walkway is at the bottom and lookouts are at the top; 16 km S. ***Tooma:*** historic town with old hotel (c. 1880); 34 km SE. ***Wineries:*** several with cellar doors including Glenburnie Vineyard; brochure from visitor centre; (02) 6948 2570.

See also RIVERINA, p. 34

Tumut

Pop. 5926

Map ref. 554 D10 | 565 A4 | 566 C2

5 Adelong Rd; (02) 6947 7025; www.tumut.nsw.gov.au

96.3 FM Sounds of the Mountains, 97.9 FM ABC Riverina

Tumut (pronounced 'Tyoomut') is located in a fertile valley, surrounded by spectacular mountain scenery. The poplar and willow trees planted by early settlers make summer and autumn particularly striking. Prior to European settlement, Tumut was the seasonal meeting place for three Aboriginal tribes. Each summer the tribes would journey to the mountains to feast on Bogong moths.

Old Butter Factory Tourist Complex: local art and craft, and visitor centre; Snowy Mountains Hwy. ***Millet Broom Factory:*** 90 per cent of the state's broom millet comes from the region and visitors can see the factory in action; open Mon–Fri; Snowy Mountains Hwy. ***Tumut Art Society Gallery:*** specialises in work by local artists; open 10am–4pm Tues–Thurs and Sat; Cnr Tumut Plains Rd and Snowy Mountains Hwy. ***Tumut Museum:*** large collection of farm and domestic items and an excellent display of Miles Franklin memorabilia (the author was born in nearby Talbingo); open 1–4pm Sat and Sun; Cnr Capper and Merrivale sts. ***River walk:*** along Tumut River; from Elm Dr. ***Historical and tree-identifying walks:*** include Alex Stockwell Memorial Gardens with European trees and a WW I memorial; brochure from visitor centre.

Tumut Show: Mar. ***Festival of the Falling Leaf:*** autumn celebration; Apr/May. ***Boxing Day Horse Races:*** Dec.

Chit Chat: cafe with homemade treats; Shop 5, Wynyard Centre, Wynyard St; (02) 6947 1187. ***The Coach House:*** cafe with homemade bread; Tumut Connection, Russell St; (02) 6947 9143.

Nimbo Fork Lodge: luxurious fly-fishing lodge; 330 Nimbo Rd; (02) 6944 9099.

Blowering Reservoir This enormous dam is an excellent centre for watersports and fishing for rainbow trout, brown trout and perch. With the dam containing the largest trout hatchery in Australia, almost everyone catches a fish. There is a spectacular lookout over the dam wall, and the Blowering Cliffs walk (19 km S) is a pleasant 5 km stroll in Kosciuszko National Park along stunning granite cliffs; 10 km S.

Air Escape: powered hang-gliding; airport, off Snowy Mountains Hwy; 6 km E.

Tumut Valley Violets: largest African violet farm in Australia with over 1000 varieties; Tumut Plains Rd; 7 km S. ***Snowy Mountains Trout Farm:*** NSW's largest trout farm with fresh trout sales; 10 km SE. ***Talbingo Dam and Reservoir:*** dam in steep, wooded country; 40 km S. ***Power station:*** Tumut Power 3, tours available; 45 km S. ***Kosciuszko National Park:*** massive alpine park to the south-east includes nearby Yarrangobilly Caves; (02) 6947 7025; 60 km S. *For more details see Adaminaby.*

See also RIVERINA, p. 34

Tweed Heads

Pop. 84 325

Map ref. 557 O1 | 645 G11 | 646 H10 | 653 N11

Wharf St; (07) 5536 6737 or 1800 674 414; www.tweedtourism.com.au

94.5 FM ABC North Coast, 96.9 FM ABC Radio National

Tweed Heads is the state's northernmost town and – along with its twin town Coolangatta over the Queensland border – is a popular holiday destination at the southern end of the Gold Coast. The region has long been celebrated for its weather, surf beaches, night-life and laid-back atmosphere.

Point Danger This lookout is on the Queensland–NSW border and overlooks Duranbah Beach, which is popular for surfing. It was named by Captain James Cook to warn of the dangerous coral reefs that lay under the waves off the coast. The world's first laser-beam lighthouse is located here. Dolphins may be seen off the coast along the pleasant cliff-edge walk. There are several picnic spots with stunning ocean views.

Tweed Maritime and Heritage Museum: four original buildings house maritime, heritage and photographic collections; Pioneer Park, Kennedy Dr, Tweed Heads West. ***Tweed Cruise Boats:*** cruises visit locations along the Tweed River; River Tce. ***Fishing and diving charters and houseboat hire:*** guided and self-guide river excursions; bookings at visitor centre. ***Tweed Snorkelling and Whale Adventures:*** whale-watching adventures and snorkelling tours to Cook Island marine reserve; bookings (07) 5536 6737. ***Catch a Crab Cruises, Birds Bay Oyster Farm, deep-sea fishing:*** these tours depart daily, bookings from visitor centre.

Craft market: Florence St; Sun. ***Tweed Harbour Fireworks Challenge:*** Apr. ***Wintersun Carnival:*** festival and music; June. ***Greenback Tailor Fishing Competition:*** June. ***Tweed River Festival:*** Oct.

Ivory Tavern: international; 156 Wharf St; (07) 5506 9988. ***Signatures Restaurant:*** seafood/modern Australian; Quality Resort Twin Towns, Wharf St; (07) 5536 2277.

Cook's Endeavour Motor Inn: traditional motel; 26–28 Frances St; (07) 5536 5399. ***Penny Ridge Winery Resort:*** luxurious bungalows, stunning location; 363 Carool Rd; (07) 5590 9033.

Minjungbal Aboriginal Cultural Centre The Aboriginal Heritage Unit of the Australian Museum is dedicated to self-determination and the importance of promoting, protecting and preserving Australian Indigenous cultures. The unit runs this museum, which features displays on all aspects of Aboriginal life on the north coast. There is also a walk encompassing a ceremonial bora ring and a mangrove and rainforest area. Located just over Boyds Bay Bridge.

Currumbin Wildlife Sanctuary: home to a huge range of Australian native wildlife; night tours available; Currumbin; (07) 5534 1266; 7 km N. ***Beaches:*** idyllic white sandy beaches for surfing and swimming, including Fingal (3 km S) and Kingscliff (14 km S). ***Melaleuca Station:*** re-created 1930s railway station in a tea-tree plantation with train rides, a tea-tree oil distillation plant and an animal nursery; Chinderah; 9 km S. ***Tropical Fruit World:*** home to the world's largest variety of tropical fruit with plantation safari, jungle riverboat cruise, fauna park and fruit tastings; 15 km S. ***John Hogan Rainforest:*** spectacular palm rainforest walks and picnic areas; 17 km SW.

See also TROPICAL NORTH COAST, p. 28

Ulladulla

Pop. 10 302
Map ref. 555 I11 | 565 G5 | 567 N4

Civic Centre, Princes Hwy; (02) 4455 1269; www.shoalhaven.nsw.gov.au

106.7 FM 2ST, 94.9 Power FM, 702 AM ABC Local

This fishing town, built around a safe harbour, is surrounded by beautiful lakes, lagoons and white sandy beaches. It is a popular holiday destination, especially for surfing and fishing, and enjoys mild weather all year-round. Visitors flock here each Easter Sunday for the Blessing of the Fleet ceremony.

Coomie Nulunga Cultural Trail This 30 min signposted walk along the headland was created by the local Aboriginal Land Council. Along the path are hand-painted and hand-carved information posts incorporating names of local plants and animals. Dawn and dusk are the best times to experience the wildlife along the walk, but visitors are advised to stay on the path for the good of the local fauna and for their own protection (from snakes). Starts Deering St opposite Lighthouse Oval carpark.

Funland Timezone: large indoor family fun park; Princes Hwy; (02) 4454 3220. ***Warden Head:*** lighthouse views and walking tracks. ***South Pacific Heathland Reserve:*** walks among native plants and birdlife; Dowling St. ***Ulladulla Wildflower Reserve:*** 12 ha with walking trails and over 100 plant types including waratah and Christmas bush. Best in spring; Warden St.

Royal Coastal Patrol Markets: harbour wharf; 2nd Sun each month. ***Blessing of the Fleet:*** Easter. ***Gamefishing Tournament:*** June.

Carmelo's: Italian; Shop 1, 2, 10 Wason St; (02) 4454 1443. ***Elizans at Ulladulla Guest House:*** modern Australian; 39 Burrill St; (02) 4455 1796. ***Hayden's Pies:*** excellent pies; 166 Princes Hwy; (02) 4455 7798. ***Bannisters:*** modern Australian; Bannisters Point Lodge, 191 Mitchell Pde, Mollymook; (02) 4455 3044.

Ulladulla Guest House: luxurious variety of accommodation styles; 39 Burrill St; (02) 4455 1796. ***Bannisters Point Lodge:*** luxury contemporary waterfront accommodation; 191 Mitchell Pde, Mollymook; (02) 4455 3044.

Pigeon House Mountain, Morton National Park The Ulladulla section of Morton National Park features this eye-catching mountain, which Captain James Cook thought looked like a square dovehouse with a dome on top, hence its name. The local Aboriginal people obviously had a different viewpoint and named it Didhol, meaning 'woman's breast'. The mountain has now been assigned a dual name by the Geographic Names Board. The area is an Aboriginal women's Dreaming area. A 5 km return walk to the summit (for the reasonably fit) provides 360-degree views taking in the ocean, Budawang Mountains and Clyde River Valley. 25 km NW. *For the Nowra section of the park see Nowra.*

Mollymook: excellent surfing and beach fishing; 2 km N. ***Narrawallee Beach:*** popular surf beach; nearby Narrawallee Inlet has calm shallow water ideal for children; 4 km N. ***Lakes:*** good swimming, fishing and waterskiing at Burrill Lake (5 km SW) and Lake Conjola (23 km NW). ***Milton:*** historic town with art galleries and outdoor cafes. Village markets are held on the highway 1st Sat each month, the Scarecrow Festival in June and the Escape Arts Festival in Sept; 7 km NW. ***Pointer Gap Lookout:*** beautiful coastal views; 20 km NW. ***Sussex Inlet:*** coastal hamlet with fishing carnival in July; 47 km N.

See also SOUTH COAST, p. 32

Uralla

Pop. 2270
Map ref. 550 B1 | 557 J8

104 Bridge St; (02) 6778 4496; ww.uralla.com

100.3 Box FM, 819 AM ABC Local

This charming New England town is famous for its connection with bushranger Captain Thunderbolt, who lived and died in the area. It is no coincidence that this was a rich goldmining district at the time. Uralla's name comes from the Anaiwan word for 'ceremonial place'. Uralla is at its most striking when the European deciduous trees change colour in autumn.

McCrossin's Mill This restored 3-storey granite and brick flour mill (1870) is now a museum of local history. The ground floor and gardens are used for functions, but the upper levels have fascinating exhibitions, including the Wool Industry, Gold Mining (featuring a replica Chinese joss house) and an Aboriginal diorama. The Thunderbolt exhibition contains a set of 9 paintings depicting the events leading up to his death, painted by Phillip Pomroy. Inquiries (02) 6778 3022; Salisbury St.

Hassett's Military Museum: large and impressive collection features displays on military history and memorabilia, including war vehicles, uniforms and a field kitchen. Many of the displays were donated by local families; Bridge St; (02) 6778 4600. ***Thunderbolt statue:*** of 'gentleman' bushranger Fred Ward dominates the corner of Bridge St and Thunderbolt's Way. ***Thunderbolt's grave:*** he was hunted by police for over 6 years before being shot dead at nearby Kentucky Creek in 1870. His grave is clearly identified among some other magnificent Victorian monumental masonary headstones; Old Uralla Cemetery, John St. ***New England Brass and Iron Lace Foundry:*** beginning in 1872, it is the oldest of its kind still operating in Australia. Visitors are welcome and tours available; East St; (02) 6778 3297. ***Self-guide heritage walk:*** easy 2 km walk that includes 30 historic buildings, most built in the late 1800s; brochure from visitor centre.

Get off your Arts: Mar/Apr. ***Thunderbolt Country Fair:*** Nov.

Stokers Restaurant and Bar: bistro in heritage building; Bushranger Motor Inn, 37 Bridge St; (02) 6778 3777. ***Thunderbolt Inn:*** bistro-style country fare; Cnr Bridge and Hill sts; (02) 6778 4048. ***White Rose Cafe:*** traditional cafe; 82 Bridge St; (02) 6778 4159.

Cruickshanks Farmstay and B&B: working cattle farm; 313 Mihi Rd, off Tourist Dr 19; (02) 6778 2148.

Mt Yarrowyck Nature Reserve This dry eucalypt reserve has plentiful wildlife including kangaroos, wallaroos and wallabies, and is ideal for bushwalking and picnics. The highlight is the 3 km Aboriginal cultural walk – about halfway along the trail is a large overhang of granite boulders under which is a set of red-ochre paintings of circles and bird tracks. 23 km NW.

Dangars Lagoon: bird sanctuary and hide; 5 km SE. ***Gold fossicking:*** gold and small precious stones can still be found; map and equipment hire from visitor centre; 5 km sw. ***Thunderbolt's Rock:*** used by the bushranger as a lookout. Climb with care; 6 km s. ***Tourist Drive 19:*** signposted drive includes historic Gostwyck Church (11 km SE) and Dangars Falls and Gorge (40 km s); brochure from visitor centre.

See also NEW ENGLAND, p. 29

Urunga

Pop. 2685
Map ref. 550 H2 | 551 D12 | 557 M8

 **Pacific Hwy; (02) 6655 5711; www.bellingen.com**

 738 AM ABC North Coast

Urunga is a sleepy, attractive town at the junction of the Bellinger and Kalang rivers and is regarded by locals as one of the best fishing spots on the north coast. Because it is bypassed by the Pacific Highway, the town has remained relatively untouched by tourism. A large percentage of the population are retirees and there are some beautiful walks around the foreshore.

Oceanview Hotel: refurbished with original furniture. Meals and accommodation are available; Morgo St. ***Urunga Museum:*** in historic building with photographs, documents and paintings from the local area; Morgo St. ***Anchor's Wharf:*** riverside restaurant and boat hire. ***The Honey Place:*** huge concrete replica of an old-style straw beehive with glass beehive display, honey-tasting, gallery and gardens; Pacific Hwy. ***Watersports and fishing:*** both on the rivers and beach; brochures from visitor centre.

Anchors Wharf Cafe: seafood/steak and boat hire; 4–6 Bellingen St; (02) 6655 5588.

Aquarelle B&B: self-contained guesthouses; 152 Osprey Rd, Hungry Head; (02) 6655 3174. ***Maino Gabuna:*** rustic accommodation options; Wollumbin Dr, Hungry Head; (02) 6655 6017.

Bongil Bongil National Park This stunning park is 10 km of unspoilt coastal beaches, pristine estuaries, wetlands, rainforest and magnificent views. The estuaries are perfect for canoeing and birdwatching, with abundant protected birdlife. The beaches provide outstanding fishing, surfing and swimming, as well as important nesting areas for a variety of wading birds and terns. There are signposted bushwalks and scenic picnic spots; (02) 6651 5946; 15 km N.

Hungry Head: beautiful beach for surfing and swimming; 3 km s. ***Raleigh:*** charming town with Prince of Peace Anglican Church (1900), winery, horseriding and a go-kart complex; 4 km N.

See also HOLIDAY COAST, p. 27

Wagga Wagga

Pop. 62 000
Map ref. 554 B9 | 559 P9 | 585 M1

Tarcutta St; (02) 6926 9621 or 1300 100 122; www.waggawaggaaustralia.com.au

 91.3 Star FM, 102.7 FM ABC Local

Wagga Wagga ('place of many crows' in the Wiradjuri language) is the largest inland city in New South Wales and is regarded as the capital of the Riverina. In 1864 Wagga Wagga received international attention when a man arrived claiming to be Roger Tichborne, a baronet who was believed drowned when his ship disappeared off South America. While Tichborne's mother believed him, the trustees of the estate were not so sure. What followed is believed to be the longest court case in England's history. The man was found to be Arthur Orton, a butcher, and sentenced to 14 years for perjury. Mark Twain found this story so fascinating that he insisted on visiting Wagga Wagga when he visited Australia in the 1890s.

Museum of the Riverina The museum is divided into 2 locations, one at the historic Council Chambers (Baylis St) and the other at the Botanic Gardens Wagga Wagga (Lord Baden Powell Dr). The chambers museum has a regular program of travelling exhibitions (inquiries on (02) 6926 9655). The museum at the botanic gardens focuses on the people, places and events that have been important to Wagga Wagga and incorporates the Sporting Hall of Fame, which features local stars such as former Australian cricket captain Mark Taylor.

Botanic Gardens: themed gardens, mini-zoo, free-flight aviary, miniature railway; picnic and barbecue facilities; Willans Hill. ***Wagga Wagga Regional Art Gallery:*** offers an extensive and changing exhibition program and includes the National Art Glass Collection; closed Mon; Cnr Baylis and Morrow sts; inquiries (02) 6926 9660. ***Lake Albert:*** watersports, fishing, bushwalking and birdwatching; Lake Albert Rd. ***Charles Sturt University Winery and Cheese Factory:*** open daily for tastings with a range of premium wines and handmade cheeses; off McKeown Dr. ***Bikeways:*** along Lake Albert, Wollundry Lagoon, Flowerdale Lagoon and the Murrumbidgee River.

Farmers market: Wollundry Lagoon; 2nd Sat each month. ***Wagga Wagga Wine and Food Festival:*** Mar. ***Gold Cup Racing Carnival:*** May. ***Jazz and Blues Festival:*** Sept.

Clancy's Restaurant: modern Australian; Quality Inn Carriage House Motel, cnr Sturt Hwy and Eunony Bridge Rd; (02) 6922 7374. ***Three Chefs Restaurant:*** modern Australian; Townhouse International Motel, 70 Morgan St; (02) 6921 5897.

Comfort Inn Prince of Wales: 32 comfortable guest rooms; 143 Fitzmaurice St; (02) 6921 1922. ***Little Bunda Cottages:*** self-contained cottages; 221 Coolamon Rd; (02) 6931 7016. ***The Manor:*** 4-star mansion; 38 Morrow St; (02) 6921 2958.

Junee This important railhead and commercial centre is located on Olympic Way. It has several historic buildings and museums. Monte Cristo Homestead is a restored colonial mansion with an impressive carriage collection. The Roundhouse Museum in Harold St contains an original workshop, locomotives, a model train and memorabilia. The visitor centre is located in Railway Sq in 19th-century railway refreshment rooms. 41 km NE.

Lake Albert: watersports, fishing, bushwalking and birdwatching; 7 km s. ***RAAF Wagga Heritage Centre:*** indoor and outdoor exhibits; open Wed, Sat, Sun 10am–4pm; 10 km E. ***The Rock:*** small town noted for its unusual scenery. Walking trails through the reserve lead to the summit of the Rock; 32 km sw. ***Aurora Clydesdale Stud and Pioneer Farm:*** encounters with Clydesdales and other farm animals on a working farm; closed Thurs and

Sun; 33 km W. ***Lockhart:*** historic town with National Trust–listed Green St for an impressive turn-of-the-century streetscape. Also several pleasant walking tracks in the area, and picnic races in Oct; 65 km W. ***Wineries:*** several in the area offering cellar-door tastings. Wagga Wagga Winery's tasting area and restaurant have an early-Australian theme; brochure from visitor centre. ***Walking tracks:*** guided and self-guided tours of the city sights; brochures from visitor centre.

See also RIVERINA, p. 34

Walcha

Pop. 1625
Map ref. 550 B3 | 557 J9

51W Fitzroy St; (02) 6774 2460; www.walchansw.com.au

 88.5 FM ABC Local, 90.1 FM ABC Radio National

Walcha (pronounced 'Wolka') is an attractive service town to the local farming regions on the eastern slopes of the Great Dividing Range. Modern sculptures are featured throughout the town and the beautiful Apsley Falls are a must-see for visitors.

Amaroo Museum and Cultural Centre This unique centre features Aboriginal art and craft with the artists working on-site; visitors are invited to watch them at work. The artists use traditional designs combined with contemporary flair to make original clothing, homewares, gifts, jewellery and art. Also on display is a collection of local Aboriginal artefacts. Open Mon–Fri; Derby St; (02) 6777 1111.

Open-Air Sculptures Situated throughout town, the works include street furniture created by local, national and international artisans; brochure from visitor centre.

Pioneer Cottage and Museum: includes a blacksmith's shop and the first Tiger Moth used for crop dusting in Australia; Derby St.

Walcha Bushmans Carnival and Campdraft: Jan. ***Timber Expo:*** May. ***Walcha Garden Festival:*** Nov.

Cafe Graze: great food and coffee; 21N Derby St; (02) 6777 2409. ***Walcha Road Hotel Restaurant:*** modern Australian; Walcha Rd; (02) 6777 5829.

Anglea House B&B: beautiful self-contained guesthouse; Cnr Thunderbolts Way and Hill St; (02) 6777 2187. ***Arran Eco-Farmstay:*** European-influenced working vineyard; Winterbourne Rd; (02) 6777 9181.

Oxley Wild Rivers National Park This national park encompasses a high plateau, deep gorges and numerous waterfalls. The Walcha section of the park features Apsley Falls (20 km E), where 7 platforms and a bridge provide access to both sides of the gorge and waterfall. Tia Falls (35 km E) has beautiful rainforest scenery. Campsites at Riverside and Youdales Hut are accessible by 4WD.

Trout fishing: several good locations; brochure from visitor centre.

See also NEW ENGLAND, p. 29

Walgett

Pop. 1734
Map ref. 556 B5

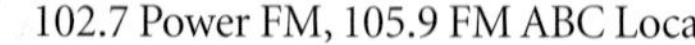

88 Fox St; (02) 6828 6139; www.walgett.nsw.gov.au

102.7 Power FM, 105.9 FM ABC Local

Walgett is the service centre to a large pastoral region. The name Walgett, which means 'the meeting of two waters', is apt, as the town is at the junction of the Barwon and Namoi rivers. The rivers provide excellent Murray cod and yellow-belly fishing. The area is rich in Aboriginal history, and archaeological digs in the shire have demonstrated that human life existed here up to 40 000 years ago.

Norman 'Tracker' Walford Track: signposted 1.5 km scenic walk includes the first European settler's grave on the banks of the Namoi River; from levee bank at end of Warrena St. ***Dharriwaa Elders Group:*** Aboriginal arts and craft with local work including paintings, carved emu eggs and wooden items; Fox St. ***Hot artesian springs:*** relaxing and therapeutic baths at swimming pool; Montekeila St.

Campdraft and rodeo: Aug. ***Bulldust to Bitumen Festival:*** Sept.

Caloola B&B: traditional working sheep and cattle farm; Gwydir Hwy; (02) 6828 1124. ***Nolans on the Barwon:*** relaxing accommodation and holiday park; cnr Brewarina and Cumborah rds; (02) 6828 1154. ***Nomads Cryon:*** picturesque working farm; Weetonga, Kamilaroi Hwy; (02) 6828 5237.

Come-by-Chance This town 'came by chance' to William Colless when all of the land in the area was thought to be allocated, but it was discovered that some had been missed. Colless came to own most of the buildings, including the police station, post office, hotel, blacksmith shop and cemetery. It is now an attractive and quiet town with riverside picnic spots, bushwalks and abundant wildlife. There are picnic races in Sept. 65 km SE.

Grawin, Glengarry and Sheepyard opal fields: go fossicking, but be warned that water is scarce so an adequate supply should be carried. Brochure from visitor centre; 70 km NW. ***Macquarie Marshes:*** 100 km SW; *see Nyngan.*

See also NEW ENGLAND, p. 29

Warialda

Pop. 1204
Map ref. 556 H5

Heritage Centre, Hope St; (02) 6729 0046.

102.9 Gem FM, 648 AM ABC New England North West

Warialda is a historic town in a rich farming district. Its location on Warialda Creek gives the town a charm and contributes to its lush greenery. The origin of Warialda's name is uncertain but is thought to mean 'place of wild honey' and is presumed to be in the language of the original inhabitants, the Weraerai people.

Carinda House: historic home, now a craft shop featuring local work; Stephen St. ***Pioneer Cemetery:*** historic graves from as early as the 1850s in a bushland setting; Queen and Stephen sts. ***Heritage Centre:*** visitor centre and Well's Family Gem and Mineral collection; Hope St. ***Koorilgur Nature Walk:*** 3.6 km stroll through areas of wildflowers and birdlife; self-guide brochure from visitor centre. ***Self-guide historical walk:*** historic town buildings in Stephen and Hope sts; brochure from visitor centre.

Agricultural Show: May. ***Warialda Off-Road 200:*** motor race; Sept. ***Honey Festival:*** Nov.

Rose Cottage Guest House: self-contained cottage; 69 Queen St; (02) 6729 1238.

Cranky Rock Nature Reserve It is rumoured that during the gold rush a 'cranky' Chinese man, after being challenged about a wrongdoing, jumped to his death from the highest of the balancing granite boulders. Today you'll find picnic spots, camping, fossicking, wildflowers and wildlife. A suspension bridge leads to an observation deck above Reedy Creek for breathtaking views. 8 km E.

See also NEW ENGLAND, p. 29

Wauchope

Pop. 5499
Map ref. 550 F6 | 557 M11

High St; (02) 6586 4055.

100.7 2MC FM, 684 AM ABC Local

Wauchope (pronounced 'Waw-hope') is the centre of the local dairy and cattle industries, and the gateway to over 40 000 hectares of national parks and state forests. Bursting with country hospitality, its popularity with visitors has vastly increased since the introduction of the fascinating Timbertown.

Historical town walk: self-guide walk past historic buildings such as the old courthouse and the bank; brochure from visitor centre.

Hastings Farmers Markets: Wauchope Showground; 4th Sat each month. ***Community Markets:*** Hastings St; 1st Sat each month. ***Lasiandra Festival:*** community festival; Mar. ***Jazz in the Vines:*** Bago Vineyards; 2nd Sun each month.

Cooking with Company: cooking classes; The Company Farm, 3470 Oxley Hwy, Gannons Creek; (02) 6585 6495.

Jaspers Peak Holiday Retreat: versatile farmstay accommodation; Oxley Hwy; (02) 6587 7155. ***Timbertown Resort and Motel:*** comfortable accommodation options; 230 High St; (02) 6585 1355.

Timbertown This re-created 1880s sawmillers' village demonstrates the struggles and achievements of early pioneers. It features steam-train rides, Cobb & Co horserides and carriage rides, blacksmith, wood turner, art gallery, farmyard patting pen, bullock demonstrations, whip cracking, boutique winery, old-fashioned lolly shop and saloon bar with bush ballads. Oxley Hwy; (02) 6586 1940; 2 km w.

Billabong Koala and Nature Park: 2.5 ha of lush parkland and waterways with exotic and native animals and birds, and koala patting 3 times daily; 10 km E. ***Bellrowan Valley Horse Riding:*** relaxing horserides through the Australian bush with experienced guides. Daily trail rides and overnight packages; Beechwood; 5 km NW. ***Old Bottlebutt:*** the largest known bloodwood tree in NSW; 6 km S. ***Bago Winery:*** cellar-door tastings and sales with regular Jazz in the Vines concerts; 8 km SW. ***Bago Bluff National Park:*** signposted bushwalks (various fitness levels) through rugged wilderness; 12 km W. ***Werrikimbie National Park:*** magnificent World Heritage–listed wilderness with rainforests, rivers and wildflowers (best viewed in spring). Also several excellent sites for camping and picnics; 80 km W. ***4WD tours and abseiling:*** bookings at visitor centre.

See also HOLIDAY COAST, p. 27

Wee Waa

Pop. 1692
Map ref. 556 E6

Narrabri Shire Visitor Information Centre, Newell Hwy, Narrabri; (02) 6799 6760 or 1800 659 931; www.weewaa.com

91.3 Max FM, 648 AM ABC New England North Coast

Wee Waa is a dynamic rural community near the Namoi River and also the base for the Namoi Cotton Cooperative, the largest grower-owned organisation in the country. Cotton has only been grown here since the 1960s, but the town claims to be the 'cotton capital of Australia'. Wee Waa, meaning 'fire for roasting', comes from the local Aboriginal language.

Wee Waa Museum: fascinating display of machinery, artefacts and documents pertaining to the history of the Wee Waa district. Open Wed–Sat at various times; Rose St; (02) 6796 1760.

Wee Waa Show: May. ***Village Fest:*** food, music and games; June. ***Rodeo:*** Sept. ***Christmas Mardi Gras:*** Dec.

Wee Waa Chinese Restaurant: traditional Chinese; Bowling Club, Alma St; (02) 6795 4108. ***Wee Waa Hot Bread Shop:*** bakery cafe with freshly baked bread; 82 Rose St; (02) 6795 4393.

Wee Waa Motel: comfortable accommodation options; 148 Rose St; (02) 6795 4522.

Guided cotton gin and farm tour First the tour visits a local cotton farm to view the picking and pressing of cotton into modules ready for transporting to the cotton gin. At the gin the cotton is transformed from modules into bales and then goes to the classing department for sorting. Runs Mar–Aug; bookings (02) 6799 6760.

Cuttabri Wine Shanty The slab-construction shanty was built in 1882 and was once a Cobb & Co coach stop between Wee Waa and Pilliga. It was issued the second liquor licence in Australia and is the only wine shanty still operating in the country. 25 km SW.

Yarrie Lake: boating, swimming and birdwatching; 24 km S. ***Cubbaroo Cellars:*** cellar-door tastings and sales; 48 km W. ***Barren Junction:*** hot artesian bore baths (over 100 years old) in a pleasant location surrounded by tamarind trees; 51 km W.

See also NEW ENGLAND, p. 29

Wellington

Pop. 4660
Map ref. 554 F2

Cameron Park, Nanima Cr; (02) 6845 1733 or 1800 621 614; www.visitwellington.com.au

89.5 Zoo FM, 549 AM ABC Central West

Wellington is a typical Australian country town with a wide main street, numerous monuments to significant local people and attractive parklands. Sitting at the foot of Mount Arthur, it is best known for the nearby Wellington Caves.

Oxley Museum: the history of Wellington is told with photographs and artefacts; in the old bank (1883); open 1.30–4.30pm Mon–Fri, other times by appt; Cnr Percy and Warne sts. ***Orana Aboriginal Corporation:*** authentic Aboriginal ceramics, paintings, clothing and artefacts; open Mon–Fri; Swift St. ***Cameron Park:*** known for its rose gardens and suspension bridge over the Bell River, it also has picnic and barbecue facilities; Nanima Cres. ***Self-guide town walk:*** taking in historic buildings including hotels and churches; brochure from visitor centre.

Market: Cnr Percy St and and Nanima Cres; last Sat each month. ***Vintage Fair:*** with street parade and swap meet; Mar. ***The Wellington Boot:*** horseraces; Mar. ***Wellington Show:*** Apr. ***Festivale:*** week of celebrations; Oct. ***Carols in the Cave:*** Dec.

Keston Rose Garden Cafe: light lunches and cakes; Mudgee Rd; (02) 6845 3508. ***Kimbells Bakery and Coffee Shop:*** cafe in heritage building; 44 Warne St; (02) 6845 4647. ***The Grange Restaurant:*** modern Australian; Hermitage Hill, 35 Maxwell St; (02) 6845 4469.

Carinya B&B: quaint traditional B&B; 111 Arthur St; (02) 6845 4320. ***Wellington Caves Holiday Complex:*** secluded, relaxed holiday park; Caves Rd; (02) 6845 2970.

Wellington Caves These fascinating limestone caves include Cathedral Cave, with a giant stalagmite, and Gaden Cave, with rare cave coral. There are guided tours through the old phosphate mine (wheelchair-accessible). Nearby is an aviary, an opal shop, Japanese gardens, picnic facilities, kiosk and the bottle house, a structure made from over 9000 wine bottles. 9 km s.

Mt Arthur Reserve: walks to the lookout at the summit of Mt Binjang; maps from visitor centre; 3 km w. ***Angora Tourist Farm:*** demonstrations of shearing angora rabbits and alpacas; 20 km sw. ***Eris Fleming Gallery:*** original oil paintings and watercolours by the artist; 26 km s. ***Nangara Gallery:*** Aboriginal art and craft with artefacts dating back over 20 000 years; 26 km sw. ***Lake Burrendong State Park:*** watersports, fishing, campsites and cabins, and spectacular lake views from the main wall. Burrendong Arboretum is a beautiful spot for birdwatching and features several pleasant walking tracks. Also excellent camping, picnic and barbecue sites at Mookerawa Waters Park; 32 km se. ***Stuart Town:*** small gold-rush town formerly known as Ironbark, made famous by Banjo Paterson's poem 'The Man from Ironbark'; 38 km se. ***Wineries:*** several in the area offering cellar-door tastings and sales. Glenfinlass Wines sell exclusively through cellar door; brochure from visitor centre.

See also Central West, p. 30

Wentworth

Pop. 1305
Map ref. 558 D5 | 588 F5

66 Darling St; (03) 5027 3624; www.wentworth.nsw.gov.au

90.7 Hot FM, 999 AM ABC Broken Hill

At the junction of the Murray and Darling rivers, Wentworth was once a busy and important town. With the introduction of the railways it became quieter and is now an attractive and peaceful holiday town with a rich history.

Old Wentworth Gaol The first Australian-designed gaol by colonial architect James Barnett. The bricks were made on-site from local clay, and bluestone was transported from Victoria. Construction took from 1879 to 1891. Closed as a gaol in 1927, the building is in remarkably good condition. Beverley St.

Pioneer World Museum: over 3000 historic artefacts including space junk, prehistoric animals and the country's largest collection of paddleboat photos; Beverley St. ***Sturt's Tree:*** tree on the riverbank marked by explorer Charles Sturt when he weighed anchor and identified the junction of the Murray and Darling rivers in 1830; Willow Bend Caravan Park, Darling St. ***Fotherby Park:*** PS *Ruby*, a historic paddlesteamer (1907), and statue of 'The Possum', a man who became a hermit during the Depression and lived in trees for 50 years; Wentworth St. ***Lock 10:*** weir and park for picnics; south-west edge of town. ***Historical town walk:*** self-guide walk includes the town courthouse (1870s) and Customs House (1 of 2 original customs houses still standing in Australia); brochure from visitor centre.

Market: Inland Botanic Gardens; 1st and 3rd Sat each month. ***National Trust Festival Week:*** Apr. ***Country Music Festival:*** Sept/Oct. ***Wentworth Cup:*** horserace; Nov.

Perry Sandhills These magnificent orange dunes are estimated to have originated during the last Ice Age, around 40 000 years ago. Skeletal remains of mega-fauna (kangaroos, wombats, emus and lions) have been found here. In WW II the area was used as a bombing range, but recently it has been used in film and television. The Music Under the Stars concerts are held here each Mar. Off Silver City Hwy; 5 km nw.

Mungo National Park: 157 km n; *see Balranald*. ***Yelta:*** former Aboriginal mission, now a Victorian town with model aircraft display; 12 km se. ***Australian Inland Botanic Gardens:*** the desert blooms with some exotic and colourful plant life. There are tractor/train tours of the gardens and a light lunch the last Sun each month; 28 km se. ***Pooncarie:*** 'outback oasis' with natural two-tier wharf, weir, museum and craft gallery; 117 km n. ***Harry Nanya Aboriginal Cultural Tours:*** travel to Mungo National Park with a Barkindji guide; brochure and bookings at visitor centre. ***Heritage and nature driving tours:*** various sites include Mildura and Lake Victoria; brochure from visitor centre. ***Houseboat hire:*** short- or long-term river holidays; brochure from visitor centre.

See also Murray, p. 35

West Wyalong

Pop. 3189
Map ref. 554 B6 | 559 P6

89–91 Main St; (02) 6972 3645; www.visitwestwyalong.com.au

99.7 Star FM, 549 AM ABC Local

John Oxley was the first European explorer to visit West Wyalong. He disliked the region, claiming 'these desolate areas would never again be visited by civilised man'. He was proved wrong when squatters moved in, and the discovery of gold in 1893 meant the town became inundated with settlers. West Wyalong is now in one of the state's most productive agricultural regions.

West Wyalong Local Aboriginal Land Council Arts and Crafts The craft shop features local Aboriginal work. Handcrafted items are on display and for sale and include boomerangs, didgeridoos, hand-woven baskets, clothing, and traditional beauty and skin-care products. The library features a collection of historic and contemporary titles relating to Aboriginal heritage, culture and modern issues. Open Mon–Fri; Main St; (02) 6972 3493.

Bland District Historical Museum: displays of goldmining including a scale model of a goldmine and records from mines such as the Black Snake, the Blue Jacket and the Shamrock and Thistle; Main St.

West Wyalong Campdraft: Sept.

Cameo Inn Motel: traditional motel accommodation; 263 Neeld St; (02) 6972 2255. ***Club Inn Resort:*** beautiful accommodation nestled on golf course; Newell Hwy; (02) 6972 2000.

Lake Cowal When it is full, this is the largest natural lake in NSW and a bird and wildlife sanctuary. There are over 180 species of waterbird living in the area, with many rare or endangered. The lake is also excellent for fishing. No visitor facilities are provided. Via Clear Ridge; 48 km ne.

Barmedman: mineral-salt pool believed to help arthritis and rheumatism; 32 km se. ***Weethalle Whistlestop:*** Devonshire teas, art and craft; Hay Rd; 65 km w.

See also Riverina, p. 34

White Cliffs

Pop. 120
Map ref. 560 F7

White Cliffs General Store, Keraro Rd; (08) 8091 6611.

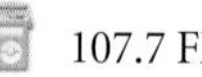

107.7 FM ABC Broken Hill, 1584 AM ABC Local

White Cliffs is first and foremost an opal town. The first mining lease was granted in 1890, and a boom followed with an influx of 4500 people. The area is still known for its opals, particularly the unique opal 'pineapples' and the opalised remains of a plesiosaur, a 2-metre-long 100-million-year-old fossil found in 1976. The intense heat has forced many people to build underground, often in the remains of old opal mines. The buildings left on the surface are surrounded by a pale and eerie moonscape with an estimated 50 000 abandoned opal digs.

Solar Power Station The country's first solar power station was established by the Australian National University in 1981 at White Cliffs because it receives the most solar radiation in NSW. The row of 14 giant mirrored dishes is a striking sight between the blue sky and red earth. Next to council depot.

Jock's Place: dugout home and museum with an opal seam along one wall; Turley's Hill. ***Otto Photography:*** gallery of outback landscape photos; Smith's Hill. ***Self-guide and guided historical walks and fossicking:*** include the old police station (1897) and school (1900) and several fossicking sites; brochures and maps from visitor centre. ***Underground accommodation:*** various standards available in dugout premises. Underground temperatures come as a relief at 22°C; details from visitor centre. ***Opal shops:*** several in town sell local gems; details from visitor centre.

PJ's Underground B&B: luxury underground B&B; Dugout 72, Turley's Hill; (08) 8091 6626. ***White Cliffs Underground Motel:*** simple underground accommodation; Smiths Hill; (08) 8091 6647.

Paroo–Darling National Park: section 20 km E of town contains magnificent Peery Lake, part of the Paroo River overflow, where there is birdlife, Aboriginal cultural sites and walking trails. Southern section of park has camping along the Darling River. ***Wilcannia:*** small town with many fine sandstone buildings, an opening bridge across the Darling River and an old paddlesteamer wharf. Also a self-guide historical walk available; brochure from council offices in Reid St; 93 km S. ***Mutawintji National Park:*** 150 km SW; *see Broken Hill.*

See also OUTBACK, p. 36

Windsor

Pop. 1899
Map ref. 542 A2 | 545 I5 | 547 P7 | 555 J6

Hawkesbury Visitor Information Centre, Ham Common, Hawkesbury Valley Way, Clarendon; (02) 4578 0233 or 1300 362 874; www.hawkesburytourism.com.au

 89.9 FM Hawkesbury Radio, 702 AM ABC Local

Windsor, located on a high bank of the Hawkesbury River, is the third oldest European settlement on mainland Australia, after Sydney Cove and Parramatta. There are still many old buildings standing and the surrounding national parks make for breathtaking scenery.

Hawkesbury Regional Museum Built as a home in the 1820s, the building became the Daniel O'Connell Inn in 1843. In the late 1800s it was used to print *The Australian*, a weekly newspaper. Today it houses the history of the local area in photographs, documents and artefacts, with special displays on riverboat history and the Richmond Royal Australian Air Force base. Thompson Sq.

Hawkesbury Regional Gallery: displays contemporary and traditional works by national and international artists; George St. ***St Matthew's Church:*** designed by convict architect Francis Greenway and built in 1817, St Matthews is the oldest Anglican church in the country. The adjacent graveyard dates back to 1810; Moses St. ***Self-guide tourist walk/drive:*** historic sites include the original courthouse and doctor's house; brochure from visitor centre.

Market: Windsor Mall; Sun. ***Bridge to Bridge Powerboat Classic:*** May. ***Bridge to Bridge Water Ski Race:*** Nov. ***Hawkesbury Canoe Classic:*** Nov. ***Hawkesbury Wine, Food and Music Affair Raceday:*** Nov.

Clydesdale Horse Drawn Restaurant: modern Australian; 61 Hawkesbury Valley Way; (02) 4577 4544. ***The Harvest Restaurant:*** modern Australian/international; The Sebel Resort & Spa, 61 Hawkesbury Valley Way; (02) 4577 4222.

Windsor Terrace Motel: rustic motel accommodation; 47 George St; (02) 4577 5999.

Cattai National Park First Fleet assistant surgeon Thomas Arndell was granted this land and today the park features his 1821 cottage. There are also grain silos and the ruins of a windmill believed to be the oldest industrial building in the country. The old farm features attractive picnic and barbecue areas and campsites. In a separate section nearby, Mitchell Park offers walking tracks and canoeing on Cattai Creek. 14 km NE.

Ebenezer: picturesque town with Australia's oldest church (1809), colonial graveyard and schoolhouse; 11 km N. ***Wollemi National Park:*** Windsor section features the spectacular Colo River and activities including abseiling, canoeing (bring your own), bushwalking and 4WD touring; via Colo; 26 km N.

See also CENTRAL COAST & HAWKESBURY, p. 25

Wingham

Pop. 4813
Map ref. 550 D8 | 555 N1 | 557 K12

Manning Valley Visitor Information Centre, 21 Manning River Dr, Taree North; (02) 6592 5444 or 1800 182 733; www.gtcc.nsw.gov.au

 107.3 MAX FM, 756 AM ABC Local

Heritage-listed Wingham is the oldest town in the Manning Valley. It has many Federation buildings surrounding the enchanting town common, which was based on a traditional English square. The wonderful Manning River and Wingham Brush bring nature to the centre of town. The Chinese Garden at the town's entrance marks the importance of the early Chinese settlers to the town's heritage. Best-selling Australian author Di Morrissey's book *The Valley* is based on the pioneering characters and places of the Manning Valley. It is largely set in and around Wingham where she was born.

Wingham Brush The unique brush is part of the last 10 ha of subtropical flood-plain rainforest in NSW. It is home to 195 species of native plants including giant Moreton Bay fig trees, a large population of endangered grey-headed flying foxes and 100 bird species. The brush includes a boardwalk, picnic and barbecue facilities and a boat-launching area on the Manning River. Farquar St.

Manning Valley Historical Museum: housed in an old general store (1880s), this has one of the most extensive collections of historical memorabilia on the north coast. Includes displays on local farming, commercial and timber history; part of an attractive square bounded by Isabella, Bent, Farquar and Wynter sts. ***Manning River:*** picturesque waterway with several locations for swimming, boating, fishing and waterskiing. ***Self-guide historical town walk:*** tour of the town's Federation buildings; brochure from museum.

Market: Wynter St; 2nd Sat each month. ***Wingham Farmers Market:*** Wingham Showground, Gloucester Rd; 1st Sat each month. ***Summertime Rodeo:*** Jan. ***Wingham Beef Week and Scottish Heritage Festival:*** May. ***Junior Rodeo and Bute Ute Show:*** July. ***All Breeds Horse Spectacular and A-Koo-Stik Festival:*** Oct. ***Killabakh Day in the Country:*** country fair; Nov.

Bent on Food: cafe and provedore; 95 Isabella St; (02) 6557 0727.

Wingham Motel: traditional motel accommodation; 13 Bent St; (02) 6553 4295.

Tourist Drive 8 This enjoyable drive begins in Taree and passes through Wingham, Comboyne and Bybarra before finishing in Wauchope (36 km is unsealed; not suitable for motorists towing caravans). The highlight is Ellenborough Falls, 40 km N, one of the highest single-drop falls in NSW, with easy walking trails, lookouts and barbecue facilities. Red Tail Wines in Marlee (13 km N) has free tastings. Brochure from visitor centre.

See also HOLIDAY COAST, p. 27

Wisemans Ferry

Pop. 80
Map ref. 543 A7 | 545 K2 | 555 K5

Hawkesbury River Tourist Information Centre; 5 Bridge St, Brooklyn; (02) 9985 7064; www.hawkesburyaustralia.com.au

89.9 FM Hawkesbury Radio, 702 AM ABC Local

Wisemans Ferry is a sleepy town built around what was once an important crossing on the Hawkesbury River. The mainland route from Sydney to Newcastle had always gone via this region, but when people started using the Castle Hill route, Solomon Wiseman, who had been granted a parcel of land and opened an inn, built a ferry to take people and cargo across the river. Today car ferries still cross at this point.

Wisemans Ferry Inn Before it was an inn, this was the home of Solomon Wiseman; he called it Cobham Hall. Wiseman later opened a section of the building as an inn and it is said to be haunted by his wife, whom he allegedly pushed down the front steps to her death. The inn provides food and accommodation. Old Northern Rd; (02) 4566 4301.

Cemetery: early settlers' graves include that of Peter Hibbs, who travelled on the HMS *Sirius* with Captain Phillip in 1788; Settlers Rd.

Riverbend Restaurant: modern Australian; Retreat at Wisemans, 5564 Old Northern Rd; (02) 4560 0593. ***Wiseman's Steakhouse Bistro:*** steakhouse, traditional pub; Wiseman's Inn Hotel, Lot 1, Old Northern Rd; (02) 4566 4739.

Retreat at Wisemans: luxury resort; 5564 Old Northern Rd; (02) 4566 4422.

Dharug National Park The multicoloured sandstone provides striking scenery on this historic land. The convict-built Old Great North Road is a great example of early 19th-century roadbuilding. Convicts quarried, dressed and shifted large sandstone blocks to build walls and bridges, but the road was abandoned before it was finished because of poor planning. Signposted walking tracks lead through beautiful bushland and to Aboriginal rock engravings. The clear-water tributaries are popular for swimming, fishing and canoeing. North side of the river.

Yengo National Park and Parr State Conservation Area: rugged land of gorges, cliffs and rocky outcrops. Discovery walks, talks and 4WD tours are conducted by NPWS; north-east side of the river (accessible by ferry); bookings (02) 4784 7301. ***Marramarra National Park:*** undeveloped park with wetlands and mangroves for canoeing, camping, bushwalking (experienced only) and birdwatching; 28 km S.

See also CENTRAL COAST & HAWKESBURY, p. 25

Woolgoolga

Pop. 4358
Map ref. 551 E11 | 557 N7

Cnr Beach and Boundary sts; (02) 6654 8080; www.coffscoast.com.au

105.5 Star FM, 738 AM ABC Radio National

'Woopi' (as it is affectionately known to locals) is a charming and relaxed seaside town with a significant Sikh population. Punjabi migrants who were working on the Queensland cane fields headed south for work on banana plantations, many settling in Woolgoolga. Today Indians make up between a quarter and a half of the town's population, providing a unique cultural mix. The beaches are popular for fishing, surfing and swimming.

Guru Nanak Sikh Temple: spectacular white temple with gold domes; River St. ***Woolgoolga Art Gallery:*** exhibits local works; Turon Pde.

Market: Beach St; 2nd Sat each month.

Possum's Cafe: light meals; Shop 4, 53 Beach St; (02) 6654 2807.

Halcyon Retreat: self-contained suites; 237 Woolgoolga Creek Rd; (02) 6654 7750. ***Waterside Cabins at Woolgoolga:*** resort-style accommodation; cnr Pacific Hwy and Hearnes Lake Rd; (02) 6654 1644.

Yarrawarra Aboriginal Cultural Centre The focus of this centre is to help Aboriginal and Islander peoples maintain their heritage while teaching others about it. Visitors are encouraged to browse through the rooms of locally produced art, craft, books and CDs. The Bush Tucker Cafe offers meals with an Indigenous twist. Tours offered through the local area explore middens, ochre quarries and campsites while teaching about bush tucker and natural medicines. Stone and tool workshops and accommodation are also offered. Red Rock Rd; (02) 6649 2669; 10 km N.

Yuraygir National Park: Woolgoolga section is excellent for bushwalking, canoeing, fishing, surfing, swimming, picnicking and camping on unspoiled coastline; 10 km N. ***Wedding Bells State Forest:*** subtropical and eucalypt forest with walking trails to Sealy Lookout and Mt Caramba Lookout; 14 km NW.

See also TROPICAL NORTH COAST, p. 28

Woy Woy

Pop. 9985
Map ref. 542 H1 | 543 D8 | 545 N4 | 555 K6

Shop 1, 18–22 The Boulevard; (02) 4341 2888 or 1300 130 708; www.visitcentralcoast.com.au

101.3 Sea FM, 702 AM ABC Local

continued on p. 110

WOLLONGONG

Pop. 263 537

Map ref. 552 | 553 F5 | 555 J8 | 565 H2

[AERIAL VIEW OF WOLLONGONG]

Southern Gateway Centre; Bulli Tops; 1800 240 737; www.tourismwollongong.com

96.5 Wave FM, 1431 AM ABC Radio National

Located just one hour's drive south of Sydney, Wollongong has an undeserved reputation for being an unattractive industrial city. In fact, Wollongong enjoys some of the best coastal scenery and beaches in the state, superbly positioned with mountains to the west and ocean to the east. It has been awarded the title of 'Australia's most liveable regional city'.

Wollongong Botanic Garden The magnificent gardens encompass 27 ha of undulating land and feature a sunken rose garden, a woodland garden, rainforests, and flora representing a range of plant communities. Guided walks and bus tours are available. Various workshops including crafts and gardening throughout the year; Murphys Ave, Keiraville; (02) 4225 2636; call for opening hours. The adjacent Gleniffer Brae Manor House is a Gothic-style 1930s house now home to the Wollongong Conservatorium of Music and used for music recitals. Northfields Ave; Keiraville.

Flagstaff Hill: 180 degree views of the ocean and historic lighthouse (1872); Endeavour Dr. ***Illawarra Historical Society Museum:*** highlights include handicraft room and Victorian parlour; open weekends and 12–3pm Thurs; Market St; (02) 4283 2854. ***Wollongong City Gallery:*** collection of 19th- and 20th-century art including Aboriginal art; cnr Burelli and Kembla sts; 10am–5pm Tue–Fri, 12–4pm Sat–Sun; (02) 4228 7500. ***Mall:*** soaring steel arches and water displays; Crown St. ***Five Islands Brewing Company:*** locally brewed beers; Crown St; (02) 4220 2854. ***Wollongong Harbour:*** home to a huge fishing fleet and Breakwater Lighthouse. ***Surfing beaches and rockpools:*** to the north and south, with excellent surfing and swimming conditions. ***Foreshore parks:*** several with superb coastal views and picnic facilities.

Market: Harbour St; Sat. ***Illawarra Folk Festival:*** Jan. ***Wings over Illawarra:*** Feb. ***Mt Kembla Mining Heritage Festival:*** July. ***Viva La Gong:*** street arts festival; Oct.

Diggies Cafe: beachside cafe; North Wollongong Beach; (02) 4226 2688. ***Lagoon Seafood Restaurant:*** seafood, incredible views; Stuart Park, cnr George Hanley Dr and Kembla St, North Wollongong; (02) 4226 1677. ***Lorenzo's Diner:*** modern Italian; 119 Keira St; (02) 4229 5633.

Above Wollongong at Pleasant Heights B&B: boutique B&B; 77 New Mount Pleasant Rd; (02) 4283 3355.

Royal National Park Established in 1879, this is the second oldest national park in the world after Yellowstone in the USA. There is much natural diversity packed into a compact parkland. Highlights include walking and cycling along Lady Carrington Dr through rich forest, swimming at Wattamolla, enjoying the Victorian-park atmosphere at Audley and walking the magnificent 26 km Coast Track. (02) 9542 0648. 35 km N. Grand Pacific Dr, a new tourist route along the coast, links Royal National Park with Wollongong. The 70 km drive takes in spectacular scenery and the Sea Cliff Bridge.

Illawarra Escarpment: forms the western backdrop to the city and has spectacular lookouts at Stanwell Tops, Sublime Pt, Mt Keira and Mt Kembla. ***Wollongong Science Centre and Planetarium:*** hands-on displays and activities for all ages; Squires Way, Fairy Meadow; open 10am–4pm daily; (02) 4286 5000; 2 km N. ***Nan Tien Temple:*** largest Buddhist temple in the Southern Hemisphere with a range of programs available; closed Mon; 5 km SW. Berkeley Rd, Berkeley. (02) 4272 0600. ***Lake Illawarra:*** stretching from the southern Pacific Ocean to the foothills of the Illawarra Range, the lake offers good prawning, fishing and sailing. Boat hire is available; 5 km S. ***Port Kembla:*** up-close view of local industry and the steel works at Australia's Industry World with tours; bookings (02) 4275 7023. ***Mt Kembla:*** site of horrific 1902 mining disaster. Also here are several historic buildings and a historical museum featuring a pioneer kitchen, a blacksmith's shop and a reconstruction of the Mt Kembla disaster. The Mt Kembla Mining Heritage Festival is held each winter in memory of those lost; 10 km W. ***Bulli Pass Scenic Reserve:*** steep scenic drive with stunning coastal views. Bulli Lookout at the top of the escarpment has great views and a walking path leads to Sublime Pt Lookout, which enjoys stunning views over Wollongong and has a restaurant; 16 km N. ***Symbio Wildlife Park:*** koalas, eagles, wombats and reptiles, as well as Sumatran tigers; Lawrence Hargrave Drive, Helensburgh; (02) 4294 1244; 32 km N. ***Just Cruisin Motorcycle Tours:*** cruise the spectacular south coast on a chauffeured Harley Davidson; (02) 4294 2598. ***Lawrence Hargrave Memorial and Lookout:*** on Bald Hill, this was the site of aviator Hargrave's first attempt at flight in the early 1900s. Now popular for hang-gliding; 36 km N. ***Illawarra Fly Treetop Walk:*** the Treetop Walk offers inspiring views; 182 Knights Hill Rd; (02) 4885 1010. ***Heathcote National Park:*** excellent for bushwalks through rugged bushland, past hidden pools and gorges; 40 km N. (02) 9542 0648. ***Dolphin Watch Cruises:*** See dolphins and whales at Jervis Bay; 50 Owen Street, Huskisson; (02) 4441 6311 or 1800 246 010.

See also SOUTHERN HIGHLANDS, p. 24

Woy Woy is the largest of the numerous holiday villages clustered around Brisbane Water, a shallow but enormous inlet. Along with nearby Broken Bay, the Hawkesbury River and Pittwater, it draws visitors looking for aquatic holidays. The nearby national parks encompass breathtaking wilderness and lookouts.

Woy Woy Hotel: historic 1897 hotel offering meals and accommodation; The Boulevard. ***Waterfront reserve:*** picnic facilities with Brisbane Water view.

Agweil: modern bistro; Everglades Country Club, Dunban Rd; (02) 4341 1866. ***Riley's Brasserie:*** international; Ettalong Beach Club, 52 The Esplanade, Ettalong Beach; (02) 4343 0111. ***Bells at Killcare:*** modern Italian; 107 The Scenic Rd, Killcare; (02) 4360 2411. ***Lizotte's:*** modern Australian, with live music; Lot 3, Avoca Dr, Kincumber; (02) 4368 2017. ***Pearls on the Beach:*** modern Australian; 1 Tourmaline Ave, Pearl Beach; (02) 4342 4400. ***Stillwaters:*** modern Australian; 1 Restella Ave, Davistown; (02) 4369 1300 ***Yum Yum Eatery:*** modern Australian; 60 Araluen Dr, Hardys Bay; (02) 4360 2999.

Glades Motor Inn: peaceful country accommodation; 15 Dunban Rd; (02) 4341 7374.

Brisbane Water National Park A beautiful park of rugged sandstone with spring wildflowers, bushwalks and birdlife. Staples Lookout has superb coastal views. Warrah Lookout enjoys a sea of colour in spring when the wildflowers bloom. A highlight is the Bulgandry Aboriginal engravings on Woy Woy Rd. 3 km sw.

Ettalong Beach: great swimming beach with seaside markets each Sat and Sun (Mon on long weekends); 3 km s. ***Milson Island:*** recreation reserve once used as an asylum and then as a gaol. ***HMAS*** Parramatta: a WW I ship, it ran aground off the northern shore and the wreck is still there today; 5 km s. ***Mt Ettalong Lookout:*** stunning coastal views; 6 km s. ***Pearl Beach:*** chic holiday spot favoured by the rich and famous, with magnificent sunsets; 12 km s. ***Boating, fishing and swimming:*** excellent conditions on Brisbane Water, Broken Bay and Hawkesbury River.

See also CENTRAL COAST & HAWKESBURY, p. 25

Wyong

Pop. 149 382
Map ref. 543 E6 | 545 P1 | 549 A12 | 555 L5

Rotary Park, Terrigal Dr, Terrigal; (02) 4343 4444 or 1300 130 708; www.visitcentralcoast.com.au

101.3 Sea FM, 1512 AM ABC Radio National

Wyong is an attractive holiday town surrounded by Tuggerah Lakes and the forests of Watagan, Olney and Ourimbah. After World War II it became a popular area for retirees and it retains a relaxed atmosphere today.

District Museum Features displays of local history, including early ferry services across the lakes and records of the logging era. It is situated in historic Alison Homestead, with picnic and barbecue facilities on 2 ha of rolling lawns. Cape Rd; (02) 4352 1886.

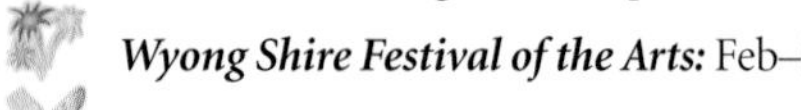

Wyong Shire Festival of the Arts: Feb–Mar.

Karinyas Restaurant and Bar: modern Australian; Kooindah Waters Residential Golf & Spa Resort, 40 Kooindah Blvd; (02) 4355 5777 or 1800 705 355.

Olney State Forest This native rainforest has several scenic walks. The Pines picnic area has an education shelter and Mandalong and Muirs lookouts have sensational views. 17 km NW.

Burbank Nursery: 20 ha of azaleas (flowering in Sept); 3 km s. ***Fowlers Lookout:*** spectacular forest views; 10 km sw. ***Yarramalong Macadamia Nut Farm:*** offers tours, talks and sales in beautiful Yarramalong Valley; 18 km w. ***Frazer Park:*** recreational park in natural bush setting; 28 km NE.

See also CENTRAL COAST & HAWKESBURY, p. 25

Yamba

Pop. 5515
Map ref. 551 F8 | 557 O5

Lower Clarence Visitor Centre, Ferry Park, Pacific Hwy, Maclean; (02) 6645 4121; www.yambansw.com.au

100.3 Yamba Radio FM, 738 AM ABC North Coast

This quiet holiday town at the mouth of the Clarence River offers excellent sea, lake and river fishing. It is the largest coastal resort in the Clarence Valley, with excellent facilities for visitors, but it manages to maintain a peaceful atmosphere. Fishing fleets from Yamba, Iluka and Maclean catch approximately 20 per cent of the state's seafood, so this is a great spot for lovers of fresh seafood.

Story House Museum This quaint museum tells the story of the development of Yamba from the time it was merely a point of entry to the Clarence River. The collection of photographs and records tells a compelling tale of early development of a typical Australian coastal town. River St.

Clarence River Lighthouse: coastal views from the base; via Pilot St. ***Yamba Boat Harbour Marina:*** departure point for daily ferry service to Iluka, river cruises, deep-sea fishing charters and whale-watching trips. Also houseboat hire; off Yamba Rd. ***arthouse australia:*** showcase of local artwork; Coldstream St. ***Whiting Beach:*** sandy river beach ideal for children. ***Coastal beaches:*** several in town with excellent swimming and surfing conditions; map from visitor centre.

River Market: Ford Park; 4th Sun each month. ***Easter Yachting Regatta:*** Easter. ***Family Fishing Festival:*** Oct. ***Hot Rod Run:*** Nov.

Beachwood Cafe: Mediterranean; 22 High St; (02) 6646 9781. ***Pacific Hotel Bistro:*** international; Pacific Hotel, 18 Pilot St; (02) 6646 2125.

Angourie Rainforest Resort: contemporary self-contained luxury; 166 Angourie Rd; (02) 6646 8600. ***Moby Dick Waterfront Resort Motel:*** waterfront accommodation; 27–29 Yamba Rd; (02) 6646 2196.

Maclean This quirky village is known as the 'Scottish town' because of the many Scots who first settled here. Some street signs are in Gaelic as well as English. Highlights in the town include Scottish Corner, Bicentennial Museum and a self-guide historical walk with a brochure available from the visitor centre. There is a market on the 2nd Sat each month and a Highland Gathering each Easter. A 24 hr ferry service crosses the river to Lawrence; 17 km w.

Lake Wooloweyah: fishing and prawning; 4 km s. ***Yuraygir National Park:*** Yamba section offers sand ridges and banksia heath, and is excellent for swimming, fishing and bushwalking; (02) 6627 0200; 5 km s. ***The Blue Pool:*** deep freshwater pool 50 m from the ocean, popular for swimming; 5 km s.

See also TROPICAL NORTH COAST, p. 28

Yass

Pop. 5330
Map ref. 554 F9 | 565 C3

Coronation Park, Comur St; (02) 6226 2557 or 1300 886 014; www.yass.nsw.gov.au

100.3 Yass FM, 549 AM ABC Central West

Yass is set in rolling countryside on the Yass River. Explorers Hume and Hovell passed through on their expedition to Port Phillip Bay. Hume returned in 1839. Yass is also the end of the Hume and Hovell Walking Track, which begins in Albury.

Cooma Cottage The National Trust has restored and now maintains this former home of explorer Hamilton Hume. He lived with his wife in the riverside house from 1839 until his death in 1873. It now operates as a museum with relics and documents telling of Hume's life and explorations. Open Thurs–Sun, closed winter; Yass Valley Way.

Yass Cemetery: contains the grave of explorer Hamilton Hume; via Rossi St. ***Yass and District Museum:*** historical displays including a war exhibit encompassing the Boer War, WW I and WW II. Open Sat and Sun, or by appt, archive open Tues; Comur St. ***Railway Museum:*** history of the Yass tramway; open Sun; Lead St. ***Self-guide town walk and drive:*** highlight is the National Trust–listed main street; brochure from visitor centre.

Picnic races: Mar. ***Yass Show and Rodeo:*** Mar. ***Wine Roses and all that Jazz:*** food, wine and music festival in the Murrumbateman area; Oct. Rodeo: Nov.

Cafe Dolcetto: cakes, sandwiches; 129 Comur St; (02) 6226 1277. ***Ewe'n Me:*** steak, seafood; Thunderbird Motel, 264 Comur St; (02) 6226 1158. ***Swaggers Restaurant:*** steakhouse; Sundowner Motor Inn, cnr Laidlaw and Castor sts; (02) 6226 3188. ***Smokehouse Cafe:*** lovely cafe in restored farm cottage, gourmet produce available for purchase; Poachers Pantry, 'Marakei', Nanima Rd, Hall; (02) 6230 2487.

Kerrowgair B&B: popular rustic B&B; 24 Grampian St; (02) 6226 4932.

Wee Jasper This picturesque village, where Banjo Paterson owned a holiday home, is set in a valley at the foot of the Brindabella Ranges. The Goodradigbee River is excellent for trout fishing. Carey's Caves are full of limestone formations and were the site of the 1957 discovery of the spine of a large extinct wombat. 50 km SW.

Bookham: village with historic cemetery, Sheep Show and Country Fair in Apr; 30 km W. ***Binalong:*** historic town with Motor Museum, Southern Cross Glass and the grave of bushranger Johnny Gilbert; 37 km NW. ***Burrinjuck Waters State Park:*** bushwalking, watersports and fishing. Burrinjuck Ski Classic is held each Nov (water level permitting); off Hume Hwy; 54 km SW. ***Brindabella National Park:*** birdwatching, camping and bushwalking in alpine surrounds. 4WD access only; via Wee Jasper; (02) 6122 3100; 61 km SW. ***Wineries:*** in Murrumbateman area with cellar-door tastings. Follow the signs on Barton Hwy; brochure from visitor centre. ***Hume and Hovell Walking Track:*** 23-day, 440 km trek from Albury to Yass. For a kit (including maps), contact Land and Property Management Authority Wagga (02) 6937 2700.

See also CAPITAL COUNTRY, p. 31

Young

Pop. 7139
Map ref. 554 E7 | 565 B1

2 Short St; (02) 6382 3394 or 1800 628 233; www.visityoung.com.au

93.9 Star FM, 96.3 FM ABC Local

Young is an attractive town in the western foothills of the Great Dividing Range with a fascinating history of goldmining. The Lambing Flat goldfields were rushed after a discovery was announced in 1860. Within a year there were an estimated 20 000 miners in town, 2000 of whom were Chinese. A combination of lawlessness and racism boiled over in the Lambing Flat riots in 1861, which gave rise to the Chinese Immigration Restriction Act, the first legislation to herald the infamous White Australia Policy. Today the town is the peaceful centre of a cherry-farming district.

Lambing Flat Folk Museum This museum is recognised as one of the finest in the country. Meticulously maintained photographs and relics tell the story of the town during the 1800s and 1900s. The full horrific story of the Lambing Flat riots is covered. Items on display include a 'roll-up' flag carried by miners during the riots. Campbell St; (02) 6382 2248.

Burrangong Art Gallery: hosts changing exhibitions from guest and local artists; at visitor centre; Short St. ***J. D.'s Jam Factory:*** tours, tastings and Devonshire teas; Henry Lawson Way; (02) 6382 4060.

Hilltops Flavours of the Harvest Festival (Young): Feb. ***Lambing Flat Festival:*** Apr. ***National Cherry Festival:*** Nov/Dec.

Doves Restaurant: modern Australian; Youngs Services Club, Cloete St; (02) 6382 1944. ***Hilltops Restaurant:*** modern Australian; Hilltops Retreat Motor Inn, Olympic Hwy; (02) 6382 3300. ***Young Cafe de Jour:*** modern Australian; 4/21 Lovell St; (02) 6382 1413.

Adrianna's B&B: romantic B&B; 4484 Olympic Hwy; (02) 6382 2231. ***Hilltops Retreat Motor Inn:*** traditional motel with impressive restaurant; 4662 Olympic Hwy; (02) 6382 3300. ***Young Federation Motor Inn:*** simple, comfortable, affordable motel; 109–119 Main St; (02) 6382 5644.

Chinaman's Dam Recreation Area: scenic walks, playground, and picnic and barbecue facilities. Includes Lambing Flat Chinese Tribute Gardens with Pool of Tranquility; Pitstone Rd; 4 km SE. ***Yandilla Mustard Seed Oil Enterprises:*** tours by appt; bookings (02) 6943 2516; 20 km S. ***Murringo:*** historic buildings, a glassblower and engraver; 21 km E. ***Hilltops:*** cool-climate wine region; cellar-door information at visitor centre. ***Cherries and stone fruit:*** sales and pick-your-own throughout the area Nov–Dec. Cherries blossom in Sept/Oct.

See also CAPITAL COUNTRY, p. 31

63

AUSTRALIAN CAPITAL TERRITORY

[HORSERIDING IN TIDBINBILLA NATURE RESERVE]

CANBERRA IS...

A picnic in the AUSTRALIAN NATIONAL BOTANIC GARDENS / A visit to PARLIAMENT HOUSE / Fine dining in one of MANUKA'S many restaurants / A wander through the NATIONAL GALLERY OF AUSTRALIA's Sculpture Garden / Fun science activities at QUESTACON – The National Science and Technology Centre / City views from the top of MOUNT AINSLIE / A trip to the NATIONAL MUSEUM OF AUSTRALIA / Viewing portraits of well-known Australians at the AUSTRALIAN PORTRAIT GALLERY

VISITOR INFORMATION
Canberra and Region Visitor Centre
→ 330 Northbourne Ave, Dickson
→ (02) 6205 0044 or 1300 554 114
www.visitcanberra.com.au

Canberra really is the bush capital – kangaroo-dotted nature reserves are scattered throughout the city, and the Brindabella mountain range bounds the south-western edge. As the national capital, Canberra claims some of Australia's most significant institutions, including magnificent art galleries and a remarkable war memorial. Grand public buildings and monuments complement the order and beauty of the city's original design, and the landmark flagpole of Parliament House can be seen from many parts of the city.

Many visitors are attracted by Canberra's impressive national collections or the experience of witnessing federal politics, but the city has much more to offer. The capital also boasts cool-climate wineries, top-class restaurants and bars, an annual balloon festival, attractions for children and a full calendar of vibrant cultural and sporting events.

With a population of about 352 000, life in Canberra moves at a comfortable pace; and with the snow and the sea both only two hours away, the locals have the best of both worlds. The city's creation solved the debate between Sydney and Melbourne over the location of Australia's capital, and Canberra is now one of the world's few completely planned cities.

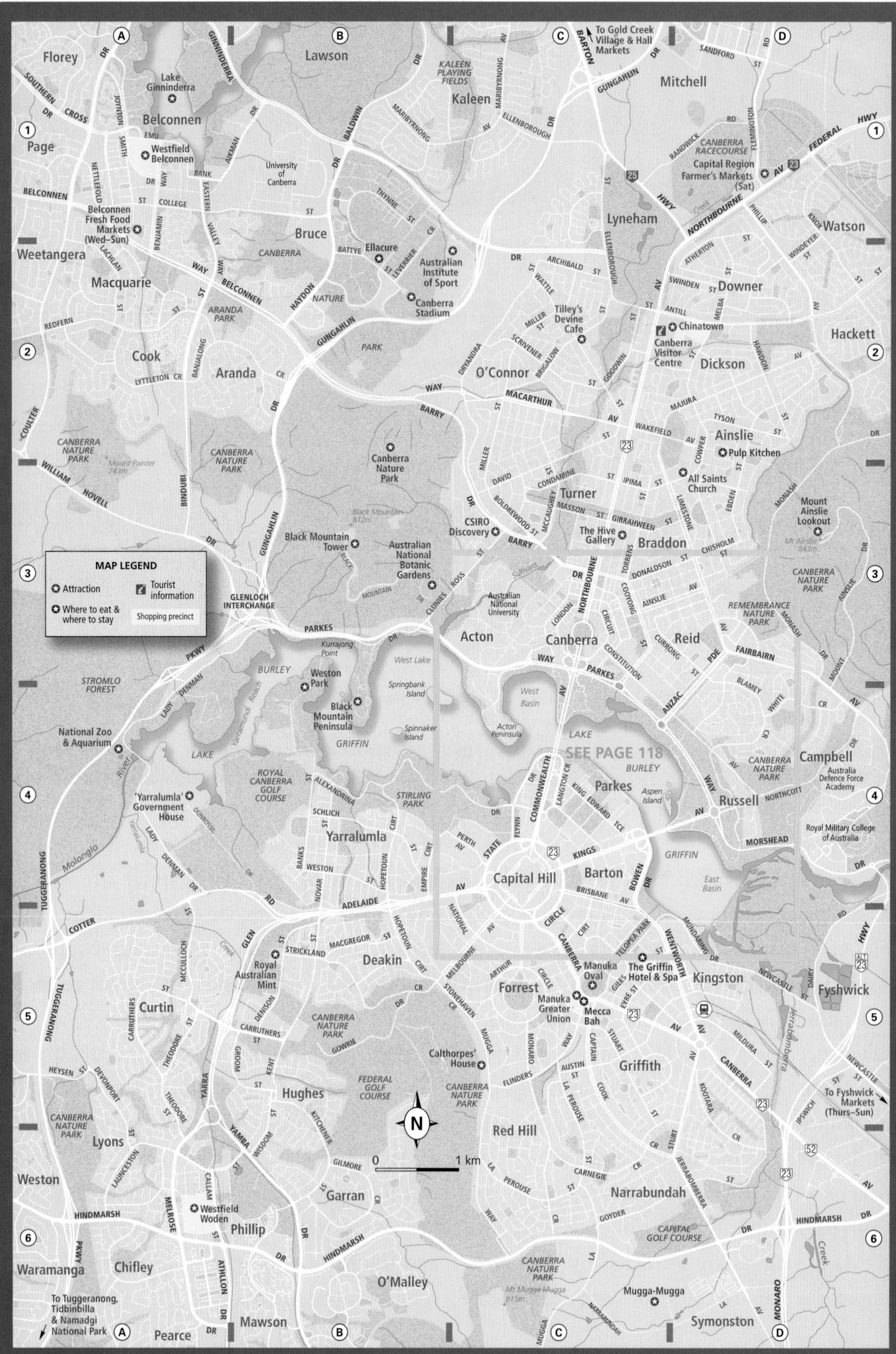

MAP LEGEND
Attraction
Tourist information
Where to eat & where to stay
Shopping precinct
Belconnen
Lake Ginninderra
Westfield Belconnen
University of Canberra
Belconnen Fresh Food Markets (Wed–Sun)
Bruce
Ellacure
Australian Institute of Sport
Canberra Stadium
Mitchell
Canberra Racecourse
Capital Region Farmer's Markets (Sat)
Lyneham
Watson
Downer
Chinatown
Canberra Visitor Centre
Dickson
Hackett
O'Connor
Tilley's Devine Cafe
Ainslie
Pulp Kitchen
All Saints Church
Turner
Braddon
The Hive Gallery
Mount Ainslie Lookout
Canberra Nature Park
Black Mountain Tower
Australian National Botanic Gardens
CSIRO Discovery
Australian National University
Acton
Canberra
Reid
Weston Park
Black Mountain Peninsula
National Zoo & Aquarium
'Yarralumla' Government House
Yarralumla
SEE PAGE 118
Parkes
Russell
Campbell
Australia Defence Force Academy
Royal Military College of Australia
Capital Hill
Barton
Kingston
Fyshwick
Royal Australian Mint
Deakin
Forrest
Manuka Oval
The Griffin Hotel & Spa
Manuka Greater Union
Mecca Bah
Curtin
Calthorpes' House
Griffith
Hughes
Red Hill
Narrabundah
Lyons
Garran
Westfield Woden
Phillip
Chifley
Waramanga
O'Malley
Mugga-Mugga
Symonston
Mawson
Pearce
Weston
Florey
Page
Lawson
Kaleen
Weetangera
Macquarie
Cook
Aranda
To Gold Creek Village & Hall Markets
To Fyshwick Markets (Thurs–Sun)
To Tuggeranong, Tidbinbilla & Namadgi National Park
0 1 km

Parliamentary Triangle & War Memorial

The vision of Canberra's architect, Walter Burley Griffin, can be seen in the tree-lined avenues, spectacular lake views and spacious parks of central Canberra. The focal point is Parliament House, atop Capital Hill, situated at the apex of the Parliamentary Triangle. An integral part of Burley Griffin's plan was the vista from Capital Hill, extending to Lake Burley Griffin, bounded by Commonwealth and Kings avenues, up the broad sweep of Anzac Parade to the Australian War Memorial.

National Capital Exhibition 118 B2

The National Capital Exhibition tells the story of Canberra through interactive displays, rare photographs, a laser model of the city and various audiovisual material. Learn about the area's Indigenous inhabitants, European settlement and Walter Burley Griffin's design for the city. *Regatta Point, Barrine Dr, Commonwealth Park; (02) 6272 2902; open 9am–5pm Mon–Fri, 10am–4pm Sat–Sun; closed public holidays except Australia Day and Canberra Day admission free.*

Parliament House 118 A5

Designed by the American-based architects Mitchell/Giurgola & Thorp, Parliament House was officially opened by the Queen in 1988. It is home to both houses of Federal Parliament (the Senate and the House of Representatives). If your visit coincides with the sitting of parliament, you can see democracy in action from the public galleries (check parliament's website, www.aph.gov.au, for sitting dates). The permanent displays include an extensive collection of Australian art, the Great Hall Tapestry and one of only four surviving 1297 copies of the Magna Carta. Once inside, take the lift to the roof for magnificent views of the city. It is worth noting that the flag flying atop the 81-metre flagpole is roughly the size of the side of a double-decker bus. Guided tours of the complex are recommended; brochures also give visitors the option of self-guide tours. Those with a keen eye may spot the fossils in the main foyer's marble floor. Those who prefer outdoor attractions can walk through the 23 hectares of landscaped gardens. *Parliament Dr, off State Circle; (02) 6277 5399; recorded information (02) 6277 2727; open 9am–5pm daily; admission free.*

Reconciliation Place 118 B4

Reconciliation Place, adjacent to Commonwealth Place on Lake Burley Griffin, symbolises the journey of reconciliation between Indigenous and white Australians, in the past, present and future. The location of Reconciliation Place within the Parliamentary Zone places the reconciliation process physically and symbolically at the heart of Australian democratic and cultural life. A series of public artworks, known as slivers, surround a central circular mound, while pathways link Reconciliation Place and Commonwealth Place, national institutions and Lake Burley Griffin.

climate

Canberra has a climate of extremes and seasons of beauty. Summer in the capital is hot and very dry, with temperatures sometimes reaching the mid to high 30s. Winter is cold and frosty, thanks to Canberra's altitude of roughly 600 metres, its inland location and its proximity to the Snowy Mountains. Don't let this stop you, though – Canberra winter days are breathtakingly crisp and usually fine, with cloudless, bright blue skies after the early morning fogs lift. Spring is Canberra at its most pleasant – the city is awash with colour and flowers abound – although some say that it is in autumn that Canberra is at its most beautiful.

J	F	M	A	M	J	J	A	S	O	N	D	
28	28	24	20	15	12	11	13	16	19	23	26	MAX °C
13	13	11	7	4	1	0	1	4	6	9	11	MIN °C
59	51	55	49	47	37	52	47	65	61	58	46	RAIN MM
5	4	4	5	5	5	6	7	7	7	7	5	RAIN DAYS

[PARLIAMENT HOUSE]

Museum of Australian Democracy at Old Parliament House 118 B4

Home to Australia's Federal Parliament from 1927 to 1988, this heritage-listed building captivates its audiences with its rich history as well as several innovative and vibrant exhibitions. It's now home to the Museum of Australian Democracy, Canberra's newest cultural attraction, which provides insights into the stories and events that have shaped Australia's democracy. Wander through the corridors of power or take a free guided tour. In summer enjoy Friday night drinks and music in the lower house courtyard. Flanking the building are the **Old Parliament House Rose Gardens**, while across the road are the **Aboriginal Tent Embassy**, established in 1972 to protest against a lack of land rights for the country's Indigenous peoples, and the **National Rose Gardens**. *18 King George Tce, Parkes; (02) 6270 8222; open 9am–5pm daily.*

National Portrait Gallery 118 B4

The National Portrait Gallery opened in December 2008, in a new building next door to the High Court of Australia and the National Gallery of Australia in Parkes. The National Portrait Gallery has a permanent display of over 450 portraits that represent the human face of Australia – those who have contributed to Australian society or whose lives have set them apart. Included are paintings, sculptures, photographs and multimedia works. The gallery also hosts changing exhibitions as well as those online. *King Edward Tce, Parkes; (02) 6102 7000; open 10am–5pm daily; general admission free.*

National Archives of Australia 118 B4

The National Archives occupy what was Canberra's first GPO, opened in 1927. The permanent exhibition, 'Memory of a Nation', highlights the extent and diversity of the treasures contained within the National Archives' collection, including ASIO surveillance photos, wooden balls used in National Service ballots to conscript young men to the Vietnam War and an 1897 draft of the Constitution complete with Australia's first prime minister Edmund Barton's edits. Learn how to search for war service and migration records to trace your family history. The Federation Gallery displays Australia's 'birth certificate'– Queen Victoria's Royal Commission of Assent and Australia's original Constitution. *Queen Victoria Tce, Parkes; (02) 6212 3600; open 9am–5pm daily; admission free.*

National Library of Australia 118 B3

Collecting since 1901, this is the country's largest reference library. The present building contains over six million books as well as newspapers, periodicals, photographs and other documents. There are free tours every hour and if you are there on Thursday take the free behind-the-scenes weekly tour. The Library cafe, bookplate, is a good spot for lunch or coffee. *Parkes Pl, Parkes; (02) 6262 1111; open 9am–9pm Mon–Thurs, 9am–5pm Fri–Sun; admission free.*

Questacon – The National Science and Technology Centre 118 B4

Making science fun and relevant for everyone, Questacon has many interactive exhibits – you can experience an earthquake and a cyclone, see lightning created and free-fall 6 metres on the vertical slide. *King Edward Tce, Parkes; (02) 6270 2800; open 9am–5pm daily.*

High Court of Australia 118 C4

Located between Questacon and the National Gallery of Australia, the High Court is notable for its glass-encased public gallery and timber courtrooms. Murals by artist Jan Senbergs reflect the history and functions of the court, and the role of the states in Federation. Visitors can explore the building, talk to the knowledgeable attendants and view a short film on the court's work. *Parkes Pl, Parkes; (02) 6270 6811; open 9.45am–4.30pm Mon–Fri, 12pm–4pm Sun; admission free.*

National Gallery of Australia 118 C4

Established in 1911, the gallery has been housed across the road from the High Court since 1982. With over 120 000 works, the collection provides a brilliant overview of Australian art. The

monuments

Anzac Parade Memorials Eleven dramatic monuments set along the striking red gravel of Anzac Parade commemorate Australian involvement in military conflicts, and the Australians who served in the various defence forces. 118 C2

Australian–American Memorial Celebrates America's World War II contribution to Australia's defence in the Pacific. 118 D4

Captain Cook Memorial Jet An impressive 150-metre water jet – hire a paddleboat to cruise under the spray. 118 B3

National Carillon The largest carillon in Australia, with 53 bells, it was a gift from the British government to celebrate Canberra's 50th anniversary. Regular recitals can be heard. 118 C4

See also Australian War Memorial, p. 120, Reconciliation Place, p. 117

getting around

Canberra is an easy city to get around if you have a car a map or GPS. The road infrastructure is probably the best in Australia, with wide, well-planned roads, and visitors will find that they can cover long distances in a short time. Certainly in the central area of the city it is very easy to travel from one attraction to the next by car or public transport. Walking between the attractions in the Parliamentary Triangle is also very pleasant.

Buses are the only public transport available in Canberra, but these services can be variable at off-peak times. For a convenient way to get around the main attractions, catch one of the sightseeing buses that depart from the Melbourne Building on Northbourne Avenue.

Canberra is a city full of cyclists and major roads have on-road cycling lanes, often marked with green where roads merge. Take care, as cyclists have right of way on the green lanes.

Public transport ACTION Buses 13 1710.

Motoring organisation NRMA 13 1122, roadside assistance 13 1111.

Car rental ACT Car Rentals (02) 6282 7272; Avis 13 6333; Budget 13 2727; Hertz 13 3039; Rumbles (02) 6280 7444; Thrifty 1300 367 227.

Taxis Canberra Cabs 13 2227.

Bicycle hire Mr Spokes Bike Hire (02) 6257 1188.

[HALL OF MEMORY, AUSTRALIAN WAR MEMORIAL]

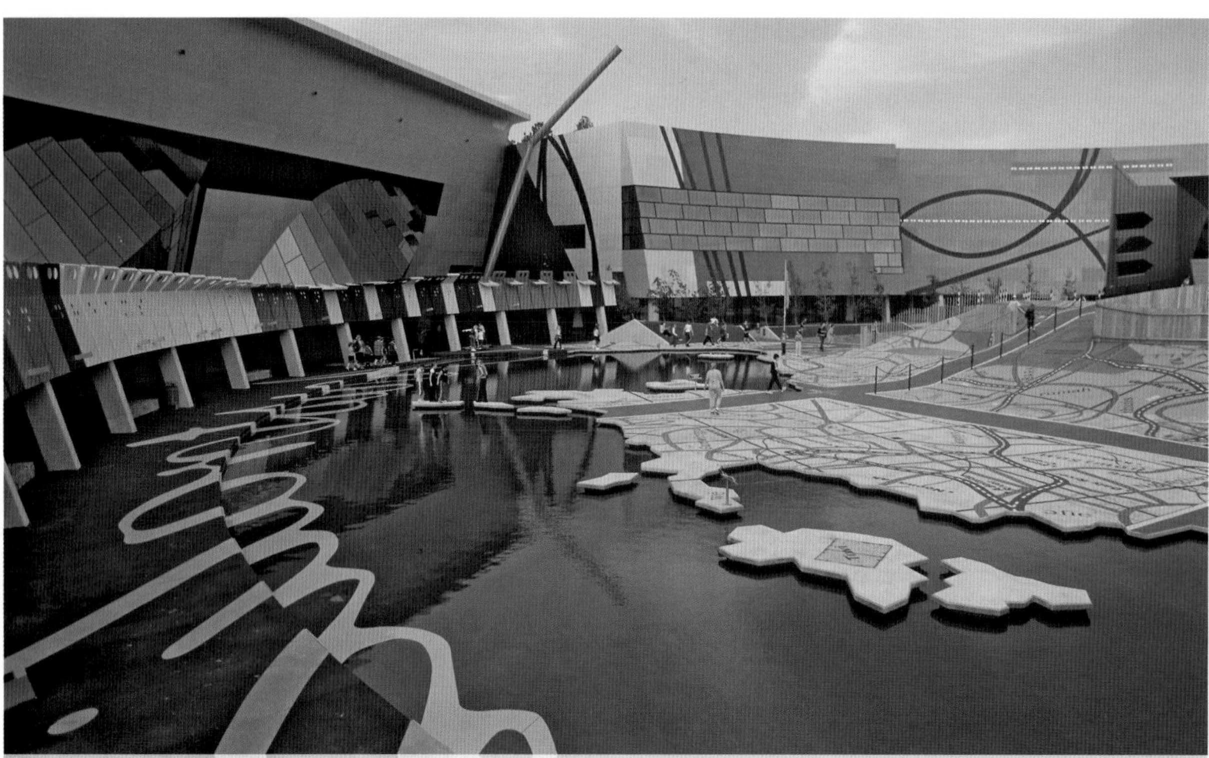

[NATIONAL MUSEUM OF AUSTRALIA]

international collection is just as impressive and includes Jackson Pollock's *Blue Poles* and Australian-born Ron Mueck's *Pregnant Woman*. Enjoy a picnic lunch in the tranquil **Sculpture Garden**, which comprises a series of native gardens and contains over 50 sculptures set between the gallery and the lake. *Parkes Pl, Parkes; (02) 6240 6411; open 10am–5pm daily; general admission free.*

Blundell's Cottage 118 C3

Built in the 1860s, the historic Blundell's Cottage was part of the Campbell family's 32 000-acre (12 800-hectare) estate and home to tenant farm workers. The cottage is now a museum that offers a close encounter with the region's early farming history. *Wendouree Dr, off Constitution Ave, Parkes; (02) 6272 2902; open 11am–4pm daily.*

Australian War Memorial 118 D1

The Australian War Memorial, set at the foot of Mount Ainslie and at the end of Anzac Parade, commemorates and honours the Australian men and women who have served in war. The memorial has undergone redevelopment, culminating in the opening of Aircraft Hall, the Second World War Gallery and ANZAC Hall. Free guided tours run daily. Visitors can browse over 20 exhibition galleries or find moments of silent contemplation at the Hall of Memory, the Tomb of the Unknown Australian Soldier, the Pool of Reflection and the Roll of Honour, which lists the names of over 102 000 Australian servicemen and servicewomen who died in war. *Treloar Cres, Campbell; (02) 6243 4211; open 10am–5pm daily; admission free.*

National Zoo & Aquarium 116 A4

This modern zoo, with naturalistic enclosures rather than cages, has a strong commitment to breeding endangered species and conservation. Animals range from the big cats to bears, otters and monkeys. Wander through the aquarium to see colourful marine life in the Great Barrier Reef exhibit and visit the shark-filled Predators of the Deep exhibit. Special tours include handfeeding the big cats and bears, meeting a cheetah and going on a ZooVenture. (These tours can cost significantly more than general admission, but include all-day zoo entry.) *Scrivener Dam, Lady Denman Dr, Yarralumla; (02) 6287 8400; open 10am–5pm daily.*

Royal Australian Mint 116 B5

Opened in 1965, the mint has the capacity to produce two million coins per day. You can learn how coins are made and see the production floor from an elevated gallery. There are displays of old coins, a video on coin production, and visitors can even

top events

Summernats Car Festival Canberra becomes a 'cruise city' full of street machines. January.

National Multicultural Festival A showcase of cultures in food, art and performance. February.

Canberra Festival Balloon Spectacular An eight-day festival with over 50 hot-air balloons launched daily from the lawns of the Museum of Australian Democracy at Old Parliament House. March.

National Folk Festival Hundreds of performers and thousands of spectators create one of Canberra's biggest parties. Easter.

Rally of Canberra Car rally as part of the Australian Rally Championship. June.

Fireside Festival Warm up winter with fireside food, wine and arts. August.

Floriade A celebration of spring, flowers and fun. September–October.

Wine, Roses and All That Jazz Open weekend of the cool-climate vineyards, with music, food and much wine-tasting. November.

make their own coins. *Denison St, Deakin; (02) 6202 6800; open 9am–4pm Mon–Fri, 10am–4pm Sat–Sun; admission free.*

City Centre & Surrounds

The modest size of Canberra's city centre, Civic, is a reminder that Canberra's population is only just over 352 000 people. There are many attractions in the centre, along with excellent shopping, the peaceful Glebe Park, great cafes and Casino Canberra.

Garema Place 118 B1

The top end of Civic, this open public space is encircled by cafes and restaurants, and features a large movie screen (which shows movies at night during summer). Garema Place is a popular meeting place and is next to the Civic bus interchange. On Friday and Saturday evenings the Irish pub King O'Malley's is usually bursting at the seams, and the outdoor cafes are full of relaxed diners. If you head south from Garema Place down the tree-lined City Walk towards the large Canberra Centre mall, you will find Civic's landmark carousel. Built in 1914 in Melbourne, its hand-carved horses and two elephants were imported from Germany.

Canberra Museum and Gallery 118 B1

CMAG's permanent collection reflects the history, environment, culture and community of the Canberra region. Dynamic exhibitions and diverse community events celebrate Canberra's social history and visual arts. It is also playing host to the Nolan Collection, works by the renowned Australian painter Sidney Nolan, while the Nolan Gallery at Lanyon is refurbished. *Cnr London Circuit and Civic Sq, Civic; (02) 6207 3968; open 10am–5pm Tues–Fri, 12–4pm Sat–Sun; admission free.*

National Film and Sound Archive 118 A2

Housed in the former Institute of Anatomy next to the landscaped grounds of the Australian National University, Australia's national screen and sound archive collects, preserves and shares Australia's film and sound heritage. The displays include film memorabilia, special exhibitions and interactive activities. *McCoy Circuit, Acton; (02) 6248 2000; open 9am–5pm Mon–Fri, 10am–5pm Sat–Sun; admission free.*

National Museum of Australia 118 A3

Opened in 2001, this thoroughly modern museum uses state-of-the-art technology and hands-on interactive displays to show stories about Australia's past, present, people, issues and future. Begin your experience in Circa, a rotating cinema, to get an overview of the three main themes of the museum: land, nation, people. The permanent displays include 'First Australians', an amazing display of Indigenous culture, experience, dance and music; and 'Nation', where you can celebrate Australian icons such as the Hills hoist, *Play School* and Vegemite. 'Old New Land' takes you through 20 000 years of environmental change, and in 'Eternity' you can share the joy and sorrow of ordinary and extraordinary Australians. Children will love 'KSpace', a virtual-reality experience, and 'Our Place', with four different cubbyhouses to explore. Fascinating temporary exhibitions change throughout the year. Guided tours available. *Lawson Cres, Acton Peninsula; (02) 6208 5000; open 9am–5pm daily; general admission free.*

CSIRO Discovery 116 C3

The centre showcases Australia's scientific research and innovation with exhibitions and interactive displays. Take part in the hands-on experiments, experience the virtual-reality theatre and find out about the latest research breakthroughs. *Off Clunies Ross St, Acton; (02) 6246 4646; open 9am–5pm Mon–Fri, 11am–3pm Sun.*

St John the Baptist Church and St John's Schoolhouse Museum 118 C2

Visit Canberra's oldest church and first schoolhouse, built in the 1840s. The buildings have been restored and now form an interesting museum. The adjoining cemetery has some of Canberra's oldest headstones. *Constitution Ave, Reid; (02) 6249 6839; open 10am–12pm Wed, 2–4pm Sat–Sun and public holidays; general admission free, museum entry by donation.*

shopping

Canberra Centre, City Recently expanded, this stylish mall has department stores, a wide variety of smaller shops and specialty stores, in the heart of the CBD. 118 C1

Kingston Old-fashioned specialty shopping and restaurants. 116 C5

Manuka Up-market fashion and homewares, streets lined with cafes, and a supermarket open until midnight. 116 C5

Westfield Woden One of Canberra's largest shopping malls with department stores, over 200 specialty shops, Hoyts 8 cinemas and plenty of parking. 116 A6

Other suburban malls These include Westfield Shopping Centre in Belconnen, the Tuggeranong Hyperdome and two factory outlet centres in Fyshwick. 116 A1, 574 D6, E5

where to eat

Ellacure Great cafe with a good, casual menu in the pizza and pasta mode. Cnr Braybrook and Battye sts, Bruce; (02) 6251 0990; open Tues–Sun for lunch and Tues–Sat for dinner. 116 B2

Iori Japanese Inner-city Japanese restaurant where the emphasis is on really exciting food. 41 East Row; (02) 6257 2334; open Mon–Fri for lunch and Mon–Sat for dinner. 118 B1

Mecca Bah Buzzing Middle Eastern bazaar with a fun menu and a great feel. Shop 25–29, Manuka Terrace, cnr Flinders Way and Franklin St, Manuka; (02) 6260 6700; open daily for lunch and dinner. 116 C5

Mezzalira Top-notch Italian restaurant in the city where you'll get serious and very good food. Melbourne Building, cnr London Circuit and West Row; (02) 6230 0025; open Mon–Fri for lunch and Mon–Sat for dinner. 118 B1

Pulp Kitchen Classy, simple brasserie-style restaurant that offers some of the best, most uncluttered food in town. Shop 1, Wakefield Gardens, Ainslie shops; (02) 6257 4334; open Tues–Fri for lunch and Tues–Sat for dinner. 116 D2

[GINNINDERRA CREEK]

Inner North

Incorporating some of the oldest suburbs in Canberra, this area is an enjoyable mix of relaxed cafes, interesting attractions and nature reserves with excellent walking and mountain-bike tracks.

markets

Belconnen Fresh Food Markets Fresh fruit, vegetables, produce and home to one of Australia's 'big things' – the Giant Mushroom. Wed–Sun. 116 A1

Fyshwick Markets Fresh produce markets with a great atmosphere and excellent value. Thurs–Sun. 564 E5

Gorman House Markets, Braddon Arts and crafts, home-baked treats, plants, clothes and a great up-beat atmosphere, near the city centre. Sat. 118 C1

Hall Markets Set in the historic showgrounds of Hall village on the outskirts of Canberra, with up to 500 stalls of crafts, home produce, plants and homemade stylish clothing. 1st Sun each month (closed Jan). 564 D3

Old Bus Depot Markets, Kingston A local favourite, showcasing the creativity of the Canberra region, with handcrafted and home-produced arts, crafts and jewellery, gourmet food, New Age therapies, kids' activities and musical entertainment. Sun. 118 C5

Capital Region Farmer's Markets, Exhibition Park Farmers and growers from the coast to the capital bring their food to these markets off Northbourne Ave. Sat. 116 D1

Mount Ainslie Lookout 116 D3

Drive to the top of Mount Ainslie for a stunning view of the city and the surrounding mountain ranges. From here you will clearly see the geometry of the capital's design. During autumn this is the best place to see the capital's stunning array of natural colours. *Mt Ainslie Dr, off Fairbairn Ave, Campbell.*

Ainslie, Braddon and Dickson 116 C3, D2

If you are in need of a good coffee, a tasty Chinese meal or a beer in the sun, head to the inner north-eastern suburbs. This is an area with an 'alternative' atmosphere, traditionally populated by Canberra's students. The shopping precincts of these suburbs offer small art galleries, interesting fashion and street cafes. **Ainslie** and **Braddon** have some of Canberra's earliest houses, and make for pleasant walking. In Braddon, browse the Hive Gallery (on Lonsdale Street) for one-of-a-kind gifts. Ainslie is also home to the historic **All Saints Church** (on Cowper Street). Further north, **Dickson** is a hive of activity and home to Canberra's small version of Chinatown (on Woolley Street).

North-west

The inner north-west of Canberra is a hub of activity, with the leafy suburbs of Turner, O'Connor and Lyneham particularly popular with students from the nearby Australian National University. Canberra Nature Park's Mount Ainslie and Mount Majura lie on the edge of the city centre. Further out is Belconnen, which is the main centre for northern Canberra. It offers a large shopping mall, a top-quality sports and aquatic centre, and many other facilities. Also in Belconnen you will find Lake Ginninderra, a favourite recreational and picnic spot.

Australian National Botanic Gardens 116 B3

At the base of Black Mountain, these magnificent gardens (roughly 90 hectares, with 40 hectares developed and the remainder bushland) have the largest collection of Australian native flora in the world. The gardens are organised into sections representing Australia's various climatic zones and ecosystems. Follow the Aboriginal Plant Use Walk for an understanding of the species used for foods and medicines by the country's Indigenous inhabitants. In summer, people flock here for evening picnics and weekend twilight jazz. *Clunies Ross St, Acton; (02) 6250 9450; open 8.30am–5pm daily, extended hours in Jan to 6pm weekdays and 8pm weekends; admission free.*

Black Mountain Tower 116 B3

One of Canberra's landmarks, the communications tower rises 195 metres above the summit of Black Mountain and offers superb views of Canberra. It has two open viewing platforms, an exhibition gallery and **Alto**, a revolving restaurant with a growing reputation. *Black Mountain Dr, Acton; (02) 6219 6111; open 9am–10pm daily.*

Australian Institute of Sport 116 C2

Providing world-class facilities across a broad range of disciplines, the AIS is the training ground for many elite athletes. The Sports Visitors Centre is the entrance point for visitors and has exhibitions and displays of sporting memorabilia. Tours of the AIS are led by athletes and give visitors an insight into the life of an elite athlete. The tours include Sportex, an interactive exhibition where you can test your sporting skills against those recorded by our Olympians and see the latest in sport technology. *Leverrier Cres, Bruce; (02) 6214 1111; open 9am–5pm daily; general admission free.*

North

In Canberra's north is Gold Creek Village, a tourist attraction that takes its name from an old property established in the area in the mid-1800s. To make matters confusing, there is no Gold Creek, and no history of goldmining in this area – Gold Creek Station was apparently named after a racehorse. The station, north of the village, is still a large working merino property and is open to visitors wanting a taste of Australian rural life (groups only; bookings (02) 6227 6586). However, Gold Creek Village is the real drawcard here, with specialty shops, galleries, cafes, historic buildings and a host of modern attractions.

walks & tours

Anzac Parade Walk Follow the war memorials set along the regal Anzac Parade on this self-guide walk (2.5 kilometres). Brochure available from the visitor centre or download a tour podcast (www.nationalcapital.gov.au).

Australians of the Year Walk On the lake foreshore in front of the National Library, this walk honours the award's recipients, with plaques for each year of the award.

Balloon Aloft! Experience the beauty of Canberra as you float above the city in a hot-air balloon. Bookings on (02) 6285 1540.

Burley Griffin Walk or Ride A self-guide walk takes in the north-eastern shores of Lake Burley Griffin, from the National Capital Exhibition to the National Carillon (4.6 kilometres). Brochure available from the visitor centre.

Canberra Tracks Drive one, or all three, of these fascinating, signed road trips. Track 1 explores the path of Canberra's original Indigenous peoples; Track 2 takes you through the pastoral era; and Track 3 gives you an overview of Canberra's layout, via lookouts. Brochures available from the visitor centre.

Lakeside Walk This self-guide walk begins at Commonwealth Place and passes the National Gallery, the High Court and the National Library (1.8 kilometres). Brochure available from the visitor centre.

where to stay

Diamant Sleek, ultra-modern hotel with bright splashes of modern art close to the CBD. 15 Edinburgh Ave; (02) 6175 2222. 118 B2

Hyatt Canberra A grand Art Deco hotel offering old-world luxury in the shadow of Parliament. Commonwealth Ave; (02) 6270 1234. 118 B3

Redbrow Garden Luxury lakeside B&B set in a wildlife haven and country property just 20 minutes from the city. 1143 Nanima Rd, Murrumbateman; (02) 6226 8166. 564 D1

The Griffin Hotel & Spa Stylish boutique apartment hotel in the trendy cafe neighbourhood of Kingston. 15 Tench St; 1800 622 637. 116 C5

Hotel Realm Canberra's newest five-star hotel. Modern luxury in the heart of the Parliamentary Triangle. 18 National Circuit; (02) 6163 1888. 118 B5

entertainment

Cinema The big suburban malls of Woden and Belconnen have Hoyts multiplexes, but for arthouse and independent cinema try the Dendy Canberra Centre in the city or Limelight Cinemas in Tuggeranong. Manuka Greater Union has a wide range of interesting new releases. See the *Canberra Times* for details of current films.

Live music Canberra has a surprisingly busy music scene and there is usually a good selection of live music around town, ranging from rock gigs at the ANU Union Bar to mellow jazz or soul at Tilley's Devine Cafe in Lyneham. Big acts play at the Canberra Theatre Centre or the Royal Theatre, both in the city centre. You will find nightclubs aplenty in Civic, Kingston, Manuka and Braddon. For gig details pick up a copy of the *Canberra Times* on Thursday for its lift-out entertainment guide 'Fly' or look for the free street magazine *BMA*.

Classical music and performing arts The main venue for performing arts is the Canberra Theatre Centre. A couple of smaller venues around town, such as Gorman House Arts Centre in Braddon and the Street Theatre on Childers Street, cater to more eclectic tastes. Classical music concerts most often take place at Llewellyn Hall, part of the ANU's Canberra School of Music, in Acton. See the *Canberra Times* for details of what's on.

If a visit to the Australian Institute of Sport leaves you wanting more, then there is plenty of sport to see in Canberra. Watch a game of **basketball** as the two Canberra WNBL teams, the Capitals and the AIS, go head-to-head or play interstate teams. The Capitals play their home games at Southern Cross Stadium in Tuggeranong and the AIS team at the AIS.

See Canberra's **Rugby League** side, the Raiders, take on the rest of the nation at Canberra Stadium. The ACT Brumbies, the capital's **Rugby Union** team, also play at Canberra Stadium, to partisan capacity crowds. Canberra doesn't have an **AFL** (Australian Football League) team of its own, but one Melbourne team plays a couple of 'home' games a year at the picturesque Manuka Oval.

If **cricket** is more your thing, visit in summer to see the Prime Minister's XI take on one of the touring international sides at Manuka Oval.

If you like dust, mud and excitement, try the Rally of Canberra **car rally** in June.

National Dinosaur Museum 564 E3

The museum has an extensive display of fossilised dinosaur remains, full skeletons and full-size replicas. During school holidays there are plenty of activities for children as well as guided tours. *Cnr Gold Creek Rd and Barton Hwy, Gold Creek Village; 1800 356 000; open 10am–5pm Sat–Thurs.*

Cockington Green 564 E3

Journey through a magical world of miniatures, from a Stonehenge replica to a village cricket match amid beautifully manicured gardens, or take a ride on a miniature steam train. A heritage rose walk links the miniatures with the steam train. *11 Gold Creek Rd, Gold Creek Village; (02) 6230 2273; open 9.30am–5pm daily.*

Canberra Reptile Sanctuary 564 E3

Dedicated to research, conservation and education, this is the place for all things reptilian. Both Australian and exotic species are on display, including a boa constrictor, the giant day gecko from Madagascar and local favourites such as the blue-tongue lizard and children's python. The collection is continually increasing so there will be something new each time you visit. *O'Hanlon Pl, Gold Creek Village; (02) 6253 8533; open 10am–5pm Mon–Sat 10–4 Sun. canberrareptilesanctuary.org.au*

The Bird Walk 564 E3

Walk among over 500 birds representing 54 species from Australia and around the world in this aviary, which measures 1000 square metres. You can photograph and handfeed the colourful creatures, which fly free in the landscaped 9-metre-high enclosure. *Federation Sq, Gold Creek Village; (02) 6230 2044; open 10am–4.30pm daily in summer, 11am–3pm daily in winter.*

Ginninderra Village 564 E3

The old village of Ginninderra predates **Hall**, the nearby town that began life as a village well before the city of Canberra was created. Ginninderra serviced the surrounding farming districts, including the property of Gold Creek Station. The Ginninderra Village that is today part of Gold Creek Village contains the old Ginninderra schoolhouse and the township's old Roman Catholic church. They can be found among giftware stores, craft studios and galleries. *O'Hanlon Pl, Gold Creek Village; open daily.*

Inner South

Drive around Canberra's leafy inner-south suburbs to see official residences, Art Deco bungalows, carefully tended gardens and the diplomatic precinct. Most of the embassies are located in the suburbs of Yarralumla and Forrest, between State Circle and Empire Circuit. The Lodge, the prime minister's Canberra residence, next to Parliament House, is noticeable by the large cream brick wall that surrounds it. It is not open to the public, but occasional open days are held. Likewise, Government House, the governor-general's residence at Yarralumla, a grand 1920s building, is only open to the public once or twice a year. There are views of the grounds from a lookout on Lady Denman Drive.

Manuka and Kingston 116 C5, D5

Over the years many of the original homes in these two suburbs have been replaced by townhouses and apartments – their proximity to both Lake Burley Griffin and the city centre has made them sought-after addresses. Both Manuka and Kingston

[HOT-AIR BALLOONS OVER LAKE BURLEY GRIFFIN]

have vibrant shopping precincts, with a fabulous variety of cafes, restaurants, nightclubs, bars and pubs. Up-market shops and public squares add to the vibe, and the lovely **Telopea Park** is only a few minutes away from both suburbs. On the Kingston Foreshore you'll find the **Canberra Glassworks**, nestled in an old powerhouse next to the weekly Old Bus Depot Markets. The nearby suburb of **Griffith** has some of Canberra's best organic and alternative shopping, along with a couple of excellent restaurants and cafes.

Calthorpes' House 116 C5

Built in 1927 in Spanish Mission style, the house contains the original furnishings and photos, offering glimpses into what domestic life was like in the then-fledgling capital. Explore the 1920s garden or hide in the World War II air-raid shelter. There are guided tours during the week (group bookings only) and open house on the weekend. *24 Mugga Way, Red Hill; (02) 6295 1945; open 1–4pm Sat–Sun.*

Mugga-Mugga 116 C6

Set on 17 hectares of grazing land, Mugga-Mugga is a collection of buildings and cultural objects dating from the 1830s to the 1970s. The highlight is the 1830s shepherd's cottage, which has been carefully preserved and furnished with household belongings from the early 1900s. The option of combined admission fees with Calthorpes' House and Lanyon Homestead *(see Day tours, below)* offers good value. *Narrabundah La, Symonston; (02) 6239 5607; open 1.30–4.30pm Sat–Sun.*

South

Tuggeranong Homestead 564 D6

Sitting on 31 hectares, this historic homestead started life as an 1830s cottage, was added to between 1890 and 1903, then was rebuilt in 1908 (part of the old cottage was incorporated into the drawing room). A stone barn built by convicts in the 1830s is still standing. The property was bought by the Commonwealth government in 1917 for military purposes. War historian Charles Bean lived here from 1919 to 1925 while he wrote the official history of the Australian involvement in World War I. A cafe and markets are held on the site every second Sunday of the month. *Johnson Dr, opposite Calwell Shops, Richardson; (02) 6292 8888; open Sat–Sun or by appt.*

Lake Tuggeranong 564 D6

Artificial Lake Tuggeranong offers waterside recreation for Canberra's inhabitants, who come here to ride the cycle path, enjoy a picnic, have a swim or go fishing, sailing or windsurfing.

waterside retreats

Black Mountain Peninsula A picnic and barbecue spot on the edge of the lake in the shadow of Black Mountain. 116 B4

Casuarina Sands Where the Cotter and Murrumbidgee rivers meet. 564 C5

Commonwealth Park Formal gardens, parkland and public art on the edge of the lake. 118 C3

Commonwealth Place Promenade on the shores of the lake, with the International Flag Display and vistas of the Australian War Memorial and Parliament House. 118 B3

Jerrabomberra Wetlands A refuge for wildlife, including 77 bird species. 118 D5

Weston Park A woodland and lakeside recreation area with play equipment for children. 116 B4

day tours

Namadgi National Park Namadgi is part of the Australian Alps and the informative visitor centre is only a 45-minute drive from the city centre. Walking the 160 kilometres of marked trails is a popular way to explore the park. Beautiful scenery can be enjoyed in the rugged Bimberi Wilderness in the western part of the park. Namadgi has a wide range of natural environments, an abundance of native wildlife and a rich Aboriginal heritage with the evidence of several rock art sites in the park. For more details see p. 31

Lanyon Homestead South of Canberra is one of Australia's most historic grazing properties and a beautiful 19th-century homestead. On the banks of the Murrumbidgee River, the homestead and its gardens provide a glimpse of the 1850s. The homestead's outbuildings (the kitchen, dairy, storerooms and workers' barracks) were built from wood and stone – the stone was cut and quarried by convict labour. Parts of the homestead have been restored and furnished in the style of the period.

Tidbinbilla Nature Reserve South-west of Canberra is the Tidbinbilla Nature Reserve, which includes dry and wet forests, subalpine areas and wetlands. Enjoy a bushwalk or picnic, play with the kids in the Playground at Tidbinbilla, or marvel at the free-ranging wildlife at The Sanctuary. Near the reserve is the Canberra Deep Space Communication Complex, one of three facilities in the world that form NASA's Deep Space Network. Here you can view a genuine piece of moon rock that is 3.8 billion years old, astronaut suits, space food, spacecraft models and photographs. Just south of the Tidbinbilla Nature Reserve is Corin Forest, which has an 800-metre bobsled alpine slide, a flying fox, a waterslide and some great bushwalks and picnic areas.

Wine district Canberra district wineries have established a sound reputation for their cool-climate wines. With 140 vineyards and 33 wineries within 35 minutes' drive from the city centre, making a winery tour is a must. You can pick up a guide to the wineries from the visitor centre or from cellar doors around the region. In addition to cellar-door sales, some wineries offer excellent dining facilities and entertainment, with the annual Wine, Roses and all that Jazz Festival proving popular each November and the Wine Harvest festival each April. For more details see p. 31

Historic towns A number of towns within easy driving distance of Canberra are noted for their heritage buildings and are filled with a sense of the area's farming and goldmining history. These towns include Captains Flat, Goulburn, Braidwood, Bungendore, Yass and Young; most are within an hour or so of the capital. Historic Bungendore, full of craft galleries and antique shops, is a great daytrip, although at only 30 minutes away it is easy enough to visit just for a fine meal or an enjoyable shopping expedition. For more details see p. 31

VICTORIA IS...

Barracking for your team at a footy match at the MCG / Watching penguins waddle from the sea to their burrows on PHILLIP ISLAND / Taking a paddlesteamer ride on the Murray River from ECHUCA / Panning for gold and enjoying traditional hard-boiled lollies at SOVEREIGN HILL in Ballarat / Soaking in a mineral spa in DAYLESFORD

[MOUNTAIN ASH TREES, YARRA RANGES NATIONAL PARK]

or HEPBURN SPRINGS / Driving along the spectacular GREAT OCEAN ROAD to see the TWELVE APOSTLES / Bushwalking in a timeless mountain landscape in GRAMPIANS NATIONAL PARK / A daytrip to the DANDENONGS, stopping off at some cellar doors in the YARRA VALLEY / Hitting the slopes at MOUNT BULLER

VICTORIA is possibly Australia's most diverse state. In a half-hour drive from Melbourne you could be taking in mist-laden mountain ranges and fern gullies. In an hour you could be lying on a sandy beach in a sheltered bay, or surfing in the rugged Southern Ocean. In around four hours you could be standing on the edge of the immense desert that stretches away into Australia's interior. In a country full of mind-numbing distances, nothing seems far away in Victoria.

More than five million people live in Victoria, with 3.8 million in Melbourne. The city was only founded in 1835, as a kind of afterthought to Sydney and Hobart, but by the 1850s Victoria was off to a racing start. A deluge of people from all corners of the world fanned out across the state in response to the madness that was gold. It brought prosperity to Victoria and it also brought the certain wildness treasured in the state's history – uprisings like the Eureka Rebellion and bushrangers like Ned Kelly.

Two centuries later, Victoria has also recognised the richness of its natural landscape. To the west of Melbourne, beyond Geelong, a tract of cool-temperate rainforest unravels on its way to

[MOUNT HOTHAM AT SUNSET, ALPINE NATIONAL PARK]

the vivid green Cape Otway, where a lighthouse stands on the cliff-top. The Great Ocean Road winds past here, en route to the state's iconic limestone stacks, the Twelve Apostles.

On the other side of Melbourne, the land falls away into a series of peninsulas, islands and isthmuses. One leads to Wilsons Promontory, an untouched landscape of forested hills, tea-brown rivers and beaches strewn with enormous rust-red boulders.

[GEELONG WATERFRONT]

The amber-hued Yarra Valley produces some of the country's finest cool-climate wines, and from here the landscape begins its gradual climb up into the High Country, which becomes a vista of snowfields in winter.

Perhaps Victoria's most cherished place is the Grampians, an offshoot of the Great Dividing Range. With a quarter of the state's flora and 80 per cent of its Aboriginal rock art, the Grampians is a living gallery and a superb place for bushwalking and camping.

fact file

Population 5 547 500
Total land area 227 010 square kilometres
People per square kilometre 22.1
Sheep per square kilometre 94
Length of coastline 1868 kilometres
Number of islands 184
Longest river Goulburn River (566 kilometres)
Largest lake Lake Corangamite (209 square kilometres)
Highest mountain Mount Bogong (1986 metres), Alpine National Park
Hottest place Mildura (77 days per year above 30°C)
Wettest place Weeaproinah (1900 millimetres per year), Otway Ranges
Oldest permanent settlement Portland (1834)
Most famous beach Bells Beach, Torquay
Tonnes of gold mined 2500 (2 per cent of world total)
Litres of milk produced on Victorian dairy farms per year 7 billion
Quirkiest festival Great Vanilla Slice Triumph, Ouyen
Famous people Germaine Greer, Barry Humphries, Kylie Minogue
Original name for the Twelve Apostles The Sow and Piglets
Best invention Bionic ear
First Ned Kelly film released *The Story of the Kelly Gang*, 1906 (also believed to be the world's first feature film)
Local beer Victoria Bitter

gift ideas

Raspberry drops (Sovereign Hill, Ballarat)
Delicious, old-fashioned lollies from the gold-rush days are still made and sold at Charles Spencer's Confectionery Shop in Main Street and at the Sovereign Hill Gift Shop. See Ballarat p. 166, 579 C10

Replica of Melbourne tram (Best of Souvenirs, Melbourne) Prince Christian of Denmark received a real tram on the occasion of his birth, but you can take home a smaller – and beautifully crafted – version of Melbourne's distinctive transportation. Melbourne Visitor Information Centre, Federation Square, cnr Flinders and Swanston sts, Melbourne. 134 C4

Football souvenirs (National Sports Museum, Richmond) Scarves, beanies and jerseys with team colours and souvenirs from the home of Australian Rules Football. See MCG and National Sports Museum p. 145, 132 E3

Bread, pastries and provisions (Phillippa's, Armadale and Brighton, Melbourne)
A small business that produces delicious baked goods. Highly recommended are the chocolate brownie and caramel, date and walnut blondie. 1030 High St, Armadale; 608 Hampton St, Brighton. 133 F5, 570 D7

Beach boxes souvenirs (The Esplanade Market St Kilda, Melbourne) Victoria's famous beach boxes are represented on fridge magnets, key holders, coasters and prints. See Markets p. 137, 132 D6

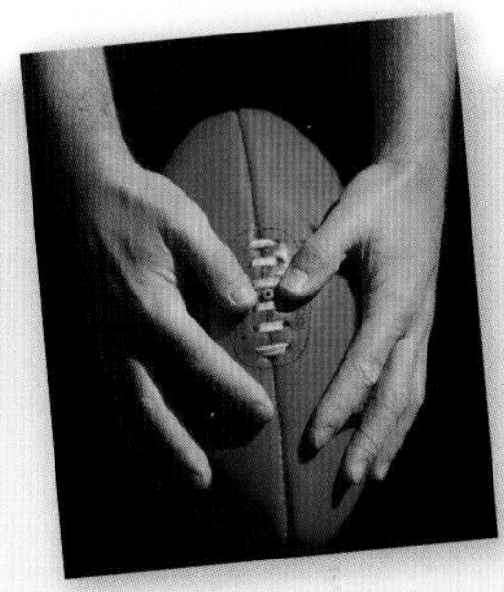

Stefano's products (Mildura) Dine at the acclaimed Stefano's restaurant and then take home jams, chutneys, pasta and pasta sauces. See Mildura p. 195, 558 E6

Wine (Yarra Valley and Mornington Peninsula) Fantastic chardonnay, pinot noir and sparkling varieties from these two wine-producing regions. See Yarra & Dandenongs p.148, 571 C6 and Mornington Peninsula p. 149, 573 I10

Red Hill muesli (Red Hill Market) This popular muesli comes in five different varieties including roast hazelnut, tropical and Wicked (with chocolate). See Flinders p. 179, 570 D12

Ned Kelly memorabilia (Glenrowan)
Victoria's famous bushranger is remembered through everything from replicas of his head armour to belt buckles and T-shirts. See Glenrowan p. 180, 586 E5

Spa and skincare products (Hepburn Springs and Daylesford) The heart of Victoria's Spa Country offers a range of locally made beauty, spa and skincare products from shops and spa centres. See Spa & Garden Country p. 151, 579 E8

MELBOURNE IS...

A footy match at the MCG / Admiring architecture and art in FEDERATION SQUARE / Shopping for produce at the QUEEN VICTORIA MARKET / A stroll along the ST KILDA foreshore / Taking a ferry trip to WILLIAMSTOWN / Moonlight Cinema in the ROYAL BOTANIC GARDENS during summer / Sweating it out in the crowd at the AUSTRALIAN OPEN / Stopping for coffee in DEGRAVES STREET, one of the city's laneways / Live music at a pub in FITZROY / Waterside dining at SOUTHGATE or NewQuay, DOCKLANDS

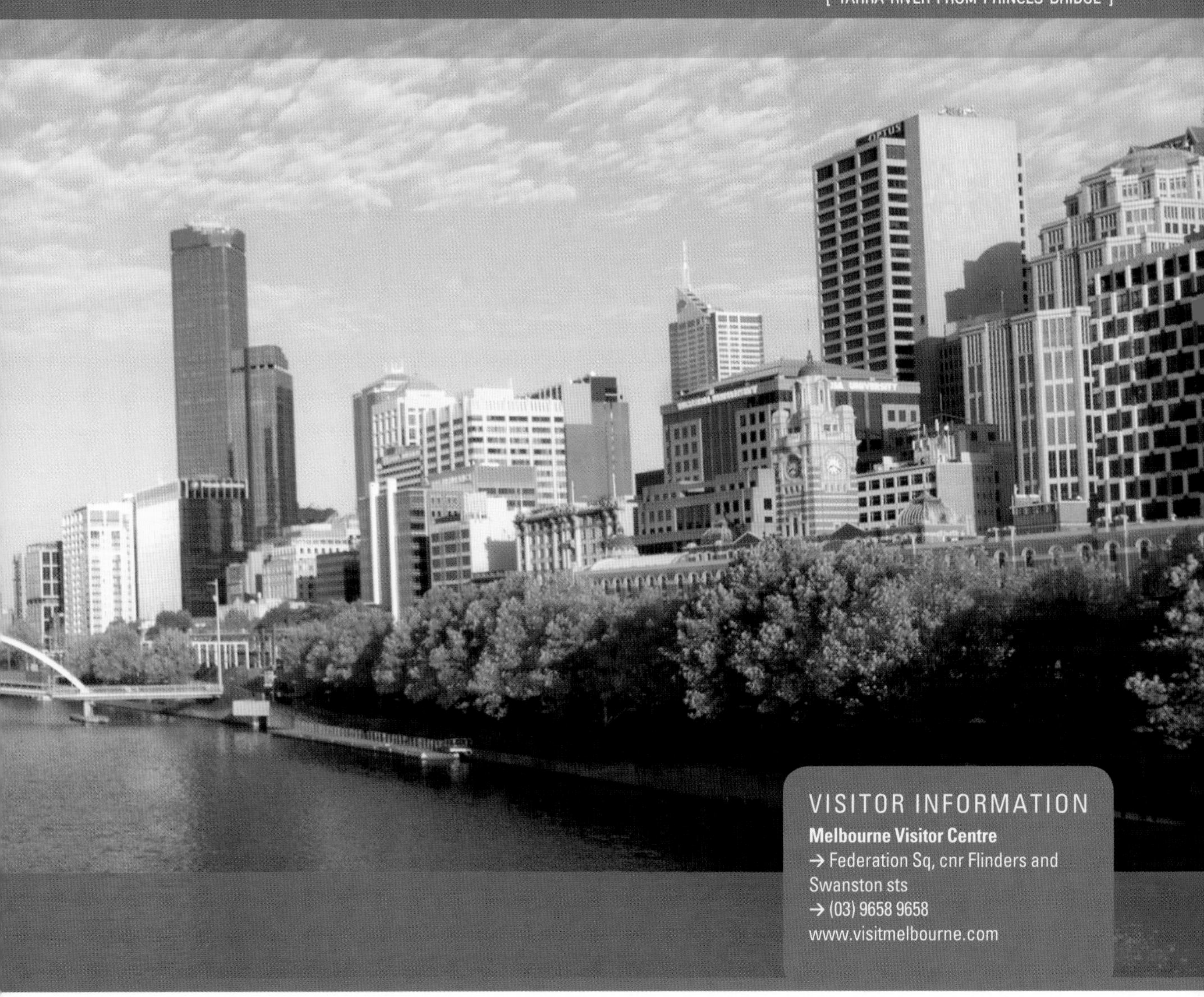

VISITOR INFORMATION

Melbourne Visitor Centre

→ Federation Sq, cnr Flinders and Swanston sts

→ (03) 9658 9658

www.visitmelbourne.com

Melbourne is renowned as Australia's cultural capital. The city has a decidedly European feel, with neo-Gothic banks and cathedrals, much-loved department stores, art galleries and theatres around every corner. And hidden among these buildings is a string of vibrant laneways given over to cafe culture and boutique shopping. Yet Melbourne wouldn't be Melbourne without sport – seeing a footy match at the MCG is a must.

Melbourne was born in 1835, and quickly became a city. With the boom of Victoria's goldfields, unbelievable wealth was poured into public buildings and tramways, grand boulevards and High Victorian masterpieces.

Today Melbourne's population of around 3 893 000 still enjoys the good life, at the very centre of which is a love of good food and fine dining. You can find comfort food in a cosy corner pub or meals with a view and a waterfront setting – a trend in so many of the country's coastal cities. Southbank, the shopping and eating precinct on the Yarra River, has become an extension of the city centre, while Docklands is the city's latest waterside area.

You might come to Melbourne for the dining and the shopping; the gardens and the architecture; the arts and music; the football, cricket and tennis. The city has as much diversity as it has suburbs, and at last check these were marching right down the Mornington Peninsula.

Maribyrnong
Homemaker City Maribyrnong
Highpoint SC
Moonee Ponds
Ascot Vale
Travancore
Brunswick
Brunswick East
East Brunswick Club
CERES Market (Wed & Sa
Flemington
Flemington Market
Flemington Racecourse
Royal Park
Royal Park Golf Course
Melbourne Zoo
Princes Hill
Princes Park
Parkville
Carlton North
Kensington
J.J. Holland Park
Footscray
Footscray Market
Seddon
North Melbourne
University of Melbourne
Ian Potter Museum of Art
Cinema Nova
Brunetti
La Mama Theatre
Carlton
Bar Open & Mario's
The Evelyn
The Vegie B
Babka
The Night Cat
Napier Ho
Fire Services Museum Victoria
St Patrick's Cathedral
Fitzroy Gardens
West Melbourne
Docklands
Medibank Icehouse
Wonderland Fun Park
Harbour Town
NewQuay
Etihad Stadium
Waterfront City
Cow Up A Tree
Victoria Harbour
Fox Classic Car Collection
Blowhole
Webb Bridge
Coode Island
Swanson Dock
Appleton Dock
Victoria Dock
RIVER
Yarraville
SEE PAGE 134
Melbourne
Southbank
Melbourne Park
MCG & National Sports Museum
Olympic Park
AAMI Park
Government House
Ornamental Lake
Ian Potter Foundation Children's Garden
Moonlight Cinema
Royal Botanic Gardens
Shrine of Remembrance
La Trobe's Cottage
Scienceworks & Melbourne Planetarium
Spotswood
Westgate Park
Port Melbourne
Murphy Reserve
South Melbourne Market (Wed & Fri–Sun)
South Melbourne
Newport
Newport Park
Sandridge Beach
Princes Pier
Station Pier
Webb Dock
Gasworks Art Park
Albert Park
Melbourne Sports & Aquatic Centre
Albert Park Lake
Albert Park Public Golf Course
Australian Grand Prix Circuit
Middle Park
Fawkner Park
The Lyall Hotel
HOBSONS BAY
PORT PHILLIP BAY
Spirit of Tasmania ferries Melbourne to Devonport
Greenwich Bay
Ferguson St Pier
Gem Pier
Gellibrand Pier
Breakwater Pier
Williamstown
Williamstown Beach
Shelley Beach
Point Gellibrand
St Kilda West
St Kilda Harbour
The Prince
St Kilda Pier
St Kilda Sea Baths
St Kilda
The Esplanade Hotel
The Espy
Market St Kilda
The Palais Theatre
Luna Park
St Kilda Beach
St Kilda Marina
Jewish Museum of Australia
Astor Theatre
Chapel Str
Elwood
To Brighton Beach
N
0
1 km
MAP LEGEND
Attraction
Where to eat & where to stay
Tourist information
Shopping precinct

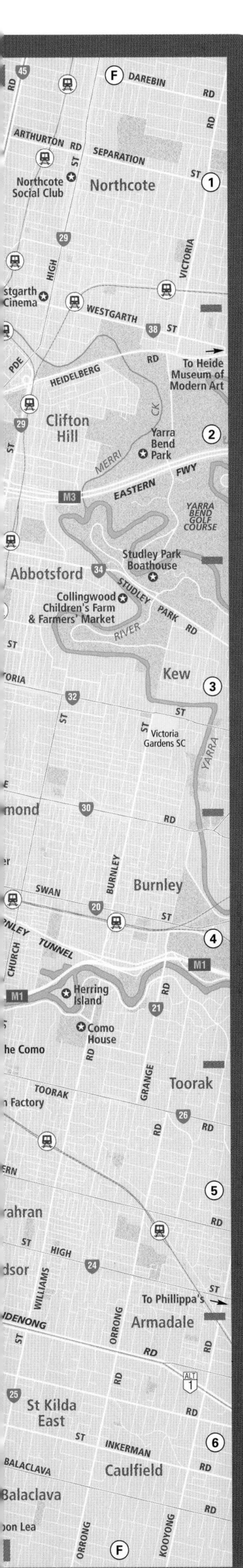

CBD Central

Melbourne's central business district (CBD) lies on the north bank of the Yarra River. The train system runs a ring around the CBD and trams amble up and down most of its main streets. Melbourne's heart is bounded by Flinders, Elizabeth, Little Bourke and Russell streets. This central core takes in the eclectic corner of Swanston and Flinders streets, and further in there's a charming network of arcades and backstreets.

A station, a pub and a cathedral 134 C4

As the major train station in the CBD, **Flinders Street Station** is the first port of call for many people travelling in from the suburbs.

On the three corners facing the station are three other landmarks: Federation Square *(see next entry)*, St Paul's Cathedral and the **Young and Jackson Hotel**. Across Swanston Street, the grandiose **St Paul's Cathedral** was built in 1891. Its mosaic interior is well worth a look.

Federation Square 134 C4

Federation Square is the biggest building project to occur in Melbourne in decades – if not in actual size, then at least in terms of its public significance and architectural ambition. The central piazza is paved with 7500 square metres of coloured Kimberley sandstone. Surrounding it are bars, cafes, restaurants and shops, many of them with unique views over Flinders Street Station and the Yarra. Down by the river, Federation Wharf is home to Rentabike *(see Getting around, p. 135)* and a departure point for river cruises *(see Walks & tours, p. 142)*. Fed Square's must-visit attractions are the Ian Potter Centre: NGV Australia and the Australian Centre for the Moving Image *(see below)*.

Ian Potter Centre: NGV Australia 134 C4

Australian art has finally found a home of its own at this gallery. On the ground floor is a large space dedicated to Indigenous art – from traditional sculptures and bark paintings to the bright and expressive works of modern Aboriginal artists. Also in the gallery are the best of the colonial artists, such as Tom Roberts and Arthur Streeton. The NGV Kids Corner is one of the city's best destinations for pre-schoolers. *Federation Sq; (03) 8620 2222; open 10am–5pm Tues–Sun; general admission free. The National Gallery of Victoria's international collection can be found on St Kilda Rd (see p. 140).*

Australian Centre for the Moving Image (ACMI) 134 C4

ACMI is a museum of the 21st century and an Australian first, exploring all current guises of the moving image. Here you will find cinemas screening films and darkened galleries with screen-based art, as well as temporary exhibitions. Check the newspapers for details of films running or visit the website (www.acmi.net.au). *Federation Sq; (03) 8663 2200, bookings (03) 8663 2583; open 10am–6pm daily, open later for scheduled film screenings; general admission free.*

Swanston Street 134 C3, C4

Your experience of Swanston Street can be totally different depending on which side of the road you walk on. One side of the street is a grand boulevard lined with trees, and dotted with significant buildings and quirky public sculptures. The other side (the west side) seems overcrowded with discount stores, souvenir shops and fast-food outlets. Swanston Street is closed to cars other than taxis.

On the corner of Swanston and Collins streets is the prominent **Melbourne Town Hall**, a venue for various public events including the Melbourne International Comedy Festival. Opposite is a statue of **Burke and Wills**, the two explorers who set out on a doomed journey to find the fabled inland sea. On the corner of Bourke Street is the quirky sculpture, *Three Businessmen Who Brought Their Own Lunch*. Swanston Street also boasts some fine historic buildings *(see Grand old buildings, p. 139)*, and further north is the State Library of Victoria *(see p. 138)*.

Collins Street 134 C4, D3

This is Melbourne's most dignified street. The top end near Spring Street has been dubbed the 'Paris end' with its European ambience and designer boutiques. Near Elizabeth Street

climate

'Four seasons in one day' is a familiar phrase to all Melburnians. It might reach 38°C in the morning then drop to 20°C in the afternoon – and the weather the next day is anyone's guess. Generally though, winter is cold – daytime temperatures of 11–12°C are not unusual – and spring is wet. January and February are hot, with temperatures anywhere between the mid-20s and high 30s. The favourite season of many locals is autumn, when the weather is usually dry and stable.

J	F	M	A	M	J	J	A	S	O	N	D	
25	25	23	20	16	14	13	14	17	19	21	24	MAX °C
14	14	13	10	8	6	5	6	7	9	11	12	MIN °C
48	47	50	57	56	49	47	50	58	66	59	59	RAIN MM
5	5	6	8	9	9	9	10	10	10	8	7	RAIN DAYS

VICTORIA

North Melbourne
Carlton
Melbourne
Southbank
South Wharf
Melbourne Museum & IMAX Theatre
Royal Exhibition Building
Carlton Gardens
Queen Victoria Market (Tues & Thurs–Sun)
City Baths
Old Melbourne Gaol
Old Melbourne Magistrates' Court
Bennetts Lane Jazz Club
Melbourne Central
State Library of Victoria
On3
QV
Chinese Museum
Princess Theatre
Parliament
Parliament of Victoria
Her Majesty's Theatre
St James' Old Cathedral
Flagstaff Gardens
Flagstaff
Former Royal Mint
Chinatown
Flower Drum
Pellegrini's Espresso Bar
Hotel Windsor
Koorie Heritage Trust Cultural Centre
Supreme Court of Victoria
Melbourne's GPO
David Jones
Three Businessmen Who Brought Their Own Lunch
Greater Union Cinema
Old Treasury Building
Myer Melbourne
Bourke St Mall
Melbourne Club
Royal Arcade
Capitol Theatre
Melbourne Town Hall
Kino Cinemas
Collins Place
Treasury Gardens
Manchester Unity Building
Regent Theatre
The Block Arcade
Flinders Lane Gallery
Craft Victoria
Australia on Collins
Burke & Wills Statue
The Westin Melbourne
Anna Schwartz Gallery
Gallery Gabrielle Pizzi
Old ANZ Bank & Banking Museum
Manchester Lane
Nicholas Building
Forum Theatre
Young & Jackson Hotel
St Paul's Cathedral
Degraves Espresso
Olderfleet Building
Vue de monde
The Rialto
Southern Cross Station
ACMI
Federation Square
Ian Potter Centre: NGV Australia
Flinders Street Station
Chocolate Buddha
Birrarung Marr
Immigration Museum
Yarra River
Princes Bridge
Sandridge Bridge
Queens Bridge
Enterprize Park
Hamer Hall
Langham Hotel
Southgate
Sunday Market
Alexandra Gardens
Melbourne Aquarium
Queensbridge Square
Victoria Police Museum
Batman Park
Kings Bridge
Eureka Tower & Skydeck 88
The Arts Centre Theatres Building
Queen Victoria Gardens
Spencer Street Bridge
Crown Entertainment Complex
NGV International
To DFO South Wharf
Polly Woodside Melbourne Maritime Museum
Melbourne Convention Centre
Melbourne Exhibition Centre
Melbourne Recital Centre
MTC Theatre
Sidney Myer Music Bowl
Burnley Tunnel
Domain Tunnel
Kings Domain
CityLink
Australian Centre for Contemporary Art (ACCA)
CUB Malthouse
Royal Botanic Gardens
Sturt St Reserve
West Gate Fwy
University Square
Lincoln Square
Piazza Italia
0 300 m

is **Australia on Collins**, where you will find some of the big names in Australian fashion as well as a ground-floor food court. Towards Spencer Street are impressive buildings, such as the Old ANZ Bank *(see Grand old buildings, p. 139)*.

Laneways and arcades 134 C3, C4

From Flinders Street Station to Bourke Street Mall, you can slip through a world of cafes, fashion boutiques and jewellers, many of them selling one-off items that you just won't find in malls and department stores. But it is also worth it for the walk alone – narrow, darkened and usually bustling, these laneways seem to be a completely separate world to the rest of the city.

The section of **Degraves Street** closest to Flinders Lane is closed to cars and is full of cafes spilling onto the paved street. Across from Degraves Street is **Centre Place**, with more cafes as well as bars and designer-fashion outlets. Towards Collins Street, Centre Place becomes a covered arcade.

Block Arcade runs between Collins and Little Collins streets. It boasts Italian mosaic floors, ornate glass ceilings, tearooms and exclusive clothing boutiques. Follow the arcade to Elizabeth Street, or to the laneway that joins it to Little Collins Street, where there are yet more cafes.

Over Little Collins Street is **Royal Arcade**, Australia's oldest surviving arcade. Above the Little Collins Street entrance stand two giants, **Gog and Magog**, of the ancient British legend. You can take Royal Arcade to either Bourke Street Mall or Elizabeth Street. (To get to the mall you can also take the adjacent, cafe-lined **Causeway**.)

Redevelopment projects such as the GPO development in Bourke Street Mall *(see next entry)*, the renovation of Melbourne Central *(see p. 139)* and the QV site *(see p. 138)*, have opened up more of the old laneways, restoring the original vision of Melbourne's designer, Robert Hoddle.

Bourke Street Mall 134 C3

Bourke Street Mall is the heart of Melbourne's shopping district, with big department stores and brand-name fashion outlets. Between Elizabeth and Swanston streets the mall is closed to cars, making it the territory of trams and pedestrians.

The **Myer** and **David Jones** department stores both have entrances on the mall. Myer, in particular, is renowned for its fine window displays at Christmas. At the west end of the mall is **Melbourne's GPO**, once the city's post office but now a smart shopping complex showcasing a who's who of fashion labels, cafes and bars. If you fancy a break from shopping, grab a seat along the mall and watch life wander by. More often than not there will be a busker performing for your entertainment.

Art spaces 134 C4

Flinders Lane is home to a terrific array of art galleries *(see Flinders Lane galleries, p. 137)*, including the **Anna Schwartz Gallery**. With exhibitions that draw on a wide variety of art forms and often push the envelope, expect to be surprised and delighted, especially if you're into conceptual art. On the corner of Flinders Lane and Swanston Street, the **Nicholas Building** is an important address in the Melbourne arts scene, home to numerous, small, artist-run spaces. *Anna Schwartz Gallery: 185 Flinders La; (03) 9654 6131; open 12–6pm Tues–Fri, 1–5pm Sat; admission free.*

CBD West

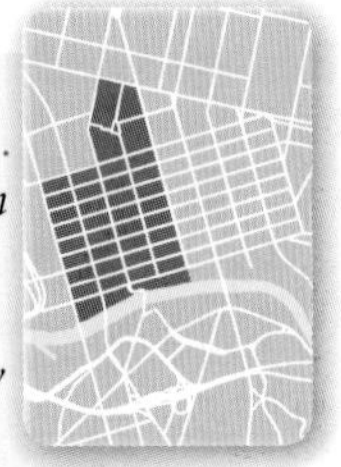

This part of Melbourne stretches from the Queen Victoria Market down to the Yarra River, taking in the city's legal district including the Supreme Court of Victoria on William Street. Bounded by Spencer Street to the west, the old Spencer Street train station was replaced in 2006 with the architecturally superb Southern Cross Railway Station. It is worth a visit even if you're not planning a train ride.

getting around

Melbourne's trams are an icon, but also a very good way of getting around the city. The City Circle tram is free and extends to Docklands. Trams depart every 12 minutes between 10am and 6pm from Sunday to Wednesday, and till 9pm from Thursday to Saturday. Other (paid) services head out into the suburbs, with especially good coverage of the eastern, south-eastern and northern suburbs. A map of the different services can be found inside most trams.

Trains are generally a faster option if there is a service that goes to your destination. Details of services can be found at each of the five stations in the CBD *(see map on p. 134)*.

Buses tend to cover the areas that trains and trams don't service. The free Melbourne City Tourist Shuttle is a hop-on, hop-off bus service stopping at 13 key city destinations. The service runs every 30 minutes between 9.30am and 4.30pm daily and includes an informative on-board commentary. Details of routes and stops can be found at www.thatsmelbourne.com.au, or pick up a brochure from the Melbourne Visitor Centre at Federation Square.

Melbourne's public transport system is progressively moving from Metcard tickets to plastic myki smart cards. While this is happening you will notice two types of ticketing equipment on the train, tram and bus networks. The price of your journey will depend on which of the two 'zones' you need to travel to. For an update on ticketing and an excellent journey planner facility, see www.metlinkmelbourne.com.au

For drivers, the much-talked-about feature of Melbourne's roads is the hook-turn, a process of moving to the left of the road in order to turn right, and therefore getting out of the way of trams. If you wish to use the tollways CityLink or EastLink, either an e-TAG or a trip pass is required (there are no tollbooths, but day passes can be purchased over the phone either before or after making a journey).

Public transport Tram, train and bus information line 13 1638.

Airport shuttle bus Skybus (03) 9335 2811.

Tollways CityLink 13 2629; EastLink 13 5465.

Motoring organisation RACV 13 7228, roadside assistance 13 1111.

Car rental Avis 13 6333; Budget 1300 362 848; Hertz 13 3039; Thrifty 1300 367 227.

Taxis 13CABS 13 2227; Silver Top 13 1008; West Suburban (03) 9689 1144.

Water taxi Melbourne Water Taxis 0416 068 655.

Tourist buses AAT Kings 1300 228 546; Australian Pacific Tours 1300 336 932.

Bicycle hire Rentabike @ Federation Square 0417 339 203; Bike Now (South Melbourne) (03) 9696 8588; St Kilda Cycles (03) 9534 3074.

top events

Australian Open One of the world's four major tennis Grand Slams. January.

Australian Grand Prix Elite motor racing and plenty of off-track entertainment. March.

Melbourne Food and Wine Festival Eat your way through Melbourne and regional Victoria. March.

Melbourne International Comedy Festival Just as many laughs as in Edinburgh and Montreal. April.

Melbourne International Film Festival Features, shorts and experimental pieces from around the world. July–August.

AFL Grand Final The whole city goes footy-mad. September.

Melbourne International Arts Festival Visual arts, theatre, dance and music in indoor and outdoor venues. The alternative Melbourne Fringe Festival usually overlaps. October.

Melbourne Cup The pinnacle of the Spring Racing Carnival. November.

Boxing Day Test Boxing Day in Melbourne wouldn't be the same without this cricket match. December.

Hardware Lane 134 B3

Sitting between Bourke and Lonsdale streets, west of Bourke Street Mall, Hardware Lane comes alive at lunchtime and in the evening when office workers and bar seekers crowd the outdoor tables. The cobblestone paving, the window boxes and the brightly painted facades of the old buildings add to the atmosphere.

Former Royal Mint 134 B3

Two blocks north-west of the Supreme Court of Victoria, the Old Royal Mint was built in Renaissance Revival style, set off by a dazzling coat of arms. Originally constructed to mint the bounty from Victoria's goldfields, it operated until 1968 and is now home to the **Hellenic Museum**, which holds items from the Byzantine period and displays related to Greek migration to Australia. *Cnr William and Latrobe sts; (03) 8615 9016; open 10am–4pm Mon–Fri.*

shopping

Bourke Street Mall, City With department stores Myer and David Jones as well as the swish Melbourne's GPO. 134 C3

Collins Street, City Glamorous shopping strip with the big names in high-end fashion. Check out the boutique-style shopping on adjacent Flinders Lane and Little Collins Street too. 134 C4, D3

Melbourne Central and QV, City These two nearby precincts are a shopper's paradise. 134 C3

Chapel Street, South Yarra Where shopping is an event to dress up for. 132 E5

DFO, South Warf Shopping mall of factory outlets for bargain hunters. 134 A5

Southgate, Southbank A classy range of clothing, art and gifts. 134 C4

Bridge Road, Richmond Back-to-back factory outlets and designer warehouses. 133 F3

Brunswick Street, Fitzroy Unique fashion boutiques and giftshops. 132 E2, E3

Melbourne Aquarium 134 B5

Beside the Yarra River, take a journey into the depths of the ocean, past rockpool and mangrove habitats and a surreal display of jellyfish, then into a tunnel and the 'fishbowl' for a close-up encounter with sharks, stingrays and multitudes of fish. The **Antarctica** exhibition, featuring King and Gentoo penguins among other Antarctic creatures, is a must-see. *Cnr King and Flinders sts; (03) 9923 5999; open 9.30am–6pm daily.*

Immigration Museum 134 B4

At first glance this might seem like a specialist museum, but no subject could be more generally relevant in Australia, where migration has been constant since the first days of European settlement. The Immigration Museum is about journeys, tumultuous new beginnings, and people coming from all corners of the world and bringing their traditions with them. It also investigates Australia's changing government policies on immigration, and how they continue to shape the country. *400 Flinders St; (03) 9927 2700; open 10am–5pm daily.*

Koorie Heritage Trust Cultural Centre 134 A3

As you walk through this centre you realise the drastic, violent and totally irreversible changes made to a culture over 40 000 years old. Displays take you through the local traditions and lifestyle, including food and crafts, as well as events that have occurred in the last two centuries. There are also changing exhibitions by local Aboriginal artists. *295 King St; (03) 8622 2600; open 10am–4pm daily; entry by donation.*

Queen Victoria Market 134 B2

This famous market is spread across 7 hectares under the shelter of a massive shed. The meat hall is at the Elizabeth Street end, while outside all manner of fruit, vegetable and herb stalls extend towards the horizon. On weekends, Saturdays in particular, the aisles are crammed with shoppers from all over Melbourne, and the wide range of clothing and souvenirs make this a hot spot for tourists as well. *Main entrance cnr Elizabeth and Victoria sts; (03) 9320 5822; open 6am–2pm Tues and Thurs, 6am–5pm Fri, 6am–3pm Sat, 9am–4pm Sun.*

On Wednesday evenings during summer, the 'Queen Vic' takes on a whole new character with the **Suzuki Night Market**. At these times it feels more like a festival than a market, with live music, international food and a healthy dose of alternative-clothing and craft stalls.

Flagstaff Gardens 134 A2

Originally known as Burial Hill – many of Melbourne's early settlers ended up here –Flagstaff Gardens were Melbourne's first public gardens and once served as a signalling station for ships arriving from Britain. With open lawns, mature trees (including several lovely Moreton Bay fig trees and avenues of elms), a rose garden, public barbecues and tennis/netball/handball/volleyball courts, it is a lovely space in which to enjoy some time out from the city bustle. *Bounded by William, Latrobe, King and Dudley sts; sports bookings (03) 9663 5888.*

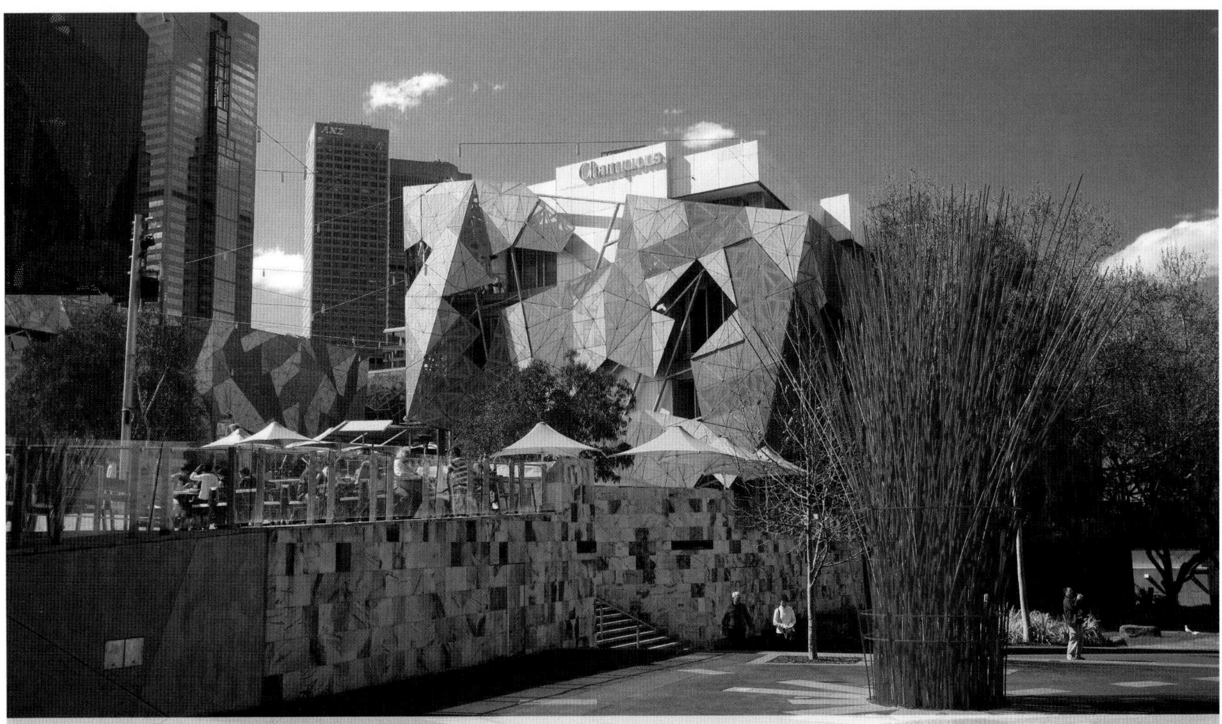

[RED CENTRE SCULPTURE, FEDERATION SQUARE]

CBD East

This is Melbourne's most distinguished quarter, taking in some fine old buildings and the government district. The Parliament of Victoria occupies a suitably prominent position on Spring Street at the top of a hill. At the top end of Collins Street is the Melbourne Club, where Melbourne's male elite have been socialising and doing business since the city's earliest days.

Parliament of Victoria 134 D3

The Parliament of Victoria is perched atop a grand run of steps, which are a popular spot for wedding photos. Built in stages between 1856 and 1929, the building remains incomplete – an ornate dome in the centre was originally supposed to be double its height. From 1901 to 1927 this was the seat of federal government before it moved to Canberra. Free tours of the building run on days when parliament is not sitting; sitting dates are published on the parliament's website (www.parliament.vic.gov.au). *Spring St, facing Bourke St; (03) 9651 8911.*

The imposing structure just south of parliament facing Collins St is the **Old Treasury Building**. It is now home to the Victorian Marriage Registry and glamorous wedding parties can often be seen posing for photos on the steps. The building also houses the Built on Gold exhibition and a program of temporary exhibitions. *Spring St; (03) 9651 2233; open 10am–4pm Sun–Fri; admission free.*

Flinders Lane galleries 134 C4, D3

This laneway boasts the highest concentration of commercial galleries in Australia, mainly in the section between Spring and Swanston streets. There is a strong focus on Indigenous and contemporary art, with standout galleries including **Flinders Lane Gallery** and **Anna Schwartz Gallery** *(see Art spaces, p. 135)*, exhibiting emerging and established Australian contemporary artists; **Gallery Gabrielle Pizzi**, with Aboriginal art from many of Australia's lesser known regions; and **Craft Victoria**, with an excellent gallery and retail space fostering creativity in craft and design. *Flinders Lane Gallery: 137 Flinders La; (03) 9654 3332; open 11am–6pm Tues–Fri, 11am–4pm Sat. Gallery Gabrielle Pizzi: Level 3, 75 Flinders La; (03) 9654 2944; open 10am–5.30pm Tues–Fri, 12–4pm Sat. Craft Victoria: 31 Flinders La; (03) 9650 7775; open 10am–5pm Mon–Sat; admission free.*

markets

Sunday Market, Southbank Crafts galore, leading from the Arts Centre. 134 C4

The Esplanade Market St Kilda Melbourne's oldest art and craft market, with over 150 artisans. Sun. 132 D6

Collingwood Children's Farm – Farmers' Market Victorian produce, from free-range eggs to fresh fruit and vegetables, in a lovely bushland setting on the Yarra. 2nd Sat each month. 133 F3

South Melbourne Market Produce, deli items, clothing and homewares. Wed and Fri–Sun. 132 D4

Flemington Racecourse Market Art, craft and regional produce. 3rd or 4th Sun each month. 132 B2

CERES Market Count your organic food miles as you shop at this community-run environmental park in Brunswick East. Cafe, nursery and inspiration for greening your life. Wed and Sat. 132 E1

Camberwell Market Melbourne's best trash and treasure event. Sun. 573 J6

St Andrews Community Market Laid-back market with alternative crafts, foods, music and clothing, an hour's drive from the city. Sat. 573 K4

See also Queen Victoria Market, p. 136, and Prahran Market, p. 142

entertainment

Cinema The major cinemas in the city are Hoyts in Melbourne Central's On3 entertainment floor and Greater Union in Russell Street. For arthouse films, try Kino Cinemas in Collins Place. Standout cinemas in the inner-city area include: Cinema Nova in Carlton for a great range of popular and arthouse films; Village Jam Factory in South Yarra or Village in the Crown Entertainment Complex for a Hollywood-style experience; the Rivoli in Camberwell or the Westgarth in Northcote for an old-world cinema experience; and the Astor in St Kilda East, where they play re-runs of the classics and recent releases. See daily newspapers for details.

Live music Melbourne is renowned for its live-music scene. Fitzroy is one of the major centres of original music, with venues like Bar Open and the Evelyn hosting bands most nights. Further north are the Northcote Social Club and East Brunswick Club, venues for local and international acts. On the south side of town is The Esplanade Hotel (The Espy) in St Kilda, one of Melbourne's best original rock venues, and in Richmond there's the Corner Hotel, showing many local acts. Other venues dabble in jazz but Bennetts Lane Jazz Club, off Little Lonsdale Street, is the real deal. For a boogie to anything from reggae to funk, try the lamplit Night Cat in Johnston Street, Fitzroy. Bigger local and international acts play at other venues around town. Pick up one of the free street publications, *Beat* or *Inpress*, or get the 'EG' lift-out from *The Age* on Fridays.

Classical music and performing arts The Theatres Building at the Arts Centre is Melbourne's premier venue for theatre, opera and ballet, and Hamer Hall, next door, is the venue for classical music concerts. The new Melbourne Recital Centre is a world-class chamber music venue. Popular musicals and theatrical productions are held at the Regent, Her Majesty's and Princess theatres. The Malthouse Theatre company and the Melbourne Theatre Company (MTC) host plays, and La Mama in Carlton is the venue for more experimental works. Check out the arts section of *The Age* for details. Most performances are booked through Ticketmaster and Ticketek.

where to eat

Chocolate Buddha Quick and delicious modern Japanese served at long communal tables overlooking Federation Square's piazza. Federation Square, cnr Flinders and Swanston sts; (03) 9654 5688. 134 C4

Degraves Espresso Excellent coffee, breakfasts and lunches served amidst stylishly 'distressed' decor or outside at a laneway table. 23–25 Degraves St; (03) 9654 1245. 134 C4

Flower Drum Melbourne's iconic Cantonese restaurant, with dishes such as crayfish with ginger and their famous Peking duck. 17 Market La; (03) 9662 3655. 134 C3

Pellegrini's Espresso Bar The city's original Italian eatery, where the pasta arrives only minutes after ordering and the ice-cold granitas are always good. 66 Bourke St; (03) 9662 1885. 134 D3

Vue de monde Watch super-chef Shannon Bennett and staff prepare some of the best food in Australia in this sky-high dining room. Level 55, Rialto South Tower, 525 Collins St; (03) 9691 3888; open Tues–Fri for lunch and Tues–Sat for dinner. 134 B4

CBD North

This part of town is occupied mainly by office buildings, but there are some interesting places among them, such as the Old Melbourne Gaol, the State Library of Victoria, Chinatown, and the QV shopping and food precinct.

QV 134 C3

Occupying the block between Swanston, Lonsdale, Russell and Little Lonsdale streets, QV is divided up into a series of laneways. You can weave through here at your leisure, maybe stopping in at one of the cafes or go shopping with a purpose: high-end fashion stores line Albert Coates Lane, while Artemis Lane is a wonderland of homewares. You can also explore Red Cape and Jane Bell lanes. (The medical-themed lane names relate to the site's history – QV was once the site of the Queen Victoria Women's Hospital, a part of which still remains.) When you've finished shopping, step into one of QV's various eateries. *(03) 9207 9200; open 10am–6pm Mon–Wed and Sat, 10am–7pm Thurs, 10am–9pm Fri, 10am–5pm Sun.*

State Library of Victoria 134 C2

The State Library of Victoria's front steps and lawn make a great spot for soaking up the sun. In fact, pre–Fed Square, this was the city centre's biggest public space and the main meeting spot for demonstrations. On the third floor of the Roman-style building is an impressive five-storey octagonal reading room, restored to its original sky-lit splendour. The library also incorporates several art galleries and is home to the Wheeler Centre: Books, Writing and Ideas, the centerpiece of Melbourne's UNESCO City of Literature Initiative. This new hub for Victoria's literary community hosts a year-round program of talks, readings and debates. *Cnr Lonsdale and Swanston sts; (03) 8664 7000; open 10am–9pm Mon–Thurs, 10am–6pm Fri–Sun.*

Old Melbourne Gaol 134 C2

Melbourne Gaol was the setting for the execution of some of early Victoria's most notorious criminals. Discover the horrifying reality of death masks and the *Particulars of Execution*, a how-to book on this gruesome subject. If you are brave enough, join a candle-lit Hangman's Night Tour (conducted four times weekly at 8.30pm, 7.30pm in winter) or a ghost hunt (held monthly); contact Ticketek (13 2849) for bookings. *Russell St, between Victoria and Latrobe sts; (03) 8663 7228; open 9.30am–5pm daily; at 12.30pm and 2pm Sat the story of Ned Kelly, 'Such a Life', is performed (free with entry).*

Chinatown 134 C3

Chinatown has prospered and flourished since the first Chinese migrated to Victoria at the beginning of the gold rush. Decorated archways herald the entrance to the Little Bourke Street strip at the Swanston, Russell and Exhibition street ends. Like Chinatowns around the world, Melbourne's Chinatown is distinctive, with lanterns decorating the street at night and exotic aromas drifting out through the doorways of small restaurants.

The **Chinese Museum**, in Cohen Place off Little Bourke Street, tells the tale of the Chinese who migrated to Australia in search of the 'New Gold Mountain', and is also the resting place of Dai

Loong (Big Dragon), which roams the streets during Chinese New Year. *22 Cohen Pl; (03) 9662 2888; open 10am–5pm daily.*

Lonsdale Street 134 C3

The section of Lonsdale Street between Russell and Swanston streets is the centre of Melbourne's Greek community. On the southern side are Greek bookshops, music stores and, of course, cafes and restaurants.

Melbourne Central 134 C3

Melbourne Central is built around the underground railway station of the same name, with arcades and laneways branching out in all directions. The visual focus of the complex is a historic 9-storey shot tower, preserved under a massive glass cone. There are over 300 shops on four levels and an entertainment zone called On3. *(03) 9686 1000; open 10am–6pm Mon–Thurs and Sat, 10am–9pm Fri, 10am–5pm Sun.*

Southbank

This inner-city suburb takes in some of Melbourne's best leisure and dining precincts, as well as a concentration of public arts institutions. Behind Flinders Street Station is Southgate, a stylish shopping and dining area on the Yarra River. On the riverbanks here are also some interesting public sculpture pieces.

Eureka Tower 134 C5

Standing 300 metres and 92 storeys high, the Eureka Tower is Melbourne's tallest building and boasts the Southern Hemisphere's highest viewing platform. This structure is a particularly prominent feature of Melbourne's city skyline due to the 24-carat, gold-plated glass across the top ten levels. On the 88th floor, **Eureka Skydeck 88** offers the ultimate 360-degree view of the city, bay and mountains. Viewfinders placed around the deck pinpoint major attractions. For a heart-stopping experience, step inside The Edge, a 3-metre glass cube that projects out of the side of the building. On the ground floor is the Serendipity Table, with information on the history of Melbourne. *7 Riverside Quay, Southbank; (03) 9693 8888; open 10am–10pm daily.*

Southgate 134 C4

Over 15 years ago, Southgate was just like Docklands *(see p. 142)* – an industrial site being slowly reinvented. Apartments, office blocks, shops, restaurants and a tree-lined promenade have been added to form what is today an essential part of Melbourne.

On the ground floor of the complex is a food court and shops. As you make your way to the top, the restaurants and bars become increasingly exclusive and the shops become boutiques selling glassware, art and jewellery. Some of Melbourne's finest restaurants are located on the top floors. *(03) 9686 1000; open 10am–6pm Mon–Wed, 10am–7pm Thurs, 10am–10pm Fri–Sun: food court breakfast–dinner.*

grand old buildings

Old ANZ Bank Known as the Gothic Bank, with an incredible, gold-leafed interior and a banking museum. 380 Collins St.

Manchester Unity Building Chicago-style building with stark vertical lines, once the city's tallest skyscraper. Cnr Collins and Swanston sts.

Capitol Theatre Designed by the architects of Canberra with a ceiling that will amaze you. 109–117 Swanston St.

St Patrick's Cathedral Victoria's largest church building, built with tonnes of Footscray bluestone. Cnr Gisborne St and Cathedral Pl, East Melbourne.

Princess Theatre The dramatic exterior culminates in three domes with cast-iron tiaras. 163 Spring St.

Forum Theatre Moorish domes, and a starry night sky on the inner ceiling of the main theatre. Cnr Russell and Flinders sts.

St Paul's Cathedral Gothic cathedral made of sandstone. Cnr Flinders and Swanston sts.

Old Melbourne Magistrates' Court The rough sandstone exterior and deeply set archways make for a grim atmosphere. Cnr Russell and Latrobe sts.

Olderfleet Building An intricate Gotham City facade. Nearby is The Rialto, designed by the same architect. 477 Collins St.

Regent Theatre Melbourne's most glamorous theatre, with an interior of Spanish-style lattice and red carpet. 191 Collins St.

St James' Old Cathedral A humble relic of Melbourne's founding years. 419–435 King St.

University of Melbourne More historic buildings than you can count. Parkville.

Hotel Windsor Layered like a wedding cake and fit for a queen. 111 Spring St.

City Baths A feast of domes on the skyline, this building dates back to the days when bathrooms were a luxury few could afford. Cnr Swanston and Victoria sts.

See also Royal Exhibition Building, p. 144

where to stay

Langham Hotel Located riverside on Southbank Promenade, this world-class hotel enjoys spectacular city and river views with a superlative in-house Chuan Spa. 1 Southgate Ave; (03) 8696 8888. 134 C5

The Como In the heart of trendy South Yarra's fashion shopping strip, this intimate and discreet hotel is regarded by visiting celebrities as their 'home away from home'. 630 Chapel St, South Yarra; (03) 9825 2222. 133 F4

The Lyall Hotel A chic intimate property of just 40 suites in tree-lined fashionable South Yarra, it boasts a champagne bar, in-house spa and a plethora of restaurants nearby. 14 Murphy St, South Yarra; (03) 9868 8222. 132 E4

The Prince This glamorous hotel in bayside St Kilda dazzles with its innovative rooms, highly acclaimed restaurant, Circa, The Prince, and adjacent pampering Aurora Spa Retreat. 2 Acland St, St Kilda; (03) 9536 1111. 132 D6

The Westin Melbourne Situated centrally on City Square, the Westin enjoys the best location in town. It also boasts the chic Martini Bar, Wellness Centre and Allegro restaurant. 205 Collins St; (03) 9635 2222. 134 C4

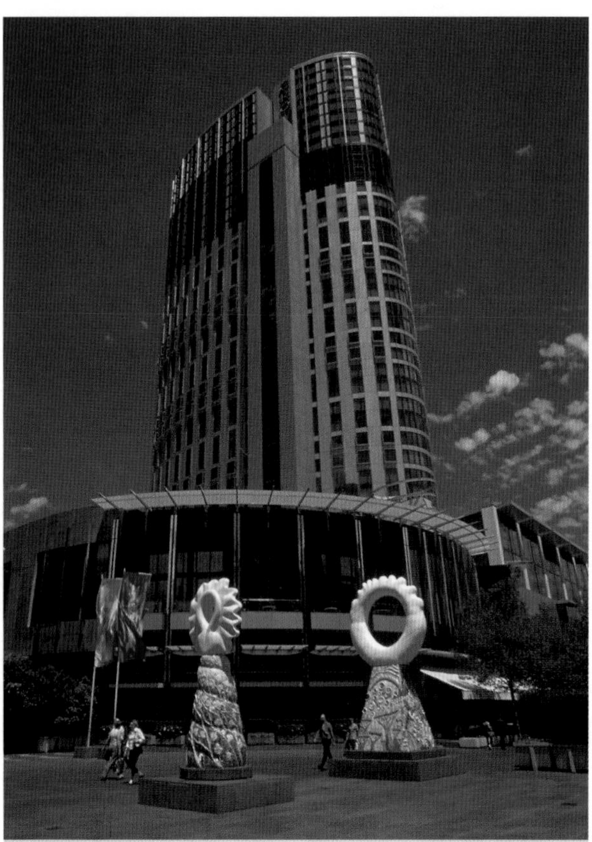

[SCULPTURES OUTSIDE CROWN CASINO]

Crown Entertainment Complex 134 B5

This huge complex begins just over Queensbridge Street. As well as a casino, Crown contains shops, a food court, nightclubs and cinemas. Restaurants include world-class dining such as Neil Perry's Rockpool Bar and Grill, Japanese restaurant Nobu and Silks for Chinese banquets. The big names in fashion such as Versace and Prada reside here too, and items regularly top the $1000 mark.

Melbourne Convention and Exhibition Centre 134 A5

Over Clarendon Street from Crown is the **Melbourne Convention and Exhibition Centre**, with its striking entrance angled upwards over the water. This is the venue for most of Melbourne's major expos, from car shows to wedding exhibitions. *2 Clarendon St, South Wharf; (03) 9235 8000.*

Next to the centre in Duke's Dock is the 1885 tall ship **Polly Woodside**. Visitors are invited to step aboard for a journey into Melbourne's maritime history. *2A Clarendon St, South Wharf; (03) 9699 9760; open 9.30am–5pm daily.*

The Arts Centre 134 C5

The Arts Centre, on St Kilda Road over Princes Bridge from Flinders Street Station, consists of two main buildings – **Hamer Hall** and the **Theatres Building**, with its distinctive lattice spire intended to resemble a ballerina's tutu. The Theatres Building plunges six levels below St Kilda Road and includes three theatres: the State Theatre, with seating for 2000 and a venue for opera, ballets and musicals; the Playhouse, for drama; and the George Fairfax Studio, a smaller drama venue.

If you are not heading to a concert or a theatre show, you can still visit **Gallery 1**, also under the spire. This is the main exhibition space for the **Performing Arts Collection**, which preserves a variety of Australian performing-arts memorabilia. *100 St Kilda Rd; 1300 182 183; open 8am–end of last performance daily; admission free.*

NGV International 134 C5

The large grey building next to the Arts Centre has been the home of the National Gallery of Victoria since 1968. After recent renovations, it now houses the gallery's highly regarded international component, including permanent and touring exhibitions. There are more than 30 galleries, a water curtain at the entrance that has become a Melbourne icon, and a magnificent stained-glass roof by Leonard French towards the back of the building. *180 St Kilda Rd; (03) 8620 2222; open 10am–5pm Wed–Mon; general admission free. NGV Australia can be found at Federation Square (see p. 133).*

museums

Jewish Museum of Australia A record of the experiences of Australia's many Jewish migrants. 26 Alma Rd, St Kilda; (03) 8534 3600; open 10am–4pm Tues–Thurs, 10am–5pm Sun.

Montsalvat An artists' colony that began in 1934; French provincial buildings, restaurant, and artworks for view in the gallery. 7 Hillcrest Ave, Eltham; (03) 9439 7712; open 9am–5pm daily.

ANZ Banking Museum Old money boxes, staff uniforms, historic displays and an interactive ATM exhibit in the glorious Gothic setting of the Old ANZ Bank. 380 Collins St; open 10am–3pm Mon–Fri; admission free.

Ian Potter Museum of Art An extensive art collection, including cultural artefacts and contemporary artworks. University of Melbourne, Swanston St, between Faraday and Elgin sts, Parkville; (03) 8344 5148; open 10am–5pm Tues–Fri, 12–5pm Sat–Sun; admission free.

Victoria Police Museum Victoria's life of crime revealed, from the capture of Ned Kelly to the Hoddle Street shootings. Lower Concourse Level, World Trade Centre, 637 Flinders St; (03) 9247 5214; open 10am–4pm Mon–Fri; admission free.

Fire Services Museum Victoria Huge collection of fire brigade memorabilia, including vintage vehicles and historic photos. 39 Gisborne St, East Melbourne; (03) 9662 2907; open 9am–3pm Thurs–Fri, 10am–4pm Sun.

See also Melbourne Museum, p. 145, Melbourne Cricket Ground (MCG) and National Sports Museum, p. 145, Sturt Street arts, p. 141, Immigration Museum, p. 136, Koorie Heritage Trust Cultural Centre, p. 136, ACCA, p. 141, Heide Museum of Modern Art, p. 146, Ian Potter Centre: NGV Australia, p. 133, NGV International, this page, Australian Centre for the Moving Image, p. 133, Chinese Museum, p. 138, Flinders Lane galleries, p. 137, and Arts Centre, this page

Sturt Street arts 134 C5, C6

Behind St Kilda Road is Sturt Street, home to four arts institutions – the CUB Malthouse, the Australian Centre for Contemporary Art (ACCA) and the new Melbourne Recital Centre (MRC)/Melbourne Theatre Company (MTC) complex.

Once a malt factory, the **CUB Malthouse** has four theatres and is home to the Malthouse Theatre company, dedicated to contemporary theatre. Next door, the **ACCA** is housed in a rusted-steel building that is intended to resemble the colour of Uluṟu against a blue sky. Inside the stark structural forms are changing contemporary Australian and international exhibitions, quite often confronting and interactive. *CUB Malthouse: 113 Sturt St; bookings on (03) 9685 5111. ACCA: 111 Sturt St; (03) 9697 9999; open 10am–5pm Tues–Fri, 11am–6pm Sat–Sun and public holidays; admission free.*

Resembling a giant beehive in parts, the stunning new **Melbourne Recital Centre** gives ACCA a run for its money in the architectural stakes. Comprising the 1001-seat Elisabeth Murdoch Hall for chamber music, the 150-seat Salon for more intimate performances and public spaces including a cafe-bar, it's an exciting addition to Melbourne's arts precinct. The adjacent new home of the **Melbourne Theatre Company** is an equally fascinating, cutting-edge structure. *Cnr Sturt St and Southbank Blvd, Southbank; MRC bookings on (03) 9699 3333, MTC bookings on (03) 8688 0800.*

CBD South-east

Sidney Myer Music Bowl 134 D5

Set in the Kings Domain gardens, the iconic Sidney Myer Music Bowl is a massive grassy amphitheatre leading down to a stage, above which floats a soaring, tent-like roof. Much-loved by Melburnians for its annual Christmas extravaganza, Carols by Candlelight, it also hosts numerous other summertime events, including free concerts by the Melbourne Symphony Orchestra. Just fabulous for a champagne picnic on a starry summer's eve. *Linlithgow Ave, Kings Domain; 1300 182 183.*

Gardens 134 C6, 132 E4

Bordered by the curve of the Yarra and the bitumen of St Kilda Road is a series of public gardens, including Kings Domain and the Royal Botanic Gardens. In **Kings Domain** is the Sidney Myer Music Bowl *(see previous entry)* and the imposing **Shrine of Remembrance**. Deep within the shrine stands a statue of two soldiers, representing the generations of Australians who have fought in various wars around the world. You access the shrine from the chambers below, through the visitor centre. From the balcony at the top there are views straight down St Kilda Road. The shrine is the centre for ANZAC Day commemorations. *Birdwood Ave, South Yarra; (03) 9661 8100; open 10am–5pm daily; admission free.*

To the east of the shrine is **La Trobe's Cottage**, a prefabricated house built in England and brought to Australia for the residence of Victoria's first lieutenant governor. Magnificent **Government House** is located nearby but is obscured by a tangle of vegetation. *Cnr Birdwood Ave and Dallas Brooks Drive; (03) 9656 9800; open 2–4pm Sun, Oct–May, and select days Jun–Sept. Tours of La Trobe's Cottage and Government House, Mon and Wed; (03) 8663 7260.*

Directly opposite the shrine, across Birdwood Avenue, is Observatory Gate, behind which are the renowned **Royal Botanic Gardens**. Inside the gate is a cafe and the historic Melbourne Observatory. The visitor centre here has information on walks and activities in the gardens, including night tours at the observatory and the Aboriginal Heritage Walk *(see Walks & tours, p. 142)*. The **Ian Potter Foundation Children's Garden**, was created as a fun and interactive space for kids, with a program of hands-on activities. In summer the botanic gardens also host **Moonlight Cinema**; programs are available around town, and the films on offer range from new releases to the classics. *Birdwood Ave; (03) 9252 2300; open 7.30am–sunset daily; Ian Potter Foundation Children's Garden open 10am–sunset, Wed–Sun in school term time and daily in school holidays, but closed eight weeks in winter.*

If you are after a riverfront picnic spot, look just outside the botanic gardens across Alexandra Avenue. Free gas barbecues dot the Yarra bank from Swan Street Bridge to Anderson Street.

Melbourne and Olympic parks 132 E4

Between the city and the suburbs of Richmond and Abbotsford are Melbourne's biggest sporting venues, scattered on either side of the rail yards like giant resting UFOs. The two major ones are the **MCG** *(p. 145)* and **Melbourne Park**, incorporating the **Rod Laver** and **Hisense** arenas, and home of the Australian Open and big-ticket concerts. Heading towards Richmond, **AAMI Park** in the **Olympic Park** precinct is the home of Melbourne Rebels (Rugby Union), Melbourne Storm (Rugby League) and Melbourne Victory (soccer). *Olympic Blvd; (03) 9286 1600.*

parks and gardens

Birrarung Marr Melbourne's newest park, with sculptural displays, a colourful playground and a bike track leading up to Federation Square. City. 134 D4

Albert Park A great spot for exercising around the lake, and the site for the Australian Formula One Grand Prix. Albert Park. 132 D5

Gasworks Arts Park Sculptures, native gardens, barbecues, a cafe and artist studios in the former South Melbourne Gasworks. Albert Park. 132 C5

Yarra Bend Park Closest bushland to the city, with boats for hire, a golf course, great views and a strong Aboriginal heritage. Kew/ Fairfield. 133 F2

Eltham Lower Park Featuring the Diamond Valley Miniature Railway, which offers rides for kids on Sunday 11am–5pm. Eltham. 570 F5

Brimbank Park With wetlands, a children's farm, a visitor centre with a cafe and walking trails along the Maribyrnong River. Keilor. 570 D5

Jells Park A haven for waterbirds and a great place for a stroll through the bush. Wheelers Hill. 570 F7

Wattle Park Native bush and birds, a nine-hole golf course and accessible by tram. Surrey Hills. 570 F6

Westerfolds Park On the Yarra and popular for canoeing and cycling, with the Mia Mia Aboriginal Art Gallery on top of the hill. Templestowe. 570 F5

See also Royal Botanic Gardens, this page, Fitzroy Gardens, p. 145

Docklands

Melbourne's CBD is once again on the move, this time west across Spencer Street and the rail yards to Docklands. Until the 1960s this was a busy shipping port; now it is being transformed into a residential, business, entertainment and retail precinct, with nine mini neighbourhoods including NewQuay, Waterfront City and Victoria Harbour.

NewQuay and Waterfront City 132 C3

At the northern end of Docklands, the adjacent waterfront neighbourhoods of NewQuay and Waterfront City comprise high-rise apartments, offices, retail outlets, galleries, marinas and the bulk of Docklands' bars and restaurants. The precinct is home to **Wonderland Fun Park**, with a host of rides for kids, and the impressive **Medibank Icehouse**, a world-class ice sports and entertainment venue designed to resemble the fault line in a glacier. It features a public skating rink with spectacular sound and lights, and a huge stadium for events such as ice hockey and figure skating. *1300 756 699; open daily; session times and events www.icehouse.com.au.*

The new **Harbour Town** shopping centre is home to a large range of brand outlets and fashion stores, while trash-and-treasure lovers will want to check out the weekly **Sunday Market** (10am–5pm) at Waterfront City Promenade.

Etihad Stadium 132 C3

A major venue for Australian Football League (AFL) games, other sporting events and concerts, Etihad Stadium is a Colosseum-like stadium with a retractable roof (very handy for when the weather is inclement). You can go on a 'Behind the Scenes' tour to inspect, among other things, the AFL players' change rooms and the coaches' box. The one-hour tours leave at 11am, 1pm and 3pm Monday to Friday from the Customer Service Centre opposite Gates 2 and 3; different times on event days. *Bounded by Bourke and Latrobe sts, Wurundjeri Way and Harbour Esplanade; tour bookings (03) 8625 7277.*

Art Journey 132 C3, C4

A self-guide urban art tour, the Art Journey takes you past more than 30 superb, large-scale public artworks, including the whimsical wind-powered **Blowhole**, the sublimely sculptural **Webb Bridge** for pedestrians and cyclists, and the amusing **Cow Up A Tree** (which is exactly that). You can also make use of the facilities along the way, including barbecues, recreation areas and a terrific children's playground. Pick up a copy of the Art Journey brochure from the Melbourne Visitor Information Centre, or download one from the Docklands website (www.docklands.com).

Fox Classic Car Collection 132 C3

With its incredible collection of vintage cars, including Bentley, Jaguar, Rolls Royce and some ultra-cool Porsche and Mercedes Benz makes, the Fox Classic Car Collection has up to 50 vehicles on display at any given time. Located in the historic Queen's Warehouse building (originally built as a customs house). *Cnr Batmans Hill Dr and Collins St; (03) 9620 4086; open 10am–2pm Tues Feb–Nov.*

Inner South-east

Beyond the gardens south-east of the CBD are some of the city's most exclusive suburbs: South Yarra, Toorak, Malvern and Armadale. South Yarra and Toorak centre around Toorak Road, where exclusive clothing and footwear stores, as well as cafes and food shops, line the street. Chapel Street runs in the other direction, from South Yarra into St Kilda, and is virtually non-stop shops for three major blocks, from Toorak Road to Dandenong Road.

Chapel and Greville streets 132 E5

The northern end of Chapel Street, between Toorak and Commercial roads, is the place to come for the latest in fashion. About halfway down this stretch is the **Jam Factory**, a shopping complex and food court inside the old premises of the Australian Jam Company. Inside is a cinema complex that screens most of the new releases as well as some arthouse titles.

Just around the corner from Chapel Street on Commercial Road is the **Prahran Market**, Australia's oldest continually running market. *163 Commercial Rd, South Yarra;*

walks & tours

Golden Mile Heritage Trail Walk with a guide or navigate this trail on your own. It leads from the Immigration Museum to Melbourne Museum, taking in historic buildings and heritage attractions along the way (walkers gain discounted entry to various places). Bookings on 1300 780 045, or buy a self-guide brochure from the Melbourne Visitor Information Centre at Federation Square. 10am daily.

Aboriginal Heritage Walk With an Aboriginal guide and a gum leaf for a ticket, stroll through the Royal Botanic Gardens and learn about the bushfoods, medicines and traditional lore of the Boonerwrung and Woiwurrung people, whose traditional lands meet here. Bookings on (03) 9252 2429; 11am Thurs–Fri and some Sundays.

MELTours Various tours focusing on the city's laneways, art, history, shopping and architecture. Bookings on 0407 380 969.

Haunted Melbourne Ghost Tour Get the adrenaline pumping as you traipse down dark alleys and enter city buildings that the ghosts of early Melbourne are known to haunt. Bookings on (03) 9670 2585.

Chocoholic Tours A range of tours to get you drooling, taking in Melbourne's best chocolatiers, candy-makers, ice-creameries and cafes. Bookings on (03) 9686 4655.

Harley Rides Take the Introduction to Melbourne Tour, exploring the bay and over the West Gate Bridge with the wind whistling through your hair. Bookings on 1800 182 282; tours daily.

Foodies' Tour Get tips on picking the best fresh produce, meet the specialist traders and taste samples from the deli at Queen Victoria Market. Bookings on (03) 9320 5822; 10am Tues and Thurs–Sat.

Carlton & United Brewery Tours Free tastings of CUB draught beers are preceded by a tour around Abbotsford Brewery, the home of Fosters. Bookings (03) 9420 6800. 10am and 2pm Mon–Fri.

River Cruises A trip down the Yarra or the Maribyrnong or across to Williamstown will give you new views of Melbourne. Melbourne River Cruises (03) 8610 2600, City River Cruises (03) 9650 2214, Williamstown Ferry (03) 9517 9444, Maribyrnong River Cruises (03) 9689 6431.

(03) 8290 8220; open dawn–5pm Tues, Thurs and Sat, dawn–6pm Fri, 10am–3pm Sun.

A little further up Chapel Street, Greville Street runs off to the west. The narrow strip is lined with cafes, bars, hip clothing shops and original occupants such as **Greville Records**.

Just south of Greville Street is **Chapel Street Bazaar**, a treasure trove for collectors. If the northern end of Chapel Street is the fashion darling, the Prahran/Windsor end, south of Commercial Road, is its more down-to-earth, homely sibling. Lined with dozens of interesting cafes, restaurants and shops, it is a fantastic stretch to browse and graze.

Como House and Herring Island 132 F4

Como House is a National Trust–listed mansion at the end of Williams Road. Stroll around the gardens and take a tour of the house, complete with the original furnishings of the Armytage family. You can visit the gorgeous cafe (open Wednesday to Sunday) without paying admission. Nearby is Como Landing, where you can take a punt across to Herring Island in the middle of the Yarra. This artificial island boasts a sculpture park and picnic/barbecue facilities. *Como House: cnr Williams Rd and Lechlade Ave, South Yarra; (03) 9827 2500; open 10am–4pm Wed, Sat–Sun (daily May–Aug). Herring Island punt: 11am–5pm Sat–Sun Dec–Apr.*

Inner North

Carlton and Fitzroy are two lively inner suburbs to the north and north-east of the city. Carlton is the heart of Victorian terrace territory, while Fitzroy is where many young Melburnians would choose to live if they could afford it. Smith Street, a major street to the east, is blossoming with cafes, health-food shops and independent fashion designers.

sport

Melbourne is possibly Australia's most sporting city, with hardly a gap in the calendar for the true sports enthusiast. **AFL** (Australian Football League) is indisputably at the top of the list. The season begins at the end of March and as it nears the finals in September, footy madness eclipses the city. Victoria has ten teams in the league, and the blockbuster matches are played at the **MCG** and **Etihad Stadium**.

After the football comes the **Spring Racing Carnival**, as much a social event as a horseracing one. October and November are packed with events at racetracks across the state, with the city events held at Caulfield, Moonee Valley, Sandown and Flemington racetracks. The Melbourne Cup, 'the race that stops the nation', is held at Flemington on the first Tuesday in November, and is a local public holiday.

Cricket takes Melbourne through the heat of summer. One Day International and Test matches are usually played at the MCG, and the popular Boxing Day Test gives Christmas in Melbourne a sporting twist.

In January Melbourne hosts the Australian Open, one of the world's four major **tennis** Grand Slams. The venue is **Melbourne Park**, home to the Rod Laver Arena and the Hisense Arena, both of which host other sporting events and concerts throughout the year.

Come March, and the **Australian Formula One Grand Prix** comes to town, attracting a large international crowd. The cars race around Albert Park Lake, which for the rest of the year is the setting for rather more low-key sporting pursuits such as jogging and rollerblading.

See also Melbourne Cricket Ground (MCG) and National Sports Museum, p. 145, and Melbourne and Olympic parks, p. 141

[ETERNAL FLAME, SHRINE OF REMEMBRANCE]

[ST KILDA PIER AT SUNSET]

other suburbs

Brunswick Sydney Road is the place to come to for bargain fabrics, authentic Turkish bread and a healthy dose of Middle Eastern culture. 132 D1

Footscray A mini-Saigon that is the lesser-known version of Victoria Street, Abbotsford, now with a growing African flavour. Jam-packed with cheap eateries and one of Melbourne's best produce markets. 132 A2

Camberwell and Canterbury Melbourne's eastern money belt, with fashion outlets lining Camberwell's Burke Road, and the elegant Maling Road shopping precinct in Canterbury. 570 E6

Yarraville A gem tucked away in a largely industrial sweep of suburbs, with cafes and a superb Art Deco cinema. 132 A3

Hawthorn With a strong student culture from the nearby university, and a strip of shops on Glenferrie Road offering everything from Asian groceries to smart fashion. 570 E6

Balaclava Kosher butchers mixed with an emerging cafe culture. 133 F6

Black Rock One of many bayside suburbs shifting from a sleepy village into sought-after real estate, fronting two of Melbourne's best beaches. 573 J7

Dandenong One of Melbourne's most culturally diverse communities – with food stores galore and a vibrant market. 573 K7

Eltham All native trees and mud-brick architecture, this suburb feels like a piece of the country only 30 minutes from the city. 570 F5

Lygon Street, Carlton 132 D2

This is Carlton's main artery and the centre of Melbourne's Italian population, with many restaurants, cafes, bookstores, clothing shops and the excellent **Cinema Nova**. Stop for authentic pasta, pizza, gelato and good coffee at places such as **Brunetti**, on Faraday Street, an institution that is always crowded with Italian pastry lovers. You can also head to **Rathdowne Street**, parallel to Lygon Street, which has more cafes, restaurants and food stores.

Brunswick Street, Fitzroy 132 E2

Brunswick Street offers an eclectic mix of cafes, pubs, live-music venues and shops. Anything goes in Brunswick Street – young professionals come here for leisurely weekend breakfasts at cafes such as **Babka** and **Mario's**.

Royal Exhibition Building 134 D1

The Royal Exhibition Building is Melbourne's most significant historic building – and arguably the country's now that it has become Australia's first man-made structure to achieve World Heritage status. The building and the adjacent Carlton Gardens were admitted to the list in 2004, joining the likes of Uluṟu and the Great Barrier Reef.

A vast hall topped with a central dome, it is considered an enduring monument to the international exhibition movement that began in the mid-19th century. No comparable 'great halls' survive from other international exhibitions held elsewhere in the world.

At dusk each night the building is illuminated, creating a vista that harks back to the heady days of 1880s Melbourne. Tours to view its interior run from the adjacent Melbourne Museum *(see next entry)* at 2pm daily whenever the building is not in use (bookings 13 1102). The area comes alive during the Melbourne International Flower & Garden Show (March–April). *11 Nicholson St, Carlton.*

Melbourne Museum 134 D1

Melbourne Museum is housed in the spaceship-like structure of metal and glass next to the Royal Exhibition Building. It is the home of Phar Lap, Australia's champion racing horse, standing proud and tall in a dimly lit room. Bunjilaka is an Aboriginal cultural centre telling the Koorie story from the Koorie perspective. Other features of the museum include dinosaur skeletons, a living rainforest and impressive displays on science, the mind and the body. Located in the same building is the **IMAX Theatre**, screening films in 2D and 3D. *Melbourne Museum: Nicholson St, Carlton; 13 1102; open 10am–5pm daily.*

Melbourne Zoo 132 D2

This is Australia's oldest zoo, and the single iron-barred enclosure that remains is a testimony to the days when animals were kept in minuscule cages. Today things are rather different – take the Trail of the Elephants, for instance, where elephants live in a re-creation of an Asian rainforest, complete with an elephant-sized plunge pool (and Asian hawker stalls for visitors). Another perennial favourite here is the Butterfly House, where butterflies are quite happy to land on you as you pass through. From mid-January to mid-March the zoo runs a popular program of open-air, evening jazz sessions called Zoo Twilights. *Elliott Ave, Parkville; (03) 9285 9300; open 9am–5pm daily, to 9.30pm for Zoo Twilights.*

Inner East

Richmond and Abbotsford lie to the east of the city, and for both food and clothing the combination of quality and price here is hard to beat. Further out is a stretch of parkland that follows the winding path of the Yarra River.

Bridge Road, Richmond 133 F3

Bridge Road is Richmond's main artery, and between Hoddle and Church streets it is a shopper's heaven. Many a tour bus pulls up here, with shoppers pouring into the designer-clothing stores and factory outlets. In the next block, between Church and Burnley streets, is a strip of reasonably priced restaurants offering various cuisines. Swan Street, south of Bridge Road, is another good spot for wining and dining, and features the **Corner Hotel**, staging local and international bands.

Victoria Street, Abbotsford 133 F3

Victoria Street, north of Bridge Road, is a living, breathing piece of Vietnam. From Hoddle Street to Church Street it overflows with Asian grocery stores and Vietnamese restaurants, where the focus is on authentic food, fast.

Melbourne Cricket Ground (MCG) and National Sports Museum 132 E3

A footy or cricket match at the MCG (or the 'G' as it is locally called) would have to be one of Melbourne's top experiences. But if you visit in the off-season or simply can't get enough sport in your system, then you can take a tour on non-event days, 10am–3pm, from Gate 3 in the Olympic Stand. Tour prices can include entry to the MCG's interactive National Sports Museum. Some of the 3500 artefacts include swimmer Ian Thorpe's swimsuit, the inaugural Brownlow Medal, legendary cricketer Sir Don Bradman's baggy green cap and Bart Cummings' 12 Melbourne Cups. Visitors can also try out their own sporting skills in the 'Game On' room. *Brunton Ave, Richmond; open 10am–5pm daily; (03) 9657 8879.*

Fitzroy Gardens 132 E3

Fitzroy Gardens, bounded by Albert, Clarendon and Lansdowne streets and Wellington Parade, are one of a handful of public gardens surrounding the CBD, but the only one that can boast **Cooks' Cottage**, a fairy tree and a model Tudor village. *Cook's Cottage: (03) 9419 4677; open 9am–5pm daily.*

Yarra Bend Park 133 F2

Yarra Bend Park is a bushland sanctuary that feels far, far away from the city even though it is, in fact, just a few minutes' drive from it. It features walking tracks and a golf course, and boat-hire facilities at the historic **Studley Park Boathouse**. Go boating on the river, then dock for a spot of Devonshire tea. *Off Studley Park Rd, Kew; (03) 9853 1828; boathouse and kiosk open 9am–5pm daily.*

Bayside

Melbourne's eastern bayside suburbs sprawl down towards the Mornington Peninsula. At the top of the bay is Port Melbourne, once the entry point for many thousands of migrants and now the docking point for Spirit of Tasmania ferries. Port Melbourne's Bay Street has a range of pubs, shops and cafes, as does Clarendon Street, South Melbourne. On Coventry Street is the South Melbourne Market and south of here is Albert Park, the venue for the Australian Formula One Grand Prix each March as well as a spot for jogging and boating. Various other sports are also on offer in Albert Park's Melbourne Sports and Aquatic Centre. Further south again is St Kilda.

St Kilda 132 D6

St Kilda began life as a seaside holiday destination, so separate from the city that on the sandy track that was then St Kilda Road, travellers ran the risk of a run-in with a bushranger.

Fitzroy Street is a long line of shoulder-to-shoulder cafes, restaurants, bars and pubs. Straight ahead is the palm-lined foreshore and the beach, and around the corner is the much-loved **St Kilda Pier** with its historic kiosk that offers great views and quality food.

The path along the foreshore goes from Port Melbourne in the north to beyond Brighton in the south and is often busy with cyclists, rollerbladers and walkers. **The Esplanade Hotel** (The Espy) is an integral part of Melbourne's live-music scene, and **The Palais Theatre**, a grand, French-style theatre, is the venue for concerts. Luna Park *(see next entry)* is next door to The Palais Theatre, and there is an arts and crafts market on The Esplanade every Sunday *(see Markets, p. 137)*.

Just around the corner from The Esplanade are the continental cake shops that have made **Acland Street** famous.

South from St Kilda is a string of swimming beaches, including **Brighton Beach** with its trademark colourful bathing boxes and views of the city.

Rippon Lea Estate 132 E6

The beautiful Rippon Lea Estate gardens are as much of a drawcard as the grand Romanesque Rippon Lea mansion. Visitors can explore the paths, discover hidden water features, and picnic on the lawns. *192 Hotham St, Elsternwick; (03) 9523 6095; gardens open 10am–5pm daily; entry to house by tour, last departure 3.30pm.*

Luna Park 132 E6

While it is now appropriately modernised, Luna Park still feels like a colourful chunk of the early 20th century, when a ride on the Scenic Railway rollercoaster was a big night out. Since it opened in 1912, many things about the park have lived on, including the huge and famous (and much-renovated) face that forms its entrance. Among the traditional rides such as the carousel, Ferris wheel and rollercoaster are the more modern and heart racing Shock Drop, Enterprise and G Force. *Lower Esplanade, St Kilda; (03) 9525 5033; see www.lunapark.com.au for monthly operating hours.*

Western Bayside

Williamstown 132 A6

Travel over the West Gate Bridge and you'll find a landscape of factories and new suburbs. But south of the bridge, on the western arm of Hobsons Bay, is Williamstown – one of Melbourne's true gems. Bobbing up and down in its harbour are a fleet of yachts, and along Nelson Place are restaurants, bars and cafes in old maritime buildings. Williamstown's main beach is around the other side of the bay. For those after an alternative to driving, there is a ferry service to Williamstown from St Kilda and Southgate, with an optional stop at Scienceworks *(see next entry)*. *For information on the ferry, see Walks and tours, p. 142.*

Scienceworks 132 A4

This is the place to come to 'push it, pull it, spin it, bang it', and inadvertently get a grasp on all things science. It is a great place to bring the kids, with interactive exhibitions that make learning about science fun. Getting involved is enouraged! Also in the complex is the **Melbourne Planetarium**, with simulated night skies and 3D adventures through space. A historic pumping station is located on site, and tours are available. *2 Booker St, Spotswood; 13 1102; open 10am–4.30pm daily.*

North-east

Heide Museum of Modern Art 573 J5

Surrounded by beautiful parklands and intriguing sculptures, this complex was formerly home to museum founders John and Sunday Reed. The Reeds emerged as patrons of the arts in the 1930s and 40s, and today the support for modern art continues in the museum, which has changing exhibitions in its three galleries. Sunday Reed's original kitchen garden lives on and supplies produce to Café Vue at Heide, operated by renowned chef Shannon Bennett of Vue de monde. *7 Templestowe Rd, Bulleen; (03) 9850 1500; open 10am–5pm Tues–Sun; restaurant open Thurs–Sat for dinner and Tues–Sun for breakfast and lunch; admission free to gardens and sculpture park.*

day tours

The Dandenongs These scenic hills at the edge of Melbourne's eastern suburbs are a popular daytrip. Native rainforests of mountain ash and giant ferns, cool-climate gardens, the popular steam train Puffing Billy, and galleries, craft shops and cafes, many serving Devonshire tea, are among the many attractions. For more details see p. 148

Yarra Valley and Healesville High-quality pinot noir and sparkling wines are produced across one of Australia's best-known wine areas. Pick up a brochure from the information centre in Healesville and map out your wine-tasting tour. Worthy of its own daytrip is Healesville Sanctuary, featuring around 200 species of native animals in a bushland setting. For more details see p. 148

Mornington Peninsula This holiday centre features fine-food producers, around 40 cool-climate vineyards, historic holiday villages, quiet coastal national parks, 18 golf courses and many attractions for children. For more details see p. 149

Phillip Island The nightly Penguin Parade on Phillip Island is one of Victoria's signature attractions. For the avid wildlife-watcher, seals and koalas are the other stars of the show, though the island also boasts magnificent coastal scenery and great surf breaks. For more details see p. 150

Sovereign Hill Ballarat's award-winning re-creation of a 19th-century goldmining village conjures up the detail and drama of life during one of the nation's most exciting periods of history. You can even stay the night at a new accommodation complex within the village, with full period costume thrown in! For more details see Ballarat, p. 154

Bellarine Peninsula The Bellarine Peninsula separates the waters of Port Phillip from the famously rugged coastline of Victoria's south-west. Beyond the historic buildings, streets and Geelong's waterfront are quaint coastal villages, excellent beaches, golf courses and wineries. For more details see p. 152

Mount Macedon and Hanging Rock Country mansions and superb 19th- and 20th-century European-style gardens sit comfortably in the native bushland. Here you'll find wineries, cafes, nurseries, galleries and the mysteriously beautiful Hanging Rock. For more details see p. 151

Spa country For a few hours of health-giving indulgence, visit the historic spa complex at Hepburn Springs. Explore Daylesford, enjoy a meal at one of the region's excellent eateries, or take a peaceful forest drive. For more details see p. 151

REGIONS
victoria

[OTWAY FLY, SOUTH-WEST COAST]

[SUNSET OVER MOUNT BUFFALO, HIGH COUNTRY]

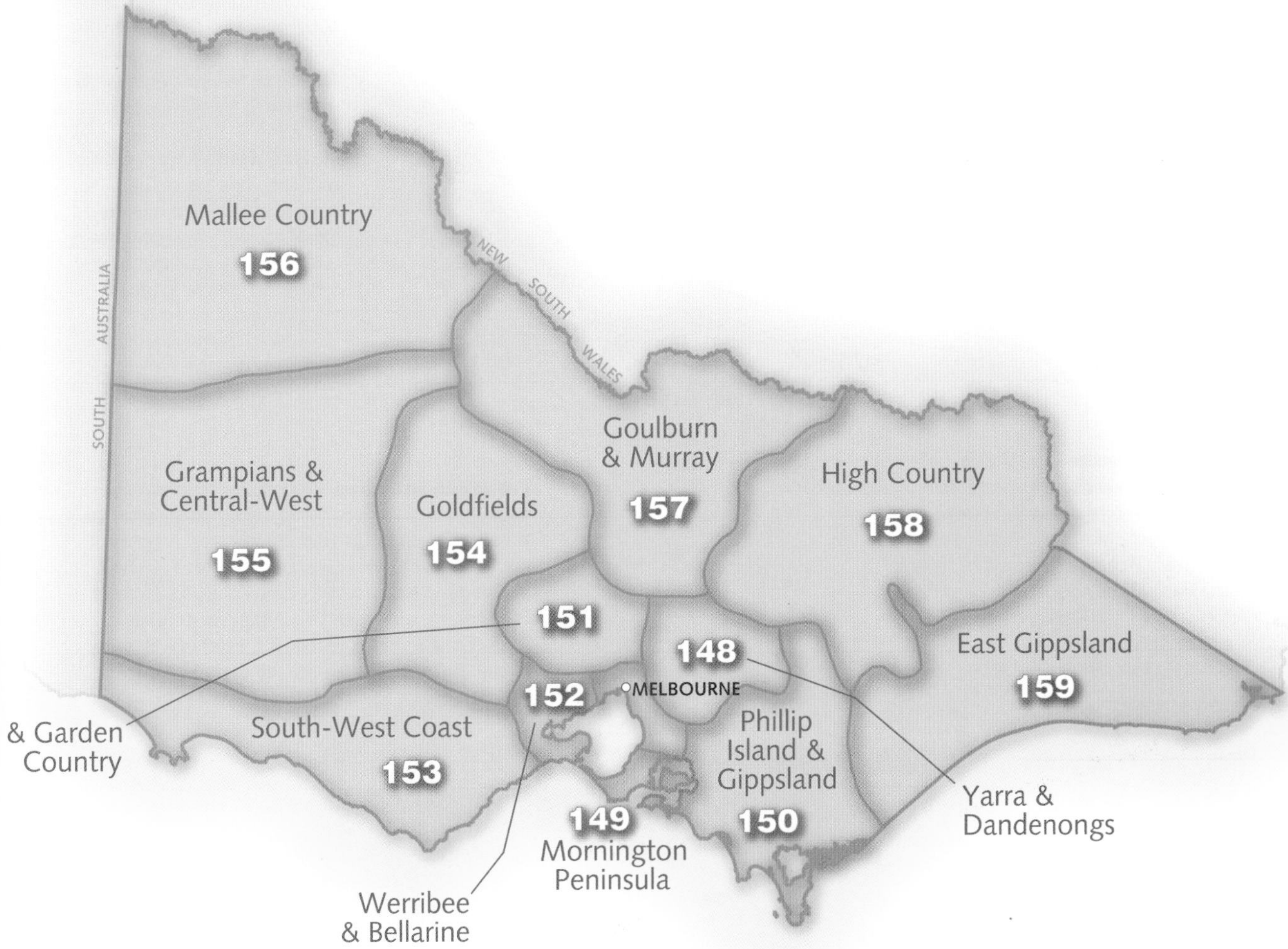

SOUTH AUSTRALIA

NEW SOUTH WALES

Mallee Country 156

Grampians & Central-West 155

Goldfields 154

Goulburn & Murray 157

High Country 158

East Gippsland 159

Spa & Garden Country 151

Yarra & Dandenongs 148

MELBOURNE

Werribee & Bellarine 152

South-West Coast 153

Mornington Peninsula 149

Phillip Island & Gippsland 150

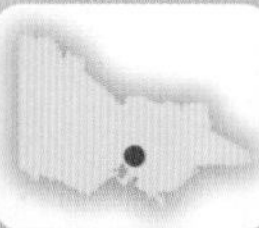

YARRA & DANDENONGS

A daytrip to the Dandenongs, stopping off at cellar doors in the Yarra Valley, combines native forest landscapes with indulgent wining and dining on Melbourne's doorstep.

[DE BORTOLI VINEYARD, YARRA VALLEY]

Healesville Sanctuary Spread across 32 hectares of bushland, this world-renowned native animal sanctuary has over 200 species, most roaming in natural settings. Special features include talks by keepers, a nocturnal viewing area, the bird-of-prey displays and the platypus exhibit.

Yarra Valley wineries The 50 or so wineries of this district produce high-quality chardonnay, cabernet sauvignon and pinot noir. Visit Domaine Chandon, built by French champagne makers Moët & Chandon; the magnificent tasting-room offers fine views across the vine-covered valley. TarraWarra Estate features an impressive art gallery with temporary exhibitions of work by artists such as Brett Whiteley, Sidney Nolan and Arthur Boyd, and De Bortoli is home to a restaurant serving fine Italian-style fare. While touring the region for wines, follow the Yarra Valley Regional Food Trail, a self-guide tour taking in the many gourmet food outlets in the region.

Marysville and Lake Mountain For 100 years, the beautiful subalpine village of Marysville was a much-frequented holiday destination for Melburnians. Providing access to the magnificent 84-metre Steavenson Falls nearby, the town charmed visitors with its cafes, galleries and popular guesthouses. In February 2009, Marysville was destroyed by bushfire, and many people lost their lives. The surviving residents have vowed to rebuild their much-loved town. Nearby Lake Mountain also suffered in the bushfires but it remains a popular tobogganing and cross-country skiing resort in winter, and a great bushwalking destination in summer.

Gardens of the Dandenongs Mountain ash forests and fern gullies frame the historic cool-climate gardens of one of Australia's best-known gardening regions. Many of the private gardens are open daily. Public gardens include the National Rhododendron Gardens and the R. J. Hamer Forest Arboretum (both near Olinda), the William Ricketts Sanctuary (Mount Dandenong) and the Alfred Nicholas Memorial Gardens (Sherbrooke).

Puffing Billy Steam Train This magical little steam train takes you on a 25-kilometre journey through lush forest and tree ferns from Belgrave to Gembrook and back again. A genuine relic of Victoria's early rail days, kids love the two-hour round trip, dangling their legs out the windows. Stopping at Emerald Lake along the way, a lovely tranquil spot ideal for picnics, paddleboating and walks, you can alight here to spend the day before returning on the train in the late afternoon.

TOP EVENTS

FEB	Grape Grazing Festival (throughout Yarra Valley wine region)
MAY	Great Train Race (Emerald)
JULY	Winterfest (Warburton)
AUG–NOV	Rhododendron and Daffodil Festivals (Olinda)
SEPT–OCT	Tesselaar's Tulip Festival (Silvan, near Olinda)

CLIMATE

HEALESVILLE

J	F	M	A	M	J	J	A	S	O	N	D	
26	26	24	19	16	12	12	14	16	19	22	24	MAX °C
11	12	11	9	7	4	4	5	6	8	9	11	MIN °C
58	68	64	91	96	82	87	98	94	106	93	86	RAIN MM
7	7	8	11	14	14	16	17	15	14	12	10	RAIN DAYS

TOURS OF THE FOREST

This area has some of the state's best forest scenery. Bushwalkers, horseriders and cyclists can travel the 38-kilometre Warburton Rail Trail, starting in Lilydale. The Beeches, an area of rainforest near Marysville, has a 5-kilometre stroll through forests of ancient beech and mountain ash. The less energetic can take a forest drive through or around Yarra Ranges National Park, choosing from the Black Spur Drive (between Healesville and Marysville), Acheron Way (from Warburton to Marysville) and the Lady Talbot Forest Drive (a shorter drive in the Marysville region). For something special, visit Mount Donna Buang Rainforest Gallery (near Warburton), which includes a viewing platform and a raised walkway through the rainforest.

→ For more detail see maps 570, 571, 573, 575 & 582. For descriptions of Ⓣ towns see Towns from A–Z (p. 160).

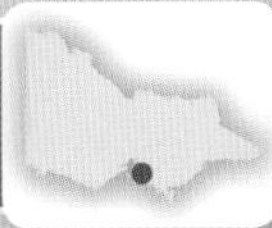

MORNINGTON PENINSULA

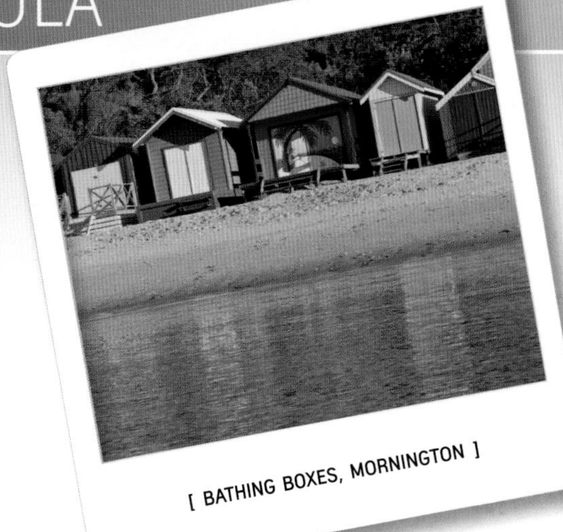

[BATHING BOXES, MORNINGTON]

This region's well-serviced seaside towns are popular during summer, while in winter wineries in the beautiful hinterland have welcoming cellar doors.

Sorrento sojourns The Queenscliff–Sorrento car ferry crosses Port Phillip several times a day, offering visitors a tour of the two peninsulas – Bellarine and Mornington – without a long drive by land. Dolphin cruises, some including a swim with Port Phillip's bottlenose dolphins, operate in summer.

Arthurs Seat Just inland from Dromana, Arthurs Seat offers superb views across Melbourne and the bay. The 300-metre summit is reached by foot or vehicle – at the top are picnic facilities and a restaurant. Nearby is Arthurs Seat Maze with a series of themed gardens and mazes, where the large Maize Maze Festival is held each autumn.

Portsea Near the north-west tip of the peninsula, this village has long been favoured by Melbourne's wealthy. It has large houses (some with private boathouses), elegant hotels, B&Bs, good restaurants and a legendary pub. Further west, don't miss Fort Nepean, once an important defence site, and London Bridge, a rock formation off Portsea Surf Beach.

Mornington Peninsula National Park This park extends along the south-west coast of the peninsula, where the Bass Strait surf pounds windswept beaches and headlands. A 32-kilometre walking track runs from Portsea Surf Beach right down to Cape Schanck, with its historic 1858 lighthouse (offering accommodation).

Red Hill Market The small town of Red Hill, in the scenic undulating landscape of the Mornington Peninsula, is famous for its market held on the first Saturday of each month (from September to May). Affectionately known as the 'grand dame' of Victoria's craft market scene, the market has featured exceptional local produce, clothing and crafts for over 30 years. A visit to this much-lauded community event is a real treat.

TOP EVENTS

JAN	Portsea Swim Classic (Portsea)
FEB	Rye Beach Sand Sculpting Championship (Rye)
FEB–APR	Maize Maze Festival (Arthurs Seat)
MAR	Colour the Sky Kite Festival (Rosebud)
OCT	Pinot Week (throughout wine region)
OCT	Mornington Food and Wine Festival (throughout wine region)

CLIMATE

MORNINGTON

J	F	M	A	M	J	J	A	S	O	N	D	
25	25	23	19	16	14	13	14	16	18	20	23	MAX °C
13	14	13	11	9	7	7	7	8	10	11	12	MIN °C
45	42	50	63	70	71	69	71	72	71	59	54	RAIN MM
7	7	8	11	14	15	15	16	14	13	10	8	RAIN DAYS

WINERIES

The grape came relatively late to the peninsula: the oldest vineyard, Elgee Park, north of Merricks, was established early in the 1970s. Viticulture exploded during the 1980s and 90s; now there are nearly 40 wineries in this cool-climate region, most clustered around Red Hill and many set in postcard-perfect landscapes. The vineyards tend to be small and concentrate on the classic varieties of pinot noir and chardonnay. Stonier, Tucks Ridge and Main Ridge Estate are a few of the names to look out for.

→ For more detail see maps 570, 572–3, 574–5 & 582. For descriptions of towns see Towns from A–Z (p. 160).

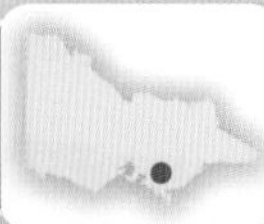

PHILLIP ISLAND & GIPPSLAND

This diverse area features historic towns, snowfields, the remote beauty of Wilsons Promontory and the famous Phillip Island penguins.

[WILSONS PROMONTORY COASTLINE]

Gourmet Deli Trail Central Gippsland is home to producers of trout, venison, cheese, berries, potatoes, herbs and wine. The trail covers the area north and south of Warragul, and annotated maps are available from visitor centres.

Korumburra's Coal Creek Heritage Village A delightful re-creation of a 19th-century coalmining village on the site of the original Coal Creek mine (1890s), this heritage village has over 30 historic buildings. Visitors can take a guided tour through a coalmine, go on a carriage ride or board a small locomotive for a meander through tranquil bushland.

Phillip Island Phillip Island is best known for its little penguins, but visitors can also walk around Cape Woolamai, drive across the bridge to Churchill Island to see a historic homestead, or immerse themselves in motor-racing history at the Phillip Island Grand Prix Circuit Visitor Centre.

Baw Baw National Park The highest part of the park, Mount Baw Baw (via Moe), has ski facilities and, unlike many other slopes, is seldom crowded. The eastern section (via Erica and Walhalla) is popular in summer with bushwalkers, wildflower enthusiasts and campers.

Walhalla The perfectly preserved former goldmining town of Walhalla is set in a steep valley. Take the signposted town walk or a 45-minute ride on the Walhalla Goldfields Railway, or inspect the Long Tunnel Mine.

Wilsons Promontory The Prom is a remote and beautiful landscape supporting diverse native flora and fauna in a near-wilderness. Around 150 kilometres of walking tracks along bays and through bush begin at Tidal River and other points along the access road.

Port Albert Today Port Albert is a quaint fishing village, but as the state's first official port, it was once the gateway to Gippsland and Victoria. About 40 old buildings survive. For anglers, plentiful snapper, whiting, flathead, bream and trevally are found in the protected waters offshore.

TOP EVENTS

FEB–MAR	World Superbike Championships (Phillip Island)
MAR	Jazz Festival (Inverloch)
MAR	Blue Rock Classic (multi-sport race, Moe)
MAR	Fishing Contest (Port Albert)
EASTER	Tarra Festival (Yarram)
OCT	Australian Motorcycle Grand Prix (Phillip Island)

CLIMATE

WARRAGUL

J	F	M	A	M	J	J	A	S	O	N	D	
26	26	24	20	16	13	13	14	16	19	21	23	MAX °C
13	13	12	9	7	5	4	5	6	8	9	11	MIN °C
62	52	69	84	94	93	91	103	104	109	89	80	RAIN MM
8	7	10	13	15	16	16	17	16	15	13	11	RAIN DAYS

PHILLIP ISLAND WILDLIFE

The Penguin Parade on Summerland Beach is a major international tourist attraction. Just after sunset, the world's smallest penguins, at 33 centimetres tall, come home to their burrows in the sand dunes after a day in the sea. To protect the penguins, visitors are restricted to designated viewing areas and cameras are prohibited. Bookings are essential during peak holiday periods. The visitor centre also offers a simulated underwater tour showing the penguins foraging for food and avoiding predators. On the island's western tip you can walk along a cliff-top boardwalk for views across to Seal Rocks, 2 kilometres offshore, where thousands of Australian fur seals live. For an up-close view of the frolicking animals, take a cruise from Cowes or watch them on cameras that you control yourself at the Nobbies Centre. Also visit the Koala Conservation Centre near Cowes, where you can see the delightful marsupials snoozing in the treetops.

→ For more detail see maps 573, 575 & 582–3. For descriptions of Ⓣ towns see Towns from A–Z (p. 160).

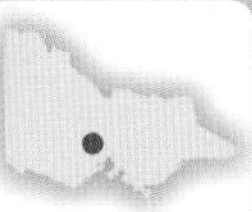

SPA & GARDEN COUNTRY

Grand European-style gardens and historic spa towns feature in this district, which retains the air of a 19th-century hill retreat.

[PARMA VILLA, HEPBURN SPRINGS]

Daylesford and Hepburn Springs The spa complex at Hepburn Springs offers heated mineral spas, flotation tanks, saunas and massages. The adjacent town of Daylesford is an attractive weekend destination with galleries, antique shops, heritage buildings, B&Bs and fantastic restaurants and cafes. The Convent Gallery, a 19th-century former convent, houses local artwork, sculpture and jewellery.

Hanging Rock This impressive rock formation north-east of Woodend was created by the erosion of solidified lava. The spot was the inspiration for Joan Lindsay's novel *Picnic at Hanging Rock*, and a setting for the subsequent film. Walking tracks lead up to a superb view, with glimpses of koalas along the way, and the Discovery Centre tells the story of the site.

Lake House Daylesford's beloved Lake House is regarded as one of regional Victoria's best restaurants, if not the best. Opened in 1984, it is partly responsible for the town's transformation into a gourmet destination. The white, light-filled restaurant overlooks picturesque Lake Daylesford, and the menu might feature local yabbies, free-range pork and dishes that tip their hat to the owner's Russian heritage. Luxury accommodation is also available, as is pampering at the on-site Salus Day Spa.

Organ Pipes National Park Lava flows have created a 20-metre wall of basalt columns in this small park near Sunbury. The 'organ pipes', within a gorge, can be seen close-up via an easy walking trail. A regeneration program is gradually bringing this formerly denuded area back to its native state.

Lerderderg State Park The Lerderderg River has cut a deep gorge through sandstone and slate in this 14 250-hectare park. Rugged ridges enclose much of the river, and there are also some interesting relics of old goldmining days. Explore on foot, or take a scenic drive via O'Briens Road (turn off south of Blackwood).

TOP EVENTS

JAN	New Year's Day and Australia Day races (Hanging Rock, near Woodend)
APR–MAY	Hepburn Springs Swiss–Italian Festival (Hepburn Springs)
JUNE	Winter Arts Festival (Woodend)
SEPT	Daffodil and Arts Festival (Kyneton)
OCT	Macedon Ranges Budburst Food and Wine Festival (throughout wine region)
DEC	Highland Gathering (Daylesford)

CLIMATE

KYNETON

J	F	M	A	M	J	J	A	S	O	N	D	
27	27	24	18	14	11	10	12	15	18	22	25	MAX °C
10	10	8	6	4	2	2	2	3	5	7	9	MIN °C
37	39	47	54	75	90	82	84	74	69	52	50	RAIN MM
5	5	6	9	12	15	16	16	13	11	9	7	RAIN DAYS

COUNTRY GARDENS

Gardens flourish in the volcanic soil and cool, moist climate of the spa country. There are botanic gardens in Daylesford and Malmsbury. Around the beautiful village of Mount Macedon, classic mountain-side gardens surround large houses; check with a visitor centre for their spring and autumn open days. Cope-Williams Winery at Romsey is well known for its English-style garden and cricket green, while at Blackwood the beautiful Garden of St Erth offers 2 hectares of exotic and native species (closed Wednesday and Thursday). For something special, visit the Lavandula Swiss Italian Farm at Shepherds Flat, where lavender sits beside other cottage-style plants.

→ For more detail see maps 572–3, 579, 582, 584 & 593. For descriptions of Ⓣ towns see Towns from A–Z (p. 160).

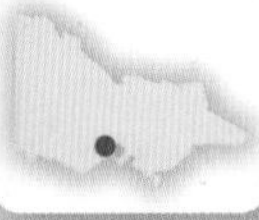

WERRIBEE & BELLARINE

Beyond the regional centre of Geelong, diverse attractions beckon: watersports, wineries, Victorian-era hotels and even an African-style safari park.

[VICTORIA STATE ROSE GARDEN, WERRIBEE PARK]

The You Yangs These granite tors rise abruptly from the Werribee Plains. There is a fairly easy walk (3.2 kilometres return) from the carpark to the top of Flinders Peak. On a clear day the view extends to Mount Macedon, Geelong and the skyscrapers of Melbourne.

Werribee Park Built in Italianate style during the 1870s, Werribee Mansion is now preserved in all its splendour, grandly furnished to re-create the lifestyle of a wealthy family in Victoria's boom years. Take a tour through the house, then visit the Victoria State Rose Garden before heading off on a safari at the Werribee Open Range Zoo.

Geelong wine region Vines were first planted around Geelong in the 1850s, but were uprooted during the 1870s phylloxera outbreak. Today's industry was established in the 1960s and there are now around 20 wineries, including eight on the Bellarine Peninsula. Scotchmans Hill, near Drysdale, has views to the coast and a growing reputation for pinot noir and chardonnay.

Geelong Victoria's second largest city is a bustling port that grew with the state's wool industry. The redeveloped Eastern Beach Waterfront district features cafes, restaurants and colourful sculptures, with the historic Cunningham Pier as a centrepiece. Take a stroll along the promenade or cool off in the restored 1930s sea baths.

See Mornington Peninsula region on page 149

TOP EVENTS

JAN	Skandia Geelong Week (Geelong)
FEB–MAR	Spray Farm Estate summer concert series (Bellarine)
MAR	Highland Gathering (Geelong)
JUNE	National Celtic Folk Festival (Geelong)
OCT	Geelong Cup (horseracing, Geelong)
NOV	Queenscliff Music Festival (Queenscliff)

CLIMATE

GEELONG

J	F	M	A	M	J	J	A	S	O	N	D	
25	26	24	20	17	14	14	15	17	19	21	24	MAX °C
14	14	13	10	8	6	5	6	7	8	10	12	MIN °C
44	38	35	39	47	43	42	47	53	63	52	48	RAIN MM
8	6	9	12	14	16	16	17	16	15	13	10	RAIN DAYS

HOLIDAY HAVENS

Just an hour or so from Melbourne, this district is enormously popular for weekend getaways. Queenscliff offers luxurious accommodation and fine dining, and the Maritime Centre and Museum in town displays the region's historic relationship with the sea. The tiny historic town of Point Lonsdale has great views of the turbulent entrance to Port Phillip. Ocean Grove, the peninsula's biggest town, is popular with retirees, surfers and scubadivers. A bridge across the estuary leads to Barwon Heads, which was the setting for the television series *SeaChange*. This is a small town with an excellent surfing beach and good accommodation and restaurants.

→ For more detail see maps 570, 572, 574, 582 & 593. For descriptions of Ⓣ towns see Towns from A–Z (p. 160).

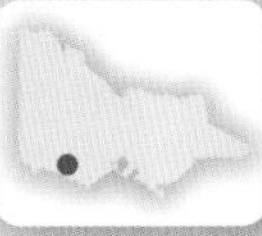

SOUTH-WEST COAST

The breathtaking coastal landscape and charming holiday towns of the Great Ocean Road are world renowned, while inland lies a volcanic landscape.

[THE TWELVE APOSTLES, PORT CAMPBELL NATIONAL PARK]

Mount Eccles National Park This park is at the far edge of the 20 000-year-old volcanic landscape that extends west from Melbourne. There are excellent walking trails, and camping is available.

Warrnambool's southern right whales Each year from June to September, southern right whales can be spotted from Warrnambool's Logans Beach; they return annually to these waters from Antarctica to give birth and raise their young. There's a purpose-built viewing platform at the beach (binoculars or a telescope are recommended), and the local visitor centre releases daily information on whale sightings.

Surf coast Torquay is Victoria's premier surfing town. Factory outlets offer great bargains on surf gear and the local Surfworld Australia museum celebrates the wonders of the wave. Bells and Jan Juc beaches are just around the corner.

Cape Bridgewater A two-hour walk leads to a viewing platform overlooking one of the country's largest Australian fur seal colonies; take a boat trip into the mouth of a cave for a closer look at these charming creatures.

Port Fairy Port Fairy is a superbly preserved whaling port, with historic bluestone buildings lining the main street. In summer it offers lazy beachside holidays, and in winter, a refuge from the cold in one of the many cosy restaurants and B&Bs.

Lorne This popular resort village could be called the capital of the south-west coast, with excellent cafes and restaurants and a lively summertime crowd. Nearby, in Great Otway National Park, beautiful forests and waterfalls provide time-out for bushwalkers and nature lovers.

The Twelve Apostles These spectacular limestone stacks were part of the cliffs until wind and water left them stranded in wild surf off the shore. Preserved in Port Campbell National Park, they are one of Australia's most photographed sights.

The Otways The contrast between rugged coastline and tranquil, temperate rainforest is at its most impressive in Great Otway National Park. In the north, the Otway Fly takes visitors on a suspended walkway to a treetop lookout; in the south, at Cape Otway, a lighthouse offers views over the sea. Another highlight is Melba Gully, where, at dusk, visitors can witness a show of twinkling lights from glow worms.

TOP EVENTS

JAN	Pier to Pub Swim and Mountain to Surf Foot Race (Lorne)
FEB	Go Country Music Festival and Truck Show (Colac)
FEB	Wunta Fiesta (Warrnambool)
MAR	Port Fairy Folk Festival (Port Fairy)
EASTER	Rip Curl Pro (Bells Beach, near Torquay)
DEC–JAN	Falls Festival (Great Otway National Park)

CLIMATE

WARRNAMBOOL

J	F	M	A	M	J	J	A	S	O	N	D	
23	24	22	19	16	14	13	14	15	17	19	21	MAX °C
11	12	11	9	7	6	5	6	7	7	9	10	MIN °C
49	35	53	69	75	101	96	107	109	82	66	63	RAIN MM
11	9	12	15	17	20	20	20	19	16	15	13	RAIN DAYS

SHIPWRECKS

There are about 80 wrecks along the vital yet treacherous south-west-coast shipping route. Victoria's Historic Shipwreck Trail, between Moonlight Head (in Port Campbell National Park) and Port Fairy, marks 25 sites with plaques telling the history of the wrecks. Not to be missed is the evocative *Loch Ard* site, near Port Campbell. On a direct route from London to Melbourne, the *Loch Ard* ran into trouble while negotiating the entrance to Bass Strait; fog and haze prevented the captain from seeing that the ship was only a short distance from the cliffs. In the struggle to change direction the ship hit the cliffs and soon sank – only two people managed to swim ashore to the now well-known Loch Ard Gorge. At Flagstaff Hill Maritime Village in Warrnambool you can see a magnificent statue of a peacock, which was being transported on the *Loch Ard* for display in Melbourne's International Exhibition of 1880 and was washed ashore after the wreck. You can also watch the sound-and-laser spectacular 'Shipwrecked', which brings the tale of the *Loch Ard* to life.

→ For more detail see maps 572, 574, 580 & 592–3. For descriptions of Ⓣ towns see Towns from A–Z (p. 160).

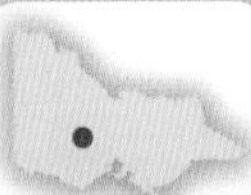

GOLDFIELDS

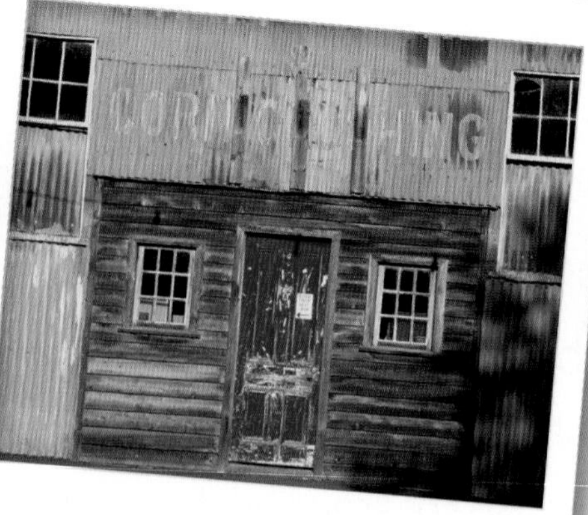

[CORRUGATED-IRON BUILDING, MALDON]

This region is an historic jewel of rural Australia with an intense concentration of Victorian architecture in grand public buildings and parks.

Backblocks of the goldfields Quieter gold towns are tucked away north-west of Bendigo. In Dunolly 126 nuggets were found – see replicas of some of the most impressive finds at the Goldfields Historical and Arts Society. St Arnaud boasts the beautiful Queen Mary Gardens and a number of old pubs and verandah-fronted shops. There are eucalyptus distilleries at Inglewood and Wedderburn.

Maryborough Old Railway Station Grand for the size of the town, the former railway station now houses a tourist complex that includes an antique emporium, a woodworking shop and a restaurant and cafe.

Maldon The 1860s streets of Maldon are shaded by European trees and lined with old buildings of local stone. Declared a Notable Town by the National Trust, Maldon has historic B&Bs and a tourist steam train.

Sovereign Hill One of Victoria's top tourist attractions, Ballarat's Sovereign Hill is a living museum. Blacksmiths, bakers and storekeepers in period dress ply their trades amid the tents, while miners pan for gold.

Bendigo's Chinese sites The restored Joss House on the city's northern outskirts and the Golden Dragon Museum in Bridge Street are reminders of the substantial presence of Chinese immigrants on the goldfields. A ceremonial archway leads from the museum to the Garden of Joy, built in 1996 to represent the Chinese landscape in miniature.

Pall Mall The tree-lined, French-style boulevard of Pall Mall in Bendigo is probably country Australia's most impressive street. Many of its buildings date back to the gold rush and vintage trams continue to operate.

Castlemaine This historic goldmining town's original market building, with its classical Roman facade, now houses visitor information and a gold-diggings interpretive centre. Visit the 1860s Buda Historic Home, with its heritage-listed garden, and the Castlemaine Art Gallery with its impressive collection of local and international art.

Ballarat Begonia Festival Every year in early March for the past 100-odd years, Ballarat's 40-hectare Botanic Gardens have been awash with colour during the city's annual Begonia Festival. Apart from stunning floral displays, the 20 000-plus people who come to the festival enjoy gardening forums, street parades, fireworks, art shows, kids' activities and much more.

TOP EVENTS

JAN	Organs of the Ballarat Goldfields (Ballarat)
MAR	Country Food and Wine Race Meeting (Avoca)
MAR	Ballarat Begonia Festival (Ballarat)
EASTER	Easter Festival (Bendigo)
OCT–NOV	Folk Festival (Maldon)
NOV	Festival of Gardens (Castlemaine)

CLIMATE

BENDIGO

J	F	M	A	M	J	J	A	S	O	N	D	
29	29	26	21	16	13	12	14	16	20	24	26	MAX °C
14	15	13	9	7	4	3	5	6	8	11	13	MIN °C
33	32	36	41	55	61	56	59	54	52	37	33	RAIN MM
5	4	5	7	10	12	13	13	11	10	7	6	RAIN DAYS

GOLD-RUSH HISTORY

Sovereign Hill in Ballarat is one of the country's best historic theme parks. It offers a complete re-creation of life on the 1850s goldfields. The nearby Gold Museum, part of the Sovereign Hill complex, features displays of gold nuggets and coins, and changing exhibits on the history of gold. The Eureka Stockade Centre offers interpretive displays on Australia's only armed insurrection, which took place in 1854. See the original Eureka Flag at the Ballarat Fine Art Gallery, which also houses an excellent collection of work by artists such as Tom Roberts, Sidney Nolan, Russell Drysdale and Fred Williams. In Eureka Street is the tiny Montrose Cottage (1856), an ex-miner's house furnished in the style of the period; here a museum display movingly recalls the lives and contribution of women in the gold-rush era. In Bendigo, the Central Deborah Gold Mine offers tours 80 metres down a reef mine, and excellent displays on goldmining techniques.

→ For more detail see maps 572, 579, 591 & 593. For descriptions of towns see Towns from A–Z (p. 160).

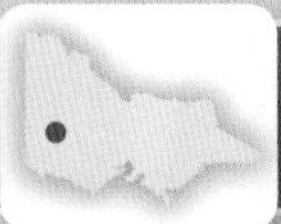

GRAMPIANS & CENTRAL WEST

Ancient mountains, semi-arid plains and farming landscapes typify this area, while Aboriginal rock-art sites are remnants of life before white settlement.

[VIEW FROM MOUNT ARAPILES]

Little Desert National Park During spring, more than 600 varieties of wildflowers and over 40 types of ground orchids flourish in this 132 647-hectare park. With nearly 600 kilometres of tracks, it is ideal for four-wheel driving, but perhaps the best way to appreciate the colourful spring display is on foot: for keen hikers, there is the 84-kilometre Desert Discovery Walk.

The olive groves of Laharum Mount Zero Olives at Laharum is the largest olive plantation in the Southern Hemisphere, with 55 000 trees on 730 hectares. The first trees were planted in 1943, after World War II stopped olive oil imports. You can buy oil, vinegar and lentils, and stay overnight.

Mount Arapiles Mount Arapiles is regarded as Australia's best rock climbing venue, attracting interstate and international enthusiasts with its 2000 rock climbing routes marked out across 365 metres of sandstone cliffs.

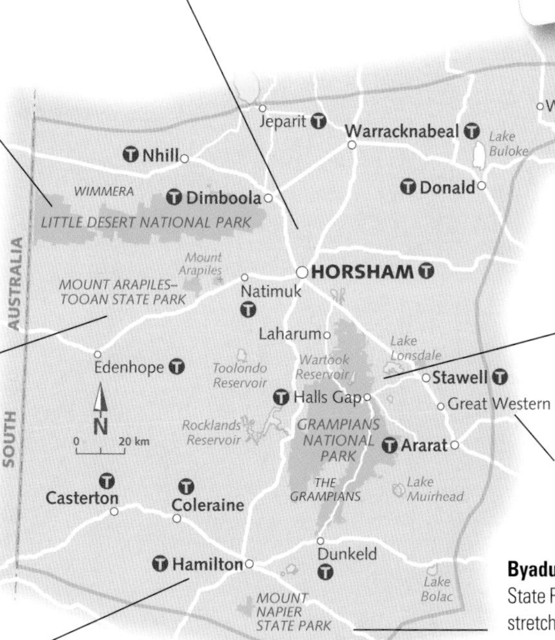

Grampians day drive and balloon flights From Halls Gap, drive to Boroka Lookout, Reed Lookout and MacKenzie Falls. Break for lunch at Zumsteins, home to a large kangaroo population. Dawn hot-air balloon flights over the Grampians leave from Stawell.

Wines of Great Western Grapevines were first planted at Seppelt's Great Western vineyards in 1865. Today the winery is best known for its red and white sparkling wines. Other wineries in the area include Best's and Garden Gully.

Hamilton Hamilton is the commercial hub of the wool-rich Western District. Gracious houses and churches on its tree-lined streets testify to over a century of prosperity. Close to town are historic homesteads in magnificent gardens.

Byaduk Caves These caves, located in Mount Napier State Park, are part of a giant, 24-kilometre lava flow stretching to Mt Eccles in the south-west, evidence of the volcanic activity that shaped the region's landscape. The caves, one of which is open, are a wonderland of ropey lava, columns, stalactites and stalagmites.

TOP EVENTS

JAN	Picnic Races (Great Western)
FEB	Grampians Jazz Festival (Halls Gap)
MAR	Jailhouse Rock Festival (Ararat)
EASTER	Stawell Easter Gift (professional footrace, Stawell)
MAY	Grampians Grape Escape (Halls Gap)
JUNE–JULY	Australian Kelpie Muster and Kelpie Working Dog Auction (Casterton)

CLIMATE

STAWELL

J	F	M	A	M	J	J	A	S	O	N	D	
27	28	25	20	16	13	12	14	16	19	22	26	MAX °C
13	13	12	9	7	5	4	5	6	8	9	11	MIN °C
36	28	33	41	57	60	67	67	64	58	36	29	RAIN MM
6	5	7	9	12	16	18	18	14	12	9	7	RAIN DAYS

ABORIGINAL CULTURE

The Djab Wurrung and Jardwadjali people shared the territory they called Gariwerd for at least 5000 years before European settlement, although some evidence suggests up to 30 000 years of habitation. Brambuk – The National Park and Cultural Centre near Halls Gap, run by five Koorie communities, is an excellent first stop for information about the region's Indigenous heritage. There is a ceremonial ground for everything from dance performances to boomerang-throwing, while bush tucker–inspired meals are served in the cafe. The region contains 100 recorded rock-art sites, representing more than 80 per cent of all sites in Victoria. A Brambuk-guided tour of some of the sites (most are in Grampians National Park) is probably the most rewarding way to experience the meaning and nature of the art. Notable sites include Gulgurn Manja, featuring over 190 kangaroo, emu and handprint motifs, and Ngamadidj, a site decorated with 16 figures painted in white clay. Bunjil's Shelter is just outside the park near Stawell, and is the only site in the area where more than one colour is used and a known figure is represented. Bunjil was a creator spirit from the Dreaming, responsible for the people, the land and the law.

→ For more detail see maps 590–1 & 592–3. For descriptions of T towns see Towns from A–Z (p. 160).

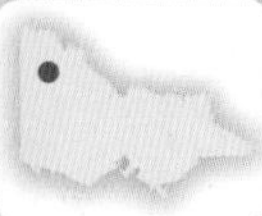

MALLEE COUNTRY

The Murray River is the lifeblood of this region, allowing the cultivation of fruit crops and the development of various riverside settlements.

[EMU AND CHICKS, HATTAH–KULKYNE NATIONAL PARK]

Mildura With its museums and galleries, excellent dining and surrounding wineries and orchards, Mildura is like a colourful Mediterranean oasis. Go to the zoo, visit the Mildura Arts Complex and the Rio Vista museum (once the home of William Chaffey), or book a day tour with Indigenous guides to Mungo National Park in New South Wales.

Stefano's The star of Mildura is Stefano's, the kind of place you dine at one night and love so much you go again the next. This may have something to do with the restaurant's beautiful setting in a historic hotel cellar – and celebrity chef Stefano de Pieri's food. The menu is degustation, with five different courses nightly, each focusing on the best local and seasonal produce; you can expect handmade pastas and dishes featuring the finest quality beef, pork and lamb.

River district wines The wine-producing areas of Mildura and Swan Hill are an unrecognised heartland of the Australian wine industry, producing 37 per cent of the total output – most for the bulk market, although prestige production is rising. The dozen or so wineries include the large Lindemans Karadoc estate, with cellar-door tastings and sales.

Swan Hill Pioneer Settlement This 7-hectare park offers a lively experience of river-port life in early Australia. Wander through the 19th-century-style streets, complete with staff in period dress. Ride on a paddlesteamer or book for the popular Sound and Light Tour. Also in the park is an Aboriginal canoe tree.

Wyperfeld National Park A park of brilliant sunsets, huge open spaces and spring wildflowers, Wyperfeld is explored via walks from Wonga Campground and Information Centre, 50 kilometres west of Hopetoun. The park is home to the endangered mallee fowl, a turkey-size bird that makes nesting mounds up to 5 metres across.

TOP EVENTS

MAR	Arts Festival (Mildura)
AUG–SEPT	Great Australian Vanilla Slice Triumph (Ouyen)
SEPT	Vintage Tractor Pull (Mildura)
SEPT–OCT	Country Music Festival (Mildura)

CLIMATE

MILDURA

J	F	M	A	M	J	J	A	S	O	N	D	
32	32	28	23	19	16	15	17	20	24	27	30	MAX °C
17	16	14	10	8	5	4	5	7	10	12	15	MIN °C
21	22	19	19	27	23	26	28	29	31	24	23	RAIN MM
4	3	4	4	7	8	9	9	8	7	6	4	RAIN DAYS

MURRAY AND MALLEE WILDLIFE

This landscape of rivers and plains is home to abundant wildlife. Most notable is the birdlife: spoonbills, herons, eagles, mallee fowl, harriers and kites are to be found in the parks and on the roadsides and riverbanks. Hattah–Kulkyne National Park protects around 200 bird species as well as the red kangaroo. Murray–Sunset National Park – true desert country in parts – includes riverine plains. It supports an array of native fauna, including mallee fowl and the rare black-eared miner. To see the kangaroos and birdlife of Wyperfeld National Park, follow the Brambruk Nature Trail.

→ For more detail see maps 588–9 & 591. For descriptions of T towns see Towns from A–Z (p. 160).

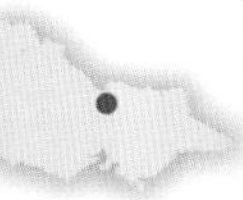

GOULBURN & MURRAY

The Goulburn Valley contains Victoria's richest farm land, while the Murray has historical attractions, particularly the paddlesteamers of Echuca.

[PADDLESTEAMER, MURRAY RIVER]

Echuca The historic port of Echuca, with its impressive red-gum wharf, recalls the late 19th century, when the Murray carried wool and other goods from farms and stations. A number of beautifully restored paddlesteamers are moored here, including some offering cruises up the river.

Barmah State Park and State Forest These neighbouring areas form the state's biggest river red gum forest. This is a popular area for bushwalking, canoeing and camping. Walking trails take in various Aboriginal sites, and the Dharnya Centre interprets the culture of the local Yorta Yorta people.

Cobram This lovely fruit-growing town is surrounded by peach, nectarine, pear and orange orchards. It also offers access to a number of wide, sandy beaches on the Murray, perfect for swimming, fishing, picnicking and watersports. Camping facilities are available, and across the river in New South Wales is the renowned 36-hole Cobram Barooga golf course.

Tahbilk winery Tahbilk was established in 1860 on the sandy loam of the Goulburn River north of Nagambie. One of Australia's most beautiful wine properties, it has a National Trust–classified cellar, which – along with other buildings – is a timepiece of the early Australian wine industry. Tahbilk wines have won over 1000 awards.

Seymour's Royal Hotel Dating from 1848, this classic old pub on Emily Street in Seymour is a great spot to stop for a beer and counter meal. If the hotel looks vaguely familiar, it is because you might have seen it before in the famous, haunting painting *The Cricketers* (1948), by one of Australia's most notable artists, Russell Drysdale.

VICTORIA

TOP EVENTS

JAN	Peaches and Cream Festival (Cobram, odd-numbered years)
FEB	Southern 80 Ski Race (Torrumbarry Weir to Echuca)
FEB	Riverboats, Jazz, Food and Wine Festival (Echuca)
MAR	Bridge to Bridge Swim (Cohuna)
APR	Barmah Muster (Barmah State Forest)
JUNE	Steam, Horse and Vintage Rally (Echuca)

CLIMATE

ECHUCA

J	F	M	A	M	J	J	A	S	O	N	D	
31	31	27	22	18	14	13	15	18	22	26	29	MAX °C
15	15	13	9	7	5	4	5	6	9	11	14	MIN °C
27	27	31	33	42	43	41	43	40	43	32	28	RAIN MM
4	4	5	6	9	10	11	11	10	9	6	5	RAIN DAYS

FOOD AND WINE

Irrigation has transformed the once-dusty Goulburn Valley into the fruit bowl of Victoria. Orchards, market gardens and farms supply the canneries in Shepparton, which are among the largest in the Southern Hemisphere. The valley is dotted with outlets for venison, poultry, smoked trout, berries, organic vegetables, honey, jams, preserves, fruit juices, mustards, pickles, vinegar and liquored truffles. The region's nine wineries make reliable reds and distinctive whites. Near Nagambie, Tahbilk Winery and Mitchelton Wines are both must-visit cellar doors, not only for their superb wines but also for their scenic surrounds. Tahbilk is set next to a wetlands and wildlife reserve, while Mitchelton offers great views of the region from its famous tower. At nearby Avenel, Plunkett Fowles is another excellent cellar door with various wines available for tasting, including their 490m range inspired by the grapes grown at high altitude in the Strathbogie Ranges.

→ For more detail see maps 584 & 586. For descriptions of towns see Towns from A–Z (p. 160).

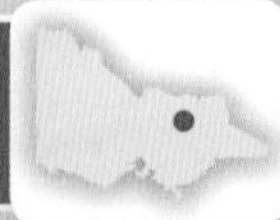

HIGH COUNTRY

[ALPINE LAKE, BOGONG HIGH PLAINS]

The Victorian Alps offer challenging skiing; beyond the foothills lie charming old gold towns and Rutherglen wineries.

Wines of Rutherglen The vines on the alluvial flats in a shallow loop of the Murray produce some of the world's great fortified wines. The region is known for tokays and muscats, big reds and – more recently – lighter reds such as gamay. There are over a dozen wineries near Rutherglen; look out for All Saints, with its historic building, and Pfeiffer, Chambers, Gehrig Estate and Campbells.

Kelly country A giant effigy of Ned Kelly greets visitors to Glenrowan. After killing three local policemen in 1878, the Kelly Gang hid for two years in the Warby Range, raiding nearby towns. Visit the Ned Kelly Memorial Museum and Homestead in Gladstone Street.

Historic Beechworth The National Trust has classified over 30 buildings in what is now one of Australia's best-preserved gold-rush towns. Dine in a stately former bank, visit the powder magazine, wander through an evocative cemetery for Chinese goldminers, and sample the delectable goods from Beechworth Bakery.

Mount Buffalo National Park This 31 000-hectare national park is the state's oldest, declared in 1898. A plateau of boulders and tors includes The Horn, the park's highest point and a great place for views at sunrise. Walking tracks are set among streams, waterfalls, stunning wildflowers, and snow gum and mountain ash forest. There is summer camping, swimming and canoeing at Lake Catani. In winter, the Mount Buffalo ski area is popular with families.

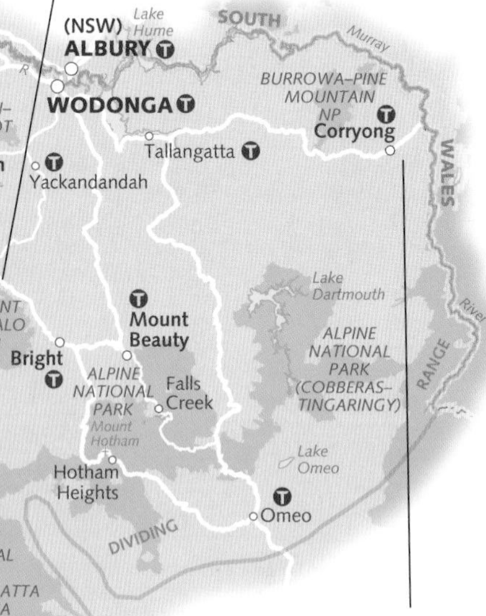

Lake Eildon Created by damming the Goulburn River in the 1950s, this lake is popular with watersports enthusiasts, anglers and houseboat holiday-makers. The surrounding Lake Eildon National Park offers bushwalking, camping and four-wheel-drive tracks through the foothills of the Victorian Alps.

Horseback riding Jack Riley, believed to be the original Man from Snowy River, is buried in Corryong's pretty hillside cemetery. Each year, at an annual festival honouring his memory, horseriding competitions are held to find his modern-day equivalent. A gentle trail ride through this stunning countryside is a great way to appreciate this High Country legend.

TOP EVENTS

MAR	The High Country Autumn Festival (Mansfield)
MAR	Tastes of Rutherglen (district wineries)
APR	The Man from Snowy River Bush Festival (Corryong)
APR–MAY	Autumn Festival (Bright)
OCT–NOV	Festival of Jazz (Wangaratta)
NOV	Celtic Festival (Beechworth)

CLIMATE

MOUNT BULLER

J	F	M	A	M	J	J	A	S	O	N	D	
16	17	14	10	7	3	2	2	5	8	12	14	MAX °C
8	8	6	4	1	–1	–3	–2	–1	1	4	6	MIN °C
83	66	79	136	161	160	190	185	156	185	155	124	RAIN MM
7	5	8	12	14	14	15	16	14	16	13	11	RAIN DAYS

SKI COUNTRY

Victoria's ski resorts are within easy reach of Melbourne. They include, in order of their distance from the city: Mount Buller (via Mansfield), Mount Buffalo (via Myrtleford), Mount Hotham (via Bright) and Falls Creek (via Mount Beauty). The ski season starts officially on the Queen's Birthday weekend, early in June, and ends on the first weekend in October. Snowsport conditions, however, depend on the weather. All resorts offer a range of skiing, from protected runs for beginners to cross-country ski trails. Mount Hotham, known as the 'powder snow capital' of Australia, has the most challenging runs for experienced downhill skiers and snowboarders. After the snow melts a range of summer activities come in to play. The adventurous can try mountain-bike riding, tandem paragliding, abseiling or caving. Mountain lakes and streams offer trout fishing, swimming, sailing and canoeing. Trails across the mountains, ablaze with wildflowers in summer, can be explored on horseback or on foot. The less energetic can just breathe the crystalline air and gaze across the hazy blue ridges.

→ For more detail see maps 573, 582–3, 584–5 & 586–7. For descriptions of towns see Towns from A–Z (p. 160).

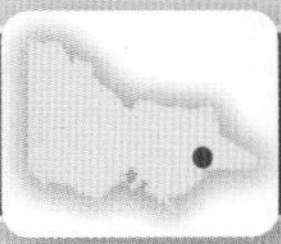

EAST GIPPSLAND

Natural wonders and outdoor activities abound in this region, which is blessed with some magnificent national parks and inland waterways.

[POINT HICKS, CROAJINGOLONG NATIONAL PARK]

Snowy River National Park The much-celebrated Snowy River begins as a trickle near Mount Kosciuszko and passes through wild limestone gorges and forest country before reaching a coastal lagoon. McKillops Bridge (via Buchan) is a beautiful area with camping and barbecue facilities, swimming spots and some good short walks.

Buchan Caves Reserve This spectacular limestone cave system comprises more than 350 caves, of which the Royal and Fairy caves are the most accessible. Fairy Cave is over 400 metres long, with impressive and elaborate stalactites and stalagmites, while Royal Cave impresses with pretty calcite-rimmed pools. Tours of both caves are run daily. After exploring the caves, you can cool off in a spring-fed swimming pool.

Mallacoota Surrounded by the remote ocean beaches, estuarine waterways and unspoiled bush of Croajingolong National Park, this old-fashioned resort offers one of the best nature-based holidays in the state, with excellent fishing, walking, boating and swimming. Hire a boat in town and explore Mallacoota Inlet.

Lakes Entrance Lakes Entrance, at the head of the lakes, is a great base for fishing and boating, and offers accommodation to suit all budgets. For a holiday afloat, book a self-drive cruiser from nearby Metung. There are several wineries in the area – Wyanga Park Winery offers a lakes cruise from town to its cellar door and restaurant.

Large lakes and a long beach The Gippsland Lakes are fed by five major rivers and contained on the coastal side by Ninety Mile Beach. At the system's centre, The Lakes National Park offers birdwatching, walking, swimming and camping. Access is via boat from Paynesville or road and foot from Loch Sport.

TOP EVENTS

JAN	Foothills Festival (Buchan)
FEB	Jazz Festival (Paynesville)
FEB	Blues and Arts Festival (Bruthen)
MAR	Marlay Point–Paynesville Overnight Yacht Race
MAR	Line Dancing Championships (Bairnsdale)
EASTER	Rodeo (Buchan)

CLIMATE

LAKES ENTRANCE

J	F	M	A	M	J	J	A	S	O	N	D	
24	24	22	20	17	15	15	16	17	19	20	22	MAX °C
15	15	13	11	9	7	6	6	8	9	11	13	MIN °C
56	41	56	63	67	65	54	51	59	64	71	70	RAIN MM
9	7	10	10	11	13	12	14	13	13	12	11	RAIN DAYS

FISHING

Fishing is a huge drawcard in East Gippsland. Fish for trout in the mountain streams and rivers – such as the Delegate River – or head for the coast. The lakes, rivers and inlets around Paynesville, Marlo, Bemm River and Mallacoota are great for bream, trevally and flathead. Boat angling is the best choice here, although land-based angling also yields results. Ninety Mile Beach and the remote beaches of The Lakes National Park provide some of the best surf-fishing in Victoria, with salmon, tailor and flathead among the prospects.

→ For more detail see maps 565, 581 & 583. For descriptions of T towns see Towns from A–Z (p. 160).

TOWNS A–Z

victoria

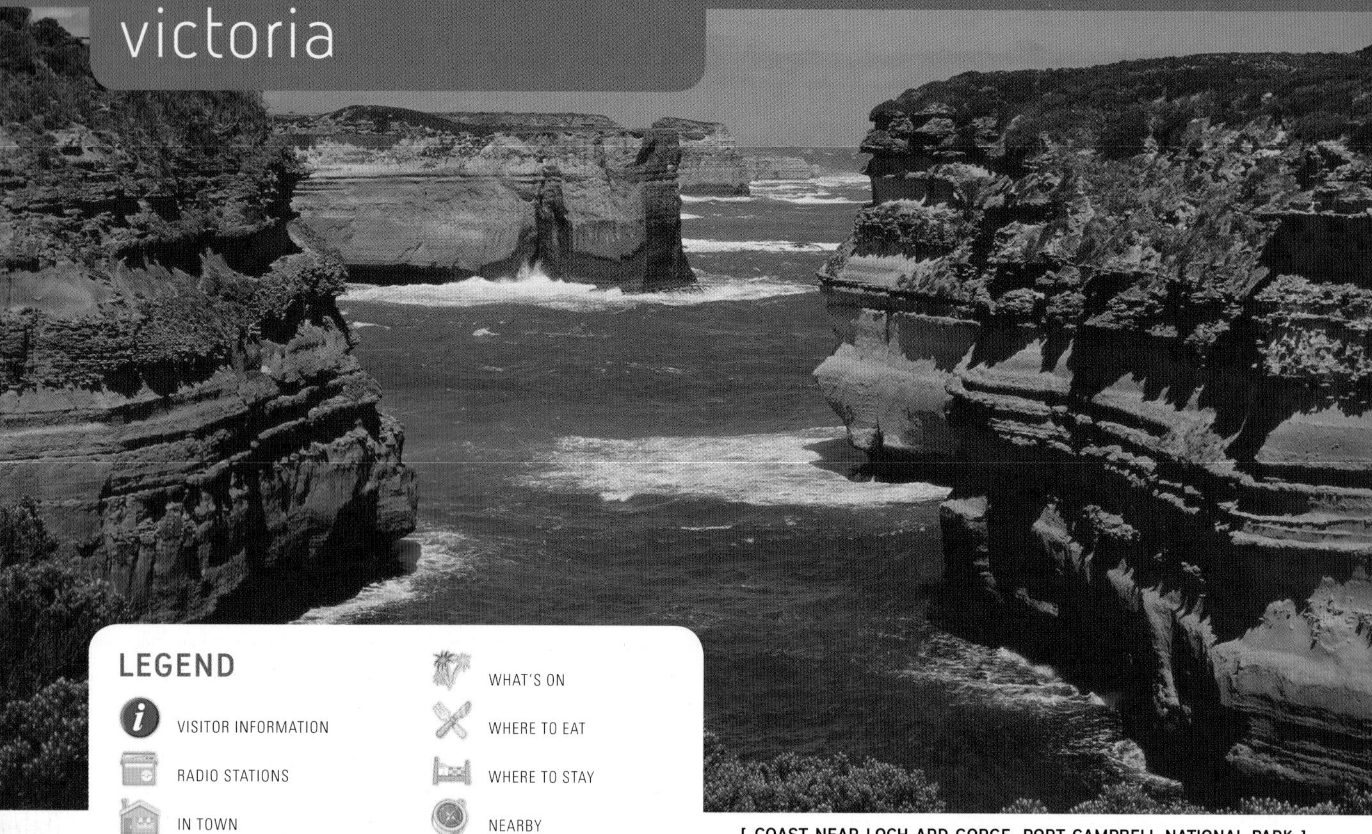

[COAST NEAR LOCH ARD GORGE, PORT CAMPBELL NATIONAL PARK]

LEGEND

VISITOR INFORMATION

RADIO STATIONS

IN TOWN

WHAT'S ON

WHERE TO EAT

WHERE TO STAY

NEARBY

Food and accommodation listings in town are ordered alphabetically with places nearby listed at the end

Alexandra

Pop. 2142
Map ref. 571 F1 | 573 N1 | 582 F1 | 584 F12 | 586 B10

36 Grant St; (03) 5772 1100 or 1800 652 298; www.alexandratourism.com

102.9 FM ABC Local Radio, 106.9 UGFM Upper Goulburn Community Radio

Alexandra was apparently named after Alexandra, Princess of Wales, although, coincidentally, three men named Alexander discovered gold here in 1866. Situated in the foothills of the Great Dividing Range, Alexandra is supported primarily by agriculture. Nearby, the Goulburn River is an important trout fishery.

Alexandra Timber Tramways: museum housed in the original railway station that offers an insight into the timber industry around Alexandra; open 2nd Sun each month; Station St. ***Art and craft galleries:*** many outlets around town displaying and selling local art, pottery and glassware.

Bush market: Perkins St; 3rd Sat each month (excluding winter). ***Picnic Races:*** Jan, Mar, Oct and Nov. ***Truck, Ute and Rod Show:*** June. ***Open Gardens Weekend:*** Oct. ***Rose Festival:*** Nov.

Mia Mia Tea Rooms: good country fare; 79 Grant St; (03) 5772 2122. ***Stonelea Country Estate:*** excellent food in a country-house setting; Connellys Creek Rd, Acheron; (03) 5772 2222. ***Tea Rooms at Yarck:*** regional Italian; 6585 Maroondah Hwy, Yarck; (03) 5773 4233.

Athlone Country Cottages: 3 well-equipped cottages; 266 UT Creek Rd; (03) 5772 2992. ***Mittagong Homestead and Cottages:*** 3 well-appointed cottages; 462 Spring Creek Rd; (03) 5772 2250. ***Stonelea Country Estate:*** cottage and homestead accommodation; Connellys Creek Rd, Acheron; (03) 5772 2222.

McKenzie Nature Reserve: in virgin bushland, with orchids and wildflowers during winter and spring. ***Self-guide tourist drives:*** the Skyline Rd from Alexandra to Eildon features lookouts along the way; information from visitor centre. ***Taggerty:*** home to the Willowbank Gallery and a bush market; open 4th Sat each month; 18 km S. ***Trout fishing:*** in the Goulburn, Acheron and Rubicon rivers. ***Lake Eildon National Park:*** excellent walking trails in the north-west section of the park; *for more details see Eildon*. ***Bonnie Doon:*** a good base for exploring the lake region. Activities include trail-riding, bushwalking, watersports and scenic drives; 37 km NE near Lake Eildon.

See also HIGH COUNTRY, **p. 158**

Anglesea

Pop. 2292
Map ref. 572 E10 | 580 F3 | 593 N9

Off Great Ocean Rd; or ring Torquay Information Centre, (03) 5261 4219 or 1300 614 219; www.visitsurfcoast.com

94.7 The Pulse FM, 774 AM ABC Local Radio

A pretty and sheltered part of the surf coast, Anglesea is one of the smaller holiday hamlets along the Great Ocean Road.

The main beaches are patrolled from Christmas through to Easter, making it a favourite destination for both swimmers and beginner surfers.

Coogoorah Reserve Set on the Anglesea River, the name of this park means 'swampy reed creek'. Coogoorah was established after the 1983 Ash Wednesday fires and now features a network of boardwalks weaving through the distinctive wetland vegetation. Keep an eye out for local birdlife, including the peregrine falcon.

Anglesea Golf Course: golfers share the greens with kangaroos; Golf Links Rd. ***Melaleuca Gallery:*** open daily 11am–5.30pm; Great Ocean Rd. ***Viewing platform:*** overlooks open-cut brown-coal mine and power station; behind town in Coalmine Rd. ***Paddleboats:*** for hire on the banks of the Anglesea River.

Markets: local crafts and produce, held over summer, Easter and Melbourne Cup weekend; by the Anglesea River. ***Rock to Ramp Swim:*** Jan. ***Anglesea Art Show:*** June. ***ANGAIR Wildflower and Art Show:*** Sept.

Locanda Del Mare: a slice of Italy; 5 Diggers Pde; (03) 5263 2904. ***Pete's Place:*** cosy seafood cafe; 113 Great Ocean Rd; (03) 5263 2500. ***A La Grecque:*** modern Greek cuisine; 60 Great Ocean Rd, Aireys Inlet; (03) 5289 6922.

Aireys-on-Aireys: 3 luxury villas; 19 Aireys St, Aireys Inlet; (03) 5289 6844.

Aireys Inlet This pretty little town is overlooked by a lighthouse built to guide passing ships along this treacherous coastline. Painkalac Creek flows out to the ocean here, creating the inlet of the town's name and a safe swimming spot. Horseriding and fishing are favourite activities along the sheltered beaches. South of town is the Great Ocean Road Memorial Arch, built in 1939. 11 km sw.

J. E. Loveridge Lookout: 1 km w. ***Pt Roadknight Beach:*** a shallow, protected beach, popular with families; 2 km sw. ***Ironbark Basin Reserve:*** features ocean views, local birdlife and good bushwalking. The Pt Addis Koorie Cultural Walk leads through the park, highlighting sites of Indigenous significance; 7 km NW, off Pt Addis Rd. ***Great Otway National Park:*** the park begins near Anglesea and stretches to the south and west. The section near Anglesea features unique heathland flora and good walking trails; access via Aireys Inlet, 11 km sw. *For more details see Apollo Bay.* ***Surf Coast Walk:*** 30 km from Torquay to Moggs Creek (south of Aireys Inlet). The track passes through Anglesea. ***Surf schools:*** learn to surf on one of the beginner courses available at nearby beaches; details from visitor centre.

See also **SOUTH-WEST COAST, p. 153**

Apollo Bay

Pop. 1373
Map ref. 572 B12, | 580 C5 | 593 L10

Great Ocean Road Visitor Information Centre, 100 Great Ocean Rd; 1300 689 297; www.visitgreatoceanroad.org.au

89.5 FM ABC Local Radio, 104.7 Otway FM

Named after a local schooner, Apollo Bay has become the resting place of many shipwrecks, yet it maintains an appeal for all lovers of the ocean. The town is situated near Otway National Park with a wonderful contrast between rugged coastline and tranquil green hills. The seaside town is popular with fishing enthusiasts and, like many other coastal towns, its population swells significantly over summer as visitors flock here for the holidays.

Old Cable Station Museum: features artefacts from Australia's telecommunications history and informative displays exploring the history of the region; open 2–5pm weekends and school and public holidays; Great Ocean Rd. ***Bass Strait Shell Museum:*** holds an array of shells and provides many facts about the marine life along the south-west coast; Noel St. ***Great Ocean Walk:*** enjoy stunning views on this 91 km walk between Apollo Bay and Glenample Homestead, near the Twelve Apostles. Walkers must register to use campgrounds en route. Further information available at www.greatoceanwalk.com.au

Foreshore Market: each Sat. ***Apollo Bay Music Festival:*** Apr.

Bay Leaf Cafe: substantial cafe food; 131 Great Ocean Rd; (03) 5237 6470. ***Chris's Beacon Point Restaurant & Villas:*** Mediterranean/Greek; 280 Skenes Creek Rd; (03) 5237 6411. ***Great Ocean Road Hotel:*** modern Australian and bar menu; 29 Great Ocean Rd; (03) 5237 6240. ***La Bimba Restaurant & Cafe:*** modern Mediterranean; 125 Great Ocean Rd; (03) 5237 7411.

Captain's at the Bay: stylish accommodation options; 21 Pascoe St; (03) 5237 6771. ***Chocolate Gannets:*** villas with views; 6180 Great Ocean Rd; 1300 500 139. ***Chris's Beacon Point Restaurant & Villas:*** villas and studios with views; 280 Skenes Creek Rd; (03) 5237 6411. ***Point of View:*** villas for couples; 165 Tuxion Rd; 0427 376 377.

Great Otway National Park The park includes some of the most rugged coastline in Victoria, particularly around Cape Otway and the stretch of coast towards Princetown. It is an ideal location for a bushwalking adventure taking in sights through the park to the sea, from the scenic Elliot River down to adjacent Shelly Beach. Maits Rest Rainforest Trail is a great little walk for all levels. Many species of wildlife inhabit the park, including koalas and the rare tiger quoll. Also look out for the historic Cape Otway Lighthouse, built in 1848. The Great Ocean Rd, west of Apollo Bay, passes through the park. Contact Parks Victoria on 13 1963; 13 km sw.

Otway Fly The consistently popular Otway Fly is a steel-trussed walkway perched high among the temperate rainforest treetops of the Otway Ranges. The 'Fly' is 25 m high and stretches for 600 m. It is accessible to all ages and levels of mobility. Get a bird's-eye view of ancient myrtle beech, blackwood and mountain ash while looking out for a variety of wildlife, including pygmy possums and the raucous yellow-tailed black cockatoo. A springboard bridge takes you over Youngs Creek, where you might spot a shy platypus. If you want an additional rush, there is a 'zip line tour' that allows you to fly on 30 m-high cables between cloud staions. Inquiries on 1800 300 477; 62 km NW via Lavers Hill.

Marriners Lookout: with views across Skenes Creek and Apollo Bay; 1.5 km NW. ***Barham Paradise Scenic Reserve:*** in the Barham River Valley, it is home to a variety of distinctive moisture-loving trees and ferns; 7 km NW. ***Forests and Waterfall Drive:*** 109 km loop drive featuring spectacular Otway Ranges scenery. Waterfalls include Beauchamp, Triplet and Houptoun falls. Drive starts at Apollo Bay, travels west to Lavers Hill and around to Skenes Creek. Map from visitor centre. ***Charter flights:*** views of the Twelve Apostles, the Bay of Islands and the 'Shipwreck Coast'; details from visitor centre.

See also **SOUTH-WEST COAST, p. 153**

Ararat

Pop. 7170
Map ref. 591 I12 | 593 I2

 **Railway Station Complex, 91 High St; (03) 5355 0281 or 1800 657 158; www.visitararat.com.au**

99.9 VoiceFM, 107.9 FM ABC Local Radio

Ararat is a city with a vibrant history. Once inhabited by the Tjapwurong Aboriginal people, the promising lands soon saw squatters move in, and the area really started to boom when gold was discovered in 1854. Thousands of prospectors arrived, and Ararat finally came into existence when Chinese immigrants rested on the town's site in 1857, after walking from South Australian ports in order to avoid Victorian poll taxes. One member of the party discovered alluvial gold, and Ararat was born. Today Ararat is a service centre to its agricultural surrounds.

J Ward The town's original gaol, 'J Ward' served as an asylum for the criminally insane for many years and offers an eerie glimpse into the history of criminal confinement. Now guided tours reveal in chilling detail what life was like for the inmates. Girdlestone St; daily tours (03) 5352 3357.

Gum San Chinese Heritage Centre Gum San means 'hill of gold', a fitting name for this impressive centre built in traditional Southern Chinese style and incorporating the principles of feng shui. The centre celebrates the contribution of the Chinese community both to Ararat, which is said to be the only goldfields town founded by Chinese prospectors, and to the surrounding Goldfields region. The experience is brought to life with interactive displays and an original Canton lead-mining tunnel, uncovered during the building of the centre. Western Hwy; (03) 5352 1078.

Alexandra Park and Botanical Gardens: an attractive formal garden featuring ornamental lakes, fountains and an orchid glasshouse; Vincent St. ***Historical self-guide tours (walking or driving):*** of particular note are the bluestone buildings in Barkly St, including the post office, town hall, civic square and war memorial; details from visitor centre. ***Ararat Art Gallery:*** a regional gallery specialising in wool and fibre pieces by local artists; Barkly St. ***Langi Morgala Museum:*** displays Aboriginal artefacts; Queen St.

Jailhouse Rock Festival: Mar. ***Australian Orchid Festival:*** Sept. ***Golden Gateway Festival:*** held over 10 days; Oct.

Nectar Ambrosia: smart eatery in old pub; 157–159 Barkly St; (03) 5352 7344.

Mt Buangor State Park The park features the Fern Tree Waterfalls and the 3 impressive peaks of Mt Buangor, Mt Sugarloaf and Cave Hill. Its diverse terrain with many varieties of eucalypts offers great sightseeing, bushwalking and picnicking. There are more than 130 species of birds, as well as eastern grey kangaroos, wallabies and echidnas. Contact Parks Victoria on 13 1963. Access to the southern section is via Ferntree Rd off the Western Hwy; 30 km E. Mt Buangor and Cave Hill can be accessed from the main Mt Cole Rd in the Mt Cole State Forest.

Garden Gully Winery: hosts a scarecrow competition each Apr, with ingenious entries from across the state scattered through the vineyard; 17 km N on Western Hwy. Many more of the region's wineries can be accessed on the Great Grape Rd, a circuit through Ballarat and St Arnaud. This region is famous for sparkling whites and traditional old shiraz varieties; brochure and map from visitor centre. ***Green Hill Lake:*** great for fishing and water activities; 4 km E. ***McDonald Park Wildflower Reserve:*** an extensive display of flora indigenous to the area, including wattles and banksias, impressive during the spring months; 5 km N on Western Hwy. ***One Tree Hill Lookout:*** 360-degree views across the region; 5 km NW. ***Langi Ghiran State Park:*** Mt Langi Ghiran and Mt Gorrin form the key features of this park. A popular walk starts at the picnic area along Easter Creek, then goes to the Old Langi Ghiran Reservoir and along the stone water race to a scenic lookout; access via Western Hwy, Kartuk Rd; 14 km E. ***Mt Cole State Forest:*** adjoins Mt Buangor State Park, with bushwalking, horseriding, 4WD tracks and trail-bike riding. The Ben Nevis Fire Tower offers spectacular views; 35 km E.

See also GRAMPIANS & CENTRAL WEST, p. 155

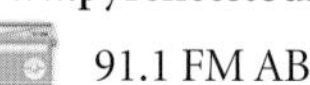

Avoca

Pop. 948
Map ref. 579 A6 | 591 L12 | 593 L1

122 High St; (03) 5465 1000 or 1800 206 622; www.pyreneestourism.com.au

91.1 FM ABC Local Radio, 96.5 Radio KLFM

Avoca was built during the gold boom of the 19th century and is renowned for its wide main street, divided by a stretch of park with trees and a war memorial. Avoca is set in the picturesque Pyrenees Ranges, with the Avoca River flowing by the town.

Historic walk: takes in the original courthouse, one of the oldest surviving courts in Victoria, as well as the powder magazine and Lalor's, one of the state's earliest pharmacies; map from visitor centre. ***Cemetery:*** Chinese burial ground from the goldmining period; on outskirts of town.

Avoca Fine Wine, Arts and Craft Market: 3rd Sun each month. ***Blue Pyrenees Pink Lamb and Purple Shiraz Race Meeting:*** country race meeting; Mar. ***Petanque Tournaments (French Bowls):*** Mar and Dec. ***Mt Avoca Anzac Day Races:*** Apr. ***Taltarni Cup Races:*** Oct.

Warrenmang Vineyard & Resort: regional food; 188 Mountain Creek Rd, Moonambel; (03) 5467 2233.

eco-luxe @ Mount Avoca: eco-friendly luxury lodges; Mount Avoca Winery, Moates La, Avoca; (03) 5465 3282. ***Warrenmang Vineyard & Resort:*** chalets and luxury suites; 188 Mountain Creek Rd, Moonambel; (03) 5467 2233.

Pyrenees Ranges State Forest Covering a large stretch of bushland, these ranges are great for bushwalking and picnics and camping. Visitors can see a variety of wildlife, including koalas, wallabies, kangaroos and goannas. Orchids and lilies can be found growing around the base of the ranges in season. An 18 km walking track starts at The Waterfall camping area and finishes at Warrenmang–Glenlofty Rd. Access via Sunraysia or Pyrenees hwys. For further information contact the Department of Sustainability and Environment Customer Sevice Centre on 13 1186.

Blue Pyrenees Estate: with underground cellar, petanque piste and gourmet lunches on weekends; 7 km W. ***Mt Lonarch Arts:*** displays and sells fine bone china made on the premises; Mt Lonarch; 10 km S. ***Warrenmang Vineyard Resort:*** with cottage-style accommodation and a restaurant specialising in regional produce. The vineyard is also the venue for A Sparkling Affair each Nov, an event celebrating the release of sparkling wines; 22 km NW. ***Wine-tasting tours:*** including self-guide Great Grape Rd; details from visitor centre.

See also GOLDFIELDS, p. 154

Bacchus Marsh

Pop. 13 258
Map ref. 572 G4 | 579 G12 | 582 A3 | 593 O5

156 Main St; (03) 5367 7488; www.discoverbacchusmarsh.org

98.5 3APL Apple FM, 774 AM ABC Local Radio

Bacchus Marsh shares part of its name with the Roman god of wine, but is actually better known for the apples that grow so well in the fertile valley region between the Werribee and Lerderderg rivers. Considered a satellite town within commuting distance of Melbourne, Bacchus Marsh retains a certain charm with stunning heritage buildings and a rural atmosphere.

Avenue of Honour Visitors to the town are greeted by the sight of the renowned Avenue of Honour, an elm-lined stretch of road built in honour of the Australian soldiers who fought in WW I. Eastern approach to town.

Big Apple Tourist Orchard: fresh produce market; Avenue of Honour. ***Historic buildings:*** include The Manor, the original home of the town's founder, Captain Bacchus (now privately owned), and Border Inn, built in 1850, thought to be the state's first stop for Cobb & Co coaches travelling to the goldfields; details from visitor centre. ***Local history museum:*** connected to the blacksmith cottage and forge; open Sat–Sun; Main St. ***Naturipe Fruits, Strawberry, Peach and Nectarine Farm:*** pick-your-own fruits and roadside sales; Avenue of Honour.

Rotary Art Show: June.

Lerderderg State Park Featuring the imposing Lerderderg Gorge, the park is a great venue for picnics, bushwalking and swimming, while the Lerderderg River is ideal for trout fishing. The area was mined during the gold rush, and remnants from the water races used for washing gold can still be found upstream from O'Brien's Crossing. Late winter and spring are good times to see wildflowers and blossoming shrubs. Look out for koalas nestled in giant manna gums and for the magnificent sulphur-crested cockatoo and the wedge-tailed eagle. Contact Parks Victoria on 13 1963. Access via Western Fwy to Bacchus Marsh–Gisborne and Lerderderg Gorge rds; 10 km N.

Werribee Gorge State Park Over time the Werribee River has carved through ancient seabed sediment and lava flows to form a spectacular gorge. The name 'Werribee' comes from the Aboriginal word 'Wearibi', meaning 'swimming place' or 'backbone', perhaps in reference to the snake-like path of the river. Rock climbing is permitted at Falcons Lookout and a popular walk follows the Werribee River from the Meikles Pt picnic area, providing views of the river and the gorge cliff-faces. Contact Parks Victoria on 13 1963. Access via Western Fwy and Pentland Hills Rd to Myers Rd, or via Ironbark Rd (the Ballan–Ingliston Rd) from the Bacchus Marsh–Anakie Rd; 10 km W.

Long Forest Flora Reserve: a great example of the distinctive mallee scrub that once covered the region; 2 km NE. ***St Anne's Vineyard:*** with a bluestone cellar built from the remains of the old Ballarat Gaol; Western Fwy; 6 km W. ***Merrimu Reservoir:*** attractive park area with picnic facilities; about 10 km NE. ***Melton:*** now virtually a satellite suburb of Melbourne, this town has a long and rich history of horse breeding and training. Visit the Willow Homestead to see exhibits detailing the life of early settlers (open Wed, Fri and Sun), picnic on the Werribee River at Melton Reservoir, or taste the fine wines in the nearby Sunbury Wine Region; 14 km E. ***Brisbane Ranges National Park:*** with good walking tracks, wildflowers during spring and the imposing, steep-sided Anakie Gorge; 16 km SW. ***Ballan:*** try the refreshing mineral-spring water at Bostock Reservoir, or join in the festivities at the Vintage Machinery and Vehicle Rally in Feb, and an Autumn Festival held each Mar; 20 km NW. ***Blackwood:*** visit the Mineral Springs Reserve and Garden of St Erth (closed Wed and Thurs). Blackwood is also the start of the 53 km return scenic drive through the Wombat State Forest; 31 km NW.

See also SPA & GARDEN COUNTRY, p. 151

Bairnsdale

Pop. 11 284
Map ref. 583 M5

240 Main St; (03) 5152 3444 or 1800 637 060; www.lakesandwilderness.com.au

100.7 FM ABC Local Radio, 105.5 3REG Radio East Gippsland FM

An attractive rural centre situated on the Mitchell River Flats and considered to be the western gateway to the lakes and wilderness region of East Gippsland. The area has a rich Koorie history brought to life through local landmarks, especially in Mitchell River National Park, where a fascinating piece of Aboriginal folklore is based around the Den of Nargun.

Aboriginal culture The Krowathunkoolong Keeping Place, on Dalmahoy St, details the cultural history of the region's Kurnai Aboriginal people and provides an insight into the impact of white settlement. To explore local Aboriginal history further, visit Howitt Park, Princes Hwy – a tree here has a 4 m scar where bark has been removed to make a canoe. The Bataluk Cultural Trail from Sale to Cann River takes in these and other Indigenous sites of East Gippsland. Details of the trail from Krowathunkoolong.

Historical Museum: built in 1891, contains relics from Bairnsdale's past; Macarthur St. ***Jolly Jumbuck Country Craft Centre:*** wool spinning and knitting mills, plus woollen products for sale; edge of town. ***Self-guide heritage walks:*** take in St Mary's Church, with wall and ceiling murals by Italian artist Francesco Floreani, and the Court House, a magnificent, castle-like construction; details from visitor centre.

Howitt Park Market: 4th Sun each month. ***Line Dancing Championships:*** Mar. ***East Gippsland Agricultural Field Days:*** popular event with family entertainment; Apr. ***Bairnsdale Easter Races:*** Easter. ***Bairnsdale Cup:*** Sept.

Paper Chase: modern bookshop cafe; 168 Main St; (03) 5152 5181.

Comfort Inn Riversleigh: spacious rooms in heritage building; 1 Nicholson St; (03) 5152 6966. ***Tara House:*** 3 ensuite rooms; 37 Day St; (03) 5153 2253. ***Waterholes Guest House:*** rooms each with own verandah; 540 Archies Rd; (03) 5157 9330.

Mitchell River National Park Set in the remnants of temperate rainforest, this park has its own piece of mythology. According to Koorie history, Nargun was a beast made all of stone except for his hands, arms and breast. The fierce creature would drag the unwary to his den, a shallow cave beneath a waterfall on the Woolshed Creek. This Den of Nargun can be found in the park, as can giant kanooka trees, wildflowers and over 150 species of birds. There is a circuit walk to Bluff Lookout and Mitchell River, and Billy Goat Bend is good for picnics. Contact Parks Victoria on 13 1963; Princes Hwy; 15 km W near Lindenow.

VICTORIA

McLeods Morass Wildlife Reserve: a boardwalk extends over the freshwater marshland, allowing a close-up view of the many species of waterbirds found here; southern outskirts of town, access via Macarthur St; 2 km S. ***Wineries:*** include Nicholson River Winery, for tastings and sales; 10 km E. ***Bruthen:*** hosts a Blues Bash each Feb; 24 km NW. ***Dargo:*** historic township in Dargo River valley and major producer of walnuts. Dargo Valley Winery has accommodation and cellar-door sales. The road beyond Dargo offers a scenic drive through high plains to Hotham Heights, stunning in spring when wattles bloom (unsealed road, check conditions); 93 km NW.

See also EAST GIPPSLAND, p. 159

Ballarat

see inset box on next page

Beechworth

Pop. 2644
Map ref. 585 J7 | 586 H4

Ford St; (03) 5728 3233 or 1300 366 321; www.beechworth.com

101.3 Oak FM, 106.5 FM ABC Local Radio

Set in the picturesque surrounds of the Australian Alps, Beechworth is one of the state's best preserved 19th-century gold towns, with over 30 buildings listed by the National Trust. The grandeur of Beechworth's buildings can be explained by the fact that during the 1850s over four million ounces of gold were mined here. There is a delightful tale about Beechworth's heyday: the story goes that Daniel Cameron, a political candidate vying for support from the Ovens Valley community, rode at the head of a procession through the town on a horse shod with golden shoes. Sceptics claim they were merely gilded, but the tale offers a glimpse into the wealth of Beechworth during the gold rush.

Historic and cultural precinct This fantastic precinct provides a snapshot of 19th-century Beechworth. Featuring fine, honey-coloured granite buildings, the area incorporates the telegraph station, gold office, Chinese prospectors' office, town hall and powder magazine. Of particular interest is the courthouse, site of many infamous trials including Ned Kelly's, and where Sir Isaac Isaacs began his legal career. Also in the precinct is the Robert O'Hara Burke Memorial Museum, with the interesting 'Strand of Time' exhibition where 19th-century Beechworth shops are brought to life.

Beechworth Gaol Built in 1859, the original wooden gates of this gaol were replaced with iron ones when it was feared prisoners would break out in sympathy with Ned Kelly during his trial. The gaol is located in William St but not presently open to the public.

Walking tours: Ned Kelly and Gold Rush walking tours operate daily, and Ghost Tours are available at the former Mayday Hills Asylum; bookings at visitor centre. ***Carriage Museum and Australian Light Horse Exhibition:*** National Trust horse-drawn carriage display and Australian Light Horse Exhibition housed at the historic Murray Breweries, which also offers turn-of-the-century gourmet cordial made to time-honoured recipes; 29 Last St. ***Beechworth Honey Experience:*** interpretive display on the history of honey; includes a glass-fronted live bee display. A wide range of premium Australian honey is on offer in the concept shop; Cnr Ford and Church sts. ***Harry Power's Cell:*** under the shire offices, where the 'gentleman bushranger' was once briefly held; Albert Rd. ***The Beechworth Pantry:*** gourmet cafe and centre for produce of the north-east; Ford St. ***Beechworth Bakery:*** famous for its pastries and cakes; Camp St.

Country Craft Market: Queen Victoria Park, 4 times a year; details from visitor centre. ***Golden Horseshoes Festival:*** a celebration of the town's past, with street parades and a variety of market stalls; Easter. ***Beechworth Harvest Celebration:*** May; details from visitor centre. ***Drive Back in Time:*** vintage car rally; May. ***Celtic Festival:*** music festival; Nov.

Provenance Restaurant: contemporary regional food; 86 Ford St; (03) 5728 1786. ***The Green Shed Bistro:*** European-influenced; 37 Camp St; (03) 5728 2360. ***The Ox and Hound:*** contemporary bistro; 52 Ford St; (03) 5728 2123. ***Wardens Food & Wine:*** modern Italian; 32 Ford St; (03) 5728 1377.

Black Springs Bakery: charming provincial-style barn; 464 Wangaratta Rd; (03) 5728 2565. ***Freeman on Ford:*** 4 Victorian-era and 2 1930s-style rooms; 97 Ford St; (03) 5728 2371. ***Provenance Restaurant & Luxury Suites:*** 4 contemporary suites; 86 Ford St; (03) 5728 1786. ***Stone Cottage:*** 1-bedroom cottage, 2-bedroom barn; 6 Tanswell St; (03) 5728 2857. ***The Stanley:*** boutique rooms at country pub; 1 Wallace St, Stanley; (03) 5728 6502.

Beechworth Cemetery This cemetery is a fascinating piece of goldfields history. More than 2000 Chinese goldminers are buried here. Twin ceremonial Chinese burning towers stand as a monument to those who died seeking their fortune far from home. Northern outskirts of town.

Beechworth Historic Park: surrounds the town and includes Woolshed Falls Historical Walk through former alluvial goldmining sites. ***Gorge Scenic Drive (5 km):*** starts north of town. ***Beechworth Forest Drive:*** takes in Fletcher Dam; 3 km SE towards Stanley. ***Kellys Lookout:*** at Woolshed Creek; about 4 km N. ***Mt Pilot Lookout:*** views of Murray Valley, plus signposted Aboriginal cave paintings nearby; 5 km N. ***Stanley:*** a historic goldmining settlement with fantastic views of the alps from the summit of Mt Stanley; 10 km SE. ***Wineries:*** 5 cellar doors in and around Beechworth for tastings and sales; map from visitor centre.

See also HIGH COUNTRY, p. 158

Benalla

Pop. 9128
Map ref. 584 G8 | 586 D5

14 Mair St; (03) 5762 1749; www.benallaonline.com.au

97.7 FM ABC Local Radio, 101.3 Oak FM

Motorists from Melbourne entering Benalla will notice the Rose Gardens positioned beside the highway a short distance before Lake Benalla – gardens for which the city has become known as the 'Rose City'. The town is Sir Edward 'Weary' Dunlop's birthplace and proudly advertises the fact with a museum display and a statue in his honour at the Benalla Botanical Gardens.

Benalla Art Gallery Set beside Lake Benalla, the gallery has an impressive collection including contemporary Australian art, works by Sidney Nolan, Arthur Streeton, Tom Roberts and Arthur Boyd, and a substantial collection of Indigenous art. Built in 1975, the gallery is a striking work of modern architecture. There is a permanent exhibition featuring the works of Laurie Ledger, a local resident, and examples of the Heidelberg School and early colonial art. Bridge St; (03) 5762 3027.

Benalla Ceramic Art Mural: a Gaudi-inspired community construction, this fascinating 3D mural is opposite the art gallery on Lake Benalla. ***The Creators Gallery:*** paintings, pottery and craft; at the information centre. ***Benalla Costume and Pioneer Museum:*** has period costumes, a Ned Kelly exhibit (including Kelly's cummerbund) and a feature display of Benalla's 'famous sons', in particular, Sir Edward 'Weary' Dunlop; Mair St. ***Lake Benalla:*** created in Broken River, it has good recreation and picnic facilities and is a haven for waterbirds. Take the self-guide walk around the lake. ***Botanical Gardens:*** features a splendid collection of roses and memorial statue of Sir Edward 'Weary' Dunlop; Bridge St. ***Aeropark:*** centre for the Gliding Club of Victoria, offering hot-air ballooning and glider flights; northern outskirts of town; bookings on (03) 5762 1058.

Lakeside Craft and Farmers Market: near the Civic Centre; 3rd Sat each month. ***Benalla Festival:*** Feb/Mar.

Benalla Gallery Cafe: contemporary local produce menu; Benalla Art Gallery, Bridge St East; (03) 5762 3777. ***Georgina's:*** modern Australian menu; 100 Bridge St East; (03) 5762 1334. ***North Eastern Hotel:*** regional contemporary dishes; 1–3 Nunn St; (03) 5762 7333. ***Raffety's Restaurant:*** homemade modern Australian dishes; 55 Nunn St; (03) 5762 4066.

Merriollia: 3 B&B rooms; 17–19 Cecil St; (03) 5762 3786. ***Nillahcootie Estate:*** 3-bedroom guesthouse; 3630 Midland Hwy, Lima South; (03) 5768 2685.

Reef Hills State Park Features grey box, river red gum, wildflowers in spring and wattle blossom in winter. The park offers scenic drives, bushwalks, picnics and horseriding. There are more than 100 species of birds, including gang-gang cockatoos and crimson rosellas, plus animals such as eastern grey kangaroos, sugar gliders, brush-tailed possums, echidnas and bats. Contact Parks Victoria on 13 1963; 4 km SW, western side of the Midland Hwy.

Lake Mokoan: depending on water levels, great for fishing, boating and waterskiing; 10 km NE. ***1950s-style cinema:*** showing classic films at Swanpool; 23 km S.

See also HIGH COUNTRY, p. 158

Bendigo

see inset box on page 168

Bright

see inset box on page 171

Buchan

Pop. 326
Map ref. 565 A11 | 581 A10 | 583 P4

General Store, Main St; (03) 5155 9202 or 1800 637 060; www.lakesandwilderness.com.au

90.7 FM 3REG Radio East Gippsland, 828 AM ABC Local Radio

Situated in East Gippsland, Buchan is primarily an agricultural town renowned for offering some of the best caving in Victoria. Although the origin of the town's name is disputed, it is said to be derived from the Aboriginal term for either 'smoke-signal expert' or 'place of the grass bag'.

Foothills Festival: Jan. ***Canni Creek Races:*** Jan. ***Rodeo:*** Easter. ***Flower Show:*** Nov.

Parks Victoria Wilderness Retreat Buchan Caves: 5 comfortable tents; Buchan Caves Reserve; 13 1963.

Buchan Caves Reserve The reserve features more than 350 limestone caves, of which the Royal and Fairy caves are the most accessible – the Fairy Cave alone is over 400 m long, with impressive stalactites. Europeans did not discover the caves until 1907, but from then on they became a popular tourist destination. Now visitors can cool off in the spring-fed swimming pool after exploring the caves. Tours of the Royal and Fairy caves run daily. Off Buchan Rd, north of town; (03) 5162 1900.

Snowy River Scenic Drive The drive takes in the Buchan and Snowy rivers junction and runs along the edge of Snowy River National Park to Gelantipy. Beyond Gelantipy is Little River Gorge, Victoria's deepest gorge. A short walking track leads to a cliff-top lookout. Near the gorge is McKillops Bridge, a safe swimming spot, a good site to launch canoes, and the starting point for 2 walking tracks. Care is required on the road beyond Gelantipy; 4WD is recommended. Details from visitor centre.

Suggan Buggan: this historic townsite, surrounded by Alpine National Park, features an 1865 schoolhouse and the Eagle Loft Gallery for local art and craft; 64 km N.

See also EAST GIPPSLAND, p. 159

Camperdown

Pop. 3164
Map ref. 580 H7 | 593 J7

Old Courthouse, Manifold St; (03) 5593 3390; www.greatoceanrd.org

104.7 Otway FM, 594 AM ABC Local Radio

Located at the foot of Mount Leura, a volcanic cone, Camperdown is more famous for its natural attractions than for the town itself, being situated on the world's third largest volcanic plain. But that should not detract from Camperdown – National Trust–listed Finlay Avenue features 2 kilometres of regal elm trees, while in the town centre the Gothic-style Manifold Clock Tower proudly stands as a tribute to the region's first European pioneers.

Manifold Clock Tower: an imposing structure built in 1896; open 1st Sun each month; Cnr Manifold and Pike sts. ***Historical Society Museum:*** displays Aboriginal artefacts, local historical photographs, and household and farming implements; Manifold St. ***Courthouse:*** built in 1886–87, described as one of the most distinctive courthouses in Australia; Manifold St. ***Buggy Museum:*** collection of 30 restored horse-drawn buggies; Ower St.

Craft market: Finlay Ave or Theatre Royal; 1st Sun each month. ***Heritage Festival:*** Nov.

Purrumbete Homestead: perfect for an indulgent holiday; 3551 Princes Hwy; (03) 5594 7374. ***Timboon House & Stables B&B:*** 3 rooms in historic house; 320 Old Geelong Rd; (03) 5593 1003.

Crater lakes Surrounding Camperdown are spectacular crater lakes that provide an interesting history of volcanic activity over the past 20 000 years. Travelling west, take the scenic drive around the rims of lakes Bullen Merri and Gnotuk, and join in the watersports and swimming at South Beach. The lakes are regularly stocked with Chinook salmon and redfin. For a scenic picnic spot, and some fishing, visit Lake Purrumbete; 15 km SE. One of the most impressive lakes is Lake Corangamite, the Southern Hemisphere's largest permanent salt lake. This lake lies 25 km E, but the best viewing spot is Red Rock Lookout; *see Colac.*

Derrinallum and Mt Elephant Mt Elephant rises to almost 200 m behind the small township of Derrinallum – it doesn't sound like a lot, but across the plains of the Western District you can see it from up to 60 km away. A gash in the elephant's western side is

continued on p. 167

BALLARAT

Pop. 85 196

Map ref. 572 D3 | 527 | 579 C10 | 593 M4

43 Lydiard St North; 1800 446 633 or 5320 5741; www.visitballarat.com.au

99.9 VoiceFM, 107.9 FM ABC Local Radio

Ballarat is Victoria's largest inland city and features grand old buildings and wide streets that create an air of splendour. Built on the wealth of the region's goldfields, Ballarat offers activities ranging from fine dining in the many restaurants to real-life experiences of the area's goldmining past. Lake Wendouree provides a beautiful backdrop for picnics. Ballarat was the site of the infamous Eureka Rebellion of 3 December 1854. When goldfields police attempted to quell the miners' anger over strict mining-licence laws, a bloody massacre eventuated. The Eureka Rebellion is viewed by many as a symbol of the Australian workers' struggle for equity and a 'fair go'. The best place to get a feel for this historic event is at Sovereign Hill.

Sovereign Hill This is the main destination for visitors to Ballarat and a good place to get a taste for what life was like on the Victorian goldfields. Spread over 60 ha, Sovereign Hill is a replica goldmining town, complete with authentically dressed townspeople. Panning for gold is a popular activity, while in the evening the 'Blood on the Southern Cross' show re-enacts the Eureka Rebellion. Bradshaw St; (03) 5337 1100.

Eureka Centre for Australian Democracy: multimillion dollar cultural centre with information about the famous battle; cnr Eureka and Rodier sts; (03) 5333 1854. ***Ballarat Botanic Gardens:*** an impressive collection of native and exotic plants; Prime Minister Ave features busts of all of Australia's prime ministers; Wendouree Pde. ***Ballarat Wildlife Park:*** houses native Australian animals such as koalas, kangaroos, quokkas and crocodiles; Cnr York & Fussell sts, Ballarat East; (03) 5333 5933. ***Art Gallery of Ballarat:*** holds a significant collection of Australian art. The original Eureka Rebellion flag is also on display; 40 Lydiard St North; (03) 5320 5791. ***Gold Museum:*** details the rich goldmining history of the area; opposite Sovereign Hill; Bradshaw St; (03) 5337 1107; admission free with Sovereign Hill Ticket. ***Historic buildings:*** include Her Majesty's Theatre, built in 1875 and Australia's oldest intact, purpose-built theatre, and Craig's Royal and the George hotels, with classic old-world surroundings; Lydiard St. ***Vintage Tramway:*** via Wendouree Pde; rides weekends, and public and school holidays. ***Avenue of Honour and Arch of Victory:*** honours those who fought in WW I; western edge of city.

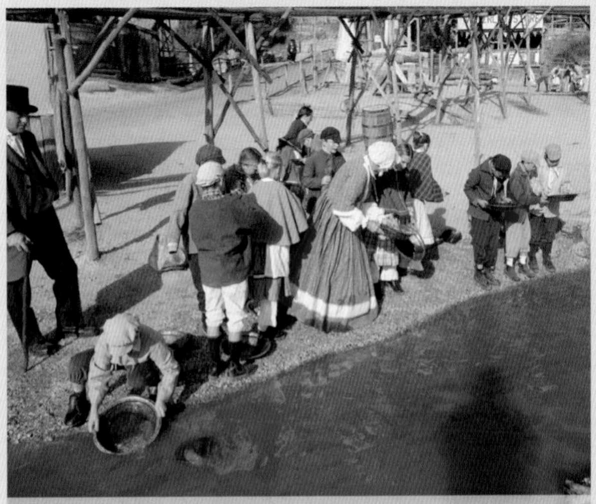

[PANNING FOR GOLD, SOVEREIGN HILL]

Ballarat Lakeside Farmers Market: Lake Wendouree foreshore; 2nd and 4th Sat each month. ***Organs of the Ballarat Goldfields:*** music festival held in historic venues; Jan. ***Begonia Festival:*** popular event for garden lovers; Mar. ***Heritage Weekend:*** May. ***Royal South Street Eisteddfod:*** music festival; Sept–Nov. ***Ballarat Cup:*** Nov.

Europa Cafe: cafe with all-day breakfast; 411 Sturt St; (03) 5338 7672. ***L'espresso:*** Italian cafe; 417 Sturt St; (03) 5333 1789. ***Masons Cafe & Foodstore:*** excellent blackboard specials; 32 Drummond St North; (03) 5333 3895. ***Boatshed Restaurant:*** Mod Oz restaurant by the lake; 27a Lake Wendouree Foreshore; (03) 5333 5533

Steinfeld's: Elegant heritage rooms in Sovereign Hill village; Magpie St; (03) 5337 1159. ***Craig's Royal Hotel:*** restored historic hotel rooms; 10 Lydiard St South; (03) 5331 1377. ***Ansonia on Lydiard:*** boutique B&B hotel near Craig's; 32 Lydiard St Sth; (03) 5332 4678. ***Montrose of Ballarat:*** heritage cottage; 111 Eureka St; 0429 439 448.

Enfield State Park Great for bushwalking or horseriding, the park has many species of orchids and numerous animals including echidnas, koalas, bats and frogs. There is a pretty picnic ground and walking tracks at Remote Long Gully. Also featured are the remnants of early goldmining settlements, including the Berringa Mines Historic Reserve. Contact Parks Victoria on 13 1963; access via Incolls and Misery Creek rds; 25 km S.

Buninyong Buninyong features many fine art and craft galleries. Ballarat Bird World, home to many species of parrots, has raised walkways through the aviaries; 408 Eddy Ave, Mt Helen; (03) 5341 3843. Buninyong Good Life Festival; 3rd weekend in Oct, and in May for the Buninyong Film Festival. The Mt Buninyong Lookout east of town offers great views. 13 km SE.

Kirks and Gong Gong reserves: ideal for picnics and bushwalking, these parks include many unique, indigenous plants; on opposite sides of Daylesford Rd; 5 km NE. ***Kryal Castle:*** replica of a medieval castle, with daily tours and family entertainment; (03) 5334 7422; 9 km E. ***Yuulong Lavender Estate:*** set in scenic landscaped gardens, the estate produces and sells lavender products; closed May–Sep; Yendon Rd, Mt Egerton; (03) 5368 9453; 25 km SE. ***Lal Lal Falls:*** plunge 30 m into the Moorabool River; 18 km SE. ***Lal Lal Blast Furnace:*** fascinating 19th-century archaeological remains; 18 km SE. ***Lake Burrumbeet:*** this 2100 ha lake is a popular fishing spot, especially for redfin in spring and summer. Watersports and family activities are available on the lake; various boat ramps provide access. Caravan parks are set on the lakeside and are popular with holiday-makers; 22 km NW. ***Skipton:*** in town is an eel factory selling smoked eel and other products; 51 km SW. South of town are the Mt Widderin Caves – one has been named the Ballroom, as it was once a venue for dances and concerts. The caves are on private property; tours by appt (03) 5340 2081. ***Beaufort:*** a small town on the shores of Lake Beaufort, an artificial lake surrounded by gardens, providing a picturesque location for picnics and leisurely walks; 54 km W. South of town is Lake Goldsmith, home of a major rally of steam-driven machinery and vehicles each May and Oct. ***Mooramong Homestead:*** built in the 1870s and then altered during the 1930s by its ex-Hollywood owners. It is surrounded by beautiful gardens and a flora and fauna reserve, and is open for tours 3rd Sun each month; 56 km NW via Skipton. ***Great Grape Rd:*** circuit through Avoca, St Arnaud and Stawell, visiting local wineries.

See also GOLDFIELDS, p. 154

the result of decades of quarrying. The mountain is actually the scoria cone of an extinct volcano, and inside is a 90 m deep crater. Now owned by the community, there is a walking trail to the top, and the Music on the Mount festival is held here in Nov. Lake Tooliorook on the other side of town offers good fishing for trout and redfin, and watersports. 40 km N.

Camperdown–Timboon Rail Trail: walking or riding track through bush, following historic railway line. ***Mt Leura:*** extinct volcano close to the perfect cone of Mt Sugarloaf. A lookout offers excellent views over crater lakes and volcanoes, and north across the plains to the Grampians; 1 km S. ***Camperdown Botanic Gardens:*** feature rare examples of Himalayan oak and a lookout over lakes Bullen Merri and Gnotuk; 3 km W. ***Cobden Miniature Trains:*** operates 3rd Sun each month; Cobden; 13 km S.

See also SOUTH-WEST COAST, p. 153

Cann River

Pop. 223
Map ref. 565 C11 | 581 E10

Parks Victoria Cann River office, (03) 5158 6351; or East Gippsland Visitor Information Centre, 1800 637 060; www.lakesandwilderness.com.au

101.7 FM 3MGB Wilderness Radio, 106.1 FM ABC Local Radio

Cann River is situated at the junction of the Princes and Cann Valley highways, and is notable for its proximity to several spectacular national parks. The area boasts excellent fishing, bushwalking and camping in the rugged hinterland, with nearby Point Hicks notable for being the first land on the east coast of Australia to be sighted by Europeans.

Point Hicks Lighthouse: 2 cottages, each sleeps 6; Point Hicks Rd; (03) 5156 0432.

Lind National Park The park includes the Euchre Valley Nature Drive through temperate rainforest gullies. It also supports open eucalypt forests with grey gum, messmate and silvertop ash. Watch for wildlife such as the pretty masked owl and the elusive long-footed potoroo. Has picnic facilities. 15 km W.

Coopracambra National Park This park is in one of the most remote sections of Victoria. Ancient fossil footprints have been found in the red sandstone gorge of the Genoa River, and the surrounding granite peaks create a spectacular scene. The 35 000 ha area protects unique ecosystems and rare flora and fauna. Only experienced and well-equipped hikers should undertake walks in the rugged and remote parts of this park. A 'trip intentions' form needs to be lodged at the Cann River or Mallacoota office of Parks Victoria prior to departure, and parks staff must be notified upon return. 30 km N near NSW border.

Croajingolong National Park: the road travelling south of Cann River leads to Pt Hicks and its historic 1890 lighthouse (daily tours offered); *for further details on the park see Mallacoota.*

See also EAST GIPPSLAND, p. 159

Casterton

Pop. 1654
Map ref. 592 C3 | 601 H11

Shiels Tce; (03) 5581 2070; www.castertonnow.org.au

94.1 FM ABC Local Radio, 99.3 Coastal FM

Casterton is a Roman name meaning 'walled city', given to the town because of the natural wall of hills surrounding the valley where it lies. These hills, and the Glenelg River that flows through town, create an idyllic rural atmosphere. The region is colloquially known as 'Kelpie Country' as it is the birthplace of this world-famous breed of working dog. In the mid-1800s a prized Scottish collie female pup from nearby Warrock Homestead was sold to a stockman named Jack Gleeson, who named her 'Kelpie' – she was bred out with various 'black and tan' dogs, and so began the long line of the working man's best friend.

Historical Museum: housed in the old railway station, the museum displays local artefacts; open by appt; Cnr Jackson and Clarke sts. ***Alma and Judith Zaadstra Fine Art Gallery:*** Henty St. ***Mickle Lookout:*** a great view across the town; Moodie St, off Robertson St on the eastern edge of town.

Vintage Car Rally: Mar. ***Polocrosse Championships:*** Mar. ***Casterton Cup:*** June. ***Australian Kelpie Muster and Kelpie Working Dog Auction:*** June/July.

Dergholm State Park The park features a great diversity of vegetation, including woodlands, open forests, heaths and swamps. An abundance of wildlife thrives, including echidnas, koalas, kangaroos, reptiles and the endangered red-tailed black cockatoo. A key attraction is Baileys Rocks, unique giant green-coloured granite boulders. Contact Parks Victoria on 13 1963; 50 km N.

Long Lead Swamp: waterbirds, kangaroos, emus and a trail-bike track; Penola Rd; 11 km W. ***Geological formations:*** in particular, The Hummocks, 12 km NE, and The Bluff, viewable from Dartmoor Rd, 20 km SW. Both rock formations are around 150 million years old. ***Warrock Homestead:*** a unique collection of 33 buildings erected by its founder, George Robertson. The homestead was built in 1843 and is National Trust–classified; open day on Easter Sun; 26 km N. ***Bilston's Tree:*** 50 m high and arguably the world's largest red gum; Glenmia Rd; 30 km N.

See also GRAMPIANS & CENTRAL WEST, p. 155

VICTORIA

BENDIGO

Pop. 81 941

Map ref. 578 | 579 F3 | 584 A9 | 591 O10

Former Post Office, 51–67 Pall Mall; (03) 5434 6060 or 1800 813 153; www.bendigotourism.com

89.5 The Fresh FM, 91.1 FM ABC Local Radio, 96.5 Radio KLFM

Bendigo was the place of one of the world's most exciting gold rushes, with more gold found here between 1850 and 1900 than anywhere else in the world. Elaborate buildings and monuments from the golden past line the main streets, offering an ever-present reminder of the riches from the goldfields. Today modern life weaves itself around this legacy with a vibrant pace. The town's new wealth can be seen in many areas including art, culture, dining, wine and shopping.

Golden Dragon Museum The museum commemorates the contribution of the Chinese community to life on the goldfields. On display are exhibitions depicting the daily life and hardships of Chinese immigrants and an impressive collection of Chinese memorabilia and processional regalia, including what is said to be the world's oldest imperial dragon, 'Loong' (which first appeared at the Bendigo Easter Fair in 1892), and the world's longest imperial dragon, 'Sun Loong'. Adjacent to the museum is the Classical Chinese Garden of Joy and newly developed Dai Gum San precinct; 1-11 Bridge St; (03) 5441 5044.

Central Deborah Gold Mine To get a feel for life in a goldmining town is to take a trip down this mine, where you can still see traces of gold in the quartz reef 20 storeys below the ground. The Central Deborah Gold Mine was the last commercial goldmine to operate in Bendigo. From 1939 to 1954 around a tonne of gold was excavated; 76 Violet St; tour details (03) 5443 8322.

Bendigo Art Gallery: well regarded for contemporary exhibitions plus an extensive permanent collection with a focus on Australian artists, including Arthur Boyd, Tom Roberts and Arthur Streeton. Guided tours daily; 42 View St; (03) 5434 6088. ***Self-guide heritage walk:*** takes in landmarks including The Hotel Shamrock, built in 1897, cnr Pall Mall and Williamson St; Sacred Heart Cathedral, the largest outside Melbourne, Wattle St; Alexandra Fountain, built in 1881, one of the largest and most ornate fountains in regional Victoria, at Charing Cross; and the Renaissance-style post office and law courts at Pall Mall; details from visitor centre. ***Bendigo Pottery:*** Australia's oldest working pottery, with potters at work, a cafe and sales; 146 Midland Hwy, Epsom; (03) 5448 4404. ***Dudley House:*** National Trust–classified building; View St. ***Vintage Trams:*** run from Central Deborah Gold Mine on 8 km city trip, including a stop at the Tram Depot Museum; taped commentary provided. ***Chinese Joss House:*** National Trust–classified temple built by Chinese miners; included on the vintage tram trip; Finn St, North Bendigo; (03) 5442 1685. ***Rosalind Park:*** majestic parklands that sit beautifully in the centre of Bendigo, offering stately gardens for leisure and relaxation; includes a lookout tower, Cascades water feature and Conservatory Gardens; Pall Mall. ***Discovery Science and Technology Centre:*** features more than 100 hands-on displays; 7 Railway Pl; (03) 5444 4400. ***Post Office Gallery:*** changing exhibitions and displays of Bendigo's history and culture; at the visitor centre; 51–67 Pall Mall.

[THE HISTORIC FORMER BENDIGO POST OFFICE, NOW HOME TO THE BENDIGO VISITOR CENTRE AND POST OFFICE GALLERY]

Bridge Street Market: features local produce and handmade arts and crafts; Bridge St near Golden Dragon Museum; 8am–1pm 3rd Sat each month. ***Bendigo Showgrounds Market:*** Prince of Wales Showgrounds, Holmes St; 8.30am–3pm each Sun. ***Bendigo Farmers Market:*** fresh regional produce; Sidney Myer Pl; 9am–1pm, 2nd Sat each month. ***Bendigo International Madison:*** major cycling event; Mar long weekend. ***Bendigo Easter Festival:*** first held in 1871, the festival spans 4 days and is a major event on the town's calendar, with free music and entertainment, craft markets, art exhibits, food, wine and the famous procession featuring 'Sun Loong'. ***Australian Sheep and Wool Show:*** showcases everything from farming to fashion; July. ***Bendigo Heritage Uncorked:*** wine event in the historic streets; 2nd weekend Oct. ***National Swap Meet:*** Australia's largest meet for vintage cars and bikes; Bendigo Showgrounds; Nov. ***Bendigo Cup:*** horseracing; Nov.

GPO Bendigo: cafe-bar; 60–64 Pall Mall; (03) 5443 4343. ***The Bridge:*** contemporary gastropub; 49 Bridge St; (03) 5443 7811. ***The Dispensary Enoteca:*** European menu; 9 Chancery La; (03) 5444 5885. ***Whirrakee Restaurant:*** contemporary French; 17 View St; (03) 5441 5557. ***Wine Bank on View:*** Mediterranean wine bar; 45 View St; (03) 5444 4655.

Anchorage by the Lake: apartments decorated in French or oriental style; 300–302 Napier St; (03) 5442 4777. ***Fountain View Suites:*** boutique suites; 10–12 View St; (03) 5435 2121. ***The Hotel Shamrock:*** historic hotel rooms; Cnr Pall Mall and Williamson St; (03) 5443 0333. ***Byronsvale Vineyard B&B:*** 3 converted-stable apartments; 51 Andrews Rd, Maiden Gully; (03) 5447 2790.

Greater Bendigo National Park The park, which extends to the north and south of town, protects some high-quality box-ironbark forest and is popular for scenic driving, cycling, walking and camping. Relics of the region's goldmining and eucalyptus-oil industries can be found within. Fauna includes over 170 species of birds including the grey shrike-thrush, a pretty songbird. In the early morning and later in the evening, look out for eastern grey kangaroos, black wallabies and echidnas. Detailed maps of the park are available at the visitor centre. Contact Parks Victoria on 13 1963; access via Loddon Valley Hwy through Eaglehawk; 8 km N.

One Tree Hill observation tower: panoramic views; 4 km S. ***Eaglehawk:*** site of the gold rush in 1852, it features remnants of goldmining days and fine examples of 19th-century architecture; details of self-guide heritage tour from visitor centre; 6.5 km NW. ***Mandurang:*** features historic wineries and is the exact centre of Victoria; 8 km S. ***Bendigo Wine Region:*** more than 30 wineries are located around Bendigo, producing award-winning wines and offering welcoming cellar-door experiences with tastings and sales; wine booklet with map available from visitor centre.

See also GOLDFIELDS, p. 154

Castlemaine

Pop. 7250
Map ref. 579 F6 | 591 O12 | 593 N1

Castlemaine Market Building, 44 Mostyn St; (03) 5471 1795 or 1800 171 888; www.maldoncastlemaine.com

91.1 FM ABC Local Radio, 106.3 Radio KLFM

Castlemaine is a classic goldmining town known for its grand old buildings and sprawling botanical gardens. This area was the site of the greatest alluvial gold rush that the world has ever seen. Now the town relies largely on agriculture and the manufacturing sectors, as well as being home to a thriving artistic community that takes inspiration from the area's red hills.

Castlemaine Art Gallery Housed in an elegant Art Deco building, the gallery was designed in 1931 by Peter Meldrum and is renowned for its collection of Australian art. Along with the permanent collection, many exhibitions appear here. Works by Rembrandt, Francisco Goya and Andy Warhol have all been displayed at this delightful gallery. 14 Lyttleton St; (03) 5472 2292.

Buda Historic Home and Garden Buda is considered to have one of the most significant examples of 19th-century gardens in Victoria. The house itself is furnished with period pieces and art and craft created by the Leviny family, who lived here for 118 years. Ernest Leviny was a silversmith and jeweller. Five of his 6 daughters never married, but remained at Buda and pursued woodwork, photography and embroidery. Open 12–5pm Wed–Sat, 10am–5pm Sun; 42 Hunter St; (03) 5472 1032.

Victorian Goldfields Railway This historic railway runs from Castlemaine to Maldon. The steam train journeys through box-ironbark forest in a region that saw some of the richest goldmining in the country. As well as the regular timetable, it also hosts special events throughout the year. Castlemaine Railway Station, Kennedy St; recorded information (03) 5475 2966, inquiries (03) 5470 6658.

Diggings Interpretive Centre: housed in the restored 19th-century Castlemaine Market building, the centre features interactive displays about the area's many goldmines as well as various exhibitions; Mostyn St. ***Theatre Royal:*** hosts live shows and films and also offers luxurious backstage accommodation; Hargraves St. ***Castlemaine Botanic Gardens:*** one of Victoria's oldest and most impressive 19th-century gardens; cnr Walker and Downes rds. ***Old Castlemaine Gaol:*** restored gaol now offers tours; Bowden St. ***Food and wine producers:*** dotted throughout the area; food and wine trail brochures from visitor centre.

Castlemaine Farmers Market: 1st Sun each month. ***Wesley Hill Market:*** each Sat; 2.5 km E. ***Castlemaine State Festival:*** odd-numbered years, Apr. ***Festival of Gardens:*** Nov.

Bold Cafe Gallery: Asian-influenced nursery cafe; 146 Duke St; (03) 5470 6038. ***Saffs Cafe:*** casual cafe food; 64 Mostyn St; (03) 5470 6722. ***The Empyre:*** modern gastropub; 68 Mostyn St; (03) 5472 5166. ***Togs Place:*** popular breakfast/lunch spot; 58 Lyttleton St; (03) 5470 5090.

The Empyre: 6 beautiful suites; 68 Mostyn St; (03) 5472 5166. ***The Hermitage B&B:*** 2nd floor of sandstone cottage; 181 Blakeley Rd; (03) 5472 2008. ***Tuckpoint:*** historic 2-bedroom cottage; 60 Kennedy St; 0439 035 382. ***Sage Cottage:*** 2-bedroom miners cottage; 25 Castlemaine St, Fryerstown; (03) 5473 4322. ***Shack 14 @ Prospect House:*** eco-friendly hideaway, sleeps 4; Hooper Rd, Chewton; (03) 5472 1677.

Castlemaine Diggings National Heritage Park The wealth on Castlemaine's streets springs from the huge hauls of gold found on the Mt Alexander Diggings, east and south of town. Thousands of miners worked the fields. Towns such as Fryerstown, Vaughan and Glenluce, now almost ghost towns, supported breweries, schools, churches and hotels. Today visitors can explore Chinese cemeteries, mineral springs, waterwheels and old townsites. Fossicking is popular. Details of self-guide walks and drives from visitor centre. 4 km S.

Chewton: historic buildings line the streets of this former gold town; 4 km E. ***Harcourt:*** this town is known for its many wineries, including Harcourt Valley Vineyard and Blackjack Vineyards, with tastings and cellar-door sales. Also at Harcourt is the Skydancers Orchid and Butterfly Gardens. The town hosts the Apple Festival in Mar and spring, and the Orchid Festival in Oct; 9 km NE. ***Big Tree:*** a giant red gum over 500 years old; Guildford; 14 km SW. ***Koala Reserve:*** Mt Alexander; 19 km NE.

See also GOLDFIELDS, p. 154

Chiltern

Pop. 1067
Map ref. 559 N12 | 585 J6 | 586 H2

30 Main St; (03) 5726 1611; www.chilternvic.com

101.3 Oak FM, 106.5 FM ABC Local Radio

Now surrounded by rich pastoral farmland, Chiltern was once at the centre of a goldmining boom and had as many as 14 suburbs. After the Indigo gold discovery in the 1850s, there was a major influx of miners and settlers, but the boom was brief and farming was soon prominent in the town's economy. Today the rich heritage of the 19th century can be seen in the streetscapes, a vision not lost on Australian filmmakers keen for that 'authentic' 1800s scene.

Athenaeum Museum: historic building with heritage display; Conness St. ***Dow's Pharmacy:*** old chemist shop with original features; Conness St. ***Star Theatre and Grapevine Museum:*** the quaint theatre still operates and the museum, formerly the Grapevine Hotel, boasts the largest grapevine in Australia, planted in 1867 and recorded in the Guinness World Records; Main St. ***Federal Standard newspaper office:*** open by appt for groups; Main St. ***Lakeview House:*** former home of author Henry Handel Richardson; open afternoons on weekends and public and school holidays; Victoria St. ***Lake Anderson:*** picnic and barbecue facilities; access via Main St.

Antique Fair: Aug. ***Ironbark Festival:*** heritage fair with woodchopping, live music and markets; Oct.

The Mulberry Tree: 2 rooms in former bank; 28 Conness St; (03) 5726 1277. ***Koendidda Country House:*** elegant B&B; 79 Pooleys Rd, Barnawartha; (02) 6026 7340.

Chiltern–Mt Pilot National Park This park stretches from around Chiltern south to Beechworth and protects remnant box-ironbark forest, which once covered much of this part of Victoria. Also featured are significant goldmining relics, including the impressive Magenta Goldmine (around 2 km E). Of the park's 21 000 ha, 7000 were exposed to bushfire in Jan 2003. But its regeneration is evidence of the hardiness of the forest, and there are now upgraded visitor facilities. An introduction to the forest scenery and goldmining history is on the 25 km scenic drive signposted from Chiltern. Other activities include canoeing and rafting, fishing, and cycling and walking trips along the many marked trails. Contact Parks Victoria on 13 1963; access via Hume Hwy and the road south to Beechworth.

See also HIGH COUNTRY, p. 158

Clunes

Pop. 1024
Map ref. 572 C1 | 579 C8 | 593 M2

Old School Complex, 70 Bailey St; (03) 5345 3896; www.visitclunes.com.au

99.9 VoiceFM, 107.9 FM ABC Local Radio

The first registered gold strike in the state was made at Clunes on 7 July 1851. The town, north of Ballarat, is said to be one of the most intact gold towns in Victoria, featuring historic buildings throughout. Surrounding the town are a number of extinct volcanoes. A view of these can be obtained 3 kilometres to the south, on the road to Ballarat. The town was used as a location for the film *Ned Kelly*, starring Heath Ledger.

Clunes Museum: local history museum featuring displays on the gold-rush era; open weekends and school and public holidays; Fraser St. ***Bottle Museum:*** in former South Clunes State School; open Wed–Sun; Bailey St. ***Queens Park:*** on the banks of Creswick Creek, the park was created over 100 years ago.

Market: Fraser St; 2nd Sun each month. ***Booktown:*** large-scale book fair; May. ***Words in Winter Celebration:*** Aug.

Talbot This delightful, historic town has many 1860–70 buildings. Attractions include the Arts and Historical Museum in the former Methodist Church; the Bull and Mouth Restaurant in an old bluestone building and a market (holds the honour of being the first farmers market in the region) selling local produce, 3rd Sun each month. 18 km NW.

Mt Beckworth Scenic Reserve: popular picnic and horseriding reserve with panoramic views from the summit; 8 km W.

See also GOLDFIELDS, p. 154

Cobram

Pop. 5061
Map ref. 559 L12 | 584 F4

Cnr Station St and Punt Rd; (03) 5872 2132 or 1800 607 607

101.3 Oak FM, 106.5 FM ABC Local Radio

At Cobram and nearby Barooga (across the New South Wales border) the Murray River is bordered by sandy beaches, making it a great spot for fishing, watersports and picnics. The stretch of land between the township and the river features river red gum forests and lush wetlands, with tracks leading to various beaches, the most accessible of which is Thompsons Beach, located near the bridge off Boorin Street. The town is supported by orchards and dairies, earning it the nickname 'peaches and cream country'.

Historic log cabin: built in Yarrawonga in 1875, then moved piece by piece to its current location; opposite the information centre on Station St. ***Station Gallery:*** at the railway station, displays a collection of art by local artists.

Market: Punt Rd; 1st Sat each month. ***Peaches and Cream Festival:*** free peaches and cream, a rodeo, fishing competitions and other activities; odd-numbered years, Jan. ***Rotary Art Show:*** May. ***Antique Fair:*** June. ***Open Gardens Display:*** Oct.

Tokemata Retreat: 4 cottages by golf range; 100 Cemetery Rd, Cobram East; (03) 5873 5332.

Quinn Island Flora and Fauna Reserve: home to abundant birdlife and Aboriginal artefacts, including scar trees, flint tools and middens, the island can be explored on a self-guide walk; on the Murray River, accessed via a pedestrian bridge off River Rd. ***Binghi Boomerang Factory:*** large manufacturer and exporter of boomerangs. Free demonstrations are offered with purchases; Tocumwal Rd, Barooga, across the river. ***Scenic Drive Strawberry Farm:*** strawberry-picking during warmer months; Torgannah Rd, Koonoomoo; 11 km NW. ***Cactus Country:*** Australia's largest cacti gardens; Strathmerton; 16 km W. ***Ulupna Island:*** part of Barmah State Park; turn-off after Strathmerton; *see Echuca for details.* ***Murray River Horse Trails:*** a fantastic way to explore the Murray River beaches; (03) 5868 2221.

See also GOULBURN & MURRAY, p. 157

Cohuna

Pop. 1891
Map ref. 559 I11 | 584 A3 | 591 O5

Gannawarra Shire Council, 49 Victoria St, Kerang; (03) 5450 9333; www.gannawarra.vic.gov.au

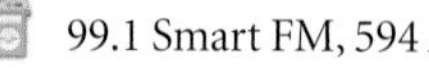
99.1 Smart FM, 594 AM ABC Local Radio

A peaceful, small service centre located on the Murray River. Cohuna's claim to fame is that its casein factory developed

continued on p. 172

BRIGHT

Pop. 2113

Map ref. 585 K9 | 587 I7

119 Gavan St; 1300 551 117; www.brightescapes.com.au

89.7 FM ABC Local Radio, 101.3 Oak FM

Bright is situated in the Ovens Valley in the foothills of the Victorian Alps. A particularly striking element of the town is the avenues of deciduous trees, at their peak during the autumn months. The Bright Autumn Festival is held annually in celebration of the spectacular seasonal changes. The Ovens River flows through the town, providing a delightful location for picnics or camping. The town also offers off-the-mountain accommodation for nearby Mount Hotham and Mount Buffalo.

Old Tobacco Sheds You could easily spend half a day here – wandering through the sheds filled with antiques and bric-a-brac and through the makeshift museums, which give an insight into the local tobacco industry and the gold rush. Also on-site is a historic hut, and the Sharefarmers Cafe serves Devonshire tea. Great Alpine Rd.

Gallery 90: local art and craft; at the visitor centre. ***Centenary Park:*** with a deep weir, children's playground and picnic facilities; Gavan St. ***Bright Art Gallery and Cultural Centre:*** community-owned gallery, displays and sells fine art and handicrafts; Mountbatten Ave. ***Bright Brewery:*** enjoy award-winning beers or brew your own; 121 Great Alpine Rd. ***Bright and District Historical Museum:*** in the old railway station building, with artefacts and photographs from the town's past; open by appt (contact visitor centre); Cnr Gavan and Anderson sts. ***Walking tracks:*** well-marked tracks around the area include Canyon Walk along the Ovens River, where remains of gold-workings can be seen; details from visitor centre. ***Murray to the Mountains Rail Trail:*** Bright sits at one end of this 94 km track suitable for cycling and walking; links several townships.

Craft market: Burke St; 3rd Sat each month. ***Autumn Festival:*** activities include craft markets and entertainment; Apr/May. ***Alpine Spring Festival:*** free entertainment, displays and open gardens, celebrating the beauty of Bright in spring; Oct.

Poplars: French-inspired menu; Shop 8, Star Rd; (03) 5755 1655. ***Simone's of Bright:*** excellent Italian cuisine; 98 Gavan St; (03) 5755 2266. ***Villa Gusto:*** Tuscan-style regional menu; 630 Buckland Valley Rd, Buckland; (03) 5756 2000.

Centenary Peaks: stylish apartments; 1 Delany Ave; (03) 5750 1433. ***The Buckland Studio Retreat:*** luxury eco-retreats; 116 McCormacks La, Buckland; 0419 133 318. ***The Odd Frog:*** architecturally designed studios; 3 McFadyens La; 0418 362 791. ***The Kilns:*** 3-bedroom corrugated-iron homestead; Cavedons La, Porepunkah; 0408 553 332. ***Villa Gusto:*** Tuscan-style suites; 630 Buckland Valley Rd, Buckland; (03) 5756 2000.

Wandiligong A National Trust–classified hamlet, the area contains well-preserved historic buildings from the town's goldmining days. The tiny village is set in a rich green valley, with an enormous hedge maze as the dominant feature and over 2 km of walkways surrounded by lush gardens. The maze is well signposted. Open 10am–5pm Wed–Sun; 6 km S.

Mt Buffalo National Park This is not a large park, but it is one of Victoria's favourites. In winter it is a haven for skiers. In summer bushwalkers and campers descend on the park, taking in the superb views from the granite peaks, the gushing waterfalls and the display of alpine wildflowers. Lake Catani is a popular spot for canoeists, and rock climbing and hang-gliding are also popular. Contact Parks Victoria on 13 1963; 10 km NW.

Tower Hill Lookout: 4 km NW. ***Boyntons/Feathertop Winery:*** open for sales and tastings; at junction of Ovens and Buckland rivers, Porepunkah; 6 km NW. ***The Red Stag Deer and Emu Farm:*** Hughes La, Eurobin; 16 km NW. ***Harrietville:*** a former goldmining village located just outside the Alpine National Park. Attractions include Pioneer Park, an open-air museum and picnic area; Tavare Park, with a swing bridge and picnic and barbecue facilities; and a lavender farm; 20 km SE. ***Alpine National Park:*** *see Mount Beauty*; to the south-east of town.

See also HIGH COUNTRY, p. 158

VICTORIA

[THE COLOURS OF AUTUMN CREATE A PICTURESQUE SETTING]

produce that became part of the diet of the astronauts flying the Apollo space missions. East of town is Gunbower Island, at the junction of the Murray River and Gunbower Creek. The island is home to abundant wildlife, including kangaroos and emus, plus breeding rookeries for birdlife during flood years.

Cohuna Historical Museum: housed in the former Scots Church, the museum features memorabilia relating to explorer Major Mitchell; Sampson St.

Murray River International Music Festival: Feb. ***Bridge to Bridge Swim:*** Mar. ***Austoberfest:*** Oct.

Gunbower Family Hotel: excellent steakhouse; Murray Valley Hwy, Gunbower; (03) 5487 1214.

Gunbower Island This island, surrounded by Gunbower Creek and the Murray River, is an internationally recognised wetland, with a great variety of waterbirds and stands of river red gum forest. A 5 km canoe trail flows through Safes Lagoon and bushwalking is another highlight.

Grove Patchwork Cottage and Tearooms: for local art and craft; Murray Valley Hwy; 4 km SE. ***Mathers Waterwheel Museum:*** features waterwheel memorabilia and outdoor aviary; Brays Rd; 9 km W. ***Murray Goulburn Factory:*** cheese factory; Leitchville; 16 km SE. ***Kow Swamp:*** bird sanctuary with picnic spots and fishing at Box Bridge; 23 km S. ***Section of Major Mitchell Trail:*** 1700 km trail that retraces this explorer's footsteps from Mildura to Wodonga via Portland. From Cohuna, follow the signposted trail along Gunbower Creek down to Mt Hope; 28 km S. ***Torrumbarry Weir:*** during winter the entire weir structure is removed, while in summer waterskiing is popular; 40 km SE.

See also GOULBURN & MURRAY, p. 157

Colac

Pop. 10 859
Map ref. 572 B9 | 580 B2 | 593 L8

Cnr Murray and Queen sts (Princes Hwy); (03) 5231 3730; www.visitotways.com

 104.7 Otway FM, 594 AM 3WV ABC Local Radio

Colac was built by the shores of Lake Colac on the volcanic plain that covers much of Victoria's Western District. The lake was once the largest freshwater body in Victoria, but harsh drought has seen it almost depleted. Still, the town acts as the gateway to the Otways. The area was once described by novelist Rolf Boldrewood as 'a scene of surpassing beauty and rural loveliness … this Colac country was the finest, the richest as to soil and pasture that I had up to that time ever looked on'.

Colac Heritage Walk: self-guide tour of the history and architectural wonders of Colac; details from visitor centre. ***Performing Arts and Cultural Centre:*** incorporates the Colac Cinema, open daily; and the Historical Centre, open 2–4pm Thurs, Fri and Sun; Cnr Gellibrand and Ray sts. ***Botanic Gardens:*** unusual in that visitors are allowed to drive through the gardens. Picnic, barbecue and playground facilities are provided; by Lake Colac. ***Barongarook Creek:*** prolific birdlife, and a walking track leading from Princes Hwy to Lake Colac; on the northern outskirts of town.

Lions Club Market: Memorial Sq, Murray St; 3rd Sun each month. ***Go Country Music Festival and Truck Show:*** Feb. ***Colac Cup:*** Feb. ***Kana Festival:*** community festival with family entertainment, music and displays; Mar. ***Garden Expo:*** Colac Showgrounds; Oct.

Old Lorne Road Olives: Mediterranean-inspired cafe; 45 Old Lorne Rd, Deans Marsh; (03) 5236 3479. ***Otway Estate Winery and Brewery:*** tasting plates and more; 10–30 Hoveys Rd, Barongarook; (03) 5233 8400.

Elliminook Heritage B&B: charming garden B&B; 585 Warncoort Rd, Birregurra; (03) 5236 2080.

Red Rock Lookout The lookout features a reserve with picnic and barbecue facilities, plus spectacular views across 30 volcanic lakes, including Lake Corangamite, Victoria's largest saltwater lake. At the base of the lookout is the Red Rock Winery. Near Alvie; 22 km N. The Volcano Discovery Trail goes from Colac to Millicent in SA, and follows the history of volcanic activity in the region; details from visitor centre.

Old Beechy Rail Trail: 45 km trail that follows one of the state's former narrow-gauge railway lines from Colac to Beech Forest, suitable for walkers and cyclists. The trail starts at Colac railway station; details from visitor centre. ***Art and craft galleries:*** at Barongarook (12 km SE); details from visitor centre. ***Burtons Lookout:*** features Otway Estate Winery and Brewery with its well-known Prickly Moses ale range, views of the Otways; 13 km S. ***Tarndwarncoort Homestead:*** wool displays and sales; off Warncoort Cemetery Rd; 15 km E. ***Birregurra:*** township located at the foot of the Otway Ranges and the edge of volcanic plains; 20 km E. ***Forrest:*** old timber and logging town in the Otway Ranges; 32 km SE. Attractions nearby include fishing, walking and picnics at the West Barwon Reservoir (2 km S), or spotting a platypus at Lake Elizabeth, formed by a landslide in 1952 (5 km SE).

See also SOUTH-WEST COAST, p. 153

Coleraine

Pop. 992
Map ref. 592 E3

Lonsdale St, Hamilton; 1800 807 056; www.sthgrampians.vic.gov.au

 94.1 FM ABC Local Radio, 99.3 Coastal FM

Situated in Victoria's Western District, Coleraine is a small, picturesque town supported by wool and beef industries. A chocolate factory, the ultimate native garden and vintage cars are just a few of the intriguing prospects that await in Coleraine.

Peter Francis Points Arboretum Two thousand species of native flora are found here, including 500 species of eucalyptus. 'The Points' sprawls up the hillside behind the town, with great views from the top, on Portland–Coleraine Rd. In town is the Eucalyptus Discovery Centre, designed to complement the arboretum and give an insight into the natural history and commercial applications of eucalypts. Whyte St.

Glenelg Fine Confectionery: immerse yourself in the rich aroma of German-style continental chocolates; tastings available; Whyte St. ***Historic Railway Station:*** also site of the visitor centre, it displays and sells local arts and crafts; Pilleau St. ***Coleraine Classic Cars:*** open by appt; Whyte St.

Tour of Southern Grampians Cycling: Apr. ***Coleraine Cup:*** Sept.

Bochara Wines: wine-tasting available Fri–Sun; Glenelg Hwy. ***Glacier Ridge Redgum:*** gallery of wooden products for sale; open by appt. ***Balmoral:*** historic township west of the Grampians; 49 km N. Nearby features include the Glendinning Homestead, just east of town, with gardens and a wildlife sanctuary. The town is also the gateway to Rocklands Reservoir, for watersports and fishing, and Black Range State Park, for bushwalking. It also holds the Balmoral Annual Show in Mar.

See also GRAMPIANS & CENTRAL WEST, p. 155

Corryong

Pop. 1229
Map ref. 566 A8 | 585 O7 | 587 O3

50 Hanson St; (02) 6076 2277; www.pureuppermurrayvalleys.com

88.7 FM Radio Upper Murray, 99.7 FM ABC Local Radio

Welcome to authentic 'Man from Snowy River' country. This district offers superb mountain scenery and excellent trout fishing in the Murray River and its tributaries, with the town being known as the home and final resting place of Jack Riley, the original 'Man from Snowy River'. A life-size statue depicting 'that terrible descent' made famous by Banjo Paterson's poem sits in the town. An annual festival honours Riley's memory with a feature event called the 'Challenge' to find his modern-day equivalent. Corryong is also the Victorian gateway to Kosciuszko National Park across the New South Wales border.

The Man from Snowy River Folk Museum Banjo Paterson's poem evoked the lives of the High Country's settlers. This charming museum proudly does the same, with local exhibits, memorabilia and photos depicting the hardships of local life, as well as a unique collection of historic skis. Hanson St.

Jack Riley's grave: Corryong cemetery. ***Man from Snowy River Statue:*** Hanson St. ***Large wooden galleon:*** Murray Valley Hwy. ***Playle's Hill Lookout:*** for a great view of the township; Donaldson St.

Towong Cup: Sat long weekend in Mar. ***The Man from Snowy River Bush Festival:*** music, art and horsemanship challenges; Apr. ***Upper Murray Challenge:*** 1st Sat in Oct. ***Corryong Pro Rodeo:*** New Year's Eve.

Burrowa–Pine Mountain National Park Pine Mountain is one of Australia's largest monoliths. Mt Burrowa is home to wet-forest plants and unique wildlife, including wombats and gliders. Both mountains provide excellent and diverse opportunities for bushwalkers, campers, climbers and birdwatchers. The Cudgewa Bluff Falls offer fabulous scenery and bushwalking. Contact Parks Victoria on 13 1963; main access is from the Cudgewa–Tintaldra Rd, which runs off Murray Valley Hwy; 27 km W.

Khancoban This NSW town was built by the Snowy Hydro for workers on the hydro-electric scheme. Its willow- and poplar-lined streets, historic rose garden and mountains give the town a European feel. Huge trout are caught in Khancoban Pondage. Nearby, Murray 1 Power Station Visitor Centre reveals the workings of this 10-turbine station. South, along Alpine Way through Kosciuszko National Park, is the spectacular Scammell's Spur Lookout and historic Geehi Hut. 32 km E.

Nariel: Nariel Creek is a good spot for trout fishing. The town hosts the Nariel Creek Folk Music Festival each Dec; 8 km SW. ***Towong:*** historic Towong Racecourse is where scenes from *Phar Lap* were filmed. Gangster Squizzy Taylor also once stole the takings; 12 km E. ***Lookouts:*** lookout with views over Kosciuszko National Park at Towong, 12 km NE; Embery's Lookout over Mt Mittamatite, 16 km N. ***Walwa:*** hire canoes and mountain bikes from Upper Murray Holiday Resort and Winery; 47 km NW.

Touring routes: Murray River Rd, Lakeside Loop, Mitta Valley Loop; details from visitor centre.

See also HIGH COUNTRY, p. 158

Cowes

see inset box on next page

Creswick

Pop. 2487
Map ref. 572 D2 | 579 C9 | 593 M3

1 Raglan St; (03) 5345 1114; www.creswick.net

99.9 VoiceFM, 107.9 FM ABC Local Radio

Creswick is an attractive and historic town, a symbol of the rich and heady life of the gold-rush days of the 1850s. Unfortunately, the goldmining also meant that the surrounding forests were decimated. Today the town is surrounded by pine plantations over 100 years old; they exist thanks to the initiative and foresight of local pioneer John La Gerche and – while they are no replacement for the Australian bush – they have given Creswick the title of 'the home of forestry'. Creswick was the birthplace of renowned Australian artist Norman Lindsay, many of whose paintings can be seen in the local historical museum.

Historic walk: self-guide tour, map from visitor centre. ***Giant Mullock Heaps:*** indicate how deep mines went; Ullina Rd. ***Creswick Museum:*** photos and memorabilia from the town's goldmining past as well as an exhibition of Lindsay paintings; open Sun, public holidays or by appt; Albert St. ***Gold Battery:*** est. 1897; Battery Cres. ***Creswick Woollen Mills:*** last coloured woollen mill of its type in Australia; offers product sales, regular demonstrations and exhibitions; Railway Pde.

Creswick Makers Market: 1st Sun each month. ***Creswick CALCAN Market:*** 3rd Sat each month. ***Forestry Fiesta:*** Oct.

Harvest 383: regional produce menu; Novotel Forest Resort Creswick, 1500 Midland Hwy; (03) 5345 9600.

Novotel Forest Resort Creswick: luxury rooms; 1500 Midland Hwy; (03) 5345 9600. ***Rossmore Cottage:*** 2-bedroom miners cottage; 12 Bald Hills Rd; (03) 5345 2759.

Creswick Regional Park After La Gerche replanted the denuded hills around Creswick in the 1890s, the state established a nursery that it continues to operate today. Further natural history can be explored on the various walking trails, including the 30 min Landcare Trail or the longer La Gerche Forest Walk. Visit St Georges Lake, once a mining dam and now popular for picnics and watersports, and Koala Park, an old breeding ground for koalas that was highly unsuccessful (they escaped over the fences). Slaty Creek is great for gold panning or picnics, with abundant birdlife. The park stretches east and south-east of town. Contact Parks Victoria on 13 1963 or the Creswick Landcare Centre, located within the park, on (03) 5345 2200.

Tangled Maze: a maze formed by climbing plants; 5 km E. ***Smeaton:*** pretty little town with the historic Smeaton House, the Tuki Trout Farm and Anderson's Mill; 16 km NE.

See also GOLDFIELDS, p. 154

Daylesford

see inset box on page 177

COWES

Pop. 4217

Map ref. 573 K11 | 575 L10 | 582 D8

895 Phillip Island Tourist Rd, Newhaven; (03) 5956 7447 or 1300 366 422; www.visitphillipisland.com

89.1 3MFM South Gippsland, 774 AM ABC Local Radio

Situated on the north side of Phillip Island, Cowes is the island's major town. It is linked to the Mornington Peninsula by a passenger ferry service to Stony Point and by road to Melbourne via the San Remo bridge. The foreshore offers coastal walks, safe swimming beaches and fishing from the jetty. Seal-watching cruises to Seal Rocks operate from the jetty and are the best way to see the fur seals close-up. There are many activities and events that attract 3.5 million visitors to Phillip Island each year. All are within easy reach of Cowes.

Seal-watching cruises: depart from the jetty to Seal Rocks; bookings on 1300 763 739.

Market: crafts and second-hand goods; Settlement Rd; each Sun. ***Farmers market:*** Churchill Island; 4th Sat each month. ***World Superbike Championships:*** Grand Prix Circuit; Feb/Mar. ***Churchill Island Heritage Farms Easter Fun:*** farm show; Easter. ***Australian Motorcycle Grand Prix:*** Grand Prix Circuit; Oct. ***V8 Supercars:*** Grand Prix Circuit; Sept.

Harry's on the Esplanade: waterfront dining; Shop 5, 17 The Esplanade; (03) 5952 6226. ***Hotel:*** casual bistro dining; 11–13 The Esplanade; (03) 5952 2060. ***Infused:*** modern Australian; 115 Thompson Ave; (03) 5952 2655. ***Foreshore Bar and Restaurant:*** seafront gastropub dining; 11 Beach Rd, Rhyll; (03) 5956 9520.

All Seasons Phillip Island Resort: villas on acreage; 2128 Phillip Island Rd; (03) 5952 8000. ***Holmwood Guesthouse:*** 3 B&B rooms, 2 cottages; 37 Chapel St; (03) 5952 3082. ***Quest Phillip Island:*** centrally located apartments; Cnr Bass Ave and Chapel St; (03) 5952 2644. ***Glen Isla House:*** boutique accommodation in 1870 homestead; 230 Church St; (03) 5952 1882.

Penguin Parade The nightly penguin parade is Phillip Island's most popular attraction. During this world-famous event, little penguins emerge from the sea after a tiring fishing expedition and cross Summerland Beach to their little homes in the dunes. Tours run at sunset each night, and the penguins can be spotted from the boardwalks and viewing platforms. The site also has an interactive visitor centre with fascinating details about these adorable creatures. Note that no cameras are allowed beyond the visitor centre. Bookings on (03) 5951 2800; 12 km sw.

The Nobbies Centre and Seal Rocks An interactive centre gives visitors an insight into local marine life, including Australia's largest colony of Australian fur seals via cameras that you can control yourself. Outside, the island features a cliff-side boardwalk with views of the fantastic natural landmark The Nobbies and out to Seal Rocks. Walk around to the Blowhole to hear the thunderous noise of huge waves and look out for the nesting sites of vast colonies of seagulls and short-tailed shearwaters that migrate to the island annually. Informative displays explain each natural attraction. Ventnor Rd; 15 km sw.

Phillip Island Wildlife Park: features native fauna, with visitors able to handfeed kangaroos and wallabies; Phillip Island Rd; (03) 5952 2038; 3 km s. ***Koala Conservation Centre:*** view these lovely creatures in their natural habitat from an elevated boardwalk; Phillip Island Rd; (03) 5951 2800; 5 km se. ***Grand Prix Circuit:*** the circuit is steeped in both old and recent history, which is detailed thoroughly in the visitor centre. You can also go go-karting on a replica of the Grand Prix track; Back Beach Rd; (03) 5952 9400; 6 km s. ***A Maze 'N Things:*** family fun park featuring a large timber maze, optical-illusion rooms and 'maxigolf'; Phillip Island Rd; 6 km se. ***Rhyll Inlet:*** wetlands of international significance, with the marshes and mangroves providing an important breeding ground for wading birds. There are various loop walks, as well as an excellent view from the Conservation Hill Observation Tower; 7 km e. ***Rhyll Trout & Bush Tucker Farm:*** fish for rainbow trout in the indoor Rainforest Pool, take a fishing lesson or wander around the self-guide bush tucker trail; open 10am–5pm; 36 Rhyll–Newhaven Rd, Rhyll; (03) 5956 9255. ***Wineries:*** Phillip Island Vineyard and Winery offers tastings, sales and casual dining; Berrys Beach Rd; (03) 5956 8465; 7 km sw. Purple Hen Wines also offers tastings and light meals; 96 McFees Rd, Rhyll; (03) 5956 9244; 9 km se. ***Churchill Island:*** a road bridge provides access to this protected parkland, which features a historic homestead, a walking track and abundant birdlife; 16 km se. ***National Vietnam Veterans Museum:*** details the history of Australian involvement in the Vietnam War, displaying around 6000 artefacts; 25 Veterans Dr, Newhaven; (03) 5956 6400; 16 km se. ***Phillip Island Chocolate Factory:*** visit Panny's Amazing World of Chocolate with information on the chocolate-making process and a model of Dame Edna made from 12 000 chocolate pieces; visitors can even make their own chocolate; Newhaven; (03) 5956 6600; 16 km se. ***Pelicans:*** see these unusual birds up close, with feeding time daily at 12pm; San Remo Pier (opposite the Fishing Co-op); 17 km se. ***Cape Woolamai:*** the beach is renowned for its fierce and exciting surf (patrolled in season). From the beach there are a number of 2–4 hr loop walks, many to the southern end of the cape and passing the Pinnacles rock formations on the way. South of Cape Woolamai township; 18 km se. ***Wildlife Wonderland:*** centre includes the Earthworm Museum, a giant earthworm and Wombat World; (03)5678 2222; 31 km e.

[LITTLE PENGUINS COMING ASHORE AT DUSK]

See also PHILLIP ISLAND & GIPPSLAND, p. 150

Dimboola

Pop. 1493
Map ref. 590 F8

Dim E-Shop, 109–111 Lloyd St; (03) 5389 1588; www.dimboola.com.au

 96.5 Triple H FM, 594 AM ABC Local Radio

Dimboola, on the Wimmera River, is a key access point to the Little Desert National Park. The area was home to the Wotjobaluk Aboriginal people until the first European settlers arrived. The district was known as 'Nine Creeks' because of the many little streams that appear when the river recedes after floods. Many of the early white settlers were German.

Historic buildings: include the mechanics institute in Lloyd St and the Victoria Hotel, a grand 2-storey structure with grapevines hanging from the verandahs (cnr Wimmera and Victoria sts). ***Walking track:*** follows a scenic stretch of the Wimmera River. The track can be followed all the way to the Horseshoe Bend camping ground in the Little Desert National Park 7 km away; details of walks from visitor centre.

Little Desert National Park This park covers 132 647 ha. The eastern block (the section nearest to Dimboola) has picnic and camping facilities and good walking tracks. The park does not resemble the typical desert – it contains extensive heathlands and, during spring, more than 600 varieties of wildflowers and over 40 types of ground orchids. The park is home to the distinctive mallee fowl, and the large ground-nests built by the male birds can be seen during breeding season. Kangaroos, possums and bearded dragons are just some of the other wildlife that inhabit the park. 6 km sw. *See also Nhill.*

Pink Lake: a salt lake that reflects a deep pinkish colour, particularly impressive at sunset, but has dried up in recent years; 9 km NW. ***Ebenezer Mission Station:*** founded in 1859 in an attempt to bring Christianity to the local Aboriginal people. The site contains fascinating ruins of the original buildings, a cemetery and a restored limestone church; off the Dimboola–Jeparit Rd; 15 km N. ***Kiata Lowan Sanctuary:*** the first part of Little Desert National Park to be reserved, in 1955. Home to the mallee fowl; Kiata; 26 km w.

See also GRAMPIANS & CENTRAL WEST, p. 155

Donald

Pop. 1429
Map ref. 558 F12 | 591 J7

Council Offices, cnr Houston and McCulloch sts; (03) 5497 1300.

 96.5 Triple H FM, 99.1 FM ABC Local Radio

Donald is on the scenic Richardson River and referred to by locals as 'Home of the Duck', owing to the many waterbirds that live in the region. The town also features Bullocks Head, a tree on the riverbank with a growth that looks like its namesake. The 'bull' is also used as a flood gauge – according to how high the waters are, the 'bull' is either dipping his feet, having a drink or, when the water is really high, going for a swim.

Bullocks Head Lookout: beside Richardson River; Byrne St. ***Steam Train Park:*** a restored steam locomotive, an adventure playground and barbecue facilities; Cnr Hammill and Walker sts. ***Historic Police Station:*** dates back to 1865; Wood St. ***Shepherds hut:*** built by early settlers; Wood St. ***Agricultural Museum:*** an impressive collection of agricultural machinery; Hammill St. ***Scilleys Island:*** reserve on the Richardson River featuring wildlife, walking tracks and picnic facilities; access by footbridge from Sunraysia Hwy. ***Kooka's Country Cookies:*** tours and sales; Sunraysia Hwy.

Scottish Dancing Country Weekend: June. ***Donald Cup:*** Nov.

Lake Buloke The lake is filled by the floodwaters of the Richardson River, so its size varies greatly with the seasons. This extensive wetland area is home to a variety of birdlife and is a popular venue for fishing, picnicking and bushwalking. The end of the park closest to town is a protected bird sanctuary. 10 km N.

Fishing There is good fishing for redfin and trout in the many waterways close to town. Good spots include Lake Cope Cope, 10 km s; Lake Batyo Catyo and Richardson River Weir, both 20 km s; Watchem Lake, 35 km N; and the Avoca River, which runs through Charlton, 43 km NE.

Mt Jeffcott: flora, kangaroos and views over Lake Buloke; 20 km NE.

See also GRAMPIANS & CENTRAL WEST, p. 155

Drysdale

Pop. 10 217
Map ref. 570 A9 | 572 G8 | 574 D4 | 582 A6 | 593 O7

Queenscliff Visitor Information Centre, 55 Hesse St, Queenscliff; (03) 5258 4843; www.visitgreatoceanroad.org.au

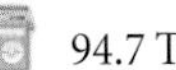 94.7 The Pulse FM, 774 AM ABC Local Radio

Drysdale, situated on the Bellarine Peninsula, is primarily a service centre for the local farming community. The town is close to the beaches of Port Phillip Bay and there are a number of wineries in the area, including the delightful Spray Farm Winery. Drysdale is now considered a satellite town of Geelong, yet retains a charming, holiday-resort atmosphere.

Old Courthouse: home of the Bellarine Historical Society; High St. ***Drysdale Community Crafts:*** High St.

Celtic Festival: June. ***Community Market:*** at the reserve on Duke St; 3rd Sun each month Sept–Apr.

Loam Restaurant: local produce menu; Lighthouse Olive Grove, 650 Andersons Rd; (03) 5251 1101. ***Port Pier Cafe:*** casual Spanish; Portarlington Foreshore, Pier St, Portarlington; (03) 5259 1080. ***The Ol' Duke Hotel:*** contemporary cuisine; 40 Newcombe St, Portarlington; (03) 5259 1250.

The Ol' Duke Hotel: 6 bedrooms, 2 self-catering apartments; 40 Newcombe St, Portarlington; (03) 5259 1250.

Wineries For over 150 years vines have been grown on the Bellarine Peninsula, and most vineyards here today remain family owned and operated. Owing to the peninsula's varying soil conditions, a range of white and red wines are produced. Many wineries in the area offer cellar-door tastings and sales. These include the historic Spray Farm Winery, which runs the summer concert series each Feb and Mar, known as A Day on the Green, in a natural amphitheatre. Great views to the sea can be had from Scotchmans Hill Winery. Winery map from visitor centre.

Bellarine Peninsula Railway: steam-train rides from Queenscliff to Drysdale and return; *see Queenscliff.* ***Lake Lorne picnic area:*** 1 km sw. ***Portarlington:*** a popular seaside resort town with a restored flour mill featuring displays of agricultural history, a safe bay for children to swim in and fresh mussels for sale near the pier. There is a market at Parks Hall, last Sun each month; 10 km NE.

St Leonards: a small beach resort, which includes Edwards Point Wildlife Reserve, a memorial commemorating the landing of Matthew Flinders in 1802 and of John Batman in 1835; 14 km E.

See also WERRIBEE & BELLARINE, p. 152

Dunkeld

Pop. 400
Map ref. 592 G4

Lonsdale St, Hamilton; 1800 807 056; www.sthgrampians.vic.gov.au

 94.1 FM ABC Local Radio, 99.3 Coastal FM

Dunkeld is considered the southern gateway to the Grampians, and its natural beauty has long been recognised since the explorer Major Thomas Mitchell camped here in 1836. It was originally named Mount Sturgeon after the mountain that towers over the town. Both Mount Sturgeon and Mount Abrupt (to the north of town) have been renamed to recognise the ancient Aboriginal heritage of the landscape; they are now known as Mount Wuragarri and Mount Murdadjoog respectively.

Dunkeld Arboretum Exotic species from all over the world have been planted. Ideal for walking, cycling, fishing and picnics. Old Ararat Rd.

Historical Museum Housed in an old church, the museum features displays on the history of the local Aboriginal people, the wool industry and the journeys of explorer Major Mitchell. It also offers dining and accommodation. Open weekends or by appt; Templeton St.

Corea Wines: open by appt. ***Varrenti Wines:*** open 12–5pm; Blackwood Rd. ***Sandra Kranz Art Studio:*** Glass St. ***Waiting Room Art Gallery:*** Parker St. ***Bushwalking:*** walking trails include Mt Abrupt, Mt Sturgeon and the Piccaninny Walk.

Dunkeld Cup: Nov. ***Dunkeld Arts Festival:*** biennial, Nov.

Gourmet Pantry: cafe and provedore; 109 Parker St (Glenelg Hwy); (03) 5577 2288. ***Royal Mail Hotel:*** outstanding contemporary menu, exceptional wine list; 98 Parker St (Glenelg Hwy); (03) 5577 2241.

Aquila Eco Lodges: 2 tree houses and 2 loft houses; Manns Rd; (03) 5577 2582. ***Griffins Hill:*** stylish yoga retreat; Victoria Valley Rd; (03) 5577 2499. ***Royal Mail Hotel:*** rooms and shearers' cottages; 98 Parker St (Glenelg Hwy); (03) 5577 2241. ***Southern Grampians Cottages:*** log cabins; 33–35 Victoria Valley Rd; (03) 5577 2457.

Grampians National Park The southern section of the park includes Victoria Valley Rd, a scenic drive that stops at Freshwater Lake Reserve (8 km N), popular for picnics. Also near Dunkeld are various hiking destinations and the Chimney Pots, a formation popular for rock climbing; access via Henty Hwy. *For further details on the park see Halls Gap.*

Grampians Pure Sheep Dairy: sample some of the sheep milk, yoghurts and cheeses while also watching how they are made; Glenelg Hwy.

See also GRAMPIANS & CENTRAL WEST, p. 155

Dunolly

Pop. 605
Map ref. 579 C4 | 591 M10

109 Broadway; (03) 5468 1205.

91.1 FM ABC Local Radio, 99.1 Goldfields FM

The towns of Dunolly, Wedderburn and Inglewood formed the rich goldfield region colloquially known in the 1850s as the 'Golden Triangle'. The district has produced more gold nuggets than any other goldfield in Australia. The 'Welcome Stranger', considered to be the largest nugget ever discovered, was found 15 kilometres north-west of Dunolly, at Moliagul.

Restored courthouse: offers a display relating to gold discoveries in the area; open Sat afternoons; Market St. ***Original lock-up and stables:*** viewable from street only; Market St. ***Gold-themed tours of the region:*** include gold panning in local creeks; details from visitor centre.

Market: with local produce, crafts and second-hand goods; Market St; 3rd Sat each month. ***Community Street Market:*** Broadway; 4th Sat each month.

Moliagul: the Welcome Stranger Discovery Walk leads to a monument where the Welcome Stranger nugget was found in 1869. Moliagul is also the birthplace of Rev. John Flynn, founder of the Royal Flying Doctor Service; 15 km NW. ***Laanecoorie Reservoir:*** a great spot for swimming, boating and waterskiing, water levels permitting, with camping and picnic facilities; 16 km E. ***Tarnagulla:*** a small mining town with splendid Victorian architecture and a flora reserve nearby; 16 km NE. ***Bealiba:*** hosts a market 2nd Sun each month; 21 km NW.

See also GOLDFIELDS, p. 154

Echuca

Pop. 12 361
Map ref. 559 J12 | 584 C5

2 Heygarth St; (03) 5480 7555 or 1800 804 446; www.echucamoama.com

 91.1 FM ABC Local Radio, 104.7 Radio EMFM

Visitors to Echuca are transported back in time by the sight of beautiful old paddleboats cruising down the Murray River. The town is at the junction of the Murray, Campaspe and Goulburn rivers. Once Australia's largest inland port, its name comes from an Aboriginal word meaning 'meeting of the waters'. A historic iron bridge joins Echuca to Moama in New South Wales.

Port of Echuca The massive red-gum wharf has been restored to the grandeur of its heyday, with huge paddlesteamers anchored here. Cruises are available on many boats, including the paddlesteamer *Pevensey*, renamed *Philadelphia* for the TV miniseries *All the Rivers Run*; the D26 logging barge; PS *Alexander Arbuthnot*; and PS *Adelaide*. Cruises are also available on PS *Canberra*, *Pride of the Murray* and PS *Emmylou*. The MV *Mary Ann* also features a fine restaurant.

Historic buildings: many along Murray Espl include the Star Hotel, with an underground bar and escape tunnel, and the Bridge Hotel, built by Henry Hopwood, the founder of Echuca, who ran the original punt service. ***Red Gum Works:*** wood-turning demonstrations; Murray Espl. ***Sharp's Magic Movie House and Penny Arcade:*** award-winning attractions; Murray Espl. ***Echuca Historical Society Museum:*** housed in former police station; open 11am–3pm daily; High St. ***Billabong Carriages:*** Murray Espl. ***National Holden Museum:*** Warren St.

Southern 80 Ski Race: from Torrumbarry Weir to Echuca; Feb. ***Riverboats, Jazz, Food and Wine Festival:*** Feb. ***Steam, Horse and Vintage Rally:*** June. ***Winter Blues Festival:*** July. ***Port of Echuca Steam Heritage Festival:*** Oct.

Ceres: appealing modern menu; 2 Nish St; (03) 5482 5599. ***Left Bank:*** modern bistro; 551 High St; (03) 5480 3772. ***Oscar W's Wharfside:*** modern Australian; 101 Murray Espl, Port of

continued on p. 178

DAYLESFORD

Pop. 3071

Map ref. 572 E1 | 579 E8 | 593 N3

98 Vincent St; (03) 5321 6123; www.visitdaylesford.com

99.9 VoiceFM, 107.9 FM ABC Local Radio

Daylesford is at the centre of Victoria's spa country. The area developed with the discovery of gold, which lured many Swiss-Italian settlers, but it was the discovery of natural mineral springs that proved a more lasting attraction. Of the 72 documented springs in the area, the most famous are nearby Hepburn Springs. The water is rich with minerals that dissolve into it as it flows from the crest of the Great Dividing Range through underground rocks, and it is known for its rejuvenating and healing qualities. Daylesford has grown as a destination in itself, complete with beautiful gardens, interesting shopping, great eating and a huge range of accommodation. The streets are lined with trees that blaze with colour in autumn, and inside the attractive old buildings are restaurants, cafes, galleries, bookshops, bakeries and chocolate shops. Overlooking the lake is one of regional Victoria's most highly regarded restaurants, the Lake House.

Convent Gallery A magnificent building surrounded by delightful cottage gardens, this former convent and girls school has been restored and features an impressive collection of artwork, sculptures and jewellery. A cafe serves local produce and Devonshire tea. Daly St.

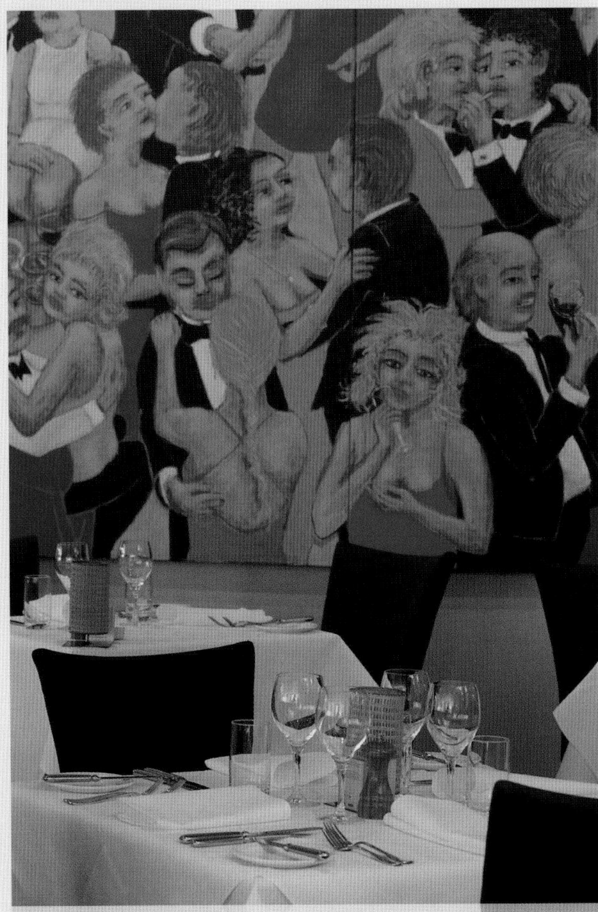

[LAKE HOUSE RESTAURANT]

Historical Museum: features a collection of photographs from the region's past and artefacts from the local Djadja Wurrung people; open weekends and public and school holidays; Vincent St. ***Lake Daylesford:*** a lovely spot for picnics, with paddleboats and rowboats for hire in the warmer months. The Tipperary walking track starts here and ends at the Mineral Springs Reserve; access to the lake is from Bleakly Rd. ***Wombat Hill Botanical Gardens:*** est. 1861, these lovely gardens are situated on the hill overlooking town. ***Daylesford Spa Country Train:*** leaves railway station for Bullarto (11 km SE) each Sun.

Market: for arts, crafts and local produce; near railway station; each Sun morning. ***Silver Streak Champagne Train:*** train journey with gourmet food; 1st Sat each month. ***Hepburn Swiss–Italian Festival:*** Apr/May. ***Highland Gathering:*** Dec.

Frangos & Frangos: city-chic European menu; 82 Vincent St; (03) 5348 2363. ***Lake House:*** exceptional regional dining; King St; (03) 5348 3329. ***Mercato @ Daylesford:*** modern Italian; 32 Raglan St; (03) 5348 4488. ***The Farmers Arms Hotel:*** hearty rustic dishes; 1 East St; (03) 5348 2091. ***Cosy Corner:*** quaint eatery; 3 Tenth St, Hepburn Springs; (03) 5348 2576.

Lake Daylesford Lodge: luxurious lakeside villas; 32 King St; (03) 5348 4422. ***Lake House:*** waterfront suites and lodge rooms; King St; (03) 5348 3329. ***Latte Cottage:*** contemporary 2-bedroom cottage; 33 Fulcher St; 0416 264 165. ***Peppers Springs Retreat & Spa:*** luxury hotel and villas; 124 Main Rd, Hepburn Springs; (03) 5348 2202.

Hepburn Springs spas The Hepburn Spa and Bathhouse Wellness Retreat offers pure mineral water spas and hydrotherapy, massage therapies and an extensive range of relaxation, health and beauty treatments. Mineral Springs Cres, Hepburn Springs; (03) 5348 8888. Dating back to 1894, the recently renovated Hepburn Bathhouse has state-of-the-art communal and private mineral bathing, spas and therapies utilising the renowned local mineral springs, plus a day spa for massage, facials and indulgent beauty treatments. Mineral Springs Reserve Rd, Hepburn Springs; (03) 5321 6000; 4 km N.

Lavandula Swiss Italian Farm A sprawling estate featuring fields of lavender, cottage gardens and sales of lavender-based products. The Lavandula Harvest Festival is a popular event with a variety of family entertainment, held in Jan. Open daily, winter months weekends only; Shepherds Flat; (03) 5476 4393; 10 km N.

Hepburn Regional Park: located around Daylesford and Hepburn Springs, this park features goldmining relics, mineral springs and the impressive Mt Franklin, an extinct volcano, with panoramic views from the summit and picnic, barbecue and camping facilities around the base. There are good walking tracks. ***Waterfalls:*** several in area, including Sailors Falls, 5 km S; Loddon Falls, 10 km NE; Trentham Falls, 21 km SE. ***Breakneck Gorge:*** early goldmining site; 5 km N. ***Glenlyon:*** small town that hosts the popular Glenlyon Sports Day on New Year's Day, and the Fine Food and Wine Fayre in July; 8 km NE. ***Lyonville Mineral Springs:*** picnic and barbecue facilities; 15 km SE. ***Yandoit:*** historic Swiss-Italian settlement; 18 km NW.

See also SPA & GARDEN COUNTRY, p. 151

VICTORIA

Echuca; (03) 5482 5133. PS ***Emmylou:*** set-price 3-course dinner; 57 Murray Espl; (03) 5482 5244.

PS *Emmylou:* 2-berth cabins; 57 Murray Espl; (03) 5482 5244. ***Quest Echuca:*** stylish serviced apartments; 25–29 Heygarth St; (03) 5481 3900. ***Bright on the Murray:*** 5 B&B homestead rooms; 93 Goldsborough Rd, Moama; (03) 5483 6264. ***Perricoota Vines Retreat:*** spa villas; 400 Perricoota Rd, Moama; (03) 5482 6655 or 1800 826 655.

Barmah State Park This park combines with Barmah State Forest to contain the largest river red gum forest in Victoria. Nearby are Barmah Lakes, a good location for fishing and swimming. Canoes and barbecue pontoons are available for hire. Ulupna Island, in the eastern section of the park (near Strathmerton), has river beaches, camping and a large population of koalas. Barmah Muster, a cattle muster, is held in the state forest in Apr. Contact Parks Victoria on 13 1963; 39 km NE.

Moama: attractions include the Silverstone Go-Kart Track and the Horseshoe Lagoon nature reserve; 2 km N. ***Mathoura:*** set among the mighty red gums, Mathoura is a charming Murray town over the NSW border. Fishing is popular, with sites including Gulpa Creek and the Edward and Murray rivers. To see the forest in its splendour, take the Moira Forest Walkway or, for that authentic Murray River experience, visit nearby Picnic Pt, popular for camping, picnics, waterskiing and fishing; 40 km N. ***Nathalia:*** a town on Broken Creek with many historic buildings. Walking tracks along the creek take in fishing spots, old homesteads and a lookout; 57 km E.

See also **Goulburn & Murray, p. 157**

Edenhope

Pop. 787
Map ref. 590 C11 | 601 H9

96 Elizabeth St; (03) 5585 1509; www.westwimmera.vic.gov.au

94.1 FM ABC Local Radio, 96.5 Triple H FM

Just 30 kilometres from the South Australian border, Edenhope is set on the shores of Lake Wallace, a haven for waterbirds. The town is renowned as the site where, in 1868, Australia's first all-Aboriginal cricket team trained – their coach was T. W. Wills, who went on to establish Australian Rules football. A cairn in Lake Street honours the achievements of this early cricket team.

Edenhope Antiques: offers an extensive variety of antique wares; Elizabeth St. ***Bennetts Bakery:*** Elizabeth St. ***Lake Wallace:*** walking tracks and birdwatching hides; Wimmera Hwy.

Henley-on-Lake Wallace Festival: with market and family entertainment; Feb. ***Races:*** Mar long weekend.

Harrow One of Victoria's oldest inland towns, Harrow has many historic buildings in Main St, including the Hermitage Hotel, the police station and an early log gaol. The Johnny Mullagh Cricket Centre, a celebration of the first Australian Aboriginal cricketer to travel overseas, is also located in town. Kelly's Garage and Transport Museum in Main St is popular with car enthusiasts, and the National Bush Billycart Championship is held here in Mar. 32 km SE.

Dergholm State Park: 26 km S; *see Casterton.* ***Naracoorte Caves National Park:*** World Heritage site of fabulous caves with extensive fossil history to explore; around 50 km W over SA border. ***Fishing:*** redfin, trout and yabbies in many lakes and swamps nearby. Availability depends on water levels; contact visitor centre for locations.

See also **Grampians & Central West, p. 155**

Eildon

Pop. 742
Map ref. 571 H1 | 573 O1 | 582 G1 | 584 G12 | 586 C11

High St; (03) 5774 2909; www.murrindinditourism.com.au

97.3 FM ABC Local Radio, 106.9 UGFM Upper Goulburn Community Radio

Eildon established itself as a town to service dam workers, and later holiday-makers, when the Goulburn River was dammed to create Lake Eildon. This is the state's largest constructed lake, irrigating a vast stretch of northern Victoria and providing hydro-electric power. In recent years low water levels have revealed homesteads that were submerged when the dam was constructed. The lake and the surrounding national park are popular summer holiday destinations, especially for watersports, fishing and boating.

Lions Club Monster Market: Easter. ***Opening of Fishing Season Festival:*** Sep.

Robyns Nest B&B: retreat for 2 couples; 13 High St; (03) 5779 1064.

Lake Eildon National Park Comprising the lake and surrounding woodlands, hills and wilderness areas, this national park provides a venue for many water- and land-based activities. When full, Lake Eildon has 6 times the capacity of Sydney Harbour. Hire a kayak, boat or houseboat from the outlets in Eildon to explore the waters, or enjoy the thrills of waterskiing with the picturesque foothills of the Australian Alps providing a backdrop. In the surrounding hills and woodlands there are various nature walks, scenic drives and panoramic lookout points. Many of the walks start at the campgrounds; details from visitor centre, or from Parks Victoria on 13 1963.

Lake Eildon Wall Lookout: 1 km N. ***Eildon Pondage and Goulburn River:*** for excellent fishing – there is no closed season for trout in Lake Eildon. ***Mt Pinniger:*** for views of Mt Buller, the alps and the lake; 3 km E. ***Freshwater Discovery Centre:*** native-fish aquariums and displays; Snobs Creek; 6 km SW. ***Waterfalls:*** include Snobs Creek Falls and Rubicon Falls; 18 km SW via Thornton. ***Eildon Trout Farm:*** towards Thornton on Back Eildon Rd.

See also **High Country, p. 158**

Emerald

Pop. 6317
Map ref. 570 H7 | 571 C10 | 573 L7 | 575 O1 | 582 E5

Dandenong Ranges Information Centre, 1211 Burwood Hwy, Upper Ferntree Gully; (03) 9758 7522; www.dandenongrangestourism.com.au

97.1 FM 3MDR Mountain District Radio, 774 AM ABC Local Radio

Emerald is a delightful little town set in the Dandenong Ranges, which lie behind Melbourne's eastern suburbs. Over the weekend people come from the city into 'the hills' to take in the scenic forests and visit the cafes, galleries, and antique and craft stores.

Emerald Lake The lake is a lovely, tranquil spot ideal for picnics and walks. Attractions include the largest model railway display in the Southern Hemisphere, paddleboats, cafe and tearooms, fishing, free wading pool (summer months) and a variety of walking trails. Picnic shelters are available for hire throughout the year. Puffing Billy stops at Emerald Lake and many passengers spend a day here before returning on the train in the late afternoon. Emerald Lake Rd.

Galleries and craft shops: a wide variety, specialising in locally made products; along Main St.

PAVE (Performing and Visual Arts in Emerald) Festival: Apr. ***Great Train Race:*** runners attempt to race Puffing Billy from Belgrave to Emerald Lake Park; 1st Sun in May.

Elevation at Emerald: casual dining with views; 374 Main Rd; (03) 5968 2911.

Fernglade on Menzies: self-contained and B&B suites; 11 Caroline Cres; (03) 5968 2228. ***Fernhem Cottages:*** 3 country-style cottages; 109 Emerald–Monbulk Rd; (03) 5968 5462.

Puffing Billy Victoria's favourite steam train runs between Belgrave and Gembrook, stopping at Emerald Lake. The views from the train are of tall trees and ferny gullies, and if you time your trip for the last Sat of the month you could catch the local craft and produce market at Gembrook station. Also at Gembrook is the Motorist Cafe and Museum. Puffing Billy operates every day of the year, except Christmas Day; 24 hr recorded timetable and fare information on 1900 937 069, all other inquiries (03) 9757 0700; Belgrave 9 km W, Gembrook 14 km E.

Menzies Creek This town is home to Cotswold House, where visitors enjoy gourmet food amid fantastic views. Nearby is Cardinia Reservoir Park, where picnic spots are shared with free-roaming kangaroos, and Lake Aura Vale, a popular spot for sailing. Belgrave–Gembrook Rd; 4 km NW.

Sherbrooke Equestrian Park: trail-rides; Wellington Rd; 3 km W. ***Australian Rainbow Trout Farm:*** Macclesfield; 8 km N. ***Sherbrooke Art Gallery:*** impressive collection of local artwork; Monbulk Rd, Belgrave; 11 km NW. ***Bimbimbie Wildlife Park:*** Mt Burnett; 12 km SE.

See also YARRA & DANDENONGS, p. 148

Euroa

Pop. 2773
Map ref. 584 F9 | 586 A7

Strathbogie Ranges Tourism Information Service, BP Service Centre, Tarcombe St; 1300 134 610; www.strathbogieregion.com.au

97.7 FM ABC Local Radio, 98.5 FM

Euroa was the scene of one of Ned Kelly's most infamous acts. In 1878 the notorious bushranger staged a daring robbery, rounding up some 50 hostages and making off with money and gold worth nearly £2000. The Strathbogie Ranges, once one of the Kelly Gang's hideouts, now provide a pretty backdrop to the town, and the region really comes to life in spring, when stunning wildflowers bloom. During this time and in autumn a number of private gardens are open to the public.

Farmers Arms Historical Museum The museum features displays explaining the history of Ned Kelly and Eliza Forlonge; Eliza and her sister are said to have imported the first merino sheep into Victoria. Open Fri–Mon afternoons; Kirkland Ave.

Walking trail: self-guide trail to see the rich history and architecture of the town, including the National Bank building and the post office, both in Binney St; brochure available from visitor centre. ***Seven Creeks Park:*** good freshwater fishing, particularly for trout; Kirkland Ave.

Miniature steam-train rides: Turnbull St; last Sun each month. ***Wool Week:*** Oct.

Ruffy Produce Store: excellent regional produce store; 26 Nolans Rd, Ruffy; (03) 5790 4387.

Forlonge B&B: 2 air-conditioned suites; 76 Anderson St; (03) 5795 2460.

Faithfull Creek Waterfall: 9 km NE. ***Longwood:*** includes the delightful White Hart Hotel and horse-drawn carriage rides; 14 km SW. ***Gooram Falls:*** a scenic drive takes in the falls and parts of the Strathbogie Ranges; 20 km SE. ***Locksley:*** popular for gliding and parachuting; 20 km SW. ***Polly McQuinns Weir:*** historic river crossing and reservoir; Strathbogie Rd; 20 km SE. ***Mt Wombat Lookout:*** spectacular views of surrounding country and the Australian Alps; 25 km SE. ***Blue Wren Lavender Farm:*** lavender products and Devonshire tea; Boho South; 28 km E. ***Avenel Maze:*** Ned Kelly–themed maze; open Thurs–Mon, school and public holidays; 37 km SW.

See also GOULBURN & MURRAY, p. 157

Flinders

Pop. 575
Map ref. 573 I11 | 575 I10 | 582 C8

Nepean Hwy, Dromana; (03) 5987 3078 or 1800 804 009.

98.7 3RPP FM, 774 AM ABC Local Radio

Flinders is set on the south coast of the Mornington Peninsula, a region famous for its wineries. During the 1880s, Flinders became known as a health and recreation resort and a number of guesthouses and hotels began to emerge. Today Flinders remains a popular holiday spot, with its renowned cliff-top golf course and gastropub. Heritage buildings have wide verandahs, often shading antique and curio shops or excellent cafes, giving the town an enchanting and historic air. This, combined with taking in the view across the bay to The Nobbies and Seal Rocks, makes it easy to understand the town's perennial appeal.

Foreshore Reserve: popular for picnics and fishing from the jetty. ***Studio @ Flinders:*** small but unique art gallery with emphasis on ceramics, also exhibits handcrafted jewellery, glass, textiles, wood and paintings; Cook St. ***Historic buildings:*** 'Bimbi', built in the 1870s, is the earliest remaining dwelling in Flinders; King St. 'Wilga' is another fine Victorian-era home; King St. ***Flinders Golf Links:*** great views across Bass Strait; West Head, Wood St.

Peninsula Piers and Pinots: the region's winemakers showcase their pinots with local food and produce at Flinders Pier; Mar long weekend.

Foxeys Hangout: casual winery cafe; 795 White Hill Rd, Red Hill; (03) 5989 2022. ***Merricks General Store:*** wine-friendly rustic fare; 3458 Frankston–Flinders Rd, Merricks; (03) 5989 8088. ***Montalto:*** slick vineyard restaurant; 33 Shoreham Rd, Red Hill South; (03) 5989 8412. ***Ten Minutes by Tractor:*** cellar-door restaurant, dishes matched with wines; 1333 Mornington–Flinders Rd, Main Ridge; (03) 5989 6080.

Flinders B&B: 1-bedroom fisherman's cottage; 94 Cook St; (03) 5989 0301. ***Cape Schanck Light Station:*** 4 cottages; 420 Cape Schanck Rd, Cape Schanck; (03) 5968 6411. ***Lindenderry:*** luxury country house; 142 Arthurs Seat Rd, Red Hill; (03) 5989 2933. ***Shoreham Beach House:*** beach house, sleeps 10; 20 Myers Dr, Shoreham; (03) 5989 8433.

Red Hill This is fine wine country, where vineyards are interspersed with noted art galleries, farm gates, cafes and restaurants. The Red Hill Market is legendary and held on the first Sat of each month from Sept to May. It specialises in local

VICTORIA

crafts, clothing and fresh produce. The town also features a number of galleries and The Cherry Farm, where you can 'pick your own' cherries and berries in a pleasant setting (in season); Arkwells La. The Mornington Peninsula Winter Wine Fest is held annually on the Queen's Birthday weekend (June) and the Cool Climate Wine Show is in Mar.

Mornington Peninsula National Park The park covers 2686 ha and features a diverse range of vegetation, from the basalt cliff-faces of Cape Schanck to banksia woodlands, coastal dune scrubs and swampland. One of the park's many attractions is the Cape Schanck Lighthouse, built in 1859, which provides accommodation in one of the lighthouse keepers' houses. Historic Pt Nepean retains its original fortifications and has information displays and soundscapes. Also available here is a 'hop-on, hop-off' tractor train with commentary, and bicycle hire. There are ocean beaches for swimming and surfing, while the Bushranger Bay Nature Walk, starting at Cape Schanck, and the Farnsworth Track at Portsea are just 2 of the many walks on offer. Contact Parks Victoria on 13 1963; access to Cape Schanck from Rosebud–Flinders Rd; 15 km W.

French Island National Park French Island once served as a prison where inmates kept themselves entertained with their own 9-hole golf course. This unique reserve features a range of environments from mangrove saltmarsh to open woodlands. During spring more than 100 varieties of orchids come into bloom. The park is home to the most significant population of koalas in Victoria. Long-nosed potoroos and majestic sea-eagles can also be spotted. There is a variety of walking tracks on the island and bicycles can be hired from the general store. There are also guesthouses, and camping and picnic facilities. Contact Parks Victoria on 13 1963; access is via a 30 min ferry trip from Stony Pt, 30 km NE of Flinders.

Ashcombe Maze and Lavender Gardens: a large hedge maze surrounded by beautifully landscaped gardens; closed Aug; Red Hill Rd, Shoreham; 6 km N. ***Ace Hi Horseriding and Wildlife Park:*** beach and bush trail-rides and a native-animal sanctuary; Cape Schanck; 11 km W. ***Main Ridge:*** Sunny Ridge Strawberry Farm – pick your own berries in season; Mornington–Flinders Rd. Also The Pig and Whistle, English-style pub, Purves Rd; 11 km NW. ***Pt Leo:*** great surf beach; 12 km NE via Shoreham. ***Balnarring:*** hosts a market specialising in handmade crafts; 3rd Sat each month Nov–May; 17 km NE. Nearby is Coolart Homestead, an impressive Victorian mansion with historical displays, gardens, wetlands and a bird-observation area.

See also MORNINGTON PENINSULA, p. 149

Foster

Pop. 1039
Map ref. 582 G9

Stockyard Gallery, Main St; 1800 630 704; www.visitpromcountry.com.au

 89.5 3MFM South Gippsland, 100.7 FM ABC Local Radio

Foster was originally a goldmining town settled in the 1870s. The town boasts close access to Wilsons Promontory – affectionately called 'the Prom' – and is a popular base for visitors. Set in the centre of a rich agricultural area, Foster is the main shopping precinct for the Prom, Corner Inlet and Waratah Bay.

Historical Museum: in old post office; Main St. ***Stockyard Gallery:*** Main St. ***Hayes Walk:*** view the site of Victory Mine, Foster's largest goldmine; starts in town behind the carpark. ***Pearl Park:*** picturesque picnic spot.

Tastes of Prom Country: Jan. ***Great Southern Portrait Prize:*** Jan. ***Prom Coast Seachange Festival:*** Apr. ***Mt Best Art Show:*** Apr. ***Prom Country Challenge:*** fun run; Aug. ***Promontory Home Produce and Craft Market:*** Nov–Apr.

Basia Mille Luxury Apartments: Tuscan-inspired accommodation; 1 Taylor Crt, Fish Creek; (03) 5687 1453

Wilsons Promontory National Park The Prom is well loved across the state for its wild and untouched scenery. Its 130 km coastline is framed by granite headlands, mountains, forests and fern gullies. Bordered on all sides by sea, it hangs from Victoria by a thin, sandy isthmus. Limited road access means opportunities for walking are plentiful. The park features dozens of walking tracks, ranging from easy strolls to more challenging overnight hikes that take visitors to one of 11 campsites only accessible by foot. Hikes range from beginner to intermediate, and permits are required. Detailed information is provided at the park's own visitor centre: the remnants of a commando training camp from WW II. Contact Parks Victoria on 13 1963; 32 km S.

Toora An internationally recognised wetland site located on Corner Inlet, it is renowned for the huge variety of migratory birds that nest in the area. Toora is also home to Agnes Falls, a wind farm, a lavender farm and the Bird Hide where you can watch migratory and indigenous birdlife. 12 km E.

Foster North Lookout: 6 km NW. ***Wineries:*** Windy Ridge Winery; 10 km S. ***Fish Creek:*** A rural village, which attracts many visitors en route to the Prom. From the novelty of the giant mullet on top of the Promontory Gate Hotel to the fish-shaped seats around town, there is more to this unusually themed town than meets the eye. Galleries and vineyards are located in the area. Access the Great Southern Trail and walk, ride or cycle your way to Foster. Nearby Mt Nicol offers a lookout with spectacular views; 13 km SW. ***Turtons Creek Reserve:*** features mountain ash, blackwood and tree ferns, and a small waterfall. Bush camping is available; 18 km N. ***Coastal towns:*** popular bases during summer months; Sandy Pt, 22 km S; Waratah Bay, 34 km SW; Walkerville, 36 km SW. ***Cape Liptrap:*** lighthouse, views over rugged coastline and Bass Strait; 46 km SW.

See also PHILLIP ISLAND & GIPPSLAND, p. 150

Geelong

see inset box on next page

Glenrowan

Pop. 324
Map ref. 584 H8 | 586 E5

Wangaratta Visitor Information Centre, 100 Murphy St, Wangaratta; 1800 801 065; www.visitwangaratta.com.au

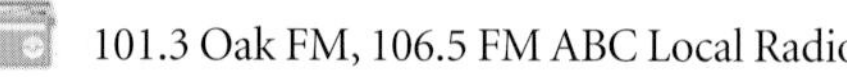 101.3 Oak FM, 106.5 FM ABC Local Radio

Glenrowan is a town well known to most Victorians as the site of Ned Kelly's final showdown with the police in 1880. Most of the attractions in Glenrowan revolve around the legends surrounding Kelly's life – a giant statue of Kelly himself towers over shops in Gladstone Street. Almost as legendary as Ned Kelly are the numerous wineries and fruit orchards in the area.

Kate's Cottage Museum: with an extensive collection of Kelly memorabilia as well as a replica of the Kelly homestead and blacksmith shop; Gladstone St. ***Cobb & Co Museum:*** an underground museum featuring notorious stories of Kelly and other bushrangers; Gladstone St. ***Kellyland:*** a computer-animated show of Kelly's capture; Gladstone St. ***Kelly Gang***

Siege Site Walk: discover the sites and history that led to the famous siege on this self-guide walk (brochure available). ***Wine and produce outlets:*** over 22 local wines are offered for tastings and sales at the Buffalo Mountain Wine Centre; Gladstone St. Gourmet jams and fruit products are also available at Smiths Orchard and The Big Cherry; Warby Range Rd. ***White Cottage Herb Garden:*** herb sales; Hill St.

Trails, Tastings and Tales wine and food event: June.

Baileys Old Block Cafe: weekend platters and grazing food; Baileys of Glenrowan, Taminick Gap Rd; (03) 5766 2392.

Warby Range State Park The 'Warbys', as they are known locally, extend for 25 km north of Glenrowan. The steep ranges provide excellent viewing points, especially from Ryans Lookout. Other lookouts include the Pangarang Lookout near the Pine Gully Picnic Area and the Mt Glenrowan Lookout, the highest point of the Warbys at 513 m. There are well-marked tracks for bushwalkers and a variety of pleasant picnic spots amid open forests and woodlands, with wildflowers blossoming during the warmer months. Access from Taminick Gap Rd.

See also HIGH COUNTRY, p. 158

Halls Gap

see inset box on page 185

Hamilton

Pop. 9379
Map ref. 592 F4

Lonsdale St; (03) 5572 3746 or 1800 807 056; www.sthgrampians.vic.gov.au

 94.1 FM ABC Local Radio, 99.3 Coastal FM

Hamilton is a prominent rural centre in the heart of a sheep-grazing district. This industry is such an important part of the town's economy that it has been dubbed the 'Wool Capital of the World'. A thriving country city, Hamilton is filled with cultural experiences, whether gazing at botanical, artistic or architectural beauty, browsing through great shops or putting in a bid as part of a 50 000-head sheep sale.

Hamilton Art Gallery This gallery is said to be one of regional Australia's finest, featuring a diverse collection of fine arts and museum pieces dating back to the earliest European settlements in Australia. Many trinkets and treasures of the region's first stately homes are on display, as well as English and European glass, ceramic and silver work. There is also a good collection of colonial art from the Western District. Guided heritage tours of the gallery and district are available. Brown St.

Botanic Gardens First planted in 1870 and classified by the National Trust in 1990, these gardens have long been regarded as one of the most impressive in rural Victoria. Designed by the curator of the Melbourne Botanic Gardens, William Guilfoyle, the gardens feature his 'signature' design elements of sweeping lawns interrupted by lakes, islands, and contrasting plant and flower beds. Keep an eye out for the free-flight aviary, enormous English oaks and historic band rotunda. French St.

Lake Hamilton: attractive landscaped man-made lake used for swimming, sailing, yachting and rowing, and featuring an excellent walking/bike track; off Ballarat Rd. ***Sir Reginald Ansett Transport Museum:*** birthplace of Ansett Airlines, the museum tells the story of Ansett and our aviation history in one of the airline's original hangars; Ballarat Rd. ***Hamilton Pastoral Museum:*** features farm equipment, tractors, engines, household items and small-town memorabilia; Ballarat Rd. ***Big Wool Bales:*** built in the shape of 5 giant wool bales and surrounded by native red gums, the Wool Bales tell the fascinating story of Australia's wool industry; Coleraine Rd. ***Mt Baimbridge Lavender:*** set on 12 acres. Wander through gardens and browse the gallery; Mt Baimbridge Rd; tours available by appt (03) 5572 4342. ***Hamilton History Centre:*** features the history of early Western District families and town settlement; Gray St.

Hamilton Farmer's Market: last Sat each month. ***Harvest Rally:*** Jan. ***Beef Expo:*** Feb. ***Hamilton Cup:*** Apr. ***Promenade of Sacred Music:*** Apr. ***Plough and Seed Rally:*** May. ***Sheepvention:*** promotes the sheep and wool industries; Aug.

Darriwill Farm: produce/wine store and cafe; 169 Grey St; (03) 5571 2088. ***Hamilton Strand Restaurant:*** modern Italian; 100 Thompson St; (03) 5571 9144. ***Roxburgh House:*** wine-bar-cum-cosy-cafe; 64 Thompson St; (03) 5572 4857. ***Cafe Catalpa:*** game-heavy menu; 7921 Hamilton Hwy, Tarrington; (03) 5572 1888.

Garland Cottage: 1-bedroom cottage; 18 Skene St; (03) 5572 1054. ***Seven Palmer Street:*** ultra-chic 2-bedroom house; 7 Palmer St; 0408 212 100.

 **Grampians National Park** *See Halls Gap.* 35 km NE.

Mt Eccles National Park A range of walks let visitors explore the scoria cones and caves formed thousands of years ago by volcanoes. The 3 main craters hold a 700 m long lake, Lake Surprise, fed by underground springs. Contact Parks Victoria on 13 1963; near Macarthur; 40 km S.

Tarrington: established by German settlers and originally named Hochkirch, this area is fast becoming a well-known 'pinot noir' grape-producing area; 12 km SE. ***Waterfalls:*** Nigretta Falls has a viewing platform; 15 km NW. Also Wannon Falls; 19 km W. ***Mt Napier State Park:*** features Byaduk Caves (lava caves) near the park's western entrance. Only 1 cave is accessible to the public; 18 km S. ***Cavendish:*** a small town en route to the Grampians, notable for the 3 beautiful private gardens open during the Southern Grampians Open Gardens Festival each Oct; 25 km N. ***Penshurst:*** a lovely historic town at the foot of Mt Rouse. Excellent views from the top of the mountain, where there is a crater lake. Country Muster each Feb; 31 km SE.

See also GRAMPIANS & CENTRAL WEST, p. 155

Healesville

Pop. 7357
Map ref. 573 L4 | 582 E4

Yarra Valley Visitor Information Centre, Old Courthouse, Harker St; (03) 5962 2600; www.visityarravalley.com.au

 99.1 Yarra Valley FM, 774 AM ABC Local Radio

To the west of Yarra Ranges National Park and within easy reach of Melbourne, Healesville has a charming rural atmosphere. There are good restaurants and cafes in town, all focusing on quality local produce, especially the world-class Yarra Valley wines. On top of this is a host of art and craft boutiques and two major attractions – TarraWarra Museum of Art and the famous Healesville Sanctuary, one of the best places in Victoria to experience Australia's unique wildlife close-up.

continued on p. 183

VICTORIA

GEELONG

Pop. 160 989

Map ref. 572 F8 | 574 B4 | 576 | 580 H1 | 582 A6 | 593 O7

Princes Hwy, Little River; (03) 5283 1735 or 1800 620 888; www.visitgreatoceanroad.org.au

94.7 The Pulse FM, 774 AM ABC Local Radio

Situated on Corio Bay, Geelong is the largest provincial city in Victoria. Geelong was traditionally a wool-processing centre, and the National Wool Museum in Moorabool Street details its early dependence upon the industry. The town was first settled by Europeans in the 1830s, but Geelong and its surrounds were originally home to the Wathaurong people, with whom the famous convict escapee William Buckley lived for many years. Buckley later described the unique culture of the Aboriginal tribes who welcomed him into their lives, and his writing is now one of the most priceless historical records of Indigenous culture in southern Australia. Geelong is a beautifully laid-out city, and a drive along the scenic Esplanade reveals magnificent old mansions built during its heyday.

Waterfront Geelong This superbly restored promenade stretches along Eastern Beach and offers a variety of attractions. Visitors can relax in the historic, 1930s-built sea-baths, enjoy fine dining in seaside restaurants and cafes or stroll along the famous Bollards Trail featuring colourful sculptures. The Waterfront district is on Eastern Beach Rd, with the beautiful old Cunningham Pier as a centrepiece.

National Wool Museum Housed in a historic bluestone woolstore, the centre features audiovisual displays plus re-created shearers' quarters and a mill-worker's cottage. There is a licensed restaurant and bar in the cellar, and a souvenir shop selling locally made wool products; 26 Moorabool St; (03) 5272 4701.

Geelong Art Gallery: this regional gallery is considered one of the finest in the state. The focus is on late-19th- and early-20th-century paintings by British artists and members of the Royal Academy, such as Tom Roberts and Arthur Streeton; Little Malop St; (03) 5229 3645. ***Historic buildings:*** there are over 100 National Trust classifications in Geelong, including Merchiston Hall, Osborne House and Corio Villa. 'The Heights' is a 14-room prefabricated timber mansion set in landscaped gardens; contact visitor centre for details of open days; Aphrasia St, Newtown. Christ Church, still in continuous use, is the oldest Anglican Church in Victoria; Moorabool St. ***Ford Discovery Centre:*** Geelong has long been a major manufacturing centre for Ford and this centre details the history of Ford cars with interactive displays; closed Tues; Cnr Brougham and Gheringhap sts; (03) 5227 8700. ***Wintergarden:*** a historic building housing a gallery, a nursery, antiques and a giftshop; 51 McKillop St. ***Botanic Gardens:*** overlooking Corio Bay and featuring a good collection of native and exotic plants; part of Eastern Park; Garden St. ***Johnstone Park:*** picnic and barbecue facilities; Cnr Mercer and Gheringhap sts. ***Queens Park:*** walks to Buckley Falls; Queens Park Rd, Newtown. ***Balyang Bird Sanctuary:*** 50 Marnock Rd, Newton, off Shannon Ave. ***Barwon River:*** extensive walking tracks and bike paths in parkland by the river. ***Norlane Water World:*** water slides; Princes Hwy, Norlane. ***Corio Bay beaches:*** popular for swimming, fishing and sailing; boat ramps provided.

[COLOURFUL BOLLARDS ALONG GEELONG'S WATERFRONT]

Steampacket Gardens Market: on foreshore at Eastern Beach; 1st Sun each month. ***Geelong Farmers Market:*** Little Malop St; 2nd Sat each month. ***Audi Victoria Week***; sailing regatta; Jan. ***Pako Festa:*** Victoria's premier multicultural event; Pakington St; last Sat in Feb. ***Highland Gathering:*** Mar. ***National Celtic Folk Festival:*** Port Arlington, June. ***Geelong Show:*** Oct. ***Geelong Cup:*** Oct. ***Christmas Carols by the Bay:*** Eastern Beach; Dec. ***Geelong New Year Waterfront Festival:*** New Year's Eve.

2 Faces: contemporary cuisine; 8 Malop St; (03) 5229 4546. ***Fishermen's Pier:*** seafood; Yarra St, Eastern Beach; (03) 5222 4100. ***Pettavel Winery & Restaurant:*** gastronomic winery; 65 Pettavel Rd, Waurn Ponds; (03) 5266 1120. ***The Beach House:*** contemporary beachside; Eastern Beach Reserve, Waterfront Bay; (03) 5221 8322. ***The Pier Geelong:*** Dock Cafe, City Quarter Bar and Baveras Restaurant; Cunningham Pier; (03) 5222 6444.

Chifley on the Esplanade: hotel rooms and apartments; 13 The Esplanade; (03) 5244 7700 or 1300 650 464. ***Four Points by Sheraton Geelong:*** waterfront rooms/studios; 10–14 Eastern Beach Rd; (03) 5223 1377. ***Haymarket Boutique Hotel:*** 6 centrally located suites; 244 Moorabool St; (03) 5221 1174. ***Mercure Hotel Geelong:*** variety of room types; Cnr Gheringhap and Myers sts; (03) 5223 6200.

You Yangs Regional Park These granite outcrops that rise 352 m above Werribee's lava plains have an ancient link to the Wathaurong people as they provided a much-needed water source – rock wells were created to catch water, and many of them can still be seen at Big Rock. The park's activities include the 12 km Great Circle Drive and the climb to Flinders Peak for fantastic views of Geelong, Corio Bay, Mt Macedon and Melbourne's skyline. Contact Parks Victoria on 13 1963; 24 km N.

Werribee Park and Open Range Zoo The key feature of Werribee Park is a beautifully preserved 1870s mansion with the interior restored to its original opulence. The mansion is surrounded by 12 ha of gardens, including a grotto and a farmyard area, complete with a blacksmith. Within the grounds is the Victoria State Rose Garden with over 500 varieties of flowers. Next to the park is the Werribee Open Range Zoo, developed around the Werribee River. The zoo covers 200 ha and has a variety of animals native to the grasslands of Africa, Asia, North America and Australia, including giraffes, rhinos, meerkats, cheetahs and vervet monkeys. Guided safaris through the replicated African savannah are a must. Access from the Princes Hwy; Werribee Park 13 1963; Open Range Zoo (03) 9731 9600; 40 km NE.

Fyansford: one of the oldest settlements in the region, with historic buildings including the Swan Inn, Balmoral Hotel and Fyansford Hotel. The Monash Bridge across the Moorabool River is thought to be one of the earliest reinforced-concrete bridges in Victoria; outskirts of Geelong; 4 km W. ***Adventure Park:*** Victoria's first waterpark, with more than 20 attractions and rides; open Oct–Apr; 1251 Bellarine Hwy; (03) 5250 2756; 15 km SE. ***Avalon Airfield:*** hosts the Australian International Air Show in odd-numbered years; off Princes Hwy; 20 km NE. ***Serendip Sanctuary:*** a wildlife research station that includes nature trails, bird hides and a visitor centre; just south of the You Yangs. ***Fairy Park:*** features miniature houses and scenes from fairytales; 2388 Ballan Rd, Anakie; (03) 5284 1262; 29 km N. ***Steiglitz:*** once a gold town, Steiglitz is now almost deserted. The restored courthouse is open on Sun; 37 km NW. ***Geelong Wine Region:*** stretching north past Anakie, around Geelong and south along the Bellarine Peninsula, this region produces a diverse range of red and white wines. Cellar doors offer tastings and sales; map from visitor centre.

See also WERRIBEE & BELLARINE, **p. 152**

Silvermist Studio Gallery: handmade gold and silver jewellery; Maroondah Hwy. ***Open-air trolley rides:*** from Healesville railway station; open Sun and public holidays. ***Giant Steps/Innocent Bystander Winery:*** thoroughly modern cellar door in the town centre with an excellent bakery, bistro serving mouth-watering pizzas, and cheese room; 336 Maroondah Hwy; (03) 5962 6111 or 1800 661 624.

Market: River St; 1st Sun each month. ***Grape Grazing Festival:*** events held throughout wine district to celebrate the harvest; Feb. ***Australian Car Rally Championship:*** Sept.

3777: contemporary with views; Mt Rael Retreat, 140 Healesville–Yarra Glen Rd; (03) 5962 1977. ***Giant Steps/ Innocent Bystander:*** modern bakery, cafe and cellar door; 336 Maroondah Hwy; (03) 5962 6111 or 1800 661 624. ***Healesville Hotel:*** country gastropub; 256 Maroondah Hwy; (03) 5962 4002. ***Bella Vedere:*** bakery, cafe-cum-restaurant; 874 Maroondah Hwy, Coldstream; (03) 5962 6161.

Healesville Hotel: 7 hotel rooms; 256 Maroondah Hwy: (03) 5962 4002. ***Mt Rael Retreat:*** modern suites with views, adult guests only; 140 Healesville–Yarra Glen Rd; (03) 5962 1977. ***RACV Healesville Country Club:*** recently renovated club with modern rooms, RACV club members and guests only; Healesville–Yarra Glen Rd; (03) 5962 4899. ***Sebel Heritage Yarra Valley:*** luxury accommodation in the Yarra Ranges; Heritage Ave, Chirnsdie Park; (03) 9760 3333.

Healesville Sanctuary Australia's unique animal species are on show at this 32 ha reserve. The sanctuary is also one of the few places in the world to have successfully bred platypus in captivity. Allow at least half a day to visit and see the animal hospital or go on a behind-the-scenes keeper tour (bookings on (03) 5957 2800). Badger Creek Rd; 4 km S.

TarraWarra Museum of Art TarraWarra Estate has been operating as a vineyard since 1983, producing a selection of fine chardonnay and pinot noir. Now there is a striking building housing an extensive private collection of modern art. The collection focuses on the 3 key themes of Australian Modernism – landscape, figuration and abstraction – and works by artists such as Howard Arkley, Arthur Boyd and Brett Whiteley can be found within. Healesville–Yarra Glen Rd; (03) 5957 3100.

VICTORIA

Hedgend Maze: giant maze and fun park; Albert Rd; 2.5 km s. ***Coranderrk Aboriginal Cemetery:*** once the burial ground for an Aboriginal mission, and the final resting place of well-known Wurundjeri leader William Barak; 3 km s. ***Maroondah Reservoir Park:*** a magnificent park set in lush forests with walking tracks and a lookout nearby; 3 km NE. ***Donnelly's Weir Park:*** starting point of the 5000 km Bicentennial National Trail to Cooktown (Queensland). The park also has short walking tracks and picnic facilities; 4 km N. ***Badger Weir Park:*** picnic area in a natural setting; 7 km SE. ***Mallesons Lookout:*** views of Yarra Valley to Melbourne; 8 km s. ***Mt St Leonard:*** good views from the summit; 14 km N. ***Toolangi:*** attractions include the Singing Gardens of C. J. Dennis, a beautiful, formal garden; the Toolangi Forest Discovery Centre, for a fascinating insight into the local forests and how they were formed; and Toolangi Pottery; 20 km NW. ***Wineries:*** around 85 in the area open for tastings and sales. Tours available; details from visitor centre. *See also Yarra Glen.*

See also YARRA & DANDENONGS, p. 148

Heathcote

Pop. 1572
Map ref. 584 B10 | 591 P11

Cnr High and Barrack sts; (03) 5433 3121; www.heathcote.org.au

 91.1 FM ABC Local Radio, 100.7 Highlands FM

Heathcote is located near the outskirts of the scenic Heathcote–Graytown National Park, with the McIvor Creek flowing by the town. Heathcote was established during the gold rush, but is now known as a prominent wine region with good red wines produced from a number of new vineyards.

Courthouse Crafts: displays relating to the gold rush, plus arts and crafts; High St. ***Pink Cliffs:*** eroded soil from gold sluices gave the cliffs their remarkable pink colour; Pink Cliffs Rd, off Hospital Rd. ***McIvor Range Reserve:*** walking tracks; off Barrack St; details of walks from visitor centre. ***Heathcote Winery:*** this winery, in the old Thomas Craven Stores building, has an art gallery and cellar-door sales; High St.

World's Longest Lunch: Mar. ***Rodeo:*** Mar long weekend. ***Heathcote Wine and Food Festival:*** 1st weekend in Oct.

Emeu Inn: regional produce; 187 High St; (03) 5433 2668.

 ***Emeu Inn:*** 6 suites, 1 cottage; 187 High St; (03) 5433 2668. ***Hut on the Hill:*** 2-bedroom luxury cottage; 720 Dairy Flat Rd; (03) 5433 2329. ***Olive Grove Retreat:*** chic guesthouse with pool; Travers La; 0403 876 988.

Heathcote–Graytown National Park Compared with many of Victoria's national parks, Heathcote–Graytown was declared quite late as part of a statewide plan to preserve box-ironbark forest. Now protecting the largest forest of this type in the state, the park is not only an important nature reserve but also has a long history of settlement. Take one of the many walks or scenic drives to explore evidence of Aboriginal, goldmining and pioneering history, or take in scenic views from the lookouts at Mt Black, Mt Ida and Viewing Rock (just near Heathcote). Contact Parks Victoria on 13 1963; access from Northern Hwy and Heathcote–Nagambie Rd.

Wineries: Heathcote shiraz makes wine lovers go weak at the knees. Its depth is the result of the dark, red Cambrian soil and the continental climate. Jasper Hill is the most exclusive name in the area. Its elegant red wines can be hard to come by, so if you manage to find a bottle, it is worth purchasing it on the spot. Other good wineries include Heathcote Winery with its cellar door located on the town's main street, Shelmerdine Vineyards, Wild Duck Creek Estate and Red Edge. To find most of the region's wines under one roof, head to Cellar and Store, which also stocks a full range of local gourmet produce.

Lake Eppalock: one of the state's largest lakes, great for fishing, watersports and picnics; 10 km W.

See also GOULBURN & MURRAY, p. 157

Hopetoun

Pop. 591
Map ref. 558 E10 | 590 H4

Gateway Beet, 75 Lascelles St; (03) 5083 3001; www.hopetounvictoria.com.au

92.9 3MBR-FM Mallee Border Radio, 594 AM ABC Local Radio

This small Mallee town, south-east of Wyperfeld National Park, was named after the first governor-general of Australia, the Earl of Hopetoun. The Earl was a friend of Edward Lascelles, who played a major role in developing the Mallee Country by eradicating vermin, developing water strategies to cope with the dry conditions, and enticing settlers to the region.

Hopetoun House: the residence of Lascelles, this majestic building is now National Trust–classified; Evelyn St. ***Mallee Mural:*** depicts history of the region; wall of Dr Pete's Memorial Park, cnr Lascelles and Austin sts. ***Lake Lascelles:*** good for boating, swimming and fishing when filled, presently dry. Camping facilities available; access from end of Austin St.

Hopetoun Bowl Club Annual Carnival: Apr. ***Hopetoun A & P Society Annual Show:*** Oct.

Wyperfeld National Park Outlet Creek connects the network of lake beds that are the main highlight for visitors to this park. They fill only when Lake Albacutya overflows, which in turn fills only when Lake Hindmarsh overflows. Eastern grey kangaroos can be seen grazing on Wirrengren and the other lake beds, and the Eastern Lookout Nature Drive is a great way to see the range of vegetation in the park – river red gums, black box, mallee and cypress pine, and wildflowers in spring. A variety of walking trails leave from the 2 campgrounds – Wonga Campground in the south and Casuarina Campground in the north. Contact Parks Victoria on 13 1963; 50 km NW.

Patchewollock: the northern gateway to the Wyperfeld National Park, and also home of the Patchewollock Outback Pub; 35 km NW.

See also MALLEE COUNTRY, p. 156

Horsham

Pop. 14 120
Map ref. 590 G9

20 O'Callaghan Pde; (03) 5382 1832 or 1800 633 218; www.grampianslittledesert.com.au

 96.5 Triple H FM, 594 AM ABC Local Radio

Horsham is an important centre for the Wimmera district. Prior to European settlement, Horsham and its surrounds were occupied by the Jardwa and Wotjobaluk Aboriginal people, who referred to the region as 'Wopetbungundilar'. This term is thought to have meant 'place of flowers', a reference to the flowers that grow along the banks of the Wimmera River. Flowers continue to play an important role in Horsham, which is

continued on p. 186

HALLS GAP

Pop. 279

Map ref. 590 H12 | 592 H1

Grampians Rd; (03) 5356 4616 or 1800 065 599; www.grampianstravel.com.au

94.1 FM ABC Local Radio, 99.9 VoiceFM

The little village of Halls Gap is set in the heart of the Grampians. It was named after Charles Browing Hall, who discovered the gap and valley in 1841. The valley was later developed by cattle-station owners, but the town really took off in the early 1900s when tourists, nature lovers and botanists caught on to the beauty and diversity of the mountain ranges that would later become Grampians National Park. The town itself has its own charm – shops, galleries and cafes lend a laid-back atmosphere that befits the location, while in the evening long-billed corellas arrive to roost opposite the shops in the main street.

Grampians Jazz Festival: Feb. ***Grampians Grape Escape:*** wine and food festival; May. ***Wildflower Exhibition:*** Oct.

Boroka Downs: 5 contemporary retreats; 51 Birdswing Rd; (03) 5356 6243. ***DULC Cabins:*** 3 environmentally friendly cabins; 9 Thryptomene Crt; (03) 5356 4711. ***Lakuna Retreat:*** architect-designed 3-bedroom home; 81 High Rd; (03) 5221 1606. ***Kangaroos in the Top Paddock:*** adults-only bushland retreat; Northern Grampians Rd, Wartook; (03) 9497 2020.

Grampians National Park Aboriginal occupation of the area known as the Grampians dates back over 5000 years (some evidence suggests up to 30 000 years). To local Koorie communities, this magnificent mountain range is known as Gariwerd. Within the 168 000 ha park is a startling array of vegetation and wildlife, including 200 bird species and a quarter of Victoria's native flora species. The heathlands abound in colourful shows of wildflowers including Grampians boronia, blue pincushion lily and Grampians parrot-pea. Twenty of the park's 800 plant species are not found anywhere else in the world. Brambuk – The National Park and Cultural Centre, a short walk or drive south of Halls Gap, is an excellent first stop in the park. It features interactive displays and written information about the park's attractions, bringing to life the culture of the local Jardwadjali and Djab Wurrung people. Natural highlights of the Grampians include MacKenzie Falls, the largest of the park's many waterfalls; Zumsteins picnic ground, a beautiful spot with tame and friendly kangaroos; and the Balconies, a rock ledge once known as the Jaws of Death, offering views over Victoria Valley. The most popular section of the park is the Wonderland Range, true to its name with features including Elephants Hide, Grand Canyon, Venus Baths and Silent Street. There are over 90 bushwalks available in the park, all varying in length and degree of difficulty. Visitors are advised to consult a ranger before embarking on one of the longer treks. For further information, contact Brambuk – The National Park and Cultural Centre on (03) 5361 4000.

The Gap Vineyard: cellar-door tastings and sales of award-winning white and red wine varieties, as well as port; closed Mon and Tues; Pomonal Rd; 2 km E. ***Grampians Adventure Golf:*** world-class minigolf, 18-hole course set on 2 acres; 481 Grampians Rd; 4 km S near Lake Bellfield. ***Halls Gap Zoo:*** explore the park's nature track and view the animals, many of which are free-range; (03) 5356 4668; 7 km SE. ***Grampians Horse Riding Adventures:*** morning and afternoon rides lasting 2 hrs, also has on-site accommodation; Brimpaen; (03) 5383 9225; 44 km W.

See also GRAMPIANS & CENTRAL WEST, p. 155

[BOROKA LOOKOUT, GRAMPIANS NATIONAL PARK]

VICTORIA

considered to be one of the prettiest regional towns in Victoria – the town prides itself on its clean streets and picturesque gardens. Although the Wimmera is a renowned wheat-growing region, Horsham is also a centre for fine wool production.

Horsham Regional Art Gallery This is one of Victoria's key regional galleries, with an extensive collection housed in a 1930s Art Deco building. Most of the artwork is centred on the Mack Jost collection of Australian art, with contemporary Australian photography another specialty. 80 Wilson St; (03) 5362 2888.

Botanic Gardens: picturesquely set on the banks of the Wimmera River; Cnr Baker and Firebrace sts. ***The Wool Factory:*** produces extra-fine wool from Saxon-Merino sheep, with tours daily; Golf Course Rd. ***Wimmera River:*** key attraction for the town, with scenic picnic spots along the river's edges. Visit the river at dusk for spectacular sunsets.

Market: showgrounds on McPherson St; 2nd Sun each month. ***Haven Recreation Market:*** 1st Sat each month. ***Wimmera Machinery Field Days:*** Longerenong; Mar. ***Art Is:*** community festival; Apr. ***Awakenings Festival:*** largest Australian festival involving disabled patrons; Oct. ***Spring Garden Festival:*** Oct. ***Horsham Show:*** Oct. ***Kannamaroo Rock 'n' Roll Festival:*** Nov. ***Karkana Strawberry Festival:*** Nov.

Cafe Chickpea: hearty breakfasts and light lunches; 30A Pynsent St; (03) 5382 3998.

Orange Grove B&B: mud-brick cottage or modern unit; 123 Keatings Rd; (03) 5382 0583. ***Sylvania Park Mohair Farm:*** 2 farm homesteads; 808 East Rd, Wimmera; (03) 5382 2811.

Murtoa This town lies on the edge of Lake Marma, which has dried out in recent times. The Water Tower Museum (open Sun) displays the history of the area as well as James Hill's 1885–1930 taxidermy collection of some 500 birds and animals. On the eastern side of town, among the grain silos, is an unusual relic called the Stick Shed. The roof of this now empty storage shed is held up with 640 unmilled tree trunks, and the interior is an evocative sight (open once a year in Oct); 31 km NE.

Jung: market on last Sat each month; 10 km NE. ***Fishing:*** redfin and trout in local lakes, depending on water levels. Reasonable levels at Taylors Lake; 18 km SE. ***Mount Zero Olives:*** the largest olive grove in the Southern Hemisphere, with tastings and sales of olive oil; Laharum; 30 km S.

See also GRAMPIANS & CENTRAL WEST, p. 155

Inglewood

Pop. 684
Map ref. 579 D1 | 591 M9

Loddon Visitor Information Centre, Wedderburn Community Centre, 24 Wilson St, Wedderburn; (03) 5494 3489; www.loddonalive.com.au

89.5 The Fresh FM, 91.1 FM ABC Local Radio

North along the Calder Highway from Bendigo is the 'Golden Triangle' town of Inglewood. Sizeable gold nuggets were found in this area during the gold rush and are still being unearthed. Inglewood is also known as Blue Eucy town, due to the once vigorous blue mallee eucalyptus oil industry. The town is also the birthplace of Australian aviator Sir Reginald Ansett.

Old eucalyptus oil distillery: not in operation but can be viewed; Calder Hwy, northern end of town. ***Old courthouse:*** local historical memorabilia; open by appt; Southey St. ***Streetscape:*** historic buildings are evidence of the town's goldmining history.

Kooyoora State Park The park sits at the northern end of the Bealiba Range and features extensive box-ironbark forests. The Eastern Walking Circuit offers a great opportunity for bushwalkers, passing through strange rock formations and giant granite slabs. The Summit Track leads to Melville Caves Lookout. The caves were once the haunt of the notorious bushranger Captain Melville. Camping is allowed around the caves. Contact Parks Victoria on 13 1963; 16 km W.

Bridgewater on Loddon: fishing and watersports, Old Loddon Vines Vineyard, Water Wheel Vineyards and horse-drawn caravans for hire; 8 km SE. ***Loddon Valley wineries:*** the warm climate and clay soils of this region are known for producing outstanding red varieties and award-winning chardonnays. Taste the wines at cellar doors like Pandalowie at Bridgewater on Loddon (8 km SE) and Kingower (11 km SW); winery map from visitor information centre.

See also GOLDFIELDS, p. 154

Inverloch

Pop. 3682
Map ref. 573 M12 | 582 F9

A'Beckett St; 1300 762 433; www.visitbasscoast.com

88.1 3MFM South Gippsland, 100.7 FM ABC Local Radio

Inverloch is a small seaside resort set on the protected waters of Anderson Inlet, east of Wonthaggi. It is characterised by long stretches of pristine beach that offer good surf and excellent fishing.

Bunurong Environment Centre: natural history displays with special focus on dinosaur diggings; also sales of natural products; The Esplanade. ***Shell Museum:*** The Esplanade.

Inverloch Food and Wine Festival: Feb. ***Annual Dinosaur Dig:*** Feb/Mar. ***Inverloch Jazz Festival:*** Mar. ***Inverloch Billy Cart Derby:*** Nov.

Tomo: sophisticated Japanese; Shop 1, 23 A'Beckett St; (03) 5674 3444. ***Vela Nine:*** modern Australian/Mediterranean; 9 A'Beckett St; (03) 5674 1188.

RACV Inverloch Resort: hotel rooms, villas and caravan sites; 70 Cape Paterson–Inverloch Rd; (03) 5674 0000.

Bunurong Coastal Drive Stretching the 14 km of coastline between Inverloch and Cape Paterson is this spectacular coastal drive with magnificent views to Venus Bay and beyond. Carparks offer access to beaches and coastal walks along the drive. The waters offshore are protected within Bunurong Marine National Park, and offer opportunities to surf, snorkel, scuba dive or simply explore the numerous rockpools that are dotted along the coast.

Anderson Inlet: the most southerly habitat for mangroves in Australia. This calm inlet is popular for windsurfing and watersports, and nearby Townsend Bluff and Maher's Landing offer good birdwatching; adjacent to town. ***Fishing:*** in nearby waterways such as the Tarwin River; 20 km SE.

See also PHILLIP ISLAND & GIPPSLAND, p. 150

Jeparit

Pop. 374
Map ref. 558 D12 | 590 F6

Wimmera–Mallee Pioneer Museum, Charles St; (03) 5397 2101.

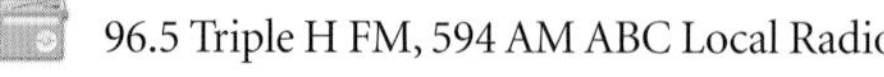

96.5 Triple H FM, 594 AM ABC Local Radio

This little town in the Wimmera is 5 kilometres south-east of Lake Hindmarsh, which was once the largest natural freshwater lake in

Victoria. Sadly, it has been empty for the past several years. Former prime minister Sir Robert Menzies was born here in 1894.

Wimmera–Mallee Pioneer Museum This unique museum details what life was like for early settlers in the Wimmera through a collection of colonial buildings furnished in the style of the period. The buildings on display are spread over a 4 ha complex and include log cabins, a church and a blacksmith's shop. The museum also features displays of restored farm machinery. Southern entrance to town, Charles St; (03) 5397 2101.

Menzies Sq: site of the dwelling where Menzies was born; Cnr Charles and Roy sts. ***Wimmera River Walk:*** 6 km return; starts at museum.

Lake Hindmarsh Victoria's largest freshwater lake has seen dire water levels for the past several years. It was fed by the Wimmera River. Boating, waterskiing and fishing were all popular pastimes (Schulzes Beach has a boat ramp), with pelicans and other waterbirds existing at the lake in breeding colonies. Picnic and camping spots are available on the lake's shores. A historic fisherman's hut can also be seen. Contact the visitor centre for an update on water levels. 5 km NW.

Rainbow: a charming little Wimmera township, with Pasco's Cash Store, an original country general store, and Yurunga Homestead, a beautiful Edwardian home with a large collection of antiques and original fittings (northern edge of town, key available); 35 km N. ***Pella:*** former German settlement with Lutheran church and old schoolhouse; 40 km NW via Rainbow. ***Lake Albacutya:*** fills only when Lake Hindmarsh overflows; 44 km N. ***Wyperfeld National Park:*** great for bushwalking; known for its birdlife, including the endangered mallee fowl, and wildflowers in spring; 60 km NW via Rainbow; *for details see Hopetoun.*

See also GRAMPIANS & CENTRAL WEST, p. 155

Kerang

Pop. 3780
Map ref. 558 H11 | 591 N4

Sir John Gorton Library, cnr Murray Valley Hwy and Shadforth St; (03) 5452 1546; www.gannawarra.vic.gov.au

99.1 Smart FM, 102.1 FM ABC Local Radio

Kerang, situated on the Loddon River just south of the New South Wales border, lies at the southern end of the Kerang wetlands and lakes. They extend from Kerang 42 kilometres north-west to Lake Boga and offer a wonderland for watersports enthusiasts and birdwatchers; the lakes contain what are reputedly the world's largest ibis breeding grounds. The town itself is a service centre for its agricultural surrounds.

Lester Lookout Tower: town views; Cnr Murray Valley Hwy and Shadforth St. ***Historical Museum:*** focuses on cars and antique farm machinery; Riverwood Dr.

Races: Easter and Boxing Day. ***Tour of the Murray River:*** cycling race; late Aug/early Sept.

Reedy Lakes: a series of 3 lakes. Apex Park, a recreation reserve for swimming, picnicking and boating, is set by the first lake, and the second features a large ibis rookery. Picnic facilities are available at the third lake; 8 km NW. ***Leaghur State Park:*** on the Loddon River flood plain, this peaceful park is a perfect spot for a leisurely walk through the black box woodlands and wetlands; 25 km SW. ***Murrabit:*** a historic timber town on the Murray surrounded by picturesque forests, with a country market 1st Sat each month; 27 km N. ***Quambatook:*** hosts the Australian Tractor Pull Championship each Easter; 40 km SW. ***Lake Boga:*** popular for watersports, with good sandy beaches; 42 km NW. ***Fishing:*** Meran, Kangaroo and Charm lakes all offer freshwater fishing; details from visitor centre.

See also GOULBURN & MURRAY, p. 157

Kilmore

Pop. 4720
Map ref. 573 J1 | 582 C1 | 584 C12

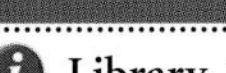

Library, 12 Sydney St; (03) 5781 1319.

97.1 OKR FM, 774 AM ABC Local Radio

Kilmore is Victoria's oldest inland town, known for its historic buildings and many horseracing events. Like many towns in the central goldfields, Kilmore was the scene of a Kelly family saga. In this case, it was Ned Kelly's father who had a run-in with the law. In 1865 John 'Red' Kelly was arrested for killing a squatter's calf to feed his family, and was locked away in the Kilmore Gaol for six months. It was a crime that Ned had actually committed. Soon after Red's release, he died of dropsy and was buried in the small town of Avenel, where the Kelly family lived for some time.

Old Kilmore Gaol An impressive bluestone building, established in 1859, that is now a privately owned auction house; Sutherland St.

Hudson Park: picnic/barbecue facilities; Cnr Sydney and Foote sts. ***Historic buildings:*** Whitburgh Cottage, Piper St, and a number of 1850s shops and hotels along Sydney St; brochure from visitor centre.

Kilmore Celtic Festival: June. ***Kilmore Cup:*** harness racing; Oct.

Bindley House B&B: 2 self-contained suites; 20–22 Powlett St; (03) 5781 1142. ***Laurel Hill Cottage:*** 2-bedroom cottage; 12 Melrose Dr; (03) 5782 1630. ***Woodbury Estate B&B:*** 2 luxury in-house rooms; 110A Butlers Rd; (03) 5782 0307.

Tramways Heritage Centre at Bylands: extensive display of cable cars and early electric trams, with tram rides available; open Sun only; just south of town. ***Broadford:*** a small town featuring a historic precinct on High St; 17 km NE. ***Mt Piper Walking Track:*** wildlife and wildflowers can be spotted along the way (1 hr return); near Broadford. ***Strath Creek:*** walks to Strath Creek Falls and a drive through the Valley of a Thousand Hills; starts at outskirts of Broadford.

See also SPA & GARDEN COUNTRY, p. 151

Koo-wee-rup

Pop. 1423
Map ref. 570 H10 | 573 L9 | 575 O6 | 582 E7

Newsagency, 277 Rossiter Rd; (03) 5997 1456.

103.1 3BBR FM West Gippsland Community Radio, 774 AM ABC Local Radio

Koo-wee-rup and the surrounding agricultural area exist on reclaimed and drained swampland. It has given rise to Australia's largest asparagus-growing district. The town's name derives from the Aboriginal name meaning 'blackfish swimming', a reference to the fish that were once plentiful in the swamp.

Historical Society Museum: local history; open Sun; Rossiter Rd.

VICTORIA

Swamp Observation Tower: views of remaining swampland and across to Western Port. A market with local produce operates regularly at the base; South Gippsland Hwy; 2 km SE. ***Bayles Fauna Reserve:*** native animals; 8 km NE. ***Harewood House:*** restored 1850s house with original furnishings; South Gippsland Hwy towards Tooradin. ***Tooradin:*** offers good boating and fishing on Sawtells Inlet; 10 km W. ***Caldermeade Farm:*** originally a premier beef cattle property but now a fully operational modern dairy farm focused on educating and entertaining visitors; 10 km SE. ***Pakenham:*** now considered a suburb of Melbourne, Pakenham is home to the Military Vehicle Museum, Army Rd, and the Berwick Pakenham Historical Society Museum, John St; 13 km N. ***Tynong:*** attractions include Victoria's Farm Shed, featuring farm animals and shearing, and Gumbaya Park, a family fun park; 20 km NE. ***Royal Botanic Gardens Cranbourne:*** renowned, wonderfully maintained native gardens; 22 km NW. ***Grantville:*** hosts a market 4th Sun each month; 30 km S.

See also **PHILLIP ISLAND & GIPPSLAND, p. 150**

Korumburra

Pop. 3145
Map ref. 573 N11 | 582 F8

Prom Country Information Centre, South Gippsland Hwy; (03) 5655 2233 or 1800 630 704; www.visitpromcountry.com.au

88.1 3MFM South Gippsland, 100.7 FM ABC Local Radio

Established in 1887, Korumburra stands firmly as the heritage centre of South Gippsland. The township was a primary producer of black coal for Victoria's rail industry until the last mine closed in 1958. Korumburra is set in the rolling green hills of South Gippsland, with scenic drives found in any direction.

Coal Creek Heritage Village Coal Creek is an open-air museum that offers all the fascination of life in a 19th-century coalmining village, including history and memorabilia of the area. The village contains beautiful picnic areas, bush tramway and cafe, and community events are held throughout the year. South Gippsland Hwy; (03) 5655 1811.

Korumburra Federation Art Gallery: South Gippsland Hwy. ***Whitelaw Antiques & Collectibles:*** 9 Mine Rd.

Coal Creek Farmers Market: Railway Siding; 2nd Sat each month. ***Korumburra Agricultural Show:*** Feb. ***Rotary Club of Korumburra Art Show:*** Feb.

Cypress Hill B&B: 3 Mediterranean-style guesthouse suites; 75 Korumburra–Bena Rd, Whitelaw; (03) 5657 2240.

South Gippsland Tourist Railway This railway travels to Leongatha, Korumburra, Loch and Nyora, and provides a scenic way to view the ever-changing South Gippsland landscape. Trains operate Sun and public holidays with a Wed service during school holidays. The grand Edwardian Railway Station behind the main street is also worth a visit. (03) 5658 1111.

Loch: a thriving art and craft village with cosy eateries, antique stores and galleries; 14 km NW. ***Poowong:*** beautiful country town nestled among the rolling hills of South Gippsland with Poowong Pioneer Chapel, est. 1878; 18 km NW.

See also **PHILLIP ISLAND & GIPPSLAND, p. 150**

Kyneton

Pop. 4286
Map ref. 572 G1 | 579 G8 | 582 A1 | 584 A11 | 593 O2

Jean Haynes Reserve, High St; (03) 5422 6110; www.visitmacedonranges.com

91.1 FM ABC Local Radio, 100.7 Highlands FM

Part of Victoria's picturesque spa and garden country, Kyneton is a well-preserved town with many attractive bluestone buildings. Caroline Chisholm, who helped many migrants find their feet in this country, lived in Kyneton, where her family owned a store and her husband worked as a magistrate. While living in the town, she established a series of affordable, overnight shelters for travellers on the Mount Alexander Road (now the Calder Highway), a road frequented by gold prospectors. Remnants of the shelters can be seen at the historic township of Carlsruhe, south-east of Kyneton.

Kyneton Museum: in a former bank building, with a drop-log cottage in the grounds; open Fri–Sun; Piper St. ***Wool on Piper:*** features a spinning mill, with yarns and handmade garments for sale; Piper St. ***Botanic Gardens:*** 8 ha area scenically located above Pipers Creek. The gardens feature rare varieties of trees; Clowes St. ***Historic buildings:*** many in town, including mechanics institute on Mollison St and old police depot, Jenning St. ***Campaspe River Walk:*** scenic walk with picnic spots; access from Piper St.

Farmers market: farmgate produce; Piper St; 2nd Sat each month. ***Kyneton Daffodil and Arts Festival:*** Sept. ***Kyneton Cup:*** Nov.

Annie Smithers' Bistrot: French Provincial–style; 72 Piper St; (03) 5422 2039. ***Little Swallow Cafe:*** European-style cafe; 58A Piper St; (03) 5422 6241. ***Pizza Verde:*** city-smart pizzeria; 62 Piper St; (03) 5422 7400. ***Royal George Hotel:*** contemporary dining room; 24 Piper St; (03) 5422 1390. ***Star Anise Bistro:*** modern bistro; 29A Piper St; (03) 5422 2777.

Rockville: chic, spacious house; 59 Ebden St; (03) 5422 2617. ***St Agnes Homestead:*** 3 rooms and 1 apartment; 30 Burton Ave (continuation of Piper St); (03) 5422 2639. ***Tilwinda B&B and Woolshed Barn:*** 2 rooms, 1 barn; 1793 Kyneton–Trentham Rd; (03) 5422 2772.

Trentham and Wombat State Forest This picturesque spa-country town has a mixed history of gold, timber and farming. It has a charming streetscape and attractions include a historic foundry and Minifie's Berry Farm, where you can pick your own berries in season. Just north-east of town is Wombat State Forest – deep within is Victoria's largest single-drop waterfall, Trentham Falls. Cascading 32 m onto a quartz gravel base, the falls are an impressive backdrop for a picnic. 22 km SW.

Reservoirs: several offering scenic locations for walks and picnics. Upper Coliban, Lauriston and Malmsbury reservoirs all nearby. ***Paramoor Farm and Winery:*** a former Clydesdale horse farm, now winery and B&B; Carlsruhe; 5 km SE. ***Malmsbury:*** a town noted for its old bluestone buildings. It features historic Botanic Gardens and The Mill, National Trust–classified, not open to the public. Also wineries in the area; 10 km NW. ***Turpins and Cascade Falls:*** with picnic area and walks; near Metcalfe; 22 km N.

See also **SPA & GARDEN COUNTRY, p. 151**

Leongatha

Pop. 4501
Map ref. 573 N11 | 582 G8

Michael Place Complex; (03) 5662 2111; www.visitpromcountry.com.au

88.1 3MFM South Gippsland, 100.7 FM ABC Local Radio

continued on p. 190

LAKES ENTRANCE

Pop. 5545

Map ref. 583 O6

Lakes and Wilderness Tourism, cnr Esplanade and Marine Pde; (03) 5155 1966 or 1800 637 060; www.discovereastgippsland.com.au

90.7 FM 3REG Radio East Gippsland, 100.7 FM ABC Local Radio

Lakes Entrance is a lovely holiday town situated at the eastern end of the Gippsland Lakes, an inland network of waterways covering more than 400 square kilometres. The artificially created 'entrance' of the town's name allows the Tasman Sea and the lakes to meet, creating a safe harbour that is home to one of the largest fishing fleets in Australia. While many of the attractions in Lakes Entrance are based around the water, there is also the opportunity for gourmets to indulge themselves with a variety of cafes and restaurants lining The Esplanade, plus sales of fresh fish and local wines.

Seashell Museum: The Esplanade.

Markets: Lakes Entrance Primary School, Myer St; 3rd Sat each month. ***Arts Festival:*** Forest Tech Living Resource Centre; Mar long weekend. ***Lakes Motor Fest:*** biannual; Apr. ***New Year's Eve Entertainment and Fireworks***: Dec.

Boathouse Restaurant: local seafood; 201 The Esplanade; (03) 5155 3055. ***Ferryman's Seafood Cafe:*** super fresh fish; Middle Boat Harbor, The Esplanade; (03) 5155 3000. ***Miriam's:*** European, local seafood; Level 1, cnr The Esplanade and Bulmer St; (03) 5155 3999. ***The Metung Galley:*** excellent local-produce-driven menu; 3/59 Metung Rd, Metung; (03) 5156 2330.

Goldsmith's in the Forest: 4-room private retreat; 191 Harrisons Track; (03) 5155 2518. ***The Lakes Apartments:*** contemporary apartments; 35 Church St; (03) 5155 6100. ***Waverley House Cottages:*** cottages with modern furnishings; 205 Palmers Rd; (03) 5155 1257. ***5 Knots:*** high-tech waterfront apartments; 42 Metung Rd, Metung; (03) 5156 2462.

Gippsland Lakes Five rivers end their journey to the sea here, forming a vast expanse of water tucked in behind Ninety Mile Beach. The lakes are a true playground for anyone with an interest in water activities, especially fishing and boating. Explore the lakes on a sightseeing cruise, including one to Wyanga Park Winery, or on the ever-popular houseboats that can be hired over summer. Details from visitor centre.

Jemmys Point: great views of the region; 1 km W. ***Lake Bunga:*** nature trail along foreshore; 3 km E. ***Lake Tyers:*** sheltered waters ideal for fishing, swimming and boating. Cruises depart from Fishermans Landing in town. Lake is 6–23 km NE, depending on access point. Lake Tyers Forest Park is great for bushwalking, wildlife-spotting, picnicking and camping; 20 km NE. ***Nyerimilang Heritage Park:*** 1920s homestead, with original farm buildings and the wonderfully maintained East Gippsland Botanic Gardens. Rose Pruning Day, with demonstrations, is held in July; 10 km NW. ***Swan Reach:*** Rosewood Pottery; Malcolm Cameron Studio Gallery, open weekends; 14 km NW. ***Metung:*** a scenic town on Lake King with boat hire, cruises and a marina regatta each Jan. Chainsaw Sculpture Gallery has chainsaw sculpture and a display of Annemieke Mein's embroidery art; 15 km W. ***Nicholson River Winery:*** 22 km NW. ***East Gippsland Carriage Co:*** restored horse-drawn carriages and tours; 30 km E. ***Bataluk Cultural Trail:*** driving tour taking in Aboriginal cultural sites in the East Gippsland region; self-guide brochure from visitor centre.

See also EAST GIPPSLAND, p. 159

[SCRUB, SAND AND SEA IN LAKES ENTRANCE]

VICTORIA

Leongatha is a thriving town, considered the commercial centre of South Gippsland. Idyllically positioned as a gateway to Gippsland destinations and attractions, any major road departing Leongatha provides access to popular attractions, all within an easy one hour's drive.

Historical Society Museum: McCartin St. ***Leongatha Gallery:*** Cnr McCartin St and Michael Pl. ***Mushroom Crafts:*** craft sales and gallery; 40 Blair St. ***Great Southern Rail Trail:*** commencement of 50 km rail trail that winds between Leongatha and Foster.

South Gippsland Golf Classic: Feb. ***Music for the People Concert:*** Mossvale Park; Feb. ***Mossvale Music Festival:*** Mar. ***Raw Vibes Youth Festival:*** Mar. ***Daffodil Festival:*** Sept. ***Garden and Lifestyle Expo:*** Nov.

Koonwarra Store: foodstore and emporium, excellent regional fare, courtyard setting; South Gippsland Hwy, Koonwarra; (03) 5664 2285.

Lamont House: 3-bedroom historic cottage; 19 Long St; (03) 5662 4510. ***Lyre Bird Hill Winery:*** 3 B&B rooms, family cottage; 370 Inverloch Rd, Koonwarra; (03) 5664 3204.

Koonwarra Situated between Leongatha and Meeniyan on the South Gippsland Hwy, Koonwarra became the first eco-wise town in Australia. The town prides itself on its commitment to sustainable lifestyles. Drop in to the Koonwarra Store to purchase local wines, cheese and pantry items. On the first Sat of each month, Koonwarra holds a farmers market. The town also boasts an organic cooking school, day spa, specialty shops, pottery and winery nearby. 8 km SE.

Meeniyan: great place to visit for the art and craft enthusiast. Places of interest include Meeniyan Art Gallery, South Gippsland Craft Merchants, Beth's Antiques and Lacy Jewellery. Meeniyan hosts an annual art and craft exhibition over the Melbourne Cup weekend; 16 km SE. ***Mossvale Park:*** tranquil setting for a picnic or barbecue. Music concerts and festivals are held here in Feb and Mar; 16 km NE. ***Mirboo North:*** situated among the picturesque Strzelecki Ranges, the township is decorated with murals depicting the history of the area. Grand Ridge Brewery, Lyre Bird Forest Walk and the Grand Ridge Rail Trail are also located here; 26 km NE.

See also PHILLIP ISLAND & GIPPSLAND, p. 150

Lorne

Pop. 971
Map ref. 572 D11 | 580 E4 | 593 M9

15 Mountjoy Pde; (03) 5289 1152; www.visitgreatoceanroad.org.au

94.7 The Pulse FM, 774 AM ABC Local Radio

Lorne is one of Victoria's most attractive and lively coastal resorts. The approach into town along the Great Ocean Road is truly spectacular, with the superb mountain scenery of the Otways on one side and the rugged Bass Strait coast on the other. The village of Lorne was established in 1871 and quickly became popular with pastoralists from inland areas, leading to its development around the picturesque Louttit Bay. When the Great Ocean Road opened in 1932 Lorne became much more accessible; however, the area has remained relatively unspoiled with good beaches, surfing, fishing and bushwalking in the hills – activities made all the more enjoyable by the area's pleasant, mild climate.

Teddys Lookout: excellent bay views; behind the town, at the end of George St. ***Shipwreck Walk:*** walk along the beach taking in sites of the numerous shipwrecks along this stretch of coast; details from visitor centre. ***Foreshore Reserve:*** great spot for a picnic, with paddleboats available for hire. ***Qdos:*** contemporary art gallery; Allenvale Rd. ***Lorne Fisheries:*** on the recently opened new pier with daily supplies.

Pier to Pub Swim: Jan. ***Mountain to Surf Foot Race:*** Jan. ***Great Ocean Road Marathon:*** May. ***Anaconda Adventure Race:*** Dec. ***Falls Festival:*** Dec–Jan.

Ba Ba Lu Bar and Restaurant: Spanish-style menu; 6A Mountjoy Pde; (03) 5289 1808. ***Bistro C Bar & Restaurant:*** mainly local seafood; Cumberland Lorne Resort, 150 Mountjoy Pde; (03) 5289 2455. ***Lorne Ovenhouse:*** modern wood-fired menu; 46A Mountjoy Pde; (03) 5289 2544. ***Qdos:*** cafe-bar next to gallery; 35 Allenvale Rd; (03) 5289 1989.

Acacia Villas: 4 bushland villas; 9 Fletcher St; (03) 5289 2066. ***Cumberland Lorne Resort:*** apartments and penthouses; 150 Mountjoy Pde; (03) 5289 4444. ***Grand Pacific Hotel:*** rooms and apartments; 268 Mountjoy Pde; (03) 5289 1609. ***Mantra Erskine Beach Resort:*** apartments and hotel rooms; 1–35 Mountjoy Pde; (03) 5228 9777.

Great Otway National Park This park covers 103 000 ha and includes a range of environments, from the timbered ridges of the eastern Otways to fern gullies, waterfalls and a coastline with tall cliffs, coves and sandy beaches. Around Lorne there are more than 100 walking tracks and the rock platforms along the coast provide ideal spots for ocean fishing. The Falls Festival, a major rock-music festival, is held over New Year's Eve at a property near the Erskine Falls, 9 km NW. These popular falls are a peaceful location and drop 30 m over moss-covered rocks. As well as driving, you can walk to the falls from Lorne along the river. The park surrounds Lorne and can be accessed from various points along the Great Ocean Rd. Contact Parks Victoria on 13 1963.

Cumberland River Valley: walking tracks and camping; 4 km sw. ***Mt Defiance Lookout:*** 10 km sw. ***Wye River:*** a small coastal village, good for rock and surf fishing, surfing and camping; 17 km sw. ***Old Lorne Road Olives:*** olive grove, cafe and gallery; closed Tues and Wed; Deans Marsh Rd; 22 km N. ***Gentle Annie Berry Gardens:*** pick your own; open Nov–Apr; 26 km NW via Deans Marsh. ***Scenic drives:*** west through the Otway Ranges, and south-west or north-east on the Great Ocean Rd.

See also SOUTH-WEST COAST, p. 153

Maffra

Pop. 4151
Map ref. 583 K6

96 Johnson St; (03) 5141 1811; www.tourismwellington.com.au

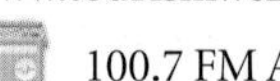
100.7 FM ABC Local Radio, 104.7 Gippsland FM

Maffra, settled in the 1840s, has the charm and old-style hospitality of another era. Named after Maffra in Portugal because many of the early Gippsland settlers had fought in that area of Europe during the Peninsula War, the town's early days were fraught with drought until a sugar beet industry established in the 1890s provided a major boost. The Glenmaggie Irrigation Scheme of 1919 also signalled a new heyday and ensured the viable and lengthy success of today's dairy industry. The sugar beet factory closed in 1948 owing to World War II's labour shortages and the competing dairy industry, but Maffra continues to support its rich agricultural surrounds. It also holds a great sense of history in its original shop verandahs and grand homesteads.

Maffra Sugar Beet Historic Museum: local history museum with special interest in the sugar beet industry; open Sun afternoon; River St. ***Mineral and gemstone display:*** large collection of rare gemstones and fossils at the information centre; Johnson St. ***All Seasons Herb Garden:*** Foster St. ***Gippsland Vehicle Collection:*** outstanding rotating display of interesting vehicles; located in a historic vegetable dehydrating factory; Maffra–Sale Rd; (03) 5147 3223. ***Gippsland Plains Rail Trail:*** recreational trail for cycling and walking that passes through town; still under development but when complete will link Stratford in the east to Traralgon in the west by traversing dairy country.

Gippsland Harvest Festival: Mar. ***Mardi Gras:*** Mar.

Stratford: the scenic Avon River flows through town. Knobs Reserve is a site where the local Aboriginal people once sharpened axe heads on sandstone grinding stones – it is part of the Bataluk Cultural Trail, which takes in significant Indigenous sites throughout East Gippsland. Stratford hosts the Shakespeare Celebration in May; 9 km E. ***Australian Wildlife Art Gallery and Sculpture:*** Princes Hwy near Munro; 25 km E. ***Robotic Dairy:*** the first Australian dairy farm to install 4 'Astronaut Milking Robots' where the cows decide when to be milked; open on public visitor days or by appt; Toongabbie Rd, Winnindoo; (03) 5199 2212; 26 km W. ***Lake Glenmaggie:*** popular watersports venue; 42 km NW via Heyfield. ***Alpine National Park:*** sprawls from Licola, 75 km NW, to the NSW border. Near Licola is Lake Tali Karng, which lies 850 m above sea level and is a popular bushwalking destination during the warmer months. ***Scenic drives:*** the Traralgon to Stratford Tourist Route highlights attractions of the area. For stunning scenery, drive north along Forest Rd, through the Macalister River Valley to Licola and Mt Tamboritha in Alpine National Park; or to Jamieson (166 km NW via Heyfield), with access to snowfields or Lake Eildon.

See also EAST GIPPSLAND, p. 159

Maldon

Pop. 1222
Map ref. 579 E5 | 591 N11 | 593 N1

High St; (03) 5475 2569; www.maldon.org.au

91.1 FM ABC Local Radio, 106.3 Radio KLFM

Maldon is one of Victoria's best-known gold towns and a popular weekend getaway for Melburnians. The town has been wonderfully preserved, with the wide, tree-lined main street featuring old buildings and shopfronts. Aside from the cafes and galleries, the town seems unchanged from the gold rush days. Maldon was declared Australia's first 'notable town' by the National Trust in 1966.

Historic town walk: grab a brochure from the visitor centre and take to the wide, old footpaths to discover the historic delights of Maldon. See preserved 19th-century shopfronts and old stone cottages. Highlights include the restored Dabb's General Store in Main St, and the Maldon Hospital in Adair St. ***Museum:*** displays on mining as well as domestic memorabilia from Maldon's past, in heritage building; open 1.30–4pm daily; High St. ***The Beehive Chimney:*** southern end of Church St. ***Anzac Hill:*** the walk to the top is rewarded with magnificent views of the area; southern end of High St.

Fair: Easter. ***Vintage Car Hill Climb:*** Oct. ***Folk Festival:*** Oct–Nov.

Penny School Gallery/Cafe: international cuisine; 11 Church St; (03) 5475 1911. ***Zen Eden Produce:*** vegetarian cafe and produce store; 6 Main St; 0408 319 188.

Clare House: 3 B&B rooms; 99 High St; (03) 5475 2229. ***Heatherlie:*** romantic cottage with pool; High St; 0413 123 650. ***Mount Hawke of Maldon:*** 3 B&B rooms in historic house; 24 Adair St; (03) 5475 1192. ***Wywurri B&B:*** in-house accommodation for a couple; 3 Templeton St; (03) 5475 2794.

Porcupine Township This award-winning tourist attraction is a reconstruction of an early 1850s goldmining town, with an array of slab, shingle and mudbrick buildings moved here from other goldfields. The village, complete with a blacksmith's, a doctor's surgery and even a bowling alley, is located in rugged bushland on the site of the original Porcupine diggings, where the first gold discovery between Castlemaine and Bendigo was made. Visitors to the township can pan for gold, handfeed emus or take a ride on the Little Toot train, which does a circuit through the diggings. Cnr Maldon–Bendigo and Allans rds; (03) 5475 1000; 3 km NE.

Mt Tarrangower Lookout Tower: town views; 2 km W. ***Carman's Tunnel Mine:*** guided mine tours feature relics from goldmining days; 2 km SW. ***Nuggetty Ranges and Mt Moorol:*** 2 km N. ***Cairn Curran Reservoir:*** great for watersports and fishing, water levels permitting; features picnic facilities and a sailing club near the spillway; 10 km SW. ***Victorian Goldfields Railway:*** historic steam trains run from Maldon Railway Station (Hornsby St) through scenic forest to Castlemaine; operates Wed, Sun and public holidays; bookings on (03) 5470 6658.

See also GOLDFIELDS, p. 154

Mallacoota

Pop. 973
Map ref. 565 D12 | 581 G11

Visitor Information Shed, Main Wharf; (03) 5158 0800; www.visitmallacoota.com

101.7 FM 3MGB Wilderness Radio, 104.9 FM ABC Local Radio

Mallacoota is a popular holiday centre in far East Gippsland, surrounded by the scenic Croajingolong National Park, which features Point Hicks, notable for being the first land on the east coast of Australia to be sighted by Europeans. There are spectacular surf beaches near the town, with Mallacoota Inlet offering great fishing.

WW II bunker and museum: restored and located at the airport. ***Information shed:*** a mural depicting Mallacoota's history is painted on the external walls; Main Wharf. ***The Spotted Dog Gold Mine:*** established in 1894, this was the most successful goldmine in the Mallacoota district.

Holiday markets: Easter and Christmas. ***Bream Fishing Classic:*** Mar (round 1) and June (round 2). ***Tour of Gippsland Cycling Event:*** July/Aug. ***Flora and Fauna weekend:*** 1st weekend in Nov.

Lucy's Noodles: authentic noodles with home-grown produce; 64 Maurice Ave; (03) 5158 0666.

Karbeethong Lodge: stylish historic guesthouse; 16 Schnapper Point Dr; (03) 5158 0411. ***Gipsy Point Lodge:*** guesthouse and 3 cottages; 35 Macdonald St, Gipsy Point; (03) 5158 8205 or 1800 063 556.

VICTORIA

Croajingolong National Park This park takes up a vast portion of what has been dubbed the Wilderness Coast. It protects remote beaches, tall forests, heathland, rainforest, estuaries and granite peaks, as well as creatures such as wallabies, possums, goannas and lyrebirds. Offshore, you might be lucky enough to spot dolphins, seals or southern right and humpback whales. Pt Hicks Lighthouse is a popular spot to visit, and Tamboon and Mallacoota inlets are good spots for canoeing. Access the park via a track west of town or various roads south of the Princes Hwy; contact Parks Victoria on 13 1963.

Surf beaches: Bastion Point, 2 km s; Bekta, 5 km s. ***Gabo Island Lightstation Reserve:*** take a scenic daytrip or stay in the Lightkeeper's Residence; 11 km E (offshore). ***Gipsy Pt:*** a quiet holiday retreat overlooking the Genoa River; 16 km NW.

See also EAST GIPPSLAND, p. 159

Mansfield

Pop. 2846
Map ref. 584 H11 | 586 D9

The Station Precinct, 173 Maroondah Hwy; 1800 039 049; www.mansfield-mtbuller.com.au

 99.7 FM Radio Mansfield, 103.7 FM ABC Local Radio

Mansfield is located in Victoria's High Country at the junction of the Midland and Maroondah highways. It is within easy reach of Lake Eildon's network of rivers, Alpine National Park and Mansfield State Forest. Activities ranging from hiking to horseriding to skiing make it an ideal destination for anyone with a love of outdoor adventure, no matter what the season.

Troopers' Monument: monument to police officers shot by Ned Kelly at Stringybark Creek; Cnr High St and Midland Hwy. ***Mansfield Mullum Wetlands Walk:*** along reclaimed railway line; starts from behind the visitor centre. ***Self-guide town walk:*** take in many buildings of historical significance.

The High Country Autumn Festival and Merrijig Rodeo: Mar long weekend. ***High Country Spring Arts Festival:*** 24 Oct – 4 Nov.

Mansfield Hotel: pizzas and DIY barbecue; 86 High St; (03) 5775 2101. ***Mansfield Regional Produce Store:*** popular local cafe, great coffee; 68 High St; (03) 5779 1404. ***The Magnolia Restaurant:*** modern Australian; 190 Mt Buller Rd; (03) 5779 1444. ***Jamieson Brewery:*** menu complements house-crafted ales; Eildon Rd, Jamieson; (03) 5777 0515.

Burnt Creek Cottages: luxurious cottages; 68 O'Hanlons Rd; (03) 5775 3067. ***Highton Manor:*** manor house plus motel-style units; 140 Highton La; (03) 5775 2700. ***Magnolia Gourmet Country House:*** 2 attic rooms; 190 Mount Buller Rd; (03) 5779 1444. ***Wombat Hills Cottages:*** 3 romantic stone cottages; 55 Lochiel Rd, Barwite; (03) 5776 9507.

Mt Buller Victoria's largest and best alpine skiing resort is Mt Buller, whose summit stands 1804 m above sea level. The 24 lifts, including the new 6-seater Holden chairlift (first of its kind in Australia), give access to 180 ha of ski trails, from gentle 'family runs' to heart-stopping double black diamond chutes. If you are a beginner, take on the friendly Bourke Street (Green Run) to find your 'ski legs', or join one of the ski schools there. There is also a half pipe at Boggy Creek and Terrain Park, or cross-country skiing at nearby Mt Stirling. Mt Buller Village offers resort accommodation, and the ski season runs between early June and late Sept. (03) 5777 6077; 47 km E.

Craig's Hut The High Country is synonymous with courageous and hardy cattlemen, transformed into Australian legends by Banjo Paterson's iconic ballad 'The Man from Snowy River'. The men would build huts on the high plains for shelter during summer cattle drives. Craig's Hut on Mt Stirling is a replica of one such shelter, used as a set on the 1983 film *The Man from Snowy River*. It burnt down in the 2006 bushfires, but was rebuilt and reopened in January 2008. The last 2 km of the track to the hut is 4WD or 1.5 km via a fairly steep walking track. 50 km E.

Delatite Winery: Stoneys Rd; 7 km SE. ***Mt Samaria State Park:*** scenic drives, camping and bushwalking; 14 km N. ***Lake Eildon:*** houseboat hire, fishing and sailing; 15 km s; *see Eildon for further details.* ***Lake Nillahcootie:*** popular for boating, fishing and watersports; 20 km NW. ***Jamieson:*** an old goldmining town on the Jamieson River with historic buildings; 37 km s. ***Alpine National Park:*** begins around 40 km SE (*see Mount Beauty*). ***Scenic drive:*** take the road over the mountains to Whitfield (62 km NE), in the King River Valley, passing through spectacular scenery, including Powers Lookout (48 km NE) for views over the valley. ***Lake William Hovell:*** for boating and fishing; 85 km NE. ***Mt Skene:*** great for bushwalking, with wildflowers in summer; 85 km SE via Jamieson. ***Fishing:*** good spots include the Delatite, Howqua, Jamieson and Goulburn rivers. ***Horse trail-riding:*** a different way to explore the region, from 2 hr rides to 10-day treks; details from visitor centre. ***Mountain-biking:*** summer months reveal an expanding network of downhill and cross-country trails at Mt Buller and Mt Stirling.

See also HIGH COUNTRY, p. 158

Maryborough

Pop. 7690
Map ref. 579 C5 | 591 M11 | 593 M1

Cnr Alma and Nolan sts; (03) 5460 4511 or 1800 356 511; www.visitmaryborough.com.au

 99.1 Goldfields FM, 107.9 FM ABC Local Radio

Maryborough is a small city set on the northern slopes of the Great Dividing Range. Its historic 19th-century buildings, particularly around the civic square, are a testament to the riches brought by the gold rush of the 1850s. Stroll through the streets to enjoy the cafes, craft shops and magnificent buildings, such as the National Trust–listed courthouse, post office and town hall.

Maryborough Railway Station So immense and impressive is this building that Mark Twain, on his visit to the town, remarked that Maryborough was 'a station with a town attached'. Rumour has it that the building was actually intended for Maryborough in Queensland. The beautifully preserved station houses the visitor centre, the extensive Antique Emporium, the Woodworkers Gallery (open weekends only), and Twains Wood and Craft Gallery. Station St.

Pioneer Memorial Tower: Bristol Hill. ***Worsley Cottage:*** a historical museum featuring local relics; open Sun; Palmerston St. ***Central Goldfields Art Gallery:*** features an impressive collection of local artworks, housed in the old fire station; Neill St. ***Phillips Gardens:*** Alma St.

Maryborough Highland Gathering: New Year's Day. ***Energy Breakthrough:*** energy expo; Nov.

Paddys Ranges State Park This park offers the chance to enjoy red ironbark and grey box vegetation on a scenic walk or drive. The majority of walks start from the picnic area – see old goldmines and relics or keep an eye out for the rare painted honeyeater and other birdlife. There is also fossicking within the park, but in designated areas only. Access to the park is just south of Maryborough. Contact Parks Victoria on 13 1963.

Aboriginal wells: impressive rock wells; 4 km S. ***Carisbrook:*** holds a popular tourist market 1st Sun each month; 7 km E.

See also GOLDFIELDS, p. 154

Marysville

Pop. 516
Map ref. 571 F5 | 573 N3 | 582 F3

49 Darwin St; 0429 000 394; www.marysvilletourism.com

98.5 UGFM Upper Goulburn Community Radio, 774 AM ABC Local Radio

The historic town of Marysville was almost totally destroyed by bushfire on 7 February 2009. With the help of the community, state and federal governments, this once idyllic township is slowly being rebuilt.

Uncle Fred & Aunty Val's Lolly Shop & Produce Store: old-fashioned candy store; 8 Murchison St; (03) 5963 3644; ***Bruno's Art & Sculpture Garden:*** gardens featuring sculptures by artist Bruno Torfs; open 10am–5pm daily; (03) 5963 3513.

Marysville Community Market: 2nd Sun each month. ***Marysville Farmers Market:*** 4th Sun each month. ***Taggerty 4 Seasons Market:*** Apr.

Fraga's Cafe: stop for lunch or a coffee; 1/19 Murchison St; (03) 5963 3216. ***Marysville Country Bakery:*** country-style bakery and cafe; 17 Murchison St; (03) 5963 3477. ***My Chef Mike:*** restaurant serving excellent pizzas; 49 Darwin St; (03) 5963 4512. ***Black Spur Inn:*** country-style restaurant; 436 Maroondah Hwy, Narbethong; (03) 5963 7121.

Delderfield: 2 private suites; 1 Darwin St; (03) 5963 4345. ***Black Spur Inn:*** country-style suites; 436 Maroondah Hwy, Narbethong; (03) 5963 7121. ***Woodlands Rainforest Retreat:*** 4 luxurious bungalows; 137 Manby Rd, Narbethong; (03) 5963 7150.

Cathedral Range State Park The word 'imposing' does not do justice to the 7 km rocky ridge that forms the backbone of this park. Challenging hikes up the ridge to lookout points offer unparalleled views to the valley below. Walks can include overnight stays at the Farmyard, so named because lyrebirds imitate the noises of the domestic animals in the farmyards below. Contact Parks Victoria on 13 1963; 15 km N.

Lake Mountain Renowned for first-rate cross-country skiing, the area is also great for tobogganing, snow tubing and sled rides. When the snow melts and the wildflowers bloom, hikers can take the Summit Walk (4 km return) over the mountain. Ski and walk brochure available from visitor centre.

Lady Talbot Forest Drive: this 46 km route begins east of town. Stop en route to enjoy picnic spots, walking tracks and lookouts. ***Buxton Trout & Salmon Farm:*** drop a line in one of the well-stocked ponds, purchase smoked fish or enjoy a barbecue lunch; open 9am–5pm daily; 2118 Maroondah Hwy, Buxton; (03) 5774 7370; 12 km SE. ***Scenic walks:*** many tracks in the area, including a short walk to Steavenson Falls, from Falls Rd; 4 km loop walk in Cumberland Memorial Scenic Reserve, 16 km E; 4 km Beeches Walk through ancient beech and mountain ash forests (accessed via Lady Talbot Forest Dr). ***Big River State Forest:*** camping, fishing and gold fossicking; 30 km E.

See also YARRA & DANDENONGS, p. 148

Milawa

Pop. 202
Map ref. 585 I8 | 586 H6

Wangaratta Visitor Information Centre, 100 Murphy St, Wangaratta; 1800 801 065; www.visitwangaratta.com.au

97.7 FM ABC Local Radio, 101.3 Oak FM

Milawa is the perfect destination for lovers of fine food and wine. The Milawa gourmet region boasts over 13 wineries, including the renowned Brown Brothers vineyard. Other fresh local produce outlets sell olives, honey, cheese, chocolates and berries.

Milawa Mustards: a wide range of locally produced mustards; set in attractive cottage gardens; Snow Rd. ***Milawa Cheese Company:*** sales and tastings of specialist, gourmet cheeses; Factory Rd. ***Milawa Muse Gallery:*** ever-changing collection of various art mediums complementing the fine quality of the region; Milawa Cheese Factory complex. ***Brown Brothers:*** cellar-door tastings and sales; Bobinawarrah Rd. ***EV Olives:*** working olive grove open for tastings and sales; Everton Rd, Markwood.

A Weekend Fit for a King: be treated like a king at the King Valley wineries festival; Queen's Birthday weekend, June. ***Beat the Winter Blues and Jazz Festival:*** July. ***King Valley Shed Show:*** Oct. ***La Dolce Vita:*** wine and food festival at the Milawa/King Valley wineries; Nov.

Restaurant Merlot: contemporary Australian; Lindenwarrah Country House Hotel, Milawa–Bobinawarrah Rd; (03) 5720 5777. ***The Ageing Frog Bistrot:*** French-influenced menu; Milawa Cheese Company, Factory Rd; (03) 5727 3589. ***The Epicurean Centre:*** rustic dishes to match wines; Brown Brothers Vineyard, 239 Milawa–Bobinawarrah Rd; (03) 5720 5540.

Lindenwarrah Country House Hotel: boutique hotel rooms; Milawa–Bobinawarrah Rd; (03) 5720 5777. ***Casa Luna Gourmet Accommodation:*** 2 stylish rooms; 1569 Boggy Creek Rd, Myrrhee; (03) 5729 7650. ***Willoaks Boutique Accommodation:*** farmstay with 2 guest rooms; 31 Tetleys La, Oxley; (03) 5727 3292.

King Valley Wine Region This region produces wines with a distinctly Italian influence. Taste cabernet sauvignon, merlot, pinot noir, riesling and chardonnay at various cellar doors, as well as the Italian varieties of sangiovese, pinot grigio and nebbiolo.

Oxley: home to many wineries as well as the Blue Ox Blueberry Farm and King River Cafe; 4 km W.

See also HIGH COUNTRY, p. 158

Mildura

see inset box on next page

Moe

Pop. 15 581
Map ref. 573 P9 | 582 H7

Latrobe Visitor Information Centre, The Old Church, Southside Central, Princes Hwy; 1800 621 409; www.visitlatrobevalley.com

100.7 FM ABC Local Radio, 104.7 Gippsland FM

Like many of the towns in this region, Moe is supported by the power industry, but it has managed to avoid becoming a grim industrial centre. Instead there is a small-town feel and a number of pretty gardens and public parks.

VICTORIA

Gippsland Heritage Park Also known as Old Gippstown, this is a re-creation of a 19th-century community with over 30 restored buildings and a fine collection of fully restored horse-drawn carriages. Lloyd St; (03) 5127 8709.

Cinderella Dolls: Andrew St. ***Race track:*** picturesque country horse track with regular meetings; Waterloo Rd.

Old Gippstown Market: at Gippsland Heritage Park, with local crafts and produce; last Sat each month Sept–May. ***Fairies in the Park:*** Feb. ***Jazz Festival:*** Mar. ***Blue Rock Classic:*** cross-country horserace; Mar. ***Moe Cup:*** horserace; Oct.

Baw Baw National Park The landscape of Baw Baw ranges from densely forested river valleys to alpine plateaus and the activities on offer are equally varied – from canoeing river rapids and fishing for trout to skiing, horseriding and bushwalking. Wildflowers carpet the alpine areas in spring. Baw Baw Alpine Resort is located 90 km north of Moe, while the popular Aberfeldy picnic and camping area is accessed via a track north of Walhalla. Contact Parks Victoria on 13 1963.

Edward Hunter Heritage Bush Reserve: 3 km s via Coalville St. ***Trafalgar Lookout and Narracan Falls:*** near Trafalgar; 10 km w. ***Old Brown Coal Mine Museum:*** explore the history and memorabilia of the original township known as 'Brown Coal Mine' and the establishment of the power industry in the Latrobe Valley; Cnr Third St and Latrobe River Rd, Yallourn North; 10 km E. ***Blue Rock Dam:*** fishing, swimming and sailing; 20 km NW. ***Thorpdale:*** a town renowned for its potatoes. A bakery sells potato bread and a potato festival is held each Mar; 22 km sw. ***Walhalla Mountain River Trail:*** leads to the picturesque old mining township of Walhalla; Tourist Route 91; details from visitor centre. *See Walhalla.*

See also PHILLIP ISLAND & GIPPSLAND, p. 150

Mornington

Pop. 20 821
Map ref. 570 D10 | 573 J9 | 575 J5 | 582 C7

Mornington Peninsula Visitor Information Centre, Point Nepean Rd, Dromana; (03) 5987 3078 or 1800 804 009; www.visitmorningtonpeninsula.org

98.7 3RPP FM, 774 AM ABC Local Radio

Mornington was once the hub of the Mornington Peninsula, which is the reason this long arm of land was eventually given the same name. Today Melbourne's urban sprawl has just about reached the town, and it has virtually become a suburb. It still retains a seaside village ambience, particularly with a historic courthouse and post office museum that provide a glimpse of the past. The Rocks restaurant on the harbour provides a stunning view over the famous yachts. In the distance, colourful bathing boxes line Mills Beach.

Historic Mornington Pier: built in the 1850s, the pier remains popular today for walks and fishing. ***Mornington Peninsula Regional Gallery:*** print and drawing collection, including works by Dobell, Drysdale and Nolan; open Tues–Sat; Dunns Rd. ***World of Motorcycles Museum:*** Tyabb Rd. ***Old post office:*** now home to a local history display; Cnr Main St and The Esplanade. ***Mornington Tourist Railway:*** 10 km journey on steam train; departs from cnr Yuilles and Watt rds; 1st, 2nd and 3rd Sun each month, with additional trips running on Thurs in Jan.

Street Market: Main St; each Wed. ***Mornington Racecourse Craft Market:*** 2nd Sun each month. ***RACV Vintage Car Rally:*** Jan. ***Mornington Food and Wine Festival:*** Oct.

Afghan Marco Polo: Afghan cuisine; 9–11 Main St; (03) 5975 5154. ***Brass Razu Wine Bar:*** modern European; 13 Main St; (03) 5975 0108. ***The Rocks:*** seafood with yachts; Mornington Yacht Club, 1 Schnapper Point Dr; (03) 5973 5599.

Glynt Manor: boutique hotel; 10 Greenslade Crt, Mount Martha; (03) 5974 8400. ***George's Boutique B&B & Culinary Retreat:*** 4 luxury suites; 776 Arthurs Seat Rd, Arthurs Seat; (03) 5981 8700. ***Morning Star Estate:*** boutique hotel; 2 Sunnyside Rd, Mount Eliza; (03) 9787 7760. ***Woodman Estate:*** boutique hotel and spa retreat; 136 Graydens Rd, Moorooduc; (03) 5978 8455.

Arthurs Seat State Park At 309 m, Arthurs Seat is the highest point on the Mornington Peninsula. The summit can be reached by foot or vehicle and offers panoramic views of the bay and surrounding bushland. Picnic facilities and a restaurant are on the summit. There are many short walks, plus the historic Seawinds Park with gardens and sculptures. The Enchanted Maze Garden is set in superb gardens, with a variety of mazes and the Maize Maze Festival in Feb–Apr. Contact Parks Victoria on 13 1963; Arthurs Seat Rd, near Dromana; 16 km sw.

Mornington Peninsula Wine Region Although the first vines were planted in 1886, the wine industry here did not truly take off until the 1970s. Comprising over 170 vineyards, the region consists predominantly of boutique wineries, with more than 50 cellar doors. While enjoying the sunshine it seems appropriate to try the crisp chardonnay that is famed in this region – or even a pinot noir or cabernet sauvignon as the chilly sea breeze kicks in of an evening. Wine festivals include Pinot Week each Mar, Queen's Birthday Wine Weekend each June and the Cool Whites Festival each Nov. Winery map from visitor centre.

Mount Martha: here is The Briars with a significant collection of Napoleonic artefacts and furniture. The town also features many gardens, plus wetlands great for birdwatching and walks; 7 km s. ***Ballam Park:*** historic, French-farmhouse-style homestead built in 1845; open Sun; Cranbourne Rd, Frankston; 14 km NE. ***Mulberry Hill:*** former home of artist Sir Daryl Lindsay and author Joan Lindsay, who wrote *Picnic at Hanging Rock*; open Sun afternoons; Golf Links Rd, Baxter; 14 km NE. ***Tyabb Packing House Antique Centre:*** Australia's largest collection of antiques and collectibles; Mornington–Tyabb Rd, Tyabb; 16 km E. ***Hastings:*** coastal town on Western Port with 2 km walking trail through wetlands and mangrove habitat; 21 km SE. ***Moonlit Sanctuary Wildlife Conservation Park:*** wildlife park featuring endangered native Australian animals; visit 11am–5pm or take an evening tour; night tour bookings on (03) 5978 7935; 550 Tyabb–Tooradin Rd, Pearcedale; 24 km E. ***Beaches:*** sandy bays popular with holiday-makers between Mornington and Mount Martha.

See also MORNINGTON PENINSULA, p. 149

Morwell

Pop. 13 398
Map ref. 582 H7

Latrobe Visitor Information Centre, The Old Church, Southside Central, Princes Hwy, Traralgon; 1800 621 409; www.visitlatrobevalley.com

100.7 FM ABC Local Radio, 104.7 Gippsland FM

Morwell is primarily an industrial town and Victoria's major producer of electricity. Nestled in the heart of the Latrobe Valley, it contains one of the world's largest deposits of brown coal. Among all the heavy machinery is the impressive Centenary Rose

continued on p. 196

MILDURA

Pop. 46 035

Map ref. 558 E6 | 588 G6

Alfred Deakin Centre, 180–190 Deakin Ave; (03) 5018 8380 or 1800 039 043; www.visitmildura.com.au

104.3 FM ABC Local Radio, 106.7 HOTFM Sunraysia Community Radio

Mildura offers a Riviera lifestyle, with the Murray River flowing by the town and sunny, mild weather throughout the year. One of Victoria's major rural cities, Mildura's development has been aided by the expansion of irrigation, which has allowed the city to become a premier fruit-growing region.

Mildura Arts Centre & Rio Vista The complex is set on the banks of the Murray River and includes an art gallery, an amphitheatre and the Rio Vista museum. Rio Vista was the home of the town's founders, the Chaffey brothers, and is now preserved as a museum displaying colonial household items. The arts centre and theatre, reopening in late 2011 after a major redevelopment, houses an impressive permanent collection including Australia's largest display of Orpen paintings, works by Brangwyn, frequent temporary exhibitions and performing arts. Outside, a delightful Sculpture Trail winds through the landscape gardens surrounding the centre; 199 Cureton Ave; (03) 5018 8330.

The Alfred Deakin Centre: interactive exhibitions and displays of the region; Deakin Ave. ***Mildura Brewery:*** produces natural and specialty beers inside the former Art Deco Astor Theatre; view the brewing process or eat at the Brewery Pub; 20 Langtree Ave; (03) 5021 5399. ***Langtree Hall:*** Mildura's first public hall now contains antiques and memorabilia; open Tues–Sat; 79 Walnut Ave; (03) 5021 3090. ***Mildura Wharf:*** paddleboats departing here for river cruises include the steam-driven *PS Melbourne*; *PV Rothbury* for day trips to Trentham Winery on Thursday and Gol Gol Hotel on Tuesday; and *PV Mundoo* for a Thursday dinner cruise. Access from Hugh King Drive; bookings (03) 5023 2200. ***Snakes and Ladders:*** fun park featuring 'dunny' collection; Seventeenth St, Cabarita; (03) 5025 3575. ***Aquacoaster waterslide:*** Cnr Seventh St and Orange Ave.

Arts Festival: Mar. ***Cup Carnival:*** May. ***Writers Festival:*** July. ***Golf Week:*** July. ***Country Music Festival:*** Sept. ***Vintage Tractor Pull:*** Oct. ***Jazz Food and Wine Festival:*** Nov.

New Spanish Bar and Grill: up-market steakhouse; Quality Hotel Mildura Grand, Seventh St; (03) 5021 2377. ***Stefano's:*** excellent Italian restaurant in the cellar of the majestic Grand Hotel, featuring the culinary skills of TV chef Stefano de Pieri; Langtree Ave; (03) 5023 0511. ***The Gol Gol Hotel:*** good pub dining; Sturt Hwy, Gol Gol; (03) 5024 8492. ***Trentham Estate Restaurant:*** modern Australian; Sturt Hwy, Trentham Cliffs; (03) 5024 8888.

Murray Haven: beautifully renovated 4-bedroom house; 118A Thirteenth St; 0419 514 861. ***Pied-à-terre:*** luxury 5-bedroom home; 97 Chaffey Ave; (03) 5022 9883. ***Quality Hotel Mildura Grand:*** refurbished rooms; Seventh St; (03) 5023 0511 or 1800 034 228. ***Adventure Houseboats:*** slick solar-powered vessels; 16 Sturt Hwy, Buronga; (03) 5023 4787.

Murray Darling Wine Region The Mediterranean-style climate mixed with irrigated lands has contributed to making this wine region Victoria's largest. The region is well regarded for its varieties of chardonnay, cabernet sauvignon and shiraz. Among the large-scale wineries such as Lindemans Karadoc, smaller boutique wineries offer specialty wines for tastings and sales. Brochure available from visitor centre.

[AERIAL VIEW OF THE MURRAY RIVER]

Orange World: tours of citrus-growing region; Silver City Hwy, Buronga; (03) 5023 5197; 6 km N. ***Australian Inland Botanic Gardens:*** unique semi-arid botanic gardens with tractor tour and lunch last Sat each month; open daily; River Rd, Buronga; (03) 5023 3612; 8km NE. ***Ornamental Lakes Park:*** farmers market held 1st and 3rd Sat each month; Hugh King Dr. ***Kings Billabong Wildlife Reserve:*** situated on the Murray River flood plain, home to river red gums and abundant birdlife. Attractions include Psyche Pump Station, Bruce's Bend Marina and Kings Billabong Lookout; 8 km SE. ***Angus Park:*** dried fruits and confectionery; 10 km SE. ***Red Cliffs:*** an important area for the citrus and dried fruit industries. The town features the 'Big Lizzie' steam traction engine; 15 km S. ***Hattah–Kulkyne National Park:*** 70 km S; *see Ouyen*. ***Murray–Sunset National Park:*** attractions near Mildura include Lindsay Island, for boating, swimming and fishing; access from Sturt Hwy, about 100 km W of Mildura. *See Murrayville for further details on park.* ***Mungo National Park:*** 104 km NE over NSW border (*see Balranald, NSW*).

See also MALLEE COUNTRY, p. 156

VICTORIA

Garden, featuring over 4000 rose bushes and regarded as one of the finest rose gardens in the Southern Hemisphere.

PowerWorks: dynamic displays on the electrical industries; tours of mines and power stations daily; Ridge Rd. ***Centenary Rose Garden:*** off Commercial Rd. ***Latrobe Regional Gallery:*** hosts outstanding works of contemporary Australian art by local and national artists; Commercial Rd. ***Gippsland Immigration Wall of Recognition:*** acknowledges all immigrants who contributed to the development of the Gippsland region.

Market: Latrobe Rd; each Sun.

Morwell National Park This park protects some of the last remnant vegetation of the Strzelecki Ranges, including pockets of rainforest and fern gullies. The area was once occupied by the Woollum Woollum people, who hunted in the ranges. In the 1840s European settlers cleared much of the surrounding land. On the Fosters Gully Nature Walk, keep your eyes peeled for orchids (over 40 species are found here) and native animals. Contact Parks Victoria on 13 1963; 16 km S.

Hazelwood Pondage: warm water ideal for year-round watersports; 5 km S. ***Arts Resource Collective:*** housed in an old butter factory; Yinnar; 12 km SW. ***Lake Narracan:*** fishing and waterskiing; 15 km NW. ***Narracan Falls:*** 27 km W. ***Scenic drives:*** routes along the Strzelecki Ranges and Baw Baw mountains offer impressive views over the Latrobe Valley.

See also **PHILLIP ISLAND & GIPPSLAND, p. 150**

Mount Beauty

Pop. 1705
Map ref. 585 L10 | 587 J7

Alpine Discovery Centre, 31 Bogong High Plains Rd; (03) 5754 1962 or 1800 111 885; www.visitalpinevictoria.com.au

92.5 Alpine Radio FM, 720 AM ABC Local Radio

At the foot of Mount Bogong, Victoria's highest mountain at 1986 metres, Mount Beauty was originally an accommodation town for workers on the Kiewa Hydro-electric Scheme. Today, the town is regarded as an adventure mecca and a focal point for a variety of adventure activities, with mountain-biking, hang-gliding and bushwalking just a few on offer. The pairing of adventure with more leisurely outdoor activities like golf, swimming and fishing makes Mount Beauty an ideal holiday destination.

Mt Beauty Pondage: for watersports and fishing; just north of Main St. ***Wineries:*** cool-climate vineyards at Annapurna Estate, Bogong Estate and Recline. ***Scenic walks:*** several scenic walking tracks; details from visitor centre.

Markets: Hollonds St; 1st Sat each month. ***Bogong Cup:*** Jan. ***MTBA National Series:*** mountain-biking championships; Jan. ***Music Muster:*** Apr. ***Mitta Mitta to Mount Beauty Mountain Bike Challenge:*** Oct.

Alpine National Park Covering 646 000 ha in 4 sections, this is Victoria's largest park, containing the highest mountains in the state. Most of Australia's south-east rivers, including the mighty Murray, have their source here. The area is known for its outstanding snowfields during winter, and bushwalking and wildflowers in summer. Other activities include horseriding, canoeing, rafting and mountain-bike riding. 30 km SE.

Falls Creek Surrounded by Alpine National Park, Falls Creek is a winter playground for downhill and cross-country skiers. When the snow is falling, the ski resort caters for skiers and snowboarders with a variety of runs and terrain, including some to suit beginners. Novelty tours are also available, such as the Snowmobile Tours. Each Aug, Falls Creek hosts the Kangaroo Hoppet cross-country ski race. In spring and summer, take a walk on the Bogong High Plains or fly-fish in one of the lakes and rivers nearby; 30 km SE.

Tawonga Gap: features a lookout over valleys; 13 km NW. ***Bogong:*** scenic walks around Lake Guy and nearby Clover Arboretum for picnics; 15 km SE. ***Scenic drives:*** to Falls Creek and the Bogong High Plains (not accessible in winter beyond Falls Creek); details from visitor centre.

See also **HIGH COUNTRY, p. 158**

Murrayville

Pop. 210
Map ref. 558 B9 | 588 C11 | 590 C1 | 603 O10

Ouyen Information Centre, 17 Oke St, Ouyen; (03) 5092 2006; www.visitmildura.com.au/murrayville

103.5 3MBR-FM Mallee Border Radio, 594 AM ABC Local Radio

Murrayville is a small town on the Mallee Highway near the South Australia border. It is near three major, remote national parks: Murray–Sunset National Park; Wyperfeld National Park; and Big Desert Wilderness Park, one of Victoria's largest wilderness zones.

Historic buildings: include the restored railway station and the old courthouse. ***Walking tracks:*** several, including the Pine Hill Walking Trail in the town.

Murray–Sunset National Park Millions of years ago this area was submerged beneath the sea. When the sea retreated, large sand ridges and dunes were left. Now there is a variety of vegetation including grasslands, saltbush and mallee eucalypts. In spring, wildflowers abound; look out for Victoria's largest flower, the Murray lily. Access roads to the park are off Mallee Hwy. The Pink Lakes saltwater lakes, with a distinctive, pinkish hue, are a key attraction and are especially remarkable at sunset. There are many good walking tracks near the lakes, as well as excellent camping facilities. Lakes access via Pink Lakes Rd (turn-off at Linga, 50 km E); contact Parks Victoria on 13 1963. *For north section of park see Mildura.*

Cowangie: a small, historic town with several 19th-century buildings, including Kow Plains Homestead; 19 km E. ***Big Desert Wilderness Park:*** a remote park with no access other than by foot. True to its name, this park has remained relatively untouched by Europeans and includes many reptile species and plants adapted to arid conditions; the track south of town takes you close to the park boundary. ***Wyperfeld National Park:*** access via Underbool or by 4WD track south of Murrayville; *for further details see Hopetoun.*

See also **MALLEE COUNTRY, p. 156**

Myrtleford

Pop. 2726
Map ref. 585 J8 | 586 H6

Post Office Complex, Great Alpine Rd; (03) 5752 1044; www.visitalpinevictoria.com.au

91.7 FM ABC Local Radio, 101.3 Oak FM

Myrtleford is a pretty town in Victoria's alpine High Country. Originally called Myrtle Creek Run in the early 1800s, it is a thriving agricultural district and gateway to Mount Hotham.

The Phoenix Tree: a sculpture created by Hans Knorr from the trunk of a red gum; Lions Park. ***The Big Tree:*** a huge old red gum; Smith St. ***Old School:*** the town's original school, now fully restored; open Thurs, Sun or by appt; Albert St. ***Swing Bridge over Myrtle Creek:*** Standish St. ***Reform Hill Lookout:*** a scenic walking track from Elgin St leads to the lookout, which has great views across town; end of Halls Rd. ***Parks:*** Rotary Park in Myrtle St and Apex Park in Standish St are both delightful picnic spots. ***Michelini Wines:*** Great Alpine Rd. ***Murray to the Mountains Rail Trail:*** the cycle touring loop from Bright to Wangaratta runs through the town.

Market: local produce; Great Alpine Rd; each Sat Jan–Apr. ***Alpine Classic Bike Ride:*** to Bright; Jan. ***Myrtleford Festival:*** Mar. ***Golden Spurs Rodeo:*** Dec.

Plump Harvest: regional produce store/cafe; 72 Great Alpine Rd; (03) 5752 2257. ***The Butter Factory:*** produce store/cafe; 15 Great Alpine Rd; (03) 5752 2300.

Motel on Alpine: range of motel rooms; 258 Great Alpine Rd; (03) 5752 1438.

Mt Buffalo National Park: 7 km S; *see Bright.* ***Wineries:*** several in the region, including Rosewhite Vineyards and Winery; open weekends, public holidays and throughout Jan; Happy Valley Rd; 8 km SE. ***Gapsted:*** home to the Victorian Alps Winery, offers tours and sales; 8 km NW. ***Eurobin:*** a number of farms near the town with sales of local produce, including Red Stag Deer Farm and Bright Berry Farm, offering homemade jams and berries Dec–Mar; 16 km SE. ***Fishing:*** in the Ovens and Buffalo rivers and Lake Buffalo (25 km S).

See also HIGH COUNTRY, p. 158

Nagambie

Pop. 1381
Map ref. 584 D9

319 High St; (03) 5794 2647 or 1800 444 647; www.nagambielakestourism.com.au

97.7 FM ABC Local Radio, 98.5 FM

Nagambie is found between Seymour and Shepparton on the Goulburn Valley Highway. The town is on the shores of Lake Nagambie, which was created by the construction of the Goulburn Weir in 1891. Activities such as waterskiing, speedboating and especially rowing are popular on this man-made lake.

The Jetty: fine dining, lakeside apartments and spa; High St. ***The Grapevine by the Lake:*** sales of nuts and local produce; High St. ***Nagambie Lakes Entertainment Centre:*** renovated 1890s building with bars, gaming lounge and non-motorised boat hire; High St.

Rowing Regatta: Jan–Mar. ***Head of the River:*** Mar. ***Nagambie On Water Festival:*** Mar long weekend. After ***Vintage Festival:*** May. ***Shiraz Challenge:*** a search for the best shiraz in the region; Nov. ***Fireworks over the Lake:*** New Year's Eve.

Mitchelton Wines: regional produce menu; Mitchells Town Rd; (03) 5736 2221. ***Tahbilk Wetlands Cafe:*** casual cafe food; Tahbilk Winery, 254 O'Neils Rd, Tabilk; (03) 5794 2555. ***The Jetty:*** modern Australian; 317 High St; (03) 5794 1964.

Blackwood Park Country House: 3-bedroom house; Mitchelton Wines, Mitchells Town Rd; (03) 5736 2221. ***Nagambie Lakes Leisure Park:*** cabins and caravan sites; 69 Loddings La; (03) 5794 2373.

Goulburn Valley Wine Region The top wine here is shiraz, but also look out for marsanne at the area's 2 main players, Tahbilk Winery and Mitchelton Wines. Tahbilk's vineyard features a 3-tiered chateau, a museum detailing the estate's history, and a wetland and wildlife reserve. Mitchelton offers a 55-metre lookout tower with views of Goulburn River, and an excellent restaurant.

Goulburn Weir The construction of this weir resulted in the creation of Lake Nagambie. It is the diversion weir on the Goulburn River for the Goulburn Valley Irrigation area and feeds water by channel and pipeline to Bendigo, among other places. A walkway runs across the weir offering views of the structure and lake. Picnic and barbecue facilities are available; 12 km N.

See also GOULBURN & MURRAY, p. 157

Natimuk

Pop. 445
Map ref. 590 F9

National Hotel, Main St; (03) 5387 1300.

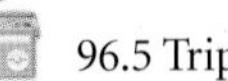
96.5 Triple H FM, 594 AM ABC Local Radio

Natimuk is popular for its proximity to Mount Arapiles, a 369-metre sandstone monolith that has been described as 'Victoria's Ayers Rock'. The mountain was first climbed by Major Mitchell in 1836, and today is a popular rock-climbing destination with over 2000 marked climbing routes.

Arapiles Historical Society Museum: housed in the old courthouse; open by appt; Main St. ***Arapiles Craft Shop:*** features local arts and crafts; Main St. ***The Goat Gallery:*** showcases works of local and regional artists; Main St. ***Self-guide heritage trail:*** details from visitor centre.

Nati Frinj: biennial arts festival; Oct/Nov.

Mt Arapiles–Tooan State Park This park is divided into 2 blocks, the larger Tooan block and the smaller Mt Arapiles block; Mt Arapiles offers rock climbing and is by far the most popular. Mitre Rock presents a smaller climbing challenge if required. Should you choose not to scale one of the various rock faces, great views are still available from the walking tracks, or you can drive to the summit. Nature study is another possibility – a huge 14% of the state's flora is represented in the Mt Arapiles section alone. Contact Parks Victoria on 13 1963; access is from the Wimmera Hwy; 12 km SW.

Banksia Hill Flower Farm: 10 km E. ***Duffholme Museum:*** 21 km W.

See also GRAMPIANS & CENTRAL WEST, p. 155

Nhill

Pop. 1919
Map ref. 558 C12 | 590 E7

Victoria St; (03) 5391 3086.

96.5 Triple H FM, 594 AM ABC Local Radio

The name of this town may be derived from the Aboriginal word 'nyell', meaning 'white mist on water'. Nhill is exactly halfway between Melbourne and Adelaide, and claims to have the largest single-bin silo in the Southern Hemisphere. The town is a good starting point for tours of Little Desert National Park.

Historical Society Museum: open Thurs, Fri or by appt; McPherson St. ***Cottage of John Shaw Neilson (lyric poet):*** open by appt; Jaypex Park, Victoria St. ***Boardwalk:*** scenic walk from Jaypex Park to Nhill Lake. ***Lowana Craft Shop:*** local crafts; Victoria St. ***Self-guide historical walk:*** details from visitor centre.

VICTORIA

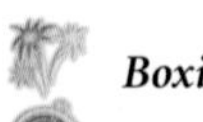 ***Boxing Day Races:*** Dec.

Little Desert National Park: the Little Desert Lodge is in the central section of the park, south of Nhill, and is a departure point for day tours and a popular place to stay. There are walking trails in the central and western sections. 15 km S. *See also Dimboola.* ***Mallee Dam:*** lately dry, once offered fantastic birdwatching with bird hides provided; 20 km SW.

See also GRAMPIANS & CENTRAL WEST, p. 155

Ocean Grove

Pop. 14 351

Map ref. 572 G9 | 574 D6 | 580 H2 | 582 A7 | 593 O8

Geelong Visitor Information Centre, Stead Park, Princes Hwy, Corio; (03) 5275 5797 or 1800 620 888; www.visitgreatoceanroad.org.au

 94.7 The Pulse FM, 774 AM ABC Local Radio

Ocean Grove is a popular summer-holiday destination near the mouth of the Barwon River. The beaches around the town offer great surfing and safe swimming, with surf patrols operating during the summer months.

Ocean Grove Nature Reserve This reserve contains the only significant example of woodland on the Bellarine Peninsula, preserved virtually as it was prior to European settlement. A bird hide lets visitors look out for any number of the 130 different species that live here. Contact Parks Victoria on 13 1963; Grubb Rd.

Ti-Tree Village: peaceful cottages, some with a spa; 34 Orton St; (03) 5255 4433.

HMAS *Canberra* Dive Site Opened to the public in December 2009, this purposely sunken warship is Victoria's first artificial site created specifically for diving. Divers can visit every area of the ship's 138 m length, from the captain's cabin to the galley. Bookings must be made with a charter boat operator or dive shop to be taken to the dive site. For details, call 0414 922 916. 4 km SE.

Barwon Heads This is a pretty seaside town on the Barwon River, made famous as the location for the television series *SeaChange*. There are several good restaurants in the area, including one right on the Barwon River that provides a scenic environment for fine dining. The Barwon Heads Golf Club is one of the top 3 public courses in the state. The Jirrahlinga Koala and Wildlife Reserve on Taits Rd lets visitors encounter the delightful, yet often elusive, koala in a natural environment. Barwon Heads is a short drive from Ocean Grove. 3 km S.

Lake Connewarre State Game Reserve: with mangrove swamps and great walks, the game reserve is home to a variety of wildlife, including wallabies; 7 km N. ***Wallington:*** the town is home to A Maze'N Games, a timber maze with minigolf, picnic/barbecue facilities and a cafe, Koombahla Park Equestrian Centre, Adventure Park and Bellarine Adventure Golf. A strawberry fair is held in Wallington in Nov; 8 km N.

See also WERRIBEE & BELLARINE, p. 152

Olinda

Pop. 1568

Map ref. 570 H6 | 571 B9 | 573 L6 | 582 E5

Dandenong Ranges Information Centre, 1211 Burwood Hwy, Upper Ferntree Gully; (03) 9758 7522; www.olindavillage.com.au

98.1 Radio Eastern FM, 774 AM ABC Local Radio

Olinda is in the centre of the Dandenong Ranges, a landscape of towering mountain ash forests, lush fern gullies, waterfalls, English gardens and picnic spots. The ranges have been a retreat for Melburnians since the 1800s. Olinda and nearby Sassafras are known for their many galleries and cafes, particularly the numerous tearooms serving traditional Devonshire teas.

Rhododendron and Daffodil Festivals: Aug–Nov.

Wild Oak Restaurant and Wine Bar: French-influenced modern Australian menu; Cnr Ridge and Mt Dandenong Tourist rds; (03) 9751 2033. ***Ladyhawke Cafe:*** Middle Eastern cafe; 1365–1367 Mt Dandenong Tourist Rd, Mount Dandenong; (03) 9751 1104. ***Miss Marple's Tea Room:*** traditional tea room fare; 382 Mt Dandenong Tourist Rd, Sassafras; (03) 9755 1610. ***Ripe – Australian Produce:*** foodie cafe/produce store; 376–378 Mt Dandenong Tourist Rd, Sassafras; (03) 9755 2100.

Folly Farm Rural Retreat: 1-bedroom country cottage; 192 Falls Rd; (03) 9751 2544. ***Lochiel Luxury Accommodation:*** contemporary cottages; 1590–1594 Mt Dandenong Tourist Rd; (03) 9751 2300. ***Woolrich Retreat:*** executive garden retreat; 7–9 Monash Ave; (03) 9751 2464 or 1300 553 011. ***Japanese Mountain Retreat:*** traditional Japanese accommodation; 14 Mountain Cres, Montrose; (03) 9737 0086.

Dandenong Ranges National Park This park offers great walking tracks and picnic facilities. Visitors may be lucky enough to spot an elusive lyrebird, a species renowned for its ability to mimic sounds – from other bird calls to human voices and even chainsaws. Most walking tracks leave from picnic grounds, such as the Thousand Steps Track from Ferntree Gully Picnic Ground (south-west via the Mt Dandenong Tourist Rd) and the walk to Sherbrooke Falls from the Sherbrooke Picnic Ground (via Sherbrooke Rd from the Mt Dandenong Tourist Rd). The park extends to the east and west of town. Contact Parks Victoria on 13 1963; 2 km NE, 6 km SW.

William Ricketts Sanctuary William Ricketts was a well-known artist and conservationist whose intricate sculptures focus on Aboriginal people and the complexities of Australia's native vegetation. Many sculptures are displayed in a bushland setting on the scenic Mt Dandenong Tourist Rd. Contact Parks Victoria on 13 1963; 3 km N.

National Rhododendron Gardens: the gardens begin just east of town and are something of a mecca for garden enthusiasts, with superb displays of rhododendrons and azaleas in season. ***R. J. Hamer Arboretum:*** good walking tracks through 100 ha of rare and exotic trees; Olinda–Monbulk Rd, shortly after turn-off to Rhododendron Gardens. ***Cloudehill Gardens:*** twilight concerts are held here in summer; south of R. J. Hamer Arboretum. ***Mt Dandenong Lookout:*** spectacular views over Melbourne; picnic/barbecue facilities; 3 km N. ***Alfred Nicholas Gardens:*** an ornamental lake with the original boathouse and the George Tindale Memorial Garden, with flowering plants beneath mountain ash trees. The original Nicholas family home (built 1920s) is here; Sherbrooke; 4 km SE. ***Kawarra Australian Plant Garden:*** an impressive collection of native plants; Kalorama; 4.5 km N. ***Markets:*** art, craft, plants and homemade goods; nearby markets include Kallista Market, 6 km S, 1st Sat each month, and Upper Ferntree Gully Market, 12 km SE, every Sat and Sun. ***Burrinja Gallery:*** a memorial to artist Lin Onus, with Aboriginal and Oceanic sculptures and paintings; Upwey; 10 km SW. ***Silvan:*** prominent flower-growing region with many tulip farms. The famous Tesselaar's Tulip Farm hosts a popular festival each Sept–Oct with sales of flowers and bulbs, and traditional Dutch music and food; Monbulk Rd; 15 km NE. ***Silvan Dam:*** an area to the north of this major Melbourne water supply has walking tracks and picnic/barbecue facilities; turn-off after Silvan. ***Mont De Lancey:*** wonderfully preserved house, built in 1882 and set in landscaped gardens; includes a museum and a chapel; open 2nd Sat each month; Wandin North; 22 km NE via Mt Evelyn.

See also YARRA & DANDENONGS, p. 148

Omeo

Pop. 226
Map ref. 583 M1 | 585 M12 | 587 M11

152 Great Alpine Rd; (03) 5159 1679; www.omeoregion.com.au

90.9 FM High Country Radio, 720 AM ABC Local Radio

Omeo is an Aboriginal word meaning mountains – appropriate for this picturesque town in the Victorian Alps. Today Omeo is a peaceful farming community, but it wasn't always so. During the 1800s gold rush, Omeo was an unruly frontier town, which early Australian novelist Rolf Boldrewood described as the roughest goldfield in Australia. Despite taking damage in the 1939 Black Friday bushfires, several historic buildings still remain.

A. M. Pearson Historical Park The park preserves a piece of Omeo's rich history in a peaceful, bushland setting. Buildings on display include the old courthouse, which now houses a museum, a log gaol, stables and a blacksmith's. Day Ave (Great Alpine Rd).

Historic buildings: many distinctive structures from the 19th century can be seen in town, including the post office, primary school and shire offices; Day Ave. ***Shops:*** several unique stores, including the German Cuckoo Clock Shop, Petersen's Gallery, High Country Paintings and Octagon Bookshop; Day Ave.

High Country Calf Sales: Mar. ***Alpine Discovery Festival and Picnic Races:*** Mar. ***Rodeo, Easter Market and Easter Egg Hunt:*** Easter. ***Cobungra Polo Match:*** Apr.

The Golden Age Hotel Motel: up-market country fare; 189 Day Ave; (03) 5159 1344.

Oriental Claims The Claims was a major goldmining area, and remains the highest alluvial goldfield in Australia. French-Canadians, Americans and Europeans all worked alongside Australians and Chinese during the gold boom. The word 'Oriental' in the mine's name may conjure an image of Chinese workers, but 'Oriental Claims' was actually the name of a European company. The Omeo Sluicing Company, however, was Chinese. There are a variety of walks around the site and visitors should look out for the variety of flora, including wild orchids. High cliffs, left by the hydraulic sluicing process, offer impressive views across town, and signs throughout the Parks Victoria–managed site explain the history of the Claims. Contact Parks Victoria on 13 1963; 1.5 km W on Great Alpine Rd.

Mt Hotham This popular downhill ski resort is suited to both budding and experienced skiers. Skiing areas range from the beginners' Big D Playground through to the more advanced slopes around Mary's Slide and the black diamond chutes of Heavenly Valley. In summer, the mountain is a popular hiking and mountain-bike-riding destination. (03) 5759 3550; 56 km W.

Livingstone Park and Creek: walking tracks and swimming area adjacent to the Oriental Claims. ***Mt Markey Winery:*** on the site of the old Cassilis Hotel, on the touring loop from Omeo to Swifts Creek; Cassilis Rd, Cassilis; 15 km S. ***Lake Omeo:*** scenic natural landscape, dry for most parts of the year; Benambra; 21 km NE. ***Benambra:*** gateway to the Alpine National Park; 24 km N. ***Anglers Rest:*** historic Blue Duck Inn, a good base for horseriding, whitewater rafting and fly-fishing; 29 km NW. ***Swifts Creek:*** this town situated at the junction of Swifts Creek and Tambo River has the Great Alpine Art Gallery; 40 km S. ***Taylors Crossing suspension bridge:*** part of the scenic Australian Alps Walking Track and also a great base for camping and fishing the Mitta Mitta River; off Tablelands Rd; 44 km NE. ***Dinner Plain:*** relaxed village surrounded by the Alpine National Park that offers many activities such as skiing, walking, horseriding and Australia's first indoor-outdoor alpine spa; 46 km W. ***Ensay:*** small but picturesque town that is home to the well-known Ensay Winery; 70 km S. ***Great Alpine Rd:*** covers over 300 km from the High Country to Gippsland Lakes and offers 6 individual self-guide touring routes with a diverse combination of scenery; details from visitor centre. Note that some drives cross state forests or alpine areas, so be alert for timber trucks and check conditions in winter. Omeo is located on this road. ***Mitta Mitta and Cobungra rivers:*** great trout fishing, waterskiing and whitewater rafting (only available in spring); details from visitor centre. ***High Country tours:*** explore the high plains around Omeo – on horseback, by 4WD or, for keen hikers, on challenging bushwalks; details from visitor centre.

See also HIGH COUNTRY, p. 158

Orbost

Pop. 2096
Map ref. 565 A12 | 581 B11

Slab Hut, 39 Nicholson St; (03) 5154 2424; www.lakesandwilderness.com.au

90.7 FM 3REG Radio East Gippsland, 97.1 FM ABC Local Radio

Situated on the banks of the legendary Snowy River, Orbost is on the Princes Highway and surrounded by spectacular coastal and mountain territory. For those who love arts and crafts, there are many shops in the area supplying and displaying local products.

Visitor Information Centre: display explains complex rainforest ecology; Slab Hut, 39 Nicholson St. ***Old Pump House:*** behind relocated 1872 slab hut; Forest Rd. ***Historical Museum:*** details local history with displays of artefacts; Ruskin St. ***Snowy River Country Craft:*** Forest Rd. ***Netherbyre Gemstone and Art Galley:*** Cnr Browning and Carlyle sts. ***Exhibition Centre:*** equipped with 2 galleries, one dedicated to the National Collection of Australian Wood Design, the other presenting monthly exhibitions; Nicholson St. ***Mirrawong Woolworks:*** sells wool and felt handmade items; 295 Nicholson St. ***Heritage walk:*** weaves its way through town with storyboards, fingerboards and plaques explaining the historic buildings; begins at Slab Hut.

Australian Wood Design Exhibition: Jan.

A Lovely Little Lunch: local produce cafe menu; 125A Nicholson St; (03) 5154 1303. ***PS Curlip II:*** lunches and snacks on a paddlesteamer; PS *Curlip* Centre, 1 Browning St; (03) 5154 1699.

Errinundra National Park The park is one of the largest remaining stands of cool temperate rainforest in Victoria, and features giant eucalypt forests. There is the rainforest boardwalk and for keen hikers there are walking tracks, as well as camping and picnic facilities. Enjoy superb views from Ellery View, Ocean

VICTORIA

View Lookout and the peak of Mt Morris. In winter, snow and rain can make access difficult. Errinundra Rd, off Princes Hwy; contact Parks Victoria on 13 1963; 54 km NE.

Marlo: a popular fishing spot also known for its galleries and Bush Races in Jan. Take a cruise on the paddle steamer *Curlip*; 14 km S. ***Cape Conran Coastal Park:*** rugged coastal scenery and excellent walks. Turn south after Cabbage Tree Creek (26 km E) or take the coastal route from Marlo. ***Cabbage Tree Palms Flora Reserve:*** 27 km E. ***Bemm River Scenic Reserve:*** a 1 km signposted rainforest walk and picnic facilities; off Princes Hwy; 40 km E. ***Snowy River National Park:*** in the south of the park is Raymond Creek Falls. A 40 min return walk leads to the falls, with a further 1 hr walk leading to the Snowy River; 42 km N; 2WD access, check road conditions. McKillops Bridge, 148 km N via Deddick, is one of the most accessible parts of this park; *for more details see Buchan*. ***Sydenham Inlet:*** a good spot for bream fishing; 58 km E. ***Tranquil Valley Tavern:*** on the banks of the Delegate River near the NSW border; about 115 km NE. ***Baldwin Spencer Trail:*** a 262 km scenic drive following the route of this explorer, taking in old mining sites and Errinundra National Park; details from visitor centre.

See also **East Gippsland, p. 159**

Ouyen

Pop. 1058
Map ref. 558 E8 | 588 H11

17 Oke St; (03) 5092 2006; www.visitmildura.com.au/ouyen

92.9 3MBR-FM Mallee Border Radio, 594 AM ABC Local Radio

Ouyen was once little more than a station on the Melbourne–Mildura train route, but it has since grown to become an important service town. Ouyen is at the centre of the Mallee region, which was developed in the early 1900s – relatively late when compared with other regions of rural Victoria. This was mainly due to the difficulties in clearing the land as well as the harsh climate. The current success of agriculture in the region, in particular wheat-growing, is a testament to the hardiness of early farmers and settlers.

Roxy Theatre: newly restored and functioning tropical-style theatre, only one of its type in southern Australia; Oke St.

Great Australian Vanilla Slice Triumph: 1st Fri in Sept.

Hattah–Kulkyne National Park This park protects an area of 48 000 ha that includes typical mallee country with both low scrub and open native pine woodlands. The freshwater Hattah Lakes are seasonally filled by creeks connected to the Murray River, which brings the area to life with plants and waterbirds. Activities within the park include bushwalking, canoeing, fishing and scenic drives. There are picnic and camping facilities at Mournpall and Lake Hattah. Contact Parks Victoria on 13 1963; off the Calder Hwy; 35 km N.

Speed: Mallee Machinery Field Days held here in Aug; 39 km S.

See also **Mallee Country, p. 156**

Paynesville

Pop. 3455
Map ref. 583 N6

Community Craft Centre, Esplanade; (03) 5156 7479; www.lakesandwilderness.com.au

90.7 FM 3REG Radio East Gippsland, 100.7 FM ABC Local Radio

Paynesville is a popular tourist resort close to the rural city of Bairnsdale, on the McMillan Straits. The town is set on the Gippsland Lakes and the beaches of the Tasman Sea, making it a favourite destination for fishing and waterskiing.

St Peter-by-the-Lake Church: built in 1961, this unique structure incorporates seafaring images in its design; The Esplanade. ***Community Craft Centre:*** displays and sells local arts and crafts; The Esplanade.

Market: Gilsenan Reserve; 2nd Sun each month. ***Jazz Festival:*** Feb. ***Marlay Pt Paynesville Overnight Yacht Race:*** Mar long weekend.

Cafe Espas: contemporary seafood; Raymond Island; (03) 5156 7275.

Captain's Cove: smart nautical units; 13 Mitchell St; (03) 5156 7223.

Gippsland Lakes This area incorporates The Lakes National Park, Gippsland Lakes Coastal Park and the famous Ninety Mile Beach – an incredible stretch of scenic coastline offering great swimming beaches. Lake cruises, boat charters and organised scenic tours of the region are all available; details from visitor centre.

Eagle Pt: a small fishing community set by Lake King. The Mitchell River empties here, where it forms curious silt jetties that stretch out into the distance. The town hosts the annual Australian Powerboat Racing Championships at Easter; 2 km NW. ***Raymond Island:*** Koala Reserve and Riviera Meadows, an animal farm that specialises in miniature breeds; the island is just east of Paynesville and can be accessed by a ferry that departs from the foreshore.

See also **East Gippsland, p. 159**

Phillip Island

see Cowes

Port Albert

Pop. 253
Map ref. 583 I10

Old Courthouse, 9 Rodgers St, Yarram; (03) 5182 6553.

100.7 FM ABC Local Radio, 104.7 Gippsland FM

Port Albert is a tranquil port on the south-east coast. Looking at this peaceful village now, it is hard to believe that it was the first established port in Victoria, with ships from Europe and America once docking at its jetty. Ships from China arrived here during the gold rush, bringing thousands of prospectors to the Gippsland goldfields. Still a commercial fishing port, the sheltered waters of Port Albert are popular with anglers and boat owners, which sees its population swell considerably during summer.

Port Albert Hotel This attractive old building has wide verandahs, and offers genuine country hospitality and a glimpse into the area's past. The hotel was first licensed in 1842, which makes it one of the oldest hotels in Victoria still operating. Wharf St.

Historic buildings: include original government offices and stores, and the Bank of Victoria, which now houses a maritime museum with photographs and relics from the town's past. Georgian and Victorian architectural styles are evident in over 40 buildings; Tarraville Rd. ***Warren Curry Art:*** a gallery featuring country-town streetscapes; Tarraville Rd.

Fishing Contest: Mar.

Wildfish: waterfront dining, freshest local seafood; 40 Wharf St; (03) 5183 2007.

Nooramunga Marine and Coastal Park Surrounding Port Albert and comprising the waters and sand islands offshore, this marine park is a fishing enthusiast's delight. Snapper, flathead and Australian salmon can be caught from the surf beaches or from a boat. The Aboriginal middens that dot the shorelines prove that fishing has been carried on here for many thousands of years. This park is an important reserve for migratory wading birds. Camping is allowed but permits must be obtained. Contact Parks Victoria on 13 1963.

Christ Church: built in 1856, this was the first church to be established in Gippsland; Tarraville; 5 km NE. ***Beaches:*** Manns, for swimming, 10 km NE; and Woodside, on Ninety Mile Beach, for good surfing, 34 km NE. Note that both beaches are patrolled during summer. ***St Margaret Island:*** a protected area featuring a wildlife sanctuary; 12 km E.

See also PHILLIP ISLAND & GIPPSLAND, p. 150

Port Campbell

Pop. 258
Map ref. 580 F10 | 593 I9

12 Apostles Visitor Information Centre, 26 Morris St; 1300 137 255; www.visit12apostles.com.au

 103.7 3WAY-FM, 774 AM ABC Local Radio

This peaceful seaside resort – the base of a small crayfishing industry – is in the centre of Port Campbell National Park on the Great Ocean Road. The Twelve Apostles, one of Victoria's most famous attractions, can be found nearby.

Historical Museum: open Wed, Thurs and Sat; Lord St. ***Fishing:*** good from rocks and pier; boat charters available.

Market: Lord St; each Sun in summer and Easter.

Room Six Cafe Restaurant: Mediterranean- and Japanese-influenced menu; 28 Lord St; (03) 5598 6242. ***The Craypot Bistro:*** great-value food, seasonal crayfish; Port Campbell Hotel, 40 Lord St; (03) 5598 6320. ***Timboon Railway Shed Distillery:*** local-produce-driven menu; The Railway Yard, Bailey St, Timboon; (03) 5598 3555.

Daysy Hill Country Cottages: cottage, suite and cabin accommodation; 2585 Cobden–Port Campbell Rd; (03) 5598 6226. ***Loch Ard Motor Inn:*** motel rooms and apartments; 18–24 Lord St; (03) 5598 6328. ***Sea Foam Villas:*** stylish apartments; 14 Lord St; (03) 5598 6413.

Port Campbell National Park The park is a major attraction on the Great Ocean Rd, with magnificent rock formations jutting out into the ocean. Particularly impressive when viewed at dusk (when penguins can be seen) and dawn, the key coastal features are The Arch, 5 km W; London Bridge, 6 km W; Loch Ard Gorge, 7 km SE; and the world-famous Twelve Apostles, which begin 12 km SE of Port Campbell and stretch along the coast. Other notable features are The Grotto, Bay of Islands and Bay of Martyrs. There are walking tracks throughout the park, and the Historic Shipwreck Trail marks 25 sites along the coast between Moonlight Head and Port Fairy (sites are also popular with divers – a charter company is based in Port Campbell; details from visitor centre). For the ultimate view of this coastline, take an ever-popular scenic flight (details from visitor centre).

Mutton Bird Island: attracts short-tailed shearwaters, best viewed at dawn and dusk Sept–Apr; just off coast. ***Great Ocean Walk:*** between Apollo Bay and the Twelve Apostles. The 91 km walk offers stunning views; walkers must register to use campgrounds en route; further information available at greatoceanwalk.com.au ***Timboon:*** a pretty town in the centre of a dairy district. Timboon Farmhouse Cheese offers tastings and sales of gourmet cheeses, while Timboon Railway Shed Distillery offers a variety of spirits. A scenic drive goes from Port Campbell to the town. It is also on one end of the Camperdown–Timboon (Crater to Coast) Rail Trail. Pick your own berries in season at nearby Berry World; 16 km N. ***Otway Deer and Wildlife Park:*** 19 km SE. ***Gourmet Food and Wine Loop:*** map from visitor centre.

See also SOUTH-WEST COAST, p. 153

Port Fairy

Pop. 2597
Map ref. 580 B8 | 592 F8

Railway Place, Bank St; (03) 5568 2682; www.visitportfairy.com

 103.7 3WAY-FM, 1602 AM ABC Local Radio

Port Fairy was once a centre for the whaling industry and one of the largest ports in Australia. Today many visitors are attracted to it for its charming old-world feel, its legacy of historic bluestone buildings, the small fleet of fishing boats that line the old wharf, and its great beach and lively atmosphere in summer. The town truly comes alive in March, when the Port Fairy Folk Festival is held. International folk and blues acts play, and tickets are best booked well in advance.

History Centre: displays relating to local history housed in the old courthouse; Gipps St. ***Battery Hill:*** old fort and signal station at the river mouth; end of Griffith St. ***Port Fairy Wharf:*** sales of fish and crayfish when in season. ***Historic buildings:*** many are National Trust–classified, including the splendid timber home of Captain Mills, Gipps St; Mott's Cottage, Sackville St; Caledonian Inn, Bank St; Seacombe House, Cox St; St John's Church of England, Regent St; and the Gazette Office, Sackville St.

Port Fairy Folk Festival: Mar. ***Spring Music Festival:*** Oct. ***Moyneyana Festival:*** family entertainment; Dec.

Merrijig Inn: innovative food using fresh produce; 1 Campbell St; (03) 5568 2324. ***Portofino on Bank:*** contemporary Mediterranean; 26 Bank St; (03) 5568 2251. ***Saltra Brasserie:*** modern Australian; 20 Bank St; (03) 5568 3058. ***Time & Tide:*** gallery-cum-cafe; 21 Thistle Pl; (03) 5568 2134.

Hearn's Beachside Villas: waterside spa villas; 13–17 Thistle Pl; (03) 5568 3150. ***Merrijig Inn:*** historic B&B rooms; 1 Campbell St; (03) 5568 2324. ***Moyne Mill Townhouse:*** riverside accommodation; Unit 4, 25 Gipps St; (03) 5568 2229. ***Oscars Waterfront Boutique Hotel:*** French Provincial–style rooms; 41B Gipps St; (03) 5568 3022.

Griffiths Island Connected to town by a causeway, this island is home to a large colony of short-tailed shearwaters. Each year they travel across the Pacific Ocean from North America to nest in the same burrows (Sept–Apr). Also on the island is a much-photographed lighthouse.

The Crags: rugged coastal rock formations; 12 km W. ***Yambuk:*** a small township centred on an old inn with Yambuk Lake, a popular recreation area, nearby; 17 km W. ***Lady Julia Percy Island:*** home to a fur seal colony; charters can be arranged from Port Fairy Wharf; 22 km off coast. ***Codrington Wind Farm:*** Victoria's first wind-power station; 27 km W. ***Mahogany Walk to Warrnambool:*** a 6–7 hr walk (one way, can return by bus) taking

VICTORIA

in a magnificent stretch of coastline; details from visitor centre. ***Historic Shipwreck Trail:*** between Port Fairy and Moonlight Head with 25 wreck sites signposted along the way.

See also South-West Coast, p. 153

Portland

Pop. 9824
Map ref. 592 D8

Lee Breakwater Rd; (03) 5523 2671 or 1800 035 567; www.visitgreatoceanroad.org.au

 96.9 AM ABC Local Radio, 99.3 Coastal FM

Portland is the most westerly of Victoria's major coastal towns and the only deep-water port between Melbourne and Adelaide. It was also the first permanent settlement in Victoria, founded in 1834 by the Hentys. The township, which features many National Trust–classified buildings, overlooks Portland Bay. The Kerrup–Tjmara people, who once numbered in the thousands, were the original inhabitants of the district and referred to it as 'Pulumbete' meaning 'Little Lake' – a reference to the scenic lake now known as Fawthorp Lagoon.

Portland Maritime Discovery Centre The centre features a 13 m sperm whale skeleton, and the lifeboat used to rescue 19 survivors from the *Admella* shipwreck in 1859. Another wreck, the *Regia*, is displayed in 2 m of water. The centre shares the building with the information centre. Lee Breakwater Rd.

Botanical Gardens: established in 1857, with both native and exotic plant life. A restored 1850s bluestone worker's cottage is within the grounds and open to the public; Cliff St. ***Historical buildings:*** more than 200 around town, many National Trust–classified. The best way to explore buildings such as the courthouse, Steam Packet Inn and Mac's Hotel is to take either a guided or self-guide walk; details from visitor centre. ***History House:*** a historical museum and family research centre in the old town hall; Charles St. ***Burswood:*** a bluestone, regency-style mansion that was once the home of pioneer settler Edward Henty. The house is set amid 5 ha of gardens; Cape Nelson Rd. ***Fawthorp Lagoon:*** prolific birdlife; Glenelg St. ***Powerhouse Car Museum:*** Percy St. ***Watertower Lookout:*** displays of WW II memorabilia on the way up the 133 steps to magnificent 360-degree views across Portland and the ocean, where whales and dolphins can sometimes be spotted; Percy St. Another good spot for whale-watching is Battery Hill.

Anzac Day Floral Display: Apr. ***Portland Bay Festival:*** Nov. ***3 Bays Marathon:*** Nov.

Clock by the Bay: contemporary Australian; Cnr Cliff and Bentinck sts; (03) 5523 4777. ***Lido Larder:*** modern cafe; 5 Julia St; (03) 5521 1741. ***Bridgewater Bay Beach Cafe:*** views and food worthy of detour; 1661 Bridgewater Rd, Cape Bridgewater; (03) 5526 7155.

Clifftop Accommodation Portland: 3 studio apartments; 13 Clifton Crt; (03) 5523 1126. ***Sheoak Tinntean:*** 2 studio apartments; 22 Cavendish St; (03) 5523 2296.

Cape Bridgewater This cape is home to a 650-strong colony of Australian fur seals. A 2 hr return walk leads to a viewing platform, or you can take a 45 min boat ride that leads into the mouth of a cave to see them up close (bookings essential, (03) 5526 7247). Across the cape towards Discovery Bay are the Petrified Forest and the Blowholes – spectacular during high seas. 21 km sw.

Lower Glenelg National Park The Glenelg River is a central feature of the park. It has cut an impressive 50 m deep gorge through a slab of limestone. Watch for platypus, water rats, moorhens and herons around the water's edge. Bushwalking, camping, fishing and canoeing are all popular, and Jones Lookout and the Bulley Ranges offer great views. Also in the park are the Princess Margaret Rose Caves on the north side of the river – you can drive there via Nelson or Dartmoor. Alternatively, boat tours operate from Nelson. Contact Parks Victoria on 13 1963; 44 km nw.

Cape Nelson: here a lighthouse perches on top of tall cliffs and lightstation tours are available; 11 km sw. ***Narrawong State Forest:*** a short walk leads to Whalers Pt, where Aboriginal people once watched for whales; 18 km ne. ***Discovery Bay Coastal Park:*** Cape Bridgewater is included in this park, though the majority of it is remote and relatively untouched. The Great South West Walk *(see below)* offers the best chance to take in the park's scenery. Behind Cape Bridgewater are the Bridgewater Lakes (19 km w) – popular for waterskiing and fishing. A walking track leads from here to the beach. ***Mt Richmond National Park:*** a 'mountain' formed by an extinct volcano. The area has abundant spring wildflowers and native fauna, including the elusive potoroo; 25 km nw. ***Heywood:*** home to the Bower Birds Nest Museum, and the Wood, Wine and Roses Festival in Feb. Budj Bim National Heritage Landscape, the traditional lands of the Gunditjmara people, is located here also. Visitors can experience the aquaculture system including stone eel traps, permanent stone houses and smoking trees; Budj Bim Tours (03) 5527 1699; 28 km n. ***Nelson:*** a charming hamlet near the mouth of the Glenelg River. There is good waterskiing in the area; 70 km nw. ***Great South West Walk:*** this epic 250 km walking trail takes in the full range of local scenery – the Glenelg River, Discovery and Bridgewater bays and Cape Nelson are some of the highlights. It is possible to do just small sections of the walk; maps and details from visitor centre.

See also South-West Coast, p. 153

Pyramid Hill

Pop. 467
Map ref. 559 I12 | 591 N6

Loddon Visitor Information Centre, Wedderburn Community Centre, 24 Wilson St, Wedderburn; (03) 5494 3489; www.pyramidhill.net.au

 91.1 FM ABC Local Radio, 104.7 Radio EMFM

Pyramid Hill's namesake is an unusually shaped, 187-metre-high hill. The town, which is located in a wheat-growing district about 30 kilometres from the New South Wales border, was a source of inspiration to notable Australian author Katherine Susannah Pritchard, who based a character in her book *Child of the Hurricane* on a woman she met while staying in Pyramid Hill during World War I.

Pyramid Hill A climb to the top of this eerily symmetrical hill reveals views of the surrounding irrigation and wheat district. There are abundant wildflowers in spring.

Historical Museum: features local story displays; open Sun afternoons or by appt; McKay St.

Terrick Terrick National Park The park is a large Murray pine forest reserve with granite outcrops, including Mitiamo Rock. There is a variety of good walking tracks, and the park is a key nesting area for the distinctive brolga. Contact Parks Victoria on 13 1963; access is via the Pyramid Hill–Kow Swamp Rd; 20 km se.

Mt Hope: named by explorer Major Mitchell, who 'hoped' he would be able to spot the sea from the mountain's peak. Now known for its wildflowers; 16 km NE. ***Boort:*** nearby lakes provide a habitat for swans, ibis, pelicans and other waterbirds, and a place for watersports, fishing and picnics; 40 km W.

See also GOULBURN & MURRAY, p. 157

Queenscliff

Pop. 3892
Map ref. 570 A10 | 572 H9 | 574 E6 | 582 B7 | 593 P8

55 Hesse St; (03) 5258 4843; www.queenscliffe.vic.gov.au

 94.7 The Pulse FM, 774 AM ABC Local Radio

Queenscliff is a charming seaside town on the Bellarine Peninsula. It began life as a resort for wealthy Victorians in the 1800s, as testified by lavish buildings such as the Queenscliff Hotel, with its ornate lattice work and plush interiors. The town's wide main street is lined with cafes and restaurants, plus an array of art galleries, and the nearby beaches become a playground for holiday-makers during summer. A ferry runs between Queenscliff and Sorrento, a resort town across Port Phillip Bay.

Queenscliff Maritime Museum The museum explores the town's long association with ships and the sea through a collection of maritime memorabilia. It features a re-created fisherman's cottage, a diving-technology display and an array of navigational equipment. Weeroona Pde; (03) 5258 3440.

Marine Discovery Centre This is a great family destination where visitors can learn all about the local marine life. It has a number of aquariums and touch-tanks. The centre also runs various tours, including boat cruises off Port Phillip and 'rockpool rambles'. Adjacent to the Maritime Museum, Weeroona Pde; (03) 5258 3344.

The Blues Train An incredibly popular attraction that provides a unique dining and entertainment experience on board a steam train. Round trips provide 4 carriages, each with a different blues musician. Guests can change carriages at stops on the journey and purchase drinks from the mobile bar at each station platform. Operates Sat nights Oct–May. Departs from and returns to Queenscliff Railway Station; bookings at Ticketek on 132 849, or contact visitor centre for more information.

Fort Queenscliff: built during the Crimean War, it includes the unique 'Black Lighthouse'. Tours of the fort run most days; details from visitor centre. ***Queenscliff Historical Museum:*** open daily 2–4pm; Hesse St. ***Bellarine Peninsula Railway:*** beautifully restored steam trains run between Queenscliff and Drysdale. There are many engines on display around the station. Trains run Sun, public holidays and other times during school holidays; Symonds St. ***Bellarine Rail Trail:*** 32.5 km track that extends from South Geelong to Queenscliff; on reaching Drysdale it runs along the Bellarine Peninsula Railway and ends in the town.

Market: with crafts and second-hand goods; Princes Park, Gellibrand St; last Sun each month except winter. ***Queenscliff Music Festival:*** major event attracting local and international music acts; Nov.

Apostle Queenscliff: modern Australian; 79 Hesse St; (03) 5258 3097. ***Athelstane House:*** contemporary Australian; 4 Hobson St; (03) 5258 1024. ***Vue Grand Hotel:*** classic menu, impressive dining room; 46 Hesse Street; (03) 5258 1544.

Athelstane House: 8 rooms and 1 apartment; 4 Hobson St; (03) 5258 1024. ***Lathamstowe:*** B&B rooms, 3-bedroom cottage; 44 Gellibrand St; (03) 5258 4110. ***Queenscliff Hotel:*** vintage ensuite bedrooms; 16 Gellibrand St; (03) 5258 1066. ***Vue Grand Hotel:*** grand hotel rooms; 46 Hesse St; (03) 5258 1544.

Pt Lonsdale This peaceful holiday town offers gorgeous beaches suitable for either surfing or swimming. A lookout from the cliff-top provides a great view of the treacherous entrance to Port Phillip known as 'The Rip'. A market is held here on the 2nd Sunday of each month. 6 km SW.

Lake Victoria: an important waterbird habitat; 7 km SW via Pt Lonsdale. ***Harold Holt Marine Reserve:*** incorporates Mud Island and coastal reserves. Guided boat tours can be arranged from the Marine Discovery Centre.

See also WERRIBEE & BELLARINE, p. 152

Robinvale

Pop. 2216
Map ref. 558 F7 | 589 J8

Kyndalyn Park Information Centre, Bromley Rd; (03) 5026 1388; www.murrayriver.com.au/html/towns/robinvaleeuston.html

 90.7 HOTFM Sunraysia Community Radio

Robinvale is set on the New South Wales border by a pretty stretch of the Murray River. The Robinswood Homestead, built in 1926, was home to the town's founder, Herbert Cuttle (you can find the homestead in River Road). Herbert's son, Robin, was killed during World War I, so he named both the homestead and the town in Robin's honour. As another form of remembrance, the town has a sister city in France, near where young Robin died.

Rural Life Museum: housed in the information centre, with locally grown almonds for sale; open by appt; Bromley Rd. ***Murray River:*** the beaches around Robinvale are popular for picnics and fishing, while in the river waterskiing and swimming are favourite summer pastimes.

Ski Race: Mar. ***Tennis Tournament:*** Easter. ***Almond Blossom Festival:*** Aug.

Euston Weir and Lock on Murray: created as an irrigation water store, it features a 'fish ladder' that enables fish to jump over the weir. Picnic and barbecue facilities are provided; Pethard Rd, south-west edge of town. ***Robinvale Organic and Bio-dynamic Wines:*** tastings and sales of these distinctive, preservative-free wines. Also a children's playground; Sea Lake Rd; 5 km S. ***Olive oil:*** this region is renowned for its award-winning olive oil. Robinvale Estate offers farmgate sales and tastings; Tol Tol Rd; 8 km SE. There is also Boundary Bend Estate; Boundary Bend; Murray Valley Hwy; 1 km S of the Murray River. ***Robinvale Indigenous Arts and Crafts:*** learn about the local bush tucker; River Rd. ***Hattah–Kulkyne National Park:*** 66 km SW; *see Ouyen*.

See also MALLEE COUNTRY, p. 156

Rochester

Pop. 2827
Map ref. 584 C7

Council offices, 43 Mackay St; (03) 5484 4500; www.rochester.org.au

 91.1 FM ABC Local Radio, 104.7 Radio EMFM

On the Campaspe River, near Echuca, Rochester is the centre of a rich dairying and tomato-growing area. There are several lakes and waterways near town, making Rochester a popular destination for freshwater fishing.

VICTORIA

The 'Oppy' Museum The museum details the history of Sir Hubert Opperman, affectionately known as Oppy, a champion cyclist who competed in the Tour de France. There is a collection of memorabilia related to Oppy's career as a cyclist, as well as artefacts from the town's past. A statue of Oppy is opposite the museum. Moore St.

Heritage walk: take in the town's attractive old buildings. ***Campaspe River Walk:*** a pleasant, signposted walk by the river.

Kyabram Fauna Park This park, owned by the Kyabram community, is home to over 140 animal species – everything from wombats to waterfowl. It has been built from the ground up on a piece of degraded farmland, and is now heavily involved in breeding programs for endangered species such as the eastern barred bandicoot. There is a walk-through aviary and Australia's first energy-efficient reptile house. (03) 5852 2883; 35 km NE.

Campaspe Siphon: an impressive engineering feat, where the Waranga–Western Main irrigation channel was redirected under the Campaspe River; 5 km N. ***Fishing:*** nearby channels, rivers and lakes are popular with anglers for redfin and carp. Lakes include Greens Lake and Lake Cooper (14 km SE), also good for picnicking and watersports. ***Elmore:*** here is the Campaspe Run Rural Discovery Centre, which explains Koorie and colonial history and heritage. Elmore Field Days are held each Oct; 17 km S.

See also GOULBURN & MURRAY, p. 157

Rushworth

Pop. 1041
Map ref. 584 D8

33 High St; (03) 5856 1117; www.campaspe.vic.gov.au/community/rushworth/main.htm

97.7 FM ABC Local Radio, 98.5 FM 98.5

Situated in central Victoria off the Goulburn Valley Highway, this delightful little town was once a goldmining settlement. The original site of the township was known as Nuggetty owing to the numerous gold nuggets found during the 19th century. Rushworth has retained much of its original character, with well-preserved early buildings lining the main street.

Historic buildings: many along High St are National Trust–classified, including the Church of England, the Band Rotunda, the former Imperial Hotel, the Glasgow Buildings and the Whistle Stop. Take the High St Heritage Walk to see these and others; map from visitor centre. ***History Museum:*** housed in the old mechanics institute with displays relating to the town's goldmining heritage; Cnr High and Parker sts. ***Growlers Hill Lookout Tower:*** views of the town, Rushworth State Forest and the surrounding Goulburn Valley; Reed St.

Rushworth State Forest The largest natural ironbark forest in the world, Rushworth State Forest is also renowned for the orchids and wildflowers that blossom here in spring. Picnics and bushwalks are popular activities in this attractive reserve where over 100 species of birds, along with echidnas and kangaroos, can be seen. Access via Whroo Rd; 3 km S.

Jones's Eucalyptus Distillery: eucalyptus oil is extracted from blue mallee gum; Parramatta Gully Rd, just south of town. ***Waranga Basin:*** an artificial diversion of the Goulburn weir constructed in 1916, now a haven for boating, fishing, swimming and watersports; 6 km NE. ***Whroo Historic Reserve:*** Balaclava Hill, an open-cut goldmine, along with camping and picnic facilities, the Whroo cemetery and an Aboriginal waterhole; 7 km S. ***Murchison:*** a small town picturesquely set on the Goulburn River. Town attractions include the Italian war memorial and chapel; Meteorite Park, the site of a meteorite fall in 1969; Longleat Winery; and Campbell's Bend Picnic Reserve; 19 km E. ***Days Mill:*** a flour mill with buildings dating from 1865; 39 km NE via Murchison. ***Town ruins:*** goldmining played a huge role in the development of this region, but not all towns survived the end of the gold rush. Ruins of Angustown, Bailieston and Graytown are all to the south of Rushworth.

See also GOULBURN & MURRAY, p. 157

Rutherglen

Pop. 1991
Map ref. 559 N12 | 585 J6 | 586 G1

57 Main St; (02) 6033 6300 or 1800 622 871; www.rutherglenvic.com

101.3 Oak FM, 106.5 FM ABC Local Radio

Rutherglen is the centre of one of the most important winegrowing districts in Victoria, with a cluster of vineyards surrounding the town. Many of the local wineries are best known for their fortified wines. Rutherglen's main street features preserved late-19th-century architecture.

Rutherglen Wine Experience: interpretive displays of Rutherglen's wine history; visitor centre, Main St. ***Common School Museum:*** local history displays and a re-creation of a Victorian-era schoolroom; behind Main St. ***Historic tours:*** take a self-guide walk, bike ride or drive, following maps provided at the visitor centre. ***Lake King:*** originally constructed in 1874 as Rutherglen's water storage, it is now a wildlife sanctuary and offers a scenic walk.

Tastes of Rutherglen: celebration of the region's gourmet food and wine; Mar. ***Rutherglen and District Art Society Show:*** Mar. ***Easter in Rutherglen:*** Easter. ***Winery Walkabout:*** June. ***Country Fair:*** June. ***Wine Show:*** Sept. ***Tour de Rutherglen:*** cycling and wine event; Oct. ***Young Bloods and Bloody Legends:*** food and wine event; Oct.

Beaumont's Cafe: contemporary fare; 84 Main St; (02) 6032 7428. ***Vintara Winery, Brewery & Cafe:*** modern menu matched to beers, wines; 105 Fraser Rd; 0447 327 517. ***Pickled Sisters Cafe:*** local produce cafe; Cofield Wines, Distillery Rd, Wahgunyah; (02) 6033 2377. ***Terrace Restaurant:*** regional vineyard menu; All Saints Estate, All Saints Rd, Wahgunyah; (02) 6035 2209.

Bank on Main: 2 B&B rooms; 80 Main St; (02) 6032 7000. ***Ready Cottage B&B:*** original 3-room family homestead; 92 High St; (02) 6032 7407. ***The House at Mount Prior:*** elegant rooms; 1194 Gooramadda Rd; (02) 6026 5256. ***Vineyards at Tuileries:*** spa units; 13–35 Drummond St; (02) 6032 9033.

Rutherglen Wine Region At many of the fantastic wineries in this region the appeal extends beyond cellar-door tastings and sales, such as at the All Saints Estate, 10 km NW, which features a National Trust–classified, castle-like building and a fine restaurant. A Day on the Green concert is held here in Feb. In the grounds of the Bullers Calliope Vineyard, 6 km W, is a bird park, with over 100 native and exotic species. Gehrig Estate, 21 km E, is Victoria's oldest continuously operating vineyard. It displays historic farming implements and has a charming restaurant.

Great Northern Mine: marked by mullock heaps associated with the first alluvial goldmine in the district. Historical details are provided on-site; Great Northern Rd, 5 km E. ***Lake Moodemere:*** found near the winery of the same name, the lake is popular for watersports and features ancient Aboriginal canoe trees by the shores; 8 km W. ***Old customs house:*** a relic from the time when a tax was payable on goods from NSW; 10 km NW.

See also HIGH COUNTRY, p. 158

St Arnaud

Pop. 2274
Map ref. 591 K9

4 Napier St; (03) 5495 1268 or 1800 014 455.

 91.1 FM ABC Local Radio, 99.1 Goldfields FM

A former goldmining town surrounded by forests and scenic hill country, St Arnaud is a service centre for the district's farming community, yet retains a peaceful rural atmosphere. The main street is lined with well-preserved historic buildings, many of which feature impressive ornate lacework. Together, these buildings form a nationally recognised historic streetscape.

Self-guide historic tour: brochure available from visitor centre. ***Queen Mary Gardens:*** great spot for a picnic; Napier St. ***Old Post Office:*** now a B&B and restaurant; Napier St. ***Police lock-up:*** built in 1862; Jennings St.

 ***Heritage Festival:*** Nov.

St Arnaud Range National Park The park protects an oasis of dense box-ironbark forest and woodland surrounded by agricultural land. Over 270 different species of native flora have been recorded here and provide a glimpse of what the area would have looked like before the land-clearing that occurred during and after the gold rush. Within the park are the Teddington Reservoirs, popular for brown trout and redfin fishing. The rugged terrain throughout provides a great opportunity for keen bushwalkers or 4WD enthusiasts. Wedge-tailed eagles can be seen soaring above the steep, forested ranges. Contact Parks Victoria on 13 1963; Sunraysia Hwy; 15 km S.

Great Grape Rd: wine-themed circuit through Stawell and Ballarat; details from visitor centre.

See also GOLDFIELDS, p. 154

Sale

Pop. 13 335
Map ref. 583 K7

Wellington Visitors Information Centre, Princes Hwy; (03) 5144 1108 or 1800 677 520; www.tourismwellington.com.au

 104.7 Gippsland FM, 828 AM ABC Local Radio

Situated by the Thomson River near the Latrobe River junction, Sale grew on the back of the gold rush and became Gippsland's first city in 1950. Although largely considered an industrial town, with the nearby Bass Strait oilfields providing a large part of the town's economy, Sale has a lot more to offer. The Port of Sale is being redeveloped and there are many good cafes and restaurants, and a number of fine-art galleries and craft outlets. The lakes near Sale are home to the unique Australian black swan – the bird that has become a symbol for the town.

Gippsland Arts Gallery The gallery promotes the work of artists and craftspeople in central Gippsland. Works range from traditional landscapes to visual statements on environmental and cultural issues, and may be in any medium from painting and photography to film and video. Foster St; (03) 5142 3372.

Lake Guthridge Parklands This major recreational area within Sale comprises the Lake Guthridge and Lake Guyatt precincts, the Sale Botanic Gardens and the Regional Aquatic Complex. The precinct showcases over 35 ha of historically significant botanic gardens, walking trails, Indigenous artworks and a contemporary fauna park. It also provides sensory gardens, abundant seating, an adventure playground for children and tennis courts. Foster St.

Historical Museum: local history memorabilia; Foster St. ***Ramahyuck Aboriginal Corporation:*** offers local arts and crafts and is part of the Bataluk Cultural Trail, which takes in sites of Aboriginal significance in the region; Foster St. ***Historical buildings:*** include Our Lady of Sion Convent in York St; Magistrates Court and Supreme Court, Foster St; St Paul's Anglican Cathedral featuring fine stained-glass windows, Cunninghame St; St Mary's Cathedral, Foster St. Bicentennial clock tower in the mall utilises the original bluestone base, ironwork and clock mechanisms; Raymond St. ***RAAF base:*** home of the famous Roulettes aerobatic team; Raglan St. ***Sale Common and State Game Refuge:*** protected wetland area with a boardwalk; south-east edge of town. ***Textile art:*** Sale is the home of internationally recognised textile artist Annemieke Mein. Her work is on permanent display in the foyer of the Port of Sale Civic Centre, ESSO BHP Billiton Wellington Entertainment Centre and St Mary's Cathedral; Foster St.

Sale Cup: Greenwattle Racecourse; Oct.

Bis Cucina: contemporary menu; 100 Foster St; (03) 5144 3388. ***Il Nido:*** modern Italian; 29 Desailly St; (03) 5144 4099. ***Relish @ the Gallery:*** modern cafe; Gippsland Art Gallery, 68–70 Foster St; (03) 5144 5044.

Minnies B&B: groovy 2-bedroom country house; 202 Gibsons Rd; (03) 5144 3344. ***Quest Sale:*** modern studios and apartments; 180–184 York St; (03) 5142 0900. ***Frog Gully Cottages:*** 2 contemporary cottages; Lot 2419 Rosedale Rd, Longford; (03) 5149 7242.

Holey Plains State Park The open eucalypt forests in this park are home to abundant wildlife, while swamps provide a habitat for many frog species. There is a good swimming lake, and a series of fascinating fossils can be seen nearby in a limestone quarry wall. Bushwalking, picnicking and camping are all popular activities, particularly around Harriers Swamp. Access from Princes Hwy; 14 km SW.

Fishing: good fishing for trout in the Avon River near Marlay Pt and also in the Macalister, Thomson and Latrobe rivers, especially at Swing Bridge; 5 km S. ***Marlay Pt:*** on the shores of Lake Wellington with boat-launching facilities provided. The yacht club here sponsors an overnight yacht race to Paynesville each Mar; 25 km E. ***Seaspray:*** a popular holiday spot on Ninety Mile Beach; offers excellent surfing and fishing; 32 km S. ***Golden and Paradise beaches:*** 2 more townships on Ninety Mile Beach with great surfing and fishing; 35 km SE. ***Loch Sport:*** set on Gippsland Lakes and popular for camping and fishing; 65 km SE. *For details on Gippsland Lakes see Lakes Entrance and Paynesville.* ***Howitt Bike Trail:*** 13-day round trip beginning and ending in Sale; details from visitor centre. ***Bataluk Cultural Trail:*** takes in sites of Indigenous significance from Sale to Cann River; brochure from visitor centre.

See also EAST GIPPSLAND, p. 159

Seymour

Pop. 6062
Map ref. 584 D10

Old Courthouse, Emily St; (03) 5799 0233.

 87.6 Seymour FM, 97.7 FM ABC Local Radio

Seymour is a commercial, industrial and agricultural town on the Goulburn River. The area was recommended for a military base by Lord Kitchener during his visit in 1909. Nearby Puckapunyal became an important training place for troops during World War II, and remains a major army base today.

Royal Hotel: featured in Russell Drysdale's famous 1941 painting *Moody's Pub*; Emily St. ***Old Courthouse:*** built in 1864, it now houses local art; Emily St. ***Fine Art Gallery:*** in the old post office; Emily St. ***Goulburn River:*** a walking track goes by the river and the Old Goulburn Bridge has been preserved as a historic relic. ***Goulburn Park:*** for picnics and swimming; Cnr Progress and Guild sts. ***Seymour Railway Heritage Centre:*** restored steam engines and carriages; open by appt; Railway Pl. ***Australian Light Horse Memorial Park:*** Goulburn Valley Hwy.

Market: Kings Park; 3rd Sat each month. ***Alternative Farming Expo:*** Feb. ***Tastes of the Goulburn:*** food and wine festival; Oct. ***Seymour Cup:*** Oct.

Plunkett Fowles Cellar Door and Cafe: local-produce-driven menu; Cnr Hume Hwy and Lambing Gully Rd, Avenel; (03) 5796 2150. ***Trawool Shed Cafe:*** slick contemporary cafe; 8447 Goulburn Valley Hwy, Trawool; (03) 5799 1595.

Rosehill Cottages: 2 cosy cottages; 8447 Goulburn Valley Hwy, Trawool; (03) 5799 1595. ***Trawool Valley Resort:*** picturesque country resort; 8150 Goulburn Valley Hwy, Trawool; (03) 5792 1444.

Tallarook State Forest Tallarook is a popular destination for bushwalking, camping, rock climbing and horseriding. The key features are Mt Hickey, the highest point in the park and the location of a fire-lookout tower, and Falls Creek Reservoir, a scenic picnic spot. Warragul Rocks offers great views over the Goulburn River. 10 km s.

Travellers note: *Lookout from Warragul Rocks can only be accessed via private property. The landowner requests that any visitor contact him first to arrange access: Ron Milanovic, 0413 402 744.*

Wineries: several in the area, including Somerset Crossing Vineyards, 2 km s; Plunkett Fowles, 21 km NE; Tahbilk, 26 km N; Mitchelton, 28 km N. ***RAAC Memorial and Army Tank Museum:*** Puckapunyal army base; 10 km w.

See also GOULBURN & MURRAY, p. 157

Shepparton

Pop. 44 598
Map ref. 584 E7

534 Wyndham St; (03) 5831 4400 or 1800 808 839; www.greatershepparton.com.au

 97.7 FM ABC Local Radio, 98.5 FM

Shepparton has recently become a popular destination for conferences and sporting events, and so has plenty of modern accommodation and good restaurants in town. Indeed, Shepparton is a thriving city and is considered the 'capital' of the Goulburn Valley. It is home to many orchards irrigated by the Goulburn Irrigation Scheme.

Art Gallery: features Australian paintings and ceramics; Welsford St. ***Bangerang Cultural Centre:*** displays and dioramas on local Aboriginal culture; Parkside Dr. ***Historical Museum:*** in the Historical Precinct; open even-dated Sun afternoons; High St. ***Emerald Bank Heritage Farm:*** displays of 1930s farming methods; Goulburn Valley Hwy. ***Victoria Park Lake:*** scenic picnic spot; Tom Collins Dr. ***Reedy Swamp Walk:*** prolific birdlife; at the end of Wanganui Rd. ***Moooving Art:*** mobile interactive public art of life-size 3-D cow sculptures; various parks in Shepparton including Monash Park, Queens Gardens and Murchison riverbank. ***Factory sales:*** Pental Soaps and Campbells Soups.

Trash and treasure market: Melbourne Rd; each Sun. ***Craft market:*** Queens Gardens, Wyndham St; 3rd Sun each month. ***International Dairy Week:*** Jan. ***Bush Market Day:*** Feb. ***Shepparton Fest:*** major local arts festival with family entertainment; Mar. ***Spring Car Nationals:*** car competitions; Nov.

Bohjass: contemporary menu with tapas; Level 1, 276B Wyndham St; (03) 5822 0237. ***Cellar 47 Restaurant & Wine Bar:*** top regional restaurant; 166–170 High St; (03) 5831 1882. ***Letizia's:*** innovative cross-cultural menu; 67 Fryers St; (03) 5831 8822. ***Teller:*** modern Australian; 108 McLennan St, Mooroopna; (03) 5825 3344.

Central Shepparton Apartments: 2-bedroom serviced apartments; 507 Wyndham St; (03) 5821 4482. ***The Carrington:*** boutique and traditional motel accommodation; 505 Wyndham St; (03) 5821 3355. ***The Churches:*** beautifully renovated century-old church; Woodlands Estate Lavender, 325 Poplar Ave, Orrvale; (03) 5829 1019.

SPC Ardmona KidsTown: a fun attraction with a maze, flying fox, enormous playground and miniature railway, and camel rides on the weekends; Midland Hwy; (03) 5831 4213; 3 km w. ***Mooroopna:*** a small town in the fruit-growing district. It hosts the Fruit Salad Day in Feb. 5 km w. ***Kialla:*** Ross Patterson Gallery, with displays and sales of local artwork. Also here is Belstack Strawberry Farm, where you can pick your own berries; Goulburn Valley Hwy; 9 km s. ***Tatura:*** a museum with displays on local WW II internment camps. Taste of Tatura is held each Mar; 17 km sw. ***Wunghnu:*** the town (pronounced 'one ewe') is centred on the well-known Institute Tavern. A tractor-pull festival is held each Easter; 32 km N.

See also GOULBURN & MURRAY, p. 157

Stawell

Pop. 5879
Map ref. 591 I11 | 593 I1

Stawell and Grampians Visitor Information Centre, 50–52 Western Hwy; (03) 5358 2314 or 1800 330 080.

 96.5 Triple H FM, 594 AM ABC Local Radio

Pastoral runs were established in the Stawell region in the 1840s, but it was the discovery of gold in 1853 by a shepherd at nearby Pleasant Creek that was the catalyst for creating a town. Stawell remains a goldmining centre with Victoria's largest mine. However, it is actually better known as the home of the Stawell Gift, Australia's richest footrace, and the gateway to the Grampians.

Stawell Gift Hall of Fame Museum In 1878 the Stawell Athletic Club was formed by local farmers and businessmen who were keen to have a sports day each Easter. The club put up the prize pool of £110, and the race was on. The annual Stawell Gift has run almost continuously since and is now one of the most prestigious races in the world. The race has been run at Central

continued on p. 208

SORRENTO

Pop. 1530

Map ref. 570 B11 | 572 H9 | 574 F7 | 593 P8

[POINT KING]

Mornington Peninsula Visitor Information Centre, 359B Pt Nepean Rd, Dromana; (03) 5987 3078 or 1800 804 009.

98.7 3RPP FM, 774 AM ABC Local Radio

Just inside Port Phillip Heads on the Mornington Peninsula, in 1803 Sorrento was the site of Victoria's first European settlement. The town is close to historic Point Nepean and major surf and bayside beaches. Its population swells significantly over summer as visitors flock to soak up the holiday-resort atmosphere. A ferry links Sorrento to Queenscliff on the Bellarine Peninsula.

Collins Settlement Historic Site: marks the state's first European settlement and includes early graves; on Sullivan Bay. ***Historic buildings:*** include Sorrento Hotel on Hotham Rd and Continental Hotel on Ocean Beach Rd. Both are fine examples of early Victorian architecture, with the latter reputed to be the largest limestone building in the Southern Hemisphere; the visitor centre can give details of self-guide historical walks. ***Nepean Historical Society Museum and Heritage Gallery:*** a collection of local artefacts and memorabilia in the National Trust–classified mechanics institute. Adjacent is Watt's Cottage and the Pioneer Memorial Garden; Melbourne Rd. ***Dolphin and seal cruises:*** depart from the pier; not in winter months.

Craft market: primary school, cnr Kerferd and Coppin rds; last Sat each month. ***Street Festival:*** Oct.

Acquolina: Italian trattoria; 26 Ocean Beach Rd; (03) 5984 0811. ***Loquat Restaurant & Bar:*** modern Mediterranean; 3183 Point Nepean Rd; (03) 5984 4444. ***Smokehouse Sorrento:*** pizzas and Mediterranean fare; 182 Ocean Beach Rd; (03) 5984 1246. ***The Baths:*** iconic beachside eatery; 3278 Point Nepean Rd; (03) 5984 1500.

Oceanic Sorrento: original guesthouse and new apartments; 231 Ocean Beach Rd; (03) 5984 4166. ***Sorrento Beach Motel:*** beach-style studios; 780 Melbourne Rd; (03) 5984 1356. ***Aquabelle Apartments:*** 3 architect-designed apartments; Level 1, 2331–2335 Point Nepean Rd, Rye; 1300 880 319. ***Lakeside Villas at Crittenden Estate:*** 3 contemporary villas; Crittenden Estate, Harrisons Rd, Dromana; (03) 5987 3275. ***Peppers Moonah Links Resort:*** modern golf resort; 55 Peter Thomson Dr, Fingal; 5988 2000.

Mornington Peninsula National Park The park incorporates Sorrento, Rye and Portsea back beaches and stretches south-east to Cape Schanck and beyond (*see Flinders*). Walks, picnics and swimming are the main attractions, but there is also the unique rock formation of London Bridge, at Portsea. The rugged coastline offers good surfing. Pt Nepean and historic Fort Nepean can be accessed by a daily transport service departing from Portsea. A former Quarantine Station on Pt Nepean has recently been opened to the public (after being closed for 150 years) and offers a self-guide walking tour through the Boiler House. Walking tracks also link up with other attractions like London Bridge. Contact Parks Victoria on 13 1963.

Portsea: an opulent holiday town with good, safe swimming beaches. It hosts the Portsea Swim Classic each Jan; 4 km NW. ***Popes Eye Marine Reserve:*** an artificially created horseshoe-shaped island and reef, now a popular spot for diving. Gannets nest here. Cruises available; details from visitor centre; 5 km offshore at Portsea. ***Rye:*** a beachside holiday spot with horseriding trips on offer and the annual Beach Sand Sculpting Championship each Feb; 8 km E. ***Peninsula Hot Springs:*** relaxing, outdoor, naturally heated pools. Private mineral pools, baths and massage therapies available; Springs La, Rye. ***Moonah Links Golf Course:*** 2 fantastic 18-hole golf courses, one designed specifically for the Australian Open; Peter Thompson Dr, Fingal. ***Rosebud:*** a bayside resort town with gorgeous, safe swimming beaches. Summer fishing trips depart from Rosebud pier. A film festival is held here each Nov; 15 km E. ***McCrae Homestead:*** National Trust–classified drop-slab property built in 1844; open afternoons; McCrae; 17 km E.

See also MORNINGTON PENINSULA, p. 149

VICTORIA

Park since 1898. Visit the museum to discover the glory and heartbreak of the race since its inception. Open 9–11am Mon–Fri; Main St; (03) 5358 1326.

Big Hill Lookout and Stawell Gold Mine viewing area: the Pioneers Lookout at the summit of this local landmark presents magnificent 360-degree views of the surrounding area. Continue down Reefs Rd to Stawell Gold Mine viewing area to hear about the daily operations of Victoria's largest gold-producing mine. ***Casper's Mini World:*** miniature tourist park with working models of famous world attractions such as the Eiffel Tower and including dioramas and commentaries; London Rd. ***Fraser Park:*** displays of mining equipment; Main St. ***Pleasant Creek Court House Museum:*** local history memorabilia; Western Hwy. ***Stawell Ironbark Forest:*** spring wildflowers, including rare orchids; northern outskirts of town, off Newington Rd.

SES Market: Drill Hall, Sloane St; 1st Sun each month. ***Farmer's Market:*** Harness Racing Club, Patrick St; last Sun each month. ***Stawell Gift:*** Easter.

Bunjil's Shelter This is Victoria's most important Aboriginal rock-art site. It depicts the creator figure, Bunjil, sitting inside a small alcove with his 2 dingoes. Bunjil created the geographical features of the land, and then created people, before disappearing into the sky to look down on the earth as a star. The site is thought to have been used for ceremonies by the local Djab Wurrung and Jardwadjali people. Off Pomonal Rd; 11 km S.

The Sisters Rocks: huge granite tors; beside Western Hwy; 3 km SE. ***Great Western:*** picturesque wine village with Seppelt Great Western Winery, est. 1865, featuring National Trust–classified underground tunnels of cellars and Champagne Picnic Races in Jan; 16 km SE. ***Tottington Woolshed:*** rare example of a 19th-century woolshed; road to St Arnaud; 55 km NE. ***Great Grape Rd:*** circuit through Ballarat and St Arnaud, stopping at wineries; details from visitor centre.

See also GRAMPIANS & CENTRAL WEST, p. 155

Swan Hill

Pop. 9684
Map ref. 558 G9 | 589 M12 | 591 L2

Swan Hill Region Information Centre, cnr McCrae and Curlewis sts; (03) 5032 3033 or 1800 625 373; www.swanhillonline.com

99.1 Smart FM, 102.1 FM ABC Local Radio

In 1836, explorer Thomas Mitchell named this spot Swan Hill because of the black swans that kept him awake all night. The town's swans remain, but there are many other attractions in this pleasant city on the Murray Valley Highway.

Swan Hill Pioneer Settlement This museum re-creates the Murray and Mallee regions from the 1830s to the 1930s. There are barber shops and chemists, and rides available on the PS *Pyap* or horse-drawn carts. There is also the Sound and Light Tour; bookings required. End of Gray St on Little Murray River.

Swan Hill Regional Art Gallery: an impressive permanent collection plus touring exhibitions; opposite the Pioneer Settlement Museum.

Market: Curlewis St; 3rd Sun each month. ***Racing Cup Carnival:*** June. ***Australian Inland Wine Show:*** Oct.

Carriages Restaurant: hospitality students prepare international menu; 423 Campbell St; (03) 5033 2796. ***Java Spice:*** South-East Asian; 17 Beveridge St; (03) 5033 0511. ***Quo Vadis:*** Italian restaurant and pizza parlour; 255–259 Campbell St; (03) 5032 4408. ***Yutaka Sawa:*** Japanese; 107 Campbell St; (03) 5032 3515.

BIG 4 Swan Hill: holiday park cabins; 186 Murray Valley Hwy; (03) 5032 4372 or 1800 990 389. ***Best's Riverbed and Breakfast:*** 5 in-house rooms; 7 Kidman Reid Dr, Murray Downs; (03) 5032 2126. ***Burrabliss B&B:*** B&B and self-catering options; 169 Lakeside Dr, Lake Boga; (03) 5037 2527. ***Murray Downs Resort:*** hotel rooms and apartments; Lot 5 Murray Downs Dr, Murray Downs; (03) 5033 1966 or 1800 807 574.

Swan Hill Wine Region The region, which starts around Tresco to the south-east and ends around Piangil to the north-west, takes advantage of the Murray River and the Mediterranean-style climate. The first vines were planted here in 1930, but the proliferation of vineyards really began when Sicilian immigrants arrived on the Murray after WW II. Today cellar doors offer tastings and sales of predominantly shiraz, colombard and chardonnay varieties. Winery map from visitor centre.

Lake Boga The town has an interesting history as an RAAF flying-boat repair depot during WW II. The depot serviced over 400 flying boats, one of which can be seen at the Flying Boat Museum. The underground museum is in the original communications bunker in Willakool Dr. At Lake Boga, the water mass is popular for watersports, fishing and camping, and is home to a variety of bird species that can be seen on the various walks. A yachting regatta is held here each Easter. 17 km SE.

Lakeside Nursery and Gardens: over 300 varieties of roses; 10 km NW. ***Tyntyndyer Homestead:*** built in 1846; open Tues and Thurs 10am–4pm or by appt; Murray Valley Hwy; 20 km NW. ***Nyah:*** good market with local produce; 2nd Sat each month; 27 km NW. ***Tooleybuc:*** situated in NSW, it has a tranquil riverside feel and good fishing, picnicking and riverside walks. The Bridgekeepers Cottage has sales and displays of dolls and crafts; 46 km N.

See also MALLEE COUNTRY, p. 156

Tallangatta

Pop. 955
Map ref. 585 L7 | 587 K3

50 Hanson St, Corryong; (02) 6076 2277; www.pureuppermurrayvalleys.com

101.3 Oak FM, 106.5 FM ABC Local Radio

When the old town of Tallangatta was going to be submerged in 1956 after the level of the Hume Weir was raised, the residents simply moved the entire township 8 kilometres west. Tallangatta now has an attractive lakeside location and sits directly north of Victoria's beautiful alpine region.

The Hub: local art and craft, and Lord's Hut, the only remaining slab hut in the district; Towong St.

Farm and Water Festival: Apr. ***Fifties Festival:*** Oct. ***Garage Sale Festival:*** Oct.

Lake Hume Tallangatta is on the shores of this enormous and attractive lake, formed when the then largest weir in the Southern Hemisphere was constructed. It is now a picturesque spot for swimming, waterskiing, windsurfing and fishing. The foreshore reserves are perfect for barbecues.

Eskdale: craft shops, and trout fishing in the Mitta Mitta River; 33 km S. ***Lake Dartmouth:*** great for trout fishing and boating; hosts the Dartmouth Cup Fishing Competition over the June long weekend. Also here is The Witches Garden featuring unique medicinal plants; 58 km SE. ***Mitta Mitta:*** remnants of a large open-cut goldmine. Also a gallery, Butcher's Hook Antiques and Bharatralia Jungle Camp. Hosts the Mitta Muster on Sun on the long weekend in Mar; 60 km S. ***Australian Alps Walking Track:*** passes over Mt Wills; 108 km S via Mitta Mitta. ***Scenic drives:*** to Cravensville, to Mitta Mitta along Omeo Hwy and to Tawonga and Mount Beauty.

See also HIGH COUNTRY, p. 158

Terang

Pop. 1830
286 C4 | 288 C4

Old Courthouse, 22 High St; (03) 5592 1984.

103.7 3WAY-FM, 774 AM ABC Local Radio

Terang is in a fertile dairy-farming district. It is a well-laid-out town with grand avenues of deciduous trees, and is known throughout the state for its horseracing carnivals.

Cottage Crafts Shop: in the old courthouse on High St. ***District Historical Museum:*** old railway station and memorabilia; open 3rd Sun each month; Princes Hwy. ***Lions Walking Track:*** 4.8 km, beside dry lake beds and majestic old trees; begins behind Civic Centre on High St. ***Historic buildings:*** many examples of early-20th-century commercial architecture. A Gothic-style Presbyterian church is in High St.

Terang Cup: Nov.

Demo Dairy: demonstrates dairy-farming practices; Princes Hwy, 4 km W. ***Lake Keilambete:*** 2.5 times saltier than the sea and reputed to have therapeutic properties; must obtain permission to visit as it is surrounded by private land; 4 km NW. ***Model Barn Australia:*** collection of model cars, boats and planes; open by appt; Robertson Rd; 5 km E. ***Noorat:*** birthplace of Alan Marshall, author of *I Can Jump Puddles*. The Alan Marshall Walking Track here involves a gentle climb to the summit of Mt Noorat, an extinct volcano, with excellent views of the crater, the surrounding district and the Grampians; 6 km N.

See also SOUTH-WEST COAST, p. 153

Torquay

Pop. 9848
Map ref. 572 F9 | 574 A7 | 580 G2 | 593 N8

Surfworld Australia, Surf City Plaza, cnr Surfcoast Hwy and Beach Rd; (03) 5261 4219 or 1300 614 219; www.visitsurfcoast.com

94.7 The Pulse FM, 774 AM ABC Local Radio

Torquay was one of the first resort towns on Victoria's coast, and remains one of the most popular today. It was named in honour of the famous English resort, but its heritage is very different. Not only does Torquay and its coast have some of the best surf beaches in the world, it was also the birthplace of world leaders in surfboards, wetsuits and other apparel, including Rip Curl and Quiksilver, founded here in the 1960s and 70s.

Surf City Plaza This modern plaza houses some of the biggest names in surfing retail alongside smaller outlets. The complex boasts the world's biggest surfing museum, Surfworld. See how board technology has developed over the last century, find out exactly what makes a good wave, and learn about the history of surfing at Bells Beach. A theatre here screens classic 1960s and 70s surf flicks and the latest surf videos. Beach Rd.

Fishermans Beach: good spot for fishing, with a sheltered swimming beach and a large sundial on the foreshore. ***Tiger Moth World:*** theme park based around the1930s Tiger Moth biplane. Joy-flights available; Blackgate Rd. ***Surf schools:*** programs available to suit all abilities, with many courses run during summer school holidays; details from visitor centre.

Cowrie Community Market: foreshore; 3rd Sun each month Sept–Apr. ***Danger 1000 Ocean Swim:*** Jan. ***Surf for Life Surfing Contest:*** Jan. ***Kustom Jetty Surf Pro:*** Jan. ***Rip Curl Pro:*** Easter. ***Hightide Festival:*** fireworks display; Dec.

Growlers: cafe-bar and restaurant; 23 The Esplanade; (03) 5264 8455. ***Scorched:*** modern Australian; 17 The Esplanade; (03) 5261 6142. ***The Surf Rider:*** modern fusion menu; 26 Bell St; (03) 5261 6477. ***Sunnybrae Restaurant and Cooking School:*** regionally inspired weekend lunches; Cnr Cape Otway and Lorne rds, Birregurra; (03) 5236 2276.

Crowne Plaza Torquay: beachfront rooms and apartments; 100 The Esplanade; (03) 5261 1500 or 1800 007 697. ***Peppers The Sands Resort:*** golf resort; 2 Sands Blvd; (03) 5264 3333.

Surf coast It is no wonder the coast that runs from Torquay through to Eastern View (past Anglesea) has dubbed itself the Surf Coast. Submerged reefs cause huge waves that are a surfer's paradise. Most famous is Bells Beach, around 5 km SW of Torquay. The clay cliffs provide a natural amphitheatre for one of the best surf beaches in the world and the longest running surf competition, the Rip Curl Pro, which started in 1973 and attracts top competitors. Other good surf beaches include Jan Juc, Anglesea and Fairhaven. To see the coast on foot, take the 30 km Surf Coast Walk, starting at Torquay and travelling south to Moggs Creek.

Hinterland: delightful towns like Bellbrae, Deans Marsh and Birregurra are dotted along the vista of the Surf Coast hinterland; starts 8 km W. ***Bicycle lane:*** runs along Surfcoast Hwy from Grovedale to Anglesea.

See also SOUTH-WEST COAST, p. 153

Traralgon

Pop. 21 960
Map ref. 583 I7

Latrobe Visitor Information Centre, The Old Church, Southside Central, Princes Hwy; 1800 621 409; www.visitlatrobevalley.com

100.7 FM ABC Local Radio, 104.7 Gippsland FM

Traralgon is one of the Latrobe Valley's largest towns; a commerical hub located on the main Gippsland rail and road routes. Primarily a service centre for neighbouring agricultural communities, timber and electricity production, it also retains a certain village atmosphere with historic buildings in its wide streets and attractive public gardens.

Historic buildings: include the old post office and courthouse; Cnr Franklin and Kay sts. ***Victory Park:*** a great spot for picnics. Also here is a band rotunda and miniature railway; Princes Hwy.

Farmers market: 4th Sat each month. ***International Junior Tennis Championships:*** Jan. ***Traralgon Cup:*** Nov.

Cafe Aura: cafe-deli-cum-restaurant; Shop 3, 19–25 Seymour St; (03) 5174 1517. ***Iimis Cafe:*** good Mediterranean/ Greek; 28 Seymour St; (03) 5174 4577. ***Neilsons:*** contemporary

Australian; 13 Seymour St; (03) 5175 0100. ***Terrace Cafe:*** modern fare; Century Inn, 5 Airfield Rd; (03) 5173 9400.

Walhalla Mountain Rivers Trail: this scenic drive (Tourist Route 91) winds through pretty hills to the north of town. ***Loy Yang power station:*** tours available; 5 km S. ***Toongabbie:*** a small town that hosts the Festival of Roses each Nov; 19 km NE. ***Hazelwood Cooling Pond:*** year-round warm water makes this a popular swimming spot; outskirts of Churchill; 20 km SW. ***Tarra–Bulga National Park:*** temperate rainforest; 30 km S.

See also PHILLIP ISLAND & GIPPSLAND, p. 150

Walhalla

Pop. 15
Map ref. 583 I5

Latrobe Visitor Information Centre, The Old Church, Southside Central, Princes Hwy, Traralgon; 1800 621 409.

100.7 FM ABC Local Radio, 104.7 Gippsland FM

This tiny goldmining town is tucked away in dense mountain country in Gippsland – in a steep, narrow valley with sides so sheer that some cemetery graves were dug lengthways into the hillside. The town has a tiny population and is a relic from a long-gone era – it was only connected to electricity in 1998.

Long Tunnel Gold Mine One of the most prosperous goldmines in the state during the 19th century with over 13 tonnes of gold extracted here. Guided tours take visitors through sites such as Cohen's reef and the original machinery chamber 150 m below the ground. Tours operate daily at 1.30pm; Main St.

Historic buildings and goldmining remains: include the old post office, bakery and Windsor House, now a B&B. ***Walks:*** excellent walks in the town area, including one to a cricket ground on top of a 200 m hill. Another walk leads to a historic cemetery with graves of early miners; details from visitor centre. ***Old Fire Station:*** with hand-operated fire engines and firefighting memorabilia; open weekends and public holidays. ***Museum and Corner Store:*** local history displays plus goldmining artefacts; Main St. ***Walhalla Goldfields Railway:*** wonderfully restored old steam engine; departs from Thomson Station on Wed, Sat, Sun and public holidays. ***Gold panning:*** try your luck along pretty Stringers Creek, which runs through town. ***Ghost tours:*** spook yourself with a night-time guided ghost tour of Walhalla using old-fashioned lanterns; details from visitor centre.

Parker's Restaurant: contemporary hotel dining room; Walhalla's Star Hotel, Main Rd; (03) 5165 6262.

Walhalla's Star Hotel: modern hotel suites; Main Rd; (03) 5165 6262.

Deloraine Gardens: terraced gardens; just north of town. ***Thomson River:*** excellent fishing and canoeing; 4 km S. ***Rawson:*** a town built to accommodate those who helped construct the nearby Thomson Dam. Mountain trail-rides are available; 8 km SW. ***Erica:*** visit this small timber town to see a timber-industry display at the Erica Hotel. The King of the Mountain Woodchop is held in town each Jan; 12 km SW. ***Baw Baw National Park:*** park areas accessible from Walhalla include the Aberfeldy River picnic and camping area; 12 km N. *See Moe.* ***Moondarra State Park:*** great for walks and picnics. Moondarra Reservoir is nearby; 30 km S. ***4WD tours:*** to gold-era 'suburbs' such as Coopers Creek and Erica. Tours can be organised through Mountain Top Experience, (03) 5134 6876. ***Australian Alps Walking Track:*** starts at Walhalla and goes for an incredible 655 km. It can be done in sections; details from visitor centre.

See also PHILLIP ISLAND & GIPPSLAND, p. 150

Wangaratta

Pop. 16 846
Map ref. 585 I7 | 586 F4

100 Murphy St; 1800 801 065; www.visitwangaratta.com.au

101.3 Oak FM, 106.5 FM ABC Local Radio

Wangaratta lies in a rich agricultural district in north-eastern Victoria that produces a diverse range of crops including kiwifruit, wine grapes, walnuts and wheat. An entry for both the Murray to the Mountains Rail Trail and the Great Alpine Road, it offers the services of a rural city while retaining a country-town warmth. A short drive in any direction will lead to world-class wineries, gourmet food and some spectacular views.

Self-guide historical walk: historic sites and buildings, such as the majestic Holy Trinity cathedral, Vine Hotel Cellar Museum and the Wangaratta Historical Museum; details from visitor centre. ***Wangaratta Cemetery:*** headless body of infamous bushranger Daniel 'Mad Dog' Morgan is buried here; Tone Rd. ***Wangaratta Exhibitions Gallery:*** changing exhibitions by national and regional artists; Ovens St. ***Brucks Textile Factory:*** a factory outlet for Sheridan sheets; Sicily Ave. ***Australian Country Spinners:*** an outlet for local wool products; Textile Ave.

Wangaratta Trash and Treasure Market: Olympic Swimming Pool carpark, Swan St; each Sun. ***Wangaratta Stitched Up Festival:*** textile displays; June/July. ***Wangaratta Show:*** agricultural show; Oct. ***Festival of Jazz:*** well-known jazz festival; Oct/Nov.

Quality Hotel Wangaratta Gateway: imaginative dining room; 29–37 Ryley St; (03) 5721 8399. ***Rinaldo's:*** Italian; 8–10 Tone Rd; (03) 5721 8800. ***Tread:*** tapas-style menu; 56–58 Faithful St; (03) 5721 4635.

King River Stables: 3 bedrooms, sleeps 8; 170 Oxley Flats Rd; (03) 5721 8195. ***Quality Hotel Wangaratta Gateway:*** refurbished suites and apartments; 29–37 Ryley St; (03) 5721 8399.

Warby Range State Park The steep ranges of the 'Warbys', as they are known locally, provide excellent viewing points, especially from Ryan's Lookout. Other lookouts include the Pangarang Lookout, near Pine Gully Picnic Area, and Mt Glenrowan Lookout, the highest point of the Warbys at 513 m. There are well-marked tracks for bushwalkers and a variety of pleasant picnic spots amid open forests and woodlands, with wildflowers blossoming during the warmer months. 12 km W.

Eldorado Eldorado is a fascinating old goldmining township named after the mythical city of gold. The main relic of the gold era is a huge dredge, the largest in the Southern Hemisphere, which was built in 1936. There is a walking track with information boards around the lake where the dredge now sits. The Eldorado Museum provides details of the town's mining past, alongside WW II relics and a gemstone collection. 20 km NE.

Wombi Toys: old-fashioned, handmade toys for sale; Whorouly; 25 km SE. ***Reids Creek:*** popular with anglers, gem fossickers and gold panners; near Beechworth; 28 km E. ***Newton's Prickle Berry Farm:*** pick your own blackberries and buy organic berry jams; Whitfield; 45 km S. ***Scenic drives:*** one goes for 307 km along the Great Alpine Rd through the alps to Bairnsdale. The road south leads through the beautiful King Valley and to Paradise Falls. A network of minor roads allows you to fully explore the area, including a number of tiny, unspoiled townships such as Whitfield, Cheshunt and Carboor. ***Murray to the Mountains Rail Trail:*** following historical railway lines with 94 km of bitumen sealed track, the trail ventures into pine forests, natural bushland and open valleys. It links several townships. Suitable for both cycling and walking; a gentle gradient makes the track appropriate for all ages and levels of fitness.

See also HIGH COUNTRY, p. 158

Warburton

Pop. 1949
Map ref. 571 F8 | 573 M5 | 582 F4

Water Wheel Visitor and Information Centre, 3400 Warburton Hwy; (03) 5966 9600; www.warburtononline.com

99.1 Yarra Valley FM, 774 AM ABC Local Radio

Warburton was established when gold was discovered in the 1880s, but its picturesque location and proximity to Melbourne meant it quickly became a popular tourist town, with many guesthouses built over the years. There are fine cafes and antique and craft shops in town.

Information Centre: local history display and an old-style, operating water wheel, 6 m in diameter. A wood-fired bakery is adjacent to the centre; Warburton Hwy. ***River Walk:*** 9 km return walk, following a pretty stretch of the Yarra River; starts at Signs Bridge on Warburton Hwy. ***Upper Yarra Arts Centre:*** cinema with regular screenings and a variety of live performances held during the year; Warburton Hwy. ***Warburton Golf Course:*** with great views across the river valley; Dammans Rd. ***O'Shannassy Aqueduct Trail:*** good walking and cycling track that follows the historic open channelled aqueduct; details from visitor centre.

Film Festival: Upper Yarra Arts Centre; June. ***Winterfest:*** wood festival; July.

Wild Thyme: bohemian cafe for music lovers; 3391 Warburton Hwy; (03) 5966 5050. ***Bulong Estate Winery & Restaurant:*** modern French menu; 70 Summerhill Rd, Yarra Junction; (03) 5967 1358.

Charnwood Cottages: 3 luxurious country cottages; 2 Wellington Rd; (03) 5966 2526. ***Forget Me Not Cottages:*** cottages and apartment for couples; 18 Brett Rd; (03) 5966 5805. ***3 Kings B&B:*** contemporary apartments; 2482 Warburton Hwy, Yarra Junction; 0409 678 046.

Yarra Ranges National Park Here, tall mountain ash trees give way to pockets of cool temperate rainforest. Mt Donna Buang, a popular daytrip destination – especially during winter, when it is often snow-covered – is 17 km NW of Warburton. The Rainforest Gallery on the southern slopes of the mountain features a treetop viewing platform and walkway. Night walk tours here reveal some of Victoria's unique nocturnal creatures. Acheron Way is a scenic 37 km drive north through the park to Marysville. Along the way are views of Mt Victoria and Ben Cairn. Drive starts 1 km E of town.

Yarra Centre: indoor sports and swimming; Yarra Junction, Warburton Hwy; 9 km SW. ***Yarra Junction Historical Museum:*** local history displays; open 1–5pm Sun or by appt; Warburton Hwy; 10 km SW. ***Upper Yarra Reservoir:*** picnic and camping facilities; 23 km NE. ***Walk into History:*** takes in the goldmining and timber region from Warburton East to Powelltown (25 km S); details from visitor centre. ***Ada Tree:*** a giant mountain ash over 300 years old; access from Powelltown. ***Yellingbo State Fauna Reserve:*** good for nature spotting. Home to the helmeted honeyeater, a state emblem; 25 km SW. ***Vineyards:*** several in the region, many with tastings and sales. They include the Yarra Burn Winery, the Five Oaks Vineyard and the Brahams Creek Winery. ***Rail trails:*** former railway tracks now used for walking, bikeriding or horseriding, the main one being the Lilydale to Warburton trail; details from visitor centre.

See also YARRA & DANDENONGS, p. 148

Warracknabeal

Pop. 2491
Map ref. 558 E12 | 590 H7

119 Scott St; (03) 5398 1632; www.wag.wimmera.com.au

96.5 Triple H FM, 594 AM ABC Local Radio

The town's Aboriginal name means 'the place of the big red gums shading the watercourse', a name that is both beautifully descriptive and accurate, especially for the part of town around Yarriambiack Creek. Warracknabeal is a major service town at the centre of a wheat-growing district.

Historical Centre: includes a pharmaceutical collection, clocks, and antique furnishings of child's nursery; open afternoons; Scott St. ***Black Arrow Tour:*** a self-guide driving tour of historic buildings. ***Walks:*** including the Yarriambiack Creek Walk; details from visitor centre. ***National Trust–classified buildings:*** include the post office, the Warracknabeal Hotel and the original log-built town lock-up. ***Lions Park:*** by the pleasant Yarriambiack Creek with picnic spots and a flora and fauna park; Craig Ave.

Y-Fest: golf, horseracing, machinery and country music; Easter.

North Western Agricultural Machinery Museum: displays of farm machinery from the last 100 years; Henty Hwy; 3 km S.

See also GRAMPIANS & CENTRAL WEST, p. 155

Warragul

Pop. 11 501
Map ref. 573 N9 | 582 G6

Gippsland Food and Wine, 123 Princes Hwy, Yarragon; 1300 133 309; www.bawbawcountry.com.au

99.1 Yarra Valley FM, 104.7 Gippsland FM

Warragul is a thriving rural town with a growing commuter population, being the dairying centre that supplies much of Melbourne's milk. An excellent base to explore the delightful countryside including the Baw Baw snowfields and 'Gippsland Gourmet Country'.

West Gippsland Arts Centre Part of the town's fantastic, architect-designed civic centre complex, the centre is a mecca for art lovers from across the state. It houses a good permanent collection of contemporary visual arts and is known for the variety of theatre productions and events held here throughout the year. Ask inside for a full program of events. Civic Pl; (03) 5624 2456.

Harvest of Gippsland: Mar. ***Gippsland Field Days:*** Mar/Apr.

The Grange Cafe and Deli: local-produce menu; 15 Palmerston St; (03) 5623 6698. ***Jack's at Jindivick:*** modern Australian; 1070 Jacksons Track, Jindivick; (03) 5628 5424. ***Sticcado Cafe:*** features local Gippsland beef; Shop 6, The Village Walk, Yarragon; (03) 5634 2101. ***The Outpost Retreat:*** modern regional fare; 38 Loch Valley Rd, Noojee; (03) 5628 9669.

continued on p. 213

WARRNAMBOOL

Pop. 30 393

Map ref. 580 C8 | 592 G8

Adjacent Flagstaff Hill Maritime Museum, Merri St; (03) 5564 7837; www.warrnamboolinfo.com.au

103.7 3WAY-FM, 594 AM ABC Local Radio

Warrnambool lies at the end of the Great Ocean Road on a notorious section of coastline that has seen over 80 shipwrecks. The best known was the *Loch Ard* in 1878, which claimed the lives of all but two of those on board. While the wreck site itself is closer to Port Campbell, impressive relics from the ship are held at the Flagstaff Hill Maritime Museum in town. Warrnambool, as Victoria's fifth largest city, offers first-rate accommodation and dining as well as a fantastic swimming beach. The southern right whales that migrate here in winter are another major drawcard.

Flagstaff Hill Maritime Village This reconstructed 19th-century maritime village is complete with a bank, hotel, schoolhouse and surgery. There are also 2 operational lighthouses and an authentic keeper's cottage, now housing the Shipwreck Museum, where relics retrieved from the *Loch Ard* – including the famous earthenware Loch Ard Peacock – are kept. On display is the Flagstaff Hill tapestry, depicting themes of Aboriginal history, sealing, whaling, exploration, immigration and settlement. At night, visitors can watch the sound-and-light show 'Shipwrecked', which details the story of the *Loch Ard*. Merri St; (03) 5559 4600.

[SHEEP OUTSIDE THE FLAGSTAFF HILL MARITIME MUSEUM]

Main beach: a safe swimming beach, with a walkway along the foreshore from the Breakwater to near the mouth of the Hopkins River. ***Lake Pertobe Adventure Playground:*** a great spot for family picnics; opposite main beach, Pertobe Rd. ***Art Gallery:*** local artwork, plus European and avant-garde collections; Timor St. ***Customs House Gallery:*** Gilles St. ***Botanic Gardens:*** pretty regional gardens designed by Guilfoyle (a curator of Melbourne's Royal Botanic Gardens) in 1879; Botanic Rd. ***Fletcher Jones Gardens/Mill Markets:*** award-winning landscaped gardens and market in front of former Fletcher Jones factory; Raglan Pde, eastern approach to town. ***History House:*** local history museum; open 1st Sun each month or by appt; Gilles St. ***Portuguese Padrao:*** monument to early Portuguese explorers; Cannon Hill, southern end of Liebig St. ***Heritage walk:*** 3 km self-guide walk taking in historic buildings around town; details from visitor centre. ***Hopkins River:*** great for fishing and boating. Blue Hole is a popular swimming spot with rockpools. Cruises are available; east of town. ***Proudfoots Boathouse:*** National Trust–classified boathouse on the Hopkins. ***Wollaston Bridge:*** an unusual bridge, built over 100 years ago; northern outskirts of town. ***Tours and charters:*** fishing, whale-watching and diving tours (including shipwreck sites); details from visitor centre.

Trash and treasure market: showgrounds on Koroit St; each Sun. ***Hillside market:*** Flagstaff Hill; operates throughout summer. ***Wunta Fiesta:*** family entertainment, food stalls and music; Feb. ***Tarerer Festival:*** Indigenous culture and music; Mar. ***Racing Carnival:*** May. ***Fun 4 Kids:*** children's festival; June. ***Melbourne–Warrnambool Cycling Classic:*** Oct. ***Flower Shows:*** held in spring.

Donnelly's Restaurant: contemporary regional fare; 78 Liebig St; (03) 5561 3188. ***Kermond's Hamburgers:*** excellent old-fashioned hamburgers; 151 Lava St; (03) 5562 4854. ***Nonna Casalinga:*** up-market Italian; 69 Liebig St; (03) 5562 2051. ***Pippies by the Bay:*** contemporary food with a view; 91 Merri St; (03) 5561 2188.

104 on Merri Apartments: luxury 2-bedroom apartments; 104 Merri St; 0448 668 738. ***Aqua Ocean Villas:*** 4 contemporary apartments; 72 Merri St; (03) 5562 5600. ***Logans Beach Spa and Fitness:*** 2 comfortable apartments; 7 Logans Beach Rd; (03) 5561 3750. ***The Sebel Deep Blue Warrnambool:*** hotel rooms and apartments; 16 Pertobe Rd; (03) 5559 2000.

Tower Hill State Game Reserve This is a beautiful piece of preserved bushland featuring an extinct volcano and a crater lake, with tiny islands. Nature walk starts at the Worn Gundidj Visitor Centre in the reserve. For further information contact (03) 5561 5315; 12 km NW, just after the turn-off to Koroit.

Logans Beach Each year in June, southern right whales return to the waters along the south coast of Australia to give birth, raise their young and start the breeding cycle again. Each female seems to have a favourite spot to give birth, which means that many familiar faces keep reappearing at Warrnambool's Logans Beach. The beach features a purpose-built viewing platform above the sand dunes (binoculars or telescopes are recommended), and the local visitor centre releases information on whale sightings daily.

Allansford Cheeseworld: for cheese tastings and sales; 10 km E. ***Hopkins Falls:*** scenic picnic spot, particularly spectacular in winter after heavy rain. In spring hundreds of baby eels migrate up the falls, creating a most unusual sight; 16 km NE. ***Cudgee Creek Wildlife Park:*** deer, crocodiles and other native fauna, plus an aviary. Picnic and barbecue facilities are provided; Cudgee; 18 km E. ***Koroit:*** National Trust–classified buildings, good local arts and crafts shops, botanic gardens and an Irish festival in Apr; 19 km NW.

See also SOUTH-WEST COAST, p. 153

Mt Worth State Park This park protects a rich variety of native flora including the silver wattle and the Victorian Christmas bush. The Giant's Circuit is a walk that takes in a massive old mountain ash that is 7 m in circumference. Other walks include the Moonlight Creek and MacDonalds tracks, both of which are easily accessible. No camping is permitted. Contact Parks Victoria on 13 1963; access via Grand Ridge Rd; 22 km SE.

Yarragon Nestled in the foothills of the Strzelecki Ranges and with views of green rolling hills, Yarragon is a wonderful destination with an abundance of delightful shops and accommodation options. It boasts one of Gippsland's leading antique stores and a unique gallery renowned for its quality original artwork, exquisite jewellery, beautiful handblown glass and much more. Sample local wines and gourmet produce, including award-winning cheeses from Tarago River and Jindi Cheese. 13 km SE.

Gippsland Gourmet Country: the renowned 'Gippsland Gourmet Country' takes in lush green pastures and state forests to reveal a diverse range of superb gourmet delights. Previously known as 'Gourmet Deli Country', Gippsland Gourmet encompasses some of the best food and wine producers in the region. Sure to tempt your tastebuds and tantalise the senses; details from visitor centre. ***Darnum Musical Village:*** a complex of buildings housing a collection of musical instruments dating back to the 1400s; Princes Hwy; 8 km E. ***Oakbank Angoras and Alpacas:*** sales of yarn and knitted goods; near Drouin, 8 km W. ***Waterfalls:*** Glen Cromie, Drouin West (10 km NW); Glen Nayook, south of Nayook; and Toorongo Falls, just north of Noojee. ***Neerim South:*** visit Tarago Cheese Company for tastings and sales of top-quality cheeses, or enjoy a picnic or barbecue at the pleasant reserve near the Tarago Reservoir. Scenic drives through mountain country start from town; 17 km N. ***Grand Ridge Road:*** 132 km drive that starts at Seaview, 17 km S, and traverses the Strzelecki Ranges to Tarra–Bulga National Park *(see Yarram for park details)*. ***Nayook:*** good fresh produce, a fruit-and-berry farm, and the Country Farm Perennials Nursery and Gardens; 29 km N. ***Childers:*** Sunny Creek Fruit and Berry Farm, and Windrush Cottage; 31 km SE. ***Noojee:*** a mountain town featuring a historic trestle bridge and the Alpine Trout Farm; 39 km N.

See also PHILLIP ISLAND & GIPPSLAND, p. 150

Wedderburn

Pop. 702
Map ref. 591 L8

Loddon Visitor Information Centre, Wedderburn Community Centre, 24 Wilson St; (03) 5494 3489; www.loddonalive.com.au

91.1 FM ABC Local Radio, 99.1 Goldfields FM

Wedderburn, part of the 'Golden Triangle', was once one of Victoria's richest goldmining towns. Many large nuggets have been unearthed here in the past and – for some lucky people – continue to be discovered today. The town's annual Detector Jamboree, with music, historical re-enactments and family entertainment, is growing every year and recognises the importance of gold in the development of so many towns.

Hard Hill Tourist Reserve Hard Hill is a fascinating former mining district with original gold diggings and Government Battery. There is a good walking track through the site, where old mining machinery can be seen. Hard Hill is in a pleasant bushland setting, and picnic facilities are provided. Nearby is a fully operational eucalyptus distillery offering tours and selling eucalyptus products. Northern outskirts of town.

Coach House Cafe and Museum: a 1910 building restored to its original appearance, with authentic, old-fashioned stock and coach-builders quarters; High St. ***Bakehouse Pottery:*** old bakery now used as a pottery, also home to gold pistachio nuts; High St. ***Nardoo Creek Walk:*** takes in the key historic buildings around town; map from visitor centre.

Wedderburn Detector Jamboree: gold festival; Mar long weekend. ***Historic Engine Exhibition:*** Sept.

Mt Korong: bushwalking; 16 km SE. ***Wychitella Forest Reserve:*** wildlife sanctuary set in mallee forest, home to mallee fowl; 16 km N. ***Kooyoora State Park:*** see Inglewood for details. ***Fossickers Drive:*** takes in goldmining sites, places of Aboriginal significance, local wineries and the Melville Caves; details from visitor centre.

See also GOLDFIELDS, p. 154

Welshpool

Pop. 200
Map ref. 582 H9

Old Courthouse, 9 Rodgers St, Yarram; (03) 5182 6553.

89.5 3MFM South Gippsland, 828 AM ABC Local Radio

Welshpool is a small dairying community in South Gippsland. On the coast nearby, Port Welshpool is a deep-sea port servicing the local fishing and oil industries. Barry Beach Marine Terminal, a short distance west of Port Welshpool, services the offshore oil rigs in Bass Strait.

Port Welshpool This popular coastal town has all the natural attractions that a seaside village could want. It is frequented by families who enjoy the safe beaches and fabulous coastal walks, and has fantastic views across to Wilsons Promontory. Fishing enthusiasts should drop a line from the historic jetty, or try from a boat. The port's long link with the sea is detailed in the

VICTORIA

Port Welshpool and District Maritime Museum, which exhibits shipping relics and local history displays as well. 2 km S.

Franklin River Reserve: great bushwalking with well-marked tracks; near Toora; 11 km W. ***Agnes Falls:*** the highest single-span falls in the state, spectacular after heavy rain; 19 km NW. ***Scenic drive:*** head west to see magnificent views from Mt Fatigue; off South Gippsland Hwy. ***Fishing and boating:*** excellent along the coast.

See also PHILLIP ISLAND & GIPPSLAND, p. 150

Winchelsea

Pop. 1336
Map ref. 572 D8 | 580 E1 | 593 M8

Old Library, Willis St (Princes Hwy); open 11am–4pm Fri–Sun; 1300 614 219; www.historicwinchelsea.com.au

 94.7 The Pulse FM, 774 AM ABC Local Radio

This charming little town on the Barwon River west of Geelong was first developed with cattle runs in the 1830s. Many of the historic buildings that grew from this development can still be seen around town, the most impressive being the nearby Barwon Park Homestead – a mansion built by famous settlers of the district, Thomas and Elizabeth Austin. Winchelsea soon became a key stopover for travellers taking the road from Colac to Geelong, and it still serves that purpose for travellers on the Princes Highway.

Barwon Bridge: an impressive arched bridge, built from stone in 1867; Princes Hwy. ***Antiques and collectibles:*** many shops in town that outline its history; Main St and Princes Hwy. ***Winchelsea Historical Trail:*** map available from visitor information centre, or check township information boards. ***Barwon Hotel:*** known locally as the 'bottom pub' of the town, offers country style; Main St. ***Winchelsea Tavern:*** recently renovated Art Deco 'top pub'; Princes Hwy. ***Old Shire Hall:*** beautifully restored bluestone building, now housing popular tearooms; Princes Hwy. ***Old Library:*** houses the visitor information centre; Princes Hwy. ***Marjorie Lawrence Trail:*** details the life of one of the world's most adored dramatic sopranos from the 1900s; details from visitor centre.

Winchelsea Festival: wool shearing, dog trials, wool classing, children's activities, local produce and market stall, plus more; Nov.

Barwon Park Homestead Only the greatest estate would satisfy Elizabeth Austin, and her husband, Thomas, acquiesced. Barwon Park, built in 1869, was the biggest mansion in the Western District. Featuring 42 rooms furnished largely with original pieces, the bluestone building is an impressive example of 19th-century design. The name Austin might be familiar: Thomas Austin reputedly imported the first of Australia's devastating rabbit population and Elizabeth Austin contributed to major charities, and established the Austin Hospital in Melbourne. Open 11am–4pm Wed and Sun; Inverleigh Rd; (03) 5267 2209; 3 km N.

Country Dahlias Gardens: beautiful gardens, best viewed during spring, with sales of dahlia plants; open Feb–Apr; Mathison Rd; 5 km S. ***Killwarrie Cottage:*** rose garden display, home-grown vegetables; open Nov–Mar; 7 km SW.

See also SOUTH-WEST COAST, p. 153

Wodonga

Pop. 29 713
Map ref. 554 A12 | 559 N12 | 585 K6 | 587 I2

Gateway Visitor Information Centre, Lincoln Causeway; 1300 796 222; www.destinationalburywodonga.com.au

 106.5 FM ABC Local Radio, 107.3 Ten-73 Border FM

Wodonga and its twin town, Albury (in New South Wales), sit astride the Murray River. There are many attractions around the Murray and nearby Lake Hume, making the region a popular holiday destination.

Gateway Village: includes woodwork shops and cafes. Also houses the visitor centre; Lincoln Causeway. ***Huon Hill Lookout:*** maps from visitor centre. ***Sumsion Gardens:*** a pretty lakeside park with walking track, picnic and barbecue facilities; Church St. ***Tennis Centre:*** the largest grass court centre of its kind in Australia; Melrose Dr.

Wodonga Craft Market: Woodland Gr; 1st Sat each month. ***Farmers market:*** Gateway Village; 2nd Sat each month. ***Wodonga Show:*** Mar. ***In My Backyard Festival:*** raises environmental awareness; Mar. ***Wine and Food Festival:*** Oct.

Zilch Food Store: friendly modern cafe; 8/1 Stanley St; (02) 6056 2400.

Mt Granya State Park This landscape contrasts between steep, rocky slopes and open eucalypt forests. Bushwalking is a popular pastime and the display of wildflowers in spring is magnificent. There is a pleasant picnic spot at Cottontree Creek, and a short walk leads to the Mt Granya summit, which offers spectacular views of the alps. Murray River Rd; 56 km E.

Military Museum: Bandiana; 4 km SE. ***Hume Weir:*** good spot for walks and picnics; 15 km E. ***Tours:*** winery and fishing tours, as well as scenic drives through the Upper Murray region, the mountain valleys of north-east Victoria and the Riverina; details from visitor centre.

See also HIGH COUNTRY, p. 158

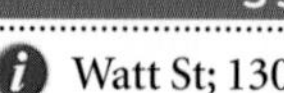

Wonthaggi

Pop. 6528
Map ref. 573 L12 | 582 E9

Watt St; 1300 854 334; www.visitbasscoast.com

88.1 3MFM South Gippsland, 100.7 FM ABC Local Radio

Once the main supplier of coal to the Victorian Railways, Wonthaggi, near the beachside town of Cape Paterson, is South Gippsland's largest town. There are good tourist facilities in town and a number of pretty beaches nearby.

Bass Coast Rail Trail: 16 km trail that runs between Wonthaggi and Anderson. Suitable for walking and cycling, it is the only coastal rail trail in Victoria with landscape that varies from flat farmland and bushland to rugged coastline.

Energy and Innovation Festival: Mar.

State Coal Mine The demand for black coal created a thriving industry in Wonthaggi from 1909 until 1968, and the mine site has been retained to show visitors the lifestyle and working conditions of the miners. Daily underground tours offer close-up views of the coalface, a short walk into the East Area Mine and a cable-hauled skip ride to the surface. Above ground, visit the museum for an introduction to the history of the mine and of Wonthaggi itself, or take a walk around the historic buildings. Inquiries (03) 5672 3053 or Parks Victoria on 13 1963; Cape Paterson Rd; 1.5 km S.

Cape Paterson: waters offshore are protected by Bunurong Marine and Coastal Park and are good for surfing, swimming, snorkelling and scuba diving; 8 km S. ***George Bass Coastal Walk:*** starts at Kilcunda; 11 km NW. Ask at visitor centre for details of other walks. ***Gippsland Gourmet Country:*** takes in central Gippsland's gourmet food and wine producers; details from visitor centre.

See also PHILLIP ISLAND & GIPPSLAND, p. 15

Woodend

Pop. 3166
Map ref. 572 G2 | 579 H9 | 582 B2 | 584 A12 | 593 P3

High St, beside Five Mile Creek; (03) 5427 2033 or 1800 244 711; www.visitmacedonranges.com

 100.7 Highlands FM, 774 AM ABC Local Radio

During the gold rushes of the 1850s, travellers sought refuge from mud, bogs and bushrangers at the 'wood's end' around Five Mile Creek, where a town eventually grew. In the late 19th century Woodend became a resort town, and its lovely gardens and proximity to spectacular natural sights, such as Hanging Rock and Mount Macedon, still make it a popular daytrip and weekend getaway for visitors from Melbourne.

Bluestone Bridge: built in 1862, the bridge crosses Five Mile Creek on the northern outskirts of town. ***Clock Tower:*** built as a WW I memorial; Calder Hwy. ***Courthouse:*** historic structure built in 1870; Forest St.

Craft market: 3rd Sun each month Sept–May. ***Woodend Winter Arts Festival:*** June. ***Macedon Ranges Budburst Festival:*** held throughout the wine district; Nov.

Campaspe Country House: lovely regional dining room; Goldies La; (03) 5427 2273. ***Holgate Brewhouse:*** modern menu; 79 High St; (03) 5427 2510.

Campaspe Country House: rooms and a cottage; Goldies La; (03) 5427 2273. ***Holgate Brewhouse:*** ensuite rooms; 79 High St; (03) 5427 2510. ***Woodbury Cottage B&B:*** charming converted stables; 18 Jason Dr; (03) 5427 1876.

Hanging Rock A massive rock formation made famous by *Picnic at Hanging Rock*, the novel by Joan Lindsay that was later made into a film. The story, about schoolgirls who mysteriously vanished while on a picnic in the reserve, became something of a legend. There is certainly something eerie about Hanging Rock with its strange rock formations and narrow tracks through dense bushland. Hanging Rock is renowned for the annual races held at its base, especially the New Year's Day and Australia Day races. Other events include a Vintage Car Rally and Harvest Picnic, both held in Feb. The reserve also has a discovery centre and cafe. Access from South Rock Rd, off Calder Hwy; 8 km NE.

Macedon: a town at the foot of Mt Macedon. Home to the Church of the Resurrection, with stained-glass windows designed by Leonard French, and excellent plant nurseries; 8 km SE. ***Mt Macedon:*** a township located higher up the mountain, 2 km from Macedon, renowned for its beautiful gardens, many open to the public in autumn and spring. ***Macedon Regional Park:*** bushwalking and scenic drives. The Camels Hump marks the start of a signposted walk to the summit of the mountain where there stands a huge WW I memorial cross. Access via turn-off after Mt Macedon township. ***Wineries:*** several in region include Hanging Rock Winery at Newham; 10 km NE; details from visitor centre. ***Gisborne:*** a variety of craft outlets. Gisborne Steam Park holds a steam-train rally each May; 16 km SE. ***Glen Erin at Romsey/Cope-Williams Winery:*** tastings and sales, surrounded by charming English-style gardens, tennis courts and a cricket green; Romsey; 19 km E. ***Lancefield:*** historic buildings and wineries. Mad Gallery and Bankart Gallery offer contemporary and fine art, respectively. The town is also home to a woodchopping competition in Mar and a farmers market, for local produce, 4th Sat each month; 25 km NE. ***Monegeetta:*** in town is the Mintaro Homestead, a smaller replica of Melbourne's Government House, but not open to the public; 27 km E via Romsey.

See also SPA & GARDEN COUNTRY, p. 151

Wycheproof

Pop. 686
Map ref. 558 G12 | 591 K6

Wycheproof Community Resource Centre, 280 Broadway; (03) 5493 7455; www.wycheproof.vic.au

 99.1 Goldfields FM, 102.1 FM ABC Local Radio

Wycheproof is renowned for the long wheat trains that travel down the middle of the main street, towing up to 60 carriages behind them. There are many historic buildings in town, as well as rare, old peppercorn trees. Mount Wycheproof, at a mere 43 metres, has been named the smallest mountain in the world.

Mt Wycheproof A walking track leads up and around the mountain. Emus and kangaroos can be seen up close in a fauna park at the mountain's base.

Willandra Museum: farm machinery, old buildings and historical memorabilia; open by appt; Calder Hwy. ***Centenary Park:*** aviaries, 2 log cabins and barbecue facilities; Calder Hwy.

 ***Music on the Mount:*** Oct. ***Racing Carnival:*** Oct/Nov.

Tchum Lakes: artificially created lakes, great for fishing and watersports dependent on water levels; 23 km W. ***Birchip:*** visitors to town are greeted by the town's beloved 'Big Red' mallee bull in the main street. Also in town is the Soldiers Memorial Park with large, shady Moreton Bay fig trees, a great spot for a picnic; 31 km W.

See also GRAMPIANS & CENTRAL WEST, p. 155

Yackandandah

Pop. 663
Map ref. 585 K7 | 587 I4

The Athenaeum, High St; (02) 6027 1988; www.uniqueyackandandah.com.au

 101.3 Oak FM, 106.5 FM ABC Local Radio

Yackandandah, with its avenues of English trees and traditional buildings, is so rich with history that the entire town is National Trust–classified. It is situated south of Wodonga in the heart of the north-east goldfields region. In fact, many of the town's creeks still yield alluvial gold.

Historic buildings: the post office, several banks and general stores, with the Bank of Victoria now preserved as a museum, open Sun and school holidays. Explore these and other buildings on a self-guide walk; details from visitor centre; High St. ***Ray Riddington's Premier Store and Gallery:*** displays and sales of local art; High St. ***The Old Stone Bridge:*** a beautiful old structure, built in 1857; High St. ***Arts and crafts:*** many outlets in town, including Yackandandah Workshop, cnr Kars and

VICTORIA

Hammond sts, and Wildon Thyme, High St. ***Antiques:*** Finders Bric-a-Brac and Old Wares, High St; Frankly Speaking, High St; and Vintage Sounds Restoration, specialising in antique gramophones, radios and telephones, Windham St. ***Rosedale Garden and Tea Rooms:*** Devonshire teas; Kars St.

Folk Festival: 3rd weekend Mar. ***Spring Migration Festival:*** Sept. ***Flower Show:*** Oct.

Star Hotel: pub classics; 30 High St; (02) 6027 1493. ***Sticky Tarts Cafe:*** charming casual cafe; 26 High St; (02) 6027 1853.

Creek Haven Cottage: cosy 3-bedroom cottage; 281 Osbornes Flat Rd; (02) 6027 1389. ***Karalilla B&B:*** 4 rooms in Victorian homestead; 271 Ben Valley La; (02) 6027 1788.

Kars Reef Goldmine: take a tour of this fascinating old goldmine, or try your hand at gold panning (licence required); details of tours from visitor centre; Kars St. ***Lavender Patch Plant Farm:*** sales of plants and lavender products; Beechworth Rd; 4 km W. ***Kirbys Flat Pottery and Gallery:*** Kirbys Flat Rd; open weekends or by appt; 4 km S. ***Indigo Valley:*** a picturesque area with a scenic drive leading along the valley floor to Barnawatha; 6 km NW. ***Allans Flat:*** a great destination for food lovers, with The Vienna Patisserie for coffee, ice-cream and delicious Austrian cakes (closed Tues). Also here are Parks Wines and Schmidt's Strawberry Winery, both with tastings and sales; 10 km NE. ***Wombat Valley Tramways:*** a small-gauge railway; open Easter or by appt for groups; Leneva; 16 km NE.

See also HIGH COUNTRY, p. 158

Yarra Glen

Pop. 1901
Map ref. 570 H4 | 571 C6 | 573 L4 | 582 E4

Yarra Valley Visitor Information Centre, Old Courthouse, Harker St, Healesville; (03) 5962 2600; www.yarraglen.com.au

99.1 Yarra Valley FM, 774 AM ABC Local Radio

Yarra Glen is in the heart of the Yarra Valley wine region, nestled between the Yarra River and the Great Dividing Range. A gorgeous area featuring lush, vine-covered hills and fertile valleys; all within easy reach of Melbourne. Fine wines and top-quality local produce; fascinating antique, specialty gift and clothing shops; and not forgetting restaurants; are all in town to entice. For the more adventurous, Yarra Glen is home to hot-air ballooning, scenic helicopter flights over the valley and even skydiving at nearby Coldstream airfield. The bushfire of February 2009 came very close to the township of Yarra Glen, and some businesses and townsfolk were directly affected.

Yarra Glen Grand Hotel: imposing heritage-listed and National Trust–classified hotel, built in 1888 with a recently refurbished restaurant, stands like a sentinel in the main street; Bell St. ***Hargreaves Hill Brewing Co:*** the old Colonial Bank building houses a fine-dining restaurant specialising in local produce and an extensive choice of Yarra Valley and European boutique beers; Bell St. ***Tea Leaves:*** a treasure chest containing Melbourne's largest range of premium-quality teas, novelty teapots and tea accessories: Bell St. ***Den of Antiquities and Yarra Valley Antique Centre:*** boasts large collections of genuine antique furniture, china and glass, vintage radios and other collectibles; Bell St. ***Yarra Glen Railway Station:*** old station on the 1888 Healesville–Lilydale railway line, rebuilt in 1915 and now being restored as part of the Yarra Valley Tourist Railway; King St.

Yarra Valley Farmers Market: Historic Barn at Yering Station; 3rd Sun each month. ***Gulf Station Farmers Market:*** Gulf Station Historic Farm; 4th Sun each month. ***Craft market:*** racecourse; 1st Sun each month Sept–June. ***Grape Grazing Festival:*** wine, food and music throughout the valley; Feb. ***Shortest Lunch:*** midwinter fine food and wine at the smaller boutique wineries; June. ***Melba Festival and Melba Wine Trail:*** arts, culture, food and wine event; Oct.

Hargreaves Hill Brewing Company Restaurant: Mediterranean-style dishes; 25 Bell St; (03) 9730 1905. ***TarraWarra Estate Wine Bar Cafe:*** rustic European fare; 311 Healesville–Yarra Glen Rd; (03) 5957 3510. ***Eleonore's Restaurant at Chateau Yering:*** elegant dining room; Chateau Yering, 42 Melba Hwy, Yering; (03) 9237 3333 or 1800 237 333. ***Yering Station Wine Bar Restaurant:*** modern Australian; 38 Melba Hwy, Yering; (03) 9730 0100.

Art at Linden Gate: 3-suite country escape; 899 Healesville–Yarra Glen Rd; (03) 9730 1861. ***Valley Guest House:*** garden suites; 319 Steels Creek Rd; (03) 9730 1822. ***Chateau Yering:*** historic luxury hotel; 42 Melba Hwy, Yering; (03) 9237 3333 or 1800 237 333.

Yarra Valley Wine Region Victoria's first vines were planted in the Yarra Valley by the Ryrie brothers in 1838 – these plantings would develop into Yering Station, today home to one of the region's finest wineries and restaurants. It produces excellent shiraz viognier as well as award-winning pinot noir. With over 55 cellar door outlets and around 100 wineries, the Yarra Valley is also home to other exceptional names. The De Bortoli Winery in Dixons Creek is Australia's oldest family-owned winery; Domaine Chandon, owned by the legendary Moët & Chandon, makes sophisticated sparkling wines that can be enjoyed in the glass-walled tasting room overlooking the vines. To combine wine with art, head to TarraWarra Estate, well known for its chardonnay and pinot noir, and its modern gallery that holds exhibitions of work by Australian greats from Arthur Boyd to Peter Booth; *see Healesville*. Excellent restaurants abound throughout the valley, many at wineries. Tour companies offer bus tours and personalised winery tours by car through the Yarra Valley region. The ever-popular Grape Grazing Festival is celebrated each Feb, and the smaller wineries' Shedfest wine festival is held in Oct. Winery map available from visitor centre.

Kinglake National Park Kinglake's beautiful messmate forests, fern gullies, panoramic lookouts and bushwalking tracks were devastated by the bushfire of February 2009, and the park is closed until further notice. Phone Parks Victoria on 13 1963 for an update on progress on the park's reopening.

Gulf Station: this National Trust–owned pastoral property, preserved as it was during pioneering days, features old-fashioned farming implements and early animal breeds; open Wed–Sun and public holidays; 2 km NE. ***Yarra Valley Dairy:*** a working dairy with sales of specialty cheeses, clotted cream and local produce; 4 km S. ***Ponyland Equestrian Centre:*** trail-rides and riding lessons; 7 km W. ***Sugarloaf Reservoir Park:*** sailing, fishing and walking, with barbecue and picnic facilities available; 10 km W. ***Yarra Valley Regional Food Trail:*** a self-guide tour, taking in the many gourmet food outlets in the region; details from visitor centre.

See also YARRA & DANDENONGS, p. 148

Yarram

Pop. 1716
Map ref. 583 I9

Old Courthouse, 9 Rodgers St; (03) 5182 6553; www.tourismwellington.com.au

 89.5 3MFM South Gippsland, 100.7 FM ABC Local Radio

Yarram is deep in the dairy country of South Gippsland, and at the heart of some of its most beautiful locales, from the splendour of Ninety Mile Beach to the refreshingly cool atmosphere of Tarra–Bulga National Park. Yarram was originally settled on a swamp, and its name is derived from an Aboriginal word meaning 'plenty of water'. In town are some notable examples of early architecture, including the recently restored Regent Theatre and the historic courthouse.

Regent Theatre: built in 1930, this theatre has been wonderfully restored. Cinemas operate on weekends and school holidays; Commercial Rd.

 ***Tarra Festival:*** Easter.

Tarra–Bulga National Park Tarra–Bulga is a tranquil park with spectacular river and mountain views. Fern Gully Walk, starting from the Bulga picnic ground, takes in the dense temperate rainforests of mountain ash, myrtle and sassafras. The walk leads across a suspension bridge high among the treetops. A walk to Cyathea or Tarra falls, surrounded by lush fern gullies, completes the rainforest experience. Keep an eye out for rosellas, lyrebirds and the occasional koala. The Tarra–Bulga Visitor Centre is on Grand Ridge Rd near Balook; from Yarram, access the park from Tarra Valley Rd; 20 km NW.

Won Wron Forest: great for walks, with wildflowers in spring; Hyland Hwy; 16 km N. ***Beaches:*** there are many attractive beaches in the region, including Manns, for fishing, 16 km SE; McLoughlins, 23 km E; and Woodside Beach, which is patrolled in summer; 29 km E. ***Tarra Valley:*** there are many great gardens, including Eilean Donan Gardens and Riverbank Nursery; located just north-west of Yarram. ***Scenic drive:*** a 46 km circuit goes from Yarram through Hiawatha and takes in Minnie Ha Ha Falls on Albert River, where picnic and camping facilities are provided.

See also PHILLIP ISLAND & GIPPSLAND, p. 150

Yarrawonga

Pop. 5726
Map ref. 559 M12 | 584 H5 | 586 D1

Irvine Pde; (03) 5744 1989 or 1800 062 260; www.yarrawongamulwala.com.au

 101.3 Oak FM, 106.5 FM ABC Local Radio

Yarrawonga and its sister town Mulwala, across the New South Wales border, are separated by a pleasant stretch of the Murray River and the attractive Lake Mulwala. The 6000-hectare lake was created in 1939 during the building of the Yarrawonga Weir, which is central to irrigation in the Murray Valley. Yarrawonga's proximity to such great water features has made it a popular holiday resort. The sandy beaches and calm waters are ideal for watersports, and are also home to abundant wildlife.

Yarrawonga and Mulwala foreshores: great locations for walks and picnics, with shady willows, water slides, barbecue facilities and boat ramps. ***Canning A.R.T.S Gallery:*** work by local artists; Belmore St. ***Tudor House Clock Museum:*** Lynch St. ***Cruises:*** daily cruises along the Murray on paddleboats *Paradise Queen* or *Lady Murray*; depart from Bank St.

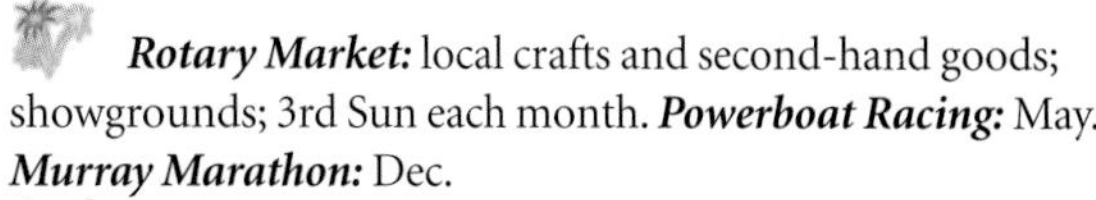

Rotary Market: local crafts and second-hand goods; showgrounds; 3rd Sun each month. ***Powerboat Racing:*** May. ***Murray Marathon:*** Dec.

Lady Murray Cruises: barbecue lunches, 3-course dinners; Yarrawonga foreshore; (03) 5744 2005.

Fyffe Field Wines: tastings and sales; Murray Valley Hwy; 19 km W. ***Fishing:*** Murray River for Murray cod and yellow-belly. ***Guided tours:*** local wineries; book at visitor centre.

See also GOULBURN & MURRAY, p. 157

Yea

Pop. 1051
Map ref. 571 D1 | 573 L1 | 582 E1 | 584 E12

Old Railway Station, Station St; (03) 5797 2663; www.murrindinditourism.com.au

88.9 UGFM Upper Goulburn Community Radio, 97.7 FM ABC Local Radio

This town sits by the Yea River, a tributary of the Goulburn River. Hume and Hovell, the first explorers through the region, discovered this wonderfully fertile area – a discovery that led in part to the settlement of the rest of Victoria. Near the Yea–Tallarook Road there are beautiful gorges and fern gullies, a reminder of what Yea looked like thousands of years ago.

Historic buildings: Beaufort Manor, High St; General Store, now a restaurant, High St. ***Wetlands Walk:*** sightings of abundant birdlife and glider possums; eastern outskirts of town.

Market: local craft and produce; Main St; 1st Sat each month Sept–May. ***Autumn Fest:*** Mar.

Marmalades: hearty country food; 20 High St; (03) 5797 2999.

Cheviot Glen Cottages: 2 country cottages; 175 Limestone Rd, Cheviot Hills; (03) 5797 2617.

Murrindindi Reserve: see the impressive Murrindindi Cascades and a variety of wildlife including wombats, platypus and lyrebirds; 11 km SE. ***Ibis Rookery:*** Kerrisdale; 17 km W. ***Flowerdale Winery:*** Whittlesea–Yea Rd; 23 km SW. ***Grotto:*** a beautiful old church set in the hillside; Caveat; 27 km N. ***Berry King Farm:*** pick your own fruit; Two Hills Rd, Glenburn; 28 km S. ***Wilhelmina Falls:*** spectacular falls and a great spot for walks and picnics; access via Melba Hwy; 32 km S. ***Kinglake National Park:*** 32 km S; *see Yarra Glen*. ***Mineral springs:*** Dropmore, off back road to Euroa; 47 km N. ***Scenic drives:*** many in the region. Best time is Aug–Sept when wattles are in bloom; maps from visitor centre.

See also GOULBURN & MURRAY, p. 157

SOUTH AUSTRALIA IS...

Seeing a performance at the ADELAIDE FESTIVAL OF ARTS / Walking along NORTH TERRACE in Adelaide / Admiring the glittering, panoramic night view from WINDY POINT in Belair / Enjoying excellent wine and gourmet food in the BAROSSA VALLEY / Birdwatching in COORONG NATIONAL PARK / Taking a scenic flight

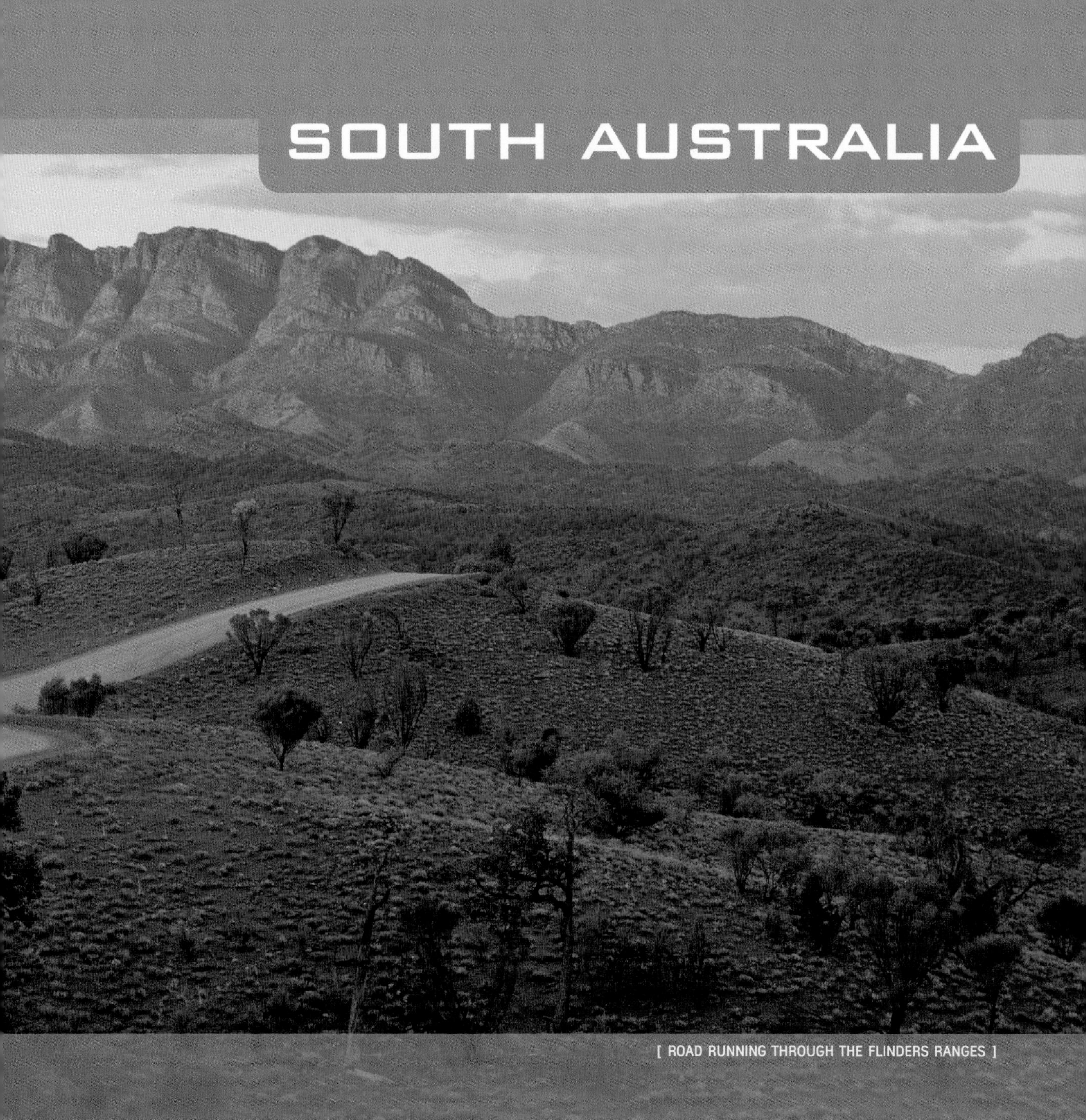

[ROAD RUNNING THROUGH THE FLINDERS RANGES]

over WILPENA POUND in the Flinders Ranges / Exploring NARACOORTE CAVES by guided tour / Visiting historic towns, wineries and galleries in the ADELAIDE HILLS / Sleeping underground and searching for opals in COOBER PEDY / Discovering native fauna on KANGAROO ISLAND / Driving across the NULLARBOR

SOUTH AUSTRALIA is a million square kilometres of ancient Dreamtime landscapes and wild coastal beauty. It is also a land of incredible contrasts: the endless desert of the north and the fertile vales of the south-east are a world apart.

In 1836 Colonel William Light chose the site for the capital of South Australia on Kaurna land beside the River Torrens. The settlement's early days were far from ideal as the first colonists huddled in squalid mud huts, perhaps regretting they had no convict labour to call on. But today the world's first planned city is a gracious capital of wide streets and generous public parks.

In the 1840s German Lutherans fleeing persecution in Europe settled in the Adelaide Hills and the Barossa Valley, bringing traditions of wine-growing and social liberty that have flourished here ever since. In the early 1970s the election of flamboyant rebel Don Dunstan as premier launched a decade of social reform unmatched in any other state. South Australians are proud of their history of social innovation, and their state has a well-earned reputation for tolerance and cultural diversity.

[BUNDA CLIFFS, HEAD OF BIGHT]

For travellers, South Australia is the perfect place to get off the beaten track. The state has the most centralised population in the country and the outback begins just an hour or two up the road from Adelaide.

The Flinders Ranges are one of the oldest mountain ranges on earth. At their centre is the natural amphitheatre Wilpena Pound (Ikara), a lost world of cypress pines and hidden creeks, its gorges created by Akurra the serpent as he travelled north with a grumbling belly full of salt water.

[STURT'S DESERT PEA, SOUTH AUSTRALIA'S FLORAL EMBLEM]

The state's Southern Ocean coastline includes the sheer cliffs of the Great Australian Bight and the sheltered wetlands of the Coorong. This refuge for native and migratory birds begs you to sit quietly with a pair of binoculars.

fact file

Population 1 644 600
Total land area 984 377 square kilometres
People per square kilometre 1.6
Sheep per square kilometre 13.25
Length of coastline 3816 kilometres
Number of islands 346
Longest river Murray River (650 kilometres)
Largest lake Lake Eyre (9500 square kilometres)
Highest mountain Mount Woodroffe (1440 metres), Musgrave Ranges, Pitjantjatjara Land
Lowest place Lake Eyre, 15 metres below sea level (Australia's lowest point)
Hottest place 50.7°C – Australia's hottest recorded temperature – was reached in Oodnadatta in 1960
Driest place Lake Eyre is Australia's driest place with a mere 125 millimetres of rainfall per year
Longest place name Nooldoonooldoona, a waterhole in the Gammon Ranges
Best surf Cactus Beach, near Ceduna
Best discoveries and inventions Penicillin by Howard Florey (he describes the discovery as 'a terrible amount of luck'); the wine cask by Tom Angove
Most dangerous coast There are 80 shipwrecks around Kangaroo Island
Best political stunt Premier Don Dunstan wore pink hot pants into parliament to campaign for gay law reform in the 1970s
Favourite takeaway food Adelaide's pie floater (a meat pie floating in a bowl of pea soup – with or without tomato sauce)
Local beer Coopers
Interesting fact Adelaide boasts a higher ratio of restaurants to residents than any other state

gift ideas

Leather boots (R. M. Williams Outback Heritage Museum, Adelaide) This iconic bushman began making saddlery and leather boots while camping in South Australia's Gammon Ranges. 5 Percy St, Prospect. 596 D4

Strawberry jam (Beerenberg, Hahndorf) Now world famous, but the original signature product can still be bought from the pioneer farm. Mt Barker Rd, Hahndorf. See Hahndorf p. 256, 596 F9

Ligurian honey (Kangaroo Island) The purest organic honey of its type in the world. See Kangaroo Island p. 236, 599 C11

Glass art (JamFactory, Adelaide) Pick up a finely crafted, unique piece of glass or ceramic art from an emerging or established artist. See JamFactory p. 229, 226 B2

Wine (Barossa and Clare valleys) Take home some cases of wine from two of Australia's eminent wine regions, particularly famous for shiraz and riesling. See Barossa Valley p. 238 and Clare p. 252, 597 F4, 603 J6

Jewellery (Zu design, Adelaide Arcade, Adelaide) Distinctive contemporary objects of adornment from Adelaide artists. 226 C2

Shiraz-filled chocolates (David Medlow, McLaren Vale) Delightfully rich shiraz liqueur encased in dark chocolate. McLarens on the Lake, Kangarilla Rd, McLaren Vale. See McLaren Vale p. 263, 597 B12

Premium oysters (Coffin Bay) Pure from the nutrient-rich water currents of the Southern Ocean. See Coffin Bay p. 252, 602 C8

King George whiting (South Australian waters) Sweet, delicate and always fresh from stores around the state.

Opals (Coober Pedy) Although available around the country, purchase these beautiful stones directly from this mining town. See Coober Pedy p. 255, 609 O10

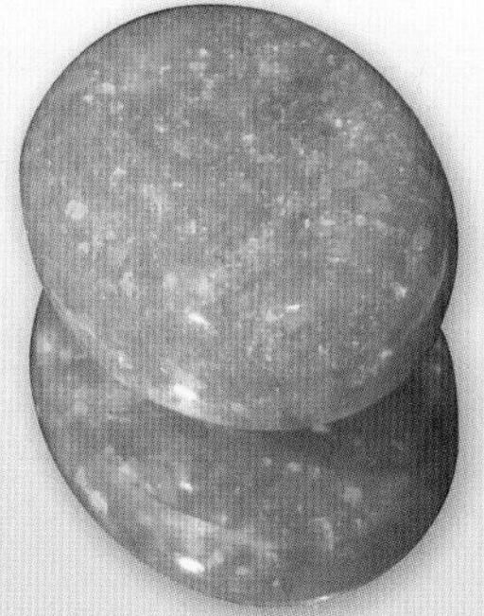

ADELAIDE is...

Coffee at Lucia's in ADELAIDE CENTRAL MARKET / A Popeye cruise along the RIVER TORRENS / A tram trip to GLENELG / Surveying the city from LIGHT'S VISION / A visit to the Australian Aboriginal Cultures Gallery at the SOUTH AUSTRALIAN MUSEUM / Shopping at Adelaide's historic RUNDLE MALL / HENLEY SQUARE on a Sunday afternoon, for lunch and a walk by the sea / A night out in EAST END / A stroll around the JAMFACTORY craft and design gallery

VISITOR INFORMATION
South Australian Visitor and Travel Centre
→ 18 King William St
→ (08) 8303 2220 or 1300 764 227
www.southaustralia.com

One of the best-planned cities in the world, Adelaide remains testament to the work of its first surveyor, Colonel William Light, whose statue stands on Montefiore Hill, overlooking Adelaide Oval.

Settled in 1836, Adelaide was Australia's first free settlement. Like other well-planned cities around the world, Adelaide has few skyscrapers and its architecture blends both heritage and contemporary styles, retaining a 'human scale'.

Since the 1970s Adelaide has been famous for food and wine. The state is the powerhouse of the booming Australian wine industry, producing almost 60 per cent of the total output, while Adelaide Central Market is possibly the finest fresh-produce market in Australia. The city is renowned for its restaurants – from the fish cafes of Gouger Street to the many gourmet eateries dotted around the CBD, and tucked away in quiet corners.

Adelaide also knows how to throw a party, and with a population of 1 200 000 the city is compact enough to generate a feeling of all-over revelry. First on the calendar is the Adelaide Festival of Arts, one of the world's great arts festivals. During February and March every second year, the festival and the now annual Adelaide Fringe take over the city.

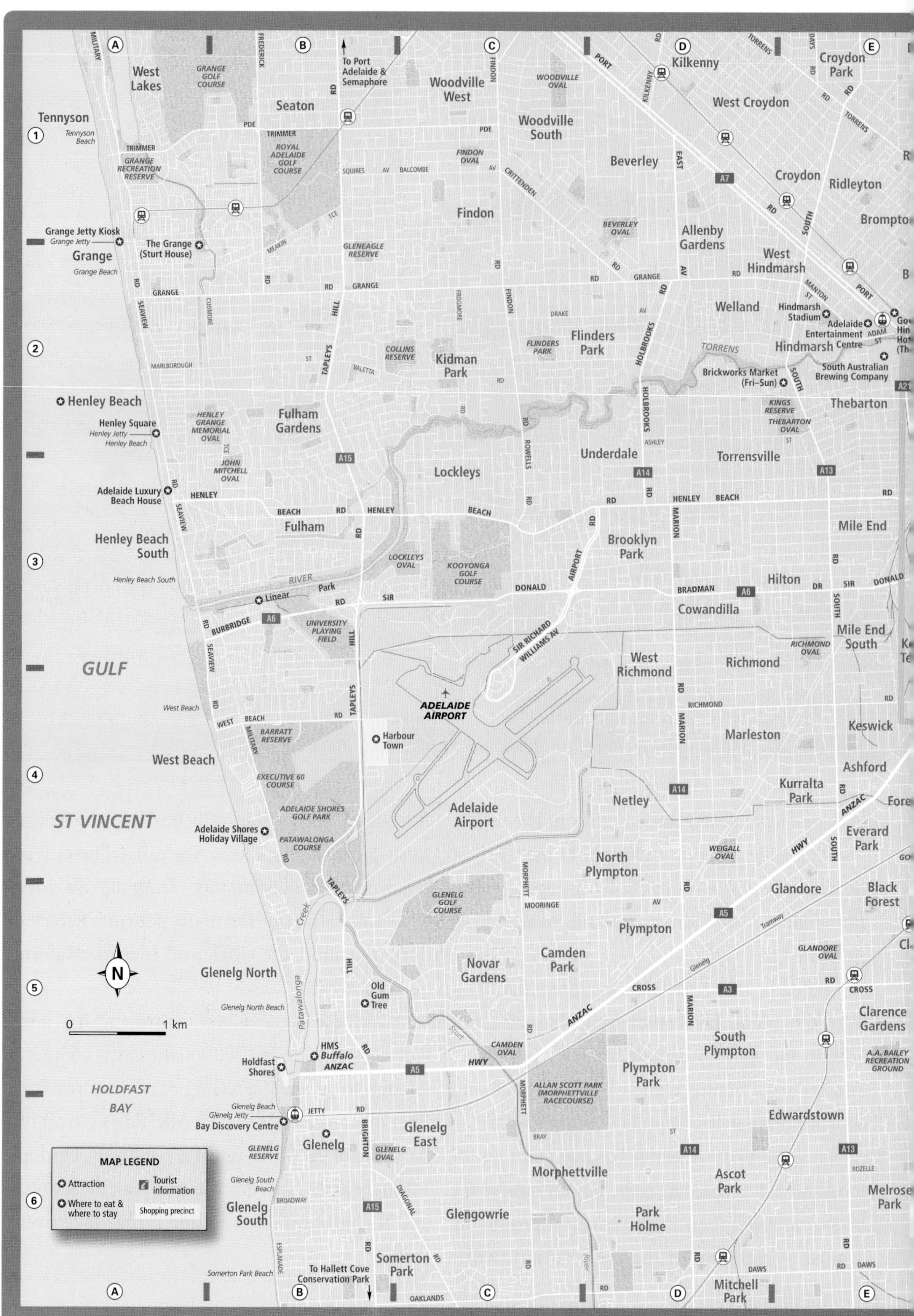
MAP LEGEND
Attraction
Tourist information
Where to eat & where to stay
Shopping precinct
GULF
ST VINCENT
HOLDFAST BAY
0
1 km
N
West Lakes
GRANGE GOLF COURSE
Tennyson
Tennyson Beach
To Port Adelaide & Semaphore
Seaton
ROYAL ADELAIDE GOLF COURSE
GRANGE RECREATION RESERVE
Woodville West
WOODVILLE OVAL
Woodville South
Kilkenny
Croydon Park
West Croydon
Beverley
Croydon
Ridleyton
Brompton
FINDON OVAL
Findon
BEVERLEY OVAL
Allenby Gardens
West Hindmarsh
Grange Jetty Kiosk
Grange Jetty
The Grange (Sturt House)
Grange
Grange Beach
GLENEAGLE RESERVE
Welland
Hindmarsh Stadium
Adelaide Entertainment Centre
Hindmarsh
Flinders Park
FLINDERS PARK
COLLINS RESERVE
Kidman Park
TORRENS
Brickworks Market (Fri–Sun)
South Australian Brewing Company
Henley Beach
Henley Square
Henley Jetty
Henley Beach
HENLEY GRANGE MEMORIAL OVAL
Fulham Gardens
KINGS RESERVE
THEBARTON OVAL
Thebarton
JOHN MITCHELL OVAL
Lockleys
Underdale
Torrensville
Adelaide Luxury Beach House
Fulham
Brooklyn Park
Mile End
Henley Beach South
Henley Beach South
LOCKLEYS OVAL
KOOYONGA GOLF COURSE
RIVER
Linear Park
Hilton
Cowandilla
UNIVERSITY PLAYING FIELD
Mile End South
RICHMOND OVAL
West Richmond
Richmond
ADELAIDE AIRPORT
West Beach
BARRATT RESERVE
Harbour Town
Marleston
Keswick
West Beach
EXECUTIVE 60 COURSE
Ashford
Kurralta Park
Netley
ADELAIDE SHORES GOLF PARK
Adelaide Airport
Adelaide Shores Holiday Village
PATAWALONGA COURSE
Everard Park
North Plympton
WEIGALL OVAL
Glandore
Black Forest
GLENELG GOLF COURSE
Creek
Plympton
Camden Park
Novar Gardens
GLANDORE OVAL
Glenelg North
Glenelg North Beach
Patawalonga
Old Gum Tree
Clarence Gardens
HMS Buffalo
CAMDEN OVAL
South Plympton
Holdfast Shores
Plympton Park
A.A. BAILEY RECREATION GROUND
ALLAN SCOTT PARK (MORPHETTVILLE RACECOURSE)
Glenelg Beach
Glenelg Jetty
Bay Discovery Centre
Glenelg
Glenelg East
Edwardstown
GLENELG RESERVE
GLENELG OVAL
Morphettville
Ascot Park
Glenelg South Beach
Glenelg South
Glengowrie
Park Holme
Melrose Park
Somerton Park Beach
To Hallett Cove Conservation Park
Somerton Park
Mitchell Park
A15
A7
A14
A13
A6
A5
A3

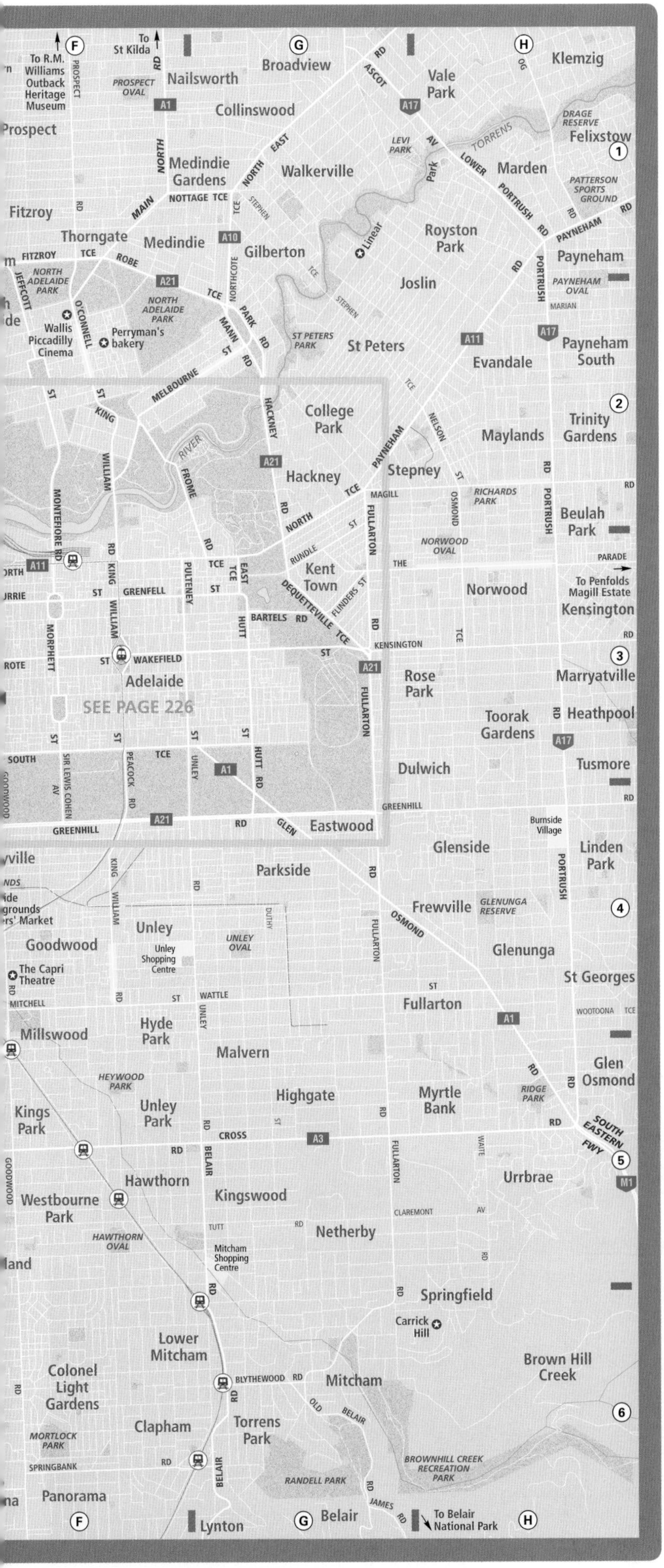

CBD North

The northern section of central Adelaide is based around North Terrace and Rundle Mall. North Terrace, a wide, tree-lined boulevard with a university located at either end, hosts three of South Australia's most important cultural institutions: the state library, the museum and the art gallery. Rundle Mall lies one block south.

Rundle Mall 226 C2

Rundle Mall was Australia's first shopping mall and forms the city's shopping heart, with major department stores, specialty clothing outlets, souvenir and craft stores, food outlets and music shops. Public sculptures dot the mall, including two silver balls colloquially known as the 'Malls Balls'. While you're here, visit **Haigh's Chocolates** on historic Beehive Corner (*you can also visit the Haigh's factory in Parkside, see p. 231*).

Rundle Mall runs parallel to Grenfell Street, and they are joined by two treasures of old Adelaide – Adelaide Arcade and Regent Arcade. Opened in 1885, **Adelaide Arcade** was the first retail establishment in the country to have electric lights. It is a thoroughly Victorian affair, lined with small shops of every kind on the ground floor and charming balcony level.

East End 226 C2

The east end of Rundle Mall, beyond the main shopping zone, is the hub of Adelaide's nightlife. Friday and Saturday nights in the East End throb with longstanding pubs – such as the Austral and the Exeter – among newer bars and clubs, all of them bursting with revellers. With over 50 cafes and restaurants, this is also a superb place to come for food, and the shopping is great too. **Mary Martin Bookshop**, 'purveyors of fine literature since 1945', is Adelaide's oldest bookseller. On Sundays Rundle Street is transformed into an exciting craft and fashion street market.

climate

Adelaide's weather is described as temperate Mediterranean, with temperatures in summer rising to around 40°C on a number of days, and falling to a minimum near zero a couple of times in winter. The in-between seasons are near-perfect, with maximum temperatures in the mid-20s for much of March, April, September and October.

J	F	M	A	M	J	J	A	S	O	N	D	
28	28	26	22	18	15	14	16	18	21	24	26	MAX °C
16	16	15	12	10	8	7	8	9	11	13	15	MIN °C
20	20	24	44	68	71	66	61	51	44	30	26	RAIN MM
2	2	3	6	10	11	12	11	9	7	5	4	RAIN DAYS

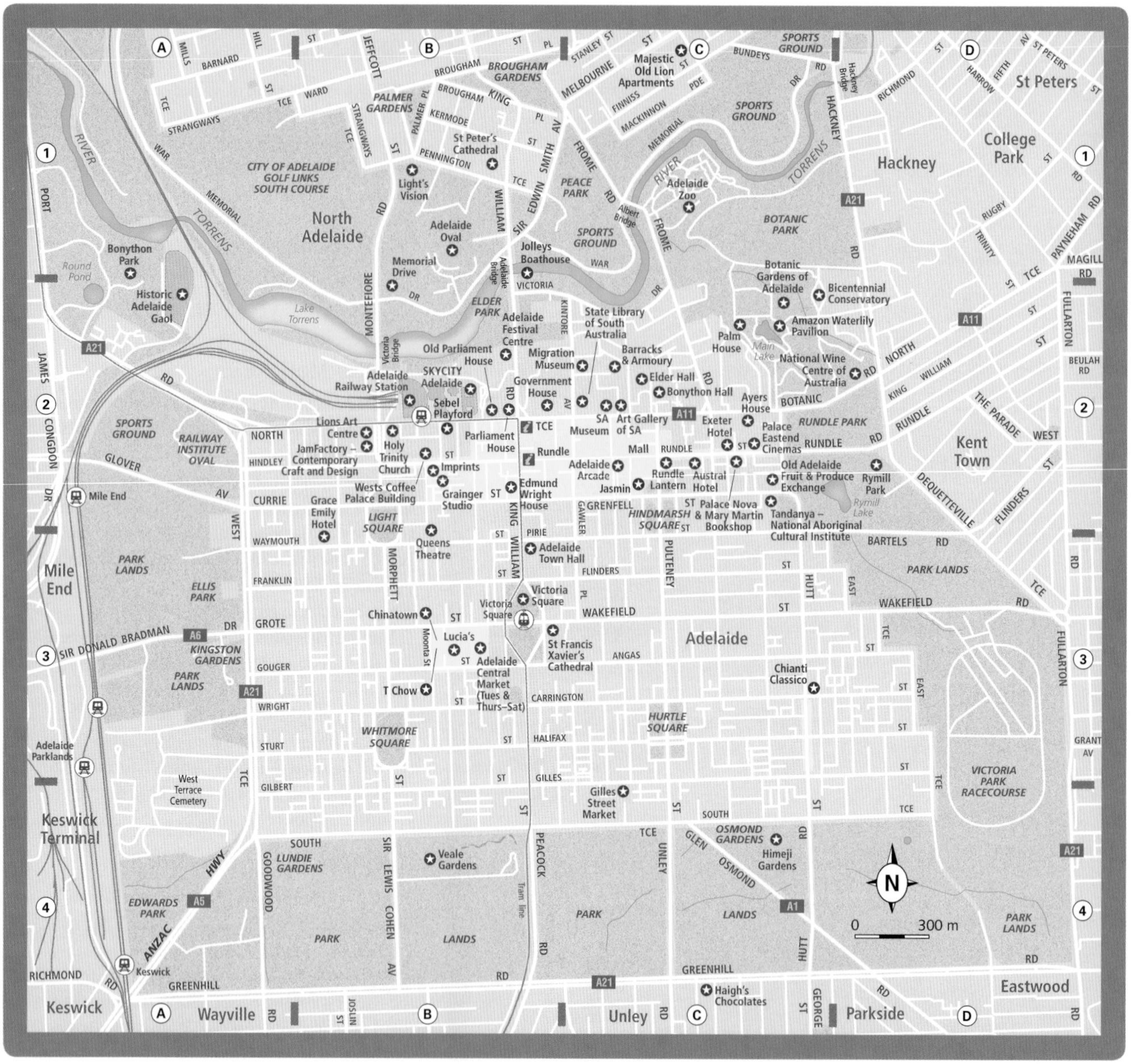

North Adelaide
CITY OF ADELAIDE GOLF LINKS SOUTH COURSE
Light's Vision
St Peter's Cathedral
Adelaide Oval
Memorial Drive
Bonython Park
Historic Adelaide Gaol
Round Pond
Lake Torrens
Jolleys Boathouse
Adelaide Zoo
BOTANIC PARK
Botanic Gardens of Adelaide
Bicentennial Conservatory
Palm House
Amazon Waterlily Pavilion
National Wine Centre of Australia
Majestic Old Lion Apartments
PALMER GARDENS
BROUGHAM GARDENS
PEACE PARK
SPORTS GROUND
ELDER PARK
Adelaide Festival Centre
State Library of South Australia
Old Parliament House
Migration Museum
Barracks & Armoury
SKYCITY Adelaide
Adelaide Railway Station
Government House
Elder Hall
Bonython Hall
Sebel Playford
SA Museum
Art Gallery of SA
Ayers House
Exeter Hotel
Palace Eastend Cinemas
RUNDLE PARK
Lions Art Centre
JamFactory – Contemporary Craft and Design
Holy Trinity Church
Parliament House
Rundle Mall
Adelaide Arcade
Rundle Lantern
Austral Hotel
Jasmin
Old Adelaide Fruit & Produce Exchange
Rymill Park
Imprints
Wests Coffee Palace Building
Grainger Studio
Edmund Wright House
Grace Emily Hotel
LIGHT SQUARE
Queens Theatre
HINDMARSH SQUARE
Palace Nova
Mary Martin Bookshop
Tandanya – National Aboriginal Cultural Institute
Adelaide Town Hall
Victoria Square
Chinatown
Lucia's
Adelaide Central Market (Tues & Thurs–Sat)
St Francis Xavier's Cathedral
T Chow
Chianti Classico
Adelaide
HURTLE SQUARE
WHITMORE SQUARE
Gilles Street Market
OSMOND GARDENS
Himeji Gardens
Veale Gardens
LUNDIE GARDENS
KINGSTON GARDENS
ELLIS PARK
RAILWAY INSTITUTE OVAL
PARK LANDS
EDWARDS PARK
West Terrace Cemetery
VICTORIA PARK RACECOURSE
Haigh's Chocolates
Mile End
Keswick Terminal
Keswick
Adelaide Parklands
Wayville
Unley
Parkside
Eastwood
Kent Town
Hackney
College Park
St Peters
0 300 m

[SOUTH AUSTRALIAN MUSEUM]

getting around

Adelaide has a wide range of public transport. First there is the city's tram service, which runs from Glenelg, all the way to the city, through Victoria Square and City West, to the Adelaide Entertainment Centre on Port Road, Hindmarsh. Then there is the Adelaide O-Bahn – the longest and fastest guided bus service in the world. It travels along Currie and Grenfell streets in the city, then heads out to Westfield Tea Tree Plaza in Modbury.

The City Loop Bus and the Terrace to Terrace Tram are two free services that operate around the city centre. The City Loop bus runs every 15 minutes and takes in North Terrace and Light, Hindmarsh and Victoria squares; the Terrace to Terrace Tram runs every ten minutes and includes Victoria Square, King William Street and the railway station. Board either the Glenelg or South Terrace tram anywhere between South Terrace and North Terrace during shopping hours.

The Adelaide Metroticket JetBus is a daily service linking the airport to the city, Glenelg, West Beach and the north-eastern suburbs.

Four train routes operate from the CBD to Adelaide's suburbs: to Gawler in the north, Outer Harbour in the north-west, Noarlunga in the south and Belair in the Adelaide Hills. There are also plenty of bus services operating around the suburbs. All public transport in Adelaide is covered by one ticketing system, and tickets can be purchased at train stations and on buses and trams, as well as newsagents and convenience stores displaying the Metroticket signage, and the Adelaide Metro InfoCentre, corner King William and Currie streets in the City centre.

Public transport Passenger Transport InfoLine 1300 311 108 or 1800 182 160.

Airport shuttle bus Skylink Airport Shuttle (08) 8413 6196.

Motoring organisation RAA (08) 8202 4600, roadside assistance 13 1111.

Car rental Avis 13 6333; Budget 13 2727; Europcar 13 1390; Hertz 13 3039; Thrifty 1300 367 227.

Taxis Independent Taxis 13 2211; Suburban Taxis 13 1008; Yellow Cabs 13 2227.

Tourist bus Adelaide Explorer (replica tram) (08) 8293 2966.

Bicycle hire Contact Bicycle SA for operators (08) 8168 9999.

The **Rundle Lantern**, on the corner of Rundle and Pulteney streets, uses computer-controlled LEDs to deliver a platform for digital art. It operates each night from dusk until midnight, with extended hours for special events.

Hindley Street 226 B2

At the other end of Rundle Mall, Hindley Street, sometimes known as the West End, is another lively part of Adelaide. The ornate Wests Coffee Palace building, constructed in 1903 as the Austral Stores and serving as a coffee palace from 1908 onwards, is now the home of **Arts SA**.

Parliament House & Government House 226 B2

These impressive buildings adorn the northern intersection of North Terrace and King William Road. Government House, the oldest in Australia, is set on a sweep of manicured lawns and was the meeting place of the first council of government. As the council expanded, new buildings were constructed, including what is now **Old Parliament House.** These buildings, along with several statues and monuments – including the **War Memorial** at the corner of Kintore Avenue – make for an interesting and informative walk. Government House normally has two open days a year; phone (08) 8203 9800 for details. Guided tours of Parliament House run on non-sitting days (Monday and Friday) at 10am and 2pm; sitting days are published on the website (www.parliament.sa.gov.au).

State Library of South Australia 226 C2

The State Library complex is a wonderful blend of charming 19th-century buildings and modern technology, bringing the history of South Australian culture to the people, as well as showcasing the library's collection. It also includes the Bradman Digital Library and Trail, based on the life of cricketer Sir Donald Bradman, who lived in Adelaide for much of his life. The archives are the ultimate research source covering his career. *North Tce; open daily; admission free.*

Migration Museum 226 C2

This museum details immigrant life from pioneering days up to today. The museum building was once Adelaide's Destitute Asylum, where many of the city's aged, homeless and underprivileged lived (and died). The stories of the women and children who lived there from the 1850s to 1918 are told in the 'Behind the Wall' exhibition. *82 Kintore Ave; open 10am–5pm Mon–Fri, 1–5pm Sat–Sun and public holidays; admission free.*

top events

Schützenfest Traditional German folk festival with music, food and frivolity in Bonython Park. January.

Jacobs Creek Tour Down Under Bike riders descend on Adelaide for this race into the hills. January.

Carnevale Adelaide Italian culture, food, song and dance in Rymill Park. February.

Adelaide Festival of Arts The city's defining event, and one of the world's highest regarded arts festivals. February–March (even-numbered years).

Adelaide Fringe Alongside the Festival of Arts, the edgy performances of this world-renowned annual festival are great entertainment. February–March.

WOMADelaide A huge festival of world music and dance in Botanic Park. March.

Clipsal 500 Adelaide V8 supercars race on city streets in what is said to be Australia's best motor event. March.

Glendi Festival Greek culture, food, song and dance in Adelaide Showgrounds, October.

Tasting Australia International event celebrating food, wine and beer as well as chefs and writers. October (odd-numbered years).

Feast Festival One of the country's top gay and lesbian events, with theatre, film, dance and more. November.

Credit Union Christmas Pageant An Adelaide institution since 1933 welcomes Father Christmas to the city streets. November.

shopping

Rundle Mall The CBD's main shopping area, with major department stores as well as individual offerings of clothing, chocolates and much more. 226 C2

Melbourne Street, North Adelaide Adelaide's most exclusive shopping strip, with designer fashion boutiques. 225 G2

The Parade, Norwood Very cosmopolitan, with an array of stores and plenty of places to stop for a coffee, a meal or a drink. 225 H3

King William Road, Hyde Park Hip fashion outlets sprinkled among cafes, and furniture and homewares stores. 225 F4

Jetty Road, Glenelg A great mix of stores, and food and drink outlets. 224 B6

Harbour Town This complex offers seconds and discount stores. 224 B4

Magill Road, Stepney For antiques and second-hand treasures. 225 H2

See also Adelaide Central Market, p. 230, and Gouger Street, p. 230

markets

Fisherman's Wharf Market, Port Adelaide Fresh seafood and other produce are the main drawcards. Sun. 597 B8

Torrens Island Open Air Market, Port Adelaide Buy fish direct from the boat and quality local produce. Sun. 597 B7

Brickworks Market, Torrensville Located in an old kiln, the market's 100 stalls offer a wide range of wares, including clothing and crafts, as well as produce. Fri–Sun. 224 E2

Junction Markets, Kilburn Popular market offering fresh produce and specialty stores. Thurs–Sun. 596 C6

Adelaide Showgrounds Farmers' Market, Goodwood Fresh produce and gourmet goods. 9am–1pm Sun. 225 F4

See also Adelaide Central Market, p. 230

Gillies Street Market. 3rd Sunday each month. Retro, new designers, vintage, food , DJs , 91 Gillies Street, city.

[ADELAIDE FESTIVAL CENTRE]

South Australian Museum 226 C2

For over a century, the South Australian Museum has played a crucial role in researching, documenting and exhibiting every facet of Aboriginal culture. The excellent **Australian Aboriginal Cultures Gallery** contains over 3000 objects and is the world's most comprehensive Aboriginal cultural exhibition.

In the **Origin Energy Fossil Gallery** is a fascinating collection of opalised fossils – look up to see the impressive model skeleton of the *Addyman plesiosaur*, a marine animal over 100 million years old.

Australia's richest natural history art exhibition, the Waterhouse Natural History Art Prize is held each year in the **ETSA Utilities Gallery** at the museum from July to September. *North Tce; open daily.*

Art Gallery of South Australia 226 C2

This gallery has grouped its permanent collection in three major categories. The Australian collection includes some fine works from distinguished South Australian artists Margaret Preston and Stella Bowen. An impressive display of contemporary British artwork is housed in the European collection, as is an array of major works by French sculptor Auguste Rodin.

The Asian collection includes delicate Japanese artworks and South-East Asian ceramics. There is also a program of changing exhibitions. *North Tce; open daily; general admission free.*

Ayers House 226 C2

North Terrace was the home of wealthy and prestigious figures such as Premier Henry Ayers, whose name identified the country's most recognisable landmark (Ayers Rock) until it was changed back to its Indigenous name, Uluṟu. After years in the hands of the Royal Adelaide Hospital, the original decorations of the building have now been painstakingly restored, and one wing has been transformed into a luxurious restaurant. *288 North Tce; open 10am–4pm Tues–Fri, 1–4pm Sat–Sun.*

Tandanya – National Aboriginal Cultural Institute 226 C2

This is a vibrant meeting place of cultures with a strong emphasis on the culture of the Kaurna people, the traditional owners of the land on which Adelaide stands. In the gallery, see the exciting work of emerging artists, or taste modern versions of traditional bush tucker in the cafe. Indigenous tours are also offered, encompassing both the centre and the grounds. *253 Grenfell St; open daily.*

Botanic Gardens of Adelaide 226 C2

The gardens were laid out in the mid-1800s and retain a northern-European style. The **Palm House**, imported in 1875, is a fine example of German engineering. The impressive **Bicentennial Conservatory** is considered to be the largest single-span conservatory in the Southern Hemisphere. Inside, walkways guide visitors through a lush rainforest environment, past endangered native and exotic plant species. **Amazon Waterlily Pavilion** is the most recent addition, opened in 2007. The fully licensed restaurant overlooking the Main Lake serves some of South Australia's finest wines and innovative cuisine. *North Tce; open daily, guided walks 10.30am.*

National Wine Centre of Australia 226 D2

Here visitors can take a Wine Discovery Journey through the different stages of wine-production, meet winemakers and try winemaking – all through virtual technology. The centre has its own vineyard and a retail centre for wine-tasting. *Cnr Botanic and Hackney rds; open daily; admission free.*

Adelaide Zoo 226 C1

The preservation of the zoo's 19th-century buildings and landscaped gardens has led many to call it the most attractive zoo in Australia. The **Elephant House** is a highlight of 1900s architecture – the enclosure's design was based on an Indian temple, and is now classified by the National Trust. The zoo's newest residents – giant pandas Wang Wang and Funi – are also very popular. *Frome Rd; open daily.*

The River Torrens 226 B1

The river provides a scenic setting for a host of leisure activities including enjoying a ride on Adelaide's **Popeye** cruises and motor launches. From the Elder Park Wharf, Popeye cruises operate every hour, on the hour, on weekdays, and every 20 minutes on weekends. Paddleboats can be hired on weekends from **Jolleys Boathouse** under the Adelaide Bridge on King William Road. You can also take a romantic Venetian gondola ride from Red Ochre restaurant day or night. Alternatively, enjoy good food and lovely views at the boathouse restaurant.

Adelaide Festival Centre 226 B2

In the 1970s, South Australia had a premier who was determined to turn the arts into a viable industry rather than just a pastime. Don Dunstan's accomplishments include developing the South Australian Film Corporation and increasing support for the state's other major arts groups. Overlooking Elder Park and the River Torrens, his Festival Centre consists of four theatres and is an architectural landmark. Visitors can take in a show or wander through the foyer to view the centre's collection of artworks as well as the changing displays of the **Performing Arts Collection of South Australia**, with its costumes, props and photographs. Visit the website (www.adelaidefestivalcentre.com.au) to find out what's on.

JamFactory – Contemporary Craft and Design 226 B2

Each year a rigorous application process sifts through hopefuls keen on becoming a Design Associate of the JamFactory. The pieces produced here are said to be some of the most innovative

where to stay

Adelaide Luxury Beach House A stunning beachfront residence with designer interiors, gulf views and outdoor living options. Esplanade, Henley Beach; 0418 675 339. 224 A3

Adelaide Shores Holiday Village Self-contained seafront bungalows offering relaxed family accommodation just metres from the beach. Military Rd, West Beach; (08) 8355 7360. 224 B4

Bishops Garden Hotel A lavish three-bedroom apartment in a historic villa with open living spaces and private garden. Molesworth St, North Adelaide; (08) 8272 1355. 225 F2

Majestic Old Lion Apartments Stylish courtyard units with self-contained comforts and close to cafes and parkland attractions. 9 Jerningham St, North Adelaide; (08) 8334 7799. 226 C1

Sebel Playford Adelaide A welcoming, boutique-style luxury hotel, acclaimed for its exceptional service and Art Nouveau flair. 120 North Tce; (08) 8213 8888. 226 B2

where to eat

Chianti Classico A welcoming family-run trattoria serving Italian regional fare with dash and rustic charm. 160 Hutt St; (08) 8232 7955; open daily for breakfast, lunch and dinner. 226 C3

Jasmin Perennial favourite for authentic Indian dishes brought to life by quality ingredients and spices. 31 Hindmarsh Sq; (08) 8223 7837; open Thurs–Fri for lunch and Tues–Sat for dinner. 226 C2

Jolleys Boathouse Smart, bistro-style dining featuring contemporary flavours in a relaxing riverside setting. Jolleys La; (08) 8223 2891; open Sun–Fri for lunch and Mon–Sat for dinner. 226 B1

Magill Estate Glorious city and vineyard views complement the artful marriage of fine cuisine with Penfolds wines. 78 Penfold Rd, Magill; (08) 8301 5551; open Fri for lunch and Tues–Sat for dinner. 597 C9

T Chow A diverse, crowd-pleasing menu of southern Chinese fare, including famed duck dishes. 68 Moonta St; (08) 8410 1413; open daily for lunch and dinner. 226 B3

entertainment

Cinema Those who want an atmospheric cinema-going experience should head for the Wallis Piccadilly Cinema in North Adelaide, which screens mainstream and arthouse films. A similar line-up is available at the Palace Nova cinemas in Rundle Street's East End. Adelaide's main alternative cinema is the Mercury at the Lion Arts Centre – it shows cult classics, foreign language and arthouse movies, and holds short film festivals. The Capri Theatre, in Goodwood, is notable for the live organ recitals that introduce film screenings on Tuesday, Saturday and Sunday nights. Check the *Adelaide Advertiser* for daily listings.

Live music The best live-music venue in Adelaide, without question, is the Governor Hindmarsh Hotel on Port Road, opposite the Adelaide Entertainment Centre. Offering everything from Irish fiddlers in the front bar to major international acts in the large, barn-like concert room at the back, The Gov (as it's known to the locals) has been the top venue in town for at least a decade. Plenty of other pubs also offer live music, like the Grace Emily on Waymouth Street, an intimate venue that also boasts an excellent beer garden. The best way to find out what's on is to pick up a copy of *Rip It Up*, the local contemporary-music paper.

Classical music and performing arts The Adelaide Festival Centre is the hub of mainstream performing arts activity in Adelaide, hosting the three major and highly regarded state companies: State Theatre SA, State Opera of South Australia and Adelaide Symphony Orchestra (ASO). The ASO also holds small recitals at its Grainger Studio in Hindley Street. Adelaide nurtures a diverse range of theatre companies – look out for performances by the Australian Dance Theatre (contemporary dance), Vitalstatistix (women's theatre) and Windmill Performing Arts (children's theatre). The Lion Arts Centre is the venue for two innovative companies: Nexus (cabaret) and Doppio Teatro (multicultural). Indigenous dance is also on display in daily performances at Tandanya. The *Adelaide Advertiser* publishes a listing of entertainment events each Thursday.

grand old buildings

Barracks and Armoury Built in 1855, these magnificent examples of colonial architecture were the local Australian Army headquarters from 1857 to 1870. Behind South Australian Museum.

Adelaide Railway Station Built in 1856, the main feature is the Great Hall, with marble floors, Corinthian columns and a domed ceiling. SKYCITY Adelaide casino now occupies much of the station's upper floors. North Tce.

Edmund Wright House Wright was the architect of some of Adelaide's grandest buildings, including Parliament House, the GPO and the Adelaide and Glenelg town halls. This elaborate Italianate creation was completed in 1878 and used as a banking chamber. 59 King William St.

Adelaide Town Hall Opened in 1866, this building is much admired for its magnificent tower and classic portico, and its equally grand interior. Free tours run on Monday at 10am; bookings essential (08) 8203 7590. Regular classical music concerts and recitals. 128 King William St.

Bonython Hall Part of the University of Adelaide, the hall has an unusual sloping floor. This was insisted upon by benefactor Langdon Bonython, a strict Methodist, to prevent dancing. North Tce.

Elder Hall This church-like building in the grounds of the University of Adelaide is in fact a concert venue with a spectacular pipe organ. Next to Bonython Hall.

Churches As well as St Francis Xavier's Cathedral near Victoria Square, there's St Peter's Cathedral in North Adelaide and the Holy Trinity Church on North Terrace.

Old Adelaide Fruit and Produce Exchange The charming facade of this old wholesale market still stands on East Terrace. The rear has been converted into apartments.

Queens Theatre For a theatre, this is a surprisingly humble affair – perhaps because it is the oldest theatre on the mainland, built in 1840. Cnr Gilles Arcade and Playhouse La.

The Grange (Sturt House) The home of Charles Sturt, the man who sailed into South Australia on a whaleboat down the Murray, which led to the settlement of the state. Many original furnishings remain. Jetty St, Grange; open Fri–Sun and public holidays.

See also Ayers House, p. 228, Migration Museum, p. 227, Parliament House & Government House, p. 227, and Carrick Hill, p. 233

and impressive craft designs on the Australian market. Changing exhibits of international and JamFactory works are showcased in the gallery, and many one-off designs are available for purchase in the store. Nearby is another Adelaide arts institution, the **Australia Experimental Art Foundation** at the Lion Arts Centre. This centre has some of the best in new media, such as video and installation art. *JamFactory: 19 Morphett St; open daily; admission free. There is also a JamFactory outlet in Rundle Plaza, open Mon–Fri.*

Historic Adelaide Gaol 226 A2

Today you can tour the cells and yards of this complex – which operated as a gaol from 1841 to 1988 – and imagine the days when Sister Mary MacKillop visited the prisoners, or take a night tour (bookings required) or a Ghost Tour. *Gaol Rd; tour bookings (08) 8231 4062; open 11am–4pm Mon–Fri and Sun.*

CBD South

The southern part of central Adelaide is home to some stellar attractions – the Adelaide Central Market is an absolute must-see, and there is superb dining on Gouger Street.

Adelaide Central Market 226 B3

Located just off Victoria Square, the market fills the block between Grote and Gouger streets. A showcase for South Australia's fresh produce, the market has evolved into a significant shopping and social centre. It's a wonderland of fruit, vegetables, fish, breads, cheeses, meats and nuts, with rows of stalls surrounded by alleys and arcades crammed with shops and cafes. Grab lunch at Zuma's, Malacca Corner or Lucia's, or just wander the aisles of the market and experience its sensory delights. *Open Tues and Thurs–Sat.*

Gouger Street 226 B3

Originally a spin-off from Central Market, Gouger Street has grown into an attraction in its own right with its restaurants, cafes, bars and clubs. Gouger Street winds down earlier than East End, but has its own character and appeal.

There are several Chinese restaurants on Gouger Street, but Adelaide's **Chinatown** is centred on Moonta Street between Gouger and Grote streets. Among the noodle bars and Chinese shops look for T Chow, an Adelaide institution.

Inner North

North Adelaide is one of the city's most exclusive residential addresses. It is also rich in history, with a mix of stately mansions and humble workers' cottages. From Light's Vision you can take a walk through Lower North Adelaide to get the full picture. Take Palmer Place (the section that runs parallel to Jeffcott Street) into Brougham Place to see old mansions with commanding views across the city.

Brougham Place continues into Stanley Street. Down the hill you'll find small stone cottages that formed part of South Australia's first subdivision. Melbourne Street is the heart of this area and is Adelaide's classiest street, with art galleries, boutiques, restaurants and the restored Lion Hotel.

To find the true cafe heart of North Adelaide, head to O'Connell Street – it has old pubs bedecked with Victorian iron lace, as well as restaurants and cafes. Tynte Street, which runs off O'Connell Street and was once North Adelaide's main street, has many fine 19th-century buildings.

R. M. Williams Outback Heritage Museum 597 A9

Reginald Murray Williams was a classic example of the 1930s battler. He made his first pair of boots in outback South Australia and later started up a 'factory' in his father's shed. Out of this grew an Australian legend. This museum, on the site of that shed, charts the story of his life in fascinating photographs, objects and displays. *5 Percy St, Prospect; open Mon–Fri; admission free.*

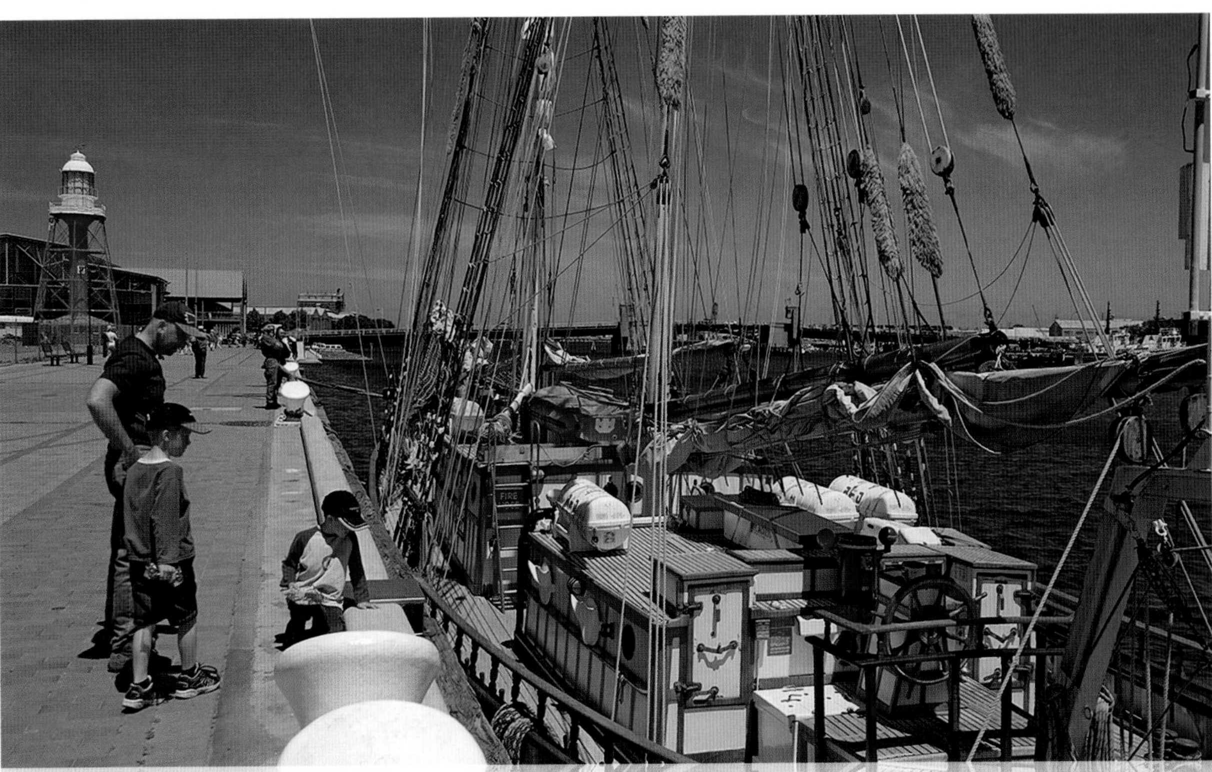

[SHIP DOCKED IN PORT ADELAIDE]

Inner South

Haigh's Chocolates 226 C4

Australia's oldest chocolate-maker is located just south of Adelaide's CBD. The visitor centre includes a viewing area, displays of original factory machinery and a delightful old-world shop with a tempting array of chocolates. You'll be sure to find your favourites after taking a tour (bookings essential). *154 Greenhill Rd, Parkside; tour bookings (08) 8372 7077; free tours 11am, 1pm and 2pm Mon–Sat.*

Inner West

South Australian Brewing Company 224 E2

Not to be outdone in Australia's premier wine state, this company, which makes a range of brands including Tooheys, XXXX, Hahn, Lion Red, Steinlager and Speight's, offers tours of its premises. Learn all about making beer by taking a tour (bookings essential), from the raw materials through to brewing, canning and bottling. *107 Port Rd, Thebarton; tour bookings (08) 8354 8888; tours 10.30am Mon–Thurs.*

Southern Bayside

Glenelg 224 B6

One of Adelaide's great features is its long frontage to Gulf St Vincent. Glenelg is South Australia's best-known beach and its most developed piece of coastal real estate. It spills with shops, restaurants, high-rise apartments and people strolling, sunbaking, cycling and rollerblading. It's around 20 minutes from the city centre, on the sweep of Holdfast Bay.

Since the late 1800s Adelaidians have flocked to Glenelg to partake of the sea air – the Glenelg tram, which was once the main mode of transport, still runs to the suburb from Victoria Square in the city. Other relics of the era have survived too, most notably the grandiose 1875 town hall. Housed within it is the **Bay Discovery Centre**, where an exhibition details the rich history of the area.

walks & tours

Adelaide Oval Tours Get an up-close view of this Adelaide icon, its magical old scoreboard and its brand new modern grandstand. Tours leave from the southern gate each weekday at 10am (no bookings required). (08) 8300 3800.

Market Adventures Follow the experts through Adelaide's famous Central Market. The same company also runs a 'Grazing on Gouger' tour, giving people the chance to eat five courses at five of Gouger Street's best restaurants. Bookings on (08) 8336 8333.

Wineries at your doorstep Visit McLaren Vale, one of Australia's greatest wine regions, just 40 minutes from Adelaide. Enjoy the local tasty foods too. Check out www.mclarenvale.info

City of Adelaide Historical Walking Trails Pick up a brochure from the visitor centre and head out on themed walks covering everything from the grand buildings of North Adelaide to the places of interest around Rundle Mall and Adelaide's historic cinemas and theatres.

Port Walks Take a walk through historic Port Adelaide. Contact the Port Adelaide Visitor Centre for details. 66 Commercial Rd; (08) 8405 6560.

Yurrebilla Trail Take in the magnificent bushland on Adelaide's doorstep on this 52-kilometre trail that links Black Hill and Morialta conservation parks in the north with Belair National Park in the south. The scenery includes unsurpassed views of the city. Walkers can try smaller sections of the hike. Contact the visitor centre for details or visit the website (www.environment.sa.gov.au/parks/sanpr/yurrebilla).

parks and gardens

Rymill Park The centrepiece is Rymill Lake, with rowboats for hire, but there is also a large rose garden. City. 226 D2

Himeji Gardens Created in conjunction with Himeji, Adelaide's sister city, this garden blends classic Japanese styles. City. 226 C4

Veale Gardens For some years the green belt around Adelaide was more paddock than parkland, but a town clerk named William Veale started beautifying the gardens in the mid-1900s. Veale Gardens were his pièce de résistance. City. 226 B4

Bonython Park A park with native style. The river attracts birdlife, and there is also a children's playground and a popular picnic area. Thebarton. 226 A1

Linear Park This park follows the curves of the Torrens all the way from West Beach to Paradise, and is great for bike riding. The sections near the city offer fantastic views, while further out the track runs through a thick corridor of bushland. 224 B3, 225 G1

sport

Adelaide has been passionate about **cricket** from its earliest beginnings. Until his death in 2001, Adelaide was the proud home of Australia's greatest cricketer, Sir Donald Bradman. The city is also home to the Adelaide Oval, regarded as one of the most beautiful sporting arenas in the world. Adelaide's cricket test, usually in December, is a great event. Drop in and visit the famed Bradman Collection, featuring many items donated by the legendary player. *Open 9.30am–4.30pm Mon–Fri.*

With two local **AFL** (Australian Football League) teams, Adelaide and Port Adelaide, football is the city's other sporting passion. Showdown is the twice-yearly match between the two teams, and the whole city stops to watch what is considered Adelaide's own grand final. AAMI Stadium at West Lakes is the city's home of AFL.

In March there is a **motor race** that literally takes over the city. Sections of the Clipsal 500 Adelaide track, centring on Victoria Park Racecourse, are run on Wakefield Road and East Terrace in the south-east of the city. This V8 event was created after Adelaide lost the Australian Formula One Grand Prix to Melbourne and, interestingly, is more popular than Formula One ever was.

Also on the calendar is the AAPT Championships, the international **tennis** tournament held at Memorial Drive near the Adelaide Oval in early January. Adelaide's United **soccer** team matches are held at Hindmarsh Stadium, and the NBL 36ers play **basketball** on their home court at Beverley.

Each Year, on a public holiday Monday early in March, Adelaide comes alive with **horseracing** fever for the SKYCITY Adelaide Cup at Allan Scott Park, Morphettville. Each January, Adelaide comes alive when the world's best **pro cyclists** take to the streets for a week of racing in the Tour Down Under.

Staged annually on the last Sunday in September, the Bay to Birdwood **vintage car rally** takes place. More than 1500 vehicles drive the 70-kilometre route from Adelaide Shores, West Lakes to the National Motor Museum Birdwood, watched by over 100 000 spectators.

History is re-created along the foreshore where a replica of the HMS *Buffalo* – the ship that carried the first South Australian settlers – now stands. Unlike the original, it has an onboard restaurant offering fine seafood dining, along with a small museum. Nearby is **Holdfast Shores**, a recent development with restaurants, boutique shopping and a massive marina.

Walking trails lead up and down the coast (for information call into the Bay Discovery Centre), and the calm water is perfect for swimming, fishing and sailing. And there is, of course, the long, iconic South Australian pier – this is the spot to dine on fish and chips at sunset. Experience swimming with dolphins from the *Temptation* catamaran (0412 811 838; www.dolphinboat.com.au), or simply enjoy a relaxing cruise.

Hallett Cove Conservation Park 597 A10

The southern suburbs of Adelaide begin to blur into the Fleurieu Peninsula just south of Brighton. Here you'll find Hallett Cove Conservation Park, which protects an incredible record of glaciation. A walking trail takes visitors on a guided geological tour to see some spectacular ochre- and sand-coloured rock formations. For the energetic, take the Marino Rocks to Hallett Cove boardwalk and reward yourself with refreshments at the Marino Rocks Café.

Western Bayside

Henley Beach to Semaphore 224 A2

At Henley Beach another long wooden jetty extends into Gulf St Vincent, flanked on either side by a long strip of sand and lapping waters. Set just back from the sea is **Henley Square**, a drawcard especially on weekends with restaurants, bars and cafes often humming with live music.

Straight up Seaview Road is the suburb of **Grange**, which has its own combination of beach and jetty. Relax and enjoy the view over coffee or a cafe meal at the Grange Jetty Kiosk.

Further along is **Semaphore**, a suburb with fine heritage buildings, many cafes and uninterrupted sea views. The Art Deco **Semaphore Palais** offers dining right on the waterfront, while the Semaphore Waterslide Complex has activities for the kids. A walking and bike-riding track heads north to Largs Bay and North Haven Marina, and visitors can also take a 40-minute ride aboard a tourist train to Fort Glanville, built in 1878 in response to fears of a Russian invasion. *Train operates 11am–4pm Sat–Sun (Oct–April), school and public holidays.*

North-west

Port Adelaide 597 B8

Without being overly commercialised, 'the Port', established in 1840, retains the charm of its maritime history and serves as an interesting destination in its own right, with weekend wharf markets, pubs and cafes, and galleries and antique shops. This was South Australia's first State Heritage Area, with historic buildings and a sense of the days when the old ports were busy trading and social centres.

This is, however, still very much a working port, with a grain terminal, a tug-base, and fishing and pleasure boats in the inner port – on the upper reaches of the Port River – and a container terminal and wharves for large ships in the outer harbour, at the mouth of the river.

The area is also famous for its dolphin population. Visitors can drive along the Dolphin Trail to the six hot spots where dolphins are regularly spotted, or take a cruise to get up close. Details at the Port Adelaide Visitor Information Centre. *66 Commercial Rd; 08 8405 6560 or 1800 629 888; open daily.*

South Australian Maritime Museum 597 B8

Behind the red-and-white lighthouse in the middle of Port Adelaide, on Queens Wharf, visitors can immerse themselves in South Australia's beach-going lifestyle, climb aboard a replica ketch, or learn about the tentative beginnings of the state's rock-lobster industry. This is a first-class centre with a program of permanent and changing exhibitions. *126 Lipson St, Port Adelaide; open daily.*

Ships' Graveyard 597 B8

Many shipwrecks and other sunken vessels around Australia provide magnificent sites for divers – but the one in Port Adelaide can be accessed from land. Around 25 wrecks were abandoned on the south side of Garden Island, including barges, sailing ships and steamers – their disintegrating skeletons create an eerie skyline above the water. Visitors can view this watery graveyard from the Garden Island Bridge, or – for a closer look – on a kayak tour with Adventure Kayaking SA, *(08) 8295 8812.*

North-east

Waterworld Aquatic Centre

The newly redeveloped Waterworld will refresh and entertain. Three new waterslides and a fun Splash Ground is a popular with the kids. *Golden Grove Road, Ridgehaven, (08) 8397 7439*

North

St Kilda 597 B7

St Kilda is more of an outer village than an outer suburb, even though it is only a short distance (as the crow flies) from Port Adelaide and the Lefevre Peninsula. The access road crosses vast salt crystallisation pans, but beyond them is an idyllic little marina and other attractions that make this an excellent daytrip destination. First is the **Mangrove Trail and Interpretive Centre**, which gives a fascinating insight into this little-understood ecosystem and its inhabitants. Then there is the **Tramway Museum**, open afternoons on Sundays and public holidays, which has 20 historic trams and provides tram rides to the nearby adventure playground. The playground is St Kilda's showcase attraction, with a range of unique play equipment that includes a wooden ship, giant slides and a monorail.

South

Carrick Hill 225 H6

This 1930s house was once the home of Edward and Ursula Hayward – Edward was the son of wealthy retailers, and Ursula was the daughter of wealthy pastoralists. On their year-long honeymoon they collected many of the fittings for their house, including 16th- to 18th-century windows, doors, staircases and panelling. When they returned, their house was built in the style of a 17th-century English manor. The gardens surrounding the house contain a citrus orchard, a pleached pear arbour and avenues of poplars. The house also has an impressive collection of art. The **Children's Storybook Trail** is an enchanting walk encompassing scenes from 10 popular children's classics throughout the grounds. *46 Carrick Hill Dr, Springfield; open Wed–Sun and public holidays.*

East

Penfolds Magill Estate 597 C9

This winery, located between the city and the hills, was the first venture of Penfolds, one of Australia's best-known winemakers. Established in 1844, the vintage cellar is still used for making shiraz, and the Still House, once used for making brandy, is now used for cellar-door tastings and sales. The Magill Estate Restaurant has views over the Adelaide Plains, fine food and an incredible wine list – including Grange by the glass! *78 Penfold Rd, Magill; (08) 8301 5551; tours 11am and 3pm daily, lunch Fri, dinner Tues–Sat.*

day tours

Adelaide Hills The hills directly east of Adelaide have long been a retreat for citysiders including, most famously, 19th-century governors. Today the attractions of this beautiful semi-rural area include cool-climate wineries, gourmet produce, forests and lookouts over the city. For more details see p. 237

Barossa Valley Australia's best-known winegrowing region is a landscape of rolling yellow hills carpeted with vines. It boasts around 50 wineries, including some of the top names in the business. The district owes much to its strong German heritage, which is also expressed in the local food, architecture and many cultural events. For more details see p. 238

Clare Valley Boutique wineries, attractive 19th-century buildings and magnificent food and accommodation make the scenic Clare Valley a favourite weekend retreat. Just east of Clare is another world altogether – the old mining region of Burra, with landmarks that recall the immense copper boom. For more details see p. 239

Fleurieu Peninsula The small seaside villages along Gulf St Vincent and the historic maritime town of Victor Harbor are irresistible seaside destinations close to the capital. En route to the peninsula, visitors can stop in at one of the cellar doors around McLaren Vale, one of the country's top wine regions. For more details see p. 235

Murraylands Before ending its long journey at Lake Alexandrina and onto the rich wetlands of the Coorong, the Murray River passes through diverse landscapes of rugged cliffs, mallee scrub, river red gum forests and pastoral lands. Visitors can relive the river's rich history as a bustling trade route and take a relaxing cruise from one of the ports of yesteryear. For more details see p. 241

REGIONS
south australia

NORTHERN TERRITORY
QUEENSLAND
WESTERN AUSTRALIA
NEW SOUTH WALES
VICTORIA

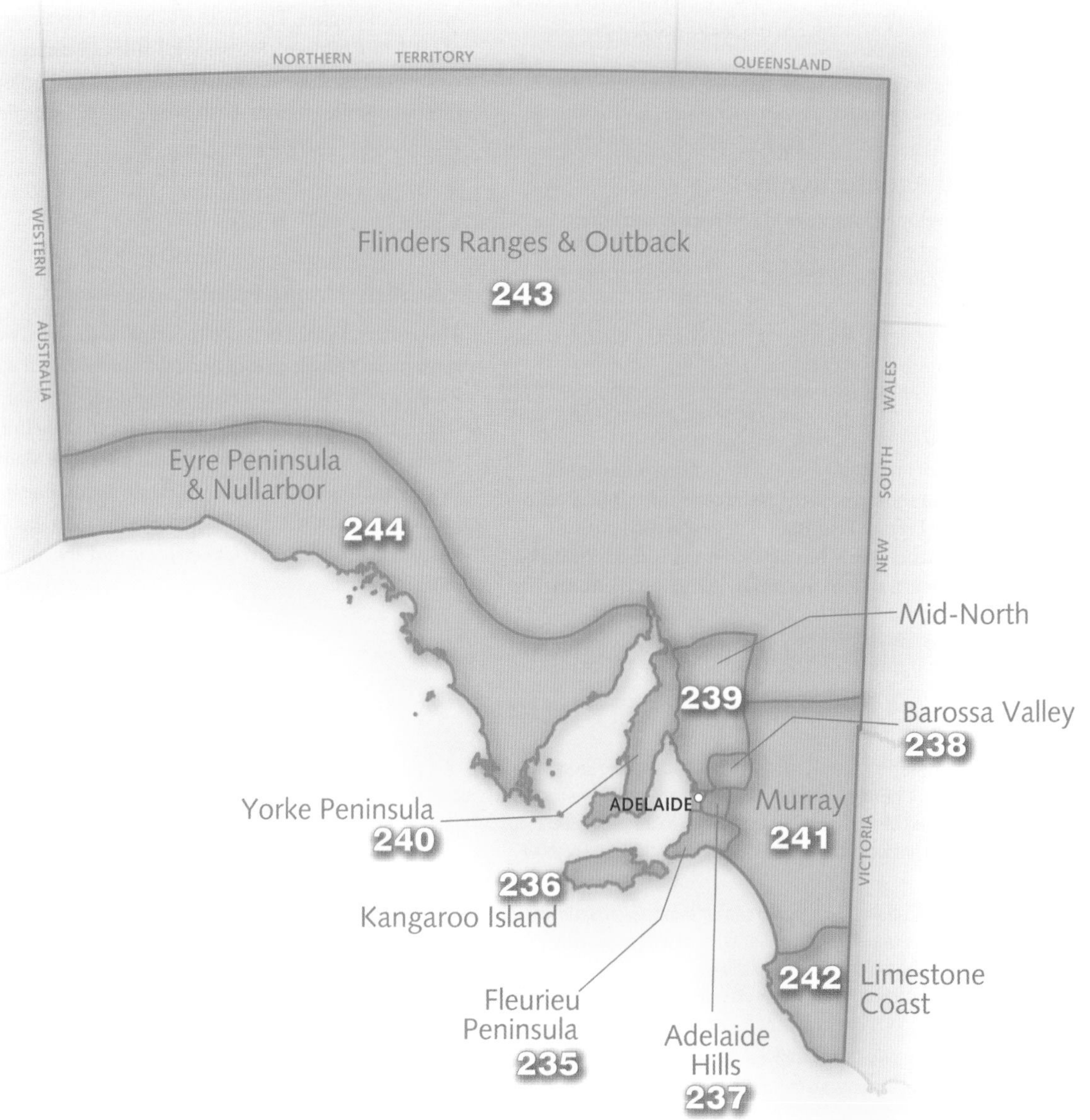

FLEURIEU PENINSULA

This region is South Australia's most popular and accessible holiday destination, and is also known for its wineries and gourmet produce.

[ALMOND BLOSSOM AT MCLAREN VALE WINERY]

McLaren Vale While wineries are a major feature of this area, vines are not the only side to agriculture here – olives, almonds, avocados and stone fruits are also grown in the fertile surrounds. In town, visit the Almond and Olive Train and the McLaren Vale Olive Grove for local produce, and head to Medlow Confectionery for high-quality gels and chocolates.

Mount Compass Gourmet Trail Mount Compass is the centre of gourmet food production in the region, with trout, berry, deer, pheasant and marron farms open for viewing and sales. Pick up a touring map from the visitor centre in McLaren Vale.

Victor Harbor Historic Victor Harbor on Encounter Bay began as a whaling port in the 1830s, later becoming the ocean port for trade along the Murray River. Today visitors can soak up the relaxed atmosphere of this popular holiday resort town. Take a walk around the many heritage sites, jump on a horse-drawn tram to Granite Island to see penguins, dolphins and maybe even whales, or ride around the bay to Goolwa on the historic Cockle Train, in operation since 1887.

Gulf St Vincent coast The Fleurieu Peninsula's western coastline includes spectacular scenery and popular holiday towns. Go snorkelling at Port Noarlunga, dive off Aldinga Beach, bathe at one of a number of family beaches or visit the heritage-listed sand dunes at Normanville.

TOP EVENTS

JAN	Milang to Goolwa Freshwater Sailing Classic
JUNE	Sea and Vines Festival (McLaren Vale)
JUNE	Playtime Festival and Whale Season Launch (Victor Harbor)
JULY–AUG	Almond Blossom Festival (Willunga)
OCT	Glenbarr Highland Gathering (Strathalbyn)
OCT	Fiesta! (McLaren Vale)

CLIMATE

VICTOR HARBOR

J	F	M	A	M	J	J	A	S	O	N	D	
25	25	23	21	19	16	15	16	18	20	22	24	MAX °C
15	15	14	12	10	8	8	8	9	10	12	14	MIN °C
21	20	23	43	62	71	75	67	56	46	28	24	RAIN MM
5	4	6	10	14	15	17	16	14	11	8	7	RAIN DAYS

WINE DISTRICTS

The Fleurieu Peninsula has two main wine-producing districts: McLaren Vale and Langhorne Creek. The McLaren Vale district is set against a landscape of weathered hills and rolling acres. With a viticulture history dating back to 1838, the region is highly regarded particularly for its full-bodied reds, most notably its shiraz. Over 50 wineries, ranging from boutique outfits to big players, are located in the district – Tourist Route 60 is an excellent way to explore both the wineries and the region. The second district is based around the highly productive Langhorne Creek area on the floodplains of the Bremer and Angas rivers. Vines have been grown here since the 1850s, but only five cellar doors offer tastings and sales. Many of the grapes grown here are sold to wine producers in other regions around the country.

→ For more detail see maps 596, 597, 598, 599, 601 & 603. For descriptions of T towns see Towns from A–Z (p. 245).

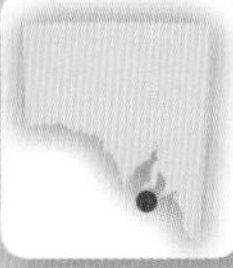

KANGAROO ISLAND

[KANGAROO ISLAND SEA LION]

Australia's third largest island is a relatively untouched paradise with a large population of native creatures that live undisturbed in pristine, natural habitats.

Gourmet produce Fast developing as an authentic, grassroots foodie destination, 'KI' is deservedly proud of its 'clean and green' status. The island offers flapping-fresh seafood, superb cheese and honey. Try Ferguson Australia in Kingscote for fresh rock lobster and crab, Island Pure Sheep Dairy in Cygnet River (for its haloumi, in particular) and Clifford's Honey Farm in MacGillivray. Cafe-style eateries showcasing local produce are dotted across the island; Rockpool Cafe in remote Stokes Bay is one favourite.

Cape Borda Lighthouse This unusually shaped lighthouse – squat and square rather than tall and round – was built in 1858 and converted to automatic operation in 1989. Guided tours are conducted regularly and accommodation is available in the old lighthouse keeper's residence.

Kingscote The island's largest town, Kingscote is situated on the Bay of Shoals. It was the site of the state's first European settlement (1836), and its long pioneering history is presented in the excellent National Trust Museum. Other attractions include tours to see the nearby colony of little penguins.

Flinders Chase National Park This park is the location of the Remarkable Rocks, enormous weathered boulders that are perched precariously on a cliff. It is also known for its wonderfully varied springtime wildflowers and for its wildlife – including the New Zealand fur seal colony at Cape du Couedic and the Cape Barren geese often seen wandering around the visitor centre. Walking trails lead into the wilderness.

Seal Bay Conservation Park Some 700 Australian sea lions feed their young and rest between fishing expeditions at Seal Bay. See the creatures from a boardwalk or up-close on a ranger-guided tour, which takes you down onto the beach.

TOP EVENTS

FEB	Racing Carnival (Kingscote)
FEB	Street Party (Kingscote)
EASTER	Art Exhibition (Penneshaw)
OCT	Kangaroo Island Art Feast (Penneshaw)

CLIMATE

KINGSCOTE

J	F	M	A	M	J	J	A	S	O	N	D	
24	24	22	20	18	15	15	15	17	19	21	22	MAX °C
15	15	14	13	11	9	8	8	9	10	12	14	MIN °C
15	17	18	35	58	73	77	65	47	37	23	19	RAIN MM
4	4	5	9	13	16	18	17	13	10	7	5	RAIN DAYS

WILDLIFE-WATCHING

As a result of its isolation from the mainland, Kangaroo Island has one of Australia's most impressive concentrations of wildlife. Seal Bay is home to a colony of Australian sea lions, while at Cape du Couedic in Flinders Chase National Park there is a 600-strong colony of New Zealand fur seals. This national park is also the best place to see land animals like kangaroos, tammar wallabies, brush-tailed possums and the occasional koala or platypus. Little penguins can be seen on tours operating from Kingscote or Penneshaw. There is a large and varied bird population (240 species) across the island, with Murray Lagoon (en route to Seal Bay) home to many waterbirds.

→ For more detail see maps 598, 599, 601, 602 & 603. For descriptions of Ⓣ towns see Towns from A–Z (p. 245).

ADELAIDE HILLS

An interesting mix of Australian bushland and European-style farmland is the setting for this region's historic villages, gardens, museums, galleries and wineries.

[GERMAN-STYLE CRAFT SHOP, HAHNDORF]

Cleland Wildlife Park Within Cleland Conservation Park, this excellent wildlife park has a large collection of everybody's favourite animals, including kangaroos, emus and wallabies. Nocturnal walks reveal some of the rarer species, including bettongs and bandicoots. Daytime visits could be combined with a picnic in the adjacent Mount Lofty Botanic Gardens.

Belair National Park South Australia's oldest national park, established in 1891, offers a natural landscape of eucalypt forests and brilliant flowering plants. Green thumbs will enjoy a stroll through the gardens of the governor's old summer residence, which dates back to 1859.

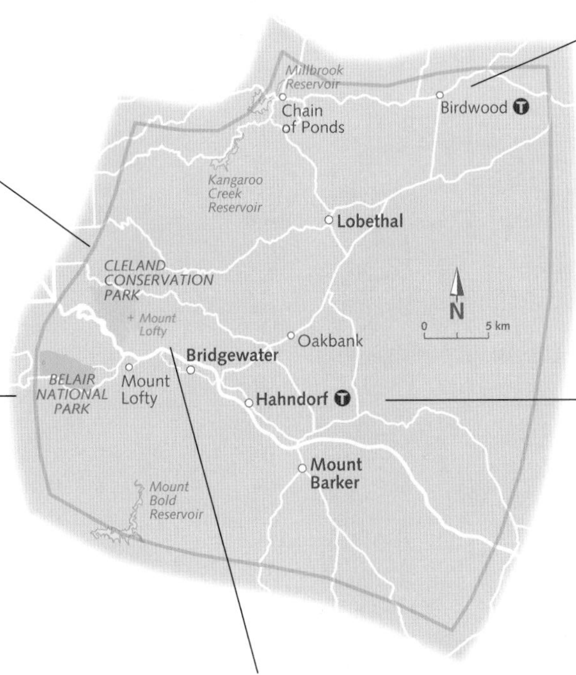

Birdwood's National Motor Museum This impressive museum has the largest collection of vintage vehicles in the Southern Hemisphere, with more than 300 classic cars, motorcycles and commercial vehicles housed in an 1852 flour mill. Learn about the history of motoring and visit the workshop complex to see the vehicles being lovingly restored by coach builders and mechanics.

Hahndorf This distinctive town was settled in the 1830s by Prussian refugees. Its heritage is preserved in the village architecture and German-style shops, museums and cafes. Visit The Cedars, former home of artist Hans Heysen, and see contemporary artworks at the Hahndorf Academy.

Mount Lofty Lookout Enjoy spectacular views of Adelaide and the hills from the lookout at the 727-metre summit of Mount Lofty. Drop in to the information centre to plan your day, and enjoy 'food with a view' in the adjacent restaurant and cafe.

TOP EVENTS

EASTER	Oakbank Easter Racing Carnival (Oakbank)
APR	Rock and Roll Rendezvous (Birdwood)
MAY	Jazz and Heritage Festival (Mount Barker)
SEPT	Bay to Birdwood Run (vintage vehicles, even-numbered years)
SEPT–OCT	Heysen Festival (Hahndorf)
DEC	Lights of Lobethal Festival (Lobethal)

CLIMATE

MOUNT BARKER

J	F	M	A	M	J	J	A	S	O	N	D	
27	27	25	20	17	14	13	14	16	19	22	25	MAX °C
12	12	10	8	7	5	5	5	6	7	9	10	MIN °C
26	26	31	59	89	100	105	103	86	68	40	34	RAIN MM
6	5	7	11	15	16	17	18	15	13	9	7	RAIN DAYS

FOOD AND WINE

The Adelaide Hills are becoming increasingly well known as a touring destination for those with gourmet inclinations. Highlights include a visit to Petaluma's Bridgewater Mill, an old flour mill, for tastings and sales of Petaluma wines or lunch at the attached restaurant. In Hahndorf try German-style produce, including breads and smallgoods, at the many outlets. Here you can also buy full-flavoured berry produce at Beerenberg Strawberry Farm or go wine-tasting at Hahndorf Hill Winery or Hillstowe Wines. There is an excellent introduction to South Australia's regional wine and cheese at the Birdwood Wine and Cheese Centre, which offers tastings. Fine boutique wineries are also found around Birdwood.

→ For more detail see maps 596, 597, 598, 601 & 603. For descriptions of Ⓣ towns see Towns from A–Z (p. 245).

BAROSSA VALLEY

Australia's best-known wine region offers a landscape of vine-covered hills dotted with historic villages, stone cottages and the grand buildings of old wine estates.

[CHATEAU DE KARLSBURG, BAROSSA VALLEY]

Seppeltsfield The Seppelts estate, established in the 1850s, is one of the grandest in the country. Elegant bluestone buildings are surrounded by superb gardens, and the property is accessed by an avenue of 2000 date palms. Do not miss the hill-top mausoleum built in the style of a Doric temple.

Barossa wineries In this famed region is a landscape of old, gnarled shiraz vines planted as early as the 1840s, adding character to the newer vines and varieties. The Barossa is best known for its shiraz, semillon and chardonnay. The biennial Barossa Vintage Festival is a time of great activity and colour.

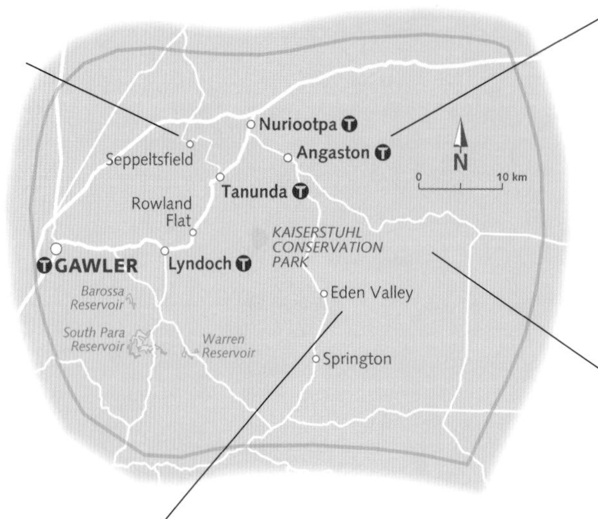

Kaiserstuhl Conservation Park This rugged mountain-side park offers a view of what the Barossa would have looked like before European settlement. A couple of excellent walking trails allow visitors to explore the varied terrain and glimpse the local wildlife.

Eden Valley wineries This elevated wine-producing area is regarded as a distinct region, with a small number of wineries open for tastings. In Springton stands the Herbig Tree, a giant, hollow red gum that was the temporary home of a German family in the 1850s.

TOP EVENTS

FEB	Barossa under the Stars (Tanunda)
EASTER	Gawler Gourmet and Heritage Festival (Gawler)
APR	Barossa Vintage Festival (throughout region, odd-numbered years)
OCT	Barossa International Music Festival (throughout region)
OCT	Barossa Band Festival (Tanunda)

CLIMATE

NURIOOTPA

J	F	M	A	M	J	J	A	S	O	N	D	
29	29	26	21	17	14	13	14	17	20	24	26	MAX °C
14	14	12	9	7	5	4	5	6	8	10	12	MIN °C
19	19	22	38	55	56	66	64	60	49	29	24	RAIN MM
5	4	5	8	12	13	16	16	13	11	8	6	RAIN DAYS

GOURMET TRADITION

While many parts of Australia have developed gourmet credentials over the last decade or so, the Barossa has a culinary heritage that dates back to the mid-1800s. German-style baking is a highlight: try the Lyndoch Bakery, or the Apex Bakery in Tanunda, established 70 years ago. Old-fashioned ice-cream is made and served at Tanunda's Nice Ice and in Angaston you'll find the shopfront for Australia's biggest processor of dried fruit – the Angas Park Fruit Company. Maggie Beer's Farm Shop at Nuriootpa sells the cook's renowned products – everything from quince paste to verjuice – while Angaston Gourmet Foods is a one-stop shop for all the best local foods.

→ For more detail see maps 596, 597, 601 & 603. For descriptions of Ⓣ towns see Towns from A–Z (p. 245).

MID-NORTH

[FARMHOUSE BESIDE THE BARRIER HIGHWAY]

Clare Valley wineries and the historically important coppermining sites of Burra make this a great place for a weekend away.

Clare Valley wineries This wine-producing district extends for 35 kilometres across the fertile valley, including big names as well as charming boutique establishments. Well-known names include Annie's Lane, Taylors, Pikes and Mount Horrocks. Although the climate is generally Mediterranean in style, many Clare Valley wines have cool-climate characteristics. Clare Valley rieslings are regarded as among the best in Australia.

Clare Valley produce The Clare Valley is famed for its gourmet food producers, who embrace the slow-food philosophy in their lovingly created offerings, such as chutneys, jams, olive oils and sausages, available at Wild Saffron in Clare and other outlets throughout the region.

Riesling Trail Running through the picturesque Clare Valley wine region between Clare and Auburn, this scenic 27-kilometre route is the perfect way to visit the surrounding villages and wineries on foot or by bicycle. One of the first rail trails to be commissioned in South Australia, it traverses the old Riverton to Spalding railway line, passing by several cellar doors on the way.

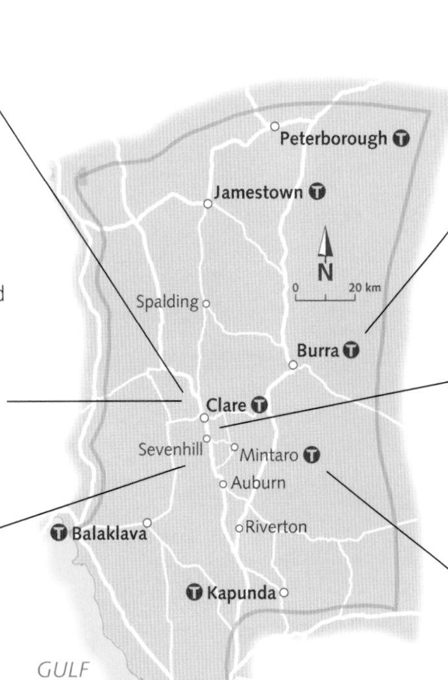

Burra Located in the sparse landscape of the Bald Hills Range, this former coppermining centre is a timepiece of the mid-19th century. Pick up a Burra Heritage Passport and visit mine shafts, museums and heritage buildings. Marvel at the creek-bed dugouts that were home to over 1800 miners, and see the working conditions on a guided mine tour.

Sevenhill Cellars Austrian Jesuit priests established the Clare Valley's first winery here in 1851 to ensure a steady supply of altar wine. The winery continues to be run by Jesuits, and now offers a range of table wines. Drop into St Aloysius Church (1875), next to the cellars.

Mintaro Mintaro is an almost-intact 19th-century village, with many attractive stone buildings incorporating the region's unique slate. It was the first town in South Australia to be declared a State Heritage Area. Just south-east is Martindale Hall, an 1879 mansion used in the film Picnic at Hanging Rock (1975) and now open to the public.

TOP EVENTS

FEB	Jailhouse Rock Festival (Burra)
MAR	Music in the Monster Mine (Burra)
EASTER	Clare Races (horseracing, Clare)
MAY	Antique and Decorating Fair (Burra)
MAY	Gourmet Weekend (Clare)
SEPT	Celtic Music Festival (Kapunda, dates vary)

CLIMATE

CLARE

J	F	M	A	M	J	J	A	S	O	N	D	
30	29	27	22	17	14	13	15	18	21	25	27	MAX °C
13	13	12	8	6	4	3	4	5	7	10	12	MIN °C
25	24	25	47	73	80	82	80	73	57	37	29	RAIN MM
4	4	5	8	12	14	15	15	13	11	7	6	RAIN DAYS

MID-NORTH HISTORY

Historically, this is one of the most interesting and well-preserved areas of rural South Australia. In 1839 Edward Eyre explored the Clare Valley and his favourable reports led quickly to pastoral settlement. Jesuit priests planted the first Clare Valley vines at Sevenhill in 1851, and a booming wine industry followed. Copper deposits found at Burra and Kapunda in the 1840s drove a huge mining industry, but this fell into decline in the 1870s. The community's perseverance in preserving Burra's history led to the town's State Heritage Area listing in 1993. Kapunda had a change of direction after mining, with cattle baron Sir Sidney Kidman initiating an agricultural industry that has since grown exponentially. This mixture of viticulture, agriculture and mining history is a Clare Valley attraction in itself – endless stories of survival and success can be found in the museums and displays throughout the region.

SOUTH AUSTRALIA

→ For more detail see maps 597, 599, 601, 603 & 605. For descriptions of towns see Towns from A–Z (p. 245).

YORKE PENINSULA

This long, boot-shaped peninsula is a popular beachside holiday destination with the surrounding waters offering excellent fishing, diving and surfing.

[WOOL BAY, NEAR STANSBURY]

Little Cornwall The Cornish miners that flocked to Wallaroo, Moonta and Kadina in the early 1860s prompted the colloquial name for this triangle of towns – Little Cornwall. The discovery of substantial copper deposits led to a prosperous mining industry, revealed in the heritage architecture of the three towns. Visit the Moonta Mines State Heritage Area, with walking trails to mining ruins.

Innes National Park At the southern tip of the peninsula, Innes National Park protects salt lakes, low mallee scrub, wildflowers, sandy beaches and rugged cliffs. Browns Beach, West Cape and Pondalowie Bay are popular for surfing, diving and fishing. A 1904 shipwreck can be glimpsed from Ethel Beach, and the interesting remains of Inneston mining town are found in the southern area of the park.

Port Victoria This town was once the main port of call for the clippers and windjammers that transported grain to the Northern Hemisphere, a period of history recorded in the local Maritime Museum. Port Victoria is now a resort town, offering access to swimming beaches and Wardang Island, a popular diving spot with an underwater heritage trail featuring eight wrecks.

Edithburgh The town's jetty was once the site of a large shipping operation, when tonnes of salt harvested from the nearby lakes were exported from here. Edithburgh is now a popular resort town. Attractions include a tidal pool for safe swimming, good diving locations and access to Troubridge Island, home to populations of little penguins, black-faced shags and crested terns.

TOP EVENTS

MAY	Kernewek Lowender (Cornish festival, odd-numbered years, Kadina, Moonta and Wallaroo)
JUNE	Winter Fun Fishing Contest (Port Broughton)
SEPT	Blessing of the Fleet (Port Pirie)
OCT	Garden Fair (Moonta)
OCT	Yorke Surfing Classic (Innes National Park)
OCT	Festival of Country Music (Port Pirie)

CLIMATE

KADINA

J	F	M	A	M	J	J	A	S	O	N	D	
31	30	28	24	19	16	15	17	20	23	26	28	MAX °C
16	16	14	11	9	7	6	6	8	10	12	14	MIN °C
15	19	19	33	46	52	49	45	39	34	23	18	RAIN MM
3	3	4	6	10	12	13	13	10	8	6	4	RAIN DAYS

FISHING

The Yorke Peninsula is one of South Australia's top fishing destinations, with jetties at Wallaroo, Moonta Bay, Edithburgh, Stansbury and Port Victoria, and rocky points and sandy coves all through the region, providing excellent opportunities for land-based anglers. Snapper, squid, tommy ruff, garfish and whiting are among the more commonly caught species. Reef-fishing is also popular. Browns Beach, on the western side of Innes National Park, is renowned for its big hauls of salmon. Near Goose Island, just north of Wardang Island on the west coast, are two reefs that offer excellent boat fishing for a variety of species.

→ For more detail see maps 599, 601, 602–3 & 604. For descriptions of Ⓣ towns see Towns from A–Z (p. 245).

MURRAY

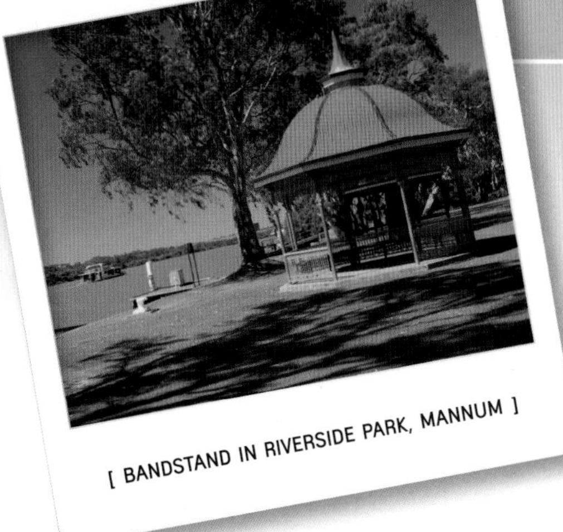
[BANDSTAND IN RIVERSIDE PARK, MANNUM]

Rich in history, this region also encompasses cliff-lined river valleys, mallee scrub, river red gum forests, lagoons, orchards and vineyards.

Morgan Morgan's days as a busy river port may be over, but a rich legacy of sites and buildings preserves something of the excitement of the 19th-century river trade. Look out for the wharves (built in 1877), the customs house and the courthouse. For a thorough look at the history of Morgan, drop in to the Port of Morgan Historical Museum in the old railway buildings.

Camping and watersports From Murray Bridge to Renmark there are caravan parks and camping grounds with river frontage and access to a wide range of watersports, including canoeing, fishing and swimming. Murray River National Park provides a couple of quiet spots at the north-east end of the river for those who like their recreation in a park setting.

Houseboat on the Murray River A terrific way to discover the secrets of the mighty Murray is to captain your own houseboat. Slow down and take in the captivating landscapes of river red gums, limestone cliffs, plains and mallee scenery; at times you may have this beautiful river all to yourself. Houseboats are for hire at numerous places along the South Australian stretches of the river, at towns such as Murray Bridge, Renmark, Mannum and Morgan.

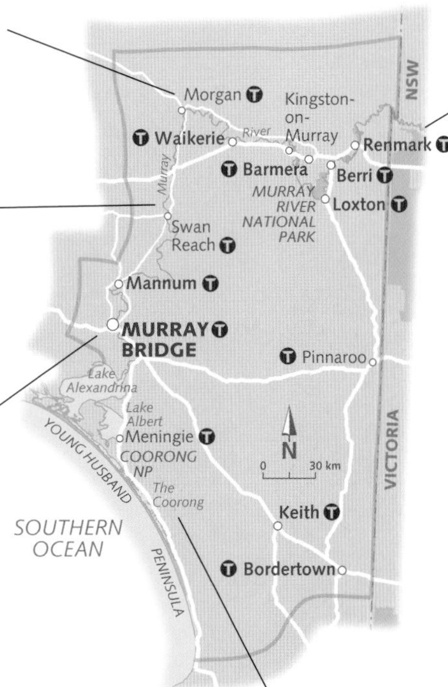

Riverland produce The Riverland is the fruit bowl of South Australia, producing over 90 per cent of the state's citrus fruit, stone fruit and nuts. Wine is also produced here; tastings and sales are available at half a dozen estates, including Angoves near Renmark, Berri Estates Winery near Berri, and Banrock Station Wine and Wetland Centre at Kingston-on-Murray.

The Coorong The Coorong, a shallow lagoon protected within Coorong National Park, is one of Australia's most significant wetlands. It stretches 135 kilometres along the coast, separated from the Southern Ocean by the dunes of the Younghusband Peninsula. The Coorong is best known for its abundant birdlife (and for birdwatching). Over 240 species have been recorded, including many migratory species.

TOP EVENTS

FEB	Riverland Greek Festival (Renmark)
FEB	Mardi Gras (Loxton)
MAR	Rotary Food Fair (Waikerie)
JUNE	South Australian Country Music Festival (Barmera)
JUNE	Riverland Balloon Fiesta (Renmark)
OCT	Rose Festival (Renmark)

CLIMATE

RENMARK

J	F	M	A	M	J	J	A	S	O	N	D	
33	32	29	24	20	17	16	18	21	24	28	30	MAX °C
17	17	14	11	8	6	5	6	8	11	13	15	MIN °C
16	19	14	18	25	25	23	25	28	28	21	18	RAIN MM
3	3	3	4	6	8	8	9	7	6	4	4	RAIN DAYS

PADDLESTEAMERS

Paddlesteamers were first used in South Australia when PS *Mary Ann* was launched at Mannum in 1853. Carrying goods and passengers, they were vital in the development of the all-important trade route that ran from the mouth of the Murray River into New South Wales and Victoria (and vice versa). Two of the original boats, PS *Mayflower* and PS *Marion*, operate as day-cruisers, departing from Morgan and Mannum respectively. Longer tours are available from Mannum on PS *River Murray Princess* – the largest paddleboat ever built in the Southern Hemisphere.

→ For more detail see maps 597, 601 & 603. For descriptions of Ⓣ towns see Towns from A–Z (p. 245).

LIMESTONE COAST

A major holiday region, the Limestone Coast features the World Heritage–listed Naracoorte Caves, Mount Gambier and the prestigious wineries of the Coonawarra.

[COASTAL SCENERY NEAR BEACHPORT]

Bool Lagoon Game Reserve One of southern Australia's largest and most diverse freshwater lagoon systems, Bool Lagoon was declared a reserve in 1967 and is on the Ramsar List of Wetlands of International Importance. These wetlands rarely dry out, creating a refuge for over 150 bird species, including brolgas, black swans, magpie geese and tiny fairy-wrens. A network of boardwalks, trails and observation hides, such as a 500-metre boardwalk to an ibis rookery, allows close-up views of the wetland activity.

Robe Settled in the 1840s, Robe is one of the state's oldest and best-preserved towns. It boasts a fine collection of stone cottages, shops, public buildings and hotels, many of them National Trust–classified. Set around Guichen Bay along a beautiful stretch of coast, Robe combines a quaint fishing-village atmosphere with excellent facilities for holiday-makers.

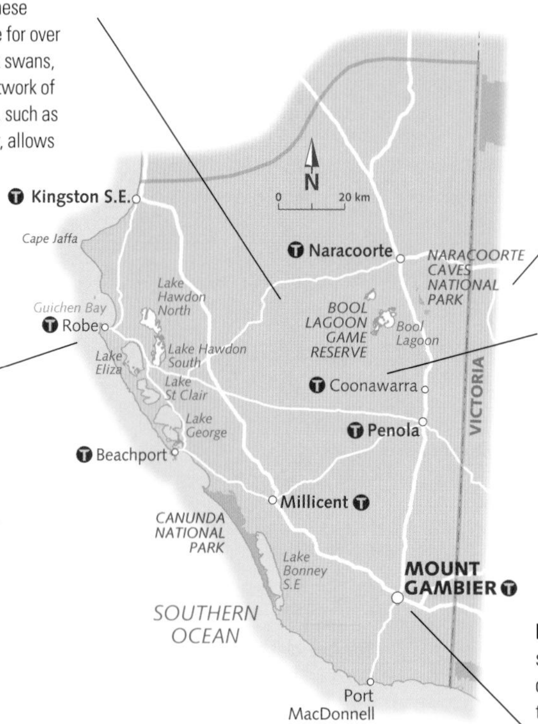

Naracoorte Caves Naracoorte Caves National Park protects one of Australia's most significant cave systems, a fact reflected by its World Heritage listing. There are 60 known caves, several of which are open to the public. At the excellent Wonambi Fossil Centre, displays show how the fossils found in these caves have played a key role in charting the continent's evolutionary history.

The Coonawarra The Coonawarra is Australia's most valuable piece of wine real estate. The first vines were planted by John Riddoch in the late-1800s. Since then this 12-kilometre-long and 2-kilometre-wide stretch of terra rossa soil has continually produced wine of the highest quality, chiefly bold red varieties. The region's cabernet sauvignon is outstanding. Over 20 wineries offer cellar-door tastings and sales.

Mount Gambier This major regional centre is set on the slopes of an extinct volcano – its crater contains the intensely coloured Blue Lake. Beneath the town, stretching from the coast all the way to Bordertown in the north, is an enormous wedge of limestone that has given rise to many caves. Nearby examples include Engelbrecht Cave (popular with divers) and the Umpherston Sinkhole.

TOP EVENTS

JAN	Cape Jaffa Seafood and Wine Festival (Cape Jaffa)
FEB	Blue Lake City Muster (country music event, Mount Gambier)
FEB	Taste the Limestone Coast (Naracoorte)
MAY	Penola Coonawarra Arts Festival (Penola)
MAY	Mount Gambier Gold Cup Carnival (Mount Gambier)
OCT	Cabernet Celebrations (Coonawarra/Penola)

CLIMATE

MOUNT GAMBIER

J	F	M	A	M	J	J	A	S	O	N	D	
25	25	23	19	16	14	13	14	16	18	20	23	MAX °C
11	12	10	9	7	6	5	5	6	7	8	10	MIN °C
26	25	35	55	72	84	99	94	73	63	47	37	RAIN MM
9	8	11	14	18	19	21	21	19	17	14	12	RAIN DAYS

SEASIDE TOWNS

In summer, the Limestone Coast's quiet fishing towns transform into busy holiday centres, offering great beaches and fantastic seafood. Visit Kingston S.E., a major lobster port at the southern end of Coorong National Park – nearby lakes and lagoons are havens for birdlife. Head further south to Robe, with its lovely combination of maritime history and windswept scenery, or to Beachport, settled as a whaling station in the 1830s. Today the town's long jetty is a popular spot for fishing. For more fishing, don't miss Port MacDonnell, where you'll find South Australia's largest lobster fleet.

→ For more detail see map 601. For descriptions of towns see Towns from A–Z (p. 245).

FLINDERS RANGES & OUTBACK

This vast region covers around 70 per cent of South Australia. Here you will find some of the country's most legendary landscapes and the quintessential outback town of Coober Pedy.

[WINDMILL IN CARRIETON, NEAR QUORN]

Coober Pedy Coober Pedy is considered the opal capital of the world, producing an incredible 80 per cent of the world's gem-quality supply. It is also Australia's most famous underground town, with homes, churches and art galleries all set into the hills to keep cool. Visit the museums and mines, and try your hand at prospecting.

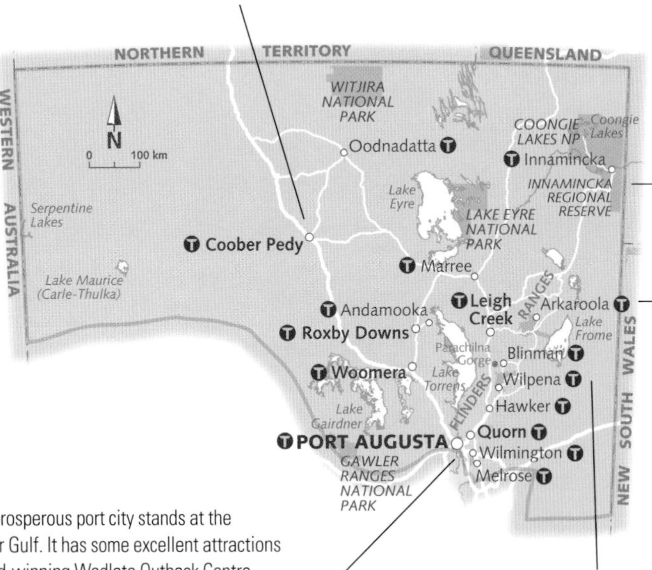

Innamincka Regional Reserve Four-wheel-drive enthusiasts flock to this remote reserve to tackle the terrain, but there is more here than just adventure driving. The area's wetlands are internationally significant, particularly Coongie Lakes, which provide a vital refuge for waterbirds. Closer to Innamincka is the Cullyamurra Waterhole on Cooper Creek, where visitors can see Aboriginal rock carvings.

Ridgetop Tour The signature attraction of the northern Flinders Ranges, this four-hour, four-wheel-drive tour in an open-top vehicle takes you along an insanely steep track, originally built for mining exploration. The guided tour, run by Arkaroola Wilderness Sanctuary, scales heart-stoppingly precipitous ridges to reward passengers with panoramic views featuring red granite mountains and the desert beyond.

Port Augusta This prosperous port city stands at the northern tip of Spencer Gulf. It has some excellent attractions for visitors – the award-winning Wadlata Outback Centre provides an Indigenous, natural and social history of the outback and ranges, while the Australian Arid Lands Botanic Garden offers an in-depth look at the country's little-understood arid environments.

Wilpena Pound The Pound, part of Flinders Ranges National Park, is one of the most extraordinary geological formations in Australia. It is a vast natural amphitheatre, surrounded by sheer cliffs and jagged rocks that change colour according to the light. An old homestead within the pound, built by the Hill family in 1902 and abandoned after floods in 1914, stands as a reminder of the difficulties of farming in this environment.

TOP EVENTS

FEB	Wilpena under the Stars (Wilpena)
EASTER	Opal Festival (Coober Pedy)
MAY	Races and gymkhana (Oodnadatta)
JULY	Pichi Richi Marathon (Quorn)
JULY	Australian Camel Cup (Marree)
NOV	Outback Surfboat Carnival (Port Augusta)

CLIMATE

HAWKER

J	F	M	A	M	J	J	A	S	O	N	D	
34	33	30	25	20	16	16	18	22	26	29	32	MAX °C
18	18	15	11	7	5	4	4	7	10	13	16	MIN °C
19	21	17	20	31	39	35	33	28	25	22	21	RAIN MM
3	3	2	3	5	7	7	7	6	5	4	3	RAIN DAYS

TRAILS AND TRACKS

Beginning (or ending) in the Flinders Ranges are two long-distance trails for bushwalkers and bikeriders. The longest track is the 1200-kilometre Heysen Trail, which begins at Cape Jervis and follows the mountain ranges to Parachilna Gorge north of Flinders Ranges National Park, taking in the best of the scenery of the Fleurieu Peninsula, Adelaide Hills and Barossa Valley en route. The 900-kilometre Mawson Trail is a bike track that runs between Adelaide and Blinman, named after explorer Sir Douglas Mawson. The trail explores the lesser known parts of the Mount Lofty and Flinders ranges, travelling along unmade and rarely used roads, farm-access tracks and national-park trails. Beyond the ranges are some of Australia's premier four-wheel-drive treks, including the Oodnadatta Track – taking in hot springs, old railway sidings and sections of the Dog Fence – and the Birdsville Track, crossing impressive sandhills and Cooper Creek.

→ For more detail see maps 604–5, 606–7 & 608–9. For descriptions of towns see Towns from A–Z (p. 245).

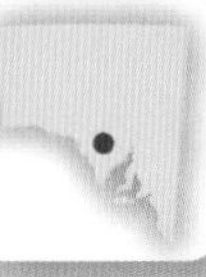

EYRE PENINSULA & NULLARBOR

This diverse region features a coastline that is ideal for surfing and fishing, as well as the vast, treeless plain of the Nullarbor.

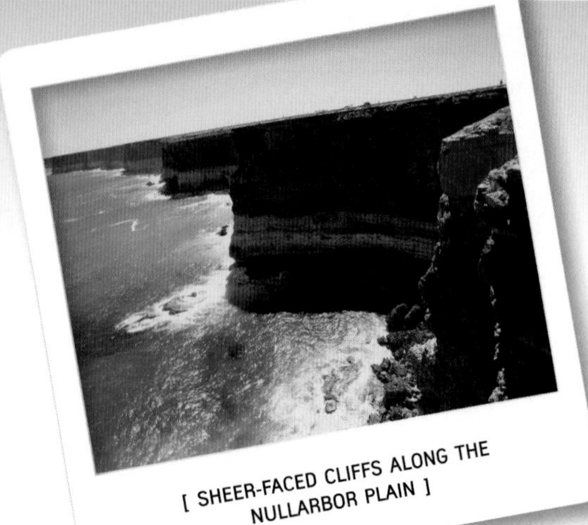

[SHEER-FACED CLIFFS ALONG THE NULLARBOR PLAIN]

Nullarbor Plain The Nullarbor, Latin for 'treeless', is a plain of 250 000 square kilometres, resting on a massive area of limestone riddled with caves. Along the coast, a long line of sheer-faced cliffs drop suddenly into the foaming waters of the Great Australian Bight. Head of Bight offers some of the country's best whale-watching, between June and October.

Murphy's Haystacks Located some 40 kilometres south-east of Streaky Bay, these ancient wind-worn granite inselbergs (from the German words for 'island' and 'mountain') have stood sentinel over the surrounding plains for nearly 34 000 years. The Stonehenge-like arrangement of pink-tinted pillars and boulders is one of the most photographed attractions on the Eyre Peninsula, and is a lovely (if unusual) location for a picnic (undercover facilities are available).

Surf coast This 73-kilometre stretch of coast between Ceduna and the tiny settlement of Penong justifiably promotes itself as a surfing paradise. There are renowned surf beaches along the remote coastline, the most famous being Cactus Beach, just south of Penong, which boasts three world-famous breaks.

WARNING: Sharks have been known to frequent these waters – seek local advice before entering.

Spencer Gulf coast Here calm waters lap against the shores of peaceful holiday villages and the northern regional centre of Whyalla. At Port Lincoln, the huge natural harbour is home to Australia's foremost tuna fleet. Beaches, museums, golf, walks and drives all feature, but the biggest drawcard is fishing – jetty-fishing or gamefishing – in some of the best waters in Australia.

Coffin Bay National Park This park protects a pristine coastal wilderness of exposed cliffs, small coves and beaches. Bush camping, surfing, walking and fishing are all popular activities. Many of the vehicle tracks are four-wheel-drive only, but for conventional vehicles there is a popular scenic tour called the Yangie Trail, beginning at Coffin Bay township.

TOP EVENTS

JAN	Tunarama Festival (Port Lincoln)
FEB–MAR	Adelaide to Lincoln Yacht Race and Lincoln Week Regatta (Port Lincoln)
EASTER	Australian Snapper Championship (Whyalla)
APR	Sculptures on the Cliff (Elliston)
JUNE–AUG	Australian Salmon Fishing Competition (Elliston)
OCT	Oysterfest (Ceduna)

CLIMATE

PORT LINCOLN

J	F	M	A	M	J	J	A	S	O	N	D	
25	26	24	22	19	17	16	17	18	20	22	24	MAX °C
15	16	15	13	11	9	8	8	9	11	12	14	MIN °C
13	15	20	37	57	75	79	69	50	36	22	19	RAIN MM
4	4	5	10	14	16	18	17	13	11	7	6	RAIN DAYS

WILDLIFE

This sparsely settled region remains a wildlife haven. Rare and endangered bird species find refuge in many of the parks and reserves; at Nullarbor National Park look out for Major Mitchell cockatoos, or see migrating birds land at Lincoln National Park after their journey from the chilly Arctic Circle. Lolling about on the sands after days of fishing far out at sea are the rare and endangered Australian sea lions of Point Labatt Conservation Park – they are Australia's only resident mainland colony. One of the most important parks in the region is the Sir Joseph Banks Group Conservation Park, on islands offshore from Tumby Bay. This park is Australia's largest breeding ground for Cape Barren geese, and is home to dolphins, New Zealand fur seals and Australian sea lions.

→ For more detail see maps 602, 604 & 610–1. For descriptions of Ⓣ towns see Towns from A–Z (p. 245).

TOWNS A–Z

south australia

LEGEND

VISITOR INFORMATION

RADIO STATIONS

IN TOWN

WHAT'S ON

WHERE TO EAT

WHERE TO STAY

NEARBY

Food and accommodation listings in town are ordered alphabetically with places nearby listed at the end

[PIPI SHELLS ON THE BEACH, THE COORONG]

Aldinga Beach

Pop. 5979

Map ref. 597 A12 | 598 E6 | 599 H9 | 601 B4 | 603 J10

McLaren Vale and Fleurieu Visitor Centre, Main Rd, McLaren Vale; (08) 8323 9944 or 1800 628 410; www.mclarenvale.info

94.7 5EFM, 639 AM ABC North and West

The rolling hills of the southern Mount Lofty Ranges form the backdrop to Aldinga Beach, a long curve of white sand facing Gulf St Vincent. One and a half kilometres off the coast is one of the state's best diving spots – the Aldinga Drop Off, an underwater cliff where divers say the marine life has to be seen to be believed. The township – to the west of the original Aldinga, which grew as a small farming centre in the mid-1800s – is a popular holiday spot.

Gnome Caves: 5 theme 'caves', great for kids; Aldinga Beach Rd.

Lake Coolangatta Cafe: casual beachfront seafood; 1 Norman Rd, Silver Sands; (08) 8557 4000.

Cockleshell Cottage: beachside, country-style comfort; 10 Boomerang Ave; 0409 702 208. ***Silver Sands B&B:*** elevated waterfront apartment with views; 277 Esplanade; (08) 8557 4002.

Star of Greece In 1888 the *Star of Greece* plunged to the ocean floor in a wild storm. The ship was only a short distance from land, but at 3am, and in gigantic swells, 17 of the 28 people on board drowned. Today a portion of the vessel can be seen from shore at low tide. A plaque lies on the seabed for the benefit of divers, but for those wanting to stay dry, pictures of the wreck line the walls of the Star of Greece Cafe. Port Willunga; 3 km N.

Aldinga Scrub Conservation Park: offers walks through remnant coastal vegetation, and a blaze of colourful wildflowers in spring; end of Dover St, off Aldinga Beach Rd; 1 km S. ***Aldinga:*** Uniting Church cemetery has the graves of those who died in the *Star of Greece* shipwreck. Community market held 1st Sat each month in Institute Hall, Old Coach Rd; 4 km NE. ***Beaches:*** many north and south of Aldinga Beach including Port Willunga Beach, with the remains of an old jetty and caves built in the cliff by anglers (3 km N); Sellicks Beach, with boat access and good fishing (8 km S); and Maslin Beach, Australia's first official nudist beach (10 km N). ***Lookouts:*** one south of Sellicks Beach (11 km S) and another over the Myponga Reservoir (23 km S).

See also FLEURIEU PENINSULA, p. 235

Andamooka

Pop. 528

Map ref. 604 F3

Dukes Bottlehouse Motel (incorporating post office), 275 Opal Creek Blvd; (08) 8672 7007; www.andamookaopal.com.au

105.9 FM ABC North and West

If Queen Elizabeth II had ever been to Andamooka, perhaps she would have thought differently about the Andamooka Opal given to her as a gift in 1954 on her first visit to Australia. The opal weighed 203 carats and glistened in blues, reds and greens. It was the result of an extensive search for the most beautiful opal in the state, yet Andamooka itself is a misshapen collection of tin sheds, dugouts and fibros in the middle of the desert. With constant water shortages, no local council and an all-consuming drive to find opals, residents have become experts in making do. The town offers old-fashioned outback hospitality to an increasing number of tourists.

Opal showrooms: showrooms in town include Andamooka Gems and Trains, attached to Dukes Bottlehouse Motel, with opals and a model railway. ***Historic miners huts:*** a handful of old semi-dugouts line the creek bed in the centre of town, complete with old tools and furnishings. Access is by tour, which includes a visit to an underground mine; details from post office. ***Cemetery:*** with miners' nicknames on the headstones.

Market and barbecue: local art and craft; Sat long weekends (Easter, June and Oct).

Fossicking: noodling in unclaimed mullock dumps surrounding town; details from post office. ***Lake Torrens:*** one of the state's largest salt lakes stretches away to the south-east; 4WD recommended for access.

See also Flinders Ranges & Outback, p. 243

Angaston

Pop. 1865
Map ref. 597 F4 | 601 C2 | 603 K8

Barossa Visitor Information Centre, 66–68 Murray St, Tanunda; (08) 8563 0600 or 1300 852 982; www.barossa.com

 89.1 Triple B FM, 1062 AM ABC Riverland

Angaston takes its name from George Fife Angas who purchased the original plot of land on which the town now stands. He was a prominent figure in the South Australian Company and one of the shareholders who used his substantial buying power to get the best plots of land and dictate the terms of purchase. Many of the town's public buildings were funded by him, even before he emigrated. In a sense, the town's strong German heritage was also funded by him as he sponsored many Lutherans to make the journey to South Australia. Angaston still has strong ties with its history. In town is a German butcher shop that has been making wursts for more than 60 years, a blacksmith shop over a century old and a cafe and specialty food shop named the South Australian Company Store. Jacarandas and Moreton Bay figs line the main street.

A. H. Doddridge Blacksmith Shop This is the town's original blacksmith, started by Cornish immigrant William Doddridge. The shop closed in 1966 and 15 years later it was purchased by local townspeople. On Sat and Sun it operates as a working smithy, complete with the original bellows that Doddridge brought out from England. Murray St.

Angas Park Fruit Co: retail outlet for Angas Park dried fruits and nuts; Murray St. ***The Abbey:*** second-hand clothing and period pieces inside the old church; open Thurs–Sun; Murray St. ***The Lego Man:*** one of Australia's largest Lego collections; Jubilee Ave. ***Food outlets:*** include Angaston Gourmet Foods, famous for their baguettes, and the Barossa Valley Cheese Company, which makes cheese on the premises; Murray St. ***Angaston Heritage Walk:*** brochure from visitor centre.

Farmers market: behind Vintners Bar and Grill, Nuriootpa Rd; Sat mornings. ***Angaston Show:*** Feb/Mar. ***Barossa Vintage Festival:*** celebration of locally produced food and wine in various locations; odd-numbered years, Apr. ***Barossa Gourmet Weekend:*** Aug.

Blond Coffee: homemade snacks and lunches; 60 Murray St; (08) 8564 3444. ***Vintners Bar & Grill:*** robust Mediterranean flavours; Nuriootpa Rd; (08) 8564 2488.

Caithness Manor: hosted heritage suites; 12 Hill St; (08) 8564 2761. ***Collingrove Homestead:*** stately historic residence; Eden Valley Rd; (08) 8564 2061. ***Naimanya Cottage:*** secluded stone cottage; Pohlner Rd; (08) 8565 3275. ***Strathlyn B&B:*** gracious hosted and self-contained suites; Nuriootpa Rd; (08) 8564 2430.

Wineries: Barossa Valley wineries surrounding Angaston include Yalumba, with its cellar door on Eden Valley Rd, (08) 8561 3200, and Saltram, Nuriootpa Rd, (08) 8561 0200. *For more information on the Barossa Valley, see Nuriootpa, Lyndoch and Tanunda.* ***Collingrove Homestead:*** the old Angas family home, now owned by the National Trust and open for tours, boutique accommodation and dining; 7 km SE. ***Mengler Hill Lookout:*** views over the Barossa Valley; 8 km SW. ***Kaiserstuhl Conservation Park:*** a small pocket of native flora and fauna, with walking trails; 10 km S. ***Butcher, Baker, Winemaker Trail:*** between Lyndoch and Angaston, taking in wineries and gourmet-food producers along the way. Smartcard available for purchase, which allows you VIP experiences and rewards points; details and brochure from visitor centre.

See also Barossa Valley, p. 238

Ardrossan

Pop. 1125
Map ref. 599 F5 | 601 A1 | 603 I7

Ardrossan Bakery, 39 First St, (08) 8837 3015; or Harvest Corner Visitor Information Centre, 29 Main St, Minlaton, (08) 8853 2600 or 1800 202 445; www.yorkepeninsula.com.au

 89.3 Gulf FM, 639 AM ABC North and West

A cluster of bright white grain silos sit atop the red clay cliffs at Ardrossan, an industrial town on Yorke Peninsula. The town has two jetties – one for the export of grain, salt and dolomite, and the other for the benefit of local anglers. Ardrossan is well known for its blue swimmer crabs that are found under the jetty or in the shallows at low tide. The best season for crabbing is between September and April.

Stump-jump Plough A lonely stump-jump plough stands in the cliff-top park opposite East Tce. Mallee scrub once covered much of this area and caused endless grief to early farmers because it was so difficult to clear. The invention of the stump-jump plough made it possible to jump over stumps left in the ground and plough on ahead. The plough's design was perfected in Ardrossan. The original factory, on Fifth St, now houses a historical museum; open 2.30–4.30pm Sun.

Kalinda Shores: beachfront family holiday home; 108 Black Point Dr, Black Point; (08) 8838 2208.

***Zanoni* wreck** South Australia's most complete shipwreck is 15 km south-east of Ardrossan off Rogues Pt. The wreck was lost for over 100 years, but was eventually rediscovered by some local fishermen. It lies virtually in one piece on the seabed. Some artefacts from the ship can be found at the Ardrossan historical museum, but divers wanting to see the wreck in situ need a permit from Heritage South Australia, (08) 8204 9245.

Walking trail: 3 km track along cliff-tops to Tiddy Widdy Beach; begins at the boat ramp in town. ***BHP Lookout:*** view of Gulf St Vincent and dolomite mines; 2 km S. ***Clinton Conservation Park:***

mangrove swamps and tidal flats with an array of birdlife; begins after Port Clinton, 25 km N, and stretches around the head of Gulf St Vincent.

See also YORKE PENINSULA, p. 240

Arkaroola

Map ref. 600 F2 | 605 K3

Arkaroola Village reception; (08) 8648 4848 or 1800 676 042; www.arkaroola.com.au

999 AM ABC Broken Hill

Arkaroola is set in an incredible landscape of ranges laced with precious minerals, waterholes nestled inside tall gorges and places with songful names like Nooldoonooldoona and Bararranna. What is more, the Flinders Ranges are still alive, rumbling with up to 200 small earthquakes a year. It was a place that geologist Reg Sprigg found fascinating, and worth conserving. He purchased the Arkaroola property in 1968 and created a wildlife sanctuary for endangered species. Today a weather station, seismograph station and observatory (tours available) add to its significance and the spectacular four-wheel-drive tracks entice many visitors. The village has excellent facilities for such a remote outpost.

Arkaroola Wilderness Sanctuary: famed resort in rugged terrain; (08) 8648 4848 or 1800 676 042. ***Grindell's Hut:*** remote, rustic bush hut; Wortupa Loop Rd, Vulkathunha–Gammon Ranges National Park, via Balcanoona; (08) 8648 0049.

Ridgetop Tour This, the signature attraction of the northern Flinders Ranges, is a 4WD tour along an insanely steep track. The original track, built for mining exploration, wound through the creek beds, but run-off from the ridges washed the road away in just a few years. The idea was formed to create a track along the ridges themselves. A few bulldozers later, the track was complete. This is a guided tour, but Arkaroola Wilderness Sanctuary also has 100 km of self-guide 4WD tracks, including the popular Echo Camp Backtrack.

Vulkathunha–Gammon Ranges National Park This park is directly south of Arkaroola, taking in much of the distinctive scenery of the northern Flinders Ranges. The Adnyamathanha people believe that the Dreamtime serpent, Arakaroo, drank adjacent Lake Frome dry and carved out Arkaroola Gorge as he dragged his body back to his resting spot, inside Mainwater Pound. His restlessness is the cause of the earthquakes. Features include the surprisingly lush Weetootla Gorge, fed by a permanent spring, and Italowie Gorge, the unlikely spot where an impoverished R. M. Williams began making shoes. Park Headquarters at Balcanoona; (08) 8648 0048; 32 km S.

Waterholes: many picturesque waterholes along Arkaroola Creek and tributaries west and north-east of the village. ***Bolla Bollana Smelter ruins:*** where the ore from surrounding mines was once treated. It includes a Cornish beehive-shaped kiln; 7 km NW. ***Paralana Hot Springs:*** the only active geyser in Australia, where water heated by radioactive minerals bubbles through the rocks. Swimming or extended exposure is not recommended; 27 km NE. ***Big Moro Gorge:*** rockpools surrounded by limestone outcrops. The gorge is on Nantawarrina Aboriginal Land; obtain permit from Nepabunna Community Council, (08) 8648 3764; 59 km S. ***Astronomical Tours:*** boasts some of the best star-watching conditions in the Southern Hemisphere at 3 magnificent observatories; (08) 8648 4848. ***Scenic flights:*** over the ranges or further afield; details from village reception.

See also FLINDERS RANGES & OUTBACK, p. 243

Balaklava

Pop. 1627
Map ref. 599 H4 | 603 J7

Council offices, Scotland Pl; (08) 8862 0800.

90.9 Flow FM, 1062 AM ABC Riverland

Balaklava is set on the Wakefield River in an area dominated by traditional wheat and sheep farms. It sprang up as a stopping point between the Burra copper fields and Port Wakefield, but a grain merchant from Adelaide, Charles Fisher, soon turned the focus to agriculture. He built grain stores here before there was any sign of grain. This proved a canny move, as it lured farmers to the area. The town features old sandstone buildings and a 'silent cop', a curious keep-left sign in the middle of a roundabout.

Courthouse Gallery The arts are alive and well in Balaklava, as shown by this community-run art gallery that has a changing program of local and visiting exhibitions, plus a popular art prize in July. Open 2–4pm Thurs, Fri and Sun; Edith Tce.

Balaklava Museum: old household items and local memorabilia; Old Centenary Hall, May Tce. ***Urlwin Park Agricultural Museum:*** old agricultural machinery, 2 old relocated banks and a working telephone exchange; open 2–4pm 2nd and 4th Sun each month or by appt; Short Tce; (08) 8862 1854. ***Walking trail:*** scenic 3 km track along the riverbank.

Adelaide Plains Cup Festival: arts and crafts exhibitions, as well as golf and clay-shooting competitions; 1st Sun in Mar. ***Balaklava Cup:*** major regional horserace; Aug. ***Balaklava Show:*** Sept.

Port Wakefield Behind the highway's long line of takeaways and petrol stations is a quiet town that began life as a cargo port to carry the copper mined in Burra's Monster Mine back to Port Adelaide. It is set on the mangrove-lined Wakefield River at the top of Gulf St Vincent. The wharf, which has a floor of mud at low tide, is now used by the local fishing industry. A historical walk brochure is available from the Port Wakefield/Rivergum Information Centre. This is a popular spot for fishing, crabbing and swimming. 25 km W.

Devils Garden: a picnic spot among river box gums, once a 'devil of a place' for bullock wagons to get through as the black soil quickly turned to mud; 7 km NE. ***Rocks Reserve:*** walking trails and unique rock formations by the river; 10 km E. ***Balaklava Gliding Club:*** offers weekend 'air experience flights' with an instructor; Whitwarta Airfield; (08) 8864 5062; 10 km NW.

See also MID-NORTH, p. 239

Barmera

Pop. 1927
Map ref. 558 A6 | 601 G1 | 603 N7

Barwell Ave; (08) 8588 2289; www.berribarmera.sa.gov.au

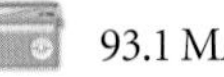
93.1 MAGIC FM, 1062 AM ABC Riverland

Barmera lies in the middle of a swooping hairpin bend of the Murray River, close to the Victorian border, but it is hard to tell where the river stops and where the flood plains and tributaries begin in the area to the west of town. The wetlands eventually

SOUTH AUSTRALIA

 RADIO STATIONS  IN TOWN   WHAT'S ON  WHERE TO EAT   WHERE TO STAY  NEARBY

flow into Lake Bonney, a large body of water to the north of Barmera. Swimming, waterskiing, sailing and fishing are some of the activities popular on the lake. The town was established in 1921 as a settlement for returned World War I soldiers, who were all promised a patch of well-irrigated farmland.

Rocky's Country Music Hall of Fame Dean 'Rocky' Page established Barmera's famous country music festival and was a well-known musician in his own right. Within the centre is an array of country music memorabilia and a display of replica guitars with the handprints of the legends who used them. The pièce de résistance is Slim Dusty's hat. Open Wed–Fri; Barwell Ave; (08) 8588 1463.

Donald Campbell Obelisk: commemorates an attempt in 1964 to break the world water-speed record, but 347.5 km/h was not quite enough to make the books; Queen Elizabeth Dr. ***Bonneyview Winery:*** cosy atmosphere with award-winning reds and light pastries; Sturt Hwy; (08) 8588 2279.

5RM Barmera Main Street Market: 1st Sun each month (Sept–Dec). ***Lake Bonney Yachting Regatta:*** Easter. ***South Australian Country Music Festival:*** June. ***Riverland Field Days:*** Sept. ***Barmera Bonnie Sheepdog Trials:*** Oct.

Banrock Station Wine & Wetland Centre: stylish eco-cafe; Holmes Rd, Kingston on Murray; (08) 8583 0299. ***The Overland Corner Hotel:*** historic bush pub; Old Coach Rd, Overland Corner; (08) 8588 7021.

Banrock Station Wine & Wetland Centre Fruity wines mix with a cacophony of birds and frogs at Banrock Station. In these new times of sensitive agriculture, Banrock is working with environmental organisations to breathe life back into a pocket of wetland that was ruined by irrigation (almost 70% of all Murray wetlands have been affected). The natural cycles of flooding and drying have seen the return of black swans, ibis and native fish, and a boardwalk gives visitors a close-up look. Kingston-on-Murray; (08) 8583 0299; 10 km W.

Overland Corner This was the first settlement in the area, a convenient stop en route for drovers and people travelling to the goldfields. By 1855 a police post had been established to deal with the odd bushranger and quell the problems flaring between drovers and the indigenous inhabitants. In 1859 the Overland Corner Hotel opened its doors. Its thick limestone walls and red gum floors have seen many floods. An 8 km walking track into the adjacent Herons Bend Reserve leaves from the hotel. 19 km NW.

Cobdogla Irrigation and Steam Museum: has the world's only working Humphrey Pump, used in the early days of irrigation. Also local memorabilia and steam-train rides; open 1–3pm Sun for train rides; inquiries (08) 8588 2323 or Barmera visitor centre; 5 km W. ***Highway Fern Haven:*** garden centre featuring an indoor rainforest; 5 km E. ***Napper's Old Accommodation House:*** ruins of a hotel built in 1850 on the shores of the lake; turn east over Napper Bridge; 10 km NW. ***Loch Luna Game Reserve:*** linking Lake Bonney and the Murray, these wetlands form an important refuge for waterbirds and one of the few inland nesting sites for sea eagles. Chambers Creek, which loops around the reserve, is popular for canoeing; turn west over Napper Bridge; 16 km NW. ***Moorook Game Reserve:*** these wetlands surround Wachtels Lagoon; 16 km SW. ***Loveday Internment Camps:*** guided tour or self-guide drive to the camps where Japanese, Italian and German POWs were held during WW II; details from visitor centre.

See also MURRAY, p. 241

Beachport

Pop. 342
Map ref. 601 E10

Millicent Rd; (08) 8735 8029; www.wattlerange.sa.gov.au

107.7 5THE FM, 1161 AM ABC South East

Beachport started out as a whaling port. Today the crayfish industry has taken over and the town has South Australia's second longest jetty, favoured by anglers young and old with regular catches including whiting, flathead and garfish. People flock here for summer holidays to relax on the beautiful sandy beaches and swim in the bay. The Bowman Scenic Drive provides stunning views over the Southern Ocean with access to sheltered coves and rocky headlands for the adventurous to explore. It is also a great place for whale-watching.

Old Wool and Grain Store Museum The old store contains a whaling and fishing display including harpoons and whaling pots, as well as relics from shipwrecks off the coastline. Upstairs rooms are furnished in the style of the day. Natural history display features local and migratory seabirds. Open 10am–4pm daily over holiday period, 10am–1pm Sun at other times; Railway Tce.

Lanky's Walk: a short walk through bushland to Lanky's Well, where the last full-blood member of the Boandik tribe camped while working as a police tracker; begins on Railway Tce North; details on this and other walks from visitor centre. ***Pool of Siloam:*** this small lake, 7 times saltier than the sea, is said to be a cure for all manner of ailments. Also a popular swimming spot; end of McCourt St. ***Lighthouse:*** the original lighthouse was located on Penguin Island, a breeding ground for seals and penguins offshore from Cape Martin, where the current lighthouse now stands. It offers good views of the island from the cape; south of town.

Market: Sat long weekends (Jan, Easter and Oct). ***Duck Race:*** Jan. ***Festival by the Sea:*** stalls and entertainment; odd-numbered years, Feb/Mar.

Beachport Harbourmasters: trendy beachfront apartment; 1 Beach Rd; (08) 8735 8197. ***Beachport Retreat:*** spacious holiday home; 10 Ethel St; (08) 8735 7211.

Beachport Conservation Park This park is a succession of white beaches, sand dunes and rugged limestone cliffs, with the southern shore of Lake George lying inland. The coast is dotted with ancient shell middens and is accessed primarily by 4WD or on foot. Five Mile Drift, a beach on Lake George, is a good base for swimming, sailing and windsurfing. Access to the coast side is via Bowman Scenic Dr, which begins at the lighthouse. Access to the Lake George side is via Railway Tce North. 4 km NW.

Woakwine Cutting: an incredible gorge, cut through Woakwine Range by one man to drain swampland and allow farming, with viewing platform, information boards and machinery exhibit; 10 km N.

See also LIMESTONE COAST, p. 242

Berri

Pop. 4009
Map ref. 558 A6 | 588 A6 | 601 G1 | 603 N7

Riverview Dr; (08) 8582 5511; www.berribarmera.sa.gov.au

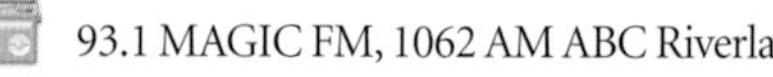

93.1 MAGIC FM, 1062 AM ABC Riverland

Orange products and wine are big business in this Riverland town, which has Australia's largest winery and is one of the major growing and manufacturing centres of the country's biggest orange-juice company. The Big Orange, on the north-west outskirts of town, makes this rather clear. The name 'Berri' has nothing to do with fruit, though. It comes from the Aboriginal 'Bery Bery', thought to mean 'bend in the river'. The town was established in 1911, the year after irrigation of the Murray began.

Berri Direct: makers of Berri fruit juice and other products, with sales and a video presentation detailing production history of the Riverland; Old Sturt Hwy. ***Riverlands Gallery:*** local and touring art exhibitions; open weekdays; Wilson St. ***Gilbert Street Gallery:*** artwork by local artists, including some amazing glassware and woodwork; Gilbert St. ***Berri Community Mural:*** enormous community-painted mural commemorating the past and present fruit industry; Old Sturt Hwy next to Berri Direct. ***Berri Lookout Tower:*** panoramic views of river, town and surrounds from a converted water tower; Cnr Fiedler St and Vaughan Tce. ***Lions Club Walking Trail:*** 4 km riverfront walk from Berri Marina to Martin Bend Reserve, a popular spot for picnics and waterskiing; an Aboriginal mural and totems are under the bridge and further along are monuments to famous Aboriginal tracker Jimmy James. ***Birdwatching safaris:*** to Bookmark Biosphere Reserve; tours offered by Jolly Goodfellows Birding; bookings (08) 8583 5530. ***Berri Air Tours:*** offering scenic flights over the Riverland; bookings (08) 8582 2799.

Farmers market: Sat. ***Speedboat Spectacular:*** Mar (subject to river conditions). ***Riverland Renaissance:*** Oct. ***Craft Fair:*** Nov.

Cragg's Creek Cafe: smart waterfront tapas and snacks; Riverview Dr; (08) 8582 4466.

Riverbush Cottages: roomy, relaxed, riverfront homes; Old Sturt Hwy; (08) 8582 3455 or 1800 088 191.

Berri Estates Winery This impressive winery was founded in 1922 and has grown to be the largest winery and distillery in Australia. The Murray River and the temperate climate have much to do with the quality of the wines here, including reds, whites and fortified wines, as well as brandy. Between Berri and Glossop.

Murray River National Park, Katarapko section In this park, see the merging of 2 distinct vegetations – of the famous Murray River flood plains and the equally renowned Mallee region. The 6 km Mallee Drive takes the visitor into the heart of the park and to the distinctive mallee terrain. There are also walking trails to see some of the park's inhabitants, including the ever-popular kangaroo. Just south of Berri; (08) 8595 2111.

Angas Park Fruit Company: dried fruits and other products; open 8.30am–5pm weekdays, 10am–12pm Sat and public holidays; Old Sturt Hwy; (08) 8561 0800; 3 km W. ***Monash:*** a small irrigation town best known for the free family attractions at the Monash Adventure Park on Morgan Rd. Enjoy delicate handmade chocolates at the Chocolates and More store opposite the park, or taste the wines at nearby Thachi Wines; 12 km NW.

See also MURRAY, p. 241

Birdwood

Pop. 733

Map ref. 596 H5 | 597 E8 | 601 C3 | 603 K9

 Shannon St; (08) 8568 5577.

 98.7 Power FM, 1062 AM ABC Riverland

The small town of Birdwood is set picturesquely in the Torrens Valley in the northern part of the popular Adelaide Hills district. The region's beauty would have been a welcome sight for German settlers escaping religious persecution in the 1840s. Like many of the German-settled towns in the area, Birdwood was originally named after a Prussian town, Blumberg. However, anti-German sentiment during World War I created a feeling of unrest and the town's name was changed to Birdwood after the commander of the ANZAC forces in Gallipoli, Sir William Birdwood.

National Motor Museum The largest in the Southern Hemisphere, this impressive collection of over 300 vintage cars, motorcycles and commercial vehicles is housed in an 1852 flour mill. The vehicles are lovingly restored, often from simply a shell. Visit the workshop complex to see the process of restoration as coach builders and mechanics work tirelessly on these old machines. The building's original history as a flour mill can be seen in the Mill Building. Shannon St; (08) 8568 5006.

Birdwood Wine and Cheese Centre: introduces visitors to boutique wines and cheeses of regional SA with tastings and sales; open Wed–Sun; Shannon St. ***Blumberg Inn:*** imposing 1865 inn harking back to German-settler days; Main St.

Rock and Roll Rendezvous: Apr. ***Bay to Birdwood Run:*** vintage motoring event attracting more than 1600 vehicles; even-numbered years, Sept.

Chain of Ponds Vineyard Cottage: large stone residence with vistas; Mannum Rd, Gumeracha; (08) 8389 1415. ***Stoneybank Settlement Cottages:*** quaint, rural heritage hideaways; Lot 100 Stoneybank La, Mount Pleasant; (08) 8568 2075.

Lobethal The quaint town of Lobethal features historic German-style cottages and an 1842 Lutheran seminary. Fairyland Village takes the visitor into the world of fairytales; open weekends. The National Costume Museum houses a collection of dresses, suits and accessories dating from 1812 (closed Mon). The town lights up each Christmas in the 'Lights of Lobethal' festival. 13 km SW.

The Toy Factory: a family business manufacturing wooden toys from a shop adjacent to an 18 m giant rocking horse; Gumeracha; 7 km W. ***Chain of Ponds Wine:*** boutique winery with tastings, sales, viewing platform, restaurant and B&B (1880s cottage); 9 km W. ***Herbig Tree:*** an extraordinary insight into Friedrich Herbig and the hollow red gum tree where he raised a family in the 1850s. School museum and pioneer cemetery also on-site; Springton; group bookings only (08) 8568 2287; 15 km NE. ***Malcolm Creek Vineyard:*** boutique winery with cellar door and friendly deer; open weekends and public holidays; Bonython Rd, Kersbrook; 20 km NW. ***Roachdale Reserve:*** self-guide nature trail with brochure; 23 km NW via Kersbrook. ***Torrens Gorge:*** spectacular cliffs and streams make this a popular spot for picnics; 25 km W. ***Samphire Wines and Pottery:*** a small boutique winery and handmade pottery shop; Cnr Watts Gully and Robertson rds; 27 km NW via Kersbrook. ***Warren Conservation Park:*** difficult trails, including part of the long-distance Heysen Trail, lead to spectacular views over countryside and Warren Gorge; adjacent to Samphire Wines and Pottery. ***Mt Crawford Forest:*** walkers, horseriders and cyclists will enjoy the forest tracks of this park, which is scattered in various locations north, west and south-west of Birdwood; map from information centre on Warren Rd (signposted turn-off between Kersbrook and Williamstown).

See also ADELAIDE HILLS, p. 237

Blinman

Pop. 151
Map ref. 600 C7 | 605 I5

General store, Mine Rd; (08) 8648 4370.

During the 19th century numerous mining townships dotted the northern Flinders Ranges. Blinman is the sole surviving town surveyed at the time. The discovery of copper here in 1859 was accidental. The story goes that a shepherd, Robert Blinman, used to watch his sheep from a boulder and one day he absentmindedly broke off a chunk and discovered it was copper. Historic buildings in the main street recall that rich, and short-lived, era.

Land Rover Jamboree: Easter. ***Cook Outback:*** food/wine festival with camp oven competition; Oct long weekend.

Blinman Hotel: rustic bush pub; Mine Rd; (08) 8648 4867. ***Prairie Hotel:*** innovative outback cuisine; Cnr High St and West Tce, Parachilna; (08) 8648 4844.

Alpana Station: historic station accommodation; Wilpena Rd; (08) 8648 4626. ***Blinman Cottage:*** cosy, heritage character; 201 Hancock St; 0417 084 003. ***Gum Creek Station:*** basic, homestead-style accommodation; Wilpena Rd; (08) 8648 4883. ***Moolooloo Station:*** rustic and remote shearers quarters; Glass Gorge Rd; (08) 8648 4861. ***Prairie Hotel:*** innovative outback luxury; Cnr High St and West Tce, Parachilna; (08) 8648 4844.

Blinman Mine Historic Site: a 1 km self-guide walk explains the history and geology of the site. Contact the caretaker on (08) 8648 4874 for a guided tour; just north-east of Blinman. ***Great Wall of China:*** impressive limestone ridge; Wilpena Rd; 10 km s. ***Angorichina Tourist Village:*** start point for 4 km walk along creek bed to Blinman Pools, permanent spring-fed pools in scenic surrounds. Accommodation ranges from tents to cabins; (08) 8648 4842; 14 km w. ***Glass and Parachilna gorges:*** 10 km NW and 15 km W of town are these 2 beautiful gorges. Parachilna Gorge is the end point of the 1200 km Heysen Trail (bushwalking trail), which begins at Cape Jervis; information from Parks SA (08) 8124 4792. ***Flinders Ranges National Park:*** *see Wilpena for more details*; 26 km s; ***Prairie Hotel:*** a historic hotel at Parachilna offering cuisine with a bush-tucker twist as well as first-rate accommodation. Hotel staff can arrange activities including 4WD tours, scenic flights and visits to nearby Nilpena Station on the edge of Lake Torrens to sample the outback life; (08) 8648 4844; 32 km w. ***Scenic drive:*** travel east through Eregunda Valley (around 20 km E), then north-east to Mt Chambers Gorge (around 75 km NE), with its rockpools and Aboriginal carvings; further north is Vulkathunha–Gammon Ranges National Park; *see Arkaroola*. ***Mawson Trail:*** this 900 km bike trail from Adelaide ends in Blinman. It is named after famous Australian explorer Sir Douglas Mawson and traverses the Mt Lofty and Flinders ranges; details from Bicycle SA, (08) 8411 0233.

See also FLINDERS RANGES & OUTBACK, p. 243

Bordertown

Pop. 2584
Map ref. 558 A12 | 590 A7 | 601 G7

81 North Tce; (08) 8752 0700; www.tatiara.sa.gov.au

106.1 5TCB FM, 1062 AM ABC Riverland

In spite of its name, Bordertown is actually 18 kilometres from the South Australia–Victoria border in the fertile country of the Tatiara district. Thought to be the Aboriginal word for 'good country', Tatiara's name is justified by the region's productive wool and grain industries. For a different native-animal experience, look out for Bordertown's famous white kangaroos, Australia's only known colony. Former Australian prime minister Robert (Bob) J. L. Hawke was born here.

 ***Robert J. L. Hawke's childhood home:*** includes memorabilia. Visit by appt only, details from visitor centre; Farquhar St. ***Bordertown Wildlife Park:*** native birds and animals, including pure-white kangaroos; Dukes Hwy. ***Hawke Gallery:*** in foyer of council chambers; Woolshed St. ***Bordertown Recreation Lake:*** popular spot for fishing, canoeing and walking, with artwork on display; northern outskirts of town.

Bordertown Show: Oct.

Dunalan Cottage: roomy farmstay residence; Dukes Hwy; (08) 8753 2323. ***Curlew Park:*** contemporary country homestead; Lot 146 Chark Rd, Mundulla; (08) 8753 4133.

Padthaway The Padthaway district has long been regarded as a leader in Australian wine production. An excellent place to start the tasting of such wines is the Padthaway Estate Winery, where meals and wine-tastings enrich the palate. Adjacent is the historic Padthaway Homestead, an 1882 building rising regally above the surrounding region. After indulging in gastronomic delights, walk among the magnificent red gums and stringybarks at nearby Padthaway Conservation Park. 42 km sw.

Clayton Farm: incorporates the Bordertown and District Agricultural Museum and features vintage farm machinery and a National Trust–classified thatched-roof building and woolshed; 3 km s. ***Mundulla:*** a historic township featuring the heritage-listed Mundulla Hotel; 10 km sw. ***Bangham Conservation Park:*** a significant habitat for the red-tailed black cockatoo; 30 km SE.

See also MURRAY, p. 241

Burra

Pop. 976
Map ref. 603 K5

2 Market Sq; (08) 8892 2154; www.visitburra.com.au

105.1 Trax FM, 1062 AM ABC Riverland

The Burra region exploded into activity when copper was found by two shepherds in 1845. Settlements were established based on the miners' country of origin: Aberdeen for the Scottish, Hampton for the English, Redruth for the Cornish and Llwchwr for the Welsh. The combined settlement grew to be the second largest in South Australia, but the miners were fickle – with riches promised on the Victorian goldfields, they did not stay for long. In 1877 the Monster Mine closed. Luckily, Burra did not turn into a ghost town. Instead, the rich heritage of its past has been carefully preserved by the community, resulting in the town being declared a State Heritage Area in 1993. Burra is in the Bald Hill Ranges, named for the 'naked' hills around the town.

Burra Heritage Passport This 'passport' allows visitors to discover the major heritage sites of Burra – armed with an unlimited-access 'key' and a brief history of each site outlined in the pamphlet *Discovering Historic Burra*. Included in the passport is the Burra Historic Mine Site (off Market St), with an ore dressing tower and powder magazine offering views of the open-cut mine and town. At the site is Morphetts Enginehouse Museum (additional entry fee), featuring an excavated 30 m entry tunnel and engine displays. Another site is the Burra Creek Miners' Dugouts (alongside Blyth St) – these dugouts cut into the creek beds housed 1800 people in the boom. Visit the Unicorn Brewery Cellars (Bridge Tce), which date back to 1873, the police lock-up and stables (Tregony St) – these were the first built outside Adelaide – and Redruth Gaol (off Tregony St), which served as a gaol, then a girls reformatory from 1856 to 1922. You can also visit the ruins of a private English township called Hampton, on the northern outskirts of town. Passports are available from the visitor centre and can be upgraded to provide access to the town's museums.

Bon Accord Mine Complex: National Trust interpretive centre with working forge and model of Burra Mine. Guided tours available; closed Mon and Fri; Railway Tce. ***Market Square Museum:*** an old-style general store, post office and family home returned to its heyday; opposite visitor centre. ***Malowen Lowarth Cottage:*** restored 1850s Cornish miner's cottage; Kingston St. ***Burra Regional Art Gallery:*** local and touring exhibitions; Market St. ***Thorogoods:*** enjoy some of Australia's best apple wine and cider; John Barker St. ***Antique shops:*** in Commercial and Market sts. ***Burra Creek:*** canoeing and picnicking.

Jailhouse Rock Festival: Feb. ***Music in the Monster Mine:*** Mar. ***Antique and Decorating Fair:*** May. ***Burra Show:*** Oct.

Burra Bakery: classic Cornish fare; 16 Commercial St; (08) 8892 2070. ***Burra Hotel:*** hearty country fare; 5 Market Sq; (08) 8892 2389.

Birch Cottage: quaint, homely country retreat; 10 Thames St; (08) 8892 2210. ***Burra Heritage Cottages:*** charming and historic miners digs; 8–18 Truro St; (08) 8892 2461.

Burra Trail Rides: horseriding adventures in Bald Hills Range; bookings 0427 808 402. ***Burra Gorge:*** picnics, camping and walking tracks around gorge and permanent springs; 23 km SE. ***Dares Hill Drive:*** scenic 90 km drive with lookout; begins 30 km N near Hallett; map from visitor centre.

See also MID-NORTH, p. 239

Ceduna

Pop. 2304
Map ref. 611 M8

58 Poynton St; (08) 8625 2780 or 1800 639 413; www.cedunatourism.com.au

94.5 5CCR FM, 693 AM ABC Eyre Peninsula and West Coast

The name Ceduna is derived from the Aboriginal word 'chedoona', meaning resting place, which is apt for those who have just traversed the Nullarbor. Ceduna is also the last major town for those about to embark on the journey west – the place to check your car and stock up on food and water. The difficulty of obtaining supplies and provisions has a long history around Ceduna. Denial Bay, where the original settlement of McKenzie was situated, was where large cargo ships brought provisions for the early pioneers. Ceduna was established later, in 1896, and is situated on the shores of Murat Bay with sandy coves, sheltered bays and offshore islands. In the 1850s there was also a whaling station on St Peter Island (visible from Thevenard).

Old Schoolhouse National Trust Museum: pioneering artefacts, including those from British atomic testing at Maralinga; closed Sun; Park Tce. ***Ceduna Arts Cultural Centre:*** original paintings, local pottery and ceramics; open weekdays; Cnr Eyre Hwy and Kuhlmann St. ***Oyster tours:*** offered to Denial Bay, Thevenard and Smoky Bay; bookings at visitor centre. ***Ceduna Oyster Bar:*** fresh oysters year-round; western outskirts, on Eyre Hwy. ***Local beaches:*** swimming, boating, waterskiing and fishing. The foreshore is an ideal spot for walks and picnics (sharks have been known to frequent these waters – seek local advice). ***Encounter Coastal Trail:*** 3.8 km interpretive trail from the foreshore to Thevenard.

Oysterfest: community festival including street parade and fireworks; Oct long weekend. ***Ceduna Races:*** horseracing; Dec and Jan.

Great Australian Bight Marine Park The park preserves the fragile ecosystem of the Great Australian Bight. It has spectacular wildlife sights, including the breeding and calving of southern right whales from June to Oct. Spend a day observing these giant creatures from the viewing platform at Head of Bight. There are also spectacular views of the Bunda Cliffs, which begin at the head and trail all the way to the WA border. Whale-watching permits are purchased from the visitor centre on-site. Interpretive centre also on-site. 300 km W.

Nullarbor National Park Aboriginal culture is closely linked with this park's network of caves, part of the largest karst landscape in the world (Murrawijinie Caves north of Nullarbor Roadhouse are the only caves accessible to the public). Vast and mainly flat, the park's most beautiful scenery is along the coast where the cliffs stretch for 200 km overlooking the Southern Ocean. Visitors should take care along the unstable cliff edges. Rare and endangered species such as the Major Mitchell cockatoo and the peregrine falcon are often sighted. Also watch out for the southern hairy-nosed wombat. 300 km W.

Thevenard: a deep-sea port that handles bulk grain, gypsum and salt, as well as a large fishing fleet noted for whiting hauls. Bill's Seafood Tours gives an insight into the commercial fishing industry; details from visitor centre. A 3.6 km interpretive trail, 'Tracks Along the Coast', runs from the Sailing Club to Pinky Pt; 4 km SW. ***Denial Bay:*** visit the McKenzie ruins to see an early pioneering home and the heritage-listed landing where cargo was brought to shore. Denial Bay jetty is good for fishing and crabbing; 14 km W. ***Davenport Creek:*** see pure-white sandhills and swim in the sheltered creek. Beyond the sandhills is excellent surfing and waterskiing; 40 km W. ***South-east towns and beaches:*** Decres Bay for swimming, snorkelling and rock-fishing (10 km SE); Laura Bay with cove-swimming near the conservation park (18 km SE); Smoky Bay for safe swimming, fishing and boating (40 km SE); Pt Brown for surf beaches, salmon fishing and coastal walks (56 km SE). ***Penong:*** more than 40 windmills draw the town's water from underground. See historical memorabilia and local crafts at the Penong Woolshed Museum. Camel day rides and safaris on offer; 73 km W. ***Cactus Beach:*** renowned for its 'perfect' surfing breaks; 94 km W. ***Fowlers Bay:*** this town, surrounded by a conservation park, offers long, sandy beaches and excellent fishing; 139 km SW. ***North-east conservation parks and reserves:*** comprising Yellabinna, Yumbarra, Pureba, Nunnyah and Koolgera, an extensive wilderness area of dunes and mallee country. Rare species of wildlife live here, including dunnarts and mallee fowl. 4WD is essential, and visitors must be experienced in outback travel; north of Ceduna. ***Googs Track:*** 4WD trek from Ceduna to the Trans-Australia railway track (154 km N) through Yumbarra Conservation Park and Yellabinna Regional Reserve; details from visitor centre.

See also EYRE PENINSULA & NULLARBOR, p. 244

Clare

Pop. 3061
Map ref. 599 H2 | 603 J6

Cnr Main North and Spring Gully rds; (08) 8842 2131 or 1800 242 131; www.clarevalley.com.au

105.1 Trax FM, 639 ABC North and West

Clare is known as the 'Garden of the North'. In the mid-1800s, Edward John Eyre reported favourably on the area and pastoral settlement followed; the town came to be known as Clare after the county in Ireland. The land has proved as favourable as Eyre claimed and Clare continues to boast a rich agricultural industry, including the famous Clare Valley wine region. The first vines were planted by Jesuit priests at Sevenhill in 1851. The Sevenhill Cellars are still operated by Jesuit brothers and the monastery buildings, including the historic St Aloysius Church, are of special interest.

Old Police Station Museum: once a prison, a casualty hospital and housing for government employees, it is now a National Trust museum with historic artefacts and photographs; open weekends and public holidays or by appt; Neagles Rock Rd; (08) 8842 2376. ***Lookouts:*** Billy Goat Hill from Wright St and Neagles Rock Lookout on Neagles Rock Rd. ***Town walk:*** self-guide trail; brochure from visitor centre.

Clare Races: horseraces among the vines; Easter. ***Spanish Festival:*** Apr. ***Clare Gourmet Weekend:*** May. ***Clare Show:*** Oct.

Salt n Vines Restaurant: seafood and steaks; Wendouree Rd; (08) 8842 1796. ***Wild Saffron Gourmet Food & Catering:*** snacks and foodie treats; 288 Main North Rd; (08) 8842 4255. ***Penna Lane Wines:*** cellar-door platters; Penna La, Skilly Hills; (08) 8843 4364. ***Sevenhill Hotel:*** sturdy pub tucker; Main North Rd, Sevenhill; (08) 8843 4217. ***Skillogalee Winery Restaurant:*** stylish country fare; Trevarrick Rd, Sevenhill; (08) 8843 4311.

Brice Hill Country Lodge: elegant country resort; 56–66 Warenda Rd; (08) 8842 2925. ***Molly's Chase:*** secluded log cabins; Leighton Rd, Sevenhill; 0413 550 225. ***Mundawora Mews:*** historic bluestone cottages; Main North Rd; (08) 8842 3762. ***Skillogalee:*** luxury winery retreat; Trevarrick Rd, Sevenhill; (08) 8843 4311. ***Thorn Park by the Vines:*** heritage flair, graciously hosted; Quarry Rd, Sevenhill; (08) 8843 4304.

Clare Valley wine region The Clare Valley wine region consists of 12 valleys of undulating hills and lowlands, home to over 30 boutique and commercial wineries. The area is most famous for producing a fine riesling, grown to perfection because of the high terrain and continental climate. The wineries are around Clare, Polish Hill River, Mintaro, Sevenhill, Penwortham, Watervale and Auburn. For those wanting to see the villages and wineries on foot or bicycle, the 27 km Riesling Trail is a good option. It starts from either Clare or Auburn and follows an old railway line. Winery map from visitor centre.

Watervale: a historic town with a self-guide-walk leaflet available from the visitor centre; 12 km s. ***Blyth:*** a little country town overlooking the western plains. Take a short walk on the interpretive botanical trail or picnic at Brooks Lookout. Medika Gallery, originally a Lutheran church, offers an art gallery and Australian craft sales; 13 km w. ***Auburn:*** the birthplace in 1876 of poet C. J. Dennis. Take a self-guide walk through the National Trust historic precinct in St Vincent St; 26 km s. ***Scenic drive:*** travel south to Spring Gully Conservation Park with its walking tracks and rare red stringybarks.

See also MID-NORTH, p. 239

Coffin Bay

Pop. 582
Map ref. 602 D8

Beachcomber Agencies, Esplanade; (08) 8685 4057; www.coffinbay.net

89.9 Magic FM, 1485 AM ABC Eyre Peninsula and West Coast

A picturesque holiday town and fishing village on the shores of a beautiful estuary, Coffin Bay is popular particularly in summer, when the population quadruples. The bay offers sailing, waterskiing, swimming and fishing. The town was originally known as Oyster Town because of the abundant natural oysters in the bay, but they were dredged to extinction last century. Today the cultivated oysters are among the best in the country. The bay – in spite of what some locals will try to tell you – was named by Matthew Flinders in 1802 in honour of his friend Sir Isaac Coffin.

Fishing: in the bay or in game fishing areas; boat hire and charters available. ***Oyster Walk:*** 12 km walkway along foreshore and bushland from a lookout (excellent view of Coffin Bay) to Long Beach; brochure from visitor centre. ***Coffin Bay Explorer:*** catamaran offering a 3 hr cruise on Coffin Bay, visiting oyster leases and greeting dolphins and other spectacular sea life; bookings essential, 0428 880 621.

The Oysterbeds Restaurant: bayside seafood; 61 Esplanade; (08) 8685 4000.

Railway Shack 1: basic waterfront cottage; The Esplanade; 0427 844 568. ***Sheoak Eastside Apartment:*** elevated family beach house; 257 Esplanade; 0427 844 568. ***Mt Dutton Bay Woolshed:*** historic waterfront stone house; 1 Woolshed Dr, Mt Dutton Bay; (08) 8685 4031.

Coffin Bay National Park A mixture of rugged coastal landscapes and calm bays and waterways makes this diverse park a pleasure to wander and drive through. Conventional vehicles can access the eastern part of the park where walks through she-oak and samphire swamps reveal incredible birdlife. The beaches and lookouts provide a different perspective on this remote wilderness. 4WD vehicles and bushwalkers can access the western part, which includes Gunyah Beach, Pt Sir Isaac and the Coffin Bay Peninsula. The park extends south and west of town; (08) 8688 3111.

Kellidie Bay Conservation Park: a limestone landscape popular for walking and canoeing; eastern outskirts of Coffin Bay. ***Yangie Trail:*** the 10 km trail starts at Coffin Bay and travels south-west via Yangie Bay Lookout, which offers magnificent views to Pt Avoid and Yangie Bay. ***Mt Dutton Bay:*** features a restored heritage-listed jetty and woolshed, the latter now a shearing and farming museum; 40 km N. ***Farm Beach:*** popular swimming spot; 50 km N. ***Gallipoli Beach:*** location for the film Gallipoli (1981); 55 km N. ***Scenic coastal drive:*** between Mt Hope and Sheringa; 105 km N.

See also EYRE PENINSULA & NULLARBOR, p. 244

Coober Pedy

see inset box on next page

Coonawarra

Pop. 310
Map ref. 590 A12 | 592 A2 | 601 G10

27 Arthur St, Penola; (08) 8737 2855; www.wattlerange.sa.gov.au

 96.1 Star FM, 1161 AM ABC South East

Unlike many other Australian wine regions, Coonawarra was a planned horticulture scheme – and a very successful one at that. John Riddoch, a Scottish immigrant, acquired extensive lands in South Australia's south-east in the late 1800s. He subdivided 800 hectares of his landholding specifically for orchards and vineyards. Prominent wine professionals such as Wolf Blass damned this region as a place that could never produce decent wine, and the original John Riddoch wine estate was nearly sold to the Department of Forestry and Lands (thankfully, David Wynn purchased the property and it is now Wynns Coonawarra). In the 1950s, large wine companies such as Penfolds and Yalumba finally began recognising the depth of the region's reds, and opinions began to change. Coonawarra's famed terra rossa (red earth), combined with the region's particular climate, is now known to create some of the best cabernet sauvignon in the country, as well as excellent shiraz, merlot, riesling and chardonnay. Wynns Coonawarra produces world-class shiraz and cabernet sauvignon, some for purchase at reasonable prices. Other excellent wineries to visit are Balnaves of Coonawarra, Brands of Coonawarra, Majella and Zema Estate.

Coonawarra Cup: Jan. ***Coonawarra After Dark:*** cellar doors open in the evenings; Apr. ***Penola Coonawarra Festival:*** arts, food and wine; May. ***Coonawarra Cellar Dwellers:*** July. ***Cabernet Celebrations:*** Oct.

Kitchen @ The Poplars: tasty breakfasts and lunches; The Poplars Winery, Riddoch Hwy; (08) 8736 3065. ***Upstairs at Hollick:*** elegant winery dining; Ravenswood La; (08) 8737 2752.

Punters Vineyard Retreat: contemporary vineyard escape; Punters Corner Vineyard, cnr Riddoch Hwy and Racecourse Rd; (08) 8737 2007. ***The Menzies Retreat:*** stylish modern vineyard accommodation; Yalumba Winery, Riddoch Hwy; (08) 8737 3603.

See also Limestone Coast, p. 242

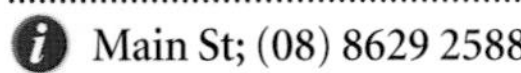

Cowell

Pop. 883
Map ref. 599 B1 | 602 G5

Main St; (08) 8629 2588.

89.9 Magic FM, 639 AM ABC North and West

This pleasant Eyre Peninsula township is on the almost land-locked Franklin Harbour – its entrance is merely 100 metres wide. Matthew Flinders sailed past here in 1802 and, understandably, mistook the harbour for a large lagoon. The sandy beach is safe for swimming and the fishing is excellent. Oyster farming is a relatively new local industry, and fresh oysters can be purchased year-round. The world's oldest and perhaps largest jade deposit is in the district. Discovered in the Minbrie Range in 1965, the deposit is believed to have been formed around 1700 million years ago by the shifting of the earth's surface.

Franklin Harbour Historical Museum: in the old post office and its attached residence (1888), now operated by the National Trust and featuring local history displays; closed Mon and Tues; Main St; (08) 8629 2686. ***Ruston Proctor Steam Tractor Museum:*** open-air agricultural museum; Lincoln Hwy. ***Cowell Jade Motel:*** showroom and sales of local jade jewellery; Lincoln Hwy. ***Turner Aquaculture:*** tours of an oyster factory; Oyster Dr. ***Foreshore and Mangrove Boardwalk***: ideal for a picnic, with barbecue area and adventure playground for the kids; The Esplanade. ***Boat hire:*** from the caravan park.

Fireworks Night: includes street party; Dec.

Franklin Harbour Hotel: old-style pub with views; 1 Main St; (08) 8629 2015 or 1800 303 449.

Scenic drive: 20 km drive south to Port Gibbon along a coast renowned for its history of wrecked and sunken ketches; interpretive signs detail the history at each site. ***Franklin Harbour Conservation Park:*** coastal peninsula park of sand dunes and mangrove habitat, popular for bush camping and fishing; 5 km s. ***May Gibbs Memorial:*** marks the location of children's author May Gibbs' first home; Cleve Rd; 10 km s. ***The Knob:*** good fishing from sheltered beach and rocks; 13 km s. ***Lucky Bay:*** safe swimming for children and the start of a 4WD track to Victoria Pt with excellent views of the harbour. Sea SA operates a ferry from here to Wallaroo on the Yorke Peninsula 4 times a day on weekdays and twice a day on weekends; bookings (08) 8823 0777; 16 km E. ***Port Gibbon:*** old shipping port with remains of original jetty. Sea lions are visible from the short walk to the point; 25 km s. ***Yeldulknie Weir and Reservoir:*** picnics and walking; 37 km w. ***Cleve:*** a service town with murals depicting its early days and an observation point at Tickleberry Hill; 42 km w. ***Arno Bay:*** a holiday town with sandy beaches and a jetty for fishing. Regular yacht races are held on Sun in summer; 44 km sw.

See also Eyre Peninsula & Nullarbor, p. 244

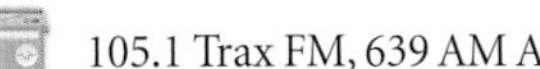

Crystal Brook

Pop. 1188
Map ref. 603 I4 | 604 H12

Port Pirie Regional Tourism and Arts Centre, 3 Mary Elie St, Port Pirie; (08) 8633 8700 or 1800 000 424.

105.1 Trax FM, 639 AM ABC North and West

Crystal Brook serves the sheep and wheat country at the southern point of the Flinders Ranges. It once formed part of a vast sheep station, Crystal Brook Run, which extended from the current town to Port Pirie in the north-west. The country feel of the town begins on entering the tree-lined main street.

National Trust Old Bakehouse Museum: local history collection in the town's first 2-storey building; open 2–4pm Sun and public holidays or by appt; Brandis St; (08) 8636 2328. ***Crystal Crafts:*** local craft; Bowman St. ***Creekside parks:*** popular spots for picnics.

Crystal Brook Show: Aug.

The Big Goanna: we love our things big, be it fruit, rocking horses, or in this case, reptiles; 3 km N. ***Bowman Fauna Park:*** enjoyable walks, including part of the Heysen Trail, around ruins of the Bowman family property, Crystal Brook Run (1847); 8 km E. ***Koolunga:*** a small community, home to the mythical bunyip – in 1883 2 attempts to capture the beast were unsuccessful. Also craft outlets and the Bunyip River Walk on the banks of Broughton River; 10 km E. ***Gladstone:*** set in rich rural country in the Rocky River Valley. Heritage-listed Gladstone Gaol offers daily tours. Try traditional soft drinks, including Old Style Ginger Beer, at the Trends Drink Factory in Sixth Ave (open weekdays) or discover the town's history on foot by picking up a map from the caravan park; 21 km NE. ***Redhill:*** riverside walk, museum, craft shop and antique shop; 25 km s. ***Laura:*** boyhood home of C. J. Dennis, author of *The Songs of a Sentimental Bloke*, known for its cottage crafts, art galleries and historic buildings. Leaflet available for self-guide walking tour from Biles Art Gallery,

Herbert St. The Folk Fair each Apr brings thousands of visitors to the town; 32 km N. West of town is the Beetaloo Valley and Reservoir, a pleasant picnic spot in cooler months. ***Snowtown:*** surrounded by large salt lakes that change colour according to weather conditions. Lake View Dr is a scenic 6 km drive around the lakes. Lochiel–Ninnes Rd Lookout provides panoramic country and lake views; 50 km S.

See also YORKE PENINSULA, p. 240

Edithburgh

Pop. 394
Map ref. 599 E8 | 602 H9

Cnr Weaver and Towler sts, Stansbury; (08) 8852 4577; www.yorkepeninsula.com.au

 98.9 Flow FM, 1044 5CS AM

Edithburgh is located on the foreshore at the south-eastern tip of Yorke Peninsula. This is an area synonymous with shipwrecks and, although reflecting tragic maritime days of old, it is a source of excitement for the diving enthusiast. Despite the construction of a lighthouse in 1856, over 26 vessels were wrecked on the coast between West Cape in Innes National Park and Troubridge Point just south of Edithburgh. Today Edithburgh is a popular coastal holiday destination overlooking Gulf St Vincent and Troubridge Island.

Edithburgh Museum: a community museum with local history of the town and region featuring a historical maritime collection; open 2–4pm Sun and public holidays or by appt; Edith St; (08) 8852 6187. ***Native Flora Park:*** walk through eucalypts and casuarinas and see a variety of birdlife; Ansty Tce. ***Bakehouse Arts and Crafts:*** local handcrafts and produce in a historic 1890 building; Blanche St. ***Town jetty:*** built in 1873 to service large shipments of salt found inland, it offers views to Troubridge Island and is popular with anglers; end of Edith St. ***Natural tidal pool:*** excellent for swimming; foreshore. ***Nature walks:*** extend south to Sultana Pt or north to Coobowie.

Gala Day: family day of entertainment and stalls; Oct.

Faversham's Restaurant: fine dining; Edithburgh House, 7 Edith St; (08) 8852 6373.

Edithburgh House: gracious heritage B&B; 7 Edith St; (08) 8852 6373. ***Tipper's Edithburgh:*** renovated rustic hideaway; 35 Blanche St; (08) 8852 6181. ***Troubridge Island Hideaway:*** offshore lighthouse keeper's cottage; Troubridge Island; (08) 8852 6290.

Dive sites Discover the south coast of Yorke Peninsula with *The Investigator Strait Maritime Heritage Trail* brochure that includes the history and maps of 26 dive sites. By far the worst recorded shipwreck was that of the *Clan Ranald*, a huge steel steamer that, through incompetence and greed, was wrecked in 1909 just west of Troubridge Hill. The disaster claimed 40 lives – 36 bodies were later buried in the Edithburgh Cemetery.

Wattle Point Wind Farm: currently Australia's largest wind farm with 55 turbines; free viewing area 3 km SW of Edithburgh. ***Sultana Pt:*** fishing and swimming; 2 km S. ***Coobowie:*** a coastal town popular for swimming; 5 km N. ***Troubridge Island Conservation Park:*** home to penguins, black-faced shags and crested terns; tours available (30 min by boat) from town. ***Scenic drive:*** west along the coast to Innes National Park; *see Yorketown*.

See also YORKE PENINSULA, p. 240

Elliston

Pop. 199
Map ref. 602 B5

Town Hall, 6 Memorial Dr; (08) 8687 9200.

89.5 Magic FM, 693 AM ABC Eyre Peninsula and West Coast

Nestled in a range of hills on the shores of picturesque Waterloo Bay is the small community of Elliston. The waters of the bay used to have rich abalone beds, but fierce exploitation in the 1960s decimated them. Thanks to a hatchery and rehabilitation, the abalone population is now back on the increase. The rugged and scenic coastline is spectacular and – with its excellent fishing and safe swimming beaches – Elliston is becoming a popular holiday destination. The 12-kilometre stretch of coast north of Elliston, known as Elliston's Great Ocean View, is said to rival the landscape of Victoria's Great Ocean Road, but on a smaller scale.

Town Hall Mural: the Southern Hemisphere's largest mural, which represents the history of the town and district; Main St. ***Jetty:*** the heritage-listed 1889 jetty has been restored and is lit by night.

Sculptures on the Cliff: Anzac Day weekend, Apr. ***Australian Salmon Fishing Competition:*** June–Aug.

Elliston's Great Ocean View: scenic cliff-top drive north of town to adjacent Anxious Bay, with fabulous coastal views. Along the way is Blackfellows, reputedly one of the best surfing beaches in Australia. ***Anxious Bay:*** good fishing from beach and ledges for King George whiting. The boat ramp here provides access to Waldegrave Island (4 km offshore) and Flinders Island (35 km offshore), both good for fishing and seal-spotting (seek local advice about conditions before departing). ***Locks Well:*** a long stairwell down to a famous salmon-fishing and surf beach with coastal lookout; 12 km SE. ***Walkers Rock:*** good beaches for swimming and rock-fishing; 15 km N. ***Lake Newland Conservation Park:*** significant dunes separate the park's salt lakes and wetlands from the sea. Walk along bush tracks or try a spot of fishing; 26 km N. ***Scenic drive:*** to Sheringa (40 km SE). From Sheringa Beach, a popular fishing spot, see whales, dolphins and seals offshore. ***Talia Caves:*** spectacular scene of caves with waves crashing on edge. Another good spot for beach-fishing; 45 km N.

See also EYRE PENINSULA & NULLARBOR, p. 244

Gawler

Pop. 20 002
Map ref. 596 E2 | 597 D5 | 601 C2 | 603 J8

2 Lyndoch Rd; (08) 8522 9260; www.gawler.sa.gov.au

89.1 BBB FM, 639 AM ABC North and West

Set in the fork of the North and South Parra rivers and surrounded by rolling hills, it is no wonder that Gawler was picked, in 1839, as the site of South Australia's first country town. The grand architecture of that era can be seen in its stately homes and buildings, especially in the Church Hill State Heritage Area. Gawler is today a major service centre to a thriving agricultural district and has a growing Adelaide commuter population.

Historical walking trail The excellent trail brochure guides the visitor past stately buildings and homes, many with original cast-iron lacework. The Church Hill State Heritage Area provides a fascinating snapshot of town planning in the 1830s, and the 2.4 km walk around it includes a look at several churches, the old school, the courthouse, and cottages of the Victorian era. Also on the way are Gawler Mill, Eagle Foundry (now a popular B&B), the

continued on p. 257

COOBER PEDY

Pop. 1470

Map ref. 606 A10 | 609 O10

Council offices, Hutchison St; (08) 8672 4617 or 1800 637 076; www.opalcapitaloftheworld.com.au

104.5 Dusty Radio FM, 106.1 FM ABC Local

On 1 February 1915 a group of gold prospectors discovered opal in the area surrounding Coober Pedy. It was to become the biggest opal field in the world, which today provides around 80 per cent of the world's gem-quality opals. The name Coober Pedy is derived from the Aboriginal phrase 'Kupa Piti', loosely translating to 'white man's hole in the ground'. The town's unique underground style of living was first established by soldiers returning from World War I who were used to trench life. Today much of the population call the 'dugouts' home – as ideal places to escape the severe summer temperatures and cold winter nights. The landscape of Coober Pedy is desolate and harsh, and dotted with thousands of mines.

Umoona Underground Mine and Museum This award-winning underground centre provides an all-round look at Coober Pedy. A detailed town history and an Aboriginal interpretive centre comprehensively document Coober Pedy's evolution. Experience 'dugout' life in an underground home or mine life on an on-site mine tour. An excellent documentary on Coober Pedy is shown in the underground cinema. Hutchison St; (08) 8762 5228.

Old Timers Mine: museum in an original 1916 mine featuring 3 large opal seams and interpretive centre with self-guide walk; Crowders Gully Rd. ***Underground churches:*** St Peter and St Pauls was the first underground church in the world; Hutchison St. Other interesting churches include the Catacomb Church on Catacomb Rd and the Serbian Orthodox Church with its 'ballroom' style. ***Desert Cave:*** international underground hotel with shopping complex and display gallery detailing the early hardships of miners in Coober Pedy; open 8am–8pm daily; Hutchison St. ***Underground Art Gallery:*** works of central Australian artists, including Aboriginal pieces; Hutchison St. ***Opal retailers:*** outlets offer jewellery and opal stone sales; many demonstrate the skill required to cut and polish opals. ***Big Winch Lookout:*** monument and lookout over town; Italian Club Rd. ***Outback scenic and charter flights:*** both local and outback tours; depart from airport; bookings (08) 8672 3067. ***Mine tours:*** to local mines; details from visitor centre.

Opal Festival: Easter. ***Coober Pedy Greek Glendi:*** celebration of Greek culture; July. ***Horseracing:*** Aug.

Tom & Mary's Greek Taverna: famous Greek cafe; Hutchinson St; (08) 8672 5622.

Desert Cave Hotel: full-service outback luxury; Hutchison St; (08) 8672 5688 or 1800 088 521. ***Down to Erth B&B:*** underground 2-bedroom apartment; Lot 1795 Wedgetail Cres; (08) 8672 5762.

Moon Plain and the Breakaways The rocky landscape of Moon Plain (15 km NE) has been the backdrop for many movies, especially those with a science-fiction bent. Likewise, the 40 sq km reserve of the Breakaways (30 km N). The stark arid landscape of flat-topped outcrops and stony gibber desert is breathtaking, as is the wildlife that has adapted to these harsh conditions. Passes to the reserve are available from the visitor centre and other outlets in town. Return via the road past part of the Dog Fence, a 5300 km fence built to protect sheep properties in the south from wild dogs.

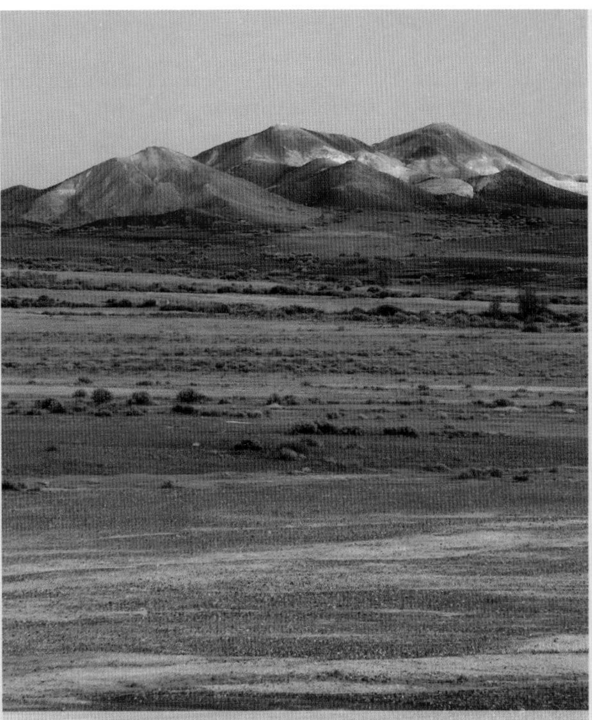

[THE BREAKAWAYS, NEAR COOBER PEDY]

The Mail Run This overland adventure is a unique way of discovering the remote outback of SA. Travelling with Coober Pedy's mailman, the tour travels past waterholes and through scenic landscapes on its delivery run to Oodnadatta, William Creek and the remote cattle stations in between. Travelling 600 km over outback roads and the renowned Oodnadatta Track, the mailman offers up fascinating stories and history of the landscape and people. Tours depart from the Underground Bookshop at 9am on Mon and Thurs; bookings (08) 8672 5226 or 1800 069 911.

Underground Pottery: handmade pottery depicting the colours and landscape of the desert; 2 km W. ***William Creek:*** the smallest town in SA, situated in Anna Creek station, the world's largest cattle station. A race meeting is held the weekend before Easter each year. Flights offered over Lake Eyre; 166 km E. ***Painted Desert:*** rich colours paint the Arckaringa Hills, also noted for flora and fauna; 234 km N.

Travellers note: *Opal fields are pocked with diggings. Beware of unprotected mine shafts. For safety reasons, visitors to the mines are advised to join a tour. Avoid entering any field area unless escorted. Trespassers on claims can be fined a minimum of $1000. Coober Pedy is the last stop for petrol between Cadney Homestead (151 km N) and Glendambo (252 km S).*

See also **FLINDERS RANGES & OUTBACK, p. 243**

HAHNDORF

Pop. 1806

Map ref. 596 F9 | 597 D10 | 598 H3 | 603 K9

Adelaide Hills Visitor Information Centre, 68 Main St; (08) 8388 1185; www.visitadelaidehills.com.au

639 AM ABC North and West

In the heart of the Adelaide Hills is Hahndorf, Australia's oldest surviving German settlement. Prussian Lutheran refugees fleeing religious persecution in their homelands settled the area in the late 1830s. The town has retained a distinctly Germanic look and many local businesses and attractions are still operated by descendants of the original German pioneers. The surrounding countryside with its rolling hills, historic villages, vineyards, gourmet-produce farms and native bushland is renowned as a tourist destination.

Main Street: a mix of historic buildings, German-style bakeries, delis, cafes, restaurants and art and craft shops. ***The Cedars:*** historic paintings, gardens and home of famous landscape artist Hans Heysen; closed Sat; Heysen Rd. ***Hahndorf Academy:*** displays work of local artists in SA's largest regional gallery; Main St. ***Clocks and Collectables:*** housed in a historic butcher shop specialising in cuckoo and grandfather clocks. ***Hahndorf Farm Barn:*** interactive farm animal shows, with petting and feeding; Mt Barker Rd. ***Hahndorf Hill Winery:*** boutique cool-climate winery with tastings and sales; open weekends; Pains Rd. ***Beerenberg Strawberry Farm:*** balcony viewing of the farm and packing area, jam and condiment sales, and berry-picking Oct–May; Mt Barker Rd. ***Historic Hahndorf:*** obtain *A Guide to Historic Hahndorf*, which lists 42 historic properties; available from visitor centre.

[THE CEDARS, HOME OF ARTIST HANS HEYSEN]

Heysen Festival: major 10-day event celebrating the life of Hans Heysen, with the Heysen Art Prize, food, wine and entertainment; Sept–Oct. ***Oktoberfest:*** Oct.

Hahndorf Hill Winery: sophisticated cellar-door lunches; Lot 10 Pains Rd; (08) 8388 7512. ***The Lane Vineyard:*** chic menu and fine views; Ravenswood La; (08) 8388 1250. ***Udder Delights Cheese Cellar:*** snacks, platters and cheese tastings; 91A Main Rd; (08) 8388 1588. ***The Organic Market & Cafe:*** famed hills fare; 5 Druids Ave, Stirling; (08) 8339 7131. ***Petaluma Restaurant:*** sublime modern cuisine; Bridgewater Mill, Mt Barker Rd, Bridgewater; (08) 8339 9200.

The Manna: sleek modern motel units; 25 Main St; (08) 8388 1000 or 1800 882 682. ***Adelaide Hills Country Cottages:*** romantic, rural hideaways; Oakwood Rd, Oakbank; (08) 8388 4193. ***Hannah's Cottage:*** secluded country haven; 'Botathan', Lot 14 Jones Rd, Balhannah; (08) 8388 4148. ***Mount Lofty House:*** gracious boutique hotel; 74 Mt Lofty Summit Rd, Crafers; (08) 8339 6777. ***Thorngrove Manor Hotel:*** authentic baronial-themed luxury; 2 Glenside La, Stirling; (08) 8339 6748.

Cleland Conservation Park This park protects a variety of vegetation, from the stringybark forests in the highlands to the woods and grasses of the lowlands. It also includes panoramic views from the Mt Lofty Summit viewing platform. A highlight is the award-winning Cleland Wildlife Park where native animals wander freely about. Kids will enjoy the daily animal-feeding shows and the guided night tours. Bookings (08) 8339 2444. For a rich cultural experience, take the guided Aboriginal tour on the Yurridla Trail. Mt Lofty Summit Rd; 14 km NW.

Belair National Park The oldest national park in SA was declared in 1891. There are plenty of activities, including 5 defined bushwalks ranging from the easy Wood Duck Walk to the more challenging 6.5 km Waterfall Walk, and cycling and horseriding tracks. Visitors can also visit Old Government House, built in 1859 to serve as the governor's summer residence; open Sun and public holidays; (08) 8278 5477. 19 km W.

Mt Barker: a historic town renowned for its gourmet outlets. The *SteamRanger* tourist train operates from here, including regular trips to Strathalbyn and on to the coastal section from Goolwa to Victor Harbor (this section is known as the Cockle Train; *see Goolwa*). For *SteamRanger* information and bookings contact 1300 655 991. Local events include the Highland Gathering and Heritage Festival in Feb; 6 km SE. ***Bridgewater:*** a historic town with excellent gardens and Petaluma's Bridgewater Mill, which has wine-tastings, sales and an up-market restaurant for lunch in the historic flour mill; 6 km W. ***Oakbank:*** craft shops and historic buildings with self-guide heritage-walk brochure available from local businesses. The Oakbank Easter Racing Carnival, held each Apr, is the Southern Hemisphere's biggest picnic race meeting and brings thousands of visitors to the town each year; 7 km N. ***Aldgate:*** a historic village featuring art and craft shops and historic sites that include the Aldgate Pump and the National Trust–listed Stangate House with its extensive camellia garden; 8 km W. ***Stirling:*** renowned for its European gardens and architecture, including the National Estate–listed Beechworth

Heritage Garden (Snows Rd). See colourful parrots in the aviaries at Stirling Parrot Farm (Milan Tce); 10 km W. ***Warrawong Sanctuary:*** large native animal reserve for reintroduced and endangered species, with walking trails and a boardwalk around Platypus Lakes. Guided dawn and evening tours are available; bookings (08) 8370 9197; 10 km SW via Mylor. ***Jupiter Creek Goldfields:*** walking trails with interpretive signs across historic fields discovered in 1852; 12 km SW. ***Woodside Heritage Village:*** includes Melba's Chocolate Factory (with guided tours and sales) as well as craft studios with artisans at work; Woodside; 13 km NE. ***Mt Lofty Botanic Gardens:*** Australia's largest botanic gardens, with walking trails past cool-climate garden species, on the eastern face of Mt Lofty; access via Summit or Piccadilly rds; 14 km NW. ***Wittunga Botanic Gardens:*** native plants; Shepherds Hill Rd; 21 km W at Blackwood.

See also ADELAIDE HILLS, p. 237

Anglican Church with its pipe organ (open Sun) and Para Para Mansion (1862), a grand mansion that features a domed ballroom and has had British royalty as its guests (closed to public).

Gawler Museum: located in the Old Telegraph Station, this National Trust museum displays the history of Gawler's pioneer past; closed Sat–Mon; Murray St. ***Historic main street walking tour:*** the town is filled with some of SA's oldest and most historic hotels, pubs and churches; brochure from visitor centre. ***The Food Forest:*** award-winning permaculture farm producing 160 varieties of organically certified food; tours available; on the Gawler River. ***Fielke Cricket Bats:*** the only cricket bat maker in SA, visit the workshop and see how the English willow bats are meticulously handcrafted; Crown St. ***Dead Man's Pass Reserve:*** so named because an early pioneer was found dead in the hollow of a tree. It has picnic facilities and a walking trail; southern end of Murray St. ***Community Art Gallery:*** local artwork on display; 23rd Street. ***Adelaide Soaring Club:*** glider flights over Gawler and region; Wells Rd; (08) 8522 1877.

Market: Gawler Railway Station; Sun mornings. ***Antique and Collectors Auction:*** Gawler Greyhound Pavilion; 1st Sun each month. ***Band Festival:*** Mar. ***Gawler Gourmet and Heritage Festival:*** even-numbered years, Apr. ***Barossa Cup:*** horseracing; May. ***Gawler Show:*** Aug. ***Christmas Street Festival:*** Dec.

The Garden Pavilion: sumptuous French Provincial–style garden rooms; AL-RU Farm, One Tree Hill Rd, One Tree Hill; (08) 8280 7353. ***The Miner's Cottage:*** secluded heritage charm; Goldfields Rd, Cockatoo Valley; (08) 8524 6213 or 1800 646 213.

Roseworthy Agricultural Museum: a dryland farming museum featuring vintage farm equipment, located at the Roseworthy campus of the University of Adelaide; open every Wed and 3rd Sun each month; (08) 8303 7739; 15 km N. ***Freeling:*** a rural town with a self-guide historical walk; 17 km NE. ***Two Wells:*** named for 2 Aboriginal wells found by original settlers, forgotten, then rediscovered in 1967. It has craft shops with local crafts; Main Rd; 24 km W. ***Stockport Astronomical Facility:*** run by the Astronomical Society of SA, featuring impressive observatory with public viewing nights; (08) 8338 1231; 30 km N.

See also BAROSSA VALLEY, p. 238

Goolwa

Pop. 5881

Map ref. 598 G8 | 601 C4 | 603 J11

Signal Point River Murray Interpretive Centre, Goolwa Wharf; (08) 8555 3488; www.alexandrina.sa.gov.au

87.6 Alex FM, 639 AM ABC North and West

Goolwa is a rapidly growing holiday town on the last big bend of the Murray River before it reaches open waters. In 2007 Goolwa became the first town in Australia to be declared a Cittaslow, or 'slow town', joining a network of other towns in Europe and throughout the world which aim to improve the quality of life in towns. Goolwa was originally surveyed as the capital of South Australia, but Adelaide was later thought to be a better option. Goolwa did, however, boom as a river port from the 1850s to the 1880s – in the golden days of the riverboats. The area is excellent for fishing, with freshwater fishing in the Murray and saltwater fishing in the Southern Ocean and the Coorong, as well as boating, surfing, watersports, birdwatching and photography.

Signal Point River Murray Interpretive Centre Learn about the river and district before European settlement and see the impact of later development. There is also a detailed history on Goolwa's river port, the paddlesteamers and local Aboriginal people. A highlight is the 3D model showing how the Murray River interacts with the SA landscape. The Wharf.

National Trust Museum: documents the history of Goolwa and early navigation of the Murray River; Porter St. ***South Coast Regional Arts Centre:*** explore the restored old court and police station and see exhibitions by local artists; Goolwa Tce. ***Goolwa Barrage:*** desalination point to prevent salt water from reaching the Murray River; open Mon–Fri; Barrage Rd. ***Armfield Slipway:*** a working exhibition of wooden boatbuilding and restoration; open Tues mornings and Fri afternoons; Barrage Rd. ***Horse-Drawn Railway Carriage:*** first carriages used in SA between Goolwa and Port Elliot from 1854; Cadell St. ***Goolwa Beach:*** popular surfing and swimming beach with large cockles to be dug up (in season); end of Beach Rd. ***Cockle Train:*** journey around Encounter Bay from Goolwa Wharf to Victor Harbor, stopping at Port Elliot. The train has been operating since 1887, and its name comes from the abundance of large cockles found on Goolwa's surf beach. It forms part of the longer *SteamRanger* tourist railway from Mount Barker; *see Hahndorf*; operates most weekends and public/school holidays; bookings (08) 8231 4366. ***Cruises:*** day tours to the mouth of the Murray, the Coorong, the Barrages and the Lower Murray; details and bookings at visitor centre. ***Scenic flights:*** over the region; contact visitor centre for details. ***Goolwa Heritage Walk:*** a self-guide tour to see the 19th-century architecture of the river port's boom days; brochure available from visitor centre.

Goolwa Wharf Markets: 1st and 3rd Sun each month. ***Milang to Goolwa Freshwater Sailing Classic:*** Jan. ***Wooden Boat Festival:*** odd-numbered years, Mar.

SOUTH AUSTRALIA

Aquacaf Gourmet Cafe: innovative riverside fare; Barrage Rd; (08) 8555 1235. ***Blues Restaurant:*** relaxed snacks and seafood; Main Goolwa Rd, Middleton; (08) 8554 1800.

Jackling Cottage: updated heritage retreat; 18 Oliver St; 0433 571 927. ***The Yellow House:*** colourful waterfront family escape; 3 Barrage Rd; 0417 861 940. ***Boatman's Cabin:*** relaxed lakefront boathouse; Island View Dr, Clayton Bay; (08) 8537 0372. ***Coorong Beach House:*** stylish waterfront family escape; 1 Goolwa Channel Dr, Hindmarsh Island; (08) 8383 0504.

Hindmarsh Island: Captain Sturt located the mouth of the Murray River from here in 1830 – visit the Captain Sturt Lookout and monument. The island is now popular for both freshwater and saltwater fishing and is home to the kid-friendly Narnu farm, which offers horseriding, cottage accommodation and the opportunity to milk a cow; east of town; farm bookings (08) 8555 2002. ***Currency Creek Game Reserve:*** feeding grounds, breeding rookeries and hides for many waterbirds; access by boat only; 3 km E. ***Currency Creek:*** Lions Park is a popular picnic spot with a walking track along the creek. Near town is an Aboriginal canoe tree (a eucalypt carved to make a canoe) and also the Currency Creek Winery with a restaurant and fauna park; 7 km N.

See also FLEURIEU PENINSULA, p. 235

Hahndorf

see inset box on previous page

Hawker

Pop. 225
Map ref. 600 A12 | 604 H8

Hawker Motors, cnr Wilpena and Cradock rds; (08) 8648 4022 or 1800 777 880; www.hawkersa.info

105.9 Magic FM, 639 AM ABC North and West

This small outback town was once a thriving railway centre, and historic buildings are still well preserved in its streets. Hawker was also once an agricultural region producing bumper crops of wheat. Serious drought sent the crops into decline and the industry died. Today Hawker is the place to begin exploring the fantastic natural attractions of the southern Flinders Ranges.

Fred Teague's Museum: local history displays; Hawker Motors, cnr Wilpena and Cradock rds. ***Jeff Morgan Gallery:*** including a 30 m painting of the view from Wilpena Pound; Cradock Rd. ***Heritage walk:*** self-guide walk on numbered path; brochure from visitor centre. ***Scenic flights and 4WD tours:*** contact visitor centre for details.

Horseracing: May. ***Art Exhibition:*** Sept/Oct.

Meaney's Rest: quaint sandstone cottage; 68 Craddock Rd; (08) 8648 4022 or 1800 777 880. ***Arkaba Station:*** stylish homestead B&B; Wilpena Rd; (02) 9571 6399 or 1300 790 561.

Jarvis Hill Lookout: walking trail with views over the countryside; 7 km SW. ***Yourambulla Caves:*** Aboriginal rock paintings in hillside caves; 12 km SW. ***Willow Waters:*** popular picnic spot with a short walk to Ochre Wall; 20 km E off Cradock Rd. ***Moralana Scenic Drive:*** 22 km drive with superb views of Wilpena Pound and the Elder Range; leaves Hawker–Wilpena Rd 23 km N. ***Cradock:*** a tiny town with National Heritage–listed St Gabriel's Church (1882); 26 km SE. ***Kanyaka Homestead Historic Site:*** ruins of the homestead, stables and woolshed once part of a large sheep run, with informative displays explaining the history of each ruin; 28 km SW. ***Kanyaka Death Rock:*** once an Aboriginal ceremonial site, it overlooks a permanent waterhole; near Kanyaka Homestead ruins. ***Long-distance trails:*** close to Hawker you can pick up sections of the Heysen (walking) and Mawson (cycling) trails; information from visitor centre.

See also FLINDERS RANGES & OUTBACK, p. 243

Innamincka

Pop. 15
Map ref. 607 N7 | 662 F8

Inland Mission; (08) 8675 9909.

1602 AM ABC North and West

This tiny settlement is built around a hotel and trading post on Cooper Creek. The first European explorer to visit the area was Charles Sturt, who discovered the Cooper in 1846 while vainly searching for an inland sea. It was also the final destination of the ill-fated Burke and Wills expedition. In 1860 all but one of Burke and Wills' party perished near the creek. John King survived owing to the outback skills of the Aboriginal people who found him. Innamincka was once a customs depot and service centre for surrounding pastoral properties, but now mainly services travellers, many of whom come from the south via the intrepid Strzelecki Track (for four-wheel-drive vehicles only).

Australian Inland Mission Built in 1928 to service the medical needs of remote pastoral properties, this mission was attended by a rotating staff of 2 nurses on horseback. Injured workers, flood victims and even fallen jockeys from the races called on their expertise. The mission was abandoned in the early 1950s when the Royal Flying Doctor Service began providing services. The restored classic outback building now houses the national parks headquarters as well as a tribute museum to the nurses who faced the trials of this isolated region.

Boat hire and tours: fishing and cruising trips; contact hotel for details on (08) 8675 9901.

Picnic Race Meeting: Aug.

Innamincka Regional Reserve This isolated but spectacular reserve is popular with 4WD enthusiasts and nature lovers. It covers 13 800 sq km and comprises important wetland areas, the most impressive being Coongie Lakes (112 km NW). This internationally significant wetland area is a haven for wildlife, particularly waterbirds. Closer to Innamincka is Cullyamurra Waterhole, on Cooper Creek (16 km NE), for bush camping and fishing. Aboriginal rock carvings and the Cullyamurra Choke are accessible by foot at the eastern end of the waterhole. 4WD is essential in parts of the park; after heavy rains roads become impassable. A Desert Parks Pass is required; visit the Inland Mission in Innamincka or phone 1800 816 078.

Memorial plaques: to Charles Sturt and Burke and Wills; 2 km N. ***Burke and Wills Dig Tree:*** famous Dig Tree where supplies were buried for their expedition; 71 km across border in Qld. ***Strzelecki Regional Reserve:*** sand-dune desert country with birdwatching and camping at Montecollina Bore; 167 km SW.

Travellers note: *Motorists intending to travel along the Strzelecki Track should ring the Northern Roads Condition Hotline on 1300 361 033 to check conditions before departure. There are no supplies or petrol between Lyndhurst and Innamincka.*

See also FLINDERS RANGES & OUTBACK, p. 243

Jamestown

Pop. 1408
Map ref. 603 J4 | 605 I11

Country Retreat Caravan Park, 103 Ayr St; (08) 8664 0077; www. clarevalley.com.au

105.1 Trax FM, 639 AM ABC North and West

Jamestown survived the demise of wheat crops in the late 1800s to become an important service town to the thriving agricultural farmlands of the Clare Valley. John Bristow Hughes took up the first pastoral lease in 1841 and the strength of stud sheep and cattle farms, cereals, dairy produce and timber grew rapidly. A look at the names of towns in South Australia will reveal that the governors, politicians and surveyors of the day were bent on commemorating themselves or people they liked. Jamestown followed this trend, named after Sir James Fergusson, then state governor.

Railway Station Museum: a National Trust museum detailing local rail and Bundaleer Forest history and featuring the Both-designed iron lung (invented at Caltowie); Mannanarie Rd; (08) 8664 0522. ***Heritage murals:*** on town buildings. ***Belalie Creek:*** parks for picnics along the banks; floodlit at night. ***Town and cemetery walks:*** self-guide tours; brochure available from caravan park.

Producers Market: 1st Sun each month. ***Bundaleer Forest Weekend:*** fine music, food and wine in the autumn; Mar. ***Jamestown Show:*** largest 1-day show in SA; Oct. ***Christmas Pageant:*** Dec.

North Bundaleer: lavish, luxurious heritage homestead; RM Williams Way; (08) 8665 4024.

Bundaleer Forest Reserve This plantation forest, established in 1876, was the first in the world. Walking tracks start from the Arboretum, Georgetown Rd and the picnic area, and range from botanic walks to historic trails past building ruins and extensive dry-stone walls. The longer Mawson (cycling) and Heysen (walking) trails also travel through the reserve, as does a scenic drive from Jamestown, which then continues towards New Campbell Hill, Mt Remarkable and The Bluff. Each Easter Sunday the Bilby Easter Egg Hunt is held in the reserve. Spalding Rd; 9 km S.

Appila Springs: scenic picnic and camping spot; 31 km NW via Appila. ***Spalding:*** a town in the Broughton River valley. Picnic areas and excellent trout fishing; 34 km S.

See also MID-NORTH, p. 239

Kadina

Pop. 4026
Map ref. 599 E3 | 602 H6

The Farm Shed Museum and Tourist Centre, 50 Moonta Rd; (08) 8821 2333 or 1800 654 991; www.yorkepeninsula.com.au

89.3 Gulf FM, 639 AM ABC North and West

Kadina exists solely as a result of the digging habits of wombats. In 1860 upturned ground from wombat diggings revealed copper. This was the starting point of coppermining on Yorke Peninsula. The wombats were commemorated by the naming of Kadina's 1862 hotel: Wombat Hotel. Kadina, along with Wallaroo and Moonta, formed part of a copper triangle colloquially named Little Cornwall because of the number of Cornish immigrants recruited to work in the mines. Kadina is now the commercial centre and largest town on Yorke Peninsula.

The Farm Shed Museum and Tourist Centre: home to the Dry Land Farming Interpretative Centre; Matta House, former home to the mining manager's family; a 1950s style schoolroom; Kadina story, a display of the town's history; and the visitor information centre; Moonta Rd; (08) 8821 2333. ***Ascot Theatre Gallery:*** cultural centre exhibiting local artists' work, with sales; closed Sun; Graves St. ***Victoria Square Park:*** historic band rotunda and Wallaroo Mine Monument; Main St. ***Heritage walk:*** self-guide walk includes historic hotels such as the Wombat and the Royal Exchange with iron-lace balconies and shady verandahs; brochure from visitor centre.

Kernewek Lowender: Cornish festival held with Wallaroo and Moonta; odd-numbered years, May. ***Kadina Show:*** Aug.

Yorke Peninsula Field Days: Australia's oldest field days, started in 1884, are held each Sept (odd-numbered years); Paskeville; 19 km SE.

See also YORKE PENINSULA, p. 240

Kangaroo Island

see Kingscote

Kapunda

Pop. 2479
Map ref. 597 E2 | 601 C1 | 603 K7

Cnr Hill and Main sts; (08) 8566 2902; www.clarevalley.com.au

99.5 Flow FM, 1062 AM ABC Riverland

Kapunda is between two wine districts – the Barossa Valley and the Clare Valley – but its history is very different. Copper was discovered here by Francis Dutton, a sheep farmer, in 1842. It was to be the highest-grade copper ore found in the world. Settlement followed, and Kapunda came into existence as Australia's first coppermining town. When the mines closed in 1878, Australia's 'cattle king' Sir Sidney Kidman moved in, eventually controlling 26 million hectares of land across Australia.

Kapunda Museum: excellent folk museum with a short Kapunda history film and displays of old agricultural machinery, an original fire engine and other vehicles in the pavilion. Detailed mining history is in Bagot's Fortune interpretive centre; open 1–4pm daily; Hill St; (08) 8566 2286. ***Community Gallery:*** significant regional gallery with local and touring art exhibitions; cnr Main and Hill sts. ***'Map Kernow':*** 8 m bronze statue commemorating early miners, many of whom migrated from Cornwall in England; end of Main St at southern entrance of town. ***High school's main building:*** former residence of Sir Sidney Kidman; West Tce. ***Gundry's Hill Lookout:*** views over township and surrounding countryside; West Tce. ***Heritage trail:*** 10 km self-guide tour through town and historic Kapunda Copper Mine; *Discovering Historic Kapunda* brochure available at visitor centre.

Farm Fair: including fashion, craft and antiques; odd-numbered years, Apr/May. ***Celtic Music Festival:*** Sept (can vary). ***Kapunda Show:*** Nov.

Wheatsheaf Hotel: revamped pub, hearty food; Burra Rd, Allendale North; (08) 8566 2198.

Anlaby Homestead: historic pastoral estate; Anlaby Rd; (08) 8566 2465. ***Oldham House:*** lavish, stately country house; 14 Jeffs St; (08) 8566 3503.

continued on p. 261

KINGSCOTE

Pop. 1691

Map ref. 597 A11 | 599 C11 | 602 G12

 **Kangaroo Island Gateway Visitor Information Centre, Howard Dr, Penneshaw; (08) 8553 1185; www.tourkangarooisland.com.au**

90.7 KIX FM

In the early days of Australia's European settlement, Kangaroo Island was a haven for some of the country's most rugged characters – escaped convicts and deserters from English and American whaleboats. These men formed gangs, hunted more than their fair share of whales, seals, kangaroos, wallabies and possums, and went on raids to the mainland to kidnap Aboriginal women. It was an island truly without law. Two centuries on, the only ruggedness to speak of is found along the island's southern coast, where the surf is Southern Ocean–style and the seals are now left in peace. The north shore is a rippling line of bays and coves, with grass-covered hills sweeping down into Investigator Strait. On the shores of the resplendent Nepean Bay is Kingscote, the island's capital and the state's first official settlement (est. 1836). Amidst the beautiful scenery and wildlife, Kinsgscote is also the island's main commercial and business hub.

Kangaroo Island Marine Centre This interpretive marine centre is a must to visit. The size and shape of the aquariums allow visitors to get up close and personal with the vast array of sea life, which includes big bellied seahorses and giant cuttles. The Penguin Tour is the ultimate highlight, where you can observe the little penguins interact with each other. The penguins' breeding season is Mar–Dec, which is the best time to visit. There are 2 tours every night; times vary during the year. Pelican feedings and aquarium tours also on offer. The Wharf; (08) 8553 3112.

Hope Cottage Museum: this National Trust museum exhibits the town's long pioneering and maritime history through memorabilia and photographs; open 1–4pm daily; Centenary Ave; (08) 8553 3017. ***St Alban's Church:*** view the impressive stained-glass windows and pioneer memorials; Osmond St. ***Kangaroo Island Gallery:*** beautiful local artworks from ceramics and glassworks to paintings and woodcrafts; Murray St. ***Cemetery***: SA's oldest cemetrey; Seaview Rd. ***Fishing:*** the jetty is a hot spot for keen anglers.

Racing Carnival: Feb. ***Ozone Street Party:*** Feb. ***Art Exhibition:*** Penneshaw; Easter. ***Kangaroo Island Art Feast:*** Penneshaw; Oct. ***Kangaroo Island Speed Shears:*** Parndana; Nov. ***Kingscote Show:*** Nov.

Marron Cafe: island fare; Harriet Rd, Parndana; (08) 8559 4114. ***Sorrento's Restaurant:*** contemporary fine dining; Kangaroo Island Seafront Resort, 49 North Tce, Penneshaw; (08) 8553 1028 or 1800 624 624. ***The Rockpool Cafe:*** laid-back beachfront seafood; Stokes Bay; (08) 8559 2277.

Blue Seas Beach House: modern oceanfront hideaway; D'Estrees Bay; (08) 8553 8268. ***Cape Borda Lighthouse:*** remote lighthouse keepers cottages; Cape Borda; (08) 8559 7235. ***Kangaroo Island Beach House:*** beachfront family retreat; Island Beach; 0403 986 007. ***Sea Dragon Lodge:*** dramatic, secluded coastal haven; Willoughby Rd, Cape Willoughby; (08) 8553 1449. ***Southern Ocean Lodge:*** stunning, stylish wilderness resort; Hanson Bay; (02) 9918 4355.

Seal Bay Conservation Park This park protects the habitat of the rare Australian sea lion, which faced extinction on the SA coast during the 1800s. Guided beach tours provide close-up encounters with the snoozing creatures. There are also views down to the beach from the 400 m boardwalk that runs through dunes to an observation deck. (08) 8559 4207; 58 km sw.

Flinders Chase National Park On the south coast of this vast park are the precariously positioned granite boulders called the Remarkable Rocks, gradually being eroded by wind and sea to form spectacular shapes. Nearby is the Cape du Couedic lighthouse, and a colony of New Zealand fur seals that can be seen from the boardwalk down to Admirals Arch, a sea cave.

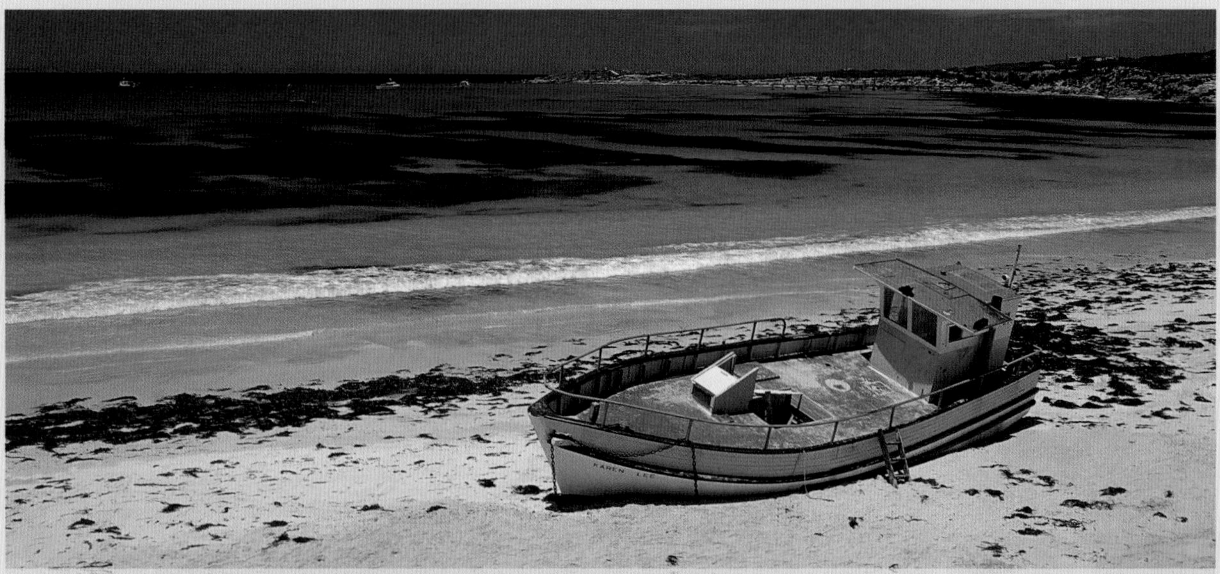

[VIVONNE BAY, POPULAR FOR SURFING AND FISHING]

There are many walking trails and a detailed map is available from the visitor centre at Rocky River. Watch for the Cape Barren geese around the visitor centre. (08) 8559 7220; 93 km sw.

Island Pure Sheep Dairy: tasting and sales of produce, and milking demonstrations; open 1–5pm daily; Cygnet River; 12 km sw. ***Emu Bay:*** excellent swimming at the beach with fishing from the jetty; 17 km sw. ***Emu Ridge Eucalyptus Oil Distillery:*** sales of eucalyptus-oil products; open 9am–2pm daily; Wilsons Rd, off South Coast Rd; 20 km s. ***Clifford's Honey Farm:*** sales and free tasting of honey produced by pure Ligurian bees; Hundred Line Rd; 30 km s. ***Prospect Hill Lookout:*** spectacular views from the spot where Matthew Flinders surveyed Kangaroo Island; on the narrow neck to Dudley Peninsula; 35 km se. ***American River:*** a fishing village overlooking Eastern Cove and Pelican Lagoon, havens for birdlife; 40 km se. ***Parndana:*** a small town known as the Place of Little Gums, featuring the Soldier Settlement Museum; 40 km sw. ***Murray Lagoon:*** well-known waterbird area with tea-tree walk; 40 km sw. ***Stokes Bay:*** natural rock tunnel leads to a rockpool, ideal for swimming; 50 km w. ***Little Sahara:*** large sand dunes surrounded by bush; 55 km sw. ***Penneshaw:*** a small town on Dudley Peninsula where the vehicular ferry arrives from Cape Jervis. The town features a National Trust folk museum and nightly penguin tours from the Penguin Interpretive Centre at Hog Bay Beach; 60 km e. ***Vivonne Bay:*** popular beach for surfing and fishing (beware of strong undertow – safe swimming near the jetty, boat ramp and Harriet River); 63 km sw. ***Antechamber Bay:*** picturesque beach and area, excellent for bushwalking, fishing and swimming; 72 km se. ***Kelly Hill Conservation Park:*** sugar gum forest walks and guided tours of limestone caves; 85 km sw. ***Western River Wilderness Protection Area:*** 2.5 km track to winter waterfall; 85 km e. ***Scott Cove:*** from here you can view SA's highest coastal cliffs, at 263 m; near Cape Borda; 100 km w. ***Cape Borda Lightstation:*** historical tours offered of 1858 lighthouse. Also here is the Cape Borda Heritage Museum. Cannon is fired at 12.30pm; 105 km w.

See also Kangaroo Island, p. 236

Pines Reserve: nature and wildlife reserve; 6 km nw on road to Tarlee. ***Anlaby Station:*** historic Dutton Homestead and gardens, once a setting for large prestigious parties. Also a coach collection and historic station buildings; open 10am–4pm Wed–Sun; Anlaby Rd; 16 km ne. ***Tarlee:*** historic local-stone buildings; 16 km nw. ***Riverton:*** a historic town in the Gilbert Valley, once a stopover point for copper-hauling bullock teams. Many historic buildings remain, including the heritage-listed railway station and Scholz Park Museum, which incorporates a cottage, blacksmith and wheelwright shop; open 1–4pm weekends; 30 km nw. ***Scenic drive:*** 28 km drive north-east through sheep, wheat and dairy country to Eudunda.

See also Mid-North, p. 239

Keith

Pop. 1089
Map ref. 601 F6

Mobil Roadhouse, cnr Riddock and Dukes hwys; (08) 8755 1700; www.tatiara.sa.gov.au

104.5 5TCB FM, 1062 AM ABC Riverland

Keith is a farming town in the area formerly known as the Ninety Mile Desert. Settlers found the original land unpromising, but the area has since been transformed from infertile pasture to productive farmland with the addition to the soil of missing trace elements and water piped from the Murray.

Congregational Church: National Trust church with 11 locally made leadlight windows depicting the town's life and pioneering history; Heritage St. ***Early Settler's Cottage:*** limestone pioneer cottage; open by appt; Heritage St; (08) 8755 1118. ***Keith Water Feature:*** water sculpture; Heritage St.

Keith Show: Oct.

McIntyre Cottage: romantic Mediterranean-style villa; 3 Tolmer Tce; (08) 8755 1126.

Ngarkat group of conservation parks Protecting 262 700 ha of sand dunes, mallee and heath are the 4 adjacent conservation parks – Ngarkat, Scorpion Springs, Mt Rescue and Mt Shaugh. The walking trails are an excellent introduction to the region's vegetation. Birdwatching is particularly good at Rabbit Island (Mt Rescue) and Comet Bore (Ngarkat). For panoramic views, try walking to the summit of Mt Rescue, Goose Hill or Mt Shaugh. Visitors can drive through the parks on the Pinaroo–Bordertown Rd. The main entrance is via Snozwells Rd near Tintinara.

Monster Conservation Park: scenic views and picnic spots; 10 km s. ***Tintinara Homestead:*** one of the first homesteads in the area, with woolshed, shearers quarters and pioneer cottage; caters for visitors; 38 km nw. ***Mt Boothby Conservation Park:*** mallee scrub, granite outcrops and wildflowers in spring; 58 km nw via Tintinara.

See also Murray, p. 241

Kimba

Pop. 636
Map ref. 602 F4 | 604 D11

Kimba Mobil Roadhouse, Eyre Hwy; (08) 8627 2040; www.kimba.sa.gov.au

89.9 Magic FM, 1485 AM ABC Eyre Peninsula and West Coast

A small town on the Eyre Peninsula, Kimba is 'halfway across Australia' according to the huge sign on the Eyre Highway. It is the gateway to the outback, a fact that explorer Edward John Eyre confirmed when he traversed the harsh landscape in 1839. Early settlers thought the country too arid for settlement and it wasn't until demand for wheat production grew, and rail services were extended to the area in 1913, that the Kimba region developed. It is now major sheep- and wheat-farming country.

Kimba and Gawler Ranges Historical Museum: a 'living' museum featuring local history and a Pioneer House (1908), a

SOUTH AUSTRALIA

blacksmith's shop and 'Clancy' the fire truck; open 1.30–4pm Sat (closed Jan) or by appt; Eyre Hwy; (08) 8627 2349. ***Halfway Across Australia Gem Shop and the Big Galah:*** standing 8 m high, the Big Galah is in front of the gem shop, which offers sales of local gemstones, carved emu eggs, opal and locally mined jade, including rare black jade; Eyre Hwy. ***Pine 'n' Pug Gallery:*** local craft; High St. ***Roora Walking Trail:*** meanders through 3 km of bushland to White Knob Lookout; starts at north-eastern outskirts of town.

Kimba Show: Sept.

Lake Gilles Conservation Park: habitat for mallee fowl; 20 km NE. ***Caralue Bluff:*** popular for rock climbing; 20 km SW. ***Carappee Hill Conservation Park:*** bush camping and walking; 25 km SW. ***Darke Peak:*** excellent views from the summit and a memorial at the base to John Charles Darke, an explorer who was speared to death in 1844; 40 km SW. ***Pinkawillinie Conservation Park:*** the largest mallee vegetation area on the peninsula and a habitat for small desert birds, emus and western grey kangaroos; turn-off 50 km W.

See also EYRE PENINSULA & NULLARBOR, p. 244

Kingscote

see inset box on previous page

Kingston S.E.

Pop. 1632
Map ref. 601 E8

BP Roadhouse, 1 Princes Hwy; (08) 8767 2404; www.thelimestonecoast.com

107.7 5THE FM, 1161 AM ABC South East

Known as the 'Gateway to the South East', Kingston S.E. is at the southern end of Coorong National Park on Lacepede Bay. The area was once home to the Ngarrindjeri, river people who mastered the waterways of the Coorong and the Murray River. This famous lobster town was established in 1858 and its shallow lakes and lagoons are a haven for birdlife and a delight for photographers.

National Trust Museum: pioneer museum; open 2–4.30pm daily during school holidays, or by appt; Cooke St; (08) 8767 2114. ***Cape Jaffa Lighthouse:*** built in the 1860s on the Margaret Brock Reef, it was dismantled and re-erected on its current site in the 1970s; open 2–4.30pm daily during school holidays, or by appt; Marine Pde; (08) 8767 2591. ***Analematic Sundial:*** an unusual sundial, 1 of only 8 in the world; on an island in the creek adjacent to Apex Park in East Tce. ***Aboriginal burial ground:*** Dowdy St. ***Power House engine:*** historic engine that produced the town's energy until 1974; Lions Park, Holland St. ***The Big Lobster:*** 17 m high 'Larry Lobster' has sales of cooked lobster; Princes Hwy.

Fishing Contest: Jan.

Kingston Shore Villas: modern 2-storey beachfront villas; 99 Marine Pde; 0417 886 329.

Butchers Gap Conservation Park: this important coastal park provides a winter refuge for bird species. Follow the walking trail from the carpark; 6 km SW. ***The Granites:*** rocky outcrops, a striking sight from the beach; 18 km N. ***Cape Jaffa:*** scenic drive south-west from Kingston S.E. leads to this small fishing village popular with anglers and divers. The Cape Jaffa Seafood and Wine Festival is held here each Jan; 18 km SW. ***Mt Scott Conservation Park:*** part of a former coastal dune system, with walks through stringybark forest; 20 km E. ***Jip Jip Conservation Park:*** features a prominent outcrop of unusually shaped granite boulders; 50 km NE.

See also LIMESTONE COAST, p. 242

Leigh Creek

Pop. 547
Map ref. 600 B4 | 604 H4

13 Black Oak Dr; (08) 8675 2723.

1602 AM ABC North and West

Located in the Flinders Ranges, Leigh Creek is a modern coalmining town that services a huge open-cut mine to the north. The original township (13 kilometres north) was unfortunately placed, as it was situated over a large coal seam. The lure of the dollar led to the town's relocation in 1982 to its current site. A tree-planting scheme has transformed the town from a barren landscape to an attractive oasis.

Copley Bush Bakery & Quandong Cafe: great pies and cakes; Railway Tce, Copley; (08) 8675 2683.

Warraweena Conservation Park: rustic, remote shearers quarters; via Beltana Roadhouse; (08) 8675 2770.

Coal mine tours: tours to the open-cut mine each Sat from Mar to late Oct and during school holidays; contact visitor centre for details. ***Aroona Dam:*** in a steep-sided valley with coloured walls; picnic area near gorge; 4 km W. ***Coalmine viewing area:*** turn-off 14 km N (area is 3 km down road to coalmine). ***Beltana:*** almost a ghost town, it has a historic reserve and holds a Picnic Race Meeting and Gymkhana each Sept; 27 km S. ***Lyndhurst:*** starting point of the famous Strzelecki Track. It also features a unique gallery of sculptures by well-known talc-stone artist 'Talc Alf'; tours by appt (08) 8675 7781; 39 km N. ***Ochre Cliffs:*** here Aboriginal people used to dig for ochre. The colours range from white to reds, yellows and browns; 44 km N via Lyndhurst. ***Sliding Rock Mine ruins:*** access track is rough in places; 60 km SE. ***Vulkathunha–Gammon Ranges National Park:*** 64 km E; *see Arkaroola.*

Travellers note: *Care must be taken on outback roads. Check road conditions with Northern Roads Condition Hotline on 1300 361 033 before departure.*

See also FLINDERS RANGES & OUTBACK, p. 243

Loxton

Pop. 3432
Map ref. 558 A6 | 588 A7 | 601 G2 | 603 N8

Bookpurnong Tce; (08) 8584 7919; www.riverland.info

93.1 Magic FM, 1062 AM ABC Riverland

Although the area around Loxton was originally settled largely by German immigrants, the town's boom began when servicemen returned from World War II. The enticement of irrigated allotments brought a great number of them to town and the success of current-day industries, such as the production of citrus fruits, wine, dried fruit, wool and wheat, was due to their skill on the land. Loxton's delightful setting on the Murray River has made the town the 'Garden City of the Riverland'.

Loxton Historical Village The Riverland's pioneering history comes to life in the 30 historic buildings, all fully furnished in the styles of late 1880s to mid-1900s. A highlight is the pine-and-pug building, Loxtons Hut, built by the town's namesake, boundary rider William Loxton. Visit on one of the Village Alive days held thrice yearly in the village. Locals dress up in period costume and the whole village steps back 100 years. Allan Hosking Dr; (08) 8584 7194.

Terrace Gallery: local art and pottery displays and sales; part of visitor centre; Bookpurnong Tce. ***The Pines Historical Home:*** resplendent gardens and historic home filled with antique furniture and fine china. Tours of home every Sun followed by afternoon tea; Henry St; (08) 8584 4646. ***Pepper tree:*** grown from a seed planted by Loxton over 110 years ago; near the historical village. ***Nature trail:*** along riverfront; canoes for hire. ***Heritage walk:*** brochure from visitor centre.

Mardi Gras: Feb. ***Nippy's Loxton Gift:*** SA's largest and richest footrace; Feb. ***Riverland Harvest Festival & Great Grape Stomp:*** Apr. ***Loxton Annual Spring Show:*** Oct. ***Loxton Lights Up:*** Christmas lights throughout town; self-guide tour map available; Nov–Dec.

Mill Cottage: inviting stone home and garden; 2 Mill Rd; 0439 866 990. ***Thiele's Heritage B&B:*** evocative cliff-top farmhouse; Casson Ave, Loxton North; 0402 837 012.

McGuigan Simeon Wines Loxton Cellars: wine-tasting and sales; Bookpurnong Rd, Loxton North; (08) 8584 7236; 4 km N. ***Torambre Nissen Hut Wines:*** boutique winery with award-winning shirazes and merlots; Balfour–Ogilvy Rd, Loxton North; (08) 8584 1530; 6 km NE. ***Lock 4:*** picnic/barbecue area; 14 km N on Murray River. ***MV Loch Luna Eco Cruise:*** relaxing 3 hr cruise around the Nockburra and Chambers creeks; departs daily from Kingston at 9am and 1.30pm; 0449 122 271. ***Banrock Station Wine and Wetland Centre:*** magnificently restored wetlands and mallee woodlands, with 250 ha of picturesque vineyards and a wine centre; Holmes Rd, Kingston-on-Murray; (08) 8583 0299; 35km NW.

See also MURRAY, p. 241

Lyndoch

Pop. 1419
Map ref. 596 G2 | 597 E5 | 601 C2 | 603 K8

Kies Family Wines, Lot 2, Barossa Valley Hwy; (08) 8524 4110; www.barossa-region.org

89.1 Triple B FM, 639 AM ABC North and West

Lyndoch is one of the oldest towns in South Australia. The first European explorers, led by Colonel Light in 1837, described the area around Lyndoch as 'a beautiful valley'. The undulating landscape and picturesque setting attracted Lutheran immigrants and English gentry, who began growing grapes here. By 1850 Johann Gramp had produced his first wine from the grapes at Jacob's Creek.

Spinifex Arts & Crafts: retail shop selling exclusive quality hand-crafted goods including pottery, quilts, folk art and paintings of local scenes. All goods are locally made in the Barossa region; Barossa Valley Way. ***Helicopter and balloon flights:*** scenic flights over the Barossa region; contact visitor centre for details. ***Historic Lyndoch Walk:*** self-guide walk featuring buildings from the mid-1800s, including many built from locally quarried hard ironstone; brochure from visitor centre.

Barossa under the Stars: food, wine and music; Tanunda; Feb. ***Barossa Vintage Festival:*** celebration of food and wine in various locations; odd-numbered years, Apr. ***Barossa Gourmet Weekend:*** Aug.

Abbotsford Country House: gracious hosted suites; Yaldara Dr; (08) 8524 4662. ***Bandicoot Nest:*** stone cottage with modern makeover; address provided on booking; (08) 8524 5353. ***Barossa Country Cottages:*** easygoing garden cottages; 55 Gilbert St; (08) 8524 4426. ***Barossa Pavilions:*** dashing, designer hilltop retreats; 461 Yaldara Dr; (08) 8524 5497 or 1800 990 055.

Para Wirra Recreation Park Para Wirra comes from the Aboriginal words for 'river with scrub'. The park has a large recreational area with extensive facilities including tennis courts, picnic and barbecue areas, and walking trails ranging from short 800 m walks to more extensive 7.5 km trails. The park consists of mostly eucalypts and is home to a large variety of native birds – including inquisitive emus that meander around the picnic areas. The historic Barossa Goldfield Trails (1.2 km or 5 km loop walks) cover the history of the old goldmines. 12 km SW.

Barossa Wine Region: wineries and cellar doors lie around Lyndoch and Rowland Flat, just north-east, with rich red varieties as well as crisp whites. At Rowland Flat is the Jacob's Creek Visitor Centre where viticulture history is on display. For a total Barossa experience, take the self-drive Butcher, Baker, Winemaker trail from Lyndoch to Angaston and Eden Valley; brochure available from visitor centre. *For more information on the region see Tanunda, Angaston and Nuriootpa.* ***Sandy Creek Conservation Park:*** on undulating sand dunes, with walking trails and birdlife. See western grey kangaroos and echidnas at dusk; 5.5 km W. ***Lyndoch Lavender Farm and Cafe:*** wander through rows of over 60 lavender varieties. The nursery and farm shop offer lavender-product sales; open daily Sept–Feb, Mon–Fri Mar–April; cnr Hoffnungsthal and Tweedies Gully rds; 6 km SE. ***Barossa Reservoir and Whispering Wall:*** acoustic phenomenon allowing whispered messages at one end to be audible at the other end, 140 m away; 8 km SW.

See also BAROSSA VALLEY, p. 238

McLaren Vale

Pop. 2907
Map ref. 596 B12 | 597 B12 | 598 E5 | 599 H9 | 601 B4 | 603 J10

Main Rd; (08) 8323 9944 or 1800 628 410; www.mclarenvale.info

94.7 5EFM, 639 AM ABC North and West

McLaren Vale is recognised as a region of vineyards, orchards and gourmet-produce farms. The first grape vines were planted here in 1838, just two years after Adelaide was settled, and McLaren Vale has remained a prominent wine region ever since. Shiraz is the most transcendent drop and is known for its overtones of dark chocolate. Cabernet sauvignon, grenache, chardonnay and sauvignon blanc are also very respectable. Coriole makes wonderful shiraz as well as olive oil, while d'Arenberg has many interesting varieties and a good restaurant, d'Arry's Verandah. Other wineries to visit include Geoff Merrill Wines and two of the oldest and biggest names in the district: Hardys Tintara and Rosemont Estate. The best time to visit the region is late winter, when almond blossoms provide a gentle pink blush. The town is also known for its coastal vistas to the west, which provide subject matter for the artists who exhibit in the galleries.

McLaren Vale Visitors Centre: picturesque landscaped grounds with a vineyard and a centre that features changing art exhibitions, sales of local craft and produce, and a cafe; Main Rd; (08) 8323 9944. ***The Old Bank Artel:*** community cooperative of local crafts including pottery, jewellery and metalwork; Main Rd. ***Almond and Olive Train:*** sales of local produce, including almonds, in a restored railway carriage; Main Rd. ***McLaren Valley Bakery:*** produces unique 'wine pies'; Central Shopping Centre, Main Rd. ***McLaren Vale Olive Grove:*** grows over 26 varieties of olives and sells olive products, arts and crafts, and local gourmet produce; Warners Rd. ***Medlow Confectionery and FruChocs Showcase:*** tastings and sales of gourmet chocolate and other confectionery. Interactive confectionery machine for kids; Main Rd. ***McLaren Vale Heritage Trail:*** self-guide trail of historic sites, including wineries, with audio CD available; starts at visitor centre.

Sea and Vines Festival: music, food and wine; June. ***Fiesta!:*** month-long food and wine festival; Oct.

d'Arry's Verandah Restaurant: glorious cellar-door views, regional cuisine; d'Arenberg Wines, Osborn Rd; (08) 8329 4848. ***Red Poles:*** arty cafe; McMurtrie Rd; (08) 8323 8994. ***Salopian Inn:*** creative contemporary cooking; Cnr McMurtrie and Main rds; (08) 8323 8769. ***The Kitchen Door:*** stylish rustic fare; Penny's Hill Winery, Main Rd; (08) 8556 4000.

Bellevue B&B: comfortable suites, impeccably hosted; 12 Chalk Hill Rd; 0432 868 402. ***McLaren Ridge Log Cabins:*** hilltop vineyard havens; Whitings Rd; (08) 8383 0504. ***Aunt Amanda's Cottage:*** cute, secluded farmhouse retreat; Peters Creek Rd, Kangarilla; (08) 8383 7122.

Onkaparinga River National Park The Onkaparinga River, SA's second longest river, travels through valleys and gorges to Gulf St Vincent. The walks in Onkaparinga Gorge are impressive, but very steep. More regulated walking trails are on the northern side of the gorge. The estuary section of the park is an altogether different environment and is best explored on the 5 km interpretive trail. Look out for the 27 species of native orchids. Access is via Main South Rd, Old Noarlunga; (08) 8278 5477; 7 km NW.

Old Noarlunga: self-guide tour of historic colonial buildings (brochure available). Walks into Onkaparinga National Park start from here; 7 km NW. ***Port Noarlunga:*** popular holiday destination with historic streetscapes. A marked underwater trail along the reef is provided for divers and snorkellers; 11 km NW. ***Coastal beaches:*** safe family beaches to the north-west include O'Sullivan, Christies and Moana.

See also FLEURIEU PENINSULA, p. 235

Maitland

Pop. 1055
Map ref. 599 E5 | 602 H7

Council offices, 8 Elizabeth St; (08) 8832 0000; www.yorkepeninsula.com.au

89.3 Gulf FM, 639 AM ABC North and West

Maitland represents a much smaller version of the city of Adelaide, with the town layout in the same pattern of radiating squares. It is in the heart of Yorke Peninsula and is central to a rich agricultural region. In recent years tourism has grown dramatically on the peninsula, but barley and wheat industries remain strong.

Maitland Museum: located in the former school, this National Trust museum documents local indigenous and settlement history; open 2–4pm Sun, public/school holidays, other times by appt; cnr Gardiner and Kilkerran tces; (08) 8832 2220. ***White Flint Olive Grove:*** range of olive oil products produced from own trees, great food and tea on offer; appt only; South Tce; (08) 8832 2874. ***St John's Anglican Church:*** stained-glass windows depict biblical stories in an Australian setting; cnr Alice and Caroline sts. ***Aboriginal cultural tours:*** a range of tours through Adjahdura Land (Yorke Peninsula), with an Aboriginal guide; bookings and inquiries 0429 367 121. ***Heritage town walk:*** interpretive walk; brochure from council in Elizabeth St.

Maitland Show: Mar. ***Maitland Art/Craft Fair:*** June and Nov.

Gregory's Wines: Yorke Peninsula's only commercial vineyard. Cellar-door tastings and sales 10am–5pm weekends or by appt; Lizard Rd; (08) 8834 1258; 13 km S. ***Balgowan:*** this town has safe, sandy beaches and is popular with anglers; 15 km W.

See also YORKE PENINSULA, p. 240

Mannum

Pop. 2037
Map ref. 597 H9 | 601 D3 | 603 L9

6 Randell St; (08) 8569 1303; www.murraylands.info

98.7 Power FM, 1062 AM ABC Riverland

Mannum is one of the oldest towns on the Murray River, at the romantic heart of the old paddlesteamer days. In 1853 the 'Father of Mannum', William Randell, built the first Murray River paddlesteamer, *Mary Ann* (named after his mother), in order to transport his flour to the Victorian goldfields. The paddlesteamer set out from Mannum in 1853 and started a boom in the river transport industry. Another first for Mannum was Australia's first steam car, built in 1894 by David Shearer.

Mannum Dock Museum This excellent museum documents the changing history of the Mannum region from ancient days through Indigenous habitation, European settlement and river history to the present day. Outside is the renowned Randell's Dry Dock, where the grand lady of the Murray, PS *Marion*, is moored. Passenger cruises on the restored paddlesteamer still operate. Randell St; (08) 8569 2733.

Mary Ann Reserve: popular recreation reserve on the riverbank with a replica of PS *Mary Ann*'s boiling engine. PS *River Murray Princess* is moored here between cruises. ***Ferry service:*** twin ferries operate to the eastern side of the river. ***River cruises and houseboat hire:*** afternoon, day and overnight cruises are available from the town wharf, including the Murray Expedition Captain's Dinner. Alternatively, hire a houseboat to discover the Murray River your own way. Contact visitor centre for details. ***Town lookout:*** off Purnong Rd to the east. ***Scenic and historical walks:*** brochures from visitor centre.

Mannum Show: Mar. ***Houseboat Hirers' Open Days:*** May. ***Christmas Pageant:*** Dec.

Baseby House: ultra-modern riverfront B&B; 152 River La; (08) 8569 2495. ***River Dream Boatel:*** opulent, architect-designed cruising; Mannum Waterfront, 16 Randell St; (08) 8223 3030. ***Riverview Rise Retreats:*** glamorous getaway for couples; Bowhill Rd; 0400 310 380. ***Unforgettable Houseboats:*** 11-strong fleet, multi-award-wining operator; 69 River La; (08) 8569 2559 or 1800 656 323.

Mannum Falls: picnics and scenic walks, best visited in winter (after rains) when the waterfall is flowing; 6 km SE. ***Kia Marina:*** the largest river marina in SA; boats and houseboats for hire; 8 km NE. ***Lowan Conservation Park:*** mallee vegetation park with varied wildlife, including fat-tailed dunnarts, mallee fowl and western grey kangaroos; turn-off 28 km E at Bowhill. ***Purnong:*** scenic drive north-east, runs parallel to excellent Halidon Bird Sanctuary; 33 km E.

See also MURRAY, p. 241

Marree

Pop. 300
Map ref. 604 H1 | 606 H12 | 662 A12

Marree Outback Roadhouse and General Store; (08) 8675 8360.

105.7 AM ABC Local

Marree is the perfect image of a tiny outback town. It is frequented by four-wheel-drive enthusiasts taking on the legendary Birdsville and Oodnadatta tracks. The settlement was established in 1872 as a camp for the Overland Telegraph Line as it was being constructed, and also became a railhead for the Great Northern Railway (which was later known as the *Ghan*). The town soon serviced all travellers and workers heading north, including the famous Afghan traders who drove their camel trains into the desert and played a significant role in opening up the outback.

Aboriginal Heritage Museum: features artefacts and cultural history; in Arabunna Aboriginal Community Centre. ***Marree Heritage Park:*** includes Tom Kruse's truck that once carried out the famous outback mail run on the Birdsville Track in the 1950s. ***Camel sculpture:*** made out of railway sleepers. ***Scenic flights:*** including over Lake Eyre and the Marree Man, a 4 km long carving in a plateau of an Aboriginal hunter. The carving, visible only from the air, appeared mysteriously in 1998, and is slowly fading; contact visitor centre for details.

 ***Australian Camel Cup:*** July.

Lake Eyre National Park Of international significance, Lake Eyre is dry for most of the time – it has filled to capacity on only 3 occasions in the last 150 years. When water does fill parts of the lake (usually due to heavy rains in Queensland funnelled south via creeks and rivers), birds flock to it. Avoid visiting in the hotter months (Nov–Mar). Lake Eyre North is accessed via the Oodnadatta Track, 195 km W of Marree. Lake Eyre South is accessed via the 94 km track north of Marree (along this track is Muloorina Station, which offers camping alongside the Frome River). Both access routes are 4WD only. Lake Eyre South also meets the Oodnadatta Track about 90 km W of Marree, where there are good views. A Desert Parks Pass is required for the park and is available from Marree Post Office or by contacting the Desert Parks Hotline on 1800 816 078. Scenic flights are perhaps the most rewarding option, from both Marree and William Creek.

Oodnadatta Track: a 600 km 4WD track from Marree to Marla. Highlights along the track include the Dog Fence (around 40 km W) and the railway-siding ruins at Curdimurka Siding and Bore (90 km W) from the original Great Northern Railway line to Alice Springs. A short distance beyond Curdimurka is Wabma Kadarbu Mound Springs Conservation Park, with a series of springs – fed by water from the Great Artesian Basin – supporting a small ecosystem of plants and animals. Between Marree and Marla, fuel is available only at William Creek (202 km NW) and Oodnadatta (405 km NW). ***Birdsville Track:*** famous 4WD track from Marree to Birdsville (in Queensland) of just over 500 km, once a major cattle run. Highlights on the track include the failed date palm plantation at Lake Harry Homestead (30 km N) and the meeting of the Tirari and Strzelecki deserts at Natterannie Sandhills (140 km N, after Cooper Creek crossing). Cooper Creek may have to be bypassed if flooded (with a 48 km detour to a ferry). Between Marree and Birdsville, fuel is available only at Mungerannie Roadhouse (204 km N).

Travellers note: *Care must be taken when attempting the Birdsville and Oodnadatta tracks. These tracks are unsealed, with sandy patches. Heavy rain in the area can cut access for several days. Motorists are advised to ring the Northern Roads Condition Hotline on 1300 361 033 before departure.*

See also FLINDERS RANGES & OUTBACK, p. 243

Melrose

Pop. 450
Map ref. 603 I3 | 604 H10

Melrose Caravan and Tourist Park, Joes Rd; (08) 8666 2060; www.mountremarkable.com.au

 105.1 Trax FM, 639 AM ABC North and West

Melrose, a quiet settlement at the foot of Mount Remarkable, is the oldest town in the Flinders Ranges. Indigenous groups who occupied the southern Flinders Ranges around Mount Remarkable and Melrose resisted European settlement in the 1840s, but after just a few decades the population was reduced to a handful. Pastoral properties were established on the mountainous slopes of the ranges, but Melrose truly took off when copper deposits were found nearby in 1846. Bushwalking through Mount Remarkable National Park is a highlight of any visit to this area. The arid north country meets the wet conditions of southern regions to provide a diverse landscape to explore.

National Trust Museum: documents local history, with particular focus on early law enforcement. Original stone buildings include the courthouse and lock-up; open 2–5pm; Stuart St. ***Bluey Blundstone Blacksmith Shop:*** a restored shop with a cafe and B&B; closed Tues; Stuart St. ***Serendipity Gallery:*** Australiana arts and crafts; Stuart St. ***War Memorial and Lookout Hill:*** views over surrounding region. ***4WD tours:*** to local landmarks; contact visitor centre for operators. ***Heritage walk:*** self-guide walk includes ruins of Jacka's Brewery and Melrose Mine; brochure from visitor centre.

 ***Melrose Show:*** Oct.

The North Star Hotel: rejuvenated landmark pub; Nott St; (08) 8666 2110. ***The Old Bakery Stone Hut:*** classic English-style pies; 1 Main North Rd, Stone Hut; (08) 8663 2165

Mount Remarkable Cottage: private, self-contained bush retreat; Diocesan Rd; (08) 8666 2173. ***The North Star Hotel:*** heritage pub with hip makeover; Nott St; (08) 8666 2110.

Mt Remarkable National Park Part of the southern Flinders Ranges and popular with bushwalkers. Marked trails through the park's gorges and ranges vary in scope from short scenic walks to long 3-day hikes. Highlights include pretty Alligator Gorge and the tough but worthwhile 5 hr return walk from Melrose to the summit of Mt Remarkable (960 m), with breathtaking views from the top. Access to the park by vehicle is via Mambray Creek or Wilmington. Foot access is from carparks and Melrose. (08) 8634 7068; 2 km W.

Cathedral Rock: an impressive rock formation on Mt Remarkable Creek; just west of Melrose. ***Murray Town:*** a farming town with nearby scenic lookouts at Box Hill, Magnus Hill and Baroota Nob. Remarkable View Wines offers tasting and sales on weekends. Starting point for a scenic drive west through Port Germein Gorge; 14 km S. ***Booleroo Centre:*** this service town to a rich farming community features the Booleroo Steam Traction Preservation Society's Museum. Annual Rally Day is held in Mar/Apr; open by appt; (08) 8667 2193; 15 km SE. ***Wirrabara:*** you could easily lose half a day in this town's bakery-cum-antiques shop, wandering through the rooms of relics and antiques just to work up an appetite for another homemade pastry; 25 km S.

See also FLINDERS RANGES & OUTBACK, p. 243

Meningie

Pop. 939
Map ref. 601 D5 | 603 L11

The Chambers, 14 Princes Hwy; (08) 8575 1770; www.thelimestonecoast.com

98.7 Power FM, 1161 AM ABC South East

Today Meningie is an attractive lakeside town, but it was once a wilderness area, home to the Ngarrindjeri people, who had a self-sufficient lifestyle on the water. They made canoes to fish on the waterways and shelters to protect themselves from the weather. However, European settlement – after Captain Charles Sturt's journey down the Murray from 1829 to 1830 – soon wiped out much of the population, largely through violence and the introduction of smallpox. Stretching south from the mouth of the Murray and located just south of Meningie, the Coorong, with its lakes, birdlife, fishing and deserted ocean beaches, attracts visitors year-round.

The Cheese Factory Museum Restaurant: a restaurant and separate community museum with special interest in the changing population of Meningie and the Coorong; closed Mon; Fiebig Rd. ***Coorong Cottage Industries:*** local craft and produce; The Chambers, Princes Hwy.

Campbell Park Cottage: stylish, exclusive homestead accommodation; Campbell Park Rd; 0412 887 910. ***Poltalloch Station:*** heritage lakefront estate; Poltalloch Rd, Narrung; (08) 8574 0043.

Coorong National Park Listed as a 'wetland of international importance', this park's waterways, islands and vast saltpans demonstrate a diverse ecological environment invaluable for the refuge and habitat of migratory and drought-stricken birds. Throughout the park are reminders of the long history of habitation by the Ngarrindjeri people. There are ancient midden heaps and burial grounds, and the Ngarrindjeri people continue to have strong links with the Coorong. There are a number of ways to see the park: take a boat, canoe or cruise on the waterways; walk one of the varied tracks offered; drive your 4WD onto the Southern Ocean beach; or simply sit and soak up the park's atmosphere. Walking is the best way to access great coastal views, birdwatching spots and historic ruins. The most comprehensive walk in the park is the Nukan Kungun Hike. This 27 km hike starts at Salt Creek and includes smaller, side trails, including the informative walk to Chinaman's Well, the ruins of a temporary settlement that sprang up en route to the goldfields. The hike ends at the 42 Mile Crossing Sand Dune Walk, which leads to the wild Southern Ocean; (08) 8575 1200; 10 km S.

Scenic drive: follows Lake Albert to the west, adjacent to Lake Alexandrina, which is the largest permanent freshwater lake in the country (50 000 ha). Ferry crossing at Narrung. ***Camp Coorong:*** cultural centre offering bush-tucker tours and other traditional Aboriginal experiences; (08) 8575 1557; 10 km S. ***Coorong Wilderness Lodge:*** operated by the Ngarrindjeri people, this accommodation lodge offers Aboriginal heritage tours of the Coorong as well as traditional bush tucker and other cultural experiences; Hacks Point; (08) 8575 6001; 25 km S. ***Poltalloch Station:*** a historic pastoral property established in the 1830s as a sheep station. Guided tours take in a heritage-listed farm village and museum, historic farm machinery, and past and present farm life. Cottage accommodation available; Poltalloch Rd near Narrung; (08) 8574 0043; 30 km NW.

See also MURRAY, p. 241

Millicent

Pop. 4768
Map ref. 601 F11

1 Mt Gambier Rd; (08) 8733 0904; www.wattlerange.sa.gov.au

107.7 5THE FM, 1161 AM ABC South East

Millicent, a prosperous, friendly and vibrant community located in the heart of the Limestone Coast region, is named after Millecent Glen, wife of one of the early pioneers and daughter of the first Anglican Bishop of Adelaide. In 1876 the barque *Geltwood* was wrecked off Canunda Beach, with debris, bodies and cargo littering the sands. Relics from the *Geltwood* can be found in the award-winning National Trust Living History Museum. The town's distinctive aroma can be attributed to the surrounding pine forests, which support a pulp mill, a paper mill and sawmill.

Visitor Centre and National Trust Living History Museum: centre provides extensive information on the region and surrounding areas. Souvenirs, arts and crafts, and local produce available. History museum includes a shipwreck room, farm machinery shed, Aboriginal and natural history rooms, a T-Class locomotive and the largest collection of restored horse-drawn vehicles in SA; Mt Gambier Rd; (08) 8733 0904. ***Lake McIntyre:*** boardwalks and bird hides to view the lake's prolific birdlife, native fish and yabbies; northern edge of town.

Nangula Market Days: 2nd Sun each month (excluding Jan). ***Tall Timbers Wood Work Exhibition:*** pine and red-gum craft; Mar–May. ***Geltwood Craft Festival:*** Mar/Apr. ***Pines Enduro Off-Road Championships:*** Sept. ***Millicent Show:*** Sept.

Canunda National Park The massive sand-dune system of the southern part of the park rises to cliffs and scrublands in the north. These 2 sections provide quite different experiences. In the north (accessed via Southend and Millicent) the walking trails pass along cliff-tops and through the scrubland. In the south (accessed via Carpenter Rocks and Millicent) the beaches and wetlands provide picturesque coastal walks. You can surf, 4WD, birdwatch, bushwalk and fish (excellent from the beaches and rocks). (08) 8735 1177; 10 km W.

Woakwine Range Wind Farms: dozens of giant turbines dominate the Woakwine Range skyline, comprising the largest wind farm development in the Southern Hemisphere. Take a drive along the Wind Farm Tourist Drive. Maps available at the Visitor Centre; 2 km S. ***Mt Muirhead Lookout:*** a large viewing platform provides views of Millicent, pine plantations and Mt Burr Range. It is also the start of the Volcanoes Discovery Trail (brochure from visitor centre); 6 km NE. ***Mount Burr:*** a historic timber town, the first to plant pines for commercial use on the Limestone Coast; 10 km NE. ***Tantanoola:*** home of the famous 'Tantanoola Tiger', a Syrian wolf shot in the 1890s, now stuffed and displayed in the Tantanoola Tiger Hotel; 20 km SE. ***Tantanoola Caves Conservation Park:*** daily tours of an imposing dolomite cavern in an ancient limestone cliff. Also walks and picnic areas; 21 km SE. ***Glencoe Woolshed:*** National Trust limestone woolshed once occupied by 38 shearers; 29 km SE.

See also LIMESTONE COAST, p. 242

Minlaton

Pop. 772
Map ref. 599 E6 | 602 H8

Harvest Corner Visitor Information Centre, 29 Main St; (08) 8853 2600 or 1800 202 445; www.yorkepeninsula.com.au

98.9 Flow FM, 639 AM ABC North and West

continued on p. 268

MOUNT GAMBIER

Pop. 23 494

Map ref. 592 A5 | 601 G11

Mount Gambier Visitor Information Centre and the Lady Nelson Discovery Centre, Jubilee Hwy East; (08) 8724 9750 or 1800 087 187; www.mountgambiertourism.com.au

96.1 Star FM, 1476 AM ABC South East, 5GTR FM 100.01

Mount Gambier is set on an extinct volcano – the area boasts a fascinating network of volcanic craters above sea level and one of limestone caves beneath. Lieutenant James Grant named the volcano in 1800 – he sighted it from HMS *Lady Nelson*, off the coast. The original settlement was known as Gambier Town. Today Mount Gambier is at the centre of the largest softwood pine plantation in the Commonwealth and is surrounded by farming, viticulture and dairy country.

Blue Lake The lakes formed in the craters of the extinct volcano have become an important recreational area for locals and visitors alike. The most spectacular is Blue Lake, so-called because the water's dull, blue-grey winter colour changes to a vibrant blue each Nov and stays that way until Mar the following year. Discover the area on the 3.6 km walking track around the shores. Aquifer Tours offers a trip in a lift down an old well shaft. The area also includes a wildlife park and an adventure playground; southern outskirts of town; (08) 8723 1199.

Old Courthouse: a National Trust dolomite building with a local history museum; open by appt; Bay Rd; (08) 8725 5284. ***Lady Nelson Visitor Information and Discovery Centre:*** full-scale replica of HMS *Lady Nelson*, interactive displays on the region's history and geography, and free local information packs; Jubilee Hwy East. ***Riddoch Art Gallery:*** changing exhibitions of local and touring art and sculpture; open Tues–Sun; Commercial St East. ***Cave Garden:*** a cave used as a water supply for early settlers, now a rose garden with a suspended viewing platform; Bay Rd. ***Umpherston Sinkhole:*** a sunken garden on the floor of a collapsed cave; floodlit at night; Jubilee Hwy East. ***Engelbrecht Cave:*** guided tour of the limestone cave system under the city; contact visitor centre for details. ***Centenary Tower:*** views of the city, the lakes area and surrounding countryside; top of Mt Gambier, 190 m above sea level. ***Heritage walk:*** self-guide tour of historic buildings, many constructed of white Mt Gambier stone; brochure from visitor centre.

Mount Gambier Market: behind Harvey Norman; Sat. ***World Series Sprintcars:*** Jan. ***Generations in Jazz:*** May. ***Mount Gambier Gold Cup Carnival:*** May. ***Mount Gambier Show:*** Oct. ***Christmas Parade:*** Nov.

Sage & Muntries Restaurant: imaginative, regional dishes; 78 Commercial St West; (08) 8724 8400. ***Sorrento's Cafe:*** relaxed all-day eatery; 6 Bay Rd; (08) 8723 0900. ***The Barn Steakhouse:*** excellent steaks and grills; Punt Rd; (08) 8726 8250.

Clarendon Chalets: self-contained garden units; Clarke Rd; (08) 8726 8306. ***Colhurst House:*** graceful, historic mansion suites; 3 Colhurst Pl; (08) 8723 1309. ***Eliza Cottage:*** cosy, restored town residence; 30 Wehl St South; 0407 422 877.

Haig's Vineyards: wine-tastings and sales; 4 km SE. ***Mt Schank:*** excellent views of the surrounding district from the summit of an extinct volcano. Note that the 2 summit walks are very steep; 17 km S. ***Nelson and Lower Glenelg National Park:*** over the Victorian border; *see Portland (Vic).*

See also LIMESTONE COAST, p. 242

[BLUE LAKE]

SOUTH AUSTRALIA

This small rural centre on Yorke Peninsula was originally called Gum Flat, because of the giant eucalypts in the area, but was later changed to Minlaton – from the Aboriginal word 'minlacowrie', thought to mean 'sweet water'. Aviator Captain Harry Butler, pilot of the *Red Devil*, a 1916 Bristol monoplane, was born here.

Butler Memorial: in a hangar-like building stands the *Red Devil*, a fighter plane – thought to be the only one of its type left in the world – that fought in France and, less romantically, flew mail between Adelaide and Minlaton; Main St. ***Minlaton Museum:*** National Trust museum in the historic general store features a local history display and Harry Butler memorabilia; open 9.30am–1pm Tues–Fri, 9.30am–12pm Sat; Main St. ***Harvest Corner Yorke Peninsula Visitor Information Centre:*** accredited visitor centre with tourist information, craft and local produce sales, a gallery featuring local artists and tearooms; Main St. ***The Creamery:*** crafts, ceramics and art exhibitions in a historic building; Maitland Rd. ***Gum Flat Gallery:*** art workshop and gallery; open 10am–2pm Wed; Main St. ***Minlaton Fauna Park:*** popular spot for picnics, with kangaroos, emus and up to 43 species of birds; Maitland Rd. ***Minlaton Walking Trail:*** a trail to the only naturally occurring stand of river red gums in the Yorke Peninsula. See a number of historic landmarks including an old horse dip and ancient Aboriginal wells, catch a local glimpse of the birdlife at the bird hide, and learn about the area through interpretive signs; contact visitor centre for details.

Minlaton Show: Oct.

Red Devil Cafe: smart pub dining; Minlaton Hotel, 26 Main St; (08) 8853 2014.

Ramsay Park: native flora and fauna park; between Minlaton and Port Vincent to the east. ***Port Rickaby:*** quiet swimming and fishing spot; 16 km NW.

See also YORKE PENINSULA, p. 240

Mintaro

Pop. 223
Map ref. 603 J6

Cnr Main North and Spring Gully rds; (08) 8842 2131 or 1800 242 131; www.clarevalley.com.au

105.1 Trax FM, 639 AM ABC North and West

Although it is in the Clare Valley region, Mintaro's prosperity is not linked with the valley's booming wine industry. Instead, its buildings date back to the 1840s and '50s, when bullock drays carried copper from the Monster Mine at Burra to Port Wakefield in the south. Many of the buildings use local slate. In 1984 Mintaro became the first town in South Australia to be classified a State Heritage Area.

Timandra Garden: one of the town's fine garden displays; enter via Timandra Nursery, Kingston St. ***Mintaro Garden Maze:*** kids will love getting lost in the maze, comprising over 800 conifers; open 10am–4pm Thurs–Mon; Jacka St. ***Reillys Wines:*** wine-tastings, cellar-door sales and a restaurant; Burra Rd. ***Mintaro Cellars:*** wine-tastings and sales; Leasingham Rd. ***Mintaro Slate Quarries:*** fine-quality slate, produced since the quarry opened in 1854, is used world-wide for billiard tables; open 7.30am–3pm Mon–Fri; viewing platform in Kadlunga Rd. ***Heritage walk:*** self-guide trail includes 18 heritage-listed colonial buildings and 2 historic cemeteries; brochure available.

Reilly's Wines & Restaurant: satisfying country meals; Cnr Burra and Hill sts; (08) 8843 9013. ***The Station Cafe:*** weekend snacks and platters; Mount Horrocks Wines, Curling St, Auburn; (08) 8849 2202.

Reilly's Country Retreat: quaint vineyard cottages; Reilly's Wines; Cnr Burra and Hill sts; (08) 8843 9013. ***William Hunt's Retreat:*** stylish converted barn; Lot 32 Burra St; 0447 008 135. ***Dennis Cottage:*** well-appointed garden haven; St Vincent St, Auburn; (08) 8843 0048. ***Martindale Hall:*** magnificent heritage mansion; Manoora Rd, Martindale; (08) 8843 9088.

Martindale Hall This 1879 mansion was built for Edmund Bowman who, the story goes, commissioned it for his bride-to-be from English high society. The lady declined his offer of marriage, and Bowman lived there on his own until 1891. Today visitors can explore the National Trust, Georgian-style home with its Italian Renaissance interior. A room in the mansion was used in the 1975 film *Picnic at Hanging Rock* as a girls' dormitory. Accommodation and dining are available. Open 11am–4pm Mon–Fri, 12–4pm weekends; Manoora Rd; (08) 8843 9088; 3 km SE.

Polish Hill River Valley: a subregion of the Clare Valley wine region, with cellar doors offering tastings and sales; between Mintaro and Sevenhill. ***Waterloo:*** features the historic Wellington Hotel, once a Cobb & Co staging post; 23 km SE.

See also MID-NORTH, p. 239

Moonta

Pop. 3353
Map ref. 599 E3 | 602 H6

Railway Station, Blanche Tce; (08) 8825 1891; www.yorkepeninsula.com.au

89.3 Gulf FM, 639 AM ABC North and West

The towns of Moonta, Kadina and Wallaroo form the 'Copper Coast' or 'Little Cornwall', so called because of abundant copper finds and the significant Cornish population. Like so many other copper discoveries in South Australia, Moonta's was made by a local shepherd – in this case, Paddy Ryan in 1861. It was to prove a fortunate find: Moonta Mining Company paid over £1 million in dividends. Thousands of miners, including experienced labourers from Cornwall, flocked to the area. The mines were abandoned in the 1920s because of the slump in copper prices and rising labour costs. Moonta has survived as an agricultural service town with an increasing tourist trade.

All Saints Church of England: features a locally constructed copper bell; cnr Blanche and Milne tces. ***Queen Square:*** park for picnics, with the imposing town hall opposite; George St. ***Heritage walks and drives:*** self-guide trails to see heritage stone buildings and historic mine sites; brochure from visitor centre.

Kernewek Lowender: prize-winning Cornish festival held with Kadina and Wallaroo; odd-numbered years, May. ***Moonta Garden Fair:*** Oct. ***Moonta Antiques and Collectible Fair:*** Nov.

Moonta Mines State Heritage Area Take a historical walk or drive from Moonta to this significant heritage area. Interpretive walking trails guide the visitor to the major sites, including the Hughes Pump House, shafts, tailing heaps and ruins of mine offices. A 50 min historical railway tour runs from the museum (tours depart Wed 2pm, Sat and Sun 1–3pm on the hour, public/school holidays daily 11–3pm on the hour). Also on the site is a historic 1880 pipe organ in the Moonta Mines Heritage Uniting Church; Cornish lifestyle history and memorabilia at the Moonta Mines Museum; and the National Trust furnished Miners Cottage and Heritage Garden. Open 1–4pm daily. Enjoy old-style sweets at the Moonta Mines Sweet Shop. Via Verran Tce; 2 km SE.

Wheal Hughes Copper Mine: underground tour of one of the modern mines; tours Wed, Sat and Sun; Wallaroo Rd;

(08) 8825 1891; 3 km N. ***Moonta Bay:*** a popular seaside town for fishing and swimming. See native animals at the Moonta Wildlife Park; 5 km W.

See also YORKE PENINSULA, p. 240

Morgan

Pop. 425
Map ref. 603 L6

Shell Morgan Roadhouse, 14–18 Fourth St; (08) 8540 2354; www.riverland.info

 93.1 Magic FM, 1062 AM ABC Riverland

This Murray River town was once a thriving port and a stop on the rail trade route to Adelaide. Settlers saw the potential of the region, and Morgan boomed as soon as it was declared a town in 1878. Now it is a quiet holiday destination, but evidence of its boom days can still be seen in the streetscapes and the historic wharf and rail precinct.

Port of Morgan Museum This comprehensive museum is dedicated to the rail- and river-trade history of Morgan. In the old railway buildings are museum exhibits focusing on the paddlesteamers and trains that were the lifeblood of the town. The Landseer Building has vintage vehicles and a 12 m mural depicting the old Murray River lifestyle. Other highlights are the restored wharf and permanently moored PS *Mayflower*. Open 2–4pm Wed, Sat and Sun; Riverfront; (08) 8540 2085.

Houseboats: for hire; contact visitor centre for details. ***Heritage walk:*** self-guide trail covers 41 historic sites, including the impressive wharves (1877) standing 12 m high, the customs house and courthouse, the sunken barge and steamer, and the rail precinct; brochure from visitor centre.

Mallyons on the Murray: home-grown and handmade delights; Morgan–Renmark Rd; (08) 8543 2263.

Bessie's of Morgan: charming, historic settler's cottage; 16 Seventh St; (08) 8541 9096.

Morgan Conservation Park: a diverse landscape of river flats, sand dunes and mallee scrub with abundant birdlife; across the Murray River from Morgan. ***White Dam Conservation Park:*** well known for red and western grey kangaroo populations; 9 km NW. ***Cadell:*** scenic 12 km drive east from Morgan via a ferry crossing (operates 24 hrs) to Cadell, a major citrus-growing region.

See also MURRAY, p. 241

Mount Gambier

see inset box on previous page

Murray Bridge

Pop. 14 048
Map ref. 597 H11 | 601 D3 | 603 L10

3 South Tce; (08) 8539 1142 or 1800 442 784; www.murraybridge.sa.gov.au

 98.7 Power FM, 1062 AM ABC Riverland

Murray Bridge, just as its name suggests, is all about bridges. The town was established in 1879 when a road bridge was built over the Murray River. The plan to make the river a major trade route from east to west and back had become a reality. In 1886 the construction of a railway line between Adelaide and Melbourne cemented the town's importance. Now watersports, river cruises and a relaxed river atmosphere make Murray Bridge South Australia's largest river town, a perfect holiday spot.

Captain's Cottage Museum: local history museum; open 10am–4pm weekends; Thomas St. ***Dundee's Wildlife Park:*** crocodiles, koalas and a bird sanctuary; Jervois Rd. ***Murray Bridge Regional Gallery:*** regular art exhibitions from local artists; closed Sat; Sixth St. ***Heritage and Cultural Community Mural:*** depicts significant aspects of Murray Bridge; Third St. ***Sturt Reserve:*** offers fishing, swimming, picnic and playground facilities, as well as the mythical Aboriginal creature, the Bunyip (coin-operated); Murray Cod Dr. ***Long Island Marina:*** houseboat hire and recreational facilities; Roper Rd. ***Thiele Reserve:*** popular for waterskiing; east of the river. ***Avoca Dell:*** a popular picnic spot with boating, waterskiing, minigolf and caravan facilities. ***Swanport Wetlands:*** recreational reserve with raised walkways and bird hides; adjacent to Swanport Bridge. ***Charter and regular cruises:*** on the Murray River; contact visitor centre. ***Town and riverside walks:*** brochure from visitor centre.

AutoFest: Australia Day weekend Jan. ***International Pedal Prix:*** novelty bikes and endurance event; Sept. ***Murray Bridge Show:*** Sept. ***Waterski Race:*** over 110 km; Nov. ***Christmas Pageant and Fireworks:*** Nov.

Balcony On Sixth Lodge: suites, basic to deluxe; 6 Sixth St; (08) 8531 1411. ***Childsdale The Udder Place:*** dairy converted to rural retreat; Long Flat Rd; (08) 8531 1153. ***Oz Houseboats:*** huge fleet, budget to luxury craft; Long Island Marina; (08) 8365 7776. ***Rainforest Retreat:*** self-contained house in garden oasis; 19A Torrens Rd; (08) 8532 6447.

Monarto Zoological Park This open-range 1000 ha zoo features Australian, African and Asian animals. It also runs a breeding program for rare and endangered species. Jump on a safari bus tour to see the animals up close. On the way you might encounter the huge giraffe herd or some cheetahs, zebras or rhinoceroses. Tours depart hourly 10.30am–3.30pm. There are also walking tracks through native bushland and mallee country. Princes Hwy; (08) 8534 4100; 10 km W.

Sunnyside Reserve Lookout: views across the wetlands and Murray River; 10 km E. ***Willow Point Winery:*** cellar-door tastings and sales of famous ports, sherries and muscats; closed Sun; Jervois Rd; 10 km S. ***Ferries–McDonald and Monarto conservation parks:*** walking trails through important mallee conservation areas, prolific birdlife, and blossom in spring; 16 km W. ***Tailem Bend:*** a historic railway town with views across the Murray River. A children's playground features an old steam locomotive, and over 90 historic buildings are displayed at the Old Tailem Town Pioneer Village; 25 km SE. ***Wellington:*** situated where Lake Alexandrina meets the Murray River. A museum is in the restored courthouse. Wellington and nearby Jervois have free 24 hr vehicle ferries; 32 km S. ***Karoonda:*** the heart of the Mallee, Karoonda is well known for the 1930 meteorite fall nearby (monument at RSL Park). Natural attractions include the limestone caves of Bakara Plains and walking trails in Pioneer Historical Park; 66 km E.

See also MURRAY, p. 241

Naracoorte

Pop. 4889
Map ref. 590 A10 | 601 G9

MacDonnell St; (08) 8762 1399 or 1800 244 421; www.naracoortelucindale.sa.gov.au

89.7 5TCB FM, 1161 AM ABC South East

continued on p. 271

SOUTH AUSTRALIA

NURIOOTPA

Pop. 4415

Map ref. 597 F4 | 601 C1 | 603 K8

Barossa Visitor Centre, 66–68 Murray St, Tanunda; (08) 8563 0600 or 1300 852 982; www.barossa.com

89.1 BBB FM, 639 AM ABC North and West

The long history of winemaking in this Barossa town is apparent when travelling down the main street. Old vines that glow red in autumn drape the verandahs of equally old buildings. Surprisingly, the town actually began life as a pub. As a trade route was being established northwards to the Kapunda coppermines, William Coulthard foresaw the demand for rest and refreshment. He built the Old Red Gum Slab Hotel in 1854 and the town developed around it. The Para River runs through Nuriootpa, its course marked by parks and picnic spots.

Coulthard Reserve: popular recreation area; off Penrice Rd. ***Barossa Trike Tours:*** provides chauffeured Barossa tours aboard an Oztrike Chopper 4 with seating for 3 as you take in the beauty and splendour of Australia's most famous wine region; South Tce; 0438 623 342.

Barossa Vintage Festival: celebration of food and wine in various locations; odd-numbered years, Apr. ***Barossa Gourmet Weekend:*** Aug. ***Barossa Farmers' Market:*** in historic Vintners Sheds, near Angaston, each Sat morning.

Maggie Beer's Farm Shop: snacks and goodies; Pheasant Farm Rd; (08) 8562 4477. ***The Branch:*** versatile country cafe; 15 Murray St; (08) 8562 4561. ***Appellation:*** benchmark regional fine dining; The Louise, cnr Seppeltsfield and Stonewall rds, Marananga; (08) 8562 4144.

The Lodge Country House: elegant, heritage homestead; Seppeltsfield Winery, Seppeltsfield; (08) 8562 8277. ***The Louise:*** opulent resort-style vineyard suites; Cnr Seppeltsfield and Stonewell rds, Marananga; (08) 8562 2722. ***Seppeltsfield Vineyard Cottage:*** exquisite, secluded vineyard retreat; Gerald Roberts Rd, Seppeltsfield; (08) 8563 4059.

Barossa Wine Region The Barossa is Australia's eminent wine region, a landscape of historic villages panning out to vine-swept hills and grand buildings on old wine estates. Shiraz is the premier drop, with semillon the star of the whites. Some of the shiraz vines date back to the 1840s, and several winemaking families, many with German backgrounds, are into their sixth generation. Senior names include Yalumba (which is officially part of the Eden Valley), Penfolds and Seppelt. You can sample the iconic Penfolds Grange at Penfolds Barossa Valley, and visit the Winemakers' Laboratory to blend your own wine to take home in a personlised bottle. Seppelt is a must-visit winery with its elegant bluestone buildings and gardens. Its range of fortified wines includes Spanish styles and classic tawny wines – the jewel is Para Liqueur, a tawny released when it is 100 years old. Peter Lehmann and Wolf Blass are other well-known wineries in the area. Smaller gems include Charles Melton, known for its Nine Popes blend of shiraz, grenache and mourvedre and for the Rose of Virginia; the nearby Rockford, with fantastic wines seldom seen in other Australian states; Torbreck Vintners, offering excellent shiraz and shiraz viognier; and Langmeil. For some history on the Barossa and winemaking, the Jacob's Creek Visitor Centre has a gallery with displays next to its wine-tasting area, where you can sample some of this well-known label's varieties; *for more information on the region see Tanunda, Lyndoch and Angaston.*

Light Pass: a small, historic township with notable Lutheran churches and Luhrs Pioneer German Cottage, displaying German artefacts; 3 km E. ***Maggie Beer's Farm Shop:*** tastings and sales of gourmet farm produce from renowned chef and writer Maggie Beer, as well as Pheasant Farm and Beer Brothers wines. Enjoy a gourmet lunch (gourmet-style picnic lunch, (08) 8562 4477); Pheasant Farm Rd; 5 km SW. ***Wolf Blass Visitor Centre:*** an opportunity to discover your own unique and memorable Wolf Blass experience, be it through learning about one of Australia's most storied winemakers or enjoying the tasting room; Sturt Hwy; (08) 8568 7311.

See also BAROSSA VALLEY, p. 238

[PENFOLDS WINERY, BAROSSA VALLEY]

Naracoorte dates from the 1840s, but its growth has been slow. In the 1850s it was a stopover for Victorian gold escorts and miners. Since then it has developed a rich agricultural industry. Today it is renowned for its natural attractions, including the parks and gardens but more significantly the Naracoorte Caves, protected within South Australia's only World Heritage area.

The Sheep's Back: a comprehensive museum in the former flour mill (1870) details the history and community of the wool industry, with a craft and souvenir shop and information centre; MacDonnell St. ***Naracoorte Art Gallery:*** local and touring exhibitions; open Tues–Fri; Ormerod St. ***Mini Jumbuk Centre:*** display gallery and sales of woollen products; Smith St. ***Pioneer Park:*** restored locomotive on display; MacDonnell St. ***Walking trail:*** starts at the town centre and winds 5 km along the Naracoorte Creek.

Taste the Limestone Coast: wine and gourmet food festival; Feb. ***Naracoorte Horse Trials:*** May. ***Swap Meet:*** May. ***Limestone Coast Children's Expo:*** Oct. ***Naracoorte Show:*** Oct. ***Christmas Pageant:*** Dec.

Sherwood Cottages Country Retreat: endearing, luxury stone cottage; McMillan Rd; (08) 8762 1652. ***Willowbrook Cottages:*** restful home and garden; 5 Jenkins Tce; (08) 8762 0259.

Naracoorte Caves National Park For thousands of years the 26 Naracoorte Caves – today protected by national park and World Heritage listing – have acted as a natural trap for animals, providing an environment that was just right for fossilisation. Twenty fossil deposits have been found – an incredible record of Australia's evolution over the last 500 000 years. Guided walking tours take in the chambers, extensive stalagmite and stalactite deposits and fossil collections. The Victoria Fossil Cave Tour is an introduction to the ancient animal history of Australia, while the natural delights of the caves, including helictites and fabulous domed ceilings, are accessed on the 30 min Alexandra Cave Tour. The world of bats is celebrated on the Bat Tour, the highlight being unhindered views of the bats' activity from infra-red cameras. Adventure caving allows visitors to see the caves in their raw state, while also providing an opportunity for exciting squeezes and crawls through some very tight spaces. For caving beginners, try the Blackberry and Stick-Tomato tours. For the more experienced cavers, enjoy the crawls and sights on the Starburst Chamber Tour. The Fox Cave Tour is the ultimate caving experience, with access to the cave system by a small entrance, leading to great fossil collections, vast speleothem development and incredible scenery. You can get details on these tours from Wonambi Fossil Centre, located within the park; (08) 8762 2340; 12 km SE.

Wrattonbully Wine Region: a recently established wine region focusing mainly on red wine varieties; 15 km SE. ***Bool Lagoon Game Reserve:*** wetland area of international significance, a haven for ibis and over 100 waterbird species. It includes boardwalks and a bird hide; 17 km S. ***Frances:*** a historic railway town that each Mar holds the Frances Folk Gathering; 38 km NE. ***Lucindale:*** a small country town featuring a Historical Society Museum and Jubilee Park with a lake, island and bird haven. It holds mammoth South East Field Days each Mar; 41 km W. ***Padthaway:*** prominent wine region; 47 km NW; *see Bordertown*.

See also LIMESTONE COAST, p. 242

Oodnadatta

Pop. 277
Map ref. 606 B6

Pink Roadhouse, Ikaturka Tce; (08) 8670 7822 or 1800 802 074.

95.3 FM ABC North and West

Oodnadatta is a gutsy outback town on the legendary Oodnadatta Track. It was once a major railway town, but the line's closure in 1981 left it largely deserted. Local Aboriginal people have successfully kept the town operating since then. Today many travellers use it to refuel and gather supplies before heading out to the major desert parks to the north. It is believed that the name Oodnadatta originated from an Aboriginal term meaning 'yellow blossom of the mulga'.

Pink Roadhouse: a town icon, and also the place to go for information on local road conditions and outback travel advice; Ikaturka Tce. ***Railway Station Museum:*** well-preserved sandstone station (1890), now a local museum; key available from roadhouse.

Races and gymkhana: May. ***Bronco Branding:*** July.

Witjira National Park This arid park is famous for the Dalhousie Springs. These thermal springs emerge from the Great Artesian Basin deep below the surface and are said to be therapeutic (visitors can swim in the main spring). They are also a habitat for many fish species that can adapt to the changing water conditions. A Desert Parks Pass is required; they are available from Mt Dare Homestead (which has fuel and supplies), the Pink Roadhouse or Parks SA (1800 816 078); 180 km N.

Oodnadatta Track: runs from Marree (404 km SE) through Oodnadatta and joins the Stuart Hwy at Marla (212 km NW); *see Marree*. ***Neales River:*** swim in permanent waterholes. ***The Painted Desert:*** superb desert scenery of richly coloured hills; 100 km SW. ***Simpson Desert Conservation Park and Regional Reserve:*** 4WD tracks across enormous dune desert east of Witjira. Travellers must be totally self-sufficient; details from visitor centre.

See also FLINDERS RANGES & OUTBACK, p. 243

Penola

Pop. 1315
Map ref. 592 A2 | 601 G10

The John Riddoch Centre, 27 Arthur St; (08) 8737 2855; www.wattlerange.sa.gov.au

107.7 5THE FM, 1161 AM ABC South East

Penola is one of the oldest towns in south-east South Australia, and has some excellent wineries nearby. The town is noted for its association with Mary MacKillop, a Josephite nun who in 1866 established Australia's first school to cater for children regardless of their family's income or social class. In 2010 she was canonised by the Vatican, making her the first Australian to be declared a saint. Penola is also noted for its literary roots – several Australian poets have been inspired by the landscape and lifestyle.

Mary MacKillop Interpretive Centre and Woods MacKillop Schoolhouse: details the lives of Mary MacKillop and Father Julian Tenison (who shared Mary's dream) through photos, memorabilia and displays in this 1860s-style schoolhouse; Portland St. ***Petticoat Lane:*** heritage area of original cottages, including Sharam Cottage, the first built in

SOUTH AUSTRALIA

town; many are now retail outlets. ***The John Riddoch Centre:*** incorporates the Local History Exhibition and Hydrocarbon Centre, featuring hands-on and static displays on natural gas; Arthur St. ***Toffee and Treats:*** old-fashioned sweet sales; Church St. ***Heritage walk:*** details from visitor centre.

Vigneron Cup: Jan. ***Petanque Festival:*** Feb. ***Penola Coonawarra Arts Festival:*** arts, food and wine; May. ***Cellar Dwellers:*** July. ***Cabernet Celebrations:*** Oct. ***Penola Show:*** Oct/Nov.

Pipers of Penola: exquisite contemporary cuisine; 58 Riddoch St; (08) 8737 3999.

Merlot and Verdelho: adjoining luxury townhouses; 14 Arthur St; (08) 8737 3035 or 0413 512 559. ***must @ Coonawarra:*** cutting-edge designer apartments; 126 Church St; (08) 8737 3444. ***Sarah's Cottage:*** intimate, refurbished cottage; 24 Julian St; (03) 5571 9030.

Yallam Park: a magnificent 2-storey Victorian home with original decorations; by appt (08) 8737 2435; 8 km W. ***Penola Conservation Park:*** signposted woodland and wetland walk; 10 km W. ***Coonawarra Wine Region:*** more than 20 wineries, most open for tastings and cellar-door sales; north to Coonawarra. ***Nangwarry Forestry & Logging Museum:*** features a fascinating array of mill machinery, firefighting equipment, photographs and other artefacts from a bygone era; 18 km S. ***Glencoe Woolshed:*** built in 1863, this shed is unique as it was never converted to mechanised shearing. Now a museum with relics of the period; via Nangwarry; 50km S.

See also LIMESTONE COAST, p. 242

Peterborough

Pop. 1689
Map ref. 603 K3 | 605 I11

Main St; (08) 8651 2708; www.peterborough.au.com

105.1 Trax FM, 639 AM ABC North and West

Peterborough is a town obsessed with the railway. Its very existence and growth can be claimed by that industry. In 1881 the line to Jamestown was opened and over the next few years the town became a key intersection between all the major South Australian towns. Locals boast about how, in a mammoth one-day effort, 105 trains travelled the Broken Hill to Port Pirie line. The rail passion continued even after many of the lines closed, and today each town entrance has a welcoming model steam train.

Steamtown With a 100-year-old rail history this dynamic museum, located around the old locomotive workshops, is a collection of historic rolling stock, including a converted Morris car that rides the tracks. Also on display is Australia's only 3-gauge roundhouse and turntable. Main St; (08) 8651 3355.

Town hall: a beautiful, ornate 1927 building with its original theatre and a Federation wall hanging in the foyer; Main St. ***The Gold Battery:*** ore-crushing machine; contact visitor centre; end Tripney Ave. ***Meldonfield:*** view a world of miniature horse-drawn carriages modelled on the style used in the 1800s; Lloyd St. ***Dragons Rest Habitat Garden:*** reptiles and exotic plant life; Watkins Rd. ***Victoria Park:*** features a lake and islands with deer and kangaroo enclosure and a playground; Queen St. ***Bus tour:*** guided tour of sights and history of town; contact visitor centre. ***Town walk and drive:*** self-guide tour; brochure from visitor centre.

Rodeo: Feb.

Terowie: an old railway town with well-preserved 19th-century main street. Self-guide drive or walk tour; brochure *A Tour of Terowie* available from tearooms; 24 km SE. ***Magnetic Hill:*** park the car, turn off the engine and watch it roll uphill; 32 km NW via Black Rock.

See also MID-NORTH, p. 239

Pinnaroo

Pop. 587
Map ref. 558 B9 | 588 B11 | 590 B1 | 601 H4 | 603 O10

Mallee Tourist and Heritage Centre, Railway Tce Sth; (08) 8577 8644.

107.5 3MBR FM, 1062 AM ABC Riverland

In the 19th century the harshness of the land prevented settlers from properly establishing a farming community here. Instead, they chose the more fertile conditions south-west. The arrival of rail in 1906 and the influx of farming families allowed the community to grow. Although conditions remained tough, the now-renowned Mallee spirit of the farmers allowed the region's agricultural industry to strengthen to what it is today.

Mallee Tourist and Heritage Centre Established in 1999, the centre dwarfs its former home in the old railway station which is now a Pioneer Women's Museum. The new building comprises the D. A. Wurfel Grain Collection, featuring the largest cereal collection in Australia (1300 varieties); working letter presses in the Printing Museum; dioramas, interpretive displays and photos depicting local history in the Heritage Museum; and a collection of restored farm machinery in the Gum Family Collection. Open 10am–1pm or by appt; Railway Tce Sth; (08) 8577 8644.

Animal Park and Aviary: South Tce.

Pinnaroo Show: Oct.

Karte Conservation Park: includes a walking trail through low scrub and 40 m high sand dunes; 30 km NW. ***Billiatt Conservation Park:*** the 1 km walk through mallee scrub and dune country ends with panoramic views from Trig Point; 37 km NW. ***Lameroo:*** Mallee town with historic 1898 Byrne pug-and-pine homestead (Yappara Rd) and railway station (Railway Tce); 40 km W. ***Ngarkat group of conservation parks:*** south-west of town; *see Keith*. ***Peebinga Conservation Park:*** important reserve for the rare western whipbird; Loxton Rd; 42 km N.

See also MURRAY, p. 241

Port Augusta

Pop. 13 255
Map ref. 602 H2 | 604 G9

Wadlata Outback Centre, 41 Flinders Tce; (08) 8641 9193 or 1800 633 060; www.wadlata.sa.gov.au

105.9 Magic FM, 639 AM ABC North and West

Port Augusta is the most northerly port in South Australia. The difficulty of land transportation in the 1800s prompted the town's establishment in 1854. It was a major wool and wheat shipping depot until its closure in 1973 – luckily the power stations built by the State Electricity Trust were already generating the city's chief income. Fuelled by coal from the huge open-cut mines at Leigh Creek, the stations generate more than a third of the state's electricity. Port Augusta is also a supply centre for outback areas, an important link on the *Indian–Pacific* railway and a stopover for the Adelaide to Darwin *Ghan* train.

Wadlata Outback Centre This award-winning complex (recently upgraded) covers the natural history of the outback and Flinders Ranges, as well as the people that have called it home throughout the ages. There are hands-on interpretive displays, audiovisual presentations and artefacts. Discover the landscape of 15 million years ago in the Tunnel of Time, and hear ancient Dreamtime stories. The centre is a place in which to learn – Wadlata is an Aboriginal word for communicating. 41 Flinders Tce.

Homestead Park Pioneer Museum: picnic areas, re-creation of a blacksmith's shop, miniature steam and diesel train rides (1st and 3rd Sun each month), and the restored 130-year-old pine-log Yudnappinna homestead; Elsie St; (08) 8642 2035. ***Fountain Gallery:*** local and touring art and cultural exhibitions; open Mon–Fri; Flinders Tce. ***Gladstone Square:*** landscaped square surrounded by historic sites, including the courthouse, barracks and Presbyterian church; cnr Jervois and Marryatt sts. ***Australian Arid Lands Botanic Garden:*** walks through 200 ha of arid-zone vegetation. Guided tours 10am weekdays; northern outskirts, on the Stuart Hwy; (08) 8641 1049. ***McLellan Lookout:*** site of Matthew Flinders' landing in 1802; Whiting Pde. ***Water Tower Lookout:*** spectacular views from the balcony of the 1882 tower; Mitchell Tce. ***Matthew Flinders Lookout:*** excellent view of Spencer Gulf and the Flinders Ranges; end of McSporran Cres. ***Boat cruises and adventure tours:*** contact visitor centre. ***Heritage walk:*** self-guide town walk includes courthouse and the magnificent stained glass in St Augustine's Church; brochure from visitor centre. ***Curdnatta Art and Pottery Gallery:*** high-quality painting, pottery and fabric art; Flinders Tce.

Cup Carnival: horseracing; June. ***Outback Surfboat Carnival:*** Nov.

Majestic Oasis Apartments: smart serviced units; Marryatt St; (08) 8648 9000 or 1800 008 648.

Spencer Gulf: watersports, yachting and fishing for King George whiting in northern waters. ***Scenic drive:*** north-east to the splendid Pichi Richi Pass, historic Quorn and Warren Gorge. See the same sights by train on Pichi Richi Railway, a 33 km round trip operating from Quorn; *see Quorn.*

See also FLINDERS RANGES & OUTBACK, p. 243

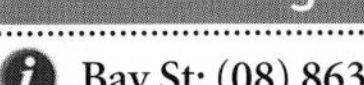

Port Broughton

Pop. 910
Map ref. 599 F1 | 603 I5 | 604 G12

Bay St; (08) 8635 2261; www.yorke peninsula.com.au

87.6 Easy FM, 639 AM ABC North and West

This Yorke Peninsula holiday town has a quiet coastal feel in winter and bustles with sun-seeking holiday-makers in summer. Set on a quiet inlet on Spencer Gulf, it has a long fishing history. In the 1900s the fishing fleets and ketches operated from the jetty. Today the town is still a major port for fishing boats and each week truckloads of blue swimmer crabs depart for city restaurants.

Heritage Centre: local history museum in the old school; Edmund St. ***Sailboat hire and fishing charters:*** from foreshore. ***Town jetty:*** popular fishing spot. ***Historical walking trail:*** grab a *Walk Around Port Broughton* booklet from the visitor centre and navigate the historical sights of the town, including the Heritage Plaques on the foreshore.

Winter Fun Fishing Competition: June. ***Rubber Duck Race:*** Oct.

Fisherman Bay: fishing, boating and holiday spot with over 400 holiday shacks; 5 km N along the coast.

See also YORKE PENINSULA, p. 240

Port Elliot

Pop. 1750
Map ref. 598 F8 | 599 H10 | 601 B5 | 603 J11

Goolwa Wharf; (08) 8555 3488; www.visitalexandrina.com

89.3 5EFM, 639 AM ABC North and West

Port Elliot is a charming historic town set on scenic Horseshoe Bay. Its popularity as a holiday destination lies in the fabulous beaches and the relaxed coastal atmosphere. The town was established in 1854, the year Australia's first public (horse-drawn) railway began operating between Goolwa and the town. Port Elliot's intended purpose as an ocean port for the Murray River was, however, unsuccessful. The bay proved less protected than was first thought and the port was moved to Victor Harbor.

National Trust Historical Display: interpretive centre detailing local history in the old railway station; The Strand. ***The Strand:*** historic street of art and craft shops, cafes and restaurants. ***Cockle Train:*** stops at the railway station on Henry St on its journey from Goolwa Wharf to Victor Harbor, so you can do a section of the journey from here; *for more details see Goolwa.* ***Freeman Nob:*** spectacular views and coastal walks; end of The Strand. ***Encounter Bikeway:*** scenic coastal route between Goolwa and Victor Harbor. ***Horseshoe Bay:*** safe family beach with fishing from jetty. ***Boomer Beach:*** popular surfing beach; western edge of town. ***Maritime Heritage Trail:*** the town's story illustrated in foreshore displays. ***Heritage walk:*** brochure from railway station.

Market: Lakala Reserve; 1st and 3rd Sat each month. ***Port Elliot Show:*** Oct.

Flying Fish Cafe: popular seafood destination; 1 The Foreshore, Horseshoe Bay; (08) 8554 3504. ***Stranded Mexican:*** versatile Mexican fare; 12 The Strand; (08) 8554 2022.

Brooklands Heritage B&B: snug, restored settler's cottage, Heysen Rd; (08) 8554 3808. ***The Summer House:*** contemporary luxury home; 7 Charlotte St; (08) 8363 4510. ***Trafalgar House:*** majestic 19th-century guesthouse; 25 The Strand; (08) 8554 3888.

Basham Beach Regional Park: scenic coastal trails with interpretive signage and southern right whale sightings during their migration season, June–Sept; just north-east of Port Elliot. ***Middleton:*** coastal town with heritage bakery, the old flour mill and fabulous beaches; 3 km NE. ***Crows Nest Lookout:*** excellent views of the coast; 6 km N.

See also FLEURIEU PENINSULA, p. 235

Port Lincoln

Pop. 13 046
Map ref. 602 D8

3 Adelaide Pl; (08) 8683 3544 or 1300 788 378; www.visitportlincoln.net

89.9 Magic FM, 1485 AM ABC Eyre Peninsula and West Coast

Each January this township on the Eyre Peninsula celebrates the life of the tuna – one of the few festivals in Australia devoted to a fish and a fair indication of the reign tuna has over this town. Lincoln Cove, the marina, is the base for Australia's largest tuna fleet and tuna-farming industry. Port Lincoln is set on attractive Boston Bay, which is three times the size of Sydney Harbour. The townsite was reached by Matthew Flinders in his expedition of 1802 and he named it in honour of his home, Lincolnshire, in

England. Sheltered waters, a Mediterranean climate and scenic coastal roads make this a popular holiday spot.

Mill Cottage: National Trust museum with early pioneering artefacts and paintings; open 10am–2pm Wed, 2–4pm Sun or by appt; Flinders Hwy. ***Railway Museum:*** relics of the railway past displayed in a historic 1926 stone building; open 1–4pm Wed, Sat and Sun; Railway Pl. ***Axel Stenross Maritime Museum:*** features original boatbuilding tools and working slipway; open 9.30–11.30am and 1–4.30pm Tues, Thurs and Sun, 1–4.30pm Sat or by appt; Lincoln Hwy. ***Settler's Cottage:*** stone cottage with early pioneer photos and documents; open 2–4.30pm Sun (closed July and Aug) or by appt; in Flinders Park, Flinders Hwy. ***Kotz Stationary Engines:*** museum collection of oil and petrol engines; Baltimore St. ***Nautilus Theatre:*** features 2 galleries of local and touring art, a gallery shop and a wine bar; Tasman Tce. ***Kuju Arts and Crafts:*** Aboriginal craft sales; closed Sat and Sun; Ravendale Rd. ***Mayne Gallery:*** local arts and crafts; open 12.30–4.30pm daily; King St. ***Lincoln Cove:*** includes marina, leisure centre with water slide, holiday charter boats and the base for the commercial fishing fleet (tastings of local catches available). Guided walking tours of the marina are available from the visitor centre; off Ravendale Rd. ***Boston Bay:*** swimming, waterskiing, yachting and excellent fishing. ***Yacht and boat charters:*** for gamefishing, diving, day fishing and for viewing sea lions, dolphins and birdlife around Sir Joseph Banks Group Conservation Park and Dangerous Reef; contact visitor centre for details. ***Aquaculture Cruise:*** offers you a chance to view the working tuna farms (when in season) and to taste some mouth-watering local sashimi; a sea lion colony can also be visted; contact visitor centre for details. ***Boston Island boat tours:*** cruises around bay and island; contact visitor centre for details. ***Adventure tours and safaris:*** offshore and land adventure offered, including close-up tuna tours, shark expeditions and 4WD safaris; contact visitor centre for details. ***Old Mill Lookout:*** panoramic views of town and bay; Dorset Pl. ***Parnkalla Walking Trail:*** 14 km trail with coastal views and abundant wildlife. It forms part of the longer Investigator Walking Trail from North Shields to Lincoln National Park; brochure from visitor centre.

Tunarama Festival: Jan. ***Adelaide to Lincoln Yacht Race and Lincoln Week Regatta:*** Feb/Mar. ***Port Lincoln Show:*** Aug.

Del Giorno's Cafe Restaurant: wonderful Italian-inspired seafood; 80 Tasman Tce; (08) 8683 0577. ***The Marina Bistro:*** seafood and grills; The Marina Hotel, 13 Jubilee Dr; (08) 8682 6141.

Bay 10 Holiday House: neatly renovated foreshore villa; 24 Lincoln Hwy; (08) 8682 1010. ***Belle Vista:*** smartly restored stone bungalow; 12 Normandy Pl; (08) 8683 0180. ***The Anchorage Holiday Apartment:*** spacious marina accommodation; 6/33 South Point Dr; (08) 8683 0992. ***Donington Cottage:*** rustic and remote lighthouse keeper's residence; Donington Rd, Lincoln National Park; (08) 8683 3544 or 1300 788 378.

Lincoln National Park This spectacular coastal park has a network of walking trails through rugged wilderness areas to fantastic coastal scenery. The park is an important sanctuary for migrating birds. To see the park from a height, take the 1.1 km return hike up Stamford Hill. At the top is the Flinders Monument and panoramic views of the coast. For a true, uninterrupted wilderness experience, grab a key and permit from the visitor centre and head on to Memory Cove, a calm bay with a fantastic beach. There is also a replica of the plaque placed by Matthew Flinders in 1802 in memory of 8 crew members lost in seas nearby. 4WD enthusiasts would enjoy the challenges of the Sleaford Bay coast. (08) 8688 3111; 20 km S.

Delacolline Estate Wines: well known for blended variety of sauvignon blanc/semillon; tastings and sales; open 1.30–4pm weekends; Whillas Rd; 1 km W. ***Winters Hill Lookout:*** views to Boston Bay, Boston Island and Port Lincoln; Flinders Hwy; 5 km NW. ***Boston Bay Winery:*** tastings and sales; closed Tues and Wed; Lincoln Hwy; 6 km N. ***Roseview Emu Park and Rose Gardens:*** picturesque gardens in bush setting, with sales of emu produce; Little Swamp La; 10 km NW. ***Glen-Forest Tourist Park:*** native animals, bird-feeding and miniature golf course; Greenpatch; 15 km NW. ***Poonindie Church:*** quaint old church built in 1850 with the unique feature of 2 chimneys; 20 km N. ***Mikkira Station and Koala Park:*** historic 1842 homestead, with bushwalks to see native wildlife. Permit required, available from Port Lincoln Visitor centre; off Fishery Bay Rd; 26 km SW. ***Constantia Designer Craftsmen:*** guided tours of world-class furniture factory and showroom; open Mon–Fri; on road to Whalers Way. ***Whalers Way:*** cliff-top drive through privately owned sanctuary inhabited by seals, ospreys, kangaroos and emus. Permit from visitor centre; 32 km S.

See also EYRE PENINSULA & NULLARBOR, p. 244

Port MacDonnell

Pop. 624
Map ref. 592 A6 | 601 G12

7 Charles St; (08) 8738 2576; www.thelimestonecoast. com.au

96.1 Star FM, 1476 AM ABC South East

Port MacDonnell is a quiet fishing town that was once a thriving port. The establishment of the breakwater in 1975 has ensured the southern rock lobster trade many more years of fruitful operation. The fleet is now the largest in Australia. While fishing is the main focus of the area, the rich maritime history, fascinating crystal pools and coastal scenery attract visitors year-round.

Port MacDonnell and District Maritime Museum The long maritime history of this stretch of coast is littered with stories of shipwrecks and bravery. Here photos and salvaged artefacts bring the old days to life. A particularly tragic story is the crash of the *Admella* on an off-coast reef in 1859. Only 24 of the 113 people aboard survived. The ship's bell and cannon are on display in the museum. There is also a focus on community history and on the rock lobster industry. Open 12.30–4.30pm Wed, Fri and Sun; Meylin St; (08) 8738 7259.

Clarke's Park: popular picnic spot with natural spring; northern outskirts. ***Fishing:*** anglers will enjoy fishing from the jetty and landing. Boat charters available for deep-sea catches of tuna; details from visitor centre. ***Heritage walk:*** includes historic cemetery with hidden headstones; contact visitor centre.

Bayside Festival: Jan.

Ewens Ponds and Piccaninnie Ponds conservation parks For a unique snorkelling or diving experience, visit the crystal-clear waters of these parks. At Ewens Ponds (7 km E) there are 3 ponds, connected via channels. Snorkel on the surface to see the amazing plant life underwater, or go diving for the ultimate experience. The deep caverns in Piccaninnie Ponds (20 km E) offer visitors an insight into the underwater world. Snorkellers can gaze into the depths of the Chasm, while divers can explore the limestone-filtered waters of the Cathedral, so named because of its regal white walls. While no experience is necessary for snorkelling, divers require qualifications. Inquiries and bookings to SA Parks and Wildlife; (08) 8735 1177.

Cape Northumberland Heritage and Nature Park: a coastal park famous for sunrises and sunsets. Other highlights include a historic lighthouse, a penguin colony and unusual rock formations; just west of town. ***Dingley Dell Conservation Park:*** the historic 1862 restored cottage that is located here was once the home of Australian poet Adam Lindsay Gordon and features displays on his life and work; tours 10am–4pm daily; 2 km W. ***Germein Reserve:*** 8 km boardwalk (loop track) through wetlands; opposite Dingley Dell. ***Southern Ocean Shipwreck Trail:*** over 89 vessels came to grief on the section of coast from the Victorian border to the Murray River mouth. The drive trail includes 10 interpretive sites; brochure from visitor centre.

See also LIMESTONE COAST, p. 242

Port Pirie

Pop. 13 204
Map ref. 603 I4 | 604 H11

Regional Tourism and Arts Centre, 3 Mary Elie St; (08) 8633 8700 or 1800 000 424; www.pirie.sa.gov.au

105.1 Trax FM, 639 ABC North and West

Industry in its splendour greets the visitor at this major industrial and commercial centre. The oil tanks, grain silos and 250-metre-high smokestack all tower over the city, while on the waterfront huge local and overseas vessels are loaded and discharged. Broken Hill Proprietary Company (BHP) began mining lead in 1889 and various South Australian ports at that time vied for BHP's smelting business. Port Pirie eventually won, and created what is today the largest lead-zinc smelter in the world. Wheat and barley from the mid-north are also exported from here. Port Pirie shows great character in its old buildings and attractive main street, and Spencer Gulf and the Port Pirie River offer swimming, waterskiing, fishing and yachting.

Regional Tourism and Arts Centre This award-winning centre comprises an eclectic mix of exhibitions, art and information. A lifelike fibreglass model of the largest white pointer shark taken from SA's waters is on display. Local and regional history is presented through a series of art pieces and on the miniature railway, Pirie Rail Express, which replicates the journey from Port Pirie to Broken Hill (runs 1st and 3rd Sun each month). There are local and touring art exhibitions in the art gallery and the centre runs tours to the Pasminco smelter. Mary Elie St; (08) 8633 8700.

National Trust Museum: located in historic town buildings, including the old customs house (1882) and the Victorian pavilion-style railway station, the museum houses a local history display and rooms furnished in early-1900s style; Ellen St. ***Memorial Park:*** features the John Pirie anchor, memorials, and the Northern Festival Centre; Memorial Dr. ***Fishing:*** good local spots include the main wharf. ***Self-guide walks:*** including National Trust Walking Tours and The Journey Landscape, a 1.6 km nature trail representing changes in vegetation from Broken Hill to Port Pirie; brochures from visitor centre.

State Masters Games: even-numbered years, Apr. ***Blessing of the Fleet:*** celebrates the role of Italians in establishing the local fishing industry; Sept. ***Festival of Country Music:*** Oct.

Sampsons Cottage: landmark shop, now self-contained B&B; 66 Ellen St; (08) 8632 3096.

Weeroona Island: good fishing and holiday area accessible by car; 13 km N. ***Port Germein:*** a quiet beachside town with a tidal beach safe for swimming. At 1.7 km, the town's jetty is one of the longest in Australia; 23 km N. ***Telowie Gorge Conservation Park:*** follow the marked Nukunu Trail from the park's entrance to the breathtaking Telowie Gorge on the south-west edge of the Flinders Ranges. Care should be taken on less-formal tracks in the park; 24 km NE.

See also YORKE PENINSULA, p. 240

Port Victoria

Pop. 344
Map ref. 599 D5 | 602 H8

Port Victoria Kiosk, Esplanade; or The Farm Shed Museum and Tourist Centre, 50 Moonta Rd, Kadina; (08) 8821 2333 or 1800 654 991; www.yorkepeninsula.com.au

89.3 Gulf FM, 639 AM ABC North and West

A tiny township on the west coast of Yorke Peninsula, Port Victoria was tipped to be a thriving port town after James Hughes travelled up the coast in 1840. Hughes, a land surveyor, studied the coastline from his schooner, *Victoria*, and reported favourably on the region. It became an important port for grain exports, with windjammers transporting wheat from here to Europe. The town still proudly proclaims that it is the 'last of the windjammer ports'.

Maritime Museum: displays, relics and artefacts of the great era of the windjammer; open 2–4pm weekends and public holidays; Main St. ***Jetty:*** original 1888 jetty with good swimming and fishing; end of Main St. ***Geology trail:*** 4 km interpretive track along the foreshore explains the coast's ancient volcanic history; brochure from visitor centre.

Goose Island Conservation Park: important breeding area for several bird species and the Australian sea lion; 13 km offshore; access by private boat. ***Wardang Island:*** this large island is an Aboriginal reserve, and permission for access is required from Goreta (Point Pearce) Aboriginal Community Council; (08) 8836 7205; near Goose Island. ***Wardang Island Maritime Heritage Trail:*** this scuba-diving and overland trail includes 8 shipwreck sites with underwater plaques around Wardang Island and 6 interpretive signs at Port Victoria; waterproof self-guide leaflet available from visitor centre.

See also YORKE PENINSULA, p. 240

Quorn

Pop. 1073
Map ref. 603 I1 | 604 H9

3 Seventh St; (08) 8648 6419; www.flindersranges.com

89.1 5UMA FM, 1242 5CS AM

Nestled in a valley in the Flinders Ranges, Quorn was established as a town on the Great Northern Railway line in 1878. The line was built by Chinese and British workers and operated for over 45 years (it closed in 1957). Part of the line through Pichi Richi Pass has been restored as a tourist railway, taking passengers on a scenic 33-kilometre round trip via Port Augusta. The town's old charm has not been lost on movie producers – the historic streetscapes and surrounding landscapes have been used in many films.

Railway Workshop Tours: guided tours of the workshop where locomotives travelling on the Pichi Richi line are maintained and restored. Tours by appt; book at visitor centre. ***Junction Art Gallery:*** local art exhibition; Railway Tce. ***Outback Colours Art Gallery:*** Seventh St. ***Town walks:*** the Walking Tour of Quorn and the Quorn Historic Buildings walk; brochures from visitor centre.

Flinders Ranges Bush Festival: Apr. ***Taste of the Outback:*** Apr. ***Race Meeting:*** June. ***Pichi Richi Marathon:*** July. ***Quorn Show:*** Sept. ***Spring Craft Fair:*** Oct. ***Christmas Pageant and Party:*** Dec.

Quandong Cafe: heritage cafe; 31 First St; (08) 8648 6155.

Endilloe Lodge: luxurious, contemporary stone cottage; 319 Schmidt Rd; (08) 8648 6536. ***Wilderness Cabin:*** remote bush retreat; Horseshoe Range via Quorn; (08) 8648 6438.

Pichi Richi Railway: historical tourist train travels through dramatic countryside from Quorn to Port Augusta and back; tours by arrangement; bookings 1800 440 101 or through visitor's centre. ***Quorn Native Flora Reserve:*** stone reserve, once the town's quarry, with informative brochure available that details the reserve's flora; Quarry Rd; 2 km NW. ***Pichi Richi Camel Tours:*** award-winning camel tours through the gorgeous native bushland, with candlelit dinners and moonlight rides on offer; Devils Peak Rd; (08) 8648 6640; 6 km SE. ***The Dutchmans Stern Conservation Park:*** colourful rocky outcrops observed on 2 trails through the park. The Ridge Top Trail (8.2 km return) offers spectacular views of the Flinders Ranges and Spencer Gulf; 8 km W. ***Devil's Peak Walking Trail:*** panoramic views up steep climb to the summit; closed Nov–Apr (fire season); 10 km S. ***Mt Brown Conservation Park:*** mixed landscape of ridges and woodland. The loop trail, starting at Waukarie Falls, offers a side climb to the Mt Brown summit; Richman Valley Rd; 15 km S. ***Warren Gorge:*** imposing red cliffs popular with climbers. Also the habitat of the rare yellow-footed rock wallaby; 23 km N. ***Buckaringa Gorge Scenic Drive:*** drive past Buckaringa Sanctuary and Proby's Grave (he was the first settler at Kanyaka Station) to a lookout accessed via a short walk; begins 35 km N.

See also FLINDERS RANGES & OUTBACK, p. 243

Robe

Pop. 1249
Map ref. 601 E9

The Robe Institute, Mundy Tce; (08) 8768 2465 or 1300 367 144; www.council.robe.sa.gov.au

 107.7 5THE FM, 1161 AM ABC South East

Guichen Bay and Robe's coastline would have been a welcome sight to the Chinese immigrants arriving in the mid-1800s. During the Victorian gold rush, around 16 500 Chinese disembarked here and travelled overland to the goldfields to avoid the Poll Tax enforced at Victorian ports. Robe had a thriving export trade before rail was introduced, which has left a legacy of historic buildings, from quaint stone cottages to the Caledonian Inn, with internal doors salvaged from shipwrecks. Today Robe is one of the state's most significant historic towns, but also a fishing port and holiday centre, famous for its crayfish and its secluded beaches.

The Robe Institute: incorporates the visitor centre, library and Historic Interpretation Centre with photographic and audiovisual displays on Robe's history; Mundy Tce. ***Robe Customs House:*** historic 1863 building, once the hub of Robe's export trade, now a museum featuring Chinese artefacts and displays; open 2–4pm Tues, Sat and daily in Jan; Royal Circus. ***Art and craft galleries:*** throughout town, especially in Smillie and Victoria sts. ***Deep Sea Fishing charter:*** sightseeing cruise also on offer; bookings (08) 8768 1807. ***Crayfish fleet:*** anchors in Lake Butler (Robe's harbour); sells fresh crayfish and fish Oct–Apr. ***Walk and scenic drive tours:*** self-guide tours available. Take the town walk past 81 historic buildings and sites; brochures from visitor centre.

Robe Easter Surfing Classic: Easter. ***Blessing of the Fleet:*** celebrates the role of Italians in establishing the local fishing industry; Sept. ***Robe Village Fair:*** last full weekend in Nov.

Caledonian Inn: historic pub, smart food; 1 Victoria St; (08) 8768 2029. ***The Gallerie Restaurant & Wine Bar:*** stylish seafood; 2 Victoria St; (08) 8768 2256.

Caledonian Inn: historic pub and seafront suites; 1 Victoria St; (08) 8768 2029. ***Cricklewood:*** tranquil, relaxed lakefront cottage; 24 Woolundry Rd; (08) 8768 2137. ***Grey Masts:*** historic seaside cottages; Smillie St; 0411 627 146. ***Honeyfield:*** French Provincial–style cottage hideaway; Robe Rd; 1300 760 629. ***The Shore:*** spacious luxury beachfront residence; 11 Wrattonbully Rd; 1300 760 629.

Lake Fellmongery: popular spot for waterskiing; 1 km SE. ***Long Beach:*** 17 km pristine beach for surfing and swimming. Cars are allowed on the sand; 2 km N. ***Little Dip Conservation Park:*** features a complex, moving sand-dune system, salt lakes, freshwater lakes and abundant wildlife. Drive or walk through native bush to beaches for surfing and beach-fishing; some areas 4WD only; 2 km S. ***Beacon Hill:*** panoramic views of Robe, lakes and coast from lookout tower; Beacon Hill Rd; 2 km SE. ***The Obelisk:*** navigational marker at Cape Dombey. Scenic access via cliff walk from the Old Gaol at Robe; 2 km W. ***Mt Benson Wine Region:*** young wine region specialising in shiraz and cabernet sauvignon. Eight cellar doors offer tastings and sales; 18 km N.

See also LIMESTONE COAST, p. 242

Roxby Downs

Pop. 3848
Map ref. 604 E4

Roxby Downs Cultural Precinct; (08) 8671 2001; www.roxbydowns.com

 102.7 AM ABC Local, 105.5 ROX FM

In 1975 Roxby Downs station was a hard-working property on the red sand dunes of central South Australia. That was until a body of copper and uranium, the largest in the world, was discovered near a dam. Roxby Downs, the township, was built to accommodate the employees of the Olympic Dam mining project and has many modern facilities.

Cultural Precinct: incorporates the visitor centre, cinema, cafe, art gallery with local and touring exhibitions, and interpretive display on town and dam history; Richardson Pl. ***Arid Discovery:*** area of native landscape with sunset tours to see reintroduced native animals, including bilbies and burrowing bettongs. A highlight is the close viewing of animals in the observation hide; contact visitor centre for tour details. ***Emu Walk:*** self-guide flora walk through town; contact visitor centre.

Market: Richardson Pl; closest Sat to the 15th each month. ***Outback Fringe Festival:*** Apr.

Olympic Dam Mining Complex: an extensive underground system of roadways and trains services the mine that produces refined copper, uranium oxide, gold and silver. The mine is 9 km N, but limited views are available at the site. Olympic Dam Tours run surface tours 3 days a week; times and bookings through the visitor centre.

See also FLINDERS RANGES & OUTBACK, p. 243

RENMARK

Pop. 4342

Map ref. 558 A5 | 588 B5 | 601 G1 | 603 O7

[ERODED RED CLIFFS ALONG THE MURRAY RIVER]

84 Murray Ave; (08) 8586 6704; www.renmarkparinga.sa.gov.au

93.1 Magic FM, 1062 AM ABC Riverland

It is hard to imagine that the lush lands around Renmark, thriving with orchards and vineyards, were once a veritable wasteland. In 1887 the Canadian-born Chaffey brothers were granted 30 000 acres (12 000 hectares) by the South Australian government to test their irrigation scheme. Theirs was the first of its type to succeed in Australia and today the farmlands are still irrigated with water piped from the Murray River.

Olivewood: National Trust historic building, formerly the Chaffey homestead, dressed in period furnishings, and with famous olive trees in the orchard; closed Wed; cnr Renmark Ave and Twenty-first St. ***PS Industry:*** 1911 grand lady of the river still operates on steam when taking visitors on her monthly cruises; 90 min tours run at 11am and 1.30pm first Sun each month; bookings at visitor centre. ***Renmark Hotel:*** historic community-owned and -run hotel; Murray Ave. ***Nuts About Fruit:*** sales of local dried fruit, nuts and other produce; closed Sun; Renmark Ave. ***Renmark Riverfront walk:*** wander along and take in great views of the town along the Murray River. ***Murray River cruises:*** houseboat hire or paddlesteamer tours to cruise the mighty Murray; contact visitor centre for details.

Dash for Cash: Feb. ***Riverland Dingy Derby:*** Feb. ***Riverland Balloon Fiesta:*** June. ***Rose Festival:*** Oct. ***Renmark Show:*** Oct. ***World Future Cycle Challenge:*** Nov.

River's Edge Restaurant: regional cuisine; Renmark Club, Murray Ave; (08) 8586 6611.

Liba Liba Houseboats: 14 paddlewheel boats, 4–12 berth; Jane Eliza Landing; (08) 8586 6734 or 1800 810 252. ***Willows & Waterbirds:*** large, convenient family home; 41 Murray Ave; (08) 8295 8836. ***Customs House Houseboats:*** secluded boating along Victorian border; Customs House Marina, via Renmark; 1300 557 706. ***Paringa House B&B Stone Cottages:*** romantic riverfront getaways; 1 Museum Dr, Paringa; (08) 8595 5217. ***Riverfun Houseboat Hire:*** 10 vessels, classic to ultra-modern; Lock 5 Rd, Paringa; (08) 8595 5537.

Bookmark Biosphere Reserve This reserve incorporates the mallee country and arid outback landscapes of Chowilla Regional Reserve and Danggali Conservation Park. In Chowilla (50 km N) are stretches of flood plains interspersed with native woodland and scrubland. Fishing, canoeing and birdwatching are popular and the history of the flood plains is explained on the Old Coach Road Vehicle Trail. Danggali (90 km N) is a vast wilderness area with interesting trails to explore. The 2 drive tours, Nanya's Pad Interpretive Drive (100 km circuit, 2WD accessible) and Tipperary Drive (100 km circuit, 4WD only), are both excellent introductions to the mallee scrub region, while the 10 km Target Mark Walking Trail passes through native vegetation to the dam.

Lock and Weir No. 5: picnic in surrounding parklands; 2 km SE. ***Paringa:*** small farming community featuring a historic suspension bridge (1927), Bert Dix Memorial Park and nearby Headings Cliffs Lookout; 4 km E. ***Angove's:*** producers of St Agnes Brandy as well as wine, with cellar-door tastings and sales; Bookmark Ave; 5 km SW. ***Bredl's Wonder World of Wildlife:*** unique fauna, particularly reptiles. Handling and feeding times between 11am and 3pm; (08) 8595 1431; 7 km SW. ***Ruston's Rose Garden:*** the Southern Hemisphere's largest rose garden with over 50 000 bushes and 4000 varieties; open Oct–May; Moorna St, off Sturt Hwy; (08) 8586 6191; 7 km SW. ***Dunlop Big Tyre:*** spans the Sturt Hwy at Yamba Roadhouse and marks the fruit-fly inspection point (no fruit allowed between Victoria and SA); 16 km SE. ***Murray River National Park, Bulyong Island section:*** popular park for water-based activities, fishing and birdwatching; just upstream from Renmark on the Murray River.

See also MURRAY, p. 241

Stansbury

Pop. 521
Map ref. 599 E7 | 602 H9

Cnr Weaver and Towler sts; (08) 8852 4577; www.stansburysa.com

 98.9 Flow FM, 639 AM ABC North and West

Situated on the lower east coast of Yorke Peninsula and with views of Gulf St Vincent, Stansbury was originally known as Oyster Bay because of its claim to the best oyster beds in the state. The town has always serviced the farms inland, but its mainstay today is tourism. The bay is excellent for fishing and watersports, including diving and waterskiing.

Schoolhouse Museum: this local history museum in Stansbury's first school features cultural and environmental displays as well as the headmaster's rooms furnished in early-1900s style; open 2–4pm Wed and Sun, daily in Jan; North Tce. ***Oyster farms:*** see daily operations of local oyster farms and try fresh oysters. ***Fishing:*** popular spots include the jetty, rocks and beach. ***Mills' Gully Lookout:*** popular picnic spot with panoramic views of bay, town and Gulf St Vincent; northern outskirts of town. ***Coastal trails:*** walking and cycling trails past reserves, lookouts and a historic cemetery; start at foreshore caravan park; brochure from visitor centre.

Stansbury Seaside Markets: monthly Oct–May; check dates with visitor centre. ***Stansbury and Port Vincent Wooden and Classic Boat Regatta:*** even-numbered years, Apr. ***Sheepdog Trials:*** odd-numbered years, May.

Musgrave Manor: extravagant, high-tech seaside home; Lot 2 Musgrave La; 0417 855 064.

Kleines Point Quarry: SA's largest limestone quarry; 5 km S. ***Lake Sundown:*** one of the many salt lakes in the area and a photographer's delight at sunset; 15 km NW. ***Port Vincent:*** popular holiday destination with good swimming, yachting and waterskiing; 17 km N.

See also YORKE PENINSULA, p. 240

Strathalbyn

Pop. 3894
Map ref. 597 E12 | 598 H5 | 601 C4 | 603 K10

Old Railway Station, South Tce; (08) 8536 3212; www.visitalexandrina.com

 94.7 5EFM, 639 AM ABC North and West

This heritage town has some of the most picturesque and historic streetscapes in country South Australia. It has a predominantly Scottish heritage, first settled by Dr John Rankine, who emigrated with 105 other Scotsmen in the late 1830s. The town is set on the Angas River, with the Soldiers Memorial Gardens following the watercourse through the town. Strathalbyn is renowned for its antique and craft shops.

National Trust Museum: history display in the courtroom, Victorian-era relics in the courthouse, and a historical room and photographic displays in the Old Police Station; open 2–5pm Wed, Thurs, Sat and Sun; Rankine St. ***Old Railway Station:*** complex includes the visitor centre, the Station Master's Gallery with local and touring art exhibitions (open Wed–Sun), and the station for the tourist railway from Mount Barker, the *SteamRanger* (*see Hahndorf*); South Tce. ***St Andrew's Church:*** impressive church with castle-like tower; Alfred Pl. ***Original Lolly Shop:*** old-fashioned lollies and fudge; High St. ***Antiques, art and craft shops:*** outlets in High St. ***Heritage walk:*** self-guide trail featuring over 30 heritage buildings and the architectural delights of Albyn Tce; brochure available from visitor centre.

Collectors, Hobbies and Antique Fair: Aug. ***Strathalbyn Show:*** Oct. ***Glenbarr Highland Gathering:*** Oct. ***Rotary Duck Race:*** plastic ducks; Nov.

Victoria Hotel: creative pub meals; 16 Albyn Tce; (08) 8536 2202.

Gasworks B&B: charming heritage cottages; 12 South Tce; (08) 8536 4291. ***Longview Lodge Apartments:*** designer vineyard retreats; Longview Vineyard, Pound Rd, Macclesfield; (08) 8388 9694. ***The Old Oak B&B:*** secluded rural homestead; Bald Hill Rd, Bull Creek; (08) 8536 6069.

Milang This old riverboat town is now a popular holiday destination on the shores of Lake Alexandrina, Australia's largest freshwater lake. The lake offers fishing, sailing and windsurfing. In town, visit the Port Milang Railway for its local history display and pick up a Heritage Trail brochure for a self-guide walk. Each Australia Day weekend the Milang–Goolwa Freshwater Classic fills the town with visitors who come to watch hundreds of yachts begin the race. 20 km SE.

Lookout: views over town and district; 7 km SW. ***Ashbourne:*** buy local produce at roadside stalls and at the country market held 3rd Sun each month; 14 km W. ***Langhorne Creek wine region:*** on the Bremer and Angas rivers flood plains, the first vines were planted in the 1850s. This winemaking region has always specialised in red varieties, particularly cabernet sauvignon and shiraz. Five cellar doors offer tastings and sales; 15 km E. ***Meadows:*** features Pottery at Paris Creek and Iris Gardens (open Oct–Mar) nearby. The Country Fair is held each Oct; 15 km NW.

See also FLEURIEU PENINSULA, p. 235

Streaky Bay

Pop. 1059
Map ref. 602 A3 | 611 N10

Rural Transaction and Visitor Information Centre, 21 Bay Rd; (08) 8626 7033; www.streakybay.sa.gov.au

99.3 Flow FM, 693 AM ABC Eyre Peninsula and West Coast

A holiday town, fishing port and agricultural centre for the cereal-growing hinterland. The bay was first sighted in 1627 by Dutch explorer Peter Nuyts, but it wasn't fully explored until 1802 by Matthew Flinders. Flinders named the bay after the 'streaky' colour of the water, caused by seaweed oils. While this town is pretty, it is the surrounding bays and coves, sandy beaches and towering cliffs that bring the visitors.

National Trust Museum: early pioneer history displays in the old school, as well as a restored pioneer cottage and a doctor's surgery; open 2–4pm Tues and Fri, or by appt; Montgomerie Tce; (08) 8626 1443. ***Powerhouse Restored Engine Centre:*** display of old working engines; open 2–5pm Tues and Fri; Alfred Tce. ***Shell Roadhouse:*** Great White Shark replica (original caught with rod and reel); Alfred Tce. ***Fishing:*** for King George whiting, southern rock lobster, salmon, mullaway, garfish, abalone and shark (check with PIRSA & Fisheries centre).

Perlubie Sports Day: Jan. ***Streaky Bay Cup:*** horseracing; Apr.

Mocean Cafe: innovative seafront dining; 34B Alfred Tce; (08) 8626 1775.

Scenic drives: include Westall Way Scenic Drive, which starts 9 km S, taking in rock formations, high cliffs, quiet pools and the Yanerbie Sand Dunes. Also the drive west of town to Cape Bauer and the Blowhole (20 km NW), for views across the

Bight. ***Calpatanna Waterhole Conservation Park:*** bushwalking in coastal park to an important Aboriginal waterhole; excellent birdwatching; 28 km SE. ***Murphy's Haystacks:*** a much-photographed cluster of pink granite boulders, with interpretive signage and paths; 40 km SE. ***Baird Bay:*** a small coastal town with an attractive beach for swimming, boating and fishing. Baird Bay Charters and Ocean Ecotours offer swims with sea lions and dolphins; (08) 8626 5017; 45 km SE. ***Point Labatt Conservation Park:*** from the cliff-top viewing platform, see the rare and endangered Australian sea lions sleeping on the beach (this colony is the only permanent one on the Australian mainland). Parts of access road unsealed; 50 km SE. ***Venus Bay Conservation Park:*** important reserve for breeding and reintroduction of native species. The park includes the peninsula and 7 islands with beach-fishing and swimming. Peninsula access is 4WD only; turn-off 50 km SE. ***Acraman Creek Conservation Park:*** this mangrove and mallee park is an important refuge for coastal birds. Popular activities include canoeing and fishing. 2WD access to beach, 4WD to Point Lindsay; turn-off 53 km N. ***Port Kenny:*** this small township on Venus Bay offers excellent fishing, boating and swimming, with sea lion and dolphin tours available; 62 km SE. ***Venus Bay:*** fishing village renowned for catches of King George whiting, trevally, garfish and many more. Its waters are safe for swimming and watersports, and nearby beaches are good for surfing. Needle Eye Lookout close by provides fantastic views, with southern right whale sightings June–Oct; 76 km S.

See also EYRE PENINSULA & NULLARBOR, p. 244

Swan Reach

Pop. 237
Map ref. 601 E2 | 603 L8

General Store, 47 Anzac Ave; (08) 8570 2036 or 1800 442 784; www.murraylands.info

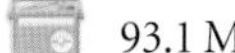

93.1 Magic FM, 1062 AM ABC Riverland

This quiet little township on the Murray River was once one of five large sheep stations; the original homestead is now the Swan Reach Hotel. Established as one of the first river ports for Murray River trade, the introduction of rail, and Morgan's rise as one of the state's busiest ports, saw the era of paddlesteamers in Swan Reach decline. Today the picturesque river scenery and excellent fishing make the town a popular holiday destination.

Swan Reach Museum: local history displays with special interest in Swan Reach's flood history, the waters having devastated the town in the early 1900s; Nildottie Rd.

Griffens Marina Houseboats: luxury 10–12 berth boats; Griffens Marina, Blanchetown; (08) 8540 5250.

Yookamurra Sanctuary This sanctuary represents an initiative to restore 1100 ha of land to its original state. Fittingly, the sanctuary is named Yookamurra after the Aboriginal word for 'yesterday'. The mallee vegetation that was found here before European habitation has been replanted; keep an eye out for the rare and endangered numbat or the bilby and woylie. Walking tours and overnight stays are available. Bookings are essential, (08) 8562 5011; Pipeline Rd, Sedan; 21 km W.

Murray Aquaculture Yabby Farm: catch your own yabbies; 1.5 km E. ***Ridley and Swan Reach conservation parks:*** both parks represent typical western Murray vegetation and protect the habitat of the hairy-nosed wombat; 7.5 km S and 10 km W respectively. ***Ngaut Ngaut Boardwalk:*** guided tours of archaeological site, established when an ancient skeleton was discovered; Nildottie; 14 km S. ***Big Bend:*** imposing Murray cliffs, the tallest found on the river, home to diverse flora and fauna. Spectacular nightly tours are available; inquiries (08) 8570 1097; 20 km downstream. ***Bakara Conservation Park:*** mallee-covered plains and sand dunes, important habitat for the mallee fowl; 32 km E. ***Brookfield Conservation Park:*** bushwalking in limestone country to see hairy-nosed wombats, red kangaroos and a variety of bird species; 40 km NW.

See also MURRAY, p. 241

Tanunda

Pop. 4152
Map ref. 597 E4 | 601 C2 | 603 K8

Barossa Visitor Information Centre, Murray St; (08) 8563 0600 or 1300 812 662; www.barossa-region.org

89.1 BBB FM, 639 AM ABC North and West

Tanunda is at the heart of the Barossa and surrounded by vineyards. The modern-day township grew out of the village of Langmeil, which was the focal point for early German settlement. The German Lutherans found it only natural to plant vines, as it was a basic part of their lifestyle. Many of the Barossa's shiraz vines date back to those early days. Tanunda has a boisterous German spirit and fine examples of Lutheran churches.

Barossa Historical Museum: situated in the former post and telegraph office (1865), its collections specialise in German heritage; Murray St. ***Gourmet produce:*** specialty stores include Tanunda Bakery for German breads (Murray St), Tanunda's Nice Ice for homemade ice-cream (Kavel Arcade) and Apex Bakery for traditional pastries (Elizabeth St). ***Heritage walk:*** includes many historic Lutheran churches; brochure available from visitor centre.

Barossa Under the Stars: Feb. ***Tanunda Show:*** Mar. ***Barossa Vintage Festival:*** celebration of locally produced food and wine in various locations; odd-numbered years, Apr. ***Barossa Gourmet Weekend:*** Aug. ***Barossa Band Festival:*** Oct.

1918 Bistro & Grill: modern Australian; 94 Murray St; (08) 8563 0405. ***Apex Bakery:*** traditional German bakery; 1A Elizabeth St, (08) 8563 2483. ***Krondorf Road Cafe:*** authentic German recipes; Krondorf Rd; (08) 8563 0889. ***Jacob's Restaurant:*** sleek cellar-door dining; Jacob's Creek Visitor Centre, Barossa Valley Way, Rowland Flat; (08) 8521 3000.

Goat Square Cottages: quaint heritage town houses; 33 John St; (08) 8524 5353 or 1800 227 677. ***Jacob's Creek Retreat:*** opulent, European-themed resort; Nitschke Rd; (08) 8563 1123. ***Lawley Farm:*** beautifully restored settler's cottages; Krondorf Rd; (08) 8563 2141. ***Stonewell Cottages:*** dreamy lakefront retreats; Stonewell Rd; 0417 848 977. ***Sonntag House:*** peaceful heritage cottage; Bethany Rd, Bethany; 0419 814 349.

Barossa Wine Region The Mediterranean-style climate, varying soils, specialised winemakers and long history (dating back to the 1840s) have created a world-renowned wine region in the Barossa Valley. Nearly all outfits offer cellar-door tastings and sales. Close to Tanunda is the Barossa Small Winemakers Centre, housed in the cellar door at Chateau Tanunda, Basedow Rd, and showcasing the rare and handmade varieties of the Barossa's small producers. At the Chateau Dorrien Winery Tourism Centre in Barossa Valley Way there is an interesting mural depicting

continued on p. 281

VICTOR HARBOR

Pop. 10 377

Map ref. 598 F9 | 599 H10 | 601 B5 | 603 J11

Causeway Building, Esplanade; (08) 8551 0777; www.tourismvictorharbor.com.au

89.3 5EFM, 99.7 Power FM, 1125 5MU, 90.1 Happy FM, 891 ABC

In the 1830s the crystal waters of Encounter Bay – and the Southern Ocean beyond – throbbed with the whalers and sealers of the south. Granite Island housed a whaling station, Victor Harbor was its port, and life revolved around the ocean slaughters. Today the whalers and sealers are gone, Granite Island is a recreation park and Victor Harbor is a holiday town. The naming of Encounter Bay comes from the unexpected meeting in the bay between explorers Matthew Flinders and Nicolas Baudin.

South Australian Whale Centre This unique centre focuses on the 25 species of whale and dolphin, and other marine life found in the southern Australian waters, with an aim to educate and conserve these species. Past atrocities are displayed alongside interactive displays and presentations that reveal the wonders of the amazing creatures. Between May and October each year, Southern Right whales mate and breed in Encounter Bay. The centre offers whale cruises and sighting information, as well as a Whale Information Hotline for the latest sightings (in season) 1900 WHALES (1900 942 537). Open 9.30am to 5pm daily (exc Christmas Day); Railway Tce; (08) 8551 0750.

Encounter Coast Discovery Centre: National Trust museum that covers Aboriginal, whaling, settler and recent local history. A museum walk finishes at the Old Customs House, which has period furnishings; open 1–4pm daily; Flinders Pde. ***Cockle Train:*** departs from Railway Tce for return journey to Goolwa; bookings 1300 655 991; *for more details see Goolwa*. ***Horse Drawn Tram:*** operates daily; (08) 8551 0720; www.horsedrawntram.com.au. ***Amusement Park:*** family fun fair on the beach with the state's historic Ferris wheel, dodgem cars and an inflatable slide; open long weekends and school holidays; the Causeway; 0418 845 540. ***SteamRanger Heritage Railway:*** heritage steam and diesel tourist trains run to Port Elliot, Goolwa, Strathalbyn and Mount Barker; 1300 655 991; www.steamranger.org.au.

[HORSE-DRAWN TRAM FROM GRANITE ISLAND]

Scenic flights: helicopter joy-flights over Victor Harbor, Granite Island, local vineyards and the mouth of the Murray; (08) 8552 8196.

Rotary Art Show: Jan. ***Coast to Coast Bike Ride and Victor Harbor Triathlon:*** Mar. ***Whale Season Opening:*** June. Whaletime Playtime Festival: July. ***Rock n Roll Festival:*** Sept. ***New Year's Eve Celebration:*** Dec.

Ocean Grill Restaurant: steaks and seafood; Anchorage Seafront Hotel, 21 Flinders Pde; (08) 8552 5970. ***Waterside Restaurant:*** contemporary bayside dining; Whalers Inn Resort, 121 Franklin Pde, Encounter Bay; (08) 8552 4400.

Anchorage Seafront Hotel: affordable foreshore suites; 21 Flinders Pde; (08) 8552 5970. ***Riverhouse:*** engaging country retreat, 447 Waggon Rd; (08) 8552 5657. ***Whalers Inn Resort:*** self-contained suites with spectacular views; 121 Franklin Pde, Encounter Bay; (08) 8552 4400.

Granite Island Recreation Park Granite Island has a long and varied history. It has significance in the Ramindjeri people's Dreamtime; in 1837 a whaling station was established; and today the island is a recreation park. This history is detailed on the Kaiki Trail, a 1.5 km walk around the island. A highlight is the Below Decks Oceanarium, just off the Screwpile Jetty, with close-up views of marine life and tours daily; at dusk, take a guided Penguin Discovery Tour to see the penguins scuttle in and out of their burrows (all tour bookings (08) 8552 7555). The island is linked to the mainland by a 630 m causeway. Walk or take the horse-drawn tram, the last one remaining in the Southern Hemisphere; tram departs from entrance to Causeway at 10am daily. Tickets available at visitor centre or on the Horse Tram.

Hindmarsh River Estuary: peaceful picnic and fishing spot with boardwalk through coastal scrub; 1 km NE. ***Greenhills Adventure Park:*** family-fun activities including go-karts, jumping castle and water slide; Waggon Rd; 3.5 km N. ***Victor Harbor Winery:*** cellar-door tastings and sales of cool-climate reds, whites and fortified wines; open Wed–Sun; Hindmarsh Valley; 4 km N. ***Urimbirra Wildlife Park:*** popular fauna park with a wetland bird sanctuary, crocodile-feeding and children's farmyard; Adelaide Rd; 5 km N. ***Big Duck Boat Tours:*** Spectacular half hour and 1 hour tours taking in Encounter Bay and coastal parts of Victor Harbor. Leaves from Granite Island causeway; 0405 125 312. ***Nangawooka Flora Reserve:*** tranquil walks through native bushlands with over 1250 native plant varieties on show; opposite Urimbirra Wildlife Park. ***The Bluff (Rosetta Head):*** 500-million-year-old mass of granite, well worth the 100 m climb for the views; 5 km SW. ***Newland Head Conservation Park:*** known for its wild surf and coastal vegetation, this park protects the headland and Waitpinga and Parsons beaches, which offer surf-fishing opportunities and beach walks; turn-off 15 km SW. ***Hindmarsh Falls:*** pleasant walks and spectacular waterfall (during winter); 15 km NW. ***Mt Billy Conservation Park:*** mallee and forest park renowned for its rare orchid species; 18 km NW. ***Inman Valley:*** features Glacier Rock, said to be the first recorded discovery of glaciation in Australia; 19 km NW.

See also FLEURIEU PENINSULA, p. 235

Barossa heritage. Winery map available from visitor centre. *For more information see Angaston, Lyndoch and Nuriootpa.*

Norm's Coolies: see performances by a unique breed of sheepdog, Norm's coolie, at 2pm Mon, Wed and Sat; just south on Barossa Valley Way. ***Bethany:*** this pretty village was the first German settlement in the Barossa. The creekside picnic area, pioneer cemetery, attractive streetscapes and walking trail along Rifle Range Rd make it well worth a visit; 3 km SE. ***The Keg Factory:*** makers of American and French oak kegs, as well as barrel furniture and wine racks; St Halletts Rd; 4 km SW.

See also BAROSSA VALLEY, p. 238

Tumby Bay

Pop. 1348
Map ref. 602 E7

Hales MiniMart, 1 Bratten Way; (08) 8688 2584; www.tumbybay.sa.gov.au

89.9 Magic FM, 1485 AM ABC Eyre Peninsula and West Coast

Tumby Bay is a pretty coastal town on the east coast of Eyre Peninsula. Its development was slow – Matthew Flinders discovered the bay in 1802, settlers arrived in the 1840s and the jetty was built in 1874 to ship the grain produce, but still there was no town. It took until the early 1900s for any official settlement to be established. Now the famous long, crescent beach, white sand and blue water attract holiday-makers.

C. L. Alexander National Trust Museum: depicts early pioneer history in an old timber schoolroom; open 10–11am Wed, 2.30–4.30pm Sun or by appt; West Tce; (08) 8688 2760. ***Rotunda Art Gallery:*** local art display and a fantastic mural on the outside of the rotunda; open 10am–12pm Mon and Wed or by appt; Tumby Tce; (08) 8688 2678. ***Excell Blacksmith and Engineering Workshop Museum:*** original workshop and equipment dating from the early 1900s; open 1.30–4.30pm 4th Sun each month, or by appt; Barraud St; (08) 8688 2101. ***Mangrove boardwalk:*** 70 m walkway with interpretive signs explaining ecology of mangroves; Berryman St. ***Fishing:*** from the recreational jetty, beach, rocks or boats (hire and charters available).

Tumby Bay Beach House: smart, revived waterfront shack; 1 Elfrieda Dr; 0429 830 328.

Koppio Smithy Museum The early 1900s come to life in this extensive National Trust museum in the Koppio Hills. Consisting of the restored Blacksmith's Shop (1903), historic log cottage 'Glenleigh' (1893) and schoolrooms, the museum houses an eclectic collection of Aboriginal artefacts, early pioneer furniture, firearms and early machinery. Closed Mon; 30 km SW.

Trinity Haven Scenic Drive: travels south from town along the coast and offers scenic coastal views and secluded beaches and bays. ***Island Lookout Tower and Reserve:*** views of town, coast and islands. Enjoy a picnic in the reserve; Harvey Dr; 3 km S. ***Lipson Cove:*** popular spot for anglers. Walk to the coastal sanctuary on Lipson Island at low tide; 10 km NE. ***Ponta and Cowleys beaches:*** fishing catches include snapper and bream; 15 km NE. ***Moody Tanks:*** State Heritage–listed water-storage tanks once used to service passing steam trains; 30 km W. ***Cummins:*** rich rail heritage celebrated each Apr at the World Championship Kalamazoo Classic; 37 km NW. ***Port Neill:*** an old port town with a safe beach for fishing and watersports. Also Ramsay Bicentennial Gardens, and vintage vehicles at Vic and Jill Fauser's Living Museum. Port Neill Lookout, nearby, provides fantastic views of the coast; 42 km NE. ***Sir Joseph Banks Group Conservation Park:*** comprising around 20 islands and reefs, this park is a breeding area for migrating coastal birds and the Australian sea lion colony at Dangerous Reef; boat access is from Tumby Bay, Port Lincoln and 250 m north of Lipson Cove.

See also EYRE PENINSULA & NULLARBOR, p. 244

Waikerie

Pop. 1744
Map ref. 601 E1 | 603 M7

Orange Tree Giftmania, Sturt Hwy; (08) 8541 2332.

93.1 Magic FM, 1062 AM ABC Riverland

Waikerie, the citrus centre of Australia, is surrounded by an oasis of irrigated orchards and vineyards in the midst of the mallee-scrub country of the Riverland. Owing to its position on cliff-tops, the area around Waikerie was not a promising settlement. However, in an experiment by the South Australian government in 1894 that attempted to alleviate unemployment and decentralise capital, 281 people were relocated from Adelaide. It was an instant town. Waikerie has beautiful views of the river gums and sandstone cliffs along the Murray River, which is a popular spot for fishing, boating and waterskiing – and the skies above are a glider's paradise due to the fantastic thermals and flat landscape.

Rain Moth Gallery: local art exhibitions; open 10.30am–2.30pm Mon–Fri, 10am–1pm Sat; Peake Tce. ***Waikerie Murray River Queen:*** unique floating motel, restaurant and cafe; moored near the ferry. ***Harts Lagoon:*** wetland area with bird hide; Ramco Rd. ***Houseboat hire:*** scenic trips along the Murray; contact visitor centre for details. ***Bush Safari:*** camel or 4WD tours to the river and outback country north-east of Waikerie; bookings (08) 8543 2280. ***Scenic walk:*** along cliff-top to lookout; northern outskirts of town.

Rotary Food Fair: Mar. ***Horse and Pony Club Easter Horse Show:*** Easter. ***Music on the Murray:*** odd numbered years, Apr. ***Riverland Rock 'n' Roll Festival:*** May. ***Hit n Miss Tractor Pull:*** odd-numbered years, Sept.

Orange Tree Giftmania: local produce sales – including citrus and dried fruits – and souvenirs. Enjoy Murray River views from the viewing platform; Sturt Hwy. ***Waikerie Gliding Club:*** offers recreational flights, beginner courses and cross-country training; Waikerie Aerodrome, off Sturt Hwy, east side of town; inquiries (08) 8541 2644. ***Maize Island Conservation Park:*** this waterbird reserve has fantastic cliffs and lagoons. Beware of strong currents when swimming; 2 km N. ***Pooginook Conservation Park:*** both dense and open mallee country, home to kangaroos, hairy-nosed wombats and the ever-busy mallee fowl; 12 km NE. ***Stockyard Plain Disposal Basin Reserve:*** varied plant and birdlife – over 130 bird species identified; key available from visitor centre; 12 km SW. ***Broken Cliffs:*** popular fishing spot; Taylorville Rd; 15 km NE. ***Birds Australia Gluepot Reserve:*** important mallee area that forms part of the Bookmark Biosphere Reserve (*see Renmark*). Also significant bird refuge, with over 17 threatened Australian species to be seen on the 14 walking trails; access key from Shell Service Station in Waikerie; 64 km N.

See also MURRAY, p. 241

SOUTH AUSTRALIA

Wallaroo

Pop. 3050
Map ref. 599 E3 | 602 H6

The Farm Shed Museum and Tourist Centre, 50 Moonta Rd, Kadina; (08) 8821 2333 or 1800 654 991; www.yorkepeninsula.com.au

89.3 Gulf FM, 639 AM ABC North and West

Vast grain silos greet visitors to Wallaroo, a coastal town and shipping port on the west coast of Yorke Peninsula. The town is an interesting mix of tourism and industry. The safe beaches and excellent fishing prove popular with holiday-makers, while the commercial port controls exports of barley and wheat. Wallaroo exists thanks to a lucky shepherd's discovery of copper in 1859. Vast deposits were uncovered and soon thousands of Cornish miners arrived. The area boomed until the 1920s, when copper prices dropped and the industry slowly died out. Wallaroo's buildings and old Cornish-style cottages are a reminder of its colourful past. Wallaroo and nearby towns Moonta and Kadina are part of the 'Copper Coast' or 'Little Cornwall'.

Wallaroo Heritage and Nautical Museum This National Trust museum in Wallaroo's original 1865 post office features shipwreck displays, maps, charts, model ships and records, as well as local cultural and religious history. Meet George, the unlucky giant squid eaten then recovered from a whale's belly 30 years ago. Open 10.30am–4pm Mon–Fri, 2–4pm Sat–Sun; Jetty Rd; (08) 8823 3015.

Yorke Peninsula Railway: historical diesel-train journey from Wallaroo to Bute; runs 2nd Sun each month and school holidays; contact visitor centre for details. ***Ausbulk:*** informative drive through grain-handling facility; Lydia Tce. ***Boat hire and charters:*** for the ultimate gulf-fishing experience. ***Self-guide historical walk:*** highlight is the 1865 Hughes chimney stack, which contains over 300 000 bricks and measures more than 7 sq m at its base; brochure available from museum or town hall.

Kernewek Lowender: Cornish festival held in conjunction with Moonta and Kadina; odd-numbered years, May.

The Boatshed Restaurant: waterfront seafood specialists; 1 Jetty Rd; (08) 8823 3455.

The Cornucopia Hotel: updated heritage hotel; 49 Owen Tce; (08) 8823 3457. ***Tipara:*** modern waterfront apartment; 7 Tipara Crt, Moonta Bay; (08) 8843 0187. ***Top Deck Cliff House:*** beachfront family retreat; 1 Queen Pl, Moonta Bay; (08) 8636 2343.

Bird Island: crabbing; 10 km S.

Travellers note: *To avoid the extra driving distance to the Eyre Peninsula, Sea SA runs a ferry service between Wallaroo and Lucky Bay 4 times a day on weekdays, and twice a day on weekends. Bookings (08) 8823 0777.*

See also YORKE PENINSULA, p. 240

Whyalla

Pop. 21 123
Map ref. 602 H3 | 604 G11

Lincoln Hwy; (08) 8645 7900 or 1800 088 589; www.whyalla.com

107.7 5YYY FM, 639 AM ABC North and West

Whyalla, northern gateway to Eyre Peninsula, has grown from the small settlement of Hummock Hill to the largest provincial city in South Australia. It has become known for its heavy industry since iron ore was found in the 1890s around Iron Knob and the steel works opened in 1964. Whyalla also has an interesting natural attraction. Each year, from May to August, an incredible number of cuttlefish spawn on the rocky coast just north – a must-see for diving and snorkelling enthusiasts. The city is modern and offers safe beaches, excellent fishing and boating.

Whyalla Maritime Museum The central attraction is HMAS *Whyalla*, a 650-tonne corvette, the largest permanently land-locked ship in Australia. It was the first ship built in the BHP shipyards. Guided tours of the ship are included in the entry price and run on the hour 11am–3pm Apr– Oct, 10am–2pm Nov–Mar. The lives of the 4 wartime corvettes built by BHP are documented, as are histories of the shipbuilding industry and maritime heritage of Spencer Gulf. Lincoln Hwy; (08) 8645 8900.

Mt Laura Homestead Museum: National Trust museum featuring the original homestead with progressive city-history displays, period furnishings in the 1914 Gay St Cottage, and the Telecommunications Museum; open 10am–12pm Mon–Fri, 2–4pm Sun; Ekblom St. ***Tanderra Craft Village:*** art and craft shops, market and tearooms; open 10am–4pm last weekend each month; next to Maritime Museum. ***Whyalla Wetlands:*** park and wetlands area with walking trails and a picnic/barbecue area; Lincoln Hwy. ***Foreshore and marina:*** safe beach, jetty for recreational fishing, picnic/barbecue area, access to Ada Ryan Gardens, and a marina with boatlaunching facilities. ***Ada Ryan Gardens:*** mini-zoo with picnic facilities under shady trees; Cudmore Tce. ***Murray Cod Tour:*** see a fully operational inland freshwater aquaculture venture, and learn about recycling water and the various growth stages of the Murray cod. Closed footwear required; book at visitor centre. ***Steelworks Tour:*** 2 hr guided tour explains steelmaking process; departs 9.30am Mon, Wed and Fri; book at visitor centre. ***Hummock Hill Lookout:*** views of city, gulf, steel works and coast from WW II observation post; Queen Elizabeth Dr. ***Flinders and Freycinet Lookout:*** Farrel St; *Whyalla Visitor Guide* from visitor centre.

Australian Snapper Championship: Easter. ***Whyalla Show:*** family activities, rides and stalls; Aug.

Whyalla Conservation Park: 30 min walking trail through typical semi-arid flora and over Wild Dog Hill; 10 km N off Lincoln Hwy. ***Port Bonython and Point Lowly:*** this area of coast offers beautiful views of Spencer Gulf, fishing from rocks, and dolphin sightings. Lowly Beach is a popular swimming beach and the Freycinet Trail is a scenic drive from just before Port Bonython along Fitzgerald Bay to Point Douglas (parts are gravel); 34 km E. ***Iron Knob:*** a mining town with museum and mine lookout tours (depart from the museum at 10am and 2pm Mon–Fri); 53 km NW.

See also EYRE PENINSULA & NULLARBOR, p. 244

Willunga

Pop. 2103
Map ref. 597 B12 | 598 E6 | 599 H9 | 601 B4 | 603 J10

McLaren Vale Visitors Centre, Main Rd, McLaren Vale; (08) 8323 9944 or 1800 628 410; www.mclarenvale.info

94.7 5EFM, 639 AM ABC North and West

The historic town of Willunga grew rapidly around the slate quarries, which drove the town's economy until the late 1800s. Fortunately, by that time Willunga already had a thriving new industry – almonds. The town sits at the southern edge of the McLaren Vale wine region and is surrounded by farmlands and olive groves. Its name is derived from the Aboriginal word 'willa-unga', meaning 'the place of green trees'.

Willunga Courthouse Museum: National Trust museum with local history displays in the original 1855 courtroom, cells

continued on p. 284

WILPENA

Map ref. 600 B10 | 605 I7

Wilpena Rd, via Hawker; (08) 8648 0048; www.wilpenapound.com.au

Wilpena consists of a resort and caravan/camping park on the edge of Wilpena Pound, in Flinders Ranges National Park. In 1902 the Hill family, wheat farmers, built a homestead inside the pound, but abandoned it after a flood washed away the access road in 1914. The pound is a vast natural amphitheatre surrounded by peaks that change colour with the light, and is a fantastic destination for bushwalking.

Wilpena Pound Resort: partly powered by the largest solar-power system in the Southern Hemisphere (viewing area accessed by a walking trail). The visitor centre at the resort has extensive information on 4WD and organised tours, self-guide drives, scenic flights, bushwalking and hiking; (08) 8648 0004.

Wilpena Under the Stars: black-tie dinner and dance to raise funds for the Royal Flying Doctor Service; Feb. ***Tastes of the Outback:*** Apr. ***Flinders Ranges Event Program:*** events run in autumn and spring, including guided walks, tours and cultural activities; details from visitor centre.

Rawnsley Park Station: multi-faceted bush resort; Wilpena Rd, (08) 8648 0030. ***Willow Springs:*** rustic station accommodation: Wilpena Rd, (08) 8648 4022. ***Wilpena Pound Resort:*** famed outback holiday destination; Wilpena Pound, (08) 8648 0004.

Flinders Ranges National Park For thousands of years the ancient landscapes of the Flinders Ranges were home to the Adnyamathanha people – the 'people of the rocks'. Their Muda (Dreamtime) stories tell of the creation of the slopes and gorges that ripple across the landscape for over 400 km, from south-east of Port Augusta to north of Arkaroola. In the 1850s stock runs were established at Arkaba, Wilpena and Aroona. Foreign plant and animal species were introduced and the natural balance of the ranges was altered. Within 50 years of European settlement many endemic animals had been pushed to extinction. Today conservationists are trying to recover the natural balance of the area, and have had success in the recovery of yellow-footed rock wallabies. The central ranges are a fabulous place for hikers. There are 17 walks and hikes to choose from and the choice is difficult. All provide a different historical, geological or scenic look at the ranges. For a look into early European settlement take the 5.4 km return Hills Homestead Walk into the extraordinary natural rock formation of Wilpena Pound. Impressive rock paintings depicting the creation of the ranges can be seen on the Arkaroo Rock Hike (3 km loop track). And for nature lovers there is the Bunyeroo Gorge Hike, a 7.5 km return trail that follows the gorge and reveals fantastic wildlife and rock formations. There are also driving tours that reveal some of the park's most spectacular scenery. The popular Brachina Gorge Geological Trail is a 20 km drive that details the long physical history of the ranges, from when the hills were layers of sediment beneath an ocean, to when they were compressed and pushed up into the shape of mountains. Look out for the yellow-footed rock wallaby in the rocky upper slopes of this beautiful gorge. All the roads north of Wilpena are unsealed, but are generally 2WD accessible. Among all of this, at Wilpena, is some of the best accommodation and facilities north of Adelaide. Drop into the Wilpena Pound Visitor Information Centre for a park guide; (08) 8648 0048.

Sacred Canyon: Aboriginal rock carvings and paintings; 19 km E. ***Rawnsley Park Station:*** camping and holiday-unit accommodation, scenic flights, horseriding and 4WD tours; (08) 8648 0030; 20 km S on Hawker Rd. ***Moralana Scenic Drive:*** 22 km route between Elder Range and south-west wall of Wilpena Pound with lookouts and picnic spots en route; drive starts 25 km S.

See also FLINDERS RANGES & OUTBACK, p. 243

[WILPENA POUND'S NATURAL AMPHITHEATRE]

and stables; guided 'Willunga Walks and Talks' tours; open 11am–4pm Tues, 1–5pm weekends; High St; (08) 8556 2195. ***Quarry:*** operated for 60 years (1842–1902), now a National Trust site; Delabole Rd. ***Historical walk:*** self-guide walk featuring historic pug cottages, colonial architecture and an Anglican church with an Elizabethan bronze bell; brochure from museum.

Willunga Farmers Market: Hill St; Sat mornings. ***Willunga Quarry Market:*** country market with local produce and crafts; Aldinga Rd; 2nd Sat each month. ***Almond Blossom Festival:*** running since 1970, celebrates the blooming of almond trees; July. ***Fleurieu Folk Festival:*** Oct.

Fino Restaurant: rustic regional cuisine; 8 Hill St; (08) 8556 4488. ***Star of Greece:*** stunning cliff-top seafood; The Esplanade, Port Willunga; (08) 8557 7420. ***Victory Hotel:*** popular pub and cellar; Main South Rd, Sellicks Beach; (08) 8556 3083.

Citrus Cottage: immaculate heritage cottage; 37 High St; (08) 8557 8516. ***Sea & Vines Cottage:*** secluded historic farmhouse; Culley Rd; (08) 8295 8659. ***Saltaire:*** easy-going seaside holiday home; 4 Marlin Rd, Port Willunga; (08) 8339 4151. ***The Blue Grape:*** vineyard hideaway; Newman Cl, Willunga South; (08) 8556 4078.

Mt Magnificent Conservation Park: explore virtually untouched rocky landscapes and vegetation popular for picnics and scenic walks. The highlight is the walk to the Mt Magnificent summit for coastal views; 12 km SE. ***Mount Compass:*** a small farming town featuring the Wetlands Boardwalk and many farms open for viewing and sales, offering both primary products and gourmet food. Australia's only Cow Race is held here each Jan or Feb; 14 km S. ***Fleurieu Big Shed:*** local produce, art and craft sales; 15 km S. ***Kyeema Conservation Park:*** completely burnt out in the 1983 Ash Wednesday fires and then again in the fires of 1994 and 2001, this park is evidence of nature's ability to constantly regenerate. It is home to over 70 species of birdlife and offers good hiking and camping. Part of the Heysen Trail passes through it; 14 km NE.

See also FLEURIEU PENINSULA, p. 235

Wilmington

Pop. 217
Map ref. 603 I2 | 604 H10

Wilmington General Store, Main North Rd; (08) 8667 5155.

105.9 Magic FM, 639 AM ABC North and West

Robert Blinman had the foresight to build an inn, called the Roundwood Hotel, at the base of Horrocks Pass in 1861, and soon the Cobb & Co coaches were stopping there on their passenger routes. The town was built around the first hotel, and before long the farming community was thriving. Originally named Beautiful Valley by European explorers, the name was changed to Wilmington in 1876, although the original name still persists in many local establishments. Today the town retains much of its old-time feel and is renowned for its stone buildings.

Wilmington Hotel: built around 1876, the hotel is one of the town's oldest buildings and was first called The Globe Hotel. Original Cobb & Co coach stables are at the rear of the building; Main North Rd. ***Mt Maria Walking Trail:*** 2 km walking trail starting from town leads to vantage point over Wilmington; brochure available from general store.

Night Rodeo: Jan. ***Wilmington Show:*** Sept.

Alligator Lodge: bushland holiday retreat; Alligator Gorge Rd, Mt Remarkable National Park; (08) 8634 7068.

Spring Creek Mine Drive: 24 km scenic loop beginning south of town, passing mountain and farm scenery and an old copper mine, now the town's water supply; brochure from general store. ***Horrocks Pass and Hancocks Lookout:*** this historic pass was named after explorer John Horrocks who traversed the pass in 1846. Hancocks Lookout, at the highest point of the pass, offers magnificent views to Spencer Gulf; 8 km W off road to Port Augusta. ***Mt Remarkable National Park:*** 13 km S; *see Melrose*. ***Winninowie Conservation Park:*** coastal park of creeks and samphire flats, home to abundant birdlife; 26 km SW. ***Hammond:*** historic ghost town; 26 km NE. ***Bruce:*** historic railway town featuring 1880s architecture; 35 km N. ***Carrieton:*** historic buildings and Yanyarrie Whim Well in town. A rodeo is held here each Dec; 56 km NE. See Aboriginal carvings a further 9 km along Belton Rd.

See also FLINDERS RANGES & OUTBACK, p. 243

Wilpena

see inset box on previous page

Woomera

Pop. 294
Map ref. 604 E5

Dewrang Ave; (08) 8673 7042; www.woomera.com.au

101.7 Flow FM, 1584 AM ABC North and West

Woomera and its testing range were established in 1947 as a site for launching British experimental rockets during the Cold War era. The town was a restricted area until 1982. The Woomera Prohibited Area remains today and is still one of the largest land-based rocket ranges in the world.

Woomera Heritage and Visitor Information Centre: provides a detailed history of the area through videos, exhibitions, rocket relics and photographic displays. It also includes a bowling alley. Tours of the Rocket Range can be booked and depart here; Dewrang Ave. ***Missile Park:*** open-air defence display of rockets, aircraft and weapons; cnr Banool and Dewrang aves. ***Baker Observatory:*** viewing the night sky through a computer-controlled telescope; contact visitor centre for details.

See also FLINDERS RANGES & OUTBACK, p. 243

Wudinna

Pop. 517
Map ref. 602 C3 | 604 B11

44 Eyre Hwy; (08) 8680 2969; www.lehunte.sa.gov.au

89.5 Magic FM, 693 AM ABC Eyre Peninsula and West Coast

The enormous silos in Wudinna are indicative of the town's major grain industry, predominantly wheat and barley, grown here since the first pastoral lease was granted in 1861. Wudinna was proclaimed a town in 1916 and has since grown as a service centre to the Eyre Peninsula. A little travelling in the surrounding countryside will reveal unusually shaped granite outcrops – the area is known as granite country.

Gawler Ranges Cultural Centre: dedicated to the exhibition of artwork with a ranges theme; Ballantyne St.

Wudinna Show: Sept.

Gawler Ranges National Park This rugged national park offers fantastic gorge and rocky-outcrop scenery, spectacular when the spring wildflowers are in bloom. There are no marked trails, but highlights of drive tours include the Organ Pipes, a large and unique formation of volcanic rhyolite, the Kolay Mirica Falls and Yandinga Gorge. Some areas are accessible by 2WD, but

4WD is generally recommended. Roads may be impassable after rain. Guided tours into the ranges are offered by 2 operators: Gawler Ranges Wilderness Safaris (1800 243 343) and Nullarbor Traveller (1800 816 858). 40 km N.

Wudinna Granite Trail: signposted 25 km tourist drive to all major rock formations in the area. ***Mt Polda Rock Recreation Reserve:*** walking trail for excellent birdwatching with views from the top of Polda Rock; 7 km NE. ***Mt Wudinna Recreation Reserve:*** the mountain is thought to be the second-largest granite outcrop in the Southern Hemisphere. At its base is a picnic area, a 30 min return interpretive walking trail, and original stone walls used as water catchments. Enjoy scenic views at the mountain's summit. On the road to the reserve look out for Turtle Rock; 10 km NE. ***Ucontitchie Hill:*** isolated and unique granite formations, similar to Kangaroo Island's Remarkable Rocks; 32 km S. ***Minnipa:*** home to the Agricultural Centre, which provides invaluable research into sustainable dryland farming. Nearby are granite formations of geological significance, including Yarwondutta Rock (2 km N), Tcharkuldu Rock (4 km E) and the wave-like formation of Pildappa Rock (15 km N); 37 km NW. ***Koongawa:*** memorial to explorer John Charles Darke; 50 km E.

See also EYRE PENINSULA & NULLARBOR, p. 244

Yankalilla

Pop. 552
Map ref. 598 D8 | 599 G10 | 603 J10

104 Main South Rd; (08) 8558 2999; www.yankalilla.sa.gov.au

89.3 5EFM, 639 AM ABC North and West

Since the first land grant in 1842, Yankalilla has been the centre of a thriving farming industry. It is a growing settlement just inland from the west coast of the Fleurieu Peninsula, but it still retains its old country flavour. It has even adopted the slogan 'Yankalilla Bay – you'll love what we haven't done to the place'. In recent times it has seen an influx of visitors keen to see the apparition at Our Lady of Yankalilla Shrine. The Blessed Virgin Mary was first sighted here in 1996.

Yankalilla District Historical Museum: local history and interpretive trail; open 1–4pm Sun; Main South Rd. ***Anglican Church:*** historic and known for apparition of Mary; Main St.

Leafy Sea Dragon Festival: arts and cultural festival; odd-numbered years, Apr. ***Yankalilla Show:*** Oct long weekend.

Lilla's Cafe: creative cafe fare; 117 Main South Rd; (08) 8558 2525.

Rattleys at Pear Tree Hollow: secluded farm cottage; 9 Nosworthy Rd, Inman Valley; (08) 8558 8234. ***Ridgetop Retreats:*** contemporary bush pavilions; Deep Creek Conservation Park; (08) 8598 4169. ***The Beach House:*** family-sized retreat; 21A Ronald St, Normanville; (08) 8558 3223.

Deep Creek Conservation Park Take one of the many walks along rugged coastal cliffs, tranquil creeks, majestic forests and scenic waterfalls. Walks range from easy and short to more challenging long-distance hikes. Keep an eye out for the western grey kangaroos at dusk on the Aaron Creek Hiking Trail or drop a line at Blowhole Creek and Boat Harbour beaches. Permits are required in the park (self-registration). 26 km SW.

Normanville: a seaside town with beach and heritage-listed sand dunes. Shipwrecks are popular with divers; 3 km W. ***Myponga Conservation Park:*** popular bushwalking and birdwatching park; 9 km NE. ***Myponga:*** a historic town with fantastic views from the Myponga Reservoir; 14 km NE. ***Second Valley:*** a peaceful picnic spot with a jetty for fishing; 17 km SW. ***Rapid Bay:*** this seaside town offers excellent fishing and diving opportunities. Sightings of the endangered leafy sea dragon in the bay make diving a must for any enthusiast; 27 km SW. ***Talisker Conservation Park:*** an interpretive trail explains the old silver-mine workings in the park; 30 km SW. ***Cape Jervis:*** breathtaking sea and coastal views on entering town. Vehicular ferries to Kangaroo Island depart from here, and it is also the starting point of the 1200 km Heysen Trail (bushwalking trail) to the Flinders Ranges. Morgan's and Fishery beaches nearby have good fishing; 35 km S.

See also FLEURIEU PENINSULA, p. 235

Yorketown

Pop. 687
Map ref. 599 E8 | 602 H9

SYP Telecentre, Yorketown or Harvest Corner Visitor Information Centre, 29 Main St; (08) 8853 2600 or 1800 202 445; www.yorkepeninsula.com.au

98.9 Flow FM, 639 AM ABC North and West

Yorketown is a small rural community at the southern end of Yorke Peninsula. The surrounding landscape is dotted with many inland salt lakes, some of which are still mined. In the late 1840s farmers were eager to take up land here as it was prime crop-producing land. The town was settled in 1872 and has remained an important service centre on the peninsula since.

Courthouse Photographic Display: photographs of the area's pioneering days; open Fri mornings or by appt.

Marion Bay Tavern: relaxed beachside meals; Stenhouse Bay Rd, Marion Bay; (08) 8854 4141.

Innes National Park Heritage Accommodation: roomy, rustic cottages; Inneston, Innes National Park; (08) 8854 3200. ***Marion Bay Holiday Villas:*** peaceful seaside cottages; Cnr Waratah Ave and Templetonia Cres, Marion Bay; (08) 8854 4142.

Innes National Park In summer, soak up the sun at beaches or bays with excellent (but challenging) surf breaks at Chinamans Reef, Pondalowie Bay and West Cape. In winter, keep an eye out at Stenhouse Bay and Cape Spencer for migrating southern right whales. Diving is popular, especially near the Gap, an eroded gap in a 60 m high cliff. Other activities include beach and jetty fishing, and walking on coastal and inland tracks. Accommodation is something special in this park – enjoy fabulous coastal camping in the mallee scrub or stay at the heritage lodge in the old mining township of Inneston. The annual Yorke Surfing Classic is held here each Oct. (08) 8854 3200; 81 km SW.

Ballywire Farm and Tearooms: Experience a fully working farm by joining in farming activities. Tearooms and a licensed restaurant on-site, as well as farm produce for sale, interactive displays and a games area for kids; open 10am–5pm Wed–Sun (daily school holidays); (08) 8852 1053; 11km S. ***Bublacowie Military Museum:*** personal stories, memorabilia and documents, and also a craft centre; closed Mon; 25 km N. ***Corny Point:*** coastal town featuring a lighthouse and lookout, and fishing and camping; 69 km NW. ***Daly Head:*** great surfing spot with nearby blowhole; 75 km W. ***Marion Bay:*** popular with surfers and visitors to nearby Innes National Park; 79 km SW.

See also YORKE PENINSULA, p. 240

WESTERN AUSTRALIA IS...

Spending a day at Perth's beaches, including COTTESLOE / Eating fish and chips on FREMANTLE'S WHARF / Seeing magnificent beaches and searching for quokkas on ROTTNEST ISLAND / Taking a camel ride along CABLE BEACH, Broome / Sampling wines at cellar doors in MARGARET RIVER / Feeding dolphins at MONKEY MIA /

WESTERN AUSTRALIA

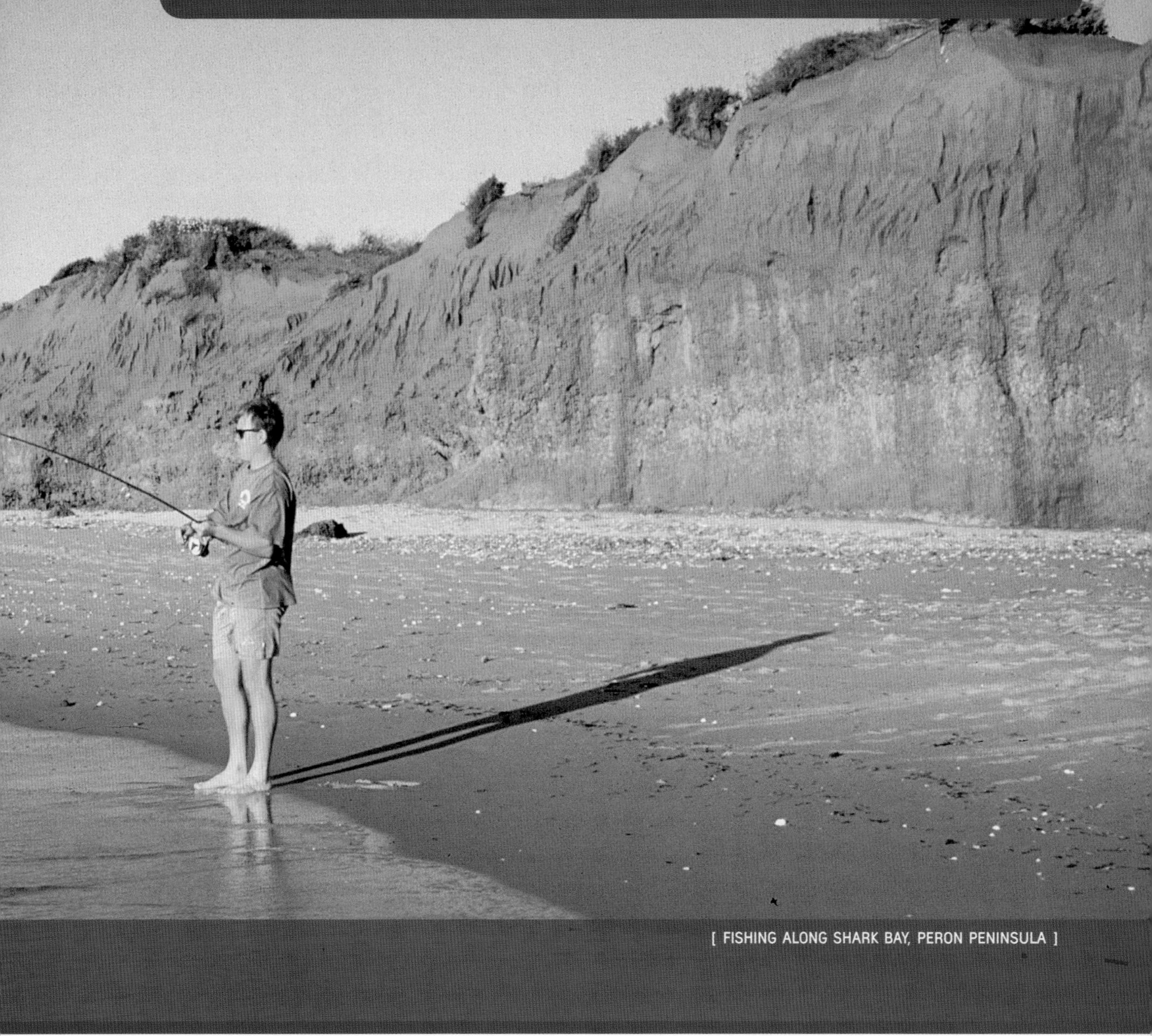

[FISHING ALONG SHARK BAY, PERON PENINSULA]

A scenic flight over the Bungle Bungles in PURNULULU NATIONAL PARK / Walking through giant stands of karri and tingle trees in the VALLEY OF THE GIANTS / A visit to NAMBUNG NATIONAL PARK, home to the moonscape of limestone pillars known as the PINNACLES / Swimming with whale sharks in NINGALOO REEF

WESTERN AUSTRALIA is defined by its size. Spanning an area of 2.5 million square kilometres, it covers one-third of the Australian continent. In dramatic contrast to its size, its population is just over two million, around one-tenth of Australia's total population. Over 72 per cent of Western Australians live in or around the capital city of Perth.

Within this great state there are incredibly diverse landscapes – an ancient terrain of rugged ranges and dramatic gorges to the north, towering forests to the south, arid deserts to the east and 12 889 kilometres of the world's most pristine coastline to the west. To match the huge variety in landscape are huge differences in climate, from the tropical humidity of the north and the dryness of the desert to the temperate Mediterranean-style climate of the South-West.

After driving for hours along empty highways, you will get a true feeling for the state's vastness. But you will be amply rewarded when you reach your destination. Western Australia boasts precious natural features, including the 350-million-year-old Bungle Bungle Range, the limestone sentinels of the Pinnacles Desert and the majestic karri forests of the South-West.

[NATURE'S WINDOW, KALBARRI NATIONAL PARK]

There is the extraordinary marine life of Ningaloo Reef, the friendly dolphins of Monkey Mia and Rottnest Island's famous quokkas.

Western Australia's historic sites are also a highlight. The Aboriginal people who first inhabited the land up to 65 000 years ago left a legacy of distinctive rock art. Albany, the site of the state's first European settlement in 1826, boasts well-preserved heritage buildings, while gracious 19th-century buildings in the capital city of Perth and its nearby port of Fremantle hark back to the days of the Swan River Colony. Remnants of great gold discoveries remain around Coolgardie and Kalgoorlie from the 1890s, which transformed Western Australia into one of the world's great producers of gold, iron ore, nickel, diamonds, mineral sands and natural gas.

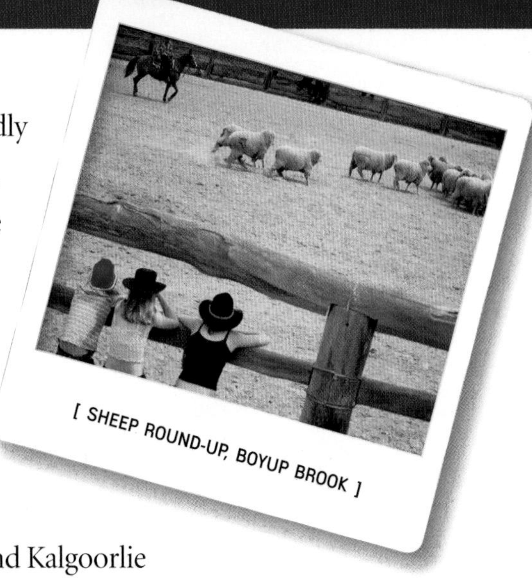

[SHEEP ROUND-UP, BOYUP BROOK]

fact file

Population 2 296 400
Total land area 2 529 875 square kilometres
People per square kilometre 0.8
Sheep and cattle per square kilometre 9.8
Nearest interstate city Adelaide, 2700 kilometres east
Length of coastline 12 889 kilometres
Number of islands 3747
Longest river Gascoyne River (760 kilometres)
Largest constructed reservoir Lake Argyle (storage volume 10 760 million cubic metres)
Highest mountain Mount Meharry (1253 metres), Karijini National Park
Highest waterfall King George Falls (80 metres), northern Kimberley
Highest town Tom Price (747 metres)
Hottest place Marble Bar (160 days a year over 37.5°C)
Coldest place Bridgetown (33 days a year begin at below 2°C)
Most remote town Warburton
Strangest place name Walkaway
Most famous person Rolf Harris
Quirkiest festival Milk Carton Regatta, Hillarys Boat Harbour
Number of wildflowers 12 500 species
Most challenging road Gibb River Road, the Kimberley
Best beach Cable Beach, Broome
Most identifiable food Pavlova (created at Perth's Esplanade Hotel)
Local beer Swan Lager

gift ideas

Gold nuggets (Kalgoorlie–Boulder) Visit the Australian Prospectors and Mining Hall of Fame to find gold nuggets in presentation cases and on chains as pendants. See Kalgoorlie–Boulder p. 338, 620 H5

Cricket merchandise (WACA, East Perth) Cricket lovers will be thrilled to receive a memento from the world-famous Western Australian Cricket Association grounds. See Museums p. 298, 292 E3

Ugg boots (Uggs-N-Rugs, Kenwick, Perth) This iconic sheepskin footwear is available around Australia, but in January 2006, Uggs-N-Rugs won a legal battle to remove an American company's trademark for 'ugh boots' from the Australian Trademarks Registry – a major victory for Australian manufacturers. 9 Royal St, Kenwick. 614 E7

Arts and crafts (Fremantle Markets, Fremantle) Over 150 stalls sell everything from local arts, crafts and clothes to fresh produce. See Fremantle Markets p. 302, 301 B2

Argyle diamonds (The Kimberley) Diamonds of unique brilliance in various colours, including the world's only intense pink diamonds. See Kimberley p. 314, 625 K5

Freshwater pearls (Broome) With its unique pearling heritage, exquisite pearl jewellery is available all over Broome. Willie Creek Pearls also have a showroom in Sorrento Quay, Hillarys Boat Harbour, north of Perth. See Broome p. 320, 624 H8

Wine (Margaret River) There is an abundance of fine wine to choose from in this top wine-producing region. See Margaret River p. 346, 616 B8

Jarrah, karri and marri woodwork (South-West region) Western Australia's unique woods are turned into handcrafted furniture, boxes, bowls, platters, salt and pepper grinders, to name but a few. See The South-West p. 307, 616

Wood-fired bread, biscuits and nut cake (New Norcia) Made by New Norcia's Benedictine monk community. Their rich, intensely flavoured, panforte-style nut cake is also available at the New Norcia Bakeries at 163 Scarborough Beach Rd, Mount Hawthorn, and Bagot Rd, Subiaco. See New Norcia p. 352, 618 C2

Quokka soft toy (Rottnest Island) Get a soft toy version of this small marsupial, which famously lives on 'Rotto', from the general store or gift shop. See Rottnest Island p. 305, 615 A6

PERTH IS...

Views across the city from KINGS PARK / SWIMMING at any of Perth's beaches / Eating fish 'n' chips on FREMANTLE'S WHARF / Sipping a coffee on the 'cappuccino strip' of SOUTH TERRACE, Fremantle / Seeing black swans at LAKE MONGER / A cricket match at the WACA / A picnic on the MATILDA BAY foreshore / A visit to the WESTERN AUSTRALIAN MUSEUM / Eating out in NORTHBRIDGE / Browsing the eclectic offerings at the FREMANTLE MARKETS / A footy match at PATERSONS STADIUM, SUBIACO / Touring the forbidding FREMANTLE PRISON

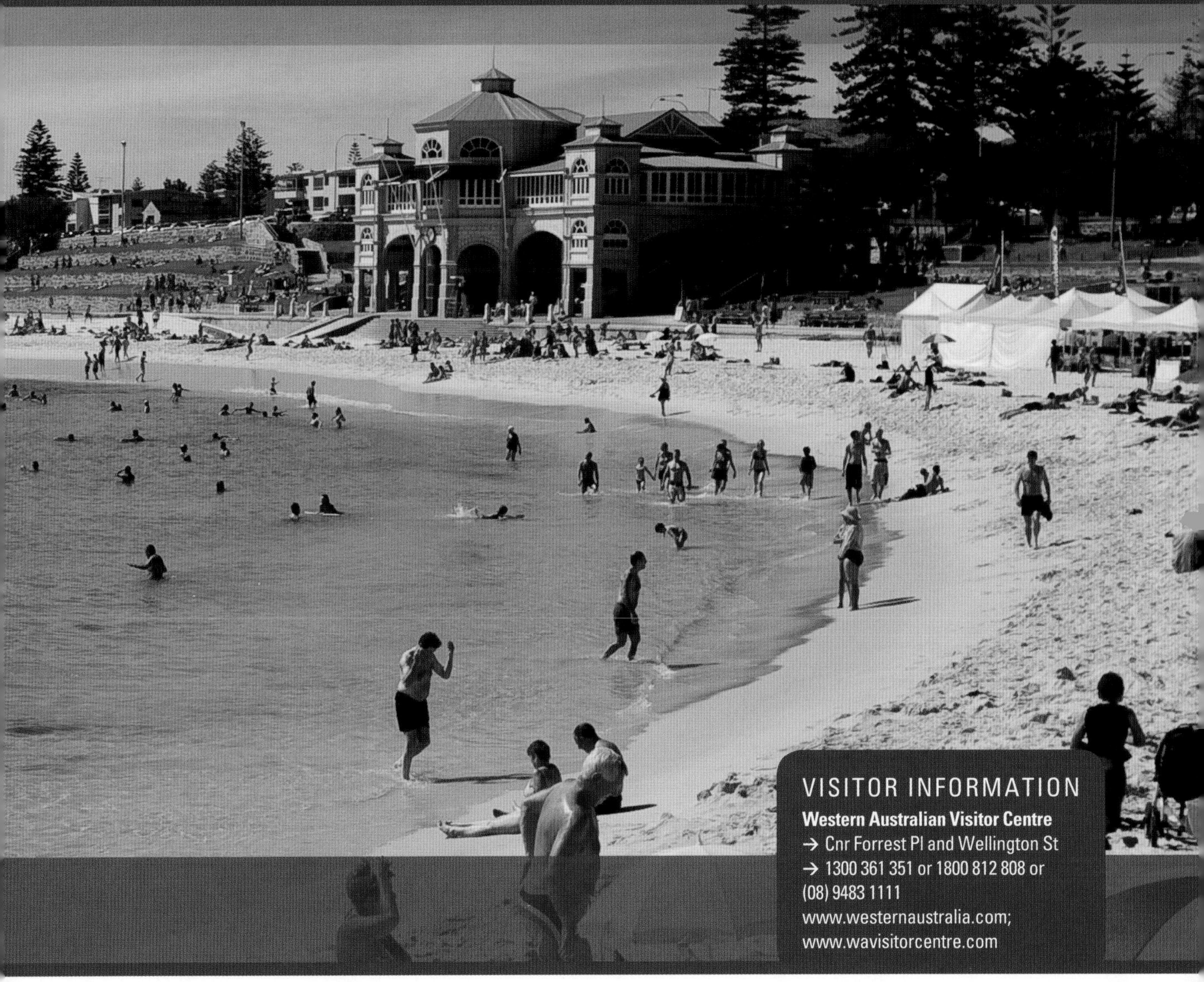

VISITOR INFORMATION

Western Australian Visitor Centre

→ Cnr Forrest Pl and Wellington St

→ 1300 361 351 or 1800 812 808 or (08) 9483 1111

www.westernaustralia.com; www.wavisitorcentre.com

Perth is the most isolated capital city in the world, closer to Singapore than it is to Sydney. Its nearest neighbour, Adelaide, is 2700 kilometres away by road. Yet it is exactly this isolation that has allowed Perth to retain a feeling of space and relaxed charm.

Claimed to be the sunniest state capital in Australia, Perth has a Mediterranean climate: hot and dry in summer, cool and wet in winter. This climate, and the city's proximity to both river and ocean, fosters a relaxed lifestyle for the population of 1 658 992. One of Perth's great attributes is that its water frontages are public land, accessible to everyone. Picnicking is a popular pastime, while cafes and bars spill their tables and chairs out onto pavements to make the most of the glorious weather.

Yet for all Perth's coastal beauty, it is the Swan River that defines the city. North of the river is Kings Park and the old-money riverside suburbs with their grand homes; further on are the beaches and the newer northern beach suburbs stretching up the coast. At the mouth of the Swan is the historic port city of Fremantle, with its rich maritime history, creative community and street-cafe culture. Upstream from Perth – where the river dwindles to a meandering waterway – is the Swan Valley, the state's oldest wine district.

INDIAN OCEAN
Trigg
Karrinyup
Gwelup
Stirling
Balcatta
Nollamara
Dianella
Innaloo
Tuart Hill
Yokine
Scarborough
Doubleview
Osborne Park
Joondanna
Coolbinia
Inglewood
Woodlands
Mount Hawthorn
Menora
Glendalough
North Perth
Mount Lawley
Wembley Downs
Herdsman
Churchlands
Leederville
Wembley
Highgate
West Leederville
Floreat
Northbridge
Jolimont
West Perth
Perth
City Beach
Daglish
Subiaco
Kings Park
Shenton Park
Mount Claremont
Karrakatta
Crawley
Swanbourne
Claremont
Nedlands
South Perth
Dalkeith
Peppermint Grove
Cottesloe
Applecross
Mosman Park
Attadale
Mount Pleasant
North Fremantle
Bicton
Ardross
Alfred Cove
East Fremantle
Melville
Booragoon
Myaree
Palmyra
SEE PAGE 294
Rendezvous Observation City Hotel
Scarborough Fair Markets (Sat–Sun)
Innaloo Megaplex (Greater Union)
Westfield Innaloo
New Norcia Bakery
Ha-Lu
Pension of Perth
Must Winebar
ME Bank Stadium
Hyde Park Hotel
Leederville Hotel
Luna Palace
Scitech Discovery Centre
Patersons Stadium
Subiaco Station St Markets (Fri–Sun)
Subiaco Pavilion Markets (Thurs–Sun)
Ace Cinemas Subiaco
Regal Theatre
Indigenart Mossenson Galleries
The Outram
Parliament House
Subiaco Arts Centre
Aspects of Kings Park
Botanic Garden
Lotterywest Federation Walkway
Moonlight Cinema
Old Swan Brewery
Old Mill
Perth Zoo
WA Basketball Centre
Challenge Stadium
Somerville Auditorium
University of Western Australia
Windsor
Cygnet Cinema
Greenhill Galleries
Bayview Centre
Claremont Museum
Ocean Beach Hotel
Cottesloe Civic Centre
Indiana Cottesloe Beach Restaurant
The Grove Shopping Centre
Point Walter Reserve
Point Heathcote Reserve
Telecommunications Museum
Wireless Hill Park
Garden City
Buckland Hill Lighthouse
To Hillarys Boat Harbour & AQWA
To Wanneroo Markets & Yanchep NP
To Fremantle
To Adventure World & Mandurah
1 km
N

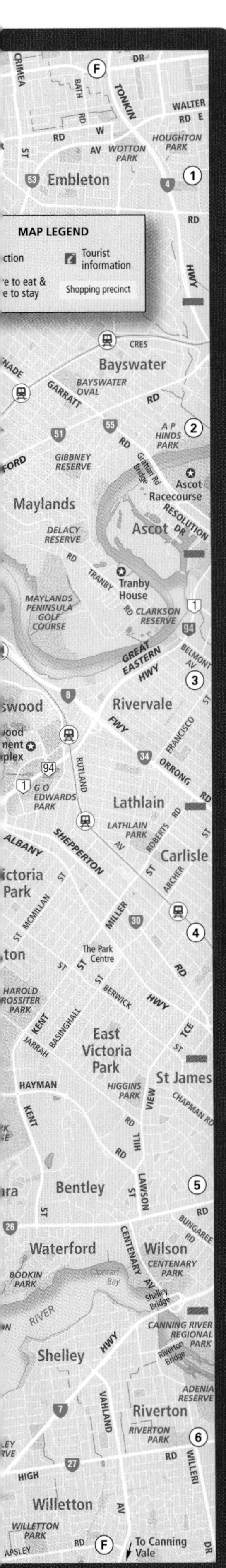

City Centre

Perth's city centre is a compact mix of towering skyscrapers and elegant colonial buildings. It is bordered to the south and east by the Swan River, with stretches of grassy parkland fringing the riverbank. Perth's central business district (CBD) harbours the city's large pedestrian-only shopping precinct, made up of a series of malls and arcades. This is connected northwards to the Perth train station by an overhead walkway across Wellington Street. To the west, the ultra-hip King Street is renowned for its gourmet cafes, galleries and fashion houses. On the south side of the city is St Georges Terrace, the main commercial street, a strip of high-rise buildings interspersed with remnants of Perth's early British heritage. Just beyond it is the main bus depot, the Esplanade Busport, and at the river end of Barrack Street is the city's jetty at Barrack Square.

Malls, arcades & a touch of old England 294 B2

Perth's central shopping precinct is in the blocks bounded by St Georges Terrace and William, Wellington and Barrack streets. These three main shopping blocks encompass the vehicle-free zones of **Hay Street Mall, Murray Street Mall** and **Forrest Place.** Between them, the two malls contain a swag of brand-name fashion outlets, bookstores and homewares shops. The big department stores of Myer and David Jones both have entrances on Murray Street Mall, and the western side of Forrest Place is home to the GPO. A series of arcades and underground walkways run from Murray Street Mall through to Hay Street Mall and on to St Georges Terrace, making it possible to shop in the city without ever crossing a street. **Carillon City** is a modern shopping centre between the malls, while **London Court**, an arcade with the appearance of a quaint Elizabethan street, runs from Hay Street Mall to St Georges Terrace, and is Perth's only open-air arcade. At the mall end, knights joust above a replica of Big Ben every 15 minutes, while St George and the Dragon do battle above the clock at the St Georges Terrace end.

King Street 294 B2

This historic precinct of commercial buildings between Hay and Wellington streets dates from the 1890s gold rush. While the street has been restored to its turn-of-the-century character, its commercial interests are entirely modern: designer fashion houses, specialist bookstores, art galleries and gourmet cafes, such as the ever-popular **No 44 King Street**, with its homemade bread and extensive wine list.

St Georges Terrace 294 A2, B2

The city's main commercial street is lined with modern office towers that overshadow a number of historic buildings. At the western end of the terrace is the **Barracks Archway**, the only remains of the Pensioners' Barracks, a structure that originally had two wings and 120 rooms. This building housed the retired British soldiers who guarded convicts in the mid-1800s. The **Central Government Building** on the corner of Barrack Street marks the spot where Perth was founded with a tree-felling ceremony in 1829. Around 50 years later, convicts and hired labour commenced work on the building that stands there today. At one stage it housed the GPO, and a plaque on the building's east corner marks the point from which all distances in the state are measured. Other historic buildings along the terrace include the Cloisters, the Old Perth Boys' School, the Deanery, St George's Cathedral, Government House and the Old Court House *(see Grand old buildings, p. 298)*. As you're walking along the Terrace, look out for the commemorative plaques inlaid in the footpath which celebrate the achievements of over 170 notable Western Australians. This was a sesquicentennial project in 1979 celebrating 150 years since the foundation of the Swan River Colony.

City gardens 294 C2, C3

Two delightful city gardens are located in the block bounded by St Georges Terrace, Riverside Drive, Barrack Street and Victoria Avenue. **Stirling Gardens** offer ornamental trees, well-kept lawns and plenty of shady spots. Along the footpath on the St Georges Terrace side there are large statues of kangaroos – a great photo opportunity. Just nearby within the gardens is an ore obelisk, a memorial acknowledging the state's role as one of the world's foremost producers of minerals. Look carefully in the garden beds from the footpath on the Barrack

climate

Perth is Australia's sunniest capital, with an annual average of eight hours of sunshine per day. All this sunshine gives Perth a Mediterranean climate of hot, dry summers and mild, wet winters. The average maximum temperature in summer is 31°C; however, heat waves of temperatures in the high 30s and low 40s are not unusual. Fortunately, an afternoon sea breeze affectionately known as 'the Fremantle Doctor' eases the heat of summer.

J	F	M	A	M	J	J	A	S	O	N	D	
30	31	29	25	22	19	18	18	20	22	26	28	MAX °C
17	17	16	13	10	8	7	8	9	11	14	16	MIN °C
8	12	19	45	123	184	173	136	80	54	21	14	RAIN MM
3	3	4	8	14	17	18	17	14	11	6	4	RAIN DAYS

Street side of the gardens and you'll find small statues of May Gibbs' 'Gumnut Babies' nestled amongst the ferns. The gardens also house the oldest public building in Perth, now the Francis Burt Law Museum *(see Museums, p. 298)*. The **Supreme Court Gardens**, further towards Riverside Drive, are a popular location for concerts on warm summer evenings, including the annual Carols by Candlelight.

Bells & a jetty 294 B3

Barrack Square, at the water's edge south of the city, is where you'll find the Bell Tower, home of the Swan Bells, and the ferry jetty. The Barrack Street Jetty is the departure point for ferry services to Fremantle, South Perth, Rottnest Island and Carnac Island, along with various leisure cruises on the Swan River and to the vineyards of the Swan Valley. Behind Jetty 6 is the Willem de Vlamingh Memorial, which features a sundial indicating Amsterdam time and many historical references.

The Swan Bells consist of 18 'change-ringing' bells, which form the largest set in the world. Twelve of the bells, given to the state in 1988 by the British government, come from London's St Martin-in-the-Fields church. The bell tower – which cost $6 million to build amid great controversy – offers galleries from which you can view the bellringers and the bells in action. A viewing platform at the top of the bell tower provides excellent views of the river and city. *Barrack Sq, cnr Barrack St and Riverside Dr; (08) 6210 0444; open from 10am daily, closing times vary seasonally (check the website, www.thebelltower.com.au); full bell ringing 12–1pm Mon, Tues, Thurs and Sat–Sun.*

Northbridge

Northbridge, which lies north of the city centre across the train line, is Perth's centre for the arts and the heart of the city's nightlife. This inner-city suburb is connected to the city centre via a walkway that crosses Perth Railway Station and leads directly to the Perth Cultural Centre. Bounded by Roe, Francis, Beaufort and William streets, the Perth Cultural Centre includes the state museum, art gallery, theatre centre, state library and Perth's institute of contemporary arts. William Street and the streets further west are packed with restaurants and bars that offer great eating, drinking and nightclubbing.

Art Gallery of Western Australia 294 C1

The state's principal public art gallery, founded in 1895, houses collections of Australian and international paintings, sculpture, prints, craft and decorative arts. The gallery's collection of Aboriginal art is one of the finest in Australia. *Perth Cultural Centre, James St Mall; (08) 9492 6622 (24-hour information line) or (08) 9492 6600; open 10am–5pm Wed–Mon; general admission free; free guided tours available.*

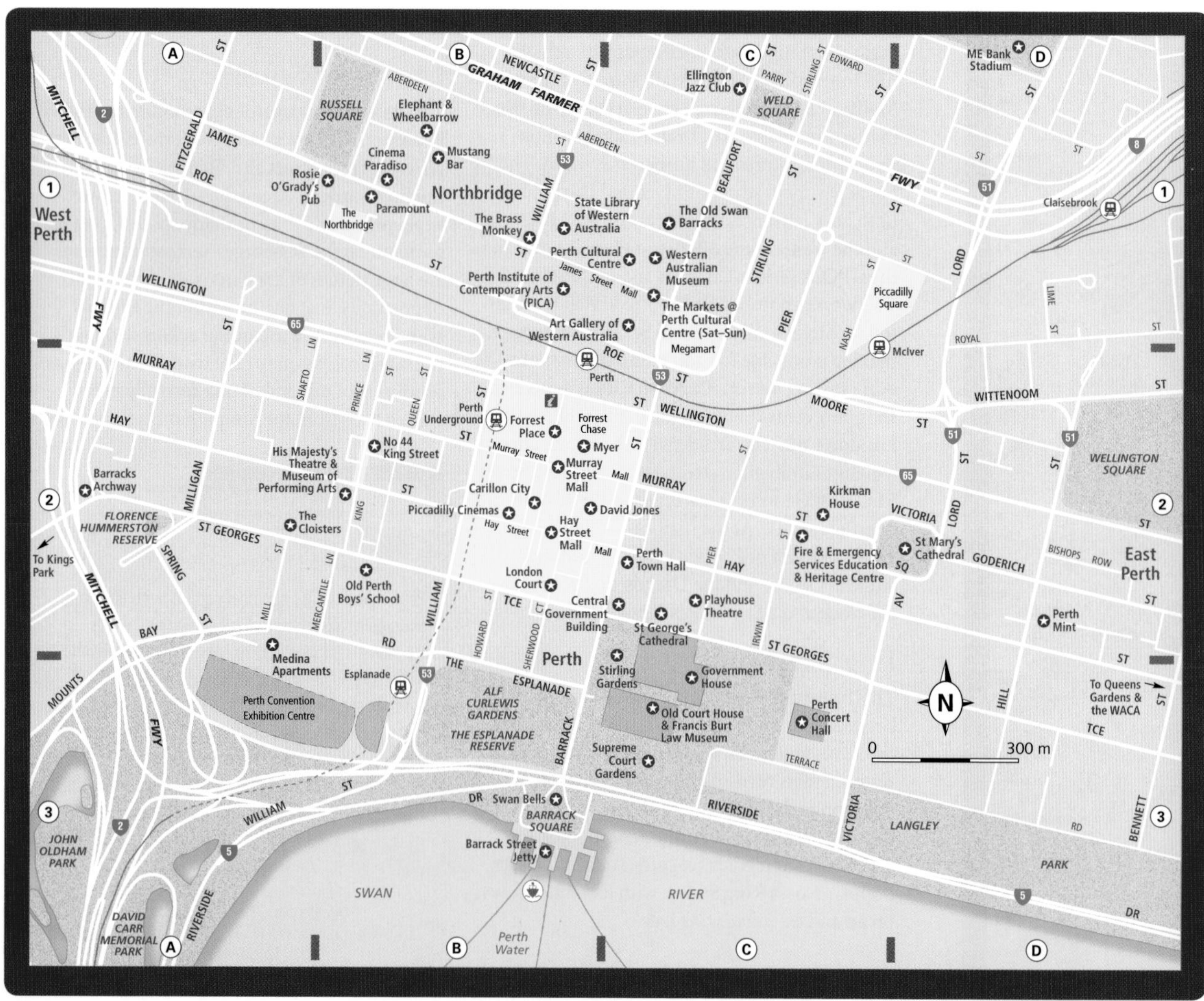

Western Australian Museum 294 C1

At this comprehensive museum you can see a 25-metre whale skeleton, the 11-tonne Mundrabilla meteorite and 'Megamouth', a rare species of shark. There are exhibitions concerning the state's Aboriginal people; the origins of the universe; and dinosaur, bird, butterfly, mammal and marine galleries. The interactive Discovery Centre is great for children. Within the museum complex is the **Old Gaol**, built in 1856 and used by the Swan River Colony until 1888. *Perth Cultural Centre, James St Mall; (08) 9212 3700; open 9.30am–5pm Thurs–Tues; admission by donation; free guided tours available.*

Perth Institute of Contemporary Arts 294 B1

Commonly referred to by its acronym, PICA, the Perth Institute of Contemporary Arts is where you can sample the latest in visual and performance art. There is an ever-changing program of exhibitions. *Perth Cultural Centre, James St Mall; (08) 9228 6300; open 11am–6pm Tues–Sun; general admission free.*

East Perth

East Perth is where you'll find the Western Australian Cricket Association oval, known colloquially as the WACA (pronounced 'Wacka'), where cricket matches entertain the crowds over the summer months. Gloucester Park Raceway, with night harness horseracing on Friday nights, is nearby. Once the dead end of town – except when night matches or race meets lit up the night sky – East Perth is now an enclave of offices and hip inner-city apartments. There are two historically significant sites in this precinct, the Perth Mint and Queens Gardens.

The Perth Mint 294 D2

The Mint's imposing facade was built in 1899 from Rottnest Island limestone, and is one of the best examples of Perth's gold-boom architecture. Here at Australia's oldest operating mint you can see the world's largest collection of natural gold specimens, including the 'Golden Beauty', a 11.5-kilogram nugget. Visitors can hold a 400-ounce gold bar, mint their own coins, take a guided tour and watch gold being poured. *310 Hay St (cnr Hill St), East Perth; (08) 9421 7277; open 9am–5pm daily.*

Queens Gardens 292 E3

The serene, English-style Queens Gardens feature a tranquil water garden complete with a replica of the famous Peter Pan statue that graces London's Kensington Gardens. These gardens were originally clay pits, where bricks were kilned for use in early colonial buildings such as the Perth Town Hall. *Cnr Hay and Plain sts.*

WACA 292 E3

Across Hale Street from Queens Gardens is the WACA, the home of the state's cricket team, the Western Warriors. True lovers of the sport should visit the WACA's cricket museum *(see Museums, p. 298).*

getting around

The city is compact and easy to explore. A free bus service known as the CAT (Central Area Transit) System operates regular services, every five–ten minutes, around central Perth. The blue CAT runs in a north–south loop, the red CAT operates in an east–west loop, and the yellow CAT travels to the city centre from East Perth. (Note that a CAT bus also services Fremantle.) You can also travel free on Transperth buses or trains within the Free Transit Zone in the city centre, but only on trips that start and finish within the zone.

Trains run from the city out to the northern suburbs and down to Fremantle while the Southern Suburbs Railway links Perth to Mandurah. Ferries and cruise boats depart regularly from Barrack Street Jetty to various destinations, including Fremantle, South Perth, Rottnest Island and the Swan Valley wine region. (Transperth runs the ferry to South Perth, while private operators travel further afield.) Perth, with its largely flat landscape, is also excellent for cycling; maps of the city's 700-kilometre bike network are available at bike shops or online at the Department of Transport website, www.transport.wa.gov.au.

Public transport Transperth (bus, train and ferry) 13 6213.

Airport shuttle bus Airport–city shuttle 1300 666 806 or (08) 9277 7958.

Swan River Cruises Captain Cook Cruises (08) 9325 3341.

Rottnest Island ferries Rottnest Express 1300 467 688: Rottnest Fast Ferries (08) 9246 1039.

Motoring organisation RAC of WA 13 1703.

Car rental Avis 13 6333; Budget 13 2727; Hertz 13 3039; Thrifty 1300 367 227.

Taxis Black and White Taxis 13 1008; Swan Taxis 13 1330.

Bicycle hire About Bike Hire (08) 9221 2665.

See also Getting to Fremantle, p. 301

top events

Hopman Cup Prestigious international tennis event. January.

Perth Cup Western Australia's premier horseracing event. January.

Australia Day Skyworks A day-long party of events, culminating in a spectacular fireworks display. January.

Perth International Arts Festival Music, theatre, opera, dance, visual arts and film. February.

Eat, Drink Perth A month-long celebration of delicious events for Perth's biggest food festival. March.

City of Perth Winter Arts Season A three-month program of locally created arts and culture. June–August.

Kings Park Wildflower Festival Australia's premier native plant and wildflower exhibition. September.

Perth Royal Show Showcases the state's primary and secondary resources. September–October.

See also Top events in Fremantle, p. 302

shopping

London Court Mock-Tudor arcade with souvenir, jewellery and antique stores. 294 B2

Hay Street Mall, Murray Street Mall and Forrest Place The CBD's main shopping precinct with brand-name fashion outlets and major department stores Myer and David Jones. 294 B2

King Street High fashion, galleries and cafes with style. 294 B2

Rokeby Road, Subiaco Funky local designers sit alongside more established labels. 292 C3

Bay View Terrace, Claremont Perth's up-market fashion hot spot. 292 B5

Napoleon Street, Cottesloe Cafes, boutiques and designer homewares. 292 B5

[THE FUTURISTIC SWAN BELLS TOWER]

markets

Subiaco Pavilion Markets Art and craft stalls with large food hall in restored warehouse adjacent to station. Cnr Rokeby and Roberts rds, Subiaco; 10am–9pm Thurs–Fri, 9am–5pm Sat–Sun. 292 C3

Subiaco Station Street Markets A colourful outdoor and undercover market with an eclectic array of goods and live entertainment. 52 Station St, Subiaco; (08) 9382 2832; 9am–5.30pm Fri–Sun. 292 C3

Scarborough Fair Markets Stalls of antiques, bric-a-brac, arts and crafts, surf gear and hippie wear, jewellery and organic produce. 9am–5.30pm Sat–Sun. 292 A1

Canning Vale Markets Also known as Market City, this is the primary fruit and vegetable whole market for the state. On Sundays, it becomes a huge undercover flea market with over 300 stalls. 280 Bannister Rd, Canning Vale; (08) 9456 9200; 7am–1pm Sun. 615 C6

Wanneroo Markets Five acres of undercover markets north of Perth boast over 120 variety stalls, a food court and locally grown fruit and vegetables. 33 Prindiville Dr, Wangara; (08) 9409 8397; 9am–5pm Fri–Sun. 615 B4

See also Fremantle Markets, p. 302

Inner East

Burswood Entertainment Complex 293 F3

Built on an artificial island on the southern banks of the Swan River, this complex includes a casino, hotel, convention centre, tennis courts, golf course and Burswood Dome indoor stadium (the venue for the Hopman Cup). The Atrium Lobby has an impressive 47-metre-high pyramid of shimmering glass containing a tropical garden and waterfall. Burswood Park, the beautifully landscaped gardens that surround the complex, offers paths for walkers, cyclists and joggers, a heritage trail and a children's playground. *Great Eastern Hwy, Burswood; (08) 9362 7777.*

Inner North

Tranby House 293 F3

Just beyond East Perth in the suburb of Maylands is historic Tranby House. Built in 1839 overlooking the picturesque Swan River, Tranby is one of the oldest surviving buildings from the early settlement of the Swan River Colony and is a unique example of colonial farmhouse architecture. A major attraction is the property's gardens which feature two century-old oak trees. The house has been beautifully restored by the National Trust. *Johnson Rd, Maylands; (08) 9272 2630; open 10am–4pm Wed–Sun (closed 24 Dec – 6 Feb).*

Inner West

Just minutes from the city centre, Kings Park is visited by millions of people each year. Beyond it at the end of Kings Park Road is the suburb of Subiaco, with its popular shopping, cafe and market precinct. Below Kings Park, Mounts Bay Road winds its way along the river's edge to the suburb of Crawley, passing the Old Swan Brewery site, now a riverside complex of up-market offices, apartments and restaurants. The distinctive clock tower of the University of Western Australia's Winthrop Hall is an easily spotted landmark. Across Matilda Bay Road from the university is the grassy Matilda Bay shoreline, with shady spots and views back up the river towards the city.

Kings Park & Botanic Garden 292 D4

The first stop for any visitor to Perth has to be Kings Park. Standing on top of Mount Eliza, you enjoy sweeping views of the city and the Swan River, with the Darling Range in the distance. Within this huge 400-hectare natural bushland reserve there are landscaped gardens and walkways, lakes, playgrounds, a restaurant and cafes.

Fraser Avenue, the main entrance road into the park, is lined with towering lemon-scented gums, honouring those who perished in war. The clock tower and bronze portrait bust at this entrance is a memorial to Edith Cowan, the first woman elected to an Australian parliament. A tireless advocate for women's rights and children's welfare, she now lends her name to one of Perth's universities.

Opened in 2003, the Lotterywest Federation Walkway is a combination of on-ground pathways, elevated walkway and spectacular steel-and-glass bridge, extending 620 metres through the Botanic Garden. It is a snapshot of Western Australia's famed flora; at ground level you'll pass boabs, boronias and tuart trees, while the walkway through the treetops takes you close to karri, marri, tingle and jarrah trees. The Botanic Garden itself is spread over 17 hectares and planted with more than 1700 native species.

The Botanic Garden's newest attraction arrived in July 2008 when a giant boab tree weighing 36 tonnes and estimated to be 750 years old was transported 3200 kilometres from the Kimberley town of Warnum and transplanted in the garden opposite Forrest Carpark. It joins long-time favourite attractions such as the State War Memorial precinct; the Pioneer Women's Memorial, with its water fountains which periodically shoot skywards; the nearby DNA Tower, offering spectacular views; the fantastic dinosaur and fossil creations at Synergy Parkland; and the child-friendly Lotterywest Family Area.

In spring the annual Kings Park Festival showcases the best of the state's wildflowers, attracting over 500 000 visitors from around the world. This month-long event showcases all aspects of the park, with myriad spectacular events and activities. During summer the park is a favourite venue for live outdoor entertainment and moonlight movies (www.bgpa.wa.gov.au). *Fraser Ave; (08) 9480 3600; open daily; free guided walks from the visitor centre at 10am and 2pm daily.*

University of Western Australia 292 C4

With its distinctive Mediterranean-style architecture and landscaped gardens, the University of Western Australia is renowned as one of Australia's most beautiful campuses. Here you'll find **Winthrop Hall**, with its majestic clock tower and reflection pond, and the **Sunken Garden**, backdrop for many a wedding photo. The **Lawrence Wilson Art Gallery**, home to the university's extensive collection of Australian art, includes works by Sidney Nolan, Arthur Boyd, Fred Williams and Rupert Bunny. *35 Stirling Hwy, Crawley; (08) 6488 3707; open 11am–5pm Tues–Fri, 12–5pm Sun; admission free.*

The university is also home to the **Berndt Museum of Anthropology**, an internationally renowned collection of traditional and contemporary Aboriginal art and artefacts. Currently in the process of relocating, the collection will be temporarily housed in the Lawrence Wilson Art Gallery until a new, purpose-built museum can be funded. *www.berndt.uwa.edu.au*

where to eat

The Old Brewery Perth's premier steakhouse in a stunning waterside venue just minutes from the CBD. 173 Mounts Bay Rd; (08) 9211 8999; open weekends for breakfast, daily for lunch and dinner. 614 E6

Ha-Lu Relatively new and already an award finalist offering Japanese cuisine served in the Japanese pub style. Shop 4/401 Oxford St, Mt Hawthorn; (08) 9444 0577; open Wed–Sun for dinner. 292 D2

Must Winebar Bistro-style French Provincial cuisine with impressive wine list and always busy. 519 Beaufort St, Highgate; (08) 9328 8255; open daily for lunch and dinner. 292 E3

New Norcia Bakeries Authentic sourdough and yeasted breads baked in a traditional wood-fired oven; also many other tempting delights. 163 Scarborough Beach Rd, Mt Hawthorn; (08) 9442 4114; open daily for breakfast and lunch. 292 D2

Restaurant Amusé Michelin-star trained chef Hadleigh Troy brings an innovative degustation menu to his awaiting fans. 64 Bronte St, East Perth; (08) 9325 4900; open Tues–Sat for dinner. 292 E3

See also Where to eat in Fremantle, p. 303

entertainment

Cinema After once reigning supreme in town, Hoyts and Greater Union have moved out to the suburbs, leaving the city centre bereft of any major cinemas. There is, however, the Piccadilly Cinema in Piccadilly Arcade in the Hay Street Mall, which screens new-release films in Perth's only surviving grand old Art Deco cinema. The most easily accessible Hoyts cinemas are in Fremantle at the Queensgate on William Street and the Millennium on Collie Street. The closest Greater Union cinema to the city is in Innaloo. Subiaco has an independent cinema, the Ace at 500 Hay Street on the corner of Alvan. There's also Cinema Paradiso in Northbridge, the Luna Palace in Leederville, the Windsor in Nedlands, the Cygnet in Como and, in Fremantle, Luna on SX (Essex Street). In summer, there are a number of outdoor cinemas that operate; favourites are the Moonlight Cinema in Kings Park and the Somerville Auditorium at the University of Western Australia. The latter, which screens films for the Perth International Arts Festival from December through March, is defined by a cathedral of Norfolk pine trees and patrons sit on deckchair-style seats under the stars. Programs and session times, including those for the open-air cinemas over summer, are listed daily in *The West Australian.*

Live music Northbridge, Leederville and Subiaco are the places to go for live music, as a healthy pub scene supports local musicians. In Northbridge popular venues include the Brass Monkey, the Mustang Bar, the Paramount, the Elephant & Wheelbarrow and Rosie O'Grady's. The Leederville Hotel on Oxford Street has a legendary Sunday afternoon session; the Ocean Beach Hotel in Cottesloe (known locally as the 'OBH') adds sunset views from the bar. In Fremantle, the premier live music venues are the Fly by Night Musicians Club on Parry Street, Kulcha on South Terrace and Mojos in North Freo. Pick up the free street publication *Xpress* for gig guides or visit www.xpressmag.com.au. Jazz venues include the Hyde Park Hotel in Bulwer Street, North Perth; the Navy Club in High Street, Fremantle; and the Ellington Jazz Club at 191 Beaufort Street, Perth. Aficionados should visit www.jazzwa.com for gig information.

Classical music and performing arts His Majesty's Theatre, Australia's only remaining Edwardian theatre, is Perth's premier venue for high-end theatre, opera and ballet. At the beginning of 2011, the State Theatre Centre of WA was opened in Northbridge. Home to both the Black Swan State Theatre Company and the Perth Theatre Company, it features three performance venues: the Heath Ledger Theatre, the Studio Underground and the outdoor Courtyard. Other theatre productions can be seen at the Subiaco Arts Centre and the Regal Theatre, while Fremantle is home to Deckchair Theatre in High Street and Spare Parts Puppet Theatre in Short Street. The Perth Concert Hall in St Georges Terrace is the fine music venue for concerts by local and international musicians. The Burswood Entertainment Complex hosts touring shows and musicals; ME Bank Stadium in East Perth is the venue for big international acts. In the warmer months, outdoor concerts are held in the Supreme Court and Queens Gardens, and in Kings Park. Check *The West Australian* for details.

grand old buildings

Government House Gothic arches and turrets reminiscent of the Tower of London. St Georges Tce (opposite Pier St).

His Majesty's Theatre 'The Maj', built in 1904, features an opulent Edwardian exterior. 825 Hay St; free foyer tours 10am–4pm Mon–Fri.

Kirkman House In front of this gracious edifice is an immense Moreton Bay fig tree, planted in the 1890s and now classified by the National Trust. 10 Murray St.

Old Court House Perth's oldest surviving building (1836), now home to the Francis Burt Law Museum. Cnr St Georges Tce and Barrack St.

Old Perth Boys' School Perth's first purpose-built school was made from sandstone ferried up the Swan River by convict labour. 139 St Georges Tce.

Perth Town Hall Built by convict labour (1867–70) in the style of an English Jacobean market hall. Cnr Hay and Barrack sts.

St George's Cathedral This 1879 Anglican church features an impressive jarrah ceiling. 38 St Georges Tce.

St Mary's Cathedral Grand Gothic-style cathedral, one end of which was built in 1865. Victoria Sq.

The Cloisters Check out the decorative brickwork of this 1858 building, originally a boys' school. The old banyan tree adjoining it is something special too. 200 St Georges Tce.

See also Fremantle's grand old buildings, p. 302

museums

Fire and Emergency Services Education and Heritage Centre Refurbished in 2009, this limestone building dating from 1900 houses exhibitions on the history of the Perth fire brigade, including fire rescue and old Big Red engines. Cnr Murray and Irwin sts; (08) 9416 3400; 10am–4pm Tues–Thurs; admission free.

Francis Burt Law Museum The history of the state's legal system is housed in the Old Court House. Stirling Gardens, cnr St Georges Tce and Barrack St; (08) 9325 4787; open 10am–2.30pm Wed–Fri; admission free.

Museum of Performing Arts Entertainment history brought to life through exhibitions of costumes and memorabilia taken from backstage archives. His Majesty's Theatre, 825 Hay St; (08) 9265 0900; open 10am–4pm Mon–Fri; admission by gold coin donation.

WACA Museum Offers cricket memorabilia for fans of the sport. Gate 2, Nelson Cres, East Perth; (08) 9265 7222; open 10am–3pm Mon–Fri except match days; tours of ground and museum run at 10am and 1pm Mon–Fri.

Army Museum of Western Australia Houses WA army memorabilia dating from colonial times to the present day. Artillery Barracks, Burt St, Fremantle; (08) 9430 2535; open 11am–4pm Wed–Sun.

See also The Perth Mint, p. 295, Western Australian Museum, p. 295, Western Australian Maritime Museum, p. 302, Shipwreck Galleries, p. 302, and Fremantle Arts Centre, p. 303

Foreshore suburbs 292 C4, C5

Extending along the river foreshore from Matilda Bay towards the ocean is a series of exclusive waterfront suburbs with charming village-style shopping areas, fashionable galleries and foreshore restaurants. **Nedlands** is the suburb closest to the University of Western Australia. **Dalkeith's** Jutland Parade takes you to Point Resolution, with magnificent views of the river to the south and west. Follow the walking paths down the hillside to White Beach on the foreshore, a popular recreational spot.

Subiaco 292 C3

Beyond West Perth is the popular shopping, cafe and market precinct of Subiaco, with its village-style main street, Rokeby Road (pronounced 'Rock-a-bee'). 'Subi', as it is known to the locals, is one of Perth's oldest suburbs, and there are some fine old homes in the back streets behind Rokeby Road. Subiaco is also where you'll find WA's home of AFL football, **Patersons Stadium**, formerly known as Subiaco Oval.

Parliament House 292 D3

Go inside the corridors of power on a free, hour-long guided tour. *Harvest Tce, West Perth; (08) 9222 7259; tours 10.30am Mon and Thurs.*

Scitech Discovery Centre 292 D3

This interactive science and technology centre has more than 160 hands-on exhibits. You can touch, switch, climb, crank and explore – all in the name of science. There's also **Horizon**, a state-of-the-art planetarium screening extraordinary journeys into space on the largest dome screen in the Southern Hemisphere. *City West, cnr Sutherland and Railway sts, West Perth; (08) 9215 0700; open 9.30am–4pm Mon–Fri, 10am–5pm Sat–Sun, school holidays and public holidays.*

Inner North-west

Lake Monger 292 C3

Lake Monger, a large urban wetland 5 kilometres north of Perth's CBD in the suburb of Wembley, is the best place in Perth to see Western Australia's famous black swans. Ever since 1697, when Dutch explorer Willem de Vlamingh named the Swan River after the numerous black swans he found on its waters, the majestic waterbirds have been inextricably linked with Western Australia. The first settlement was named the Swan River Colony, and the black swan became the official State Bird Emblem of Western Australia in 1973.

South-western Beachside

With pristine white-sand beaches stretching northwards up the coast, swimming and surfing are a way of life in Perth. Several stunning beaches, including Cottesloe, are close to the city. Scattered along these beach-fronts are lovely cafes, restaurants and bars. Enjoy the sublime Perth experience of sipping a glass of Western Australian wine as you watch the sun sink into the Indian Ocean.

Cottesloe 292 A5

Cottesloe is distinguished by its towering Norfolk Island pines and the **Indiana Cottesloe Beach Restaurant**, a neo-colonial

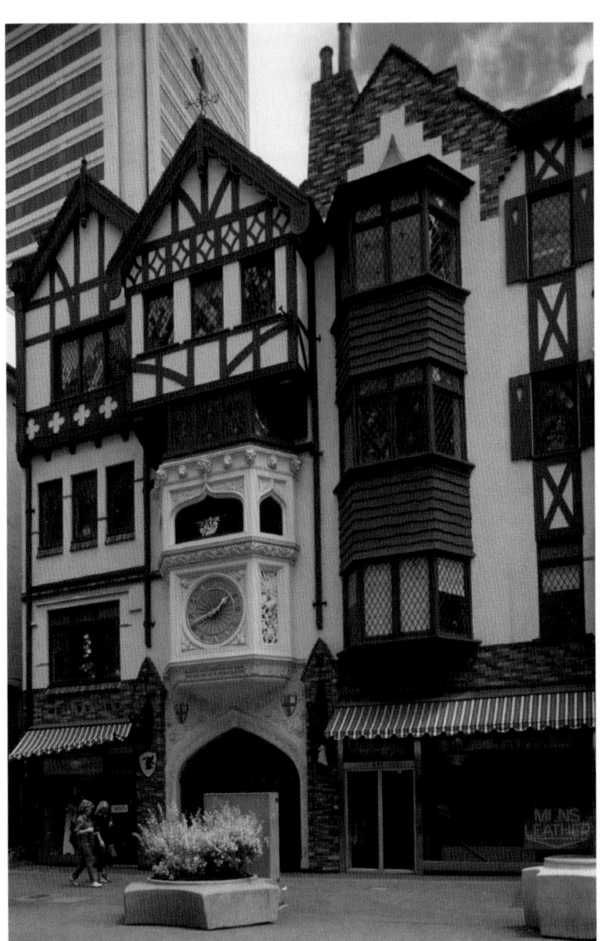

[LONDON COURT]

building of grand proportions that sits right on the beach. *99 Marine Pde, Cottesloe; (08) 9385 5005.*

The Spanish-style **Cottesloe Civic Centre**, once the private mansion of flamboyant millionaire Claude de Bernales, boasts magnificent gardens with sweeping views of the Indian Ocean. *109 Broome St; (08) 9285 5000; gardens open daily; admission free.*

Claremont 292 B4

This up-market suburb is home to Perth's swankiest shopping area, centred around Bay View Terrace. The **Claremont Museum**, on Victoria Avenue, was built in 1862 by convicts and the Pensioner Guards as the Freshwater Bay School. It now offers an interesting social history display of more than 9000 items related to the district, including artefacts, photographs, oral history tapes and documents. The main exhibitions are housed in the schoolhouse, which has been carefully preserved to provide a glimpse into the colonial education system. Nearby is a children's playground and picnic area. *66 Victoria Ave, Claremont; (08) 9340 6983; open 12–4pm Mon–Fri; admission free.*

Peppermint Grove 292 B5

Peppermint Grove boasts some of Perth's grandest homes. A drive along The Esplanade takes you past grass-backed riverside beaches, natural bushland and shady picnic areas. Follow the road around to Bay View Park – another great place for picnicking – in Mosman Park, and up the hill for sweeping views of Mosman Bay, the Swan River and Perth city skyline.

galleries

Aspects of Kings Park Contemporary Australian craft and design, with a focus on local artists. Fraser Ave, Kings Park and Botanic Garden, West Perth; (08) 9480 3900; open 9am–5pm daily.

Indigenart Mossenson Galleries One of Australia's foremost Aboriginal art galleries.115 Hay St, Subiaco; (08) 9388 2899; open 10am–5pm Mon–Fri, 11am–4pm Sat.

Greenhill Galleries New York–style gallery features works of leading Australian artists. 6 Gugeri St, Claremont; (08) 9383 4433; open 10am–5pm Tues–Fri, 10am–4pm Sat.

Holmes à Court Gallery Changing display of works from Australia's finest private art collection. Level 1, 11 Brown St, East Perth; (08) 9218 4540; open 12–5pm Wed–Sun; general admission free.

Kailis Jewellery Perfectly matched strands and handcrafted pieces made from exquisite cultured and seedless pearls. 29 King St; (08) 9422 3888; cnr Marine Tce and Collie St, Fremantle; (08) 9239 9330.

See also Art Gallery of Western Australia, p. 294, and Perth Institute of Contemporary Arts, p. 295

North-western Beachside

Scarborough 292 A2

Renowned for its beachside cafe society, the beach here is a top spot for surfers and sailboarders. The five-star Rendezvous Observation City Hotel dominates the landscape, reminiscent of the Gold Coast resorts. On weekends the Scarborough Fair Markets set up stalls just opposite the hotel, where you can shop for antiques, bric-a-brac, arts and crafts plus – of course – surf gear and hippie wear. *Cnr Scarborough Beach Rd and West Coast Hwy; open 9am–5.30pm Sat–Sun.*

where to stay

Burswood Entertainment Complex Choose from the upmarket InterContinental Perth Burswood hotel or the more budget-friendly Holiday Inn. Great Eastern Hwy, Burswood; (08) 9362 7777. 293 F3

Medina Apartments Situated in two locations in the CDB, both complexes include self-contained studios, one- and two-bedroom apartments. 33 Mounts Bay Rd and 138 Barrack St; 1300 633 462. 294 A2

Pension of Perth Romantic and historic B&B, each room has been restored to its former glory and includes an en-suite. 3 Throssell St; (08) 9228 9049. 292 D3

The Old Swan Barracks Built in 1896, the converted barracks offer budget accommodation close to nightlife and the CBD. 2–8 Francis St, Northbridge; (08) 9428 0000. 294 C1

The Outram One of the city's most luxurious places to stay and situated in the elite West Perth district. 32 Outram St; (08) 9322 4888. 292 D3

See also Where to stay in Fremantle, p. 303

sport

AFL (Australian Football League) is the most popular spectator sport in Perth, with crowds flocking to Patersons Stadium in Subiaco from April through September to support their local teams, the West Coast Eagles and the Fremantle Dockers. Patersons Stadium, formerly known as Subiaco Oval is also home to Western Australia's **Rugby Union** team, the Western Force, which plays in the Super 14 international rugby union competition.

The **cricket** season takes up where the footy leaves off, with the famous WACA hosting both interstate and international test matches over the summer months.

Perth's **soccer** club, the Perth Glory, play at ME Bank Stadium (formerly Members Equity Stadium) in East Perth. **Basketball** fans can catch the popular Perth Wildcats from September to February at Challenge Stadium in Mount Claremont, while their female counterparts, the Perth Lynx, play nearby at the new $44 million WA Basketball Centre.

In January, Perth's Burswood Dome hosts the Hopman Cup, a prestigious international **tennis** event.

Horseracing is a year-round event, split between two venues: Ascot racecourse in summer and Belmont Park in winter. Events such as the Perth Cup (held on New Year's Day), the Easter Racing Carnival and the Opening Day at Ascot draw huge crowds. Night harness racing can be seen at Gloucester Park every Friday night.

Hillarys Boat Harbour 615 B4

This ocean-side complex houses a marina, a world-class aquarium *(see next entry)*, and Sorrento Quay, a 'village' of shops, cafes, restaurants and resort apartments. Specialty shops include Margaret River on the Boardwalk, offering a range of gourmet products from the famous Margaret River region; Willie Creek Pearls, showcasing pearl jewellery from Broome's popular tourist attraction, the award-winning Willie Creek Pearl Farm; The British Lolly Shop, a feast of 'old-country' chocolates, sweets, soft drinks, crisps and snacks; and The Jarrah Joint, specialising in fine jarrah souvenirs and gifts. If the sandy beaches are too tame for you, try The Great Escape, a leisure park with water slides, miniature golf and trampolines. Ferries to Rottnest Island run from Hillarys, and there are whale-watching cruises from September to November. *Southside Dr, Hillarys; Boat Harbour, (08) 9448 7544; Sorrento Quay, (08) 9246 9788.*

Aquarium of Western Australia (AQWA) 615 B4

The highlight of this attraction, housing five different aquariums, is an incredible 98-metre underwater tunnel aquarium, surrounded by approximately 3 million litres of the Indian Ocean. Here you'll see thousands of marine creatures including sharks, stingrays, seals, crocodiles, sea dragons and turtles. For the ultimate underwater sensation, visitors can snorkel or swim with sharks. *Hillarys Boat Harbour, 91 Southside Dr, Hillarys; (08) 9447 7500; open 10am–5pm daily.*

Inner South

Perth Zoo 292 D4

Perth Zoo, with its shady gardens, walkways and picnic areas, is just a short ferry ride across the river from the city. Visit the Australian Walkabout for a close-up look at native animals in a bush setting. Other exhibits include the Penguin Plunge, African Savannah, Reptile Encounter and Rainforest Retreat. *20 Labouchere Rd, South Perth; (08) 9474 0444 or (08) 9474 3551 (recorded information line); open 9am–5pm daily.*

Old Mill 292 D4

This picturesque whitewashed windmill at the southern end of the Narrows Bridge is one of the earliest buildings in the Swan River Colony – the foundation stone was laid by Captain James Stirling, the first governor, in 1835. The mill, the adjacent miller's cottage and the recently established Education Centre review the development of flour milling in the early days of the colony. Coloured lights turn the huge Norfolk Island pine next to the Old Mill into a giant Christmas tree every December, making a festive display in the summer night sky. *Mill Point Rd, South Perth; (08) 9367 5788; open 10am–4pm Tues–Fri, 1–4pm Sat–Sun.*

South

Wireless Hill Park 292 D6

This natural bushland area boasts a magnificent springtime wildflower display. Three lookout towers provide views of

walks & tours

Perth Walking Tours Take a free city orientation tour at 11am Monday to Saturday and noon on Sunday, or learn about Perth's history and culture on a free guided tour at 2pm weekdays. City of Perth Information Kiosk, Murray St Mall (near Forrest Pl).

Two Feet & A Heartbeat Walking Tours Combining Perth's history with quirky tales and emerging culture, these walks include the Perth Urban Adventure and Eat/Drink/Walk Perth… there's even a Perth Shopping Tour. www.twofeet.com.au

Swan River Cruises Choose from full- or half-day cruises up the Swan River to the Swan Valley, with wine tastings included. Captain Cook Cruises, bookings on (08) 9325 3341.

Kings Park Walks Free guided walks, including the Botanic Garden Discovery, Bushland Nature Trail and the Memorials Walk, among others. (08) 9480 3600; 10am and 2pm daily.

Kings Park Indigenous Heritage Tour Learn about bush medicines, bush tucker and Indigenous history in this 1.5-hour tour. Bookings on (08) 9483 1106.

Fremantle Prison Tours Choose from a range of tours including the fascinating 'Doing Time Tour', every 30 minutes between 10am and 5pm daily, and the spooky 'Torchlight Tour' on Wednesdays and Fridays at 7pm. 1 The Terrace, Fremantle; (08) 9336 9200.

His Majesty's Theatre Tours Theatre Historian Ivan King will take you on a gossipy and anecdotal wander through ten decades of colourful show business history on the 2-hour Grand Historical or 1-hour Behind the Scenes tours. (08) 9265 0900.

Rock 'n' Roll Mountain Biking Tours Enjoy a scenic tour of the bushlands of Western Australia on a safe, controlled mountain biking experience. 0410 949 182 or 0428 263 668; www.rockandrollmountainbiking.com.au.

the Swan River and the city skyline. A **Telecommunications Museum** is housed in the original Wireless Station and is open by appointment only. *Almondbury Rd, Ardross; (08) 9364 0158.*

Point Walter Reserve 292 B6

This recreation area on the river is a pleasant spot for a picnic. Pick a shady peppermint tree by the water's edge to sit under and watch children paddling in the river. A kiosk, cafe, free barbecue facilities and children's playground area are nearby. Stroll out on the jetty to see what's biting, and walk out on the huge sandbar for great views both up and down river. *Honour Ave, Bicton.*

Point Heathcote Reserve 292 D5

Cross the Canning River – a large tributary of the Swan River – via Canning Bridge and wind your way through the affluent suburb of Applecross to Point Heathcote Reserve on Duncraig Road. This hilltop playground and parkland area, which includes a local museum, art gallery, restaurant, kiosk and free barbecue facilities, is the best spot south of the Swan for sweeping views of the river and city.

Outer South

Adventure World 615 C6

The state's biggest amusement park offers over 30 thrill rides and water slides, including the Turbo Mountain Rollercoaster, the Power Surge, the Shotgun and the Tunnel of Terror that corkscrews its way down Water Mountain. Nearby Bibra Lake is a great place to see black swans and other waterbirds. *179 Progress Dr, Bibra Lake; (08) 9417 9666; open 10am–5pm Thurs–Mon, daily in Jan and school holidays.*

fremantle

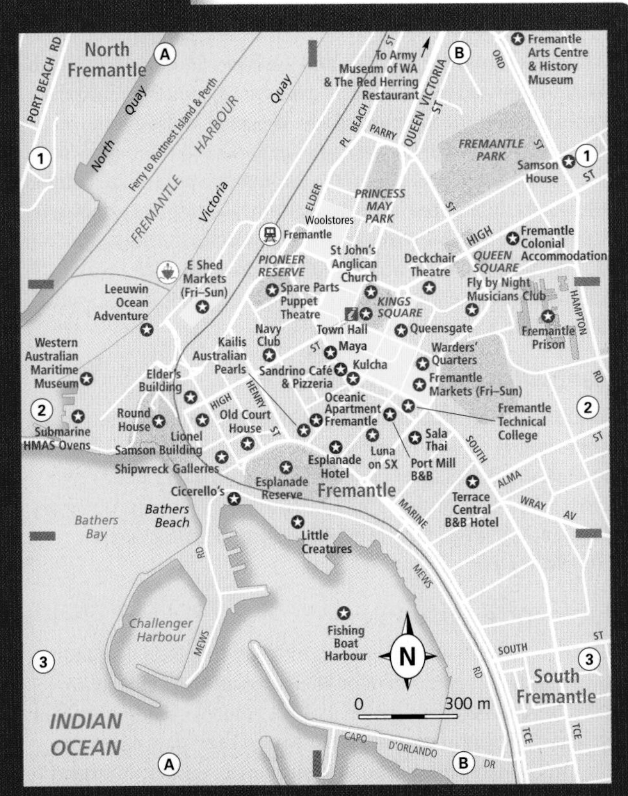

Fremantle

VISITOR INFORMATION

Fremantle Visitor Centre
→ Town Hall, Kings Sq, High St
→ (08) 9431 7878
www.fremantlewa.com.au

Although now linked to Perth by a sprawl of suburbs, Fremantle ('Freo' to the locals) has a feel that is quite different in both architecture and atmosphere. Today it is a major boat and fishing centre at the mouth of the Swan River, but it also has the streetscape of a 19th-century port. It is a place to stay, unwind and watch the world go by. You can shop at the famous Fremantle Markets, or rest at a cafe on **South Terrace** and wait for the arrival of 'the Fremantle Doctor', the refreshing afternoon wind that blows in off the Indian Ocean.

European settlers arrived at Fremantle in 1829. The settlement developed gradually, its existence dependent on whaling and fishing. The population was boosted with the arrival in 1850 of British convicts, who constructed the forbidding Fremantle Prison (now open to the public) and the imposing lunatic asylum, now the Fremantle Arts Centre. Many heritage houses and terraces with cast-iron balconies from this period have survived.

Fremantle was at the centre of the world stage in 1987 when it hosted the America's Cup series of yacht races, following the win by *Australia II* – a Fremantle yacht – in 1983. Preparations for this huge event included the restoration of many old buildings in Fremantle, and the boost to its tourist economy has lasted to the present day.

getting to fremantle

By car A drive of 20–30 minutes, either via Stirling Highway on the north bank of the Swan River or via Canning Highway on the south.

By train A 30-minute journey from Perth Railway Station, Wellington Street. Trains depart every 15 minutes on weekdays and less frequently on weekends.

By bus Many buses and routes link both cities. Timetables and route details from Transperth 13 6213.

By ferry Various ferry operators travel twice daily between Perth and Fremantle, departing Barrack Street Jetty, Perth. *See Getting around, p. 295*

Combined travel packages For combined ferry, train and 'tram' tours of Perth and Fremantle, contact either Fremantle Tram Tours (08) 9433 6674 or Perth Tram Company (08) 9322 2006.

Once you've arrived Look out for the free orange CAT bus, which runs regular services from Victoria Quay, through the city centre and down to South Fremantle. 'Tram' tours are another option, departing hourly between 10am and 5pm from the town hall.

top events in fremantle

Araluen's Fremantle Chilli Festival A weekend celebration of everything chilli. March.

Fremantle Street Arts Festival Local, national and international buskers perform on the streets. Easter.

West Coast Blues 'n Roots Festival Day-long celebration of blues, roots, reggae and rock, with many big-name acts. April.

Heritage Festival Exhibitions, talks, dinners, concerts and film screenings promoting Freo's diverse cultural heritage. May–June.

Blessing of the Fleet Traditional Italian blessing of the fishing fleet. October.

Fremantle Festival Performing arts and community activities culminating in a street parade and dance party. November.

Shipwreck Galleries 301 A2

Part of the Western Australian Maritime Museum, the galleries are one of three sites showcasing Western Australia's maritime history. The state's treacherous coastline has doomed many ships to a watery grave. The Shipwreck Galleries, housed in the restored convict-built Commissariat building, document this chapter of maritime history. The most popular exhibit is the stern of the *Batavia*, wrecked in 1629, which has been reconstructed from recovered timbers. *Cliff St; www.museum.wa.gov.au/oursites/shipwreckgalleries; open 9.30am–5pm Thurs–Tues; admission by donation.*

Grand old buildings

Fremantle has a compact cluster of lovely old buildings, beginning with the Georgian-style **Elder's Building** at 11 Cliff Street. Made of brick and Donnybrook stone, it was once the hub of Fremantle's overseas trade. At 31–35 Cliff Street, the **Lionel Samson Building**'s rich facade epitomises the opulent style of gold-rush architecture. The **Esplanade Hotel**, on the corner of Marine Terrace and Essex Street, dates from the 1890s and its newer extension blends in seamlessly with the original facade. Now the Challenger TAFE e-Tech, the **Fremantle Technical College** displays Art Nouveau decorative touches and boasts Donnybrook stone facings and plinth. It's on the corner of South Terrace and Essex Street. On the corner of Ellen and Ord streets, the grand old **Samson House** dates from 1900 and was originally built for Michael Samson, who later became mayor of Fremantle. Finally, there's the stone bell-tower and large stained-glass window of **St John's Anglican Church**, on the corner of Adelaide and Queen streets, and Henderson Street's **Warders' Quarters**, a row of convict-built cottages built in 1851, and until recently used to house warders from Fremantle Prison. *(See also the Round House and Fremantle Prison, on this page.)*

where to stay in fremantle

Esplanade Hotel Impressive heritage building is centrally located to the town centre and each room is elegantly decorated. Cnr Marine Tce and Essex St; (08) 9432 4000. 301 B2

Fremantle Colonial Accommodation Step back in time with a stay in a two-storey terrace house or a historic prison cottage. 214 High St; (08) 9430 6568. 301 B1

Oceanic Apartment Fremantle Luxurious ground-floor 3-bedroom apartment in the converted 1898 Oceanic Hotel. 3/8 Collie St; 0420 309 030. 301 B2

Port Mill B&B Award-winning ensuite rooms with balconies in converted historic flour mill. 3/17 Essex St; (08) 9433 3832. 301 B2

Terrace Central B&B Hotel Refurbished heritage hotel located on South Terrace's vibrant 'Cappuccino Strip'.79–85 South Tce; (08) 9335 6600. 614 C8

Fremantle Markets 301 B2

One of the city's most popular attractions are these National Trust–classified markets, with their ornate gold-rush era architecture, which were opened in 1897. They offer a diversity of stalls: fresh produce, food, books, clothes, bric-a-brac, pottery and crafts. There is also a great tavern bar where buskers often perform. *Cnr South Tce and Henderson St; (08) 9335 2515; open 9am–8pm Fri, 9am–6pm Sat–Sun.*

The Round House 301 A2

Its name a misnomer, the Round House is actually a dodecahedron, with its 12 sides erected around a central yard. Built in 1831 by the first settlers as a prison, it is the oldest public building in the state. At 1pm each day, the Round House's signal station fires a cannon – the time gun – and this activates a time ball (an instrument once used to give accurate time readings to vessels out at sea). The Whalers' Tunnel underneath the Round House was cut by a whaling company in 1837 for access to Bathers Bay. *10 Arthur Head; (08) 9336 6897; open 10.30am–3.30pm daily; admission by gold coin donation.*

Western Australian Maritime Museum 301 A2

This stunning, nautically inspired building perched on the waterfront has six galleries, each of which explores a different theme in the state's maritime history. Highlights of the collection of historic and significant boats include the yacht that won the America's Cup – *Australia II* – and the *Parry Endeavour*, in which Western Australian Jon Sanders circumnavigated the world three times. Outside the museum are the Welcome Walls, a series of engraved panels that list the names of thousands of migrants who arrived in Western Australia through the port of Fremantle, while next door to the museum is another of its prize exhibits, the submarine HMAS *Ovens*, which was in service during World War II. Take the tour to see what conditions are like inside a real submarine. *Victoria Quay; (08) 9431 8444; open 9.30am–5pm Thurs–Tues.*

Leeuwin Ocean Adventure 301 A2

The *Leeuwin II* – a 55-metre, three-masted barquentine – is rated as the largest ocean-going tall ship in Australia. It is available for half-day or twilight sails and for longer voyages along the coast of Western Australia and the Northern Territory. When in Fremantle, it is open to the public. *B Berth, Victoria Quay; (08) 9430 4105.*

Fremantle Prison 301 B2

The first convicts arrived in Fremantle in 1850. Built with their own hands, this complex of buildings was initially used as a barracks and became a prison in 1867. Huge, forbidding and full of history, it was in use until 1991. Now visitors can experience the atmosphere on a guided tour, running every half-hour and taking in the isolation chamber and the gallows. The entrance can be reached via steps and a walkway around Fremantle Oval from Parry Street. *1 The Terrace; (08) 9336 9200; open 10am–5pm daily; Tunnels Tours daily and Torchlight Tours Wed and Fri (bookings essential).*

Fremantle Arts Centre 301 B1

This magnificent limestone building, with its steeply pitched roofs and Gothic arches, was also built by convicts. The colony's first lunatic asylum, it now offers an interesting display on the history of Fremantle, contemporary art exhibitions, a craft shop with a range of wares made by Western Australian artists, a ghost walk and a garden area with a cafe. *1 Finnerty St; (08) 9432 9550 (event info line) or (08) 9432 9555; open 10am–5pm daily; admission free.*

Fishing Boat Harbour 301 B3

This popular restaurant strip for locals and tourists alike overlooks the boats of the local fishermen. Sample the fresh catch of the day at restaurants such as **Cicerello's**, famous for its fish and chips. On the footpath in front of Cicerello's is a statue of AC/DC's Bon Scott, unveiled in October 2008. Across the railway line is the **Esplanade Reserve**, an ideal spot to relax under the giant Norfolk Pines.

where to eat in fremantle

Cicerello's The epitome of fish and chip dining on Fremantle's Fishing Boat Harbour. 2/1 Howard St; (08) 9335 1911; open daily for lunch and dinner. 301 A2

Little Creatures Harbour-side brewery, bar and restaurant serves hearty fare to its enormous following. 40 Mews Rd; (08) 9430 5555; open daily for lunch and dinner, and Sun for breakfast. 301 A2

Sandrino Café and Pizzeria Rated 'the best pizzas in town' in the 2011 *Good Food Guide*. 95 Market St; (08) 9335 4487; open 11am–9.30pm daily. 301 B2

Sala Thai A local favourite, serves Thai cuisine comparable to its country of origin. 22 Norfolk St; (08) 9335 7749; open daily for dinner. 301 B2

The Red Herring Restaurant Up-market seafood restaurant boasts a magical setting right on the river. 26 Riverside Rd, East Fremantle; (08) 9339 1611; open daily for lunch and dinner, and Sun for breakfast.

Maya Widely considered to be the best Indian food in town. 75 Market St; (08) 9335 2796; open Fri for lunch, Tues–Sun for dinner. 301 B2

day tours

Rottnest Island Just off the coast of Perth, the low-key island resort of Rottnest makes for a perfect day tour. Access is via ferry from Fremantle, Perth or Hillarys Boat Harbour. No private cars are permitted: island transport is by foot, bicycle or bus. Visitors to Rottnest can divide their time between the beach and the scenic and historic attractions of the island. For more details see p. 305

Darling Range Follow the Great Eastern Highway for a tour of the Darling Range and its 80 000 hectares of escarpment and jarrah forest in the Hills Forest area. Highlights include a scenic drive through John Forrest National Park and a visit to the huge, forest-fringed Mundaring Weir. For more details see p. 306

Swan Valley A premier winegrowing district, with vineyards along the scenic Swan River. Other attractions include the historic town of Guildford; Woodbridge House, a Victorian mansion in West Midland; Walyunga National Park; and Whiteman Park, a 2500-hectare area that includes Caversham Wildlife Park. For more details see p. 306

Yanchep National Park On the coast north of Perth, Yanchep has long been one of the city's favourite recreation areas. Have your photo taken with a koala; see didgeridoo and dance performances; or take a guided tour of Crystal Cave, where stalactites hang above the inky waters of an underground pool.

[WESTERN AUSTRALIAN MARITIME MUSEUM]

REGIONS
western australia

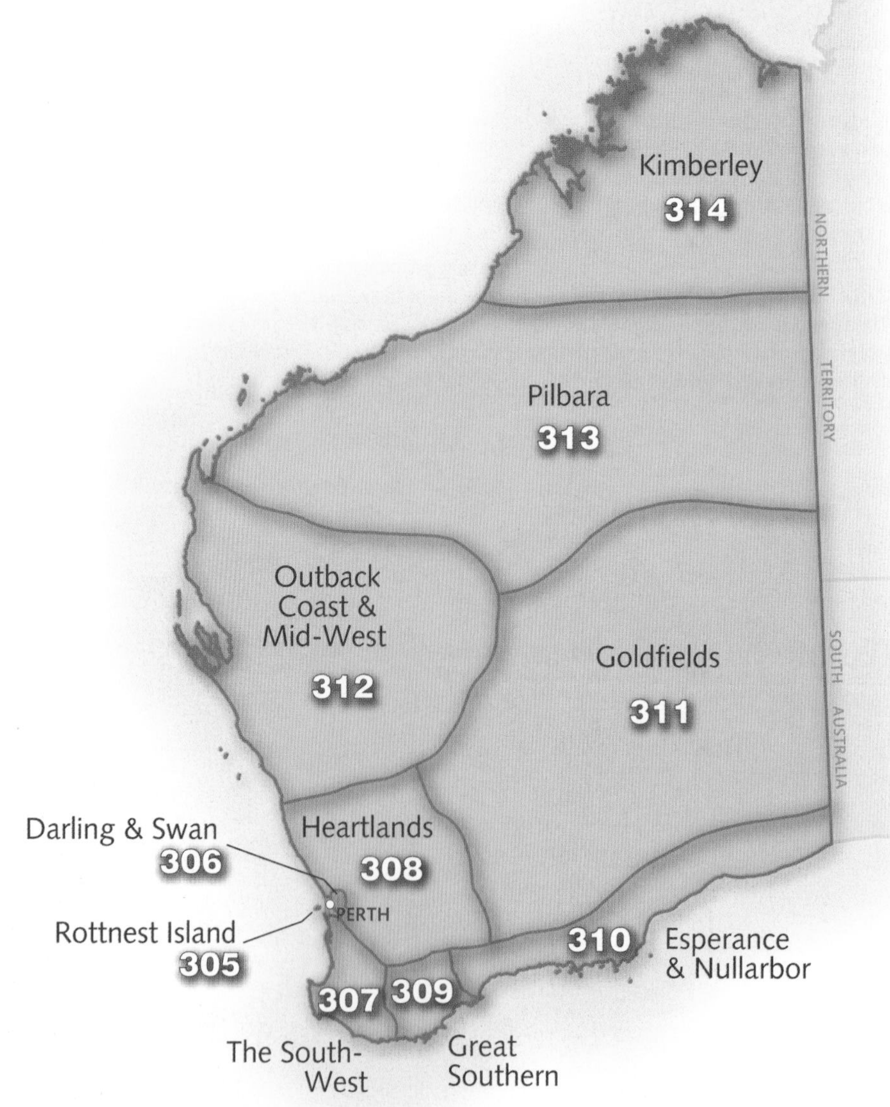

ROTTNEST ISLAND

In azure waters only 18 kilometres from Perth, Rottnest Island's white, sandy beaches are perfect for a range of aquatic pleasures. The island is also home to the famous quokka.

[ROTTNEST ISLAND QUOKKA]

The Basin An outer reef surrounds Rottnest, protecting the clear waters and creating calm conditions for family swimming. The Basin is one of a number of beautiful sandy beaches on the eastern end of the island. It is within easy walking distance of Thomson Bay and has basic facilities.

Underwater wonderland The diversity of fish and coral species, and the numerous shipwrecks found around the island, make Rottnest a favourite site for scubadivers and snorkellers. Dive charters and snorkelling tours are popular; if you prefer not to get wet, enjoy the underwater scenery aboard the glass-bottomed Underwater Explorer.

Quokka country The quokka is a native marsupial found primarily on Rottnest Island. It is semi-nocturnal and furry, and grows to the size of a hare. Find the interpretive signs about a kilometre south of Thomson Bay, just before Kingstown Barracks; if you don't see one of the 10 000 quokkas on the island here, then there are good viewing spots along the boardwalk at Garden Lake.

West End The 'West End' of Rottnest can be reached on an 11-kilometre bike ride along a sealed road, or on a bus tour. There are stunning ocean views from Cape Vlamingh (where you may also spot a humpback whale in winter) and a 1-kilometre heritage trail that affords sightings of wedge-tailed shearwaters, fairy terns, quokkas and bottlenose dolphins.

TOP EVENTS

FEB	Rottnest Channel Swim
MAR	The Big Splash (watersports race, Fremantle–Rottnest Island)
JUNE	The State Open Surfing Series
OCT	Marathon and Fun Run
DEC	Rottnest Swim Thru

CLIMATE

ROTTNEST ISLAND

J	F	M	A	M	J	J	A	S	O	N	D	
26	27	25	23	20	18	17	17	18	20	22	24	MAX °C
18	19	18	16	14	13	12	12	12	13	15	17	MIN °C
7	13	14	37	106	156	149	104	61	39	17	10	RAIN MM
2	2	4	8	15	18	20	18	14	11	6	3	RAIN DAYS

ISLAND HERITAGE

Known as Wadjemup to the Nyungar people, Rottnest Island was unoccupied when Europeans arrived, although there is evidence of Aboriginal occupation around 7000 years ago when the island was linked to the mainland. The island was given the unflattering name Rotte-nest, meaning 'rats' nest', in 1696 by explorer Willem de Vlamingh, who mistook the island's quokkas for large rats. Europeans settled on the island in 1831. From 1838 to 1903 it was used as a prison for Aboriginal people. During World War I it became an internment camp and in 1917 it was declared an A-class reserve. World War II saw it used as a military post. Many heritage sites can be found, including an original 1840s streetscape – Thomson Bay's Vincent Way. Other interesting sites are the Chapel (1858), the octagonal prison building known as the Quad (1864), the Oliver Hill Gun Battery (1930s) and Rottnest Lighthouse (1859) on Wadjemup Hill.

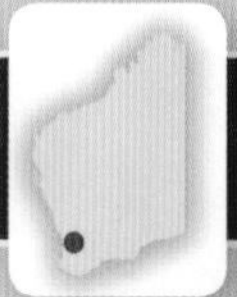

DARLING & SWAN

Barely half an hour from the centre of Perth are two distinct country landscapes perfect for picnicking, walking, sampling local produce and looking for wildflowers.

[SWAN VALLEY VINEYARD]

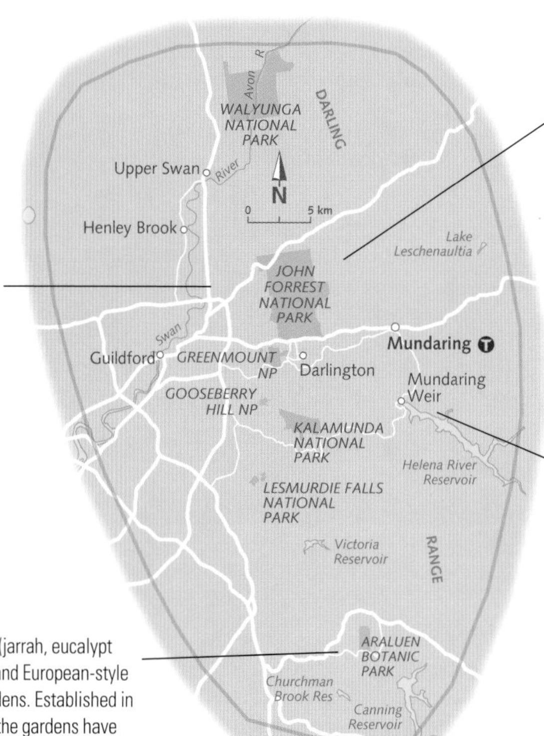

Swan Valley wineries The Swan Valley region is renowned for its chardonnay, shiraz, chenin blanc and verdelho. It boasts over 40 wineries; of these, Houghton and Sandalford are the best known. For a treat that offers scenery as well as wine-tasting, take a Swan River wineries cruise from Perth.

John Forrest National Park Declared a national park in 1947, John Forrest is one of the oldest and best-loved picnic spots in the Perth Hills. A drive through the park has vantage points with superb views across Perth and the coastal plain. A popular walk is the Heritage Trail on the western edge, past waterfalls and an old rail tunnel.

Mundaring Weir The rolling lawns and bush-clad surrounds of this weir reserve make it ideal for picnics. At the foot of the weir, the Number 1 Pump Station commemorates C. Y. O'Connor's extraordinary engineering feat in piping water to the goldfields some 600 kilometres east. Nearby, at the Hills Forest Discovery Centre, you can sign up for nature-based activities such as bushcraft.

Araluen Botanic Park Tall forest trees (jarrah, eucalypt and marri) frame the rockpools, cascades and European-style terraces of these beautiful 59-hectare gardens. Established in the 1930s by the Young Australia League, the gardens have picturesque walking trails, picnic and barbecue areas, and, in spring, magnificent tulip displays.

TOP EVENTS

MAR–APR	Taste of the Valley (food, wine and art festival, throughout Swan Valley)
AUG	Avon Descent (whitewater race down Avon and Swan rivers, from Northam to Perth)
OCT	Spring in the Valley (wine festival, throughout Swan Valley)
NOV	Darlington Arts Festival (Darlington)

CLIMATE

GUILDFORD

J	F	M	A	M	J	J	A	S	O	N	D	
32	32	30	27	22	19	18	19	21	23	27	30	MAX °C
17	17	15	13	10	8	7	8	9	10	13	15	MIN °C
8	10	17	43	122	177	172	139	86	56	20	13	RAIN MM
2	2	4	7	14	17	19	17	14	11	6	4	RAIN DAYS

HERITAGE SITES

In historic Guildford, roam among yesteryear's farm tools, fashions and household items at the Old Courthouse, Gaol and Museum. Then enjoy a drink at the nearby Rose and Crown Hotel (1841), the oldest trading hotel in the state. The faithfully restored Woodbridge House (1885), picturesquely located on the banks of the Swan River, is a fine example of late-Victorian architecture. And just south of Upper Swan, at Henley Brook on the western bank of the river, All Saints Church (1839–41) is the state's oldest church.

→ For more detail see maps 614, 615, 618 & 620. For descriptions of Ⓣ towns see Towns from A–Z (p. 315).

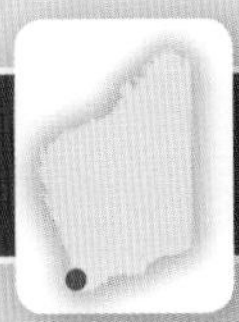

THE SOUTH-WEST

This region is renowned for its world-class wines, excellent surf breaks, towering old-growth forests and the 963-kilometre Bibbulmun Track.

[MEELUP BEACH, DUNSBOROUGH]

Swim with dolphins In Bunbury's Koombana Bay you can swim under ranger guidance with wild bottlenose dolphins. Learn about the dolphins and other marine life at the Dolphin Discovery Centre's interpretive museum and theatre.

Busselton Jetty Stretching a graceful 2 kilometres into Geographe Bay, this wooden jetty is the longest in the Southern Hemisphere. Built in 1865, the jetty originally serviced American whaling ships. Today a small tourist train takes you from the Interpretive Centre at one end to the Underwater Observatory at the other.

Lake Cave One of hundreds of limestone caves in Leeuwin–Naturaliste National Park, magnificent Lake Cave is centred on the incredibly still, eerie waters of an underground lake, which reflects the cave's stunning crystal formations. Its other highlight is the Suspended Table, a huge column hanging precariously from the cave's ceiling.

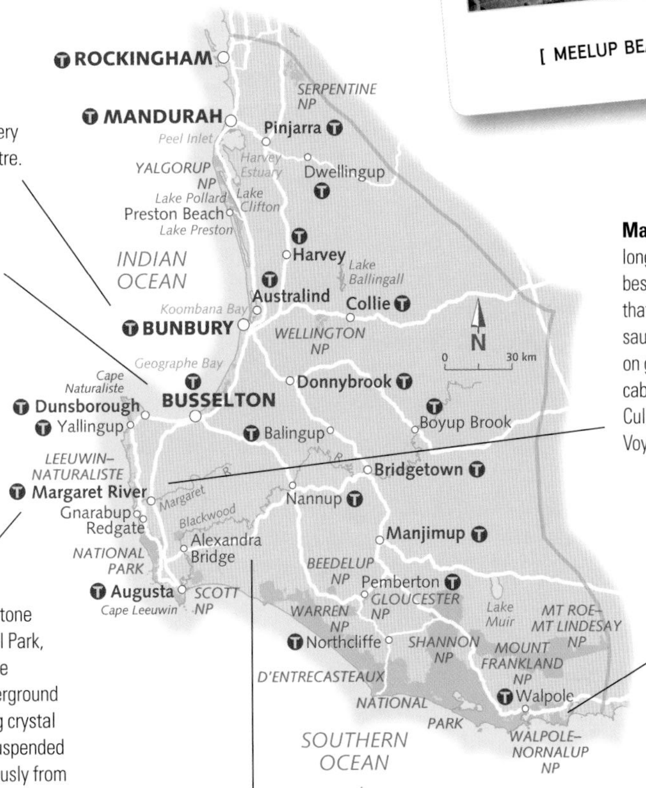

Margaret River This beautiful area has long had a reputation as one of Australia's best wine-producing regions, a reputation that rests principally on its cabernet sauvignon and chardonnay grapes grown on grey-brown, gravelly sandy soils. Try the cabernet from Vasse Felix, Moss Wood and Cullen, and the chardonnay from Leeuwin, Voyager and Ashbrook.

Valley of the Giants Gain a unique perspective on the majestic southern forests as you walk through the upper branches of giant tingle and karri trees. East of Walpole, this walkway, 38 metres above the forest floor, is one of the highest and longest of its kind in the world.

Blackwood Valley The Blackwood River meanders for 500 kilometres through wheat-belt plains and forested valleys to its broad estuary at Augusta. Secluded spots along the river between Nannup and Alexandra Bridge offer tranquil camping, fishing, swimming and canoeing. The Sheoak Walk is a one-hour loop through the forest close to Nannup.

TOP EVENTS

FEB–MAR	Leeuwin Estate Concert (near Margaret River)
MAR–APR	Margaret River Pro (surfing competition)
MAY	Margaret River Wine Region Festival (Margaret River)
NOV	Blues at Bridgetown Festival (Bridgetown)
DEC	Ironman Western Australia Triathlon (Busselton)
DEC	Cherry Harmony Festival (Manjimup)

CLIMATE

BUSSELTON

J	F	M	A	M	J	J	A	S	O	N	D	
29	28	26	23	19	17	16	17	18	20	24	27	MAX °C
14	14	13	11	9	8	8	8	8	9	11	13	MIN °C
10	11	22	42	118	175	167	117	75	52	24	13	RAIN MM
3	2	4	8	15	19	22	19	16	13	7	4	RAIN DAYS

JARRAH, KARRI, TUART AND TINGLE

Western Australia's only forests are in the cool, well-watered South-West. The grey-barked tuart tree grows only on coastal limestone; just outside Busselton is the largest remaining pure tuart forest in the world. Jarrah, a beautifully grained, deep-red hardwood, flourishes between Dwellingup and Collie. The Forest Heritage Centre at Dwellingup has interpretive displays on the issues of forest management, and a treetop walk. Forests of karri – one of the world's tallest trees, that can reach 90 metres in 100 years – are found in the wetter areas, from Manjimup to Walpole. Near Pemberton, 4000 hectares of old-growth karri forest are protected by Warren and Beedelup national parks. In Gloucester National Park, right outside the town, is the 61-metre Gloucester Tree, with a spiral ladder to the top. The Valley of the Giants, east of Walpole, is home to towering red tingle trees. Nearby is the 25-metre-wide Giant Red Tingle, a huge fire-hollowed tree that is regarded as one of the ten largest living things on the planet.

→ For more detail see maps 614, 615, 616, 617, 618 & 620. For descriptions of towns see Towns from A–Z (p. 315).

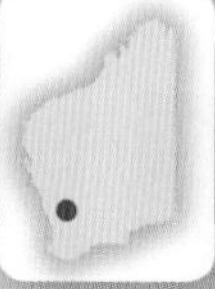

HEARTLANDS

[THE PINNACLES, NAMBUNG NATIONAL PARK]

Dominated by the wheat belt, this region features the Avon Valley, New Norcia and the extraordinary Pinnacles Desert.

The Pinnacles Thousands of limestone pillars, the eroded remnants of what was once a thick bed of limestone, create a weirdly beautiful landscape in Nambung National Park. Other park attractions include a beautiful coastline with superb beaches where visitors can fish, swim, snorkel, walk or picnic.

New Norcia This town, with its Spanish Colonial architecture, was built in 1846 by Benedictine monks who established a mission for the local Indigenous population. It remains Australia's only monastic town. Visitors can tour the buildings and visit the fascinating museum and art gallery.

Avon Valley In the 1860s, bushranger Moondyne Joe hid in the forests, caves and wildflower fields of this lush valley. Now Avon Valley National Park preserves much of the landscape. In the valley's heart, and marking the start of the wheat belt, are the historic towns of Northam and York.

Wave Rock and Mulka's Cave Wave Rock, east of Hyden, is a 2.7-billion-year-old piece of granite, 15 metres high and 100 metres long. It looks like a giant wave frozen at the moment of breaking and has vertical bands of colour created by algal growth. To the north is Mulka's Cave, decorated with Aboriginal rock art.

TOP EVENTS

JAN	Lancelin Ocean Classic (windsurfing competition)
MAY	Moondyne Festival (Toodyay)
AUG	Avon Descent (whitewater race down Avon and Swan rivers, from Northam to Perth)
SEPT–OCT	Jazz Festival (York)
OCT	Kulin Bush Races (Kulin)
NOV	Blessing of the Fleet (Jurien Bay)

CLIMATE

NORTHAM

J	F	M	A	M	J	J	A	S	O	N	D	
34	34	31	26	21	18	17	18	20	24	28	32	MAX °C
17	17	15	12	9	7	5	6	7	9	12	15	MIN °C
10	13	19	23	57	83	84	62	37	25	12	9	RAIN MM
2	2	3	6	11	15	16	14	11	7	4	2	RAIN DAYS

WILDFLOWERS

Western Australia is home to some 12 500 kinds of wildflowers, and the Heartlands is one of the most accessible areas to see magnificent wildflower displays. Top spots include the Chittering Valley near Gingin, Lesueur and Badgingarra national parks near Jurien Bay, and the sand plains around Southern Cross. Up to 20 species of native orchids flourish around Hyden in spring; yellow wattles light up the countryside around Dalwallinu; and massed displays of white, pink and yellow everlastings line the road from Moora to Wubin.

→ For more detail see maps 614, 615, 618 & 620. For descriptions of ⓘ towns see Towns from A–Z (p. 315).

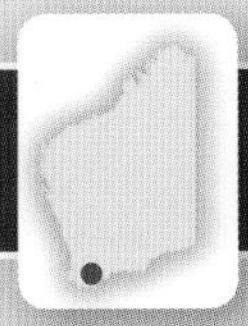

GREAT SOUTHERN

This varied region encompasses sheep country, parts of the wheat belt, wineries, surf beaches and the historic town of Albany.

[LILY WINDMILL, STIRLING RANGE]

Great Southern wineries This region has over 40 wineries, extending from Frankland east through Mount Barker to Porongurup. Since the first plantings in 1967, the region has gained an international reputation for its aromatic rieslings. More recently, premium chardonnay, shiraz, cabernet sauvignon and pinot noir are also proving popular. Many of the wineries are open for tastings and sales.

Stirling Range National Park The Stirling Range rises abruptly above the surrounding plains; its rock-faces are composed of the sands and silts laid down in the delta of an ancient river. The highest peaks can be veiled in swirling mists, which creates a cool and humid environment that supports a proliferation of flowering plants. Bluff Knoll, at 1073 metres, is one of the state's premier hiking challenges.

The Old Farm at Strawberry Hill The Old Farm, near Albany, is a delightful stone cottage on Western Australia's oldest cultivated farm. Originally home to the first government resident in the state, the 1836 building at Strawberry Hill, as it was known, was the centre of Albany's social life in its day. These days the property welcomes swarms of visitors to its picturesque heritage buildings and stunning gardens.

Albany Whaleworld The bloody realities of whaling are displayed at the old Cheynes Beach Whaling Station on Frenchman Bay, 25 kilometres south-east of Albany. Visitors can explore *Cheynes IV*, a restored whalechaser, and relive the sights and sounds of the hunt. A 3D theatrette occupies one of the old whale-oil storage tanks.

Torndirrup National Park Located 15 minutes from Albany, this park features the Blowholes, the Gap and Natural Bridge, all sculpted into their current form by the treacherous Southern Ocean. Granite outcrops and cliffs alternate with dunes, and sandy heath supports peppermint, banksia and karri. Easy walking trails take in some of the spectacular sights.

TOP EVENTS

JAN	Mount Barker D'Vine Wine Festival (Mount Barker)
JAN	Vintage Blues Festival (Albany)
MAR	Wine Summer Festival (Porongurup)
EASTER	Brave New Works (new performance art, Denmark)
SEPT	Wildflower Display (Cranbrook)
SEPT–OCT	Great Southern Wine Festival (Albany)

CLIMATE

ALBANY

J	F	M	A	M	J	J	A	S	O	N	D	
25	25	24	22	19	17	16	16	17	19	21	24	MAX °C
14	14	13	12	10	8	8	7	8	9	11	12	MIN °C
27	24	28	63	102	103	124	106	82	78	48	25	RAIN MM
8	9	11	14	18	19	21	21	18	15	13	10	RAIN DAYS

ALBANY HERITAGE

Albany, the oldest white settlement in Western Australia, was officially founded on 21 January 1827. The founders were a party of soldiers and convicts, under the command of Major Edmund Lockyer, who had arrived on the *Amity* a month earlier. Albany's magnificent harbour, once commanding the sea lanes running between Europe, Asia and eastern Australia, became a whaling station and later a coaling port for steamships. Museums now occupy several of the town's historic buildings, and a full-size replica of the *Amity* stands next to one of them, the Residency Museum. Stirling Terrace has some evocative Victorian shopfronts, while Princess Royal Fortress on Mount Adelaide has restored buildings and gun emplacements, and offers fine views.

→ For more detail see maps 617, 618 & 620. For descriptions of towns see Towns from A–Z (p. 315).

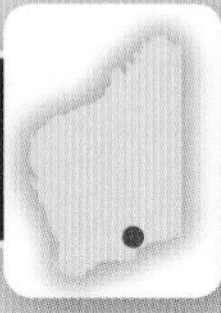

ESPERANCE & NULLARBOR

Esperance's beaches are famed for their white sand and turquoise waters, while to the north-east lies the vast Nullarbor Plain.

[TURQUOISE WATERS AND WHITE SAND, CAPE ARID NATIONAL PARK]

90-mile straight Four kilometres west of Caiguna, be sure to have your photo taken beside the signpost marking the eastern end of the longest straight stretch of road in Australia, which runs for 90 miles (146.6 kilometres) between Caiguna and a point east of Balladonia.

Cape Le Grand National Park Swimming beaches, sheltered coves, heathlands, sand plains and the Whistling Rock are all features of this park, 56 kilometres east of Esperance. There are easy walking trails and two camping areas. Scenic spots include Thistle Cove, Hellfire Bay and Lucky Bay (where luck might have you spot a kangaroo on the beach).

Eucla This isolated outpost was established in 1877 as a telegraph station; today coastal dunes partially obscure the station's ruins. On the beach, a lonely jetty stretches out into startlingly blue waters, and in Eucla National Park you can wander among the rippling Delisser Sandhills.

Great Ocean Drive A 38-kilometre circuit drive explores the coast west of Esperance. Attractions include Australia's first wind farm, sheltered swimming at Twilight Cove, and Pink Lake, rendered lipstick-colour by algae. There are coastal lookouts, and sightings of southern right whales from June to October.

Archipelago of the Recherche This group of 105 granite islands and 1500 islets stretches for 250 kilometres along the Esperance coast. Boat tours, available from Esperance, may provide sightings of fur seals, sea lions, dolphins and, in season, southern right whales. Visitors can stay overnight in safari huts or camp on Woody Island.

TOP EVENTS

JAN	Summer Festival (Hopetoun)
SEPT	Wildflower Show (Ravensthorpe)
SEPT	Wildflower Show (Esperance)
OCT	Agricultural Show (Esperance)

CLIMATE

ESPERANCE

J	F	M	A	M	J	J	A	S	O	N	D	
26	26	25	23	20	18	17	18	19	21	23	25	MAX °C
16	16	15	13	11	9	8	9	10	11	13	14	MIN °C
22	27	31	43	76	82	98	84	58	50	36	17	RAIN MM
6	6	8	11	14	16	17	17	14	12	10	7	RAIN DAYS

THE NULLARBOR

The Nullarbor is one of the country's essential touring experiences. This 250 000-square-kilometre treeless limestone slab was initially part of the seabed, and has been formed in part by deposits of marine fossils. The terrain is riddled with sinkholes, caverns and caves, only some of which are open to the public. Murrawijinie Caves include Koonalda Cave, which is open to the public and contains rock art that dates back 20 000 years. Although it can seem featureless, the country is far from monotonous, particularly where the highway veers to the coast for a view of dramatic cliffs and the wild Southern Ocean, and perhaps a lucky sighting of migrating southern right whales.

→ For more detail see maps 620–1. For descriptions of T towns see Towns from A–Z (p. 315).

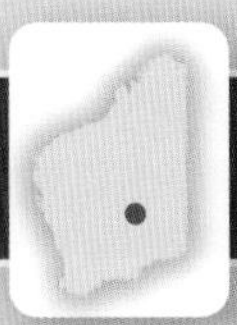

GOLDFIELDS

[GWALIA MINE SITE]

Gold continues to be mined in this region, while sheep stations the size of small nations produce fine wool. The area offers an excellent opportunity to delve into its fascinating heritage.

North of Kalgoorlie The town of Menzies has 130 people and several intact old buildings, while Kookynie has retained its spacious 1894 Grand Hotel. Gwalia, almost a ghost town, has a museum, the restored State Hotel and tin houses preserved in their lived-in state. Laverton, 100 kilometres east, has historic buildings saved by the nickel industry.

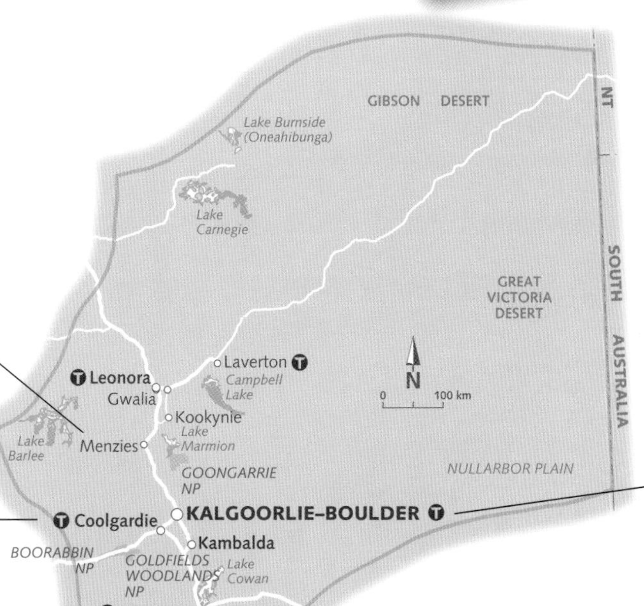

Coolgardie The 100-plus street markers scattered through Coolgardie are a good introduction to this town, the first settlement in the eastern goldfields. Coolgardie's splendid historic buildings include the Marble Bar Hotel, now the RSL, and the 1898 Warden's Court. The latter is an architectural treasure and houses the comprehensive Goldfields Exhibition Museum and the visitor centre.

Kalgoorlie–Boulder
Kalgoorlie–Boulder produces half of Australia's gold. Take a circuit of the open-cut Superpit on the Loopline Railway, see the gold vault at the Museum of the Goldfields, and sample the rigours of 1890s mining at the Australian Prospectors and Miners Hall of Fame.

Peak Charles National Park In this park south-west of Norseman, granite mountains rise in wave-cut platforms to a height of 651 metres. Walking trails to the summit of Peak Charles and its twin, Peak Eleanora, should only be attempted in favourable weather, but at the top are fantastic views across saltpans, sand plains and dry woodlands.

TOP EVENTS

FEB	Undies 500 Car Rally (Kalgoorlie–Boulder)
MAR	Norseman Cup (horseracing, Norseman)
MAY–JUNE	Leonora Golden Gift and Festival (Leonora)
SEPT	Race Round (horseracing, Kalgoorlie–Boulder)
SEPT–OCT	Metal Detecting Championships (Coolgardie, odd-numbered years)
OCT	Balzano Barrow Race (Kanowna to Kalgoorlie–Boulder)

CLIMATE

KALGOORLIE–BOULDER

J	F	M	A	M	J	J	A	S	O	N	D	
34	32	30	25	20	18	17	18	22	26	29	32	MAX °C
18	18	16	12	8	6	5	5	8	11	14	17	MIN °C
22	28	19	19	28	31	26	20	15	16	18	15	RAIN MM
3	4	4	5	7	8	9	7	5	4	4	3	RAIN DAYS

GOLDFIELDS HISTORY

The discovery of gold in the region in 1892 secured the economic success of Western Australia. Since then, goldmines from Norseman to Laverton have yielded well over 1000 tonnes. A railway from Perth in 1896 and a water pipeline in 1903 helped Kalgoorlie and Boulder sustain a population of 30 000, the liquor requirements of which were met by 93 hotels. By 1900, surface gold was exhausted and big companies went underground. Exhausted mines have left a belt of ghost towns north of Kalgoorlie, while nickel mining since the 1960s has allowed towns such as Kambalda, Leonora and Laverton to survive.

→ For more detail see maps 620–1 & 623. For descriptions of 🅣 towns see Towns from A–Z (p. 315).

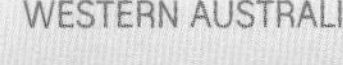

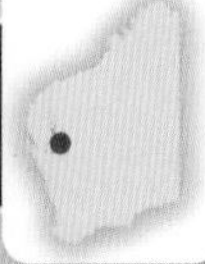

OUTBACK COAST & MID-WEST

This vast area is known for the richness and rarity of its natural features, from colourful Ningaloo Reef to the eroded cliffs and gorges of Kalbarri National Park.

[AN UNDERWATER LANDSCAPE OF FISH AND CORAL, NINGALOO REEF]

Ningaloo Reef Ningaloo Marine Park protects the state's largest reef, a stunning underwater landscape of fish and coral located directly off the beach. For a quintessential Ningaloo experience, take a swim with the whale sharks – this is one of the few places in the world where they come close to shore. These gentle giants migrate to the reef between March and June; tours can be arranged in Exmouth.

Monkey Mia The wild bottlenose dolphins of Monkey Mia are world famous for their daily morning ritual of swimming into the shallows to be handfed with fish. Under the guidance of rangers, visitors can wander into the water and witness this rare event. For a total marine encounter, dugong-watching cruises can also be arranged from here.

Kalbarri National Park This park is best known for its 80 kilometres of gorges carved out by the Murchison River. Watersports are popular on the river's lower reaches. The park is also one of the world's richest wildflower areas. Dolphins, whale sharks and whales frequent the coastal waters, and the fishing is excellent.

Geraldton Geralton is surrounded by superb swimming and surfing beaches, but if you prefer architecture to aquatics follow the Hawes Heritage Trail, which highlights the remarkable church buildings (1915–39) of architect-priest Monsignor John Cyril Hawes. In Geraldton, the Byzantine-styled St Francis Xavier Cathedral is considered to be one of Hawes' masterpieces.

TOP EVENTS

MAR	Sport Fishing Classic (Kalbarri)
MAR	Gamex (gamefishing competition, Exmouth)
APR–MAY	Whale Shark Festival (Exmouth)
AUG–SEPT	Mullewa Wildflower Show (Mullewa)
OCT	Airing of the Quilts (Northampton)
OCT	Sunshine Festival (Geraldton)

CLIMATE

CARNARVON

J	F	M	A	M	J	J	A	S	O	N	D	
31	33	31	29	26	23	22	23	24	26	27	29	MAX °C
22	23	22	19	15	12	11	12	14	16	19	21	MIN °C
12	21	16	14	38	48	47	19	6	6	4	2	RAIN MM
2	3	2	3	5	7	7	5	3	2	1	1	RAIN DAYS

SHARK BAY

World Heritage–listed Shark Bay is a sunny paradise of bays, inlets and shallow azure waters, blessed with a great number of unusual features. It boasts the world's most diverse and abundant examples of stromatolites – the world's oldest living fossils – which dot the shores of Hamelin Pool in rocky lumps. The bay region supports the largest number and greatest area of seagrass species in the world. Covering 4000 square kilometres, these vast underwater meadows are home to around 14 000 dugongs, 10 per cent of the world's total number. The bay's extraordinary marine population also includes humpback whales resting on their long migrations, manta rays, green and loggerhead turtles and, most famously, the dolphins that regularly visit Monkey Mia.

→ For more detail see maps 619, 620 & 622. For descriptions of towns see Towns from A–Z (p. 315).

PILBARA

Home to the ochre-hued Hamersley Range, this region is characterised by rust-red landscapes and vast tidal flats broken by mangroves.

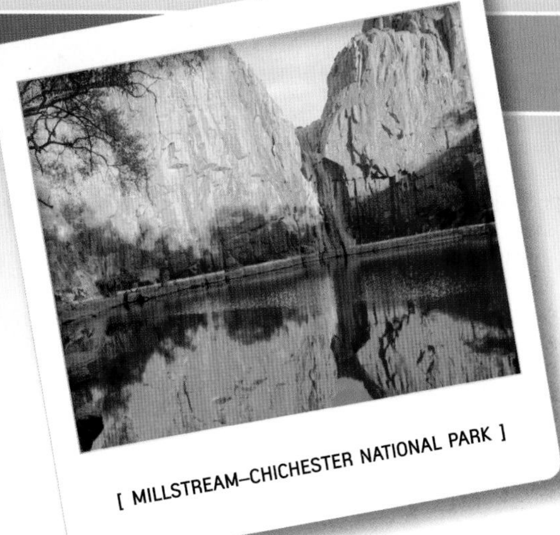

[MILLSTREAM–CHICHESTER NATIONAL PARK]

Pilbara islands The Dampier Archipelago and the Montebello and Mackerel islands seem a far-flung beach paradise from the industrial ports of the Pilbara. In reality, the nearest islands – those of the Dampier Archipelago – are just 20 minutes by boat from Dampier. These islands are a haven for marine life, including turtles, dolphins and migrating humpback whales, and are a renowned location for fishing.

Cossack This first port in the north-west was built between 1870 and 1898, and is now a ghost town. Many buildings have been restored. The old post office houses a gallery and the courthouse has a museum, while the police barracks offer budget accommodation.

Marble Bar The mining town of Marble Bar has gained the dubious reputation as the hottest town in Australia. For 161 consecutive days in 1923–24, the temperature in town did not drop below 37.8° Celsius. The heart of the town is the Ironclad Hotel, a classic outback pub where you can enjoy a counter meal and that cold beer you're going to need. You can also experience Marble Bar's goldmining past at Comet Gold Mine, a museum and tourist centre.

Millstream–Chichester National Park Located within the arid beauty of Millstream–Chichester National Park, spring-fed Chinderwarriner Pool has an almost mirage-like quality. Encircled by remnant rainforest, the oasis is a haven for a range of plants and animals, many of them rare. Visit historic Millstream homestead, and take a refreshing dip in the pool.

Karijini National Park Karijini was the name given to this area by the original inhabitants, the Banjima. It is renowned for extraordinary gorges, multicoloured walls, and hidden pools and waterfalls. Brilliant wildflowers carpet the rust-red hills in spring. Camping is available inside the park.

TOP EVENTS

MAR	Campdraft and Rodeo (Newman)
MAY	Welcome to Hedland Night (Port Hedland)
JULY	Roebourne Cup (horseracing, Roebourne)
AUG	FeNaCING Festival (Karratha)
SEPT	Pilbara Music Festival (Port Hedland)

CLIMATE

ROEBOURNE

J	F	M	A	M	J	J	A	S	O	N	D	
39	38	38	35	30	27	27	29	32	35	38	39	MAX °C
26	26	25	22	18	15	14	15	17	20	23	25	MIN °C
59	67	63	30	29	30	14	5	1	1	1	10	RAIN MM
3	5	3	1	3	3	2	1	0	0	0	1	RAIN DAYS

HAMERSLEY RANGE RESOURCES

One of the world's richest deposits of iron ore was discovered in Hamersley Range in 1962, spearheading the Hamersley Iron Project. Towns with swimming pools, gardens and golf courses then sprang up in this landscape of mulga scrub, spinifex and red mountains. Visitors can inspect open-cut mines at Tom Price and Newman. At Dampier and Port Hedland there are iron-ore shipping ports; the latter boasts the largest iron-ore export centre in Australia. Offshore from Karratha is the massive North West Shelf Gas Project; a visitor centre at the processing plant on Burrup Peninsula explains its operations.

→ For more detail see maps 619, 622, 624 & 626–7. For descriptions of T towns see Towns from A–Z (p. 315).

KIMBERLEY

[BUNGLE BUNGLE RANGE, PURNULULU NATIONAL PARK]

Covering more than 420 000 square kilometres, this is an ancient landscape of mighty ranges, spectacular gorges and arid desert.

Gibb River Road One of Australia's premier four-wheel-drive destinations, the Gibb River Road is a 649-kilometre outback adventure between Derby and Kununurra, traversing some of the most spectacular gorge country of the Kimberley. Highlights of the rugged and diverse landscape include the many gorges, where waterfalls and crystal-clear pools are fringed by palms and pandanus.

Port of pearls Broome's attractions include its tropical climate, cosmopolitan character and world-famous pearling industry. Enjoy a camel ride at sunset along Cable Beach, renowned as one of the most beautiful beaches in the world.

Lake Argyle Lake Argyle was formed in the 1960s as part of the Ord River Scheme, the success of which is evident in the lush crops within its irrigation area. It has transformed a dusty, million-acre cattle station into a habitat for waterbirds, fish and crocodiles; the hills and ridges of the former station have become islands.

Purnululu National Park A rough track off the Great Northern Highway leads to the spectacular Bungle Bungle Range in Purnululu National Park, on the Ord River. A fantastic landscape of huge black-and-orange sandstone domes is intersected by narrow, palm-lined gorges where pools reflect sunlight off sheer walls.

Geikie Gorge The Fitzroy River cuts through the Geikie Range to create a 7-kilometre gorge just north-east of Fitzroy Crossing. The riverbanks are inhabited by freshwater crocodiles, fruit bats and many bird species, and the only way to see the gorge is by boat – during the Dry.

Wolfe Creek Crater Two hours south of Halls Creek by unsealed road is the world's second largest meteorite crater. It is 850 metres across and was probably formed by a meteorite weighing at least several thousand tonnes crashing to earth a million years ago. It is most impressive from the air; scenic flights run from Halls Creek.

TOP EVENTS

APR–MAY	King Tide Day (celebrating Australia's highest tide, Derby)
MAY	Ord Valley Muster (Kununurra)
JULY	Boab Festival (mardi gras, mud football and Stockmen and Bushies weekend, Derby)
JULY	Rodeo (Fitzroy Crossing)
AUG	Opera Under the Stars (Broome)
AUG–SEPT	Shinju Matsuri (Festival of the Pearl, Broome)

CLIMATE

HALLS CREEK

J	F	M	A	M	J	J	A	S	O	N	D	
37	36	36	34	30	27	27	30	34	37	38	38	MAX °C
24	24	23	20	17	14	13	15	19	23	25	25	MIN °C
153	137	74	22	13	5	6	2	4	17	37	77	RAIN MM
13	13	8	3	2	1	1	1	1	3	6	11	RAIN DAYS

ABORIGINAL ART

The Kimberley is one of Australia's most important regions for Aboriginal rock art. It is renowned for two styles – the Bradshaw and the Wandjina. The Bradshaw figures, as they are known, are painted in red ochre. According to one Aboriginal legend, birds drew the figures using their beaks. One rock-face frieze shows figures dancing and swaying; another depicts figures elaborately decorated with headdresses, tassels, skirts and epaulets. Significant Bradshaw sites have been found on the Drysdale River. The more recent Wandjina figures, depictions of ancestor spirits from the sky and sea who brought rain and fertility, are in solid red or black, outlined in red ochre, and sometimes on a white background. Wandjina figures are typically human-like, with pallid faces and wide, staring eyes, halos around their heads and, for reasons of religious belief, no mouths. Good examples of Wandjina art have been found near Kalumburu on the King Edward River and at the burial site known as Panda-Goornnya on the Drysdale River.

→ For more detail see maps 624–5 & 628–9. For descriptions of towns see Towns from A–Z (p. 315).

TOWNS A–Z
western australia

[FITZROY BLUFF, THE KIMBERLEY]

LEGEND

VISITOR INFORMATION
RADIO STATIONS
IN TOWN
WHAT'S ON
WHERE TO EAT
WHERE TO STAY
NEARBY

Food and accommodation listings in town are ordered alphabetically with places nearby listed at the end

Albany

see inset box on next page

Augusta

Pop. 1072
Map ref. 616 C11 | 618 B11 | 620 B11

Blackwood Ave; (08) 9758 0166; www.margaretriver.com

98.3 FM ABC South West Radio, 99.1 FM ABC Radio National

The town of Augusta lies in the south-west corner of Western Australia. The state's third-oldest settlement sits high on the slopes of the Hardy Inlet, overlooking the mouth of the Blackwood River and the waters of Flinders Bay. Just beyond it lies Cape Leeuwin with its unforgettable signpost dividing the oceans: the Southern Ocean to the south and the Indian Ocean to the west.

Augusta Historical Museum Augusta's difficult beginning in 1830 is documented in this collection of artefacts and photographs. An exhibit details the 1986 rescue of whales that beached themselves near the town. Blackwood Ave; (08) 9758 0465.

Crafters Croft: locally made handcrafts, jams, emu-oil products; Ellis St.

Augusta River Festival: Mar. ***Spring Flower Show:*** Sept/Oct.

The Colourpatch Cafe: modern Australian; 98 Albany Tce; (08) 9758 1295.

Augusta Sheoak Chalets: chalets with panoramic views; 298 Hillview Rd; 0419 555 072. ***Baywatch Manor:*** budget B&B; 9 Heppingstone View (rear of 88 Blackwood Ave); (08) 9758 1290. ***Blackwood River Houseboats:*** houseboats along the Blackwood River; Lot 450 Bussell Hwy, Westbay; (08) 9758 0181. ***Brilea Cottages and B&B:*** self-contained cottages and B&B; Bussell Hwy, Karridale; (08) 9758 5001. ***Molloy Island Hideaway:*** self-contained house; Fairlawn Pl, Molloy Island; 0403 338 813.

Leeuwin–Naturaliste National Park This park extends 120 km from Cape Naturaliste in the north to Cape Leeuwin in the south. Close to Augusta are three major attractions: Cape Leeuwin, Jewel Cave and Hamelin Bay. Cape Leeuwin (8 km sw) marks the most south-westerly point of Australia. Climb 176 steps to the top of the limestone lighthouse, mainland Australia's tallest. Nearby is the Old Water Wheel, built in 1895 from timber that has since calcified, giving it the appearance of stone. Jewel Cave (8 km NW on Caves Rd) is renowned for its limestone formations, including the longest straw stalactite found in any tourist cave. At Hamelin Bay (18 km NW) a windswept beach and the skeleton of an old jetty give little indication of the massive amounts of jarrah and karri that were once transported from here. In the heyday of the local timber industry, the port's exposure to the treacherous north-west winds resulted in 11 wrecks. These now form the state's most unusual Heritage Trail: the Hamelin Bay Wreck Trail, for experienced divers. *See also Margaret River and Dunsborough.*

The Landing Place: where the first European settlers landed in 1830; 3 km s. ***Whale Rescue Memorial:*** commemorates the 1986 rescue of beached pilot whales; 4 km s. ***Matthew Flinders Memorial:*** Flinders began mapping the Australian coastline from

continued on p. 317

ALBANY

Pop. 25 197

Map ref. 617 D11 | 618 G11 | 620 E11

Old Railway Station, Proudlove Pde; (08) 9841 9290 or 1800 644 088; www.albanytourist.com.au

100.9 FM Albany Community Radio, 630 AM ABC Local Radio

Albany, a picturesque city on Western Australia's south coast, is the site of the state's first European settlement. On Boxing Day 1826, Major Edmund Lockyer, with a party of soldiers and convicts from New South Wales, came ashore to establish a military and penal outpost. Ninety years later, Albany was the embarkation point for Australian troops during World War I and, for many, their last view of the continent. A whaling industry, which began in the 1940s, defined the town until the Cheynes Beach Whaling Company closed in 1978. Nowadays, whale-watching has taken its place. Lying within the protected shelter of the Princess Royal Harbour on the edge of King George Sound, Albany is one of the state's most popular tourist destinations.

Historic buildings As WA's oldest town, Albany boasts more than 50 buildings of historical significance dating back to the early years of the settlement. Two of the oldest were built in the 1830s: Patrick Taylor Cottage on Duke St, which houses an extensive collection of period costumes and household goods, and the Old Farm at Strawberry Hill on Middleton Rd, site of the first government farm in WA. Other heritage buildings include the Old Gaol (1851), with its collection of social history artefacts, and The Residency (1850s), a showcase of historical and environmental exhibits, both in Residency Rd. There are self-guide walks available, including the Colonial Buildings Historical Walk; brochures from visitor centre.

The Amity: full-scale replica of the brig that brought Albany's first settlers from Sydney in 1826; Princess Royal Dr. ***St. John's Church:*** 1848 Anglican church is the oldest in the state; York St. ***Albany Entertainment Centre:*** architecturally stunning new performing arts centre on foreshore with breathtaking harbour views; Toll Place. ***Vancouver Arts Centre:*** gallery, craft shop, studio and workshop complex, originally the Albany Cottage Hospital (1887); Vancouver St. ***House of Gems:*** extensive range of gemstones and jewellery; Cnr York St and Stirling Tce. ***Dog Rock:*** granite outcrop resembling the head of an enormous labrador is a photo opportunity not to be missed; Middleton Rd. ***Princess Royal Fortress:*** Albany's first federal fortress, commissioned in 1893 and fully operational until the 1950s, now houses military museums and war memorials; off Forts Rd. ***The White Star Hotel:*** charming historic hotel with microbrewery; Stirling Tce. ***Spectacular views:*** lookouts at the peaks of Mt Clarence and Mt Melville have 360-degree views. Near the top of Mt Clarence is the Desert Mounted Corps Memorial statue, a recast of the original statue erected at Suez in 1932; Apex Dr. John Barnesby Memorial Lookout at the top of Mt Melville is 23 m high, with observation decks; Melville Dr. ***Mt Clarence Downhill:*** new downhill mountain bike trail adjacent to the peak; Apex Dr. ***Bibbulmun Track:*** 963 km walking track to Perth begins at Albany's Old Railway Station in Proudlove Pde; *see below*. ***Whale-watching:*** cruises daily from town jetty to see southern rights; June–Oct.

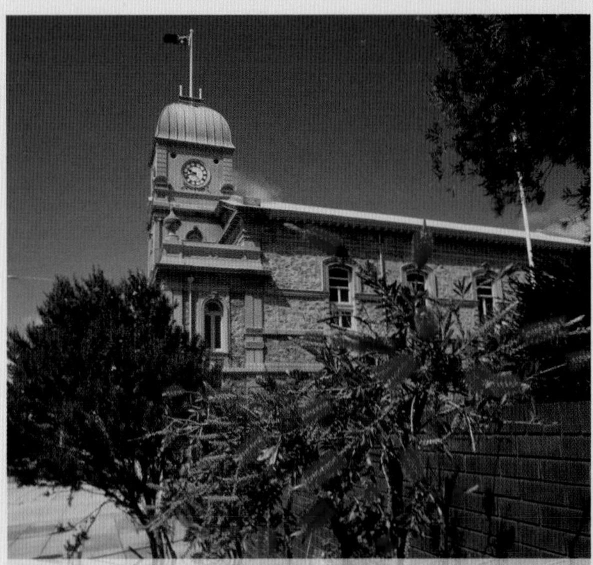

[ALBANY'S TOWN HALL]

Albany Farmers' Market: Collie St; 8am–12pm Sat. ***Albany Boatshed Markets:*** Princess Royal Dr; 10am–1pm Sun. ***Vintage Blues Festival:*** Jan. ***Taste Great Southern Food and Wine Festival:*** Feb/Mar.

Earl of Spencer: pub fare in historic tavern; Cnr Earl and Spencer sts; (08) 9841 1322. ***Lime 303:*** contemporary, creative restaurant; The Dog Rock Motel, 303 Middleton Rd; (08) 9845 7298. ***The Naked Bean:*** for coffee addicts; 21 Sanford Rd; (08) 9841 4225. ***The Wild Duck Restaurant:*** sophisticated dining; 112 York St; (08) 9842 2554.

Foreshore Apartments: harbour-view apartments; 81–89 Proudlove Pde; (08) 9841 1506. ***Middleton Beach Holiday Park:*** beachside holiday park; 28 Flinders Pde; (08) 9841 3593 or 1800 644 674. ***Old Surrey:*** self-contained manor; 55–59 Burt St, Mount Clarence; 0413 015 145. ***The Priory:*** elegant B&B; 55–59 Burt St, Mount Clarence; 0413 015 145. ***The Rocks Albany:*** stately B&B; 182–188 Grey St West; (08) 9842 5969. ***Cape Howe Cottages:*** self-contained cottages halfway between Albany and Denmark; 322 Tennessee Rd South, Lowlands Beach; (08) 9845 1295. ***The Lily Dutch Windmill:*** Nordic-style self-contained accommodation 94 km north of Albany with views to Stirling Range; 9793 Chester Pass Rd, Amelup; (08) 9827 9205 or 1800 980 002.

Bibbulmun Track At 963 km, this is WA's only long-distance walking trail and one of the longest continuously marked trails in Australia. It stretches from Kalamunda, a suburb on the outskirts of Perth, to Albany. On the way it passes through some of the state's most picturesque southern towns including Dwellingup, Collie, Balingup, Pemberton, Northcliffe, Walpole and Denmark. Named after a local Aboriginal language group, the track is marked by a stylised image of the 'Waugal' (rainbow serpent), a spirit being from the Aboriginal Dreaming. Whether taking a short walk or a 5-day hike, easy access points enable walkers of all ages and fitness levels to experience the Bibbulmun Track. Walk the track in springtime and see the bush at its best with WA's amazing array of wildflowers. Near Walpole you'll encounter the massive red tingle trees of the Valley of the Giants. Other well-known natural attractions on the track include Mt Cook, the highest point in the Darling Range, Beedelup Falls and the Gloucester Tree lookout. Details and maps from the Bibbulmun Track Office of DEC (08) 9334 0265 or visit www.bibbulmuntrack.org.au

Torndirrup National Park Torndirrup is one of the most visited parks in the state, featuring abundant wildflowers, wildlife and bushwalking trails. The park is renowned for its rugged coastal scenery, including such features as the Gap, a chasm with a 24 m drop to the sea, and the Natural Bridge, a span of granite eroded by huge seas to form a giant arch. Exercise extreme caution on this dangerous coastline; king waves can rush in unexpectedly. 17 km S.

Whale World Even before the Cheynes Beach Whaling Company closed in 1978, Albany's oldest industry was a major tourist attraction. In its heyday, the company's chasers took up to 850 whales a season. View the restored whale-chaser *Cheynes IV*, whale skeletons, the old processing factory, an aircraft display and the world's largest collection of marine mammal paintings. This is the only whaling museum in the world created from a working whaling station. Free guided tours are available on (08) 9844 4021. 25 km SE.

Albany Marron & Bird Park: tour the aquaculture production tanks full of marron, see over 250 native and exotic birds, and enjoy a meal at Nippers Cafe; 304 Two Peoples Bay Rd; 23 km E. ***Deer-O-Dome:*** showcases the Australian deer industry; 6 km N. ***Mt Romance Sandalwood Factory:*** skincare products, perfumes, therapeutics and free guided tours; (08) 9845 6888; 12 km N. ***Albany Wind Farm:*** 12 giant turbines, each 100 m high; 12 km SW. ***Pt Possession Heritage Trail:*** views and interpretive plaques; Vancouver Peninsula; 20 km SE. ***Fishing:*** Emu Pt (8 km NE), Oyster Harbour (15 km NE), Jimmy Newhill's Harbour (20 km S) and Frenchman Bay (25 km SE). ***Diving:*** former HMAS *Perth* was scuttled in 2001 as an artificial dive reef; Frenchman Bay; 25 km SE. ***West Cape Howe National Park:*** walking, fishing, swimming and hang-gliding; 30 km W. ***Two Peoples Bay Nature Reserve:*** sanctuary for the noisy scrub bird, thought to be extinct but rediscovered in 1961; 40 km E. ***Tours:*** include sailing, wineries, eco-tours, 4WD driving and national parks tours; details from visitor centre.

See also GREAT SOUTHERN, p. 309

Cape Leeuwin in December 1801; 5 km S. ***Alexandra Bridge:*** picnic and camping spot with towering jarrah trees and beautiful wildflowers in season; 10 km N. ***Boranup Maze and Lookout:*** the maze offers a short walking track under trellis, while the lookout provides a picnic area with panoramic views towards the coast; 18 km N. ***Augusta–Busselton Heritage Trail:*** 100 km trail traces the history of the area through the pioneering Bussell and Molloy families, who settled in Augusta only to move further up the coast looking for suitable agricultural land; maps from visitor centre. ***Cruises:*** Blackwood River and Hardy Inlet. ***Marron in season:*** fishing licence required and available at the post office; Blackwood Ave. ***Whale-watching:*** charter boats and coastal vantage points offer sightings of migrating humpback whales (June–Aug) and southern right whales (June–Oct), plus pods of dolphins and fur seals; details from visitor centre.

See also THE SOUTH-WEST, p. 307

Australind

Pop. 8716
Map ref. 616 E3 | 618 C8 | 620 C9

Henton Cottage, cnr Old Coast and Paris rds; (08) 9796 0122.

95.7 Hot FM, 1224 AM ABC Radio National

Lying on the Leschenault Estuary and bordered by the Collie River, Australind offers a multitude of aquatic pleasures including fishing, crabbing, prawning, swimming, boating, sailing and windsurfing. The town's unusual name is a contraction of Australia and India, coined by its founders in the hope of a prosperous trade in horses between the two countries.

St Nicholas Church: built in 1840, reputedly the smallest church in Australia; at only 3.6 m wide and 8.2 m long; Paris Rd. ***Henton Cottage:*** early 1840s heritage building now houses the visitor centre and an art and craft gallery; Cnr Old Coast and Paris rds. ***Featured Wood Gallery:*** fine furniture and craft made from the local timbers of jarrah, she-oak, marri, banksia and blackbutt. Also includes a museum of Australian and American West history; Piggott Dr; (08) 9797 2411. ***Pioneer Memorial:*** site of the first settlers' landing in 1840; Old Coast Rd. ***Cathedral Ave:*** scenic 2 km drive through arching paperbark trees with sightings of kangaroos and black swans, especially at sunset; off Old Coast Rd.

Carols in the Park: Dec.

Australind Tourist Park: ideal crabbing location; Lot 9 Old Coast Rd; (08) 9725 1206. ***Cook's Park on Australind Waters:*** 2-bedroom house; 474 Cathedral Ave; (08) 9796 0505. ***Leschenault Inlet Caravan Park:*** shady caravan park; Lot 52 Cathedral Ave; (08) 9797 1095.

Leschenault Inlet Offers recreational attractions from the simple pleasure of fishing from the Leschenault Inlet Fishing Groyne to picnicking, camping and bushwalking in the Peninsula Conservation Park. The park is a haven for native wildlife with over 60 species of birds recorded. Only walking or cycling is permitted in the park except for 4WD beach access from Buffalo Rd (1 km S). The Leschenault Waterway Discovery Centre has an interpretive gazebo with information on the estuary environment. Old Coast Rd; 2 km S.

Pioneer Cemetery: graves dating back to 1842 and beautiful wildflowers in season; Old Coast Rd; 2 km N. ***Binningup and Myalup:*** pleasant beach towns north of Leschenault. ***Australind–Bunbury Tourist Drive:*** coastal scenery, excellent crabbing and picnic spots; brochures from visitor centre.

See also THE SOUTH-WEST, p. 307

Balingup

Pop. 443
Map ref. 616 G6 | 618 C9 | 620 C10

Brockman St; (08) 9764 1818; www.balinguptourism.com.au

93.3 ABC Classic FM, 1044 AM ABC Local Radio

This small town, nestled in the Blackwood River Valley, is surrounded by rolling hills, forests and orchards. Balingup is renowned for its glowing summer sunsets, amazing autumn colours and misty winter mornings.

Birdwood Park Fruit Winery: unique award-winning fruit wines, chutneys, jams and fruits; Brockman St; (08) 9764 1172. ***Tinderbox:*** herbal and natural products; South Western Hwy. ***Old Cheese Factory Craft Centre:*** the largest art and craft centre in WA, including pottery and timber products; Balingup–Nannup Rd; (08) 9764 1018.

Opera in the Valley: Jan. ***Small Farm Field Day:*** festival and roadside scarecrows; Apr. ***Medieval Carnivale:*** Aug. ***Festival of Country Gardens:*** Oct/Nov. ***Balingup Jalbrook Concert:*** Nov.

Balingup Bronze Cafe: healthy and tasty; Balingup Bronze Gallery, South Western Hwy; (08) 9764 1843. ***Fre-Jac French Restaurant:*** French cuisine; Forrest St; (08) 9764 1883.

Balingup Heights Hilltop Forest Cottages: timber cottages; 65 Nannup Rd; (08) 9764 1283. ***Balingup Jalbrook Cottages & Alpacas:*** range of cottages; Lot 1 Jayes Rd; (08) 9764 1616. ***Balingup Rose B&B:*** historic B&B; 208 Jayes Rd; (08) 9764 1205.

Golden Valley Tree Park This 60 ha arboretum boasts a superb collection of exotic and native trees. Other attractions include a tree information gazebo, walk trails, lookout and the historic Golden Valley Homestead. Old Padbury Rd; 2 km S.

Jalbrook Alpacas and Knitwear Gallery: alpacas to feed and alpaca knitwear; accommodation also available; (08) 9764 1190; 2 km E. ***Lavender Farm:*** oil-producing lavender farm with open gardens, picnic area, art gallery and giftshop. Take a distillation tour; open Sept–Apr, Balingup–Nannup Rd; (08) 9764 1436; 2.5 km W. ***Balingup Heights Scenic Lookout:*** stunning views of town and orchards; off Balingup–Nannup Rd; 2.5 km W. ***Greenbushes:*** boasts WA's first metal-producing mine (1888), still in production and now the world's largest tantalum producer. The Discovery Centre has interactive displays and walking trails, and there is an excellent lookout at the mine; 10 km W. ***Heritage Country Cheese:*** cheese-producing factory with viewing window and tastings; 16 km W. ***Wineries:*** several in area; details from visitor centre. ***Balingup–Nannup Rd:*** enjoy wonderful scenery, interesting and historic landmarks and great marroning, fishing and picnic sites. ***Bibbulmun Track:*** sections of this trail pass through Balingup; *see Albany*.

See also THE SOUTH-WEST, p. 307

Balladonia

Pop. 20
Map ref. 621 K7

Balladonia Roadhouse, Eyre Hwy; (08) 9039 3453.

Balladonia lies on the Eyre Highway on the western edge of the Nullarbor Plain. Its closest towns are Norseman, 191 kilometres to the west, and Caiguna, 182 kilometres to the east. This arid desert shrubland is one of the world's oldest landscapes, containing seashells millions of years old from when the area was ocean floor. Balladonia made world headlines in 1979 when space debris from NASA's *Skylab* landed 40 kilometres east on Woorlba Station.

Cultural Heritage Museum Learn about the crash-landing of *Skylab*, local Indigenous culture, early explorers, Afghan cameleers and other chapters in the area's history. Balladonia Roadhouse, Eyre Hwy.

Balladonia Roadhouse: light meals and hearty fare; Eyre Hwy; (08) 9039 3453.

Balladonia Caravan Facility: caravan sites, dorms, motel rooms; Eyre Hwy; (08) 9039 3453.

90-Mile Straight Have your photo taken beside the signpost marking the western end of the longest straight stretch of road in Australia, which runs for 90 miles (146.6 km) between Balladonia and Caiguna. Begins 35 km E.

Newman Rocks: superb views from rocky outcrop, with picnic and camping areas on-site; 50 km W. ***Cape Arid National Park and Israelite Bay:*** great birdwatching and fishing; access via 4WD track, south of town; check track conditions at roadhouse.

See also ESPERANCE & NULLARBOR, p. 310

Beverley

Pop. 850
Map ref. 618 E5 | 620 D7

Aeronautical Museum, 139 Vincent St; (08) 9646 1555; www.beverleywa.com

99.7 FM ABC News Radio, 531 AM ABC Local Radio

Beverley is a small town set on the banks of the Avon River 130 kilometres east of Perth. Its main street boasts some beautifully preserved buildings, representing Federation to Art Deco architectural styles. This farming community, while having long been associated with wheat and wool, also produces grapes, olives, emus, deer and yabbies.

Aeronautical Museum This museum presents a comprehensive display of early aviation in WA. The museum's star attraction is the *Silver Centenary*, a biplane built between 1928 and 1930 by local man Selby Ford and his cousin Tom Shackles. Ford designed the plane in chalk on the floor where he worked. The plane first flew in July 1930, but was never licensed because of the lack of design blueprints. Vincent St.

Station Gallery: art exhibitions and sales in the Tudor-style (1889) railway station; Vincent St. ***Dead Finish Museum:*** the oldest building in town (1872) houses memorabilia and historic items from wooden cotton wheels to washing boards; open 11am–3pm Sun Mar–Nov or by appt through visitor centre; Hunt Rd.

Yabbie Races: Apr. ***Annual Quick Shear:*** Aug.

Freemasons Hotel: classic Australian pub; 104 Vincent St; (08) 9646 1094.

Beverley B&B: heritage B&B; 131 Forrest St; (08) 9646 0073.

Avondale Discovery Farm Avondale is an agricultural research station with displays of historic farming machinery and tools. The 1850s homestead is furnished in period style and set in traditional gardens. There is also an animal nursery, Clydesdale horses and a picnic area with barbecues and a children's playground. A land-care education centre houses interactive displays. The farm hosts the Clydesdale and Vintage Day in June. Waterhatch Rd; (08) 9646 1004; 6 km W.

Brookton: attractions of this nearby town include the Old Police Station Museum and the Brookton Pioneer Heritage Trail, which highlights places significant to the local Aboriginal people; 32 km S. ***County Peak Lookout:*** spectacular views from the summit; 35 km SE. ***The Avon Ascent:*** take a self-drive tour of the Avon Valley; maps from visitor centre.

See also HEARTLANDS, p. 308

Boyup Brook

Pop. 531
Map ref. 618 D9 | 620 C10

Cnr Bridge and Abel sts; (08) 9765 1444; www.bbvisitor.mysouthwest.com.au

 100.5 Hot FM, 1044 AM ABC Local Radio

Boyup Brook is on the tranquil Blackwood River in the heart of Western Australia's grass-tree country. The town's name is thought to derive from the Aboriginal word 'booyup', meaning 'place of big stones' or 'place of much smoke', which was given to the nearby Boyup Pool. Seasonal wildflowers are abundant during September and October.

Carnaby Beetle and Butterfly Collection Keith Carnaby was such a leading light in the field of entomology that beetles have been named after him. His collection of Jewel beetles, part of which is on display at the Boyup Brook Tourist Information Centre, is regarded as the best outside the British Museum of Natural History. Cnr Bridge and Abel sts.

Pioneers' Museum: displays of historic agricultural, commercial and domestic equipment; open 2–5pm Mon, Wed, Fri or by appt; Jayes Rd. ***Sandakan War Memorial:*** honours 1500 Australian POWs sent to Sandakan to build an airfield for the Japanese; Sandakan Park. ***The Flax Mill:*** built during WW II for processing flax needed for war materials. At its peak it operated 24 hrs a day and employed over 400 people. A scale model of the mill can be viewed on-site, which is now the caravan park; off Barron St. ***Heritage walk:*** follows 23 plaques around town centre; self-guide pamphlet from visitor centre. ***Bicentennial Walking Trail:*** pleasant walk around town and beside the Blackwood River.

Country Music Festival and Ute Muster: Feb. ***Blackwood Marathon:*** running, canoeing, horseriding, cycling and swimming a 58 km course to Bridgetown; Oct.

Chudacud Estate: versatile vineyard restaurant; Lot 22 Wade Rd; (08) 9764 4053.

Scotts Brook: tranquil B&B; 201B Scotts Brook Rd; (08) 9765 3014. ***Tulip Cottage:*** restored cottage; 30 Bridge St; (08) 9765 1223.

Harvey Dickson's Country Music Centre This entertainment shed is decorated wall-to-wall and floor-to-rafter with music memorabilia spanning 100 years. The 'record room' containing hundreds of records also has Elvis memorabilia. There is a music show in Sept and a rodeo in Oct, with basic bush camping facilities. Open by appt; Arthur River Rd; (08) 9765 1125; 5 km N.

Roo Gully Wildlife Sanctuary: for injured and orphaned Australian wildlife, with a special focus on raising unfurred marsupial young; 1 km N. ***Gregory Tree:*** remaining stump of a tree blazed by explorer Augustus Gregory in 1845; Gibbs Rd; 15 km NE. ***Norlup Homestead:*** built in 1874, this is one of the district's first farms; to view contact (08) 9767 3034; off Norlup Rd; 27 km SE. ***Wineries:*** Scotts Brook Winery (20 km SE) and Blackwood Crest Winery (at Kilikup, 40 km E); both open daily. ***Haddleton Flora Reserve:*** displays of brown and pink boronia in season. Not suitable for campers or caravans; 50 km NE. ***Boyup Brook flora drives:*** self-guide maps from visitor centre.

See also THE SOUTH-WEST, p. 307

Bremer Bay

Pop. 239
Map ref. 620 G10

Community Resource Centre and Library, Mary St; (08) 9837 4171; www.bremerbay.com

 103.5 WA FM, 531 AM ABC Local Radio

Bremer Bay on the south coast is a wide expanse of crystal-clear blue water and striking white sand. The main beach, only a ten-minute walk from the town, has a sheltered cove for swimming and fishing. Just north of Bremer Bay is the magnificent Fitzgerald River National Park with its four rivers, dramatic gorges, wide sand plains, rugged cliffs, pebbly beaches and spectacular displays of wildflowers between August and October.

Watersports: fishing, boating, swimming, surfing, waterskiing, scuba diving, bay cruises and seasonal whale-watching are the town's main attractions. ***Rammed-earth buildings:*** the Bremer Bay Hotel/Motel on Frantom Way and Catholic Church on Mary St are excellent examples of rammed-earth construction.

Mount Barren Restaurant: local produce; Bremer Bay Resort, 1 Frantom Way; (08) 9837 4133.

Bremer Bay Beaches Tourist Resort Caravan Park: 4-star holiday park; Wellstead Rd; (08) 9837 4290. ***Bremer Bay Resort:*** well-appointed resort; 1 Frantom Way; (08) 9837 4133. ***Quaalup Homestead Wilderness Retreat:*** range of simple accommodation; Fitzgerald River National Park; (08) 9837 4124.

Fitzgerald River National Park This huge 242 739 ha park, lying between Bremer Bay and Hopetoun to the east, is renowned for its scenery and flora. A staggering 1800 species of flowering plants have been recorded. Royal hakea, endemic to this region, is one of the most striking. Quaalup Homestead (1858), restored as a museum, offers meals and accommodation in the park. Pt Ann has a viewing platform for whale-watching (southern rights, June–Oct). Campgrounds, barbecues and picnic areas available. 17 km N.

Wellstead Homestead Museum: the first residence in the area, now incorporating a gallery and museum with family heirlooms, historic farm equipment and vintage cars; Peppermint Grove, Wellstead Rd; (08) 9837 4448; 9 km SW. ***Surfing:*** nearby beaches include Native Dog Beach, Dillon Bay, Fosters Beach and Trigelow Beach; directions from visitor centre.

See also GREAT SOUTHERN, p. 309

Bridgetown

Pop. 2321
Map ref. 616 H8 | 618 D9 | 620 C10

154 Hampton St; (08) 9761 1740 or 1800 777 140; www.bridgetown.com.au

 100.5 Hot FM, 1044 AM ABC Local Radio

Bridgetown is a picturesque timber town nestled among rolling hills on the banks of the Blackwood River. Crossing the river, Bridgetown boasts the longest wooden bridge in the state, made of the area's famous jarrah. In addition to tourism, timber milling and mining (lithium, tantalum and tin) are now the largest industries in the area.

Brierley Jigsaw Gallery The only public jigsaw gallery in the Southern Hemisphere, Brierley has over 170 jigsaws ranging from the world's smallest wooden puzzle to a huge 9000-piece jigsaw. A highlight is an 8000-piece jigsaw of the Sistine Chapel. Back of visitor centre, Hampton St.

continued on p. 321

BROOME

Pop. 11 547

Map ref. 624 H8 | 628 A11

1 Hamersley Street; (08) 9195 2200; www.broomevisitorcentre.com.au

101.3 WA FM, Spirit 102.9 FM, 675 AM ABC Local Radio

Broome is distinguished by its pearling history, cosmopolitan character and startling natural assets: white sandy beaches, turquoise water and red soils. The discovery of pearling grounds off the coast in the 1880s led to the foundation of the Broome township in 1883. A melting pot of nationalities flocked to its shores in the hope of making a fortune. Japanese, Malays and Koepangers joined the Aboriginal pearl divers, while the Chinese became the shopkeepers in town. By 1910 Broome was the world's leading pearling centre. In those early, heady days, over 400 pearling luggers operated out of Broome. The industry suffered when world markets collapsed in 1914, but stabilised in the 1970s as cultured-pearl farming developed. Today remnants of Broome's exotic past are everywhere, with the town's multicultural society ensuring a dynamic array of cultural influences. Broome's beaches are ideal for swimming and there is good fishing year-round.

Pearl Luggers Experience Broome's pearling heritage by visiting two restored pearling luggers in Chinatown. Tours daily. Dampier Tce; (08) 9192 2059.

Japanese Cemetery The largest Japanese cemetery in Australia contains the graves of over 900 Japanese pearl divers, dating back to 1896. This is a sobering reminder of the perils of the early pearling days when the bends, cyclones and sharks claimed many lives. Cnr Port Dr and Savannah Way.

Staircase to the Moon This beautiful optical illusion is caused by a full moon reflecting off the exposed mudflats of Roebuck Bay at extremely low tides. Town Beach; 3 nights monthly from Mar–Oct; check dates and times at visitor centre.

Chinatown: an extraordinary mix of colonial and Asian influences, Chinatown was once the bustling hub of Broome where pearl sheds, billiard saloons and Chinese eateries flourished; now it is home to some of the world's finest pearl showrooms. ***Buildings on Hamersley Street:*** distinctive Broome-style architecture including the courthouse, made of teak inside and corrugated iron outside; Captain Gregory's House, a classic old pearling master's house, built in 1915, now an art gallery; and Matso's Broome Brewery, once the Union Bank Building. ***Historical Museum:*** pearling display and collection of photographs and literature on Broome's past; Robinson St. ***Bedford Park:*** war memorial, replica of explorer William Dampier's sea chest and an old train coach; Hamersley St. ***Shell House:*** one of the largest shell collections in Australia; Dampier Tce. ***Sun Pictures:*** the world's oldest operating outdoor cinema, opened in 1916; Carnarvon St. ***Sisters of St John of God Convent:*** built in 1926 by a Japanese shipbuilder using traditional methods that emphasise the external framing of the building; Cnr Barker and Weld sts. ***Deep Water Jetty:*** good for fishing; Port Dr. ***Heritage trail:*** 2 km walk introduces places of interest; self-guide pamphlet from visitor centre.

Courthouse Markets: Hamersley St; Sat and Sun Apr–Oct Sat only Nov–Mar. ***Town Beach Markets:*** Robinson St; 1st 2 nights of the Staircase to the Moon; check with visitor centre for dates and times. ***Race Round:*** horseracing; June–Aug. ***Opera Under the Stars:*** Aug. ***Shinju Matsuri:*** Festival of the Pearl, recalls Broome's heyday and includes Dragon Boat Regatta; Aug/Sept. ***Mango Festival:*** Nov.

Matso's Broome Brewery: laid-back brewery; 60 Hammersley St; (08) 9193 5811. ***The Club Restaurant:*** fine dining; Cable Beach Club Resort & Spa, Cable Beach Rd; 1800 199 099. ***The Old Zoo Cafe:*** alfresco cafe; 2 Challenor Dr; (08) 9193 6200. ***The Wharf Restaurant:*** casual seafood; On the Wharf, Port Dr; (08) 9192 5700.

Cable Beach Club Resort & Spa: outstanding resort; Cable Beach Rd; 1800 199 099. ***Kimberley Klub Broome:*** centrally located hostel; 62 Fredrick St; (08) 9192 3233 or 1800 004 345. ***McAlpine House:*** historic restored house; 55 Herbert St; (08) 9192 0510

[CAMEL RIDE ON CABLE BEACH]

or 1800 746 282. ***The Bungalow-Broome:*** romantic bungalow; 3 McKenzie Rd; 0417 918 420. ***The Courthouse B&B:*** luxurious B&B; 10 Stewart St; (08) 9192 2733. ***Kooljaman at Cape Leveque:*** beach safari retreat; Cape Leveque; (08) 9192 4970.

Cable Beach With its 22 km of pristine white sands fringing the turquoise waters of the Indian Ocean, Cable Beach is one of the most stunning beaches in the world. Every day the beach is washed clean by high tides ranging from 4 m to 10 m. It takes its name from the telegraph cable laid between Broome and Java in 1889. Why not do that quintessential Broome activity and ride a camel along this famous beach. Details from visitor centre. 7 km NW.

Gantheaume Point Dinosaur footprints believed to be 130 million years old can be seen at very low tide. A plaster cast of the tracks has been embedded at the top of the cliff. Nearby, view the almost perfectly round Anastasia's Pool, built by a lighthouse keeper for his wife. 5 km NE.

Crocodile Park: home to some of Australia's biggest crocodiles; Cable Beach Rd; 7 km NW. ***Riddell Beach:*** enjoy the dramatic sight of Broome's distinctive red soils, known as 'pindan', meeting white sands and brilliant blue water; 7 km SW. ***Buccaneer Rock:*** at entrance to Dampier Creek, this landmark commemorates Captain William Dampier and HMS *Roebuck*; 1 km SE. ***Broome Bird Observatory:*** see some of the 310 species of migratory wader birds that arrive each year from Siberia; 17 km E. ***Willie Creek Pearl Farm:*** the Kimberley's only pearl farm open to the public, with daily tours; (08) 9192 0000; 35 km N. ***Dampier Peninsula:*** this remote area north of Broome boasts unspoiled coastline (4WD access only). Record-breaking game fish have been caught in the surrounding waters. Charters and tours leave from Dampier. The Sacred Heart Church at Beagle Bay (118 km NE) was built by Pallotine monks in 1917 and boasts a magnificent pearl-shell altar. Lombadina (200 km NE) is a former mission now home to an Aboriginal community that offers sightseeing, fishing and mudcrabbing tours; contact (08) 9192 4936. On the eastern side of the peninsula is Cygnet Bay Pearl Farm, the oldest Australian and family-owned pearl farm, which offers 1-hour and 1-day tours; (08) 9192 4283. Cape Leveque, at the northern end of the peninsula, is well known for its pristine beaches and rugged pindan cliffs; 220 km NE. ***Buccaneer Archipelago:*** in Broome you can arrange scenic flights over this magnificent landscape that stretches north-east of the Dampier Peninsula. Also known as the Thousand Islands, this is a dramatic coastal area of rugged red cliffs, spectacular waterfalls and secluded white sandy beaches. Here you'll find whirlpools created by massive 11 m tides and the amazing horizontal two-way waterfall of Talbot Bay. ***Hovercraft Spirit of Broome:*** tours of Roebuck Bay; details from visitor centre.

Travellers note: *Poisonous jellyfish frequent this stretch of coast, especially Nov–May. Pay attention to warning signs on the beaches and wear protective clothing if swimming.*

See also KIMBERLEY, p. 314

Bridgedale: historic house owned by John Blechynden, one of the area's first European settlers, constructed in 1862 of local timber and bricks made from riverbank clay; South Western Hwy. ***Memorial Park:*** picnic area with a giant chessboard, 3 ft high pieces for hire from visitor centre; South Western Hwy.

Blackwood River Park Markets: Sun mornings each fortnight. ***State Downriver Kayaking Championships:*** Aug. ***Blackwood Classic Powerboat Race:*** Sept. ***Blackwood Marathon:*** between Boyup Brook and Bridgetown; Oct. ***Blues at Bridgetown Festival:*** Nov. ***Festival of Country Gardens:*** Nov.

Bridgetown Pottery Tearooms & Gallery: home-style cooking; 81 Hampton St; (08) 9761 1038. ***Nelson's of Bridgetown:*** country charm; 38 Hampton St; (08) 9761 1641 or 1800 635 565. ***The 1896 Cafe:*** wholesome cafe; 145–151 Hampton St; (08) 9761 1699. ***The Bridgetown Hotel:*** pub grub; 157 Hampton St; (08) 9761 1034.

Ford House Retreat: romantic B&B; Eedle Tce; (08) 9761 1816. ***Maranup Ford:*** range of accommodation; Maranup Ford Rd; (08) 9761 1200. ***Nelson's of Bridgetown:*** range of accommodation; 38 Hampton St; (08) 9761 1641 or 1800 635 565. ***Tortoiseshell Farm:*** cottages and B&B; Polina Rd; (08) 9761 1089.

The Cidery Discover the history of Bridgetown's apple industry and sample fresh juice, cider or award-winning beers. The orchard contains over 80 varieties of apples. Closed Tues; Cnr Forrest St and Gifford Rd; (08) 9761 2204; 2 km N.

Geegelup Heritage Trail: 52 km walk retraces history of agriculture, mining and timber in the region. It starts at Blackwood River Park. ***Scenic drives:*** choose from 8 scenic drives in the district through green hills, orchards and valleys into karri and jarrah timber country; self-guide maps from visitor centre. ***Excellent views:*** Sutton's Lookout, off Phillips St and Hester Hill, 5 km N. ***Bridgetown Jarrah Park:*** ideal place for a picnic or bushwalk. The Tree Fallers and Shield Tree trails commemorate the early timber history of the town; Brockman Hwy; 20 km W. ***Karri Gully:*** bushwalking and picnicking; 20 km W.

See also THE SOUTH-WEST, p. 307

Bunbury

see inset box on next page

Busselton

Pop. 15 385
Map ref. 616 C6 | 618 B9 | 620 B10

38 Peel Tce; (08) 9752 1288; www.geographebay.com

96.5 FM Western Tourist Radio, 684 AM ABC Local Radio

First settled by Europeans in 1834, Busselton is one of the oldest towns in Western Australia. It is situated on the shores of Geographe Bay and the picturesque Vasse River. Sheltered from most prevailing winds, the tranquil waters of the bay are an aquatic playground edged with 30 kilometres of white sand beaches. Over the past three decades, the traditional industries of timber, dairying, cattle and sheep have been joined by grape-growing and winemaking. Fishing is also important, particularly crayfish and salmon in season. In spring, the wildflowers are magnificent.

Busselton Jetty The longest timber jetty in the Southern Hemisphere was built over a 95-year period, beginning in 1865, principally for the export of timber. Over 5000 ships from all over the world docked here through the ages of sail, steam and diesel, before the port closed in 1972. The jetty stretches a graceful 1.8 km into Geographe Bay and has always been a popular spot for fishing, snorkelling and scuba diving because of the variety of marine life. An Interpretive Centre at the base of the jetty displays historical and environmental exhibits. At the seaward end is an Underwater Observatory featuring an observation chamber with viewing windows 8 m beneath the surface revealing vividly coloured corals, sponges and fish. Tours are available, bookings essential. End of Queen St.

Ballarat Engine: first steam locomotive in WA; Pries Ave. ***St Mary's Anglican Church:*** built in 1844 of limestone and jarrah, with a she-oak shingle roof. The churchyard has many pioneer graves, including John Garrett Bussell's, after whom Busselton was named; Peel Tce. ***Nautical Lady Entertainment World:*** family fun park with giant water slide, flying fox, minigolf, skate park, racing cars, lookout tower and nautical museum; on beachfront at end of Queen St; (08) 9752 3473. ***Old Courthouse:*** restored gaol cells and arts complex; Queen St. ***Busselton Historical Museum:*** originally a creamery, now houses historic domestic equipment; closed Tues; Peel Tce. ***Vasse River Parkland:*** barbecue facilities; Peel Tce.

Markets: Barnard Park, 1st and 3rd Sat each month; Railway Building Park, Causeway Rd, 2nd and 4th Sun each month. ***Southbound:*** music festival; Jan. ***Festival of Busselton:*** Jan. ***Beach Festival:*** Jan. ***Busselton Jetty Swim:*** Feb. ***Geographe Bay Race Week:*** yachting; Feb. ***Bluewater Fishing Classic:*** Mar. ***Great Escapade:*** cycling; Mar. ***Busselton Agricultural Show:*** one of the oldest and largest country shows in WA; Oct/Nov. ***Smell the Roses, Taste the Wine:*** Nov. ***Ironman Western Australia Triathlon:*** Dec.

Newtown House Restaurant: fine dining; 737 Bussell Hwy; (08) 9755 4485. ***The Equinox:*** beachside dining; 343 Queen St; (08) 9752 4641. ***The Goose:*** jetty-side longtime favourite; Geographe Bay Rd; (08) 9754 7700. ***Vasse Bar Cafe:*** funky Mediterranean; 44 Queen St; (08) 9754 8560.

Beachlands Holiday Park: well-appointed holiday park; 10 Earnshaw Rd; (08) 9752 2107 or 1800 622 107. ***Mandalay Holiday Resort and Tourist Park:*** beachside holiday park; off Bussell Hwy at Lockhart St, Broadwater; (08) 9752 1328 or 1800 248 231. ***Martin Fields Beach Retreat:*** beachside retreat; 24 Lockville Rd, Geographe; (08) 9754 2001.

Tuart Forest National Park The majestic tuart tree grows only on coastal limestone 200 km either side of Perth. Known locally as the Ludlow Tuart Forest, this 2049 ha park protects the largest natural tuart forest in the world. It also has the tallest and largest specimens of tuart trees on the Swan Coastal Plain, up to 33 m high and 10 m wide. Enjoy scenic drives, forest walks and picnics in a magnificent setting. 12 km SE.

Wonnerup House: built in 1859, now a National Trust museum and fine example of colonial architecture, furnished in period style; 10 km N. ***Bunyip Craft Centre:*** Ludlow; 15 km E. ***Wineries:*** numerous vineyards and wineries in the area. Many are open for cellar-door tastings; maps from visitor centre. ***Augusta–Busselton Heritage Trail:*** maps from visitor centre.

See also THE SOUTH-WEST, p. 307

Caiguna

Pop. 10
Map ref. 621 L7

Caiguna Roadhouse; (08) 9039 3459.

The small community of Caiguna, on the Nullarbor Plain, consists of a 24-hour roadhouse, caravan park, motel, restaurant and service station. The nearest towns are Balladonia, 182 kilometres west, and Cocklebiddy, 65 kilometres east. To the south is the coastal wilderness of Nuytsland Nature Reserve. From immediately east of Caiguna until Border Village, locals operate on Central Western Time, 45 minutes ahead of the rest of Western Australia.

John Eyre Motel: comfortable motel rooms; Eyre Hwy; (08) 9039 3459.

John Baxter Memorial In 1841, the explorer John Baxter, together with an Aboriginal guide known as Wylie and two other unnamed Aboriginal men, accompanied Edward John Eyre on his epic journey across the Nullarbor Plain. The party left Fowlers Bay in SA on 25 Feb and reached the site of modern-day Eucla on 12 Mar. Later, the two unnamed Aboriginal men killed Baxter and, taking most of the supplies, fled into the desert. Eyre and Wylie walked for another month and eventually reached Thistle Cove (near Esperance), where they were rescued by a French whaler. The Baxter memorial is on the Baxter Cliffs overlooking the Great Australian Bight; 4WD access only. 38 km S.

90-Mile Straight: have your photo taken beside the signpost marking the eastern end of the longest straight stretch of road in Australia, which runs for 90 miles (146.6 km) between Caiguna and a point east of Balladonia; 4 km W. ***Caiguna Blowhole:*** a hole in the flat limestone landscape where the earth seemingly breathes in and out; 5 km W.

See also ESPERANCE & NULLARBOR, p. 310

Carnamah

Pop. 358
Map ref. 620 B4

Council offices, Macpherson St; (08) 9951 7000; www.carnamah.wa.gov.au

 101.9 WA FM, 612 AM ABC Radio National

Carnamah is a typical wheat-belt town servicing the surrounding wheat and sheep properties. From late July through to December the shire of Carnamah and the rest of the wheat belt blossoms into a wildflower wonderland. This is one of Western Australia's richest areas of flowering plants, with more than 600 species.

Historical Society Museum: displays historic domestic equipment and old farm machinery; Macpherson St.

North Midlands Agricultural Show, Rodeo and Ute Parade: Sept.

Three Springs Tourist Lodge: B&B accommodation; Three Springs; (08) 9954 1065.

Tathra National Park This park, with its diverse range of spring wildflowers, is named after the Nyungar word for 'beautiful place'. 25 km SW.

Macpherson's Homestead: an excellent example of pioneering architecture (1869), once the home of Duncan Macpherson, the first settler in the area; open by appt; Bunjil Rd; (08) 9951 1690; 1 km E. ***Yarra Yarra Lake:*** this salt lake changes from pink in summer to deep blue in winter. View it from the Lakes Lookout; 16 km S. ***Eneabba:*** spectacular wildflowers surround this mining town; 74 km W. ***Lake Indoon:*** a freshwater lake popular for sailing,

continued on p. 325

BUNBURY

Pop. 54 967

Map ref. 616 D4 | 618 C8 | 620 C9

[VIEW OF BUNBURY FROM LESCHENAULT INLET]

Old Railway Station, Carmody Pl; (08) 9792 7205 or 1800 286 287; www.visitbunbury.com.au

95.7 Hot FM, 684 AM ABC Local Radio

Bunbury is known as the 'city of three waters', surrounded by the Indian Ocean, Koombana Bay and the Leschenault Inlet. This is a water-lover's paradise with fishing, crabbing, diving, white sandy beaches, sailing and kayaking. Bunbury is also known for its wild dolphins that come close to the beach at Koombana Bay. Bunbury was settled by Europeans in 1838 and the Koombana Bay whalers were a source of initial prosperity. Today the port is the main outlet for the timber and mining industries.

Dolphin Discovery Centre Wild bottlenose dolphins regularly visit Koombana Bay. The centre has interpretative displays on dolphins and other marine life, and offers visitors the chance to swim with dolphins under ranger guidance. Dolphin visits usually occur in the mornings; however, times and days of visits are unpredictable. If you prefer not to get wet, take a dolphin-spotting cruise on the bay. Open daily 8am–5pm Sept–May, 9am–3pm June–Aug; Koombana Dr; (08) 9791 3088.

Historic buildings: many date back to the early decades of the settlement, including the 1865 Rose Hotel; Cnr Victoria and Wellington sts; details from visitor centre. ***King Cottage:*** this cottage was built in 1880 by Henry King using homemade bricks. It now displays items of domestic life from the early 20th century; open 2–4pm daily; Forrest Ave. ***Sir John Forrest Monument:*** born in Picton on the outskirts of Bunbury in 1847, Sir John Forrest was elected the first Premier of WA in 1890 and entered Federal Parliament in 1901; Cnr Victoria and Stephen sts. ***Victoria Street:*** a 'cappuccino strip' of sidewalk cafes and restaurants. ***Bunbury Regional Art Galleries:*** built in 1887, formerly a convent for the Sisters of Mercy and now the largest art gallery in the South-West; Wittenoom St. ***Miniature Railway Track:*** take a ride on this 800 m track through the trees at Forrest Park; 3rd Sun each month; Blair St. ***Lookouts:*** Boulter's Heights, Haig Cres and Marlston Hill; Apex Dr. ***Lighthouse:*** painted in black-and-white check, this striking landmark has a lookout at the base; end of Ocean Dr. ***Basaltic rock:*** formed by volcanic lava flow 150 million years ago; foreshore at end of Clifton St, off Ocean Dr. ***Mangrove boardwalk:*** 200 m elevated boardwalk lets you view the southernmost mangrove colony in WA, estimated to be 20 000 years old; Koombana Dr. ***Big Swamp Wildlife Park:*** handfeed kangaroos, see bettongs, wombats, swamp wallabies and more, and enjoy the South-West's largest walk-through aviary; Prince Phillip Dr. ***Heritage trail:*** 12 km walk from the Old Railway Station; brochures from visitor centre.

Bunbury Carnaval: Mar. ***Bunbury International Jazz Festival:*** May. ***Geographe Crush:*** Food and Wine Festival; Nov.

Alexanders Restaurant: modern Australian; Quality Hotel Lord Forrest, 20 Symmons St; (08) 9726 5777. ***Aristos Waterfront:*** casual seafood; 2/15 Bonnefoi Blvd; (08) 9791 6477. ***Boardwalk Bar & Bistro:*** local seafood; The Parade Hotel, 1 Austral Pde; (08) 9721 2933. ***Mojo's Restaurant:*** contemporary Australian; Grand Cinema Complex, Victoria St; (08) 9792 5900. ***Vat 2:*** harbour-side dining; 2 Jetty Rd; (08) 9791 8833. ***Carlaminda Wines Bistro:*** French fare; Carlaminda Wines, Richards Rd, Ferguson Valley; (08) 9728 3002. ***Hackersley:*** seasonal menu; Ferguson Rd, Ferguson Valley; (08) 9728 3033. ***Truffles Restaurant:*** fresh local produce; Meadowbrooke Estate, 33 Turner St, Boyanup; (08) 9731 5550.

All Seasons Sanctuary Golf Resort: golfing enthusiast resort; Cnr Old Coast Rd and Australind Bypass, Pelican Pt; (08) 9725 2777 or 1800 677 309. ***Boathouse B&B:*** harbour-side B&B; 11 Austral Pde; (08) 9721 4140. ***Bunbury Silo Accommodation:*** converted silo apartments; off Casurina Dr, Marlston Waterfront; 0439 973 285. ***Quality Hotel Lord Forrest:*** premier town accommodation; 20 Symmons St; (08) 9726 5777 or 1800 097 811.

St Marks Anglican Church: built in 1842, this is the second oldest church in WA. The churchyard contains the graves of many early Bunbury settlers; 5 km SE at Picton. **Lena *Dive Wreck:*** apprehended by the navy in 2002 for illegal fishing, the *Lena* was sunk three nautical miles from Bunbury as a dive wreck; suitable for snorkelling and diving for all levels of experience. ***South West Gemstone Museum:*** over 2000 gemstones; 12 km S. ***Wineries:*** at the heart of the Geographe Wine Region, many in the area offer cellar-door tastings, including Killerby Wines (10 km S) and Capel Vale Wines (27 km S); details from visitor centre. ***Abseiling tours:*** on the quarry face of the Wellington Dam; details from visitor centre. ***Scenic flights:*** over Bunbury and surrounds.

See also THE SOUTH-WEST, p. 307

CARNARVON

Pop. 5283

Map ref. 619 B7

Civic Centre, 21 Robinson St; (08) 9941 1146; www.carnarvon.org.au

99.7 Hot Hits FM, 846 AM ABC Local Radio

Carnarvon is a large coastal town at the mouth of the Gascoyne River. The river and the fertile red earth surrounding it are crucial to the town's thriving agricultural industry. Plantations stretching for 15 kilometres along the riverbanks draw water from the aquifer of the river basin to grow a host of tropical fruits such as bananas, mangoes, avocados, pineapples, pawpaws and melons. Carnarvon gained national prominence when a NASA tracking station operated nearby at Browns Range from 1964 to 1974.

Robinson Street In 1876 the region's founding fathers, Aubrey Brown, John Monger and C. S. Brockman, overlanded 4000 sheep from York. Carnarvon was gazetted in 1883 and developed into the centre of an efficient wool-producing area. Camel teams, driven by Afghan camel drivers, brought the wool to Carnarvon from the outlying sheep stations. This is the reason for the extraordinary width of the town's main street, which, at 40 m, gave the camel teams enough room to turn around.

Pioneer Park: good picnic spot; Olivia Tce. ***Murals:*** up to 15 buildings in the town, including the Civic Centre, are adorned with murals painted by local artists. ***Heritage walking trail:*** 20 historic landmarks around the town; maps from visitor centre. ***Gwoonwardu Mia:*** the Gascoyne Aboriginal Heritage and Cultural Centre celebrates the history and culture of the five indigenous language groups of the Gascoyne region. It houses an exhibition gallery, local indigenous art and crafts for sale, and a cafe that serves lunches with a regional bush tucker flavour. ***Gascoyne Food Trail:*** follow this self-drive tour of the plantations, orchards and fresh produce outlets in Carnarvon; contact Visitor Centre for brochure. ***Bumbak's Plantation Tours:*** learn just what it takes to grow bananas, grapes and mangoes at one of the oldest family-run plantations in Carnarvon. The shop sells delicious homemade preserves, jams, fruit icecreams and other treats. 449 North River Rd, tours at 10am daily.

Gascoyne Growers Markets and Courtyard Markets: Civic Centre; Sat mornings May–Oct. ***Fremantle-Carnarvon Bluewater Classic:*** odd-numbered years; Apr. ***Carnafin:*** fishing competition, May/June. ***Taste the Gascoyne:*** Aug/Sept. ***Carnarvon Festival:*** Aug/Sept. ***Carnarvon Cup:*** Sept. ***Kickstarters Gascoyne Dash:*** Oct.

[POINT QUOBBA]

Hacienda Crab Shack: the freshest seafood lunch in town; Small Boat Harbour; (08) 9941 4078. ***Old Post Office Café:*** great value Italian meals and pizza; Robinson St; (08) 9941 1800. ***Sails Restaurant:*** international menu with alfresco dining option for hot nights; Best Western Hospitality Inn, 6 West St; (08) 9941 1600.

Carnarvon Central Apartments: 2-bedroom apartments; 120 Robinson St; (08) 9941 1317. ***The Carnarvon Hotel:*** motel and backpacker accommodation; 121 Olivia Tce; (08) 9941 1181. ***Wintersun Caravan Park:*** range of accommodation; 546 Robinson St; (08) 9941 8150 or 1300 555 585. ***Quobba Station:*** working sheep station; Gnarloo Rd, 80 km north-west of Carnarvon; (08) 9948 5098.

Carnarvon Heritage Precinct On Babbage Island and connected to the township by a causeway, this heritage precinct incorporates the One Mile Jetty. Built in 1897, this is the longest jetty in WA's north, stretching for 1493 m into the Indian Ocean. It offers excellent fishing and a jetty train runs its length. Other attractions include the Lighthouse Keeper's Cottage museum, prawning factory at the old whaling station (tours in season, check times at visitor centre) and Pelican Pt, for picnics, swimming and fishing.

Blowholes Jets of water shoot up to 20 m in the air after being forced through holes in the coastal rock. When you arrive at the Blowholes, you are greeted by a huge sign declaring 'KING WAVES KILL' – a cautionary reminder that this picturesque coastline has claimed the lives of over 30 people in freak waves. 73 km N. Nearby, a sheltered lagoon provides good swimming and snorkelling (1 km S). A further 7 km north of the blowholes is a cairn commemorating the loss of HMAS *Sydney* in 1941.

'The Big Dish': a huge 29 m wide reflector, part of the old NASA station, with views of town and plantations from the base; 8 km E. ***Bibbawarra Artesian Bore:*** hot water surfaces at 65° C and picnic area nearby; 16 km N. ***Bibbawarra Trough:*** 180 m long, believed to be the longest in the Southern Hemisphere; adjacent to bore; 16 km N. ***Miaboolya Beach:*** good fishing, crabbing and swimming; 22 km N. ***Rocky Pool:*** picnic area and deep billabong ideal for swimming (after rains) and wildlife watching; Gascoyne Rd; 55 km E. ***Red Bluff:*** world-renowned surfing spot with waves 1–6 m, depending on the time of the year; 143 km N. ***Mt Augustus:*** considered the biggest rock in the world, twice the size of Uluṟu. It is known as Burringurrah to the local Aboriginal people. This 'monocline' is over 1750 million years old, cloaked in thick scrub, and offers many interesting rock formations, caves and Indigenous rock art. Camping and powered sites are available at Mt Augustus Outback Tourist Resort; (08) 9943 0527; road conditions vary; 450 km E; *see also Gascoyne Junction.* ***Fishing:*** excellent fishing for snapper or groper and game fishing for marlin or sailfish; charter boats available from Williams St. Also excellent fishing off One Mile Jetty.

See also OUTBACK COAST & MID-WEST, p. 312

boating, camping, picnics and barbecues (swimming is forbidden due to poor water quality); 85 km w.

See also Heartlands, p. 308

Cervantes

Pop. 506
Map ref. 618 A2 | 620 A5

Pinnacles Visitor Centre, Cadiz St; (08) 9652 7672 or 1800 610 660; www.visitpinnaclescountry.com.au

 99.9 WA FM, 612 AM ABC Radio National

This small but thriving fishing town was established in 1962 and named after the American whaling ship *Cervantes*, which sank off the coast in 1844. The town's fishing fleet nearly doubles in rock lobster season, and in spring the town is surrounded by spectacular displays of wildflowers with vistas of wattles stretching from horizon to horizon. Not far from Cervantes is one of Australia's best-known landscapes, the Pinnacles Desert, lying at the heart of Nambung National Park.

Pinnacle Wildflowers: displays of native WA flora, dried flower arrangements, souvenirs. Flowers are visible year-round, but at their peak in Aug and Sept; Bradley Loop. ***Thirsty Pt:*** lookout has superb views of the bay and Cervantes islands. A trail connects the lookouts between Thirsty Pt and Hansen Bay. Popular in wildflower season; off Seville St.

Ronsard Bay Tavern: bistro dining; 219 Cadiz St; (08) 9652 7009. ***The Europa Anchor Restaurant:*** local seafood; Cervantes Pinnacles Motel, 7 Aragon St; (08) 9652 7145.

Cervantes Pinnacles Motel: comfortable motel; 7 Aragon St; (08) 9652 7145. ***Pinnacles Caravan Park:*** beachfront caravan park; 35 Aragon St; (08) 9652 7060.

Nambung National Park In the Pinnacles Desert, thousands of limestone pillars rise out of a stark landscape of yellow sand, reaching over 3 m in places. They are the eroded remnants of a bed of limestone, created from sea-shells breaking down into lime-rich sands. See formations like the Indian Chief, Garden Wall and Milk Bottles. The loop drive is one-way and not suitable for caravans. The park allows day visits only; tours departing morning and sunset can be arranged at visitor centre. 17 km s.

Lake Thetis Stromatolites: one of WA's six known locations of stromatolites, the oldest living organism on earth; 5 km s. ***Kangaroo Pt:*** good picnic spot; 9 km s. ***Hangover Bay:*** a stunning white sandy beach ideal for swimming, snorkelling, windsurfing and surfing; 13 km s.

See also Heartlands, p. 308

Cocklebiddy

Pop. 75
Map ref. 621 M6

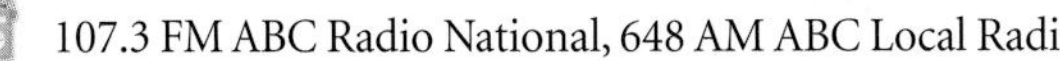

Cocklebiddy Roadhouse; (08) 9039 3462.

107.3 FM ABC Radio National, 648 AM ABC Local Radio

This tiny settlement, comprising a roadhouse with motel units, caravan sites and camping facilities, lies between Madura and Caiguna on the Nullarbor Plain. Nuytsland Nature Reserve extends southwards, a 400 000-hectare strip running along the Great Australian Bight. Locals operate on Central Western Time, 45 minutes ahead of the rest of the state.

Cocklebiddy Wedgetail Inn: convenient motel; Eyre Hwy; (08) 9039 3462.

Eyre Bird Observatory Housed in the fully restored 1897 Eyre Telegraph Station, Australia's first bird observatory offers birdwatching, bushwalking and beachcombing in Nuytsland Nature Reserve. Over 240 species of birds have been recorded at Eyre, including Major Mitchell cockatoos, brush bronzewings, honeyeaters and mallee fowl. It is near the site where Edward John Eyre found water and rested during his Nullarbor journey in February 1841. Courses, tours and whale-watching (June–Oct) as well as accommodation can be arranged on (08) 9039 3450. 4WD access only. 50 km se.

Chapel Rock: picnic area; 4 km e. ***Twilight Cove:*** fishing and whale-watching spot with views of 70 m high limestone cliffs overlooking the Great Australian Bight; 4WD access only; 32 km s.

See also Esperance & Nullarbor, p. 310

Collie

Pop. 7084
Map ref. 616 G3 | 618 D8 | 620 C9

Old Collie Post Office, 63 Throssell St; (08) 9734 2051; www.collierivervalley.org.au

 95.7 Hot FM, 684 AM ABC Local Radio

Collie is Western Australia's only coalmining town. The surrounding area was first explored in 1829 when Captain James Stirling led a reconnaissance party to the land south of Perth. The region was originally considered ideal for timber production and as pasturelands. However, the discovery of coal along the Collie River in 1883 changed the region's fortunes. In dense jarrah forest, near the winding Collie River, the town has many parks and gardens. The drive into Collie on the Coalfields Highway along the top of the Darling Scarp offers spectacular views of the surrounding forests, rolling hills and farms.

Tourist Coal Mine Step back in time and gain an insight into the mining industry and the working conditions in underground mines. This replica mine was constructed in 1983 to commemorate the 100-year anniversary of coal discovery. Tours by appt only; details from visitor centre. Throssell St.

Coalfields Museum: displays of historic photographs, coalmining equipment, rocks and minerals, woodwork by local miner Fred Kohler, a doll house and art housed in the historic Roads Board building; Throssell St. ***Collie Railway Station:*** the rebuilt station houses railway memorabilia, a scale model of the Collie township with model trains, tearooms and a giftshop; Throssell St. ***Soldiers' Park:*** bordering the Collie River, features include a war memorial, rose garden, gazebo and childrens' playground; Steere St. ***All Saints Anglican Church:*** impressive Norman-style church distinctive for its unusual stained-glass windows, extensive use of jarrah timbers and elaborate mural, which in 1922 took renowned stage artist Philip Goatcher 8 months to complete. Tours by appt; contact visitor centre; Venn St. ***Old Collie Goods Shed:*** restoration of rolling stock; Forrest St. ***Central Precinct Historic Walk:*** self-guide walk of historic buildings; maps from visitor centre. ***Collie River Walk:*** pleasant walk along riverbank; maps from visitor centre.

Market: Old Goods Shed, Forrest St; 1st and 3rd Sun each month (except winter). ***Collie Rock and Coal Music Festival:*** Mar. ***Collie–Donnybrook Cycle Race:*** Aug. ***Collie River Valley Marathon:*** Sept. ***Griffin Festival:*** Sept.

The Ridge: contemporary menu; Collie Ridge Motel, 185–195 Throssell St; (08) 9734 6666.

Peppermint Lane Lodge: charming B&B; 351 Wellington Mill Rd, Wellington Mill; (08) 9728 3138.

Wellington National Park Covering 4000 ha, this park is characterised by jarrah forest. Picnic, swim, canoe or camp at Honeymoon Pool or Potters Gorge, or go rafting in winter on the rapids below the Wellington Dam wall (note that work is being carried out on the wall until 2010, when the Quarry picnic area will reopen). 18 km W.

Minninup Pool: where the Collie River is at its widest, ideal for swimming, canoeing or picnicking; off Mungalup Rd; 3 km S. ***Stockton Lake:*** camping and waterskiing; 8 km E. ***Brew 42:*** microbrewery producing 6 different beers, especially traditional Irish and English ales. Tastings and sales Thurs–Sun or by arrangement; Allanson; (08) 97344784; 8km W. ***Harris Dam:*** beautiful picnic area; 14 km N. ***Collie River Scenic Drive:*** views of jarrah forest and wildflowers in season; maps from visitor centre. ***Munda Biddi Trail:*** starting in the hills near Perth, this bike trail winds through scenic river valleys and forests south to Collie; details from visitor centre. ***Bibbulmun Track:*** sections of this trail pass through Collie; *see Albany.*

See also THE SOUTH-WEST, p. 307

Coolgardie

Pop. 801
Map ref. 620 H5

Goldfields Exhibition Building, Bayley St; (08) 9026 6090; www.coolgardie.wa.gov.au

97.9 Hot FM, 648 AM ABC Local Radio

After alluvial gold was found in 1892, Coolgardie grew in ten years to a town of 15 000 people, 23 hotels, six banks and two stock exchanges. The main street, lined with some magnificent buildings, was made wide enough for camel trains to turn around in. As in many outback towns, the heat and the isolation led to innovation, in this case that of the Coolgardie safe, which used water and a breeze to keep food cool before the days of electricity.

Historic buildings There are 23 buildings in the town centre that have been listed on the National Estate register, many of them on the main street, Bayley St. Over 100 markers are positioned at buildings and historic sites across the town, using stories and photographs to recapture the gold-rush days. The index to markers is in Bayley St next to the visitor centre.

Goldfields Exhibition Museum Local photographs and displays inside the old Warden's Court including a display on the famous Varischetti mine rescue. In 1907 Modesto Varischetti was trapped underground in a flooded mine for 9 days. Varischetti survived in an air pocket until divers eventually found him. The dramatic rescue captured world attention. Bayley St.

Ben Prior's Open-Air Museum: unusual collection of machinery and memorabilia; Cnr Bayley and Hunt sts. ***Warden Finnerty's House:*** striking 1895 example of early Australian architecture and furnishings; open 11am–4pm daily except Wed; McKenzie St. ***C. Y. O'Connor Dedication:*** fountain and water course in memory of O'Connor, who masterminded the Goldfields Water Supply Scheme; McKenzie St. ***Gaol tree:*** used for prisoners in early gold-rush days, before a gaol was built; Hunt St. ***Lindsay's Pit Lookout:*** over open-cut goldmine; Ford St.

Coolgardie Day: Sept. ***Metal Detecting Championships:*** odd-numbered years, Sept/Oct.

Coolgardie Motel: comfortable, versatile motel; 49–53 Bayley St; (08) 9026 6080.

Coolgardie Cemetery The town cemetery gives you an inkling of the harshness of the early gold-rush years. The register of burials records that of the first 32 burials, the names of 15 were unknown, and many entries for 'male child' and 'female child' note 'fever' as the cause of death. One of the most significant graves is that of Ernest Giles, an Englishman whose name is associated with the exploration of inland Australia. 1 km W.

Coolgardie Camel Farm: offers rides on the 'ships of the desert'; (08) 9026 6159; 4 km W. ***Gnarlbine Rock:*** originally an Aboriginal well, then one of the few water sources for the early prospectors; 30 km SW. ***Kunanalling Hotel:*** once a town of over 800 people, the ruins of the hotel are all that remain; 32 km N. ***Victoria Rock:*** camping, and spectacular views from the summit; 55 km SW. ***Burra Rock:*** popular camping and picnic area (55 km S). Cave Hill, a similar destination, lies a further 40 km S (4WD only). ***Rowles Lagoon Conservation Park:*** picnicking and camping spots available although recently there has been no water; 65 km N. ***Wallaroo Rocks:*** three dams with scenic views and good bushwalking; 90 km W. ***Golden Quest Discovery Trail:*** Coolgardie forms part of this 965 km self-guide drive trail of the goldfields; book, map and CD from visitor centre.

See also GOLDFIELDS, p. 311

Coral Bay

Pop. 192
Map ref. 619 B5

Coastal Adventure Tours, Coral Bay Arcade, Robinson St; (08) 9948 5190; www.coralbaytours.com.au

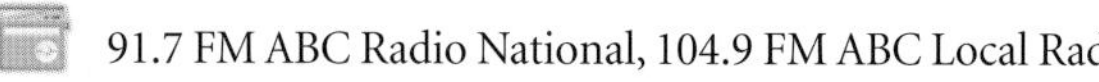
91.7 FM ABC Radio National, 104.9 FM ABC Local Radio

Coral Bay is famous for one thing: its proximity to Ningaloo Marine Park. Ningaloo Reef boasts an incredible diversity of marine life and beautiful coral formations. At Coral Bay the coral gardens lie close to the shore, which makes access to the reef as easy as a gentle swim. Lying at the southern end of Ningaloo Marine Park, Coral Bay has pristine beaches and a near-perfect climate: it is warm and dry regardless of the season, and the water temperature only varies from 18°C to 28°C degrees. Swimming, snorkelling, scuba diving, and beach, reef and deep-sea fishing (outside sanctuary areas) are available year-round.

Fin's Cafe: beachside cafe; People's Park Shopping Village; (08) 9942 5900. ***Shades:*** casual dining; Ningaloo Reef Resort, 1 Robinson St; (08) 9942 5934 or 1800 795 522. ***The Reef Cafe:*** Italian and seafood; Robinson St; (08) 9942 5882.

Bayview Coral Bay: range of accommodation options; Robinson St; (08) 9385 6655. ***Ningaloo Reef Resort:*** range of accommodation options; 1 Robinson St; (08) 9942 5934 or 1800 795 522. ***The Ningaloo Club:*** hostel; 46 Robinson St; (08) 9948 5100 or (08) 9385 6655.

Ningaloo Marine Park This park protects the 260 km long Ningaloo Reef, the longest fringing coral reef in Australia. It is the only large reef in the world found so close to a continental land mass: about 100 m offshore at its nearest point and less than 7 km at its furthest. This means that even novice snorkellers and children can access the coral gardens. The reef is home to over 500 species of fish, 250 species of coral, manta rays, turtles and a variety of other marine creatures, with seasonal visits from humpback whales, dolphins and whale sharks. Ningaloo Reef is famous for the latter, and from Apr to June visitors from around the world visit the reef to swim with these gentle giants.

Pt Cloates: the wrecks of the *Zvir*, *Fin*, *Perth* and *Rapid* lie on the reef just off the point; 4WD access only; 8 km N. ***Tours:*** glass-bottomed boat cruises, snorkel and dive tours, kayak tours, fishing charters, scenic flights and marine wildlife-watching tours to see whale sharks (Apr–June), humpback whales (June–Nov) and manta rays (all year); details from visitor centre.

See also OUTBACK COAST & MID-WEST, p. 312

Corrigin

Pop. 687
Map ref. 618 F5 | 620 E8

Corrigin Resource Centre, Larke Cres; (08) 9063 2778; www.corrigin.wa.gov.au

92.5 ABC Classic FM, 100.5 Hot FM

Corrigin was established in the early 1900s and was one of the last wheat-belt towns to be settled. Today the town has a healthy obsession with dogs, as demonstrated by its Dog Cemetery and its national record for lining up 1527 utes with dogs in the back.

Corrigin Pioneer Museum Superb collection of old agricultural equipment including an original Sunshine harvester and some early steam-driven farm machinery. A small working steam train carries passengers on a short circuit around the museum and local rest area. Open Sun and by appt; Kunjin St.

RSL Monument: a Turkish mountain gun from Gallipoli; McAndrew Ave.

Dog in a Ute event: held in varying years in Apr; dates from visitor centre.

Corrigin Windmill Motel: motel rooms plus family lodge; Brookton Hwy; (08) 9063 2390.

Dog Cemetery Loving dog owners have gone to the considerable expense of having elaborate headstones placed over the remains of their faithful four-footed friends. There are over 80 dogs buried in the cemetery, with gravestones dedicated to Dusty, Rover, Spot et al. There is even one statue of a dog almost 2 m high. Brookton Hwy; 7 km W.

Wildflower scenic drive: signposted with lookout; 3 km W. ***Gorge Rock:*** large granite outcrop with picnic area; 20 km SE.

See also HEARTLANDS, p. 308

Cranbrook

Pop. 279
Map ref. 618 F10 | 620 E10

Council offices, Gathorne St; (08) 9826 1008; www.cranbrook.wa.gov.au

95.3 Hot FM, 630 AM ABC Local Radio

The small town of Cranbrook greets travellers with a large sign announcing that it is the 'Gateway to the Stirlings'. A mere 10 kilometres away is Stirling Range National Park, a mecca for bushwalkers and climbers. The nearby Frankland area has gained a national reputation for its premium-quality wines.

Station House Museum: restored and furnished 1930s-style; Gathorne St. ***Wildflower walk:*** 300 m walk to Stirling Gateway with displays of orchids in spring; Salt River Rd.

Cranbrook Shire on Show: Apr. ***Wildflower Display:*** Sept. ***Art trail and Photographic Competition:*** Oct.

Stirling Range National Park Surrounded by a flat, sandy plain, the Stirling Range rises abruptly to over 1000 m, its jagged peaks veiled in swirling mists. The cool, humid environment created by these low clouds supports 1500 flowering plant species, many unique to the area, earning the park recognition as one of the top-10 biodiversity hot spots in the world. This National Heritage–listed park is one of WA's premier destinations for bushwalking. Best time to visit is Oct–Dec. 10 km SE.

Sukey Hill Lookout: expansive views of farmland, salt lakes and Stirling Range; off Salt River Rd; 5 km E. ***Lake Poorrarecup:*** swimming and waterskiing; 40 km SW. ***Wineries:*** the nearby Frankland River region boasts several wineries, including Alkoomi, Frankland Estate and Ferngrove; 50 km W. ***Wildflower drive and heritage trail:*** brochures from visitor centre.

See also GREAT SOUTHERN, p. 309

Cue

Pop. 273
Map ref. 619 H10 | 620 D1 | 622 D11

Golden Art Shop and Tourist Information Centre, Austin St; (08) 9963 1936; www.cue.wa.gov.au

102.9 WA FM, 106.1 FM ABC Local Radio

This town was once known as the 'Queen of the Murchison'. In 1891 Mick Fitzgerald and Ed Heffernan found large nuggets of gold not far from what was to become the main street. It was their prospecting mate, Tom Cue, who registered the claim on their behalf and when the town was officially proclaimed in 1894, it bore his name. Within ten years the population of this boom town had exploded to about 10 000 people. While Cue's population has dwindled, the legacy of those heady gold-rush days is evident in the town's remarkably grandiose buildings.

Heritage buildings Many early buildings still stand and are classified by the National Trust. A stroll up the main street takes in the elegant band rotunda, the former Gentleman's Club (now the shire offices, housing a photographic display of the region's history), the Old Gaol, the courthouse, the post office and the police station. One block west in Dowley St is the former Masonic Lodge built in 1899 and reputed to be the largest corrugated-iron structure in the Southern Hemisphere.

Murchison Club Hotel: hotel and motel accommodation; Austin St; (08) 9963 1020.

Walga Rock This monolith is 1.5 km long and 5 km around the base. It has several Aboriginal rock paintings. One of the most extraordinary paintings, considering that Cue is over 300 km from the sea, is of a white, square-rigged sailing ship. It is believed to depict one of the Dutch ships that visited WA's mid-west shores in the 17th century. 50 km W.

Day Dawn: once Cue's twin town, thanks to the fabulous wealth of the Great Fingall Mine. The mine office, a magnificent century-old stone building now perched precariously on the edge of a new open-cut mine, is all that remains of the town; 5 km W. ***Milly Soak:*** popular picnic spot for early Cue residents. A tent hospital was set up nearby during the typhoid epidemic; three lone graves are the only reminder of the thousands who died; 16km N. ***Heritage trail:*** includes the abandoned towns Big Bell and Day Dawn; brochures from visitor centre. ***Fossicking:*** areas surrounding the town; details from visitor centre.

See also OUTBACK COAST & MID-WEST, p. 312

Denham

Pop. 609
Map ref. 619 B9

Knight Tce; (08) 9948 1590 or 1300 135 887; www.sharkbaywa.com.au

105.3 Hot Hits FM, 107.5 FM ABC Radio National

On the middle peninsula of Shark Bay, Denham is the most westerly town in Australia. Dirk Hartog, the Dutch navigator, landed on an island at the bay's entrance in 1616, the first known European to land on the continent. Centuries later, in 1858, Captain H. M. Denham surveyed the area and a town bearing his name was established. The Shark Bay region was once known for its pearling and fishing and the streets of Denham were literally paved with pearl shells. In the 1960s, however, the local roads board poured bitumen over the pearl shells, and so destroyed what could have been a unique tourist attraction. Fortunately, several buildings made from coquina shell block still stand in the town. Today Shark Bay is renowned for the wild dolphins that come inshore at Monkey Mia (pronounced 'my-a'). As a World Heritage area, it also protects dugongs, humpback whales, green and loggerhead turtles, important seagrass feeding grounds and a colony of stromatolites, the world's oldest living fossils.

Shell block buildings: St Andrews Anglican Church, cnr Hughes and Brockman sts, and the Old Pearlers Restaurant, cnr Knight Tce and Durlacher St, were both built from coquina shell block. ***Town Bluff:*** popular walk for beachcombers; from town along beach to bluff. ***Pioneer Park:*** contains the stone on which Captain Denham carved his name in 1858; Hughes St.

Old Pearler Restaurant: maritime-themed seafood restaurant; 71 Knight Tce; (08) 9948 1373. ***The Boughshed Restaurant:*** international cuisine; Monkey Mia Dolphin Resort, Monkey Mia; (08) 9948 1320 or 1800 653 611.

Bay Lodge Economy Beachfront: beachfront budget accommodation; 113 Knight Tce; (08) 9948 1278 or 1800 812 780. ***Denham Villas:*** self-contained accommodation; 8 Durlacher St; 9948 1264. ***Heritage Resort Shark Bay:*** beachfront elegance; Cnr Knight Tce and Durlacher St; (08) 9948 1133. ***Monkey Mia Dolphin Resort:*** wide range of accommodation; Monkey Mia; (08) 9948 1320 or 1800 653 611.

Monkey Mia The daily shore visits by the wild bottlenose dolphins at Monkey Mia are a world-famous phenomenon. The dolphins swim into the shallows, providing a unique opportunity for humans to make contact with them. It began in the 1960s when a local woman started feeding the dolphins that followed her husband's fishing boat to the shoreline. Feeding still occurs, although now it is carefully monitored by rangers to ensure that the dolphins maintain their hunting and survival skills. Visiting times, and the number of dolphins, vary. 26 km NE.

Dirk Hartog Island The state's largest and most historically significant island, named after Dutchman Dirk Hartog who landed here in 1616 – 154 years before Captain Cook. Hartog left behind an inscribed pewter plate, which was removed in 1697 by his countryman Willem de Vlamingh and replaced with another plate. The original was returned to Holland; Vlamingh's plate is now housed in the Maritime Museum in Fremantle. Flights and cruises depart daily; bookings at visitor centre. 30 km W.

Hamelin Pool stromatolites The shores of Hamelin Pool are dotted with stromatolites, the world's largest and oldest living fossils. These colonies of micro-organisms resemble the oldest and simplest forms of life on earth, dated at around 3.5 million years old. The Hamelin Pool stromatolites are relatively new colonies however, about 3000 years old. They thrive here because of the extreme salinity of the water, the occurrence of calcium bicarbonate and the limited water circulation. Visitors can view these extraordinary life forms from a boardwalk. Close by is the Flint Cliff Telegraph Station and Post Office Museum (1884) with a history of the region. 88 km SE.

Dugongs The Shark Bay World Heritage Area has the largest seagrass meadows in the world, covering about 4000 sq km. These meadows are home to around 10 000 dugongs, 10% of the world's remaining population. An endangered species, the dugong is nature's only vegetarian sea mammal. Also known as a sea cow, the dugong can live for up to 70 years and grow up to 3 m long. Tours are available offering visitors a unique opportunity to see dugongs in the wild. Details from visitor centre.

Little Lagoon: ideal fishing and picnic spot; 3 km N. ***Francois Peron National Park:*** Peron Homestead with its 'hot tub' of artesian water; 4WD access only; 7 km N. ***Ocean Park:*** marine park with aquarium and touch pool; 9 km S. ***Eagle Bluff:*** habitat of sea eagle and a good viewing spot for sharks and stingrays; 20 km S. ***Blue Lagoon Pearl Farm:*** working platform where black pearls are harvested; Monkey Mia; 26 km NE. ***Shell Beach:*** 120 km of unique coastline comprising countless tiny coquina shells; 45 km SE. ***Steep Pt:*** western-most point on mainland with spectacular scenery; 4WD access only; 260 km W. ***Zuytdorp Cliffs:*** extend from beneath Shark Bay region south to Kalbarri; 4WD access only. ***Tours:*** boat trips and charter flights to historic Dirk Hartog Island, catamaran cruises, safaris and coach tours; details from visitor centre.

See also OUTBACK COAST & MID-WEST, p. 312

Denmark

Pop. 2735
Map ref. 617 A11 | 618 F11 | 620 E11

73 South Coast Hwy; (08) 9848 2055; www.denmark.com.au

92.1 FM ABC News Radio, 630 AM ABC Local Radio

Denmark lies at the foot of Mt Shadforth, overlooking the tranquil Denmark River and Wilson Inlet. It is surrounded by forests of towering karri trees that sweep down to meet the Southern Ocean. The Aboriginal name for the Denmark River is 'koorabup', meaning 'place of the black swan'. Originally a timber town, Denmark's economy is today sustained by a combination of dairying, beef cattle, fishing, timber and tourism. The town is close to some of the most beautiful coastline in the state.

Bert Bolle Barometer Huge water barometer that visitors can view by climbing the surrounding tower. Reputedly the world's largest, housed in the visitor centre.

Historical Museum: in old police station; Mitchell St. ***Bandstand:*** located on the riverbank with seating for the audience on the other side of the river; Holling Rd. ***Arts and crafts:*** galleries abound, including the Old Butter Factory in North St; details from visitor centre. ***Mt Shadforth Lookout:*** magnificent views; Mohr Dr. ***Berridge and Thornton parks:*** shaded picnic areas; along riverbank in Holling Rd.

Craft Market: Berridge Park; Jan, Easter and Dec. ***Pantomime:*** Civic Centre; Jan. ***Brave New Works:*** new performance art; Easter.

Che Sera Sera: modern Australian, water views; The Denmark Waterfront, 63 Inlet Dr; (08) 9848 3314. ***Greenpool Restaurant:*** slick fine dining; Forest Hill Vineyard, cnr Myers Rd and South Coast Hwy; (08) 9848 0091. ***The Southern End***

Restaurant: local produce; Denmark Observatory Resort, 427 Mt Shadforth Rd; (08) 9848 2600.

Chimes Spa Retreat: luxurious pampering; Mt Shadforth Scenic Dr; (08) 9848 2255. ***Misty Valley Country Cottages:*** cosy country cottages; 52 Hovea Rd; (08) 9840 9239. ***Mt Lindesay View B&B:*** well appointed with spectacular views; Cnr Mt Shadforth Tourist Dr and McNabb Rd; (08) 9848 1933. ***Pensione Verde:*** intimate guesthouse affiliated with organic cooking school; 31 Strickland St; (08) 9848 1700. ***Tree Tops:*** romantic getaway; 21 Payne Rd, Weedon Hill; (08) 9848 2055 or (08) 9364 1594.

William Bay National Park This relatively small 1867 ha park protects stunning coastline and forest between Walpole and Denmark on WA's south coast. It is renowned for its primeval windswept granite tors. Green's Pool, a natural rockpool in the park, remains calm and safe for swimming and snorkelling all year-round. Nearby are the Elephant Rocks, massive rounded boulders resembling elephants; Madfish Bay, a good fishing spot; and Waterfall Beach for swimming. 17 km SW.

Ocean Beach: one of the finest surfing beaches in WA; 8 km S. ***Monkey Rock:*** lookout with panoramic views; 10 km SW. ***Bartholomew's Meadery:*** honey, honey wines, gourmet honey ice-cream and other bee products, as well as a live beehive display; 20 km W. ***Pentland Alpaca Stud and Tourist Farm:*** diverse collection of animals, including alpacas, koalas, kangaroos, bison, water buffalo, llamas and many more; Cnr McLeod and Scotsdale rds; (08) 9840 9262; 20 km W. ***Eden Gate Blueberry Farm:*** spray-free fruit, a range of blueberry products and blueberry wines; open Thurs–Mon Dec–Apr; 25 km E. ***Whale-watching:*** viewing platform above Lowlands Beach (southern rights June–Oct); 28 km E. ***Fishing:*** at Wilson Inlet, Ocean Beach (8 km S) and Parry Beach (25 km W). ***West Cape Howe National Park:*** Torbay Head, WA's most southerly point, and Cosy Corner, a protected beach perfect for swimming; 30 km SW. ***Wineries:*** many wineries open for cellar-door tastings, including Howard Park Winery, West Cape Howe and Tinglewood Wines; maps from visitor centre. ***Scenic drives:*** the 25 km Mt Shadforth Scenic Drive and the 34 km Scotsdale Tourist Drive both feature lush forests, ocean views, wineries and galleries; maps from visitor centre. ***Heritage trails:*** 3 km Mokare trail, 5 km Karri Walk or 9 km Wilson Inlet trail; maps from visitor centre. ***Bibbulmun Track:*** a section of this world-class 963 km long-distance trail passes through Denmark; *see Albany*. ***Valley of the Giants Tree Top Walk:*** *see Walpole*; 65 km W.

See also GREAT SOUTHERN, p. 309

Derby

Pop. 3091
Map ref. 625 I7 | 628 D9

2 Clarendon St; (08) 9191 1426 or 1800 621 426; www.derbytourism.com.au

102.7 WA FM, 873 AM ABC Local Radio

It is said that Derby, known as the 'Gateway to the Gorges', is where the real Kimberley region begins. The first town settled in the Kimberley, it features some spectacular natural attractions nearby: the Devonian Reef Gorges of Windjana and Tunnel Creek are only a few hours' drive along the Gibb River Road, and the magnificent islands of the Buccaneer Archipelago are just a short cruise away. Although King Sound was first explored in 1688, it wasn't until the early 1880s that the Port of Derby was established as a landing point for wool shipments and Derby was proclaimed a townsite. The first jetty was built in 1885, the same year that gold was discovered at Halls Creek. Miners and prospectors poured into the port on their way to the goldfields but by the 1890s, as gold fever died, the port was used almost exclusively for the export of live cattle and sheep. In 1951 iron-ore mining began at Cockatoo Island, which revitalised the town. Derby is now a service centre for the region's rich pastoral and mining industries. Rain closes some roads in the area from November to March, so check conditions before setting out on any excursion.

Old Derby Gaol: built in 1906, this is the oldest building in town; Loch St. ***Wharfinger House Museum:*** built in the 1920s for the local harbourmaster, the design is typical of the tropics. It now houses an extensive collection of historical memorabilia and Aboriginal artefacts. Key from visitor centre; Loch St. ***Derby Jetty:*** some of the highest tides in Australia, up to 12 m, can be seen from the jetty. It is now used to export ore from various local mines.

Market: Clarendon St; each Sat May–Sept. ***King Tide Day:*** festival celebrating highest tide in Australia; Apr/May. ***Moonrise Rock Festival:*** June. ***Derby Races:*** June/July. ***Mowanjum Festival:*** indigenous art and culture; July. ***Boab Festival:*** Mardi Gras, mud football, mud crab races and bush poets; July. ***Derby Rodeo:*** Aug. ***Boxing Day Sports:*** Dec.

Derby Boab Inn: international cuisine; 100 Loch St; (08) 9191 1044. ***Oasis Bistro:*** bistro dining; King Sound Resort Hotel, 112 Loch St; (08) 9193 1044. ***The Point Restaurant:*** seaside seafood; 1 Jetty Rd; (08) 9191 1195.

Jila Gallery Apartment: centrally located apartment; 18 Clarendon St; (08) 9193 2560. ***Kimberley Cottages:*** self-contained cottages; 18 Windjana Rd; (08) 9191 1114. ***Kimberley Entrance Caravan Park:*** shady caravan park; 2–12 Rowan St; (08) 9193 1055. ***King Sound Resort Hotel:*** central hotel accommodation; 112 Loch St; (08) 9193 1044.

Windjana Gorge National Park A 350-million-year-old Devonian reef rises majestically above the surrounding plains. An easy walking trail winds through the gorge, taking in primeval life forms fossilised within the gorge walls. 145 km E.

Tunnel Creek National Park Wear sandshoes, carry a torch and be prepared to get wet as you explore the 750 m long cave that runs through the Napier Range. Nearby Pigeon's Cave was the hideout of an 1890s Aboriginal outlaw, Jandamarra, also known as 'Pigeon'. Tour details from visitor centre. 184 km E.

Prison tree: 1000-year-old boab tree formerly used as a prison; 7 km S. ***Myall's Bore:*** beside the bore stands a 120 m long cattle trough reputed to be the longest in the Southern Hemisphere; 7 km S. ***Gorges:*** Lennard Gorge (190 km E), Bell Gorge (214 km E), Manning Gorge (306 km E), Barnett River Gorge (340 km NE) and Sir John Gorge (350 km E); 4WD access only. ***Mitchell Plateau:*** highlights include the Wandjina rock art and spectacular Mitchell Falls, King Edward River and Surveyor's Pool. In this remote region, visitors must be entirely self-sufficient; via Gibb River Rd and Kalumburu Rd; 580 km NE. Scenic flights can also be arranged from Drysdale River Station and Kununurra. ***Pigeon Heritage Trail:*** follow the story of the Aboriginal outlaw Jandamarra, nicknamed 'Pigeon', and his people, the Bunuba; maps from visitor centre. ***Gibb River Rd:*** 4WD road between Derby and Wyndham traverses some of the most spectacular gorge country of the Kimberley; guidebook and current road conditions from visitor centre. ***Buccaneer Archipelago:*** in Derby you can arrange a scenic flight or cruise around this archipelago which begins north of King Sound; *see Broome.*

See also **Kimberley, p. 314**

Dongara–Denison

Pop. 3052
Map ref. 620 A4

9 Waldeck St; (08) 9927 1404; www.irwin.wa.gov.au

96.5 WA FM, 828 AM ABC Local Radio

Dongara and its nearby twin town of Port Denison lie on the coast 359 kilometres north of Perth. Dongara–Denison is the self-proclaimed 'Lobster Capital' of the state, with its offshore reefs supporting a profitable industry. Dongara's main street is lined with magnificent Moreton Bay fig trees while Port Denison provides local anglers with a large marina and harbour.

Irwin District Museum Housed in Dongara's Old Police Station, Courthouse and Gaol (1870), the museum features exhibits on the history of the buildings, the invasion of rabbits into WA and the Irwin Coast shipwrecks. Open 10am–4pm Mon–Fri; Waldeck St; (08) 9927 1323.

Russ Cottage: a beautifully restored farm-worker's cottage (1870). The hard-packed material of the kitchen floor was made from scores of anthills, and the flood-level marker near the front door indicates how high the nearby Irwin River rose during the record flood of 1971; open 10am–12pm Sun or by appt; St Dominics Rd, Dongara. ***The Priory Hotel:*** this 1881 building has been an inn, a priory and a boarding college for girls and is now once again a hotel; St Dominics Rd, Dongara. ***Church of St John the Baptist:*** (1884) its pews were made from the driftwood of shipwrecks and its church bell is said to have come from Fremantle Gaol; Cnr Waldeck and Church sts, Dongara. ***The Royal Steam Flour Mill:*** (1894) it served the local wheat-growing community until its closure in 1935; northern end of Waldeck St, Dongara. ***Cemetery:*** headstones dating from 1874 and a wall of remembrance to Dominican sisters; brochure from visitor centre; Dodd St, Dongara. ***Town heritage trail:*** 1.6 km walk that features 28 historic Dongara sites; maps from visitor centre. ***Fisherman's Lookout:*** 1 remaining of 2 obelisks built in 1869, with panoramic views of Port Denison; Pt Leander Dr, Port Denison.

Monthly Market: Priory Gardens; 1st Sat each month. ***Craft Market:*** old police station, Dongara; Easter and Christmas. ***Dongara Races:*** Easter. ***Larry Lobster Community Festival and Blessing of the Fleet:*** at the start of each rock lobster season; Nov.

The Season Tree: rustic cafe; 8 Moreton Tce; (08) 9927 1400.

Getaway Beach: contemporary beach houses; off Brand Hwy; (08) 9927 2458. ***Port Denison Holiday Units:*** self-contained holiday units; 14 Carnarvon St; (08) 9927 1104. ***Sea Spray Beach Holiday Park:*** beachside apartments; 79–81 Church St; (08) 9927 1165.

Silverdale Olive Orchards: olive oil products and tastings; open Sat Apr–Nov or by appt; 10 km N. ***Mingenew:*** small town in agricultural surrounds. Nearby is Fossil Cliff, filled with marine fossils over 250 million years old; 47 km E.

See also **Outback Coast & Mid-West, p. 312**

Donnybrook

Pop. 1932
Map ref. 616 F5 | 618 C9 | 620 C10

Old Railway Station, South Western Hwy; (08) 9731 1720; www.donnybrook-balingup.wa.gov.au

 95.7 Hot FM, 1224 AM ABC Radio National

Donnybrook is the centre of the oldest and largest apple-growing area in Western Australia. This is the home of the Granny Smith apple and where Lady William apples were developed. Gold was found here in 1897 but mined for only four years. Donnybrook is famous for its sandstone, which has been used in construction statewide since the early 1900s. In Perth, the GPO, St Mary's Cathedral and the University of Western Australia buildings have all been faced with Donnybrook stone. The quarry can be seen from the Upper Capel Road out of town.

Memorial Hall: built of Donnybrook stone; Bentley St. ***Anchor and Hope Inn:*** (1862) the oldest homestead in the district, now a private property; view outside from South Western Hwy. ***Trigwell Place:*** picnic and barbecue facilities, and canoeing on nearby Preston River; South Western Hwy.

Gourmet Wine and Food Fest: Feb. ***Apple Festival Ball:*** even-numbered years, Easter. ***Marathon Relay:*** Nov.

Old Goldfields Orchard & Cider Factory: country cosiness and international cuisine; 75 Goldfields Rd; (08) 9731 0311.

Boronia Farm: self-contained cottage on organic farm; 47 Williams Rd; (08) 9731 7154. ***Country Charm Retreat B&B:*** country B&B; 629 Hurst Rd; (08) 9731 2010. ***Jarragon B&B:*** centrally located cottage; 9 Collins St; (08) 9731 1930. ***Kirup Kabins Farmstay:*** self-contained cottages; Lot 3 Mailman Rd; (08) 9731 6272.

Old Goldfields Orchard and Cider Factory Combines goldfield history with a working orchard and restaurant. Climb the reconstructed poppet head over the mine, study the history of gold on the property and try your hand at gold prospecting. The orchard provides seasonal fruit for sale and you can enjoy tastings of cider, fruit juice and wines. Open 9.30am–4.30pm Wed–Sun and public/school holidays; Goldfields Rd; (08) 9731 0322; 6 km S.

Boyanup: features a transport museum; 12 km NW. ***Ironstone Gully Falls:*** barbecue area en route to Capel; 19 km W. ***Gnomesville:*** surprising roadside collection of garden gnomes; by the side of the Wellington Mills roundabout on the road between Dardanup and Lowden; 25 km SE.

See also **The South-West, p. 307**

Dunsborough

Pop. 3373
Map ref. 616 B6 | 618 B9 | 620 B10

Seymour Blvd; (08) 9752 1288; www.geographebay.com

98.4 FM Western Tourist Radio, 1224 AM ABC Radio National

Dunsborough is a picturesque coastal town on the south-western tip of Geographe Bay. Just west of the town is Leeuwin–Naturaliste National Park with its dramatic coastline and seasonal wildflower displays. Many of the wineries of the South-West region are only a short drive from the town.

Market: Dunsborough Hall, cnr Gibney St and Gifford Rd; 1st Sat each month. ***Margaret River Wine Festival:*** throughout region; Apr/May.

Dunsborough Bakery: renowned bakery; 243 Naturaliste Tce; (08) 9755 3137. ***Other Side of the Moon Restaurant:*** swanky fine dining; Quay West Resort, Bunker Bay Rd, Bunker Bay; (08) 9756 9159. ***Wise Vineyard Restaurant:*** seasonal international cuisine; 80 Eagle Bay Rd; (08) 9755 3331.

Dunsborough Rail Carriages & Farm Cottages: restored rail carriages; 123 Commonage Rd; (08) 9755 3865. ***Newberry Manor:*** character-filled B&B; 16 Newberry Rd; (08) 9756 7542. ***Quay West Resort Bunker Bay:*** sublime resort; Bunker Bay Rd, Bunker Bay; (08) 9756 9100 or 1800 010 449. ***Whalers Cove:*** contemporary self-contained accommodation; 3 Lecaille Crt; (08) 9755 3699.

Leeuwin–Naturaliste National Park Close to Dunsborough at the northern end of the park is Cape Naturaliste, with its lighthouse, museum and whale-watching platform (humpback whales linger offshore Sept–Nov). Walking tracks offer spectacular views of the coastline. Sugarloaf Rock is a dramatic formation just south of the lighthouse – it is also a habitat of the endangered red-tailed tropic bird. 13 km NW. *See also Margaret River and Augusta.*

Country Life Farm: animals galore, plus merry-go-round, giant slide and bouncing castles; Caves Rd; (08) 9755 3707; 1 km W. ***Simmo's Icecreamery:*** 39 flavours of homemade ice-cream made fresh daily; Commonage Rd; 5 km SE. ***Quindalup Fauna Park:*** specialises in birds, fish, tropical butterflies and baby animals; (08) 9755 3933; 5 km E. ***Wreck of HMAS Swan:*** the largest accessible dive-wreck site in the Southern Hemisphere; tour bookings and permits at visitor centre; off Pt Picquet, just south of Eagle Bay; 8 km NW. ***Beaches:*** to the north-west, popular for fishing, swimming and snorkelling, include Meelup (5 km), Eagle Bay (8 km) and Bunker Bay (12 km). ***Wineries:*** as part of the Margaret River wine region, there are many wineries nearby; details from visitor centre. ***Tours and activities:*** whale-watching charters (Sept–Nov); deep-sea fishing charters; scuba diving, snorkelling and canoeing; wildflower displays in season.

See also **The South-West, p. 307**

Dwellingup

Pop. 344
Map ref. 615 E10 | 618 C7 | 620 C8

Marrinup St; (08) 9538 1108; www.murray.wa.gov.au

97.3 Coast FM, 684 AM ABC Local Radio

Set among pristine jarrah forest, this is a thriving timber town that was virtually destroyed in 1961 when lightning started a bushfire that lasted for five days, burnt 140 000 hectares of forest and destroyed several nearby towns. Dwellingup was the only town to be rebuilt, and is now a forest-management centre. The Hotham Valley Tourist Railway operates here.

Forest Heritage Centre This centre records WA's jarrah forest heritage and promotes fine wood design, training and education. The building is formed from rammed earth and designed to represent three jarrah leaves on a bough. It includes an Interpretive Centre, a School of Wood and a Forest Heritage Gallery. Learn about conservation and walk among the treetops on an 11 m high canopy walkway. Acacia Rd; (08) 9538 1395.

Historical Centre: includes a photographic display depicting early 1900s life in the mill towns. Also a 1939 Mack Fire Truck, the only one in WA; visitor centre, Marrinup St. ***Community Hotel:*** last community hotel in WA; Marrinup St.

Log Chop and Community Fair: Feb. ***Giant Pumpkin Competition:*** Apr.

Dwellingup Community Hotel Motel: counter meals; Marrinup St; (08) 9538 1056. ***Dwellingup Millhouse Restaurant:*** cosy cafe; 41 McLarty St; (08) 9538 1122. ***Lake Navarino Restaurant:*** cosy restaurant, international cuisine; Lake Navarino Forest Resort, 147 Invarell Rd, Waroona; (08) 9733 3000 or 1800 650 626. ***Newbliss Vineyard Cafe:*** rustic lunches; Newbliss Winery, Lot 20 Irwin Rd; (08) 9538 1665.

Banksia Springs Cottages: well-appointed cottages; Banksiadale Rd; (08) 9538 1880. ***Dwellingup Bunkhouses Outdoor Adventure Camp:*** perfect for group adventurists; Lot 1379 Vandals Rd; (08) 9538 1314. ***Dwellingup Chalet & Caravan Park:*** peaceful accommodation; 23 Del Park Rd; (08) 9538 1157. ***Lake Navarino Forest Resort:*** variety of accommodation in country setting; 147 Invarell Rd, Waroona; (08) 9733 3000 or 1800 650 626.

Lane–Poole Reserve Provides opportunities for picnicking, swimming, canoeing, rafting, fishing, camping and walking. Walk trails include sections of the Bibbulmun Track, the 18 km King Jarrah Track from Nanga Mill, the 17 km Nanga Circuit and a 1.5 km loop from Island Pool. 10 km S.

Marrinup Forest Tour: unique 16 km vehicle and walk tour that features many aspects of the Darling Scarp including the Marrinup POW camp and remnants of old mills and towns of days gone by; maps from visitor centre. ***Hotham Valley Tourist Railway:*** travel from Perth via Pinjarra to Dwellingup by train, taking in lush green dairy country before climbing the Darling Range, WA's steepest and most spectacular section of railway, and finishing in the heart of the jarrah forest; steam-hauled May–Oct, diesel-hauled Nov–Apr; bookings (08) 9221 4444; Dwellingup Railway Station. ***Etmilyn Forest Tramway:*** takes visitors 8 km through farms and old-growth jarrah forest to the pioneer settlement of Etmilyn. ***Bibbulmun Track:*** long-distance walk trail runs through the middle of the town; *see Albany*. ***Munda Biddi Trail:*** WA's first long-distance off-road bike track begins in Mundaring near Perth and winds 182 km through native forest to Dwellingup. It will eventually be extended to Albany; details from visitor centre.

See also **The South-West, p. 307**

Esperance

Pop. 9536
Map ref. 621 I9

Museum Village, Dempster St; (08) 9083 1555 or 1300 664 455; www.visitesperance.com

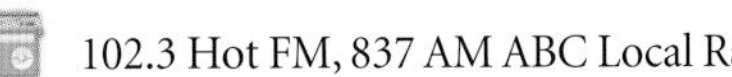

102.3 Hot FM, 837 AM ABC Local Radio

Esperance was a sleepy backwater until, in the 1950s, it was found that adding trace elements to the sandy soil made farming feasible. The town became a port and service centre for the agricultural and pastoral hinterland. However, it is the magnificent scenery, the pristine beaches and the proximity of many national parks that draw visitors to this town. Take the Great Ocean Drive, 38 kilometres of postcard-perfect scenery, and you will understand why Esperance is a popular holiday spot.

Municipal Museum Visit one of WA's outstanding regional museums. See exhibits about shipwrecks, including the famous *Sanko Harvest*, and learn of Australia's only recorded pirate, the bloodthirsty Black Jack Anderson, who roamed the Recherche

Archipelago. There is also a comprehensive display about *Skylab*, which crashed and spread debris through the area in 1979. Open 1.30–4.30pm daily; James St; (08) 9071 1579.

Museum Village: collection of historic buildings housing craft shops, pottery shops, art gallery, cafe and visitor centre; Dempster St. ***Cannery Arts Centre:*** local exhibitions with wind garden and views behind; Norseman Rd. ***Mermaid Leather:*** unique range of leather products made from fish and shark skins; Wood St. ***Aquarium:*** 14 aquariums and touch pool; the Esplanade.

Market: Dempster St; Sun mornings. ***Wildflower Show:*** Sept. ***Agricultural Show:*** Oct.

Bonaparte Seafood Restaurant: traditional town favourite; 51 The Esplanade; (08) 9071 7727. ***Loose Goose:*** vibrant modern restaurant; 9A Andrews St; (08) 9071 2320. ***Ocean Blues Restaurant:*** simple seaside fare; 19 The Esplanade; (08) 9071 7107. ***Taylor Street Cafe:*** beachside cafe; Taylor St Jetty; (08) 9071 4317. ***The Deck:*** a favourite for ice-creams; Cnr Clarke and Veal sts, Hopetoun; (08) 9838 3303.

Bay of Islands B&B: seaview B&B; 73 Twilight Beach Rd; (08) 9072 1995. ***Seascape Beach House:*** holiday house set in bush surrounds; 11 Cornell St; (08) 9071 3150. ***The Jetty Resort:*** range of apartments; 1 The Esplanade; (08) 9071 3333. ***Woody Island Ecostays:*** safari huts, eco accommodation; via Mackenzie's Island Cruises, 71 The Esplanade; (08) 9071 5757.

Great Ocean Drive One of Australia's most spectacular scenic drives, this 38 km loop road passes wind farms, which supply 30% of the town's electricity and some of the region's best-known natural attractions. Maps from visitor centre.

Cape Le Grand National Park This spectacular coastline is lined with pristine beaches, including Hellfire Bay and Thistle Cove. At Lucky Bay, kangaroos can often be spotted lying on the beach. Visit Whistling Rock, which 'whistles' under certain wind conditions, and climb Frenchman's Peak for breathtaking views. There are magnificent displays of wildflowers in spring, and many bushwalks. Camping at Cape Le Grand and Lucky Bay. 56 km E.

Recherche Archipelago The Esperance region is known as the Bay of Isles because of this collection of 110 islands dotted along the coast that provide a haven for seals and sea lions. Cruises (3 hrs 30 min, subject to numbers and weather) take you around Cull, Button, Charlie, Woody and other islands; landing is permitted only on Woody Island. For an extraordinary camping experience, try a safari hut on Woody Island (open Sept–Apr). These canvas huts set high on timber decking overlook an idyllic turquoise bay framed by eucalyptus trees. Woody Island also has an interpretive centre to provide information to visitors.

Rotary Lookout: views of bay, town and archipelago; Wireless Hill; 2 km W. ***Pink Lake:*** a pink saltwater lake; 5 km W. ***Twilight Cove:*** sheltered swimming; 12 km W. ***Observatory Pt and Lookout:*** dramatic views of bay and islands; 17 km W. ***Monjingup Lake Nature Reserve:*** walk trails, birdwatching and wildflowers in spring; 20 km W. ***Dalyup River Wines:*** the most isolated winery in WA; open weekends in summer; (08) 9076 5027; 42 km W. ***Stokes National Park:*** beautiful coastal and inlet scenery; 80 km W. ***Cape Arid National Park:*** birdwatching, fishing, camping and 4WD routes; 120 km E. ***Whale-watching:*** southern right whales visit bays and protected waters to calve (June–Oct); along the Great Ocean Drive and at Cape Arid. ***Great Country Drive:*** takes visitors 92 km inland; maps from visitor centre.

See also ESPERANCE & NULLARBOR, p. 310

Eucla

Pop. 50
Map ref. 610 C7 | 621 P6

Eucla Motel; (08) 9039 3468.

97.1 FM ABC Radio National, 531 AM ABC Local Radio

Eucla is the largest settlement on the Nullarbor Plain, located just near the South Australian border. The ruins of a telegraph station exist at the original townsite and beyond the ruins are the remains of a jetty, a reminder of pioneering days when supplies were transported by boat. Eucla is today located on the Hampton Tableland and operates on Central Western Time, 45 minutes ahead of the rest of Western Australia.

Telegraph station ruins Opened in 1877 (just 33 years after Samuel Morse invented the telegraph), the Eucla Telegraph Station helped link WA with the rest of Australia and the world, often sending over 20 000 messages a year. The first message, sent to Perth in December 1877, stated simply, 'Eucla line opened. Hurrah.' 4 km S.

Eucla Museum: local history, including exhibits of the telegraph station, told through newspaper clippings and old photographs; Eucla Motel. ***Travellers' Cross:*** dedicated to travellers and illuminated at night; on the escarpment, west of town. ***Bureau of Meteorology:*** visitors welcome; east of town; (08) 9039 3444. ***9-hole golf course:*** site of the Golf Classic in May; north of town.

Eucla Amber Motor Hotel: counter meals; Eyre Hwy; (08) 9039 3468.

Eucla Amber Motor Hotel: hotel rooms, caravan park; Eyre Hwy; (08) 9039 3468.

Eucla National Park This small park extends between Eucla and Border Village. On the coast near the SA border is Wilsons Bluff Lookout, with views to the east following the Bunda Cliffs into the distance. Closer to Eucla are the enormous sculptural shapes of the Delisser Sandhills. Mark your footprints in the dunes.

Border Village: quarantine checkpoint for people entering WA (travellers should ensure they are not carrying fruit, vegetables, honey, used fruit and produce containers, plants or seeds). The Border Dash starts here every Oct; 13 km E.

See also ESPERANCE & NULLARBOR, p. 310

Exmouth

Pop. 1845
Map ref. 619 B3

Murat Rd; (08) 9949 1176 or 1800 287 328; www.exmouthwa.com.au

107.7 ABC Radio National, 1188 AM ABC Local Radio

One of the newest towns in Australia, Exmouth was founded in 1967 as a support town for the Harold E. Holt US Naval Communications Station, the main source of local employment. Excellent year-round fishing and proximity to Cape Range National Park and Ningaloo Reef have since made Exmouth a major tourist destination. The town is the nearest point in Australia to the continental shelf.

Mall Market: each Sun Apr–Sept. ***Gamex:*** world-class game-fishing competition; Mar. ***Whale Shark Festival:*** Apr/May. ***Art Quest:*** July. ***Bill Fish Bonanza:*** Oct/ Nov.

Mantaray's Restaurant: modern brasserie; Novotel Ningaloo Resort, Madaffari Dr; (08) 9949 0000. ***Potshot Resort Restaurant:*** relaxed dining; Potshot Resort, Murat Rd; (08) 9949 1200. ***Sailfish Bar and Restaurant:*** alfresco dining and beer garden; Seabreeze Resort, 116 North C St; (08) 9949 1800. ***Whaler's Restaurant:*** local seafood; 5 Kennedy St; (08) 9949 2416.

Best Western Sea Breeze Resort: close access to Ningaloo Marine Park; 116 North C St; (08) 9949 1800. ***Exmouth Cape Holiday Park:*** range of accommodation; 3 Truscott Cres; (08) 9949 1101 or 1800 621 101. ***Novotel Ningaloo Resort:*** ideal for exploring Ningaloo Marine Park; Madaffari Dr; (08) 9949 0000. ***Sal Salis Ningaloo Reef:*** camping in style; North West Cape, Cape Range National Park; (02) 9571 6399 or 1300 790 561.

Cape Range National Park This rugged landscape of arid rocky gorges is edged by the stunning coastline of Ningaloo Marine Park. Wildlife is abundant, with emus, euros, rock wallabies and red kangaroos often sighted. In late winter there is a beautiful array of wildflowers including the Sturt's desert pea and the superb bird flower. Attractions within the park include Shothole Canyon, an impressive gorge; Mangrove Bay, a sanctuary zone with a bird hide overlooking a lagoon; and Mandu Mandu Gorge where you can walk along an ancient river bed. Yardie Creek is the only gorge with permanent water. Turquoise Bay is a popular beach for swimming and snorkelling (watch for currents). The Milyering Visitor Centre (54 km sw), made of rammed earth and run by solar power, is 52 km from Exmouth on the western side of the park and offers information on both Cape Range and Ningaloo. Contact (08) 9949 2808.

Ningaloo Marine Park: *see Coral Bay*. ***Naval Communication Station:*** the centre tower in its antenna field, at 388 m, is one of the tallest structures in the Southern Hemisphere; not open to public; 5 km N. ***Vlamingh Head Lighthouse and Lookout:*** built in 1912, Australia's only kerosene-burning lighthouse served as a beacon to mariners until 1967. The lookout offers panoramic 360-degree views; 19 km N. ***Learmonth Jetty:*** popular fishing spot, rebuilt after Cyclone Vance; 33 km S. ***Wildlife-watching:*** turtle-nesting (Nov–Jan); coral-spawning (Mar–Apr); boat cruises and air flights to see whale sharks (Mar–June); humpback whales (Aug–Nov) from lighthouse (17 km N) and from whale-watching boat tours. Snorkellers can swim with whale sharks and manta rays located by cruise boats. Coral-viewing boat cruises also available; details from visitor centre.

See also OUTBACK COAST & MID-WEST, p. 312

Fitzroy Crossing

Pop. 925
Map ref. 625 K8 | 628 H11

Cnr Great Northern Hwy and Flynn Dr; (08) 9191 5355.

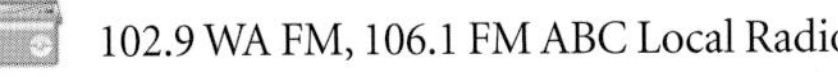
102.9 WA FM, 106.1 FM ABC Local Radio

Fitzroy Crossing is in the heart of the Kimberley region. As its name suggests, the original townsite was chosen as the best place to ford the mighty Fitzroy River. In the wet season, the river can rise over 20 metres and spread out up to 15 kilometres from its banks. Fitzroy Crossing's main attraction is its proximity to the magnificent 30-metre-deep Geikie Gorge with its sheer yellow, orange and grey walls. Check road conditions before any excursions from December to March, as this area is prone to flooding.

Crossing Inn First established in the 1890s as a shanty inn and trade store for passing stockmen, prospectors and drovers, it has operated on the same site ever since and is one of the very few hotels in the state to retain a true outback atmosphere. A stop-off and drink are a must for all travellers passing by. Skuthorp Rd; (08) 9191 5080.

Rodeo: July. ***Garnduwa Festival:*** sporting events; Oct.

Riverside Restaurant: traditional Australian fare; Fitzroy River Lodge, Great Northern Hwy; (08) 9191 5141.

Fitzroy River Lodge: range of accommodation; Great Northern Hwy; (08) 9191 5141. ***Imintji Wilderness Camp:*** wilderness camp with amazing views; Gibb River Rd; (08) 9277 8444 or 1800 889 389. ***Mornington Wilderness Camp:*** safari tents; Gibb River Rd; (08) 9191 7406 or 1800 631 946.

Geikie Gorge National Park Geikie Gorge has cliffs and sculptured rock formations carved by water through an ancient limestone reef. The Fitzroy River is home to sharks, sawfish and stingrays that have, over centuries, adapted to the fresh water. Freshwater crocodiles up to 3 m long and barramundi are plentiful, best seen on a guided boat tour. Aboriginal heritage and cultural tours are run by guides from the local Bunuba tribe; bookings essential. DEC rangers run tours on the geology, wildlife and history of the area. Entry to park is restricted during wet season (Dec–Mar). Details from visitor centre. 18 km NE.

Causeway Crossing: a concrete crossing that was the only way across the river until the new bridge was built in the 1970s; Geikie Gorge Rd; 4 km NE. ***Tunnel Creek National Park:*** unique formation created by waters from the creek cutting a 750 m tunnel through the ancient reef; 4WD access only; 110 km NW. ***Windjana Gorge National Park:*** 350-million-year-old Devonian reef rising majestically above the surrounding plains. An easy walking trail takes you past primeval life forms fossilised within the gorge walls; 4WD access only; 145 km NW. ***4WD tours:*** to Tunnel Creek and Windjana Gorge; bookings essential, details from visitor centre.

See also KIMBERLEY, p. 314

Gascoyne Junction

Pop. 46
Map ref. 613 D7 | 622 A8

Shire offices, 4 Scott St; (08) 9943 0988; www.gascoyneonline.com.au

Lying at the junction of the Lyons and Gascoyne rivers, Gascoyne Junction is a small administration centre for the pastoral industry. Sheep stations in the area, ranging in size from around 36 000 to 400 000 hectares, produce a wool clip exceeding 1.5 million kilograms annually.

Junction Hotel: see the high-water mark from the 1982 floods on the wall of this Aussie pub; Carnarvon–Mullewa Rd. ***Old Roads Board Museum:*** memorabilia of the area; Scott St.

Bush Races: Sept/Oct. ***Gascoyne Dash:*** cross-country endurance; Oct.

The Junction Hotel: counter meals; Lot 27, Carnarvon Meekatharra Rd; (08) 9943 0504.

Bidgemia Station: working cattle station; Carnarvon Rd; (08) 9943 0501.

Kennedy Range National Park Along with spectacular scenery, the park is home to fossils of the earliest known species of banksia in Australia and marine fossils that reflect the history of the region as an ocean bed. Ideal for sightseeing, hiking and bush camping, trails start from the camping area and pass through gorges where you can see honeycomb-like rock formations. 60 km N.

Mt Augustus National Park Mt Augustus is the world's largest monolith, twice the size of Uluru. It is also known as Burringurrah, named after a boy who, in Aboriginal legend, broke tribal law by running away from his initiation. On capture, he was speared in the upper right leg. The spear broke as the boy fell to the ground, leaving a section protruding from his leg. It is said that, as you look at Mt Augustus, you can see the shape of the boy's body with the stump of the spear being the small peak at the eastern end called Edney's Lookout. There are several walking and driving trails; maps from visitor centre. 294 km NE; *see Carnarvon*.

See also OUTBACK COAST & MID-WEST, p. 312

Geraldton

Pop. 31 550
Map ref. 619 D12 | 620 A3

Bill Sewell Complex, cnr Chapman Rd and Bayley St; (08) 9921 3999 or 1800 818 881; www.geraldtontourist.com.au

 94.9 ABC Classic FM, 96.5 WA FM

Situated on the spectacular Batavia Coast, Geraldton is the largest town in the mid-west region. As a port city, it is the major centre for the wheat belt and is renowned for its rock lobster industry. Geraldton is also regarded as one of the best windsurfing locations in the world. The nearby Houtman Abrolhos Islands are the site of 16 known shipwrecks. The most infamous is that of the Dutch ship *Batavia*, which foundered on a reef in 1629.

HMAS *Sydney* Memorial Built on Mt Scott overlooking the town to commemorate the loss of 645 men from HMAS *Sydney* on 19 November 1941. The ship sank after an encounter with the German raider HSK *Kormoran*. The wrecks of both ships were found in March 2008. Seven pillars representing the seven seas hold aloft a 9 m high domed roof formed of 645 interlocking figures of seagulls. At night an eternal flame lights the cupola. Near the memorial is the bronze sculpture of a woman looking out to sea, representing the women left behind waiting for those who would not return. Tours of the memorial site are conducted daily at 10.30am. Cnr George Rd and Brede St.

WA Museum Geraldton Exhibits focus on the cultural and natural heritage of the Geraldton region. Maritime displays include finds from Australia's oldest shipwrecks, notably the original stone portico destined to adorn the castle gateway in the city of Batavia and lost to the sea when the *Batavia* sank in 1629. Museum Pl, Batavia Coast Marina; (08) 9921 5080. Adjacent in the Geraldton Marina is a replica of the *Batavia* longboat.

Historic buildings: explore the town's historic architecture dating back to the mid-1800s, with works by noted architect Monsignor John Cyril Hawes a highlight. Many of the buildings have been restored and are open to the public, including the Old Geraldton Gaol (1858), which is now a craft centre and the Bill Sewell Complex (1884), which was built as a hospital and subsequently became a prison. In Cathedral Ave, St Francis Xavier Cathedral offers tours (10am Mon, Wed and Fri), and the Cathedral of the Holy Cross has one of the largest areas of stained glass in Australia. Self-guide walks are available, including the Heritage Trail; details from visitor centre. ***Geraldton Regional Art Gallery:*** the original Geraldton Town Hall (1907) converted to house art exhibitions and workshops; closed Mon; Chapman Rd. ***Leon Baker Jewellers:*** international jeweller works with Abrolhos pearls and Argyle diamonds. Workshop tours are available; Marine Tce Mall. ***Rock Lobster Factory:*** take a tour and follow the journey of Geraldton's most famous export, the western rock lobster, from processor to plate; covered shoes required; tours 9.30am Mon–Fri Nov–June; Willcock Dr, Fisherman's Wharf. ***Pt Moore Lighthouse:*** assembled in 1878 from steel sections prefabricated in England, and standing 34 m tall, this is the only lighthouse of its kind in Australia; Willcock Dr.

 ***Sunshine Festival:*** Oct.

Boatshed Restaurant: modern Australian and seafood; 359 Marine Tce; (08) 9921 5500. ***The Freemason's Hotel:*** brasserie style; cnr Durlacher St and Marine Terrace Mall; (08) 9964 3457. ***Tides of Geraldton:*** contemporary dining with harbour views; 103 Marine Tce; (08) 9965 4999. ***Chapman Valley Wines Restaurant:*** vineyard platters; Chapman Valley Wines, Lot 14 Howatharra Rd, Nanson; (08) 9920 5148.

Drummond Cove Holiday Park: well-equipped caravan park; North West Coastal Hwy; (08) 9938 2524. ***Mantra Geraldton:*** self-contained studios and apartments; 221 Foreshore Dr; (08) 9956 1300 or 1300 987 604. ***Marina Views:*** harbour-view apartments; Foreshore Dr; (08) 9938 3848. ***Ocean Centre Hotel:*** boutique hotel; Cnr Foreshore Dr and Cathedral Ave; (08) 9921 7777.

Houtman Abrolhos Islands These 122 reef islands with a fascinating history span 100 km of ocean and are the main source of rock lobster for the local lobster fishing industry. There are 16 known shipwrecks in the Abrolhos Islands, the most infamous of which is that of the Dutch ship *Batavia* from 1629. Captain Pelsaert and 47 of the survivors sailed north to Batavia (modern-day Jakarta) for help. When they returned three and a half months later, they discovered that a mutiny had taken place and 125 of the remaining survivors had been massacred. All of the mutineers were hanged, except for two who were marooned on the mainland, becoming Australia's first white inhabitants. There is no record of their subsequent fate. The wreck was discovered in 1963 and some skeletons of victims of the mutiny have been found on Beacon Island. The islands now offer diving, snorkelling, surfing, windsurfing, fishing and birdwatching. Access is via boat or plane; tours and charters are available. Details from and bookings at visitor centre. 65 km W.

Fishing: good fishing spots at Sunset Beach (6 km N) and Drummond Cove (10 km N). ***Mill's Park Lookout:*** excellent views over Moresby Range and coastal plain; 10 km NE. ***Oakabella Homestead:*** one of the region's oldest pioneering homesteads with a rare buttressed barn; tours available; 30 km N. ***Chapman Valley:*** an area of scenic drives and spectacular scenery, once home to the first coffee bean plantation in Australia. Also a huge diversity of wildflowers on display July–Oct. Wineries include Chapman Valley Wines, the northernmost winery in WA; enjoy free tastings; 35 km NE. ***Scenic flights:*** tours over nearby Abrolhos Islands, Murchison Gorges or the coastal cliffs of Kalbarri; details from visitor centre.

See also OUTBACK COAST & MID-WEST, p. 312

Gingin

Pop. 527
Map ref. 615 C1 | 618 B3 | 620 C7

Council offices, Brockman St; (08) 9575 2211; www.gingin.wa.gov.au

 720 AM ABC Local Radio, 810 AM ABC Radio National

Gingin is one of the oldest towns in Western Australia, having been settled in 1832, only two years after the establishment of the Swan River Colony. For tourists, it has the charm of old original stone buildings within a picturesque natural setting. Situated 84 kilometres north of Perth, it is an ideal destination for a daytrip from the city.

Historic buildings Enjoy a pleasant self-guide stroll around the town on the Gingin Walkabout Trail, which features many fine examples of early architecture including Philbey's Cottage and St Luke's Anglican Church, both made from local stone. Maps from visitor centre.

Granville Park: in the heart of the town with free barbecue facilities, playground and picnic area. ***Self-guide walks:*** stroll along the Gingin Brook on the Jim Gordon VC Trail or try the Three Bridges Recreation Trail, rebuilt after being destroyed by fire in Dec 2002; maps from visitor centre.

Horticultural Expo: Apr. ***British Car Club Day:*** May. ***Market Day Festival:*** Sept.

Amirage Restaurant: European fare; 1654 Gingin Brook Rd; (08) 9575 7646. ***Kyotmunga Estate:*** platter lunches; 287 Chittering Valley Rd, Lower Chittering; (08) 9571 8001. ***Stringybark Winery & Restaurant:*** international cuisine; 2060 Chittering Rd, Chittering; (08) 9571 8069. ***Willowbrook Farm Tearooms:*** country tearooms; 1679 Gingin Brook Rd; (08) 9575 7566.

Brookside: chalets and shared accommodation; 1010 Chitna Rd, West Gingin; (08) 9575 7585. ***Bindoon's Windmill Farm:*** chalet and homestead accommodation; Great Northern Hwy, Bindoon; (08) 9576 1136. ***Orchard Glory Farm Resort:*** self-contained chalets; 41 Mooliabeenee Rd, Bindoon; (08) 9576 2888. ***The Orchard Villa:*** European-inspired home; Great Northern Hwy, Bindoon; (08) 9271 2270.

Gravity Discovery Centre Opened in 2003, this $4 million centre offers hands-on and static scientific displays on gravity, magnetism and electricity. It includes the biggest public astronomy centre in the Southern Hemisphere and the largest telescope in WA. Visitors can take a high-tech look at heavenly bodies in an evening presentation (bookings essential) and see a number of WA inventions relating to physics. Military Rd; (08) 9575 7577; 15 km SW.

Cemetery: with a spectacular display of kangaroo paws in early spring; northern outskirts of town. ***Jylland Winery:*** open to public, wine-tastings and cellar-door sales; 2 km S. ***West Coast Honey:*** live bee display, honey extraction, tastings, sales of honey and bee products; open 9am–4pm Wed–Sun, or Mon and Tues by appt; Gingin Brook Rd; 3 km W. ***Moore River National Park:*** special area for conservation featuring banksia woodlands and wildflower displays in spring; 20 km NW.

See also HEARTLANDS, p. 308

Greenough

Pop. 15 394
Map ref. 620 A3

Cnr Chapman Rd and Bayly St, Geraldton; (08) 9921 3999 or 1800 818 881; www.cgg.wa.gov.au

96.5 WA FM, 99.7 FM ABC Radio National

Lying 24 kilometres south of Geraldton, the Greenough Flats form a flood plain close to the mouth of the Greenough River. At its peak in the 1860s and '70s, Greenough (pronounced 'Grennuff') was a highly successful wheat-growing area. However, the combined effects of drought, crop disease and floods led to the area's decline and from 1900 the population dropped dramatically. The historic hamlet that was once the centre of this farming community has been extensively restored and is classified by the National Trust.

Central Greenough Historic Settlement Precinct of 11 restored stone buildings dating from the 1860s including a school, police station, courthouse, gaol and churches. Fully re-created interior furnishings. Self-guide maps are available, or tours by appt. Cnr Brand Hwy and McCartney Rd; (08) 9926 1084.

Pioneer Museum: folk display located in an original limestone cottage; tours available; Brand Hwy. ***Leaning trees:*** these trees are a unique sight, having grown sideways in response to the harsh salt-laden winds that blow from the Indian Ocean; seen from Brand Hwy on the Greenough Flats. ***Hampton Arms Inn:*** fully restored historic inn (1863); Company Rd.

Abrolhos Restaurant: seafood; Greenough River Resort, Greenough River Rd; (08) 9921 5888.

Rock of Ages Cottage: charming heritage B&B; Phillips Rd; (08) 9926 1154.

Walkaway Railway Station: built in the style of a traditional British railway station, now housing a railway and heritage museum; closed Mon; Evans Rd; 10 km E. ***Greenough River mouth:*** ideal for swimming, canoeing, beach and rock fishing, birdwatching and photography; 14 km N. ***Flat Rocks:*** surfing, swimming and rock-fishing. A round of the State Surfing Championships is held here in June every year; 10 km S. ***Ellendale Pool:*** this deep, freshwater swimming hole beneath spectacular sandstone cliffs is an ideal picnic area; 23 km E. ***Greenough River Nature Trail:*** self-guide walk; brochures from visitor centre. ***The Greenough/Walkaway Heritage Trail:*** 57 km self-drive tour of the area; maps from visitor centre.

See also OUTBACK COAST & MID-WEST, p. 312

Halls Creek

Pop. 1209
Map ref. 625 N7 | 629 L11

Hall St; (08) 9168 6262; www.hallscreek.wa.gov.au

102.9 WA FM, 106.1 FM ABC Local Radio

In the heart of the Kimberley region and on the edge of the Great Sandy Desert, Halls Creek is the site of the first payable gold discovery in Western Australia. In 1885 Jack Slattery and Charlie Hall (after whom the town is named) discovered gold, thereby sparking a gold rush that brought over 15 000 people to the area. In 1917 a seriously injured stockman named James 'Jimmy' Darcy was taken into Halls Creek. With neither doctor nor hospital in the town, the local postmaster carried out an emergency operation using a penknife as instructions were telegraphed by morse code from Perth. The Perth doctor then set out on the ten-day journey to Halls Creek via cattle boat, model-T Ford, horse-drawn sulky and finally, on foot, only to discover that the patient had died the day before his arrival. The event inspired Reverend John Flynn to establish the Royal Flying Doctor Service in 1928, a development that helped to encourage settlement throughout the outback.

Russian Jack Memorial: tribute to a prospector who pushed his sick friend in a wheelbarrow to Wyndham for medical help; Thomas St. ***Trackers Hut:*** restored original hut of Aboriginal trackers; Robert St, behind police station.

Rodeo: July. ***Picnic Races:*** Oct.

Russian Jack's Restaurant: hearty fare; Best Western Halls Creek Motel, 194 Great Northern Hwy; (08) 9168 9600.

Best Western Halls Creek Motel: central Kimberley location; 194 Great Northern Hwy; (08) 9168 9600. ***Kimberley Hotel:*** motel units; Roberta Ave; (08) 9168 6101.

Wolfe Creek Crater National Park Wolfe Creek Crater is the second-largest meteorite crater in the world. Named after Robert Wolfe, a Halls Creek prospector, it is 870–950 m across and in Aboriginal legend, said to be the site of the emergence of a powerful rainbow serpent from the earth. Scenic flights afford magnificent views; details from visitor centre.148 km s.

Purnululu National Park This World Heritage Area in the outback of the east Kimberley is home to the Bungle Bungle Range, a remarkable landscape of tiger-striped, beehive-shaped rock domes. A scenic flight is the best way to gain a perspective of the Bungle Bungles' massive size and spectacular scenery (details from visitor centre). The most visited site in Purnululu is Cathedral Gorge, a fairly easy walk. A couple of days and a backpack allow you to explore nearby Piccaninny Creek and Gorge, camping overnight. On the northern side of the park is Echidna Chasm, a narrow gorge totally different from those on the southern side. Purnululu is also rich in Aboriginal art, and there are many traditional burial sites within its boundaries. Purnululu is open to visitors (Apr–Dec) and is accessible by 4WD. There are few facilities and no accommodation; visitors must carry in all food and water and notify a ranger. 160 km E.

China Wall: white quartz formation said to resemble Great Wall of China; 6 km E. ***Caroline's Pool:*** deep pool ideal for swimming (in wet season) and picnicking; 15 km E. ***Old Halls Creek:*** remnants of original town including graveyard where James Darcy is buried; prospecting available; 16 km E. ***Palm Springs:*** fishing, swimming and picnicking; 45 km E. ***Sawpit Gorge:*** fishing, swimming, picnicking; 52 km E. ***Billiluna Aboriginal Community:*** fishing, swimming, camping, birdwatching and bushwalking, and bush tucker and cultural tours; 180 km s.

See also KIMBERLEY, p. 314

Harvey

Pop. 2602
Map ref. 616 F2 | 618 C7 | 620 C9

James Stirling Pl; (08) 9729 1122; www.harveytourism.com

96.5 FM Harvey Community Radio, 810 AM ABC Radio National

On the Harvey River, 18 kilometres from the coast, the thriving town of Harvey is surrounded by fertile, irrigated plains. Beef production, citrus orchards and viticulture flourish in the region and intensive dairy farming provides the bulk of Western Australia's milk supply. Bordered by the Darling Range, Harvey offers a wealth of natural attractions, from the magnificent scenic drives through the escarpment to the pristine white beaches with excellent sunsets and fishing on the coast.

Tourist and Interpretive Centre: tourist information and display of local industries and May Gibbs characters; James Stirling Pl. ***Big Orange:*** lookout, one of Australia's big icons; Third St. ***Harvey Museum:*** memorabilia housed in renovated railway station; open 2–4pm Sun or by appt; Harper St. ***Stirling Cottage:*** replica of the home of Governor Stirling, which later became the home of May Gibbs, author of *Snugglepot and Cuddlepie*; James Stirling Pl tourist precinct. ***Heritage Gardens:*** picturesque country gardens on the banks of the Harvey River; James Stirling Pl tourist precinct. ***Internment Camp Memorial Shrine:*** the only roadside shrine of its type in the world, built by prisoners of war in the 1940s; collect key from visitor centre; South Western Hwy. ***Heritage trail:*** 6.2 km self-guide walk includes historic buildings and sights of town; maps from visitor centre. ***Mosaics and murals:*** unique collection throughout the region. See Uduc Rd, South Western Hwy, and entrances to Myalup and Binningup.

Summer Series Concerts: Jan–Mar. ***Harvest Festival:*** Mar. ***Spring Fair:*** Sept.

Old Coast Road Brewery: tasting plates and pub fare; West Break Rd, Myalup; 1300 792 106.

Bluehills Farmstay Harvey: stylish chalets; Weir Rd; 0439 313 898. ***Harvey Hills Farmstay Chalets:*** family-friendly accommodation; Weir Rd; (08) 9729 1434. ***Harvey Rainbow Caravan Park:*** range of accommodation; 199 Kennedy St; (08) 9729 2239. ***Highland Valley Homestead:*** country luxury; 402 Collie River Rd, Burekup; (08) 9726 3080. ***Top Paddock Cottage:*** contemporary 4-star cottage; Vista Ridge Estate, 7 Newell St; (08) 9729 3240. ***Indian Ocean Retreat & Caravan Park:*** beachfront chalets; Myalup Beach Rd, Myalup; (08) 9720 1113.

Harvey Dam: landscaped park with viewing platform, amphitheatre, barbecues and playground. Fishing is allowed in season with permit; 3 km E. ***HaVe Cheese:*** tours and gourmet cheese tasting; (08) 9729 3949; 3 km s. ***White Rocks Museum and Dairy:*** founded in 1887. Compare current technology with display of machinery from the past; open 2–4pm or by appt for groups; 15 km s. ***Beaches:*** Myalup Beach provides good swimming, surfing and beach fishing; 21 km w. Binningup Beach is protected by a reef that runs parallel to shore and is ideal for sheltered swimming, snorkelling, beach fishing and boating; 25 km sw. ***Wineries:*** more than 10 wineries open to the public, only a short distance from town; details from visitor centre.

See also THE SOUTH-WEST, p. 307

Hyden

Pop. 281
Map ref. 618 H5 | 620 F8

Wave Rock; (08) 9880 5182; www.waverock.com.au

648 AM ABC Local Radio, 1296 AM ABC Radio National

The small wheat-belt town of Hyden is synonymous with its famous nearby attraction, Wave Rock, originally known as Hyde's Rock in honour of a sandalwood cutter who lived in the area. A typing error by the Lands Department made it Hyden Rock, and the emerging town soon became known as Hyden. The area around the town boasts beautiful wildflowers in spring, including a wide variety of native orchids.

Wave Rock Weekender: music festival; Sept/Oct.

Wave Rock Motel Homestead: 3 restaurants on-site; 2 Lynch St; (08) 9880 5052.

Wave Rock Cabins & Caravan Park: 3-star park; Wave Rock Rd; (08) 9880 5022. ***Wave Rock Lakeside Resort:*** lakeside 2-bedroom cottages; Wave Rock Rd; (08) 9880 5022. ***Wave Rock Motel Homestead:*** units and suites; 2 Lynch St; (08) 9880 5052.

Wave Rock Resembling a breaking wave, this 100 m long and 15 m high granite cliff owes its shape to wind action over the past 2.7 billion years. Vertical bands of colour are caused by streaks of algae and chemical staining from run-off waters (4 km E). At Wave Rock Visitor Centre see the largest lace collection in the Southern Hemisphere with fine examples of antique lace, including lace worn by Queen Victoria. There are local wildflower species on display, an Australiana collection at the Pioneer Town, fauna in a natural bush environment and a walking trail.

Hippo's Yawn: rock formation; 5 km E via Wave Rock. ***The Humps and Mulka's Cave:*** Aboriginal wall paintings; 22 km N via Wave Rock. ***Rabbit-Proof Fence:*** see the fence where it meets the road; 56 km E.

See also HEARTLANDS, p. 308

Jurien Bay

Pop. 1175
Map ref. 620 A5

Council offices, 110 Bashford St; (08) 9652 1020; www.dandaragan.wa.gov.au

 103.1 WA FM, 107.9 FM ABC Radio National

Jurien Bay, settled in the mid-1850s, is the centre of a lobster fishing industry. The jetty was constructed in 1885 to enable a more efficient route to markets for locally produced wool and hides. Located within a sheltered bay protected by reefs and islands, the town has wide beaches and sparkling waters ideal for swimming, waterskiing, windsurfing, snorkelling, diving and surfing. The Jurien Bay boat harbour services the fishing fleet and has facilities for holiday boating and fishing. Anglers can fish from boat, jetty and beach.

Jurien Bay Charters: boat and fishing charters, scuba diving, sea lion tours; bookings at dive shop; Carmella St. ***Old jetty site:*** plaque commemorates site of original jetty. Remains of the jetty's timber piles have been discovered 65 m inland from high-water mark, which indicates the gradual build-up of coastline over time; Hastings St.

Market: Bashford St; usually last Sat each month. ***Blessing of the Fleet:*** Nov.

Leuseurs Gallery Cafe: relaxed cafe; 36 Bashford St; (08) 9652 2113. ***Sandpiper Bar & Grill:*** international cuisine; Cnr Roberts and Sandpiper sts; (08) 9652 1229.

Apex Camp Jurien: peaceful holiday park; 15 Bashford St; (08) 9652 1010. ***Jurien Bay Tourist Park:*** deluxe chalets, caravans and camping; Roberts St; (08) 9652 1595. ***Ocean View Retreat:*** 5-bedroom beach house; 6 Coubrough Pl; (08) 9255 2653. ***Seafront Estate:*** modern villa; Heaton St; (08) 9652 2055.

Jurien Bay Marine Park Established in August 2003, this marine park extends from Wedge Island to Green Head and encompasses major sea lion and seabird breeding areas. The reefs within the park are populated by a wide range of plants and animals including the rare Australian sea lion, and the seagrass meadows are a breeding ground for western rock lobsters.

Lions Lookout: spectacular views of town and surrounds; 5 km E. ***Drovers Cave National Park:*** rough limestone country with numerous caves, all of which have secured entrances limiting public access; 4WD access only; 7 km E. ***Grigsons Lookout:*** panoramic views of ocean and hinterland. Also wildflowers July–Nov; 15 km N. ***Lesueur National Park:*** with over 900 species of flora, representing 10% of the state's known flora, Lesueur is an important area for flora conservation. Enjoy coastal views from a lookout; 23 km E. ***Stockyard Gully National Park:*** walk through 300 m Stockyard Gully Tunnel along winding underground creek; 4WD access only; 50 km N.

See also HEARTLANDS, p. 308

Kalbarri

Pop. 1329
Map ref. 619 C11

Grey St; (08) 9937 1104 or 1800 639 468; www.kalbarriwa.info

 102.9 WA FM, 106.1 FM ABC Local Radio

Kalbarri lies at the mouth of the Murchison River, flanked by Kalbarri National Park. Established in 1951, the town is a popular holiday resort, famous for the magnificent gorges up to 130 metres deep along the river. Just south of the township a cairn marks the spot where in 1629 Captain Pelsaert of the Dutch East India Company marooned two crew members implicated in the *Batavia* shipwreck and massacre. These were the first, albeit unwilling, white inhabitants of Australia.

Oceanarium: large aquariums and touch pools; Grey St; (08) 9937 2027. ***Pelican feeding:*** daily feeding by volunteers on the river foreshore; starts 8.45am; off Grey St. ***Family Entertainment Centre:*** trampolines, minigolf, bicycle hire; Magee Cres; (08) 9937 1105.

Sport Fishing Classic: Mar. ***Canoe and Cray Carnival:*** June.

Black Rock Cafe: cafe with emphasis on seafood; 80 Grey St; (08) 9937 1062. ***The Grass Tree Cafe & Restaurant:*** seafood with an Asian twist; 94–96 Grey St; (08) 9937 2288. ***The Jetty Seafood Shack:*** seafood restaurant with marina views; Shop 1, Marina Shopping Centre, 365 Grey St; (08) 9937 1067. ***Zuytdorp Restaurant:*** international cuisine; Kalbarri Beach Resort, Clotworthy St; (08) 9937 1061 or 1800 096 002.

Gecko Lodge: luxury B&B; 9 Glass St; (08) 9937 1900. ***Kalbarri Palm Resort:*** hotel and apartment accommodation; 8 Porter St; (08) 9937 2333 or 1800 819 029. ***Kalbarri Seafront Villas:*** spacious villas; 108 Grey St; (08) 9937 1025. ***Kalbarri Tudor Holiday Park:*** centrally located holiday park; 10 Porter St; (08) 9937 1077 or 1800 681 077.

Kalbarri National Park Gazetted in 1963, this park has dramatic coastal cliffs along its western boundary, towering river gorges and seasonal wildflowers, many of which are unique to the park. The Murchison River has carved a gorge through sedimentary rock known as Tumblagooda sandstone, creating a striking contrast of brownish red and purple against white bands of stone. Embedded in these layers are some of the earliest signs of animal life on earth. There are many lookouts including Nature's Window at the Loop, which overlooks the Murchison Gorge, and the breathtaking scenery at Z Bend Lookout. Along the 20 km coastal section of the park, lookouts such as Mushroom Rock, Pot Alley and Eagle Gorge offer panoramic views and whale-watching sites. Bushwalking, rock climbing, abseiling, canoeing tours, rafting, cruises, camping safaris, coach and wilderness tours, and barbecue facilities are all available. Details from visitor centre. 57 km E.

continued on p. 340

KALGOORLIE–BOULDER

Pop. 28 243

Map ref. 620 H5

Cnr Hannan and Wilson sts; 1800 004 653 or (08) 9021 1966; www.kalgoorlietourism.com

97.9 Hot FM, 648 AM ABC Local Radio

Kalgoorlie is the centre of Western Australia's goldmining industry. It was once known as Hannan's Find in honour of Paddy Hannan, the first prospector to discover gold in the area. In June 1893 Hannan was among a party of about 150 men who set out from Coolgardie to search for some lost prospectors. After a stop at Mount Charlotte, Hannan and two others were left behind, as one of their horses had lost a shoe. Here they stumbled on a rich goldfield – it soon grew to encompass the 'Golden Mile', which is reputedly the world's richest square mile of gold-bearing ore. Rapid development of Kalgoorlie and nearby Boulder followed, with thousands of men travelling to the field from all over the world. The shortage of water was always a problem, but in 1903 engineer C. Y. O'Connor opened a pipeline that pumped water 560 kilometres from Perth. In its heyday Kalgoorlie and Boulder boasted eight breweries and 93 hotels. The two towns amalgamated in 1989 to form Kalgoorlie–Boulder, and today the population is again close to peak levels due to the region's second gold boom. Unfortunatley, an earthquake in 2010 damaged many of the town's heritage buildings.

Historic buildings Although only a few kilometres apart, Kalgoorlie and Boulder developed independently for many years. The amalgamated towns now form a city with two main streets, each lined with impressive hotels and civic buildings. Built at the turn of the century, when people were flocking to the area, many of these buildings display ornamentation and fittings that reflect the confidence and wealth of the mining interests and are fine examples of early Australian architecture. The Kalgoorlie Town Hall (1908), on the corner of Hannan and Wilson sts, has hosted many a famous performer on its stage, including Dame Nellie Melba and Percy Grainger. It now displays a collection of memorabilia; tours 1.30pm Wed. The offices of newspapers the *Kalgoorlie Miner* and *Western Argus* was the first 3-storey building in town. Burt St in Boulder is regarded as one of the most significant streetscapes in WA. Buildings to see include the Grand Hotel (1897), the Old Chemist (1900), which now houses a pharmaceutical museum, and the post office (1899), which was once so busy it employed 49 staff. The Boulder Town Hall, built in 1902, is currently closed due to extensive damage caused by a 5.2 magnitude earthquake which struck Kalgoorlie-Boulder in April 2010. The Town Hall is home to one of the world's last remaining Goatcher stage curtains. Phillip Goatcher was one of the greatest scenic painters of Victorian times and his hand-painted drop curtains graced theatres in London, Paris and New York. Another building damaged in the earthquake was the Goldfields War Museum; a small part of the exhibition has been moved to the Kalgoorlie Town Hall (open 9am–4pm Mon–Fri). The self-guided Kalgoorlie Inner City Trail will take you on a tour of 40 historic buildings; contact Visitor Centre for brochure.

Paddy Hannan's Statue: a monument to the first man to discover gold in Kalgoorlie; Hannan St. ***St Mary's Church:*** built in 1902 of Coolgardie pressed bricks, many of which are believed to contain gold; Cnr Brookman and Porter sts, Kalgoorlie. ***WA School of Mines Mineral Museum:*** displays include over 3000 mineral and ore specimens and many gold nuggets; open 8.30am–12.30pm Mon–Fri, closed on school holidays; Egan St, Kalgoorlie. ***WA Museum Kalgoorlie–Boulder:*** panoramic views of the city from the massive mining headframe at the entrance. Known locally as the Museum of the Goldfields, displays include a million-dollar gold collection, nuggets and jewellery. See the narrowest pub in the Southern Hemisphere, a re-created 1930s miner's cottage and other heritage buildings. Guided tours available; open 10am-4.30pm daily; 17 Hannan St, Kalgoorlie. ***Goldfields Arts Centre:*** art gallery and theatre; Cassidy St, Kalgoorlie. ***Goldfields Aboriginal Art Gallery:*** examples of local Aboriginal art and artefacts for sale; open 9am–5pm Mon–Fri or by appt; Dugan St, Kalgoorlie; (08) 9021 8533. ***Paddy Hannan's Tree:*** a plaque marks the spot where Paddy Hannan first discovered gold; Brown Avenue, Kalgoorlie. ***Red-light district:*** view the few remaining 'starting stalls', in which women once posed as prospective clients walked by, and visit the only working brothel in the world that visitors can tour – Langtrees; tours by appt; (08) 9026 2181. ***Super Pit Lookout:*** underground mining on the Golden Mile became singly owned in the 1980s and '90s and was converted into an open-cut operation – what is now known as the Superpit. Peer into its depths from the lookout off Goldfields Hwy, Boulder (can coincide visit with blasting; check times at visitor centre); daily tours (free on market day) and scenic flights are also available. ***Mt Charlotte:*** the reservoir holds water pumped from the Mundaring Weir in Perth via the pipeline of C. Y. O'Connor. A lookout provides good views of the city; off Goldfields Hwy, Kalgoorlie. ***Hammond Park:*** miniature Bavarian castle made from thousands of local gemstones. There is also a sanctuary for kangaroos and emus, and aviaries for a variety of birdlife; Lyall St, Kalgoorlie. ***Arboretum:*** a living museum of species of the semi-arid zone and adjacent desert areas, this 26.5 ha parkland has interpretive walking trails and recreation facilities; Hawkins St, adjacent Hammond Park. ***Miners' Monument:*** tribute to mine workers; Burt St, Boulder. ***WMC Nickel Pots:*** massive nickel pots and interpretive panels describe the story of the development of the nickel industry in the region; Goldfields Hwy. ***Loopline Railway Museum:*** celebrates what was once one of the busiest railways in Australia, which connected Kalgoorlie and Boulder to outlying towns in the region; Boulder City Railway Station, cnr Burt and Hamilton sts; open 9am–1pm daily. ***Royal Flying Doctor Visitor Centre:*** climb on board an authentic RFDS plane, 45-minute tours available, on the hour; open 10am–3pm Mon–Fri; Airport, Hart Kerspien Dr. ***Karlkurla Bushland:*** pronounced 'gullgirla', this natural regrowth area of bushland offers a 4 km signposted walk trail, picnic areas and lookout over the city and nearby mining areas; Riverina Way. ***Walks:*** guided and self-guide heritage walks; details and maps from visitor centre.

Kalgoorlie Market Day: St. Barbara Sq; 1st Sun each month. ***Boulder Market Day:*** Burt St; 3rd Sun each month. ***Community Fair:*** Apr. ***Menzies to Kalgoorlie Cycle Race:*** June. ***Diggers and Dealers Mining Forum:*** Aug. ***Kalgoorlie Cup and Boulder Cup:*** Sept. ***Back to Boulder Festival:*** Oct. ***Balzano Barrow Race:*** Oct. ***St Barbara's Mining and Community Festival:*** Dec.

Barista 202: funky cafe; 202 Hannan St, Kalgoorlie; (08) 9022 2228. ***Saltimbocca:*** Italian restaurant; 90 Egan St; (08) 9022 8028. ***The Cornwall:*** pub and bistro; 25 Hopkins St, Boulder; (08) 9093 0900. ***The Exchange Hotel:*** traditional Irish and Australian food; 135 Hannan St, Kalgoorlie; (08) 9021 2833.

All Seasons Kalgoorlie Plaza Hotel: centrally located hotel; 45 Egan St, Kalgoorlie; (08) 9080 5900. ***Kalgoorlie Overland Motel:*** range of motel rooms; 566 Hannan St, Kalgoorlie; (08) 9021 1433. ***Palace Hotel:*** centrally located historic pub; 137 Hannan St, Kalgoorlie; (08) 9021 2788. ***Quest Yelverton Kalgoorlie:*** premier serviced apartments; 210 Egan St, Kalgoorlie; (08) 9022 8181. ***Rydges Kalgoorlie:*** studios and apartments; 21 Davidson St, Kalgoorlie; (08) 9080 0800 or 1300 857 922.

Australian Prospectors and Miners Hall of Fame Tour a historic underground mine, watch a gold pour or visit the Exploration Zone, which is designed specifically for young people. You will find interactive exhibits on exploration, mineral discoveries and surface and underground mining, including panning for gold. The Environmental Garden details the stages and techniques involved in mine-site rehabilitation. At Hannan's North Tourist Mine, historic buildings re-create an early gold-rush town and visitors can take a first-hand look at the cramped and difficult working conditions of the miners. Goldfields Hwy; (08) 9026 2700; 8.30am–4.30pm daily; 7 km N.

Bush 2-Up: visit the original corrugated-iron shack and bush ring where Australia's only legal bush 2-up school used to operate; off Goldfields Hwy; 8 km E. ***Kanowna Belle Gold Mine lookouts:*** wander the ghost town remains of Kanowna and see day-to-day mining activities from 2 lookouts over a previously mined open pit and processing plant; 20 km E. ***Broad Arrow:*** see the pub where scenes from the Googie Withers movie *Nickel Queen* were shot in the 1970s. Every wall is autographed by visitors; 38 km N. ***Ora Banda:*** recently restored inn; 54 km NW. ***Kambalda:*** nickel-mining town on Lake Lefroy (salt). Head to Red Hill Lookout for views across the expanse; 55 km S. ***Prospecting:*** visitors to the area may obtain a miner's right from the Dept of Mineral and Petroleum Resources in Brookman St, Kalgoorlie; strict conditions apply; details from the visitor centre. ***Lake Ballard:*** a collection of 51 black steel sculptures created by world-renowned artist Antony Gormley are scattered through the vast 10 square km white salt plain of Lake Ballard. Made using scans of locals from the nearby town of Menzies, these taut, stick-like body-forms are best seen at sunrise and sunset. 80 km NW. ***Golden Quest Discovery Trail:*** 965 km drive that traces the gold rushes of the 1890s through Coolgardie, Kalgoorlie–Boulder, Menzies, Kookynie, Gwalia, Leonora and Laverton; pick up the map, book and CD from visitor centre. ***Golden Pipeline Heritage Trail:*** follow the course of the pipeline from Mundaring Weir to Mt Charlotte. Finding a reliable water supply to support the eastern goldfields' booming population became imperative after the 1890s gold rush. C. Y. O'Connor's solution was radically brilliant: the construction of a reservoir at Mundaring in the hills outside Perth and a 556 km water pipeline to Kalgoorlie. His project was criticised relentlessly by the press and public, which affected O'Connor deeply. On 19 Mar 1902 he went for his usual morning ride along the beach in Fremantle. As he neared Robb Jetty, he rode his horse into the sea and shot himself. The pipeline was a success, delivering as promised 22 million litres of water a day to Kalgoorlie, and continues to operate today. On the coast just south of Fremantle, a half-submerged statue of a man on a horse is a poignant tribute to this man of genius; guidebook from visitor centre. ***Tours:*** self-drive or guided 4WD tours available to many attractions. Also fossicking, prospecting, camping and museum tours, and self-guide wildflower tours; details from visitor centre. ***Scenic flights:*** flights over Coolgardie, the Superpit, Lake Lefroy; details from visitor centre.

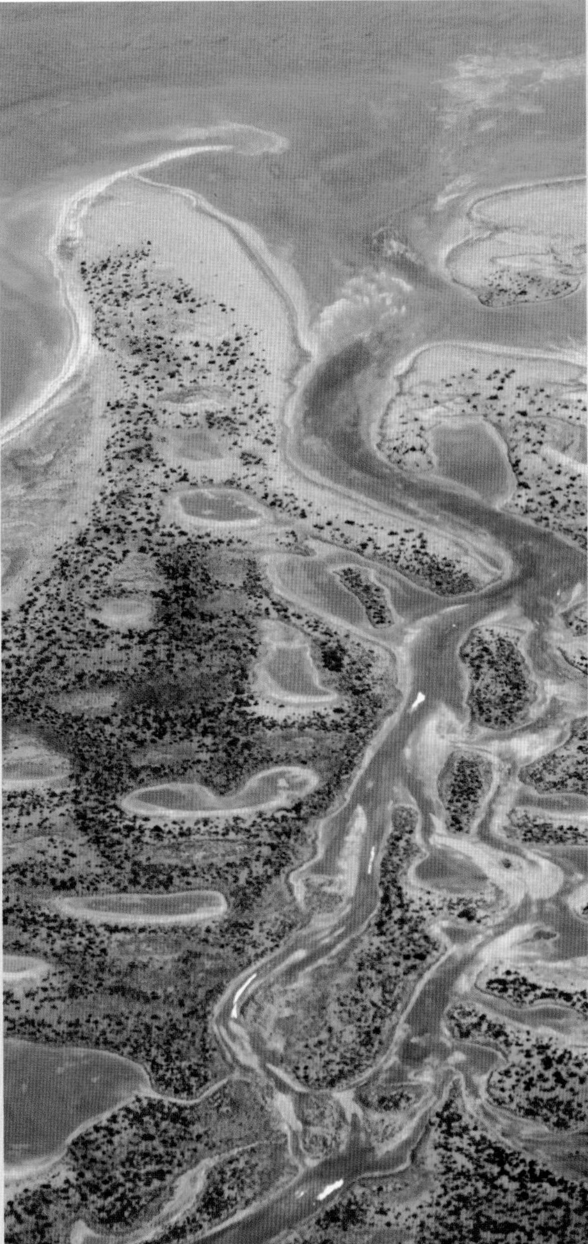
[AERIAL VIEW OF SALT LAKE NORTH-EAST OF KALGOORLIE]

See also GOLDFIELDS, p. 311

Rainbow Jungle: breeding centre for rare and endangered species of parrots, cockatoos and exotic birds set in landscaped tropical gardens. Also here is the largest walk-in parrot free-flight area in Australia and an outdoor cinema featuring the latest movies; Red Bluff Rd; 3 km S. ***Seahorse Sanctuary:*** aquaculture centre focused on the conservation of seahorses and other tropical marine fish; Red Bluff Rd; 3 km S. ***Wildflower Centre:*** view over 200 species on display along a 1.8 km walking trail. Visit the plant nursery and herbarium, where you can purchase seeds and souvenirs; open July–Nov; Ajana–Kalbarri Rd; 3km E. ***Murchison House Station:*** tours available of one of the oldest and largest stations in WA, which includes historic buildings and cemetery, display of local arts and crafts, and wildflowers; seasonal, check with visitor centre; Ajana–Kalbarri Rd; 4 km E. ***Big River Ranch:*** enjoy horseriding through the spectacular countryside; Ajana–Kalbarri Rd; 4 km E. ***Wittecarra Creek:*** cairn marking the site where 2 of the mutineers from the Dutch ship *Batavia* were left as punishment for their participation in the murders of 125 survivors of the wreck; 4 km S. ***Hutt River Province:*** independant sovereign state founded in 1970; 50 km SE; *see Northampton.*

See also OUTBACK COAST & MID-WEST, p. 312

Kalgoorlie–Boulder

see inset box on previous page

Karratha

Pop. 11 727
Map ref. 619 E1 | 622 B2 | 624 B12 | 626 D5

4548 Karratha Rd; (08) 9144 4600.

104.1 FM ABC News Radio, 106.5 WA FM

Karratha is the Aboriginal word for 'good country'. Founded in 1968 as a result of expansion of the iron-ore industry, Karratha was originally established for workers on the huge industrial projects nearby. For visitors, Karratha is an ideal centre from which to explore the fascinating Pilbara region.

TV Hill Lookout: excellent views over town centre and beyond; off Millstream Rd. ***Jaburara Heritage Trail:*** 3 hr walk features Aboriginal rock carvings; pamphlet from visitor centre.

FeNaClNG Festival: this celebration takes its name from the town's mining roots (Fe is the chemical symbol for iron ore, NaCl is the symbol for salt, and NG is an abbreviation for natural gas); Aug. ***Gamefishing Classic:*** Aug.

Etcetera Brasserie: fine dining, Pilbara-style; Karratha International Hotel, cnr Hillview and Millstream rds; (08) 9187 3333. ***Hearson's Bistro:*** poolside fine dining; All Seasons Karratha, Lot 1079 Searipple Rd; (08) 9185 1155. ***Whim Creek Pub:*** counter meals; North West Coastal Hwy, Whim Creek; (08) 9176 4914.

Balmoral Holiday Park: long-term accommodation; Balmoral Rd; (08) 9815 3628. ***Best Western Karratha Central Apartments:*** modern self-contained accommodation; 27 Warambie Rd; (08) 9143 9888. ***Pilbara Holiday Park:*** variety of accommodation options; Rosemary Rd; (08) 9185 1855 or 1800 451 855. ***Dampier Archipelago:*** luxury yacht; contact Karratha Visitor Centre; (08) 9144 4600.

Millstream–Chichester National Park Rolling hills, spectacular escarpments and tree-lined watercourses with hidden rockpools characterise this park. The remarkable oasis of Millstream is an area of tropical palm-fringed freshwater springs, well known to the Afghan cameleers of Pilbara's past. Other notably scenic spots are Python, Deepreach and Circular pools, and Cliff Lookout. The Millstream Homestead Visitor Centre, housed in the Gordon family homestead (1919), has displays dedicated to the local Aboriginal people, early settlers and the natural environment. Popular activities include bushwalking, picnicking, camping, fishing, swimming and boating. Tours are available; details from visitor centre. 124 km S.

Salt Harvest Ponds: Australia's largest evaporative salt fields. Tours are available; details from visitor centre; 15 km N. ***Dampier:*** port facility servicing the iron-ore operations at Tom Price and Paraburdoo. Watersports and boat hire are available; 22 km N. ***Hamersley Iron Port Facilities:*** 3 hr tour and audiovisual presentation daily; bookings essential at visitor centre; 22 km N. ***Cleaverville Beach:*** scenic spot ideal for camping, boating, fishing and swimming; 26 km NE. ***North-West Shelf Gas Project Visitor Centre:*** displays on the history and technology of Australia's largest natural resource development, with panoramic views over the massive onshore gas plant; open 10am–4pm Mon–Fri Apr–Oct, 10am–1pm Mon–Fri Nov–Mar; Burrup Peninsula; 30 km N. ***Aboriginal rock carvings:*** there are more than 10 000 engravings on the Burrup Peninsula alone, including some of the earliest examples of art in Australia. A debate is currently raging over the damage being done to this magnificent outdoor gallery by the adjacent gas project; check with visitor centre for locations. ***Dampier Archipelago:*** 42 islands and islets ideal for swimming, snorkelling, boating, whale-watching and fishing; take a boat tour from Dampier. ***Montebello Islands:*** site of Australia's first shipwreck, the *Tryal*, which ran aground and sank in 1622. It is now a good spot for snorkelling, beachcombing, fishing and diving; beyond Dampier Archipelago. ***Scenic flights:*** over the Pilbara outback; details from visitor centre.

See also PILBARA, p. 313

Katanning

Pop. 3806
Map ref. 618 F8 | 620 E10

Old Mill, cnr Austral Tce and Clive St; (08) 9821 4390; www.katanningwa.com

94.9 Hot FM, 612 AM ABC Radio National

Katanning lies in the middle of a prosperous grain-growing and pastoral area. A significant development in the town's history was the 1889 completion of the Great Southern Railway, which linked Perth and Albany. Construction was undertaken at both ends, and a cairn north of town marks the spot where the lines were joined.

Old Mill Museum: built in 1889, it features an outstanding display of vintage roller flour-milling processes; Cnr Clive St and Austral Tce. ***All Ages Playground and Miniature Steam Railway:*** scenic grounds with playground equipment for all ages. Covered shoes required to ride the train, which runs on the 2nd and 4th Sun of each month; Cnr Great Southern Hwy and Clive St. ***Kobeelya:*** a majestic residence (1902) with seven bedrooms, ballroom, billiard room, tennis courts and croquet lawn, now a conference centre; Brownie St. ***Old Winery Ruins:*** inspect the ruins of the original turreted distillery and brick vats, with old ploughs and machinery on display; Andrews Rd. ***Historical Museum:*** the original school building has been converted into a museum of local memorabilia; open 2–4pm Sun, or by appt; Taylor St. ***Sale Yards:*** one of the biggest yards in Australia, sheep sales every Wed at 8am; viewing platform for visitors; Daping St. ***Heritage Rose Garden:*** with roses dating from 1830; Austral Tce. ***Piesse Memorial Statue:*** unveiled in 1916, this statue of Frederick H. Piesse, the founder of Katanning, was sculpted by P. C. Porcelli, a well-known artist in the early days of WA; Austral

Tce. ***Art Gallery:*** a changing display and local collection; closed Sun; Austral Tce.

Farmers markets: Pemble St; 3rd Sat each month. ***Spring Lamb Festival:*** Oct.

Kimberley Restaurant: intimate dining, Kimberley-style; New Lodge Motel, 170 Clive St; (08) 9821 1788. ***Loretta @ Feddy:*** seasonal menu; Federal Hotel, 111 Clive St; (08) 9821 7128.

New Lodge Motel: comfortable motel accommodation; 170 Clive St; (08) 9821 1788. ***Woodchester B&B:*** colonial B&B; 19 Clive St; (08) 9821 7007.

Police Pools (Twonkwillingup): site of the original camp for the district's first police officers. Enjoy swimming, picnicking, birdwatching and bushwalking; 3 km s. ***Lake Ewlyamartup:*** picturesque freshwater lake ideal for picnicking, swimming, boating and waterskiing, particularly in early summer when the water level is high; 22 km e. ***Katanning-Piesse Heritage Trail:*** 20 km self-drive/walk trail; maps from visitor centre. ***Watersports:*** the lakes surrounding the town are excellent for recreational boating, waterskiing and swimming; details from visitor centre.

See also Great Southern, p. 309

Kellerberrin

Pop. 868
Map ref. 618 F4 | 620 D7

Shire offices, 110 Massingham St; (08) 9045 4006; www.kellerberrin.wa.gov.au

 107.3 FM ABC Radio National, 1215 AM ABC Local Radio

Centrally located in the wheat belt, Kellerberrin is 200 kilometres east of Perth. In springtime, magnificent displays of wildflowers adorn the roadsides, hills and plains around the town.

International Art Space Kellerberrin Australia This art gallery, built in 1998, is home to an ambitious art project. International artists are given the opportunity to live and work within the local community for a 3-month period. Workshops and mentoring programs provide collaboration between these established artists and emerging Australian talent. Many of the exhibitions created are then displayed in larger venues throughout Australia and the world. Massingham St; (08) 9045 4739.

Pioneer Park and Folk Museum: located in the old Agricultural Hall, displays include local artefacts, farming machinery and photographic records. Pick up the key from tourist information or Dryandra building next door; Cnr Leake and Bedford sts. ***Centenary Park:*** children's playground, in-line skate and BMX track, maze, heritage walkway and barbecue facilities all in the centre of town; Leake St. ***Golden Pipeline Lookout:*** interpretive information at viewing platform with views of the countryside and pipeline; via Moore St. ***Heritage trail:*** self-guide town walk that includes historic buildings and churches; brochures from visitor centre.

Keela Dreaming Cultural Festival: odd-numbered years, Mar. ***Central Wheatbelt Harness Racing Cup:*** May.

Kellerberrin Motor Hotel: hearty counter meals; 108 Massingham St; (08) 9045 5000.

Kellerberrin Caravan Park: 6 powered sites; Connelly St; 0428 138 474. ***Kellerberrin Motor Hotel:*** centrally located hotel; 108 Massingham St; (08) 9045 5000. ***The Prev:*** B&B with conference facilities; 1 George St; 0427 063 638.

Durokoppin Reserve: take a self-guide scenic drive through this woodland area, which is beautiful in the wildflower season; maps from visitor centre; 27 km n. ***Kokerbin Wave Rock:*** the third largest monolith in WA. The Devil's Marbles and a historic well are also at the site. Restricted vehicle access to the summit, but the walk will reward with panoramic views; 30 km s. ***Cunderdin:*** museum housed in the No 3 pumping station has displays on the pipeline, wheat-belt farming and the Meckering earthquake; 45 km w. ***Golden Pipeline Heritage Trail:*** one of the main stops along the trail, which follows the water pipeline of C. Y. O'Connor from Mundaring Weir to the goldfields; guidebook from visitor centre. *See Kalgoorlie–Boulder.*

See also Heartlands, p. 308

Kojonup

Pop. 1124
Map ref. 618 F9 | 620 D10

143 Albany Hwy; (08) 9831 0500; www.kojonupvisitors.com

558 AM ABC Local Radio, 612 AM ABC Radio National

A freshwater spring first attracted white settlement of the town now known as Kojonup. In 1837 Alfred Hillman arrived in the area after being sent by Governor Stirling to survey a road between Albany and the Swan River Colony. He was guided to the freshwater spring by local Aboriginal people and his promising report back to Governor Stirling resulted in a military outpost being established. The Shire of Kojonup was the first shire in Western Australia to have a million sheep within its boundaries.

Kodja Place Visitor and Interpretative Centre: fascinating and fun displays about the land and its people, with stories of Aboriginal heritage and white settlement. It also includes the Australian Rose Maze, the only rose garden in the world growing exclusively Australian roses; 143 Albany Hwy. ***Kojonup Museum:*** in historic schoolhouse building with displays of local memorabilia; open by appt; Spring St. ***A. W. Potts Kokoda Track Memorial:*** a life-size statue of the brigadier facing towards his beloved farm 'Barrule'; Albany Hwy. ***Centenary of Federation Wool Wagon:*** commemorates the significance of the sheep industry to the Kojonup community; Albany Hwy. ***Kojonup Spring:*** grassy picnic area; Spring St. ***Military Barracks:*** built in 1845, this is one of the oldest surviving military buildings in WA and features historical information about the building; open by appt; Spring St. ***Elverd's Cottage:*** display of pioneer tools and farm machinery; open by appt; Soldier Rd. ***Kodja Place Bush Tucker Walk:*** follows the old railway line east where 3000 trees and shrubs indigenous to the area have been planted; maps from visitor centre. ***Town walk trail:*** self-guide signposted walk of historic sights; maps from visitor centre.

Wildflower Week: Sept. ***Kojonup Show:*** Oct.

Commercial Hotel: homely pub; 118 Albany Hwy; (08) 9831 1044.

Jacaranda Heights B&B: secluded residence; 14 Stock Rd; (08) 9831 1200. ***Kemminup Farm Homestay:*** country-style B&B; Kemminup Rd; (08) 9831 1286. ***Kojonup B&B:*** converted barn; 47 Newstead Rd; (08) 9831 1119.

Myrtle Benn Memorial Flora and Fauna Sanctuary: walk one of the numerous trails among local flora and fauna including many protected species; Tunney Rd; 1 km w. ***Farrar Reserve:*** scenic bushland and spectacular wildflower display in season; Blackwood Rd; 8 km W. ***Australian Bush Heritage Block:*** natural woodland featuring wandoo and species unique to the South-West; 16 km N. ***Lake Towerinning:*** boating, waterskiing, horseriding, camping; 40 km NW. ***Aboriginal guided tours:*** tours of Aboriginal heritage sites; details and bookings at visitor centre.

See also GREAT SOUTHERN, p. 309

Kulin

Pop. 354
Map ref. 618 G6 | 620 E8

Resource Centre, Johnston St; (08) 9880 1021; www.kulin.wa.gov.au

720 AM ABC Local Radio

The sheep- and grain-farming districts surrounding Kulin provide spectacular wildflower displays in season. The flowering gum, *Eucalyptus macrocarpa*, is the town's floral emblem. A stand of jarrah trees, not native to the area and not known to occur elsewhere in the wheat belt, grows near the town. According to Aboriginal legend, two tribal groups met at the site and, as a sign of friendship, drove their spears into the ground. From these spears, the jarrah trees grew. The Kulin Bush Races event has expanded from horseracing to a major attraction including a weekend of live music, an art and craft show, foot races and Clydesdale horserides. In the months prior to the Kulin Races, tin horses appear in the paddocks lining the road on the way to the racetrack. These, along with the tin horses from past years, create an unusual spectacle.

Tin Horse Highway: starting in town and heading to the Jilakin racetrack, the highway is lined with horses made from a wide variety of materials. ***Kulin Herbarium:*** specialising in local flora; open by appt; Johnston St. ***Butlers Garage:*** built in the 1930s, this restored garage houses a museum of cars and machinery; open by appt; Cnr Johnston and Stewart sts. ***Memorial Slide and Swimming Pool:*** the longest water slide in regional WA; pool open 12–7pm Tues–Fri, 10am–7pm weekends/ public holidays, summer months; check for opening hours of water slide; Holt Rock Rd; (08) 9880 1222.

Charity Car Rally: Sept. ***Kulin Bush Races:*** Oct. ***Longneck Roughneck Des Cook Memorial Quick Shears Shearing Competition:*** Oct.

Kulin Hotel & Motel: Australian pub; Johnston St; (08) 9880 1201.

Kulin Caravan Park: includes railway carriages; Rankin St; (08) 9880 1204.

Macrocarpa Walk Trail: 1 km self-guide signposted walk trail through natural bush; brochure from visitor centre; 1 km W. ***Jilakin Rock and Lake:*** granite monolith overlooking a 1214 ha lake; 16 km E. ***Hopkins Nature Reserve:*** important flora conservation area; 20 km E. ***Buckley's Breakaways:*** unusual pink and white rock formations; 70 km E. ***Dragon Rocks Nature Reserve:*** wildflower reserve with orchids and wildlife; 75 km E.

See also HEARTLANDS, p. 308

Kununurra

Pop. 3745
Map ref. 625 D4 | 629 M5 | 634 A11 | 636 A2

75 Coolibah Dr; (08) 9168 1177 or 1800 586 868; www.kununurratourism.com

102.5 WA FM, 819 AM ABC Local Radio

Kununurra lies in the East Kimberley region not far from the Northern Territory border. It was established in the 1960s alongside Lake Kununurra on the Ord River at the centre of the massive Ord River Irrigation Scheme. Adjacent is the magnificent Mirima National Park. Lake Argyle to the south, in the Carr Boyd Range, was created by the damming of the Ord River and is the largest body of fresh water in Australia. Islands in the lake were once mountain peaks. The word 'Kununurra' means 'meeting of big waters' in the language of the local Aboriginal people. The climate in Kununurra and the East Kimberley is divided into two seasons, the Dry and the Wet. The Dry extends from April to October and is characterised by blue skies, clear days and cool nights. The Wet, from November to March, is a time of hot, humid days, when frequent thunderstorms deliver most of the annual rainfall to the region.

Historical Society Museum: artefacts and photos of the development of the town; Coolibah Dr. ***Lovell Gallery:*** art gallery exhibiting Kimberley artworks for sale; Konkerberry Dr. ***Waringarri Aboriginal Arts:*** large and varied display of Aboriginal art and artefacts for sale; open daily, weekends by appt; Speargrass Rd. ***Red Rock Art:*** a gallery and studio for Indigenous painters from across the Kimberley; open Mon–Fri; Coolibah Dr. ***Kelly's Knob Lookout:*** panoramic view of town and Ord Valley; off Speargrass Rd. ***Celebrity Tree Park:*** arboretum on the shore of Lake Kununurra where celebrities, including John Farnham, HRH Princess Anne, Harry Butler and Rolf Harris, have planted trees. Lily Creek Lagoon at the edge of the park is a good spot for birdwatching. The boat ramp was once part of the road to Darwin; off Victoria Hwy. ***Historical Society walk trails:*** choose between two trails of different lengths; maps from historical society.

Ord Valley Muster: May. ***Kununurra Races:*** Aug. ***Rodeo and Campdraft:*** Aug. ***Night Rodeo:*** Sept.

Gulliver's Tavern: typical pub fare; 186 Cottontree Ave; (08) 9168 1666. ***Ivanhoes Gallery Restaurant:*** international cuisine; All Seasons Kununurra, Messmate Way; (08) 9168 4000. ***Kelly's Bar & Grill:*** relaxed dining; Kununurra Country Club Resort, 47 Coolibah Dr; (08) 9168 1024 or 1800 808 999.

Diversion Cruises & Hire: houseboat on Lake Kununurra; 1 Lily Creek Dr; (08) 9168 3333. ***Hidden Valley Tourist Park:*** ideally located tourist park; Weaber Plains Rd; (08) 9168 1790. ***Hotel Kununurra:*** central and comfortable hotel; 37 Messmate Way; (08) 9168 0400. ***Kununurra Country Club Resort:*** centrally located resort; 47 Coolibah Dr; (08) 9168 1024 or 1800 808 999. ***El Questro Wilderness Park:*** featuring Aboriginal rock art, hot springs, spectacular scenery, camping, boating, fishing and a range of accommodation options; Gibb River Rd; (08) 9169 1777.

Mirima National Park Known by locals as the 'mini-Bungle Bungles', a striking feature of this park is the boab trees that grow on the rock faces, the seeds having been carried there by rock wallabies and left in their dung. There are walking trails within the park, and between May and Aug guided walks are available. Details from visitor centre; 2 km E.

Ivanhoe Farm: tastings and sales of melons and other local produce; open May–Sept; 1 km N. ***Lake Kununurra:*** formed after the completion of the Diversion Dam as part of the Ord River Scheme, the lake is home to a large variety of flora and fauna and is ideal for sailing, rowing, waterskiing and boat tours; details from visitor centre; 2 km S. ***City of Ruins:*** unusual sandstone formation of pinnacles and outcrops that resemble the ruins of an ancient city; off Weaber Plains Rd; 6 km N. ***Ord River and Diversion Dam:*** abundance of wildlife and spectacular scenery and a variety of watersports and cruises available; details from visitor centre;

7 km w. ***Top Rockz Gallery:*** exhibits gemstones and precious metals; open May–Sept; 10 km N. ***Ivanhoe Crossing:*** permanently flooded causeway is an ideal fishing spot; Ivanhoe Rd; 12 km N. ***Hoochery Distillery:*** visit a traditional old country and western saloon bar or take a tour of the only licensed distillery in WA; closed Sun; Weaber Plains Rd; 15 km N. ***Zebra Rock Gallery:*** view the amazing display of zebra rock, nearly 600 million years old and believed to be unique to the Kimberley, or feed fish from the lakeside jetty; Packsaddle Rd; 16 km S. ***Middle Springs:*** picturesque spot with diverse birdlife; 4WD access only; 30 km N. ***Black Rock Falls:*** spectacular waterfall during the wet season that spills over rocks stained by the minerals in the water; 4WD access only; Apr–Oct (subject to road conditions); 32 km N. ***Parry Lagoons Nature Reserve:*** enjoy birdwatching from a shaded bird hide at Marlgu Billabong or scan the wide vistas of the flood plain and distant hills afforded from the lookout at Telegraph Hill; 65 km NW. ***The Grotto:*** ideal swimming hole (in the wet season) at the base of 140 stone steps; 70 km NW. ***Argyle Downs Homestead Museum:*** built in 1884 and relocated when the lake was formed, the building is a fine example of an early station homestead; open 7am–4pm dry season, wet season by appt; Parker Rd; 70 km S. ***Lake Argyle:*** the view of the hills that pop out of the main body of water is said to resemble a crocodile basking in the sun and is known locally as Crocodile Ridge. Fishing, birdwatching, camping, bushwalking, sailing, canoeing and lake cruises are all available; 72 km S. ***Argyle Diamond Mine:*** the largest producing diamond mine in the world. Access is via tour only; details from visitor centre; 120 km S. ***Purnululu National Park:*** scenic flights available in town; 375 km S; *see Halls Creek*. **Mitchell River National Park:** one of the Kimberley's newest national parks protects this scenic and biologically important area. Mitchell Falls and Surveyor's Pool are the two main attractions for visitors. The area is remote with 4WD access to the park only, and is about 16 hours' drive from Kununurra; 680 km NW. ***Scenic flights:*** flights from town take visitors over the remarkable Bungle Bungles, Argyle Diamond Mine, Mitchell Plateau or Kalumburu; details from and bookings at visitor centre. ***Tours:*** bushwalks, safaris, camping, canoeing, 4WD or coach; details from visitor centre.

See also KIMBERLEY, **p. 314**

Lake Grace

Pop. 503
Map ref. 618 H7 | 620 F9

Stationmaster's house, Stubbs St; (08) 9865 2140; www.lakegrace.wa.gov.au

91.7 WA FM, 531 AM ABC Local Radio

The area around Lake Grace is a major grain-growing region for the state, producing wheat, canola, oats, barley, lupins and legumes. Sandy plains nearby are transformed into a sea of colour at the height of the wildflower season in September and October.

Inland Mission Hospital Museum: the only remaining inland mission hospital in WA, this fully restored building (est. 1926) is now a fascinating medical museum. Approach the building via Apex Park along the interpretive walkway; open daily by appt; Stubbs St. ***Mural:*** artwork depicting pioneering women was begun in 1912; Stubbs St. ***Memorial Swimming Pool:*** includes water playground for children; open 11am–6pm daily Oct–Apr; Bishop St.

Market: visitor centre; every Sat Oct–Mar. ***Art Exhibition:*** Oct.

Lake Grace Hotel: pub fare; Stubbs St; (08) 9865 1219.

Darean Farm Country Retreat: self-contained timber cottage; Lot 4126 South Rd; (08) 9865 1068.

Wildflower walk: easy walk through natural bushland with informative signage; details from visitor centre; 3 km E. ***Lake Grace:*** combination of two shallow salt lakes that gives the town its name; 9 km w. ***Lake Grace Lookout:*** ideal spot to view the north and south lakes system; 12 km w. ***White Cliffs:*** unusual rock formation and picnic spot on private property; details from visitor centre; 17 km S. ***Holland Track:*** in 1893 John Holland and his partners cut a track from Broomehill through bushland to Coolgardie in the goldfields. Hundreds of prospectors and their families trudged along this track in search of fortune, and cartwheel ruts are still evident today. A plaque marks the place where the track crosses the road; Newdegate Rd; 23 km E. ***Dingo Rock:*** now on private property, this reservoir for water run-off was built by labourers from Fremantle Gaol. Wildflowers are beautiful in season; details from visitor centre; 25 km NE. ***Newdegate:*** small town with a pioneer museum in the heritage-listed Hainsworth building. One of WA's major agricultural events, the Machinery Field Days, is held here in Sept each year; 52 km E.

See also HEARTLANDS, **p. 308**

Lake King

Pop. 219
Map ref. 620 G8

Lake's Breaks, Church Ave; (08) 9874 4007; www.lakeking.com.au

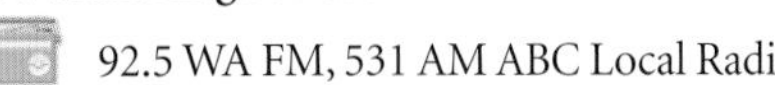
92.5 WA FM, 531 AM ABC Local Radio

This small rural town lies on the fringe of sheep- and grain-farming country. With a tavern and several stores, Lake King is a stopping place for visitors travelling across the arid country around Frank Hann National Park to Norseman (adequate preparations must be made as there are no stops en route). Outstanding wildflowers in late spring include rare and endangered species.

Self-guide walks: signposted walk trails; maps from visitor centre.

Lake King Tavern & Motel: traditional meals; 165 Varley Rd; (08) 9874 4048.

Lake King Caravan Park: powered and unpowered sites; Critchley Ave; (08) 9838 0052. ***Lake King Tavern & Motel:*** rammed-earth motel; 165 Varley Rd; (08) 9874 4048.

Lake King and Causeway: 9 km road across the salt lake studded with native scrub and wildflowers. Lookout at eastern end; 5 km w. ***Pallarup Reserve:*** pioneer well and lake with abundant wildflowers in season; 15 km S. ***Mt Madden:*** cairn and lookout with picnic area that forms part of the Roe Heritage Trail; 25 km SE. ***Frank Hann National Park:*** good example of inland sand plain heath flora with seasonal wildflowers. The rabbit-proof fence forms a boundary to the park. Access is subject to weather conditions; 32 km E. ***Roe Heritage Drive Trail:*** begins south of Lake King and covers natural reserves and historic sites. It offers panoramic views from the Roe Hill lookout and retraces part of J. S. Roe's explorations in 1848; maps from visitor centre.

See also HEARTLANDS, **p. 308**

Laverton

Pop. 314
Map ref. 621 I2 | 622 I12

Great Beyond Explorers Hall of Fame, Augusta St; (08) 9031 1361; www.laverton.wa.gov.au

 102.1 WA FM, 106.1 FM ABC Local Radio

Surrounded by old mine workings and modern mines, Laverton is on the edge of the Great Victoria Desert. In 1900 Laverton was a booming district of gold strikes and mines, yet gold price fluctuations in the late 1950s made it almost a ghost town. In 1969 nickel was discovered at Mount Windarra, which sparked a nickel boom. Early in 1995 a cyclone blew through Laverton, leaving it flooded and isolated for three months. Mines closed down and supplies were brought in by air. During this time, locals held a 'wheelie bin' race from the pub to the sports club, which is now an annual event. Today the town has two major gold mines and one of the world's largest nickel-mining operations. Wildflowers are brilliant in season.

Historic buildings Restored buildings include the courthouse, the Old Police Station and Gaol, and the Mt Crawford Homestead. The original police sergeant's House is now the local museum with displays of local memorabilia. Details from visitor centre.

Great Beyond Explorers Hall of Fame: uses cutting-edge technology to bring to life the characters and stories of the early explorers of the region. It also houses the visitor centre; Augusta St. ***Cross-Cultural Centre:*** houses the Laverton Outback Gallery, a collection of art and artefacts made and sold by the local Wongi people; Augusta St.

Race Day: June. ***Laverton Day and Wheelie Bin Race:*** Nov.

The Desert Inn Hotel: traditional meals; 2 Laver St; (08) 9031 1188.

The Desert Inn Hotel & Motel: self-contained units; 2 Laver St; (08) 9031 1188.

Giles Breakaway: scenic area with interesting rock formation; 25 km E. ***Lake Carey:*** swimming and picnic spot exhibiting starkly contrasting scenery; 26 km W. ***Windarra Heritage Trail:*** walk includes rehabilitated mine site and interpretive plaques; Windarra Minesite Rd; 28 km NW. ***Empress Springs:*** discovered in 1896 by explorer David Carnegie and named after Queen Victoria. The spring is in limestone at the end of a tunnel that runs from the base of a 7 m deep cave. A chain ladder allows access to the cave. Enclosed shoes and torch required; 305 km NE. ***Warburton:*** the Tjulyuru Arts and Culture Regional Gallery showcases the art and culture of the Ngaanyatjarraku people; 565 km NE. ***Golden Quest Discovery Trail:*** Laverton forms part of the 965 km self-guide drive trail of the goldfields; book, map and CD available. ***Outback Highway:*** travel the 1200 km to Uluṟu from Laverton via the Great Central Rd. All roads are unsealed but regularly maintained.

Travellers note: *Permits are required to travel through Aboriginal reserves and communities – they can be obtained from the Dept of Indigenous Affairs in Perth, and the Central Lands Council in Alice Springs. Water is scarce. Fuel, supplies and accommodation are available at the Tjukayirla, Warburton and Warakurna roadhouses. Check road conditions before departure at the Laverton Police Station or Laverton Shire Offices; road can be closed due to heavy rain.*

See also GOLDFIELDS, p. 311

Leonora

Pop. 400
Map ref. 620 H2

Rural Transaction Centre, cnr Tower and Trump sts; (08) 9037 7016; www.leonora.wa.gov.au

 102.5 WA FM, 105.7 FM ABC Local Radio

Leonora is the busy railhead for the north-eastern goldfields and the surrounding pastoral region. Mount Leonora was discovered in 1869 by John Forrest. Gold was later found in the area and, in 1896, the first claims were pegged. By 1908 Leonora boasted seven hotels and was the largest centre on the north-eastern goldfields. Many of the original buildings were constructed of corrugated iron and hessian, as these were versatile materials and light to transport. You can get a glimpse of these structures at nearby Gwalia, a town once linked to Leonora by tram.

Historic buildings: buildings from the turn of the century include the police station, courthouse, fire station and Masonic Lodge; details from visitor centre. ***Tank Hill:*** excellent view over town; Queen Victoria St. ***Miners Cottages:*** miners camps, auctioned and restored by locals; around town.

Leonora Golden Gift and Festival: May/June.

White House Hotel: traditional pub; 120 Tower St; (08) 9037 6030.

Hoover House B&B: 3 antique-decorated rooms; Tower St; (08) 9037 7122.

Gwalia Gwalia is a mining ghost town that has been restored to show visitors what life was like in the pioneering gold-rush days. The original mine manager's office now houses the Gwalia Historical Museum with displays of memorabilia that include the largest steam winding engine in the world and a headframe designed by Herbert Hoover, the first mine manager of the Sons of Gwalia mine and eventually the 31st President of the United States. Self-guide heritage walk available. 4 km S.

Mt Leonora: sweeping views of the surrounding plains and mining operations; 4 km S. ***Malcolm Dam:*** picnic spot at the dam, which was built in 1902 to provide water for the railway; 15 km E. ***Kookynie:*** tiny township with restored shopfronts, and historic memorabilia on display at the Grand Hotel. The nearby Niagara Dam was built in 1897 with cement carried by camel from Coolgardie; 92 km SE. ***Menzies:*** small goldmining town with an interesting historic cemetery. View 'stick figure' sculptures, for which locals posed, about 55 km west of town on bed of Lake Ballard; 110 km S. ***Golden Quest Discovery Trail:*** self-guide drive trail of 965 km takes in many of the towns of the northern goldfields; maps, book and CD from visitor centre, or visit www.goldenquesttrail.com

See also GOLDFIELDS, p. 311

Madura

Pop. 9
Map ref. 621 N6

Madura Pass Oasis Motel and Roadhouse; (08) 9039 3464.

Madura, comprising a roadhouse, motel and caravan park on the Eyre Highway, lies midway between Adelaide and Perth on the Nullarbor Plain. It is remarkable, given the isolation of the area, that Madura Station was settled in 1876 to breed horses, which were then shipped across to India for use by the British Army. Now Madura is surrounded by private sheep stations. Locals operate on Central Western Time, 45 minutes ahead of the rest of Western Australia.

WESTERN AUSTRALIA

Blowholes: smaller versions of the one found at Caiguna. Look for the red marker beside the track; 1 km N. ***Madura Pass Lookout:*** spectacular views of the Roe Plains and Southern Ocean; 1 km W on highway.

See also ESPERANCE & NULLARBOR, p. 310

Mandurah

Pop. 67 783
Map ref. 615 C9 | 618 B6 | 620 C8

75 Mandurah Tce; (08) 9550 3999; www.visitmandurah.com

97.3 Coast FM, 810 AM ABC Radio National

Mandurah has long been a popular holiday destination for Perth residents. The Murray, Serpentine and Harvey rivers meet at the town to form the vast inland waterway of Peel Inlet and the Harvey Estuary. This river junction was once a meeting site for Aboriginal groups who travelled here to barter. The town's name is derived from the Aboriginal word 'mandjar', meaning trading place. The river and the Indian Ocean offer a variety of watersports and excellent fishing and prawning. But, the aquatic activity for which Mandurah is perhaps best known is crabbing. It brings thousands during summer weekends, wading the shallows with scoop nets and stout shoes.

Christ's Church Built in 1870, this Anglican church has hand-worked pews believed to be the work of early settler Joseph Cooper. Many of the district's pioneers are buried in the churchyard, including Thomas Peel, the founder of Mandurah. In 1994 the church was extended and a bell-tower added to house eight bells from England. Cnr Pinjarra Rd and Sholl St.

Australian Sailing Museum: displays and models of vessels from the 1860s to Olympic yachts; Ormsby Tce. ***Hall's Cottage:*** (1832) restored home of one of the original settlers, Henry Hall; Leighton Rd. ***Museum:*** (1898) originally a school then a police station, now houses displays on Mandurah's social, fishing and canning histories; open Tues–Sun; Pinjarra Rd. ***King Carnival Amusement Park:*** fun-fair attractions including giant water slide, Ferris wheel and more; Leighton Rd. ***Estuary Drive:*** scenic drive along Peel Inlet and Harvey Estuary.

Smart Street Markets: off Mandurah Tce; each Sat. ***Crab Fest:*** Mar. ***Boat and Fishing Show:*** Oct.

Cafe Pronto: international cuisine; Cnr Pinjarra Rd and Mandurah Tce; (08) 9535 1004. ***Red Manna Waterfront Restaurant:*** waterfront seafood; upstairs, 5/9 Mandurah Tce; (08) 9581 1248. ***Scusi:*** fine dining; Shop 6, Lot 4 Old Coast Rd, Halls Head; (08) 9586 3479. ***The Miami Bakehouse:*** bakery cafe; Falcon Grove Shopping Centre, Old Coast Rd; (08) 9534 2705. ***Millbrook Winery Restaurant:*** innovative fresh produce; Millbrook Winery, Old Chestnut La, Jarrahdale; (08) 9525 5796. ***Cafe on the Dam:*** relaxed meals; Serpentine Dam, Kingsbury Dr, Jarrahdale; (08) 9525 9920.

Dolphin House Boats: variety of houseboats; Mandurah Ocean Marina; (08) 9535 9898. ***Lakeside Holiday Apartments:*** 2-bedroom apartments; 1 Lakes Cres, South Yunderup; (08) 9537 7634. ***Peel Manor House:*** elegant hotel; 164 Fletcher Rd, Karnup; (08) 9524 2838. ***Seashells Resort Mandurah:*** well-appointed apartments; 16 Dolphin Dr; (08) 9550 3000 or 1800 800 850.

Coopers Mill: the first flour mill in the Murray region, located on Cooleenup Island near the mouth of the Serpentine River. Joseph Cooper built it by collecting limestone rocks and every morning, sailing them across to the island. Accessible only by water; details from visitor centre. ***Abingdon Miniature Village:*** display of miniature heritage buildings and gardens from UK and Australia with maze, picnic area and children's playground; Husband Rd; 3 km E. ***Marapana Wildlife Park:*** handfeed and touch native and exotic animals; 14 km N at Karnup. ***Yalgorup National Park:*** swamps, woodlands and coastal lakes abounding with birdlife. Lake Clifton is one of only three places in Australia where the living fossils called thrombolites survive. A boardwalk allows close-up viewing; 45 km S. ***Wineries:*** several in the area; details from visitor centre. ***Tours:*** including estuary cruises and dolphin interaction tours; details from visitor centre.

See also THE SOUTH-WEST, p. 307

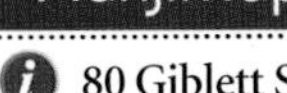

Manjimup

Pop. 4236
Map ref. 616 H10 | 617 B1 | 618 D10 | 620 C10

80 Giblett St; (08) 9771 1831; www.manjimupwa.com

100.5 Hot FM, 738 AM ABC Local Radio

Manjimup is the gateway to the South-West region's tall-timber country. Magnificent karri forests and rich farmlands surround the town. While timber is the town's main industry, Manjimup is also the centre of a thriving fruit and vegetable industry that supplies both local and Asian markets. The area is well known for its apples and is the birthplace of the delicious Pink Lady apple.

Timber and Heritage Park A must-see for any visitor to Manjimup, this 10 ha park includes the state's only timber museum, an exhibition of old steam engines, an 18 m climbable fire-lookout tower and a historic village with an early settler's cottage, blacksmith's shop, old police station and lock-up, one-teacher school and early mill house. Set in natural bush and parkland, there are many delightful spots for picnics or barbecues. Cnr Rose and Edwards sts.

Manjimup Motocross 15000: June. ***Festival of Country Gardens:*** Oct/Nov. ***Cherry Harmony Festival:*** Dec.

Cabernet Restaurant: international cuisine; Kingsley Motel, 74 Chopping St; (08) 9771 1177 or 1800 359 177. ***Déjà vu Cafe:*** family-friendly cafe; 43A Giblett St; (08) 9771 2978. ***Slice of Heaven Cafe:*** home-style cooking; 31B Rose St; (08) 9777 1331.

Dingup House B&B: heritage homestead; 114 Dingup Rd; (08) 9772 4206. ***Fonty's Pool & Caravan Park:*** cottage, cabins and camping; Seven Day Rd; (08) 9771 2105. ***Karri Glade Chalets:*** contemporary chalets in lovely bush setting; Graphite Rd; (08) 9772 1120. ***Diamond Tree Angus Stud:*** cattle farmstay; Channybearup Rd; (08) 9776 1348.

One Tree Bridge In 1904 a single enormous karri tree was felled so that it dropped across the 25 m wide Donnelly River, forming the basis of a bridge. Winter floods in 1966 swept most of the bridge away; the 17 m piece salvaged is displayed near the original site with information boards. Nearby is Glenoran Pool, a scenic spot for catching rainbow trout and marron in season, with walking trails and picnic areas. Graphite Rd; 21 km W.

King Jarrah: estimated to be 600 years old, this massive tree is the centrepiece for several forest walks; Perup Rd; 3 km E. ***Wine and Truffle Company:*** tastings and information about

continued on p. 347

MARGARET RIVER

Pop. 4413

Map ref. 616 B8 | 618 B10 | 620 B10

[CAPE MENTELLE VINEYARD]

100 Bussell Hwy (cnr Tunbridge St); (08) 9780 5911; www.margaretriver.com

100.3 Hot FM, 1224 AM ABC Radio National

One of the best-known towns in Western Australia, Margaret River is synonymous with world-class wines, magnificent coastal scenery, excellent surf breaks and spectacular cave formations. In addition, the region boasts a thriving arts scene, boutique breweries, gourmet food outlets, and restaurants with views of sweeping vineyards or the sparkling ocean. The bustling township lies on the Margaret River near the coast, 280 kilometres south-west of Perth.

Wine Tourism Showroom Although the first grapevines were only planted in the area in 1967, Margaret River is now considered to be one of the top wine-producing regions in Australia. A huge 20% of Australia's premium wines are made here. The 'terroir' is perfect for grape-growing: cool frost-free winters, good moisture-retaining soils and low summer rainfall provide a long, slow ripening period. These conditions produce greater intensity of fruit flavour, the starting point for all great wines. There are now more than 120 wine producers in the region, many of which are open to the public for tastings and cellar-door sales. Others have restaurants and tours of their premises. A good place to start a wine tour is the Margaret River Wine Tourism Showroom at the visitor centre. It provides information about the regional wineries and vineyards, an interactive wineries screen, videos, a sensory display and wine-making paraphernalia. 100 Bussell Hwy.

Rotary Park: picnic area on the riverbank with a display steam engine; Bussell Hwy. ***St Thomas More Catholic Church:*** one of the first modern buildings built of rammed earth; Wallcliffe Rd. ***Fudge Factory:*** fudge and chocolate made before your eyes; 152 Bussell Hwy; (08) 9758 8881. ***Arts and crafts:*** many in town, including Margaret River Gallery on Bussell Hwy; details from visitor centre.

Town Square Markets: 8am–12pm Sun in summer, 2nd Sun each month rest of year. ***Margaret River Wine Region Festival:*** Mar. ***Great WA Bike Ride:*** 560 km week-long bike ride; Mar. ***Margaret River Pro:*** surfing competition; Apr. ***Vintage Stomp:*** end-of-harvest festivities; May. ***Flourish Margaret River:*** celebration of nature; Oct. ***Margaret River Classic:*** surfing competition; Oct/Nov.

The Berry Farm Cottage Cafe: charming cafe; 43 Bessell Rd; (08) 9757 5054. ***Leeuwin Restaurant:*** superb international cuisine; Leeuwin Estate Winery, Stevens Rd; (08) 9759 0000. ***The Margaret River Chocolate Company:*** chocolate-inspired cafe; cnr Harman's Mill and Harman's South rds, Metricup; (08) 9755 6555. ***Brookwood Estate:*** ploughman's lunches; 430 Treeton Rd, Cowaramup; (08) 9755 5604. ***Cullen Restaurant:*** seasonal menu in a relaxed vineyard setting; Cullen Wines, Caves Rd, Cowaramup; (08) 9755 5656. ***Sea Gardens Cafe:*** tasty menu and ocean views; 9 Mitchell Dr, Prevelly; (08) 9757 3074. ***Udderly Divine Cafe:*** dairy-themed cafe; 22 Bussell Hwy, Cowaramup; (08) 9755 5519.

Basildene Manor: old-world grandeur; Wallcliffe Rd; (08) 9757 3140. ***Gilgara Retreat:*** old station homestead replica and self-contained garden suites; 3 Old Ellen Brook Rd, Gracetown; (08) 9757 2705. ***Studio 9:*** chic apartment opposite visitor centre; Bussell Hwy; (08) 9757 2871. ***Beach Barnacle:*** traditional beach house; 25 Wooredah Cres, Prevelly; (08) 9757 3519. ***Beckett's Flat Vineyard Cottage:*** romantic cottage; 49 Beckett Rd, Metricup; (08) 9755 7402. ***Craythorne Country House:*** peaceful B&B; 180 Worgan Rd, Metricup; (08) 9755 7477. ***Island Brook Estate:*** cedar chalets in lush jarrah forest; 7388 Bussell Hwy, Metricup; (08) 9755 7501. ***Lake View Siesta:*** motel-style accommodation; 30 Lake View Cres, Prevelly; (08) 9757 2579. ***Redgate Farmstay:*** pet-friendly cottages; 1032 Redgate Rd, Witchcliffe; (08) 9757 6400. ***The Grove Vineyard:*** well-appointed chalets; Cnr Metricup and Carter rds, Wilyabrup; (08) 9755 7458. ***The Noble Grape Guest House:*** charming B&B; 29 Bussell Hwy, Cowaramup; (08) 9755 5538. ***The Roozen Residence:*** architecturally designed minimalist beach house; 4 Chapel Pl, Prevelly; 0407 479 004. ***Villa Crisafina:*** Mediterranean-style house; 45 Baudin Dr, Gnarabup; (08) 9757 1050.

Mammoth and Lake caves, Leeuwin–Naturaliste National Park Lying beneath the Leeuwin–Naturaliste Ridge that separates the hinterland from the coast is one of the world's most extensive and beautiful limestone cave systems. Mammoth Cave is home to the fossil remains of prehistoric animals. An audio self-guide tour lets you travel through the cave at your own pace (21 km sw). Only a few kilometres away is Lake Cave, with its famous reflective lake and delicate formations. Book your guided tour at the adjacent interpretive centre, CaveWorks, which also features a walk-through cave model, interactive displays and a boardwalk with spectacular views of a collapsed cavern; (08) 9757 7411; 25 km sw. Further south, also in the national park, is Jewel Cave; *see Augusta and Dunsborough.*

Amaze'n Margaret River: family-friendly venue with giant hedge maze, ground puzzles, outdoor games and picnic area; cnr Bussell Hwy and Gnarawary Rd; (08) 9758 7439; 4 km s. ***Eagles Heritage:*** the largest collection of birds of prey in Australia, with free-flight displays at 11am and 1.30pm daily; Boodjidup Rd; (08) 9757 2960; 5 km sw. ***Surfer's Pt:*** the centre of surfing in Margaret River and home to the Margaret River Pro; 8 km w, just north of Prevelley. ***Leeuwin Estate Winery:*** the picturesque venue for the world-renowned open-air Leeuwin Estate Concert, held each Feb/Mar; Stevens Rd; (08) 9759 0000; 8 km s. ***Margaret River Regional Wine Centre:*** a one-stop centre for tastings and sales, offering in-depth information on local wines and wineries; 9 Bussell Hwy, Cowaramup; (08) 9755 5501; 13 km N. ***Candy Cow:*** free tastings of fudge, nougat and honeycomb, with demonstrations at 11am Wed–Sun; 3 Bottrill St, Cowaramup; (08) 9755 9155; 13 km N. ***Miller's Icecream:*** operating dairy makes fresh icecream, 'from cow to cone'; 314 Wirring Rd, Cowaramup; (08) 9755 9850; 13 km N. ***Gnarabup Beach:*** safe swimming beach for all the family; 13 km w. ***Margaret River Dairy Factory:*** free tastings of cheese and yoghurt; Bussell Hwy, Cowarumup; (08) 9755 7588; 15 km N. ***The Berry Farm:*** jams, pickles, naturally fermented vinegars, fruit and berry wines; 43 Bessell Rd; (08) 9757 5054; 15 km SE. ***Ellensbrook Homestead:*** this wattle-and-daub homestead (1857) was once the home of Alfred and Ellen Bussell, the district's first pioneers; open weekends; (08) 9755 5173. Nearby is the beautiful Meekadarabee Waterfall, a place steeped in Aboriginal legend; 15 km NW. ***Olio Bello:*** boutique, handmade olive oil, with tastings available; Armstrong Rd, Cowaramup; (08) 9755 9771; 17 km NW. ***Grove Vineyard:*** Margaret River's only liqueur factory also boasts a winery, distillery, nano-brewery and cafe; tastings available; cnr Metricup and Carter rds, Wilyabrup; (08) 9755 7458; 20 km N. ***Arts and crafts:*** many in area, including Boranup Galleries, 20 km s on Caves Rd; details from visitor centre. ***Margaret River Chocolate Company:*** free chocolate tastings, interactive displays and viewing windows to watch the chocolate products being made; Cnr Harmans Mill and Harmans South rds, Metricup; (08) 9755 6555; 30 km NW. ***Bootleg Brewery:*** enjoy naturally brewed boutique beers in a picturesque setting; Puzey Rd, Wilyabrup; (08) 9755 6300; 45 km NW. ***Bushtucker River and Winery Tours:*** experience the Margaret River region through its wine, wilderness and food; (08) 9757 9084. ***Land-based activities:*** abseiling, caving, canoeing, coastal treks, horseriding and hiking; details from visitor centre. ***Heritage trails:*** including the Rails-to-Trails and Margaret River heritage trails; maps from visitor centre.

See also THE SOUTH-WEST, p. 307

truffle farming and dog training; (08) 9777 2474; 7 km sw. ***Fonty's Pool:*** dammed in 1925 by Archie Fontanini for the irrigation of vegetables, it is now a popular swimming pool and picnic area in landscaped grounds; Seven Day Rd; 7 km s. ***Dingup Church:*** built in 1896 by the pioneer Giblett family and doubling as the school, this church is one of the few remaining local soapstone buildings; Balbarrup Rd; 8 km E. ***Pioneer Cemetery:*** poignant descriptions on headstones testify to the hardships faced by first settlers; Perup Rd; 8 km E. ***Diamond Tree Lookout:*** one of 8 tree towers constructed from the late 1930s as fire lookouts. Climb to the wooden cabin atop this 51 m karri, used as a fire lookout from 1941 to 1947; 9 km s. ***Fontanini's Nut Farm:*** gather chestnuts, walnuts, hazelnuts and fruit in season; open Apr–June; Seven Day Rd; 10 km s. ***Nyamup:*** old mill town redeveloped as a tourist village; 20 km SE. ***Four Aces:*** four giant karri trees 220–250 years old and 67–79 m high stand in Indian file; Graphite Rd; 23 km w. ***Great Forest Trees Drive:*** self-guide drive through Shannon National Park; maps from visitor centre; 45 km s. ***Perup Forest Ecology Centre:*** night spotlight walks to see rare, endangered and common native animals; contact DEC on (08) 9771 7988 for details; 50 km E. ***Lake Muir Lookout/Bird Observatory:*** boardwalk over salt lake to bird hide; 55 km E. ***Wineries:*** several in area; maps from visitor centre.

See also THE SOUTH-WEST, p. 307

Marble Bar

Pop. 194
Map ref. 622 E2 | 624 E12 | 627 K5

11 Francis St; (08) 9176 1375.

102.7 WA FM, 107.5 FM ABC Radio National

Marble Bar has gained a dubious reputation as the hottest town in Australia. For 161 consecutive days in 1923–24 the temperature in Marble Bar did not drop below 37.8°C (100°F). This mining town was named after a bar of mineral deposit that crosses the nearby Coongan River and was originally mistaken for marble. It proved to be jasper, a coloured variety of quartz.

Government buildings: built of local stone in 1895, now National Trust listed; General St.

Marble Bar Races: July.

Marble Bar Travellers Stop: seafood; Halse Rd; (08) 9176 1166.

Ironclad Hotel: heritage hotel; 15 Francis St; (08) 9176 1066. ***Marble Bar Travellers Stop:*** simple double rooms; Halse Rd; (08) 9176 1166.

Comet Gold Mine This mine operated from 1936 to 1955. The Comet is now a museum and tourist centre with displays of gemstones, rocks, minerals and local history. Also here is a 75 m high smoke stack, reputed to be the tallest in the Southern Hemisphere. Underground mine tours occur twice daily. (08) 9176 1015; 10 km s.

Marble Bar Pool: site of the famous jasper bar (splash water on it to reveal its colours) and a popular swimming spot; 4 km W. ***Corunna Downs RAAF Base:*** built in 1943 as a base for long-range attacks on the Japanese-occupied islands of the Indonesian archipelago; 40 km SE. ***Doolena Gorge:*** watch the cliff-face glow bright red as the sun sets; 45 km NW.

See also PILBARA, p. 313

Margaret River

see inset box on page 346

Meekatharra

Pop. 799
Map ref. 619 H9 | 622 D10

Shire offices, 54 Main St; (08) 9981 1002; www.meekashire.wa.gov.au

 103.1 WA FM, 106.3 FM ABC Local Radio

The name Meekatharra is believed to be an Aboriginal word meaning 'place of little water' – an apt description for a town sitting on the edge of a desert. Meekatharra is now the centre of a vast mining and pastoral area. It came into existence in the 1880s when gold was discovered in the area. However, the gold rush was short-lived and it was only the arrival of the railway in 1910 that ensured its survival. The town became the railhead at the end of the Canning Stock Route, a series of 54 wells stretching from the East Kimberleys to the Murchison. The railway was closed in 1978, but the town continues to provide necessary links to remote outback areas through its Royal Flying Doctor Service.

Royal Flying Doctor Service: operates an important base in Meekatharra; open to public 9am–2pm daily; Main St. ***Old Courthouse:*** National Trust building; Darlot St. ***Meekatharra Museum:*** photographic display and items of memorabilia from Meekatharra's past; open 8am–4.30pm Mon–Fri; shire offices, Main St. ***State Battery:*** relocated to the town centre in recognition of the early prospectors and miners; Main St. ***Meeka Rangelands Discovery Trail:*** walk or drive this trail for insight into the town's mining past, Aboriginal heritage and landscapes; maps from visitor centre.

The Royal Mail Hotel: counter meals; Main St; (08) 9981 1148.

Auski Inland Motel: best motel in town; Main St; (08) 9981 1433.

Peace Gorge: this area of granite formations is an ideal picnic spot; 5 km N. ***Meteorological Office:*** watch the launching of weather balloons twice daily at the airport; tours available on (08) 9981 1191; 5 km SE. ***Bilyuin Pool:*** swimming (but check water level in summer); 88 km NW. ***Old Police Station:*** remains of the first police station in the Murchison; Mt Gould; 156 km NW.

See also OUTBACK COAST & MID-WEST, p. 312

Merredin

Pop. 2556
Map ref. 618 G3 | 620 E6

Barrack St; (08) 9041 1666; www.wheatbelttourism.com

105.1 Hot FM, 107.3 FM ABC Radio National

Merredin started as a shanty town where miners stopped on their way to the goldfields. In 1893 the railway reached the town and a water catchment was established on Merredin Peak, guaranteeing the town's importance to the surrounding region. An incredible 40 per cent of the state's wheat is grown within a 100-kilometre radius of the town.

Cummins Theatre Heritage-listed theatre that was totally recycled from Coolgardie in 1928. Used regularly for local productions, events and visiting artists. Open 9am–3pm Mon–Fri; Bates St.

Military Museum: significant collection of restored military vehicles and equipment; closed Sat; Great Eastern Hwy. ***Old Railway Station Museum:*** prize exhibits are the 1897 locomotive that once hauled the *Kalgoorlie Express* and the old signal box with 95 switching and signal levers; open 9am–3pm daily; Great Eastern Hwy. ***Pioneer Park:*** picnic area adjacent to the highway including a historic water tower that once supplied the steam trains; Great Eastern Hwy. ***Merredin Peak Heritage Walk:*** self-guide walk that retraces the early history of Merredin and its links with the goldfields and the railway; maps from visitor centre. ***Merredin Peak Heritage Trail:*** leads to great views of the countryside, and a rock catchment channel and dam from the 1890s. Adjacent to the peak is the interpretation site of a WW II field hospital that had over 600 patients in 1942; off Benson Rd. ***Tamma Parkland Trail:*** a 1.2 km walk around this 23 ha of bushland will give an insight into the flora and fauna of the area. Wildflowers in spring; South Ave.

Community Show: Oct.

Denzil's Restaurant: international cuisine; Merredin Olympic Motel, Lot 5 Great Eastern Hwy; (08) 9041 1588. ***Merredin Motel & Gumtree Restaurant:*** very decent meals; 10 Gamenya Ave; (08) 9041 1886.

Merredin B&B: converted bank residence; 30 Bates St; (08) 9041 4358. ***Merredin Oasis Hotel Motel:*** range of rooms; 8 Great Eastern Hwy; (08) 9041 1133. ***Merredin Plaza All Suites:*** town's premier accommodation; 149 Great Eastern Hwy; (08) 9041 1755. ***Merredin Tourist Park:*** self-contained units and camping; 2 Oats St; (08) 9041 1535.

Pumping Station No 4: built in 1902 but closed in 1960 to make way for electrically driven stations, this fine example of early industrial architecture was designed by C. Y. O'Connor; 3 km W. ***Hunt's Dam:*** one of several wells sunk by convicts under the direction of Charles Hunt in 1866. It is now a good spot for picnics and bushwalking; 5 km N. ***Totadgin Conservation Park:*** interpretive walk, Hunt's Well and mini rock formation similar to Wave Rock. Picnic tables and wildflowers in spring; 16 km SW. ***Rabbit-Proof Fence:*** roadside display gives an insight into the history of this feature; 25 km E. ***Mangowine Homestead:*** now a restored National Trust property, in the 1880s this was a wayside stop en route to the Yilgam goldfields. Nearby is the Billyacatting Conservation Park with interpretative signage; open 1–4pm Mon–Sat, 10am–4pm Sun (closed Jan); Nungarin; 40 km NW. ***Shackelton:*** Australia's smallest bank; 85 km SW. ***Kokerbin Rock:*** superb views from summit; 90 km SW. ***Koorda:*** museum and several wildlife reserves in area; 140 km NW.

See also HEARTLANDS, p. 308

Moora

Pop. 1606
Map ref. 618 B2 | 620 C6

34 Padbury St; (08) 9651 1401; www.moora.wa.gov.au

90.9 WA FM, 612 AM ABC Radio National

On the banks of the Moore River, Moora is the largest town between Perth and Geraldton. The area in its virgin state was a large salmon gum forest. Many of these attractive trees can still be seen.

Historical Society Genealogical Records and Photo Display: open by appt; Clinch St. ***Painted Roads Initiative:*** murals by community artists on and in town buildings.

Moora Races: Oct.

Moora Hotel: traditional pub; 1 Gardiner St; (08) 9651 1177.

The Drovers Inn: hotel and motel accommodation; Cnr Dandaragan and Padbury sts; (08) 9651 1108.

The Berkshire Valley Folk Museum James Clinch, who came from England in 1839, created a village in the dry countryside of WA based on his home town in Berkshire. Over a 25-year period from 1847, Clinch built a homestead, barn, manager's cottage, stables, shearing shed and bridge. The elaborate buildings were made from adobe, pise, handmade bricks and unworked stone. Open by appt; 19 km E.

Western Wildflower Farm: one of the largest exporters of dried wildflowers in WA, with dried flowers, seeds and souvenirs for sale and an interpretive education centre; open 9am–5pm daily Easter–Christmas; Midlands Rd, Coomberdale; (08) 9651 8010; 19 km N. ***Watheroo National Park:*** site of Jingamia Cave; 50 km N. ***Moora Wildflower Drive:*** from Moora to Watheroo National Park, identifying flowers on the way; maps from visitor centre.

See also HEARTLANDS, p. 308

Morawa

Pop. 594
Map ref. 620 B4

Winfield St (May–Oct), (08) 9971 1421; or council offices, cnr Dreghorn and Prater sts, (08) 9971 1204; www.morawa.wa.gov.au

103.1 WA FM, 531 AM ABC Local Radio

Morawa, a small wheat-belt town, has the distinction of being home to the first commercial iron ore to be exported from Australia. In springtime the area around Morawa is ablaze with wildflowers.

Church of the Holy Cross and Old Presbytery From 1915 to 1939, the famous WA architect-priest Monsignor John C. Hawes designed a large number of churches and church buildings in WA's mid-west region. Morawa boasts 2 of them: the Church of the Holy Cross and an unusually small stone hermitage known as the Old Presbytery. The latter, which Hawes used when visiting the town, is reputed to be the smallest presbytery in the world with only enough room for a bed, table and chair. Both buildings are part of the Monsignor Hawes Heritage Trail. Church is usually open; if not, contact council offices; Davis St.

Historical Museum: housed in the old police station and gaol with displays of farm machinery, household items and a collection of windmills; open by appt; Cnr Prater and Gill sts.

Everlastings Restaurant: buffet food; 10 Evan St; (08) 9971 1771.

Everlastings Guest Homes: modern hotel and cottage; 10 Evan St; (08) 9971 1771. ***Morawa Chalets:*** well-appointed chalets; White Ave; (08) 9971 1204. ***Morawa Marian Convent B&B:*** comfortable accommodation; Davis St; (08) 9971 1555.

Koolanooka Mine Site and Lookout: scenic views and a delightful wildflower walk in season; 9 km E. ***Perenjori:*** nearby town has historic St Joseph's Church, designed by Monsignor Hawes, and the Perenjori–Rothsay Heritage Trail, a 180 km self-drive tour taking in Rothsay, a goldmining ghost town; 18 km SE. ***Bilya Rock Reserve:*** with a large cairn, reportedly placed there by John Forrest in the 1870s as a trigonometrical survey point; 20 km N. ***Koolanooka Springs Reserve:*** ideal for picnics; 26 km E.

See also OUTBACK COAST & MID-WEST, p. 312

Mount Barker

Pop. 1760
Map ref. 617 C9 | 618 G11 | 620 E11

Old Railway Station, Albany Hwy; (08) 9851 1163; www.mountbarkertourismwa.com.au

92.1 FM ABC News Radio, 95.3 Hot FM

Mount Barker lies in the Great Southern region of Western Australia, with the Stirling Range to the north and the Porongurups to the east. The area was settled by Europeans in the 1830s. Vineyards were first established here in the late 1960s. Today, Mount Barker is a major wine-producing area.

Old Police Station and Gaol: built by convicts in 1867–68, it is now a museum of memorabilia; open 10am–4pm Sat and Sun, daily during school holidays, or by appt; Albany Hwy, north of town. ***Banksia Farm:*** complete collection of banksia and dryandra species; guided tours daily, closed July; Pearce Rd; (08) 9851 1770.

Mount Barker D'Vine Wine Festival: Jan. ***Mount Barker Wildflower Festival:*** Sept/Oct.

Bluff Knoll Cafe: hillside cafe; Cnr Bluff Knoll and Chester Pass rds, Borden; (08) 9827 9293. ***Galafrey Wines:*** vineyard platters; 432 Quangellup Rd; (08) 9851 2022. ***Porongurup Tearooms:*** traditional tea rooms; 1972 Porongurup Rd, Porongurup; (08) 9853 1110. ***The Vineyard Cafe:*** hilltop cafe; Windrush Wines, cnr St Werburgh's and Hay River rds; (08) 9851 1353.

Karribank Country Retreat: range of accommodation options; 1983 Porongurup Rd, Porongurup; (08) 9853 1022. ***Kendenup Lodge & Cottages:*** charming cottages and lodge rooms; 217 Moorilup Rd, Kendenup; (08) 9851 4233. ***The Sleeping Lady:*** quality B&B; 2658 Porongurup Rd, Porongurup; (08) 9853 1113. ***Cloud Nine Spa Chalets:*** luxury chalets; 278 Moorialup Rd, East Porongurup; (08) 9853 1111. ***Stirling Range Retreat:*** range of accommodation and activities; Chester Pass Rd, Borden; (08) 9827 9229.

Porongurup National Park This is a park of dramatic contrasts, from stark granite outcrops and peaks to lush forests of magnificent karri trees. Many unusual rock formations, such as Castle Rock and Balancing Rock, make the range a fascinating place for bush rambles. The Tree in the Rock, a mature karri, extends its roots down through a crevice in a granite boulder. 24 km E.

Lookout and TV tower: easily pinpointed on the summit of Mt Barker by 168 m high television tower, it offers panoramic views of the area from the Stirling Ranges to Albany; 5 km SW. ***St Werburgh's Chapel:*** small mud-walled chapel (1872) overlooking Hay River Valley; 12 km SW. ***Kendenup:*** historic town, location of WA's first gold find; 16 km E. ***Porongurup:*** hosts Wine Summer Festival each Mar and boasts many small wineries; 24 km E. ***Wineries:*** several in the area, including Goundrey, Marribrook and Plantagenet; maps from visitor centre. ***Mt Barker Heritage Trail:*** 30 km drive tracing the development of the Mt Barker farming district; maps from visitor centre. ***Stirling Range National Park:*** 80 km NE; *see Cranbrook*.

See also GREAT SOUTHERN, p. 309

Mount Magnet

Pop. 422
Map ref. 619 H11 | 620 D2 | 622 D12

Hepburn St; (08) 9963 4172; www.mtmagnet.wa.gov.au

 102.5 WA FM, 105.7 FM ABC Local Radio

In 1854 the hill that rises above this Murchison goldmining town was named West Mount Magnet by surveyor Robert Austin after he noticed that its magnetic ironstone was playing havoc with his compass. Now known by its Aboriginal name, Warramboo Hill, it affords a remarkable view over the town and mines. Located 562 kilometres north-east of Perth on the Great Northern Highway, Mount Magnet offers visitors a rich mining history, rugged granite breakaway countryside and breathtaking wildflowers in season.

Mining and Pastoral Museum: collection of mining and pioneering artefacts includes a Crossley engine from the original State Battery; Hepburn St. ***Heritage trail:*** see the surviving historic buildings and sites of the gold-rush era on this 1.4 km walk; maps from visitor centre.

Mount Magnet Hotel: traditional pub; 36 Hepburn St; (08) 9963 4002.

Wondinong Pastoral Station: working outback property; 72 km north-east of Mount Magnet; (08) 9963 5823.

The Granites: rocky outcrop with picnic area and Aboriginal rock paintings; 7 km N. ***Heritage drive:*** 37 km drive of local historic and natural sights, including views of old open-cut goldmine. Also takes in the Granites and various ghost towns; maps from visitor centre. ***Fossicking for gemstones:*** take care as there are dangerous old mine shafts in the area.

See also OUTBACK COAST & MID-WEST, p. 312

Mullewa

Pop. 419
Map ref. 619 E12 | 620 B3

Jose St; (08) 9961 1500 ; www.mullewatourism.com.au

 107.5 FM ABC Radio National, 828 AM ABC Local Radio

Mullewa, 100 kilometres north-east of Geraldton, is in the heart of wildflower country. In spring, the countryside surrounding the town bursts forth with one of the finest displays of wildflowers in Western Australia. The wreath flower is the star attraction of the annual Mullewa Wildflower Show.

Our Lady of Mt Carmel Church This small church is widely considered to be the crowning achievement of noted priest-architect Monsignor John C. Hawes. Built of local stone, this gem of Romanesque design took seven years to build with Hawes as architect, stonemason, carpenter, modeller and moulder. The adjoining Priest House is now a museum in honour of Hawes, housing his personal belongings, books, furniture and drawings (Cnr Doney and Bowes sts). Both of these buildings are part of the Monsignor Hawes Heritage Trail, which also features the Pioneer Cemetery (Mullewa–Carnarvon Rd; 1 km N) and a site at the old showground just outside town, where a rock carved by Monsignor Hawes was once a simple altar where he held mass for the local Aboriginal people (Mt Magnet Rd; 1.5 km E). Details and maps from visitor centre.

 ***Mullewa Wildflower Show:*** Aug/Sept.

Butterabby Gravesite This gravesite is a grim reminder of the harsh pioneering days. A stone monument recalls the spearing to death in 1864 of a convict labourer and the hanging in 1865 of five Aboriginal people accused of the crime. Mullewa–Mingenew Rd; 18 km S.

Tenindewa Pioneer Well: example of the art of stone pitching that was common at the time of construction, reputedly built by Chinese labourers en route to the Murchison goldfields. Also walking trails; 18 km W. ***Bindoo Hill:*** glacial moraine where ice-smoothed rocks dropped as the face of the glacier melted around 225 million years ago; 40 km NW. ***Coalseam Conservation Park:*** remnants of the state's first coal shafts, now a picnic ground; 45 km SW. ***Tallering Peak and Gorges:*** ideal picnic spot; Mullewa–Carnarvon Rd; check accessibility at visitor centre; 59 km N. ***Wooleen Homestead:*** stay on a working sheep and cattle station in the central Murchison district. Visit Boodra Rock and Aboriginal sites, and experience station life; (08) 9963 7973; 194 km N.

See also OUTBACK COAST & MID-WEST, p. 312

Mundaring

Pop. 3004
Map ref. 614 G5 | 615 E4 | 618 C5 | 620 C7

The Old School, 7225 Great Eastern Hwy; (08) 9295 0202; www.mundaringtourism.com.au

 94.5 Mix FM, 720 AM ABC Local Radio

Mundaring is virtually an outer suburb of Perth, being only 34 kilometres east. Nearby, the picturesque Mundaring Weir is the water source for the goldfields 500 kilometres further east. The original dam opened in 1903. The hilly bush setting makes the weir a popular picnic spot.

Mundaring Arts Centre: contemporary WA fine art and design with comprehensive exhibition program; Great Eastern Hwy; (08) 9295 3991. ***Mundaring District Museum:*** displays on the diverse history of the shire; Great Eastern Hwy. ***Mundaring Sculpture Park:*** collection of sculptures by WA artists, set in natural bush park with grassed areas for picnics and children's playground; Jacoby St.

Market: Nichol St, 2nd Sun each month. ***Truffle Festival:*** Aug. ***Perth Hills Wine Show:*** Aug. ***Trek the Trail:*** Sept. ***Darlington Arts Festival:*** Nov.

Little Caesars Pizzeria: award-winning pizzeria; Shop 7, 7125 Great Eastern Hwy; (08) 9295 6611. ***The Loose Box:*** French cuisine at its best; 6825 Great Eastern Hwy; (08) 9295 1787. ***Black Swan Winery & Restaurant:*** up-market international cuisine; 8600 West Swan Rd, Henley Brook; (08) 9296 6090. ***Darlington Estate Restaurant:*** modern Australian; Darlington Estate Winery, 1495 Nelson Rd, Darlington; (08) 9299 6268. ***Elmar's in the Valley:*** German fare; 8731 West Swan Rd, Henley Brook; (08) 9296 6354. ***Le Paris Brest Cafe & Patisserie:*** French fare; Shop 9, Kalamunda Village Shopping Centre, 22 Haynes St, Kalamunda; (08) 9293 2752. ***Little River Winery Restaurant:*** provincial French fare; 2 Forest Rd, Henley Brook; (08) 9296 4462. ***Sandalford Restaurant:*** European food, vineyard setting; Sandalford Wines, 3210 West Swan Rd, Caversham; (08) 9374 9301. ***Sittella Restaurant:*** international cuisine with picturesque views; Sittella Winery, 100 Barrett St; Herne Hill; (08) 9296 2600. ***Stewart's Restaurant:*** award-winning European cuisine; Brookleigh Estate, 1235 Great Northern Hwy, Upper Swan; (08) 9296 6966.

The Loose Box: country chic–style cottages; 6825 Great Eastern Hwy; (08) 9295 1787. ***Catton Hall Country Homestead:***

luxury cottage; Wilkins Rd, Mt Helena; (08) 9572 1375. ***Chapel Farm Getaway:*** uniquely designed rooms; 231 Toodyay Rd, Middle Swan; (08) 9250 4755. ***Grandis Cottages:*** self-contained cottages; 45 Casuarina Pl, Henley Brook; (08) 9296 3400. ***Grandview B&B:*** B&B with stunning views; 30 Girrawheen Dr, Gooseberry Hill; (08) 9293 2518. ***Hidden Valley Eco Spa Lodges:*** country luxury; 85 Carinyah Rd, Pickering Brook; (08) 9293 7337. ***Settlers Rest Farmstay:*** character 3-bedroom cottage; 90 George St, West Swan; (08) 9250 4540. ***Strelley Brook Cottage:*** historic mud-brick farmhouse; 90 Lefroy St, Herne Hill; (08) 9296 1876. ***Tannamurra:*** luxury home; off Great Northern Hwy, via Lynward Park Estate; (08) 9430 9933. ***Vines Resort & Country Club:*** resort for golf enthusiasts; Verdelho Dr, The Vines; (08) 9297 3000.

Mundaring Weir The Number 1 Pump Station, formerly known as the C. Y. O'Connor Museum, houses an exhibition on the mammoth project of connecting the weir to the goldfields (open 10am–4pm Mon–Sun and public holidays; Mundaring Weir Rd; 8 km s). This is also the starting point of the 560 km Golden Pipeline Heritage Trail to Kalgoorlie, which follows the route of O'Connor's water pipeline, taking in towns and heritage sites. Nearby, the Perth Hills National Park Centre includes Nearer to Nature who provide hands-on activities including bush craft, animal encounters, bush walks and information about Aboriginal culture; contact (08) 9295 2244. The Mundaring Weir Gallery, built in 1908 as a Mechanics Institute Hall, showcases the work of local craftspeople; open 10am–4pm Thurs, Sat, Sun and public holidays; Cnr Hall and Weir Village rds.

John Forrest National Park: on a high point of the Darling Range with sensational views and a lovely picnic spot beside a natural pool at Rocky Pool; 6 km w. ***Karakamia Sanctuary:*** native wildlife sanctuary with guided dusk walks; bookings essential (08) 9572 3169; Lilydale Rd, Chidlow; 8 km NE. ***Calamunnda Camel Farm:*** camel rides; open Thurs–Sun; 361 Paulls Valley Rd; (08) 9293 1156; 10 km s. ***Lake Leschenaultia:*** swimming, canoeing, bushwalks and camping with picnic/barbecue facilities; 12 km NW. ***Kalamunda National Park:*** walking trails through jarrah forest, including the first section of the 963 km Bibbulmun Track; 23 km s; *see Albany*. ***Kalamunda History Village:*** collection of historic buildings; open Sat–Thurs; 23 km s. ***Lesmurdie Falls National Park:*** good views of Perth and Rottnest Island near spectacular falls over the Darling Escarpment; 29 km s. ***Walyunga National Park:*** beautiful bushland and wildflowers and venue for the Avon Descent, a major whitewater canoeing event held each Aug; 30 km NW. ***Wineries:*** several in the area; *Heart of the Hills Wine Trail* map from visitor centre. ***Munda Biddi Bike Trail:*** passes through Mundaring; maps from visitor centre.

See also DARLING & SWAN, p. 306

Nannup

Pop. 503
Map ref. 616 F8 | 618 C10 | 620 C10

4 Brockman St; (08) 9756 1211; www.nannupwa.com

98.9 FM ABC Radio National, 684 AM ABC Local Radio

Nannup is a historic mill town in the Blackwood Valley south of Perth. Known as 'The Garden Village', it has beautiful private and public gardens, tulip farms, daffodils and wildflowers. The countryside is a series of lush, rolling pastures alongside jarrah forests and pine plantations.

Old Police Station: now a visitor centre, original cell block open for viewing; Brockman St. ***Town Arboretum:*** fine collection of old trees planted in 1926; Brockman St. ***Kealley's Gemstone Museum:*** displays of rocks, shells, gemstones, bottles and stamps; closed Wed; Warren Rd. ***Marinko Tomas Memorial:*** memorial to the local boy who was the first serviceman from WA killed in the Vietnam War; Warren Rd. ***Blackwood Wines:*** beautiful winery and restaurant overlooking an artificial lake; closed Wed; Kearney St; (08) 9756 0077. ***Arts, crafts and antiques:*** many in town, including Crafty Creations for quality timber goods on Warren Rd (closed June). ***Heritage trail:*** in 2 sections, with a 2.5 km town walk highlighting historic buildings, and a 9 km scenic drive; maps from visitor centre.

Market: Warren Rd; 2nd Sat each month. ***Art and Photography Exhibition:*** Jan. ***Music Festival:*** Feb/Mar. ***Forest Car Rally:*** Mar. ***Nannup Cup Boat Race:*** June. ***Flower and Garden Festival:*** tulips; Aug. ***Rose Festival:*** Nov.

Blackwood Bistro: modern Australian; Blackwood Wines, Kearney St; (08) 9756 0077. ***Mulberry Tree Restaurant:*** home-style cooking; 62 Warren Rd; (08) 9756 3038. ***Nannup Bridge Cafe:*** cosy eatery; 1 Warren Rd; (08) 9756 1287. ***Tathra Restaurant:*** specialises in local produce; Tathra Winery, Blackwood River Tourist Dr (Route 251); (08) 9756 2040.

Beyonderup Falls: spa chalets; Balingup Rd; (08) 9756 2034. ***Crabapple Lane B&B:*** tasteful B&B; Barrabup Rd; (08) 9756 0017. ***Redgum Hill Country Retreat:*** B&B and self-contained accommodation; Balingup Rd; (08) 9756 2056. ***Tathra Hill Top Retreat:*** country adult retreat; Blackwood River Tourist Dr (Route 251); (08) 9756 2040.

Kondil Park: bushwalks and wildflowers in season; 3 km w. ***Barrabup Pool:*** largest of several pools, ideal for swimming, fishing and camping. Also has barbecue facilities; 10 km w. ***Cambray Sheep Cheese:*** award-winning cheeses, visitors can watch the milking and cheese-making; samples available. There are also guest cottages; 12 km N. ***Carlotta Crustaceans:*** marron farm; 14 km s off Vasse Hwy. ***Tathra:*** fruit winery and restaurant; open 11am–4.30pm daily; 14 km NE. ***Mythic Mazes:*** mazes including sculptures of myths from around the world; 20 km N. ***Donnelly River Wines:*** open daily for cellar-door tastings; (08) 9301 5555; 45 km s. ***Blackwood River:*** camping, swimming, canoeing and trout fishing. ***Self-guide walks:*** wildflower (in spring), waterfall (in winter) and forest walks; maps from visitor centre. ***Scenic drives:*** through jarrah forest and pine plantations, including 40 km Blackwood Scenic Drive; maps from visitor centre.

See also THE SOUTH-WEST, p. 307

Narrogin

Pop. 4240
Map ref. 618 E7 | 620 D9

Dryandra Country Visitor Centre, cnr Park and Fairway sts; (08) 9881 2064; www.dryandratourism.org.au

100.5 Hot FM, 918 AM Radio West

Narrogin, 192 kilometres south-east of Perth on the Great Southern Highway, is the commercial hub of a prosperous agricultural area. Sheep, pigs and cereal farms are the major industries. First settled in the 1870s, the town's name is derived from an Aboriginal word 'gnarojin', meaning waterhole.

Gnarojin Park This park is a national award winner for its original designs and artworks portraying local history and culture, which include the Centenary Pathway, marked with 100 locally designed commemorative tiles, Newton House Barbecue and Noongar Cultural Sites. Gordon St.

History Hall: local history collection; Egerton St. ***Old Courthouse Museum:*** built in 1894 as a school, it later became the district courthouse; open Mon–Sat; Egerton St. ***Narrogin Art Gallery:*** exhibitions; open 10am–4pm Tues–Fri, 10am–12pm Sat; Federal St. ***Lions Lookout:*** excellent views; Kipling St. ***Heritage trail:*** self-guide walk around the town's historic buildings; maps from visitor centre.

State Gliding Championships: Jan. ***Spring Festival:*** Oct. ***Rev Heads:*** car rally; Nov/Dec.

Albert's Restaurant: international cuisine; Albert Facey Motor Inn, 78 Williams Rd; (08) 9881 1899.

Albert Facey Motor Inn: hotel accommodation; 78 Williams Rd; (08) 9881 1899. ***Chuckem Farm B&B:*** homestead B&B and self-contained cottage; 1481 Tarwonga Rd; (08) 9881 1188. ***Eden Valley Farmstay:*** 2-bedroom cottage; 3733 Williams–Kondinin Rd; (08) 9881 5864.

Dryandra Woodland One of the few remaining areas of virgin forest in the wheat belt, Dryandra is a paradise for birdwatchers and bushwalkers. The open, graceful eucalypt woodlands of white-barked wandoo, powderbark and thickets of rock she-oak support many species of flora and fauna including numbats (the state's animal emblem), woylies, tammar wallabies, brush-tailed possums and many others. Over 100 species of birds have been identified, including the mound-building mallee fowl. Tune your radio to 100FM for 'Sounds of Dryandra', a 25 km radio drive trail with 6 stops featuring tales of the local Nyungar people, early forestry days, bush railways and Dryandra's unique wildlife. There are day-visitor facilities and accommodation, walk trails, a weekend Ecology Course (runs in autumn and spring; try your hand at radio-tracking, trapping and spotlighting) and school holiday programs. 22 km NW.

Barna Mia Animal Sanctuary The sanctuary, within the Dryandra Woodland, has guided spotlight walks at night that reveal threatened marsupials, including the bilby and boodie. Bookings essential; contact DEC on (08) 9881 9200 or visitor centre which also has maps; Narrogin–Wandering Rd; 26 km NW.

Foxes Lair Nature Reserve: 60 ha of bushland with good walking trails, wildflowers in spring and 40 species of birds; maps and brochures from visitor centre; Williams Rd; 1 km SW. ***Yilliminning and Birdwhistle Rocks:*** unusual rock formations; 11 km E. ***District heritage trail:*** maps from visitor centre.

See also HEARTLANDS, p. 308

New Norcia

Pop. 70
Map ref. 618 C2 | 620 C6

Museum and art gallery, Great Northern Hwy; (08) 9654 8056; www.newnorcia.wa.edu.au

612 AM ABC Radio National, 720 AM ABC Local Radio

In 1846 Spanish Benedictine monks established a mission 132 kilometres north of Perth in the secluded Moore Valley in an attempt to help the local Aboriginal population. They named their mission after the Italian town of Norcia, the birthplace of the order's founder, St Benedict. The imposing Spanish-inspired buildings of New Norcia, surrounded by the gum trees and dry grasses of the wheat belt, provide a most unexpected vista. The town still operates as a monastery and is Australia's only monastic town. Visitors may join the monks at daily prayers.

Abbey Church This fine example of bush architecture was built using a combination of stones, mud plaster, rough-hewn trees and wooden shingles. It is the oldest Catholic church still in use in WA and contains the tomb of Dom Rosendo Salvado, the founder of New Norcia and its first Abbot. Hanging on a wall is the painting of Our Lady of Good Counsel, given to Salvado before he left for Australia in 1845 by Bishop (later Saint) Vincent Palotti. One of New Norcia's most famous stories relates how, in 1847, Salvado placed this revered painting in the path of a bushfire threatening the mission's crops. The wind suddenly changed direction and drove the flames back to the part already burnt, and the danger was averted.

Museum and art gallery: the museum tells the story of New Norcia's history as an Aboriginal mission, while the art gallery houses priceless religious art from Australia and Europe as well as Spanish artefacts, many of which were gifts from Queen Isabella of Spain. The Museum Gift Shop features New Norcia's own produce including bread, nutcake, pan chocolatti, biscotti, wine, honey and olive oil. ***Monastery:*** daily tours of the interior. A guesthouse allows visitors to experience the monastic life for a few days. ***New Norcia Hotel:*** this magnificent building, featuring a massive divided central staircase and high, moulded pressed-metal ceilings, was opened in 1927 as a hostel to accommodate parents of the children who were boarding at the town's colleges. ***Old Flour Mill:*** the oldest surviving building in New Norcia dates from the 1850s. ***Heritage trail:*** 2 km self-guide walk highlights New Norcia's historic and cultural significance and the role of the Benedictine monks in colonial history; maps from visitor centre. ***Guided tour:*** 2 hr tour of the town with an experienced guide takes you inside buildings not otherwise open to the public; 11am and 1.30pm daily from museum.

New Norcia Hotel: town landmark; Great Northern Hwy; (08) 9654 8034.

Monastery Guesthouse: Benedictine hospitality; Great Northern Hwy; (08) 9654 8002. ***New Norcia Hotel:*** simple hotel accommodation; Great Northern Hwy; (08) 9654 8034. ***Nirranda Farmstay:*** spacious 4-bedroom homestead; 9236 Great Northern Hwy, Wannamal; (08) 9655 9046.

Mogumber: town with one of the state's highest timber-and-concrete bridges; 24 km SW. ***Piawaning:*** magnificent stand of eucalypts north of town; 31 km NE. ***Bolgart:*** site of historic hotel; 49 km SE. ***Wyening Mission:*** former mission, now a historic site; open by appt; 50 km SE.

See also HEARTLANDS, p. 308

Newman

Pop. 4247
Map ref. 622 E5 | 627 L11

Cnr Fortescue Ave and Newman Dr; (08) 9175 2888; www.newman-wa.org

88.9 Red FM, 567 AM ABC Local Radio

Located in the heart of the Pilbara, Newman was built in the late 1960s by the Mount Newman Mining Company to house the workforce required at nearby Mount Whaleback, the largest open-cut iron-ore mine in the world. At the same time a 426-kilometre railway was constructed between Newman and Port Hedland to transport the ore for export to Japan.

Mt Whaleback Mine Tours: the mine produces over 100 million tonnes of iron ore every year. Tours (minimum 4 people) run Mon–Sat Apr–Sept, Tues and Thurs Oct–Mar; bookings at visitor centre. ***Mining and Pastoral Museum:*** interesting display of relics from the town's short history, including the first Haulpak (giant iron-ore truck) used at Mt Whaleback and outback station life; located at visitor centre. ***Radio Hill Lookout:*** panoramic view of town and surrounding area; off Newman Dr.

Campdraft and Rodeo: Mar. ***Fortescue Festival:*** Aug.

Newman Hotel Motel: traditional pub food; Newman Dr; (08) 9175 1101.

Newman Hotel Motel: centrally located motel rooms; Newman Dr; (08) 9175 1101. ***Seasons Hotel Newman:*** variety of modern rooms; Newman Dr; (08) 9177 8666.

Ophthalmia Dam: swimming, sailing, barbecues and picnics (no camping); 20 km E. ***Mt Newman:*** excellent views; 25 km NW. ***Eagle Rock Falls:*** permanent pools and picnic spots nearby; 4WD access only; 80 km NW. ***Wanna Munna:*** site of Aboriginal rock carvings; 74 km W. ***Rockpools and waterholes:*** at Kalgans Pool (65 km NE), Three Pools (75 km N) and Weeli Wolli (99 km W). ***Newman Waterholes and Art Sites Tour:*** maps from visitor centre. ***Karijini National Park:*** 196 km N; *see Tom Price.*

See also PILBARA, p. 313

Norseman

Pop. 861
Map ref. 621 I7

68 Roberts St; (08) 9039 1071; www.norseman.info

105.7 FM ABC Goldfields, 107.3 FM ABC Radio National

Norseman is the last large town on the Eyre Highway for travellers heading east towards South Australia. Gold put Norseman on the map in the 1890s with one of the richest quartz reefs in Australia. The town is steeped in goldmining history, reflected in its colossal tailings dump. If visitors could stand atop this dump they could have up to $50 million in gold underfoot (although the rock has been processed, much residual gold remains). The story behind the town's name has become folklore. The settlement sprang up in 1894 when a horse owned by prospector Laurie Sinclair pawed the ground and unearthed a nugget of gold; the site proved to be a substantial reef. The horse's name was Norseman.

Historical Collection: mining tools and household items; open 10am–1pm Mon–Sat; Battery Rd. ***Phoenix Park:*** open-plan park with displays, stream and picnic facilities; Prinsep St. ***Statue of Norseman:*** bronze statue by Robert Hitchcock commemorates Norseman, the horse; Cnr Roberts and Ramsay sts. ***Camel Train:*** corrugated-iron sculptures represent the camel trains of the pioneer days; Prinsep St. ***Gem fossicking:*** gemstone permits from visitor centre.

Norseman Cup: Mar.

Norseman Hotel: traditional pub food; 90 Roberts St; (08) 9039 1023.

Norseman Great Western Motel and Great Western Travel Village: Norseman's premier motel; Prinsep St; (08) 9039 1633.

Beacon Hill Lookout: spectacular at sunrise and sunset, this lookout offers an outstanding 360-degree panorama of the salt lakes, Mt Jimberlana, the township, tailings dump and surrounding hills and valleys; Mines Rd; 2 km E. ***Mt Jimberlana:*** reputed to be one of the oldest geological areas in the world. Take the walking trail to the summit for great views; 5 km E. ***Dundas Rocks:*** barbecue and picnic area amid granite outcrops near old Dundas townsite, where the lonely grave of a seven-month-old child is one of the only signs that the area was once inhabited. Travel here via the highway (22 km S) or along the 33 km heritage trail that follows an original Cobb & Co route (maps from visitor centre). ***Buldania Rocks:*** picnic area with beautiful spring wildflowers; 28 km E. ***Bromus Dam:*** freshwater dam with picnic area; 32 km S. ***Peak Charles National Park:*** good-weather track for experienced walkers and climbers to Peak Eleanora, with a magnificent view from top; 50 km S, then 40 km W off hwy. ***Cave Hill Nature Reserve:*** a granite outcrop with a cave set in its side and a dam nearby. Popular spot for camping, picknicking and rock climbing; 4WD access only; 55 km N, then 50 km W. ***Fraser Range Station:*** working pastoral property that specialises in Damara sheep. The station has a range of available accommodation and opportunities to experience a remote working pastoral property; (08) 9039 3210; 100 km E.

See also GOLDFIELDS, p. 311

Northam

Pop. 6007
Map ref. 615 G2 | 618 D4 | 620 C7

Avon Valley Visitor Centre, 2 Grey St; (08) 9622 2100; www.visitnorthamwa.com.au

96.5 Hot FM, 1215 AM ABC Local Radio

Northam lies in the heart of the fertile Avon Valley. The Avon River winds its way through the town and on its waters you'll find white swans, a most unusual sight in a state where the emblem is a black swan. White swans were brought to Northam from England in the 1900s and have flourished here. Northam is also synonymous with hot-air ballooning as it is one of the few areas in Western Australia ideally suited to this pastime. Northam is home to the famous Avon Descent, a 133-kilometre whitewater race down the Avon and Swan rivers to Perth.

Historic buildings Of the many historic buildings in Northam, two are particularly noteworthy: Morby Cottage (1836) on Avon Dr, the home of Northam's first settler, John Morrell, now a museum and open 10.30am–4pm Sun or by appt; and the National Trust–classified Sir James Mitchell House (1905), cnr Duke and Hawes sts, with its elaborate Italianate architecture. Take the 90 min self-guide walk for the full tour of the town. Maps from visitor centre.

Old Railway Station Museum: displays include a steam engine and renovated carriages, plus numerous artefacts from the early 1900s; open 10am–4pm Sun or by appt; Fitzgerald St West. ***Visitor centre:*** exhibition showcasing the area's significant postwar migrant history; Grey St. ***Suspension bridge:*** the longest pedestrian suspension bridge in Australia crosses the Avon River adjacent to the visitor centre.

Vintage Car Swap Meet: Feb. ***Avon Descent:*** Aug. ***Northam Cup:*** Oct. ***Motorcycle Festival:*** Nov. ***Wheatbelt Cultural Festival:*** Dec.

3twotwo: charming country pub; Avon Bridge Hotel, 322 Fitzgerald St; (08) 9622 1023. ***Cafe Yasou:*** Mediterranean flavours of Cyprus and Greece; 175 Fitzgerald St; (08) 9622 3128.

Avon Bridge Hotel: hotel and motel rooms; 322 Fitzgerald St; (08) 9622 1023. ***Brackson House B&B:*** heritage B&B; 7 Katrine Rd; (08) 9622 5262. ***Egoline Reflections:*** heritage homestead; Northam–Toodyay Rd; (08) 9622 5811. ***Uralia Cottage:*** 1-bedroom cottage with pool; Cnr 59 Gordon St and Uralia Tce; (08) 9622 1742. ***Mystique Maison:*** ideal for group retreats; 10 Forrest St, Goomalling; (08) 9629 1673.

Mt Ommanney Lookout: excellent views of the township and agricultural areas beyond; 1.5 km W. ***Hot-air ballooning:*** Northam Airfield; Mar–Nov; bookings essential; 2 km NE. ***Meckering:*** small town made famous in 1968 when an earthquake left a huge fault line in its wake; 35 km E. ***Cunderdin Museum:*** The museum housed in the No 3 pumping station has displays on the pipeline, wheat-belt farming and the Meckering earthquake; 50 km E.

See also HEARTLANDS, p. 308

Northampton

Pop. 814
Map ref. 619 D12 | 620 A3

Old police station, Hampton Rd; (08) 9934 1488; www.northamptonwa.com.au

 96.5 WA FM

Northampton, nestled in the valley of Nokarena Brook, 51 kilometres north of Geraldton, was awarded Historic Town status by the National Trust in 1993. It was declared a townsite in 1864 and is one of the oldest settlements in Western Australia. A former lead-mining centre, its prosperity is now based on sheep and wheat-farming.

Chiverton House Historical Museum: unusual memorabilia housed in what was originally the home of Captain Samuel Mitchell, mine manager and geologist. Surrounding gardens include herbarium and restored farm machinery; open 10am–12pm and 2–4pm Fri–Mon; Hampton Rd. ***Mary Street Railway Precinct:*** railway memorabilia at the site of the town's 2nd railway station, built 1913; Eastern end. ***Church of Our Lady in Ara Coeli:*** designed in 1936 by Monsignor John Hawes, WA's famous architect-priest; Hampton Rd. ***Gwalla Church and Cemetery:*** ruins of town's first church (1864); Gwalla St. ***Hampton Road Heritage Walk:*** 2 km walk includes 37 buildings of historical interest including the Miners Arms Hotel (1868) and the Old Railway Station (1879); maps from visitor centre.

Market: Kings Park, cnr Essex St and Hampton Rd; 1st Sat each month. ***Airing of the Quilts:*** quilts hung in main street; Oct.

Miners Arms Hotel: counter meals; Hampton Rd; (08) 9934 1281. ***The Railway Tavern:*** great pizzas, history displays; 71 North West Coastal Hwy; (08) 9934 1120.

Old Miners' Cottages B&B: restored cottages; Brook St; (08) 9934 1864.

Alma School House: built in 1916 as a one-teacher school; 12 km N. ***Aboriginal cave paintings:*** at the mouth of the Bowes River; 17 km W. ***Oakabella Homestead:*** one of the first farms in WA to plant canola, or rapeseed as it was then known. Take a guided tour of the historic homestead and outbuildings; open daily Mar–Nov; 18 km S. ***Horrocks Beach:*** beautiful bays, sandy beaches, good swimming, fishing and surfing; 20 km W. ***Lynton Station:*** ruins of labour-hiring depot for convicts, used in 1853–56; 35 km NW. ***Lynton House:*** squat building with slits for windows, probably designed as protection from hostile Aboriginal people; 35 km NW. ***Principality of Hutt River:*** visitors are given a tour by the royals of this 75 sq km principality. With their own government, money and postage stamps, Hutt River exists as an independent sovereign state, seceded from Australia in 1970; 35 km NW. ***Hutt Lagoon:*** appears pink in midday sun; 45 km NW. ***Port Gregory:*** beach settlement, ideal for swimming, fishing and windsurfing; 47 km NW. ***Warribano Chimney:*** Australia's first lead smelter; 60 km N.

See also OUTBACK COAST & MID-WEST, p. 312

Northcliffe

Pop. 299
Map ref. 616 H12 | 617 C4 | 618 D11 | 620 C11

Muirillup Rd; (08) 9776 7203; www.northcliffe.org.au

102.7 WA FM, 105.9 FM ABC Local Radio

Magnificent virgin karri forests surround the township of Northcliffe, 31 kilometres south of Pemberton in the state's South-West. Just a kilometre from the town centre is Northcliffe Forest Park, where you can see purple-crowned lorikeets, scarlet robins and in spring, a profusion of wildflowers. Not far away is the coastal settlement of Windy Harbour, a popular swimming beach.

Pioneer Museum Northcliffe came into existence as a result of the Group Settlement Scheme, a WA government plan to resettle returned WW I soldiers and immigrants by offering them rural land to farm. The scheme was enthusiastically backed by English newspaper magnate Lord Northcliffe (hence the town's name). Unfortunately, by the 1920s, when the scheme began, all the good land in the state had already been settled. The group settlers were left to contend with inhospitable country and with only crosscut saws and axes, they were faced with the daunting task of clearing some of the world's biggest trees from their land. It is not surprising that by the mid-1930s all of the Group Settlement projects in the South-West timber country had failed. A visit to the Pioneer Museum with its excellent displays is the best way to understand the hardships the group settlers experienced. Open 10am–2pm daily Sept–May, 10am–2pm Sat, Sun and school holidays June–Aug; Wheatley Coast Rd; (08) 9775 1022.

Canoe and mountain-bike hire: details from visitor centre.

Great Karri Ride: Mar. ***Mountain Bike Championship:*** May/June. ***Night-time Mountain Bike Race:*** Nov.

The Dairy Lounge Cafe: dairy-inspired eatery; Bannister Downs Farm, Muirillup Rd; (08) 9776 6300.

Northcliffe Hotel: traditional hotel; Lot 8 Wheatley Coast Rd; (08) 9776 7089. ***Riverway Chalets:*** relaxing rammed-earth chalets; Riverway Rd; (08) 9776 7183. ***Watermark Kilns:*** kilns converted to self-contained accommodation; Karri Hill Rd; (08) 9776 7349.

Northcliffe Forest Park: follow the Hollow Butt Karri and Twin Karri walking trails or enjoy a picnic; Wheatley Coast Rd. ***Warren River:*** trout fishing and sandy beaches; 8 km N. ***Mt Chudalup:*** spectacular views of the surrounding D'Entrecasteaux National Park and coastline from the summit of this giant granite outcrop; 10 km S. ***Moon's Crossing:*** delightful picnic spot; 13 km NW. ***Lane Poole Falls and Boorara Tree:*** 3 km walking trail leads to the falls, passing the Boorara Tree with 50 m high fire-lookout cabin; 18 km SE. ***Pt D'Entrecasteaux:*** limestone cliffs, popular with rock climbers, rise 150 m above the sea where 4 viewing platforms provide superb views; 27 km S. ***Windy Harbour:*** swimming, snorkelling, fishing, camping and whale-watching (from platform, best times Sept–Nov); 27 km S.

Cathedral Rocks: watch seals and dolphins; 27 km s. ***Salmon Beach:*** surf beach offers salmon fishing Apr–June; 27 km s. ***The Great Forest Trees Drive:*** 48 km self-guide scenic drive takes in the karri giants at Snake Gully Lookout, the Boardwalk and Big Tree Grove; maps from visitor centre. ***Bibbulmun Track:*** section of this long-distance walking trail links the 3 national parks around Northcliffe: D'Entrecasteaux (5 km s), Warren (20 km NW) and Shannon (30 km E); *see Albany*. ***Mountain-bike trails:*** 4 permanent trails have been established around Northcliffe; details from visitor centre.

See also THE SOUTH-WEST, p. 307

Onslow

Pop. 574
Map ref. 619 C2 | 626 A8

Second Ave (May–Oct), (08) 9184 6644; or council offices, Second Ave, (08) 9184 6001; www.ashburton.wa.gov.au

 106.7 WA FM, 107.5 FM ABC Radio National

Onslow, on the north-west coast between Exmouth and Karratha, is the supply base for offshore gas and oil fields. This part of the coast is among the north's most cyclone-prone and Onslow has often suffered severe damage. The town was originally at the Ashburton River mouth and a bustling pearling centre. In the 1890s gold was discovered nearby. In 1925 the townsite was moved to Beadon Bay after cyclones caused the river to silt up. During World War II, submarines refuelled here and the town was bombed twice. In the 1950s it was the mainland base for Britain's nuclear experiments at Montebello Islands. In 1963 Onslow was almost completely destroyed by a cyclone. It is now an attractive tree-shaded town.

Goods Shed Museum: memorabilia from the town's long history and collections of old bottles, shells and rocks; in visitor centre; Second Ave. ***Beadon Creek and Groyne:*** popular fishing spot. ***Ian Blair Memorial Walkway:*** 1 km scenic walk; starts at Beadon Pt and finishes at Sunset Beach. ***Heritage trail:*** covers sites of interest in town; maps from visitor centre.

Beadon Bay Hotel: counter meals; Second Ave; (08) 9184 6002. ***Nikki's Licensed Restaurant:*** steak and seafood; 17 First Ave; (08) 9184 6121.

Onslow Mackerel Motel: 4-star motel; Cnr Second Ave and Third St; (08) 9184 6586. ***Onslow Sun Chalets:*** beachfront motel rooms and chalets; Second Ave; (08) 9184 6058. ***Club Thevenard:*** variety of accommodation; Thevenard Island; (08) 9184 6444. ***Direction Island:*** secluded beach cabin; Direction Island; (08) 9184 6444.

Termite mounds: with interpretive display; 10 km s. ***Mackerel Islands:*** excellent fishing destination. Charter boats are available for daytrips or extended fishing safaris; details from visitor centre; 22 km off the coast. ***Ashburton River:*** swimming, camping and picnicking; 45 km sw. ***Old townsite heritage trail:*** self-guide walk around original townsite including old gaol; maps from visitor centre; 45 km sw.

See also PILBARA, p. 313

Pemberton

Pop. 757
Map ref. 616 H11 | 617 B2 | 618 D11 | 620 C11

Brockman St; (08) 9776 1133 or 1800 671 133; www.pembertontourist. com.au

97.3 WA FM, 558 AM ABC Local Radio

Pemberton sits in a quiet valley surrounded by some of the tallest trees in the world and, in spring, brilliant wildflowers. This is the heart of karri country, with 4000 hectares of protected virgin karri forest in the nearby Warren and Beedelup national parks. Pemberton is a centre for high-quality woodcraft and is renowned for its excellent rainbow trout and marron fishing.

Karri Forest Discovery Centre: interpretive centre includes museum with collection of historic photographs and forestry equipment; at visitor centre. ***Pioneer Museum:*** utensils, tools and other memorabilia from pioneer days plus a full-scale settler's hut; at visitor centre. ***Craft galleries:*** many in town, including the Fine Woodcraft Gallery in Dickinson St and the Peter Kovacsy Studio in Jamieson St.

Mill Hall Markets: Brockman St; 2nd Sat each month and public holidays. ***CWA Markets:*** Brockman St; 4th Sat each month. ***Autumn Festival:*** May.

Hidden River Estate Restaurant: superb international cuisine; Mullineaux Rd; (08) 9776 1437. ***Jarrah Jacks Brewery:*** innovative brewery menu; Lot 2 Kemp Rd; (08) 9776 1333. ***King Trout Restaurant and Marron Farm:*** seafood; Cnr Northcliffe and Old Vasse rds; (08) 9776 1352. ***The Shamrock Restaurant:*** local fare; 18 Brockman St; (08) 9776 1186.

Clover Cottage: self-contained cottages; 251 Wheatley Coast Rd; (08) 9773 1262. ***Peppermint Grove Retreat:*** elegant chalets; Lot 5198 Channybearup Rd; (08) 9776 0056. ***Pump Hill Farm Cottages:*** rammed-earth cottages; Pump Hill Rd; (08) 9776 1379. ***Salitage Suites:*** adults-only suites; Salitage Winery, Vasse Hwy; (08) 9776 1195. ***Warren River Resort:*** cottages set in the forest; 713 Pemberton–Northcliffe Rd; (08) 9776 1400.

Gloucester National Park In this park is the town's most popular tourist attraction, the Gloucester Tree. With its fire lookout teetering 61 m above the ground and a spine-tingling 153 rungs spiralling upwards, this is not a climb for the faint-hearted. The Gloucester Tree is one of eight tree towers constructed from the late 1930s as fire lookouts. As the extremely tall trees in the southern forests offered few vantage points for fire-lookout towers, it was decided to simply build a cabin high enough in one of the taller trees to serve the purpose. Also within the park are the Cascades, a scenic spot for picnicking, bushwalking and fishing. 1 km s.

Warren National Park This park boasts some of the most easily accessible virgin karri forest. The Dave Evans Bicentennial Tree has another fire lookout with picnic facilities and walking tracks nearby. 9 km sw.

Beedelup National Park Here you'll find the Walk Through Tree, a 75 m, 400-year-old karri with a hole cut in it big enough for people to walk through. The Beedelup Falls, a total drop of 106 m, are rocky cascades best seen after heavy rain. Nearby are walk trails and a suspension bridge. 18 km w.

Lavender & Berry Farm: enjoy berry scones, lavender biscuits and other unusual produce; Browns Rd; 4 km N. ***Big Brook Dam:*** the dam has its own beach, picnic and barbecue facilities, trout and marron fishing in season, and walking trails; 7 km N. ***Big Brook Arboretum:*** established in 1928 to study the growth of imported trees from around the world; 7 km N. ***King Trout Restaurant and Marron Farm:*** catch and cook your own trout;

closed Thurs; 8 km sw. ***Founder's Forest:*** part of the 100-Year-Old Forest, with karri regrowth trees over 120 years old; 10 km N. ***Wineries:*** more than 28 wineries in the area, many offering tours, tastings and sales; details from visitor centre. ***Pemberton Tramway:*** tramcars based on 1907 Fremantle trams operate daily through tall-forest country to the Warren River Bridge; (08) 9776 1322. ***Fishing:*** in rivers, inland fishing licence is required for trout and marron; contact post office for details and permits. ***Tours:*** river tours, scenic bus tours, 4WD adventure tours, self-guide forest drives, walking trails and eco-tours; details from visitor centre. ***Drive trails:*** include the Heartbreak Trail, a one-way drive through the karri forest of Warren National Park; maps from visitor centre. ***Walk trails:*** include the 1 hr return Rainbow Trail; maps from visitor centre. ***Bibbulmun Track:*** walking trail passes through Pemberton; *see Albany*.

See also THE SOUTH-WEST, p. 307

Pingelly

Pop. 817
Map ref. 618 E6 | 620 D8

Council offices, 17 Queen St; (08) 9887 1066; www.pingelly.wa.gov.au

99.7 FM ABC News Radio, 612 AM ABC Radio National

Located 158 kilometres south-east of Perth on the Great Southern Highway, Pingelly is part of the central-southern farming district. Sandalwood was once a local industry, but today sheep and wheat are the major produce.

Courthouse Museum: built in 1907, now houses historic memorabilia and photographs; Parade St. ***Apex Lookout:*** fine views of town and surrounding countryside; Stone St.

Autumn Country Show and Ute Muster: Mar.

The Exchange Tavern: counter meals; 1 Pasture St; (08) 9887 0180.

Pingelly Roadhouse and Motel: town's premier accommodation; 8 Quadrant St; (08) 9887 1015.

Pingelly Heights Observatory: audio tour and telescope viewing of the stars and constellations; bookings essential (08) 9887 0088; 5 km NE. ***Moorumbine Heritage Trail:*** walk or drive through this old townsite featuring early settlers' cottages and St Patrick's Church, built in 1873; maps from council office; 8 km E. ***Tutanning Flora and Fauna Reserve:*** botanist Guy Shorteridge collected over 400 species of plants from here for the British Museum between 1903 and 1906; 22 km E. ***Boyagin Nature Reserve:*** widely recognised as one of the few areas of original fauna and flora left in the wheat belt, this picnic reserve has important stands of powderbark, jarrah and marri trees and is home to numbats and tammar wallabies; 26 km NW.

See also HEARTLANDS, p. 308

Pinjarra

Pop. 3295
Map ref. 614 D10 | 618 C6 | 620 C8

Pinjarra Heritage Train Station, Fimmel La; (08) 9531 1438; www.pinjarravisitorcentre.com.au

97.3 Coast FM, 585 AM ABC News Radio

A pleasant 84-kilometre drive south of Perth along the shaded South Western Highway brings you to Pinjarra, picturesquely set on the Murray River. Predominantly a dairying, cattle-farming and timber-producing area, Pinjarra was also once known as the horse capital of Western Australia when horses were bred for the British Army in India. Today horseracing, pacing and equestrian events are a major part of Pinjarra culture. The Alcoa Refinery north-east of town is the largest alumina refinery in Australia.

Edenvale Complex Built in 1888 with locally fired clay bricks, Edenvale was the home of Edward McLarty, member of the state's Legislative Council for 22 years. Nearby is Liveringa (1874), the original residence of the McLarty family, now an art gallery. There is a Heritage Rose Garden featuring 364 varieties of old-fashioned roses, a quilters' display in the Old School House, and a machinery museum. Cnr George and Henry sts.

Suspension bridge: across the Murray River, with picnic areas at both ends; George St. ***Heritage trail:*** 30 min river walk follows series of tiles explaining the heritage of the area; maps from visitor centre.

Railway Markets: Lions Park; 2nd Sun morning each month Sept–June. ***Community Markets:*** Fimmel La; 4th Sat morning each month Sept–May. ***Pinjarra Cup:*** Mar. ***Pinjarra Festival:*** June. ***Murray Arts and Craft Open Day:*** Nov.

Edenvale Homestead & Heritage Tearooms: light meals; 1 George St; (08) 9531 2223. ***Raven Wines:*** revolving menu concept; 41 Wilson Rd; (08) 9531 2774. ***Redcliffe on the Murray:*** modern Australian; 13 Sutton St; (08) 9531 3894.

Lazy River Boutique B&B: luxurious spa suites; 9 Wilson Rd; (08) 9531 4550. ***Pinjarra Cabins & Caravan Park:*** peaceful caravan park; 1716 Pinjarra Rd; (08) 9531 1374. ***Nautica Lodge:*** luxury marine-themed B&B; 203 Culeenup Rd, North Yunderup; (08) 9537 8000.

Fairbridge Established by Kingsley Fairbridge in 1912 as a farm school for British children, many of them orphans. Over the years more than 8000 English children were educated here. The boarding houses, which are today used as holiday cottages, have famous British names such as Clive, Shakespeare, Nightingale, Exeter, Evelyn and Raleigh. South Western Hwy; 1800 440 770; 6 km N.

Peel Zoo: set in lush native flora, includes opportunities to feed and interact with the animals and has picnic area and barbecues; (08) 9531 4322; 2 km N. ***Old Blythewood:*** beautiful National Trust property built in 1859 by John McLarty, who arrived in Australia in 1839; open Sat 10.30am–3.30pm, Sun 12.30pm–3.30pm, or by appt; 4 km S. ***Alcoa Mine and Refinery Tours:*** includes the mining process, and the world's biggest bulldozer. Tours are free Wed and the last Fri of each month (except during the summer school holidays); bookings essential (08) 9531 6752; 6km NE. ***Alcoa Scarp Lookout:*** good views of coastal plain, surrounding farming area and Alcoa Refinery; 14 km E. ***North Dandalup Dam:*** recreation lake, picnic area and coastal views from lookout; 22 km NE. ***South Dandalup Dam:*** barbecues and picnic areas; 30 km E. ***Lake Navarino:*** formerly known as Waroona Dam, it is good for watersports, fishing, walking and horseriding; 33 km S. ***Coopers Mill:*** first flour mill in the Murray region, located on Cooleenup Island near the mouth of the Serpentine River. It is accessible only by water; details from visitor centre. ***Hotham Valley Tourist Railway:*** travel from Pinjarra to Dwellingup by train, taking in lush green dairy country before climbing the steep and spectacular Darling Range and finishing in the heart of the jarrah forest. The train is steam-hauled May–Oct and diesel-hauled Nov–Apr; check times with visitor centre. ***Ravenswood Adventures:*** explore the Murray River by kayak, dinghy, canoe or pedal boat; Ravenswood; (08) 9537 7173.

See also THE SOUTH-WEST, p. 307

Port Hedland

Pop. 11 557
Map ref. 622 D1 | 624 D11 | 626 H3

13 Wedge St; (08) 9173 1711.

 91.7 WA FM, 94.9 FM ABC News Radio

Port Hedland was named after Captain Peter Hedland, who reached this deep-water harbour in 1863. An iron-ore boom that began in the early 1960s saw the town grow at a remarkable rate. Today, Port Hedland handles the largest iron-ore export tonnage of any Australian port. Iron ore is loaded onto huge ore carriers; the 2.6-kilometre trains operated by BHP Iron Ore are hard to miss. Salt production is another major industry with about 2 million tonnes exported per annum.

Don Rhodes Mining Museum: open-air museum with displays of historic railway and mining machinery; Wilson St. ***Pioneer and Pearlers Cemetery:*** used between 1912 and 1968, it has graves of early gold prospectors and Japanese pearl divers; off Stevens St. ***Town tour:*** visit the town's many attractions; 11am Mon, Wed and Fri June–Oct; bookings at visitor centre. ***BHP Iron Ore Tour:*** see the enormous machinery required to run this industrial giant; departs 9.30am Mon–Fri from visitor centre. ***Heritage trail:*** 1.8 km self-guide walk around the town; maps from visitor centre.

Australia Day Festival: Jan. ***Port Hedland Cup:*** Aug. ***Pilbara Music Festival:*** Sept. ***Welcome to Hedland Night:*** every few months.

Esplanade Hotel Port Hedland: hearty counter meals; 2–4 Anderson St; (08) 9173 2783. ***Heddy's Bar & Bistro:*** bistro by the sea; All Seasons Port Hedland, cnr Lukis and McGregor sts; (08) 9173 1511. ***The Pilbara Room Restaurant:*** best dining in town; Hospitality Inn Port Hedland, Webster St; (08) 9173 1044. ***Wedge Street Coffee Shop:*** light meals; 12A Wedge St; (08) 9173 2128.

All Seasons Port Hedland: hotel with ocean views; Cnr Lukis and McGregor sts; (08) 9173 1511. ***Cooke Point Holiday Park:*** family-friendly holiday park; 2 Taylor St; (08) 9173 1271 or 1800 459 999. ***Hospitality Inn Port Hedland:*** comfortable motel accommodation; Webster St; (08) 9173 1044. ***The Lodge Motel:*** hotel rooms and self-contained apartments; 5–13 Hawke Pl, South Hedlands; (08) 9172 2188.

Stairway to the Moon Like the Broome version, this beautiful illusion is created when a full moon rises over the ocean at low tide. The moon's rays hit pools of water left by the receding tide, creating the image of a stairway leading up to the moon. It lasts for about 15 minutes. Check with visitor centre for dates and times. Coastal side of Goode St; 7 km E.

Pretty Pool: picnic, fish and swim at this scenic tidal pool; 8 km NE. ***Dampier Salt:*** see giant cone-shaped mounds of salt awaiting export; 8 km S. ***Royal Flying Doctor Service:*** operates an important base in Port Hedland at the airport; open to public 8–11am Mon–Fri; closed public holidays, school holidays and weekends; 15 km S. ***School of the Air:*** experience schooling outback-style at the airport; open to public 8–11am Mon–Fri; closed public and school holidays; 15 km S. ***Turtle-watching:*** flatback turtles nest in the area Oct–Mar at Pretty Pool, Cooke Pt and Cemetery Beach. ***Cruises:*** scenic harbour and sunset cruises; details from visitor centre.

Travellers note: *Poisonous stonefish frequent this stretch of coast, especially Nov–Mar, so wear strong shoes when walking on rocky reef areas and make local inquiries before swimming in the sea.*

See also PILBARA, p. 313

Ravensthorpe

Pop. 438
Map ref. 620 G9

Morgans St; (08) 9838 1277; www.ravensthorpe.wa.gov.au

101.9 WA FM, 105.9 FM ABC Local Radio

Ravensthorpe is encircled by the Ravensthorpe Range. This unspoiled bushland is home to many plants unique to the area such as the Qualup bell, warted yate and Ravensthorpe bottlebrush. Gold was discovered here in 1898 and by 1909 the population had increased to around 3000. Coppermining reached a peak in the late 1960s; the last coppermine closed in 1972. Many old mine shafts can be seen around the district and fossicking is a favourite pastime.

Historical Society Museum: local history memorabilia; in Dance Cottage near visitor centre, Morgans St. ***Historic buildings:*** many in town including the impressive Palace Hotel (1907) and the restored Commercial Hotel (now a community centre); both in Morgans St. ***Rangeview Park:*** local plant species, picnic and barbecue facilities; Morgans St.

Wildflower Show: Sept.

Ravensthorpe Palace Motor Hotel: traditional pub food; 28 Morgan St; (08) 9838 1005.

Ravensthorpe Palace Motor Hotel: historic building; 28 Morgan St; (08) 9838 1005.

WA Time Meridian: plaque on a boulder marks the WA time meridian; at First Rest Bay west of town. ***Eremia Camel Treks:*** offers rides along the beach; open by appt Dec–Apr; Hopetoun Rd; (08) 9838 1092; 2 km SE. ***Old Copper Smelter:*** in operation 1906–18, now site of tailings dumps and old equipment; 2 km SE. ***Archer Drive Lookout:*** extensive views over farms and hills; 3 km N in Ravensthorpe Range. ***Mt Desmond Lookout:*** magnificent views in all directions; Ethel Daw Dr; 17 km SE in Ravensthorpe Range. ***Hopetown:*** seaside village with pristine beaches ideal for swimming, surfing, windsurfing, fishing and boating. Summer Festival each June. Walk on the Hopetoun Trail Head Loop (part of the Hopetoun–Ravensthorpe Heritage Walk) or visit Fitzgerald River National Park to the west; *see Bremer Bay*; 49 km S. ***Scenic drives:*** include the 170 km circular Hamersley Drive Heritage Trail; maps from visitor centre. ***Rock-collecting:*** check locally to avoid trespass.

See also ESPERANCE & NULLARBOR, p. 310

Rockingham

Pop. 67 521
Map ref. 614 C11 | 615 B7 | 618 B6 | 620 B8

43 Kent St; (08) 9592 3464; www.rockinghamvisitorcentre.com.au

 97.7 FM ABC Classic Radio, 720 AM ABC Local Radio

Lying on the edge of Cockburn Sound just 47 kilometres south of Perth, the coastal city of Rockingham offers sheltered waters ideal for swimming, snorkelling, sailing, windsurfing, fishing and crabbing. Established in 1872 to ship timber from Jarrahdale to England, Rockingham was the busiest port in Western Australia until the end of the 19th century, after which all port activities were shifted north to Fremantle. It was only because of the industrial area nearby at Kwinana in the 1950s and the development of the HMAS *Stirling* Naval Base on Garden Island in the 1970s that the town was revitalised. Today, its magnificent beaches and proximity to Perth are Rockingham's main attractions.

Rockingham Museum: folk museum featuring local history exhibits including displays on the Group Settlement farms, the timber industry, domestic items and antique photographic equipment; open 1–4pm Tues, Wed, Thurs and Sat, 10am–4pm Sun; Cnr Flinders La and Kent St; (08) 9592 3455. ***Art Gallery and Craft Centre:*** features local artists, with market every 3rd Sun; Civic Blvd. ***The Granary:*** museum of artefacts celebrating the history of WA's grain industry; open for tours (minimum group size 4) by appt; northern end of Rockingham Rd; (08) 9599 6333. ***Mersey Pt Jetty:*** departure point for cruises and island tours; Shoalwater. ***Kwinana Beach:*** hull of wrecked SS *Kwinana*. ***Cape Peron:*** the lookout was once the main observation post for a WW II coastal battery; Pt Peron Rd. ***Bell and Churchill Park:*** family picnics and barbecues in shaded grounds; Rockingham Rd.

Swap Mart: opposite Churchill Park; each Sun. ***Sunset Jazz Festival:*** Mar. ***Musslefest:*** Oct. ***Spring Festival:*** Nov.

Bettyblue Bistro: generous portions, steak and seafood; Shop 3, 1 Railway Tce; (08) 9528 4228. ***Emma's on the Boardwalk:*** innovative with bay views; Shop 7–8, 1–3 Railway Tce; (08) 9592 8881. ***Sunsets Cafe Bistro:*** bay-side bistro; The Boardwalk, Palm Beach; (08) 9528 1910. ***Y2K Cafe & Restaurant:*** relaxed dining; 57B Rockingham Beach Rd; (08) 9529 1044.

5 Star @ Aria: beachfront penthouse; Rockingham Beach Rd; 0407 419 194. ***Luxury @ Nautilus:*** beachside apartment; 18–24 Kent St; Rockingham Visitor Centre (08) 9592 3464. ***Manuel Towers:*** spectacular Mediterranean-style B&B; 32A Arcadia Dr, Shoalwater; (08) 9592 2698. ***Payne's Find:*** tri-level waterfront house; 171 Rockingham Beach Rd; Rockingham Visitor Centre (08) 9592 3464.

Penguin Island Take a trip to this offshore island, which is home to a colony of little penguins. The Discovery Centre allows you to see the penguins up close in an environment similar to their natural habitat and to learn about them through daily feedings, commentaries and displays. The island also provides picnic areas, lookouts and a network of boardwalks, and you can swim, snorkel or scuba dive at any of the pristine beaches. The island is open to the public in daylight hours Sept–June. Ferries to the island leave regularly from Mersey Pt, south of Rockingham. The ferry also provides bay cruises and snorkelling tours.

Sloan's Cottage: restored pioneer cottage; open Mon–Fri; Leda; 2 km W. ***Lake Richmond:*** walks, flora and fauna, and thrombolites (domed rock-like structures like the famous stromatolites of Hamelin Pool near Denham, built by ancient micro-organisms); 4 km SW. ***Marapana Wildlife Park:*** handfeed and touch native and exotic animals; Karnup; 15 km SE. ***Wineries:*** in the area include Baldivis Estate (15 km SE) and Peel Estate (17 km SE); details from visitor centre. ***Secret Harbour:*** surfing, snorkelling and windsurfing; 20 km S. ***Serpentine Dam:*** major water storage with brilliant wildflowers in spring, bushland and the nearby Serpentine Falls; 48 km SE. ***Garden Island:*** home to HMAS *Stirling* Naval Base, two-thirds of the island is open to the public but is accessible only by private boat during daylight hours. ***Shoalwater Bay Islands Marine Park:*** extends from just south of Garden Island to Becher Pt. Cruises of the park are available; details from visitor centre. ***Dolphin Watch Cruises:*** swim with dolphins between Pt Peron and Garden Island; daily Sept–May; details from visitor centre. ***Scenic drives:*** including Old Rockingham Heritage Trail, a 30 km drive that takes in 23 points of interest in the Rockingham–Kwinana area, and Rockingham–Jarrahdale Timber Heritage Trail, a 36 km drive retracing the route of the 1872 timber railway; maps from visitor centre.

See also THE SOUTH-WEST, p. 307

Roebourne

Pop. 853

Map ref. 619 F1 | 622 B2 | 624 B12 | 626 E5

Old Gaol, Queens St; (08) 9182 1060; www.roebourne.wa.gov.au

107.5 FM ABC Radio National, 702 AM ABC Local Radio

Named after John Septimus Roe, Western Australia's first surveyor-general, Roebourne was established in 1866 and is the oldest town on the north-west coast. As the centre for early mining and pastoral industries in the Pilbara, it was connected by tramway to the pearling port of Cossack and later to Point Samson. Now Cossack is a ghost town and Point Samson is known for its beachside pleasures.

Historic buildings: some original stone buildings remain, many of which have been classified by the National Trust. The Old Gaol, designed by the well-known colonial architect George Temple Poole, now operates as the visitor centre and museum; Queen St. ***Mt Welcome:*** offers views of the coastal plains and rugged hills surrounding town. Spot the railroad from Cape Lambert to Pannawonica, and the pipeline carrying water from Millstream to Wickham and Cape Lambert; Fisher Dr.

Roebourne Cup: July.

Moby's Kitchen: beachside dining; Bartley Crt, Point Samson; (08) 9187 1435. ***TaTa's Restaurant:*** international cuisine; Point Samson Resort, 56 Samson Rd, Point Samson; (08) 9187 1052. ***Trawlers Tavern:*** ocean views; Roebourne Port Samson Rd; (08) 9187 1503. ***Red Rock Cafe:*** steak, pizza, seafood; Mulga Way, Wickham; (08) 9187 1303.

Amani Cottage: self-contained cottage; 1 McLeod St, Point Samson; (08) 9187 1085. ***Point Samson Resort:*** seaside resort; 56 Point Samson Rd, Point Samson; (08) 9187 1052. ***Samson Beach Chalets:*** architect-designed chalets; 44 Bartley Crt, Point Samson; (08) 9187 0202. ***Samson Hideaway:*** self-contained apartment; 49 Meares Dr, Point Samson; (08) 9187 0330.

Cossack Originally named Tien Tsin after the boat that brought the first settlers there in 1863, Cossack was the first port in the north-west region. During its days as a pearling centre in the late 1800s the population increased dramatically. Although now a ghost town, the beautiful stone buildings have been restored and nine are classified by the National Trust. Cruises are available from the wharf. 14 km N.

Wickham: the company town for Robe River Iron Ore offers a spectacular view from Tank Hill lookout; tours available; 12 km N. ***Pt Samson:*** good fishing, swimming, snorkelling and diving. Boat hire, whale-watching and fishing charters are available; 19 km N. ***Cleaverville:*** camping and fishing; 25 km N. ***Harding Dam:*** ideal picnic spot; 27 km S. ***Millstream–Chichester National Park:*** 150 km S; *see Karratha*. ***Emma Withnell Heritage Trail:*** 52 km historic self-drive trail, named after the first European woman in the north-west, takes in Roebourne, Cossack, Wickham and Pt Samson; maps from visitor centre. ***Tours:*** include historic

Pearls and Past tour and trips to Jarman Island; details from visitor centre.

See also PILBARA, p. 313

Southern Cross

Pop. 709
Map ref. 620 F6

Council offices, Antares St; (08) 9049 1001; www.yilgarn.wa.gov.au

 100.7 WA FM, 106.3 FM ABC Local Radio

A small, flourishing town on the Great Eastern Highway, Southern Cross is the centre of a prosperous agricultural and pastoral region. Its claim to fame is as the site of the first major gold discovery in the huge eastern goldfields. Although it never matched the fever pitch of Kalgoorlie and Coolgardie, Southern Cross remains the centre for a significant gold-producing area. The town's wide streets, like the town itself, were named after stars and constellations.

Yilgarn History Museum: originally the town courthouse and mining registrar's office, it now houses displays on mining, agriculture, water supply and military involvement; Antares St. ***Historic buildings:*** including the post office in Antares St, the Railway Tavern in Spica St and the restored Palace Hotel in Orion St.

King of the Cross: 2 days of motorcycle races; Aug.

Club Hotel: serves lunch and dinner; 21 Antares St; (08) 9049 1202. ***Palace Hotel:*** traditional pub; Great Eastern Hwy; (08) 9049 1555. ***Southern Cross Motel:*** international cuisine; Canopus St; (08) 9049 1144.

Southern Cross Caravan Park & Motor Lodge: sites, cabins, motel rooms; Great Eastern Hwy; (08) 9049 1212.

Hunt's Soak: once an important water source, now a picnic area; 7 km N. ***Frog Rock:*** large rock with wave-like formations. Popular picnic spot; 34 km S. ***Baladjie Rock:*** granite outcrop with spectacular views; 50 km NW. ***Karalee Rock and Dam:*** this dam was built to provide water for steam trains and is now popular for swimming and picnics; 52 km E.

See also HEARTLANDS, p. 308

Tom Price

Pop. 2721
Map ref. 619 G3 | 622 C5 | 626 G10

Central Rd; (08) 9188 1112; www.tompricewa.com.au

103.3 WA FM, 567 AM ABC Local Radio

The huge iron-ore deposit now known as Mount Tom Price was discovered in 1962 in the heart of the Pilbara, after which the Hamersley Iron Project was established. A mine, two towns (Dampier and Tom Price) and a railway line between them all followed. Today, the town is an oasis in a dry countryside. On the edge of the Hamersley Range at an altitude of 747 metres, this is the state's highest town, hence its nickname of 'Top Town in WA'.

Tom Price Hotel Motel: international cuisine; Central Rd; (08) 9189 1101.

Karijini Eco Retreat: luxury eco-tents and traditional campsites; off Weano Rd, Karijini National Park; (08) 9425 5591.

Karijini National Park The second largest national park in Western Australia, this park features ochre-coloured rock faces with bright-white snappy gums, bundles of spinifex dotting the red earth and chasms up to 100 m deep. The waterfalls and rockpools of Karijini offer some of the best swimming in the state. The park protects the many different wildlife habitats, plants and animals of the Pilbara. The landscape is dotted with huge termite mounds and the rock piles of the rare pebble mouse; other species include red kangaroos and rock wallabies, and reptiles from legless lizards to pythons. Kalamina Gorge and Pool is the most accessible gorge, while at Hamersley Gorge a wave of tectonic rock acts as a backdrop to a swimming hole and natural spa. Oxer Lookout reveals where the Joffre, Hancock, Weano and Red gorges meet. Mt Bruce, the second-tallest peak in the state, offers spectacular views and interpretive signs along the trail to the top. The Karijini Visitor Centre is located off the road to Dales Gorge and has information on camping. 50 km E.

Kings Lake: constructed lake with nearby park offering picnic and barbecue facilities (no swimming); 2 km W. ***Mt Nameless Lookout:*** stunning views of district around Tom Price; 6 km W via walking trail or 4WD track. ***Aboriginal carvings:*** thought to be 35 000 years old; 10 km S. ***Mine tours:*** marvel at the sheer enormity of Hamersley Iron's open-cut iron-ore mine; bookings essential at visitor centre. ***Hamersley Iron Access Road:*** this private road is the most direct route between Tom Price and Karratha via Karijini and Millstream national parks. It requires a permit to travel along it; available from visitor centre.

Travellers note: *To the north-east is Wittenoom, an old asbestos-mining town. Although the mine was closed in 1966, there is still a health risk from microscopic asbestos fibres present in the abandoned mine tailings in and around Wittenoom. If disturbed and inhaled, blue asbestos dust may cause cancer. The Ashburton Shire Council advocates avoidance of the Wittenoom area.*

See also PILBARA, p. 313

Toodyay

Pop. 1068
Map ref. 615 F2 | 618 D4 | 620 C7

7 Piesse St; (08) 9574 2435; www.toodyay.com

558 AM ABC Local Radio, 612 AM ABC Radio National

This National Trust–classified town is nestled in the Avon Valley surrounded by picturesque farming country and bushland. The name originates from 'Duidgee', which means the 'place of plenty'. Founded in 1836, Toodyay was one of the first inland towns to be established in the colony. It was a favourite haunt of Western Australia's most famous bushranger, Joseph Bolitho Johns, who was more commonly known as 'Moondyne Joe'.

Historic buildings Some original buildings from the early settlement of Toodyay still stand, including Stirling House (1908), Connor's Mill (1870s) with displays of working flour-milling equipment, the Old Newcastle Gaol (1865) built by convict labour and where Moondyne Joe was imprisoned, and the Police Stables (1870) built by convict labour from random rubblestone. Self-guide walk available; pamphlet from visitor centre.

Duidgee Park: popular picnic spot on the banks of the river has a miniature railway and a walking track; check at visitor centre for running times; Harper Rd. ***Newcastle Park:*** contains a unique stone monument of Charlotte Davies, the first white female to set foot on the soil of the Swan River Colony. Also has a children's playground; Stirling Tce. ***Pelham Reserve and Lookout:*** nature walks, a lookout with views over the town and a memorial to James Drummond, the town's first resident botanist; Duke St.

Moondyne Festival: May. ***Avon Descent:*** whitewater rafting; Aug. ***Kombi Konnection:*** Sept. ***Music Festival:*** Sept.

Cola Cafe and Museum: with a huge selection of Coca-Cola memorabilia on display and music from the 1950s playing, this is a cafe with a difference; 128 Stirling Tce; (08) 9574 4407. ***Victoria Hotel:*** international cuisine; 116 Stirling Tce; (08) 9574 2206.

Avondale Estate: cosy country retreat; 9 Railway Rd; (08) 9574 4033. ***Ipswich View Homestead:*** heritage B&B; 45 Folewood Rd; (08) 9574 4038. ***Mountain Park Retreat:*** large Queenslander-style home; Cnr Dumbarton Rd and Nairn Dr; (08) 9255 2653. ***Toodyay Farmstay:*** 2-bedroom cottage; 51 Leeder St, West Toodyay; (08) 9385 8824.

Avon Valley National Park The park offers spectacular scenery with abundant wildflowers in season. Being at the northern limit of the jarrah forests, the jarrah and marri trees mingle with wandoo woodland. This mix of trees creates diverse habitats for fauna, including a wide variety of birdlife. The Avon River, which in summer and autumn is a series of pools, swells to become impressive rapids during winter and spring. These rapids provide the backdrop for the Avon Descent, a well-known annual whitewater race held every Aug, which begins in Northam and passes through the park. The park is ideal for camping, bushwalking, canoeing and picnicking, although all roads are unsealed. Whitewater rafting tours are available. 25 km SW.

Pecan Hill Tearoom Museum: tearoom and museum set in a pecan nut orchard with a lookout offering spectacular views over the Avon Valley; Julimar Rd; 4 km NW. ***Coorinja Winery:*** dating from the 1870s; open for tastings and sales; closed Sun; 4 km SW. ***Ringa Railway Bridge:*** constructed in 1888, this timber bridge has 18 spans, but is not readily accessible; details from visitor centre; 6 km SW. ***Windmill Hill Cutting:*** the deepest railway cutting in Australia; 6 km SE. ***Avonlea Park Alpaca Farm:*** alpacas and other farm animals, and sales of alpaca wool products; 12 km SE. ***Emu Farm:*** one of the oldest in Australia. Crafts and emu products for sale; 15 km SW. ***Cartref Park Country Garden:*** 2 ha park of English-style landscaped gardens and native plants, with prolific birdlife; 16 km NW. ***Oliomio Farm:*** olive and lavendar farm offering tastings and sales; Parkland Dr; 20 km NW. ***Wyening Mission:*** this Benedictine farm and winery runs tours of their mission, cellars and grounds; by appt; (08) 9364 6463; 38 km N. ***Toodyay Pioneer Heritage Trail:*** honouring the pioneering spirit in the Avon Valley, this 20 km self-drive trail retraces the route of the first settlers; maps from visitor centre. ***Avon Valley Tourist Drive:*** 95 km scenic drive includes Toodyay, Northam, York and Beverley; maps from visitor centre. ***Hotham Valley Steam Railway:*** the famous steam-train service runs special trips including a monthly 'murder-mystery' night; details from visitor centre.

See also HEARTLANDS, p. 308

Wagin

Pop. 1424
Map ref. 618 F8 | 620 D9

Historical Village, Kitchener St; (08) 9861 1232; www.wagintouristinfo.com.au

96.3 FM ABC News Radio, 558 AM ABC Local Radio, 142.2 AM, 161.1 AM Radio Great Southern

Wagin, 177 kilometres east of Bunbury, is the sheep capital of Western Australia. The importance of the wool industry to the district is celebrated in its annual Wagin Woolorama, one of the largest rural shows in the state, and its Giant Ram, an enormous structure that visitors from around the country come to photograph.

Wagin Historical Village Explore 24 relocated or re-created historic buildings and machinery providing a glimpse of pioneering rural life. The buildings are furnished with original pieces, and audio commentaries are available. Open 10am–4pm; Kitchener St.

Giant Ram and Ram Park: 9 m high statue provides a photo opportunity not to miss; Arthur River Rd. ***Wagin Heritage Trail:*** self-guide walk around the town; maps from visitor centre.

Wagin Woolorama: Mar. ***Foundation Day:*** June.

The Palace Hotel: traditional pub; 51 Tudhoe St; (08) 9861 1003.

Morans' Wagin Hotel: family-run hotel; 77 Tudor St; (08) 9861 1017.

Puntapin Rock: spectacular views over the town and surrounding farmlands from the top of the rock. Enjoy the picnic and barbecue facilities nearby; Bullock Hill Rd; 4 km SE. ***Mt Latham:*** interesting rock formation with walk trails, a lookout and abundant wildflowers in season; Arthur River Rd; 8 km W. ***Lake Norring:*** swimming, sailing and waterskiing; picnic and barbecue facilities; water levels vary considerably; check conditions at visitor centre; 17 km SW. ***Lake Dumbleyung:*** where Donald Campbell established a world water-speed record in 1964. Swimming, boating and birdwatching are subject to water levels that vary considerably; check conditions at visitor centre; 18 km E. ***Wait-jen Trail:*** self-guide 10.5 km signposted walk that follows ancient Aboriginal Dreaming. The word 'wait-jen' means 'emu footprint' in the language of the local Aboriginal people; maps from visitor centre. ***Wheat Belt Wildflower Drive:*** self-guide drive that includes the Tarin Rock Nature Reserve; maps from visitor centre.

See also HEARTLANDS, p. 308

Walpole

Pop. 322
Map ref. 617 F6 | 618 E12 | 620 D11

Pioneer Cottage, South Coast Hwy; (08) 9840 1111; www.walpole.com.au

102.9 WA FM, 106.1 FM ABC Local Radio

Walpole is entirely surrounded by national park and is the only place in the South-West where the forest meets the sea. The area is renowned for its striking ocean and forest scenery, which provides an idyllic setting for outdoor activities. The town of Walpole was established in 1930 through the Nornalup Land Settlement Scheme for city families hit by the Great Depression.

Pioneer Cottage: re-creation of a historic cottage to commemorate the district's pioneer settlers; South Coast Hwy.

Easter Markets: Apr.

Tree Top Restaurant: local produce; Tree Top Walk Motel, Nockolds St; (08) 9840 1444. ***Slow Food Cafe:*** local produce; Old Kent River Wines, Kent River, South Coast Hwy; (08) 9855 1589. ***The Nornalup Teahouse Restaurant:*** cottage cafe; 6684 South Coast Hwy, Nornalup; (08) 9840 1422. ***Thurlby Herb Farm Cafe:*** fresh produce; Lot 3 Gardiner Rd; (08) 9840 1249.

Ayr Sailean: self-contained farmstay accommodation; 1 Tindale Rd, Bow Bridge; (08) 9840 8098. ***Che Sara Sara Chalets:*** tranquil riverside chalets; 92 Nunn Rd, Hazelvale; (08) 9840 8004. ***Houseboat Holidays:*** well-appointed houseboats; Lot 660 Boronia St; (08) 9840 1310. ***Tree Elle Retreat:*** superbly appointed self-contained retreat; Lot 4, South Coast Hwy; (08) 9840 8471. ***Walpole Wilderness Resort:*** self-contained cottages; Gardiner Rd; (08) 9840 1481.

Walpole–Nornalup National Park The many forest attractions include the Valley of the Giants, the Hilltop Drive and Lookout, Circular Pool and the Knoll. The park is probably best known for its huge, buttressed red tingle trees, some more than 400 years old, which are unique to the Walpole area. The world-class Bibbulmun Track, between Perth and Albany, passes through the park. *See Albany*; 4 km SE.

Valley of the Giants Here visitors can wander over a walkway suspended 38 m above the forest floor, the highest and longest tree-top walkway of its kind in the world. The Ancient Empire interpretive boardwalk weaves its way through the veteran tingle trees. Twilight walks are available in holiday season. Contact (08) 9840 8263; 16 km E.

Giant Red Tingle: a 25 m circumference defines this tree as one of the 10 largest living things on the planet; 2 km E. ***Hilltop Lookout:*** views over the Frankland River out to the Southern Ocean; 2 km W. ***John Rate Lookout:*** panoramic views over the mouth of the Deep River and of the nearby coastline and forests; 4 km W. ***Thurlby Herb Farm:*** herb garden display with sales of herbal products; Gardiner Rd; 13 km N. ***Mandalay Beach:*** site of the 1911 shipwreck of the Norwegian *Mandalay*. A boardwalk has descriptive notes about the wreck. Also popular for fishing; 20 km W. ***Dinosaur World:*** a collection of native birds and reptiles, and exotic birds; 25 km E. ***Mt Frankland National Park:*** noted for its exceptional variety of birdlife, it also offers breathtaking views from the top of Mt Frankland, known as 'Caldyanup' to the local Aboriginal people; 29 km N. ***Peaceful Bay:*** small fishing village with an excellent beach for swimming; 35 km E. ***Fernhook Falls:*** ideal picnic spot with boardwalk, at its best in winter when it is popular for canoeing and kayaking; 36 km NW. ***Walk trails:*** many trails in the area, including self-guide Horseyard Hill Walk Trail through the karri forest and the signposted Coalmine Beach Heritage Trail from the coastal heathland to the inlets; maps from visitor centre. ***Tours:*** take a guided cruise through the inlets and rivers, hire a boat or canoe, or go on a forest tour or wilderness eco-cruise; details from visitor centre.

See also THE SOUTH-WEST, p. 307

Wickepin

Pop. 245
Map ref. 618 F6 | 620 D8

District Resource and Telecentre, 24 Wogolin Rd; (08) 9888 1500; www.wickepin.wa.au

100.5 Hot FM, 612 AM ABC Radio National

The first settlers arrived in the Wickepin area in the 1890s. Albert Facey's internationally acclaimed autobiography, *A Fortunate Life*, details these pioneering times. However, Facey is not the only major literary figure to feature in Wickepin's history. The poet and playwright, Dorothy Hewett, was born in Wickepin in 1923, and much of her work deals with life in the area.

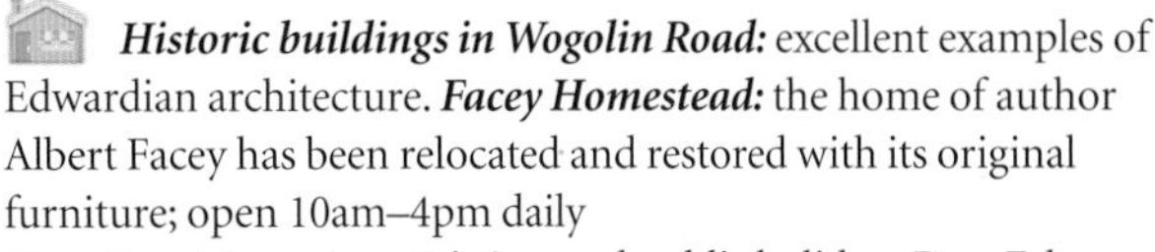

Historic buildings in Wogolin Road: excellent examples of Edwardian architecture. ***Facey Homestead:*** the home of author Albert Facey has been relocated and restored with its original furniture; open 10am–4pm daily Mar–Nov, 10am–4pm Fri–Sun and public holidays Dec–Feb.

Wickepin Hotel: traditional pub; 34 Wogolin Rd; (08) 9888 1192.

Wickepin Hotel: standard hotel rooms; 34 Wogolin Rd; (08) 9888 1192.

Tarling Well: the circular stone well marks the original intended site for the town; 8 km W. ***Malyalling Rock:*** unusual rock formation; 15 km NE. ***Toolibin Lake Reserve:*** see a wide variety of waterfowl while you enjoy a barbecue; 20 km S. ***Yealering and Yealering Lake:*** historic photographs on display in town, and swimming, boating, windsurfing and birdwatching at lake; 30 km NE. ***Harrismith Walk Path:*** self-guide trail through wildflowers in season, including orchids and some species unique to the area; brochures from visitor centre. ***Albert Facey Heritage Trail:*** 86 km self-drive trail brings to life the story of Albert Facey and the harshness of life in the early pioneering days of the wheat belt; maps from visitor centre.

See also HEARTLANDS, p. 308

Wyndham

Pop. 672
Map ref. 625 N4 | 629 C5

Kimberley Motors, Great Northern Hwy; (08) 9161 1281; www.wyndham.wa.au

102.9 WA FM, 1017 AM ABC Local Radio

Wyndham, in the Kimberley region, is the most northerly town and safe port in Western Australia. The entrance to the town is guarded by the 'Big Croc', a 20-metre-long concrete crocodile.

Historical Society Museum: in the old courthouse building, its displays include a photographic record of the town's history, artefacts and machinery; open Apr–Oct; O'Donnell St. ***Warriu Dreamtime Park:*** bronze statues representing the Aboriginal heritage of the area; Koolama St. ***Zoological Gardens and Crocodile Park:*** daily feeding of crocodiles and alligators; Barytes Rd; (08) 9161 1124. ***Durack's Old Store:*** has an informative plaque with details of its history; O'Donnell St. ***Pioneer Cemetery:*** gravestones of some of the area's original settlers; Great Northern Hwy. ***Boat charters:*** scenic, fishing and camping cruises; details from visitor centre.

Races: Aug. ***Art and Craft Show:*** Aug/Sept.

Wyndham Town Hotel: traditional pub; O'Donnell St; (08) 9161 1003.

Kimberley Coastal Camp: wilderness retreat; Admiralty Gulf, access by air; 0417 902 006.

Three Mile Valley On offer is spectacular scenery typical of the Kimberley region, with rough red gorges and pools of clear, cold water during the wet season. Walk trails lead the visitor through the brilliant displays of wildflowers in season. Three Mile Valley is the home of the 'Trial Tree', a sacred Aboriginal site into which, when a person died of unnatural causes, the body was placed. Rocks were placed around the base of the tree, each rock representing a relative of the deceased. When the body started to decompose, the first rock to be marked by the decomposition indicated the name of the person responsible for the death. This person was then banished from the tribe. 3 km N.

Afghan Cemetery: containing the graves of Afghan camel drivers who carried supplies throughout the Kimberley region. All the gravestones face towards Mecca; 1 km E. ***Koolama Wreck Site:*** the *Koolama* was hit by Japanese bombs near Darwin in 1942. After limping along the coast to Wyndham, it sank just 40 m from the jetty. The spot is marked by unusual swirling in the water; 5 km NW. ***Five Rivers Lookout:*** spectacular views of the Kimberley landscape from the highest point of the Bastion Range, particularly good for viewing the striking sunsets. Also a good picnic area with barbecue facilities; 5 km N. ***Moochalabra Dam:*** completed in 1971, the dam was constructed to provide an assured water supply to the Wyndham area. The construction is unique in Australia, designed to allow overflow to pass through the rock on the crest of the hill. 4WD access only; King River Rd; 18 km SW. ***Aboriginal rock paintings:*** 4WD access only; well signposted off the King River Rd; 18 km SW. ***Parry Lagoons Nature Reserve:*** visitors can enjoy birdwatching from a shaded bird hide at Marlgu Billabong or scan the wide vistas of the flood plain and distant hills afforded from the lookout at Telegraph Hill; 20 km SE. ***Prison Tree:*** 2000–4000-year-old boab tree once used by local police as a lock-up; King River Rd; 22 km SW. ***The Grotto:*** this rock-edged waterhole, estimated to be 100 m deep, is a cool, shaded oasis offering year-round swimming; 36 km E.

See also KIMBERLEY, p. 314

Yalgoo

Pop. 164
Map ref. 619 F11 | 620 C2

Old Railway Station, Geraldton–Mt Magnet Rd; (08) 9962 8157; www.yalgoo.wa.gov.au

104.5 WA FM, 106.1 FM ABC Local Radio

Alluvial gold was discovered in the 1890s in Yalgoo, which lies 216 kilometres east of Geraldton. Today, gold is still found in the district and visitors are encouraged to try their luck fossicking in the area. The name Yalgoo is from the Aboriginal word meaning 'blood', a rather odd fact given that in 1993 a Yalgoo resident was the first person in Australia to be the victim of a parcel bomb.

 ***Courthouse Museum:*** exhibits of local artefacts; Gibbons St. ***Gaol:*** built in 1896 and recently relocated to the museum precinct, it has photographs illustrating the town's history; Gibbons St. ***Chapel of St Hyacinth:*** designed by Monsignor Hawes and built in 1919 for the Dominican Sisters who lived in a wooden convent school near the chapel; Henty St. ***Heritage walk:*** self-guide town walk; pamphlet from visitor centre.

Tardie and Yuin Stations: outback homestead; (08) 9963 7980.

Cemetery: the history of Yalgoo as told through headstones; 5 km W. ***Joker's Tunnel:*** a tunnel carved through solid rock by early prospectors and named after the Joker's mining syndicate, it has panoramic views near the entrance; 10 km SE. ***Meteorite crater:*** discovered in 1961, a portion of the meteorite is held at the WA Museum in Perth; 100 km N. ***Gascoyne Murchison Outback Pathways:*** three self-drive trips exploring outback history: the Wool Wagon Pathway, the Kingsford Smith Mail Run and the Miners Pathway; maps from visitor centre or downloadable from website.

See also OUTBACK COAST & MID-WEST, p. 312

Yallingup

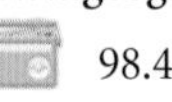

Pop. 300
Map ref. 616 A6

Seymour Blvd, Dunsborough; (08) 9752 1288; www.geographebay.com

98.4 FM Western Tourist Radio, 684 AM ABC Local Radio

Yallingup, known for its limestone caves and world-class surf breaks, is also an ideal location for swimming, fishing and beachcombing. Nearby Leeuwin–Naturaliste National Park offers spectacular scenery, interesting bushwalks and beautiful wildflowers in season. Art and craft galleries abound and many of the wineries of the South-West region are only a short drive from the town. Yallingup continues to live up to its Aboriginal meaning of 'place of love', with nearby Caves House Hotel being a favourite destination for generations of honeymooners and holiday-makers.

Wardan Aboriginal Centre: houses an interpretive display and a gallery that sells local Indigenous arts. Book ahead for workshops, including boomerang throwing classes; Injidup Springs Rd; (08) 9756 6566. ***Caves House Hotel:*** originally built in 1903 as a holiday hotel, the building was damaged by fire in 1938 and rebuilt; off Caves Rd; (08) 9755 2131. ***Yallingup Beach:*** surfing, scuba diving, whale-watching and salmon fishing in season.

Amberley Estate Seafood and Semillon Weekend: Feb. ***Yallingup Malibu Surfing Classic:*** Dec.

Amberley Restaurant: picturesque vineyard setting; Amberley Estate, cnr Thornton and Wildwood rds; (08) 9750 1112. ***Cape Lodge Restaurant:*** superb fine dining; 3341 Caves Rd; (08) 9755 6311. ***Flutes Restaurant:*** Asian-influenced menu in charming surrounds; Brookland Valley Winery, Caves Rd; Wilyabrup; (08) 9755 6250. ***Lamont's Margaret River:*** dishes made with love; Gunyulgup Valley Dr; (08) 9755 2434. ***Caves House Hotel:*** local produce; Yallingup Beach Rd; (08) 9750 1500.

Cape Lodge: sublime world-class hotel; 3341 Caves Rd; (08) 9755 6311. ***Fourwells:*** 4-bedroom house; address confirmed with booking; (08) 9385 5611. ***Injidup Spa Retreat:*** luxury 2-bedroom suites; Cape Clairault Rd, Injidup; (08) 9750 1300. ***Seashells Resort Yallingup:*** Art Deco–style beachside resort; Yallingup Beach Rd; (08) 9750 1500. ***Smiths Beach Resort:*** stylish beachfront resort; Smiths Beach Rd; (08) 9750 1200.

Leeuwin–Naturaliste National Park: stretches north and south of town; *see Dunsborough, Margaret River and Augusta.* ***Ngilgi Cave:*** (pronounced 'Nillgee') semi-guided tours are available for this stunning display of stalactite, stalagmite and shawl rock formations. An interpretive area details the history of the cave; 2 km E. ***Canal Rocks and Smith's Beach:*** interesting rock formation plus fishing, surfing, swimming, snorkelling and diving; 5 km SW. ***Gunyulgup Gallery:*** over 120 artists and craftspeople are represented, with paintings, prints, ceramics and sculpture on display and for sale; (08) 9755 2177; 9 km SW. ***Yallingup Gallery:*** specialises in custom-built furniture; 9 km SW. ***Quinninup Falls:*** falls that are particularly attractive in winter and can be reached only by 4WD or on foot; 10 km S. ***Shearing Shed:*** sales of wool products, and shearing demonstrations at 11am; closed Fri; Wildwood Rd; (08) 9755 2309; 10 km E. ***Wineries and breweries:*** many wineries and boutique breweries in the area offer tastings and sales; details from visitor centre.

See also THE SOUTH-WEST, p. 307

Yanchep

Pop. 2481
Map ref. 615 A2 | 618 B4 | 620 B7

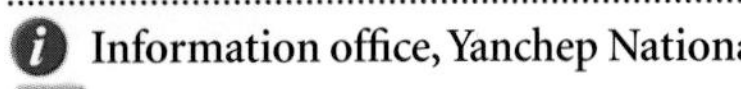

Information office, Yanchep National Park; (08) 9561 1004

94.5 Mix FM, 720 AM ABC Local Radio

Only 58 kilometres north of Perth, Yanchep is a rapidly developing recreational area and popular tourist destination. It provides safe, sandy beaches and good fishing areas, as well as natural attractions such as the series of caves found within Yanchep National Park. The town derives its name from the Aboriginal word 'yanjet', which means bullrushes, a feature of the area.

Yanchep Lagoon: good swimming and fishing beach; off Lagoon Dr.

Blue Dolphin Restaurant: relaxed setting; Two Rocks Shopping Centre, Two Rocks; (08) 9561 1469. ***Chocolate Drops:*** handmade chocolates, light meals; Yanchep National Park, Wanneroo Rd; (08) 9561 6699. ***Lindsay's Restaurant:*** buffet and à la carte; Club Capricorn, Two Rocks Rd; (08) 9561 1106. ***The Tudor Manor Restaurant:*** popular seafood platters; Yanchep Inn, Yanchep National Park; (08) 9561 1001.

Casa Del Mar: architecturally designed villa; 10B Arney Crt; (08) 9298 9367. ***Oceanview Retreat:*** modern holiday home; address confirmed with bookings; (08) 9276 4257. ***Yanchep Holiday Village:*** self-contained apartments in natural setting; St Andrews Dr; (08) 9561 2244. ***Yanchep Inn:*** hotel- and motel-style accommodation; Yanchep National Park; (08) 9561 1001.

Yanchep National Park On a belt of coastal limestone, this 2842 ha park has forests of massive tuart trees, underground caves and spring wildflowers. Within the park, attractions include the historic Tudor-style Yanchep Inn; the Crystal Cave featuring magnificent limestone formations (daily tours available); a koala boardwalk; rowing-boat hire on freshwater Loch McNess; self-guide walk trails; and Aboriginal cultural tours (available on weekends and public holidays). Boomerang Gorge follows an ancient collapsed cave system and has an interpretive nature trail with access for people with disabilities. Grassy areas with barbecues and picnic tables provide a perfect setting for a family outing. 5 km E.

Marina: charter fishing boat hire; 6 km NW at Two Rocks. ***Guilderton:*** peaceful town at the mouth of the Moore River. Estuary provides safe swimming, and upper reaches of the river can be explored by boat or canoe (hire on river foreshore). Also good fishing; 37 km N. ***Ledge Pt:*** great destination for diving, with dive trail to 14 shipwrecks. Also the starting point for Lancelin Ocean Classic, a major windsurfing race each Jan; 62 km N. ***Lancelin:*** great base for fishing and boating because of a natural breakwater offshore. White sandy beaches provide safe swimming, and sand dunes at the edge of town have designated areas for off-road vehicles and sand-boarding; 71 km N.

See also HEARTLANDS, p. 308

York

Pop. 2091
Map ref. 615 H4 | 618 D4 | 620 C7

Town Hall, 81 Avon Tce; (08) 9641 1301; www.yorkwa.org

101.3 York FM, 612 AM ABC Radio National

On the banks of the Avon River in the fertile Avon Valley, York is one of the best preserved and restored 19th-century towns in Australia. It is now classified by the National Trust as 'York Historic Town'. Settled in 1831, only two years after the establishment of the Swan River Colony, York was the first inland European settlement in Western Australia.

Historic buildings There are a significant number of carefully preserved historic buildings in York. The three remaining hotels are fine examples of early coaching inns. The Romanesque-style Town Hall (1911) reflects gold-rush wealth. Details from visitor centre.

Avon Park: on the banks of the river with playground and barbecue facilities; Low St. ***York Motor Museum:*** vehicles on display represent the development of motor transport; Avon Tce. ***Mill Gallery:*** display and sales of unique and award-winning recycled jarrah furniture and craft; Broome St. ***Residency Museum:*** personal possessions, ceramics and silverware reflect aspects of civic and religious life in early York; open 1–3pm Tues–Thurs, 11am–3.30pm Sat, Sun and public holidays; Brook St. ***Suspension bridge and walk trail:*** built in 1906, the bridge crosses the Avon River at Avon Park. A 1.5 km nature and heritage walk starts at the bridge; Low St.

Antique Fair: Apr. ***York Society Photographic Awards:*** Apr. ***Olive Festival:*** June. ***Jazz Festival:*** Sept/Oct. ***Spring Garden Festival:*** Oct.

Cafe Bugatti: town favourite; 104 Avon Tce; (08) 9641 1583. ***Restaurant Eboracum:*** international cuisine; Imperial Hotel York, 83 Avon Tce; (08) 9641 1255. ***York Mill Bakehouse:*** arty cafe; 7–13 Broome St; (08) 9641 2447.

Old Albion B&B: heritage cottages; 19 Avon Tce; (08) 9641 2608. ***Imperial Hotel York:*** romantic heritage hotel; 83 Avon Tce; (08) 9641 1255. ***Quellington School House Farmstay:*** 1880s school converted to cottage; Sees Rd, Quellington; (08) 9641 1343. ***Tipperary Church:*** church converted into B&B; 2092 Northam Rd; 0439 965 275.

Mt Brown Lookout: provides 360-degree views over town; 3 km SE. ***Gwambygine Park:*** overlooking the river with boardwalk and viewing platform; 10 km S. ***Skydive Express:*** award-winning centre; (08) 9641 2905; 10 km N. ***Toapin Weir and Mt Stirling:*** panoramic views; 64 km E. ***Self-drive trails:*** eight different routes including the Avon Ascent through Perth's scenic hinterland to a series of special places in the Avon Valley; maps from visitor centre.

See also HEARTLANDS, p. 308

THE NORTHERN TERRITORY IS...

Shopping at the MINDIL BEACH SUNSET MARKETS / Swimming in LITCHFIELD NATIONAL PARK / Exploring billabongs and Aboriginal rock-art galleries in KAKADU NATIONAL PARK / Seeing spectacular ULURU at sunrise and sunset / Visiting the ARALUEN CULTURAL PRECINCT / Fishing for barramundi

NORTHERN TERRITORY

[NOURLANGIE ROCK, KAKADU NATIONAL PARK]

in the TERRITORY'S WEST COAST RIVERS / Boarding an ADELAIDE RIVER JUMPING CROCODILES CRUISE / Learning about the Aboriginal legend behind the DEVILS MARBLES (KARLU KARLU) / Enjoying a drink at the historic DALY WATERS PUB / Climbing Kings Canyon in WATARRKA NATIONAL PARK

THE NORTHERN TERRITORY is Australia's least settled state or territory, with vast tracts of desert and tropical woodlands. But to regard this country as empty is to do it a disservice; Aboriginal people have lived and travelled across the territory for thousands of years, and still do. Many non-Aboriginal Australians see it as the last great frontier because of its remoteness, spectacular landscapes and hardy outback characters.

Desert regions lie towards central Australia, while the tropical Top End is lapped by the Timor and Arafura seas. Although the diversity of landscape and wildlife makes it one of Australia's most inspiring destinations, visitors should expect to cover a lot of distance between highlights.

The coastline and offshore islands are places of special beauty – pearly white beaches interspersed with rocky red cliffs and rich mangrove habitats. The coastal rivers are home to thousands of bird and marine species, and their flood plains carry the annual wet season deluge into the Timor and Arafura seas, and the Gulf of Carpentaria. The rivers are also spawning grounds for barramundi, which attract anglers from around the world.

[KINGS CANYON, WATARRKA NATIONAL PARK]

The north-east includes Arnhem Land, the largest Aboriginal reserve in Australia, where trade and mingling of cultures occurred between Yolngu people and Indonesian seafarers from the 1600s. Today it is home to many groups who still live a semi-traditional lifestyle. It is also the custodial land of Australia's most famous Indigenous instrument, the didgeridoo. Here visitors can explore parts of the Gove and Cobourg peninsulas, with their green vegetation, turquoise waters and great fishing.

[SALTWATER CROCODILE]

The Red Centre is ancient and breathtaking, a land of intense as well as muted tones created by beautiful gorges, rock holes and vistas. While many travellers are drawn to Uluru and Kata Tjuta, the surrounding countryside is equally impressive – from the rolling red sandhills of the Simpson Desert to the undulating grasslands west of Glen Helen. North of Alice Springs, the Tanami Desert is incredibly remote and vastly interesting.

fact file

Population 229 700
Total land area 1 335 742 square kilometres
People per square kilometre 0.15
Beef cattle per square kilometre 1.3
Length of coastline 5437 kilometres
Number of islands 887
Longest river Victoria River (560 kilometres)
Highest mountain Mount Zeil (1531 metres), West MacDonnell National Park
Highest waterfall Jim Jim Falls (160 metres), Kakadu National Park
Highest town Areyonga (700 metres), west of Hermannsburg
Hottest place Aputula (Finke), 48.3°C in 1960
Strangest place name Humpty Doo
Quirkiest festival Henley-on-Todd Regatta with boat races in the dry riverbed, Alice Springs
Longest road Stuart Highway (approximately 2000 kilometres)
Most scenic road Larapinta Drive, from Alice Springs to Hermannsburg
Most famous pub Daly Waters Pub
Most impressive gorge Nitmiluk (Katherine) Gorge, Nitmiluk National Park
Most identifiable trees Pandanus palm and desert she-oak
Most impressive sight Electrical storms in the build-up to the wet season, Darwin
Favourite food Barramundi
Local beer NT Draught
Interesting fact Some 50 per cent of the Northern Territory is either Aboriginal land or land under claim

gift ideas

Aboriginal art and handmade baskets (Arnhem Land, Kakadu National Park & Darwin) Authentic pieces are best purchased from Aboriginal craft centres in places like Maningrida and Gunbalunya, Arnhem Land. Alternatively, visit Marrawuddi Gallery, Kakadu National Park, or Framed – The Darwin Gallery, 55 Stuart Hwy, Darwin. See Kakadu & Arnhem Land p. 379, 634–5

Barramundi-skin wallets and shoes (Barra Shack, Humpty Doo) Stop at this popular shop on the way to Kakadu National Park to pick up items that are true evidence of an outback experience. 41 Acacia Rd, Humpty Doo. 632 E4

Didgeridoos (The Didgeridoo Hut, Humpty Doo) Purchase an iconic Aboriginal item from this store owned and run by Indigenous people. 1 Arnhem Hwy, Humpty Doo. 632 E4

Handmade whips, crocodile-skin and kangaroo-skin products (Mindil Beach Sunset Markets, Darwin) A range of fantastic Top End products are on offer, from buffalo horns to necklaces with a crocodile tooth. In particular, visit Mick's Whips for various whips and handcrafted crocodile or kangaroo leather goods. See Mindil Beach Sunset Markets p. 374, 370 B5

Pukumani Poles (Bathurst & Melville islands) Aboriginal burial poles found only on the Tiwi Islands can be purchased at several arts and crafts centres. See Tiwi Islands p. 377, 634 D2

South Sea pearls (Paspaley & Bynoe Harbour Pearl Company, Darwin) Found in Australia's northern waters, the largest pearls in the world are made into elegant jewellery pieces. Paspaley, 19 The Mall; Bynoe Harbour Pearl Company, Shop 11, Darwin Wharf Precinct. See Shopping p. 375, 372 C2

Yothu Yindi CD (Arnhem Land) The most famous Indigenous band comes from the Yolngu people of Arnhem Land, and their well-known song 'Treaty' catapulted Indigenous music into the mainstream Australian music scene. The yearly Garma Festival of Traditional Culture, held during October in Gulkula, Arnhem Land, is organised by the Yothu Yindi Foundation. See Kakadu & Arnhem Land p. 379, 635

***Walkabout Chefs* book (Darwin)** With recipes by renowned outback chef Steve Sunk, available from any bookshop in Darwin.

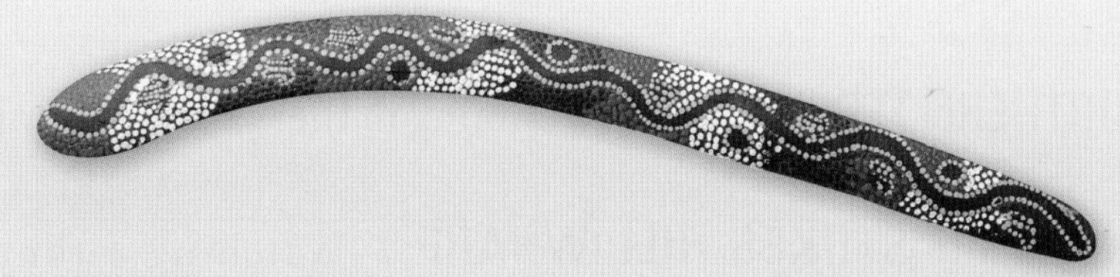

DARWIN IS...

Window-shopping for pearls in SMITH STREET MALL / A stroll along THE ESPLANADE / Watching a movie at the DECKCHAIR CINEMA / A safe encounter with reptiles at CROCOSAURUS COVE / A picnic at FANNIE BAY / Sunset drinks at the DARWIN SAILING CLUB / Shopping for arts and crafts at the MINDIL BEACH SUNSET MARKETS / Handfeeding fish at AQUASCENE / Exploring ABORIGINAL ART GALLERIES / Learning about cyclones at the MUSEUM AND ART GALLERY OF THE NORTHERN TERRITORY / Swimming in the wave lagoon at the DARWIN WATERFRONT / Fishing off STOKES HILL WHARF / Watching a thunderstorm in the evening sky

[SUNSET OVER CULLEN BAY MARINA]

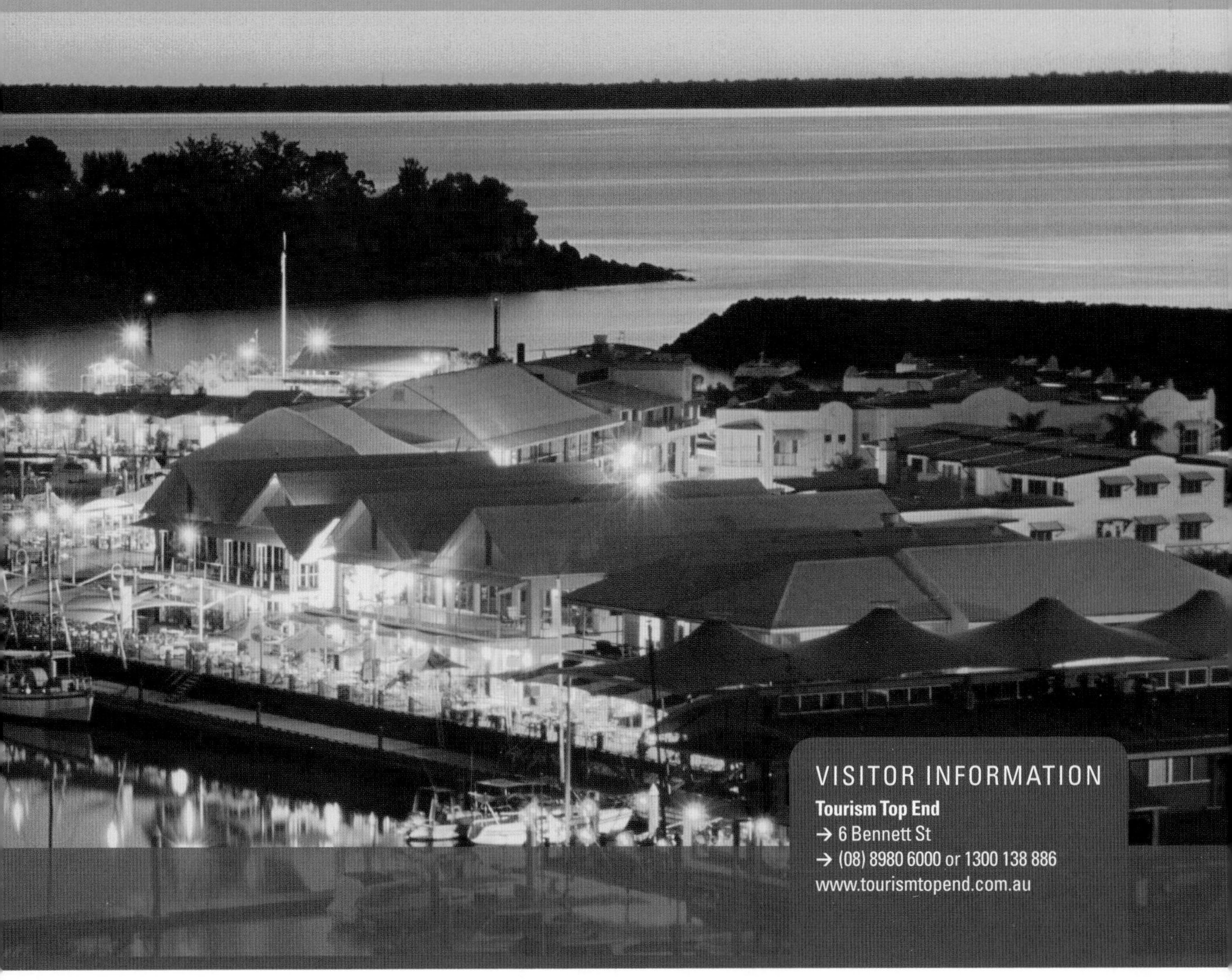

VISITOR INFORMATION

Tourism Top End

→ 6 Bennett St

→ (08) 8980 6000 or 1300 138 886

www.tourismtopend.com.au

Regarded as Australia's northern outpost, Darwin's proximity to Asia and its immersion in Aboriginal culture make it one of the world's most interesting cities. It retains a tropical, colonial feel despite having been largely rebuilt after the devastation wreaked by cyclone Tracy over 30 years ago.

The founding fathers of Darwin laid out the city centre on a small peninsula that juts into one of the finest harbours in northern Australia. Their names live on in the wide streets of the city centre, which is easy to get around and lacks the winding lanes of older Australian cities.

Built on the land of the Larrakia Aboriginal people, Darwin is a beautiful green city. Manicured lawns, and hedges of bougainvillea and frangipani adorn parks and roadways, while the waters of the Timor Sea lap three sides of the city.

The population of 125 000 comprises over 50 ethnic groups who live together harmoniously here. This diversity stretches back to the early days of Darwin's development when Aboriginal, European and Chinese people worked side by side. More recent arrivals include migrants from Greece, Timor, Indonesia and Africa.

Evidence of the early days remain, but Darwin is also a modern city. With so much natural beauty around its harbour, along the beaches and in its tropical parks, it remains one of the most fascinating cities in the world.

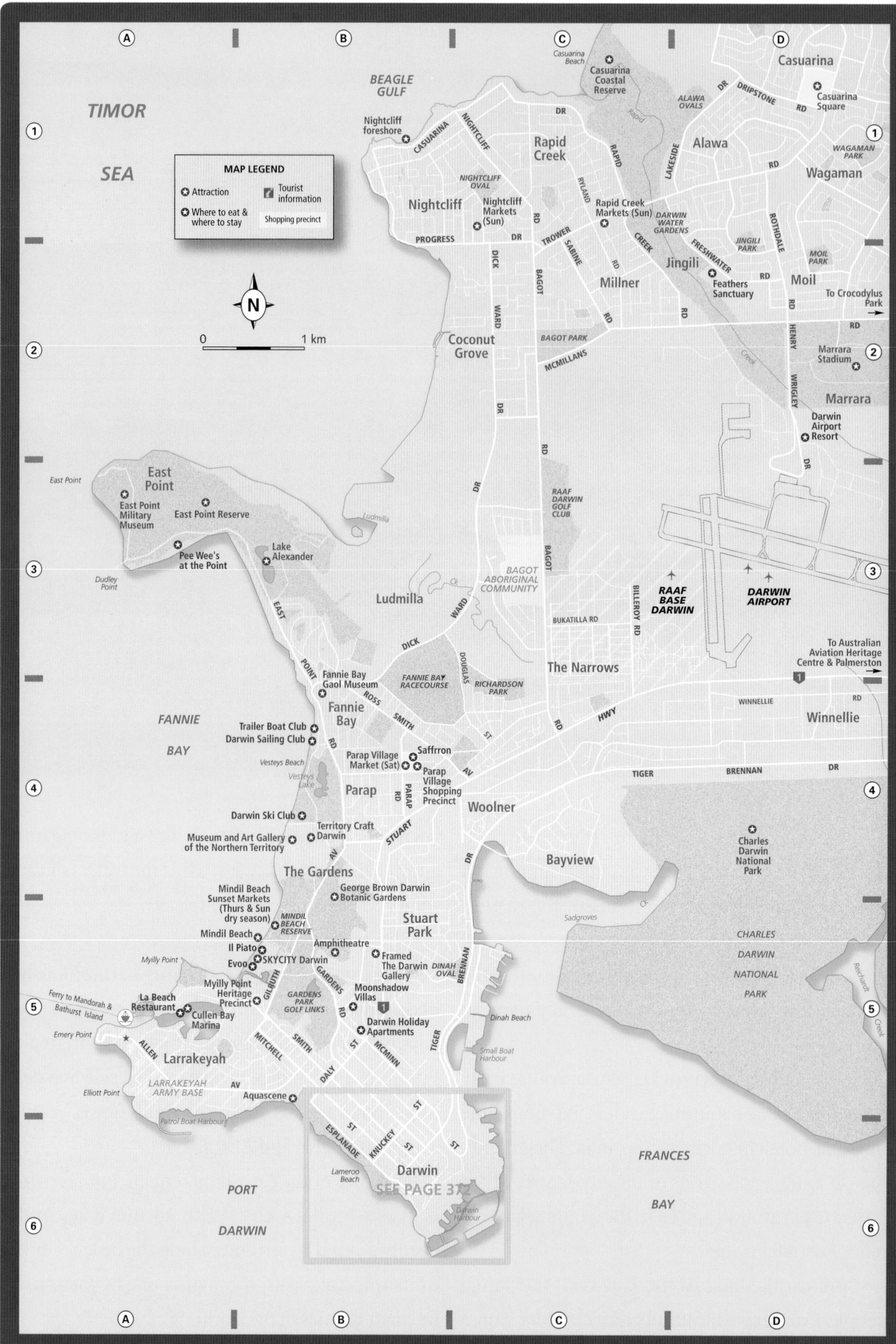

TIMOR
SEA
BEAGLE
GULF
MAP LEGEND
Attraction
Tourist information
Where to eat & where to stay
Shopping precinct
0
1 km
Casuarina
Casuarina Beach
Casuarina Coastal Reserve
Casuarina Square
Nightcliff foreshore
Rapid Creek
Alawa
Wagaman
Nightcliff
Nightcliff Markets (Sun)
Rapid Creek Markets (Sun)
Jingili
Millner
Moil
Feathers Sanctuary
To Crocodylus Park
Coconut Grove
Marrara Stadium
Marrara
Darwin Airport Resort
East Point
East Point Military Museum
East Point Reserve
Pee Wee's at the Point
Lake Alexander
Ludmilla
BAGOT ABORIGINAL COMMUNITY
RAAF DARWIN GOLF CLUB
RAAF BASE DARWIN
DARWIN AIRPORT
The Narrows
To Australian Aviation Heritage Centre & Palmerston
Fannie Bay Gaol Museum
Fannie Bay
FANNIE BAY RACECOURSE
RICHARDSON PARK
Winnellie
FANNIE BAY
Trailer Boat Club
Darwin Sailing Club
Parap Village Market (Sat)
Saffron
Parap Village Shopping Precinct
Parap
Woolner
Darwin Ski Club
Territory Craft Darwin
Museum and Art Gallery of the Northern Territory
The Gardens
Bayview
Charles Darwin National Park
George Brown Darwin Botanic Gardens
Mindil Beach Sunset Markets (Thurs & Sun dry season)
MINDIL BEACH RESERVE
Stuart Park
Mindil Beach
Il Piato
Evoo
SKYCITY Darwin
Amphitheatre
Framed The Darwin Gallery
DINAH OVAL
CHARLES DARWIN NATIONAL PARK
Myilly Point Heritage Precinct
Moonshadow Villas
La Beach Restaurant
Cullen Bay Marina
GARDENS PARK GOLF LINKS
Ferry to Mandorah & Bathurst Island
Darwin Holiday Apartments
Dinah Beach
Larrakeyah
LARRAKEYAH ARMY BASE
Aquascene
Darwin
SEE PAGE 372
PORT
DARWIN
FRANCES
BAY

City Centre

Like most of Darwin, the city centre is open and vibrant with wide streets, leafy parks, a shady mall and outdoor dining. Arching shade trees and towering palms are features of the lush parks and reserves, while stunning Aboriginal art and artefacts are characteristic of the retail areas. Tall, modern structures have begun to replace the old colonial style buildings, but they still retain a fresh, tropical flavour with overhanging eaves, corrugated iron and lush vegetation.

Smith Street 372 B2, C2

Smith Street Mall is the retail heart of the central business district (CBD). Shady **Raintree Park** at the northern end is popular with tourists and locals at lunchtime, and a walk down the mall between May and the end of August often reveals buskers from all over Australia who have travelled north to beat the southern winter. Plazas and small arcades reveal shops where visitors can buy Aboriginal art, locally made jewellery and tropical clothing. Halfway down the mall is the **Victoria Hotel** a Darwin landmark and one of its oldest pubs; *(08) 8981 4011*. Almost directly opposite is the **Star Arcade**, which used to house the old open-air Star Theatre. Today the arcade has a quirky collection of boutiques and cafes. At the southern end of the mall are some of the city's grandest buildings where it is possible to look at or buy South Sea pearls and authentic crocodile products.

Across Bennett Street, Smith Street continues south towards the harbour past **Brown's Mart Theatre** which was built in 1883 and is now home to the Darwin Theatre Company; *(08) 8981 5522*.

Behind Brown's Mart are the Darwin City Council Chambers, where an ancient banyan tree known as the **Tree of Knowledge** casts a huge umbrella of shade. Planted at the end of the 19th century, the tree has been a meeting place, dormitory and soapbox for generations. At the southern end of Smith Street is **Christ Church Cathedral**, built in 1902 and restored after damage from both Japanese bombers and cyclone Tracy.

Mitchell Street 372 A1, B2

One street to the west of Smith Street is Mitchell Street, a popular dining and entertainment area specialising in outdoor eateries. With backpacker accommodation, pubs and outdoor dining areas, Mitchell Street is the party precinct of Darwin. You'll find cinemas and the **Darwin Entertainment Centre** which hosts many exciting theatrical and musical performances; *(08) 8980 3366 (see Top events, p. 373)*. At **Crocosaurus Cove** visitors can swim with crocodiles – safely protected inside a perspex cage! *(08) 98981 7522*.

Cavenagh Street 372 B1, C1

Cavenagh Street is one street east of Smith Street. Among the commercial buildings and government departments, it boasts a few art galleries and cafes.

This street was Darwin's original Chinatown. In the late 1800s the southern end was full of ramshackle huts and shops with the occasional opium den. While some of the original stone buildings remain up near Darwin Post Office, a reminder of its Asian history is at nearby Litchfield Street where a modern **Chinese Temple** is built on the site of an older temple that was constructed in 1887. The **Northern Territory Chinese Museum** next door has displays taking visitors through the history of Chinese people in Darwin from the establishment of Chinatown and market gardens to the bombings in World War II. *Open during the dry season; admission by gold coin donation.*

The Esplanade 372 B2

Much of Darwin's accommodation is built along The Esplanade, with balconies and windows looking out over Darwin Harbour.

climate

Darwin has a constant temperature of around 30°C and the weather is always warm and humid – just how humid depends on the time of year. Most people visit during the dry season, which extends from May to the end of September. The 'build-up' period (between October and December) is famous for hot, stifling weather and massive electrical storms. During the wet season (which can last until late April), the Asian monsoon drops over Darwin, often bringing days of cleansing rain. Occasionally a cyclone comes and the streets of the city and surrounding landscape become waterlogged and flood-bound. The wet season may be uncomfortable, but it is the time when the landscape is green and lush, and native trees burst into flower and attract thousands of birds.

J	F	M	A	M	J	J	A	S	O	N	D	
32	32	32	33	32	31	30	31	33	34	34	33	MAX °C
25	25	24	24	22	20	19	20	23	25	25	25	MIN °C
393	329	258	102	14	3	1	1	12	52	124	241	RAIN MM
15	15	13	5	1	0	0	0	1	3	8	11	RAIN DAYS

getting around

Darwin is easy to get around – city streets are laid out in a grid, most attractions are within walking distance, and traffic is rarely heavy.

A regular public bus service covers many of the suburbs as well as the satellite town of Palmerston (the city terminus is on Harry Chan Avenue). Taxis are located at the Knuckey Street end of The Mall and (weekends only) outside the Cinema Complex on Mitchell Street.

Darwin's network of bicycle paths extends from the city out to the northern suburbs.

Bicycle Hire At many of the backpacker lodges (08) 8981 0227.

Public transport Bus Service (08) 8924 7666.

Airport shuttle bus Darwin Airport Shuttle Bus 1800 358 945; Metro Minibus (08) 8983 0577.

Police road report 1800 246 199 (a good source of information for travel outside Darwin, particularly in the wet season).

Motoring organisation AANT (08) 8925 5901, roadside assistance 13 1111.

Car rental Avis (08) 8945 0662; Britz Campervan Rentals 1800 331 454; Budget 1800 22 55 88; Hertz 1800 22 55 88; Sargent's four-wheel drive hire service (08) 8947 2736.

Taxis Radio Taxis 131 008.

Bus tours Tour Tub is an open-air bus service with a 'hop-on hop-off' bus to the city's top sights, departing every hour from the corner of Smith Street Mall and Knuckey Street. $40 pp. (08) 8985 6322.

Bicycle and scooter hire Darwin Scooter Hire (08) 8941 2434.

Oil rigs can often be seen being towed out to the offshore fields of the Timor Sea, and other cargo ships lie at anchor, awaiting shipments. There are many Australian and American wrecks at the bottom of the harbour, sunk by the Japanese bombers that struck without warning in February 1942. There are memorial sites all around the harbour recording the hundreds of bombing raids that were made on the city during World War II including the **Cenotaph** honouring service men and women.

Beautiful **Bicentennial Park** runs the length of The Esplanade and a walking/cycling track goes from Doctors Gully in the north to the Wharf Precinct in the south. The park is a great place to relax under a shady tree, enjoy lunch or simply watch the day go by; in the mornings and evenings, this is also a popular area for joggers and walkers. Midway along The Esplanade there's a pathway down to **Lameroo Beach** and at the southern end of the track, steps go down to the **Deckchair Cinema** at the end of Jervois Road off Kitchener Drive. *(08) 8941 4377*. The open-air cinema shows current and popular films 'under the stars' every night during the dry season.

On the corner of Knuckey Street, at the southern end of The Esplanade, is **Old Admiralty House**, built in 1937. It shows off tropical design and living standards before the city was devastated by cyclone Tracy in 1974, and is now incorporated into **Char Restaurant** a beautiful outdoor dining space. *(08) 8981 4544*. On the opposite corner is **Lyons Cottage**, the former British Australian Telegraph (BAT) headquarters for the Overland Telegraph. Built in 1925 for BAT staff, it now houses Aboriginal Bush Traders which promotes Indigenous tourism expeirence as well as arts, crafts. *0448 329 933*.

State Square

State Square is one of Darwin's most outstanding precincts. The thoroughly modern Parliament House and Supreme Court buildings are either loved or hated by locals, but most people agree that Government House, one of the few intact reminders of the city's colonial days, is utterly charming. The wide green lawns of Liberty Square and surrounding shade trees make this an ideal area in which to eat a sandwich or simply relax.

Palmerston Town Hall 372 C2

Darwin was initially named Palmerston, and the now-ruined Palmerston Town Hall was built in 1883. The building was partially destroyed by cyclone Tracy in 1974, and now stands as a memorial to the city's early colonial days and to the ferocity of the cyclone. The ruins occasionally host outdoor performances such as *A Midsummer Night's Dream*.

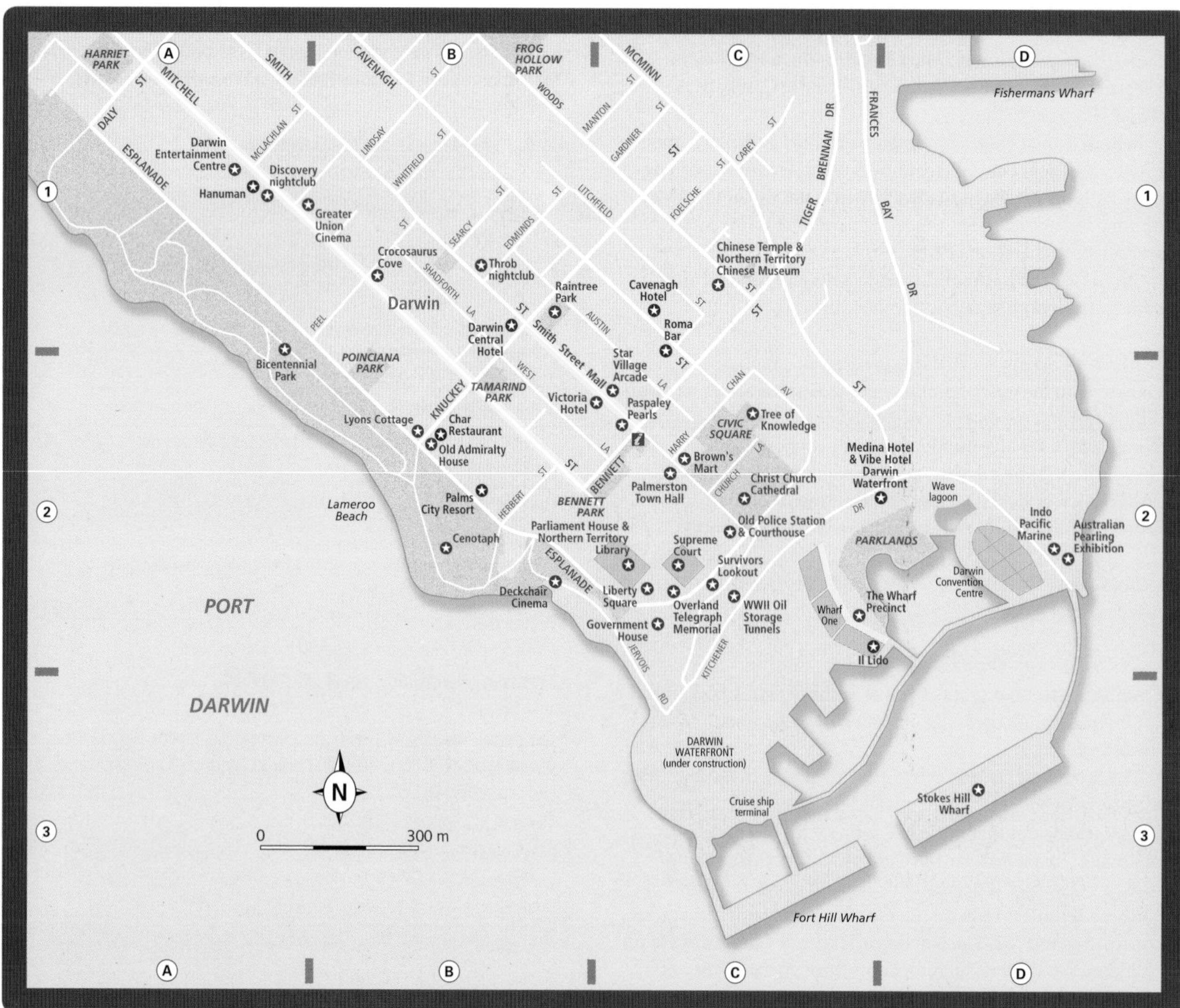

Parliament House, the Supreme Court & Liberty Square 372 C2

Parliament House is a large white rectangular building with one of the finest views of Darwin Harbour. Opened in 1994, this modern, imposing edifice also houses the **Northern Territory Library** and **Speaker's Corner Cafe** that opens out on to an area of lawn with a great view. *(08) 8946 1439.* Visitors are encouraged to look through the grand hall that displays art and photographic exhibitions. *Free guided tours (08) 8946 1434.*

Across the wide courtyard at the front of Parliament House is the Supreme Court building, which was built in 1990. High ceilings and an atmosphere of modern grandeur are also a feature of this building – it has a spectacular foyer with a giant mosaic floor designed by Aboriginal artist Norah Napaljarri Nelson, and a permanent exhibit of Arnhem Land burial poles. At the southern side of Parliament House, next to the Supreme Court, is an open, grassed area called Liberty Square. Edged by spreading rain trees, this is the place where unionists met early last century to protest against the administration of Dr John Gilruth, who was eventually forced to flee the city as a result of a popular uprising against him. *Free guided tours are available Sat and Wed. Contact the Visitor Centre for times.*

Government House, Old Police Station & Courthouse 372 C2

Across the road from Liberty Square is Government House, an elegant, gabled, colonial-style building built in 1879, which survived both cyclones and World War II bombs. The building is open to the public once a year and is the venue for formal government occasions and ceremonies. Further along The Esplanade, on the corner of Smith Street, is the Old Police Station and Courthouse. They are now home to the offices for the Northern Territory administrator and staff. Directly opposite these buildings is **Survivors Lookout**. It surveys the Wharf Precinct and Darwin Harbour to East Arm Port and the terminus of the *Ghan*, a historic train that runs across the continent from Adelaide. The lookout tells the story of the WWII battles that took place over the harbour.

The Wharf Precinct

Steeped in history and perched upon one of the largest natural harbours in Australia, this precinct is the original base of the city of Darwin. Stokes Hill and Fort Hill wharves date back more than a century to a time when clippers and steam ships used to call in at Darwin to load exotic cargoes such as crocodile skins, buffalo hides and pearls. Today the port is used for luxury liners and warships, as a venue for outdoor dining, and as the home of Darwin's convention centre and new Waterfront recreation development.

World War II Oil Storage Tunnels 372 C2

A set of stairs leads from Survivors Lookout to the Wharf Precinct. At the bottom of the stairs are storage tunnels that were built underground after the above-ground tanks were bombed in early war time raids on Darwin. One of the five tunnels is open to the public and there are historical displays of the war years. *Kitchener Dr; (08) 8985 6333; open daily.*

top events

Sky City Triple Crown V8 Supercars Championship Three days of V8 supercars at Hidden Valley Raceway. June.

Royal Darwin Show Three days of rides, and equestrian and agricultural displays at the Darwin Showgrounds. July.

Darwin Beer Can Regatta Darwin-style boat races in vessels built from beer cans. June/July.

Darwin Fringe Festival Theatre, dance, music, visual arts, film and comedy. June

Darwin Cup Carnival Premier horseracing. July/August.

Festival of Darwin A feast of visual and performing arts that attracts people from around the world. August.

Telstra National Aboriginal and Torres Strait Islander Art Awards Exquisite Aboriginal art at the Museum and Art Gallery of the Northern Territory. August–October.

Indo Pacific Marine 372 D2

Find out what lies in Darwin Harbour – from deadly stone fish that can inflict terrible pain and even death, to the beautiful coral that lies hidden in the Top End's sometimes murky waters. Indo Pacific Marine is one of the few places in the world that has been able to transfer a living ecosystem from the water into a land-based exhibition. Also on show are creatures endemic to the Top End, such as the deadly box jellyfish. *Kitchener Dr; (08) 8981 1294; open daily 10am–4pm.*

Australian Pearling Exhibition 372 D2

Housed in the same building as Indo Pacific Marine, this exhibition gives a detailed history of pearling in northern Australia and highlights the important contribution Aboriginal, Islander and Japanese people made to this multimillion-dollar industry. Also on display are some wonderful examples of South Sea pearls, which are found in northern waters and are the largest pearls in the world. Pearling makes a huge contribution to the northern Australian economy and is still an important source of employment. *(08) 8999 6573; open daily.*

swimming

Swimming in waters around Darwin is not recommended, particularly during the build-up to the wet season (October to December) and during the wet season (up until the end of April), because of the hidden dangers of box jellyfish and crocodiles. Although Darwin Harbour and the foreshores are patrolled regularly by Parks & Wildlife Commission officers, and saltwater crocodiles are relocated to farms, the seas are not completely free of dangerous creatures. Box jellyfish pose the biggest problem because they are small, almost transparent, and deadly. Some people do swim in the sea during the dry season, when box jellyfish are least threatening, but most locals do not. Preferred swimming spots include community pools at Parap, Nightcliff and Casuarina, Lake Alexander at East Point Reserve *(see p. 376)*, the lagoon and wave pool at the Waterfront and natural springs outside the city.

markets

Parap Village Market This market year-round is about waking up to the splendours of Asia. Colourful flowers, frozen-fruit ice-cream, silk-screened sarongs, Asian food, relaxing massages and exotic blended drinks are all on offer. Parap Shopping Centre; Sat mornings. 370 B4

Rapid Creek Markets This little slice of Asia is a favourite for locals and visitors. Stock up on exotic fresh fruit and vegetables. Rapid Creek Shopping Centre; Sun mornings. 370 C1

Nightcliff Markets A relaxing morning market in an outdoor setting. Nightcliff Shopping Centre; Sun mornings. 370 C1

See also Mindil Beach Sunset Markets, on this page.

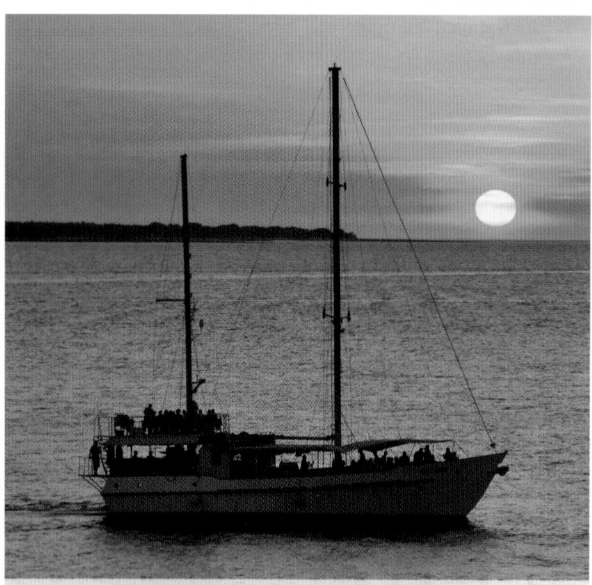

[SUNSET SAIL, PORT DARWIN]

where to stay

Darwin Airport Resort Modern resort. 1 Henry Wrigley Dr, Marrara; (08) 8920 3333 or 1800 600 975. 370 D2

Darwin Central Hotel Good value and very central. Knuckey St; (08) 8944 9000. 372 B1

Darwin Holiday Apartments Offer a range of self-contained apartments. 0410 368 475. 370 B5

Feathers Sanctuary Luxury B&B retreat for lovers of wildlife, particularly birds. 49A Freshwater Rd, Jingili; (08) 8985 2144; open dry season. 370 D2

Hidden Valley (08) 8984 2888 and **Shady Camp** (08) 8984 3330 caravan parks are the closest to the city.

Medina and Vibe Apartments or rooms at the Darwin Waterfront. Kitchener Dr; (08) 8982 9998. 372 C2

Moonshadow Villas Stylish indoor/ outdoor living, within walking distance of the CBD. 6 Gardens Hill Cres; (08) 8981 8850. 370 B5

Palms City Resort Conveniently located on The Esplanade. 64 The Esplanade; (08) 8982 9200. 372 B2

Stokes Hill Wharf 372 D3

Once northern Australia's most important port catering to sailing vessels and steamers, Stokes Hill Wharf is now a place for restaurants, bars and sunset cruises. When the fish are biting, the wharf is one of the best places in Darwin to fish. During the build-up to the wet season, head here to watch one of the city's magnificent storms as it gathers across the harbour. Neighbouring Fort Hill Wharf is the defence vessel facility and incorporates a new cruise ship terminal.

The new Darwin Waterfront development has transformed the Wharf Precinct into a lively recreation and commercial zone, with a waterfront promenade lined with restaurants, cafes and boutiques. Swim in the lagoon or ride the waves in the wave pool. *Open daily 10am–6pm.*

Inner North

The beaches of the inner north are idyllic in the dry season, but can be deadly during the build-up to the wet season when box jellyfish float in shallow waters along the shoreline (see Swimming, p. 373). Most visitors content themselves with sunbathing or walking the picturesque foreshores.

Aquascene 370 B5

This is one of Darwin's most popular attractions, where visitors can handfeed fish that live in Darwin Harbour. Opening times depend on the high tide, but Aquascene publishes feeding times every day in the *NT News* and weekly timetables can be found at Tourism Top End. Aquascene not only offers the chance to interact with the marine life of the harbour, but also an opportunity to learn about northern Australian creatures through informative talks. *Doctors Gully Rd; (08) 8981 7837.*

Cullen Bay Marina 370 A5

Cullen Bay is a popular dining and recreational area where waterside mansions and berths for yachts are built alongside holiday apartments. A wide walking path fringes a gently sloping sandy beach, and a variety of restaurants and cafes overlook the blue waters of the marina. A regular ferry service crosses Darwin Harbour to Mandorah, departing from the front of the lock at Cullen Bay, while sunset and evening charters are based within the marina itself. At about a kilometre north of central Darwin, the road down to the marina (Kahlin Avenue) passes **Myilly Point Heritage Precinct**. This small group of pre–World War II houses was constructed for senior public servants and survived cyclone Tracy; they are considered excellent examples of tropical residential architecture. High Teas are held every Sunday afternoon in the gardens of **Burnett House**. *Confirm times with the National Trust (08) 8981 2848.*

Mindil Beach Sunset Markets 370 B5

The Mindil Beach Sunset Markets are Darwin's most popular tourist attraction. Located off Gilruth Avenue, the markets operate on Thursday and Sunday evenings from the end of April to the end of October, taking advantage of the superb dry-season weather. Up to 10 000 people enjoy the food of more than 30 nations and wander between stalls that offer everything from Aboriginal arts and Asian crafts to tarot-card readings, massages and kangaroo sausages. Live performances by theatrical and singing troupes are spiced with whip-cracking and poetry readings. Bands play at night and there is an occasional offshore fireworks display. At sunset, people often walk over the sand dunes to the beach to experience a quintessential Darwin moment.

SKYCITY 370 B5

Also on Mindil Beach, SKYCITY casino lies next to the market strips. The casino frequently holds concerts on its lawns. There are several restaurants in the casino, including **Evoo** (see where to eat below), one of Darwin's finest dining venues. SKYCITY is also the hub of the horseracing scene in August when a gala ball is held on the lawns. *Gilruth Ave, Mindil Beach; (08) 8943 8888.*

George Brown Darwin Botanic Gardens 370 B5

The botanic gardens have paths that wind through one of the best collections of tropical plants in Australia. Established in the 1870s, the gardens cover an area of 42 hectares and contain attractions such as an extensive collection of tropical orchids and palms, and a self-guide Aboriginal plant-use trail. The gardens are popular with family groups and wedding parties. A wonderful tree house incorporated into a huge African mahogany is a hit with kids. *Geranium St, The Gardens.*

Museum and Art Gallery of the Northern Territory 370 B4

The Museum and Art Gallery of the Northern Territory is one of Darwin's main cultural icons. This institution houses one of the finest Aboriginal art collections in Australia, which is enhanced every year with the work of entrants in the Aboriginal and Torres Strait Islander Art Award. A spine-tingling cyclone Tracy gallery details what happened during and after that fateful Christmas in 1974. The museum also possesses an excellent natural history display of fauna and flora of the Top End and South-East Asia. The Maritime Boatshed, a vast room filled with all sorts of vessels that have travelled to northern Australia over the years, is impressive in its size and in the diversity of its exhibits from tiny jukung, with their slender outriggers and woven sails, to the large fishing vessels that limped to Australian shores overloaded with refugees. The nearby **Territory Craft Darwin** exhibits the work of local artists and craft producers. *Conacher St, Bullocky Point; open daily; general admission free.*

Sunset dining 370 B4

One of the pleasures of visiting Darwin is being able to enjoy a meal by the beach as the sun sets over Fannie Bay. A special treat is **Pee Wees**, at East Point Reserve. *(08) 8981 6868.* There are several other popular dining venues, such as the **Darwin Ski Club** (08) 8981 6630, on the beach at Fannie Bay next to the Museum and Art Gallery of the Northern Territory, and the popular **Darwin Sailing Club**, just 200 metres further along the coast on Atkins Drive. *(08) 8981 1700.* Or pack a picnic and head up to **East Point Reserve**.

Fannie Bay Gaol Museum 370 B4

One of Darwin's most interesting destinations is Fannie Bay Gaol Museum, which served as Darwin's prison between 1883 and 1979. Located barely 300 metres from the Sailing Club, the gaol housed some of Darwin's most desperate criminals. The cells and gallows provide a sobering display for visitors, but are sometimes used as a backdrop for dinner parties and social events. *East Point Rd; open daily; admission free; (08) 8941 2260.*

shopping

Smith Street Mall, City Darwin's major shopping precinct with a wide range of interesting shops, including Aboriginal art and artefacts all interspersed with outdoor eateries, intriguing arcades and a galleria. Paspaley Pearls is a good place to buy pearls and unique crocodile products can be found next door. Territory Colours, 21 Knuckey Street, has great examples of local art and craft, while Animale, in the Galleria, is a terrific frock shop. 372 B2, C2

Casuarina Square Every popular department and chain store is under one roof here, and there is also a huge eatery with more than 30 restaurants and cafes. 370 D1

Framed – The Darwin Gallery, City One of the oldest art galleries in Darwin, with some of the best Aboriginal and Islander art in northern Australia. Stuart Highway (08) 8981 2994. 370 B5

Cullen Bay and The Wharf Precinct Both precincts offer excellent outdoor dining venues with several unique boutiques and gift shops.

Parap With its village atmosphere (and Saturday Markets), Parap is a unique shopping destination especially for Aboriginal art and crafts.

North

Darwin is blessed with a particularly attractive coastline, lush parks and excellent walking paths that meander through a beautiful urban environment. The suburbs of Fannie Bay, Nightcliff and Casuarina all front onto the Timor Sea which is flat, calm and often like liquid gold at sunset. Tall palm trees sway in the breeze that blows onshore every afternoon, while parks full of fig trees provide shelter and food for the huge variety of birds that inhabit Darwin. Parks teem with wildlife. Wallabies can be seen grazing along the foreshore and sea turtles come ashore at a nearby island to nest.

where to eat

Char Restaurant Best steaks in a garden setting. 70 The Esplanade; (08) 8981 4544. 372 B2

Cullen Bay Marina Boatshed Coffee House (08) 8981 0200 and Buzz Café (08) 8941 1141, both overlook the yachts and marina.

Evoo Premier dining in Darwin. SKYCITY; (08) 8943 8940. 370 B5

Hanuman Special Thai dining experience. 93 Mitchell St; (08) 8941 3500. 372 A1

Il Lido Cocktails and dinner on the water. 19 Kitchener Dr; (08) 8941 0900. 372 C2

Il Piato Upmarket food. SKYCITY; (08) 8943 8940. 370 B5

La Beach Restaurant Views over the ocean, great for sunset. Cullen Bay Marina; (08) 8941 7400. 370 A5

Pee Wee's at the Point Fine views and 'must do' dining. East Point; (08) 8981 6868. 370 A3

Roma Bar Traditional breakfast or Asian-inspired dinner. 9–11 Cavenagh St; (08) 8981 6729. 372 C1

Saffrron Indian with a difference. 34 Parap Rd; (08) 8981 2383. 370 B4

entertainment

Cinema One of the first examples of entertainment infrastructure ever established in Darwin was the outdoor cinema that ran during the dry season – the nights were clear and balmy and the temperature perfect. That tradition continues to this day at the Deckchair Cinema (on the shores of Darwin Harbour just around from the new Waterfront Development; (08) 8981 0700).

There are also cinema complexes on Mitchell Street (08) 8981 5999, Casuarina Shopping Square (08) 8945 7777and The Hub in Palmerston (08) 8931 2555. Check the *NT News* for daily showings.

Live music Darwin has a young population and is a regular stopover for naval and cruise vessels from many countries. There are plenty of clubs, pubs and restaurants, mainly in Mitchell Street that provide live music. SKYCITY casino has sunset jazz on its lawns during the dry season and The Wharf Precinct has occasional outdoor entertainment. Nightclubs such as Discovery and Throb have djs interspersed with live music, while the Amphitheatre in the Botanic Gardens is a popular venue for touring musicians especially during the Darwin Festival. Friday's edition of the *NT News* publishes a round-up of what's on in the Gig Guide, or visitors can check with Top End Tourism, Bennett Street (08) 8980 6000.

Classical music and performing arts Darwin has a vibrant arts community that has been deeply influenced by Aboriginal and Asian cultures. Local playwrights, dancers and artists are always producing some tropical gem that can be viewed at interesting venues such as Brown's Mart, the Palmerston Town Hall ruins in Smith Street or the Darwin Performing Arts Centre (DPAC) on Mitchell Street.

During the dry season the Darwin Festival, held in August, is a fantastic conglomeration of local and international artists and performances – all held outdoors in a purpose built festival 'marquee'. Check with Top End Tourism for the dates or see the *NT News* for weekly programs.

walks & tours

Darwin City Heritage Walk Visitors can either organise a guided tour with Darwin Walking Tours (08) 8981 0227 or pick up a map from the visitors centre.

Batji Experience Darwin through the perspective of its Traditional Owners, the Larrakia people 1300 881 186.

George Brown Darwin Botanic Gardens Walks Self-guide walks through different habitats; pamphlets available from the visitor centre at the Geranium Street entrance. A guide can be organised if you book ahead. (08) 8981 1958.

Sunset Cruises Take a cruise on one of the charter boats at Cullen Bay Marina or Stokes Wharf to see the sun slip below the horizon over the Timor Sea. City of Darwin Cruises 0417 855 829; Darwin Harbour Cruises (08) 8942 3131; Spirit of Darwin 0417 381 977.

Ferry to Mandorah Cast a line off the Jetty or have lunch at the Mandorah Beach Hotel (08) 8978 5044. It's 130 kilometres by road, but only a 20-minute ferry ride from Cullen Bay Marina. Mandorah Ferry Service (08) 8978 5044.

Tour Tub Hop-on, hop-off bus. (08) 8985 6322.

Fishing Tours Darwin fishing tours travel agency specialises in Darwin fishing adventures. (08) 8945 7686.

[THE CAGE OF DEATH, CROCOSAURUS COVE]

East Point Reserve 370 A3

The cliff-top paths into and along East Point Reserve are spectacular and very popular with joggers, roller bladers and cyclists. Early morning and sunset find a range of people taking a constitutional along these paths that meander between palm trees, parks and barbecue areas. The waters of Fannie Bay are enticing but, like all beaches in Darwin, are unsafe for swimming, particularly late in the year because of the presence of box jellyfish. **Lake Alexander**, at East Point Reserve, is an alternative swimming area, along with public baths at Parap, Nightcliff and Casuarina and the lagoon at the Waterfront. These are definitely the options preferred by the locals. A raised boardwalk on the northern side of the lake gives access to mangroves that are teeming with bird, fish and plant life.

East Point Military Museum, in the north of the reserve, sits in the shadow of two huge cement gun emplacements that were manned during World War II to protect the city. The museum highlights the role of Darwin in World War II and the parts played by service personnel in the defence of northern Australia. *Alec Fong Lim Dr, East Point; open daily; (08) 8981 9702.*

Nightcliff foreshore 370 B1

The Nightcliff foreshore provides an excellent cliff-top walkway that runs from Nightcliff boat ramp to a bridge over Rapid Creek, on to Casuarina Beach. A beautiful swimming pool, parkland, play areas for children and fishing spots can be found all along here.

North-east

Australian Aviation Heritage Centre 632 D3

Darwin was the first port of call for many early aviators – Sir Charles Kingsford Smith touched down in present-day Parap on his historic flight between Britain and Australia. The Australian Aviation Heritage Centre houses an American B-52 bomber and the wreckage of a Japanese Zero. The city's rich aviation history is detailed at the centre, along with superbly restored exhibits. *557 Stuart Highway; (08) 8947 214; open daily.*

Casuarina Coastal Reserve 370 C1

This reserve boasts a magnificent white sandy beach that backs on to dunes and thickets of native she-oaks. Several hundred metres from Casuarina Beach is Free Beach, Darwin's only recognised nudist bathing area, popular with sunbathers and swimmers during the dry season.

Leanyer Recreation Park 632 D2

This popular family park offers bike-riding, skateboarding and basketball facilities; an all-abilities playground; shaded barbecues and picnic areas; and a water fun park with paddling pools, water cannons and a water playground. *Off Vanderlin Dr; open 8am–8pm daily; admission free.*

Outer North-east

Holmes Jungle Nature Park 632 D3

Off Vanderlin Drive in Karama, this nature park features a monsoonal rainforest full of native fauna and flora. Excellent walking trails meander through the jungle, and the reserve is home to many species of birds, mammals and reptiles. *Open 7am–7pm daily.*

South-east

Charles Darwin National Park 370 D4

Mangroves merge with tropical woodland at this national park, off Tiger Brennan Drive east of the city. There is an excellent view of the city from a lookout on the escarpment, and the park contains a number of Aboriginal shell middens and some World War II bunkers. It also incorporates some of the most pristine mangrove areas in northern Australia, and if accessed by boat there is excellent fishing.

East

Crocodylus Park 632 D3

Crocodylus Park is a research centre as well as a tourist attraction. Crocodiles emerge from the murky waters at feeding times (10am, 12pm and 2pm) to take pieces of meat that are dangled over the side of their enclosures. Well-trained guides allow children – under supervision – to handle baby crocodiles that have their jaws taped shut to avoid any sharp teeth.

Work by scientists at Crocodylus Park helped establish the Northern Territory's ground-breaking policies on crocodile preservation. There are other animals on show including monkeys and large cats, such as tigers. *End of McMillans Rd, past the airport, near the Police Centre; open daily; (08) 8922 4500.*

sport

Darwinites love their sport. **AFL** (Australian Football League) is like a second religion in Darwin and the surrounding communities, particularly for Aboriginal and Torres Strait Islander players, many of whom go on to play for big clubs. In March the NTFL grand final is held at Marrara Stadium and the Tiwi Islands Annual Football Art Sale is held on Bathurst Island – grand-final day is the only time Bathurst Island is open to visitors without a permit. Darwin and Territory teams often play a curtain-raiser to the Australian AFL season. Any match between the Aboriginal All Stars and an AFL team is a must-see game.

Rugby League is well supported and matches are played at Richardson Park in Fannie Bay; **Rugby Union** is played at the headquarters at Marrara.

The Marrara sporting complex is also home to **hockey**, **basketball** and **gymnastics**. Fannie Bay and Darwin Harbour are great for **sailing**. The foreshore is also popular for **recreational fishing**.

The annual Darwin Cup **horseracing** carnival in August draws people from around the country to Darwin's Fannie Bay track.

The Hidden Valley leg of the V8 supercar series in June brings **motorsport** to town. Three days of car racing is turned into a week of celebrations as Top Enders turn out in tens of thousands for this event.

The Arafura Games are held bi-annually with **athletes** competing from all around the world.

NORTHERN TERRITORY

day tours

Howard Springs Nature Park A pleasant, shaded swimming area south of Darwin with many native birds and animals and a crocodile-free pool. For more details see p. 380

Territory Wildlife Park and Berry Springs Nature Park This award-winning park showcases the Top End's flora and fauna. Many exhibits are connected by a 4-kilometre walk or shuttle train route. Berry Springs is next door with a swimming pool and picnic facilities. For more details see p. 380

Window on the Wetlands Looking out over the Adelaide River flood plains, this centre not only has excellent interpretive displays of Top End ecology, but also has excellent views. Nearby, popular jumping crocodile cruises run on the Adelaide River. For more details see p. 380

Mary River The river is a popular fishing area, with places such as Corroboree Billabong and Shady Camp popular for boating (houseboats can be hired at Corroboree) and fishing. But visitors should be careful: crocodiles are plentiful in these areas. For more details see p. 380

Litchfield National Park A two-hour drive south of Darwin, this park is a wonderful place to visit or camp, with bubbling streams and gushing waterfalls that flow year-round. Visitors can take walks, admire the scenic views or take advantage of the picnic areas. While generally crocodile-free, you should still read the signs before venturing into any pools. For more details see p. 380

Kakadu National Park In three hours visitors can be in Jabiru, the centre for exploring the aboriginal rock art, gorges and wetlands of Kakadu National Park. For more details see p. 379

Katherine Gorge Further afield (about a 4-hour drive) are the famous gorges of Katherine River. For more details see p. 381

The Tiwi Islands Bathurst and Melville islands lie 80 kilometres north of Darwin and are the traditional homes of the Tiwi Aboriginal people. 'Tiwi' is said to mean 'the people' or 'we, the chosen people'. Renowned artists, the Tiwi people produce beautiful silk-screened clothing, woven bangles, painted conch shells, carvings, pottery and elaborately decorated Pukumani burial poles. Tiwi Tours run one- and two-day tours to Bathurst Island from Darwin and arrange permits, accommodation and transport via light aircraft. (08) 8922 2720. For more details see p. 380

REGIONS
northern territory

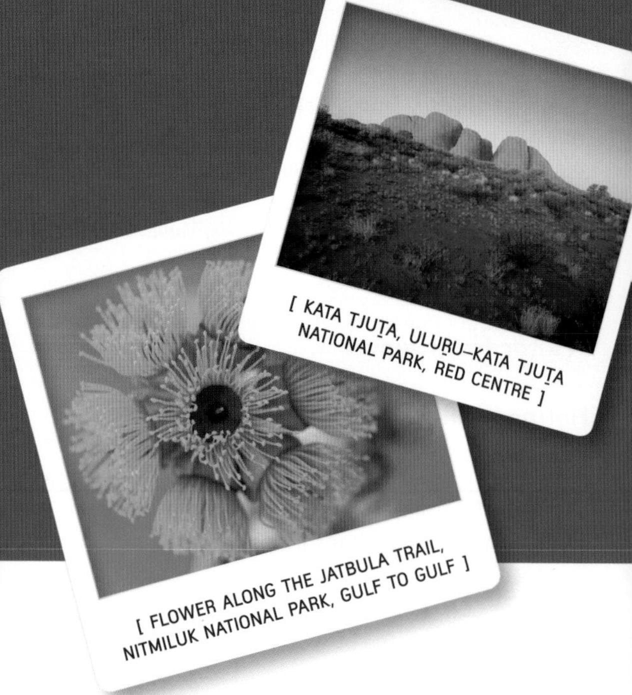
[KATA TJUṮA, ULURU–KATA TJUṮA NATIONAL PARK, RED CENTRE]

[FLOWER ALONG THE JATBULA TRAIL, NITMILUK NATIONAL PARK, GULF TO GULF]

KAKADU & ARNHEM LAND

World Heritage–listed Kakadu National Park protects one-third of the country's bird species and is bordered by Aboriginal-owned Arnhem Land – visitors must apply for a permit before entering.

[JACANA SEARCHING FOR FOOD IN A BILLABONG, KAKADU NATIONAL PARK]

NORTHERN TERRITORY

Cobourg Peninsula Four Iwaidja clan groups are custodians of this peninsula, which is protected by Garig Gunak Barlu National Park. Here visitors can fish and explore the ruins of Victoria Settlement, set up by the British in Port Essington in 1838 to defend the north.

Walks at South Alligator River Towards the end of the dry season, thousands of waterbirds congregate to feed in the Mamukala Wetlands. A short nature trail through the wetlands starts just east of South Alligator River crossing on the Arnhem Highway. The Gungarre Monsoon Rainforest Walk passes through a forest environment.

Yellow Water Yellow Water (Ngurrungurrudjba) is a spectacular wetland area with prolific birdlife, especially in the dry season. Boat tours give visitors a close-up view of the birdlife and the Territory's crocodiles.

Nourlangie Rock This is one of Kakadu's main Aboriginal rock-art areas. The Nourlangie Art Site walk takes visitors through a variety of rock-art styles, including prime examples of Kakadu X-ray art, which shows the anatomy of humans and animals in rich detail.

Sunset at Ubirr Aboriginal people once camped in the rock shelters at Ubirr in Kakadu National Park, leaving the legacy of a spectacular rock-art site. A 1-kilometre circuit covers the main natural galleries, which feature portrayals of extinct animals and animated figures in motion. Another highlight is a shorter side track that climbs steeply to Nardab Lookout, a rocky vantage point with sensational views.

Port Essington
COBOURG PENINSULA
GARIG GUNAK BARLU NP
ARAFURA SEA
VAN DIEMEN GULF
Nhulunbuy
Ubirr
South Alligator R
KAKADU NATIONAL PARK
Jabiru
Nourlangie Rock
Cooinda
ARNHEM LAND ABORIGINAL LAND TRUST
GOVE PENINSULA
GULF OF CARPENTARIA
N
0 50 km
GROOTE EYLANDT

Jim Jim Falls and Twin Falls Jim Jim Falls (Barrkmalam) and Twin Falls (Gungkurdul) are reached via a four-wheel-drive track off the Kakadu Highway. An incredible volume of water cascades over Jim Jim Falls in the wet season, when the falls can only be seen from the air.

TOP EVENTS

MAR	Aurora Kakadu Klash fishing competition (Kakadu National Park)
AUG	Garma Festival of Traditional Culture (Nhulunbuy)
SEPT	Mahbilil Festival (Jabiru)

CLIMATE

JABIRU

J	F	M	A	M	J	J	A	S	O	N	D	
34	33	33	35	33	32	32	34	36	38	37	35	MAX °C
25	24	24	24	22	19	18	19	21	24	25	25	MIN °C
347	332	318	66	11	1	3	4	9	27	158	211	RAIN MM
22	21	20	7	2	0	0	0	1	3	12	16	RAIN DAYS

YOLNGU CULTURE

The Yolngu people have lived in north-east Arnhem Land for over 40 000 years and, despite Australia's European settlement, have managed to keep their culture relatively intact. It was only from the 1930s that Yolngu people had steady contact with Balanda (white people). They were not truly affected by this contact until 1963, when a proposal to develop part of Arnhem Land for mining was put forth. In a famous response, the Yolngu sent their opposing petition to the House of Representatives on a piece of bark. While mining went ahead, the petition now hangs in Parliament House in Canberra, seen as having been instrumental in the fight for Aboriginal land rights. The Yolngu people are well known for their arts and crafts – particularly for their bark paintings – and for their skill at playing the didgeridoo. Most of the members of Australia's most prominent Indigenous band, Yothu Yindi, are of Yolngu descent.

→ For more detail see maps 633 & 634–5. For descriptions of towns see Towns from A–Z (p. 383).

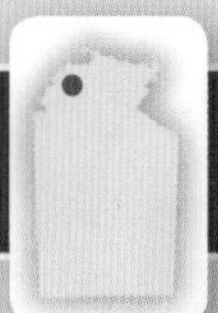

AROUND DARWIN

The attractions within easy reach of Darwin showcase extraordinary geological features and offer a glimpse of the local culture.

[WANGI FALLS, LITCHFIELD NATIONAL PARK]

Tiwi Islands Bathurst and Melville islands, 80 kilometres offshore from Darwin, belong to the Tiwi people, whose unique culture results from their long isolation. The landscape includes escarpments, lakes, waterfalls, beaches and forests, but perhaps the biggest attraction is the renowned Tiwi art and craft.

Territory Wildlife Park This award-winning park showcases northern Australia's native fauna and flora. Features include aviaries, raptor displays, an aquarium tunnel and a large nocturnal house. Nearby, Berry Springs Nature Park has a spring-fed pool – safe for swimming – in natural bushland.

Litchfield National Park Two hours from Darwin, this is a world of waterfalls, gorges, pockets of rainforest, giant termite mounds and rock formations that resemble lost civilisations. Enjoy the scenic pools beneath the waterfalls – most offer crocodile-free swimming, but read the signs before you take the plunge.

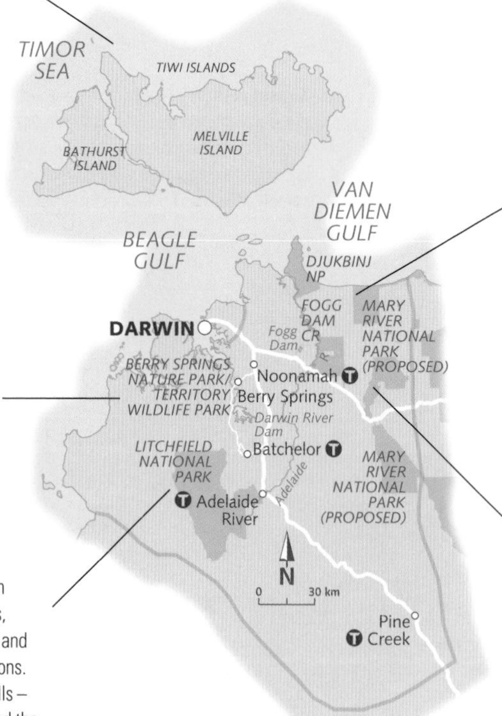

Fogg Dam Conservation Reserve
Fogg Dam was built in the 1950s to provide irrigation to the Humpty Doo Rice Project; the agricultural scheme failed, but the dam became a dry-season refuge for an amazing range of wildlife, especially waterbirds. It's best to visit at sunrise or sunset, when there is the most activity. There are a few lovely walks through the park, such as the Monsoon Forest Walk, a 2.7-kilometre trail that traverses a variety of fascinating habitats; also inquire about ranger-guided nocturnal walks.

Window on the Wetlands Visitor Centre
With superb views over the Adelaide River flood plains, this interpretive centre offers an insight into the Top End's fascinating and unique wetland environments. Interactive displays detail the ecology of the wetlands, as well as local Aboriginal and European history. Time your visit for early morning or late afternoon, when the vistas are at their best.

TOP EVENTS

MAY	Races (horseracing, Pine Creek)
JUNE	Bush Race Meeting (horseracing, Adelaide River)
JUNE	Northern Territory Gold-panning Championships and Didgeridoo Festival (Pine Creek)
JUNE	Gold Rush Festival (Pine Creek)
JUNE	Rodeo and Campdraft (Adelaide River)
JUNE	Rodeo (Pine Creek)

CLIMATE

ADELAIDE RIVER

J	F	M	A	M	J	J	A	S	O	N	D	
32	31	32	33	32	31	30	31	33	33	33	33	MAX °C
25	25	25	24	22	20	19	21	23	25	25	25	MIN °C
429	353	322	103	21	1	1	6	16	73	141	250	RAIN MM
21	20	19	9	2	1	1	1	2	7	12	16	RAIN DAYS

CROCODILES

Crocodiles are both compelling and deadly creatures. Of the two types in northern Australia, the most dangerous is the estuarine crocodile ('saltie'). This well-camouflaged reptile is found out at sea, along the coastline, in creeks, and in rivers up to 100 kilometres from the coast. Never go swimming where salties have been seen. The freshwater crocodile ('freshie') inhabits rivers and lagoons; it is smaller, but can still inflict a serious wound. Both types nest and sun themselves near the water's edge. Always seek local advice before swimming, camping or boating. View these reptiles safely at Darwin Crocodile Farm, 40 kilometres south of the city; Crocodylus Park, an education and research centre in Darwin's north-eastern suburbs; and Crocosaurus Cove in the city centre.

→ For more detail see maps 632–3 & 634. For descriptions of Ⓣ towns see Towns from A–Z (p. 383).

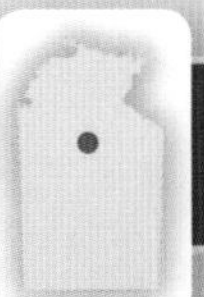

GULF TO GULF

Experience the natural riches of the Territory away from the crowds. Some of this region's attractions are well known, while others are scarcely on the map.

[RUGGED SANDSTONE FORMATIONS, KEEP RIVER NATIONAL PARK]

NORTHERN TERRITORY

Nitmiluk (Katherine Gorge) National Park
The traditional home of the Jawoyn people, this park is world-renowned for its 13 stunning gorges, carved from red sandstone over a period of 20 million years. Visitors can navigate the gorges in canoes, take a boat tour with a Jawoyn guide, swim in the pools or explore the 100-kilometre network of walking tracks.

Never Never country Jeannie Gunn wrote We of the Never Never (1908) after living at Elsey Station. Now you can visit a faithful replica of Gunn's home on the property of Mataranka Homestead. Nearby in Elsey National Park visitors can enjoy the Mataranka thermal pool, which pumps out 20 million litres of water daily.

JOSEPH BONAPARTE GULF
NITMILUK (KATHERINE GORGE) NATIONAL PARK
Nitmiluk (Katherine) Gorge
GULF OF CARPENTARIA
Daly River
Katherine
Roper R
Roper Bar
Mataranka
ELSEY NATIONAL PARK
KEEP RIVER NATIONAL PARK
GREGORY NATIONAL PARK
LIMMEN NATIONAL PARK (PROPOSED)
Timber Creek
Daly Waters
GREGORY NATIONAL PARK
Victoria River
River
Borroloola
McArthur
Cape Crawford
Elliott
Lake Woods
Tarrabool Lake
Corella Lake
Lake Sylvester
WESTERN AUSTRALIA
QUEENSLAND
N
0 80 km

Keep River National Park
This remote park includes the traditional land of the Miriwoong and Kadjerong peoples and contains many important rock-art sites, including the accessible Nganalam site. A major attraction is the rugged sandstone formations, similar to the Bungle Bungles over the Western Australian border. There are designated camping areas and good walking tracks.

Eastern frontiers The area adjoining the Gulf of Carpentaria is popular with four-wheel-drive travellers seeking new frontiers. The main settlement, Borroloola, is a popular spot for barramundi anglers and offers access to the waters around Barranyi (North Island) National Park.

Lost City The Lost City is an eerie landscape of free-standing sandstone pillars rising from the plains and suggesting a long forgotten civilisation. While there is a great walk among the towering formations, access is for experienced four-wheel drivers only, but there's always the much more civilised option of a helicopter ride from Cape Crawford.

TOP EVENTS

EASTER	Barra Classic (Borroloola)
APR–MAY	Fishing competitions (Timber Creek)
MAY	Katherine Cup (horseracing, Katherine)
MAY	Back to the Never Never Festival (Mataranka)
AUG	Flying Fox Festival (Katherine)
AUG	Rodeo (Mataranka)

CLIMATE

KATHERINE

J	F	M	A	M	J	J	A	S	O	N	D	
35	34	35	34	32	30	30	33	35	38	38	37	MAX °C
24	24	23	20	17	14	13	16	20	24	25	24	MIN °C
234	215	161	33	6	2	1	1	6	29	89	195	RAIN MM
15	14	10	2	1	0	0	0	1	3	8	12	RAIN DAYS

BARRAMUNDI FISHING

Barramundi, Australia's premier native sport fish, is nowhere as prolific or accessible to anglers as it is in the Northern Territory. There is good barra fishing along both east and west coasts. The rugged west features tropical wetlands and a network of waterways. Popular spots include the Daly River (known for the size of its barramundi and its fishing lodges), Nitmiluk (Katherine Gorge) National Park and Victoria River via Timber Creek. The numerous eastern rivers are smaller, with minimal tides but very clear water. Top spots east are Borroloola on the McArthur River and Roper Bar on the Roper River.

→ For more detail see maps 634–5 & 636–7. For descriptions of Ⓣ towns see Towns from A–Z (p. 383).

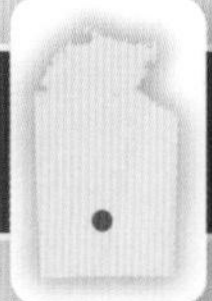

RED CENTRE

The Red Centre is Australia's geographical, scenic and mythical heart, with spectacular landforms, deserts, blue skies and a monumental sense of scale.

[DEVILS MARBLES (KARLU KARLU)]

Devils Marbles (Karlu Karlu) This is one of the most iconic outback destinations, where massive red and orange boulders sit precariously atop each other in a broad desert playground; legend has it that they are the eggs of the Rainbow Serpent. A network of informal tracks meanders through the rock formations.

Kings Canyon Within Watarrka National Park, spectacular Kings Canyon has sandstone walls rising to 270 metres. A 6-kilometre return trail scales the side of the canyon and leads past beehive formations to the Garden of Eden.

Kata Tjut**a** Uluṟu's sister formation; comprising 36 magnificently domed and coloured shapes, Kata Tjuṯa means 'many heads'. The three-hour Valley of the Winds walk winds through the crevices and gorges of the rock system.

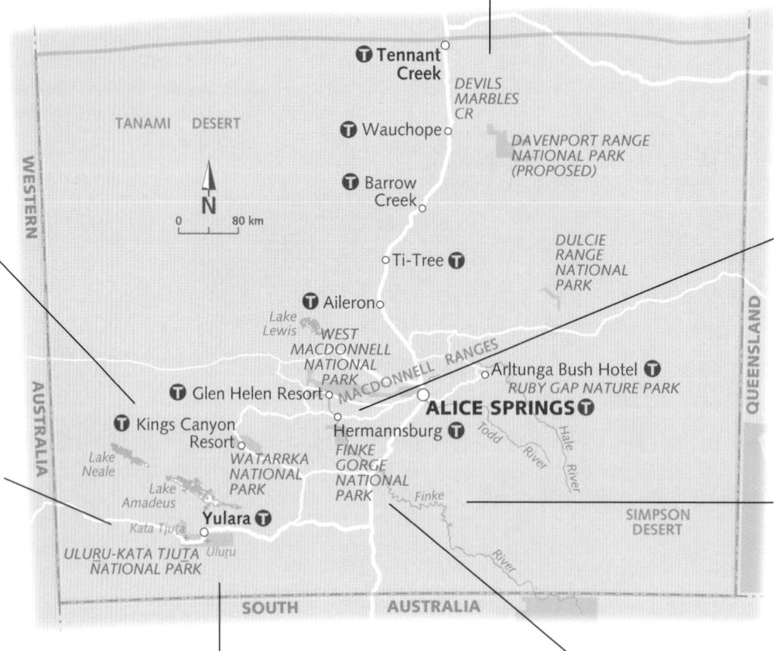

Hermannsburg (Ntaria) This Aboriginal-owned site was established as a German Lutheran mission in 1877. Today the historic precinct displays preserved buildings from that time, as well as works of art from the acclaimed Hermannsburg School, of which Albert Namatjira is the most famous artist.

Simpson Desert and Chambers Pillar The world's largest sand-dune desert was formed around 18 000 years ago after the continent's central lakes dried up. The main attraction is Chambers Pillar, a sandstone obelisk towering 50 metres above the plain.

Exploring Uluṟu The A<u>n</u>angu prefer tourists not to climb Uluṟu. There are four guided walks as an alternative: a 9.4-kilometre walk around the base; the Mala walk to art sites; the Kuniya walk, which introduces creation stories; and the Liru walk, explaining the use of bush foods and materials, such as quandong fruit.

Finke Gorge National Park This park has four-wheel-drive access only. Its unique feature is Palm Valley, a 10 000-year-old oasis with 3000 red fan palms. The Finke River has maintained its course for over 100 million years and is possibly the world's oldest river.

TOP EVENTS

MAY	Bangtail Muster (Alice Springs)
JULY	Camel Cup (Alice Springs)
JULY	Alice Springs Show (Alice Springs)
AUG	Tennant Creek Cup (horseracing, Tennant Creek)
AUG	Henley-on-Todd Regatta (Alice Springs)
AUG–SEPT	Desert Harmony Festival (Tennant Creek)

CLIMATE

ALICE SPRINGS

J	F	M	A	M	J	J	A	S	O	N	D	
36	35	33	28	23	20	20	23	27	31	34	35	MAX °C
21	21	17	13	8	5	4	6	10	15	18	20	MIN °C
38	44	33	18	19	15	14	10	9	22	28	38	RAIN MM
5	5	3	2	3	3	3	2	2	5	6	6	RAIN DAYS

THE STORY OF ULUṞU

Uluṟu lies in the territory of the A<u>n</u>angu people. European explorer William Gosse named it Ayers Rock in 1873. Along with The Olgas (now Kata Tjuṯa) and surrounding land it became a national park in 1958, and was returned to its traditional owners and gazetted as Uluṟu. The rock is Australia's most identifiable natural icon. It is a massive, red, rounded monolith rising 348 metres above the plain, sitting 863 metres above sea level, and reaching 6 kilometres below the earth's surface. Uluṟu's circumference measures 9.4 kilometres. It has no joints – so despite its valleys, fissures and caves, it is a true monolith. Uluṟu attracts tourists because of its size and singularity. For the A<u>n</u>angu, however, the rock is not a single spiritual object but a thing of many parts; Uluṟu and Kata Tjuṯa were laid down during the Tjukurpa (creation period) and are joined by the iwara (tracks) of the ancestral beings that created all the land.

→ For more detail see maps 636–7, 638–9 & 640–1. For descriptions of T towns see Towns from A–Z (p. 383).

TOWNS A–Z
northern territory

LEGEND

- VISITOR INFORMATION
- RADIO STATIONS
- IN TOWN
- WHAT'S ON
- WHERE TO EAT
- WHERE TO STAY
- NEARBY

Food and accommodation listings in town are ordered alphabetically with places nearby listed at the end

[LOST CITY, CAPE CRAWFORD]

Adelaide River

Pop. 188
Map ref. 632 E8 | 634 D6

Adelaide River Inn, Stuart Hwy; (08) 8976 7047.

 98.9 FM ABC Territory Radio

Adelaide River is a small town located near the headwaters of the river of the same name, known for its large population of saltwater crocodiles. A settlement was established here as a base for workers on the Overland Telegraph Line in the early 1870s. The population was boosted by gold discoveries at Pine Creek and the subsequent building of the Northern Australia Railway, which operated from 1888 to 1976. During World War II the relatively sheltered Adelaide River was a major military base for 30 000 Australian and US soldiers.

Adelaide River Inn The hotel was a favourite watering hole of soldiers during WW II and still has war photographs and memorabilia adorning the walls. Hard to ignore is the main attraction, Charlie the Buffalo, suitably stuffed, from the film *Crocodile Dundee*. Meals and accommodation are available. Stuart Hwy.

Adelaide River Railway Station: National Trust–classified station (1888), now a museum featuring relics of local history including the railway construction and WW II; open Apr–Sept, other times by appt; Stuart Hwy; (08) 8976 7101.

Rodeo and Campdraft: June. ***Bush Race Meeting:*** June and Aug.

Adelaide River Inn: good pub food; 106 Stuart Hwy; (08) 8976 7047.

Mount Bundy Station: historic cattle station; Haynes Rd; (08) 8976 7009. ***Daly River Mango Farm:*** rustic accommodation; off Port Keats Rd, Daly River; (08) 8978 2464 or 1800 000 576. ***Perry's on the Daly:*** camping ground with 1 room and a house; Mayo Park, via Daly River Crossing; (08) 8978 2452. ***Wooliana on the Daly:*** group accommodation; Woolianna Rd, Daly River; (08) 8978 2478.

Daly River Copper was discovered here in 1883 and mining began in 1884, which led to a bloody conflict between miners and Aboriginal people. Mining ceased about 1909. Daly River is now a sleepy riverside town with a roadside inn and local Aboriginal arts and crafts. It is said that the country's best barramundi are caught in the Daly River and the area teems with crocodiles and other reptiles, buffalo, cockatoos and wild pigs. The Merrepen Arts Festival is held here each June. 102 km SW.

War Cemetery This is Australia's only war cemetery on native soil. There are graves of 434 military personnel and 54 civilians killed in the 1942 Japanese air raids, along with memorials for those lost in Timor and other northern regions. 1 km E.

Mt Bundy Station: rural experience with fishing, walking, swimming and a wide range of accommodation; 3 km NE. ***Robin Falls:*** pleasant 15 min walk to falls that flow most of the year; 15 km S. ***Snake Creek Armament Depot:*** wartime military base and weapon storage area with 40 buildings still intact; 17 km N. ***Grove Hill Heritage Hotel and Museum:*** example of outback ingenuity made from recycled materials in a fashion designed to withstand the harsh weather of the Top End. Hotel offers meals and accommodation; Goldfields Rd; around 70 km SE.

See also AROUND DARWIN, p. 380

ALICE SPRINGS

Pop. 21 619

Map ref. 639 I7 | 641 J3

Central Australian Tourism Industry Association, 60 Gregory Tce; (08) 8952 5800 or 1800 645 199; www.centralaustraliantourism.com

96.9 Sun FM, 783 AM ABC Territory Radio

Located on the Todd River in the MacDonnell Ranges, Alice Springs is almost 1500 kilometres from the nearest capital city. In 1871 Overland Telegraph Line surveyor William Whitfield Mills discovered a permanent waterhole just north of today's city. Mills named the water source after Alice Todd, wife of South Australian Superintendent of Telegraphs Sir Charles Todd. A repeater station was built on the site. In 1888 the South Australian government gazetted a town 3 kilometres to the south. It was called Stuart until 1933, when the name Alice Springs was adopted. Supplies came to the slow-growing settlement by camel train from Port Augusta. The railway line from Adelaide, known as the Ghan after the original Afghan camel drivers, was completed in 1929. Today 'the Alice' is an oasis of modern civilisation in the middle of a vast and largely uninhabited desert, made all the more likeable by not taking itself too seriously (as some of its annual events testify).

Araluen Cultural Precinct This precinct includes some of Alice Springs' best cultural attractions. At the Araluen Centre are 3 galleries of Aboriginal art, with an emphasis on work from the central desert, and a magnificent stained-glass window by local artist Wenten Rubuntja. The Museum of Central Australia offers an insight into the geological and natural history of the Red Centre, with interpretive displays and impressive fossils. Also in the precinct are the Strehlow Research Centre, Craft Central, the Central Australian Aviation Museum, several memorials and the Yeperenye Sculpture, depicting a Dreamtime caterpillar. One ticket covers the entry to each attraction. Cnr Larapinta Dr and Memorial Ave; (08) 8951 1120.

Alice Springs Desert Park Sir David Attenborough was so impressed by this desert park that he proclaimed, 'there is no museum or wildlife park in the world that could match it'. The park invites visitors to explore the arid lands and the relationship between its plants, animals and people. A walking trail leads through 3 habitats: Desert Rivers, Sand Country and Woodland. There are films, interactive displays, free audio-guides, guided day tours, nocturnal tours to experience the Central Australian desert at night (optionally with dinner) and talks about flora, fauna and the ability of Aboriginal people to survive in such harsh conditions. The park has a spectacular desert nocturnal house with native marsupials such as the bilby and mala, and there are free-flying birds of prey in twice daily shows. Larapinta Dr; (08) 8951 8788.

John Flynn Memorial Uniting Church: in memory of the founder of Royal Flying Doctor Service and Australian Inland Mission; Todd Mall. (Flynn's grave can be found 1 km beyond the entrance to Alice Springs Desert Park.) ***Adelaide House:*** originally a hospital designed by Rev Flynn, now a museum displaying pedal-radio equipment he used and other artefacts and photographs; open 10am–4pm Mon–Fri and 10am–12pm Sat Mar–Nov; Todd Mall. ***Sounds of Starlight Theatre:*** musical journey through Central Australia; open Apr–Nov; Todd Mall; (08) 8953 0826. ***Alice Springs Reptile Centre:*** houses the largest reptile display in Central Australia, including goannas, lizards, pythons and some of the world's most venomous snakes. There's also Terry, the saltwater crocodile; and a new Cave Room featuring a live gecko exhibit; open daily; Stuart Tce; (08) 8952 8700. ***Royal Flying Doctor Service Base:*** operational base since 1939, with daily tours and presentations and an interactive museum including a pedal radio; Stuart Tce; (08) 8952 1129. ***National Pioneer Women's Hall of Fame:*** national project dedicated to preserving women's place in Australia's history; open Feb–mid-Dec; Old Alice Springs Gaol, 2 Stuart Tce. ***Old Stuart Town Gaol:*** the harsh but functional design of stone and timber reflects the community attitude to prisoners at the time of construction (1907); open 10am–12.30pm Mon–Sat; Parsons St; (08) 8952 4516. ***Anzac Hill:*** the most visited landmark in Alice Springs, this war memorial features panoramic views of town and the MacDonnell Ranges; Anzac Hill Rd. ***Stuart Town Cemetery:*** original town cemetery with graves of the earliest pioneers dating from 1889; George Cres. ***Alice Springs School of the Air:*** the first of its kind in Australia with a classroom size of 1.3 million sq km. See interpretive displays and hear lessons being broadcast; Head St; (08) 8951 6834. ***Old Hartley Street School:*** originally constructed in 1930 and closed in 1965, the school was added to several times as the population surged, reflecting the changing styles and requirements of school design through different periods; open 10am–2.30pm Mon–Fri; Hartley St; (08) 8952 4516. ***Olive Pink Botanic Gardens:*** named after an anthropologist who worked with Central Desert people, this is Australia's only arid-zone botanic garden. Covering 16 ha, it contains over 300 Central Australian plant species in simulated habitats such as sand dunes, woodlands and creeks; Cnr of Barrett Dr and Tuncks St; (08) 8952 2514. ***The Residency:*** grand home completed in 1927 and housing Central Australia's regional administrator until 1973, it has welcomed many VIPs, including Queen Elizabeth II, and now showcases local history; Cnr Parsons and Hartley St. ***Aboriginal cultural experiences, art galleries and artefacts:*** Alice Springs is renowned for its rich aboriginal culture and art. Tours and galleries plus ceremonial dances and didgeridoo playing are all available. Obtain more information from the visitor centre.

Alice Springs Council Market: Todd Mall; contact city council (08) 8950 0505 for dates and times. ***Heritage Festival:*** Apr. ***Racing Carnival:*** Apr/May. ***Bangtail Muster:*** street parade and sports day; May. ***Finke Desert Race:*** car and motorbike racing; June. ***Beanie Festival:*** June. ***Camel Cup:*** July. ***Show:*** cooking, crafts and camels; July. ***Henley-on-Todd Regatta:*** boats wheeled or carried along a dry riverbed; Aug. ***Masters Games:*** mature-age athletics carnival, even-numbered years; Oct. ***Desert Festival:*** Arts festival; Sept. ***Christmas Carnival:*** Dec.

The Juicy Rump: part of Lasseters Casino complex, this bar offers 17 beers on tap and an impressive menu of surf and turf; Barrett Dr; (08) 8950 7777. ***Alice Vietnamese Restaurant:*** authentic Vietnamese; Heffernan Rd; (08) 8952 0745. ***Casa Nostra Pizza Bar & Spaghetti House:*** authentic Italian; cnr Undoolya Rd and Sturt Tce; (08) 8952 6749. ***Hanuman Alice Springs:*** excellent modern Asian; Crowne Plaza, 82 Barrett Dr;

[BUSH TUCKER DISPLAY, ALICE SPRINGS DESERT PARK]

NORTHERN TERRITORY

(08) 8953 7188. ***Red Ochre Grill Restaurant:*** contemporary and native cuisine; Leichhardt Tce; (08) 8952 9614. ***Bojangles Saloon and Dining Room:*** good atmosphere, in the centre of town; Todd St; (08) 8952 2873. ***Oscar's Cafe & Restaurant:*** Portuguese and Spanish; 86 Todd Mall; (08) 8953 0930. ***Barra on Todd Restaurant and Bar:*** part of the Chifley Alice Springs Resort, this stylish, modern Australian restaurant usually has two or three barramundi dishes on the menu as well as a wide choice of meat and vegetarian dishes; Stott Tce; (08) 8951 4545. ***The Overlanders Steakhouse:*** dares diners to try the 'Drover's Blowout', a multi-course meal of soup, emu, crocodile, camel, barramundi, beef, kangaroo and dessert; 72 Hartley Tce; (08) 8952 2159.

Alice on Todd: modern self-contained apartments; Cnr South Tce and Strehlow St; (08) 8953 8033. ***Chifley Alice Springs Resort:*** low-set quality accommodation; 34 Stott Tce; (08) 8951 4545. ***Cavenagh Lodge B&B:*** private, comfortable B&B; 4 Cavenagh Cres; (08) 8952 2257. ***Lasseters Hotel Casino:*** modern rooms in great setting; 93 Barrett Dr; (08) 8950 7777 or 1800 808 975. ***The Swagmans Rest Motel:*** budget self-contained rooms; 67–69 Gap Rd; (08) 8953 1333 or 1800 089 612. ***White Gum Motel:*** comfortable rooms close to CBD; 17 Gap Rd; (08) 8952 5144 or 1800 624 110. ***Bond Springs Station:*** historic cattle station; Bond Springs; (08) 8952 9888. ***Ooraminna Station Homestead:*** historic cattle station; Old South Rd; (08) 8953 0170. ***Ross River Resort:*** caravan park; Ross Hwy; (08) 8956 9711 or 1800 241 711.

West MacDonnell National Park The majestic MacDonnell Ranges, once higher than the Himalayas, were formed over 800 million years ago. Over time the ancient peaks have been dramatically eroded so that what remains is a spectacular environment of rugged gorges, hidden waterholes, remnant rainforest and an unexpectedly large number of animal species and diversity of flora. The jewel of the ranges is Ormiston Gorge, one of the most beautiful in Australia, while the Ochre Pits are a natural quarry once mined by Aboriginal people for ochre which they used in rock art, painting and ceremonial decoration. The Larapinta Trail is a huge 230 km walk divided into 12 sections covering the major sites of the ranges and can be walked separately. Contact the Parks & Wildlife Commission of the Northern Territory for maps and tour options on (08) 8951 8250. 17 km W

Emily and Jessie Gaps Nature Park Located in the East MacDonnells, Emily Gap (13 km E) and Jessie Gap (18 km E) both contain Aboriginal rock art and are important spiritual sites to the Eastern Arrernte people. The caterpillar beings of Mparntwe (Alice Springs) originated where Emily Gap lies today. They formed Emily Gap and other topographical features around Alice Springs and then spread across to the edge of the Simpson Desert. Both gaps are popular barbecue and picnic places. (08) 8951 8250.

Alice Springs Telegraph Station Historical Reserve: area protecting original stone buildings and equipment, with historical display, guided tours, bushwalking and wildlife (Alice

ALICE SPRINGS

continued from previous page

Springs waterhole is located here); 3 km N. ***Old Timers Traeger Museum:*** unique museum set in a retirement home and run by volunteer residents. It displays photographs and memorabilia from the early days of white settlement in Central Australia; 5 km S. ***Pyndan Camel Tracks:*** ride a camel through the picturesque Ilparpa Valley alongside the Western MacDonnell Ranges; 0416 170 164; 15 kms. ***National Road Transport Hall of Fame:*** impressive collection of old trucks, cars and motorbikes; 8 km S. ***Ghan Railway Museum:*** re-creation of a 1930s railway siding featuring the Old Ghan, which runs on 8 km of private line between MacDonnell Siding and Mt Ertiva; 10 km S. ***Tropic of Capricorn Marker:*** bicentennial project marking the Tropic of Capricorn; 30 km N. ***Ewaninga Rock Carvings Conservation Reserve:*** soft sandstone outcrops form natural galleries of sacred Aboriginal paintings. Custodians request that visitors do not climb on rocks or interfere with the paintings; 35 km SE. ***Corroboree Rock Conservation Reserve:*** significant Eastern Arrernte site for ceremonial activities with a short walk and information signs; 43 km E. ***Standley Chasm:*** spectacular narrow, sheer-sided gorge that is particularly striking at midday when sun lights up the rocks. An attractive 1 km creek walk leads to the chasm and picnic facilities; 50 km W. ***Rainbow Valley Conservation Reserve:*** stunning freestanding sandstone cliffs that change colour at sunrise and sunset. Access by 4WD only; 101 km S. ***Mud Tank zircon field:*** prospecting for zircons, guided fossicking tours and gem-cutting at the caravan park; 135 km NE. ***Henbury Meteorites Conservation Reserve:*** contains 12 craters created when meteorites (comprising 90% iron) crashed to earth 4700 years ago. The largest of the meteorites was over 100 kg and is now at the Museum of Central Australia in Alice Springs; 147 km SW. ***Chambers Pillar Historical Reserve:*** Chambers Pillar, named by John McDouall Stuart in 1860, is a 40 m high solitary rock pillar left standing on a Simpson Desert plain after 340 million years of erosion. It served as a landmark for early pioneers and explorers and is best viewed at dawn or dusk; 149 km S. ***Mereenie Loop:*** this unsealed road links the West MacDonnell Ranges, Hermannsburg, Glen Helen and Palm Valley with Watarrka National Park (Kings Canyon). A permit is required because a section of the route passes through Aboriginal land; obtain a map and permit from the visitor centre. ***Tours:*** experience scenic attractions and Aboriginal culture by foot, bus or coach, train, limousine, 4WD safari, Harley-Davidson motorcycle, camel, horse, aeroplane, helicopter or hot-air balloon; brochures at visitor centre.

See also **RED CENTRE, p. 382**

Aileron

Map ref. 638 H5

Aileron Roadhouse, Stuart Hwy; (08) 8956 9703.

The highlight of the year at this popular rest stop on the Stuart Highway is the annual cricket match between local grape farmers and government officials from the Department of Primary Industries. There are even attendant seagulls, which are made by sticking parts of packing-cases onto sticks and painting them white.

Aileron Hotel Roadhouse The roadhouse is a welcome oasis with a swimming pool, Aboriginal art, playground and picnic and barbecue facilities. A 17 m sculpture of Charlie Quartpot, a local rainmaker, stands beside the roadhouse which offers meals, including a roast for Sunday lunch, and accommodation. There is also an adjoining campsite. Stuart Hwy.

Cricket Match: dates from visitor centre.

Aileron Hotel and Roadhouse: modern roadhouse rooms; Stuart Hwy; (08) 8956 9703.

Ryans Well Historical Reserve The reserve is named after Ned Ryan, a 19th-century stonemason and bushman who was an expert at sinking wells. He accompanied John McKinlay on his ill-fated exploration of Arnhem Land in 1866. When they became trapped on the East Alligator River during the wet season, Ryan and another bushman fashioned a raft out of the skins of 27 pack horses to negotiate their escape. Ryans Well was hand-dug in 1889 as part of an attempt to encourage settlement in the NT. Today there is a plaque beside the well explaining the process of raising water. 7 km SE.

Glen Maggie Homestead: ruins of 1914 homestead, once used as a telegraph office but abandoned in 1934; 7 km SE. ***Native Gap Conservation Reservation:*** sacred Aboriginal site with picnic area surrounded by cypress pines and magnificent views of the Hahn Range; 12 km SE.

See also **RED CENTRE, p. 382**

Arltunga Bush Hotel

Map ref. 639 K6 | 641 N1

Arltunga Historical Reserve Visitor Centre, (08) 8956 9770; or Arltunga Bush Hotel, (08) 8956 9797.

Arltunga was named after a subgroup of the Arrernte people who had lived in the area for at least 22 000 years before Europeans arrived.When gold was discovered in 1887, prospectors travelled 600 kilometres from the Oodnadatta railhead, often on foot, to get there. When the gold ran out, the people left, and the only remaining signs of life today are the hotel (closed 1 December to 1 March) and adjoining campground. Ruins of the town have been well preserved by the dry climate.

N'Dhala Gorge Nature Park The shady gorge features extensive walking tracks that provide access to a large number of rock-art sites with carvings and paintings, including more than 6000 petroglyphs. The carvings are of 2 distinct types – finely pecked and heavily pounded – and are thought to represent 2 different periods. There are also rare plants in the park such as the peach-leafed poison bush and the undoolya wattle. Access by 4WD only. 53 km SW.

Arltunga Historical Reserve: ruins of Arltunga, some of which have been restored, including police station and gaol, mines and the Government Battery and Cyanide Works (1896). The visitor centre has displays on the history of the town; behind the hotel. ***Ruby Gap Nature Park:*** the site of Central Australia's first mining rush, in 1886, when rubies were thought to have been found. After the stones proved to be relatively valueless garnets the boom went bust and people left. The stunning gorges are now popular sites for bushwalking and camping. Track closures possible Nov–Apr due to wet conditions; 37 km E. ***Trephina Gorge Nature Park:*** known for its sheer quartzite cliffs, red river gums and sandy creek bed. Several waterholes attract native wildlife and provide a beautiful setting for bushwalking, swimming, camping and picnicking; 43 km W.

See also RED CENTRE, p. 382

Barrow Creek

Pop. 11
Map ref. 639 I2

 **Barrow Creek Hotel; (08) 8956 9753.**

 107.3 FM ABC Territory Radio

Although its appearance today is that of a small wayside stop on the highway, Barrow Creek was originally an important telegraph station. It was also the site of an 1874 punitive expedition against the Kaytej people by police after a telegraph station master and linesman were killed during an assault by 20 Kaytej men. This attack (thought to have been a reaction to the abuse of an Aboriginal woman) is the only known planned attack on staff of the Overland Telegraph Line.

Old Telegraph Station This restored stone building (1872) is set against the breathtaking backdrop of the Forster Ranges. It is 1 of 15 telegraph stations that formed the original network from Port Augusta to Port Darwin. A key is available from the hotel for those who wish to look inside. Stuart Hwy.

Barrow Creek Hotel: outback pub (1932), with original bar, cellar and tin ceilings, and memorabilia of the area on display; Stuart Hwy. ***Cemetery:*** graves of early settlers and local characters; information available at the hotel.

See also RED CENTRE, p. 382

Batchelor

Pop. 477
Map ref. 632 E7 | 634 D5

 **Information booth, Tarkarri Rd; (08) 8976 0045.**

 93.7 8KIN FM, 98.9 FM ABC Territory Radio

Established as a large air force base during World War II, Batchelor became prominent with the discovery of uranium at nearby Rum Jungle in 1949. Today Batchelor thrives on tourism, in particular, because of its close proximity to the increasingly popular Litchfield National Park.

Karlstein Castle Czech immigrant Bernie Havlik worked at Rum Jungle and was a gardener there until his retirement in 1977. As a gardener he had been frustrated by a rocky outcrop that was too large to move and too difficult to keep tidy, so he decided to build over it. Havlik spent 5 years constructing a mini replica of the original Karlstein Castle that still stands in Bohemia. He added finishing touches, despite serious illness, and died just after completion. Rum Jungle Rd.

Coomalie Cultural Centre: bush tucker garden plus display of Aboriginal works and culture; Batchelor Institute, cnr Awilla Rd and Nurudina St.

Lingalonga Festival: cultural festival; Aug/Sept.

Batchelor Butterfly Cafe: good barra and buffalo; 8 Meneling Rd; (08) 8976 0199. ***Monsoon Cafe:*** tasty burgers and cheesecake; Litchfield Park Rd; (08) 8978 2077.

Historic Retreat: restored 1950s guesthouse; 19 Pinaroo Cres; (08) 8976 0554. ***Lake Bennett Wilderness Resort:*** stunning resort set in 125 ha of tropical bushland on the shore of an 81 ha freshwater lake; Chinner Rd, Winnellie; (08) 8976 0960 or 1800 999 089. ***Rum Jungle Bungalows:*** bush cabins; 10 Meneling Rd; (08) 8976 0555.

Litchfield National Park Sandstone formations and monsoon rainforest feature in this easily accessible park. There are fantastic swimming spots: plunge into the rainforest-fringed pool at Wangi Falls, swim beneath the cascading Florence Falls or relax in the waters of Buley Rockhole. Do not miss the Lost City (access by 4WD only), a series of windswept formations that eerily resemble the ruins of an ancient civilisation, and the magnetic termite mounds, so called because they all align north–south. (08) 8976 0282; 20 km W.

Lake Bennett Wilderness Resort: this resort has a plethora of birdlife and native fauna that can be enjoyed in the many serene picnic areas. A range of activities are available including abseiling, swimming and barramundi fishing. Dining and accommodation also on-site; Chinner Rd; (08) 8976 0960; 18 km NE. ***Batchelor Air Charter:*** scenic flights, parachuting and gliding; Batchelor Aerodrome; 1 km W. ***Rum Jungle Lake:*** canoeing, kayaking, diving and swimming; 10 km W.

See also AROUND DARWIN, p. 380

Borroloola

Pop. 775
Map ref. 635 M11 | 637 M2

Lot 384, Robinson Rd; (08) 8975 8799.

102.9 8MAB FM, 106.1 FM ABC Territory Radio

Located beside the McArthur River, Borroloola is a small settlement with a chequered past. As a frontier town and port in the 19th century it was a base for rum smuggling from Thursday Island. It became known as the home of criminals, murderers and alcoholics, a reputation it lost only when it was virtually deserted in the 1930s. Today Borroloola's fortunes have vastly improved and it is popular with fishing and four-wheel-drive enthusiasts. Although it lies on Aboriginal land, no permit is required to enter.

Police Station Museum: built in 1886, the museum is the oldest surviving outpost in the NT and displays memorabilia and photographs that illustrate the town's history; Robinson Rd. ***Cemetery:*** pioneer graves; Searcy St and other scattered sites; map from visitor centre.

Barra Classic: Easter. ***Rodeo:*** Aug.

Heartbreak Hotel Cape Crawford: budget rooms with pub food; Cnr Carpentaria and Tablelands hwys, Cape Crawford; (08) 8975 9928. ***King Ash Bay Lodge:*** stopover between WA and Queensland; Batten Point Rd, King Ash Bay; (08) 8975 8998 or (08) 8975 9650.

Cape Crawford This charming town is an excellent base for exploring Bukalara rocks (60 km E) and the Lost City (20 km SE). Bukalara rocks are a mass of chasms winding through ancient sandstone structures (the area is very remote and a guide is recommended). The Lost City is a collection of sandstone turrets, domes and arches formed by water seeping through cracks and eroding the sandstone. It is an important Aboriginal ceremonial site and accessible only by air or 4WD. Flights can be arranged from Cape Crawford. 110 km SW.

King Ash Bay: popular fishing spot year-round; 40 km NE. ***Caranbirini Conservation Reserve:*** protects weathered rock escarpments and a semi-permanent waterhole rimmed by riverine vegetation and open woodland, along with many native birds. Surrounding the waterhole are 25 m sandstone spires providing a vivid contrast in colour and shape to the surrounding countryside; 46 km SW. ***Barranyi (North Island) National Park:*** sun-drenched wilderness with long sandy beaches and excellent angling. No permit is required, but visitors are requested to register with the Borroloola Ranger Station. Offshore; (08) 8975 8792; 70 km NE. ***Lorella Springs:*** campground and caravan park at thermal springs. The road is unsealed and accessible only during the dry season; 170 km NW. ***Limmen Bight River Fishing Camp:*** ideal conditions for barramundi and mud crabs. Accommodation is available but check road access in wet season; 250 km NW. ***Fishing charters:*** river and offshore fishing trips; bookings at visitor centre. ***Scenic flights:*** over town and the islands of the Sir Edward Pellew Group; brochure from visitor centre.

See also **GULF TO GULF, p. 381**

Daly Waters

Pop. 20
Map ref. 634 H12 | 636 H3

 **Daly Waters Pub, Stuart St; (08) 8975 9927.**

 106.1 FM ABC Territory Radio

The nearby springs, from which this tiny settlement takes its name, were named after the governor of South Australia by John McDouall Stuart during his south–north crossing of Australia in 1862. The town's size belies its historical importance as the first international refuelling stop for Qantas in 1935.

Daly Waters Pub Known as one of the great authentic Australian pubs with characters to match. The knickers donated by patrons and stapled to the beam above the bar have all been removed. Accommodation and meals are available with the menu including 'bumnuts on toast' and 'ambuggers'. Stuart St.

Campdraft: Sept.

Daly Waters Aviation Complex: display on local aviation in the oldest hangar in the NT (1930). Key available from Daly Waters Pub; off Stuart Hwy; 1 km NE. ***Tree:*** marked with an 'S', reputedly by explorer John McDouall Stuart; 1 km N. ***Dunmarra:*** roadhouse and accommodation; 44 km S. ***Historic marker:*** commemorates the joining of north and south sections of the Overland Telegraph Line; 79 km S. ***Larrimah:*** remains of historic WW II building at Gorrie Airfield and museum with relics of the local transport industry and WW II; 92 km N.

See also **GULF TO GULF, p. 381**

Elliott

Pop. 353
Map ref. 637 I5

Elliott Hotel, Stuart Hwy; (08) 8969 2069.

 102.9 8KIN FM, 105.3 FM ABC Territory Radio

Elliott is a shady one-street town that is used as a cattle service stop. Stock being transported from the north are given a chemical tick bath here to prevent infection of herds in the south. The town was named after Lieutenant Snow Elliott, the officer in charge of an army camp on this site during World War II.

Nature Walk: interpretive signs introduce native flora and its traditional Aboriginal uses; details from visitor centre.

Mardi Gras: Apr.

Renner Springs This roadside stop on the Stuart Hwy is thought of as the place where the tropical Top End gives way to the dry Red Centre. It was named after Frederick Renner, doctor to workers on the Overland Telegraph Line. Dr Renner discovered the springs when he noticed flocks of birds gathering in the area. Fuel, supplies and meals are available and there are pleasant picnic and barbecue facilities. 91 km S.

Newcastle Waters: once-thriving old droving town featuring historic buildings and a bronze statue, The Drover; 19 km NW.

See also **GULF TO GULF, p. 381**

Glen Helen Resort

Map ref. 638 G7 | 640 C3

Namatjira Dr; (08) 8956 7489; www.glenhelen.com.au

106.1 8ACR FM, 783 AM ABC Territory Radio

This small homestead-style resort is an excellent base for exploring the superb scenery of Glen Helen Gorge and other attractions in West MacDonnell National Park. Its facilities include accommodation, restaurant, bar, entertainment, tour information and an attractive natural swimming hole.

Glen Helen Gorge This breathtaking sandstone formation was created by the erosive action of the Finke River over thousands of years. There is a beautiful walk along the Finke River with towering cliffs providing a habitat for black-footed wallabies. Helicopter flights provide awe-inspiring views over Mt Sonder. 300 m E.

West MacDonnell National Park: majestic mountain range and wilderness surrounding Glen Helen Resort and featuring Ormiston Gorge (12 km NE), Ochre Pits (21 km E) and the Larapinta Trail; see *Alice Springs*. ***Tnorala (Gosse Bluff) Conservation Reserve:*** huge crater, 25 km in diameter, formed when a comet struck earth over 130 million years ago, with excellent views from Tylers Pass. Access permit, from visitor centre, is included with Mereenie Loop permit (Mereenie Loop links West MacDonnells with Kings Canyon Resort in the south-west); 50 km SW.

See also **RED CENTRE, p. 382**

Hermannsburg

Pop. 555
Map ref. 638 H7 | 640 D5

Ntaria Supermarket; (08) 8956 7480; www.hermannsburg.com.au

106.1 8ACR FM, 783 AM ABC Territory Radio

This Aboriginal community lives on the site of a former mission station established by German Lutherans in 1877. During the first 14 years, they recorded the Arrernte language and culture, compiled an Arrernte dictionary and translated the New Testament into the local language. From 1894 to 1922 the mission was run by Pastor Carl Strehlow, who restored and constructed most of the extant buildings. His son, T. G. H. Strehlow, assembled a vast collection of anthropological items relating to the Arrernte way of life. Renowned artist Albert Namatjira,

JABIRU

Pop. 1140

Map ref. 633 O4 | 634 G5

 6 Tasman Plaza; (08) 8979 2548.

 747 AM ABC Territory Radio

Located deep within Kakadu National Park, Jabiru was first established because of the nearby uranium mine and, although this still operates, the town is now a major centre for the thousands who come to explore Kakadu each year. Facilities include a supermarket, a bakery and a large town swimming pool. Swimming pools are the only safe place to swim in Kakadu. Note that takeaway alcohol cannot be purchased.

Gagudju Crocodile Holiday Inn This NT icon is a 250 m building that resembles a crocodile from the air. The entrance (the mouth of the crocodile) leads to a cool marble reception area, designed to represent a billabong, and accommodation is in the belly of the beast. The design was approved by the Gagudju people, to whom the crocodile is a totem, and was a finalist at the prestigious Quaternario Architectural Awards in Venice. Flinders St.

Aurora Kakadu Lodge and Caravan Park: accommodation laid out in traditional Aboriginal circular motif; Jabiru Dr; (08) 8979 2422.

Gunbalunya (Oenpelli) Open Day: the Aboriginal community of Gunbalunya opens its doors to visitors, with sports, art and Aboriginal culture; Aug. ***Mahbilil Festival:*** music, dance and artistic expression celebrating Kakadu; Sept.

Escarpment Restaurant: stylish, modern Australian; Gagudju Crocodile Holiday Inn, Flinders St; (08) 8979 9000. ***Jabiru Golf Club:*** good, well-priced food; Jabiru Dr; (08) 8979 2575. ***Jabiru Sports & Social Club:*** steak, barra and Sunday roast; Lakeside Dr; (08) 8979 2326. ***Aurora Kakadu Lodge and Caravan Park:*** a bistro alongside the pool is open to non-residents. See above. ***The Bark Hut Inn:*** good pub fare; Arnhem Hwy, Annaburroo; 135 km w; (08) 8978 8988. ***Mimi Restaurant:*** relaxed, outdoor seating; Gagudju Lodge Cooinda, Kakadu Hwy, Jim Jim; (08) 8979 0145; 50 km s. ***The Wetlands Restaurant:*** buffet breakfast and dinner; Aurora Kakadu, Arnhem Hwy, South Alligator; (08) 8979 0166; 40 km w.

Gagudju Crocodile Holiday Inn: luxury rooms in a crocodile; Flinders St; (08) 8979 9000. ***Aurora Kakadu:*** comfortable mid-range accommodation; Arnhem Hwy, South Alligator; (08) 8979 0166 or 1800 818 845; 40 km w. ***Gagudju Lodge Cooinda:*** diverse range of accommodation near Yellow Waters wetland; Kakadu Hwy, Cooinda; 50 km s; (08) 8979 0145. ***Mount Borradaile Safari Camp:*** wilderness safari camp; Mt Borradaile, Arnhem Land; (08) 8927 5240. ***Peppers Seven Spirit Bay Wilderness Resort:*** luxury coastal wilderness stay; Garig Gunak Barlu National Park, Cobourg Peninsula; (08) 8979 0281.

Kakadu National Park A World Heritage site listed for both its natural value and cultural significance, Kakadu is a place of rare beauty and grand landscapes, abundant flora and fauna, impressive rock art and ancient mythology. The largest national park in Australia, it encompasses the flood plain of the South Alligator River system and is bordered to the east by the massive escarpment of the Arnhem Land Plateau. The wide-ranging habitats, from arid sandstone hills, savannah woodlands and monsoon forests to freshwater flood plains and tidal mudflats, support an immense variety of wildlife, some rare, endangered or endemic. There are over 50 species of mammal, including kangaroos, wallabies, quolls, bandicoots, bats and dugong, as well as a plethora of reptiles and birdlife. With around 5000 rock-art sites, the park has the world's largest and possibly oldest collection of rock art, which reveals the complex culture of the Aboriginal people since the Creation Time, when their ancestors are believed to have created all landforms and living things. Nourlangie Rock shows a variety of styles including X-ray art, which illustrates the anatomy of humans and animals in rich detail. Other highlights of the park include Jim Jim Falls and Twin Falls, the Yellow Water wetland area at Cooinda and sunset at Ubirr rock, with free ranger tours at 5pm each day from May–Nov. Also stop in at the Warradjan Aboriginal Cultural Centre in Cooinda, which features educational displays and exhibitions. ***Travellers note:*** *Kakadu is a special place. It is important to respect the land and its people and refrain from entering restricted areas such as sacred sites, ceremonial sites and burial grounds. For more information contact (08) 8938 1120.*

[FLOOD PLAIN OF THE SOUTH ALLIGATOR RIVER, KAKADU NATIONAL PARK]

Tourist walk: 1.5 km stroll from town centre through bush to Bowali Visitor Centre, which features an audiovisual presentation and interpretive displays on Kakadu. ***Scenic flights:*** over Kakadu parklands to see inaccessible sandstone formations standing 300 m above vast flood plains, seasonal waterfalls, wetland wilderness, remote beaches and ancient Aboriginal rock-art sites; 6 km E. ***Ranger Uranium Mine:*** open-cut mine with educational tours available; bookings through tours office 1800 089 113; 9 km E. ***Aboriginal cultural tours:*** day tours and cruises; contact Bowali Visitor Centre (08) 8938 1120. Note that a permit is required to enter Kakadu National Park. Details from the Visitors Centre.

See also KAKADU & ARNHEM LAND, p. 379

NORTHERN TERRITORY

KATHERINE

Pop. 5848

Map ref. 634 F8

[ENTRANCE TO KATHERINE GORGE SYSTEM]

Cnr Lindsay St and Katherine Tce. (08) 8972 2650 or 1800 653 142; www.visitkatherine.com.au

101.3 Katherine FM, 106.1 FM ABC Territory Radio

Katherine, on the south side of the Katherine River, has always been a busy and important area for local Aboriginal groups, who used the river and gorge as meeting places. Explorer John McDouall Stuart named the river on his way through in 1862 after the second daughter of James Chambers, one of his expedition sponsors. The town grew up around an Overland Telegraph station and was named after the river. Today its economic mainstays are pastural, tourism and the Tindal RAAF airbase.

Katherine Museum: built as an air terminal in 1944–45, the museum houses artefacts, maps, photographs and farming displays. There is also memorabilia relating to Dr Clyde Fenton, who as a pioneer medical aviator and Katherine's Medical Officer between 1934 and 1937 serviced an area of 8 000 000 sq km in a second-hand Gypsy Moth he bought for £500. The plane is on display. Check opening times; Gorge Rd; (08) 8972 3945. ***Railway Station Museum:*** displays of the area's railway history with an old steam engine in nearby Ryan Park; check at the Visitors Centre for opening times; Railway Tce. ***School of the Air:*** see displays about the history of the school and listen in on lessons; open Apr–Oct, closed on weekends; Giles St. ***O'Keefe House:*** built during WW II and used as the officers' mess, it became home to Sister Olive O'Keefe, whose work with the Flying Doctor and Katherine Hospital from the 1930s to 1950s made her a NT identity. The house is one of the oldest in town and is a classic example of bush ingenuity, using local cypress pine, corrugated iron and flywire; open Mon–Fri May–Sept; Riverbank Dr. ***NT Rare Rocks:*** unusual rock and gem displays; Zimmin Dr; (08) 8971 0889.

Tick Markets: Lindsay St; 1st Sat each month Apr–Sept. ***Katherine Community Markets and Farmers' Market:*** 8am–12pm every Sat, Mar–Dec; Ryan Park. ***Katherine Show:*** Rodeo, Exhibitions; July. ***Katherine Cup:*** horseracing; May. ***Katherine Festival:*** community festival, theatre and music; Aug.

Katherine Country Club: traditional Australian; Pearce St, off Victoria Hwy; (08) 8972 1276. ***Katie's Bistro:*** good cooking; Knotts Crossing Resort, cnr Giles and Cameron sts; (08) 8972 2511. ***Kumbidgee Lodge Tea Rooms:*** barramundi a specialty; 4739 Gorge Rd. ***The Carriage Stonegrill Restaurant & Bar:*** hot-rock cooking; Paraway Motel, 35 First St; (08) 8972 2644. ***The Katherine Club:*** value-for-money; Cnr Second St and O'Shea Tce; (08) 8972 1250. ***Silver Screen:*** casual dining; 20 First St; (08) 8972 3140. ***Coffe Club:*** cafe, bar, restaurant; 23 Katherine Tce; (08) 8972 3990.

Hotel All Seasons Katherine: modern tropical setting; Stuart Hwy; (08) 8972 1744. ***Knotts Crossing Resort:*** tropical landscape, rooms and cabins; Cnr Giles and Cameron sts; (08) 8972 2511. ***Maud Creek Lodge:*** private rustic retreat; Gorge Rd; (08) 8971 0877. ***Low Level Caravan Park:*** near Nitmiluk Gorge; 20 Shadforth Rd; (08) 8972 3962. ***Shady Lane Tourist Park:*** close to essentials; 1828 Gorge Rd; (08) 8971 0491. ***Nitmiluk NP Campgrounds:*** 1300 146 743.

Nitmiluk National Park Nitmiluk National Park This 292 800 ha wilderness is owned by the Jawoyn people and managed jointly with NT Parks & Wildlife Commission. The Katherine River flows through a broad valley that narrows dramatically between the high sandstone cliffs of the magnificent Nitmiluk Gorge; 29 km NE. High above the floodline, on the overhangs of the ancient rock walls, are Aboriginal paintings thousands of years old. The best way to explore the gorge is by flat-bottomed boat (daily cruises; bookings at visitor centre), but canoe hire is available. Another highlight of the park is Edith Falls (Leliyn); 66 kms NE, which drops into a paperbark- and pandanus-fringed natural pool that is a popular swimming spot. There are signposted bushwalks, picnic areas and campsites. Fauna in Nitmiluk includes many reptile and amphibian species, kangaroos and wallabies in higher reaches and rare birds such as the hooded parrot and Gouldian finch. The countryside is at its best Nov–Mar although the weather is hot. Magnificent scenery can be enjoyed from helicopter tours, scenic flights, boat cruises etc.; brochures at visitor centre.

Knotts Crossing: site of region's first Overland Telegraph station (1870s) around which the original township of Katherine developed. By 1888 there was a hotel, store and police station. The hotel lost its licence in 1916 and the store was given a Gallon Licence. The Gallon Licence Store (now a private residence) operated until 1942; Giles St; 2 km E. ***Natural hot springs:*** on the banks of the Katherine River with picnic facilities and pleasant walking trails; Victoria Hwy; 3 km S. ***Low Level Nature Reserve:*** weir built by US soldiers during WW II, now a popular waterhole for picnicking and canoeing (bring own canoe). The river is teeming with barramundi, black bream and northern snapping turtles so fishing is a common pastime; 3 km S. ***Springvale Homestead:*** originally the home of Alfred Giles, this is the oldest remaining homestead in NT (1879). Accommodation, restaurant, swimming pool and camping; Shadforth Rd; 8 km W. ***Cutta Cutta Caves Nature Park:*** 1499 ha of the only accessible tropical limestone caves in the NT with fascinating formations 15 m underground; regular tours each day; (08) 8972 1940; 27 km SW. ***Manyallaluk Aboriginal Community:*** camping and Aboriginal cultural tours; bookings essential, (08) 8971 0877; 100 km NE. ***Flora River Nature Park:*** great campsites, interesting mineral formations, pools and cascades along the river; 122 km SW, of which 46 km is unsealed.

See also GULF TO GULF, p. 381

the first Aboriginal to paint landscapes in a European style, was born at the mission in 1902. In 1982 the mission and its land was returned to the Arrernte. Visitors are restricted to the shop, petrol station and historic precinct.

Historic precinct The National Heritage–listed mission site comprises about 13 main buildings, mostly stone, the earliest dating from 1877 but generally from the period 1897–1910. They include Strehlow's House (1879), home of the pastor in charge of the mission and now Kata-Anga Tea Rooms with a reputation for delicious apple strudel; old manse (1888), currently a watercolour gallery housing work by Aboriginal artists of the Hermannsburg school (guided tours available); old schoolhouse (1896); tannery (1941); and Old Colonists House (1885), now a museum displaying historic items from the missionary era.

Finke Gorge National Park For millions of years the Finke River has carved its way through the weathered ranges, creating red-hued gorges and wide valleys. There are astonishing rock formations and dry creek beds that wind through sandstone ravines, where rare flora flourishes. The park's most famous feature, Palm Valley, is a refuge for about 3000 red cabbage palms (*Livistona mariae*), which are found nowhere else in the world. The 46 000 ha area is great for bushwalking: a 1.5 km climb leads to Kalarranga Lookout for views of the amazing Amphitheatre rock; the 5 km Mpaara Walk with informative signs explains the mythology of Western Arrernte culture; and the 2 km Arankaia Walk passes through the lush oasis of Palm Valley. Park access by 4WD only. (08) 8951 8290; 20 km S.

Monument to Albert Namatjira: a 6 m red sandstone memorial to the legendary artist; Larapinta Dr; 2 km E. ***Wallace Rockhole Aboriginal community:*** cultural tours and camping; 46 km SE.

See also RED CENTRE, p. 382

Jabiru

see inset box on previous page

Kings Canyon Resort

Map ref. 638 F8

Luritja Rd; (08) 8956 7442; www.kingscanyonresort.com.au

Kings Canyon Resort, in Watarrka National Park, is an excellent base from which to explore the region. It has various standards of accommodation, a petrol station, supermarket, laundry, tennis courts and swimming pools.

Watarrka National Park This park's many rock holes and gorges provide refuge from the harsh conditions for many species of plants and animals. The great attraction is Kings Canyon, an enormous amphitheatre with sheer sandstone walls rising to 270 m. The 870 m Carmichael Crag is known for its majestic colours, which are particularly vibrant at sunset. There are well-signed trails. The Rim Walk is a boardwalk through prehistoric cycads in the lush Garden of Eden and takes in unusual rock formations such as the Lost City. There are wonderful views across the canyon. The Kings Creek Walk is a 1 hr return walk along the canyon floor. The 6 km (3–4 hr) Kings Canyon Walk is rough going and recommended only for experienced walkers. Tours of the park include Aboriginal tours and scenic flights; brochures available from visitor centre. (08) 8951 8211.

Kings Creek Station: working cattle and camel station with campsites and accommodation. Quad (4-wheeled motorcycle), helicopter and camel tours of the area; 36 km SE. ***Mereenie Loop:*** links Alice Springs, Kings Canyon, Uluṟu and Kata Tjuṯa via the West MacDonnell Ranges and Glen Helen. Permit required because a section of the route passes through Aboriginal land; map and permit from visitor centre.

See also RED CENTRE, p. 382

Mataranka

Pop. 249
Map ref. 634 H9

Stockyard Gallery, Stuart Hwy; (08) 8975 4530.

106.1 FM ABC Territory Radio

Visitors are lured to Mataranka for its thermal springs and its sense of literary history. Jeannie Gunn, author of *We of the Never Never*, lived at nearby Elsey Station in the early 20th century. The town has adopted the term 'Never Never', using it to name a museum and a festival. Generally the phrase now refers to the vast remote area of inland northern Australia.

Stockyard Gallery: showcases NT artists' work, including leather sculpture; Devonshire teas available; Stuart Hwy. ***Territory Manor:*** restaurant and accommodation with daily feeding of barramundi at 9.30am and 1pm (open to the public); Stuart Hwy. ***We of the Never Never Museum:*** outdoor displays of pioneer life, railway and military history and the Overland Telegraph; open Mon–Fri; Stuart Hwy. ***Giant termite mound:*** sculpture with recorded information; Stuart Hwy.

Back to the Never Never Festival: celebration of the outback including a cattle muster; May. ***Art show:*** May. ***Rodeo:*** Aug.

Mataranka Homestead Tourist Resort: good food in bush setting; Homestead Rd; (08) 8975 4544. ***Territory Manor Holiday Park:*** barra and beef; Martin Rd, off Stuart Hwy; (08) 8975 4516.

Coodardie Station Stay: family station stay; Stuart Hwy; (08) 8975 4460. ***Territory Manor Motel & Caravan Park:*** surrounded by lush gardens; Martins Rd, off Stuart Hwy; (08) 8975 4516

Elsey National Park The park encircles the Roper River, with rainforest, paperbark woodlands, and tufa limestone formations at Mataranka Falls. The thermal pools at Mataranka Springs and nearby Bitter Springs are believed to have therapeutic powers. At Mataranka Thermal Pool water (34°C) rises from underground at the rate of 30.5 million litres per day to provide a beautiful swimming spot surrounded by palm trees. The Roper River offers excellent canoeing (hire available) and barramundi fishing. There are scenic walking tracks through pockets of rainforest with wildlife observation points and camping areas. The walk from Twelve Mile Yards leads to the small but beautiful Mataranka Falls. Mataranka Homestead Tourist Resort is a replica of Jeannie Gunn's home, Elsey Homestead, and offers accommodation and camping. The replica was used in the 1982 film *We of the Never Never*. (08) 8975 4560; 5 km E.

Elsey Cemetery: graves of outback pioneers immortalised by Jeannie Gunn, who lived on Elsey Station from 1902 to 1903; 20 km SE. ***We of the Never Never:*** a cairn marks the site of the original homestead near the cemetery; 20 km SE.

See also GULF TO GULF, p. 381

NORTHERN TERRITORY

Nhulunbuy

Pop. 4111
Map ref. 635 N4

Chamber of Commerce, Endeavour Sq; (08) 8987 1985.

106.9 Gove FM, 990 AM ABC Territory Radio

Located on the north-eastern tip of Arnhem Land on the Gove Peninsula, Nhulunbuy and its surrounds are held freehold by the Yolngu people. Originally a service town for the bauxite-mining industry, it is now the administrative centre for the Arnhem region and a pleasant and relaxed hideaway. Access is by a year-round daily air service from Darwin or Cairns or, with a permit, by four-wheel drive through Arnhem Land.

Gayngaru Wetlands Interpretive Walk The path surrounds an attractive lagoon that is home to around 200 bird species. There are 2 viewing platforms and a bird hide, as well as signs near local flora explaining their uses in Aboriginal food and bush medicine. Visitors can take the Winter walk or Tropical Summer walk, which is shorter as a result of higher water levels. Centre of Nhulunbuy.

Garma Festival: celebration of the arts and culture; Aug. ***Beach Volleyball Day:*** Sept. ***Annual Ball:*** Sept. ***John Jones Memorial Billfish Classic:*** Oct/Nov.

Katie's Kitchen: home-style cooking; Drimmie Head Rd; (08) 8987 3077. ***Macassans Restaurant:*** modern Australian; The Arnhem Club, 1 Franklyn St; (08) 8987 0600. ***The Walkabout Lodge Restaurant:*** modern Australian with good seafood; 12 Westal St; (08) 8987 1777. ***Bistro Sea:*** steak, seafood and Asian dishes; 35 Bougainvillea Dr, Alyangula; (08) 8987 7100. ***Golf Shed Bistro:*** Australian and Asian dishes; Alyangula Golf Club, 1 Alebewa Rd, Alyangula; (08) 8987 6060. ***Seagrass Restaurant:*** modern Australian with Asian influences; Dugong Beach Resort, 1 Bougainvillea Dr, Alyangula; (08) 8987 7077 or 1800 877 077.

Dugong Beach Resort: modern luxury cabins and rooms; 1 Bougainvillea Dr, Alyangula; (08) 8987 7077 or 1800 877 077. ***Gove Peninsula Motel:*** modern motel rooms; 1 Matthew Flinders Way; (08) 8987 0700. ***Walkabout Lodge:*** modern hotel; 12 Westal St; (08) 8987 1777.

Buku-Larrnggay Mulka This renowned community-based Aboriginal art museum was set up to educate visitors in the ways of local law and culture, and to share the art of the Yolngu people. A permit is not required for a museum visit. Open Mon–Fri and Sat morning; (08) 8987 1701; 20 km SE.

Dhamitjinya (East Woody Island) and Galaru (East Woody Beach): island with long sandy beaches, tropical clear blue water and amazing sunset views; 3 km N. ***Nambara Arts and Crafts:*** traditional and contemporary Aboriginal art and craft; 15 km NW. ***Tours:*** include boat charters for outstanding game, reef and barramundi fishing, eco and cultural tours with Yolngu guides, 4WD tours, birdwatching and croc-spotting tours and bauxite-mine tours. Details from visitor centre.

Travellers note: *Visitors intending to drive to Nhulunbuy must obtain a permit from the Northern Land Council beforehand, (08) 8987 8500; conditions apply. Allow 2 weeks for processing. A recreation permit is also required for travel outside the Nhulunbuy Town Lease, available from Dhimurru Land Management, (08) 8987 3992.*

See also KAKADU & ARNHEM LAND, p. 379

Noonamah

Pop. 485
Map ref. 632 E4

Noonamah Tavern, Stuart Hwy; (08) 8988 1054.

106.1 FM ABC Territory Radio

Noonamah is a tiny town outside Darwin at the centre of numerous parks and reserves. It is a great base to experience wildlife, native bushland and safe swimming spots (something of a rarity, considering the Northern Territory's crocodile population).

Bamurru Plains: luxury lodges; Swim Creek Station, Harold Knowles Rd; (02) 9571 6399 or 1300 790 561. ***Eden at Fogg Dam:*** tropical B&B; 530 Anzac Pde, Middle Point; (08) 8988 5599. ***Mary River Houseboats:*** explore the waterway by houseboat; Corroboree Billabong, off Arnhem Hwy; (08) 8978 8925. ***Mary River Park:*** bush cabins; Mary River Crossing, Arnhem Hwy; (08) 8978 8877. ***Point Stuart Wilderness Lodge:*** comfortable wilderness stay; Point Stuart Rd, Point Stuart; (08) 8978 8914.

Territory Wildlife Park This 400 ha park provides an easy way to view native animals in their natural habitat. A tunnel leads through an extensive aquarium that represents a Top End river system, where visitors can come face to face with a 3.7 m crocodile. Other highlights include a bird walk, nocturnal house and daily show of birds of prey. 14 km SW.

Didgeridoo Hut: see Aboriginal craftspeople make didgeridoos and weave baskets and dillybags. Works are for sale in the gallery and there is an emu farm on-site; 8 km N. ***Lakes Resort:*** great facilities for watersports including waterski and jetski hire; accommodation available; 10 km SW. ***Berry Springs Nature Park:*** safe swimming in spring-fed pools in a monsoon forest with pleasant walking trails and picnic areas; 13 km SW next to Territory Wildlife Park. ***Jenny's Orchid Garden:*** huge and colourful collection of tropical orchids, many on sale in the nursery; 22 km NW. ***Howard Springs Nature Park:*** swim in the freshwater springs and see barramundi and turtles or relax with a picnic in the shade by the children's pool. 23 km NW. ***Manton Dam:*** safe swimming, fishing, watersports and shady picnic spots; 25 km S. ***Tumbling Waters Holiday Park:*** fauna enclosure and caravan park; 26 km SW. ***Fogg Dam:*** wetland with prolific wildlife that is best viewed at sunrise or sunset; 41 km NE. ***Window on the Wetlands:*** interpretive centre offering insight into the Top End's fascinating wetland environments plus superb views over the Adelaide River flood plains; 40 km NE (7 km beyond Fogg Dam turn-off). ***Adelaide River Jumping Crocodile Cruises:*** see crocodiles jump out of the water with the lure of food. Several tours each day; bookings (08) 8988 8144; 41 km NE (next to wetlands centre). ***Fishing:*** excellent barramundi fishing at Mary River Crossing (45 km E of Fogg Dam turn-off); several 4WD tracks lead north from here to prime fishing spots in Mary River National Park such as Corroboree Billabong (houseboat hire available).

See also AROUND DARWIN, p. 380

Pine Creek

Pop. 256
Map ref. 633 I12 | 634 F7

Diggers Rest Motel, 32 Main Tce, (08) 8976 1442; or Shire office, (08) 8976 1391.

96.7 Mix FM, 106.1 FM ABC Territory Radio

The town was named by Overland Telegraph workers because of the prolific pine trees growing along the banks of the tiny creek. Today the town benefits from the reopened goldmine, one of the largest open-cut goldmines in the Northern Territory.

Railway Station Museum The station (1888) is at the terminus of the uncompleted 19th-century transcontinental railway system and now houses photographs and memorabilia. The adjacent Miners Park points to an important visible link between the railway and the mines, which depended on the railway for survival. There are interpretive signs and displays of old mining machinery that reflect life on the goldfields. Main Tce.

Water Gardens: ponds with walking trails, birdlife and picnic spots; Main Tce. ***National Trust Museum:*** once a doctor's residence, military hospital then post office, building now houses a historical collection, including a display on the Overland Telegraph; open 11am–5pm Mon–Fri and 11am–1pm Sat Apr–Oct, 1–5pm Mon–Fri Nov–Mar; Railway Tce. ***Enterprise Pit Mine Lookout:*** panoramic views of open-cut goldmine that was once Enterprise Hill, but is now a water-filled pit; Enterprise Pit. ***Gun Alley Gold Mining:*** features restored steam ore crusher and gold-panning tours; Gun Alley Rd. ***Bird Park:*** tropical birds in lush garden setting; Gun Alley Rd. ***Old Timers Rock Hut:*** rock and mineral display; Jenson St. ***Town walk:*** takes in historic buildings and mining sites; brochure from visitor centre.

Races: horseracing; May. ***Gold Rush Festival and Gold-panning Championships:*** June. ***Rodeo:*** June.

Mayse's Cafe: good Australian homemade food; 40 Moule St; (08) 8976 1241.

Grove Hill Heritage Hotel: rustic, comfortable hotel; Goldfields Rd, Grove Hill Siding, via Adelaide River; (08) 8978 2489.

Butterfly Gorge Nature Park This park is a wilderness of sheer cliff-faces, dense vegetation and scenic shady river walks. Butterfly Gorge was named for the butterflies that settle in its rock crevices and is a beautiful and safe swimming and picnic spot. Access is by 4WD only. Open only during the dry season (May–Sept); 113 km NW.

Copperfield Recreation Reserve: safe swimming in deep-water lake and picnicking on foreshore plus a Didgeridoo Jam held each May; 6 km SW. ***Bonrook Lodge and Station:*** wild horse sanctuary with trail rides and overnight camps; 6 km SW. ***Umbrawarra Gorge Nature Park:*** good swimming, rock climbing and walking trails; 22 km SW. ***Tjuwaliyn (Douglas) Hot Springs Park:*** sacred place for Wagiman women with hot springs; camping available; off Stuart Hwy; 64 km NW. ***The Rock Hole:*** attractive secluded waterhole; 4WD access only (via Kakadu Hwy); 65 km NE. ***Gold fossicking:*** several locations; licence required (available along with maps from visitor centre). ***Kakadu National Park:*** best-known park in the Top End, this is a massive tropical savannah woodland and freshwater wetland. Highlights nearby include the spectacular views at Bukbukluk Lookout (87 km NE) and the beautiful falls and permanent waterhole at Waterfall Creek (113 km NE). See *Jabiru*.

See also AROUND DARWIN, p. 380

Tennant Creek

Pop. 2922
Map ref. 637 J9

Peko Rd; 1800 500 879; www.barklytourism.com.au

102.1 8CCC FM, 106.1 FM ABC Territory Radio

Tennant Creek is midway between Alice Springs and Katherine and according to legend, emerged when a beer truck broke down here. Gold was found in the area in the early 1930s and the town grew rapidly in the wake of Australia's last great gold rush. Gold exploration still continues in the Barkly Tableland, an area larger than Victoria.

Nyinkka Nyunyu Cultural Centre This centre was built near a Warumungu sacred site and its name means 'home of the spiky-tailed goanna'. Dioramas illustrate the history of the area, an Aboriginal art gallery showcases the Tennant Creek art movement, and bush tucker, dance performances and displays explain the Aboriginal people's relationship with the land. The centre is set in landscaped gardens featuring plants used for bush tucker and medicine. Paterson St; (08) 8962 2221.

Tuxworth Fullwood Museum: housed in an old WW II army hospital (1942) and listed by the National Trust, this museum has a photographic collection and displays of early mine buildings and equipment, a 1930s police cell and a steam tractor engine; open May–Sept, check times with the visitor centre; Schmidt St. ***Purkiss Reserve:*** pleasant picnic area with swimming pool nearby; Ambrose St. ***Julalikari Arts:*** Aboriginal art centre; (08) 8962 2163. ***Winanjjikari Music Centre:*** resident aboriginal musicians; (08) 8962 3282. ***Self-guided tours:*** scenic drives and heritage walks including an old Australian Inland Mission, built in 1934 of prefabricated corrugated iron, and the catholic church, built in 1911 and relocated from Pine Creek in 1936. Brochures and maps from the visitor centre.

Drag Racing: May. ***Show:*** cooking, craft, exhibitions and a big dog show; July. ***Cup Day:*** horseracing; May. ***Desert Harmony Festival:*** arts and culture; Aug/Sept. ***Brunette Downs Races:*** campdrafting, bronco branding and more; June; 360 kms NE.

Anna's Restaurant: steak, chicken and barra; Bluestone Motor Inn, 1 Paterson St; (08) 8962 2617. ***Fernanda's Restaurant:*** Mediterranean cuisine; 1 Noble St; (08) 8962 3999. ***Jajjikari Cafe:*** cafe food in terrific setting; Nyinkka Nyunyu, Paterson St; (08) 8962 2221. ***Memories Restaurant:*** home-style cooking; Tennant Creek Memorial Club, 48 Schmidt St; (08) 8962 2474. ***Eldorado Restaurant:*** in or outdoor dining; 1 Paterson St; (08) 8962 2617. ***Wok's Up:*** Chinese at the Sporties Club; Ambrose St; (08) 8962 3888.

Bluestone Motor Inn: modern comfortable rooms; 1 Paterson St; (08) 8962 2617. ***Eldorado Motor Inn:*** refurbished modern rooms; 195 Paterson St; (08) 8962 2402. ***Safari Motor Lodge:*** Davidson St; (08) 8962 2207. ***Desert Sands:*** self–contained appartments; Paterson St; (08) 8962 1346. ***Goldfields Hotel Motel:*** standard accommodation; 113 Paterson St; (08) 8962 2030. ***Outback Caravan Park:*** nice sites; Peko Rd; (08) 8962 2459. ***Barkly Homestead:*** comfortable cool rooms; 210 kms E and the only place to stay between Tennant Creek and Camooweal, QLD; (08) 8964 4549. ***Three Ways Roadhouse:*** camping, rooms and good roadhouse meals at the junction of Stuart and Barkly Hwys, with memorial to John Flynn, founder of the Royal Australian Flying Doctor Service, nearby; 24 km N; cnr Stuart and Barkly Hwys; (08) 8962 2744.

YULARA

Pop. 983

Map ref. 638 E10 | 640 D9

Alice Springs Visitor Information Centre; Gregory Tce; (08) 8952 5800. Ayers Rock Resort; (08) 8957 7888.

99.7 FM ABC Territory Radio, 100.5 8HA FM

Yulara is a resort town that was built specifically to cater for visitors coming to see Uluṟu and Kata Tjuṯa. It offers excellent visitor facilities and food and accommodation for all budgets. Advance bookings for accommodation are essential.

Information Centre: displays of geology, history, flora and fauna; spectacular photographic collection.

All dining options are in ***Ayres Rock Resort:*** Yulara Dr; 1300 134 044 for bookings. ***Sounds of Silence:*** ultimate dining experience, alfresco dining under the stars; (08) 8956 2229. ***Pioneer BBQ and Bar:*** casual open-air dining; (08) 8957 7606. ***Gecko's Cafe:*** gourmet pizzas and pasta; (08) 8956 2562. ***Red Rock Deli***: salads and wraps; (08) 8956 2229. ***Kuniya Restaurant:*** modern Australian; Sails in the Desert Hotel; (08) 8957 7888.

Ayres Rock Resort: a collection of hotels, apartments and a campground; Yulara Dr; Contact Voyages Travel Centre 1300 134 044 or (02) 8296 8010 for all properties. ***Desert Gardens Hotel:*** modern rooms with views. ***Emu Walk Apartments:*** self-contained apartments. ***Longitude 131°:*** luxury eco-camping. ***Outback Pioneer Hotel and Lodge:*** budget priced rooms and dormitories. ***Sails in the Desert Hotel:*** five-star accommodation. ***The Lost Camel Hotel:*** modern studio-style apartments. ***Voyages Ayres Rock Campground:*** tent sites, powered sites and a/c cabins. ***Curtin Springs Cattle Station & Wayside Inn:*** rustic outback rooms; working cattle station Curtin Springs, Lasseter Hwy; (08) 8956 2906. 85km E. ***Erldunda Station B&B:*** cottage on working cattle station; Erldunda Station; (08) 8956 0997. 200km E on the junction of the Stuart and Lassiter Hwy.

Uluṟu Australia's most recognisable natural landmark and the largest monolith in the world, Uluṟu features stunning Aboriginal rock-art sites that can be viewed on guided walks and tours around the base. These sites highlight the rock's significance for Aboriginal people. The traditional owners prefer visitors not to climb Uluṟu and there are countless tales of a curse befalling those who take a piece of the rock home as a souvenir. The spectacular changing colours of Uluṟu at sunrise and sunset are not to be missed. 20 km SE.

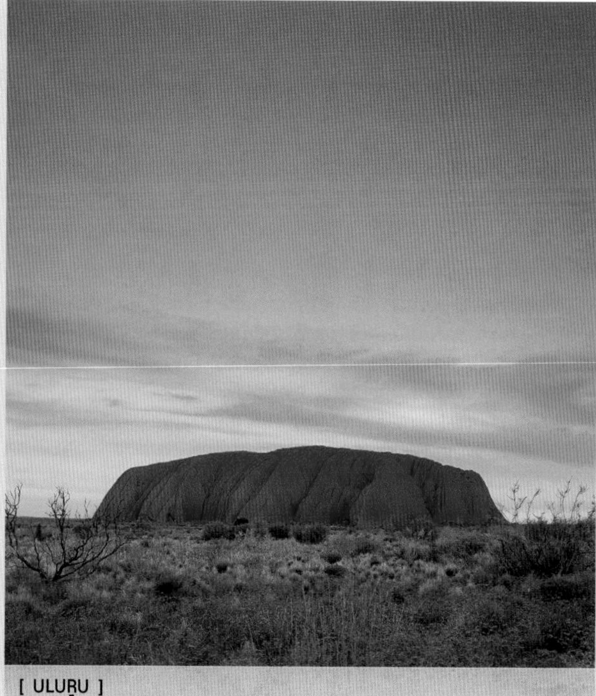

[ULUṞU]

Uluṟu-Kata Tjuṯa Cultural Centre: designed in the shape of two snakes, with displays of Aboriginal culture and sales of artwork; on approach road to Uluṟu; (08) 8956 2214; 17 km S. ***Kata Tjuṯa:*** remarkable rock formations, with spectacular Valley of the Winds walk, fantastic views and a variety of flora and fauna; 50 km W. ***Tours:*** helicopter, coach and safari tours; call information centre for bookings.

See also RED CENTRE, p. 382

Battery Hill This comprises an underground mine, a 10-head stamp battery and two museums. There are daily tours of the underground mine with working machinery highlighting gold extraction. Bill Allen Lookout, just past the battery, offers panoramic views, with plaques identifying significant sites. Peko Rd; 3.5 km E.

Tennant Creek Cemetery: pioneer graves with plaques reminding us of the hardships of pioneer life; 2 km S. ***Lake Mary Ann:*** artificial dam ideal for swimming, canoeing, windsurfing and picnics; 6 km NE. ***Juno Bush Camp:*** bush camping; contact visitor centre for details; 10 km E. ***Telegraph Station:*** restored stone buildings (1872), once the domain of telegraph workers whose isolated lives are revealed by interpretive signs; key is available from Tennant Creek Visitor Centre; 11 km N. ***The Pebbles:*** spectacular in their quantity, these are miniature versions of the Devils Marbles (huge balancing boulders found north of Wauchope). The site is sacred to the Munga Munga women; 16 km NW (6km of unsealed road). ***Memorial Attack Creek Historical Reserve:*** memorial marks the encounter between John McDouall Stuart and local Aboriginal people; 72 km N. ***Karlu Karlu Devils Marbles Conservation Reserve:*** a must see phenomenon of huge, red granite boulders; bush camping is available. 106 km S and a 2 km sealed loop.

See also RED CENTRE, p. 382

Ti Tree

Map ref. 639 I3

 **Ti Tree Roadhouse, Stuart Hwy; (08) 8956 9741.**

 105.9 8ACR FM, 107.7 FM ABC Territory Radio

This rest stop on the Stuart Highway took its name from nearby Ti Tree Wells, the source of plentiful sweet water in the 1800s. Today the desert region supports remarkably successful fruit and vegetable industries.

Red Sand Art Gallery: Pmara Jutunta art with exhibitions and sales; food available; Stuart Hwy. ***Ti Tree Park:*** picnic area and playground; Stuart Hwy.

Central Mt Stuart Historical Reserve: the sandstone peak was noted by John McDouall Stuart as the geographical centre of Australia; no facilities, but a monument at the base; 18 km N.

See also RED CENTRE, p. 382

Timber Creek

Pop. 227
Map ref. 634 D10 | 636 D2

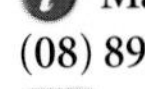 **Max's Victoria River Boat Tours, Victoria Hwy; (08) 8975 0850.**

 106.9 FM Territory Radio

In 1855 explorer A. C. Gregory was the first European to visit this area as he followed the Victoria River south from the Timor Sea. His boat was wrecked at the site of Timber Creek, where he found the timber he needed to make repairs. The town today is growing in importance as a stop on the journey from the Kimberley in Western Australia to the major centres of the Northern Territory and as a gateway to Gregory National Park.

Timber Creek Police Station Museum: displays of historic artefacts in restored police station (1908); open 10am–12pm Mon–Fri; off Victoria Hwy. ***Office for Parks & Wildlife Commission of the Northern Territory:*** information for travellers to Gregory and Keep River national parks; Victoria Hwy. ***Tours:*** boat and fishing tours, river cruises (with abundant crocodiles) and scenic flights; brochures from visitor centre.

Fishing competitions: Apr–May. ***Rodeo:*** May. ***Campdraft and Gymkhana:*** Sept.

Bullo River Station: great outback experience; Bullo River Station via Timber Creek; (08) 9168 7375.

Gregory National Park This is the NT's second largest national park with 2 sections covering 13 000 sq km of ranges, gorges, sandstone escarpments, remnant rainforest, eucalypts and boab trees. There are opportunities for boating, canoeing and bushwalking, scenic flights and cruises, as well as Aboriginal and European heritage sites. Gregory's Tree (in the Victoria River section to the east) stands at Gregory's campsite with historic inscriptions and audio presentation. The tree also has special significance for the Ngarinman people and is a registered sacred site. The spectacular dolomite blocks and huge cliffs of Limestone Gorge can be found in the Bullita section to the west. Bullita station has traditional timber stockyards, an old homestead, interpretive displays and shady camping spots in summer. Check with Parks & Wildlife Commission for current access details; (08) 8975 0888; 31 km W and 92 km E.

Jasper Gorge: scenic gorge with permanent waterhole and Aboriginal rock art; 48 km SE. ***Victoria River Roadhouse:*** rest stop where the Victoria Hwy crosses the Victoria River; cruises, fishing trips and accommodation; several scenic walks in the area including the Joe Creek Walk (10 km W); 92 km E. ***Keep River National Park:*** rugged scenery, Aboriginal rock art and wildlife. Most trails are 4WD only; check with Parks & Wildlife Commission for current access details; (08) 9167 8827; 180 km W.

See also GULF TO GULF, p. 381

Wauchope

Pop. 10
Map ref. 637 J11 | 639 J1

Wauchope Hotel, Stuart Hwy; (08) 8964 1963.

106.1 FM ABC Territory Radio

Wauchope (pronounced 'walk-up') was established to cater for the wolfram-mining and cattle-farming communities nearby. Today it is a service town and tourist destination thanks to its proximity to the popular Devils Marbles.

Wauchope Hotel This desert oasis offers fuel, meals (licensed restaurant and takeaway), various standards of accommodation and an adjacent campground. The walls of the pub are adorned with signed bank notes (a 'bush bank' where customers pin deposits to the wall to be retrieved on a later visit) and photographs of patrons. Stuart Hwy.

Wauchope Hotel and Roadhouse: rustic outback pub; Stuart Hwy; (08) 8964 1963. ***Wycliffe Well Holiday Park:*** variety of cabins and camping; Wycliffe Well, Stuart Hwy; (08) 8964 1966 or 1800 222 195.

Wycliffe Well Once a market garden supplying troops during WW II, this roadhouse now features pleasant picnic lawns surrounding Wycliffe Lake. The area is ranked 5th in the world for the number of UFO sightings. Wycliffe Well also claims to have Australia's largest range of beers, but any connection between that and the UFO sightings is yet to be proven. Accommodation available. 17 km S.

Devils Marbles: large, precariously balanced granite boulders; 8 km N. ***Davenport Range National Park:*** isolated area with important Aboriginal heritage and waterhole ecology sites; high clearance vehicles or 4WD access only; advise travel plans at the Wauchope Hotel; (08) 8962 4599; 118 km E.

See also RED CENTRE, p. 382

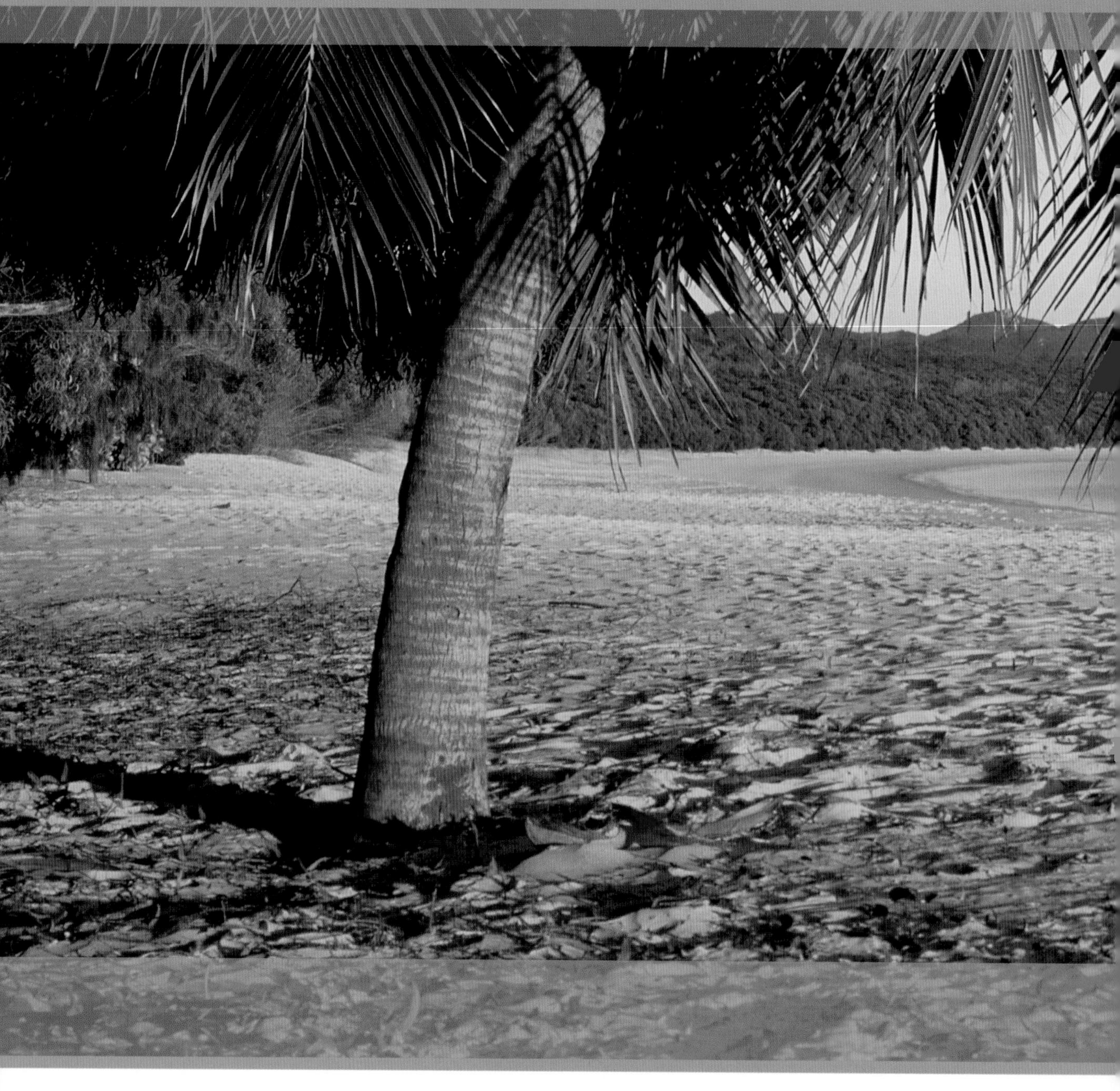

QUEENSLAND is...

Snorkelling or scuba diving among fish and coral in the GREAT BARRIER REEF / A daytrip to one – or more – of the GOLD COAST'S many theme parks / Four-wheel driving on FRASER ISLAND'S BEACHES / Fossicking for gemstones around ANAKIE, RUBYVALE, SAPPHIRE or WILLOWS GEMFIELDS / Discovering tropical rainforest in

QUEENSLAND

[WHITEHAVEN BEACH, WHITSUNDAY ISLANDS NATIONAL PARK]

DAINTREE NATIONAL PARK / Taking a ride on the KURANDA SCENIC RAILWAY, returning via the Skyrail Rainforest Cableway / Sailing in the WHITSUNDAYS / Walking through NOOSA NATIONAL PARK, followed by a coffee and shopping on HASTINGS STREET / Travelling on the GULFLANDER train between Normanton and the old goldmining town of Croydon

QUEENSLAND is Australia's second largest state and offers numerous idyllic holiday destinations. Myriad islands, cays and atolls are scattered along its 6973-kilometre coastline. The Great Barrier Reef offers the ultimate in diving, and there are 2000 species of fish, dugongs, turtles and extensive coral gardens, all protected by World Heritage–listing.

By contrast, the arid west gives visitors a chance to experience some of Australia's unique outback in towns such as Winton, established by those searching for the lost Burke and Wills expedition. Winton also has a special place in Australian folklore as the location of Dagworth woolshed where Banjo Paterson wrote the iconic *Waltzing Matilda* in 1895.

Two-thirds of Queensland lies above the Tropic of Capricorn. In the monsoonal Far North, visitors can venture into magnificent ancient rainforests, like those of the Daintree National Park, where cool respite lies in places such as the boulder-strewn Mossman Gorge.

South of Brisbane is the famous Gold Coast. With more waterways than Venice and 300 days of sunshine each year, it is the perfect place for swimming and surfing. The theme parks here will terrify and astound, while in the hinterland, an emerald-green paradise allows visitors to soak up magnificent views among waterfalls and rainforest trees.

[PANDANUS PALM, NOOSA NATIONAL PARK]

Captain James Cook and his crew were the first Europeans to unexpectedly enjoy the Queensland coast after they ran aground on a reef near Cape Tribulation in 1770. Dutch explorer Willem Jansz had sailed along the western side of Cape York 164 years earlier, but received a hostile reception from the local Aboriginal people.

European settlement of Queensland occurred quite late compared with the rest of Australia. In 1824 a convict station was built near Moreton Bay to cater for the most intractable prisoners from southern gaols, but after a year of active resistance by Aboriginal tribes it was abandoned and relocated to where Brisbane stands today.

In recent years Queensland has shaken off its reputation as a quiet backwater. This modern state is fast becoming the envy of the rest of the country with its stunning natural features, relaxed pace and languid lifestyle, all enhanced by a climate close to perfect.

[A PAIR OF RAINBOW LORIKEETS]

fact file

Population 4 516 400
Total land area 1 722 000 square kilometres
People per square kilometre 2.3
Sheep and cattle per square kilometre 12
Length of coastline 6973 kilometres
Number of islands 1955
Longest river Flinders River (840 kilometres)
Largest lake Lake Dalrymple (dam), 220 square kilometres
Highest mountain Mount Bartle Frere (1622 metres), Wooroonooran National Park
Highest waterfall Wallaman Falls (305 metres), Girringun (Lumholtz) National Park
Highest town Ravenshoe (930 metres)
Hottest place Cloncurry (average 37°C in summer)
Coldest place Stanthorpe (46 days per year begin at below 0°C)
Wettest place Tully gets 4300 millimetres per year, and has a giant gumboot to prove it
Sunniest town Townsville (average 300 days of sunshine per year)
Most remote town Birdsville
Major industries Sugar and mining
Most famous person Steve Irwin, the 'Crocodile Hunter'
Number of 'big things' 39
Best beach Whitehaven Beach, Whitsunday Island
Local beer XXXX

gift ideas

Arts and crafts (The Valley Markets, Brisbane) Quirky, unusual and interesting handmade arts, crafts and fashion. See Markets p. 408, 404 D1

Beer glasses (XXXX Brewery, Brisbane) A little something from Queensland's iconic XXXX brewery. See Milton p. 411, 402 D4

Pineapple souvenirs (Big Pineapple, Nambour) Take home your own slice of the Sunshine State, to eat or keep. See Nambour p. 467, 647 G4

Ginger products (Buderim Ginger Factory, Yandina) All Queensland grown and made, including everything from ginger chocolate to ginger perfume. See Yandina p. 482, 647 F3

'Crocodile Hunter' souvenirs (Australia Zoo, Beerwah) Choose from an incredible range of souvenirs from this internationally renowned zoo. See Landsborough p. 457, 647 F5

Macadamia nuts (Nutworks, Yandina) Delicious and unusual produce, as well as skin products from this working nut factory. See Yandina p. 482, 647 F3

Great Barrier Reef mementos (Reef HQ, Townsville) Beautiful photos, books and souvenirs of the Great Barrier Reef. See Townsville p. 480, 650 G10

Tropical fruit (Cairns & The Tropics region) Black sapote, soursop and carambola are just some of the Dr Seuss–sounding tropical fruits you can taste and purchase. Visit Cape Tribulation Exotic Fruit Farm at Lot 5, Nicole Dr, Cape Tribulation, and other producers around Far North Queensland. See Cairns & The Tropics p. 425, 657

Australian Arabica coffee (The Australian Coffee Centre, Mareeba) The Skybury Coffee company makes pure Arabica coffee from beans grown near the Great Dividing Range. See Mareeba p. 459, 650 D2

Vodka and liqueurs (Tamborine Mountain Distillery, North Tamborine) Award-winning vodka, schnapps and other liqueurs in hand-painted bottles from Australia's smallest pot still distillery. 87–91 Beacon Rd, North Tamborine. See North Tamborine p. 469, 645 E8

BRISBANE IS...

A bike ride through the CITY BOTANIC GARDENS / Breakfast by the river at the POWERHOUSE arts centre / People-watching on fashionable JAMES STREET / City views from MOUNT COOT-THA / The latest blockbuster expo at the GALLERY OF MODERN ART / Riverside dining at EAGLE STREET PIER / An adventure climb on the STORY BRIDGE / Hugging a koala at LONE PINE KOALA SANCTUARY / A stroll along the RAINFOREST WALK and a swim at STREETS BEACH in South Bank / A CityCat ferry cruise on the BRISBANE RIVER

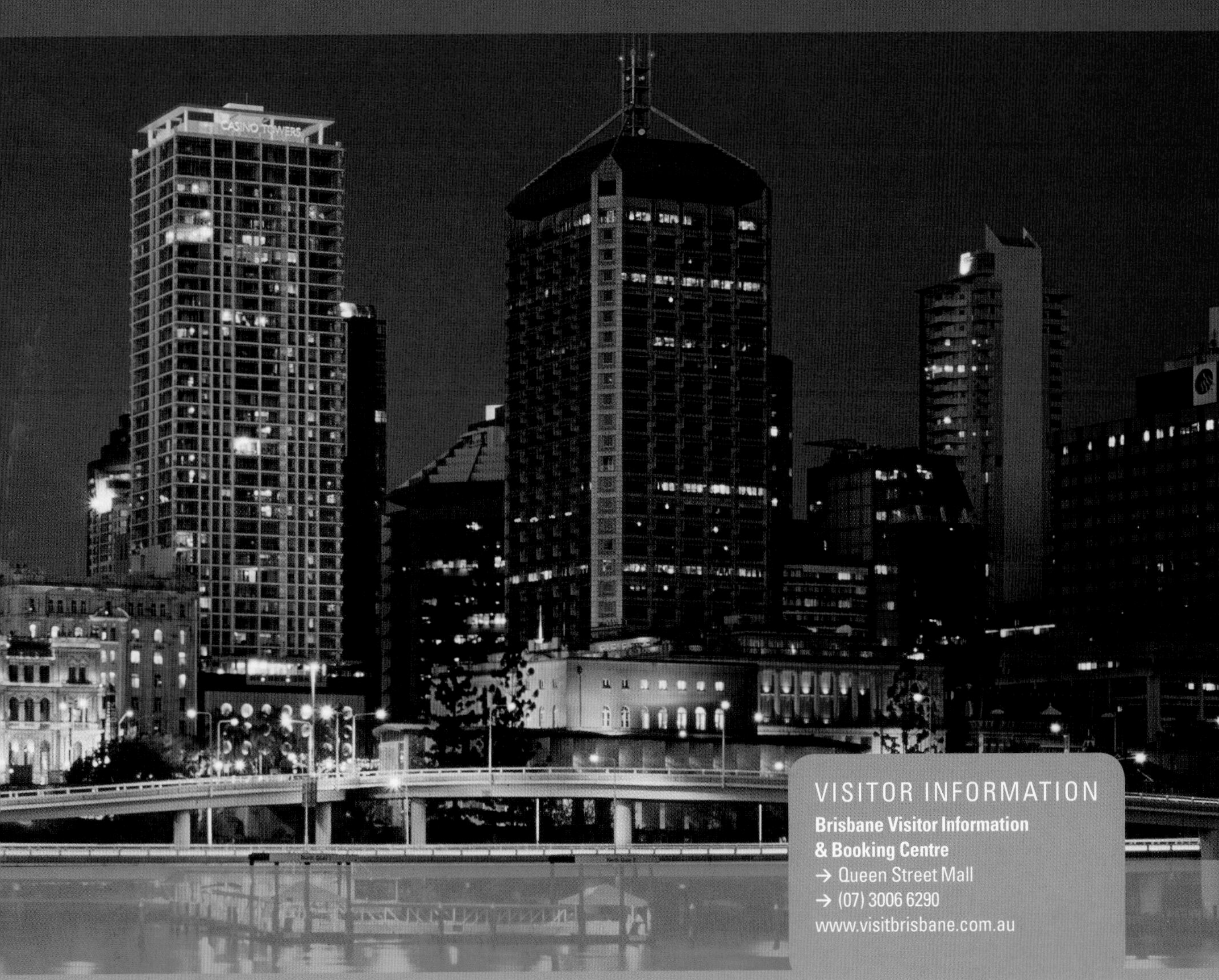

VISITOR INFORMATION

Brisbane Visitor Information & Booking Centre

→ Queen Street Mall

→ (07) 3006 6290

www.visitbrisbane.com.au

The subtropical climate may be warm, but Brisbane is decidedly cool. The population has doubled in the last two decades, and as a result, Brisbane has been busily reinventing itself. There is a lively young arts scene and the city has been ranked among the hottest places in the world for new music. Its young fashion designers are making a name for themselves, and the Gallery of Modern Art is bringing blockbuster exhibitions down under. But while the CBD skyline may be spiked by glittering high-rise buildings, and the river lined with big boats, thankfully, the city has lost little of its friendliness in the make-over. Life is as relaxed as ever, and firmly focused on the outdoors.

Brisbane's hilly terrain provides breathing space and a beautiful backdrop to the CBD. Step into the nearby suburbs and you will find stately Moreton Bay fig trees standing sentinel in the suburban streets and mango trees blooming in the backyards of those distinctive weatherboard houses on stilts known as 'Queenslanders'. With their shady verandahs and tin roofs just made for the patter of summer rain, you can still find them within walking distance of the CBD.

From the coast to the suburbs, the year-round warm climate means that Brisbane is tops for a holiday; whether you want city parks or national parks, markets or museums, nightlife or wildlife, you will find it all here.

MAP LEGEND
Attraction
Tourist information
Where to eat & where to stay
Shopping precinct
0
1 km
N
Upper Kedron
Keperra
KEPERRA COUNTRY GOLF CLUB
KEPERRA BUSHLAND
Mitchelton
MITCHELTON PARK
TERALBA PARK
Everton Park
ENOGGERA MEMORIAL PARK
Gaythorne
FERGUSON PARK
Stafford Heights
Stafford
Kedron
KEONG PARK
GIBSON PARK
GRINSTEAD PARK
GRANGE FOREST PARK
HICKEY PARK
Gordon Park
ALDERLEY GROVE
EMERSON PARK
Grange
LANHAM PARK
Lutwyche
ENOGGERA MILITARY CAMP (GALLIPOLI BARRACKS)
Enoggera
Alderley
SEDGELEY PARK
BANKS STREET RESERVE
Wilston
Windsor
The Gap
ASHGROVE GOLF CLUB
ASHGROVE SPORTS GROUND
DONNINGTON PARK
Ashgrove
Newmarket
SPENCER PARK
FINSBURY PARK
DOWNEY PARK
Northey Street City Farm Organic Growers Markets (Sun)
BALLYMORE PARK
RASEY PARK
Herston
PATEN PARK
JUBILEE PARK
GILBERT PARK
WOOLCOCK PARK
Kelvin Grove
Red Hill
VICTORIA PARK GOLF COMPLEX
Fortitude Valley
The Tivoli
Roundhouse Theatre
BOWMAN PARK
Bardon
Spring Hill
Paddington
Montrachet
Suncorp Stadium
Milton
XXXX Brewery
Mount Coot-tha Forest
JC Slaughter Falls
Mount Coot-tha
Brisbane Botanic Gardens (Mt Coot-tha)
Sir Thomas Brisbane Planetarium
Toowong Cemetery
Auchenflower
FREW PARK
MILTON PARK
Go Between Bridge
SEE PAGE 404
Brisbane
Mt Coot-tha Lookout
BRISBANE FOREST PARK
TOOWONG MEMORIAL PARK
DUNMORE PARK
Davies Park Markets West End (Sat)
DAVIES PARK
Toowong
Regatta Hotel
Toowong Village Shopping Centre
West End
Thomas Dixon Centre
ORLEIGH PARK
Taringa
Green Hill Reservoir
PERRIN PARK
Highgate Hill
Chapel Hill
MOORE PARK
Indooroopilly Shopping Centre
JACK COOK PARK
IRONSIDE PARK
St Lucia
Dutton Park
The Gabba
Woolloongabba
JACK SPEARE PARK
ROBERTSON PARK
University of Queensland
Walter Taylor Bridge
THE JACK BOWERS OVAL
Indooroopilly
ST. LUCIA GOLF LINKS
Kenmore
CUBBERLA CREEK RESERVE
GORDON THOMSON PARK
THOMAS PARK BOUGAINVILLEA GARDENS
GOODWIN PARK
Fairfield
FAIRVIEW PARK
RAINFOREST PARK
Chelmer
LEYSHON PARK
ROBINSON PARK
Greenslopes
EKIBIN PARK
GRACEVILLE MEMORIAL PARK
FAULKNER PARK
Indooroopilly Island
Graceville
INDOOROOPILLY GOLF CLUB
Yeronga
Annerley
Fig Tree Pocket
To Lone Pine Koala Sanctuary
Sherwood
BERT ST. CLAIR OVAL
YERONGA PARK
Yeerongpilly
BRISBANE RIVER

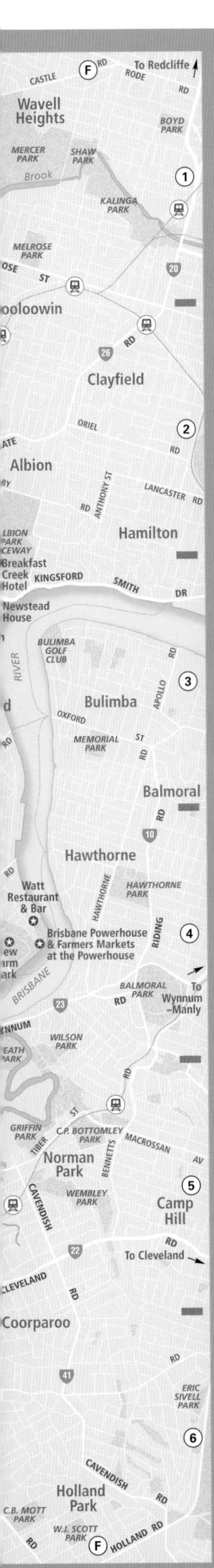

City Centre

Despite the fact that it is the commercial and retail heart of Brisbane, the city centre retains the buoyant holiday spirit that pervades the entire Sunshine State. If you lose your way in the CBD grid, remember that the streets named after royal women all run north-east and parallel to each other, while the king-inspired streets run north-west.

Queen Street Mall 404 B4

With QueensPlaza, Myer Centre, Broadway on the Mall, the heritage-listed Brisbane Arcade and MacArthur Central (adjoining the GPO), Queen Street Mall offers something for everyone and is the hub of the CBD. Catering only to pedestrians (but with the central bus station conveniently located beneath), it is packed with people seven days a week. Along with a public stage, there are outdoor cafes and restaurants dotted down its centre. At the top (the George Street end) is the Treasury Casino *(see next entry)*, the Brisbane Square Library and surrounding eateries, and the Victoria Bridge leading to the South Bank parklands. At the other end, the mall leads onto ANZAC Square *(see below)* and Post Office Square. Cutting through the middle is Albert Street, leading to more shopping on Elizabeth Street, including the Elizabeth Arcade – a focus for edgy fashion – and Adelaide Street with the Brisbane City Hall *(see below)* and the recently redeveloped King George Square. Get a literary insight into Brisbane by following the **Albert Street Literary Walk** – look for the 32 brass plaques in the pavement.

Treasury Casino 404 B4

One of Brisbane's spectacular buildings, the Treasury Building, built between 1885 and 1928, is now the Treasury Casino. With restaurants, bars and live bands nightly, the casino is a top nightspot even if you don't want to play the tables. *George St; open 24 hours daily; admission free.*

ANZAC Square 404 C3

Popular with lunchtime office workers and foraging pigeons and Ibis alike, ANZAC Square, between Ann and Adelaide streets, is a peaceful retreat. The square's **Shrine of Remembrance**, built in 1930, honours the Australian soldiers who died in World War I with its eternal flame. In the pedestrian tunnel behind the square is the World War II **Shrine of Memories** (open Monday to Friday 9am–2pm only) where you can see unit plaques, honour rolls and a mosaic made from hand-cut glass enamels and soils from official World War II cemeteries. From ANZAC Square take the steps to the walkways over Adelaide Street to reach **Post Office Square**, another of Brisbane's grassy public squares. Opposite the square is the **General Post Office**, built in the 1870s; even with the busy post office crowds, you can still get an impression of its history.

Brisbane City Hall 404 B3

The newly revamped **King George Square** is a popular place for public gatherings, and it's here you will find the historic Brisbane City Hall. Built throughout the 1920s and opened in 1930, this impressive sandstone building is topped by a soaring 92-metre clock tower. Unfortunately there are no tours inside – City Hall is closed for repairs until late 2012. *Between Ann and Adelaide sts; (07) 3403 8888.*

Museum of Brisbane 404 B3

With City Hall closed for restoration, the Museum of Brisbane has moved to a temporary location just around the corner from King George Square. Also known as MoB, the museum has several exhibition spaces and celebrates the history, culture and people of Brisbane. Displays incorporate design, craft and visual arts. The MoB Store is a good spot to pick up something created by one of Brisbane's talented writers, artists or musicians. *157 Ann St; (07) 3403 8888; open daily; admission free.*

St John's Cathedral 404 C2

This striking example of Gothic-Revival architecture was designed in 1888 and construction began 17 years later. St John's has the distinction of being the last medieval construction project of its kind in the world.

climate

Queensland isn't called the 'Sunshine State' for nothing, and with an average of 300 days of sunshine each year, the south-east corner has (according to many) the most liveable climate in the state and possibly the country. Summers are hot and steamy; winters are mild and dry. Summer days in Brisbane average in the high 20s or low 30s. The hot days often build up to spectacular evening thunderstorms, and Brisbane's annual average rainfall of 970 mm occurs mostly in summer. In winter the average temperature is 21°C – a very pleasant 'winter' for anyone from down south! Winter evenings can be a bit cool though, so make sure you pack at least one jumper.

J	F	M	A	M	J	J	A	S	O	N	D	
30	30	29	27	25	22	22	23	26	27	28	29	MAX °C
21	21	20	17	14	12	10	11	14	16	19	20	MIN °C
114	128	89	56	64	60	23	42	33	84	111	158	RAIN MM
8	11	8	7	6	6	3	4	4	7	9	10	RAIN DAYS

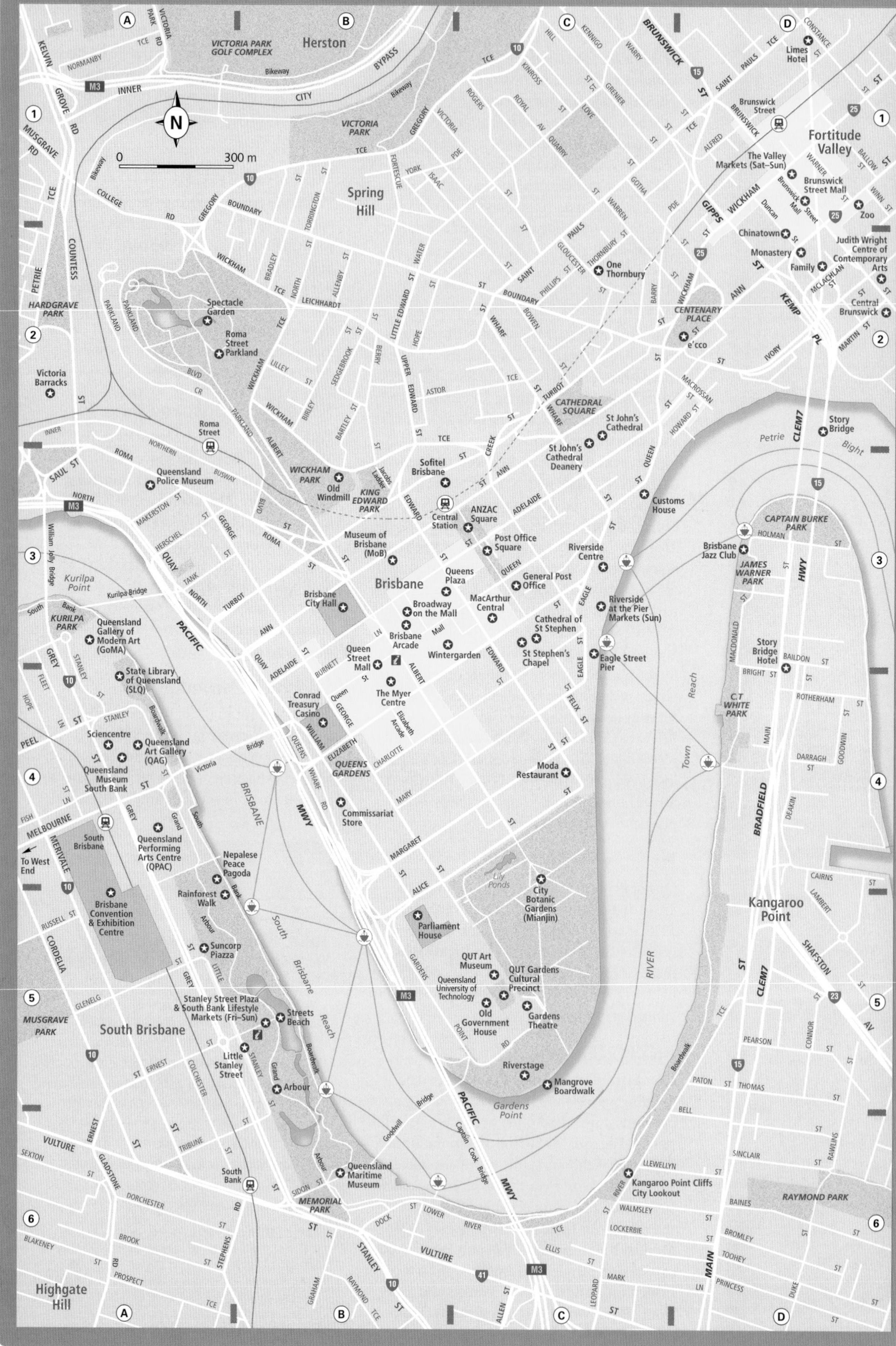

Herston
Victoria Park Golf Complex
Victoria Park
Spring Hill
Fortitude Valley
Brisbane
Kangaroo Point
South Brisbane
Highgate Hill
300 m
Limes Hotel
Brunswick Street
The Valley Markets (Sat–Sun)
Brunswick Street Mall
Zoo
Chinatown
Monastery
Family
Judith Wright Centre of Contemporary Arts
Central Brunswick
One Thornbury
Centenary Place
e'cco
Spectacle Garden
Roma Street Parkland
Victoria Barracks
Hardgrave Park
Roma Street
Queensland Police Museum
Wickham Park
Old Windmill
King Edward Park
Sofitel Brisbane
Central Station
ANZAC Square
Cathedral Square
St John's Cathedral
St John's Cathedral Deanery
Customs House
Story Bridge
Captain Burke Park
Brisbane Jazz Club
James Warner Park
Museum of Brisbane (MoB)
Post Office Square
Riverside Centre
Queens Plaza
General Post Office
MacArthur Central
Riverside at the Pier Markets (Sun)
Brisbane City Hall
Broadway on the Mall
Brisbane Arcade
Wintergarden
Cathedral of St Stephen
St Stephen's Chapel
Eagle Street Pier
Story Bridge Hotel
C.T. White Park
Queen Street Mall
The Myer Centre
Conrad Treasury Casino
Elizabeth Arcade
Queens Gardens
Moda Restaurant
Commissariat Store
Kurilpa Point
Kurilpa Park
Queensland Gallery of Modern Art (GoMA)
State Library of Queensland (SLQ)
Sciencentre
Queensland Art Gallery (QAG)
Queensland Museum South Bank
South Brisbane
Queensland Performing Arts Centre (QPAC)
To West End
Brisbane Convention & Exhibition Centre
Nepalese Peace Pagoda
Rainforest Walk
Suncorp Piazza
City Botanic Gardens (Mianjin)
Parliament House
QUT Art Museum
QUT Gardens Cultural Precinct
Queensland University of Technology
Old Government House
Gardens Theatre
Riverstage
Mangrove Boardwalk
Gardens Point
Musgrave Park
Stanley Street Plaza & South Bank Lifestyle Markets (Fri–Sun)
Streets Beach
Little Stanley Street
Arbour
South Bank
Queensland Maritime Museum
Memorial Park
Kangaroo Point Cliffs City Lookout
Raymond Park
Brisbane River
South Brisbane Reach
Town Reach
Petrie Bight
Pacific Mwy
Bradfield Hwy
Inner City Bypass

The original project ran out of funds and remained unfinished until, thanks to a recent injection of capital, it was finally completed 103 years after building started. It has the only fully stone-vaulted ceiling in Australia, as well as extensive woodcarvings by Queensland artists, sandstone arches and beautiful stained-glass windows. Next to the cathedral is the **Deanery**, built in 1850 and formerly the residence of Queensland's first governor, Sir George Bowen. Free tours are conducted daily. *373 Ann St; (07) 3835 2222.*

Cathedral of St Stephen 404 C3

This magnificent cathedral is a quiet place of worship amid the hustle and bustle of the city. The cathedral grounds include **St Stephen's Chapel**, the oldest surviving church in Queensland. Guided tours are available. *249 Elizabeth St; (07) 3336 9111.*

Commissariat Store 404 B4

Built by convicts in 1829 from Brisbane tuff – a local stone quarried at nearby Kangaroo Point – this is one of the oldest buildings in Brisbane (with newer additions). Today it is home to the offices, library and museum of the Royal Historical Society of Queensland. There is a convict display, and tours are available. *115 William St; (07) 3221 4198; open 10am–4pm Tues–Fri.*

Foreshore

The serpentine Brisbane River curves its way through the whole city and seems to be everywhere at once. It is the place for award-winning restaurants, cafes and spectacular vistas. Ferries and CityCats stop at the Riverside Wharf and near the Botanic Gardens. Hire a bike in the Botanic Gardens – it is a great way to see everything without having to mix it with the road traffic.

Customs House 404 C3

Built in 1889 and beautifully restored, this magnificent building on the Brisbane River served as the city's Customs House for almost a century until port activities shifted closer to the mouth of the river. Now a cultural and educational facility of the University of Queensland, it has function rooms and a brasserie with waterfront tables. On Sundays there are free guided tours. *399 Queen St; (07) 3365 8999; open daily; admission free.*

Riverside Centre & Eagle Street Pier 404 C3

These two neighbouring office precincts dominate the CBD reach of the river. By day they are the busy hub of corporate Brisbane, and their riverside cafes, bars and restaurants are packed with city professionals. At night the precincts turn on the glamour, and with the city lights twinkling on the water it's here you will find some of Brisbane's best-known and most-awarded restaurants. On Sundays they transform themselves once again, into the popular **Riverside at the Pier Markets** *(see Markets, p. 408)*. It's worth having a coffee on the boardwalk just to enjoy the views of the landmark **Story Bridge**, the largest steel cantilever bridge in Australia.

City Botanic Gardens 404 C5

These beautiful historic gardens, established in 1855 right in the heart of the city, are Queensland's oldest public gardens and recognised for both their natural and historic heritage. You can spend hours strolling along the avenues lined with majestic bunya pines and Moreton Bay figs, exploring the rainforest glade or taking the **Mangrove Boardwalk** along the bank of the river. Hidden gardens and shady nooks provide secluded picnic spots and, if you take the riverside path, you can enjoy stunning views of the river and Kangaroo Point Cliffs. Take a free guided tour, hire a bike or, if exercise isn't on the agenda, simply sit on the grassy foreshore and enjoy the views of the river and Kangaroo Point Cliffs. *Gardens Point, Alice St.*

getting around

Brisbane has well-signed, well-maintained roads, but it's not an easy city for first-time visitors to negotiate. The region's growth has resulted in a crisscrossing network of major motorways and some significant new roadwork projects. In the centre itself, there are many one-way streets. To make matters more confusing, the Brisbane River twists its way through the city and suburbs. An up-to-date road map and some careful route planning at the beginning of each day is a good idea. As the city has grown more crowded, cycling has become a viable and healthy option for getting around and more locals are taking to their bikes. Brisbane boasts an ever-expanding network of generally picturesque bikeways and pedestrian paths. The Cycle2City facility in the King George Square Bus Station is a first for Australia and offers showers, lockers, laundry and bike security for local city workers and visitors alike.

Trains, buses and ferries cater for all needs and a couple of bus routes are designed specifically for visitors *(see below)*. A boat trip on the Brisbane River is a must. Plenty of tours are available to riverside tourist attractions *(see Walks & tours, p. 411)* and there is an excellent commuter ferry and catamaran (CityCat) service. The Cats travel at high speed and standing on the deck and feeling the cool wind in your hair is the best way to see Brisbane. The Go Card can be used on all transport – ferries, trains and buses – and offers discounts for frequent users. The Off Peak Daily offers cheaper travel between 9am and 3.30pm and on all weekends and gazetted public holidays.

Public transport Translink (bus, ferry, CityCat and rail) 13 1230.

Airtrain Train from the airport to the city and Gold Coast 1800 119 091.

Airport shuttle bus Coachtrans Airport Service (07) 3358 9700.

Motoring organisation RACQ 13 1905, roadside assistance 13 1111.

Car rental Avis 13 6333; Budget 1300 362 848; Europcar 131 390; Hertz 13 3039; Thrifty 1300 367 227.

Taxis Black and White Cabs 13 1008; Yellow Cabs 13 1924.

Bus tours City Sights bus tours (07) 3403 8888; The Loop (free bus circling the CBD) 13 1230.

Bicycle hire Bicycle Revolution (07) 3342 7829; Gardens Cycle Hire, Botanic Gardens 0408 003 198.

shopping

Queen Street Mall Brisbane's premier shopping precinct, with major department stores, malls and arcades including QueensPlaza, Myer Centre, Brisbane Arcade, MacArthur Central and Wintergarden. Also young labels in Elizabeth Street. Everything you'll ever need for retail therapy. 404 B4

Fortitude Valley For up-and-coming designer fashion and innovative chic, collectibles and books, art and trendy homewares, adventure gear and trinkets. 402 E3

Emporium More than 35 specialty retailers offering top-end fashion, jewellery, art, books, wine and food. 402 E3

Little Stanley Street, South Bank Edgy designer fashion, boutique homewares and gifts. 404 B5

Paddington For antiques, fashion and boutique homewares on Latrobe and Given terraces. 402 C3

Logan Rd, South Brisbane A quirky street of antiques, edgy fashion, accessories and homewares. 402 E5

Westfield Chermside Shopping Centre Huge mall on Brisbane's north side with 350 specialty stores, all the major department stores and a 16-cinema complex including a Gold Class theatre. 644 E6

Indooroopilly Shopping Centre A mall in Brisbane's western suburbs with more than 250 specialty shops, major department stores, a gym and a 16-cinema complex. Fabulous shopping. 402 B5

Direct Factory Outlet (DFO) Brisbane Next to Brisbane Airport, this centre offers up to 120 discounted brands. 644 F6

Stones Corner Factory outlets and fashion seconds only 15 minutes south of the CBD. 402 E5

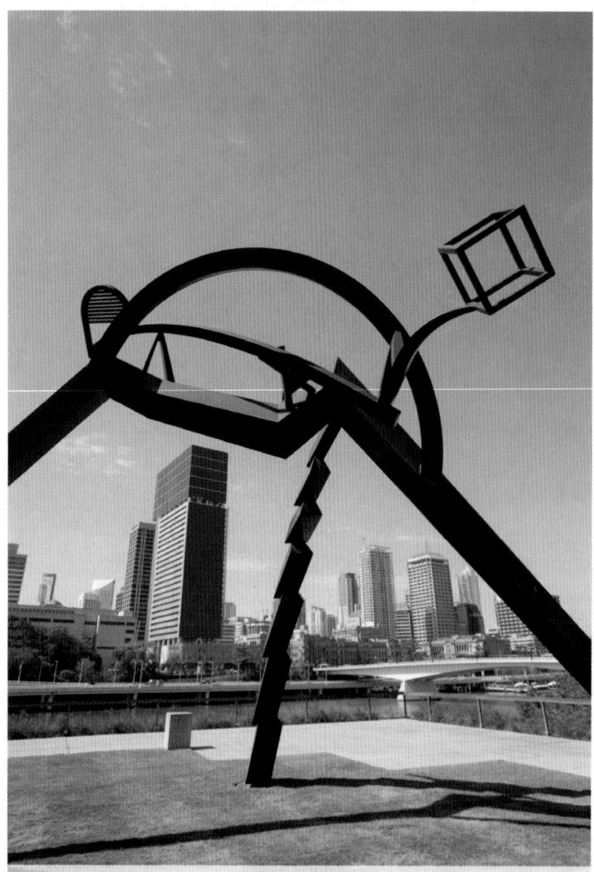

[SCULPTURE AT THE QUEENSLAND GALLERY OF MODERN ART]

QUT Gardens Cultural Precinct 404 C5

Launched in 2000, Queensland University of Technology's Gardens Cultural Precinct, at Gardens Point next to the City Botanic Gardens, encompasses **Old Government House**, the QUT Art Museum and the Gardens Theatre. Now run by QUT, the graceful sandstone Old Government House, built in 1860, was home to the Queensland governor for most of the state's first 50 years. *(07) 3138 8005; open 10am–5pm Sun–Fri; admission free.*

QUT Art Museum is housed in a 1930s neoclassical building and shows QUT's art collections, work by students and diverse contemporary exhibitions. *(07) 3138 5370; open 10am–5pm Tues–Fri, 10am–8pm Wed, 12–4pm Sat–Sun; admission free.*

The cultural venue of the **Gardens Theatre** offers shows by QUT students and visiting international and Australian theatre companies.

Parliament House 404 B5

Overlooking the City Botanic Gardens, this grand old seat of government was built in 1868 (new buildings have since been added to the precinct). The two sandstone wings hold majestic staircases, stained-glass windows and ornate chandeliers. Fringed with palms, the parliamentary precinct is a showcase of Queensland's history. Watch the Queensland State Parliament in action from the visitors' gallery when the House is sitting, or when it's not in session take one of the 30-minute tours that are run on demand. *Cnr George and Alice sts; (07) 3406 7562; open daily; admission free.*

Roma Street Parkland & Surrounds

Formerly rail yards and engine sheds, this area has been transformed by the creation of the Roma Street Parkland. Pedestrian links give access from the city centre, so it is an easy stroll when you need a break from Queen Street shopping. If you are coming into town via train, hop off at Roma Street Station and head to the Queensland Police Museum nearby. The museum has displays on themes such as police heritage, the Dog Squad and police investigative techniques. Open 9am–4pm Mon–Fri, 10am–3pm last Sun of the month (Feb–Nov). The military museum at Victoria Barracks has a collection of weapons and medals from 1860 to 1980. Open 12–4pm Sun (or by appt).

Old Windmill 404 B3

Brisbane's oldest convict-built structure on Wickham Terrace is also Australia's oldest extant windmill. Built in 1828 to grind flour, it was also used as a treadmill for punishment. In 1861 it was converted into a signal station. Since then it has been used as both an observatory and a fire tower, and in 1934 the first successful experimental television transmittal was made from it. It no longer has its original sails and treadmill, but it remains as an evocative reminder of the convict past. An information board lists its fascinating history and it is a pleasant walk from the CBD up the Jacobs Ladder steps through the leafy King Edward Park.

Roma Street Parkland 404 A2

This huge subtropical garden is a wonderful haven in Brisbane's city centre and an easy walk from Queen Street through King George Square. The 16 hectares of parkland include landscaped gardens, Queensland's largest public art collection and hundreds of unique plants. Meander along the network of pathways or simply escape from the summer heat under the shade of one of the many mature trees. Kids can spend time in the playground or play on the Celebration Lawn. There is a lake precinct and a subtropical rainforest, and the **Spectacle Garden** is the parkland's horticultural heart. Enjoy a barbecue in one of the many picnic areas or have someone else do the cooking at the parkland's cafe. There are brochures for self-guide themed walks, or you can take a guided tour or hop on the trackless train that travels the paths.

South Bank

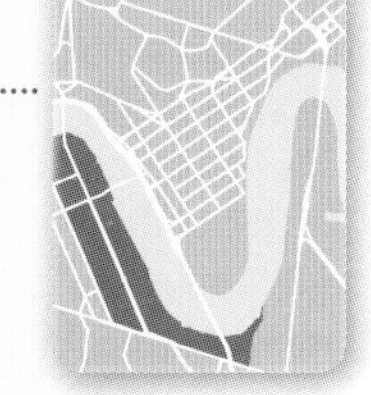

Set aside a day to spend at South Bank; with its large areas of parklands and some of Queensland's major cultural institutions, it's one of Brisbane's favourite places to play and celebrate. On the opposite side of the river to the CBD, it is an easy walk across Victoria Bridge from Queen Street, or the pedestrian-and-cycle-only Goodwill Bridge. Throughout the year thousands gather here for Christmas and New Year parties, and events such as the Brisbane Festival with its Riverfire pyrotechnic extravaganza. There are also regular concerts and multicultural events.

The Arbour

The beautiful kilometre-long Arbour – created by tendril-like metal columns covered in vibrant magenta bougainvillea – winds through the length of the South Bank precinct. Along the way, stop to watch some of the free entertainment – from buskers and street performers to international acts in the 2600-seat **Suncorp Piazza**. For something more peaceful, discover the ornate **Nepalese Peace Pagoda** set among tropical rainforest trees or take a detour and stroll along the shady **Rainforest Walk**, a raised boardwalk through the sort of subtropical vegetation for which Queensland is famed.

Kangaroo Point

You can walk east along the **Cliffs Boardwalk** to watch rock climbers and abseilers do their stuff on **Kangaroo Point Cliffs**, located on the bend of the river beyond the Captain Cook Bridge. If you're keen, you can learn to abseil on the spot. From the Kangaroo Point Cliffs City Lookout there are great views of the city.

Streets Beach 404 B5

A sandy beach right in the centre of Brisbane, Streets Beach has a lagoon with enough water to fill five Olympic swimming pools. Patrolled by lifesavers and with views across the Brisbane River to the CBD, the lagoon is popular with kids and sunbathers alike. Children also love South Bank's water playground, **Aquativity**, with its many fountains and water themed sculptures, and the ultra-modern playground at Picnic Island Green.

Clem Jones Promenade

Walking or cycling along the riverside promenade is a great way to see the sights of South Bank. This is also a popular eating spot, with more than 50 cafes, restaurants and bars including the new **Riverbend** riverfront dining precinct. There are plenty of shady spots to sit back, relax and enjoy a good meal or a coffee among sculptures, tropical foliage and water features. Plus you'll find barbecue areas ideal for families or groups.

Queensland Performing Arts Centre 404 A4

This is the place for theatre, music and opera. The Cremorne Theatre is dedicated to experimental theatre; the Playhouse offers cutting-edge technology in stage design; the Lyric Theatre is the flagship of the centre, hosting everything from Opera Queensland performances to the latest blockbuster musical; and the Concert Hall is home to regular performances by the Queensland Orchestra as well as international classical, jazz and pop artists. At the ground-floor entrance of the Cremorne Theatre is the Tony Gould Gallery, the exhibition space for the **QPAC Museum**, which has free exhibitions related to the performing arts. There are cafes and a restaurant, perfect for post-theatre suppers. Guided tours of the centre must be booked in advance. *Cnr Grey and Melbourne sts; guided tours (07) 3840 7444; general admission free.*

Queensland Museum South Bank 404 A4

This excellent museum has an extensive natural history collection, including a fascinating endangered species exhibit, and displays on Queensland's Indigenous and European history. Interactive displays for kids sit alongside the skeleton of a humpback whale and dinosaur displays. Plan to spend a few hours here and take in **Sciencentre**, an interactive science experience. *Cnr Grey and Melbourne sts; open daily; general admission free.*

Queensland Art Gallery

Overlooking the Brisbane River, Brisbane's premier cultural attraction houses the state's permanent art collection as well as regular special exhibitions. Not to be missed is the Watermall, an internal feature that runs the entire length of the building and is flooded with natural light from a huge skylight. Free guided tours are available. The **Queensland Gallery of Modern Art (GoMA)**, which opened at the end of 2006, showcases cutting-edge contemporary art and is home to the highly successful Asia–Pacific Triennial of Contemporary Art. GoMA's architecture is equally impressive, encouraging a connection between the

top events

Paniyiri Greek Festival Enjoy Greek food, dance, music, cooking classes and live entertainment as Musgrave Park, South Brisbane, is transformed with markets, tavernas and all things Greek. May.

Brisbane Winter Racing Carnival Excitement on and off the racecourse. May–June.

Ekka (Royal Queensland Show) A Brisbane institution where the city meets the country for ten days, with fireworks and wild rides. August.

Brisbane Festival Queensland's largest celebration of the performing and visual arts, featuring fireworks, fantastic food and live music. September.

Brisbane Jazz Festival at the Powerhouse Some of the world's most acclaimed musicians perform innovative jazz. October.

Brisbane International Film Festival Superb showcase of the latest in Australian and overseas film, with an international atmosphere. November.

Woodford Folk Festival Huge award-winning folk festival, just out of Brisbane, with local, national and international musicians and artists, and lots of stalls. December.

markets

Riverside at the Pier Markets Open-air markets on the city reach of the Brisbane River at Waterfront Place, with bric-a-brac, craft, fashion, food. Sun. 404 C3

South Bank Lifestyle Markets Street performances, art and craft, jewellery and live music. Fri night and Sat–Sun. 404 B5

The Valley Markets Brisbane's alternative markets, with vintage clothing, tarot readings, gifts, old books and an exciting atmosphere. Brunswick St Mall and Chinatown Mall; Sat–Sun. 404 D1

Powerhouse Farmers Markets Mix in with the locals combing the stalls for locally grown, farm-fresh produce, gourmet food, cut flowers and fresh seafood, all in the atmospheric surrounds of the Brisbane Powerhouse. New Farm, 2nd and 4th Sat each month. 403 F4

Davies Park Market West End Friendly, cosmopolitan markets on the river at Davies Park, West End, with fruit and veg, plenty of organics, bargains, art and craft, and free entertainment. Sat. 402 C4

Northey Street City Farm Organic Growers Markets Brisbane's only completely organic market, with organic produce and a nursery, set in a permaculture garden at Northey Street City Farm, Windsor. Sun. 402 E3

art and the public, and offering river views. *Stanley Pl; open 10am–5pm Mon–Fri, 9am–5pm Sat–Sun; general admission free.*

State Library of Queensland 404 A4

The State Library of Queensland must rank as one of the most beautiful repositories of books, archives, manuscripts, images, prints and maps you'll find anywhere. There are regular exhibitions and events (visit www.slq.qld.gov.au for details). *(07) 3840 7666; open 10am–8pm Mon–Thurs, 10am–5pm Fri–Sun.*

Queensland Maritime Museum 404 B6

No maritime enthusiast should miss this local institution, which is self-funded and run by volunteers. There are three galleries displaying relics, memorabilia and exhibits, such as a wall map of vessels lost in Queensland seas – more than1500 since the first recorded wreck in 1791. Look for displays on sailing, old diving equipment and the navigation instruments used by early explorers of Queensland's coastline; there's also an impressive collection of historic seagoing craft, ocean-liner replicas and nautical machinery. In the Dry Dock explore Australia's only remaining World War II frigate, HMAS *Diamantina*. You can also see a coal-fired, steam-powered 1925 tug, the SS *Forceful. Sidon St at South Bank (next to Goodwill Bridge); open daily from 9.30am (last entry 3.30pm).*

where to eat

Anise Bistro & Wine Bar Exciting small bar and equally exciting food at this tiny French treasure. 697 Brunswick St, New Farm; (07) 3358 1558; open Fri–Sat for lunch and Mon–Sat for dinner. 402 E4

e'cco Fabulous modern Australian bistro; one of the very best in town. 100 Boundary St; (07) 3831 8344; open Tues–Fri for lunch and Tues–Sat for dinner. 404 D2

Moda Contemporary Australian cuisine, with influences from Spain, France and Italy. An innovative and changing menu uses local and seasonal produce. 12 Edward St; (07) 322 7655; open Tue–Fri for lunch and Mon–Sat for dinner. 404 C4

Montrachet Local pinnacle of French cuisine; fun and relaxed with superb seafood. 224 Given Tce, Paddington; (07) 3367 0030; open Mon–Fri for lunch and Mon–Thurs for dinner. 402 D3

Watt Restaurant + Bar Fusion fare in a riverside setting at New Farm Park. 119 Lamington St, New Farm; (07) 3358 5464; open Tues–Sun for lunch, Tues–Sat for dinner and weekends for breakfast. 403 F4

Grey Street precinct 404 B5

The success of South Bank has spilled over into the adjoining area, especially Little Stanley and Grey streets, where you will find cafes and shops and people out and about enjoying the sunshine. **Little Stanley Street** is home to designer fashion stores, gourmet food shops and great cafes, bars and restaurants. Closed to traffic, **Stanley Street Plaza** also has eateries aplenty and buzzes day and night. **Grey Street**, a grand tree-lined boulevard, has shops, cafes, a hotel and a cinema complex as well as the nearby **Brisbane Convention and Exhibition Centre**, a 7.5-hectare complex used for regular trade and retail exhibitions.

Fortitude Valley

Known by the locals simply as 'the Valley', this cosmopolitan area of inner-city Brisbane is a fascinating mix of seedy history, stylish restaurants and alternative-lifestylers. These days it's more hip than hippie, and still dances to a rock 'n' roll beat as the seeding ground for young bands making their way onto the world stage. First settled in 1849 by 256 free settlers who arrived in Moreton Bay aboard the SS Fortitude, *the Valley still retains much of its 19th-century heritage. Its old buildings are home to many private art galleries and artists' studios. The creative hub is the Judith Wright Centre of Contemporary Arts.*

Brunswick Street Mall 404 D1

Dominated for almost a century by the McWhirters building, formerly a department store and now stylish apartments with shops at street level, Brunswick Street is packed with shoppers at weekends cruising through the **Valley Markets** where the vibe is busy but always relaxed *(see Markets, this page)*. Free bands play in the mall on Saturdays and Sundays, and at night crowds flock to the restaurants and nightclubs. Enjoy the alfresco dining at any time of day. Further up Brunswick Street, you will find another foodies' delight at **Central Brunswick**. For cutting-edge fashion go to Ann Street which runs across Brunswick Street, or to the TCB – formerly the TC Beirne Building – which links Chinatown with Brunswick Street Mall. Further down Ann Street there are more pubs and bars, including the GPO, in the old Fortitude Valley Post Office.

Chinatown 404 D2

Brisbane's Chinatown is a sensory delight. Restaurants offer excellent value and even better yum cha, supermarkets have traditional Chinese treasures and the busy Valley Markets extend into Chinatown at weekends. Elements of feng shui were used in the design of this mall on Duncan Street and you can expect to see peaceful practitioners of tai chi there on weekends. If you are in Brisbane in late January/early February, head to Chinatown for Chinese New Year celebrations.

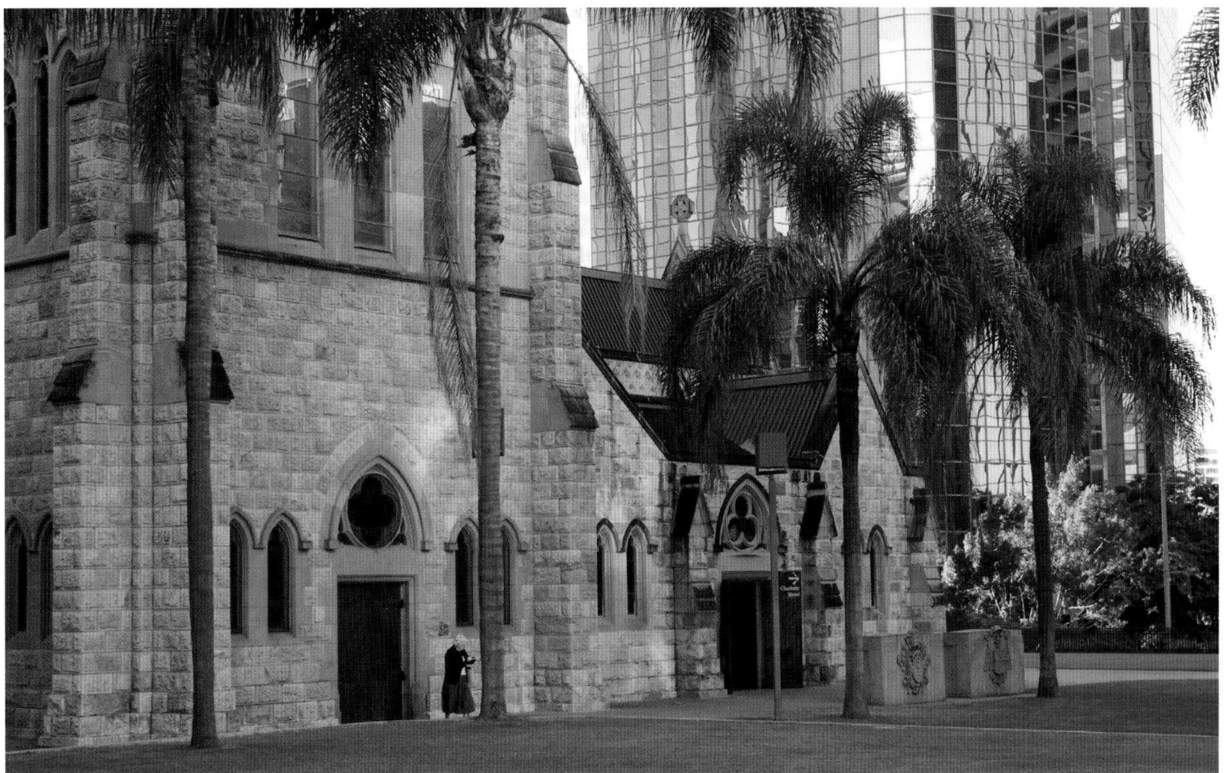

[CATHEDRAL OF ST STEPHEN]

James Street precinct 402 E3

Lined with shade trees, James Street is Brisbane's hottest new lifestyle precinct, where homewares and furniture stores cater to the cashed-up new inner-city residents, while outdoor cafes, restaurants and bars satisfy appetites for good coffee, fine dining and cocktails. The **James Street Market** is an excellent spot for picking up fresh food and produce along with flowers, nuts and juices. The **Centro** development includes a cinema complex. A short walk across Ann Street, **Emporium** has a big 'village square' of outdoor cafes, restaurants, high-end fashion shops, gifts and bookshops. A luxury hotel is here, along with the adjoining French restaurant and patisserie Belle Epoque.

New Farm & Newstead

Adjoining Fortitude Valley are other lively suburbs well worth a visit. New Farm is located on a peninsula created by the deep curves of the Brisbane River. Further along the river, Newstead and Teneriffe's historic wool stores and warehouses have been converted into stylish apartments, cafes and shops.

New Farm Park 403 F4

Originally an 1846 racecourse until it was bought by the Brisbane City Council in 1913, New Farm Park is a favourite spot for locals at any time, in any season. Stroll through the rose garden, picnic under the trees, listen to a band in the Rotunda, contemplate the river, and let the kids loose in the much-prized playground. First settled in the 1800s, the inner-city suburb of New Farm has a growing population and changing dynamic. The strongly Italian local community has now been joined by fashion-conscious young inner-urbans. There is still a village feel, however, with outdoor cafes, fashion shops, art galleries and bookshops.

Newstead & Teneriffe 403 F3

Between New Farm and Breakfast Creek, these suburbs once encompassed one of Brisbane's major industrial precincts. Its deep-water frontage made it the perfect dock for ships and in the late 19th century a railway was built, bringing export wool from the booming inland sheep properties. In 1987 development zoning was changed from industrial to high-density residential, and these days the rugged industrial architecture of the old wool stores, warehouses, wharves and laneways provides the framework for stylish restaurants, cafes and bars, and antiques and homewares shops.

where to stay

Emporium Hotel Modern, luxury boutique hotel in a precinct of up-market shops and restaurants, just a short walk to Fortitude Valley. 1000 Ann St, Fortitude Valley; (07) 3253 6999 or 1300 883 611. 402 E3

Limes Hotel Australia's first Design Hotel; each of the 21 rooms has a balcony or courtyard with hammock, plus there's a rooftop bar. 142 Constance St, Fortitude Valley; (07) 3852 9000. 404 D1

One Thornbury Boutique B&B in a character building built in 1886, close to CBD and Fortitude Valley. 1 Thornbury St, Spring Hill; (07) 3839 5334. 404 C2

Sofitel Brisbane Luxury in an unbeatable location with proximity to the CBD and adjacent Central Railway. 249 Turbot St; (07) 3835 3535. 404 B3

QUEENSLAND

entertainment

Cinema Major cinemas can be found at the Myer Centre in the city (a Birch Carroll & Coyle complex) and at most of the big suburban malls. See the *Courier-Mail* for movie times. If you're in town in November you can catch the Brisbane International Film Festival (BIFF). Brisbane is a city devoted to film festivals. Look out for them throughout the year.

Live music The Valley is the centre of Brisbane's live music scene and has been the birthplace of the 'Valley Sound' – a collective name for some of Australia's best and most innovative bands. Crowded with partying people most evenings, Brunswick and Ann streets are lined with fantastic bars, pubs, nightclubs – including the award-winning multilevel Family, and the monastically imbued Monastery – and alternative music venues such as the Zoo, a Brisbane institution. Kangaroo Point has some good live-music venues, including the renowned Story Bridge Hotel and the Brisbane Jazz Club. For jazz you could also head to West End. Popular alternative acts play gigs at the Tivoli in the Valley; international acts play the Brisbane Entertainment Centre at Boondall. Buy a copy of the *Courier-Mail* on Thursdays and Fridays for music listings, or pick up one of the free street papers such as *Time Off* for details of what's on.

Classical music and performing arts Brisbane has a pulsing creative life with theatres of all sizes spread throughout the city. For high-end performing arts and music, the premier venue is the Queensland Performing Arts Centre at South Bank, which includes the Concert Hall and the Lyric Theatre. The Powerhouse, on the river at New Farm, is the place to go for world-ranking contemporary music, dance and theatre. The Judith Wright Centre of Contemporary Arts in the Valley focuses on cutting-edge contemporary performance. West End is the address of the Queensland Ballet, in the Thomas Dixon Centre. The Valley has a number of smaller venues for experimental and alternative theatre, and there are often live street performances taking place in Brunswick Street Mall. The Roundhouse Theatre in the Kelvin Grove Urban Village is the venue for the La Boite Theatre Company and some brilliant productions. See the *Courier-Mail* for details of what's on.

Brisbane Powerhouse 403 F4

This live arts precinct, in the restored 100-year-old New Farm Powerhouse, is the place to go for contemporary music, theatre, dance and art. Go to www.brisbanepowerhouse.org for details of upcoming shows. The landmark building itself is worth a visit and it is a great spot for lunch, dinner or just a coffee by the river. Every second and fourth Saturday its forecourts are packed with locals foraging for the best in fresh food at the Farmers Markets. It's also popular with dog owners and the markets run a doggie cafe *(see Markets, p. 408). 119 Lamington St, New Farm; (07) 3358 8622; open 9am–5pm Mon–Fri, 10am–4pm Sat–Sun.*

Newstead House 403 F3

Brisbane's oldest surviving residence, Newstead House was built in 1846 by Patrick Leslie, one of the early pioneers of the city. Beautifully restored and furnished in the style of the late Victorian era, with its spacious verandahs, formal gardens and lawns running down to the river, it recreates a quintessential Australian homestead. *Newstead Park, Breakfast Creek Rd, Newstead; (07) 3216 1846; open 10am–4pm Mon–Thurs, 2pm–5pm Sun.*

Breakfast Creek 403 F3

Named by explorer John Oxley in 1826 when he stopped one morning to eat while charting the Brisbane River, Breakfast Creek sits on a wide, open stretch of water. Here you will find the historic **Breakfast Creek Hotel** – or the 'Brekkie Creek', as it is affectionately dubbed by the locals. Built in 1889, but given a stunning contemporary renewal in 2003, this Queensland institution is the ideal place for top-class steak and a cold beer. Nearby is **Breakfast Creek Wharf** with restaurants, cafes and river boat tours.

Inner North-east

Miegunyah House Museum 403 E3

Built in 1886 in traditional Queensland fashion with long verandahs and ornate ironwork, Miegunyah has been restored to its original style. It is now home to the Queensland Women's Historical Association and serves as a memorial to the state's pioneering women. Take a guided tour or enjoy morning or afternoon tea on the wide verandah overlooking the garden. *35 Jordan Tce, Bowen Hills; (07) 3252 2979; open 10.30am–3pm Wed, 10.30am–4pm Sat–Sun.*

Racecourse Road & Hamilton 403 F2

Drive through Hamilton and Ascot if you want to see some stunning examples of grand 19th-century Queensland homes. And if you feel like a flutter and a day out at the races, **Eagle Farm** is the state's premier track. Racecourse Road, which runs from the river to the racecourse, is the place to go for smart outdoor restaurants and cafes – and if you need to pick up a hat for the occasion, there are some excellent high-end fashion shops. On Kingsford-Smith Drive, **Brett's Wharf** has an acclaimed seafood restaurant and a ferry stop. Further along the river, the newer **Portside** development is Brisbane's international cruise-ship terminal and offers smart restaurants, quality shopping and cinemas.

Inner East

Bulimba & Woolloongabba 402 E5

The renewal and expansion of Brisbane has seen the proliferation of cafe and shopping strips right across the inner suburbs from north to south. It's in the suburbs that you will find some of the best coffee and most innovative little gift, homewares and fashion shops. On the southside, Oxford Street at **Bulimba** offers a leafy strip that is a short ferry ride from the CBD, and provides a pleasant break from the city bustle. **Logan Road** is one of Brisbane's main thoroughfares, but running south from the Story Bridge through Woolloongabba there is a little dog-leg behind the **Gabba** cricket ground that's a 19th-century hideaway of antique shops, tiny cafes, top restaurants, young designer fashion and quality homewares.

Inner South

West End 402 C4

Close to the CBD but a million miles away in mood, this is the suburb of choice for artists, students and those seeking a low-key lifestyle. If you want buildings with character, a multicultural mix of cafes and restaurants, organic provedores and alternative fashion, this is the place to find them. You can see19th-century houses throughout the suburb and the 1912 Gas Stripping Tower in Davies Park is a local landmark. The park is also home to organic markets on Saturdays *(see Markets, p. 408)*.

Inner West

Milton 402 D4

Dominated by the iconic XXXX Brewery *(see next entry)* and the legendary sportsground the Suncorp Stadium, (formerly Lang Park), Milton is also known for the bustling Park Road. Developed 20 years ago and endowed with a mock mini-Eiffel Tower, this is a lively street of restaurants and cafes, fashion, jewellery and bookshops. There are more restaurants, cafes and delis at nearby **Baroona Road** and **Nash St, Rosalie**.

XXXX Brewery 402 D4

Brisbanites are familiar with the pungent smell of brewing beer wafting from the Castlemaine Perkins XXXX Brewery. For beer-loving visitors, the brewery's Ale House is a must. Bookings are recommended for tailored tours that finish with a beer sampling. *Cnr Black and Paten sts (just off Milton Rd), Milton; (07) 3361 7597; open 10am–5.30pm Mon–Tues and Thurs–Fri, 10am–9.30pm Wed, 11.30am–3pm Sat, 11.30am–3pm Sun.*

Paddington 402 C3

Known for its distinctive style, the streets of this hilly inner-city suburb are lined with old timber cottages and on **Latrobe** and **Given terraces** many have been converted into small shops, cafes and businesses. Together, the two terraces make for a very long shopping strip that changes in character from one end to the other. At the Latrobe end there are some excellent antique shops and small art galleries along with quirky cafes and recycled fashion shops. Closer to the city, Given Terrace offers quality fashion, an organic provedore and some of Brisbane's best restaurants. Also here is the popular Paddington Tavern, better known to locals as the 'Paddo'. **Caxton Street**, at the Petrie Terrace end of Paddington, is close to the Suncorp Stadium and offers nightlife and a variety of bars and restaurants.

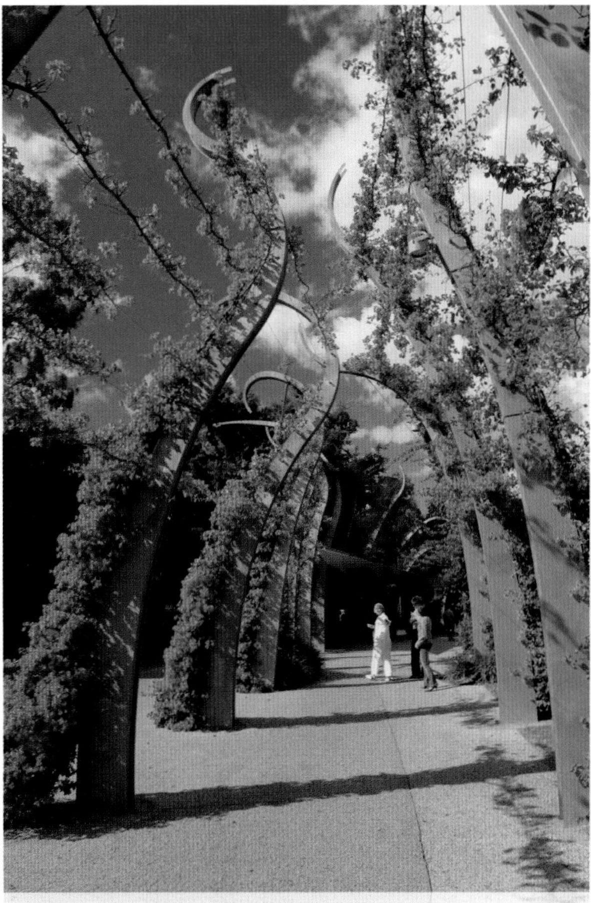

[COLOURFUL ARBOUR IN SOUTH BANK]

QUEENSLAND

walks & tours

Balloons Over Brisbane See the sun rise over Brisbane city at dawn – take a hot-air balloon flight, followed by a gourmet champagne breakfast. Bookings on (07) 3844 6671.

Brisbane CityWalk Explore Brisbane's green heart on this leisurely, self-guide walking tour that takes in the CBD's highlights via its three main parkland areas: the City Botanic Gardens, South Bank and Roma Street Parkland. Brochure available from the visitor centre in the Queen Street Mall.

Brisbane's Living Heritage Network Grab a copy of the Network's free guide, *Cultural & Heritage Places...of Greater Brisbane*, and discover the historic places dotted around the city and its surrounds. Booklets are available from the visitor centre and Brisbane City Council. (07) 3006 6290.

City Sights Bus Tours See the cultural and historic attractions of Brisbane in comfort. Set your own pace – you can hop off and on the clearly signed buses at any time. Brochure and tickets available from the visitor centre or call (07) 3404 8888 for more information.

Gonewalking Discover Brisbane on foot with the help of experienced Brisbane City Council volunteers in this free community walking program. Walks take about an hour and cover between 3.5km and 6.5km, depending on your pace. (07) 3403 8888.

Ghost Tours Scare yourself silly on one of a variety of serious ghost tours exploring Brisbane's haunted history. You can also pick up a copy of the Haunted Brisbane guidebook. Bookings on (07) 3344 7265.

Moreton Bay Cruises Explore beautiful Moreton Bay on one of the many cruises on offer. Details from the visitor information centre.

River tours Travel the river up to Lone Pine with Mirimar Cruises (1300 729 742) or enjoy a fine meal with stunning views on a Kookaburra River Queen cruise ((07) 3221 1300).

Story Bridge Adventure Climb Take in spectacular 360-degree views of the city on this 2.5-hour climb up Brisbane's iconic Story Bridge. Bookings on 1300 254 627.

Wynnum Manly Heritage Trail Once you've explored historic Brisbane, do the same in the bayside area. Brochures available from the Wynnum Manly visitor centre.

West

Toowong 402 C4

Toowong Cemetery has more than 100 000 graves, some dating back to 1875. Its headstones tell the story of the trials, tribulations and triumphs of Queensland's early settlers. This is also the final resting place of one prime minister, two Queensland governors, 13 Queensland premiers, 11 Queensland Labor leaders, at least 15 Brisbane mayors and many other prominent political, religious, sports, arts and business figures. Also built at the end of the 19th century, but very much a part of the living, the **Regatta Hotel** enjoys self-proclaimed 'oarsome' river views. A stunning revamp in 2001 has made it a dining and drinking spot that well deserves its popularity.

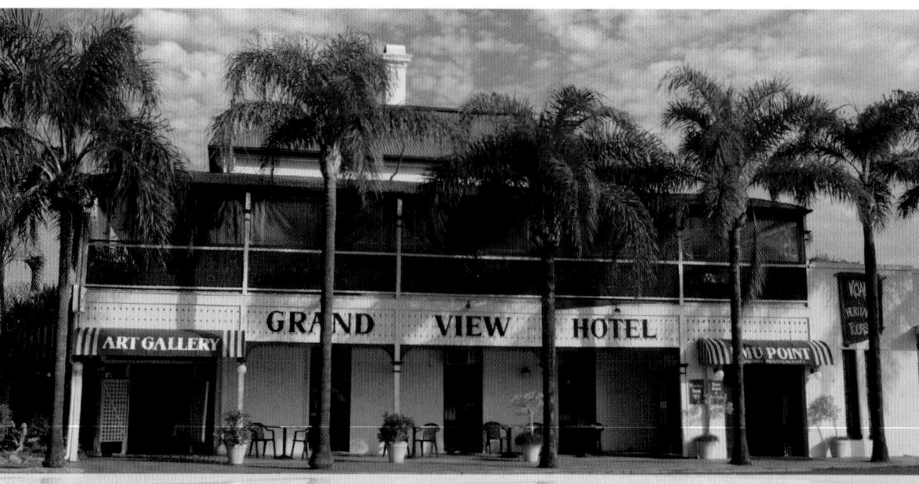

[HISTORIC GRAND VIEW HOTEL, CLEVELAND]

Mount Coot-tha Forest 402 A4

To get a bearing on Brisbane's geography, drive up **Sir Samuel Griffith Drive** to the **Mount Coot-tha Lookout**. You will see the Brisbane River snaking through leafy suburbs, impressive views of the CBD and, unless it's very hazy, the islands in Moreton Bay. To the west are rugged bush-covered ranges. Look down at your feet and you will see that the lookout's paving reflects the layout of the city, the river and the ocean. There are beautiful walks and picnic areas in the park, including the local favourite, **J. C. Slaughter Falls.**

sport

Brisbane's weather is perfect for sport, and Brisbane's stadiums are state of the art. If **AFL** (Australian Football League) is your passion, you can't miss the Brisbane Lions at their home ground, the Gabba. The Gabba is also home to Queensland **cricket** – watch the Bulls defend the state's cricketing honour. You will find the Gabba (known formally as the Brisbane Cricket Ground) in the suburb of Woolloongabba, south of Kangaroo Point.

Rugby League is a way of life in Queensland, culminating in State of Origin. You can watch the Brisbane Broncos, the city's rugby league team, at the redeveloped Lang Park (also known as Suncorp Stadium), a Brisbane sporting institution in Milton. Ticket-holders enjoy free public transport on match days. Or show your true colours by supporting the Reds, Queensland's **Rugby Union** team, at their matches, which are also at Lang Park.

The Brisbane International **tennis** tournament attracts the world's leading players in the first week of January. Matches are held at the Queensland Tennis Centre in Tennyson.

If you prefer the sport of thoroughbreds, head to Brisbane's **horseracing** venues at Doomben and Eagle Farm. Or for car racing at its most thrilling, head south for the **Gold Coast 600 Supercarnivale** in October.

Brisbane Botanic Gardens – Mount Coot-tha 402 B4

At the foot of Mount Coot-tha, these botanic gardens are well worth a visit for the Bonsai House, Fern House, Japanese Garden, the Tropical Dome and the Australian Plant Communities section, which showcases native species. Free guided walks are available. Located in the gardens is the **Sir Thomas Brisbane Planetarium**, which holds a range of regular and special events, some involving planetarium astronomers. *Mt Coot-tha Rd, Toowong; (07) 3403 2535; open daily; admission free.*

Lone Pine Koala Sanctuary 645 D5

The world's largest and oldest koala sanctuary, Lone Pine has more than 130 koalas. Visitors can hold koalas, handfeed kangaroos and get up close to many other species in a natural environment. The sanctuary is set on 20 hectares next to the Brisbane River. You can get to Lone Pine by road (11 kilometres from the city) or catch Mirimar's Koala and River Cruise *(see Walks & tours, p. 411)*. There are informative talks daily. *Jesmond Rd, Fig Tree Pocket; (07) 3378 1366; open daily.*

Outer North

The northern bayside suburbs of Scarborough, Redcliffe, Brighton, Margate, Sandgate and Shorncliffe were popular holiday places for Brisbanites prior to World War II, and the legacy lives on in the form of beautiful old boarding houses, mature gardens, the Redcliffe Jetty (although now rebuilt) and Shorncliffe Pier. Together they form a pleasant coastal drive (between 25 minutes and 45 minutes from Brisbane), with plenty of worthwhile places to stop.

Northern beaches 644 F3

Eating fish and chips at Morgans Seafood, a renowned local institution in **Scarborough**, while watching pelicans glide and seagulls wheel over the bobbing trawlers against a background of the Glass House Mountains, is a must-do. Travel south to **Redcliffe**, the site of Queensland's first European settlement in 1824, and follow one of the heritage trails (ask at the information centre for brochures). Redcliffe has good beaches and a landscaped foreshore lined with little cafes – great for coffee with a view over to the Moreton Island sandhills. Its jetty is a popular

day tours

Mount Glorious The sleepy settlement of Mount Glorious lies to the north-west of the city in Brisbane Forest Park, a 28 500-hectare reserve of subtropical forests and hills. Mount Glorious is the base for a number of enjoyable walking tracks. Nearby Wivenhoe Lookout offers panoramic views of the surrounding country. For more details see p. 415

Daisy Hill Conservation Park Close to Brisbane's south-eastern suburbs is Daisy Hill Conservation Park, best known for its large colony of koalas. The Daisy Hill Koala Centre has a variety of displays and, from a tower, you can see koalas in their favourite place – the treetops. (07) 3299 1032. For more details see p. 415

Bribie Island Connected to the mainland by a bridge near Caboolture, Bribie is the most accessible of the Moreton Bay islands. Fishing, boating and crabbing are popular activities. For a quiet picnic and walk, visit Buckleys Hole Conservation Park at the southern end. For more details see p. 416

Moreton Island This impossibly beautiful sand island is mostly national park, reached by passenger ferry from Eagle Farm or vehicular ferry from the Port of Brisbane. Vehicle access is four-wheel drive only, but guided tours are available. Walking, swimming, fishing and dolphin-watching are some of the activities on offer. The big sand hills are a must-visit attraction. For more details see p. 416

North Stradbroke Island A favourite getaway for Brisbanites, Straddie, as it's affectionately known, is the most developed of the Moreton Bay islands, with small townships at Point Lookout, Dunwich and Amity. Visit Blue Lake National Park, at the centre of the island, for swimming and walking (access by four-wheel drive or a 45-minute walk), or enjoy ocean views along the North Gorge Headlands Walk. The island is reached by vehicular ferry from Cleveland. For more details see p. 416

The Gold Coast An hour's drive from Brisbane, the Gold Coast is arguably Australia's most famous and busiest holiday region, with beautiful surf beaches and huge theme parks – perfect for kids of all ages. All activities on offer, from deep-sea fishing and golf to dining and shopping, are of international-resort standard. For more details see p. 417

Gold Coast Hinterland Dubbed the 'green behind the gold', the region offers a peaceful retreat from the bustle of the coast. Follow the winding scenic road up to Lamington National Park, part of a World Heritage area and Queensland's most visited park. It preserves a rainforest environment and a large wildlife population with many bird species, including bowerbirds and lyrebirds. For more details see p. 417

Toowoomba Travel to Toowoomba from Brisbane and you will find yourself climbing the Great Dividing Range. This grand old lady of the Darling Downs is perched on the edge of the escarpment at 800 metres above sea level. It's not dubbed the 'Garden City' for nothing – Toowoomba is famous for its parks, gardens and tree-lined streets, and each September it celebrates the Carnival of the Flowers. The city has also long been known for its antique shops and tea parlours, and these days also offers good coffee and innovative restaurants. For more details see p. 418

fishing spot and the little cafe here is decorated with historical photos. There's also a memorial to John Oxley, who landed in 1823, and a plaque dedicated to Mathew Flinders, who came ashore here in 1799. **Margate** beach has a string of shaded picnic spots with lovely bay views. At **Woody Point**, you will find the historic Belvedere Hotel, built in the 1890s.

Alma Park Zoo 644 E3

Set in award-winning rainforest gardens, the zoo is 30 minutes north of the city, inland from Redcliffe. It displays both Australian and exotic animals, and offers the chance to get up close to many species. You can touch koalas, and feed kangaroos and deer in the walk-through enclosures – all great for kids. *Alma Rd, Dakabin; open daily.*

Eastern Bayside

Wynnum & Manly 644 G7

Half an hour east of the city, Wynnum and Manly offer an easy-reach bayside experience. The **Wynnum Mangrove Boardwalk** is worth a look, especially for wildlife – watch out for fish and migrating birds. If you're feeling active, hire a bike to ride the 1.5-kilometre **Esplanade** from Manly to Lota. The **Manly Boat Harbour** is a vista of yachts, and charters and boat hire for both fishing and sailing are available. Tours to the island of St Helena also depart from here. This island is two-thirds national park, and served as a penal settlement between 1867 and 1932. Prison life is recalled in the stark but strangely beautiful structures that remain. North of Wynnum, at the mouth of the Brisbane River, is the 19th-century **Fort Lytton**. A museum at the fort (open only on Sundays and public holidays) interprets Queensland's military and social history from 1879. Further north is the **Port of Brisbane**, which offers public tours and a visitor centre. *Wynnum–Manly Visitor Information Centre: 43A Cambridge Pde, Manly; (07) 3348 3524.*

Outer Eastern Bayside

Cleveland 644 H8

About 40 minutes south-east of Brisbane, leafy Cleveland is the centre of the Redland Bay district. The area is rich in history, so follow the heritage trail from the Old Courthouse to explore it. Enjoy the colourful Cleveland Bayside Markets on Sundays or try one of the delicious counter meals at Brisbane's oldest pub, the 1851 **Grand View Hotel** – with Queensland's oldest banyan tree out front. Another historic building is **Ormiston House**, built in 1862 and now restored and open to visitors. *(07) 3286 1425; open Sun afternoons Mar–Nov.*

It is at Cleveland that you can catch a ferry or water taxi to North Stradbroke Island *(see Day tours, on this page)*. South of Cleveland, Victoria Point and Redland Bay offer good fishing. Boats can be hired along this section of the coast.

REGIONS
queensland

[WHITSUNDAY ISLANDS NATIONAL PARK, GREAT BARRIER REEF]

[CORKSCREW ROLLERCOASTER AT SEA WORLD, GOLD COAST & HINTERLAND]

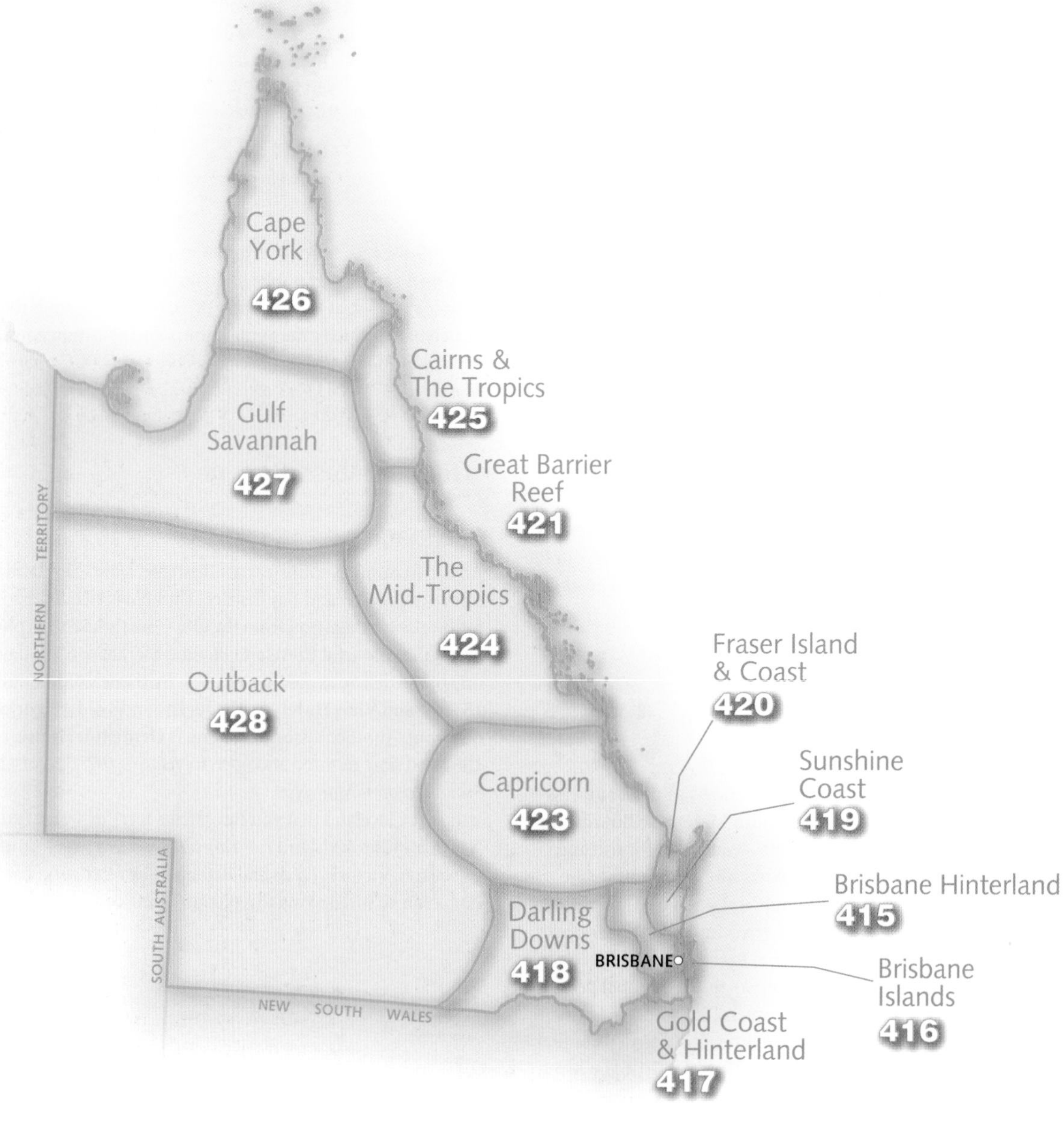

BRISBANE

NORTHERN TERRITORY

SOUTH AUSTRALIA

NEW SOUTH WALES

BRISBANE HINTERLAND

Heavily forested hills create a subtropical haven just a stone's throw away from Brisbane's hustle and bustle.

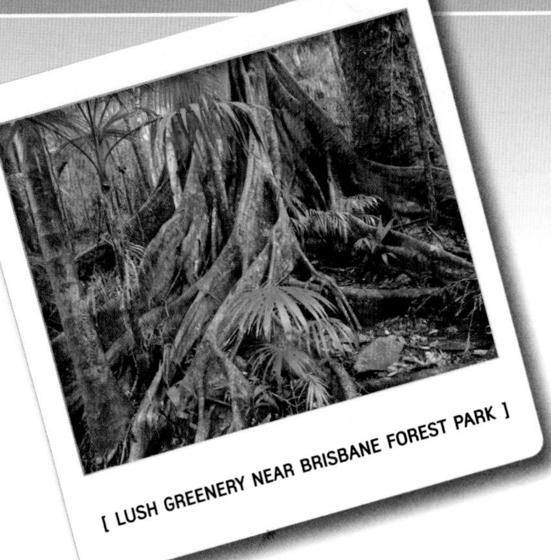

[LUSH GREENERY NEAR BRISBANE FOREST PARK]

Booubyjan Homestead Two Irish brothers, the Clements, took up this run near Goomeri in 1847, beginning with sheep and then moving to cattle in the 1880s. Many generations later, the property is still in the family. The homestead, open daily, provides a glimpse of pioneering life in Queensland's early years.

South Burnett This comfortable slice of rural Queensland invites you to the historic timber towns of Blackbutt and Yarraman, and to wineries along the scenic Barambah Wine Trail. Go to Nanango or Kilkivan to fossick for gold and bed down for the night in a B&B or farmstay.

Bunya Mountains This isolated spur of the Great Dividing Range is a cool, moist region of waterfalls, green and scarlet king parrots, and the remaining stands of bunya pine, a species much depleted by early timber-getters. Walk the easy 4-kilometre Scenic Circuit from the Dandabah camping area, which winds through rainforest to Pine Gorge Lookout.

Brisbane Forest Park Few cities have an attraction such as Brisbane Forest Park on their doorstep, with its pristine rainforest, towering trees, cascading waterfalls, deep pools, mountain streams and incredible wildlife. The small settlement of Mount Glorious is a base for forest walking tracks. Wivenhoe Lookout, 10 kilometres further on, has superb views west to Lake Wivenhoe.

TOP EVENTS

MAR	Wine and Food in the Park Festival (Kingaroy)
MAY	Pine Rivers Festival (Strathpine)
MAY	Pumpkin Festival (Goomeri)
AUG	Peanut Festival (Kingaroy)
OCT	Annual Fishing Competition (Bjelke-Petersen Dam, south-east of Murgon)
OCT	Funfest (Nanango)

ANIMAL ANTICS

Walkabout Creek Wildlife Centre, in Brisbane Forest Park, features a freshwater creek environment populated with water dragons, frogs, platypus, pythons and fish. Daisy Hill State Forest is a pocket of eucalypt forest and acacia scrub 25 kilometres south of Brisbane, where visitors can scan the canopy for koalas from a treetop tower. The Daisy Hill Koala Centre in the central picnic area has information about koalas and their habitats.

CLIMATE

MOUNT GLORIOUS

J	F	M	A	M	J	J	A	S	O	N	D	
25	24	24	21	18	16	15	17	20	22	24	25	MAX °C
18	17	17	15	12	10	9	9	11	14	15	17	MIN °C
238	252	222	129	126	84	86	56	57	114	123	167	RAIN MM
15	16	16	12	11	8	8	7	7	10	11	13	RAIN DAYS

→ For more detail see maps 644, 645, 647 & 653. For descriptions of towns see Towns from A–Z (p. 429).

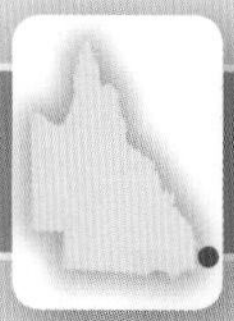

BRISBANE ISLANDS

The islands of Moreton, North Stradbroke and Bribie are holiday destinations with endless white beaches, lakes and pockets of forest, while St Helena draws visitors keen to discover remnants of its penal settlement.

[CYLINDER BEACH, NORTH STRADBROKE ISLAND]

St Helena Island This low sandy island, 8 kilometres from the mouth of the Brisbane River, was used as a prison from 1867 to 1932. It was dubbed 'the hell-hole of the South Pacific'. Historic ruins remain and are protected in the island's national park. Tours of the island depart from the Brisbane suburbs of Manly and Breakfast Creek.

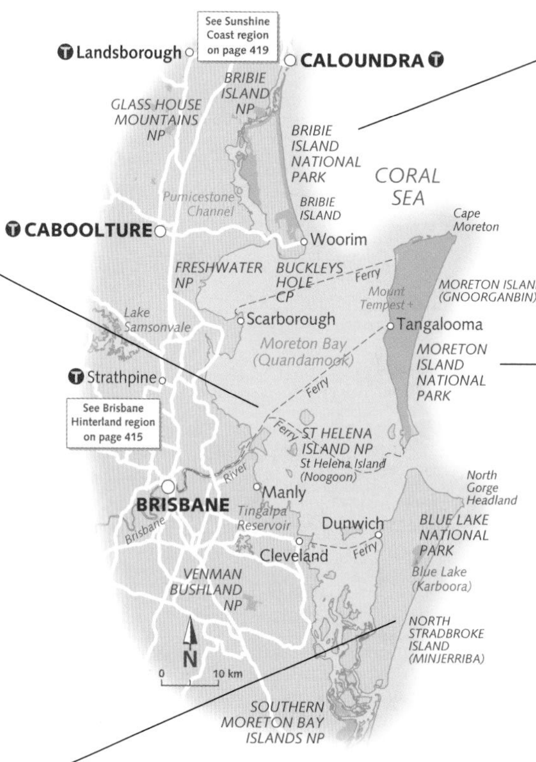

Bribie Island About a third of Bribie is protected by national park. See the magnificent birdlife and spectacular wildflower displays on one of the Bicentennial bushwalks, or go fishing, boating or crabbing. Woorim, in the south-east, is an old-fashioned resort with great surfing beaches. Nearby, Buckleys Hole Conservation Park has good picnic spots for daytrippers and walking tracks to the beach. The island is connected to the mainland by a bridge east of Caboolture.

Moreton Island Almost all of this large island is national park. It is also mainly sand, with 280-metre Mount Tempest possibly the world's highest stable sandhill. On the east coast is an unbroken 36-kilometre surf beach, with calmer beaches on the west coast. Get to the island by passenger or vehicular ferry from Scarborough or the Brisbane River. A four-wheel drive and a permit are required for self-drive touring.

North Stradbroke Island 'Straddie' is a coastal and bushland paradise, with contained pockets of development. Blue Lake National Park is an ecologically significant wetland; access is by four-wheel drive or a 45-minute walk. Other island walking trails include the popular North Gorge Headland Walk. Travel to North Stradbroke by vehicular ferry from Cleveland. The ferry arrives at Dunwich, the site of a 19th-century quarantine and penal centre.

TOP EVENTS

AUG	Bribie Island Greek Festival (Bribie Island)
SEPT	Festival in the Ruins (celebrates the history and ecology of St Helena Island)

CLIMATE

MORETON ISLAND

J	F	M	A	M	J	J	A	S	O	N	D	
29	29	28	26	23	21	20	22	24	26	28	29	MAX °C
21	21	19	17	13	11	10	10	13	16	18	20	MIN °C
159	158	141	92	74	68	57	46	46	75	97	133	RAIN MM
8	9	9	7	6	5	4	4	4	6	6	7	RAIN DAYS

MARINE LIFE

The marine population of Moreton Bay includes dolphins, whales, dugongs and turtles. Visitors to Moreton Island can see dolphins at the Tangalooma Wild Dolphin Resort, where a care program has been developed, or at several spots along the western shore. Migrating humpback whales can be seen between July and November from Cape Moreton and from North Gorge Headland on North Stradbroke Island. Pumicestone Channel, between Bribie and the mainland, is a haven for turtles, dolphins and dugongs. Diving and snorkelling facilities are available on all three islands, allowing visitors to explore the crystal waters and rich underwater life of this magnificent bay.

→ For more detail see maps 644, 645, 647 & 653. For descriptions of Ⓣ towns see Towns from A–Z (p. 429).

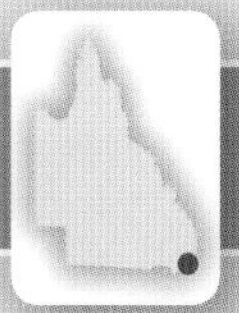

GOLD COAST & HINTERLAND

With unbelievably good weather and no less than 35 beautiful beaches, the Gold Coast has become Australia's biggest holiday destination.

[BURLEIGH HEADS ARTS AND CRAFTS MARKET]

Warner Bros. Movie World This popular theme park south of Oxenford offers visitors the chance to 'meet' their favourite Hollywood characters and see the business of movie-making up close. Rides include the Wild West Falls Adventure and the Road Runner Rollercoaster (for toddlers).

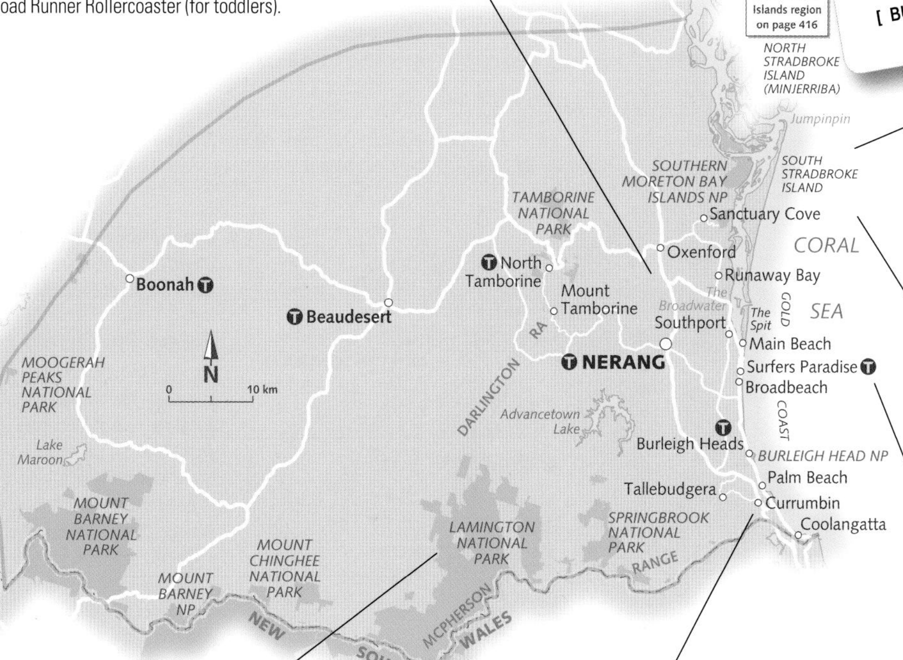

South Stradbroke Island
South Stradbroke, separated from North Stradbroke by the popular fishing channel Jumpinpin, is a peaceful alternative to the Gold Coast. Access is by launch from Runaway Bay. Cars are not permitted; once on the island visitors must walk or cycle.

Golf at Sanctuary Cove
With 40 courses, the Gold Coast is one of the Southern Hemisphere's great golfing destinations. Sanctuary Cove boasts two championship courses: the exclusive Pines and the immaculate Palms.

Surfers Paradise
Surfers Paradise is the Gold Coast's signature settlement – high-rise apartments fronting one of the state's most beautiful beaches. The first big hotel was built here in the 1930s, among little more than a clutch of shacks.

Lamington National Park Part of a World Heritage area, this popular park preserves a wonderland of rainforest and volcanic ridges, crisscrossed by 160 kilometres of walking tracks. The main picnic, camping and walking areas are at Binna Burra and Green Mountains, sites of the award-winning Binna Burra Mountain Lodge and O'Reilly's Rainforest Retreat. Don't miss the Tree Top Walk for an up-close look at the canopy's flora.

Currumbin Wildlife Sanctuary
Be captivated by the wild and roaming animals in this National Trust reserve. The crocodile wetlands let you get up close, and the daily feedings in the aviary make for a loud and exciting experience.

TOP EVENTS

JAN	Magic Millions Racing Carnival (Gold Coast Turf Club)
MAY–JUNE	Blues on Broadbeach Music Festival (Broadbeach)
JUNE	Country and Horse Festival (Beaudesert)
AUG–SEPT	Gold Coast Show (Southport)
OCT	Springtime on the Mountain (Tamborine Mountain)
OCT	Gold Coast SuperGP (Surfers Paradise)

CLIMATE

COOLANGATTA

J	F	M	A	M	J	J	A	S	O	N	D	
28	28	27	25	23	21	20	21	22	24	26	26	MAX °C
20	20	19	17	13	11	9	10	12	15	17	19	MIN °C
184	181	213	114	124	122	96	103	49	108	137	166	RAIN MM
14	15	16	14	10	9	7	9	9	11	11	13	RAIN DAYS

GOLDEN BEACHES

The Gold Coast is a surfers' mecca and, mixed with the fabulous weather, also the perfect destination for a relaxing beach holiday. Best for surfers are the southern beaches, including Currumbin and Kirra Point (in Coolangatta), said to have one of the ten best breaks in the world. Greenmount Beach, in Coolangatta, is great for families, as is Tallebudgera, which offers both estuary and ocean swimming. Towards Surfers Paradise the crowds get larger, but the beaches are still spectacular. Surfers Paradise is renowned for its beachfront – try out your volleyball skills in the free competition near Cavill Avenue. There is an open-ocean beach on The Spit, near Main Beach, with the marine park Sea World close by. For a more remote option, try the 22 kilometres of beach on the east coast of South Stradbroke Island. Once you've soaked up enough magnificent Gold Coast sunshine, you could explore the island's wetlands and rainforest.

→ For more detail see maps 645, 646 & 653. For descriptions of Ⓣ towns see Towns from A–Z (p. 429).

DARLING DOWNS

This region's rich volcanic soil yields some of the country's most magnificent gardens. Winter, a popular time for touring, is known locally as 'Brass Monkey Season'.

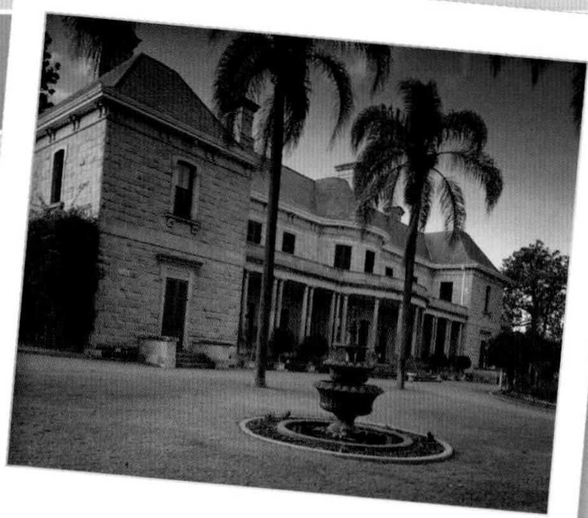

[JIMBOUR HOUSE WINERY, NEAR DALBY]

Jondaryan Woolshed
This 1859 woolshed on historic Jondaryan Station is the centrepiece of a complex of old farm buildings. Sheepdog and shearing demonstrations are held daily, and for a taste of yesteryear, visitors can sit down for a yarn over the billy tea and damper.

Toowoomba Japanese garden A thousand visitors a week stroll the 3 kilometres of paths at Ju Raku En, a Japanese garden at the University of Southern Queensland. Opened in 1989, it showcases the harmony and beauty of ancient Japanese garden design with its lake, willowy beeches, islands, bridges, stream and pavilion.

Queen Mary Falls
Stretching for 55 kilometres along the Great Dividing Range, World Heritage–listed Main Range National Park boasts impressive peaks and escarpments, and the delightful Queen Mary Falls. At the falls, which are part of the Murray-Darling, one of Australia's longest river systems, there's a 400-metre cliff walk that takes you to the top, as well as a 2-kilometre circuit trail revealing distinct changes in the vegetation.

Allora This historic town lies just off the highway between Toowoomba and Warwick. Victorian verandahed shopfronts and three old timber hotels line the main street. St David's Anglican Church (1888) is one of Queensland's finest timber churches. Glengallan Homestead, south of town, was built in 1867, during the golden age of pastoralism.

Granite Belt wineries Queensland's only significant wine region is on an 800-metre plateau in the Great Dividing Range, around Ballandean and Stanthorpe. Over 40 boutique wineries, many with tastings and sales, grow major grape varieties on the well-drained granite soils, favouring soft reds made from shiraz and merlot grapes.

TOP EVENTS

FEB	Melon Festival (Chinchilla, odd-numbered years)
MAR	Cotton on to Energy Festival (Dalby)
MAY	Opera in the Vineyards (Ballandean Estate Winery)
AUG	Australian Heritage Festival (Jondaryan Woolshed)
SEPT	Carnival of Flowers (Toowoomba)
OCT	Campdraft and Rodeo (Warwick)

CLIMATE

TOOWOOMBA

J	F	M	A	M	J	J	A	S	O	N	D	
28	27	26	23	20	17	16	18	21	24	26	28	MAX °C
17	17	15	12	9	6	5	6	9	12	14	16	MIN °C
135	122	95	63	60	58	54	40	48	73	89	120	RAIN MM
12	11	11	8	8	8	7	6	7	8	10	11	RAIN DAYS

GARDENS OF THE DOWNS

The climate and soils of the Darling Downs have created one of Australia's great gardening districts. Toowoomba has 150 public parks and gardens, including the Scented Garden – for visually impaired people – and the 6-hectare mountainside Boyce Gardens, with 700 species of trees, shrubs and perennials. Warwick is known for its roses, particularly the red 'City of Warwick', best seen in the Jubilee Gardens. There are superb private gardens throughout the region – some open daily, some seasonally and some as part of the Open Garden Scheme; check with an information centre for details.

→ For more detail see maps 652–3. For descriptions of Ⓣ towns see Towns from A–Z (p. 429).

SUNSHINE COAST

[DRIVING NEAR THE GLASS HOUSE MOUNTAINS]

Bathed by the blue Coral Sea and fringed by native bush, the Sunshine Coast has near-perfect weather, with winter temperatures around 25°C.

Mountain villages The 70-kilometre scenic drive here is one of Queensland's best. Starting on the Bruce Highway near Landsborough, it passes the antique shops, B&Bs, galleries and cafes of the pretty mountain villages of Maleny, Montville, Flaxton and Mapleton, offering beautiful coastal and mountain views along the way. The drive ends near Nambour.

Glass House Mountains These 20-million-year-old crags, the giant cores of extinct volcanoes, mark the southern entrance to the Sunshine Coast. Glass House Mountains Road leads to sealed and unsealed routes through the mountains, with some spectacular lookouts along the way. There are walking trails, picnic grounds, and challenges aplenty for rockclimbers.

Tin Can Inlet
Rainbow Beach
Tin Can Bay
Noosa River
GREAT SANDY NATIONAL PARK
GYMPIE
N
0 10 km
Lake Cootharaba
Teewah
Pomona
Lake Cooroibah
CORAL
Noosa Heads
Lake Weyba
NOOSA NP
Eumundi
SEA
Kenilworth
Yandina
MOUNT COOLUM NP
CONONDALE NATIONAL PARK
Mapleton
Flaxton
NAMBOUR
MAROOCHYDORE
Montville
Lake Baroon
BUDERIM
MOOLOOLABA
Maleny
MOOLOOLAH RIVER NP
DULARCHA NP
Landsborough
CALOUNDRA
GLASS HOUSE MOUNTAINS NP
Glass House Mountains
BRIBIE ISLAND

See Brisbane Islands region on page 416

The coloured sands of Teewah
Located in the Cooloola section of Great Sandy National Park, these coloured sands rise in 40 000-year-old cliffs. It is thought that oxidisation or decaying vegetation has caused the colouring; Aboriginal legend attributes it to the slaying of a rainbow serpent.

Gondola ride on the Noosa River
A very special way to appreciate Noosa's surrounds is on a gondola cruise with Gondolas of Noosa. It doesn't get much better than gliding through Noosa Sound's calm waters listening to the strains of Italian opera as the sun goes down. This is also a very romantic mode of transport when eating at Ricky's River Bar and Restaurant – catch the gondola at the Sheraton wharf, but be sure to book beforehand.

Coastal towns The southern towns of Caloundra, Mooloolaba and Maroochydore make pleasant daytrips from Brisbane or good spots for an extended family holiday, with patrolled surfing beaches and protected lakes and rivers for boating and fishing. Central Noosa Heads offers luxury hotels, top restaurants, hip bars and stylish boutiques, as well as the pandanus-fringed beaches of Noosa National Park.

TOP EVENTS

JAN	Ginger Flower Festival (Yandina)
JULY	Family Fishing Classic (Rainbow Beach)
SEPT	Scarecrow Festival (Maleny)
SEPT	Jazz Festival (Noosa Heads)
SEPT	Seafood Festival (Tin Can Bay)
OCT	Gold Rush Festival (Gympie)

CLIMATE

NAMBOUR

J	F	M	A	M	J	J	A	S	O	N	D	
30	29	28	26	24	22	21	22	25	27	28	29	MAX °C
19	20	18	15	12	9	8	8	10	14	16	18	MIN °C
242	262	236	149	143	91	92	53	48	105	141	176	RAIN MM
16	18	18	13	13	9	9	8	9	12	12	13	RAIN DAYS

TROPICAL PRODUCE

The Sunshine Coast hinterland, with its subtropical climate and volcanic soils, is renowned for its produce. Nambour's Big Pineapple symbolises the importance of food as an industry and a tourist attraction. Visitors can take a train, trolley and boat through a plantation growing pineapples and other fruit, macadamia nuts, spices and flowers. Yandina's Ginger Factory, the world's largest, sells ginger products including ginger ice-cream. For freshly picked local fruit and vegetables, visit the Wednesday- and Saturday-morning markets at Eumundi, north of Yandina.

→ For more detail see maps 647, 648 & 653. For descriptions of T towns see Towns from A–Z (p. 429).

FRASER ISLAND & COAST

This region features all of Queensland's signature attractions: spectacular white beaches, coloured sand cliffs, wildflower heathland and rainforest, and whale-watching in Hervey Bay.

[FOUR-WHEEL DRIVING, FRASER ISLAND]

Fraser Island's dingoes Roaming around on the world's largest sand island is what is thought to be the purest strain of dingoes in eastern Australia. By the time these native dogs arrived on the mainland – they came with Asian seafarers around 5000 years ago – Fraser Island was already disconnected from the continent, and the dingoes swam the few kilometres across Great Sandy Strait. Unlike most mainland dingoes, they have not been hybridised by contact with domestic dogs.

WARNING: Dingoes have been known to attack. Stay with small children and never feed or coax animals.

Hervey Bay This once sleepy settlement is now a booming resort town. The bay itself is a large, calm body of water warmed by tropical currents. The area's protected beaches are perfect for family swimming. Other popular activities include sailing, diving, windsurfing, fishing, kayaking, whale-watching and skydiving.

Maryborough Heritage Walk and Drive Established in 1847, Maryborough is one of Queensland's oldest and best-preserved provincial cities. A self-guide brochure leads visitors through tree-lined streets, past heritage sites and well-restored Queenslander houses, and along the historic streetscape of Wharf Street.

Great Sandy Strait This narrow strait between the mainland and Fraser Island makes for good boating; there are houseboats and other vessels for hire. Drop into the Kingfisher Bay Resort and Village on Fraser Island, facing the strait. Look out for dugongs, the world's only plant-eating marine mammals, and fish at the mouth of the Mary River, around River Heads.

Fraser Island's east coast Also Fraser Island's surf coast this takes in the beautiful Seventy Five Mile Beach; The Cathedrals, 15-metre sheer cliffs composed of different-coloured sands; the wreck of the *Maheno*, a trans-Tasman luxury liner; and Eli Creek, a freshwater creek filtered through the dunes, where visitors can float beneath the pandanus trees.

Lake McKenzie World Heritage–listed Fraser Island is distinguished by its 40 perched dune lakes, formed by water collecting on an impermeable layer of decaying matter. Idyllic Lake McKenzie is definitely one of the most beautiful of these freshwater lakes – its shallow water is dazzling aquamarine and ringed by white sandy beaches backed by paperbark trees. To swim here, in some of the purest drinking water in the world, is a wonderful experience.

TOP EVENTS

EASTER	Amateur Fishing Classic (Burrum Heads)
APR–MAY	Bay to Bay Yacht Race (Tin Can Bay to Hervey Bay)
AUG	Whale Festival (Hervey Bay)

CLIMATE

MARYBOROUGH

J	F	M	A	M	J	J	A	S	O	N	D	
31	30	29	27	25	22	22	23	26	28	29	31	MAX °C
21	21	19	17	13	10	9	9	12	15	18	20	MIN °C
166	173	159	90	80	67	54	40	43	75	85	128	RAIN MM
13	14	14	12	11	8	7	6	6	8	9	11	RAIN DAYS

WHALE-WATCHING

At the start of winter humpback whales begin migrating to the warmer waters of Hervey Bay. Visitors are never far behind them, many making their way up the coast for the chance to see these amazing creatures offshore. Each year around 2000 humpback whales migrate from the Antarctic to Australia's eastern subtropical coast. Between July and November, up to 400 rest and regroup in Hervey Bay. For an up-close view of the majestic whales, various tours operate from the boat harbour in town.

→ For more detail see maps 648 & 653. For descriptions of Ⓣ towns see Towns from A–Z (p. 429).

GREAT BARRIER REEF

Australia's most prized and visited natural destination extends over 2000 kilometres, taking in tropical islands, rare and brilliantly coloured corals, exotic fish, and sea-going mammals and birds.

[SNORKELLING, GREAT BARRIER REEF]

Tropical North Islands

Lizard Island This northernmost island has tourist facilities, gamefishing, snorkelling, diving and bushwalking. The small, luxurious resort has bungalow-style lodgings (maximum 80 people) or camping (maximum 20 people); camping permits from Cairns parks office.

Fitzroy Island A low-key destination with a national park, white coral beaches and magnificent flora and fauna for bushwalking, diving and snorkelling. Cabins and hostel-style accommodation (maximum 160 people) and camping (maximum 20 people) are available.

Bedarra Island This island of untouched tropical beauty is off-limits to day visitors and children under 15. There's bushwalking, snorkelling, fishing, swimming, windsurfing, sailing and tennis, and an exclusive resort (maximum 30 people).

Orpheus Island A small island surrounded by coral reefs and protected by national park, Orpheus offers birdwatching, watersports, glass-bottomed boat tours, island walks and fishing. There's a five-star resort (maximum 74 people) or bush camping (maximum 54 people); camping permits from Ingham parks office.

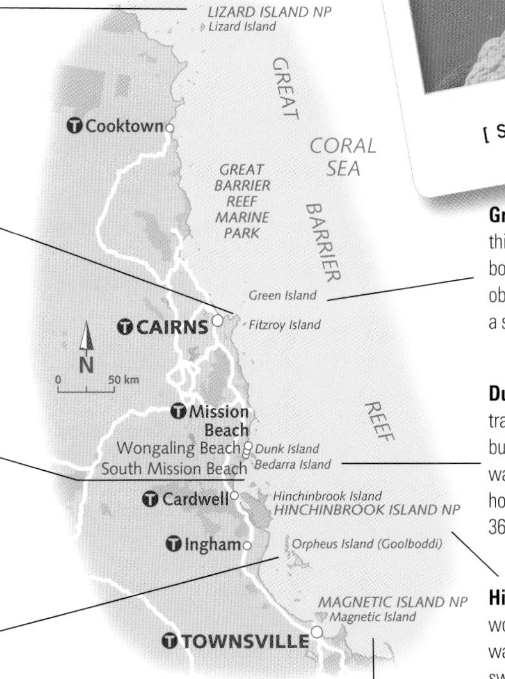

Green Island A true coral cay covered with thick tropical vegetation, with glass-bottomed boats for reef viewing and an underwater observatory. This popular daytrip destination has a small resort (maximum 90 people).

Dunk Island National park with walking tracks through rainforest, and prolific birdlife, butterflies and wild orchids. Go parasailing, waterskiing, sailing, clay-target shooting and horseriding. Resort accommodation (maximum 360 people) and camping (maximum 30 people).

Hinchinbrook Island National park with a wonderland of mountains, tropical vegetation, waterfalls and sandy beaches for snorkelling, swimming, fishing and bushwalking. Small, low-key resort (maximum 45 people).

Magnetic Island With a national park and beautiful beaches, activities include horseriding, bushwalking, snorkelling, parasailing, swimming, fishing, sea-kayaking and reef excursions. There's a permanent population and accommodation from budget to deluxe.

QUEENSLAND

ISLAND ACCESS

Magnetic Island 8 kilometres NE of Townsville.
From Townsville, by vehicular ferry, catamaran or water taxi.
Orpheus Island 80 kilometres N of Townsville.
From Townsville or Cairns, by sea plane.
Hinchinbrook Island 5 kilometres E of Cardwell.
From Cardwell, by launch.
Bedarra Island 35 kilometres NE of Cardwell.
From Dunk Island, by launch.
Dunk Island 5 kilometres SE of Mission Beach.
From Cairns, by plane. From Clump Point near Mission Beach, by launch.
From Wongaling Beach and South Mission Beach, by water taxi.
Fitzroy Island 30 kilometres SE of Cairns.
From Cairns, by catamaran.
Green Island 27 kilometres NE of Cairns.
From Cairns, by catamaran, sea plane or helicopter.
Lizard Island 93 kilometres NE of Cooktown.
From Cairns or Cooktown, by plane or sea plane.

CLIMATE

FITZROY ISLAND

J	F	M	A	M	J	J	A	S	O	N	D	
31	30	29	28	26	24	24	25	27	29	30	31	MAX °C
12	12	10	8	7	5	4	5	6	7	9	10	MIN °C
27	25	31	60	89	100	106	103	86	68	40	35	RAIN MM
6	5	7	11	5	18	17	18	15	13	9	7	RAIN DAYS

→ For more detail see maps 649, 650, 651, 653, 655 & 657. For descriptions of Ⓣ towns see Towns from A–Z (p. 429).

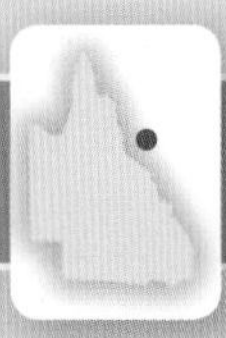

GREAT BARRIER REEF

Whitsunday Islands

Daydream Island A small island of volcanic rock, coral and dense tropical foliage. Daydream has a Kids Club, tennis, outdoor cinema, watersports centre, snorkelling, diving, and reef and island trips. Luxurious resort (maximum 900 people).

South Molle Island Numerous inlets and splendid views of Whitsunday Passage, plus great wildflowers in spring and early summer. There's golf, bushwalking, snorkelling, scuba diving, windsurfing and sailing. Medium-size resort (maximum 520 people).

Hayman Island Close to the outer reef, with fishing, sightseeing trips, scenic flights, diving, watersports, Kids Club and whale-watching excursions. Luxury resort (maximum 450 people).

Whitsunday Island Entirely uninhabited national park, with a beautiful 7-kilometre white silica beach and complex mangrove system. Camping only (maximum 40 people); details from Airlie Beach parks office.

Hamilton Island A large island with a wide range of facilities and activities, shops, marina and fauna park. There's windsurfing, sailing, fishing, scuba diving, parasailing, helicopter rides, tennis, squash, and reef and inter-island trips. Resort (maximum 1500 people).

Lindeman Island Secluded beaches, national park, and prolific birds and butterflies, a golf course, and full range of watersports and other island activities. Club Med resort (maximum 460 people).

Brampton Island National park and wildlife sanctuary with fine golden beaches, snorkelling trail, bushwalking, sea-plane trips and watersports. Resort-style accommodation (maximum 280 people).

ISLAND ACCESS

Whitsunday Islands

Brampton Island 32 kilometres NE of Mackay.
From Mackay, by light plane or launch. From Hamilton Island, by plane.

Lindeman Island 67 kilometres N of Mackay.
From Airlie Beach or Shute Harbour, by light plane or boat. From Mackay, by plane. From Hamilton Island, by boat or plane.

Hamilton Island 16 kilometres SE of Shute Harbour.
Direct flight from Sydney, Brisbane and Melbourne; connections from all major cities. From Shute Harbour, by launch.

South Molle Island 8 kilometres E of Shute Harbour.
From Shute Harbour or Hamilton Island, by launch.

Daydream Island 5 kilometres NE Shute Harbour.
From Shute Harbour or Hamilton Island, by launch or helicopter.

Whitsunday Island 25 kilometres E of Shute Harbour.
From Shute Harbour or Airlie Beach, by boat.

Hayman Island 25 kilometres NE of Shute Harbour.
Direct flight to Hamilton Island from Sydney and Brisbane (connections from all major cities), then by launch to Hayman Island. From Airlie Beach, by water taxi.

Southern Reef Islands

Lady Elliot Island 80 kilometres NE of Bundaberg.
From Bundaberg or Hervey Bay, by plane.

Heron Island 72 kilometres NE of Gladstone.
From Gladstone, by catamaran or charter helicopter.

Great Keppel Island 48 kilometres NE of Rockhampton.
From Rockhampton, by light plane. From Yeppoon, by launch.

CLIMATE

HAMILTON ISLAND

J	F	M	A	M	J	J	A	S	O	N	D	
30	30	29	27	25	23	22	23	25	28	29	30	MAX °C
25	25	24	23	21	19	18	18	20	22	23	24	MIN °C
13	322	262	242	159	100	80	59	23	52	89	215	RAIN MM
15	18	19	19	18	12	10	11	7	8	8	13	RAIN DAYS

Southern Reef Islands

Great Keppel Island With white sandy beaches and unspoiled tropical island scenery, there's tennis, waterskiing, diving, snorkelling, fishing, sea-kayaking, golf, parasailing, coral viewing and island cruises. Kids Club during holidays. Camping, cabins and lodge-style accommodation (maximum 650 people).

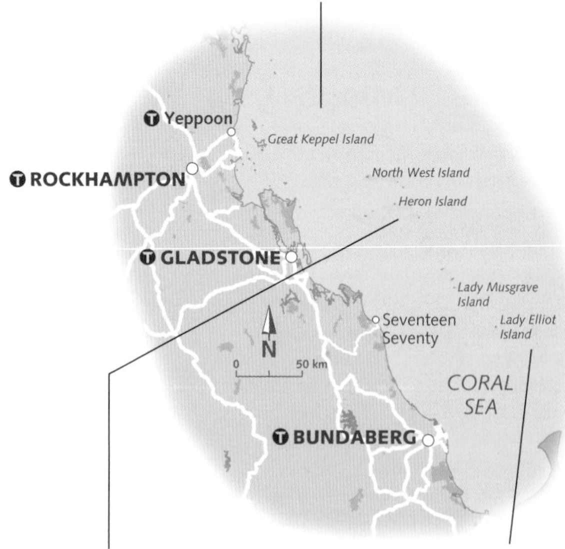

Heron Island A small coral cay, entirely national park, with turtle-nesting site, birdwatching and prolific flora, diving and snorkelling, and reef and ecology walks. Resort-style accommodation (maximum 250 people).

Lady Elliot Island A small coral cay with 19 major dive sites, bird rookeries, turtle-nesting site and whale-watching opportunities. Low-key resort (maximum 140 people), ranging from budget to island suites.

CAPRICORN

Relatively untouched by commercial development, this region features remote beaches and unspoiled bushland.

[CARNARVON NATIONAL PARK]

Gemfields Some of the world's richest sapphire fields are found around the tiny, ramshackle settlements of Anakie, Sapphire, Rubyvale and Willows Gemfields, approximately 50 kilometres west of Emerald. Fossicking licences can be bought on the gemfields for a small fee.

Mount Morgan This goldmining town south of Rockhampton has hardly changed in a century. Take a tour to the Southern Hemisphere's largest excavation, or see the ancient dinosaur footprints in nearby caves.

Rockhampton The town has many heritage buildings – particularly grand are those in Quay Street. There is also an excellent discovery centre at the Customs House, and historic botanic gardens.

Capricorn Coast Thirteen beaches stretch out along Keppel Bay, taking in Yeppoon, Emu Park and Keppel Sands. Elsewhere, picturesque bays are framed by rocky headlands, pockets of rainforest, peaceful estuarine waters and wetlands.

Carnarvon National Park Carnarvon Gorge, the signature attraction of Carnarvon National Park, bends and twists its way around 30 kilometres of semi-arid terrain. Within its sandstone walls exists a cool, green world of delicate ferns and mosses fed by Carnarvon Creek. The 298 000-hectare park also contains some magnificent Aboriginal rock-art sites.

Rum town On the southern coast of the Capricorn region is the city of Bundaberg, renowned for its parks and gardens, in particular the Botanical Gardens, with the excellent Hinkler House Memorial Museum.

Mon Repos turtle rookery Mon Repos Conservation Park is one of Australia's most important turtle rookeries. From November to March there is an on-site interpretative centre and guided night tours.

TOP EVENTS

MAR	1770 Longboard Classic (surfing competion, Agnes Water)
MAR	Harbour Festival (Gladstone, includes finish of Brisbane–Gladstone Yacht Race)
JUNE	Orange Festival (Gayndah, odd-numbered years)
SEPT	Bundy in Bloom Festival (Bundaberg)
SEPT	Seafood Festival (Gladstone)
OCT	Rocky Barra Bounty (Rockhampton)

CLIMATE

ROCKHAMPTON

J	F	M	A	M	J	J	A	S	O	N	D	
32	31	30	29	26	23	23	25	27	30	31	32	MAX °C
22	22	21	18	14	11	9	11	14	17	19	21	MIN °C
136	141	103	47	52	35	31	29	24	48	68	105	RAIN MM
11	12	10	7	7	5	5	4	4	7	8	10	RAIN DAYS

DISCOVERY COAST

Seventeen Seventy, a small town on a narrow, hilly peninsula above an estuary, was named to mark Captain Cook's landing at Bustard Bay on 24 May 1770. Today's visitors come for the views from the headland north across the bay, and for fishing, mudcrabbing and boating. Agnes Water, a few kilometres south, has Queensland's northernmost surfing beach; rolling surf and a balmy climate attract visitors all year-round. Eurimbula National Park, just across Round Hill Inlet from Seventeen Seventy, has dunes, mangroves, salt marshes and eucalypt forests. From Agnes Water, an 8-kilometre track south to Deepwater National Park is suitable for four-wheel drives only. The long beaches of this park form a breeding ground for loggerhead turtles.

→ For more detail see maps 648, 653 & 654–5. For descriptions of Ⓣ towns see Towns from A–Z (p. 429).

THE MID-TROPICS

An alternative to some of Queensland's busier coastal areas, The Mid-Tropics has stretches of sandy shoreline, warm tropical waters and rainforest.

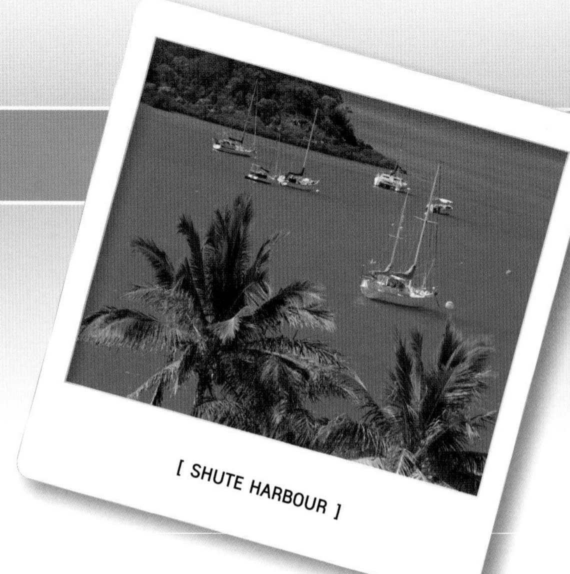

[SHUTE HARBOUR]

Marine attractions At Reef HQ in Townsville, touch-tanks and underwater-viewing tunnels reveal some of the Great Barrier Reef's ecological mysteries. Next door, the Museum of Tropical Queensland features a full-scale reproduction of the bow of HMS *Pandora*, a British vessel wrecked on the reef in 1791.

Charters Towers Charters Towers was Queensland's second largest city during the 1870s gold rush; at that time it was known as 'The World' because of its cosmopolitan population. Today it is a showpiece, with many beautifully preserved buildings including the Bank of Commerce, now restored as the New World Theatre Complex. The 'Ghosts of Gold' Heritage Trail allows visitors to experience the town's glamorous history.

Wreck diving A must for divers is the SS *Yongala* wreck, which sank near Cape Bowling Green during a cyclone in 1911, and lay undiscovered for half a century. Despite being under water for almost 100 years, details such as the engine room, toilets, portholes and most of the ship's name are still evident. The marine life is excellent, with beautiful corals, giant groupers and trevally, barracuda, stingrays, turtles and hundreds of other sea creatures.

Hibiscus Coast Steep rainforest-clad hills plunge to rocky headlands and white sandy beaches in this lovely and surprisingly peaceful district north of Mackay. Cape Hillsborough National Park offers the most pristine scenery, with kangaroos often seen hopping along the deserted beaches.

TOP EVENTS

APR–MAY	Ten Days in the Towers (Charters Towers)
MAY	Australian Italian Festival (Ingham)
JULY	Festival of the Arts (Mackay)
JULY–AUG	Australian Festival of Chamber Music (Townsville)
SEPT	Burdekin Water Festival (Ayr)
OCT	Greek Fest (Townsville)

CLIMATE

MACKAY

J	F	M	A	M	J	J	A	S	O	N	D	
30	29	28	27	24	22	21	22	25	27	29	30	MAX °C
23	23	22	20	17	14	13	14	16	20	22	23	MIN °C
293	311	303	134	104	59	47	30	15	38	87	175	RAIN MM
16	17	17	15	13	7	7	6	5	7	9	12	RAIN DAYS

HERITAGE

This region preserves some interesting pockets of heritage. Ravenswood, east of Charters Towers, is now almost a ghost town. It flourished in the second half of the 19th century as a centre for the surrounding goldfields, and many of its buildings from this period are in a near-original state. West of Mackay is Greenmount Historic Homestead – now a museum – built in 1915 on the Cook family's grazing property. Bowen, established in 1861, is north Queensland's oldest town. The diverse history here is recorded in 25 street murals detailing the stories, personalities and events of the towns past.

→ For more detail see maps 649, 650, 654–5 & 657. For descriptions of T towns see Towns from A–Z (p. 429).

CAIRNS & THE TROPICS

Tourists flock to this region to experience its ancient rainforest and coastline fronting part of the world's most spectacular reef. The national parks here offer camping, wildlife-watching and four-wheel driving.

[THE ELUSIVE SOUTHERN CASSOWARY, DAINTREE NATIONAL PARK]

Cooktown Captain Cook beached the *Endeavour* near the site of Cooktown in 1770. The town was built a century later when gold was discovered on the Palmer River to the south-west. Cooktown has excellent botanic gardens dating from the 1880s, and cruises travel along the river and out to the reef.

Atherton Tableland This 900-metre-high tableland south-west of Cairns is a productive farming district, thanks to the high rainfall and rich volcanic soil. Near Yungaburra is the remarkable Curtain Fig Tree, a strangler fig that has subsumed its host, sending down a curtain of roots. Volcanic lakes and spectacular waterfalls, including Millaa Millaa Falls and Zillie Falls, are among the other scenic attractions.

Whitewater rafting on the Tully River The Tully River, which descends from the Atherton Tableland through rainforest gorges, is one of Queensland's three main rafting rivers (the other two are the Barron and North Johnstone). It receives one of the highest annual rainfalls in Australia, explaining its rapids, which are pretty ferocious in parts, and has more than 45 individual rapids up to grade 4. In short, it's perfect for whitewater rafting.

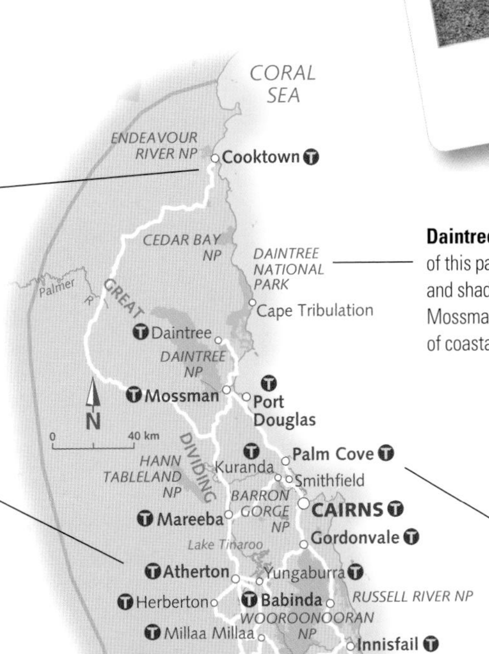

Daintree National Park The Mossman Gorge section of this park takes visitors into the rainforest's green and shady heart via an easy 2.7-kilometre walk to the Mossman River. The Cape Tribulation section is a rich mix of coastal rainforest, mangroves, swamp and heath.

Tjapukai Aboriginal Cultural Park An Aboriginal group, originally a dance company, set up this park in Smithfield to make their unique Far North Queensland culture more accessible. In the theatres you can hear Aboriginal creation stories and watch Aboriginal dance, and in the Camp Village you can learn to throw a boomerang.

QUEENSLAND

TOP EVENTS

MAY	Village Carnivale (Port Douglas)
JUNE	Discovery Festival (Cooktown)
AUG–SEPT	Festival Cairns (Cairns)
OCT	Walkamin Country Music Festival (Mareeba)
OCT	Yungaburra Folk Festival (Yungaburra)
OCT	Palm Cove Fiesta (Palm Cove)

CLIMATE

CAIRNS

J	F	M	A	M	J	J	A	S	O	N	D	
31	31	30	29	28	26	26	27	28	29	31	31	MAX °C
24	24	23	22	20	18	17	18	19	21	22	23	MIN °C
413	435	442	191	94	49	28	27	36	38	90	175	RAIN MM
18	19	20	17	14	10	9	8	8	8	10	13	RAIN DAYS

TROPICAL RAINFORESTS

The World Heritage Wet Tropics Area covers 894 000 hectares along the eastern escarpment of the Great Dividing Range between Townsville and Cooktown, and features rainforest, mountains, gorges, fast-flowing rivers and countless waterfalls. The rainforest here is one of the oldest continually existing rainforests on earth, with an estimated age of more than 100 million years. It represents major stages in the earth's evolutionary history and has the world's greatest concentration of primitive flowering plants, or 'green dinosaurs', as they are known. Its biological diversity is astounding; while the Wet Tropics covers just 0.1 per cent of the continent, it contains 30 per cent of Australia's marsupial species, 60 per cent of its bats and 62 per cent of its butterflies.

→ For more detail see maps 650, 651 & 657. For descriptions of T towns see Towns from A–Z (p. 429).

CAPE YORK

One of Australia's last undeveloped frontiers, Cape York showcases some of the country's finest Aboriginal rock art, spectacular wildlife and amazing landscapes.

[FISHING BOAT WINDING THROUGH MANGROVES, CAPE YORK PENINSULA]

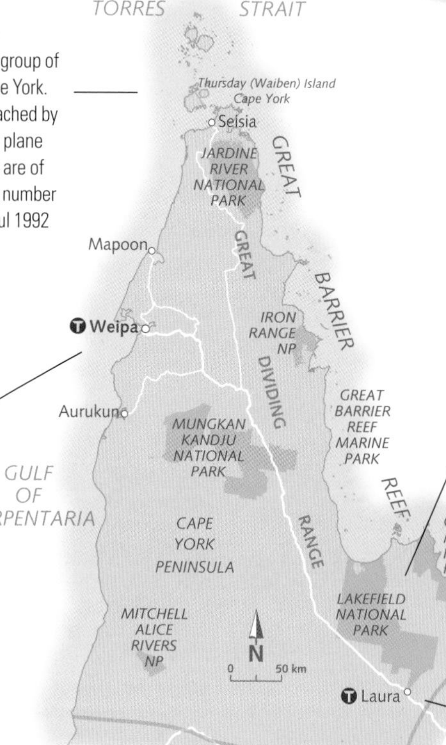

Torres Strait Islands Australia's only non-Aboriginal indigenous people come from this group of around 100 islands off the northern tip of Cape York. The commercial centre is Thursday Island, reached by ferry from Seisia or Punsand Bay or by ship or plane from Cairns. The Torres Strait Islander people are of Melanesian descent and include among their number the late Eddie Mabo, famous for his successful 1992 land claim in Australia's High Court.

Weipa The world's largest bauxite deposits are found around Weipa. In 1961 Comalco built a modern mining town, with all facilities, on the site of an Aboriginal mission station and reserve. Today the company offers tours of its mining operations. The fishing around Weipa (and indeed the whole of the cape) is excellent.

Lakefield National Park Lakefield is Queensland's second largest national park, and the most accessible on the Cape York Peninsula. A near-wilderness of grassland, woodland, swamp and mangroves is cut by three major rivers and their tributaries. Access for conventional vehicles is via the township of Laura (during the Dry only). Visit the Old Laura Homestead, fish for barramundi and watch the wildlife.

Quinkan country To the south-east of Laura are the spectacular Aboriginal rock-art sites of Split Rock and Gu Gu Yalangi. Encompassing the area between Laura and Cooktown, these form just a small part of one of the largest collections of prehistoric rock art in the world. The most distinctive works are the Quinkan figures, stick-like figures representing spirits that emerge suddenly from rock crevices.

TOP EVENTS

JUNE	Laura Aboriginal Dance and Cultural Festival (Laura, odd-numbered years)
JUNE	Mabo Day (Torres Strait Islands)
JULY	Coming of the Light Festival (Torres Strait Islands)
SEPT	Torres Strait Cultural Festival (Torres Strait Islands, even-numbered years)

CLIMATE

WEIPA

J	F	M	A	M	J	J	A	S	O	N	D	
32	31	32	32	32	31	30	32	33	35	35	33	MAX °C
24	24	24	23	21	20	19	19	20	21	23	24	MIN °C
421	401	338	82	6	2	1	1	1	14	98	256	RAIN MM
22	21	20	10	3	2	1	1	1	3	8	16	RAIN DAYS

ABORIGINAL CAPE YORK

Before white people arrived, the Aboriginal people of Cape York were divided into two main groups – East Cape and West Cape – with many language and cultural groups within this broad division. The distinctive art of the East Cape people survives in the rock-art galleries around Laura, some of the most extensive and unusual in Australia. The Aboriginal groups of West Cape, an area that proved largely resistant to European expansion, maintained sizeable pockets of land around Mapoon and Aurukun. Native title legislation since the 1950s has handed much of the West Cape back to its Aboriginal owners.

→ For more detail see maps 656–7 & 658. For descriptions of T towns see Towns from A–Z (p. 429).

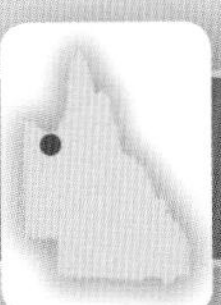

GULF SAVANNAH

A remote and far-flung area of Queensland, this region is characterised by vast tracts of savannah grasslands, lagoons and mangrove-lined estuaries.

[CANOEING IN BOODJAMULLA (LAWN HILL) NATIONAL PARK]

Tourist train Every Wednesday the *Gulflander* leaves Normanton on a 153-kilometre journey to the historic goldmining town of Croydon. With stops at points of interest along the line, the trip takes four hours. Travellers can explore Croydon with a local guide and return to Normanton on Thursday.

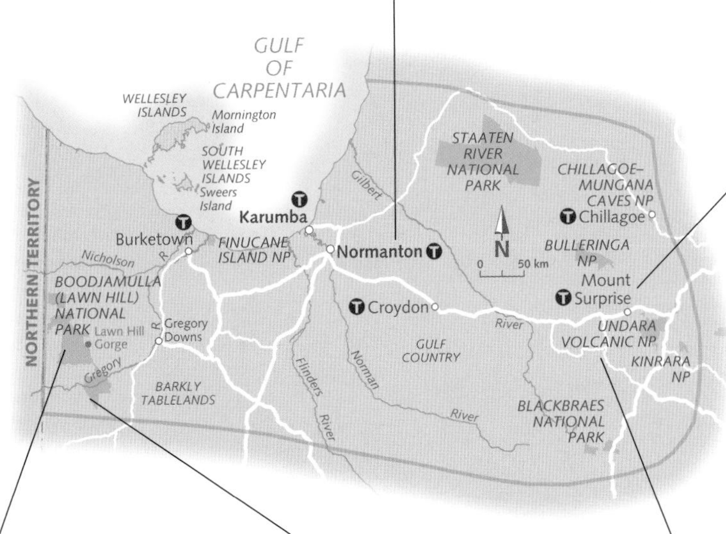

Fossicking around Mount Surprise The historic rail town of Mount Surprise has excellent gemfields in its vicinity that are perfect for fossicking, especially for topaz, quartz and aquamarine. Mount Surprise Gems in Garland Street (which doubles as the visitor centre) runs tours to nearby fossicking spots, so you don't need to worry about licences. If you're out of luck on the field, you can always purchase the stunning gemstones in town.

Boodjamulla (Lawn Hill) National Park Lawn Hill Gorge, about 205 kilometres from Burketown, is this remote park's main attraction. The gorge has given life to an oasis of lush rainforest. Canoeing, swimming and walking are the main activities, and Aboriginal rock-art sites are accessible at two locations.

Riversleigh Fossil Site The fossils in this World Heritage–listed part of Boodjamulla (Lawn Hill) National Park record the evolution of mammals over 20 million years, as the vegetation changed from rainforest to semi-arid grassland. Guided tours provide an insight into the ancient world, and there is a self-guide interpretive trail.

Lava tubes Undara, an Aboriginal word for long, accurately describes the tubes in Undara Volcanic National Park east of Mount Surprise. At 160 kilometres, one of the lava tubes here is the longest on earth. The caves in nearby Chillagoe–Mungana Caves National Park are a feast for the senses, with bat colonies and richly decorative stalactites and stalagmites. Tours are necessary to visit both the tubes and the caves. The incredible geological history of the area is detailed at The Hub in Chillagoe.

QUEENSLAND

TOP EVENTS

EASTER	World Barramundi Championships (Burketown)
EASTER	Barra Classic (Normanton)
APR	Fishing Competition (Karumba)
APR–MAY	Gregory River Canoe Marathon (Gregory)
MAY	Chillagoe Big Weekend (Chillagoe)
JUNE	Normanton Show and Rodeo (Normanton)

CLIMATE

NORMANTON

J	F	M	A	M	J	J	A	S	O	N	D	
35	34	34	34	32	29	29	31	34	36	37	36	MAX °C
25	25	24	22	19	16	15	17	20	23	25	25	MIN °C
260	249	158	31	8	9	3	2	3	10	44	143	RAIN MM
14	14	9	2	1	1	1	0	0	1	4	9	RAIN DAYS

FISHING THE GULF

The gulf is one of Australia's true fishing frontiers. Anglers can fish the rivers – the Nicholson, Albert, Flinders, Norman and Gilbert – as well as the coastal beaches and the offshore waters and reefs of the gulf, accessible via the island resorts. Karumba, on the Norman River estuary, is a popular base for both river and offshore anglers. Sweers Island, in the Wellesley group, has a fishing resort offering access to thousands of hectares of reef, where coral trout, parrotfish, sweetlip and sea perch (and in winter, pelagics such as mackerel and tuna) are plentiful. Mornington Island is home to the Birri Fishing Resort, offering crabbing and sport- and bottom-fishing, all with professional masters. You can also stay at Escott Barramundi Lodge on the Nicholson River, reached via Burketown. Fishing charters take you along the nearby lagoons and rivers, where you can catch barramundi, catfish and mangrove jack.

→ For more detail see maps 656–7 & 659. For descriptions of T towns see Towns from A–Z (p. 429).

OUTBACK

Outback Queensland is remote, sparsely populated and – in parts – beautiful. Many legends of Australia's pioneering days evolved here, and are celebrated in the region's museums and monuments.

[OLD MAILBOX AT CRAVENS PEAK, BOULIA]

Mount Isa This is Queensland's largest outback town. For a quintessential outback experience, visit Outback at Isa, complete with underground mining tunnels and mining machinery. Nearby is the Kalkadoon Tribal Centre and Cultural Keeping Place, with artefacts of the fierce and proud Kalkadoon people.

Channel Country Monsoon rains in the tropical north flood the hundreds of inland river channels that meander through Queensland's south-west. Here cattle graze on huge semi-desert pastoral holdings. Spectacular red sandhills are found in the area, particularly in Simpson Desert National Park, beyond Birdsville. On certain nights under certain conditions, the town of Boulia, capital of the Channel Country, is handed over to the realm of the Min Min reputedly a strange ball of white light that hovers above the ground, disappearing, reappearing and moving around at will. The Min Min Encounter in town is an impressive centre where you can soak up the stories surrounding this outback legend.

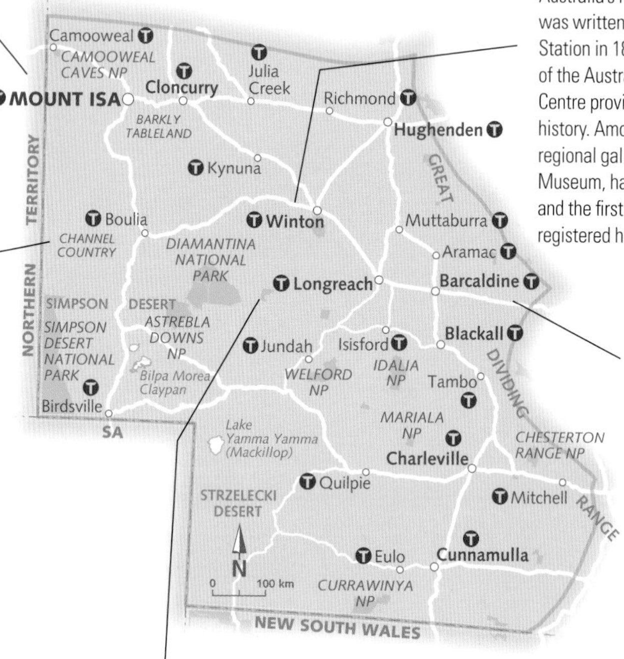

Waltzing Matilda Centre The area surrounding Winton is known as Matilda Country, because Australia's most famous song, *Waltzing Matilda*, was written by Banjo Paterson at nearby Dagworth Station in 1895. A tribute to the character and life of the Australian swagman, the Waltzing Matilda Centre provides an interesting look at the nation's history. Among other things, it incorporates the regional gallery, interactive exhibits and the Matilda Museum, harking back to Winton's pioneering days and the first days of Qantas (whose first office was registered here in 1920).

Barcaldine This 'Garden City of the West' was the first Australian town to tap the waters of the Great Artesian Basin, an event commemorated by the town's giant windmill. The Australian Workers Heritage Centre, in town, recollects the 1891 Shearers' Strike that led to the formation of the Australian Workers Party, forerunner of the Australian Labor Party. It also features re-created work precincts in tribute to all the workers who helped shape early Australia.

Australian Stockman's Hall of Fame and Outback Heritage Centre This impressive institution is a lasting tribute to the outback people – Indigenous, European settler and current-day inhabitants alike. The imaginative displays show the development of the outback. Don't miss the nearby Qantas Founders Outback Heritage Museum, which tells the story of the oldest airline in the English-speaking world.

TOP EVENTS

APR	Dirt and Dust Festival (Julia Creek)
APR	Surf Carnival (Kynuna)
AUG	Isa Rodeo (Mount Isa)
AUG	Drover's Camp Festival (Camooweal)
SEPT	Birdsville Cup Racing Carnival (Birdsville)
SEPT	Bilby Festival (Charleville)

CLIMATE

LONGREACH

J	F	M	A	M	J	J	A	S	O	N	D	
37	36	35	31	27	24	23	26	30	34	36	37	MAX °C
23	23	20	16	12	8	7	8	12	17	20	22	MIN °C
58	68	43	10	13	5	4	3	3	16	16	44	RAIN MM
7	7	5	3	3	2	2	2	2	4	4	6	RAIN DAYS

ANCIENT ANIMALS

Queensland's outback is a veritable feast of fossils that document its changing ecological history. There is a fossil collection at Flinders Discovery Centre in Hughenden, its main exhibit being the replica skeleton of *Muttaburrasaurus langdoni* (the real thing was found in 1963 in a creek near Muttaburra). Kronosaurus Korner in Richmond is well known for its vertebrate fossils, and the Riversleigh Fossil Centre in Mount Isa showcases the tremendous findings of the Riversleigh Fossil Site. At Lark Quarry Conservation Park south-west of Winton, the preserved tracks of a dinosaur stampede lie protected under a shelter.

→ For more detail see maps 652, 654, 659, 660–1 & 662–3. For descriptions of Ⓣ towns see Towns from A–Z (p. 429).

[RAINFOREST CANOPY, DAINTREE NATIONAL PARK]

LEGEND

VISITOR INFORMATION
RADIO STATIONS
IN TOWN
WHAT'S ON
WHERE TO EAT
WHERE TO STAY
NEARBY

Food and accommodation listings in town are ordered alphabetically with places nearby listed at the end

Airlie Beach

Pop. 2752
Map ref. 649 F5 | 655 J3

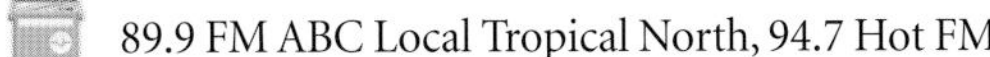

277 Shute Harbour Rd; (07) 9496 6665 or 1800 819 366; www.destinationwhitsundays.com.au

89.9 FM ABC Local Tropical North, 94.7 Hot FM

Airlie Beach is at the centre of the thriving Whitsunday coast. This tropical holiday town offers a cosmopolitan blend of bars, restaurants and shops just metres from the beach. From Abel Point Marina, daytrips to the outer Great Barrier Reef and Whitsunday islands are on offer. Watersports available include sailing, snorkelling, diving and fishing. Nearby Shute Harbour is one of the largest marine passenger terminals in Australia and, along with Airlie Beach, services the majority of the Whitsunday islands.

Airlie Beach Lagoon: safe, year-round swimming in landscaped environment; foreshore. ***Whale watching:*** tours July–Sept depart Abel Point Marina; details from visitor centre. ***Sailing:*** be it traditional sailing, adventure or luxury crewed; details from visitor centre. ***Skydiving:*** tandem skydive and take in the stunning views from very high above; (07) 4946 9115.

Community market: Airlie Beach Esplanade; Sat mornings. ***Meridien Marinas Race Week:*** sailing; Aug. ***Whitsunday Fun Race:*** competitions for cruising yachts; Aug/Sept. ***Festival of Sport:*** includes triathlon; Sept.

Déjà Vu: modern Australian; Golden Orchid Dr; (07) 4948 4309. ***Air Whitsunday Peppers Palm Bay daytrip:*** seaplane tour with lunch; Air Whitsunday, Whitsunday Airport; (07) 4946 9111. ***Daydream Island Day Tripper:*** lunch cruise; Fantasea Cruises, 11 Shute Harbour; (07) 4946 5111 or 1800 650 851.

Airlie Waterfront B&B: waterfront location; Cnr Broadwater Ave and Mazlin St; (07) 4946 7631. ***Pinnacles Resort and Spa:*** luxury resort; 16 Golden Orchid Dr; (07) 4948 4800. ***Whitsunday Terraces:*** great views from these self-contained apartments; 5 Golden Orchid Dr; (07) 4946 6788 or 1800 075 062.

Conway National Park Covering 35 km of coastline, this park is renowned for its natural beauty and as the habitat of the Proserpine rock wallaby (endangered species). Walks start in Airlie Beach and Shute Harbour. The Mt Rooper Lookout is a highlight, featuring a panoramic view over Hamilton, Dent, Long and Henning islands. Access the park off the road to Shute Harbour; (07) 4967 7355; 3 km SE.

Islands of Whitsunday Passage These islands offer an abundance of sights and activities, whether they be the sports and social activities at the resorts on Hamilton, Daydream and Hayman islands, or the more secluded island experience on Whitsunday, Hook and South Molle. Whitsunday Island is famous for the white sand and clear water of Whitehaven Beach. The waterfalls on Hook Island are impressive, as are the butterflies on the shores of the island's aptly named Butterfly Bay. Big-game enthusiasts strive to catch black marlin off Hayman Island Sept–Nov.

Crocodile safaris and fishing trips: to nearby coastal wetlands; details from visitor centre. ***Scenic flights:*** various tours over the 74 Whitsunday islands; details from visitor centre.

See also THE MID-TROPICS, p. 424

Allora

Pop. 920
Map ref. 653 L10

49 Albion St (New England Hwy), Warwick; (07) 4661 3122; www.southerndownsholidays.com.au

89.3 Rainbow FM, 747 AM ABC Southern QLD

Allora is a charming town in the Darling Downs, central to its rich agricultural surrounds. Explored and settled with stud farms in the 1840s, Allora's main street is noted for its well-preserved historic buildings and old-time country feel.

Allora Museum: noted for its replica of the Talgai Skull, an Aboriginal cranium dating back 15 000 years; open 1.30–4pm Sun; old courthouse, Drayton St. ***St David's Anglican Church:*** built in 1888 and said to be one of the finest timber churches in country Queensland; Church St. ***The Gnomery:*** over 100 different moveable handcrafted toys that feature in a theatre performance; New England Hwy.

Glengallan Farmers Markets: 1st Sun of each season. ***Heritage Weekend:*** Jan. ***Allora Show:*** Feb. ***Allora Auction:*** cars to farm machinery; June.

Goomburra Forest Retreat: peaceful retreat; 268 Forestry Reserve Rd; (07) 4666 6058.

Goomburra, Main Range National Park Located in the western foothills of the Great Dividing Range, this park has short walks around Dalrymple Creek and spectacular views from Mt Castle and Sylvesters lookouts. Take the Inverramsay Rd 40 km E to the forest. The last 6 km is unsealed and may be impassable following heavy rain. (07) 4666 1133; 49 km E.

Glengallan Homestead and Heritage Centre: restored 1867 sandstone mansion. Documents and photos chronicle its history as a pastoral station; open Thurs–Sun; New England Hwy; 11 km S.

See also **Darling Downs, p. 418**

Aramac

Pop. 340
Map ref. 654 D9 | 661 P9

Post office, 22 Gordon St; (07) 4651 3147; www.aramac.qld.gov.au

540 AM ABC Western QLD

Aramac is a small service town west of the Great Dividing Range. The town was named by explorer William Landsborough; the name is an acronym for the name of Queensland premier Sir Robert Ramsay Mackenzie (RRMac). The town's sole water supply is from two bores that tap into the Great Artesian Basin.

Harry Redford Interpretive Centre: photographic exhibition of cattle drives, also local arts and crafts; Gordon St. ***White Bull replica:*** commemorating Captain Starlight's arrest for cattle stealing; Gordon St. ***Tramway Museum:*** with old rail motor and historical exhibits; McWhannell St.

Harry Redford Cattle Drive: May–June. ***Ballyneety Rodeo:*** Sept.

Lake Dunn This freshwater lake and its surrounds have greatest appeal to birdwatchers. It is also popular for swimming and fishing. Follow signs to 'The Lake'; 68 km NE.

Forest Den National Park This remote park is an important wildlife sanctuary due to its semi-permanent waterholes. Have a picnic next to Torrens Creek and go birdwatching at dusk. 4WD recommended. Torrens Creek Rd; (07) 4652 7333; 110 km N.

Gray Rock: large sandstone rock engraved with the names of hundreds of Cobb & Co travellers. This was once the site of a hotel – a nearby cave was used as the hotel's cellar. 35 km E. ***Lake Galilee:*** 15 000 ha saltwater lake with large waterfowl population; some of access road unsealed; 100 km NE.

See also **Outback, p. 428**

Atherton

Pop. 6249
Map ref. 650 D3 | 651 D9 | 657 L7

Cnr Silo and Main sts; (07) 4091 4222; www.athertontableland.com

97.9 Hot FM, 720 AM ABC Far North QLD

Originally called Prior's Pocket and renamed in 1885, this town is the commercial hub of the Atherton Tableland. It is an area renowned for its volcanic crater lakes, waterfalls and fertile farmlands. Surrounding the town is dense rainforest that abounds in birdlife, and the nearby parks and forests offer a variety of watersports, bushwalking and outdoor activities.

Chinese Interpretive Centre and Old Post Office Gallery Atherton once had a large population of Chinese working for local timber cutters. This centre exhibits photos of these days and has artefacts and works by local artists and potters. Tours of the nearby Hou Wang Temple, built in 1903 and recently restored, depart from here. Open Wed–Sun; Herberton Rd.

Hallorans Hill Conservation Park: walk to the rim of this extinct volcanic cone on the Atherton Tableland, where there is a spectacular lookout and informative displays; off Kennedy Hwy. ***Crystal Caves:*** explore underground tunnels and chambers lined with crystals, fossils and fluorescent minerals. The above-ground Fascinating Facets shop sells a range of jewellery and gemstones; Main St. ***Birds of Prey:*** interactive bird show featuring some amazing aerial predators; shows 11am and 2pm Wed–Sun (closed Feb, Mar and Nov); Herberton Rd; (07) 4091 6945.

VP60 Celebrations: military parade; Aug. ***Maize Festival:*** Aug.

Tolga Woodworks Cafe and Gallery: gourmet cafe food; Kennedy Hwy, Tolga; (07) 4095 4488.

Atherton Rainforest Motor Inn: garden adjoins rainforest; Cnr Simms Rd and Kennedy Hwy; (07) 4095 4141. ***The Summit Rainforest Retreat:*** Land for Wildlife property; 22 Twelfth Ave; (07) 4091 7300. ***Barking Owl Retreat:*** 2 secluded cottages; 409 Hough Rd, Kairi; (07) 4095 8455.

Lake Tinaroo With 200 km of shoreline, Lake Tinaroo is ideal for fishing, waterskiing and sailing. Walking tracks circle the lake, and dinghies and houseboats are available for hire. The Danbulla Forest Drive is a scenic 28 km drive. 15 km NE via Kairi.

Hasties Swamp National Park: local and migratory birds visit this swamp, including whistling ducks and magpie geese; (07) 4091 1844; 3 km S. ***Tolga:*** this town has a railway museum and craft outlets; 5 km N. ***Wongabel State Forest:*** important wildlife refuge in Wet Tropics World Heritage Area. An informative heritage trail gives an insight into Aboriginal culture and history; Kennedy Hwy; 8 km S.

See also **Cairns & The Tropics, p. 425**

Ayr

Pop. 8094
Map ref. 649 B2 | 654 H2 | 657 O12

Burdekin Visitor Information Centre, Bruce Hwy, Plantation Creek Park; (07) 4783 5988; www.burdekintourism.com.au

 97.1 Sweet FM, 630 AM ABC North QLD

This busy town south-east of Townsville is surrounded by sugarcane fields – the most productive in Australia – and is the largest mango-growing area in the country. On the north side of the Burdekin River, it is linked to Home Hill to the south by the 1097-metre Silver Link Bridge, which ensures the towns are not cut off when the river floods.

Burdekin Cultural Complex: 530-seat theatre, library and activities centre. The forecourt has distinctive Living Lagoon water feature; Queen St. ***Gubulla Munda:*** 60 m carpet snake sculpture is the totem for the Juru Tribe, who were the original inhabitants of the area. Nearby Juru Walk passes through remnant dry tropical rainforest; Bruce Hwy, southern entrance to Ayr.

Market: Plantation Creek Park; Bruce Hwy; Sun (not 4th Sun of the month). ***Barra Rush:*** fishing competition; Feb. ***Auto Fest:*** car show; Apr. ***Water Festival:*** Sept.

Home Hill This small town is just south of Ayr over the Silver Link. The towering tea trees along the main street bear plaques commemorating the town's pioneering families. The Silver Link Interpretive Centre in Eighth Ave gives a photographic history of the Burdekin River Bridge. Ashworth's Tourist Centre houses Ashworth's Jewellers, the Rock Shop and the impressive Treasures of the Earth Display, while Zaro's Cultural Gallery has original islander artworks. The Canefield Ashes cricket competition is held in Apr, the Burdekin Grower Race Day in May and the Harvest Festival in Nov; 12 km S.

Hutchings Lagoon: watersports; 5 km NW. ***Brandon Heritage Precinct:*** includes district's oldest church, St Patrick's (1897), now a local history museum. Ye Olde Machinery Place has an antique farm machinery display; 6 km N. ***Lynchs Beach:*** beach walks, birdwatching, swimming and fishing. Departure point for dive trips to world-famous shipwreck, SS *Yongala*; 18 km N. ***Mt Kelly:*** Great views of surrounding farmlands; 18 km SW. ***Charlies Hill:*** WW II historic site; 24 km S. ***Groper Creek:*** great fishing spot with camping available; 24 km SE. ***Horseshoe Lagoon:*** birdwatching; 35 km N. ***Mt Inkerman:*** good views at top, plus picnic and barbecue facilities; 30 km S. ***Burdekin Dam:*** biggest dam in Queensland, holding the equivalent of 4 Sydney Harbours. Fishing and camping on offer; 180 km NW.

See also THE MID-TROPICS, p. 424

Babinda

Pop. 1168
Map ref. 650 E4 | 651 G10 | 657 M7

 **Cnr Bruce Hwy and Munro St; (07) 4067 1008.**

94.9 Kool FM, 720 AM ABC Far North QLD

A small sugar town south of Cairns, Babinda boasts abundant wildlife, secluded swimming holes and untouched rainforest in its surrounds. In adjacent Wooroonooran National Park are the state's two highest mountains, Mount Bartle Frere (1622 metres) and Mount Bellenden Ker (1592 metres).

M&J Aboriginality: a family business specialising in traditional Aboriginal artefacts. Make your own didgeridoo or learn how to throw a boomerang; Howard Kennedy Dr; (07) 4067 1660.

Harvestfest: celebrate the start of the cane season; May.

Wooroonooran National Park Part of the Wet Tropics World Heritage Area, the park has endemic species of plants and animals and spectacular walks through tropical rainforest. Swim in the watering hole at Josephine Falls, located at the base of Mt Bartle Frere, or see the Boulders, a large group of rocks worn smooth by tropical rains. Also visit the Mamu Canopy Rainforest Walk, which has spectacular views of the North Johnstone Gorge. Access off Bruce Hwy, west of Babinda; (07) 4061 5900.

Deeral: departure point for cruises through rainforest and saltwater crocodile haunts of the Mulgrave and Russell rivers. Deeral Cooperative makes footwear and Aboriginal artefacts; Nelson Rd; 14 km N. ***Bramston Beach:*** small community behind long palm-lined beach; Bruce Hwy S to Miriwinni, then 12 km E. ***Russell River National Park:*** small park on the coast with good birdwatching, swimming and canoeing; no facilities, 4WD access; (07) 4046 6600; 6 km N of Brampton Beach.

See also CAIRNS & THE TROPICS, p. 425

Barcaldine

Pop. 1338
Map ref. 654 D10 | 661 P10

Oak St; (07) 4651 1724; www.barcaldine.qld.gov.au

100.9 West FM, 540 AM ABC Western QLD

After the 1891 Shearers' Strike, the Australian Labor Party was born in this pastoral and rail town located some 100 kilometres east of Longreach. Barcaldine's supply of artesian water ensures its status as 'Garden City of the West'; the streets are named after trees.

Australian Workers Heritage Centre This centre is a tribute to the working men and women of Australia – the shearers, teachers, policemen and other workers who helped build the nation. The interpretive displays also cover the events leading to the formation of the Labor Party. Open 9am–5pm Mon–Sat, 10am–4pm Sun and public holidays; Ash St; (07) 4651 2422.

Tree of Knowledge: was a large ghost gum, which stood outside the railway station. It was fatally poisoned in 2006, but is being preserved and will be placed inside an $8 million memorial; Oak St. ***Folk Museum:*** display of historical memorabilia from the area; Cnr Gidyea and Beech sts. ***Roses 'n' Things:*** sit among over 800 roses of various varieties while enjoying Devonshire tea; Coolibah St. ***National Trust–classified buildings:*** Masonic lodge, Beech St; Anglican church, Elm St; shire hall, Ash St. ***Murals and musical instruments:*** Barcaldine is home to several murals including one painted by the late D'Arcy Doyle. There are also 2 musical instruments in the parks, which visitors are free to play; Oak St. ***Between the Bougainvilleas Heritage Trail:*** this award-winning heritage trail showcases Barcaldine's varied and colourful history; details from visitor centre. ***Artesian Country tours:*** including some to Aboriginal carvings and caves; details from visitor centre. ***Bike hire:*** a gold coin donation will allow you to cycle around the town at your own leisure and take in all the sites.

Mini steam-train rides: depart Folk Museum; last Sun each month (Mar–Oct). ***Easter in the Outback:*** Easter. ***Tree of Knowledge Festival:*** festival includes the Revfest, goat races, May Day parade and markets; May.

Bicentennial Park: has botanical walk through bushland; Blackall Rd; 9 km S. ***Lloyd Jones Weir:*** great fishing and birdwatching venue; 15 km SW.

See also OUTBACK, p. 428

Beaudesert

Pop. 5386
Map ref. 551 F1 | 557 N1 | 645 C9 | 653 N10

Historical Museum, 54 Brisbane St; (07) 5541 3740; www.bsc.qld.gov.au

 101.5 Beau FM, 747 AM ABC Southern QLD

Beaudesert lies in the valley of the Logan River, in the Gold Coast hinterland. The town was built up around the homestead of Edward Hawkins – his property was immense, comprising land from the coast to the Logan River. Those origins continue with the area noted for its dairying, beef cattle, and fruit and vegetable produce, making the country markets a great attraction.

Centre for Arts and Culture: the region's newest attraction and hub for musicians, artists, comedians and film buffs. State-of-the-art auditorium for large performances as well as intimate spaces; Brisbane St. ***Historical Museum:*** displays of old machinery and tools; Brisbane St. ***Community Arts Centre:*** art gallery, teahouse and craft shop; Enterprise Dr.

Market: Westerman Park; 1st Sat each month. ***Bikes and Bands Charity Fesitval:*** food and music; Feb. ***Rodeo:*** May. ***Equine Expo:*** June. ***Country and Horse Festival:*** June. ***Beaudesert Show:*** Sept.

Barney Creek Vineyard Cottages: 3 private cottages; 198 Seidenspinner Rd, Barney View; (07) 5544 3285. ***Cedar Glen Farmstay:*** homestead and private cottages; 3338 Kerry Rd, Lost World Valley; (07) 5544 8170. ***Katandra Mountain Farm House:*** pets welcome, including horses; 3213 Kerry Rd, Lost World Valley; 1800 503 475. ***Worendo Cottages:*** also has cooking school; 97 Cedar Glen Rd, Darlington; (07) 5544 8104.

Mt Barney National Park A remote park where the rugged peaks of Barney, Maroon, May and Lindesay mountains stand as remnants of the ancient Focal Peak Shield Volcano. The walks are not for the inexperienced, but picnicking at Yellow Pinch at the base of Mt Barney is an alternative. The challenging 10 hr ascent to Mt Barney's summit on the Logan's Ridge track rewards walkers with spectacular views; 55 km SW.

Darlington Park: recreation area with picnic/barbecue facilities; 12 km S. ***Tamrookum:*** has a fine example of a timber church; tours by appt; 24 km SW. ***Bigriggen Park:*** recreation area with picnic/barbecue facilities; 30 km SW. ***Rathdowney:*** great viewpoint from Captain Logan's Lookout in John St; 32 km S. ***Lamington National Park:*** 40 km SW; *see Nerang*.

See also GOLD COAST & HINTERLAND, p. 417

Biggenden

Pop. 644
Map ref. 648 B5 | 653 L4

Rose's Gallery Cafe Shoppe, George St; (07) 4127 1901; www.biggenden.qld.gov.au

 102.5 Breeze FM, 855 AM ABC Wide Bay

This agricultural town south-west of the city of Bundaberg, known as the 'Rose of the Burnett', is proud of its impressive range of roses in the main street. Situated in a valley, the majestic ranges of Mount Walsh National Park tower over the town.

Historical Museum: exhibits history of shire and life of the early pioneers; open Thurs and 2nd Sat each month, or by appt; Brisbane St; (07) 4127 5137. ***Blacksmith Shop:*** established in 1891, this recently restored shop contains displays and relics; open by appt; George St; (07) 4127 1298.

Market: Lions Park; 3rd Sat each month. ***Biggenden Show:*** May. ***Dallarnil Sports Day:*** June. ***Auto Spectacular:*** Aug. ***Rodeo:*** Nov.

Mt Walsh National Park Featuring the impressive Bluff Range, this wilderness park commands the skyline. Walks take in rugged granite outcrops and gullies and are for the experienced bushwalker only. From the picnic area the views are still commanding; Maryborough Rd; (07) 4121 1800; 8 km S.

Mt Woowoonga: bushwalking in a forestry reserve; picnic and barbecue facilities; 10 km NW. ***Coalstoun Lakes National Park:*** protects 2 volcanic crater lakes. Walk up the northern crater for a view over the rim; 20 km SW. ***Coongara Rock:*** a volcanic core surrounded by rainforest; 4WD access only; 20 km S. ***Chowey Bridge:*** 1905 concrete arch railway bridge, one of 2 surviving in the country; 20 km NW. ***Brooweena:*** small town with Pioneer Museum; Biggenden Rd; 30 km SE. ***Paradise Dam:*** this 30 000-megalitre dam took 4 years to build and has a tourist centre. Walking track and fishing areas nearby; barbecue facilities; 35 km NW.

See also CAPRICORN, p. 423

Biloela

Pop. 5369
Map ref. 653 J2

Queensland Heritage Park, Exhibition Ave; (07) 4992 2400; www.biloela.com

 96.5 Sea FM, 837 AM ABC Capricornia

This thriving town in the fertile Callide Valley is part of the Banana Shire, but do not expect to find any bananas grown here. The area was actually named after a bullock called 'Banana', whose job was to lure wild cattle into enclosures, a difficult feat that was much applauded by local stockmen.

Queensland Heritage Park Exhibition Originally an exhibition at the Expo '88 in Brisbane, the park is home to the famous silo, which stands 28 m tall. Inside are exhibitions on the history of the Callide and Dawson valleys as well as scenes of rural life. Also in the complex is Pioneer Place, home to Biloela's first church and railway station, where photographs and memorabilia document the area's past. Exhibition Ave.

Spirit of the Land Mural: amazing mural depicting the history of women in the shire; State Farm Rd. ***Greycliffe Homestead:*** original slab hut converted to a museum showcases the area's pioneering heritage; open by appt; Gladstone Rd; (07) 4992 1572.

Rotary Car, Ute and Bike Show: Mar. ***Biloela Rockfest:*** Mar. ***Callide Fishing Competition:*** Mar. ***Callide Valley Show:*** May. ***Old Wheels in Motion (Callide):*** festival celebrating farm machinery and vintage cars; July. ***Arts Festival (Brigalow):*** Oct. ***Food and Comedy Festival:*** Oct. ***Coal and Country Festival (Moura):*** Oct.

Mt Scoria Known locally as the 'Musical Mountain' because of the basalt columns at the top that ring when hit with another rock. Walks and trails around the mountain. 14 km S.

Thangool: renowned for its race days; 10 km SE. ***Callide Dam:*** excellent for boating, fishing and swimming; 12 km NE. ***Callide Power Station:*** near Callide Dam. ***Callide Mine Lookout:*** view over Biloela, the mine and the dam; 18 km NE, past dam. ***Kroombit Tops National Park:*** 25 km E; *see Gladstone*. ***Baralaba:*** historic village; watersports on the Dawson River; 100 km NW.

See also CAPRICORN, p. 423

Birdsville

Pop. 115
Map ref. 607 K2 | 662 D3

 **Wirrari Centre, Billabong Blvd; (07) 4656 3300.**

 540 AM ABC Western QLD

Birdsville is a tiny town at the northern end of the Birdsville Track, a major cattle route developed in the 1880s. In the 1870s the first European settlers arrived in the area. By 1900 the town was flourishing, boasting three hotels, several stores, a customs office and a cordial factory. When, after Federation in 1901, the toll on cattle crossing the border was abolished, the town's prosperity slowly declined. John Flynn, the famous 'Flynn of the Inland', opened an Australian Inland Mission at Birdsville in 1923. Cattle remains a major trade, as well as the tourism accompanying four-wheel-drive enthusiasts keen to take on the Birdsville Track and Simpson Desert National Park.

Working Museum: housed in 6 buildings, the museum showcases an array of old relics from tools to pottery and farming equipment. Join proprietor John Menzies for a tour and demonstration in harness making and coach building; tours daily at 9am, 11am and 3pm; Waddie Dr. ***Blue Poles Gallery:*** art by local painter Wolfgang John. Cafe on-site; Graham St. ***Artesian bore:*** water comes out at near boiling point from this 1219 m deep bore; behind the bore is a geothermal power plant; Graham St. ***Adelaide Street:*** ruins of Royal Hotel (1883), a reminder of Birdsville's boom days; Birdsville Hotel (1884), still an important overnight stop for travellers; cemetery, housing the grave sites of early pioneers.

Rodeo and Bronco Branding: May. ***Birdsville Gift:*** footrace; June. ***Birdsville Cup Racing Carnival:*** the 1st meeting of this annual event was held in 1882 and the tradition continues on the claypan track south-east of town. Held 1st Fri and Sat in Sept, when the population swells to over 6000.

Birdsville Caravan Park: centre of town; 1 Florence St; (07) 4656 3214. ***Birdsville Hotel:*** legendary pub; Adelaide St; (07) 4656 3244.

Simpson Desert National Park West of Birdsville, this arid national park is the largest in Queensland. The parallel windblown sand dunes are enormous – up to 90 m high, about 1 km apart, and can extend up to 200 km. A self-guide drive includes 10 sites, starting at the eastern park boundary and following the track to Poeppel's Corner. Walking any distance is not recommended and a 4WD is essential. Visit only between Apr and Oct, *see note below*; (07) 4656 3249; 65 km W.

Waddi Trees and Dingo Cave Lookout: 14 km N. ***Big Red:*** huge sand dune; 35 km W. ***Bedourie:*** Eyre Creek runs through town providing waterholes that are home to the endangered bilby and peregrine falcons; 191 km N.

Travellers note: *Travel in this area can be hazardous, especially in the hotter months (approximately Oct–Mar). Motorists are advised to check the RACQ Road Conditions Report on 1300 130 595 (or* www.racq.com.au*) for information before departing down the Birdsville Track and to advise police if heading west to Simpson Desert National Park. There is no hotel or fuel at Betoota, 164 km* E, *but fuel is available at Windorah, 375 km* E.

See also OUTBACK, p. 428

Blackall

Pop. 1160
Map ref. 652 A1 | 654 D12 | 661 P12 | 663 O1

Shamrock St; (07) 4657 4637; www.blackall.qld.gov.au

 95.1 West FM, 540 AM ABC Western QLD

Blackall is west of the Great Dividing Range in sheep and cattle country, and was home to the legendary sheep-shearer Jackie Howe. In 1892 he set the record of shearing 321 sheep with blade shears in less than eight hours at Alice Downs Station.

Fossilised tree stump: preserved tree stump estimated to be possibsly 225 million years old; Shamrock St. ***Major Mitchell Memorial Clock:*** commemorates the founding of Blackall in 1846; Shamrock St. ***The Black Stump:*** the reference point used when the area was surveyed in 1886. Beautiful mural painting of the stump at the site; Thistle St. ***Jackie Howe Memorial:*** statue of the legendary shearer who holds the record for shearing 321 sheep in 7 hrs 40 min; Shamrock St. ***Historic Ram Park:*** incorporates the living history of the Blackall district, with shearing occuring all year-round; Shamrock St. ***Pioneer Bore:*** first artesian bore sunk in Queensland, with display of replica drilling plant; Aqua St.

Blackall Show: May. ***Campdraft:*** May. ***Heartland Festival:*** May. ***Christmas in July:*** July. ***Flower Show:*** Sept. ***Springtime Affair:*** celebration of the season; Sept.

Idalia National Park Renowned habitat of the yellow-footed rock-wallaby, which can be spotted at Emmet Pocket Lookout (which also has amazing panoramic views) or along the Bullock Gorge walking track. A self-guide drive begins at the information centre, 12 km beyond the park entrance. (07) 4652 7333; 70 km SW on Yaraka Rd; at Benlidi siding turn south.

Blackall Wool Scour: restored steam-driven wool-processing plant with demonstrations of machinery (steam operating May–Sept only); Evora Rd; 4 km N.

See also OUTBACK, p. 428

Blackwater

Pop. 5030
Map ref. 655 J11

Central Highlands Regional Council, McKenzie St; (07) 4980 5555.

 92.7 4BCB FM, 1548 AM ABC Capricornia

Blackwater is west of Rockhampton in the Capricorn region and is known as the coal capital of Queensland. The coal is transported directly from coalmines south of town to Gladstone by train. The name 'Blackwater' is not a reference to the effects of mining operations, however, but comes from the discolouration of local waterholes caused by tea trees.

Japanese Garden complex This ornate traditional Japanese garden was constructed over 8 months in 1998. It symbolises the relationship Blackwater shares with its sister town Fujisawa in Japan. The complex also houses the newly opened Blackwater International Coal Centre, which has interactive and interpretive touch-screen displays, a cinema screening films on coal mining and a cafe. There is also a craft shop and information centre that was once the town station. Capricorn Hwy.

Lions Park: displays the flags of 37 nations to commemorate the nationality of every worker on the coalmines. In terms of size and variety, the display is second only to that of the United Nations'

QUEENSLAND

building in New York; Capricorn Hwy to the west of town. ***Helicopter flights over Blackwater Coal Mine:*** see the mine in action; details from visitor centre.

May Day Festival: May. ***World Dingo Trap Throwing Competition:*** July. ***Rodeo:*** Oct. ***Craft Fair and Art Exhibition:*** regional arts and crafts exhibit; Oct.

Blackdown Tableland National Park This national park offers spectacular scenery over mountains and lowlands, including some beautiful waterfalls. It is the traditional home of the Ghungalu people, whose stencil art can be seen on the 2.8 km Mimosa Culture Track. Walk through to Rainbow Falls Gorge and swim in rockpools. (07) 4986 1964; 30 km E to turnoff.

Bedford Weir: dam excellent for fishing; 20 km N. ***Comet:*** in town is the Leichhardt Dig Tree, where the explorer buried letters and marked the tree 'dig'; 30 km W.

See also CAPRICORN, p. 423

Boonah

Pop. 2282
Map ref. 551 D1 | 557 N1 | 645 A9 | 653 M10

Boonah–Fassifern Rd; (07) 5463 2233; www.scenicrim.qld.gov.au

100.1 Rim FM, 747 AM ABC Southern QLD

Boonah is set in the picturesque Fassifern Valley, surrounded by hills. Once noted as a 'beautiful vale' by 19th-century explorers, a little expedition in the surrounding region will reveal the beauty and ruggedness of the area. West of town is Main Range National Park, part of the Scenic Rim.

Cultural Centre: incorporates regional art gallery; open Mon–Fri, gallery open Wed–Sun; High St. ***Art and Soul:*** local art and photography; Walter St. ***Gliding and ultralight tours:*** flights over the Scenic Rim; details from visitor centre.

Country market: Springleigh Park; 2nd and 4th Sat each month. ***Rodeo:*** Apr. ***Country Show:*** May. ***SPAR Arts Festival:*** Sept. ***Orchid Show:*** Oct.

Stark House B&B: 3 rooms with ensuites; 143 Kalbar Connection Rd, Kalbar; (07) 5463 9365. ***Wiss House B&B:*** heritage-listed luxury; 7 Ann St, Kalbar; (07) 5463 9030. ***Zengarra Country House:*** music pavilion enchants guests; 2225 Lake Moogerah Rd, Moogerah; (07) 5463 5600.

Main Range National Park A World Heritage–listed park of rugged mountains and landscapes with spectacular lookouts. There are walks starting at the Cunninghams Gap and Spicers Gap campsites. See the varied birdlife, including the satin bowerbird, on the 8.4 km return Box Forest track. Access park from Cunningham Hwy; (07) 4666 1133; 40 km W. In the south of the park is Queen Mary Falls; *see Killarney*.

Templin: has historical museum chronicling the history of the area; 5 km NW. ***Kalbar:*** historic German town with magnificent buildings including the heritage-listed Wiss Emporium; 10 km NW. ***Moogerah Peaks National Park:*** excellent for birdwatching, and with lookouts over the Fassifern Valley. The Frog Buttress at Mt French is one of the best rock-climbing sites in Queensland; contact EPA on 1300 130 372 for more information; 12 km W. ***Lakes Maroon and Moogerah:*** ideal for camping and watersports; 20 km S and SW.

See also GOLD COAST & HINTERLAND, p. 417

Boulia

Pop. 206
Map ref. 660 F8

Min Min Encounter, Herbert St; (07) 4746 3386; www.outbackholidays.tq.com.au

102.5 Hot FM, 106.1 ABC FM North West QLD

Boulia is the capital of the Channel Country and is on the Burke River, named after the ill-fated explorer Robert O'Hara Burke. The town is famous for random appearances of the mysterious Min Min light, a ball of light that sometimes reveals itself to travellers at night. The isolated Diamantina National Park nearby is a haven for threatened species.

Stonehouse Museum: built in 1888, this National Trust–listed site was one of the first houses built in western Queensland. It is now a museum housing the history of the Jones family, as well as Aboriginal artefacts and photographs; Pituri St. ***Min Min Encounter:*** high-tech re-creation of the Min Min light, with outback characters as your guide; Herbert St. ***Red Stump:*** warns travellers of the dangers of the Simpson Desert; Herbert St. ***Corroboree Tree:*** last known of the Pitta Pitta community; near Boulia State School.

Rodeo Races and Campdraft: Easter. ***Camel Race:*** July. ***Back to Boulia:*** traditional games weekend; Sept.

Desert Sands Motel: 11 motel units, 1 self-contained unit and real Australian hospitality; Herbert St; (07) 4746 3000. ***The Australian Hotel:*** iconic outback pub; Herbert St; (07) 4746 3144.

Diamantina National Park This remote park south-east of Boulia is rich in colours and landscapes. Follow the 157 km Warracoota self-guide circuit drive to view the spectacular sand dunes, claypans and ranges and many rare and threatened species in their native habitat, including the greater bilby, kowari and peregrine falcon. Canoe or fish in the winding creeks and rivers. 4WD access only. Roads may become impassable after rain; check road conditions before travelling; (07) 4652 7333; 147 km SE.

Ruins of Police Barracks: 19 km NE. ***Cawnpore Hills:*** good views from summit; 108 km E. ***Burke and Wills Tree:*** on the west bank of the Burke River; 110 km NE. ***Ruins of Min Min Hotel:*** burned down in 1918, where the Min Min light was first sighted; 130 km E.

See also OUTBACK, p. 428

Bowen

Pop. 7483
Map ref. 649 E4 | 655 J3

Bruce Hwy next to Big Mango, Mount Gordon; (07) 4786 4222; www.tourismbowen.com.au

88.0 Explore FM, 630 AM ABC North QLD

At the north of the Whitsundays, Bowen is positioned within 5 kilometres of eight pristine beaches and bays. Named after the state's first governor, Bowen was established in 1861 – the first settlement in north Queensland. The town and surrounding area is well known for its mangoes, the Big Mango being testimony to the fact. Bowen has more recently gained fame as the film set for Baz Luhrmann's epic movie, *Australia*.

Historical murals Around the buildings and streets of Bowen's town centre are 25 murals by local and national artists, each illustrating an aspect of the region's history. The mural by Australian artist Ken Done was displayed at Expo '88. A new mural is commissioned every 2 years.

Historical Museum: covers history of area; open 9.30am–3.30pm Mon–Fri, 10am–12pm Sun Sept–Apr, closed Feb and Mar; Gordon St. ***Summergarden Theatre:*** styled on the classic movie houses of Southern California. Used now to screen films and stage performances; Murroona Rd.

Bowen Show: June. ***Polocrosse Carnivale:*** June. ***Bowen River Rodeo (Collinsville):*** July. ***Bowen Fest:*** Aug. ***Bowen Family Fishing Classic:*** Sept. ***Coral Coast Festival:*** Oct. ***Bowen Cup:*** Oct.

Rose Bay Resort: apartments on the bay; 2 Pandanus St, Rose Bay; (07) 4786 9000.

Bays and beaches Choose from 8 excellent spots for swimming, snorkelling and fishing in spectacular surrounds. Rose and Horseshoe bays are connected by a walking track with panoramic views over the ocean; a sidetrack leads to Murray Bay. Impressive corals and fish can be found at Grays, Horseshoe, Murray and Rose bays. Diving enthusiasts should head to Horseshoe and Murray bays. For a more exclusive swim, visit secluded Coral Bay. Details and directions from visitor centre.

Big Mango: tribute to the local Kensington Mango, grown since the 1880s. A shop sells all things mango-related – the mango ice-cream is a highlight; Bruce Hwy, Mt Gordon; 7 km S. ***Gloucester Island National Park:*** group of secluded islands 23 km offshore, part of the Great Barrier Reef, boasting beaches and rainforest. Campers must be self-sufficient and obtain a permit. Access via private boat from Dingo Beach, Hideaway Bay, Bowen or Airlie Beach; (07) 4967 7355. ***Cape Upstart National Park:*** remote granite headland flanked with sandy beaches; self-sufficient visitors only; access by boat, ramps at Molongle Bay and Elliot River; (07) 4967 7355; 50 km NW. ***Collinsville:*** coalmine tours can be arranged at Bowen visitor centre; 82 km SW.

See also THE MID-TROPICS, p. 424

Buderim

Pop. 34 454
Map ref. 647 G4 | 648 F12 | 653 N7

Old Post Office, Burnett St; (07) 5477 0944; www.buderim.com

90.3 FM ABC Coast, 91.1 Hot 91 FM

Buderim is just inland from the Sunshine Coast, high on the fertile red soil of Buderim Mountain, a plateau overlooking the surrounding bushland and ocean. With its wide streets and abundance of small-scale art and craft galleries, it escapes the crush of nearby towns like Maroochydore and Mooloolaba.

Pioneer Cottage This restored 1876 National Trust timber cottage is one of Buderim's earliest houses and retains many of its original furnishings. Now home to the local historical society, it has exhibits on the history of the town and its surrounds. Open 11am–3pm daily; Ballinger Cres; (07) 5450 1966.

Buderim Forest Park: subtropical rainforest reserve and a great place for a picnic or barbecue. In the south, via Quorn Cl, is the Edna Walling Memorial Garden and Serenity Falls; in the north, via Lindsay Rd, is Harry's Restaurant and a boardwalk along Martins Creek. ***Foote Sanctuary:*** rainforest walks and more than 80 bird species; car entry via Foote St. ***Arts and crafts galleries:*** various shops selling locally made items; Main St. ***Ginger Shoppe:*** while the Buderim Ginger Factory may have relocated to Yandina, you can purchase products from the shop; Burnett St.

Australia Day celebrations: parade and fair; Jan.

Parle on Buderim B&B: guest privacy is paramount; 6 Parle Crt; (07) 5450 1413.

Mooloolah River National Park: 6 km SE; *see Mooloolaba.*

See also SUNSHINE COAST, p. 419

Bundaberg

see inset box on next page

Burketown

Pop. 176
Map ref. 656 A7 | 659 D4

Old Post Office (Apr–Sept); or council offices (Oct–Mar), Musgrave St, (07) 4745 5100; www.burkeshirecouncil.com

567 AM ABC North West QLD

Burketown is on the edge of the Gulf of Carpentaria, on the dividing line between the wetlands to the north and the Gulf Savannah plains to the south. It was named after the ill-fated explorer Robert Burke, who was the first European (with partner William John Wills) to arrive in the area. Regularly in spring, the natural phenomenon known as Morning Glory takes over the horizon at dawn between Burketown and Sweers Island, offshore. The clouds appear as rolling tube-like formations and can extend for more than 1000 kilometres.

Museum and Information Centre: in the original post office, with displays on the history of the area plus local arts and crafts. Open Apr–Oct; Musgrave St. ***Artesian bore:*** operating for over 100 years and quite a sight to see due to the build up of minerals; The Great Top Rd.

World Barramundi Championships: Easter. ***May Day Weekend (Gregory):*** May. ***Arts and Crafts Festival:*** Aug. ***Campdraft:*** Sept.

Burketown Caravan Park: pet friendly; Sloman St; (07) 4745 5118. ***Burketown Pub:*** ensuite units and hotel rooms; Cnr Beames and Musgrave sts; (07) 4745 5104. ***Adels Grove:*** range of accommodation; Lawn Hill; (07) 4748 5502.

Boodjamulla (Lawn Hill) National Park Approximately 90 km W of the town of Gregory, park highlights include canoeing in lush Lawn Hill Gorge, home to the Waanyi rock-art sites of Wild Dog Dreaming and Rainbow Dreaming, and the early-morning climb to Island Stack. The park's Riversleigh section contains some of the world's most significant mammalian fossils; public access is restricted to D site. Access to the park by conventional vehicle is best from Cloncurry; 4WD access from Burketown, Mount Isa and Camooweal; 200 km SW.

Original Gulf Meatworks: machinery relics of this once-thriving industry; just north of town. ***Colonial Flat:*** site of the Landsborough Tree. Blazed by the explorer in 1862 on his search for Burke and Wills, it became the depot camp for search parties and the resting place of Landsborough's ship Firefly – the first ship to enter the Albert River; 5 km E. ***Nicholson River wetlands:*** breeding ground for crocodiles, fish and birds; 17 km W. ***Bluebush Swamp:*** large wetland area ideal for birdwatchers; 30 km SW. ***Sweers Island:*** excellent spot for lure and fly fishing, plus golden beaches and over 100 species of birds; access by aircraft or boat, details from visitor centre; 30 km N. ***Leichhardt Falls:*** picturesque flowing falls in rainy months; 71 km SE.

continued on p. 437

QUEENSLAND

BUNDABERG

Pop. 59 766

Map ref. 648 D2 | 653 M3

271 Bourbong St; (07) 4153 8888 or 1300 722 099; www.bundabergregion.info

93.9 Hitz FM, 855 AM ABC Wide Bay

Bundaberg, the southernmost access point to the Great Barrier Reef, is proud of its parks and gardens. Even more recognisable is its world-famous amber spirit, Bundaberg Rum, and the surrounding fields of sugarcane. In harvest season, from July to November, the cane fires give the area a smoky haze. As the sugar industry was being developed in the 1880s, and Australian labour costs were rising, South Sea Islanders were placed under 'contract' to work on the cane fields as a cheap alternative. In fact, the majority of these labourers had been lured from their island homes onto boats under the pretence of trading goods. In 1901, when Australia's Commonwealth government was established, the Kanakas (as the labourers were known) were allowed to return home. A Kanaka-built basalt stone wall can still be seen near Bargara, a short distance north-east of town. Bundaberg more proudly claims the aviator Bert Hinkler as one of its own. In 1928 Hinkler was the first to successfully fly from England to Australia in a flight of just over 15 days.

Bundaberg Botanical Gardens In this picturesque setting stand many buildings from Bundaberg's past. Hinkler House Memorial Museum, inside Bert Hinkler's relocated Southampton home, is a tribute to him and to aviation history. Fairymead House Sugar Museum, a restored plantation house, recalls years of sugar production, and the nearby Historical Museum chronicles the general history of the area. To see more of the grounds, take the restored steam train around the lakes, which runs every Sun (also Wed on school holidays); Mt Perry Rd.

Bundaberg Rum Distillery: discover the distillation process first-hand; tours daily; Avenue St. ***Schmeider's Cooperage and Craft Centre:*** demonstrations of barrel making and glass-blowing, and sales of local crafts and handmade crystal jewellery; Alexandra St. ***Arts Centre:*** 3 galleries devoted to local and visiting art exhibitions; cnr Barolin and Quay sts. ***Baldwin Swamp Conservation Park:*** boardwalks and pathways, waterlily lagoons, abundant birdlife and native fauna; Steindl St. ***Tropical Wines:*** taste unique wines made from local fruit; Mt Perry Rd. ***Alexandra Park and Zoo:*** historic band rotunda, cactus garden, zoo (free admission); riverbank, Quay St. ***Whaling Wall:*** a 7-storey-high whale mural; Bourbong St. ***Heritage city walk:*** self-guide walking tour of 28 significant sites and buildings; starts at historic post office, Barolin St. ***Bundaberg Barrel:*** giant barrel and home to Bundaberg Brewed Drinks. With interactive tours, holographic 3D adventure and free sampling; Bargara Rd.

Shalom College Markets: local crafts; Fitzgerald St; every Sun. ***PCYC Markets:*** Maryborough St; 2nd Sun each month. ***Big House Piano:*** music at Fairymead House; 1st Sun each month. ***Aussie Country Muster:*** music festival; May. ***Bundaberg Show:*** May. ***International Air Show:*** July. ***Multicultural Food and Wine Festival:*** Aug. ***Bundy in Bloom Festival:*** celebration of spring; Sept. ***Arts Festival:*** Oct. ***Bundy Thunder:*** power boat spectacular; Nov. ***Port2Port Yacht Rally:*** Nov.

[LOGGERHEAD TURTLE AT MON REPOS CONSERVATION PARK]

Reef Gateway Motor Inn: Best Western member; 11 Takalvan St; (07) 4132 6999. ***Kacy's Bargara Beach Motel:*** on the waterfront; Cnr Bauer St and Esplanade, Bargara; (07) 4130 1100. ***Kelly's Beach Resort:*** self-contained villas; 6 Trevors Rd, Bargara; (07) 4154 7200 or 1800 246 141.

Mon Repos Conservation Park This park contains the largest and most accessible mainland loggerhead turtle rookery in eastern Australia. Between Nov and Mar these giant sea turtles come ashore to lay their eggs. Hatchlings leave their nests for the sea from mid-Jan to late Mar. Access to the park is restricted during these times – guided night tours depart from the park information centre for viewing turtles up-close. When turtles are not hatching, snorkelling and exploring the rockpools are popular. (07) 4159 1652; 15 km NE.

Burrum Coast National Park Broken into 2 sections, this national park offers a variety of landscapes and activities. In the northern Kinkuna section it is relatively undeveloped. The vegetation along the beach is rugged and spectacular, and birdwatching in the wallum heath is a highlight. Access is via Palm Beach Road; 14 km SW; 4WD and sand-driving experience are necessary. The southern Woodgate section is more developed, with boardwalks and established tracks that allow the visitor to see abundant wildlife from every vantage. Access is via Woodgate, a small town with a magnificent ocean beach; (07) 4131 1600; 57 km S.

Hummock Lookout: panoramic view of Bundaberg, cane fields and coast; 7 km NE. ***Meadowvale Nature Park:*** rainforest and walkway to Splitters Creek; 10 km W. ***Sharon Nature Park:*** rainforest, native fauna and walkway to Burnett River; 12 km SW. ***Bargara:*** coastal town with a popular surf beach and year-round fishing on man-made reef. Turtles often nest at nearby Neilson Park, Kelly's and Rifle Range beaches; 13 km NE. ***Fishing spots:*** area renowned for its wide variety of fishing. Excellent spots at Burnett Heads (15 km NE), Elliott Heads (18 km SE) and Moore Park (21 km NW). ***Mystery craters:*** 35 small craters in sandstone slab, the origin of which causes much debate, but confirmed to be over 25 million years old; 25 km SW. ***Littabella National Park:*** lagoons and billabongs surrounded by tea tree forest. Many sand tracks for the 4WD enthusiast can be found at nearby Norval Park Beach; (07) 4131 1600; 38 km NW. ***Lady Elliot and Lady Musgrave islands:*** excellent spots for snorkelling and fishing. Wilderness camping on Lady Musgrave; seabird nesting on Lady Elliot during summer; sea access from Bundaberg Port plus air access to Lady Elliot from Hinkler Airport.

See also CAPRICORN, p. 423

Gregory: small outback town that holds the Gregory River Canoe Marathon in May; 113 km S. ***Fishing and boat tours:*** to nearby estuaries and Wellesley Islands; details from visitor centre.

See also GULF SAVANNAH, p. 427

Burleigh Heads

Pop. 7606

Map ref. 551 G1 | 557 O1 | 645 G10 | 646 F8 | 653 N11

Shop 14B, Coolangatta Pl, Cnr Griffith and Warner sts, Coolangatta; 1300 309 440; www.verygoldcoast.com.au

 89.3 4CRB FM, 91.7 FM ABC Gold Cast

Burleigh Heads is a suburb of the Gold Coast and is situated between Coolangatta and the tourist mecca Surfers Paradise. It is known for its breathtaking scenery, highlighted by the stunning Burleigh Head National Park. The famed south-easterly swells and surrounding parklands make the beaches on Burleigh Heads some of the best in the world, attracting international surfing tournaments. The relaxed charm of Burleigh Heads can be enjoyed from under the beachside pines and pandanus palms or in a nearby restaurant overlooking the stunning Pacific Ocean.

Burleigh Head National Park: take the 2.8 km Ocean View circuit to experience the coastal vegetation, rainforest and mangroves or go to Tumgun Lookout to watch for dolphins and humpback whales (seasonal); access from Goodwin Tce; 1300 130 372. ***David Fleay Wildlife Park:*** displays Queensland's native animals in a natural setting with the only display of Lumholtz's tree kangaroo and mahogany gliders in the world. The park also has crocodile feeding in summer and Aboriginal heritage programs; West Burleigh Rd. ***Paramount Adventure Centre:*** enjoy a rush of adrenaline and adventure at the Gold Coast's largest adventure centre. Choose from indoor rock climbing, kayaking, mountain-biking or take part in a learn-to-surf program; Hutchinson St.

Burleigh Art and Craft Market: beachfront; last Sun each month. ***Coolangatta Markets:*** beachfront; 2nd Sun each month. ***Burleigh Car Boot Sale:*** Stocklands Shopping Centre Car Park; 2nd Sun each month. ***Quiksilver Roxy Pro Surfing:*** Feb/Mar. ***Gold Coast Cup Outrigger Canoe Marathon:*** Apr. ***Wintersun Festival:*** Coolangatta; May/June. ***Coolangatta Gold:*** famous ironman and ironwoman race; Oct.

Vecchia Roma: traditional Italian family restaurant; 1748 Gold Coast Hwy, cnr West Burleigh Rd; (07) 5535 5988. ***Fish & Wine:*** very fine fish dishes; Relections Tower 2, 110 Marine Pde, Coolangatta; (07) 5536 7775.

Hillhaven Holiday Apartments: great ocean-view apartments; 2 Goodwin Tce; (07) 5535 1055. ***Sands Mediterranean Resort:*** good-value apartments; 220 The Esplanade, North Burleigh; (07) 5535 7188 or 1300 302 733. ***Bella Mare Holiday Apartments:*** near several beaches; 5 Hill St, Coolangatta; (07) 5599 2755. ***Sanctuary Beach Resort:*** good family resort; 47 Teemangum St, Currumbin; (07) 5598 2524. ***Sanctuary Lake Apartments:*** near wildlife sanctuary; 1/40 Teemangum St, Currumbin; (07) 5534 3344.

Currumbin Wildlife Sanctuary This 20 ha reserve is owned by the National Trust. There are free-ranging animals in open areas, the Crocodile Wetlands with raised walkways over pools of freshwater and saltwater crocodiles, a walk-through rainforest aviary and a miniature railway. A highlight is the twice-daily feeding of wild rainbow lorikeets. The 'Wildnight Tours' are interactive tours to see the nocturnal wildlife. 8 km SE.

Springbrook National Park, Mt Cougal section This small section of the park contains a subtropical rainforest remnant and is part of the Gondwana Rainforests of Australia World Heritage Area. Mt Cougal's twin peaks and the Currumbin Valley are an interesting and diverse landscape. There is a scenic drive through the valley and a walking track through rainforest, past cascades, to the remains of an old bush sawmill. End of Currumbin Creek Rd; 27 km sw. Springbrook and Natural Bridge sections: *see Nerang.*

Palm Beach: popular golden-sands beach that has won Queensland's Cleanest Beach Award in previous years; 4 km SE. ***Superbee Honeyworld:*** live displays, Walks with Bees tour, honey making and sales; opposite sanctuary; 8 km SE. ***Olson's Bird Gardens:*** large landscaped aviaries in subtropical setting with a lilly pilly hedge maze; Currumbin Creek Rd; 16 km sw. ***Greenmount and Coolangatta beaches:*** sheltered white-sand beaches with beautiful views of the coast; 13 km SE. ***Rainbow Bay:*** sheltered beach excellent for swimming. Walk along the coast to Snapper Rocks; 15 km SE. ***Snapper Rocks:*** top surf area with the 'Superbank', one of the world's longest point breaks; 15 km SE. ***Point Danger:*** named by Captain Cook as he sailed by. It offers excellent panoramic views over the ocean and coast. Catch a glimpse of dolphins from the Captain Cook Memorial Lighthouse; 15 km SE. ***Tom Beaston Outlook (Razorback Lookout):*** excellent views; behind Tweed Heads; 16 km SE.

See also GOLD COAST & HINTERLAND, p. 417

Caboolture

Pop. 17 739

Map ref. 644 E1 | 645 D1 | 647 F7 | 653 N8

Bruce Hwy, Burpengary; 1800 833 100; www.moretonbay.qld.gov.au

 105.1 4OUR FM, 612 AM ABC Brisbane

continued on p. 439

QUEENSLAND

CAIRNS

Pop. 122 732

Map ref. 650 E2 | 651 F7 | 657 M6

Cairns & Tropical North Visitor Information Centre, 51 The Esplanade; (07) 4051 3588; www.cairnsgreatbarrierreef.org.au

102.7 Zinc, 801 AM ABC Far North QLD

This modern, colourful city is the capital of the Tropical North and the gateway to the Great Barrier Reef, but was once a service town for the sugar plantations to the south. Tourism boomed with the airport's upgrade in 1984 and the influx of visitors and commercial enterprises resulted in the unusual mix of modern architecture and original Queenslander homes that can be seen today. The Esplanade traces the bay foreshore and blends city life with the natural attractions of the Coral Sea. With Cairns' superb location – the Great Barrier Reef to the east, mountain rainforests of the Wet Tropics and plains of the Atherton Tableland to the west, and palm-fringed beaches to the north and south – it is a good base for many activities. For fishing enthusiasts, Cairns is famous for its black marlin.

Flecker Botanic Gardens Established in 1886 as a recreational reserve, they are now the only wet tropical botanic gardens in Australia. The gardens display tropical plants from around the world, including a number of endangered species and over 200 species of palm. Follow the boardwalks through remnant lowland swamp to adjacent Centenary Lakes to see turtles and mangrove birds. Access gardens via Collins Ave, Edge Hill.

Mt Whitfield Conservation Park: 2 major walking tracks through forested mountain range to summit for views of Cairns and Coral Sea; located behind botanic gardens, access via Collins Ave, Edge Hill. ***McLeod Street Pioneer Cemetery:*** honours local pioneers. ***Tanks Art Centre:*** multipurpose centre in revamped WW II oil storage tanks, including gallery with local art; call for event listings; 46 Collins Ave, Edge Hill; (07) 4032 6600. ***Cairns Regional Gallery:*** local artists exhibit in this National Trust–classified building; Mon–Sat 10am–5pm, Sun 1pm–5pm; cnr Shields and Abbott sts; (07) 4046 4800. ***Cairns Museum:*** displays of Aboriginal, gold rush, timber and sugarcane history; Mon–Sat 10am–4pm; cnr Lake and Shields sts. ***Bulk Sugar Terminal:*** guided tours by arrangement during crushing season (June–Dec); Cook St; (07) 4051 3533. ***Foreshore Promenade:*** landscaped area with safe swimming lagoon; pool closed Wed mornings. ***Kuranda Scenic Railway:*** daily trips through Barron Gorge to rainforest village of Kuranda; leaves Cairns Railway Station in Bunda St; (07) 4036 9333; *see also Kuranda*. ***Game fishing:*** contact visitor centre for details. ***Dive schools:*** contact visitor centre for details.

Rusty's Market: Grafton and Sheridan Sts; Fri 5am–6pm, Sat 6am–3pm and Sun 6am–2pm. ***Esplanade Market:*** Sat 8am–5pm. ***Cairns Show:*** July. ***Cairns Cup:*** Aug. ***Cairns Festival:*** Aug–Sept.

Bayleaf Balinese Restaurant: Balinese food with flair; Bay Village Tropical Retreat, cnr Lake and Gatton Sts; (07) 4047 7955. ***Blue Sky Brewery:*** hearty meals and beers; 34–42 Lake St; (07) 4057 0500. ***Bushfire Flame Grill:*** world menu with Brazilian flame grill and Japanese barbecue; cnr The Esplanade and Spence St; (07) 4044 1879. ***Food Trail Tours:*** tastes of tropical produce; pick-ups from Cairns and northern beaches accommodation; (07) 4032 0322.

Hides Hotel: affordable heritage pub; 87 Lake St; (07) 4051 1266 or 1800 079 266. ***Pacific International Hotel:*** spacious rooms and friendly staff; cnr The Esplanade and Spence St; (07) 4051 7888. ***Pullman Reef Hotel Casino:*** luxurious upmarket hotel; 35–41 Wharf St; (07) 4030 8888.

[SWIMMING LAGOON, CAIRNS ESPLANADE]

Great Barrier Reef and islands Take a tour, charter a boat or fly to see some of the spectacular sights just offshore. Rainforest-covered Fitzroy Island to the east has impressive snorkelling sites at Welcome and Sharkfin bays. Green Island to the north-east is a true coral cay and the surrounding reef is home to magnificent tropical fish; they can be seen from a glass-bottom boat, in the underwater observatory or by snorkelling or helmet diving. Smaller Michaelmas and Upolo cays to the north-east are important sites for ground-nesting seabirds. The surrounding waters are excellent for reef swimming. You can also take a trip to the outer Barrier Reef, which is known for its spectacular underwater scenes and huge variety of marine life. For tours contact the visitor centre.

Tjapukai Aboriginal Cultural Park The group began as an Aboriginal dance company in 1987, but the demand for more cultural information prompted the move to this large park. Four theatres, both live and film, illustrate the history and culture of the rainforest people of Tropical North Queensland. Learn skills such as bush medicine and spear throwing on a day tour or experience the theatre of an Aboriginal ceremony in the 'Tjapukai By Night' experience; open daily 9am–5pm, evening shows 7pm–10pm; Cairns Western Arterial Rd, Caravonica; (07) 4042 9999; 11 km NW.

Beaches: incredible 26 km of beaches extending from Machans Beach on the north bank of Barron River, 9 km N, north to Ellis Beach. ***Skyrail Rainforest Cableway:*** spectacular gondola ride through rainforest to Kuranda; open daily; cnr Cook Hwy and Cairns Western Arterial Rd, Smithfield; (07) 4038 1555; 11 km N. ***Bungy jumping:*** choose from a variety of jumps and other thrills in the rainforest; daily 10am–5pm; 5 McGregor Rd, Smithfield, or free transport from Cairns; (07) 4057 7188; 13 km N. ***Crystal Cascades:*** walks by cascades and a secluded freshwater swimming hole; end of Redlynch Valley; 18 km SW. ***Lake Morris and Copperlode Dam:*** walking tracks; 19 km SW. ***Barron and Freshwater valleys:*** bushwalking, hiking, whitewater rafting and camping to the west and north-west of Cairns; contact visitor centre for details. ***Safaris:*** 4WD to Cape York and the Gulf Country; contact visitor centre for details.

See also CAIRNS & THE TROPICS, p. 425

At the northern edge of Greater Brisbane and the southern opening to the Sunshine Coast, Caboolture is surrounded by subtropical fruit farms. The town was settled in 1842 after the restricted land around Moreton Bay penal colony was opened up. The historical village north of town exhibits much of this history. Bribie Island to the east has spectacular aquatic and wildlife attractions, which bring many visitors to the region.

Trail of Reflections: self-guide trail of open-air artwork and sculptures around town that illustrate the history of the area; starts in King St; details from visitor centre.

Market: showgrounds, Beerburrum Rd; Sun mornings. ***Country Music Festival:*** Apr/May. ***Medieval Tournament:*** July. ***Caboolture Show:*** Aug. ***Rodeo:*** Nov/Dec.

Caboolture Riverlakes Motel: overlooks riverside parklands; 14 Morayfield Rd; (07) 5499 1766. ***Beachmere Palms Motel:*** 6 quiet waterside units; 30 Biggs Ave, Beachmere; (07) 5496 8577.

Bribie Island This island park is separated from the mainland by Pumicestone Passage, where mangroves flourish and dugongs, dolphins, turtles and over 350 species of birds live. National park covers about a third of the island and offers secluded pristine white beaches. Follow the Bicentennial Bushwalks to discover the park on foot, boat along Pumicestone Passage or 4WD along the ocean beach (permit required). Fishing and surfing are popular at Woorim Beach, just north of which are old WW II bunkers. See migratory birds in summer at Buckleys Hole Conservation Park on the south-west tip of the island. Bridge access to island; 21 km E.

Ferryman cruises: cruise the waters of Pumicestone Passage; (07) 3408 7124. ***Caboolture Historical Village:*** over 50 restored buildings of historical importance house museums, with themes including maritime and transport; open 9.30am–3.30pm daily; Beerburrum Rd; 2 km N. ***Airfield:*** Warplane and Flight Heritage Museum with displays of WW II memorabilia and restored fighter planes. Tiger Moth and Mustang flights and gliding on offer; McNaught Rd; 2 km E. ***Sheep Station Creek Conservation Park:*** walks through open forest. See remains of the old bridge on original road leading from Brisbane to Gympie; 6 km SW. ***Abbey Museum:*** traces growth of western civilisation with displays of art and antiques; open 10am–4pm Mon–Sat; just off road to Bribie Island; 9 km E. ***Woodford:*** the town has one of the largest narrow-gauge steam locomotive collections in Australia; Margaret St; 22 km NW. ***Mt Mee State Forest:*** boardwalks through subtropical rainforest and lookouts over Neurum Valley, Moreton Bay and surrounds; 23 km W. ***Donnybrook, Toorbul and Beachmere:*** coastal fishing towns to the east.

See also BRISBANE ISLANDS, p. 416

Caloundra

Pop. 87 596

Map ref. 647 H5 | 648 F12 | 653 N8

7 Caloundra Rd; (07) 5420 6240 or 1800 644 969; www.caloundratourism.com.au

90.3 FM ABC Coast

This popular holiday spot at the southern tip of the Sunshine Coast was once a retirement haven. It now boasts a diverse population of retirees and young Brisbane commuters keen on the seaside lifestyle. The nearby beaches offer a variety of watersports – the calm waters of Golden Beach are especially popular with windsurfers. The fishing between Bribie Island and the mainland in Pumicestone Passage is excellent.

Queensland Air Museum Founded by members of the Aviation Historical Society of Australia in 1973, this museum collects important relics of Queensland's aviation heritage. Memorabilia on display includes old fighter planes and bombers. Airport, Pathfinder Dr; (07) 5492 5930.

Caloundra Regional Art Gallery: local and touring art exhibitions; open 10am–4pm Wed–Sun; Omrah Ave. ***Ben Bennet Botanical Park:*** easy walks through natural bushland; Queen St. ***Suncoast Helicopter Flights:*** over Glass House Mountains and Sunshine Coast; bookings on (07) 5499 6900. ***Caloundra Cruises:*** morning, lunchtime and afternoon cruises, as well as sunset charters. Scenic Pumicestone Passage cruises have spectacular views of Bribie Island and Moreton Bay; bookings on (07) 5492 8280. ***Blue Water Kayak Tours:*** paddle in the tranquil Moreton Bay Marine Park.

Country Market: Bulcock St; Sun. ***City Show:*** May. ***Cairns Cup:*** June. ***Open Cockpit Weekend:*** July. ***Taste of the Coast:*** food and wine festival; July. ***Bowls Carnival:*** July–Aug. ***Classic Boat Regatta:*** Aug. ***Art and Craft Festival:*** Oct. ***Caloundra Music Festival:*** Oct.

Alfie's Mooo Char and Bar: steak from paddock to plate; Cnr Otranto Tce and The Esplanade, Bulcock Beach; (07) 5492 8155. ***Between the Flags Restaurant:*** Lifesaving Club bistro; Ormonde Tce; (07) 5491 8418. ***Cafe by the Beach:*** popular surf cafe; 12 Seaview Tce, Moffat Beach; (07) 5491 9505. ***The Moorings Cafe:*** masters of big breakfasts; 84 The Esplanade, Golden Beach; (07) 5492 2466.

Bluewater Point and Deepwater Point Resort: quiet spot on the river; 13 Nicklin Way, Minyama; (07) 5477 9900. ***Currimundi Lakeside B&B:*** gourmet breakfasts; 1 Rosea Crt; (07) 5493 9123. ***Dicky Beach Family Holiday Park:*** on a patrolled beach; Beerburrum St; (07) 5491 3342.

Currimundi Lake Conservation Park This unspoiled coastal park offers quiet walks beside the lake and through to the beach. Canoe and swim in the lake or see the finches and friarbirds in the remnant wallum heath. In spring the wildflowers are spectacular. Access from Coongara Espl; 4 km N.

Opals Down Under: opal-cutting demonstrations and 'scratch patch' where visitors fossick for their own gemstones; 14 km NW. ***Aussie World:*** family fun-park in native garden setting, with over 30 rides and games, Side Show Alley and one of the country's growing number of Ettamogah Pubs, based on Ken Maynard's cartoon; Palmview; 18 km NW. ***Surrounding beaches:*** include patrolled beaches of Bulcock, Kings, and Dicky with the wreck of SS *Dicky* (1893); excellent fishing at Moffat and Shelly beaches. ***Scenic drives:*** taking in the beaches to the north, the Blackall Range with art galleries and views of the Sunshine Coast, and the Glass House Mountains with magnificent walks and scenery; details from visitor centre.

See also **Brisbane Islands, p. 416**

Camooweal

Pop. 197
Map ref. 637 P10 | 659 A9 | 660 C1

Drovers Camp, Camooweal; (07) 4748 2022.

102.5 Hot FM, 567 AM ABC North West QLD

North-west of Mount Isa, Camooweal is the last Queensland town before the Northern Territory border. It was once the centre for enormous cattle drives travelling south. Some say that the town is a suburb of Mount Isa, which would make the 188 kilometres of Barkly Highway between Mount Isa and Camooweal one of the longest main streets in the world. To the south of town are the incredible Camooweal Caves, a series of sinkhole caves that have evolved over millions of years.

Drover's Camp: memorabilia shed and info centre; tours May–Sept 10am–5pm; Barkly Hwy; (07) 4748 2022.

Cricket tournament: Apr. ***Campdraft:*** June. ***Horseracing:*** Aug. ***Drovers Camp Festival:*** Aug.

Camooweal Caves National Park On the Barkly Tableland, this national park is still evolving as water continues to filter through the soluble dolomite to create and transform the extensive cave system. The underground caves are linked by vertical shafts and only the experienced caver should attempt them. The Great Nowranie Cave is excellent to explore with an 18 m drop at the opening (climbing gear is essential). Caves may flood during wet season. If exploring the caves, inform local police or ranger beforehand. 4WD access is recommended. (07) 4722 5224; 24 km S.

Cemetery: headstones tell local history; 1 km E. ***Boodjamulla (Lawn Hill) National Park:*** around 300 km N via Gregory; *see Burketown.*

See also **Outback, p. 428**

Cardwell

Pop. 1251
Map ref. 650 E7 | 657 M9

Rainforest and Reef Information Centre, 142 Victoria St; (07) 4066 8601; www.gspeak.com.au/cardwell

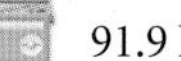

91.9 Kool FM, 630 AM ABC North QLD

Cardwell is a coastal town overlooking Rockingham Bay and the nearby islands of the Great Barrier Reef. Ferries transport visitors to nearby Hinchinbrook Island, the largest island national park in Australia. Between the island and the mainland is Hinchinbrook Channel (Cardwell is at the northern edge), a popular spot for fishing and a sheltered area for houseboats.

Bush Telegraph Heritage Centre This complex comprises the old post office and telegraph station, in operation 1870–1982, and the original magistrates court and gaol cells. An informative history of communications and the region is provided through interpretive displays. Open 10am–1pm Mon–Fri, 9am–12pm Sat, other times by appt; Bruce Hwy; (07) 4066 2412.

Rainforest and Reef Information Centre: interpretive centre that acquaints visitors with landscape, flora and fauna of northern Queensland; Bruce Hwy, near jetty. ***Coral Sea Battle Memorial Park:*** large war memorial that commemorates the WW II battle off the coast between Australian/US forces and the Japanese; beachfront. ***Boat hire and cruises:*** explore the tropical waters and islands to the east at the helm of a yacht, houseboat or cruiser, or travel with an organised cruise; details from visitor centre. ***Snorkelling and scuba-diving tours:*** details from visitor centre.

Market: Cardwell Espl; 1st Sun each month. ***Coral Sea Battle Memorial Commemoration:*** May. ***Seafest:*** Aug. ***Fishing Classic (Port Hinchinbrook):*** Sept.

Cardwell Beachfront Motel and Holiday Units: 6 units, close to beach; Cnr Bruce Hwy and Scott St; (07) 4066 8776. ***Mudbrick Manor:*** 5 rooms and suites; Lot 13 Stoney Creek Rd; (07) 4066 2299. ***Port Hinchinbrook Resort:*** cabins and apartments; 1 Front St; (07) 4066 2000 or 1800 220 077.

Hinchinbrook Island National Park An amazing variety of vegetation covers this island park, including rainforest, wetlands, forests and woodlands. The 32 km Thorsborne Trail on the east coast is renowned for its spectacular scenery as it winds past waterfalls and along pristine beaches. Many people allow 4 days or more for the walk, camping on a different beach each night.

Hikers must be self-sufficient and bookings are essential (limited number of walkers allowed). Shorter walks are from the camping areas at Macushla and The Haven. Access the island via ferry from Cardwell or Lucinda; (07) 4066 8601.

Scenic drive in Cardwell State Forest: this 26 km circuit from Cardwell takes in a lookout, waterfalls, swimming holes and picnic spots; begins on Braesnose St. ***Edmund Kennedy National Park:*** boardwalk through extensive mangrove forests and variety of other vegetation to beach, with spectacular view of islands. This park is a habitat of the endangered mahogany glider; (07) 4066 8601; 4 km N. ***Five Mile Swimming Hole:*** attractive picnic and swimming spot safe from crocodiles, sharks and stingers; 7 km S. ***Dalrymple's Gap:*** original service path and stone bridge through range; 15 km S. ***Brook Islands:*** nesting area for Torresian imperial pigeons (Sept–Feb). Excellent snorkelling on reef of northern 3 islands. Sea access only; 30 km NE. ***Murray Falls:*** climb the steep 1 km path to viewing platform over falls and surrounds; 42 km NW. ***Girringun (Lumholtz) National Park:*** travel through World Heritage rainforest on road (dry weather only) to the 3-tier 91 m Blencoe Falls; (07) 4066 8601; 71 km W; *see Ingham for southern parts of park.*

See also THE MID-TROPICS, p. 424

Charleville

Pop. 3275
Map ref. 652 B5 | 663 P5

Matilda Hwy; (07) 4654 3057; www.murweh.qld.gov.au

101.7 Triple C FM, 603 AM ABC Western QLD

Charleville is in the heart of mulga country on the banks of the Warrego River and at the centre of a rich sheep and cattle district. By the late 1890s the town had its own brewery, ten hotels and 500 registered bullock teams. Cobb & Co recognised the value of Charleville's location on a major stock route and opened a coach-building factory in 1893. There are also strong links with aviation: the first London–Sydney flight landed here in 1919, Qantas' first fare-paying service took off in 1922 and record-setting aviator Amy Johnson landed nearby in 1930. Charleville marks the terminus of the Westlander rail service from Brisbane.

Cosmos Centre This centre explores the Australian night sky and its significance to Aboriginal culture. There are multimedia displays, nightly shows and interactive areas where the wonders of the sky are observed through powerful Meade Telescopes. The outback night sky has never looked so beautiful; Matilda Hwy; (07) 4654 7771.

Royal Flying Doctor Service Visitor Centre: museum displaying memorabilia from the past and present. View the documentary entitled A Day in the Life of the Flying Doctor; Old Cunnamulla Rd. ***Historic House Museum:*** a wonderful example of early Queensland architecture. Machinery displays including steam engine and a rail ambulance in restored Queensland National Bank building; open 9am–3pm Mon–Fri, 9am–12pm Sat, other times by appt; Alfred St. ***Vortex Gun:*** in 1902 this 5 m long gun was used in an unsuccessful rain-making experiment; Bicentennial Park, Matilda Hwy. ***Heritage trail:*** self-guide walk past heritage buildings; brochure from visitor centre. ***Bilby Tours:*** one of Australia's most rare and endangered animals, bilbies are captively bred in Charleville. Get up-close and personal with these marsupials on the night tour; details from visitor centre.

Market: Historic House Museum; 1st Sat each month. ***Charleville Show:*** May. ***Bilby Festival:*** food, music and celebration of National Bilby Day; Sept.

Evening Star Tourist Park: on a cattle station; Adavale Rd; (07) 4654 2430. ***Hotel Corones:*** famous Queensland outback hotel; 33 Wills St; (07) 4654 1022. ***Mulga Country Motor Inn:*** luxury in the outback; Cunnamulla Rd; (07) 4654 3255. ***Waltzing Matilda Motor Inn:*** comfort close to town; 125 Alfred St; (07) 4654 1720.

Tregole National Park This semi-arid national park has a vulnerable and fragile ecosystem. It is largely made up of ooline forest – rainforest species dating back to the Ice Age. Follow the 2.1 km circuit track to see the diverse vegetation and spectacular birds of the park. (07) 4654 1255; 99 km E via Morven.

Monument: marks the spot where Ross and Keith Smith landed with engine trouble on the first London–Sydney flight; 19 km NW.

See also OUTBACK, p. 428

Charters Towers

Pop. 7978
Map ref. 654 F3

74 Mosman St; (07) 4761 5533; www.charterstowers.qld.gov.au

95.9 Hot FM, 630 AM ABC North QLD

Charters Towers is in the Burdekin Basin south-west of Townsville. The town's gold rush began on 25 December 1871 when Aboriginal horse-boy Jupiter discovered gold while looking for lost horses. He brought gold-laden quartz to his employer, Hugh Mosman, who rode to Ravenswood to register the claim, and the gold rush was on. Between 1872 and 1916 Charters Towers produced ore worth £25 million. At the height of the gold rush it was Queensland's second largest city and was commonly referred to as 'The World' because of its cosmopolitan population. This rich history can be seen in the preserved streetscapes of 19th- and 20th-century architecture. To the north-west of Charters Towers is the 120-kilometre Great Basalt Wall, a lava wall created from the Toomba basalt flow.

Ghosts of Gold Heritage Trail This informative tour reveals the rich history of the town in the district known as 'One Square Mile'. Over 60 heritage-listed buildings are in the precinct. Of particular interest are the re-created workings of the Stock Exchange in Mosman St; the once heavily mined Towers Hill (1.5 km W) with interpretive walking trails and a film screening at night; and the Venus Gold Battery in Millchester Rd, an old gold-processing plant where the 'ghosts' come alive. Starts in the orientation centre behind the visitor centre.

Charters Towers Folk Museum: local historical memorabilia; Mosman St. ***Civic Club:*** once a gentlemen's club, this remarkable building (1900) still contains the original billiard tables; Ryan St. ***Rotary Lookout:*** panoramic views over region; Fraser St.

National Trust Markets: Stock Exchange Arcade; 1st and 3rd Sun each month. ***Showgrounds Markets:*** Cnr Mary and Show sts; 2nd Sun each month. ***Goldfield Ashes Cricket Carnival:*** Jan. ***Rodeo:*** Easter. ***Ten Days in the Towers:*** festival including music and bush poets; Apr–May. ***Country Music Festival:*** May Day weekend. ***Charters Towers Show:*** July.

Bluff Downs Historic Station: experience station life; 29 km off Lynd Hwy; (07) 4770 4084.

QUEENSLAND

Dalrymple National Park This small national park on the Burdekin River comprises mainly woodland and is an important area for native animals including rock wallabies and sugar gliders. A highlight is the 4 million-year-old solidified lava wall, the Great Basalt Wall, parts of which are accessible from this park. The site of the old Dalrymple township is also of interest. Contact the ranger before setting out on any walks. (07) 4722 5224; 46 km N.

Ravenswood: another gold-rush centre, but now 'not quite a ghost town'. Two hotels remain as magnificent examples of the prosperity of this town's wealthy beginnings. It's worth taking a stroll down Main St, which resembles a set from an old movie; 85 km E. ***Greenvale:*** near the Gregory Development Rd with a roadhouse, caravan park, golf course and the well-known Three Rivers Hotel, made famous by Slim Dusty's song of the same name; 209 km NW. ***Burdekin Falls Dam:*** recreation area with barramundi fishing; 165 km SE via Ravenswood. ***Blackwood National Park:*** a woodland park of undulating hills and stony ridges. See the Belyando blackwood trees that give the park its name, and walk on fire trails to discover the park's interesting birdlife including squatter pigeons and speckled warblers; (07) 4722 5224; 180 km S.

See also THE MID-TOPICS, p. 424

Childers

Pop. 1352
Map ref. 648 C4 | 653 M4

Palace Hotel (Childers Memorial), 72 Churchill St; (07) 4126 1994.

 102.5 Breeze FM, 855 AM ABC Wide Bay

Childers is a picturesque National Trust town south of Bundaberg, part of the state's sugarcane belt. With leafy streets and a lovely outlook over the surrounding valleys, the town's history has been blighted by fire. One ravaged the town in 1902 (though many of the heritage buildings survived) and another engulfed a backpacker's hostel in 2000, tragically killing 15 people and making international news. A memorial to those lost in this fire stands in Churchill Street. Thankfully life has returned to normal in Childers and the flow of backpackers on the fruit-picking trail is as strong as ever.

Pharmaceutical Museum: collection of memorabilia including leather-bound prescription books and Aboriginal wares; open 8.45am–4.30pm Mon–Fri, 8.30am–12pm Sat; Churchill St. ***Historical complex:*** area of historic buildings including school, cottage and locomotive; Taylor St. ***Baker's Military and Memorabilia Museum:*** 16 000 items on display covering all the major wars, including uniforms and communications equipment; closed Sun (open by appt); Ashby La. ***Childers Art Gallery:*** Churchill St. ***Snakes Downunder:*** informative exhibits on native snakes; open 9am–3pm, closed Wed; Lucketts Rd. ***Historic Childers:*** self-guide town walk past historic buildings; highlight is the Old Butcher's Shop (1896) in Churchill St.

Village Market: historical society; 3rd Sun each month. ***Festival of Cultures:*** a week of celebrations also includes workshops by visiting artists and performers; July.

Gateway Motor Inn: quiet and comfortable; Cnr Bruce Hwy and Butchers Rd; (07) 4126 1288 or 1800 100 603.

Flying High Bird Habitat The native vegetation here is home to 600 bird species. Follow the boardwalk tracks to see the many Australian parrots and finches. Cnr Bruce Hwy and Old Creek Rd, Apple Creek; 5 km N.

Buxton and Walkers Point: unspoiled fishing villages to the east. ***Woodgate:*** coastal town to the north-east with 16 km beach. Access to Burrum Coast National Park is from here; *see Bundaberg*. ***Goodnight Scrub National Park:*** 27 km W; *see Gin Gin*.

See also CAPRICORN, p. 423

Chillagoe

Pop. 223
Map ref. 650 A3 | 657 J6

The Hub, 23 Queen St; (07) 4094 7111; www.chillagoehub.com.au

 720 AM ABC Far North QLD

Chillagoe is a small outback town west of Cairns and the Atherton Tableland. Once a thriving mining town after silver and copper deposits were found in 1887, the town was practically deserted after the smelter closed in 1940. Chillagoe's history and the well-preserved Aboriginal rock art and limestone caves in the area make it a popular spot with visitors.

The Hub This major interpretive centre is constructed from local materials including marble and copper. Informative displays cover the geographical history of the local landscape (dating back 2 billion years), the town's mining and pioneering past and the region's Aboriginal heritage. Queen St; (07) 4094 7111.

Heritage Museum: displays on local history and relics of old mining days; Hill St. ***Historic cemetery:*** headstones from early settlement; Railway Line Rd. ***Historical walks:*** self-guide or guided walks taking in the old State Smelter and disused marble quarry just south of town; details from visitor centre.

Big Weekend and Rodeo: May. ***Annual Wheelbarrow Race:*** May. ***Country Music Festival:*** July.

Chillagoe–Mungana Caves National Park The impressive rugged limestone outcrops and magnificent caves of this park, south of town, are studied by scientists world-wide. The cave system was originally an ancient coral reef and is home to a wide variety of bats. Fossilised bones, including those of a giant kangaroo, have been discovered in the caves. Guided tours only; tickets from The Hub. Above ground are magnificent Aboriginal rock paintings at Balancing Rock. (07) 4046 6600; 7 km S.

Mungana: Aboriginal rock paintings at the Archways. Also a historic cemetery; 16 km W. ***Almaden:*** small town where cattle own the main street; 30 km SE. ***4WD self-guide adventure trek:*** mud maps and clues provided; details from visitor centre. ***Tag-along tours:*** full- and half-day tours visiting Aboriginal sites, marble quarries and other places; details from visitor centre.

See also GULF SAVANNAH, p. 427

Chinchilla

Pop. 3684
Map ref. 653 J7

Warrego Hwy; (07) 4668 9564; www.chinchilla.org.au

97.1 Breeze FM, 747 AM ABC Southern QLD

Chinchilla is a prosperous town in the western Darling Downs. Explorer Ludwig Leichhardt named the area in 1844 after the local Aboriginal name for the native cypress pines, 'jinchilla' – there are still many in town. Today Chinchilla is known as the 'melon capital' of Australia as it produces around 25 per cent of the country's watermelons.

Historical Museum: a varied collection of memorabilia including steam engines, a replica 1910 sawmill and a slab cottage. Also an excellent display of local petrified wood known as 'Chinchilla Red'; open 9am–4pm Wed–Sun; Villiers St. ***Pioneer Cemetery:*** headstones tell early history. Also a monument to Ludwig Leichhardt; Warrego Hwy.

Melon Festival: odd-numbered years, Feb. ***Market:*** Warrego Hwy; Easter Sat. ***Rotary May Day Celebrations:*** May. ***Chinchilla Show:*** May. ***Polocrosse Carnival:*** Sept. ***Grand Father Clock Campdraft:*** Oct. ***Mardi Gras:*** Nov. ***Chinchilla Cup:*** Dec.

Cactoblastis Memorial Hall: dedicated to the insect introduced from South America to eradicate the prickly pear cactus; Boonarga, 8 km E. ***Fossicking:*** for petrified wood at nearby properties; licences and details from visitor centre. ***Fishing:*** good spots include Charleys Creek and the Condamine River; details from visitor centre. ***Chinchilla Weir:*** popular spot for boating and waterskiing, plus freshwater fishing for golden perch and jewfish; 10 km S.

See also DARLING DOWNS, p. 418

Clermont

Pop. 1851
Map ref. 649 B12 | 654 H9

Cnr Herschel and Karmoo sts; (07) 4983 4755; www.isaac.qld.gov.au

102.1 4HI FM, 1548 AM ABC Capricornia

Clermont is in the central highlands south-west of Mackay. It was established over 130 years ago after the discovery of gold at Nelson's Gully. At first the settlement was at Hood's Lagoon, but was moved to higher ground after a major flood in 1916 in which 63 people died. To the east of Clermont are the prominent cone-shaped mountains of Peak Range National Park. The Wolfgang Peak between Clermont and Mackay is particularly spectacular.

Hood's Lagoon and Centenary Park Walk the boardwalks in this picturesque setting to see the colourful birdlife. The park has interesting memorials and monuments that include the Sister Mary MacKillop grotto, an Aboriginal monument and a war memorial. The tree marker at the flood memorial plaque demonstrates how high the water rose. Access via Lime St.

Railway Wagon Murals: paintings on 4 original wagons depict industries within the Belyando Shire; Hershel St. ***The Stump:*** memorial to the 1916 flood; Cnr Drummond and Capricorn sts.

Rodeo: Apr. ***Clermont Show:*** May. ***Campdraft:*** June. ***Gold and Coal Festival:*** Aug. ***Country Music Festival:*** Oct.

Cemetery: headstones dating back to 1860s and mass grave of 1916 flood victims; 2.5 km NE. ***Clermont Museum and Heritage Park:*** museum exhibits historic artefacts and machinery, including the steam engine used to shift the town after flood and historic buildings, such as the old Masonic lodge, with displays of local family histories; open daily, times vary; (07) 4983 3311; 4 km NW. ***Remnants of Copperfield:*** old coppermining town with museum in original Copperfield store, chimneystack, and cemetery containing 19th-century graves of copperminers; 5 km S. ***Theresa Creek Dam:*** popular spot for waterskiing, sailboarding and fishing (permit required). Bushwalks nearby; off Peakvale Rd; 17 km SW. ***Blair Athol open-cut mine:*** free half-day tour to see the largest seam of steaming coal in the world; departs from Clermont Visitor Centre; Blair Athol; 23 km NW. ***Blackwood National Park:*** 187 km NW; *see Charters Towers*. ***Fossicking for gold:*** obtain licence and fossicking kit to start the search in nearby area; details from visitor centre.

See also THE MID-TROPICS, p. 424

Clifton

Pop. 1065
Map ref. 653 L10

Council offices, 70 King St; (07) 4697 4222; www.clifton.qld.gov.au

91.9 CFM, 747 AM ABC Southern QLD

Clifton is south of Toowoomba in the fertile agricultural lands of the Darling Downs. The town's charming street facades have been a popular backdrop for many Australian movies, including the epic film adaptation of The *Thorn Birds*. The town is a vision of colour during the renowned annual Rose and Iris Show.

Historical Museum: in the old butter factory, with displays of early implements and farm life; open 1st and 3rd Sun each month; King St. ***Alister Clark Rose Garden:*** largest Queensland collection of these roses; Edward St.

Clifton Show: Apr. ***Country Week:*** includes the Rose and Iris Show and horseracing; Oct.

Nobby This small town provided inspiration to writer Arthur Hoey Davis (Steele Rudd), author of *On Our Selection* and creator of the famous Dad and Dave characters. Rudd's Pub (1893) has a museum of pioneering memorabilia. Nearby is the burial site of Sister Kenny along with a memorial and museum dedicated to her – she was renowned for her unorthodox method of treating poliomyelitis. 8 km N.

Darling Downs Zoo Queensland's newest zoo is situated in the serene countryside where you can view native and exotic animals from Africa, South America and Asia. The zoo has a shady picnic area as well as barbecue facilities and a small kiosk. Open weekends, public and school holidays, or by appt (for large groups); Gatton–Clifton Rd; (07) 4696 4107; 10 km E.

Leyburn: holds the Historic Race Car Sprints in Aug; 30 km W.

See also DARLING DOWNS, p. 418

Cloncurry

Pop. 2385
Map ref. 659 F11 | 660 H3

Mary Kathleen Park, McIlwraith St; (07) 4742 1361; www.cloncurry.qld.gov.au

102.5 Hot FM, 567 AM ABC North West QLD

Cloncurry is an important mining town in the Gulf Savannah region. In 1861 John McKinlay, leading a search for Burke and Wills, reported traces of copper in the area. Six years later, pastoralist Ernest Henry discovered the first copper lodes. During World War I Cloncurry was the centre of a copper boom and the largest source of the mineral in Australia. Copper prices slumped postwar and a pastoral industry was developed. Today main industries include grazing, coppermining and goldmining. The town's interesting history extends to aviation as well. Qantas was conceived here – the original hangar can still be seen at the airport – and the town became the first base for the famous Royal Flying Doctor Service in 1928.

John Flynn Place This complex includes the RFDS Museum, with history on the service and memorabilia including

QUEENSLAND

the first RFDS aircraft and an original Traegar pedal wireless. There is also the Fred McKay Art Gallery, with changing exhibits of local art, the Alfred Traegar Cultural Centre and the Alan Vickers Outdoor Theatre. Open 8am–4.30pm Mon–Fri, 9am–3pm Sat–Sun (May–Sept); Cnr Daintree and King sts.

Mary Kathleen Memorial Park: with buildings from abandoned uranium mining town of Mary Kathleen including Old Police Station and Town Office – now a museum with historic items such as Robert O'Hara Burke's water bottle, local Aboriginal artefacts and a comprehensive rock, mineral and gem collection. Visitor information centre is also on-site; McIlwraith St. ***Shire hall:*** restored 1939 building; Scarr St. ***Cemeteries:*** varied cultural background of Cloncurry can be seen in the 3 cemeteries: Cloncurry Old Cemetery, including the grave of Dame Mary Gilmore (Sir Hudson Fyshe Dr); Afghan Cemetery (part of Cloncurry Old Cemetery); and Chinese Cemetery (Flinders Hwy). ***Original Qantas hangar:*** airport; Sir Hudson Fyshe Dr. ***Historic buildings:*** including the post office, Brodie Hardware Store and the Post Office Hotel; brochure from visitor centre.

Markets: Florence Park, Scarr St; 1st Sat each month. Mary Kathleen Park, McIlwraith St; 4th Sun each month. ***Cloncurry Show:*** June. ***Stockman's Challenge and Campdraft:*** night rodeo; July. ***Rockhana:*** mineral and gem festival; July. ***Merry Muster Rodeo:*** July/Aug. ***Art Show:*** Sept. ***Battle of the Mines:*** rugby league carnival; Oct. ***Cloncurry Cup:*** Oct.

Gidgee Inn Motel: rammed-earth motel; Matilda Hwy; (07) 4742 1599.

Rotary Lookout: over Cloncurry and the river; Mt Isa Hwy; 2 km w. ***Chinamen Creek Dam:*** peaceful area with abundant birdlife; 3 km w. ***Ernest Henry Copper and Gold Mine:*** tours available. Departs from Mary Kathleen Park; 29 km NE. ***Burke and Wills cairn:*** near Corella River; 43 km w. ***Kajabbi:*** town holds Yabby Races in May; 77 km NW. ***Kuridala:*** this one-time mining town is now a ghost town. Explore the ruins including the old cemetery; 88 km s. ***Fossicking:*** for amethysts and other gemstones; details from visitor centre.

See also OUTBACK, p. 428

Cooktown

Pop. 1339
Map ref. 657 L3

Nature's Powerhouse, Cooktown Botanic Gardens, Finch Bay Rd; (07) 4069 6004 or 1800 174 895; www.cook.qld.gov.au

105.7 FM ABC Far North QLD

Cooktown is the last main town before the wilderness that is Cape York. In 1770, Captain Cook beached the *Endeavour* here for repairs after running aground on the Great Barrier Reef. Cooktown was founded more than 100 years later after gold was discovered on the Palmer River. It became the gold-rush port with 37 hotels and a transient population of some 18 000, including 6000 Chinese. Nearby are some of the most rugged and remote national parks in Australia. They form part of the Wet Tropics World Heritage Area and are a special experience for today's intrepid explorers.

James Cook Museum Housed in the old convent school (1888), this museum documents Aboriginal life prior to European settlement, Cook's voyages and the 1870s gold-rush past. Relics include the anchor from the *Endeavour*. Helen St.

Botanic Gardens: these are the oldest botanic gardens in Queensland, with native, European and exotic plants. The 'Cooktown Interpretive Centre: Nature's Powerhouse' has botanical and wildlife illustrations, and there are also walking trails to Cherry Tree and Finch bays; Walker St. ***Historic cemetery:*** documents the varied cultural heritage of Cooktown and includes the grave of tutor, early immigrant and heroine Mary Watson; Boundary Rd. ***Chinese shrine:*** to many who died on the goldfields; near cemetery. ***Cooktown Wharf:*** dates back to 1880s and is an excellent spot for fishing. ***The Milibi Wall:*** collage of traditional art by local Aboriginal people; near wharf. ***Fishing:*** spanish mackerel and barramundi at the wharf. ***Walking trails:*** plenty around the town and outlying areas, including the Scenic Rim and Wharf and Foreshore Walk; details from visitor centre.

Market: Lions Park; Sat. ***Discovery Festival:*** re-enactment of Cook's landing; June. ***Cooktown Races:*** July. ***Art Festival:*** Aug. ***Agricultural Show:*** Aug. ***Wallaby Creek Festival (Rossville):*** music festival; Sept.

The Sovereign Resort Hotel: quality hotel in colonial building; 128 Charlotte St; (07) 4043 0500. ***Mungumby Lodge:*** 10 ensuite bungalows; Mungumby Rd, Helenvale; (07) 4060 3158.

Lizard Island National Park Comprises 6 islands to the north-east of Cooktown surrounded by the blue waters and coral reefs of the northern Great Barrier Reef. Four of the islands are important seabird nesting sites. Lizard Island is a resort island popular for sailing and fishing. Also on offer is snorkelling in the giant clam gardens of Watsons Bay. Walk to Cooks Look for a spectacular view. There are 11 species of lizard here, and green and loggerhead turtles nest in late spring. Regular flights depart from Cairns; charter flights depart from Cairns and Cooktown; charter or private boat hire is available at Cooktown. (07) 4069 5777 (use same contact number for all national parks below).

Endeavour River National Park: just north of town is this park of diverse landscapes, including coastal dunes, mangrove forests and catchment areas of the Endeavour River. Most of the park is accessible only by boat (ramps at Cooktown); southern vehicle access is via Starcke St, Marton. ***Black Mountain National Park:*** impressive mountain range of granite boulders and a refuge for varied and threatened wildlife; Cooktown Developmental Rd; (07) 4069 5777; 25 km s. ***Helenvale:*** small town with historic Lions Den Hotel (est. 1875) and rodeo in June; 30 km s. ***Rossville:*** town with markets every 2nd Sat and Wallaby Creek Folk Festival at Home Rule Rainforest Lodge in Sept; 38 km s. ***Cedar Bay (Mangkal-Mangkalba) National Park:*** this remote coastal park is an attractive mix of rainforest, beaches and fringing reefs, with a variety of wildlife including the rare Bennett's tree kangaroo. Walk on the old donkey track once used by tin miners (remains of tin workings can be seen). Access by boat or by the walking track, which starts at Home Rule Rainforest Lodge, Rossville, 38 km s. ***Elim Beach and the Coloured Sands:*** spectacular beach with white silica sandhills and surrounding heathlands. The Coloured Sands are found 400 m along the beach (Aboriginal land, permit required from Hope Vale Community Centre); 65 km N. ***Cape Melville National Park:*** this rugged park on the Cape York Peninsula has spectacular coastal scenery. Much of the plant life is rare, including the foxtail palm. Visitors must be self-sufficient; 4WD access only in dry weather; southern access via Starcke homestead, western access via Kalpower Crossing in Lakefield National Park; 140 km NW. ***Flinders Group National Park:*** comprising 7 continental islands in Princess Charlotte Bay. There are 2 self-guide trails taking in bushfoods and Aboriginal rock-art sites on Stanley Island. Access by charter or private boat or sea plane; 195 km NW. ***Sportfishing safaris:*** to nearby waterways; details from visitor centre. ***Guurbi Aboriginal Tours:*** full- and half-day tours of Aboriginal rock-art sites and informative history of bush tucker and traditional medicine; details from visitor centre. ***Bicentennial National Trail:*** 5000 km trail that runs from Cooktown to Healesville in Victoria, for walkers, bike riders and horseriders. It is possible to do just a section of trail; details from visitor centre.

Travellers note: *Before driving to remote areas check road conditions and restrictions with Queensland Parks & Wildlife Service, 5 Webber Espl, Cooktown, (07) 4069 5777.*

See also CAIRNS & THE TROPICS, p. 425

Crows Nest

Pop. 1443
Map ref. 653 L8

Hampton Visitor Information Centre, 8623 New England Hwy, Hampton (12 km S); (07) 4697 9066 or 1800 009 066; www.crowsnest.info

 100.7 C FM, 747 AM ABC Southern QLD

On the western slopes of the Great Dividing Range north of Toowoomba is the small town of Crows Nest. It was named after Jimmy Crow, a Kabi-Kabi Aboriginal man who made his home in a hollow tree near the present police station. He was an invaluable source of directions for passing bullock teams staying overnight in the area. A memorial to Crow can be found in Centenary Park.

Carbethon Folk Museum and Pioneer Village: many interesting old buildings and over 20 000 items of memorabilia documenting the history of the shire; open 10am–3pm Thurs–Sun; Thallon St. ***Bullocky's Rest and Applegum Walk:*** a 1.5 km track follows the creeks to Hartmann Park and a lookout over Pump Hole. Visit in late winter to see the beautiful wildflowers; entry from New England Hwy.

Village Markets: arts, crafts and local produce; 1st Sun each month. ***Crows Nest Day:*** children's worm races, crow calling competitions and battle of the band concert; Oct.

Bunnyconnellen Olive Grove and Vineyard: self-contained studio; Swain Rd; (07) 4697 9555.

Crows Nest National Park This popular park features a variety of landscapes, including granite outcrops and eucalypt forest. The wildlife is spectacular: see the platypus in the creek and the brush-tailed rock wallabies on the rocky cliffs. A steep track from the creek leads to an excellent lookout over Crows Nest Falls; follow this further to Koonin Lookout for spectacular views over the gorge, known locally as the Valley of Diamonds. (07) 4699 4333; 6 km E (look for sign to Valley of Diamonds).

Holland Wines: offers a vast collection of exquisite wines: 5 km N. ***Bunnyconnellen Olive Grove and Vineyard:*** perfect getaway for couples with rooms available. Enjoy the magnificent wines and gourmet food; open weekends; Swain Rd, off New England Hwy, 10 km N. ***Lake Cressbrook:*** set among picturesque hills is this excellent spot for windsurfing and boating. Fish the lake for silver perch; 17 km E. ***Ravensbourne National Park:*** a small park comprising remnant rainforest and wet eucalypt forest with over 80 species of birds, including the black-breasted button-quail that can be seen feeding on the rainforest floor on the Cedar Block track. Many bushwalks start at Blackbean picnic area; Esk–Hampton Rd; (07) 4699 4333; 25 km SE. ***Beutel's Lookout:*** picnic area with scenic views across the Brisbane Valley; adjacent to Ravensbourne National Park; 25 km SE. ***Goombungee:*** historic town with museum. Famous for running rural ironman and ironwoman competitions on Australia Day; 32 km W.

See also DARLING DOWNS, p. 418

Croydon

Pop. 255
Map ref. 656 F8

 **Samwell St; (07) 4745 6125; www.croydon.qld.gov.au**

 105.9 FM ABC Far North QLD

Croydon is a small town on the grassland plains of the Gulf Savannah. It marks the eastern terminus for the Gulflander train service, which leaves each Thursday morning for Normanton. The train line was established in the late 1800s to service Croydon's booming gold industry. Many original buildings dating from 1887 to 1897, and classified by the National Trust and Australian Heritage Commission, have been restored to the splendour of the town's goldmining days.

Historical precinct: many restored buildings bring the rich history of Croydon to life. Highlights include the old courthouse with original furniture, and the old hospital featuring original hospital documents. ***Outdoor Museum:*** displays mining machinery from the age of steam; Samwell St. ***Old Police Precinct and Gaol:*** historic documents and access to gaol cells; Samwell St. ***General Store and Museum:*** restored store in the old ironmongers shop; Sircom St.

 ***Poddy Dodger Festival:*** June.

Mining Museum: interesting display of early mining machinery including battery stamper; 1 km N. ***Historic cemetery:*** includes old Chinese gravestones; 2 km S. ***Chinese temple site:*** Heritage-listed archaeological site preserving 50 years of Chinese settlement that followed the gold discoveries of the 1880s. Take the Heritage Trail to see how the Chinese lived; on road to Lake Belmore. ***Lake Belmore:*** one of many sites for birdwatching, along with beautiful picnic areas and barbecue facilities. Also swimming, waterskiing and fishing (limits apply); 5 km N.

See also GULF SAVANNAH, p. 427

Cunnamulla

Pop. 1218
Map ref. 655 A9 | 663 N9

 **Centenary Park, Jane St; (07) 4655 8470; www.paroo.info**

 96.5 Triple C FM, 603 AM ABC Western QLD

Cunnamulla is on the Warrego River, north of the New South Wales border. It is the biggest wool-loading station on the Queensland railway network, with two million sheep in the area. Explorers Sir Thomas Mitchell and Edmund Kennedy were the first European visitors, arriving in 1846 and 1847 respectively, and by 1879 the town was thriving with regular Cobb & Co services to the west. In 1880, Joseph Wells held up the local bank but could not find his escape horse. Locals bailed him up in a tree that is now known as the Robber's Tree and can be found in Stockyard Street. The wetland birdlife in the area, particularly the black swans, brolgas and pelicans, is spectacular.

Cunnamulla Fella Centre Art Gallery and Museum: gallery showcases different artist exhibitions throughout the year, while the musem displays the rich history and heritage of the town and wider Paroo shire; Jane St. ***Yupunyah Tree:*** planted by Princess Anne; Cnr Louise and Stockyard sts. ***Outback Botanic***

QUEENSLAND

Gardens and Herbarium: Matilda Hwy. ***Outback Dreaming Arts and Crafts:*** exhibits and sales of local Aboriginal art; Stockyard St. ***The Heritage Trail:*** discover the days of the late 1880s on this self-guide trail; details from visitor centre.

Cunnamulla Outback Masters Games: odd-numbered years, Apr. ***Horseracing:*** Aug. ***Cunnamulla Festival and Bullride Championships:*** Nov.

Nardoo Station: ensuite rooms and camp sites; Matilda Hwy; (07) 4655 4833.

Wyandra: small town featuring Powerhouse Museum and Heritage Trail; 100 km N. ***Noorama:*** remote sheep station that holds picnic races each Apr; 110 km SE. ***Culgoa Floodplain National Park:*** 195 km SE; *see St George.*

See also OUTBACK, p. 428

Daintree

Pop. 78
Map ref. 651 D2 | 657 L4

5 Stewart St; (07) 4098 6133; www.daintreevillage.asn.au

639 AM ABC Far North QLD

The unspoiled township of Daintree lies in the tropical rainforest of Far North Queensland, at the heart of the Daintree River catchment basin surrounded by the McDowall Ranges. Daintree began as a logging town in the 1870s. Now tourism is the major industry with the Daintree River and World Heritage rainforest nearby. Saltwater crocodiles can be seen in the mangrove-lined creeks and tributaries of the Daintree River.

Daintree Timber Gallery: hand-crafted pieces from local woods exhibited and sold. Also displays showcasing local history, including logging past; Stewart St. ***Eliza's Rainforest Gallery:*** handmade pottery of pixies, elves and gnomes; Stewart St.

Julaymba: elegant, contemporary bush flavours; Daintree Eco Lodge & Spa, 20 Daintree Rd; (07) 4098 6100 or 1800 808 010. ***Papaya Cafe:*** try the crocodile dishes; 3–5 Stewart St; (07) 4098 6173.

Daintree Eco Lodge & Spa: Daintree Valley Haven: 3 self-catering cabins; Stewart Creek Rd; (07) 4098 6206. ***Daintree Wild B&B:*** 4 rooms in wildlife park; 2054 Daintree Rd; (07) 4098 7272.

Daintree National Park Arguably Australia's most beautiful and famous rainforest. The lush tangle of green protected within it is an incredible remnant from the days of Gondwana, forming part of the Wet Tropics World Heritage Area. This park, which is split into 3 distinct sections – Mossman Gorge, Cape Tribulation and Snapper Island – will dazzle visitors with its diverse landscapes, which, apart from beautiful rainforest, include canopies of sprawling fan palms, deserted mangrove-lined beaches and boulder-strewn gorges.

Daintree Discovery Centre: informative displays for visitors. A boardwalk and aerial walkway passes through the rainforest canopy; 11 km SE. ***Wonga Belle Orchid Gardens:*** 3.5 ha of lush gardens; 17 km SE. ***Cape Tribulation:*** spectacular beaches and reefs. Reef tours and horseriding on beach can be arranged at visitor centre or Mason's Shop in Cape Tribulation; for details call (07) 4098 0070; access via Daintree River cable ferry (car ferry runs 6am to 12am). ***River cruises:*** to see the saltwater crocodiles and plentiful birdlife; details from visitor centre. Horseriding tours: in Cape Tribulation; (07) 4041 3244.

Travellers note: *Beware of estuarine crocodiles that live in the sea and estuaries; never cross tidal creeks at high tide, swim in creeks, prepare food at water's edge or camp close to deep waterholes. Beware of marine stingers between Oct and Mar.*

See also CAIRNS & THE TROPICS, p. 425

Dalby

Pop. 9775
Map ref. 653 K8

Thomas Jack Park, Cnr Drayton and Condamine sts; (07) 4662 1066; www.dalby.info

89.9 FM Dalby Community Radio, 747 AM ABC Southern QLD

Sitting at the crossroads of the Warrego, Moonie, Condamine and Bunya highways, Dalby is conveniently linked to all of Australia's capital cities. It was a small rural town until the soldier resettlement program after World War II in which the population influx allowed the surrounding agricultural industry to thrive. It is now a relaxed country town with uncluttered landscapes, charismatic local pubs and home-grown produce.

Thomas Jack Park: beautifully landscaped park with playground equipment and tranquil lagoon; Cnr Drayton and Condamine sts. ***Cultural and Administration Centre:*** regional art gallery and cinema; Drayton St. ***Pioneer Park Museum:*** comprises historic buildings, household and agricultural items and a craft shop; Black St. ***The Crossing:*** an obelisk marks the spot where explorer Henry Dennis camped in 1841; Edward St. ***Historic cairn:*** pays homage to the cactoblastis, the Argentinean caterpillar that controlled prickly pear in the 1920s; Myall Creek picnic area, Marble St. ***Myall Creek Walk:*** walk along banks of Myall Creek to see varied birdlife. ***Heritage walk:*** self-guide walk provides insight into town's history; brochure from visitor centre.

Cotton on to Energy Festival: 4-day festival; Mar. ***Country Race Days:*** Newmarket Races; Mar. Picnic Races: May. Saints Race Day: July. Plough Inn Cup: Aug. ***Dalby Show:*** Apr. ***Spring Art Show:*** Sept. ***Spring Garden Week:*** Sept.

Lake Birdwater Conservation Park The 350 ha lake is an important breeding ground for waterfowl. There are over 240 species of bird that can be seen from the short walks around the lake. Waterskiing and boating are popular on the main body of the lake when it is full (permit required); 29 km SW.

Historic Jimbour House: attractive French homestead, formal gardens and boutique winery; 29 km N. ***Bell:*** small town at the base of Bunya Mountains with traditional arts and crafts stores; 41 km NW. ***Cecil Plains:*** cotton town with historic murals and Cecil Plains Homestead. Also a popular spot for canoeing down Condamine River; 42 km S. ***Rimfire Vineyards and Winery:*** boutique winery offering tastings; 47 km NE. ***Bunya Mountains National Park:*** 63 km NE; *see Kingaroy.*

See also DARLING DOWNS, p. 418

Emerald

Pop. 10 999
Map ref. 655 I10

Clermont St; (07) 4982 4142; www.emerald.qld.gov.au

96.3 4EEE FM, 1548 AM ABC Capricornia

Shady Moreton Bay fig trees line the main street in Emerald, an attractive town at the hub of the Central Highlands. Emerald was established in 1879 as a service town while the railway from Rockhampton to the west was being constructed. Several fires ravaged the town in the mid-1900s, destroying much of this early history. The largest sapphire fields in the Southern Hemisphere are nearby, where visitors can fossick for their own gems.

Pioneer Cottage Complex: historic cottage and lock-up gaol with padded cells. There is also a church and a communications museum; Morton Park, Clermont St. ***Botanic Gardens:*** walk around native display, herb garden and melaleuca maze. You can also visit a traditional bush chapel and ride the monorail; banks of Nogoa River. ***Railway station:*** restored 1900 National Trust–classified station with attractive lacework and pillared portico; Clermont St. ***Fossilised tree:*** 250 million years old; in front of town hall. ***Mosaic pathway:*** 21 pictures depict 100 years of Emerald history; next to visitor centre. ***Van Gogh Sunflower Painting:*** largest one (on an easel) in the world; Morton Park, Clermont St.

Sunflower Festival: Easter. ***Gemfest:*** Aug.

Sapphire gemfields To the west of Emerald is the largest sapphire-producing area in the Southern Hemisphere, incorporating Rubyvale, Sapphire, Anakie and Willows gemfields. The towns feature walk-in mines, fossicking parks, gem-faceting demonstrations, jewellers and museums. Obtain a licence and map of the mining areas from local stores or the Department of Natural Resources in Emerald to start fossicking. Rubyvale offers 4WD tours of gemfields including Tomahawk Creek; 43 km W.

Lake Maraboon/Fairbairn Dam: popular spot for watersports and fishing, especially for the red claw crayfish; 18 km S. ***Capella:*** first town settled in the area. See its history at the Capella Pioneer Village. The Crafts Fair and Vintage Machinery Rally is held in Sept; 51 km NW. ***Local cattle stations and farm stays:*** day tours in nearby area; details from visitor centre.

See also CAPRICORN, p. 423

Esk

Pop. 1163
Map ref. 647 A8 | 653 M8

2 Redbank St; (07) 5424 4000; www.somerset.qld.gov.a

 95.9 Valley FM, 747 AM ABC Southern QLD

Esk is a heritage town in the Upper Brisbane Valley renowned for its beautiful lakes and dams where watersports are popular. Deer roam in the grazing country north of town, progeny of a small herd presented to the state by Queen Victoria in 1873.

Antiques and local crafts: numerous shops in town.

Market: Highland St; Sat. ***Multicultural Festival:*** July. ***Picnic Races:*** July. ***Rail Trail Fun Run:*** July. ***Esk Show:*** Aug.

Lakes and dams Known as the Valley of the Lakes, this region is popular for swimming, fishing and boating. Lake Wivenhoe (25 km E) is the source of Brisbane's main water supply. Walk to the Fig Tree Lookout for a panoramic view or ride a horse around the lake. Lake Somerset (25 km NE) is on the Stanley River and is a popular waterskiing spot. Atkinson Dam (30 km S) also attracts watersports enthusiasts.

Caboonbah Homestead: museum that houses the Brisbane Valley Historical Society, with superb views over Lake Wivenhoe; closed Thurs; 19 km NE. ***Coominya:*** small historic town; 22 km SE. ***Ravensbourne National Park:*** 33 km W; *see Crows Nest.* ***Skydiving:*** take a tandem dive over the valley; contact visitor centre for details. ***Brisbane Valley Rail Trail:*** picturesque 148 km rural trail for biking enthusiasts. Trail runs between Fervale and Lowood, and Moore and Blackbutt. ***Kilcoy:*** small and unassuming farming town that claims to be the home of the Yowie, Australia's equivalent of the Bigfoot or Yeti. There is a large wooden statue in town of the supposed creature; 60 km N.

See also BRISBANE HINTERLAND, p. 415

Eulo

Pop. 108
Map ref. 663 M9

Centenary Park, Jane St, Cunnamulla; (07) 4655 2481; www.paroo.info

Eulo is on the banks of the Paroo River in south-west Queensland and was once a centre for opal mining. The town was originally much closer to the river but, after severe flooding, moved to where it currently stands. Eulo's population is variable as beekeepers travel from the south every winter so their bees can feed on the precious eucalypts in the area.

Eulo Queen Hotel: owes its name to Isobel Robinson who ran the hotel and reigned over the opal fields in the early 1900s; Leo St. ***Eulo Date Farm:*** taste the famous date wine; open Mar–Oct; western outskirts. ***WW II air-raid shelter:*** part of Paroo Pioneer Pathways; self-guide brochure available from visitor centre. ***Destructo Cockroach Monument:*** commemorates the death of a champion racing cockroach; near Paroo Track, where annual lizard races are held. ***Eulo Queen Opal Centre:*** opals, Aboriginal art and jewellery on display and for purchase; Leo St. ***Lizard Lounge:*** beautiful picnic area inspired by the frill-necked lizard; Cunnamulla Rd. ***Fishing:*** on the Paroo River; details from visitor centre.

Currawinya National Park The lakes and waterholes of Paroo River form an important refuge for the abundant birdlife of this park. The 85 km circuit track for vehicles starts at the Ranger Station and is the best way to view the park. See the black swans and grebes at Lake Wyara or go canoeing at Lake Numulla. For an excellent outlook over the park, climb the Granites. 4WD necessary to reach the lakes. (07) 4655 4143; 60 km SW.

Mud springs: natural pressure valve to artesian basin that is currently inactive; 9 km W. ***Yowah:*** small opal-mining town where visitors can fossick for opals or take a tour of the minefields. The 'Bluff' and 'Castles' provide excellent views over the minefields. Yowah holds a craft day in June and the Opal Festival in July, which includes an international opal jewellery competition; 87 km NW. ***Lake Bindegolly National Park:*** walk to the lakes to see pelicans and swans. The 9.2 km lake circuit track may flood after rain. No vehicles are allowed in the park; 100 km W. ***Thargomindah:*** pick up a mud map at visitor centre for Burke and Wills Dig Tree site; (07) 4655 3173; 130 km W. ***Noccundra Waterhole:*** good fishing spot on Wilson River; 260 km W.

See also OUTBACK, p. 428

Gatton

Pop. 5296
Map ref. 653 L9

34 Lake Apex Dr; (07) 5466 3450; www.lockyervalley.qld.gov.au

100.7 C FM, 747 AM ABC Southern QLD

In the Lockyer Valley west of Brisbane, Gatton was one of the first rural settlements in Queensland. The much recorded and debated 'Gatton Murders' that occurred in 1898 were the subject of Australian writer Rodney Hall's novel *Captivity Captive.* The graves of the three murdered siblings can still be seen in the cemetery. The Great Dividing Range provides a scenic

QUEENSLAND

backdrop to the surrounding fertile black soil, excellent for vegetable produce.

Gatton & District Historical Society Museum: complex of 11 buildings depicting the social history of the region; open 1–4pm or by appt; Freemans Rd. ***Lake Apex:*** park and complex include the historic village with preserved heritage buildings, memorabilia and Aboriginal carvings. Follow the walking tracks to see the diverse birdlife; Old Warrego Hwy.

Markets: Rotary Park, 1st Sat each month; Showgrounds, 2nd Sat each month. ***Clydesdale and Heavy Horse field days:*** Apr/May. ***Gatton Show:*** July. ***Multicultural Festival:*** Aug.

Agricultural College: opened in 1897. Drive through the grounds that are now part of the University of Queensland; 5 km E. ***Helidon:*** town noted for its sandstone – used in many Brisbane buildings – and spa water; 16 km W. ***Glen Rock Regional Park:*** at the head of East Haldon Valley, this park boasts rainforest gorges, creeks and excellent valley views; 40 km S. ***Tourist drive:*** 82 km circuit through surrounding countryside that includes farm visits; contact visitor centre for details.

See also BRISBANE HINTERLAND, p. 415

Gayndah

Pop. 1747
Map ref. 653 L5

Historical Museum, 3 Simon St; (07) 4161 2226; www.gayndah.qld.gov.au

 91.5 FM Burnett River Radio, 855 AM ABC Wide Bay

Gayndah, in the Capricorn region, is one of Queensland's oldest towns. Founded in 1849, it was once competing with Brisbane and Ipswich to be the state's capital. Main Street's heritage buildings and landscaped gardens illustrate the long history of the town, which is now central to a rich citrus-growing industry.

Historical Museum An award-winning museum with several historic buildings, displays, photographs and memorabilia that illustrate the town's changing history from small settlement to thriving agricultural centre – a highlight is a restored 1884 cottage. There is also an interesting display on the lungfish and its link between sea and land animals. Simon St.

Market: Jaycees Park; 1st Sun each month. ***Gayndah Show:*** Mar. ***Orange Festival:*** odd-numbered years, June. ***Triathlon:*** Oct.

Lookouts: several in area offering views over Burnett Valley; closest to town is Archers Lookout, atop the twin hills 'Duke and Duchess' overlooking town. ***Claude Warton Weir:*** excellent spot for fishing and picnics; 3 km W. ***Ban Ban Springs:*** natural springs and picnic area; 26 km S.

See also CAPRICORN, p. 423

Gin Gin

Pop. 888
Map ref. 648 B2 | 653 L3

Mulgrave St; (07) 4157 3060.

93.1 Sea FM, 855 AM ABC Wide Bay

Some of Queensland's oldest cattle properties surround this pastoral town south-west of Bundaberg. The district is known as 'Wild Scotchman Country' after James McPherson, Queensland's only authentic bushranger. His antics are re-enacted every March at the Wild Scotchman Festival.

Historical Society Museum: displays memorabilia of pioneering past in 'The Residence' – a former police sergeant's house. The old sugarcane locomotive The Bunyip forms part of the historic railway display; open 8.30am–3.30pm Mon–Fri, 8.30am–12pm Sat, other times by appt; Mulgrave St. ***Courthouse Gallery:*** fine-arts gallery in refurbished old courthouse; Mulgrave St.

Market: Historical Museum; Sat. ***Jazz and Shiraz:*** Wonbah Winery; Aug. ***Auto Machinery Show:*** Aug. ***Cane Burning and Whipcracking Festival:*** Aug. ***Didgeridoo Festival:*** Aug/Sept. ***Santa Fair:*** Dec.

Goodnight Scrub National Park A dense remnant hoop pine rainforest in the Burnett Valley, this park is home to over 60 species of butterfly. Have a bush picnic at historic Kalliwa Hut, used during the logging days of the park. Drive up to One Tree Hill (4WD only) for a spectacular panoramic view over the area, on a clear day, all the way to Bundaberg. Turn-off is 10 km south of Gin Gin; (07) 4131 1600.

Lake Monduran: an excellent spot for watersports and fishing (permit from kiosk). Catch a barramundi or Australian bass, or walk the 6 km of tracks in the bush surrounds; 24 km NW. ***Boolboonda Tunnel:*** longest non-supported tunnel in the Southern Hemisphere. It forms part of a scenic tourist drive; brochure available from visitor centre; 27 km W. ***Mount Perry:*** small mining town, home to mountain-bike racing in June; 55 km SW.

See also CAPRICORN, p. 423

Gladstone

Pop. 42 902
Map ref. 653 L1 | 655 N12

Marina Ferry Terminal, Bryan Jordan Dr; (07) 4972 4000; www.gladstoneregion.org.au

 95.1 Sea FM, 837 AM ABC Capricornia

Gladstone is a modern city on the central coast of Queensland. Matthew Flinders discovered Port Curtis, Gladstone's deep-water harbour, in 1802, but the town did not truly develop until the 1960s. Today it is an outlet for central Queensland's mineral and agricultural wealth – a prosperous seaboard city with one of Australia's busiest harbours. Set among hills with natural lookouts over the harbour and southern end of the Great Barrier Reef, Gladstone is popular for swimming, surfing and fishing – especially for mud crabs and prawns.

Tondoon Botanic Gardens: displays of all-native species of the Port Curtis region with free guided tours on weekends. Also offers a recreational lake and Mt Biondello bushwalk; Glenlyon Rd. ***Gladstone Regional Art Gallery and Museum:*** local and regional art with history exhibitions; closed Sun; Cnr Goondoon and Bramston sts. ***Maritime Museum:*** artefacts and memorabilia document 200 years of port history; open Thurs and Sun; Auckland Pt. ***Potters Place:*** fine-art gallery and craft shop; Dawson Hwy. ***Gecko Valley Winery:*** tastings and sales of award-winning wines; closed Mon; Bailiff Rd. ***Barney Point Beach:*** historic beach including Friend Park; Barney St. ***Waterfall:*** spectacular at night when floodlit; Flinders Pde.

Markets: Calliope River Historical Village; selected dates (6 times a year). ***Sunfest:*** youth sports and arts program; Jan. ***Harbour Festival:*** includes the finish of the Brisbane to Gladstone Yacht Race; Mar. ***Woodworker's Art and Crafts Weekend (Calliope):*** July. ***Multicultural Festival:*** Aug. ***Seafood Festival:*** Sept.

Auckland Hill B&B: heritage house in centre of town; 15 Yarroon St; (07) 4972 4907. ***Rydges Gladstone Hotel:*** spacious rooms; 100 Goondoon St; (07) 4970 0000 or 1300 857 922.

Xenia Central Studio Accommodation: self-catering rooms and suites; 166 Auckland St; (07) 4972 2022. ***Glassford Creek Station Farmstay:*** working cattle station; Boxvale, Builyan; (07) 4974 1185.

Capricornia Cays National Park This park, 60–100 km offshore from Gladstone, protects the 9 coral islands and cays that form the southern end of the Great Barrier Reef. The islands are important nesting sites for seabirds and loggerhead turtles. North West Island has walking tracks through forests dominated by palms and she-oaks. The most popular activities are reef walking, diving and snorkelling in the spectacular reefs or visiting the renowned dive sites on Heron Island. Access is by private boat or charter from Gladstone; air access to Heron Island. There is seasonal closure to protect nesting wildlife. (07) 4971 6500 (same number for all national parks in the area).

Boyne Island: with beautiful foreshore parks and beaches. Home to the Boyne–Tannum Hookup Fishing Competition in June. The Boyne Aluminium Smelter is Australia's largest and has an information centre and tours every Fri. The island and its twin town of Tannum Sands are linked by bridge; 25 km SE. ***Tannum Sands:*** small community offering sandy beaches with year-round swimming, picturesque Millennium Way along the beach and 15 km of scenic walkways known as the Turtle Way. Wild Cattle Island, an uninhabited national park at the southern end of the beach, can be reached on foot at low tide; 25 km SE. ***Calliope:*** small rural community with excellent fishing in nearby Calliope River with abundant mud crabs, salmon and flathead. The Calliope River Historical Village documents Port Curtis history in restored buildings and holds regular art and craft markets; 26 km SW. ***Lake Awoonga:*** a popular spot for swimming, skiing (permit required) and fishing. It has walking tracks and recreational facilities, and holds the Lions Lake Awoonga Family Fishing Festival each Sept; 30 km S. ***Mt Larcom:*** spectacular views from the summit; 33 km W. ***Castle Tower National Park:*** a rugged park of granite cliffs and the outcrops of Castle Tower and Stanley mountains. Only experienced walkers should attempt the climb to Mt Castle Tower summit, where there are superb views over the Boyne Valley and Gladstone. Access by foot or boat from Lake Awoonga; access by car from Bruce Hwy; 40 km S. ***Kroombit Tops National Park:*** this mountain park is on a plateau with sandstone cliffs and gorges, waterfalls and creeks. Drive the 90 min return loop road to explore the landscapes and walk to the site of a WW II bomber crash; 4WD recommended; 75 km SW via Calliope. ***Curtis Island National Park:*** at the north-east end of the island is this small park with a variety of vegetation and excellent spots for birdwatching. There are no walking tracks, but the 3–4-day hike along the east coast is worthwhile; access by boat from Gladstone or the Narrows. ***Great Barrier Reef tours:*** cruises depart from the Gladstone Marina; Bryan Jordan Dr.

See also CAPRICORN, p. 423

Goondiwindi

Pop. 5031
Map ref. 556 H1 | 653 I11

Cnr McLean and Bowen sts; (07) 4671 2653; www.goondiwindi.qld.gov.au

 101.9 Coast FM, 711 AM ABC Western QLD

Beside the picturesque MacIntyre River in the western Darling Downs is this border town. Explored by Allan Cunningham in 1827 and settled by pastoralists in the 1830s, the town derives its name from the Aboriginal word 'gonnawinna', meaning 'resting place of the birds'.

Customs House Museum: explore local history in the restored 1850 customs house; closed Tues; McLean St. ***Victoria Hotel:*** renowned historic pub with tower; Marshall St. ***Goondiwindi Cotton and Gin:*** one of the largest cotton manufacturers in Australia; bus tours available, contact visitor information centre for details. ***River Walk:*** watch abundant birdlife and wildlife on 2 km walk along MacIntyre River; starts at Riddles Oval, Lagoon St. ***'Goondiwindi Grey' Statue:*** tribute to famous racehorse Gunsynd; Apex Park, MacIntyre St. ***Fishing:*** some of Queensland's best fishing in and around the town, Murray cod and yellow-belly in abdundance.

Market: Town Park, Marshall St; 2nd Sun each month. ***Hell of the West Triathlon:*** Feb. ***Gourmet in Gundy:*** Sept. ***Spring Festival:*** coincides with flowering of jacarandas and silky oaks; Oct.

Southwood National Park This brigalow-belah forest park was once known as 'Wild Horse Paradise'. Have a bush picnic and look for the black cockatoos in the belah trees, or visit at night and go spotlighting for feathertail gliders. 4WD is recommended. Access from Moonie Hwy; (07) 4699 4355; 125 km NW.

Botanic Gardens of Western Woodlands: 25 ha of native plants of the Darling Basin. Also here is a lake popular for swimming and canoeing; access from Brennans Rd; 1 km W. ***Toobeah:*** small town famous for its horse events; 48 km W.

See also DARLING DOWNS, p. 418

Gordonvale

Pop. 4421
Map ref. 650 E3 | 651 F8 | 657 M6

Cnr Bruce Hwy and Munro St, Babinda; (07) 4067 1008.

 103.5 Hot FM, 801 AM ABC Far North QLD

Gordonvale is a sugar-milling town just south of Cairns with well-preserved streetscapes, historic buildings and the 922-metre-high Walsh's Pyramid that forms the backdrop to the town. People flock to the mountain in August for a race to the top. The town's less glorious claim to fame is that cane toads were released here in 1935 in an attempt to eradicate sugarcane pests.

Settlers Museum: with displays and dioramas depicting early pioneer life in the shire. Includes blacksmith's shop, old store and a Chinese display; open 10am–2pm Mon–Sat (closed Dec–Feb); Gordon St.

Goldsborough Valley State Forest The lowland rainforest along the Goldsborough Valley is protected here. Walk to the falls along the 1.6 km Kearneys Falls track and learn about the local Aboriginal culture and customs from the informative displays. The 18 km historic Goldfields Trail travels through nearby Wooroonooran National Park to the Boulders near Babinda; 25 km SW via Gillies Hwy.

Mulgrave River: runs next to Gordonvale and is popular for swimming, canoeing, kayaking and bushwalking. ***Wooroonooran National Park:*** waterfalls, walking tracks and Walsh's Pyramid; 10 km S; *For the southern sections see Babinda and Millaa Millaa.* ***Orchid Valley Nursery and Gardens:*** tours of tropical gardens; 15 km SW. ***Cairns Crocodile Farm:*** crocodiles and other native wildlife with daily tours; around 5 min north on road to Yarrabah.

See also CAIRNS & THE TROPICS, p. 425

QUEENSLAND

Gympie

Pop. 10 933
Map ref. 648 D9 | 653 M6

Mary Street Information Booth, (07) 5483 6656; or Cooloola Regional Information Centre, Matilda's Roadhouse Complex, Bruce Hwy, Kybong (15 km s), (07) 5483 5554 or 1800 444 222; www.cooloola.org.au

 96.1 Zinc FM, 1566 AM ABC Wide Bay

On the banks of the Mary River on the Sunshine Coast is the major heritage town of Gympie. It was established when James Nash discovered gold in the area in 1867 and started Queensland's first gold rush to save the state from near bankruptcy. The field proved extremely rich – four million ounces had been found by the 1920s. The gold slowed to a trickle soon after, but the dairy and agricultural industries were already well established. See the attractive jacarandas, silky oaks, cassias, poincianas and flame trees that line the streets.

Mary Valley Heritage Railway Known locally as the 'Valley Rattler', this restored 1923 steam train takes the visitor on a 40 km journey through the picturesque Mary Valley. The train departs Gympie every Wed, Sat and Sun on its way to Imbil, where it stops before returning. Special tours run each Sat. Tickets and information from visitor centre; inquiries (07) 5482 2750.

Woodworks, Forestry and Timber Museum: exhibits memorabilia from old logging days including a steam-driven sawmill; closed Sat; Fraser Rd. ***Deep Creek:*** gold-fossicking area; permits from visitor centre; Counter St. ***Public gallery:*** local and visiting art exhibitions in heritage building; Nash St. ***Heritage walk:*** self-guide walk includes the Stock Exchange and Town Hall; details from visitor centre. ***Trail rides:*** horseriding through Kiah Park and Mary Valley; details from visitor centre.

Market: Gympie South State School; 2nd and 4th Sun each month. ***Gympie Show:*** May. ***Race the Rattler:*** a race against the historic steam train; June. ***Fishing Classic:*** Rainbow Beach; July. ***Rodeo and Woodchop (Mary Valley):*** Aug. ***Country Music Muster:*** Amamoor State Forest; Aug. ***Art Festival (Mary Valley):*** Sept. ***Gold Rush Festival:*** Oct.

Amamoor State Forest Over 120 native animal species find shelter in this protected forest. See the platypus in Amamoor Creek at dusk or take the Wonga walk or Cascade circuit track starting across the road from Amama. The renowned outdoor music festival, the Country Music Muster, is held in the forest on the last weekend in Aug. 30 km s.

Gold Mining Museum: delve into the area's goldmining history. It includes Andrew Fisher House (Fisher was the first Queenslander to become prime minister); 5 km s. ***Mothar Mountain:*** rockpools and forested area for bushwalking and excellent views; 20 km SE. ***Imbil:*** picturesque town with excellent valley views. There is a market every Sun, and the nearby Lake Borumba offers great conditions for watersports and fishing – especially for golden perch and saratoga. Take the 14 km Imbil Forest Drive through scenic pine plantations just south of town; 36 km s. ***Mary Valley Scenic Way:*** enjoy this scenic route through towns of the valley, pineapple plantations and grazing farms; it runs south between Gympie and Maleny, via Kenilworth.

See also SUNSHINE COAST, p. 419

Herberton

Pop. 974
Map ref. 650 C4 | 651 C10 | 657 L7

Herberton Mining Museum, 1 Jacks Rd; (07) 4096 3474; www.herbertonvisitorcentre.com.au

Known as the 'Village in the Hills', Herberton sits about 1000 metres above sea level on the south-west ranges of the Atherton Tableland. The first settlement on the tableland, it was established in 1880 when two prospectors discovered tin in the area. It was a thriving tin-mining town, the most important in the Herbert River field, until the mine's closure in 1978.

Mining Museum: explore local mining history in this richly informative museum, with displays of antique machinery, an array of minerals and rocks, and multimedia shows. A short walking track takes you past mining relics that made up the Great Northern Claim, now a heritage-listed site; 1 Jacks Rd. ***Spy and Camera Museum:*** houses some of the world's rarest, oldest and smallest cameras. See Russian spy cameras and cameras from Hitler's Germany; Grace St. ***Herberton Historic Village:*** contains over 30 original buildings with plenty of memorabilia; (07) 4096 2002; 6 Broadway.

Ghost Walks: 3 times a year the ghosts of former miners and Herberton residents 'materialise' to retell their stories; dates vary, details from visitor centre.

Mt Hypipamee National Park Set on the Evelyn Tableland, this park boasts high-altitude rainforests and a climate that attracts birdlife and possums, including the green and lemuroid ringtail possums. The park is known locally as 'The Crater' because of its sheer-sided volcanic explosion crater 70 m wide. Walk to the viewing deck for the best vantage point, or see the Dinner Falls cascade down the narrow gorge. (07) 4091 1844; 25 km s.

Herberton Range State Forest: the temperate climate attracts a variety of wildlife in this rainforest park, including the attractive golden bowerbird. Walk to the summit of Mt Baldy for panoramic views over the tableland (the steep ascent should be attempted only by experienced walkers); Rifle Range Rd, between Atherton and Herberton. ***Irvinebank:*** many heritage-listed buildings including Mango House, Queensland National Bank Building and Loudoun House Museum; 26 km w on unsealed road. ***Emuford:*** small town featuring a historic stamper battery and museum; 51 km w.

See also CAIRNS & THE TROPICS, p. 425

Hughenden

Pop. 1154
Map ref. 654 B4 | 661 N4

Flinders Discovery Centre, Gray St; (07) 4741 1021; www.hughenden.com

 1485 AM ABC North West QLD

Hughenden is on the banks of the Flinders River, Queensland's longest river, west of the Great Dividing Range. The first recorded Europeans to pass here were members of Frederick Walker's 1861 expedition to search for explorers Burke and Wills. Two years later Ernest Henry selected a cattle station and Hughenden came into existence. The black volcanic soil in the region is rich with fossilised bones, particularly those of dinosaurs.

Flinders Discovery Centre Learn about the Flinders Shire history in this complex. The fossil exhibition's centrepiece is the 7 m replica skeleton of *Muttaburrasaurus langdoni*, a dinosaur found in Muttaburra (206 km s) – it was the first entire fossil skeleton found in Australia. The Historical Society also documents the shire history in their display. Gray St; (07) 4741 1021.

continued on p. 452

HERVEY BAY

Pop. 48 156

Map ref. 648 F4 | 653 N4

Cnr Urraween and Maryborough–Hervey Bay rds; (07) 4125 9855 or 1800 811 728; www.herveybay.qld.gov.au

107.5 FM Fraser Coast Community Radio, 855 AM ABC Wide Bay

Hervey (pronounced 'Harvey') Bay is the large area of water between Maryborough and Bundaberg, protected by Fraser Island. It is also the name of a thriving city spread out along the bay's southern shore. The climate is ideal and during winter there is an influx of visitors, not only for the weather, but also for the migrating humpback whales that frolic in the bay's warm waters between July and November. Hervey Bay is promoted as 'Australia's family aquatic playground' – there is no surf and swimming is safe, even for children. Fishing is another popular recreational activity, especially off the town's kilometre-long pier.

Botanic Gardens: interesting orchid conservatory; Elizabeth St. ***Historical Museum:*** recalls pioneer days in 19 historic buildings, including an 1898 slab cottage; open Fri–Sun; Zephyr St. ***Sea Shell Museum:*** shell creations and displays, including 100-million-year-old shell; Esplanade. ***Neptune's Reefworld:*** animals of the ocean, including performing seals; Pulgul St. ***M & K Model Railways:*** award-winning miniature gardens and model trains. Ride the replica diesel train; open Tues–Fri; Old Maryborough Rd. ***Dayman Park:*** memorial commemorates landing of Matthew Flinders in 1799 and the Z-Force commandos who trained there on the Krait in WW II. ***Regional Gallery:*** includes regularly changing exhibitions of local artists plus touring exhibitions from state and national galleries; open Tues–Sat; Old Maryborough Rd. ***Scenic walkway:*** cycle or walk 15 km along the waterfront. ***Whale-watching tours:*** half- and full-day tours to see the migratory whales off the coast, departing from Boat Harbour; contact visitor centre for details.

Nikenbah Markets: Nikenbah Animal Refuge; 1st and 3rd Sun each month. ***Koala Markets:*** Elizabeth St; 2nd and 4th Sun each month. ***Fraser Coast Multicultural Festival:*** Mar. ***Fraser Coast Show:*** May. ***World's Biggest Pub Crawl:*** June. ***Mary Poppins Festival:*** July. ***Whale Festival:*** Aug. ***Seafood Festival:*** Aug. ***Hervey Bay Jazz and Blues Festival:*** Nov.

Bayswater: seafood the specialty; 26–34 Hibiscus St; (07) 4194 9700. ***Enzo's on the Beach:*** seaside cafe, meals can be served on the beach; 351A The Esplanade, Scarness; (07) 4124 6375. ***Gatakers Landing Restaurant:*** outdoor dining, delicious seafood; Cnr The Esplanade and Mant St, Point Vernon, Gatakers Bay; (07) 4124 2470. ***Moonlight and Day Away Cruises:*** leave the mainland for lunch; from Urangan Boat Harbour; 1800 072 555.

Breakfree Great Sandy Straits: on the water's edge; 17 Buccaneer Dr, Urangan; (07) 4128 9999 or 1300 553 800. ***Emeraldene Inn & Eco-Lodge:*** rooms and apartments; 166 Urraween Rd; (07) 4124 5500. ***Peppers Pier Resort:*** 129 units on 6 floors; 26–34 Hibiscus St; (07) 4194 9700.

Fraser Island World Heritage–listed Fraser Island is the largest sand island in the world. It is an ecological wonder with lakes and forests existing purely on sand. Protected within Great Sandy National Park, the island is an oasis of beaches, beautifully coloured sand cliffs, more than 40 freshwater lakes and the spectacular tall rainforests. The island is also home to a variety of wildlife, including migratory birds and rare animals such as the ground parrot and Illidge's ant-blue butterfly. Fraser Island's dingoes are one of the purest strains in the country. Offshore, see the turtles, dugong and dolphins soak up the warm waters and, between Aug and Nov, look out for migrating humpback whales. There are a variety of walks around the island, as well as swimming spots and scenic drives. Care should be taken around the island's dingo population. Stay with children, walk in groups, never feed or coax the dingoes and keep all food and rubbish in vehicles or campground lockers. Vehicle and camping permits from Queensland Parks & Wildlife Service, 13 1304.

Great Sandy Strait: the Mary and Susan rivers to the south of Hervey Bay run into this strait where the visitor can see spectacular migratory birds, including the comb-crested jacana. Hire a houseboat to travel down the strait; contact visitor centre for details. ***Toogoom:*** quiet seaside resort town. Feed the pelicans on the boardwalk; 15 km W. ***Burrum Heads:*** pleasant holiday resort at the mouth of the Burrum River with excellent beaches and fishing. Visit at Easter for the Amateur Fishing Classic; 20 km NW. ***Burrum Coast National Park:*** 34 km NW; *see Bundaberg*. ***Brooklyn House:*** historic old Queenslander pioneer house; Howard; 36 km W. ***Lady Elliot Island:*** fly to the island for a day to snorkel on the fabulous reef; contact visitor centre for details. ***Scenic flights:*** over Hervey Bay and Fraser Island in a Tiger Moth or other small plane; contact visitor centre for details.

See also FRASER ISLAND & COAST, p. 420

[VIEWING A HUMPBACK WHALE ON A WHALE-WATCHING TOUR]

QUEENSLAND

Historic Coolibah Tree: blazed by Walker in 1861 and again by William Landsborough in 1862 when he was also searching for Burke and Wills, and their 2 companions; east bank of Station Creek, Stansfield St East. ***Flinders Poppy Art and Craft Cottage:*** unique range of handmade art and craft items for sale; open 10am–2pm daily (hours can vary); Gray St.

Market: Lions Rotary Park; last Sun each month. ***Country Music Festival:*** Apr. ***Horseracing:*** Apr and Sept. ***Hughenden Show:*** June. ***Porcupine Gorge Challenge:*** fun run; June. ***Arid Lands Festival:*** camel endurance race; Aug. ***Outback Scrap:*** Aug. ***Bullride:*** Sept.

Porcupine Gorge National Park The coloured sandstone cliffs of this park are a delight and contrast with the greenery surrounding Porcupine Creek. The gorge, known locally as the 'mini Grand Canyon', has been formed over millions of years and can be seen from the lookout just off Kennedy Development Rd. Walk down into the gorge on the 1.2 km track, but be warned, the steep walk back up is strenuous. (07) 4722 5224; 61 km N.

Basalt Byways: discover the Flinders Shire landscapes on these 4WD tracks. Cross the Flinders River and see the Flinders poppy in the valleys. The longest track is 156 km; access on Hann Hwy; 7.3 km N. ***Prairie:*** small town with mini-museum and historic relics at Cobb & Co yards; 44 km E. ***Mt Emu Goldfields:*** fossicking and bushwalking; 85 km N. ***Torrens Creek:*** the town, a major explosives dump during WW II, was nearly wiped out by 12 explosions when firebreaks accidentally hit the dump. Visit the Exchange Hotel, home of the 'dinosaur steaks'; 88 km E. ***Kooroorinya Falls:*** the small falls cascade into a natural waterhole, excellent for swimming. Walk and go birdwatching in surrounding bushland; 109 km SE via Prairie. ***White Mountains National Park:*** rugged park with white sandstone outcrops and varied vegetation. Burra Range Lookout on Flinders Hwy has excellent views over the park. There are no walking tracks; (07) 4722 5224; 111 km E. ***Chudleigh Park Gemfields:*** gem-quality peridot found in fossicking area (licence required); 155 km N. ***Moorrinya National Park:*** important conservation park protecting 18 different land types of the Lake Eyre Basin. The park is home to iconic Australian animals, such as koalas, kangaroos and dingoes, and includes remains of the old sheep-grazing property Shirley Station. Walking is for experienced bushwalkers only; (07) 4722 5224; 178 km SE via Torrens Creek.

See also OUTBACK, p. 428

Ingham

Pop. 4603
Map ref. 650 E8 | 657 M10

Tyto Wetlands Information Centre, Bruce Hwy (Townsville Rd); (07) 4776 5211; www.hinchinbrooknq.com.au

91.9 FM Kool, 630 AM ABC North QLD

Ingham is a major sugar town near the Hinchinbrook Channel. Originally, Kanaka labourers were employed in the surrounding sugarcane fields, but after their repatriation at the beginning of the 20th century, Italian immigrants took their place (the first Italians arrived in the 1890s). This strong Italian heritage is celebrated in the Australian Italian Festival each May. Ingham is at the centre of a splendid range of national parks. The Wallaman Falls in Girringun (Lumholtz) National Park is a highlight as the largest single-drop falls in Australia.

Hinchinbrook Heritage Walk and Drive: displays at each historic site in Ingham and the nearby township of Halifax illustrate the dynamic history of the shire. It starts at the Shire Hall in Ingham; brochure available from visitor centre. ***Memorial Gardens:*** picturesque waterlily lakes and native tropical vegetation. They include Bicentennial Bush House with displays of orchids and tropical plants (open weekdays); Palm Tce.

Conroy Hall Markets: McIlwraith St; 2nd Sat each month. ***Raintree Market:*** Herbert St; 3rd Sat each month. ***Ingham Arts Festival:*** Apr. ***Australian Italian Festival:*** May. ***Bullride:*** June. ***Ingham Show:*** July. ***Horseracing:*** Aug. ***Family Fishing Classic:*** Sept. ***Maraka Festival:*** Oct.

Noorla Heritage Resort: tiffin, tapas, cakes and curry; 5–9 Warren St; (07) 4776 1100 or 1800 238 077.

Noorla Heritage Resort: Art Deco villa; 5–9 Warren St; (07) 4776 1100 or 1800 238 077.

Girringun (Lumholtz) National Park The Wallaman Falls section (50 km W) is in the Herbert River Valley and features waterfalls, gorges and tropical rainforest. See the crimson rosellas on the 4 km return walk from the Falls Lookout to the base of Wallaman Falls – look out for platypus and water dragons. Take a scenic drive around the Mt Fox section of the park (75 km SW) or walk the 4 km return ascent to the dormant volcano crater. There are no formal tracks and the ascent is for experienced walkers only. 4WD is recommended in both sections during the wet season; *see Cardwell for northern parts of park*; (07) 4066 8601 (same number for other national parks in area); 20 km NW.

Paluma Range National Park The Jourama Falls section of the park (24 km S) is at the foothills of the Seaview Range. Walk the 1.5 km track through rainforest and dry forest to the Jourama Falls Lookout (take care crossing the creek) to see the vibrant birdlife including azure kingfishers and kookaburras. The Mt Spec section of the park (40 km S) features casuarina-fringed creeks in the lowlands and rainforest in the cooler mountain areas. Drive to McClelland's Lookout for a spectacular view and take the 2 short walks from there to see the varied park landscapes; 49 km S.

Tyto Wetlands: 90 ha of wetlands with birdlife and wallabies. See the rare grass owl from the viewing platform; outskirts of town. ***Cemetery:*** interesting Italian tile mausoleums; 5 km E. ***Forrest Beach:*** sandy 16 km beach overlooking Palm Islands. Stinger-net swimming enclosures are installed in summer; 20 km SE. ***Taylors Beach:*** popular family seaside spot for sailing with excellent fishing and crabbing in nearby Victoria Creek; 24 km E. ***Hinchinbrook Island National Park:*** resort island to the north-east; sea access from Lucinda; 25 km NE; *see Cardwell*. ***Lucinda:*** coastal village on the banks of the Herbert River at the southern end of the Hinchinbrook Channel. Take a safari or fishing tour through the channel. Stinger-net swimming enclosures are installed in summer; 27 km NE. ***Broadwater State Forest:*** swimming holes, walking tracks and birdwatching; 45 km W. ***Orpheus Island National Park:*** this rainforest and woodland park is a continental resort island in the Palm Islands surrounded by a marine park and fringing reefs. Snorkel and dive off the beaches or take in the wildlife on the short track from Little Pioneer Bay to Old Shepherds Hut; access is by private or charter boat from Lucinda or Taylors Beach.

See also THE MID-TROPICS, p. 424

Innisfail

Pop. 8260
Map ref. 650 E4 | 651 H11 | 657 M7

24 Bruce Hwy, Mourilyan; (07) 4063 2655; www.innisfailtourism.com.au

 98.3 Kool FM, 720 AM ABC Far North QLD

Innisfail has seen destruction by cyclone more than once – most recently cyclone Larry in 2006, which destroyed homes and much of the town's banana plantations. But today Innisfail seems to have mostly recovered and travellers come to work during fruit-picking season. The parks and walks on the riverside, as well as the classic Art Deco buildings in the town centre, add to the charm of this town. The area is also renowned for its sugar industry and the excellent fishing in nearby rivers, beaches and estuaries.

Innisfail and District Historical Society Museum: documents local history; Edith St. ***Chinese Joss House:*** reminder of Chinese presence during gold-rush days; Owen St. ***Warrina Lakes and Botanical Gardens:*** recreational facilities and walks; Charles St. ***Historical town walk:*** see classic Art Deco architecture of the town and historic Shire Hall on self-guide or guided town walk; details from visitor centre.

Market: ANZAC Memorial Park; 3rd Sat each month. ***Innisfail Show:*** July. ***Game Fishing Tournament:*** Sept. ***Karnivale:*** Oct. ***Innisfail Cup:*** Oct.

Eubenangee Swamp National Park This important park protects the last of the remnant coastal lowland rainforest around Alice River, which is part of the Wet Tropics Region. See the rainforest birds on the walk from Alice River to the swamp where jabirus and spoonbills feed. (07) 4067 6304; 13 km NW.

Flying Fish Point: popular spot for swimming, camping and fishing; 5 km NE. ***Johnstone River Crocodile Farm:*** daily feeding shows at 11am and 3pm; 8 km NE. ***Ella Bay National Park:*** small coastal park with beach and picnic spot; (07) 4067 6304; 8 km N. ***North Johnstone River Gorge:*** walking tracks to several picturesque waterfalls; 18 km W via Palmerston Hwy. ***Wooroonooran National Park:*** 33 km W; *see Millaa Millaa.* ***Crawford Lookout:*** for spectacular views of North Johnstone River; 38 km W off Palmerston Hwy.

See also CAIRNS & THE TROPICS, p. 425

Ipswich

Pop. 155 000
Map ref. 644 A9 | 645 B5 | 647 D12 | 653 M9

14 Queen Victoria Pde; (07) 3281 0555; www.ipswichtourism.com.au

 94.9 FM River, 747 AM ABC Southern QLD

Ipswich is Queensland's oldest provincial city with diverse heritage buildings throughout its streets. In 1827 a convict settlement was established alongside the Bremer River to quarry the nearby limestone deposits used in Brisbane's stone buildings. In 1842 the settlement, simply called 'Limestone', opened to free settlers and was in 1843 renamed Ipswich. The impressive Workshops Rail Museum honours Ipswich as the birthplace of Queensland Railways. Australia's largest RAAF base is in the suburb of Amberley.

Workshops Rail Museum In the North Ipswich railyards, this museum offers diverse historical displays, interactive exhibitions and an impressive variety of machinery. Watch workers restore old steam trains, or look into the future of rail by taking a simulated ride in the high-speed tilt train. North St.

Global Art Links Gallery This art gallery and social history museum merges heritage and the present day at the Old Town Hall. See the local and visiting exhibitions in the Gallery, partake in the interactive displays of Ipswich history in the Hall of Time, walk through the Indigenous installation in the Return to Kabool section, or try electronic finger-painting in the Children's Gallery. D'Arcy Doyle Pl.

Queens Park Nature Centre: native flora, animals and a bird aviary; Goleby Ave. ***Ipswich Heritage Model Railway Club:*** miniature trains; activity days on the 1st Sat and 3rd Sun of each month, visits at other times by appt; via Workshops Rail Museum. ***Historical walk:*** self-guide walk to see the renowned heritage buildings, churches and excellent domestic architecture of Ipswich, including the Uniting Church (1858) and Gooloowan (1864); outstanding colour brochure available from visitor centre.

Showground Market: Salisbury Rd; Sun. ***Farmers Market:*** Leichhardt Community Centre; 1st Sat each month (June–Nov). ***Handmade Expo:*** over 50 stalls of handmade arts and crafts; 3rd Sat each month (Aug–Dec). ***Ipswich Cup:*** horseracing; June. ***Circus Train:*** month-long event with circus performers and train rides; Sept–Oct. ***Jacaranda Festival (Goodna):*** Oct. ***Rodeo:*** Nov.

Cotton's Restaurant: modern Australian steakhouse; Spicers Hidden Vale, 617 Mt Mort Rd, Grandchester; (07) 5465 5900. ***Woodlands of Marburg:*** modern Australian in heritage house; 174 Seminary Rd, Marburg; (07) 5464 4777.

Parkview Colonials B&B: colonial houses with ensuite rooms; 72 Chermside Rd; (07) 3812 3266. ***Spicers Hidden Vale:*** luxury resort; 617 Mt Mort Rd, Grandchester; 1300 881 435. ***Woodlands of Marburg:*** heritage mansion; 174 Seminary Rd, Marburg; (07) 5464 4777.

Queensland Raceway: home to the Queensland 500 and host of the V8 Supercar Series. Winternationals Drag-racing Championship runs every June; 15 km W. ***Wolston House:*** historic home at Wacol; open weekends and public holidays; 16 km E. ***Rosewood:*** heritage town with St Brigid's Church, the largest wooden church in the South Pacific. The Railway Museum, with restored carriages and wagons and displays of area's industrial heritage, runs scenic steam-train rides on the last Sun of each month; museum open every Sun; 20 km W. ***Recreational reserves:*** popular picnic and leisure spots to the north-east of Ipswich include College's Crossing (7 km), Mt Crosby (12 km) and Lake Manchester (22 km).

See also BRISBANE HINTERLAND, p. 415

Isisford

Pop. 262
Map ref. 654 B11 | 661 N11

Council offices, St Mary St; (07) 4658 8900; www.isisford.qld.gov.au

 104.5 FM West, 603 AM ABC Western QLD

Isisford is a small outback community south of Longreach. In the mid-1800s large stations were established in the area. This brought hawkers, keen on trading their goods to the landowners. Two such hawkers were brothers William and James Whitman who, after their axle broke trying to cross the Alice River, decided

QUEENSLAND

to stay and established Isisford in 1877. First called Wittown, after the brothers, it was renamed in 1880 to recall the nearby Barcoo River ford and Isis Downs Station. The town provided inspiration to iconic Australian poet Banjo Paterson, in particular his poems Bush Christening and Clancy of the Overflow.

Outer Barcoo Interpretation Centre: features the world-class fossil exhibit of *Isisfordia duncani,* reported to be the ancestor of all modern crocodilians. Also local arts and crafts, town relics and theatrette; St Mary St. ***Whitman's Museum:*** photographic exhibition documents Isisford's history; St Mary St. ***Barcoo Weir:*** excellent spot for fishing and bush camping.

Sheep and Wool Show: May. ***Fishing Competition and Festival:*** July. ***Horse and Motorbike Gymkhana:*** June/July. ***Isisford Ross Cup:*** horseracing; Oct.

Oma Waterhole: popular for fishing and watersports, and home to a fishing competition and festival mid-year (subject to rains); 15 km SW. ***Idalia National Park:*** 62 km SE; *see Blackall.*

See also OUTBACK, p. 428

Julia Creek

Pop. 368
Map ref. 659 H11 | 661 J3

Cnr Burke and Quarrell sts; (07) 4746 7690; www.mckinlay.qld.gov.au

 99.5 FM Rebel, 567 AM ABC North West QLD

Julia Creek became known as 'The Gateway to the Gulf' after the road to Normanton was sealed in 1964. Pioneer Donald MacIntyre was the first to settle when he established Dalgonally Station in 1862. Julia Creek is actually named after his niece. A monument to Donald and his brother Duncan (an explorer who led a search for Leichhardt) can be seen by the grave site on the station boundary. The district's main industries are cattle, sheep and mining. It is also home to a rare and endangered marsupial, the Julia Creek dunnart, a tiny nocturnal hunter found only within a 100-kilometre radius of town.

Duncan MacIntyre Museum: local pioneering and cattle history; Burke St. ***Opera House:*** photographic display of McKinlay Shire; closed weekends; Julia St. ***Historical walk:*** stroll the 38 signposted historical sites around the town; map from visitor centre.

Dirt and Dust Festival: Apr. ***Horseracing:*** Apr–Aug. ***Campdraft and Rodeo:*** May, June and Aug. ***Cultural Capers:*** Oct.

WW II bunkers: remains of concrete bunkers used to assist navigation of allied aircraft; western outskirts near airport. ***Punchbowl waterhole:*** popular area for swimming and fishing on the Flinders River; 45 km NE. ***PROA Redclaw Farm:*** 12 ponds with over 16 000 redclaw; 75 km SE. ***Sedan Dip:*** popular swimming, fishing and picnicking spot; 100 km N.

See also OUTBACK, p. 428

Jundah

Pop. 93
Map ref. 663 J1

Dickson St; (07) 4658 6930; www.outbackholidays.tq.com.au

101.7 FM Triple C, 540 AM ABC Western QLD

Jundah is at the centre of the Channel Country and its name comes from the Aboriginal word for women. Gazetted as a town in 1880, for 20 years the area was important for opal mining, but lack of water caused the mines to close. The waterholes and channels of the Thomson River are filled with yabbies and fish, including yellow-belly and bream. The spectacular rock holes, red sand dunes and beauty of Welford National Park are the natural attractions of this outback town.

Jundah Museum: documents area's early pioneer heritage; Perkins St. ***Post Office:*** beautiful shopfront mural; Dickson St.

Race Carnival: Oct. ***Wooly Caulfield Cup:*** sheep races; Oct.

Welford National Park This park protects mulga lands, Channel Country and Mitchell grass downs – 3 types of natural vegetation in Queensland. See the rare earth homestead (1882) that is now listed by the National Trust (not open to the public) or go wildlife-watching to see pelicans and whistling kites at the many waterholes of the Barcoo River. There are 2 self-guide drives that start at the campground: one through the mulga vegetation to the scenic Sawyers Creek; the other a desert drive past the impressive red sand dunes. 4WD is recommended. Roads are impassable in wet weather; call 000 or Jundah police station on (07) 4658 6120 in emergency. (07) 4652 7333; 20 km S.

Stonehenge: named not for the ancient English rock formation, but for the old stone hut built for visiting bullock teams. Nearby on the Thomson River are brolgas and wild budgerigars; 68 km NE. ***Windorah:*** holds the International Yabby Race in Sept; 95 km S.

See also OUTBACK, p. 428

Karumba

Pop. 522
Map ref. 656 C6 | 659 G4

Karumba Library, Walker St, (07) 4745 9582; or call Normanton Visitor Information Centre, (07) 4745 1065; www.gulf-savannah.com.au

98.1 FM Rebel, 567 AM ABC North West QLD

Karumba is at the mouth of the Norman River in the Gulf Savannah. It is the easiest access point for the Gulf of Carpentaria, the key reason the town is the centre for the Gulf's booming prawn and barramundi industries. During the 1930s, the town was an important refuelling depot for the airships of the Empire Flying Boats, which travelled from Sydney to England. Fishing enthusiasts will enjoy the untouched waters of the gulf and nearby rivers that offer an abundance of fish.

Barramundi Discovery Centre: barramundi display and information; Riverside Dr. ***Ferryman cruises:*** tours include birdwatching, gulf sunset and night crocodile spotting; depart Karumba boat ramp; (07) 4745 9155. ***Charters and dinghy hire:*** discover the renowned fishing spots in the Gulf or on the Norman River on a charter or hire a dinghy; Karumba Port. ***Heritage walk:*** self-guide walk; brochure from Karumba library.

Market: Sunset Tavern; Sun. ***Fishing Competition:*** follows Easter event at Normanton; Apr. ***Norman River Duck Race:*** June.

End of the Road Motel: where the outback meets the ocean; Karumba Point; (07) 4745 9599.

Wetland region: extending 30 km inland from Karumba are the wetlands, habitat of the saltwater crocodile and several species of birds, including brolgas and cranes. ***Cemetery:*** early-settlement cemetery when Karumba was known as Norman Mouth telegraph station; 2 km N on road to Karumba Pt. ***Karumba Pt:*** boat hire available; note presence of saltwater crocodiles; 3 km N. ***Sweers Island:*** island in the Gulf with beaches, abundant birdlife, excellent fishing and caves; access is by boat or air; details from visitor centre.

See also GULF SAVANNAH, p. 427

Kenilworth

Pop. 238
Map ref. 647 E3 | 648 D11 | 653 M7

4 Elizabeth St; (07) 5446 0122; www.kenilworthguide.org.au

91.1 Hot FM, 612 AM ABC Brisbane

West of the Blackall Range in the Sunshine Coast hinterland is Kenilworth. This charming town is known for its handcrafted cheeses and excellent bushwalking. The spectacular gorges, waterfalls, creeks and scenic lookouts make Kenilworth State Forest a popular spot for bushwalking, camping and picnics.

Kenilworth Cheese Factory: tastings and sales of local cheeses; Charles St. ***Historical Museum:*** machinery and dairy display and audiovisual show; open 10am–3pm Sun or by appt; Alexandra St; (07) 5446 0581. ***Lasting Impressions Gallery:*** fine-art gallery; Elizabeth St. ***Artspace Gallery:*** local art on display; Elizabeth St.

Kenilworth Cheese, Wine and Food Festival: Easter. ***Kenilworth Celebrates:*** arts festival; Sept–Oct.

Another Time Another Place: 2 cabins beside a billabong; 2576 Eumundi–Kenilworth Rd; (07) 5472 3134.

Kenilworth State Forest This diverse park is in the rugged Conondale Ranges. The rainforest, tall open forest and exotic pines are home to birds and wildlife, including the threatened yellow-bellied glider. There are walks signposted, but a highlight is the steep 4 km return hike from Booloumba Creek to the summit of Mt Allan, where the forest and gorge views are breathtaking. Visit Booloumba Falls from the Gorge picnic area (3 km return) or picnic in the riverine rainforest at Peters Creek. Turn-off to the park is 6 km SW.

Kenilworth wineries: tastings and sales at boutique wineries; 4 km N. ***Kenilworth Bluff:*** steep walking track to lookout point; 6 km N. ***Conondale National Park:*** this small forest reserve west of the Mary River is suitable only for experienced walkers. Take the 37 km scenic drive, starting in the adjacent Kenilworth Forest, to enjoy the rugged delights of the park; (07) 5446 0925. ***Lake Borumba:*** picnics and watersports, and home to a fishing competition each Mar; 32 km NW.

See also SUNSHINE COAST, p. 419

Killarney

Pop. 833
Map ref. 551 C2 | 557 L1 | 653 L11

49 Albion St (New England Hwy), Warwick; (07) 4661 3122; www.southerndownsholidays.com.au

The attractive small town of Killarney is on the banks of the Condamine River close to the New South Wales border. It is appealingly situated at the foothills of the Great Dividing Range and is surrounded by beautiful mountain scenery.

Spring Creek Mountain Cafe & Cottages: local produce, great views; 1503 Spring Creek Rd; (07) 4664 7101.

Adjinbilly Rainforest Retreat Cabins: true eco-friendly accommodation; Lot 10 Adjinbilly Rd; (07) 4664 1599. ***Spring Creek Mountain Cafe & Cottages:*** cooking classes add interest; 1503 Spring Creek Rd; (07) 4664 7101.

Main Range National Park, Queen Mary Falls section The Queen Mary Falls section of this World Heritage–listed park is right on the NSW border. Most of the vegetation is open eucalypt, but in the gorge below the falls is subtropical rainforest. Follow the 2 km Queen Mary Falls circuit to the lookout for a stunning view of the 40 m falls and continue down to the rockpools at the base. If you are lucky, the rare Albert's lyrebird might be seen on the walk or the endangered brush-tailed rock-wallaby on the cliffs. 11 km E. *See also Boonah.*

Dagg's and Brown's waterfalls: stand behind Brown's Falls and see the 38 m Dagg's Falls; 4 km S. ***Cherrabah Homestead Resort:*** offers horseriding and fabulous bushwalking; 7 km S.

See also DARLING DOWNS, p. 418

Kingaroy

Pop. 7619
Map ref. 648 A10 | 653 L7

128 Haly St (opposite silos); (07) 4162 6272; www.kingaroy.qld.gov.au

96.3 FM Kingaroy, 747 AM ABC Southern QLD

Kingaroy is a large and prosperous town in the South Burnett region. The town's name derives from the Aboriginal word 'Kingaroori', meaning red ant. Found in the area, this unique ant has gradually adapted its colour to resemble the red soil plains of Kingaroy. The town is the centre for Queensland's peanut and navy-bean industries, and its giant peanut silos are landmarks. The region's relatively new wine industry is thriving with excellent boutique wineries close by. Kingaroy was also the home of Sir Johannes (Joh) Bjelke-Petersen, former premier of Queensland.

Heritage Museum: formerly the Kingaroy Power House, the museum depicts the history of Kingaroy under the themes of people, power and peanuts. Historical displays include machinery, photos and videos on the peanut and navy-bean industries; Haly St. ***Art Gallery:*** local and regional artists; open 10am–3pm Mon–Fri; Civic Sq, Glendon St. ***The Peanut Van:*** sales of local 'jumbo' peanuts; Kingaroy St. ***Apex Lookout:*** panoramic views of town; Carroll Nature Reserve, Fisher St.

Wine and Food in the Park Festival: Mar. ***Peanut Festival:*** includes the Strong Man Games; Aug.

Belltower Restaurant: modern Australian, adjoins distillery; Cnr Schellbachs and Haydens rds; (07) 4162 7000.

Bethany Cottages: Bjelke-Petersen family farm; 218 Peterson Dr; (07) 4162 7046. ***Deshon's Retreat:*** pet-friendly bush cottage; 164 Haydens Rd; (07) 4163 6688. ***Taabinga Homestead:*** a National Estate homestead; 7 Old Taabinga Rd; (07) 4164 5531. ***The Quilter's Rest:*** accommodation and quilting classes; 562 Weens Rd; (07) 4162 3987.

Bunya Mountains National Park This important park is part of the Great Dividing Range and protects the world's largest natural Bunya-pine forest. It is a significant Aboriginal site as many feasts were held here with the bunya nuts as the main fare. There are many walking trails for beginners and the experienced alike, including the short Bunya Bunya Track or the 8.4 km return Cherry Plain Track. Have a bush picnic and see the many butterflies or go spotlighting to glimpse owls and mountain possums. (07) 4668 3127; 58 km SW.

Aboriginal Bora Ring: preserved site; Cnr Reagan and Coolabunia rds;17.5 km SW. ***Mt Wooroolin Lookout:*** excellent views over Kingaroy's farmlands; 3 km W. ***Bethany:*** tour the Bjelke-Petersen's property and taste the famous Bjelke-Petersen pumpkin scones; open Wed and Sat, bookings essential; (07) 4162 7046; off Goodger Rd; 9 km SE. ***Wooroolin:*** quaint town with many heritage buildings including the Grant Hotel (1916). The Gordonbrook Dam is an excellent spot for picnics and birdwatching from the hides; 18 km N. ***Scenic aeroplane flights:*** over the Burnett region daily; (07) 4162 2629. ***Scenic glider flights:*** over surrounding area; (07) 4162 2191.

See also BRISBANE HINTERLAND, p. 415

Kuranda

Pop. 1612
Map ref. 650 D2 | 657 L6

 **Therwine St; (07) 4093 9311; www.kuranda.org**

99.5 FM Sea, 801 AM ABC Far North QLD

This small village is set in tropical rainforest on the banks of the Barron River north-west of Cairns. Its beautiful setting attracted a strong hippie culture in the 1960s and 1970s and, while it still has a bohemian feel, tourism is now the order of the day. There are many nature parks and eco-tourism experiences on offer, along with plenty of art and craft workshops, cafes and a daily market. Even transport to and from the town has been developed into an attraction – the Scenic Railway and Skyrail, both with jaw-dropping views over World Heritage–listed rainforest.

Scenic Railway and Skyrail The Scenic Railway is an engineering feat over 100 years old with tunnels, bridges and incredible views of Barron Falls. It begins in Cairns and ends 34 km later in the lush garden setting of Kuranda Station. Travel by rail on the way up and take the Skyrail on the way back (or vice versa), a journey via gondola across the treetops (ends at Caravonica Lakes, 11 km N of Cairns). (07) 4038 1555.

Butterfly Sanctuary: large enclosure, home to over 1500 tropical butterflies, including the blue Ulysses and the Australian birdwing, the country's largest butterfly; Rod Veivers Dr. ***Birdworld:*** over 50 species of birds, including the flightless cassowary and some endangered species; close to the heritage markets. ***Koala Gardens:*** Australian animals in a natural setting; close to the heritage markets. ***The Aviary:*** birds, frogs, snakes and crocodiles; Thongon St. ***Arts and crafts shops:*** including Kuranda Arts Cooperative, next to Butterfly Sanctuary, and Doongal Aboriginal Arts and Crafts on Coondoo St. ***Emu Ridge Gallery and Museum:*** unique dinosaur skeleton, fossil and gemstone museum; Therwine St. ***Australian Venom Zoo:*** visit the most venomous snakes and spiders in the world; Coondoo St. ***World Famous Honey House:*** free tastings and live bee displays; open 9am–3pm; Therwine St. ***River cruises:*** depart daily from the riverside landing below the railway station.

Kuranda Heritage Markets: Rod Veivers Dr; daily. ***Kuranda Original Markets:*** behind the Kuranda Market Arcade; daily. ***New Kuranda Markets:*** 40 shops and stalls on Coondoo St; daily. ***Kuranda Spring Festival:*** Sept.

Cedar Park Rainforest Resort: very fine affordable food; 250 Cedar Park Rd; (07) 4093 7892.

Cedar Park Rainforest Resort: 6 individually decorated rooms; 250 Cedar Park Rd; (07) 4093 7892. ***Kuranda Rainforest Accommodation Park:*** several accommodation options; 88 Kuranda Heights Rd; (07) 4093 7316.

Barron Gorge National Park Most people experience this park via the Scenic Railway or Skyrail, but those who want to get away from the crowds could set out on one of the park's bushwalking tracks leading into pockets of World Heritage wilderness. Perhaps you'll spot a Ulysses butterfly, cassowary or tree kangaroo. If you haven't already seen Barron Falls from the train or Skyrail, make your way to Barron Falls lookout 3 km from Kuranda. The falls are spectacular after rain. Access to the national park is via Cairns or Kuranda; (07) 4046 6600.

Kuranda Nature Park: wildlife, canoeing, swimming and rainforest tours aboard 4WD Hummers; 2 km W. ***Rainforestation Nature Park:*** rainforest tours, tropical fruit orchard, Pamagirri Aboriginal Dance troupe, Dreamtime walk, and a koala and wildlife park; 35 km E. ***Carrowong Wildlife Eco-tours:*** tours to see nocturnal, rare and endangered rainforest creatures; (07) 4093 7287.

See also CAIRNS & THE TROPICS, p. 425

Kynuna

Pop. 20
Map ref. 661 J5

Roadhouse, Matilda Hwy; (07) 4746 8683.

95.9 FM West, 540 AM ABC Western QLD

Kynuna is a tiny outback town famous for inspiring Banjo Paterson to write his iconic tune, Waltzing Matilda. It is said that Samuel Hoffmeister, at the time of the Shearers' Strike, drank his last drink at the Blue Heeler Hotel and then killed himself at Combo waterhole (south of town). This story stirred Paterson to write the now famous ballad.

Blue Heeler Hotel: famous hotel with illuminated blue heeler statue on the roof. ***Kynuna Roadhouse:*** obtain a mudmap from the roadhouse to find the original Swagman's Billabong.

 ***Surf Carnival:*** Apr.

Combo Waterhole Conservation Park The events described in *Waltzing Matilda* occurred in this park, which has waterholes lined with coolibahs. On the Diamantina River, the park is an important dry-weather wildlife refuge. See the Chinese-labour-constructed historic stone causeways from the 1880s or take the 40 min return waterhole walk. The turn-off to the park is 16 km E.

McKinlay Although the town is tiny, its famous Walkabout Creek Hotel was originally known as McKinlay Hotel and is the local watering hole in the film *Crocodile Dundee*, which starred famed Australian larrikin Paul Hogan. 74 km NW.

Swagman's Billabong: made famous in our national folk song; 2 km E.

See also OUTBACK, p. 428

Laidley

Pop. 2387
Map ref. 647 A12 | 653 M9

Lockyer Valley Tourist Information Centre, Jumbo's Fruit Barn Complex, Warrego Hwy, Hatton Vale (8 km N); (07) 5465 7642; www.laidley.qld.gov.au

 100.7 C FM, 747 AM ABC Southern QLD

Laidley is in the Lockyer Valley in the Brisbane hinterland. The surrounding agricultural farmland produces an abundance of fresh fruit and vegetables, which can be taste-tested at the country market. The quality of Laidley's produce comes from the region's fertile soil, which explains why the town is known as 'Queensland's country garden'.

Das Neumann Haus Restored and refurbished in 1930s style, this 1893 historic home is the oldest in the shire. It was built by a German immigrant, whose carpentry skills can be seen in the excellent details of the building. It houses a local history museum and exhibits local art and craft. Open Thurs–Sun; William St.

Historical walk: self-guide walk to heritage sites; contact visitor centre for details.

Village market: Main St; Fri. ***Country market:*** Ferrari Park; last Sat each month. ***Rodeo:*** Apr. ***Heritage Day:*** Apr. ***Laidley Show:*** July. ***Spring Festival:*** flower and art show, craft expo, street parade and markets; Sept. ***Art Exhibition:*** Oct. ***Christmas Carnival:*** Dec.

Branell Homestead B&B: luxurious homestead accommodation; 12 Paroz Rd; (07) 5465 1788. ***Frog Hollow Barnstay:*** guests enjoy the orchard; 19 Hughes Rd, Plainland; (07) 5465 6616.

Laidley Pioneer Village: original buildings from old township including blacksmith shop and slab hut; 1 km S. ***Narda Lagoon:*** flora and fauna sanctuary with picturesque suspension footbridge over lagoon; adjacent to pioneer village. ***Lake Dyer:*** beautiful spot for fishing, picnics and camping; 1 km W. ***Lake Clarendon:*** birdwatching area; 17 km NW. ***Laidley Valley Scenic Drive:*** attractive drive to the south of Laidley through Thornton.

See also BRISBANE HINTERLAND, p. 415

Landsborough

Pop. 2807
Map ref. 647 F5 | 648 E12 | 653 N8

Historical Museum, Maleny St; (07) 5494 1755; www.landsboroughtown.com.au

 90.3 FM ABC Coast QLD, 91.1 Hot FM

Landsborough is just north of the magnificent Glass House Mountains in the Sunshine Coast hinterland. It was named after the explorer William Landsborough and was originally a logging town for the rich woodlands of the Blackall Ranges.

Historical Museum: this excellent local museum documents the history of the shire through memorabilia, photographs and artefacts; open 9am–3pm Thurs–Mon; Maleny St. ***De Maine Pottery:*** award-winning clay pottery by Joanna De Maine; open Thurs–Mon; Maleny St.

 ***Market:*** School of Arts Memorial Hall; Sat.

Glass House Mountains National Park This park protects 8 rugged volcanic mountain peaks. The open eucalypt and mountain heath landscape is a haven for many threatened and endangered animals. Three tracks lead to mountain lookouts that provide panoramic views of the Sunshine Coast hinterland. Only experienced walkers should attempt climbing to any of the summits. (07) 5494 3983; 13 km SW.

Australia Zoo Once a small park, this zoo was made famous by 'The Crocodile Hunter', Steve Irwin, who was tragically killed by a stingray in 2006. Originally developed by Irwin's parents, the complex is now over 20 ha and home to a wide range of animals. See the otters catching fish, the birds of prey tackling the skies, the ever-popular crocodile demonstrations or feed the kangaroos by hand in the Kids Zoo. The complex also has important breeding programs for threatened and endangered species. 4 km S.

Dularcha National Park: scenic park with excellent walks. 'Dularcha' is an Aboriginal word describing blackbutt eucalyptus country; (07) 5494 3983; 1 km NE. ***Big Kart Track:*** largest outdoor go-kart track in Australia, open for day and night racing and includes the 'Bungee Bullet'; 5 km N. ***Beerburrum State Forest:*** short walks and scenic drives to lookouts; access from Beerburrum; 11 km S.

See also SUNSHINE COAST, p. 419

Laura

Pop. 225
Map ref. 657 J3

Quinkan Centre, Peninsula Development Rd; (07) 4060 3457; www.quinkancc.com.au

 92.5 FM 4CA, 106.1 FM ABC Far North QLD

Laura is a tiny town in Far North Queensland that boasts only a few buildings, including the quaint old Quinkan Pub nestled in the shade of mango trees. The area to the south-east of town is known as Quinkan country after the Aboriginal spirits depicted at the incredible Split Rock and Gu Gu Yalangi rock-art sites. Every two years Laura hosts possibly the biggest Indigenous event on Australia's calendar – the Laura Dance and Cultural Festival.

Quinkan Centre: photos and relics are on display here. Guided tours, which include rock-art sites, can also be booked from the centre; Peninsula Development Rd; (07) 4060 3457.

Laura Aboriginal Dance and Cultural Festival: around 25 Cape York and Gulf communities gather at a traditional meeting ground by the Laura River. Traditional dance, music, and art and craft feature at the 3-day event; odd-numbered years, June. ***Laura Annual Race Meeting:*** race meeting on river flat that has run since 1897, and includes dances and rodeo; June.

Split Rock and Gu Gu Yalangi rock-art sites These are Queensland's most important Aboriginal art sites. Hidden behind a tangle of trees in the chasms and crevices of the sandstone escarpment, a diorama of Aboriginal lore and culture unfolds. The Quinkan spirits – the reptile-like Imgin and stick-like Timara – can be found hiding in dark places. There are also dingoes, flying foxes, kangaroos, men, women and many hundreds of other things, both obvious and mysterious. More sites exist nearby, though only these 2 are accessible to the public. Bookings through the Quinkan Centre; 12 km S.

Lakefield National Park This park is Queensland's second largest and a highlight of any visit to the Cape. The large rivers and waterholes are excellent for fishing and boating in the dry season, but become inaccessible wetlands in the wet season. In the south is the Old Laura Homestead, once en route to the Palmer River Goldfields, and in the north are plains dotted with spectacular anthills. See the threatened gold-shouldered parrot and spectacled hare-wallaby in the rainforest fringes of the Normanby and Kennedy rivers. 4WD is recommended. Access only in the dry season (Apr–Nov); (07) 4069 5777; entrance is 27 km N.

Lakeland Downs: bananas and coffee are the main fare cultivated from the area's rich volcanic soil; 62 km S.

See also CAPE YORK, p. 426

Longreach

see inset box on next page

Mackay

see inset box on page 461

Maleny

Pop. 1299
Map ref. 647 E4 | 648 E12 | 653 N7

25 Maple St; (07) 5499 9033; www.brbta.com

 91.1 Hot FM, 612 AM ABC Brisbane

continued on p. 459

LONGREACH

Pop. 2977

Map ref. 654 B10 | 661 N10

 **Qantas Park, Eagle St; (07) 4658 3555; www.longreach.qld.gov.au**

104.5 FM West, 540 AM ABC Western QLD

On the Thomson River, Longreach is the largest town in central-west Queensland. It epitomises the outback and features the renowned Australian Stockman's Hall of Fame, devoted to the outback hero. In 1922, Longreach became the operational base for Qantas and remained so until 1934. The world's first Flying Surgeon Service started from Longreach in 1959.

Australian Stockman's Hall of Fame and Outback Heritage Centre This impressive centre was developed as a tribute to the men and women who opened up outback Australia for settlement, industry and agriculture. It deals with everything from Australia's Indigenous heritage and the challenges of outback education and communication to the life of the modern-day stockman. Highlights include a photo gallery, an old blacksmith's shop and a 1920s kitchen. Open 9am–5pm daily; Landsborough Hwy; (07) 4658 2166.

Qantas Founders Outback Heritage Museum This modern museum details the commercial flight history of the second-oldest airline in the world. Explore the restored original 1922 hangar with displays on early flights and a replica Avro 504K. Or visit the exhibition hall to see how flying has evolved over the last century. Longreach Airport, Landsborough Hwy; (07) 4658 3737.

Powerhouse Museum: displays of old agricultural machinery, power station and local history museum; open 2–5pm Apr–Oct; Swan St. ***Botanical Gardens:*** walking and cycling trails; Landsborough Hwy. ***Banjo's Outback Theatre and Woolshed:*** bush poetry, songs and shearing in shearing shed; near airport. ***Qantas Park:*** replica of original Qantas booking office, it now houses the information centre; Eagle St. ***Heritage buildings:*** highlights include the courthouse (1892) in Galah St and the railway station (1916) in Sir Hudson Fysh Dr. ***Kinnon and Co:*** offers horse-drawn adventure tours; bookings through the Station Store; (07) 4658 2006.

Easter in the Outback: Easter. ***Outback Muster and Drovers Reunion:*** May. ***Rodeo:*** July. ***Fishing Competition:*** Aug. ***Horseracing:*** Nov.

Albert Park Motor Inn: 4-star rooms, garden setting; Cnr Stork and Ilfracombe rds; (07) 4658 2411.

Lochern National Park The Thomson River plays an important role in this national park as its many waterholes are a sanctuary for a variety of birds. Camp beside the billabong in true outback style at Broadwater Hole and watch the brolgas and pelicans. For visitors passing through, drive the 16 km Bluebush Lagoon circuit to see the natural attractions of this park and glimpse Australian favourites – the kangaroo and emu. Visitors must be self-sufficient. Check road conditions before departing – roads may be impassable in wet weather; (07) 4652 7333. Turn-off 100 km S onto Longreach–Jundah Rd, then further 40 km.

Thomson River: fishing, bushwalking and swimming; 4.6 km NW. ***Ilfracombe:*** once a transport nucleus for the large Wellshot Station to the south, now an interesting outdoor machinery museum runs the length of this town on the highway. Another feature of town is the historic Wellshot Hotel, with a wool press bar and local memorabilia; 27 km E. ***Starlight's Lookout:*** said to be a resting spot of Captain Starlight. Enjoy the scenic view; 56 km NW. ***Sheep and cattle station tours:*** visit local stations to try out mustering and shearing; contact visitor centre for details. ***Thomson River cruises:*** contact visitor centre for details.

See also OUTBACK, p. 428

[CATTLE MUSTERING NEAR LONGREACH]

A steep road climbs from the coast to Maleny, at the southern end of the Blackall Range. The surrounding area is lush dairy country, although farmland is increasingly being sold for residential development. The town's peaceful community lifestyle and picturesque position, with views to the coast and Glass House Mountains, make it popular with artists.

Arts and crafts galleries: excellent quality galleries throughout town.

Handcraft Markets: Community Centre, Maple St; Sun. ***Chainsaw to Fine Furniture Expo:*** May. ***Spring Fair and Flower Show:*** Sept. ***Scarecrow Festival:*** Sept. ***Fine Art Show:*** Sept. ***Festival of Colour:*** Oct. ***Christmas Show:*** Nov.

Terrace of Maleny: seafood with style; Cnr Mountain View and Landsborough Maleny rds; (07) 5494 3700. ***Le Relais Bressan:*** French Provincial; 344 Flaxton Dr, Flaxton; (07) 5445 7157. ***Tree Houses Restaurant:*** imaginative international food; 5 Kondalilla Falls Rd, Montville; (07) 5445 7650 or 1800 444 350.

Bendles Cottages: 5 cottages with rural views; 937 Montville Rd; (07) 5494 2400. ***Eyrie Escape:*** 3 suites and resort facilities; 316 Brandenburg Rd, Bald Knob; 0414 308 666. ***The Frog House:*** luxury ensuite rooms with views; 73 Mountain View Rd; (07) 5499 9055.

Kondalilla National Park This park has scenic walks, subtropical rainforest and the spectacular Kondalilla Falls. The 4.6 km return Falls Circuit track passes rockpools to the falls, which drop 90 m. (07) 5494 3983; 21 km N via Montville.

Mary Cairncross Park: this beautiful park was donated to the community in 1941 as protected rainforest after the fierce logging days of the early 1900s. Walk through the rainforest to see superb panoramic views; 7 km SE. ***Baroon Pocket Dam:*** popular spot for fishing and boating. Follow the boardwalks through rainforest; North Maleny Rd; 8 km N. ***Montville:*** main street lined with cafes, giftshops, potteries and art and craft galleries. The town also has a growing wine industry; 16 km NE. ***Flaxton:*** charming tiny village surrounded by avocado orchards. Visit the quality art and craft galleries, Flaxton Grove Winery and miniature English village; 19 km NE. ***Flaxton Barn:*** model Swiss and German railway with tearooms; 21 km N via Flaxton. ***Maleny–Blackall Range Tourist Drive:*** this 28 km scenic drive is one of the best in south-east Queensland. Drive north-east from Maleny through to Mapleton, stopping off at museums, antique shops, fruit stalls and tearooms along the way, as well as taking in spectacular views. The drive can be extended to Nambour.

See also SUNSHINE COAST, p. 419

Mareeba

Pop. 6806
Map ref. 650 D2 | 651 D7 | 657 L6

Heritage Museum, Centenary Park, 345 Byrnes St; (07) 4092 5674; www.mareebaheritagecentre.com.au

92.3 FM Rhema, 720 AM ABC Far North QLD

Mareeba was the first town settled on the Atherton Tableland by pastoralist John Atherton in 1877. Tobacco production began in the 1950s, but was deregulated only a few years ago. Today the area produces mangoes, coffee and sugarcane. The morning balloon flights over the tableland are spectacular.

Heritage Museum: local history exhibits and information centre; Centenary Park, Byrnes St. ***Art Society Gallery:*** Centenary Park. ***Barron River Walk:*** walk along the banks to swimming hole. ***Bicentennial Lakes:*** park with plantings to encourage wildlife. Explore the park on the walking tracks and bridges; Rankine St. ***The Coffee Works:*** see the production of coffee and taste-test the results; 136 Mason St. ***The Australian Coffee Centre:*** this complex offers a 54-seat cinema, coffee labratory and restaurant. Daily tours take you through the coffee plantation and include tastings. Open 9am–5pm daily; 136 Ivicevic Rd.

Market: Centenary Park; 2nd and 5th Sat each month (if applicable). ***Great Wheelbarrow Race:*** Mareeba to Chillagoe; May. ***Rodeo:*** July. ***Multicultural Festival:*** Aug. ***Walkamin Country Music Festival:*** Oct.

Biboohra Bush Retreat B&B: homestead near wetland reserve; 326 Pickford Rd, Biboohra; (07) 4093 2787. ***Jabiru Safari Lodge:*** exceptional nature experience; Pickford Rd, Biboohra; (07) 4093 2514 or 1800 788 755.

Mareeba Tropical Savanna and Wetland Reserve This conservation reserve of over 2400 ha and 12 lagoons is home to birds, mammals, fish and freshwater crocodiles. Take a self-guide trail by hiring a timber canoe, a tour cruise or the guided 'Twilight Reserve Safari' (with cheese and wine afterwards). Open Apr–Dec (dry season); tour bookings on (07) 4093 2514; turn-off at Biboohra (7 km N).

Mako Trac International Racetrack: Australia's best track and fastest go karts. Also 18-hole minigolf course; Springs Rd, 2 km W. ***de Brueys Boutique Wines:*** taste world-class tropical fruit wines, liqueurs and ports; Fichera Rd, 4 km E. ***The Beck Museum:*** aviation and military collection; Kennedy Hwy; 5 km S. ***Warbird Adventures Adventure flights:*** experience the thrill of a flight in a real Warbird; museum open Wed–Sun; Kennedy Hwy, 5 km S. ***Mango Winery:*** produces white wine from 'Kensington Red' mangoes; Bilwon Rd, Biboohra; 7 km N. ***Granite Gorge:*** impressive boulder and rock formation; 12 km SW off Chewko Rd. ***Emerald Creek Falls:*** walk 1.9 km and see the water as it tumbles down the mountain between massive boulders; Cobra Rd; 12 km SE. ***Davies Creek National Park:*** walk the 1.1 km Davies Creek Falls circuit to see the falls crashing over boulders; (07) 4091 1844; 22 km E. ***Dimbulah:*** small town with museum in restored railway station; 47 km W. ***Tyrconnell:*** historic mining town with tours of goldmine; 68 km NW via Dimbulah. ***Scenic balloon flights:*** over Atherton Tableland; details available from visitor centre.

See also CAIRNS & THE TROPICS, p. 425

Maroochydore

Pop. 17 500
Map ref. 647 G4 | 648 F11 | 653 N7

Cnr Sixth Ave and Melrose Pde; (07) 5459 9050 or 1800 882 032; www.discovermaroochy.com.au

91.9 FM Sea, 612 AM ABC Brisbane

A popular beach resort, Maroochydore is also the business centre of the Sunshine Coast. The parklands and birdlife on the Maroochy River and the excellent surf beaches began to attract a growing tourist interest in the 1960s, which has only increased since. An incredible range of watersports is available.

Maroochy River: enjoy diverse birdlife and parklands on the southern bank with safe swimming. ***Endeavour Replica:*** replica of Captain Cook's ship; David Low Way. ***The Esplanade:*** atmospheric strip of cafes, restaurants and clothing stores.

QUEENSLAND

Market: Cnr Fishermans Rd and Bradman Ave; Sun mornings. ***Cotton Tree Street Markets:*** King St; Sun mornings.

Ebb: riverside location, contemporary menu; 6 Wharf St; (07) 5452 7771. ***Boat Shed:*** laid-back eatery with day beds; The Esplanade at Cotton Tree; (07) 5443 3808. ***The Wine Bar Restaurant:*** 40 wines by the glass, modern Australian; 4–8 Duporth Ave; (07) 5479 0188.

Amytis Gardens Retreat & Spa: 4 luxury mountain cabins; 51 Malones Rd; (07) 5450 0115. ***Novotel Twin Waters Resort:*** many types of accommodation; Ocean Dr, Twin Waters; (07) 5448 8000 or 1800 072 277.

Mt Coolum National Park Located above the surrounding sugarcane fields, the mountain offers cascading waterfalls after rain. The park is generally undeveloped, but take the rough 800 m trail to the summit to be rewarded with panoramic views of the coast. (07) 5447 3243; 13 km N via Marcoola.

Nostalgia Town: emphasises humour in history on a train ride through fantasy settings. Markets every Fri and Sat; Pacific Paradise; 7 km NW. ***Bli Bli:*** attractions include medieval Bli Bli Castle, with dungeon, torture chamber and doll museum, the 'Ski 'n Surf' waterski park and the Aussie Fishing Park. Take a cruise through Maroochy River wetlands; 9 km NW. ***Marcoola:*** coastal town with quiet beach; 11 km N. ***Coolum Beach:*** coastal resort town with long sandy beach; 17 km N. ***Eco-Cruise Maroochy:*** cruises to Dunethin Rock through sugarcane fields; details from visitor centre.

See also SUNSHINE COAST, p. 419

Maryborough

Pop. 21 499
Map ref. 648 E6 | 653 N5

City Hall, Kent St; (07) 4190 5742 or 1800 214 789; www.maryborough.qld.gov.au

103.5 FM Mix, 855 AM ABC Wide Bay

Maryborough is an attractive city on the banks of the Mary River. Its fine heritage buildings and famous timber Queenslander architecture date back to the early years of settlement, when Maryborough was a village and port. Maryborough Port was an important destination in the mid-1800s as over 22 000 immigrants arrived from Europe. Today the Mary River is a popular spot for relaxed boating and some good fishing.

Bond Store Museum: located in the Portside Centre, this museum documents the history of the region from a river port to the present day; open 9am–4pm Mon–Fri, 10am–1pm Sat–Sun; Wharf St. ***Customs House Museum:*** also in the Portside Centre, this important cultural heritage museum depicts the area's industries and early immigration and Kanaka history; Wharf St. ***Military and Colonial Museum:*** fascinating look at local military and colonial history; open 9am–3pm daily; Wharf St. ***Mary Poppins statue:*** statue of the famed and magical nanny in honour of author Pamela Lyndon Travers, who was born in the town in 1899; Cnr Kent and Richmond sts. ***Central Railway Station, Mary Ann:*** replica of Queensland's first steam engine with rides in Queens Park every Thurs, market day, and last Sun each month; Lennox St. ***Queens Park:*** overlooks the Mary River; Cnr Lennox and Bazaar sts. ***Elizabeth Park:*** extensive rose gardens; Kent St. ***ANZAC Park:*** includes Ululah Lagoon, a scenic waterbird sanctuary where black swans, wild geese, ducks and waterhens may be handfed; Cnr Cheapside and Alice sts. ***Heritage walk:*** self-guide walk past 22 historic buildings, including the impressive City Hall, St Paul's bell tower (with pealing bells) and National Trust–listed Brennan and Geraghty's Store. The walk starts at City Hall, Kent St; brochure available from visitor centre. ***Heritage drive:*** a highlight is the original site of Maryborough (until 1885), where a series of plaques document its history. Drive starts at City Hall; brochure available from visitor centre. ***Ghost tours:*** discover the town's ghostly past on tours that run once a month; details from visitor centre.

Heritage markets: Cnr Adelaide and Ellena sts; Thurs. ***World's Greatest Pub Crawl:*** June. ***Mary Poppins Festival:*** July. ***Technology Challenge:*** Sept. ***Maryborough Masters Games:*** Sept/Oct.

Teddington Weir: popular for watersports; 15 km S. ***Tiaro:*** excellent fishing for Mary River cod. See the historic Dickabram Bridge over the river, and nearby Mt Bauple National Park; 24 km SW. ***Tuan Forest:*** bushwalking; 24 km SE. ***Fraser Island:*** World Heritage–listed sand island to the east of town; *see Hervey Bay.*

See also FRASER ISLAND & COAST, p. 420

Miles

Pop. 1164
Map ref. 653 I7

Historical Village, Murilla St; (07) 4627 1492; www.murilla.qld.gov.au

94.5 FM Rebel, 747 AM ABC Southern QLD

Ludwig Leichhardt passed through this district and named the place Dogwood Crossing. The town was later renamed Miles in honour of a local member of parliament. After spring rains, this pocket of the Darling Downs is ablaze with wildflowers.

Dogwood Crossing @ Miles This modern cultural centre combines history and art in an innovative space. The excellent Wall of Water displays imagery and stories of the local people – a novel way of discovering the personal history of Miles. The art gallery has changing exhibitions from local and regional artists. There is an IT centre and library also on-site. Murilla St.

Historical Village: a pioneer settlement with all types of early buildings, a war museum, shell display and lapidary exhibition; Murilla St. ***Wildflower excursions:*** some of the most beautiful wildflowers in Australia; details from visitor centre.

St Luke's market: Dawson St; 2nd Sat each month. ***Miles Show:*** May. ***Back to the Bush:*** includes the Wildflower Festival; Sept. ***Beef, Bells and Bottle Tree Festival:*** massive month-long celebration including country music bash, bush poetry and fishing competition; even-numbered years, Sept.

Condamine: small town known for inventing the Condamine Bell, a bullfrog bell that, hung around bullocks, can be heard up to 4 km away. A replica and history display are in Bell Park. There is excellent fishing on the Condamine River, and the town holds a famous rodeo in Oct; 33 km S. ***The Gums:*** tiny settlement with historic church and nature reserve; 79 km S. ***Glenmorgan:*** the Myall Park Botanical Gardens; 134 km SW.

See also DARLING DOWNS, p. 418

Millaa Millaa

Pop. 289
Map ref. 650 D4 | 651 E11 | 657 L7

Millaa Millaa Tourist Park, Millaa Millaa–Malanda Rd; (07) 4097 2290; www.tablelands.org/millaa-millaa.html

continued on p. 462

MACKAY

Pop. 72 848

Map ref. 649 G8 | 655 K5

Mackay Tourism, Old Town Hall, 320 Nebo Rd; (07) 4952 2677 or 1300 130 001; www.mackayregion.com

101.9 FM 4MK, 630 AM ABC North QLD

The city of Mackay is an intriguing blend of 1800s and Art Deco heritage buildings mixed with modern-day architecture. Sugar was first grown here in 1865, and Mackay now produces around one-third of Australia's sugar crop and has the world's largest bulk-sugar loading terminal.

Artspace Mackay This modern art gallery forms part of the Queensland Heritage Trails Network. The museum documents Mackay's history with an interesting permanent exhibition entitled 'Spirit and Place: Mementoes of Mackay'. The gallery section has changing exhibits of both local and international artwork. Closed Mon; Gordon St.

Old Town Hall: houses the Heritage Interpretive Centre with displays on Mackay's history and visitor information; Sydney St. ***Regional Botanic Gardens:*** follow the picturesque boardwalk over a lagoon and explore specialised gardens of plants of the central coast; Lagoon St. ***Queens Park:*** includes the Orchid House that displays over 3000 orchids; Goldsmith St. ***Heritage walk:*** self-guide walk of 22 heritage buildings; brochure available from visitor centre. ***City cemetery:*** 1.5 km heritage walk; Greenmount Rd. ***Horizons Mosaic:*** locally constructed mosaics of the Mackay region's natural attractions; Victoria St. ***Mackay Museum:*** local artefacts with a research area and souvenirs for sale; open 10am–2pm every Thurs and 1st Sun each month; Casey Ave. ***Great Barrier Reef tours:*** snorkelling, diving, reef fishing and sailing tours; details at visitor centre.

Market: showgrounds, Milton St; Sat mornings. ***City Centre Markets:*** Victoria St; Sun mornings. ***Walkers Markets:*** Harbour Rd; weekends. ***Seaforth Markets***: Palm Ave; Sun mornings. ***Troppo Treasure Markets:*** showgrounds; 2nd Sun each month. ***Seaforth Fishing Classic:*** Apr. ***Festival of the Arts:*** July. ***River to Reef Festival:*** Aug. ***Sugartime Festival:*** Sept.

Church on Palmer: international cuisine; 15 Palmer St; (07) 4944 1466. ***George's Thai on the Marina:*** top-notch Thai food; Mulherin Dr; (07) 4955 5778. ***Sorbello's Italian Restaurant:*** affordable Italian; 166 Victoria St; (07) 4957 8300.

Clarion Hotel Mackay Marina: 79 suites with views; Mulherin Dr; (07) 4955 9400 or 1800 386 386. ***Lantern Motor Inn and Restaurant:*** 16 units in tropical gardens; 149–151 Nebo Rd; (07) 4951 2188. ***Ocean International:*** beachfront hotel; 1 Bridge Rd; (07) 4957 2044 or 1800 635 104. ***Comfort Resort Blue Pacific:*** waterfront location; 26 Bourke St, Blacks Beach; (07) 4954 9090 or 1800 808 386. ***Broken River Mountain Resort:*** a platypus haven; Eungella Dam Rd, Eungella National Park; (07) 4958 4000.

Eungella National Park This ecologically diverse park is home to some unusual plants and animals including the Eungella gastric brooding frog and the Mackay tulip oak. A highlight of the visit is seeing the platypus in Broken River from the viewing

[SUGARCANE HARVESTING NEAR MACKAY]

deck. Visit the Finch Hatton Gorge section of the park where the waterfalls, swimming holes and walking tracks are breathtaking. For the more adventurous, try sailing through the rainforest canopy on the eco-tour Forest Flying. Finch Hatton Gorge turn-off is just before Eungella. Broken River is 6 km S of Eungella; (07) 4944 7800 (same phone number for national parks below).

Northern beaches: visit fabulous beaches including Harbour (patrolled, fishing), Town, Blacks (area's longest beach), Bucasia, Illawong, Eimeo, Lamberts (excellent lookout) and Shoal Point. ***Farleigh Sugar Mill:*** tours during crushing season, July–Oct; 15 km N. ***That Sapphire Place:*** sapphire display and gem-cutting demonstrations; 20 km W. ***Greenmount Homestead:*** restored historic home with pioneering history museum; open 9.30am–12.30pm, closed Sun; Walkerston; 20 km W. ***Homebush:*** small town offers art and craft gallery, orchid farms and self-drive tour through historic area; 25 km S. ***Melba House:*** the home where Dame Nellie Melba spent the 1st year of her married life. It is now home to Pioneer Valley Visitor Centre; Marian; 28 km W. ***Mirani:*** small town in the Pioneer Valley with museum and Illawong Fauna and Flora Sanctuary; 33 km W. ***Kinchant Dam:*** popular spot for watersports and fishing; 40 km W. ***Hibiscus Coast:*** comprises the quaint coastal towns of Seaforth, Ball Bay and Cape Hillsborough; 43 km NW. ***Cape Hillsborough National Park:*** scenic coastal park with sandy beaches, tidal rockpools and walking trails; access from Hibiscus Coast, 44 km NW. ***Smith Islands National Park:*** the largest island of this group is Goldsmith, with long sandy beaches and snorkelling in surrounding reefs; access is by private boat or water taxi; 70 km NE via Seaforth. ***Cumberland Island National Park:*** this group of islands off Mackay coast is popular for boating and also an important rookery for flat back and green turtles; access is by private boat or water taxi. ***Brampton Island:*** resort island at southern end of Whitsunday Passage with pristine beaches and coral reef; sea and air access. ***Nebo:*** historic town of Mackay region, with local artefacts and pioneering history at the Nebo Shire Museum. Nebo holds a Campdraft in Sept; 93 km SW.

See also THE MID-TROPICS, p. 424

QUEENSLAND

 97.9 FM Hot, 720 AM ABC Far North QLD

Millaa Millaa is at the southern edge of the Atherton Tableland, and is central to a thriving dairy industry. The 17-kilometre Waterfall Circuit and rainforest-clad Wooroonooran National Park are just two of the natural attractions that bring visitors to the mild climate of this town.

Eacham Historical Society Museum: documents history of local area, with special interest in dairy and timber industries; opening times vary, contact museum for more information; Main St; (07) 4097 2147.

Wooroonooran National Park, Palmerston section More than 500 types of rainforest trees means the landscape in this park is both diverse and breathtaking. Walks include a 5 km return track leading to spectacular gorge views at Crawford's Lookout and a short 800 m track to glimpse the Tchupala Falls. A popular activity in the park is whitewater rafting on the North Johnstone River; companies include RnR Adventures; bookings on (07) 4041 9444. Access via Palmerston Hwy; (07) 4061 5900; 25 km SW.

Waterfall Circuit This 17 km circuit road includes the Zillie, Ellinjaa and Mungalli falls, as well as the popular Millaa Millaa Falls, a great spot for swimming, with walks leading to other waterfalls. The circuit road is mostly sealed and the route leaves and rejoins Palmerston Hwy east of town.

Millaa Millaa Lookout: panoramic views of tablelands and national parks; 6 km W. ***Misty Mountains walking trails:*** short- and long-distance tracks through World Heritage–listed Wet Tropics, many of which follow traditional Aboriginal paths of the Jirrbal and Mamu people; details from visitor centre. ***Hillside Eden Garden:*** award-winning garden featured on the television program *Better Homes and Gardens*, with stunning camellias and rhododendrons; Tarzali, 20 km N.

See also CAIRNS & THE TROPICS, p. 425

Miriam Vale

Pop. 362
Map ref. 653 L2

Discovery Coast Information Centre, Roe St (Bruce Hwy); (07) 4974 5428; www.gladstonerc.qld.gov.au

 93.5 FM Hot, 855 AM ABC Wide Bay

Miriam Vale lies south-east of Gladstone, in the hinterland of the 'Discovery Coast'. The town is renowned for its charming hospitality, its historic fig trees in the main street and its mud-crab sandwiches. The hinterland and coastal national parks are ideal places for bushwalking, four-wheel driving and horseriding.

Edge on Beaches Resort: 4 pools, many beaches; Cnr Captain Cook Dr and Beaches Village Circuit, Agnes Water; (07) 4902 1200. ***The Beach Shacks:*** totally tropical and relaxed; 578 Captain Cook Dr, Seventeen Seventy; (07) 4974 9463.

Deepwater National Park A diverse vegetation covers this coastal park, including paperbark forests, swamp mahogany, Moreton Bay ash and subtropical rainforest. Walk along the sandy beaches or enjoy the birdlife of the freshwater stream, Deepwater Creek. Bush camp at Wreck Rock, and explore the rockpools. 4WD access only; (07) 4131 1600; 63 km NE via Agnes Water.

Eurimbula National Park: rugged coastal park with walks along the beach, canoeing on Eurimbula Creek, fishing, and scenic views from Ganoonga Noonga Lookout. 4WD recommended; (07) 4131 1600; access between Miriam Vale and Agnes Water; 50 km NE. ***Agnes Water:*** this coastal town has the most northerly surfing beach in Queensland. A local history museum includes documents of Cook's voyage in 1770. The town holds the surfing competition 1770 Longboard Classic each Mar; 57 km NE. ***Seventeen Seventy:*** Captain Cook, while on his discovery voyage, made his second landing on Australian soil at the town site. Today the seaside village has the Joseph Banks Environmental Park and is the departure point for daytrips and fishing charters to the Great Barrier Reef and Lady Musgrave Island; 63 km NE.

See also CAPRICORN, p. 423

Mission Beach

Pop. 517
Map ref. 650 E5 | 657 M8

Porter Promenade; (07) 4068 7099; www.missionbeachtourism.com

 88.5 FM 4KZ, 630 AM ABC North QLD

Mission Beach is named for the Aboriginal mission established in the area in 1914. The beach that features in the town's name is a 14-kilometre-long strip of golden sand fringed by coconut palms and World Heritage–listed wet tropical rainforest. Artists, potters, sculptors and jewellers have settled in the area, now reliant on the strong tourism industry of the coast.

Porter Promenade: woodcarving exhibition and rainforest arboretum; next to visitor centre. ***Ulysses Link Walking Trail:*** this 1.2 km pathway along the foreshore features local history, sculptures and mosaics. ***Great Barrier Reef tours:*** cruises and day tours to islands and reefs depart from Clump Point Jetty daily. ***Boat, catamaran and jetski hire:*** details from visitor centre.

Market: Porter Promenade; 1st Sat and 3rd Sun each month. ***Monster Markets:*** Recreation Centre, Cassowary Dr; last Sun each month (Easter–Nov). ***Banana Festival:*** Aug. ***Aquatic Festival:*** Oct.

Lillypads: 2 tropical garden chalets; 1375 Cassowary Dr; (07) 4088 6133. ***Mission Beach Resort:*** all rooms lead to a pool; Wongaling Beach Rd; (07) 4088 8288 or 1800 079 024. ***Sejala on the Beach:*** luxuriously rustic beach huts; Pacific Pde; (07) 4088 6699.

Family Islands The Family Islands National Park protects this 14 km stretch of islands. The most northerly of the group, Dunk Island, is a popular holiday destination with spectacular forest, rainforest and 14 km of walking tracks. The resort is private but camping is available. The less-developed islands of Wheeler and Coombe are perfect for bush camping (visitors must be self-sufficient). There is air service to Dunk Island from Cairns and Townsville or ferry from Clump Point. Islands are accessible by water taxi from Wongaling Beach.

Clump Mountain National Park: this scenic park boasts remnant lowland rainforest, an important habitat for the southern cassowary. A highlight is the 4 km Bicton Hill Track to the summit lookout over Mission Beach and coast; just north of Mission Beach on Bingil Bay Rd; (07) 4722 5224. ***Historic cairn:*** commemorates the ill-fated 1848 Cape York expedition of Edmund Kennedy; South Mission Beach Espl. ***Aboriginal Midju:*** display of Aboriginal culture; adjacent to cairn. ***Wet Tropics walking trails:*** the area around Mission Beach offers spectacular rainforest walks, including the Lacy Creek Forest circuit (1.2 km) in the major cassowary habitat of Tam O'Shanter National Park, the Kennedy Trail (7 km) past lookouts and along beaches, and the trails in Licuala State Forest; brochure available from visitor centre. ***Adventure activities:*** include tandem parachuting, kayak trips and whitewater rafting; details from visitor centre.

See also CAIRNS & THE TROPICS, p. 425

Mitchell

Pop. 942
Map ref. 652 E6

Great Artesian Spa Complex, 2 Cambridge St; (07) 4623 8171; www.visitmitchell.com.au

 102.9 FM Outback Radio 4VL, 711 AM ABC Western QLD

Mitchell, a gateway to the outback, is on the banks of the Maranoa River at the western edge of the Darling Downs. But, located as it is on the Great Artesian Basin, it does not suffer the dry heat or exhibit the arid landscape typical of the region. The town was named after Sir Thomas Mitchell, explorer and Surveyor-General of New South Wales, who visited the region in 1846. Its long pastoral history is shown in the fine examples of heritage buildings on the main street.

Kenniff Courthouse This courthouse was in use from 1882 to 1965. It held the murder trials for the Kenniff Brothers, infamous bushrangers who killed a policeman and station manager in 1902. The courthouse is now a museum with a bushranger exhibition, visual display, and art and craft sales. The landscaped grounds incorporate a community mosaic, an operating artesian windmill and a small billabong. Cambridge St.

Great Artesian Spa Complex: with relaxing waters in a garden setting, this is Australia's largest open-air spa; Cambridge St. ***Graffiti murals:*** depict the past, present and future of the Booringa Shire; around town. ***Art galleries:*** including the Maranoa Art Gallery and Nalingu Contemporary Indigenous Gallery. ***Horse-drawn wagon tours:*** in season; details from visitor centre.

Mitchell Rodeo: Mar/Apr. ***Maranoa Diggers Races:*** Apr. ***Campdraft:*** Mar. ***Agricultural Show:*** May. ***Mitchell Show:*** May. ***Fire & Water Festival:*** Sept/Oct. ***Bushstock Contemporary Musical Festival:*** Oct. ***Christmas in the Park:*** Dec.

Berkeley Lodge Motor Inn: near Great Artesian Spa centre; 20–30 Cambridge St; (07) 4623 1666. ***Bonus Downs Farmstay:*** fully catered options available; Bollon Rd; (07) 4623 1573.

Carnarvon National Park, Mt Moffatt section This section of the park is mainly for driving, with short walks to scenic spots. See the sandstone sculptures of Cathedral Rock, Marlong Arch and Lot's Wife, or visit the Tombs for the ancient stencil art of the Nuri and Bidjara people. The high-country woodlands and forest are home to a variety of wildlife, and birdwatching for raptors and lorikeets is exceptional. 4WD recommended; (07) 4984 4505; 256 km N. Gorge section: *see Roma*. Ka Ka Mundi section: *see Springsure*. Salvator Rosa section: *see Tambo*.

Neil Turner Weir: birdwatching and picnics; 3.5 km NW. ***Fisherman's Rest:*** good fishing spot; 6 km W. ***Maranoa River Nature Walk:*** informative 1.8 km circuit walk starting at Fisherman's Rest. ***Kenniff Statues:*** depict the story of the brothers at their last stand; 7 km S. ***Major Mitchell Cruises:*** cruises down Maranoa River departing from Rotary Park, Neil Turner Weir; bookings at visitor centre. ***Aboriginal tours:*** guided tours run by Nalingu Aboriginal Corporation; details from visitor centre.

See also OUTBACK, p. 428

Monto

Pop. 1155
Map ref. 653 K3

Touch screen in Newton St; or council offices, 51A Newton St, (07) 4166 9999; www.monto.qld.gov.au

 105.1 FM Rebel, 855 AM ABC Wide Bay

Monto, one of the most recent towns in the Capricorn region, was settled in 1924. It is the centre of a rich dairy and beef cattle district and is set picturesquely on a plateau surrounded by rolling hills.

Monto History Centre: local history displays and videos; closed Sat–Sun; Cnr Kelvin and Lister sts. ***Historical and Cultural Complex:*** variety of historic artefacts and a mineral display; Flinders St.

Country Craft Market: Mulgildie Hall; 1st Sun every month. ***Campdraft:*** Mar. ***Fishing Classic:*** Mar. ***Horseracing:*** Mar. ***Monto Show:*** Apr. ***Monto Festival:*** June. ***Annual Cattle Drive and Trail Ride:*** Aug. ***Monto Garden Expo:*** Oct.

Cania Gorge Caravan and Tourist Park: campsites, cabins and villas; Phil Marshall Dr; (07) 4167 8188.

Cania Gorge National Park Part of Queensland's sandstone belt, this park has cliffs, gorges and caves of spectacular colours. The freehand Aboriginal art around the park is a reminder of the area's ancient heritage. There are over 10 walks of varying length and difficulty. See the park's goldmining history on the 1.2 km return Shamrock Mine track or experience breathtaking park views from the Giant's Chair Lookout, reached on the longer Fern Tree Pool and Giant's Chair circuit. (07) 4167 8162; 25 km NW.

Mungungo: small town with boutique Waratah Winery; 13 km N. ***Lake Cania:*** excellent spot for watersports, fishing (permit required) and walking to lookout. Annual Lake Cania Fishing Classic is held here every Mar; 11 km N via Cania Gorge Picnic Area. ***Kalpower State Forest:*** hoop pine and rainforest vegetation. 4WD or walk rugged tracks to scenic lookouts; 40 km NE. ***Wuruma Dam:*** swimming, sailing and waterskiing; 50 km S.

See also CAPRICORN, p. 423

Mooloolaba

Pop. 7376
Map ref. 647 H4 | 648 F11 | 653 N7

Cnr First Ave and Brisbane Rd; (07) 5478 2233; www.mooloolabatourism.com.au

 90.3 FM ABC Sunshine Coast, 104.9 FM Sunshine

Mooloolaba is a popular holiday destination on the Sunshine Coast. Its fabulous beaches, restaurants, nightlife and resort-style shopping contribute to the constant influx of families and young people eager for the sun. Mooloolaba Harbour is one of the safest anchorages on the east coast and the base for a major prawning and fishing fleet.

UnderWater World This award-winning complex has a fantastic 80 m walkway through seawater 'ocean', with displays of the Great Barrier Reef and underwater creatures. There are daily shows, including the seal show and crocodile feeding. Spend 15 min swimming with a seal or dive in with sharks (bookings essential). The Touch Tank is a less daunting alternative to get up-close to the animals of the sea; Wharf Complex, Parkyn Pde.

Mooloolaba Harbour: popular spot for parasailing, scuba diving and cruises; contact visitor centre for tour operators.

Triathlon: Mar. ***Etchells Australian Winter Championship:*** yacht race; June.

Aegean Apartments: 5-storey block with views; 14 River Espl; (07) 5444 1255 or 1800 644 155. ***Breakfree Seamark:*** 12-storey apartment block; 29 First Ave; (02) 8248 2350 or

QUEENSLAND

1300 130 601. ***Mantra Zanzibar:*** family-friendly high-rise; 47–51 The Esplanade; (07) 5444 5633 or 1300 553 800.

Alexandra Headland: popular coastal town with views to the Maroochy River and Mudjimba Island. Extensive beaches and parklands on the foreshore. Surf lessons and board hire are available from Mooloolaba Wharf; just north of Mooloolaba. ***Mooloolah River National Park:*** take a canoe down Mooloolah River, ride along the bike trail or walk on the fire trails in this remnant wallum heath park; straddles Sunshine Motorway; (07) 5494 3983; 6 km sw. ***Yachting and game fishing:*** trips to nearby offshore reefs; details from visitor centre.

See also SUNSHINE COAST, p. 419

Moranbah

Pop. 7131
Map ref. 649 C10 | 655 I7

Library, town square; (07) 4941 4500; www.isaac.qld.gov.au

96.9 FM Rock, 104.9 FM ABC Tropical North QLD

Moranbah is a modern mining town south-west of Mackay. It was established in 1969 to support the huge open-cut coalmines of the expanding Bowen Coal Basin. Coking coal is railed to the Hay Point export terminal just south of Mackay.

Federation Walk: 1 km scenic walk starts at Grosvenor Park; Peak Downs Hwy. ***Historical walk:*** self-guide trail past interesting sites and heritage buildings of town; brochure from visitor centre.

Australia Day Street Party: Jan. ***May Day Union Parade and Fireworks:*** May.

Tours to Peak Downs Mine: leaves from town square the last Wed each month; bookings at visitor centre. ***Isaacs River:*** recreational area with historic monuments and a hiking trail in dry weather; 13 km s. ***Lake Elphinstone:*** camping, recreation activities, waterskiing, boating and fishing; 70 km N.

See also THE MID-TROPICS, p. 424

Mossman

Pop. 1740
Map ref. 650 D1 | 651 D3 | 657 L5

Call council offices, (07) 4099 9444; or Cairns & Tropical North Visitor Information Centre, 51 The Esplanade, Cairns, (07) 4051 3588; www.cairnsgreatbarrierreef.org.au

92.5 FM 4CA, 639 AM ABC Far North QLD

The town of Mossman, in Far North Queensland, is set among green mountains and fields of sugarcane. Originally named after the explorer Hugh Mosman, the town changed the spelling of its name from Mosman to Mossman to avoid being confused with the Sydney suburb.

Market: Foxton Ave; Sat. ***Christmas Party:*** Dec.

Papillon B&B: 2 rooms overlooking pool; 36 Coral Sea Dr; (07) 4098 2760. ***Silky Oaks Lodge & Healing Waters Spa:*** fantastic treehouse and rainforest rooms; Finlayvale Rd; (07) 4098 1666.

Cooya Beach and Newell: coastal towns with popular beaches to the north-east. ***Daintree National Park:*** including the magnificent Mossman Gorge; 5 km w; *see also Daintree*. ***Karnak Rainforest Sanctuary:*** amphitheatre and rainforest; 8 km N. ***High Falls Farm:*** tropical fruit orchard, market garden and open-air restaurant in rainforest setting; 9 km N. ***Wonga:*** small town with an excellent beach and orchid gardens; 18 km N.

See also CAIRNS & THE TROPICS, p. 425

Mount Isa

Pop. 18 856
Map ref. 659 D11 | 660 F3

Outback at Isa, 19 Marian St; (07) 4749 1555 or 1300 659 660; www.mountisa.qld.gov.au

102.5 Hot FM, 567 AM ABC North West QLD

The city of Mount Isa is the most important industrial, commercial and administrative centre in north-west Queensland, an oasis of civilisation in outback spinifex and cattle country. But before John Campbell Miles first discovered a rich silver-lead deposit in 1923, the area was undeveloped. Today Mount Isa Mines operates one of the world's largest silver-lead mines. Mount Isa is also one of the world leaders in rodeos, holding the third largest, which attracts rough-riders from all over Queensland and almost doubles the town's population.

Outback at Isa This modern complex shows the splendours of Queensland's outback country. Don a hard hat and take a guided tour of the 1.2 km of underground tunnels in the Hard Times Mine. The Indigenous, pioneering and mining history of Mount Isa is explored in the Sir James Foots building, and the Outback Park offers a scenic lagoon and informative walking trail. Marian St; 1300 659 660.

Riversleigh Fossil Centre This award-winning centre explores the significant discoveries of the Riversleigh Fossil Site (267 km NW). Through colourful displays, the ancient animals and landscapes come alive. The theatrette shows an excellent film on the fossil story so far and a visit to the laboratory with a working palaeontologist brings fossil discovery up-close as precious material is extracted from rocks. Adjacent to Outback at Isa.

Kalkadoon Tribal Centre: preserves the heritage and culture of the Indigenous Kalkadoon people. There are displays of artefacts and guided tours by Kalkadoon descendants; open weekdays; Centenary Park, Marian St. ***National Royal Flying Doctor Service Visitor Centre:*** informative video, historic and modern memorabilia and photo display; open weekdays; Barkly Hwy. ***Trust Tent House:*** an example of the housing provided for miners in the 1930s and 1940s that was designed for good ventilation in extreme conditions; open by appt only; Fourth Ave; (07) 3229 1788. ***School of the Air:*** discover how distance education works in the outback; Abel Smith Pde. ***City Lookout:*** overview of city and mine area; Hilary St. ***Underground Hospital:*** tours of hospital built in WW II; Deighton St. ***Rodeo Walk of Fame:*** walk the road paved with the names of Mount Isa rodeo legends; Rodeo Dr. ***Mount Isa Mine:*** the mine's lead smelter stack is Australia's tallest free-standing structure (265 m); informative surface tours available; details from visitor centre. ***Donaldson Memorial Lookout:*** lookout and walking track; off Marian St.

Market: Library carpark, Sat. ***Mining Expo:*** Apr. ***Mount Isa Cup:*** horseracing; June. ***Mount Isa Show:*** June. ***Isa Rodeo:*** Aug. ***Mardi Gras:*** Aug. ***Fishing Classic:*** Sept.

Lake Moondarra: artificial lake for picnics and barbecues, swimming, watersports and birdwatching. Home of the Fishing Classic in Sept; 20 km N. ***Mt Frosty:*** old limestone mine and swimming hole (not recommended for children as hole is 9 m deep with no shallow areas); 53 km E. ***Lake Julius:*** canoe at the lake or see the Aboriginal cave paintings and old goldmine on the nature trails; 110 km NE. ***Station visits:*** feel the outback spirit at one of the stations in the area; details from visitor centre. ***Safari tours:*** to Boodjamulla (Lawn Hill) National Park and Riversleigh Fossil Fields; details from visitor centre. ***Air-charter flights:*** to barramundi fishing spots on Mornington and Sweers islands in the Gulf of Carpentaria; details from visitor centre.

See also OUTBACK, p. 428

Mount Morgan

Pop. 2444
Map ref. 655 M11

Heritage Railway Station, 1 Railway Pde; (07) 4938 2312; www.mountmorgan.com

 98.5 FM 4YOU, 837 AM ABC Capricornia

Located in the Capricorn region south-west of Rockhampton is the quaint historic mining town of Mount Morgan. Said to be the largest single mountain of gold in the world, Mount Morgan's gold supply was discovered and mined from the late 1800s. What was a big mountain is now a big crater – the largest excavation in the Southern Hemisphere. In the mine's heyday, around 1910, the town was home to about 14 000 people.

Railway Station: historic station with tearooms, rail museum and a restored 1904 Hunslett Steam Engine that operates regularly; Railway Pde. ***Historical Museum:*** varied collection of memorabilia traces history of this mining town; Morgan St. ***Historic suspension bridge:*** built in the 1890s and spans the Dee River. ***Historic cemetery:*** features the Chinese Heung Lew (prayer oven) and the Linda Memorial to men killed in underground mines (1894–1909); off Coronation Dr. ***The Big Stack:*** 76 m high 1905 brick chimney used to disperse mining fumes; at the mine site.

Lions Club Market: Apex Park Cemetery; 3rd Sat each month. ***Golden Mount Festival:*** May. ***Mount Morgan Show:*** Aug.

Leichhardt Hotel: 12 hotel rooms, shared bathrooms; 52 Morgan St; (07) 4938 1851.

The Big Dam: good boating and fishing; 2.7 km N via William St. ***Wowan:*** town featuring the Scrub Turkey Museum in old butter factory; 40 km SW. ***Tours:*** tour to the open-cut Mount Morgan Mine and ancient clay caves with dinosaur footprints; details available from visitor centre.

See also CAPRICORN, p. 423

Mount Surprise

Pop. 162
Map ref. 657 J8

Bedrock Village Caravan Park, Garland St (Savannah Way), (07) 4062 3193; or Mount Surprise Gems, Garland St, (07) 4062 3055; www.gulf-savannah.com.au

 97.9 FM Hot, 105.3 FM ABC Far North QLD

Mount Surprise is a historic rail town in the Gulf Savannah. Its name comes from the surprise the Aboriginal people felt when they were resting at the base of the mountain and the loud white people of Ezra Firth's pioneer party arrived in 1864. The region has excellent gemfields for fossicking, especially for topaz, quartz and aquamarine. The town is on the edge of the Undara lava field, formed from the craters of the McBride Plateau. The lava caves can be explored in Undara Volcanic National Park.

Savannahlander This unique train journey shows the rugged delights and beautiful landscapes between Mount Surprise and Cairns. Departing from Mount Surprise Station each Sat, the train stops at various towns in the Gulf Savannah region. There are also trips between Mount Surprise and Forsayth. Bookings on (07) 4053 6848.

Old Post Office Museum: documents the bush history of the region in a historic 1870 building; opposite railway station. ***Mount Surprise Gems:*** runs tours to nearby fossicking spots and provides licences; Garland St.

Campdraft: Aug.

Undara Experience: accommodation in railway carriages; Savannah Way; (07) 4097 1900 or 1800 990 992.

Undara Volcanic National Park The cooling molten lava of an erupted volcano formed the 90 km of hollow underground lava tubes, the longest in the world. You can only explore the tubes by guided tour; bookings on 1800 990 992; day tours also run from Bedrock Village. See the eggcup-shaped crater on the 2.5 km Kalkani Crater circuit or go birdwatching to see some of the park's 120 bird species. (07) 4046 6600; 42 km E.

O'Brien's Creek: renowned for quality topaz; obtain a licence before fossicking; 37 km NW. ***Tallaroo Hot Springs:*** 5 natural springs created over centuries; open Easter–Sept; 48 km NW. ***Forty Mile Scrub National Park:*** vine-thicket park on the McBride Plateau with informative short circuit track from day area; (07) 4046 6600; 56 km E. ***Georgetown:*** small town once one of many small goldmining settlements on the Etheridge Goldfields; noted for its gemstones, especially agate, and gold nuggets; 82 km W. ***Forsayth:*** old mining town; 132 km SW. ***Cobbold Gorge:*** guided boat tours through sandstone gorge; bookings essential (07) 4062 5470. A full-day tour to Forsayth and the gorge also runs from Bedrock Village; 167 km SW via Forsayth. ***Agate Creek:*** fossick for gemstones; 187 km SW via Forsayth.

See also GULF SAVANNAH, p. 427

Mourilyan

Pop. 423
Map ref. 650 E5 | 651 H12 | 657 M7

Cnr Eslick St and Bruce Hwy, Innisfail; (07) 4061 7422.

98.3 FM Kool, 720 AM ABC Far North QLD

Mourilyan is a small town in Far North Queensland. It is the bulk-sugar outlet for the Innisfail area. The history of this thriving industry can be seen at the Australian Sugar Industry Museum.

Australian Sugar Industry Museum This large museum was opened in 1977. Its permanent displays include a museum collection of photographs, books, documents and an incredible display of machinery that includes a steam engine, reputedly one of the largest ever built. See the audiovisual display on the history of Australia's sugar industry and tour the art gallery. Cnr Bruce Hwy and Peregrine St; (07) 4063 2656.

Paronella Park This 5 ha park was the vision of immigrant sugarcane worker José Paronella. After making his fortune, he started building a mansion on the site (1930–1946). Floods and other natural disasters have ruined the grand buildings, despite rebuilding efforts. Now the visitor can walk through the ruins, admire the spectacular rainforest and birdlife, swim near the falls and have Devonshire tea at the cafe. 17 km SW.

Etty Bay: quiet tropical beach with caravan and camping facilities; 9 km E.

See also CAIRNS & THE TROPICS, p. 425

Mundubbera

Pop. 1055
Map ref. 653 K4

Historical Museum, Dugong St; or call council offices, (07) 4165 4101 or (07) 4165 4549.

 93.1 FM Sea, 855 AM ABC Wide Bay

QUEENSLAND

Mundubbera is on the banks of the Burnett River and is the main citrus-growing area in Queensland. Fruit pickers flock to the town in the cooler months. The unusual company 'Bugs for Bugs' produces a group of predatory bugs that reduces the need for pesticides. The rare Neoceratodus, or lungfish, is found in the Burnett and Mary rivers.

Historical Museum: local history; Leichhardt St. ***Heritage Centre:*** several displays of early settlement history; Durong Rd. ***Jones Weir:*** popular spot for fishing; Bauer St. ***'Meeting Place of the Waters':*** 360-degree town mural; Cnr Strathdy and Stuart–Russell sts. ***Bugs for Bugs:*** company that breeds insects; tours by appt only; Bowen St; (07) 4165 4663. ***Pioneer Place:*** gorgeous park with picnic facilities right in the middle of town.

Morning Market: Uniting Church: 3rd Sat each month. ***Mundubbera Show:*** May.

Auburn River National Park In this small park the Auburn River flows through a sheer-sided gorge and over granite boulders in the riverbed. Dry rainforest grows on the upper part of the tough track that leads down the side of the gorge to the river. An easier walk is the 150 m trail to the lookout above the Auburn River. Opposite the campsite catch a glimpse of the nesting peregrine falcons. 4WD is recommended in wet weather. (07) 4165 3905; 40 km SW.

Enormous Ellendale: the big mandarin, another addition to the 'big' monuments of Queensland; outskirts of town at the Big Mandarin Caravan Park. ***Golden Mile Orchard:*** impressive orchard with tours; 5 km S. ***Eidsvold:*** town at the centre of beef cattle country featuring the Historical Museum Complex and Tolderodden Environmental Park; 37 km NW.

See also CAPRICORN, p. 423

Murgon

Pop. 2133
Map ref. 648 A9 | 653 L6

Queen Elizabeth Park; (07) 4168 3864; www.murgon.qld.gov.au

90.7 FM Crow, 855 AM ABC Wide Bay

Murgon, known as 'the beef capital of the Burnett', is one of the most attractive towns in southern Queensland. Settlement dates from 1843, but the town did not really develop until after 1904, when the railway arrived and the large stations of the area were divided up. The town's name comes from a lily pond, found on Barambah Station, which was the site of the first pastoral property in the area.

Queensland Dairy Museum: static and interactive displays illustrating the history of the dairy industry, with special interest in the development of butter; Gayndah Rd.

Murgon Show: Mar. ***Barambah Blowout Bull Ride:*** Mar/Apr. ***Kids Kapers:*** fun day for the kids including a treasure hunt; May. ***Dairy Heritage Festival:*** June. ***Rotary Arts Festival:*** Sept/Oct. ***Barambah Shakin' Grape Festival:*** Oct. ***Christmas Carnival:*** Dec.

Boat Mountain Conservation Park The flat-topped crest in this park looks like an upturned boat, hence the name. The views from the top are panoramic and take in the surrounding agricultural valley. There are 2 lookout walks and an excellent 1.8 km circuit track. Watch for the bandicoot digs along the way. 20 km NE via Boat Mountain Rd.

Jack Smith Conservation Park This park comprises valuable remnant dry rainforest that used to cover the region before clearing for agriculture began. Have a picnic overlooking the South Burnett Valley before taking the 20 min return track through scrub to see the abundant birdlife of the park; 13 km N.

Cherbourg: small Aboriginal community featuring the Ration Shed Precinct displaying a pictorial history of the area; 5 km SE. ***Wondai:*** attractions include the Regional Art Gallery, Heritage Museum and South Burnett Timber Industry Museum. Town hosts Garden Expo in September; 13 km S. ***Bjelke-Petersen Dam:*** popular spot for watersports and fishing (boat hire available) with various accommodation styles at Yallakool Tourist Park. The dam is home to the Annual Fishing Competition in Oct; 15 km SE. ***Goomeri:*** known as 'clock town' for its unique memorial clock in the town centre. It has numerous antique stores and holds the Pumpkin Festival in May; 19 km NE. ***Booubyjan Homestead:*** historic home (1847) open to public; 43 km N via Goomeri. ***Kilkivan:*** Queensland's first discovery of gold was here in 1852. Try fossicking for gold or visit the lavender farm and historical museum. The town holds the Great Horse Ride in Apr; 44 km NE. ***Proston:*** small community featuring Sidcup Castle and Crafts Museum (open Wed–Mon); 54 km W. ***Lake Boondooma:*** watersports, fishing and the Fishing Competition in Feb; 74 km NW via Proston. ***Barambah Wine Trail:*** within 15 km of Murgon, in the Moffatdale and Redgate areas, are 7 excellent wineries with tastings and sales; visit by car or bus tour. ***Bicentennial National Trail:*** a 5330 km trail for walkers, bike riders and horseriders – you can do just a part of the trail. It runs through Kilkivan. ***Fossicking:*** semi-precious stones in Cloyna and Windera region; details from visitor centre.

See also BRISBANE HINTERLAND, p. 415

Muttaburra

Pop. 106
Map ref. 654 C8 | 661 O8

Post office, Sword St; (07) 4658 7147; www.muttaburra.com

104.5 FM West, 540 AM ABC Western QLD

Despite being a tiny outback community, Muttaburra has much to promote in its history. Famous cattle duffer Harry Redford (Captain Starlight) planned his daring robbery at nearby Bowen Downs Station in 1870. He stole 1000 head of cattle (thinking they wouldn't be missed) and drove them 2400 kilometres into South Australia. He was arrested and tried; the jury acquitted him, probably because his daring was so admired. These events were the basis for Rolf Boldrewood's novel *Robbery Under Arms*. More recently, a skeleton of an unknown dinosaur was found in 1963 in a creek close to the Thomson River. It was named Muttaburrasaurus. A replica can be seen in Bruford Street.

Dr Arratta Memorial Museum: this museum has medical and hospital displays with original operating theatres and wards; tours by appt; (07) 4658 7287. ***Cassimatis General Store and Cottage:*** restored store depicts the original family business in early 1900s. The adjacent cottage was home to the Cassimatis family; tours by appt; (07) 4658 7287.

Anzac Day Celebrations: Apr. ***Landsborough Flock Ewe Show:*** June. ***Horseracing:*** Aug. ***Christmas Celebrations:*** Dec.

Muttaburra Cemetery: outskirts of town; 1 km W. ***Agate fossicking:*** 5 km W. ***Pump Hole:*** swimming and fishing for golden perch and black bream; 5 km E. ***Broadwater:*** part of Landsborough River for fishing, birdwatching, bushwalking and camping; 6 km S.

See also OUTBACK, p. 428

Nambour

Pop. 13 532
Map ref. 647 F3 | 648 E11 | 653 N7

Yandina Historic House, 3 Pioneer Rd (at the roundabout); (07) 5472 7181; www.yandinahistorichouse.com.au

 90.3 FM ABC Coast, 91.1 Hot FM

Nambour is a large, unpretentious service town in the Sunshine Coast hinterland approximately 100 km north of Brisbane. Development began in the 1860s and sugar has been the main crop since the 1890s. The town's name is derived from the Aboriginal word for the local red-flowered tea tree.

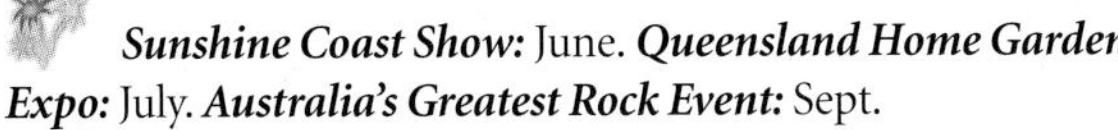

Sunshine Coast Show: June. ***Queensland Home Garden Expo:*** July. ***Australia's Greatest Rock Event:*** Sept.

Chez Claude: tiny French treasure; 4 Pine Grove Rd, Woombye; (07) 5442 1511.

Mapleton Falls National Park Volcanic columns jut out of Pencil Creek just before the creek's water falls 120 m to the valley floor. Walk to the falls lookout or see panoramic views of the Obi Obi Valley from Peregrine Lookout. Birdwatchers will delight at the early morning and dusk flights of the park's numerous bird species. (07) 5494 3983; 3 km SW of Mapleton.

The Big Pineapple The 16 m fibreglass pineapple makes this landmark hard to miss. At the time of going to press, this property was up for sale and was no longer open to the public. 7 km S.

Mapleton: attractive arts and crafts town in the Blackall Range; holds the Yarn Festival in Oct; 15 km W; *see also Maleny*. ***Mapleton Forest Reserve:*** with excellent drive through bunya pines and blackbutt forests starting just north of Mapleton. Along the drive, walk to the top of the waterfall from Poole's Dam and take the short Piccabeen Palm Groves Walk. The drive ends with spectacular views from Point Glorious.

See also SUNSHINE COAST, p. 419

Nanango

Pop. 3085
Map ref. 648 A11 | 653 L7

Henry St; (07) 4171 6871; www.nanango.qld.gov.au

 89.1 C FM, 747 AM ABC Southern QLD

Nanango is one of the oldest towns in Queensland. Gold was mined here from 1850 to 1900 and fossickers still try their luck today. The industrial Tarong Power Station and Meandu Coal Mine are nearby, yet Nanango still retains a welcoming country atmosphere.

Historic Ringsfield House: excellent example of colonial architecture of the 1900s. It houses the historical society and period furnishings; open weekdays, Sat–Sun by appt; Cnr Alfred and Cairns sts; (07) 4163 3345. ***Tarong Power Station Display:*** models and displays; adjacent to visitor centre.

Market: showgrounds; 1st Sat each month. ***Nanango Show:*** Apr. ***Nanart:*** art show; Apr/May. ***Country Music Muster:*** Sept. ***Funfest:*** Oct.

The Palms National Park Located at the Brisbane headwaters is this vine forest and subtropical rainforest park. Have a bush picnic and then take the 20 min Palm Circuit track through natural vegetation and along boardwalks. (07) 4699 4355; 42 km SW.

Tipperary Flat: tribute park to early pioneers with old goldmining camp, displays and walking track; 2 km E. ***Seven-Mile Diggings:*** gold- and gem-fossicking; permit from visitor centre; 11 km SE. ***Yarraman:*** historic timber town with heritage centre and 'mud maps' for region; 21 km S. ***Maidenwell:*** small town with Astronomical Observatory and Coomba Falls nearby; 28 km SW. ***Blackbutt:*** picturesque timber town with country markets 3rd Sun each month; 41 km SE.

See also BRISBANE HINTERLAND, p. 415

Nerang

Pop. 16 066
Map ref. 551 G1 | 645 F9 | 646 D6 | 653 N10

Cavill Walk, Surfers Paradise; (07) 5538 4419.

 92.1 FM Breeze, 612 AM ABC Brisbane

Nerang is in the Gold Coast hinterland. Today the town is much more similar in character to that dense urban strip of coast than to the small rural centre it started out as in the mid-1800s.

Nerang Forest Reserve On the north-west fringe of Nerang is this hilly rainforest and open eucalypt reserve. An excellent way of exploring the landscape is on the 2.8 km return Casuarina Grove Track through rainforest and along the creek. Look out for the black cockatoos. The reserve is a popular spot for horseriders and cyclists (permit required).

Nerang River: this gorgeous river flows through town, providing a popular spot for picnics, boating, fishing and watersports.

 ***Farmers market:*** Lavelle St; Sun.

Riviera B&B: French food and friendly hosts; 53 Evanita Dr; (07) 5533 2499. ***Binna Burra Mountain Lodge:*** cabins, tents and guesthouse accommodation; Binna Burra Rd, Beechmont; (07) 5533 3622 or 1300 246 622. ***O'Reilly's:*** great location and accommodation; Lamington National Park Rd, via Canungra; (07) 5502 4911 or 1800 688 722.

Springbrook National Park This rainforest park forms part of the Scenic Rim of mountains. The Springbrook section has an information centre, from which a short walk leads to a spectacular lookout over the Gold Coast. Access is via Springbrook (39 km SW). The Natural Bridge section features a unique rock arch over Cave Creek. Take the 1 km rainforest walk to see the natural bridge where Cave Creek plunges through an eroded hole to a cavern below. There are tours to see the glow worms or a 3 km night trail through rainforest; bookings on (07) 5533 5239. Park information (07) 5533 5147; 38 km SW. Mt Cougal section: *see Burleigh Heads.*

Lamington National Park This beautiful park to the south-west of Nerang protects the world's largest subtropical rainforest. It is a popular bushwalking and scenic park with spectacular waterfalls and mountain landscapes. Have a picnic with the rosellas and brush turkeys or go on a bushwalk – there are over 20. Access to the Green Mountains section is via Canungra; (07) 5544 0634. Binna Burra is accessed via Beechmont; (07) 5533 3584.

Carrara: nearby town offers scenic balloon flights over the Gold Coast, (07) 5578 2244, and holds weekend markets; 5 km SE. ***Hinze Dam on Advancetown Lake:*** sailing and bass-fishing;

continued on p. 469

NOOSA HEADS

Pop. 3658

Map ref. 647 H1 | 648 F10 | 653 N7

61 Hastings St; or Noosa Marina (Tewantin); 1300 066 672 or (07) 5430 5000; www.visitnoosa.com.au

90.3 FM ABC Sunshine Coast, 101.3 FM Noosa Community Radio

Noosa Heads, commonly known as Noosa, is a coastal resort town on Laguna Bay on the Sunshine Coast. The relaxed lifestyle, the weather and the safe year-round swimming make this a popular holiday destination. Cosmopolitan Hastings Street offers a relaxed cafe lifestyle, and within walking distance are the natural attractions of superb coastal scenery and the protected coves, surfing beaches and seascapes of Noosa National Park.

Noosa Main Beach: safe family swimming; beginners' surfing lessons available; 0418 787 577. ***Boutique shopping:*** browse clothing stores, gift shops and art galleries on stylish Hastings St. ***Adventure sports:*** on spectacular coastal waters of the Coral Sea and inland waterways. Activities include kite surfing, high-speed boating, surfing lessons and kayaking; book at the visitor centre. ***Camel and horse safaris:*** beach and bushland rides on Noosa's North Shore; book at the visitor centre. ***Scenic flights:*** contact visitor centre for details.

Farmers Market: Sun 7am–12pm; AFL ground, Weyba Rd, Noosaville. ***Festival of Surfing:*** Mar. ***Noosa Mardi Gras Recovery Week:*** a week-long festival following the Sydney Gay and Lesbian Mardi Gras; Mar. ***Food and Wine Festival:*** May. ***Winter Festival:*** sporting carnival; May. ***Noosa Longweekend:*** arts festival; June. ***Jazz Festival:*** Sept. ***Beach Car Classic:*** Sept. ***Triathlon Multisport Festival:*** Oct/Nov.

Berardo's Bistro on the Beach: inspired by French bistros; 8/49 Hastings St; (07) 5448 0888. ***Sails:*** delicious food, overlooking Main Beach; 75 Hastings St; (07) 5447 4235. ***Thomas Corner Eatery:*** casual dining with first class food, overlooking the Noosa River; 201 Gympie Tce, Noosaville; (07) 5470 2224. ***Café Le Monde:*** one of Noosa's longest-running eateries, with alfresco dining; 52 Hastings St; (07) 5449 2366. ***Wasabi:*** award-winning Japanese; 2 Quamby Pl, Noosa Sound; (07) 5449 2443.

Lake Weyba Cottages: lakeside cottages and pavilions; 79 Clarendon Rd, Peregian Beach; (07) 5448 2285. ***Offshore Noosa:*** modern apartments on the riverfront; 287 Gympie Tce, Noosaville; (07) 5474 4244. ***Netanya Noosa:*** beachfront luxury boutique hotel; 75 Hastings St; (07) 5447 4722. ***Sheraton Noosa Resort & Spa:*** sophisticated and relaxed; 14 Hastings St; (07) 5449 4888. ***Montpellier Boutique Resort:*** stylish two-bedroom apartments in the Noosaville restaurant precinct; 7 James St, Noosaville; (07) 5455 5033.

Noosa National Park, Headland Section This largely untouched rocky coastline park offers walks of varying length through rainforest and heathland. Escape the summer crowds of Noosa by taking the Tanglewood Track across the headland to Hells Gate, a popular lookout and whale-viewing spot. Return via the coastal track for scenic ocean views. Access the park via Park Rd in Noosa Heads, the coastal boardwalk from Hastings St, or Sunshine Beach; info and maps from the visitor centre.

Great Sandy National Park, Cooloola Section This park has stunning coloured sands, beaches, lakes, forests and sand dunes, all of which are protected. Many rare and threatened species call it home. There are walks for all ranges of fitness and stamina, from short circuit walks to overnight hikes. For the serious walker, there is the 2–4-day Cooloola Wilderness Trail with bush camping. For less strenuous activity, picnic in the rainforest at Bymien Picnic Area or see the Teewah Coloured Sands (multi-coloured sand cliffs). The park is separated from Noosa by the Noosa River and is accessed via vehicle ferry from Tewantin. Access from the north is via Rainbow Beach; 4WD recommended; (07) 5449 7792; 14 km N. Tours combining the coastline sights with the mirrored waterways of the Noosa Everglades are operated by Noosa Everglades Discovery; (07) 5449 0393.

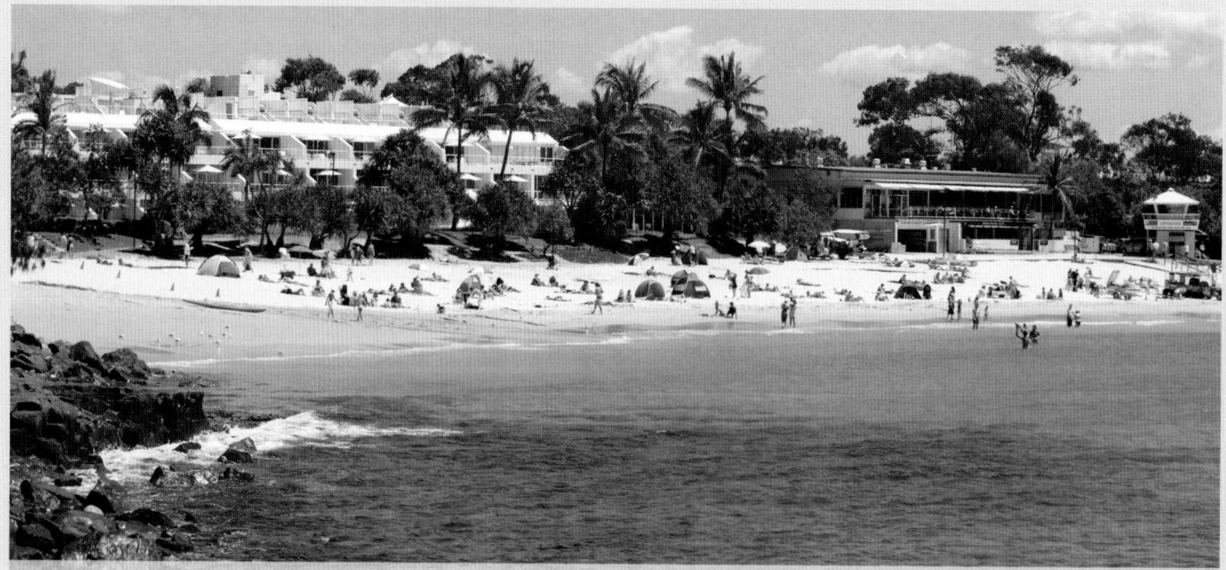

[NOOSA MAIN BEACH]

Laguna Lookout: views of Noosa River, lakes and hinterland; on Noosa Hill, access via Viewland Dr. ***Sunshine Beach:*** golden beach popular for surfing; 3 km SE. ***Noosaville:*** family-style area with Noosa River as focal point; departure point for river cruises; 5 km SW. ***Peregian Beach:*** beachside village with alfresco cafes, restaurants and boutique shops; 13km S. ***Lake Cooroibah:*** ideal for boating, sailing and windsurfing; access by car or boat from Noosaville. ***Tewantin:*** Noosa Marina in Parkyn Court has restaurants, boat hire and cruises; 7 km W. ***Tewantin State Forest:*** hilly rainforest and eucalypt forest reserve with 10 min walk to Mt Tinbeerwah Lookout offering a panoramic view over Noosa River, lakes and hinterland; 10 km W via Tewantin. ***Noosa River and the Everglades:*** the river extends over 40 km north into Great Sandy National Park. Take a cruise into the mirrored Everglades and to Harry's Hut, a relic of timber-cutting days. Kayak and canoe hire on offer, as well as camping facilities; contact visitor centre for details. ***Eumundi:*** famed markets; Wed 8am–1.30pm and Sat 6.30am–2pm (also Thurs evenings Dec/Jan); (07) 5442 7106; *see Yandina*; 21 km SW.

See also SUNSHINE COAST, p. 419

10 km SW. ***Historic River Mill:*** 1910 arrowroot mill; 10 km W. ***Mudgeeraba:*** holds the Somerset Celebration of Literature in Mar. Nearby is the Gold Coast War Museum with militia memorabilia and skirmish paintball; 12 km S. ***Paradise Country:*** this 12 ha Aussie farm offers half-day tours with camel rides, whip cracking, boomerang throwing, sheep shearing demonstrations, and damper and billy tea by a campfire; Entertainment Rd, Oxenford; (07) 5573 8270; 15 km N. ***Outback Spectacular:*** night-time show that brings Australia's outback to life with stunt riders, stockmen and a barbecue dinner; show starts at 7.30pm Tues–Sun; Pacific Mwy, Oxenford; (07) 5573 3999; 15 km N.

See also GOLD COAST & HINTERLAND, p. 417

Normanton

Pop. 1098
Map ref. 656 D7 | 659 G4

Cnr Landsborough and Caroline sts; (07) 4745 1065; www.gulf-savannah.com.au

 101.7 Rebel FM, 567 AM ABC North West QLD

Normanton thrived as a port town in the late 1800s when the gold rush was on in Croydon. The Normanton-to-Croydon railway line, established at that time, today runs the award-winning *Gulflander* tourist train. More recently, Australia's largest saltwater crocodile, known as Krys the Savannah King, was shot at nearby Archer's Creek in 1957. A life-size replica of his body, over 8 metres long, can be seen in the council park.

Normanton Railway Station: National Trust–listed Victorian building; Matilda St. ***Original well:*** settlers used it for drawing water; Landsborough St. ***Giant Barramundi:*** big monument to the fish; Landsborough St. ***Scenic walk and drive:*** self-guide tours to historic buildings including the Old Gaol in Haig St and the restored Bank of NSW building in Little Brown St; brochure available from information centre in the Burns Philip Building. **Gulflander:** this historical 140 km railway journey from Normanton to Croydon reveals the remote beauty of the Gulf Savannah; departs 8.30am Wed and returns Thurs afternoons. *Gulflander* also runs regular sunset tours with billy tea and damper, and other short trips. ***'Croc Spot' Cruise:*** at sunset on the Norman River; departs Norman Boat Ramp 5pm Mon–Sat. ***Norman River fishing tours:*** fishing trips and boat hire for barramundi and estuary fish; details from visitor centre.

 ***Barra Classic:*** Easter. ***Normanton Show and Rodeo:*** June.

Purple Pub & Brolga Palms Motel: character-filled pub, comfortable motel; Landsborough St; (07) 4745 1324.

Lakes: attract jabirus, brolgas, herons and other birds; on the outskirts of Normanton. ***Shady Lagoon:*** bush camping, birdwatching and wildlife; 18 km SE. ***Fishing:*** catching barramundi is very popular in the area – try the spots at Norman River in Glenore (25 km SE), Walkers Creek (32 km NW) or off the bridges or banks in Normanton. ***Burke and Wills cairn:*** last and most northerly camp of Burke and Wills (Camp 119) before their fatal return journey; off the Savannah Hwy; 30 km SW. ***Bang Bang Jump Up rock formation:*** a solitary hill on the surrounding flat plains with excellent views; Matilda Hwy; 106 km SW. ***Dorunda Station:*** cattle station offering barramundi and saratoga fishing in lake and rivers; accommodation is available; 197 km NE. ***Kowanyama Aboriginal community:*** excellent barramundi fishing, guesthouse and camping; permit to visit required from Kowanyama Community Council; 359 km NE.

See also GULF SAVANNAH, p. 427

North Tamborine

Pop. 1272
Map ref. 645 E9 | 646 B4 | 653 N10

Doughty Park, Main Western Rd; (07) 5545 3200; www.tamborinemtncc.org.au

 88.9 Breeze FM, 612 AM ABC Brisbane

North Tamborine is one of the towns on the Tamborine Mountain ridge in the Gold Coast hinterland. Numerous galleries, arts and crafts shops and boutique wineries make it and the nearby towns of Tamborine Village, Eagle Heights and Mount Tamborine popular weekend getaways. Tamborine National Park covers most of the mountain – the Witches Falls section was Queensland's first national park, listed in 1908.

Tamborine Mountain Distillery: award-winning distillery with a range of liqueurs, schnapps, vodkas and spirits; Beacon Rd. ***Mt Tamborine Winery and Cedar Creek Estate:*** tastings and sales; Hartley Rd. ***Hang-gliding:*** off Tamborine Mountain; Main Western Rd; (07) 5543 5631 or contact visitor centre. ***Tamborine Trolley Co:*** provides tours of mountain, including weekend hop-on hop-off tour to major sights; (07) 5545 2782.

QUEENSLAND

Produce market: showgrounds (green shed); Sun mornings. ***Tamborine Mountain Markets:*** showgrounds, Main Western Rd; 2nd Sun each month. ***Scarecrow Festival:*** May. ***Mountain Show:*** Sept. ***Springtime on the Mountain:*** open gardens and mini-markets; Oct. ***Craft Extravaganza:*** Oct. ***Tamborine Mountain Classic:*** bike, run or walk; Nov.

Songbirds in the Forest: modern Australian, lovely setting; Songbirds Rainforest Retreat, Tamborine Mountain Rd; (07) 5545 2563. ***The Polish Place:*** hearty Polish food and drink; 333 Main Western Rd; (07) 5545 1603. ***Cork 'n Fork Winery Tours:*** local wineries and lunch; Tamborine Mountain Plateau; (07) 5543 6584.

Pethers Rainforest Retreat: adults-only luxury retreat; 28B Geissmann St; (07) 5545 4577. ***Songbirds Rainforest Retreat:*** villas in the forest; Tamborine Mountain Rd; (07) 5545 2563. ***The Polish Place:*** European-style chalets; 333 Main Western Rd; (07) 5545 1603.

Tamborine National Park This picturesque mountain park protects remnant subtropical rainforest. Waterfalls, cliffs and beautiful walks make it a popular spot for visitors. The scenic drive visits the major waterfalls and lookouts and there are 22 km of walks on offer. Walk highlights are the 5.4 km Jenyns Falls circuit track and the 3 km Witches Falls circuit track, both leading to spectacular waterfalls. Park access points are on Tamborine–Oxenford Rd; (07) 5576 0271.

Eagle Heights: pretty village to the north-east with the Gallery Walk on Long Rd featuring excellent local crafts; Botanical Gardens in Forsythia Dr are set on 9 ha of rainforest with a variety of plants; historic buildings are on show at the Heritage Centre, Wongawallen Rd. ***Thunderbird Park:*** wildlife sanctuary, horseriding, laser skirmish, ropes course and fossicking for 'thunder eggs'; Tamborine Mountain Rd. ***Heritage and Mud Brick wineries:*** tastings and sales; Mt Tamborine; 5 km S.

See also GOLD COAST & HINTERLAND, p. 417

Oakey

Pop. 3653
Map ref. 653 K9

Library, 64 Campbell St; (07) 4692 0154.

102.7 FM 4DDB, 747 AM ABC Southern QLD

Oakey is an agricultural town on the Darling Downs, surrounded by beautiful rolling hills and black-soil plains. It is also the base for the aviation division of the Australian army.

Bernborough: bronze statue of famous local racehorse; Campbell St. ***Oakey Historical Museum:*** local memorabilia; Warrego Hwy. ***Australian Army Aviation Museum:*** every aircraft flown by the Australian army since WW II is represented at the museum through originals and replicas. Other aircraft can also be viewed, from the early wood Box Kite to the hi-tech Blackhawk Helicopter; at army base on Corfe Rd.

Jondaryan Woolshed Built in 1859 and still shearing under steam power, this woolshed is a memorial to pioneers of the wool industry. The complex includes the huge woolshed, historic buildings, and machinery and equipment collections. Visitors can see the shearing and sheepdog demonstrations or sit down to some billy tea and damper. Events are held throughout the year, including the Working Draft-horse Expo in June and the Australian Heritage Festival in Aug. Off Warego Hwy; 22 km NW.

See also DARLING DOWNS, p. 418

Palm Cove

Pop. 1215
Map ref. 650 D2 | 651 F6 | 657 L5

Paradise Village Shopping Centre, Williams Espl; (07) 4055 3433; www.palmcove.net

99.5 Sea FM, 639 AM ABC Far North QLD

Serene Palm Cove, north-west of Cairns, has white, sandy beaches and streets lined with palm trees. It is essentially a resort village, with locals and tourists alike soaking up the relaxed Far North atmosphere. The lifestyle is based around the water – fishing off the jetty, horseriding along the beach and swimming in the Coral Sea's crystal-clear blue water.

Reef tours: to Green Island and the Great Barrier Reef; depart Palm Cove jetty daily. ***Day tours:*** to Atherton Tableland and surrounds; details from visitor centre.

Palm Cove Fiesta: Oct.

Nu Nu Restaurant: award-winning Asian and Mediterranean fusion cuisine; 123 Williams Espl; (07) 4059 1880. ***Reef House Restaurant:*** modern Australian; The Sebel Reef House and Spa, 99 Williams Espl; (07) 4055 3633. ***Vivo Bar and Grill:*** colonial-style eatery, Italian influences; 45 Williams Espl; (07) 4059 0944.

Angsana Great Barrier Reef: world-renowned spa, absolute beach frontage; 1 Veivers Rd; (07) 4055 3000. ***Peppers Beach Club & Spa:*** luxury resort with rainforest lagoon pool; 123 Williams Espl; (07) 4059 9200 or 1300 987 600. ***Sea Temple Resort & Spa:*** contemporary Asian-style spa resort; 5 Triton St; (07) 4059 9600 or 1800 010 241. ***The Sebel Reef House & Spa:*** colonial-style architecture; 99 Williams Espl; (07) 4055 3633 or 1800 079 052.

Clifton Beach: resort village with park-lined beach and attractions such as Wild World, an interactive animal zoo, and Outback Opal Mine, with simulated mine and displays of Australia's most famous stone; 3 km S. ***Hartley's Crocodile Adventures:*** prides itself on the range of habitats that can be seen from extensive boardwalks and river cruises. Presentations include crocodile and snake shows and koala feeding; 15 km N. ***Rex Lookout:*** stunning coastal views; 17 km N.

See also CAIRNS & THE TROPICS, p. 425

Pomona

Pop. 1003
Map ref. 647 F1 | 648 E10 | 653 N7

Railway Station Gallery, Station St; (07) 5485 2950.

92.7 Mix FM, 612 AM ABC Brisbane

This small and relaxed farming centre is in the northern hinterland of the Sunshine Coast. Mount Cooroora rises 439 metres above the town. Each July mountain runners from around the world flock to attempt the base–summit and back again race, the winner being crowned King of the Mountain.

Majestic Theatre: authentic silent movie theatre with cinema museum and regular screenings; Factory St. ***Noosa Shire Museum:*** tribute to shire's past in old council chambers; Factory St. ***Railway Station Gallery:*** converted station featuring local art; open Tues–Sat; Station St.

Country market: Stan Topper Park; 2nd and 4th Sat each month. ***King of the Mountain:*** July. ***Silent Movie Festival:*** at the Majestic Theatre; Nov.

Noosa Avalon Cottages: peaceful rural retreat; 292 Pomona–Kin Kin Rd; (07) 5485 1959.

Boreen Point This sleepy town is on the shores of Lake Cootharba. The town features a 2.4 km walk from Teewah Land to Noosa's north shore beaches, and boardwalks into the surrounding wetlands. There are holiday cottages, and boat hire is available on Lake Cootharba. Try the many watersports on offer, including canoeing and kayaking. Windsurfing and yachting competitions are held here throughout the year. The lake is near where Mrs Eliza Fraser spent time with Aboriginal people after the wreck of Stirling Castle on Fraser Island in 1836; 19 km NE.

Cooroy: large residential area with excellent art gallery and cultural centre in the Old Butter Factory. The Noosa Botanical Gardens and Lake Macdonald are nearby; 10 km SE.

See also SUNSHINE COAST, p. 419

Port Douglas

Pop. 951
Map ref. 650 D1 | 651 D4 | 657 L5

Call council offices, (07) 4099 9444; or Cairns & Tropical North Visitor Information Centre, 51 The Esplanade, Cairns, (07) 4051 3588; www.cairnsgreatbarrierreef.org.au

106.3 Velvet Radio, 639 AM ABC Far North QLD

Port Douglas lies on the serene waters of a natural harbour in tropical North Queensland. Once a small village, it is now an international tourist destination. The town is surrounded by lush vegetation and pristine rainforests, and offers a village lifestyle with shops, galleries and restaurants. Its tropical mountain setting, the pristine Four Mile Beach and its proximity to the Great Barrier Reef make Port Douglas an ideal holiday destination. The drive from Cairns to Port Douglas is one of the most scenic coastal drives in Australia.

Rainforest Habitat Wildlife Sanctuary This sanctuary covers an area of 2 ha and is home to over 1600 animals. Walk along the boardwalks through the 4 habitats of North Queensland – rainforest, wetlands, woodlands and grasslands. There are regular guided tours of the sanctuary, and the special daily events of Breakfast with the Birds and Habitat After Dark (runs July–Oct). Cnr Captain Cook Hwy and Port Douglas Rd.

Flagstaff Hill: commands excellent views of Four Mile Beach and Low Isles; end of Island Point Rd. ***Dive schools:*** details from visitor centre. ***Great Barrier Reef tours:*** over 100 operators offer reef tours to outer Great Barrier Reef and Low Isles; details from visitor centre. ***Cane-toad racing:*** nightly at 8.15pm; Ironbar pub, 5 Macrossan St.

Market: Anzac Park; Sun. ***Village Carnivale:*** May.

Balé: modern Italian with style; Peppers Balé Resort, 1 Balé Dr; (07) 4084 3000. ***Nautilus Restaurant:*** magical forest setting, whole coral trout is a highlight; 17 Murphy St; (07) 4099 5330. ***Port O'Call Bar & Bistro:*** affordable international fare; Port O'Call Eco Lodge, cnr Port St and Craven Cl; (07) 4099 5422. ***Salsa Bar and Grill:*** excellent food in a relaxed setting; 26 Wharf St; (07) 4099 4922.

Macrossan House Boutique Holiday Apartments: comfortable apartments near good restaurants; 19 Macrossan St; (07) 4099 4366. ***Peppers Balé Resort:*** pavilion-style resort; 1 Balé Dr; (07) 4084 3000. ***Port O'Call Eco Lodge:*** high-standard budget accommodation; Cnr Port St and Craven Cl; (07) 4099 5422. ***Sheraton Mirage Port Douglas Resort:*** luxury on the beach; Davidson St; (07) 4099 5888 or 1800 073 535. ***Thala Beach Lodge:*** secluded deluxe bungalows; Private Rd, Oak Beach; (07) 4098 5700.

Tours: horseriding, sea-kayaking, rainforest tours, Lady Douglas paddlewheel cruises, 4WD safaris and coach tours to surrounding areas; details available from visitor centre.

See also CAIRNS & THE TROPICS, p. 425

Proserpine

Pop. 3314
Map ref. 649 E5 | 655 J4

Whitsunday Information Centre, Bruce Hwy; (07) 4945 3711 or 1300 717 407; www.tourismwhitsundays.com.au

Proserpine is the inland sugar town and service centre of the Whitsunday Shire. It was named after the Roman goddess of fertility, Proserpina, for the rich and fertile surrounding lands.

Historical Museum: local history dating back to settlement; closed weekends, open by appt; Bruce Hwy. ***Cultural Centre:*** displays local Aboriginal artefacts and has regular cinema screenings; Main St. ***Pioneer Park and Mill Street Park:*** beautiful spots with shady trees perfect for a picnic.

Proserpine Show: June. ***Harvest Festival:*** includes the World Championship Cane Cutting event; Oct.

Lake Proserpine at Peter Faust Dam: boat hire, waterskiing, fishing and swimming. The Cedar Creek Falls are nearby; 20 km NW on Crystalbrook Rd. ***Conway National Park:*** 28 km E; *see Airlie Beach*. ***Midge Point:*** coastal community and an ideal spot for bushwalking, fishing, crabbing and swimming; 41 km SE. ***Crocodile safaris:*** take a nature tour on an open-air tractor and cruise through mangrove river system on ***Proserpine River***; bookings on (07) 4946 5111.

See also THE MID-TROPICS, p. 424

Quilpie

Pop. 563
Map ref. 663 L5

Brolga St; (07) 4656 2166; www.quilpie.qld.gov.au

104.5 FM Outback Radio 4VL, 603 AM ABC Western QLD

Quilpie is on the banks of the Bullo River in the outback's famous Channel Country. The town was established as a rail centre for the area's large sheep and cattle properties. Today it is better known as an opal town and, in particular, for the 'Boulder Opal'. The world's largest concentration of this opal is found in the area surrounding Quilpie. The town takes its name from the Aboriginal word 'quilpeta', meaning 'stone curlew'.

Museum and gallery: historical and modern exhibitions; closed Sat–Sun from Oct–Mar; visitor centre, Brolga St. ***St Finbarr's Catholic Church:*** unique altar, font and lectern made from opal-bearing rock; Buln Buln St. ***Opal sales:*** various town outlets. ***Lyn Barnes Gallery:*** outback-inspired paintings; Brolga St.

Kangaranga Doo: Aug/Sept. ***Wool and Flower Show:*** includes show and rodeo; Sept.

Mariala National Park The park was formerly used to breed Cobb & Co horses in the early 1900s. It is remote with spectacular contrasts – the rich red earth mixed with green vegetation of mulga trees and shrubs. The threatened yellow-footed rock wallaby and pink cockatoo find refuge in this park. There are no formal walking trails. You can bush camp, but visitors must be self-sufficient. 4WD is recommended; roads may become impassable in the wet season. (07) 4654 1255; 130 km NE.

Lake Houdraman: popular watersports and recreation area; river road to Adavale; 6 km NE. ***Baldy Top:*** large geological formation with spectacular views; 6 km S. ***Opal fields:*** guided tours (no general access); details from visitor centre; 75 km W. ***Toompine:*** historic hotel, cemetery and designated opal-fossicking areas nearby; 76 km S. ***Eromanga:*** this place is reputedly Australia's furthest town from the sea. It features the Royal Hotel, once a Cobb & Co staging post, and holds a rodeo at Easter; 103 km W.

See also OUTBACK, p. 428

Ravenshoe

Pop. 914
Map ref. 650 D4 | 651 D12 | 657 L7

24 Moore St; (07) 4097 7700; www.ravenshoevisitorcentre.com.au

 97.9 Hot FM, 720 AM ABC Far North QLD

At 930 metres, Ravenshoe is the highest town in Queensland. Situated on the Atherton Tableland, it is surrounded by World Heritage rainforest, with 350 species of birds, 14 species of kangaroos and 12 species of possums. Once a town with a thriving logging industry, a new, alternative-lifestlye population has now developed.

Nganyaji Interpretive Centre: showcases the lifestyle of the local Jirrbal people, including hunting techniques and community life; Moore St. ***Scenic train-ride:*** heritage steam-train ride to nearby Tumoulin; details available from visitor centre.

Market: railway station; 4th Sun each month. ***Torimba Garden Party:*** Sept. ***Torimba Festival:*** includes Festival of the Forest and mardi gras; Oct.

Innot Hot Springs These natural thermal springs reputedly have healing powers. The spring water was originally bottled and sent to Europe as a healing remedy until the 1900s. 3 km E of the town is the Windy Hill Wind Farm. 32 km SW.

Millstream Falls National Park: enjoy the 1 km return walk past falls and rockpools to the Millstream Falls, the widest single-drop waterfall in Australia; (07) 4091 1844; 3 km SW. ***Tully Falls:*** walk 300 m to Tully Falls (in wet season) and the gorge; 25 km S. ***Lake Koombooloomba:*** popular spot for swimming, watersports, camping and fishing for barramundi; 34 km S. ***Mount Garnet:*** old tin-mining town with prospecting sites nearby; 47 km W.

See also CAIRNS & THE TROPICS, p. 425

Richmond

Pop. 554
Map ref. 661 L3

Kronosaurus Korner, 91–93 Goldring St; (07) 4741 3429 or 1300 576 665; www.richmond.qld.gov.au

 99.7 Rebel FM, 540 AM ABC Western QLD

This small town on the Flinders River in the Gulf Country serves the surrounding sheep and cattle properties. The town's main street is lined with beautiful bougainvilleas. In recent years Richmond has become the centre of attention as an area rich in marine fossils dating back around 100 million years, when outback Queensland was submerged under an inland sea.

Kronosaurus Korner This marine fossil museum has a renowned collection of vertebrate fossils, all found in the Richmond Shire. The museum and exhibition space holds more than 200 exhibits, including the 100-million-year-old armoured dinosaur Minmi, Australia's best-preserved dinosaur. There is also an activity centre, children's discovery area and fossil preparation area where visitors can watch the palaeontologist at work. Guided museum tours, as well as tours to nearby fossicking sites with a palaeontologist, are available (groups of 10 or more, bookings essential). Goldring St; (07) 4741 3429.

Cobb & Co coach: beautifully restored coach with informative history display; Lions Park, Goldring St. ***Gidgee Wheel Arts and Crafts:*** local craft; Harris St. ***Lake Fred Tritton:*** recreational lake for waterskiing, picnics and walks. Enjoy the Richmond Community Bush Tucker Garden, which showcases beautiful native plants and was a finalist for the Banksia Environmental Awards; eastern outskirts. ***Cambridge Store:*** replica of the old Cambridge Downs Homestead from the late 1800s; contains memorablia and machinery from pioneer days; Goldring St. ***Heritage walk:*** follow the self-guide trail with informative history at each stop. It includes the historic flagstone and adobe building, Richmond Hotel, St John the Baptist Church and the Pioneer Cemetery; brochure available from visitor centre.

Fossil Festival: even-numbered years, Apr/May. ***Richmond Great Outback Challenge:*** sports carnival; odd-numbered years, Oct.

Fossicking sites: guided tours to nearby areas; details from visitor centre.

See also OUTBACK, p. 428

Rockhampton

Pop. 68 832
Map ref. 655 M11

Customs House, 208 Quay St; (07) 4922 5339; capricorntourism.com.au

 98.5 4YOU FM, 837 AM ABC Capricornia

Rockhampton is a prosperous city that straddles the Tropic of Capricorn. It is known as the beef capital of Australia, with some 2.5 million cattle in the region. Quay Street is Australia's longest National Trust–classified streetscape, with over 20 heritage buildings set off by flowering bauhinia and brilliant bougainvilleas. Watch out for the bent-wing bat exodus in summer at Mount Etna, just north of the town, and the summer solstice light spectacular in early December to mid-January at the Capricorn Caves.

Mt Archer National Park On the north-east outskirts of Rockhampton in the Berserker Ranges, this park provides a backdrop to the city. Take a scenic drive up the mountain to Frazer Park for panoramic views. Explore the open forest and subtropical vegetation of the mountain on the 11 km walk from top to bottom. Be advised that the return trip is quite strenuous. Other shorter walks lead to scenic lookouts. Access to the summit is via Frenchville Rd; to the base via German St; (07) 4936 0511.

Botanic Gardens Set on the Athelstane Range, these heritage-listed gardens are over 130 years old. There are tropical displays, an orchid and fern house, a Japanese-style garden and the bird haven of Murray Lagoon. The city zoo has free entry and koala and lorikeet feedings at 3pm daily. Access via Spencer and Ann sts.

Customs House: this heritage building (1901) houses the visitor centre and 'Rockhampton Discovery Centre' exhibition, which introduces the visitor to the history, culture and lifestyle of Rockhampton; Quay St. ***Archer Park Station and Steam Tram Museum:*** interactive displays document the history of rail transport in Rockhampton. A fully restored Purrey Steam Tram operates 10am–1pm every Sun; closed Sat; Cnr Denison and Cambridge sts. ***Kershaw Gardens:*** follow the Australian native flora Braille Trail; Bruce Hwy. ***Rockhampton City Art Gallery:*** changing exhibitions featuring various well-known Australian artists such as Sidney Nolan; Victoria Pde. ***Great Western Hotel:*** operates weekly rodeos at an indoor arena; Stanley St. ***Fitzroy River:*** a great spot for watersports, rowing and fishing. A barrage in Savage St that separates tidal salt water from upstream fresh water provides opportunities for barramundi fishing. ***Capricorn Spire:*** marks the line of the Tropic of Capricorn; Curtis Park, Gladstone Rd. ***Heritage walk:*** self-guide trail around city centre; brochure available from visitor centre.

Garden Expo: Apr. ***Beef Australia:*** every 3 years, May. ***Rocky Swap Meet:*** Aug. ***Big River Jazz:*** Sept. ***Rocky Barra Bounty:*** Oct.

Coffee House Luxury Apartment Motel: right in the centre of town; Cnr William and Bolsover sts; (07) 4927 5722. ***Henderson Park Farm Retreat:*** 2 cabins beside a creek; 88 Barretts Rd, Barmoya; (07) 4934 2794.

Dreamtime Cultural Centre: displays on Aboriginal and Torres Strait Islander culture set on ancient tribal site. Guided tours available; open 10am–3.30pm Mon–Fri; 7 km N on Bruce Hwy. ***Old Glenmore Historic Homestead:*** National Trust–classified complex of historic buildings and displays; open by appt only; 8 km N. ***Rockhampton Heritage Village:*** heritage buildings with unusual Time after Time clock collection and Life before Electricity exhibition; Parkhurst; 9 km N. ***St Christopher's Chapel:*** open-air chapel built by American servicemen; Emu Park Rd; 20 km E. ***Capricorn Caves:*** guided tours and wild caving adventures through limestone cave system. Visit in Dec for Carols in the Caverns; 23 km N. ***Mt Hay Gemstone Tourist Park:*** thunder-egg fossicking and sales; 38 km W. ***Capricorn Coast Scenic Loop tourist drive:*** through coast and hinterland. ***Great Keppel Island:*** catch the cruise boat from Rosslyn Bay to visit the beautiful white sandy beaches and clear blue waters of the island; perfect spot for snorkelling and watersports.

See also CAPRICORN, p. 423

Roma

Pop. 5983
Map ref. 652 G6

Big Rig Visitor Information Centre, 2 Riggers Rd; (07) 4622 8676 or 1800 222 399; www.visitmaranoa.com.au

101.7 4RRR FM, 711 AM ABC Western QLD

Roma is in the Western Downs region and was named after the wife of Queensland's first governor. Roma boasts a few historic 'firsts' for Queensland and Australia. In 1863 Samuel Symons Bassett brought vine cuttings to Roma and Queensland's first wine-making enterprise began – Romavilla Winery is still running today. In the same year Captain Starlight faced trial in Roma for cattle stealing. Australia's first natural gas strike was at Hospital Hill in 1900. The excellent complex in town, The Big Rig, documents the oil and gas industry since this discovery.

The Big Rig This unique complex is set on an old oil derrick and features historic oil rigs and machinery displays. Photographs, memorabilia and multimedia displays provide a comprehensive history of oil and gas discovery and usage in Australia from 1900 to the present day. A highlight is the sound and light show on summer nights. Adjacent to the complex is a historic slab hut, recreational area and 1915 mini steam train that travels on a 1.4 km circuit. Open daily; 2 Riggers Rd; (07) 4622 4355.

Roma–Bungil Cultural Centre: large 3D clay mural by a local artist, depicting Roma's history; cnr Bungil and Quintin Sts; (07) 4622 1266. ***Heroes Avenue:*** heritage–listed street of bottle trees commemorating local soldiers who died in WWI; Wyndham and Bungil Sts. ***Romavilla Winery:*** Queensland's oldest winery; 8am–5pm weekdays, 9am–12pm and 2pm–4pm Sat; 77–83 Northern Rd; (07) 4622 1822.

RSL Market: RSL Hall; 1st Sun each month, from early morning. ***Easter in the Country:*** Easter. ***Picnic Races:*** Mar. ***Polocrosse Club Annual Carnival:*** Aug. ***Roma Cup:*** Nov. ***Christmas In The Park:*** Big Rig Parklands; Dec.

Roma Explorers Inn Motel: 4-star accommodation with excellent facilities, close to town centre; Warrego Hwy; (07) 4620 1400. ***Carnarvon Gorge Wilderness Lodge:*** safari cabins near national park; via Rolleston; (07) 4984 4503.

Carnarvon National Park, Gorge section The Carnarvon Creek winds through the steep-sided Carnarvon Gorge, flanked by white sandstone cliffs. There are over 21 km of walks through rainforest to waterfalls and caves, and Aboriginal rock art can be found throughout the park – see rock engravings, stencils and paintings at Cathedral Cave and the Art Gallery. The turn-off to the park is 199 km N on Carnarvon Developmental Rd; (07) 4984 4505. Mt Moffatt section: *see Mitchell.* Ka Ka Mundi section: *see Springsure.* Salvator Rosa section: *see Tambo.*

Roma Salesyard: largest inland cattle market in Australia with sales on Tues and Thurs from 8am; 4 km E. ***Meadowbank Museum:*** historic vehicle and machinery collection, and farm animals; by appt; (07) 4622 3836; 12 km W. ***Surat:*** small town featuring the Cobb & Co Changing Station Complex with museum, art gallery and aquarium. Try fishing for Murray cod on Balonne River; 78 km SE.

See also DARLING DOWNS, p. 418

St George

Pop. 2411
Map ref. 652 F9

Cnr The Terrace and Roe St; (07) 4620 8877; www.balonne.qld.gov.au

102.9 FM 4ROM, 711 AM ABC Western QLD

On the banks of the Balonne River is the Western Downs town of St George. It is the last post in southern Queensland before the heavily populated east coast finally gives way to the sparseness of the outback. The river crossing was discovered by explorer Sir Thomas Mitchell on St George's Day, 1846, giving the town its name. St George is often referred to as the inland fishing capital of Queensland with lakes and rivers nearby, which also support the area's rich cotton, grape and grain industry.

Heritage Centre: historic buildings and a local history museum featuring Aboriginal display; Victoria St. ***The Unique Egg:*** carved, illuminated emu eggs; 9am–5pm weekdays, 9am–12pm Sat; Balonne Sports Store; Victoria St; (07) 4625 3490. ***Town murals:*** around town, depicting scenes of St George's history. ***Riversands Vineyard:*** boutique winery on the banks of the Balonne River; open daily but closed Sun in Feb; Whytes Rd; (07) 4625 3643. ***Jack Taylor Weir:*** Sir Thomas Mitchell cairn commemorating the explorer's landmark crossing in 1846; western outskirts.

St George Show: with rodeo; May. ***Boolba Wool and Craft Show:*** May. ***Horseracing:*** St George Jockey Club; July.

QUEENSLAND

Nindigully Campdraft: Sept. ***Nindigully Country Music Spectacular:*** includes pig races; Nov.

Beardmore Dam: popular spot for watersports and picnics in surrounding parklands. It offers excellent fishing for yellowbelly and Murray cod; 21 km N. ***Ancient rock well:*** hand-hewn by Aboriginal people, possibly thousands of years ago; 37 km E. ***Nindigully:*** town where the film *Paperback Hero* was filmed. Features Nindigully Pub, which has the oldest continual licence in Queensland, since 1864; George Rd; (07) 4625 9637; 44 km SE. ***Boolba:*** holds the impressive Boolba Wool and Craft Show each May; 50 km W. ***Begonia Historical Homesetad:*** grazing property with farmstay and museum; (07) 4625 7415; 73 km N. ***Thallon:*** small town with excellent swimming and fishing at Barney's Beach on the Moonie River. The nearby Bullamon homestead (1860), mentioned in Steele Rudd's *Memoirs of Corporal Keeley*, has original shingle roof and canvas ceilings; tours by appt; (07) 4625 9217; 76 km SE. ***Bollon:*** large koala population in river red gums along Wallan Creek and a heritage and craft centre in George St; 112 km W. ***Thrushton National Park:*** undeveloped park of mulga scrub, sand plains and woodlands; access in dry weather only and 4WD recommended; (07) 4624 3535; 132 km NW. ***Culgoa Floodplain National Park:*** in the Murray–Darling basin, this park, with over 150 species of birds, is excellent for birdwatchers. There are no formal walking tracks and visitors must be self-sufficient. 4WD recommended; access via Brenda Rd, Goodooga; (07) 4654 1255; 200 km SW.

See also DARLING DOWNS, p. 418

Sarina

Pop. 3284
Map ref. 649 G8 | 655 K6

Sarina Tourist Art and Craft Centre, Bruce Hwy; (07) 4956 2251; www.sarina.qld.gov.au

93.5 Kids FM

Sarina is in the hinterland of what has been dubbed the Serenity Coast, at the base of the Connor Range. It is central to the Queensland sugar belt. To the east and south are fine beaches, many of which are renowned fishing spots.

Sarina Sugar Shed: Australia's only fully operational miniature sugar mill and distillery. Take a tour and watch the complex process of turning sugar canes into granules. There are also multimedia presentations and a novelty shop on-site; closed Sun; Railway Sq. ***Sarina Tourist Art and Craft Centre:*** excellent variety of local art and craft as well as visitor centre; Bruce Hwy. ***'Field of Dreams' Historical Centre:*** local industry history and memorabilia; open 10am–2pm Tues, Wed and Fri; Railway Sq.

Market: showgrounds; last Sun each month. ***Mud Trials:*** buggy racing on a mud track; May. ***Scope Visual Arts Competition:*** May. ***Discover Sarina Festival:*** May.

Cape Palmerston National Park This remote park features rugged coastal landscapes of headlands, swamps and sand dunes. Watch for the soaring sea eagles overhead or the birdlife around the swamp. There are spots for bush camping. The park has no official walking tracks, but the outlook from the cape is spectacular. 4WD recommended. (07) 4944 7800; 46 km SE via Ilbilbie.

Beaches: including Grasstree and Half Tide beaches to the north-east (Grasstree hosts annual bike race; date depends on tidal conditions); also Sarina Beach (east), popular for boating, fishing and swimming, with a lookout over the coast, and Armstrong Beach (south), great for swimming, prawning and fishing. ***Hay Point Lookout:*** viewing gallery at Hay Point and Dalrymple Bay coal terminal complex – informative video and excellent views; 12 km N. ***Salonika Beach:*** attractive beach with amazing wildlife, including loggerhead turtles and whales in season; adjacent to Hay Point Lookout. ***Lake Barfield:*** picnic area and bird sanctuary; 12 km N. ***Carmila:*** small town with beach just to the east. Visit the nearby Flaggy Rock Exotic Fruit Garden for delicious ice-cream; 65 km S. ***Clairview:*** popular spot for beach fishing and crabbing; 73 km S. ***St Lawrence:*** once a major port, this town is now a historical tribute to past days with many historic buildings and the remains of the wharf; 110 km S.

See also THE MID-TROPICS, p. 424

Springsure

Pop. 828
Map ref. 655 I12

Information shed, Rolleston Rd; or call council offices, (07) 4984 1166.

94.7 Hot FM, 1548 AM ABC Capricornia

Mount Zamia towers over Springsure, a small valley town in the Central Highlands, settled in the 1860s. Springsure's early history is dominated by conflicts between local Aboriginal groups and the intruding European settlers. The 1861 massacre of 19 Europeans at Cullin-la-ringo is commemorated at Old Rainworth Fort.

Aboriginal Yumba-Burin (resting place): in Cemetery Reserve, containing 3 bark burials (around 600 years old). ***Rich Park Memorials:*** includes cattleyard displays and a Dakota engine from a plane that crashed during WW II. ***Historic Hospital:*** heritage-listed building (1868) includes museum.

Horseracing: Mar. ***Springsure Show:*** May.

Minerva Hills National Park The park includes the Boorambool and Zamia mountains and has unusual wildlife, such as the fawn-footed melomys, on a 2.2 km walking track to a spectacular lookout, or have a bush picnic at Fred's Gorge. (07) 4984 1716; 4 km W of Springsure, part of road unsealed.

Carnarvon National Park, Ka Ka Mundi section This remote section of the park is in Queensland's brigalow belt and features undulating plains and sandstone cliffs. See the king parrots and fig birds around the springs and creeks or the area's pastoral history at the old cattleyards near the springs. (07) 4984 1716; west on Springsure–Tambo Rd for 50 km, then south on Buckland Rd. Mt Moffatt section: *see Mitchell*. Gorge section: *see Roma*. Salvator Rosa section: *see Tambo*.

Old Rainworth Historical Complex: National Trust–listed buildings of old storehouse built after 1861 massacre; open 9am–2pm weekdays (closed Thurs), 9am–5pm Sat–Sun; 10 km S.

See also CAPRICORN, p. 423

Stanthorpe

Pop. 4268
Map ref. 551 A3 | 557 L2 | 653 K11

26 Leslie Pde; (07) 4681 2057 or 1800 060 877; www.granitebeltwinecountry.com.au

90.1 Breeze FM, 747 AM ABC Southern QLD

Stanthorpe is the main town in the Granite Belt and mountain ranges along the border between Queensland and New South Wales. The town came into being after the discovery of tin at Quartpot Creek in 1872, but the mineral boom did not last. The climate is cool, said to be the coldest in Queensland, but the numerous wineries in the vicinity are welcoming. Visit in spring to see the fruit trees, wattles and wildflowers in bloom.

Heritage Museum: displays the region's past in historic buildings, such as a schoolroom, gaol and shepherd's hut; closed Mon–Tues; High St. ***Regional Art Gallery:*** touring and local exhibitions; open Mon–Fri, Sat–Sun afternoons; Weeroona Park, Marsh St.

Market in the Mountains: Cnr Marsh and Lock sts; 2nd and 4th Sun each month. ***Apple and Grape Harvest Festival:*** even-numbered years, 1st weekend in Mar. ***Rodeo:*** Mar. ***Brass Monkey Season: winter festival;*** June–Aug. ***Primavera:*** Sept–Nov. ***Australian Small Winemakers Show:*** Oct.

Vineyard Cafe: regional food and wine; New England Hwy, Ballandean; (07) 4684 1270.

Beverley Bush Cottages: 2 cottages and licensed restaurant; 25 Turner La, Severnlea; (07) 4683 5100. ***Granite Gardens:*** 3 cottages amid 2000 roses; 90 Nicholson Rd, Thorndale; (07) 4683 5161.

Granite Belt Wine Region Around Stanthorpe are over 40 boutique wineries open for tastings and sales. To the north, among others, are Old Caves Winery and Heritage Wines at Cottonvale; 12 km N. High-quality wineries are near Glen Aplin; 11 km SW. Ballandean, 21 km SW, features Golden Grove, Winewood, Bungawarra and Robinson's Family wineries. The Ballandean Estate holds Opera in the Vineyard in May and Jazz in the Vineyard in Aug/Sept. Brochure from visitor centre; bus tours available.

Girraween National Park The stunning granite and creek scenery is popular with bushwalkers. It is a remarkable Queensland park, not only for its cooler weather, but also because it is home to animals that are rarely seen in the tropical state, such as the common wombat. There is a variety of walks, from short to overnight hikes (taking in adjacent Bald Rock National Park). Of particular note is the 3 km return track to the Pyramid, which has panoramic views. Visit in spring to see the wildflower display. (07) 4684 5157 (same number for national parks below); turn-off 40 km S of Stanthorpe; further 8 km to park.

Mt Marlay: excellent views; 1 km E. ***Storm King Dam:*** popular spot for picnics, canoeing, waterskiing, and fishing for Murray cod and silver perch; 26 km SE. ***Boonoo Boonoo National Park (NSW):*** spectacular waterfall; 60 km SE. ***Bald Rock National Park (NSW):*** incredible granite rock formation (second biggest monolith in the world); 65 km SE. ***Sundown National Park:*** rugged national park of gorges and high peaks. Go birdwatching to see the herons and azure kingfishers along the river or take the short Red Rock Gorge Lookout Track for spectacular views; 80 km SW. ***Heritage trail:*** a historical drive tour of surrounding towns; brochure from visitor centre.

See also DARLING DOWNS, p. 418

Strathpine

Pop. 9534
Map ref. 644 D5 | 645 D3 | 647 F10 | 653 N9

Pine Rivers Tourism Centre, Daisy Cottage, cnr Gympie and South Pine rds; (07) 3205 4793; www.brisbanehinterland.com

97.3 4B FM, 612 AM ABC Brisbane

Strathpine is north of Brisbane in the Pine Rivers region, a district that includes the forested areas and national parks closest to the capital. Taking advantage of this rural setting so close to the city are a number of art and craft industries.

Pine Rivers Festival: May.

Brisbane Forest Park This natural bushland forest park has a scenic drive from the south-east corner to lookouts, mountain towns and attractive landscapes, ending at Lake Wivenhoe. There are many picnic spots – Jollys Lookout is a highlight, with views over Brisbane, the valley and north to the Glass House Mountains. A guide to the park's walks is available from the park headquarters at 60 Mt Nebo Rd, The Gap. Access via Ferny Hills.

Lake Samsonvale: fishing, watersports and bushwalking; 8 km NW. ***Old Petrie Town:*** heritage park that holds markets each Sun and the popular Twilight Markets 1st Fri each month (Jan–Oct) then every Fri (Nov–Dec). Catch the free bus from Petrie Railway Station; 9 km N. ***Alma Park Zoo:*** palm garden and subtropical rainforest zoo with native and exotic animals. Feed the koalas and explore the Friendship Farm for children; Alma Rd, Dakabin; 14 km N. ***Australian Woolshed:*** demonstrations of shearing, spinning and working sheepdogs; bush dances with bush band; Ferny Hills; 16 km SW. ***Osprey House:*** environmental centre; Dohles Rocks Rd; 18 km N. ***Brisbane's Vineyard:*** wine tastings and sales; Mt Nebo Rd, Mount Nebo; 46 km SW.

See also BRISBANE HINTERLAND, p. 415

Surfers Paradise

see inset box on next page

Tambo

Pop. 346
Map ref. 652 B2 | 663 P2

Council offices, Arthur St; (07) 4654 6133; www.tambo.qld.gov.au

100.3 Outback Radio 4LG FM, 603 AM ABC Western QLD

Tambo is the oldest town in central-western Queensland. It was established in the mid-1860s to service the surrounding pastoral properties, which it continues to do today. This long history can be seen in the heritage buildings on Arthur Street.

Old Post Office Museum: display of historic photographs; Arthur St. ***Tambo Teddies Workshop:*** produces popular all-wool teddies; Arthur St. ***Coolibah Walk:*** nature walk along banks of the Barcoo River.

Tambo Stock Show: Apr. ***Day/Night Rodeo:*** Oct. ***Market Day:*** Dec.

Carnarvon National Park, Salvator Rosa section One of the more remote sections of the park, Salvator Rosa is a perfect spot to escape the crowds. The attractive Nogoa River and Louisa Creek flow through the valley. See the spectacular rock formations, Belinda Springs and other natural attractions on the self-guide trail, which starts at the Nogoa River camping area. (07) 4984 4505; 130 km E. Mt Moffatt section: *see Mitchell*. Gorge section: *see Roma*. Ka Ka Mundi section: *see Springsure*.

Wilderness Way: a 420 km self-guide drive including Aboriginal rock art, historic European settlement sites and the Salvator Rosa section of the Carnarvon National Park. Check road conditions before departing; brochure available from council office.

See also OUTBACK, p. 428

QUEENSLAND

SURFERS PARADISE

Pop. 18 501

Map ref. 551 G1 | 557 O1 | 645 G10 | 646 F6 | 653 N10

Gold Coast Tourism Bureau, Cavill Walk; (07) 5538 4419 or 1300 309 440; www.verygoldcoast.com.au

89.3 4CRB FM, 91.7 FM ABC Coast

Surfers Paradise is the Gold Coast's signature settlement – high-rise apartments fronting one of the state's most beautiful beaches. The first big hotel was built here in the 1930s, among little more than a clutch of shacks. Since then the area has become an international holiday metropolis, attracting every kind of visitor from backpacker to jetsetter.

Circle on Cavill: exciting new-generation shopping and leisure precinct with Circle Big Screen entertainment; access via Surfers Paradise Blvd, Cavill and Ferny Aves. ***Orchid Avenue:*** famous strip with the best live music in town. ***Ripley's Believe It or Not:*** the 12 galleries of amazing feats, facts and figures will surprise and amaze. There are interactive displays and movies that bring events to life; Raptis Plaza, Cavill Ave. ***Haunted House:*** from the creators of Dracula's world-famous cabaret, this walk-through attraction offers 5 levels of fright and a shop of horrors; Surfers Paradise Blvd. ***Q1 Observation Deck:*** the tallest residential building in the world, the Q1 Observation Deck gives breathtaking panoramic views from Brisbane to Byron Bay at 230m above sea level; Hamilton Ave. ***Adventure activities and tours:*** try surfing lessons or scenic flights to see Surfers Paradise in a new light or take a tour to the Gold Coast hinterland; contact visitor centre for details. ***Minus 5 Ice Lounge:*** Circle on Cavill. ***Flycoaster and Bungee Rocket:*** thrill rides; Cypress Ave. ***Aquaduck Tours:*** take an adventure on both land and water in this amphibious vehicle; departs Orchid Ave.

Surfers Paradise Markets: The Esplanade, between Hanlan St and Elkhorn Ave; every Wed and Fri evening. ***Broadbeach Markets:*** Kurrawa Park; 1st and 3rd Sun each month. ***Lantern Market:*** Broadbeach Mall; Fri nights May–Sept. ***Farmers Market:*** Marina Mirage Shopping Centre; 1st Sat each month. ***Magic Millions Racing Carnival:*** Jan. ***Gold Coast Big Day Out:*** Music festival; Jan. ***Australian Beach Volleyball Championships:*** Mar. ***Blues on Broadbeach Music Festival:*** May–June. ***GC Bazaar:*** month-long festival of fashion, food and fun; June. ***Gold Coast Marathon:*** July. ***Tastes of the Gold Coast Festival:*** Aug. ***Tastes of Broadbeach Festival Food, Wine and Jazz Festival:*** Aug. ***Gold Coast Show:*** Aug–Sept. ***Gold Coast Eisteddfod:*** Aug–Sept. ***Go Xtreme Festival:*** actions sports at the Mudgeeraba Showgrounds: Sept. ***Gold Coast Indy 300:*** Oct. ***New Year's Eve fireworks:*** Dec.

Absynthe: innovative French and Australian cuisine; Q1 Building, Surfers Paradise Blvd; (07) 5504 6466. ***Chill on Tedder:*** great seafood, delicious desserts; Shop 10/26 Tedder Ave, Main Beach; (07) 5528 0388. ***Kurrawa Surf Club:*** modern bistro; Old Burleigh Rd, Broadbeach; (07) 5538 0806. ***Moo Moo The Wine Bar and Grill:*** modern steakhouse; Broadbeach on the Park Resort, 2685 Gold Coast Hwy, Broadbeach; (07) 5539 9952. ***Ristorante Fellini:*** classic Italian fine dining; Marina Mirage, Sea World Dr, Main Beach; (07) 5531 0300.

Q1 Resort & Spa: world's tallest residential tower; Hamilton Ave; (07) 5630 4524 or 1300 792 008. ***Surfers Paradise Marriott Resort & Spa:*** excellent facilities throughout; 158 Ferny Ave; (07) 5592 9800. ***Chevron Palms Resort:*** good-value low-rise; 50 Stanhill Dr, Chevron Island; (07) 5538 7933. ***Harbour Side Resort:*** good value close to attractions; 132 Marine Pde, Southport; (07) 5591 6666. ***Palazzo Versace:*** world's first designer hotel; Sea World Dr, Main Beach; (07) 5509 8000 or 1800 098 000. ***Sea World Resort:*** park entry with accommodation packages; Sea World Dr, Main Beach; (07) 5591 0000.

[HIGH-RISE BUILDINGS ALONG SURFERS PARADISE'S BEACHFRONT]

Sofitel Gold Coast: child-friendly and luxurious; 81 Surf Pde, Broadbeach; (07) 5592 2250 or 1800 074 465. ***Surfers Paradise YHA:*** dorms and doubles, no under 18s; 70 Sea World Dr, Main Beach; (07) 5571 1776. ***The Ritz Resort:*** desirable northern aspect; 8 Philip Ave, Broadbeach; (07) 5531 5185.

Broadbeach The cosmopolitan heart of the Gold Coast has sophisticated wine bars, chic cafes, sun-drenched beaches and a vibrant nightlife. Visit Pacific Fair, Australia's largest shopping centre, hit the famous Conrad Jupiters Casino and try your luck at some blackjack, or stroll down to Kurrawa Beach, home of Australia's major surf lifesaving competition. 3 km S.

Main Beach This wealthy area just north of Surfers Paradise is awash with trendy boutiques and chic eateries. The Southport Spit Jetty, or The Spit as it's commonly known, is home to Sea World and the famous Palazzo Versace Hotel, as well as many specialty shops, restaurants, outdoor cafes and weekend entertainment at Marina Mirage and Mariner's Cove. It is also a hot spot for fishing while diving can be done at the nearby wreck of the *Scottish Prince*. 3 km N.

South Stradbroke Island This resort island boasts peaceful coves to the west and lively ocean beaches to the east, separated by wetland and remnant rainforest. See the abundant bird and butterfly species and discover the pleasures of windsurfing and sailing on either a daytrip or longer stay. Access by ferry or private boat from Runaway Bay Marina; 11 km N.

Theme parks The Gold Coast is the theme park capital of Australia, with 4 major theme parks in the Oxenford and Coomera areas. Warner Bros. Movie World is 'Hollywood on the Gold Coast' with thrilling rides, stunts and shows that will interest all ages. Just down the road is Wet 'n Wild Water World. Enjoy the thrill-seeking rides or relax on Calypso Beach. Further north is Dreamworld, which has a diverse range of attractions from The Giant Drop (the tallest free-fall in the world) to the famous *Big Brother* House. Next to Dreamworld is Whitewater World, which has the latest technology in water theme-park slides and is home to the 'world's best' water slide.

Gold Coast Arts Centre: art gallery incorporating contemporary and historical Australian art. Evandale sculpture walk nearby; Bundall Rd; 2 km W. ***Miami and Nobby beaches:*** both beaches are separated by a headland known as Magic Mountain. Miami is home to many diehard surfers; 6 km S. ***The Broadwater:*** sheltered waterways excellent for boating (hire available), watersports and shore walks; access from Labrador and Southport; 5 km N. ***Surfers Riverwalk:*** scenic 9 km walk from Sundale Bridge at Southport to Pacific Fair at Broadbeach. ***Sanctuary Cove:*** famous area for championship golf courses. Hire a houseboat or take a cruise; 23 km NW. ***High-speed racing:*** 3 professional circuit tracks to test the visitor's driving skills in V8 Supercars, Commodores and WRXs. Courses at Ormeau and Pimpama; Pacific Hwy towards Brisbane; 33 km NW.

See also GOLD COAST & HINTERLAND, p. 417

Taroom

Pop. 626
Map ref. 653 I4

 **17 Kelman St; (07) 4628 6113.**

 102.1 Rebel FM

Taroom is on the banks of the Dawson River in the Capricorn region. Since settlers took up land in 1845, cattle raising has been the main local industry. The coolibah tree in Taroom's main street was marked 'L. L.' by explorer Ludwig Leichhardt on his 1844 trip from Jimbour House near Dalby to Port Essington, north of Darwin.

Museum: old telephone-exchange equipment, farm machinery and items of local history; open 9am–3pm Mon and Fri or by appt; Kelman St; (07) 4627 3231.

Campdraft: Apr and Oct. ***Taroom Show:*** May. ***Rodeo:*** July.

Expedition National Park This remote park features the 14 km Robinson Gorge. The 4 km return track leads to an excellent lookout. Only experienced walkers should attempt the rough track down into the gorge. The Cattle Dip Lookout overlooks a permanent waterhole reached via the 8 km return Shepherd's Peak Trail or from the carpark at the lookout. 4WD recommended; some of the road is gravel. (07) 4624 3535; 128 km NW.

Lake Palm Tree Creek: rare Livistona palms; 15 km N. ***Murphy Conservation Park:*** pristine lake with birdlife, and picnic and camping spots. It was the site of Leichhardt's campsite in 1844; 30 km N. ***Glebe Weir:*** waterskiing and fishing; 40 km NE.

Wandoan: the local heritage trail in the town visits all the major sights, including the Waterloo Plain Environmental Park; brochure from visitor centre in Royd St; 59 km S. ***Scenic and historic drives:*** self-drive tours to nearby sights; brochure from visitor centre.

See also CAPRICORN, p. 423

Texas

Pop. 694
Map ref. 557 J3 | 653 J12

Newsagency, 19 High St; (07) 4653 1384; texasqld4385.com.au

 90.1 Breeze FM, 711 AM ABC Western QLD

Texas lies alongside the Dumaresq River (pronounced Du-meric) and the Queensland–New South Wales border. Its name comes from the similarity between an 1850s land dispute in the area to a dispute between the Republic of Texas and Mexico. The town was originally on the river flat, 2 kilometres from its current position. Severe floods forced the move. Remains of the original town can be seen on the river off Schwenke Street.

Heritage Centre and Tobacco Museum: located in the old police building (1893), with memorabilia that shows 100 years of the tobacco industry, as well as horse-drawn vehicles, mini shearing shed and the gaol; open Sat or by appt; Fleming St; (07) 4653 1410. ***Art Gallery:*** local and touring art exhibitions; open Tues–Sat; High St. ***Riverside Freezing and Rabbit Works:***

interpretive display on the rabbit skins factory and Aboriginal artefacts; open by appt; Mingoola Rd; (07) 4653 1453.

Heritage Centre Markets: Heritage Centre; 1st Sat in June and Dec. ***Texas Show:*** July. ***Country Music Roundup:*** Sept. ***Horseracing:*** Dec.

Beacon Lookout: regional views; 3 km SE on Stanthorpe Rd. ***Cunningham Weir:*** site where Allan Cunningham crossed the Dumaresq River in 1827; 31 km W off Texas–Yelarbon Rd. ***Glenlyon Dam:*** excellent fishing spot; 45 km SE. ***Inglewood:*** Texas's twin town at the centre of Australia's olive industry. Visit Inglewood Heritage Centre for local memorabilia (open 2nd and last Sat each month or by appt). Tour the local olive groves and follow the scenic drives in area; brochure available; 55 km N. ***Coolmunda Reservoir:*** picnics, boating and fishing; 75 km NE via Inglewood. ***Dumaresq River:*** winding river popular for canoeing and fishing, and hiking through wilderness areas along its banks.

See also DARLING DOWNS, p. 418

Theodore

Pop. 442
Map ref. 653 I3

The Boulevard; (07) 4993 1900.

96.5 Sea FM, 855 AM ABC Wide Bay

Theodore, first called Castle Creek, was the site of Queensland's first irrigation project, opened in 1924. With its rich black soils, the town is surrounded by pastoral and grazing properties, with sheep, cattle, sorghum, wheat and cotton. Located on the Dawson River, irrigation has resulted in palm-lined streets and a tropical air. The town was named after Edward (Red Ted) Theodore, union leader then Queensland premier from 1919 to 1925.

Theodore Hotel: only cooperative hotel in Queensland; The Boulevard. ***Dawson Folk Museum:*** provides local history; open by appt; Second Ave; (07) 4993 1686.

Fishing Competition: Mar. ***Theodore Show:*** May.

Isla Gorge National Park This highland park has gorges and spectacular rock formations. The camping area overlooks the gorge and a 2 km return walk leads to the Isla Gorge Lookout. If staying overnight, watch the changing colours of the sandstone cliffs as the sun sets. (07) 4627 3358; turn-off 35 km S.

Theodore Weir: popular spot for fishing; southern outskirts of town. ***Moura:*** major cattle town that holds the annual Coal and Country Festival in Aug; 48 km NW. ***Cracow:*** where gold was produced from famous Golden Plateau mine 1932–76; 49 km SE.

See also CAPRICORN, p. 423

Tin Can Bay

Pop. 1920
Map ref. 648 F8 | 653 N6

Cooloola Regional Information Centre, Matilda's Roadhouse Complex, Bruce Hwy, Kybong (15 km S of Gympie); (07) 5483 5554 or 1800 444 222; www.tincanbaytourism.org.au

96.1 Zinc FM, 855 AM ABC Wide Bay

Tin Can Bay is a well-known fishing and prawning region north-east of Gympie. It was originally known as Tuncanbar to the local Aboriginal people. The town is a relaxing hamlet offering watersports on the quiet waters of Tin Can Bay inlet.

Environmental Walkway: a 9.5 km trail for birdwatching on the Tin Can Bay foreshore. ***Boat and yacht hire:*** cruise the inlet and Sandy Strait; Tin Can Bay Marina. ***Canoeing:*** eco-tours down estuaries; (07) 5486 4417. ***Norman Point boat ramp:*** see the dolphins up-close before 10am; access point to waterways.

Yacht Race: May. ***War Birds:*** model planes and helicopters; May. ***Seafood Festival:*** Sept.

Rainbow Beach This relaxing coastal town to the east across the inlet offers a pristine sandy beach popular with surfers. The Family Fishing Classic is held here each July. For adventure, try paragliding from the Carlo Sand Blow; bookings on (07) 5486 3048. Tours include dolphin ferry cruises, safaris and 4WD tours. The road south (4WD) leads to the coloured sands and beaches of the Cooloola section of Great Sandy National Park; *see Noosa Heads*. Rainbow Beach is 41 km E by road.

Carlo Point: great for fishing and swimming. There is also boat access to the inlet, with houseboats and yachts available for hire; 43 km E via Rainbow Beach. ***Inskip Point:*** camp along the point or take the car ferry to Fraser Island; 53 km NE via Rainbow Beach.

See also SUNSHINE COAST, p. 419

Townsville

see inset box on next page

Tully

Pop. 2459
Map ref. 650 E6 | 657 M8

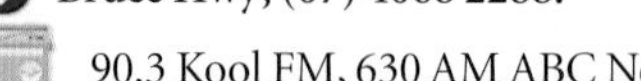

Bruce Hwy; (07) 4068 2288.

90.3 Kool FM, 630 AM ABC North QLD

At the foot of Mount Tyson, Tully receives one of the highest annual rainfalls in Australia – around 4200 millimetres. This abundance of rain supports swift rapids on the Tully River – an attraction for any whitewater-rafting enthusiast. The area was settled in the 1870s by a family keen on growing sugarcane, and the town grew when the government decided to build a sugar mill in 1925. Sugarcane remains a major industry.

Tully Sugar Mill: informative tours in the crushing season (approximately June–Nov); tickets from the visitor centre. ***Golden Gumboot:*** 7.9 m high gumboot erected to celebrate Tully as Australia's wettest town.

Tully Show: July. ***Golden Gumboot Festival:*** Sept.

Tully Gorge Alcock State Forest This state forest incorporates the Tully Gorge and the raging waters of the Tully River. Visit the Frank Roberts Lookout for gorge views, take the Rainforest Butterfly Walk and visit the Cardstone Weir boardwalk in the afternoons to watch rafters negotiate the rapids. Head to the top reaches of the river for superb scenery and swimming. Visit in dry season only (May–Dec); 40 km W.

Alligator's Nest: beautiful rainforest with swimming in stream; 10 km S. ***Tully Heads:*** estuary and beachside fishing; separated from Hull Heads by Googorra Beach; 22 km SE. ***Echo Creek Walking Trail:*** take this guided trail through rainforest, walking a traditional Aboriginal trading route with Jirrba guides; turn-off after Euramo; 30 km SW. ***Murray Upper State Forest:*** rainforest walks to cascades, rockpools and Murray Falls; turn-off 38 km S. ***Misty Mountains walking trails:*** day walks or longer (up to 44 km) in Wet Tropics World Heritage Area; details from visitor centre. ***Whitewater rafting:*** operators run from Tully to the renowned rapids of the Tully River; details from visitor centre.

See also CAIRNS & THE TROPICS, p. 425

TOOWOOMBA

Pop. 114 479

Map ref. 653 L9

86 James St; (07) 4639 3797 or 1800 331 155; www.toowoombaholidays.info

100.7 CFM, 747 AM ABC Southern QLD

Toowoomba is a city with a distinctive charm and graciousness in its wide, tree-lined streets, colonial architecture and over 240 parks and gardens. The city is perched 700 metres above sea level on the rim of the Great Dividing Range. It began in 1849 as a small village near an important staging post. It is now the commercial centre for the fertile Darling Downs where visitors are captivated by the mild summers, crisp leafy autumns and vibrant springs.

Cobb & Co Museum The museum traces the history of horse drawn vehicles in the Darling Downs region. It has been recently expanded to include the National Carriage Factory, featuring an open-plan training centre that hosts workshops for heritage skills including blacksmithing, saddlery, silversmithing and glass art, with a viewing area for visitors. There is also an interactive discovery centre for children. Open 10am–4pm daily; 27 Lindsay St; (07) 4659 4900.

Parks and gardens No trip to the 'Garden City' would be complete without a visit to some of the superb parks and gardens. Lake Annand is a popular recreation spot with boardwalks, bridges and ducks; Mackenzie St. There are imposing European trees in Queen's Park, which also includes the Botanic Gardens; Lindsay and Margaret Sts. Laurel Bank Park features the unique Scented Garden, designed for the visually impaired; cnr Herries and West Sts. Birdwatchers should visit Waterbird Habitat where native birds can be watched from observation platforms and floating islands; Mackenzie and Alderley Sts. The impressive Japanese Garden, at the University of Southern Queensland, is the largest in Australia; off West St.

Regional Art Gallery: changing exhibitions in Queensland's oldest gallery; open 10am–4pm Tues–Sat, 1–4pm Sun; 531 Ruthven St; (07) 4688 6652. ***Royal Bull's Head Inn:*** National Trust–listed building (1859) with small museum; open 10am–4pm Fri–Sun; Brisbane St, Drayton; (07) 4637 2278. ***Empire Theatre:*** live theatre in this restored Art Deco building, opened in 1911; 56 Neil St; 1300 655 299. ***Picnic Point:*** offers views of Lockyer Valley, mountains and waterfall. Enjoy the recreational facilities and walks through bushland. Perfect venue for Summer Tunes concerts during summer; Tourist Dr, eastern outskirts. ***City tour:*** Stonestreets Coaches offers a two-hour tour of the city; book at the visitor centre or (07) 4687 5555.

Markets: PCYC Markets; 7am–12pm Sun; 219a James St. Darling Downs Farmers Markets; 1st and 3rd Sun each month; 7am–12pm; Victoria St. ***Toowoomba Royal Show:*** Mar/Apr. Easterfest: music festival; Easter. ***Carnival of Flowers:*** Sept.

Spotted Cow Hotel: European beers and food; Ruthven St (cnr Campbell St); (07) 4639 3264. ***Veraison:*** modern Australian, good wine list; 205 Margaret St; (07) 4638 5909. ***Weis:*** legendary seafood smorgasbord; 2 Margaret St; (07) 4632 7666.

Beccles on Margaret B&B: 1938 Queensland bungalow; 25 Margaret St, East Toowoomba; (07) 4638 5254. ***Quality Hotel Platinum International:*** modern hotel; 326 James St; (07) 4634 0400. ***Vacy Hall Historic Guesthouse:*** National Trust mansion; 135 Russell St; (07) 4639 2055.

[COBB & CO MUSEUM]

Highfields: growing town featuring Orchid Park, Danish Flower Art Centre, quaint shopping complexes and historical village with vintage machinery and buildings; 12 km N. ***Cabarlah:*** small community that has the Black Forest Hill Cuckoo Clock Centre and holds excellent country markets on the last Sun each month; 19 km N. ***Southbrook:*** nearby Prestbury Farmstay offers accommodation and the experience of rural life; (07) 4691 0195; 34 km SW. ***Lake Cooby:*** fishing, walking and picnic facilities; 35 km N. ***Pittsworth:*** in town is the Pioneer Historical Village featuring a single-teacher school, early farming machinery and a display commemorating Arthur Postle, or the 'Crimson Flash', once the fastest man in the world; village open 10am–1pm Wed–Fri and 10am–4pm Sun or by appt; Pioneer Way; (07) 4619 8000; 46 km SW. ***Yandilla:*** here you'll find the quaint, steeped All Saints Anglican Church (1877); 77 km SW. ***Millmerran:*** colourful murals illustrate the history of this industrial town, including the dairy industry mural at the Old Butter Factory, and the mural on the water reservoir showing the development of transport. The town's historical museum is open 10am–3pm Tues–Thurs; (07) 4695 2560; 87 km SW. ***Scenic drives:*** take in places such as Spring Bluff, with old railway station and superb gardens (16 km N); Murphy's Creek and Heifer Creek.

See also DARLING DOWNS, p. 418

TOWNSVILLE

Pop. 143 329

Map ref. 649 A1 | 650 G10 | 654 G1 | 657 N11

Bruce Hwy, (07) 4778 3555 or 1800 801 902; or The Mall Information Centre, Flinders Mall, (07) 4721 3660; www.townsvilleholidays.info

106.3 FM Townsville's Best Mix, 630 AM ABC North QLD

Townsville is a bustling and vibrant tropical city boasting an enviable lifestyle, diverse landscapes, activities for all ages and a certain charm that comes with a long-standing history. Established in 1864, Townsville was set up to service a new cattle industry. The many historic buildings found around the city are a reminder of this heritage. These days, the city is better known for its cosmopolitan tastes and fast growth. Its iconic foreshore, The Strand, offers an excellent choice of cafes and restaurants, combined with tropical parks and spectacular views to Magnetic Island.

Reef HQ Home to the headquarters for the Great Barrier Reef Marine Park Authority, this underwater observatory is both informative and visually breathtaking. Its living reef is the largest 'captive' reef in the world. Get up-close at the touch pools or see marine feeding and dive shows. Open 9.30am–5pm daily; Flinders St East; (07) 4750 0800.

Museum of Tropical Queensland: featuring artefacts from the wreck of HMAS Pandora; open 9.30am–5pm daily; 70-102 Flinders St; (07) 4726 0600. ***Maritime Museum:*** maritime items of historical significance; open 10am–4pm weekdays, 12pm–4pm weekends; 42-68 Palmer St, South Townsville; (07) 4721 5759. ***Rock Pool Baths:*** year-round swimming; The Strand. ***Perc Tucker Regional Gallery:*** local and touring exhibitions; open 10am–5pm weekdays, 10am–2pm weekends; Flinders Mall; (07) 4727 9000. ***Botanic Gardens:*** Anderson Park, Kings Rd.

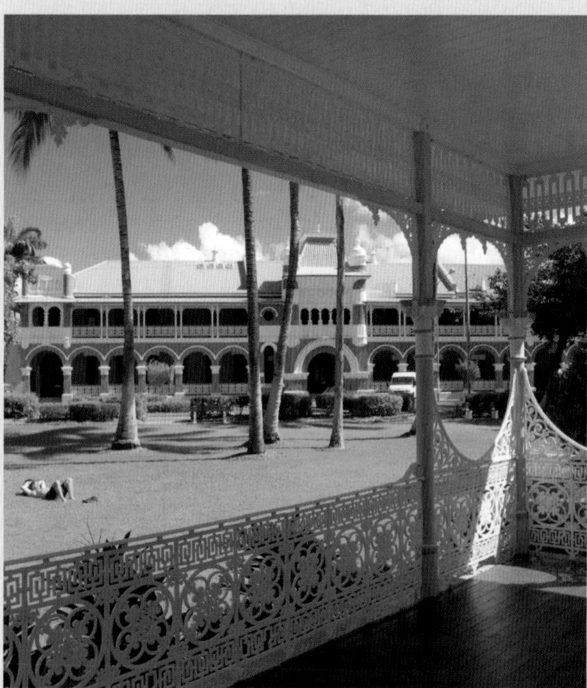

[TROPICAL GARDENS OUTSIDE THE OLD QUEENS HOTEL, THE STRAND]

IMAX Dome Theatre: excellent viewing experience in this theatre, the first of its kind in the Southern Hemisphere; call for viewing times; Reef HQ Building, Flinders St East; (07) 4721 1481. ***Town Common Conservation Park:*** a coastline park with prolific birdlife, Aboriginal Plant Trail and forest walks; Cape Pallarenda Rd. ***Castle Hill Lookout:*** off Stanley St. ***Jupiters Townsville Hotel and Casino:*** Sir Leslie Thiess Dr; (07) 4722 2333. ***Ferry Terminal:*** for Magnetic Island, daytrips and dive cruises to the Great Barrier Reef, and extended cruises to the Whitsundays; Sir Leslie Thiess Dr. ***Coral Princess:*** cruise to Cairns via resort islands and reef on a luxury catamaran; contact visitor centre for details. ***Adventure activities:*** numerous tour operators offer scuba diving, jetskiing, waterskiing, abseiling, whitewater rafting and more; contact visitor centre for details.

Showground Market: Ingham Rd, West End; 6am–2pm Sun. ***Cotters Market:*** Flinders Mall; 8.30am–1pm Sun. ***Horseshoe Bay Market:*** Beachside Park, Magnetic Island; 9am–1pm last Sun each month. ***Opal Fashion Bash:*** Apr. ***Italian Festival:*** May. ***Townsville Show:*** July. ***Australian Festival of Chamber Music:*** July/Aug. ***Townsville Cup:*** horseracing; July. ***Magnetic Island Race Week:*** sailing; Sept. ***Greek Fest:*** Oct.

Michel's Cafe & Bar: international fare; 7 Palmer St, South Townsville; (07) 4724 1460. ***SugarTrain:*** modern Australian; 14 Palmer St; (07) 4753 2000. ***Masala:*** Indian; 79 Palmer St, South Townsville; (07) 4721 3388. ***Watermark:*** seafood and steaks; 72 The Strand, North Ward; (07) 4724 4281.

Jupiters Townsville Hotel and Casino: 4.5-star hotel; Sir Leslie Thiess Dr; (07) 4722 2333. ***Balgal Beach Holiday Units:*** simple beachside units with lovely outdoor pool; 284 Ocean Pde; Balgal Beach; (07) 4770 7296. ***Hidden Valley Cabins:*** carbon-neutral resort about 90 min northwest of the city; Pine Creek Rd, Hidden Valley; (07) 4770 8088.

Magnetic Island More than half of this beautiful island is covered by the Magnetic Island National Park. The sandy beaches, granite headlands and hoop pine rainforest make this an attractive daytrip, or stay overnight in the accommodation on offer. See the natural attractions by taking some of the many walks around the island or hire a bike for a different view. Snorkel and swim at Alma Bay and see spectacular views from the old WW II forts. Access from Townsville by fast cat, passenger ferry or vehicle barge. 8 km NE.

Bowling Green Bay National Park This coastal park offers much for the self-sufficient visitor. Granite mountains blend with a variety of landscapes including saltpans and mangrove country. Walk along Alligator Creek to see cascades and waterfalls. Stay overnight at the Alligator Creek campsite and go spotlighting to glimpse brush-tail possums and sugar gliders; call 13 74 68 for camping bookings. Turn-off Bruce Hwy 28 km SE; park is further 6 km.

Billabong Sanctuary: covering 10 ha of rainforest, eucalypt forest and wetlands. See koala feeding and crocodile shows; Bruce Hwy, Nome; (07) 4778 8344; 17 km SE. ***Giru:*** small community with waterfalls, bushwalks and swimming nearby; 50 km SE.

See also THE MID-TROPICS, p. 424

Warwick

Pop. 12 564
Map ref. 551 B1 | 557 L1 | 653 L10

49 Albion St (New England Hwy); (07) 4661 3122; www.southerndownsholidays.com.au

 89.3 Rainbow FM, 747 AM ABC Southern QLD

Warwick is an attractive city set alongside the willow-shaded Condamine River. It is known as the 'Rose and Rodeo City', as the Warwick Rodeo dates back to the 1850s, and the parks and gardens have an abundance of roses. There is even a red rose cultivated especially for Warwick – the City of Warwick Rose (or Arofuto Rose). The area was explored by Allan Cunningham in 1827, and in 1840 the Leslie brothers established a sheep station at Canning Downs. Warwick was eventually established in 1849 on the site that Patrick Leslie selected. The surrounding pastures support famous horse and cattle studs.

Australian Rodeo Heritage Centre: follow the history and relive the glory of the Australian Professional Rodeo Association's greatest champions; open 10am–3pm Mon–Sat or by appt; Alice St. ***Pringle Cottage:*** historic home (1870) housing large historic photo collection, vehicles and machinery; closed Mon–Tues; Dragon St. ***Warwick Regional Art Gallery:*** local and touring exhibitions; closed Mon; Albion St. ***Jubilee Gardens:*** see the displays of roses that Warwick is famous for; Cnr Palmerin and Fitzroy sts. ***Lookout:*** viewing platform for regional views; Glen Rd. ***Historic walk or drive:*** self-guide tour of historic sandstone buildings dating from the 1880s and 1890s; brochure from visitor centre.

Rock Swap Festival: Easter. ***Jumpers and Jazz:*** July. ***Campdraft and Rodeo:*** Oct.

Spicers Peak Lodge Restaurant: up-market modern Australian; Wilkinsons Rd, Maryvale; (07) 4666 1083 or 1300 773 452.

Coachman's Inn Warwick: modern suites and apartments; 91 Wood St; (07) 4660 2100. ***Spicers Peak Lodge:*** small luxury resort; Wilkinsons Rd, Maryvale; (07) 4666 1083 or 1300 773 452.

Leslie Dam: watersports, fishing and swimming; 15 km W. ***Main Range National Park:*** 61 km NE; *see Boonah and Killarney.* ***Heritage drive:*** 80 km cultural drive in region; brochure from visitor centre.

See also DARLING DOWNS, p. 418

Weipa

Pop. 2832
Map ref. 658 B7

Evans Landing; (07) 4069 7566.

 100.9 KIG FM, 1044 AM ABC Far North QLD

This coastal part of Cape York was reputedly the first area in Australia to be explored by Europeans (1605). The town of Weipa was built in 1961 on the site of a mission station and Aboriginal reserve, and is now home to the world's largest bauxite mine. Although the town is remote, it offers a full range of services for travellers.

Western Cape Cultural Centre This centre was established to introduce the visitor to the culture of western Cape York. The range of artefacts and photos bring the area's Indigenous and European history alive. A highlight is the ceramic wall mural depicting sacred images of the local Aboriginal people. There is also information about the landscapes and ecosystems of the Cape, and sales of local arts and crafts. Open 10am–3pm daily; Evans Landing; (07) 4069 7566.

Tours of bauxite mine: guided tours provide insight into the mining process at Weipa; details from visitor centre. ***Fishing tours:*** Weipa's fishing spots can be explored on tours; details from visitor centre. ***Boat and houseboat hire:*** details from visitor centre.

 ***Fishing Competition:*** June. ***Bullride:*** Aug.

Thursday Island and the Torres Strait The islands of the Torres Strait stretch from the tip of Cape York Peninsula to Papua New Guinea and comprise 17 inhabited islands. The first Europeans passed through the Islands in the 1600s, and by the late 1800s a pearling industry was established, which continues today along with crayfishing, prawning and trochus industries. Thursday Island is the administrative centre. Visit the Torres Strait Island Cultural Centre – it preserves the cultural heritage of the islands and documents their art, culture, geography and history in an excellent interpretive display. Other surrounding islands worth a visit are Friday Island where you can see pearls being cultivated at Kazu; Horn Island, which was an important posting for Australian Troops in WW II; and Badu Island, where you can enjoy traditional dances, arts and crafts and food. Getting to the Islands involves either a flight from Cairns, a trip on a cargo vessel from Cairns (through Seaswift, (07) 4035 1234), or a ferry ride from Cape York (through Peddells, (07) 4069 1551).

Jardine River National Park This remote park is on the north-east tip of Cape York Peninsula. It was known to early explorers as the 'wet desert' because of its abundant waterways, but lack of food. These waters attract varied birdlife including the rare palm cockatoo. See the Fruit Bat Falls from the boardwalk or fish in restricted areas. 4WD access only. Visit between May–Oct; off Peninsula Development Rd, south of Bamaga; (07) 4069 5777.

Fishing and camping: a number of areas developed for the well-equipped visitor. ***Mungkan Kandju National Park:*** wilderness park of open forests, swamps and dense rainforest. There is excellent birdlife around lagoons and bushwalking along Archer River. 4WD access only; visit between May–Nov; (07) 4060 1137; turn-off 29 km N of Coen. ***Mapoon:*** camping and scenery; permit required; 85 km N. ***Iron Range National Park:*** this important lowland tropical rainforest park is a haven for wildlife. There is good fishing at Chili Beach, and bush camping for self-sufficient visitors only. 4WD recommended; visit only between Apr–Sept; (07) 4060 7170; 216 km E.

Travellers note: *Roads to the Cape may become impassable during the wet season (Nov–Apr). Motorists are advised to check the RACQ Road Conditions Report on 1300 130 595 (or www.racq.com.au) before departing. Permits for travel over Aboriginal land can be sought in Weipa; details from visitor centre. Beware of crocodiles in rivers, estuaries and coastal areas. Also beware of marine stingers in coastal areas (Oct–Apr) and swim within enclosures where possible.*

See also CAPE YORK, p. 426

Winton

Pop. 983
Map ref. 661 L7

Waltzing Matilda Centre, 50 Elderslie St; (07) 4657 1466; www.matildacentre.com.au

 95.9 West FM, 540 AM ABC Western QLD

The area surrounding Winton is known as Matilda Country, as Australia's most famous song, *Waltzing Matilda*, was written by Banjo Paterson at nearby Dagworth Station in 1895. Combo Waterhole (near Kynuna) was the setting for the ballad and the tune had its first airing in Winton. A less auspicious event in Winton's history was the declaration of martial law in the 1890s following the Shearers' Strike. In 1920 the first office of Qantas was registered here. The town's water supply comes from deep artesian bores at a temperature of 83°C. The movie *The Proposition* was filmed in the town in 2004.

Waltzing Matilda Centre Created as a tribute to the life of the swagman. The centre incorporates the 'Billabong Courtyard' with its sound and light show, the regional art gallery, interactive exhibits showcasing the swagmans life in 'Home of the Legend' hall, and the Matilda Museum that harks back to Winton's pioneering days and the first days of Qantas. Elderslie St; (07) 4657 1466.

Royal Theatre: historic open-air movie theatre and museum, one of the oldest still operating in Australia; Elderslie St. ***Corfield and Fitzmaurice Store:*** charming National Trust–listed store with diorama of Lark Quarry dinosaur stampede; Elderslie St. ***Gift and Gem Centre:*** displays and sales; the 'Opal Walk' leads to the theatre museum; Elderslie St. ***Arno's Wall:*** ongoing concrete-wall creation proudly containing 'every item imaginable'; Vindex St. ***North Gregory Hotel:*** built in 1878, this historical structure has been ravaged by 3 separate fires. Artwork and opal collection on display; Elderslie St.

Winton Show: June. ***Opal Expo:*** July. ***Camel Races:*** July. ***Rodeo:*** Aug. ***Outback Festival:*** odd-numbered years, Sept.

North Gregory Hotel Motel: family friendly; 67 Elderslie St; (07) 4657 1375.

Lark Quarry Conservation Park This park features the preserved tracks of a dinosaur stampede from 93 million years ago – the only track of this type known in the world. The stampede occurred when a Therapod chased a group of smaller dinosaurs across the mud flats. The 'trackways' are sheltered and can be visited on a tour (details from visitor centre). The park also offers a short walk past ancient rock formations, known as the Winton Formation, to a lookout over the region. 110 km sw.

Bladensburg National Park The vast plains and ridges of this park provide an important sanctuary for a variety of wildlife, including kangaroos, dunnarts and emus. Skull Hole (40 km s) has Aboriginal paintings and bora ceremonial grounds and is believed to be the site of a late-1880s Aboriginal massacre. Walking should only be attempted by experienced bushwalkers. 'Route of the River Gums', a self-drive tour, shows the region's varied landscapes. (07) 4652 7333; drive starts 8 km s.

Carisbrooke Station: a working sheep station with Aboriginal cave paintings and scenic drives in the surrounds; day tours and accommodation available; 85 km sw. ***Opalton:*** see the remains of the historic town or try fossicking for opals in one of the oldest fields in Queensland; licence available from visitor centre; 115 km s. ***Air charters and ground tours:*** to major regional sights; details from visitor centre.

See also OUTBACK, p. 428

Yandina

Pop. 1078
Map ref. 647 F3 | 648 E11 | 653 N7

Yandina Historic House, 3 Pioneer Rd (at the roundabout); (07) 5472 7181; www.yandinahistorichouse.com.au

90.3 AM ABC Coast FM, 91.1 Hot FM

Yandina is in the Sunshine Coast hinterland north of Nambour. The first land claims in the area were made here in 1868. It is now home to The Ginger Factory, the largest of its kind anywhere, giving rise to Yandina's title of 'Ginger Capital of the World'.

The Ginger Factory This award-winning complex is devoted to everything ginger. Visitors can see Gingertown, watch ginger-cooking demonstrations, see ginger being processed and ride on the historic Queensland Cane Train through subtropical gardens with acres of tropical plants, water features and a plant nursery – a highlight are the stunning flowering gingers. Pioneer Rd.

Yandina Historic House: local history display, arts and crafts, art gallery and visitor centre; Pioneer Rd. ***Nutworks Macadamia Processes:*** see processing of macadamia nuts and taste-test the results; opposite The Ginger Factory. ***Fairhill Native Plants and Botanic Gardens:*** fabulous nursery and gardens – a must for any native-plant buff; also includes the excellent Elements Restaurant, with breakfast, lunch and afternoon tea in botanic garden setting; Fairhill Rd. ***Heritage trail:*** self-guide trail around town sights; brochure from visitor centre.

Market: town centre; Sat. ***Ginger Flower Festival:*** Jan.

Spirit House: superb Thai food; 20 Ninderry Rd, Yandina; (07) 5446 8994.

Ninderry House: 3 ensuite rooms; 8 Karnu Dr; (07) 5446 8556. ***Ninderry Manor Luxury Retreat:*** 3 guest rooms, hosts are qualified masseurs; 12 Karnu Dr; (07) 5472 7255.

Eumundi This historic town has a variety of excellent galleries to visit, including some with Indigenous Australian art. Also sample exquisite handcrafted chocolates at Cocoa Chocolate, all made on-site (Etheridge St); or browse through the maze of shelves at Berkelouw Books in the main street. The impressive country markets are renowned for their size with over 600 stalls. The quality of the fresh produce, art and craft and cut flowers, along with the wonderful atmosphere, brings visitors to the town each Wed and Sat, and on Thurs night Dec–Jan; 10 km N.

Wappa Dam: popular picnic area; west of Yandina. ***Yandina Speedway:*** offers a variety of motor races; call (07) 5446 7552 for details; just west on Wappa Falls Rd.

See also SUNSHINE COAST, p. 419

Yeppoon

Pop. 13 282
Map ref. 655 M10

Capricorn Coast Tourist Information Centre, Scenic Hwy; (07) 4939 4888 or 1800 675 785; www.capricorncoast.com.au

91.3 FM Radio Nag

The popular coastal resort of Yeppoon lies on the shores of Keppel Bay. Yeppoon and the beaches to its south – Cooee Bay, Rosslyn Bay, Causeway Lake, Emu Park and Keppel Sands – are known as the Capricorn Coast. Great Keppel Island Resort lies 13 kilometres offshore and is a popular holiday destination offering great swimming, snorkelling and diving.

The Esplanade: attractive strip of shops, galleries and cafes overlooking parkland and crystal-clear water. ***Doll and Antiquity Museum:*** Hidden Valley Rd.

Market: showgrounds; Sat mornings. ***Fig Tree Markets:*** 1st Sun each month. ***Australia Day Celebrations:*** Jan. ***Yeppoon Show:*** June. ***Ozfest:*** World Cooeeing Competition; July. ***Village Arts Festival:*** Aug. ***Rodeo:*** Sept. ***Yeppoon Tropical Pinefest:*** Sept/Oct.

Capricorn Resort Yeppoon: family-friendly golf resort; Farnborough Rd; (07) 4939 5111 or 1300 857 922.

Great Keppel Island Popular island holiday destination with over 15 beaches to explore, swim and relax on. There is snorkelling and diving offshore and walks exploring the island's centre, including an interesting Aboriginal culture trail. Various styles of accommodation are offered. Access is by ferry from Keppel Bay Marina at Rosslyn Bay Harbour. 7 km s.

Byfield National Park This coastal park offers uninterrupted views of the ocean from its long beaches. Explore the open woodlands and forest or take in coastal views from the headlands at Five Rock and Stockyard Point. Fishing and boating are popular at Sandy Point at the south of the park. 4WD only; experience in sand driving is essential. (07) 4936 0511; 32 km N.

Boating: bareboat and fishing charters, sea access to Great Keppel Island and nearby underwater observatory, and water taxis to Keppel Bay islands; all from Keppel Bay Marina; 7 km s. ***Wetland Tour:*** Australian nature tour at Rydges Capricorn International Resort; 9 km N. ***Cooberrie Park:*** noted flora and fauna reserve; 15 km N. ***Emu Park:*** small village community with historical museum and interesting 'singing ship' memorial to Captain Cook – the sea breezes cause hidden organ pipes to make sounds. There is a Service of Remembrance to American troops each July, and Octoberfest is held each Oct; 19 km s. ***Byfield State Forest:*** the extremely rare Byfield fern is harvested here. Walks include the 4.3 km Stony Creek circuit track through rainforest and the boardwalk along Waterpark Creek; adjacent to Byfield National Park. ***Keppel Sands:*** this popular spot for fishing and crabbing is home to the excellent emerging Joskeleigh South Sea Island Museum, with the Koorana Crocodile Farm nearby; 38 km sw. ***Keppel Bay Islands National Park:*** this scenic group of islands is popular for walks, snorkelling, reef-walking and swimming; private boat or water taxi from Keppel Bay Marina; (07) 4933 6595. ***Capricorn Coast Coffee:*** Australia's largest coffee plantation; tours available; details from visitor centre. ***Scenic flights:*** over islands and surrounds from Hedlow Airport (between Yeppoon and Rockhampton); details from visitor centre.

See also CAPRICORN, p. 423

Yungaburra

Pop. 932

Map ref. 650 D3 | 651 E9 | 657 L7

Allumbah Pocket Cottages, Gillies Hwy, (07) 4095 3023; or Nick's Swiss–Italian Restaurant, Gillies Hwy, (07) 4095 3330; www.athertontableland.com

102.5 4KZ FM, 720 AM ABC Far North QLD

Yungaburra is a historic town on the edge of the Atherton Tableland. Originally a resting spot for miners, it was slow to develop. The tourism boom did not hit until the coastal road opened from Cairns in 1926. Today the town offers craft shops, galleries, cafes and restaurants.

Historical precinct walk: take this self-guide walk past heritage buildings, including the popular Lake Eacham Hotel in Cedar St; brochure available from visitor centre. ***Platypus viewing platform:*** see the elusive animal at sunrise and sunset; Peterson Creek, Gillies Hwy. ***Galleries, craft and gem shops:*** various outlets in town.

Market: renowned produce and craft market; Gillies Hwy; 4th Sat each month. ***Tour de Tableland:*** bike race; May. ***Malanda Show:*** July. ***Folk Festival:*** Oct.

Eden House: modern Australian; 20 Gillies Hwy; (07) 4089 7000. ***Flynn's Licensed Restaurant:*** European-influenced fare; 17 Eacham Rd; (07) 4095 2235. ***Nick's Restaurant and Yodeller's Bar:*** Swiss-Italian; 33 Gillies Hwy; (07) 4095 3330.

Eden House Retreat & Mountain Spa: cottages and villas; 20 Gillies Hwy; (07) 4089 7000. ***Mt Quincan Crater Retreat:*** romantic retreat; Peeramon Rd; (07) 4095 2255. ***Williams Lodge:*** luxury suites in historic family home; Cedar St; (07) 4095 3449.

Crater Lakes National Park The 2 volcanic lakes, Lake Eacham (8 km E) and Lake Barrine (12 km NE), are surrounded by rainforest and offer watersports, bushwalking and birdwatching. Look for the eastern water dragons along the 3 km track around Lake Barrine or take a wildlife cruise on the lake. There is a children's pool at Lake Eacham, a self-guide trail through the rainforest and a 3 km circuit shore track. Both lakes are popular recreation areas. (07) 4091 1844.

Malanda Malanda is a small town in the middle of rich dairy-farming country. It claims the longest milk run in the world (to Alice Springs) and boasts the still-operating 19th-century Majestic Theatre, and the Gourmet Food Factory specialising in local produce. On the southern outskirts of town is the Malanda Falls Conservation Park, with signposted rainforest walks, and the Malanda Environmental Centre, which has displays on local history, vulcanology, flora and fauna. The Malanda Falls actually flow into the local swimming pool. 20 km s.

Curtain Fig Tree: spectacular example of strangler fig with aerial roots in curtain-like formation; 2.5 km sw. ***Tinaburra:*** a great spot for swimming and watersports on Lake Tinaroo; boat ramp provides access. Nearby Seven Sisters are 7 volcanic cinder cones; 3 km N. ***Lake Tinaroo:*** watersports and fishing; travel around on a houseboat or dinghy – hire available; north of Yungaburra. ***Heales Outlook:*** spectacular views over Gillies Range; 16 km NE. ***Tinaroo Falls Dam outlet:*** views over lake; 23 km N.

See also CAIRNS & THE TROPICS, p. 425

TASMANIA IS...

Shopping for fresh produce and crafts at Hobart's SALAMANCA MARKET / Walking through a forest canopy at the TAHUNE FOREST AIRWALK / A cruise down the GORDON RIVER from Strahan / Bushwalking in CRADLE MOUNTAIN–LAKE ST CLAIR NATIONAL PARK / Learning the chilling history behind PORT

TASMANIA

[TROUSERS BAY, FLINDERS ISLAND]

ARTHUR HISTORIC SITE / Visiting cellar doors along the TAMAR VALLEY WINE ROUTE / Taking the one-hour walk to the lookout over WINEGLASS BAY in Freycinet National Park / A chairlift ride to the top of THE NUT, Stanley / Discovering CATARACT GORGE and the nearby landscaped gardens in Launceston

TASMANIA is bursting with wonderful surprises. A winding country road can suddenly reveal a colonial village, a boutique vineyard or a breathtaking ocean view like that of Wineglass Bay. From landscape and history to food and culture, Australia's island state is a feast for travellers.

Although it is Australia's smallest state – only 296 kilometres from south to north and 315 kilometres east to west – Tasmania's territory also includes the Bass Strait islands and subantarctic Macquarie Island. A population of just over 500 000 is eclipsed by over half a million visitors each year, and Tasmania is famous for its friendly, welcoming and relaxed pace of life.

Tasmania's Indigenous peoples have been here for around 35 000 years, and despite the terrible impact of white settlement, they are a large and increasingly influential community today. Middens are common around the coastline, showing where generations of Aboriginal people cooked shellfish meals.

Abel Tasman sighted and named Van Diemen's Land in 1642, closely followed by French and British explorers. The British – never keen to be outdone by the French – acted in 1803 to establish a presence on the River Derwent. With the arrival of the British, white settlement

[SLEEPY BAY, FREYCINET PENINSULA]

got off to a rollicking and violent start as a penal colony for the first 50 years.

In more recent history, Tasmania is the home of the world's first 'green' political party. Local environmental politics captured international attention in the 1980s when the No Dams campaign saved the Franklin River from being flooded for a hydro-electric scheme.

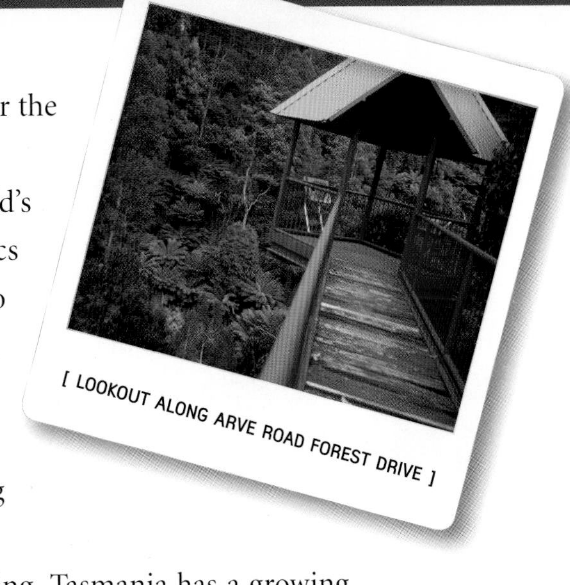

[LOOKOUT ALONG ARVE ROAD FOREST DRIVE]

The island's spirited cultural life includes the renowned Tasmanian Symphony Orchestra, David Walsh's astounding Museum of Old and New Art, and a full diary of festivals.

Although two-thirds of the land is too harsh for farming, Tasmania has a growing reputation for boutique agriculture and aquaculture. A gastronomic circumnavigation of the island offers as much diversity as the landscape and a chance to discover Tasmania's cool-climate wines, fresh seafood, fruits and fine cheeses.

fact file

Population 507 600
Total land area 68 102 square kilometres
People per square kilometre 7.1
Sheep per square kilometre 49.6
Length of coastline 2833 kilometres
Number of islands 1000
Longest river South Esk River (252 kilometres)
Largest lake Lake Gordon (hydro-electric impoundment) (271 square kilometres)
Highest mountain Mount Ossa (1614 metres), Cradle Mountain–Lake St Clair National Park
Coldest place At Liawenee in August the average minimum temperature is – 1.8°C
Longest place name Teebelebbberrer Mennapeboneyer (Aboriginal name for Little Swanport River)
Best beach Wineglass Bay, Freycinet National Park
Biggest surfable wave Shipstern Bluff near Port Arthur
Tallest tree At 99.6 m, the Arve Valley's giant swamp gum 'Centurion' is the world's tallest eucalypt
Most famous son Ricky Ponting, captain of the Australian cricket team
Most famous daughter Danish Crown Princess Mary Donaldson
Quirkiest festival National Penny Farthing Championships, Evandale
Most expensive gourmet produce Stigmas of *Crocus sativus*, grown on a saffron farm, which can attract a price of $30 000 per kilogram
Local beers Cascade in the south; Boags in the north

gift ideas

Arts and crafts (Salamanca Market, Hobart) A range of fantastic products on offer, from Jemma Clements' hand-blown perfume bottles to handcrafted vegetable soaps. See Salamanca Place p. 492, 492 C3

Huon pine box (Handmark Gallery, Hobart) Buttery aromatic wood endemic to Tasmania and treasured by woodworkers. 77 Salamanca Pl, Hobart. 492 C3

Peter Dombrovskis poster (Wilderness Society Shop, Hobart) Haunting wilderness images by the man whose photographs influenced Tasmanian conservation history, including the protection of the Franklin River against flooding. Galleria building, 33 Salamanca Pl, Hobart. 492 C3

Leatherwood honey in decorative tin (Tasmanian Honey Company, Perth) Full-flavoured honey from the nectar of trees found in Tasmania's rainforest wilderness. See Longford p. 519, 675 H11

Tasmanian Devil soft toy (Tasmanian Devil Conservation Park, Taranna) You can't snuggle up with the real thing, so cuddle a stuffed version of this famous marsupial. See Eaglehawk Neck p. 513, 671 M8

Shell necklace (Queen Victoria Museum and Art Gallery, Launceston) Tasmania's Aboriginal women string traditional necklaces of iridescent blue-green maireener shells. See Launceston p. 520, 675 G10

Sweets (Richmond) Enter the town's old-fashioned lolly shop to find giant Tasmania-shaped freckles or uniquely Tasmanian Esmeraldas (coconut ice covered in toffee). See Richmond p. 524, 669 J3

Ghosts of Port Arthur book or DVD (Port Arthur) Scare your friends witless with tales of strange apparitions, available from the giftshop. See Port Arthur p. 525, 669 M9

Oysters (Barilla Bay, Cambridge) Tasmania's deep, cool and flowing waters are the key to large and juicy oysters. Barilla Bay will package them for air travel. 1388 Tasman Hwy, Cambridge. 669 J5

Cheese (King Island Dairy, King Island) King Island's world-famous, handcrafted cheeses can be bought directly from the Fromagerie Tasting Room next to the factory in Loorana. The company's range covers everything from soft white cheeses to aged cheddars, yoghurts and desserts. See Currie p. 510, 671 O11

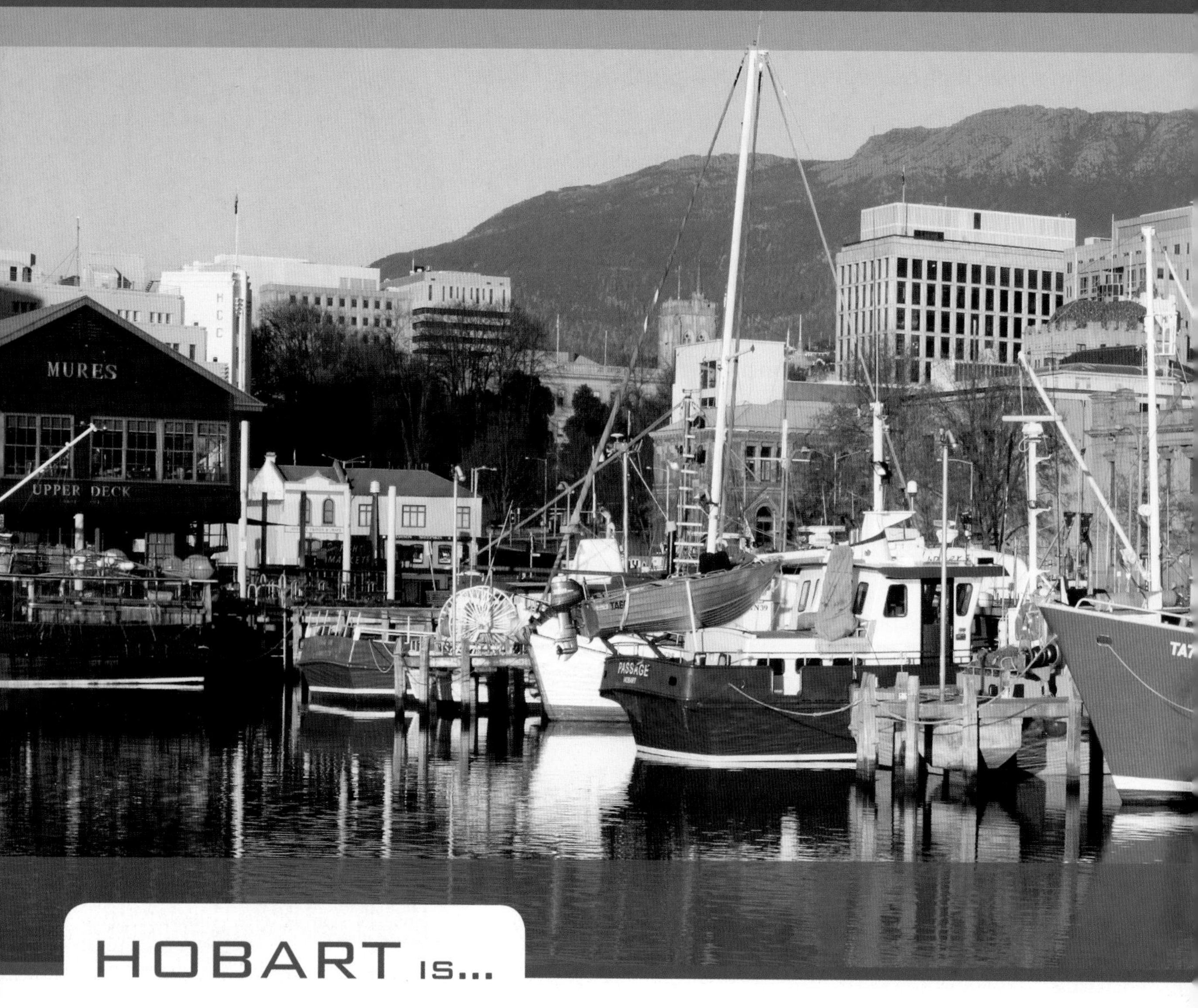

HOBART IS...

Shopping for fresh produce and crafts at SALAMANCA MARKET / Fish and chips at CONSTITUTION DOCK / A drive to the summit of MOUNT WELLINGTON / Exploring Hobart's WATERFRONT / An afternoon at the ROYAL TASMANIAN BOTANICAL GARDENS / Wandering around historic BATTERY POINT / A visit to the TASMANIAN MUSEUM AND ART GALLERY / A counter meal in a warm, colonial-era pub / Having breakfast and a browse around THE HENRY JONES ART HOTEL / Watching beer production at CASCADE BREWERY / Coffee and a movie at North Hobart's independent STATE CINEMA

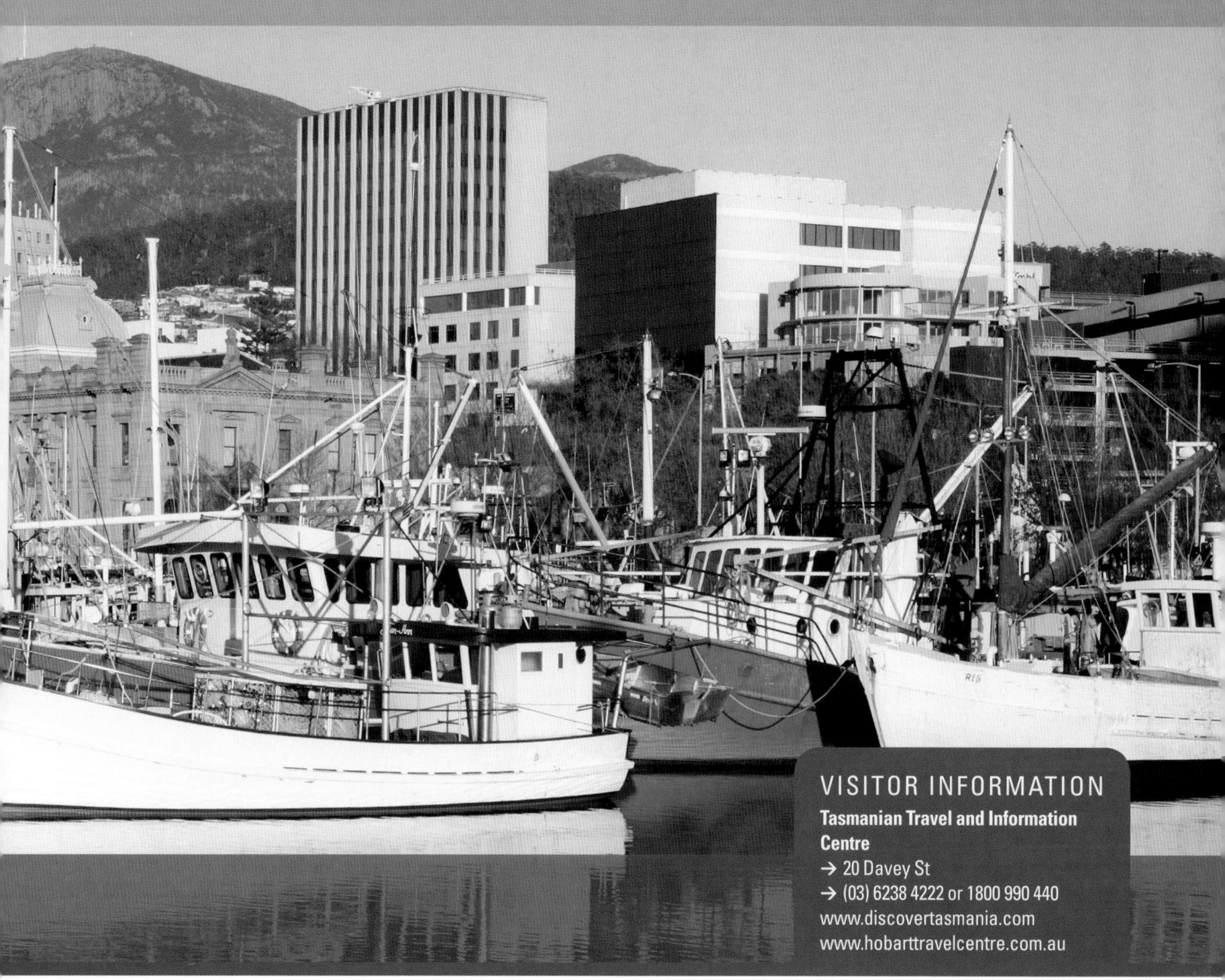

VISITOR INFORMATION

Tasmanian Travel and Information Centre

→ 20 Davey St

→ (03) 6238 4222 or 1800 990 440

www.discovertasmania.com

www.hobarttravelcentre.com.au

At the southern tip of Australia, Hobart lies nestled between the slopes of Mount Wellington and the Derwent estuary. It was the second city after Sydney to be established, yet today it is the smallest of the capitals with just 210 000 people enjoying its glorious location and unhurried, easy-going lifestyle.

Impelled to set up colonies in the face of French exploration, the English established Hobart at Sullivans Cove in 1804 on land known to its Aboriginal inhabitants as Nibberloonne. Whaling and sealing brought wealth to the town, its dockside soon dotted with taverns doing a brisk trade among seafarers and traders. Hobart's fortunes still centre on its deep-water harbour, but these days fishing trawlers and freighters moor alongside tourist ferries, Antarctic research vessels and luxury ocean liners.

Boasting internationally recognised temperate wilderness on its doorstep, Hobart's abundance of natural beauty propelled it to the forefront of environmental politics in 1972, becoming home to the world's first 'green' political party. With a cosmopolitan literary and arts culture centred on a handful of hip galleries and cafes around the waterfront, but without the hustle and bustle of bigger cities, Hobart has turned small-city attributes to its advantage with its laid-back friendly vibe.

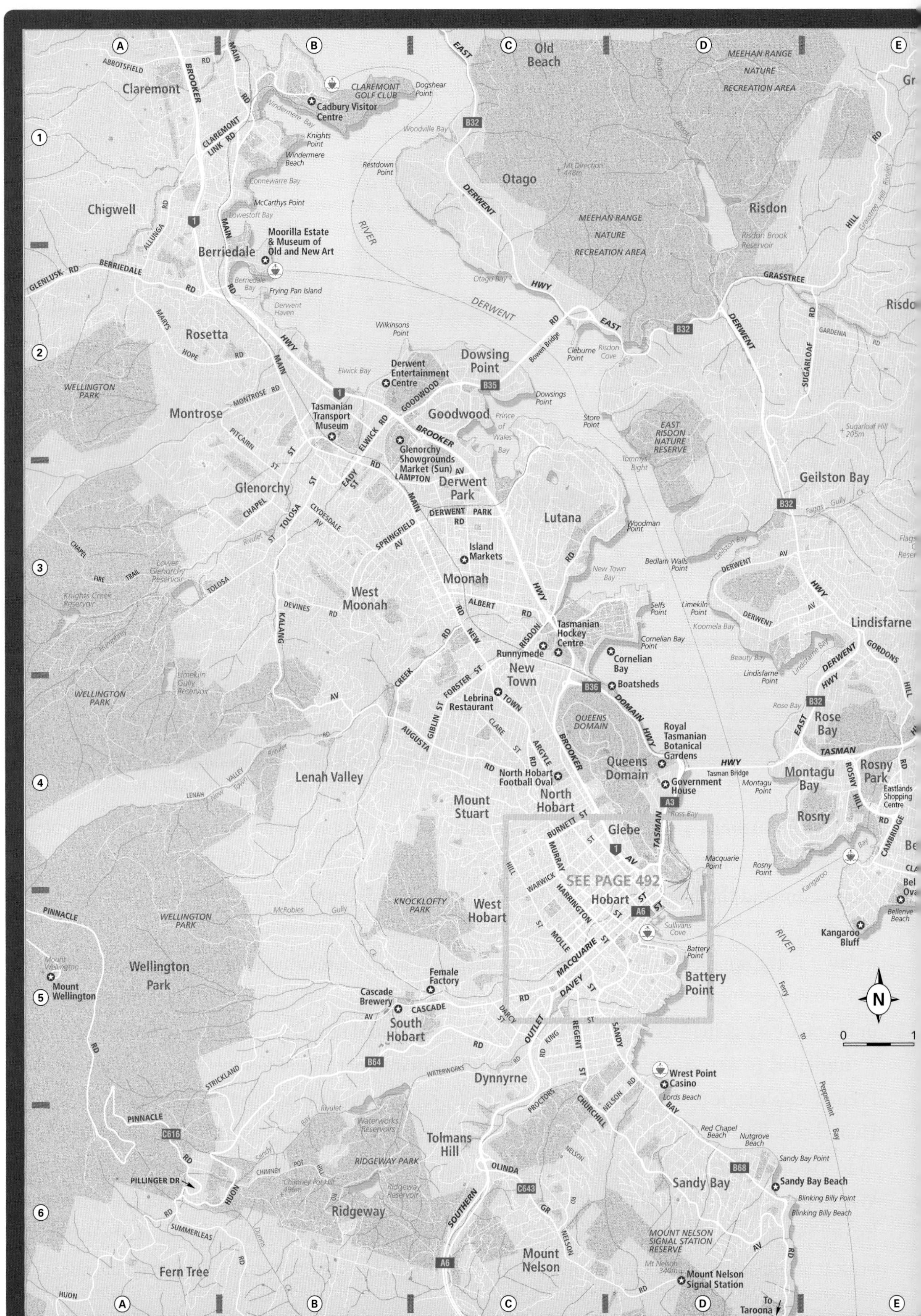
Claremont
Cadbury Visitor Centre
CLAREMONT GOLF CLUB
Dogshear Point
Knights Point
Windermere Beach
Windermere Bay
Woodville Bay
Restdown Point
Old Beach
Otago
Mt Direction 448m
MEEHAN RANGE NATURE RECREATION AREA
Risdon
Risdon Brook Reservoir
Chigwell
Connewarre Bay
McCarthys Point
Lowestoft Bay
Moorilla Estate & Museum of Old and New Art
Berriedale
Berriedale Bay
Frying Pan Island
Derwent Haven
RIVER
DERWENT
Otago Bay
Rosetta
Wilkinsons Point
Elwick Bay
Derwent Entertainment Centre
Dowsing Point
Dowsings Point
Bowen Bridge
Cleburne Point
Risdon Cove
WELLINGTON PARK
Montrose
Tasmanian Transport Museum
Goodwood
Prince of Wales Bay
Store Point
EAST RISDON NATURE RESERVE
Tommys Bight
Sugarloaf Hill 205m
Geilston Bay
Glenorchy Showgrounds Market (Sun)
Glenorchy
Derwent Park
Lutana
Woodman Point
Island Markets
Moonah
West Moonah
Lower Glenorchy Reservoir
Knights Creek Reservoir
New Town Bay
Bedlam Walls Point
Geilston Bay
Limekiln Point
Koomela Bay
Selfs Point
Tasmanian Hockey Centre
Runnymede
Cornelian Bay Point
Cornelian Bay
Beauty Bay
Lindisfarne
Lindisfarne Point
Lindisfarne Bay
Boatsheds
New Town
Lebrina Restaurant
Limekiln Gully Reservoir
WELLINGTON PARK
QUEENS DOMAIN
Royal Tasmanian Botanical Gardens
Rose Bay
Rose Bay
Queens Domain
Tasman Bridge
Montagu Point
Montagu Bay
Rosny Park
Eastlands Shopping Centre
Lenah Valley
North Hobart Football Oval
Government House
North Hobart
Mount Stuart
Ross Bay
Rosny
Glebe
Macquarie Point
Rosny Point
SEE PAGE 492
Hobart
KNOCKLOFTY PARK
West Hobart
Sullivans Cove
Kangaroo Bluff
Bellerive Beach
WELLINGTON PARK
Wellington Park
Mount Wellington
Mount Wellington
Female Factory
Cascade Brewery
South Hobart
Battery Point
Battery Point
Dynnyrne
Wrest Point Casino
Lords Beach
Red Chapel Beach
Nutgrove Beach
Tolmans Hill
Waterworks Reservoirs
RIDGEWAY PARK
Chimney Pot Hill 496m
Ridgeway Reservoir
Ridgeway
Sandy Bay
Sandy Bay Point
Sandy Bay Beach
Blinking Billy Point
Blinking Billy Beach
MOUNT NELSON SIGNAL STATION RESERVE
Mt Nelson 340m
Mount Nelson Signal Station
Mount Nelson
Fern Tree
PILLINGER DR
To Taroona
Ferry to Peppermint Bay
N
0 1

Downtown Hobart

Downtown Hobart is centred on the intersection of Elizabeth and Liverpool streets. There are fashion and department stores, specialty shops, banks, cafes and restaurants in the surrounding streets and arcades. Shops are open Monday to Saturday until 5pm, and on Sundays the larger stores open 10am to 4pm.

Elizabeth Mall 492 C2

Surrounded by shops and cafes, this busy meeting place is a favourite with shoppers, young people and buskers. A visitor information booth is open in the mall on weekdays until 4pm and from 10am until 2pm on Saturdays.

Cat and Fiddle Arcade 492 C2

There are several shopping arcades in the main retail area of Hobart, but the Cat and Fiddle Arcade's nursery rhyme musical clock lends it a certain charm. Much to the delight of children, it plays 'Hey Diddle Diddle' on the hour as moving parts provide accompanying action. The arcade runs between the Elizabeth Mall and Murray Street.

Liverpool Street 492 B3, C2

This is Hobart's main downtown shopping strip, with interesting arcades and side streets. **Bank Arcade** (between Argyle and Elizabeth streets) will tempt you with breads, spices and tiny cafes. **Mathers Lane** and **Criterion Street**, further along towards Murray Street, will interest those with a taste for good coffee and all things retro.

Elizabeth Street 492 B2

North of the mall, Elizabeth Street continues up a gentle slope towards North Hobart. In the first few blocks there is a cluster of shops selling polar fleece and outdoor equipment, followed by interesting cafes and antique shops.

Northern Waterfront

The northern waterfront of Sullivans Cove is the busiest, smelliest and most fascinating part of the working dock, both on and off the water. Along the Hunter Street waterfront contemporary apartments sit next to recycled 19th-century buildings, with Mount Wellington as the backdrop.

Constitution and Victoria docks 492 C2

Constitution Dock and neighbouring Victoria Dock are the heart of Sullivans Cove. There are several excellent seafood restaurants nearby as well as dockside punts offering some of the best takeaway fish and chips in Australia. Every New Year, Constitution Dock comes alive when the annual Sydney and Melbourne to Hobart yachts cross the finish line of these gruelling ocean races. Every two years **Franklin Wharf** hosts the Australian Wooden Boat Festival, the largest gathering of historic craft in the Southern Hemisphere *(see Top events p. 494)*.

Federation Concert Hall 492 C2

This concert hall is the home of the Tasmanian Symphony Orchestra and was designed to give priority to quality acoustics. Described as both intimate and grand, it seats over 1000 people. *1 Davey St; TSO Box Office 1800 001 190.*

The Henry Jones Art Hotel 492 D2

Facing Victoria Dock, this multi-award-winning hotel complex is established in what was once the IXL jam factory, owned by Hobart's first and most famous entrepreneur, Henry Jones. Alongside are galleries of fine Tasmanian furniture and Aboriginal art, cafes and restaurants. *25 Hunter St; (03) 6210 7700.*

Tasmanian Museum and Art Gallery 492 C2

This historic museum and gallery complex is home to Australia's finest collection of colonial art and Huon pine furniture. The zoology galleries will introduce you to Tasmania's unique wildlife including a glimpse of the extinct Tasmanian tiger or *thylacine*. There's also a gallery dedicated to Tasmanian Aboriginal culture. *40 Macquarie St; (03) 6211 4177; open 10am–5pm daily; tours 2.30pm Wed–Sun; admission free.*

Maritime Museum of Tasmania 492 C2

Located inside the impressive Neoclassical Carnegie Building, this museum brings to life Tasmania's rich maritime history. It displays

climate

Hobart's weather is changeable at any time of the year, so be prepared! Summer is mild with temperatures in the mid-20s, but sometimes the forecast highs are only reached for a short time as a cool sea breeze kicks in most afternoons. There are windless, sun-drenched days and chilly nights in autumn, but winter brings temperatures in the low teens and blustery cold conditions. Snow is rare in Hobart, but often settles on Mount Wellington's summit.

J	F	M	A	M	J	J	A	S	O	N	D	
23	22	20	17	13	11	10	12	14	17	19	21	MAX °C
11	11	10	8	6	4	3	4	6	7	9	10	MIN °C
44	39	42	47	42	49	48	46	47	55	54	51	RAIN MM
3	3	3	4	4	5	5	5	5	6	4	4	RAIN DAYS

pictures and equipment from the whaling era, models of various ships and fascinating relics from shipwrecks. *Cnr Davey and Argyle sts; (03) 6234 1427; open 9am–5pm daily.*

Southern Waterfront

The docks on the southern side of Sullivans Cove are surrounded by remarkable historic buildings. The work of the docks continues amid cafes, restaurants and pubs, contributing to the area's lively atmosphere.

Salamanca Place 492 C3

The sandstone warehouses along Salamanca Place are undoubtedly Australia's finest row of Georgian dockside buildings, packed with interesting shops and galleries.

Hobart's famous **Salamanca Market** is held here every Saturday, offering an amazing variety of tempting produce, arts and crafts, and the chance to meet local artisans. *(03) 6238 2843; 8.30am–3pm Sat.*

For 150 years, merchants conducted their business from the Salamanca warehouses alongside some notoriously rowdy pubs and hotels. **Knopwoods Retreat** still does a brisk trade and is popular for Friday night drinks. Nearby, at the **Salamanca Arts Centre** a free street party featuring local musicians is held every Friday evening.

Parliament House 492 C3

Situated behind the trees in Parliament Square, Hobart's majestic Parliament House faces out to the waterfront. Designed by the colonial architect John Lee Archer and built by convicts as the first Customs House, it became the home of Tasmania's parliament in 1856. *Between Salamanca Pl and Murray St; bookings (03) 6233 2200; visitors gallery open on sitting days; tours 10am and 2pm Mon–Fri, except sitting days.*

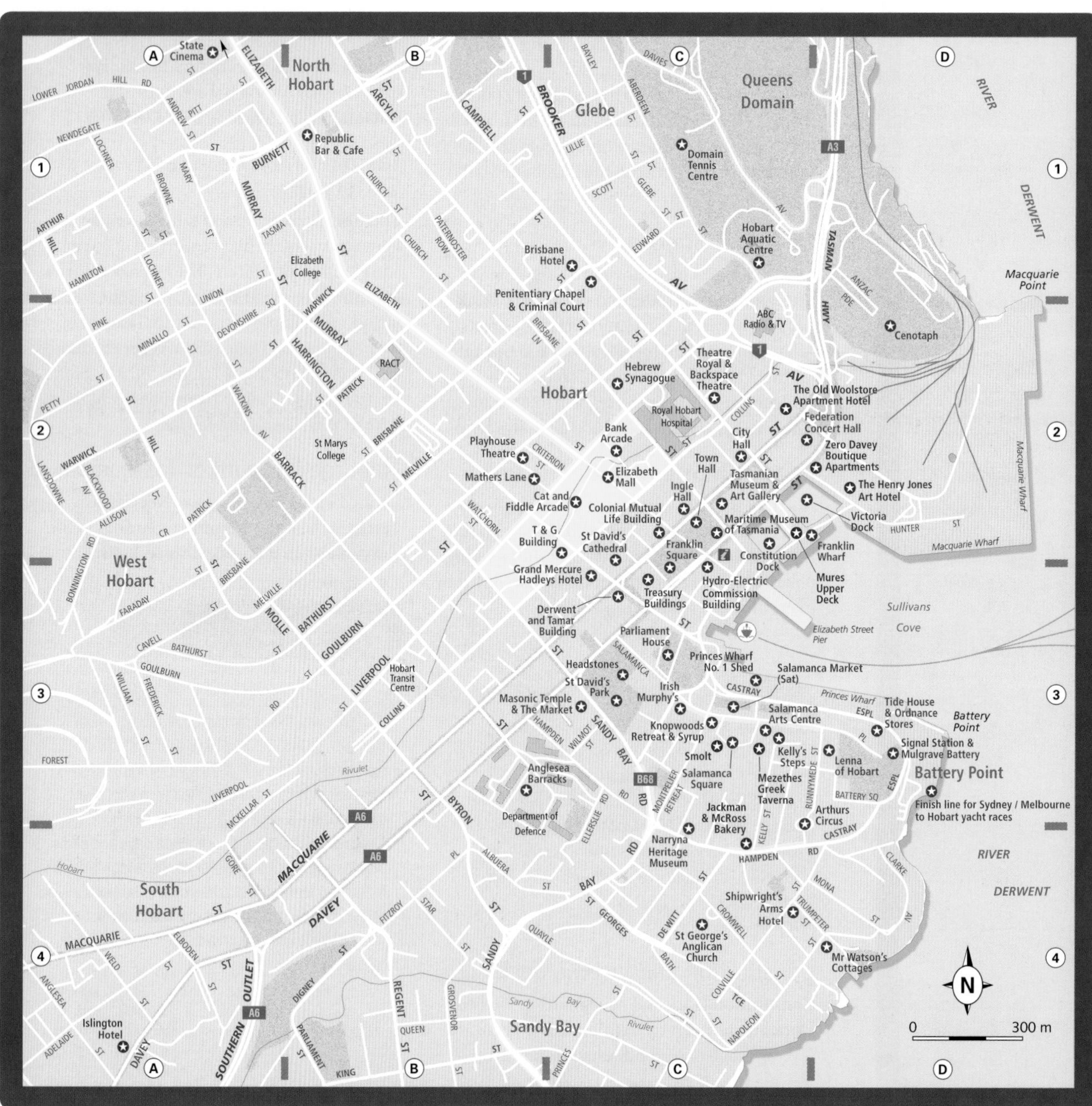

Castray Esplanade and Princes Wharf 492 C3

Built in 1870 to provide a promenade for the people of Hobart, Castray Esplanade runs behind Princes Wharf. At the city end of the esplanade are restaurants specialising in Tasmanian seafood. The Taste of Tasmania festival takes place in **Princes Wharf No. 1 Shed** at the beginning of January.

For most of the year, Princes Wharf is a working dock, berthing Antarctic research and supply vessels. The CSIRO Marine Research Laboratories are situated at the Battery Point end. On the hill are the original **Signal Station** and **Mulgrave Battery**, which was built in 1818 during panic about a rumoured Russian invasion. They are now part of Princes Park.

The octagonal **Tide House**, next to the **Ordnance Stores**, is the point from which all distances are measured in Tasmania.

Battery Point

Battery Point, built to house the workers and merchants of the port, is on the hill behind Salamanca Place. Its compact size and village atmosphere make it a perfect place to explore on foot.

Kelly's Steps 492 C3

At the end of Kelly Street, historic Kelly's Steps connect Salamanca Place to Battery Point. They were built in 1839 by Captain James Kelly, an adventurer who made a comfortable living from sealing and whaling. Kelly's good fortune was short-lived, however, and by 1842 he was bankrupt and destitute following the death of his wife and seven of his children. After a run-in with the law over another man's wife, he died at age 67.

Arthurs Circus 492 C4

Exploring Arthurs Circus, just off Hampden Road at the top of Runnymede Street, is like walking into an intact Georgian streetscape – so be sure to bring your camera. The 16 cottages were built around a circular village green between 1847 and 1852, and they are still private residences today.

Hampden Road 492 C4

Hampden Road begins at a cluster of antique shops near Sandy Bay Road and heads downhill to Castray Esplanade, which winds through the heart of Battery Point.

One of the earliest grand houses in Battery Point is **Narryna Heritage Museum**, built in 1836 by one of Hobart's early sea captains and now set up as a heritage museum. Visitors can wander through the house and see how a comfortable life would have been lived in early Hobart. The kitchen is especially worth a look. *103 Hampden Rd; (03) 6234 2791; open 10.30am–5pm Mon–Fri, 12.30–5pm Sat–Sun in summer, 10.30am–5pm Tues–Fri, 2–5pm Sat–Sun in winter.*

Trumpeter Street 492 C4

A few streets away, at the top end of Trumpeter Street is the **Shipwright's Arms Hotel** which has traded under this name since 1846. The front bar has a distinctive nautical atmosphere with every inch of wall space covered with photographs of vessels that have sailed on the River Derwent. A few doors down the hill, across Napoleon Street, are **Mr Watson's Cottages**, a row of simple Georgian brick dwellings built in 1850. Behind are the boat building slip yards that date back to the 1830s.

getting around

Metro buses regularly service the city and suburbs at peak times and less frequently during weekends. Timetables are displayed at most bus stops and are available from the Metro shop in the Hobart Bus Terminal at the Macquarie Street end of Elizabeth Street. A Day Rover ticket allows you to catch any number of buses after 9am Monday to Friday, and anytime on weekends.

On the river you can order a water taxi from Watermans Dock to Bellerive, Lindisfarne, Wrest Point or the Botanical Gardens. There are also cruise boats operating from Franklin Wharf. Information is available harbourside and from the Tasmanian Travel and Information Centre, (03) 6238 4222.

The 15-kilometre Inter-City Cycleway runs alongside a rail track between Hobart's waterfront and the northern suburb of Claremont. With a paved surface and no hills, it's popular with commuters and recreational riders alike. Bikes of all types can be hired from the Hobart end of the cycleway.

Public transport Metro bus information line 13 2201.

Airport shuttle bus Airporter City Hotels Shuttle 1300 385 511.

Motoring organisation RACT 13 2722, roadside assistance 13 1111.

Car rental Autorent Hertz (03) 6237 1111; Avis 13 6333 or (03) 6234 4222; Bargain Car Rentals (03) 6234 6959; Budget 13 2727 or (03) 6234 5222; Europcar 1800 030 118; Lo-cost Autorent (03) 6231 0550; Thrifty 1300 367 227.

Campervan and 4WD rental Britz 1800 038 171; Cruisin' Tasmania 1300 664 485; Tasmania Campers 1800 627 074; Tasmanian Campervan Hire 1800 807 119.

Taxis Australian Taxi Service 0411 286 780; Executive Taxi Service 0411 488 734; Taxi Combined Services 13 2227; United Taxis 13 1008; Yellow Cabs 13 1924.

Ferries Hobart Yellow Water Taxi 0407 036 268.

Bicycle hire Derwent Bike Hire (daily weather permitting); Cenotaph, Regatta Ground 0428 899 169.

Macquarie Street

Macquarie Street has always been Hobart's main thoroughfare, its classical buildings adding an air of sophistication. The area is now populated by government offices, lawyers' premises and consulting rooms.

Franklin Square 492 C2

Franklin Square is a peaceful city park and bus terminal. On sunny days it's a popular spot for city workers to eat lunch and a favourite meeting place for young people. Hobart's first Government House was built in Franklin Square but by 1858 the timber and thatch building was almost collapsing, so it was replaced by this park. At its centre a formal fountain supports a statue of Sir John Franklin, Arctic explorer and Governor of Tasmania (1837–43).

top events

Hobart Cup Day Join punters and picnickers for a day of racing and fashion. February.

King of the Derwent Maxi yachts battle to capture the crown. January.

MONA FOMA Hobart grooves with two weeks of eclectic jazz, rock, hip-hop and indie music. January.

Royal Hobart Regatta A family regatta and fireworks display since 1838. February.

Australian Wooden Boat Festival Biennial dockside celebration of maritime history. February (odd-numbered years).

Ten Days on the Island International island culture comes to Tasmania. March–April (odd-numbered years).

Targa Tasmania State-wide classic car rally. April–May.

Festival of Voices For four days winter is warmed by singers and choirs from all over Australia gathering to sing their hearts out. July.

Royal Hobart Show Four days of competitions and displays bringing country life to town. October.

Sydney to Hobart and Melbourne to Hobart yacht races Gruelling races end with a dockside party, and a crowd of people to welcome the yachts no matter what time of the day or night. December.

The Taste Festival Hobart sparkles with fun activities and waterfront gourmet indulgence. December–January.

Spring Community Festival Hobart's Botanical Gardens at their blooming best. October.

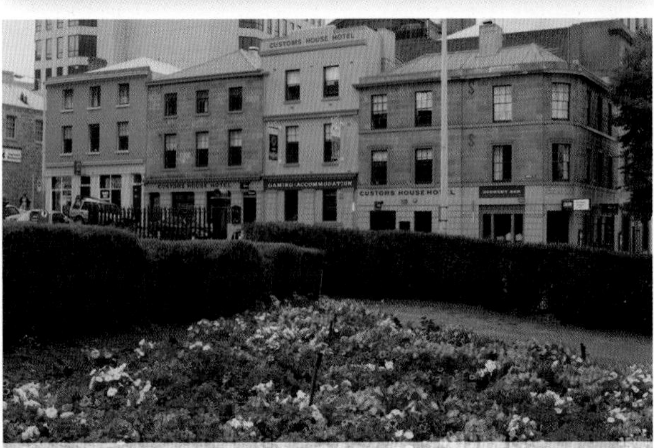

[STATE PARLIAMENT HOUSE GARDENS AND HERITAGE BUILDINGS]

shopping

Salamanca Place, City Fine craft and Tasmanian art. 492 C3

Elizabeth Street, City Gifts, outdoor clothing and antiques. 492 B2

Liverpool Street, City Fashion and jewellery. 492 C2

Cat and Fiddle Arcade, City Bargain fashion and sportswear. 492 C2

Bathurst Street, City Fine furniture and antiques. 492 C2

Sandy Bay Road, Battery Point Antique furniture, china and books. 492 C3

Sandy Bay Road, Sandy Bay Stylish fashion stores. 492 B4

Eastlands Shopping Centre, Rosny Park Hobart's largest undercover suburban mall. 490 E4

Corner of Macquarie & Murray streets 492 C3

A short walk up Macquarie Street from Franklin Square, this remarkable intersection has a historic sandstone building on every corner. It was once the administrative centre of Hobart and today offers panoramic views down Macquarie Street's elegant heritage streetscape past the Georgian **Ingle Hall**, built in 1814, and the Art Deco *Mercury* newspaper building, to the **Cenotaph** in the distance.

On one corner **St David's Cathedral**, rebuilt several times, stands where there have been churches since 1817. The present cathedral was substantially rebuilt in 1909 and its bell tower was completed in 1936. Across Murray Street, **Hadleys Hotel** stands on the site of a hotel dating back to the 1830s. Dating to the same era, the **Treasury Buildings** are still used as government offices today. The 1875 **Derwent and Tamar Building** used to have a set of punishment stocks outside the front door as the Hobart Gaol was once its neighbour.

St David's Park 492 C3

On Davey Street, at the city end of Salamanca Place, this formal English-style walled park still has old headstones displayed in the lower section – a reminder that this was the colony's first burial ground. It was made into a park in 1926, with a rotunda and sweeping lawns shaded by huge English trees. *Cnr Davey St and Sandy Bay Rd.*

Anglesea Barracks 492 B3

On the hill at the top of Davey Street is Australia's oldest military establishment still in use. There is a military museum to explore, some elegant Georgian buildings and a pair of 1770s bronze cannons. *Davey St; (03) 6237 7160; grounds open daily; museum open 9am–1pm Tues and Thurs; tours 11am Tues.*

Queens Domain

Queens Domain is a bush reserve on the hill overlooking the Derwent River, to the north-east of the city centre. It has soccer fields, a tennis centre, a cricket oval and an aquatic centre. In Glebe ('The Glebe' to locals), there are magnificent weatherboard Federation houses built on some of the steepest streets you will ever find.

Government House 490 D4

Next door to the Botanical Gardens, Tasmania's Neo-Gothic Government House is one of the finest viceregal residences in the Commonwealth. Built in 1857 from local timbers and sandstone excavated on-site, it dominates the riverside slopes between Hobart and the Tasman Bridge. *Lower Domain Rd; open to the public one Sunday each year; phone (03) 6234 2611 for the exact date; admission free.*

Royal Tasmanian Botanical Gardens 490 D4

These superb botanical gardens were established in 1818, and today they are one of the best in Australia. There is a restaurant, a beautiful old-world conservatory, a sub-Antarctic house and, of course, ABC gardener Peter Cundall's famous veggie patch. *Lower Domain Rd; (03) 6236 3076; open daily; admission free; entry to conservatory and discovery centre by gold coin donation.*

Inner North

Heading north out of Hobart city along Elizabeth Street is a strip of hip cafes, bars and eateries in North Hobart. Here you can linger over coffee and cake during the day or kick up your heels and take in the local music scene at night.

Elizabeth Street, North Hobart 492 B1

This cosmopolitan strip is busy day and night. At the Federal Street end, the independent **State Cinema** shows arthouse and foreign films in Hobart's only old cinema. Further down the street there is a wide choice of casual restaurants, bistros, cafes and takeaway outlets. *State Cinema, 375 Elizabeth St; (03) 6234 6318.*

On the Burnett Street corner, the **Republic Bar & Cafe** is a popular meeting place for an eclectic mix of musos, artists, writers, activists and students. There is live music every night of the week, complemented by excellent counter meals.

Cornelian Bay 490 D3

Cornelian Bay is tucked away north of the Botanical Gardens and easily accessible from the Inter-City Cycleway. A popular swimming beach for families in the 1950s and 1960s, it is now a tranquil waterside park reminiscent of a bygone era with barbecue facilities, an all-access playground and a row of historic Edwardian boatsheds. There is also a contemporary restaurant and a kiosk selling gourmet ice-cream and takeaway coffee.

Around both sides of the bay there are walking trails offering excellent views of the Tasman Bridge. Along these tracks, a careful eye will discover the charcoal and shell-midden evidence of past Aboriginal occupation.

Runnymede 490 C3

Built in 1836, this Georgian villa has had a number of distinguished owners including a bishop, a captain and the first lawyer to qualify in the colony. It's now in the hands of the National Trust and has been restored and refurbished to its former glory as an 1860s mansion. *61 Bay Rd, New Town; open 10am–4.30pm Mon, 12–4.30pm Sun; tours Tues, Wed, Thurs; bookings (03) 6278 1269.*

North

Between the Derwent River and Mount Wellington, Hobart's northern suburbs sweep over gently sloping hills and around the meandering foreshore. There are suburban shopping strips at Moonah and Glenorchy, catering to the retail needs of nearby suburban areas.

Tasmanian Transport Museum 490 B2

This museum contains a restored and working collection of steam engines, locomotives and carriages from Tasmania's unique and fascinating railway history. There are also some of the old trams and trolley buses that used to run on Hobart's streets. Visitors can browse through displays in the relocated New Town suburban train station, a relic from a bygone era. Occasional main-line trips are scheduled. *Anfield St, Glenorchy; (03) 6272 7721; open 1–4pm Sat–Sun and public holidays.*

markets

Glenorchy Showgrounds Market Trash and treasure, craft, produce and occasionally livestock. Royal Hobart Showgrounds, Glenorchy; 8am–2pm Sun. 490 B2

Tasmanian Farm Gate Fresh seasonal fruit and veggies plus artisan cheeses, breads and oils from local producers. Melville St car park; 9am–1pm Sun.

Sorell Market Produce, craft, and trash and treasure. Sorell Memorial Hall, Cole St; most Sun and public holidays. 669 K4

The Market Unique Tasmanian-designed and made jewellery, books, cards, craft and food. Masonic Temple, 3 Sandy Bay Rd; 10am–3pm 1st Sun each month. 492 C3

See also Salamanca Market, p. 492

where to stay

Islington Hotel Hobart's most glamorous boutique hotel in a fine Regency mansion. 321 Davey St; (03) 6220 2123. 492 A4

The Henry Jones Art Hotel Super-stylish luxury waterfront accommodation in Hobart's historic sandstone warehouses. 25 Hunter St; (03) 6210 7700. 492 D2

The Old Woolstore Apartment Hotel Multi-level apartment complex behind a heritage street facade, just a short stroll from Hobart's picturesque waterfront. 1 Macquarie St; (03) 6235 5355 or 1800 814 676. 492 C2

Wrest Point Fantastic views from Wrest Point's landmark tower and temptations in the casino below. 410 Sandy Bay Rd, Sandy Bay; 1800 703 006. 490 D5

Zero Davey Boutique Apartments Groovy waterfront apartment complex offering fabulous luxury, hip contemporary design and cool modern decor. 15 Hunter St; (03) 6270 1444 or 1300 733 422. 492 D2

where to eat

Jackman & McRoss Bakery Vibrant bakery cafe with very good coffee and heavenly sweet and savoury pastries. 57–59 Hampden Rd, Battery Point; (03) 6223 3186; open daily for breakfast, lunch and afternoon tea. 492 C4

Lebrina Restaurant Award-winning, sophisticated dining with French style and Italian flavour in an elegant historic cottage. 155 New Town Rd, New Town; (03) 6228 7775; open Tues–Sat for dinner. 490 C4

Mezethes Greek Taverna Lively crowded taverna for those with a passion for authentic Greek chargrilled seafood and meats. Salamanca Arts Centre; (03) 6224 4601; open daily for breakfast, lunch and dinner. 492 C3

Mures Upper Deck Perched above the dockside sights, sounds and smells, this is Hobart's best-loved seafood restaurant. Victoria Dock; (03) 6231 1999; open daily for lunch and dinner. 492 C2

Smolt Mediterranean-style dining and late-night 'small plates' with a piazza outlook and hip modern interior. 2 Salamanca Sq; (03) 6224 2554; open daily for breakfast, lunch and dinner. 492 C3

entertainment

Cinema There are multiscreen cinemas at Glenorchy in the northern suburbs, in Rosny Park opposite Eastlands Shopping Centre on the eastern shore and in Collins Street in the city. These are ideal for mainstream and latest-release movies. If your taste leans towards arthouse films, try the independent State Cinema in Elizabeth Street, North Hobart. See the *Mercury* for details of films being shown.

Live music Hobart has a small but thriving live-music scene based around a handful of pub venues. For a night of great blues you can't beat the Republic Bar & Cafe in North Hobart, and for original rock and roots head to the Alley Cat Bar in North Hobart or the Brisbane Hotel in the city. There are plenty of late-night pubs and clubs around Sullivans Cove including Syrup, Isobar, Bar Celona, Irish Murphy's, Lark Distillery and the Telegraph Hotel. Check out Thursday's gig guide in the *Mercury*.

Classical music and performing arts Hobart's unique Theatre Royal is the venue for visiting stage companies and local productions, such as the annual University Review. Classical music is usually performed at the Federation Concert Hall, and the Derwent Entertainment Centre hosts large events. A livelier theatre scene is supported by smaller venues such as the intimate Backspace Theatre behind the Theatre Royal. The Playhouse in Bathurst Street is home to an amateur theatrical society and the Peacock Theatre specialises in contemporary works in its small space at the Salamanca Arts Centre. See the *Mercury* for details of what's on.

Moorilla Estate and Museum of Old and New Art 490 B2

Embedded into riverside cliffs at Moorilla Estate 12 km upstream from Hobart, David Walsh's vast labyrinthine MONA houses his edgy and confronting private art collection celebrating the twin themes of sex and death. Moorilla's vines were planted by another visionary – Caludio Alcorso – in 1955 and the estate's superb wines are showcased in its modern retaurant alongside Moorilla's tasty Moo Brew beers. Take a high speed catamaran from Brooke St pier or a carbon-neutral liesurely pedal along the Scenic Inter City cycleway. *655 Main Rd, Berriedale; (03) 6277 9900; The Source restaurant open daily for lunch, dinner Fri–Sat.*

Cadbury Visitor Centre 490 B1

Among their many other chocolate delights, Cadbury's factory at Claremont makes the 90 million Freddo frogs eaten every year in Australia. Visitors can watch an informative DVD, see a demonstration of chocolate being moulded, taste the raw materials and indulge themselves in the factory shop and cafe (which makes excellent hot chocolate). *100 Cadbury Rd, Claremont; 1800 627 367; open 8am–4pm (last presentation at 3pm) Mon–Fri Sept–May, 9am–3pm (last presentation at 2pm) Mon–Fri June–Aug.*

Inner South

The prosperous southern suburbs of Hobart extend from Battery Point south along the river to Taroona and Kingston, and up into the foothills of Mount Wellington. Around the shopping area on Sandy Bay Road there are narrow streets of cottages, providing digs for students from the nearby University of Tasmania campus. Davey Street and Sandy Bay Road are lined with impressive mansions while the hills above are dotted with contemporary houses with spectacular outlooks.

Wrest Point Casino 490 D5

A Hobart landmark since it opened in 1973, the iconic Wrest Point tower houses Australia's first legal casino. From its rotating top-floor restaurant (open daily for dinner and lunch on Fridays) there are views of the Derwent River from the Tasman Bridge to Storm Bay, and to Mount Wellington behind the city. Be there at sunset for the most amazing views. *410 Sandy Bay Rd; 1800 703 006.*

Sandy Bay Beach 490 D6

Sandy Bay Beach has been a popular swimming beach for decades. Lawns beside the beach provide a relaxing place for a family barbecue, to kick a footy with the kids or take a leisurely

grand old buildings

Lenna of Hobart When Alexander McGregor made a fortune from whaling he built this rich, Italianate mansion on a cliff overlooking the cove so he could keep an eye on shipping movements. Now it's a stylish boutique hotel. 20 Runnymede St, Battery Point.

Hebrew Synagogue Australia's first synagogue and a rare example of Egyptian Revival architecture. Argyle St (between Liverpool and Bathurst sts), City.

Theatre Royal Australia's oldest theatre, with a highly decorated interior and a resident ghost. Laurence Olivier called it 'the best little theatre in the world'. 29 Campbell St, City *(see Walks & tours, p. 497)*.

Town Hall Classical Revival design by Henry Hunter, it stands where Hobart's founder, David Collins, pitched the first tent. Macquarie St (between Elizabeth and Argyle sts), City.

Penitentiary Chapel and Criminal Court Underground passages, solitary cells and an execution yard. Cnr Brisbane and Campbell sts, City *(see Walks & tours, p. 497)*.

City Hall Built from a competition-winning design in 1915, it is perhaps Hobart's most underrated public building. Macquarie St (between Market and Campbell sts), City.

T & G Building Built for an insurance company, it has an Egyptian-inspired clock tower. Cnr Collins and Murray sts, City.

Hydro-Electric Commission Building The design brief said that it should represent the new age of electricity, and its Art Deco facade suggests energy and modernity. Cnr Elizabeth and Davey sts, City.

Colonial Mutual Life Building Inter-war building with Gothic gargoyles, Moorish balconies, Art Deco chevrons and multicoloured roofing tiles. Cnr Elizabeth and Macquarie sts, City.

St George's Anglican Church Built by two noted colonial architects – the body in 1836–38 by John Lee Archer, and the spire in 1847 by James Blackburn – this is Australia's finest Classical Revival church. 28 Cromwell St, Battery Point.

stroll. There are a number of nearby eateries, including a fine-dining seafood restaurant in the old regatta pavilion, a tiny wood-fired pizza restaurant on the foreshore, a beachside contemporary wine bar and a Chinese restaurant on Sandy Bay Road.

Female Factory 490 C5

Serving at various times as a rum distillery, contagious diseases hospital, home for 'imperial lunatics' and, most recently, as a fudge factory, this historic site was also a female prison for 50 years. The nearby Remembrance Garden was created in 1995. Visitors can take a tour of the site and have an 1830s-style 'Morning Tea with the Matron' in her refurbished cottage. *16 Degraves St, South Hobart; Morning Tea with the Matron bookings (03) 6233 1559; Mon, Wed and Fri; tours weekdays; bookings (03) 6233 6656.*

Cascade Brewery 490 B5

Cascade beers have been brewed here since 1824, making it Australia's oldest brewery. Locally grown hops, Tasmanian barley and mountain water combine to produce top beers. There are several brewery tours daily with samples included. *140 Cascade Rd, South Hobart; bookings (03) 6224 1117; open daily.*

Mount Nelson Signal Station 490 D6

In 1811 on a visit to Van Diemen's Land, Governor Macquarie ordered that a signal station be erected on 350-metre 'Nelson Hill' to relay sightings of boats to Hobart. The site affords sweeping views of the Derwent Valley and Storm Bay, and there is a restaurant with a spectacular outlook – especially at night.

Taroona Shot Tower 667 E12

Towering above the winding Channel Highway south of Taroona, this 58-metre-tall sandstone tower was erected to make lead shot

[SALAMANCA MARKET]

walks & tours

Hobart Historic Walks Guided walks of Hobart or Battery Point with great stories of the early days, an informative tour discovering Hobart's polar links, or a pub crawl with colourful tales thrown in. Bookings on (03) 6278 3338.

Art Deco in Hobart Self-guide tour brings Hobart's collection of Art Deco buildings to life. Brochure available from the Tasmanian Travel and Information Centre.

Hobart Waterways Tour Peek into Hobart's convict-built aqueducts while keeping your feet dry above ground. Bookings on (03) 6238 4222 or 1800 990 440.

River cruises Combine a river cruise with delicious food and wine; step back into history aboard a square rigger; or cruise to Cadbury Schweppes Chocolate Factory, MONA, Peppermint Bay or Port Arthur, or take a jet boat out into Storm Bay for a thrilling ride. For information and timetables head to the Brooke Street Pier area in Sullivans Cove, or phone Captain Fells (03) 6223 5893, Navigators (03) 6223 1914, Peppermint Bay Cruises 1300 137 919, Hobart Yellow Water Taxi 0407 036 268, Wild Thing Adventures 1800 751 229, Hobart Harbour Jet 0404 078 687 or Windeward Bound Sailing Adventures 0418 120 243.

Kayak tours Get a sea-level perspective on Hobart's waterfront with a daytime or evening paddle around the docks. Blackaby's Sea Kayaks 0418 124 072; Freycinet Adventures (03) 6257 0500; Hobart Urban Adventures (03) 6227 1388.

Ghost Tours of Hobart and Battery Point Nerves of steel are needed for this sunset tour of Hobart's spooky past. Bookings on 0439 335 696.

Old Rokeby Historic Trail A self-guide tour of the outer suburb of Rokeby, one of Hobart's earliest rural districts. Brochures available at the trailhead in Hawthorn Place, Rokeby. (03) 6247 6925.

Theatre Royal Tour A guided tour backstage. Bookings on (03) 6233 2299; 11am Mon, Wed and Fri.

Penitentiary Chapel Historic Site Grim history of gallows, cells and tunnels illustrated by lamplight if you dare. Bookings on 0417 361 392.

Sullivans Cove Walks Guided evening and day walks around the waterfront unveiling Hobart's past. Bookings on (03) 6245 1208 or 0409 252 318.

Louisa's Walk Follow the bleak life of convict Louisa Ryan to the Female Factory where she was imprisoned. Bookings on (03) 6229 8959 or 0437 276 417.

Mount Wellington Walks Guided walks delving into the history, botany and ancient past of this intriguing mountain. Bookings on 0439 551 197.

Mount Wellington Descent Plummet down from Mount Wellington's 1270-metre summit to sea level on a mountain bike. Gloves and ear warmers supplied. Bookings on (03) 6274 1880.

The Grand Hobart Walk. An all-day ramble on foot, bus and boat around Hobart's highlights. Bookings on (03) 6227 1388.

sport

The **Sydney to Hobart Yacht Race**, held in December, is Hobart's premier sporting event. Other twilight and weekend sailing events take place on the Derwent throughout the year.

A state-wide Australian Rules **football** league is up and running, with games played at the North Hobart Football Oval. In summer the action shifts to Bellerive Oval – Tasmania's premier **cricket** ground *(see entry on this page)*.

Hobart has an international-standard **hockey** centre at New Town, and national and international games attract large crowds. In January the Domain Tennis Centre hosts the Moorilla Hobart International **tennis** tournament, and Hobart's **horseracing** calendar is dominated by the AAMI Hobart Cup, run at Tattersall's Park in February.

Tasmania lays claim to Australia's very first **golf** course at Bothwell. Around Hobart there is a nine-hole course on the eastern shore at Rosny Park, while 18-hole courses are at Kingston Beach in the southern suburbs and Claremont In the north.

in 1870. The tower summit and its remarkably good views of the Derwent are reached via hundreds of steps. There is a tearooms on-site to help you recover from the climb. *234 Channel Hwy, Taroona; (03) 6227 8885; open daily.*

Inner East

The 'eastern shore', as it is locally known, is a residential area spreading along the River Derwent south to the beachside suburbs of Howrah, Rokeby and Lauderdale. It is reached by travelling over the arching Tasman Bridge.

Bellerive 490 E4

Bellerive was settled in the 1820s. Today it's a quiet suburban village overlooking the river, bordered by a marina to the north and sandy beaches to the south. Back in 1885, in a nervous response to reported sightings of Russian ships in the Derwent, a military fort was built on **Kangaroo Bluff.** Its guns were never fired, but can still be seen today mounted over a network of tunnels and bunkers. *(03) 6248 4053; open daily; admission free.*

Bellerive's best-known landmark is the **Bellerive Oval**, home to the **Tasmanian Cricket Museum** featuring the Ricky Ponting Corner unveiled by Ricky in 2010. This picturesque oval hosts Tests and one-day matches, and is the home ground of the state cricket team – the Tasmanian Tigers – and the site of their 2007 inaugural Pura Cup victory. *Derwent St, Bellerive; tour bookings (03) 6282 0433; museum open 10am–3pm Tues–Thurs, 10am–12pm Fri; oval tours 10am Tues, or by arrangement.*

East

Beyond the eastern shore's built-up areas, the peninsula between Storm Bay and Frederick Henry Bay is a favourite haunt for surfers, within an easy drive from the city. Few tourists head out this way so there are crowd-free surf breaks and plenty of long deserted dune-backed beaches.

Rokeby 667 H9

Rokeby is a 20-minute drive out of Hobart along the South Arm Highway. It was settled in 1809 and quickly became an important food-producing area for the growing colony, with Tasmania's first crops of wheat and apples harvested here.

St Matthew's Church houses a collection of notable items including its organ, which was the first keyboard brought to Australia. Tasmania's first chaplain, the notorious Reverend Robert Knopwood, spent his last days in the Rokeby area and was buried in a plain coffin, with no nameplate, in St Matthew's churchyard. *Cnr King St and North Pde.*

Storm Bay beaches 669 K7, L5

Surrounding Hobart's outer-eastern suburbs there are some lovely sheltered beaches that are well worth a visit. The beaches facing south into Storm Bay are excellent surfing spots and usually uncrowded. **Clifton Beach** is a popular surfing location reached via a turn-off from the South Arm Highway, while a further 8 kilometres south the **Goat Bluff** cliff-top lookout offers spectacular views of wild and deserted surf beaches on either side of the bluff.

For a swim in the waves at a safe, patrolled beach, head to **Carlton** (take the turn-off to Dodges Ferry from the Arthur Highway, after Sorell).

day tours

Richmond This small settlement just north of Hobart is probably Australia's best-preserved Georgian colonial village. Highlights include the convict-built Richmond Bridge, Australia's oldest bridge; the gaol, which predates Port Arthur; and galleries and cafes housed in historic shopfronts and cottages. For more details see p. 501

Tasman Peninsula The stunning setting of Port Arthur – lawns, gardens, cliffs – and the beauty of its sandstone buildings belie the site's tragic history. Other sites on the peninsula worth a look include the spectacular rock formations and blow holes around Eaglehawk Neck. For more details see p. 501

Derwent Valley The Derwent Valley, with its neat agricultural landscape and historic buildings, forms one of the loveliest rural areas of Australia, reminiscent of England. Visit the trout hatchery of Salmon Ponds, the National Trust–classified New Norfolk, and the hop museum at Oast House. For more details see p. 501

D'Entrecasteaux Channel The beauty and intricacy of Tasmania's south-eastern coastline can be experienced on a leisurely drive south from Hobart. There are stunning water views, particularly at Tinderbox (via Kingston), and Verona Sands at the Huon River entrance. At Kettering, a car ferry goes to remote Bruny Island. For more details see p. 501

Huon Valley The Huon Valley is the centre of a growing gourmet food industry. The signposted Huon Trail follows the valley between rows of apple trees, with a backdrop of forested mountains. In the far south, at Hastings, visitors can tour a dolomite cave and swim in a thermal pool. For more details see p. 501

REGIONS
tasmania

[TASMANIAN DEVIL]

[SCOPARIA IN BLOOM, WALLS OF JERUSALEM NATIONAL PARK]

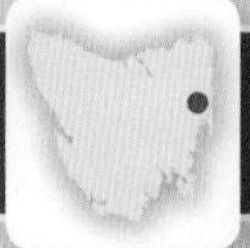

EAST COAST

[WINEGLASS BAY, FREYCINET NATIONAL PARK]

Tasmania's premier seaside destination boasts exquisite coastal scenery, historic sites, gourmet produce, a lush hinterland and, offshore, a rich marine environment.

Fishing The fishing is excellent along this coast, particularly around St Helens, which is the base for those planning to fish the East Australian Current for tuna, marlin and shark. For land-based and inshore anglers there is the long, narrow estuary of Georges Bay. And for those who prefer the eating to the catching, fish and chip shops like Captain's Catch in St Helens and restaurants like Kabuki by the Sea near Swansea specialise in only the freshest east-coast seafood.

Penguin tours in Bicheno Every night at dusk, Bicheno Penguin Tours allows people to get very close to these beautiful little creatures as they return from their day's fishing. At the peak of the season, close to 600 birds inhabit the rookery. It's a popular and magical wildlife experience, run in an environmentally sensitive way.

Mount William National Park This fairly remote park, created in 1973, protects Tasmania's forester kangaroo and many bird species. The view from Mount William takes in the sandy beaches and coastal heath of the state's north-east corner and extends north to the Furneaux Islands and south to St Marys.

Underwater wonders The underwater landscape of the east coast is an unsung wonder. Offshore from Bicheno is Governor Island Marine Reserve. Here, beneath clear waters, granite outcrops create cliffs, caves and deep fissures, which provide a home for diverse marine communities. Glass-bottomed boat and diving tours are available.

Freycinet Peninsula This long, narrow paradise features forests, cliffs, beaches and walking trails. The beautiful Peninsula Walking Track ends at Wineglass Bay, regarded by many as one of the world's best beaches. Coles Bay, which now has an up-market resort, is a peaceful base from which to explore the area.

TOP EVENTS

JAN	St Helens Regatta and Seafood Festival
MAR	Fingal Valley Festival (Fingal)
MAR	Tasmanian Game Fishing Classic (St Helens)
EASTER	Jazz Concert (Freycinet Vineyard, near Bicheno)
JUNE	Winter Solstice Festival (St Marys)

CLIMATE

BICHENO

J	F	M	A	M	J	J	A	S	O	N	D	
21	21	20	19	16	14	14	15	16	17	18	20	MAX °C
13	13	12	10	9	7	6	6	7	8	10	11	MIN °C
52	57	59	60	58	61	53	51	44	56	58	69	RAIN MM
8	8	8	9	9	9	9	10	9	10	10	10	RAIN DAYS

EAST COAST GOURMET TRAIL

Much of this region's appeal lies in its magnificent seafood. Visitors can taste and buy oysters at Freycinet Marine Farm near Coles Bay, and buy crays, oysters and scallops from Wardlaw's Cray Store at Chain of Lagoons. The area is also known for its dairy produce – try the cheddar at Pyengana Dairy Company in the tiny river town of Pyengana. Kate's Berry Farm near Swansea is a gourmet institution, specialising in fresh berries and berry products. At Swansea Wine and Wool Centre visitors can taste the pinots and chardonnays of the small local wine industry.

→ For more detail see maps 671 & 673. For descriptions of T towns see Towns from A–Z (p. 506).

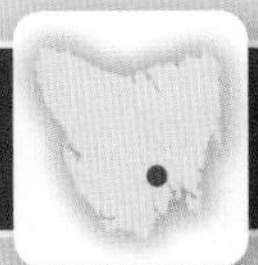

SOUTH-EAST

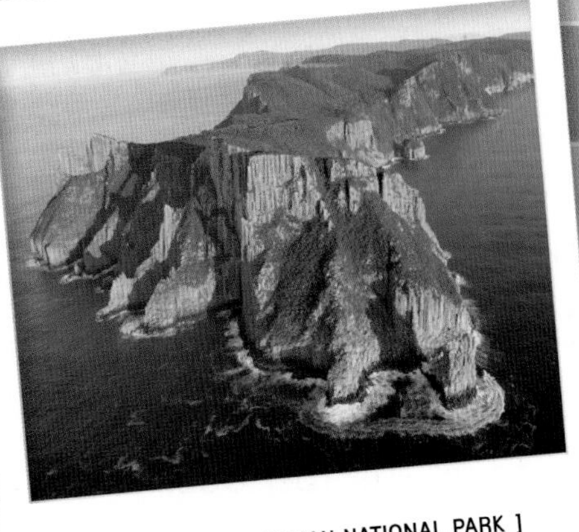

[CAPE PILLAR, TASMAN NATIONAL PARK]

Little development has ensured the preservation of many colonial sites in this region, including Port Arthur Historic Site.

Trout fishing The region's lakes and rivers boast some of Australia's best freshwater fishing. A favourite among trout anglers is the Ouse River, which joins the Derwent north of Hamilton. Other spots include the Clyde, Jordan and Coal rivers east of the Derwent, and the Tyenna, Styx and Plenty rivers west of the Derwent.

Tahune Forest AirWalk Tahune Forest Reserve's most famous attraction is its treetop walk, the award-winning AirWalk. The last section of the suspended steel path is cantilevered and hangs 48 metres above the Huon and Picton rivers that converge below – not for the faint of heart.

Hastings Caves This cave and thermal spring complex near Dover features the magnificent Newdegate Cave, with 245 steps leading to its vast chambers of formations created over more than 40 million years. Tours of its stalactites, stalagmites and helictites (distorted stalactites) run daily.

Southern Tasmanian Wine Route Tasmania's first vineyard was established at New Town (now a suburb of Hobart) in 1821. The modern industry began in 1958 when Claudio Alcorso set up the now acclaimed Moorilla Estate on the Derwent. Today a number of wineries fan out from Hobart, with most offering cellar-door tastings.

Tasmanian Devil Conservation Park The feature of this park is undeniably the Tasmanian devils, fierce-looking black creatures the size of small dogs. The park also has displays on the thylacine (Tasmanian tiger), regarded as extinct despite unconfirmed sightings.

Port Arthur Historic Site Port Arthur is one of Australia's most significant historic sites. Imposing sandstone prison buildings are set in 40 hectares of spectacular landscaping. Highlights are the ghost tours and summer boat trips to the Isle of the Dead, the final resting place for convicts and prison personnel alike.

TOP EVENTS

JAN	Cygnet Folk Festival (Cygnet)
FEB–MAR	International Highland Spin-In (wool-spinning competition, Bothwell, odd-numbered years)
MAR	Taste of the Huon (different town each year)
MAR	Village Colonial Fair (Richmond)
OCT	Olie Bollen Festival (Dutch community festival, Kingston)
NOV	Huon Agricultural Show (Huonville)

CLIMATE

NEW NORFOLK

J	F	M	A	M	J	J	A	S	O	N	D	
24	24	21	18	14	11	11	13	15	17	19	21	MAX °C
11	11	10	7	5	3	2	3	5	6	8	10	MIN °C
40	35	39	48	45	49	48	47	49	55	47	50	RAIN MM
8	7	9	10	11	12	13	14	13	14	12	11	RAIN DAYS

THE CONVICT SYSTEM

Reminders of Australia's convict years are a feature of Tasmania's south-east. The island was colonised in 1803 as a penal settlement, and over the next 50 years 52 227 males and 12 595 females – 46 per cent of the entire Australian convict consignment – were transported to its remote shores. Many prisoners worked on public buildings and infrastructure until the early 1820s, after which most were assigned to free settlers. Reoffending convicts from other prison colonies were sent to Port Arthur from 1830, and in the following years, a penal settlement for re-convicted criminals was built there. Conditions were harsh and escape almost impossible. Some of the state's most emotive and colourful stories stem from its convict past and can be appreciated at the Coal Mines Historic Site (north of Port Arthur) and Eaglehawk Neck, while the fruits of convict labour can be seen in many early structures, such as those in the heritage town of Richmond.

→ For more detail see maps 667, 668–9 & 670–1. For descriptions of Ⓣ towns see Towns from A–Z (p. 506).

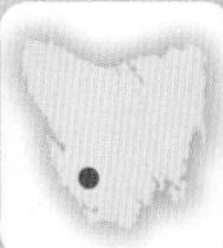

SOUTH-WEST WILDERNESS

This almost uninhabited landscape of mountains, rivers, waterfalls, gorges and 1000-year-old trees contains the famous Franklin River.

[WEST COAST WILDERNESS RAILWAY]

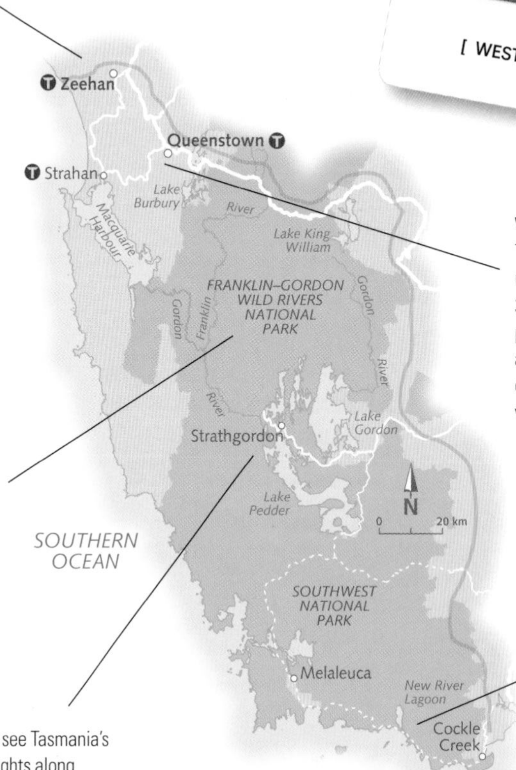

Zeehan A walk around National Trust–classified Zeehan reveals something of its heyday in the late 19th century, when it was a booming silver-mining town known as Silver City. Back then it had 26 hotels; the few that remain include the Grand Hotel, which encompasses Gaiety Theatre (once the largest concert hall in Australia). Other buildings of historical interest include the post office, bank and St Luke's Church, and the West Coast Pioneers' Memorial Museum is great for learning about the mines.

West Coast Wilderness Railway One of Tasmania's most recent tourist attractions is a restored 1896 rack-and-pinion railway that travels 34 kilometres across wild rivers and through pristine forests. The journey begins at Queenstown, a mining town huddled beneath some huge and ominous hills, and finishes at Strahan, a charming village on Macquarie Harbour.

Franklin–Gordon Wild Rivers National Park One way to visit this grand wilderness is by boat from Strahan. Cruises run up the Gordon to Heritage Landing, where there is a short walk to a 2000-year-old Huon pine. For a little more adventure, guided whitewater rafting trips begin near the Lyell Highway east of Queenstown and run down the Franklin River.

Strathgordon Strathgordon is the place to see Tasmania's massive hydro-electricity industry at work. Sights along Gordon River Road include the huge lakes Pedder and Gordon, the Gordon Dam and the underground Gordon Power Station. Bushwalkers can enter Southwest National Park via the Creepy Crawly Nature Trail.

South Coast Track This six-to-nine-day walk along the entirely uninhabited south coast has become a mecca for experienced trekkers. The walk starts at Cockle Creek and finishes at Melaleuca, where you can either continue walking north or take a prearranged flight out. The track crosses an unforgettable landscape of mountain ranges, rivers, swampy plains and wild beaches.

TOP EVENTS

JAN	Mount Lyell Picnic (Strahan)

CLIMATE

STRATHGORDON

J	F	M	A	M	J	J	A	S	O	N	D	
19	20	17	14	12	9	9	10	12	13	16	17	MAX °C
10	10	9	7	5	4	3	3	4	5	7	8	MIN °C
150	111	150	208	235	218	270	283	270	253	187	199	RAIN MM
17	14	18	20	22	22	25	25	24	23	20	20	RAIN DAYS

PRESERVING THE WILDERNESS

The 1972 flooding of Lake Pedder for the Gordon River Hydro-Electric Scheme sparked a campaign to preserve the south-west wilderness from further inroads. Despite the region's 1982 World Heritage listing, the Tasmanian government pressed on. Conservationists blockaded a proposed dam site from December 1982 until the election of a new federal Labor government in March 1983. Arrests and clashes with police made headlines and earned the movement support from mainstream Australia and abroad. Finally, federal legislation to stop the project survived a High Court challenge. The historical and ecological significance of the south-west wilderness is imaginatively presented at the Strahan Visitor Centre.

→ For more detail see maps 668, 670, 672 & 674. For descriptions of Ⓣ towns see Towns from A–Z (p. 506).

NORTH-WEST

This region features unparalleled views of Bass Strait, farmland and the beauty of Cradle Mountain–Lake St Clair National Park.

[DOVE LAKE, CRADLE MOUNTAIN–LAKE ST CLAIR NATIONAL PARK]

Woolnorth The Van Diemen's Land Company, which was granted tracts of north-west Tasmania in the 1820s, still owns this sheep, cattle and plantation-timber property. Tours from Smithton include a visit to the wind farm and to spectacular Cape Grim, where the air is reputedly the world's cleanest.

The Nut Historic Stanley is dominated by the 152-metre-high volcanic Circular Head (known as The Nut), Tasmania's version of Uluṟu. A steep stairway and a chairlift go to the cliff-top, where a 40-minute circuit walk offers spectacular views.

Bass Highway The spectacular scenery on the Ulverstone to Stanley section of this highway recalls Victoria's Great Ocean Road. The route's highlights include the little penguins at the village of Penguin, the Lactos Cheese Tasting Centre in Burnie, the colourful fields of Table Cape Tulip Farm near Wynyard, and the picturesque town of Boat Harbour.

Pieman River cruise Corinna is a delightful spot with a goldmining heritage. Once a bustling town and river port, today it is a popular launching pad for cruises (or canoeing trips) down the Pieman River, compared by many to the mighty Gordon. Daily four-hour return trips aboard Pieman River Cruises' *Arcadia II* pass lush rainforest incorporating stands of majestic Huon pine on the way to rugged Pieman Head, where you can go ashore and explore untouched wilderness.

Cradle Mountain–Lake St Clair National Park This magnificent glaciated landscape is part of the Tasmanian Wilderness World Heritage Area. While the park has become synonymous with the Overland Track, it also includes some fantastic short walks. Take the three-hour circuit around Dove Lake in the north for views of the majestic Cradle Mountain from all angles.

TOP EVENTS

JAN	Henley-on-the-Mersey Regatta (Latrobe)
FEB	Festival in the Park (Ulverstone)
MAR	Taste the Harvest Festival (Devonport)
MAR	Rosebery Miners, Axemen, Bush and Blarney Festival (Rosebery)
MAR	Rip Curl West Coast Classic (near Marrawah)
OCT	Burnie Shines (month-long festival, Burnie)

CLIMATE

WARATAH

J	F	M	A	M	J	J	A	S	O	N	D	
18	18	16	12	10	8	7	8	10	12	14	16	MAX °C
6	7	6	4	3	2	1	1	2	3	4	5	MIN °C
110	95	123	174	213	229	251	251	226	204	168	142	RAIN MM
16	14	18	21	23	23	25	25	24	23	20	18	RAIN DAYS

OVERLAND TRACK

The 85-kilometre Overland Track, one of Australia's most awe-inspiring treks, takes walkers into the heart of Tasmania's alpine wilderness, past lakes and tarns, buttongrass heath, wildflowers, woodland and rainforest. It runs the length of Cradle Mountain–Lake St Clair National Park and passes Mount Ossa, Tasmania's highest peak. Each year about 9000 people attempt the five-to-eight-day trek. While you can choose to do the walk with a group and a guide, staying in well-appointed private huts along the way, most people do the walk unguided. There are 12 basic, unattended huts for overnight stays, but these quickly fill up, so be prepared to camp. Summer is the best season for walking here, but even then walkers need to prepare for all conditions.

→ For more detail see maps 670, 672–3, 674 & 675. For descriptions of Ⓣ towns see Towns from A–Z (p. 506).

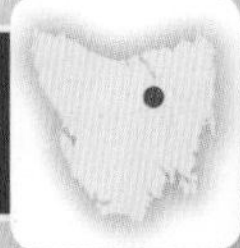

MIDLANDS & THE NORTH

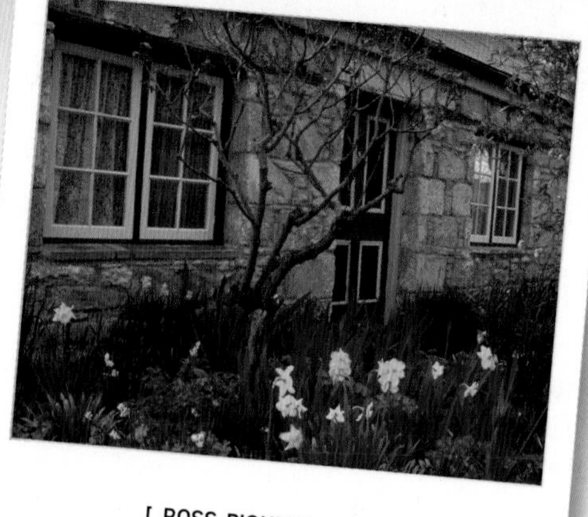

[ROSS PIONEER HOME]

This area, reminiscent of rural England, features elegant mansions and small cottages. It's also home to some excellent wineries and gourmet food spots.

A stroll through Launceston Begin at Princes Square, a serene park surrounded by Georgian churches. Walk north up John Street to one of Australia's oldest synagogues, and then to the town hall and some imposing Victorian bank buildings. Continue past a restored Georgian warehouse to the elegant customs house facing the river. Lastly, walk south along the river to Launceston's star attraction, Cataract Gorge, where you can hike, swim, take the chairlift and cross the suspension bridge; at night it's lit in spectacular fashion.

BASS STRAIT
NARAWNTAPU NATIONAL PARK
George Town
Bridport
Gladstone
Pipers Brook
Beauty Point
Beaconsfield
Scottsdale
Derby
Exeter
Tamar River
Lilydale
LAUNCESTON
Deloraine
MOLE CREEK KARST NATIONAL PARK
Mole Creek
Hadspen
Westbury
Longford
Evandale
BEN LOMOND NATIONAL PARK
Legges Tor
Lake Rowallan
CENTRAL PLATEAU CONSERVATION AREA
WALLS OF JERUSALEM NATIONAL PARK
N
0 20 km
Great Lake
Arthurs Lake
Campbell Town
Miena
Lake Sorell
Bronte Lagoon
Lake Echo
Ross
Lake Crescent
Oatlands

Ben Lomond National Park Tasmania's best skiing is to be found on Ben Lomond Range, a plateau rising to over 1575 metres. The scenery is magnificent, especially the view from Legges Tor, Tasmania's second highest peak. The park also offers easy bushwalking and is ablaze with wildflowers in spring and early summer.

Trout fishing Tasmania's Central Plateau Conservation Area is famous for the wild trout that inhabit literally thousands of lakes, tarns and connecting streams. Although bushwalking is the only option to reach many lakes in this wilderness, plenty of the most famous waters, such as Arthurs Lake and Great Lake, can be reached by road; these two lakes are the very best and most popular trout waters, producing consistent quality fish. Bronte Lagoon is also highly regarded, particularly for fly-fishing.

Woolmers Estate Regarded as Australia's most significant colonial property, Woolmers (near Longford) has buildings, antique cars, photographs, art and furniture that reflect the life of six generations of one family, from 1817 to the present day. The Servants Kitchen Restaurant serves morning and afternoon teas.

Ross This tiny village (founded in 1812) boasts the decoratively carved Ross Bridge, a fine example of convict construction. Female convicts were held at Ross Female Factory nearby, which visitors can tour today. The wool centre explains the other focus of this charming town.

TOP EVENTS

JAN	Tamar Valley Folk Festival (George Town)
FEB	Launceston Cup (horseracing, Launceston)
FEB	Festivale (food and wine, Launceston)
FEB	Village Fair and National Penny Farthing Championships (Evandale)
FEB	Exeter Show (Exeter)
NOV	Tasmanian Craft Fair (Deloraine)

CLIMATE

LAUNCESTON

J	F	M	A	M	J	J	A	S	O	N	D	
23	23	21	17	14	11	11	12	14	16	19	21	MAX °C
10	10	9	7	5	3	2	3	4	6	7	9	MIN °C
44	39	38	55	62	62	78	78	64	62	51	51	RAIN MM
8	7	8	10	12	13	15	16	13	13	11	10	RAIN DAYS

TASMANIAN WINE ROUTE

The wineries of northern Tasmania hug the banks of the River Tamar, extending east out to Pipers Brook and Lilydale. Pipers Brook Vineyard, established in 1974, is the largest and best-known producer in the region. Cool-climate varieties ripen well in northern Tasmania, and the superb quality of the wines has much to do with the dry, warm autumn. Local riesling, pinot noir and sparkling wines are gaining international recognition, while cabernet sauvignon, pinot gris, sauvignon blanc, chardonnay and gewurztraminer are also grown. Start a self-guide driving tour by picking up a Tasmanian Wine Route brochure at a visitor centre.

→ For more detail see maps 670–1, 672–3, 674 & 675. For descriptions of towns see Towns from A–Z (p. 506).

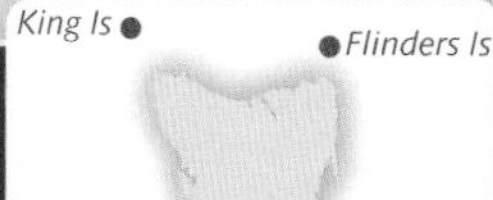

BASS STRAIT ISLANDS

These islands are known for their gourmet produce, fauna and windswept beauty, and offer low-key holidays with a range of activities available.

[CATARAQUI POINT, KING ISLAND]

Killiecrankie diamond Flinders Island is home to the elusive Killiecrankie diamond, which is actually topaz, a semi-precious stone. You can try your luck fossicking around Killiecrankie Bay or nearby Mines Creek, designated as official fossicking areas. The stone you're looking for is usually colourless, but sometimes pale blue or amber gems can be found.

King Island produce King Island has, in recent years, developed a strong reputation for quality produce, for everything from cheese to beef. The island's name is synonymous with award-winning cheeses and creams made from unpasteurised milk. The peerless double brie and camembert can be sampled at King Island Dairy, north of Currie.

Wybalenna Wybalenna ('black man's home') was set up on Flinders Island in 1834 to house around 160 Aboriginal people, the few survivors of Tasmania's pre-European population of over 4000. Less than a third of the people held there survived the appalling living conditions. Located near Emita, it is one of the most important – albeit tragic – historic sites in Tasmania, and includes a National Trust–restored church and cemetery.

Strzelecki National Park The granite Strzelecki Range occupies the south-west corner of Flinders Island. On a clear day, the highlight of this largely undeveloped park is a five-hour-return walk to the summit of Strzelecki Peaks, offering spectacular views across Franklin Sound.

Wreck diving In the storm-lashed waters around King Island lie almost 60 shipwrecks. The best known is the *Cataraqui*, which sank in 1845 off the coast south of Currie with the loss of 399 immigrants and crew, making it Australia's worst peacetime disaster. A number of wrecks around the island are accessible to scubadivers; diving tours are available.

TOP EVENTS

MAR	King Island Imperial 20 (marathon from Naracoopa to Currie, King Island)
EASTER	Three Peaks Race (sailing and hiking event visiting Lady Barron, Flinders Island)
OCT	Flinders Island Show (Flinders Island)
NOV–DEC	King Island Horseracing Carnival (Currie)

CLIMATE

CURRIE

J	F	M	A	M	J	J	A	S	O	N	D	
20	21	20	17	15	14	13	13	14	16	17	19	MAX °C
13	13	13	11	10	9	8	8	8	9	10	11	MIN °C
36	39	48	68	99	102	124	115	84	75	60	52	RAIN MM
11	10	14	17	21	22	24	24	21	19	15	13	RAIN DAYS

FISHING

The Bass Strait islands offer superb coastal fishing. On Flinders Island Australian salmon, flathead, gummy shark, silver trevally, pike and squid can be caught from the rocks and beaches. From Lady Barron, Emita and Killiecrankie several charter boats take anglers offshore for catches of all of the above as well as snapper, yellowtail kingfish, trumpeter and various species of tuna. On King Island there is excellent fishing for Australian salmon, flathead and whiting from the beaches along the east coast. South of Currie is British Admiral Reef, where boat-anglers can try for morwong, warehou, yellowtail kingfish and squid.

TASMANIA

→ For more detail see maps 670 (inset), 671 (inset) & 673. For descriptions of Ⓣ towns see Towns from A–Z (p. 506).

TOWNS A–Z
tasmania

LEGEND

 VISITOR INFORMATION
 RADIO STATIONS
 IN TOWN
 WHAT'S ON
 WHERE TO EAT
WHERE TO STAY
NEARBY

Food and accommodation listings in town are ordered alphabetically with places nearby listed at the end

[CRADLE MOUNTAIN–LAKE ST CLAIR NATIONAL PARK]

Adventure Bay

Pop. 620
Map ref. 669 I11 | 671 K10

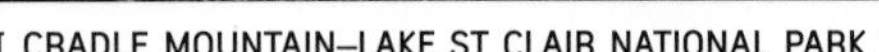

Bruny D'Entrecasteaux Visitor Centre, 81 Ferry Rd, Kettering; (03) 6267 4494; www.brunyisland.org.au

95.3 Huon FM, 936 AM ABC Local Radio

Adventure Bay is the primary town on largely undeveloped Bruny Island, a peaceful retreat for many Tasmanians with its striking landscape untainted by excessive tourism. The island is more accurately two islands joined by a narrow isthmus. On one side the coast is pounded by the waves of the Pacific, and on the other it is gently lapped by the waters of the D'Entrecasteaux Channel. After a long Aboriginal occupation, Europeans discovered the islands. Captain Cook's interaction with the Aboriginal people was largely amicable, but the sealers who subsequently came to Bruny decimated the population. Of those who survived, Truganini is the most famous. Memorials dot the island, standing as stark reminders of the grim past amid the spectacular scenery.

Eco-cruise Arguably one of Tasmania's greatest attractions, this 3 hr cruise takes in the seal colonies, penguins, dolphins, humpback whales and abundant birdlife in the area. It focuses on the ecological and historical nature of the coastline. The tour leaves from Adventure Bay daily; bookings on (03) 6293 1465.

Morella Island Retreat: hothouse, cafe and gum tree maze. ***Bligh Museum of Pacific Exploration:*** constructed from handmade convict bricks and housing displays of island history.

Bruny Island Cheese Co: cafe with cheese tastings; 1807 Bruny Island Main Rd, Great Bay; (03) 6260 6353. ***Get Shucked Oyster Farm:*** fresh Pacific oysters; 1650 Bruny Island Main Rd, Great Bay; 0428 606 250. ***The Hothouse Cafe:*** home-cooked fare in casual surrounds; 46 Adventure Bay Rd; (03) 6293 1131.

Adventure Bay Eco Village: old-fashioned beachside shacks; 1005 Main Rd, Cookville; (03) 6293 2096 or 1300 889 557. ***Bruny Beach House:*** modern beach house; Nebraska Rd, Dennes Point; 0419 315 626. ***Poppies on the Beach:*** secluded beachside A-frame cabin; Main Rd, Great Bay; (03) 6293 1131. ***The Tree House:*** luxurious waterfront pole-house; 66 Matthew Flinders Dr, Alonnah; 0405 192 892.

Bushwalking Bruny Island is a bushwalker's delight. At the southern tip is South Bruny National Park with stunning scenery and several tracks. Labillardiere State Reserve has a vast range of vegetation including eucalypt woodlands, shrublands, herblands and wildflowers. The rainforest at Mavista Falls is another breathtaking spot, as is Cape Queen Elizabeth. Check with visitor centre for details.

Truganini memorial: boardwalk and lookout; on isthmus between islands; 11 km N. ***Birdlife:*** little penguins and short-tailed shearwaters on Bruny's ocean beaches. ***Cape Bruny Lighthouse:*** second oldest lighthouse in Australia (1836); South Bruny National Park via Lunawanna; 29 km SW. ***Beaches:*** many good swimming beaches (details from visitor centre); Dennes Point Beach on North Bruny has picnic facilities and Cloudy Bay (turn left at Lunawanna) is a magnificent surf beach.

See also SOUTH-EAST, p. 501

Beaconsfield

Pop. 1055
Map ref. 673 J6 | 675 E7

Tamar Visitor Centre, Main Rd, Exeter; (03) 6394 4454 or 1800 637 989; www.tamarvalley.com.au

91.7 FM ABC Local Radio, 103.7 FM City Park Radio

Now a modest apple-growing centre in Tasmania's north, Beaconsfield was once the wealthiest gold town in Tasmania. Gold rush–era relics include two massive Romanesque arches at the old pithead of the Tasmania Gold Mine. The Gold Festival is held each December and draws a large crowd. Tourism increased significantly in Beaconsfield following the dramatic rescue of two trapped miners in 2006.

Beaconsfield Mine and Heritage Centre This mine constantly struggled with water problems, but still managed to produce 26 tonnes of gold before it was closed in 1914. There's now a museum complex along with a miner's cottage.

Beaconsfield Walk of Gold: self-guide historical walk that starts at the Heritage Centre; 1.8 km round trip with signage; West St. ***Van Diemen's Land Gallery:*** art and craft; Weld St. ***Gem and Stone Creations:*** gallery and giftshop; Weld St.

Tamar Valley Wholefoods & Coffee Shop: cafe and organic groceries; 110B Weld St; (03) 6383 1120.

Exchange Hotel: friendly heritage pub; 141 Weld St; (03) 6383 1113. ***Tamar River Retreat:*** riverside cottages; 123 Kayena Rd, Kayena; (03) 6394 7030.

Holwell Gorge This is a fern-covered gorge with beautiful waterfalls in the Dazzler Range to the west of town. There are basic picnic facilities at the eastern entrance and a 3 hr return hiking track. Contact Parks & Wildlife Service on 1300 135 513; access is via Holwell or Greens Beach rds; 8 km w.

Auld Kirk: convict-built church (1843) at Sidmouth with views of Batman Bridge, which features a 100 m A-frame across the River Tamar; 13 km SE. ***Wineries:*** cellar-door tastings and sales to the east of town around Rowella, Kayena and Sidmouth; Sidmouth is 13 km E. ***Lavender House:*** exhibition fields of 70 varieties of lavender, plus tearooms with specialty lavender scones and giftshop with 50 health and beauty products; 15 km E.

See also MIDLANDS & THE NORTH, p. 504

Beauty Point

Pop. 1113
Map ref. 673 J6 | 675 E6

Tamar Visitor Centre, Main Rd, Exeter; (03) 6394 4454 or 1800 637 989; www.tamarvalley.com.au

91.7 FM ABC Local Radio, 95.3 Tamar FM

Beauty Point is the base for the Australian Maritime College and is also a good spot for fishing and yachting. The town's main attractions are Seahorse World and Platypus House.

Seahorse World The centre acts as a breeding farm for aquariums and the Chinese medicine market, but also studies endangered species to ensure their survival. Take a guided tour to see how seahorses are bred – newborns are the size of a child's fingernail. There's also a touch pool and expo centre for local wines and craft. Inspection Head Wharf, Flinders St; (03) 6383 4111.

Platypus House This centre has 3 live displays: one with platypus and native beaver rats, another with Tasmanian butterflies, and a third with many Tasmanian leeches, blood-sucking worms and spiders. Scientists here are currently conducting research into a disease attacking the Tasmanian platypus population. Inspection Head Wharf, Flinders St; (03) 6383 4884.

Sandy Beach: safe swimming spot.

Three Peaks Race: 3-day sailing and mountain-climbing event that begins in town each Easter, heading north to Flinders Island, then south down the east coast to Hobart.

Carbones Cafe: Italian cuisine with a view; 225 Flinders St; (03) 6383 4099. ***Tamar Cove Restaurant:*** casual, family-friendly restaurant; 4421 West Tamar Hwy; (03) 6383 4375.

Beauty Point Cottages: riverbank cottages or homestead suites; 14 Flinders St; (03) 6383 4556. ***Kateland Manor Estate B&B:*** elegant luxury B&B; 170 West Arm Rd; 0418 128 742. ***Pomona Spa Cottages:*** 3 Federation-style cottages; 77 Flinders St; (03) 6383 4073. ***Greens Beach Luxury Escape:*** glass-fronted beach house; 19 Pars Rd, Greens Beach; 0408 376 211.

Narawntapu National Park This park is a popular spot for horseriding and waterskiing (conditions apply), and has excellent walks. Contact Parks & Wildlife Service on 1300 135 513; access via Greens Beach or Badger Head or Bakers Beach rds.

York Town monument: site marks the first settlement in northern Tasmania (1804); 10 km w. ***Kelso and Greens Beach:*** popular holiday towns to the north-west.

See also MIDLANDS & THE NORTH, p. 504

Bicheno

see inset box on next page

Bothwell

Pop. 377
Map ref. 671 J4

Australasian Golf Museum, Market Pl; (03) 6259 4033; www.bothwell.com.au

95.7 Heart FM, 936 AM ABC Local Radio

Australia's first golf course was created in Bothwell in 1837, making golf a point of pride. There are many craft shops and art galleries in town, as well as over 50 National Trust–classified buildings.

Heritage Walk Bothwell has 53 colonial cottages, churches, houses and official buildings around Queens Park. The best way to appreciate the historic buildings is by a self-guide walking tour; pamphlet from visitor centre.

Australasian Golf Museum: displays of golfing memorabilia and history. The museum also sells award-winning Tasmanian Highland Cheese and local tartan; Market Pl. ***Bothwell Grange:*** guesthouse with tearooms and art gallery; Alexander St.

International Highland Spin-In: odd-numbered years, Feb–Mar.

Bothwell Fat Doe Bakery & Coffee Shop: excellent pies and pastries; 12 Patrick St; (03) 6259 5551. ***Castle Hotel:*** great counter meals; 14 Patrick St; (03) 6259 5502.

The Priory Country Lodge: exclusive fishing lodge; 2 Wentworth St; (03) 6259 4012.

Ratho Home to the first game of golf in Australia, Ratho was the elegant 'gentleman's residence' of Alexander Reid in the early 1800s. It is a stone house with wooden Ionic columns. The famous golf course is still intact and in use. Lake Hwy; 3 km w.

continued on p. 509

BICHENO

Pop. 639

Map ref. 671 O1 | 673 O11

 **Foster St; (03) 6375 1500; www.gsbc.tas.gov.au**

 89.7 FM ABC Local Radio, 98.5 Star FM

Bicheno (pronounced 'bish-eno') was set up as a whaling and sealing centre in 1803, predating the official settlement of Van Diemen's Land by a few months. A local Aboriginal women, Waubedebar, became a heroine after saving two white men from drowning in a storm. Landmarks in town bear her name and her grave is in Lions Park. After a short stint as a coal port, during which time the population increased considerably, Bicheno relaxed back into what it does best – fishing – and is known for its abundant seafood. Situated on the east coast, the town has the mildest climate in Tasmania. It is blessed with sandy beaches and popular dive spots, and draws holiday-makers from far afield. Native rock orchids, unique to the east coast, bloom in October and November.

Bicheno Penguin Tours Join a tour at dusk to watch little penguins return home from the sea. On most nights they waddle right past you as they make their way through the coastal vegetation to their burrows. Tours depart from operator's office on Burgess St; bookings on (03) 6375 1333.

Scuba diving Bicheno has one of the best temperate dive locations in the world. The Bicheno Dive Centre runs tours to over 20 locations, including Governor Island Marine Reserve. In winter, divers have been known to swim among pods of dolphins or near migrating whales. Bookings on (03) 6375 1138.

Bicentennial Foreshore Walk: a 3 km track with great views, this walk starts at Redhill Beach and continues to the blowhole. ***The Gulch:*** natural harbour; foreshore. ***Whalers Lookout:*** off Foster St. ***Sea Life Centre:*** licensed restaurant and giftshop; 1 Tasman Hwy; (03) 6375 1121. ***The Glass Bottom Boat:*** stay warm and dry while you drift above the watery world of Bicheno's marine life in Tasmania's only glass-bottom boat; bookings on (03) 6375 1294. ***Fishing tours:*** rock lobster and premium gamefishing in particular; details from visitor centre.

Cyrano Restaurant: casual French cuisine; 77 Burgess St; (03) 6375 1137. ***Facets Restaurant and Bar:*** modern Australian; Diamond Island Resort, 69 Tasman Hwy; (03) 6375 0100. ***Pasini's Cafe Wine Bar Deli:*** gourmet deli, takeaway and bistro serving local wines; Shop 2 70 Burgess St; (03) 6375 1076.

Bicheno Gaol Cottages: historic self-contained cottages; Cnr Burgess and James sts; (03) 6375 1266. ***Bicheno Hideaway:*** oceanfront chalets; 179 Harvey's Farm Rd; (03) 6375 1312. ***Diamond Island Resort:*** award-winning beachfront apartments; 69 Tasman Hwy; (03) 6375 0100. ***Aurora Beach Cottage:*** timber and stone cottages; 207 Champ St, Seymour; (03) 6375 1774.

East Coast Natureworld This park offers encounters with the region's diverse fauna. The often-misunderstood Tasmanian devil is here, as are Forester kangaroos, Cape Barren geese and Bennett's wallabies. (03) 6375 1311; 8 km N.

Douglas–Apsley National Park This national park is Tasmania's last largely undisturbed area of dry eucalypt forest. It encompasses the catchments of the Denison, Douglas and Apsley rivers and has stunning gorges and waterfalls. There is a viewing platform at Apsley Gorge Lookout and Waterhole, and safe swimming spots. Contact Parks & Wildlife Service on 1300 135 513; 14 km NW.

Freycinet National Park: begins directly south of town; *see Coles Bay*. ***Vineyards:*** to the south-west of town, including Springbrook Vineyard and adjacent Freycinet Vineyard, which hosts a jazz concert each Easter; 18 km SW on Tasman Hwy. ***Scenic flights:*** offered by Freycinet Air over the region; bookings on (03) 6375 1694.

See also EAST COAST, p. 500

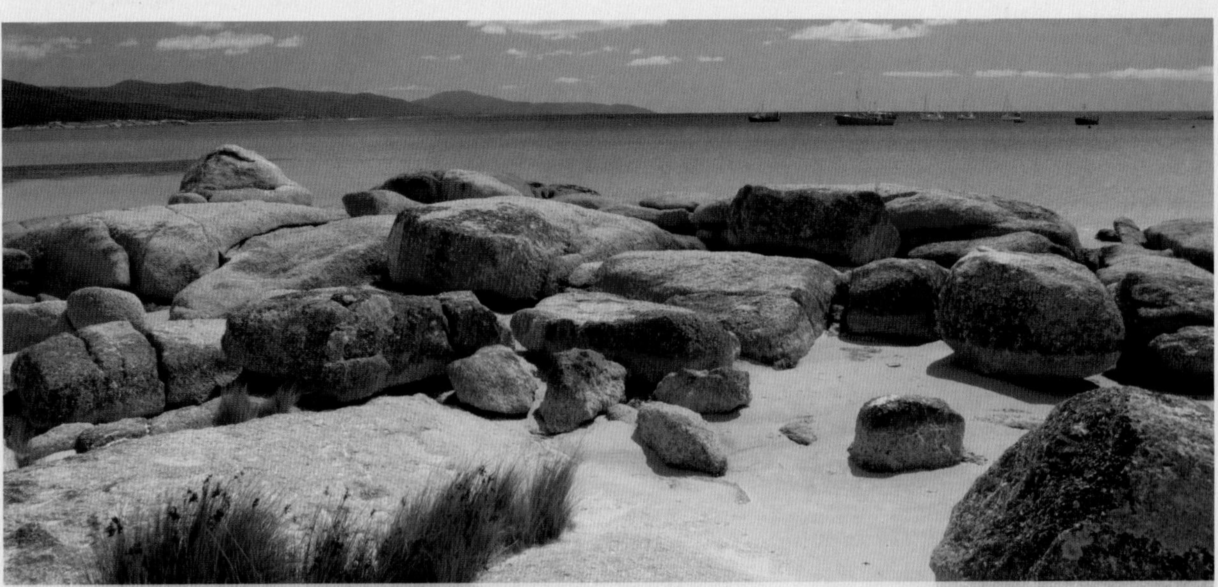

[ORANGE LICHEN–COVERED ROCKS ALONG BICHENO'S COASTLINE]

Trout fishing: in nearby lakes, rivers and streams; contact visitor centre for details.

See also SOUTH-EAST, p. 501

Bridport

Pop. 1325
Map ref. 673 L5

Main Rd; (03) 6356 1881; www.northeasttasmania.com.au

91.7 FM ABC Local Radio, 103.7 FM City Park Radio

Bridport is a popular seaside holiday retreat for Launceston and Scottsdale residents. It has many safe swimming beaches and fishing.

Bridport Wildflower Reserve Best during Sept and Oct, this wildflower reserve spans 50 ha of coastal heath and woodland. There is a 2.2 km walking track that covers the length of the reserve and takes in scenic Adams Beach. Access via Main St.

Tom's Turnery: functional and souvenir woodturning; 31 Edward St.

Bridport Triathlon: Jan.

Bridport Bay Inn Bistro & Woodfired Pizza: seafood and tasty pizzas; 105 Main St; (03) 6356 1238. ***Bridport Cafe:*** environmentally aware gourmet Mediterranean cafe; 89 Main St; (03) 6356 0057. ***Flying Teapot Cafe & Gallery:*** garden cafe; 1800 Bridport Rd; (03) 6356 1918.

Barnbougle Cottages: golf course and cottages; 425 Waterhouse Rd, Barnbougle Dunes; (03) 6356 0094. ***Bridairre:*** old-fashioned B&B with ocean views; 32 Frances St; (03) 6356 1438. ***Bridport Resort:*** seaside spa villas; 35 Main St; (03) 6356 1789. ***Platypus Park Country Retreat:*** self-contained farmstay; 20 Ada St; (03) 6356 1873.

Waterhouse Protected Area: offers 6700 ha of coastal bush camping, rockpools, sand dunes and beaches. Access via the old goldmining village, Waterhouse; 26 km E. ***Wineries:*** south-west of town around Pipers Brook and include Delamere, Dalrymple and Pipers Brook; cellar-door tastings and sales.

See also MIDLANDS & THE NORTH, p. 504

Bruny Island

see Adventure Bay

Burnie

Pop. 77 408
Map ref. 672 F5

Tasmanian Travel and Information Centre, Civic Centre Plaza, off Little Alexander St; (03) 6434 6111; www.tasmaniantravelcentre.com.au

102.5 FM ABC Local Radio, 106.1 Coast FM

The first European pioneers believed the Burnie area to be agriculturally rich, but high rainfall and dense forests covering the surrounding hills made farming virtually impossible. The deep waters in Emu Bay, however, rescued the community by providing an ideal port for the local industries of tin and timber. Today Burnie, Tasmania's fourth largest city, is a vibrant city with beautiful parklands and charming heritage buildings.

Burnie Park This park has lawns, shaded walkways, diverse native flora and animal enclosures with ducks, swans, wallabies, emus, peacocks and rabbits. Burnie Inn, the city's oldest building, is in the park and has been restored as a teahouse. A brochure is available from the park information centre. Bass Hwy.

Pioneer Village Museum: re-creation of old Burnie town that houses almost 20 000 items from the late 1800s and early 1900s; Civic Centre Plaza. Little Penguin Observation Centre: free guided tours Oct–Feb; Parsonage Point; 0437 436 803. ***Creative Paper:*** recycled-paper art with demonstrations and activities; tours available daily; Old Surrey Rd. ***Walking track:*** 17 km track that skirts the city; start at boardwalk. ***Australian Paper:*** mill tours held 2pm Mon–Fri; Bass Hwy; bookings on (03) 6430 7882. ***The Cheese Tasting Centre:*** dairy samples and other specialty produce; Old Surrey Rd. ***Hellyers Road Distillery:*** makers of single malt whisky; tours, tastings, sales and licensed eatery; open 9.30am–5.30pm daily; Old Surrey Rd, adjacent to Cheese Tasting Centre. ***Burnie Regional Art Gallery:*** impressive collection of Australian contemporary prints; open 10am–4.30pm Mon–Fri, 1.30–4.30pm Sat, Sun and public holidays; Wilmot St.

Burnie Farmers Market: Wivenhoe Showgrounds; 1st and 3rd Sat morning each month. ***Burnie Ten:*** 10 km road race; Oct. ***Burnie Shines:*** month-long community festival; Oct.

Bayviews Restaurant and Lounge Bar: smart seafront bistro; 1st floor, North Tce; (03) 6431 7999. ***Cafe Europa:*** Mediterranean fare; Cnr Cattley and Wilson sts; (03) 6431 1897. ***Rialto Gallery Restaurant:*** traditional Italian; 46 Wilmot St; (03) 6431 7718.

Apartments Down Town: opulent Art Deco apartments; 52 Alexander St; (03) 6432 3219. ***Glen Osborne House:*** glorious heritage mansion; 9 Aileen Cres; (03) 6431 9866. ***Seabreeze Cottages:*** cottages from bygone eras; 6 Mollison St and 82 Bass Hwy, Cooee; (03) 6435 3424.

Emu Valley Rhododendron Gardens Considered the city's floral emblem, the rhododendron has pride of place in Burnie. These gardens have over 9000 wild and hybrid rhododendrons on display in a natural 12 ha amphitheatre, and host the floral festival in Oct. Open Aug–Feb; off Cascade Rd; 6 km S.

Fern Glade: tranquil reserve on Emu River with walking tracks and picnic areas; off Old Surrey Rd; 5 km W. ***Annsleigh Garden and Tearooms:*** voted one of the 10 best gardens in Australia and comprising 2 ha of beautiful gardens and novelty buildings, plus souvenirs and food; open Sept–May; Mount Rd; 9 km S. ***Upper Natone Forest Reserve:*** popular picnic spot; Upper Natone Rd; 30 km S. ***Lake Kara:*** good trout fishing; signposted from Hampshire; 30 km S. ***Bushwalks and waterfalls:*** many in area, but Guide Falls (near Ridgley, 17 km S) is most accessible.

See also NORTH-WEST, p. 503

Campbell Town

Pop. 776
Map ref. 671 L1 | 673 L11

Heritage Highway Museum and Visitor Centre, 103 High St; (03) 6381 1353; www.campbelltowntasmania.com

95.7 Heart FM, 1161 AM ABC Local Radio

This small town has been prominent in Tasmania's history: the first telephone call in the Southern Hemisphere was made from here to Launceston; the British Commonwealth's first agricultural show was held here in 1839, and the event is still held today; and it is the birthplace of Harold Gatty, the first person to fly around the world.

TASMANIA

Heritage buildings Of the 35 heritage buildings listed on the National Estate, The Grange, an old manor house built in the centre of town in 1840, is possibly the grandest. Others include the Fox Hunter's Return, Campbell Town Inn and Red Bridge, a 3-arched structure built over the Elizabeth River by convicts in the 1830s. There is a map in Grange Park at the northern end of High Street that lists the main historic buildings.

Heritage Highway Museum: displays on local history and the first round-the-world flight navigation; 103 High St.

Market: Town Hall; 4th Sun each month. ***Great Painting Race:*** Jan. ***Agricultural show:*** Australia's oldest; June.

Caledonia Bistro: hearty pub meals; 118 High St; (03) 6381 1158. ***Zeps Cafe:*** travellers cafe with city vibe; 92 High St; (03) 6381 1344. ***St Andrews Inn:*** international menu in historic surrounds; 12819 Midland Hwy, Cleveland; (03) 6391 5525.

The Broadwater on Macquarie: secluded fishing cabin; 'Barton', 2464 Macquarie Rd; (03) 6398 5114. ***Foxhunters Return:*** convict-built coaching inn; 132 High St; (03) 6381 1602. ***The Gables:*** Art Deco B&B and cottages; 35 High St; (03) 6381 1347. ***St Andrews Inn:*** restored coaching inn; 12819 Midland Hwy, Cleveland; (03) 6391 5525.

Fishing: Macquarie River, just west of town, and Lake Leake, 30 km E, are 2 good trout-fishing spots in the area.

See also MIDLANDS & THE NORTH, p. 504

Coles Bay

Pop. 473
Map ref. 671 O3

Freycinet National Park; (03) 6256 7000; www.freycinetcolesbay.com

98.5 Star FM, 106.1 FM ABC Local Radio

Coles Bay is a tiny town with a few stores and facilities. It is a gateway to the spectacular scenery of The Hazards, Wineglass Bay and Freycinet Peninsula.

Fishing competition: Southern Game Fishing Club; Mar. ***Three Peaks Race:*** sailing and mountain-climbing event; Easter. ***Freycinet Challenge:*** running, sea-kayaking, and road and mountain-bike racing at Freycinet Lodge; Oct. ***Mark Webber Challenge:*** Nov.

Madge Malloys: super fresh seafood; 3 Garnet Ave; (03) 6257 0399. ***The Bay Restaurant:*** tranquil, intimate dining; Freycinet Lodge, Freycinet National Park; (03) 6225 7016 or 1800 420 155. ***The Edge Restaurant:*** fine seafood and gorgeous views; Edge of the Bay Resort, 2308 Main Rd; (03) 6257 0102. ***The Oystercatcher:*** rustic seafood cafe and takeaway; 6 Garnet Ave; (03) 6257 0033.

Cove Beach Apartments: colourful modern apartments; 97 The Esplanade; 0417 609 151. ***Freycinet Beach Apartments:*** architect-designed minimalist apartments; Meika Crt, Swanwick; 0428 245 336. ***Freycinet Lodge:*** beautiful bushland cabins; Freycinet National Park; 1800 084 620. ***Iluka Holiday Centre:*** family-friendly budget cabins; The Esplanade; (03) 6257 0115 or 1800 786 512.

Freycinet National Park This park is world renowned for its stunning coastal scenery, challenging rock climbs, abundant wildlife, and range of walking tracks. The park is covered in wildflowers, including 60 varieties of ground orchid. Visitors can take the 1 hr walk to the lookout over stunning Wineglass Bay, or the 2.5 hr return walk to the beach. There are many safe swimming beaches and waterskiing, scuba diving, canoeing and sailing facilities. Check with visitor centre for details.

Freycinet Marine Farm: working oyster farm with guided tours and sampling; open 5–6pm Mon–Fri; tour bookings on (03) 6257 0140; 9 km NW. ***Moulting Lagoon Game Reserve:*** wetlands of international importance with many bird species; 12 km NW. ***Fishing:*** Great Oyster Bay is a renowned fishing spot – species include flathead and Australian salmon. Coles Bay is also a base for big-game fishing, particularly southern bluefin tuna in autumn; contact visitor centre for tour details.

See also EAST COAST, p. 500

Currie

Pop. 745
Map ref. 671 O11

The Trend, Edward St, Currie; (03) 6462 1355 or 1800 645 014; www.kingisland.org.au

88.5 FM ABC Local Radio, 88.9 Coast FM

Located in Bass Strait, Currie is the main town on King Island. The island is well known for its gourmet produce, including superb soft cheeses and grass-fed beef. The Roaring Forties are responsible for many of the 57 shipwrecks scattered around the island. The island now has five lighthouses (to avoid any further disaster), including one at Cape Wickham, which at 48 metres is the tallest lighthouse in the Southern Hemisphere.

King Island Cultural Centre: local arts, community projects, resident artists and personal stories of King Island residents; Currie Harbour. ***King Island Historical Society Museum:*** memorabilia of the island; open 2–4pm daily in summer, closed in winter; Lighthouse St. ***The Boathouse Gallery:*** displays and sales of pottery, sculptures and paintings, plus barbecue facilities; Edward St. ***Observatory:*** tours when sky is clear; Rifle Range Rd. ***King Island Maritime Trail:*** track with interpretive signs takes in shipwreck sites; details from visitor centre. ***Tours:*** full- and half-day tours of the island run by King Island Coaches; bookings on (03) 6462 1138.

King Island Imperial 20: marathon from Naracoopa to Currie; Mar. ***King Island Show:*** Mar. ***King Island Dramatic Society Play:*** Mar. ***Queenscliff to Grassy Yacht Race:*** Mar. ***King Island Horseracing Carnival:*** Nov/Dec.

Boomerang by the Sea: perfect seafood, panoramic views; Golf Club Rd; (03) 6462 1288. ***King Island Bakery:*** gourmet pies and breads; 5 Main St; (03) 6462 1337. ***Parers King Island Hotel Bistro:*** hearty pub-style meals; Main St; (03) 6462 1633. ***Kings Cuisine at the Bold Head Brasserie:*** local-produce-driven menu; Grassy Club, 10 Main Rd, Grassy; (03) 6461 1003.

Boomerang by the Sea: comfy seaside motel with views; Golf Club Rd; (03) 6462 1288 or 1800 221 288. ***Devil's Gap Retreat:*** private coastal cottages; Charles St; (03) 6462 1180. ***Naracoopa Holiday Units:*** beachside cottages; Beach Rd, Naracoopa; (03) 6461 1326. ***Portside Links:*** modern units with harbour views; Grassy Harbour Rd, Grassy; (03) 6461 1134.

King Island Dairy Internationally recognised for its cheeses, King Island Dairy has a Fromagerie Tasting Room with an excellent range of brie, camembert, cheddar, washed rind, triple cream and many others. The cows graze year-round, which makes their milk better for cheese production. Loorana; 9 km N.

Fishing: excellent salmon, flathead and whiting fishing from east-coast beaches; morwong, warehou, yellowtail kingfish and squid fishing at British Admiral Reef; 3 km S of Currie. ***Penguins:*** see them return to shore at dusk at Grassy in the south-east; 25 km SE. ***King Island Kelp Craft:*** specialising in the rare form of bull kelp handcrafts, with displays and sales; 6 Currie Rd, Grassy; 25 km SE. ***Naracoopa:*** seaside town with growing tourist industry; 25 km E. ***Seal Rocks State Reserve:*** covers 800 ha, with calcified forest and stunning cliffs at Seal Rocks; 30 km S. ***Lavinia Nature Reserve:*** 6400 ha of ocean beaches, heath, dunes, wetland bird habitats, lagoons and a rare suspended lake formation. Lavinia Point has a popular surfing beach; 40 km NE. ***Cape Wickham:*** lighthouse on rugged cliffs; north of Egg Lagoon; 45 km NE. ***Surfing:*** King Island is an ideal location for surfing.

See also BASS STRAIT ISLANDS, p. 505

Cygnet

Pop. 840
Map ref. 668 G8 | 671 J9

Huon River Jet Boats and Visitor Centre, Esplanade, Huonville; (03) 6264 1838; www.huonjet.com

 98.5 Huon FM, 936 AM ABC Local Radio

Cygnet was originally named Port de Cygne Noir by French admiral Bruni D'Entrecasteaux – after the black swans in the bay. However, in 1915 two local troopers were assaulted in a pub brawl and, in an attempt to avoid eternal infamy, the town name was changed to Cygnet. Now it is a fruit-growing and alternative-lifestyle community. In spring, the surrounding area blooms with magnificent wattle, apple and pear blossom.

Living History Museum: memorabilia, historic photos; open Thurs–Sun; Mary St. ***Trading Post:*** antiques, second-hand treasures and collectibles; Mary St. ***Town square:*** Presbytery, Catholic church (with stained-glass windows dedicated to lost miners) and convent; popular photo spot. ***Phoenix Creations:*** turns recycled materials into one-off furniture and crafts; Mary St. ***Near and Far:*** handcrafts and gifts; Mary St.

Cygnet Folk Festival: music, dance and art festival; Jan.

Lotus Eaters Cafe: worldly cuisine and relaxed surrounds; 10 Mary St; (03) 6295 1996. ***The School House Coffee Shop:*** casual cafe in heritage building; 23A Mary St; (03) 6295 1206. ***The Red Velvet Lounge:*** popular, eclectic cafe/restaurant; 24 Mary St; (03) 6295 0466.

Cygnet Hotel: budget-priced heritage rooms; 77 Mary St; (03) 6295 1267. ***Hartzview Vineyard:*** private vineyard homestead; 70 Dillons Rd, Gardners Bay; (03) 6295 1623. ***Riverside:*** serene and stylish riverside house; 35 Graces Rd, Glaziers Bay; (03) 6295 1952.

Birdwatching: black swans and other species inhabit Port Cygnet; viewing areas off Channel Hwy. ***Wineries:*** many including Hartzview Wine Centre, known for pinot noir, ports and liqueurs. There are cellar-door tastings and sales; Gardners Bay; 10 km SE. ***Beaches:*** good boat-launching facilities and beaches at Randalls Bay (14 km S), Egg and Bacon Bay (15 km S) and Verona Sands (18 km S). ***Nine Pin Point Marine Nature Reserve:*** reef; near Verona Sands; 18 km S. ***Fishing:*** sea-run trout and brown trout from the Huon River; west and south of town. ***Berry orchards:*** orchards with pick-your-own sales.

See also SOUTH-EAST, p. 501

Deloraine

Pop. 2242
Map ref. 673 I8 | 675 C10

Great Western Tiers Visitor Centre, 98–100 Emu Bay Rd; (03) 6362 3471; www.greatwesterntiers.net.au

 91.7 FM ABC Local Radio, 103.7 FM City Park Radio

Deloraine has become the artistic hub of northern Tasmania. Artists are inspired by the magnificent scenery, working farms and hedgerows, and the Great Western Tiers nearby. Situated on the Meander River, Deloraine is Tasmania's largest inland town and a busy regional centre that has been classified by the National Trust as a town of historical significance.

Yarns Artwork in Silk Created by more than 300 people and taking 10 000 hours to complete, Yarns is a 200 m reflection of the Great Western Tiers of Tasmania in 4 large panels. Accompanied by an audio presentation and sound and light effects, the Yarns presentation operates every half-hour. 98 Emu Bay Rd.

Jahadi Tours and Art Gallery Jahadi Indigenous Experience runs half- and full-day tours in and around Deloraine that take in the natural landscape and focus on Aboriginal sites such as caves, middens and rock shelters. Tours are organised for a maximum of 8 people to ensure a personal experience; bookings on (03) 6363 6172. The Jahadi Art Gallery exhibits artwork from local Indigenous artists; 900 Mole Creek Rd.

Deloraine Folk Museum: showcases the life of a country publican with exhibition gallery, garden, dairy, blacksmith's shop and family history room; 98 Emu Bay Rd. ***Galleries:*** sales and exhibits of local artwork from paintings to furniture. Venues include Gallery B on Emu Bay Rd, Artifakt on Emu Bay Rd, Artist's Garret on West Church St and Bowerbank Mill on the Bass Hwy.

Market: showgrounds, Lake Hwy; 1st Sat each month plus 3rd Sat Feb–May. ***Tasmanian Craft Fair:*** largest working craft fair in Australia; Nov.

Christmas Hills Raspberry Farm Cafe: 'go straight to dessert'; Bass Hwy, Elizabeth Town; (03) 6362 2186. ***etc (Elizabeth Town Bakery Cafe):*** super roadside licensed cafe; 5783 Bass Hwy, Elizabeth Town; (03) 6368 1350.

Bonney's Inn: award-winning historic B&B; 19 West Pde; (03) 6362 2974. ***Bowerbank Mill B&B:*** 4-storey restored flour mill; 4455 Meander Valley Rd; (03) 6362 2628. ***Peppers Calstock:*** exclusive 19th-century boutique hotel; Highland Lakes Rd; (03) 6362 2642. ***Three Willows Vineyard B&B:*** organic vineyard homestead; 46 Montana Rd, Red Hills; 0438 507 069.

Liffey Falls State Reserve Liffey Falls is surrounded by cool–temperate rainforest species of sassafras, myrtle and leatherwood. There is a 45 min return nature walk from the picnic area through lush tree ferns, taking in smaller falls along the way. Contact Parks & Wildlife Service on 1300 135 513; 29 km S.

Exton: tiny township full of antique shops and charming old cottages; 5 km E. ***41 Degrees South Aquaculture:*** salmon farm, wetlands and ginseng nursery, with tastings and cafe; Montana Rd; 8 km W. ***Ashgrove Farm Cheeses:*** sales and tastings of English-style cheeses; Elizabeth Town; 10 km NW. ***Lobster Falls:*** 2 hr return walk from roadside through riverside forest with lookout and wide variety of local birdlife; 11 km W. ***Quamby Bluff:*** solitary mountain behind town, with 6 hr return walking track to summit starting near Lake Hwy; 20 km S. ***Meander Falls:*** stunning falls reached by 5–6 hr return trek for experienced

TASMANIA

walkers that also takes in the Tiers. Shorter walks are outlined at information booth in forest reserve carpark; 28 km S. ***Fishing:*** excellent trout fishing in Meander River (north and south of town) and Mersey River (around 20 km E). ***Scenic drive:*** through Central Highlands area via Golden Valley to Great Lake, one of the largest high-water lakes in Australia. Check road conditions with Parks & Wildlife Service; (03) 6259 8348.

See also MIDLANDS & THE NORTH, p. 504

Derby

Pop. 302
Map ref. 673 N6

Derby Tin Mine Centre, Main St; (03) 6354 1062; www.northeasttasmania.com.au

 91.7 FM ABC Local Radio, 94.5 Sea FM

This small north-eastern town was born when tin was discovered in 1874 and the 'Brothers Mine' opened two years later. The Cascade Dam was built and the mine prospered until 1929, when the dam flooded and swept through the town, killing 14 people. The mine closed and, although it eventually reopened, the town never fully recovered. Tourists stroll through the charming streets to view the old buildings and tin mine memorabilia.

Derby Tin Mine Centre The town's major attraction, the centre is a reconstruction of this old mining village. It includes a miner's cottage, general store, butcher's shop, huge sluice and the historic Derby gaol. The old Derby School houses a comprehensive museum with history displays, gemstones, minerals and tin-panning apparatus. Main St.

Red Door Gallerae: paintings, carvings and furniture; Main St. ***Bankhouse Manor:*** arts, crafts and collectibles; Main St.

Derby River Derby: raft race down Ringarooma River with markets, exhibitions and children's activities; Oct.

Berries Cafe: cute country-style cafe; 72 Main St; (03) 6354 2520.

Cobblers Accommodation: rustic miner's cottage; 63 Main St; (03) 6354 2145. ***Tin Dragon Trail Cottages:*** peaceful rural cottages; 3 Coxs La, Branxholm; (03) 6354 6210. ***Winnaleah Hotel:*** comfortable pub rooms; 12 Main St, Winnaleah; (03) 6354 2331.

Ralphs Falls The longest single-drop waterfall in Australia. A 20 min return walk under a myrtle rainforest canopy arrives at Norms Lookout at the top of the falls. From here views of the Ringarooma Valley, Bass Strait and the Furneaux Islands. Picnic and barbecue facilities are available. 15 km SE of Ringarooma.

Gemstone fossicking: Weld River in Moorina, where the largest sapphire in Tasmania was discovered; Tasman Hwy; 8 km NE. ***Miners cemetery:*** historic cemetery where early tin miners, including some Chinese, were buried; Moorina; 8 km NE. ***Fishing:*** excellent trout fishing along Ringarooma River north of town.

See also MIDLANDS & THE NORTH, p. 504

Devonport

Pop. 22 317
Map ref. 672 H6 | 675 A6

92 Formby Rd; (03) 6424 4466; www.devonporttasmania.travel

 100.5 FM ABC Local Radio, 104.7 Coast FM

Devonport, on the banks of the Mersey River, is the home port of the *Spirit of Tasmania I* and *II*. This vibrant seaport city is framed by the dramatic headland of Mersey Bluff, beautiful coastal reserves and parklands, and Bass Strait. The community is fuelled by farming, manufacturing and, of course, tourism.

Home Hill A National Trust property, this was home to prime minister Joseph Lyons and Dame Enid Lyons, along with their 12 children. The rich collection of personal material provides insight into Australian political life and international relations during the momentous events of the mid-20th century. Joseph Lyons was Tasmania's premier from 1923 to 1928 and became prime minister in 1932. He is the only Tasmanian to have achieved the 'top job'. Dame Enid became the first female member of the House of Representatives in 1943 and was sworn in as the first female federal minister of the Crown in 1949. The Home Hill property comprises a beautiful old building in well-maintained gardens of wisteria and imposing trees. 77 Middle Rd; (03) 6424 3028.

Tiagarra Aboriginal Culture Centre and Museum Meaning 'keep' or 'keeping place', Tiagarra is one of the few preserved Aboriginal rock carvings sites in Tasmania. A 1 km circuit walk takes in the carvings. An adjoining art centre exhibits over 2000 artefacts. Dioramas depict the lifestyle of the original inhabitants. Bluff Rd, Mersey Bluff; (03) 6424 8250.

Don River Railway Owned and operated by the Van Diemen Light Railway Society; the train offers short rides along Don River to Coles Beach. On-site is a museum housing the largest collection of old steam locomotives and passenger carriages in Tasmania. Train rides hourly. Forth Main Rd, Don; (03) 6424 6335.

Devonport Maritime Museum & Historical Society: features detailed models and displays from the days of sail, through the age of steam, to the present seagoing passenger ferries. 6 Gloucester Ave; (03) 6424 7100. ***Imaginarium Science Centre:*** learn and enjoy everyday science through hands-on, interactive exhibitions, displays, science shows and special events. Something for all age groups. Entrance off the Wenvoe St carpark; (03) 6423 1466. ***Devonport Regional Gallery:*** displays the work of established and emerging Tasmanian artists in a century-old building; 45–47 Stewart St; (03) 6424 8296. ***Mersey River Cruises:*** enjoy the scenery and history of this busy port; operates Nov–Apr; departs from pontoon opposite visitor centre, Formby Rd. ***Australian Weaving Mills Factory Outlet:*** factory seconds for Dickies and Dri-Glo products; 45 Tasman St. ***Devonport Lighthouse:*** striking icon completed in 1899 and part of the National Estate; Mersey Bluff. ***Scenic flights:*** through Tasair from Devonport airport; bookings on (03) 6427 9777.

Devonport Cup: Jan. ***Apex Regatta:*** Mar. ***Taste the Harvest Festival:*** Mar. ***Devonport Jazz Weekend:*** July. ***Devonport Show:*** Nov. ***Athletic and Cycling Carnivals:*** Dec.

Essence Food and Wine: fine dining, licensed; 28 Forbes St; (03) 6424 6431. ***Hawley's Gingerbread House:*** hearty fare close to ferry terminal; 71 Wright St, East Devonport; (03) 6427 0477. ***High Tide Waterfront Restaurant:*** stylish riverside dining; 17 Devonport Rd; (03) 6424 6200. ***Taco Villa:*** popular Mexican restaurant; Shop 4, 1 Kempling St; (03) 6424 6762.

Birchmore of Devonport: central historic B&B; 8–10 Oldaker St; (03) 6423 1336. ***Mersey Bank Apartments:*** luxury apartments above various restaurants; 153–159 Rooke St; (03) 6423 5141. ***Glencoe Rural Retreat:*** French-style luxury B&B; 1468 Sheffield Rd, Barrington; (03) 6492 3267.

Tasmanian Arboretum 58 ha reserve of rare native and exotic plants and trees. Picnic areas, walking tracks and a lake where you can watch waterbirds and platypus. Eugenana; 10 km S.

House of Anvers Chocolate Factory: the total chocolate experience – manufacturing viewing room, chocolate museum and tasting centre. Bass Hwy; 10 km sw. ***Braddon's Lookout:*** panoramic view of coastline; near Forth; 16 km w. ***Kelcey Tier Nature Walk:*** 160 ha of native bushland and 3.6 km circuit walk with superb views of Devonport and Mersey River.

See also North-West, p. 503

Dover

Pop. 461
Map ref. 668 F10 | 671 J10

Forest and Heritage Centre, Church St, Geeveston; (03) 6297 1836; www.forestandheritagecentre.com

 95.3 Huon FM, 936 AM ABC Local Radio

Dover lies beside the waters of Esperance Bay and the D'Entrecasteaux Channel, with the imposing figure of Adamson's Peak in the background. The three islands directly offshore, Faith, Hope and Charity, were named perhaps to inspire the convicts held at the original probation station. The town is a popular destination for yachting enthusiasts.

Commandant's Office: well-preserved remnant of Dover's penal history; Beach Rd.

Dover Hotel: local fare in homely pub; Main Rd; (03) 6298 1210. ***Gingerbread House Bakery Cafe:*** breakfasts and light lunches; Cnr Station St and Main Rd; (03) 6298 1502.

Driftwood Cottages: bayside cottages; Bayview Rd; (03) 6298 1441. ***Riseley Cottage B&B:*** spacious, elegant accommodation; 170 Narrows Rd, Strathblane; (03) 6298 1630.

Faith Island: several historic graves; access by boat. ***Walking trails:*** epic Tasmanian Trail to Devonport, as well as Dover Coast and Duckhole Lake tracks; details from visitor centre. ***Beaches:*** safe swimming beaches surround town.

See also South-East, p. 501

Dunalley

Pop. 310
Map ref. 669 M6 | 671 M7

Dunalley Hotel, 210 Arthur Hwy; (03) 6253 5101.

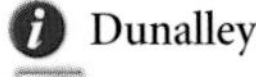
 97.7 Tasman FM, 936 AM ABC Local Radio

A quaint fishing village, Dunalley is on the isthmus separating the Forestier and Tasman peninsulas from the mainland. Nearby is the Denison Canal, which is Australia's only purpose-built, hand-dug sea canal. The swing bridge that spans the canal has become quite a spectacle for visitors.

Tasman Monument Erected in 1942, this monument commemorates the landing of Abel Tasman and his crew. The actual landing occurred to the north-east on the Forestier Peninsula, near the fairly inaccessible Cape Paul Lamanon. Imlay St.

Dunalley Fish Market: fishmonger and cafe; 11 Fulham Rd; (03) 6253 5428. ***Dunalley Hotel:*** pub tucker, especially seafood; 210 Arthur Hwy; (03) 6253 5101. ***Dunalley Waterfront Cafe:*** contemporary cafe and collectibles; 4 Imlay St; (03) 6253 5122. ***Murdunna Store:*** roadside tea room; 4050 Arthur Hwy, Murdunna; (03) 6253 5196.

Potters Croft: loft rooms and gallery; 1 Arthur Hwy; (03) 6253 5469. ***Beachbreaks:*** glass-fronted beach house; 357 Marion Bay Rd, Bream Creek; (03) 6253 5476.

Copping Colonial and Convict Exhibition Housing a vast array of antiques and memorabilia from the convict era, it has the added highlight of containing one of only 3 cars manufactured in Australia in the 19th century. Arthur Hwy; (03) 6253 5373; 11 km n.

Marian Bay: popular swimming beach; 14 km ne.

See also South-East, p. 501

Eaglehawk Neck

Pop. 267
Map ref. 669 N7 | 671 M8

Port Arthur Historic Site, Arthur Hwy, Port Arthur; (03) 6251 2310 or 1800 659 101.

97.7 Tasman FM, 936 AM ABC Local Radio

Present-day Eaglehawk Neck is a pleasant fishing destination, speckled with small holiday retreats and striking scenery. Situated on the narrow isthmus between the Forestier and Tasman peninsulas, Eaglehawk Neck was the perfect natural prison gate for the convict settlement at Port Arthur. Few prisoners escaped by sea, so Eaglehawk Neck was essentially the only viable way out. The isthmus was guarded by soldiers and a line of ferocious tethered dogs. Most convicts knew not to bother, but William Hunt, convict and former strolling actor, tackled the isthmus in a kangaroo skin. As two guards took aim with their muskets, their efforts were cut short by a plaintive shout coming from the kangaroo, 'Don't shoot! It's only me – Billy Hunt!'

Bronze dog sculpture: marks the infamous dogline; access by short walking track off Arthur Hwy. ***Scuba diving:*** the area has a huge diversity of dive sites including the spectacular formations of Sisters Rocks, the 25 m high giant kelp forests, the seal colony at Hippolyte Rock, the SS *Nord* wreck and amazing sea-cave systems; Eaglehawk Dive Centre; bookings on (03) 6250 3566. ***Surfing:*** good surf beaches at Eaglehawk Neck and Pirates Bay.

Eaglehawk Cafe & Guesthouse: local seasonal menu; 5131 Arthur Hwy; (03) 6250 3331. ***Lufra Hotel:*** counter meals; 380 Pirates Bay Dr; (03) 6250 3262 or 1800 639 532. ***The Mussel Boys:*** oysters, mussels, seafood and meats to delight; 5927 Arthur Hwy, Taranna; (03) 6250 3088.

Eaglehawk Cafe & Guesthouse: 2-storey weatherboard guesthouse; 5131 Arthur Hwy; (03) 6250 3331. ***Lufra Hotel:*** iconic Art Deco hotel; 380 Pirates Bay Dr; (03) 6250 3262 or 1800 639 532. ***Four Seasons Holiday Cottages:*** modern A-frame cabins; 5732 Arthur Hwy, Taranna; 0407 044 483. ***Norfolk Bay Convict Station:*** historic waterfront homestead; 5862 Arthur Hwy, Taranna; (03) 6250 3487.

Tasman National Park Tasman Blowhole, Tasmans Arch and Devils Kitchen are the key attractions here, occurring in rocks that are Permian in age (about 250 million years old). There are numerous walks, the full track reaching from Eaglehawk Neck to Fortescue Bay, and there are shorter walks to Tasmans Arch, Waterfall Bay and Patersons Arch. Check with visitor centre for details or contact Parks & Wildlife Service on 1300 135 513; 4 km se.

Tasmanian Devil Conservation Park This animal-rescue centre features Tasmanian devils, quolls, eagles, wallabies, owls and wombats. There is a 1.5 km bird trail and free flight bird show, Kings of the Wind (11.15am and 3.30pm). The Tasmanian devil feeding time is worth waiting for (10am, 11am, 1.30pm and 5pm; 4.30pm in winter). Arthur Hwy, Taranna; (03) 6250 3230; 12 km s.

Tessellated Pavement: these rocks appear to have been neatly tiled, but their formation is entirely natural. Earth movements have fractured the pavement over the years. 1 km N. ***Pirates Bay Lookout:*** views across the bay, past the eastern side of Eaglehawk Neck to the massive coastal cliffs of the Tasman Peninsula; 1.5 km N. ***Doo Town:*** holiday town in which most of the houses bear names with variations of 'doo'; 3 km S. ***Federation Chocolate:*** chocolate factory with free tastings and historical museum; South St, Taranna; 10 km SW.

See also SOUTH-EAST, p. 501

Evandale

Pop. 1055
Map ref. 673 K9 | 675 H11

Tourism and History Centre, 18 High St; (03) 6391 8128; www.evandaletasmania.com

91.7 FM ABC Local Radio, 103.7 FM City Park Radio

Just south of Launceston is this classified town, with beautiful buildings of historical and architectural importance. Cyclists come from as far as the Czech Republic for the annual Penny Farthing Championships, a race along the triangular circuit in the centre of the town.

Heritage walk Evandale is best appreciated with a copy of the brochure *Let's Talk About Evandale*, which lists over 35 historic buildings and sites in the town and many more in the district. Among them are Blenheim (1832), which was once a hotel, St Andrews Uniting Church (1840) with its classic bell-tower and Doric columns, and the former Presbyterian Manse (1840). Brochure available from visitor centre.

Miniature railway: steam railway; open Sun; adjacent to market, Logan Rd.

Market: over 100 stalls; Falls Park, Russell St; Sun. ***Village Fair and National Penny Farthing Championships:*** largest annual event in the world devoted to racing antique bicycles; Feb. ***Railex:*** model railway exhibition; Mar. ***Glover Prize:*** Australia's second-richest art prize; Mar.

Clarendon Arms Hotel: hearty pub food; 11 Russell St; (03) 6391 8181. ***Ingleside Bakery:*** cafe and bakery; 4 Russell St; (03) 6391 8682. ***Josef Chromy Cellar Door Cafe:*** modern Australian vineyard restaurant; 370 Relbia Rd, Relbia; (03) 6335 8700.

Grandma's House: cute cottage in heritage garden; 10 Rogers La; (03) 6391 8444. ***Greg and Gill's Place:*** 2 self-contained apartments; 35 Collins St; (03) 6391 8248. ***Solomon Cottage:*** B&B in old bakery; 1 High St; (03) 6391 8331. ***Wesleyan Chapel:*** antique church to yourself; 28 Russell St; (03) 6331 9337.

Clarendon House Just north of Nile is the stunning National Trust residence, Clarendon House. It was built in 1836 by James Cox, a wealthy grazier and merchant, and has been restored by the National Trust. Clarendon's high-ceilinged rooms, extensive formal gardens and range of connected buildings (dairy, bakehouse, gardener's cottage and stable) make it one of the most impressive Georgian houses in Australia. (03) 6398 6220; 8 km S.

Ben Lomond National Park Site of Tasmania's largest alpine area and premier ski resort, this park has both downhill and cross-country skiing, with ski tows and ski hire. The park also offers walking tracks and picnic areas in summer. Legges Tor, the second highest point in Tasmania, has spectacular views. The area blooms with alpine wildflowers in summer. Contact Parks & Wildlife Service on 1300 135 513; 47 km E.

Symmons Plains International Raceway: venue for national V8 Supercars meeting in Nov. A track is open for conditional public use; bookings on (03) 6398 2952; 10 km S. ***John Glover's grave:*** burial site of prominent Tasmanian artist beside church designed by Glover; Deddington; 24 km SE. ***Trout fishing:*** in North Esk and South Esk rivers.

See also MIDLANDS & THE NORTH, p. 504

Exeter

Pop. 339
Map ref. 673 J7 | 675 F8

Tamar Visitor Centre, Main Rd, Exeter; (03) 6394 4454 or 1800 637 989; www.tamarvalley.com.au

91.7 FM ABC Local Radio, 95.3 Tamar FM

Exeter is a small community in Tasmania's north-east, best known for its scenic surroundings. It lies just north of Launceston, in the centre of Tamar Valley wine country, which, along with cold-climate wines, has a variety of orchards.

John Temple Gallery: beautiful, framed Tasmanian photographs.

Show: Feb.

Exeter Bakery: great selection of pastries, slices and gourmet rolls including their famous scallop pies; 104 Main Rd: (03) 6394 4069. ***Koukla's:*** home-style Greek; 285 Gravelly Beach Rd; (03) 6394 4013. ***Daniel Alps at Strathlynn:*** award-winning regional fine dining; 95 Rosevears Dr, Rosevears; (03) 6330 2388. ***Estelle Restaurant at Rosevears Estate:*** hip winery restaurant; Rosevears Vineyard, 1A Waldhorn Dr, Rosevears; (03) 6330 1800.

Viewenmore Villa for Two: elevated octagonal house; 312 Rosevears Dr, Rosevears; 0421 422 779. ***Conmel Cottage:*** old-world 1-bedroom cottage; 125 Rosevears Dr, Rosevears; (03) 6330 1466. ***Rosevears Vineyard Retreat:*** modern chalets with vineyard views; 1A Waldhorn Dr, Rosevears; (03) 6330 1800. ***Aspect Tamar Valley Resort:*** quirky Swiss-village chalets; 7 Waldhorn Dr, Grindelwald; (03) 6330 0400 or 18000 TAMAR.

Brady's Lookout This scenic lookout was once the hideout of bushranger Matthew Brady, who used the high vantage point to find prospective victims on the road below. Today the site is more reputable, but retains its magnificent view of the Tamar Valley and is an ideal picnic spot. 5 km SE.

Glengarry Bush Maze: excellent family venue with maze and cafe; closed in winter, except Wed; Jay Dee Rd, Glengarry; 8 km SW. ***Paper Beach:*** 5 km return walking track to Supply River, where there are ruins of the first water-driven flour mill in Tasmania; 9 km E. ***Artisan Gallery and Wine Centre:*** displays and sales of Tasmanian arts, crafts and wines from smaller, independent vineyards in the area; Robigana; 10 km N. ***Tamar Valley Resort at Grindelwald:*** resort in Swiss architectural style with Swiss bakery, chocolatier, crafts, souvenirs and a world-class minigolf course; 10 km SE. ***Notley Fern Gorge:*** 11 ha wildlife and rainforest sanctuary with giant man-ferns and moss-covered forest. A 2 hr return walk leads to Gowans Creek; Notley Hills; 11 km SW. ***Tasmanian Wine Route:*** a string of wineries with cellar-door tastings and sales runs on either side of the Tamar, north and south of town. Closest to Exeter are wineries around Rosevears, 10 km S; brochure available from visitor centre.

See also MIDLANDS & THE NORTH, p. 504

Fingal

Pop. 340
Map ref. 673 N9

Old Tasmanian Hotel Community Centre, Main Rd; (03) 6374 2344.

 100.3 Star FM, 1161 AM ABC Local Radio

Poet James McAuley wrote of Fingal's 'blonding summer grasses', 'mauve thistledown' and the river that 'winds in silence through wide blue hours, days'. Indeed, the crags of Ben Lomond National Park and the lush valley make Fingal a quiet inspiration for many writers. The town was established in 1827 as a convict station and distinguished itself by becoming the headquarters of the state's coal industry. Just north of Fingal, at Mangana, Tasmania's first payable gold was discovered in 1852.

Historic buildings There are many heritage buildings throughout the township, particularly in Talbot St, including the Holder Brothers General Store (1859), St Peter's Church (1867) and Fingal Hotel (1840s), which claims to stock the largest collection of Scotch whiskies in the Southern Hemisphere.

Fingal Valley Festival: incorporates the World Coal Shovelling Championships and Roof Bolting Championships; Mar.

Glenesk Holiday Cottage: country cottage with brass beds; 9 Talbot St; (03) 6374 2195. ***Mayfield Manor:*** 100-year-old country manor; 18 Talbot St; (03) 6374 2285. ***The Fingal Hotel:*** pub-style accommodation; 4 Talbot St; (03) 6374 2121. ***St Pauls River Cabins:*** modern cabins with access to trout fishing; 1207 Royal George Rd, Avoca; (03) 6384 2211.

Evercreech Forest Reserve This reserve is home to the impressive White Knights, the tallest white gums in the world, including a specimen 89 m high. A 20 min circuit walk passes through a man-fern grove and blackwoods, then up a hill for a superb view. There is also a 45 min return walk to Evercreech Falls, and many picnic and barbecue spots. Contact Forestry Tasmania on 1800 367 378; on the road to Mathinna; 30 km N.

Avoca: small township with many historical buildings; 27 km SW. ***Mathinna Falls:*** magnificent 4-tier waterfall over a drop of 80 m, with an easy 30 min return walk to falls base; 36 km N. ***Ben Lomond National Park:*** *see Evandale*; 72 km NW.

See also EAST COAST, p. 500

Flinders Island

see Whitemark

Geeveston

Pop. 761
Map ref. 668 F8 | 671 I9

Forest and Heritage Centre, Church St; (03) 6297 1836; www.forestandheritagecentre.com

 95.3 Huon FM, 936 AM ABC Local Radio

Geeveston, on the cusp of enormous Southwest National Park, is driven by thriving timber and forestry industries, and slow-moving timber trucks frequent the roads. Swamp Gum, the trunk of a logged eucalypt 15.8 metres in length and weighing 57 tonnes, stands on the highway as the town's mascot. The other principal industry, apple farming, is responsible for the magnificent apple blossom in late September.

Forest and Heritage Centre Comprising 4 different sections, including the Forest Room and Hartz Gallery, the centre offers a comprehensive look at forest practices with computer games, timber species exhibits and a woodturning viewing area. Church St; (03) 6297 1836.

Southern Design Centre Tasmanian timber furniture is lovingly crafted on-site here. Watch resident artisans demonstrate skills in various mediums and browse the showrooms for art and craft, exquisite giftware and home furnishings. 11 School Rd.

Geeveston Highlands Salmon and Trout Fishery: world's first catch-and-release Atlantic salmon fishery with a 1.6 ha salmon lake and 0.4 ha trout lake. Fly-fishing tuition available; bookings on (03) 6297 0030; 172 Kermandie Rd. ***Bears Went Over the Mountain:*** sales of teddy bears and assorted antiques; 2 Church St.

 ***Tasmanian Forest Festival:*** biennial event in Mar.

Apple Core Cafe: friendly eatery; 13 George St; (03) 6297 1100.

Bears over the Mountain: teddy-bear-themed B&B; 2 Church St; (03) 6297 0110. ***Cambridge House:*** grand Edwardian B&B; 2 School Rd; (03) 6297 1561.

Hartz Mountains National Park The Huon Valley used to be wholly glaciated and this national park displays some remarkable glacial features and morainal deposits. Lake Hartz is the largest of the glacial lakes that surround the 1255 m high Hartz Mountain. There are walking tracks through forests of Tasmanian waratah, snow gums, yellow gum and alpine heath, and Waratah Lookout affords fantastic views. Self-guide brochure and park pass are available from the visitor centre. Off Arve Rd; 23 km SW.

Tahune Forest AirWalk Opened in 2001, this is the longest and highest forest canopy walk in the world. It stretches 597 m through the treetops of the Tahune Forest Reserve and, at its highest point, is 48 m above the forest floor. It provides a bird's-eye view of wet eucalypt forest and the Huon and Picton rivers. Within the forest, visitors can go fishing, rafting and camping. Tickets from visitor centre; 28 km W.

Arve Forest Drive: scenic drive following the Arve River Valley that takes in the Look-In Lookout (an information booth and lookout perch), the Big Tree Lookout (remarkable, large swamp gum), picnic areas and the Keoghs Creek Walk (a great short streamside walk); 10 km W. ***Southwest National Park:*** Tasmania's largest national park offers walking tracks of varying difficulty, beautiful scenic drives and plentiful fishing. Contact Parks & Wildlife Service in Huonville for up-to-date information on track and weather conditions; (03) 6264 8460; 60 km S.

See also SOUTH-EAST, p. 501

George Town

Pop. 4265
Map ref. 673 J6 | 675 E6

Cnr Victoria St and Main Rd; (03) 6382 1700; www.georgetown.tas.gov.au

 91.7 FM ABC Local Radio, 95.3 Tamar FM

George Town is Australia's third oldest settlement, after Sydney and Hobart, and Tasmania's oldest town. European settlement can be traced back to 1804 when William Paterson camped here after running his ship, HMS *Buffalo*, aground at York Cove. Ignoring the disaster, he ran up the flag, fired three shots in the air and played the national anthem. A memorial stands at Windmill Point to honour this optimism.

The Grove: this classic Georgian stone house (c. 1838) was the home of Mathew Friend, the port officer and magistrate of the settlement; 25 Cimitiere St. ***York Cove:*** scenic cove where George Town's centre was built, with mooring and pontoon facilities, and restaurants. ***Self-guide Discovery Trail:*** walking route through town; brochure from visitor centre. ***George Town Watch House:*** community history room and female factory display; Macquarie St.

Tamar Valley Folk Festival: Jan.

Cove Restaurant: Mediterranean-influenced modern Tasmanian; York Cove, 2 Ferry Blvd; (03) 6382 9990. ***The Pier Hotel Bistro:*** bistro with pizzas; 5 Elizabeth St; (03) 6382 1300.

Ben Hyrons Cottage: historic cottage B&B; 40 Anne St; (03) 6382 1399. ***Charles Robbins:*** stylish modern foreshore apartments; 3 Charles Robbins Pl; (03) 6382 4448. ***Tam O'Shanter Views:*** isolated glass-fronted beach house; 29 Seascape Dr, Tam O'Shanter; 0412 146 166

Low Head This popular holiday retreat has safe swimming and surf beaches. The Maritime Museum is housed in Australia's oldest continuously used pilot station and has fascinating displays of memorabilia discovered in nearby shipwrecks. At the Tamar River's entrance stands a 12 m high lighthouse, built in 1888, behind which lies a little penguin colony. Penguin-watching tours start around sunset; bookings on 0418 361 860; 5 km N.

Mt George Lookout: scenic views of George Town, the north coast, and south to the Western Tiers. The lookout has a replica of a Tamar Valley semaphore mast used to relay messages in the 1800s; 1 km E. ***Fishing:*** excellent fishing at Lake Lauriston and Curries River Dam; 13 km E. ***Lefroy:*** old goldmining settlement, now a ghost town, with ruins of old buildings; 16 km E. ***Hillwood Strawberry Farm:*** sales and pick-your-own patch, along with sampling of local fruit wines and cheeses; 24 km SE. ***Seal tours:*** cruises to Tenth Island fur seal colony, a short distance offshore in Bass Strait; bookings on (03) 6382 3452. ***Tasmanian Wine Route:*** many wineries to the east of town with cellar-door sales and tastings; brochure from visitor centre. ***Beaches:*** the area has many beautiful beaches including East Beach (facing Bass Strait and ideal for walking, swimming and surfing) and Lagoon Beach on the Tamar River (for family swimming).

See also MIDLANDS & THE NORTH, p. 504

Gladstone

Pop. 42
Map ref. 673 O5

Gladstone Hotel, 37 Chaffey St; (03) 6357 2143.

91.7 FM ABC Local Radio, 93.7 Star FM

The north-eastern district surrounding Gladstone was once a thriving mining area, yielding both tin and gold. Today many of the once-substantial townships nearby are ghost towns. Gladstone has survived, but its successful mining days have long since given way to tourism. It acts as a tiny service centre for surrounding dairy, sheep and cattle farms, as well as for visitors to Mount William National Park, and has the distinction of being Tasmania's most north-easterly town.

Gladstone cemetery: A historic reminder of the miners, including many Chinese, who were drawn to the area; Carr St.

Bay of Fires Lodge: walk-in luxury lodge; Mt William National Park; (03) 6392 2211. ***Gladstone Hotel:*** basic pub accommodation; Chaffey St; (03) 6357 2143.

Mt William National Park With rolling hills, rugged headlands and pristine beaches this park offers swimming, fishing, diving and bushwalking. Georges Rocks and Eddystone Pt are favoured diving spots, while Ansons Bay is well known for bream and Australian bass fishing. Walks vary in difficulty and are signposted. At Eddystone Pt, at the southern end of the park, stands a historic, pink-granite lighthouse. Contact Parks & Wildlife Service on 1300 135 513; 25 km E.

Bay of Fires Walk This 4-day guided walk takes in the best of Mt William National Park while offering first-class accommodation and catering. The highlight (along with the scenery) is kayaking in Ansons Bay and, of course, bedding down for the night in the architecturally superb Bay of Fires Lodge, surrounded by bush. (03) 6391 9339.

Blue Lake: disused tin mine filled with brilliant blue water (coloured by pyrites) and safe for swimming and waterskiing; South Mt Cameron; 8 km S. ***Cube Rock:*** large granite monolith on an outcrop, reached by 3 hr return climb; South Mt Cameron; 8 km S. ***Beaches:*** magnificent beaches to the north, including Petal Pt; 25 km N. ***Geological formations:*** impressive granite formations between Gladstone and South Mt Cameron. ***Gem fossicking:*** smoky quartz, topaz and amethyst can be found in the district; contact visitor centre for details.

See also MIDLANDS & THE NORTH, p. 504

Hadspen

Pop. 1929
Map ref. 673 K8 | 675 G10

Launceston Travel and Information Centre, Cornwall Square Transit Centre, 12–16 St John St, Launceston; (03) 6336 3133 or 1800 651 827; www.ltvtasmania.com.au

93.7 FM ABC Local Radio, 103.7 FM City Park Radio

Hadspen's best-known resident, Thomas Reibey III, became premier of Tasmania after being fired as archdeacon of Launceston's Church of England. Reibey was prepared to fund construction of Hadspen's Church of the Good Shepherd, but withdrew his offer after a dispute with the bishop, who allegedly discovered Reibey's unorthodox sexual preferences and refused the 'tainted' money. As a result, the church only reached completion in 1961, more than 100 years later.

Historic buildings: Hadspen's Main Rd is lined with heritage buildings, including the Red Feather Inn (c. 1844) and Hadspen Gaol (c. 1840).

Carrick Inn: traditional pub fare; 46 Meander Valley Hwy, Carrick; (03) 6393 6143. ***Rutherglen Bistro:*** family dining; Rutherglen Holiday Village, Bass Hwy; (03) 6393 6307.

Red Feather Inn: coaching inn turned French Provincial hotel; 42 Main St; (03) 6393 6506. ***The Stables at Hawthorn Villa:*** self-contained units in historic gardens; Cnr Bass Hwy and Bishopsbourne Rd, Carrick; (03) 6393 6150.

Entally House Thomas Reibey's original abode, built in 1819, is one of the most impressive heritage homes in the state. Entally has sprawling gardens (maintained by Parks & Wildlife Service), Regency furniture and other antiques, as well as a stunning riverside location on the South Esk River. 1 km W.

Carrick: neighbouring town with historical buildings. It hosts Agfest, one of Australia's biggest agricultural field days, in May; 10 km SW. ***Tasmanian Copper Gallery:*** sales and exhibitions of original copper artworks; 1 Church St, Carrick; 10 km SW.

See also MIDLANDS & THE NORTH, p. 504

Hamilton

Pop. 300
Map ref. 668 F1 | 671 I5

Council offices, Tarleton St; (03) 6286 3202.

97.1 Mid FM, 936 AM ABC Local Radio

Hamilton, an unspoilt National Trust–classified town, has avoided the commercialisation found in some parts of the state. Its buildings and tranquil lifestyle just outside Hobart conjure up an image of 1830s Tasmania.

Hamilton Sheep Centre Demonstrates sheepshearing methods and the use of working farm dogs. Farm tours can be arranged, with meals provided. Bookings on (03) 6286 3332.

Glen Clyde House: convict-built c. 1840, this place now houses an award-winning craft gallery and tearooms; Grace St. ***Hamilton Heritage Museum:*** small museum with artefacts and memorabilia of the area; Old Warder's Cottage, Cumberland St. ***Jackson's Emporium:*** sells local products and wines; Lyell Hwy. ***Heritage buildings:*** many buildings of historical importance include the Old Schoolhouse (1856) and St Peter's Church (1837), which is notable for having just 1 door to prevent the once largely convict congregation from escaping.

The Highlander Arms: hearty pub-style meals; Tarraleah Estate, 5 Oldina Dr; (03) 6289 0111. ***Wildside Restaurant:*** gourmet dining; Tarraleah Estate, 5 Oldina Dr, Tarraleah; (03) 6289 0111.

Hamilton's Cottage Collection: convict-built cottages; 'Uralla', Main Rd; (03) 6286 3270. ***McCauley's Cottage:*** 1846 garden cottage; 21 Franklin Pl; (03) 6286 3232. ***The Old Schoolhouse:*** 2-bedroom cottages; Lyell Hwy; (03) 6286 3292. ***Tarraleah:*** award-winning Art Deco mountain luxury; 5 Oldina Dr, Tarraleah; (03) 6289 0111.

Lake Meadowbank: popular venue for picnics, boating, waterskiing and trout fishing; 10 km NW.

See also SOUTH-EAST, p. 501

Hastings

Pop. 35
Map ref. 668 F11 | 671 I10

Hastings Caves and Thermal Springs Centre, Hastings Cave Rd; (03) 6298 3209.

95.3 Huon FM, 936 AM ABC Local Radio

Hastings lies in Tasmania's far south and is known for the stunning dolomite caves to the west of town. The caves were discovered in 1917 by a group of timber workers who, among others, flocked to the small town in more prosperous days.

Southport Hotel Bar & Restaurant: Australia's southernmost counter meals; 8777 Huon Hwy, Southport; (03) 6298 3144.

The Jetty House: eco-friendly bayside homestead: 8848 Huon Hwy, Southport; (03) 6298 3139. ***Lune River Cottage:*** 3 affordable rooms; Lot 2, Lune River Rd, Lune River; (03) 6298 3302. ***Southern Forest B&B:*** 3 modern rooms; 30 Jager Rd, Southport; (03) 6298 3306.

Hastings Caves and Thermal Springs This is the only cave system in Tasmania occurring in dolomite rather than limestone. Newdegate Cave, which began forming more than 40 million years ago, has stalactites, stalagmites, columns, shawls, flowstone and the more unusual helictites, making it – and especially Titania's Palace within it – one of Australia's most beautiful caves. Tours run throughout the day. Near Newdegate Cave is a thermal pool surrounded by native bushland. It remains at 28°C year-round and is an extremely popular swimming and picnic spot. The Sensory Trail, an easy walk through magnificent forest, starts near the pool. 8 km NW.

Ida Bay Railway: originally built to carry dolomite, the train now carries passengers to Deep Hole Bay along a scenic section of track. Train times change, so ring for details (03) 6298 3110; Lune River Rd; 5 km S. ***Lunaris:*** gemstone display and shop; Lune River Rd; 5 km S. ***Southport:*** seaside resort town and one of the oldest settlements in the area; good fishing, swimming, surfing and bushwalking; 6 km SE. ***Cockle Creek:*** the southernmost point of Australia that can be reached by car, the town is surrounded by beautiful beaches and mountainous terrain and is the start of the 10-day South Coast walking track; 25 km S. ***Hastings Forest Tour:*** this self-drive tour begins off Hastings Rd west of town, leads north to the Esperance River, then heads to Dover. Short walks and picnic spots en route; map available from visitor centre.

See also SOUTH-EAST, p. 501

Huonville

Pop. 1814
Map ref. 668 G7 | 671 J8

Huon River Jet Boats and Visitor Centre; Esplanade; (03) 6264 1838; www.huonjet.com

98.5 Huon FM, 936 AM ABC Local Radio

Huonville produces more than half of Tasmania's apples and is surrounded by blossoming fields of apples, cherries, plums, pears, berries and hops. Although relatively small, it is a prosperous community and the largest town in the Huon Valley.

Apple and Heritage Centre This centre houses 3 attractions: the museum, providing a comprehensive insight into the lives of Huon Valley's early settlers; Apple Blossom's Gifts, with a selection of gifts, souvenirs and samples of local products; and The Starving Artist, an art gallery and heritage apple orchard. 2064 Huon Hwy; (03) 6266 4345.

Huon River Jet Boats: exciting 35 min jet-boat rides along Huon River rapids; river cruises also on offer, and aqua-bikes and pedal boats for hire; bookings on (03) 6264 1838; Esplanade.

Market: Websters Car Park, Cool Store Rd; 10am–2pm, 2nd and 4th Sun each month. ***A Taste of the Huon:*** festival venue changes towns yearly; check visitor centre for details; Mar. ***Huon Agricultural Show:*** biggest 1-day agricultural show in state; Nov.

Huon Manor: classy seafood dishes; 1 Short St; (03) 6264 1311. ***Home Hill Winery Restaurant:*** dramatic award-winning vineyard restaurant; 38 Nairn St, Ranelagh; (03) 6264 1200. ***Petty Sessions:*** exciting Tasmanian fare; 3445 Huon Hwy, Franklin; (03) 6266 3488.

Huon Bush Retreats: eco-friendly forest cabins; 300 Browns Rd, Ranelagh; (03) 6264 2233. ***Kay Creek Cottage:*** storybook garden cottage hideaway; 17 Kay St, Franklin; (03) 6266 3524.

Ranelagh Almost an outer suburb of Huonville, Ranelagh has the atmosphere of a charming English village, complete with an old oast house to process the hops. 5 km NW.

Wooden Boat Centre – Tasmania: workshop and interpretive centre; Main St, Franklin; 8 km sw. ***Glen Huon Model Village:*** features doll and rock display including the 'crooked man'; open Sept–May; Glen Huon; 8 km w. ***Studio Karma:*** studio, gallery and home of Terry Choi-Lundberg; original paintings, quality cards and prints of the artist's work; open 10am–4pm Thurs–Sun, by appt May–Sept; Glen Huon; 8 km w. ***Pelverata Falls:*** stunning waterfall with medium-to-difficult walk over scree slope; 14 km SE. ***Huon Bushventures:*** 4WD tours of the region; bookings on (03) 6264 1838. ***Fishing:*** good trout fishing in Huon River and tributaries.

See also SOUTH-EAST, p. 501

Kettering

Pop. 389
Map ref. 668 H8 | 671 K8

Bruny D'Entrecasteaux Visitor Centre, 81 Ferry Rd; (03) 6267 4494; www.tasmaniaholiday.com

 95.3 Huon FM, 936 AM ABC Local Radio

The area around Kettering, in the state's south-east, was first explored in 1792 by Bruni D'Entrecasteaux, after whom the surrounding channel is named. The town was settled in the 1800s by timber cutters, sealers and whalers, and the community was a transient one. Kettering is now principally the launching point to Bruny Island, but is charming in its own right with a sheltered harbour full of yachts and fishing vessels.

Oyster Cove Marina: well-known marina with boats for hire, skippered cruises and fishing charters; Ferry Rd. ***Ocean-kayaking:*** guided day tours with Roaring 40s Ocean Kayaking company; bookings on (03) 6267 5000. ***Herons Rise Vineyard:*** picturesque vineyard specialising in cool-climate white and pinot noir wines; sales; 120 Saddle Rd.

Farm Gate Cafe: gourmet provedore and cafe; Cnr Channel Hwy and Saddle Rd; (03) 6267 4497. ***Fleurtys Cafe:*** uniquely Tasmanian gourmet meals; 3866 Channel Hwy, Birchs Bay; (03) 6267 4078. ***Peppermint Bay Dining Room:*** top Tasmanian gourmet food; 3435 Channel Hwy, Woodbridge; (03) 6267 4088.

Anstey Barton: ultra-luxurious modern mansion; 82 Ferry Rd; (03) 6267 4199. ***Herons Rise Vineyard Cottages:*** rustic vineyard cottages; 100 Saddle Rd; (03) 6267 4339. ***Tulendena:*** peaceful garden cottage; 29 Bloomsbury La; (03) 6267 4348.

Peppermint Bay This is one of Tasmania's top gastronomic experiences, with a first-class restaurant and a store full of local products. There is also an arts and crafts gallery and fine views across the waterfront. The Peppermint Bay Cruise, departing from Hobart and touring the D'Entrecasteaux Channel along the way, is another way to visit this spectacular location; bookings on 1800 751 229. 3435 Channel Hwy, Woodbridge; 5 km s.

Woodbridge Hill Handweaving Studio: weaving tuition and sales of products woven from silk, cotton, linen, wool, alpaca, mohair and collie-dog hair; Woodbridge Hill Rd; 4 km s. ***Channel Historical and Folk Museum:*** displays of historical memorabilia of the D'Entrecasteaux Channel region; closed Wed; 2361 Channel Hwy, Lower Snug; 5 km N. ***Conningham:*** good swimming and boating beaches; 6 km N. ***Snug Falls:*** pleasant 1.5 hr return walk to falls; 8 km N. ***Grandvewe Sheep Cheesery:*** 15 different types of cheese at Tasmania's only sheep cheesery; Birchs Bay; 9 km s. ***Bruny Island ferry:*** trips throughout the day from Ferry Rd Terminal; for information on the island, *see Adventure Bay*.

See also SOUTH-EAST, p. 501

King Island

see Currie

Kingston

Pop. 17 286
Map ref. 669 I6 | 671 K8

Council offices, Civic Centre, 15 Channel Hwy; (03) 6211 8200; www.kingborough.tas.gov.au

92.1 Hobart FM, 936 AM ABC Local Radio

Almost an outer suburb of Hobart, this pleasant seaside town sits just beyond the city limits. Its literary claim to fame is that Nobel Laureate Patrick White holidayed at Kingston Beach as a child.

Australian Antarctic Headquarters This centre houses displays on the dramatic events of Antarctic exploration, including historic photographs and items such as Sir Douglas Mawson's sledge. Open Mon–Fri; Channel Hwy; (03) 6232 3209.

Rotary Market: library carpark; Sun. ***Olie Bollen Festival:*** Dutch festival; Oct.

Citrus Moon Cafe: wholefoods cafe; 23 Beach Rd; (03) 6229 2388. ***Doms Asian Teahouse:*** South-East Asian; Shop 9A Opal Drive Shopping Centre, Blackmans Bay; (03) 6229 7633. ***The Beach:*** popular beachside bistro; 14 Ocean Espl, Blackmans Bay; (03) 6229 7600. ***Pear Ridge:*** French-style food from a potager garden; 1683 Channel Hwy, Margate; (03) 6267 1811.

Anubha Mountain Health Retreat: pampering resort with magnificent views; 680 Summerleas Rd; (03) 6239 1573. ***D'Entrecasteaux:*** charming provincial boutique hotel; 77 Howden Rd, Howden; (03) 6267 1161. ***Kingston Beach Motel:*** budget rooms in great location; Cnr Beach Rd and Osbourne Espl, Kingston Beach; (03) 6229 8969. ***Tranquilla:*** sunny seaside B&B; 30 Osbourne Espl, Kingston Beach; (03) 6229 6282.

Shot Tower This tower, one of the state's most historic industrial buildings (with a National Trust 'A' classification), was completed in 1870. There are wonderful views of the Derwent Estuary from the top of the 66 m structure and at the base there is a museum and a craft shop. Channel Hwy, Taroona; 4 km NE.

Kingston Beach: safe swimming beach at the mouth of the River Derwent; 3 km SE. ***Kingston Beach Golf Course:*** well-regarded course with specific holes picked by international players as their favourites; Channel Hwy. ***Boronia Hill Flora Trail:*** 2 km track follows ridge line between Kingston and Blackmans Bay through remnant bush; begins at end of Jindabyne Rd and finishes at Peter Murrell Reserve, well known for its many native orchids; Kingston. ***Blackmans Bay blowhole:*** small blowhole at the northern end of the beach that is spectacular in stormy weather; Blowhole Rd; 7 km s. ***Alum Cliff walk:*** great walk from Browns River area along the coastal cliffs to Taroona. ***Train:*** 1950s passenger train (non-operational) with tearooms in the buffet car, adjacent antique sales and Sun market; Margate; 9 km s. ***Tinderbox Marine Reserve:*** follow the underwater snorkel trail; Tinderbox; 11 km s. ***Scenic drives:*** south through Blackmans Bay, Tinderbox and Howden, with magnificent views of Droughty Pt, South Arm Peninsula, Storm Bay and Bruny Island from Piersons Pt. ***Fishing:*** good fishing at Browns River. Red Tag Trout runs tours from Kingston; bookings on (03) 6229 5896. ***Horseriding:*** trail rides at Cheval Equitation; bookings on (03) 6229 4303.

See also SOUTH-EAST, p. 501

Latrobe

Pop. 2846
Map ref. 672 H6 | 674 H2 | 675 B7

48 Gilbert St; (03) 6421 4699; www.latrobetasmania.com.au

 100.5 FM ABC Local Radio, 104.7 Coast FM

Latrobe was once Tasmania's third largest settlement with inns, hotels, a hospital and no less than three newspapers in circulation. As it was the best place to cross the Mersey River, it became the highest-profile town on the north coast. Since the early 19th century, however, Latrobe has ceded its importance as a port town and relaxed into a gentler pace.

Australian Axeman's Hall of Fame This attraction honours the region's renowned axemen and details the role of the town in the creation of woodchop competition. Also featured at the facility is the Platypus and Trout Experience, including an interactive display of Tasmanian wildlife heritage. There is also a cafe and giftshop selling local crafts and souvenirs. 1 Bells Pde.

Platypus tours: with a LandCare guide; check at visitor centre for departure times and cost; bookings on (03) 6426 1774. ***Court House Museum:*** located in a heritage building under the National Trust register with over 600 prints and photographs; open 1–4pm Tues–Fri, other times by appt; Gilbert St; (03) 6426 2777. ***Sherwood Hall Museum:*** original home of pioneering couple Thomas Johnson and Dolly Dalrymple; open 10am–4pm daily Oct–Mar, Tues and Thurs or by appt in other months; Bells Parade. ***Anvers Confectionery:*** factory of well-known Belgian chocolatier, this outlet has tastings and sales of premium chocolates, fudges and pralines; 9025 Bass Hwy. ***Sheean Memorial Walk:*** 3 km return walk commemorating local soldiers and WW II hero; Gilbert St. ***Bells Parade Reserve:*** beautiful picnic ground along riverbank where town's docks were once located; River Rd. ***Historical walk:*** starts at western end of Gilbert St and turns into Hamilton St.

Markets: Gilbert St; 7am–2pm Sun. ***Henley-on-the-Mersey regatta:*** Jan. ***Chocolate Winterfest:*** July. ***Latrobe Wheel Race:*** prestige cycling event; Dec. ***Latrobe Gift footrace:*** Dec.

House of Anvers: European cafe; 9025 Bass Hwy; (03) 6426 2703. ***Lucas Hotel:*** excellent bistro meals; 46 Gilbert St; (03) 6426 1101.

Lucas Hotel: quiet ensuite rooms; 46 Gilbert St; (03) 6426 1101. ***Lucinda B&B:*** Victorian mansion in parkland gardens; 17 Forth St; (03) 6426 2285. ***Sherwood View:*** rural B&B; 298 Coal Hill Rd; (03) 6426 2797.

Warrawee Forest Reserve An excellent place for platypus viewing, swimming, bushwalking and barbecues. A 5 km walking track winds through sclerophyll forest and a boardwalk has been installed around the lake to allow disabled access to trout fishing. Tours are run by LandCare; bookings on (03) 6426 2877; 3 km s.

Henry Somerset Orchid Reserve: over 40 native orchids and other rare flora; Railton Rd; 7 km s.

See also NORTH-WEST, p. 503

Launceston

see inset box on next page

Lilydale

Pop. 292
Map ref. 673 K7 | 675 H8

Launceston Travel and Information Centre, Cornwall Square Transit Centre, 12–16 St John St, Launceston; (03) 6336 3133 or 1800 651 827; www.ltvtasmania.com.au

 91.7 FM ABC Local Radio, 95.3 Tamar FM

Originally called Germantown, this small north-eastern township is better known for the 'Englishness' of its gardens and the almost French quality of its countryside. Yet the bushwalks through surrounding reserves and waterfalls are distinctly Australian, with native temperate rainforests lining the trails.

Painted Poles: 15 hydro poles painted by professional and community artists to show the local history.

Lilydale Tavern: good pub staples; 1983 Main Rd; (03) 6395 1230. ***Pipers Brook Winery Cafe:*** acclaimed vineyard cafe; 1216 Pipers Brook Rd, Pipers Brook; (03) 6382 7527.

Eagle Park and Cherry Top: passive-solar mudbrick house; 81 Lalla Rd, Bottom Downie; (03) 6395 1167.

Lilydale Falls The waterfall is situated within the temperate rainforest of Lilydale Park. A picnic area and playground are on-site, as well as 2 oak trees planted in 1937 from acorns picked near Windsor Castle in England to commemorate the coronation of King George VI. 3 km N.

Tamar Valley Wine Route: numerous vineyards between Lilydale and Pipers Brook include Providence (4 km N), Clover Hill (12 km N) and Brook Eden (15 km N); *see also Launceston*; brochure from visitor centre. ***Walker Rhododendron Reserve:*** 12 ha park reputed to have the best rhododendron display in Australia. Other species are also on display; Lalla; 4 km W. ***Appleshed:*** teahouse with local art and craft; Lalla; 4 km W. ***Hollybank Forest Reserve:*** 140 ha forest reserve with Australia's first continuous cable treetop tour, arboretum, picnic facilities and information centre with details on walking tracks; marked turn-off near Underwood; 5 km S. ***Mt Arthur:*** 3 hr return scenic walk to summit (1187 m); 20 km SE. ***Bridestowe Estate Lavender Farm:*** this is one of the world's largest single commercial lavender farms; 296 Gillespies Rd, Nabowla; 26 km N.

See also MIDLANDS & THE NORTH, p. 504

Longford

Pop. 3030
Map ref. 673 K9 | 675 G12

JJ's Bakery at the Old Mill Cafe, 52 Wellington St; (03) 6391 2364

 91.7 FM ABC Local Radio, 103.7 FM City Park Radio

Longford was established when numerous settlers from Norfolk Island were given land grants in the area in 1813. Fittingly, the district became known as Norfolk Plains, while the settlement itself was called Latour. Today Longford is classified as a historic town and serves the rich agricultural district just south of Launceston.

Historic buildings These include the Queen's Arms (1835) in Wellington St, Longford House (1839) in Catherine St and the Racecourse Hotel (1840s) in Wellington St, which was originally built as a railway station then used as a hospital and later a pub, and in which a patron was murdered after stealing

continued on p. 521

LAUNCESTON

Pop. 99 676
Map ref. 673 K8 | 675 G10

Cornwall Square Transit Centre, 12–16 St John St; (03) 6336 3133 or 1800 651 827; www.visitlauncestontamar.com.au.com.au

91.7 FM ABC Local Radio, 103.7 FM City Park Radio

Tasmania's second largest city and a busy tourist centre, Launceston lies nestled in northern hilly country where the Tamar, North Esk and South Esk rivers meet. An elegantly laid-back place, Launceston has the highest concentration of 19th-century buildings in Australia. Yet it is also a city of contrasts, where modern marinas meet graceful Georgian and Victorian streetscapes and parks – and you're seldom without a view of the Tamar River or surrounding valley.

Cataract Gorge This is one of Launceston's outstanding natural attractions. Historic Kings Bridge spans the Tamar River at the gorge entrance. Above the cliffs on the north side is an elegant Victorian park with lawns, European trees, peacocks and a restaurant. The world's longest single-span chairlift and a suspension bridge link this area to First Basin's lawns on the south side, which has a swimming pool and a kiosk. Walks and self-guide nature trails run on both sides of the gorge.

Queen Victoria Museum and Art Gallery This is considered one of the best regional museums in Australia, with permanent exhibits on Aboriginal and convict history, Tasmanian flora and fauna, colonial art, a Chinese joss house, a blacksmith shop and many temporary exhibitions. Other features include Phenomena Factory – an interactive science centre for kids of all ages – as well as the Launceston Planetarium and a railway museum. Renovations are currently being done on the museum's original Royal Park building, which should reopen in 2011. In the meantime, all exhibits are housed in the Inveresk building; 2 Invermay Rd, Inveresk; (03) 6323 3777.

Boags Brewery J Boag and Son Brewery was where James Boag commenced his brewing tradition on the banks of the Esk River in 1852. Full tours of the brewery start and finish at the Boags Centre for Beer Lovers. The centre houses a museum and a retail store. Tours Mon–Fri; tasting class Sat; William St; (03) 6332 6300.

City Park: magnificent 5 ha park with European deciduous trees, it features a Japanese Macaque monkey enclosure, the John Hart conservatory, annual display beds, senses garden, and monuments; Cnr Tamar and Brisbane sts. ***Seaport:*** new riverside complex with restaurants, shops, marina and hotel at the head of the Tamar River. ***Design Centre of Tasmania:*** houses Australia's only museum collection of contemporary wood design. Runs national and international exhibitions and tours of crafts, design and art; City Park. ***National Automobile Museum of Tasmania:*** displays over 40 fully restored classic vehicles spanning 100 years of style and technical achievement; Cimitiere St. ***Ritchies Mill & Stillwater River Cafe & Restaurant:*** a 4-storey grain mill from the 1800s, now home to one of Launceston's most renowned restaurants. The award-winning Stillwater restaurant opened on the ground floor of the Mill in 2000, while The Mill Providore & Gallery is located on the 1st and 2nd floor of the mill, showcasing local Tasmanian produce, art and design; Paterson St. ***Aurora Stadium:*** the home of AFL football in Tasmania, the stadium hosts 5 rostered matches per year and a variety of other events.

[BOAGS CENTRE FOR BEER LOVERS]

Walking tour: self-guide tour takes in 25 places of historical importance including Morton House, Milton Hall and Princes Sq, where the fountain was changed from a half-naked nymph to a pineapple after locals objected; brochure from visitor centre. ***Waverley Woollen Mills Factory Outlet Shop:*** woollen products from Australia's oldest mill; open 10am–4pm Mon–Fri; George St. ***Old Umbrella Shop:*** unique 1860s shop preserved by the National Trust and housing a giftshop and information centre; George St. ***Cocobean Chocolate:*** a chocolate boutique for connoisseurs of handcrafted chocolates, drinks and desserts; open Mon–Fri, 9.30am–2pm Sat; 82 George St; (03) 6331 7016. ***Scenic flights:*** Heli Adventures Tasmania in Launceston; bookings on (03) 6334 0444.

Launceston Cup: state's biggest race day; Feb. ***Festivale:*** food and wine; Feb. ***MS Fest:*** major youth music festival; Feb. ***Targa Tasmania:*** tarmac road rally; Apr–May. ***National Trust Tasmanian Heritage Festival:*** Apr. ***Agfest:*** Tasmania's largest agricultural show; May. ***Royal Launceston Show:*** Oct. ***Tamar Valley Classic:*** yacht race; Nov.

Black Cow Bistro: beef connoisseur's heaven; Cnr George and Paterson sts; (03) 6331 9333. ***Calabrisella:*** traditional Italian; 56 Wellington St; (03) 6331 1958. ***Flip:*** creative burgers; cnr York and Bathurst sts; (03) 6334 6844. ***Smokey Joe's Creole Cafe:*** Caribbean and Cajun Creole; 20 Lawrence St; (03) 6331 0530. ***Stillwater River Cafe & Restaurant:*** fine Asian-inspired dining: Ritchies Mill, 2 Bridge Rd; (03) 6331 4153.

Fiona's of Launceston: old-style central B&B; 141A George St; (03) 6334 5965. ***Peppers Seaport:*** swish riverside hotel; 28 Seaport Blvd; (03) 6345 3333. ***The Lido Apartments:*** 1930s Art Deco apartments; 47–49 Elphin Rd; (03) 6337 3000. ***TWOFOURTWO:*** hip, contemporary luxury apartments; 242 Charles St; (03) 6331 9242. ***Werona Heritage Accommodation:*** opulent Federation hillside mansion; 33 Trevallyn Rd, Trevallyn; (03) 6334 2272. ***Hotel Charles:*** stylish modern accommodation with fabulous views; 287 Charles St; 1300 703 284.

Tamar Valley Wine Route With around 30 wineries in the area, Pipers Brook Vineyard is one of the state's biggest names, and makes beautiful gewurztraminer and pinot noir. Its cellar also offers tastings of its Ninth Island Label and Kreglinger sparkling wines. Other wineries worth visiting include Bay of Fires and Tamar Ridge. For masterpieces in sparkling, visit Clover Hill, whose picturesque cellar door is perfect for a picnic lunch, and Jansz, with its modern, architectural wine room featuring an interpretive centre. Brochures are available from the visitor centre.

Franklin House A National Trust–listed Georgian building, Franklin House was built by convicts in 1838 for a Launceston brewer. It is furnished elaborately with period pieces and is a popular historical attraction. 6 km s.

Punchbowl Reserve: spectacular park with rhododendron plantation, a small gorge and native and European fauna in natural surroundings; 5 km sw. ***Trevallyn State Reserve:*** neighbour to the Cataract Gorge, good picnic spot with trail rides, kayaking, walking tracks and water activities at Trevallyn Dam; 6 km w. ***Hollybrook Treetops Adventure:*** flying fox adventure park with a kilometre of cable strung between treetop 'cloud stations'; open daily; Hoillybrook Rd, Underwood. ***Tamar Island Wetlands:*** urban wetlands reserve; a haven for birdlife, with a 3 km boardwalk to Tamar Island, pleasing views from the middle of the river and an outstanding Interpretation Centre that offers visitors the opportunity to learn about the great value of the wetlands; 9 km NW. ***Tasmanian Zoo:*** Tasmanian devils, emus, wallabies and other native fauna. Good fly-fishing in the lakes with tuition available; 17 km w. ***Grindelwald:*** Swiss village with chalet-themed Tamar Valley Resort and Swiss-style shopping square selling Swiss chocolate, cakes, crafts and souvenirs; 19 km NW.

See also MIDLANDS & THE NORTH, p. 504

and swallowing 2 gold sovereigns from local farmhands. The Racecourse Hotel is now a guesthouse and restaurant. Also on Wellington St is Christ Church, an 1839 sandstone building with outstanding stained-glass windows, pioneer gravestones and a clock and bell presented by George VI. Path of History is the self-guide walk; brochure available from visitor centre.

Walk: track along the South Esk River. ***The Village Green:*** originally the site of the town market, now a picnic and barbecue spot; Cnr Wellington and Archer sts.

Longford Picnic Day Races: held at the oldest operating racecourse in Australia; Jan. ***Blessing of the Harvest Festival:*** celebration of the rural tradition with parade and Sheaf Tossing Championships; Mar.

JJs Bakery and Old Mill Cafe: historic bakery cafe; 52 Wellington St; (03) 6391 2364. ***The Racecourse Inn Restaurant:*** elegant modern Australian; 114 Marlborough St; (03) 6391 2352. ***The River's Edge Cafe:*** Italian-inspired garden cafe; The River's Edge Venue, 38 Tannery Rd; (03) 6391 2559.

Longford Boutique Accommodation: luxury heritage B&B; 6 Marlborough St; (03) 6391 2126. ***The Racecourse Inn:*** rose-covered convict-brick B&B; 114 Marlborough St; (03) 6391 2352. ***Woolmers Estate:*** timeless colonial cottages; Woolmers La; (03) 6391 2230.

Woolmers Estate Woolmers was built c. 1817 by the Archer family, who lived there for 6 generations. Tours of the house, the outbuildings, the gardens and the National Rose Garden are conducted daily and the Servants Kitchen restaurant serves morning and afternoon teas. Lyell Hwy; (03) 6286 3332; 5 km s.

Brickendon: historic Georgian homestead built in 1824, now a working farm and historic farm village; open Wed–Sun June–July, Tues–Sun Sept–May; 2 km s. ***Perth:*** small town with historic buildings, including Eskleigh and Jolly Farmer Inn, and market on Sun mornings; 5 km NE. ***Tasmanian Honey Company:*** tastings and sales of excellent range of honeys including leatherwood and flavoured varieties; 25A Main Rd, Perth; 5 km NE. ***Woodstock Lagoon Wildlife Sanctuary:*** 150 ha sanctuary for nesting waterfowl; 9 km w. ***Cressy:*** good fly-fishing at Brumby's Creek, especially in Nov when mayflies hatch; 10 km s.

See also MIDLANDS & THE NORTH, p. 504

Miena

Pop. 104
Map ref. 671 I1 | 673 I11

Great Lake Hotel, Great Lake Hwy; (03) 6259 8163.

91.7 FM ABC Local Radio, 95.7 Heart FM

Miena is on the shores of Great Lake on Tasmania's Central Plateau and has been popular with anglers since brown trout were released into the lake in 1870. The Aboriginal name (pronounced 'my-enna') translates to 'lagoon-like'. The surrounding region, known as the Lake Country, can become very cold, with snow and road closures, even in summer.

Great Lake Hotel: fireside counter meals; 3096 Marlborough Hwy; (03) 6259 8163.

Central Highlands Lodge: lodge for fly-fishing devotees; Haddens Bay; (03) 6259 8179. ***Great Lake Hotel:*** lakeside rooms and apartments; 3096 Marlborough Hwy; (03) 6259 8163. ***Shannon Rise Lodge:*** self-contained lodge; 52 Lake Hwy, Haddens Bay; (03) 5334 3851 ***Skittleball Plain Homestead:*** isolated rustic cabin; Little Pine Lagoon; (03) 6424 1288.

Fishing The 22 km Great Lake is the second largest freshwater lake in Australia and has excellent trout fishing. But Arthurs Lake, 23 km E, is said to be even better. The Highland Dun mayflies that hatch in summer generate an abundance of speckled brown trout, drawing fly-fishing enthusiasts. West of Liawenee (about 7 km N) are more locations for fly-fishing in isolated lakes and tarns of the Central Plateau Conservation Area (4WD recommended for several lakes, while some are accessible only to experienced bushwalkers). Lakes are closed over winter.

Bushwalking: along the shores of Great Lake. ***Circle of Life:*** bronze sculptures by Steven Walker, each representing an aspect of the region's history; Steppes; 27 km SE. ***Waddamana Power Museum:*** housed in the first station built by the Hydro-Electric Corporation, it includes history of hydro-electricity in Tasmania; 33 km S. ***Lake St Clair:*** a boat service from Cynthia Bay provides access to the north of the lake and to the renowned 85 km Overland Track, which passes through Cradle Mountain–Lake St Clair National Park; *see Sheffield*; 63 km W.

See also **MIDLANDS & THE NORTH, p. 504**

Mole Creek

Pop. 223
Map ref. 672 H8 | 674 H5 | 675 A11

46 Pioneer Dr; (03) 6363 1487; www.greatwesterntiers.org.au

91.7 FM ABC Local Radio

Mole Creek is named after the nearby creek that 'burrows' underground. Most visitors come to explore the limestone caves in Mole Creek Karst National Park. The unique honey from the leatherwood tree, which grows only in the west-coast rainforests of Tasmania, is also a drawcard. Each summer, apiarists transport hives to the nearby leatherwood forests.

Stephens Leatherwood Honey Factory At the home of Tasmania's unique aromatic honey, visitors can see clover and leatherwood honey being extracted and bottled. Tastings and sales are available. Open Mon–Fri; 25 Pioneer Dr; (03) 6363 1170.

Mole Creek Tiger Bar: local hotel with information and memorabilia on the Tasmanian tiger; Pioneer Dr.

Laurel Berry Restaurant: local produce and flavours; Mole Creek Guest House, 100 Pioneer Dr; (03) 6363 1399.

Blackwood Park Cottages: delightfully romantic garden cottages; 445 Mersey Hill Rd; (03) 6363 1208. ***Old Wesley Dale Heritage Accommodation:*** cosy stone cottage; 1970 Mole Creek Rd; (03) 6363 1212.

Mole Creek Karst National Park Set in the forests of the Western Tiers, this national park protects the Marakoopa and King Solomons caves, spectacular caverns of calcite formations created by underground streams. Marakoopa Cave has a magnificent glow-worm display, while King Solomons Cave offers coloured stalagmites and stalactites and sparkling calcite crystals. Guided tours daily; details (03) 6363 5182; 13 km W.

Alum Cliffs Gorge: spectacular 30 min return walk; 3 km NE. ***Trowunna Wildlife Park:*** see Tasmanian devils and other native fauna; 4 km E. ***The Honey Farm:*** over 50 flavours of honey and an interactive bee display; open Sun–Fri; Chudleigh; 8 km E. ***Devils Gullet State Reserve:*** World Heritage area with natural lookout on a 600 m high cliff, reached by 30 min return walking track; 40 km SE. ***Wild Cave Tours:*** adventure tours to caves that are closed to the general public; bookings on (03) 6367 8142. ***Cradle Wilderness Tours:*** guided 4WD tours; bookings on (03) 6363 1173. ***Walls of Jerusalem National Park:*** the park is accessible on foot only, from the end of the road that turns off around 15 km W of Mole Creek and heads down to Lake Rowallan. With glacial lakes, pencil pines and dolerite peaks, it is a wonderland for self-sufficient and experienced bushwalkers.

See also **MIDLANDS & THE NORTH, p. 504**

New Norfolk

Pop;. 5232
Map ref. 668 G4 | 671 J6

Derwent Valley Visitor Information Centre, Circle St; (03) 6261 3700; www.newnorfolk.org

96.1 Hobart FM, 936 AM ABC Local Radio

This National Trust–classified town has a look similar to that of Kent in England. Located on the River Derwent in Tasmania's south-east, the town was named for the European settlers from the abandoned Norfolk Island penal settlement who were granted land here. The district is also renowned for its hops industry, and has more recently become a mecca for those seeking antiques and collectibles.

Oast House A working oast house from 1867 to 1969, the building has since been converted into a museum that showcases the growing and processing of hops. A cafe and craft market are on-site. Tynwald Park Reserve, Lyell Hwy.

Tynwald House: charming rural residence next to the Oast House, with restaurant; Tynwald Park Reserve. ***Old Colony Inn:*** heritage house (1835) with folk museum, restaurant and craft shop, and Australia's largest antique doll's house; Montagu St. ***Church of St Matthew:*** reputedly the oldest church in Tasmania (1823), with striking stained-glass windows; Bathurst St. ***Rosedown:*** beautiful rose and daffodil gardens along riverbank. Bookings preferred, (03) 6261 2030; closed May–Sept; 134 Hamilton Rd. ***Antique and collectibles outlets:*** over 15 in the town; details from visitor centre. ***Bush Inn Hotel:*** longest continually licensed hotel in Tasmania. ***Jet-boat rides:*** on River Derwent rapids; bookings on (03) 6261 3460. ***River walk:*** from Esplanade to Tynwald Park Wetlands Conservation Area. ***Scenic lookout:*** Peppermint Hill, off Blair St. ***Self-guide historical walks:*** brochure from visitor centre; Circle St.

Drill Hall Market: collectibles and antiques; Stephen St; daily.

Possum Shed Cafe: riverside cafe; 1654 Gordon River Rd, Westerway; (03) 6288 1364. ***The Bush Inn:*** generous pub meals; 49–51 Montagu St; (03) 6261 2256. ***Tynwald:*** fine Provençal food; 1 Tynwald Rd; (03) 6261 2667.

Explorers Lodge: elegant 1940s guesthouse; 105 Derwent Tce; (03) 6261 1255. ***Rosie's Inn Private Hotel:*** lacy, floral B&B in pretty gardens; 5 Oast St; (03) 6261 1171. ***Tynwald:*** fairytale mansion in riverside gardens; 1 Tynwald Rd; (03) 6261 2667. ***Woodbridge on the Derwent:*** exclusive and world-class luxury hotel; 6 Bridge St; (03) 6261 5566.

Mt Field National Park Tasmania's first national park, Mt Field is best known for the impressive Russell and Lady Barron falls. Most walks pass through lush ferns and rainforests, while the Pandani Grove walk traverses the glaciated landscapes of the mountain country to Lake Dobson. A visitor centre on-site offers interpretive displays. Contact Parks & Wildlife Service on 1300 135 513; 40 km NW.

Norske Scog Boyer Mill: the first in the world to manufacture newsprint from hardwoods; tours available Tues and Thurs; Boyer; 5 km E. ***Salmon Ponds:*** first rainbow and brown trout farm in Australia, in operation since 1864. There is also a Museum of Trout Fishing and a restaurant; Plenty; 11 km NW. ***Possum Shed:*** locally made crafts and collectibles, and cafe with delectable coffee; Gordon River Rd, Westerway; 31 km NW. ***Styx Tall Tree Reserve:*** small reserve with the tallest hardwood trees in the world, the giant swamp gums, which grow to 92 m; 73 km W via Maydena.

See also **SOUTH-EAST, p. 501**

Oatlands

Pop. 540
Map ref. 671 K3

Heritage Highway Visitors Centre, 1 Mill La; (03) 6254 1212; www.heritagehighwaytasmania.com.au

 97.1 Mid FM, 936 AM ABC Local Radio

Approaching Oatlands from the north, look out for the topiary and striking metal sculptures by the roadside. The topiaries are a local tradition from the 1960s, while the recently created metal sculptures depict earlier times in the district. The town is on the shores of Lake Dulverton and has the largest collection of Georgian sandstone buildings in a village environment – with much of the stone from local quarries.

Callington Mill Built in 1836, the old mill is a feature of Oatlands that was fully operational until 1892. After being battered by the elements and gutted by fire in the early 1900s, it was finally restored as part of Australia's bicentenary. The view from the top floor takes in Lake Dulverton. Old Mill La. The Visitors Centre is located in the mill and runs tours 10am–4pm daily.

Lake Dulverton: the lake is stocked with trout and onshore is a wildlife sanctuary protecting many bird species. Popular picnic spot; Esplanade. ***Fielding's Ghost Tours:*** tours run by local historian Peter Fielding; bookings on (03) 6254 1135. ***Historical walk:*** takes in the many Georgian buildings, including the convict-built Court House (1829); *'Welcome to Historic Oatlands'* brochure from visitor centre. ***Scottish, Irish and Welsh Shop:*** stocks over 500 tartans, clan crests, badges, pins, and arts and crafts; open Sun–Mon and Thurs–Fri; 64 High St. ***Skulduggery:*** mystery tour game following true crime clues around town; available from visitor centre.

 ***Oatlands Open Day:*** Oct.

Casaveen: country-style lunches; 44 High St; (03) 6254 0044.

Lakeview Cottage: shared cottage accommodation; 3 Lake St; (03) 6245 1212. ***Oatlands Lodge:*** convict-built Georgian manor; 92 High St; (03) 6254 1444. ***Waverley Cottage Collection:*** heritage cottages; Bow Hill Rd; (03) 6254 1264.

Convict-built mud walls: 13 km S on Jericho Rd. ***Fishing:*** excellent trout fishing in Lake Sorell and Lake Crescent; 29 km NW.

See also MIDLANDS & THE NORTH, p. 504

Penguin

Pop. 2949
Map ref. 672 G5

78 Main Rd; (03) 6437 1421; www.coasttocanyon.com.au

 91.7 FM ABC Local Radio, 106.1 Coast FM

This northern seaside town was named after the little penguins that shuffle up the beaches, and images of the iconic bird are peppered around town. The largest example is the much-photographed Big Penguin, which stands 3 metres tall on the beachfront and is the town's premier attraction.

Dutch Windmill The windmill was presented to the town during Australia's bicentenary to commemorate the Dutch explorers and settlers. There is also a colourful tulip display in spring and play equipment on-site. Hiscutt Park, off Crescent St.

Penguins: each evening penguins come ashore; check with visitor centre for tour details; Sept–Mar. ***Johnsons Beach Reef:*** popular walking spot at low tide when reef is exposed. ***Penguin Roadside Gardens:*** originally a labour of love for 2 town residents, now a flourishing garden beside the road; Old Coast Rd to Ulverstone. ***Chocolate Lovers:*** quality European and Australian chocolates of all varieties; 100 Main Rd.

 ***Market:*** over 300 stalls undercover; Arnold St; Sun.

Neptune Grand Bistro: beachside pub; 84 Main Rd; (03) 6437 2406. ***The Groovy Penguin Cafe:*** arty retro cafe; 74 Main Rd; (03) 6437 2101. ***Wild:*** Asian-inspired fine Tasmanian fare; 87 Main Rd; (03) 6437 2000.

The Madsen: classic seaside boutique hotel; 64 Main Rd; (03) 6437 2588.

Dial Range Walks Excellent tracks include the walk up Mt Montgomery (5 km S) with magnificent views from the summit, and Ferndene Gorge (6 km S). The Ferndene Bush Walk takes in an old silver-mine shaft, Thorsby's Tunnel. Brochures on walks from visitor centre highlight more trails.

Penguin Cradle Trail This 80 km trail heads inland from the coast to the world-famous Cradle Mountain. Sections of the walk vary in difficulty. It takes 6 days to complete, but access roads mean that sections can be done as day or overnight trips.

Pioneer Park: beautiful gardens with picnic facilities and walks; Riana; 10 km SW. ***Scenic drive:*** along coast to Ulverstone.

See also NORTH-WEST, p. 503

Pontville

Pop. 2170
Map ref. 669 I3 | 671 K6

Council offices, Tivoli Rd, Gagebrook; (03) 6268 7000.

 96.1 Hobart FM, 936 AM ABC Local Radio

The area around Pontville (declared a town in 1830) was first explored by Europeans when Hobart experienced severe food shortages in the early 1800s. Soldiers were sent north to kill emus and kangaroos. One of them, Private Hugh Germain, is allegedly responsible for the unusual names found in the region, such as Bagdad, Jericho, Lake Tiberius and Jordan River. Legend has it that Germain carried copies of *Arabian Nights* and *The Bible* and found his inspiration within.

Historic buildings Many buildings remain from Pontville's early days, including the Romanesque St Mark's Church (1841), The Sheiling (1819) and The Row, thought to have been built in 1824 as soldiers quarters. On or adjacent to Midland Hwy.

The Crown Inn Bistro: friendly country pub; 256 Midland Hwy; (03) 6268 1235.

Lythgo's Row Cottages: whitewashed Georgian terrace; 253 Midland Hwy; (03) 6268 1665. ***Armytage House:*** 1830s barn meets stylish luxury; 1702 Midland Hwy, Bagdad; (03) 6268 6354.

Bonorong Wildlife Park: popular attraction showing Tasmanian devils, quolls, echidnas, koalas, wombats and Forester kangaroos; Briggs Rd, Brighton; 5 km S. ***Historic towns:*** Brighton (3 km S), Tea Tree (7 km E), Bagdad (8 km N), Broadmarsh (10 km W) and Kempton (19 km N).

See also SOUTH-EAST, p. 501

Port Arthur

see inset box on next page

  RADIO STATIONS  IN TOWN  WHAT'S ON   WHERE TO EAT  WHERE TO STAY  NEARBY

Port Sorell

Pop. 2209
Map 673 I6 | Map 675 C6

Latrobe Visitor Centre, River Rd, Bells Pde, Latrobe; (03) 6421 4699; www.latrobetasmania.com.au

 100.5 FM ABC Local Radio, 104.7 Coast FM

This holiday town on the Rubicon River enjoys a mild and sunny climate nearly all year-round. It was established in the early 1820s, but sadly, many of its oldest buildings were destroyed by bushfires. The port is now a fast-developing coastal retreat for retirees and beach lovers.

The jetty: good fishing for cocky salmon, mullet, flathead, cod and bream. There are views to Bakers Beach and Narawntapu National Park. Behind caravan park on Meredith St. ***Port Sorell Conservation Area:*** 70 ha of coastal reserve with much flora and fauna. Guided tours available; bookings on (03) 6428 6072; Park Espl. ***Estuary:*** boating and safe swimming areas.

Market: Memorial Hall; 1st and 3rd Sat each month Sept–May. ***Port Sorell Regatta:*** Jan.

Hawley House: elegant candlelit dinners; Hawley Beach; (03) 6428 6221.

Hawley House: peacocks, pets and a rooftop bath; Hawley Beach; (03) 6428 6221. ***Sails on Port Sorell:*** modern apartments on an unspoilt beach; 54 Rice St; (03) 6428 7580. ***Tranquilles:*** romantic luxury B&B; 9 Gumbowie Dr; (03) 6428 7555. ***Shearwater Country Club:*** self-contained resort accommodation; The Boulevard, Shearwater; (03) 6428 6205.

Shearwater: holiday town with shopping centre and good beach access; 5 km N. ***Hawley Beach:*** safe swimming, good fishing and historic Hawley House (1878) offering meals; 6 km N. ***Walk:*** 6 km return track from Port Sorell to Hawley Beach, offering excellent views of Narawntapu National Park and coastline. Starts at beach end of Rice St.

See also North-West, p. 503

Queenstown

Pop. 2120
Map ref. 670 D1 | 672 D11 | 674 A10

Queenstown Galley Museum, Cnr Sticht and Driffield sts; (03) 6471 1483.

90.5 FM ABC Local Radio, 92.1 FM West Coast 7XS

The discovery of gold and other mineral resources in the Mount Lyell field in the 1880s led to the rapid emergence of Queenstown. Continuous mining here from 1893 to 1994 produced over 670 000 tonnes of copper, 510 000 kilograms of silver and 20 000 kilograms of gold. Operations began again in 1995, and are now owned by the Indian company Sterlite Industries. The town has modern facilities, but its wide streets and remaining historic buildings give it an old-mining-town flavour. In certain lights, multicoloured boulders on the hillsides, denuded through a combination of felling, wildfire, erosion and poisonous fumes from the smelter, reflect the sun's rays and turn to amazing shades of pink and gold. However, many Tasmanians view the place as a haunting reminder of the devastating impact humans can have on their environment.

Galley Museum The museum is housed in the Imperial (1898), Queenstown's first brick hotel, and displays over 800 photographs and general memorabilia of the history of the west coast. Cnr Sticht and Driffield sts.

West Coast Wilderness Railway This restored 1896 rack-and-pinion railway travels over 34 km of river and forest track to Strahan. It crosses 40 bridges and passes through pristine wilderness areas. Bookings on 1800 628 288.

Spion Kop Lookout: views of Queenstown and surrounding mountains; off Bowes St. ***Paragon Theatre:*** cinema; McNamara St. ***Historical walk:*** takes in 25 locations of historical importance; *'The Walkabout Queenstown'* brochure from visitor centre.

Penghana B&B: à la carte traditional dining; 32 The Esplanade; (03) 6471 2560; dinner for non-guests by arrangement.

Empire Hotel: beautifully ornate, regal hotel; 2 Orr St; (03) 6471 1699. ***Mount Lyell Anchorage:*** Edwardian bungalow; 17 Cutten St; (03) 6471 1900. ***Penghana B&B:*** magnificent stately mansion; 32 The Esplanade; (03) 6471 2560. ***Derwent Bridge Chalets:*** cosy forest chalets; 15478 Lyell Hwy, Derwent Bridge; (03) 6289 1000.

Mt Lyell Mines: guided tours of the mines north of town, either surface (1 hr) or underground (3.5 hrs); bookings on (03) 6471 1472. ***Iron Blow:*** original open-cut mine where gold was discovered in 1883; Gormanston; 6 km E. ***Linda:*** ghost town; 7 km E. ***Mt Jukes Lookout:*** superb panoramic views; 7 km S. ***Lake Burbury:*** excellent brown and rainbow trout fishing; picnic areas; 8 km E. ***Nelson Falls:*** short walk through temperate rainforest leads to falls; 23 km E. ***Valley views:*** spectacular views from Lyell Hwy as it climbs steeply out of town.

See also South-West Wilderness, p. 502

Richmond

Pop. 879
Map ref. 669 J3 | 671 K6

Old Hobart Town model village, 21A Bridge St, opposite Henry St; (03) 6260 2502; www.richmondvillage.com.au

97.7 Tasman FM, 936 AM ABC Local Radio

Richmond, just north of Hobart, is one of the most important historic towns in Tasmania. Richmond Bridge is the oldest surviving freestone bridge in Australia, built by convicts under appalling conditions in 1823. The situation was so bad that one convict committed suicide by throwing himself off the bridge. Other convicts beat and killed an overseer who was known for his cruelty; legend has it that his ghost still haunts the bridge.

Old Hobart Town Taking 3 years to build, this is an intricate model of Hobart in the 1820s. A remarkable feat, it is also historically accurate. Bridge St.

Richmond Gaol One of Australia's best preserved convict prisons, built in 1825 and once the abode of convict Ikey Solomon, said to be the inspiration for Dickens' Fagin. Self-guide tours. Bathurst St; (03) 6260 2127.

Richmond Maze and Tearooms: 2-stage maze with surprise ending, puzzle corner, gardens and Devonshire tea; Bridge St. ***Historical walk:*** many heritage buildings throughout town include Ivy Cottage, Village Store and Richmond Arms; *'Let's Talk About Richmond'* brochure from visitor centre. ***Art Galleries:*** local arts and crafts in heritage buildings include Saddlers Court (c. 1848); Bridge St. ***Olde Time Portraits:*** photographs of people in period costume; Bridge St. ***Prospect House:*** historic Georgian mansion built in 1830s, supposedly haunted by the ghost of its past owner; restaurant; Richmond Rd.

 ***Richmond Village Colonial Fair:*** Mar.

PORT ARTHUR

Pop. 499
Map ref. 669 M9 | 671 M9

Port Arthur Historic Site, Arthur Hwy; (03) 6251 2300 or 1800 659 101; www.portarthur.org.au

97.7 Tasman FM, 936 AM ABC Local Radio

This historic settlement on the scenic Tasman Peninsula was one of Australia's most infamous penal settlements from the 1830s to the 1870s. Over 12 000 convicts from Britain, some of whom did nothing more than steal some food to survive, were shipped to Port Arthur, dubbed 'Hell on Earth'. They lived under threat of the lash and experimental punitive measures that often drove them to madness. This grim past is offset by the stark beauty of the sandstone buildings overlooking the tranquil, often misty, waters of the bay. Port Arthur is Tasmania's number one tourist attraction.

Port Arthur Historic Site Over 30 buildings and restored ruins sit on 40 ha of land, illuminating the life of the convicts and their guards. Day entry tickets include a guided historical walking tour, access to the visitor centre, interpretation gallery and museum, and a harbour cruise. Lantern-lit ghost tours depart at dusk and tours of the Isle of the Dead, the cemetery for the colony, unravel emotive stories of the convicts. The site's heritage gardens have been restored and replanted to reflect the style of the original plantings; guided and self-guide walks are available. Cruises and guided tours operate daily to the site of incarceration of juvenile male convicts at nearby Point Puer. Details from visitor centre; Arthur Hwy.

Port Arthur Memorial Garden: dedicated to the victims of the 1996 tragedy at the site in which 35 people were killed by a gunman. ***Convict Trail drive:*** scenic drive from Port Arthur to the Coal Mines Historic Sight.

Eucalypt: modern Australian cafe; 6962 Arthur Hwy; (03) 6250 2555. ***Felons Bistro:*** quality regional menu; Port Arthur Historic Site Visitor Centre; (03) 6251 2310 or 1800 659 101. ***The Commandant's Table:*** friendly country restaurant; 29 Safety Cove Rd; (03) 6268 1235. ***The Fox & Hounds Inn:*** traditional fireside pub fare; 6789 Arthur Hwy; (03) 6250 2217.

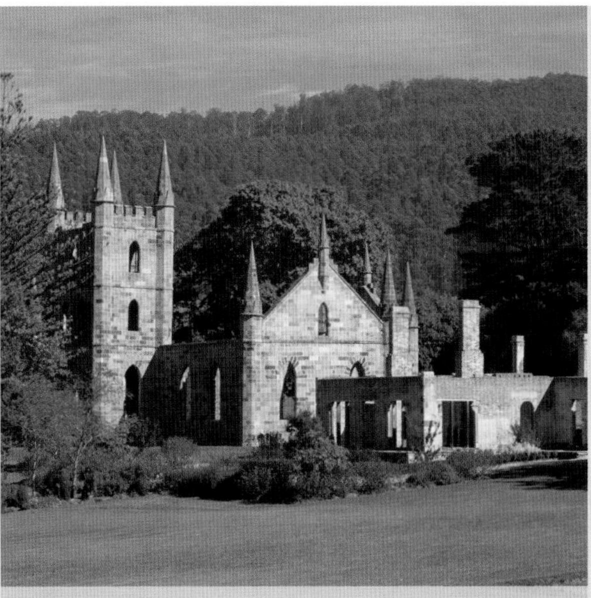

[PORT ARTHUR HISTORIC SITE]

Port Arthur Villas: villas next to the historic site; 52 Safety Cove Rd; (03) 6250 2239 or 1800 815 775. ***Sea Change Safety Cove:*** 2-storey guesthouse with views; 425 Safety Cove Rd; (03) 6250 2719. ***Stewarts Bay Lodge:*** bayside timber cottages; 6955 Arthur Hwy; (03) 6250 2888. ***Brick Point Cottage:*** waterside retro shack; 241 Safety Cove Rd, Safety Cove; 0438 070 498.

Remarkable Cave Created by wave erosion, Remarkable Cave affords spectacular views along the coastline to Cape Raoul. It is in Tasman National Park and is the starting point of a 4–5 hr return walk to Crescent Bay. 6 km S.

Coal Mines Historic Site: Tasmania's first operational mine, with self-guide tours; 30 km NW. ***Tasman Island Cruises:*** 2 hr cruise exploring the beautiful coastal scenery of Tasman National Park; bookings on (03) 6250 2200.

See also SOUTH-EAST, p. 501

Abby's Restaurant: classical dining by candlelight; Prospect House, 1384 Richmond Rd; (03) 6260 2207. ***Ashmore on Bridge Street:*** light meals, excellent coffee; 34 Bridge St; (03) 6260 2238. ***Richmond Wine Centre:*** licensed modern Australian cafe; 27 Bridge St; (03) 6260 2619. ***Meadowbank Estate:*** fine modern European dishes; 699 Richmond Rd, Cambridge; (03) 6248 4484.

Hatcher's Manor: themed farmstay accommodation complex; 73 Prossers Rd; (03) 6260 2622. ***Mrs Currie's House:*** 2-storey Georgian B&B; 4 Franklin St; (03) 6260 2766. ***Mulberry Cottage B&B:*** B&B with heritage show included in the tariff; 23A Franklin St; (03) 6260 2664.

ZooDoo This Wildlife Park is on a 330 ha farm with native fauna such as Tasmanian devils, galahs, emus and the rare albino pademelon wallaby. Middle Tea Tree Rd; (03) 6260 2444; 7 km W.

Southern Tasmanian Wine Route: vineyards north of town include Stoney (6 km N) and Crosswinds (10 km NW); brochure from visitor centre. ***Scenic drive:*** north through Campania (8 km) and Colebrook (19 km).

See also SOUTH-EAST, p. 501

Rosebery

Pop. 1032
Map ref. 672 D9 | 674 A7

Hay's Bus Services, 10–12 Esplanade; (03) 6473 1247.

92.1 FM West Coast 7XS, 106.3 FM ABC Local Radio

Like the nearby towns of Queenstown, Strahan and Zeehan, Rosebery found its economic niche in mining. The region is also known for its ancient rainforests, home to unique fauna and Huon pine, one of the oldest living things on earth.

Mine tours: surface tours of the Pasminco mine; bookings on (03) 6473 1247.

Rosebery Miners, Axemen, Bush and Blarney Festival: Irish music festival; Mar.

Tullah Lakeside Chalet Restaurant: pub fare beside scenic lake; Tullah Lakeside Resort, Farrell St, Tullah; (03) 6473 4121.

Mount Black Lodge: wilderness lodge amid stunning scenery; Hospital Rd; (03) 6473 1039.

Montezuma Falls The highest in Tasmania, the falls plummet 113 m and are accessed from Williamsford in a 3 hr return walk. The track winds through beautiful rainforest and there is a viewing platform at the falls. 11 km SW.

Wee Georgie Wood Steam Train: a 2 km ride along a scenic track on the fully restored steam train runs 1st Sun afternoon each month Sept–Apr; Tullah; 12 km NE. ***Mt Murchison:*** difficult but worthwhile 4 hr return walk rising 1275 m through ancient alpine forests; 14 km SE. ***Lake Johnson Reserve Tour:*** alpine nature tour offering a close-up look at a stand of 10 000-year-old Huon pines; bookings on (03) 6473 1247. ***Hays Trout Tours:*** fishing tours of west-coast lakes and rivers, Aug–Apr; bookings on (03) 6473 1247. ***Fishing:*** good trout fishing in nearby lakes Rosebery, Mackintosh and Murchison; north and east of town.

See also NORTH-WEST, p. 503

Ross

Pop. 270
Map ref. 671 L2 | 673 L12

Tasmanian Wool Centre, Church St; (03) 6381 5466; www.taswoolcentre.com.au

 97.1 Mid FM, 936 AM ABC Local Radio

One of the oldest and most beautiful bridges in Australia spans the Macquarie River in Ross. Completed in 1836, the bridge was designed by colonial architect John Lee Archer and constructed by convicts, one of whom, Daniel Herbert, was given a pardon for his efforts. Herbert was responsible for 186 beautiful stone carvings along the side of the bridge, comprising images of animals, plants, Celtic gods and goddesses, and even the governor of the time, George Arthur. Ross Bridge is a point of pride in this Midlands town and it complements the many old sandstone buildings that adorn the main street. The town's central junction reveals the different aspects of Ross's history and, perhaps, its potential. The four corners are known as Temptation (hotel), Recreation (town hall), Salvation (church) and Damnation (gaol).

Tasmanian Wool Centre Regional attraction housing a museum, wool exhibition and retail area – all illustrating the national importance of the wool industry. Church St.

Ross Female Factory Site Archaeologically the most intact female convict site in Australia, this operated as a probation station for female convicts and their babies in the 19th century. The women were trained as domestic help and hired out to landowners in the area. The Overseer's Cottage has a historical display and model. Off Bond St; (03) 6223 1559.

Heritage walk: takes in 40 historic buildings in town, including Uniting Church (1885); booklet from visitor centre. ***Old Ross General Store and Tearoom:*** Devonshire tea and bakery, home of the famous Tassie scallop pie; Church St. ***Skulduggery:*** mystery tour game, following true crime clues around town; available from visitor centre. ***T-Spot:*** has an array of tea and coffee produce, homewares, gifts and alpaca products; Bridge St.

Old Ross Bakery and Tearooms: stone-walled tearooms; 31 Church St; (03) 6381 5422. ***Ross Village Bakery:*** treats from the wood-fired oven; 15 Church St; (03) 6381 5246.

Colonial Cottages and Ross B&B: central heritage accommodation; 12 Church St; (03) 6381 5354. ***Man O'Ross Hotel:*** sandstone country hotel; 35 Church St; (03) 6381 5445. ***Ross Village Bakery Inn:*** wake to freshly baked bread; 15 Church St; (03) 6381 5246. ***Somercotes:*** colonial property immersed in history; Mona Vale Rd; (03) 6381 5231.

Fishing There is world-class fly-fishing for brown trout in Macquarie River and some of the state's best trout-fishing lakes (Sorell, Crescent, Tooms and Leake) are within an hour's drive.

See also MIDLANDS & THE NORTH, p. 504

St Helens

Pop. 2051
Map ref. 673 O8

St Helens History Room, 61 Cecilia St; (03) 6376 1744.

93.7 Star FM, 1584 AM ABC Local Radio

This popular resort on the shores of Georges Bay is renowned for its crayfish and scalefish, which help maintain a thriving restaurant industry. The nearby beaches have pristine white sand dunes and a mild climate year-round.

St Helens History Room: thematic display of local history, with wooden interactive models; Cecilia St. ***Beaches:*** bay beaches ideal for swimming and coastal beaches for surfing. ***Fishing:*** Scamander River has excellent bream and trout fishing. Charter boats and tours are available for deep-sea fishing and dinghy hire is available for bay fishing. Check with visitor centre for details.

St Helens Regatta and Seafood Festival: Jan. ***Athletic Carnival:*** Jan. ***Tasmanian Game Fishing Classic:*** Mar. ***Seafood and Symphony:*** Mar.

Latris Restaurant: seafood straight from the boat; Marine Pde; (03) 6376 1170. ***Margot:*** French-style bistro; 15 Pendrigh Pl; (03) 6376 2594. ***Ocean View Restaurant:*** Asian-flavoured seafood dishes; Tidal Waters Resort, 1 Quail St; (03) 6376 1999. ***Angasi:*** acclaimed modern restaurant in top spot; 64A Main Rd, Binalong Bay; (03) 6376 8222. ***Holy Cow Cafe:*** cheesemaker's licensed cafe; Pyengana Dairy Company, St Columba Falls Rd, Pyengana; (03) 6373 6157. ***Pub in the Paddock:*** good-value counter meals, home to 2 pigs that guzzle beer; 250 St Columba Falls Rd, Pyengana; (03) 6373 6121.

The Old Headmasters House: grand antique-furnished B&B; 74 Cecilia St; (03) 6376 1125. ***Tidal Waters Resort:*** stylish foreshore resort; 1 Quail St; (03) 6376 1999. ***Bay of Fires Character Cottages:*** beachside cottages with postcard views; 64–74 Main Rd, Binalong Bay; (03) 6376 8262. ***Bed in the Treetops:*** treetop pole-house B&B; 701 Binalong Bay Rd, Binalong Bay; (03) 6376 1318.

Binalong Bay: this small holiday town with gorgeous beaches is renowned for its excellent surf and rock-fishing. 11 km NE. ***St Helens Point:*** walks with stunning coastal scenery along Maurouard Beach; brochure from visitor centre; 9 km S. ***Bay of Fires Conservation Area:*** named by Captain Furneaux for the Aboriginal campfires burning along the shore as he sailed past in 1773, this area features magnificent coastal scenery, good beach fishing and camping. Aboriginal middens are found in the area. A guided walk takes in the scenery to the north (*see Gladstone for details*); begins 9 km N. ***Beaumaris:*** good beaches and lagoons;

12 km s. ***Blue Tier Forest Reserve:*** walks with wheelchair access and interpretive sites; 27 km NW. ***St Columba Falls:*** dropping nearly 90 m, the falls flow down a granite cliff-face to the South George River. Walks through sassafras and myrtle lead to a viewing platform; 38 km W.

See also EAST COAST, p. 500

St Marys

Pop. 524
Map ref. 673 O9

St Marys Coach House Restaurant, 34 Main St; (03) 6372 2529.

 100.3 Star FM, 102.7 FM ABC Local Radio

Despite being the chief centre of the Break O'Day Plains, St Marys is a small town overshadowed by the magnificent St Patricks Head in the eastern highlands. The roads to and from St Marys venture through picturesque forests or down through the valley.

Rivulet Park: platypus, Tasmanian native hens and picnic and barbecue facilities; Main St. ***St Marys Railway Station:*** restored with cafe, Woodcraft Guild and Men's Shed giftshop; Esk Main Rd. ***St Marys Hotel:*** iconic building; Main St.

New Year's Day Races: Jan. ***Winter Solstice Festival:*** June. ***Car Show:*** June. ***St Marys to Fingal Bicycle Race:*** Oct.

eScApe Cafe: licensed cafe and gallery; 21 Main St; (03) 6372 2444. ***Mt Elephant Pancake Barn:*** well-known creperie set in rainforest overlooking Chain of Lagoons; Tasman Hwy, Elephant Pass; (03) 6372 2263. ***White Sands Resort Restaurant:*** French-accented fine dining; 21554 Tasman Hwy, Ironhouse Point; (03) 6372 2228.

Rainbow Retreat: carbon-neutral eco-cabins; 1 Gillies Rd; (03) 6372 2168. ***Ivory Fields Spa Retreat and Herb Farm:*** revitalising country retreat; 56 Davis Gully Rd, Four Mile Creek; (03) 6372 2759. ***White Sands Resort:*** sun-soaked coastal resort; 21554 Tasman Hwy, Ironhouse Point; (03) 6372 2228.

St Patricks Head: challenging 1 hr 40 min return walk to top of rocky outcrop for great 360-degree views of coast and valley; 1.5 km E. ***South Sister:*** an easier lookout alternative to St Patricks Head, a 10–15 min walk leads through stringybarks and silver wattles to spectacular views of Fingal Valley; 3 km NW. ***Cornwall:*** miners wall and Cornwall Collectables shop; 6.5 km W. ***Falmouth:*** small coastal township with several convict-built structures, fine beaches and good fishing; 14 km NE. ***Scamander:*** holiday town with sea- and river-fishing, good swimming, and walks and drives through forest plantations; 17 km N. ***Douglas–Apsley National Park:*** *see Bicheno;* 25 km SE.

See also EAST COAST, p. 500

Scottsdale

Pop. 1967
Map ref. 673 M6

Forest EcoCentre, 96 King St; (03) 6352 6520; www.northeasttasmania.com.au

 91.7 FM ABC Local Radio, 103.7 FM City Park Radio

Scottsdale, the major town in Tasmania's north-east, serves some of the richest agricultural and forestry country in the state. Indeed, as Government Surveyor James Scott observed in 1852, it has 'the best soil in the island'. A visitor to the town in 1868 noted with some surprise that the town had 'neither police station nor public house, but the people appear to get on harmoniously enough without them'. Present-day Scottsdale (now with both of those establishments) still retains a sense of harmony.

Forest EcoCentre A state-of-the-art centre built to principles of ecological sustainability, this place showcases the story of the forests through greenhouse-style interpretive displays and interactive features. King St.

Anabel's of Scottsdale: National Trust building now used as a restaurant and quality motel, set in exquisite gardens; King St. ***Trail of the Tin Dragon:*** discover the history of Chinese miners who came to the region more than 100 years ago; brochure from visitor centre.

The Steak Out: surf 'n turf meals; 13–15 King St; (03) 6352 2248.

Anabel's of Scottsdale: motel-style garden units; 46 King St; (03) 6352 3277. ***Beulah Heritage B&B:*** antique-furnished Victorian house; 9 King St; (03) 6352 3723. ***Willow Lodge:*** grand country home; 119 King St; (03) 6352 2552.

Bridestowe Estate Lavender Farm One of the word's largest lavender-oil producers, Bridestowe is renowned for growing lavender that is not contaminated by cross-pollination. Tours are available in the flowering season (Dec–Jan) and there are lavender products for sale. Nabowla; (03) 6352 8182; 13 km W.

Mt Stronach: very popular 45 min climb to views of the north-east; 4 km S. ***Cuckoo Falls:*** uphill 2–3 hr return walk to falls; Tonganah; 8 km SE. ***Springfield Forest Reserve:*** popular picnic spot and 20 min loop walk through Californian redwoods, English poplars, pines and native flora; 12 km SW. ***Sidling Range Lookout:*** views of town and surrounding countryside; 16 km W. ***Mt Maurice Forest Reserve:*** walks incorporating Ralph Falls; 30 km S.

See also MIDLANDS & THE NORTH, p. 504

Sheffield

Pop. 1035
Map ref. 672 H7 | 674 H3 | 675 A8

5 Pioneer Cres; (03) 6491 1036; www.sheffieldcradleinfo.com.au

 91.7 FM ABC Local Radio, 104.7 Coast FM

Inspired by the Canadian town Chemainus, the people of Sheffield devised an attraction to encourage tourism, commissioning an artist to cover the town in murals depicting local history and scenery. Since then, murals have popped up all over town. As a result, there are many arts and crafts places in town, including marble-making and an open art space.

Story of the Sheffield Murals: audio tour that explains how the town went from rural decline to thriving outdoor art gallery; available from visitor centre. ***Kentish Museum:*** exhibits on local history and hydro-electricity; open 10am–3pm daily; Main St. ***Blue Gum Gallery:*** arts and crafts; at visitor centre. ***Steam train:*** Red Water Creek Steam and Heritage Society runs a train 1st weekend each month; Cnr Spring and Main sts. ***Badgers Range Kimberley's Lookout:*** 90 min return, with short, steep sections. Views along the walk; access from High St, Sheffield North.

Market: Claude Road Hall; 3rd Sun every 3rd month. ***Steam Fest:*** Mar. ***Mural Fest:*** Apr. ***Daffodil Show:*** Sept.

Highlander Restaurant & Scottish Scone Shop: daytime tea rooms, evening restaurant; 60 Main St; (03) 6491 1077.

Yvette's: excellent coffee, gourmet meals; 43 Main St; (03) 6491 1893. ***Highland Restaurant:*** fine Tasmanian cuisine; Cradle Mountain Lodge, 4038 Cradle Mountain Rd, Cradle Mountain; (03) 6492 2100. ***Weindorfers:*** country-style cafe; 1447 Claude Rd, Gowrie Park via Sheffield; (03) 6491 1385.

Cradle Mountain Highlanders Cottages: 12 rustic timber cabins; 3876 Cradle Mountain Rd, Cradle Mountain; (03) 6492 1116. ***Cradle Mountain Lodge:*** award-winning wilderness lodge; 4038 Cradle Mountain Rd, Cradle Mountain; (03) 6492 2100. ***Cradle Mountain Wilderness Village:*** secluded modern chalets; Cradle Mountain Rd, Cradle Mountain; (03) 6492 1500. ***Discovery Holiday Park:*** cottages, cabins and camping; 3832 Cradle Mountain Rd, Cradle Mountain; (03) 6492 1395. ***Eagles Nest Retreat:*** ultra-luxury amid spectacular scenery; 3 Browns Rd, West Kentish; (03) 6491 1511. ***Waldheim Cabins:*** rustic cabins in national park; Parks and Wildlife Service, Cradle Mountain–Lake St Clair National Park; (03) 6492 1110.

Cradle Mountain–Lake St Clair National Park Covering 124 942 ha, this national park has over 25 major peaks, including the state's highest, Mt Ossa, and possibly its most spectacular, Cradle Mountain. The terrain is marked by pristine waterfalls, U-shaped valleys, dolerite formations, forests of deciduous beech, Tasmanian myrtle, pandani and King Billy pine, and swathes of wildflowers. At the southern end of the park lies Lake St Clair. Cruises operate daily. There are many tracks for all levels of experience; check with the rangers for up-to-date information on conditions. The 3 major tracks are the Overland Track, covering 85 km and one of Australia's best-known walks for experienced, well-equipped walkers; Dove Lake Loop Track, suitable for everyone; and Summit Walk, an 8 hr return walk to the top of Cradle Mountain. The Overland Track can also be done in style, staying in well-appointed huts and with all meals provided. Contact (03) 6391 9339 for more information. ***The Wilderness Gallery***: unique showcase of environmental photography; 3718 Cradle Mountain Rd; (03) 6492 1402. For other general information, contact one of the visitor centres: Cradle Mountain (03) 6492 1110; Lake St Clair (03) 6289 1172; 61 km sw.

Tasmazia Claiming to be the world's largest maze complex, Tasmazia has a variety of mazes, model village, honey boutique, pancake parlour and lavender farm. Lake Barrington Nature Reserve; (03) 6491 1934; 14 km sw.

Stoodley Forest Reserve: experimental tree-farm forest, with a 40 min loop walk through European beech, Douglas fir, radiata pine and Tasmanian blue gum. Picnic areas; 7 km NE. ***Lake Barrington Estate Vineyard:*** tastings and sales; open Wed–Sun Nov–Apr; West Kentish; 10 km w. ***Railton:*** known as the 'Town of Topiary', Railton has over 100 living sculptures; 11 km NE. ***Mt Roland:*** well-marked bushwalks to the summit, which rises to 1234 m, with access from Claude Rd or Gowrie Park; 11 km sw. ***Lake Barrington:*** internationally recognised rowing course created by Devils Gate Dam. Picnic and barbecue facilities along the shore; 14 km sw. ***Wilmot:*** this 'Valley of Views' has magnificent views of the mountains of Cradle country and Bass Strait, novelty letterbox trail, and the original Coles store and family homestead; 38 km w.

See also NORTH-WEST, p. 503

Smithton

Pop. 3363
Map ref. 672 C4

Smithton Tourist Information Site, 29 Smith St, Smithton; (03) 6458 1330 or 1300 138 229.

88.9 Coast FM, 91.3 FM ABC Local Radio

This substantial town is renowned for its unique blackwood-swamp forests. Smithton services the most productive dairying and vegetable-growing area in the state and, with several large sawmills, is also the centre of significant forestry reserves and the gateway to The Tarkine.

Circular Head Heritage Centre: artefacts and memorabilia detailing the area's pioneering history; 8 King St. ***Lookout tower:*** excellent views from Tier Hill; Massey St. ***Western Esplanade Community Park:*** popular picnic spot overlooking the mouth of the Duck River, with fishing and walks; Western Espl. ***Britton Brothers Timber Mill:*** tours of the mill by appt (03) 6452 2522; Brittons Rd, southern outskirts of town.

Bridge Bistro: pub grub in modernised surrounds; Bridge Hotel/Motel, 2A Montagu St; (03) 6452 1389. ***Kauri Bistro:*** bistro fare in vast log cabin; Tall Timbers Hotel, Scotchtown Rd; (03) 6452 2755 or 1800 628 476. ***Marrawah Tavern:*** counter meals; Comeback Rd, Marrawah; (03) 6457 1102. ***Tarkine Forest Adventure:*** licensed cafe; Bass Hwy; (03) 6456 7199.

Rosebank Cottages Collection: country-style cottages; 46 Brooks Rd and 40 Goldie St; (03) 6452 2660. ***Tall Timbers Hotel:*** hotel complex showcasing Tasmania's timber; Scotchtown Rd; (03) 6452 2755 or 1800 628 476.

Allendale Gardens These gardens are linked to a rainforest area by 3 easy loop walks that take in the Fairy Glades, a section constructed for children, and towering 500-year-old trees. Peacocks roam the gardens, where Devonshire tea is available. Open Oct–Apr; Edith Creek; (03) 6456 4216; 13 km s.

River and Duck Bay: good fishing and boating; 2 km N. ***Sumac Lookout:*** amazing views over Arthur River and surrounding eucalypt forest; 4 km s on Sumac Rd. ***Milkshake Hills Forest Reserve:*** exquisite picnic spot among eucalypts and rainforest; 10 min loop walk or 1 hr return walk to hill-top; 40 km s. ***Surfing:*** excellent, but turbulent surf beaches near Marrawah where Rip Curl West Coast Classic is held in Mar; 51 km sw. ***Arthur River cruises:*** trips into pristine wilderness of Arthur Pieman Protected Area; run Sept–June; 70 km sw. ***Gardiner Point:*** 'Edge of the World' plaque; 70 km sw. ***Woolnorth tours:*** coach tours visit spectacular Cape Grim cliffs, wind farm, and Woolnorth sheep, cattle and timber property; bookings on (03) 6452 1493.

See also NORTH-WEST, p. 503

Sorell

Pop. 4179
Map ref. 669 K4 | 671 L6

Gordon St; (03) 6269 2924.

97.7 Tasman FM, 936 AM ABC Local Radio

The south-eastern town of Sorell was founded in 1821, and was important in early colonial times for providing most of the state's grain. It was named after Governor Sorell who attempted to curb bushranging in Tasmania, but ironically the town was later targeted by bushranger Matthew Brady, who released the prisoners from gaol and left the soldiers imprisoned in their place.

Historic buildings There are 3 churches listed in the National Estate, most notably St George's Anglican Church (1826). There are also many heritage buildings including the Old Rectory (c. 1826), Old Post Office (c. 1850) and Bluebell Inn (c. 1864).

Orielton Lagoon: important habitat for migratory wading birds on western shore of town. ***Pioneer Park:*** popular picnic spot with barbecue facilities; Parsonage Pl.

STANLEY

Pop. 459
Map ref. 672 D3

 **45 Main Rd; 1300 138 229; www.stanley.com.au**

88.9 Coast FM, 91.3 FM ABC Local Radio

This quaint village nestles at the base of an ancient outcrop called The Nut, which rises 152 metres with sheer cliffs on three sides. Matthew Flinders, upon seeing The Nut in 1798, commented that it looked like a 'cliffy round lump resembling a Christmas cake'. Today visitors come to see this 'lump' and the historic township.

The Nut The remains of volcanic rock, The Nut looms above the surrounding sea and provides spectacular coastal views. Visitors can either walk to the summit along a steep and challenging track or take the chairlift from Browns Rd. There is a 40 min circuit walk along cliffs at the summit.

Historic buildings: many in the wharf area, including the bluestone former Van Diemen's Land (VDL) Company store in Marine Park. ***Lyons Cottage:*** birthplace of former prime minister J. A. Lyons, with interesting memorabilia; Alexander Tce. ***Hearts and Craft:*** fine Tasmanian craft; shop closes during quieter months; 12 Church St. ***Discovery Centre Folk Museum:*** displays of local history; Church St. ***Touchwood:*** quality craft and woodwork; 31 Church St. ***Cemetery:*** graves of colonial architect John Lee Archer and explorer Henry Hellyer; Browns Rd. ***Stanley Seaquarium:*** Tasmanian marine species including huge rock lobsters and giant crabs; Fishermans Dock; (03) 6458 2052. ***Penguins & Platypus Tours:*** evening tours to see little penguins and platypus in their natural environment Sept–Apr; bookings on 0448 916 153. ***Seal cruises:*** daily trips to see offshore seal colonies; bookings on 0419 550 134.

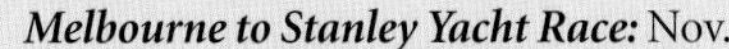

Melbourne to Stanley Yacht Race: Nov.

Hursey Seafoods: freshest seafood ever; 2 Alexander Tce; (03) 6458 1103. ***Old Cable Station Restaurant:*** intimate fine dining; 435 Greenhills Rd, West Beach; (03) 6458 1312. ***Stanley's on the Bay:*** nautical-themed seafood fine dining; 15 Wharf Rd; (03) 6458 1404.

[VIEW OF THE NUT]

@ VDL Stanley: luxury modernised bluestone warehouse; 16 Wharf Rd; (03) 6458 2032. ***Abbey's Cottages:*** 7 heritage cottages; various locations in Stanley; 1800 222 397. ***Beachside Retreat West Inlet:*** secluded futuristic eco-cabins; 253 Stanley Hwy; (03) 6458 1350. ***Horizon Deluxe Apartments:*** contemporary glass-fronted style; 88 Dovecote Rd; 0448 521 115.

Highfield Historic Site The headquarters of VDL Co, Highfield contains a homestead with 12 rooms, a chapel, convict barracks, a schoolhouse, stables, a barn, workers' cottages and large gardens. Green Hills Rd; 2 km N.

Dip Falls: double waterfall in dense rainforest and eucalypts. There is a picnic area nearby; 40 km SE off hwy, via Mawbanna.

See also NORTH-WEST, p. 503

Market: Cole St; Sun (weekly in summer, fortnightly in winter; check with visitor centre for current schedule).

Barilla Bay: oysters, seafood and ... oysters; 1388 Tasman Hwy, Cambridge; (03) 6248 5454.

Cherry Park Estate: fruit farm B&B; 114 Pawleena Rd; (03) 6265 2271. ***Sorell Barracks:*** Sorell's oldest buildings; 31 Walker St; (03) 6265 1572. ***Steeles Island Beach Cottages:*** rustic homestead and cabins; Steeles Island via River St, Carlton Beach; (03) 6265 8077.

Sorell Fruit Farm Offers pick-your-own patches of fruit as well as ready-picked. Fruits available in season include berries, apricots, cherries, pears, nashi, nectarines and peaches. Thirst quenching Devonshire tea at the cafe. Open Oct–May; Pawleena Rd; (03) 6265 2744; 2 km E.

Southern Tasmanian Wine Route: many vineyards east of town include Orani (3 km E) and Bream Creek (22 km E); brochure from visitor centre. ***Beaches:*** around Dodges Ferry and Carlton; 18 km S.

See also SOUTH-EAST, p. 501

Strahan

see inset box on next page

Swansea

Pop. 558
Map ref. 671 N3 | 673 N12

Swansea Bark Mill and East Coast Museum, 96 Tasman Hwy; (03) 6257 8094.

98.5 Star FM, 106.1 FM ABC Local Radio

Perched on Great Oyster Bay, Swansea looks out to the Freycinet Peninsula. It is part of the Glamorgan/Spring Bay shire, the oldest rural municipality in the country, and many fine old buildings testify to its age. Today it is a popular holiday destination.

TASMANIA

STRAHAN

Pop. 635
Map ref. 670 C2 | 672 C12

The Esplanade; (03) 6472 6800.

92.1 FM West Coast 7XS, 107.5 FM ABC Local Radio

This pretty little port on Macquarie Harbour, on Tasmania's west coast, is the last stop before a long stretch of ocean to Patagonia. Sometimes considered the loneliest place on earth, it was dubbed 'The Best Little Town in the World' by the *Chicago Tribune* and continues to attract visitors. Strahan (pronounced 'strawn') came into being as a penal colony operating from the isolated station of Sarah Island. Known as a particularly cruel environment, the station was shut down in 1833, but not before convict Alexander Pearce had managed to escape. Pearce and seven others set off for Hobart but found the terrain too tough to overcome. Pearce, alone when discovered, was suspected of cannibalism. The following year he again escaped and again killed and ate his cohort. Pearce finally made it to Hobart, where he was executed.

Visitor centre The centre has an impressive historical display on Tasmania's south-west, including Aboriginal history, European settlement, and events such as the fight to save the Franklin River from being dammed in the early 1980s. In the amphitheatre there is an audiovisual slideshow and a nightly performance of 'The Ship That Never Was', about convict escapes. The Esplanade.

Morrison's Mill: one of 4 remaining Huon pine sawmills in Tasmania; tours available; The Esplanade. ***Strahan Woodworks:*** woodturning, arts and crafts; The Esplanade. ***Tuts Whittle Wonders:*** carvings from forest wood; Reid St. ***Ormiston House:*** built in 1899 and a fine example of Federation architecture surrounded by magnolia trees and expansive gardens. Morning and afternoon teas are served; Bay St. ***Water Tower Hill Lookout:*** views of township and harbour; Esk St. ***West Coast Wilderness Railway:*** *see Queenstown.*

Mount Lyell Picnic: Australia Day Jan, West Strahan Beach.

Franklin Manor: distinguished fine dining; The Esplanade; (03) 6471 7311. ***Hamers Hotel:*** bistro meals; The Esplanade; (03) 6471 7191. ***Risby Cove:*** contemporary à la carte dining; The Esplanade; (03) 6471 7572. ***Schwochs Seafoods:*** excellent little seafood cafe; The Esplanade; (03) 6471 7500.

Aloft Boutique Accommodation: central contemporary apartments; 15 Reid St; (03) 6471 8095. ***Ormiston House:*** grand Federation mansion; The Esplanade; (03) 6471 7077. ***Risby Cove:*** waterline views and sublime reflections; The Esplanade; (03) 6471 7572. ***Wheelhouse Luxury Apartments:*** glass-fronted cliff-top luxury; 4 Frazer St; (03) 6471 7777.

[CRUISE BOAT ON THE GORDON RIVER]

Franklin–Gordon Wild Rivers National Park This park now has World Heritage listing after an earlier state government tried to dam the Franklin River. Protests were so heated and widespread that the federal government and High Court stepped in and vetoed the proposal, saving the dense temperate rainforest and wild rivers that make up the park. Visitors can go canoeing and whitewater rafting, and there are many bush trails for experienced walkers. A 4-day walk to Frenchmans Cap takes in magnificent alpine scenery. The 40 min return walk to Donaghys Hill is easier and overlooks the Franklin and Collingwood rivers. Contact Parks & Wildlife Service on 1300 135 513. Gordon River cruises depart from Strahan; bookings on 1800 420 500 and (03) 6471 7174. 36 km SE.

Peoples Park: popular picnic spot in botanical gardens setting with a 45 min return walk to Hogarth Falls through marked rainforest; 2 km E. ***Ocean Beach:*** Tasmania's longest beach (36 km) offers horseriding, beach fishing and the opportunity to see short-tailed shearwaters in their burrows Oct–Mar; 6 km W. ***Henty Sand Dunes:*** vast sand dunes with sandboards and toboggans for hire; 12 km N. ***Cape Sorell Lighthouse:*** 40 m high lighthouse built in 1899; 23 km SW. ***Sarah Island:*** ruins of convict station with tours available; check with visitor centre; 29 km SE. ***Scenic flights:*** bird's-eye views of Gordon River, Sir John Falls and Franklin River valley; bookings on (03) 6461 7718.

See also SOUTH-WEST WILDERNESS, p. 502

Swansea Bark Mill and East Coast Museum The only restored black-wattle bark mill in Australia has machinery that still processes the bark for tanning leather. The museum has comprehensive displays on life in the 1820s, and the adjoining Wine and Wool Centre has wines from over 50 Tasmanian vineyards and textiles from around the state. A cafe and tavern complete the complex. Tasman Hwy; (03) 6257 8094.

Historical walk: self-guide tour takes in charming heritage buildings including Morris' General Store (1838), run by the Morris family for over 100 years, and Community Centre (c. 1860), featuring the unusually large slate billiard table made for the 1880 World Exhibition; brochure from visitor centre. ***Waterloo Point:*** 1 km walking track leads to viewpoint to see short-tailed shearwaters at dusk and Aboriginal middens; Esplanade Rd.

Kate's Berry Farm: sinfully lavish berry desserts; Addison St; (03) 6257 8428. ***The Banc:*** modern Australian fine dining; Cnr Franklin and Maria sts; (03) 6257 8896. ***The Ugly Duck Out:*** international fare; 2 Franklin St; (03) 6257 8850. ***Kabuki by the Sea:*** cliff-top Japanese restaurant; 12164 Tasman Hwy, Rocky Hills; (03) 6257 8588.

Amos House and Swansea Ocean Villas: motel rooms and self-catering villas; 3 Maria St; (03) 6257 8656. ***Meredith House:*** colonial B&B; 15 Noyes St; (03) 6257 8119. ***Swansea Cottages and Sherbourne Lodge:*** country-style cottages and lodge suites; 43 Franklin St; (03) 6257 8328. ***Avalon Coastal Retreat:*** ultra-chic cliff-top villa; 11922 Tasman Hwy, Rocky Hills; 1300 361 136.

Spiky Bridge Built by convicts in 1843, the bridge was pieced together without the aid of mortar or cement. The spikes – vertical fieldstones – prevented cattle from falling over the sides. The beach nearby has a picnic area and good rock-fishing. 7.5 km s.

Coswell Beach: good spot for viewing little penguins at dusk; 1 km s along coast from Waterloo Pt. ***Duncombes Lookout:*** splendid views; 3 km s. ***Mayfield Beach:*** safe swimming beach with walking track from camping area to Three Arch Bridge. There is also great rock- and beach-fishing; 14 km s. ***Vineyards:*** Springvale, Coombend, Freycinet and Craigie Knowe vineyards all have cellar-door sales on weekends and holidays; 15 km N. ***Lost and Meetus Falls:*** bushwalks past beautiful waterfalls in dry eucalypt forest. Sheltered picnic area is nearby; 50 km NW.

See also EAST COAST, p. 500

Triabunna

Pop. 798
Map ref. 669 N1 | 671 M5

Cnr Charles St and Esplanade West; (03) 6257 4772.

90.5 FM ABC Local Radio, 97.7 Tasman FM

When Maria (pronounced 'mar-eye-ah') Island was a penal settlement, Triabunna (pronounced 'try-a-bunnah') was a garrison town and whaling base. After an initial boom, this small, south-east coast town settled into relative obscurity, content with its role as a centre for the scallop and abalone industries.

Tasmanian Seafarers' Memorial: commemorates those who lost their lives at sea; Esplanade. ***Girraween Gardens and Tearooms:*** with large lily pond, day lilies, roses and agapanthus; Henry St.

Spring Bay Seafest: Easter.

Girraween Gardens and Tearooms: light meals and snacks; 4 Henry St; (03) 6257 3458. ***Scorchers by the River:*** casual gourmet cafe; 1 Esplanade, Orford; (03) 6257 1033. ***Gateway Cafe:*** great lunches and pancakes; 1 Charles St, Orford; (03) 6257 1539.

Barton Retreat: modern luxury with 'wow' factor; 53 Barton Ave, Double Creek; 1300 763 132. ***Sanda House Colonial B&B:*** historic stone cottage; 33 Walpole St, Orford; (03) 6257 1527.

Maria Island National Park After the convicts were moved to Port Arthur, the island was leased to Italian merchant Diego Bernacchi, who envisaged first a Mediterranean idyll and then a cement works. Both projects were short-lived. The extensive fossil deposits of the Painted Cliffs are magnificent and the historic penal settlement of Darlington is also of interest. ***Maria Island Walk:*** gentle journey of 30 km over four days with gourmet dining and comfy beds at night; (03) 6234 2999. Daily ferry to the island from town. Contact Parks & Wildlife Service on 1300 135 513; 7 km s.

Orford: pleasant walk along Old Convict Rd following Prosser River; 7 km sw. ***Thumbs Lookout:*** stunning views of Maria Island; 9 km sw. ***Church of St John the Baptist:*** heritage church (1846) with a stained-glass window (taken from England's Battle Abbey) depicting John the Baptist's life; Buckland; 25 km sw. ***Beaches:*** safe swimming beaches around area; contact visitor centre for details.

See also SOUTH-EAST, p. 501

Ulverstone

Pop. 9761
Map ref. 672 G6 | 674 G1

13–15 Alexandra Rd; (03) 6425 2839; www.coasttocanyon.com.au

91.74 FM ABC Local Radio, 106.1 Coast FM

At the mouth of the Leven River on the north central coast, Ulverstone has some of Tasmania's finest tourist beaches. Relax along these sweeping beaches, cycle through waterfront parklands, take coastal and forest walks, or explore spectacular Leven Canyon and the nearby caves.

Discover the Leven Cruise A fun 1.5 hr guided tour of the upper Leven River in Ulverstone. Tours leave daily, in the cutest little passenger vessel in Tasmania. Tasma Pde; (03) 6425 2839.

Riverside Anzac Park: amazing children's playground, great picnic facilities and an interesting fountain. The park is named after a pine tree grown from a seed taken from Gallipoli; Beach Rd. ***Fairway Park:*** has a giant waterslide; Beach Rd. ***Shropshire Park:*** has a footpath inscribed with details from 75-year history of the Royal Australian Navy; Dial St. ***Legion Park:*** magnificent coastal setting; Esplanade. ***Tobruk Park:*** includes Boer War Memorial; Hobbs Pde. ***Ulverstone Local History Museum:*** interesting display of old business facades and other memorabilia; Main St. ***Shrine of Remembrance:*** clock-tower memorial designed and built by European immigrants in 1953; Reibey St. ***Ulverstone Lookout:*** views over town; Upper Maud St. ***Woodcraft Gallery and Workshop:*** demonstrations, private tutoring and sales; open Tues, Thurs and Sat; Reibey St.

Market: The Quadrant Carpark; every 2nd Sat. ***Hobbies and craft expo:*** Jan. ***Festival in the Park:*** Feb. ***Twilight Rodeo:*** Jan/Feb. ***Carnival of the Grasshopper:*** Mar. ***Forth Valley Blues Festival:*** Mar. ***Doll, Bear and Miniature Extravaganza:*** July. ***NW Woodcraft Exhibition:*** Nov. ***Cradle Coast Rotary Art Exhibition:*** Nov. ***Ulverstone Agricultural Show:*** Nov. ***Christmas Mardi Gras:*** Dec.

The Atrium: family-oriented restaurant; The Lighthouse Hotel, 33 Victoria St; (03) 6425 1197. ***Furners Hotel:*** good pub meals; 42 Reibey St; (03) 6425 1488. ***Pedro's the Restaurant:*** casual waterside seafood; Ulverstone Wharf; (03) 6425 6663.

Boscobel of Ulverstone: restored French Provincial–style mansion; 27 South Rd, West Ulverstone; (03) 6425 1727. ***Moonlight Bay Guesthouse:*** homely seaside B&B; 141 Penguin Rd, West Ulverstone; (03) 6425 1074. ***Westella:*** ornate Gothic guesthouse; 68 Westella Dr; (03) 6425 6222. ***Winterbrook:*** elegant B&B; 28 Eastland Dr; (03) 6425 6324.

Leven Canyon There are walking tracks to a viewing platform with spectacular views down this 250 m ravine. Picnic and barbecue facilities are nearby. 41 km s.

Gunns Plains Caves: these limestone caves are well-lit for visitors. Daily tours take in an underground river and glow worms. Check visitor centre for tour departure times. 24 km sw. ***Wing's Wildlife Park:*** wildlife exhibits, trout and reptile display and animal nursery are all to be found here – including Tasmanian devils.

TASMANIA

Gunns Plains; (03) 6429 1151; 24 km SW. ***Miniature railway:*** 3rd Sun each month; 2 km E. ***Goat Island Sanctuary:*** cave and good fishing, but walking access to island at low tide only; 5 km W. ***Penguins:*** view little penguins at dusk; Leith; 12 km E. ***Preston Falls:*** scenic views; 19 km S. ***Winterbrook Falls:*** 5 hr return walk to the falls for fit bushwalkers; 46 km S. ***Fishing:*** good fishing on beach (especially Turners Beach), river and estuary; contact visitor centre for details. ***Beaches:*** safe swimming beaches to east and west of town; contact visitor centre for details.

See also NORTH-WEST, p. 503

Waratah

Pop. 223
Map ref. 672 E7 | 674 A3

Wynyard Visitor Centre, 8 Exhibition Link, Wynyard; (03) 6443 8330; www.wowtas.com.au

103.3 FM ABC Local Radio, 106.1 Coast FM

Waratah was the site of the first mining boom in Tasmania. Tin deposits were discovered in 1871 by James 'Philosopher' Smith, and by the late 1800s the Mount Bischoff operation was the richest tin mine in the world. The mine closed in 1947.

Waratah Museum and Giftshop: displays early photographs and artefacts of the area and provides brochure for self-drive tour of town; Smith St. ***Waratah Waterfall:*** in the centre of town; Smith St. ***Philosopher Smith's Hut:*** replica of late-19th-century miner's hut; Smith St. ***St James Anglican Church:*** first church in Tasmania to be lit by hydro-power; Smith St. ***Lake Waratah:*** pleasant picnic and barbecue area with rhododendron garden and walks to Waratah Falls; English St.

Bischoff Hotel: Queen Anne–style hotel; 20 Main St; (03) 6439 1188. ***Waratah's O'Connor Hall Guest House and B&B:*** ornate Victorian heritage homestead; 2 Smith St; (03) 6439 1472.

Savage River National Park The rainforest here is the largest contiguous area of cool–temperate rainforest surviving in Australia. Excellent for self-sufficient bushwalking as well as fishing, 4WD and kayaking in the adjacent regional reserve. Contact Parks & Wildlife Service on 1300 135 513; access tracks lead off the main road out of Waratah; around 20 km W.

Fishing: excellent trout fishing in rivers and lakes in the area, including Talbots Lagoon; 20 km E. ***Old mines:*** walks and drives to old mining sites; brochure from visitor centre.

See also NORTH-WEST, p. 503

Westbury

Pop. 1357
Map ref. 673 J8 | 675 E11

Great Western Tiers Visitor Centre, 98–100 Emu Bay Rd, Deloraine; (03) 6362 3471; www.greatwesterntiers.org.au

91.7 FM ABC Local Radio, 103.7 FM City Park Radio

Westbury's village green gives the town an English air. Just west of Launceston, the town was surveyed in 1823 and laid out in 1828, the assumption being that it would be the main stop between Hobart and the north-west coast. Originally planned as a city, it never grew beyond the charming country town it is today.

The village green Used for parades and fairs in the 1830s, the village green is still the focal point and fairground of Westbury – with one small difference: prisoners are no longer put in the stocks for all to see. King St.

White House This collection of buildings, enclosing a courtyard, was built in 1841. Later additions include a coach depot, a bakery and a flour mill. The house has an excellent collection of 17th- and 18th-century furniture and memorabilia, and a magnificent doll's house. The outbuildings house a toy museum as well as early bicycles, vintage cars and horse-drawn vehicles. The restored bakery serves refreshments. Open Sept–June; King St.

Pearn's Steam World: said to be the largest collection of working steam traction engines in Australia; Bass Hwy. ***Westbury Maze and Tearoom:*** hedge maze composed of 3000 privet bushes; open Oct–June; Bass Hwy. ***Culzean:*** historic home with beautifully maintained temperate-climate gardens; open Sept–May; William St. ***Westbury Mineral and Tractor Shed:*** museum of vintage tractors and farm machinery featuring a scale-model tractor exhibition; Veterans Row.

Maypole Festival: Mar. ***St Patrick's Day Festival:*** Mar.

Andy's Bakery Cafe: pastries and gelato; 45 Meander Valley Rd; (03) 6393 1846. ***Fitzpatricks Inn:*** fine European-style dishes; 56 Meander Valley Hwy; (03) 6393 1153. ***Hobnobs Restaurant:*** sophisticated provincial-style food; 47 William St; (03) 6393 2007.

Elm Wood: gabled cottages in English gardens; 10 Lonsdale Prom; (03) 6393 2169. ***Fitzpatricks Inn:*** stylish renovated coaching inn; 56 Meander Valley Hwy; (03) 6393 1153. ***Westbury Gingerbread Cottages:*** central period cottages; 52 William St; (03) 6393 1140.

Fishing: good trout fishing at Four Springs Creek (15 km NE) and Brushy Lagoon (15 km NW).

See also MIDLANDS & THE NORTH, p. 504

Whitemark

Pop. 598
Map ref. 670 B10

4 Davies St, (03) 6359 2380 or 1800 994 477; www.flindersislandonline.com.au

91.7 FM ABC Local Radio

Flinders Island in Bass Strait is the largest of the 52 islands in the Furneaux Group, once part of a land bridge that joined Tasmania to the mainland. It's a beautiful place, but with a tragic history. In 1831 Tasmania's Aboriginal people, depleted to fewer than 160, were isolated on Flinders Island. A lack of good food and water meant that by 1847, when the settlement was finally abandoned, only 46 Aboriginal people remained. The Wybalenna Historic Site at Settlement Point stands as a reminder of the doomed community. Whitemark is the island's largest town.

Bowman History Room: displays of memorabilia show Whitemark since the 1920s; rear of E. M. Bowman & Co, 2 Patrick St. ***Diving and snorkelling:*** tours to shipwreck sites, limestone reefs and granite-boulder formations including Chalky Island Caves and Port Davies Reef; bookings on (03) 6359 8429. ***Fishing tours:*** from Port Davies to Prime Seal Island for pike and salmon, and Wybalenna Island for couta; bookings on (03) 6359 8429.

Three Peaks Race: annual sailing and hiking race from Beauty Pt to Flinders Island, then down the coast to Hobart. It coincides with local produce markets and children's activities at Lady Barron; Easter. ***Bass Strait Golf Classic:*** June. ***Flinders Island Show:*** local and off-shore exhibitors; showgrounds; Oct.

Sweet Surprises: coffee and light meals; 5 Lagoon Rd; (03) 6359 2138. ***Furneaux Tavern and Shearwater Restaurant:*** seafood with a view; 11 Franklin Pde, Lady Barron; (03) 6359 3521.

Interstate Hotel: budget heritage rooms; Patrick St; (03) 6359 2114. ***Partridge Farm:*** luxurious bush hideaways; Badger Corner; (03) 6359 3554. ***Echo Hills:*** farm cottage; Madeleys Rd, Lackrana; (03) 6359 6509. ***Palana Beach House:*** secluded timber beach house; 4758 Palana Rd, Palana; griggsys@gmail.com.

Strzelecki National Park The only national park in the Furneaux Group, featuring the Peaks of Flinders and Mt Strzelecki, and wetlands, heathland and lagoons. A 5 hr return walk to the summit of Mt Strzelecki is steep, but affords excellent views of Franklin Sound and its islands. Trousers Pt, featuring magnificent rust-red boulders and clear waters, is located just outside the park. Contact Parks & Wildlife Service on 1300 135 513; 15 km S.

Emita Museum This museum, run by the Historical Society, has a wide range of memorabilia and houses displays on the short-tailed shearwater industry, the War Service Land Settlement and the nautical and natural histories of Flinders Island. Open 1–4pm Sat–Sun; Settlement Point Rd, Emita; 18 km NW.

Patriarchs Wildlife Sanctuary: privately owned sanctuary with a vast range of birdlife and wallabies, which can be handfed; access via Lees Rd, Memana; 30 km NE. ***Logan Lagoon Wildlife Sanctuary:*** houses a great diversity of birdlife in winter including the red-necked stint, common greenshank and eastern curlew; east of Lady Barron; 30 km SE. ***Mt Tanner:*** lookout with stunning views of the northern end of the island and Marshall Bay; off West End Rd; Killiecrankie; 40 km N. ***Port Davies:*** from the viewing platform see short-tailed shearwaters fly into their burrows at dusk. An enormous colony of these birds breed here between Sept and Apr and then set out on an annual migration to the Northern Hemisphere. West of Emita; 20 km NW. ***Fossicking:*** Killiecrankie diamonds, a form of topaz released from decomposing granite, are found along the beach at Killiecrankie Bay; brochure at visitor centre; 43 km NW. ***Beachcombing:*** rare paper-nautilus shells wash up along the island's western beaches.

See also BASS STRAIT ISLANDS, p. 505

Wynyard

Pop. 4810
Map ref. 672 E5

8 Exhibition Link; (03) 6443 8330; www.wowtas.com.au

102.5 FM ABC Local Radio, 106.1 Coast FM

This small centre at the mouth of the Inglis River has charming timber buildings and is located on a stunning stretch of coastline in an extremely fertile pocket of the state.

Gutteridge Gardens: riverside gardens in the heart of town; Goldie St. ***Nature walks:*** include boardwalk along Inglis River; brochure at visitor centre. ***Wonders of Wynyard:*** an exhibition of veteran vehicles that includes one of the oldest Ford models in the world; Exhibition Link.

Wynyard Farmers' Market: 2nd and 4th Sat each month. ***Bloomin' Tulips Festival:*** Oct.

Buckaneers: popular seafood eatery; 4 Inglis St; (03) 6442 4101. ***The Riverview:*** international fare; The Waterfront, 1 Goldie St; (03) 6442 2531. ***Jolly Rogers on the Beach:*** modern beachside bistro; 1 Port Rd, Boat Harbour Beach; (03) 6445 1710.

Alexandria: classical, stylish Federation B&B; 1 Table Cape Rd; (03) 6442 4411. ***Cape Cottage:*** quaint bush cottage; 8A Elfrida Ave, Sisters Beach; (02) 4784 1248. ***The Harbour House:*** funky beachfront pads; Port Rd, Boat Harbour Beach; (03) 6442 2135. ***The Winged House:*** cantilevered masterpiece perched cliff-side; 400 Tollymore Rd, Table Cape; (02) 9906 3224.

Boat Harbour This is a picturesque village ideal for diving and spearfishing. Boat Harbour Beach (4 km NW of town) has safe swimming, marine life in pools at low tide, fishing and waterskiing. 10 km NW.

Rocky Cape National Park With some of the best-preserved Aboriginal rock shelters and middens in Tasmania, the park incorporates the pristine Sisters Beach. Contact Parks & Wildlife Service on 1300 135 513. 29 km NW.

Fossil Bluff: scenic views from an unusual geological structure where the oldest marsupial fossil in Australia was found; 3 km N. ***Table Cape Lookout:*** brilliant views of the coast from 190 m above sea level. En route to lookout is Table Cape Tulip Farm, with tulips, daffodils and Dutch irises blooming Sept–Oct. From lookout a short walk leads to an old lighthouse; 5 km N. ***Flowerdale Freshwater Lobster Haven:*** restaurant and artificial lakes stocked with lobster; 6.5 km NW. ***Fishing:*** Inglis and Flowerdale rivers provide excellent trout fishing; also good sea-fishing around Table Cape. ***Scenic flights:*** the best way to appreciate the patchwork colours of the fields in the area; bookings at Wynyard Airport (03) 6442 1111. ***Scenic walks and drives:*** brochure from visitor centre.

See also NORTH-WEST, p. 503

Zeehan

Pop. 846
Map ref. 672 D10

West Coast Pioneers Memorial Museum, Main St; (03) 6471 6225.

90.5 FM ABC Local Radio, 92.1 FM West Coast 7XS

After silver-lead deposits were discovered here in 1882, Zeehan boomed and between 1893 and 1908, the mine yielded ore worth $8 million, which led to its nickname, the 'Silver City of the West'. However, from 1910 the mine started to slow and Zeehan declined, threatening to become a ghost town. Fortunately, the nearby Renison Bell tin mine has drawn workers back to the area.

West Coast Pioneers Memorial Museum The old Zeehan School of Mines building (1894) has been converted into a comprehensive museum outlining the local history with extensive mineral, geological and locomotive collections. Main St.

Historic buildings: boom-era buildings on Main St include Gaiety Theatre, once Australia's largest theatre. ***Scenic drives:*** self-guide drives in town and nearby; brochure from visitor centre.

Maud's Restaurant: pub-style dishes; Hotel Cecil; 99–101 Main St; (03) 6471 6221.

Hotel Cecil and Old Miners Cottages: grand old hotel; 99–101 Main St; (03) 6471 6221. ***Corinna Wilderness Experience:*** patch of civilisation amid rainforest wilderness; Pieman River, Corinna; (03) 6446 1170.

Fishing: Trial Harbour (20 km W) and Granville Harbour (35 km NW) are popular fishing spots. ***Lake Pieman:*** boating and good trout fishing; 42 km NW. ***Cruises:*** tour the Pieman River, leaving from fascinating former goldmining town, Corinna; bookings on (03) 6446 1170; 48 km NW.

See also SOUTH-WEST WILDERNESS, p. 502

TASMANIA

ROAD ATLAS

Inter-City Route Maps

The inter-city route maps and distance charts will help you plan your route between major cities. As well, you can use the maps during your journey, since they provide information on distances between towns along the route, roadside rest areas and road conditions. The table below provides an overview of the routes mapped. The inter-city route maps can be found on pages 536–8.

INTER CITY ROUTES		DISTANCE	TIME
Sydney–Melbourne via Hume Hwy/Fwy	31 M31	879 km	12 hrs
Sydney–Melbourne via Princes Hwy/Fwy	1 A1 M1	1039 km	15 hrs
Sydney–Brisbane via New England Hwy	1 15	995 km	14 hrs
Melbourne–Adelaide via Western & Dukes hwys	M8 A8 M1	729 km	8 hrs
Melbourne–Adelaide via Princes Hwy	M1 A1 B1 M1	911 km	11 hrs
Melbourne–Brisbane via Newell Hwy	M31 A39 39 A39 A2	1676 km	20 hrs
Darwin–Adelaide via Stuart Hwy	1 87 A87 A1	3026 km	31 hrs
Adelaide–Perth via Eyre & Great Eastern hwys	A1 1 94	2700 km	32 hrs
Adelaide–Sydney via Sturt & Hume hwys	A20 20 31	1415 km	19 hrs
Perth–Darwin via Great Northern Hwy	95 1	4032 km	46 hrs
Sydney–Brisbane via Pacific Hwy	1 1 M1	936 km	14 hrs
Brisbane–Darwin via Warrego Hwy	A2 66 87 1	3406 km	39 hrs
Brisbane–Cairns via Bruce Hwy	M1 A1	1703 km	20 hrs
Hobart–Launceston via Midland Hwy	1	197 km	3 hrs
Hobart–Devonport via Midland & Bass hwys	1 B52 1	279 km	4 hrs

Freeway, with toll
M31 31 Highway, sealed, with National Highway Route Marker
A1 1 Highway, sealed, with National Route Marker
5 Highway, sealed, with Metroad Route Marker
Highway, unsealed
C141 26 Main road, sealed, with State Route Marker
Main road, unsealed
Connector road, on central city maps only
Other road, with traffic direction, on central city maps only
Other road, sealed
Other road, unsealed
Vehicle track
Walking track
Mall, on central city maps only
Railway, with station
Underground railway, with station
114 Total kilometres between two points
45 Intermediate kilometres
State border
Fruit fly exclusion zone boundary
River, with waterfall
Lake, reservoir
Intermittent lake
Coastline, with reefs and rocks

SYDNEY State capital city
GEELONG Major city / town
Deniliquin Town
Caldwell Other population centres / localities
Rorruwuy Aboriginal community
Karoonda Roadhouse Roadhouse
Nullagong Pastoral station homestead
ESSENDON Suburb, on suburbs maps only
Unley Suburb, on central city maps only
THE TWELVE APOSTLES Place of interest
Airport
Landing ground
Lighthouse
Hill, mountain, peak
Gorge, gap, pass, cave or saddle
Waterhole
Mine site
National park
Other reserve
Aboriginal / Torres Strait Islander land
Other named area
Prohibited area
Text entry in A to Z listing

Maps are in a Lamberts Conformal Conic Projection
Geocentric Datum Australia, 1994 (GDA94)

ARAFURA SEA
TORRES STRAIT
CORAL SEA
GULF OF CARPENTARIA
SOUTH PACIFIC OCEAN
GREAT BARRIER REEF
GREAT DIVIDING RANGE
NORTHERN TERRITORY
QUEENSLAND
SOUTH AUSTRALIA
NEW SOUTH WALES
VICTORIA
TASMANIA
ACT
DARWIN
Howard Springs
Jabiru
Katherine
Ngukurr
Kununurra
Wyndham
Halls Creek
Borroloola
Tennant Creek
Alice Springs
Yulara
Coober Pedy
Woomera
Port Augusta
Whyalla
Port Pirie
Ceduna
Eucla
Port Lincoln
ADELAIDE
Kingscote
Mount Isa
Cloncurry
Normanton
Hughenden
Winton
Longreach
Barcaldine
Emerald
Blackall
Charleville
Cunnamulla
Mossman
Port Douglas
Cairns
Mareeba
Atherton
Innisfail
Tully
Cardwell
Ingham
Townsville
Ayr
Bowen
Airlie Beach
Proserpine
Mackay
Moranbah
Clermont
Blackwater
Rockhampton
Gladstone
Biloela
Moura
Bundaberg
Childers
Hervey Bay
Maryborough
Gympie
Noosa Heads
Maroochydore
Caloundra
Caboolture
BRISBANE
Kingaroy
Nambour
Dalby
Toowoomba
Roma
Miles
Moonie
St George
Goondiwindi
Warwick
Tweed Heads
Murwillumbah
Lismore
Casino
Ballina
Moree
Bourke
Walgett
Narrabri
Glen Innes
Grafton
Coffs Harbour
Armidale
Gunnedah
Tamworth
Port Macquarie
Wilcannia
Broken Hill
Nyngan
Dubbo
Forster-Tuncurry
Singleton
Maitland
Raymond Terrace
Newcastle
Gosford
SYDNEY
Parkes
Orange
Bathurst
Lithgow
Cowra
Wollongong
Goulburn
Wagga Wagga
CANBERRA
Nowra
Ulladulla
Batemans Bay
Cooma
Bega
Merimbula
Renmark
Mildura
Burra
Gawler
Murray Bridge
Hay
Deniliquin
Echuca
Shepparton
Albury
Wodonga
Bendigo
Horsham
Bordertown
Kingston SE
Mount Gambier
Hamilton
Ballarat
MELBOURNE
Portland
Warrnambool
Colac
Geelong
Sale
Traralgon
Bairnsdale
BASS STRAIT
SOUTHERN OCEAN
TASMAN SEA
GREAT AUSTRALIAN BIGHT
Smithton
Burnie
George Town
Devonport
Launceston
Queenstown
Swansea
New Norfolk
HOBART
TANAMI DESERT
SIMPSON DESERT
GREAT VICTORIA DESERT
NULLARBOR PLAIN
FLINDERS RANGES
GULF SAVANNAH
CHANNEL COUNTRY
ARNHEM LAND PLATEAU
Joseph Bonaparte Gulf
Van Diemen Gulf
Spencer Gulf
TROPIC OF CAPRICORN

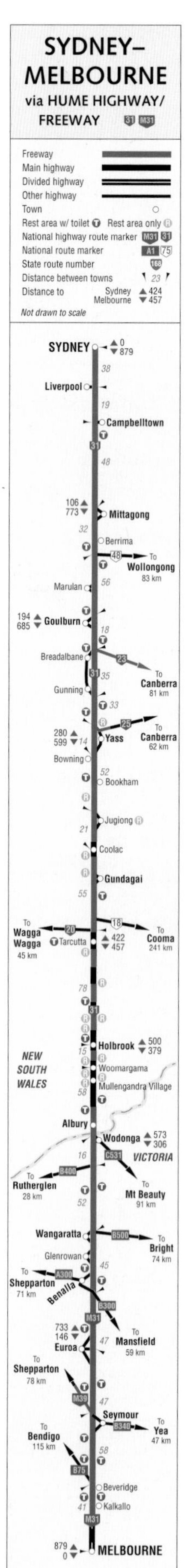
SYDNEY–MELBOURNE
via HUME HIGHWAY/ FREEWAY
Freeway
Main highway
Divided highway
Other highway
Town
Rest area w/ toilet
Rest area only
National highway route marker
National route marker
State route number
Distance between towns
Distance to Sydney Melbourne
Not drawn to scale
SYDNEY
Liverpool
Campbelltown
Mittagong
Berrima
To Wollongong 83 km
Marulan
Goulburn
Breadalbane
To Canberra 81 km
Gunning
Yass
To Canberra 62 km
Bowning
Bookham
Jugiong
Coolac
Gundagai
To Wagga Wagga 45 km
Tarcutta
To Cooma 241 km
Holbrook
Woomargama
Mullengandra Village
NEW SOUTH WALES
Albury
Wodonga
VICTORIA
To Rutherglen 28 km
To Mt Beauty 91 km
Wangaratta
To Bright 74 km
Glenrowan
To Shepparton 71 km
Benalla
Euroa
To Mansfield 59 km
To Shepparton 78 km
Seymour
To Yea 47 km
To Bendigo 115 km
Beveridge
Kalkallo
MELBOURNE

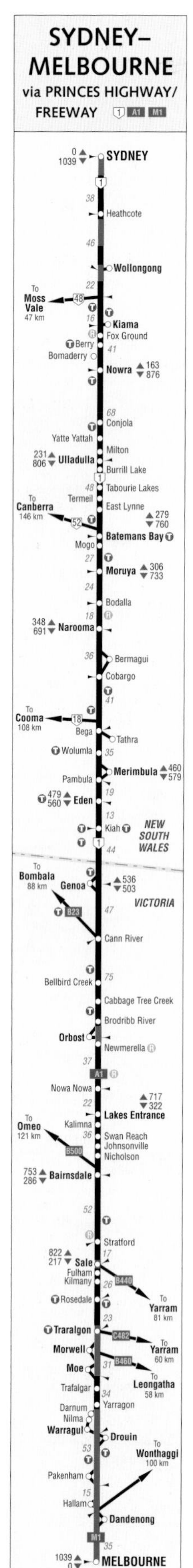
SYDNEY–MELBOURNE
via PRINCES HIGHWAY/ FREEWAY
SYDNEY
Heathcote
Wollongong
To Moss Vale 47 km
Kiama
Fox Ground
Berry
Bomaderry
Nowra
Conjola
Yatte Yattah
Milton
Ulladulla
Burrill Lake
Tabourie Lakes
Termeil
East Lynne
To Canberra 146 km
Batemans Bay
Mogo
Moruya
Bodalla
Narooma
Bermagui
Cobargo
To Cooma 108 km
Bega
Tathra
Wolumla
Merimbula
Pambula
Eden
Kiah
NEW SOUTH WALES
To Bombala 88 km
Genoa
VICTORIA
Cann River
Bellbird Creek
Cabbage Tree Creek
Brodribb River
Orbost
Newmerella
Nowa Nowa
Lakes Entrance
To Omeo 121 km
Kalimna
Swan Reach
Johnsonville
Nicholson
Bairnsdale
Stratford
Sale
Fulham
Kilmany
Rosedale
To Yarram 81 km
Traralgon
Morwell
To Yarram 60 km
Moe
To Leongatha 58 km
Trafalgar
Yarragon
Darnum
Nilma
Warragul
Drouin
To Wonthaggi 100 km
Pakenham
Hallam
Dandenong
MELBOURNE

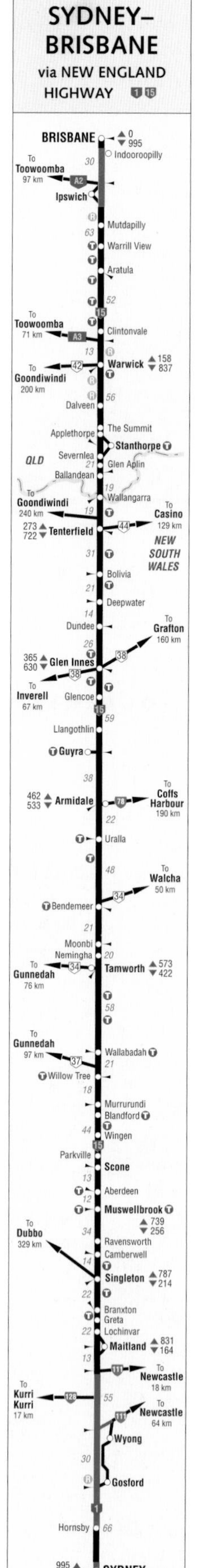
SYDNEY–BRISBANE
via NEW ENGLAND HIGHWAY
BRISBANE
Indooroopilly
To Toowoomba 97 km
Ipswich
Mutdapilly
Warrill View
Aratula
To Toowoomba 71 km
Clintonvale
To Goondiwindi 200 km
Warwick
Dalveen
Applethorpe
The Summit
Stanthorpe
QLD
Severnlea
Glen Aplin
Ballandean
Wallangarra
To Goondiwindi 240 km
To Casino 129 km
Tenterfield
NEW SOUTH WALES
Bolivia
Deepwater
Dundee
To Grafton 160 km
Glen Innes
To Inverell 67 km
Glencoe
Llangothlin
Guyra
Armidale
To Coffs Harbour 190 km
Uralla
To Walcha 50 km
Bendemeer
Moonbi
Nemingha
To Gunnedah 76 km
Tamworth
To Gunnedah 97 km
Wallabadah
Willow Tree
Murrurundi
Blandford
Wingen
Parkville
Scone
Aberdeen
Muswellbrook
To Dubbo 329 km
Ravensworth
Camberwell
Singleton
Branxton
Greta
Lochinvar
Maitland
To Newcastle 18 km
To Kurri Kurri 17 km
To Newcastle 64 km
Wyong
Gosford
Hornsby
SYDNEY

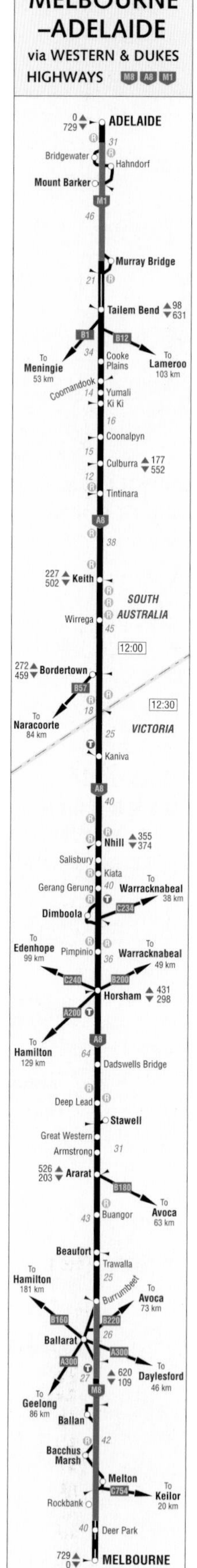
MELBOURNE –ADELAIDE
via WESTERN & DUKES HIGHWAYS
ADELAIDE
Bridgewater
Hahndorf
Mount Barker
Murray Bridge
Tailem Bend
To Meningie 53 km
Cooke Plains
To Lameroo 103 km
Coomandook
Yumali
Ki Ki
Coonalpyn
Culburra
Tintinara
Keith
SOUTH AUSTRALIA
Wirrega
12:00
Bordertown
To Naracoorte 84 km
12:30
VICTORIA
Kaniva
Nhill
Salisbury
Kiata
Gerang Gerung
To Warracknabeal 38 km
Dimboola
To Edenhope 99 km
Pimpinio
To Warracknabeal 49 km
Horsham
To Hamilton 129 km
Dadswells Bridge
Deep Lead
Stawell
Great Western
Armstrong
Ararat
To Avoca 63 km
Buangor
Beaufort
Trawalla
To Hamilton 181 km
Burrumbeet
To Avoca 73 km
Ballarat
To Daylesford 46 km
To Geelong 86 km
Ballan
Bacchus Marsh
Melton
Rockbank
To Keilor 20 km
Deer Park
MELBOURNE

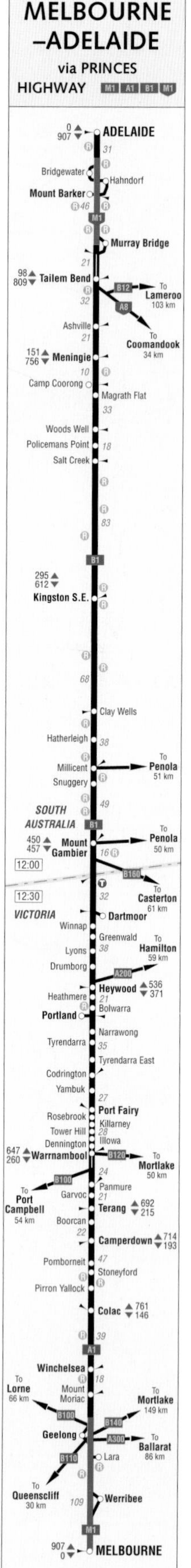
MELBOURNE –ADELAIDE
via PRINCES HIGHWAY
ADELAIDE
Bridgewater
Hahndorf
Mount Barker
Murray Bridge
Tailem Bend
To Lameroo 103 km
To Coomandook 34 km
Ashville
Meningie
Camp Coorong
Magrath Flat
Woods Well
Policemans Point
Salt Creek
Kingston S.E.
Clay Wells
Hatherleigh
Millicent
To Penola 51 km
Snuggery
SOUTH AUSTRALIA
Mount Gambier
To Penola 50 km
12:00
To Casterton 61 km
12:30
Dartmoor
VICTORIA
Winnap
Greenwald
Lyons
To Hamilton 59 km
Drumborg
Heywood
Heathmere
Bolwarra
Portland
Narrawong
Tyrendarra
Tyrendarra East
Codrington
Yambuk
Rosebrook
Port Fairy
Killarney
Tower Hill
Dennington
Illowa
Warrnambool
To Mortlake 50 km
To Port Campbell 54 km
Garvoc
Panmure
Terang
Boorcan
Camperdown
Pomborneit
Stoneyford
Pirron Yallock
Colac
Winchelsea
To Lorne 66 km
Mount Moriac
To Mortlake 149 km
Geelong
To Ballarat 86 km
Lara
To Queenscliff 30 km
Werribee
MELBOURNE

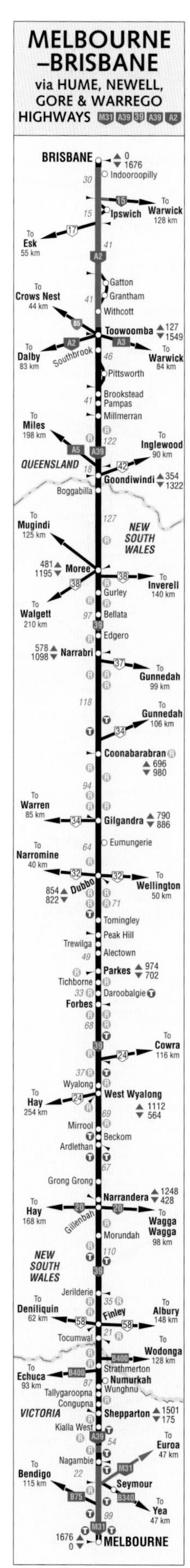
MELBOURNE –BRISBANE
via HUME, NEWELL, GORE & WARREGO HIGHWAYS
BRISBANE
Indooroopilly
Ipswich
To Warwick 128 km
To Esk 55 km
Gatton
Grantham
Withcott
To Crows Nest 44 km
Toowoomba
To Dalby 83 km
Southbrook
To Warwick 84 km
Pittsworth
Brookstead
Pampas
Millmerran
To Miles 198 km
To Inglewood 90 km
QUEENSLAND
Goondiwindi
Boggabilla
To Mugindi 125 km
NEW SOUTH WALES
Moree
To Inverell 140 km
Gurley
To Walgett 210 km
Bellata
Edgero
Narrabri
To Gunnedah 99 km
To Gunnedah 106 km
Coonabarabran
To Warren 85 km
Gilgandra
Eumungerie
To Narromine 40 km
Dubbo
To Wellington 50 km
Tomingley
Peak Hill
Trewilga
Alectown
Parkes
Tichborne
Daroobalgie
Forbes
To Cowra 116 km
Wyalong
West Wyalong
To Hay 254 km
Mirrool
Beckom
Ardlethan
Grong Grong
Narrandera
To Hay 168 km
Gillenbah
To Wagga Wagga 98 km
Morundah
NEW SOUTH WALES
Jerilderie
To Deniliquin 62 km
Finley
To Albury 148 km
Tocumwal
To Wodonga 128 km
To Echuca 93 km
Strathmerton
Numurkah
Wunghnu
Tallygaroopna
Congupna
VICTORIA
Shepparton
Kialla West
To Euroa 47 km
To Bendigo 115 km
Nagambie
Seymour
To Yea 47 km
MELBOURNE

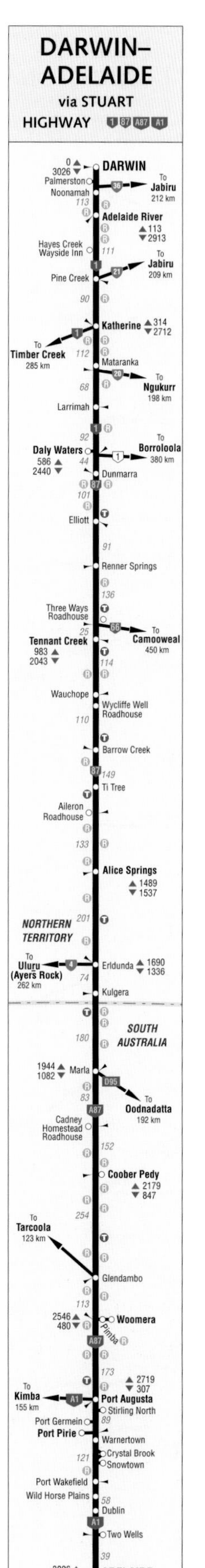
DARWIN– ADELAIDE
via STUART HIGHWAY
DARWIN
Palmerston
Noonamah
To Jabiru 212 km
Adelaide River
Hayes Creek Wayside Inn
To Jabiru 209 km
Pine Creek
Katherine
To Timber Creek 285 km
Mataranka
To Ngukurr 198 km
Larrimah
Daly Waters
To Borroloola 380 km
Dunmarra
Elliott
Renner Springs
Three Ways Roadhouse
To Camooweal 450 km
Tennant Creek
Wauchope
Wycliffe Well Roadhouse
Barrow Creek
Ti Tree
Aileron Roadhouse
Alice Springs
NORTHERN TERRITORY
To Uluru (Ayers Rock) 262 km
Erldunda
Kulgera
SOUTH AUSTRALIA
Marla
To Oodnadatta 192 km
Cadney Homestead Roadhouse
Coober Pedy
To Tarcoola 123 km
Glendambo
Woomera
Pimba
To Kimba 155 km
Port Augusta
Stirling North
Port Germein
Port Pirie
Warnertown
Crystal Brook
Snowtown
Port Wakefield
Wild Horse Plains
Dublin
Two Wells
ADELAIDE

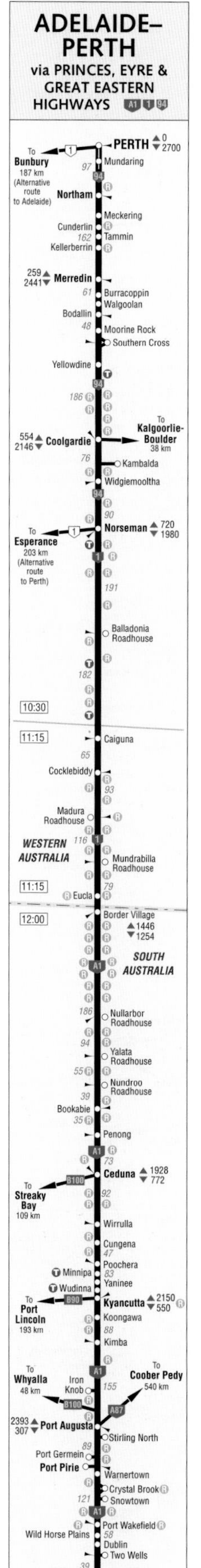
ADELAIDE– PERTH
via PRINCES, EYRE & GREAT EASTERN HIGHWAYS
PERTH
To Bunbury 187 km (Alternative route to Adelaide)
Mundaring
Northam
Meckering
Cunderlin
Tammin
Kellerberrin
Merredin
Burracoppin
Walgoolan
Bodallin
Moorine Rock
Southern Cross
Yellowdine
Coolgardie
To Kalgoorlie-Boulder 38 km
Kambalda
Widgiemooltha
Norseman
To Esperance 203 km (Alternative route to Perth)
Balladonia Roadhouse
10:30
11:15
Caiguna
Cocklebiddy
Madura Roadhouse
WESTERN AUSTRALIA
Mundrabilla Roadhouse
11:15
Eucla
12:00
Border Village
SOUTH AUSTRALIA
Nullarbor Roadhouse
Yalata Roadhouse
Nundroo Roadhouse
Bookabie
Penong
Ceduna
To Streaky Bay 109 km
Wirrulla
Cungena
Poochera
Minnipa
Wudinna
Yaninee
To Port Lincoln 193 km
Kyancutta
Koongawa
Kimba
To Whyalla 48 km
Iron Knob
To Coober Pedy 540 km
Port Augusta
Stirling North
Port Germein
Port Pirie
Warnertown
Crystal Brook
Snowtown
Port Wakefield
Wild Horse Plains
Dublin
Two Wells
ADELAIDE

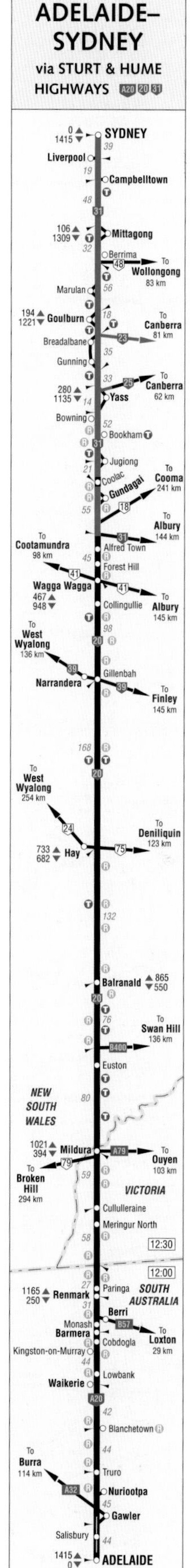
ADELAIDE– SYDNEY
via STURT & HUME HIGHWAYS
SYDNEY
Liverpool
Campbelltown
Mittagong
Berrima
To Wollongong 83 km
Marulan
Goulburn
To Canberra 81 km
Breadalbane
Gunning
To Canberra 62 km
Yass
Bowning
Bookham
Jugiong
Coolac
Gundagai
To Cooma 241 km
To Albury 144 km
Alfred Town
To Cootamundra 98 km
Forest Hill
Wagga Wagga
Collingullie
To Albury 145 km
To West Wyalong 136 km
Narrandera
Gillenbah
To Finley 145 km
To West Wyalong 254 km
To Deniliquin 123 km
Hay
Balranald
To Swan Hill 136 km
Euston
NEW SOUTH WALES
Mildura
To Ouyen 103 km
To Broken Hill 294 km
VICTORIA
Cullulleraine
Meringur North
12:30
12:00
Renmark
Paringa
SOUTH AUSTRALIA
Berri
Monash
Barmera
To Loxton 29 km
Cobdogla
Kingston-on-Murray
Waikerie
Lowbank
Blanchetown
To Burra 114 km
Truro
Nuriootpa
Gawler
Salisbury
ADELAIDE

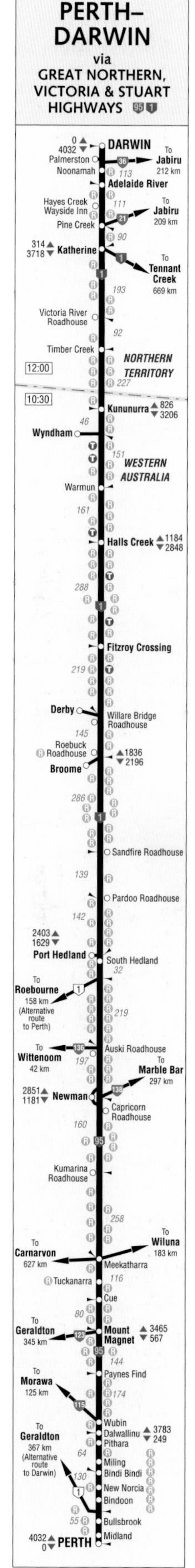
PERTH– DARWIN
via GREAT NORTHERN, VICTORIA & STUART HIGHWAYS
DARWIN
Palmerston
Noonamah
To Jabiru 212 km
Adelaide River
Hayes Creek Wayside Inn
To Jabiru 209 km
Pine Creek
Katherine
To Tennant Creek 669 km
Victoria River Roadhouse
Timber Creek
NORTHERN TERRITORY
12:00
10:30
Kununurra
Wyndham
WESTERN AUSTRALIA
Warmun
Halls Creek
Fitzroy Crossing
Derby
Willare Bridge Roadhouse
Roebuck Roadhouse
Broome
Sandfire Roadhouse
Pardoo Roadhouse
Port Hedland
South Hedland
To Roebourne 158 km (Alternative route to Perth)
Auski Roadhouse
To Wittenoom 42 km
To Marble Bar 297 km
Newman
Capricorn Roadhouse
Kumarina Roadhouse
To Carnarvon 627 km
To Wiluna 183 km
Meekatharra
Tuckanarra
Cue
To Geraldton 345 km
Mount Magnet
Paynes Find
To Morawa 125 km
Wubin
Dalwallinu
Pithara
To Geraldton 367 km (Alternative route to Darwin)
Miling
Bindi Bindi
New Norcia
Bindoon
Bullsbrook
Midland
PERTH

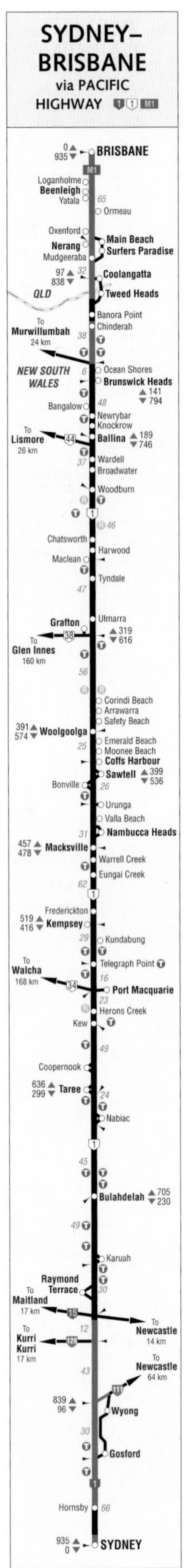
SYDNEY–BRISBANE
via PACIFIC HIGHWAY
0 935 BRISBANE
Loganholme
Beenleigh
Yatala
65
Ormeau
Oxenford
Main Beach
Nerang
Surfers Paradise
Mudgeeraba
97 32 838
Coolangatta
QLD
Tweed Heads
Banora Point
Chinderah
To Murwillumbah 24 km
38
NEW SOUTH WALES
6
Ocean Shores
Brunswick Heads
141 794
Bangalow
48
Newrybar
Knockrow
To Lismore 26 km
Ballina 189 746
Wardell
37
Broadwater
Woodburn
46
Chatsworth
Harwood
Maclean
Tyndale
47
Ulmarra
Grafton
319 616
To Glen Innes 160 km
56
Corindi Beach
Arrawarra
Safety Beach
391 574 Woolgoolga
Emerald Beach
25
Moonee Beach
Coffs Harbour
Sawtell 399 536
Bonville
26
Urunga
Valla Beach
31
Nambucca Heads
457 478 Macksville
Warrell Creek
Eungai Creek
62
Frederickton
519 416 Kempsey
29
Kundabung
Telegraph Point
To Walcha 168 km
16
Port Macquarie
23
Herons Creek
Kew
49
Coopernook
636 299 Taree
24
Nabiac
45
Bulahdelah 705 230
49
Karuah
Raymond Terrace
To Maitland 17 km
30
12
To Newcastle 14 km
To Kurri Kurri 17 km
43
To Newcastle 64 km
839 96
Wyong
30
Gosford
Hornsby
66
935 0 SYDNEY

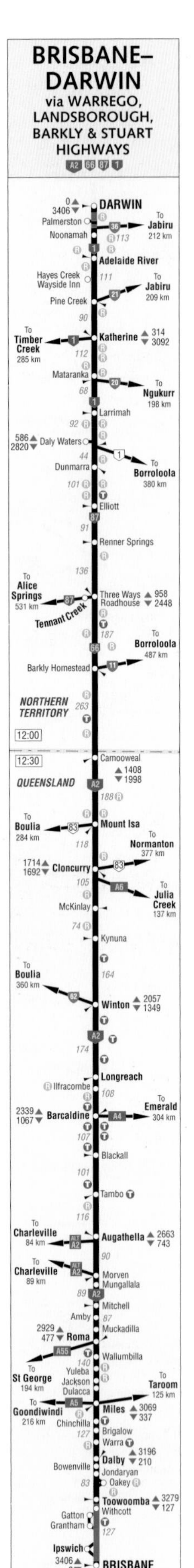
BRISBANE–DARWIN
via WARREGO, LANDSBOROUGH, BARKLY & STUART HIGHWAYS
0 3406 DARWIN
Palmerston
Noonamah
113
To Jabiru 212 km
Adelaide River
Hayes Creek Wayside Inn
111
Pine Creek
To Jabiru 209 km
90
To Timber Creek 285 km
Katherine 314 3092
112
Mataranka
To Ngukurr 198 km
68
Larrimah
92
586 2820 Daly Waters
44
Dunmarra
To Borroloola 380 km
101
Elliott
91
Renner Springs
136
To Alice Springs 531 km
Tennant Creek
Three Ways Roadhouse 958 2448
187
To Borroloola 487 km
Barkly Homestead
NORTHERN TERRITORY
263
12:00
12:30
Camooweal
QUEENSLAND
1408 1998
188
Mount Isa
To Boulia 284 km
118
To Normanton 377 km
1714 1692 Cloncurry
105
To Julia Creek 137 km
McKinlay
74
Kynuna
To Boulia 360 km
164
Winton 2057 1349
174
Longreach
Ilfracombe
108
2339 1067 Barcaldine
To Emerald 304 km
107
Blackall
101
Tambo
116
To Charleville 84 km
Augathella 2663 743
90
To Charleville 89 km
Morven
Mungallala
Mitchell
Amby
87
Muckadilla
2929 477 Roma
Wallumbilla
140
Yuleba
Jackson
Dulacca
To St George 194 km
To Taroom 125 km
To Goondiwindi 216 km
Miles 3069 337
Chinchilla
127
Brigalow
Warra
Dalby 3196 210
Bowenville
Jondaryan
83
Oakey
Toowoomba 3279 127
Withcott
Gatton
Grantham
127
Ipswich
3406 0 BRISBANE

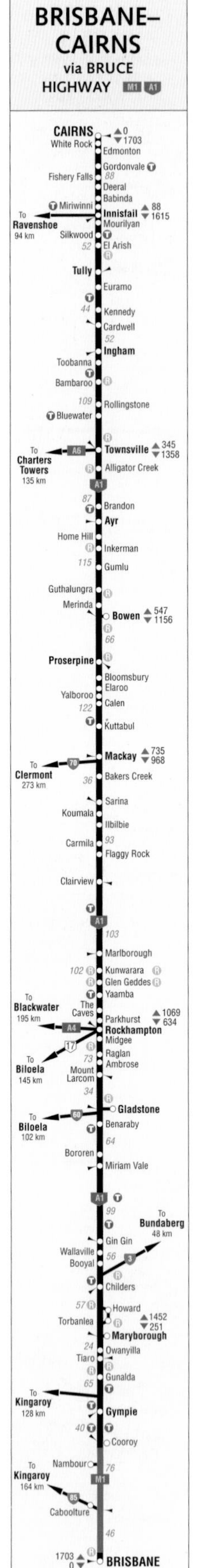
BRISBANE–CAIRNS
via BRUCE HIGHWAY
CAIRNS 0 1703
White Rock
Edmonton
Gordonvale
Fishery Falls
88
Deeral
Babinda
Miriwinni
Innisfail 88 1615
Mourilyan
To Ravenshoe 94 km
Silkwood
52
El Arish
Tully
Euramo
44
Kennedy
Cardwell
52
Ingham
Toobanna
Bambaroo
109
Rollingstone
Bluewater
To Charters Towers 135 km
Townsville 345 1358
Alligator Creek
87
Brandon
Ayr
Home Hill
Inkerman
115
Gumlu
Guthalungra
Merinda
Bowen 547 1156
66
Proserpine
Bloomsbury
Elaroo
Yalboroo
122
Calen
Kuttabul
To Clermont 273 km
Mackay 735 968
36
Bakers Creek
Sarina
Koumala
Ilbilbie
Carmila
93
Flaggy Rock
Clairview
103
Marlborough
102
Kunwarara
Glen Geddes
Yaamba
To Blackwater 195 km
The Caves
Parkhurst 1069 634
Rockhampton
Midgee
To Biloela 145 km
73
Raglan
Ambrose
Mount Larcom
34
Gladstone
To Biloela 102 km
Benaraby
64
Bororen
Miriam Vale
99
To Bundaberg 48 km
Gin Gin
Wallaville
56
Booyal
Childers
57
Howard
Torbanlea
1452 251
Maryborough
24
Owanyilla
Tiaro
Gunalda
65
To Kingaroy 128 km
Gympie
40
Cooroy
Nambour
76
To Kingaroy 164 km
Caboolture
46
1703 0 BRISBANE

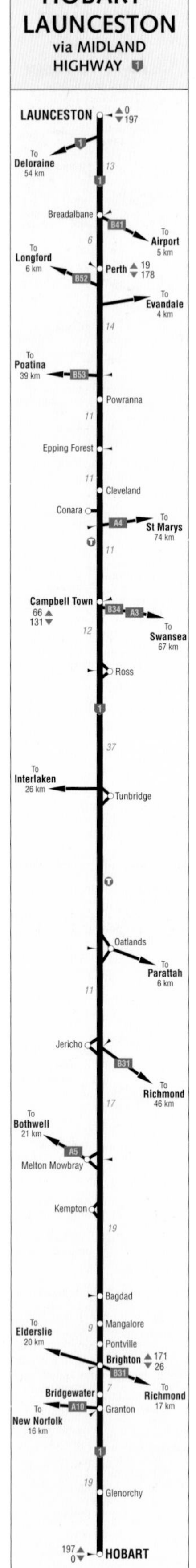
HOBART–LAUNCESTON
via MIDLAND HIGHWAY
LAUNCESTON 0 197
To Deloraine 54 km
13
Breadalbane
To Airport 5 km
6
To Longford 6 km
Perth 19 178
To Evandale 4 km
14
To Poatina 39 km
Powranna
11
Epping Forest
11
Cleveland
Conara
To St Marys 74 km
11
Campbell Town
66 131
12
To Swansea 67 km
Ross
37
To Interlaken 26 km
Tunbridge
Oatlands
To Parattah 6 km
11
Jericho
To Richmond 46 km
17
To Bothwell 21 km
Melton Mowbray
Kempton
19
Bagdad
To Elderslie 20 km
9
Mangalore
Pontville
Brighton 171 26
7
Bridgewater
Granton
To Richmond 17 km
To New Norfolk 16 km
19
Glenorchy
197 0 HOBART

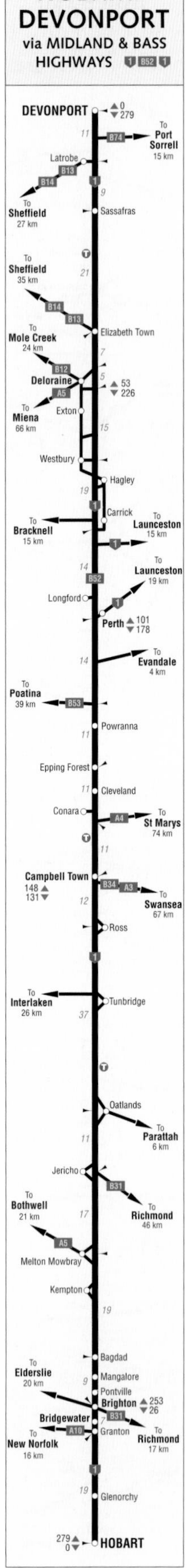
HOBART–DEVONPORT
via MIDLAND & BASS HIGHWAYS
DEVONPORT 0 279
11
To Port Sorell 15 km
Latrobe
9
To Sheffield 27 km
Sassafras
To Sheffield 35 km
21
Elizabeth Town
To Mole Creek 24 km
7
Deloraine
5
53 226
To Miena 66 km
Exton
15
Westbury
Hagley
19
Carrick
To Bracknell 15 km
To Launceston 15 km
14
To Launceston 19 km
Longford
Perth 101 178
14
To Evandale 4 km
To Poatina 39 km
11
Powranna
Epping Forest
11
Cleveland
Conara
To St Marys 74 km
11
Campbell Town
148 131
12
To Swansea 67 km
Ross
To Interlaken 26 km
37
Tunbridge
Oatlands
To Parattah 6 km
11
Jericho
To Bothwell 21 km
17
To Richmond 46 km
Melton Mowbray
Kempton
19
Bagdad
To Elderslie 20 km
9
Mangalore
Pontville
Brighton 253 26
Bridgewater
7
Granton
To New Norfolk 16 km
To Richmond 17 km
19
Glenorchy
279 0 HOBART

MAPS

NEW SOUTH WALES and AUSTRALIAN CAPITAL TERRITORY

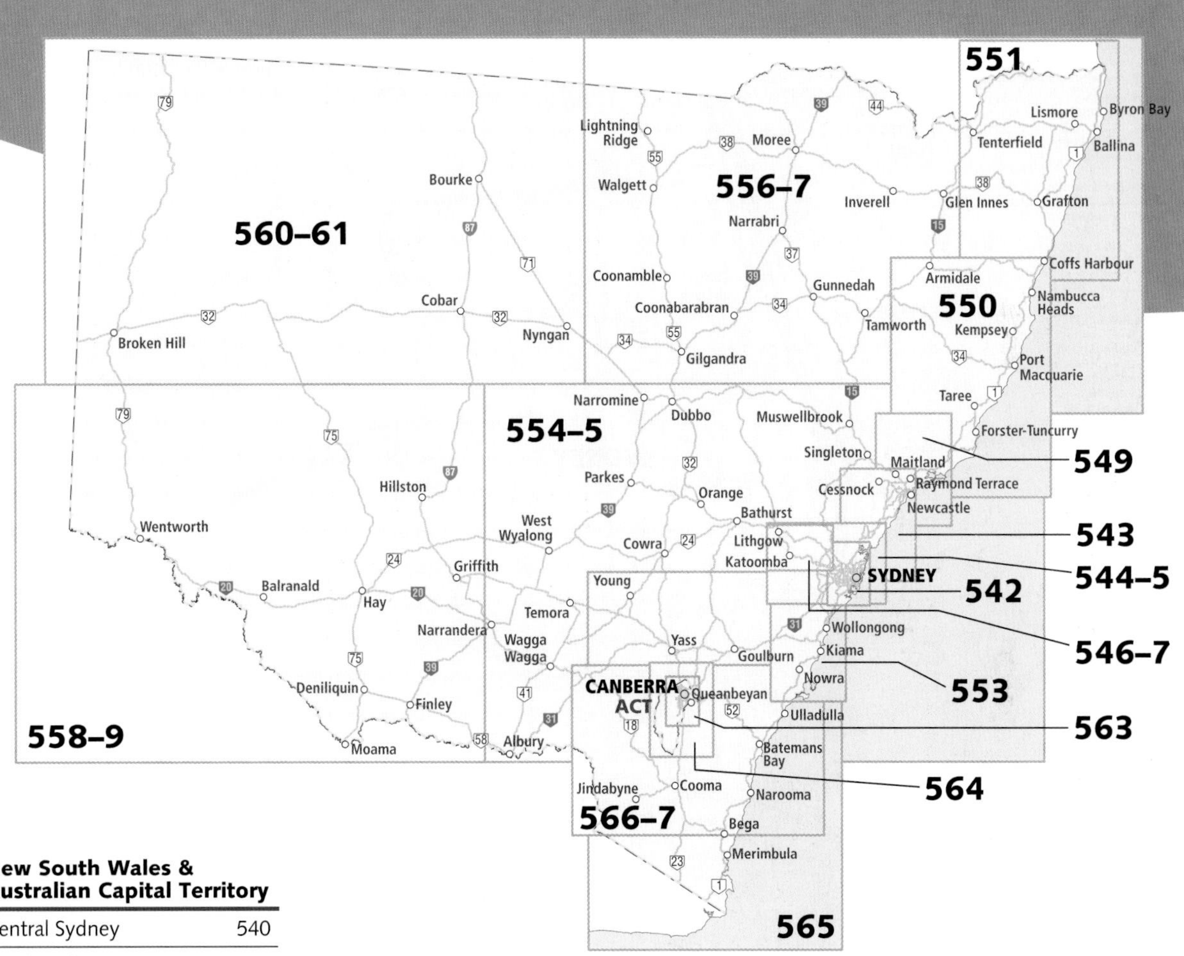

New South Wales & Australian Capital Territory

INTER-CITY ROUTES		DISTANCE
Sydney–Melbourne via Hume Hwy/Fwy	31 M31	879 km
Sydney–Melbourne via Princes Hwy/Fwy	1 A1 M1	1039 km
Sydney–Brisbane via New England Hwy	1 15	995 km
Sydney–Brisbane via Pacific Hwy	1 1 M1	935 km
Sydney–Adelaide via Hume & Sturt hwys	31 20 A20	1415 km

0 0.25 0.5 0.75 1 km

Accommodation
- Establishment Hotel 1 D5
- Four Points by Sheraton 2 C7
- Four Seasons Hotel Sydney 3 D4
- InterContinental Sydney 4 E5
- Novotel Sydney on Darling Harbour 5 A8
- The Observatory Hotel 6 C4
- Park Hyatt Sydney 7 D2
- Sebel Pier One Sydney 8 D2
- Shangri-La Hotel Sydney 9 D4
- Sydney Central YHA Hostel 10 D11
- Sydney Harbour Marriott Hotel 11 E5
- The Westin Sydney 12 D7
- The York Suites & Residences 13 C5

Note: Only a sample range of accommodation is listed; inclusion is not necessarily a recommendation.

General Information
- Cadmans Cottage (NPWS Info. Centre) 14 D3
- Central Railway Station 15 D12
- City Central Police Station 16 C9
- Coach Tour Departures 17 E3
- General Post Office 18 D6
- Interstate & Country Coach Terminal 19 D12
- Motoring Organisation (NRMA) 20 D7
- Qantas Travel Centre 21 D5
- Sydney Ferries 22 E4
- Sydney Visitor Centre, Darling Harbour 23 B9
- Sydney Visitor Centre, The Rocks 24 D3

Places of Interest
- The Art Gallery of NSW 25 G7
- Australian Museum 26 F9
- Australian National Maritime Museum 27 B7
- Campbell's Storehouse 28 D3
- Chinatown 29 C10
- Chinese Garden of Friendship 30 C10
- Government House 31 F4
- Hyde Park Barracks Museum 32 F7
- IMAX Theatre 33 C9
- Justice & Police Museum 34 E4
- Museum of Contemporary Art 35 D4
- Museum of Sydney 36 E5
- Parliament of NSW 37 F6
- Powerhouse Museum 38 B10
- Queen Victoria Building 39 D8
- Royal Botanic Gardens 40 G6
- St Mary's Cathedral 41 F8
- Star City (casino, hotel & theatres) 42 A7
- State Library of NSW 43 F6
- Susannah Place Museum 44 D4
- Sydney Aquarium 45 C7
- Sydney Harbour Bridge 46 E1
- Sydney Harbour Bridge Visitor Centre 47 D3
- Sydney Observatory Museum 48
- Sydney Opera House 49 F3
- Sydney Tower 50 E7
- Sydney Town Hall 51 D8
- Victoria Barracks 52 H12

Joins map 541
Joins map 542
Joins map 542
Joins map 542

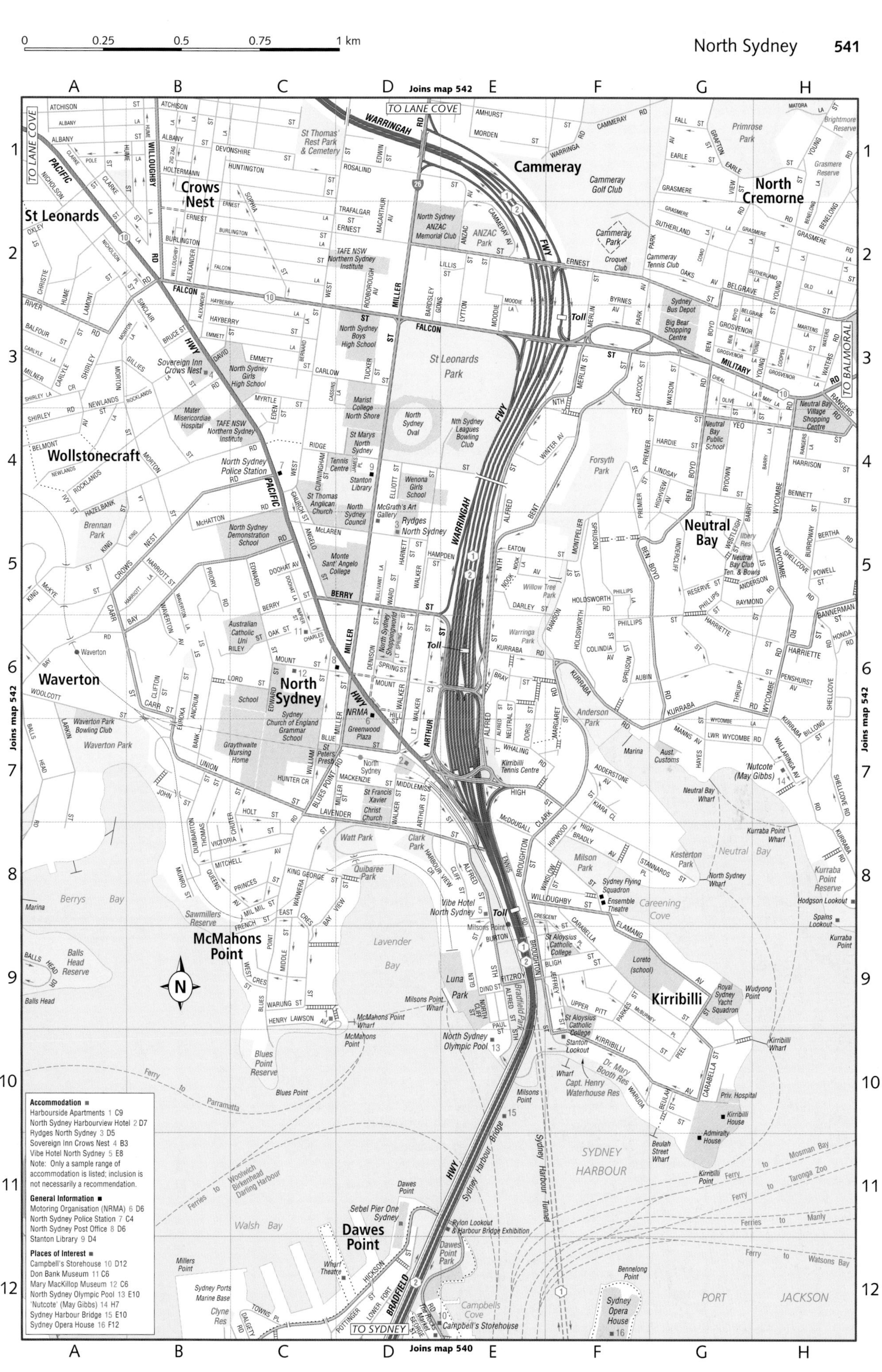
0 0.25 0.5 0.75 1 km
Joins map 542
TO LANE COVE
Joins map 540
TO SYDNEY
TO BALMORAL
St Leonards
Crows Nest
Cammeray
North Cremorne
Wollstonecraft
Waverton
North Sydney
Neutral Bay
McMahons Point
Kirribilli
Dawes Point
St Leonards Park
Berrys Bay
Lavender Bay
Neutral Bay
Careening Cove
Walsh Bay
SYDNEY HARBOUR
PORT JACKSON
Sydney Harbour Bridge
Sydney Harbour Tunnel
Accommodation
Harbourside Apartments 1 C9
North Sydney Harbourview Hotel 2 D7
Rydges North Sydney 3 D5
Sovereign Inn Crows Nest 4 B3
Vibe Hotel North Sydney 5 E8
Note: Only a sample range of accommodation is listed; inclusion is not necessarily a recommendation.
General Information
Motoring Organisation (NRMA) 6 D6
North Sydney Police Station 7 C4
North Sydney Post Office 8 D6
Stanton Library 9 D4
Places of Interest
Campbell's Storehouse 10 D12
Don Bank Museum 11 C6
Mary MacKillop Museum 12 C6
North Sydney Olympic Pool 13 E10
'Nutcote' (May Gibbs) 14 H7
Sydney Harbour Bridge 15 E10
Sydney Opera House 16 F12

0 2 4 6 8 10 km

Joins map 545

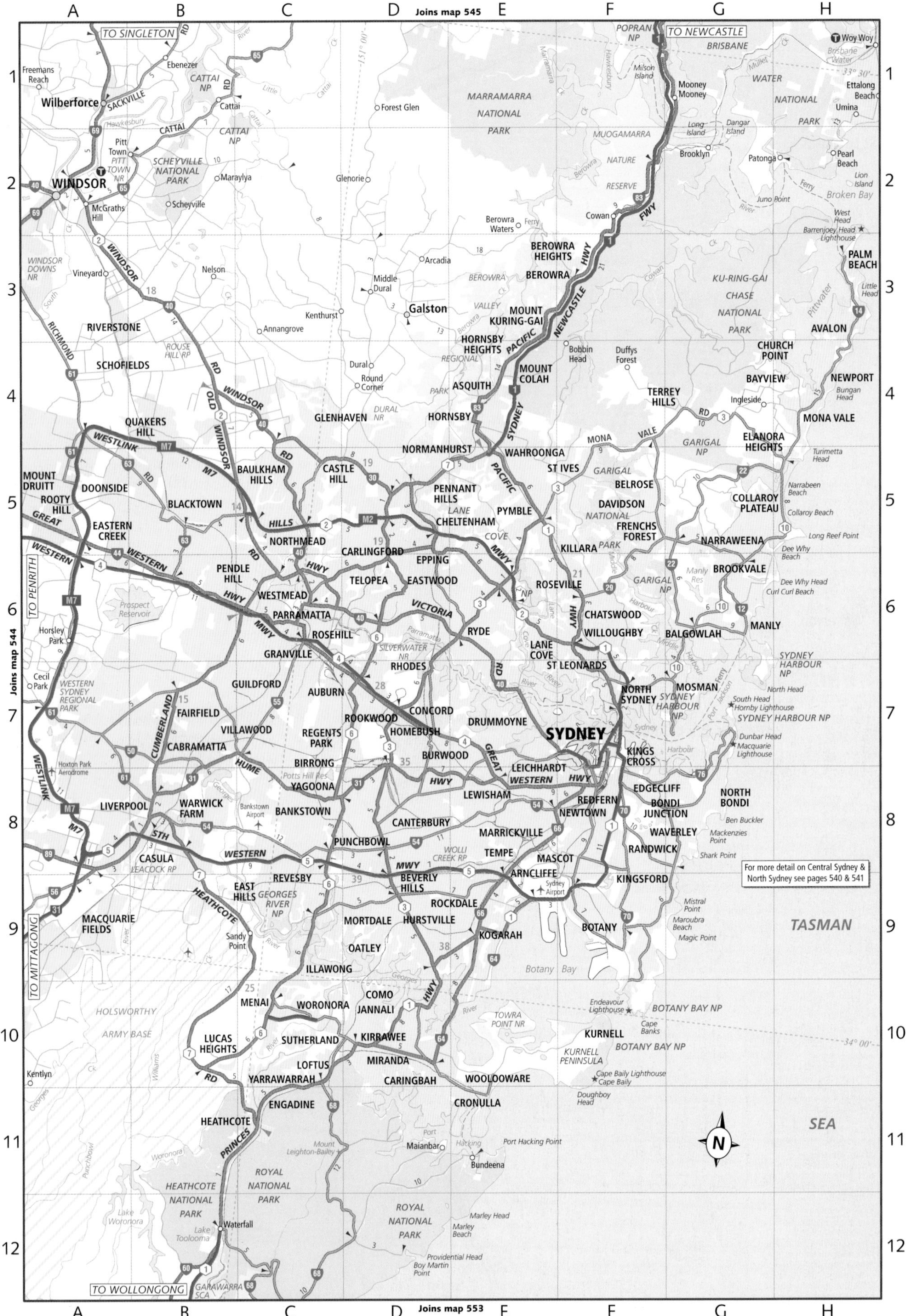

Joins map 553

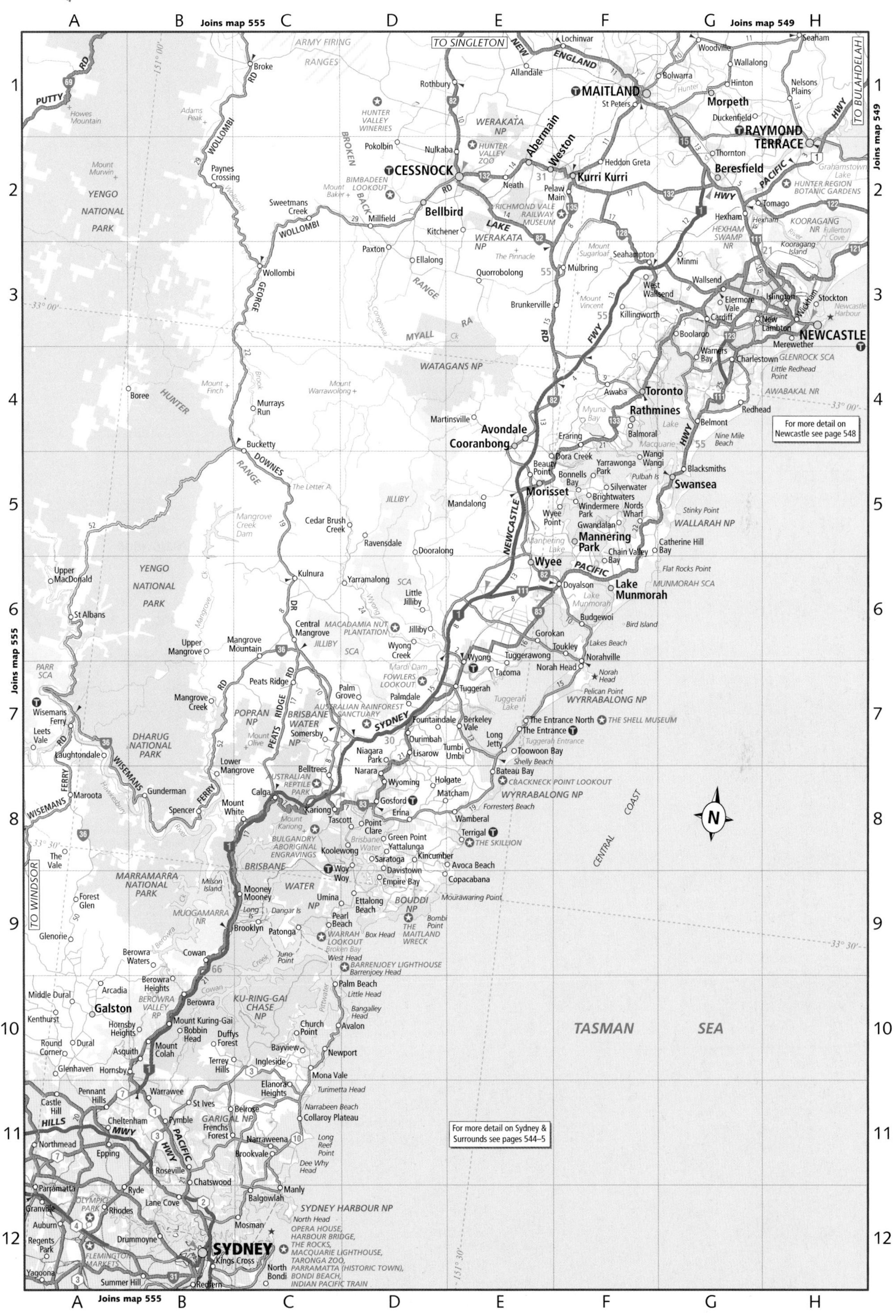
0 5 10 15 20 km
Joins map 555
Joins map 549
Joins map 555
Joins map 549
TO SINGLETON
TO BULAHDELAH
TO WINDSOR
Lochinvar
NEW ENGLAND
Seaham
Woodville
Wallalong
Allandale
Rothbury
Bolwarra
Hinton
Nelsons Plains
MAITLAND
St Peters
Morpeth
Duckenfield
RAYMOND TERRACE
HUNTER VALLEY WINERIES
WERAKATA NP
Abermain
Weston
Thornton
Pokolbin
Nulkaba
HUNTER VALLEY ZOO
Heddon Greta
Beresfield
CESSNOCK
Kurri Kurri
Neath
Pelaw Main
Grahamstown Lake
HUNTER REGION BOTANIC GARDENS
Tomago
BIMBADEEN LOOKOUT
Mount Baker
BROKEN BACK RD
RICHMOND VALE RAILWAY MUSEUM
Hexham
HEXHAM SWAMP NR
KOORAGANG NR
Kooragang Island
Fullerton Cove
Sweetmans Creek
Bellbird
Millfield
Kitchener
LAKE
WERAKATA NP
The Pinnacle
Paxton
Ellalong
Mount Sugarloaf
Seahampton
Minmi
Quorrobolong
Mulbring
Wallsend
West Wallsend
Mount Vincent
Brunkerville
Killingworth
Elermore Vale
Islington
Stockton
Newcastle Harbour
Cardiff
New Lambton
Wickham
NEWCASTLE
Boolaroo
Merewether
Warners Bay
Charlestown
GLENROCK SCA
Little Redhead Point
AWABAKAL NR
Redhead
Awaba
Toronto
Rathmines
Belmont
Nine Mile Beach
For more detail on Newcastle see page 548
MYALL
WATAGANS NP
Mount Warrawolong
Mount Finch
Martinsville
Avondale
Cooranbong
Myuna Bay
Eraring
Balmoral
Lake Macquarie
Dora Creek
Wangi Wangi
Yarrawonga Park
Blacksmiths
Swansea
Pulbah Is
Beauty Point
Bonnells Bay
Silverwater
Morisset
Brightwaters
Windermere Park
Nords Wharf
Stinky Point
WALLARAH NP
Wyee Point
Gwandalan
Mannering Park
Mannering Lake
Catherine Hill Bay
Chain Valley Bay
Wyee
PACIFIC
Flat Rocks Point
Doyalson
Lake Munmorah
MUNMORAH SCA
Budgewoi
Bird Island
Gorokan
Toukley
Lakes Beach
Norahville
Norah Head
Tuggerawong
Wyong
Tacoma
Tuggerah
Tuggerah Lake
Pelican Point
WYRRABALONG NP
The Entrance North
THE SHELL MUSEUM
The Entrance
Tuggerah Entrance
Long Jetty
Toowoon Bay
Shelly Beach
Bateau Bay
CRACKNECK POINT LOOKOUT
WYRRABALONG NP
Forresters Beach
Wamberal
Terrigal
THE SKILLION
CENTRAL COAST
N
Broke
ARMY FIRING RANGES
PUTTY RD
Howes Mountain
Adams Peak
WOLLOMBI RD
Mount Murwin
Paynes Crossing
YENGO NATIONAL PARK
WOLLOMBI
Wollombi
GEORGE
Boree
HUNTER
Murrays Run
Bucketty
DOWNES RANGE
The Letter A
JILLIBY
Mangrove Creek Dam
Cedar Brush Creek
Ravensdale
Mandalong
Dooralong
Upper MacDonald
YENGO NATIONAL PARK
Kulnura
Yarramalong
SCA
Little Jilliby
St Albans
Central Mangrove
MACADAMIA NUT PLANTATION
Jilliby
JILLIBY SCA
Wyong Creek
Upper Mangrove
Mangrove Mountain
Mardi Dam
FOWLERS LOOKOUT
PARR SCA
Peats Ridge
Palm Grove
Palmdale
Mangrove Creek
AUSTRALIAN RAINFOREST SANCTUARY
Wisemans Ferry
POPRAN NP
BRISBANE WATER NP
SYDNEY
Fountaindale
Berkeley Vale
Leets Vale
DHARUG NATIONAL PARK
Mount Olive
Somersby
Ourimbah
Tumbi Umbi
Laughtondale
Niagara Park
Lisarow
Lower Mangrove
Belltrees
Narara
Holgate
AUSTRALIAN REPTILE PARK
Wyoming
Matcham
Gunderman
Maroota
Spencer
Calga
Kariong
Gosford
Erina
WISEMANS FERRY
Mount White
Tascott
Point Clare
Mount Kariong
BULGANDRY ABORIGINAL ENGRAVINGS
Green Point
Yattalunga
Kincumber
The Vale
Koolewong
Saratoga
Davistown
Avoca Beach
Copacabana
MARRAMARRA NATIONAL PARK
BRISBANE WATER NP
Woy Woy
Empire Bay
Milson Island
Mooney Mooney
Umina
Ettalong Beach
BOUDDI NP
Mourawaring Point
Forest Glen
Long Is
Dangar Is
MUOGAMARRA NR
Bombi Point
Brooklyn
Patonga
Pearl Beach
THE MAITLAND WRECK
WARRAH LOOKOUT
Box Head
Glenorie
Broken Bay
Berowra Waters
Cowan
Juno Point
West Head
BARRENJOEY LIGHTHOUSE
Barrenjoey Head
Berowra Heights
Arcadia
Palm Beach
Middle Dural
BEROWRA VALLEY RP
KU-RING-GAI CHASE NP
Little Head
Galston
Berowra
Bangalley Head
Kenthurst
Mount Kuring-Gai
Hornsby Heights
Bobbin Head
Church Point
Avalon
TASMAN SEA
Round Corner
Dural
Asquith
Duffys Forest
Mount Colah
Bayview
Newport
Glenhaven
Hornsby
Terrey Hills
Ingleside
Mona Vale
Castle Hill
Pennant Hills
Warrawee
Elanora Heights
Turimetta Head
St Ives
Narrabeen Beach
HILLS MWY
Cheltenham
Pymble
Belrose
GARIGAL NP
Collaroy Plateau
Frenchs Forest
For more detail on Sydney & Surrounds see pages 544–5
Northmead
Epping
PACIFIC HWY
Narraweena
Long Reef Point
Brookvale
Dee Why Head
Roseville
Chatswood
Parramatta
Ryde
Manly
OLYMPIC PARK
Lane Cove
Balgowlah
SYDNEY HARBOUR NP
Granville
Rhodes
North Head
Auburn
Mosman
OPERA HOUSE, HARBOUR BRIDGE, THE ROCKS, MACQUARIE LIGHTHOUSE, TARONGA ZOO, PARRAMATTA (HISTORIC TOWN), BONDI BEACH, INDIAN PACIFIC TRAIN
Regents Park
Drummoyne
FLEMINGTON MARKETS
SYDNEY
Kings Cross
Yagoona
Summer Hill
Redfern
North Bondi
33° 00'
33° 30'
151° 00'
151° 30'
A B C D E F G H
1 2 3 4 5 6 7 8 9 10 11 12

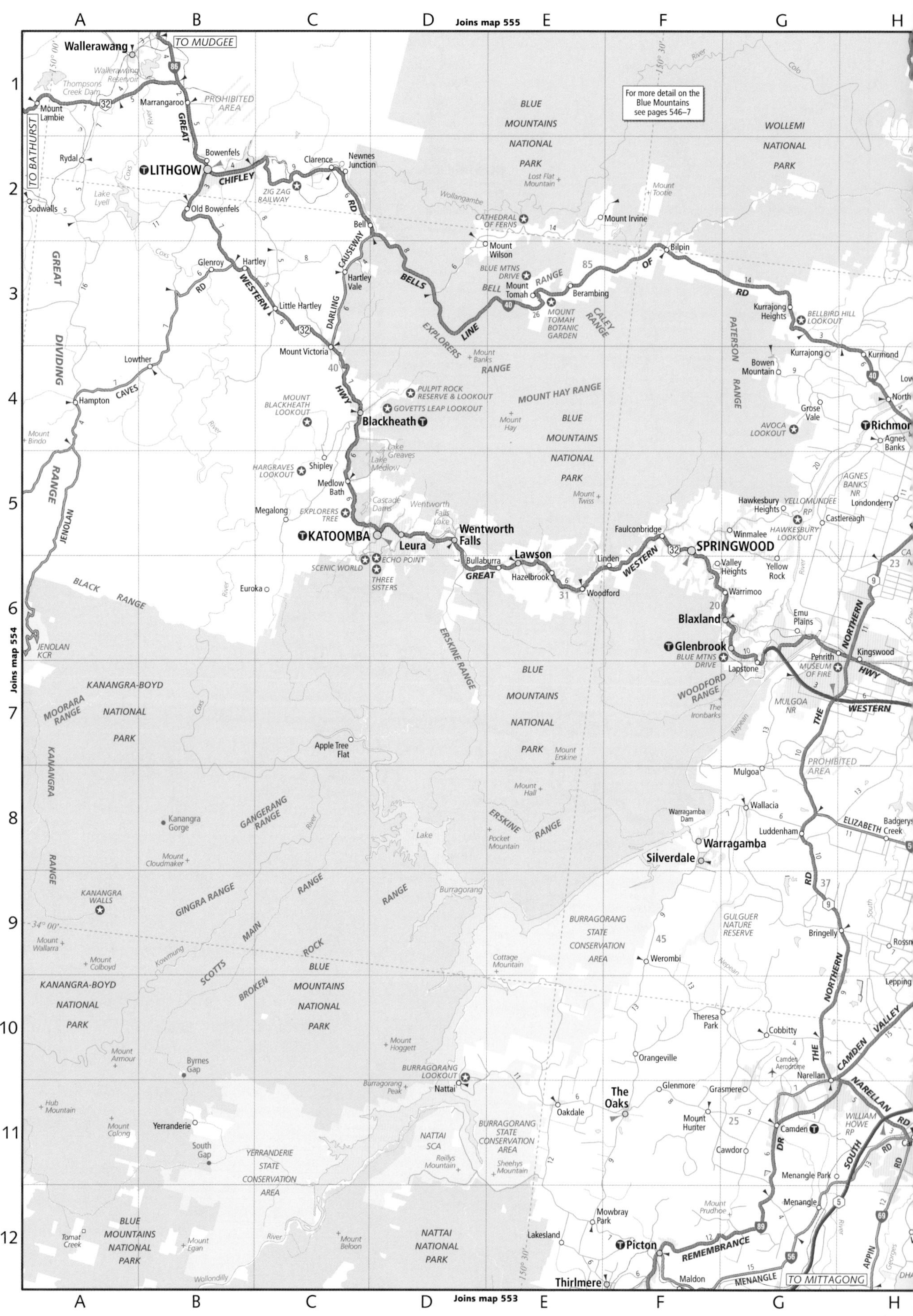
Joins map 555
Joins map 553
Joins map 554
TO MUDGEE
TO BATHURST
TO MITTAGONG
For more detail on the Blue Mountains see pages 546–7
Wallerawang
Mount Lambie
Marrangaroo
Rydal
Bowenfels
LITHGOW
Old Bowenfels
Sodwalls
Clarence
Newnes Junction
ZIG ZAG RAILWAY
Bell
Glenroy
Hartley
Hartley Vale
Little Hartley
Mount Victoria
Lowther
Hampton
Blackheath
Shipley
Medlow Bath
Megalong
KATOOMBA
Leura
Wentworth Falls
Bullaburra
Lawson
Hazelbrook
Linden
Woodford
Faulconbridge
SPRINGWOOD
Valley Heights
Warrimoo
Blaxland
Glenbrook
Lapstone
Euroka
Mount Wilson
Mount Irvine
Mount Tomah
Berambing
Bilpin
Kurrajong Heights
Kurrajong
Kurmond
Bowen Mountain
Grose Vale
Richmond
Agnes Banks
Hawkesbury Heights
Winmalee
Yellow Rock
Castlereagh
Londonderry
Emu Plains
Penrith
Kingswood
Mulgoa
Wallacia
Luddenham
Warragamba
Silverdale
Werombi
Bringelly
Theresa Park
Cobbitty
Orangeville
Narellan
Glenmore
Grasmere
Mount Hunter
The Oaks
Oakdale
Camden
Cawdor
Menangle Park
Menangle
Mowbray Park
Lakesland
Picton
Thirlmere
Maldon
Apple Tree Flat
Nattai
Yerranderie
Kanangra Gorge
Byrnes Gap
South Gap
BLUE MOUNTAINS NATIONAL PARK
WOLLEMI NATIONAL PARK
KANANGRA-BOYD NATIONAL PARK
NATTAI NATIONAL PARK
BURRAGORANG STATE CONSERVATION AREA
YERRANDERIE STATE CONSERVATION AREA
NATTAI SCA
GULGUER NATURE RESERVE
MULGOA NR
AGNES BANKS NR
YELLOMUNDEE RP
WILLIAM HOWE RP
GREAT DIVIDING RANGE
JENOLAN RANGE
BLACK RANGE
KANANGRA RANGE
MOORARA RANGE
GANGERANG RANGE
GINGRA RANGE
SCOTTS MAIN RANGE
BROKEN ROCK RANGE
ERSKINE RANGE
MOUNT HAY RANGE
CALEY RANGE
PATERSON RANGE
WOODFORD RANGE
GREAT WESTERN HWY
CHIFLEY RD
BELLS LINE OF RD
DARLING CAUSEWAY
CAVES RD
THE NORTHERN RD
ELIZABETH DR
CAMDEN VALLEY WAY
NARELLAN RD
REMEMBRANCE DR
MENANGLE RD
APPIN RD
SOUTH WESTERN MWY
CATHEDRAL OF FERNS
BLUE MTNS DRIVE
MOUNT TOMAH BOTANIC GARDEN
BELLBIRD HILL LOOKOUT
PULPIT ROCK RESERVE & LOOKOUT
GOVETTS LEAP LOOKOUT
MOUNT BLACKHEATH LOOKOUT
HARGRAVES LOOKOUT
EXPLORERS TREE
SCENIC WORLD
ECHO POINT
THREE SISTERS
AVOCA LOOKOUT
HAWKESBURY LOOKOUT
MUSEUM OF FIRE
BURRAGORANG LOOKOUT
KANANGRA WALLS
JENOLAN KCR
PROHIBITED AREA
Camden Aerodrome
Warragamba Dam
Lake Burragorang
Lake Lyell
Wallerawang Reservoir
Thompsons Creek Dam
Mount Banks
Mount Hay
Mount Twiss
Mount Erskine
Mount Hall
Pocket Mountain
Mount Cloudmaker
Mount Wallarra
Mount Colboyd
Mount Armour
Mount Colong
Hub Mountain
Mount Egan
Mount Beloon
Mount Hoggett
Burragorang Peak
Cottage Mountain
Reillys Mountain
Sheehys Mountain
Mount Prudhoe
Mount Bindo
Lost Flat Mountain
Mount Tootie
Tomat Creek
150° 00′
150° 30′
34° 00′

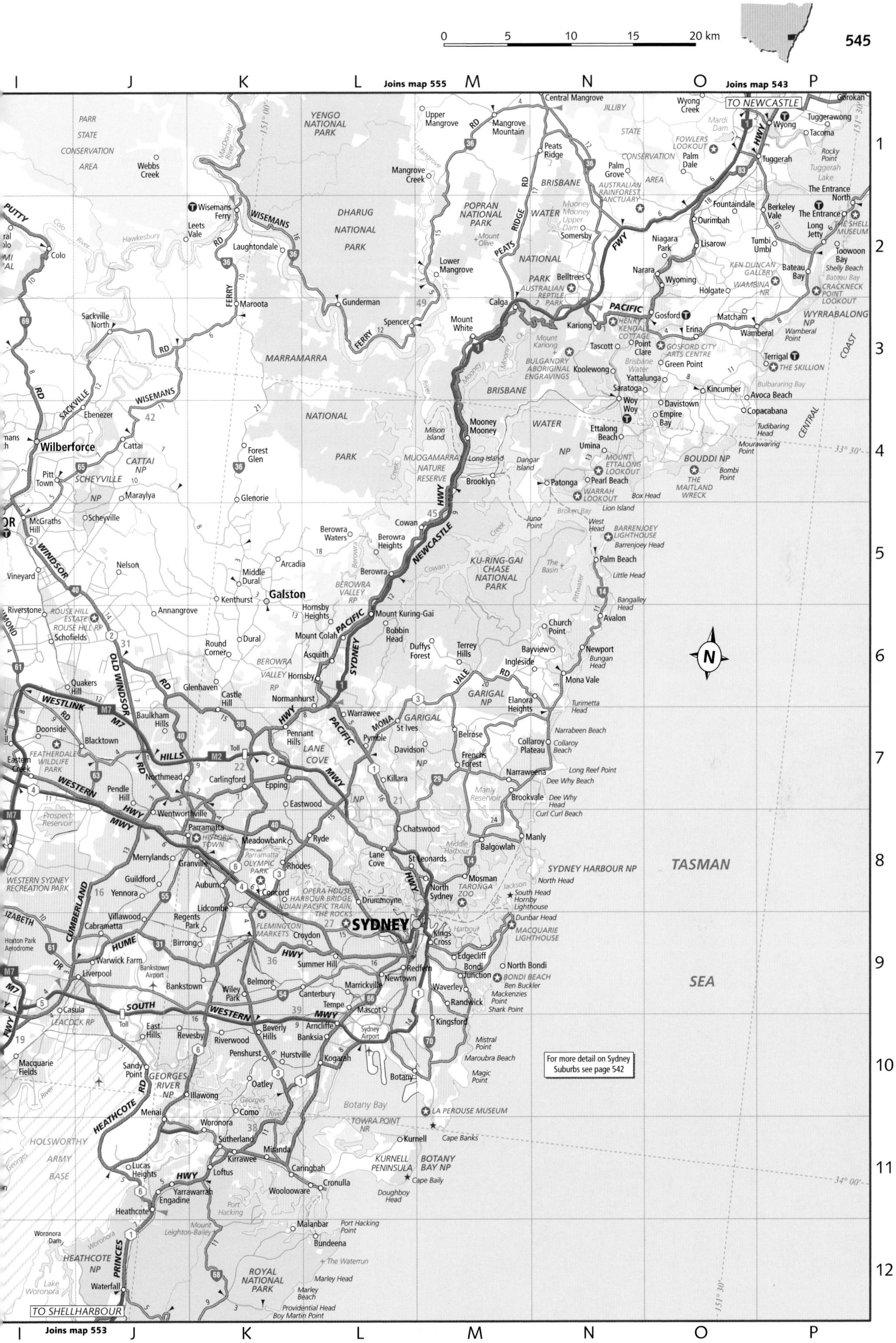

0 5 10 15 20 km

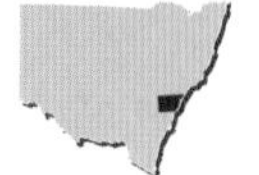

Joins map 555

Joins map 543

TO NEWCASTLE

For more detail on Sydney Suburbs see page 542

TO SHELLHARBOUR

Joins map 553

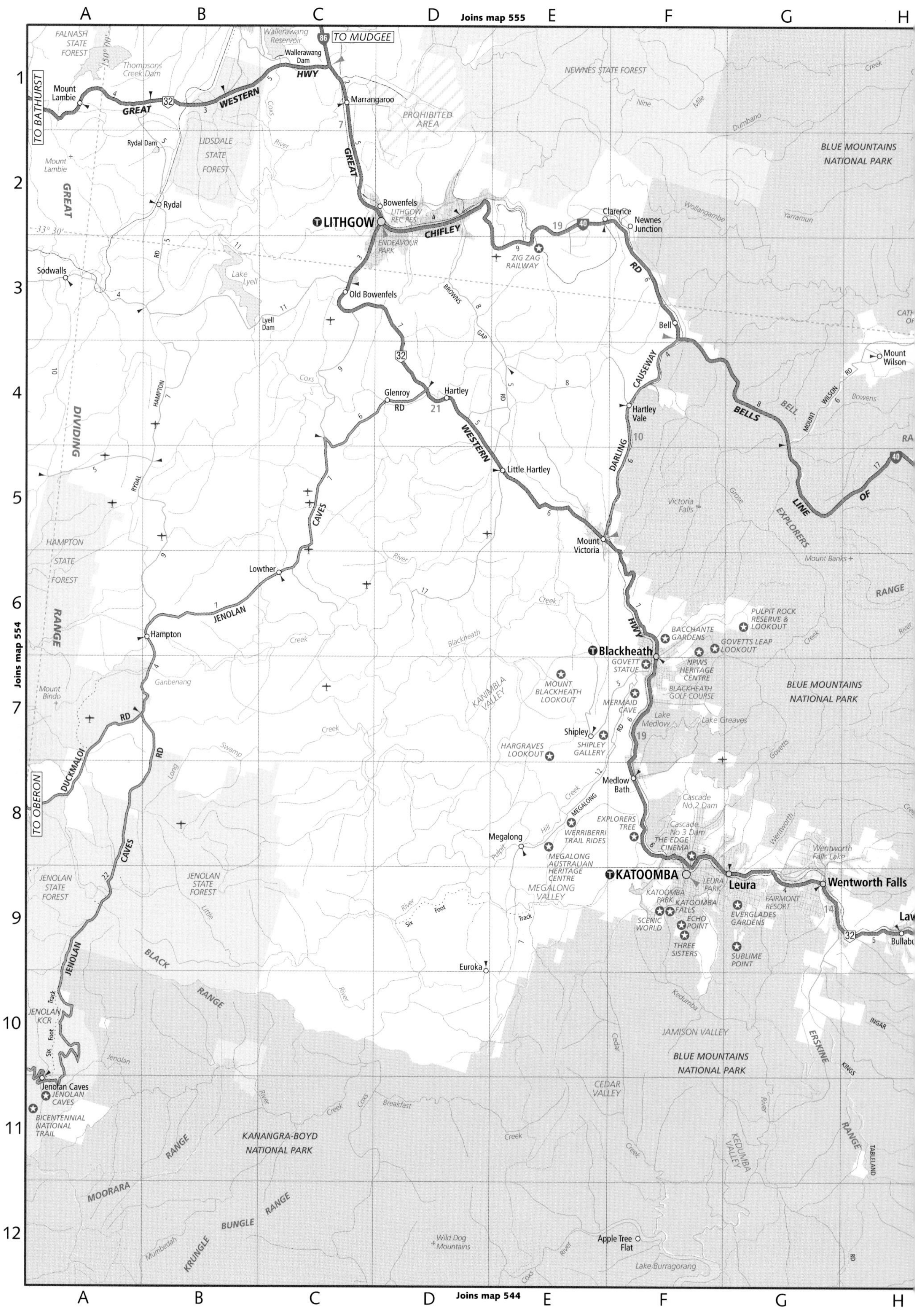
Joins map 555
TO MUDGEE
TO BATHURST
TO OBERON
Joins map 554
Joins map 544
LITHGOW
KATOOMBA
Blackheath
Leura
Wentworth Falls
Mount Victoria
Medlow Bath
Hartley
Little Hartley
Hartley Vale
Bell
Clarence
Newnes Junction
Bowenfels
Old Bowenfels
Marrangaroo
Wallerawang Dam
Mount Lambie
Rydal
Rydal Dam
Sodwalls
Lyell Dam
Glenroy
Lowther
Hampton
Megalong
Shipley
Euroka
Jenolan Caves
Apple Tree Flat
Mount Wilson
GREAT WESTERN HWY
CHIFLEY RD
DARLING CAUSEWAY
BELLS LINE OF RD
JENOLAN CAVES RD
DUCKMALOI RD
NEWNES STATE FOREST
LIDSDALE STATE FOREST
HAMPTON STATE FOREST
JENOLAN STATE FOREST
FALNASH STATE FOREST
BLUE MOUNTAINS NATIONAL PARK
KANANGRA-BOYD NATIONAL PARK
PROHIBITED AREA
ZIG ZAG RAILWAY
MOUNT BLACKHEATH LOOKOUT
HARGRAVES LOOKOUT
SHIPLEY GALLERY
MERMAID CAVE
GOVETT STATUE
NPWS HERITAGE CENTRE
BLACKHEATH GOLF COURSE
BACCHANTE GARDENS
GOVETTS LEAP LOOKOUT
PULPIT ROCK RESERVE & LOOKOUT
EXPLORERS TREE
WERRIBERRI TRAIL RIDES
MEGALONG AUSTRALIAN HERITAGE CENTRE
THE EDGE CINEMA
KATOOMBA PARK
KATOOMBA FALLS
SCENIC WORLD
ECHO POINT
THREE SISTERS
EVERGLADES GARDENS
SUBLIME POINT
FAIRMONT RESORT
LEURA PARK
JENOLAN CAVES
BICENTENNIAL NATIONAL TRAIL
JAMISON VALLEY
MEGALONG VALLEY
KANIMBLA VALLEY
CEDAR VALLEY
KEDUMBA VALLEY
GREAT DIVIDING RANGE
BLACK RANGE
MOORARA RANGE
KRUNGLE BUNGLE RANGE
ERSKINE RANGE
Lake Lyell
Lake Burragorang
Lake Medlow
Lake Greaves
Victoria Falls
Mount Banks
Wild Dog Mountains
Six Foot Track

0 5 10 15 km
Joins map 555
Joins map 544
Joins map 545
TO SINGLETON
TO PICTON
TO SYDNEY
WOLLEMI NATIONAL PARK
BLUE MOUNTAINS NATIONAL PARK
PARR STATE CONSERVATION AREA
Colo Heights
Upper Colo
Central Colo
Colo
Mount Irvine
Bilpin
Berambing
Kurrajong Heights
BELLBIRD HILL LOOKOUT
Kurrajong
Kurmond
Tennyson
Glossodia
East Kurrajong
Blaxlands Ridge
Freemans Reach
Wilberforce
Bowen Mountain
Grose Vale
North Richmond
Lowlands
Richmond
Agnes Banks
Clarendon
WINDSOR
Pitt Town
McGraths Hill
AVOCA LOOKOUT
HAWKESBURY SHOWGROUNDS
AGNES BANKS NATURE RESERVE
Londonderry
Hawkesbury Heights
HAWKESBURY LOOKOUT
Castlereagh
Winmalee
Faulconbridge
SPRINGWOOD
Valley Heights
Yellow Rock
Linden
Hazelbrook
Woodford
Warrimoo
Blaxland
Glenbrook
Emu Plains
Lapstone
Penrith
Kingswood
St Marys
Mount Druitt
Rooty Hill
Doonside
Eastern Creek
Erskine Park
Mulgoa
Vineyard
Riverstone
Schofields
CASTLEREAGH NATURE RESERVE
WINDSOR DOWNS NR
Penrith Lakes
MUSEUM OF FIRE
RED HANDS CAVE
BLUE MTNS DRIVE
WASCOE SIDING MINIATURE RAILWAY
BLAXLAND PUBLIC GARDENS
MULGOA NR
FEATHERDALE WILDLIFE PARK
Prospect Reservoir
CALEY RANGE
PATERSON RANGE
WOODFORD RANGE
HAY RANGE
BELLS LINE OF RD
PUTTY RD
KURRAJONG RD
THE NORTHERN RD
GREAT WESTERN HWY
WESTERN MWY
WESTLINK M7

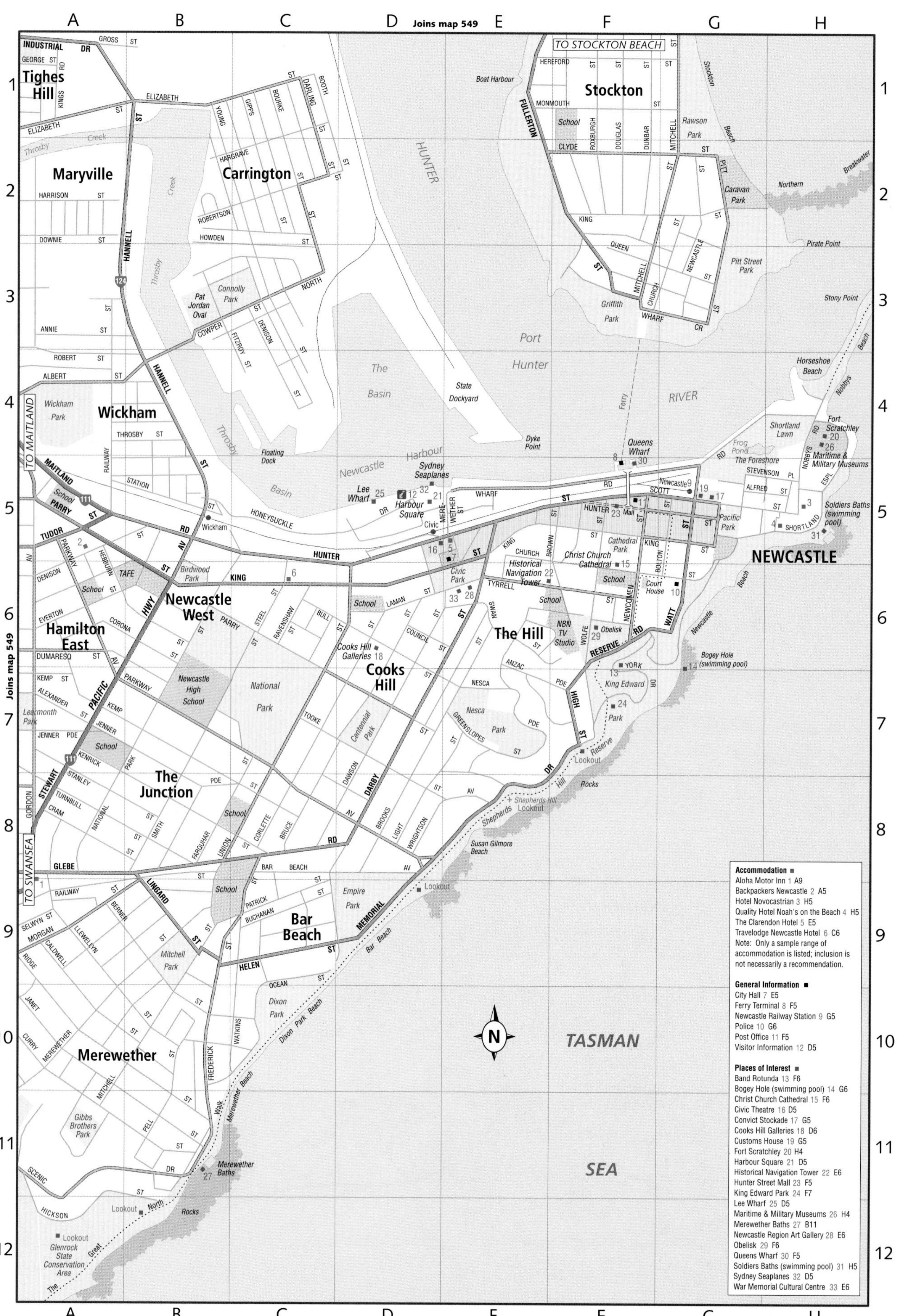
0 0.25 0.5 0.75 1 km
Joins map 549
TO STOCKTON BEACH
TO MAITLAND
TO SWANSEA
Tighes Hill
Maryville
Carrington
Stockton
Wickham
Newcastle West
Hamilton East
Cooks Hill
The Hill
NEWCASTLE
The Junction
Bar Beach
Merewether
HUNTER RIVER
Port Hunter
The Basin
Newcastle Harbour
Basin
TASMAN SEA
Accommodation
Aloha Motor Inn 1 A9
Backpackers Newcastle 2 A5
Hotel Novocastrian 3 H5
Quality Hotel Noah's on the Beach 4 H5
The Clarendon Hotel 5 E5
Travelodge Newcastle Hotel 6 C6
Note: Only a sample range of accommodation is listed; inclusion is not necessarily a recommendation.
General Information
City Hall 7 E5
Ferry Terminal 8 F5
Newcastle Railway Station 9 G5
Police 10 G6
Post Office 11 F5
Visitor Information 12 D5
Places of Interest
Band Rotunda 13 F6
Bogey Hole (swimming pool) 14 G6
Christ Church Cathedral 15 F6
Civic Theatre 16 D5
Convict Stockade 17 G5
Cooks Hill Galleries 18 D6
Customs House 19 G5
Fort Scratchley 20 H4
Harbour Square 21 D5
Historical Navigation Tower 22 E6
Hunter Street Mall 23 F5
King Edward Park 24 F7
Lee Wharf 25 D5
Maritime & Military Museums 26 H4
Merewether Baths 27 B11
Newcastle Region Art Gallery 28 E6
Obelisk 29 F6
Queens Wharf 30 F5
Soldiers Baths (swimming pool) 31 H5
Sydney Seaplanes 32 D5
War Memorial Cultural Centre 33 E6

0 5 10 15 20 km

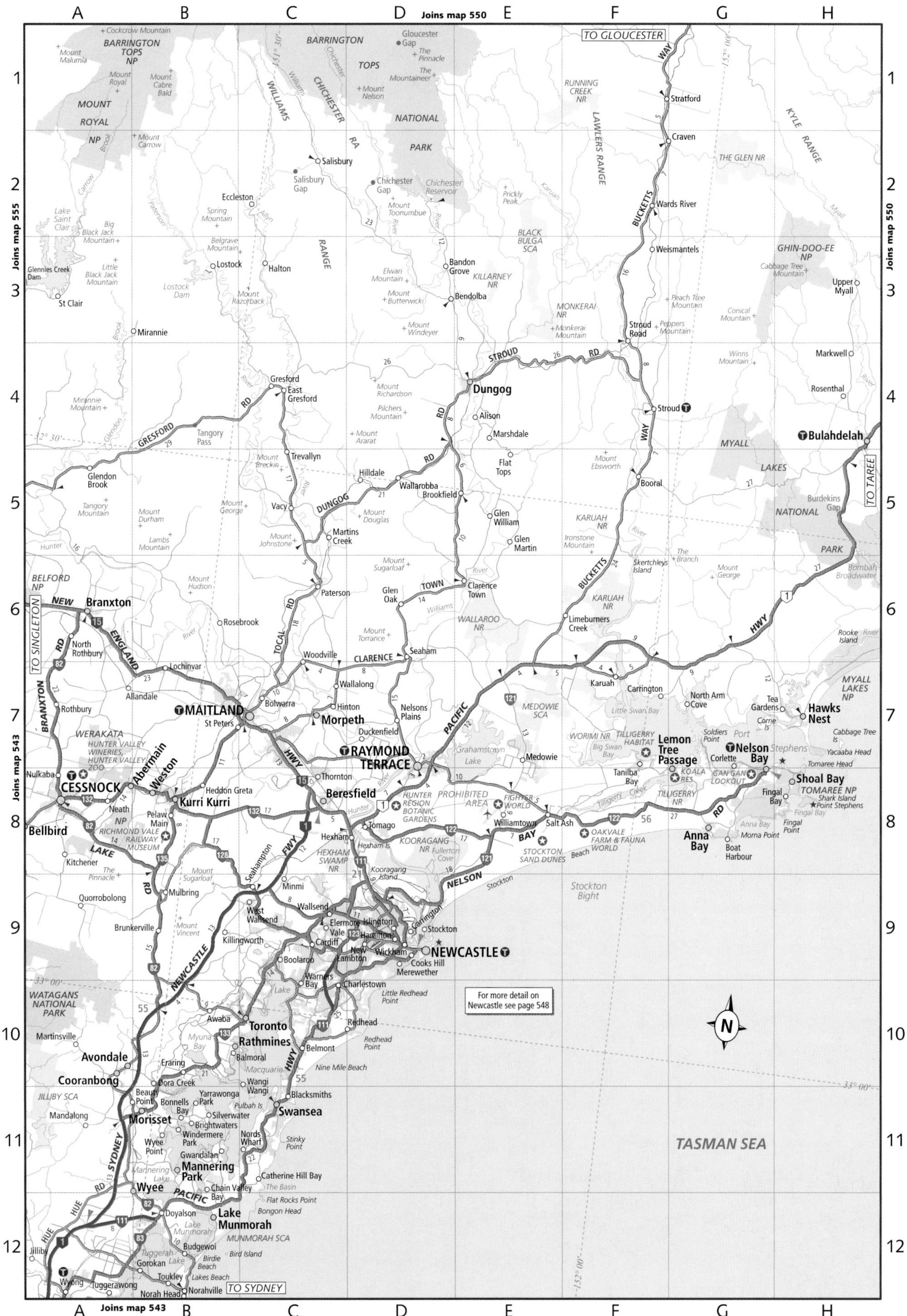
A B C D E F G H
Joins map 550
Joins map 555
Joins map 543
TO GLOUCESTER
TO TAREE
TO SINGLETON
TO SYDNEY
BARRINGTON TOPS NP
BARRINGTON TOPS NATIONAL PARK
MOUNT ROYAL NP
Cockcrow Mountain
Mount Malumla
Mount Royal
Mount Cabre Bald
Mount Carrow
Gloucester Gap
The Pinnacle
The Mountaineer
Mount Nelson
WILLIAMS
CHICHESTER RA
RUNNING CREEK NR
LAWLERS RANGE
KYLE RANGE
THE GLEN NR
Stratford
Craven
Salisbury
Salisbury Gap
Chichester Gap
Chichester Reservoir
Mount Toonumbue
Prickly Peak
Eccleston
Spring Mountain
Lake Saint Clair
Big Black Jack Mountain
Little Black Jack Mountain
Glennies Creek Dam
St Clair
Belgrave Mountain
Lostock
Lostock Dam
Halton
RANGE
Mount Razorback
Elwari Mountain
Mount Butterwicki
Bandon Grove
Bendolba
KILLARNEY NR
BLACK BULGA SCA
BUCKETTS WAY
Wards River
Weismantels
GHIN-DOO-EE NP
Cabbage Tree Mountain
Upper Myall
Peach Tree Mountain
Conical Mountain
Peppers Mountain
MONKERAI NR
Monkerai Mountain
Stroud Road
Mirannie
Mount Windeyer
STROUD RD
Winns Mountain
Markwell
Rosenthal
Gresford
East Gresford
Mount Richardson
Pilchers Mountain
Dungog
Alison
Marshdale
Stroud
Bulahdelah
Mirannie Mountain
GRESFORD RD
Tangory Pass
Mount Ararat
Mount Ebsworth
MYALL LAKES NATIONAL PARK
Trevallyn
Mount Breckin
Flat Tops
Glendon Brook
Hilldale
Wallarobba
Brookfield
Booral
Burdekins Gap
Tangory Mountain
Mount Durham
Mount George
Vacy
DUNGOG RD
Mount Douglas
Glen William
KARUAH NR
Lambs Mountain
Mount Johnstone
Martins Creek
Glen Martin
Ironstone Mountain
The Branch
Mount Sugarloaf
Skertchleys Island
Mount George
Bombah Broadwater
BELFORD NP
Mount Hudson
Paterson
Glen Oak
TOWN
Clarence Town
NEW ENGLAND HWY
Branxton
WALLAROO NR
Limeburners Creek
PACIFIC HWY
Rosebrook
TOCAL RD
Mount Torrance
Rooke Island
North Rothbury
Woodville
CLARENCE
Seaham
Lochinvar
Wallalong
Karuah
Carrington
North Arm Cove
MYALL LAKES NP
Allandale
Bolwarra
Hinton
Nelsons Plains
MEDOWIE SCA
Little Swan Bay
Tea Gardens
Hawks Nest
Corrie Is
Rothbury
BRANXTON RD
MAITLAND
St Peters
Morpeth
Duckenfield
WORIMI NR
TILLIGERRY HABITAT
Lemon Tree Passage
Soldiers Point
Port Stephens
Cabbage Tree Is
WERAKATA
HUNTER VALLEY WINERIES
HUNTER VALLEY ZOO
Abermain
Weston
RAYMOND TERRACE
Grahamstown Lake
Medowie
Big Swan Bay
KOALA RES
Nelson Bay
Corlette
Yacaaba Head
Tomaree Head
Nulkaba
CESSNOCK
Heddon Greta
Kurri Kurri
Thornton
Beresfield
PROHIBITED AREA
FIGHTER WORLD
Tanilba Bay
GAN GAN LOOKOUT
Shoal Bay
TOMAREE NP
Shark Island
Point Stephens
Fingal Bay
Neath
NP
Pelaw Main
HUNTER REGION BOTANIC GARDENS
Williamtown
Salt Ash
TILLIGERRY NR
Tilligerry Creek
Bellbird
RICHMOND VALE RAILWAY MUSEUM
LAKE RD
Tomago
Hexham
KOORAGANG NR
Fullerton Cove
NELSON BAY RD
OAKVALE FARM & FAUNA WORLD
STOCKTON SAND DUNES
Anna Bay
Morna Point
Fingal Point
Boat Harbour
Kitchener
The Pinnacle
Mount Sugarloaf
HEXHAM SWAMP NR
Kooragang Island
Stockton Beach
Stockton Bight
Quorrobolong
Mulbring
Seahampton
Minmi
FWY
West Wallsend
Wallsend
Carrington
Stockton
Brunkerville
Mount Vincent
Killingworth
Elermore Vale
Islington
Hamilton
Cardiff
New Lambton
Wickham
NEWCASTLE
Boolaroo
Cooks Hill
Merewether
NEWCASTLE FWY
Warners Bay
Charlestown
Lake
Little Redhead Point
For more detail on Newcastle see page 548
N
WATAGANS NATIONAL PARK
Awaba
Toronto
Redhead
Redhead Point
Martinsville
Myuna Bay
Rathmines
Balmoral
Belmont
Avondale
Eraring
Lake Macquarie
Nine Mile Beach
Cooranbong
Dora Creek
Wangi Wangi
HWY
JILLIBY SCA
Beauty Point
Bonnells Bay
Yarrawonga Park
Pulbah Is
Blacksmiths
Swansea
Mandalong
Morisset
Silverwater
Brightwaters
Windermere Park
Nords Wharf
Stinky Point
TASMAN SEA
Wyee Point
Gwandalan
Mannering Park
Mannering Lake
SYDNEY
Catherine Hill Bay
The Basin
Wyee
Chain Valley Bay
PACIFIC
Flat Rocks Point
HUE HUE RD
Doyalson
Lake Munmorah
Bongon Head
Lake Munmorah
MUNMORAH SCA
Budgewoi
Jilliby
Tuggerah Lake
Gorokan
Birdie Beach
Bird Island
Wyong
Tuggerawong
Toukley
Lakes Beach
Norah Head
Norahville
33° 00'
32° 30'
151° 30'
152° 00'

0 20 40 60 km

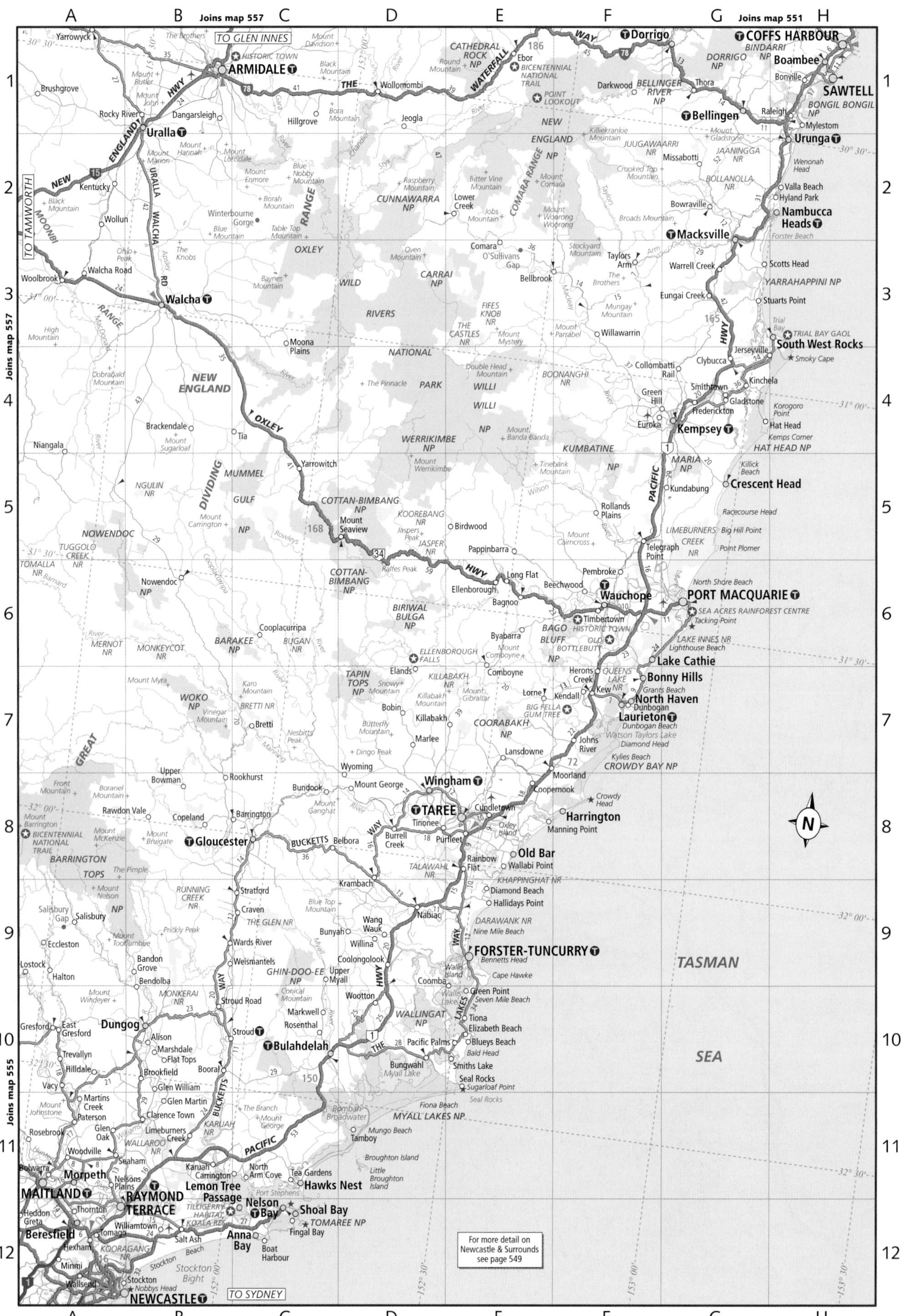
A
B
Joins map 557
C
D
E
F
G
Joins map 551
H
1
2
3
4
5
6
7
8
9
10
11
12
Joins map 557
Joins map 555
Yarrowyck
The Brothers
TO GLEN INNES
Mount Davidson
CATHEDRAL ROCK NP
186
Dorrigo
COFFS HARBOUR
BINDARRI NP
DORRIGO NP
Boambee
Bonville
SAWTELL
BONGIL BONGIL NP
HISTORIC TOWN
ARMIDALE
Black Mountain
Round Mountain
Ebor
BICENTENNIAL NATIONAL TRAIL
Brushgrove
Mount Butler
Mount John
THE
Wollomombi
WATERFALL
WAY
POINT LOOKOUT
Darkwood
BELLINGER RIVER NP
Thora
Raleigh
Mylestom
Rocky River
Dangarsleigh
Hillgrove
Bora Mountain
Jeogla
Bellingen
Urunga
Uralla
NEW ENGLAND
HWY
Mount Hannah
Mount Marion
Mount Lonsdale
NEW ENGLAND NP
Killiekrankie Mountain
JUUGAWAARRI NR
Missabotti
Mount Gladstone
JAANINGGA NR
Wenonah Head
NEW
Kentucky
URALLA
WALCHA
Mount Enmore
Blue Nobby Mountain
Raspberry Mountain
Bitter Vine Mountain
Mount Comara
Crooked Top Mountain
BOLLANOLLA NR
Valla Beach
Hyland Park
Black Mountain
Wollun
Borah Mountain
Winterbourne Gorge
Blue Mountain
Table Top Mountain
CUNNAWARRA NP
Lower Creek
Jobs Mountain
COMARA RANGE
Mount Woorong Woorong
Broads Mountain
Bowraville
Nambucca Heads
Forster Beach
Macksville
MOONBI
TO TAMWORTH
Ohio Peak
The Knobs
OXLEY
RANGE
Oven Mountain
Comara
O'Sullivans Gap
Stockyard Mountain
Taylors Arm
Warrell Creek
Scotts Head
Woolbrook
Walcha Road
Walcha
Baynes Mountain
WILD
CARRAI NP
Bellbrook
The Brothers
Eungai Creek
YARRAHAPPINI NP
Stuarts Point
RANGE
RIVERS
FIFES KNOB NR
Mungay Mountain
High Mountain
THE CASTLES NR
Mount Mystery
Mount Parrabel
Willawarrin
Trial Bay
TRIAL BAY GAOL
South West Rocks
Smoky Cape
Moona Plains
NATIONAL
Jerseyville
Clybucca
NEW ENGLAND
Cobrabald Mountain
The Pinnacle
PARK
Double Head Mountain
WILLI
WILLI
BOONANGHI NR
Collombatti Rail
Green Hill
Smithtown
Kinchela
Gladstone
Fredickton
Korogoro Point
Euroka
Kempsey
Hat Head
Kemps Corner
HAT HEAD NP
Brackendale
Mount Sugarloaf
Tia
OXLEY
NP
Mount Banda Banda
WERRIKIMBE NP
KUMBATINE NP
Niangala
Yarrowitch
Mount Werrikimbe
Tinebank Mountain
MARIA NP
Killick Beach
Crescent Head
NGULIN NR
DIVIDING
MUMMEL
GULF
NP
PACIFIC
Kundabung
Wilson
COTTAN-BIMBANG NP
KOOREBANG NR
Rollands Plains
Racecourse Head
Mount Carrington
NOWENDOC
Mount Seaview
168
Jaspers Peak
Birdwood
JASPER NR
Mount Cairncross
LIMEBURNERS CREEK NR
Big Hill Point
Point Plomer
Telegraph Point
TUGGOLO CREEK NR
TOMALLA NR
Barnard
Rowleys
34
Ralfes Peak
Pappinbarra
HWY
Pembroke
North Shore Beach
Nowendoc NP
COTTAN-BIMBANG NP
Ellenborough
Long Flat
Bagnoo
Beechwood
Wauchope
PORT MACQUARIE
SEA ACRES RAINFOREST CENTRE
Tacking Point
BIRIWAL BULGA NP
Timbertown HISTORIC TOWN
Cooplacurripa
BARAKEE NP
BUGAN NR
Byabarra
BAGO BLUFF NP
OLD BOTTLEBUTT
LAKE INNES NR
Lighthouse Beach
MERNOT NR
MONKEYCOT NR
ELLENBOROUGH FALLS
Mount Comboyne
Lake Cathie
Elands
KILLABAKH NR
Comboyne
Herons Creek
QUEENS LAKE NR
Bonny Hills
Grants Beach
Mount Myra
TAPIN TOPS NP
Karo Mountain
Snowy Mountain
Killabakh Mountain
Mount Gibraltar
Lorne
Kendall
Kew
North Haven
Dunbogan
WOKO NP
BRETTI NR
Vinegar Mountain
Bobin
Killabakh
BIG FELLA GUM TREE
Laurieton
Dunbogan Beach
Bretti
Nesbitts Peak
Butterfly Mountain
Marlee
COORABAKH NP
Watson Taylors Lake
Diamond Head
GREAT
Dingo Peak
Johns River
Lansdowne
Kylies Beach
CROWDY BAY NP
Upper Bowman
Rookhurst
Wyoming
Moorland
Front Mountain
Boranel Mountain
Bundook
Mount George
Wingham
Coopernook
Crowdy Head
Mount Barrington
Rawdon Vale
Copeland
Barrington
Mount Ganghat
TAREE
Cundletown
Harrington
Manning Point
BICENTENNIAL NATIONAL TRAIL
Mount McKenzie
Mount Bruigate
Gloucester
BUCKETTS
Belbora
Burrell Creek
Tinonee
Purfleet
Oxley Island
Old Bar
BARRINGTON TOPS
The Pimple
WAY
TALAWAHL NR
Rainbow Flat
Wallabi Point
N
KHAPPINGHAT NR
Mount Nelson
RUNNING CREEK NR
Stratford
Krambach
Diamond Beach
Hallidays Point
Salisbury Gap
Salisbury
NP
Craven
Blue Top Mountain
THE GLEN NR
Wang Wauk
Nabiac
DARAWANK NR
Nine Mile Beach
Prickly Peak
Bunyah
Willina
Mount Toolumbie
Eccleston
Wards River
FORSTER-TUNCURRY
Bennetts Head
TASMAN
Lostock
Bandon Grove
Weismantels
Coolongolook
Cape Hawke
Halton
Bendolba
GHIN-DOO-EE NP
Upper Myall
Wallis Island
Coomba
Mount Windeyer
MONKERAI NR
Conical Mountain
Wootton
Green Point
Seven Mile Beach
Stroud Road
Markwell
WALLINGAT NP
Wallis Lake
Gresford
East Gresford
Dungog
Rosenthal
LAKES
Tiona
Elizabeth Beach
Alison
Stroud
Bulahdelah
Pacific Palms
Blueys Beach
Bald Head
SEA
Trevallyn
Marshdale
Flat Tops
Hilldale
Brookfield
Booral
Bungwahl
Myall Lake
Smiths Lake
Seal Rocks
Sugarloaf Point
Vacy
Glen William
150
Seal Rocks
Mount Johnstone
Martins Creek
Paterson
Glen Martin
The Branch
Fiona Beach
Clarence Town
Bombah Broadwater
MYALL LAKES NP
Rosebrook
Glen Oak
Limeburners Creek
KARUAH NR
Mount George
Mungo Beach
Tamboy
Woodville
WALLAROO NR
PACIFIC
Broughton Island
Bolwarra
Seaham
Karuah
North Arm Cove
Little Broughton Island
Morpeth
Nelsons Plains
Carrington
Tea Gardens
Hawks Nest
MAITLAND
RAYMOND TERRACE
Lemon Tree Passage
Port Stephens
Heddon Greta
Thornton
TILLIGERRY HABITAT
Nelson Bay
Shoal Bay
KOALA RES
TOMAREE NP
Beresfield
Williamtown
Fingal Bay
Tomago
Salt Ash
Anna Bay
Hexham
Boat Harbour
KOORAGANG NR
Stockton Beach
For more detail on Newcastle & Surrounds see page 549
Minmi
Stockton Bight
Wallsend
Stockton
Nobbys Head
NEWCASTLE
TO SYDNEY
30° 30'
31° 00'
31° 30'
32° 00'
32° 30'
152° 00'
152° 30'
153° 00'
153° 30'

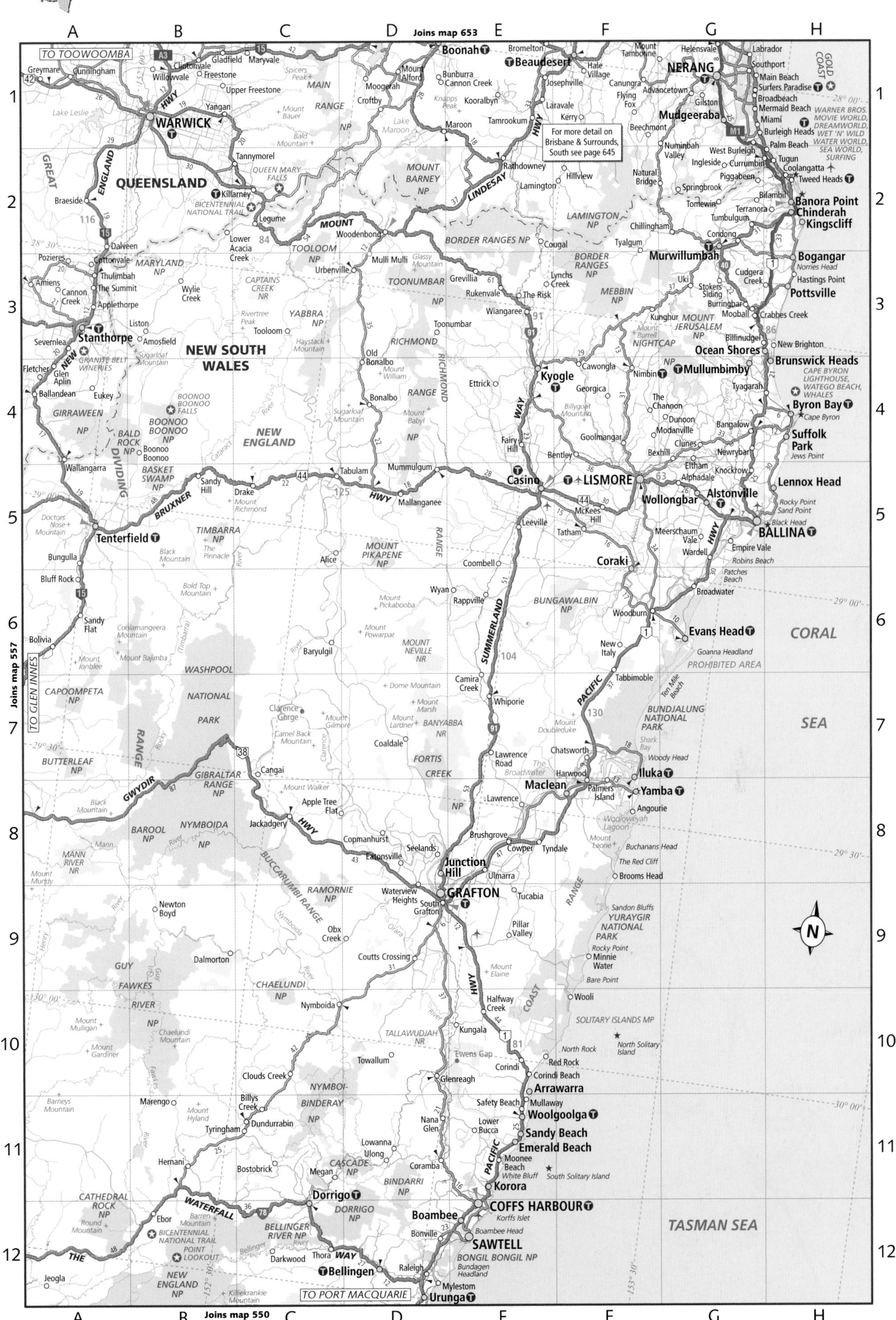

0
20
40
60 km
Joins map 653
Joins map 557
Joins map 550
TO TOOWOOMBA
TO GLEN INNES
TO PORT MACQUARIE
For more detail on Brisbane & Surrounds, South see page 645
QUEENSLAND
NEW SOUTH WALES
CORAL SEA
TASMAN SEA
Greymare
Cunningham
Clintonvale
Willowvale
Gladfield
Maryvale
Freestone
Upper Freestone
Yangan
WARWICK
Lake Leslie
Tannymorel
QUEEN MARY FALLS
Killarney
BICENTENNIAL NATIONAL TRAIL
Legume
Braeside
Lower Acacia Creek
Dalveen
Cottonvale
Thulimbah
The Summit
Applethorpe
Pozieres
Amiens
Cannon Creek
MARYLAND NP
Wylie Creek
CAPTAINS CREEK NR
Rivertree Peak
Tooloom
YABBRA NP
Liston
Amosfield
Stanthorpe
Severnlea
Fletcher
Glen Aplin
Ballandean
Eukey
GRANITE BELT WINERIES
Sugarloaf Mountain
GIRRAWEEN NP
BALD ROCK NP
BOONOO BOONOO NP
BOONOO BOONOO FALLS
Boonoo Boonoo
NEW ENGLAND
BASKET SWAMP NP
Wallangarra
Sandy Hill
Drake
Mount Richmond
Tenterfield
Doctors Nose Mountain
TIMBARRA NP
Black Mountain
The Pinnacle
Bungulla
Bluff Rock
Bold Top Mountain
Sandy Flat
Bolivia
Coolamangeera Mountain
Mount Bajimba
Mount Jonblee
WASHPOOL NATIONAL PARK
CAPOOMPETA NP
BUTTERLEAF NP
GIBRALTAR RANGE NP
Black Mountain
BAROOL NP
NYMBOIDA NP
MANN RIVER NR
Mount Mundy
Newton Boyd
Dalmorton
GUY FAWKES RIVER NP
Mount Mulligan
Chaelundi Mountain
Mount Gardiner
CHAELUNDI NP
Nymboida
Barneys Mountain
Marengo
Mount Hyland
Tyringham
Dundurrabin
Billys Creek
Clouds Creek
Hernani
Bostobrick
CATHEDRAL ROCK NP
Round Mountain
Ebor
Barren Mountain
BICENTENNIAL NATIONAL TRAIL
POINT LOOKOUT
NEW ENGLAND NP
Jeogla
Killiekrankie Mountain
Darkwood
Thora
Bellingen
BELLINGER RIVER NP
DORRIGO NP
Dorrigo
Megan
Ulong
Lowanna
CASCADE NP
Coramba
BINDARRI NP
Towallum
NYMBOI-BINDERAY NP
TALLAWUDJAH NR
Glenreagh
Nana Glen
Kungala
Ewens Gap
Lower Bucca
Boambee
Bonville
Raleigh
Mylestom
Urunga
SAWTELL
Boambee Head
BONGIL BONGIL NP
Bundagen Headland
COFFS HARBOUR
Korffs Islet
Korora
Moonee Beach
White Bluff
South Solitary Island
Emerald Beach
Sandy Beach
Woolgoolga
Mullaway
Safety Beach
Arrawarra
Corindi Beach
Corindi
Red Rock
North Rock
North Solitary Island
SOLITARY ISLANDS MP
Halfway Creek
Wooli
Bare Point
Minnie Water
Rocky Point
YURAYGIR NATIONAL PARK
Sandon Bluffs
Mount Elaine
Coutts Crossing
Pillar Valley
Obx Creek
Tucabia
Ulmarra
GRAFTON
South Grafton
Junction Hill
Waterview Heights
Seelands
Eatonsville
Copmanhurst
Cowper
Tyndale
Brushgrove
Lawrence
Maclean
Harwood
Palmers Island
Yamba
Angourie
Iluka
Woody Head
Shark Bay
Chatsworth
Lawrence Road
The Broadwater
Wooloweyah Lagoon
Mount Leone
Buchanans Head
The Red Cliff
Brooms Head
RAMORNIE NP
Jackadgery
Apple Tree Flat
Mount Walker
Cangai
Camel Back Mountain
Clarence Gorge
Mount Gilmore
Coaldale
Mount Lardner
BANYABBA NR
Mount Marsh
FORTIS CREEK NP
Dome Mountain
Camira Creek
Whiporie
Mount Doubleduke
BUNDJALUNG NATIONAL PARK
Ten Mile Beach
Tabbimoble
New Italy
Goanna Headland
PROHIBITED AREA
Evans Head
Woodburn
Broadwater
Patches Beach
Robins Beach
Empire Vale
Wardell
Meerschaum Vale
BALLINA
Black Head
Coraki
BUNGAWALBIN NP
Rappville
Wyan
Coombell
Mount Pickabooba
Mount Powarpar
MOUNT NEVILLE NR
Baryulgil
MOUNT PIKAPENE NP
Alice
Leeville
Tatham
McKees Hill
Casino
LISMORE
Wollongbar
Alstonville
Lennox Head
Rocky Point
Sand Point
Knockrow
Newrybar
Alphadale
Eltham
Bexhill
Clunes
Bentley
Goolmangar
Modanville
Dunoon
The Channon
Bangalow
Suffolk Park
Jews Point
Byron Bay
Cape Byron
CAPE BYRON LIGHTHOUSE, WATEGO BEACH, WHALES
Tyagarah
Brunswick Heads
Mullumbimby
Nimbin
Georgica
Billygoat Mountain
Kyogle
Cawongla
Etrick
Fairy Hill
Mummulgum
Mallanganee
Tabulam
Bonalbo
Old Bonalbo
Mount William
Sugarloaf Mountain
Mount Babyl
RICHMOND RANGE NP
Toonumbar
TOONUMBAR NP
Wiangaree
The Risk
Rukenvale
Grevillia
Lynchs Creek
BORDER RANGES NP
MEBBIN NP
Cougal
Urbenville
Mulli Mulli
Glassy Mountain
Woodenbong
TOOLOOM NP
Haystack Mountain
Mount Burrell
NIGHTCAP NP
Kunghur
MOUNT JERUSALEM NP
Uki
Murwillumbah
Stokers Siding
Burringbar
Mooball
Crabbes Creek
Billinudgel
Ocean Shores
New Brighton
Pottsville
Hastings Point
Cudgera Creek
Bogangar
Norries Head
Kingscliff
Chinderah
Banora Point
Terranora
Tumbulgum
Condong
Bilambil
Tomewin
Chillingham
Tyalgum
LAMINGTON NP
Natural Bridge
Springbrook
Numinbah Valley
Beechmont
Mudgeeraba
NERANG
Advancetown
Gilston
Canungra
Flying Fox
Mount Tamborine
Helensvale
Labrador
Southport
Main Beach
Surfers Paradise
Broadbeach
Mermaid Beach
Miami
Burleigh Heads
Palm Beach
West Burleigh
Ingleside
Currumbin
Piggabeen
Tugun
Coolangatta
Tweed Heads
GOLD COAST
WARNER BROS. MOVIE WORLD, DREAMWORLD, WET 'N' WILD WATER WORLD, SEA WORLD, SURFING
Hale Village
Bromelton
Beaudesert
Josephville
Laravale
Kerry
Tamrookum
Rathdowney
Lamington
Hillview
Boonah
Bunburra
Cannon Creek
Kooralbyn
Knapps Peak
Maroon
Lake Maroon
Mount Alford
Moogerah
Croftby
Spicers Peak
Mount Bauer
Bald Mountain
MAIN RANGE NP
MOUNT BARNEY NP
GREAT DIVIDING RANGE
NEW ENGLAND HWY
CUNNINGHAM HWY
MOUNT LINDESAY HWY
BRUXNER HWY
SUMMERLAND WAY
PACIFIC HWY
GWYDIR HWY
WATERFALL WAY
COAST RANGE
BUCCARUMBI RANGE
28° 00'
28° 30'
29° 00'
29° 30'
30° 00'
152° 30'
153° 30'

0 0.5 1 1.5 2 km

Joins map 553

TO SYDNEY

TO KIAMA

TASMAN SEA

WOLLONGONG

Accommodation ■
- Belmore All Suite Hotel 1 F9
- Boat Harbour Motel 2 F8
- The City Beach Motel 3 F9
- Downtown Motel 4 F9
- Hotel Ibis Wollongong 5 E9
- Novotel Wollongong Northbeach 6 F7
- Park Street Serviced Apartments 7 F7

Note: Only a sample range of accommodation is listed; inclusion is not necessarily a recommendation.

General Information ■
- Coach Terminal 8 E8
- Motoring Organisation (NRMA) 9 E9
- Police Headquarters 10 E9
- Post Office 11 F9
- WIN Sports & Entertainment Centres 12 F9
- Wollongong Railway Station 13 D9
- Wollongong Visitor Centre 14 F9

Places of Interest ■
- Illawarra Historical Society Museum 15 F9
- Illawarra Performing Arts Centre 16 F9
- International Centre 17 F9
- Wollongong Botanic Garden 18 C6
- Wollongong City Gallery 19 F9

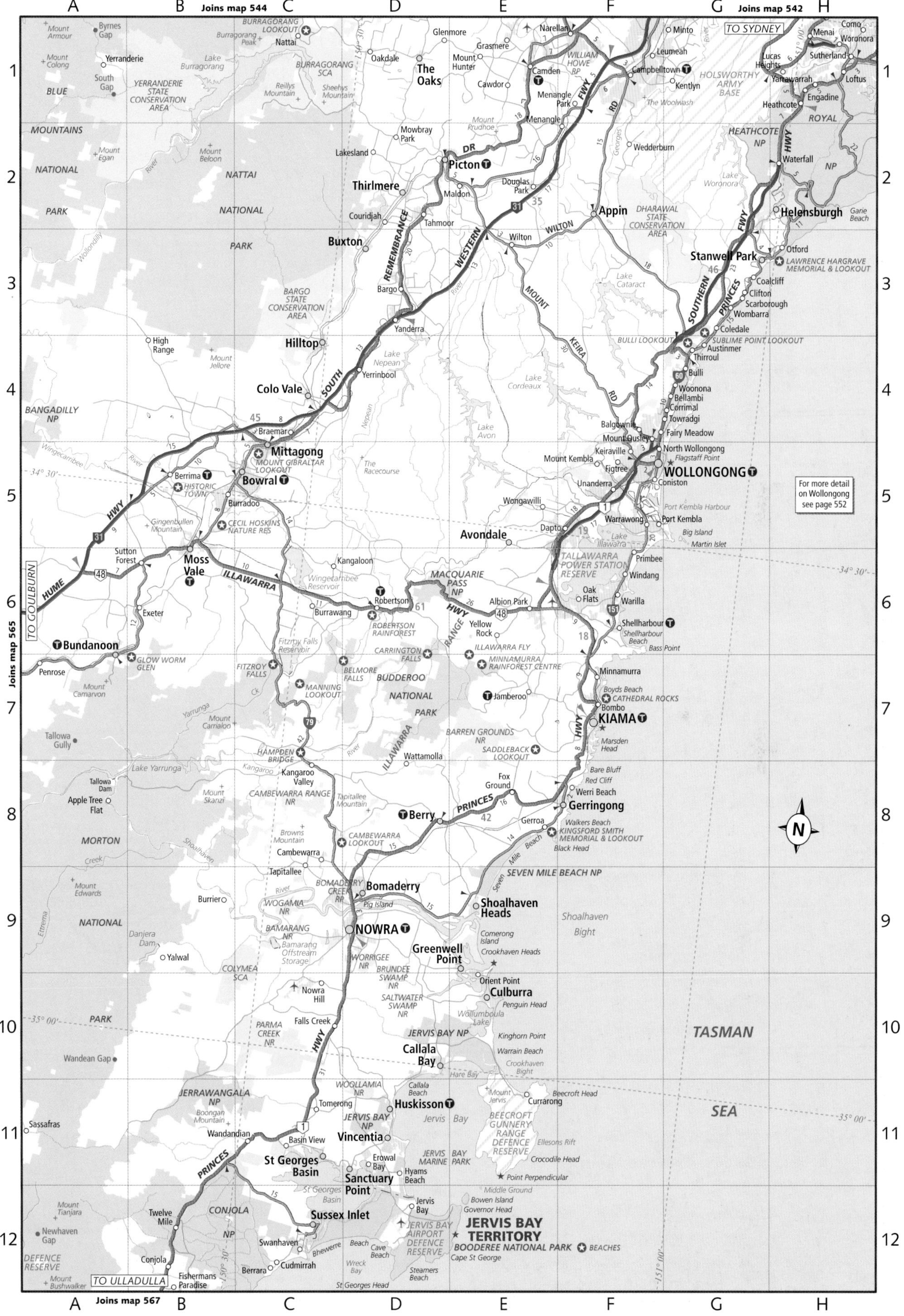

0 5 10 15 20 km
Joins map 544
Joins map 542
Joins map 565
Joins map 567
TO SYDNEY
TO GOULBURN
TO ULLADULLA
For more detail on Wollongong see page 552
Picton
Thirlmere
The Oaks
Camden
Campbelltown
Appin
Helensburgh
Stanwell Park
Hilltop
Colo Vale
Mittagong
Bowral
Moss Vale
Bundanoon
Robertson
Avondale
WOLLONGONG
KIAMA
Gerringong
Berry
Bomaderry
NOWRA
Shoalhaven Heads
Greenwell Point
Culburra
Callala Bay
Huskisson
Vincentia
St Georges Basin
Sanctuary Point
Sussex Inlet
JERVIS BAY TERRITORY
BLUE MOUNTAINS NATIONAL PARK
NATTAI NATIONAL PARK
BUDDEROO NATIONAL PARK
MORTON NATIONAL PARK
ROYAL NP
HEATHCOTE NP
BOODEREE NATIONAL PARK
TASMAN SEA

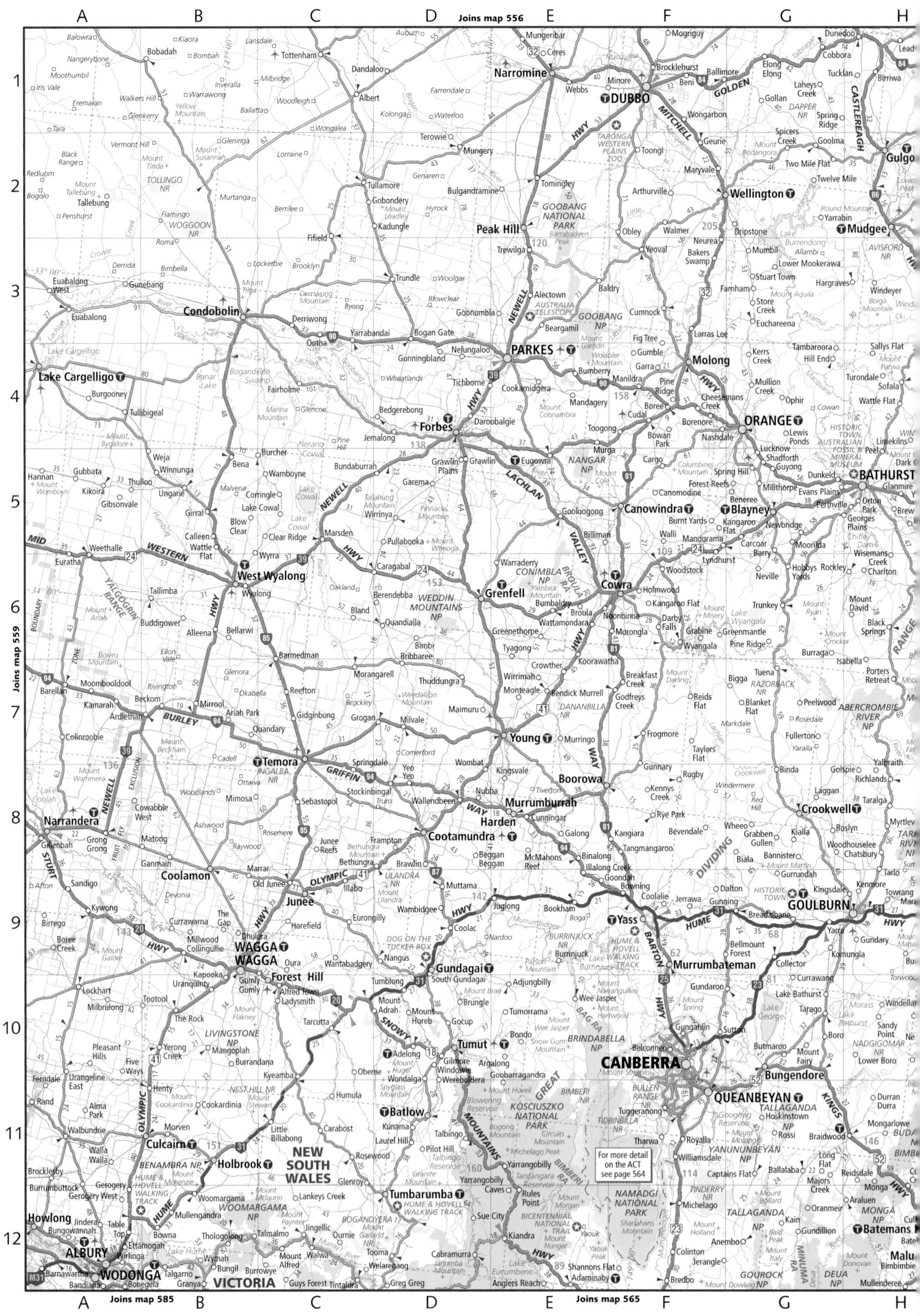
A
B
C
D
Joins map 556
E
F
G
H
1
2
3
4
5
6
7
8
9
10
11
12
Joins map 559
Joins map 585
Joins map 565
Narromine
DUBBO
Wellington
Mudgee
Gulgong
Peak Hill
GOOBANG NATIONAL PARK
Condobolin
Lake Cargelligo
PARKES
Forbes
Molong
ORANGE
BATHURST
Canowindra
Blayney
Cowra
Grenfell
West Wyalong
WEDDIN MOUNTAINS NP
Young
Boorowa
Temora
Murrumburrah
Harden
Cootamundra
Crookwell
Narrandera
Coolamon
Junee
WAGGA WAGGA
Gundagai
Yass
Murrumbateman
GOULBURN
Tumut
Batlow
CANBERRA
QUEANBEYAN
Bungendore
KOSCIUSZKO NATIONAL PARK
NAMADGI NATIONAL PARK
Tumbarumba
Holbrook
Culcairn
NEW SOUTH WALES
VICTORIA
ALBURY
WODONGA
Braidwood
Batemans Bay
For more detail on the ACT see page 564

0 20 40 60 80 100 km

I J K L M N O P

Joins map 557

1 2 3 4 5 6 7 8 9 10 11 12

For more detail on the Mid North Coast see page 550

For more detail on Newcastle & Surrounds see page 549

For more detail on the Central Coast see page 543

SYDNEY

OPERA HOUSE, HARBOUR BRIDGE, THE ROCKS, MACQUARIE LIGHTHOUSE, TARONGA ZOO, PARRAMATTA (HISTORIC TOWN), BONDI BEACH, INDIAN PACIFIC TRAIN

For more detail on Sydney & Surrounds see pages 544–5

TASMAN SEA

For more detail on the Southern Highlands see page 553

For more detail on the Snowy Mountains & The South Coast see pages 566–7

I J K L M N O P

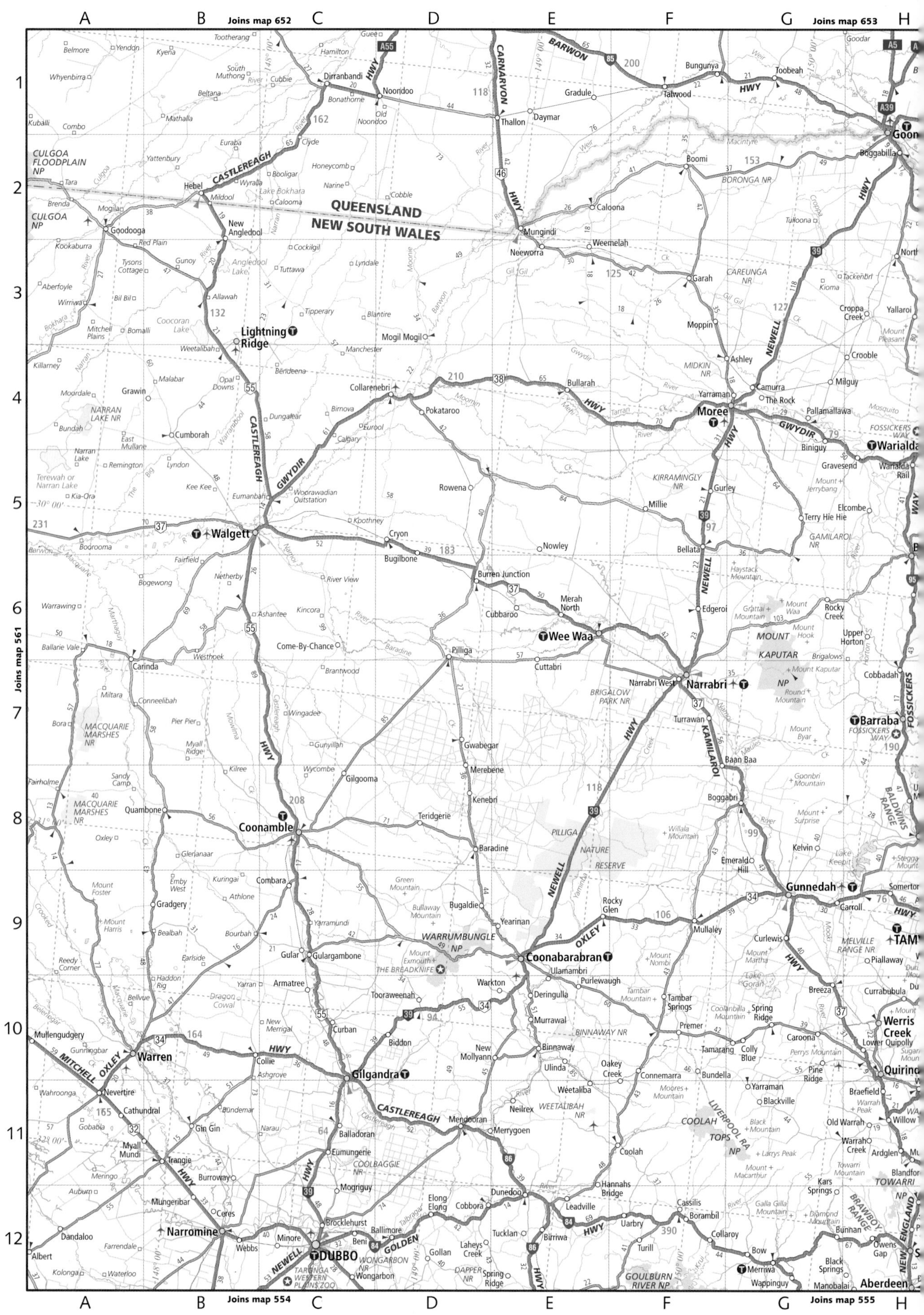

A B C D E F G H
Joins map 652
Joins map 653
Joins map 554
Joins map 555
Joins map 561
1 2 3 4 5 6 7 8 9 10 11 12
QUEENSLAND
NEW SOUTH WALES
Belmore
Yendon
Kyena
Tootherang
Guee
Hamilton
Whyenbirra
South Muthong
Cubbie
Dirranbandi
Noondoo
Bonathorne
Old Noondoo
Beltana
Mathalla
Kuballi
Combo
CULGOA FLOODPLAIN NP
Yattenbury
Euraba
Clyde
Honeycomb
Tara
Hebel
Wyralla
Booligar
Narine
Lake Bokhara
Caloomа
Cobble
Brenda
Mogila
Mildool
CULGOA NP
Goodooga
New Angledool
Kookaburra
Red Plain
Cockilgil
Tysons Cottage
Gunoy
Angledool Lake
Tuttawa
Lyndale
Aberfoyle
Wirriwa
Bil Bil
Allawah
Tipperary
Blantire
Mitchell Plains
Bomalli
Coocoran Lake
Lightning Ridge
Mogil Mogil
Weetalibah
Manchester
Killarney
Malabar
Opal Downs
Bendeena
Moordale
Grawin
Collarenebri
NARRAN LAKE NR
Dungalear
Birnova
Pokataroo
Bundah
East Mullane
Cumborah
Eurool
Calgary
Narran Lake
Remington
Lyndon
Terewah or Narran Lake
Kee Kee
Eumanbah
Woorawadian Outstation
Rowena
Kia-Ora
Walgett
Koothney
Cryon
Bugilbone
Nowley
Boorooma
Fairfield
Barwon
Burren Junction
Bogewong
Netherby
River View
Merah North
Warrawing
Ashantee
Kincora
Cubbaroo
Wee Waa
Ballarie Vale
Come-By-Chance
Pilliga
Carinda
Westhoek
Cuttabri
Brantwood
Narrabri West
Narrabri
Miltara
Conneelibah
BRIGALOW PARK NR
Wingadee
Turrawan
Bora
Pier Pier
MACQUARIE MARSHES NR
Myall Ridge
Gwabegar
Gunyillah
Baan Baa
Kilree
Merebene
Fairholme
Sandy Camp
Wycombe
Gilgooma
Kenebri
Boggabri
Quambone
Coonamble
Teridgerie
PILLIGA NATURE RESERVE
Oxley
Glenanaar
Baradine
Emerald Hill
Mount Foster
Kuringai
Combara
Green Mountain
Emby West
Athlone
Gunnedah
Gradgery
Bugaldie
Rocky Glen
Yearinan
Bullaway Mountain
Mount Harris
Yarramundi
Mullaley
Bealbah
Bourbah
WARRUMBUNGLE NP
Curlewis
Reedy Corner
Earlside
Gular
Gulargambone
Mount Exmouth
THE BREADKNIFE
Coonabarabran
Ulamambri
Haddon Rig
Yarran
Armatree
Warkton
Purlewaugh
Deringulla
Tambar Mountain
Tambar Springs
Bellvue
Dragon Cowal
Tooraweenah
Murrawal
Spring Ridge
New Merrigal
Curban
BINNAWAY NR
Premer
Mullengudgery
Warren
Collie
Biddon
New Mollyann
Binnaway
Tamarang
Colly Blue
Gunningbar
Ashgrove
Gilgandra
Ulinda
Oakey Creek
Connemarra
Bundella
Nevertire
Weetaliba
Moores Mountain
Wahroonga
Cathundral
Bundemar
Neilrex
WEETALIBAH NR
Yarraman
Blackville
Gobabla
Gin Gin
Narau
Balladoran
Mendooran
Merrygoen
COOLAH TOPS NP
LIVERPOOL RA
Black Mountain
Myall Mundi
Trangie
Eumungerie
Coolah
COOLBAGGIE NR
Larrys Peak
Meringo
Burroway
Mount Macarthur
Auburn
Mogriguy
Dunedoo
Hannahs Bridge
Mungeribar
Ceres
Elong Elong
Cobbora
Leadville
Cassilis
Galla Galla Mountain
Diamond Mountain
Dandaloo
Brocklehurst
Borambil
Narromine
Ballimore
Uarbry
Farrendale
Webbs
Minore
Beni
Tucklan
Laheys Creek
Birriwa
Turill
Collaroy
Albert
DUBBO
Gollan
Bow
Kolonga
Waterloo
WONGARBON
Wongarbon
DAPPER NR
Spring Ridge
Merriwa
Wappinguy
TARONGA WESTERN PLAINS ZOO
GOULBURN RIVER NP
Thallon
Daymar
Gradule
Talwood
Bungunya
Toobeah
Goodar
Goondiwindi
Boggabilla
Boomi
BORONGA NR
Caloona
Mungindi
Neeworra
Weemelah
Garah
CAREUNGA NR
Tulloona
Tackenbri
Kioma
Croppa Creek
Yallaroi
Mount Pleasant
Moppin
Ashley
MIDKIN NR
Crooble
Bullarah
Yarraman
Camurra
The Rock
Milguy
Moree
Pallamallawa
Mosquito
FOSSICKERS WAY
Biniguy
Warialda
Gravesend
Warialda Rail
KIRRAMINGLY NR
Gurley
Mount Jerrybang
Millie
Terry Hie Hie
Elcombe
Bellata
GAMILAROI NR
Haystack Mountain
Edgeroi
Grattai Mountain
Mount Waa
Rocky Creek
MOUNT KAPUTAR NP
Mount Hook
Upper Horton
Brigalows
Mount Kaputar
Cobbadah
Round Mountain
Barraba
Mount Byar
Goonbri Mountain
Mount Surprise
Willala Mountain
Kelvin
Lake Keepit
Somerton
Carroll
Mount Martha
MELVILLE RANGE NR
Piallaway
Mount Nombi
Lake Goran
Breeza
Currabubula
Coolanbilla Mountain
Werris Creek
Caroona
Lower Quipolly
Perrys Mountain
Pine Ridge
Quirindi
Braefield
Warrah Peak
Willow Tree
Old Warrah
Warrah Creek
Ardglen
Towarri Mountain
Kars Springs
Bunnan
Owens Gap
Black Springs
Manobalai
Aberdeen
BALDWINS RANGE
BRAWBOY RANGE
TOWARRI NP
CARNARVON HWY
BARWON HWY
CASTLEREAGH HWY
GWYDIR HWY
NEWELL HWY
KAMILAROI HWY
OXLEY HWY
MITCHELL HWY
GOLDEN HWY

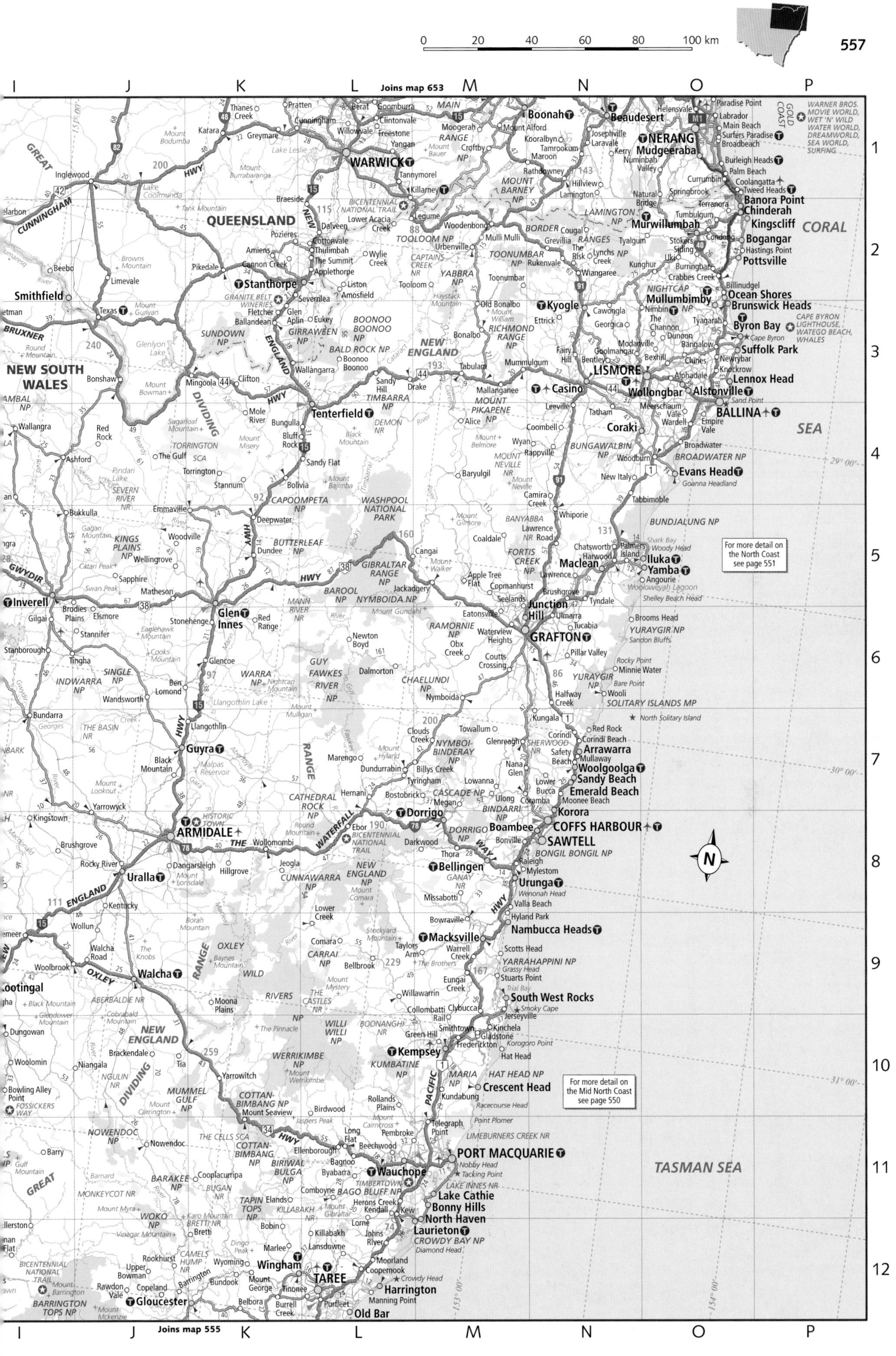

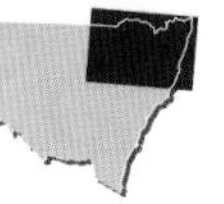

0 20 40 60 80 100 km

I J K L M N O P

Joins map 653

1 2 3 4 5 6 7 8 9 10 11 12

For more detail on the North Coast see page 551

For more detail on the Mid North Coast see page 550

WARNER BROS. MOVIE WORLD, WET 'N' WILD WATER WORLD, DREAMWORLD, SEA WORLD, SURFING

CAPE BYRON LIGHTHOUSE, WATEGO BEACH, WHALES

QUEENSLAND

NEW SOUTH WALES

CORAL SEA

TASMAN SEA

Joins map 555

I J K L M N O P

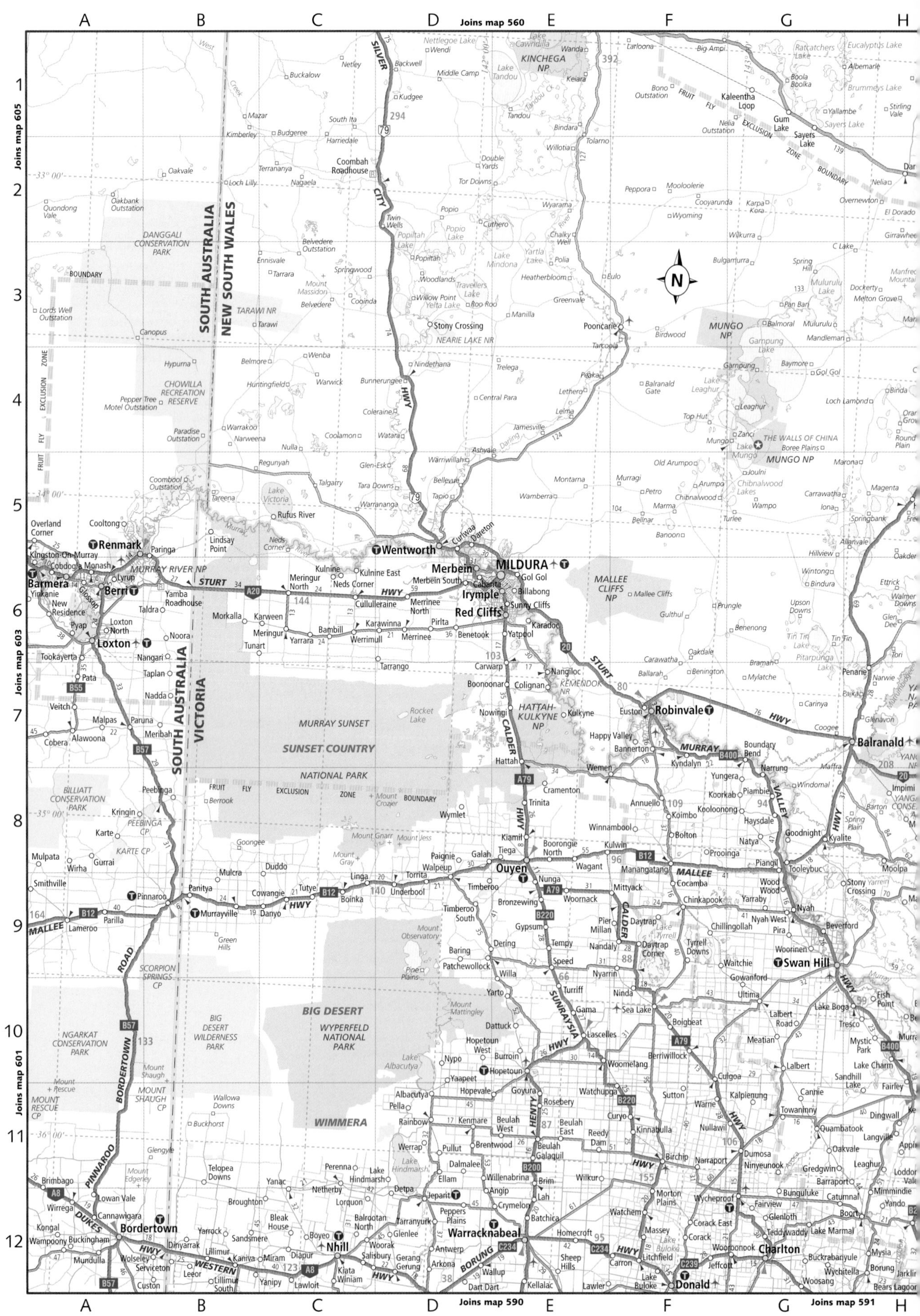
Joins map 560
Joins map 605
Joins map 603
Joins map 601
Joins map 590
Joins map 591
SOUTH AUSTRALIA
NEW SOUTH WALES
VICTORIA
MILDURA
Wentworth
Merbein
Red Cliffs
Irymple
Renmark
Berri
Loxton
Barmera
Robinvale
Balranald
Ouyen
Swan Hill
Charlton
Donald
Warracknabeal
Nhill
Bordertown
KINCHEGA NP
MUNGO NP
MURRAY SUNSET NATIONAL PARK
SUNSET COUNTRY
BIG DESERT
WYPERFELD NATIONAL PARK
WIMMERA
HATTAH-KULKYNE NP
MALLEE CLIFFS NP
STURT HWY
CALDER HWY
SILVER CITY HWY
MALLEE HWY
SUNRAYSIA HWY
HENTY HWY
BORDERTOWN PINNAROO ROAD
WESTERN HWY
DUKES HWY
BORUNG HWY
MURRAY VALLEY HWY
FRUIT FLY EXCLUSION ZONE BOUNDARY

0 20 40 60 80 100 km

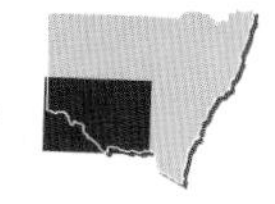

I J K L M N O P

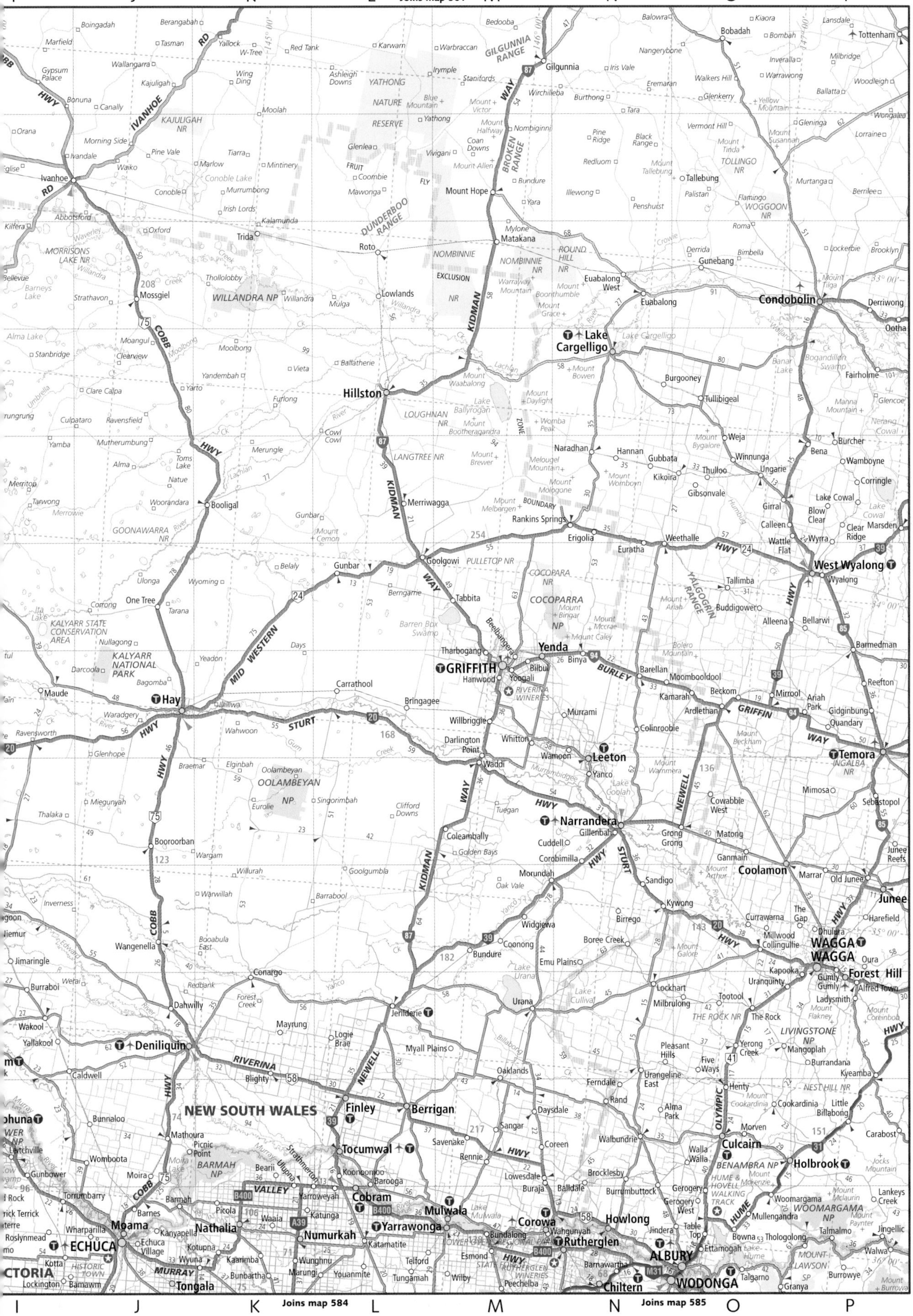

Joins map 561

1
Boingadah Berangabah Marfield Tasman Yallock W-Tree Red Tank Karwarn Warbraccan Bedooba GILGUNNIA RANGE Gilgunnia Balowra Kiaora Bobadah Bombah Lansdale Tottenham
Gypsum Palace Wallangarra Wing Ding Ashleigh Downs Irymple Staniforts Iris Vale Nangerybone Inveralla Milbridge Kajuligah YATHONG NATURE RESERVE Blue Mountain Eremaran Walkers Hill Warrawong Woodleigh Bonuna Canally Moolah Mount Victor Wirchilleba Burthong Glenkerry Ballatta Yellow Mountain IVANHOE RD HWY

2
Orana KAJULIGAH NR Yathong Mount Halfway Nombiginni Tara Vermont Hill Gleninga Wongalea Morning Side Pine Vale Tiarra Coan Downs BROKEN RANGE Pine Ridge Black Range Mount Susannah Lorraine Ivandale Marlow Mintinery Glenlea Vivigani Mount Allen Redluom Mount Tinda TOLLINGO NR Ivanhoe Waiko Conoble Lake Coombie FRUIT FLY Bundure Mount Tallebung Tallebung Murtanga Conoble Murrumbong Mawonga Mount Hope Illewong Palistan Berrilee Irish Lords Yara Penshurst Flamingo WOGGOON NR Kalamunda DUNDERBOO RANGE Roma

3
Kilfera Abbotsford Oxford Trida Mylone Matakana ROUND HILL NR Derrida Bimbella Lockerbie Brooklyn MORRISONS LAKE NR Waverley Roto NOMBINNIE NOMBINNIE NR Gunebang Willandra Crowie Bellevue Barneys Lake 208 Thollolobby EXCLUSION Warraway Mountain Mount Boorithumble Euabalong West Mount Tilga Strathavon Mossgiel WILLANDRA NP Willandra Mulga Lowlands Euabalong 91 Condobolin Derriwong Mount Grace KIDMAN WAY 58 33° 00' Alma Lake Moangul Lake Cargelligo Lake Cargelligo Ootha Stanbridge Clearview Moolbong 99 Ballatherie Banar Lake Bogandillon Swamp COBB

4
Yandembah Vieta Lachlan 58 Mount Bowen 80 Fairholme Yarto Mount Waabalong Burgooney Clare Calpa Furlong Hillston Mount Daylight Tullibigeal Glencoe Umbrella Lake Ballyrogan Manna Mountain Culpataro Ravensfield LOUGHNAN NR Mount Bootheragandra Womba Peak 73 Nerang Cowal Yamba Mutherumbung Cowl Cowl 87 Mount Bygalore Weja Burcher Merungle LANGTREE NR Naradhan Hannan Gubbata Winnunga Bena Toms Lake 94 Mount Brewer Melougel Mountain Kikoira Thulloo Ungarie Wamboyne Alma Natue Mount Wombyn Corringle

5
Merriwagga Mount Mologone Gibsonvale Lake Cowal Merritop Tarwong Woorandara Booligal Mount Melbergen BOUNDARY Girral Blow Clear Lake Cowal Merrowie Gunbar Rankins Springs Calleen Clear Ridge Marsden GOONAWARRA NR Mount Cemon 254 Erigolia Weethalle Wattle Flat Wyrra Euratha HWY 24 PULLETOP NR 39 Belaly Gunbar Goolgowi COCOPARA NR Tallimba YALGOGRIN RANGE West Wyalong Wyalong Ulonga Wyoming Berngarrie Mount Ariah Buddigower Corrong One Tree Tabbita COCOPARRA NP Mount Bingar Mount Mccrae 34° 00' Ita Lake KALYARR STATE CONSERVATION AREA Tarana Barren Box Swamp Mount Caley Alleena Bellarwi

6
Nullagong MID WESTERN Days Beelbangera Bolero Mountain Barmedman KALYARR NATIONAL PARK Yeadon Tharbogang Yenda Binya BURLEY Barellan Darcoola Bagomba GRIFFITH Hanwood Bilbul Yoogali Moombooldool Reefton Maude Carrathool RIVERINA WINERIES Beckom Mirrool Hay Bringagee Kamarah Ariah Park Gidginbung Waradgery Illilliwa Murrami Ardlethan GRIFFIN Quandary Ravensworth Wahwoon STURT 20 168 Willbriggie Colinroobie Mount Beckham WAY Glenhope Darlington Point Whitton Wamoon Leeton Temora INGALBA NR

7
Braemar Elginbah Oolambeyan OOLAMBEYAN NP Waddi Yanco Mount Wammera 136 Eurolie Singorimbah Clifford Downs Murrumbidgee Lake Coolah Cowabbie West Mimosa Sebastopol Miegunyah Tuegan NEWELL Thalaka 75 Narrandera Gillenbah Grong Grong Matong 85 Booroorban 123 Coleambally Golden Bays Cuddell Corobimilla STURT Ganmain Junee Reefs Wargam Goolgumbla Morundah Mount Arthur Coolamon Marrar Old Junee Willurah Oak Vale Sandigo Junee

8
Inverness Warwillah Barrabool Yanco Kywong Currawarna The Gap Harefield Boabula East Widgiewa Birrego 143 Dhulura Wangenella 39 Coonong Boree Creek Millwood Collingullie WAGGA WAGGA Oura Jimaringle Bundure 182 Emu Plains Mount Galore Kapooka Forest Hill Conargo Lake Urana Uranquinty Gumly Gumly Alfred Town Burraboi Wehal Redbank Lake Cullival Lockhart Tootool Ladysmith

9
Forest Creek Dahwilly Urana Milbrulong THE ROCK NR The Rock Mount Flakney Mount Coreinbob Wakool Mayrung Jerilderie Livingstone NP Yallakool Deniliquin Logie Brae Myall Plains Pleasant Hills Yerong Creek Mangoplah Caldwell RIVERINA Blighty 58 Oaklands Five Ways Burrandana Kyeamba Ferndale Urangeline East Henty NEST HILL NR

10
NEW SOUTH WALES Finley Berrigan Daysdale Rand Alma Park Mount Cookardinia Cookardinia Little Billabong Bunnaloo Mathoura 217 Sangar Walbundrie Morven Carabost Picnic Point Tocumwal Savernake Coreen Culcairn Holbrook Wombota BARMAH NP Rennie Walla Walla BENAMBRA NP Jocks Mountain Gunbower Moira Bearii Ulupna Strathmerton Koonoomoo Barooga Lowesdale Brocklesby Gerogery Mount Mckenzie Lankeys Creek Torrumbarry Barmah VALLEY Yarroweyah Cobram Mulwala Burraja Balldale Burrumbuttock Gerogery West HUME & HOVELL WALKING TRACK Woomargama WOOMARGAMA NP Mount Mclaurin Mount Paynter

11
Barnes Picola Waaia Katunga Lake Mulwala Corowa Howlong Jindera Table Top Mullengandra Talmalmo Jingellic Moama Kanyapella Nathalia Numurkah Yarrawonga Bundalong Wahgunyah Bowna Thologolong Echuca Echuca Village Kotupna Wunghnu Katamatite Esmond Rutherglen Ettamogah Lake Hume Walwa Kotta Historic Town Wyuna Kaarimba Telford Barnawartha Albury MOUNT LAWSON SP Lockington Bamawm Tongala Bunbartha Marungi Youanmite Tungamah Wilby Peechelba Chiltern Wodonga Talgarno Burrowye Granya Mount Burrowa VICTORIA RUTHERGLEN WINERIES

1 2 3 4 5 6 7 8 9 10 11 12

Joins map 554

I J K L M N O P

Joins map 584

Joins map 585

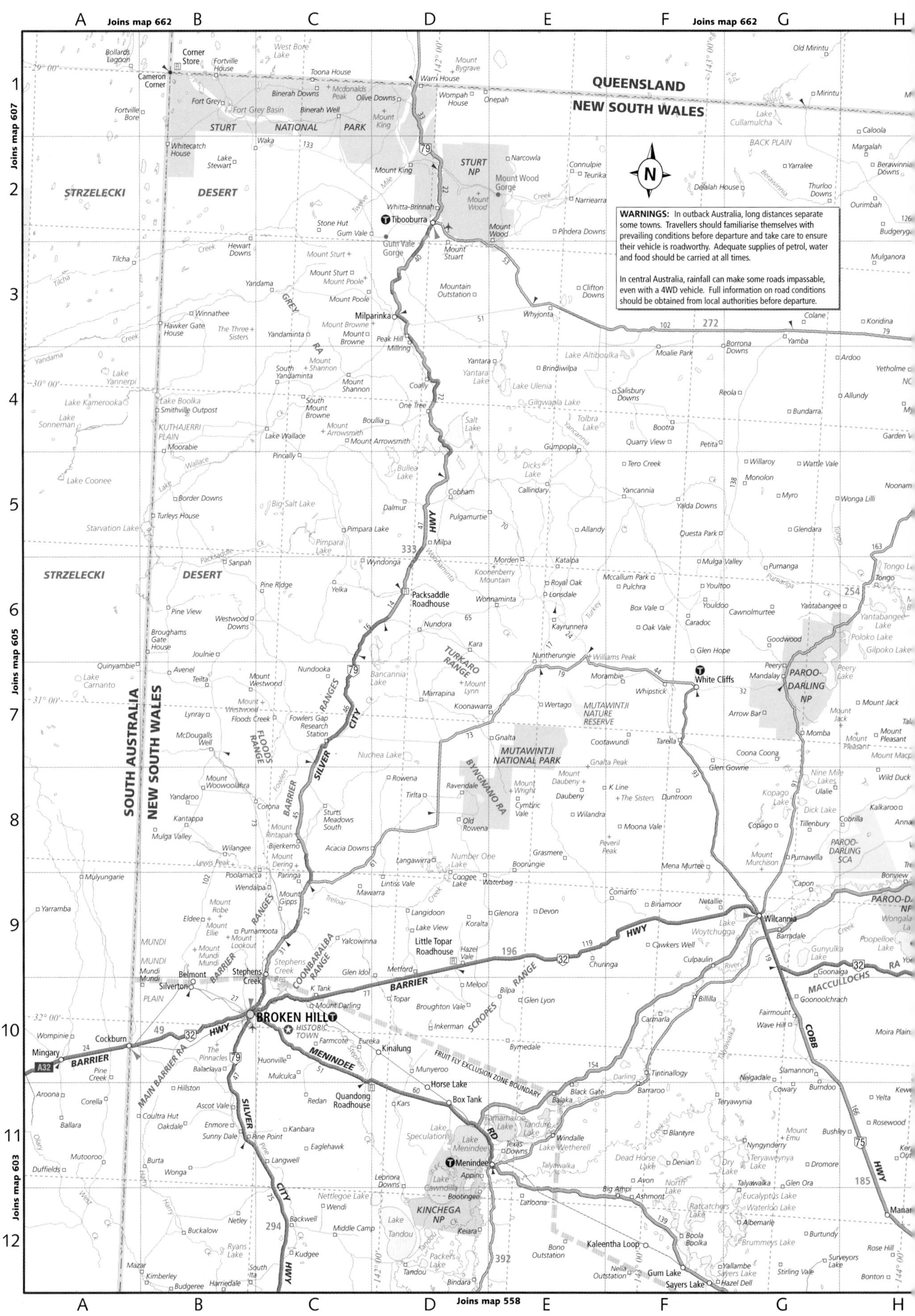

Joins map 662
Joins map 607
Joins map 605
Joins map 603
Joins map 558
A
B
C
D
E
F
G
H
1
2
3
4
5
6
7
8
9
10
11
12
QUEENSLAND
NEW SOUTH WALES
SOUTH AUSTRALIA
WARNINGS: In outback Australia, long distances separate some towns. Travellers should familiarise themselves with prevailing conditions before departure and take care to ensure their vehicle is roadworthy. Adequate supplies of petrol, water and food should be carried at all times.
In central Australia, rainfall can make some roads impassable, even with a 4WD vehicle. Full information on road conditions should be obtained from local authorities before departure.
N
Cameron Corner
Corner Store
STURT NATIONAL PARK
STURT NP
STRZELECKI
DESERT
Tibooburra
Milparinka
Packsaddle Roadhouse
White Cliffs
MUTAWINTJI NATIONAL PARK
MUTAWINTJI NATURE RESERVE
PAROO-DARLING NP
Wilcannia
BROKEN HILL
Silverton
Cockburn
Mingary
Little Topar Roadhouse
Menindee
KINCHEGA NP
Kinalung
Quandong Roadhouse
Horse Lake
Box Tank
Kaleentha Loop
Gum Lake
Sayers Lake
BARRIER HWY
SILVER CITY HWY
COBB HWY
MENINDEE RD
FRUIT FLY EXCLUSION ZONE BOUNDARY
BARRIER RANGE
GREY RA
FLOODS RANGE
COONBARALBA RANGE
SCROPES RANGE
TURKARO RANGE
BYNGNANO RA
HISTORIC TOWN

0 20 40 60 80 100 km

I J K Joins map 663 L M N O Joins map 652 P

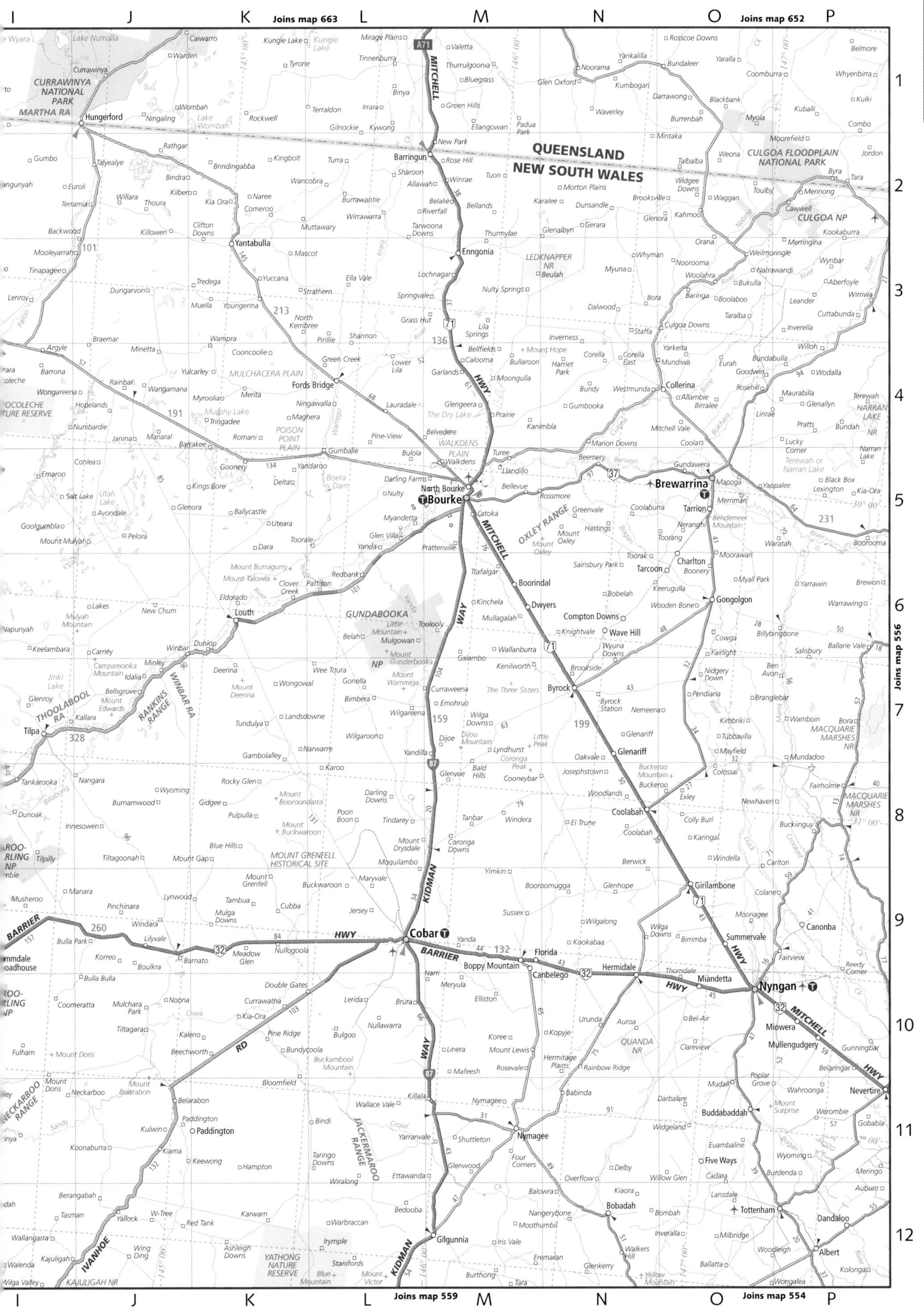

I J K L Joins map 559 M N O Joins map 554 P

Joins map 556

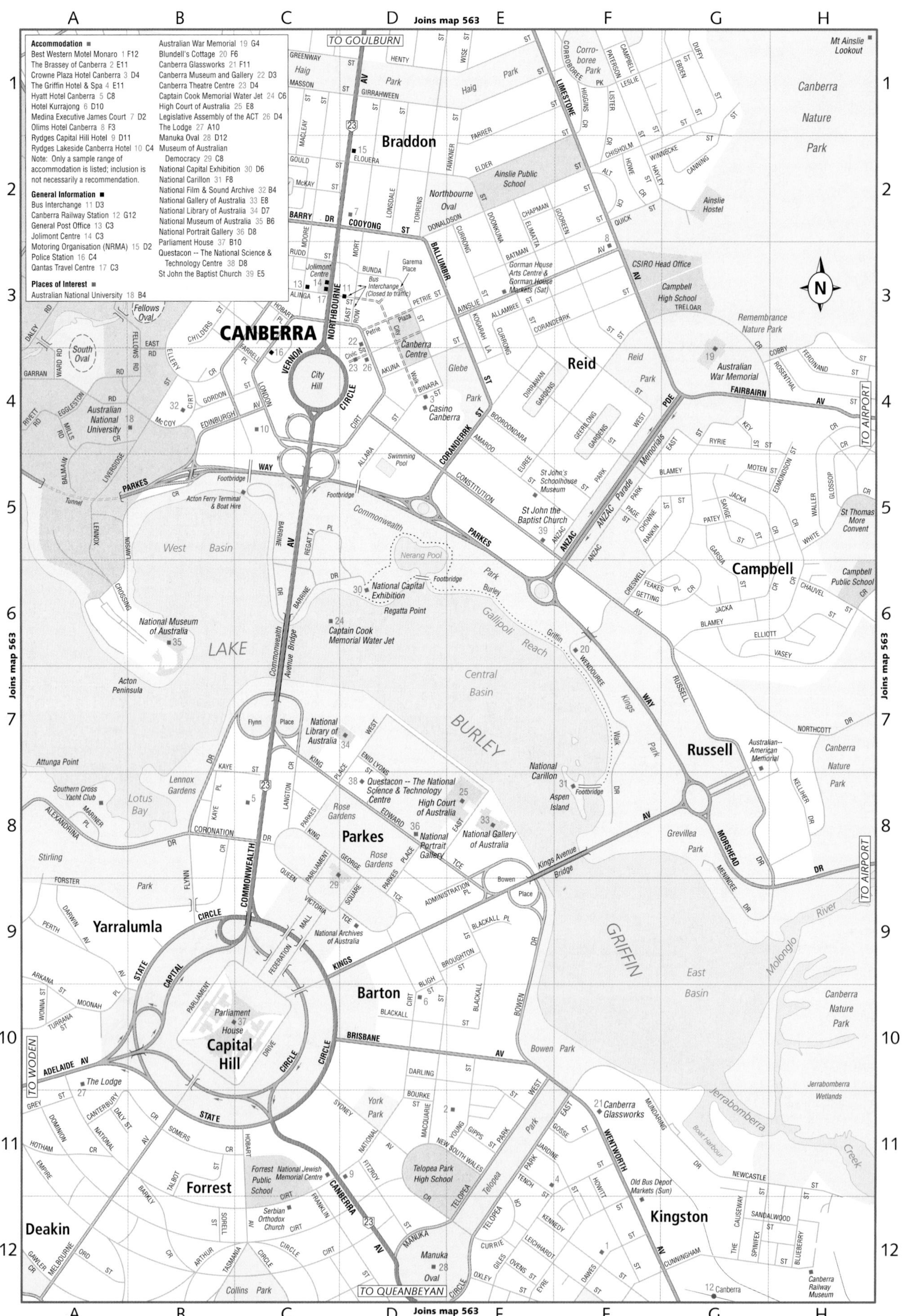
0 0.25 0.50 0.75 1 km
A B C D E F G H
Joins map 563
TO GOULBURN
Accommodation
Best Western Motel Monaro 1 F12
The Brassey of Canberra 2 E11
Crowne Plaza Hotel Canberra 3 D4
The Griffin Hotel & Spa 4 E11
Hyatt Hotel Canberra 5 C8
Hotel Kurrajong 6 D10
Medina Executive James Court 7 D2
Olims Hotel Canberra 8 F3
Rydges Capital Hill Hotel 9 D11
Rydges Lakeside Canberra Hotel 10 C4
Note: Only a sample range of accommodation is listed; inclusion is not necessarily a recommendation.
General Information
Bus Interchange 11 D3
Canberra Railway Station 12 G12
General Post Office 13 C3
Jolimont Centre 14 C3
Motoring Organisation (NRMA) 15 D2
Police Station 16 C4
Qantas Travel Centre 17 C3
Places of Interest
Australian National University 18 B4
Australian War Memorial 19 G4
Blundell's Cottage 20 F6
Canberra Glassworks 21 F11
Canberra Museum and Gallery 22 D3
Canberra Theatre Centre 23 D4
Captain Cook Memorial Water Jet 24 C6
High Court of Australia 25 E8
Legislative Assembly of the ACT 26 D4
The Lodge 27 A10
Manuka Oval 28 D12
Museum of Australian Democracy 29 C8
National Capital Exhibition 30 D6
National Carillon 31 F8
National Film & Sound Archive 32 B4
National Gallery of Australia 33 E8
National Library of Australia 34 D7
National Museum of Australia 35 B6
National Portrait Gallery 36 D8
Parliament House 37 B10
Questacon -- The National Science & Technology Centre 38 D8
St John the Baptist Church 39 E5
Braddon
CANBERRA
City Hill
Civic
Reid
Campbell
Russell
Parkes
Barton
Yarralumla
Capital Hill
Forrest
Deakin
Kingston
Mt Ainslie Lookout
Canberra Nature Park
Haig Park
Northbourne Oval
Ainslie Public School
Ainslie Hostel
CSIRO Head Office
Campbell High School
Remembrance Nature Park
Australian War Memorial
St Thomas More Convent
Campbell Public School
Australian National University
South Oval
Fellows Oval
West Basin
LAKE BURLEY GRIFFIN
Central Basin
East Basin
Acton Peninsula
National Museum of Australia
Captain Cook Memorial Water Jet
National Capital Exhibition
Regatta Point
Nerang Pool
Commonwealth Park
Gallipoli Reach
Kings Park
National Carillon
Aspen Island
Australian-American Memorial
Grevillea Park
Commonwealth Avenue Bridge
Kings Avenue Bridge
National Library of Australia
Questacon -- The National Science & Technology Centre
High Court of Australia
National Portrait Gallery
National Gallery of Australia
Rose Gardens
Lennox Gardens
Lotus Bay
Attunga Point
Southern Cross Yacht Club
Stirling Park
National Archives of Australia
Parliament House
The Lodge
Bowen Park
York Park
Telopea Park High School
Forrest Public School
National Jewish Memorial Centre
Serbian Orthodox Church
Manuka Oval
Collins Park
Canberra Glassworks
Old Bus Depot Markets (Sun)
Jerrabomberra Wetlands
Molonglo River
Jerrabomberra Creek
Canberra Railway Museum
Canberra
Casino Canberra
Glebe Park
Canberra Centre
Gorman House Arts Centre & Gorman House Markets (Sat)
St John's Schoolhouse Museum
St John the Baptist Church
Acton Ferry Terminal & Boat Hire
Swimming Pool
TO AIRPORT
TO WODEN
TO QUEANBEYAN

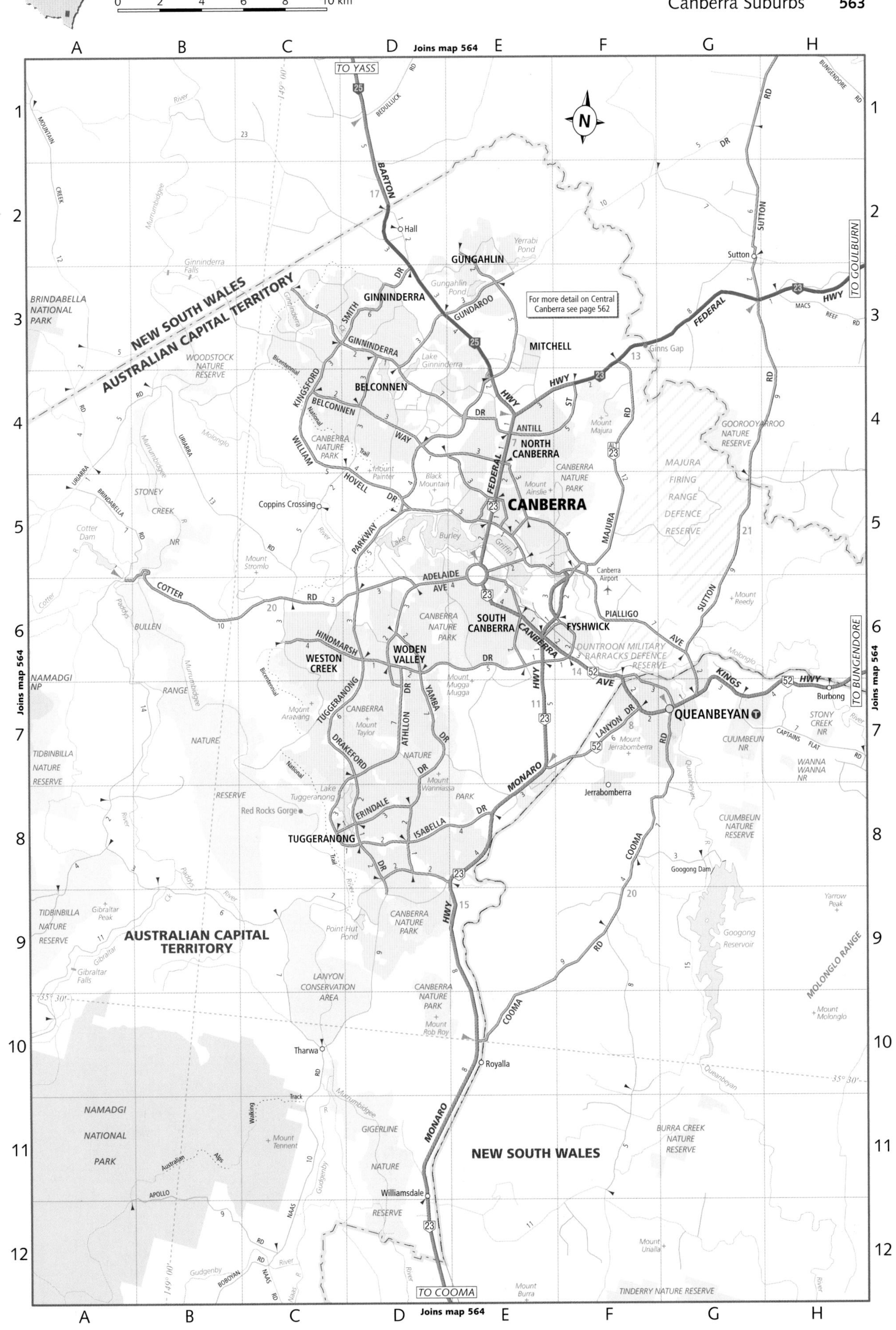

0 2 4 6 8 10 km
Joins map 564
TO YASS
TO COOMA
TO GOULBURN
TO BUNGENDORE
For more detail on Central Canberra see page 562
CANBERRA
NORTH CANBERRA
SOUTH CANBERRA
GUNGAHLIN
GINNINDERRA
BELCONNEN
MITCHELL
WESTON CREEK
WODEN VALLEY
FYSHWICK
TUGGERANONG
QUEANBEYAN
NEW SOUTH WALES
AUSTRALIAN CAPITAL TERRITORY
BRINDABELLA NATIONAL PARK
WOODSTOCK NATURE RESERVE
NAMADGI NATIONAL PARK
TIDBINBILLA NATURE RESERVE
MAJURA FIRING RANGE DEFENCE RESERVE
DUNTROON MILITARY BARRACKS DEFENCE RESERVE
GOOROOYARROO NATURE RESERVE
CUUMBEUN NATURE RESERVE
BURRA CREEK NATURE RESERVE
TINDERRY NATURE RESERVE
LANYON CONSERVATION AREA
GIGERLINE NATURE RESERVE
Hall
Sutton
Ginns Gap
Coppins Crossing
Canberra Airport
Jerrabomberra
Red Rocks Gorge
Googong Dam
Googong Reservoir
Tharwa
Royalla
Williamsdale
Burbong
BARTON HWY
FEDERAL HWY
MONARO HWY
KINGS HWY
CANBERRA AVE
COOMA RD
LANYON DR
ADELAIDE AVE
HINDMARSH DR
COTTER RD
PARKWAY
PIALLIGO AVE
MAJURA RD
Lake Burley Griffin
Lake Ginninderra
Lake Tuggeranong
Mount Ainslie
Mount Majura
Black Mountain
Mount Stromlo
Mount Taylor
Mount Mugga Mugga
Mount Tennent
Mount Rob Roy
Mount Jerrabomberra
Mount Molonglo
MOLONGLO RANGE

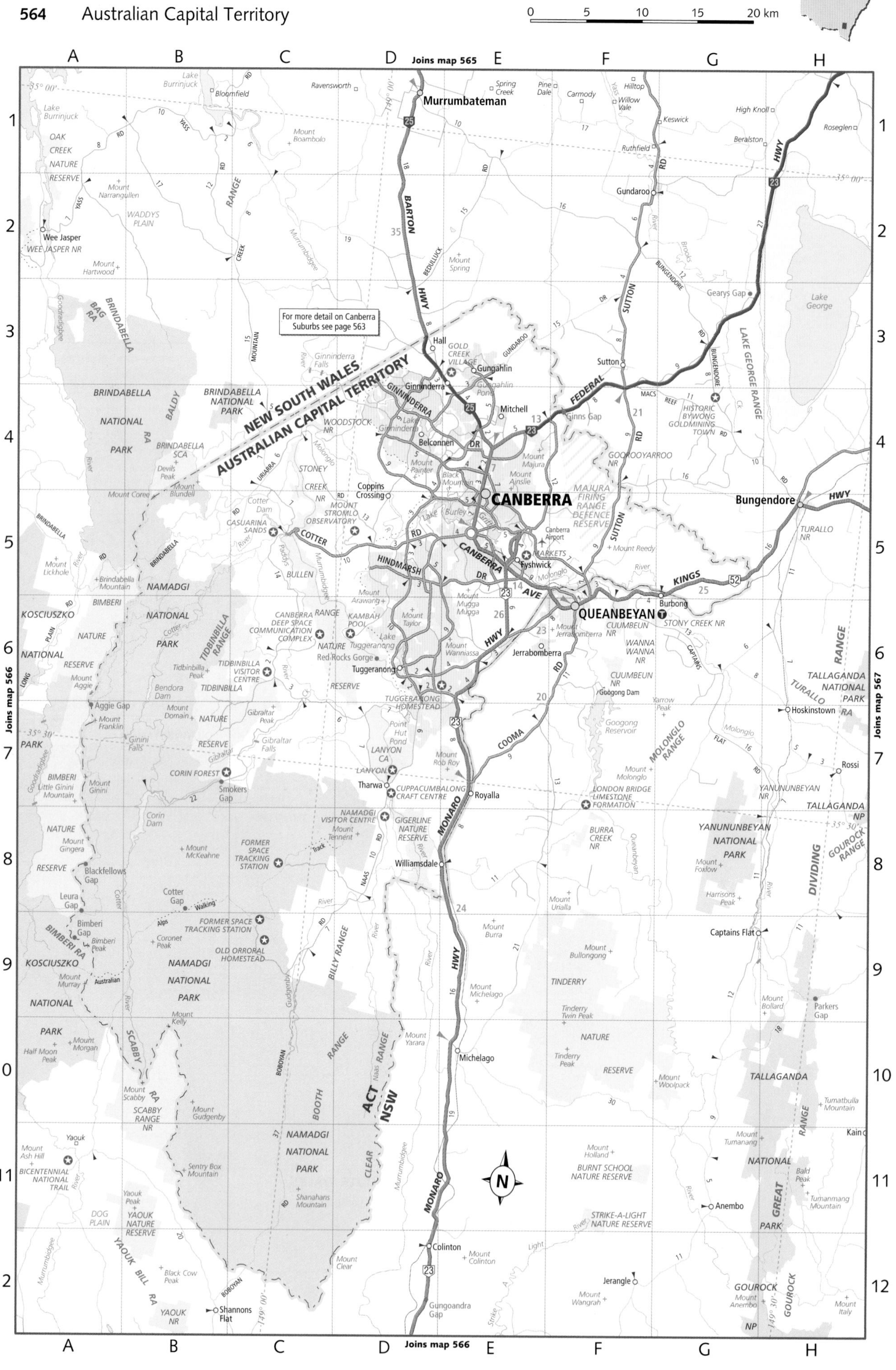
0 5 10 15 20 km
A B C D E F G H
1 2 3 4 5 6 7 8 9 10 11 12
Joins map 565
Joins map 566
Joins map 567
35° 00'
35° 30'
149° 00'
149° 30'
For more detail on Canberra Suburbs see page 563
NEW SOUTH WALES
AUSTRALIAN CAPITAL TERRITORY
ACT
NSW
CANBERRA
QUEANBEYAN
Murrumbateman
Bungendore
Lake Burrinjuck
Bloomfield
Ravensworth
Spring Creek
Pine Dale
Carmody
Hilltop
Willow Vale
Keswick
High Knoll
Roseglen
Beralston
Ruthfield
OAK CREEK NATURE RESERVE
Mount Narrangullen
Mount Boambolo
WADDYS PLAIN
Wee Jasper
WEE JASPER NR
Mount Hartwood
Mount Spring
Gundaroo
Gearys Gap
Lake George
LAKE GEORGE RANGE
BRINDABELLA
BAG RA
BALDY RA
BRINDABELLA NATIONAL PARK
BRINDABELLA SCA
Devils Peak
Mount Blundell
Mount Coree
Hall
GOLD CREEK VILLAGE
Gungahlin
Gungahlin Pond
Ginninderra
Ginninderra Falls
Lake Ginninderra
Mitchell
Sutton
Ginns Gap
MACS REEF
HISTORIC BYWONG GOLDMINING TOWN
WOODSTOCK NR
Belconnen
Mount Painter
Black Mountain
Mount Majura
Mount Ainslie
GOOROOYARROO NR
MAJURA FIRING RANGE DEFENCE RESERVE
STONEY CREEK NR
Coppins Crossing
Cotter Dam
MOUNT STROMLO OBSERVATORY
CASUARINA SANDS
Lake Burley Griffin
Canberra Airport
Mount Reedy
MARKETS
Fyshwick
Burbong
STONY CREEK NR
TURALLO NR
Mount Lickhole
Brindabella Mountain
BULLEN
Mount Arawang
Mount Mugga Mugga
KOSCIUSZKO NATIONAL PARK
BIMBERI NATURE RESERVE
NAMADGI NATIONAL PARK
TIDBINBILLA RANGE
CANBERRA DEEP SPACE COMMUNICATION COMPLEX
KAMBAH POOL
Mount Taylor
Lake Tuggeranong
Mount Jerrabomberra
CUUMBEUN NR
WANNA WANNA NR
Tidbinbilla Peak
TIDBINBILLA VISITOR CENTRE
TIDBINBILLA NATURE RESERVE
Red Rocks Gorge
Tuggeranong
Mount Wanniassa
Jerrabomberra
Googong Dam
Yarrow Peak
TALLAGANDA NATIONAL PARK
Mount Aggie
Aggie Gap
Mount Franklin
Bendora Dam
Mount Domain
Gibraltar Peak
Gibraltar Falls
TUGGERANONG HOMESTEAD
Point Hut Pond
Googong Reservoir
MOLONGLO RANGE
Hoskinstown
TURALLO RA
Ginini Falls
LANYON CA
LANYON
Tharwa
CUPPACUMBALONG CRAFT CENTRE
Mount Rob Roy
Royalla
Mount Molonglo
LONDON BRIDGE LIMESTONE FORMATION
Rossi
YANUNUNBEYAN NR
CORIN FOREST
Smokers Gap
Little Ginini Mountain
Mount Ginini
Corin Dam
NAMADGI VISITOR CENTRE
Mount Tennent
GIGERLINE NATURE RESERVE
BURRA CREEK NR
YANUNUNBEYAN NATIONAL PARK
TALLAGANDA NP
GOUROCK RANGE
DIVIDING RANGE
Mount Gingera
Blackfellows Gap
Mount McKeahne
FORMER SPACE TRACKING STATION
Williamsdale
Mount Foxlow
Leura Gap
Cotter Gap
Walking
Alps
Australian
Mount Urialla
Harrisons Peak
Bimberi Gap
Bimberi Peak
BIMBERI RA
Coronet Peak
OLD ORRORAL HOMESTEAD
BILLY RANGE
Mount Burra
Captains Flat
Mount Bullongong
Mount Murray
TINDERRY NATURE RESERVE
Mount Michelago
Tinderry Twin Peak
Mount Bollard
Parkers Gap
Half Moon Peak
Mount Morgan
Mount Kelly
SCABBY RA
Mount Yarara
Michelago
Tinderry Peak
Mount Woolpack
TALLAGANDA
Mount Scabby
SCABBY RANGE NR
Mount Gudgenby
BOOTH RANGE
Naas RANGE
CLEAR RANGE
Tumatbulla Mountain
Kain
Yaouk
Mount Ash Hill
BICENTENNIAL NATIONAL TRAIL
Sentry Box Mountain
Shanahans Mountain
Mount Holland
BURNT SCHOOL NATURE RESERVE
Mount Tumanang
NATIONAL
Bald Peak
Tumanmang Mountain
GREAT
PARK
Anembo
Yaouk Peak
YAOUK NATURE RESERVE
DOG PLAIN
STRIKE-A-LIGHT NATURE RESERVE
Colinton
Mount Colinton
Mount Clear
Black Cow Peak
YAOUK BILL RA
YAOUK NR
Shannons Flat
Gungoandra Gap
Jerangle
Mount Wangrah
GOUROCK
Mount Anembo
NP
Mount Italy
BARTON HWY
FEDERAL HWY
KINGS HWY
MONARO HWY
CANBERRA AVE
HINDMARSH DR
COTTER RD
COOMA RD
SUTTON RD
BUNGENDORE RD
GUNDAROO DR
BEDULLUCK
YASS RD
MOUNTAIN CREEK RD
URIARRA RD
BRINDABELLA RD
LONG PLAIN RD
NAAS RD
BOBOYAN RD
CAPTAINS FLAT RD
TURALLO HWY
GINNINDERRA DR
Murrumbidgee River
Molonglo River
Queanbeyan River
Goodradigbee River
Cotter River
Gudgenby River
Naas River
Paddys River
N

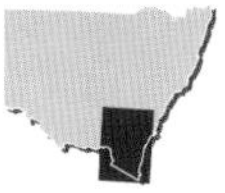

0 20 40 60 80 100 km

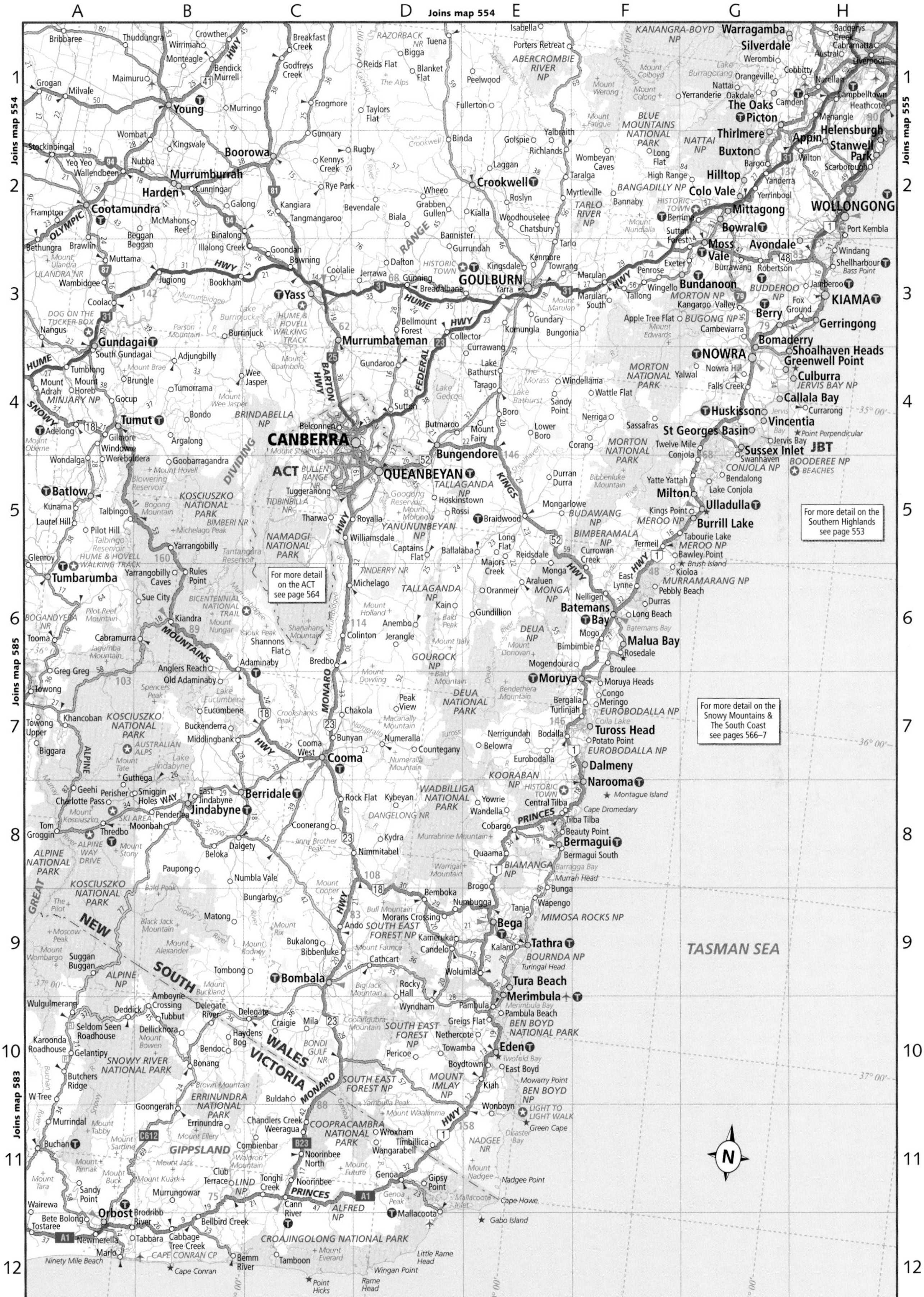

A B C D E F G H

1 2 3 4 5 6 7 8 9 10 11 12

Joins map 554

Joins map 555

Joins map 585

Joins map 583

For more detail on the ACT see page 564

For more detail on the Southern Highlands see page 553

For more detail on the Snowy Mountains & The South Coast see pages 566–7

TASMAN SEA

CANBERRA

ACT

NEW SOUTH WALES

VICTORIA

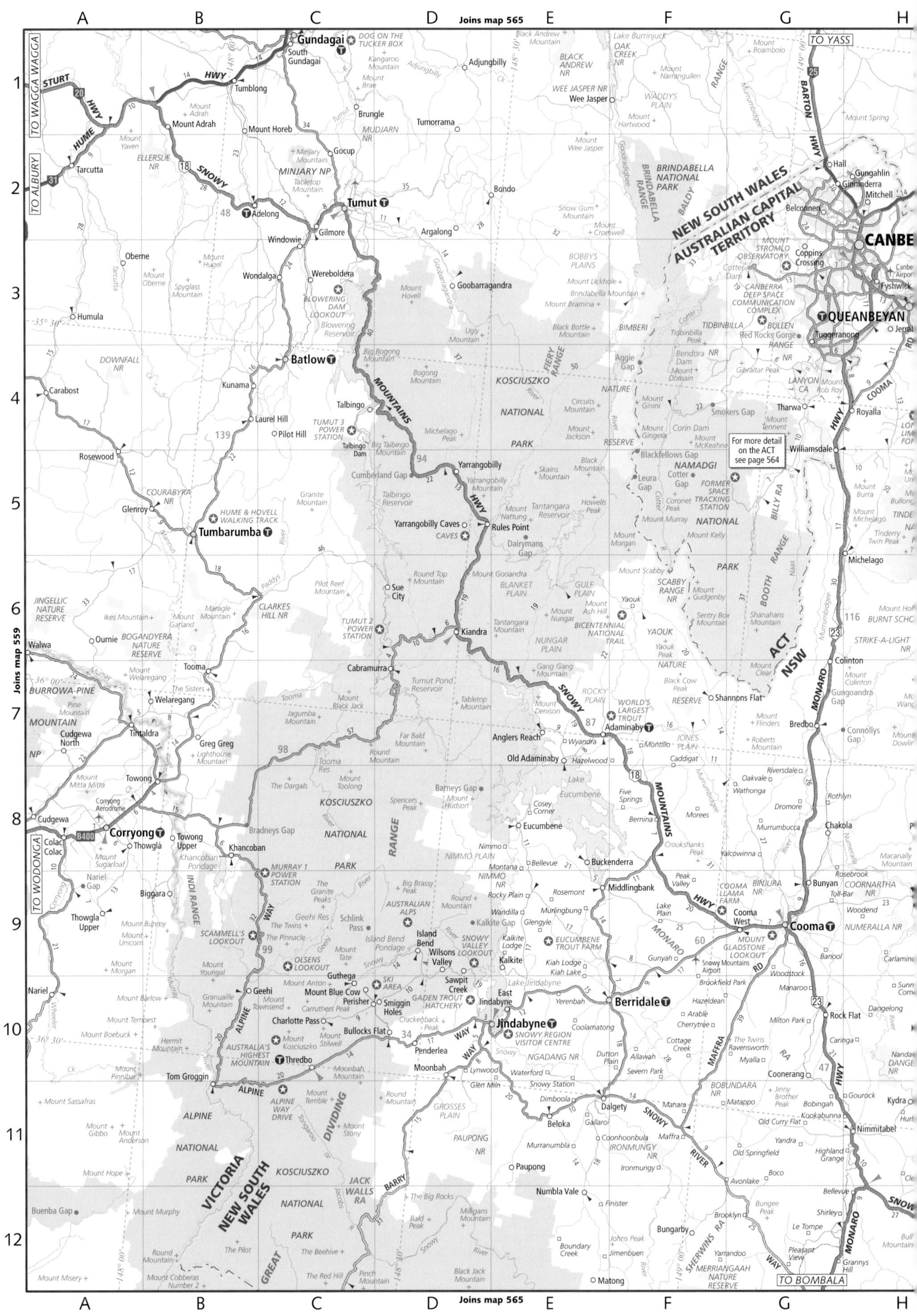
Joins map 565
Joins map 559
TO WAGGA WAGGA
TO ALBURY
TO WODONGA
TO YASS
TO BOMBALA
Gundagai
Tumut
Batlow
Tumbarumba
Adelong
Corryong
Jindabyne
Berridale
Cooma
Adaminaby
Thredbo
Khancoban
Kiandra
Yarrangobilly Caves
Queanbeyan
Canberra
Nimmitabel
Michelago
Bredbo
Tharwa
Tarcutta
Tumblong
Mount Adrah
Gilmore
Talbingo
Cabramurra
Tooma
Towong
Walwa
Dalgety
Bombala
SNOWY MOUNTAINS HWY
MONARO HWY
ALPINE WAY
BARTON HWY
HUME HWY
STURT HWY
KOSCIUSZKO NATIONAL PARK
NAMADGI NATIONAL PARK
BRINDABELLA NATIONAL PARK
ALPINE NATIONAL PARK
NEW SOUTH WALES
AUSTRALIAN CAPITAL TERRITORY
VICTORIA
ACT
NSW
For more detail on the ACT see page 564
MURRAY 1 POWER STATION
TUMUT 3 POWER STATION
TUMUT 2 POWER STATION
AUSTRALIA'S HIGHEST MOUNTAIN
SNOWY REGION VISITOR CENTRE
HUME & HOVELL WALKING TRACK
BLOWERING DAM LOOKOUT
WORLD'S LARGEST TROUT
MOUNT STROMLO OBSERVATORY
CANBERRA DEEP SPACE COMMUNICATION COMPLEX

I J K L Joins map 565 M N O P

GOULBURN
23 HWY
The Hermitage
Trentham
Currawang
Quialigo
Torwood
Kooringaroo
Mapleville
Bomaderry
TO KIAMA
SEVEN MILE BEACH NP
Shoalhaven Heads
Burrier
NOWRA
Danjera Dam
Benduck
Rogara
Westridge
Allawah
Timberton
Mount Baby
Lake Bathurst
Forest Grove
Shoalhaven Gorge
MORTON NATIONAL PARK
Yalwal
COLYMEA SCA
Nowra Hill
Greenwell Point
Shoalhaven Bight
Crookhaven Lighthouse
Orient Point
Culburra
Penguin Head
Wollumboula Lake
Kinghorn Point
Warrain Beach
Crookhaven Bight
Windellama
Lake Bathurst
Lake George
RANGE
Tarago
Sandy Point
Wattle Flat
BEES NEST NR
PIGEON HOUSE RANGE
Wandean Gap
PARMA CREEK NR
Falls Creek
JERVIS BAY NP
Callala Bay
Hare Bay
Currarong
Beecroft Head
Snake Inlet
GUNNERY RANGE DEFENCE RES
Ellesons Rift
Crocodile Head
Point Perpendicular
JERRALONG NR
Rolfes Gap
Nerriga
Cons Gap
Gilberts Gap
Sassafras
JERRAWANGALA NP
Boongan Mountain
Tomerong
Huskisson
BEECROFT
JERVIS BAY NP
Vincentia
Jervis Bay
Erowal Bay
Wandandian
Basin View
St Georges Basin
Hyams Beach
Jervis Bay
Bowen Island
Governor Head
Sussex Inlet
Swanhaven
Wreck Bay
JBT
JERVIS BAY AIRPORT DEFENCE RESERVE
Cape St George
BOODEREE NP
BEACHES
St Georges Head
JERVIS BAY MP
For more detail on the Southern Highlands see page 553
RIC MINING
Boro
BRAIDWOOD
Mount Coghill
NADGIGOMAR NATURE RESERVE
Sunset Mountain
Mount Fairy
Butmaroo
Lower Boro
Corang
Round Mountain
Mount Tianjara
Twelve Mile
CONJOLA NATIONAL PARK
Cudmirrah
Berrara
PRINCES
KINGS
SCOTT NR
DIVIDING
Square Top Mountain
Fosters Mountain
Quiltys Mountain
Newhaven Gap
DEFENCE RESERVE
Mount Bushwalker
Conjola
Fishermans Paradise
Bendalong
Red Head
Sturgiss Mountain
Clyde Gorge
NADGIGOMAR NR
DURRAN DURRA RANGE
Durran Durra
Wog Wog Mountain
Bibbenluke Mountain
Mount Cole
MORTON
Yatte Yattah
Manyana
Cunjurong
Lake Conjola
Byangee Mountain
Talaterang Mountain
POINTER GAP LOOKOUT
NARRAWALLEE CREEK NR
Narrawallee Inlet
TALLAGANDA NATIONAL PARK
BUDAWANG RANGE
Wirritin Mountain
NATIONAL PARK
Narrawallee
Bannisters Point
Milton
Mollymook
Mount Palerang
Mountain Ash Hill
Pigeon House Mountain
Kings Point
Ulladulla
Ulladulla Head
Warden Head Lighthouse
HISTORIC TOWN
Braidwood
BUDAWANG NATIONAL
Currockbilly Mountain
Rossi
Misery Mountain
RA
Mount Gillamatong
Mongarlowe
MEROO NP
Burrill Lake
Lagoon Head
Wairo Beach
Tabourie Lake
Tabourie Point
Crampton Island
MOUNT ELRINGTON PLAIN
GOUROCK
Mount Budawang
BIMBERAMALA NP
Mount Budawanga
PARK
Termeil
MEROO NP
Meroo Head
Ballalaba
Long Flat
GREAT
Majors Creek
Monga Mountain
Bawley Point
Bawley Point
Wallaces Gap
Bells Mountain
Reidsdale
Monga
Currowan Creek
Brush Island
Kioloa
Belowla Island
O'Hara Head
Snapper Point
MURRAMARANG RANGE
Parkers Gap
Frogs Hole Mountain
Araluen North
Araluen
DEUA
ARALUEN NR
MONGA NATIONAL PARK
MURRAMARANG
East Lynne
Pebbly Beach
Tranquillity Bay
Grasshopper Island
Point Upright
Oranmeir
Nelligen
Benandarah
Beagle Bay
Durras
Tumatbulla Mountain
NP
THE BIG HOLE AND MARBLE ARCH
Kain
CLYDE RIVER NP
Cullendulla
NATIONAL PARK
Tumanmang Mountain
SHELL MUSEUM
Batemans Bay
Long Beach
North Head
Gundillion
Gollarribee Mountain
Batehaven
Batemans Bay
Tollgate Islands
BOMBALAWA RA
Appletree Mountain
Bier Mountain
Surf Beach
Circuit Beach
Fairfield
Mount Berowry
Winbenby Mountain
Wandera Mountain
Mogo
MOGO ZOO
Lilli Pilli
Malua Bay
Mount Grungola
Mongamula Mountain
Larrys Mountain
Bimbimbie
Tomakin
Rosedale
Mount Italy
MINUMA RANGE
Mount Donovan
Mossy Point
Broulee Bay
Burrewarra Point
Burrewarra Point Lighthouse
GOUROCK NP
DEUA
Mogendoura
Broulee
North Beach
Moruya Aerodrome
Bald Mountain
Middle Mountain
Bendethera Mountain
NATIONAL PARK
Mullenderee
Moruya
Moruya Heads
Moruya Heads
Kiora
Pedro (Yowaga) Point
Euranbene Mountain
PARK
BADJA SWAMPS NR
Congo
Congo Point
EUROBODALLA NP
Badja Mill
Bergalia
Meringo
Mullimburra Point
TASMAN
Bingie Bingie Point
Turlinjah
Wongaburra
Loch Lomond
Little Badja
Coila Lake
Mount Coman
WEDGET PLAIN
Tuross Head
Tuross Inlet
Spring Mountain
Comerang Mountain
Old Bodalla
Bodalla
Tuross Lake
Blackfellows Point
Potato Point
Jemisons Point
Belowra
Nerrigundah
EUROBODALLA NP
Jillicambra Mountain
Mount Utopia
Eurobodalla
Brookwood
Brou Beach
Cadgee Mountain
Cadgee
Tally Ho
Wadbilliga Gap
Belowra
DSS FALLS
Belowra Mountain
Barren Jumbo Mountain
Cobra Mountain
Dalmeny
Kianga
Wagonga Head
Wadbilliga
Wagonga Inlet
Conways Gap
KOORABAN
Gembrook
Narooma
TWO RIVER PLAIN
Jeffers Mountain
GULAGA NP
Corunna
Barunga Point
Bogola Head
MONTAGUE ISLAND NR
Montague Island
SEA
Yowrie
Wandellow
NATIONAL
Image Farm
HISTORIC TOWN
Corunna Point
Cape Dromedary
WADBILLIGA
Oakside
Wandella
Central Tilba
Tilba Tilba
EUROBODALLA NP
NATIONAL
Mount Dumpling
Narira Mountain
PARK
Narira
Little Dromedary Mountain
GULAGA
Wallaga Lake
Cobargo
Glencraig
NP
Beauty Point
Haywards Beach
PARK
Murrabrine Gap
Caroon
Murrabrine Mountain
Grevillia
Cadjangarry Mountain
Mossybanks
Bermagui
Monk
Bermagui South
Brogo
Quaama
BERMAGUI NR
Jerimbut Point
Baragoot Point
Brogo Reservoir
Waterloo
BIAMANGA NATIONAL PARK
Milton Park
Cuttagee Point
Warrigal Mountain
Margaret Park
Lynwood
Greystones
Barragga Bay
Armands Bay
Pigeon Box Mountain
Brogo Pass
Mumbulla Mountain
Murrah Head
SOUTH EAST FOREST NP
Brogo
Brinawa
Glenbogen
BERMAGUI
Bunga
Goalen Head
MIMOSA ROCKS
Bunga Head
Aragunnu Bay
Irvington
Hawkshead
Warrick Farm
TATHRA
Wapengo
Bemboka
Milion Park
Bengunnu Point
Picnic Point
Numbugga
HWY
Limegrove
Tanja
Bithry Inlet
Morans Crossing
Gourlay
Hillside
TO EDEN
Bega
NATIONAL PARK

1 2 3 4 5 6 7 8 9 10 11 12

35° 00' 35° 30' 36° 00' 36° 30' 149° 30' 150° 00' 150° 30' 151° 00'

I Joins map 565 J K L M N O P

MAPS

VICTORIA

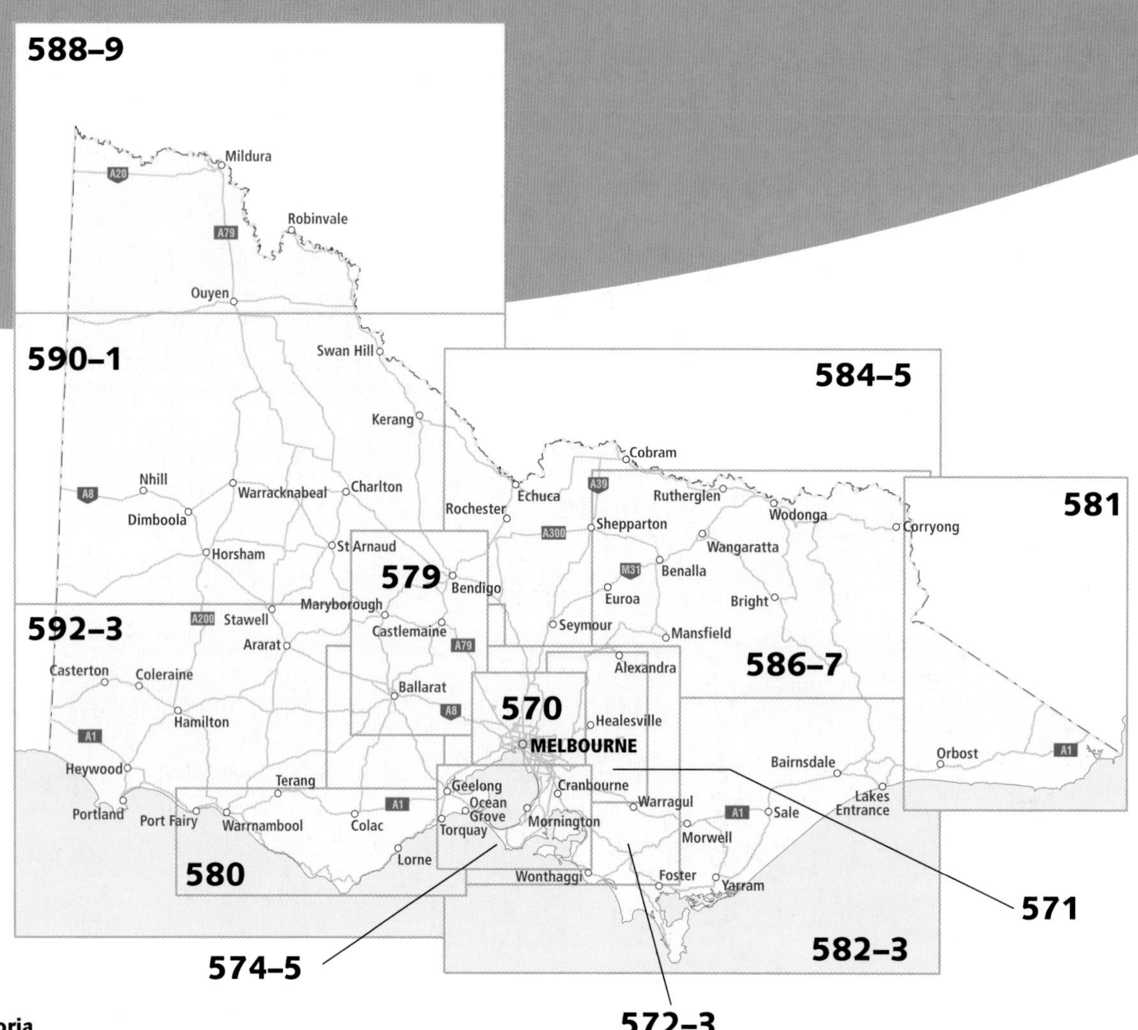

INTER-CITY ROUTES		DISTANCE
Melbourne–Sydney via Hume Hwy/Fwy	M31 31	879 km
Melbourne–Sydney via Princes Hwy/Fwy	M1 A1 1	1039 km
Melbourne–Adelaide via Western & Dukes hwys	M8 A8 M1	729 km
Melbourne–Adelaide via Princes Hwy	M1 A1 B1 M1	907 km
Melbourne–Brisbane via Newell Hwy	M31 A39 39 A39 A2	1676 km

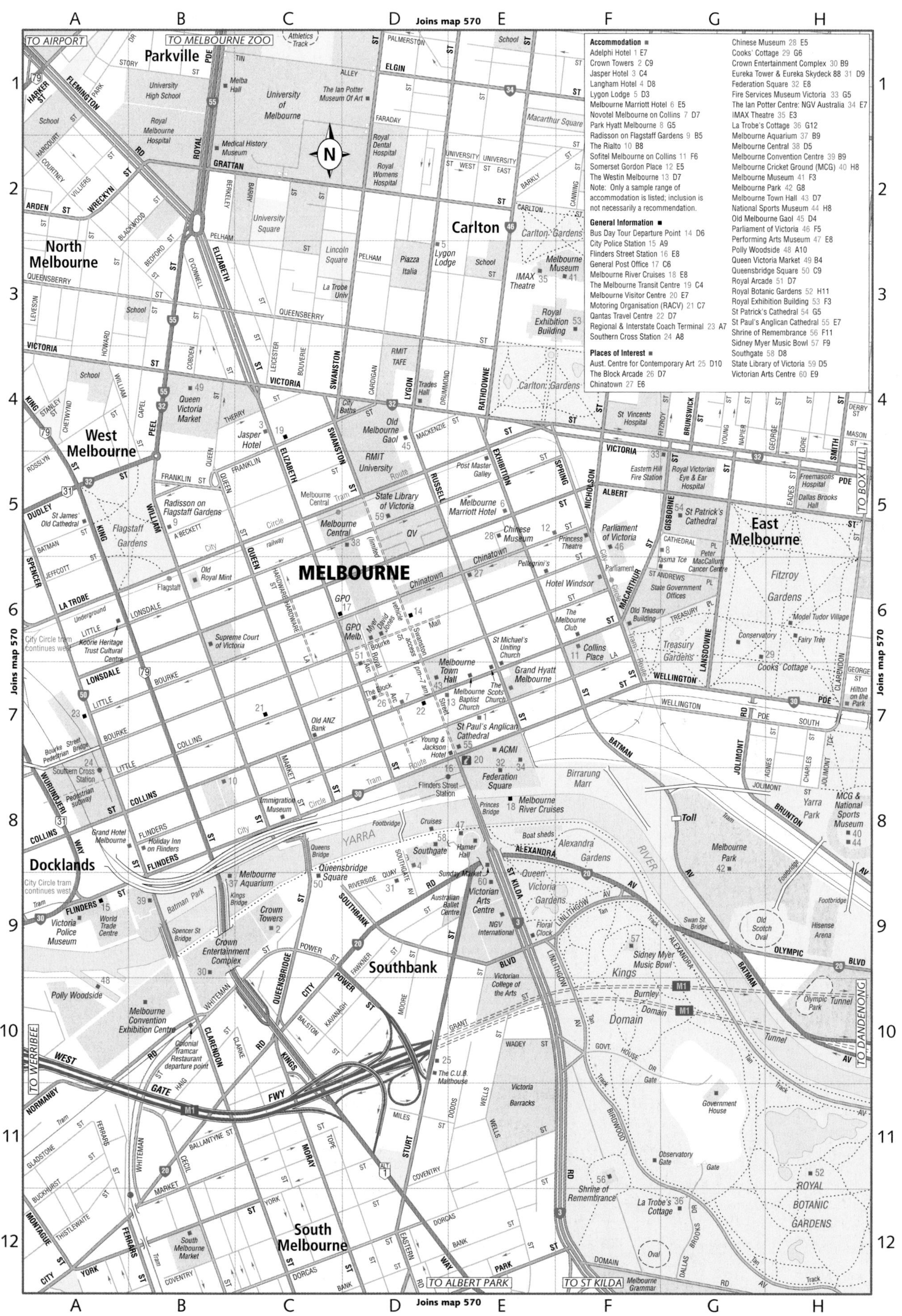
Joins map 570
TO AIRPORT
TO MELBOURNE ZOO
Parkville
North Melbourne
West Melbourne
Carlton
East Melbourne
MELBOURNE
Docklands
Southbank
South Melbourne
Accommodation
Adelphi Hotel 1 E7
Crown Towers 2 C9
Jasper Hotel 3 C4
Langham Hotel 4 D8
Lygon Lodge 5 D3
Melbourne Marriott Hotel 6 E5
Novotel Melbourne on Collins 7 D7
Park Hyatt Melbourne 8 G5
Radisson on Flagstaff Gardens 9 B5
The Rialto 10 B8
Sofitel Melbourne on Collins 11 F6
Somerset Gordon Place 12 E5
The Westin Melbourne 13 D7
Note: Only a sample range of accommodation is listed; inclusion is not necessarily a recommendation.
General Information
Bus Day Tour Departure Point 14 D6
City Police Station 15 A9
Flinders Street Station 16 E8
General Post Office 17 C6
Melbourne River Cruises 18 E8
The Melbourne Transit Centre 19 C4
Melbourne Visitor Centre 20 E7
Motoring Organisation (RACV) 21 C7
Qantas Travel Centre 22 D7
Regional & Interstate Coach Terminal 23 A7
Southern Cross Station 24 A8
Places of Interest
Aust. Centre for Contemporary Art 25 D10
The Block Arcade 26 D7
Chinatown 27 E6
Chinese Museum 28 E5
Cooks' Cottage 29 G6
Crown Entertainment Complex 30 B9
Eureka Tower & Eureka Skydeck 88 31 D9
Federation Square 32 E8
Fire Services Museum Victoria 33 G5
The Ian Potter Centre: NGV Australia 34 E7
IMAX Theatre 35 E3
La Trobe's Cottage 36 G12
Melbourne Aquarium 37 B9
Melbourne Central 38 D5
Melbourne Convention Centre 39 B9
Melbourne Cricket Ground (MCG) 40 H8
Melbourne Museum 41 F3
Melbourne Park 42 G8
Melbourne Town Hall 43 D7
National Sports Museum 44 H8
Old Melbourne Gaol 45 D4
Parliament of Victoria 46 F5
Performing Arts Museum 47 E8
Polly Woodside 48 A10
Queen Victoria Market 49 B4
Queensbridge Square 50 C9
Royal Arcade 51 D7
Royal Botanic Gardens 52 H11
Royal Exhibition Building 53 F3
St Patrick's Cathedral 54 G5
St Paul's Anglican Cathedral 55 E7
Shrine of Remembrance 56 F11
Sidney Myer Music Bowl 57 F9
Southgate 58 D8
State Library of Victoria 59 D5
Victorian Arts Centre 60 E9
TO WERRIBEE
TO ALBERT PARK
TO ST KILDA
TO DANDENONG
TO BOX HILL

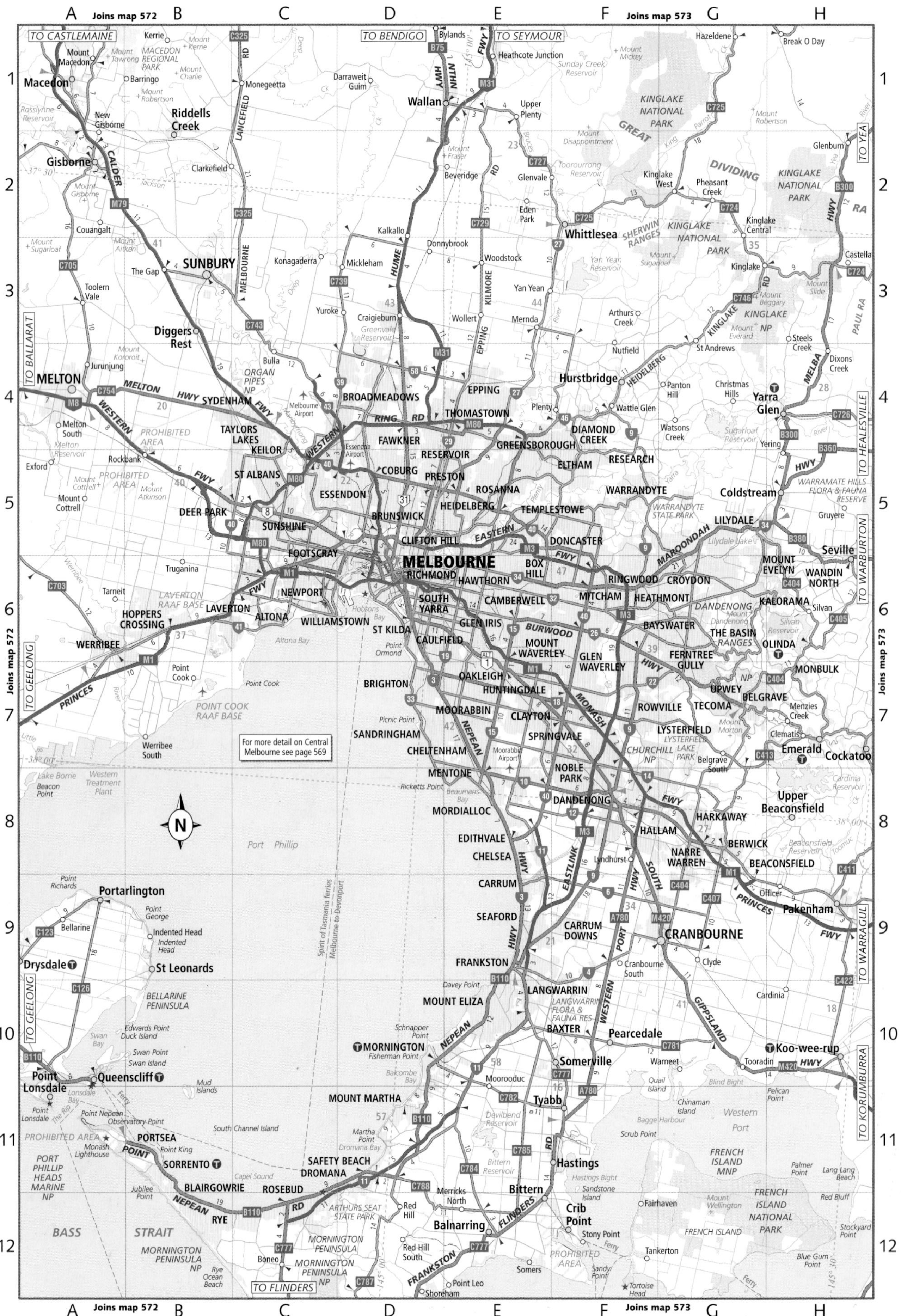
0 5 10 15 20 km
Joins map 572
Joins map 573
TO CASTLEMAINE
TO BENDIGO
TO SEYMOUR
TO BALLARAT
TO GEELONG
TO FLINDERS
TO YEA
TO HEALESVILLE
TO WARBURTON
TO WARRAGUL
TO KORUMBURRA
Macedon
Mount Macedon
Riddells Creek
Gisborne
New Gisborne
Kerrie
Barringo
Monegeetta
Darraweit Guim
Bylands
Heathcote Junction
Wallan
Upper Plenty
Clarkefield
Beveridge
Glenvale
Eden Park
Kalkallo
Donnybrook
Woodstock
Whittlesea
Couangalt
SUNBURY
The Gap
Konagaderra
Mickleham
Yan Yean
Toolern Vale
Diggers Rest
Yuroke
Craigieburn
Wollert
Mernda
Arthurs Creek
Nutfield
Bulla
ORGAN PIPES NP
Jurunjung
MELTON
Melton South
Exford
Rockbank
Mount Cottrell
DEER PARK
Truganina
Tarneit
HOPPERS CROSSING
WERRIBEE
Point Cook
Werribee South
TAYLORS LAKES
KEILOR
ST ALBANS
SUNSHINE
FOOTSCRAY
BROADMEADOWS
ESSENDON
COBURG
BRUNSWICK
FAWKNER
RESERVOIR
PRESTON
EPPING
THOMASTOWN
GREENSBOROUGH
HEIDELBERG
ROSANNA
CLIFTON HILL
MELBOURNE
RICHMOND
HAWTHORN
BOX HILL
DONCASTER
TEMPLESTOWE
ELTHAM
DIAMOND CREEK
Hurstbridge
Plenty
RESEARCH
WARRANDYTE
St Andrews
Panton Hill
Christmas Hills
Yarra Glen
Kinglake
Kinglake West
Kinglake Central
Pheasant Creek
Hazeldene
Break O Day
Glenburn
Castella
Steels Creek
Dixons Creek
Yering
Coldstream
LILYDALE
Seville
WANDIN NORTH
MOUNT EVELYN
KALORAMA
Silvan
Gruyere
RINGWOOD
CROYDON
MITCHAM
HEATHMONT
BAYSWATER
THE BASIN
OLINDA
FERNTREE GULLY
MONBULK
UPWEY
TECOMA
BELGRAVE
Menzies Creek
Clematis
Emerald
Cockatoo
Belgrave South
Upper Beaconsfield
HARKAWAY
BERWICK
BEACONSFIELD
Officer
Pakenham
NEWPORT
LAVERTON
ALTONA
WILLIAMSTOWN
ST KILDA
SOUTH YARRA
CAMBERWELL
GLEN IRIS
CAULFIELD
MOUNT WAVERLEY
GLEN WAVERLEY
OAKLEIGH
HUNTINGDALE
CLAYTON
BRIGHTON
MOORABBIN
SANDRINGHAM
CHELTENHAM
MENTONE
MORDIALLOC
SPRINGVALE
NOBLE PARK
DANDENONG
ROWVILLE
LYSTERFIELD
HALLAM
NARRE WARREN
EDITHVALE
CHELSEA
CARRUM
Lyndhurst
SEAFORD
CARRUM DOWNS
CRANBOURNE
Cranbourne South
Clyde
FRANKSTON
LANGWARRIN
MOUNT ELIZA
BAXTER
Pearcedale
Cardinia
Koo-wee-rup
Tooradin
Warneet
Somerville
Tyabb
Moorooduc
MORNINGTON
MOUNT MARTHA
Hastings
Bittern
Crib Point
Stony Point
Balnarring
Merricks North
Red Hill
Red Hill South
Somers
Point Leo
Shoreham
SAFETY BEACH
DROMANA
ROSEBUD
RYE
BLAIRGOWRIE
SORRENTO
PORTSEA
Boneo
Portarlington
Bellarine
Indented Head
St Leonards
Drysdale
Point Lonsdale
Queenscliff
BELLARINE PENINSULA
Port Phillip
Western Port
BASS STRAIT
FRENCH ISLAND NATIONAL PARK
Tankerton
Fairhaven
KINGLAKE NATIONAL PARK
MORNINGTON PENINSULA NP
PORT PHILLIP HEADS MARINE NP
For more detail on Central Melbourne see page 569

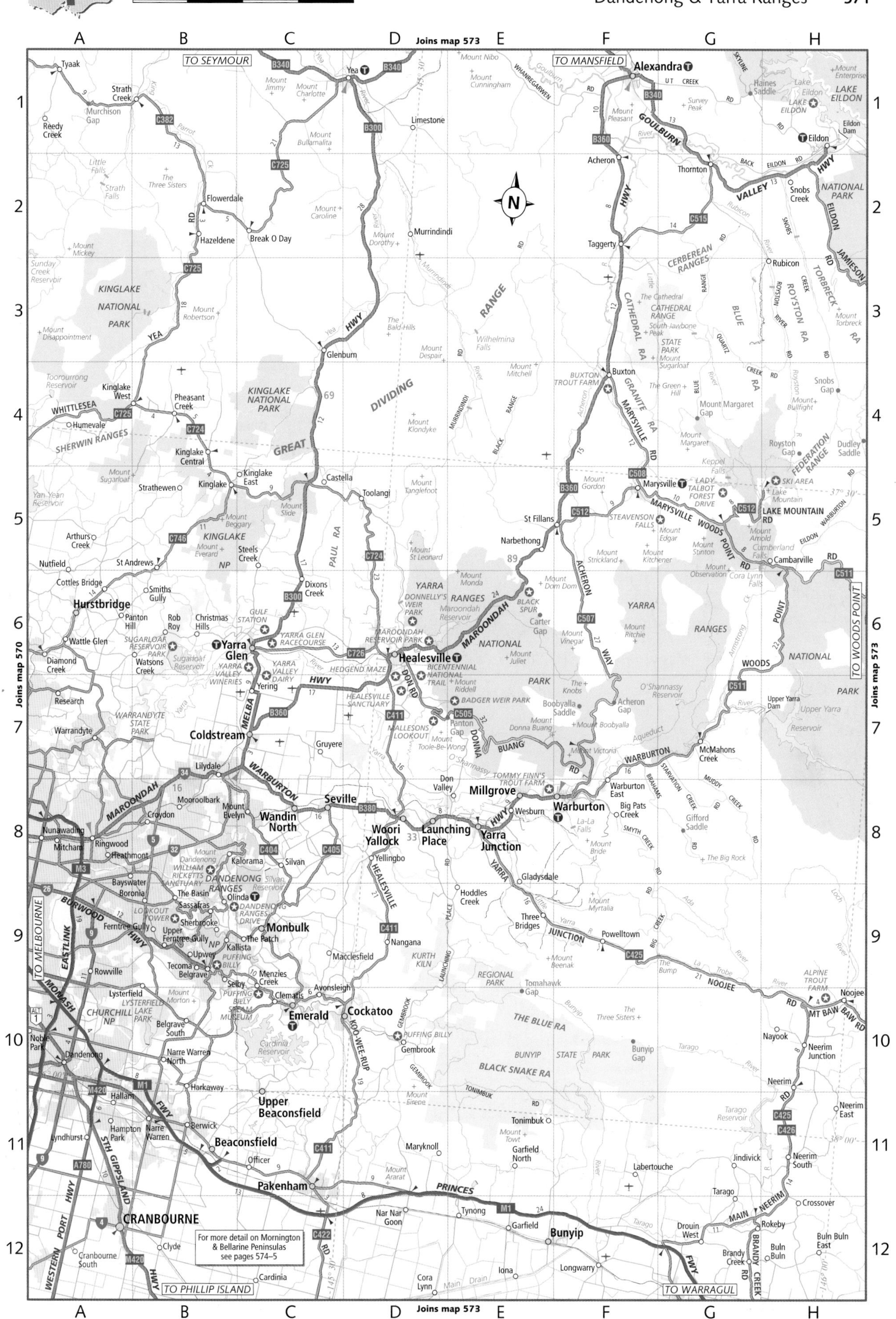
0 5 10 15 20 km
Joins map 573
TO SEYMOUR
TO MANSFIELD
Alexandra
Yea
Kinglake National Park
Healesville
Yarra Glen
Coldstream
Lilydale
Seville
Wandin North
Woori Yallock
Launching Place
Yarra Junction
Warburton
Millgrove
Marysville
Eildon
Lake Eildon
Buxton
Taggerty
Acheron
Thornton
Glenburn
Toolangi
Kinglake
Hurstbridge
Yarra Ranges National Park
Dandenong Ranges NP
Monbulk
Emerald
Cockatoo
Gembrook
Upper Beaconsfield
Beaconsfield
Pakenham
Cranbourne
Bunyip
Garfield
Neerim South
Noojee
Bunyip State Park
TO MELBOURNE
TO PHILLIP ISLAND
TO WARAGUL
TO WOODS POINT
For more detail on Mornington & Bellarine Peninsulas see pages 574–5
Joins map 570
Joins map 573

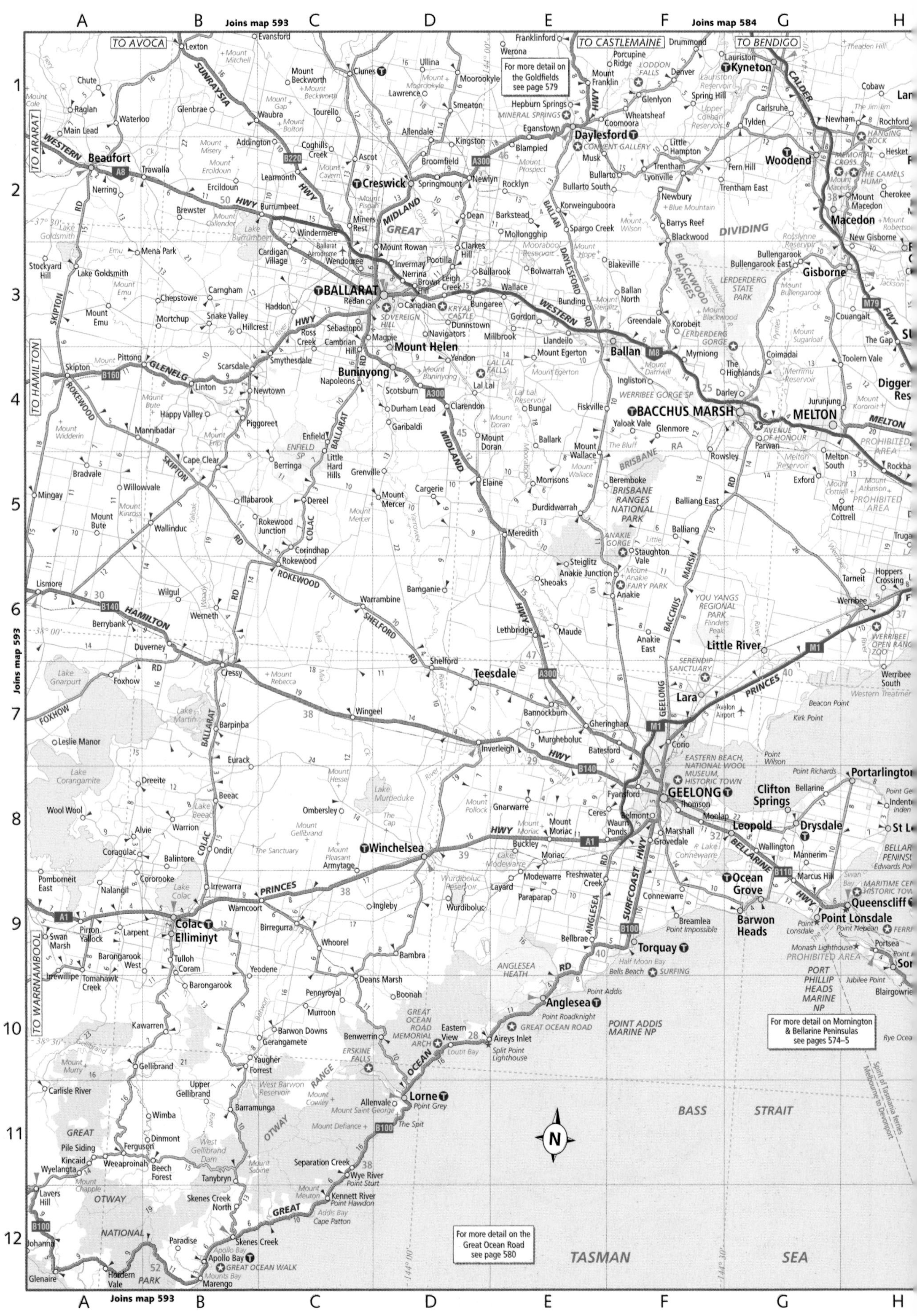
Joins map 593
Joins map 584
TO AVOCA
TO CASTLEMAINE
TO BENDIGO
TO ARARAT
TO HAMILTON
TO WARRNAMBOOL
For more detail on the Goldfields see page 579
For more detail on Mornington & Bellarine Peninsulas see pages 574–5
For more detail on the Great Ocean Road see page 580
BALLARAT
Daylesford
Kyneton
Woodend
Gisborne
Macedon
Ballan
BACCHUS MARSH
MELTON
GEELONG
Lara
Little River
Torquay
Anglesea
Lorne
Apollo Bay
Colac
Elliminyt
Winchelsea
Teesdale
Beaufort
Creswick
Buninyong
Mount Helen
Queenscliff
Point Lonsdale
Portarlington
Drysdale
Leopold
Clifton Springs
Ocean Grove
Barwon Heads
BASS STRAIT
TASMAN SEA
PORT PHILLIP HEADS MARINE NP
POINT ADDIS MARINE NP
GREAT OTWAY NATIONAL PARK
BRISBANE RANGES NATIONAL PARK
LERDERDERG STATE PARK
YOU YANGS REGIONAL PARK
WESTERN HWY
MIDLAND HWY
HAMILTON HWY
PRINCES HWY
SURFCOAST HWY
GREAT OCEAN RD
CALDER FWY
N

0 10 20 30 km

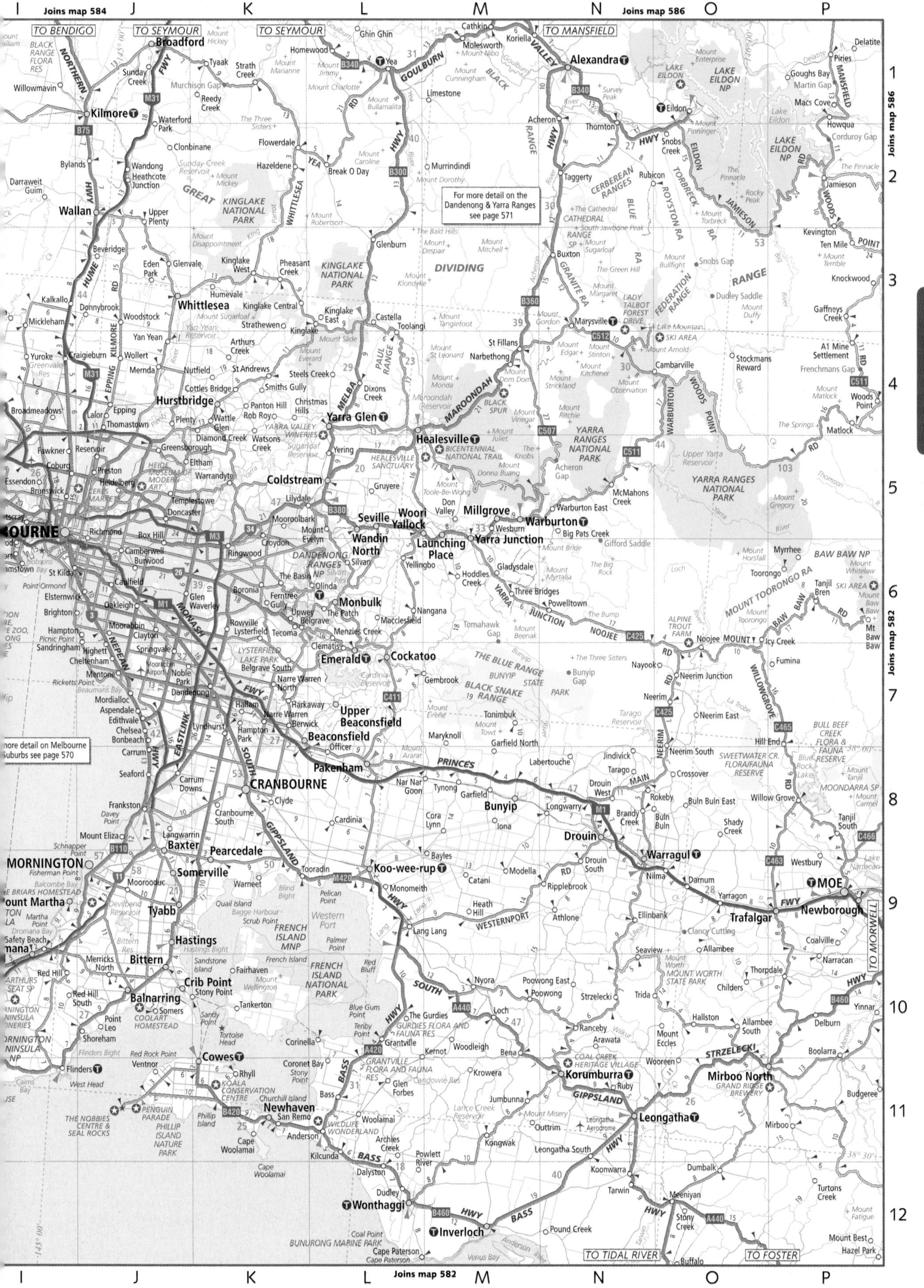

Joins map 584
Joins map 586
Joins map 582
I J K L M N O P
TO BENDIGO
TO SEYMOUR
TO SEYMOUR
TO MANSFIELD
TO MORWELL
TO TIDAL RIVER
TO FOSTER
Broadford
Kilmore
Wallan
Whittlesea
Kinglake
Yea
Alexandra
Eildon
LAKE EILDON NP
Molesworth
Cathkin
Koriella
Ghin Ghin
Homewood
Tyaak
Strath Creek
Reedy Creek
Waterford Park
Clonbinane
Wandong
Heathcote Junction
Darraweit Guim
Bylands
Willowmavin
Sunday Creek
Flowerdale
Hazeldene
Break O Day
Murrindindi
Limestone
Acheron
Thornton
Taggerty
Rubicon
Jamieson
Kevington
Ten Mile
Knockwood
Gaffneys Creek
A1 Mine Settlement
Frenchmans Gap
Woods Point
Matlock
Delatite
Piries
Goughs Bay
Martin Gap
Macs Cove
Howqua
Corduroy Gap
For more detail on the Dandenong & Yarra Ranges see page 571
GREAT DIVIDING RANGE
KINGLAKE NATIONAL PARK
Glenburn
Buxton
Marysville
LADY TALBOT FOREST DRIVE
Lake Mountain SKI AREA
Cambarville
Stockmans Reward
Upper Plenty
Beveridge
Eden Park
Glenvale
Kinglake West
Pheasant Creek
Humevale
Kinglake Central
Strathewen
Kinglake East
Castella
Toolangi
St Fillans
Narbethong
Kalkallo
Donnybrook
Woodstock
Mickleham
Yan Yean
Arthurs Creek
Yuroke
Craigieburn
Wollert
Mernda
Nutfield
St Andrews
Steels Creek
Dixons Creek
Hurstbridge
Cottles Bridge
Smiths Gully
Panton Hill
Christmas Hills
Yarra Glen
Healesville
BLACK SPUR
HEALESVILLE SANCTUARY
Broadmeadows
Lalor
Epping
Thomastown
Plenty
Wattle Glen
Rob Roy
Diamond Creek
Watsons Creek
Fawkner
Reservoir
Greensborough
Eltham
Coburg
Preston
Heidelberg
Essendon
Brunswick
Warrandyte
Coldstream
Yering
Gruyere
YARRA RANGES NATIONAL PARK
Acheron Gap
McMahons Creek
Warburton East
Templestowe
Doncaster
Lilydale
Mooroolbark
Mount Evelyn
Seville
Woori Yallock
Wandin North
Launching Place
Don Valley
Millgrove
Warburton
Wesburn
Yarra Junction
Big Pats Creek
Gifford Saddle
MELBOURNE
Richmond
Box Hill
Camberwell
Burwood
Ringwood
Croydon
DANDENONG RANGES NP
Silvan
Yellingbo
Hoddles Creek
Gladysdale
Three Bridges
Powelltown
St Kilda
Caulfield
Elsternwick
Oakleigh
Glen Waverley
Boronia
Ferntree Gully
Olinda
Monbulk
The Patch
Upwey
Belgrave
Nangana
Macclesfield
Menzies Creek
Tomahawk Gap
Brighton
Moorabbin
Clayton
Rowville
Lysterfield
Tecoma
Clematis
Emerald
Cockatoo
Gembrook
Hampton
Sandringham
Highett
Cheltenham
Springvale
Mentone
Noble Park
Dandenong
Belgrave South
Narre Warren North
Mordialloc
Aspendale
Edithvale
Chelsea
Bonbeach
Carrum
Seaford
Hallam
Narre Warren
Harkaway
Berwick
Upper Beaconsfield
Beaconsfield
Officer
Pakenham
Lyndhurst
Hampton Park
Carrum Downs
CRANBOURNE
Clyde
Cranbourne South
Frankston
Langwarrin
Baxter
Pearcedale
Mount Eliza
MORNINGTON
Somerville
Mooroodüc
Tyabb
Hastings
Bittern
Mount Martha
Safety Beach
Dromana
Merricks North
Red Hill
Red Hill South
Balnarring
Somers
Point Leo
Shoreham
Flinders
West Head
Crib Point
Stony Point
Tankerton
Cowes
Rhyll
Ventnor
KOALA CONSERVATION CENTRE
PENGUIN PARADE
THE NOBBIES CENTRE & SEAL ROCKS
PHILLIP ISLAND NATURE PARK
Newhaven
San Remo
Cape Woolamai
Anderson
Kilcunda
Dalyston
Wonthaggi
Dudley
Inverloch
Cape Paterson
BUNURONG MARINE PARK
Venus Bay
FRENCH ISLAND NATIONAL PARK
FRENCH ISLAND MNP
Tooradin
Koo-wee-rup
Monomeith
Lang Lang
Cardinia
Nar Nar Goon
Tynong
Garfield
Bunyip
Iona
Cora Lynn
Bayles
Catani
Modella
Heath Hill
Nyora
Loch
Poowong
Poowong East
The Gurdies
Grantville
Corinella
Coronet Bay
Bass
Woolamai
Glen Forbes
Kernot
Woodleigh
Bena
Korumburra
Ruby
Jumbunna
Kongwak
Outtrim
Archies Creek
Powlett River
Leongatha
Leongatha South
Koonwarra
Tarwin
Meeniyan
Stony Creek
Buffalo
Dumbalk
Pound Creek
Mirboo North
GRAND RIDGE BREWERY
Mirboo
Turtons Creek
Budgeree
Boolarra
Yinnar
Delburn
Allambee South
Hallston
Mount Eccles
Wooreen
Arawata
Ranceby
Strzelecki
Trida
Thorpdale
Childers
Narracan
Coalville
Allambee
Clanoy Cutting
Seaview
Ellinbank
Trafalgar
Yarragon
Darnum
Nilma
Warragul
Drouin
Drouin South
Drouin West
Longwarry
Labertouche
Jindivick
Tarago
Rokeby
Buln Buln
Buln Buln East
Brandy Creek
Shady Creek
Crossover
Neerim South
Neerim
Neerim East
Neerim Junction
Nayook
Noojee
Icy Creek
Fumina
Tanjil Bren
Toorongo
Hill End
Willow Grove
Tanjil South
Westbury
MOE
Newborough
Ripplebrook
Athlone
Maryknoll
Garfield North
Tonimbuk
BUNYIP STATE PARK
Bunyip Gap
BAW BAW NP
Mt Baw Baw

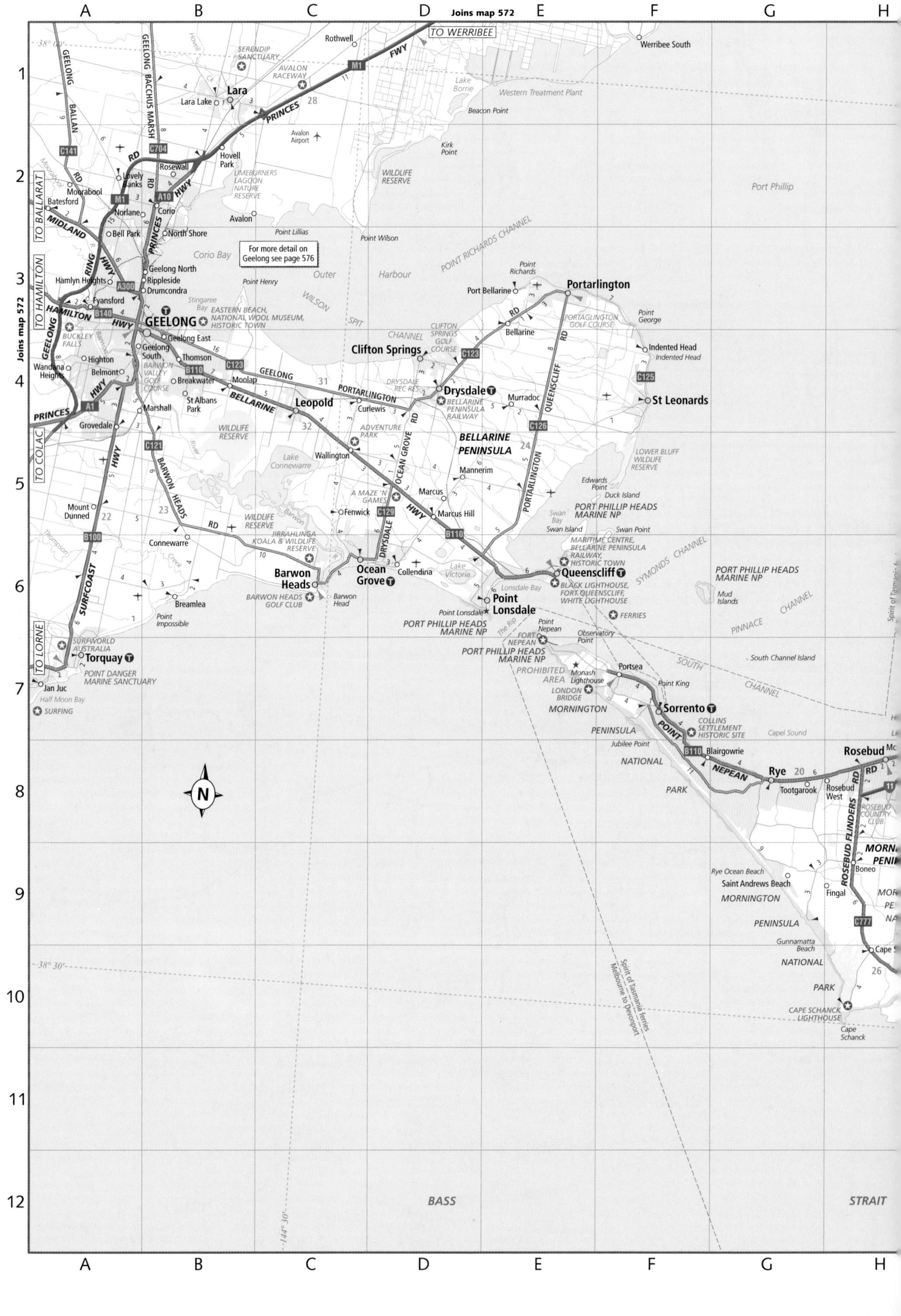

Joins map 572
TO WERRIBEE
Rothwell
Werribee South
SERENDIP SANCTUARY
AVALON RACEWAY
Lara
Lara Lake
PRINCES
FWY
Lake Borrie
Western Treatment Plant
Beacon Point
Avalon Airport
Kirk Point
Hovell Park
Rosewall
Lovely Banks
Moorabool
Batesford
Norlane
Corio
Bell Park
North Shore
LIMEBURNERS LAGOON NATURE RESERVE
WILDLIFE RESERVE
Avalon
Point Lillias
Point Wilson
Port Phillip
For more detail on Geelong see page 576
Corio Bay
Outer
Harbour
POINT RICHARDS CHANNEL
Point Richards
Geelong North
Rippleside
Drumcondra
Hamlyn Heights
Fyansford
Point Henry
WILSON
SPIT
Port Bellarine
Portarlington
PORTARLINGTON GOLF COURSE
Point George
Stingaree Bay
EASTERN BEACH, NATIONAL WOOL MUSEUM, HISTORIC TOWN
GEELONG
Geelong East
Geelong South
BUCKLEY FALLS
Highton
Belmont
Wandana Heights
Thomson
Breakwater
Moolap
CHANNEL
CLIFTON SPRINGS GOLF COURSE
Clifton Springs
Bellarine
Indented Head
DRYSDALE REC RES
Drysdale
BELLARINE PENINSULA RAILWAY
Murradoc
St Leonards
Marshall
St Albans Park
Leopold
Curlewis
Grovedale
WILDLIFE RESERVE
ADVENTURE PARK
BELLARINE PENINSULA
LOWER BLUFF WILDLIFE RESERVE
Wallington
Lake Connewarre
Mannerim
Edwards Point
Duck Island
Mount Dunned
A MAZE 'N' GAMES
Fenwick
Marcus
Marcus Hill
PORT PHILLIP HEADS MARINE NP
Swan Bay
Swan Island
Swan Point
WILDLIFE RESERVE
JIRRAHLINGA KOALA & WILDLIFE RESERVE
Connewarre
MARITIME CENTRE, BELLARINE PENINSULA RAILWAY, HISTORIC TOWN
Barwon Heads
Ocean Grove
Collendina
Lake Victoria
Queenscliff
SYMONDS CHANNEL
BLACK LIGHTHOUSE, FORT QUEENSCLIFF, WHITE LIGHTHOUSE
Lonsdale Bay
Mud Islands
BARWON HEADS GOLF CLUB
Barwon Head
Breamlea
Point Lonsdale
Point Lonsdale
FERRIES
Point Impossible
PORT PHILLIP HEADS MARINE NP
The Rip
Point Nepean
FORT NEPEAN
Observatory Point
PINNACE CHANNEL
SURFWORLD AUSTRALIA
Torquay
PORT PHILLIP HEADS MARINE NP
South Channel Island
POINT DANGER MARINE SANCTUARY
PROHIBITED AREA
Monash Lighthouse
Portsea
SOUTH CHANNEL
Jan Juc
Half Moon Bay
SURFING
LONDON BRIDGE
Point King
MORNINGTON
Sorrento
PENINSULA
COLLINS SETTLEMENT HISTORIC SITE
Capel Sound
Jubilee Point
Blairgowrie
NATIONAL
NEPEAN
Rye
Rosebud
Tootgarook
Rosebud West
PARK
ROSEBUD COUNTRY CLUB
ROSEBUD FLINDERS RD
Boneo
Rye Ocean Beach
Saint Andrews Beach
Fingal
MORNINGTON
PENINSULA
Gunnamatta Beach
NATIONAL
Cape Schanck
PARK
CAPE SCHANCK LIGHTHOUSE
Spirit of Tasmania ferries Melbourne to Devonport
BASS
STRAIT
TO BALLARAT
TO HAMILTON
Joins map 572
TO COLAC
TO LORNE
GEELONG BALLAN RD
GEELONG BACCHUS MARSH RD
MIDLAND HWY
RING RD
HAMILTON HWY
PRINCES HWY
SURFCOAST HWY
BARWON HEADS RD
BELLARINE HWY
PORTARLINGTON RD
OCEAN GROVE RD
DRYSDALE HWY
QUEENSCLIFF RD
PORTARLINGTON RD
POINT NEPEAN RD
38° 30′
144° 30′

0 5 10 15 km

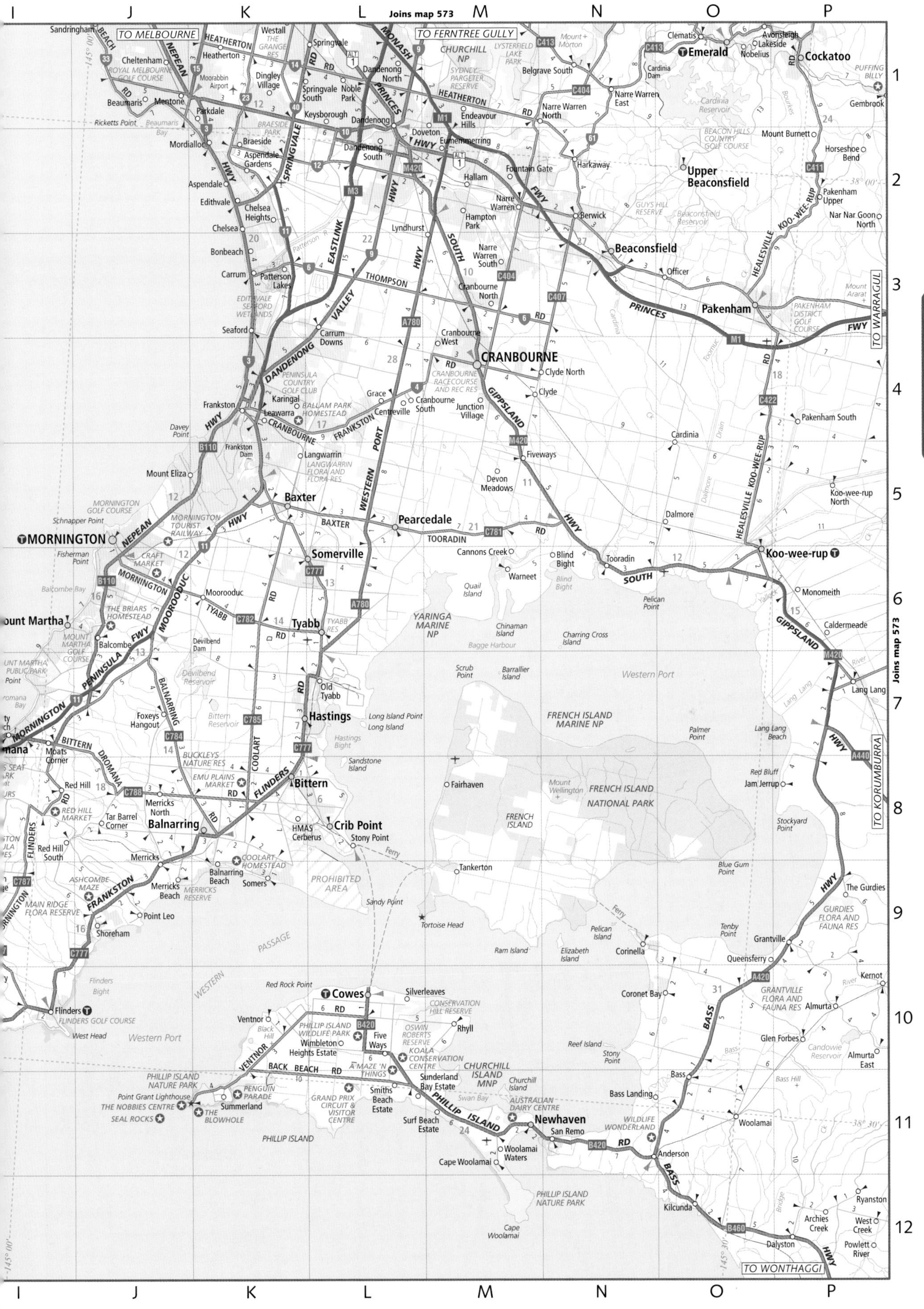
Joins map 573
TO MELBOURNE
TO FERNTREE GULLY
TO WARRAGUL
TO KORUMBURRA
TO WONTHAGGI
Sandringham
Cheltenham
Heatherton
Westall
Springvale
Dingley Village
Springvale South
Noble Park
Dandenong North
Dandenong
Dandenong South
Keysborough
Moorabbin Airport
Beaumaris
Mentone
Parkdale
Mordialloc
Braeside
Aspendale Gardens
Aspendale
Edithvale
Chelsea Heights
Chelsea
Bonbeach
Carrum
Patterson Lakes
Seaford
Carrum Downs
Frankston
Karingal
Leawarra
Langwarrin
Mount Eliza
Baxter
Somerville
MORNINGTON
Mooroduc
Tyabb
Old Tyabb
Hastings
Bittern
Crib Point
Stony Point
HMAS Cerberus
Balnarring
Balnarring Beach
Somers
Merricks
Merricks North
Merricks Beach
Point Leo
Shoreham
Flinders
Red Hill
Red Hill South
Moats Corner
Tar Barrel Corner
Foxeys Hangout
Balcombe
Pearcedale
Cannons Creek
Warneet
Blind Bight
Tooradin
Koo-wee-rup
Koo-wee-rup North
Monomeith
Caldermeade
Lang Lang
Lang Lang Beach
Jam Jerrup
Grantville
Queensferry
Corinella
Coronet Bay
Bass
Bass Landing
Anderson
Kilcunda
Dalyston
Woolamai
Glen Forbes
Almurta
Almurta East
Kernot
The Gurdies
Archies Creek
Ryanston
West Creek
Powlett River
Cowes
Silverleaves
Ventnor
Rhyll
Wimbleton Heights Estate
Five Ways
Sunderland Bay Estate
Smiths Beach Estate
Surf Beach Estate
Summerland
Newhaven
San Remo
Woolamai Waters
Cape Woolamai
Tankerton
Fairhaven
FRENCH ISLAND
FRENCH ISLAND NATIONAL PARK
FRENCH ISLAND MARINE NP
YARINGA MARINE NP
Western Port
PHILLIP ISLAND
CHURCHILL ISLAND MNP
PHILLIP ISLAND NATURE PARK
Lyndhurst
Hampton Park
Hallam
Endeavour Hills
Doveton
Eumemmerring
Narre Warren
Narre Warren North
Narre Warren South
Narre Warren East
Fountain Gate
Harkaway
Berwick
Beaconsfield
Upper Beaconsfield
Officer
Pakenham
Pakenham South
Pakenham Upper
Nar Nar Goon North
Cardinia
Dalmore
Emerald
Clematis
Avonsleigh
Lakeside
Nobelius
Cockatoo
Gembrook
Mount Burnett
Horseshoe Bend
Belgrave South
CHURCHILL NP
Cranbourne North
Cranbourne West
CRANBOURNE
Cranbourne South
Clyde North
Clyde
Junction Village
Devon Meadows
Fiveways
Grace
Centreville
Joins map 573

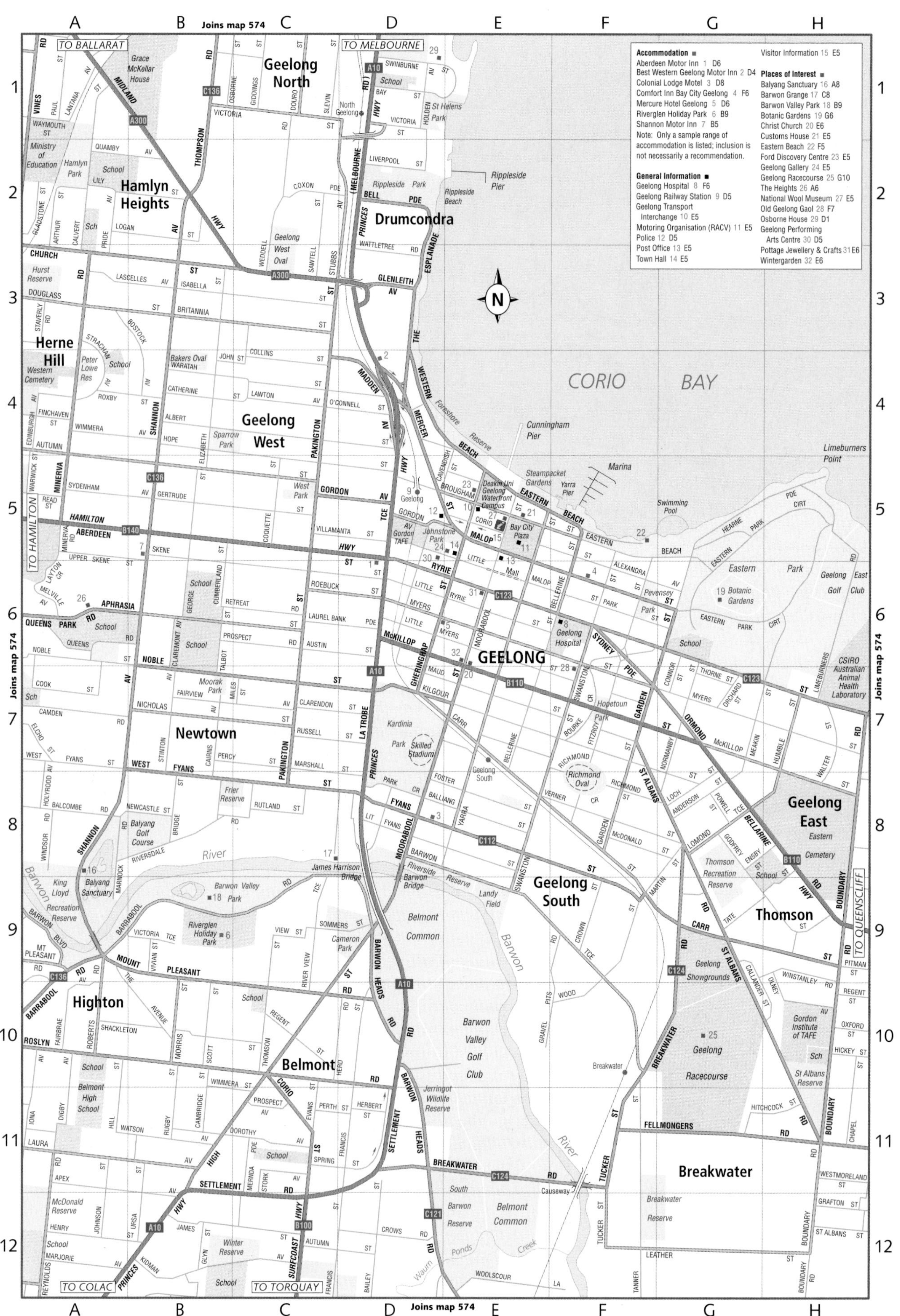
0 0.25 0.5 0.75 1 km
Joins map 574
TO BALLARAT
TO MELBOURNE
TO HAMILTON
TO COLAC
TO TORQUAY
TO QUEENSCLIFF
Accommodation
Aberdeen Motor Inn 1 D6
Best Western Geelong Motor Inn 2 D4
Colonial Lodge Motel 3 D8
Comfort Inn Bay City Geelong 4 F6
Mercure Hotel Geelong 5 D6
Riverglen Holiday Park 6 B9
Shannon Motor Inn 7 B5
Note: Only a sample range of accommodation is listed; inclusion is not necessarily a recommendation.
General Information
Geelong Hospital 8 F6
Geelong Railway Station 9 D5
Geelong Transport Interchange 10 E5
Motoring Organisation (RACV) 11 E5
Police 12 D5
Post Office 13 E5
Town Hall 14 E5
Visitor Information 15 E5
Places of Interest
Balyang Sanctuary 16 A8
Barwon Grange 17 C8
Barwon Valley Park 18 B9
Botanic Gardens 19 G6
Christ Church 20 E6
Customs House 21 E5
Eastern Beach 22 F5
Ford Discovery Centre 23 E5
Geelong Gallery 24 E5
Geelong Racecourse 25 G10
The Heights 26 A6
National Wool Museum 27 E5
Old Geelong Gaol 28 F7
Osborne House 29 D1
Geelong Performing Arts Centre 30 D5
Pottage Jewellery & Crafts 31 E6
Wintergarden 32 E6
CORIO BAY
Geelong North
Hamlyn Heights
Drumcondra
Herne Hill
Geelong West
GEELONG
Newtown
Geelong East
Geelong South
Thomson
Highton
Belmont
Breakwater
Rippleside Pier
Cunningham Pier
Limeburners Point
Eastern Park
Kardinia Park
Belmont Common
Barwon Valley Golf Club
Geelong Racecourse
Barwon River

0 0.5 1 1.5 2 km

Joins map 579

Accommodation

- Ballarat Mid City Accommodation & Conference Centre 1 D6
- Best Western Bakery Hill Motel 2 F6
- Best Western Ballarat 3 F8
- Central City Motor Inn 4 F6
- Comfort Inn Main Lead 5 F7
- Craig's Royal Hotel 6 E6
- Eureka Lodge Motel 7 H6
- George Hotel 8 E6
- The Lake View 9 C5
- Mercure Ballarat Hotel & Convention Centre 10 F8
- Miners Retreat Motel 11 H6
- Peppinella Motel 12 A10
- Ravenswood Cottage 13 E5
- Reid's Guest House 14 E6
- Sovereign Hill Lodge 15 F8
- Sovereign Park Motor Inn 16 F7
- Sovereign Views Apartments 17 F8
- Victoriana Motor Inn 18 H5

Note: Only a sample range of accommodation is listed; inclusion is not necessarily a recommendation.

General Information

- Ballarat Base Hospital 19 D6
- Ballarat Railway Station 20 E6
- Motoring Organisation (RACV) 21 E6
- Police Station 22 E6
- Post Office 23 E6
- Visitor Information 24 E6, E7

Places of Interest

- Adam Lindsay Gordon Craft Cottage 25 A5
- Aquatic & Hockey Centre 26 A5
- Art Gallery of Ballarat 27 E6
- Australian Centre for Democracy at Eureka 28 H6
- Ballarat Tramway Museum 29 A6
- Ballarat Wildlife Park 30 H7
- Gold Museum 31 F8
- Her Majesty's Theatre 32 E6
- The Mining Exchange Gold Shop 33 E6
- Montrose Cottage and Museum 34 F7
- The Robert Clark Conservatory 35 A5
- Sovereign Hill Historical Park 36 F8
- Town Hall 37 E6

Joins map 579

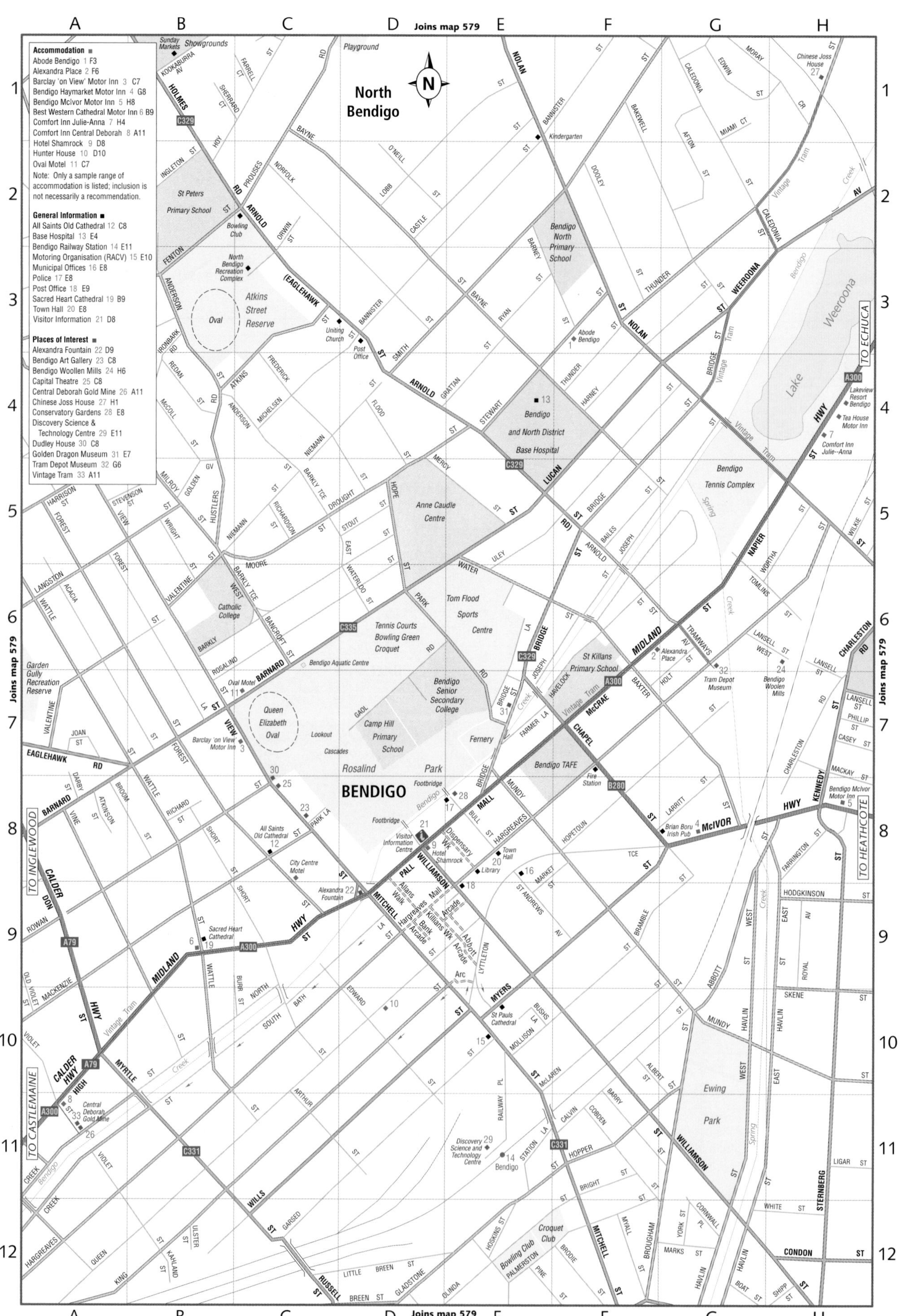
0 200 400 600 m
Joins map 579
North Bendigo
BENDIGO
Accommodation
Abode Bendigo 1 F3
Alexandra Place 2 F6
Barclay 'on View' Motor Inn 3 C7
Bendigo Haymarket Motor Inn 4 G8
Bendigo McIvor Motor Inn 5 H8
Best Western Cathedral Motor Inn 6 B9
Comfort Inn Julie-Anna 7 H4
Comfort Inn Central Deborah 8 A11
Hotel Shamrock 9 D8
Hunter House 10 D10
Oval Motel 11 C7
Note: Only a sample range of accommodation is listed; inclusion is not necessarily a recommendation.
General Information
All Saints Old Cathedral 12 C8
Base Hospital 13 E4
Bendigo Railway Station 14 E11
Motoring Organisation (RACV) 15 E10
Municipal Offices 16 E8
Police 17 E8
Post Office 18 E9
Sacred Heart Cathedral 19 B9
Town Hall 20 E8
Visitor Information 21 D8
Places of Interest
Alexandra Fountain 22 D9
Bendigo Art Gallery 23 C8
Bendigo Woollen Mills 24 H6
Capital Theatre 25 C8
Central Deborah Gold Mine 26 A11
Chinese Joss House 27 H1
Conservatory Gardens 28 E8
Discovery Science & Technology Centre 29 E11
Dudley House 30 C8
Golden Dragon Museum 31 E7
Tram Depot Museum 32 G6
Vintage Tram 33 A11
Showgrounds
Sunday Markets
Playground
Kindergarten
St Peters Primary School
Bowling Club
North Bendigo Recreation Complex
Atkins Street Reserve
Oval
Uniting Church
Post Office
Bendigo North Primary School
Bendigo and North District Base Hospital
Bendigo Tennis Complex
Lake Weeroona
Lakeview Resort Bendigo
Tea House Motor Inn
Anne Caudle Centre
Tom Flood Sports Centre
Catholic College
Tennis Courts Bowling Green Croquet
Bendigo Aquatic Centre
Bendigo Senior Secondary College
Garden Gully Recreation Reserve
Queen Elizabeth Oval
Camp Hill Primary School
Lookout
Cascades
Fernery
Rosalind Park
St Killans Primary School
Bendigo TAFE
Fire Station
Tram Depot Museum
Bendigo Woolen Mills
Brian Boru Irish Pub
Footbridge
Visitor Information Centre
City Centre Motel
Library
St Pauls Cathedral
Ewing Park
Discovery Science and Technology Centre
Croquet Club
Bowling Club
Central Deborah Gold Mine
TO INGLEWOOD
TO CASTLEMAINE
TO ECHUCA
TO HEATHCOTE
MIDLAND HWY
CALDER HWY
McIVOR HWY
PALL MALL
WILLIAMSON ST
MITCHELL ST
HARGREAVES ST
BARNARD ST
VIEW ST
NAPIER ST
HIGH ST

0 5 10 15 20 km

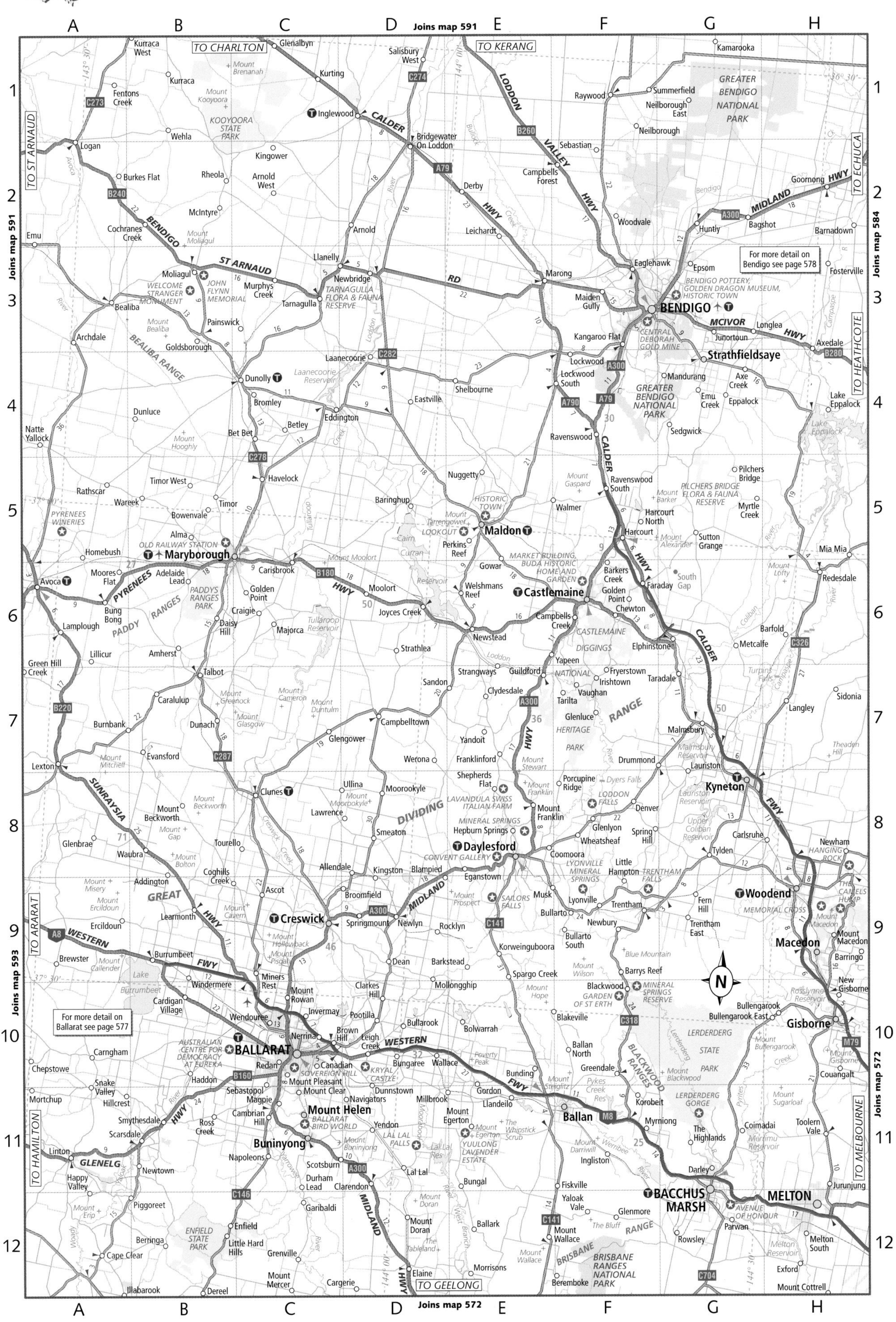
A B C D E F G H
1 2 3 4 5 6 7 8 9 10 11 12
Joins map 591
Joins map 572
Joins map 584
Joins map 593
TO CHARLTON
TO KERANG
TO ST ARNAUD
TO ECHUCA
TO HEATHCOTE
TO ARARAT
TO HAMILTON
TO MELBOURNE
TO GEELONG
For more detail on Bendigo see page 578
For more detail on Ballarat see page 577
Kurraca West
Kurraca
Fentons Creek
Glenalbyn
Mount Brenanah
Salisbury West
Kurting
Kamarooka
Raywood
Summerfield
Neilborough East
Neilborough
GREATER BENDIGO NATIONAL PARK
Mount Kooyoora
KOOYOORA STATE PARK
Inglewood
CALDER
LODDON VALLEY HWY
Logan
Wehla
Kingower
Bridgewater On Loddon
Sebastian
Burkes Flat
Rheola
Arnold West
Derby
Campbells Forest
Goornong
McIntyre
Woodvale
MIDLAND HWY
Cochranes Creek
BENDIGO
Mount Moliagul
Arnold
Leichardt
Huntly
Bagshot
Barnadown
Emu
Llanelly
Eaglehawk
Epsom
Moliagul
ST ARNAUD
Newbridge
RD
Marong
Fosterville
WELCOME STRANGER MONUMENT
JOHN FLYNN MEMORIAL
Murphys Creek
TARNAGULLA FLORA & FAUNA RESERVE
Tarnagulla
Maiden Gully
BENDIGO
BENDIGO POTTERY, GOLDEN DRAGON MUSEUM, HISTORIC TOWN
Bealiba
Mount Bealiba
Painswick
Goldsborough
MCIVOR HWY
Longlea
Junortoun
Axedale
Archdale
BEALIBA RANGE
Kangaroo Flat
CENTRAL DEBORAH GOLD MINE
Laanecoorie
Laanecoorie Reservoir
Lockwood
Lockwood South
Strathfieldsaye
Dunolly
Shelbourne
Mandurang
Axe Creek
Bromley
Eastville
GREATER BENDIGO NATIONAL PARK
Emu Creek
Eppalock
Lake Eppalock
Dunluce
Eddington
Betley
Bet Bet
Ravenswood
Sedgwick
Natte Yallock
Mount Hooghly
Havelock
Nuggetty
Ravenswood South
Pilchers Bridge
PILCHERS BRIDGE FLORA & FAUNA RESERVE
Mount Gaspard
Mount Barker
Timor West
Rathscar
Wareek
Timor
Baringhup
HISTORIC TOWN
Walmer
Harcourt North
Myrtle Creek
PYRENEES WINERIES
Bowenvale
Mount Tarrengower LOOKOUT
Maldon
Harcourt
Mount Alexander
Sutton Grange
Alma
OLD RAILWAY STATION
Maryborough
Cairn Curran
Perkins Reef
MARKET BUILDING, BUDA HISTORIC HOME AND GARDEN
Mia Mia
Homebush
Mount Moolort
Carisbrook
Gowar
Barkers Creek
South Gap
Moores Flat
PYRENEES
Adelaide Lead
PADDYS RANGES PARK
Golden Point
Moolort
Welshmans Reef
Castlemaine
Faraday
Mount Lofty
Redesdale
Avoca
Bung Bong
PADDY RANGES
Craigie
HWY
Joyces Creek
Chewton
Campbells Creek
Lamplough
Daisy Hill
Majorca
Tullaroop Reservoir
Newstead
CASTLEMAINE DIGGINGS
Elphinstone
CALDER
Barfold
Metcalfe
Lillicur
Amherst
Strathlea
Green Hill Creek
Talbot
Strangways
Guildford
NATIONAL
Yapeen
Fryerstown
Irishtown
Vaughan
Taradale
Turpins Falls
Caralulup
Mount Greenock
Mount Cameron
Mount Duntulm
Sandon
Clydesdale
Tarilta
RANGE
Langley
Sidonia
Mount Glasgow
Campbelltown
Glenluce
HERITAGE PARK
Malmsbury
Malmsbury Reservoir
Burnbank
Dunach
Glengower
Yandoit
Theaden Hill
Evansford
Werona
Franklinford
Drummond
Lauriston
Lexton
Mount Mitchell
Mount Stewart
Shepherds Flat
Dyers Falls
Kyneton
SUNRAYSIA
Clunes
Ullina
Mount Moorookyle
Moorookyle
LAVANDULA SWISS ITALIAN FARM
Mount Franklin
Porcupine Ridge
LODDON FALLS
Lauriston Reservoir
FWY
Mount Beckworth
Lawrence
DIVIDING
MINERAL SPRINGS
Hepburn Springs
Mount Franklin
Denver
Mount Gap
Smeaton
Glenlyon
Spring Hill
Upper Coliban Reservoir
Carlsruhe
Glenbrae
Waubra
Tourello
Daylesford
CONVENT GALLERY
Wheatsheaf
Newham
HANGING ROCK
Mount Bolton
Coghills Creek
Allendale
Kingston
Blampied
Eganstown
Coomoora
LYONVILLE MINERAL SPRINGS
Little Hampton
TRENTHAM FALLS
Tylden
Mount Misery
Addington
Ascot
Broomfield
Mount Prospect
SAILORS FALLS
Musk
Lyonville
Fern Hill
Woodend
THE CAMELS HUMP
Mount Ercildoun
GREAT HWY
Mount Cavern
MIDLAND
Bullarto
Trentham
MEMORIAL CROSS
Mount Macedon
Erciidoun
Learmonth
Creswick
Springmount
Newlyn
Rocklyn
Newbury
Trentham East
Macedon
WESTERN
Mount Hollowback
Mount Pisgah
Bullarto South
Korweinguboora
Blue Mountain
Brewster
Mount Callender
Burrumbeet
Dean
Barkstead
Mount Wilson
Barrys Reef
Barringo
Lake Burrumbeet
Windermere
Miners Rest
Mount Rowan
Clarkes Hill
Mollongghip
Spargo Creek
Mount Hope
Blackwood
MINERAL SPRINGS RESERVE
GARDEN OF ST ERTH
New Gisborne
Rosslynne Reservoir
Cardigan Village
Invermay
Pootilla
Bullarook
Blakeville
Bullengarook
Bullengarook East
Gisborne
Wendouree
Nerrina
Brown Hill
Leigh Creek
WESTERN
Bolwarrah
LERDERDERG STATE PARK
Mount Bullengarook
AUSTRALIAN CENTRE FOR DEMOCRACY AT EUREKA
BALLARAT
Carngham
Redan
Canadian
SOVEREIGN HILL
KRYAL CASTLE
Bungaree
Wallace
Poverty Peak
Ballan North
Greendale
BLACKWOOD RANGES
Mount Gisborne
Couangalt
Chepstowe
Snake Valley
Haddon
Mount Pleasant
Mount Clear
Dunnstown
Gordon
Bunding
Mount Steiglitz
Pykes Creek Res
Mount Blackwood
Mortchup
Hillcrest
Sebastopol
Magpie
Navigators
Millbrook
Llandeilo
Korobeit
LERDERDERG GORGE
Mount Sugarloaf
Smythesdale
HWY
Ross Creek
Cambrian Hill
Mount Helen
BALLARAT BIRD WORLD
Mount Egerton
Egerton
The Whipstick Scrub
Ballan
Myrniong
The Highlands
Coimadai
Merrimu Reservoir
Toolern Vale
Scarsdale
Buninyong
Yendon
Mount Buninyong
LAL LAL FALLS
Lal Lal Res
Mount Egerton
YUULONG LAVENDER ESTATE
Mount Darriwill
Werribee River
Ingliston
Linton
GLENELG
Napoleons
Scotsburn
Lal Lal
Darley
Happy Valley
Newtown
Durham Lead
Clarendon
Bungal
Fiskville
Jurunjung
Mount Erip
Piggoreet
Garibaldi
MIDLAND
Mount Doran
Yaloak Vale
BACCHUS MARSH
AVENUE OF HONOUR
MELTON
Glenmore
Parwan
ENFIELD STATE PARK
Enfield
Mount Doran
The Tableland
Ballark
West Branch
The Bluff
RANGE
Mount Wallace
Rowsley
Melton South
Melton Reservoir
Berringa
Little Hard Hills
Grenville
Mount Wallace
BRISBANE
BRISBANE RANGES NATIONAL PARK
Cape Clear
Elaine
Morrisons
Exford
Mount Mercer
Cargerie
Beremboke
Illabarook
Dereel
Mount Cottrell
A300 A79 A790 A8 A300 B260 B240 B180 B220 B160 B280 C273 C274 C278 C282 C287 C326 C141 C318 C146 C704 M8 M79
37° 00' 37° 30' 36° 30' 143° 30' 144° 00' 144° 30'
N

0 10 20 30 40 km

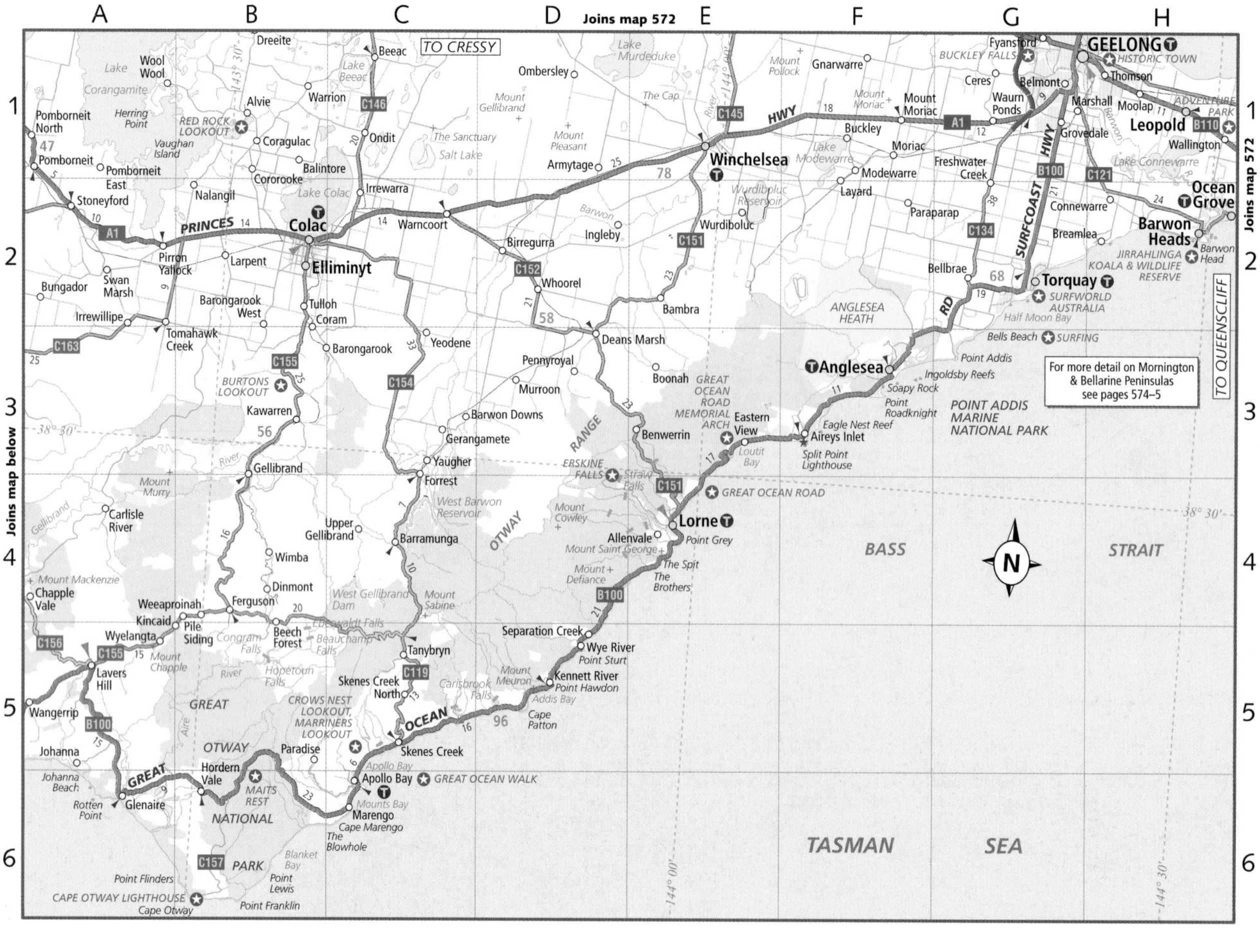

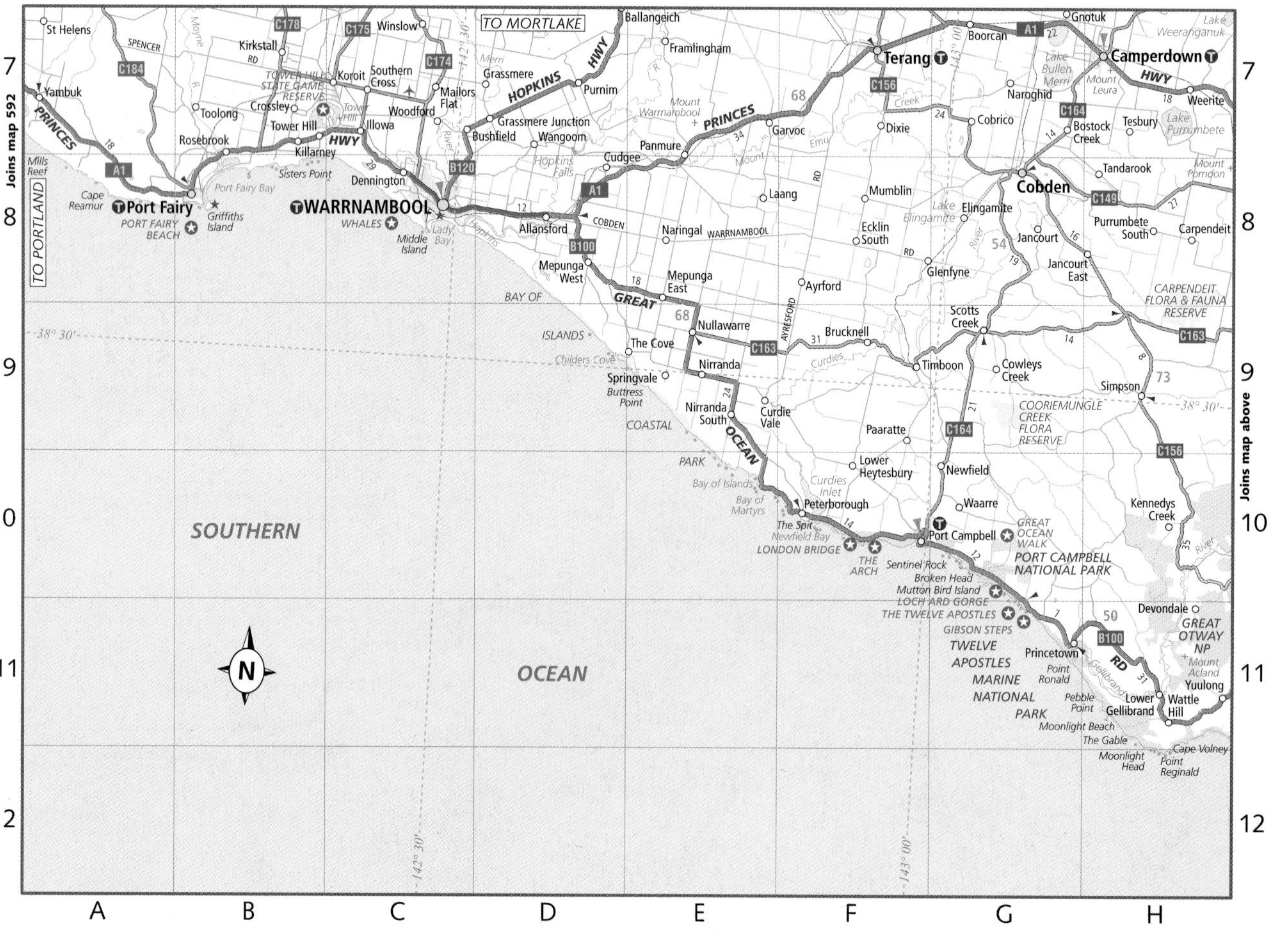

A B C D E F G H

Joins map 572

TO CRESSY

Dreeite
Beeac
Lake Beeac
Wool Wool
Lake Corangamite
Alvie
Warrion
Pomborneit North
Herring Point
RED ROCK LOOKOUT
Coragulac
Ondit
Vaughan Island
Pomborneit
Pomborneit East
Balintore
Cororooke
Irrewarra
Stoneyford
Nalangil
Lake Colac
Colac
PRINCES
Warncoort
Pirron Yallock
Larpent
Elliminyt
Bungador
Swan Marsh
Barongarook West
Tulloh
Coram
Irrewillipe
Tomahawk Creek
Barongarook
Yeodene
BURTONS LOOKOUT
Kawarren
Gellibrand
Mount Murry
Carlisle River
Yaugher
Forrest
Upper Gellibrand
West Barwon Reservoir
Barramunga
Wimba
Dinmont
Mount Mackenzie
Chapple Vale
Weeaproinah
Ferguson
Kincaid
Pile Siding
Wyelangta
Beech Forest
Congram Falls
Lavers Hill
Mount Chapple
Hopetoun Falls
West Gellibrand Dam
Mount Sabine
Tanybryn
Skenes Creek North
Wangerrip
GREAT OTWAY NATIONAL PARK
CROWS NEST LOOKOUT, MARINERS LOOKOUT
Johanna
Johanna Beach
Hordern Vale
Paradise
Rotten Point
Glenaire
MAITS REST
Skenes Creek
Apollo Bay
Marengo
Mounts Bay
Cape Marengo
The Blowhole
GREAT OCEAN WALK
Blanket Bay
Point Flinders
Point Lewis
CAPE OTWAY LIGHTHOUSE
Cape Otway
Point Franklin
Ombersley
Lake Murdeduke
The Cap
Mount Gellibrand
The Sanctuary
Salt Lake
Mount Pleasant
Armytage
Winchelsea
Birregurra
Ingleby
Whoorel
Wurdiboluc
Wurdiboluc Reservoir
Bambra
Deans Marsh
Pennyroyal
Murroon
Barwon Downs
Gerangamete
Boonah
OTWAY RANGE
Benwerrin
ERSKINE FALLS
GREAT OCEAN ROAD MEMORIAL ARCH
Eastern View
Loutit Bay
Mount Cowley
Allenvale
Mount Saint George
Lorne
Point Grey
Mount Defiance
The Spit
The Brothers
Separation Creek
Wye River
Point Sturt
Kennett River
Point Hawdon
Mount Meuron
Carisbrook Falls
Addis Bay
Cape Patton
OCEAN
Mount Pollock
Gnarwarre
Mount Moriac
Mount Moriac
Buckley
Moriac
Lake Modewarre
Modewarre
Layard
Paraparap
BUCKLEY FALLS
Fyansford
Ceres
Waurn Ponds
Belmont
GEELONG
HISTORIC TOWN
Thomson
Marshall
Moolap
Grovedale
Leopold
Wallington
Freshwater Creek
Lake Connewarre
Connewarre
Ocean Grove
Breamlea
Barwon Heads
Barwon Head
JIRRAHLINGA KOALA & WILDLIFE RESERVE
SURFCOAST HWY
Bellbrae
Torquay
SURFWORLD AUSTRALIA
Half Moon Bay
ANGLESEA HEATH
Bells Beach
SURFING
Point Addis
Anglesea
Ingoldsby Reefs
Soapy Rock
Point Roadknight
POINT ADDIS MARINE NATIONAL PARK
Eagle Nest Reef
Aireys Inlet
Split Point Lighthouse
GREAT OCEAN ROAD
BASS STRAIT
TASMAN SEA

For more detail on Mornington & Bellarine Peninsulas see pages 574–5

TO QUEENSCLIFF

Joins map 572

Joins map below

1 2 3 4 5 6

TO MORTLAKE

St Helens
Kirkstall
Winslow
Ballangeich
Framlingham
Yambuk
TOWER HILL STATE GAME RESERVE
Koroit
Southern Cross
Grassmere
Mailors Flat
HOPKINS HWY
Purnim
Toolong
Crossley
Tower Hill
Illowa
Woodford
Grassmere Junction
Bushfield
Wangoom
Mount Warrnambool
Rosebrook
Killarney
Sisters Point
Port Fairy Bay
Dennington
Hopkins Falls
Cudgee
Panmure
Garvoc
Terang
Dixie
Mills Reef
Cape Reamur
Port Fairy
Griffiths Island
PORT FAIRY BEACH
WARRNAMBOOL
WHALES
Lady Bay
Middle Island
Allansford
Naringal
Laang
Mumblin
Ecklin South
Lake Elingamite
Elingamite
Mepunga West
Mepunga East
Ayrford
Glenfyne
BAY OF ISLANDS
GREAT OCEAN RD
Nullawarre
The Cove
Childers Cove
Brucknell
Curdies
Nirranda
Springvale
Buttress Point
Timboon
COASTAL PARK
Nirranda South
Curdie Vale
Paaratte
Lower Heytesbury
Bay of Islands
Bay of Martyrs
Curdies Inlet
Newfield
Peterborough
Waarre
The Spit
Newfield Bay
LONDON BRIDGE
THE ARCH
Port Campbell
GREAT OCEAN WALK
PORT CAMPBELL NATIONAL PARK
Sentinel Rock
Broken Head
Mutton Bird Island
LOCH ARD GORGE
THE TWELVE APOSTLES
GIBSON STEPS
TWELVE APOSTLES MARINE NATIONAL PARK
Princetown
Point Ronald
Pebble Point
Moonlight Beach
The Gable
Moonlight Head
Gnotuk
Boorcan
Lake Weeranganuk
Camperdown
Lake Bullen Merri
Mount Leura
Naroghid
Weerite
Cobrico
Bostock Creek
Tesbury
Lake Purrumbete
Tandarook
Mount Porndon
Cobden
Purrumbete South
Carpendeit
Jancourt
Jancourt East
CARPENDEIT FLORA & FAUNA RESERVE
Scotts Creek
Cowleys Creek
Simpson
COORIEMUNGLE CREEK FLORA RESERVE
Kennedys Creek
Devondale
GREAT OTWAY NP
Mount Acland
Yuulong
Lower Gellibrand
Wattle Hill
Cape Volney
Point Reginald
SOUTHERN OCEAN

Joins map 592

TO PORTLAND

Joins map above

7 8 9 10 11 12

A B C D E F G H

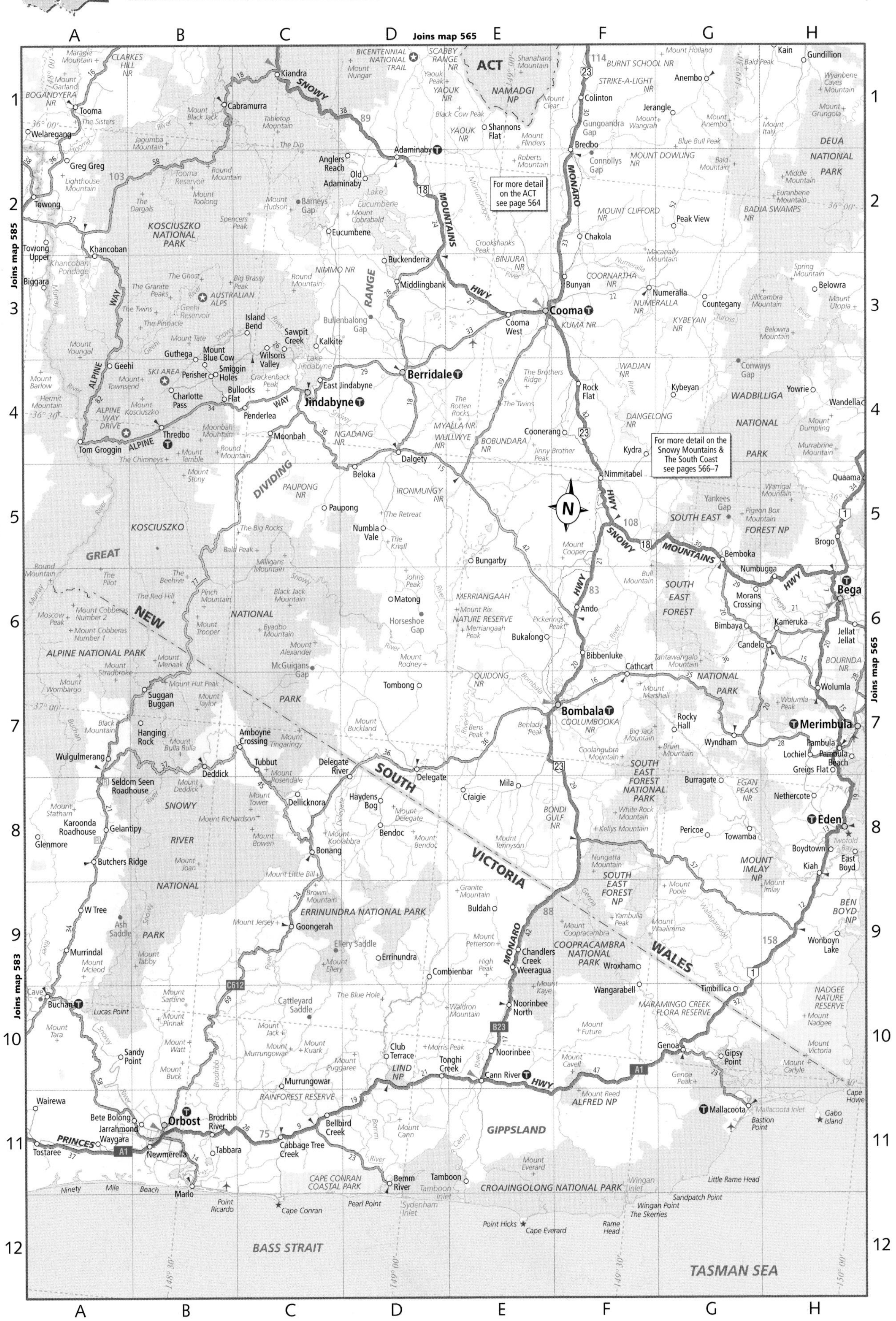
Joins map 565
Joins map 585
Joins map 583
Joins map 565
ACT
NAMADGI NP
KOSCIUSZKO NATIONAL PARK
ALPINE NATIONAL PARK
SNOWY RIVER NATIONAL PARK
ERRINUNDRA NATIONAL PARK
COOPRACAMBRA NATIONAL PARK
CROAJINGOLONG NATIONAL PARK
SOUTH EAST FOREST NATIONAL PARK
WADBILLIGA NATIONAL PARK
DEUA NATIONAL PARK
MOUNT IMLAY NP
BEN BOYD NP
ALFRED NP
LIND NP
NEW SOUTH WALES
VICTORIA
GREAT DIVIDING RANGE
SNOWY MOUNTAINS HWY
MONARO HWY
PRINCES HWY
ALPINE WAY
BASS STRAIT
TASMAN SEA
GIPPSLAND
For more detail on the ACT see page 564
For more detail on the Snowy Mountains & The South Coast see pages 566–7
Cooma
Jindabyne
Berridale
Thredbo
Bombala
Bega
Merimbula
Eden
Orbost
Buchan
Mallacoota
Cann River
Adaminaby
Kiandra
Khancoban
Tooma
Delegate
Genoa
Marlo
Cape Conran
Cape Howe
Gabo Island

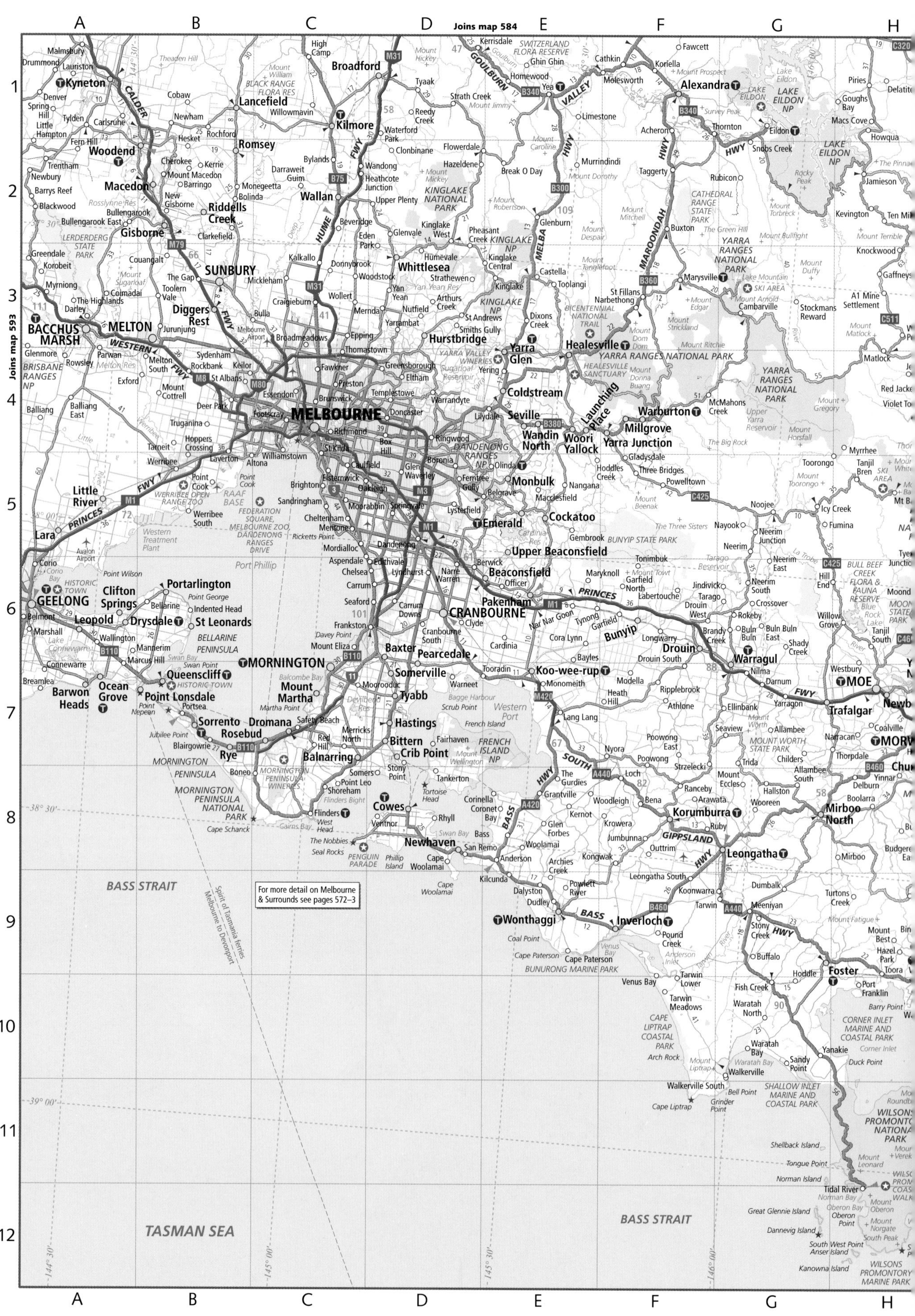
A
B
C
D
Joins map 584
E
F
G
H
1
2
3
4
5
6
7
8
9
10
11
12
Joins map 593
Malmsbury
Drummond
Lauriston
Kyneton
CALDER
Denver
Spring Hill
Tylden
Carlsruhe
Little Hampton
Fern Hill
Woodend
Trentham
Newbury
Barrys Reef
Blackwood
Macedon
Mount Macedon
Cherokee
Kerrie
Barringo
New Gisborne
Riddells Creek
Bullengarook
Bullengarook East
Gisborne
LERDERBERG STATE PARK
Clarkefield
Greendale
Korobeit
Couangalt
Myrniong
Coimadai
The Highlands
Darley
BACCHUS MARSH
MELTON
Jurunjung
The Gap
Toolern Vale
SUNBURY
Diggers Rest
Glenmore
Rowsley
Parwan
Melton South
WESTERN FWY
Sydenham
Rockbank
Keilor
St Albans
Exford
BRISBANE RANGES NP
Balliang
Balliang East
Mount Cottrell
Deer Park
Truganina
Hoppers Crossing
Tarneit
Werribee
Laverton
Altona
Williamstown
Point Cook
Little River
Lara
PRINCES FWY
WERRIBEE OPEN RANGE ZOO
RAAF BASE
Werribee South
Western Treatment Plant
Avalon Airport
Corio
Corio Bay
Point Wilson
Port Phillip
GEELONG
HISTORIC TOWN
Clifton Springs
Portarlington
Point George
Bellarine
Indented Head
St Leonards
Drysdale
Leopold
Belmont
Marshall
Wallington
Lake Connewarre
BELLARINE PENINSULA
Mannerim
Marcus Hill
Connewarre
Swan Point
Queenscliff
HISTORIC TOWN
Breamlea
Barwon Heads
Ocean Grove
Point Lonsdale
Point Nepean
Portsea
Sorrento
Dromana
Safety Beach
Rosebud
Rye
Blairgowrie
Jubilee Point
MORNINGTON PENINSULA
MORNINGTON PENINSULA NATIONAL PARK
Boneo
Cape Schanck
BASS STRAIT
Spirit of Tasmania ferries Melbourne to Devonport
For more detail on Melbourne & Surrounds see pages 572–3
TASMAN SEA
38° 30'
39° 00'
144° 30'
145° 00'
145° 30'
146° 00'
Theaden Hill
Cobaw
Newham
Hesket
Rochford
Lancefield
Romsey
Monegeetta
Bolinda
Willowmavin
Mount William
BLACK RANGE FLORA RES
High Camp
Broadford
Kilmore
Bylands
Darraweit Guim
Wandong
Heathcote Junction
Wallan
Waterford Park
Clonbinane
HUME FWY
Beveridge
Eden Park
Kalkallo
Donnybrook
Woodstock
Mickleham
Wollert
Craigieburn
Bulla
Melbourne Airport
Broadmeadows
Epping
Mernda
Thomastown
Fawkner
Preston
Essendon
Brunswick
Footscray
MELBOURNE
Richmond
St Kilda
Caulfield
Elsternwick
Oakleigh
Brighton
Sandringham
Moorabbin
Cheltenham
Mentone
Mordialloc
Aspendale
Chelsea
Carrum
Seaford
Frankston
Edithvale
Springvale
Dandenong
Lyndhurst
Carrum Downs
Langwarrin
FEDERATION SQUARE, MELBOURNE ZOO, DANDENONG RANGES DRIVE
Ricketts Point
Davey Point
Mount Eliza
MORNINGTON
Balcombe Bay
Mount Martha
Martha Point
Moorooduc
Tyabb
Hastings
Merricks North
Red Hill
Balnarring
Bittern
Crib Point
Somers
Point Leo
Shoreham
Flinders Bight
Flinders
West Head
Gairns Bay
Cowes
Ventnor
Rhyll
The Nobbies
Seal Rocks
PENGUIN PARADE
Phillip Island
Newhaven
Cape Woolamai
San Remo
Tortoise Head
Tankerton
Corinella
Coronet Bay
Swan Bay
Stony Point
French Island
FRENCH ISLAND NP
Mount Wellington
Western Port
Bagge Harbour
Scrub Point
Fairhaven
Baxter
Somerville
Pearcedale
Warneet
Tooradin
CRANBOURNE
Cranbourne South
Clyde
Cardinia
Narre Warren
Berwick
Beaconsfield
Officer
Pakenham
Nar Nar Goon
Tynong
Garfield
Bunyip
Koo-wee-rup
Monomeith
Lang Lang
Heath Hill
Modella
Mount Hickey
Tyaak
Reedy Creek
Strath Creek
Mount Jimmy
Flowerdale
Hazeldene
Mount Mickey
Upper Plenty
KINGLAKE NATIONAL PARK
Mount Robertson
Glenvale
Kinglake West
Pheasant Creek
KINGLAKE NP
Humevale
Whittlesea
Strathewen
Kinglake Central
Kinglake
Yan Yean
Yan Yean Res
Arthurs Creek
Nutfield
Yarrambat
St Andrews
Smiths Gully
Hurstbridge
Greensborough
Eltham
Templestowe
Doncaster
Warrandyte
YARRA VALLEY WINERIES
Yering
Yarra Glen
Sugarloaf Reservoir
Lilydale
Coldstream
Box Hill
Ringwood
Boronia
DANDENONG RANGES NP
Glen Waverley
Ferntree Gully
Belgrave
Olinda
Monbulk
Lysterfield
Emerald
Cockatoo
Gembrook
Upper Beaconsfield
Cardinia Res
Kerrisdale
SWITZERLAND FLORA RESERVE
Ghin Ghin
Homewood
GOULBURN VALLEY HWY
Yea
Cathkin
Molesworth
Limestone
Mount Caroline
Murrindindi
Mount Dorothy
Break O Day
Glenburn
MELBA HWY
Mount Despair
Mount Tanglefoot
Castella
Toolangi
Dixons Creek
BICENTENNIAL NATIONAL TRAIL
Healesville
HEALESVILLE SANCTUARY
YARRA RANGES NATIONAL PARK
Mount Donna Buang
Launching Place
Seville
Wandin North
Woori Yallock
Yarra Junction
Millgrove
Warburton
Gladysdale
Hoddles Creek
Three Bridges
Powelltown
Nangana
Macclesfield
Mount Beenak
The Three Sisters
BUNYIP STATE PARK
Tonimbuk
Maryknoll
Mount Toby
Garfield North
Labertouche
Tarago Reservoir
Cora Lynn
Bayles
Longwarry
Drouin
Drouin South
Drouin West
Jindivick
Ripplebrook
Athlone
Nyora
Poowong
Poowong East
Loch
The Gurdies
Grantville
Woodleigh
Kernot
Glen Forbes
Woolamai
Anderson
Kilcunda
Archies Creek
Dalyston
Wonthaggi
Coal Point
Cape Paterson
BUNURONG MARINE PARK
Inverloch
Pound Creek
Anderson Inlet
Venus Bay
Tarwin Lower
Tarwin Meadows
CAPE LIPTRAP COASTAL PARK
Arch Rock
Mount Liptrap
Walkerville
Walkerville South
Cape Liptrap
Bell Point
Grinder Point
Waratah Bay
Kongwak
Jumbunna
Krowera
Outtrim
Leongatha South
Koonwarra
Leongatha
Korumburra
Ruby
Arawata
Bena
Strzelecki
Ranceby
GIPPSLAND HWY
Fawcett
Koriella
Mount Prospect
Alexandra
Lake Eildon
LAKE EILDON NP
Thornton
Acheron
Taggerty
Snobs Creek
Eildon
Rubicon
CATHEDRAL RANGE STATE PARK
MAROONDAH HWY
Buxton
The Green Hill
Mount Bullfight
YARRA RANGES NATIONAL PARK
Marysville
Lake Mountain
SKI AREA
Mount Arnold
Cambarville
Narbethong
St Fillans
Mount Edgar
Mount Strickland
Mount Ritchie
McMahons Creek
Upper Yarra Reservoir
Mount Horsfall
The Big Rock
Noojee
Nayook
Neerim
Neerim Junction
Neerim East
Neerim South
Tarago
Crossover
Rokeby
Brandy Creek
Buln Buln
Buln Buln East
Shady Creek
Warragul
Nilma
Darnum
Yarragon
Ellinbank
Mount Worth
Seaview
MOUNT WORTH STATE PARK
Allambee
Childers
Trida
Mount Eccles
Hallston
Wooreen
Mirboo North
Mirboo
Dumbalk
Meeniyan
Stony Creek
Buffalo
Fish Creek
Waratah North
Hoddle
Foster
Port Franklin
Barry Point
CORNER INLET MARINE AND COASTAL PARK
Yanakie
Duck Point
SHALLOW INLET MARINE AND COASTAL PARK
WILSONS PROMONTORY NATIONAL PARK
Shellback Island
Tongue Point
Norman Island
Tidal River
Norman Bay
Oberon Bay
Oberon Point
Mount Oberon
Mount Norgate
South Peak
Great Glennie Island
Dannevig Island
South West Point
Anser Island
Kanowna Island
WILSONS PROMONTORY MARINE PARK
BASS STRAIT
Piries
Goughs Bay
Macs Cove
Howqua
Jamieson
Kevington
Knockwood
Gaffneys
A1 Mine Settlement
Stockmans Reward
Mount Matlock
Matlock
Red Jacket
Violet Town
Myrrhee
Toorongo
Tanjil Bren
Mount Toorongo
Icy Creek
Fumina
BULL BEEF CREEK FLORA & FAUNA RESERVE
Hill End
Willow Grove
Tanjil South
Westbury
MOE
Trafalgar
Narracan
Thorpdale
Allambee South
Yinnar
Delburn
Boolarra
Turtons Creek
Mount Best
Hazel Park
Toora
PRINCES FWY
WESTERN FWY
SOUTH GIPPSLAND HWY
BASS HWY
M1
M8
M80
M31
M79
M3
M420
B110
B75
B300
B340
B360
B380
B460
A420
A440
C425
C511

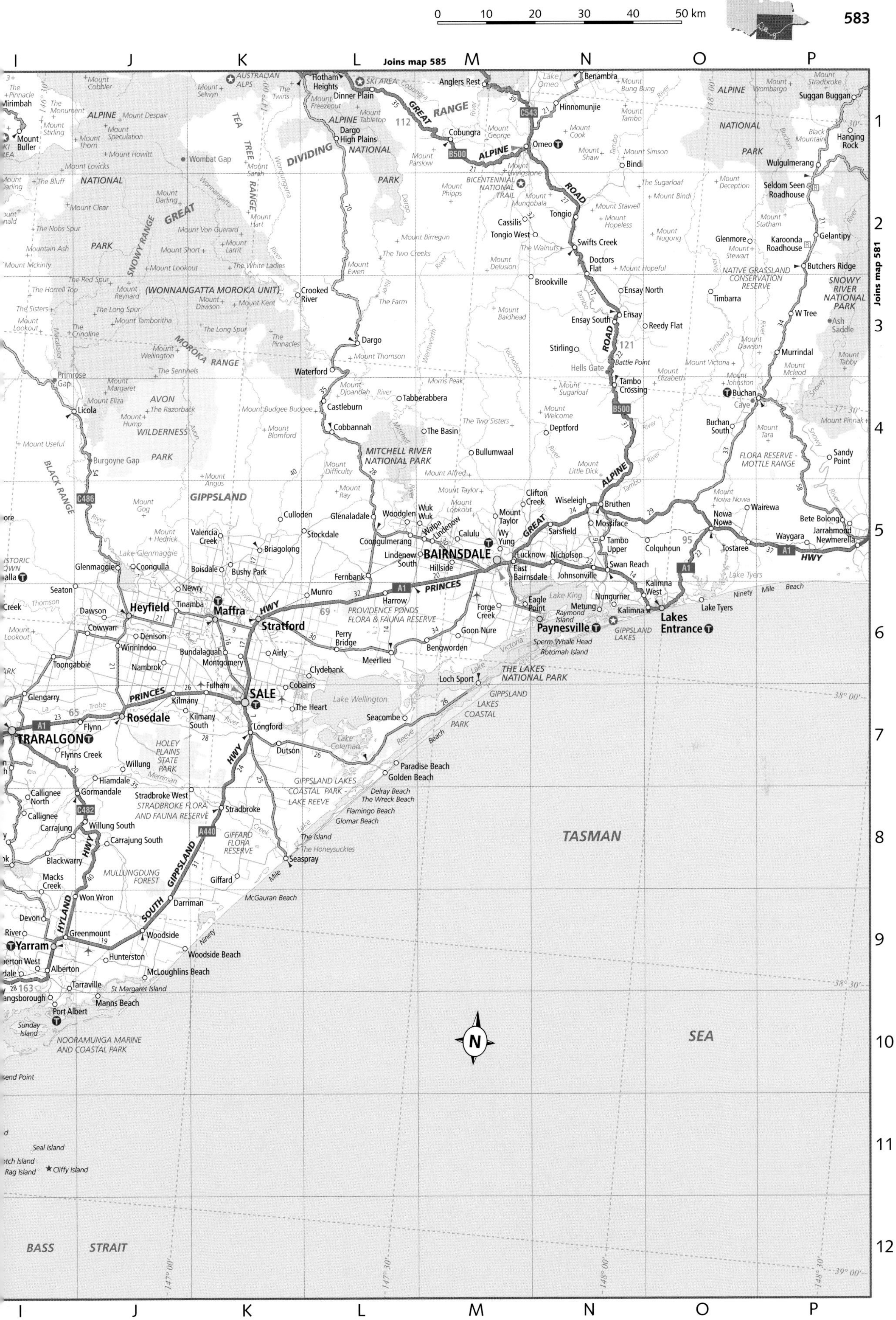

0 10 20 30 40 50 km

I J K L M N O P

Joins map 585

Joins map 581

1
The Pinnacle, Mirimbah, The Monument, Mount Cobbler, Mount Buller, Mount Stirling, ALPINE, Mount Despair, Mount Speculation, Mount Thorn, Mount Howitt, Mount Selwyn, AUSTRALIAN ALPS, The Twins, Wombat Gap, TEA TREE RANGE, Hotham Heights, SKI AREA, Dinner Plain, Mount Freezeout, Mount Tabletop, ALPINE, Dargo High Plains, DIVIDING, NATIONAL, 112, GREAT RANGE, Anglers Rest, Cobungra, Mount George, B500, ALPINE, C543, Lake Omeo, Benambra, Mount Bung Bung, Hinnomunjie, Mount Tambo, Mount Cook, Omeo, Mount Shaw, Mount Simson, Bindi, ALPINE, NATIONAL, PARK, Mount Wombargo, Mount Stradbroke, Suggan Buggan, Black Mountain, Hanging Rock, Wulgulmerang

Mount Lovicks, NATIONAL, The Bluff, Mount Darling, Mount Sarah, Mount Parslow, Mount Livingstone, BICENTENNIAL NATIONAL TRAIL, PARK, Mount Phipps, Mount Mungobala, ROAD, The Sugarloaf, Mount Bindi, Mount Deception, Seldom Seen Roadhouse

2
Mount Clear, GREAT, Mount Hart, Mount Von Guerard, The Nobs Spur, Mountain Ash, PARK, Mount Short, Mount Larrit, SNOWY RANGE, Mount Lookout, The White Ladies, Mount Birregun, Cassilis, Tongio, Tongio West, Mount Stawell, Mount Hopeless, Mount Nugong, Swifts Creek, The Walnuts, The Two Creeks, Mount Mckinty, Mount Ewen, Mount Delusion, Doctors Flat, Mount Hopeful, Mount Statham, Glenmore, Mount Stewart, Karoonda Roadhouse, Gelantipy, Butchers Ridge, NATIVE GRASSLAND CONSERVATION RESERVE, SNOWY RIVER NATIONAL PARK

The Red Spur, The Horrell Top, Mount Reynard, (WONNANGATTA MOROKA UNIT), Crooked River, Brookville, Ensay North, Timbarra

3
The Sisters, The Long Spur, Mount Dawson, Mount Kent, The Farm, Mount Baldhead, Ensay South, Ensay, Reedy Flat, W Tree, Ash Saddle, Mount Lookout, The Crinoline, Mount Tamboritha, The Long Spur, The Pinnacles, Dargo, Stirling, ROAD, 121, Battle Point, Mount Dawson, Murrindal, Mount Tabby, Mount Wellington, MOROKA RANGE, Mount Thomson, Hells Gate, Mount Victoria, Mount Mcleod, The Sentinels, Waterford, Morris Peak, Mount Elizabeth, Mount Johnston, Primrose Gap, Mount Margaret, Mount Djoandah, Mount Sugarloaf, Tambo Crossing, Buchan, Buchan Caves

4
Mount Eliza, AVON, The Razorback, Mount Budgee Budgee, Castleburn, Tabberabbera, B500, Licola, Mount Hump, WILDERNESS, Cobbannah, The Two Sisters, Mount Welcome, Deptford, Buchan South, Mount Tara, Mount Pinnacle, Mount Useful, PARK, Mount Blomford, The Basin, Bullumwaal, FLORA RESERVE - MOTTLE RANGE, Sandy Point, BLACK RANGE, Burgoyne Gap, MITCHELL RIVER NATIONAL PARK, Mount Difficulty, Mount Alfred, Mount Little Dick, Mount Angus, GIPPSLAND, Mount Ray, Mount Taylor, Clifton Creek, ALPINE, C486, Mount Gog, Mount Lookout, Wiseleigh, Bruthen, Mount Nowa Nowa, Wairewa

5
Culloden, Glenaladale, Woodglen, Wuk Wuk, Mount Taylor, Nowa Nowa, Bete Bolong, Mount Hedrick, Valencia Creek, Stockdale, Walpa, Lindenow, Calulu, GREAT, Mossiface, Sarsfield, Tambo Upper, Colquhoun, 95, Waygara, Jarrahmond, Newmerella, Lake Glenmaggie, Briagolong, Coongulmerang, Wy Yung, Tostaree, HWY, Glenmaggie, Coongulla, Boisdale, Bushy Park, Lindenow South, BAIRNSDALE, Lucknow, Nicholson, HISTORIC TOWN Walhalla, Hillside, East Bairnsdale, Johnsonville, Swan Reach, A1, Lake Tyers, Seaton, Newry, Fernbank, Kalimna West, Ninety Mile Beach, Thomson, Tinamba, Munro, PRINCES, Nungurner, Heyfield, Maffra, HWY, Harrow, Eagle Point, Lake King, Metung, Kalimna, Lake Tyers, Dawson, 69, PROVIDENCE PONDS FLORA & FAUNA RESERVE, Forge Creek, Raymond Island, Lakes Entrance

6
Cowwarr, Stratford, Perry Bridge, Goon Nure, Paynesville, GIPPSLAND LAKES, Mount Lookout, Denison, Bengworden, Sperm Whale Head, Rotamah Island, Winnindoo, Bundalaguah, Airly, Meerlieu, Toongabbie, Nambrok, Montgomery, Clydebank, THE LAKES NATIONAL PARK, Fulham, Cobains, Loch Sport, Glengarry, PRINCES, Kilmany, SALE, GIPPSLAND LAKES COASTAL PARK, Lake Wellington, The Heart, 65, Trobe, Seacombe

7
TRARALGON, Flynn, Rosedale, Kilmany South, Longford, 28, HOLEY PLAINS STATE PARK, Dutson, Lake Coleman, Flynns Creek, Reeve, Beach, Willung, Paradise Beach, Golden Beach, Hiamdale, Merriman, Callignee North, Gormandale, Stradbroke West, GIPPSLAND LAKES COASTAL PARK - LAKE REEVE, Delray Beach, The Wreck Beach, Flamingo Beach, Callignee, C482, STRADBROKE FLORA AND FAUNA RESERVE, Stradbroke, Glomar Beach, Carrajung, Willung South

8
Carrajung South, A440, GIFFARD FLORA RESERVE, The Island, The Honeysuckles, TASMAN, Blackwarry, HWY, Seaspray, MULLUNGDUNG FOREST, Giffard, Mile, Macks Creek, GIPPSLAND, Won Wron, McGauran Beach, Darriman, SOUTH

9
Devon, HYLAND, River, Greenmount, Woodside, Yarram, Ninety, Hunterston, Woodside Beach, Alberton West, Alberton, McLoughlins Beach, Tarraville, St Margaret Island, 163, Langsborough, Manns Beach, Port Albert

10
Sunday Island, NOORAMUNGA MARINE AND COASTAL PARK, N, SEA, Townsend Point

11
Seal Island, Notch Island, Rag Island, Cliffy Island

12
BASS, STRAIT

37° 30′, 38° 00′, 38° 30′, 39° 00′, 146° 30′, 147° 00′, 147° 30′, 148° 00′, 148° 30′

I J K L M N O P

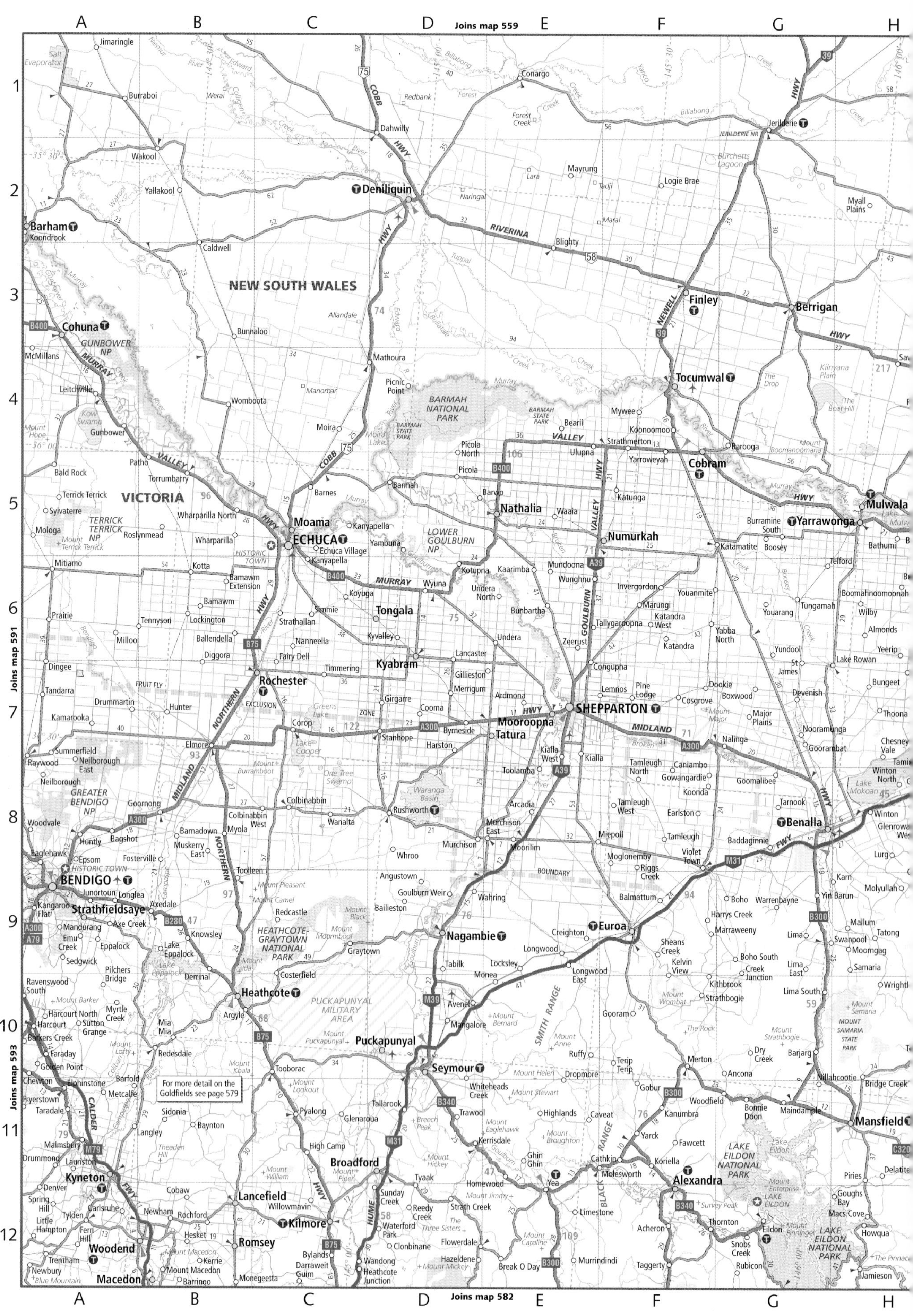
Joins map 559
Joins map 582
Joins map 591
Joins map 593
A
B
C
D
E
F
G
H
1
2
3
4
5
6
7
8
9
10
11
12
NEW SOUTH WALES
VICTORIA
Jimaringle
Burraboi
Wakool
Yallakool
Barham
Koondrook
Caldwell
Conargo
Redbank
Forest Creek
Dahwilly
Deniliquin
Naringal
Mayrung
Lara
Tadji
Maral
Logie Brae
Jerilderie
JERILDERIE NR
Burchetts Lagoon
Myall Plains
Blighty
RIVERINA
Finley
Berrigan
Bunnaloo
Allandale
Mathoura
Picnic Point
Manorbar
Womboota
Cohuna
GUNBOWER NP
McMillans
Leitchville
Kow Swamp
Mount Hope
Gunbower
Patho
Torrumbarry
Bald Rock
Terrick Terrick
Sylvaterre
Mologa
TERRICK TERRICK NP
Mount Terrick Terrick
Roslynmead
Wharparilla North
Wharparilla
BARMAH NATIONAL PARK
BARMAH STATE PARK
Moira
Moira Lake
Picola North
Picola
Barmah
Barnes
Moama
ECHUCA
Echuca Village
Kanyapella
HISTORIC TOWN
Yambuna
LOWER GOULBURN NP
Tocumwal
The Drop
Kilnyana Plain
The Boat Hill
Mywee
Bearii
Koonoomoo
Strathmerton
Ulupna
Yarroweyah
Cobram
Barooga
Mount Boomanoomana
Katunga
Barwo
Nathalia
Waaia
Numurkah
Katamatite
Burramine South
Boosey
Yarrawonga
Mulwala
Bathumi
Telford
Mitiamo
Kotta
Bamawm Extension
Bamawm
Prairie
Tennyson
Lockington
Milloo
Ballendella
Diggora
Koyuga
Tongala
Simmie
Strathallan
Nanneella
Kyvalley
Fairy Dell
Timmering
Kyabram
Lancaster
Gillieston
Merrigum
Kotupna
Kaarimba
Undera North
Wyuna
Mundoona
Wunghnu
Invergordon
Marungi
Youanmite
Katandra West
Katandra
Tallygaroopna
Bunbartha
Undera
Zeerust
Congupna
Yabba North
Youarang
Tungamah
Boomahnoomoonah
Wilby
Almonds
Yundool
St James
Yeerip
Lake Rowan
Bungeet
Dingee
Tandarra
FRUIT FLY
EXCLUSION
ZONE
Rochester
Drummartin
Hunter
Kamarooka
Girgarre
Cooma
Ardmona
Mooroopna
Tatura
SHEPPARTON
Lemnos
Pine Lodge
Dookie
Cosgrove
Boxwood
Mount Major
Major Plains
Devenish
Thoona
Corop
Greens Lake
Lake Cooper
Stanhope
Byrneside
Harston
Kialla West
Kialla
Toolamba
Tamleugh North
Caniambo
Gowangardie
Koonda
MIDLAND
Nalinga
Goomalibee
Nooramunga
Goorambat
Chesney Vale
Winton North
Lake Mokoan
Summerfield
Raywood
Neilborough East
Neilborough
Elmore
Mount Burramboot
One Tree Swamp
GREATER BENDIGO NP
Goornong
Colbinabbin
Colbinabbin West
Waranga Basin
Rushworth
Arcadia
Tamleugh West
Earlston
Tarnook
Woodvale
Huntly
Bagshot
Barnadown
Myola
Wanalta
Murchison East
Murchison
Moorilim
Miepoll
Tamleugh
Violet Town
Baddaginnie
Benalla
Winton
Glenrowan West
Eaglehawk
Epsom
HISTORIC TOWN
Fosterville
Muskerry East
Whroo
BOUNDARY
Moglonemby
Riggs Creek
Lurg
BENDIGO
Junortoun
Longlea
Toolleen
Mount Pleasant
Angustown
Goulburn Weir
Wahring
Balmattum
Boho
Warrenbayne
Karn
Molyullah
Yin Barun
Kangaroo Flat
Strathfieldsaye
Axedale
Mount Camel
Redcastle
Baillieston
Mount Black
Harrys Creek
Mandurang
Axe Creek
Knowsley
HEATHCOTE-GRAYTOWN NATIONAL PARK
Mount Moormbool
Nagambie
Creighton
Euroa
Marraweeny
Mallum
Tatong
Emu Creek
Eppalock
Lake Eppalock
Graytown
Longwood
Sheans Creek
Lima
Swanpool
Moorngag
Sedgwick
Mount Ida
Tabilk
Locksley
Longwood East
Kelvin View
Boho South
Creek Junction
Lima East
Samaria
Ravenswood South
Pilchers Bridge
Derrinal
Costerfield
Monea
Kithbrook
Strathbogie
Lima South
Wrightley
Mount Barker
Harcourt North
Harcourt
Sutton Grange
Myrtle Creek
Heathcote
Argyle
PUCKAPUNYAL MILITARY AREA
Avenel
Mount Bernard
Mangalore
SMITH RANGE
Mount Wombat
Gooram
Mount Samaria
MOUNT SAMARIA STATE PARK
Barkers Creek
Mia Mia
Mount Puckapunyal
Puckapunyal
Mount Anne
The Rock
Mount Strathbogie
Faraday
Golden Point
Mount Lofty
Redesdale
Mount Koala
Tooborac
Seymour
Ruffy
Terip Terip
Dry Creek
Barjarg
Chewton
Elphinstone
Barfold
Metcalfe
For more detail on the Goldfields see page 579
Mount Lookout
Whiteheads Creek
Mount Helen
Mount Stewart
Dropmore
Gobur
Merton
Ancona
Willahcootie
Bridge Creek
Fryerstown
Taradale
CALDER
Sidonia
Baynton
Pyalong
Glenaroua
Tallarook
Breech Peak
Trawool
Highlands
Caveat
Woodfield
Kanumbra
Bonnie Doon
Maindample
Mansfield
Malmsbury
Langley
Theaden Hill
High Camp
Mount Eaglehawk
Kerrisdale
Mount Broughton
BLACK RANGE
Yarck
Fawcett
LAKE EILDON NATIONAL PARK
Lake Eildon
Drummond
Lauriston
Mount William
Broadford
Mount Piper
Mount Hickey
Ghin Ghin
Yea
Cathkin
Koriella
Molesworth
Alexandra
Mount Enterprise
LAKE EILDON
Piries
Delatite
Kyneton
Denver
Cobaw
Lancefield
Willowmavin
Tyaak
Sunday Creek
Homewood
Mount Jimmy
Survey Peak
Goughs Bay
Macs Cove
Spring Hill
Carlsruhe
Tylden
Newham
Rochford
Kilmore
Reedy Creek
Strath Creek
Limestone
Thornton
Eildon
Mount Pinninger
Howqua
Little Hampton
Ferri Hill
Hesket
Romsey
Waterford Park
The Three Sisters
Flowerdale
Mount Caroline
Acheron
Snobs Creek
Woodend
Trentham
Mount Macedon
Kerrie
Bylands
Clonbinane
Hazeldene
Mount Mickey
Break O Day
Murrindindi
Taggerty
Rubicon
The Pinnacle
Jamieson
Newbury
Blue Mountain
Macedon
Barringo
Monegeetta
Darraweit Guim
Wandong
Heathcote Junction
HUME
HWY
FWY
COBB
NEWELL
MURRAY
VALLEY
GOULBURN
NORTHERN
MIDLAND
Salt Evaporator
35° 30'
36° 00'
36° 30'
144° 30'
145° 00'
145° 30'
146° 00'

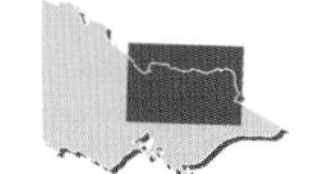

0 10 20 30 40 50 km

Joins map 559

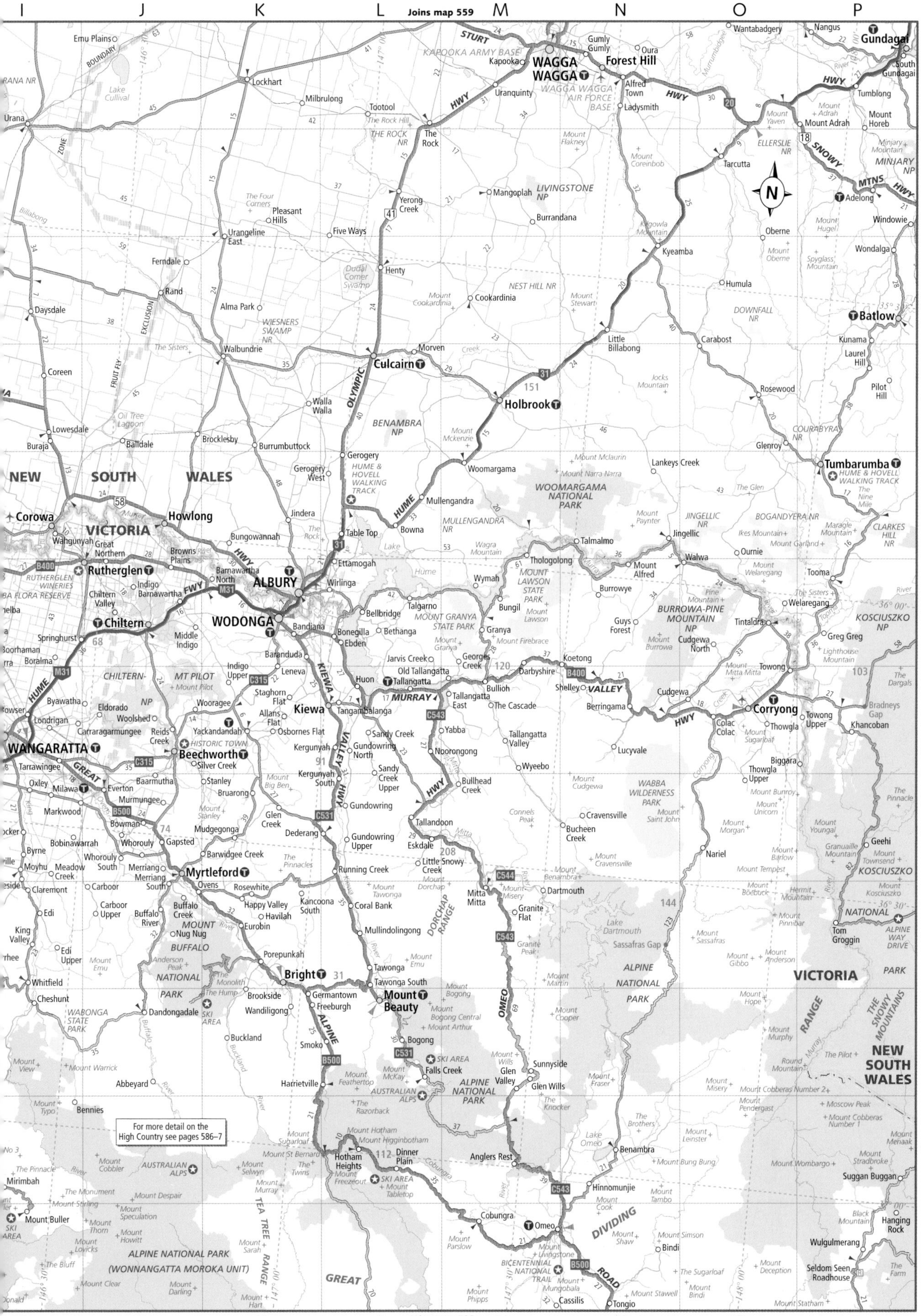

WAGGA WAGGA
ALBURY
WODONGA
Wangaratta
Corowa
Rutherglen
Chiltern
Beechworth
Myrtleford
Bright
Mount Beauty
Holbrook
Culcairn
Tumbarumba
Corryong
Tallangatta
Batlow
Gundagai
Omeo

NEW SOUTH WALES
VICTORIA

ALPINE NATIONAL PARK
KOSCIUSZKO NP
MOUNT BUFFALO NATIONAL PARK
WOOMARGAMA NATIONAL PARK
BURROWA-PINE MOUNTAIN NP

For more detail on the High Country see pages 586–7

Joins map 581

Joins map 583

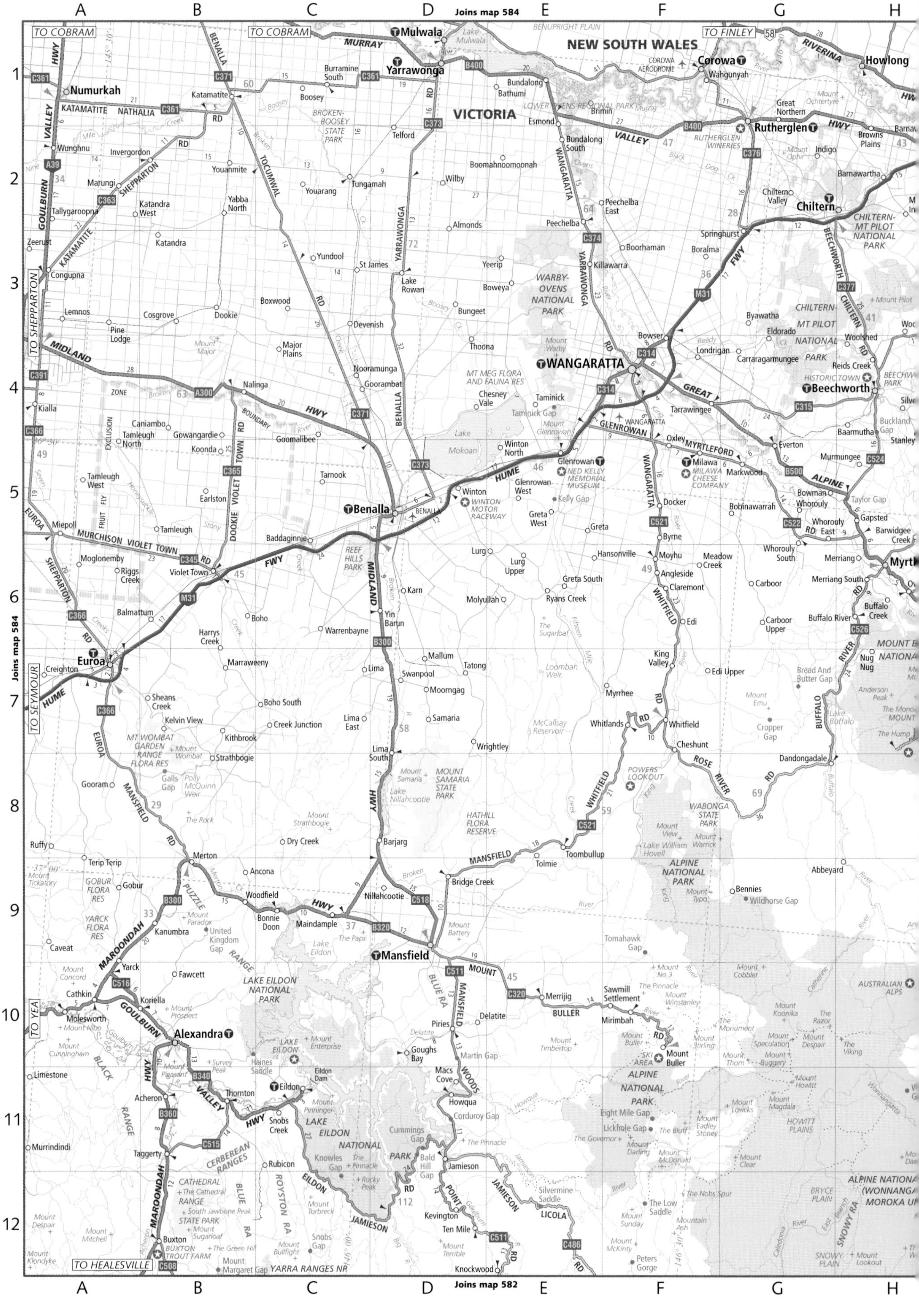

Joins map 584
Joins map 582
Joins map 584
A B C D E F G H
1 2 3 4 5 6 7 8 9 10 11 12
TO COBRAM
TO COBRAM
TO FINLEY
TO SHEPPARTON
TO SEYMOUR
TO YEA
TO HEALESVILLE
NEW SOUTH WALES
VICTORIA
Mulwala
Lake Mulwala
MURRAY
Yarrawonga
Burramine South
Boosey
Corowa
COROWA AERODROME
Wahgunyah
Howlong
RIVERINA
Bundalong
Bathumi
Esmond
LOWER OVENS REGIONAL PARK
Brimin
Rutherglen
RUTHERGLEN WINERIES
Great Northern
Browns Plains
Barnawartha
Indigo
Mount Ophir
Chiltern Valley
Chiltern
CHILTERN-MT PILOT NATIONAL PARK
Numurkah
KATAMATITE
NATHALIA
Katamatite
Wunghnu
Invergordon
Marungi
Tallygaroopna
Youanmite
Katandra West
Katandra
Yabba North
Zeerust
Congupna
GOULBURN VALLEY HWY
SHEPPARTON RD
KATAMATITE RD
BROKEN-BOOSEY STATE PARK
Telford
Tungamah
Youarang
Wilby
Boomahnoomoonah
Bundalong South
Peechelba
Peechelba East
Boorhaman
Boralma
Springhurst
Almonds
Yundool
St James
Lake Rowan
Yeerip
Killawarra
Boweya
WARBY-OVENS NATIONAL PARK
Bungeet
Lemnos
Pine Lodge
Cosgrove
Dookie
Boxwood
Devenish
Major Plains
Thoona
Mount Major
Mount Warby
Bowser
Londrigan
Carraragarmungee
Byawatha
Eldorado
Woolshed
Reids Creek
HISTORIC TOWN
Beechworth
Mount Pilot
MIDLAND
Nalinga
Noorangunga
Goorambat
MT MEG FLORA AND FAUNA RES
WANGARATTA
GREAT
Tarrawingee
Kialla
ZONE
EXCLUSION
Caniambo
Tamleugh North
Gowangardie
Koonda
Goomalibee
BOUNDARY
Chesney Vale
Taminick
Taminick Gap
Mount Glenrowan
GLENROWAN
Oxley
MYRTLEFORD
Milawa
MILAWA CHEESE COMPANY
Markwood
Baarmutha
Buckland Gap
Stanley
Everton
Murmungee
Lake Mokoan
Winton North
Glenrowan
NED KELLY MEMORIAL MUSEUM
Tamleugh West
Earlston
Tarnook
Winton
WINTON MOTOR RACEWAY
Glenrowan West
Kelly Gap
Docker
Bobinawarrah
Bowman
Whorouly
ALPINE
Taylor Gap
Gapsted
Whorouly East
Barwidgee Creek
Benalla
Miepoll
Tamleugh
FRUIT FLY
MURCHISON
VIOLET TOWN
DOOKIE VIOLET TOWN RD
Baddaginnie
REEF HILLS PARK
Greta West
Greta
Byrne
Moglonemby
Riggs Creek
Violet Town
FWY
Lurg
Lurg Upper
Hansonville
Moyhu
Meadow Creek
Whorouly South
Merriang
Angleside
Greta South
Claremont
Ryans Creek
Molyullah
Karn
Carboor
Merriang South
Buffalo Creek
Balmattum
Boho
Harrys Creek
Yin Barun
Warrenbayne
The Sugarloaf
Edi
Carboor Upper
Buffalo River
Euroa
Creighton
HUME
Marraweeny
Mallum
Tatong
Lima
Swanpool
King Valley
Edi Upper
Bread And Butter Gap
Nug Nug
Moorngag
Loombah Weir
Myrrhee
Mount Emu
Anderson Peak
Sheans Creek
Boho South
Samaria
Whitlands
Whitfield
Cropper Gap
Lake Buffalo
Kelvin View
Creek Junction
Lima East
McCallsay Reservoir
Cheshunt
The Hump
MT WOMBAT GARDEN RANGE FLORA RES
Mount Wombat
Kithbrook
Strathbogie
Wrightley
Lima South
Dandongadale
ROSE RIVER RD
Gooram
Galls Gap
Polly McQuinn Weir
Mount Samaria
MOUNT SAMARIA STATE PARK
Lake Nillahcootie
POWERS LOOKOUT
MANSFIELD
The Rock
Mount Strathbogie
HATHILL FLORA RESERVE
WHITFIELD
WABONGA STATE PARK
Mount View
Mount Warrick
Lake William Hovell
Ruffy
Dry Creek
Barjarg
Toombullup
Tolmie
Terip Terip
Merton
Ancona
ALPINE NATIONAL PARK
Abbeyard
Mount Tickatory
GOBUR FLORA RES
Gobur
PUZZLE
Woodfield
Bridge Creek
Nillahcootie
Mount Typo
Bennies
Wildhorse Gap
YARCK FLORA RES
Mount Paradox
Kanumbra
Bonnie Doon
Maindample
HWY
Mount Battery
United Kingdom Gap
RANGE
The Paps
Tomahawk Gap
Caveat
MAROONDAH
Lake Eildon
Mansfield
Yarck
Fawcett
LAKE EILDON NATIONAL PARK
MOUNT
Mount No 3
The Pinnacle
Mount Cobbler
AUSTRALIAN ALPS
Mount Concord
Cathkin
Koriella
Mount Prospect
BLUE RA
Merrijig
Sawmill Settlement
Mount Winstanley
BULLER
Molesworth
Mount Nibo
GOULBURN
Alexandra
Piries
Delatite
Mirimbah
Mount Koonika
The Razor
Mount Cunningham
LAKE EILDON
Mount Enterprise
Goughs Bay
Mount Timbertop
The Monument
Mount Buller
SKI AREA
Mount Stirling
Mount Speculation
Mount Despair
The Viking
Limestone
BLACK
Mount Pleasant
Survey Peak
Haines Saddle
Eildon Dam
Eildon
Martin Gap
Macs Cove
WOODS
Mount Thorn
Mount Buggery
Mount Howitt
Acheron
Thornton
VALLEY
Mount Pinninger
Howqua
Corduroy Gap
Eight Mile Gap
Mount Lovicks
Mount Magdala
HOWITT PLAINS
Snobs Creek
LAKE EILDON NATIONAL PARK
Cummings Gap
Lickhole Gap
The Bluff
Mount Eadley Stoney
The Governor
Mount Darling
Murrindindi
Taggerty
CERBEREAN RANGES
Knowles Gap
The Pinnacle
PARK
Bald Hill Gap
Jamieson
Mount McDonald
Mount Clear
Rubicon
Rocky Peak
Silvermine Saddle
CATHEDRAL RANGE STATE PARK
The Cathedral
South Jawbone Peak
Mount Sugarloaf
BLUE RA
ROYSTON RA
EILDON
JAMIESON
POINT
LICOLA
The Nobs Spur
The Low Saddle
Mount Sunday
Mountain Ash
BRYCE PLAIN
ALPINE NATIONAL PARK (WONNANGATTA MOROKA UNIT)
SNOWY RA
Mount Despair
Mount Mitchell
Buxton
BUXTON TROUT FARM
The Green Hill
Mount Torbreck
Mount Bullfight
Snobs Gap
Kevington
Ten Mile
Mount Terrible
Mount McKinty
Peters Gorge
Mount Klondyke
Mount Margaret Gap
YARRA RANGES NR
Knockwood
SNOWY PLAIN
Mount Lookout
37° 00'
145° 30'
146° 00'
146° 30'

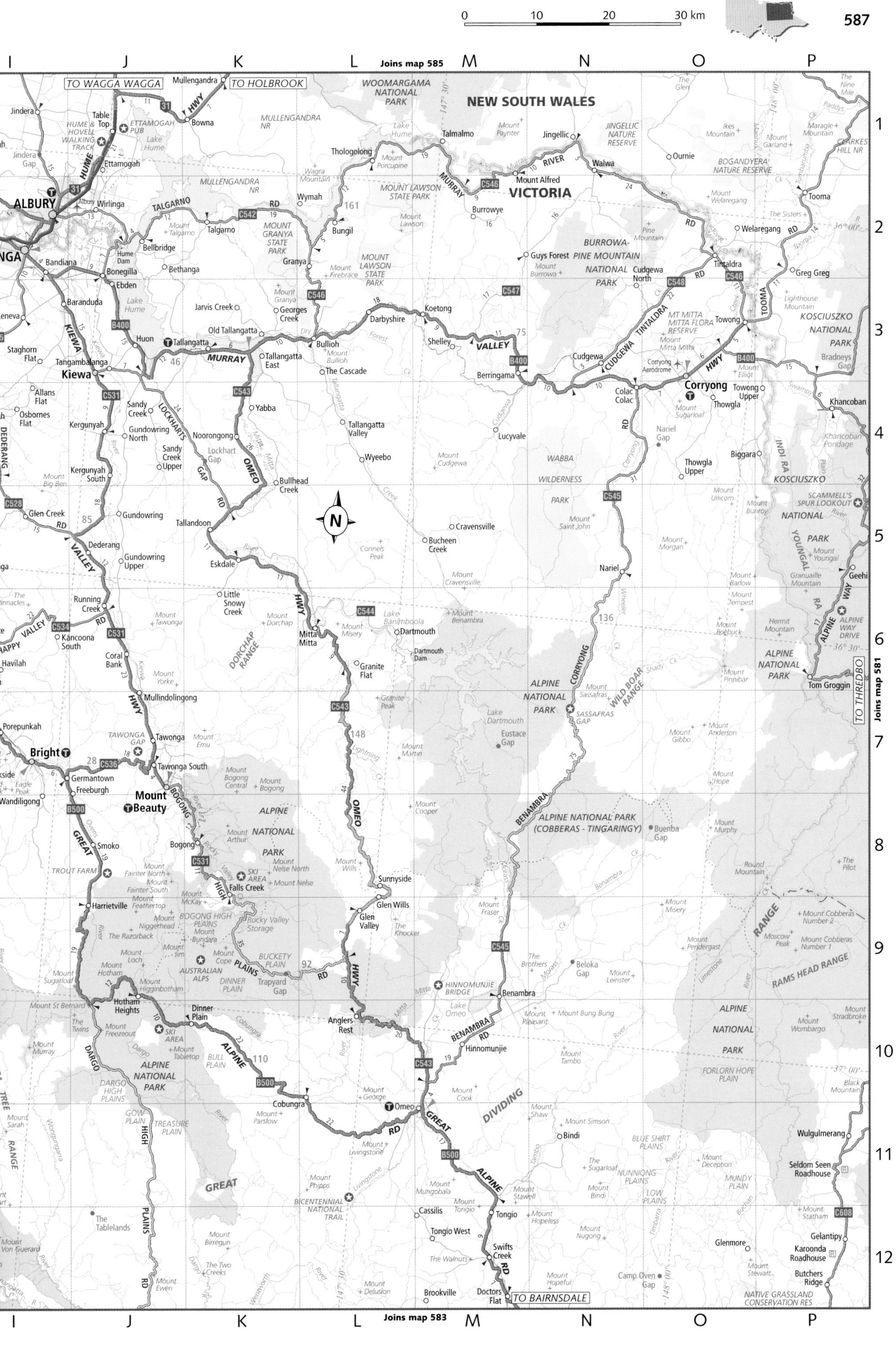

0 10 20 30 km
I J K L M N O P
1 2 3 4 5 6 7 8 9 10 11 12
Joins map 585
Joins map 583
Joins map 581
TO WAGGA WAGGA
TO HOLBROOK
TO THREDBO
TO BAIRNSDALE
NEW SOUTH WALES
VICTORIA
WOOMARGAMA NATIONAL PARK
MULLENGANDRA NR
MULLENGANDRA NR
JINGELLIC NATURE RESERVE
BOGANDYERA NATURE RESERVE
CARKES HILL NR
MOUNT GRANYA STATE PARK
MOUNT LAWSON STATE PARK
MOUNT LAWSON STATE PARK
BURROWA-PINE MOUNTAIN NATIONAL PARK
MT MITTA MITTA FLORA RESERVE
KOSCIUSZKO NATIONAL PARK
KOSCIUSZKO NATIONAL PARK
WABBA WILDERNESS PARK
ALPINE NATIONAL PARK
ALPINE NATIONAL PARK (COBBERAS - TINGARINGY)
ALPINE NATIONAL PARK
ALPINE NATIONAL PARK
ALPINE NATIONAL PARK
NATIVE GRASSLAND CONSERVATION RES
Mullengandra
Jindera
Table Top
Bowna
HUME & HOVELL WALKING TRACK
ETTAMOGAH PUB
Lake Hume
Jindera Gap
Ettamogah
ALBURY
Wirlinga
Talgarno
Bellbridge
Bethanga
Hume Dam
Bandiana
Bonegilla
Ebden
Baranduda
Leneva
Staghorn Flat
Tangambalanga
Kiewa
Huon
Tallangatta
Old Tallangatta
Tallangatta East
Jarvis Creek
Georges Creek
Granya
Wymah
Bungil
Tholoogolong
Talmalmo
Mount Alfred
Jingellic
Walwa
Ournie
Burrowye
Guys Forest
Cudgewa North
Tintaldra
Welaregang
Tooma
Greg Greg
Towong
Cudgewa
Berringama
Colac Colac
Corryong
Towong Upper
Thowgla
Khancoban
Biggara
Thowgla Upper
Nariel Gap
Koetong
Darbyshire
Shelley
Bullioh
The Cascade
Allans Flat
Osbornes Flat
Kergunyah
Sandy Creek
Gundowring North
Kergunyah South
Sandy Creek Upper
Noorongong
Lockhart Gap
Yabba
Tallangatta Valley
Wyeebo
Lucyvale
Bullhead Creek
Glen Creek
Gundowring
Tallandoon
Dederang
Gundowring Upper
Eskdale
Cravensville
Bucheen Creek
Nariel
Geehi
Running Creek
Little Snowy Creek
Kancoona South
Havilah
Coral Bank
Mitta Mitta
Dartmouth
Dartmouth Dam
Lake Banimboola
Granite Flat
Mullindolingong
Tom Groggin
Porepunkah
Bright
Tawonga
Tawonga South
TAWONGA GAP
Germantown
Freeburgh
Wandiligong
Mount Beauty
Bogong
Smoko
TROUT FARM
Harrietville
Falls Creek
SKI AREA
Sunnyside
Glen Wills
Glen Valley
SASSAFRAS GAP
Lake Dartmouth
Eustace Gap
Buenba Gap
Benambra
Beloka Gap
HINNOMUNJIE BRIDGE
Lake Omeo
Hinnomunjie
Hotham Heights
Dinner Plain
Anglers Rest
AUSTRALIAN ALPS
Cobungra
Omeo
Bindi
Wulgulmerang
Seldom Seen Roadhouse
Gelantipy
Karoonda Roadhouse
Butchers Ridge
Glenmore
Camp Oven Gap
Cassilis
Tongio
Tongio West
Swifts Creek
Brookville
Doctors Flat
The Tablelands
BICENTENNIAL NATIONAL TRAIL
SCAMMELL'S SPUR LOOKOUT
ALPINE WAY DRIVE
BOGONG HIGH PLAINS
BUCKETY PLAIN
DINNER PLAIN
BULL PLAIN
DARGO HIGH PLAINS
GOW PLAIN
TREASURE PLAIN
BLUE SHIRT PLAINS
NUNNIONG PLAINS
LOW PLAINS
MUNDY PLAIN
FORLORN HOPE PLAIN
DORCHAP RANGE
WILD BOAR RANGE
INDI RA
YOUNGAL RA
RAMS HEAD RANGE
DIVIDING RANGE
GREAT DIVIDING RANGE
HUME HWY
MURRAY VALLEY HWY
MURRAY RIVER RD
KIEWA VALLEY HWY
TALGARNO RD
LOCKHARTS GAP RD
OMEO HWY
CUDGEWA TINTALDRA RD
CORRYONG RD
BENAMBRA CORRYONG RD
BENAMBRA RD
TOOMA RD
ALPINE WAY
GREAT ALPINE RD
BOGONG HIGH PLAINS RD
HAPPY VALLEY RD
DARGO HIGH PLAINS RD
C546 C542 C547 C548 C531 C543 C544 C545 C528 C534 C536 C608 B400 B500
Mount Bogong
Mount Nelse North
Mount Nelse
Mount Feathertop
Mount Hotham
Mount Cope
Mount Jim
Mount Loch
Mount Cobberas Number 2
Mount Cobberas Number 1
Mount Pinnibar
Mount Gibbo
Mount Anderson
Mount Tambo
Mount Elliot
36° 00'
36° 30'
37° 00'
147° 30'
148° 00'

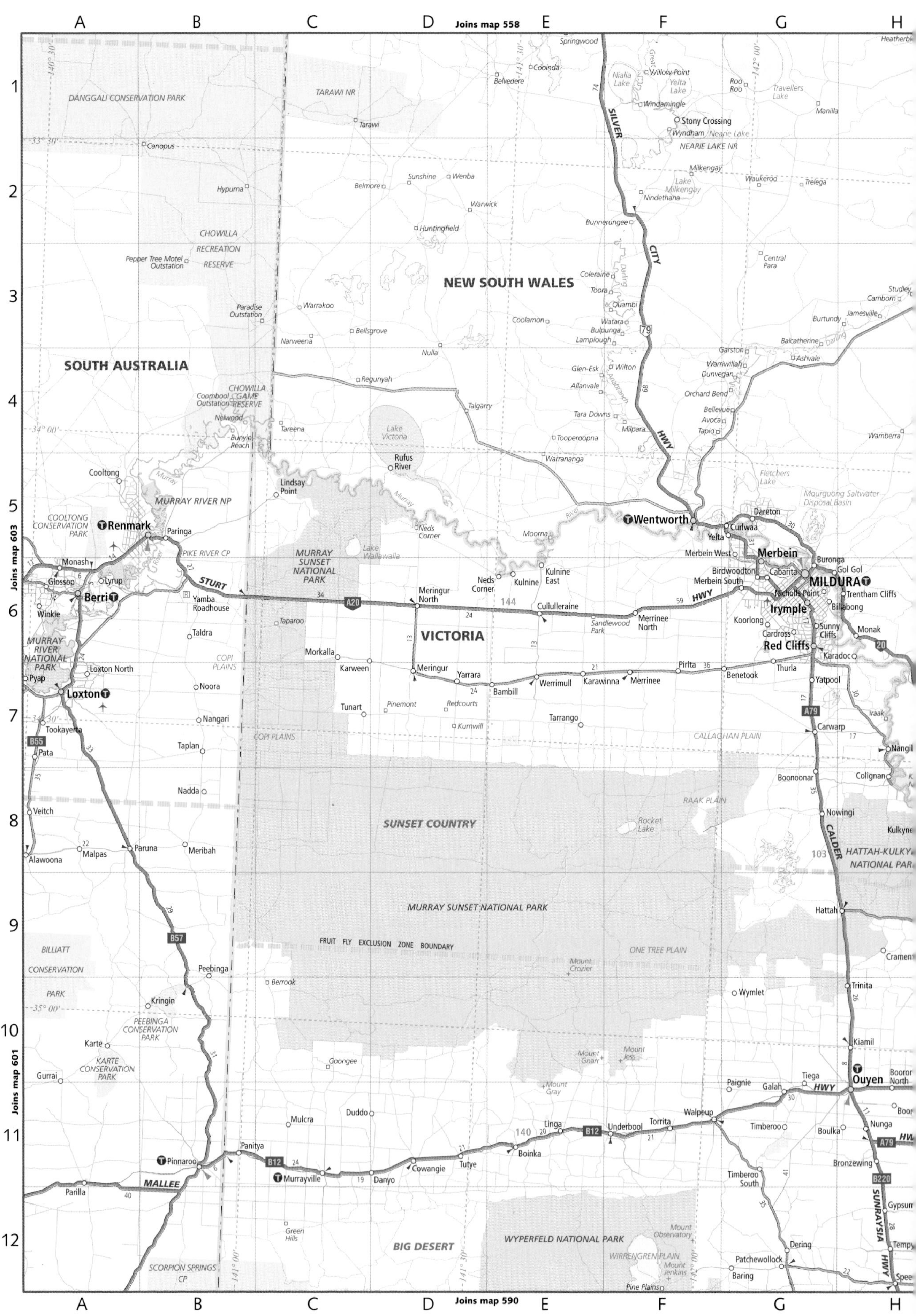

Joins map 558
Joins map 590
Joins map 603
Joins map 601
NEW SOUTH WALES
SOUTH AUSTRALIA
VICTORIA
MILDURA
Mildura
Merbein
Irymple
Red Cliffs
Wentworth
Renmark
Berri
Loxton
Ouyen
Pinnaroo
Murrayville
DANGGALI CONSERVATION PARK
TARAWI NR
CHOWILLA RECREATION RESERVE
CHOWILLA GAME RESERVE
MURRAY RIVER NP
MURRAY SUNSET NATIONAL PARK
SUNSET COUNTRY
MURRAY SUNSET NATIONAL PARK
FRUIT FLY EXCLUSION ZONE BOUNDARY
HATTAH-KULKYNE NATIONAL PARK
WYPERFELD NATIONAL PARK
BIG DESERT
SILVER CITY HWY
STURT HWY
CALDER HWY
MALLEE HWY
SUNRAYSIA HWY

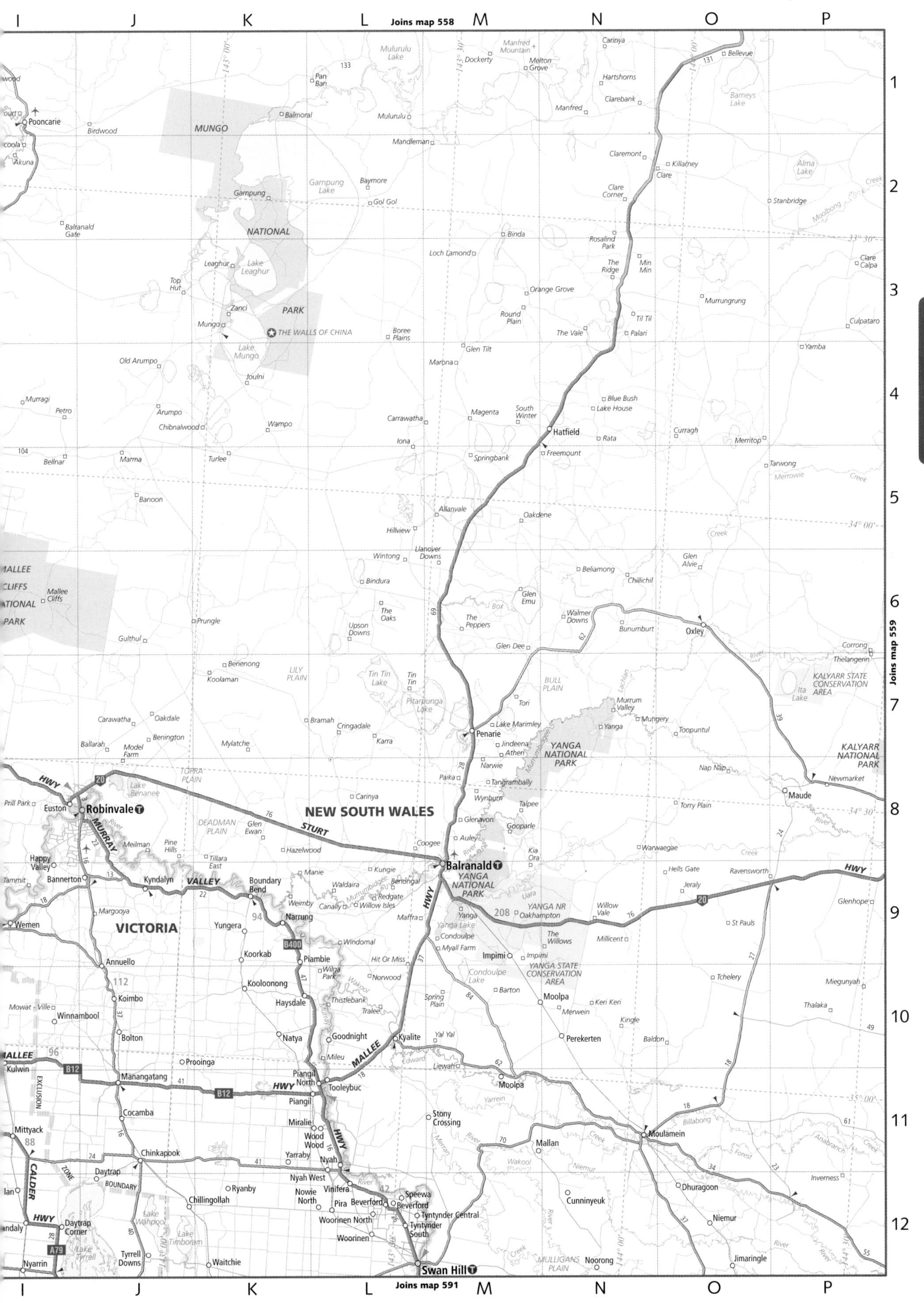
Joins map 558
Joins map 559
Joins map 591
NEW SOUTH WALES
VICTORIA
MUNGO NATIONAL PARK
YANGA NATIONAL PARK
KALYARR NATIONAL PARK
KALYARR STATE CONSERVATION AREA
YANGA STATE CONSERVATION AREA
Balranald
Robinvale
Swan Hill
Hatfield
Penarie
Oxley
Maude
Moulamein
Euston
STURT HWY
MALLEE HWY
MURRAY VALLEY HWY
CALDER HWY
THE WALLS OF CHINA

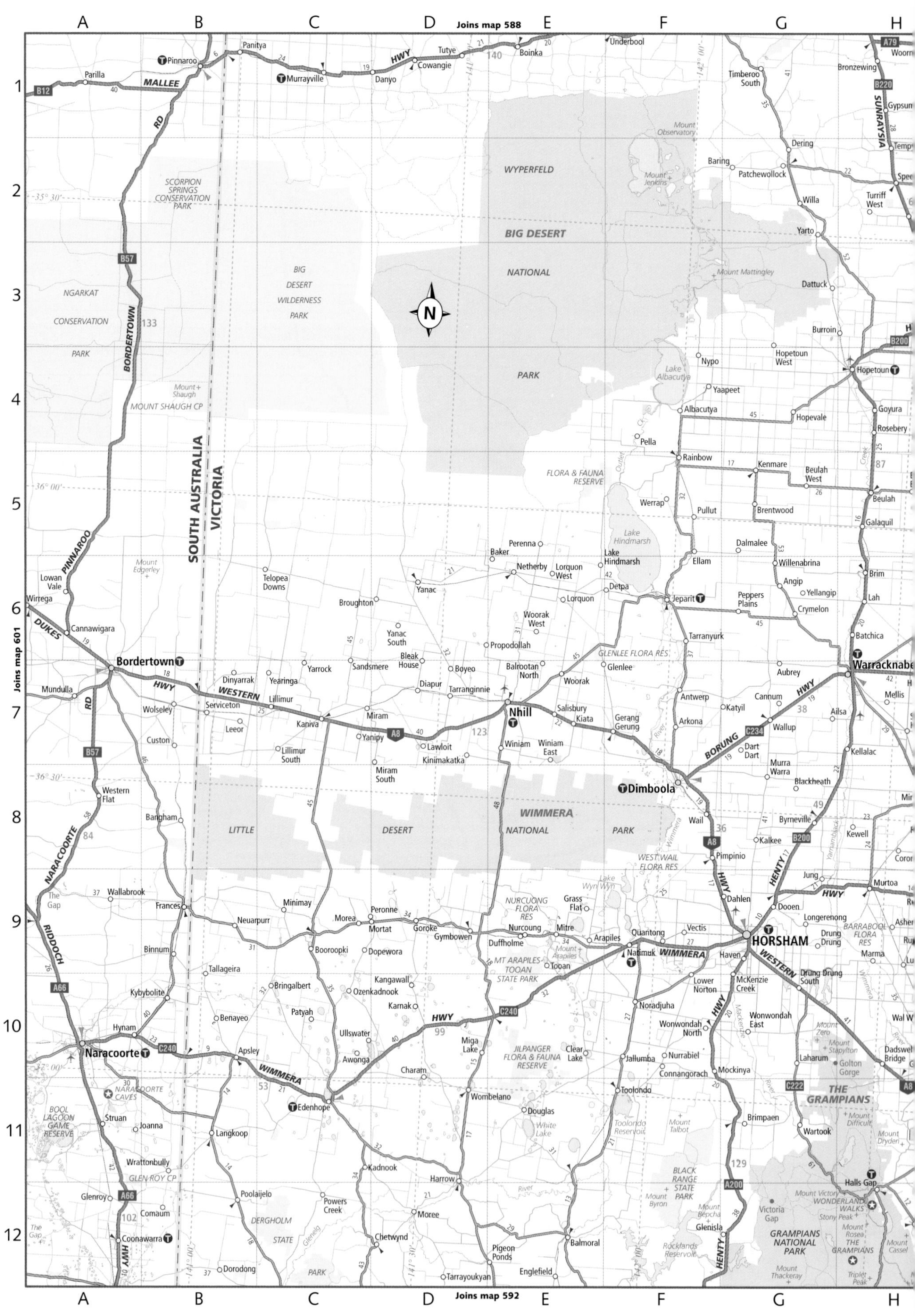

Joins map 588
Joins map 592
Joins map 601
A
B
C
D
E
F
G
H
1
2
3
4
5
6
7
8
9
10
11
12
Pinnaroo
Panitya
Parilla
MALLEE
RD
Murrayville
Danyo
HWY
Tutye
Cowangie
Boinka
Underbool
Timberoo South
Bronzewing
Woorn
SUNRAYSIA
Gypsum
Temps
Spee
Dering
Baring
Patchewollock
Mount Observatory
Mount Jenkins
Willa
Turriff West
Yarto
Dattuck
Burroin
Hopetoun West
Hopetoun
Nypo
Lake Albacutya
Yaapeet
Albacutya
Hopevale
Goyura
Rosebery
Pella
Rainbow
Kenmare
Beulah West
Beulah
Werrap
Pullut
Brentwood
Galaquil
Dalmalee
Willenabrina
Brim
Ellam
Angip
Yellangip
Lah
Jeparit
Peppers Plains
Crymelon
Tarranyurk
Batchica
Warracknabeal
Aubrey
Mellis
Antwerp
Katyil
Cannum
Ailsa
Arkona
Wallup
Kellalac
BORUNG
HWY
Dart Dart
Murra Warra
Blackheath
SCORPION SPRINGS CONSERVATION PARK
WYPERFELD
BIG DESERT
NATIONAL
PARK
BIG DESERT WILDERNESS PARK
NGARKAT CONSERVATION PARK
BORDERTOWN
Mount Shaugh
MOUNT SHAUGH CP
Mount Mattingley
FLORA & FAUNA RESERVE
Lake Hindmarsh
Lake Hindmarsh
SOUTH AUSTRALIA
VICTORIA
PINNAROO
Mount Edgerley
Lowan Vale
Wirrega
Cannawigara
DUKES
HWY
Bordertown
Telopea Downs
Perenna
Baker
Netherby
Lorquon West
Yanac
Lorquon
Detpa
Broughton
Woorak West
Yanac South
Propodollah
Bleak House
Boyeo
Balrootan North
Woorak
GLENLEE FLORA RES
Glenlee
Dinyarrak
Yearinga
Yarrock
Sandsmere
Diapur
Tarranginnie
Mundulla
WESTERN
Lillimur
Serviceton
Wolseley
Kaniva
Miram
Nhill
Salisbury
Kiata
Gerang Gerung
Leeor
Yanipy
Lawloit
Winiam
Winiam East
Custon
Lillimur South
Kinimakatka
Miram South
Dimboola
Western Flat
WIMMERA
LITTLE
DESERT
NATIONAL
PARK
Bangham
NARACOORTE
Wail
Pimpinio
Byrneville
Kewell
Kalkee
HENTY
WEST WAIL FLORA RES
Lake Wyn Wyn
Jung
Murtoa
Wallabrook
The Gap
Frances
Minimay
Neuarpurr
Morea
Peronne
Mortat
Goroke
Gymbowen
NURCUONG FLORA RES
Nurcoung
Grass Flat
Mitre
Quantong
Vectis
Dahlen
Dooen
Longerenong
Drung Drung
BARRABOOL FLORA RES
Asher
Marma
RIDDOCH
Binnum
Booroopki
Dopewora
Duffholme
Arapiles
Natimuk
WIMMERA
HORSHAM
Haven
WESTERN
Tallageira
MT ARAPILES-TOOAN STATE PARK
Mount Arapiles
Tooan
Lower Norton
McKenzie Creek
Drung Drung South
Bringalbert
Kangawall
Ozenkadnook
Kybybolite
Karnak
Noradjuha
Patyah
Benayeo
HWY
Wonwondah North
Wonwondah East
Wonwondah
Mount Zero
Mount Stapylton
Hynam
Naracoorte
Ullswater
Apsley
Miga Lake
JILPANGER FLORA & FAUNA RESERVE
Clear Lake
Jallumba
Nurrabiel
Mockinya
Laharum
Dadswell Bridge
Golton Gorge
Awonga
Charam
Connangorach
NARACOORTE CAVES
WIMMERA
THE GRAMPIANS
Edenhope
Wombelano
Toolondo
BOOL LAGOON GAME RESERVE
Struan
Joanna
Douglas
White Lake
Toolondo Reservoir
Mount Talbot
Brimpaen
Mount Difficult
Langkoop
Wartook
Mount Dryden
Wrattonbully
GLEN ROY CP
Kadnook
BLACK RANGE STATE PARK
Harrow
Halls Gap
Glenroy
Poolaijelo
Powers Creek
Mount Byron
Mount Bepcha
Mount Victory
WONDERLAND WALKS
Stony Peak
Victoria Gap
Comaum
DERGHOLM
STATE
PARK
Moree
Glenisla
Mount Rosea
THE PINNACLE
Coonawarra
Chetwynd
Balmoral
Rocklands Reservoir
GRAMPIANS NATIONAL PARK
Mount Cassell
Pigeon Ponds
HENTY
Dorodong
Englefield
Tarrayoukyan
Mount Thackeray
Triplet Peak

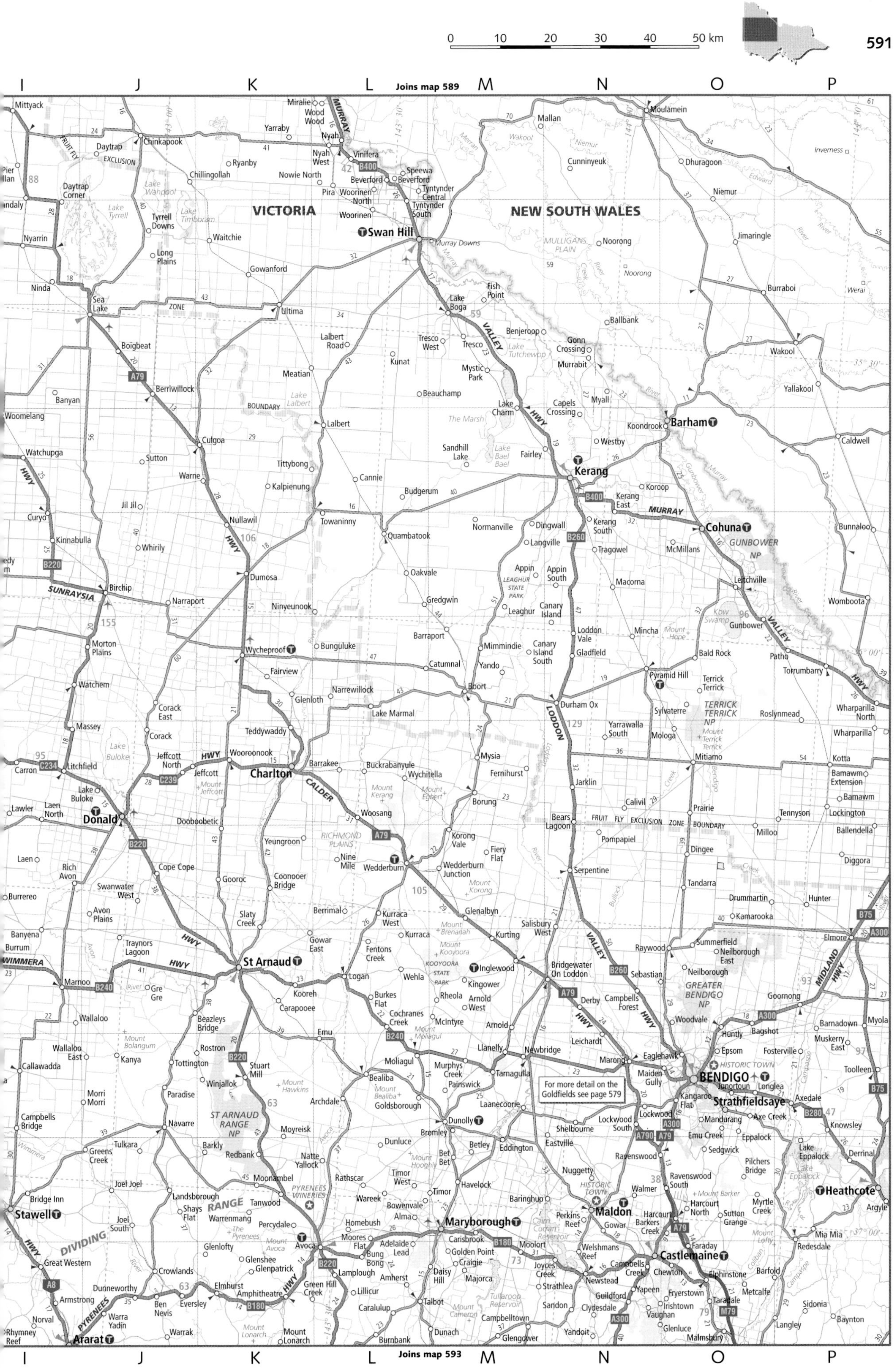

0 10 20 30 40 50 km

I J K L M N O P

Joins map 589

VICTORIA

NEW SOUTH WALES

Mittyack, Miralie, Wood Wood, Yarraby, Nyah, Nyah West, Vinifera, Mallan, Moulamein, Chinkapook, Daytrap, Ryanby, Chillingollah, Nowie North, Speewa, Beverford, Tyntynder Central, Tyntynder South, Pira, Woorinen North, Woorinen, Cunninyeuk, Dhuragoon, Inverness, Daytrap Corner, Lake Wahpool, Lake Tyrrell, Tyrrell Downs, Lake Timboram, Niemur, Wakool, Edward River, Nyarrin, Waitchie, Swan Hill, Murray Downs, Mulligans Plain, Noorong, Jimaringle, Long Plains, Gowanford, Ninda, Sea Lake, Ultima, Fish Point, Lake Boga, Burraboi, Werai, Ballbank, Boigbeat, Lalbert Road, Tresco West, Tresco, Benjeroop, Gonn Crossing, Murrabit, Kunat, Mystic Park, Lake Tutchewop, Wakool, Yallakool, Meatian, Berriwillock, Banyan, Beauchamp, Lake Charm, Capels Crossing, Myall, Woomelang, Lake Lalbert, Lalbert, The Marsh, Koondrook, Barham, Caldwell, Culgoa, Watchupga, Sutton, Tittybong, Sandhill Lake, Lake Bael Bael, Fairley, Westby, Kerang, Warne, Kalpienung, Cannie, Budgerum, Koroop, Kerang East, Jil Jil, Nullawil, Towaninny, Normanville, Dingwall, Kerang South, Cohuna, Bunnaloo, Curyo, Kinnabulla, Whirily, Quambatook, Langville, Tragowel, McMillans, Gunbower NP, Appin, Appin South, Leaghur State Park, Dumosa, Oakvale, Macorna, Leitchville, Birchip, Narraport, Ninyeunook, Gredgwin, Leaghur, Canary Island, Kow Swamp, Gunbower, Wombootа, Morton Plains, Barraport, Mimmindie, Loddon Vale, Mincha, Mount Hope, Wycheproof, Bunguluke, Catumnal, Canary Island South, Gladfield, Bald Rock, Patho, Torrumbarry, Watchem, Fairview, Glenloth, Narrewillock, Boort, Yando, Pyramid Hill, Terrick Terrick, Durham Ox, Terrick Terrick NP, Mount Terrick Terrick, Roslynmead, Wharparilla North, Wharparilla, Corack East, Corack, Massey, Lake Marmal, Sylvaterre, Yarrawalla South, Mologa, Teddywaddy, Lake Buloke, Jeffcott North, Jeffcott, Mount Jeffcott, Wooroonook, Charlton, Barrakee, Buckrabanyule, Wychitella, Mount Kerang, Mount Egbert, Mysia, Fernihurst, Mitiamo, Kotta, Bamawm Extension, Bamawm, Carron, Litchfield, Lake Buloke, Jarklin, Borung, Calivil, Prairie, Tennyson, Lockington, Lawler, Laen North, Donald, Dooboobetic, Woosang, Bears Lagoon, Pompapiel, Milloo, Ballendella, Yeungroon, Richmond Plains, Korong Vale, Fiery Flat, Dingee, Diggora, Laen, Rich Avon, Cope Cope, Nine Mile, Wedderburn, Wedderburn Junction, Serpentine, Tandarra, Burrereo, Swanwater West, Gooroc, Coonooer Bridge, Mount Korong, Drummartin, Hunter, Avon Plains, Slaty Creek, Berrimal, Kurraca West, Glenalbyn, Salisbury West, Kamarooka, Elmore, Banyena, Traynors Lagoon, Gowar East, Kurraca, Mount Brenanah, Kurting, Summerfield, Neilborough East, Burrum, Fentons Creek, Mount Kooyoora, Raywood, Neilborough, Marnoo, Gre Gre, St Arnaud, Logan, Wehla, Kooyoora State Park, Inglewood, Bridgewater On Loddon, Sebastian, Greater Bendigo NP, Kooreh, Kingower, Carapooee, Burkes Flat, Rheola, Arnold West, Derby, Campbells Forest, Goornong, Wallaloo, Beazleys Bridge, Cochranes Creek, McIntyre, Arnold, Woodvale, Huntly, Bagshot, Barnadown, Myola, Mount Bolangum, Emu, Mount Moliagul, Leichardt, Wallaloo East, Kanya, Rostron, Moliagul, Llanelly, Newbridge, Marong, Eaglehawk, Epsom, Fosterville, Muskerry East, Callawadda, Tottington, Stuart Mill, Beaiba, Murphys Creek, Tarnagulla, Maiden Gully, Bendigo, Historic Town, Toolleen, Paradise, Winjallok, Mount Hawkins, Mount Beaiba, Painswick, Junortoun, Longlea, Morri Morri, St Arnaud Range NP, Archdale, Goldsborough, Laanecoorie, Kangaroo Flat, Strathfieldsaye, Axedale, Campbells Bridge, Navarre, Moyreisk, Dunolly, Shelbourne, Lockwood South, Lockwood, Mandurang, Axe Creek, Knowsley, Tulkara, Barkly, Bromley, Betley, Eddington, Eastville, Emu Creek, Eppalock, Greens Creek, Redbank, Dunluce, Natte Yallock, Mount Hooghly, Bet Bet, Sedgwick, Lake Eppalock, Derrinal, Ravenswood, Pilchers Bridge, Moonambel, Rathscar, Timor West, Havelock, Nuggetty, Ravenswood South, Heathcote, Joel Joel, Pyrenees Wineries, Timor, Walmer, Mount Barker, Bridge Inn, Landsborough, Tanwood, Wareek, Bowenvale, Baringhup, Maldon, Harcourt North, Myrtle Creek, Argyle, Stawell, Shays Flat, Range, Warrenmang, Alma, Perkins Reef, Harcourt, Sutton Grange, Joel South, Percydale, Homebush, Maryborough, Cairn Curran Reservoir, Gowar, Barkers Creek, Mount Lofty, Dividing, Mount Avoca, The Pyrenees, Avoca, Moores Flat, Adelaide Lead, Carisbrook, Moolort, Welshmans Reef, Faraday, Mia Mia, Redesdale, Glenlofty, Bung Bong, Golden Point, Castlemaine, Great Western, Glenshee, Glenpatrick, Craigie, Joyces Creek, Newstead, Campbells Creek, Chewton, Elphinstone, Barfold, Crowlands, Lamplough, Daisy Hill, Majorca, Strathlea, Metcalfe, Dunneworthy, Elmhurst, Amphitheatre, Green Hill Creek, Amherst, Tullaroop Reservoir, Guildford, Yapeen, Fryerstown, Taradale, Sidonia, Armstrong, Ben Nevis, Eversley, Lillicur, Talbot, Mount Cameron, Sandon, Clydesdale, Irishtown, Vaughan, Langley, Baynton, Norval, Warra Yadin, Caralulup, Campbelltown, Glenluce, Rhymney Reef, Warrak, Mount Lonarch, Mount Lonarch, Dunach, Yandoit, Malmsbury, Ararat, Burnbank, Glengower

MURRAY VALLEY HWY, LODDON VALLEY HWY, CALDER HWY, SUNRAYSIA HWY, WIMMERA HWY, PYRENEES HWY, MIDLAND HWY, HENTY HWY, FRUIT FLY EXCLUSION ZONE BOUNDARY, MURRAY

For more detail on the Goldfields see page 579

Joins map 584

1 2 3 4 5 6 7 8 9 10 11 12

I J K L M N O P

Joins map 593

Joins map 590
Joins map 601
A
B
C
D
E
F
G
H
1
2
3
4
5
6
7
8
9
10
11
12
BOOL LAGOON GAME RESERVE
Wrattonbully
GLEN ROY CP
Glenroy
Comaum
Penola Swamp
Coonawarra
CAVE RANGE
Penola
Krongart
Kalangadoo
Nangwarry
Tarpeena
Wandilo
RIDDOCH
Mil Lel
MOUNT GAMBIER
BLUE LAKE
Glenburnie
Yahl
Caroline
Ob Flat
Mount Schank
Allendale East
Ewens Ponds
Port MacDonnell
Donovans Landing
SOUTH AUSTRALIA
VICTORIA
Langkoop
Poolaijelo
Dorodong
Dergholm
Mosquito Swamp
Lake Mundi
Strathdownie
Puralka
Mumbannar
PRINCES
LOWER GLENELG
Nelson
GREAT SOUTH WEST WALK
DISCOVERY
Discovery Bay
Lake Mombeong
BAY
Wanwin
GLENELG CANOEING
DERGHOLM STATE PARK
Kadnook
Powers Creek
Moree
Chetwynd
Tarrayoukyan
Brimboal
Wando Bridge
Wando Vale
Dunrobin
Casterton
Sandford
Carapook
Hilgay
Paschendale
Henty
Merino
WILKIN PARK FLORA & FAUNA RESERVE
GLENELG
Marp
Dartmoor
Winnap
Greenwald
Drik Drik
Lyons
Mount Vandyke
Mount Deception
COBBOBOONEE NP
Kentbruck
Mount Kincaid
Mt Richmond
MOUNT RICHMOND NP
COASTAL PARK
Tarragal
Cashmore
Trewalla
Cape Duquesne
Cape Bridgewater
Bridgewater Bay
CAPE NELSON STATE PARK
Cape Nelson
CAPE NELSON LIGHTHOUSE
Gorae West
Gorae
Bolwarra
Portland
Portland Bay
HISTORIC TOWN, GREAT SOUTH WEST WALK
Blacknose Point
Danger Point
Grant Bay
Cape Sir William Grant
Heathmere
Heywood
Drumborg
HENTY
Digby
Hotspur
Harrow
Pigeon Ponds
Coojar
Nareen
Konong Wootong North
Konong Wootong
Coleraine
Tahara Bridge
Tahara
Grassdale
Branxholme
Wallacedale
Condah
Myamyn
Milltown
Lake Condah
MOUNT ECCLES NP
Homerton
Narrawong
Tyrendarra
Tyrendarra East
Balmoral
Englefield
Vasey
Gatum
Brit Brit
Gringegalgona
Wootong Vale
Gritjurk
Melville Forest
Bulart
Wannon
Yulecart
Hamilton
Byaduk North
Byaduk
MOUNT NAPIER STATE PARK
Macarthur
Broadwater
Bessiebelle
Orford
Codrington
St Helens
Yambuk
Lady Julia Percy Island
Cape Reamur
Port Fairy
PORT FAIRY BEACH
BLACK RANGE STATE PARK
Mount Byron
Glenisla
Rocklands Reservoir
Mooralla
Cavendish
Kyup
Hensley Park
Karabeal
Moutajup
Strathkellar
Tarrington
Yatchaw
Tabor
Croxton East
Penshurst
Warrayure
Dunkeld
Glenthompson
HAMILTON HWY
Minhamite
Caramut
Moffat
Hawkesdale
Willatook
Warrong
Woolsthorpe
Kirkstall
Koroit
Southern Cross
Winslow
Mailors Flat
Toolong
Rosebrook
Killarney
Crossley
Tower Hill
Illowa
Woodford
Bushfield
Dennington
WARRNAMBOOL
WHALES
Allansford
Mepunga West
BAY OF ISLANDS
THE GRAMPIANS
Halls Gap
GRAMPIANS NATIONAL PARK
Victoria Gap
Mirranatwa
Wannon
Mafek
Ballangeich
Wangoom
Cudgee
The Cove
SOUTHERN
OCEAN
38° 30'
39° 00'
141° 00'
141° 30'
142° 00'
142° 30'

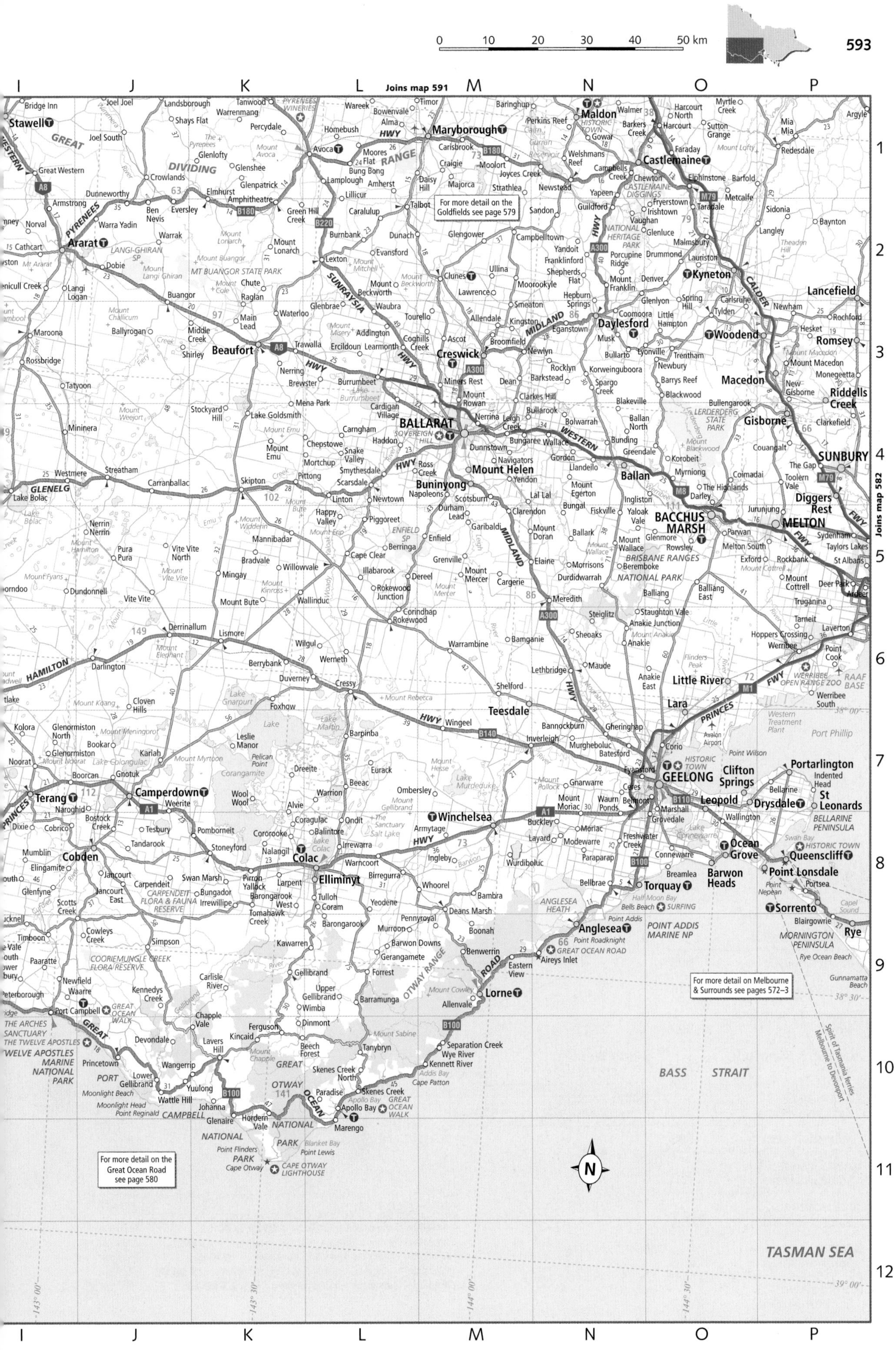

0 10 20 30 40 50 km

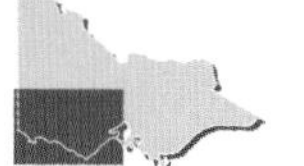

I J K L M N O P

Joins map 591

Joins map 582

For more detail on the Goldfields see page 579

For more detail on Melbourne & Surrounds see pages 572–3

For more detail on the Great Ocean Road see page 580

Stawell, Ararat, Great Western, Beaufort, Avoca, Maryborough, Maldon, Castlemaine, Kyneton, Lancefield, Daylesford, Creswick, Woodend, Macedon, Romsey, Riddells Creek, Gisborne, Sunbury, Ballarat, Buninyong, Mount Helen, Ballan, Bacchus Marsh, Melton, Diggers Rest, Lismore, Derrinallum, Skipton, Linton, Teesdale, Lara, Little River, Geelong, Portarlington, Clifton Springs, Leopold, Drysdale, St Leonards, Queenscliff, Point Lonsdale, Ocean Grove, Barwon Heads, Torquay, Anglesea, Aireys Inlet, Lorne, Apollo Bay, Colac, Elliminyt, Winchelsea, Camperdown, Terang, Cobden, Port Campbell, Princetown, Sorrento, Rye

GREAT DIVIDING RANGE · MIDLAND HWY · WESTERN HWY · CALDER FWY · PRINCES HWY · HAMILTON HWY · GLENELG HWY · SUNRAYSIA HWY · PYRENEES HWY · GREAT OCEAN ROAD

GREAT OTWAY NATIONAL PARK · PORT CAMPBELL NATIONAL PARK · BRISBANE RANGES NATIONAL PARK · LERDERDERG STATE PARK · MT BUANGOR STATE PARK · TWELVE APOSTLES MARINE NATIONAL PARK · POINT ADDIS MARINE NP

Port Phillip · BASS STRAIT · TASMAN SEA · MORNINGTON PENINSULA · BELLARINE PENINSULA

1 2 3 4 5 6 7 8 9 10 11 12

MAPS

SOUTH AUSTRALIA

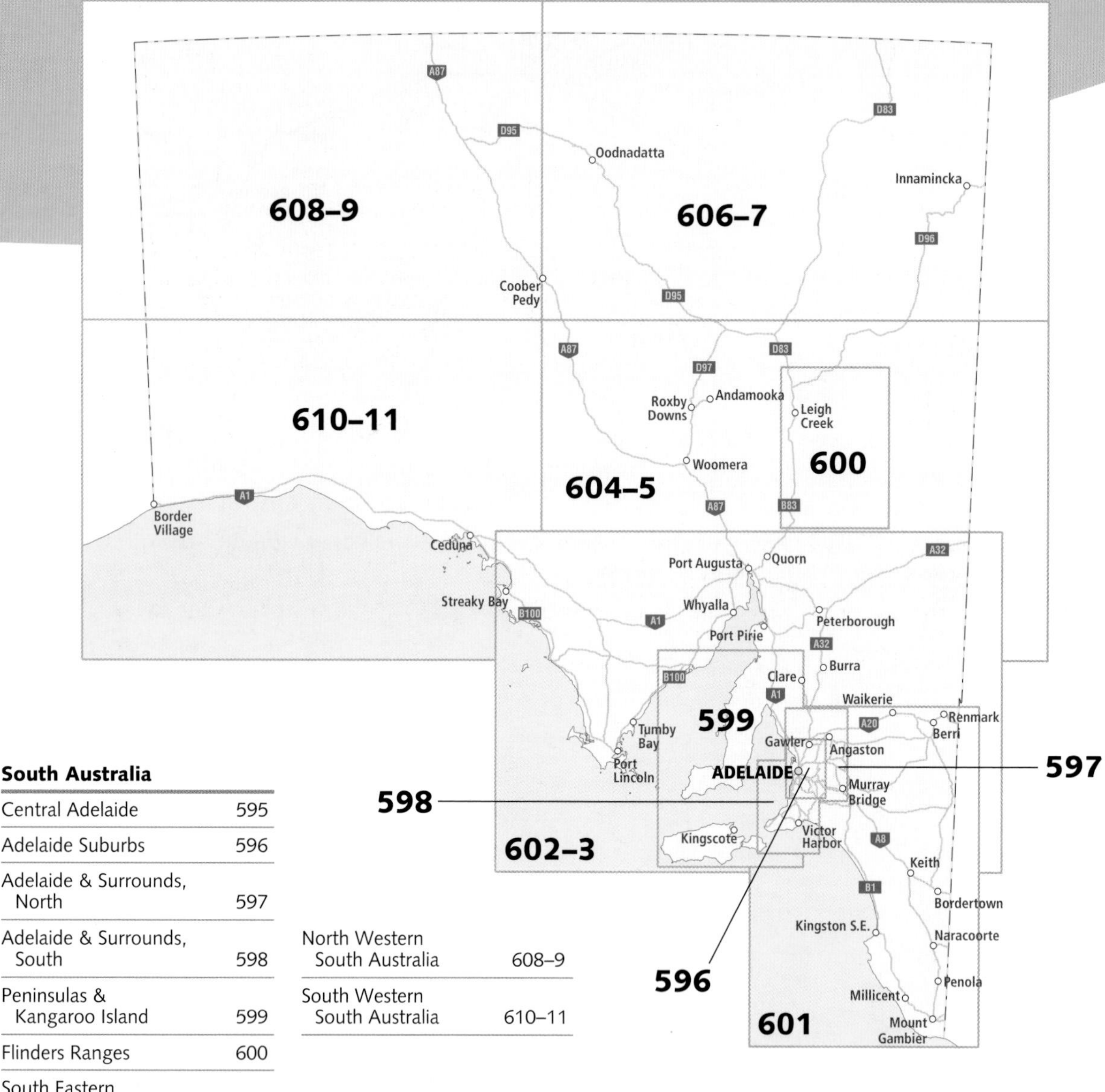

South Australia

INTER-CITY ROUTES		DISTANCE
Adelaide–Darwin via Stuart Hwy	A1 A87 87 1	3026 km
Adelaide–Perth via Eyre & Great Eastern hwys	A1 1 94	2700 km
Adelaide–Sydney via Sturt & Hume hwys	A20 20 31	1415 km
Adelaide–Melbourne via Dukes & Western hwys	M1 A8 M8	729 km
Adelaide–Melbourne via Princes Hwy	M1 B1 A1 M1	907 km

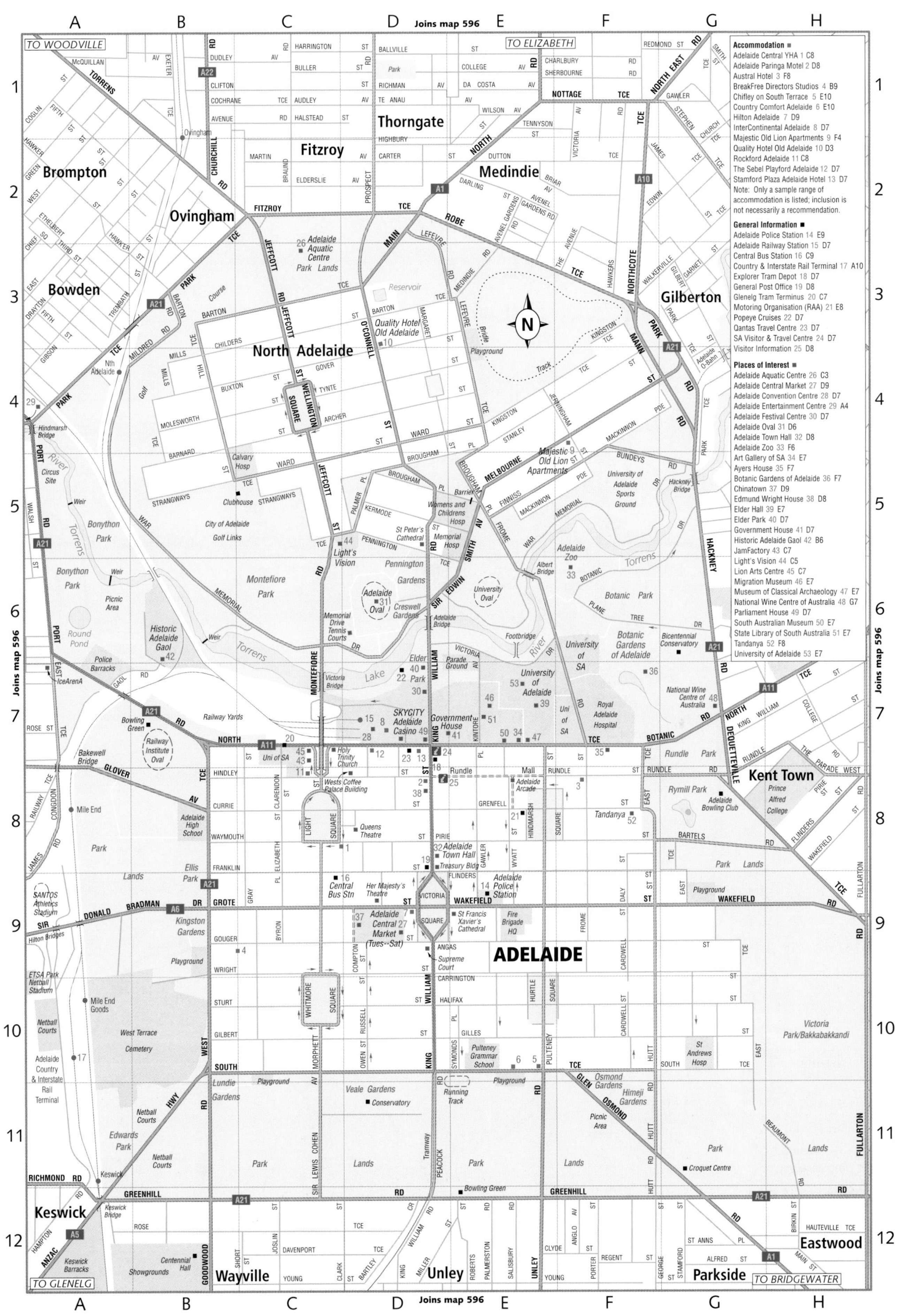
0 0.5 1 1.5 km
Joins map 596
TO WOODVILLE
TO ELIZABETH
TO GLENELG
TO BRIDGEWATER
Brompton
Bowden
Ovingham
Fitzroy
Thorngate
Medindie
Gilberton
North Adelaide
Kent Town
ADELAIDE
Keswick
Wayville
Unley
Parkside
Eastwood
Accommodation
Adelaide Central YHA 1 C8
Adelaide Paringa Motel 2 D8
Austral Hotel 3 F8
BreakFree Directors Studios 4 B9
Chifley on South Terrace 5 E10
Country Comfort Adelaide 6 E10
Hilton Adelaide 7 D9
InterContinental Adelaide 8 D7
Majestic Old Lion Apartments 9 F4
Quality Hotel Old Adelaide 10 D3
Rockford Adelaide 11 C8
The Sebel Playford Adelaide 12 D7
Stamford Plaza Adelaide Hotel 13 D7
Note: Only a sample range of accommodation is listed; inclusion is not necessarily a recommendation.
General Information
Adelaide Police Station 14 E9
Adelaide Railway Station 15 D7
Central Bus Station 16 C9
Country & Interstate Rail Terminal 17 A10
Explorer Tram Depot 18 D7
General Post Office 19 D8
Glenelg Tram Terminus 20 C7
Motoring Organisation (RAA) 21 E8
Popeye Cruises 22 D7
Qantas Travel Centre 23 D7
SA Visitor & Travel Centre 24 D7
Visitor Information 25 D8
Places of Interest
Adelaide Aquatic Centre 26 C3
Adelaide Central Market 27 D9
Adelaide Convention Centre 28 D7
Adelaide Entertainment Centre 29 A4
Adelaide Festival Centre 30 D7
Adelaide Oval 31 D6
Adelaide Town Hall 32 D8
Adelaide Zoo 33 F6
Art Gallery of SA 34 E7
Ayers House 35 F7
Botanic Gardens of Adelaide 36 F7
Chinatown 37 D9
Edmund Wright House 38 D8
Elder Hall 39 E7
Elder Park 40 D7
Government House 41 D7
Historic Adelaide Gaol 42 B6
JamFactory 43 C7
Light's Vision 44 C5
Lion Arts Centre 45 C7
Migration Museum 46 E7
Museum of Classical Archaeology 47 E7
National Wine Centre of Australia 48 G7
Parliament House 49 D7
South Australian Museum 50 E7
State Library of South Australia 51 E7
Tandanya 52 F8
University of Adelaide 53 E7

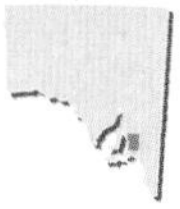

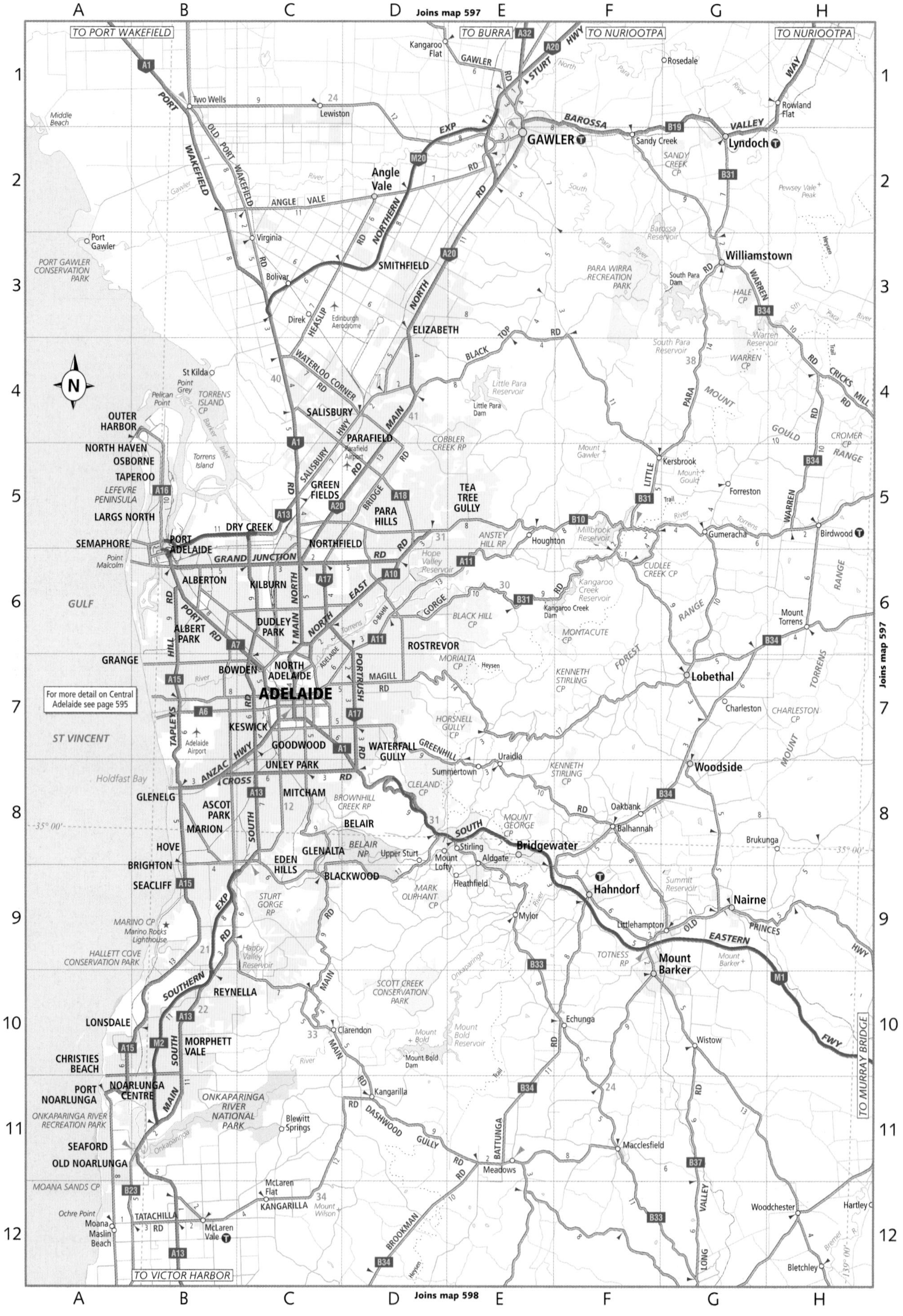

Joins map 597

Joins map 598

For more detail on Central Adelaide see page 595

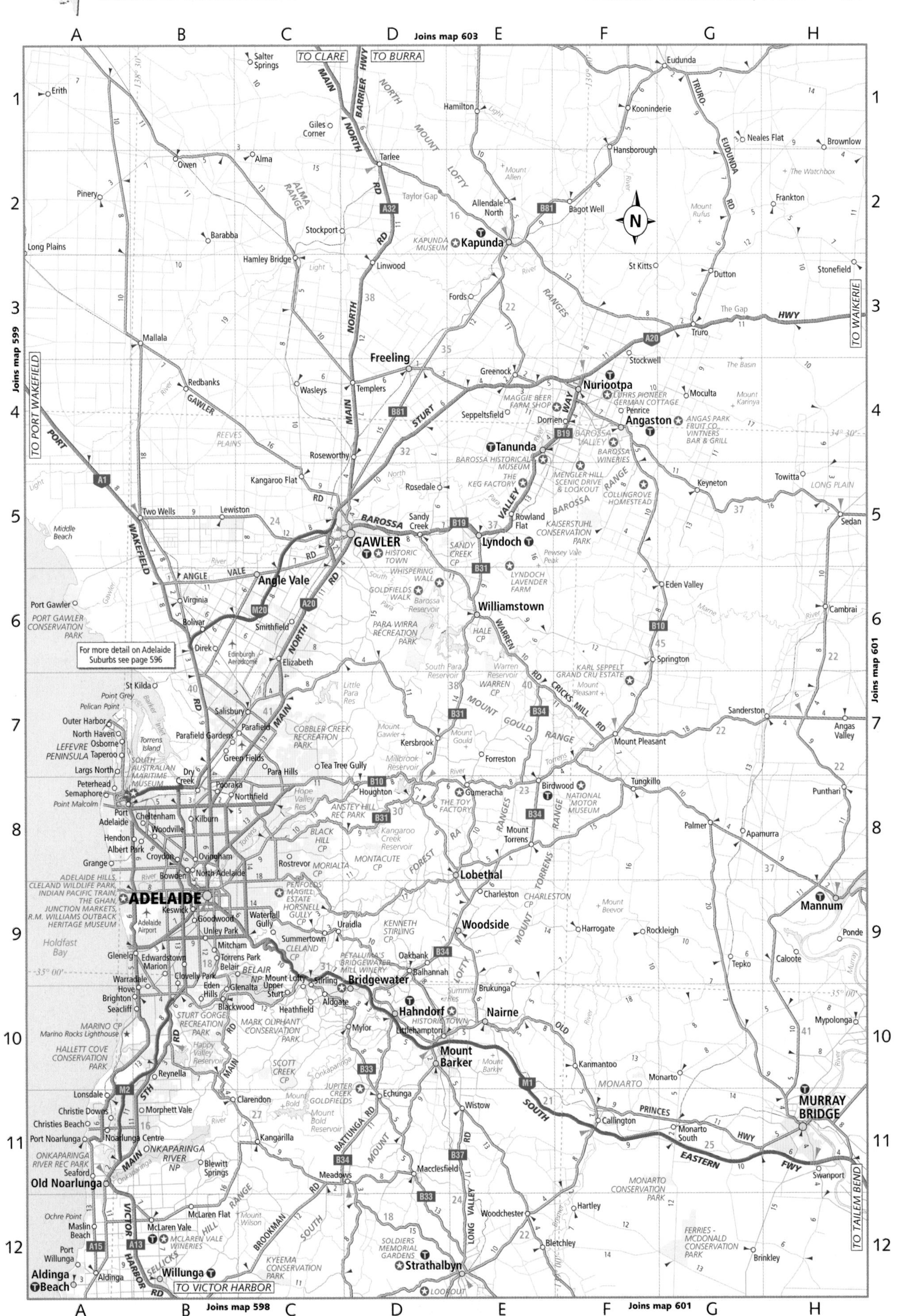
0 5 10 15 20 km
Joins map 603
Joins map 599
Joins map 601
Joins map 598
TO CLARE
TO BURRA
TO WAKERIE
TO PORT WAKEFIELD
TO VICTOR HARBOR
TO TAILEM BEND
Erith
Salter Springs
Giles Corner
Tarlee
Hamilton
Eudunda
Koonunderie
Hansborough
Neales Flat
Brownlow
The Watchbox
Owen
Alma
Pinery
Taylor Gap
Allendale North
Bagot Well
Frankton
Barabba
Stockport
KAPUNDA MUSEUM
Kapunda
Long Plains
Hamley Bridge
Linwood
St Kitts
Dutton
Stonefield
Fords
The Gap
Truro
Mallala
Freeling
Greenock
Stockwell
The Basin
Redbanks
Wasleys
Templers
Nuriootpa
MAGGIE BEER FARM SHOP
LUHRS PIONEER GERMAN COTTAGE
Penrice
Moculta
Seppeltsfield
Dorrien
Angaston
ANGAS PARK FRUIT CO.
VINTNERS BAR & GRILL
REEVES PLAINS
Tanunda
BAROSSA VALLEY
BAROSSA WINERIES
BAROSSA HISTORICAL MUSEUM
Roseworthy
THE KEG FACTORY
MENGLER HILL SCENIC DRIVE & LOOKOUT
COLLINGROVE HOMESTEAD
Keyneton
Towitta
LONG PLAIN
Kangaroo Flat
Rosedale
Two Wells
Lewiston
Sandy Creek
Rowland Flat
KAISERSTUHL CONSERVATION PARK
Sedan
Middle Beach
GAWLER
Lyndoch
HISTORIC TOWN
SANDY CREEK CP
Pewsey Vale Peak
Angle Vale
WHISPERING WALL
LYNDOCH LAVENDER FARM
Eden Valley
Virginia
GOLDFIELDS WALK
Barossa Reservoir
Port Gawler
PORT GAWLER CONSERVATION PARK
Williamstown
Cambrai
Bolivar
Smithfield
PARA WIRRA RECREATION PARK
HALE CP
Direk
Edinburgh Aerodrome
Elizabeth
Springton
For more detail on Adelaide Suburbs see page 596
South Para Reservoir
Warren Reservoir
WARREN CP
KARL SEPPELT GRAND CRU ESTATE
Mount Pleasant
St Kilda
Point Grey
Pelican Point
Little Para Res
Salisbury
Parafield
Sanderston
Outer Harbor
North Haven
Osborne
LEFEVRE PENINSULA
Taperoo
Torrens Island
Parafield Gardens
COBBLER CREEK RECREATION PARK
Mount Gawler
Kersbrook
Mount Gould
Angas Valley
Largs North
SOUTH AUSTRALIAN MARITIME MUSEUM
Green Fields
Para Hills
Tea Tree Gully
Millbrook Reservoir
Forreston
Peterhead
Semaphore
Point Malcolm
Dry Creek
Pooraka
Northfield
Houghton
Gumeracha
Birdwood
NATIONAL MOTOR MUSEUM
Tungkillo
Punthari
Port Adelaide
Cheltenham
Kilburn
Hope Valley Res
ANSTEY HILL REC PARK
THE TOY FACTORY
Hendon
Woodville
Albert Park
Croydon
BLACK HILL CP
Kangaroo Creek Reservoir
Mount Torrens
Palmer
Apamurra
Grange
Ovingham
Rostrevor
MORIALTA CP
MONTACUTE CP
Lobethal
ADELAIDE HILLS, CLELAND WILDLIFE PARK, INDIAN PACIFIC TRAIN, THE GHAN, JUNCTION MARKETS, R.M. WILLIAMS OUTBACK HERITAGE MUSEUM
Bowden
North Adelaide
PENFOLDS MAGILL ESTATE
Charleston
CHARLESTON CP
Mount Beevor
ADELAIDE
Keswick
Adelaide Airport
Goodwood
HORSNELL GULLY CP
Waterfall Gully
Mannum
Holdfast Bay
Unley Park
Uraidla
KENNETH STIRLING CP
Woodside
Harrogate
Rockleigh
Ponde
Glenelg
Edwardstown
Mitcham
Summertown
CLELAND CP
Oakbank
Tepko
Caloote
Marion
Torrens Park
Belair
PETALUMA'S BRIDGEWATER MILL WINERY
Balhannah
Warradale
Clovelly Park
BELAIR NP
Mount Lofty
Stirling
Bridgewater
Brukunga
Hove
Eden Hills
Glenalta
Upper Sturt
Brighton
Seacliff
Blackwood
Heathfield
Aldgate
Hahndorf
Nairne
Mypolonga
MARINO CP
Marino Rocks Lighthouse
STURT GORGE RECREATION PARK
MARK OLIPHANT CONSERVATION PARK
Mylor
HISTORIC TOWN
Littlehampton
HALLETT COVE CONSERVATION PARK
Happy Valley Reservoir
SCOTT CREEK CP
Mount Barker
Kanmantoo
Monarto
Reynella
JUPITER CREEK GOLDFIELDS
Echunga
MONARTO
Lonsdale
Clarendon
Mount Bold
Mount Bold Reservoir
Wistow
MURRAY BRIDGE
Christie Downs
Morphett Vale
Christies Beach
Callington
Monarto South
Port Noarlunga
Noarlunga Centre
Kangarilla
ONKAPARINGA RIVER REC PARK
ONKAPARINGA RIVER NP
Blewitt Springs
Seaford
Macclesfield
Swanport
Old Noarlunga
Meadows
MONARTO CONSERVATION PARK
Ochre Point
McLaren Flat
Mount Wilson
Woodchester
Hartley
Maslin Beach
McLaren Vale
MCLAREN VALE WINERIES
SOLDIERS MEMORIAL GARDENS
Bletchley
FERRIES-MCDONALD CONSERVATION PARK
Brinkley
Port Willunga
Aldinga Beach
Aldinga
Willunga
KYEEMA CONSERVATION PARK
Strathalbyn
LOOKOUT

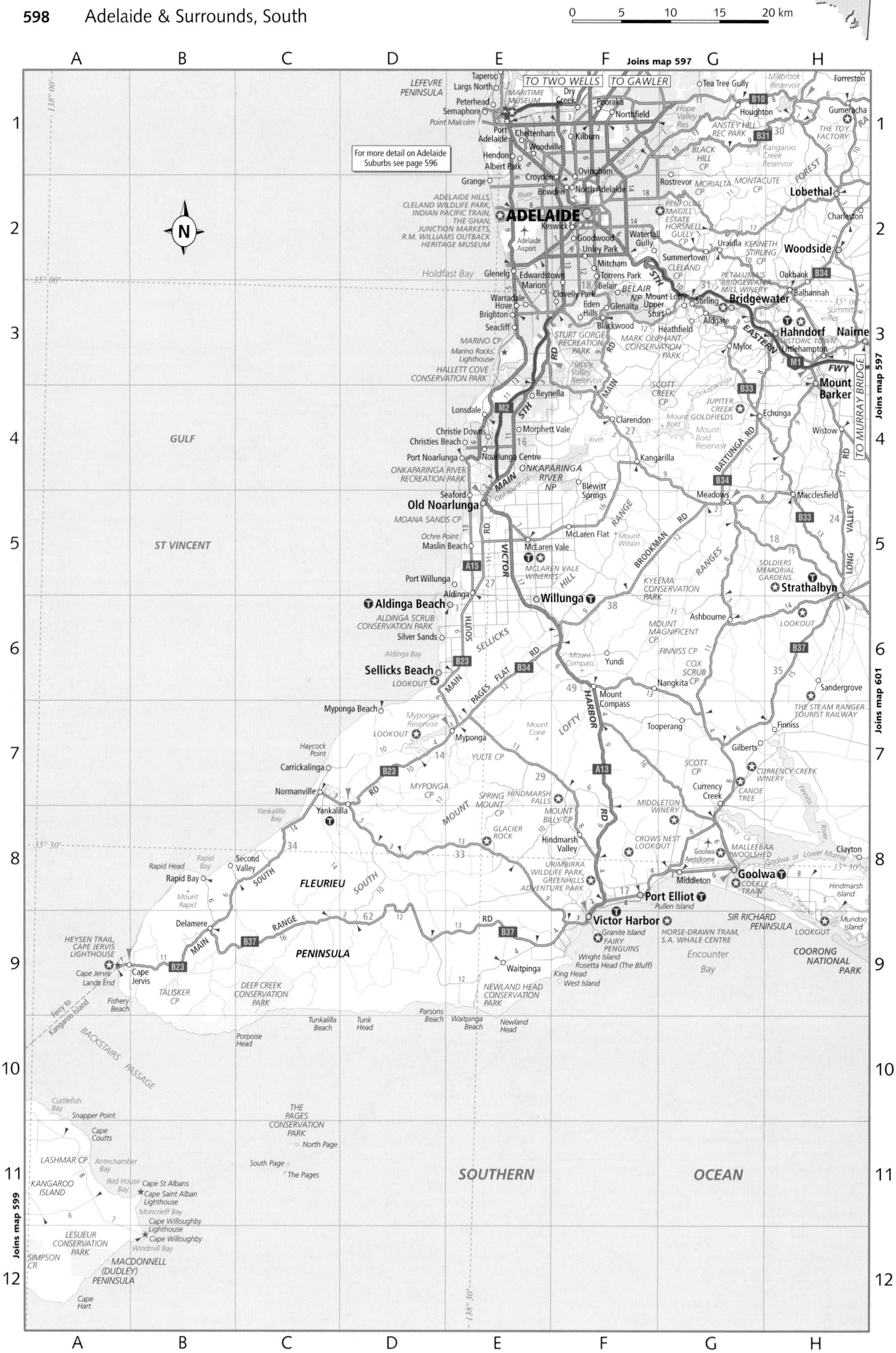
0 5 10 15 20 km
Joins map 597
Joins map 599
Joins map 601
For more detail on Adelaide Suburbs see page 596
TO TWO WELLS
TO GAWLER
TO MURRAY BRIDGE
LEFEVRE PENINSULA
Taperoo
Largs North
Peterhead
Semaphore
Point Malcolm
MARITIME MUSEUM
Port Adelaide
Dry Creek
Pooraka
Northfield
Cheltenham
Kilburn
Woodville
Hendon
Albert Park
Grange
Croydon
Bowden
Ovingham
North Adelaide
ADELAIDE
Keswick
Adelaide Airport
Goodwood
Unley Park
Mitcham
Torrens Park
Glenelg
Edwardstown
Marion
Holdfast Bay
Warradale
Hove
Brighton
Seacliff
Clovelly Park
Belair
BELAIR NP
Eden Hills
Glenalta
Blackwood
ADELAIDE HILLS, CLELAND WILDLIFE PARK, INDIAN PACIFIC TRAIN, THE GHAN, JUNCTION MARKETS, R.M. WILLIAMS OUTBACK HERITAGE MUSEUM
MARINO CP
Marino Rocks Lighthouse
HALLETT COVE CONSERVATION PARK
STURT GORGE RECREATION PARK
Happy Valley Reservoir
Reynella
Lonsdale
Christie Downs
Christies Beach
Morphett Vale
Port Noarlunga
Noarlunga Centre
ONKAPARINGA RIVER RECREATION PARK
ONKAPARINGA RIVER NP
Seaford
Old Noarlunga
MOANA SANDS CP
Ochre Point
Maslin Beach
McLaren Flat
McLaren Vale
MCLAREN VALE WINERIES
Port Willunga
Aldinga
Aldinga Beach
ALDINGA SCRUB CONSERVATION PARK
Silver Sands
Aldinga Bay
Willunga
Sellicks Beach
LOOKOUT
Myponga Beach
Myponga Reservoir
Myponga
Haycock Point
Carrickalinga
Normanville
Yankalilla
Yankalilla Bay
MYPONGA CP
YULTE CP
SPRING MOUNT CP
HINDMARSH FALLS
MOUNT BILLY CP
GLACIER ROCK
Hindmarsh Valley
Rapid Head
Rapid Bay
Mount Rapid
Second Valley
FLEURIEU PENINSULA
Delamere
HEYSEN TRAIL, CAPE JERVIS LIGHTHOUSE
Cape Jervis Lands End
Cape Jervis
TALISKER CP
Fishery Beach
DEEP CREEK CONSERVATION PARK
Tunkalilla Beach
Tunk Head
Porpoise Head
Parsons Beach
Waitpinga Beach
Waitpinga
NEWLAND HEAD CONSERVATION PARK
Newland Head
Ferry to Kangaroo Island
BACKSTAIRS PASSAGE
GULF ST VINCENT
SOUTHERN OCEAN
Encounter Bay
Victor Harbor
Granite Island
FAIRY PENGUINS
Wright Island
Rosetta Head (The Bluff)
King Head
West Island
Port Elliot
Pullen Island
HORSE-DRAWN TRAM, S.A. WHALE CENTRE
URIMBIRRA WILDLIFE PARK, GREENHILLS ADVENTURE PARK
CROWS NEST LOOKOUT
MIDDLETON WINERY
Middleton
Goolwa
Goolwa Aerodrome
COCKLE TRAIN
MALLEEBAA WOOLSHED
SIR RICHARD PENINSULA
Hindmarsh Island
Mundoo Island
COORONG NATIONAL PARK
Clayton
Lower Murray
Currency Creek
CURRENCY CREEK WINERY
CANOE TREE
SCOTT CP
Gilberts
Finniss
THE STEAM RANGER TOURIST RAILWAY
Sandergrove
Tooperang
Mount Compass
Nangkita
COX SCRUB CP
FINNISS CP
Yundi
MOUNT MAGNIFICENT CP
KYEEMA CONSERVATION PARK
Ashbourne
Strathalbyn
SOLDIERS MEMORIAL GARDENS
Macclesfield
Meadows
Blewitt Springs
Kangarilla
Clarendon
SCOTT CREEK CP
JUPITER CREEK GOLDFIELDS
Mount Bold Reservoir
Echunga
Wistow
Mount Barker
Nairne
Hahndorf
HISTORIC TOWN
Littlehampton
Mylor
Aldgate
Heathfield
MARK OLIPHANT CONSERVATION PARK
Stirling
Bridgewater
PETALUMA'S BRIDGEWATER MILL WINERY
Mount Lofty
Upper Sturt
Summit Res
Balhannah
Oakbank
Woodside
Charleston
Lobethal
Summertown
CLELAND CP
Uraidla
KENNETH STIRLING CP
Waterfall Gully
PENFOLDS MAGILL ESTATE
HORSNELL GULLY CP
Rostrevor
MORIALTA CP
MONTACUTE CP
BLACK HILL CP
ANSTEY HILL REC PARK
Hope Valley Res
Houghton
Tea Tree Gully
Millbrook Reservoir
Kangaroo Creek Reservoir
Forreston
Gumeracha
THE TOY FACTORY
Cuttlefish Bay
Snapper Point
Cape Coutts
LASHMAR CP
Antechamber Bay
KANGAROO ISLAND
Red House Bay
Cape St Albans
Cape Saint Alban Lighthouse
Moncrieff Bay
Cape Willoughby Lighthouse
Cape Willoughby
Windmill Bay
LESUEUR CONSERVATION PARK
SIMPSON CR
MACDONNELL (DUDLEY) PENINSULA
Cape Hart
THE PAGES CONSERVATION PARK
North Page
South Page
The Pages
138° 00'
138° 30'
35° 00'
35° 30'

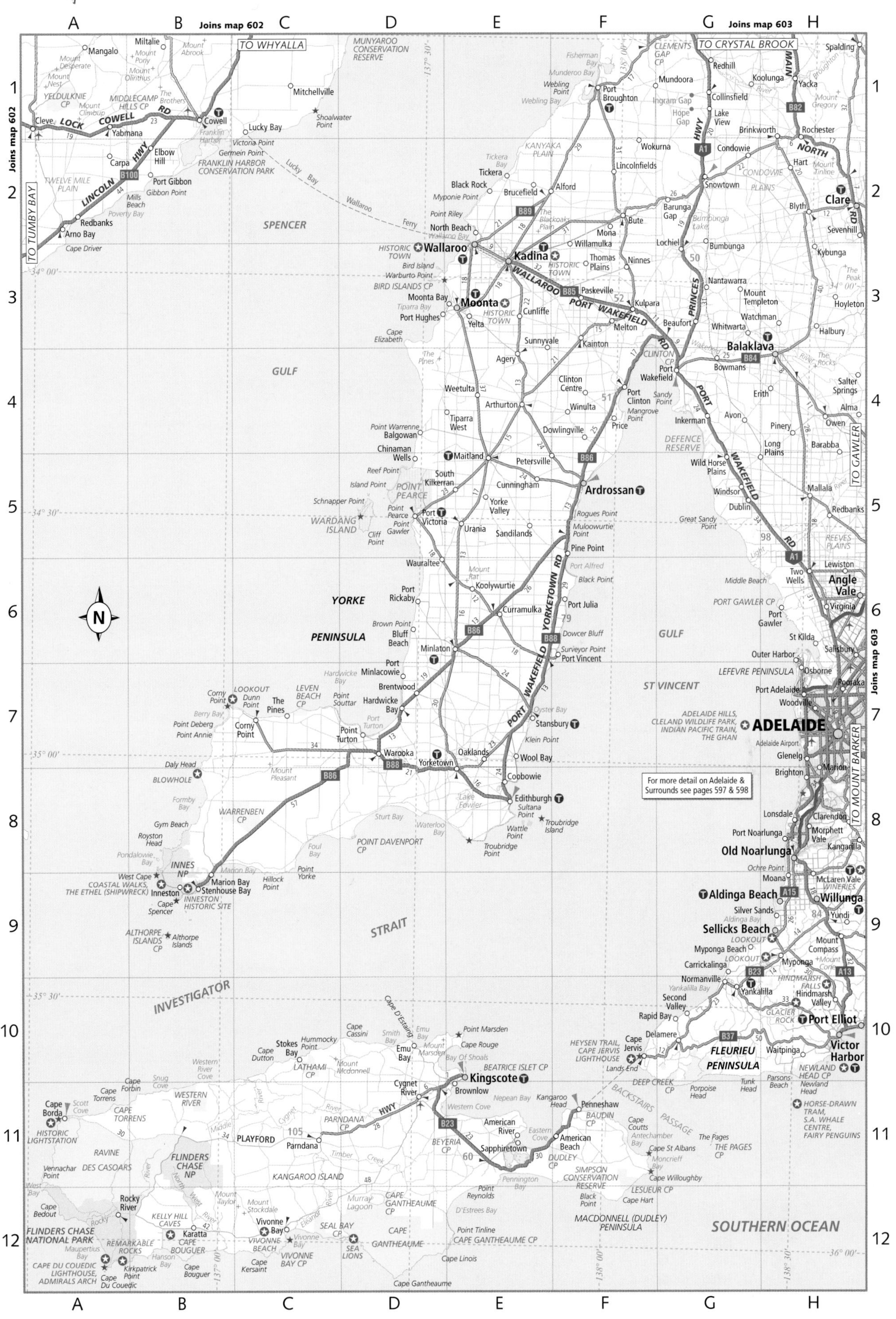

Joins map 602
Joins map 603
TO WHYALLA
TO CRYSTAL BROOK
TO TUMBY BAY
TO GAWLER
TO MOUNT BARKER
SPENCER GULF
GULF ST VINCENT
YORKE PENINSULA
INVESTIGATOR STRAIT
SOUTHERN OCEAN
KANGAROO ISLAND
FLEURIEU PENINSULA
ADELAIDE
Wallaroo
Kadina
Moonta
Ardrossan
Port Wakefield
Balaklava
Clare
Cowell
Kingscote
Victor Harbor
Port Elliot
Edithburgh
Yorketown
Minlaton
Maitland
Port Broughton
Old Noarlunga
Aldinga Beach
Sellicks Beach
Willunga
For more detail on Adelaide & Surrounds see pages 597 & 598

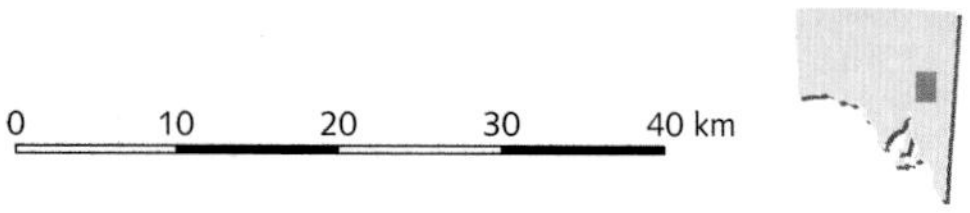

0 10 20 30 40 km

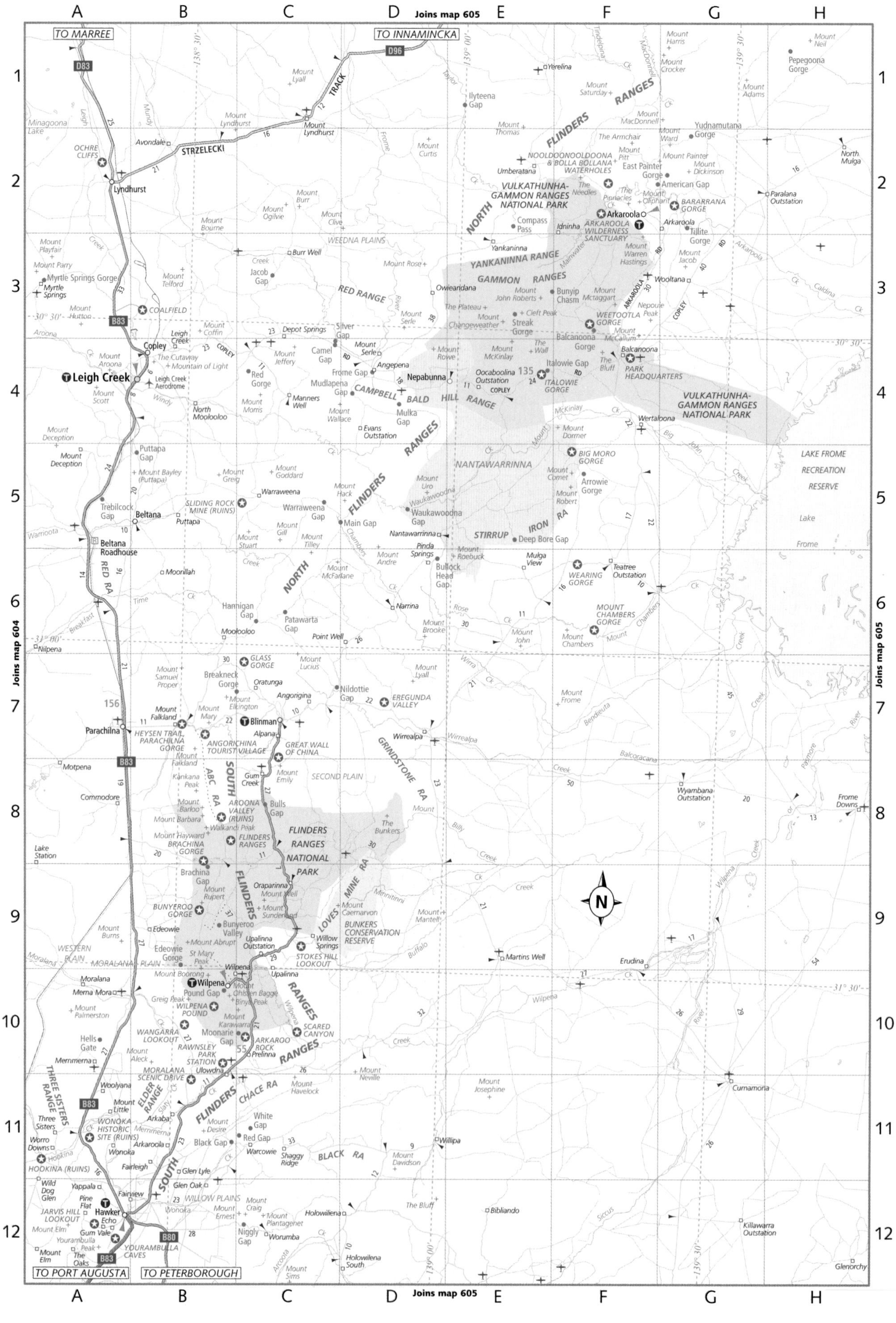

Joins map 605

TO MARREE

TO INNAMINCKA

TO PORT AUGUSTA

TO PETERBOROUGH

Joins map 604

Lyndhurst
Leigh Creek
Copley
Beltana
Beltana Roadhouse
Parachilna
Blinman
Wilpena
Hawker
Arkaroola

VULKATHUNHA-GAMMON RANGES NATIONAL PARK

FLINDERS RANGES NATIONAL PARK

NORTH FLINDERS RANGES

SOUTH FLINDERS RANGES

LAKE FROME RECREATION RESERVE

BUNKERS CONSERVATION RESERVE

Lake Frome

0 20 40 60 80 100 km

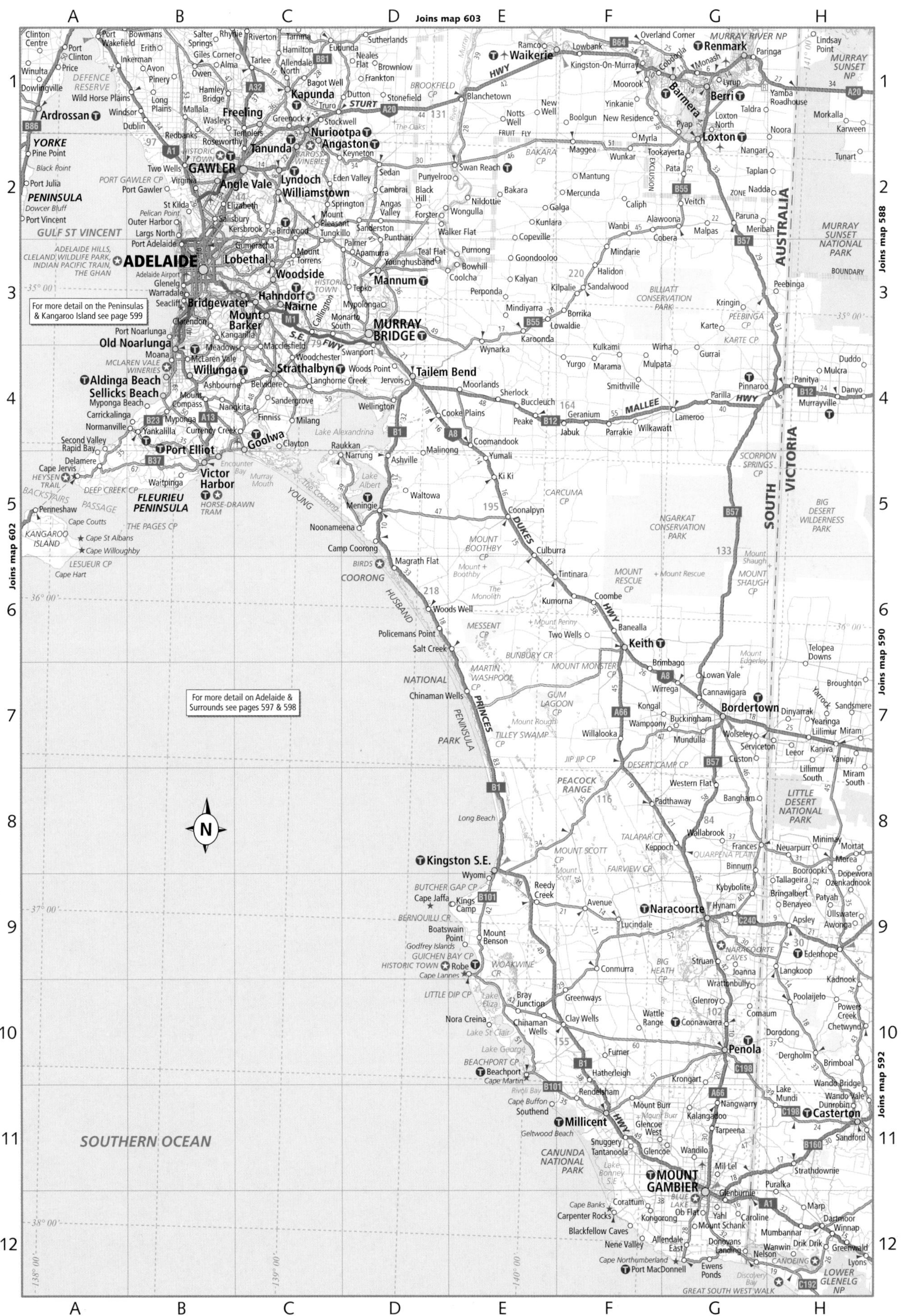

Joins map 604
Joins map 611
GREAT AUSTRALIAN BIGHT
SOUTHERN OCEAN
SPENCER GULF
EYRE PENINSULA
YORKE PENINSULA
GAWLER RANGES NATIONAL PARK
LAKE GAIRDNER NATIONAL PARK
PINKAWILLINIE CONSERVATION PARK
HAMBIDGE CONSERVATION PARK
HINCKS CONSERVATION PARK
COFFIN BAY NATIONAL PARK
LINCOLN NP
SIR JOSEPH BANKS GROUP CONSERVATION PARK
NEPTUNE ISLANDS CONSERVATION PARK
INNES NP
FLINDERS CHASE NATIONAL PARK
KANGAROO ISLAND
Streaky Bay
Ceduna
Wirrulla
Minnipa
Wudinna
Kyancutta
Kimba
Cleve
Cowell
Arno Bay
Elliston
Lock
Cummins
Tumby Bay
Coffin Bay
PORT LINCOLN
Port Augusta
Whyalla
Wallaroo
Minlaton
Yorketown
Warooka
Parndana
Kingscote
For more detail on the Peninsula & Kangaroo Island see p

0 20 40 60 80 100 km

Joins map 605

Joins map 558

Joins map 588

Joins map 601

For more detail on Adelaide & Surrounds see pages 597 & 598

ADELAIDE · MURRAY BRIDGE · BROKEN HILL · GAWLER · Peterborough · Burra · Clare · Renmark · Berri · Loxton · Waikerie · Victor Harbor · Goolwa · Tailem Bend · Mount Barker

SOUTH AUSTRALIA · NEW SOUTH WALES · VICTORIA

GULF ST VINCENT · SOUTHERN OCEAN · FLEURIEU PENINSULA

MURRAY SUNSET NATIONAL PARK · NGARKAT CONSERVATION PARK · BILLIATT CONSERVATION PARK · DANGGALI CONSERVATION PARK · CHOWILLA REGIONAL RESERVE · WYPERFELD NATIONAL PARK · BIG DESERT WILDERNESS PARK · COORONG NATIONAL PARK

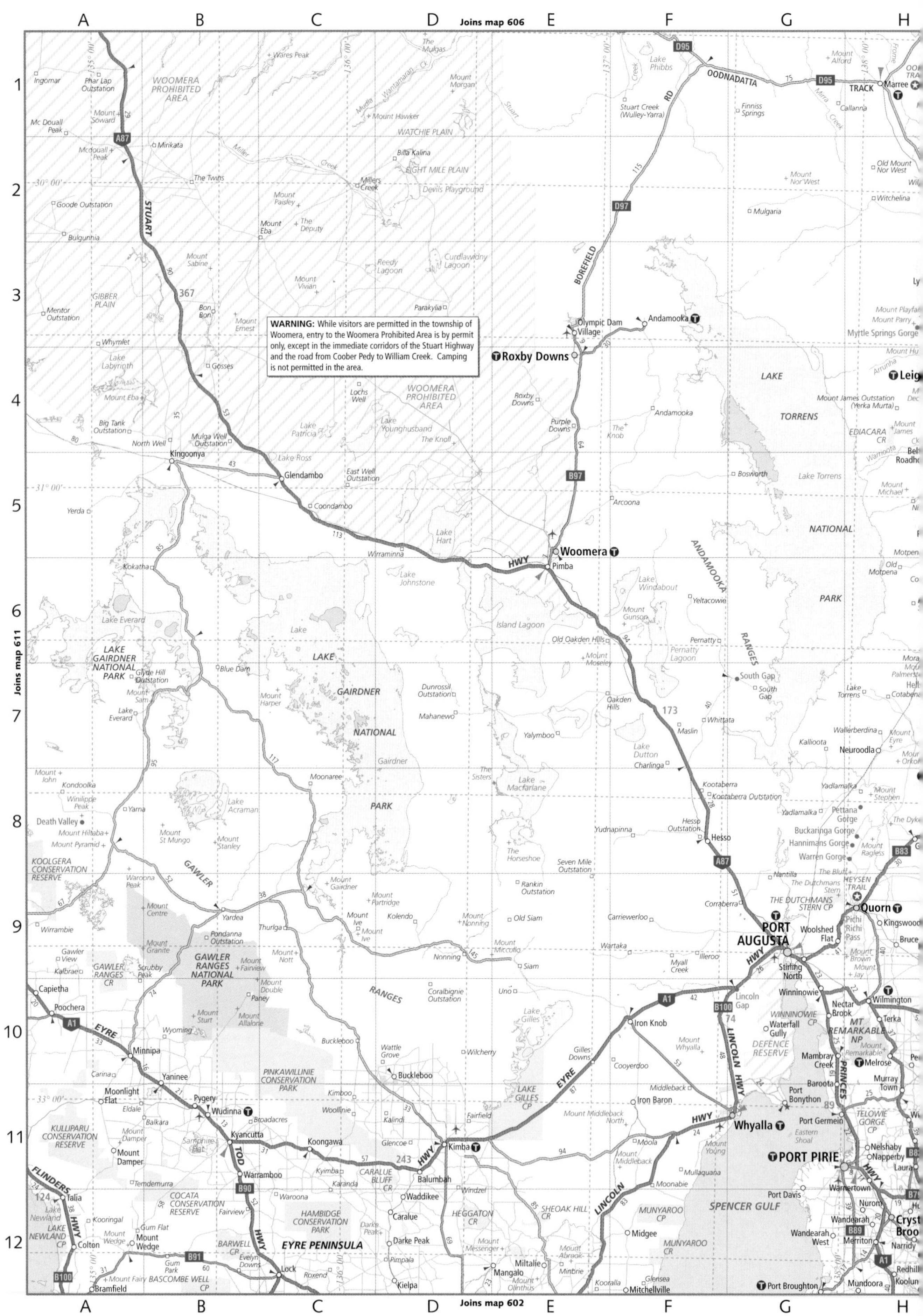

A
B
C
D
E
F
G
H
Joins map 606
Joins map 602
Joins map 611
1
2
3
4
5
6
7
8
9
10
11
12
WARNING: While visitors are permitted in the township of Woomera, entry to the Woomera Prohibited Area is by permit only, except in the immediate corridors of the Stuart Highway and the road from Coober Pedy to William Creek. Camping is not permitted in the area.
WOOMERA PROHIBITED AREA
STUART
HWY
OODNADATTA
TRACK
BOREFIELD
RD
Marree
Olympic Dam Village
Andamooka
Roxby Downs
Woomera
Pimba
Kingoonya
Glendambo
LAKE TORRENS
NATIONAL
PARK
ANDAMOOKA
RANGES
LAKE GAIRDNER NATIONAL PARK
GAWLER RANGES NATIONAL PARK
GAWLER
RANGES
PORT AUGUSTA
Quorn
Whyalla
PORT PIRIE
Iron Knob
Kimba
Wudinna
Minnipa
Kyancutta
Lock
Port Germein
Wilmington
SPENCER GULF
EYRE PENINSULA
EYRE
LINCOLN HWY
PRINCES HWY
TOD HWY
FLINDERS HWY
Port Broughton
Mitchellville
Mangalo
Darke Peak
Cleve
Buckleboo
Yardea
Wirrulla
Colton
Talia
Bramfield
Hesso
Kootaberra
Pernatty
South Gap
Whittata
Maslin
Moonabie
Iron Baron
Moola
Gilles Downs
LAKE GILLES CP
PINKAWILLINIE CONSERVATION PARK
HAMBIDGE CONSERVATION PARK
KULLIPARU CONSERVATION RESERVE
KOOLGERA CONSERVATION RESERVE
GAWLER RANGES CP
THE DUTCHMANS STERN CP
WINNINOWIE CP
MT REMARKABLE NP
TELOWIE GORGE CP
MUNYAROO CP
MUNYAROO CR

Joins map 607

Joins map 560

Joins map 558

Joins map 603

I J K L M N O P

1 2 3 4 5 6 7 8 9 10 11 12

WARNINGS: In outback Australia, long distances separate some towns. Travellers should familiarise themselves with prevailing conditions before departure and take care to ensure their vehicle is roadworthy. Adequate supplies of petrol, water and food should be carried at all times.

In central Australia, rainfall can make some roads impassable, even with a 4WD vehicle. Full information on road conditions should be obtained from local authorities before departure.

If visitors intend diverting off public roads within Aboriginal Land areas, a permit is required from the relevant Aboriginal authority.

For more detail on the Flinders Ranges see page 600

STRZELECKI DESERT

STRZELECKI REGIONAL RESERVE

VULKATHUNHA-GAMMON RANGES NATIONAL PARK

LAKE FROME REGIONAL RESERVE

NANTAWARRINNA

BUNKERS CONSERVATION RESERVE

ARID FLORA RESEARCH STATION

MINBURRA PLAIN

PANDAPPA CONSERVATION PARK

DANGGALI CONSERVATION PARK

TARAWI NATURE RESERVE

KINCHEGA NP

MUNDI MUNDI PLAIN

KOPI PLAIN

SOUTH AUSTRALIA

NEW SOUTH WALES

BROKEN HILL

HISTORIC TOWN

Peterborough

Jamestown

Blinman

Arkaroola

Nepabunna

Olary

Yunta

Mannahill

Cockburn

Mingary

Silverton

Milparinka

Packsaddle Roadhouse

Coombah Roadhouse

Quandong Roadhouse

Tibooburra

Lake Frome

Lake Callabonna

Lake Blanche

STRZELECKI TRACK

BARRIER HWY

SILVER CITY HWY

MENINDEE RD

FLINDERS RANGES

BARRIER RANGES

GREY RANGE

EXCLUSION ZONE BOUNDARY

DOG FENCE

FRUIT FLY

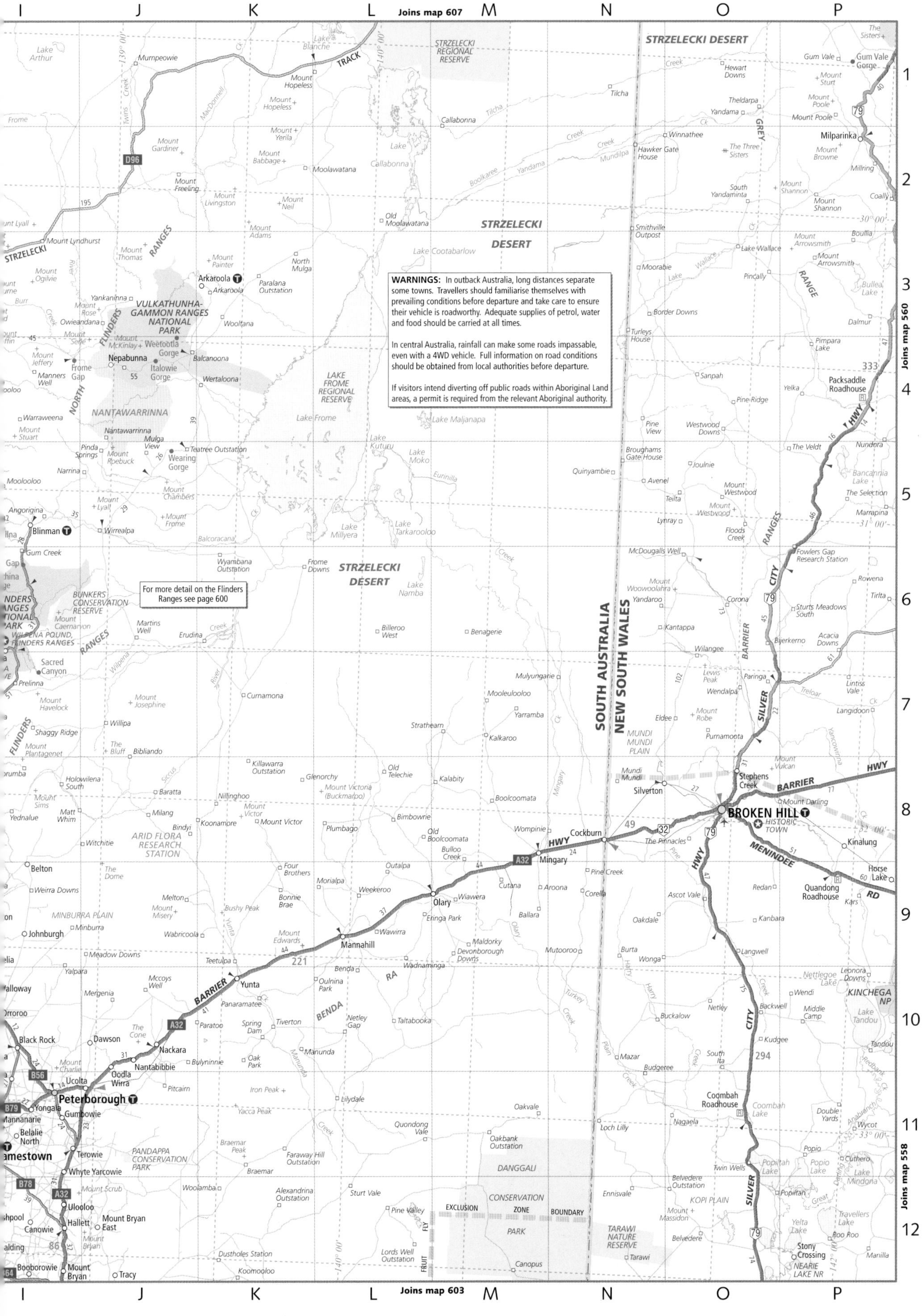

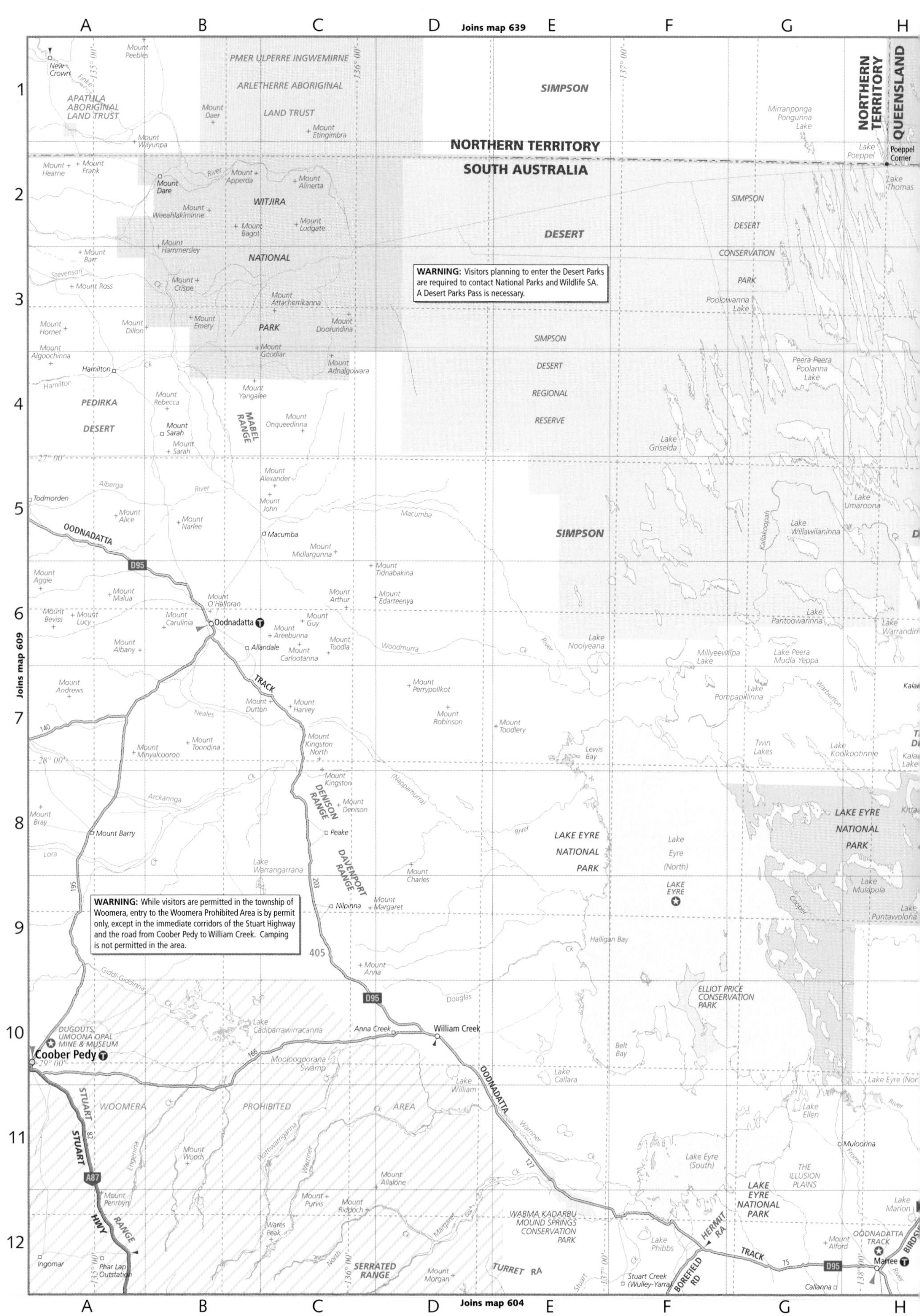

Joins map 639
Joins map 604
Joins map 609
NORTHERN TERRITORY
SOUTH AUSTRALIA
QUEENSLAND
WARNING: Visitors planning to enter the Desert Parks are required to contact National Parks and Wildlife SA. A Desert Parks Pass is necessary.
WARNING: While visitors are permitted in the township of Woomera, entry to the Woomera Prohibited Area is by permit only, except in the immediate corridors of the Stuart Highway and the road from Coober Pedy to William Creek. Camping is not permitted in the area.
PMER ULPERRE INGWEMIRNE ARLETHERRE ABORIGINAL LAND TRUST
APATULA ABORIGINAL LAND TRUST
WITJIRA NATIONAL PARK
SIMPSON DESERT
SIMPSON DESERT CONSERVATION PARK
SIMPSON DESERT REGIONAL RESERVE
PEDIRKA DESERT
LAKE EYRE NATIONAL PARK
ELLIOT PRICE CONSERVATION PARK
WABMA KADARBU MOUND SPRINGS CONSERVATION PARK
WOOMERA PROHIBITED AREA
Oodnadatta
Coober Pedy
William Creek
Marree
Poeppel Corner
OODNADATTA TRACK
STUART HWY
D95
A87
Lake Eyre (North)
Lake Eyre (South)
DUGOUTS, UMOONA OPAL MINE & MUSEUM
MABEL RANGE
DENISON RANGE
DAVENPORT RANGE
SERRATED RANGE
STUART RANGE
HERMIT RA
TURRET RA
BOREFIELD RD
THE ILLUSION PLAINS

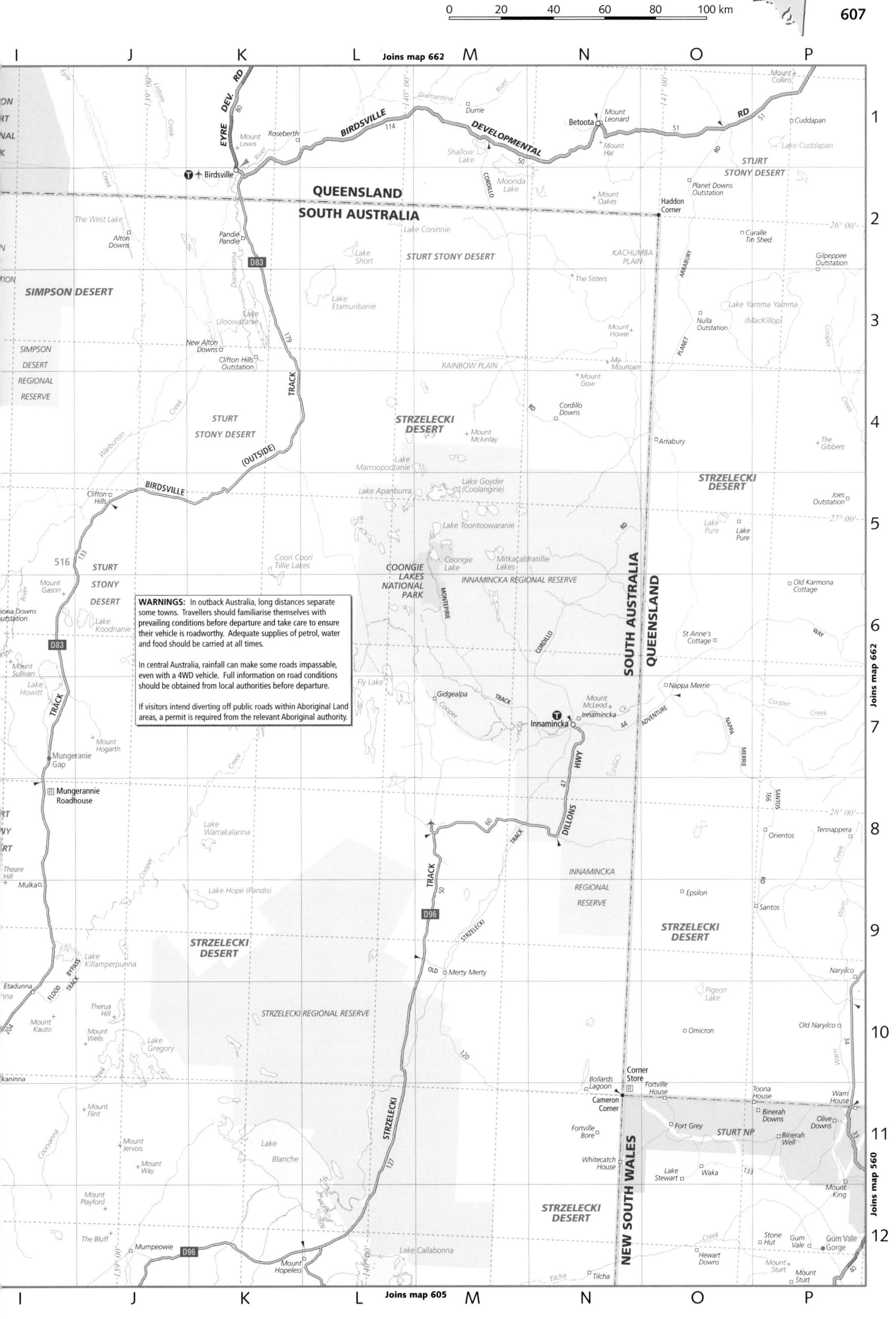

WARNINGS: In outback Australia, long distances separate some towns. Travellers should familiarise themselves with prevailing conditions before departure and take care to ensure their vehicle is roadworthy. Adequate supplies of petrol, water and food should be carried at all times.

In central Australia, rainfall can make some roads impassable, even with a 4WD vehicle. Full information on road conditions should be obtained from local authorities before departure.

If visitors intend diverting off public roads within Aboriginal Land areas, a permit is required from the relevant Aboriginal authority.

Joins map 638
Joins map 623
Joins map 621
Joins map 610
NORTHERN TERRITORY
SOUTH AUSTRALIA
WESTERN AUSTRALIA
NGAANYATJARRA CENTRAL RESERVE
PETERMANN ABORIGINAL LAND TRUST
KATITI ABORIGINAL LAND TRUST
ANANGU PITJANTJATJARA YANKUNYTJATJARA LANDS
GREAT VICTORIA DESERT
MAMUNGARI CONSERVATION PARK
MARALINGA TJARUTJA LANDS
GREAT VICTORIA DESERT NATURE RESERVE
Surveyor Generals Corner
Kalka
Pipalyatjara
Aparawatatja
Kanypi
Alpara
Amata
Pirrilyungka
Anumarrapirti
Warlpapuka
Pirntirri Mulari
Arnold Creek
MURRAY RANGE
MORGAN RANGE
TOMKINSON RA
MANN RANGES
BIRKSGATE RA
COMET RANGE
LIZZIE LIGHTFOOT RANGE
GUNBARREL
ANNE BEADELL HWY
ABORIGINAL BUSINESS RD
LAKE DEY-DEY RD
Mount Woltarlinna
Mount Lindsay
Mount Poondinna
Mount Tietkens
Mount Crombie (Ulpara)
Mount Kintore
Mount Caroline (Ulkiyanya)
Wanna Lakes
Serpentine Lakes
Forrest Lakes
Nurrari Lakes
Wyola Lake
Halinor Lake
Lake Dey-Dey
Lake Maurice (Carle-Thulka)
WARNINGS: In outback Australia, long distances separate some towns. Travellers should familiarise themselves with prevailing conditions before departure and take care to ensure their vehicle is roadworthy. Adequate supplies of petrol, water and food should be carried at all times.
In central Australia, rainfall can make some roads impassable, even with a 4WD vehicle. Full information on road conditions should be obtained from local authorities before departure.
If visitors intend diverting off public roads within Aboriginal Land areas, a permit is required from the relevant Aboriginal authority.

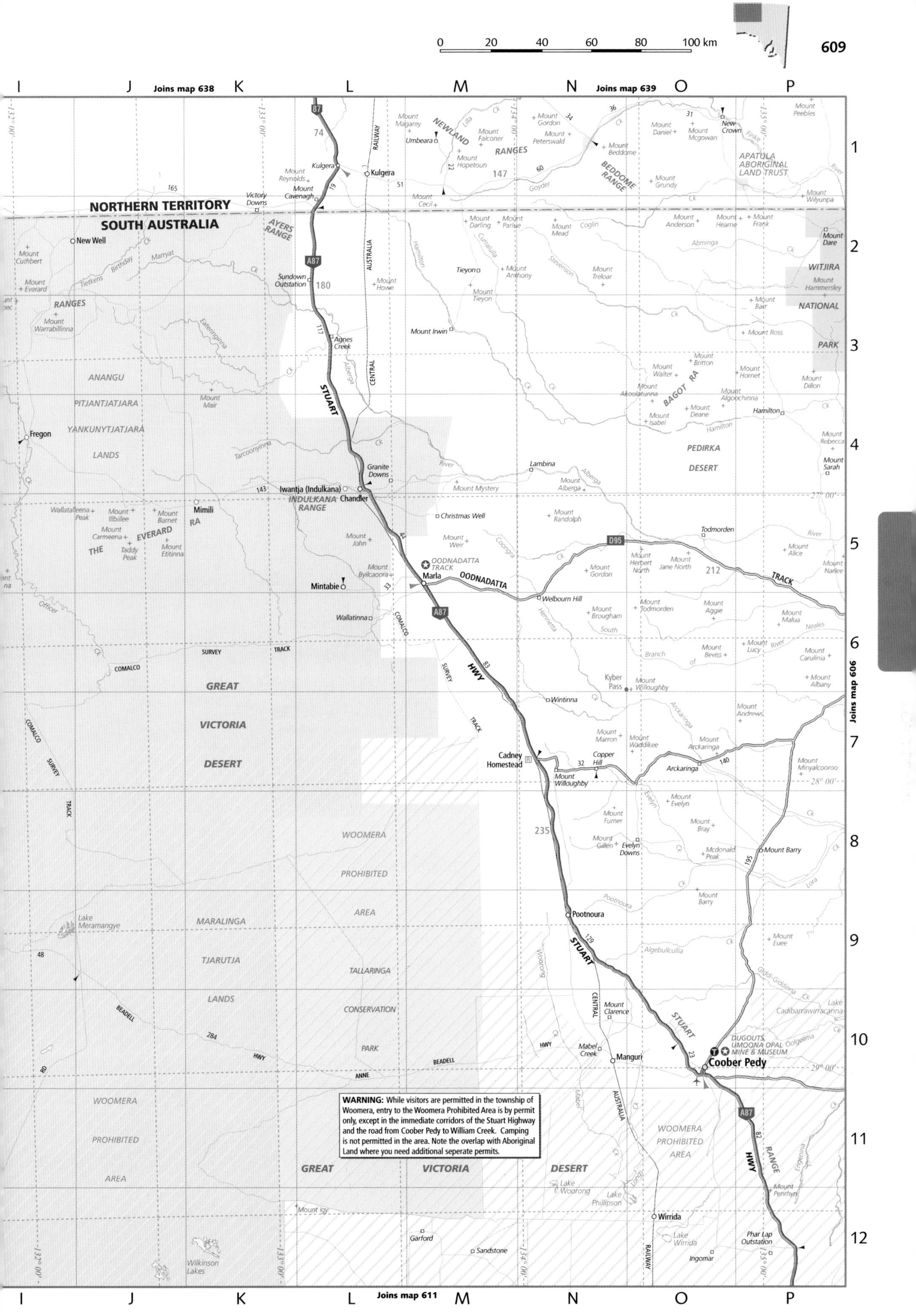

0 20 40 60 80 100 km

I J Joins map 638 K L M N Joins map 639 O P

1 2 3 4 5 6 7 8 9 10 11 12

NORTHERN TERRITORY
SOUTH AUSTRALIA

Kulgera
Mount Reynolds
Mount Cavenagh
Victory Downs
165
74
19
51
RAILWAY
Mount Magarey
Umbeara
NEWLAND RANGES
Mount Falconer
Mount Hopetoun
22
147
Mount Gordon
Mount Peterswald
34
36
60
Goyder
Mount Cecil
Mount Beddome
BEDDOME RANGE
Mount Grundy
Mount Daniel
Mount Mcgowan
31
New Crown
Finke
APATULA ABORIGINAL LAND TRUST
Mount Peebles
River
Mount Wilyunpa

AYERS RANGE
New Well
Mount Cuthbert
Mount Everard
Tietkens
Birthday
Marryat
RANGES
Mount Warrabillinna
A87
Sundown Outstation
180
AUSTRALIA
Mount Howe
Hamilton
Mount Darling
Mount Parlue
Mount Mead
Coglin
Curralulla
Tieyon
Mount Anthony
Mount Tieyon
Stevenson
Mount Treloar
Abminga
Mount Anderson
Mount Hearne
Mount Frank
Mount Dare
WITJIRA NATIONAL PARK
Mount Hammersley
Mount Barr
Mount Ross
Eateringinna
117
Agnes Creek
Alberga
CENTRAL
Mount Irwin
Mount Walter
Mount Britton
Mount Hornet
Mount Dillon
Mount Akoolatunna
BAGOT RA
Mount Algoochinna
Mount Isabel
Mount Deane
Hamilton
ANANGU PITJANTJATJARA YANKUNYTJATJARA LANDS
Mount Mair
STUART
Fregon
Tarcoonyinna
River
Granite Downs
PEDIRKA DESERT
Mount Rebecca
Mount Sarah
Lambina
Mount Alberga
Alberga
143
Iwantja (Indulkana)
Chandler
INDULKANA RANGE
Mount Mystery
27° 00'
Wallatalleena Peak
Mount Illbillee
Mount Barnet
Mimili
RA
Mount Carmeena
EVERARD
THE
Taddy Peak
Mount Etitinna
Christmas Well
Mount Randolph
Todmorden
River
Mount John
Mount Weir
Coongra
44
D95
Mount Herbert North
Mount Jane North
212
TRACK
Mount Alice
Mount Narlee
OODNADATTA TRACK
Marla
OODNADATTA
Mount Byilcaoora
33
Mintabie
Mount Gordon
Welbourn Hill
Mount Brougham
Mount Todmorden
Mount Aggie
Mount Malua
Neales
Officer
Wallatinna
COMALCO
Henrietta
South
Mount Lucy
River
Branch
of
Mount Bevis
Mount Carulinia
SURVEY TRACK
COMALCO
GREAT VICTORIA DESERT
SURVEY
HWY
83
Kyber Pass
Mount Willoughby
Mount Albany
Wintinna
Arckaringa
Mount Andrews
COMALCO SURVEY TRACK
TRACK
Mount Marron
Mount Waddikee
Mount Arckaringa
Cadney Homestead
32
Copper Hill
Arckaringa
140
Mount Minyalcooroo
Mount Willoughby
28° 00'
Evelyn
Mount Evelyn
Mount Furner
235
Mount Bray
Mount Gillen
Evelyn Downs
Mcdonald Peak
Mount Barry
WOOMERA PROHIBITED AREA
195
Lora
Mount Barry
Lake Meramangye
MARALINGA TJARUTJA LANDS
Pootnoura
Pootnoura
129
STUART
Woorong
Algebullcullia
Mount Euee
48
TALLARINGA CONSERVATION PARK
CENTRAL
Mount Clarence
Giddi-Giddinna
Lake Cadibarrawirracanna
BEADELL
STUART
284
HWY
Mabel Creek
Manguri
DUGOUTS, UMOONA OPAL MINE & MUSEUM
Oolgelima
Coober Pedy
23
29° 00'
BEADELL
ANNE
RD
Mabel
AUSTRALIA
WOOMERA PROHIBITED AREA
A87
82
RANGE
HWY
Engenina
WOOMERA PROHIBITED AREA
GREAT VICTORIA DESERT
Lake Woorong
Lake Phillipson
Mount Penrhyn
Mount Igy
Wirrida
Garford
Sandstone
RAILWAY
Lake Wirrida
Ingomar
Phar Lap Outstation
Wilkinson Lakes
132° 00'
133° 00'
134° 00'
135° 00'

WARNING: While visitors are permitted in the township of Woomera, entry to the Woomera Prohibited Area is by permit only, except in the immediate corridors of the Stuart Highway and the road from Coober Pedy to William Creek. Camping is not permitted in the area. Note the overlap with Aboriginal Land where you need additional seperate permits.

Joins map 606

I J K L Joins map 611 M N O P

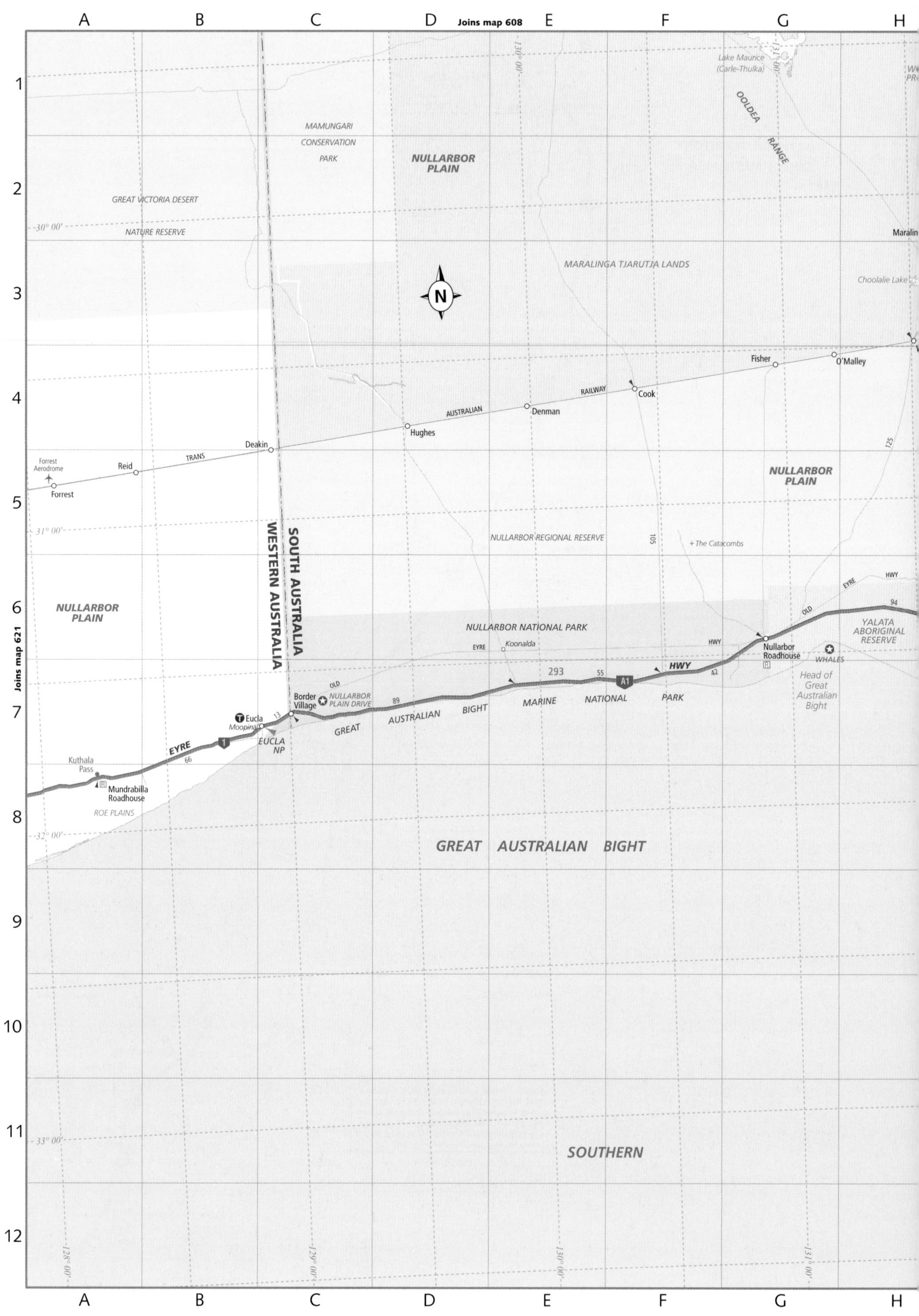

A
B
C
D
Joins map 608
E
F
G
H
1
2
3
4
5
6
7
8
9
10
11
12
Joins map 621
Lake Maurice
(Carle-Thulka)
OOLDEA
RANGE
MAMUNGARI
CONSERVATION
PARK
NULLARBOR
PLAIN
GREAT VICTORIA DESERT
NATURE RESERVE
30° 00'
130° 00'
131° 00'
Maralin
MARALINGA TJARUTJA LANDS
Choolalie Lake
N
Fisher
O'Malley
RAILWAY
Cook
AUSTRALIAN
Denman
Hughes
Deakin
TRANS
Reid
Forrest
Aerodrome
Forrest
NULLARBOR
PLAIN
31° 00'
SOUTH AUSTRALIA
WESTERN AUSTRALIA
NULLARBOR REGIONAL RESERVE
105
+ The Catacombs
125
EYRE
HWY
OLD
94
NULLARBOR
PLAIN
NULLARBOR NATIONAL PARK
YALATA
ABORIGINAL
RESERVE
EYRE
Koonalda
HWY
Nullarbor
Roadhouse
WHALES
293
55
A1
HWY
42
Head of
Great
Australian
Bight
OLD
Border
Village
NULLARBOR
PLAIN DRIVE
89
BIGHT
MARINE
NATIONAL
PARK
Eucla
Moopina
13
AUSTRALIAN
GREAT
EUCLA
NP
EYRE
1
66
Kuthala
Pass
Mundrabilla
Roadhouse
ROE PLAINS
32° 00'
GREAT AUSTRALIAN BIGHT
33° 00'
SOUTHERN
128° 00'
129° 00'
130° 00'
131° 00'

0 20 40 60 80 100 km
Joins map 609
Joins map 604
Joins map 602
WARNING: While visitors are permitted in the township of Woomera, entry to the Woomera Prohibited Area is by permit only, except in the immediate corridors of the Stuart Highway and the road from Coober Pedy to William Creek. Camping is not permitted in the area. Note the overlap with Aboriginal Land where you need additional seperate permits.
Wilkinson Lakes
MARALINGA TJARUTJA LANDS
WOOMERA PROHIBITED AREA
Lake Bring
Indooroopilly Outstation
Lake Anthony
Half Moon Lake
Jumbuck
Irria Outstation
Mount Christie
Durkin Outstation
Mulgathing
Warrior Outstation
Comet
Commonwealth Hill
Muckanippie Outstation
Bradman Outstation
Carne Outstation
Gibraltar Outstation
Ambrosia Outstation
Wirrida
Lake Wirrida
Ingomar
Phar Lap Outstation
Mount Sandy
Mc Douall Peak
Mount Soward
Gina Outstation
Mirikata
Goode Outstation
Bulgunnia
Ooraminna Outstation
Johns Outstation
Ealbara Outstation
Mentor Outstation
Whymlet
AUSTRALIAN RAILWAY
CENTRAL AUSTRALIAN RAILWAY
STUART HWY
A87
367
Doldea
Bates
TRANS AUSTRALIAN RAILWAY
Wynbring
Lyons Camp
Malbooma Outstation
Carnding Road Outstation
Tarcoola
Lake Moolkra
Wilgena
Lake Labyrinth
Mount Eba
Big Tank Outstation
North Well
Kingoonya
Lake Ifould
Lake Tallacootra
Mount Finke
Yerda
Lake Harris
YELLABINNA REGIONAL RESERVE
Kokatha
LAKE GAIRDNER NATIONAL PARK
YALATA ABORIGINAL RESERVE
Lake Everard
Yalata
Glyde Hill Outstation
YUMBARRA CR
YUMBARRA CONSERVATION PARK
Lake Everard
Nundroo
Nundroo Roadhouse
EYRE HWY
202
Pintumba
Coorabie
Wookata
FOWLERS BAY CR
Fowlers Bay
Fowlers Bay
Point Fowler
Cape Adieu
Cheetima Beach
Cape Nuyts
NUYTS REEF CP
Bookabie
CHADINGA CR
Cundilippy
Northedge
Penong
YUMBARRA CR
Koonibba
A1
PUREBA CR
PUREBA CONSERVATION PARK
Mount John
Kondoolka
Lake Acraman
CHADINGA CR
SURFING
Cactus Beach
Point Sinclair
Lake MacDonnell
Black Peak
Marbra
Corrong
NULLARBOR PLAIN DRIVE
Ceduna
Thevenard
Denial Bay
Watchbrae
Mudamuckla
NUNNYAH CR
Oak Valley
Chinbingina
Nunjikompita
KOOLGERA CR
Mount Wallaby
Mount Pollard
Winilippe Peak
Yarna
Mount Hiltaba
Mount St Mungo
Mount Pyramid
POINT BELL CR
Point Bell
Purdie Islands
Point Peter
St Peter Island
Goat Island
Cape D'Estrees
Smoky Bay
NUYTS ARCHIPELAGO CP
Lacy Island
Evans Island
Eyre Island
FLINDERS HWY
Kara-Pine
Carawa
Wallala
Waroona Peak
Mount Centre
GAWLER RANGES NP
Smoky Bay
Franklin Islands
ACRAMAN CREEK CP
Point Dillon
St Mary Bay
Point Brown
ISLES OF ST. FRANCIS CP
St Francis Island
Flagstaff
Gascoigne Bay
Point Collinson
Haslam
B100
109
222
Wirrulla
Petina
Wirrambie
Yantanabie
Gawler View
Cungena
GAWLER RANGES CR
Kolbrae
Scrubby Peak
Streaky Bay
Eba Island
Cape Bauer
Chilpanunda
Mount Jane
Coolgrana
The Bald Hills
Capietha
Poochera
Chandada
Corvisart Bay
Streaky Bay
Parla Peak
Tootla
Wyoming
Minnipa
Point Westall
Yanerbie Beach
Maryvale
SCEALE BAY CR
CALPATANNA WATERHOLE CP
Yandra
Conglima
Carina
Moonlight Flat
Yaninee
Lake Yaninee
Sceale Bay
Calca
Mount Cooper
Slade Point
Searcy Bay
Mount Hall
Colley
Mount Misery
KULLIPARU CP
Mount Damper
Point Labatt
Baird Bay
Cape Radstock
Port Kenny
124
VENUS BAY CP
Venus Bay
Anxious Bay
Talia
Talia Beach
Lake Newland
LAKE NEWLAND CP
Colton
Kooringal
Mount Wedge
COCATA CP
B91
WALDEGRAVE ISLANDS CP
Mount Fairy
Bramfield
OCEAN
I J K L M N O P
1 2 3 4 5 6 7 8 9 10 11 12

MAPS

WESTERN AUSTRALIA

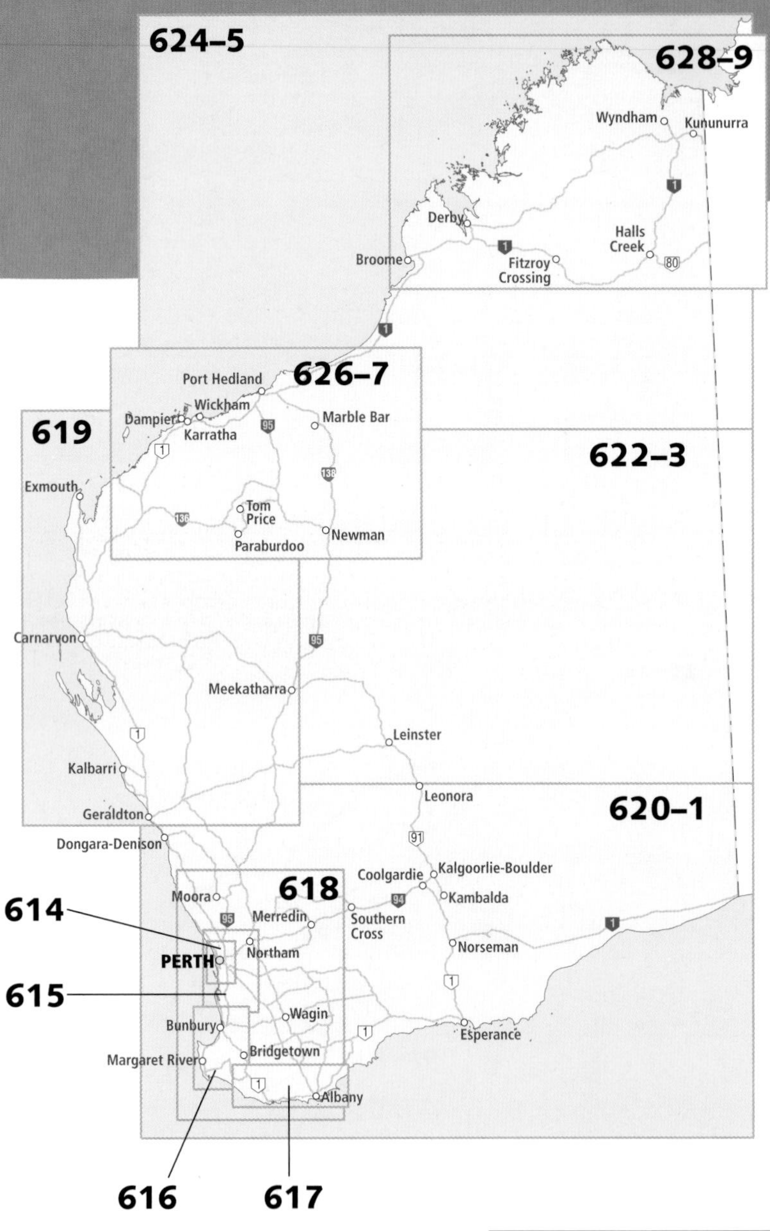

INTER-CITY ROUTES		DISTANCE
Perth–Adelaide via Great Eastern & Eyre hwys	94 1 A1	2700 km
Perth–Darwin via Great Northern Hwy	1 95	4032 km

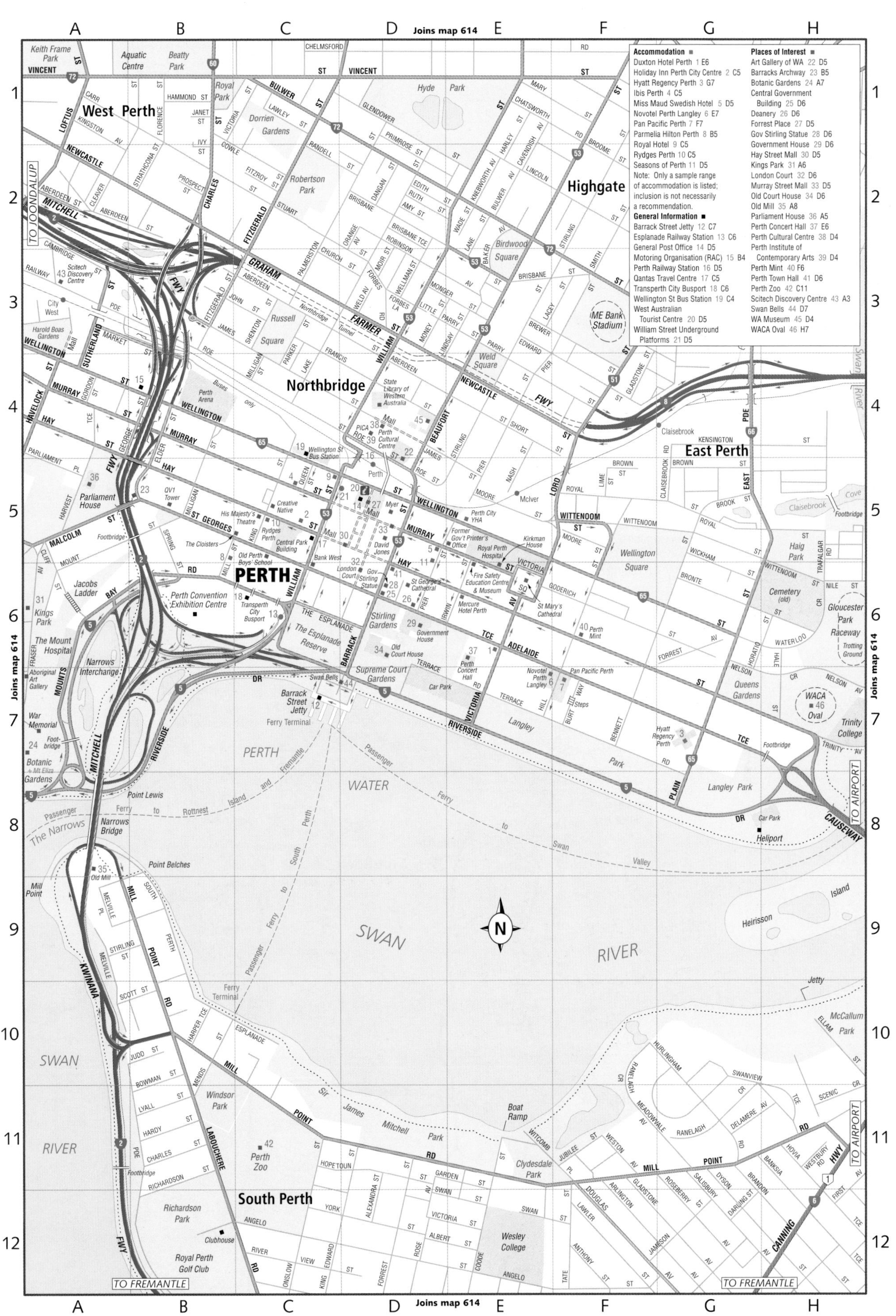
0 0.25 0.5 0.75 1 km
Joins map 614
Accommodation
Duxton Hotel Perth 1 E6
Holiday Inn Perth City Centre 2 C5
Hyatt Regency Perth 3 G7
Ibis Perth 4 C5
Miss Maud Swedish Hotel 5 D5
Novotel Perth Langley 6 E7
Pan Pacific Perth 7 F7
Parmelia Hilton Perth 8 B5
Royal Hotel 9 C5
Rydges Perth 10 C5
Seasons of Perth 11 D5
Note: Only a sample range of accommodation is listed; inclusion is not necessarily a recommendation.
General Information
Barrack Street Jetty 12 C7
Esplanade Railway Station 13 C6
General Post Office 14 D5
Motoring Organisation (RAC) 15 B4
Perth Railway Station 16 D5
Qantas Travel Centre 17 C5
Transperth City Busport 18 C6
Wellington St Bus Station 19 C4
West Australian Tourist Centre 20 D5
William Street Underground Platforms 21 D5
Places of Interest
Art Gallery of WA 22 D5
Barracks Archway 23 B5
Botanic Gardens 24 A7
Central Government Building 25 D6
Deanery 26 D6
Forrest Place 27 D5
Gov Stirling Statue 28 D6
Government House 29 D6
Hay Street Mall 30 D5
Kings Park 31 A6
London Court 32 D6
Murray Street Mall 33 D5
Old Court House 34 D6
Old Mill 35 A8
Parliament House 36 A5
Perth Concert Hall 37 E6
Perth Cultural Centre 38 D4
Perth Institute of Contemporary Arts 39 D4
Perth Mint 40 F6
Perth Town Hall 41 D6
Perth Zoo 42 C11
Scitech Discovery Centre 43 A3
Swan Bells 44 D7
WA Museum 45 D4
WACA Oval 46 H7
West Perth
Northbridge
Highgate
East Perth
PERTH
South Perth
SWAN RIVER
PERTH WATER
Kings Park
Langley Park
Heirisson Island
TO JOONDALUP
TO AIRPORT
TO FREMANTLE
Joins map 614

0 5 10 15 20 km

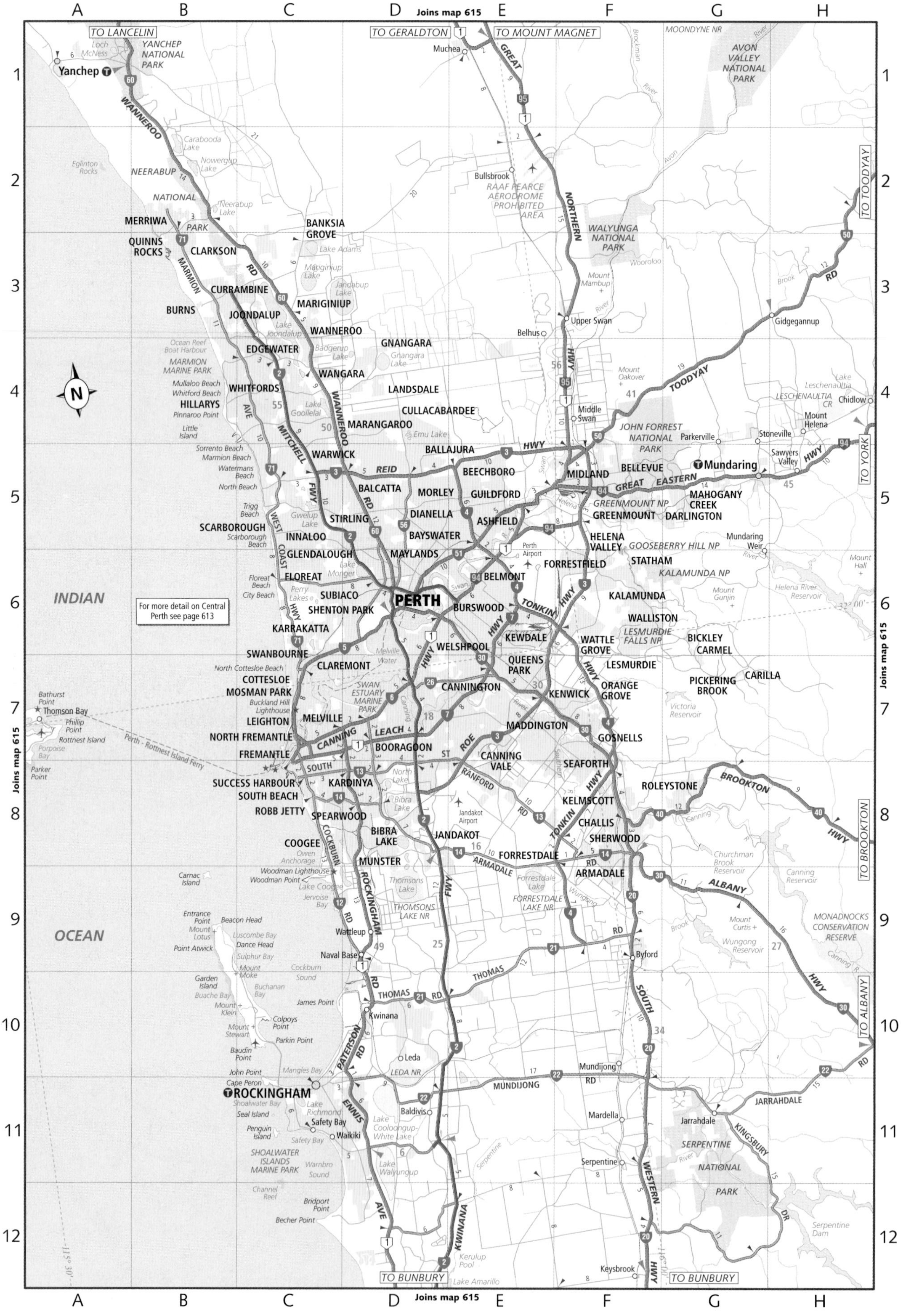

A B C D E F G H

Joins map 615

TO LANCELIN
TO GERALDTON
TO MOUNT MAGNET
MOONDYNE NR
AVON VALLEY NATIONAL PARK
Yanchep
YANCHEP NATIONAL PARK
Loch McNess
Muchea
GREAT NORTHERN HWY
WANNEROO RD
Carabooda Lake
Nowergup Lake
Eglinton Rocks
NEERABUP NATIONAL PARK
Neerabup Lake
Bullsbrook
RAAF PEARCE AERODROME PROHIBITED AREA
Brockman River
Avon
TO TOODYAY
MERRIWA
QUINNS ROCKS
CLARKSON
BANKSIA GROVE
Lake Adams
WALYUNGA NATIONAL PARK
Wooroloo
Mariginiup Lake
Jandabup Lake
CURRAMBINE
MARMION AVE
BURNS
JOONDALUP
MARIGINIUP
Lake Joondalup
WANNEROO
Mount Mambup
Upper Swan
Belhus
Gidgegannup
Brook
RD
Ocean Reef Boat Harbour
EDGEWATER
Badgerup Lake
GNANGARA
Gnangara Lake
MARMION MARINE PARK
WANGARA
Mount Oakover
TOODYAY
Lake Leschenaultia
LESCHENAULTIA CR
Chidlow
Mullaloo Beach
Whitford Beach
WHITFORDS
HILLARYS
Pinnaroo Point
LANDSDALE
CULLACABARDEE
Lake Goollelal
MARANGAROO
Middle Swan
JOHN FORREST NATIONAL PARK
Parkerville
Stoneville
Mount Helena
Little Island
Emu Lake
Sorrento Beach
Marmion Beach
Watermans Beach
North Beach
MITCHELL FWY
WARWICK
BALLAJURA
REID HWY
BEECHBORO
MIDLAND
BELLEVUE
Mundaring
Sawyers Valley
GREAT EASTERN HWY
TO YORK
BALCATTA
MORLEY
GUILDFORD
Helena
GREENMOUNT NP
MAHOGANY CREEK
Trigg Beach
Gwelup Lake
STIRLING
DIANELLA
ASHFIELD
GREENMOUNT
DARLINGTON
SCARBOROUGH
Scarborough Beach
WEST COAST HWY
INNALOO
BAYSWATER
HELENA VALLEY
GOOSEBERRY HILL NP
Mundaring Weir
GLENDALOUGH
MAYLANDS
Perth Airport
FORRESTFIELD
STATHAM
Mount Hall
Floreat Beach
City Beach
FLOREAT
Lake Monger
BELMONT
KALAMUNDA NP
Perry Lakes
SUBIACO
PERTH
Swan
TONKIN HWY
KALAMUNDA
Mount Gunjin
Helena River Reservoir
INDIAN
For more detail on Central Perth see page 613
SHENTON PARK
BURSWOOD
WALLISTON
32° 00'
KARRAKATTA
LESMURDIE FALLS NP
WELSHPOOL
KEWDALE
WATTLE GROVE
BICKLEY
CARMEL
SWANBOURNE
Melville Water
QUEENS PARK
LESMURDIE
North Cottesloe Beach
CLAREMONT
CARILLA
PICKERING BROOK
COTTESLOE
MOSMAN PARK
SWAN ESTUARY MARINE PARK
CANNINGTON
ORANGE GROVE
KENWICK
Bathurst Point
Thomson Bay
Phillip Point
Rottnest Island
Buckland Hill Lighthouse
Canning
Victoria Reservoir
LEIGHTON
MELVILLE
MADDINGTON
GOSNELLS
NORTH FREMANTLE
CANNING HWY
LEACH HWY
ROE HWY
Porpoise Bay
Perth - Rottnest Island Ferry
FREMANTLE
BOORAGOON
CANNING VALE
SEAFORTH
Parker Point
SOUTH ST
North Lake
RANFORD RD
Southern River
SUCCESS HARBOUR
KARDINYA
ROLEYSTONE
BROOKTON HWY
SOUTH BEACH
Bibra Lake
KELMSCOTT
ROBB JETTY
SPEARWOOD
Jandakot Airport
CHALLIS
TO BROOKTON
BIBRA LAKE
JANDAKOT
SHERWOOD
COOGEE
COCKBURN RD
Owen Anchorage
MUNSTER
FORRESTDALE
ARMADALE RD
ARMADALE
Churchman Brook Reservoir
Canning Reservoir
Woodman Lighthouse
Woodman Point
Lake Coogee
Thomsons Lake
Forrestdale Lake
FORRESTDALE LAKE NR
Wungong
ALBANY HWY
Carnac Island
Jervoise Bay
ROCKINGHAM RD
THOMSONS LAKE NR
KWINANA FWY
Mount Curtis
MONADNOCKS CONSERVATION RESERVE
Entrance Point
Beacon Head
Mount Lotus
Luscombe Bay
Wattleup
Wungong Reservoir
OCEAN
Dance Head
Point Atwick
Sulphur Bay
Naval Base
Byford
Garden Island
Mount Moke
Cockburn Sound
THOMAS RD
Buache Bay
Buchanan Bay
James Point
SOUTH WESTERN HWY
TO ALBANY
Mount Klein
Kwinana
Mount Stewart
Colpoys Point
Parkin Point
PATERSON RD
Baudin Point
John Point
Leda
LEDA NR
Mundijong
Cape Peron
Mangles Bay
MUNDIJONG RD
ROCKINGHAM
Shoalwater Bay
Lake Richmond
ENNIS AVE
Seal Island
Safety Bay
Baldivis
Mardella
JARRAHDALE
Jarrahdale
Penguin Island
Waikiki
Lake Cooloongup-White Lake
KINGSBURY DR
SHOALWATER ISLANDS MARINE PARK
Warnbro Sound
Lake Walyungup
SERPENTINE NATIONAL PARK
Serpentine
Channel Reef
Bridport Point
Becher Point
KWINANA
Serpentine Dam
Kerulup Pool
Lake Amarillo
Keysbrook
115° 30'
116° 00'
TO BUNBURY
TO BUNBURY
Joins map 615

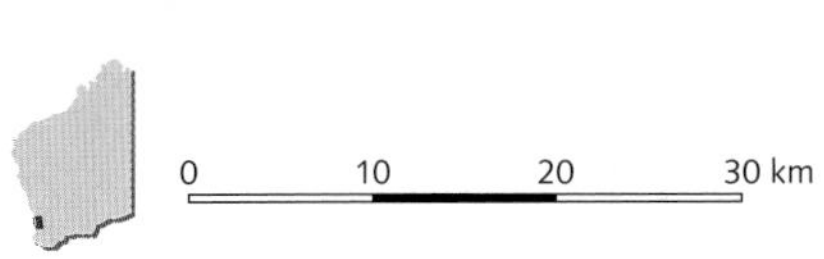

0 10 20 30 km

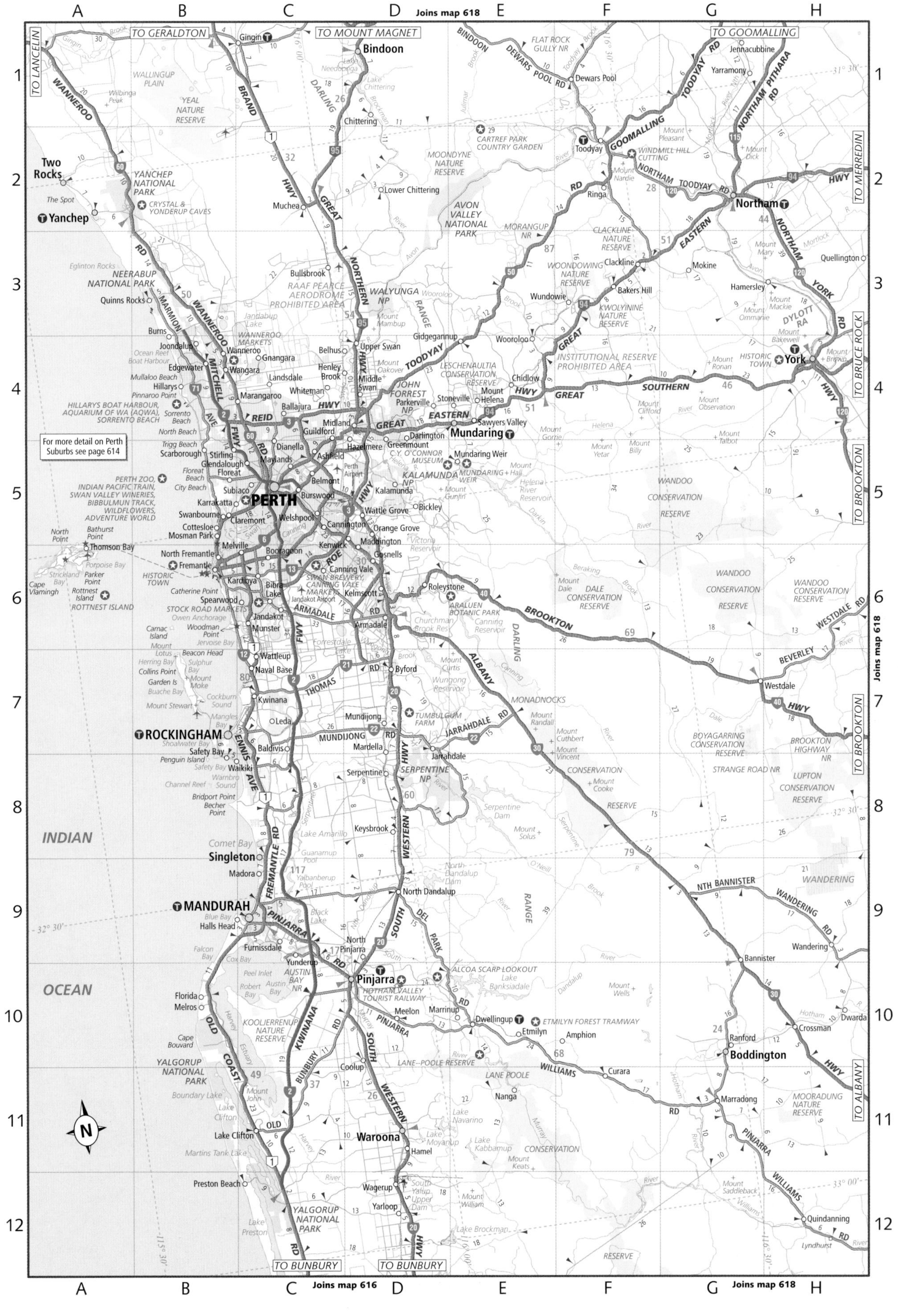

Joins map 618

TO GERALDTON
TO MOUNT MAGNET
TO GOOMALLING
TO LANCELIN
TO MERREDIN
TO BRUCE ROCK
TO BROOKTON
TO ALBANY
TO BUNBURY

For more detail on Perth Suburbs see page 614

Gingin
Bindoon
Chittering
Lower Chittering
Muchea
Dewars Pool
Toodyay
Jennacubbine
Yarramony
Northam
Two Rocks
Yanchep
YANCHEP NATIONAL PARK
CRYSTAL & YONDERUP CAVES
AVON VALLEY NATIONAL PARK
MOONDYNE NATURE RESERVE
CARTREF PARK COUNTRY GARDEN
WINDMILL HILL CUTTING
Ringa
Clackline
Bakers Hill
Wundowie
Mokine
Hamersley
Quellington
Bullsbrook
RAAF PEARCE AERODROME PROHIBITED AREA
WALYUNGA NP
NEERABUP NATIONAL PARK
Quinns Rocks
Burns
Joondalup
Wanneroo
WANNEROO MARKETS
Gnangara
Wangara
Edgewater
Hillarys
Marangaroo
Landsdale
Whiteman
Upper Swan
Belhus
Henley Brook
Middle Swan
Gidgegannup
Wooroloo
Chidlow
York
HISTORIC TOWN
INSTITUTIONAL RESERVE PROHIBITED AREA
HILLARYS BOAT HARBOUR, AQUARIUM OF WA (AQWA), SORRENTO BEACH
Sorrento Beach
North Beach
Trigg Beach
Scarborough
Stirling
Glendalough
Floreat
Ballajura
Midland
Guildford
Dianella
Maylands
Ashfield
Hazelmere
Greenmount
Darlington
Mundaring
Mundaring Weir
Sawyers Valley
Mount Helena
Stoneville
Parkerville
JOHN FORREST NP
C.Y. O'CONNOR MUSEUM
KALAMUNDA NP
Kalamunda
Bickley
PERTH ZOO, INDIAN PACIFIC TRAIN, SWAN VALLEY WINERIES, BIBBULMUN TRACK, WILDFLOWERS, ADVENTURE WORLD
Subiaco
PERTH
Karrakatta
Swanbourne
Claremont
Cottesloe
Mosman Park
Belmont
Burswood
Welshpool
Cannington
Wattle Grove
Orange Grove
Kenwick
Maddington
Gosnells
North Fremantle
Fremantle
Melville
Booragoon
Thomson Bay
ROTTNEST ISLAND
HISTORIC TOWN
Kardinya
Bibra Lake
Canning Vale
SWAN BREWERY, CANNING VALE MARKETS
Jandakot Airport
Kelmscott
Roleystone
ARALUEN BOTANIC PARK
Spearwood
STOCK ROAD MARKETS
Jandakot
Munster
Armadale
DALE CONSERVATION RESERVE
WANDOO CONSERVATION RESERVE
Wattleup
Naval Base
Byford
Kwinana
Leda
Mundijong
TUMBULGUM FARM
Mardella
Jarrahdale
Westdale
ROCKINGHAM
Safety Bay
Waikiki
Baldivis
Serpentine
SERPENTINE NP
MONADNOCKS CONSERVATION RESERVE
BOYAGARRING CONSERVATION RESERVE
BROOKTON HIGHWAY NR
STRANGE ROAD NR
LUPTON CONSERVATION RESERVE
INDIAN OCEAN
Keysbrook
Singleton
Madora
MANDURAH
Halls Head
North Dandalup
Furnissdale
North Pinjarra
Yunderup
Pinjarra
AUSTIN BAY NR
HOTHAM VALLEY TOURIST RAILWAY
ALCOA SCARP LOOKOUT
Bannister
Wandering
Dwarda
Crossman
Florida
Melros
KOOLJERRENUP NATURE RESERVE
Meelon
Marrinup
Dwellingup
ETMILYN FOREST TRAMWAY
Etmilyn
Amphion
Ranford
Boddington
YALGORUP NATIONAL PARK
Coolup
LANE-POOLE RESERVE
LANE POOLE CONSERVATION RESERVE
Nanga
Curara
Marradong
MOORADUNG NATURE RESERVE
Lake Clifton
Waroona
Hamel
Preston Beach
Wagerup
Yarloop
Quindanning
Lyndhurst

Joins map 616
Joins map 618

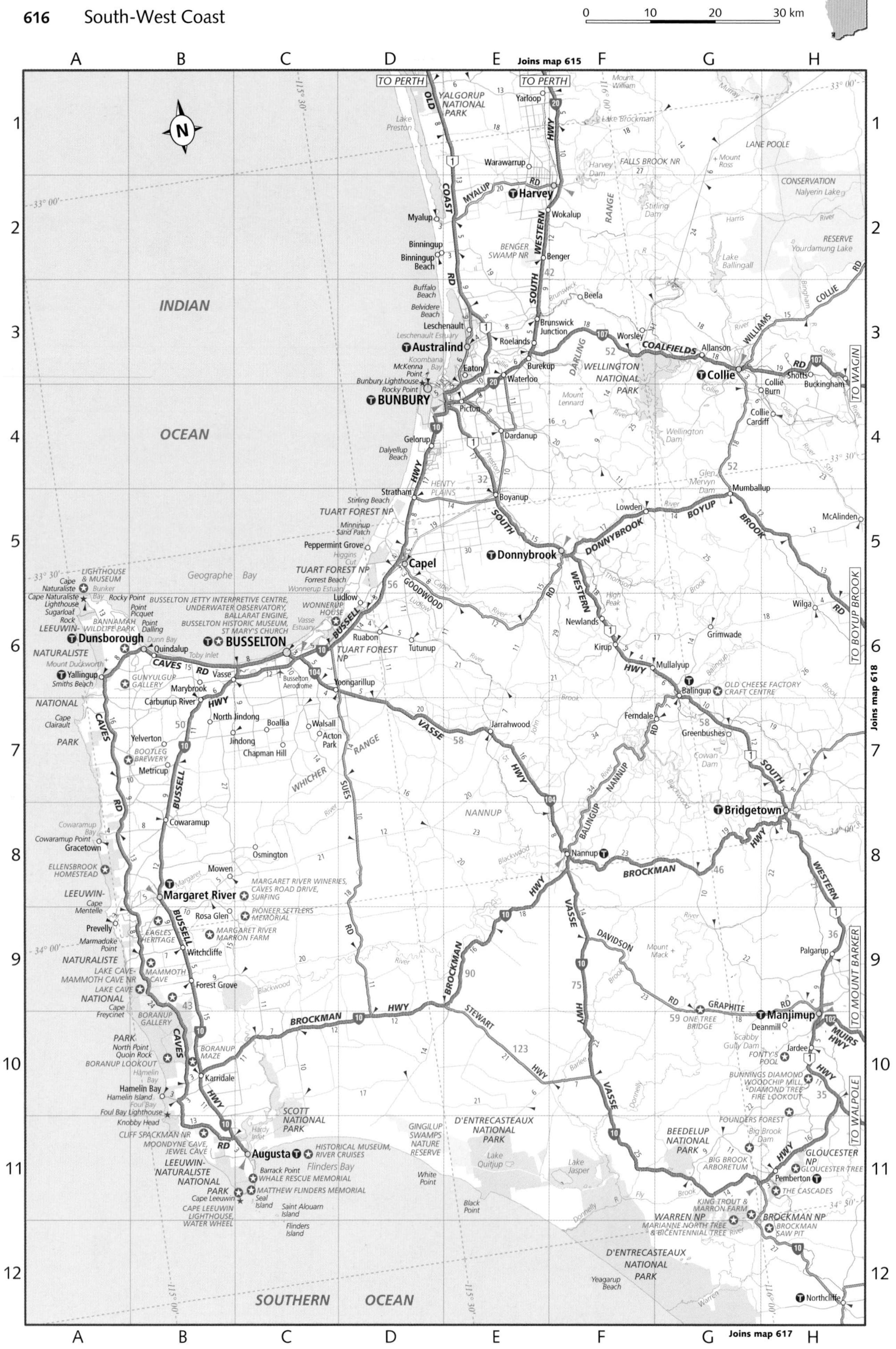

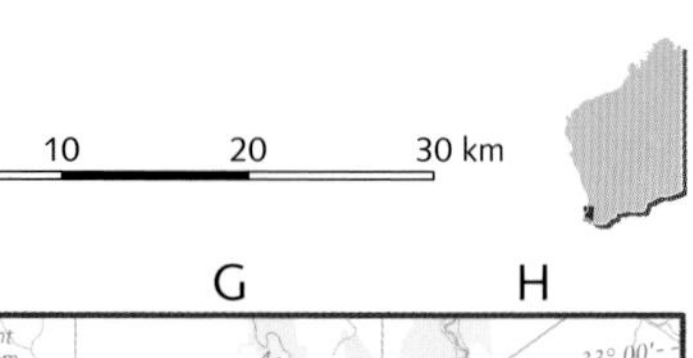

Joins map 615

TO PERTH

INDIAN

OCEAN

SOUTHERN OCEAN

Geographe Bay

Flinders Bay

Yalgorup National Park
Lake Preston
Yarloop
Lake Brockman
Mount William
Warawarrup
Harvey
Harvey Dam
Falls Brook NR
Mount Ross
Lane Poole Conservation Reserve
Nalyerin Lake
Yourdamung Lake
Myalup
Wokalup
Stirling Dam
Binningup
Binningup Beach
Benger Swamp NR
Benger
Lake Ballingall
Buffalo Beach
Belvidere Beach
Beela
Leschenault
Leschenault Estuary
Brunswick Junction
Worsley
Australind
Roelands
Koombana Bay
McKenna Point
Bunbury Lighthouse
Rocky Point
Eaton
Burekup
Waterloo
Wellington National Park
Allanson
Collie
Shotts
Buckingham
Collie Burn
Collie Cardiff
BUNBURY
Picton
Mount Lennard
Wellington Dam
Gelorup
Dalyellup Beach
Dardanup
Glen Mervyn Dam
Mumballup
Stratham
Stirling Beach
Henty Plains
Boyanup
Tuart Forest NP
Lowden
McAlinden
Minninup Sand Patch
Peppermint Grove
Higgins Cut
Capel
Donnybrook
Cape Naturaliste Lighthouse & Museum
Bunker Bay
Cape Naturaliste Lighthouse
Sugarloaf Rock
Rocky Point
Point Picquet
Point Dalling
Forrest Beach
Wonnerup Estuary
Ludlow
Wonnerup House
Busselton Jetty Interpretive Centre, Underwater Observatory, Ballarat Engine, Busselton Historic Museum, St Mary's Church
Bannamah Wildlife Park
Vasse Estuary
Mount Duckworth
Dunsborough
Dunn Bay
Quindalup
Toby Inlet
BUSSELTON
Ruabon
Tutunup
High Peak
Newlands
Wilga
Grimwade
Kirup
Mullalyup
Yallingup
Smiths Beach
Gunyulgup Gallery
Vasse
Busselton Aerodrome
Yoongarillup
Balingup
Old Cheese Factory Craft Centre
Marybrook
Carbunup River
Leeuwin-Naturaliste National Park
Cape Clairault
North Jindong
Boallia
Walsall
Acton Park
Jindong
Chapman Hill
Jarrahwood
Ferndale
Greenbushes
Yelverton
Bootleg Brewery
Metricup
Whicher Range
Cowan Dam
Nannup
Bridgetown
Cowaramup Bay
Cowaramup Point
Cowaramup
Gracetown
Osmington
Ellensbrook Homestead
Mowen
Margaret River Wineries, Caves Road Drive, Surfing
Margaret River
Cape Mentelle
Rosa Glen
Pioneer Settlers Memorial
Prevelly
Eagles Heritage
Margaret River Marron Farm
Marmaduke Point
Witchcliffe
Mount Mack
Lake Cave
Mammoth Cave NR
Mammoth Cave
Forest Grove
Palgarup
Cape Freycinet
Boranup Gallery
Graphite
One Tree Bridge
Manjimup
Deanmill
Scabby Gully Dam
Fonty's Pool
Jardee
North Point
Quoin Rock
Boranup Lookout
Boranup Maze
Hamelin Bay
Karridale
Bunnings Diamond Woodchip Mill, Diamond Tree Fire Lookout
Hamelin Island
Foul Bay
Foul Bay Lighthouse
Knobby Head
Cliff Spackman NR
Scott National Park
Hardy Inlet
Gingilup Swamps Nature Reserve
D'Entrecasteaux National Park
Founders Forest
Big Brook Dam
Beedelup National Park
Moondyne Cave, Jewel Cave
Augusta
Historical Museum, River Cruises
Barrack Point
Whale Rescue Memorial
Lake Quitjup
Lake Jasper
Big Brook Arboretum
Gloucester NP
Gloucester Tree
Pemberton
The Cascades
Cape Leeuwin
Matthew Flinders Memorial
Seal Island
Saint Alouarn Island
White Point
Black Point
King Trout & Marron Farm
Warren NP
Marianne North Tree & Bicentennial Tree
Brockman NP
Brockman Saw Pit
Cape Leeuwin Lighthouse, Water Wheel
Flinders Island
Yeagarup Beach
Northcliffe

Old Coast Rd
Coast Rd
Myalup Rd
South Western Hwy
Darling Range
Coalfields Rd
Williams
Collie
Bussell Hwy
Goodwood
Boyup Brook
Caves Rd
Vasse Hwy
Sues Rd
Balingup Nannup Rd
Brockman Hwy
Davidson Rd
Graphite Rd
Stewart Hwy
Muirs Hwy

TO WAGIN

TO BOYUP BROOK

Joins map 618

TO MOUNT BARKER

TO WALPOLE

Joins map 617

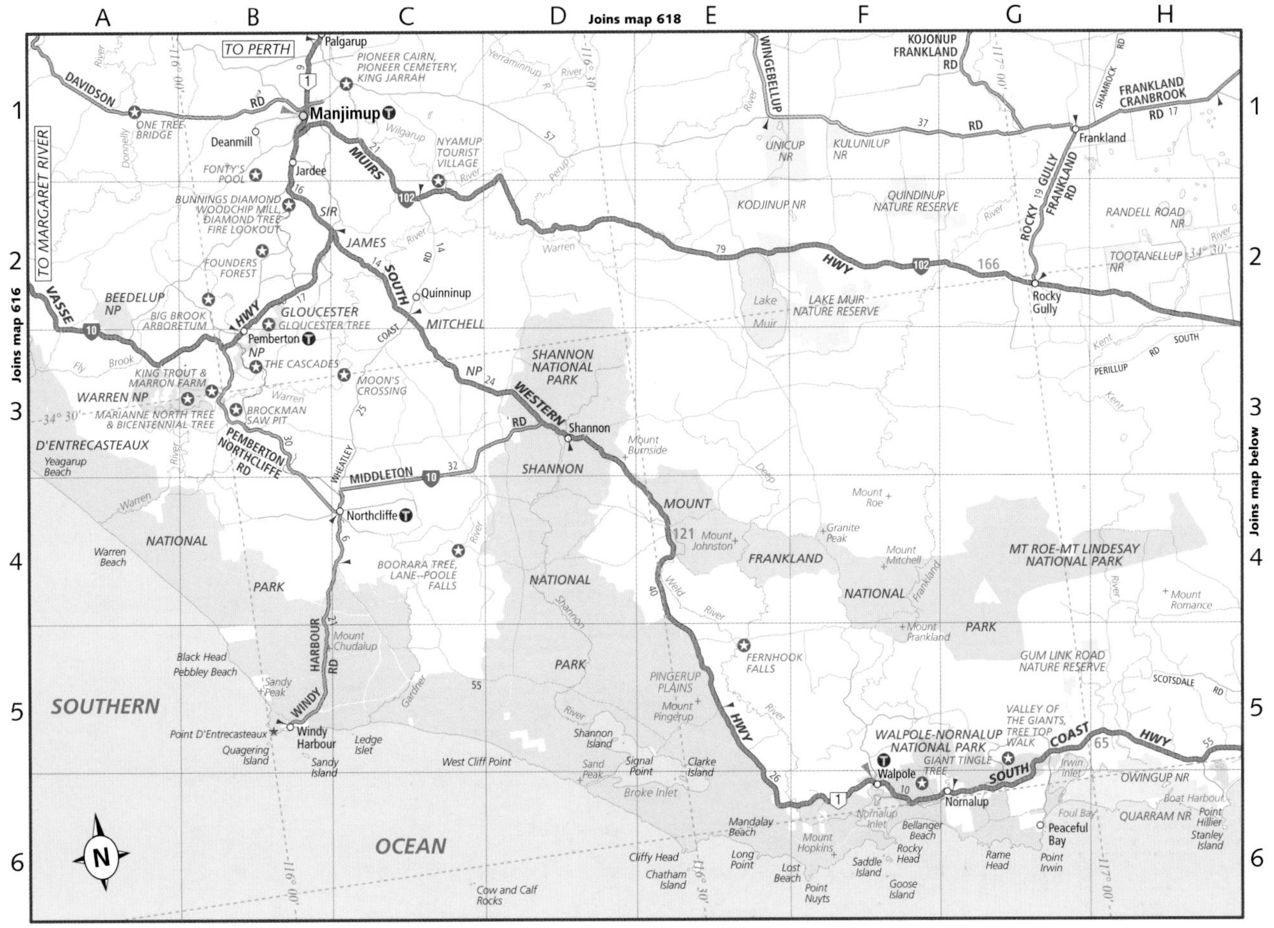

0 10 20 30 40 km

A B C D E F G H

Joins map 618

Joins map 616

Joins map below

TO PERTH
TO MARGARET RIVER
Palgarup
PIONEER CAIRN, PIONEER CEMETERY, KING JARRAH
DAVIDSON RD
ONE TREE BRIDGE
Manjimup
Deanmill
Jardee
FONTY'S POOL
NYAMUP TOURIST VILLAGE
MUIRS
BUNNINGS DIAMOND WOODCHIP MILL, DIAMOND TREE FIRE LOOKOUT
SIR JAMES SOUTH
FOUNDERS FOREST
Quinninup
BEEDELUP NP
BIG BROOK ARBORETUM
GLOUCESTER NP
GLOUCESTER TREE
Pemberton
VASSE HWY
MITCHELL
THE CASCADES
KING TROUT & MARRON FARM
MOON'S CROSSING
WARREN NP
MARIANNE NORTH TREE & BICENTENNIAL TREE
BROCKMAN SAW PIT
SHANNON NATIONAL PARK
WESTERN
Shannon
D'ENTRECASTEAUX
Yeagarup Beach
PEMBERTON NORTHCLIFFE RD
WHEATLEY
MIDDLETON RD
SHANNON
Mount Burnside
Northcliffe
Warren Beach
NATIONAL
PARK
BOORARA TREE, LANE-POOLE FALLS
NATIONAL
MOUNT
Mount Johnston
FRANKLAND
NATIONAL
PARK
Mount Roe
Granite Peak
Mount Mitchell
MT ROE-MT LINDESAY NATIONAL PARK
Mount Romance
Mount Frankland
Mount Chudalup
HARBOUR RD
WINDY
Black Head
Pebbley Beach
Sandy Peak
FERNHOOK FALLS
PINGERUP PLAINS
Mount Pingerup
GUM LINK ROAD NATURE RESERVE
SCOTSDALE RD
SOUTHERN
Point D'Entrecasteaux
Windy Harbour
Quagering Island
Sandy Island
Ledge Islet
West Cliff Point
Shannon Island
Sand Peak
Signal Point
Clarke Island
Broke Inlet
VALLEY OF THE GIANTS, TREE TOP WALK
WALPOLE-NORNALUP NATIONAL PARK
GIANT TINGLE TREE
SOUTH COAST HWY
Walpole
Nornalup
Nornalup Inlet
Irwin Inlet
OWINGUP NR
Boat Harbour
QUARRAM NR
Point Hillier
Stanley Island
OCEAN
Mandalay Beach
Cliffy Head
Long Point
Mount Hopkins
Bellanger Beach
Saddle Island
Rocky Head
Chatham Island
Lost Beach
Point Nuyts
Goose Island
Rame Head
Peaceful Bay
Foul Bay
Point Irwin
Cow and Calf Rocks
WINGEBELLUP RD
KOJONUP FRANKLAND RD
FRANKLAND CRANBROOK RD
UNICUP NR
KULUNILUP NR
Frankland
ROCKY GULLY FRANKLAND RD
KODJINUP NR
QUINDINUP NATURE RESERVE
RANDELL ROAD NR
TOOTANELLUP NR
HWY
Lake Muir
LAKE MUIR NATURE RESERVE
Rocky Gully
PERILLUP RD SOUTH

Joins map above

Joins map 618

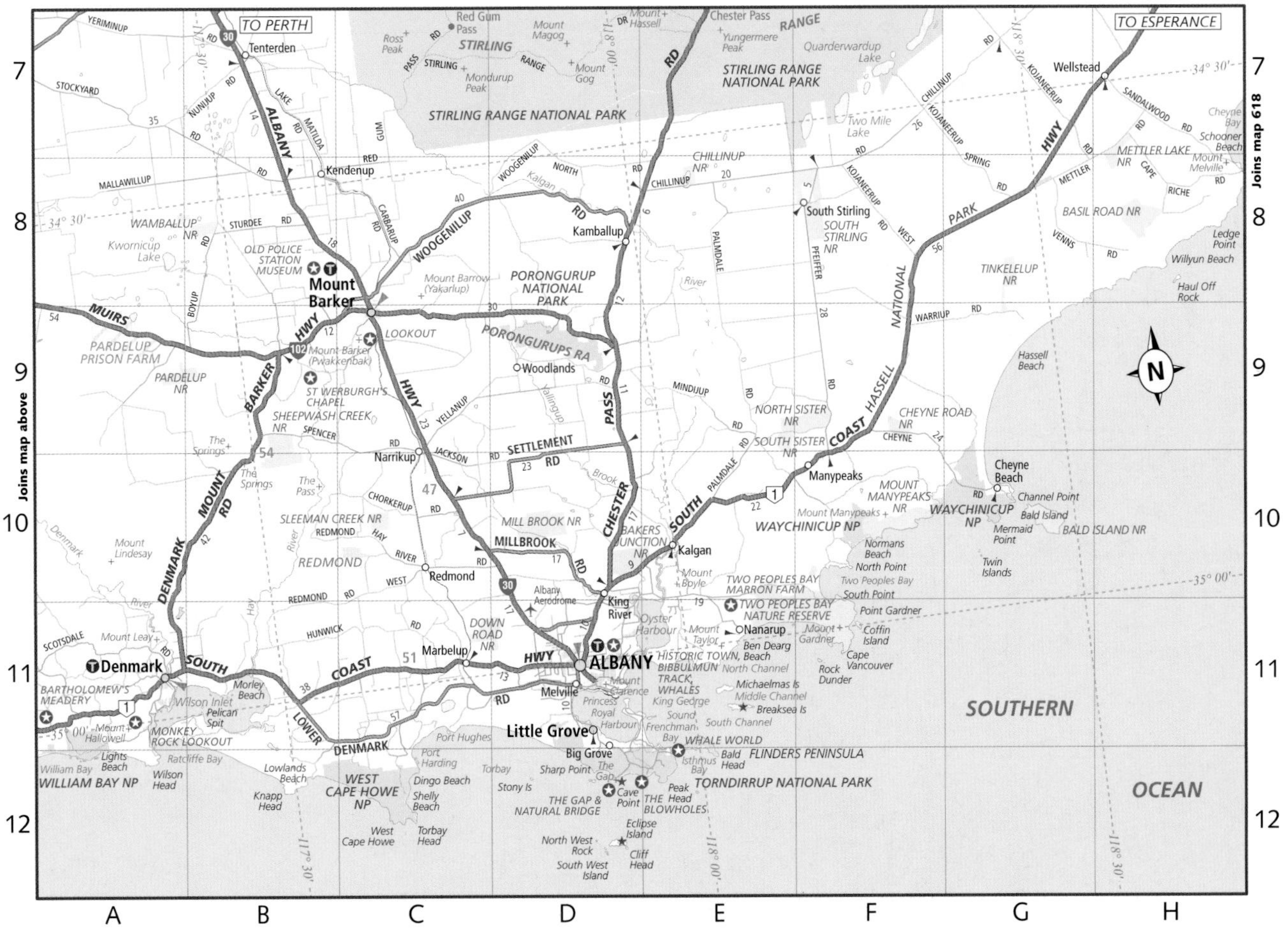

TO PERTH
TO ESPERANCE
YERIMINUP
Tenterden
Red Gum Pass
Ross Peak
Mount Magog
Mount Hassell
Chester Pass
Yungermere Peak
STIRLING RANGE
Mount Gog
Mondurup Peak
STIRLING RANGE NATIONAL PARK
Quarderwardup Lake
Wellstead
STOCKYARD
NUNIJUP
ALBANY
LAKE MATILDA
STIRLING RANGE NATIONAL PARK
Two Mile Lake
CHILLINUP
KOJANEERUP
SANDALWOOD
Cheyne Bay
Schooner Beach
MALLAWILLUP
Kendenup
WOOGENILUP
NORTH
CHILLINUP NR
SPRING
METTLER LAKE NR
Mount Melville
WAMBALLUP NR
STURDEE
Kworricup Lake
CARBARUP
Kamballup
South Stirling
SOUTH STIRLING NR
BASIL ROAD NR
RICHE
VENNS
Ledge Point
Willyun Beach
OLD POLICE STATION MUSEUM
Mount Barker
Mount Barrow (Yakarlup)
PORONGURUP NATIONAL PARK
PALMDALE
PEIFFER
NATIONAL PARK
TINKELELUP NR
Haul Off Rock
MUIRS
PARDELUP PRISON FARM
LOOKOUT
Mount Barker (Pwakkenbak)
PORONGURUPS RA
WARRIUP RD
Hassell Beach
PARDELUP NR
ST WERBURGH'S CHAPEL
Woodlands
YELLANUP
PASS
MINDIJUP
SHEEPWASH CREEK NR
SPENCER
NORTH SISTER NR
CHEYNE ROAD NR
The Springs
Narrikup
JACKSON
SETTLEMENT RD
SOUTH SISTER NR
COAST HASSELL
CHEYNE
Cheyne Beach
Manypeaks
MOUNT MANYPEAKS NR
Channel Point
Mount Manypeaks
WAYCHINICUP NP
Bald Island
BALD ISLAND NR
Mermaid Point
WAYCHINICUP NR
Denmark
Mount Lindesay
SLEEMAN CREEK NR
REDMOND
CHORKERUP
MILL BROOK NR
MILLBROOK RD
BAKERS JUNCTION NR
Kalgan
REDMOND
HAY RIVER
WEST
Redmond
CHESTER
Mount Boyle
TWO PEOPLES BAY MARRON FARM
Normans Beach
North Point
Two Peoples Bay
South Point
Point Gardner
Twin Islands
Albany Aerodrome
King River
Oyster Harbour
TWO PEOPLES BAY NATURE RESERVE
Mount Taylor
Nanarup
Mount Gardner
Coffin Island
SCOTSDALE
Mount Lea
DENMARK
HUNWICK
DOWN ROAD NP
Marbelup
ALBANY
Ben Dearg Beach
Cape Vancouver
Rock Dunder
BARTHOLOMEW'S MEADERY
SOUTH COAST
HWY
HISTORIC TOWN, BIBBULMUN TRACK, WHALES
North Channel
Michaelmas Is
Middle Channel
Breaksea Is
SOUTHERN
Mount Hallowell
Wilson Inlet
Pelican Spit
Morley Beach
Melville
Mount Clarence
Princess Royal Harbour
King George Sound
MONKEY ROCK LOOKOUT
LOWER DENMARK RD
Port Hughes
Little Grove
Frenchman Bay
South Channel
WHALE WORLD
Lights Beach
Ratcliffe Bay
Port Harding
Big Grove
Bald Head
FLINDERS PENINSULA
William Bay
WILLIAM BAY NP
Wilson Head
Lowlands Beach
WEST CAPE HOWE NP
Torbay
Dingo Beach
Sharp Point
The Gap
Isthmus Bay
TORNDIRRUP NATIONAL PARK
Knapp Head
Shelly Beach
Stony Is
THE GAP & NATURAL BRIDGE
Cave Point
Peak Head
THE BLOWHOLES
OCEAN
West Cape Howe
Torbay Head
Eclipse Island
North West Rock
South West Island
Cliff Head

A B C D E F G H

Joins map 620

1 Halfway Mill Roadhouse, ALEXANDER MORRISON NP, LESUEUR NP, WATHEROO NP, Watheroo, Dalwallinu, Pithara, Kalannie, Lake Moore, Lake Hillman, Lake Harvey, Lake O'Grady, Beacon, Wialki, Bonnie Rock, WALYAHMONING NATURE RESERVE, Hamersley Lakes, Lake Deborah West, Badgingarra, BADGINGARRA NP, Coomberdale, Miling, Ballidu, HWY

2 Cervantes, NAMBUNG NP, THE PINNACLES, BRAND, Moora, Dandaragan, Bindi Bindi, Cadoux, Koorda, Bencubbin, Mount Marshall, Mukinbudin, Lake Baladjie, Bullfinch, Warralakin, Manmanning, Lake Wallambin, Lake Brown, Piawaning, Wongan Hills, Mount Grey, WANAGARREN NR, Cataby Roadhouse, Yerecoin, Ejanding, Trayning, Kununoppin, Nungarin

3 Gillingarra, New Norcia, HISTORIC TOWN, Windmill Roadhouse, MOGUMBER, Mogumber, Regans Ford, MOORE RIVER NP, NORTHERN, Calingiri, Konnongorring, Minnivale, Wyalkatchem, Dowerin, Mount Moore, Westonia, Walgoolan, Burracoppin, Carrabin, Bodallin, Lancelin, Ledge Point, Wyening, Bolgart, Goomalling, Merredin, Nangeenan, Hines Hill, Mount Mackintosh, Seabird, Gingin, Bindoon, Dewars Pool, Baandee, Ulva, Doodlakine, Guilderton, Wilbinga Peak, Chittering, Bungulla, Tammin, Kellerberrin, Korbel

4 Two Rocks, YANCHEP NP, Yanchep, AVON VALLEY NP, Toodyay, Meckering, Waeel, Cunderdin, EASTERN, Belka, Muntadgin, Mount Camphorne, Muchea, Lower Chittering, Northam, Quellington, Mount Caroline, KOKERBIN ROCK, Jura, Bruce Rock, Quinns Rocks, Bullsbrook, Ringa, Clackline, Youndegin, Eujinyn, Joondalup, Upper Swan, Bakers Hill, York, HISTORIC TOWN, Kwolyin, Shackleton, Mount Walker, Wanneroo, Gidgegannup, Wooroloo, Greenhills, Belmunging, Yoting, Pantapin, Ardath, Narembeen, Hillarys, Midland, Mawson, Badjaling, Quairading, Mount Bebb, Babakin, Mount Arrowsmith

5 PERTH ZOO, INDIAN PACIFIC TRAIN, SWAN VALLEY WINERIES, BIBBULMUN TRACK, WILDFLOWERS, Scarborough, PERTH, Mundaring, Mount Talbot, Jacobs Well, Dangin, South Kumminin, Claremont, Helena River Reservoir, Mount Stirling, ROTTNEST ISLAND, Rottnest Island, Cape Vlamingh, Fremantle, HISTORIC TOWN, Canning Vale, AVONDALE DISCOVERY FARM, Beverley, YENYENING LAKES NR, Bilbarin, Nornakin, NORTH KARLGARIN NR, WAVE ROCK, Hyden, Garden Island, Jandakot, Armadale, BROOKTON, COUNTRY PEAK LOOKOUT, Kunjin, Corrigin, GORGE ROCK, Kwinana, Byford, Brookton, Nalya, Aldersyde, Kweda, Bulyee, Notting, Kondinin, Karlgarin, ROCKINGHAM, Westdale, Bullaring, Mount Mallet

6 For more detail on Perth & Surrounds see page 615

Waikiki, Serpentine, Jarrahdale, Mount Cuthbert, Mount Cooke, BOYAGIN ROCK, BOYAGIN NR, Pingelly, Gnarming, JILAKIN ROCK, Singleton, Keysbrook, Serpentine Dam, WANDERING, Yealering, Pingaring, Madora, North Dandalup, DARLING RA, ALBANY, Popanyinning, Malyalling, MALYALLING ROCK, Kulin, MANDURAH, Furnissdale, North Pinjarra, Wandering, Yornaning, Wickepin, Jitarning, Florida, Peel Inlet, Pinjarra, Bannister, DRYANDRA WOODLAND, Cuballing, Dudinin, Dwellingup, Crossman, Harrismith

7 YALGORUP, Coolup, Curara, Boddington, Minnigin, Ockley, Toolibin, Nanga, Marradong, Narrogin, Boundain, YILLIMINNING ROCK, Tincurrin, Lake Grace, INDIAN, Lake Clifton, Waroona, Hamel, Mount Keats, Geeralying, Dumberning, Highbury, Preston Beach, Wagerup, Williams, Lake Grace North, NATIONAL, Yarloop, Josbury, Kukerin, Quindanning, Tarwonga, Moulyinning, Warawarrup, Culbin, Piesseville, Gundaring, Nippering, Wishbone, Lake Grace South, PARK, Harvey, SOUTHERN

8 Myalup, Wokalup, Lake Ballingall, Dardadine, Arthur River, Wagin, Ballaying, Dumbleyung, CHINOCUP NR, Chinocup Lake, Kuringup, Binningup, Benger, Brunswick Junction, Hillman, Warup, HISTORIC VILLAGE, Pingrup, Boolading, Darkan, COBLININE NATURE RESERVE, Nyabing, OCEAN, Australind, Allanson, Buckingham, BUNBURY, Burekup, Waterloo, Collie, Bowelling, Dalyellup, Collie Cardiff, Cordering, Duranillin, Woodanilling, Coyrecup, Dardanup, For more detail on the South-West see page 616, Stratham, Boyanup, Mumballup, McAlinden, Boscabel, Katanning, Geographe, Lowden

9 Peppermint Grove, Bay, Ludlow, Capel, Donnybrook, Holly, Broomehill, Cape Naturaliste, Newlands, HISTORIC SPRING, Dunsborough, BUSSELTON, Kirup, Wilga, Kulikup, Muradup, Kojonup, Gnowangerup, Ongerup, Yallingup, Vasse, Yoongarillup, Mullalyup, Balingup, Boyup Brook, Dinninup, Jarrahwood, Greenbushes, Chapman Hill, Jingalup, Tambellup, Borden, CORACKERUP NR, LEEUWIN-NATURALISTE, Cowaramup, BUSSELL, Mayanup, Bridgetown, Osmington, Gracetown, Mowen, Nannup

10 Margaret River, Rosa Glen, HWY, Mount Mack, Palgarup, Cranbrook, STIRLING, RANGE, MARGARET RIVER WINERIES, CAVES ROAD DRIVE, SURFING, Prevelly, Witchcliffe, Forest Grove, BROCKMAN, Manjimup, NYAMUP TOURIST VILLAGE, Deanmill, Jardee, Tenterden, NATIONAL, PARK, Wellstead, Hamelin Bay, Karridale, SCOTT NP, SOUTH, MUIRS, Frankland, Kendenup, NATIONAL, PARK, Augusta, BEEDELUP NP, Quinninup, Mount Barker, Kamballup, South Stirling, COAST

11 Flinders Bay, Cape Leeuwin, CAPE LEEUWIN LIGHTHOUSE, D'ENTRECASTEAUX, Pemberton, WARREN NP, Lake Muir, LAKE MUIR NR, Rocky Gully, Shannon, Woodlands, NATIONAL, Northcliffe, MOUNT FRANKLAND, FERNHOOK FALLS, MT ROE-MT LINDESAY NP, Mount Lindesay, Narrikup, Cheyne Beach, Manypeaks, Bald Island, PARK, SHANNON NP, Redmond, King River, Kalgan, Two Peoples Bay, Point D'Entrecasteaux, WESTERN, VALLEY OF THE GIANTS, TREE TOP WALK, Denmark, Marbelup, Nanarup, ALBANY, HISTORIC TOWN, BIBBULMUN TRACK, WHALES, Windy Harbour, Broke Inlet, Walpole, Little Grove, Big Grove, Nornalup, Peaceful Bay, William Bay, Point Hillier, Torbay Bay, Isthmus Bay, Cliffy Head, WALPOLE-NORNALUP NP, West Cape Howe, Torbay Head, THE BLOWHOLES, Eclipse Island, Point Nuyts

12 SOUTHERN, OCEAN

For more detail on the South Coast see page 617

A B C D E F G H

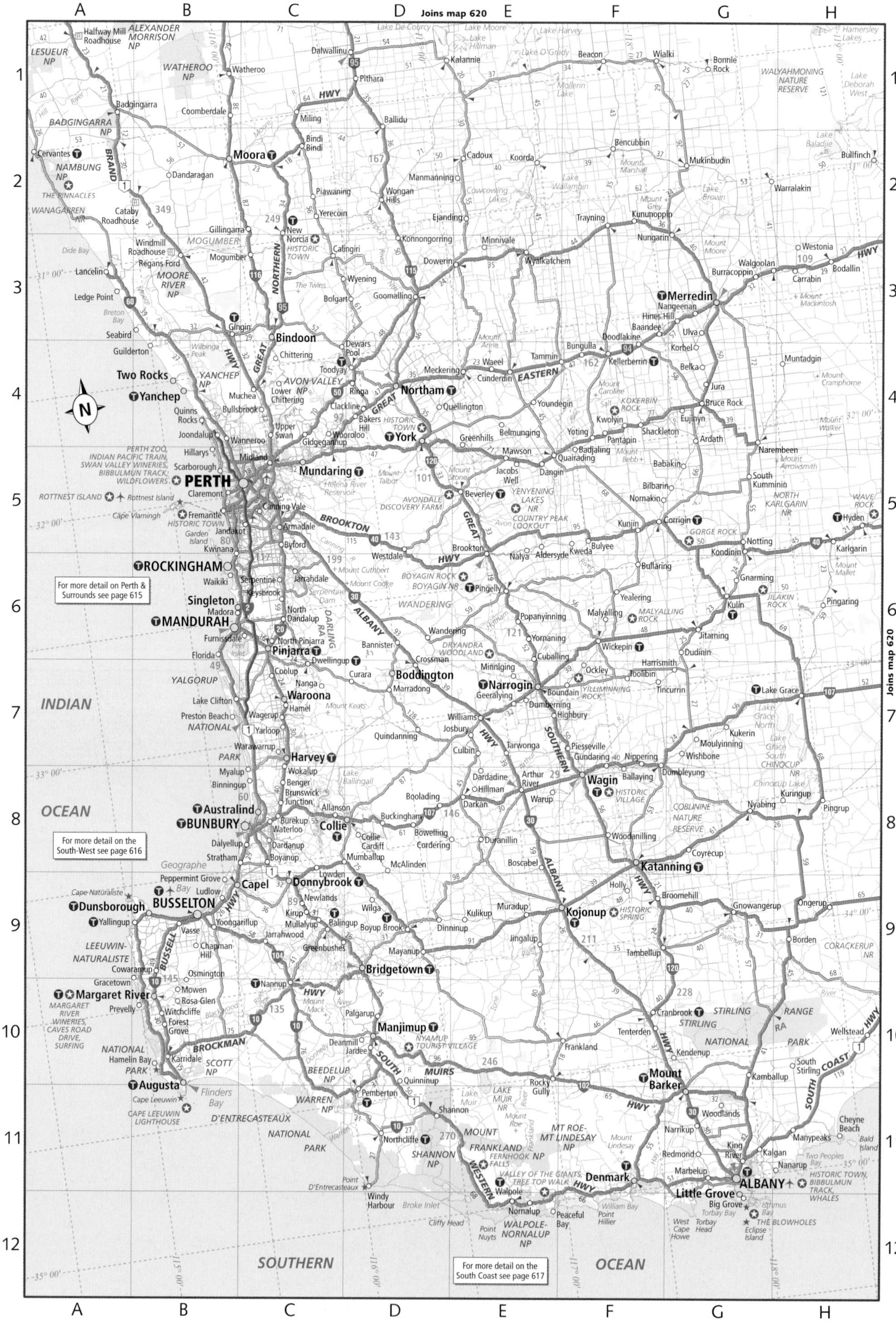

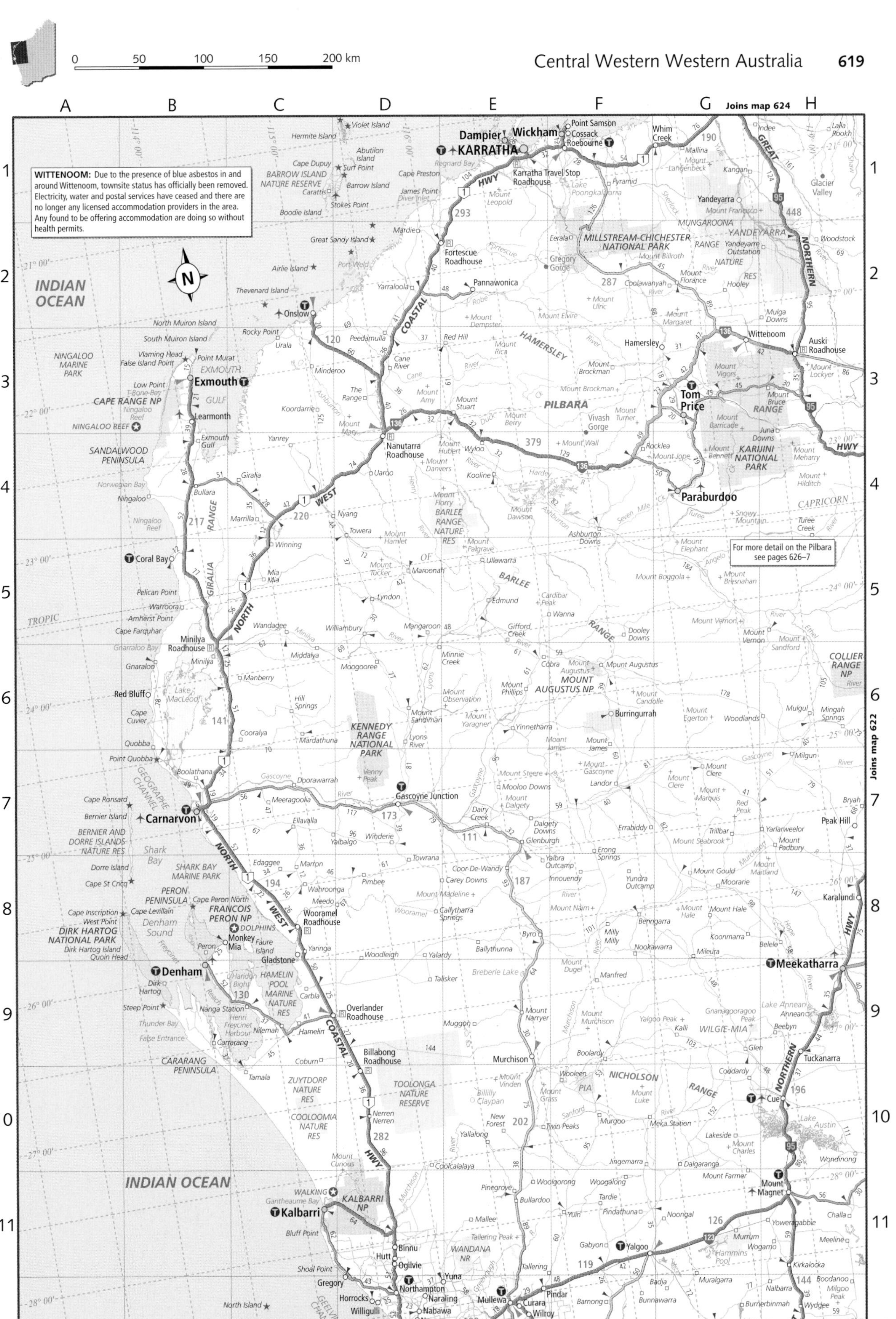
0 50 100 150 200 km
A B C D E F G H
1 2 3 4 5 6 7 8 9 10 11 12
Joins map 624
Joins map 620
Joins map 622
WITTENOOM: Due to the presence of blue asbestos in and around Wittenoom, townsite status has officially been removed. Electricity, water and postal services have ceased and there are no longer any licensed accommodation providers in the area. Any found to be offering accommodation are doing so without health permits.
For more detail on the Pilbara see pages 626–7
INDIAN OCEAN
N
Dampier
Wickham
KARRATHA
Point Samson
Cossack
Roebourne
Karratha Travel Stop Roadhouse
Whim Creek
Hermite Island
Violet Island
Abutilon Island
Cape Dupuy
Surf Point
BARROW ISLAND NATURE RESERVE
Barrow Island
Cape Preston
James Point
Stokes Point
Boodie Island
Mardie
Great Sandy Island
Fortescue Roadhouse
Airlie Island
Pannawonica
Thevenard Island
Onslow
North Muiron Island
South Muiron Island
Rocky Point
NINGALOO MARINE PARK
Vlaming Head
Point Murat
False Island Point
Exmouth
EXMOUTH GULF
Low Point
CAPE RANGE NP
Learmonth
NINGALOO REEF
SANDALWOOD PENINSULA
Ningaloo
Bullara
Giralia
Coral Bay
Pelican Point
Warroora
Amherst Point
Cape Farquhar
TROPIC OF CAPRICORN
Minilya Roadhouse
Gnaraloo
Red Bluff
Cape Cuvier
Quobba
Point Quobba
Cape Ronsard
Bernier Island
Carnarvon
BERNIER AND DORRE ISLANDS NATURE RES
Dorre Island
Shark Bay
SHARK BAY MARINE PARK
Cape St Cricq
PERON PENINSULA
Cape Peron North
FRANCOIS PERON NP
DIRK HARTOG NATIONAL PARK
Cape Inscription
Denham Sound
Monkey Mia
Denham
Steep Point
HAMELIN POOL MARINE NATURE RES
Nanga Station
Carrarang
CARARANG PENINSULA
Tamala
ZUYTDORP NATURE RES
COOLOOMIA NATURE RES
Overlander Roadhouse
Billabong Roadhouse
TOOLONGA NATURE RESERVE
Nerren Nerren
KALBARRI NP
Kalbarri
Bluff Point
Binnu
Hutt
Ogilvie
Shoal Point
Gregory
Horrocks
Northampton
Yuna
Naraling
Nabawa
Nanson
Coronation Beach
Drummond Cove
GERALDTON
Cape Burney
Walkaway
Mount Pleasant
North Island
WALLABI GROUP
HOUTMAN ABROLHOS
EASTER GROUP
WANDANA NR
Mullewa
Pindar
Tardun
Canna
Wilroy
Tallering
Gabyon
Yalgoo
Barnong
Mellenbye
Nanutarra Roadhouse
Wooramel Roadhouse
Gascoyne Junction
Minnie Creek
Mangaroon
KENNEDY RANGE NATIONAL PARK
BARLEE RANGE NATURE RES
Mount Augustus
MOUNT AUGUSTUS NP
Burringurrah
Cobra
Dalgety Downs
Glenburgh
Byro
Murchison
MILLSTREAM-CHICHESTER NATIONAL PARK
Hamersley
Wittenoom
Auski Roadhouse
Tom Price
Paraburdoo
KARIJINI NATIONAL PARK
HAMERSLEY RANGE
PILBARA
Yandeyarra
Mount Vernon
COLLIER RANGE NP
Meekatharra
Cue
Mount Magnet
Lake Austin
WILGIE-MIA
NICHOLSON RANGE
Paynes Find
Karalundi
Peak Hill
NORTH WEST COASTAL HWY
GREAT NORTHERN HWY

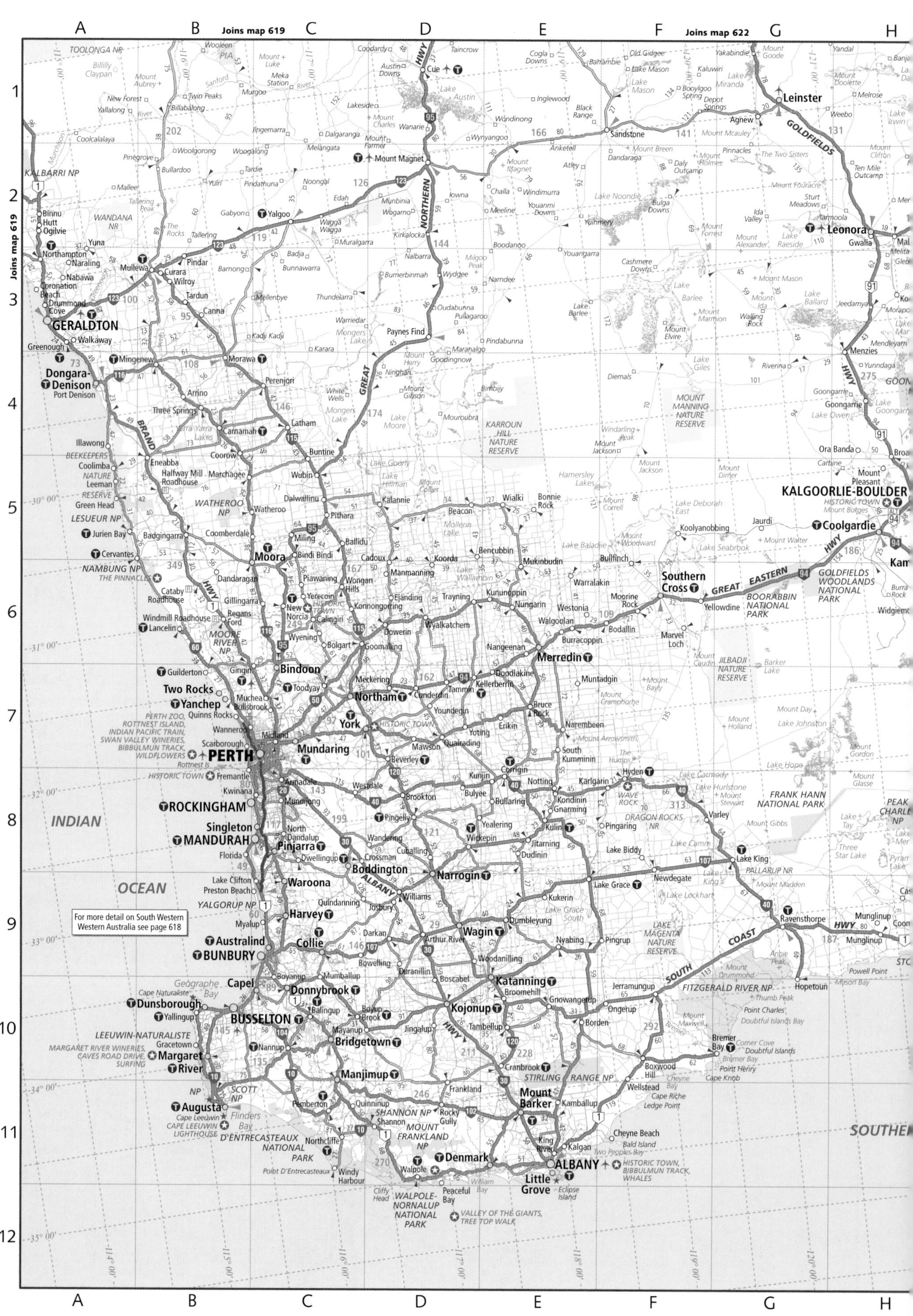

Joins map 619
Joins map 622
KALBARRI NP
WANDANA NR
GERALDTON
Northampton
Dongara-Denison
Port Denison
Mullewa
Mingenew
Morawa
Perenjori
Three Springs
Carnamah
Coorow
Eneabba
BRAND
BEEKEEPERS NATURE RESERVE
LESUEUR NP
Jurien Bay
Cervantes
NAMBUNG NP
THE PINNACLES
WATHEROO NP
Moora
New Norcia
Dalwallinu
Wubin
Latham
Yalgoo
Mount Magnet
Cue
Paynes Find
GREAT NORTHERN HWY
Sandstone
Leinster
GOLDFIELDS
Leonora
Menzies
KARROUN HILL NATURE RESERVE
MOUNT MANNING NATURE RESERVE
KALGOORLIE-BOULDER
HISTORIC TOWN
Coolgardie
GREAT EASTERN HWY
Southern Cross
BOORABBIN NATIONAL PARK
GOLDFIELDS WOODLANDS NATIONAL PARK
Merredin
Kellerberrin
Northam
York
Toodyay
Bindoon
Two Rocks
Yanchep
PERTH ZOO, ROTTNEST ISLAND, INDIAN PACIFIC TRAIN, SWAN VALLEY WINERIES, BIBBULMUN TRACK, WILDFLOWERS
PERTH
Fremantle
Mundaring
ROCKINGHAM
MANDURAH
Pinjarra
INDIAN OCEAN
YALGORUP NP
Waroona
Harvey
Australind
BUNBURY
Collie
Narrogin
Wagin
Katanning
Kojonup
Boddington
Williams
Corrigin
Kondinin
Hyden
WAVE ROCK
DRAGON ROCKS NR
Lake King
Ravensthorpe
FRANK HANN NATIONAL PARK
LAKE MAGENTA NATURE RESERVE
SOUTH COAST HWY
FITZGERALD RIVER NP
Hopetoun
Bremer Bay
For more detail on South Western Western Australia see page 618
Dunsborough
BUSSELTON
Capel
Donnybrook
Bridgetown
Nannup
LEEUWIN-NATURALISTE
MARGARET RIVER WINERIES, CAVES ROAD DRIVE, SURFING
Margaret River
Augusta
CAPE LEEUWIN LIGHTHOUSE
D'ENTRECASTEAUX NATIONAL PARK
Pemberton
Manjimup
SHANNON NP
MOUNT FRANKLAND NP
Northcliffe
Walpole
WALPOLE-NORNALUP NATIONAL PARK
VALLEY OF THE GIANTS, TREE TOP WALK
Denmark
Mount Barker
STIRLING RANGE NP
ALBANY
HISTORIC TOWN, BIBBULMUN TRACK, WHALES
Little Grove
Cranbrook
Frankland

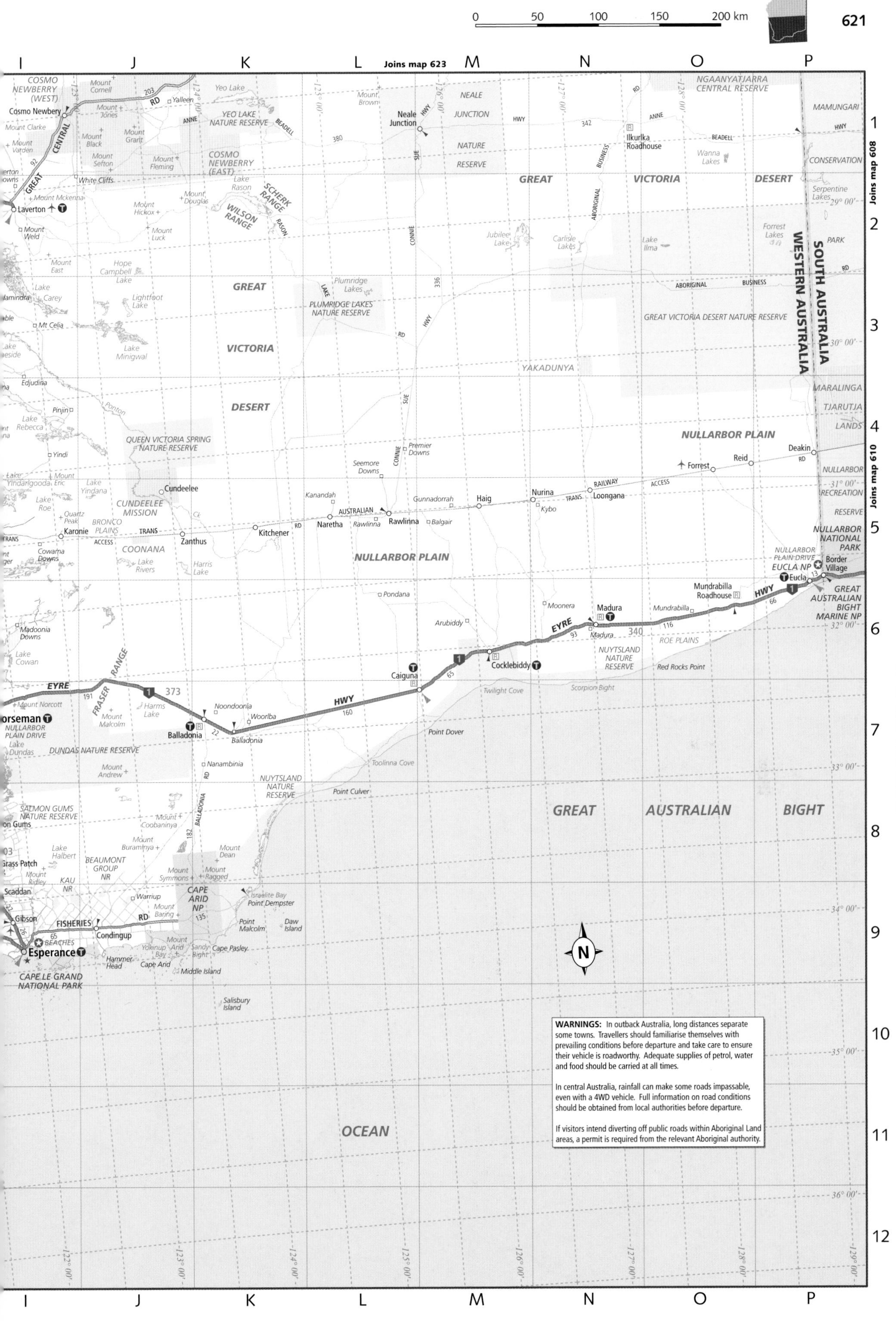
0 50 100 150 200 km
Joins map 623
Joins map 608
Joins map 610
I J K L M N O P
1 2 3 4 5 6 7 8 9 10 11 12
COSMO NEWBERRY (WEST)
Cosmo Newbery
Mount Cornell
Mount Jones
Mount Clarke
Mount Varden
Mount Black
Mount Grant
Mount Sefton
Mount Fleming
White Cliffs
Yeo Lake
YEO LAKE NATURE RESERVE
COSMO NEWBERRY (EAST)
Yelleen
GREAT CENTRAL RD
ANNE BEADELL
Lake Rason
SCHERK RANGE
WILSON RANGE
Mount Douglas
Mount Hickox
Mount Luck
Laverton
Mount Mckenna
Mount Weld
Mount East
Hope Campbell Lake
Lake Carey
Lightfoot Lake
Mt Celia
Lake Minigwal
Edjudina
Pinjin
Lake Rebecca
Yindi
Mount Eric
Lake Yindana
Lake Roe
Quartz Peak
Cundeelee
CUNDEELEE MISSION
BRONCO PLAINS
Karonie
Zanthus
COONANA
Cowarna Downs
Lake Rivers
Harris Lake
Madoonia Downs
Lake Cowan
GREAT VICTORIA DESERT
QUEEN VICTORIA SPRING NATURE RESERVE
Ponton
Mount Brown
Neale Junction
NEALE JUNCTION NATURE RESERVE
CONNIE SUE HWY
380
336
342
Plumridge Lakes
PLUMRIDGE LAKES NATURE RESERVE
Jubilee Lake
Carlisle Lakes
ABORIGINAL BUSINESS RD
Ilkurlka Roadhouse
Wanna Lakes
NGAANYATJARRA CENTRAL RESERVE
MAMUNGARI CONSERVATION PARK
Serpentine Lakes
Forrest Lakes
Lake Ilma
GREAT VICTORIA DESERT NATURE RESERVE
WESTERN AUSTRALIA
SOUTH AUSTRALIA
YAKADUNYA
MARALINGA TJARUTJA LANDS
NULLARBOR PLAIN
Premier Downs
Seemore Downs
Deakin
Reid
Forrest
NULLARBOR RECREATION RESERVE
Kanandah
AUSTRALIAN TRANS ACCESS RAILWAY
Gunnadorrah
Haig
Nurina
Loongana
Kybo
Naretha
Rawlinna
Balgair
Kitchener
NULLARBOR NATIONAL PARK
NULLARBOR PLAIN DRIVE
EUCLA NP
Border Village
Eucla
Pondana
Mundrabilla Roadhouse
Mundrabilla
Moonera
Madura
Arubiddy
ROE PLAINS
NUYTSLAND NATURE RESERVE
Red Rocks Point
GREAT AUSTRALIAN BIGHT MARINE NP
Cocklebiddy
Caiguna
EYRE HWY
Twilight Cove
Scorpion Bight
FRASER RANGE
Mount Norcott
Norseman
NULLARBOR PLAIN DRIVE
Harms Lake
Mount Malcolm
Noondoonia
Woorlba
Balladonia
Point Dover
Lake Dundas
DUNDAS NATURE RESERVE
Nanambinia
BALLADONIA RD
Mount Andrew
Toolinna Cove
Point Culver
SALMON GUMS NATURE RESERVE
Salmon Gums
Mount Coobaninya
Mount Buraminya
Lake Halbert
Grass Patch
BEAUMONT GROUP NR
Mount Ridley
KAU NR
Mount Dean
Mount Symmons
Mount Ragged
CAPE ARID NP
Israelite Bay
Point Dempster
Scaddan
Warrinup
Mount Baring
Gibson
FISHERIES RD
Condingup
Point Malcolm
Daw Island
BEACHES
Esperance
Mount Arid
Yokinup Bay
Sandy Bight
Cape Pasley
Hammer Head
Cape Arid
Middle Island
CAPE LE GRAND NATIONAL PARK
Salisbury Island
GREAT AUSTRALIAN BIGHT
OCEAN
N
373 191 22 160 65 93 340 116 66 13 182 135
WARNINGS: In outback Australia, long distances separate some towns. Travellers should familiarise themselves with prevailing conditions before departure and take care to ensure their vehicle is roadworthy. Adequate supplies of petrol, water and food should be carried at all times.
In central Australia, rainfall can make some roads impassable, even with a 4WD vehicle. Full information on road conditions should be obtained from local authorities before departure.
If visitors intend diverting off public roads within Aboriginal Land areas, a permit is required from the relevant Aboriginal authority.

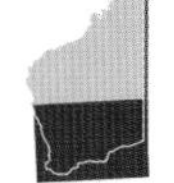

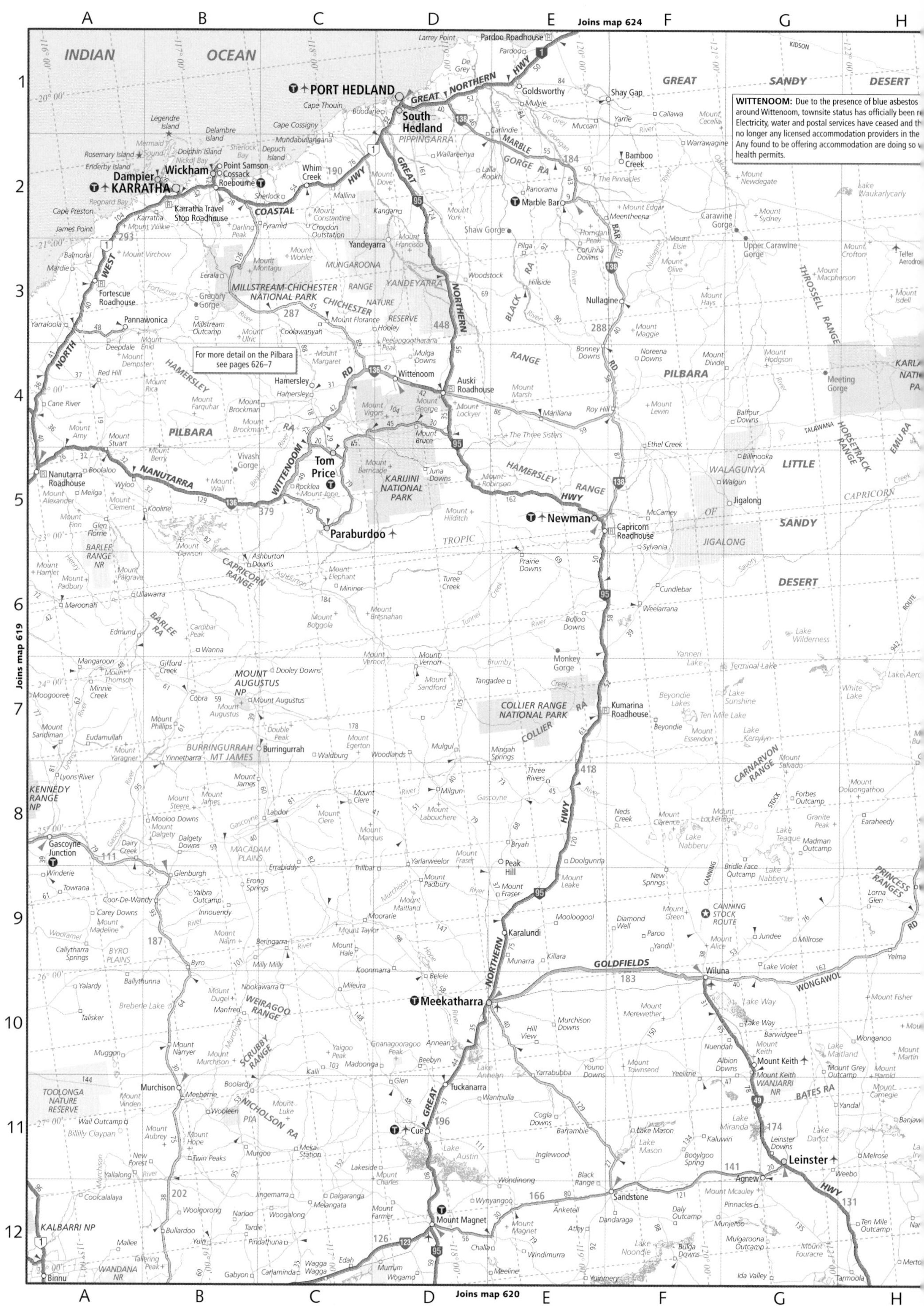

A B C D E F G H
1 2 3 4 5 6 7 8 9 10 11 12
Joins map 624
Joins map 619
Joins map 620
WITTENOOM: Due to the presence of blue asbestos around Wittenoom, townsite status has officially been re
Electricity, water and postal services have ceased and th
no longer any licensed accommodation providers in the
Any found to be offering accommodation are doing so v
health permits.
For more detail on the Pilbara see pages 626–7
INDIAN OCEAN
PORT HEDLAND
South Hedland
Dampier
KARRATHA
Wickham
Point Samson
Cossack
Roebourne
Karratha Travel Stop Roadhouse
Whim Creek
Pardoo Roadhouse
Goldsworthy
Shay Gap
Marble Bar
Nullagine
Fortescue Roadhouse
Pannawonica
Millstream-Chichester National Park
Wittenoom
Auski Roadhouse
Tom Price
Paraburdoo
Karijini National Park
Newman
Capricorn Roadhouse
Nanutarra Roadhouse
Jigalong
Mount Augustus NP
Burringurrah
Collier Range National Park
Kumarina Roadhouse
Gascoyne Junction
Glenburgh
Byro
Meekatharra
Karalundi
Wiluna
Cue
Tuckanarra
Mount Magnet
Sandstone
Leinster
Agnew
Mount Keith
Murchison
Kalbarri NP
Binnu
Kennedy Range NP
Toolonga Nature Reserve
Wanjarri NR
Barlee Range NR
Canning Stock Route
GREAT NORTHERN HWY
NORTH WEST COASTAL HWY
GOLDFIELDS
WONGAWOL
PILBARA
GREAT SANDY DESERT
LITTLE SANDY DESERT
HAMERSLEY RANGE
TROPIC OF CAPRICORN

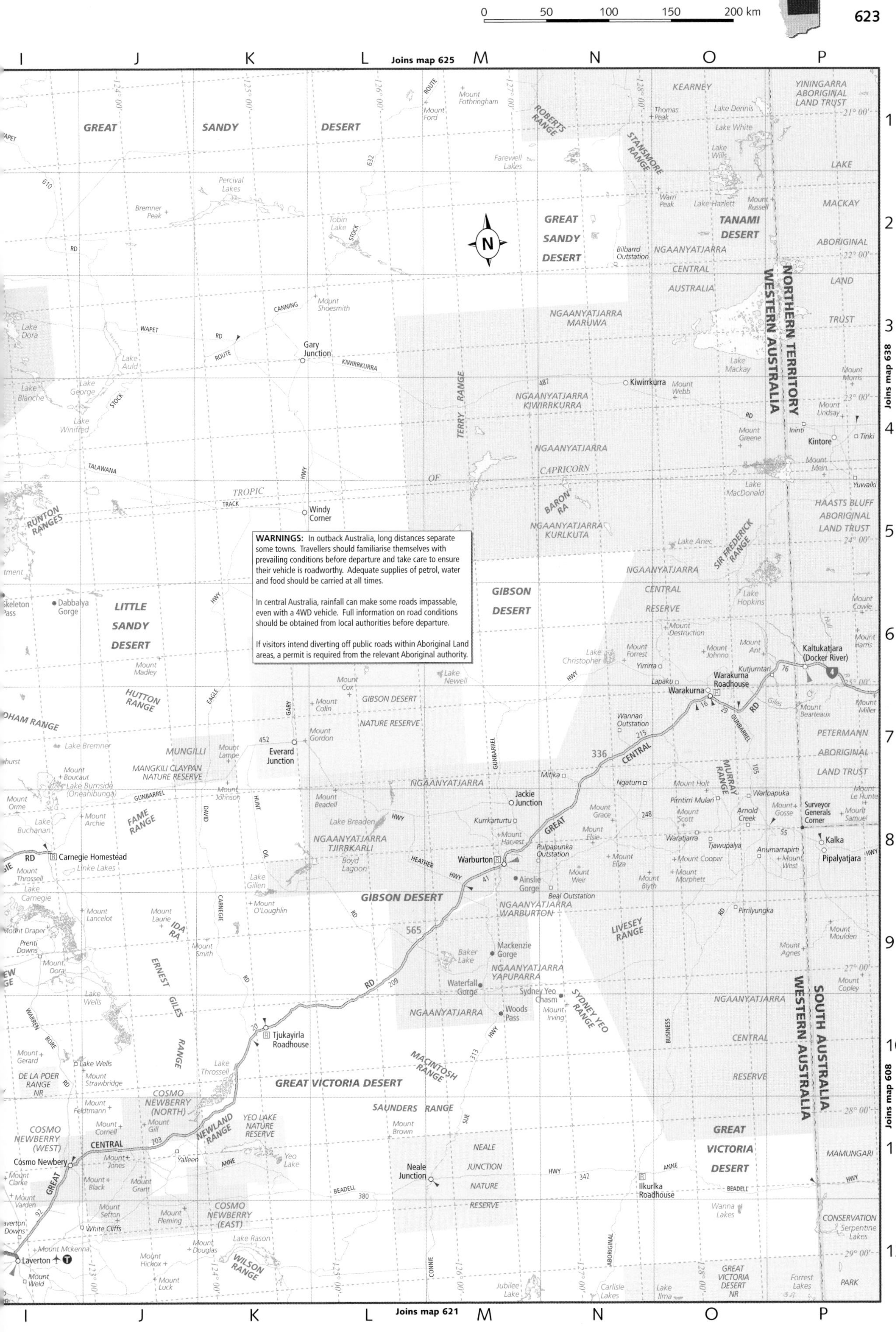

0 50 100 150 200 km

Joins map 625

I J K L M N O P

WARNINGS: In outback Australia, long distances separate some towns. Travellers should familiarise themselves with prevailing conditions before departure and take care to ensure their vehicle is roadworthy. Adequate supplies of petrol, water and food should be carried at all times.

In central Australia, rainfall can make some roads impassable, even with a 4WD vehicle. Full information on road conditions should be obtained from local authorities before departure.

If visitors intend diverting off public roads within Aboriginal Land areas, a permit is required from the relevant Aboriginal authority.

GREAT SANDY DESERT
LITTLE SANDY DESERT
GIBSON DESERT
GREAT VICTORIA DESERT
TANAMI DESERT
NORTHERN TERRITORY
WESTERN AUSTRALIA
SOUTH AUSTRALIA
YININGARRA ABORIGINAL LAND TRUST
LAKE MACKAY ABORIGINAL LAND TRUST
HAASTS BLUFF ABORIGINAL LAND TRUST
PETERMANN ABORIGINAL LAND TRUST
NGAANYATJARRA CENTRAL AUSTRALIA
NGAANYATJARRA MARUWA
NGAANYATJARRA KIWIRRKURRA
NGAANYATJARRA KURLKUTA
NGAANYATJARRA CENTRAL RESERVE
NGAANYATJARRA TJIRRKARLI
NGAANYATJARRA WARBURTON
NGAANYATJARRA YAPUPARRA
GIBSON DESERT NATURE RESERVE
MANGKILI CLAYPAN NATURE RESERVE
YEO LAKE NATURE RESERVE
NEALE JUNCTION NATURE RESERVE
COSMO NEWBERRY (NORTH)
COSMO NEWBERRY (WEST)
COSMO NEWBERRY (EAST)
DE LA POER RANGE NR
GREAT VICTORIA DESERT NR
MAMUNGARI CONSERVATION PARK
TROPIC OF CAPRICORN
Gary Junction
Windy Corner
Kiwirrkurra
Kintore
Kaltukatjara (Docker River)
Warakurna Roadhouse
Warakurna
Jackie Junction
Everard Junction
Warburton
Tjukayirla Roadhouse
Cosmo Newbery
Laverton
Neale Junction
Ilkurlka Roadhouse
Carnegie Homestead
Surveyor Generals Corner
Kalka
Pipalyatjara
Dabbalya Gorge
Mackenzie Gorge
Ainslie Gorge
Waterfall Gorge
Woods Pass
Sydney Yeo Chasm
Bilbarrd Outstation
Wannan Outstation
Pulpapunka Outstation
Beal Outstation
GREAT CENTRAL RD
GUNBARREL HWY
CANNING STOCK ROUTE
TALAWANA TRACK
HEATHER HWY
ANNE BEADELL HWY
WARREN BORE RD
Lake Dora
Lake Auld
Lake Blanche
Lake George
Lake Winifred
Lake Mackay
Lake MacDonald
Lake Hopkins
Lake Christopher
Lake Newell
Lake Breaden
Lake Gillen
Lake Throssell
Lake Wells
Lake Carnegie
Lake Rason
Yeo Lake
Baker Lake
Boyd Lagoon
Jubilee Lake
Carlisle Lakes
Lake Ilma
Forrest Lakes
Serpentine Lakes
Wanna Lakes
ROBERTS RANGE
STANSMORE RANGE
TERRY RANGE
BARON RA
SIR FREDERICK RANGE
HUTTON RANGE
RUNTON RANGES
FAME RANGE
IDA RA
ERNEST GILES RANGE
MURRAY RANGE
LIVESEY RANGE
SYDNEY YEO RANGE
MACINTOSH RANGE
SAUNDERS RANGE
NEWLAND RANGE
WILSON RANGE

Joins map 638

Joins map 608

I J K L M N O P

Joins map 621

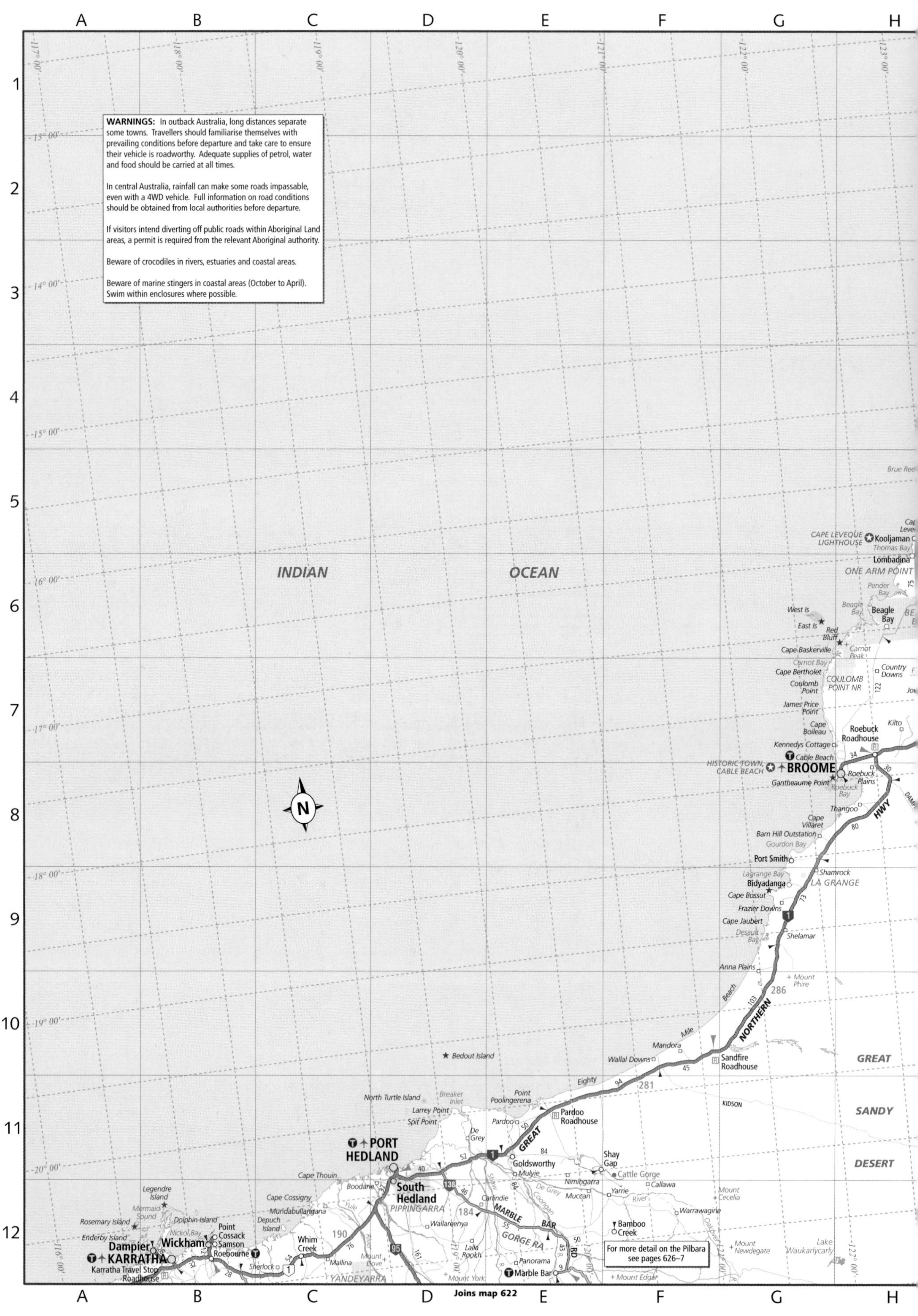
A
B
C
D
E
F
G
H
1
2
3
4
5
6
7
8
9
10
11
12
WARNINGS: In outback Australia, long distances separate some towns. Travellers should familiarise themselves with prevailing conditions before departure and take care to ensure their vehicle is roadworthy. Adequate supplies of petrol, water and food should be carried at all times.
In central Australia, rainfall can make some roads impassable, even with a 4WD vehicle. Full information on road conditions should be obtained from local authorities before departure.
If visitors intend diverting off public roads within Aboriginal Land areas, a permit is required from the relevant Aboriginal authority.
Beware of crocodiles in rivers, estuaries and coastal areas.
Beware of marine stingers in coastal areas (October to April). Swim within enclosures where possible.
117° 00'
118° 00'
119° 00'
120° 00'
121° 00'
122° 00'
123° 00'
13° 00'
14° 00'
15° 00'
16° 00'
17° 00'
18° 00'
19° 00'
20° 00'
INDIAN
OCEAN
N
Brue Reef
CAPE LEVEQUE LIGHTHOUSE
Kooljaman
Thomas Bay
Lombadina
ONE ARM POINT
Pender Bay
West Is
East Is
Red Bluff
Beagle Bay
Beagle Bay
Cape Baskerville
Carnot Peak
Carnot Bay
Cape Bertholet
Coulomb Point
COULOMB POINT NR
Country Downs
James Price Point
Cape Boileau
Roebuck Roadhouse
Kilto
Kennedys Cottage
Cable Beach
HISTORIC TOWN, CABLE BEACH
BROOME
Roebuck Plains
Gantheaume Point
Roebuck Bay
Thangoo
Cape Villaret
Barn Hill Outstation
Gourdon Bay
Port Smith
HWY
Shamrock
Lagrange Bay
Bidyadanga
LA GRANGE
Cape Bossut
Frazier Downs
Cape Jaubert
Desault Bay
Shelamar
Anna Plains
Mount Phire
286
Beach
NORTHERN
Mile
Mandora
Sandfire Roadhouse
Wallal Downs
Eighty
GREAT
SANDY
DESERT
281
KIDSON
Bedout Island
North Turtle Island
Breaker Inlet
Point Poolingerena
Larrey Point
Spit Point
Pardoo
Pardoo Roadhouse
De Grey
GREAT
PORT HEDLAND
Goldsworthy
Shay Gap
Mulyie
Cattle Gorge
Cape Thouin
Nimingarra
Callawa
Boodarie
South Hedland
PIPPINGARRA
Muccan
Yarrie
Mount Cecelia
Legendre Island
Cape Cossigny
Mundabullangana
Warrawagine
Mermaid Sound
Rosemary Island
Dolphin Island
Depuch Island
Wallareenya
184
MARBLE
BAR
Bamboo Creek
Enderby Island
Nickol Bay
Point Samson
Cossack
Whim Creek
190
GORGE RA
RD
Dampier
Wickham
Roebourne
Lalla Rookh
Mount Newdegate
Lake Waukarlycarly
KARRATHA
Karratha Travel Stop Roadhouse
Sherlock
Mallina
Mount Dove
Panorama
Marble Bar
For more detail on the Pilbara see pages 626–7
YANDEYARRA
Mount York
Mount Edgar
Joins map 622

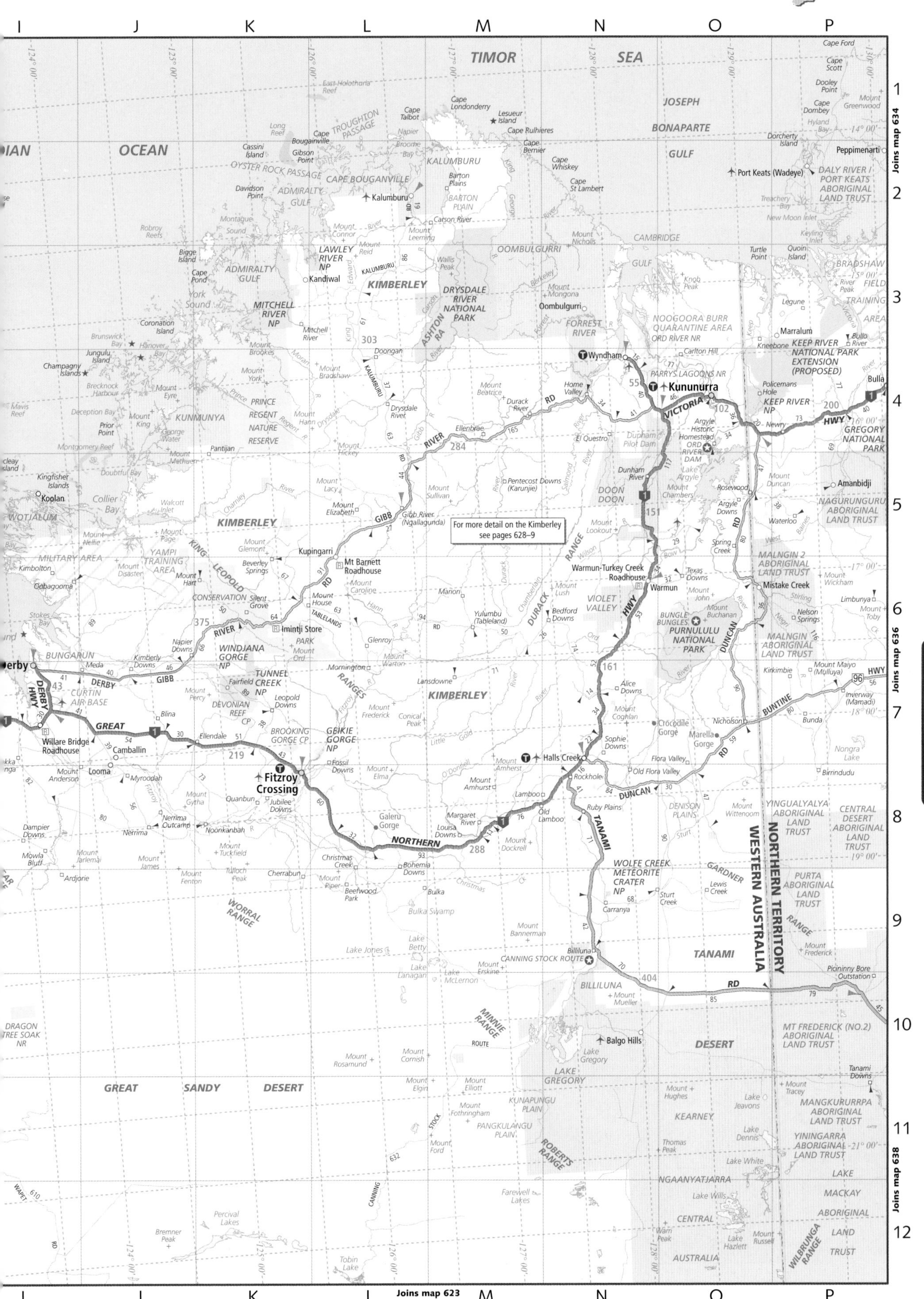

Joins map 634
Joins map 636
Joins map 638
Joins map 623

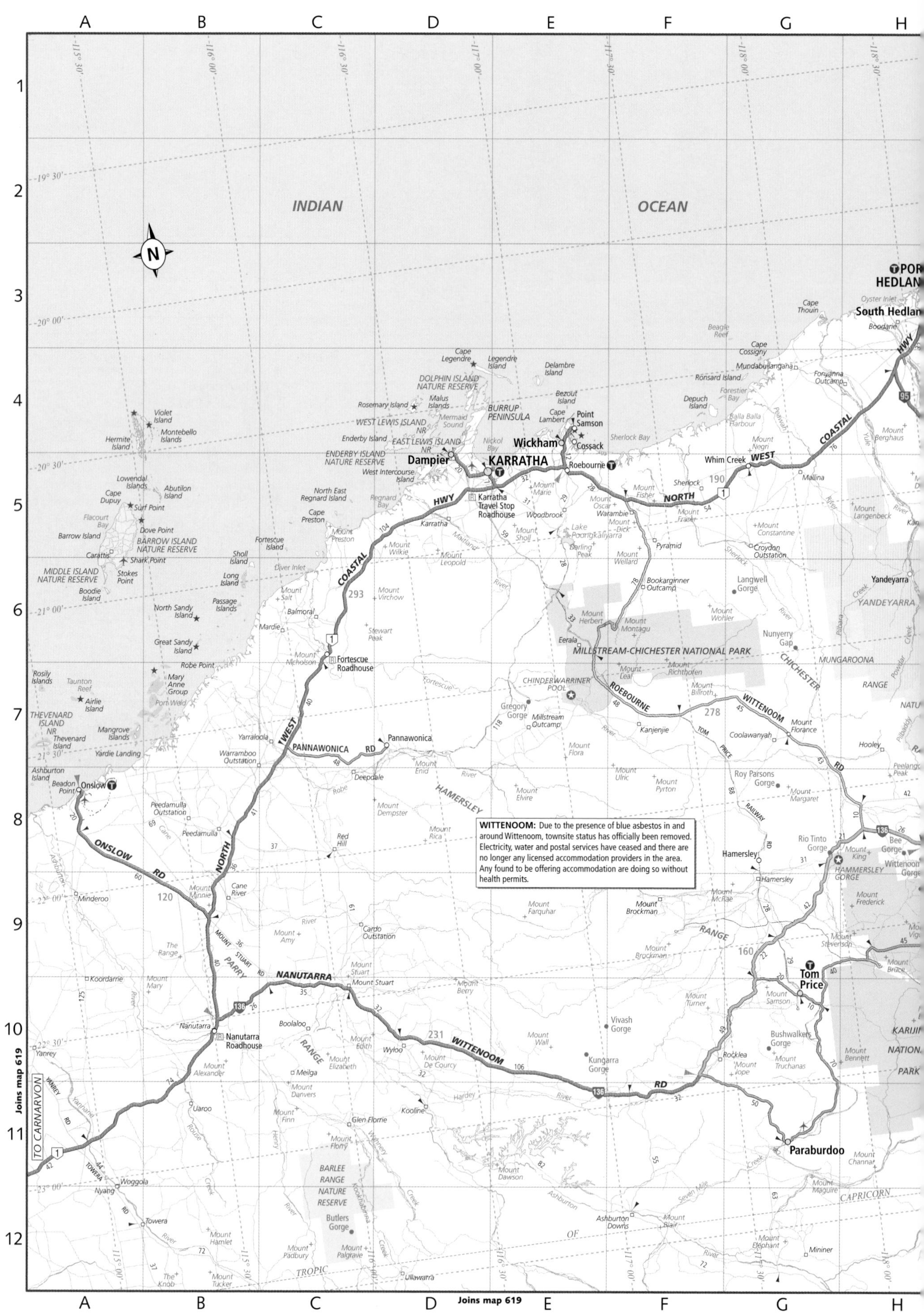
INDIAN
OCEAN
KARRATHA
Dampier
Wickham
Roebourne
Point Samson
Cossack
PORT HEDLAND
South Hedland
Whim Creek
Onslow
Tom Price
Paraburdoo
Pannawonica
Fortescue Roadhouse
Nanutarra Roadhouse
Karratha Travel Stop Roadhouse
BURRUP PENINSULA
DOLPHIN ISLAND NATURE RESERVE
WEST LEWIS ISLAND NR
EAST LEWIS ISLAND NR
ENDERBY ISLAND NATURE RESERVE
BARROW ISLAND NATURE RESERVE
MIDDLE ISLAND NATURE RESERVE
THEVENARD ISLAND NR
Montebello Islands
Barrow Island
MILLSTREAM-CHICHESTER NATIONAL PARK
CHINDERWARRINER POOL
HAMERSLEY RANGE
CHICHESTER RANGE
BARLEE RANGE NATURE RESERVE
KARIJINI NATIONAL PARK
YANDEYARRA
MUNGAROONA RANGE
NORTH WEST COASTAL HWY
ONSLOW RD
NANUTARRA WITTENOOM RD
ROEBOURNE WITTENOOM RD
PANNAWONICA RD
MOUNT STUART RD
TOM PRICE RAILWAY RD
TROPIC OF CAPRICORN
TO CARNARVON
Joins map 619
WITTENOOM: Due to the presence of blue asbestos in and around Wittenoom, townsite status has officially been removed. Electricity, water and postal services have ceased and there are no longer any licensed accommodation providers in the area. Any found to be offering accommodation are doing so without health permits.

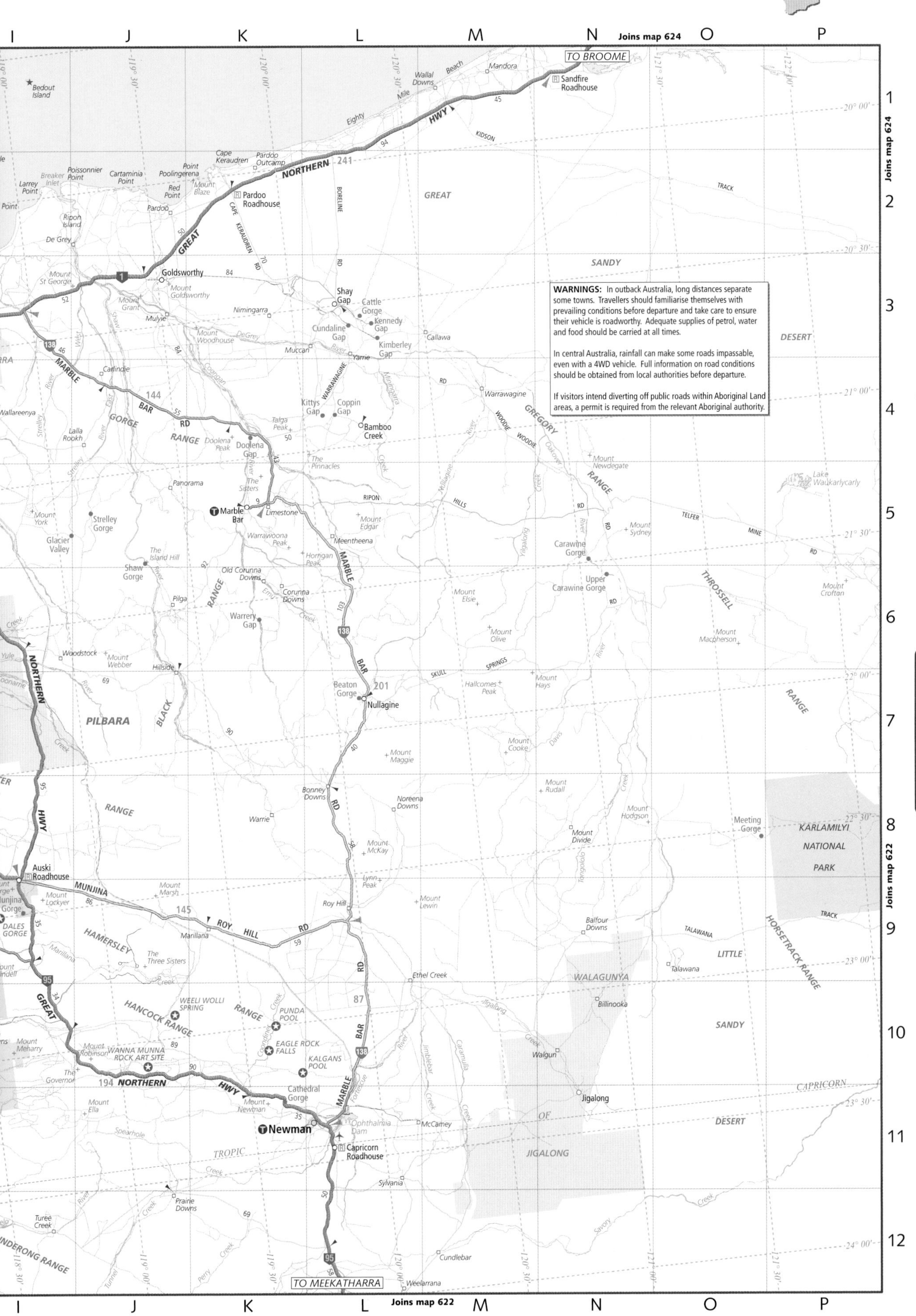
0 20 40 60 80 100 km
I J K L M N O P
Joins map 624
TO BROOME
Sandfire Roadhouse
Mandora
Wallal Downs
Eighty Mile Beach
HWY
45
94
KIDSON
TRACK
Bedout Island
Cape Keraudren
Pardoo Outcamp
NORTHERN 241
Poissonnier Point
Breaker Inlet
Larrey Point
Cartaminia Point
Point Poolingerena
Red Point
Mount Blaze
Pardoo Roadhouse
Pardoo
BORELINE RD
GREAT
Ripon Island
De Grey
CAPE KERAUDREN RD
70
50
GREAT
SANDY
Mount St George
Goldsworthy
84
Mount Goldsworthy
Shay Gap
Cattle Gorge
Kennedy Gap
52
Mount Grant
Mulyie
Nimingarra
Cundaline Gap
Kimberley Gap
Callawa
Mount Woodhouse
De Grey
Muccan
Yarrie
138
46
Carlindie
MARBLE
144
BAR
55
RD
GORGE
RANGE
WARRAWAGINE
Warrawagine
RD
Wallareenya
Lalla Rookh
Talga Peak
Kittys Gap
Coppin Gap
Bamboo Creek
GREGORY
WOODIE WOODIE
Doolena Peak
Doolena Gap
43
The Pinnacles
Mount Newdegate
RANGE
The Sisters
Panorama
RIPON
HILLS
RD
Marble Bar
9
Limestone
Mount Edgar
Mount York
Strelley Gorge
Glacier Valley
Warrawoona Peak
Meentheena
Carawine Gorge
Mount Sydney
TELFER
MINE
RD
DESERT
Lake Waukarlycarly
The Island Hill
Horrigan Peak
Shaw Gorge
Old Corunna Downs
Corunna Downs
MARBLE
RANGE
Pilga
Warrery Gap
103
Mount Elsie
Upper Carawine Gorge
THROSSELL
Mount Crofton
Mount Olive
Mount Macpherson
138
Woodstock
Mount Webber
Hillside
69
NORTHERN
BAR
SKULL SPRINGS
Hallcomes Peak
Mount Hays
Beaton Gorge
201
Nullagine
RANGE
PILBARA
BLACK
90
40
Mount Cooke
Mount Maggie
Davis
95
Bonney Downs
RD
Noreena Downs
Mount Rudall
RANGE
Warrie
Mount Hodgson
Meeting Gorge
KARLAMILYI NATIONAL PARK
HWY
Mount McKay
Mount Divide
Auski Roadhouse
MUNJINA
Mount Marsh
Lynn Peak
Munjina Gorge
Mount Lockyer
86
145
Roy Hill
Mount Lewin
DALES GORGE
35
Marillana
ROY HILL
RD
59
Balfour Downs
TALAWANA
TRACK
HORSETRACK RANGE
HAMERSLEY
The Three Sisters
LITTLE
Talawana
Ethel Creek
WALAGUNYA
95
34
WEELI WOLLI SPRING
RANGE
PUNDA POOL
87
Billinooka
GREAT
HANCOCK RANGE
SANDY
Mount Meharry
Mount Robinson
WANNA MUNNA ROCK ART SITE
EAGLE ROCK FALLS
89
KALGANS POOL
BAR
Walgun
The Governor
194 NORTHERN
90
HWY
Cathedral Gorge
MARBLE
Jigalong
CAPRICORN
Mount Ella
Mount Newman
35
Newman
Ophthalmia Dam
McCamey
OF
DESERT
Spearhole
TROPIC
Capricorn Roadhouse
JIGALONG
Sylvania
Prairie Downs
69
50
Turee Creek
ANDERONG RANGE
Cundlebar
95
TO MEEKATHARRA
Weelarrana
Joins map 622
20° 00'
20° 30'
21° 00'
21° 30'
22° 00'
22° 30'
23° 00'
23° 30'
24° 00'
119° 00'
119° 30'
120° 00'
120° 30'
121° 00'
121° 30'
122° 00'
118° 30'
1 2 3 4 5 6 7 8 9 10 11 12
WARNINGS: In outback Australia, long distances separate some towns. Travellers should familiarise themselves with prevailing conditions before departure and take care to ensure their vehicle is roadworthy. Adequate supplies of petrol, water and food should be carried at all times.
In central Australia, rainfall can make some roads impassable, even with a 4WD vehicle. Full information on road conditions should be obtained from local authorities before departure.
If visitors intend diverting off public roads within Aboriginal Land areas, a permit is required from the relevant Aboriginal authority.

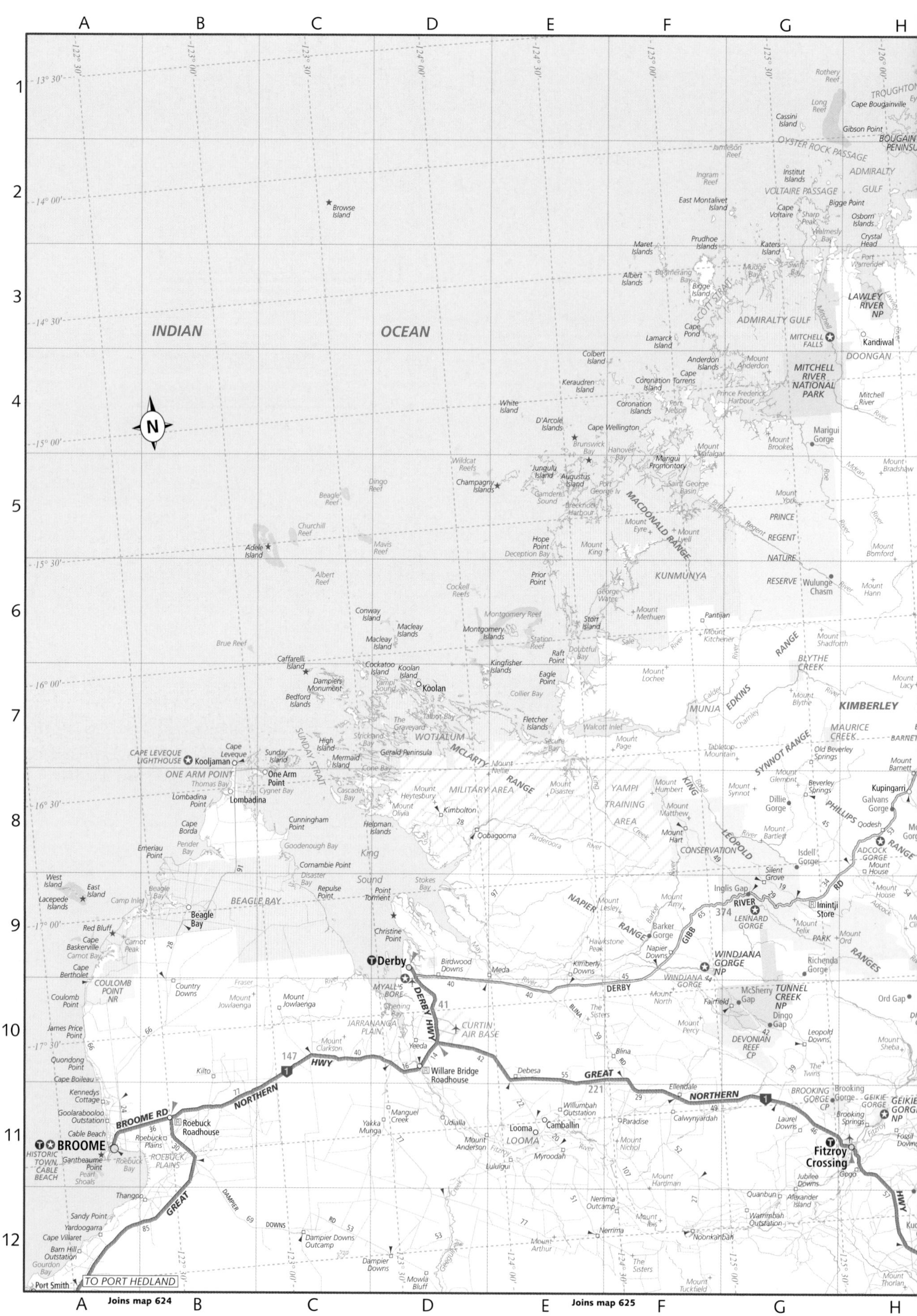
INDIAN
OCEAN
BROOME
Derby
Fitzroy Crossing
Willare Bridge Roadhouse
Roebuck Roadhouse
Beagle Bay
Lombadina
One Arm Point
Kooljaman
CAPE LEVEQUE LIGHTHOUSE
Koolan
Kimbolton
Oobagooma
Looma
Camballin
Debesa
Ellendale
Calwynyardah
Imintji Store
Barker Gorge
Kupingarri
Mitchell Falls
MITCHELL RIVER NATIONAL PARK
LAWLEY RIVER NP
Kandiwal
WINDJANA GORGE NP
TUNNEL CREEK NP
GEIKIE GORGE NP
BROOKING GORGE CP
DEVONIAN REEF CP
COULOMB POINT NR
PRINCE REGENT NATURE RESERVE
KING LEOPOLD CONSERVATION PARK
YAMPI TRAINING AREA
WOTJALUM MILITARY AREA
CURTIN AIR BASE
KIMBERLEY
MACDONALD RANGE
GREAT NORTHERN HWY
DERBY HWY
GIBB RIVER RD
BROOME RD
DAMPIER DOWNS RD
TO PORT HEDLAND
Joins map 624
Joins map 625

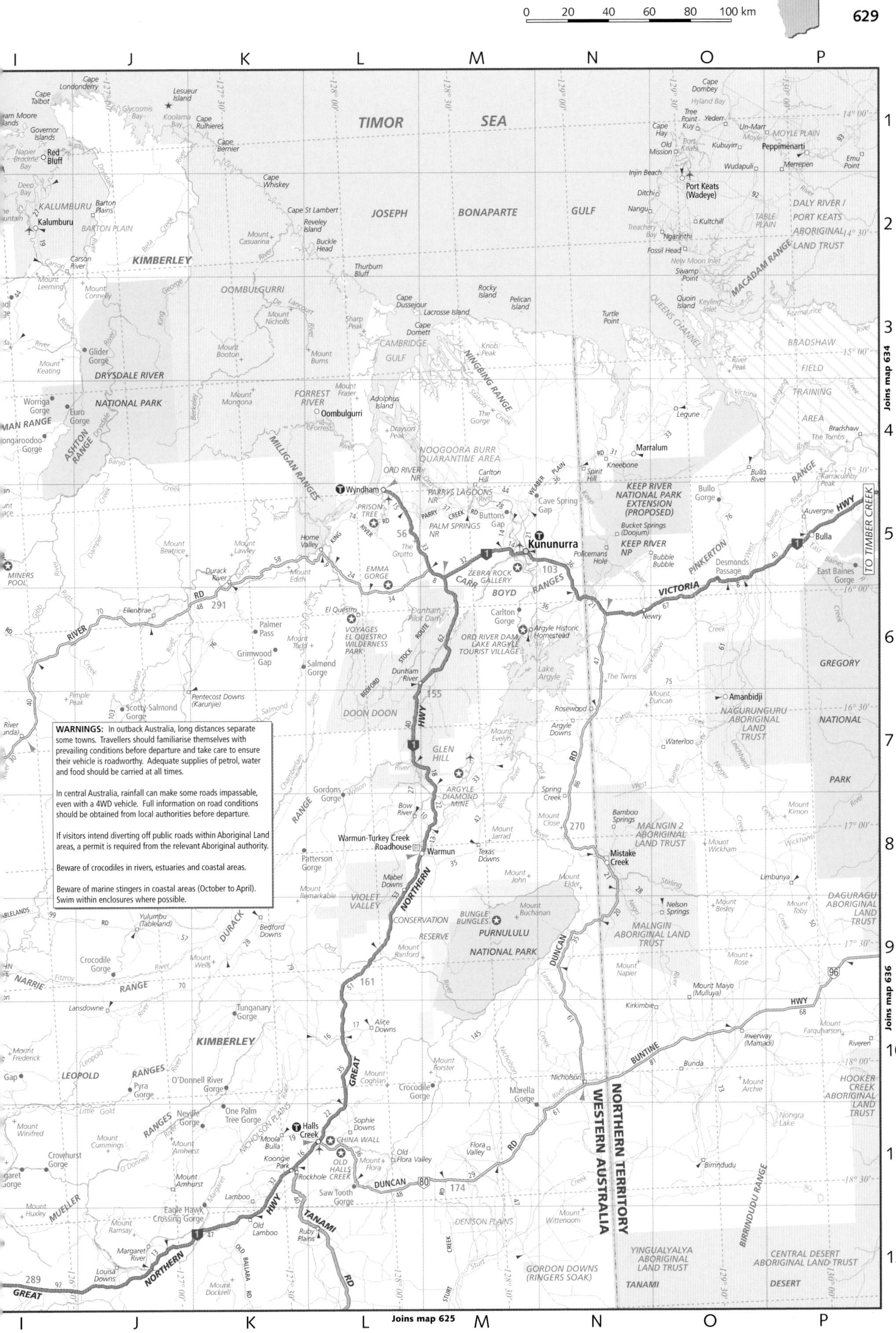

0 20 40 60 80 100 km

I J K L M N O P

TIMOR SEA

JOSEPH BONAPARTE GULF

WARNINGS: In outback Australia, long distances separate some towns. Travellers should familiarise themselves with prevailing conditions before departure and take care to ensure their vehicle is roadworthy. Adequate supplies of petrol, water and food should be carried at all times.

In central Australia, rainfall can make some roads impassable, even with a 4WD vehicle. Full information on road conditions should be obtained from local authorities before departure.

If visitors intend diverting off public roads within Aboriginal Land areas, a permit is required from the relevant Aboriginal authority.

Beware of crocodiles in rivers, estuaries and coastal areas.

Beware of marine stingers in coastal areas (October to April). Swim within enclosures where possible.

NORTHERN TERRITORY

WESTERN AUSTRALIA

Joins map 634

Joins map 636

Joins map 625

I J K L M N O P

MAPS

NORTHERN TERRITORY

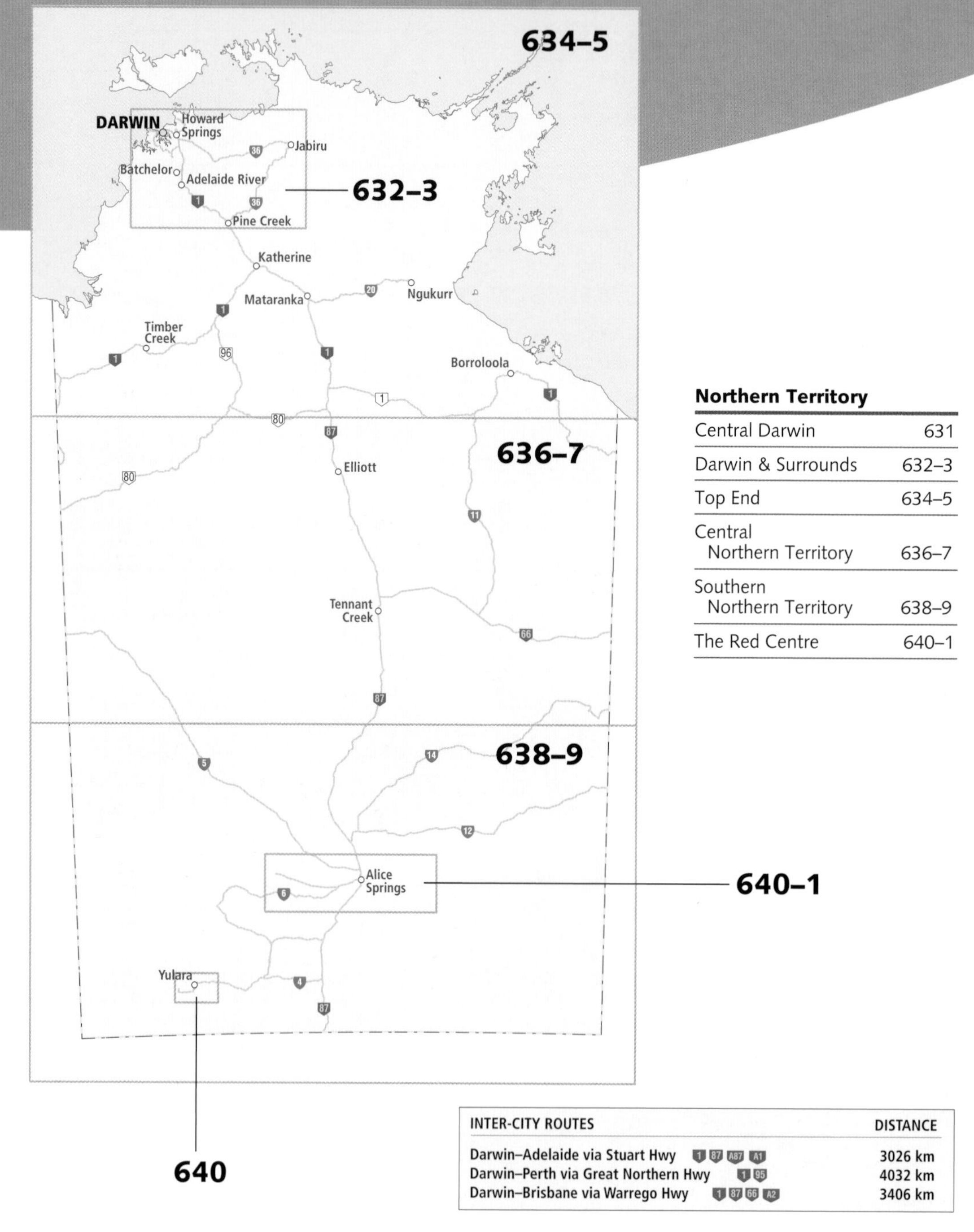

Northern Territory

INTER-CITY ROUTES	DISTANCE
Darwin–Adelaide via Stuart Hwy	3026 km
Darwin–Perth via Great Northern Hwy	4032 km
Darwin–Brisbane via Warrego Hwy	3406 km

Accommodation

- Banyan View Lodge 1 B6
- Crowne Plaza Hotel Darwin 2 E9
- Darwin YHA 3 C7
- Holiday Inn Esplanade Darwin 4 C7
- Medina Grand Darwin Waterfront 5 G9
- Novotel Darwin Atrium 6 D8
- Poinciana Inn 7 C7
- SKYCITY Darwin 8 A3
- Travelodge Mirambeena Resort Darwin 9 D7
- Vibe Hotel Darwin Waterfront 10 G10
- YMCA Top End Hostel 11 B6

Note: Only a sample range of accommodation is listed; inclusion is not necessarily a recommendation.

General Information

- Bus Terminal 12 F9
- Darwin Civic Centre & Library 13 F9
- Darwin Convention Centre 14 H10
- Darwin Memorial Uniting Church 15 D8
- Darwin Tourist Precinct 16 D8
- Garuda Indonesia Airlines 17 F8
- GPO 18 E8
- Greyhound Coaches 19 D8
- Motoring Organisation (AANT) 20 D7
- Police Station 21 E9
- Qantas Travel Centre 22 F9
- Visitor Information (Tourism Top End) 23 E9

Places of Interest

- Aquascene 24 B7
- Australian Pearling Exhibition 25 H10
- Bicentennial Park 26 D9
- Brown's Mart 27 F9
- Chinese Temple & Northern Territory Chinese Museum 28 F8
- Christ Church Cathedral 29 F9
- Darwin Performing Arts Centre 30 C7
- Deckchair Cinema (Apr–Nov) 31 E10
- George Brown Darwin Botanic Gardens 32 C1
- Government House 33 F10
- Indo Pacific Marine 34 H10
- Lyons Cottage 35 D9
- Mindil Beach Sunset Market 36 A1
- Myilly Point Heritage Precinct 37 A4
- Old Admiralty House 38 D9
- Old Police Station & Courthouse 39 F10
- Overland Telegraph Memorial 40 F10
- Palmerston Town Hall ruins 41 F9
- Parliament House & NT Library 42 E9
- Smith Street Mall 43 E9
- Stokes Hill Wharf 44 H11
- Supreme Court 45 F10
- Survivors Lookout 46 F10
- Tour Tub 47 E8
- Tree of Knowledge 48 F9
- Victoria Hotel 49 E9
- WWII Oil Storage Tunnels 50 F10

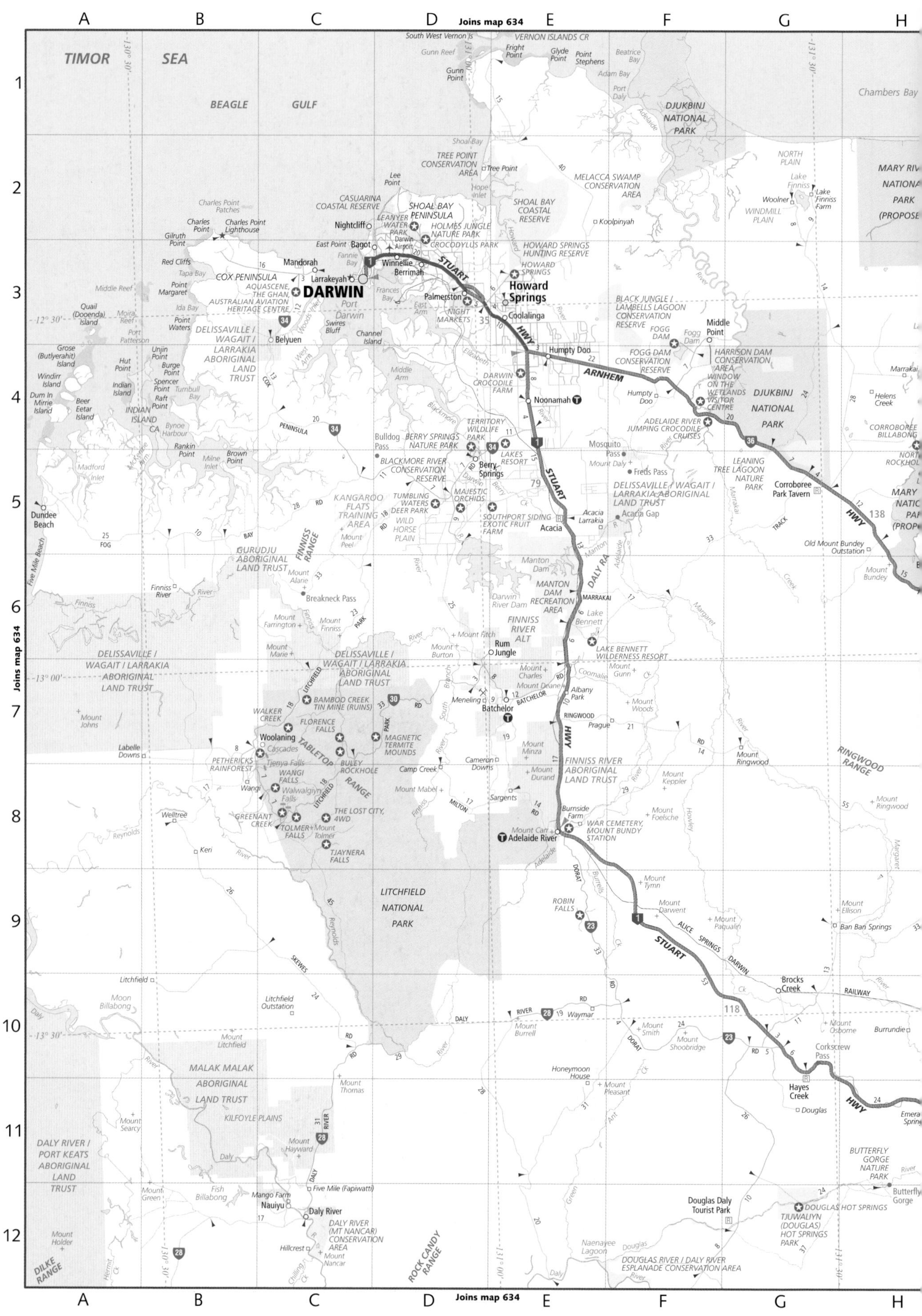
Joins map 634
TIMOR SEA
BEAGLE GULF
DARWIN
Howard Springs
Palmerston
Humpty Doo
Noonamah
Berry Springs
Batchelor
Adelaide River
Acacia
Rum Jungle
Daly River
Nauiyu
Mandorah
Belyuen
Dundee Beach
Coolalinga
Nightcliff
Bagot
Winnellie
Berrimah
Larrakeyah
DJUKBINJ NATIONAL PARK
LITCHFIELD NATIONAL PARK
MARY RIVER NATIONAL PARK (PROPOSED)
STUART HWY
ARNHEM HWY
DALY RIVER
DELISSAVILLE / WAGAIT / LARRAKIA ABORIGINAL LAND TRUST
MALAK MALAK ABORIGINAL LAND TRUST
DALY RIVER / PORT KEATS ABORIGINAL LAND TRUST
FINNISS RIVER ABORIGINAL LAND TRUST
GURUDJU ABORIGINAL LAND TRUST
TABLETOP RANGE
RINGWOOD RANGE
ROCK CANDY RANGE
DILKE RANGE
FINNISS RANGE
Hayes Creek
Brocks Creek
Douglas Daly Tourist Park
DOUGLAS HOT SPRINGS
BUTTERFLY GORGE NATURE PARK
Corroboree Park Tavern
Joins map 634

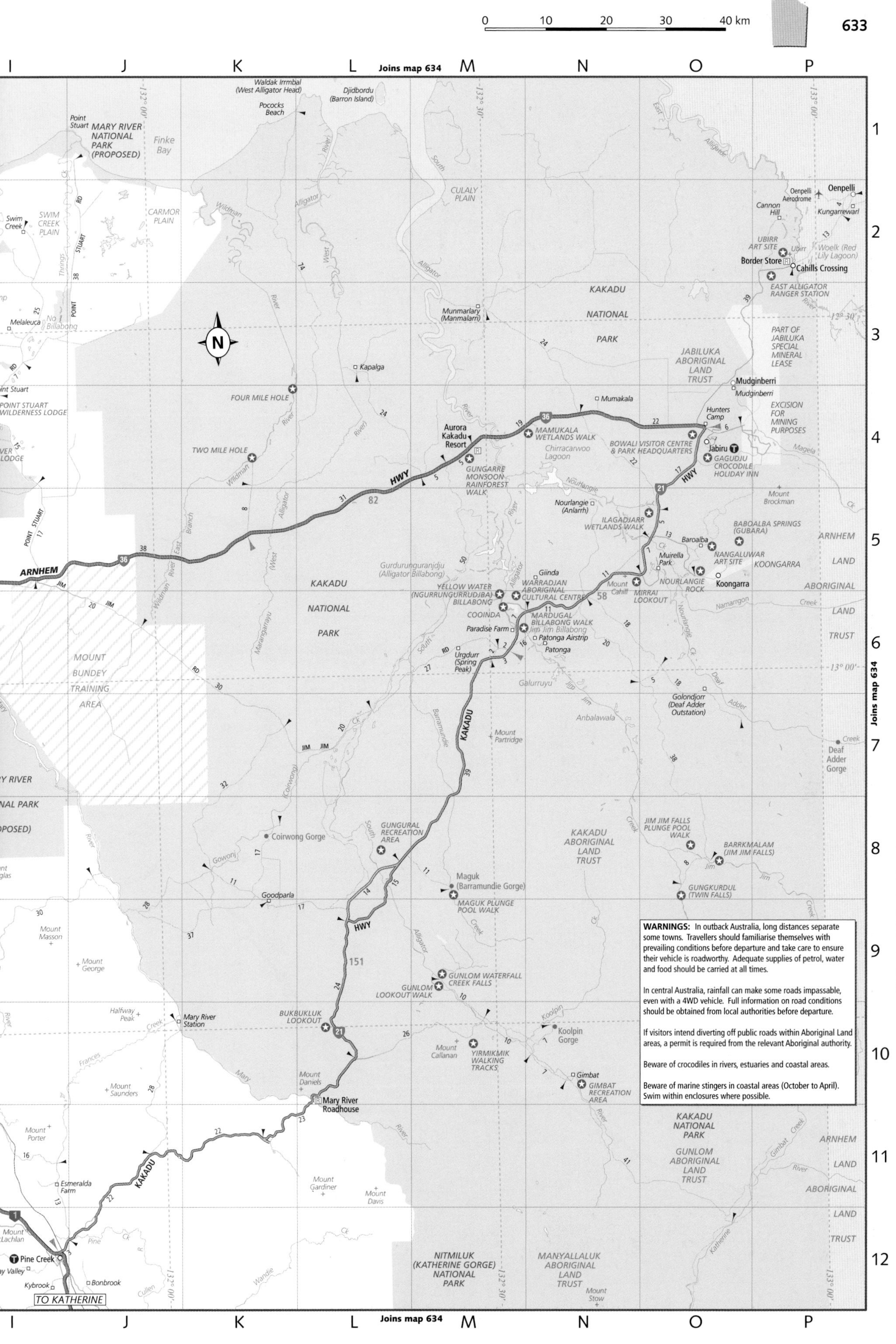
0 10 20 30 40 km
Joins map 634
Waldak Irrmbal (West Alligator Head)
Djidbordu (Barron Island)
Pococks Beach
Point Stuart
MARY RIVER NATIONAL PARK (PROPOSED)
Finke Bay
SWIM CREEK PLAIN
Swim Creek
CARMOR PLAIN
CULALY PLAIN
Oenpelli
Oenpelli Aerodrome
Cannon Hill
Kungarrewarl
UBIRR ART SITE
Ubirr
Woolk (Red Lily Lagoon)
Border Store
Cahills Crossing
EAST ALLIGATOR RANGER STATION
KAKADU NATIONAL PARK
Melaleuca
No 1 Billabong
Munmarlary (Manmalarri)
PART OF JABILUKA SPECIAL MINERAL LEASE
JABILUKA ABORIGINAL LAND TRUST
Mudginberri
Kapalga
Point Stuart
POINT STUART WILDERNESS LODGE
FOUR MILE HOLE
Mumakala
Hunters Camp
EXCISION FOR MINING PURPOSES
Aurora Kakadu Resort
MAMUKALA WETLANDS WALK
BOWALI VISITOR CENTRE & PARK HEADQUARTERS
Jabiru
GAGUDJU CROCODILE HOLIDAY INN
Chirracarwoo Lagoon
TWO MILE HOLE
GUNGARRE MONSOON RAINFOREST WALK
HWY
Mount Brockman
Nourlangie (Anlarrh)
ILAGADJARR WETLANDS WALK
BABOALBA SPRINGS (GUBARA)
Baroalba
NANGALUWAR ART SITE
Muirella Park
NOURLANGIE ROCK
Koongarra
KOONGARRA
ARNHEM LAND ABORIGINAL LAND TRUST
ARNHEM
Gurdurungurandju (Alligator Billabong)
Giinda
WARRADJAN ABORIGINAL CULTURAL CENTRE
YELLOW WATER (NGURRUNGURRUDJBA) BILLABONG
Mount Cahill
MIRRAI LOOKOUT
KAKADU NATIONAL PARK
COOINDA
MARDUGAL BILLABONG WALK
Jim Jim Billabong
Paradise Farm
Patonga Airstrip
Patonga
MOUNT BUNDEY TRAINING AREA
Urgdurr (Spring Peak)
Galurruyu
Golondjorr (Deaf Adder Outstation)
Anbalawala
Mount Partridge
KAKADU
Deaf Adder Gorge
MARY RIVER NATIONAL PARK (PROPOSED)
JIM JIM FALLS PLUNGE POOL WALK
KAKADU ABORIGINAL LAND TRUST
GUNGURAL RECREATION AREA
Coirwong Gorge
BARRKMALAM (JIM JIM FALLS)
Maguk (Barramundie Gorge)
MAGUK PLUNGE POOL WALK
GUNGKURDUL (TWIN FALLS)
Goodparla
Mount Masson
Mount George
GUNLOM WATERFALL CREEK FALLS
GUNLOM LOOKOUT WALK
Halfway Peak
Mary River Station
BUKBUKLUK LOOKOUT
Koolpin Gorge
Mount Callanan
YIRMIKMIK WALKING TRACKS
Gimbat
GIMBAT RECREATION AREA
Mount Saunders
Mount Daniels
Mary River Roadhouse
Mount Porter
KAKADU NATIONAL PARK
GUNLOM ABORIGINAL LAND TRUST
Esmeralda Farm
Mount Gardiner
Mount Davis
Mount McLachlan
Pine Creek
Kybrook
Bonbrook
TO KATHERINE
NITMILUK (KATHERINE GORGE) NATIONAL PARK
MANYALLALUK ABORIGINAL LAND TRUST
Mount Stow
Joins map 634
WARNINGS: In outback Australia, long distances separate some towns. Travellers should familiarise themselves with prevailing conditions before departure and take care to ensure their vehicle is roadworthy. Adequate supplies of petrol, water and food should be carried at all times.
In central Australia, rainfall can make some roads impassable, even with a 4WD vehicle. Full information on road conditions should be obtained from local authorities before departure.
If visitors intend diverting off public roads within Aboriginal Land areas, a permit is required from the relevant Aboriginal authority.
Beware of crocodiles in rivers, estuaries and coastal areas.
Beware of marine stingers in coastal areas (October to April). Swim within enclosures where possible.

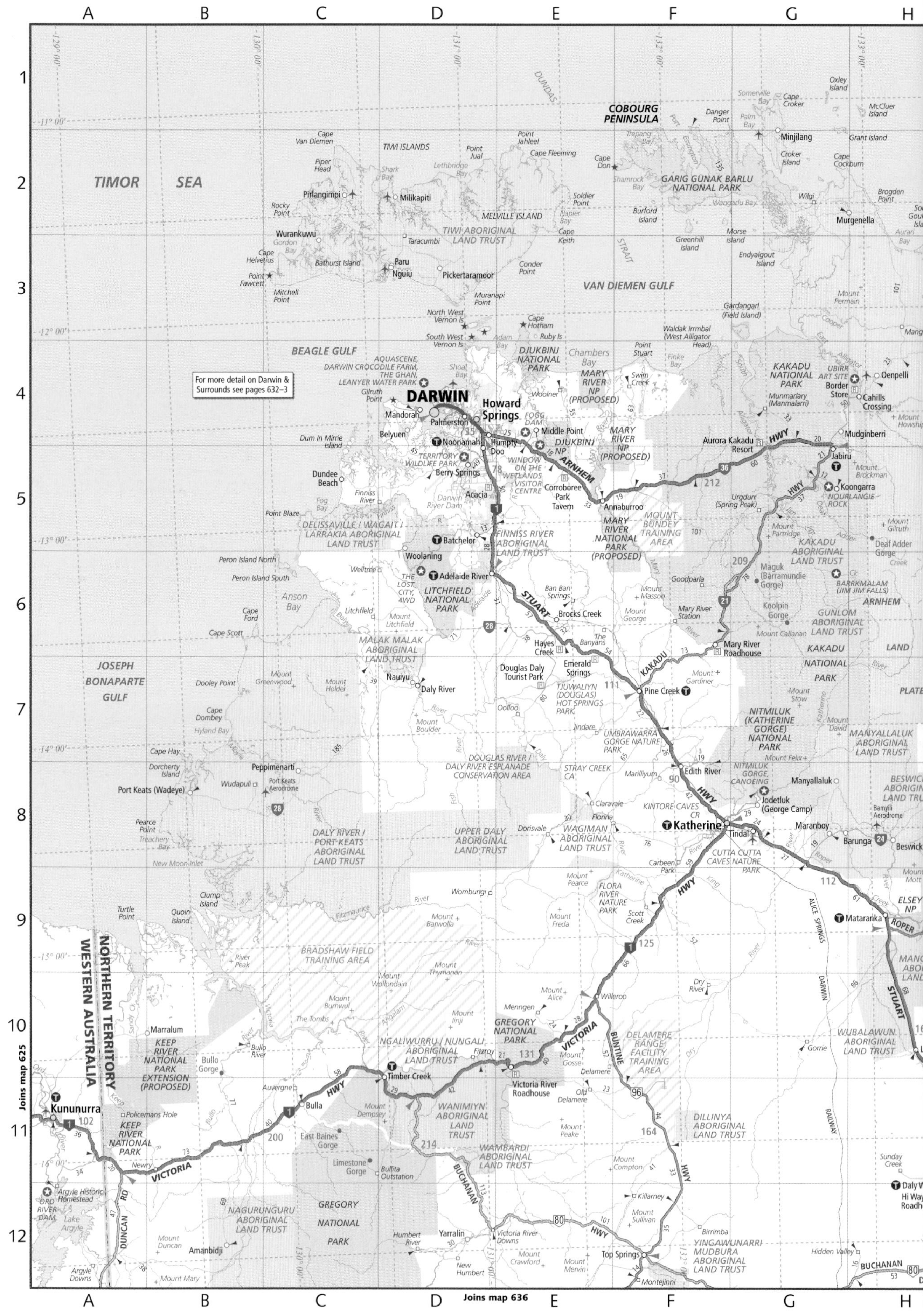

A B C D E F G H
1 2 3 4 5 6 7 8 9 10 11 12
TIMOR SEA
COBOURG PENINSULA
GARIG GUNAK BARLU NATIONAL PARK
TIWI ISLANDS
MELVILLE ISLAND
Bathurst Island
TIWI ABORIGINAL LAND TRUST
Pirlangimpi
Milikapiti
Wurankuwu
Taracumbi
Paru
Nguiu
Pickertaramoor
VAN DIEMEN GULF
BEAGLE GULF
Minjilang
Murgenella
Oenpelli
For more detail on Darwin & Surrounds see pages 632–3
AQUASCENE, DARWIN CROCODILE FARM, THE GHAN, LEANYER WATER PARK
DARWIN
Howard Springs
Palmerston
Mandorah
Belyuen
Noonamah
Humpty Doo
TERRITORY WILDLIFE PARK
Berry Springs
FOGG DAM
Middle Point
DJUKBINJ NATIONAL PARK
DJUKBINJ NP
MARY RIVER NP (PROPOSED)
WINDOW ON THE WETLANDS VISITOR CENTRE
Corroboree Park Tavern
Annaburroo
ARNHEM
Aurora Kakadu Resort
HWY
Jabiru
Koongarra
NOURLANGIE ROCK
KAKADU NATIONAL PARK
UBIRR ART SITE
Border Store
Cahills Crossing
Mudginberri
Dundee Beach
Finniss River
Acacia
Darwin River Dam
DELISSAVILLE / WAGAIT / LARRAKIA ABORIGINAL LAND TRUST
Batchelor
FINNISS RIVER ABORIGINAL LAND TRUST
MARY RIVER NATIONAL PARK (PROPOSED)
MOUNT BUNDEY TRAINING AREA
Woolaning
THE LOST CITY, 4WD
Adelaide River
LITCHFIELD NATIONAL PARK
Urgdurr (Spring Peak)
Maguk (Barramundie Gorge)
KAKADU ABORIGINAL LAND TRUST
BARRKMALAM (JIM JIM FALLS)
Goodparla
Mary River Station
Mary River Roadhouse
Koolpin Gorge
GUNLOM ABORIGINAL LAND TRUST
ARNHEM LAND
STUART
Brocks Creek
Hayes Creek
Emerald Springs
Douglas Daly Tourist Park
TIJUWALIYN (DOUGLAS) HOT SPRINGS PARK
KAKADU
Pine Creek
MALAK MALAK ABORIGINAL LAND TRUST
Nauiyu
Daly River
JOSEPH BONAPARTE GULF
Anson Bay
Oolloo
Jindare
UMBRAWARRA GORGE NATURE PARK
NITMILUK (KATHERINE GORGE) NATIONAL PARK
MANYALLALUK ABORIGINAL LAND TRUST
Peppimenarti
Port Keats (Wadeye)
Wudapuli
DOUGLAS RIVER / DALY RIVER ESPLANADE CONSERVATION AREA
STRAY CREEK CA
Edith River
NITMILUK GORGE, CANOEING
Manyallaluk
Jodetluk (George Camp)
KINTORE CAVES CR
Katherine
Tindal
Maranboy
Barunga
Beswick
DALY RIVER / PORT KEATS ABORIGINAL LAND TRUST
UPPER DALY ABORIGINAL LAND TRUST
WAGIMAN ABORIGINAL LAND TRUST
CUTTA CUTTA CAVES NATURE PARK
FLORA RIVER NATURE PARK
Mataranka
ELSEY NP
ROPER
BRADSHAW FIELD TRAINING AREA
NORTHERN TERRITORY
WESTERN AUSTRALIA
Marralum
KEEP RIVER NATIONAL PARK EXTENSION (PROPOSED)
Bullo River
GREGORY NATIONAL PARK
NGALIWURRU / NUNGALI ABORIGINAL LAND TRUST
Timber Creek
Victoria River Roadhouse
VICTORIA
Willeroo
BUNTINE
DELAMERE RANGE FACILITY TRAINING AREA
WUBALAWUN ABORIGINAL LAND TRUST
Kununurra
KEEP RIVER NATIONAL PARK
Bulla
WANIMIYN ABORIGINAL LAND TRUST
DILLINYA ABORIGINAL LAND TRUST
WAMBARDI ABORIGINAL LAND TRUST
BUCHANAN
East Baines Gorge
Limestone Gorge
Bullita Outstation
DUNCAN RD
Argyle Historic Homestead
Lake Argyle
NAGURUNGURU ABORIGINAL LAND TRUST
Amanbidji
Yarralin
Victoria River Downs
Top Springs
YINGAWUNARRI MUDBURA ABORIGINAL LAND TRUST
DARWIN
ALICE SPRINGS
RAILWAY
Joins map 625
Joins map 636

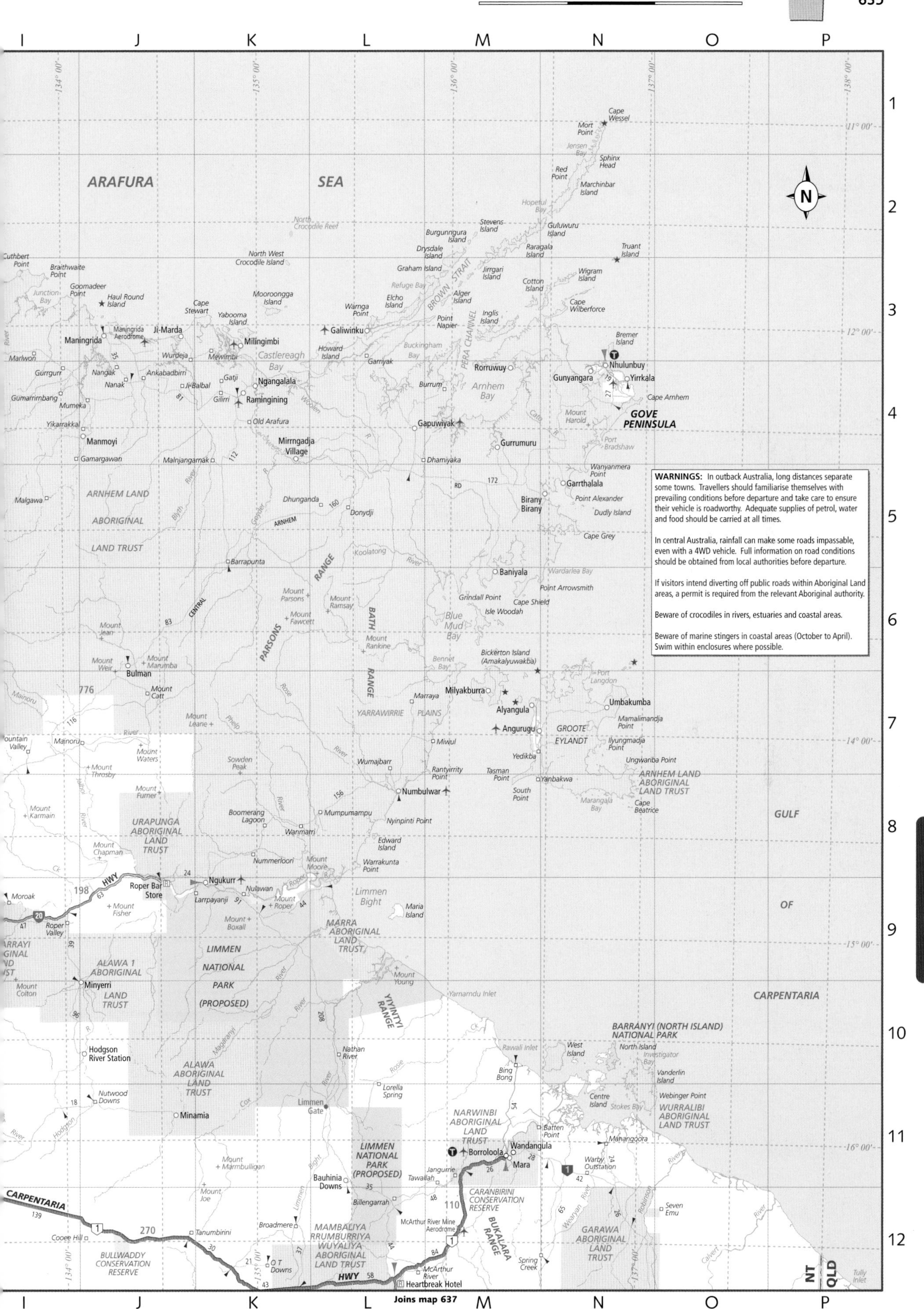

WARNINGS: In outback Australia, long distances separate some towns. Travellers should familiarise themselves with prevailing conditions before departure and take care to ensure their vehicle is roadworthy. Adequate supplies of petrol, water and food should be carried at all times.

In central Australia, rainfall can make some roads impassable, even with a 4WD vehicle. Full information on road conditions should be obtained from local authorities before departure.

If visitors intend diverting off public roads within Aboriginal Land areas, a permit is required from the relevant Aboriginal authority.

Beware of crocodiles in rivers, estuaries and coastal areas.

Beware of marine stingers in coastal areas (October to April). Swim within enclosures where possible.

Joins map 637

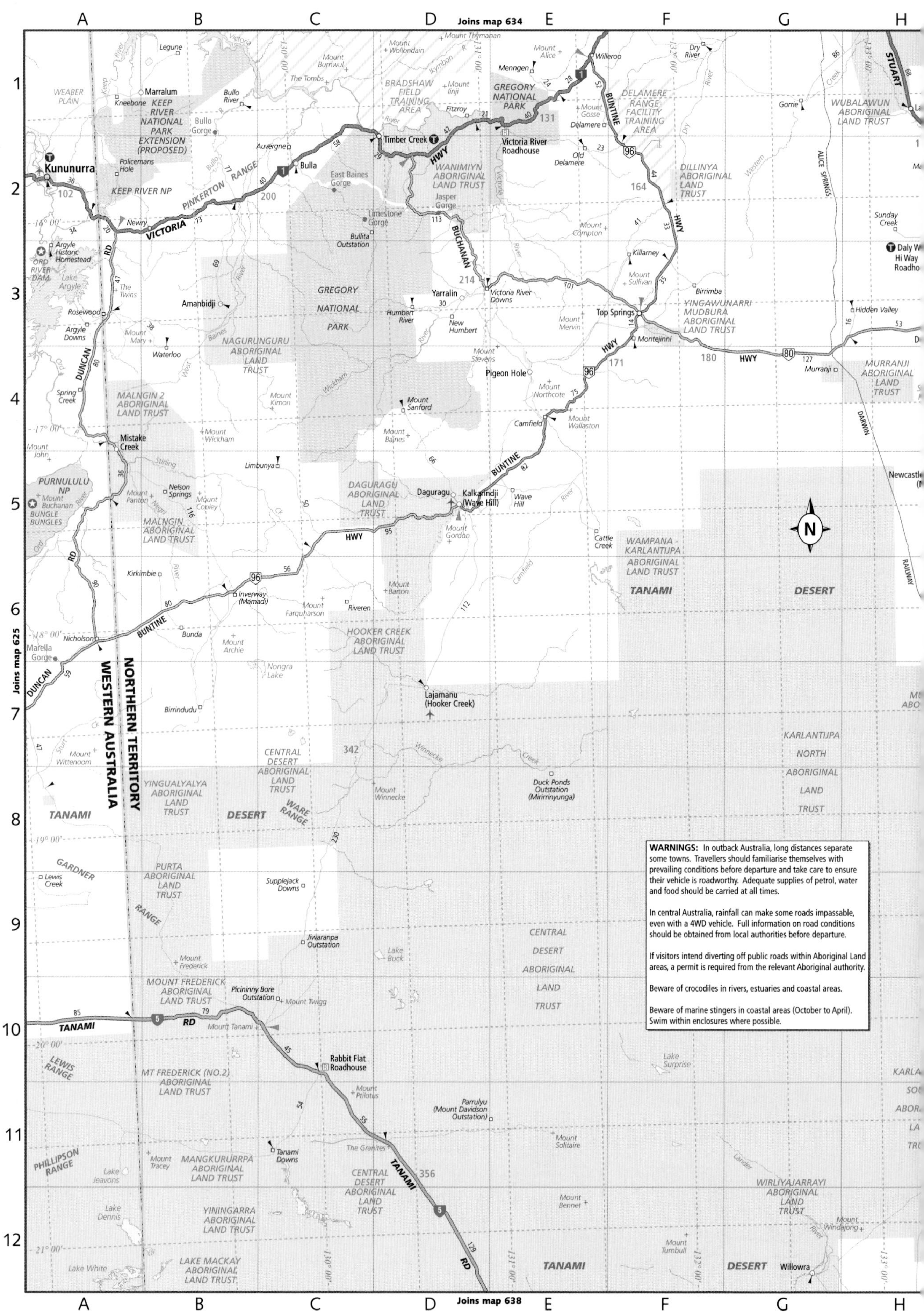
A
B
C
D
E
F
G
H
Joins map 634
Joins map 638
Joins map 625
1
2
3
4
5
6
7
8
9
10
11
12
Legune
WEABER PLAIN
Marralum
Kneebone
KEEP RIVER NATIONAL PARK EXTENSION (PROPOSED)
Bullo Gorge
Bullo River
Mount Burrwul
The Tombs
Mount Wollondain
Mount Thirmanan
BRADSHAW FIELD TRAINING AREA
Mount Iinji
Fitzroy
Mount Alice
Menngen
Willeroo
GREGORY NATIONAL PARK
Dry River
STUART
DELAMERE RANGE FACILITY TRAINING AREA
Mount Gosse
Delamere
Old Delamere
Gorrie
WUBALAWUN ABORIGINAL LAND TRUST
Kununurra
Policemans Hole
KEEP RIVER NP
PINKERTON RANGE
Auvergne
Bulla
Timber Creek
Victoria River Roadhouse
East Baines Gorge
WANIMIYN ABORIGINAL LAND TRUST
DILLINYA ABORIGINAL LAND TRUST
ALICE SPRINGS
Jasper Gorge
Limestone Gorge
Bullita Outstation
Newry
VICTORIA
HWY
BUNTINE
BUCHANAN
Mount Compton
Killarney
Sunday Creek
Daly Waters Hi Way Roadhouse
Argyle Historic Homestead
ORD RIVER DAM
Lake Argyle
The Twins
Rosewood
Argyle Downs
Amanbidji
GREGORY NATIONAL PARK
Yarralin
Victoria River Downs
Humbert River
New Humbert
Mount Sullivan
Birrimba
Top Springs
YINGAWUNARRI MUDBURA ABORIGINAL LAND TRUST
Mount Mervin
Montejinni
Hidden Valley
Mount Mary
Waterloo
NAGURUNGURU ABORIGINAL LAND TRUST
Mount Stevens
Pigeon Hole
Murranji
MURRANJI ABORIGINAL LAND TRUST
DUNCAN
RD
Spring Creek
MALNGIN 2 ABORIGINAL LAND TRUST
Mount Kimon
Mount Sanford
Mount Northcote
Camfield
Mount Wallaston
DARWIN
Mount Wickham
Mistake Creek
Mount John
Mount Baines
Limbunya
PURNULULU NP
Mount Buchanan
BUNGLE BUNGLES
Mount Panton
Nelson Springs
Mount Copley
MALNGIN ABORIGINAL LAND TRUST
DAGURAGU ABORIGINAL LAND TRUST
Daguragu
Kalkarindji (Wave Hill)
Wave Hill
Newcastle
Mount Gordon
Cattle Creek
WAMPANA - KARLANTIJPA ABORIGINAL LAND TRUST
TANAMI
DESERT
RAILWAY
Kirkimbie
Inverway (Mamadi)
Mount Barton
Riveren
Mount Farquharson
Nicholson
Bunda
Mount Archie
HOOKER CREEK ABORIGINAL LAND TRUST
Marella Gorge
NORTHERN TERRITORY
WESTERN AUSTRALIA
Nongra Lake
Lajamanu (Hooker Creek)
Birrindudu
KARLANTIJPA NORTH ABORIGINAL LAND TRUST
Mount Wittenoom
CENTRAL DESERT ABORIGINAL LAND TRUST
Winnecke Creek
Mount Winnecke
Duck Ponds Outstation (Mirirrinyunga)
YINGUALYALYA ABORIGINAL LAND TRUST
WARE RANGE
GARDNER RANGE
Lewis Creek
PURTA ABORIGINAL LAND TRUST
Supplejack Downs
Jiwaranpa Outstation
Lake Buck
Mount Frederick
MOUNT FREDERICK ABORIGINAL LAND TRUST
Picininny Bore Outstation
Mount Twigg
CENTRAL DESERT ABORIGINAL LAND TRUST
Mount Tanami
TANAMI RD
LEWIS RANGE
Rabbit Flat Roadhouse
MT FREDERICK (NO.2) ABORIGINAL LAND TRUST
Mount Pilotus
Parrulyu (Mount Davidson Outstation)
Lake Surprise
Mount Solitaire
PHILLIPSON RANGE
Mount Tracey
MANGKURURRPA ABORIGINAL LAND TRUST
Tanami Downs
The Granites
Lake Jeavons
WIRLIYAJARRAYI ABORIGINAL LAND TRUST
Mount Bennet
Lake Dennis
YININGARRA ABORIGINAL LAND TRUST
Mount Windajong
Mount Turnbull
Lake White
LAKE MACKAY ABORIGINAL LAND TRUST
Willowra
16° 00′
17° 00′
18° 00′
19° 00′
20° 00′
21° 00′
130° 00′
131° 00′
132° 00′
133° 00′
N
WARNINGS: In outback Australia, long distances separate some towns. Travellers should familiarise themselves with prevailing conditions before departure and take care to ensure their vehicle is roadworthy. Adequate supplies of petrol, water and food should be carried at all times.
In central Australia, rainfall can make some roads impassable, even with a 4WD vehicle. Full information on road conditions should be obtained from local authorities before departure.
If visitors intend diverting off public roads within Aboriginal Land areas, a permit is required from the relevant Aboriginal authority.
Beware of crocodiles in rivers, estuaries and coastal areas.
Beware of marine stingers in coastal areas (October to April). Swim within enclosures where possible.

0 50 100 150 km

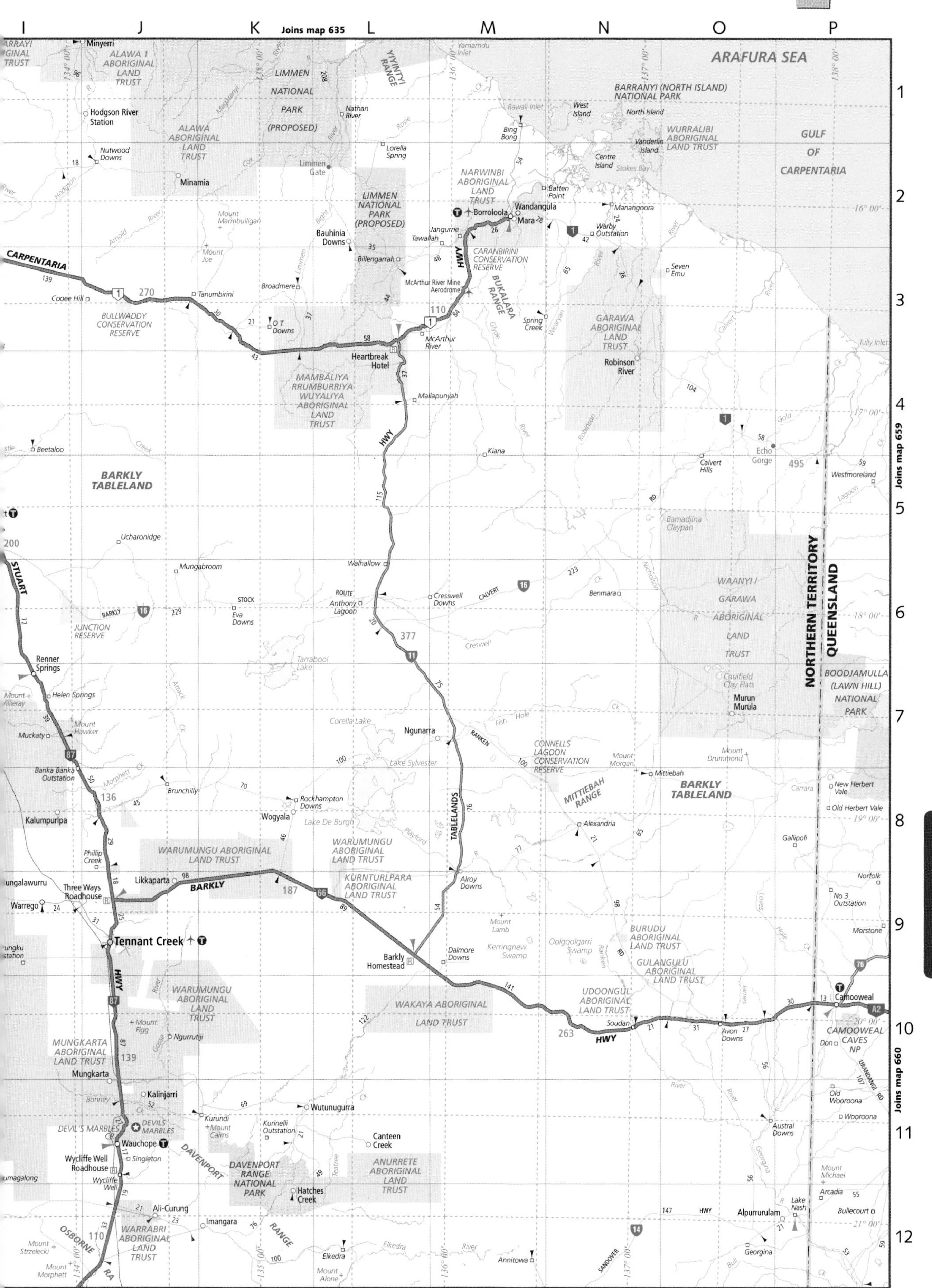

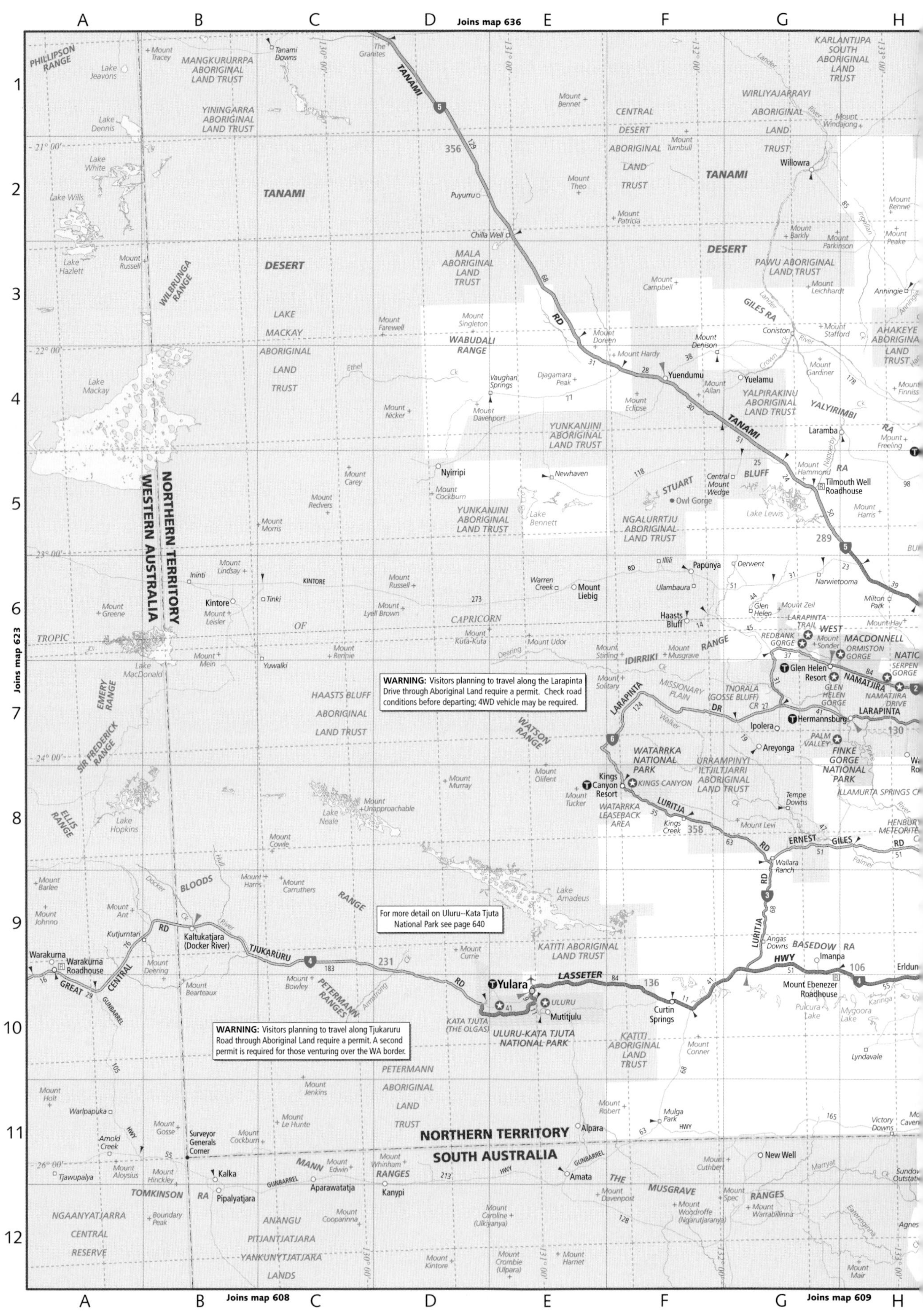
Joins map 636
Joins map 623
Joins map 608
Joins map 609
WARNING: Visitors planning to travel along the Larapinta Drive through Aboriginal Land require a permit. Check road conditions before departing; 4WD vehicle may be required.
WARNING: Visitors planning to travel along Tjukaruru Road through Aboriginal Land require a permit. A second permit is required for those venturing over the WA border.
For more detail on Uluru–Kata Tjuta National Park see page 640
NORTHERN TERRITORY
WESTERN AUSTRALIA
NORTHERN TERRITORY
SOUTH AUSTRALIA
TROPIC OF CAPRICORN
TANAMI DESERT
PHILLIPSON RANGE
MANGKURURRPA ABORIGINAL LAND TRUST
YININGARRA ABORIGINAL LAND TRUST
Tanami Downs
The Granites
TANAMI RD
Lake Jeavons
Lake Dennis
Lake White
Lake Wills
Lake Hazlett
Lake Mackay
LAKE MACKAY ABORIGINAL LAND TRUST
WILBRUNGA RANGE
CENTRAL DESERT ABORIGINAL LAND TRUST
KARLANTIJPA SOUTH ABORIGINAL LAND TRUST
WIRLIYAJARRAYI ABORIGINAL LAND TRUST
Willowra
Puyurru
Chilla Well
MALA ABORIGINAL LAND TRUST
PAWU ABORIGINAL LAND TRUST
GILES RA
Coniston
WABUDALI RANGE
Vaughan Springs
Djagamara Peak
Yuendumu
Yuelamu
YALPIRAKINU ABORIGINAL LAND TRUST
YALYIRIMBI RA
Laramba
AHAKEYE ABORIGINAL LAND TRUST
Anningie
YUNKANJINI ABORIGINAL LAND TRUST
Nyirripi
Newhaven
STUART BLUFF RA
Central Mount Wedge
Owl Gorge
Lake Lewis
Tilmouth Well Roadhouse
Lake Bennett
NGALURRTJU ABORIGINAL LAND TRUST
Ininti
Kintore
Tinki
KINTORE RD
Warren Creek
Mount Liebig
Illili
Papunya
Derwent
Ulambaura
Narwietooma
Milton Park
Haasts Bluff
Glen Helen
LARAPINTA TRAIL
WEST MACDONNELL
REDBANK GORGE
ORMISTON GORGE
SERPENTINE GORGE
Glen Helen Resort
NAMATJIRA DRIVE
IDIRRIKI RANGE
MISSIONARY PLAIN
TNORALA (GOSSE BLUFF) CR
LARAPINTA DR
Hermannsburg
Ipolera
Areyonga
PALM VALLEY
FINKE GORGE NATIONAL PARK
Lake MacDonald
Yuwalki
HAASTS BLUFF ABORIGINAL LAND TRUST
EMERY RANGE
SIR FREDERICK RANGE
WATSON RANGE
WATARRKA NATIONAL PARK
URRAMPINYI ILTJILTJARRI ABORIGINAL LAND TRUST
Kings Canyon Resort
KINGS CANYON
WATARRKA LEASEBACK AREA
LURITJA RD
Kings Creek
Tempe Downs
ILLAMURTA SPRINGS CR
ERNEST GILES RD
HENBURY METEORITES
Wallara Ranch
Lake Neale
Lake Hopkins
ELLIS RANGE
BLOODS RANGE
Lake Amadeus
Kaltukatjara (Docker River)
Kutjuntari
Warakurna
Warakurna Roadhouse
GREAT CENTRAL RD
TJUKARURU RD
PETERMANN RANGES
KATITI ABORIGINAL LAND TRUST
Yulara
LASSETER HWY
ULURU
Mutitjulu
KATA TJUTA (THE OLGAS)
ULURU-KATA TJUTA NATIONAL PARK
Curtin Springs
Angas Downs
BASEDOW RA
Imanpa
Erldunda
Mount Ebenezer Roadhouse
Pulcura Lake
Mygoora Lake
Lyndavale
PETERMANN ABORIGINAL LAND TRUST
Warlpapuka
Arnold Creek
Surveyor Generals Corner
Tjawupalya
Kalka
Pipalyatjara
Aparawatatja
Kanypi
MANN RANGES
Alpara
Amata
Mulga Park
New Well
THE MUSGRAVE RANGES
TOMKINSON RA
NGAANYATJARRA CENTRAL RESERVE
ANANGU PITJANTJATJARA YANKUNYTJATJARA LANDS
Victory Downs
Agnes
Mount Kintore
Mount Crombie (Ulpara)
Mount Harriet
Mount Caroline (Ulkiyanya)
Mount Woodroffe (Ngarutjaranya)
Mount Davenport
Mount Mair

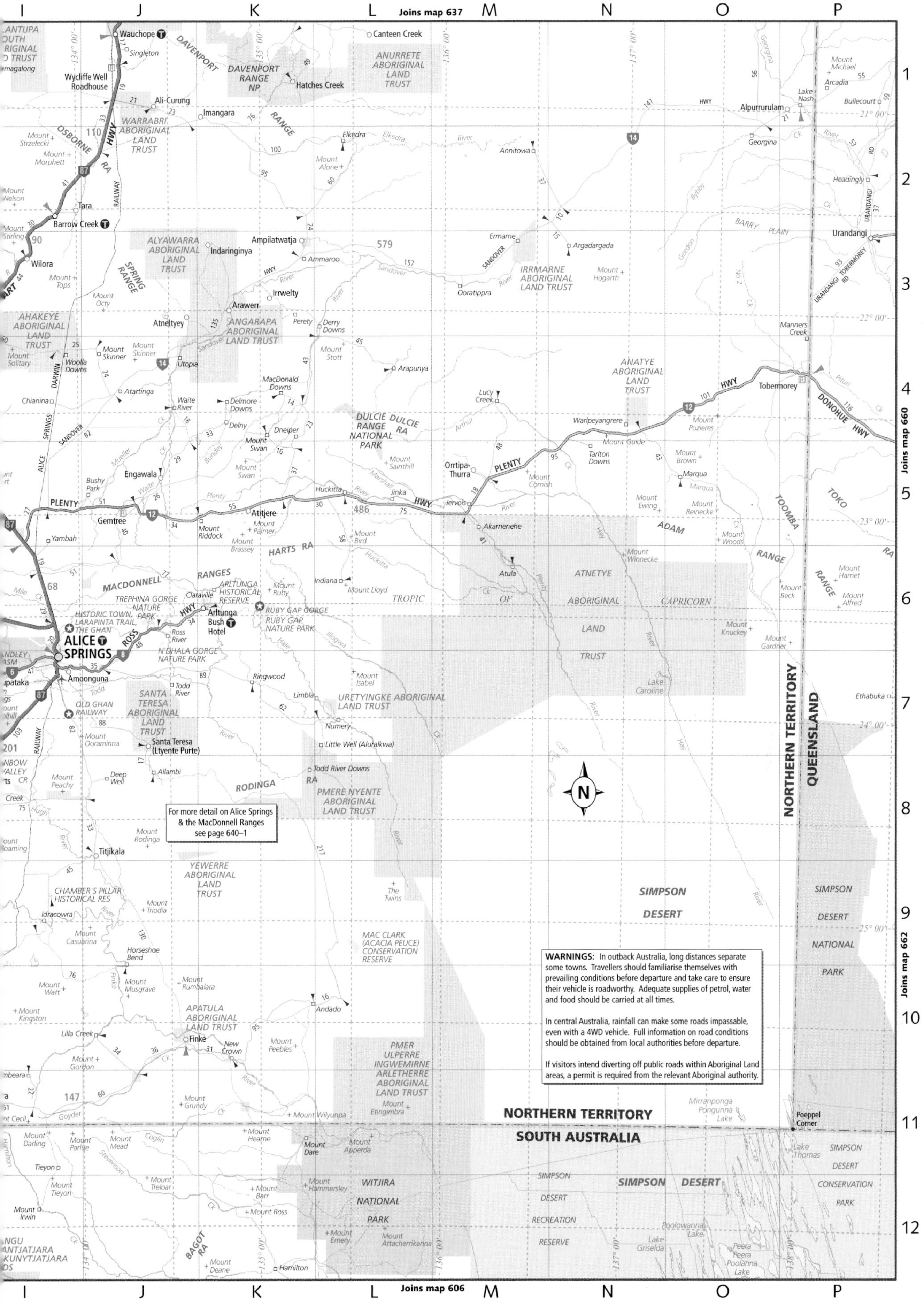
Joins map 637
Joins map 660
Joins map 662
Joins map 606
I J K L M N O P
1 2 3 4 5 6 7 8 9 10 11 12
Wauchope
Singleton
Wycliffe Well Roadhouse
DAVENPORT
DAVENPORT RANGE NP
Hatches Creek
Canteen Creek
ANURRETE ABORIGINAL LAND TRUST
Ali-Curung
Imangara
WARRABRI ABORIGINAL LAND TRUST
OSBORNE RA
Elkedra
Annitowa
Alpurrurulam
Lake Nash
Georgina
Mount Michael
Arcadia
Bullecourt
Headingly
Urandangi
Tara
Barrow Creek
ALYAWARRA ABORIGINAL LAND TRUST
Indaringinya
Ampilatwatja
Ammaroo
579
Ermarne
Argadargada
IRRMARNE ABORIGINAL LAND TRUST
Ooratippra
Wilora
SPRING RANGE
Irrwelty
Arawerr
Atneltyey
ANGARAPA ABORIGINAL LAND TRUST
Derry Downs
Perety
AHAKEYE ABORIGINAL LAND TRUST
Mount Skinner
Woolla Downs
Utopia
Arapunya
Manners Creek
ANATYE ABORIGINAL LAND TRUST
Tobermorey
DONOHUE HWY
Atartinga
MacDonald Downs
Delmore Downs
Waite River
Chianina
Delny
Dneiper
Mount Swan
DULCIE RANGE NATIONAL PARK
Lucy Creek
Warlpeyangrere
Mount Guide
Tarlton Downs
Mount Pozieres
Engawala
Bushy Park
Orrtipa-Thurra
PLENTY HWY
Marqua
Huckitta
Jinka
Jervois
486
Atitjere
Gemtree
Yambah
Mount Riddock
Akarnenehe
ADAM
TOOMBA RANGE
TOKO RA
HARTS RA
Mount Bird
MACDONNELL RANGES
ARLTUNGA HISTORICAL RESERVE
Claraville
Atula
ATNETYE ABORIGINAL LAND TRUST
TROPIC OF CAPRICORN
Indiana
Mount Lloyd
TREPHINA GORGE NATURE PARK
HISTORIC TOWN, LARAPINTA TRAIL, THE GHAN
Arltunga Bush Hotel
RUBY GAP GORGE
RUBY GAP NATURE PARK
Ross River
ALICE SPRINGS
N'DHALA GORGE NATURE PARK
Amoonguna
Ringwood
Todd River
Lake Caroline
Ethabuka
OLD GHAN RAILWAY
SANTA TERESA ABORIGINAL LAND TRUST
Limbla
URETYINGKE ABORIGINAL LAND TRUST
Numery
Santa Teresa (Ltyente Purte)
Little Well (Aluralkwa)
Todd River Downs
Allambi
RODINGA RA
PMERE NYENTE ABORIGINAL LAND TRUST
NORTHERN TERRITORY
QUEENSLAND
For more detail on Alice Springs & the MacDonnell Ranges see page 640–1
Titjikala
YEWERRE ABORIGINAL LAND TRUST
SIMPSON DESERT
SIMPSON DESERT NATIONAL PARK
CHAMBER'S PILLAR HISTORICAL RES
Idracowra
The Twins
MAC CLARK (ACACIA PEUCE) CONSERVATION RESERVE
Horseshoe Bend
WARNINGS: In outback Australia, long distances separate some towns. Travellers should familiarise themselves with prevailing conditions before departure and take care to ensure their vehicle is roadworthy. Adequate supplies of petrol, water and food should be carried at all times.
In central Australia, rainfall can make some roads impassable, even with a 4WD vehicle. Full information on road conditions should be obtained from local authorities before departure.
If visitors intend diverting off public roads within Aboriginal Land areas, a permit is required from the relevant Aboriginal authority.
APATULA ABORIGINAL LAND TRUST
Andado
Lilla Creek
Finke
New Crown
PMER ULPERRE INGWEMIRNE ARLETHERRE ABORIGINAL LAND TRUST
Mirranponga Pongunna Lake
Poeppel Corner
NORTHERN TERRITORY
SOUTH AUSTRALIA
Mount Dare
Tieyon
Lake Thomas
SIMPSON DESERT CONSERVATION PARK
WITJIRA NATIONAL PARK
SIMPSON DESERT RECREATION RESERVE
Poolowanna Lake
Lake Griselda
Peera Peera Poolanna Lake
BAGOT RA
Hamilton

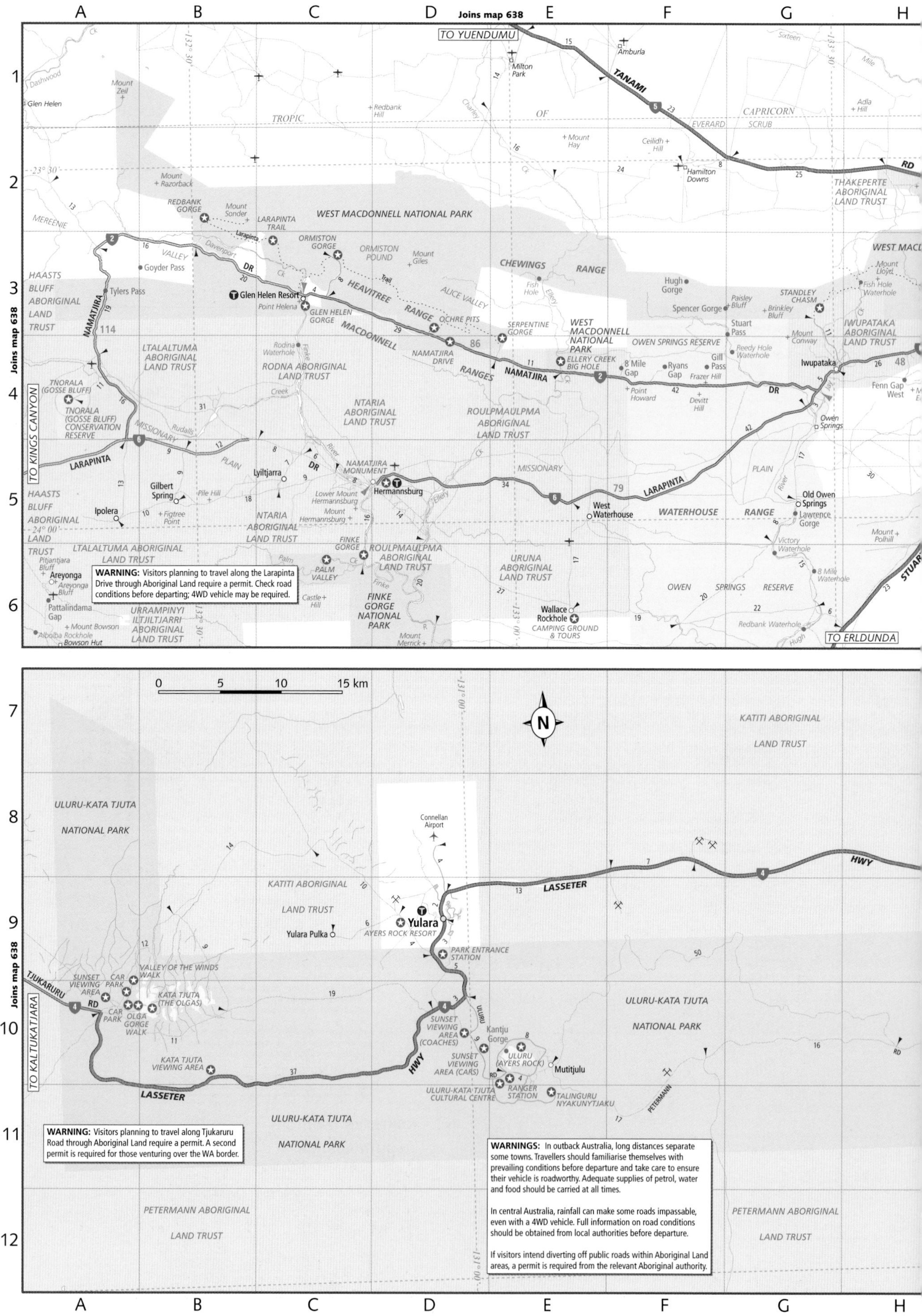

Joins map 638
TO YUENDUMU
Amburla
Milton Park
TANAMI
TROPIC OF CAPRICORN
EVERARD SCRUB
Glen Helen
Mount Zeil
Redbank Hill
Mount Hay
Ceilidh Hill
Hamilton Downs
RD
THAKEPERTE ABORIGINAL LAND TRUST
Mount Razorback
REDBANK GORGE
Mount Sonder
LARAPINTA TRAIL
WEST MACDONNELL NATIONAL PARK
MEREENIE
ORMISTON GORGE
ORMISTON POUND
Mount Giles
VALLEY
Davenport
DR
Goyder Pass
CHEWINGS RANGE
Fish Hole
Hugh Gorge
WEST MACDONNELL
Mount Lloyd
Fish Hole Waterhole
HAASTS BLUFF ABORIGINAL LAND TRUST
Tylers Pass
Glen Helen Resort
Point Helena
HEAVITREE
Trail
ALICE VALLEY
Spencer Gorge
Paisley Bluff
Brinkley Bluff
STANDLEY CHASM
NAMATJIRA
GLEN HELEN GORGE
RANGE
OCHRE PITS
SERPENTINE GORGE
WEST MACDONNELL NATIONAL PARK
Stuart Pass
Mount Conway
IWUPATAKA ABORIGINAL LAND TRUST
LTALALTUMA ABORIGINAL LAND TRUST
MACDONNELL
Rodina Waterhole
RODNA ABORIGINAL LAND TRUST
NAMATJIRA DRIVE
RANGES
OWEN SPRINGS RESERVE
Reedy Hole Waterhole
ELLERY CREEK BIG HOLE
8 Mile Gap
Ryans Gap
Gill Pass
Frazer Hill
Iwupataka
TNORALA (GOSSE BLUFF)
Creek
NTARIA ABORIGINAL LAND TRUST
ROULPMAULPMA ABORIGINAL LAND TRUST
Point Howard
Devitt Hill
Fenn Gap West
TNORALA (GOSSE BLUFF) CONSERVATION RESERVE
MISSIONARY
Rudalls
Owen Springs
TO KINGS CANYON
LARAPINTA
PLAIN
River
NAMATJIRA MONUMENT
MISSIONARY
PLAIN
Lyiltjarra
Hermannsburg
LARAPINTA
HAASTS BLUFF ABORIGINAL LAND
Gilbert Spring
Pile Hill
Lower Mount Hermannsburg
Ellery
West Waterhouse
WATERHOUSE RANGE
Old Owen Springs
Lawrence Gorge
Ipolera
Figtree Point
Mount Hermannsburg
NTARIA ABORIGINAL LAND TRUST
TRUST
LTALALTUMA ABORIGINAL LAND TRUST
FINKE GORGE
ROULPMAULPMA ABORIGINAL LAND TRUST
URUNA ABORIGINAL LAND TRUST
Victory Waterhole
Mount Polhill
Pitjantjara Bluff
Areyonga
Areyonga Bluff
WARNING: Visitors planning to travel along the Larapinta Drive through Aboriginal Land require a permit. Check road conditions before departing; 4WD vehicle may be required.
Palm
PALM VALLEY
Finke
8 Mile Waterhole
STUART
Pattalindama Gap
Castle Hill
FINKE GORGE NATIONAL PARK
OWEN SPRINGS RESERVE
URRAMPINYI ILTJILTJARRI ABORIGINAL LAND TRUST
Mount Bowson
Albolba Rockhole
Bowson Hut
Mount Merrick
Wallace Rockhole
CAMPING GROUND & TOURS
Redbank Waterhole
Hugh
TO ERLDUNDA
0 5 10 15 km
KATITI ABORIGINAL LAND TRUST
ULURU-KATA TJUTA NATIONAL PARK
Connellan Airport
HWY
LASSETER
KATITI ABORIGINAL LAND TRUST
Yulara
AYERS ROCK RESORT
Yulara Pulka
PARK ENTRANCE STATION
TJUKARURU RD
VALLEY OF THE WINDS WALK
SUNSET VIEWING AREA
CAR PARK
KATA TJUTA (THE OLGAS)
OLGA GORGE WALK
ULURU-KATA TJUTA NATIONAL PARK
TO KALTUKATJARA
ULURU
SUNSET VIEWING AREA (COACHES)
Kantju Gorge
SUNSET VIEWING AREA (CARS)
ULURU (AYERS ROCK)
Mutitjulu
KATA TJUTA VIEWING AREA
HWY
RANGER STATION
ULURU-KATA TJUTA CULTURAL CENTRE
TALINGURU NYAKUNYTJAKU
PETERMANN
LASSETER
ULURU-KATA TJUTA NATIONAL PARK
WARNING: Visitors planning to travel along Tjukaruru Road through Aboriginal Land require a permit. A second permit is required for those venturing over the WA border.
WARNINGS: In outback Australia, long distances separate some towns. Travellers should familiarise themselves with prevailing conditions before departure and take care to ensure their vehicle is roadworthy. Adequate supplies of petrol, water and food should be carried at all times.
In central Australia, rainfall can make some roads impassable, even with a 4WD vehicle. Full information on road conditions should be obtained from local authorities before departure.
If visitors intend diverting off public roads within Aboriginal Land areas, a permit is required from the relevant Aboriginal authority.
PETERMANN ABORIGINAL LAND TRUST
PETERMANN ABORIGINAL LAND TRUST

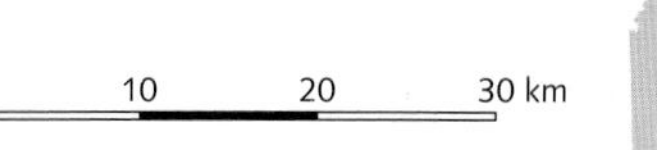

I J K L M N O P

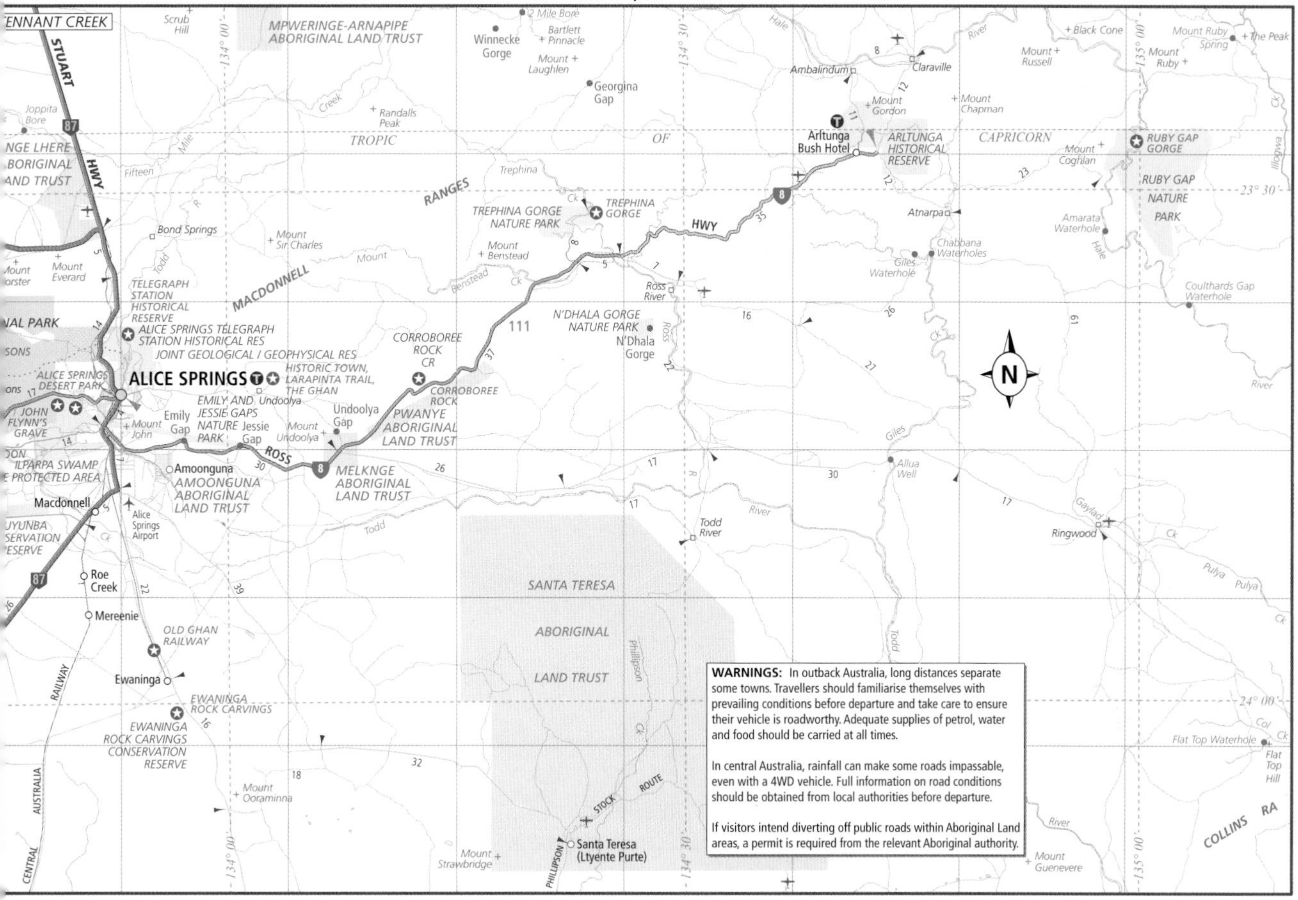

Joins map 639

WARNINGS: In outback Australia, long distances separate some towns. Travellers should familiarise themselves with prevailing conditions before departure and take care to ensure their vehicle is roadworthy. Adequate supplies of petrol, water and food should be carried at all times.

In central Australia, rainfall can make some roads impassable, even with a 4WD vehicle. Full information on road conditions should be obtained from local authorities before departure.

If visitors intend diverting off public roads within Aboriginal Land areas, a permit is required from the relevant Aboriginal authority.

Joins map 639

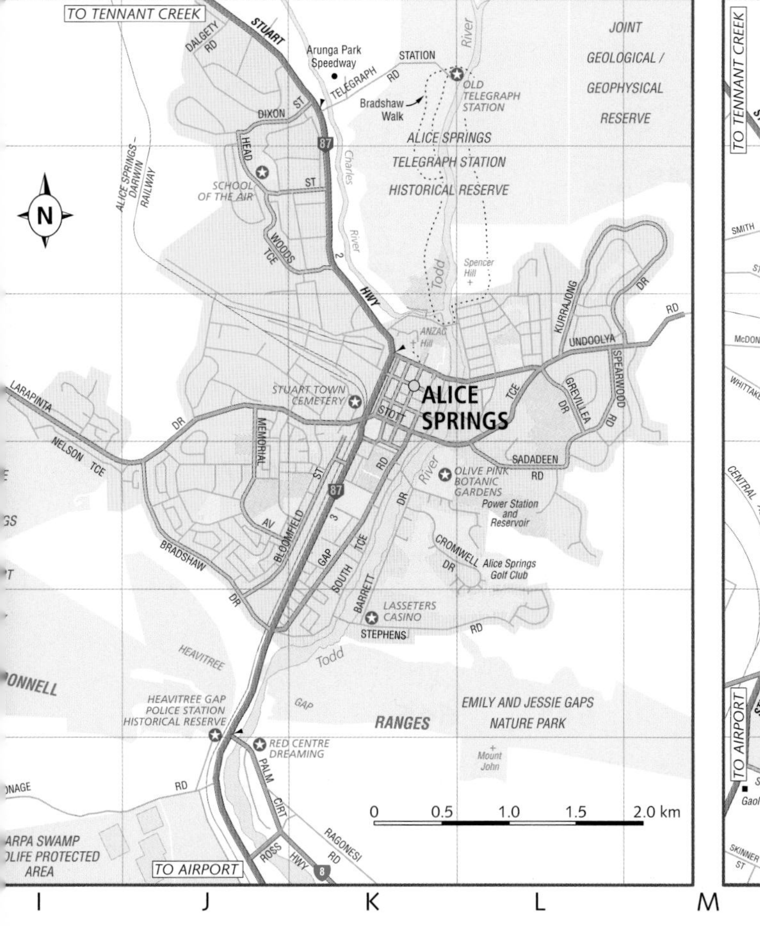

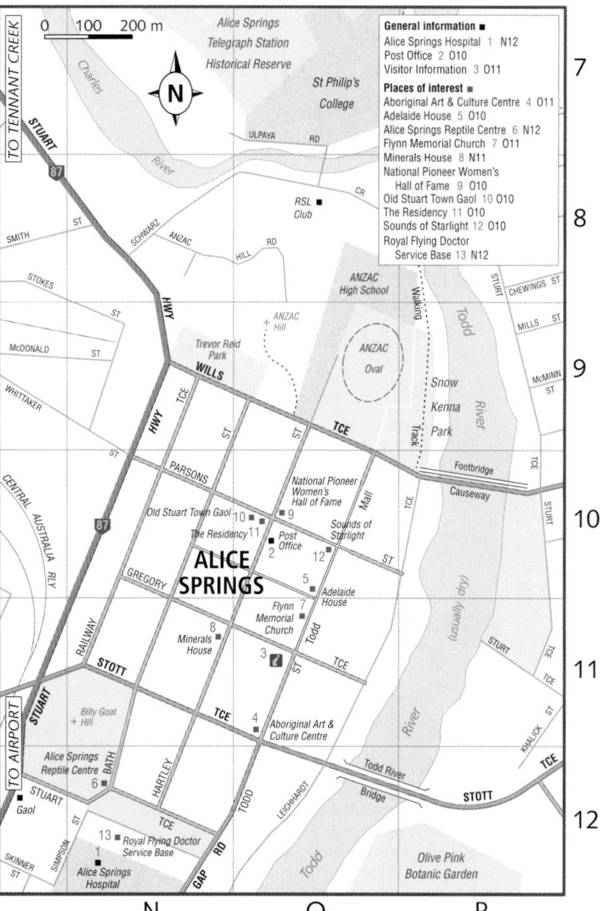

I J K L M N O P

MAPS

QUEENSLAND

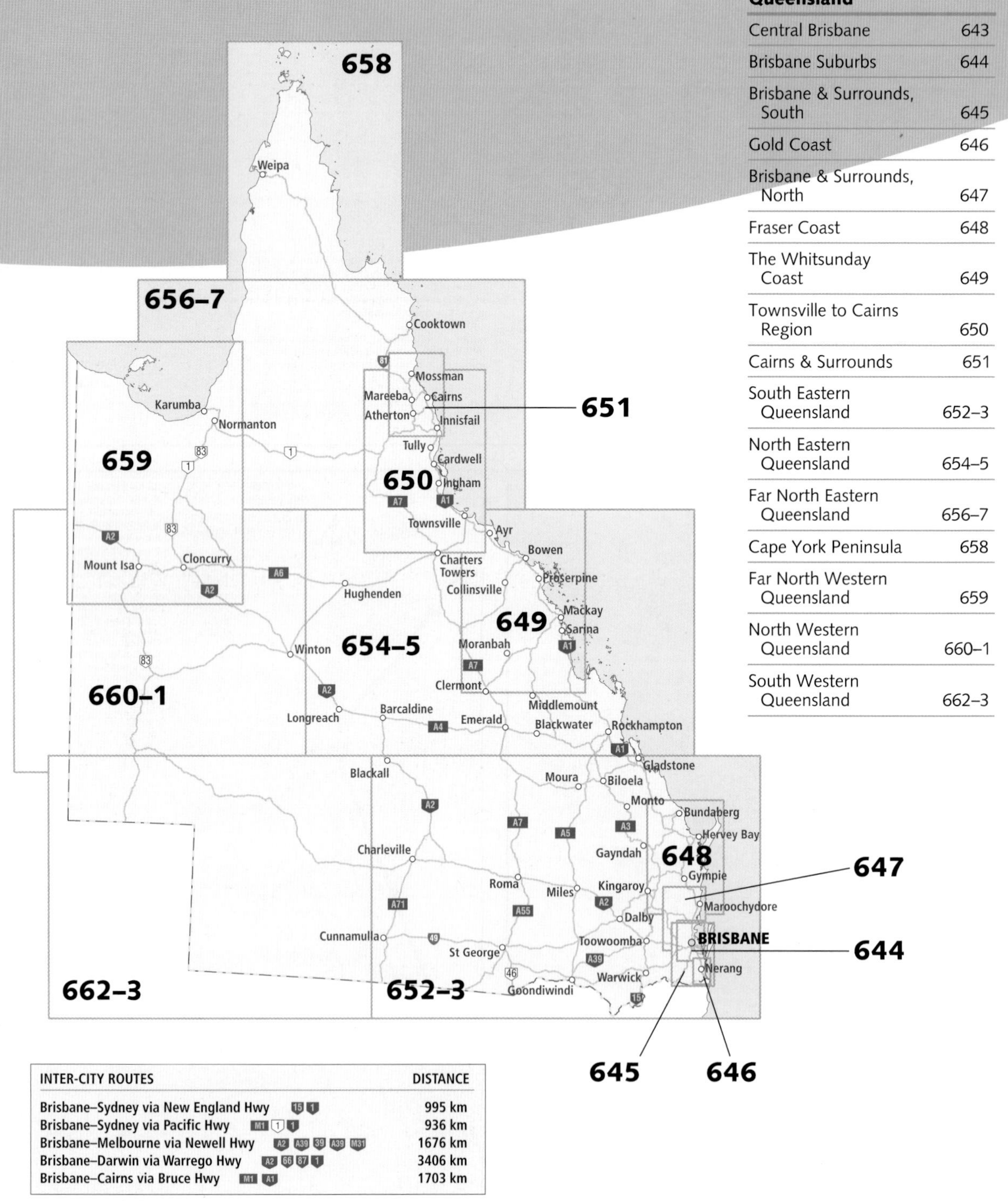

Queensland

INTER-CITY ROUTES	DISTANCE
Brisbane–Sydney via New England Hwy 15 1	995 km
Brisbane–Sydney via Pacific Hwy M1 1 1	936 km
Brisbane–Melbourne via Newell Hwy A2 A39 39 A39 M31	1676 km
Brisbane–Darwin via Warrego Hwy A2 66 87 1	3406 km
Brisbane–Cairns via Bruce Hwy M1 A1	1703 km

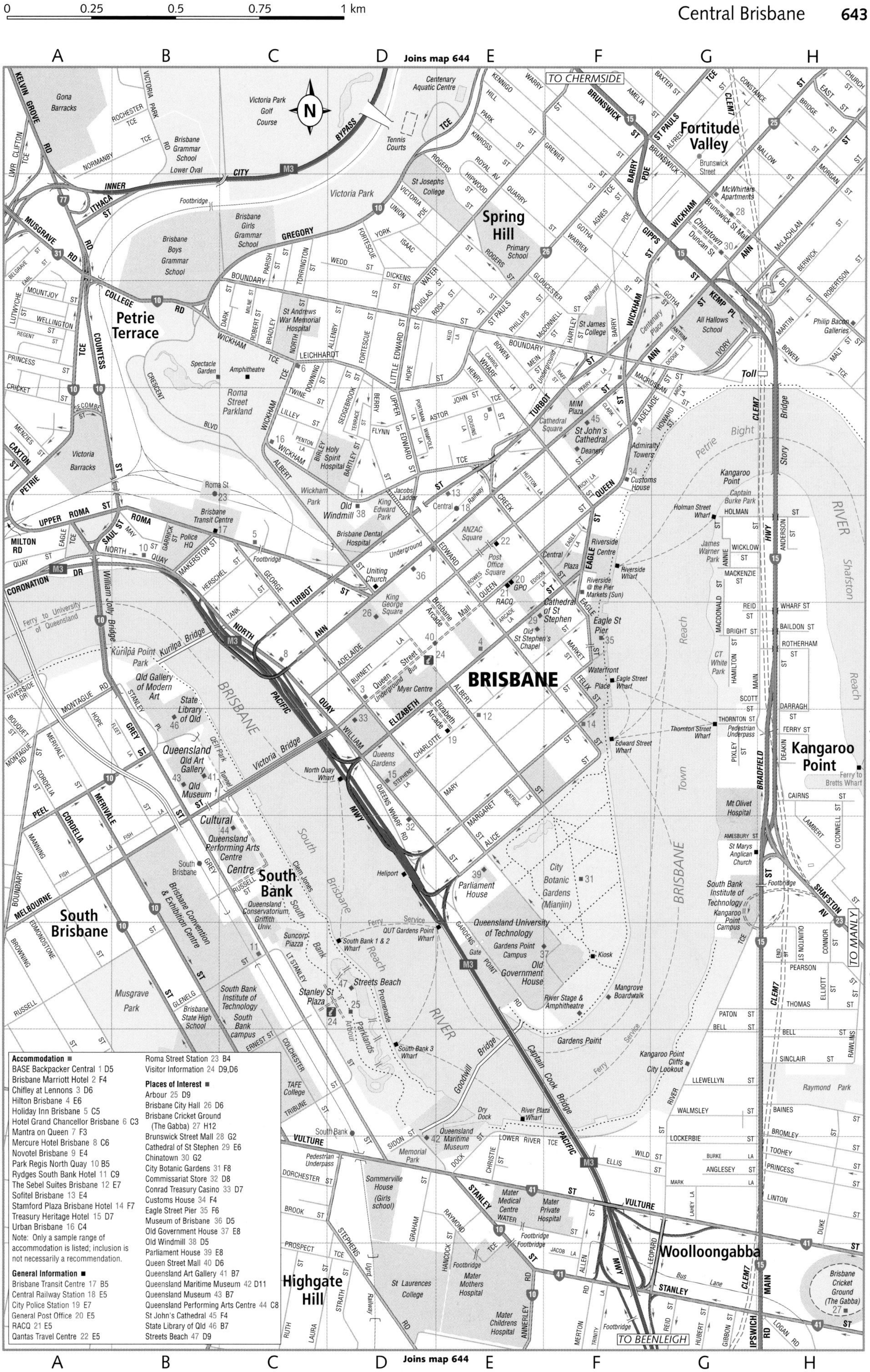

0 0.25 0.5 0.75 1 km

Joins map 644

Accommodation
- BASE Backpacker Central 1 D5
- Brisbane Marriott Hotel 2 F4
- Chifley at Lennons 3 D6
- Hilton Brisbane 4 E6
- Holiday Inn Brisbane 5 C5
- Hotel Grand Chancellor Brisbane 6 C3
- Mantra on Queen 7 F3
- Mercure Hotel Brisbane 8 C6
- Novotel Brisbane 9 E4
- Park Regis North Quay 10 B5
- Rydges South Bank Hotel 11 C9
- The Sebel Suites Brisbane 12 E7
- Sofitel Brisbane 13 E4
- Stamford Plaza Brisbane Hotel 14 F7
- Treasury Heritage Hotel 15 D7
- Urban Brisbane 16 C4

Note: Only a sample range of accommodation is listed; inclusion is not necessarily a recommendation.

General Information
- Brisbane Transit Centre 17 B5
- Central Railway Station 18 E5
- City Police Station 19 E7
- General Post Office 20 E5
- RACQ 21 E5
- Qantas Travel Centre 22 E5
- Roma Street Station 23 B4
- Visitor Information 24 D9,D6

Places of Interest
- Arbour 25 D9
- Brisbane City Hall 26 D6
- Brisbane Cricket Ground (The Gabba) 27 H12
- Brunswick Street Mall 28 G2
- Cathedral of St Stephen 29 E6
- Chinatown 30 G2
- City Botanic Gardens 31 F8
- Commissariat Store 32 D8
- Conrad Treasury Casino 33 D7
- Customs House 34 F4
- Eagle Street Pier 35 F6
- Museum of Brisbane 36 D5
- Old Government House 37 E8
- Old Windmill 38 D5
- Parliament House 39 E8
- Queen Street Mall 40 D6
- Queensland Art Gallery 41 B7
- Queensland Maritime Museum 42 D11
- Queensland Museum 43 B7
- Queensland Performing Arts Centre 44 C8
- St John's Cathedral 45 F4
- State Library of Qld 46 B7
- Streets Beach 47 D9

Joins map 644

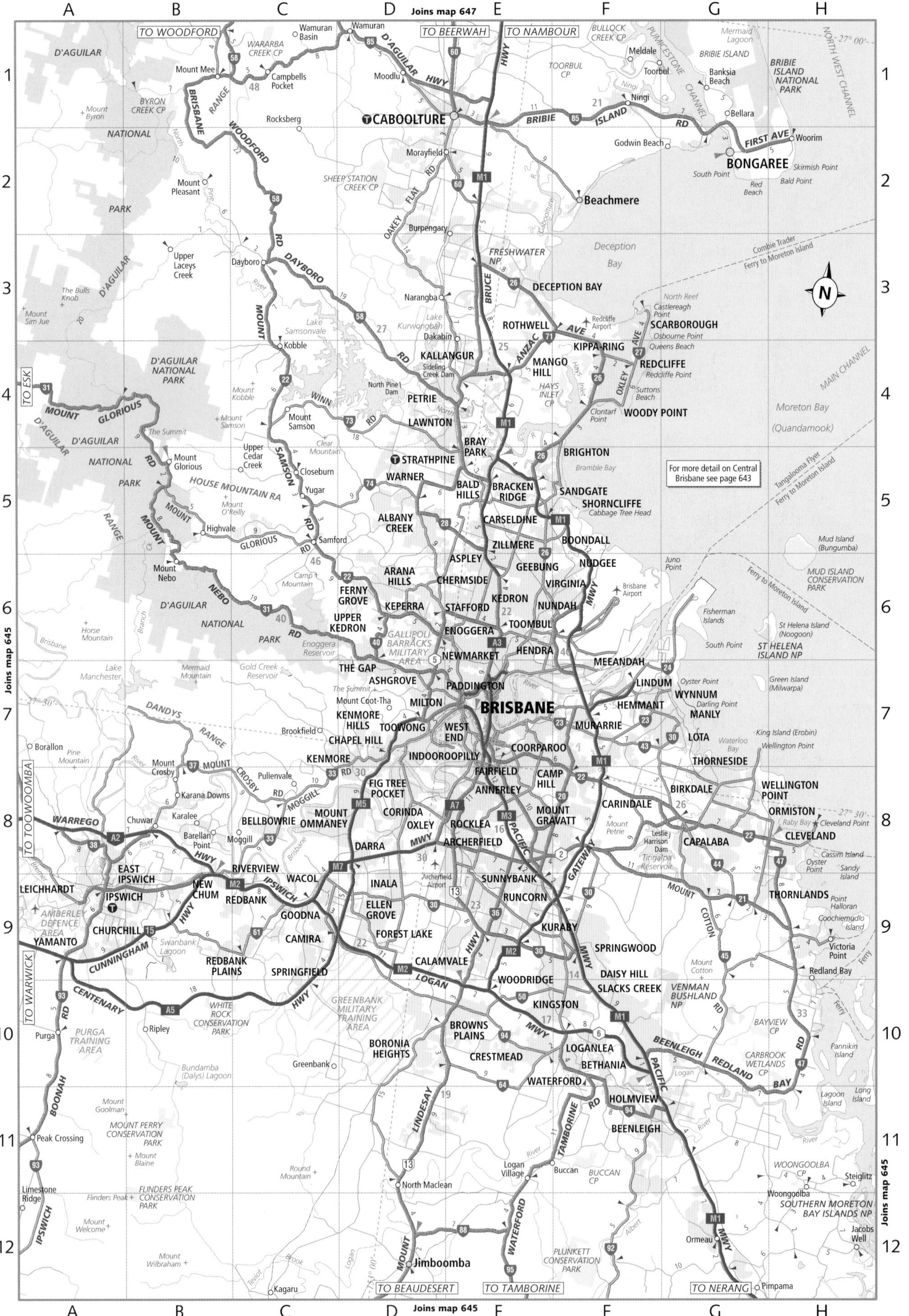
0 5 10 15 20 km
Joins map 647
Joins map 645
TO WOODFORD
TO BEERWAH
TO NAMBOUR
TO ESK
TO TOOWOOMBA
TO WARWICK
TO BEAUDESERT
TO TAMBORINE
TO NERANG
D'AGUILAR NATIONAL PARK
Mount Mee
Wamuran
Wamuran Basin
WARARBA CREEK CP
Campbells Pocket
BYRON CREEK CP
Mount Byron
Moodlu
D'AGUILAR HWY
CABOOLTURE
Rocksberg
BRISBANE RANGE
WOODFORD RD
Morayfield
SHEEP STATION CREEK CP
Mount Pleasant
OAKEY FLAT RD
Burpengary
Upper Laceys Creek
Dayboro
DAYBORO RD
Narangba
The Bulls Knob
Mount Sim Jue
Lake Samsonvale
Lake Kurwongbah
Dakabin
Kobble
KALLANGUR
Sideling Creek Dam
North Pine Dam
PETRIE
LAWNTON
MOUNT GLORIOUS
WINN RD
Mount Kobble
Mount Samson
The Summit
Mount Glorious
Upper Cedar Creek
Clear Mountain
STRATHPINE
BRAY PARK
Closeburn
SAMSON RD
HOUSE MOUNTAIN RA
Yugar
WARNER
BALD HILLS
Mount O'Reilly
Highvale
GLORIOUS RD
Samford
ALBANY CREEK
Mount Nebo
Camp Mountain
ARANA HILLS
FERNY GROVE
KEPERRA
UPPER KEDRON
NEBO RD
Horse Mountain
GALLIPOLI BARRACKS MILITARY AREA
Enoggera Reservoir
Lake Manchester
Mermaid Mountain
Gold Creek Reservoir
THE GAP
ASHGROVE
Mount Coot-Tha
KENMORE HILLS
Brookfield
CHAPEL HILL
KENMORE
DANDYS RANGE
Borallon
Pine Mountain
Mount Crosby
MOUNT CROSBY RD
Pullenvale
MOGGILL RD
Karana Downs
Karalee
Chuwar
Barellan Point
Moggill
BELLBOWRIE
WARREGO HWY
EAST IPSWICH
RIVERVIEW
LEICHHARDT
IPSWICH
NEW CHUM
REDBANK
IPSWICH
WACOL
AMBERLEY DEFENCE AREA
CHURCHILL
YAMANTO
GOODNA
CAMIRA
Swanbank Lagoon
CUNNINGHAM HWY
REDBANK PLAINS
SPRINGFIELD
CENTENARY HWY
WHITE ROCK CONSERVATION PARK
Ripley
Purga
PURGA TRAINING AREA
Bundamba (Dalys) Lagoon
BOONAH RD
Mount Goolman
MOUNT PERRY CONSERVATION PARK
Mount Blaine
Peak Crossing
Limestone Ridge
FLINDERS PEAK CONSERVATION PARK
Flinders Peak
Mount Welcome
IPSWICH
Mount Wilbraham
Kagaru
Round Mountain
Greenbank
GREENBANK MILITARY TRAINING AREA
BORONIA HEIGHTS
LINDESAY
North Maclean
MOUNT
Jimboomba
Meldale
Toorbul
TOORBUL CP
BULLOCK CREEK CP
Ningi
BRIBIE ISLAND RD
BRIBIE ISLAND
PUMICESTONE CHANNEL
Mermaid Lagoon
Banksia Beach
Bellara
BRIBIE ISLAND NATIONAL PARK
NORTH WEST CHANNEL
FIRST AVE
Woorim
BONGAREE
Godwin Beach
South Point
Red Beach
Skirmish Point
Bald Point
Beachmere
Deception Bay
FRESHWATER NP
DECEPTION BAY
BRUCE HWY
Combie Trader Ferry to Moreton Island
North Reef
Castlereagh Point
SCARBOROUGH
Osbourne Point
Queens Beach
Redcliffe Airport
ROTHWELL
ANZAC AVE
KIPPA-RING
REDCLIFFE
Redcliffe Point
MANGO HILL
HAYS INLET CP
Suttons Beach
OXLEY AVE
Clontarf Point
WOODY POINT
MAIN CHANNEL
Moreton Bay
(Quandamook)
BRIGHTON
Bramble Bay
For more detail on Central Brisbane see page 643
Tangalooma Flyer Ferry to Moreton Island
BRACKEN RIDGE
SANDGATE
SHORNCLIFFE
Cabbage Tree Head
CARSELDINE
BOONDALL
ZILLMERE
ASPLEY
GEEBUNG
NUDGEE
Juno Point
Mud Island (Bungumba)
MUD ISLAND CONSERVATION PARK
Ferry to Moreton Island
CHERMSIDE
VIRGINIA
Brisbane Airport
KEDRON
STAFFORD
NUNDAH
Fisherman Islands
St Helena Island (Noogoon)
ST HELENA ISLAND NP
ENOGGERA
TOOMBUL
South Point
NEWMARKET
HENDRA
MEEANDAH
PADDINGTON
LINDUM
Oyster Point
WYNNUM
Green Island (Milwarpa)
MILTON
BRISBANE
HEMMANT
Darling Point
MANLY
TOOWONG
WEST END
MURARRIE
LOTA
King Island (Erobin)
Waterloo Bay
Wellington Point
COORPAROO
THORNESIDE
INDOOROOPILLY
FAIRFIELD
CAMP HILL
BIRKDALE
WELLINGTON POINT
ANNERLEY
ORMISTON
FIG TREE POCKET
CARINDALE
Raby Bay
Cleveland Point
CORINDA
MOUNT OMMANEY
OXLEY
ROCKLEA
MOUNT GRAVATT
Mount Petrie
Leslie Harrison Dam
Tingalpa Reservoir
CLEVELAND
CAPALABA
ARCHERFIELD
PACIFIC MWY
DARRA
Cassim Island
Oyster Point
Sandy Island
Archerfield Airport
GATEWAY MWY
SUNNYBANK
INALA
MOUNT COTTON RD
THORNLANDS
Point Halloran
ELLEN GROVE
RUNCORN
Coochiemudlo Island
FOREST LAKE
KURABY
Victoria Point
SPRINGWOOD
CALAMVALE
LOGAN MWY
Mount Cotton
Redland Bay
DAISY HILL
WOODRIDGE
SLACKS CREEK
VENMAN BUSHLAND NP
KINGSTON
BAYVIEW CP
BROWNS PLAINS
LOGANLEA
BEENLEIGH REDLAND BAY RD
CARBROOK WETLANDS CP
Pannikin Island
CRESTMEAD
BETHANIA
WATERFORD
HOLMVIEW
Lagoon Island
Long Island
BEENLEIGH
TAMBORINE RD
Logan Village
Buccan
BUCCAN CP
WOONGOOLBA CP
Steiglitz
Woongoolba
SOUTHERN MORETON BAY ISLANDS NP
Jacobs Well
WATERFORD
PLUNKETT CONSERVATION PARK
Ormeau
Pimpama
27° 00'
27° 30'

0 10 20 30 km

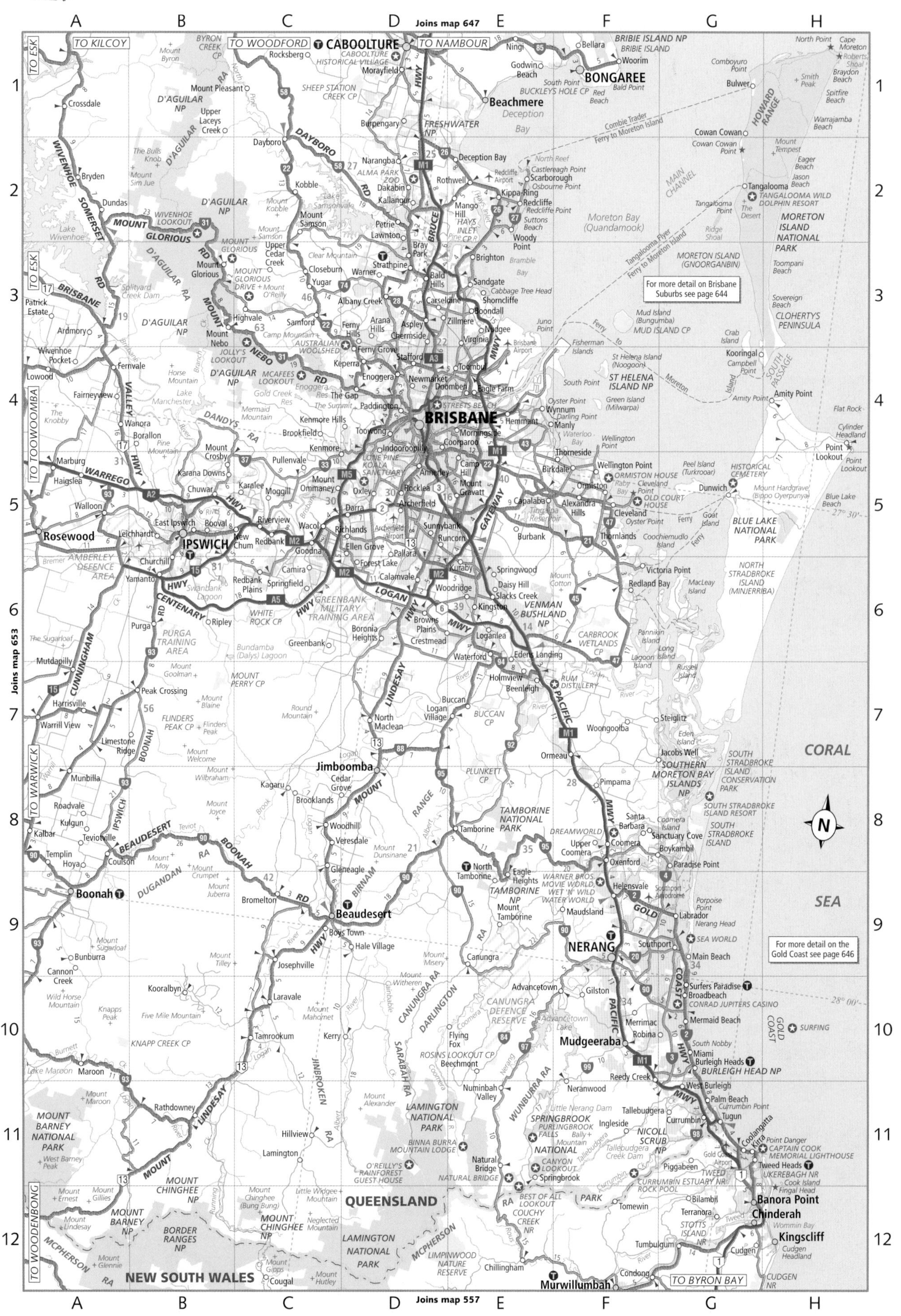

Joins map 647
Joins map 653
Joins map 557
TO KILCOY
TO WOODFORD
TO NAMBOUR
TO ESK
TO TOOWOOMBA
TO WARWICK
TO WOODENBONG
TO BYRON BAY
For more detail on Brisbane Suburbs see page 644
For more detail on the Gold Coast see page 646
CABOOLTURE
BONGAREE
Beachmere
Deception Bay
Redcliffe
Woody Point
Sandgate
Brighton
BRISBANE
Wynnum
Manly
Cleveland
Capalaba
Redland Bay
IPSWICH
Rosewood
Springfield
Logan
Beenleigh
Jimboomba
Beaudesert
Boonah
Tamborine
NERANG
Southport
Surfers Paradise
Broadbeach
Mudgeeraba
Burleigh Heads
Coolangatta
Tweed Heads
Banora Point
Chinderah
Kingscliff
Murwillumbah
Moreton Bay (Quandamook)
MORETON ISLAND NATIONAL PARK
BLUE LAKE NATIONAL PARK
NORTH STRADBROKE ISLAND (MINJERRIBA)
SOUTH STRADBROKE ISLAND
D'AGUILAR NP
TAMBORINE NATIONAL PARK
LAMINGTON NATIONAL PARK
SPRINGBROOK NATIONAL PARK
MOUNT BARNEY NATIONAL PARK
MOUNT CHINGHEE NP
BORDER RANGES NP
QUEENSLAND
NEW SOUTH WALES
CORAL SEA

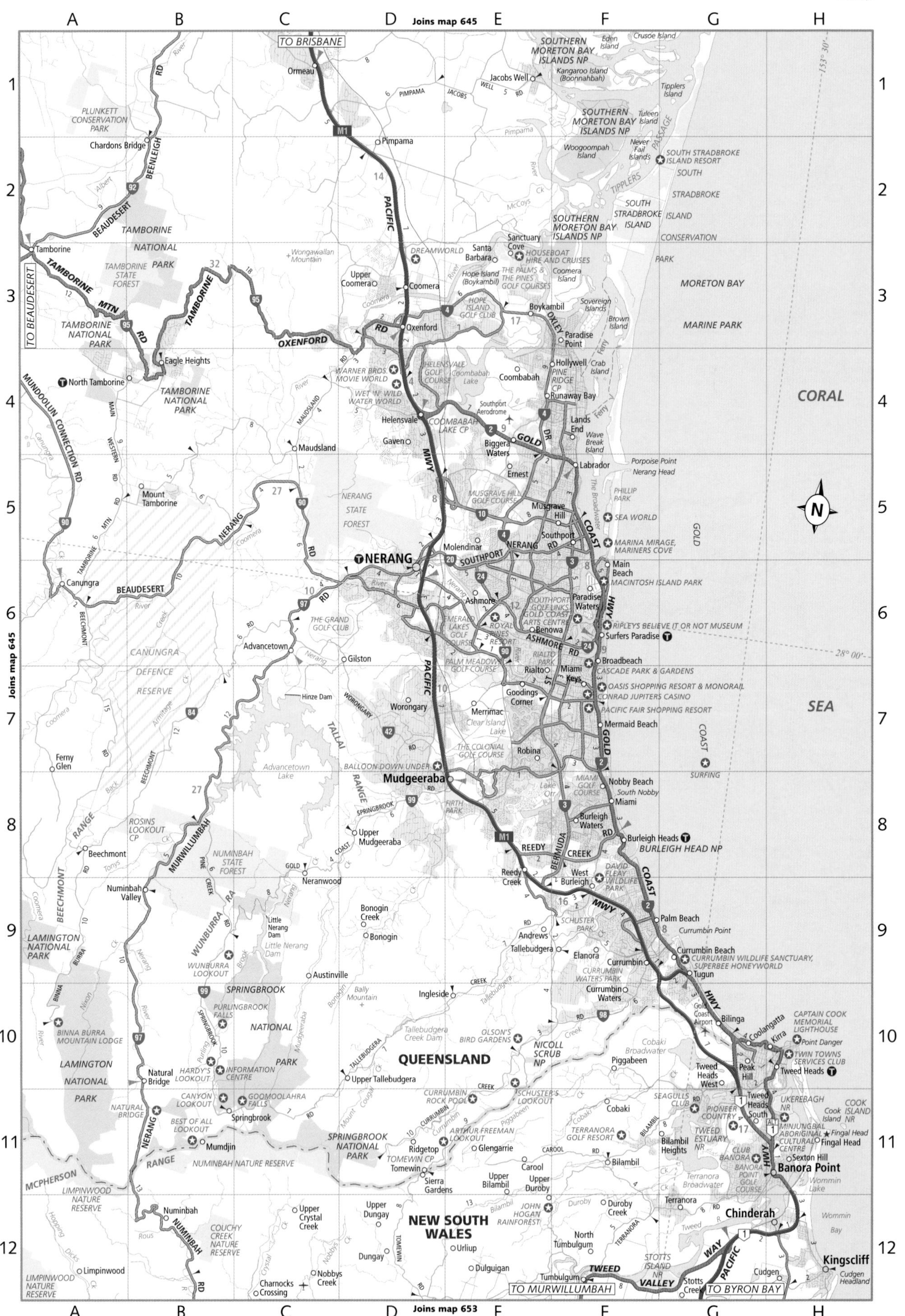
0 2 4 6 8 10 km
Joins map 645
Joins map 653
TO BRISBANE
TO BEAUDESERT
TO MURWILLUMBAH
TO BYRON BAY
CORAL
SEA
QUEENSLAND
NEW SOUTH WALES
NERANG
Southport
Surfers Paradise
Mudgeeraba
Burleigh Heads
Coolangatta
Tweed Heads
Banora Point
Chinderah
Kingscliff
MORETON BAY
MARINE PARK
SOUTH STRADBROKE ISLAND
TAMBORINE NATIONAL PARK
SPRINGBROOK NATIONAL PARK
LAMINGTON NATIONAL PARK
CANUNGRA DEFENCE RESERVE
NERANG STATE FOREST
BURLEIGH HEAD NP
SEA WORLD
DREAMWORLD
WARNER BROS. MOVIE WORLD
WET 'N' WILD WATER WORLD
CURRUMBIN WILDLIFE SANCTUARY
Gold Coast Airport

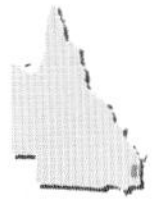

0 10 20 30 km

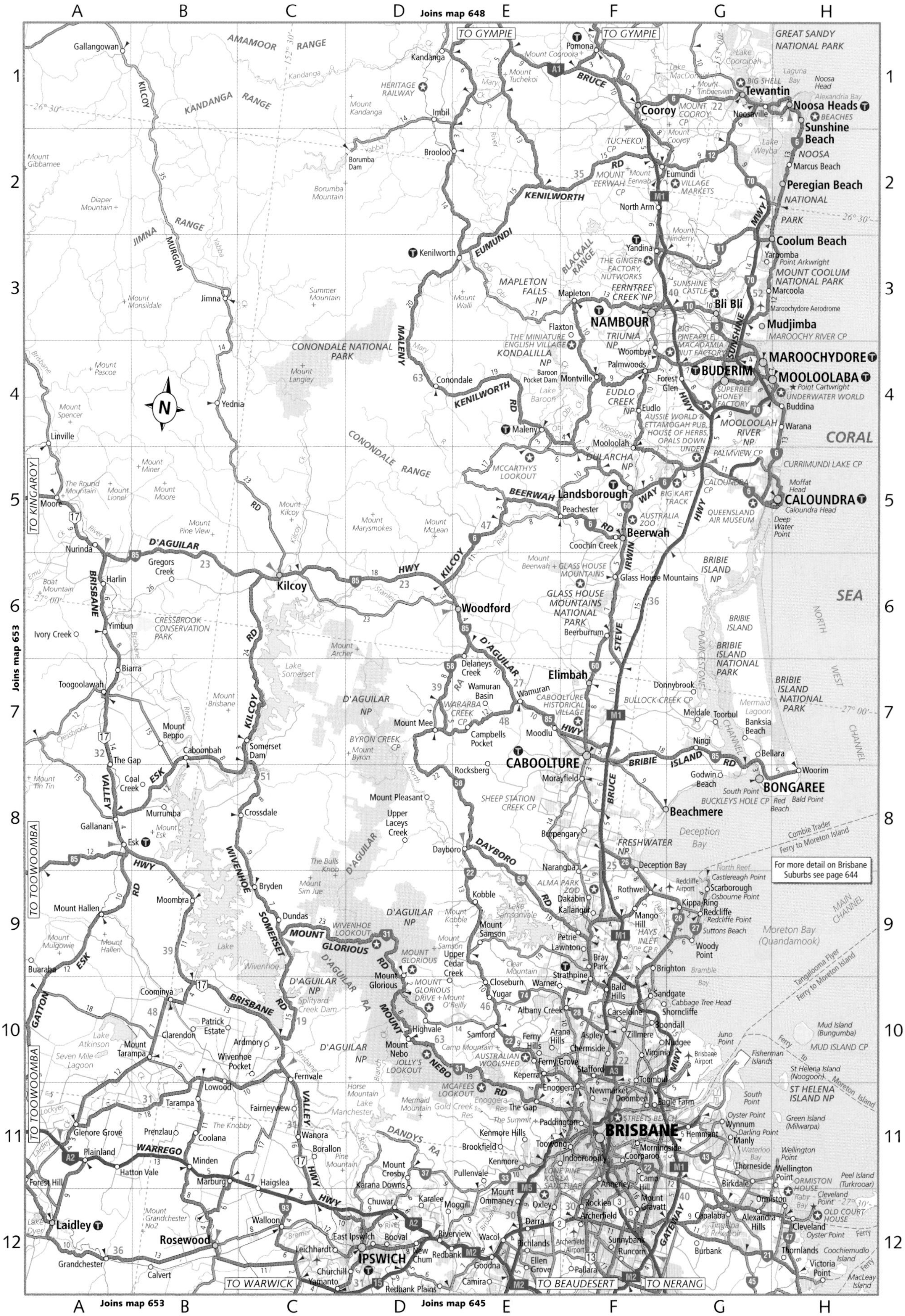

A B C D E F G H
Joins map 648
Joins map 653
Joins map 645
1 2 3 4 5 6 7 8 9 10 11 12
TO GYMPIE
TO KINGAROY
TO TOOWOOMBA
TO WARWICK
TO BEAUDESERT
TO NERANG
Gallangowan
AMAMOOR RANGE
KANDANGA RANGE
Kandanga
HERITAGE RAILWAY
Mount Kandanga
Imbil
Borumba Dam
Borumba Mountain
Brooloo
Mount Gibbarnee
Diaper Mountain
JIMNA RANGE
MURGON
KILCOY
Jimna
Mount Monsildale
Summer Mountain
Kenilworth
EUMUNDI
KENILWORTH
MAPLETON FALLS NP
Mapleton
Mount Walli
CONONDALE NATIONAL PARK
Mount Langley
MALENY
Conondale
Mount Pascoe
Yednia
Mount Spencer
Linville
CONONDALE RANGE
Maleny
MCCARTHYS LOOKOUT
Mount Miner
The Round Mountain
Mount Lionel
Mount Moore
Moore
Mount Kilcoy
Mount Marysmokes
Mount McLean
Mount Pine View
Nurinda
D'AGUILAR
Gregors Creek
Harlin
Boat Mountain
Kilcoy
HWY
Woodford
Ivory Creek
Yimbun
CRESSBROOK CONSERVATION PARK
Mount Archer
Lake Somerset
Toogoolawah
Biarra
Mount Brisbane
Mount Beppo
Delaneys Creek
Wamuran Basin
WARARBA CREEK CP
Wamuran
D'AGUILAR NP
Mount Mee
Campbells Pocket
BYRON CREEK CP
Mount Byron
The Gap
Caboonbah
Somerset Dam
Mount Tin Tin
Coal Creek
ESK
VALLEY
Murrumba
Crossdale
Rocksberg
Mount Pleasant
Upper Laceys Creek
SHEEP STATION CREEK CP
Gallanani
Esk
Mount Esk
WIVENHOE
Dayboro
DAYBORO
The Bulls Knob
Mount Sim Jue
Bryden
Kobble
Mount Hallen
Moombra
Dundas
WIVENHOE LOOKOUT
MOUNT GLORIOUS
Mount Samson
Mount Mulgowie
Buaraba
Lake Wivenhoe
SOMERSET
MOUNT GLORIOUS CP
Mount Glorious
Upper Cedar Creek
Closeburn
Clear Mountain
Yugar
Coominya
BRISBANE
Splityard Creek Dam
MOUNT GLORIOUS DRIVE
Mount O'Reilly
Samford
GATTON
Lake Atkinson
Patrick Estate
Clarendon
Ardmory
Mount Tarampa
Seven Mile Lagoon
Wivenhoe Pocket
Highvale
Camp Mountain
Mount Nebo
JOLLY'S LOOKOUT
NEBO
AUSTRALIAN WOOLSHED
Fernvale
Lowood
Horse Mountain
Lake Manchester
Mermaid Mountain
MCAFEES LOOKOUT
Gold Creek Res
Enoggera Res
The Summit
Tarampa
Fairneyview
The Knobby
Glenore Grove
Prenzlau
Wanora
Coolana
Plainland
WARREGO
Minden
Hatton Vale
Borallon
Pine Mountain
DANDYS RA
Kenmore Hills
Brookfield
Kenmore
Forest Hill
Marburg
Haigslea
Mount Crosby
Karana Downs
Pullenvale
Mount Grandchester No2
Chuwar
Karalee
Moggill
Mount Ommaney
Laidley
Lake Dyer
Rosewood
Walloon
East Ipswich
Booval
Riverview
Wacol
Leichhardt
Redbank
IPSWICH
New Chum
Grandchester
Calvert
Churchill
Yamanto
Redbank Plains
Goodna
Camira
Pomona
Mount Cooroora
Mount Tuchekoi
BRUCE
Cooroy
MOUNT COOROY CP
Mount Cooroy
TUCHEKOI CP
Eumundi
VILLAGE MARKETS
MOUNT EERWAH CP
Mount Eerwah
North Arm
Mount Ninderry
Yandina
THE GINGER FACTORY, NUTWORKS
BLACKALL RANGE
SUNSHINE CASTLE
FERNTREE CREEK NP
NAMBOUR
Bli Bli
Flaxton
THE MINIATURE ENGLISH VILLAGE
KONDALILLA NP
TRIUNIA NP
BIG PINEAPPLE, MACADAMIA NUT FACTORY
Woombye
Palmwoods
Baroon Pocket Dam
Lake Baroon
Montville
BUDERIM
Forest Glen
SUPERBEE HONEY FACTORY
EUDLO CREEK NP
Eudlo
AUSSIE WORLD & ETTAMOGAH PUB, HOUSE OF HERBS, OPALS DOWN UNDER
Mooloolah
MOOLOOLAH RIVER NP
DULARCHA NP
PALMVIEW CP
BEERWAH
Landsborough
BIG KART TRACK
CALOUNDRA CP
Peachester
AUSTRALIA ZOO
Beerwah
Coochin Creek
QUEENSLAND AIR MUSEUM
IRWIN
Mount Beerwah
GLASS HOUSE MOUNTAINS
Glass House Mountains
GLASS HOUSE MOUNTAINS NATIONAL PARK
Beerburrum
STEVE
Elimbah
CABOOLTURE HISTORICAL VILLAGE
Moodlu
CABOOLTURE
Morayfield
Donnybrook
BULLOCK CREEK CP
Meldale
Toorbul
Ningi
BRIBIE ISLAND
Godwin Beach
Beachmere
Burpengary
FRESHWATER NP
Narangba
Deception Bay
Deception Bay
ALMA PARK ZOO
Dakabin
Kallangur
Rothwell
Mango Hill
HAYS INLET CP
Petrie
Lawnton
Bray Park
Strathpine
Warner
Bald Hills
Albany Creek
Carseldine
Arana Hills
Aspley
Ferny Hills
Chermside
Ferny Grove
Keperra
Stafford
Enoggera
Newmarket
Paddington
The Gap
Toowong
Indooroopilly
KOALA SANCTUARY
Oxley
Darra
Richlands
Ellen Grove
Archerfield
Rocklea
Pallara
Sunnybank
Runcorn
Great Sandy National Park
Lake Cooroibah
Lake MacDonald
Mount Tinbeerwah
BIG SHELL
Tewantin
Noosaville
Noosa Head
Noosa Heads
Laguna Bay
Alexandria Bay
BEACHES
Sunshine Beach
Lake Weyba
NOOSA
Marcus Beach
Peregian Beach
NATIONAL PARK
Coolum Beach
Yaroomba
Point Arkwright
MOUNT COOLUM NATIONAL PARK
Marcoola
Maroochydore Aerodrome
Mudjimba
MAROOCHY RIVER CP
MAROOCHYDORE
MOOLOOLABA
Point Cartwright
UNDERWATER WORLD
Buddina
Warana
CORAL
CURRIMUNDI LAKE CP
Moffat Head
CALOUNDRA
Caloundra Head
Deep Water Point
BRIBIE ISLAND NP
SEA
NORTH
BRIBIE ISLAND NATIONAL PARK
PUMICESTONE CHANNEL
Mermaid Lagoon
Banksia Beach
Bellara
Woorim
BONGAREE
South Point
BUCKLEYS HOLE CP
Red Beach
Bald Point
Combie Trader Ferry to Moreton Island
For more detail on Brisbane Suburbs see page 644
Redcliffe Airport
North Reef
Castlereagh Point
Scarborough
Osbourne Point
Kippa-Ring
Redcliffe
Redcliffe Point
Suttons Beach
Woody Point
Moreton Bay (Quandamooka)
MAIN CHANNEL
Tangalooma Flyer Ferry to Moreton Island
Brighton
Sandgate
Cabbage Tree Head
Shorncliffe
Boondall
Zillmere
Nudgee
Juno Point
Mud Island (Bungumba)
MUD ISLAND CP
Virginia
Brisbane Airport
Fisherman Islands
St Helena Island (Noogoon)
ST HELENA ISLAND NP
Toombul
Doomben
Eagle Farm
South Point
STREETS BEACH
BRISBANE
Hemmant
Oyster Point
Wynnum
Darling Point
Manly
Green Island (Milwarpa)
Morningside
Coorparoo
Waterloo Bay
Thorneside
Wellington Point
Wellington Point
Annerley
Camp Hill
Birkdale
ORMISTON HOUSE
Raby Bay
Ormiston
Cleveland Point
Peel Island (Turkrooar)
Mount Gravatt
Capalaba
Alexandra Hills
OLD COURT HOUSE
Cleveland
Leslie Harrison Reservoir
Oyster Point
Burbank
Thornlands
Coochiemudlo Island
Victoria Point
MacLeay Island
GATEWAY
MWY
Ferry

0 10 20 30 40 50 km

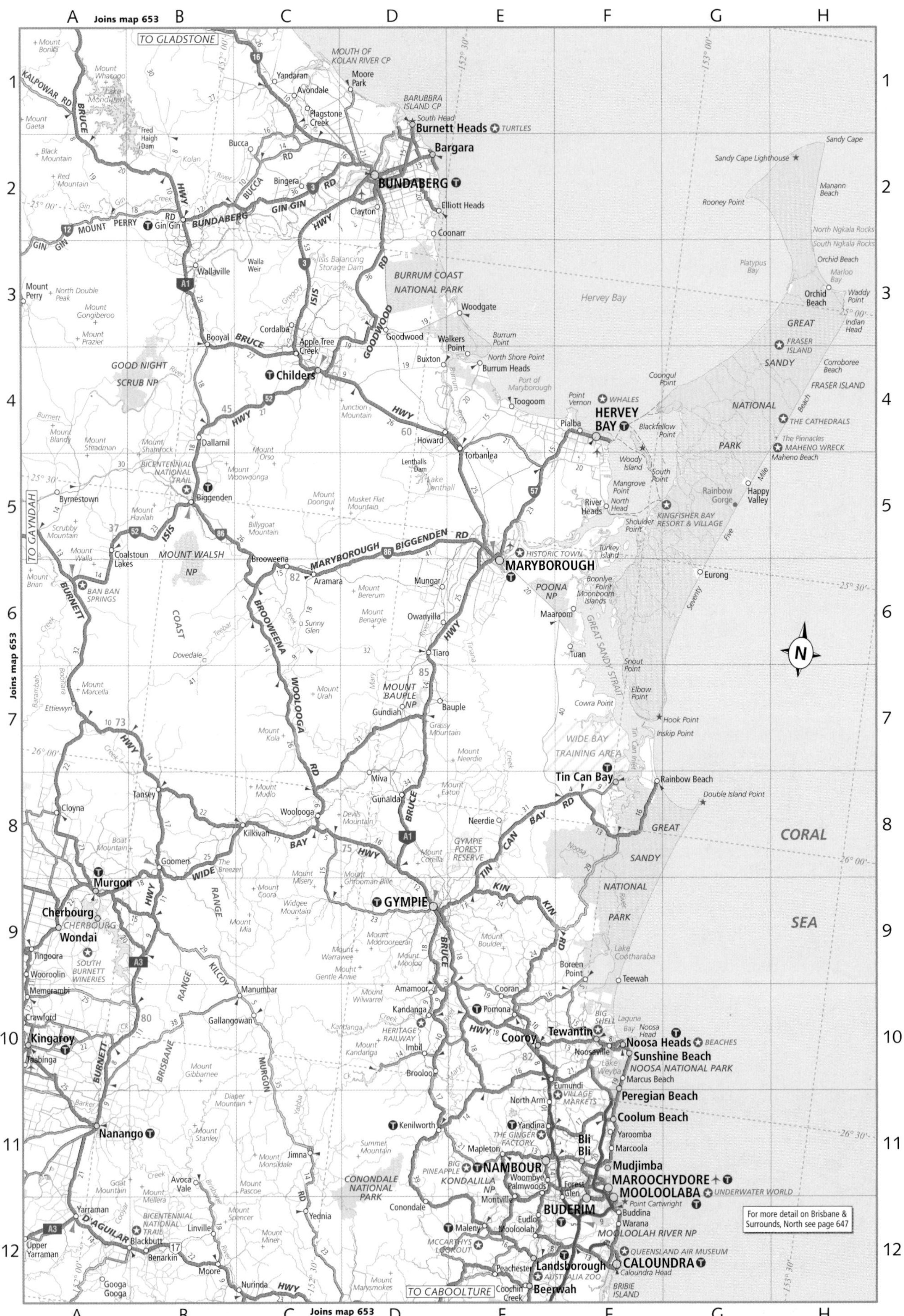

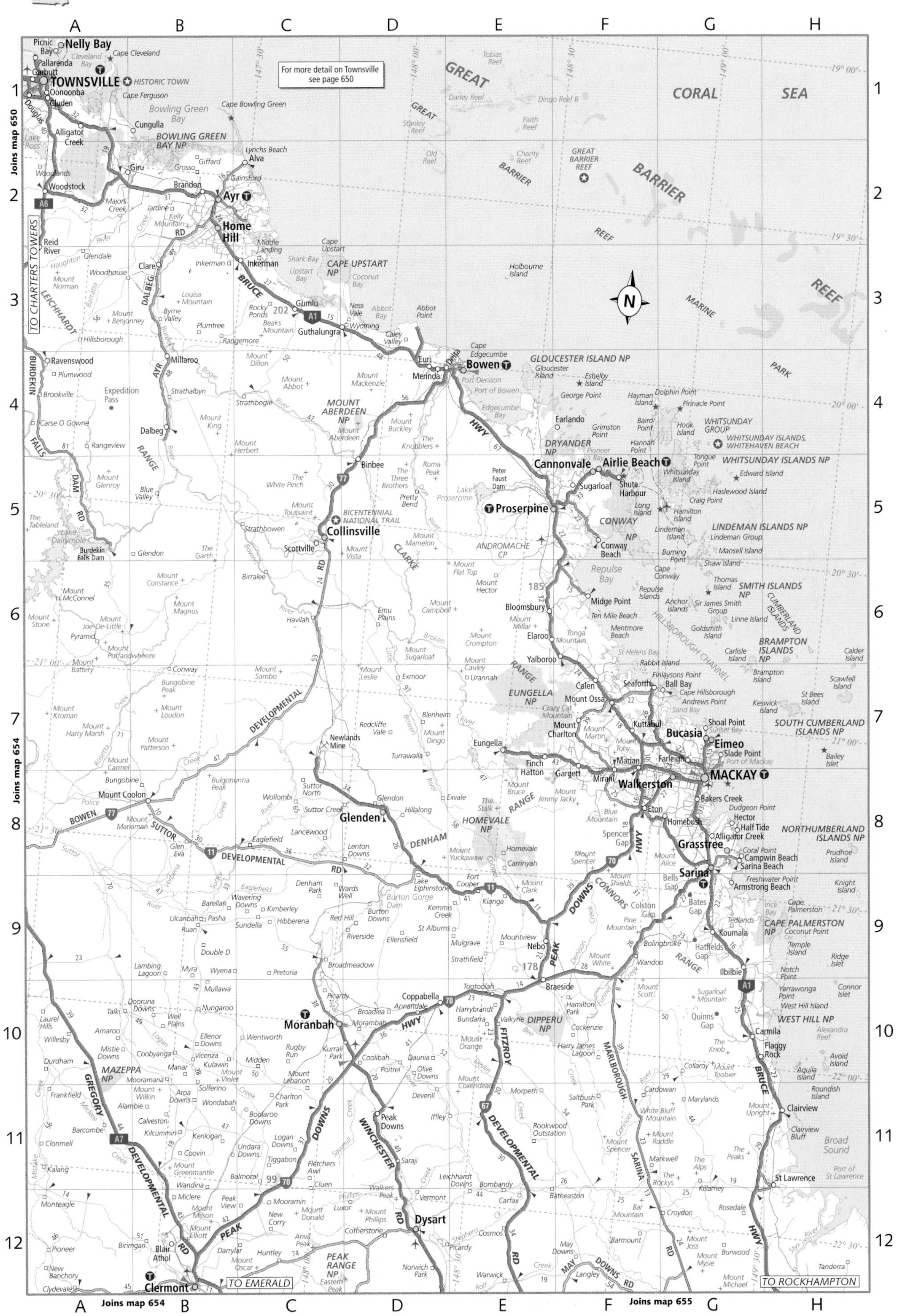
0 20 40 60 80 100 km
For more detail on Townsville see page 650
Joins map 650
Joins map 654
Joins map 655
TOWNSVILLE
Ayr
Home Hill
Bowen
Collinsville
Proserpine
Airlie Beach
Cannonvale
MACKAY
Walkerston
Sarina
Moranbah
Glenden
Dysart
Clermont
GREAT BARRIER REEF
CORAL SEA
TO CHARTERS TOWERS
TO EMERALD
TO ROCKHAMPTON

0 20 40 60 80 100 km

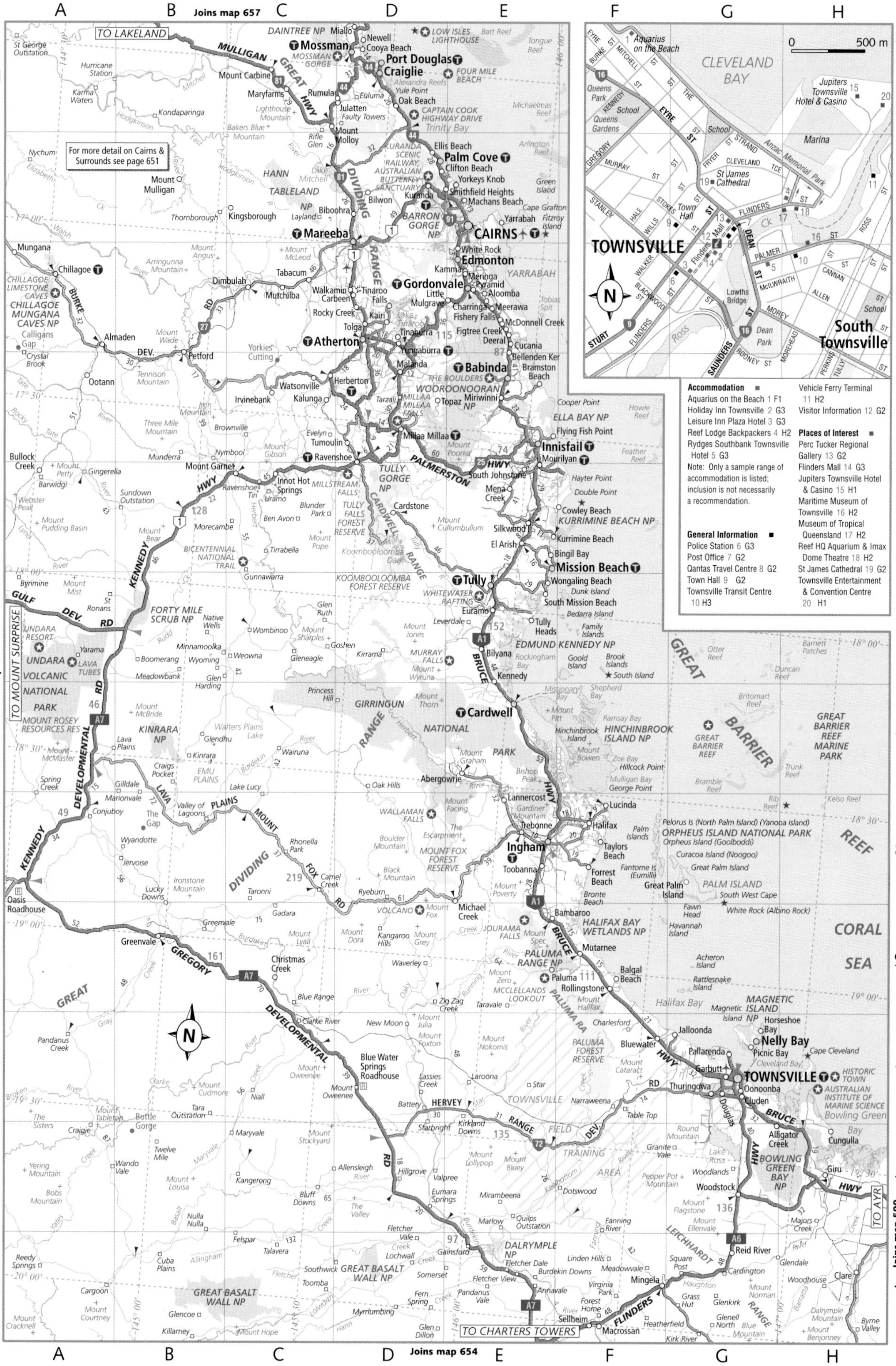
Joins map 657
TO LAKELAND
MULLIGAN HWY
Mossman
Port Douglas
Craiglie
Daintree NP
Low Isles Lighthouse
Captain Cook Highway Drive
Trinity Bay
For more detail on Cairns & Surrounds see page 651
Palm Cove
Kuranda
Mareeba
CAIRNS
Edmonton
Gordonvale
Atherton
Yungaburra
Malanda
Babinda
Herberton
Millaa Millaa
Innisfail
Mourilyan
Ravenshoe
Mount Garnet
Tully
Mission Beach
Cardwell
Ingham
Lucinda
Halifax
Paluma
TOWNSVILLE
Thuringowa
Alligator Creek
Nelly Bay
Cape Cleveland
Australian Institute of Marine Science
Mingela
Reid River
Greenvale
Oasis Roadhouse
Chillagoe
Dimbulah
Almaden
Petford
Mungana
Chillagoe Mungana Caves NP
Undara Volcanic National Park
Forty Mile Scrub NP
Kinrara NP
Great Basalt Wall NP
Dalrymple NP
Paluma Range NP
Edmund Kennedy NP
Hinchinbrook Island NP
Orpheus Island National Park
Palm Island
Halifax Bay Wetlands NP
Bowling Green Bay NP
Magnetic Island NP
Girringun National Park
Dividing Range
Great Barrier Reef
Great Barrier Reef Marine Park
Coral Sea
Kennedy Developmental Rd
Gregory Developmental Rd
Bruce Hwy
Palmerston Hwy
Hervey Range Dev. Rd
Flinders Hwy
Gulf Dev. Rd
Burke Dev. Rd
Kennedy Hwy
TO MOUNT SURPRISE
Joins map 657
TO CHARTERS TOWERS
Joins map 654
TO AYR
Joins map 599
TOWNSVILLE
South Townsville
Cleveland Bay
0 500 m
Accommodation
Aquarius on the Beach 1 F1
Holiday Inn Townsville 2 G3
Leisure Inn Plaza Hotel 3 G3
Reef Lodge Backpackers 4 H2
Rydges Southbank Townsville Hotel 5 G3
Note: Only a sample range of accommodation is listed; inclusion is not necessarily a recommendation.
General Information
Police Station 6 G3
Post Office 7 G2
Qantas Travel Centre 8 G2
Town Hall 9 G2
Townsville Transit Centre 10 H3
Vehicle Ferry Terminal 11 H2
Visitor Information 12 G2
Places of Interest
Perc Tucker Regional Gallery 13 G2
Flinders Mall 14 G3
Jupiters Townsville Hotel & Casino 15 H1
Maritime Museum of Townsville 16 H2
Museum of Tropical Queensland 17 H2
Reef HQ Aquarium & Imax Dome Theatre 18 H2
St James Cathedral 19 G2
Townsville Entertainment & Convention Centre 20 H1

0 10 20 30 40 km

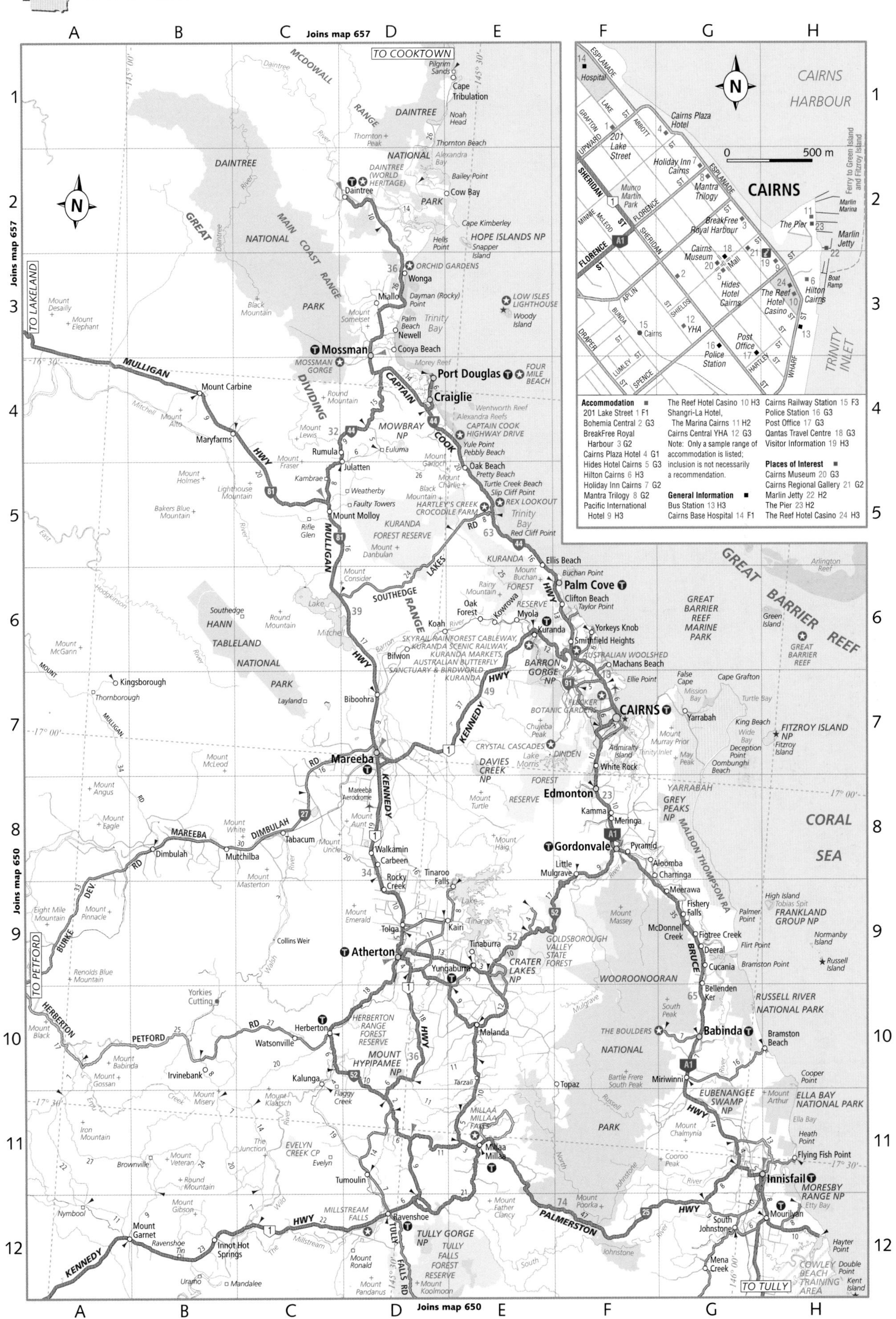

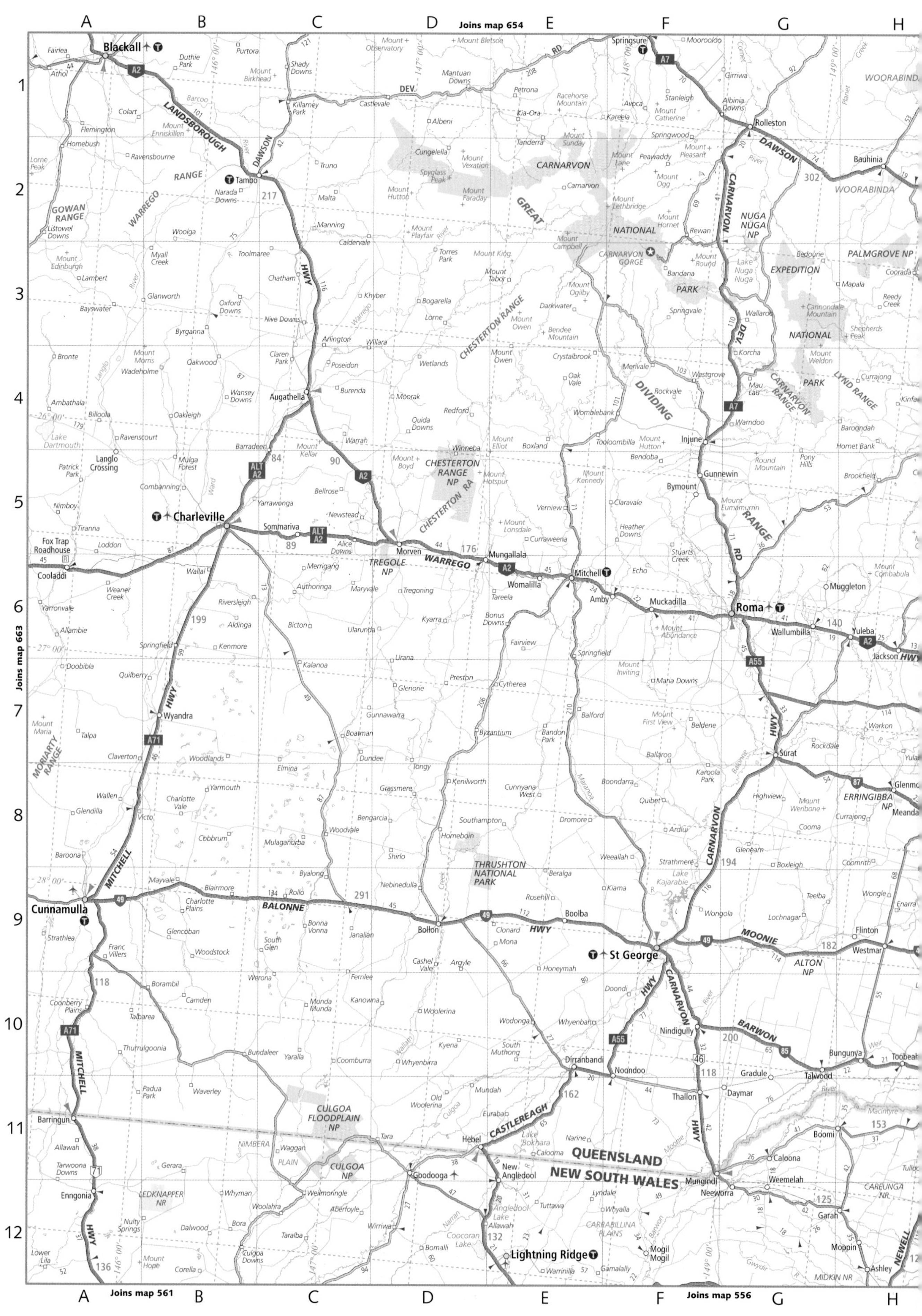
Joins map 654
Joins map 663
Joins map 561
Joins map 556
Blackall
Tambo
Augathella
Charleville
Cunnamulla
Morven
Mungallala
Mitchell
Roma
Injune
Rolleston
Springsure
St George
Dirranbandi
Hebel
Lightning Ridge
Goodooga
Mungindi
Surat
Yuleba
Wallumbilla
Bollon
Wyandra
Barringun
Enngonia
Cooladdi
Langlo Crossing
Bauhinia
Nindigully
Thallon
Bungunya
Talwood
Boomi
Garah
QUEENSLAND
NEW SOUTH WALES
CARNARVON NATIONAL PARK
CARNARVON GORGE
NUGA NUGA NP
EXPEDITION NATIONAL PARK
PALMGROVE NP
CHESTERTON RANGE NP
TREGOLE NP
THRUSHTON NATIONAL PARK
CULGOA FLOODPLAIN NP
CULGOA NP
ALTON NP
ERRINGIBBA NP
LEDKNAPPER NR
WOORABINDA
GREAT DIVIDING RANGE
WARREGO RANGE
GOWAN RANGE
MORIARTY RANGE
LANDSBOROUGH HWY
DAWSON HWY
WARREGO HWY
MITCHELL HWY
BALONNE HWY
CARNARVON HWY
MOONIE HWY
BARWON
CASTLEREAGH HWY
NEWELL

CORAL SEA
GLADSTONE
BUNDABERG
HERVEY BAY
MARYBOROUGH
GYMPIE
NAMBOUR
MAROOCHYDORE
MOOLOOLABA
CALOUNDRA
CABOOLTURE
BRISBANE
IPSWICH
TOOWOOMBA
WARWICK
NERANG
LISMORE
BALLINA
NEW SOUTH WALES
For more detail on Fraser Coast see page 648
For more detail on Brisbane & Surrounds see pages 645 & 647
Joins map 655
Joins map 557
0 25 50 75 100 km

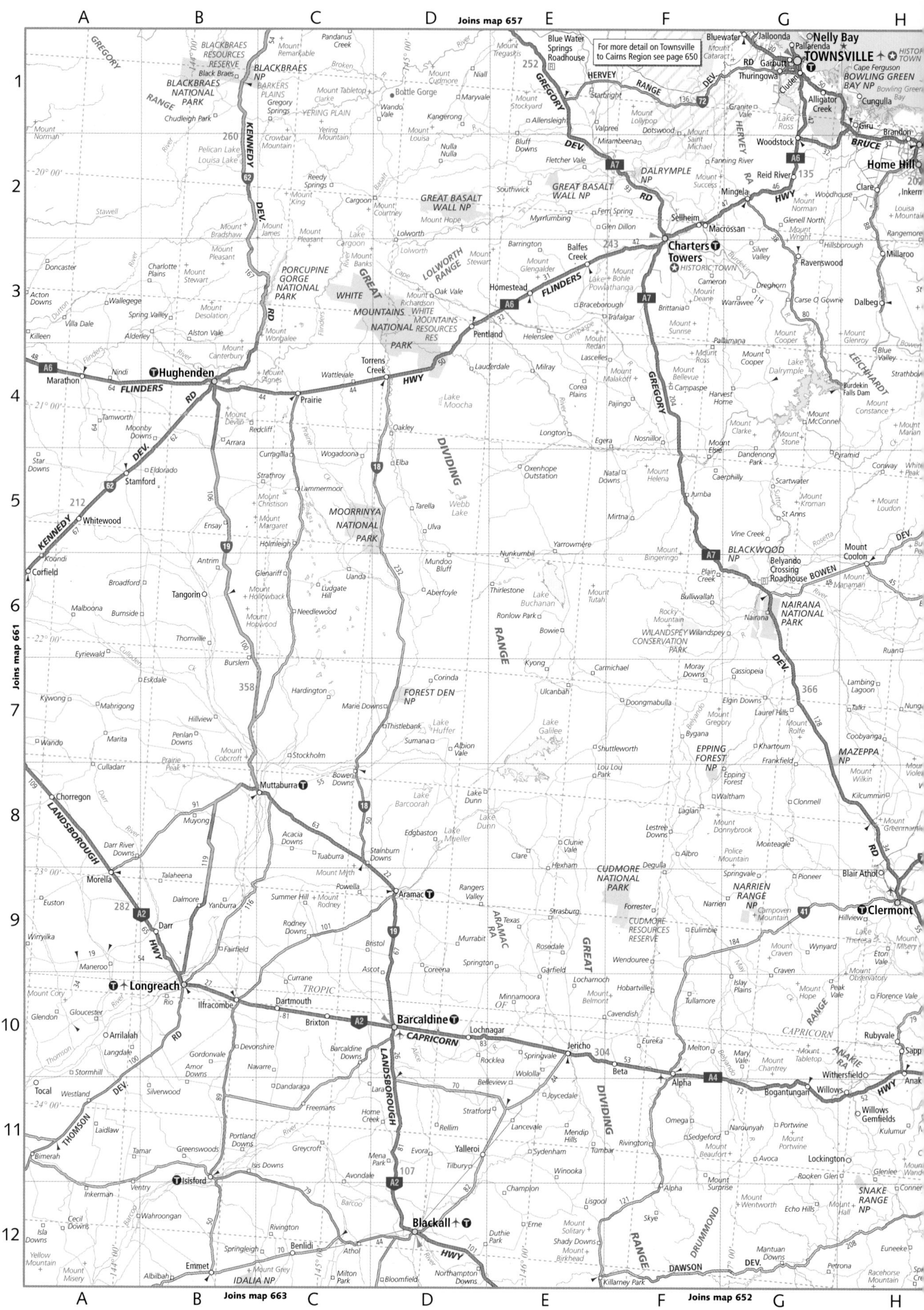
A
B
C
D
Joins map 657
E
F
G
H
1
2
3
4
5
6
7
8
9
10
11
12
Joins map 661
Joins map 663
Joins map 652
For more detail on Townsville to Cairns Region see page 650
BLACKBRAES RESOURCES RESERVE
Black Braes
BLACKBRAES NP
BLACKBRAES NATIONAL PARK
GREGORY RANGE
Chudleigh Park
Mount Remarkable
Pandanus Creek
Broken
BARKERS PLAINS
Gregory Springs
YERING PLAIN
Mount Tabletop
Clarke
Bottle Gorge
Wando Vale
Mount Cudmore
Niall
Maryvale
Kangerong
Mount Louisa
Nulla Nulla
Mount Tregaskis
Mount Stockyard
Blue Water Springs Roadhouse
252
GREGORY DEV.
Allensleigh
Bluff Downs
HERVEY RANGE
Starbright
Valpree
Mirambeena
Fletcher Vale
Mount Lollypop
Dotswood
Mount Saint Michael
Fanning River
Bluewater
Mount Cataract
Jalloonda
Nelly Bay
Pallarenda
TOWNSVILLE
HISTORIC TOWN
Garbutt
RD
Thuringowa
Cluden
Cape Ferguson
BOWLING GREEN BAY NP
Bowling Green Bay
Cungulla
Alligator Creek
Granite Vale
Lake Ross
Giru
Brandon
Woodstock
BRUCE
Home Hill
Mount Norman
260
Crowbar Mountain
Yering Mountain
Pelican Lake
Louisa Lake
KENNEDY
DEV.
RD
20° 00'
Stawell
Reedy Springs
Mount King
Cargoon
Mount Courtney
GREAT BASALT WALL NP
Mount Hope
Southwick
GREAT BASALT WALL NP
DALRYMPLE NP
Myrrlumbing
Fern Spring
Glen Dillon
Mount Success
Mingela
Reid River
135
HWY
Mount Norman
Woodhouse
Clare
Inkerman
Louisa Mountain
Sellheim
Macrossan
Glenell North
Mount Wright
Rangemore
Mount Bradshaw
Mount James
Mount Pleasant
Lake Cargoon
Lolworth
LOLWORTH RANGE
Mount Stewart
Barrington
Mount Glengalder
Balfes Creek
243
Charters Towers
HISTORIC TOWN
Silver Valley
Ravenswood
Hillsborough
Millaroo
Doncaster
Charlotte Plains
Mount Stewart
Mount Pleasant
PORCUPINE GORGE NATIONAL PARK
Mount Banks
WHITE MOUNTAINS NATIONAL PARK
GREAT WHITE MOUNTAINS
Mount Richardson
WHITE MOUNTAINS RESOURCES RES
Oak Vale
Homestead
FLINDERS
Mount Bohle
Lake Powlathanga
Cameron
Mount Deane
Dreghorn
Warrawee
Carse O Gowrie
Dalbeg
Acton Downs
Wallegege
Spring Valley
Mount Desolation
Villa Dale
Killeen
Alderley
Alston Vale
Mount Wongalee
Pentland
Helenslee
Braceborough
Trafalgar
Brittania
Mount Sunrise
Mount Redan
Pallamana
Mount Cooper
Mount Glenroy
Blue Valley
Mount Canterbury
Torrens Creek
Marathon
Nindi
Hughenden
FLINDERS
Mount Agnes
Wattlevale
Prairie
Lauderdale
Milray
Lascelles
Mount Malakoff
Mount Bellevue
Campaspe
Mount Ross
Mount Cooper
Lake Dalrymple
Burdekin Falls Dam
LEICHHARDT
Strathbogie
21° 00'
Tamworth
Moonby Downs
Mount Devlin
Redcliff
Arrara
Lake Moocha
Corea Plains
Pajingo
Harvest Home
Mount Constance
Mount McConnel
Mount Marian
Oakley
Longton
Egera
Nosnillor
Mount Clarke
Mount Else
Mount Stone
Dandenong Park
Pyramid
Star Downs
Eldorado
Stamford
Curragilla
Wogadoona
Elba
DIVIDING
Oxenhope Outstation
Natal Downs
Mount Helena
Caerphilly
Scartwater
Conway
White Peak
Strathroy
Lammermoor
212
Whitewood
Mount Christison
MOORRINYA NATIONAL PARK
Tarella
Webb Lake
Jumba
Mount Kroman
Mount Loudon
Ensay
Mount Margaret
Ulva
Mirtna
St Anns
Holmleigh
Vine Creek
Rosetta
Koondi
Corfield
Antrim
Nunkumbil
Yarrowmere
Mount Bingeringo
BLACKWOOD NP
Belyando Crossing Roadhouse
BOWEN
Mount Coolon
Mount Manaman
Broadford
Glenariff
Uanda
Mundoo Bluff
Ludgate Hill
Plain Creek
Tangorin
Mount Hollowback
Aberfoyle
Thirlestone
Lake Buchanan
Mount Tutah
Bulliwallah
NAIRANA NATIONAL PARK
Nairana
Malboona
Burnside
Mount Hopwood
Needlewood
Ronlow Park
Rocky Mountain
WILANDSPEY CONSERVATION PARK
Wilandspey
22° 00'
Thornville
Bowie
RANGE
Eyriewald
Burslem
Kyong
Carmichael
Moray Downs
Cassiopeia
Ruan
Eskdale
358
Hardington
Corinda
FOREST DEN NP
Ulcanbah
366
Lambing Lagoon
Kywong
Mahrigong
Marie Downs
Doongmabulla
Elgin Downs
Talki
Hillview
Thistlebank
Lake Huffer
Lake Galilee
Mount Gregory
Laurel Hills
Mount Rolfe
Wando
Marita
Penlan Downs
Sumana
Albion Vale
Bygana
Coobyanga
Culladarr
Prairie Peak
Mount Cobcroft
Stockholm
Shuttleworth
EPPING FOREST NP
Khartoum
Frankfield
MAZEPPA NP
Mount Wilkin
Chorregon
Muttaburra
Bowen Downs
Lou Lou Park
Epping Forest
Waltham
Clonmell
Kilcummin
LANDSBOROUGH
Muyong
Lake Barcoorah
Lake Dunn
Laglan
Mount Donnybrook
Lake Dunn
Lestree Downs
Mount Greenmantle
Darr River Downs
Acacia Downs
Edgbaston
Lake Mueller
Clunie Vale
Albro
Monteagle
Tuaburra
Stainburn Downs
Clare
Police Mountain
Morella
Talaheena
23° 00'
Mount Myth
Hexham
CUDMORE NATIONAL PARK
Degulla
Springvale
Pioneer
Blair Athol
Powella
Aramac
Rangers Valley
NARRIEN RANGE NP
Euston
282
Dalmore
Yanburra
Summer Hill
Mount Rodney
Forrester
Narrien
Campoven Mountain
Clermont
Darr
Rodney Downs
ARAMAC RA
Texas
Strasburg
CUDMORE RESOURCES RESERVE
Eulimbie
Hillview
Lake Theresa
Mount Misery
Wirryilka
Fairfield
Bristol
Murrabit
Rosedale
GREAT
Wendouree
Mount Craven
Wynyard
Eton Vale
Maneroo
Curranne
Ascot
Coreena
Springton
Garfield
Craven
Mount Observatory
Longreach
Rio
TROPIC
Minnamoora OF
Lochamoch
Hobartville
Islay Plains
Mount Hope
Peak Vale
Florence Vale
Mount Cory
Ilfracombe
Dartmouth
Mount Belmont
Tullamore
RANGE
Glendon
Gloucester
Brixton
Barcaldine
Cavendish
Arrilalah
CAPRICORN
Lochnagar
Eureka
CAPRICORN
Rubyvale
Langdale
Devonshire
Barcaldine Downs
Springvale
Rocklea
Jericho
304
Melton
Mary Vale
Mount Tabletop
ANAKIE RA
Sapphire
Gordonvale
Amor Downs
Navarre
Wololla
Beta
Mount Chantrey
Withersfield
Stormhill
Anakie
Tocal
Westland
Silverwood
Dandaraga
Lara
Belleview
Alpha
Bogantungan
Willows
24° 00'
Joycedale
DIVIDING
Willows Gemfields
Freemans
Home Creek
Stratford
Omega
Narounyah
Portwine
Laidlaw
Rellim
Lancevale
Mendip Hills
Rivington
Sedgeford
Mount Portwine
Kulumur
Bimerah
Tamar
Greenswoods
Portland Downs
Greycroft
Mena Park
Evora
Yalleroi
Sydenham
Tumbar
Mount Beaufort
Avoca
Lockington
Isis Downs
Tilbury
Glenlee
Isisford
Avondale
107
Winooka
Alpha
Mount Surprise
Rooken Glen
Mount Wanda
Ventry
Inkerman
Barcoo
Champion
Lisgool
Mount Wentworth
Echo Hills
Mount Hall
SNAKE RANGE NP
Conner
Wahroongan
Cecil Downs
Rivington
Blackall
Duthie Park
Erne
Mount Solitary
Skye
DRUMMOND
Isla Downs
Springleigh
Benlidi
Shady Downs
Mount Birkhead
Mantuan Downs
Euneeke
Yellow Mountain
Mount Misery
Emmet
Mount Grey
Athol
HWY
RANGE
DAWSON
DEV.
Petrona
Albilbah
IDALIA NP
Milton Park
Bloomfield
Northampton Downs
Killarney Park
Racehorse Mountain

I J K L M N O P

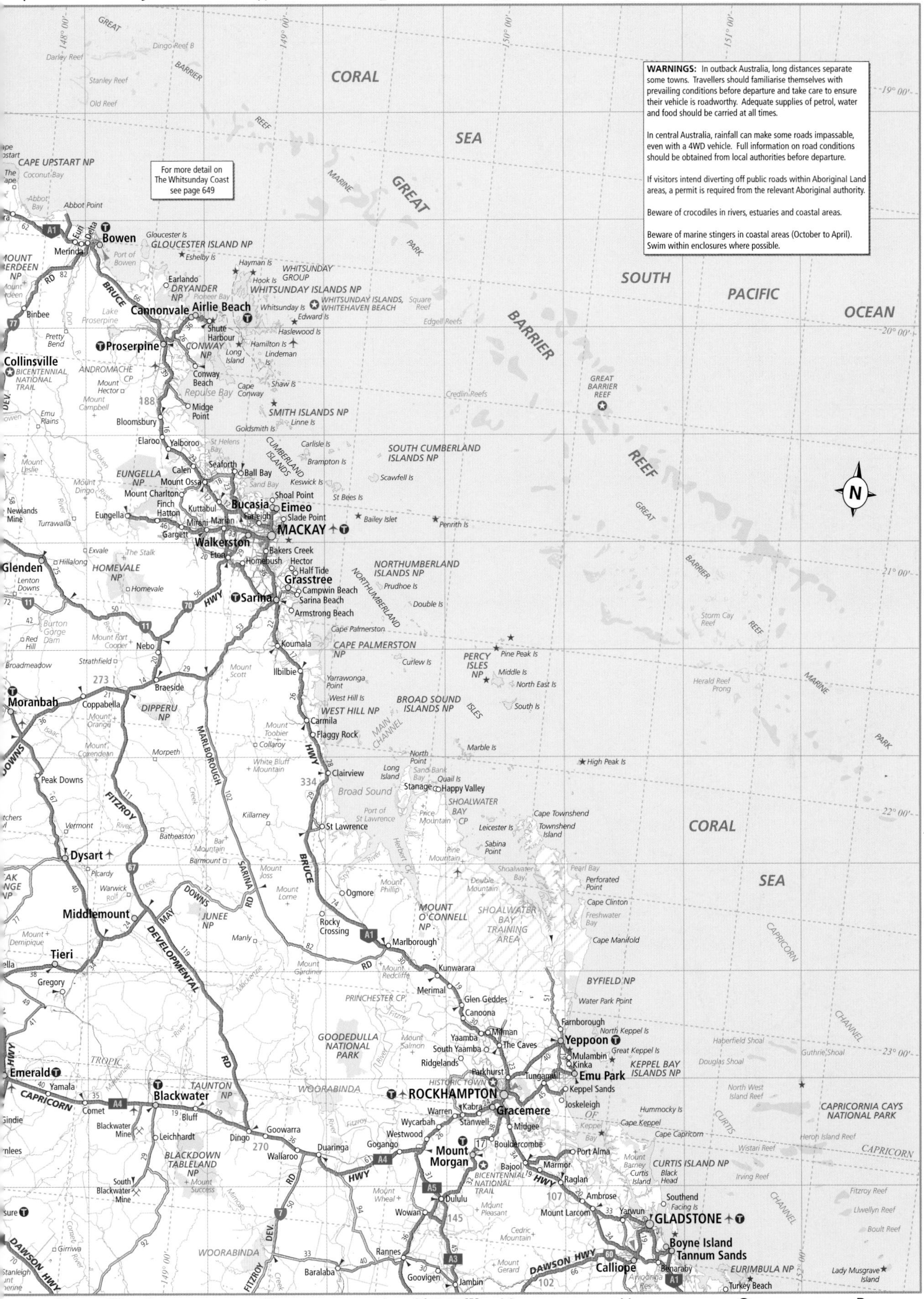
WARNINGS: In outback Australia, long distances separate some towns. Travellers should familiarise themselves with prevailing conditions before departure and take care to ensure their vehicle is roadworthy. Adequate supplies of petrol, water and food should be carried at all times.
In central Australia, rainfall can make some roads impassable, even with a 4WD vehicle. Full information on road conditions should be obtained from local authorities before departure.
If visitors intend diverting off public roads within Aboriginal Land areas, a permit is required from the relevant Aboriginal authority.
Beware of crocodiles in rivers, estuaries and coastal areas.
Beware of marine stingers in coastal areas (October to April). Swim within enclosures where possible.
For more detail on The Whitsunday Coast see page 649
CORAL SEA
SOUTH PACIFIC OCEAN
GREAT BARRIER REEF MARINE PARK
Bowen
Airlie Beach
Cannonvale
Proserpine
Collinsville
WHITSUNDAY ISLANDS NP
CAPE UPSTART NP
GLOUCESTER ISLAND NP
SMITH ISLANDS NP
SOUTH CUMBERLAND ISLANDS NP
EUNGELLA NP
MACKAY
Walkerston
Bucasia
Eimeo
Sarina
Grasstree
NORTHUMBERLAND ISLANDS NP
CAPE PALMERSTON NP
PERCY ISLES NP
BROAD SOUND ISLANDS NP
WEST HILL NP
HOMEVALE NP
Glenden
Moranbah
Nebo
Dysart
Middlemount
Tieri
Emerald
Blackwater
St Lawrence
Marlborough
Yeppoon
Emu Park
KEPPEL BAY ISLANDS NP
ROCKHAMPTON
Gracemere
Mount Morgan
GLADSTONE
Boyne Island
Tannum Sands
Calliope
CURTIS ISLAND NP
CAPRICORNIA CAYS NATIONAL PARK
EURIMBULA NP
BLACKDOWN TABLELAND NP
GOODEDULLA NATIONAL PARK
BYFIELD NP
MOUNT O'CONNELL NP
DIPPERU NP
JUNEE NP
BRUCE HWY
PEAK DOWNS HWY
MAY DOWNS RD
FITZROY DEVELOPMENTAL RD
CAPRICORN HWY
DAWSON HWY
TROPIC OF CAPRICORN

1 2 3 4 5 6 7 8 9 10 11 12

I J K L M N O P

Joins map 652
Joins map 653

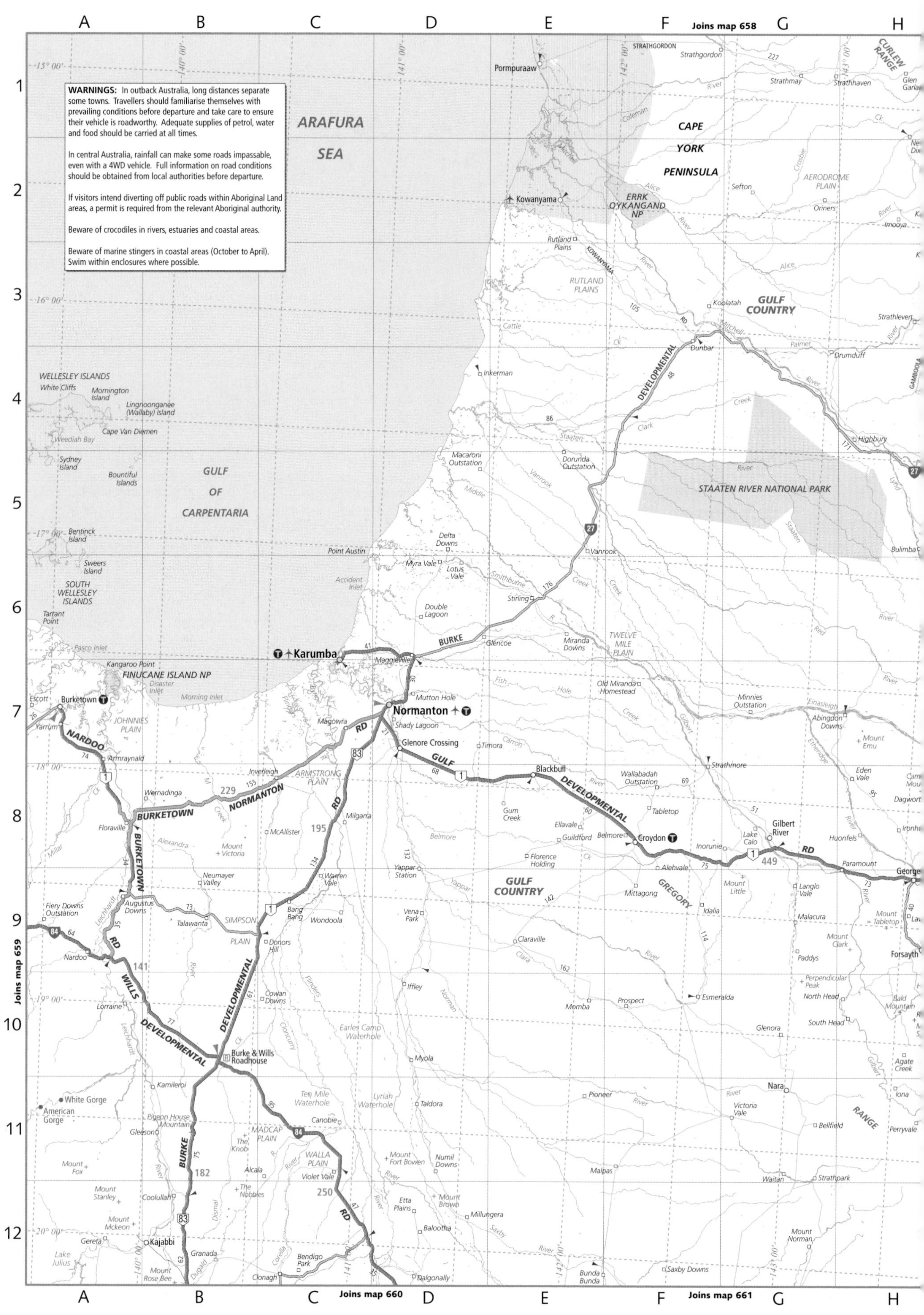
WARNINGS: In outback Australia, long distances separate some towns. Travellers should familiarise themselves with prevailing conditions before departure and take care to ensure their vehicle is roadworthy. Adequate supplies of petrol, water and food should be carried at all times.
In central Australia, rainfall can make some roads impassable, even with a 4WD vehicle. Full information on road conditions should be obtained from local authorities before departure.
If visitors intend diverting off public roads within Aboriginal Land areas, a permit is required from the relevant Aboriginal authority.
Beware of crocodiles in rivers, estuaries and coastal areas.
Beware of marine stingers in coastal areas (October to April). Swim within enclosures where possible.
Joins map 658
Joins map 659
Joins map 660
Joins map 661
ARAFURA SEA
GULF OF CARPENTARIA
CAPE YORK PENINSULA
GULF COUNTRY
STAATEN RIVER NATIONAL PARK
ERRK OYKANGAND NP
FINUCANE ISLAND NP
WELLESLEY ISLANDS
SOUTH WELLESLEY ISLANDS
Mornington Island
Bentinck Island
Sweers Island
Pormpuraaw
Kowanyama
Dunbar
Karumba
Normanton
Burketown
Croydon
Georgetown
Forsayth
Gilbert River
Blackbull
Glenore Crossing
Burke & Wills Roadhouse
Kajabbi
Strathgordon
Highbury
Inkerman
Delta Downs
Stirling
Glencoe
Esmeralda
Iffley
Myola
Taldora
Canobie
Gregory
Augustus Downs
Floraville
Nardoo
Lorraine
Gleeson
Coolullah
Granada
Clonagh
Dalgonally
Saxby Downs
Bunda Bunda
Mount Norman
Nara
Victoria Vale
Bellfield
Perryvale
Strathpark
Waitan
Glenora
Malpas
Pioneer
Momba
Prospect
Claraville
Idalia
Mittagong
Abingdon Downs
Strathmore
Minnies Outstation
Paramount
Langlo Vale
Malacura
Paddys
BURKE DEVELOPMENTAL RD
GULF DEVELOPMENTAL RD
BURKETOWN NORMANTON RD
WILLS DEVELOPMENTAL RD
NARDOO BURKETOWN RD

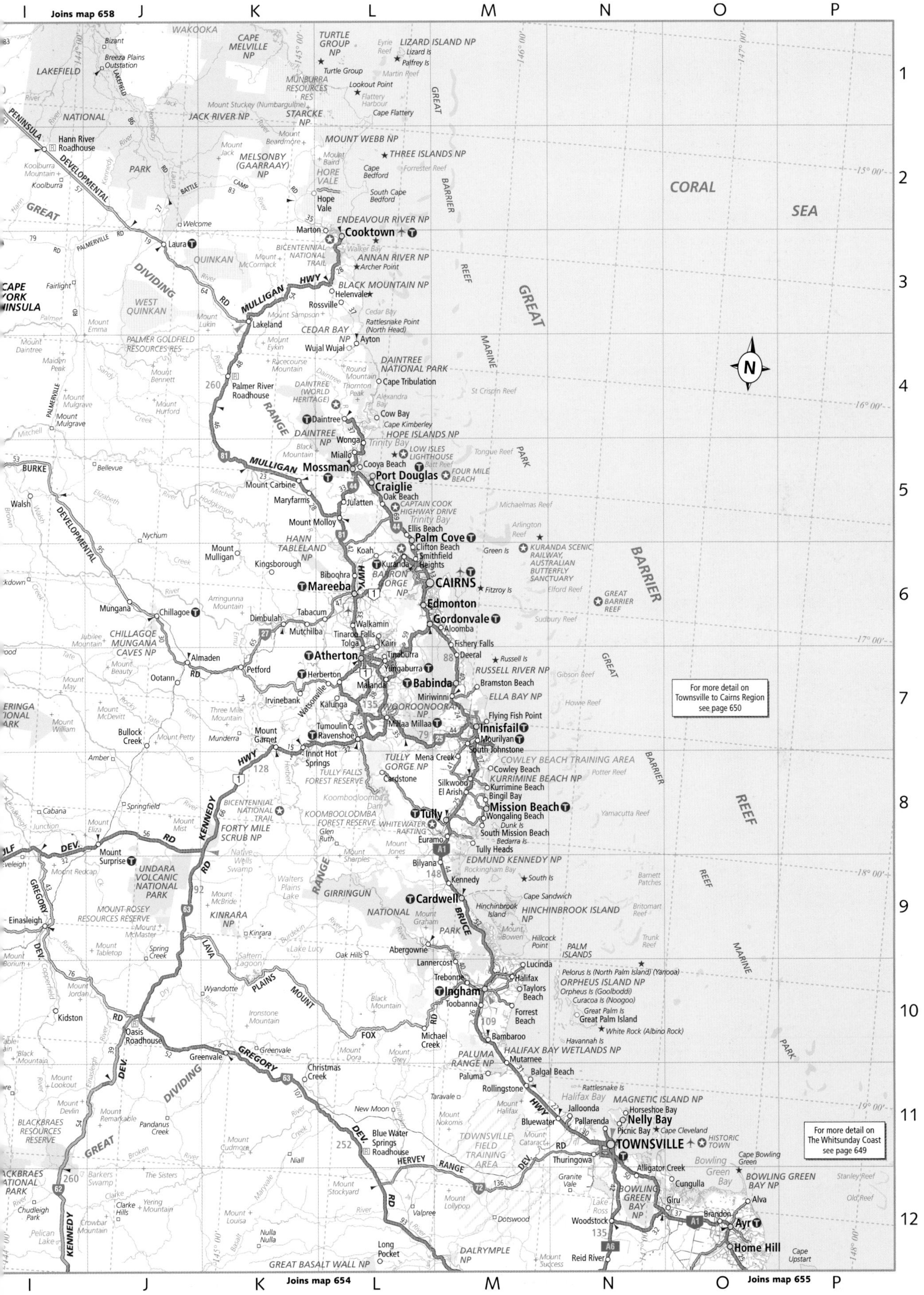

0 25 50 75 100 km

Joins map 658

I J K L M N O P

For more detail on Townsville to Cairns Region see page 650

For more detail on The Whitsunday Coast see page 649

Joins map 654

Joins map 655

0 25 50 75 100 km

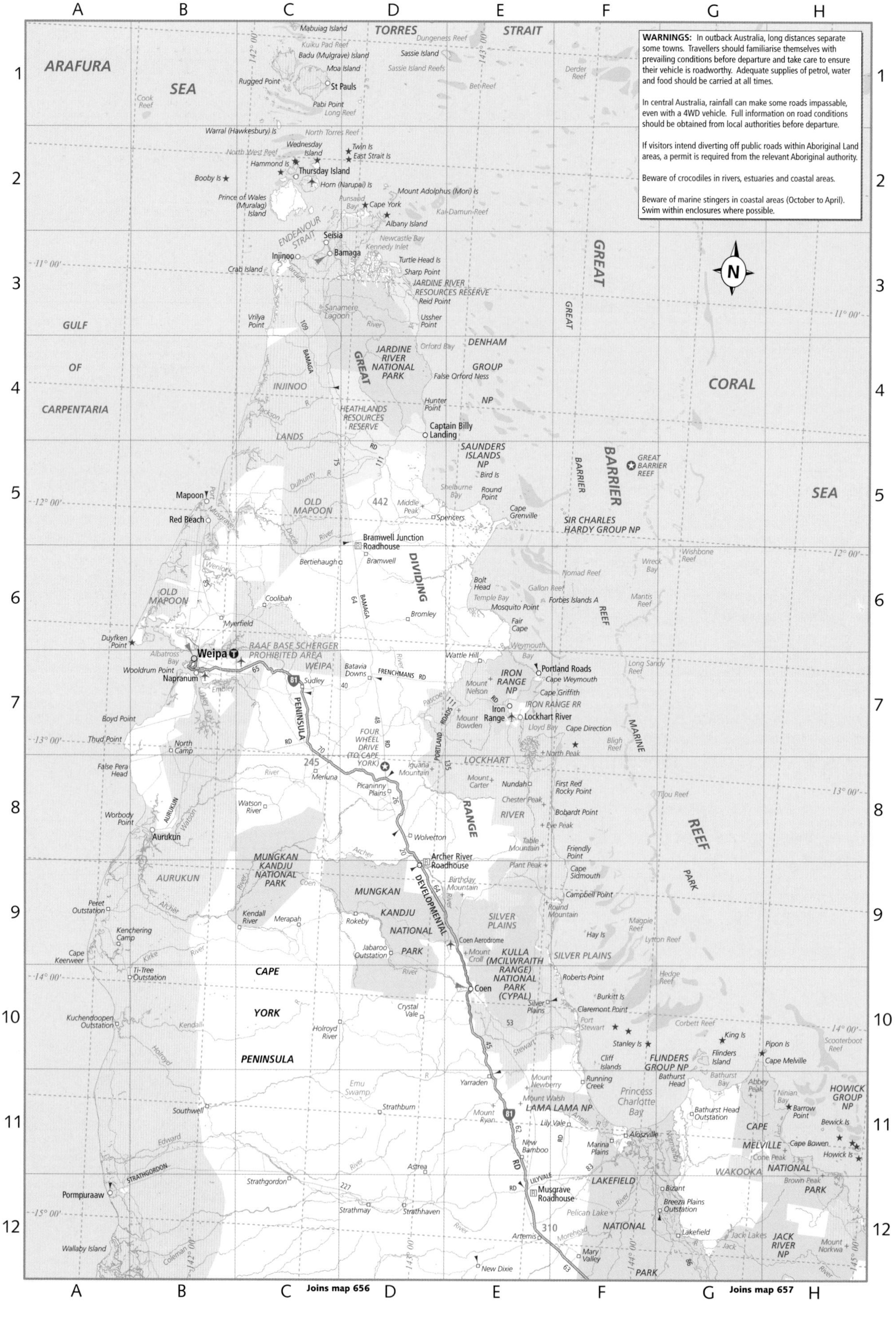

A B C D E F G H

WARNINGS: In outback Australia, long distances separate some towns. Travellers should familiarise themselves with prevailing conditions before departure and take care to ensure their vehicle is roadworthy. Adequate supplies of petrol, water and food should be carried at all times.

In central Australia, rainfall can make some roads impassable, even with a 4WD vehicle. Full information on road conditions should be obtained from local authorities before departure.

If visitors intend diverting off public roads within Aboriginal Land areas, a permit is required from the relevant Aboriginal authority.

Beware of crocodiles in rivers, estuaries and coastal areas.

Beware of marine stingers in coastal areas (October to April). Swim within enclosures where possible.

1 ARAFURA SEA
Mabuiag Island
Kuiku Pad Reef
Badu (Mulgrave) Island
Moa Island
Rugged Point
St Pauls
Pabi Point
Long Reef
Cook Reef
TORRES STRAIT
Dungeness Reef
Sassie Island
Sassie Island Reefs
Bet Reef
Derder Reef
142° 00'
143° 00'

2 Warral (Hawkesbury) Is
North Torres Reef
North West Reef
Wednesday Island
Twin Is
East Strait Is
Hammond Is
Thursday Island
Booby Is
Horn (Narupai) Is
Prince of Wales (Muralag) Island
Punsand Bay
Cape York
Mount Adolphus (Mori) Is
Albany Island
Kai-Damun Reef

3 ENDEAVOUR STRAIT
Seisia
Injinoo
Bamaga
Newcastle Bay
Kennedy Inlet
Turtle Head Is
Sharp Point
Crab Island
Jardine
JARDINE RIVER RESOURCES RESERVE
Reid Point
11° 00'
GREAT
GREAT
Sanamere Lagoon
Vrilya Point
Ussher Point
River

GULF OF CARPENTARIA
4 BAMAGA
GREAT
JARDINE RIVER NATIONAL PARK
Orford Bay
DENHAM GROUP NP
False Orford Ness
INJINOO
CORAL
Jackson R
HEATHLANDS RESOURCES RESERVE
Hunter Point
Captain Billy Landing
LANDS
RD
111
75

5 SAUNDERS ISLANDS NP
Bird Is
BARRIER
GREAT BARRIER REEF
SEA
Mapoon
12° 00'
Port Musgrave
Dulhunty R
OLD MAPOON
442
Middle Peak
Shelburne Bay
Round Point
Spencers
Cape Grenville
SIR CHARLES HARDY GROUP NP
Red Beach
Ducie River
Bramwell Junction Roadhouse
Bramwell
Bertiehaugh
DIVIDING
Wishbone Reef
12° 00'
Wreck Bay

6 Wenlock
85
OLD MAPOON
Coolibah
64
BAMAGA
Bolt Head
Temple Bay
Nomad Reef
Gallon Reef
Forbes Islands A
Mosquito Point
REEF
Mantis Reef
Bromley
Myerfield
Fair Cape
Weymouth Bay
Duyfken Point
Albatross Bay
Weipa
RAAF BASE SCHERGER PROHIBITED AREA
WEIPA
Wattle Hill
River
Long Sandy Reef

7 Wooldrum Point
Napranum
65
81
Embley
Sudley
Batavia Downs
FRENCHMANS RD
40
Portland Roads
IRON RANGE NP
Cape Weymouth
Mount Nelson
Cape Griffith
IRON RANGE RR
Iron Range
Lockhart River
PENINSULA
Pascoe
111
PORTLAND ROADS
Boyd Point
Mount Bowden
Lloyd Bay
Cape Direction
MARINE
Thud Point
13° 00'
North Camp
RD
48
RD
FOUR WHEEL DRIVE (TO CAPE YORK)
Bligh Reef
North Peak
70
245
LOCKHART
False Pera Head
Merluna
Iguana Mountain
135
River
Mount Carter
Nundah
First Red Rocky Point

8 Picaninny Plains
26
Chester Peak
Tijou Reef
13° 00'
AURUKUN
Watson
Watson River
RANGE
RIVER
Bobardt Point
Worbody Point
Eve Peak
REEF
Aurukun
Wolverton
Table Mountain
Friendly Point
Archer
MUNGKAN KANDJU NATIONAL PARK
20
Archer River Roadhouse
AURUKUN
Coen
DEVELOPMENTAL
Birthday Mountain
Plant Peak
Cape Sidmouth
PARK
River
MUNGKAN
64
River
Campbell Point

9 Peret Outstation
Archer
KANDJU
SILVER PLAINS
Round Mountain
Kendall River
Merapah
Rokeby
Magpie Reef
NATIONAL
Hay Is
Kenchering Camp
Coen Aerodrome
Lytton Reef
Kirke
River
Jabaroo Outstation
PARK
Mount Croll
KULLA (MCILWRAITH RANGE) NATIONAL PARK (CYPAL)
SILVER PLAINS
Cape Keerweer
Ti-Tree Outstation
CAPE
River
Roberts Point
Hedge Reef
14° 00'
Coen
Burkitt Is

10 Kuchendoopen Outstation
Kendall
YORK
Crystal Vale
Silver Plains
Claremont Point
Holroyd River
53
Port Stewart
Corbett Reef
14° 00'
Holroyd
Stewart
Stanley Is
King Is
Scooterboot Reef
Pipon Is
PENINSULA
45
Cliff Islands
FLINDERS GROUP NP
Flinders Island
Cape Melville
Emu Swamp
Yarraden
Mount Newberry
Running Creek
Bathurst Head
Bathurst Bay
Abbey Peak
HOWICK GROUP NP

11 Southwell
Strathburn
Mount Walsh
LAMA LAMA NP
Princess Charlotte Bay
Ninian Bay
Barrow Point
Mount Ryan
81
Annie R
Bathurst Head Outstation
Lily Vale
Edward
62
RD
CAPE
Bewick Is
MELVILLE
Cape Bowen
New Bamboo
Marina Plains
Aloszville
Normanby
Howick Is
Cone Peak
River
Astrea
RD
83
WAKOOKA NATIONAL
STRATHGORDON
LILYVALE
LAKEFIELD
Brown Peak
Pormpuraaw
Strathgordon
227
RD
Musgrave Roadhouse
Bizant
PARK

12 15° 00'
Strathmay
Strathhaven
Pelican Lake
River
Breeza Plains Outstation
River
NATIONAL
310
Morehead
Lakefield
Jack Lakes
JACK RIVER NP
Mount Norkwa
Wallaby Island
142° 00'
143° 00'
144° 00'
145° 00'
Coleman
Artemis
Mary Valley
Jack
86
New Dixie
63
PARK
River

A B C Joins map 656 D E F G Joins map 657 H

0 25 50 75 100 km

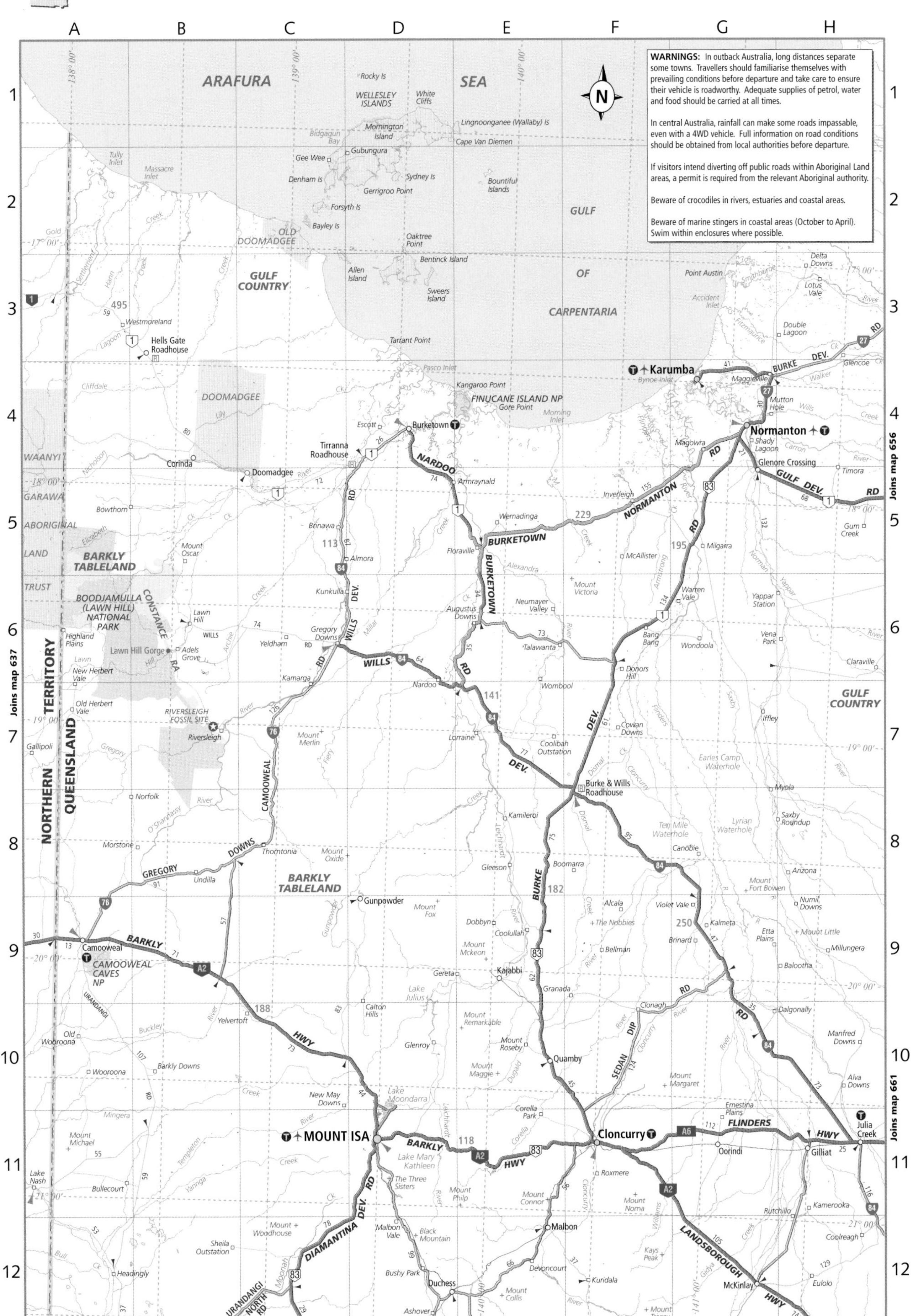

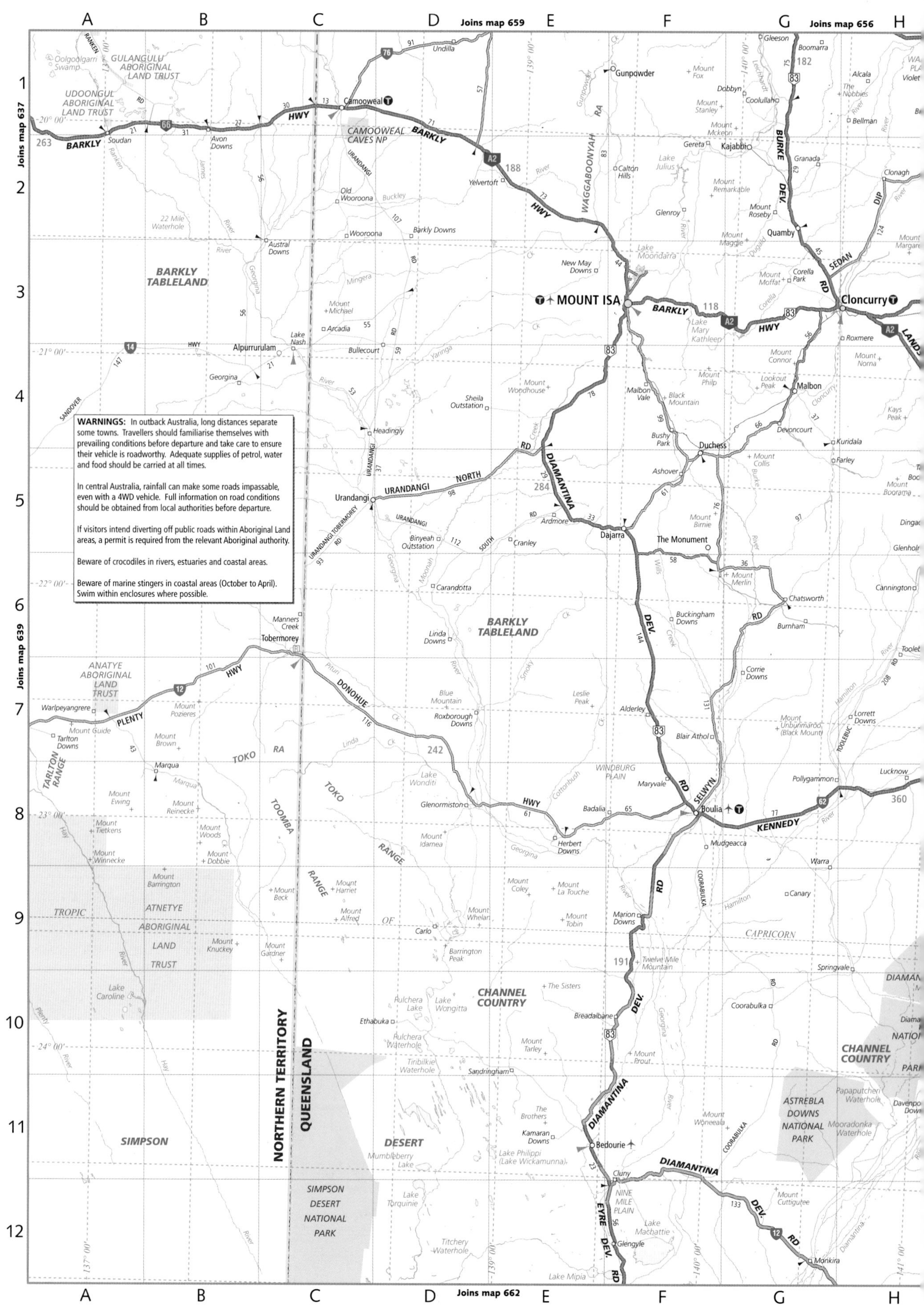

A B C D E F G H
Joins map 659
Joins map 656
Joins map 637
Joins map 639
Joins map 662
1 2 3 4 5 6 7 8 9 10 11 12
WARNINGS: In outback Australia, long distances separate some towns. Travellers should familiarise themselves with prevailing conditions before departure and take care to ensure their vehicle is roadworthy. Adequate supplies of petrol, water and food should be carried at all times.
In central Australia, rainfall can make some roads impassable, even with a 4WD vehicle. Full information on road conditions should be obtained from local authorities before departure.
If visitors intend diverting off public roads within Aboriginal Land areas, a permit is required from the relevant Aboriginal authority.
Beware of crocodiles in rivers, estuaries and coastal areas.
Beware of marine stingers in coastal areas (October to April). Swim within enclosures where possible.
NORTHERN TERRITORY
QUEENSLAND
MOUNT ISA
Cloncurry
Camooweal
Urandangi
Dajarra
Duchess
Boulia
Bedourie
Tobermorey
Alpurrurulam
Soudan
Avon Downs
Undilla
Gunpowder
Kajabbi
Quamby
Malbon
Mount Isa
BARKLY TABLELAND
BARKLY HWY
CAMOOWEAL CAVES NP
UDOONGUL ABORIGINAL LAND TRUST
GULANGULU ABORIGINAL LAND TRUST
ANATYE ABORIGINAL LAND TRUST
ATNETYE ABORIGINAL LAND TRUST
DONOHUE HWY
PLENTY HWY
KENNEDY
DIAMANTINA DEV.
BURKE DEV.
WAGGABOONYAH RA
URANDANGI NORTH RD
SELWYN RD
EYRE DEV. RD
TOKO RA
TOOMBA RANGE
TOKO RANGE
TARLTON RANGE
CHANNEL COUNTRY
SIMPSON DESERT
SIMPSON DESERT NATIONAL PARK
ASTREBLA DOWNS NATIONAL PARK
TROPIC OF CAPRICORN
WINDBURG PLAIN
NINE MILE PLAIN
Lake Moondarra
Lake Mary Kathleen
Lake Wonditi
Lake Caroline
Lake Machattie
Lake Mipia
Lake Philippi (Lake Wickamunna)
Mumblebery Lake
Lake Torquinie
Lake Wongitta
Pulchera Lake
Headingly
Carandotta
Glenormiston
Herbert Downs
Marion Downs
Breadalbane
Sandringham
Ethabuka
Carlo
Glengyle
Monkira
Cluny
Alderley
Blair Athol
Maryvale
Badalia
Mudgeacca
Chatsworth
Buckingham Downs
The Monument
Lucknow
Springvale
Coorabulka

0 25 50 75 100 km

I Joins map 656 J K L M N O Joins map 657 P

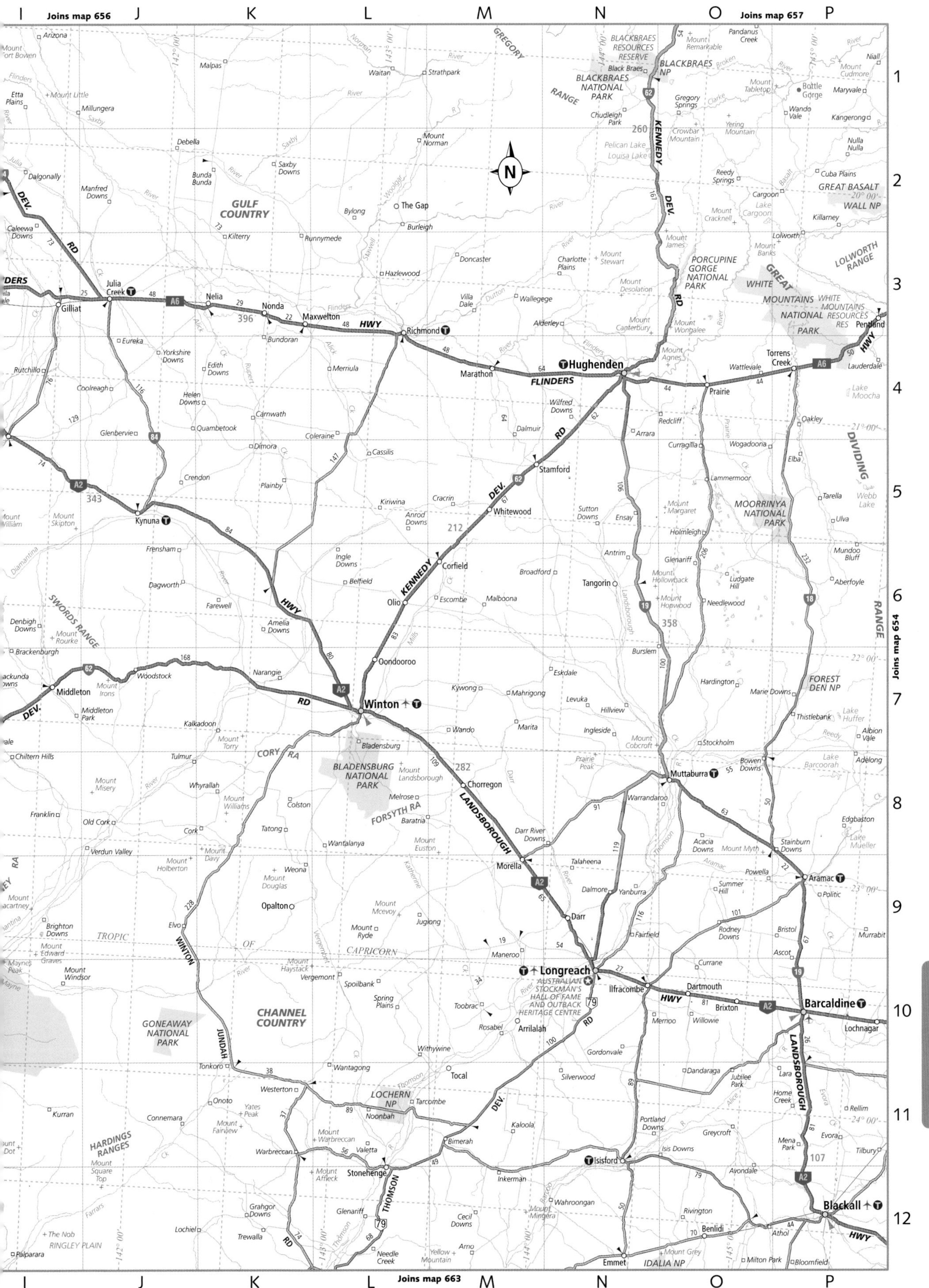

I J K L Joins map 663 M N O P

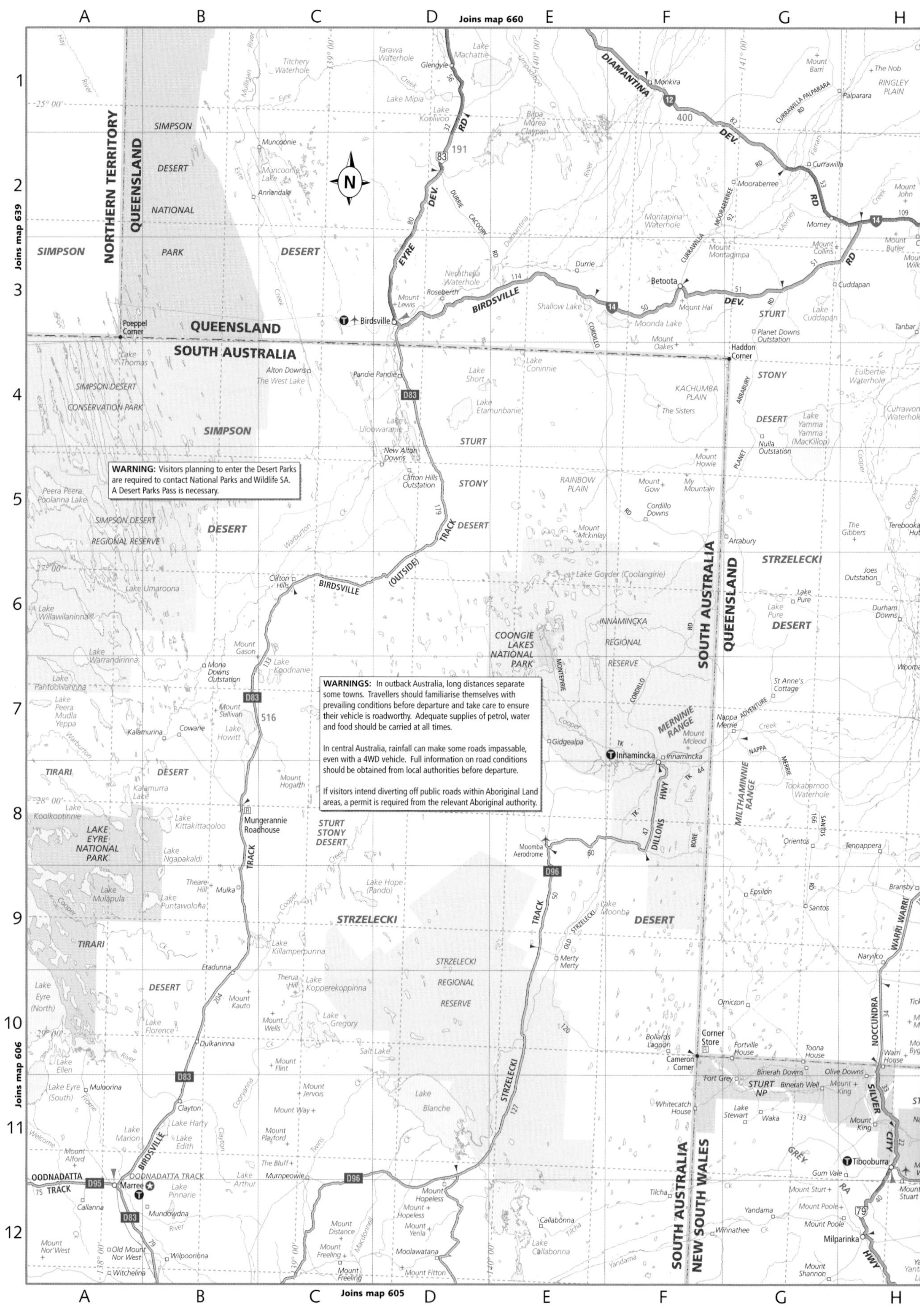

Joins map 660
Joins map 639
Joins map 606
Joins map 605
A
B
C
D
E
F
G
H
1
2
3
4
5
6
7
8
9
10
11
12
NORTHERN TERRITORY
QUEENSLAND
SIMPSON DESERT NATIONAL PARK
SIMPSON
DESERT
QUEENSLAND
SOUTH AUSTRALIA
SIMPSON DESERT CONSERVATION PARK
SIMPSON
SIMPSON DESERT REGIONAL RESERVE
DESERT
WARNING: Visitors planning to enter the Desert Parks are required to contact National Parks and Wildlife SA. A Desert Parks Pass is necessary.
WARNINGS: In outback Australia, long distances separate some towns. Travellers should familiarise themselves with prevailing conditions before departure and take care to ensure their vehicle is roadworthy. Adequate supplies of petrol, water and food should be carried at all times.
In central Australia, rainfall can make some roads impassable, even with a 4WD vehicle. Full information on road conditions should be obtained from local authorities before departure.
If visitors intend diverting off public roads within Aboriginal Land areas, a permit is required from the relevant Aboriginal authority.
Poeppel Corner
Birdsville
Glengyle
Muncoonie
Annandale
Mount Lewis
Roseberth
EYRE DEV.
BIRDSVILLE DEV.
DIAMANTINA DEV.
Monkira
Betoota
Durrie
Mooraberree
Currawilla
Morney
Cuddapan
Palparara
Haddon Corner
STURT STONY DESERT
Pandie Pandie
Alton Downs
Clifton Hills Outstation
New Alton Downs
BIRDSVILLE TRACK
Clifton Hills
Mount Gason
Mona Downs Outstation
Mount Sullivan
Cowarie
Kalamurina
TIRARI DESERT
LAKE EYRE NATIONAL PARK
Mungerannie Roadhouse
Mulka
Etadunna
Dulkaninna
Clayton
Marree
Mundowdna
Wilpoorinna
Callanna
OODNADATTA TRACK
Lake Eyre (North)
Lake Eyre (South)
Muloorina
COONGIE LAKES NATIONAL PARK
INNAMINCKA REGIONAL RESERVE
Lake Goyder (Coolangirie)
Innamincka
Gidgealpa
MERNINIE RANGE
Moomba Aerodrome
DILLONS HWY
STRZELECKI TRACK
STRZELECKI DESERT
STRZELECKI REGIONAL RESERVE
Merty Merty
Lake Blanche
Lake Callabonna
Callabonna
Mount Hopeless
Moolawatana
Mount Fitton
Murnpeowie
Lake Hope (Pando)
Lake Gregory
Arrabury
STRZELECKI DESERT
Nappa Merrie
ADVENTURE WAY
Orientos
Tennappera
Santos
Epsilon
Omicron
Cameron Corner
Corner Store
Fortville House
STURT NP
Fort Grey
Tibooburra
Milparinka
SILVER CITY HWY
NOCCUNDRA RD
WARRI WARRI RD
SOUTH AUSTRALIA
QUEENSLAND
NEW SOUTH WALES
Tilcha
Yandama
Wompah
Waka
Mount Poole
Mount Shannon

0 25 50 75 100 km

I J K L M N O P

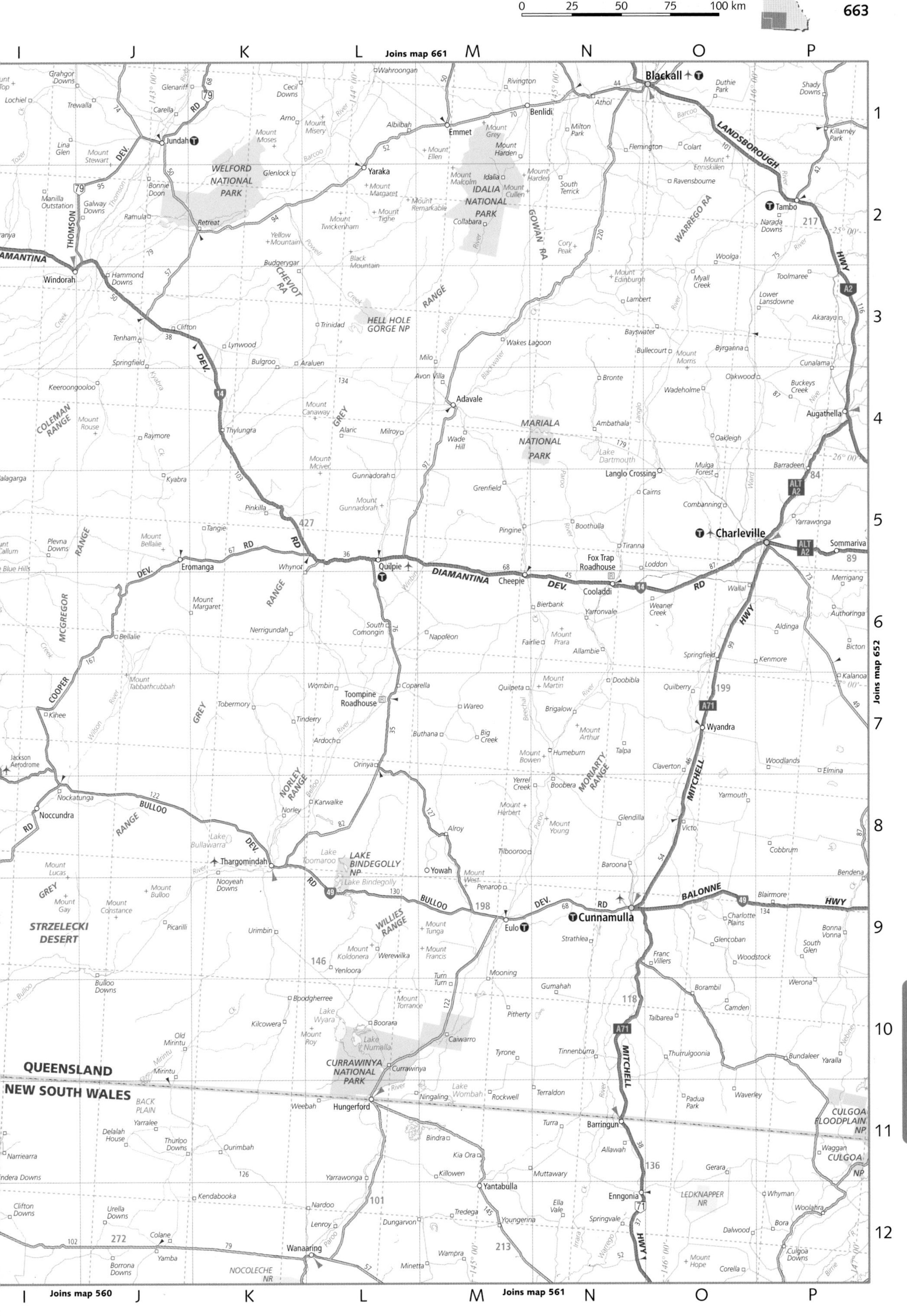

MAPS

TASMANIA

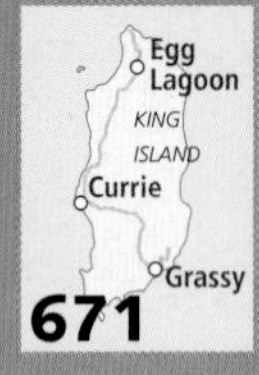

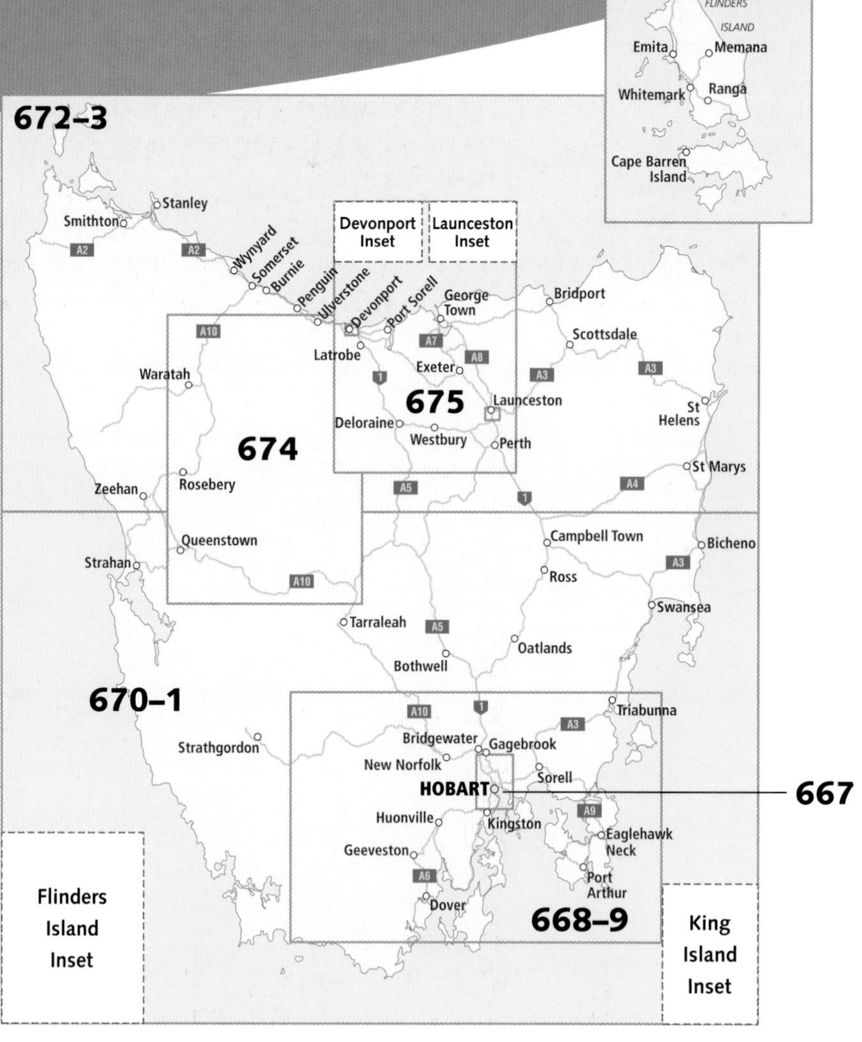

Tasmania

INTER-CITY ROUTES		DISTANCE
Hobart–Launceston via Midland Hwy	1	200 km
Hobart–Devonport via Midland & Bass hwys	1 B52 1	286 km

Approximate Distances TASMANIA	Burnie	Campbell Town	Deloraine	Devonport	Geeveston	George Town	Hobart	Launceston	New Norfolk	Oatlands	Port Arthur	Queenstown	Richmond	Rosebery	St Helens	St Marys	Scottsdale	Smithton	Sorell	Strahan	Swansea	Ulverstone
Burnie		200	101	50	391	204	333	152	328	247	432	163	304	110	300	263	222	88	318	185	267	28
Campbell Town	200		99	150	191	119	133	67	128	47	232	304	104	357	122	85	137	288	118	344	67	172
Deloraine	101	99		51	290	103	232	51	227	146	331	207	203	211	199	162	121	189	217	247	166	73
Devonport	50	150	51		341	154	283	102	278	197	382	213	254	160	250	213	172	138	268	235	217	22
Geeveston	391	191	290	341		310	58	258	95	144	157	308	85	361	313	276	328	479	84	348	197	363
George Town	204	119	103	154	310		252	52	247	166	351	310	223	314	182	182	83	292	237	350	186	176
Hobart	333	133	232	283	58	252		200	37	86	99	250	27	303	265	228	270	421	26	290	139	305
Launceston	152	67	51	102	258	52	200		195	114	299	258	171	262	167	130	70	240	185	298	134	124
New Norfolk	328	128	227	278	95	247	37	195		81	136	213	64	266	250	213	265	416	63	253	176	300
Oatlands	247	47	146	197	144	166	86	114	81		175	257	57	310	169	132	184	335	71	297	125	219
Port Arthur	432	232	331	382	157	351	99	299	136	175		349	87	402	312	275	369	520	73	389	186	404
Queenstown	163	304	207	213	308	310	250	258	213	257	349		277	53	426	389	328	253	276	40	389	191
Richmond	304	104	203	254	85	223	27	171	64	57	87	277		330	226	189	241	392	14	317	123	276
Rosebery	110	357	211	160	361	314	303	262	266	310	402	53	330		410	373	332	222	329	75	442	138
St Helens	300	122	199	250	313	182	265	167	250	169	312	426	226	410		37	99	388	240	466	126	272
St Marys	263	85	162	213	276	182	228	130	213	132	275	389	189	373	37		136	351	203	429	89	235
Scottsdale	222	137	121	172	328	83	270	70	265	184	369	328	241	332	99	136		310	255	368	204	194
Smithton	88	288	189	138	479	292	421	240	416	335	520	253	392	222	388	351	310		406	275	355	116
Sorell	318	118	217	268	84	237	26	185	63	71	73	276	14	329	240	203	255	406		316	113	290
Strahan	185	344	247	235	348	350	290	298	253	297	389	40	317	75	466	429	368	275	316		429	213
Swansea	267	67	166	217	197	186	139	134	176	125	186	389	123	442	126	89	204	355	113	429		239
Ulverstone	28	172	73	22	363	176	305	124	300	219	404	191	276	138	272	235	194	116	290	213	239	

Distances on this chart have been calculated over main roads and do not necessarily reflect the shortest route between towns.

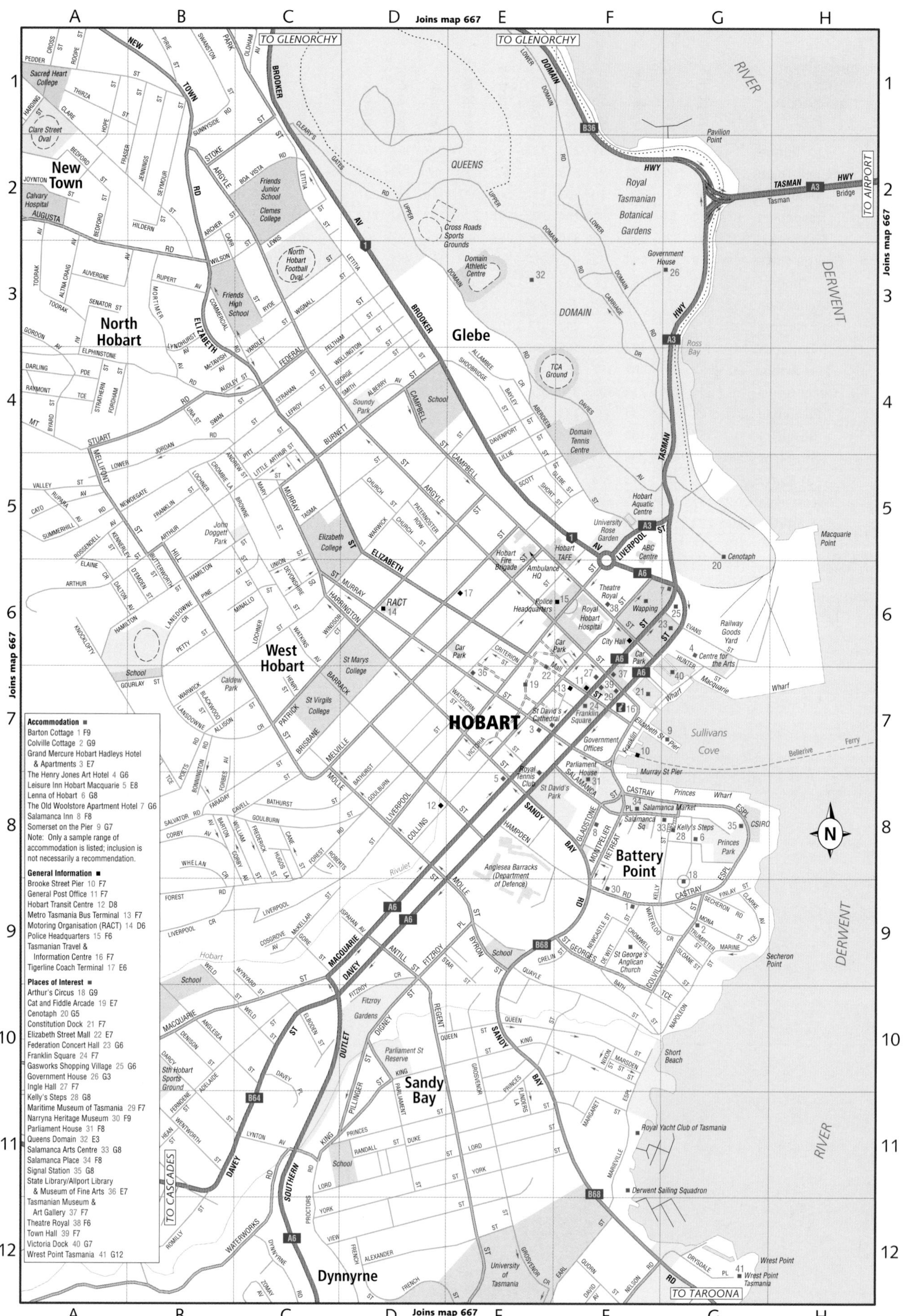
0 0.25 0.5 0.75 1 km
Joins map 667
TO GLENORCHY
TO AIRPORT
TO CASCADES
TO TAROONA
New Town
North Hobart
West Hobart
Glebe
HOBART
Battery Point
Sandy Bay
Dynnyrne
QUEENS DOMAIN
Royal Tasmanian Botanical Gardens
DERWENT RIVER
Sullivans Cove
Macquarie Point
Accommodation
Barton Cottage 1 F9
Colville Cottage 2 G9
Grand Mercure Hobart Hadleys Hotel & Apartments 3 E7
The Henry Jones Art Hotel 4 G6
Leisure Inn Hobart Macquarie 5 E8
Lenna of Hobart 6 G8
The Old Woolstore Apartment Hotel 7 G6
Salamanca Inn 8 F8
Somerset on the Pier 9 G7
Note: Only a sample range of accommodation is listed; inclusion is not necessarily a recommendation.
General Information
Brooke Street Pier 10 F7
General Post Office 11 F7
Hobart Transit Centre 12 D8
Metro Tasmania Bus Terminal 13 F7
Motoring Organisation (RACT) 14 D6
Police Headquarters 15 F6
Tasmanian Travel & Information Centre 16 F7
Tigerline Coach Terminal 17 E6
Places of Interest
Arthur's Circus 18 G9
Cat and Fiddle Arcade 19 E7
Cenotaph 20 G5
Constitution Dock 21 F7
Elizabeth Street Mall 22 E7
Federation Concert Hall 23 G6
Franklin Square 24 F7
Gasworks Shopping Village 25 G6
Government House 26 G3
Ingle Hall 27 F7
Kelly's Steps 28 G8
Maritime Museum of Tasmania 29 F7
Narryna Heritage Museum 30 F9
Parliament House 31 F8
Queens Domain 32 E3
Salamanca Arts Centre 33 G8
Salamanca Place 34 F8
Signal Station 35 G8
State Library/Allport Library & Museum of Fine Arts 36 E7
Tasmanian Museum & Art Gallery 37 F7
Theatre Royal 38 F6
Town Hall 39 F7
Victoria Dock 40 G7
Wrest Point Tasmania 41 G12

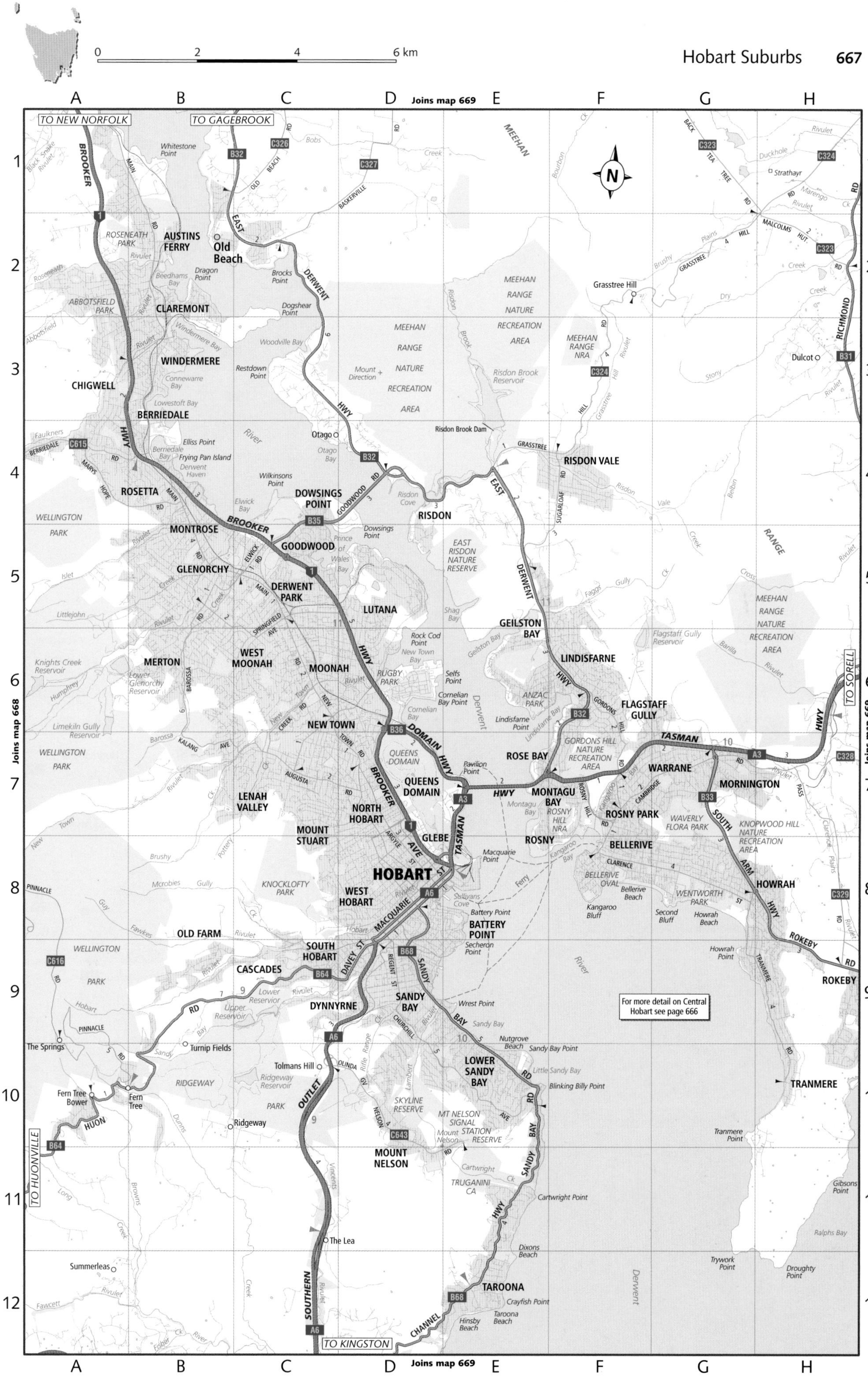

0 2 4 6 km

A B C D E F G H

Joins map 669
Joins map 668
TO NEW NORFOLK
TO GAGEBROOK
TO SORELL
TO HUONVILLE
TO KINGSTON
For more detail on Central Hobart see page 666

1 2 3 4 5 6 7 8 9 10 11 12

AUSTINS FERRY
Old Beach
CLAREMONT
WINDERMERE
CHIGWELL
BERRIEDALE
ROSETTA
MONTROSE
GLENORCHY
GOODWOOD
DOWSINGS POINT
DERWENT PARK
LUTANA
RISDON
RISDON VALE
GEILSTON BAY
LINDISFARNE
FLAGSTAFF GULLY
WEST MOONAH
MOONAH
MERTON
NEW TOWN
ROSE BAY
WARRANE
MORNINGTON
QUEENS DOMAIN
MONTAGU BAY
ROSNY PARK
ROSNY
BELLERIVE
LENAH VALLEY
NORTH HOBART
MOUNT STUART
GLEBE
HOBART
WEST HOBART
BATTERY POINT
HOWRAH
OLD FARM
SOUTH HOBART
CASCADES
DYNNYRNE
SANDY BAY
ROKEBY
LOWER SANDY BAY
TRANMERE
MOUNT NELSON
TAROONA

MEEHAN RANGE NATURE RECREATION AREA
EAST RISDON NATURE RESERVE
GORDONS HILL NATURE RECREATION AREA
KNOPWOOD HILL NATURE RECREATION AREA
WELLINGTON PARK
KNOCKLOFTY PARK
RIDGEWAY PARK
SKYLINE RESERVE
MT NELSON SIGNAL STATION RESERVE
TRUGANINI CA

Grasstree Hill
Dulcot
Strathayr
Otago
Tolmans Hill
The Springs
Turnip Fields
Fern Tree Bower
Fern Tree
Ridgeway
The Lea
Summerleas

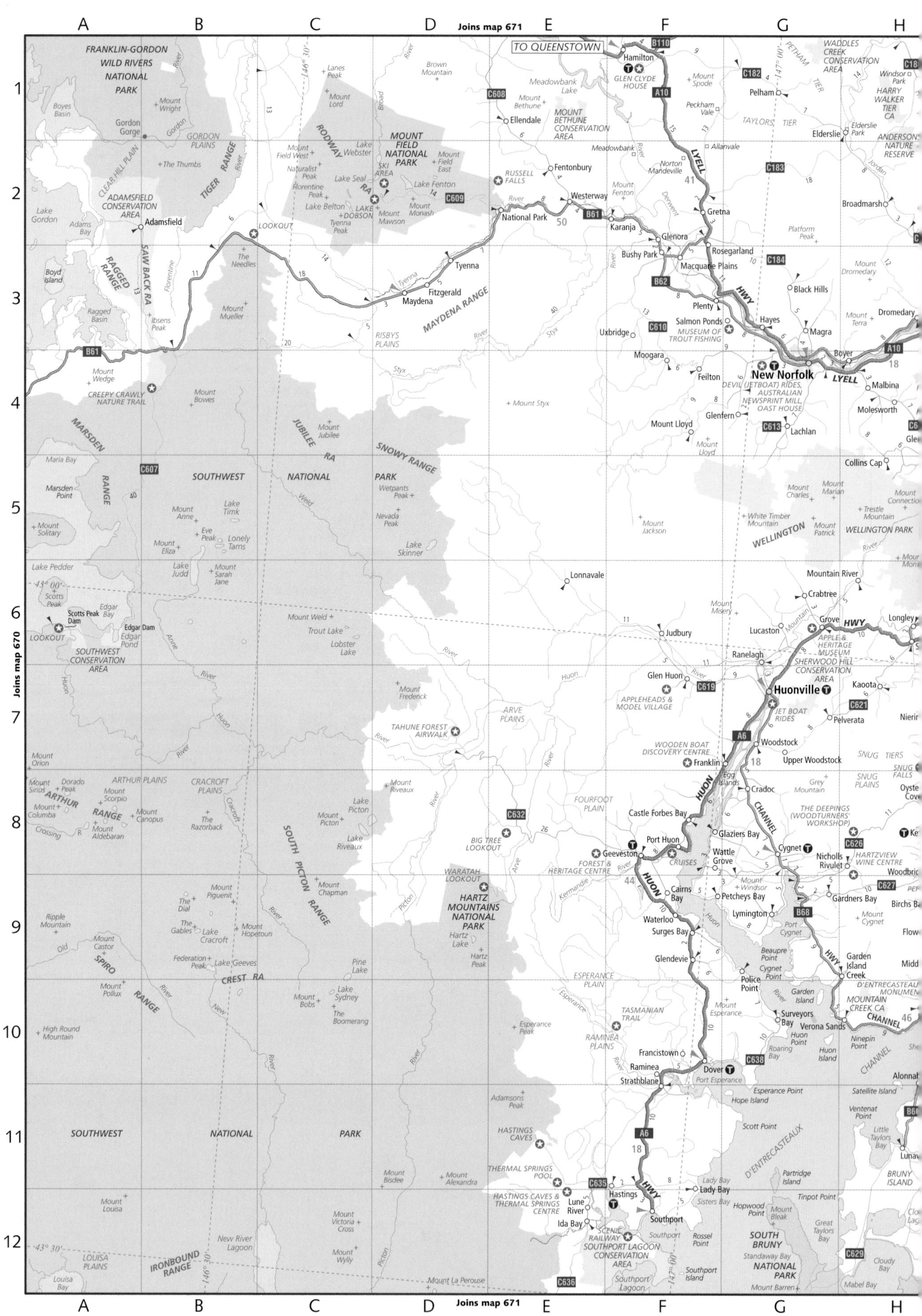

A
B
C
D
Joins map 671
E
F
G
H
1
2
3
4
5
6
7
8
9
10
11
12
Joins map 670
TO QUEENSTOWN
FRANKLIN-GORDON WILD RIVERS NATIONAL PARK
Boyes Basin
Mount Wright
Gordon Gorge
GORDON PLAINS
The Thumbs
TIGER RANGE
CLEAR HILL PLAIN
ADAMSFIELD CONSERVATION AREA
Lake Gordon
Adams Bay
Adamsfield
Boyd Island
RAGGED RANGE
SAW BACK RA
Ragged Basin
Ibsens Peak
Florentine
LOOKOUT
The Needles
Mount Mueller
Lanes Peak
Mount Lord
RODWAY RA
Mount Field West
Naturalist Peak
Florentine Peak
Lake Belton
Lake Seal
SKI AREA
LAKE DOBSON
Tyenna Peak
Lake Webster
MOUNT FIELD NATIONAL PARK
Mount Field East
Lake Fenton
Mount Mawson
Mount Monash
C609
Brown Mountain
Broad
Tyenna
Maydena
Fitzgerald
MAYDENA RANGE
RISBYS PLAINS
B61
Meadowbank Lake
C608
Mount Bethune
Ellendale
MOUNT BETHUNE CONSERVATION AREA
RUSSELL FALLS
Fentonbury
Westerway
National Park
Karanja
Meadowbank
Mount Fenton
Hamilton
GLEN CLYDE HOUSE
B110
A10
Mount Spode
C182
Pelham
Peckham Vale
TAYLORS TIER
PETHAM TIER
Allanvale
Norton Mandeville
LYELL
C183
Gretna
Glenora
Bushy Park
Rosegarland
Macquarie Plains
C184
B62
Plenty
Black Hills
Platform Peak
Mount Dromedary
Dromedary
Mount Terra
HWY
Salmon Ponds
MUSEUM OF TROUT FISHING
C610
Uxbridge
Hayes
Magra
Moogara
Feilton
New Norfolk
DEVIL (JETBOAT) RIDES, AUSTRALIAN NEWSPRINT MILL, OAST HOUSE
Boyer
LYELL
A10
Malbina
Molesworth
Glenfern
Mount Lloyd
C613
Lachlan
WADDLES CREEK CONSERVATION AREA
Windsor Park
HARRY WALKER TIER CA
Elderslie
Elderslie Park
ANDERSONS NATURE RESERVE
Broadmarsh
Jordan
Collins Cap
Mount Wedge
CREEPY CRAWLY NATURE TRAIL
Mount Bowes
MARSDEN RANGE
Maria Bay
Marsden Point
C607
SOUTHWEST NATIONAL PARK
JUBILEE RA
Mount Jubilee
SNOWY RANGE
Wetpants Peak
Nevada Peak
Lake Skinner
Mount Anne
Eve Peak
Lake Timk
Lonely Tarns
Mount Eliza
Mount Solitary
Lake Pedder
Lake Judd
Mount Sarah Jane
Mount Styx
Styx
Mount Jackson
Mount Charles
Mount Marian
White Timber Mountain
Mount Patrick
WELLINGTON
WELLINGTON PARK
Trestle Mountain
Mount Connection
Lonnavale
Mountain River
Crabtree
Mount Misery
Judbury
Lucaston
Grove
HWY
APPLE & HERITAGE MUSEUM
Longley
Ranelagh
SHERWOOD HILL CONSERVATION AREA
43° 00'
Scotts Peak
Scotts Peak Dam
Edgar Bay
Edgar Dam
Edgar Pond
LOOKOUT
SOUTHWEST CONSERVATION AREA
Mount Weld
Trout Lake
Lobster Lake
Glen Huon
C619
APPLEHEADS & MODEL VILLAGE
Huonville
Kaoota
C621
JET BOAT RIDES
Pelverata
Mount Frederick
ARVE PLAINS
TAHUNE FOREST AIRWALK
A6
Woodstock
Upper Woodstock
WOODEN BOAT DISCOVERY CENTRE
Franklin
SNUG TIERS
SNUG FALLS
SNUG PLAINS
Egg Islands
Cradoc
Grey Mountain
Mount Orion
Mount Sirius
Dorado Peak
ARTHUR PLAINS
Mount Scorpio
ARTHUR RANGE
Mount Columba
Mount Canopus
Mount Aldebaran
CRACROFT PLAINS
The Razorback
Mount Riveaux
Lake Picton
Mount Picton
Lake Riveaux
FOURFOOT PLAIN
C632
Castle Forbes Bay
HUON
CHANNEL
THE DEEPINGS (WOODTURNERS' WORKSHOP)
BIG TREE LOOKOUT
Port Huon
Glaziers Bay
Geeveston
FOREST & HERITAGE CENTRE
CRUISES
Wattle Grove
Cygnet
C626
Nicholls Rivulet
HARTZVIEW WINE CENTRE
Woodbridge
WARATAH LOOKOUT
HARTZ MOUNTAINS NATIONAL PARK
Hartz Lake
Hartz Peak
SOUTH PICTON RANGE
Mount Chapman
Mount Piguenit
The Dial
The Gables
Lake Cracroft
Mount Hopetoun
Cairns Bay
Petcheys Bay
Mount Windsor
C627
Gardners Bay
B68
Lymington
Birchs Bay
Waterloo
Surges Bay
Port Cygnet
Mount Cygnet
Glendevie
Beaupre Point
Cygnet Point
Police Point
Garden Island Creek
Ripple Mountain
Mount Castor
SPIRO RANGE
Federation Peak
Lake Geeves
CREST RA
Pine Lake
Lake Sydney
Mount Bobs
The Boomerang
Mount Pollux
ESPERANCE PLAIN
Garden Island
D'ENTRECASTEAUX MONUMENT
MOUNTAIN CREEK CA
Surveyors Bay
Verona Sands
CHANNEL
High Round Mountain
Esperance Peak
TASMANIAN TRAIL
RAMINEA PLAINS
Mount Esperance
Huon Point
Roaring Bay
Huon Island
Ninepin Point
Francistown
Raminea
Dover
Strathblane
C638
Port Esperance
Esperance Point
Hope Island
Alonnah
Satellite Island
Adamsons Peak
Scott Point
Ventenat Point
Little Taylors Bay
Lunawanna
BRUNY ISLAND
SOUTHWEST NATIONAL PARK
HASTINGS CAVES
A6
D'ENTRECASTEAUX
Mount Bisdee
Mount Alexandra
THERMAL SPRINGS POOL
C635
Hastings
Lady Bay
Partridge Island
HASTINGS CAVES & THERMAL SPRINGS CENTRE
Lune River
Sisters Bay
Tinpot Point
Mount Louisa
Hopwood Point
Mount Bleak
Great Taylors Bay
Mount Victoria Cross
Ida Bay
Southport
SCENIC RAILWAY
Rossel Point
SOUTH BRUNY NATIONAL PARK
Standaway Bay
43° 30'
LOUISA PLAINS
IRONBOUND RANGE
New River Lagoon
Mount Wylly
SOUTHPORT LAGOON CONSERVATION AREA
Southport Island
C629
Cloudy Bay
Louisa Bay
Mount La Perouse
C636
Southport Lagoon
Mount Barren
Mabel Bay
146° 30'
147° 00'

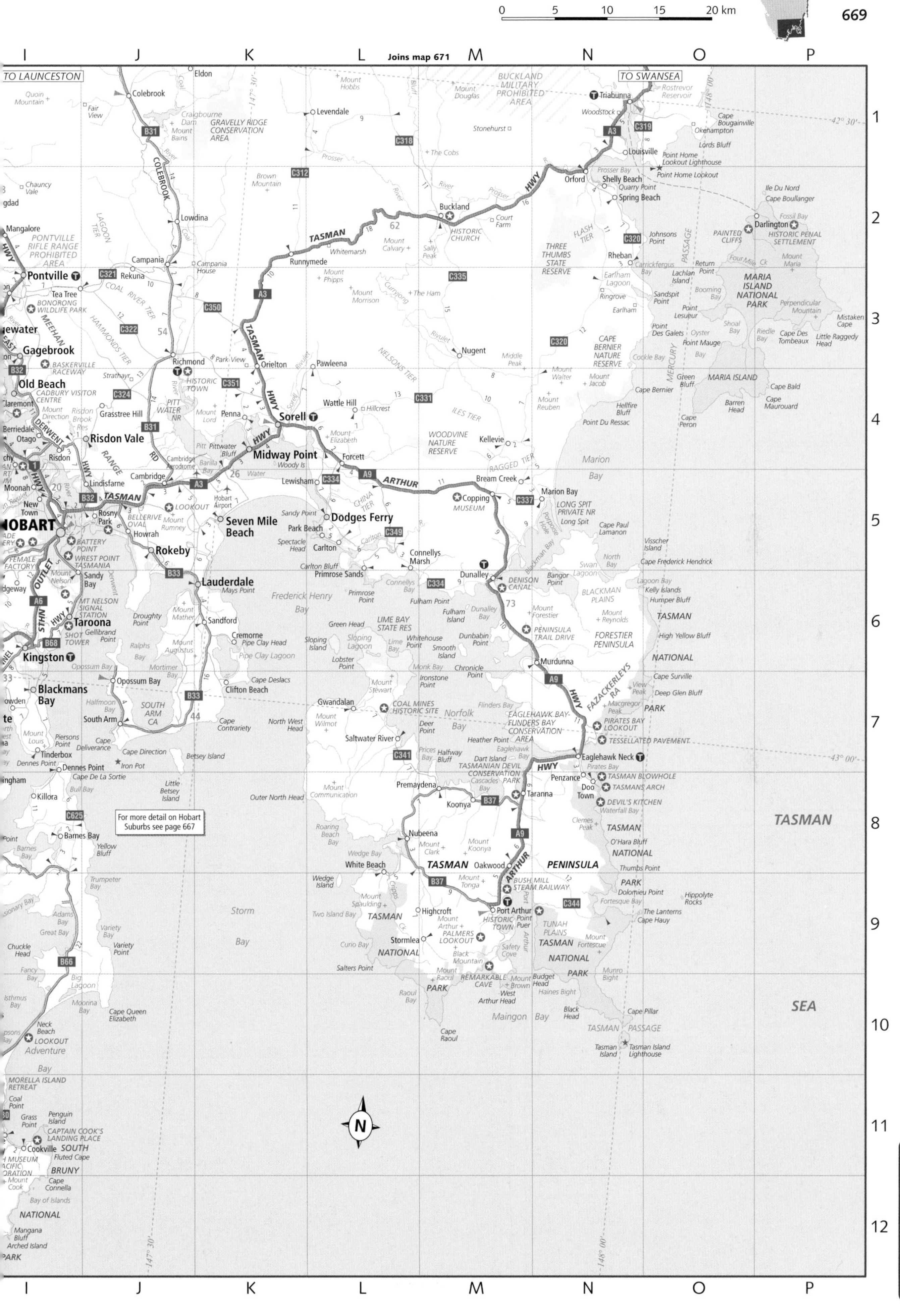

0 5 10 15 20 km
Joins map 671
TO LAUNCESTON
TO SWANSEA
HOBART
Sorell
Midway Point
Dodges Ferry
Seven Mile Beach
Lauderdale
Rokeby
Taroona
Kingston
Blackmans Bay
Risdon Vale
Old Beach
Gagebrook
Pontville
Richmond
Triabunna
Orford
Buckland
Dunalley
Eaglehawk Neck
Port Arthur
Nubeena
Taranna
Murdunna
Maria Island
MARIA ISLAND NATIONAL PARK
TASMAN NATIONAL PARK
TASMAN PENINSULA
FORESTIER PENINSULA
Storm Bay
Frederick Henry Bay
Norfolk Bay
Marion Bay
TASMAN SEA
For more detail on Hobart Suburbs see page 667

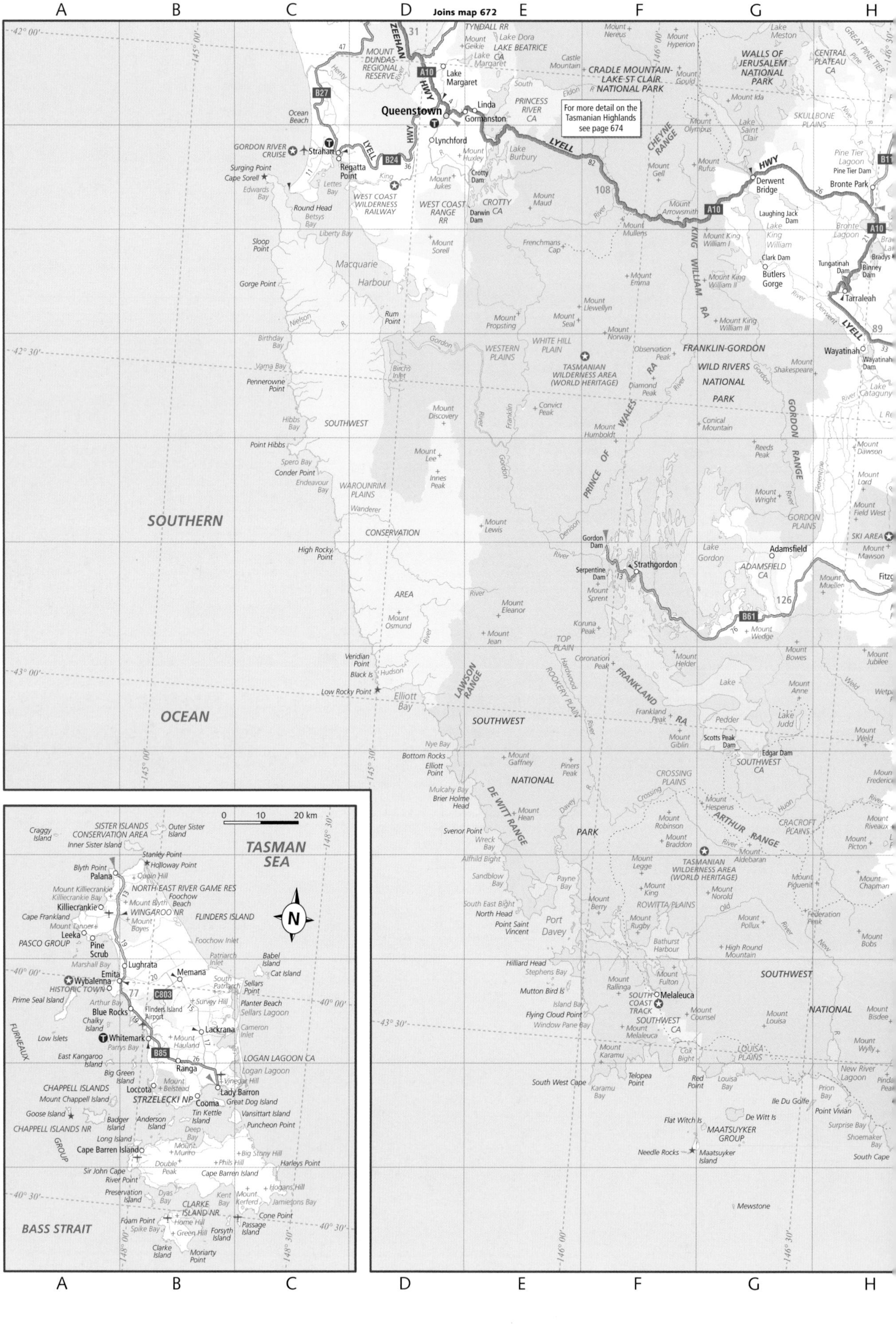

Joins map 672
A
B
C
D
E
F
G
H
1
2
3
4
5
6
7
8
9
10
11
12
42° 00'
42° 30'
43° 00'
43° 30'
145° 00'
145° 30'
146° 00'
146° 30'
TYNDALL RR
Mount Geikie
Lake Dora
LAKE BEATRICE
Lake Margaret
MOUNT DUNDAS REGIONAL RESERVE
Henty
ZEEHAN HWY
A10
Lake Margaret
Castle Mountain
CRADLE MOUNTAIN-LAKE ST CLAIR NATIONAL PARK
Mount Nereus
Mount Hyperion
Mount Gould
WALLS OF JERUSALEM NATIONAL PARK
Lake Meston
CENTRAL PLATEAU CA
GREAT PINE TIER
B27
Ocean Beach
Queenstown
Linda
Gormanston
PRINCESS RIVER CA
Eldon
South
For more detail on the Tasmanian Highlands see page 674
Mount Ida
Mount Olympus
Lake Saint Clair
SKULLBONE PLAINS
Nive
GORDON RIVER CRUISE
Strahan
LYELL
Lynchford
HWY
B24
Regatta Point
Surging Point
Cape Sorell
Lettes Bay
King
Mount Huxley
Lake Burbury
LYELL
CHEYNE RANGE
Mount Gell
Mount Rufus
HWY
Derwent Bridge
Pine Tier Lagoon
Pine Tier Dam
Bronte Park
Edwards Bay
WEST COAST WILDERNESS RAILWAY
Mount Jukes
Crotty Dam
WEST COAST RANGE RR
CROTTY CA
Darwin Dam
Mount Maud
108
Mount Arrowsmith
A10
Laughing Jack Dam
Lake King William
Bronte Lagoon
A10
Round Head
Betsys Bay
Liberty Bay
Mount Sorell
Frenchmans Cap
Mount Mullens
KING WILLIAM RA
Mount King William I
Clark Dam
Butlers Gorge
Tungatinah Dam
Binney Dam
Bradys
Sloop Point
Macquarie Harbour
Gorge Point
Mount Emma
Mount King William II
Tarraleah
Nielson
Rum Point
Mount Llewellyn
Mount Propsting
Mount Seal
Mount King William III
Derwent
LYELL
89
Birthday Bay
Gordon
WESTERN PLAINS
WHITE HILL PLAIN
Mount Norway
Observation Peak
FRANKLIN-GORDON WILD RIVERS NATIONAL PARK
Wayatinah
Wayatinah Dam
Varna Bay
Birches Inlet
TASMANIAN WILDERNESS AREA (WORLD HERITAGE)
Diamond Peak
Mount Shakespeare
Lake Catagunya
Pennerowne Point
Mount Discovery
Convict Peak
Franklin
PRINCE OF WALES RA
GORDON RANGE
Hibbs Bay
SOUTHWEST
Mount Humboldt
Conical Mountain
Point Hibbs
Mount Lee
Reeds Peak
Mount Dawson
Spero Bay
Conder Point
Innes Peak
Endeavour Bay
WAROUNRIM PLAINS
Mount Wright
Florentine
Mount Lord
Wanderer
Mount Field West
SOUTHERN
CONSERVATION
Mount Lewis
Denison
GORDON PLAINS
Gordon Dam
SKI AREA
High Rocky Point
Lake Gordon
Adamsfield
Mount Mawson
Serpentine Dam
Strathgordon
ADAMSFIELD CA
Mount Mueller
Fitzg
AREA
Mount Sprent
126
Mount Eleanor
Mount Osmund
Koruna Peak
B61
Mount Wedge
Mount Jean
TOP PLAIN
Mount Bowes
Mount Jubilee
Veridian Point
Black Is
Hudson
Coronation Peak
Mount Helder
Lake
Mount Anne
Weld
Low Rocky Point
Elliott Bay
LAWSON RANGE
ROOKERY PLAIN
FRANKLAND RA
Frankland Peak
Pedder
Lake Judd
OCEAN
SOUTHWEST
Mount Giblin
Scotts Peak Dam
Mount Weld
Nye Bay
Bottom Rocks
Elliott Point
Mount Gaffney
Piners Peak
Edgar Dam
SOUTHWEST CA
NATIONAL
CROSSING PLAINS
Mulcahy Bay
Brier Holme Head
DE WITT RANGE
Mount Hean
Mount Hesperus
Mount Robinson
ARTHUR RANGE
Huon
CRACROFT PLAINS
Mount Frederick
Mount Riveaux
Svenor Point
Wreck Bay
PARK
Mount Braddon
Mount Picton
Alfhild Bight
Mount Legge
TASMANIAN WILDERNESS AREA (WORLD HERITAGE)
Mount Aldebaran
Sandblow Bay
Payne Bay
Mount King
Mount Norold
Mount Piguenit
Mount Chapman
South East Bight
North Head
Point Saint Vincent
Mount Berry
Port Davey
ROWITTA PLAINS
Old
Federation Peak
Mount Rugby
Mount Pollux
Mount Bobs
Bathurst Harbour
High Round Mountain
Hilliard Head
Stephens Bay
Mount Rallinga
Mount Fulton
SOUTHWEST
Mutton Bird Is
SOUTH COAST TRACK
Melaleuca
Island Bay
Flying Cloud Point
Window Pane Bay
Mount Counsel
SOUTHWEST CA
Mount Louisa
NATIONAL
Mount Bisdee
Mount Melaleuca
Mount Karamu
Cox Bight
LOUISA PLAINS
Mount Wylly
New River Lagoon
Pindar Peak
South West Cape
Karamu Bay
Telopea Point
Red Point
Louisa Bay
Prion Bay
Ile Du Golfe
Point Vivian
De Witt Is
Flat Witch Is
MAATSUYKER GROUP
Surprise Bay
Shoemaker Bay
Needle Rocks
Maatsuyker Island
South Cape
Mewstone
0
10
20 km
SISTER ISLANDS CONSERVATION AREA
Outer Sister Island
Craggy Island
Inner Sister Island
TASMAN SEA
Stanley Point
Blyth Point
Holloway Point
Palana
Quoin Hill
NORTH EAST RIVER GAME RES
Mount Killiecrankie
Killiecrankie Bay
Foochow Beach
Mount Blyth
Killiecrankie
WINGAROO NR
Cape Frankland
Mount Tanner
Mount Boyes
FLINDERS ISLAND
N
Leeka
PASCO GROUP
Pine Scrub
Foochow Inlet
Marshall Bay
Lughrata
Patriarch Inlet
Babel Island
Emita
Memana
Cat Island
Wybalenna HISTORIC TOWN
77
C803
South Patriarch
Sellars Point
Prime Seal Island
Arthur Bay
Survey Hill
Planter Beach
Sellars Lagoon
40° 00'
Blue Rocks
Flinders Island Airport
Chalky Island
FURNEAUX
Cameron Inlet
Low Islets
Whitemark
Mount Hauland
Lackrana
Parrys Bay
B85
East Kangaroo Island
26
LOGAN LAGOON CA
Ranga
Logan Lagoon
Big Green Island
Mount Belstead
Vinegar Hill
CHAPPELL ISLANDS
Loccota
Lady Barron
Mount Chappell Island
STRZELECKI NP
Cooma
Great Dog Island
Goose Island
Tin Kettle Island
Vansittart Island
Badger Island
Anderson Island
Puncheon Point
CHAPPELL ISLANDS NR
GROUP
Deep Bay
Long Island
Mount Munro
Cape Barren Island
Big Stony Hill
Double Peak
Phils Hill
Harleys Point
Sir John Cape
River Point
Cape Barren Island
Hogans Hill
Preservation Island
Dyas Bay
Kent Bay
Mount Kerford
Jamiesons Bay
40° 30'
CLARKE ISLAND NR
Cone Point
BASS STRAIT
Foam Point
Spike Bay
Home Hill
Green Hill
Forsyth Island
Passage Island
148° 00'
148° 30'
Clarke Island
Moriarty Point

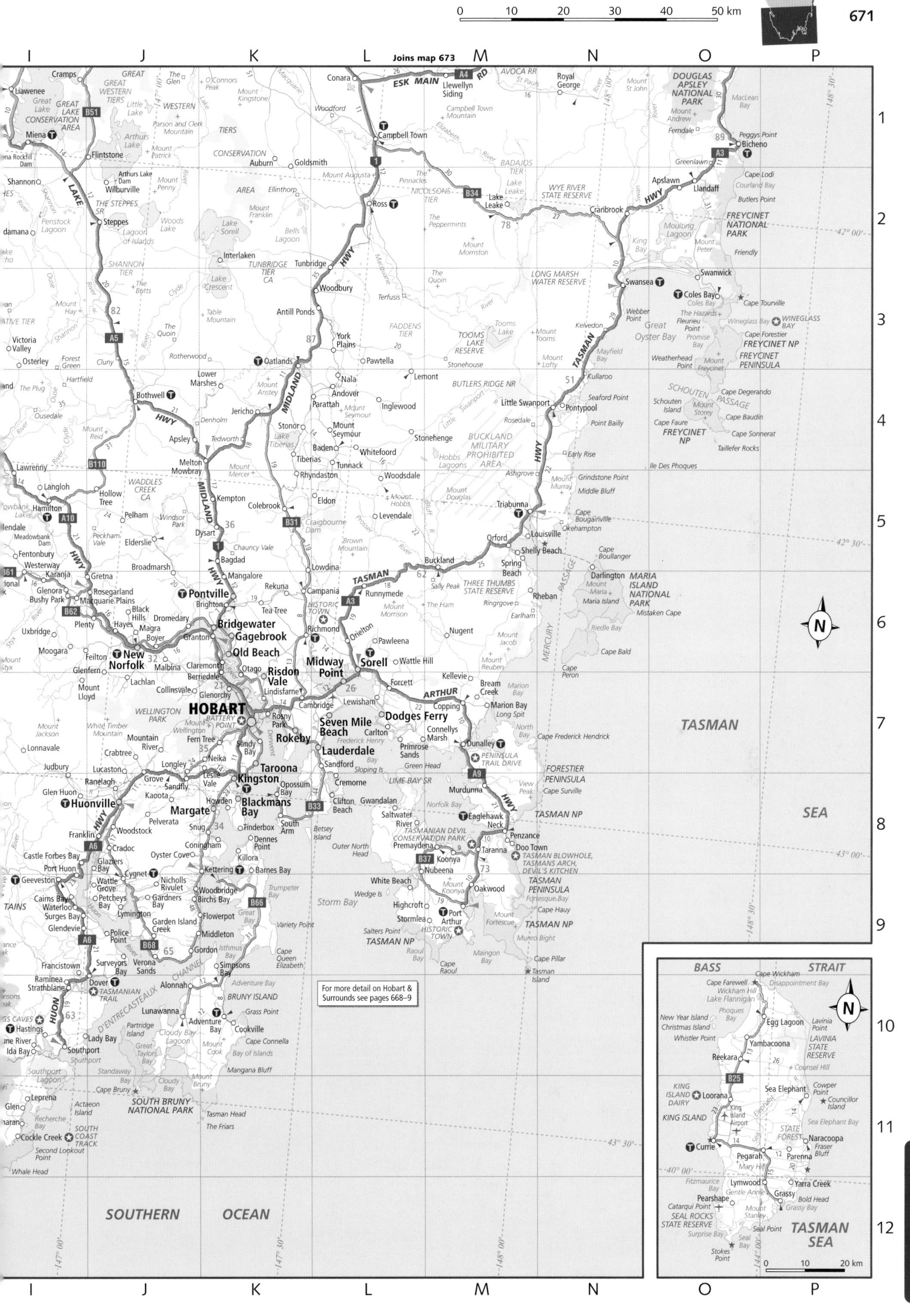

For more detail on Hobart & Surrounds see pages 668–9

A B C D E F G H
1 2 3 4 5 6 7 8 9 10 11 12
For detail on King Island see page 671
For more detail on the Tasmanian Highlands see page 674
Joins map 670
TASMAN
SEA
BASS
STRAIT
SOUTHERN
OCEAN
Spirit of Tasmania ferries Melbourne to Devonport
Albatross Is
Cape Keraudren
North West Cape
Coulomb Bay
Three Hummock Island
THREE HUMMOCK ISLAND STATE RESERVE
Cape Adamson
Cuvier Bay
Cuvier Point
Wallaby Point
HUNTER ISLAND CONSERVATION AREA
Hunter Island
Cave Bay
Steep Is
Petrel Islands
Cape Buache
Walker Island
WALKER CHANNEL
Bird Is
Trefoil Is
Woolnorth Point
Boullanger Bay
The Doughboys
Cape Grim
Valley Bay
Woolnorth
Ransonnet Bay
Robbins Island
Guyton Point
Cape Elie
Bluff Point
Studland Bay
Dodgers Point
West Montagu
Montagu
Big Bay
Shipwreck Point
Perkins Island
Perkins Bay
North Point
Half Moon Bay
Stanley
THE NUT
The Nut
Circular Head
East Inlet
Sawyer Bay
Duck Bay
Mount Cameron West
Ann Bay
Green Point
Marrawah
Redpa
Togari
HAYS TIER
Christmas Hills
Mella
Smithton
Forest
Wiltshire Junction
Port Latta
Crayfish Creek
Hellyer
Rocky Cape
ROCKY CAPE NP
South Forest
Black River
Edgcumbe Beach
WEST POINT STATE RESERVE
BASS HWY
Brittons Swamp
Irishtown
Mengha
Sisters Beach
Boat Harbour Beach
Boat Harbour
Table Cape
ARTHUR-PIEMAN CONSERVATION AREA
Mawson Bay
Alcomie
Edith Creek
Lileah
Mawbanna
Montumana
Sisters Creek
Bluff Hill Point
Nabageena
Myalla
Flowerdale
Wynyard
Roger River West
Roger River
Trowutta
DIP RANGE RR
Moorleah
Gardiner Point
Arthur River
Milabena
Lapoinya
Somerset
Camdale
Cooee
BURNIE
Wivenhoe
Mount Dipwood
Calder
Lower Mount Hicks
Sundown Point
Nelson Bay
Meunna
Preolenna
Oldina
Chasm Creek
Heybridge
Howth
Sulphur Creek
Couta Rocks
Phantom Peak
Kellatier
Yolla
Upper Mount Hicks
Elliott
West Ridgley
Stowport
Penguin
Ulverstone
Turners Beach
DEVONPORT
Temma
Kaywood
Blue Peak
Henrietta
Takone
East Yolla
Ridgley
Upper Stowport
Cuprona
West Pine
Natone
Ferndene
West Takone
Highclere
Gawler
Leith
Don
Forth
Quoiba
Spreyton
Hazard Bay
Balfour
Mount Balfour
Mount Frankland
Mount Bertha
Tewkesbury
Oonah
South Riana
Riana
North Motton
Abbotsham
Kindred
Eugenana
Ordnance Point
Hellyer Gorge
Hampshire
Upper Natone
Spalford
Gunns Plains
Sprent
Paloona
Melrose
ARTHUR-PIEMAN
SAVAGE RIVER NATIONAL PARK
Parrawe
HELLYER GORGE SR
Lake Kara
Loyetea
Heka
Preston
Central Castra
Paloona Dam
Kenneth Bay
Mount Hazelton
NORFOLK RANGE
Companion Reservoir
Mount Everett
Loyetea Peak
Warringa
Lower Barrington
Sandy Cape
LEVEN CANYON RR
Upper Castra
Barrington
Mount Norfolk
Mount Hadmar
Mount Bolton
SAVAGE RIVER REGIONAL RESERVE
Mount Cleveland
Waratah
Guildford
Loongana
Nietta
Narrawa
Wilmot
Railton
Sheffield
CONSERVATION
Mount Judith
MOUNT RD
Waratah Reservoir
Talbots Lagoon
Mount Tor
South Nietta
West Kentish
Stoodley
Roland
Lake Barrington
Mount Sunday
Savage River
Wilmot Dam
Lake Gairdner
Erriba
Staverton
Claude Road
Paradise
Beulah
AREA
Mount Meredith
Mount Ramsay
Prospect Mountain
Cethana
Moina
Cethana Dam
Gowrie Park
MOUNT ROLAND RR
Lower Beulah
VALE OF BELVOIR CA
Lorinna
ALUM CLIFFS SR
Rupert Point
Pieman Head
Hardwicke Bay
TIKKAWOPPA PLATEAU RR
MEREDITH RANGE REGIONAL RESERVE
Mount Cripps
Daisy Dell
Liena
Mole Creek
Lake Cethana
Mayberry
Corinna
Bulgobac
Mount Charter
Back Peak
Mount Kate
MOLE CREEK KARST NP
Conical Rocks Point
BERNAFAI RIDGE CA
PIEMAN RIVER STATE RESERVE
Mount Lindsay
Cradle Valley
Devils Ravine
Dove Lake
Mount Campbell
Cradle Mountain
Parangana Dam
Lake Parangana
GREAT WESTERN
Ahrberg Bay
Reece Dam
Mount Kershaw
MacKintosh Dam
MOUNT FARRELL
GRANITE TOR CONSERVATION AREA
Mount Parmeener
DEVILS GULLET SR
Lake MacKenzie
MacKenzie Dam
Bastyan Dam
Lake Rosebery
Tullah
Lake Will
CENTRAL
Granville Harbour
SURFING
Rosebery
Lake Pieman
Lake MacKintosh
Murchison Dam
Mount Murchison
CRADLE MOUNTAIN-LAKE ST CLAIR NATIONAL PARK
TASMANIAN WILDERNESS AREA (WORLD HERITAGE)
OVERLAND TRACK
Rowallan Dam
PLATEAU
PARTING CREEK RR
Renison Bell
MURCHISON
Williamsford
Lake Murchison
Mount Pelion West
Mount Oakleigh
Lake Rowallan
Mount Pillinger
Pillans Lake
MOUNT HEEMSKIRK REGIONAL RESERVE
Mount Heemskirk
Dundas
Zeehan
Mount Selina
Lake Plimsoll
Canning Peak
LAKE ST CLAIR
Mount Jerusalem
Trial Harbour
Remine
Mount Zeehan
ZEEHAN
Mount Dundas
RANGE
Mount Ossa
Mount Nereus
Mount Hyperion
Cathedral Mountain
Lake Meston
Lake Ada
GREAT PINE TIER
TYNDALL REGIONAL RESERVE
LAKE BEATRICE CA
ELDON RANGE
Eldon Peak
HENTY
MOUNT DUNDAS REGIONAL RESERVE
Lake Margaret
Castle Mountain
NATIONAL
WALLS OF JERUSALEM NATIONAL PARK
Mount Gould
WEST COAST RANGE
Mount Lyell
Lake Burbury
Linda
PRINCESS RIVER CONSERVATION AREA
PARK
Mount Ida
Queenstown
Gormanston
Pyramid Mountain
CHEYNE RANGE
Mount Olympus
Lake Saint Clair
Clarence Lagoon
GORDON RIVER CRUISE
Strahan
LYELL HWY
Lynchford
Mount Huxley
CROTTY CA
Crotty Dam
Mount Rufus
Mount Gell
Pine Tier Dam
Derwent Bridge
Bronte Park
Cape Sorell
Macquarie Heads
Regatta Point
Lettes Bay
WEST COAST WILDERNESS RAILWAY
WEST COAST RANGE RR
FRANKLIN-GORDON WILD RIVERS NATIONAL PARK
Mount Arrowsmith
Laughing Jack Dam
SOUTHWEST CONSERVATION AREA
Betsys Bay
Macquarie Harbour
Mount Darwin
Darwin Dam
Mount Mullens
Mount King William I
Lake King William
Bronte Lagoon
Sloop Point
Mount Sorell

0 10 20 30 40 50 km

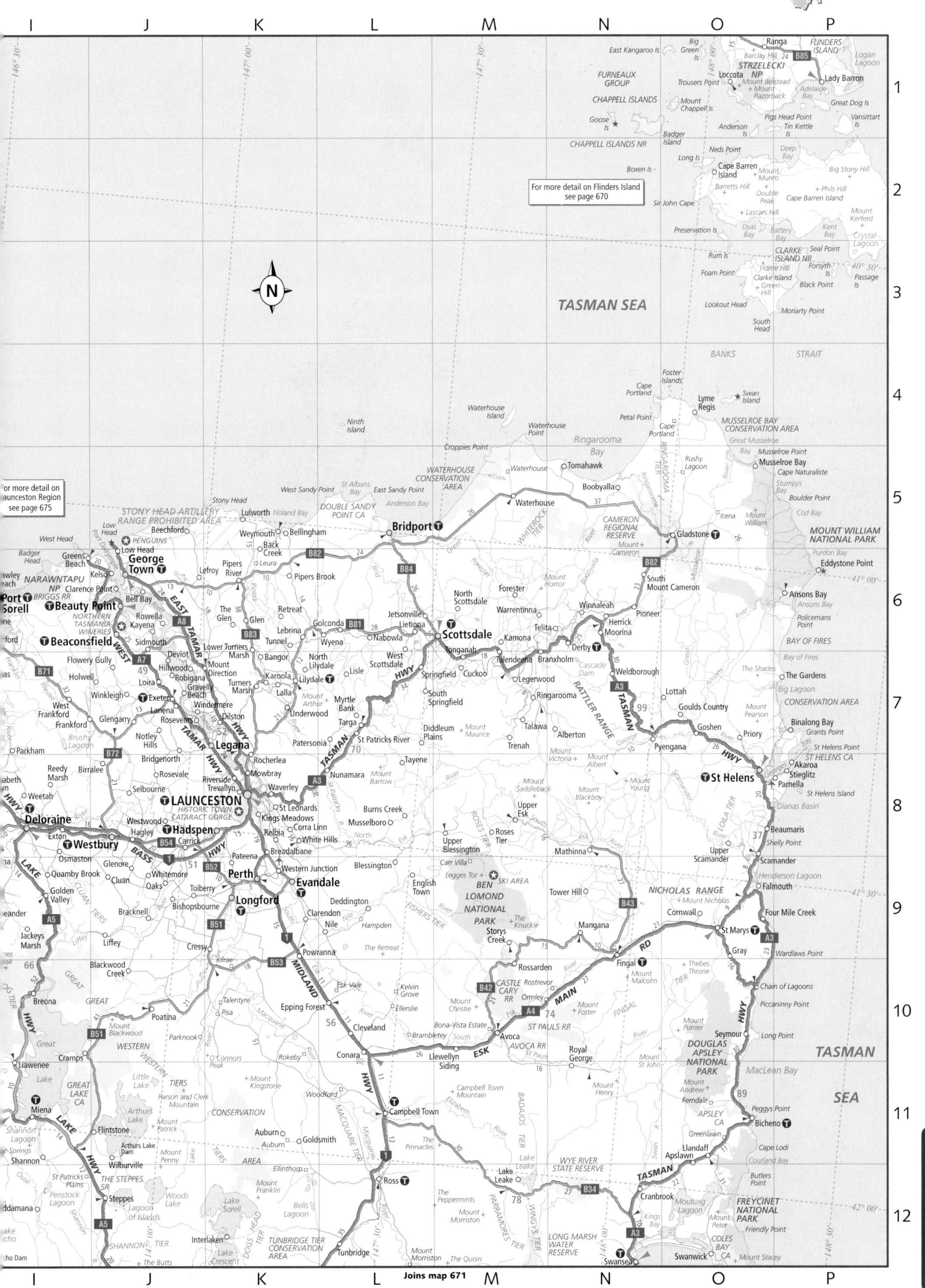

Joins map 671

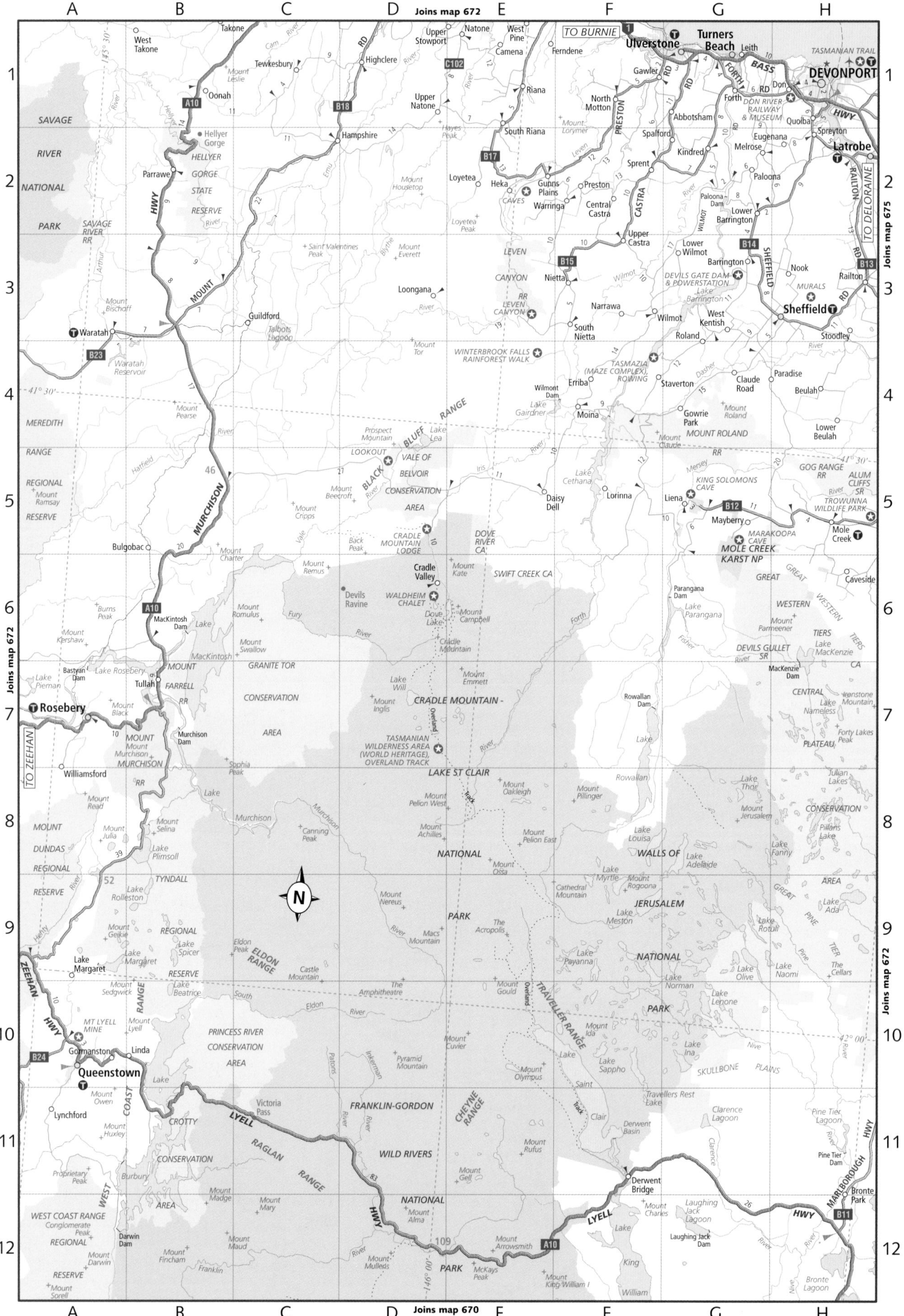
Joins map 672
Joins map 670
Joins map 675
TO BURNIE
TO DELORAINE
TO ZEEHAN
Ulverstone
Turners Beach
DEVONPORT
Latrobe
Sheffield
Queenstown
Rosebery
Tullah
Waratah
Cradle Valley
Derwent Bridge
SAVAGE RIVER NATIONAL PARK
CRADLE MOUNTAIN - LAKE ST CLAIR NATIONAL PARK
WALLS OF JERUSALEM NATIONAL PARK
FRANKLIN-GORDON WILD RIVERS NATIONAL PARK
MOLE CREEK KARST NP
MURCHISON HWY
LYELL HWY
BASS HWY
MARLBOROUGH HWY
ZEEHAN HWY
TASMANIAN WILDERNESS AREA (WORLD HERITAGE), OVERLAND TRACK

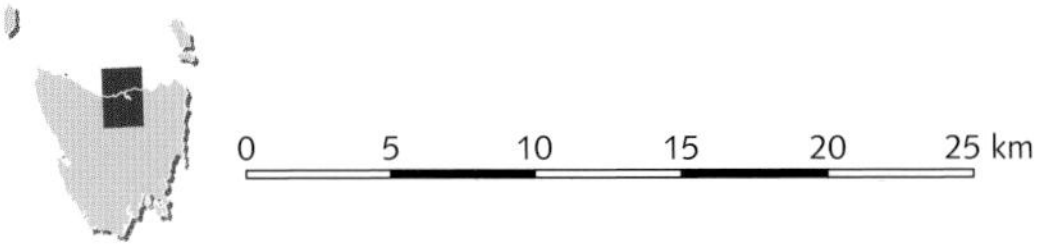

0 5 10 15 20 25 km

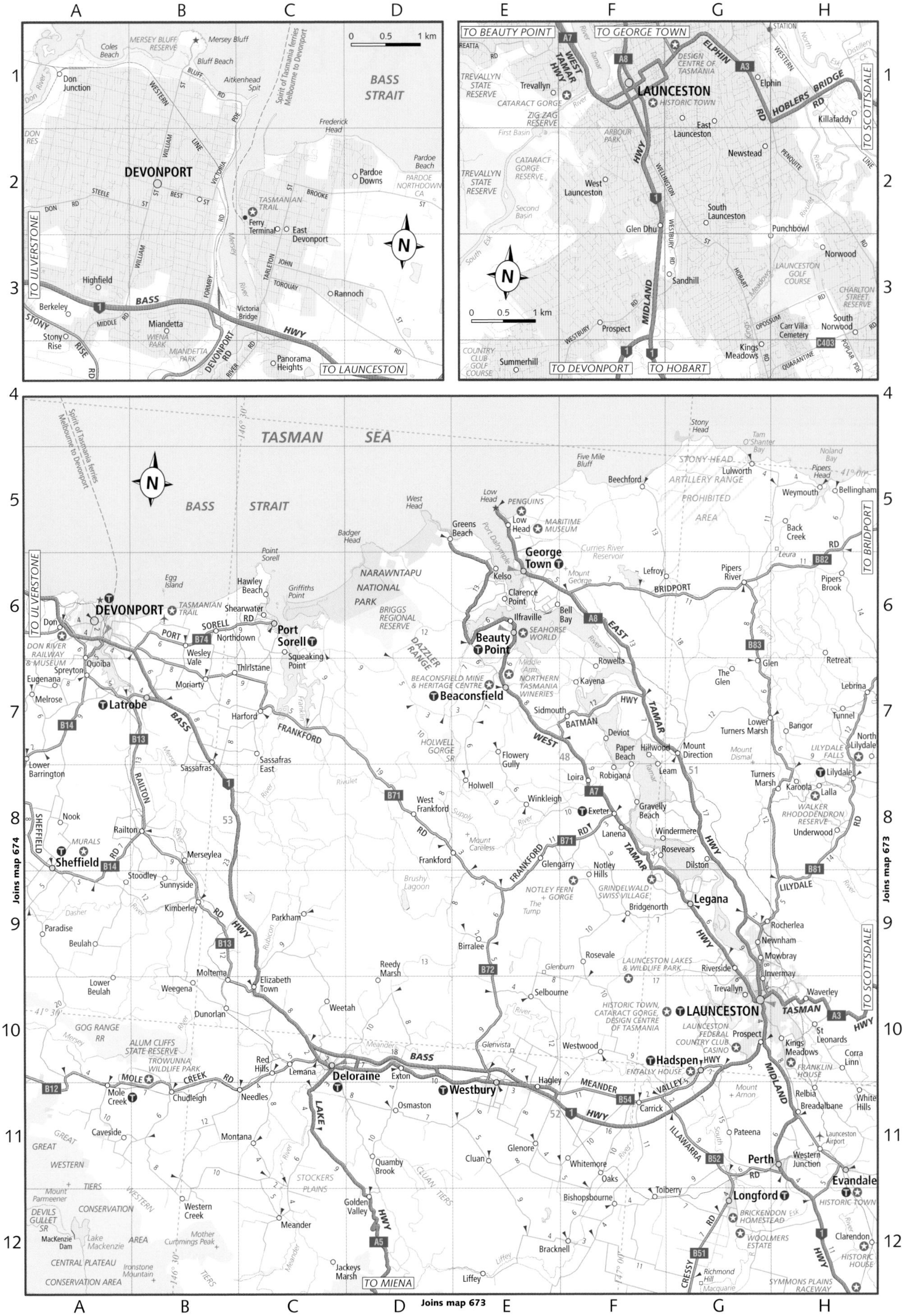

A B C D E F G H

DEVONPORT

MERSEY BLUFF RESERVE
Mersey Bluff
Coles Beach
Bluff Beach
Don Junction
Aitkenhead Spit
Spirit of Tasmania ferries Melbourne to Devonport
BASS STRAIT
DON RES
Frederick Head
Pardoe Beach
Pardoe Downs
PARDOE NORTHDOWN CA
DEVONPORT
TASMANIAN TRAIL
Ferry Terminal
East Devonport
TO ULVERSTONE
Highfield
Berkeley
Stony Rise
Miandetta
WIENA PARK
MIANDETTA PARK
Victoria Bridge
Rannoch
Panorama Heights
TO LAUNCESTON
BASS HWY
STONY RISE RD
MIDDLE RD
DEVONPORT RD
RIVER RD
WESTERN LINE
WILLIAM ST
BLUFF ST
PDE
VICTORIA
DON RD
STEELE ST
BEST ST
FORMBY RD
TARLETON ST
JOHN ST
TORQUAY
BROOKE ST
0 0.5 1 km

LAUNCESTON

TO BEAUTY POINT
TO GEORGE TOWN
TREVALLYN STATE RESERVE
Trevallyn
WEST TAMAR HWY
CATARACT GORGE
ZIG ZAG RESERVE
First Basin
ARBOUR PARK
CATARACT GORGE RESERVE
Second Basin
West Launceston
DESIGN CENTRE OF TASMANIA
LAUNCESTON
HISTORIC TOWN
East Launceston
Newstead
South Launceston
Glen Dhu
Sandhill
Prospect
Summerhill
COUNTRY CLUB GOLF COURSE
TO DEVONPORT
TO HOBART
ELPHIN RD
Elphin
HOBLERS BRIDGE RD
Killafaddy
TO SCOTTSDALE
PENQUITE RD
Punchbowl
Norwood
LAUNCESTON GOLF COURSE
CHARLTON STREET RESERVE
South Norwood
Carr Villa Cemetery
Kings Meadows
WELLINGTON ST
WESTBURY RD
MIDLAND HWY
HOBART RD
OPOSSUM RD
QUARANTINE RD
POPLAR PDE
STATION
0 0.5 1 km

TASMAN SEA
BASS STRAIT
Spirit of Tasmania ferries Melbourne to Devonport
Egg Island
Point Sorell
Hawley Beach
Griffiths Point
Badger Head
West Head
NARAWNTAPU NATIONAL PARK
BRIGGS REGIONAL RESERVE
DAZZLER RANGE
Greens Beach
Low Head
PENGUINS
Low Head
MARITIME MUSEUM
George Town
Kelso
Clarence Point
Bell Bay
Ilfraville
Beauty Point
SEAHORSE WORLD
BEACONSFIELD MINE & HERITAGE CENTRE
Beaconsfield
NORTHERN TASMANIA WINERIES
Five Mile Bluff
Stony Head
Tam O'Shanter Bay
STONY HEAD ARTILLERY RANGE PROHIBITED AREA
Beechford
Lulworth
Noland Bay
Pipers Head
Weymouth
Bellingham
Back Creek
TO BRIDPORT
Curries River Reservoir
Mount George
Lefroy
BRIDPORT RD
Pipers River
Pipers Brook
Leura
Retreat
Glen
The Glen
Lebrina
Tunnel
Bangor
Lower Turners Marsh
North Lilydale
LILYDALE FALLS
Lilydale
Turners Marsh
Mount Dismal
Karoola
Lalla
WALKER RHODODENDRON RESERVE
Underwood
LILYDALE RD
Rocherlea
Newnham
Mowbray
Invermay
Waverley
TO SCOTTSDALE
TASMAN HWY
St Leonards
Corra Linn
Kings Meadows
FRANKLIN HOUSE
Relbia
White Hills
Breadalbane
Launceston Airport
Western Junction
Evandale
HISTORIC TOWN
Clarendon
HISTORIC HOUSE
SYMMONS PLAINS RACEWAY
DEVONPORT
TASMANIAN TRAIL
TO ULVERSTONE
Don
DON RIVER RAILWAY & MUSEUM
Quoiba
Spreyton
Eugenana
Melrose
Latrobe
Shearwater
Port Sorell
Northdown
Wesley Vale
Squeaking Point
Thirlstane
Moriarty
Harford
FRANKFORD RD
Sassafras
Sassafras East
PORT SOREL RD
BASS HWY
RAILTON
Lower Barrington
Nook
Railton
SHEFFIELD
MURALS
Sheffield
Stoodley
Merseylea
Sunnyside
Kimberley
Paradise
Beulah
Lower Beulah
Weegena
Moltema
Elizabeth Town
Dunorlan
Parkham
GOG RANGE RR
ALUM CLIFFS STATE RESERVE
TROWUNNA WILDLIFE PARK
MOLE CREEK RD
Mole Creek
Chudleigh
Needles
Red Hills
Lemana
Deloraine
Exton
Caveside
Montana
GREAT WESTERN TIERS
Mount Parmeener
DEVILS GULLET SR
MacKenzie Dam
Lake Mackenzie
CENTRAL PLATEAU CONSERVATION AREA
Western Creek
Mother Cummings Peak
Irontone Mountain
Meander
STOCKERS PLAINS
LAKE HWY
Golden Valley
Jackeys Marsh
TO MIENA
Quamby Brook
Osmaston
Reedy Marsh
Weetah
HOLWELL GORGE SR
West Frankford
Frankford
Mount Careless
Brushy Lagoon
Flowery Gully
Holwell
Winkleigh
Birralee
Glenburn
Selbourne
Glenvista
Westbury
Hagley
MEANDER VALLEY HWY
Cluan
CLUAN TIERS
Glenore
Whitemore
Oaks
Bishopsbourne
Bracknell
Liffey
Kayena
Rowella
Sidmouth
BATMAN HWY
Deviot
Paper Beach
Hillwood
Mount Direction
Robigana
Loira
Leam
Exeter
Lanena
Gravelly Beach
Windermere
Rosevears
Dilston
Glengarry
Notley Hills
NOTLEY FERN GORGE
The Tump
GRINDELWALD SWISS VILLAGE
Bridgenorth
Legana
Rosevale
LAUNCESTON LAKES & WILDLIFE PARK
Riverside
Trevallyn
HISTORIC TOWN, CATARACT GORGE, DESIGN CENTRE OF TASMANIA
LAUNCESTON
LAUNCESTON FEDERAL COUNTRY CLUB CASINO
Prospect
Westwood
Hadspen
ENTALLY HOUSE
Mount Arnon
Carrick
ILLAWARRA RD
Pateena
Perth
Longford
Tolberry
BRICKENDON HOMESTEAD
WOOLMERS ESTATE
Richmond Hill
Macquarie
CRESSY RD
WEST TAMAR HWY
EAST TAMAR HWY
MIDLAND HWY
BASS HWY
41°30'
146°30'
147°00'
41°00'

Joins map 674
Joins map 673
Joins map 673

INDEX OF PLACE NAMES

This index includes all towns, localities, roadhouses and national parks shown on the maps and mentioned in the text. In addition, it includes major places of interest, landforms and water features.

Place names are followed by a map page number and grid reference, and/or the text page number on which that place is mentioned. A page number set in **bold** type indicates the main text entry for that place. For example:

Barcaldine Qld 654 D10, 661 P10, 428, **431**

Barcaldine	– Place Name
Qld	– State
654 D10, 661 P10	– Barcaldine appears on these map pages
428, **431**	– Barcaldine is mentioned on these pages
431	– Main entry for Barcaldine

The alphabetical order followed in the index is that of 'word-by-word' – a space is considered to come before 'A' in the alphabet, and the index has been ordered accordingly. For example:

White Hills
White Mountains
White Rock
Whitefoord
Whiteheads Creek
Whiteman

Names beginning with Mc are indexed as Mac and those beginning with St as Saint.

The following abbreviations and contractions are used in the index:

ACT	– Australian Capital Territory
JBT	– Jervis Bay Territory
NSW	– New South Wales
NP	– National Park
NT	– Northern Territory
Qld	– Queensland
SA	– South Australia
Tas.	– Tasmania
Vic.	– Victoria
WA	– Western Australia

A

B

C

D

F

G

H

I

J

K

L

M

O

P

Q

R

S

T

U

V

W

Y

Z

Publications manager
Astrid Browne

Project manager
Melissa Krafchek

Editors
Janet Austin, Juliette Elfick, Elizabeth Watson, Emma Adams, Melissa Krafchek, Tracy O'Shaughnessy, Stephanie Pearson, Louise McGregor

Researchers
Anthony Roberts, Emily Hewitt, Eddie Pavuna

Cover design
Jane Pennells

Internal design
Erika Budiman

Layout
Megan Ellis, Mike Kuszla

Cartographers
Paul de Leur, Claire Johnston, Bruce McGurty, Emily Maffei, Jason Sankovic

Indexer
Fay Donlevy

Photo selection
Melissa Krafchek

Pre-press
PageSet Digital Print & Pre-press

Writers
NEW SOUTH WALES Introduction by Carolyn Tate; Sydney by Ruth Ward, Jill Varley, Nick Dent, Ken Eastwood; Towns by Carolyn Tate, Ken Eastwood; Where to Eat and Where to Stay by Frances Bruce, Ken Eastwood; AUSTRALIAN CAPITAL TERRITORY Canberra by Alexandra Payne, Natasha Rudra, Margaret Wade; Where to Eat and Where to Stay by Kirsten Lawson, Natasha Rudra, Margaret Wade; VICTORIA Introduction by Rachel Pitts, Melissa Krafchek; Melbourne by Rachel Pitts, Melissa Krafchek, Michelle Bennett, Heidi Marfurt; Towns by Karina Biggs, Antonia Semler, Paul Harding; Where to Eat and Where to Stay by Tricia Welsh; SOUTH AUSTRALIA Introduction by Sue Medlock; Adelaide by Terry Plane, Rachel Pitts, Ingrid Ohlsson, Peter Dyson, Janice Finlayson, Jenny Turner; Towns by Karina Biggs, Rachel Pitts, Jenny Turner; Where to Eat and Where to Stay by Quentin Chester; WESTERN AUSTRALIA Introduction, Perth and towns by Heather Pearson; Where to Eat and Where to Stay by Carmen Jenner; NORTHERN TERRITORY Introduction and Darwin by David Hancock, Gillian Hutchison, Sue Moffitt; Towns by Carolyn Tate, Helen Duffy, David Hancock, Gillian Hutchison, Sue Moffitt; Where to Eat and Where to Stay by David Hancock, Sue Moffitt; QUEENSLAND Introduction by Sue Medlock; Brisbane by Alexandra Payne, Pamela Robson, Rowan Roebig, Lee Mylne; Towns by Karina Biggs, Jane E. Fraser; Where to Eat and Where to Stay by Liz Johnston; TASMANIA Introduction and Hobart by Sue Medlock; Towns by Emma Schwarcz; Where to Eat and Where to Stay by Sue Medlock

Explore Australia Publishing Pty Ltd
Ground floor, Building 1, 658 Church Street, Richmond, VIC 3121

Explore Australia Publishing Pty Ltd is a division of Hardie Grant Publishing Pty Ltd

hardie grant publishing

This thirtieth edition published by Explore Australia Publishing Pty Ltd, 2011

First published by George Phillip & O'Neil Pty Ltd, 1980
Second edition 1981
Third edition 1983
Reprinted 1984
Fourth edition 1985
Fifth edition 1986
Sixth edition published by Penguin Books Australia Ltd, 1987
Seventh edition 1988
Eighth edition 1989
Ninth edition 1990
Tenth edition 1991
Eleventh edition 1992
Twelfth edition 1993
Thirteenth edition 1994
Fourteenth edition 1995
Fifteenth edition 1996
Sixteenth edition 1997
Seventeenth edition 1998
Eighteenth edition 1999
Nineteenth edition 2000
Twentieth edition 2001
Twenty-first edition 2002
Twenty-second edition published by Explore Australia Publishing Pty Ltd, 2003
Twenty-third edition 2004
Twenty-fourth edition 2005
Twenty-fifth edition 2006
Twenty-sixth edition 2007
Twenty-seventh edition 2008
Twenty-eighth edition 2009
Twenty-ninth edition 2010

ISBN 9781741173673

10 9 8 7 6 5 4 3 2 1

Printed and bound in China by
1010 Printing International Ltd

Publisher's Note: Every effort has been made to ensure that the information in this book is accurate at the time of going to press. The publisher welcomes information and suggestions for correction or improvement. Write to the Publications Manager, Explore Australia Publishing, Ground floor, Building 1, 658 Church Street, Richmond, VIC 3121, Australia, or email info@exploreaustralia.net.au

Assistance with research

The publisher would like to thank the following organisations for assistance with information:

Australian Bureau of Statistics
Bureau of Meteorology
National Road Transport Commission
St John Ambulance
Tristate Fruit Fly Committee

New South Wales
Roads and Traffic Authority
New South Wales National Parks & Wildlife Service
Tourism New South Wales

Australian Capital Territory
Australian Capital Territory Land Information Centre
Australian Capital Tourism Corporation

Victoria
VicRoads
Parks Victoria
Tourism Victoria

South Australia
Transport SA
Primary Industries and Resources South Australia
National Parks & Wildlife South Australia, Department of Environment & Heritage
South Australian Tourism Commission

Western Australia
Main Roads Western Australia
Department of Indigenous Affairs Western Australia
Aboriginal Lands Trust
Department of Conservation & Land Management Western Australia
Western Australia Tourism Commission

Northern Territory
Department of Transport and Infrastructure
Northern and Central land councils
Northern Territory Department of Infrastructure, Planning and the Environment
Parks Australia
Northern Territory Tourist Commission

Queensland
Department of Main Roads
Queensland Department of Environment & Resources Management
Queensland Parks & Wildlife Service
Tourism Queensland

Tasmania
Department of Infrastructure, Energy & Resources
Parks and Wildlife Service
Tourism Tasmania

Photography credits

Cover
Idyllic Australian beach (Shutterstock)

Back cover
Uluru rock-face, Uluru–Kata Tjuta National Park, Northern Territory (Watagan View Photographics)

Front endpaper
A school of yellowstripe goatfish in the Great Barrier Reef, Queensland (© photolibrary. All rights reserved.)

Title page
Sunrise over Seven Mile Beach, Tasmania (© photolibrary. All rights reserved.)

Contents
Ormiston Gorge, West MacDonnell National Park, Northern Territory (Julie Fletcher)

Other images (left to right, top to bottom where more than one image appears on a page):
Pages iv–v (a) TNT (b) JS (c) Darren Jew/TQ (d) TV (e) © photolibrary. All rights reserved; vi–vii (a) Francisco Pelsaert or Jan Jansz from *The Voyage of the Batavia*/Wikimedia Commons (b) Ignaz Sebastian Klauber/National Library of Australia (c) Samuel Thomas Gill/National Library of Australia (d) Nicholas Chevalier/National Library of Australia (e) John Meredith/National Library of Australia (f) TV (g) John Flynn/National Library of Australia (h) National Library of Australia (i) TNSW; viii–1 AS; 2 JD; 3 (a) CK (b) ©iStockphoto.com/titelio; 4–5 JL/LT; 11 JD; 13 NR; 15 JB; 17 Mike Langford/AUS; 18 JF/AUS; 22 (a) TNSW (b) Peter & Margy Nicholas/LT; 23–24 TNSW; 25 JB; 26 JD; 27 JB; 28 Thom Stovern; 29 JL/LT; 30 GM; 31 JB; 32 Marie Lochman/LT; 33 GD; 34 JB; 35 Wayne Lawler/AUS; 36 AG; 37 Peter & Margy Nicholas/LT; 38 GD; 42 AS; 50 JLR/AUS; 53 AG; 54 JD; 57 JB; 75 JB; 86 City of Newcastle; 93 JB; 99 Courtesy of Thredbo Media; 109 Dee Kramer; 112–3 Australian Capital Tourism; 114–5 JB; 117 BP; 119 AC; 120 Peter Marsack/LT; 122, 124 & 126–127 JB; 128 AusGeo; 129 (a) AC (b) ©iStockphoto.com/Kolbz; 130–131 JD; 137 JB; 140 & 143 BB/LT; 144 Courtesy of St Kilda Pier Kiosk; 147 (a) GM (b) BP; 148 BP; 149 Courtesy of Mornington Peninsula Tourism Inc.; 150 BP; 151 Mike Leonard/AUS; 152 AS; 153 BB/LT; 154 AusGeo; 155 & 156 AC; 157 Dallas & John Heaton/AUS; 158 Hans & Judy Beste/LT; 159 AS; 160 JD; 166 TV; 168 AC; 171 JLR/AUS; 174 Gary Lewis/EAP; 177 Simon Griffiths; 182 Jon Barter/AUS; 185 AusGeo; 189 Hans & Judy Beste/LT; 195 AC; 207 AS; 212 AS; 218–219 JD; 220 JD; 221 (a) JL/LT (b) ©iStockphoto.com/Wilshire Images (c) ©iStockphoto.com/alicat; 222–223 Paul Rohal/AUS; 226 LS/LT; 228 & 231 Paul Rohal/AUS; 234 (a) NR/EAP (b) BP; 235 Stewart Roper/LT; 236 Chris Groenhout; 237 LS/LT; 238 JF/AUS; 239 RL; 240 AS; 241 JB; 242 BP; 243 AC; 244 JL/LT; 245 BP; 255 RL; 256 AusGeo; 260 GD; 267 SATC; 270 JB; 277 JLR/AUS; 280 City of Victor Harbor; 283 AusGeo; 286–287 CK; 288 JL/LT; 289 (a) CK (b) Courtesy of Cape Mentelle; 290–291 NR; 296 JL/LT; 299 LS/LT; 303 JL/LT; 304 (a) CK (b) JD; 305 JL/LT; 306 & 307 Andrew Davoll/LT; 308 JB/EAP; 309 JL/LT; 310 BB/LT; 311 DS/LT; 312 Clay Bryce/LT; 313 NR; 314 JD; 315 NR; 316 JL/LT; 320 JD; 323 DS/LT; 324 AS; 339 Tim Acker/AUS; 346 Courtesy of Cape Mentelle; 364–365 NR; 366 GM; 367 (a) AusGeo (b) TNT; 368–369 TNT; 374 GM; 376 Courtesy of Crocosaurus Cove; 378 (a) NR (b) NR/EAP; 379 BP; 380 TNT; 381 JL/LT; 382 TM; 383 DS/LT; 385 Davo Blair/AUS; 389 Stuart Grant; 390 AusGeo; 394 AS; 396–397 TM; 398 AS; 399 (a) BP (b) TQ; 400–401 JD; 406 AC; 409 NR; 411 & 412 JB; 414 (a) TQ (b) AG; 415 NR; 416 TQ; 417 JB; 418 JM; 419 GM; 420 JD; 421 TQ; 423 Wayne Lawler/AUS; 424 NR; 425 Paul Dymond; 426 JF/AUS; 427 TQ; 428 NR; 429 JM; 436 BP; 438 TQ; 451 John Carnemolla/AUS; 458 NR; 461 AS; 468 BP; 476 JD; 479 TQ; 480 NR; 484–485 TM; 486 JM; 487 (a) LS/LT (b) Sue Medlock; 488–489 JS/SFP; 494 JB; 497 RL; 499 (a) GD (b) JL/LT; 500 GD; 501 JS/SFP; 502 Courtesy of Pure Tasmania; 503 JS/SFP; 504 LS/LT; 505 JS/SFP; 506 TM; 508 & 520 JB; 525 AusGeo; 529 Courtesy of Stanley Information Centre; 530 JS/SFP.

Abbreviations

AC Andrew Chapman
AG Andrew Gregory
AUS Auscape International
AusGeo Australian Geographic
AS Australian Scenics
BB Bill Belson
BP Bruce Postle
CK Colin Kerr
DS Dennis Sarson
EAP Explore Australia Publishing
GD Grant Dixon
GM Geoff Murray
JD Jeff Drewitz
JB John Baker
JF Jean-Paul Ferrero
JL Jiri Lochman
JM John Meier
JLR Jean-Marc La Roque
JS Joe Shemesh
LS Len Stewart
LT Lochman Transparencies
NR Nick Rains
RL Rachel Lewis
SATC South Australian Tourism Commission
SFP Storm Front Productions
TM Ted Mead
TNSW Tourism New South Wales
TNT Tourism NT
TQ Tourism Queensland
TV Tourism Victoria